P9-BJF-549

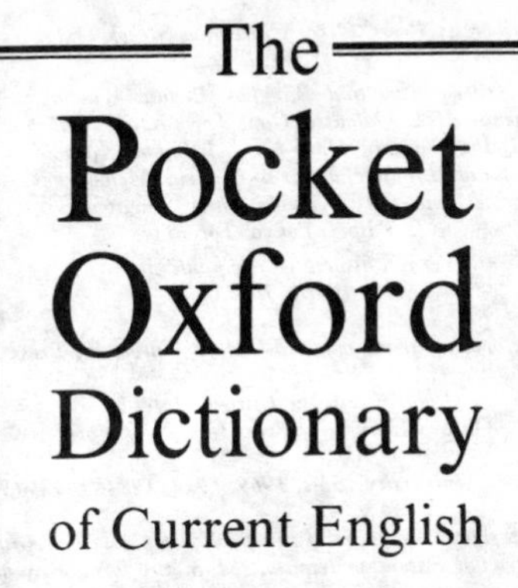

The Pocket Oxford Dictionary of Current English

First edited by
F. G. and H. W. Fowler

REVISED EIGHTH EDITION

Edited by
DELLA THOMPSON

CLARENDON PRESS · OXFORD

Oxford University Press, Great Clarendon Street, Oxford OX2 6DP
Oxford New York
Athens Auckland Bangkok Bogota Bombay
Buenos Aires Calcutta Cape Town Dar es Salaam
Delhi Florence Hong Kong Istanbul Karachi
Kuala Lumpur Madras Madrid Melbourne
Mexico City Nairobi Paris Singapore
Taipei Tokyo Toronto
and associated companies in
Berlin Ibadan

Published in the United States by
Oxford University Press Inc., New York

First edition 1924
Second edition 1934
Third edition 1939
Fourth edition 1942
Fifth edition 1969
Sixth edition 1978
Seventh edition 1984
Eighth edition 1992
Third revised edition 1996

British Library Cataloguing in Publication Data
Data available

Library of Congress Cataloging in Publication Data
Data available
ISBN 0-19-860045-3

3 5 7 9 10 8 6 4 2

Printed in the UK
by Clays p.l.c.
Bungay, Suffolk

Contents

Editorial Staff

Managing Editor	Dr Della Thompson
General Editors	Mr Andrew Hodgson Mr Chris Stewart
Science Editor	Dr Alexandra Clayton
Freelance Editors	Mrs Barbara Burge Mrs Anna Howes Ms Louise Jones
Critical Readers	Dr I. Grafe Mr M. W. Grose
Keyboarding Manager	Mrs Anne Whear
Keyboarding Assistants	Mrs Pam Marjara Mrs Kay Pepler Miss Doreen Stanbury
Proofreaders and General Contributors	Mrs Deirdre Arnold Mr Keith Harrison Mr Richard Lindsay Ms Elizabeth McIlvanney Dr Bernadette Paton Ms Lucy Reford Dr Anne St John-Hall Ms Joanna Thompson Mr F. Roland Whear

This edition owes much in its content and style to the eighth edition of the *Concise Oxford Dictionary* edited by Dr Robert Allen, who also revised the etymologies for this dictionary.

Preface to the Revised Eighth Edition

The eighth edition of the *Pocket Oxford Dictionary* takes even further the aims of the seventh edition towards making information easier to find and easier to understand. This is evident firstly in the continuation of the policy of denesting, i.e. the listing of words in a single alphabetical sequence of main entries with fewer items 'nested' at the end of entries. For example, in this edition, compound nouns written as two words (e.g. *German measles*) as well as some phrases (e.g. *long in the tooth*) are now main entries, while phrasal verbs and idioms remain listed under their first or main element (e.g. *look up to, look smart*). This makes words easier to find and reduces very long entries to a more manageable length.

Secondly, this edition no longer contains lists of undefined entries as were previously found for words with prefixes such as *over-*, *re-*, and *un-*. Instead, each is now treated as a main entry.

Thirdly, the use of sense numbers within each part of speech section (noun, verb, adjective, etc.) makes it easier to distinguish senses which in previous editions were divided only by semicolons.

The amount of information given in the dictionary about the forms of nouns, adjectives, and verbs has been greatly increased. Many more difficult plural forms of nouns are specified (e.g. for *callus*, *larynx*, and *racoon*), the treatment of verb conjugations is fuller (see, for example, *cry, dye*, and *veto*), and more comparative and superlative forms of adjectives (e.g. for *dry, noble*, and *unwieldy*) are provided.

More prominence has been given in this edition to guidance on disputed and controversial usage, and special usage notes have been introduced to this end. These are based on the current norms of standard English, i.e. the form of written and spoken English most generally accepted as a normal basis of communication in everyday life in the United Kingdom. Similarly, words that may cause great offence have been marked with the label *offens.* (offensive), in order to warn the reader of the likely effect of their use. Their inclusion in the dictionary does not in any way imply that their use is generally acceptable.

The *Pocket* has now been compiled for the first time online. This has greatly aided processes such as cross-reference checking and has allowed updating throughout the text right up to the time of printing.

D. J. T.

January 1996

Guide to the Use of the Dictionary

1. Headword

1.1 The headword is printed in bold type, or in bold italic type if the word is not naturalized in English and is usually found in italics in printed matter.

1.2 Variant spellings are given before the definition (e.g. **cabby** *n.* (also **cabbie**)); in all such cases the form given as the headword is the preferred form.

1.3 Words that are different but spelt the same way (homographs) are distinguished by superior figures (e.g. **bat**[1] and **bat**[2]).

1.4 Variant American spellings are indicated by the designation *US* (e.g. **favour**. . . *US* **favor**).

2. Pronunciation

Guidance on pronunciation follows the system of the International Phonetic Alphabet (IPA). Only the pronunciation standard in southern England is given.

3. Part-of-speech label

3.1 This is given for all main entries and derivatives.

3.2 Different parts of speech of a single word are listed separately.

3.3 Verbs, whether transitive, intransitive, or both, are given the simple designation *v.* The designation *absol.* (absolute) denotes use with an implied object (as at **abdicate**).

4. Inflexion

4.1 *Plurals of Nouns*:
Regular plurals are not given. Plural forms of those ending in *-o* (preceded by any letter other than another *o*) are always given. Other irregular forms are also given, except when the word is a compound of obvious formation (e.g. **footman**, **schoolchild**).

4.2 *Forms of Verbs*:

4.2.1 Regular forms receive no comment.

4.2.2 A doubled consonant in verbal inflexions (e.g. *rubbed, rubbing, sinned, sinning*) is shown in the form (**-bb-**, **-nn-**, etc.). Where practice differs in American usage this is noted (as at **cavil**).

4.3 *Comparative and Superlative of Adjectives and Adverbs*: Regular forms receive no comment.

5. Definition

5.1 Definitions are listed in order of comparative familiarity and importance, with the most current and important senses first.

5.2 They are separated by a number, or by a letter when the two senses are more closely related.

5.3 Round brackets enclose letters or words that are optional (as at **crash** *v.* where '(cause to) make a loud smashing noise' can mean either 'make a loud smashing noise' or 'cause to make a loud smashing noise'), and indicate typical objects of transitive verbs (such as '*milk*' and '*the skin*' in two senses of **cream** *v.*).

6. Subject and Usage labels

6.1 These are used to clarify the particular context in which a word or phrase is normally used.

6.2 Words and phrases more common in informal spoken English than in formal written English are labelled *colloq.* (colloquial) or *slang* as appropriate.

6.3 Some subject labels are used to indicate the particular relevance of a term or subject with which it is associated (e.g. *Mus.*, *Law*, *Physics*). They are not used when this is sufficiently clear from the definition itself.

6.4 Two categories of deprecated usage are indicated by special markings: *coarse slang* indicates a word that, although widely found, is still unacceptable to many people; *offens.* (offensive) indicates a use that is regarded as offensive by members of a particular ethnic, religious, or other group.

6.5 Usage notes found at the end of entries give guidance on the current norms of standard English. Some of the rules given may legitimately be broken in less formal English, and especially in conversation.

7. Compounds

7.1 Compound terms forming one word (e.g. **bathroom**,

jellyfish) are listed as main entries, as are those consisting of two or more words (e.g. **chain reaction**) or joined by a hyphen (e.g. **chain-gang**).

7.2 When a hyphenated compound in bold type is divided at the end of a line the hyphen is repeated at the beginning of the next line to show that it is a permanent feature of the spelling and not just an end-of-line hyphen.

8. Derivatives

8.1 Words formed by adding a suffix to another word are in many cases listed at the end of the entry for the main word (e.g. **chalkiness** and **chalky** at **chalk**). In this position they are not defined since they can be understood from the sense of the main word and that given at the suffix concerned; when further definition is called for they are given main entries in their own right (e.g. **changeable**).

8.2 For reasons of space words formed by certain suffixes are not always included except when some special feature of spelling or pronunciation or meaning is involved. These suffixes are -ABLE, -ER[1] (in the sense '. . .that does'), -ER[2] and -EST (see also 4.3), -ISH, -LESS, -LIKE, -LY[2], and -NESS.

9. Etymology

9.1 This is given in square brackets [ ] at the end of the entry. In the space available it can only give the direct line of derivation in outline. Forms in other languages are not given if they are exactly or nearly the same as the English form.

9.2 'Old English' is used for words that are known to have been used before AD 1150.

9.3 'Anglo-French' denotes the variety of French current in England in the Middle Ages after the Norman Conquest.

9.4 'Latin' denotes classical and Late Latin up to about AD 600; 'medieval Latin' that of the period about 600–1500; 'Anglo-Latin' denotes Latin as used in medieval England.

9.5 Where the origin of a word cannot be reliably ascertained, the form 'origin uncertain' or 'origin unknown' is used.

9.6 Names of the rarer languages that have contributed to English (such as Balti at **polo**, and Cree at **wapiti**) are given in full without explanation; they may be found explained in larger dictionaries or in encyclopedias.

10. Prefixes and Suffixes

10.1 A large selection of these is given in the main body of the text; prefixes are given in the form **ex-**, **re-**, etc., and suffixes in the form **-ion**, **-ness**, etc. These entries should be consulted to explain the many derivatives given at the end of entries (see 8.1).

10.2 Prefixes and suffixes are not normally given a pronunciation since this can change considerably when they form part of a word.

11. Cross-Reference

11.1 Cross-reference to main entries is indicated by small capitals (e.g. **calk** *US* var. of CAULK; **put a person wise** see WISE).

11.2 Cross-reference in italics to a defined phrase or compound refers to the entry for the first word unless another is specified.

Note on Proprietary Status

This dictionary includes some words which are, or are asserted to be, proprietary names or trade marks. Their inclusion does not imply that they have acquired for legal purposes a non-proprietary or general significance, nor is any other judgement implied concerning their legal status. In cases where the editor has some evidence that a word is used as a proprietary name or trade mark this is indicated by the designation *propr.*, but no judgement concerning the legal status of such words is made or implied thereby.

Abbreviations used in the Dictionary

Abbreviations in general use (such as etc., i.e.) are explained in the Dictionary itself.

abbr.	abbreviation
absol.	absolute(ly)
adj.	adjective
adv.	adverb
Aeron.	Aeronautics
Anat.	Anatomy
Anglo-Ind.	Anglo-Indian
Antiq.	Antiquity
Archaeol.	Archaeology
Archit.	Architecture
assim.	assimilated
Astrol.	Astrology
Astron.	Astronomy
Astronaut.	Astronautics
attrib.	attributive(ly)
attrib. adj.	attributive adjective
Austral.	Australian
aux.	auxiliary
Bibl.	Biblical
Biochem.	Biochemistry
Biol.	Biology
Bot.	Botany
Chem.	Chemistry
Cinematog.	Cinematography
collect.	collective(ly)
colloq.	colloquial(ly)
comb.	combination; combining
compar.	comparative
compl.	complement
conj.	conjunction
contr.	contraction
demons. adj.	demonstrative adjective
demons. pron.	demonstrative pronoun
derog.	derogatory
dial.	dialect
Eccl.	Ecclesiastical
Econ.	Economics
Electr.	Electricity
ellipt.	elliptical(ly)
emphat.	emphatic
esp.	especially
euphem.	euphemism
Exod.	Exodus
fem.	feminine
foll.	followed
Gen.	Genesis
Geol.	Geology
Geom.	Geometry
Gk	Greek
Gram.	Grammar
Hist.	History
hist.	with historical reference
imper.	imperative
Ind.	of the subcontinent comprising India, Pakistan, and Bangladesh
infin.	infinitive
int.	interjection
interrog.	interrogative
interrog. adj.	interrogative adjective
interrog. adv.	interrogative adverb
interrog. pron.	interrogative pronoun
Ir.	Irish
iron.	ironical
joc.	jocular
Judg.	Judges

Lev.	Leviticus
masc.	masculine
Math.	Mathematics
Matt.	Matthew
Mech.	Mechanics
Med.	Medicine
Meteorol.	Meteorology
Mil.	Military
Mineral.	Mineralogy
Mus.	Music
Mythol.	Mythology
n.	noun
Naut.	Nautical
neg.	negative
N.Engl.	Northern English
n.pl.	noun plural
NZ	New Zealand
obj.	objective (case)
offens.	offensive
opp.	as opposed to
orig.	originally
Parl.	Parliament(ary)
part.	participle
past part.	past participle
Pharm.	Pharmacy; Pharmacology
Philos.	Philosophy
Phonet.	Phonetics
Photog.	Photography
phr.	phrase
Physiol.	Physiology
pl.	plural
poet.	poetical
Polit.	Politics
poss.	possessive (case)
poss. pron.	possessive pronoun
prec.	preceded
predic.	predicative(ly)
predic. adj.	predicative adjective
prep.	preposition
pres.	present
pres. part.	present participle
pron.	pronoun
pronunc.	pronunciation
propr.	proprietary term
Psychol.	Psychology
RC Ch.	Roman Catholic Church
ref.	reference
refl.	reflexive
rel. adj.	relative adjective
rel. adv.	relative adverb
rel. pron.	relative pronoun
Relig.	Religion
Rev.	Revelation
rhet.	rhetorical
Rom.	Roman
S.Afr.	South African
Sc.	Scottish
Sci.	Science
sing.	singular
Stock Exch.	Stock Exchange
superl.	superlative
symb.	symbol
Theatr.	Theatre
Theol.	Theology
US	American, in American use
usu.	usually
v.	verb
var.	variant(s)
v.aux.	auxiliary verb
v.refl.	reflexive verb
Zool.	Zoology

Pronunciation Symbols

Consonants

b	*b*ut	n	*n*o	ʃ	*sh*e
d	*d*og	p	*p*en	ʒ	vi*s*ion
f	*f*ew	r	*r*ed	θ	*th*in
g	*g*et	s	*s*it	ð	*th*is
h	*h*e	t	*t*op	ŋ	ri*ng*
j	*y*es	v	*v*oice	x	lo*ch*
k	*c*at	w	*w*e	tʃ	*ch*ip
l	*l*eg	z	*z*oo	dʒ	*j*ar
m	*m*an				

Vowels

æ	c*a*t	ʌ	r*u*n	əʊ	n*o*
ɑ:	*ar*m	ʊ	p*u*t	eə	h*air*
e	b*e*d	u:	t*oo*	ɪə	n*ear*
ɜ:	h*er*	ə	*a*go	ɔɪ	b*oy*
ɪ	s*i*t	aɪ	m*y*	ʊə	p*oor*
i:	s*ee*	aʊ	h*ow*	aɪə	f*ire*
ɒ	h*o*t	eɪ	d*ay*	aʊə	s*our*
ɔ:	s*aw*				

(ə) signifies the indeterminate sound as in gard*e*n, carn*a*l, and rhyth*m*.

(r) at the end of a word indicates an r that is sounded when a word beginning with a vowel follows, as in *clutter up* and *an acre of land*.

The mark ˜ indicates a nasalized sound, as in the following sounds that are not natural in English:

æ̃ (t*i*mbre)
ɑ̃ (él*an*)
ɔ̃ (b*on* voyage)

The main or primary stress of a word is shown by ˈ preceding the relevant syllable; any secondary stress in words or phrases of three or more syllables is shown by ˌ preceding the relevant syllable.

A

A[1] /eɪ/ *n.* (*pl.* **As** or **A's**) **1** (also **a**) first letter of the alphabet. **2** *Mus.* sixth note of the diatonic scale of C major. **3** first hypothetical person or example. **4** highest category (of roads, marks, etc.). **5** (usu. **a**) *Algebra* first known quantity. □ **A1** /eɪ 'wʌn/ *colloq.* first-rate, excellent. **from A to B** from one place to another. **from A to Z** from beginning to end.

A[2] *abbr.* (also **A.**) **1** ampere(s). **2** answer.

a[1] /ə, eɪ/ *adj.* (also **an** /æn, ən/ before a vowel sound) (called the indefinite article) **1** one, some, any. **2** one like (*a Judas*). **3** one single (*not a chance*). **4** the same (*all of a size*). **5** per (*twice a year*; *seven a side*). [Old English *ān* one]

a[2] /ə/ *prep.* (usu. as *prefix*) **1** to, towards (*ashore*; *aside*). **2** (with verb in pres. part. or infin.) doing or being (*a-hunting*; *abuzz*). **3** on (*afire*). **4** in (*nowadays*). [Old English *an*, *on*, ON]

Å *abbr.* ångström(s).

a- *prefix* (also **an-** before a vowel sound) not, without (*amoral*). [Greek]

AA *abbr.* **1** Automobile Association. **2** Alcoholics Anonymous. **3** anti-aircraft.

aardvark /'ɑ:dvɑ:k/ *n.* mammal with a tubular snout and a long tongue, feeding on termites. [Afrikaans]

ab- *prefix* off, away, from (*abduct*). [Latin]

aback /ə'bæk/ *adv.* □ **take aback** surprise, disconcert. [Old English: related to A[2]]

abacus /'æbəkəs/ *n.* (*pl.* **-cuses**) **1** frame with wires along which beads are slid for calculating. **2** *Archit.* flat slab on top of a capital. [Latin from Greek from Hebrew]

abaft /ə'bɑ:ft/ *Naut.* *–adv.* in the stern half of a ship. *–prep.* nearer the stern than. [from A[2], *-baft*: see AFT]

abandon /ə'bænd(ə)n/ *–v.* **1** give up. **2** forsake, desert. **3** (often foll by *to*; often *refl.*) yield to a passion, another's control, etc. *–n.* freedom from inhibitions. □ **abandonment** *n.* [French: related to AD-, BAN]

abandoned *adj.* **1** deserted, forsaken. **2** unrestrained, profligate.

abase /ə'beɪs/ *v.* (**-sing**) (also *refl.*) humiliate, degrade. □ **abasement** *n.* [French: related to AD-, BASE[2]]

abashed /ə'bæʃt/ *predic. adj.* embarrassed, disconcerted. [French *es-* EX-[1], *baïr* astound]

abate /ə'beɪt/ *v.* (**-ting**) make or become less strong etc.; diminish. □ **abatement** *n.* [French *abatre* from Latin *batt(u)o* beat]

abattoir /'æbə,twɑ:(r)/ *n.* slaughterhouse. [French *abatre* fell, as ABATE]

abbacy /'æbəsɪ/ *n.* (*pl.* **-ies**) office or jurisdiction of an abbot or abbess. [Latin: related to ABBOT]

abbé /'æbeɪ/ *n.* (in France) abbot or priest. [French from Latin: related to ABBOT]

abbess /'æbɪs/ *n.* head of a community of nuns.

abbey /'æbɪ/ *n.* (*pl.* **-s**) **1** building(s) occupied by a community of monks or nuns. **2** the community itself. **3** building that was once an abbey.

abbot /'æbət/ *n.* head of a community of monks. [Old English from Latin *abbas*]

abbreviate /ə'bri:vɪ,eɪt/ *v.* (**-ting**) shorten, esp. represent (a word etc.) by a part of it. □ **abbreviation** /-'eɪʃ(ə)n/ *n.* [Latin: related to BRIEF]

ABC /,eɪbi:'si:/ *n.* **1** the alphabet. **2** rudiments of a subject. **3** alphabetical guide.

abdicate /'æbdɪ,keɪt/ *v.* (**-ting**) **1** (usu. *absol.*) give up or renounce (the throne). **2** renounce (a duty, right, etc.). □ **abdication** /-'keɪʃ(ə)n/ *n.* [Latin *dico* declare]

abdomen /'æbdəmən/ *n.* **1** the belly, including the stomach, bowels, etc. **2** the hinder part of an insect etc. □ **abdominal** /æb'dɒmɪn(ə)l/ *adj.* [Latin]

abduct /əb'dʌkt/ *v.* carry off or kidnap illegally. □ **abduction** *n.* **abductor** *n.* [Latin *duco* lead]

abeam /ə'bi:m/ *adv.* at right angles to a ship's or an aircraft's length.

Aberdeen Angus /,æbədi:n 'æŋgəs/ *n.* animal of a Scottish breed of hornless black cattle. [*Aberdeen* in Scotland]

Aberdonian /,æbə'dəʊnɪən/ *–adj.* of Aberdeen. *–n.* native or citizen of Aberdeen. [medieval Latin]

aberrant /ə'berənt/ *adj.* deviating from what is normal or accepted. [Latin: related to ERR]

aberration /,æbə'reɪʃ(ə)n/ *n.* **1** aberrant behaviour; moral or mental lapse. **2** *Biol.* deviation from a normal type. **3** distortion of an image because of a defect in a lens or mirror. **4** *Astron.* apparent displacement of a celestial body.

abet /ə'bet/ *v.* **(-tt-)** (usu. in **aid and abet**) encourage or assist (an offender or offence). [French: related to AD-, BAIT]

abeyance /ə'beɪəns/ *n.* (usu. prec. by *in*, *into*) temporary disuse. [French: related to AD-, *beer* gape]

abhor /əb'hɔ:(r)/ *v.* **(-rr-)** detest; regard with disgust. [Latin: related to HORROR]

abhorrence /əb'hɒrəns/ *n.* disgust; detestation.

abhorrent /əb'hɒrənt/ *adj.* (often foll. by *to*) disgusting or hateful.

abide /ə'baɪd/ *v.* **(-ding**; *past* and *past part.* **abided** or rarely **abode) 1** (usu. in *neg.*) tolerate, endure (*can't abide him*). **2** (foll. by *by*) **a** act in accordance with (*abide by the rules*). **b** keep (a promise). **3** *archaic* remain, continue. [Old English *a*-intensive prefix, BIDE]

abiding *adj.* enduring, permanent.

ability /ə'bɪlɪtɪ/ *n.* (*pl.* **-ies**) **1** (often foll. by *to* + infin.) capacity or power. **2** cleverness, talent. [French: related to ABLE]

-ability *suffix* forming nouns of quality from, or corresponding to, adjectives in *-able*.

ab initio /,æb ɪ'nɪʃɪəʊ/ *adv.* from the beginning. [Latin]

abject /'æbdʒekt/ *adj.* miserable, wretched; degraded; despicable. □ **abjection** /æb'dʒekʃ(ə)n/ *n.* [Latin *jacio -ject-* throw]

abjure /əb'dʒʊə(r)/ *v.* **(-ring)** renounce on oath (an opinion, cause, etc.). □ **abjuration** /,æbdʒʊ'reɪʃ(ə)n/ *n.* [Latin *juro* swear]

ablative /'æblətɪv/ *Gram.* *–n.* case (in Latin) of nouns and pronouns indicating an agent, instrument, or location. *–adj.* of or in the ablative. [Latin *ablatus* taken away]

ablaze /ə'bleɪz/ *predic. adj.* & *adv.* **1** on fire. **2** glittering, glowing. **3** greatly excited.

able /'eɪb(ə)l/ *adj.* **(abler, ablest) 1** (often foll. by *to* + infin.; used esp. in *is able*, *will be able*, etc., replacing tenses of *can*) having the capacity or power (*not able to come*). **2** talented, clever. □ **ably** *adv.* [Latin *habilis*]

-able *suffix* forming adjectives meaning: **1** that may or must be (*eatable*; *payable*). **2** that can be made the subject of (*dutiable*; *objectionable*). **3** relevant to or in accordance with (*fashionable*; *seasonable*). [Latin *-abilis*]

able-bodied *adj.* fit, healthy.

able-bodied seaman *n.* ordinary trained seaman.

ablution /ə'blu:ʃ(ə)n/ *n.* (usu. in *pl.*) **1** ceremonial washing of the hands, sacred vessels, etc. **2** *colloq.* **a** ordinary bodily washing. **b** place for this. [Latin *ablutio* from *luo lut-* wash]

-ably *suffix* forming adverbs corresponding to adjectives in *-able*.

ABM *abbr.* anti-ballistic missile.

abnegate /'æbnɪ,geɪt/ *v.* **(-ting)** give up or renounce (a pleasure or right etc.). [Latin *nego* deny]

abnegation /,æbnɪ'geɪʃ(ə)n/ *n.* denial; renunciation of a doctrine.

abnormal /æb'nɔ:m(ə)l/ *adj.* deviating from the norm; exceptional. □ **abnormality** /-'mælɪtɪ/ *n.* (*pl.* **-ies**). **abnormally** *adv.* [French: related to ANOMALOUS]

Abo /'æbəʊ/ (also **abo**) *Austral. slang* usu. *offens.* *–n.* (*pl.* **-s**) Aboriginal. *–adj.* Aboriginal. [abbreviation]

aboard /ə'bɔ:d/ *adv.* & *prep.* on or into (a ship, aircraft, etc.). [from A[2]]

abode[1] /ə'bəʊd/ *n.* dwelling-place. [related to ABIDE]

abode[2] see ABIDE.

abolish /ə'bɒlɪʃ/ *v.* put an end to (esp. a custom or institution). [Latin *aboleo* destroy]

abolition /,æbə'lɪʃ(ə)n/ *n.* abolishing or being abolished. □ **abolitionist** *n.*

A-bomb /'eɪbɒm/ *n.* = ATOMIC BOMB. [*A* for ATOMIC]

abominable /ə'bɒmɪnəb(ə)l/ *adj.* **1** detestable, loathsome. **2** *colloq.* very unpleasant (*abominable weather*). □ **abominably** *adv.* [Latin *abominor* deprecate]

Abominable Snowman *n.* supposed manlike or bearlike Himalayan animal; yeti.

abominate /ə'bɒmɪ,neɪt/ *v.* **(-ting)** detest, loathe. □ **abomination** /-'neɪʃ(ə)n/ *n.* [Latin: related to ABOMINABLE]

aboriginal /,æbə'rɪdʒɪn(ə)l/ *–adj.* **1** indigenous, inhabiting a land from the earliest times, esp. before the arrival of colonists. **2** (usu. **Aboriginal**) of Australian Aborigines. *–n.* **1** aboriginal inhabitant. **2** (usu. **Aboriginal**) aboriginal inhabitant of Australia. [Latin: related to ORIGIN]

aborigine /,æbə'rɪdʒɪnɪ/ *n.* (usu. in *pl.*) **1** aboriginal inhabitant. **2** (usu. **Aborigine**) aboriginal inhabitant of Australia.

■ **Usage** When referring to the people, *Aboriginal* is preferred for the singular form and *Aborigines* for the plural, although *Aboriginals* is also acceptable.

abort /ə'bɔ:t/ *v.* **1** miscarry. **2 a** effect abortion of (a foetus). **b** effect abortion in (a mother). **3** end or cause (a project etc.) to end before completion. [Latin *orior* be born]

abortion /əˈbɔːʃ(ə)n/ *n.* **1** natural or (esp.) induced expulsion of a foetus from the womb before it is able to survive independently. **2** stunted or deformed creature or thing. **3** failed project or action. □ **abortionist** *n.*

abortive /əˈbɔːtɪv/ *adj.* fruitless, unsuccessful.

abound /əˈbaʊnd/ *v.* **1** be plentiful. **2** (foll. by *in, with*) be rich; teem. [Latin *unda* wave]

about /əˈbaʊt/ *–prep.* **1 a** on the subject of (*a book about birds*). **b** relating to (*glad about it*). **c** in relation to (*symmetry about a plane*). **2** at a time near to (*about six*). **3 a** in, round (*walked about the town; a scarf about her neck*). **b** all round from a centre (*look about you*). **4** at points in (*strewn about the house*). **5** carried with (*no money about me*). **6** occupied with (*about her business*). *–adv.* **1 a** approximately (*about ten miles*). **b** *colloq.* in an understatement (*just about had enough*). **2** nearby (*a lot of flu about*). **3** in every direction (*look about*). **4** on the move; in action (*out and about*). **5** in rotation or succession (*turn and turn about*). □ **be about** (or **all about**) *colloq.* have as its essential nature (*life is all about having fun*). **be about to** be on the point of (*was about to laugh*). [Old English]

about-face *n.* & *int.* = ABOUT-TURN, ABOUT TURN.

about-turn *–n.* **1** turn made so as to face the opposite direction. **2** change of opinion or policy etc. *–int.* (**about turn**) *Mil.* command to make an about-turn.

above /əˈbʌv/ *–prep.* **1** over; on the top of; higher than; over the surface of (*head above water; above the din*). **2** more than (*above twenty people*). **3** higher in rank, importance, etc., than. **4 a** too great or good for (*not above cheating*). **b** beyond the reach of (*above my understanding; above suspicion*). *–adv.* **1** at or to a higher point; overhead (*the floor above; the sky above*). **2** earlier on a page or in a book (*as noted above*). *–adj.* preceding (*the above argument*). *–n.* (prec. by *the*) preceding text (*the above shows*). □ **above all** most of all, more than anything else. **above oneself** conceited, arrogant. [Old English: related to A²]

above-board *adj.* & *adv.* without concealment; open or openly.

abracadabra /ˌæbrəkəˈdæbrə/ *–int.* supposedly magic word used in conjuring. *–n.* spell or charm. [Latin from Greek]

abrade /əˈbreɪd/ *v.* (**-ding**) scrape or wear away (skin, rock, etc.) by rubbing. [Latin *rado* scrape]

abrasion /əˈbreɪʒ(ə)n/ *n.* **1** scraping or wearing away (of skin, rock, etc.). **2** resulting damaged area.

abrasive /əˈbreɪsɪv/ *–adj.* **1 a** tending to rub or graze. **b** capable of polishing by rubbing or grinding. **2** harsh or hurtful in manner. *–n.* abrasive substance.

abreast /əˈbrest/ *adv.* **1** side by side and facing the same way. **2** (foll. by *of*) up to date with.

abridge /əˈbrɪdʒ/ *v.* (**-ging**) shorten (a book, film, etc.). □ **abridgement** *n.* [Latin: related to ABBREVIATE]

abroad /əˈbrɔːd/ *adv.* **1** in or to a foreign country or countries. **2** widely (*scatter abroad*). **3** in circulation (*rumour abroad*).

abrogate /ˈæbrəˌgeɪt/ *v.* (**-ting**) repeal, abolish (a law etc.). □ **abrogation** /-ˈgeɪʃ(ə)n/ *n.* [Latin *rogo* propose a law]

abrupt /əˈbrʌpt/ *adj.* **1** sudden, hasty (*abrupt end*). **2** (of manner etc.) curt. **3** steep, precipitous. □ **abruptly** *adv.* **abruptness** *n.* [Latin: related to RUPTURE]

abscess /ˈæbsɪs/ *n.* (*pl.* **abscesses**) swelling containing pus. [Latin: related to AB-, CEDE]

abscissa /əbˈsɪsə/ *n.* (*pl.* **abscissae** /-siː/ or **-s**) *Math.* (in a system of coordinates) shortest distance from a point to the vertical or *y*-axis. [Latin *abscindo* cut off]

abscond /əbˈskɒnd/ *v.* depart hurriedly and furtively, esp. to avoid arrest; escape. [Latin *abscondo* secrete]

abseil /ˈæbseɪl/ *–v.* descend by using a doubled rope coiled round the body and fixed at a higher point. *–n.* descent made by abseiling. [German *ab* down, *Seil* rope]

absence /ˈæbs(ə)ns/ *n.* **1** being away. **2** time of this. **3** (foll. by *of*) lack of. □ **absence of mind** inattentiveness. [Latin *absentia*]

absent *–adj.* /ˈæbs(ə)nt/ **1** not present. **2** not existing; lacking. **3** inattentive. *–v.refl.* /æbˈsent/ go, or stay, away. □ **absently** *adv.* (in sense 3 of *adj.*).

absentee /ˌæbsənˈtiː/ *n.* person not present.

absenteeism /ˌæbsənˈtiːɪz(ə)m/ *n.* absenting oneself from work or school etc., esp. frequently or illicitly.

absentee landlord *n.* one who lets a property while living elsewhere.

absent-minded *adj.* forgetful or inattentive. □ **absent-mindedly** *adv.* **absent-mindedness** *n.*

absinth /ˈæbsɪnθ/ *n.* **1** wormwood. **2** (usu. **absinthe**) aniseed-flavoured liqueur based on this. [French from Latin]

absolute /'æbsə,lu:t/ *–adj.* **1** complete, utter (*absolute bliss*). **2** unconditional (*absolute authority*). **3** despotic (*absolute monarch*). **4** not relative or comparative (*absolute standard*). **5** *Gram.* **a** (of a construction) syntactically independent of the rest of the sentence, as in *dinner being over, we left the table.* **b** (of an adjective or transitive verb) without an expressed noun or object (e.g. *the deaf, guns kill*). **6** (of a legal decree etc.) final. *–n. Philos.* (prec. by *the*) that which can exist independently of anything else. [Latin: related to ABSOLVE]

absolutely *adv.* **1** completely, utterly. **2** in an absolute sense (*God exists absolutely*). **3** /-'lu:tlı/ *colloq.* (used in reply) quite so; yes.

absolute majority *n.* majority over all rivals combined.

absolute pitch *n.* ability to recognize or sound any given note.

absolute temperature *n.* one measured from absolute zero.

absolute zero *n.* theoretical lowest possible temperature calculated as –273.15° C (or 0° K).

absolution /,æbsə'lu:ʃ(ə)n/ *n.* formal forgiveness of sins.

absolutism /'æbsəlu:,tız(ə)m/ *n.* principle or practice of absolute government. □ **absolutist** *n.*

absolve /əb'zɒlv/ *v.* **(-ving)** (often foll. by *from, of*) set or pronounce free from blame or obligation etc. [Latin: related to SOLVE]

absorb /əb'sɔ:b/ *v.* **1** incorporate as part of itself or oneself. **2** take in, suck up (liquid, heat, knowledge, etc.). **3** reduce the effect or intensity of; deal easily with (an impact, sound, difficulty, etc.). **4** consume (resources etc.). **5** (often as **absorbing** *adj.*) engross the attention of. [Latin *sorbeo* suck in]

absorbent *–adj.* tending to absorb. *–n.* absorbent substance or organ.

absorption /əb'sɔ:pʃ(ə)n/ *n.* **1** absorbing or being absorbed. **2** mental engrossment. □ **absorptive** *adj.*

abstain /əb'steın/ *v.* **1** (usu. foll. by *from*) refrain from indulging (*abstained from smoking*). **2** decline to vote. [Latin *teneo tent-* hold]

abstemious /əb'sti:mıəs/ *adj.* moderate or ascetic, esp. in eating and drinking. □ **abstemiously** *adv.* [Latin: related to AB-, *temetum* strong drink]

abstention /əb'stenʃ(ə)n/ *n.* abstaining, esp. from voting. [Latin: related to ABSTAIN]

abstinence /'æbstınəns/ *n.* abstaining, esp. from food or alcohol. □ **abstinent** *adj.* [French: related to ABSTAIN]

abstract *–adj.* /'æbstrækt/ **1 a** of or existing in thought or theory rather than matter or practice; not concrete. **b** (of a word, esp. a noun) denoting a quality, condition, etc., not a concrete object. **2** (of art) achieving its effect by form and colour rather than by realism. *–v.* /əb'strækt/ **1** (often foll. by *from*) extract, remove. **2** summarize. *–n.* /'æbstrækt/ **1** summary. **2** abstract work of art. **3** abstraction or abstract term. [Latin: related to TRACT[1]]

abstracted /əb'stræktıd/ *adj.* inattentive, distracted. □ **abstractedly** *adv.*

abstraction /əb'strækʃ(ə)n/ *n.* **1** abstracting or taking away. **2** abstract or visionary idea. **3** abstract qualities (esp. in art). **4** absent-mindedness.

abstruse /əb'stru:s/ *adj.* hard to understand, profound. □ **abstruseness** *n.* [Latin *abstrudo -trus-* conceal]

absurd /əb'sɜ:d/ *adj.* wildly illogical or inappropriate; ridiculous. □ **absurdity** *n.* (*pl.* **-ies**). **absurdly** *adv.* [Latin: related to SURD]

ABTA /'æbtə/ *abbr.* Association of British Travel Agents.

abundance /ə'bʌnd(ə)ns/ *n.* **1** plenty; more than enough; a lot. **2** wealth. [Latin: related to ABOUND]

abundant *adj.* **1** plentiful. **2** (foll. by *in*) rich (*abundant in fruit*). □ **abundantly** *adv.*

abuse *–v.* /ə'bju:z/ **(-sing) 1** use improperly, misuse. **2** insult verbally. **3** maltreat. *–n.* /ə'bju:s/ **1** misuse. **2** insulting language. **3** unjust or corrupt practice. **4** maltreatment (*child abuse*). □ **abuser** /ə'bju:zə(r)/ *n.* [Latin: related to USE]

abusive /ə'bju:sıv/ *adj.* insulting, offensive. □ **abusively** *adv.*

abut /ə'bʌt/ *v.* **(-tt-) 1** (foll. by *on*) (of land) border on. **2** (foll. by *on, against*) (of a building) touch or lean upon (another). [Anglo-Latin *butta* strip of land: related to BUTT[1]]

abutment *n.* lateral supporting structure of a bridge, arch, etc.

abuzz /ə'bʌz/ *adv.* & *adj.* in a state of excitement or activity.

abysmal /ə'bızm(ə)l/ *adj.* **1** *colloq.* extremely bad (*abysmal food*). **2** profound, utter (*abysmal ignorance*). □ **abysmally** *adv.* [Latin: related to ABYSS]

abyss /ə'bıs/ *n.* **1** deep chasm. **2** immeasurable depth (*abyss of despair*). [Latin from Greek, = bottomless]

AC *abbr.* **1** (also **ac**) alternating current. **2** aircraftman.

Ac *symb.* actinium.

a/c *abbr.* account. [*account current*]

-ac *suffix* forming adjectives often (or only) used as nouns (*cardiac*; *maniac*). [Latin *-acus*, Greek *-akos*]

acacia /ə'keɪʃə/ *n.* tree with yellow or white flowers, esp. one yielding gum arabic. [Latin from Greek]

academia /,ækə'di:mɪə/ *n.* the academic world; scholastic life.

academic /,ækə'demɪk/ *–adj.* **1** scholarly, of learning. **2** of no practical relevance; theoretical. *–n.* teacher or scholar in a university etc. □ **academically** *adv.*

academician /ə,kædə'mɪʃ(ə)n/ *n.* member of an Academy. [French *académicien*].

academy /ə'kædəmɪ/ *n.* (*pl.* **-ies**) **1** place of specialized training (*military academy*). **2** (usu. **Academy**) society or institution of distinguished scholars, artists, scientists, etc. (*Royal Academy*). **3** *Scot.* secondary school. [Greek *akadēmeia* the place in Athens where Plato taught]

acanthus /ə'kænθəs/ *n.* (*pl.* **-thuses**) **1** herbaceous plant with spiny leaves. **2** *Archit.* representation of its leaf. [Latin from Greek]

a cappella /,ɑ: kə'pelə, ,æ-/ *adj.* & *adv.* (of choral music) unaccompanied. [Italian, = in church style]

ACAS /'eɪkæs/ *abbr.* Advisory, Conciliation, and Arbitration Service.

accede /ək'si:d/ *v.* (**-ding**) (foll. by *to*) **1** take office, esp. as monarch. **2** assent or agree. [Latin: related to CEDE]

accelerate /ək'selə,reɪt/ *v.* (**-ting**) move or cause to move or happen more quickly. □ **acceleration** /-'reɪʃ(ə)n/ *n.* [Latin: related to CELERITY]

accelerator *n.* **1** device for increasing speed, esp. the pedal controlling the speed of a vehicle's engine. **2** *Physics* apparatus for imparting high speeds to charged particles.

accent *–n.* /'æks(ə)nt/ **1** particular (esp. local or national) mode of pronunciation. **2** distinctive feature or emphasis (*accent on speed*). **3** prominence given to a syllable by stress or pitch. **4** mark on a letter or word to indicate pitch, stress, or vowel quality. *–v.* /æk'sent/ **1** emphasize (a word or syllable etc.). **2** write or print accents on (words etc.). **3** accentuate. [Latin *cantus* song]

accentuate /ək'sentʃʊ,eɪt/ *v.* (**-ting**) emphasize, make prominent. □ **accentuation** /-'eɪʃ(ə)n/ *n.* [medieval Latin: related to ACCENT]

accept /ək'sept/ *v.* **1** (also *absol.*) willingly receive (a thing offered). **2** (also *absol.*) answer affirmatively (an offer etc.). **3** regard favourably; treat as welcome (*felt accepted*). **4** believe, receive (an opinion, explanation, etc.) as adequate or valid. **5** take as suitable (*does accept cheques*). **6** undertake (an office or duty). [Latin *capio* take]

acceptable *adj.* **1** worth accepting, welcome. **2** tolerable. □ **acceptability** /-'bɪlɪtɪ/ *n.* **acceptably** *adv.* [French: related to ACCEPT]

acceptance *n.* **1** willingness to accept. **2** affirmative answer to an invitation etc. **3** approval, belief (*found wide acceptance*).

access /'ækses/ *–n.* **1** way of approach or entry (*shop with rear access*). **2 a** right or opportunity to reach or use or visit; admittance (*access to secret files, to the prisoner*). **b** accessibility. **3** *archaic* outburst (*an access of anger*). *–v.* **1** *Computing* gain access to (data etc.). **2** accession. [French: related to ACCEDE]

accessible /ək'sesɪb(ə)l/ *adj.* (often foll. by *to*) **1** reachable or obtainable; readily available. **2** easy to understand. □ **accessibility** /-'bɪlɪtɪ/ *n.*

accession /ək'seʃ(ə)n/ *–n.* **1** taking office, esp. as monarch. **2** thing added. *–v.* record the addition of (a new item) to a library etc.

accessory /ək'sesərɪ/ *n.* (*pl.* **-ies**) **1** additional or extra thing. **2** (usu. in *pl.*) small attachment, fitting, or subsidiary item of dress (e.g. shoes, gloves). **3** (often foll. by *to*) person who abets or is privy to an (esp. illegal) act. [medieval Latin: related to ACCEDE]

access road *n.* road giving access only to the properties along it.

access time *n.* *Computing* time taken to retrieve data from storage.

accident /'æksɪd(ə)nt/ *n.* **1** unfortunate esp. harmful event, caused unintentionally. **2** event that is unexpected or without apparent cause. □ **by accident** unintentionally. [Latin *cado* fall]

accidental /,æksɪ'dent(ə)l/ *–adj.* happening by chance or accident. *–n.* *Mus.* sign indicating a note's momentary departure from the key signature. □ **accidentally** *adv.*

accident-prone *adj.* clumsy.

acclaim /ə'kleɪm/ *–v.* **1** welcome or applaud enthusiastically. **2** hail as (*acclaimed him king*). *–n.* applause, welcome, public praise. [Latin *acclamo*: related to CLAIM]

acclamation /,æklə'meɪʃ(ə)n/ *n.* **1** loud and eager assent. **2** (usu. in *pl.*) shouting in a person's honour.

acclimatize /ə'klaɪmə,taɪz/ *v.* (also **-ise**) (**-zing** or **-sing**) adapt to a new climate or conditions. □ **acclimatization** /-'zeɪʃ(ə)n/ *n.* [French *acclimater*: related to CLIMATE]

accolade /'ækə,leɪd/ *n.* **1** praise given. **2** touch made with a sword at the conferring of a knighthood. [Latin *collum* neck]

accommodate /ə'kɒmə,deɪt/ *v.* (**-ting**) **1** provide lodging or room for (*flat accommodates two*). **2** adapt, harmonize, reconcile (*must accommodate himself to new ideas*). **3 a** do favour to, oblige (a person). **b** (foll. by *with*) supply (a person) with. [Latin: related to COMMODE]

accommodating *adj.* obliging, compliant.

accommodation /ə,kɒmə'deɪʃ(ə)n/ *n.* **1** lodgings. **2** adjustment, adaptation. **3** convenient arrangement; settlement, compromise.

accommodation address *n.* postal address used by a person unable or unwilling to give a permanent address.

accompaniment /ə'kʌmpənɪmənt/ *n.* **1** instrumental or orchestral support for a solo instrument, voice, or group. **2** accompanying thing. □ **accompanist** *n.* (in sense 1).

accompany /ə'kʌmpənɪ/ *v.* (**-ies, -ied**) **1** go with; escort. **2** (usu. in *passive*; foll. by *with, by*) be done or found with; supplement. **3** *Mus.* partner with accompaniment. [French: related to COMPANION]

accomplice /ə'kʌmplɪs/ *n.* partner in a crime etc. [Latin: related to COMPLEX]

accomplish /ə'kʌmplɪʃ/ *v.* succeed in doing; achieve, complete. [Latin: related to COMPLETE]

accomplished *adj.* clever, skilled.

accomplishment *n.* **1** completion (of a task etc.). **2** acquired, esp. social, skill. **3** thing achieved.

accord /ə'kɔ:d/ *—v.* **1** (often foll. by *with*) be consistent or in harmony. **2** grant (permission, a request, etc.); give (a welcome etc.). *—n.* **1** agreement, consent. **2** *Mus.* & *Art* etc. harmony. □ **of one's own accord** on one's own initiative; voluntarily. **with one accord** unanimously. [Latin *cor cord-* heart]

accordance *n.* □ **in accordance with** in conformity to. □ **accordant** *adj.*

according *adv.* **1** (foll. by *to*) **a** as stated by (*according to Mary*). **b** in proportion to (*lives according to his means*). **2** (foll. by *as* + clause) in a manner or to a degree that varies as (*pays according as he is able*).

accordingly *adv.* **1** as circumstances suggest or require (*please act accordingly*). **2** consequently (*accordingly, he left the room*).

accordion /ə'kɔ:dɪən/ *n.* musical reed instrument with concertina-like bellows, keys, and buttons. □ **accordionist** *n.* [Italian *accordare* to tune]

accost /ə'kɒst/ *v.* **1** approach and address (a person), esp. boldly. **2** (of a prostitute) solicit. [Latin *costa* rib]

account /ə'kaʊnt/ *—n.* **1** narration, description (*an account of his trip*). **2** arrangement at a bank etc. for depositing and withdrawing money, credit, etc. (*open an account*). **3** record or statement of financial transactions with the balance (*kept detailed accounts*). *—v.* consider as (*account him wise, a fool*). □ **account for 1** serve as or provide an explanation for (*that accounts for his mood*). **2** answer for (money etc. entrusted, one's conduct, etc.). **3** kill, destroy, defeat. **4** make up a specified amount of (*rent accounts for 50% of expenditure*). **by all accounts** in everyone's opinion. **call to account** require an explanation from. **give a good** (or **bad**) **account of oneself** impress (or fail to impress); be successful (or unsuccessful). **keep account of** keep a record of; follow closely. **of no account** unimportant. **of some account** important. **on account 1** (of goods) to be paid for later. **2** (of money) in part payment. **on one's account** on one's behalf (*not on my account*). **on account of** because of. **on no account** under no circumstances. **take account of** (or **take into account**) consider (*took their age into account*). **turn to account** (or **good account**) turn to one's advantage. [French: related to COUNT[1]]

accountable *adj.* **1** responsible; required to account for one's conduct. **2** explicable, understandable. □ **accountability** /-'bɪlɪtɪ/ *n.*

accountant *n.* professional keeper or verifier of accounts. □ **accountancy** *n.* **accounting** *n.*

accoutrements /ə'ku:trəmənts/ *n.pl.* (*US* **accouterments** /-təmənts/) **1** equipment, trappings. **2** soldier's equipment excluding weapons and clothes. [French]

accredit /ə'kredɪt/ *v.* (**-t-**) **1** (foll. by *to*) attribute (a saying etc.) to (a person). **2** (foll. by *with*) credit (a person) with (a saying etc.). **3** (usu. foll. by *to* or *at*) send (an ambassador etc.) with credentials. **4** gain influence for or make credible (an adviser, a statement, etc.). [French: related to CREDIT]

accredited *adj.* **1** officially recognized. **2** generally accepted.

accretion /ə'kri:ʃ(ə)n/ *n.* **1** growth or increase by accumulation, addition, or organic enlargement. **2** the resulting whole. **3 a** matter so added. **b** adhesion of this to the core matter. [Latin *cresco cret-* grow]

accrue /ə'kru:/ *v.* (**-ues, -ued, -uing**) (often foll. by *to*) come as a natural increase or advantage, esp. financial. [Latin: related to ACCRETION]

accumulate /ə'kju:mjʊ,leɪt/ *v.* (**-ting**) **1** acquire an increasing number or quantity of; amass, collect. **2** grow numerous; increase. [Latin: related to CUMULUS]

accumulation /ə,kju:mjʊ'leɪʃ(ə)n/ *n.* **1** accumulating or being accumulated. **2** accumulated mass. **3** growth of capital by continued interest. □ **accumulative** /ə'kju:mjʊlətɪv/ *adj.*

accumulator *n.* **1** rechargeable electric cell. **2** bet placed on a sequence of events, with the winnings and stake from each placed on the next.

accuracy /'ækjʊrəsɪ/ *n.* exactness or careful precision. [Latin *cura* care]

accurate /'ækjʊrət/ *adj.* careful, precise; conforming exactly with the truth or a standard. □ **accurately** *adv.*

accursed /ə'kɜ:sɪd/ *adj.* **1** under a curse. **2** *colloq.* detestable, annoying. [Old English *a-* intensive prefix, CURSE]

accusation /,ækju:'zeɪʃ(ə)n/ *n.* accusing or being accused. [French: related to ACCUSE]

accusative /ə'kju:zətɪv/ *Gram.* *–n.* case expressing the object of an action. *–adj.* of or in this case.

accusatory /ə'kju:zətərɪ/ *adj.* of or implying accusation.

accuse /ə'kju:z/ *v.* (**-sing**) (often foll. by *of*) charge with a fault or crime; blame. [Latin *accusare*: related to CAUSE]

accustom /ə'kʌstəm/ *v.* (foll. by *to*) make used to (*accustomed him to hardship*). [French: related to CUSTOM]

accustomed *adj.* **1** (usu. foll. by *to*) used to a thing. **2** customary, usual.

ace *–n.* **1** playing-card etc. with a single spot and generally signifying 'one'. **2 a** person who excels in some activity. **b** pilot who has shot down many enemy aircraft. **3** (in tennis) unreturnable stroke (esp. a service). *–adj. slang* excellent. □ **within an ace of** on the verge of. [Latin *as* unity]

acellular /eɪ'seljʊlə(r)/ *adj.* having no cells; not consisting of cells.

-aceous *suffix* forming adjectives in the sense 'of the nature of', esp. in the natural sciences (*herbaceous*). [Latin *-aceus*]

acerbic /ə'sɜ:bɪk/ *adj.* harsh and sharp, esp. in speech or manner. □ **acerbity** *n.* (*pl.* **-ies**). [Latin *acerbus* sour]

acetaldehyde /,æsɪ'tældɪ,haɪd/ *n.* colourless volatile liquid aldehyde. [from ACETIC, ALDEHYDE]

acetate /'æsɪ,teɪt/ *n.* **1** salt or ester of acetic acid, esp. the cellulose ester. **2** fabric made from this.

acetic /ə'si:tɪk/ *adj.* of or like vinegar. [Latin *acetum* vinegar]

acetic acid *n.* clear liquid acid giving vinegar its characteristic taste.

acetone /'æsɪ,təʊn/ *n.* colourless volatile liquid that dissolves organic compounds, esp. paints, varnishes, etc.

acetylene /ə'setɪ,li:n/ *n.* hydrocarbon gas burning with a bright flame, used esp. in welding.

ache /eɪk/ *–n.* **1** continuous dull pain. **2** mental distress. *–v.* (**-ching**) suffer from or be the source of an ache. [Old English]

achieve /ə'tʃi:v/ *v.* (**-ving**) **1** reach or attain, esp. by effort (*achieved victory; achieved notoriety*). **2** accomplish (a feat or task). [French *achever*: related to CHIEF]

achievement *n.* **1** something achieved. **2** act of achieving.

Achilles' heel /ə'kɪli:z/ *n.* person's weak or vulnerable point. [*Achilles*, Greek hero in the *Iliad*]

Achilles' tendon *n.* tendon connecting the heel with the calf muscles.

achromatic /,ækrəʊ'mætɪk/ *adj. Optics* **1** transmitting light without separation into constituent colours (*achromatic lens*). **2** without colour. □ **achromatically** *adv.* [French: related to A-, CHROME]

achy /'eɪkɪ/ *adj.* (**-ier, -iest**) full of or suffering from aches.

acid /'æsɪd/ *–n.* **1 a** any of a class of substances that liberate hydrogen ions in water, are usu. sour and corrosive, turn litmus red, and have a pH of less than 7. **b** any compound or atom donating protons. **2** any sour substance. **3** *slang* the drug LSD. *–adj.* **1** sour. **2** biting, sharp (*an acid wit*). **3** *Chem.* having the essential properties of an acid. □ **acidic** /ə'sɪdɪk/ *adj.* **acidify** /ə'sɪdɪ,faɪ/ *v.* (**-ies, -ied**). **acidity** /ə'sɪdɪtɪ/ *n.* **acidly** *adv.* [Latin *aceo* be sour]

acid house *n.* a type of synthesized music with a simple repetitive beat, often associated with hallucinogenic drugs.

acid rain *n.* acid, esp. from industrial waste gases, falling with rain.

acid test *n.* severe or conclusive test.

acidulous /ə'sɪdjʊləs/ *adj.* somewhat acid.

ack-ack /æ'kæk/ *colloq.* *–adj.* anti-aircraft. *–n.* anti-aircraft gun etc. [formerly signallers' term for *AA*]

acknowledge /ək'nɒlɪdʒ/ *v.* (**-ging**) **1** recognize; accept the truth of (*acknowledged its failure*). **2** confirm the receipt of (a letter etc.). **3 a** show that one has noticed (*acknowledged my arrival with* *a grunt*). **b** express appreciation of (a service etc.). **4** recognize the validity of, own (*the acknowledged king*). [from AD-, KNOWLEDGE]

acknowledgement *n.* **1** act of acknowledging. **2 a** thing given or done in gratitude. **b** letter confirming receipt of something. **3** (usu. in *pl.*) author's statement of gratitude, prefacing a book.

acme /'ækmɪ/ *n.* highest point (of achievement etc.). [Greek]

acne /'æknɪ/ *n.* skin condition with red pimples. [Latin]

acolyte /'ækə,laɪt/ *n.* **1** person assisting a priest. **2** assistant; beginner. [Greek *akolouthos* follower]

aconite /'ækə,naɪt/ *n.* **1** any of various poisonous plants, esp. monkshood. **2** drug from these. [Greek *akoniton*]

acorn /'eɪkɔːn/ *n.* fruit of the oak, with a smooth nut in a cuplike base. [Old English]

acoustic /ə'kuːstɪk/ *adj.* **1** of sound or the sense of hearing. **2** (of a musical instrument etc.) without electrical amplification (*acoustic guitar*). □ **acoustically** *adv.* [Greek *akouō* hear]

acoustics *n.pl.* **1** properties or qualities (of a room etc.) in transmitting sound. **2** (usu. as *sing.*) science of sound.

acquaint /ə'kweɪnt/ *v.* (usu. foll. by *with*) make aware of or familiar with (*acquaint me with the facts*). □ **be acquainted with** have personal knowledge of; know slightly. [Latin: related to AD-, COGNIZANCE]

acquaintance *n.* **1** being acquainted. **2** person one knows slightly. □ **acquaintanceship** *n.*

acquiesce /,ækwɪ'es/ *v.* (**-cing**) **1** agree, esp. by default. **2** (foll. by *in*) accept (an arrangement etc.). □ **acquiescence** *n.* **acquiescent** *adj.* [Latin: related to AD-, QUIET]

acquire /ə'kwaɪə(r)/ *v.* (**-ring**) gain for oneself; come into possession of. [Latin: related to AD-, *quaero quisit-* seek]

acquired immune deficiency syndrome see AIDS.

acquired taste *n.* **1** liking developed by experience. **2** object of this.

acquirement *n.* thing acquired, esp. a mental attainment.

acquisition /,ækwɪ'zɪʃ(ə)n/ *n.* **1** thing acquired, esp. when useful. **2** acquiring or being acquired. [Latin: related to ACQUIRE]

acquisitive /ə'kwɪzɪtɪv/ *adj.* keen to acquire things.

acquit /ə'kwɪt/ *v.* (**-tt-**) **1** (often foll. by *of*) declare not guilty. **2** *refl.* **a** behave or perform in a specified way (*acquitted herself well*). **b** (foll. by *of*) discharge (a duty or responsibility). □ **acquittal** *n.* [Latin: related to AD-, QUIT]

acre /'eɪkə(r)/ *n.* measure of land, 4,840 sq. yds., 0.405 ha. [Old English]

acreage /'eɪkərɪdʒ/ *n.* a number of acres; extent of land.

acrid /'ækrɪd/ *adj.* (**-er**, **-est**) bitterly pungent. □ **acridity** /ə'krɪdɪtɪ/ *n.* [Latin *acer* keen, pungent]

acrimonious /,ækrɪ'məʊnɪəs/ *adj.* bitter in manner or temper. □ **acrimony** /'ækrɪmənɪ/ *n.*

acrobat /'ækrə,bæt/ *n.* entertainer performing gymnastic feats. □ **acrobatic** /,ækrə'bætɪk/ *adj.* **acrobatically** /,ækrə'bætɪkəlɪ/ *adv.* [Greek *akrobatēs* from *akron* summit, *bainō* walk]

acrobatics /,ækrə'bætɪks/ *n.pl.* **1** acrobatic feats. **2** (as *sing.*) art of performing these.

acronym /'ækrənɪm/ *n.* word formed from the initial letters of other words (e.g. *laser*, *Nato*). [Greek *akron* end, *onoma* name]

acropolis /ə'krɒpəlɪs/ *n.* citadel of an ancient Greek city. [Greek *akron* summit, *polis* city]

across /ə'krɒs/ *—prep.* **1** to or on the other side of (*across the river*). **2** from one side to another side of (*spread across the floor*). **3** at or forming an angle with (*a stripe across the flag*). *—adv.* **1** to or on the other side (*ran across*). **2** from one side to another (*stretched across*). □ **across the board** applying to all. [French *à*, *en*, *croix*: related to CROSS]

acrostic /ə'krɒstɪk/ *n.* poem etc. in which certain letters (usu. the first and last in each line) form a word or words. [Greek *akron* end, *stikhos* row]

acrylic /ə'krɪlɪk/ *—adj.* of synthetic material made from acrylic acid. *—n.* acrylic fibre or fabric. [Latin *acer* pungent, *oleo* to smell]

acrylic acid *n.* a pungent liquid organic acid.

act *—n.* **1** something done; a deed. **2** process of doing (*caught in the act*). **3** item of entertainment. **4** pretence (*all an act*). **5** main division of a play etc. **6 a** decree of a legislative body. **b** document attesting a legal transaction. *—v.* **1** behave (*acted wisely*). **2** perform an action or function; take action (*act as referee*; *brakes failed to act*; *he acted quickly*). **3** (also foll. by *on*) have an effect (*alcohol acts on the brain*). **4 a** perform a part in a play, film, etc. **b** pretend. **5 a** play the part of (*acted Othello*; *acts the fool*). **b** perform (a play etc.). **c** portray (an incident) by actions. □ **act for** be the (esp. legal) representative of. **act of God** natural event, e.g. an earthquake. **act**

up *colloq.* misbehave; give trouble (*car is acting up*). **get one's act together** *slang* become properly organized; prepare. **put on an act** *colloq.* make a pretence. [Latin *ago act-* do]

acting –*n.* art or occupation of an actor. –*attrib. adj.* serving temporarily or as a substitute (*acting manager*).

actinism /'æktɪˌnɪz(ə)m/ *n.* property of short-wave radiation that produces chemical changes, as in photography. [Greek *aktis* ray]

actinium /æk'tɪnɪəm/ *n. Chem.* radioactive metallic element found in pitchblende. [as ACTINISM]

action /'ækʃ(ə)n/ *n.* **1** process of doing or acting (*demanded action*). **2** forcefulness or energy. **3** exertion of energy or influence (*action of acid on metal*). **4** deed, act (*not aware of his actions*). **5** (**the action**) **a** series of events in a story, play, etc. **b** *slang* exciting activity (*missed the action*). **6** battle, fighting (*killed in action*). **7 a** mechanism of an instrument. **b** style of movement of an animal or human. **8** lawsuit. □ **out of action** not working. [Latin: related to ACT]

actionable *adj.* giving cause for legal action.

action-packed *adj.* full of action or excitement.

action point *n.* proposal for action.

action replay *n.* playback of part of a television broadcast, esp. a sporting event, often in slow motion.

action stations *n.pl.* positions taken up by troops etc. ready for battle.

activate /'æktɪˌveɪt/ *v.* (**-ting**) **1** make active. **2** *Chem.* cause reaction in. **3** make radioactive.

active /'æktɪv/ –*adj.* **1** marked by action; energetic; diligent (*an active life*). **2** working, operative (*active volcano*). **3** not merely passive or inert; positive (*active support*; *active ingredients*). **4** radioactive. **5** *Gram.* designating the form of a verb whose subject performs the action (e.g. *saw* in *he saw a film*). –*n. Gram.* active form or voice of a verb. □ **actively** *adv.* [Latin: related to ACT]

active service *n.* military service in wartime.

activism *n.* policy of vigorous action, esp. for a political cause. □ **activist** *n.*

activity /æk'tɪvɪtɪ/ *n.* (*pl.* **-ies**) **1** being active; busy or energetic action. **2** (often in *pl.*) occupation or pursuit (*outdoor activities*). **3** = RADIOACTIVITY.

actor *n.* person who acts in a play, film, etc. [Latin: related to ACT]

actress *n.* female actor.

actual /'æktʃʊəl/ *adj.* (usu. *attrib.*) **1** existing in fact; real. **2** current. [Latin: related to ACT]

actuality /ˌæktʃʊ'ælɪtɪ/ *n.* (*pl.* **-ies**) **1** reality. **2** (in *pl.*) existing conditions.

actually /'æktʃʊəlɪ/ *adv.* **1** as a fact, really (*not actually very rich*). **2** strange as it may seem (*he actually refused!*).

actuary /'æktʃʊərɪ/ *n.* (*pl.* **-ies**) statistician, esp. one calculating insurance risks and premiums. □ **actuarial** /-'eərɪəl/ *adj.* [Latin *actuarius* bookkeeper]

actuate /'æktʃʊˌeɪt/ *v.* (**-ting**) **1** cause (a machine etc.) to move or function. **2** cause (a person) to act. [Latin]

acuity /ə'kju:ɪtɪ/ *n.* sharpness, acuteness. [medieval Latin: related to ACUTE]

acumen /'ækjʊmən/ *n.* keen insight or discernment. [Latin, = ACUTE thing]

acupuncture /'ækju:ˌpʌŋktʃə(r)/ *n.* medical treatment using needles in parts of the body. □ **acupuncturist** *n.* [Latin *acu* with needle]

acute /ə'kju:t/ –*adj.* (**acuter, acutest**) **1** serious, severe (*acute hardship*). **2** (of senses etc.) keen, penetrating. **3** shrewd. **4** (of a disease) coming quickly to a crisis. **5** (of an angle) less than 90°. **6** (of a sound) high, shrill. –*n.* = ACUTE ACCENT. □ **acutely** *adv.* [Latin *acutus* pointed]

acute accent *n.* diacritical mark (´) placed over certain letters in French etc., esp. to show pronunciation.

-acy *suffix* forming nouns of state or quality (*accuracy*; *piracy*), or an instance of it (*conspiracy*; *fallacy*). [French *-acie*, Latin *-acia*, *-atia*, Greek *-ateia*]

AD *abbr.* of the Christian era. [ANNO DOMINI]

ad *n. colloq.* advertisement. [abbreviation]

ad- *prefix* (altered or assimilated before some letters) implying motion or direction to, reduction or change into, addition, adherence, increase, or intensification. [Latin]

adage /'ædɪdʒ/ *n.* traditional maxim, proverb. [French from Latin]

adagio /ə'dɑ:ʒɪəʊ/ *Mus.* –*adv.* & *adj.* in slow time. –*n.* (*pl.* **-s**) such a movement or passage. [Italian]

Adam /'ædəm/ *n.* the first man. □ **not know a person from Adam** be unable to recognize a person. [Hebrew, = man]

adamant /'ædəmənt/ *adj.* stubbornly resolute; unyielding. □ **adamantly** *adv.* [Greek *adamas adamant-* untameable]

Adam's apple *n.* projection of cartilage at the front of the neck.

adapt /ə'dæpt/ *v.* **1 a** (foll. by *to*) fit, adjust (one thing to another). **b** (foll. by *to, for*) make suitable for a purpose. **c** modify (esp. a text for broadcasting etc.). **2** (also *refl.*, usu. foll. by *to*) adjust to new conditions. □ **adaptable** *adj.* **adaptation** /ˌædæp'teɪʃ(ə)n/ *n.* [Latin: related to AD-, APT]

adaptor *n.* **1** device for making equipment compatible. **2** device for connecting several electrical plugs to one socket.

add *v.* **1** join (one thing to another) as an increase or supplement. **2** put together (numbers) to find their total. **3** say further. □ **add in** include. **add up 1** find the total of. **2** (foll. by *to*) amount to. **3** *colloq.* make sense. [Latin *addo*]

addendum /ə'dendəm/ *n.* (*pl.* **-da**) **1** thing to be added. **2** material added at the end of a book.

adder *n.* small venomous snake, esp. the common viper. [Old English, originally *nadder*]

addict /'ædɪkt/ *n.* **1** person addicted, esp. to a drug. **2** *colloq.* devotee (*film addict*). [Latin: related to AD-, *dico* say]

addicted /ə'dɪktɪd/ *adj.* **1** (usu. foll. by *to*) dependent on a drug etc. as a habit. **2** devoted to an interest. □ **addiction** /ə'dɪkʃ(ə)n/ *n.*

addictive *adj.* causing addiction.

addition /ə'dɪʃ(ə)n/ *n.* **1** adding. **2** person or thing added. □ **in addition** (often foll. by *to*) also, as well (as). [Latin: related to ADD]

additional *adj.* added, extra, supplementary. □ **additionally** *adv.*

additive /'ædɪtɪv/ *n.* substance added to improve another, esp. to colour, flavour, or preserve food. [Latin: related to ADD]

addle /'æd(ə)l/ *v.* (**-ling**) **1** muddle, confuse. **2** (usu. as **addled** *adj.*) (of an egg) become rotten. [Old English, = filth]

address /ə'dres/ **–***n.* **1 a** place where a person lives or an organization is situated. **b** particulars of this, esp. for postal purposes. **c** *Computing* location of an item of stored information. **2** discourse to an audience. **–***v.* **1** write postal directions on (an envelope etc.). **2** direct (remarks etc.). **3** speak or write to, esp. formally. **4** direct one's attention to. **5** *Golf* take aim at (the ball). □ **address oneself to 1** speak or write to. **2** attend to. [French: related to AD-, DIRECT]

addressee /ˌædre'si:/ *n.* person to whom a letter etc. is addressed.

adduce /ə'dju:s/ *v.* (**-cing**) cite as an instance or as proof or evidence. □ **adducible** *adj.* [Latin: related to AD-, *duco* lead]

adenoids /'ædɪˌnɔɪdz/ *n.pl.* area of enlarged lymphatic tissue between the nose and the throat, often hindering breathing in the young. □ **adenoidal** /-'nɔɪd(ə)l/ *adj.* [Greek *adēn* gland]

adept /'ædept/ **–***adj.* (foll. by *at, in*) skilful. **–***n.* adept person. [Latin *adipiscor adept-* attain]

adequate /'ædɪkwət/ *adj.* sufficient, satisfactory. □ **adequacy** *n.* **adequately** *adv.* [Latin: related to AD-, EQUATE]

à deux /ɑ: 'dɜ:/ *adv.* & *adj.* for or between two. [French]

adhere /əd'hɪə(r)/ *v.* (**-ring**) **1** (usu. foll. by *to*) stick fast to a substance etc. **2** (foll. by *to*) behave according to (a rule, undertaking, etc.). **3** (foll. by *to*) give allegiance. [Latin *haereo* stick]

adherent **–***n.* supporter. **–***adj.* sticking, adhering. □ **adherence** *n.*

adhesion /əd'hi:ʒ(ə)n/ *n.* **1** adhering. **2** unnatural union of body tissues due to inflammation.

adhesive /əd'hi:sɪv/ **–***adj.* sticky, causing adhesion. **–***n.* adhesive substance. □ **adhesiveness** *n.*

ad hoc /æd 'hɒk/ *adv.* & *adj.* for one particular occasion or use. [Latin]

adieu /ə'dju:/ *int.* goodbye. [French, = to God]

ad infinitum /æd ˌɪnfɪ'naɪtəm/ *adv.* without limit; for ever. [Latin]

adipose /'ædɪˌpəʊz/ *adj.* of fat; fatty (*adipose tissue*). □ **adiposity** /-'pɒsɪtɪ/ *n.* [Latin *adeps* fat]

adjacent /ə'dʒeɪs(ə)nt/ *adj.* (often foll. by *to*) lying near; adjoining. □ **adjacency** *n.* [Latin *jaceo* lie]

adjective /'ædʒɪktɪv/ *n.* word used to describe or modify a noun or pronoun. □ **adjectival** /ˌædʒɪk'taɪv(ə)l/ *adj.* [Latin *jaceo* lie]

adjoin /ə'dʒɔɪn/ *v.* be next to and joined with. [Latin *jungo* join]

adjourn /ə'dʒɜ:n/ *v.* **1** put off, postpone; break off (a meeting etc.) temporarily. **2** (of a meeting) break and disperse or (foll. by *to*) transfer to another place (*adjourned to the pub*). □ **adjournment** *n.* [Latin: related to AD-, *diurnum* day]

adjudge /ə'dʒʌdʒ/ *v.* (**-ging**) **1** pronounce judgement on (a matter). **2** pronounce or award judicially. □ **adjudgement** *n.* (also **adjudgment**). [Latin *judex* judge]

adjudicate /ə'dʒu:dɪˌkeɪt/ *v.* (**-ting**) **1** act as judge in a competition, court, etc. **2** adjudge. □ **adjudication** /-'keɪʃ(ə)n/ *n.* **adjudicative** *adj.* **adjudicator** *n.*

adjunct /'ædʒʌŋkt/ *n.* **1** (foll. by *to, of*) subordinate or incidental thing. **2** *Gram.* word or phrase used to explain or amplify the predicate, subject, etc. [Latin: related to ADJOIN]

adjure /ə'dʒʊə(r)/ *v.* **(-ring)** (usu. foll. by *to* + infin.) beg or command. □ **adjuration** /,ædʒʊə'reɪʃ(ə)n/ *n.* [Latin *adjuro* put to oath: related to JURY]

adjust /ə'dʒʌst/ *v.* **1** order or position; regulate; arrange. **2** (usu. foll. by *to*) become or make suited; adapt. **3** harmonize (discrepancies). **4** assess (loss or damages). □ **adjustable** *adj.* **adjustment** *n.* [Latin *juxta* near]

adjutant /'ædʒʊt(ə)nt/ *n.* **1 a** army officer assisting a superior in administrative duties. **b** assistant. **2** (in full **adjutant bird**) giant Indian stork. [Latin: related to AD-, *juvo jut-* help]

ad lib /æd 'lɪb/ *–v.* **(-bb-)** improvise. *–adj.* improvised. *–adv.* as one pleases, to any desired extent. [abbreviation of Latin *ad libitum* according to pleasure]

admin /'ædmɪn/ *n. colloq.* administration. [abbreviation]

administer /əd'mɪnɪstə(r)/ *v.* **1** manage (business affairs etc.). **2 a** deliver or dispense, esp. formally (a punishment, sacrament, etc.). **b** (usu. foll. by *to*) direct the taking of (an oath). [Latin: related to AD-, MINISTER]

administrate /əd'mɪnɪ,streɪt/ *v.* **(-ting)** administer (esp. business affairs); act as an administrator.

administration /əd,mɪnɪ'streɪʃ(ə)n/ *n.* **1** administering, esp. public affairs. **2** government in power.

administrative /əd'mɪnɪstrətɪv/ *adj.* of the management of affairs.

administrator /əd'mɪnɪ,streɪtə(r)/ *n.* manager of a business, public affairs, or a person's estate.

admirable /'ædmərəb(ə)l/ *adj.* deserving admiration; excellent. □ **admirably** *adv.* [Latin: related to ADMIRE]

admiral /'ædmər(ə)l/ *n.* **1 a** commander-in-chief of a navy. **b** high-ranking naval officer, commander. **2** any of various butterflies. [Arabic: related to AMIR]

Admiralty *n.* (*pl.* **-ies**) (in full **Admiralty Board**) *hist.* committee superintending the Royal Navy.

admiration /,ædmə'reɪʃ(ə)n/ *n.* **1** respect; warm approval or pleasure. **2** object of this.

admire /əd'maɪə(r)/ *v.* **(-ring) 1** regard with approval, respect, or satisfaction. **2** express admiration of. □ **admirer** *n.* **admiring** *adj.* **admiringly** *adv.* [Latin: related to AD-, *miror* wonder at]

admissible /əd'mɪsɪb(ə)l/ *adj.* **1** (of an idea etc.) worth accepting or considering. **2** *Law* allowable as evidence. [Latin: related to ADMIT]

admission /əd'mɪʃ(ə)n/ *n.* **1** acknowledgement (*admission of error*). **2 a** process or right of entering. **b** charge for this (*admission is £5*).

admit /əd'mɪt/ *v.* **(-tt-) 1** (often foll. by *to be*, or *that* + clause) acknowledge; recognize as true. **2** (foll. by *to*) confess to (a deed, fault, etc.). **3** allow (a person) entrance, access, etc. **4** take (a patient) into hospital. **5** (of an enclosed space) accommodate. **6** (foll. by *of*) allow as possible. [Latin *mitto miss-* send]

admittance *n.* admitting or being admitted, usu. to a place.

admittedly *adv.* as must be admitted.

admixture /æd'mɪkstʃə(r)/ *n.* **1** thing added, esp. a minor ingredient. **2** adding of this.

admonish /əd'mɒnɪʃ/ *v.* **1** reprove. **2** urge, advise. **3** (foll. by *of*) warn. □ **admonishment** *n.* **admonition** /,ædmə'nɪʃ(ə)n/ *n.* **admonitory** *adj.* [Latin *moneo* warn]

ad nauseam /æd 'nɔ:zɪ,æm/ *adv.* excessively; disgustingly. [Latin, = to sickness]

ado /ə'du:/ *n.* fuss, busy activity; trouble. [from AT, DO[1]: originally in *much ado* = much to do]

adobe /ə'dəʊbɪ/ *n.* **1** sun-dried brick. **2** clay for making these. [Spanish]

adolescent /,ædə'les(ə)nt/ *–adj.* between childhood and adulthood. *–n.* adolescent person. □ **adolescence** *n.* [Latin *adolesco* grow up]

Adonis /ə'dəʊnɪs/ *n.* handsome young man. [Latin, name of a youth loved by Venus]

adopt /ə'dɒpt/ *v.* **1** legally take (a person) into a relationship, esp. another's child as one's own. **2** choose (a course of action etc.). **3** take over (another's idea etc.). **4** choose as a candidate for office. **5** accept responsibility for the maintenance of (a road etc.). **6** accept or approve (a report, accounts, etc.). □ **adoption** *n.* [Latin: related to AD-, OPT]

adoptive *adj.* because of adoption (*adoptive son*). [Latin: related to ADOPT]

adorable /ə'dɔ:rəb(ə)l/ *adj.* **1** deserving adoration. **2** *colloq.* delightful, charming.

adore /ə'dɔ:(r)/ *v.* **(-ring) 1** love intensely. **2** worship as divine. **3** *colloq.* like very much. □ **adoration** /,ædə'reɪʃ(ə)n/ *n.* **adorer** *n.* [Latin *adoro* worship]

adorn /ə'dɔ:n/ *v.* add beauty to; decorate. □ **adornment** *n.* [Latin: related to AD-, *orno* decorate]

adrenal /ə'dri:n(ə)l/ *–adj.* **1** at or near the kidneys. **2** of the adrenal glands. *–n.* (in full **adrenal gland**) either of two ductless glands above the kidneys, secreting adrenalin. [from AD-, RENAL]

adrenalin /ə'drenəlɪn/ *n.* (also **adrenaline**) **1** stimulative hormone secreted by the adrenal glands. **2** this extracted or synthesized for medicinal use.

adrift /ə'drɪft/ *adv.* & *predic.adj.* **1** drifting. **2** powerless; aimless. **3** *colloq.* **a** unfastened. **b** out of order, wrong (*plans went adrift*).

adroit /ə'drɔɪt/ *adj.* dexterous, skilful. [French *à droit* according to right]

adsorb /əd'sɔ:b/ *v.* (usu. of a solid) hold (molecules of a gas or liquid etc.) to its surface, forming a thin film. □ **adsorbent** *adj.* & *n.* **adsorption** *n.* [from AD-, ABSORB]

adulation /,ædjʊ'leɪʃ(ə)n/ *n.* obsequious flattery. [Latin *adulor* fawn on]

adult /'ædʌlt/ *–adj.* **1** mature, grown-up. **2** (*attrib.*) of or for adults (*adult education*). *–n.* adult person. □ **adulthood** *n.* [Latin *adolesco adultus* grow up]

adulterate /ə'dʌltə,reɪt/ *v.* (**-ting**) debase (esp. foods) by adding other substances. □ **adulterant** *adj.* & *n.* **adulteration** /-'reɪʃ(ə)n/ *n.* [Latin *adultero* corrupt]

adulterer /ə'dʌltərə(r)/ *n.* (*fem.* **adulteress**) person who commits adultery.

adultery *n.* voluntary sexual intercourse between a married person and a person other than his or her spouse. □ **adulterous** *adj.*

adumbrate /'ædʌm,breɪt/ *v.* (**-ting**) **1** indicate faintly or in outline. **2** foreshadow. **3** overshadow. □ **adumbration** /-'breɪʃ(ə)n/ *n.* [Latin: related to AD-, *umbra* shade]

advance /əd'vɑ:ns/ *–v.* (**-cing**) **1** move or put forward; progress. **2** pay or lend (money) beforehand. **3** promote (a person, cause, etc.). **4** present (a suggestion etc.). **5** (as **advanced** *adj.*) **a** well ahead. **b** socially progressive. *–n.* **1** going forward; progress. **2** prepayment; loan. **3** (in *pl.*) amorous approaches. **4** rise in price. *–attrib. adj.* done or supplied beforehand (*advance warning*). □ **advance on** approach threateningly. **in advance** ahead in place or time. [Latin: related to AB-, *ante* before]

advanced level *n.* high level of GCE examination.

advancement *n.* promotion of a person, cause, or plan.

advantage /əd'vɑ:ntɪdʒ/ *–n.* **1** beneficial feature. **2** benefit, profit. **3** (often foll. by *over*) superiority. **4** (in tennis) the next point after deuce. *–v.* (**-ging**) benefit, favour. □ **take advantage of 1** make good use of. **2** exploit, esp. unfairly. **3** *euphem.* seduce. □ **advantageous** /,ædvən'teɪdʒəs/ *adj.* [French: related to ADVANCE]

Advent /'ædvent/ *n.* **1** season before Christmas. **2** coming of Christ. **3** (**advent**) important arrival. [Latin *adventus* from *venio* come]

Adventist *n.* member of a Christian sect believing in the imminent second coming of Christ.

adventitious /,ædven'tɪʃəs/ *adj.* **1** accidental, casual. **2** added from outside. **3** *Biol.* formed accidentally or under unusual conditions. [Latin: related to ADVENT]

adventure /əd'ventʃə(r)/ *–n.* **1** unusual and exciting experience. **2** enterprise (*spirit of adventure*). *–v.* (**-ring**) dare, venture; engage in adventure. [Latin: related to ADVENT]

adventure playground *n.* playground with climbing-frames, building blocks, etc.

adventurer *n.* (*fem.* **adventuress**) **1** person who seeks adventure, esp. for personal gain or enjoyment. **2** financial speculator.

adventurous *adj.* venturesome, enterprising.

adverb /'ædvɜ:b/ *n.* word indicating manner, degree, circumstance, etc., used to modify an adjective, verb, or other adverb (e.g. *gently*, *quite*, *then*). □ **adverbial** /əd'vɜ:bɪəl/ *adj.* [Latin: related to AD-, *verbum* word, VERB]

adversary /'ædvəsərɪ/ *n.* (*pl.* **-ies**) enemy, opponent. □ **adversarial** /-'seərɪəl/ *adj.*

adverse /'ædvɜ:s/ *adj.* unfavourable; harmful. □ **adversely** *adv.* [Latin: related to AD-, *verto vers-* turn]

adversity /əd'vɜ:sɪtɪ/ *n.* misfortune, distress.

advert /'ædvɜ:t/ *n. colloq.* advertisement. [abbreviation]

advertise /'ædvə,taɪz/ *v.* (**-sing**) **1** promote (goods or services) publicly to increase sales. **2** make generally known. **3** (often foll. by *for*) seek by a notice in a newspaper etc. to buy, employ, sell, etc. [French *avertir*: related to ADVERSE]

advertisement /əd'vɜ:tɪsmənt/ *n.* **1** public announcement, esp. of goods etc. for sale or wanted, vacancies, etc. **2** advertising. [French *avertissement*: related to ADVERSE]

advice /əd'vaɪs/ *n.* **1** recommendation on how to act. **2** information given; news. **3** formal notice of a transaction.

advisable /əd'vaɪzəb(ə)l/ *adj.* to be recommended, expedient. □ **advisability** /-'bɪlɪtɪ/ *n.*

advise /əd'vaɪz/ *v.* (**-sing**) **1** (also *absol.*) give advice to. **2** recommend (*advised me to rest*). **3** (usu. foll. by *of*, or *that* + clause) inform. [Latin: related to AD-, *video vis-* see]

advisedly /əd'vaɪzɪdlɪ/ *adv.* after due consideration; deliberately.

adviser *n.* (also **advisor**) person who advises, esp. officially.

■ **Usage** The variant *advisor* is fairly common, but is considered incorrect by many people.

advisory *adj.* giving advice (*advisory body*).
advocaat /,ædvə'kɑ:t/ *n.* liqueur of eggs, sugar, and brandy. [Dutch, = ADVOCATE]
advocacy /'ædvəkəsɪ/ *n.* support or argument for a cause, policy, etc.
advocate –*n.* /'ædvəkət/ **1** (foll. by *of*) person who supports or speaks in favour. **2** person who pleads for another, esp. in a lawcourt. –*v.* /'ædvə,keɪt/ (**-ting**) recommend by argument. [Latin: related to AD-, *voco* call]
adze /ædz/ *n.* (*US* **adz**) tool like an axe, with an arched blade at right angles to the handle. [Old English]
aegis /'i:dʒɪs/ *n.* protection; support. [Greek *aigis* shield of Zeus or Athene]
aeolian harp /i:'əʊlɪən/ *n.* stringed instrument or toy sounding when the wind passes through it. [Latin *Aeolus* wind-god, from Greek]
aeon /'i:ɒn/ *n.* (also **eon**) **1** long or indefinite period. **2** an age. [Latin from Greek]
aerate /'eəreɪt/ *v.* (**-ting**) **1** charge (a liquid) with carbon dioxide. **2** expose to air. □ **aeration** /-'reɪʃ(ə)n/ *n.* [Latin *aer* AIR]
aerial /'eərɪəl/ –*n.* device for transmitting or receiving radio waves. –*adj.* **1** by or from the air; involving aircraft (*aerial attack*). **2** existing in the air. **3** of or like air. [Greek: related to AIR]
aero- *comb. form* air; aircraft. [Greek *aero-* from *aēr* air]
aerobatics /,eərə'bætɪks/ *n.pl.* **1** spectacular flying of aircraft, esp. to entertain. **2** (as *sing.*) performance of these. [from AERO-, after ACROBATICS]
aerobics /eə'rəʊbɪks/ *n.pl.* vigorous exercises designed to increase oxygen intake. □ **aerobic** *adj.* [from AERO-, Greek *bios* life]
aerodrome /'eərə,drəʊm/ *n.* small airport or airfield. [from AERO-, Greek *dromos* course]
aerodynamics /,eərəʊdaɪ'næmɪks/ *n.pl.* (usu. treated as *sing.*) dynamics of solid bodies moving through air. □ **aerodynamic** *adj.*
aerofoil /'eərə,fɔɪl/ *n.* structure with curved surfaces (e.g. a wing, fin, or tailplane) designed to give lift in flight.
aeronautics /,eərəʊ'nɔ:tɪks/ *n.pl.* (usu. treated as *sing.*) science or practice of motion in the air. □ **aeronautic** *adj.* **aeronautical** *adj.* [from AERO-, NAUTICAL]
aeroplane /'eərə,pleɪn/ *n.* powered heavier-than-air flying vehicle with fixed wings. [French: related to AERO-, PLANE[1]]
aerosol /'eərə,sɒl/ *n.* **1** pressurized container releasing a substance as a fine spray. **2** system of minute particles suspended in gas (e.g. fog or smoke). [from AERO-, SOL]
aerospace /'eərəʊ,speɪs/ *n.* **1** earth's atmosphere and outer space. **2** aviation in this.
aesthete /'i:sθi:t/ *n.* person who has or professes a special appreciation of beauty. [Greek *aisthanomai* perceive]
aesthetic /i:s'θetɪk/ –*adj.* **1** of or sensitive to beauty. **2** artistic, tasteful. –*n.* (in *pl.*) philosophy of beauty, esp. in art. □ **aesthetically** *adv.* **aestheticism** /-,sɪz(ə)m/ *n.*
aetiology /,i:tɪ'ɒlədʒɪ/ *n.* (*US* **etiology**) study of causation or of the causes of disease. □ **aetiological** /-ə'lɒdʒɪk(ə)l/ *adj.* [Greek *aitia* cause]
AF *abbr.* audio frequency.
afar /ə'fɑ:(r)/ *adv.* at or to a distance.
affable /'æfəb(ə)l/ *adj.* **1** friendly. **2** courteous. □ **affability** /-'bɪlɪtɪ/ *n.* **affably** *adv.* [Latin *affabilis*]
affair /ə'feə(r)/ *n.* **1** matter, concern, or thing to be attended to (*that is my affair*). **2 a** celebrated or notorious happening. **b** *colloq.* thing or event (*puzzling affair*). **3** = LOVE AFFAIR. **4** (in *pl.*) public or private business. [French *à faire* to do]
affect /ə'fekt/ *v.* **1 a** produce an effect on. **b** (of disease etc.) attack. **2** move emotionally. **3** pretend (*affected ignorance*). **4** pose as or use for effect (*affects the aesthete*; *affects fancy hats*). □ **affecting** *adj.* **affectingly** *adv.* [Latin *afficio affect-* influence]

■ **Usage** *Affect* should not be confused with *effect*, meaning 'to bring about'. Note also that *effect* is used as a noun as well as a verb.

affectation /,æfek'teɪʃ(ə)n/ *n.* **1** artificial manner. **2** (foll. by *of*) studied display. **3** pretence.
affected /ə'fektɪd/ *adj.* **1** pretended, artificial. **2** full of affectation.
affection /ə'fekʃ(ə)n/ *n.* **1** goodwill, fond feeling. **2** disease; diseased condition.
affectionate /ə'fekʃənət/ *adj.* loving, fond. □ **affectionately** *adv.*
affidavit /,æfɪ'deɪvɪt/ *n.* written statement confirmed by oath. [Latin, = has stated on oath]
affiliate /ə'fɪlɪ,eɪt/ –*v.* (**-ting**) (foll. by *to*, *with*) attach, adopt, or connect as a member or branch. –*n.* affiliated person etc. [Latin: related to FILIAL]

affiliation /əˌfɪlɪ'eɪʃ(ə)n/ *n.* affiliating or being affiliated.
affiliation order *n.* legal order against the supposed father of an illegitimate child for support.
affinity /ə'fɪnɪtɪ/ *n.* (*pl.* **-ies**) **1** liking or attraction; feeling of kinship. **2** relationship, esp. by marriage. **3** similarity of structure or character suggesting a relationship. **4** *Chem.* the tendency of certain substances to combine with others. [Latin *finis* border]
affirm /ə'fɜ:m/ *v.* **1** assert, state as a fact. **2** *Law* make a solemn declaration in place of an oath. □ **affirmation** /ˌæfə'meɪʃ(ə)n/ *n.* [Latin: related to FIRM[1]]
affirmative /ə'fɜ:mətɪv/ *–adj.* affirming; expressing approval. *–n.* affirmative statement or word etc.
affix *–v.* /ə'fɪks/ **1** attach, fasten. **2** add in writing. *–n.* /'æfɪks/ **1** addition. **2** *Gram.* prefix or suffix. [Latin: related to FIX]
afflict /ə'flɪkt/ *v.* distress physically or mentally. [Latin *fligo flict-* strike down]
affliction /ə'flɪkʃ(ə)n/ *n.* **1** distress, suffering. **2** cause of this.
affluent /'æflʊənt/ *adj.* wealthy, rich. □ **affluence** *n.* [Latin: related to FLUENT]
afford /ə'fɔ:d/ *v.* **1** (prec. by *can* or *be able to*) **a** have enough money, time, etc., for; be able to spare. **b** be in a position (*can't afford to be critical*). **2** provide (*affords a view of the sea*). [Old English *ge-* prefix implying completeness, FORTH]
afforest /ə'fɒrɪst/ *v.* **1** convert into forest. **2** plant with trees. □ **afforestation** /-'steɪʃ(ə)n/ *n.* [Latin: related to FOREST]
affray /ə'freɪ/ *n.* breach of the peace by fighting or rioting in public. [Anglo-French = 'remove from peace']
affront /ə'frʌnt/ *–n.* open insult. *–v.* insult openly; offend, embarrass. [Latin: related to FRONT]
Afghan /'æfgæn/ *–n.* **1 a** native or national of Afghanistan. **b** person of Afghan descent. **2** official language of Afghanistan. *–adj.* of Afghanistan. [Pashto]
Afghan hound *n.* tall hunting dog with long silky hair.
aficionado /əˌfɪsjə'nɑ:dəʊ/ *n.* (*pl.* **-s**) devotee of a sport or pastime. [Spanish]
afield /ə'fi:ld/ *adv.* to or at a distance (esp. *far afield*). [Old English: related to A[2]]
aflame /ə'fleɪm/ *adv.* & *predic.adj.* **1** in flames. **2** very excited.
afloat /ə'fləʊt/ *adv.* & *predic.adj.* **1** floating. **2** at sea. **3** out of debt or difficulty. **4** current. [Old English: related to A[2]]
afoot /ə'fʊt/ *adv.* & *predic.adj.* in operation; progressing.
afore /ə'fɔ:(r)/ *prep.* & *adv. archaic* before; previously; in front (of). [Old English: related to A[2]]
afore- *comb. form* before, previously (*aforementioned*; *aforesaid*).
aforethought *adj.* premeditated (following a noun: *malice aforethought*).
afraid /ə'freɪd/ *predic. adj.* alarmed, frightened. □ **be afraid** *colloq.* politely regret (*I'm afraid we're late*). [originally past part. of AFFRAY]
afresh /ə'freʃ/ *adv.* anew; with a fresh beginning. [earlier *of fresh*]
African /'æfrɪkən/ *–n.* **1** native (esp. dark-skinned) of Africa. **2** person of African descent. *–adj.* of Africa. [Latin]
African elephant *n.* the elephant of Africa, larger than that of India.
African violet *n.* house-plant with velvety leaves and blue, purple, or pink flowers.
Afrikaans /ˌæfrɪ'kɑ:ns/ *n.* language derived from Dutch, used in S. Africa. [Dutch, = 'African']
Afrikaner /ˌæfrɪ'kɑ:nə(r)/ *n.* Afrikaans-speaking White person in S. Africa, esp. of Dutch descent.
Afro /'æfrəʊ/ *–adj.* (of hair) tightly-curled and bushy. *–n.* (*pl.* **-s**) Afro hair-style.
Afro- *comb. form* African.
Afro-American /ˌæfrəʊə'merɪkən/ *–adj.* of American Blacks or their culture. *–n.* American Black.
Afro-Caribbean /ˌæfrəʊˌkærɪ'bi:ən/ *–n.* Caribbean person of African descent. *–adj.* of Afro-Caribbeans.
aft /ɑ:ft/ *adv. Naut.* & *Aeron.* at or towards the stern or tail. [earlier *baft*]
after /'ɑ:ftə(r)/ *–prep.* **1** following in time; later than (*after a week*). **2** in view of, in spite of (*after what you did what do you expect?*; *after all my efforts I still lost*). **3** behind (*shut the door after you*). **4** in pursuit or quest of (*run after them*). **5** about, concerning (*asked after her*). **6** in allusion to (*named after the prince*). **7** in imitation of (*a painting after Rubens*). **8** next in importance to (*best one after mine*). *–conj.* later than (*left after they arrived*). *–adv.* **1** later (*soon after*). **2** behind (*followed on after*). *–adj.* **1** later, following (*in after years*). **2** *Naut.* nearer the stern (*after cabins*). □ **after all** in spite of everything (*after all, what does it matter?*). **after one's own heart** to one's taste. [Old English]
afterbirth *n.* placenta etc. discharged from the womb after childbirth.
after-care *n.* attention after leaving hospital etc.
after-effect *n.* delayed effect following an accident, trauma, etc.

afterglow *n.* glow remaining after its source has disappeared.

afterlife *n.* life after death.

aftermath /'ɑ:ftə,mæθ/ *n.* **1** consequences, esp. unpleasant (*aftermath of war*). **2** new grass growing after mowing. [from AFTER, *math* mowing]

afternoon /,ɑ:ftə'nu:n/ *n.* time from noon or lunch-time to evening.

afterpains *n.pl.* pains caused by contraction of the womb after childbirth.

afters *n.pl. colloq.* = DESSERT 1.

aftershave *n.* lotion used after shaving.

aftertaste *n.* taste after eating or drinking.

afterthought *n.* thing thought of or added later.

afterwards /'ɑ:ftəwədz/ *adv.* (*US* **afterward**) later, subsequently. [Old English: related to AFTER, -WARD]

Ag *symb.* silver. [Latin *argentum*]

again /ə'gen/ *adv.* **1** another time; once more. **2** as previously (*home again*; *well again*). **3** in addition (*as much again*). **4** further, besides (*again, what about you?*). **5** on the other hand (*I might, and again I might not*). □ **again and again** repeatedly. [Old English]

against /ə'genst/ *prep.* **1** in opposition to (*fight against crime*). **2** into collision or in contact with (*lean against the wall*). **3** to the disadvantage of (*my age is against me*). **4** in contrast to (*against a dark background*). **5** in anticipation of (*against his coming*; *against the cold*). **6** as a compensating factor to (*income against expenditure*). **7** in return for (*issued against payment of the fee*). □ **against the grain** see GRAIN. **against time** see TIME. [from AGAIN, with inflectional *-s*]

agape /ə'geɪp/ *predic. adj.* gaping, open-mouthed. [from A[2]]

agaric /'ægərɪk/ *n.* fungus with a cap and stalk, e.g. the common mushroom. [Greek *agarikon*]

agate /'ægət/ *n.* hard usu. streaked chalcedony. [Greek *akhatēs*]

agave /ə'geɪvɪ/ *n.* plant with rosettes of narrow spiny leaves and flowers on tall stem. [*Agave*, name of a woman in Greek mythology]

age –*n.* **1** length of time that a person or thing has existed. **2 a** *colloq.* (often in *pl.*) a long time (*waited for ages*). **b** distinct historical period (*Bronze Age*). **3** old age. –*v.* **(ageing) 1** show or cause to show signs of advancing age. **2** grow old. **3** mature. □ **come of age** reach adult status (esp. *Law* at 18, formerly 21). [Latin *aetas*]

-age *suffix* forming nouns denoting: **1** action (*breakage*). **2** condition (*bondage*). **3** aggregate or number (*coverage*; *acreage*). **4** cost (*postage*). **5** result (*wreckage*). **6** place or abode (*anchorage*; *orphanage*). [Latin *-aticus*]

aged *adj.* **1** /eɪdʒd/ (*predic.*) of the age of (*aged 3*). **2** /'eɪdʒɪd/ old.

ageism *n.* prejudice or discrimination on grounds of age. □ **ageist** *adj.* & *n.*

ageless *adj.* **1** never growing or appearing old. **2** eternal.

agelong *adj.* existing for a very long time.

agency /'eɪdʒənsɪ/ *n.* (*pl.* **-ies**) **1** business or premises of an agent. **2** action; intervention (*free agency*; *by the agency of God*). [Latin: related to ACT]

agenda /ə'dʒendə/ *n.* (*pl.* **-s**) **1** list of items to be considered at a meeting. **2** things to be done.

agent /'eɪdʒ(ə)nt/ *n.* **1 a** person who acts for another in business etc. **b** spy. **2** person or thing that exerts power or produces an effect.

agent provocateur /,ɑ:ʒɑ̃ prə,vɒkə'tɜ:(r)/ *n.* (*pl.* ***agents provocateurs*** pronunc. same) person used to tempt suspected offenders to self-incriminating action. [French, = provocative agent]

age of consent *n.* age at which consent to sexual intercourse is valid in law.

age-old *adj.* very long-standing.

agglomerate –*v.* **(-ting)** /ə'glɒmə,reɪt/ collect into a mass. –*n.* /ə'glɒmərət/ mass, esp. of fused volcanic fragments. –*adj.* /ə'glɒmərət/ collected into a mass. □ **agglomeration** /-'reɪʃ(ə)n/ *n.* [Latin *glomus -meris* ball]

agglutinate /ə'glu:tɪ,neɪt/ *v.* **(-ting)** stick as with glue. □ **agglutination** /-'neɪʃ(ə)n/ *n.* **agglutinative** /-nətɪv/ *adj.* [Latin: related to GLUTEN]

aggrandize /ə'grændaɪz/ *v.* (also **-ise**) **(-zing** or **-sing) 1** increase the power, rank, or wealth of. **2** make seem greater. □ **aggrandizement** /-dɪzmənt/ *n.* [French: related to GRAND]

aggravate /'ægrə,veɪt/ *v.* **(-ting) 1** make worse or more serious. **2** annoy. □ **aggravation** /-'veɪʃ(ə)n/ *n.* [Latin *gravis* heavy]

■ **Usage** The use of *aggravate* in sense 2 is regarded by some people as incorrect, but it is common in informal use.

aggregate –*n.* /'ægrɪgət/ **1** sum total, amount assembled. **2** crushed stone etc. used in making concrete. **3** rock formed of a mass of different particles or minerals. –*adj.* /'ægrɪgət/ combined, collective, total. –*v.* /'ægrɪ,geɪt/ **(-ting) 1** collect, combine into one mass. **2** *colloq.*

amount to. **3** unite. □ **in the aggregate** as a whole. □ **aggregation** /-'geɪʃ(ə)n/ *n.* **aggregative** /'ægrɪ,geɪtɪv/ *adj.* [Latin *grex greg-* flock]

aggression /ə'greʃ(ə)n/ *n.* **1** unprovoked attacking or attack. **2** hostile or destructive behaviour. [Latin *gradior gress-* walk]

aggressive /ə'gresɪv/ *adj.* **1** given to aggression; hostile. **2** forceful, self-assertive. □ **aggressively** *adv.*

aggressor *n.* person or party that attacks without provocation.

aggrieved /ə'gri:vd/ *adj.* having a grievance. [French: related to GRIEF]

aggro /'ægrəʊ/ *n. slang* **1** aggressive hostility. **2** trouble, difficulty. [abbreviation of *aggravation* or *aggression*]

aghast /ə'gɑ:st/ *predic. adj.* filled with dismay or consternation. [past part. of obsolete *(a)gast* frighten]

agile /'ædʒaɪl/ *adj.* quick-moving, nimble, active. □ **agility** /ə'dʒɪlɪtɪ/ *n.* [Latin *agilis*: related to ACT]

agitate /'ædʒɪ,teɪt/ *v.* (**-ting**) **1** disturb or excite (a person or feelings). **2** (often foll. by *for*, *against*) campaign, esp. politically (*agitated for tax reform*). **3** shake briskly. □ **agitation** /-'teɪʃ(ə)n/ *n.* **agitator** *n.* [Latin *agito*: related to ACT]

aglow /ə'gləʊ/ *predic. adj.* glowing.

AGM *abbr.* annual general meeting.

agnail /'ægneɪl/ *n.* piece of torn skin at the root of a fingernail; resulting soreness. [Old English, = tight (metal) nail, hard excrescence in flesh]

agnostic /æg'nɒstɪk/ *—n.* person who believes that the existence of God is not provable. *—adj.* of agnosticism. □ **agnosticism** /-,sɪz(ə)m/ *n.* [from A-, GNOSTIC]

ago /ə'gəʊ/ *adv.* (prec. by duration) earlier, in the past. [originally *agone* = gone by]

agog /ə'gɒg/ *predic. adj.* eager, expectant. [French *gogue* fun]

agonize /'ægə,naɪz/ *v.* (also **-ise**) (**-zing** or **-sing**) **1** undergo (esp. mental) anguish; suffer or cause to suffer agony. **2** (as **agonized** *adj.*) expressing agony (*an agonized look*).

agony /'ægənɪ/ *n.* (*pl.* **-ies**) **1** extreme mental or physical suffering. **2** severe struggle. [Greek *agōn* struggle]

agony aunt *n. colloq.* person (esp. a woman) who answers letters in an agony column.

agony column *n. colloq.* **1** column in a magazine etc. offering personal advice to correspondents. **2** = PERSONAL COLUMN.

agoraphobia /,ægərə'fəʊbɪə/ *n.* abnormal fear of open spaces or public places. □ **agoraphobic** *adj.* & *n.* [Greek *agora* market-place]

agrarian /ə'greərɪən/ *—adj.* **1** of the land or its cultivation. **2** of landed property. *—n.* advocate of the redistribution of land. [Latin *ager* field]

agree /ə'gri:/ *v.* (**-ees**, **-eed**, **-eeing**) **1** hold the same opinion (*I agree with you*). **2** consent (*agreed to go*). **3** (often foll. by *with*) **a** become or be in harmony. **b** suit (*fish didn't agree with him*). **c** *Gram.* have the same number, gender, case, or person as. **4** reach agreement about (*agreed a price*). **5** (foll. by *on*) decide mutually on (*agreed on a compromise*). □ **be agreed** be of one opinion. [Latin: related to AD-, *gratus* pleasing]

agreeable *adj.* **1** pleasing, pleasant. **2** willing to agree. □ **agreeably** *adv.*

agreement *n.* **1** act or state of agreeing. **2** arrangement or contract.

agriculture /'ægrɪ,kʌltʃə(r)/ *n.* cultivation of the soil and rearing of animals. □ **agricultural** /-'kʌltʃər(ə)l/ *adj.* **agriculturalist** /-'kʌltʃərəlɪst/ *n.* [Latin *ager* field]

agrimony /'ægrɪmənɪ/ *n.* (*pl.* **-ies**) perennial plant with small yellow flowers. [Greek *argemōnē* poppy]

agronomy /ə'grɒnəmɪ/ *n.* science of soil management and crop production. □ **agronomist** *n.* [Greek *agros* land]

aground /ə'graʊnd/ *predic. adj.* & *adv.* on or on to the bottom of shallow water (*run aground*).

ague /'eɪgju:/ *n.* **1** *hist.* malarial fever. **2** shivering fit. [Latin: related to ACUTE]

AH *abbr.* in the year of the Hegira (AD 622); of the Muslim era. [Latin *anno Hegirae*]

ah /ɑ:/ *int.* expressing surprise, pleasure, realization, etc. [French *a*]

aha /ɑ:'hɑ:/ *int.* expressing surprise, triumph, mockery, etc. [from AH, HA[1]]

ahead /ə'hed/ *adv.* **1** further forward in space or time. **2** in the lead (*ahead on points*).

ahem /ə'həm/ *int.* used to attract attention, gain time, etc. [from HEM[2]]

ahoy /ə'hɔɪ/ *int. Naut.* call used in hailing. [from AH, HOY]

AI *abbr.* **1** artificial insemination. **2** artificial intelligence.

AID *abbr.* artificial insemination by donor.

aid *—n.* **1** help. **2** person or thing that helps. *—v.* **1** help. **2** promote (*sleep will aid recovery*). □ **in aid of 1** in support of. **2** *colloq.* for the purpose of (*what's it all in aid of?*). [Latin: related to AD-, *juvo* help]

aide /eɪd/ *n.* **1** aide-de-camp. **2** esp. *US* assistant. [French]

aide-de-camp /ˌeɪd də 'kɑ̃/ *n.* (*pl.* **aides-de-camp** pronunc. same) officer assisting a senior officer. [French]

Aids *n.* (also **AIDS**) acquired immune deficiency syndrome, an often fatal viral syndrome marked by severe loss of resistance to infection. [abbreviation]

ail *v.* **1** *archaic* (only in 3rd person interrog. or indefinite constructions) trouble or afflict (***what ails him?***). **2** (usu. **be ailing**) be ill. [Old English]

aileron /'eɪləˌrɒn/ *n.* hinged flap on an aeroplane wing. [French *aile* wing]

ailing *adj.* **1** ill. **2** in poor condition.

ailment *n.* minor illness or disorder.

aim –*v.* **1** intend or try; attempt (***aim at winning***; ***aim to win***). **2** (usu. foll. by *at*) direct or point (a weapon, remark, etc.). **3** take aim. –*n.* **1** purpose or object. **2** the directing of a weapon etc. at an object. □ **take aim** direct a weapon etc. at a target. [Latin *aestimare* reckon]

aimless *adj.* without aim or purpose. □ **aimlessly** *adv.*

ain't /eɪnt/ *contr. colloq.* **1** am, is, or are not. **2** have or has not.

■ **Usage** The use of *ain't* is usually regarded as unacceptable in spoken and written English.

air /eə(r)/ –*n.* **1** mixture mainly of oxygen and nitrogen surrounding the earth. **2** earth's atmosphere; open space in it; this as a place for flying aircraft. **3 a** distinctive impression or manner (*air of mystery*). **b** (esp. in *pl.*) pretentiousness (***gave himself airs***). **4** tune. **5** light wind. –*v.* **1** expose (clothes, a room, etc.) to fresh air or warmth to remove damp. **2** express and discuss publicly (an opinion, question, grievance, etc.). □ **by air** by or in an aircraft. **in the air 1** (of opinions etc.) prevalent. **2** (of plans etc.) uncertain. **on** (or **off**) **the air** being (or not being) broadcast. [Greek *aēr*]

airbase *n.* base for military aircraft.

air-bed *n.* inflatable mattress.

airborne *adj.* **1** transported by air. **2** (of aircraft) in the air after taking off.

air-brick *n.* perforated brick used for ventilation.

Airbus *n. propr.* short-haul passenger aircraft.

Air Chief Marshal *n.* RAF officer of high rank, above Air Marshal.

Air Commodore *n.* RAF officer next above Group Captain.

air-conditioning *n.* **1** system for regulating the humidity, ventilation, and temperature in a building. **2** apparatus for this. □ **air-conditioned** *adj.*

aircraft *n.* (*pl.* same) machine capable of flight, esp. an aeroplane or helicopter.

aircraft-carrier *n.* warship carrying and used as a base for aircraft.

aircraftman *n.* lowest rank in the RAF.

aircraftwoman *n.* lowest rank in the WRAF.

aircrew *n.* crew of an aircraft.

air-cushion *n.* **1** inflatable cushion. **2** layer of air supporting a hovercraft etc.

Airedale /'eədeɪl/ *n.* large terrier of a rough-coated breed. [*Airedale* in Yorkshire]

airer *n.* stand for airing or drying clothes etc.

airfield *n.* area with runway(s) for aircraft.

air force *n.* branch of the armed forces fighting in the air.

airgun *n.* gun using compressed air to fire pellets.

airhead *n. slang* stupid or foolish person.

air hostess *n.* stewardess in a passenger aircraft.

airless *adj.* stuffy; still, calm.

air letter *n.* sheet of light paper forming a letter for sending by airmail.

airlift –*n.* emergency transport of supplies etc. by air. –*v.* transport thus.

airline *n.* public air transport system or company.

airliner *n.* large passenger aircraft.

airlock *n.* **1** stoppage of the flow by an air bubble in a pump or pipe. **2** compartment permitting movement between areas at different pressures.

airmail *n.* **1** system of transporting mail by air. **2** mail carried by air.

airman *n.* pilot or member of an aircraft crew, esp. in an air force.

Air Marshal *n.* RAF officer of high rank, above Air Vice-Marshal.

airplane *n. US* = AEROPLANE.

air pocket *n.* apparent vacuum causing an aircraft to drop suddenly.

airport *n.* airfield with facilities for passengers and goods.

air raid *n.* attack by aircraft on ground targets.

air rifle *n.* rifle using compressed air to fire pellets.

airs and graces *n.pl.* affected manner.

airscrew *n.* aircraft propeller.

airship *n.* power-driven aircraft lighter than air.

airsick *adj.* nauseous from air travel.

airspace *n.* air above a country and subject to its jurisdiction.

air speed *n.* aircraft's speed relative to the air.

airstrip *n.* strip of ground for the take-off and landing of aircraft.

air terminal *n.* building with transport to and from an airport.

airtight *adj.* impermeable to air.

air traffic controller *n.* official who controls air traffic by radio.

Air Vice-Marshal *n.* RAF officer of high rank, just below Air Marshal.

airwaves *n.pl. colloq.* radio waves used in broadcasting.

airway *n.* recognized route of aircraft.

airwoman *n.* woman pilot or member of an aircraft crew, esp. in an air force.

airworthy *adj.* (of an aircraft) fit to fly.

airy *adj.* (**-ier**, **-iest**) **1** well-ventilated, breezy. **2** flippant, superficial. **3** light as air. **4** ethereal. □ **airily** *adv.*

airy-fairy *adj. colloq.* unrealistic, impractical.

aisle /aɪl/ *n.* **1** the part of a church on either side of the nave, divided from it by pillars. **2** passage between rows of pews, seats, etc. [Latin *ala* wing]

aitch /eɪtʃ/ *n.* the letter H. [French *ache*]

aitchbone *n.* **1** rump bone of an animal. **2** cut of beef over this. [originally *nache-bone* from Latin *natis* buttock]

ajar /əˈdʒɑː(r)/ *adv.* & *predic.adj.* (of a door) slightly open. [from A[2], obsolete *char* from Old English *cerr* a turn]

Akela /ɑːˈkeɪlə/ *n.* adult leader of Cub Scouts. [name of the leader of the wolf-pack in Kipling's *Jungle Book*]

akimbo /əˈkɪmbəʊ/ *adv.* (of the arms) with hands on the hips and elbows turned outwards. [originally *in kenebowe*, probably from Old Norse]

akin /əˈkɪn/ *predic. adj.* **1** related by blood. **2** similar.

Al *symb.* aluminium.

-al *suffix* **1** (also **-ial**) forming adjectives meaning 'relating to, of the kind of' (*central*; *tidal*; *dictatorial*). **2** forming nouns, esp. of verbal action (*removal*). [Latin *-alis*]

à la /ˈæ lɑː/ *prep.* in the manner of (*à la russe*). [French from À LA MODE]

alabaster /ˈæləˌbɑːstə(r)/ *–n.* translucent usu. white form of gypsum, used for carving etc. *–adj.* **1** of alabaster. **2** white or smooth. [Greek *alabastros*]

à la carte /ˌæ lɑː ˈkɑːt/ *adv.* & *adj.* with individually priced dishes. [French]

alacrity /əˈlækrɪtɪ/ *n.* briskness; cheerful readiness. [Latin *alacer* brisk]

à la mode /ˌæ lɑː ˈməʊd/ *adv.* & *adj.* in fashion; fashionable. [French]

alarm /əˈlɑːm/ *–n.* **1** warning of danger etc. **2 a** warning sound or device. **b** = ALARM CLOCK. **3** apprehension (*filled with alarm*). *–v.* **1** frighten or disturb. **2** warn. □ **alarming** *adj.* **alarmingly** *adv.* [Italian *all'arme!* to arms]

alarm clock *n.* clock that rings at a set time.

alarmist *n.* person stirring up alarm.

alas /əˈlæs/ *int.* expressing grief, pity, or concern. [French: related to AH, Latin *lassus* weary]

alb *n.* long white vestment worn by Christian priests. [Latin *albus* white]

albatross /ˈælbəˌtrɒs/ *n.* **1 a** long-winged, stout-bodied bird related to the petrel. **b** encumbrance. **2** *Golf* score of three strokes under par at any hole. [alteration of *alcatras*, from Spanish and Portuguese *alcatraz* from Arabic, = the jug]

albeit /ɔːlˈbiːɪt/ *conj. literary* though. [*all be it*]

albino /ælˈbiːnəʊ/ *n.* (*pl.* **-s**) **1** person or animal lacking pigment in the skin and hair (which are white), and the eyes (usu. pink). **2** plant lacking normal colouring. □ **albinism** /ˈælbɪˌnɪz(ə)m/ *n.* [Spanish and Portuguese: related to ALB]

album /ˈælbəm/ *n.* **1** book for photographs, stamps, etc. **2 a** long-playing gramophone record. **b** set of these. [Latin, = blank tablet, from *albus* white]

albumen /ˈælbjʊmɪn/ *n.* **1** egg-white. **2** substance found between the skin and germ of many seeds, usu. the edible part. [Latin: related to ALBUM]

albumin /ˈælbjʊmɪn/ *n.* water-soluble protein found in egg-white, milk, blood, etc. □ **albuminous** /ælˈbjuːmɪnəs/ *adj.*

alchemy /ˈælkəmɪ/ *n.* medieval chemistry, esp. seeking to turn base metals into gold. □ **alchemist** *n.* [Arabic]

alcohol /ˈælkəˌhɒl/ *n.* **1** (in full **ethyl alcohol**) colourless volatile inflammable liquid, esp. as the intoxicant in wine, beer, spirits, etc., and as a solvent, fuel, etc. **2** liquor containing this. **3** *Chem.* any of many organic compounds containing one or more hydroxyl groups attached to carbon atoms. [Arabic: related to KOHL]

alcoholic /ˌælkəˈhɒlɪk/ *–adj.* of, like, containing, or caused by alcohol. *–n.* person suffering from alcoholism.

alcoholism /ˈælkəhɒˌlɪz(ə)m/ *n.* condition resulting from addiction to alcohol.

alcove /ˈælkəʊv/ *n.* recess, esp. in the wall of a room. [Arabic, = the vault]

aldehyde /ˈældɪˌhaɪd/ *n. Chem.* any of a class of compounds formed by the oxidation of alcohols. [from ALCOHOL, DE-, HYDROGEN]

al dente /æl ˈdentɪ, -teɪ/ *adj.* (of pasta etc.) cooked so as to be still firm when bitten. [Italian, = 'to the tooth']

alder /ˈɔːldə(r)/ *n.* tree related to the birch. [Old English]

alderman /'ɔ:ldəmən/ *n.* esp. *hist.* co-opted member of an English county or borough council, next in dignity to the mayor. [Old English *aldor* chief, MAN]
ale *n.* beer. [Old English]
aleatory /'eɪlɪətərɪ/ *adj.* depending on chance. [Latin *alea* DIE[2]]
alehouse *n. hist.* tavern.
alembic /ə'lembɪk/ *n.* **1** *hist.* apparatus formerly used in distilling. **2** means of refining or extracting. [Greek *ambix, -ikos* cap of a still]
alert /ə'lɜ:t/ *–adj.* **1** watchful, vigilant. **2** nimble, attentive. *–n.* **1** warning call or alarm. **2** state or period of special vigilance. *–v.* (often foll. by *to*) warn. [French *alerte* from Italian *all'erta* to the watch-tower]
A level *n.* = ADVANCED LEVEL.
Alexander technique /,ælɪg'zɑ:n də(r)/ *n.* technique for controlling posture as an aid to well-being. [*Alexander*, name of a physiotherapist]
alexandrine /,ælɪg'zændraɪn/ *–adj.* (of a line of verse) having six iambic feet. *–n.* alexandrine line. [French *Alexandre*, title of a romance using this metre]
alfalfa /æl'fælfə/ *n.* clover-like plant used for fodder. [Arabic, = a green fodder]
alfresco /æl'freskəʊ/ *adv. & adj.* in the open air. [Italian]
alga /'ælgə/ *n.* (*pl.* **algae** /-dʒi:/) (usu. in *pl.*) non-flowering stemless water-plant, esp. seaweed and plankton. [Latin]
algebra /'ældʒɪbrə/ *n.* branch of mathematics that uses letters etc. to represent numbers and quantities. □ **algebraic** /-'breɪɪk/ *adj.* [ultimately from Arabic *al-jabr*, = reunion of broken parts]
Algol /'ælgɒl/ *n.* high-level computer programming language. [from ALGO(RITHM), L(ANGUAGE)]
algorithm /'ælgə,rɪð(ə)m/ *n.* process or set of rules used for calculation etc., esp. with a computer. □ **algorithmic** /-'rɪðmɪk/ *adj.* [Persian, name of a 9th-c. mathematician *al-Kuwārizmī*]
alias /'eɪlɪəs/ *–adv.* also named or known as. *–n.* assumed name. [Latin, = at another time]
alibi /'ælɪ,baɪ/ *n.* (*pl.* **-s**) **1** claim or proof that one was elsewhere when a crime etc. was committed. **2** *informal* excuse. [Latin, = elsewhere]

■ **Usage** The use of *alibi* in sense 2 is considered incorrect by some people.

alien /'eɪlɪən/ *–adj.* **1** (often foll. by *to*) unfamiliar; unacceptable or repugnant. **2** foreign. **3** of beings from other worlds. *–n.* **1** foreign-born resident who is not naturalized. **2** a being from another world. [Latin *alius* other]
alienable *adj. Law* able to be transferred to new ownership.
alienate *v.* **(-ting) 1** estrange, make hostile. **2** transfer ownership of. □ **alienation** /-'neɪʃ(ə)n/ *n.*
alight[1] /ə'laɪt/ *predic. adj.* **1** on fire. **2** lit up; excited. [*on a light* (= lighted) *fire*]
alight[2] /ə'laɪt/ *v.* **1** (often foll. by *from*) descend from a vehicle. **2** come to earth, settle. [Old English]
align /ə'laɪn/ *v.* **1** put or bring into line. **2** (usu. foll. by *with*) ally (oneself etc.) with (a cause, party, etc.). □ **alignment** *n.* [French *à ligne* into line]
alike /ə'laɪk/ *–adj.* (usu. *predic.*) similar, like. *–adv.* in a similar way.
alimentary /,ælɪ'mentərɪ/ *adj.* of or providing food or nourishment. [Latin *alo* nourish]
alimentary canal *n.* passage along which food passes during digestion.
alimony /'ælɪmənɪ/ *n.* money payable to a spouse or former spouse after separation or divorce.

■ **Usage** In UK usage this term has been replaced by *maintenance*.

aliphatic /,ælɪ'fætɪk/ *adj. Chem.* of organic compounds in which carbon atoms form open chains, not aromatic rings. [Greek *aleiphar -phat-* fat]
aliquot /'ælɪ,kwɒt/ *–adj.* (of a part or portion) contained by the whole an integral number of times (*4 is an aliquot part of 12*). *–n.* **1** aliquot part. **2** (in general use) any known fraction of a whole; sample. [Latin, = several]
alive /ə'laɪv/ *adj.* (usu. *predic.*) **1** living. **2** lively, active. **3** (foll. by *to*) aware of; alert. **4** (foll. by *with*) swarming or teeming with. [Old English: related to A[2], LIFE]
alkali /'ælkə,laɪ/ *n.* (*pl.* **-s**) **1 a** any of a class of substances that liberate hydroxide ions in water, usu. form caustic or corrosive solutions, turn litmus blue, and have a pH of more than 7, e.g. caustic soda. **b** similar but weaker substance, e.g. sodium carbonate. **2** *Chem.* any substance that reacts with or neutralizes hydrogen ions. □ **alkaline** *adj.* **alkalinity** /-'lɪnɪtɪ/ *n.* [Arabic, = the calcined ashes]
alkaloid /'ælkə,lɔɪd/ *n.* nitrogenous organic compound of plant origin, e.g. morphine, quinine.
alkane /'ælkeɪn/ *n. Chem.* saturated aliphatic hydrocarbon having the general formula C_nH_{2n+2}, including methane and ethane.
alkene /'ælki:n/ *n. Chem.* unsaturated aliphatic hydrocarbon containing a

double bond and having the general formula C_nH_{2n}, including ethylene.

alkyne /'ælkaɪn/ *n. Chem.* unsaturated aliphatic hydrocarbon containing a triple bond and having the general formula C_nH_{2n-2}, including acetylene.

all /ɔ:l/ *–adj.* **1** whole amount, quantity, or extent of (*all day*; *all his life*; *take it all*). **2** any whatever (*beyond all doubt*). **3** greatest possible (*with all speed*). *–n.* **1** all persons or things concerned; everything (*all were present*; *all is lost*). **2** (foll. by *of*) **a** the whole of (*take all of it*). **b** every one of (*all of us*). **c** *colloq.* as much as (*all of six feet*). **d** *colloq.* in a state of (*all of a dither*). **3** one's whole strength or resources (prec. by *my*, *your*, etc.). **4** (in games) each (*two goals all*). *–adv.* **1 a** entirely, quite (*was all in black*). **b** as an intensifier (*stop all this crying*). **2** *colloq.* very (*went all shy*). **3** (foll. by *the* + compar.) to that, or the utmost, extent (*if they go, all the better*; *made it all the worse*). □ **all along** from the beginning. **all and sundry** everyone. **all but** very nearly. **all for** *colloq.* strongly in favour of. **all found** with board and lodging provided free. **all in** *colloq.* exhausted. **all in all** everything considered. **all manner of** every kind of. **all of a sudden** suddenly. **all one** (or **the same**) (usu. foll. by *to*) a matter of indifference. **all out** using all one's strength (also (with hyphen) *attrib.*: *all-out effort*). **all over 1** completely finished. **2** in or on all parts of (*mud all over the car*). **3** *colloq.* typically (*you all over*). **4** *slang* effusively attentive to (a person). **all right** (*predic.*) **1** satisfactory; safe and sound; in good condition. **2** satisfactorily (*it worked out all right*). **3 a** expressing consent. **b** as an intensifier (*that's the one all right*). **all round 1** in all respects. **2** for each person. **all the same** nevertheless. **all there** *colloq.* mentally alert or normal. **all the time** throughout (despite some contrary expectation etc.). **all together** all at once; all in one place or in a group (*came all together*) (cf. ALTOGETHER). **all up with** hopeless for (a person). **at all** (with *neg.* or *interrog.*) in any way; to any extent (*did not swim at all*; *did you like it at all?*). **in all** in total; altogether. [Old English]

■ **Usage** Note the differences in meaning between *all together* and *altogether*: see note at *altogether*.

Allah /'ælə/ *n.* the Muslim and Arab name of God. [Arabic]

allay /ə'leɪ/ *v.* **1** diminish (fear, suspicion, etc.). **2** alleviate (pain etc.). [Old English *a-* intensive prefix, LAY[1]]

all-clear *n.* signal that danger etc. is over.

all comers *n.pl.* anyone who applies, takes up a challenge, etc.

allegation /,ælɪ'geɪʃ(ə)n/ *n.* **1** assertion, esp. unproved. **2** alleging. [Latin *allego* adduce]

allege /ə'ledʒ/ *v.* **(-ging) 1** declare, esp. without proof. **2** advance as an argument or excuse. [Latin *lis lit-* lawsuit]

allegedly /ə'ledʒɪdlɪ/ *adv.* as is alleged.

allegiance /ə'li:dʒ(ə)ns/ *n.* **1** loyalty (to a person or cause etc.). **2** the duty of a subject. [French: related to LIEGE]

allegory /'ælɪgərɪ/ *n.* (*pl.* **-ies**) story whose moral is represented symbolically. □ **allegorical** /,ælɪ'gɒrɪk(ə)l/ *adj.* **allegorize** /'ælɪgə,raɪz/ *v.* (also **-ise**) **(-zing** or **-sing)**. [Greek *allēgoria* other speaking]

allegretto /,ælɪ'gretəʊ/ *Mus. –adv.* & *adj.* in a fairly brisk tempo. *–n.* (*pl.* **-s**) such a passage or movement. [Italian, diminutive of ALLEGRO]

allegro /ə'legrəʊ/ *Mus. –adv.* & *adj.* in a brisk tempo. *–n.* (*pl.* **-s**) such a passage or movement. [Italian, = lively]

alleluia /,ælɪ'lu:jə/ (also **hallelujah** /,hæl-/) *–int.* God be praised. *–n.* song or shout of praise to God. [Hebrew]

Allen key /'ælən/ *n. propr.* spanner designed to turn an Allen screw. [*Allen*, name of the US manufacturer]

Allen screw /'ælən/ *n. propr.* screw with a hexagonal socket in the head.

allergic /ə'lɜ:dʒɪk/ *adj.* **1** (foll. by *to*) **a** having an allergy to. **b** *colloq.* having a strong dislike for. **2** caused by an allergy.

allergy /'ælədʒɪ/ *n.* (*pl.* **-ies**) **1** adverse reaction to certain substances, esp. particular foods, pollen, fur, or dust. **2** *colloq.* antipathy. [Greek *allos* other]

alleviate /ə'li:vɪ,eɪt/ *v.* **(-ting)** make (pain etc.) less severe. □ **alleviation** /-'eɪʃ(ə)n/ *n.* [Latin *levo* raise]

alley /'ælɪ/ *n.* (*pl.* **-s**) **1** narrow street or passageway. **2** enclosure for skittles, bowling, etc. **3** walk or lane in a park etc. [French *aller* go]

alliance /ə'laɪəns/ *n.* **1** union or agreement to cooperate, esp. of States by treaty or families by marriage. **2** (**Alliance**) political coalition party. **3** relationship; friendship. [French: related to ALLY]

allied /'ælaɪd/ *adj.* **1** (also **Allied**) associated in an alliance. **2** connected or related.

alligator /'ælɪ,geɪtə(r)/ *n.* large reptile of the crocodile family with a head broader and shorter than a crocodile's. [Spanish *el lagarto* the lizard]

all-in *attrib. adj.* inclusive of all.

all-in wrestling *n.* wrestling with few or no restrictions.
alliteration /ə,lɪtə'reɪʃ(ə)n/ *n.* repetition of the same letter or sound at the beginning of adjacent or closely connected words (e.g. *cool, calm, and collected*). □ **alliterate** /ə'lɪtə,reɪt/ *v.* (**-ting**). **alliterative** /ə'lɪtərətɪv/ *adj.* [Latin: related to LETTER]
allocate /'ælə,keɪt/ *v.* (**-ting**) (usu. foll. by *to*) assign or devote to (a purpose, person, or place). □ **allocation** /-'keɪʃ(ə)n/ *n.* [Latin: related to LOCAL]
allot /ə'lɒt/ *v.* (**-tt-**) apportion or distribute to (a person), esp. as a share or task (*they were allotted equal sums*). [French *a* to, LOT]
allotment *n.* **1** small piece of land rented by a local authority for cultivation. **2** share. **3** allotting.
allotropy /ə'lɒtrəpɪ/ *n.* existence of two or more different physical forms of a chemical element. □ **allotropic** /,ælə'trɒpɪk/ *adj.* [Greek *allos* different, *tropos* manner]
allow /ə'laʊ/ *v.* **1** (often foll. by *to* + infin.) permit. **2** assign a limited amount etc. (*was allowed £500*). **3** (usu. foll. by *for*) provide or set aside for a purpose; add or deduct in consideration (*allow £50 for expenses; allow for wastage*). [originally = commend, from French: related to AD-, Latin *laudo* praise, *loco* place]
allowance *n.* **1** amount or sum allowed, esp. regularly for a stated purpose. **2** amount allowed in reckoning. **3** deduction or discount. □ **make allowances** (often foll. by *for*) **1** consider (mitigating circumstances). **2** make excuses for (a person, bad behaviour, etc.).
alloy *–n.* /'ælɔɪ/ **1** mixture of two or more metals. **2** inferior metal mixed esp. with gold or silver. *–v.* /ə'lɔɪ/ **1** mix (metals). **2** debase by admixture. **3** moderate (*pleasure alloyed with pain*). [French: related to ALLY]
all-purpose *attrib. adj.* having many uses.
all-right *attrib. adj. colloq.* acceptable (*an all-right guy*).
all-round *attrib. adj.* (of a person) versatile.
all-rounder *n.* versatile person.
All Saints' Day *n.* 1 Nov., Christian festival in honour of saints.
All Souls' Day *n.* 2 Nov., Roman Catholic festival with prayers for the souls of the dead.
allspice *n.* **1** aromatic spice obtained from the berry of the pimento plant. **2** the berry.
all-time *attrib. adj.* (of a record etc.) unsurpassed.
allude /ə'lu:d/ *v.* (**-ding**) (foll. by *to*) refer to, esp. indirectly or briefly. [Latin: related to AD-, *ludo* play]
allure /ə'ljʊə(r)/ *–v.* (**-ring**) attract, charm, or entice. *–n.* attractiveness, personal charm, fascination. □ **allurement** *n.* [French: related to AD-, LURE]
allusion /ə'lu:ʒ(ə)n/ *n.* (often foll. by *to*) passing or indirect reference. □ **allusive** /ə'lu:sɪv/ *adj.* [Latin: related to ALLUDE]
alluvial /ə'lu:vɪəl/ *–adj.* of alluvium. *–n.* alluvium, esp. containing a precious metal.
alluvium /ə'lu:vɪəm/ *n.* (*pl.* **-via**) deposit of usu. fine fertile soil left behind by a flood, esp. in a river valley. [Latin *luo* wash]
ally /'ælaɪ/ *–n.* (*pl.* **-ies**) State, person, etc., formally cooperating or united with another, esp. (also **Ally**) in war. *–v.* also /ə'laɪ/ (**-ies, -ied**) (often *refl.* and foll. by *with*) combine in alliance. [Latin *alligo* bind]
Alma Mater /,ælmə 'mɑ:tə(r)/ *n.* one's university, school, or college. [Latin, = bounteous mother]
almanac /'ɔ:lmə,næk/ *n.* (also **almanack**) calendar, usu. with astronomical data. [medieval Latin from Greek]
almighty /ɔ:l'maɪtɪ/ *adj.* **1** having complete power. **2** (**the Almighty**) God. **3** *slang* very great (*almighty crash*). [Old English: related to ALL, MIGHTY]
almond /'ɑ:mənd/ *n.* **1** nutlike kernel of a fruit allied to the peach and plum. **2** tree bearing this. [Greek *amugdalē*]
almoner /'ɑ:mənə(r)/ *n.* social worker attached to a hospital. [French: related to ALMS]

■ **Usage** The usual term now is *medical social worker*.

almost /'ɔ:lməʊst/ *adv.* all but; very nearly. [Old English: related to ALL, MOST]
alms /ɑ:mz/ *n.pl. hist.* donation of money or food to the poor. [Greek *eleēmosunē* pity]
almshouse *n. hist.* charitable institution for the poor.
aloe /'æləʊ/ *n.* **1** plant of the lily family with toothed fleshy leaves. **2** (in *pl.*) (in full **bitter aloes**) strong laxative from aloe juice. [Old English from Greek]
aloft /ə'lɒft/ *predic. adj.* & *adv.* **1** high up, overhead. **2** upwards. [Old Norse *á lopti* in air]
alone /ə'ləʊn/ *–predic. adj.* **1** without the presence or help of others. **2** lonely (*felt alone*). *–adv.* only, exclusively. [earlier *al one*: related to ALL, ONE]
along /ə'lɒŋ/ *–prep.* beside or through (part of) the length of. *–adv.* **1** onward,

into a more advanced state (*come along*; *getting along nicely*). **2** with oneself or others (*bring a book along*). **3** beside or through part or the whole length of a thing. □ **along with** in addition to; together with. [Old English, originally adj. = facing against]

alongside /əlɒŋ'saɪd/ *–adv.* at or to the side. *–prep.* close to the side of.

aloof /ə'lu:f/ *–adj.* distant, unsympathetic. *–adv.* away, apart (*he kept aloof*). [originally *Naut.*, from A[2] + LUFF]

aloud /ə'laʊd/ *adv.* audibly.

alp *n.* **1 a** high mountain. **b** (**the Alps**) high range of mountains in Switzerland and adjoining countries. **2** pasture land on a Swiss mountainside. [originally *alps*, from Greek *alpeis*]

alpaca /æl'pækə/ *n.* **1** shaggy S. American mammal related to the llama. **2** its wool; fabric made from this. [Spanish from Quechua]

alpha /'ælfə/ *n.* **1** first letter of the Greek alphabet (Α, α). **2** first-class mark for a piece of work etc. □ **alpha and omega** beginning and end. [Latin from Greek]

alphabet /'ælfə,bet/ *n.* **1** set of letters used in writing a language. **2** symbols or signs for these. □ **alphabetical** /-'betɪk(ə)l/ *adj.* [Greek ALPHA, BETA]

alphanumeric /,ælfənju:'merɪk/ *adj.* containing both letters and numbers.

alpha particle *n.* helium nucleus emitted by a radioactive substance.

alpine /'ælpaɪn/ *–adj.* of mountainous regions or (**Alpine**) the Alps. *–n.* **1** plant growing in mountainous regions. **2** = ROCK-PLANT. [Latin: related to ALP]

already /ɔ:l'redɪ/ *adv.* **1** before the time in question (*I knew that already*). **2** as early or as soon as this (*is back already*). [from ALL, READY]

alright /ɔ:l'raɪt/ *adv.* = *all right* (see ALL).

■ **Usage** Although widely used, *alright* is still non-standard and is considered incorrect by many people.

Alsatian /æl'seɪʃ(ə)n/ *n.* large dog of a breed of wolfhound (also called GERMAN SHEPHERD). [Latin *Alsatia* Alsace]

also /'ɔ:lsəʊ/ *adv.* in addition, besides. [Old English: related to ALL, SO[1]]

also-ran *n.* **1** loser in a race. **2** undistinguished person.

altar /'ɔ:ltə(r)/ *n.* **1** table or flat block for sacrifice or offering to a deity. **2** Communion table. [Latin *altus* high]

altarpiece *n.* painting etc. above or behind an altar.

alter /'ɔ:ltə(r)/ *v.* make or become different; change. □ **alteration** /-'reɪʃ(ə)n/ *n.* [Latin *alter* other]

altercate /'ɔ:ltə,keɪt/ *v.* (**-ting**) (often foll. by *with*) dispute, wrangle. □ **altercation** /-'keɪʃ(ə)n/ *n.* [Latin]

alter ego /,ɔ:ltər 'i:gəʊ/ *n.* (*pl.* *-s*) **1** one's hidden or second self. **2** intimate friend. [Latin, = other self]

alternate *–v.* /'ɔ:ltə,neɪt/ (**-ting**) **1** (often foll. by *with*) occur or cause to occur by turns. **2** (foll. by *between*) go repeatedly from one to another (*alternated between hope and fear*). *–adj.* /ɔ:l'tɜ:nət/ **1** (with noun in *pl.*) every other (*on alternate days*). **2** (of things of two kinds) alternating (*alternate joy and misery*). □ **alternately** /-'tɜ:nətlɪ/ *adv.* **alternation** /-'neɪʃ(ə)n/ *n.* [Latin *alterno* do by turns: related to ALTER]

■ **Usage** See note at *alternative*.

alternate angles *n.pl.* two angles formed alternately on two sides of a line.

alternating current *n.* electric current reversing its direction at regular intervals.

alternative /ɔ:l'tɜ:nətɪv/ *–adj.* **1** available as another choice (*alternative route*). **2** unconventional (*alternative medicine*). *–n.* **1** any of two or more possibilities. **2** choice (*had no alternative but to go*). □ **alternatively** *adv.*

■ **Usage** The adjective *alternative* should not be confused with *alternate*, as in 'there will be a dance on alternate Saturdays'.

alternator /'ɔ:ltə,neɪtə(r)/ *n.* dynamo that generates an alternating current.

although /ɔ:l'ðəʊ/ *conj.* = THOUGH. [from ALL, THOUGH]

altimeter /'æltɪ,mi:tə(r)/ *n.* instrument indicating altitude reached.

altitude /'æltɪ,tju:d/ *n.* height, esp. of an object above sea level or above the horizon. [Latin *altus* high]

alto /'æltəʊ/ *n.* (*pl.* **-s**) **1** = CONTRALTO. **2 a** highest adult male singing-voice, above tenor. **b** singer with this voice. **3** instrument pitched second- or third-highest in its family. [Italian *alto* (*canto*) high (singing)]

altogether /,ɔ:ltə'geðə(r)/ *adv.* **1** totally, completely. **2** on the whole. **3** in total. □ **in the altogether** *colloq.* naked. [from ALL, TOGETHER]

■ **Usage** Note that *altogether* means 'in total', whereas *all together* means 'all at once' or 'all in one place'. The phrases *six rooms altogether* (in total) and *six rooms all together* (in one place) illustrate the difference.

altruism /'æltru:,ɪz(ə)m/ *n.* unselfishness as a principle of action. □ **altruist**

n. **altruistic** /ˌæltru:'ɪstɪk/ *adj.* [Italian *altrui* somebody else]
alum /'æləm/ *n.* double sulphate of aluminium and potassium. [Latin *alumen -min-*]
alumina /ə'lu:mɪnə/ *n.* aluminium oxide occurring naturally as corundum and emery.
aluminium /ˌæljʊ'mɪnɪəm/ *n.* (*US* **aluminum** /ə'lu:mɪnəm/) silvery light and malleable metallic element resistant to tarnishing by air.
aluminize /ə'lu:mɪˌnaɪz/ *v.* (also **-ise**) (**-zing** or **-sing**) coat with aluminium.
alumnus /ə'lʌmnəs/ *n.* (*pl.* **alumni** /-naɪ/; *fem.* **alumna**, *pl.* **alumnae** /-ni:/) former pupil or student. [Latin, = nursling, pupil]
always /'ɔ:lweɪz/ *adv.* **1** at all times; on all occasions. **2** whatever the circumstances. **3** repeatedly, often. [from ALL, WAY]
alyssum /'ælɪsəm/ *n.* plant with small usu. yellow or white flowers. [Greek, = curing madness]
Alzheimer's disease /'ælts,haɪməz/ *n.* brain disorder causing senility. [*Alzheimer*, name of a neurologist]
AM *abbr.* amplitude modulation.
Am *symb.* americium.
am *1st person sing. present* of BE.
a.m. *abbr.* before noon. [Latin *ante meridiem*]
amalgam /ə'mælgəm/ *n.* **1** mixture or blend. **2** alloy of mercury and another metal, used esp. in dentistry. [Greek *malagma* an emollient]
amalgamate /ə'mælgəˌmeɪt/ *v.* (**-ting**) **1** mix, unite. **2** (of metals) alloy with mercury. □ **amalgamation** /-'meɪʃ(ə)n/ *n.* [medieval Latin: related to AMALGAM]
amanuensis /əˌmænju:'ensɪs/ *n.* (*pl.* **-enses** /-si:z/) literary assistant, esp. writing from dictation. [Latin *a manu* 'at hand']
amaranth /'æməˌrænθ/ *n.* **1** plant with small green, red, or purple tinted flowers. **2** imaginary unfading flower. **3** purple colour. □ **amaranthine** /-'rænθaɪn/ *adj.* [Greek *amarantos* unfading]
amaryllis /ˌæmə'rɪlɪs/ *n.* bulbous plant with lily-like flowers. [Greek, a girl's name]
amass /ə'mæs/ *v.* heap together; accumulate. [French: related to AD-, MASS[1]]
amateur /'æmətə(r)/ *n.* person who engages in a pursuit as a pastime rather than a profession, or performs with limited skill. □ **amateurish** *adj.* **amateurism** *n.* [Latin *amator* lover: related to AMATORY]
amatory /'æmətərɪ/ *adj.* of sexual love. [Latin *amo* love]
amaze /ə'meɪz/ *v.* (**-zing**) surprise greatly, fill with wonder. □ **amazement** *n.* **amazing** *adj.* [earlier *amase* from Old English *āmasod*]
Amazon /'æməz(ə)n/ *n.* **1** female warrior of a mythical race in the Black Sea area. **2** (**amazon**) large, strong, or athletic woman. □ **Amazonian** /-'zəʊnɪən/ *adj.* [Latin from Greek]
ambassador /æm'bæsədə(r)/ *n.* **1** diplomat sent to live abroad to represent his or her country's interests. **2** promoter (*ambassador of peace*). □ **ambassadorial** /-'dɔ:rɪəl/ *adj.* [Latin *ambactus* servant]
amber —*n.* **1 a** yellow translucent fossilized resin used in jewellery. **b** colour of this. **2** yellow traffic-light meaning caution. —*adj.* of or like amber. [French from Arabic]
ambergris /'æmbəgrɪs/ *n.* waxlike secretion of the sperm whale, found floating in tropical seas and used in perfumes. [French, = grey amber]
ambidextrous /ˌæmbɪ'dekstrəs/ *adj.* able to use either hand equally well. [Latin *ambi-* on both sides, DEXTER]
ambience /'æmbɪəns/ *n.* surroundings or atmosphere. [Latin *ambio* go round]
ambient *adj.* surrounding.
ambiguous /æm'bɪgjʊəs/ *adj.* **1** having an obscure or double meaning. **2** difficult to classify. □ **ambiguity** /-'gju:ɪtɪ/ *n.* (*pl.* **-ies**). [Latin *ambi-* both ways, *ago* drive]
ambit /'æmbɪt/ *n.* scope, extent, or bounds. [Latin: related to AMBIENCE]
ambition /æm'bɪʃ(ə)n/ *n.* **1** determination to succeed. **2** object of this. [Latin, = canvassing: related to AMBIENCE]
ambitious *adj.* **1** full of ambition or high aims. **2** (foll. by *of*, or *to* + infin.) strongly determined.
ambivalence /æm'bɪvələns/ *n.* coexistence of opposing feelings. □ **ambivalent** *adj.* [Latin *ambo* both, EQUIVALENCE]
amble /'æmb(ə)l/ —*v.* (**-ling**) move at an easy pace. —*n.* such a pace. [Latin *ambulo* walk]
ambrosia /æm'brəʊzɪə/ *n.* **1** (in classical mythology) the food of the gods. **2** sublimely delicious food etc. [Greek, = elixir of life]
ambulance /'æmbjʊləns/ *n.* **1** vehicle equipped for conveying patients to hospital. **2** mobile hospital serving an army. [Latin: related to AMBLE]
ambulatory /'æmbjʊlətərɪ/ —*adj.* **1** of or for walking. **2** movable. —*n.* (*pl.* **-ies**)

arcade or cloister. [Latin: related to AMBLE]

ambuscade /ˌæmbəˈskeɪd/ *n.* & *v.* (**-ding**) = AMBUSH.

ambush /ˈæmbʊʃ/ –*n.* **1** surprise attack by persons hiding. **2** hiding-place for this. –*v.* attack from an ambush; waylay. [French: related to IN-[1], BUSH[1]]

ameliorate /əˈmiːlɪəˌreɪt/ *v.* (**-ting**) make or become better. □ **amelioration** /əˌmiːlɪəˈreɪʃ(ə)n/ *n.* **ameliorative** *adj.* [from AD-, Latin *melior* better]

amen /ɑːˈmen/ *int.* (esp. at the end of a prayer etc.) so be it. [Church Latin from Hebrew, = certainly]

amenable /əˈmiːnəb(ə)l/ *adj.* **1** responsive, docile. **2** (often foll. by *to*) answerable to law etc. [French: related to AD-, Latin *mino* drive animals]

amend /əˈmend/ *v.* **1** make minor alterations in to improve. **2** correct an error in (a document etc.). [Latin: related to EMEND]

■ **Usage** *Amend* is often confused with *emend*, a more technical word used in the context of textual correction.

amendment *n.* minor alteration or addition in a document, resolution, etc.

amends *n.* □ **make amends** (often foll. by *for*) compensate (for).

amenity /əˈmiːnɪtɪ/ *n.* (*pl.* **-ies**) **1** pleasant or useful feature or facility. **2** pleasantness (of a place etc.). [Latin *amoenus* pleasant]

American /əˈmerɪkən/ –*adj.* of America, esp. the United States. –*n.* **1** native, citizen, or inhabitant of America, esp. the US. **2** English as used in the US. □ **Americanize** *v.* (also **-ise**) (**-zing** or **-sing**). [name of navigator *Amerigo* Vespucci]

American dream *n.* ideal of democracy and prosperity.

American football *n.* football evolved from Rugby.

American Indian see INDIAN.

Americanism *n.* word etc. of US origin or usage.

americium /ˌæməˈrɪsɪəm/ *n.* artificial radioactive metallic element. [*America*, where first made]

Amerind /ˈæmərɪnd/ *adj.* & *n.* (also **Amerindian** /-ˈrɪndɪən/) = AMERICAN INDIAN (see INDIAN).

amethyst /ˈæməθɪst/ *n.* semiprecious stone of a violet or purple variety of quartz. [Greek, = preventing drunkenness]

Amharic /æmˈhærɪk/ –*n.* official and commercial language of Ethiopia. –*adj.* of this language. [*Amhara*, region of Ethiopia]

amiable /ˈeɪmɪəb(ə)l/ *adj.* (esp. of a person) friendly and pleasant, likeable. □ **amiably** *adv.* [Latin: related to AMICABLE]

amicable /ˈæmɪkəb(ə)l/ *adj.* (esp. of an arrangement, relations, etc.) friendly. □ **amicably** *adv.* [Latin *amicus* friend]

amid /əˈmɪd/ *prep.* in the middle of, among. [Old English: related to ON, MID]

amidships *adv.* in or into the middle of a ship. [from AMID, alternative form *midships*]

amidst var. of AMID.

amine /ˈeɪmiːn/ *n.* compound formed from ammonia by replacement of one or more hydrogen atoms by an organic radical or radicals.

amino acid /əˈmiːnəʊ/ *n. Biochem.* any of a group of nitrogenous organic acids occurring naturally in plant and animal tissues and forming the basic constituents of proteins. [from AMINE, ACID]

amir var. of EMIR.

amiss /əˈmɪs/ –*predic. adj.* wrong, out of order. –*adv.* wrong(ly), inappropriately (*everything went amiss*). □ **take amiss** be offended by. [Old Norse *à mis* so as to miss]

amity /ˈæmɪtɪ/ *n.* friendship. [Latin *amicus* friend]

ammeter /ˈæmɪtə(r)/ *n.* instrument for measuring electric current in amperes. [from AMPERE, -METER]

ammo /ˈæməʊ/ *n. slang* ammunition. [abbreviation]

ammonia /əˈməʊnɪə/ *n.* **1** pungent strongly alkaline gas. **2** (in general use) solution of ammonia in water. [as SAL AMMONIAC]

ammonite /ˈæməˌnaɪt/ *n.* coil-shaped fossil shell. [Latin, = horn of Jupiter Ammon]

ammunition /ˌæmjʊˈnɪʃ(ə)n/ *n.* **1** supply of bullets, shells, grenades, etc. **2** information usable in an argument. [French *la* MUNITION taken as *l'ammu-*]

amnesia /æmˈniːzɪə/ *n.* loss of memory. □ **amnesiac** /-zɪˌæk/ *n.* [Latin from Greek]

amnesty /ˈæmnɪstɪ/ –*n.* (*pl.* **-ies**) general pardon, esp. for political offences. –*v.* (**-ies, -ied**) grant an amnesty to. [Greek *amnēstia* oblivion]

amniocentesis /ˌæmnɪəʊsenˈtiːsɪs/ *n.* (*pl.* **-teses** /-siːz/) sampling of amniotic fluid to detect foetal abnormality. [from AMNION, Greek *kentēsis* pricking]

amnion /ˈæmnɪən/ *n.* (*pl.* **amnia**) innermost membrane enclosing an embryo. □ **amniotic** /-ˈɒtɪk/ *adj.* [Greek, = caul]

amoeba /əˈmiːbə/ *n.* (*pl.* **-s**) microscopic aquatic amorphous one-celled organism. □ **amoebic** *adj.* [Greek, = change]

amok /ə'mɒk/ *adv.* □ **run amok** (or **amuck**) run wild. [Malay]

among /ə'mʌŋ/ *prep.* (also **amongst**) **1** surrounded by, with (*lived among the trees; be among friends*). **2** included in (*among us were dissidents*). **3** in the category of (*among his best works*). **4 a** between; shared by (*divide it among you*). **b** from the joint resources of (*among us we can manage it*). **5** with one another (*talked among themselves*). [Old English, = in a crowd]

amoral /eɪ'mɒr(ə)l/ *adj.* **1** beyond morality. **2** without moral principles.

amorous /'æmərəs/ *adj.* of, showing, or feeling sexual love. [Latin *amor* love]

amorphous /ə'mɔ:fəs/ *adj.* **1** of no definite shape. **2** vague. **3** *Mineral.* & *Chem.* non-crystalline. [Greek *a-* not, *morphē* form]

amortize /ə'mɔ:taɪz/ *v.* (also **-ise**) (**-zing** or **-sing**) gradually extinguish (a debt) by regular instalments. [Latin *ad mortem* to death]

amount /ə'maʊnt/ *–n.* quantity, esp. a total in number, size, value, extent, etc. *–v.* (foll. by *to*) be equivalent to in number, significance, etc. [Latin *ad montem* upward]

amour /ə'mʊə(r)/ *n.* (esp. secret) love affair. [French, = love]

amour propre /æ,mʊə 'prɒpr/ *n.* self-respect. [French]

amp[1] *n.* ampere. [abbreviation]

amp[2] *n. colloq.* amplifier. [abbreviation]

ampelopsis /,æmpɪ'lɒpsɪs/ *n.* (*pl.* same) climbing plant related to the vine. [Greek *ampelos* vine, *opsis* appearance]

amperage /'æmpərɪdʒ/ *n.* strength of an electric current in amperes.

ampere /'æmpeə(r)/ *n.* SI base unit of electric current. [*Ampère*, name of a physicist]

ampersand /'æmpə,sænd/ *n.* the sign '&' (= *and*). [corruption of *and* PER SE *and*]

amphetamine /æm'fetə,mi:n/ *n.* synthetic drug used esp. as a stimulant. [abbreviation of chemical name]

amphibian /æm'fɪbɪən/ *–adj.* of a class of vertebrates (e.g. frogs) with an aquatic larval stage followed by a terrestrial adult stage. *–n.* **1** vertebrate of this class. **2** vehicle able to operate both on land and in water. [Greek *amphi-* both, *bios* life]

amphibious *adj.* **1** living or operating on land and in water. **2** involving military forces landed from the sea.

amphitheatre /'æmfɪ,θɪətə(r)/ *n.* esp. circular unroofed building with tiers of seats surrounding a central space. [Greek *amphi-* round]

amphora /'æmfərə/ *n.* (*pl.* **-phorae** /-,ri:/) narrow-necked Greek or Roman vessel with two handles. [Greek *amphoreus*]

ample /'æmp(ə)l/ *adj.* (**ampler, amplest**) **1 a** plentiful, abundant, extensive. **b** *euphem.* large, stout. **2** more than enough. □ **amply** *adv.* [Latin *amplus*]

amplifier /'æmplɪ,faɪə(r)/ *n.* electronic device for increasing the strength of electrical signals, esp. for conversion into sound.

amplify /'æmplɪ,faɪ/ *v.* (**-ies, -ied**) **1** increase the strength of (sound, electrical signals, etc.). **2** add detail to, expand (a story etc.). □ **amplification** /-fɪ'keɪʃ(ə)n/ *n.* [Latin: related to AMPLE]

amplitude /'æmplɪ,tju:d/ *n.* **1** maximum departure from average of an oscillation, alternating current, etc. **2** spaciousness; abundance. [Latin: related to AMPLE]

amplitude modulation *n.* modulation of a wave by variation of its amplitude.

ampoule /'æmpu:l/ *n.* small sealed capsule holding a solution for injection. [French: related to AMPULLA]

ampulla /æm'pʊlə/ *n.* (*pl.* **-pullae** /-li:/) **1** Roman globular flask with two handles. **2** ecclesiastical vessel. [Latin]

amputate /'æmpjʊ,teɪt/ *v.* (**-ting**) cut off surgically (a limb etc.). □ **amputation** /-'teɪʃ(ə)n/ *n.* **amputee** /-'ti:/ *n.* [Latin *amb-* about, *puto* prune]

amuck var. of AMOK.

amulet /'æmjʊlɪt/ *n.* charm worn against evil. [Latin]

amuse /ə'mju:z/ *v.* (**-sing**) **1** cause to laugh or smile. **2** interest or occupy. □ **amusing** *adj.* [French *a* cause to, *muser* stare]

amusement *n.* **1** thing that amuses. **2** being amused. **3** mechanical device (e.g. a roundabout) for entertainment at a fairground etc.

amusement arcade *n.* indoor area with slot-machines.

an see A[1].

an- see A-.

-an *suffix* (also **-ian**) forming adjectives and nouns, esp. from names of places, systems, classes, etc. (*Mexican; Anglican; crustacean*). [French *-ain*, Latin *-anus*]

Anabaptist /,ænə'bæptɪst/ *n.* member of a religious group believing in baptism only of adults. [Greek *ana* again]

anabolic steroid /,ænə'bɒlɪk/ *n.* synthetic steroid hormone used to increase muscle size.

anabolism /ə'næbə,lɪz(ə)m/ *n.* synthesis of complex molecules in living organisms from simpler ones together with the storage of energy. [Greek *anabolē* ascent]

anachronism /ə'nækrə,nız(ə)m/ *n.* **1 a** attribution of a custom, event, etc., to the wrong period. **b** thing thus attributed. **2** out-of-date person or thing. □ **anachronistic** /-'nıstık/ *adj.* [Greek *ana-* against, *khronos* time]

anaconda /,ænə'kɒndə/ *n.* large non-poisonous snake killing its prey by constriction. [Sinhalese]

anaemia /ə'ni:mıə/ *n.* (*US* **anemia**) deficiency of red blood cells or their haemoglobin, causing pallor and weariness. [Greek, = want of blood]

anaemic /ə'ni:mık/ *adj.* (*US* **anemic**) **1** of or suffering from anaemia. **2** pale, listless.

anaesthesia /,ænıs'θi:zıə/ *n.* (*US* **anes-**) absence of sensation, esp. artificially induced before surgery. [Greek]

anaesthetic /,ænıs'θetık/ (*US* **anes-**) –*n.* substance producing anaesthesia. –*adj.* producing anaesthesia.

anaesthetist /ə'ni:sθətıst/ *n.* (*US* **anes-**) specialist in the administration of anaesthetics.

anaesthetize /ə'ni:sθə,taız/ *v.* (*US* **anes-**) (also **-ise**) (**-zing** or **-sing**) administer an anaesthetic to.

anagram /'ænə,græm/ *n.* word or phrase formed by transposing the letters of another. [Greek *ana* again, *gramma* letter]

anal /'eın(ə)l/ *adj.* of the anus.

analgesia /,ænəl'dʒi:zıə/ *n.* absence or relief of pain. [Greek]

analgesic –*adj.* relieving pain. –*n.* analgesic drug.

analog *US* var. of ANALOGUE.

analogize /ə'nælə,dʒaız/ *v.* (also **-ise**) (**-zing** or **-sing**) use, or represent or explain by, analogy.

analogous /ə'næləgəs/ *adj.* (usu. foll. by *to*) partially similar or parallel. [Greek *analogos* proportionate]

analogue /'ænə,lɒg/ *n.* (*US* **analog**) **1** analogous thing. **2** (*attrib.*) (usu. **analog**) (of a computer etc.) using physical variables, e.g. voltage, to represent numbers (cf. DIGITAL).

analogy /ə'nælədʒı/ *n.* (*pl.* **-ies**) **1** correspondence; partial similarity. **2** arguing or reasoning from parallel cases. □ **analogical** /,ænə'lɒdʒık(ə)l/ *adj.* [Greek *analogia* proportion]

analyse /'ænə,laız/ *v.* (*US* **analyze**) (**-sing** or **-zing**) **1** examine in detail; ascertain the constituents of (a substance, sentence, etc.). **2** psychoanalyse.

analysis /ə'nælısıs/ *n.* (*pl.* **-lyses** /-,si:z/) **1 a** detailed examination of elements or structure. **b** statement of the result of this. **2** *Chem.* determination of the constituent parts of a mixture or compound. **3** psychoanalysis. [Greek *ana* up, *luō* loose]

analyst /'ænəlıst/ *n.* **1** person skilled in (esp. chemical or computer) analysis. **2** psychoanalyst.

analytical /,ænə'lıtık(ə)l/ *adj.* (also **analytic**) of or using analysis.

analyze *US* var. of ANALYSE.

anapaest /'ænə,pi:st/ *n.* metrical foot consisting of two short syllables followed by one long syllable (∪∪–). [Greek *anapaistos* reversed (dactyl)]

anarchism /'ænə,kız(ə)m/ *n.* political theory that all government and laws should be abolished. [French: related to ANARCHY]

anarchist /'ænəkıst/ *n.* advocate of anarchism. □ **anarchistic** /-'kıstık/ *adj.*

anarchy /'ænəkı/ *n.* disorder, esp. political. □ **anarchic** /ə'nɑ:kık/ *adj.* [Greek *an-* without, *arkhē* rule]

anathema /ə'næθəmə/ *n.* (*pl.* **-s**) **1** detested thing (*is anathema to me*). **2** ecclesiastical curse. [Greek, = thing devoted (i.e. to evil)]

anathematize *v.* (also **-ise**) (**-zing** or **-sing**) curse.

anatomy /ə'nætəmı/ *n.* (*pl.* **-ies**) **1** science of animal or plant structure. **2** such a structure. **3** analysis. □ **anatomical** /,ænə'tɒmık(ə)l/ *adj.* **anatomist** *n.* [Greek *ana-* up, *temnō* cut]

anatto var. of ANNATTO.

ANC *abbr.* African National Congress.

-ance *suffix* forming nouns expressing: **1** quality or state or an instance of one (*arrogance*; *resemblance*). **2** action (*assistance*). [French *-ance*, Latin *-antia*]

ancestor /'ænsestə(r)/ *n.* **1** person, animal, or plant from which another has descended or evolved. **2** prototype or forerunner. [Latin *ante-* before, *cedo* go]

ancestral /æn'sestr(ə)l/ *adj.* belonging to or inherited from one's ancestors.

ancestry /'ænsestrı/ *n.* (*pl.* **-ies**) **1** family descent, lineage. **2** ancestors collectively.

anchor /'æŋkə(r)/ –*n.* **1** heavy metal weight used to moor a ship or a balloon. **2** stabilizing thing. –*v.* **1** secure with an anchor. **2** fix firmly. **3** cast anchor. **4** be moored by an anchor. [Greek *agkura*]

anchorage *n.* **1** place for anchoring. **2** anchoring or lying at anchor.

anchorite /'æŋkə,raıt/ *n.* hermit; religious recluse. [Greek *anakhōreō* retire]

anchorman *n.* coordinator, esp. as compère in a broadcast.

anchovy /'æntʃəvı/ *n.* (*pl.* **-ies**) small strong-flavoured fish of the herring family. [Spanish and Portuguese *anchova*]

ancien régime /ɑ̃,sjæ̃ re'ʒi:m/ *n.* (*pl.* ***anciens régimes*** pronunc. same) **1** political and social system of pre-Revolutionary (before 1787) France. **2** any superseded regime. [French, = old rule]

ancient /'eɪnʃ(ə)nt/ *adj.* **1** of long ago, esp. before the fall of the Roman Empire in the West. **2** having lived or existed long. □ **the ancients** people of ancient times, esp. the Greeks and Romans. [Latin *ante* before]

ancillary /æn'sɪlərɪ/ *–adj.* **1** (esp. of health workers) providing essential support. **2** (often foll. by *to*) subordinate, subservient. *–n.* (*pl.* **-ies**) **1** ancillary worker. **2** auxiliary or accessory. [Latin *ancilla* handmaid]

-ancy *suffix* forming nouns denoting a quality (*constancy*) or state (*infancy*). [Latin *-antia*]

and /ænd, ənd/ *conj.* **1 a** connecting words, clauses, or sentences, to be taken jointly (*you and I*). **b** implying progression (*better and better*). **c** implying causation (*she hit him and he cried*). **d** implying great duration (*cried and cried*). **e** implying a great number (*miles and miles*). **f** implying addition (*two and two*). **g** implying variety (*there are books and books*). **2** *colloq.* to (*try and come*). **3** in relation to (*Britain and the EC*). □ **and/or** either or both of two stated alternatives. [Old English]

andante /æn'dæntɪ/ *Mus. –adv.* & *adj.* in a moderately slow tempo. *–n.* such a passage or movement. [Italian, = going]

andiron /'ænd,aɪən/ *n.* metal stand (usu. one of a pair) for supporting logs in a fireplace. [French *andier*]

androgynous /æn'drɒdʒɪnəs/ *adj.* **1** hermaphrodite. **2** *Bot.* with stamens and pistils in the same flower. [Greek *anēr andr-* man, *gunē* woman]

android /'ændrɔɪd/ *n.* robot with a human appearance. [Greek *anēr andr-* man, -OID]

anecdote /'ænɪk,dəʊt/ *n.* short, esp. true, account or story. □ **anecdotal** /-'dəʊt(ə)l/ *adj.* [Greek *anekdota* things unpublished]

anemia *US* var. of ANAEMIA.

anemic *US* var. of ANAEMIC.

anemometer /,ænɪ'mɒmɪtə(r)/ *n.* instrument for measuring wind force. [Greek *anemos* wind]

anemone /ə'nemənɪ/ *n.* plant of the buttercup family, with vividly-coloured flowers. [Greek, = wind-flower]

aneroid /'ænə,rɔɪd/ *–adj.* (of a barometer) measuring air-pressure by its action on the lid of a box containing a vacuum. *–n.* aneroid barometer. [Greek *a-* not, *nēros* water]

anesthesia etc. *US* var. of ANAESTHESIA etc.

aneurysm /'ænjʊ,rɪz(ə)m/ *n.* (also **aneurism**) excessive localized enlargement of an artery. [Greek *aneurunō* widen]

anew /ə'nju:/ *adv.* **1** again. **2** in a different way. [earlier *of newe*]

angel /'eɪndʒ(ə)l/ *n.* **1 a** attendant or messenger of God. **b** representation of this in human form with wings. **2** virtuous or obliging person. **3** *slang* financial backer of a play etc. [Greek *aggelos* messenger]

angel cake *n.* light sponge cake.

angel-fish *n.* fish with winglike fins.

angelic /æn'dʒelɪk/ *adj.* of or like an angel. □ **angelically** *adv.*

angelica /æn'dʒelɪkə/ *n.* aromatic plant or its candied stalks. [medieval Latin, = angelic (herb)]

angelus /'ændʒɪləs/ *n.* **1** Roman Catholic prayers commemorating the Incarnation, said at morning, noon, and sunset. **2** bell announcing this. [Latin *Angelus domini* (= the angel of the Lord), opening words]

anger /'æŋgə(r)/ *–n.* extreme or passionate displeasure. *–v.* make angry. [Old Norse *angr* grief]

angina /æn'dʒaɪnə/ *n.* (in full **angina pectoris** /'pektərɪs/) chest pain brought on by exertion, caused by an inadequate blood supply to the heart. [Greek *agkhonē* strangling]

angiosperm /'ændʒɪə,spɜ:m/ *n.* plant producing flowers and reproducing by seeds enclosed within a carpel, including herbaceous plants, grasses, and most trees. [Greek *aggeion* vessel]

Angle /'æŋg(ə)l/ *n.* (usu. in *pl.*) member of a N. German tribe that settled in E. Britain in the 5th c. [Latin *Anglus*, from the name *Angul* in Germany]

angle[1] /'æŋg(ə)l/ *–n.* **1** space between two meeting lines or surfaces, esp. as measured in degrees. **2** corner. **3** point of view. *–v.* (**-ling**) **1** move or place obliquely. **2** present (information) in a biased way. [Latin *angulus*]

angle[2] /'æŋg(ə)l/ *v.* (**-ling**) **1** fish with hook and line. **2** (foll. by *for*) seek an objective indirectly (*angled for a loan*). □ **angler** *n.* [Old English]

Anglican /'æŋglɪkən/ *–adj.* of the Church of England. *–n.* member of the Anglican Church. □ **Anglicanism** *n.* [Latin *Anglicanus*: related to ANGLE]

Anglicism /'æŋglɪ,sɪz(ə)m/ *n.* peculiarly English word or custom. [Latin *Anglicus*: related to ANGLE]

Anglicize /'æŋglɪ,saɪz/ *v.* (also **-ise**) (**-zing** or **-sing**) make English in character etc.

Anglo- *comb. form* **1** English. **2** of English origin. **3** English or British and. [Latin: related to ANGLE]

Anglo-Catholic /ˌæŋgləʊˈkæθəlɪk/ *–adj.* of a High Church Anglican wing emphasizing its Catholic tradition. *–n.* member of this group.

Anglo-French /ˌæŋgləʊˈfrentʃ/ *–adj.* English (or British) and French. *–n.* French language as developed in England after the Norman Conquest.

Anglo-Indian /ˌæŋgləʊˈɪndɪən/ *–adj.* **1** of England and India. **2** of British descent but Indian residence. *–n.* Anglo-Indian person.

Anglo-Norman /ˌæŋgləʊˈnɔːmən/ *–adj.* English and Norman. *–n.* Norman dialect used in England after the Norman Conquest.

Anglophile /ˈæŋgləʊˌfaɪl/ *n.* person who greatly admires England or the English.

Anglo-Saxon /ˌæŋgləʊˈsæks(ə)n/ *–adj.* **1** of the English Saxons before the Norman Conquest. **2** of English descent. *–n.* **1** Anglo-Saxon person. **2** Old English. **3** *colloq.* plain (esp. crude) English.

angora /æŋˈgɔːrə/ *n.* **1** fabric or wool from the hair of the angora goat or rabbit. **2** long-haired variety of cat, goat, or rabbit. [*Angora* (= Ankara) in Turkey]

angostura /ˌæŋgəˈstjʊərə/ *n.* aromatic bitter bark used as a flavouring. [*Angostura* (= Ciudad Bolívar) in Venezuela]

angry /ˈæŋgrɪ/ *adj.* (**-ier**, **-iest**) **1** feeling, showing, or suggesting anger (*angry sky*). **2** (of a wound etc.) inflamed, painful. □ **angrily** *adv.*

angst /æŋst/ *n.* anxiety, neurotic fear; guilt, remorse. [German]

angstrom /ˈæŋstrəm/ *n.* unit of length equal to 10^{-10} metre. [*Ångström*, name of a physicist]

anguish /ˈæŋgwɪʃ/ *n.* **1** severe mental suffering. **2** pain, agony. □ **anguished** *adj.* [Latin *angustia* tightness]

angular /ˈæŋgjʊlə(r)/ *adj.* **1** having sharp corners or (of a person) features. **2** forming an angle. **3** measured by angle (*angular distance*). □ **angularity** /-ˈlærɪtɪ/ *n.* [Latin: related to ANGLE[1]]

anhydrous /ænˈhaɪdrəs/ *adj. Chem.* without water, esp. water of crystallization. [Greek *an-* without, *hudōr* water]

aniline /ˈænɪˌliːn/ *n.* colourless oily liquid used in making dyes, drugs, and plastics. [German *Anil* indigo, former source]

animadvert /ˌænɪmædˈvɜːt/ *v.* (foll. by *on*) *literary* criticize, censure. □ **animadversion** *n.* [Latin *animus* mind, ADVERSE]

animal /ˈænɪm(ə)l/ *–n.* **1** living organism, esp. other than man, which feeds and usu. has sense-organs and a nervous system and can move quickly. **2** brutish person. *–adj.* **1** of or like an animal. **2** bestial; carnal. [Latin *animalis* having breath]

animalism *n.* **1** nature and activity of animals. **2** belief that humans are mere animals.

animality /ˌænɪˈmælɪtɪ/ *n.* **1** the animal world. **2** animal behaviour.

animalize /ˈænɪməˌlaɪz/ *v.* (also **-ise**) (**-zing** or **-sing**) make (a person) bestial, sensualize.

animate *–adj.* /ˈænɪmət/ **1** having life. **2** lively. *–v.* /ˈænɪˌmeɪt/ (**-ting**) **1** enliven. **2** give life to. [Latin *anima* breath]

animated /ˈænɪˌmeɪtɪd/ *adj.* **1** lively, vigorous. **2** having life. **3** (of a film etc.) using animation.

animation /ˌænɪˈmeɪʃ(ə)n/ *n.* **1** vivacity, ardour. **2** being alive. **3** technique of producing a moving picture from a sequence of drawings or puppet poses etc.

animism /ˈænɪˌmɪz(ə)m/ *n.* belief that inanimate and natural phenomena have souls. □ **animist** *n.* **animistic** /-ˈmɪstɪk/ *adj.*

animosity /ˌænɪˈmɒsɪtɪ/ *n.* (*pl.* **-ies**) spirit or feeling of hostility. [Latin: related to ANIMUS]

animus /ˈænɪməs/ *n.* animosity, ill feeling. [Latin, = spirit, mind]

anion /ˈænˌaɪən/ *n.* negatively charged ion. □ **anionic** /-ˈɒnɪk/ *adj.* [Greek *ana* up, ION]

anise /ˈænɪs/ *n.* plant with aromatic seeds. [Greek *anison*]

aniseed /ˈænɪˌsiːd/ *n.* seed of the anise, used for flavouring.

ankle /ˈæŋk(ə)l/ *n.* **1** joint connecting the foot with the leg. **2** this part of the leg. [Old Norse]

anklet /ˈæŋklɪt/ *n.* ornament or fetter worn round the ankle.

ankylosis /ˌæŋkɪˈləʊsɪs/ *n.* stiffening of a joint by fusion of the bones. [Greek *agkulos* crooked]

annals /ˈæn(ə)lz/ *n.pl.* **1** narrative of events year by year. **2** historical records. □ **annalist** *n.* [Latin *annus* year]

annatto /əˈnætəʊ/ *n.* (also **anatto**) orange-red dye from the pulp of a tropical fruit, used for colouring foods. [Carib name of the fruit-tree]

anneal /əˈniːl/ *v.* heat (metal or glass) and cool slowly, esp. to toughen it. [Old English *ǣlan* bake]

annelid /ˈænəlɪd/ *n.* segmented worm, e.g. the earthworm. [Latin *anulus* ring]

annex /æˈneks/ *v.* **1** (often foll. by *to*) add as a subordinate part. **2** incorporate (territory) into one's own. **3** add as a condition or consequence. **4** *colloq.* take without right. □ **annexation** /-'seɪʃ(ə)n/ *n.* [Latin *necto* bind]

annexe /'æneks/ *n.* **1** separate or added building. **2** addition to a document.

annihilate /ə'naɪə,leɪt/ *v.* (**-ting**) completely destroy or defeat. □ **annihilation** /-'leɪʃ(ə)n/ *n.* [Latin *nihil* nothing]

anniversary /,ænɪ'vɜ:sərɪ/ *n.* (*pl.* **-ies**) **1** date of an event in a previous year. **2** celebration of this. [Latin *annus* year, *verto vers-* turn]

Anno Domini /,ænəʊ 'dɒmɪ,naɪ/ *adv.* years after Christ's birth. [Latin, = in the year of the Lord]

annotate /'ænə,teɪt/ *v.* (**-ting**) add explanatory notes to. □ **annotation** /-'teɪʃ(ə)n/ *n.* [Latin *nota* mark]

announce /ə'naʊns/ *v.* (**-cing**) **1** make publicly known. **2** make known the arrival or imminence of (a guest, dinner, etc.). **3** be a sign of. □ **announcement** *n.* [Latin *nuntius* messenger]

announcer *n.* person who announces, esp. in broadcasting.

annoy /ə'nɔɪ/ *v.* **1** (often in *passive*) anger or distress slightly (*am annoyed with you*). **2** molest, harass. □ **annoyance** *n.* [Latin *in odio* hateful]

annual /'ænjʊəl/ *–adj.* **1** reckoned by the year. **2** occurring yearly. **3** living or lasting (only) a year. *–n.* **1** book etc. published yearly. **2** plant that lives only a year. □ **annually** *adv.* [Latin *annus* year]

annualized *adj.* (of rates of interest etc.) calculated on an annual basis, as a projection from figures obtained for a shorter period.

annuity /ə'nju:ɪtɪ/ *n.* (*pl.* **-ies**) **1** yearly grant or allowance. **2** investment yielding a fixed annual sum.

annul /ə'nʌl/ *v.* (**-ll-**) **1** declare invalid. **2** cancel, abolish. □ **annulment** *n.* [Latin *nullus* none]

annular /'ænjʊlə(r)/ *adj.* ring-shaped. [Latin *anulus* ring]

annular eclipse *n.* solar eclipse in which a ring of light remains visible.

annulate /'ænjʊlət/ *adj.* marked with or formed of rings.

annunciation /ə,nʌnsɪ'eɪʃ(ə)n/ *n.* **1** announcement, esp. (**Annunciation**) that made by the angel Gabriel to Mary. **2** festival of this. [Latin: related to ANNOUNCE]

anode /'ænəʊd/ *n.* positive electrode in an electrolytic cell etc. [Greek *anodos* way up]

anodize /'ænə,daɪz/ *v.* (also **-ise**) (**-zing** or **-sing**) coat (metal) with a protective layer by electrolysis.

anodyne /'ænə,daɪn/ *–adj.* **1** pain-relieving. **2** mentally soothing. *–n.* anodyne drug etc. [Greek *an-* without, *odunē* pain]

anoint /ə'nɔɪnt/ *v.* **1** apply oil or ointment to, esp. ritually. **2** (usu. foll. by *with*) smear. [Latin *inungo* anoint]

anomalous /ə'nɒmələs/ *adj.* irregular, deviant, abnormal. [Greek *an-* not, *homalos* even]

anomaly /ə'nɒməlɪ/ *n.* (*pl.* **-ies**) anomalous thing; irregularity.

anon /ə'nɒn/ *adv. archaic* soon, shortly. [Old English *on ān* into one]

anon. /ə'nɒn/ *abbr.* anonymous.

anonymous /ə'nɒnɪməs/ *adj.* **1** of unknown name or authorship. **2** without character; featureless. □ **anonymity** /,ænə'nɪmɪtɪ/ *n.* [Greek *an-* without, *onoma* name]

anorak /'ænə,ræk/ *n.* waterproof usu. hooded jacket. [Eskimo]

anorexia /,ænə'reksɪə/ *n.* lack of appetite, esp. (in full **anorexia nervosa** /nɜ:'vəʊsə/) an obsessive desire to lose weight by refusing to eat. □ **anorexic** *adj.* & *n.* [Greek *an-* without, *orexis* appetite]

another /ə'nʌðə(r)/ *–adj.* **1** an additional; one more (*another cake*). **2** person like (*another Hitler*). **3** a different (*another matter*). **4** some other (*another man's work*). *–pron.* additional, other, or different person or thing. [earlier *an other*]

answer /'ɑ:nsə(r)/ *–n.* **1** something said or done in reaction to a question, statement, or circumstance. **2** solution to a problem. *–v.* **1** make an answer or response (to) (*answer the door*). **2** suit (a purpose or need). **3** (foll. by *to, for*) be responsible (*you will answer to me for your conduct*). **4** (foll. by *to*) correspond to (esp. a description). □ **answer back** answer insolently. [Old English, = swear against (a charge)]

answerable *adj.* **1** (usu. foll. by *to, for*) responsible (*answerable to them for any accident*). **2** that can be answered.

answering machine *n.* tape recorder which answers telephone calls and takes messages.

answerphone *n.* = ANSWERING MACHINE.

ant *n.* small usu. wingless insect living in complex social colonies and proverbial for industry. [Old English]

-ant *suffix* **1** forming adjectives denoting attribution of an action (*repentant*) or state (*arrogant*). **2** forming agent nouns (*assistant*). [Latin *-ant-*, present participial stem of verbs]

antacid /ænt'æsɪd/ *–adj.* preventing or correcting acidity. *–n.* antacid agent.

antagonism /æn'tægə,nɪz(ə)m/ *n.* active hostility. [French: related to AGONY]

antagonist *n.* opponent or adversary. □ **antagonistic** /-'nɪstɪk/ *adj.*

antagonize /æn'tægə,naɪz/ *v.* (also **-ise**) (**-zing** or **-sing**) make hostile; provoke.

Antarctic /ænt'ɑ:ktɪk/ *–adj.* of the south polar regions. *–n.* this region. [Latin: related to ARCTIC]

Antarctic Circle *n.* parallel of latitude 66° 32′ S., forming an imaginary line round the Antarctic region.

ante /'æntɪ/ *–n.* **1** stake put up by a player in poker etc. before receiving cards. **2** amount payable in advance. *–v.* (**-tes, -ted**) **1** put up as an ante. **2** *US* **a** bet, stake. **b** (foll. by *up*) pay.

ante- *prefix* before, preceding. [Latin, = before]

anteater *n.* any of various mammals feeding on ants and termites.

antecedent /,æntɪ'si:d(ə)nt/ *–n.* **1** preceding thing or circumstance. **2** *Gram.* word or phrase etc. to which another word (esp. a relative pronoun) refers. **3** (in *pl.*) person's past history or ancestors. *–adj.* previous. [Latin *cedo* go]

antechamber /'æntɪ,tʃeɪmbə(r)/ *n.* ante-room.

antedate /,æntɪ'deɪt/ *v.* (**-ting**) **1** precede in time. **2** assign an earlier than actual date to.

antediluvian /,æntɪdɪ'lu:vɪən/ *adj.* **1** of the time before the Flood. **2** *colloq.* very old or out of date. [from ANTE-, Latin *diluvium* deluge]

antelope /'æntɪ,ləʊp/ *n.* (*pl.* same or **-s**) swift-moving deerlike ruminant, e.g. the gazelle and gnu. [Greek *antholops*]

antenatal /,æntɪ'neɪt(ə)l/ *adj.* **1** before birth. **2** of pregnancy.

antenna /æn'tenə/ *n.* **1** (*pl.* **-tennae** /-ni:/) each of a pair of feelers on the heads of insects, crustaceans, etc. **2** (*pl.* **-s**) = AERIAL *n.* [Latin, = sail-yard]

antepenultimate /,æntɪpɪ'nʌltɪmət/ *adj.* last but two.

ante-post /,æntɪ'pəʊst/ *adj.* (of betting) done at odds determined at the time of betting, in advance of the event concerned. [from ANTE-, POST[1]]

anterior /æn'tɪərɪə(r)/ *adj.* **1** nearer the front. **2** (often foll. by *to*) prior. [Latin from *ante* before]

ante-room /'æntɪ,ru:m/ *n.* small room leading to a main one.

anthem /'ænθəm/ *n.* **1** elaborate choral composition usu. based on a passage of scripture. **2** solemn hymn of praise etc., esp. = NATIONAL ANTHEM. [Latin: related to ANTIPHON]

anther /'ænθə(r)/ *n.* part of a stamen containing pollen. [Greek *anthos* flower]

anthill *n.* moundlike nest built by ants or termites.

anthology /æn'θɒlədʒɪ/ *n.* (*pl.* **-ies**) collection of poems, essays, stories, etc. □ **anthologist** *n.* [Greek *anthos* flower, *-logia* collection]

anthracite /'ænθrə,saɪt/ *n.* hard type of coal burning with little flame and smoke. [Greek: related to ANTHRAX]

anthrax /'ænθræks/ *n.* disease of sheep and cattle transmissible to humans. [Greek, = coal, carbuncle]

anthropocentric /,ænθrəpəʊ'sentrɪk/ *adj.* regarding mankind as the centre of existence. [Greek *anthrōpos* man]

anthropoid /'ænθrə,pɔɪd/ *–adj.* human in form. *–n.* anthropoid ape.

anthropology /,ænθrə'pɒlədʒɪ/ *n.* study of mankind, esp. its societies and customs. □ **anthropological** /-pə'lɒdʒɪk(ə)l/ *adj.* **anthropologist** *n.*

anthropomorphism /,ænθrəpə'mɔ:fɪz(ə)m/ *n.* attribution of human characteristics to a god, animal, or thing. □ **anthropomorphic** *adj.* [Greek *morphē* form]

anthropomorphous /,ænθrəpə'mɔ:fəs/ *adj.* human in form.

anti /'æntɪ/ *–prep.* opposed to. *–n.* (*pl.* **-s**) person opposed to a policy etc.

anti- *prefix* **1** opposed to (*anticlerical*). **2** preventing (*antifreeze*). **3** opposite of (*anticlimax*). **4** unconventional (*anti-hero*). [Greek]

anti-abortion /,æntɪə'bɔ:ʃ(ə)n/ *adj.* opposing abortion. □ **anti-abortionist** *n.*

anti-aircraft /,æntɪ'eəkrɑ:ft/ *adj.* (of a gun or missile) used to attack enemy aircraft.

antibiotic /,æntɪbaɪ'ɒtɪk/ *–n.* substance (e.g. penicillin) that can inhibit or destroy susceptible micro-organisms. *–adj.* functioning as an antibiotic. [Greek *bios* life]

antibody /'æntɪ,bɒdɪ/ *n.* (*pl.* **-ies**) a blood protein produced in response to and then counteracting antigens. [translation of German *Antikörper*]

antic /'æntɪk/ *n.* (usu. in *pl.*) foolish behaviour or action. [Italian *antico* ANTIQUE]

Antichrist /'æntɪ,kraɪst/ *n.* enemy of Christ. □ **antichristian** /-'krɪstʃ(ə)n/ *adj.*

anticipate /æn'tɪsɪ,peɪt/ *v.* (**-ting**) **1** deal with or use before the proper time. **2** expect, foresee (*did not anticipate a problem*). **3** forestall (a person or thing). **4** look forward to. □ **anticipation** /-'peɪʃ(ə)n/ *n.* **anticipatory** *adj.* [Latin *anti-* before, *capio* take]

■ **Usage** The use of *anticipate* in sense 2, 'expect', 'foresee', is well-established in informal use, but is regarded as incorrect by some people.

anticlerical /ˌæntɪ'klerɪk(ə)l/ *adj.* opposed to clerical influence, esp. in politics.

anticlimax /ˌæntɪ'klaɪmæks/ *n.* disappointingly trivial conclusion to something significant.

anticlockwise /ˌæntɪ'klɒkwaɪz/ *adj. & adv.* moving in a curve opposite in direction to the hands of a clock.

anticyclone /ˌæntɪ'saɪkləʊn/ *n.* system of winds rotating outwards from an area of high pressure, producing fine weather.

antidepressant /ˌæntɪdɪ'pres(ə)nt/ *–n.* drug etc. that alleviates depression. *–adj.* alleviating depression.

antidote /'æntɪˌdəʊt/ *n.* **1** medicine etc. used to counteract poison. **2** anything counteracting something unpleasant. [Greek *antidotos* given against]

antifreeze /'æntɪˌfri:z/ *n.* substance added to water to lower its freezing-point, esp. in a vehicle's radiator.

antigen /'æntɪdʒ(ə)n/ *n.* foreign substance (e.g. toxin) which causes the body to produce antibodies. [Greek *-genēs* of a kind]

anti-hero /'æntɪˌhɪərəʊ/ *n.* (*pl.* **-es**) central character in a story, lacking conventional heroic qualities.

antihistamine /ˌæntɪ'hɪstəˌmi:n/ *n.* drug that counteracts the effects of histamine, used esp. in treating allergies.

antiknock /'æntɪˌnɒk/ *n.* substance added to motor fuel to prevent premature combustion.

anti-lock /'æntɪˌlɒk/ *attrib. adj.* (of brakes) set up so as to prevent locking and skidding when applied suddenly.

antilog /'æntɪˌlɒg/ *n. colloq.* = ANTILOGARITHM. [abbreviation]

antilogarithm /ˌæntɪ'lɒgəˌrɪð(ə)m/ *n.* number to which a logarithm belongs.

antimacassar /ˌæntɪmə'kæsə(r)/ *n.* detachable protective cloth for the back of a chair etc.

antimatter /'æntɪˌmætə(r)/ *n.* matter composed solely of antiparticles.

antimony /'æntɪmənɪ/ *n.* brittle silvery metallic element used esp. in alloys. [medieval Latin]

antinomian /ˌæntɪ'nəʊmɪən/ *–adj.* believing that Christians need not obey the moral law. *–n.* **(Antinomian)** *hist.* person believing this. [Greek *nomos* law]

antinomy /æn'tɪnəmɪ/ *n.* (*pl.* **-ies**) contradiction between two reasonable beliefs or conclusions.

antinovel /'æntɪˌnɒv(ə)l/ *n.* novel avoiding the conventions of the form.

anti-nuclear /ˌæntɪ'nju:klɪə(r)/ *adj.* opposed to the development of nuclear weapons or power.

antiparticle /'æntɪˌpɑ:tɪk(ə)l/ *n.* elementary particle with the same mass but opposite charge etc. to another particle.

antipathy /æn'tɪpəθɪ/ *n.* (*pl.* **-ies**) (often foll. by *to, for, between*) strong aversion or dislike. □ **antipathetic** /-'θetɪk/ *adj.* [Greek: related to PATHETIC]

antiperspirant /ˌæntɪ'pɜ:spərənt/ *n.* substance preventing or reducing perspiration.

antiphon /'æntɪf(ə)n/ *n.* **1** hymn sung alternately by two groups. **2** versicle or phrase from this. □ **antiphonal** /-'tɪfən(ə)l/ *adj.* [Greek *phōnē* sound]

antipodes /æn'tɪpəˌdi:z/ *n.pl.* places diametrically opposite to one another on the earth, esp. (also **Antipodes**) Australasia in relation to Europe. □ **antipodean** /-'di:ən/ *adj. & n.* [Greek, = having the feet opposite]

antipope /'æntɪˌpəʊp/ *n.* pope set up in opposition to one chosen by canon law.

antipyretic /ˌæntɪpaɪə'retɪk/ *–adj.* preventing or reducing fever. *–n.* antipyretic drug.

antiquarian /ˌæntɪ'kweərɪən/ *–adj.* of or dealing in antiques or rare books. *–n.* antiquary. □ **antiquarianism** *n.*

antiquary /'æntɪkwərɪ/ *n.* (*pl.* **-ies**) student or collector of antiques etc. [Latin: related to ANTIQUE]

antiquated /'æntɪˌkweɪtɪd/ *adj.* old-fashioned.

antique /æn'ti:k/ *–n.* old object, esp. a piece of furniture, of high value. *–adj.* **1** of or from an early date. **2** old-fashioned. [Latin *antiquus*]

antiquity /æn'tɪkwətɪ/ *n.* (*pl.* **-ies**) **1** ancient times, esp. before the Middle Ages. **2** great age. **3** (usu. in *pl.*) relics from ancient times. [Latin: related to ANTIQUE]

antirrhinum /ˌæntɪ'raɪnəm/ *n.* plant with two-lipped flowers, esp. the snapdragon. [Greek, = snout]

anti-Semite /ˌæntɪ'si:maɪt/ *n.* person who is prejudiced against Jews. □ **anti-Semitic** /-sɪ'mɪtɪk/ *adj.* **anti-Semitism** /-'semɪˌtɪz(ə)m/ *n.*

antiseptic /ˌæntɪ'septɪk/ *–adj.* **1** counteracting sepsis, esp. by destroying germs. **2** sterile, uncontaminated. **3** lacking character. *–n.* antiseptic agent.

antiserum /'æntɪˌsɪərəm/ *n.* serum with a high antibody content.

antisocial /ˌæntɪ'səʊʃ(ə)l/ *adj.* **1** opposed or harmful to society. **2** not sociable.

■ **Usage** *Antisocial* is sometimes used mistakenly instead of *unsocial* in the phrase *unsocial hours*. This should be avoided.

antistatic /,æntɪ'stætɪk/ *adj.* counteracting the effects of static electricity.
anti-tank /,æntɪ'tæŋk/ *attrib. adj.* used against tanks.
antitetanus /,æntɪ'tetənəs/ *adj.* effective against tetanus.
antithesis /æn'tɪθəsɪs/ *n.* (*pl.* **-theses** /-,si:z/) **1** (foll. by *of, to*) direct opposite. **2** contrast. **3** rhetorical use of strongly contrasted words. □ **antithetical** /-'θetɪk(ə)l/ *adj.* [Greek *antitithēmi* set against]
antitoxin /,æntɪ'tɒksɪn/ *n.* antibody counteracting a toxin. □ **antitoxic** *adj.*
antitrades /'æntɪ,treɪdz/ *n.pl.* winds blowing in the opposite direction to (and usu. above) trade winds.
antiviral /,æntɪ'vaɪər(ə)l/ *adj.* effective against viruses.
antler *n.* branched horn of a stag or other deer. □ **antlered** *adj.* [French]
antonym /'æntənɪm/ *n.* word opposite in meaning to another. [Greek *onoma* name]
antrum /'æntrəm/ *n.* (*pl.* **antra**) natural cavity in the body, esp. in a bone. [Greek, = cave]
anus /'eɪnəs/ *n.* (*pl.* **anuses**) excretory opening at the end of the alimentary canal. [Latin]
anvil /'ænvɪl/ *n.* iron block on which metals are worked. [Old English]
anxiety /æŋ'zaɪətɪ/ *n.* (*pl.* **-ies**) **1** being anxious. **2** worry or concern. **3** eagerness, troubled desire. [Latin *anxietas* from *ango* choke]
anxious /'æŋkʃəs/ *adj.* **1** mentally troubled. **2** causing or marked by anxiety (*anxious moment*). **3** eager, uneasily wanting (*anxious to please*). □ **anxiously** *adv.* [Latin *anxius*]
any /'enɪ/ –*adj.* **1 a** one, no matter which, of several (*cannot find any answer*). **b** some, no matter how much or many or of what sort (*if any books arrive*; *have you any sugar?*). **2** a minimal amount of (*hardly any difference*). **3** whichever is chosen (*any fool knows*). **4** an appreciable or significant (*did not stay for any length of time*; *has any amount of money*). –*pron.* **1** any one (*did not know any of them*). **2** any number or amount (*are any of them yours?*). –*adv.* (usu. with *neg.* or *interrog.*) at all (*is that any good?*). [Old English *ǣnig*: related to ONE, -Y[1]]
anybody *n.* & *pron.* **1** any person. **2** person of importance (*is he anybody?*).
anyhow *adv.* **1** anyway. **2** in a disorderly manner or state (*does his work anyhow*).
anyone *pron.* anybody.

■ **Usage** *Anyone* is written as two words to emphasize a numerical sense, as in *any one of us can do it*.

anything *pron.* any thing; thing of any sort. □ **anything but** not at all.
anyway *adv.* **1** in any way or manner. **2** at any rate. **3** to resume (*anyway, as I was saying*).
anywhere –*adv.* in or to any place. –*pron.* any place (*anywhere will do*).
AOB *abbr.* any other business.
aorta /eɪ'ɔ:tə/ *n.* (*pl.* **-s**) main artery, giving rise to the arterial network carrying oxygenated blood to the body from the heart. □ **aortic** *adj.* [Greek *aeirō* raise]
apace /ə'peɪs/ *adv. literary* swiftly. [French *à pas*]
Apache /ə'pætʃɪ/ *n.* member of a N. American Indian tribe. [Mexican Spanish]
apart /ə'pɑ:t/ *adv.* **1** separately, not together (*keep your feet apart*). **2** into pieces (*came apart*). **3** to or on one side. **4** to or at a distance. □ **apart from 1** excepting, not considering. **2** in addition to (*apart from roses we grow irises*). [French *à part* to one side]
apartheid /ə'pɑ:teɪt/ *n.* (esp. in S. Africa) racial segregation or discrimination. [Afrikaans]
apartment /ə'pɑ:tmənt/ *n.* **1** (in *pl.*) suite of rooms. **2** single room. **3** *US* flat. [Italian *a parte*, apart]
apathy /'æpəθɪ/ *n.* lack of interest; indifference. □ **apathetic** /-'θetɪk/ *adj.* [Greek *a-* without, PATHOS]
ape –*n.* **1** tailless monkey-like primate, e.g. the gorilla, chimpanzee, orangutan, or gibbon. **2** imitator. –*v.* (**-ping**) imitate, mimic. [Old English]
apeman *n.* extinct primate held to be the forerunner of present-day man.
aperient /ə'pɪərɪənt/ –*adj.* laxative. –*n.* laxative medicine. [Latin *aperio* open]
aperitif /ə,perɪ'ti:f/ *n.* alcoholic drink taken before a meal. [Latin *aperio* open]
aperture /'æpə,tʃə(r)/ *n.* opening or gap, esp. a variable opening in a camera for admitting light. [Latin *aperio* open]
Apex /'eɪpeks/ *n.* (also **APEX**) (often *attrib.*) system of reduced fares for scheduled flights. [*A*dvance *P*urchase *Ex*cursion]
apex /'eɪpeks/ *n.* (*pl.* **-es**) **1** highest point. **2** tip or pointed end. [Latin]
aphasia /ə'feɪzɪə/ *n.* loss of verbal understanding or expression, owing to

brain damage. [Greek *aphatos* speechless]

aphelion /ə'fi:lɪən/ *n.* (*pl.* **-lia**) point in a celestial body's orbit where it is furthest from the sun. [Greek *aph'hēliou* from the sun]

aphid /'eɪfɪd/ *n.* small insect infesting and damaging plants, e.g. the greenfly.

aphis /'eɪfɪs/ *n.* (*pl.* **aphides** /-,di:z/) aphid. [invented by Linnaeus: perhaps a misreading of Greek *koris* bug]

aphorism /'æfə,rɪz(ə)m/ *n.* short pithy maxim. □ **aphoristic** /-'rɪstɪk/ *adj.* [Greek *aphorismos* definition]

aphrodisiac /,æfrə'dɪzɪ,æk/ *—adj.* arousing sexual desire. *—n.* aphrodisiac substance. [Greek *Aphroditē* goddess of love]

apiary /'eɪpɪərɪ/ *n.* (*pl.* **-ies**) place where bees are kept. □ **apiarist** *n.* [Latin *apis* bee]

apical /'eɪpɪk(ə)l/ *adj.* of, at, or forming an apex.

apiculture /'eɪpɪ,kʌltʃə(r)/ *n.* bee-keeping. □ **apiculturist** /-'kʌltʃərɪst/ *n.* [Latin *apis* bee, CULTURE]

apiece /ə'pi:s/ *adv.* for each one; severally (*five pounds apiece*). [originally *a piece*]

apish /'eɪpɪʃ/ *adj.* **1** of or like an ape. **2** foolishly imitating.

aplomb /ə'plɒm/ *n.* skilful self-assurance. [French, = straight as a plummet]

apocalypse /ə'pɒkəlɪps/ *n.* **1** violent or destructive event. **2** (**the Apocalypse**) Revelation, the last book of the New Testament. **3** revelation, esp. about the end of the world. □ **apocalyptic** /-'lɪptɪk/ *adj.* [Greek *apokaluptō* reveal]

Apocrypha /ə'pɒkrɪfə/ *n.pl.* **1** books included in the Septuagint and Vulgate versions of the Old Testament but not in the Hebrew Bible. **2** (**apocrypha**) writings etc. not considered genuine. [Greek *apokruptō* hide away]

apocryphal *adj.* of doubtful authenticity.

apogee /'æpə,dʒi:/ *n.* **1** highest point; climax. **2** point in a celestial body's orbit where it is furthest from the earth. [Greek *apogeion*]

apolitical /,eɪpə'lɪtɪk(ə)l/ *adj.* not interested in or concerned with politics.

apologetic /ə,pɒlə'dʒetɪk/ *—adj.* **1** showing or expressing regret. **2** of apologetics. *—n.* (usu. in *pl.*) reasoned defence, esp. of Christianity. □ **apologetically** *adv.*

apologia /,æpə'ləʊdʒɪə/ *n.* formal defence of opinions or conduct. [Greek: see APOLOGY]

apologist /ə'pɒlədʒɪst/ *n.* person who defends something by argument.

apologize /ə'pɒlə,dʒaɪz/ *v.* (also **-ise**) (**-zing** or **-sing**) make an apology, express regret.

apology /ə'pɒlədʒɪ/ *n.* (*pl.* **-ies**) **1** statement of regret for an offence or failure. **2** explanation or defence. **3** (foll. by *for*) poor specimen of. [Greek *apologia* from *apologeomai* speak in defence]

apophthegm /'æpə,θem/ *n.* = APHORISM. [Latin from Greek]

apoplectic /,æpə'plektɪk/ *adj.* **1** of or causing apoplexy. **2** *colloq.* enraged.

apoplexy /'æpə,pleksɪ/ *n.* sudden paralysis caused by blockage or rupture of a brain artery; stroke. [Greek *apoplēssō* disable by stroke]

apostasy /ə'pɒstəsɪ/ *n.* (*pl.* **-ies**) renunciation of a belief or faith, abandoning of principles, etc. [Greek, = defection]

apostate /ə'pɒsteɪt/ *n.* person who renounces a belief etc. □ **apostatize** *v.* (also **-ise**) (**-zing** or **-sing**).

a posteriori /,eɪ pɒ,sterɪ'ɔ:raɪ/ *—adj.* (of reasoning) proceeding from effects to causes; inductive. *—adv.* inductively. [Latin, = from what comes after]

apostle /ə'pɒs(ə)l/ *n.* **1** (**Apostle**) any of the twelve men sent out by Christ to preach the gospel. **2** leader, esp. of a new movement. [Greek *apostolos* messenger]

apostolate /ə'pɒstələt/ *n.* **1** position or authority of an Apostle. **2** leadership in reform.

apostolic /,æpə'stɒlɪk/ *adj.* **1** of the Apostles or their teaching. **2** of the Pope.

apostolic succession *n.* supposed uninterrupted transmission of spiritual authority from the Apostles through popes and bishops.

apostrophe /ə'pɒstrəfɪ/ *n.* **1** punctuation mark (') indicating: **a** omission of letters or numbers (e.g. *can't*; *May '92*). **b** possessive case (e.g. *Harry's book*; *boys' coats*). **2** exclamatory passage addressed to (an often absent) person or thing. □ **apostrophize** *v.* (also **-ise**) (**-zing** or **-sing**) (in sense 2). [Greek, = turning away]

apothecaries' measure *n.* (also **apothecaries' weight**) units formerly used in pharmacy.

apothecary /ə'pɒθəkərɪ/ *n.* (*pl.* **-ies**) *archaic* dispensing chemist. [Greek *apothēkē* storehouse]

apotheosis /ə,pɒθɪ'əʊsɪs/ *n.* (*pl.* **-theoses** /-si:z/) **1** elevation to divine status, deification. **2** glorification of a thing; sublime example (*apotheosis of chivalry*). [Greek *theos* god]

appal /ə'pɔ:l/ *v.* (**-ll-**) **1** greatly dismay or horrify. **2** (as **appalling** *adj.*) *colloq.* very bad, shocking. [French *apalir* grow pale: related to PALE[1]]

apparatus /ˌæpəˈreɪtəs/ *n.* **1** equipment for a particular function, esp. scientific or technical. **2** political or other complex organization. [Latin *paro* prepare]

apparel /əˈpær(ə)l/ *n. formal* clothing, dress. □ **apparelled** *adj.* [Romanic, = make fit, from Latin *par* equal]

apparent /əˈpærənt/ *adj.* **1** readily visible; obvious. **2** seeming. □ **apparently** *adv.* [Latin: related to APPEAR]

apparition /ˌæpəˈrɪʃ(ə)n/ *n.* remarkable or unexpected thing that appears; ghost or phantom.

appeal /əˈpiːl/ *–v.* **1** request earnestly or formally; plead. **2** (usu. foll. by *to*) attract, be of interest. **3** (foll. by *to*) resort to for support. **4** *Law* **a** (often foll. by *to*) apply (to a higher court) for reconsideration of a legal decision. **b** refer (a case) to a higher court. **5** *Cricket* call on the umpire to declare whether a batsman is out. *–n.* **1** act of appealing. **2** request for public support, esp. financial. **3** *Law* referral of a case to a higher court. **4** attractiveness. [Latin *appello* address]

appear /əˈpɪə(r)/ *v.* **1** become or be visible. **2** seem (*appeared unwell*). **3** present oneself publicly or formally. **4** be published. [Latin *appareo*]

appearance /əˈpɪərəns/ *n.* **1** act of appearing. **2** outward form as perceived (*appearance of prosperity*). **3** semblance. □ **keep up appearances** maintain an impression or pretence of virtue, affluence, etc. **make** (or **put in**) **an appearance** be present, esp. briefly.

appease /əˈpiːz/ *v.* (**-sing**) **1** make calm or quiet, esp. conciliate (a potential aggressor) by making concessions. **2** satisfy (an appetite, scruples). □ **appeasement** *n.* [French *à* to, *pais* PEACE]

appellant /əˈpelənt/ *n.* person who appeals to a higher court. [Latin *appello* address]

appellate /əˈpelət/ *attrib. adj.* (esp. of a court) concerned with appeals.

appellation /ˌæpəˈleɪʃ(ə)n/ *n. formal* name or title; nomenclature.

appellative /əˈpelətɪv/ *adj.* **1** naming. **2** *Gram.* (of a noun) designating a class, common.

append /əˈpend/ *v.* (usu. foll. by *to*) attach, affix, add, esp. to a written document. [Latin *appendo* hang]

appendage /əˈpendɪdʒ/ *n.* thing attached; addition.

appendectomy /ˌæpenˈdektəmɪ/ *n.* (also **appendicectomy** /-dɪˈsektəmɪ/) (*pl.* **-ies**) surgical removal of the appendix. [from APPENDIX, -ECTOMY]

appendicitis /əˌpendɪˈsaɪtɪs/ *n.* inflammation of the appendix.

appendix /əˈpendɪks/ *n.* (*pl.* **-dices** /-ˌsiːz/) **1** tissue forming a tube-shaped sac attached to the large intestine. **2** addition to a book etc. [Latin: related to APPEND]

appertain /ˌæpəˈteɪn/ *v.* (foll. by *to*) relate, belong, or be appropriate. [Latin: related to PERTAIN]

appetite /ˈæpɪˌtaɪt/ *n.* **1** natural craving, esp. for food or sexual activity. **2** (usu. foll. by *for*) inclination or desire. [Latin *peto* seek]

appetizer /ˈæpɪˌtaɪzə(r)/ *n.* (also **-iser**) small amount, esp. of food or drink, to stimulate the appetite.

appetizing *adj.* (also **-ising**) stimulating the appetite, esp. for food; tasty.

applaud /əˈplɔːd/ *v.* **1** express strong approval (of), esp. by clapping. **2** commend, approve (a person or action). [Latin *applaudo* clap hands]

applause /əˈplɔːz/ *n.* **1** approval shown by clapping the hands. **2** warm approval.

apple /ˈæp(ə)l/ *n.* **1** roundish firm fruit with crisp flesh. **2** tree bearing this. □ **apple of one's eye** cherished person or thing. [Old English]

apple-pie bed *n.* bed made (as a joke) with sheets folded so as to prevent a person lying flat.

apple-pie order *n.* extreme neatness.

appliance /əˈplaɪəns/ *n.* device etc. for a specific task. [related to APPLY]

applicable /ˈæplɪkəb(ə)l/ *adj.* (often foll. by *to*) that may be applied; relevant; appropriate. □ **applicability** /-ˈbɪlɪtɪ/ *n.* [medieval Latin: related to APPLY]

applicant /ˈæplɪkənt/ *n.* person who applies for something, esp. a job.

application /ˌæplɪˈkeɪʃ(ə)n/ *n.* **1** formal request. **2** act of applying. **3** substance applied. **4 a** relevance. **b** use (*has many applications*). **5** diligence.

applicator /ˈæplɪˌkeɪtə(r)/ *n.* device for applying ointment etc.

applied /əˈplaɪd/ *adj.* practical, not merely theoretical (*applied science*).

appliqué /əˈpliːkeɪ/ *–n.* cutting out of fabric patterns and attaching them to another fabric. *–v.* (**-qués**, **-quéd**, **-quéing**) decorate with appliqué. [French, = applied]

apply /əˈplaɪ/ *v.* (**-ies**, **-ied**) **1** (often foll. by *for*, *to*, or *to* + infin.) formally request. **2** (often foll. by *to*) be relevant. **3 a** make use of; employ (*apply the rules*; *apply common sense*). **b** operate (*apply the brakes*). **4** (often foll. by *to*) put or spread on. **5** *refl.* (often foll. by *to*) devote oneself. [Latin *applico* fasten to]

appoint /əˈpɔɪnt/ *v.* **1** assign a job or office to. **2** (often foll. by *for*) fix (a time, place, etc.). **3** (as **appointed** *adj.*)

equipped, furnished (*well-appointed*). □ **appointee** /-'ti:/ *n.* [French *à point* to a point]**appointment** *n.* **1** appointing or being appointed. **2** arrangement for meeting or consultation. **3 a** post or office open to applicants. **b** person appointed. **4** (usu. in *pl.*) furniture, fittings; equipment.

apportion /ə'pɔ:ʃ(ə)n/ *v.* (often foll. by *to*) share out; assign as a share. □ **apportionment** *n.* [medieval Latin: related to PORTION]

apposite /'æpəzɪt/ *adj.* (often foll. by *to*) apt, appropriate; well expressed. [Latin *appono* apply]

apposition /,æpə'zɪʃ(ə)n/ *n.* juxtaposition, esp. *Gram.* of elements sharing a syntactic function (e.g. *William the Conqueror*; *my friend Sue*).

appraisal /ə'preɪz(ə)l/ *n.* appraising or being appraised.

appraise /ə'preɪz/ *v.* **(-sing) 1** estimate the value or quality of. **2** set a price on (esp. officially). [earlier *apprize*, assimilated to PRAISE]

appreciable /ə'pri:ʃəb(ə)l/ *adj.* significant, considerable. [French: related to APPRECIATE]

appreciate /ə'pri:ʃɪ,eɪt/ *v.* **(-ting) 1 a** esteem highly; value. **b** be grateful for. **2** understand, recognize (*appreciate the danger*). **3** rise or raise in value. □ **appreciative** /-ʃətɪv/ *adj.* **appreciatory** /-ʃjətərɪ/ *adj.* [Latin *pretium* price]

appreciation /ə,pri:ʃɪ'eɪʃ(ə)n/ *n.* **1** favourable or grateful recognition. **2** sensitive estimation or judgement. **3** rise in value. [French: related to APPRECIATE]

apprehend /,æprɪ'hend/ *v.* **1** seize, arrest. **2** understand, perceive. [Latin *prehendo* grasp]

apprehension /,æprɪ'henʃ(ə)n/ *n.* **1** uneasiness, dread. **2** understanding. **3** arrest, capture.

apprehensive /,æprɪ'hensɪv/ *adj.* uneasily fearful. □ **apprehensively** *adv.*

apprentice /ə'prentɪs/ *–n.* **1** person learning a trade by working in it for an agreed period at low wages. **2** novice. *–v.* **(-cing)** (usu. foll. by *to*) engage as an apprentice (*apprenticed to a builder*). □ **apprenticeship** *n.* [French *apprendre* learn]

apprise /ə'praɪz/ *v.* **(-sing)** *formal* inform. [French *appris(e)* learnt, taught]

appro /'æprəʊ/ *n. colloq.* □ **on appro** = *on approval* (see APPROVAL). [abbreviation]

approach /ə'prəʊtʃ/ *–v.* **1** come near or nearer (to) in space or time. **2** tentatively make a proposal to. **3** be similar or approximate to (*approaching 5 million*). **4** set about (a task etc.). *–n.* **1** act or means of approaching. **2** approximation. **3** technique (*try a new approach*). **4** *Golf* stroke from the fairway to the green. **5** *Aeron.* part of a flight before landing. [Latin *prope* near]

approachable *adj.* **1** friendly, easy to talk to. **2** able to be approached.

approbation /,æprə'beɪʃ(ə)n/ *n.* approval, consent. [Latin *probo* test]

appropriate *–adj.* /ə'prəʊprɪət/ suitable, proper. *–v.* /ə'prəʊprɪ,eɪt/ **(-ting) 1** take, esp. without authority. **2** devote (money etc.) to special purposes. □ **appropriately** *adv.* **appropriation** /-'eɪʃ(ə)n/ *n.* [Latin *proprius* own]

approval /ə'pru:v(ə)l/ *n.* **1** approving. **2** consent; favourable opinion. □ **on approval** (of goods supplied) returnable if not satisfactory.

approve /ə'pru:v/ *v.* **(-ving) 1** confirm; sanction. **2** (often foll. by *of*) regard with favour. [Latin *probo* test]

approx. *abbr.* approximate(ly).

approximate *–adj.* /ə'prɒksɪmət/ fairly correct, near to the actual (*approximate price*). *–v.* /ə'prɒksɪ,meɪt/ **(-ting)** (often foll. by *to*) bring or come near (esp. in quality, number, etc.). □ **approximately** *adv.* **approximation** /-'meɪʃ(ə)n/ *n.* [Latin *proximus* nearest]

appurtenance /ə'pɜ:tɪnəns/ *n.* (usu. in *pl.*) belonging; accessory. [Latin *pertineo* belong to]

APR *abbr.* annual or annualized percentage rate (esp. of interest on loans or credit).

Apr. *abbr.* April.

après-ski /,æpreɪ'ski:/ *–n.* social activities following a day's skiing. *–attrib. adj.* (of clothes, drinks, etc.) suitable for these. [French]

apricot /'eɪprɪ,kɒt/ *–n.* **1 a** small juicy soft orange-yellow peachlike fruit. **b** tree bearing it. **2** its colour. *–adj.* orange-yellow. [Portuguese and Spanish from Arabic, ultimately from Latin *praecox* early-ripe]

April /'eɪpr(ə)l/ *n.* fourth month of the year. [Latin]

April Fool *n.* person successfully tricked on 1 April.

a priori /,eɪ praɪ'ɔ:raɪ/ *–adj.* **1** (of reasoning) from causes to effects; deductive. **2** (of concepts etc.) logically independent of experience; not derived from experience. **3** assumed without investigation (*an a priori conjecture*). *–adv.* **1** deductively. **2** as far as one knows. [Latin, = from what is before]

apron /'eɪprən/ *n.* **1** garment for covering and protecting the front of the clothes. **2** *Theatr.* part of a stage in front of the curtain. **3** area on an airfield for manoeuvring or loading. □ **tied to a person's apron-strings** dominated by or dependent on that person (usu. a woman). [originally *napron*, from French *nape* table-cloth]

apropos /ˌæprəˈpəʊ/ *–adj.* **1** appropriate. **2** *colloq.* (often foll. by *of*) in respect of. *–adv.* **1** appropriately. **2** (*absol.*) incidentally. [French *à propos*]

apse /æps/ *n.* large arched or domed recess, esp. at the end of a church. [related to APSIS]

apsis /ˈæpsɪs/ *n.* (*pl.* **apsides** /-ˌdi:z/) either of two points on the orbit of a planet etc. nearest to or furthest from the body round which it moves. [Greek (*h*)*apsis* arch, vault]

apt *adj.* **1** appropriate, suitable. **2** tending (*apt to break down*). **3** clever; quick to learn. [Latin *aptus* fitted]

aptitude /ˈæptɪˌtju:d/ *n.* **1** natural talent. **2** ability or fitness, esp. specified. [French: related to APT]

aqua /ˈækwə/ *n.* the colour aquamarine. [abbreviation]

aqua fortis /ˌækwə ˈfɔ:tɪs/ *n.* nitric acid. [Latin, = strong water]

aqualung /ˈækwəˌlʌŋ/ *n.* portable breathing-apparatus for divers. [Latin *aqua* water]

aquamarine /ˌækwəməˈri:n/ *–n.* **1** bluish-green beryl. **2** its colour. *–adj.* bluish-green. [Latin *aqua marina* sea water]

aquaplane /ˈækwəˌpleɪn/ *–n.* board for riding on water, pulled by a speedboat. *–v.* (**-ning**) **1** ride on this. **2** (of a vehicle) glide uncontrollably on a wet surface. [Latin *aqua* water, PLANE[1]]

aqua regia /ˌækwə ˈri:dʒɪə/ *n.* highly corrosive mixture of acids, attacking many substances unaffected by other reagents. [Latin, = royal water]

aquarelle /ˌækwəˈrel/ *n.* painting in thin usu. transparent water-colours. [French from Italian]

aquarium /əˈkweərɪəm/ *n.* (*pl.* **-s**) tank of water for keeping and showing fish etc. [Latin *aquarius* of water]

Aquarius /əˈkweərɪəs/ *n.* (*pl.* **-es**) **1** constellation and eleventh sign of the zodiac (the Water-carrier). **2** person born when the sun is in this sign. [Latin: related to AQUARIUM]

aquatic /əˈkwætɪk/ *–adj.* **1** growing or living in water. **2** (of a sport) played in or on water. *–n.* **1** aquatic plant or animal. **2** (in *pl.*) aquatic sports. [Latin *aqua* water]

aquatint /ˈækwətɪnt/ *n.* etched print resembling a water-colour. [Italian *acqua tinta* coloured water]

aqua vitae /ˌækwə ˈvi:taɪ/ *n.* strong alcoholic spirit, esp. brandy. [Latin, = water of life]

aqueduct /ˈækwɪˌdʌkt/ *n.* water channel, esp. a bridge on columns across a valley. [Latin *aquae ductus* conduit]

aqueous /ˈeɪkwɪəs/ *adj.* of or like water. [Latin *aqua* water]

aqueous humour *n.* clear fluid in the eye between the lens and the cornea.

aquilegia /ˌækwɪˈli:dʒə/ *n.* (usu. blue-flowered) columbine. [Latin]

aquiline /ˈækwɪˌlaɪn/ *adj.* **1** of or like an eagle. **2** (of a nose) curved. [Latin *aquila* eagle]

Ar *symb.* argon.

-ar *suffix* forming adjectives (*angular*; *linear*). [Latin *-aris*]

Arab /ˈærəb/ *–n.* **1** member of a Semitic people originating in Saudi Arabia and neighbouring countries, now widespread throughout the Middle East. **2** horse of a breed orig. native to Arabia. *–adj.* of Arabia or the Arabs (esp. with ethnic reference). [Arabic *araps*]

arabesque /ˌærəˈbesk/ *n.* **1** *Ballet* posture with one leg extended horizontally backwards and arms outstretched. **2** design of intertwined leaves, scrolls, etc. **3** *Mus.* florid piece. [French from Italian from *arabo* Arab]

Arabian /əˈreɪbɪən/ *–adj.* of or relating to Arabia (esp. in geographical contexts) (*Arabian desert*). *–n.* native of Arabia.

■ **Usage** In the sense 'native of Arabia', the usual term is now *Arab*.

Arabic /ˈærəbɪk/ *–n.* Semitic language of the Arabs. *–adj.* of the Arabs (esp. their language or literature).

arabic numeral *n.* any of the numerals 0-9.

arable /ˈærəb(ə)l/ *adj.* (of land) suitable for crop production. [Latin *aro* to plough]

arachnid /əˈræknɪd/ *n.* arthropod of a class comprising spiders, scorpions, etc. [Greek *arakhnē* spider]

arak var. of ARRACK.

Araldite /ˈærəlˌdaɪt/ *n. propr.* epoxy resin for mending china etc. [origin unknown]

Aramaic /ˌærəˈmeɪɪk/ *–n.* branch of the Semitic family of languages, esp. the language of Syria used as a lingua franca in the Near East from the sixth century BC. *–adj.* of or in Aramaic. [Greek *Aramaios* of Aram (Hebrew name of Syria)]

arbiter /ˈɑ:bɪtə(r)/ *n.* **1** arbitrator in a dispute. **2** person influential in a specific field (*arbiter of taste*). [Latin from *arbitror* to judge]

arbitrary /ˈɑ:bɪtrərɪ/ *adj.* **1** random. **2** capricious; despotic. □ **arbitrarily** *adv.*

arbitrate /ˈɑ:bɪˌtreɪt/ *v.* **(-ting)** decide by arbitration.

arbitration /ˌɑ:bɪˈtreɪʃ(ə)n/ *n.* settlement of a dispute by an impartial third party.

arbitrator *n.* person appointed to arbitrate.

arbor[1] /ˈɑ:bə(r)/ *n.* axle or spindle. [Latin, = tree]

arbor[2] *US* var. of ARBOUR.

arboreal /ɑ:ˈbɔ:rɪəl/ *adj.* of or living in trees. [Latin *arbor* tree]

arborescent /ˌɑ:bəˈres(ə)nt/ *adj.* tree-like in growth or form.

arboretum /ˌɑ:bəˈri:təm/ *n.* (*pl.* **-ta**) place cultivating and displaying rare trees.

arboriculture /ˈɑ:bərɪˌkʌltʃə(r)/ *n.* cultivation of trees and shrubs. [Latin *arbor* tree, after *agriculture*]

arbor vitae /ˌɑ:bə ˈvi:taɪ/ *n.* any of various evergreen conifers. [Latin, = tree of life]

arbour /ˈɑ:bə(r)/ *n.* (*US* **arbor**) shady garden alcove enclosed by trees etc. [Latin *herba* herb: assimilated to Latin *arbor* tree]

arbutus /ɑ:ˈbju:təs/ *n.* tree or shrub with clusters of flowers and strawberry-like berries. [Latin]

arc –*n.* **1** part of the circumference of a circle or other curve. **2** *Electr.* luminous discharge between two electrodes. –*v.* (**arced**; **arcing** /ˈɑ:kɪŋ/) form an arc; move in a curve. [Latin *arcus* bow]

arcade /ɑ:ˈkeɪd/ *n.* **1** covered walk, esp. lined with shops. **2** series of arches supporting or set along a wall. [Romanic: related to ARC]

Arcadian /ɑ:ˈkeɪdɪən/ –*n.* idealized country dweller. –*adj.* poetically rural. [Greek *Arkadia* in the Peloponnese]

arcane /ɑ:ˈkeɪn/ *adj.* mysterious, secret. [Latin *arceo* shut up]

arch[1] –*n.* **1** curved structure as an opening, as a support for a bridge, floor, etc., or as an ornament. **2** any arch-shaped curve. –*v.* **1** provide with or form into an arch. **2** span like an arch. **3** form an arch. [Latin *arcus* arc]

arch[2] *adj.* self-consciously or affectedly playful. □ **archly** *adv.* [from ARCH-, originally in *arch rogue* etc.]

arch- *comb. form* **1** chief, superior (*archbishop*). **2** pre-eminent, esp. unfavourably (*arch-enemy*). [Greek *arkhos* chief]

Archaean /ɑ:ˈki:ən/ (*US* **Archean**) –*adj.* of the earliest geological era. –*n.* this time. [Greek *arkhaios* ancient]

archaeology /ˌɑ:kɪˈɒlədʒɪ/ *n.* (*US* **archeology**) study of ancient cultures, esp. by the excavation and analysis of physical remains. □ **archaeological** /-əˈlɒdʒɪk(ə)l/ *adj.* **archaeologist** *n.* [Greek *arkhaiologia* ancient history]

archaeopteryx /ˌɑ:kɪˈɒptərɪks/ *n.* fossil bird with teeth, feathers, and a reptilian tail. [Greek *arkhaios* ancient, *pterux* wing]

archaic /ɑ:ˈkeɪɪk/ *adj.* **1 a** antiquated. **b** (of a word etc.) no longer in ordinary use. **2** of an early period of culture. □ **archaically** *adv.* [Greek *arkhē* beginning]

archaism /ˈɑ:keɪˌɪz(ə)m/ *n.* **1** use of the archaic esp. in language or art. **2** archaic word or expression. □ **archaistic** /-ˈɪstɪk/ *adj.*

archangel /ˈɑ:kˌeɪndʒ(ə)l/ *n.* angel of the highest rank.

archbishop /ɑ:tʃˈbɪʃəp/ *n.* chief bishop of a province.

archbishopric *n.* office or diocese of an archbishop.

archdeacon /ɑ:tʃˈdi:kən/ *n.* church dignitary next below a bishop. □ **archdeaconry** *n.* (*pl.* **-ies**).

archdiocese /ɑ:tʃˈdaɪəsɪs/ *n.* diocese of an archbishop. □ **archdiocesan** /-daɪˈɒsɪs(ə)n/ *adj.*

archduke /ɑ:tʃˈdju:k/ *n. hist.* chief duke (esp. as the title of a son of the Emperor of Austria). □ **archduchy** /-ˈdʌtʃɪ/ *n.* (*pl.* **-ies**). [medieval Latin *archidux*]

Archean *US* var. of ARCHAEAN.

arch-enemy /ɑ:tʃˈenəmɪ/ *n.* (*pl.* **-ies**) **1** chief enemy. **2** the Devil.

archeology *US* var. of ARCHAEOLOGY.

archer *n.* **1** person who shoots with a bow and arrows. **2** (**the Archer**) zodiacal sign or constellation Sagittarius. [Latin *arcus* bow]

archery *n.* shooting with a bow and arrows, esp. as a sport.

archetype /ˈɑ:kɪˌtaɪp/ *n.* **1** original model; prototype. **2** typical specimen. □ **archetypal** /-ˈtaɪp(ə)l/ *adj.* [Greek *tupon* stamp]

archidiaconal /ˌɑ:kɪdaɪˈækən(ə)l/ *adj.* of an archdeacon. [medieval Latin]

archiepiscopal /ˌɑ:kɪɪˈpɪskəp(ə)l/ *adj.* of an archbishop. [Church Latin from Greek]

archimandrite /ˌɑ:kɪˈmændraɪt/ *n.* **1** superior of a large monastery in the Orthodox Church. **2** honorary title of a monastic priest. [Greek *arkhi-* chief, *mandritēs* monk]

archipelago /ˌɑ:kɪˈpeləˌgəʊ/ *n.* (*pl.* **-s**) **1** group of islands. **2** sea with many islands. [Greek *arkhi-* chief, *pelagos* sea]

architect /ˈɑ:kɪˌtekt/ *n.* **1** designer of buildings etc., supervising their construction. **2** (foll. by *of*) person who brings about a specified thing (*architect of peace*). [Greek *arkhi-* chief, *tektōn* builder]

architectonic /,ɑ:kɪtek'tɒnɪk/ *adj.* **1** of architecture. **2** of the systematization of knowledge.

architecture /'ɑ:kɪ,tektʃə(r)/ *n.* **1** design and construction of buildings. **2** style of a building. **3** buildings etc. collectively. □ **architectural** /-'tektʃər(ə)l/ *adj.*

architrave /'ɑ:kɪ,treɪv/ *n.* **1** (in classical architecture) main beam resting across the tops of columns. **2** moulded frame around a doorway or window. [Italian *archi-* ARCH-, Latin *trabs* beam]

archive /'ɑ:kaɪv/ *–n.* (usu. in *pl.*) **1** collection of documents or records. **2** store for these. *–v.* **(-ving) 1** place or store in an archive. **2** *Computing* transfer (data) to a less frequently used file. [Greek *arkheia* public records]

archivist /'ɑ:kɪvɪst/ *n.* keeper of archives.

archway *n.* arched entrance or passage.

arc lamp *n.* (also **arc light**) light using an electric arc.

Arctic /'ɑ:ktɪk/ *–adj.* **1** of the north polar regions. **2** (**arctic**) *colloq.* very cold. *–n.* Arctic regions. [Greek *arktos* Great Bear]

Arctic Circle *n.* parallel of latitude 66° 33′ N, forming an imaginary line round the Arctic region.

arc welding *n.* use of an electric arc to melt metals to be welded.

ardent /'ɑ:d(ə)nt/ *adj.* **1** eager, fervent, passionate. **2** burning. □ **ardently** *adv.* [Latin *ardeo* burn]

ardour /'ɑ:də(r)/ *n.* (*US* **ardor**) zeal, enthusiasm, passion.

arduous /'ɑ:dju:əs/ *adj.* hard to accomplish; laborious, strenuous. [Latin, = steep]

are[1] *2nd sing. present & 1st, 2nd, 3rd pl. present* of BE.

are[2] /ɑ:(r)/ *n.* metric unit of measure, 100 square metres. [Latin: related to AREA]

area /'eərɪə/ *n.* **1** extent or measure of a surface (*over a large area*). **2** region (*southern area*). **3** space for a specific purpose (*dining area*). **4** scope or range. **5** space in front of the basement of a building. [Latin, = vacant space]

arena /ə'ri:nə/ *n.* **1** central part of an amphitheatre etc. **2** scene of conflict; sphere of action. [Latin, = sand]

aren't /ɑ:nt/ *contr.* **1** are not. **2** (in *interrog.*) am not (*aren't I coming too?*).

areola /æ'ri:ələ/ *n.* (*pl.* **-lae** /-li:/) circular pigmented area, esp. around a nipple. □ **areolar** *adj.* [Latin diminutive of AREA]

arête /æ'reɪt/ *n.* sharp mountain ridge. [French from Latin *arista* spine]

argent /'ɑ:dʒ(ə)nt/ *n.* & *adj. Heraldry* silver; silvery-white. [Latin *argentum*]

argon /'ɑ:gɒn/ *n.* inert gaseous element. [Greek *argos* idle]

argosy /'ɑ:gəsɪ/ *n.* (*pl.* **-ies**) *poet.* large merchant ship. [Italian *Ragusea nave* ship of Ragusa (in Dalmatia)]

argot /'ɑ:gəʊ/ *n.* jargon of a group or class. [French]

argue /'ɑ:gju:/ *v.* **(-ues, -ued, -uing) 1** (often foll. by *with*, *about*, etc.) exchange views forcefully or contentiously. **2** (often foll. by *that*) maintain by reasoning; indicate. **3** (foll. by *for*, *against*) reason. **4** treat (a matter) by reasoning. **5** (foll. by *into*, *out of*) persuade. □ **argue the toss** *colloq.* dispute a choice already made. □ **arguable** *adj.* **arguably** *adv.* [Latin *arguo* make clear, prove]

argument /'ɑ:gjʊmənt/ *n.* **1** (esp. contentious) exchange of views; dispute. **2** (often foll. by *for*, *against*) reason given; reasoning process. **3** summary of a book etc.

argumentation /,ɑ:gjʊmen'teɪʃ(ə)n/ *n.* methodical reasoning; arguing.

argumentative /,ɑ:gjʊ'mentətɪv/ *adj.* given to arguing.

Argus /'ɑ:gəs/ *n.* watchful guardian. [Greek *Argos* mythical giant with 100 eyes]

argy-bargy /,ɑ:dʒɪ'bɑ:dʒɪ/ *n.* (*pl.* **-ies**) *joc.* dispute, wrangle. [originally Scots]

aria /'ɑ:rɪə/ *n.* long accompanied solo song in an opera etc. [Italian]

arid /'ærɪd/ *adj.* **1** dry, parched. **2** uninteresting. □ **aridity** /ə'rɪdɪtɪ/ *n.* [Latin *areo* be dry]

Aries /'eəri:z/ *n.* (*pl.* same) **1** constellation and first sign of the zodiac (the Ram). **2** person born when the sun is in this sign. [Latin, = ram]

aright /ə'raɪt/ *adv.* rightly.

arise /ə'raɪz/ *v.* **(-sing;** *past* **arose;** *past part.* **arisen** /ə'rɪz(ə)n/) **1** originate. **2** (usu. foll. by *from*, *out of*) result. **3** come to one's notice; emerge. **4** rise, esp. from the dead or from kneeling. [Old English *a-* intensive prefix]

aristocracy /,ærɪ'stɒkrəsɪ/ *n.* (*pl.* **-ies**) **1** ruling class or élite; nobility. **2 a** government by an élite. **b** State so governed. **3** (often foll. by *of*) best representatives. [Greek *aristokratia* rule by the best]

aristocrat /'ærɪstə,kræt/ *n.* member of the aristocracy.

aristocratic /,ærɪstə'krætɪk/ *adj.* **1** of or like the aristocracy. **2 a** distinguished. **b** grand, stylish.

Aristotelian /,ærɪstə'ti:lɪən/ *–n.* disciple or student of Aristotle. *–adj.* of

Aristotle or his ideas. [Greek *Aristotelēs* (4th c. BC), name of a Greek philosopher]
arithmetic –*n.* /ə'rɪθmətɪk/ **1** science of numbers. **2** use of numbers; computation. –*adj.* /,ærɪθ'metɪk/ (also **arithmetical**) of arithmetic. [Greek *arithmos* number]
arithmetic mean *n.* = AVERAGE 2.
arithmetic progression *n.* sequence of numbers with constant intervals (e.g. 9, 7, 5, 3, etc.).
ark *n.* ship in which Noah escaped the Flood with his family and animals. [Old English from Latin *arca*]
Ark of the Covenant *n.* chest or cupboard containing the tables of Jewish Law.
arm[1] *n.* **1** upper limb of the human body from shoulder to hand. **2** forelimb or tentacle of an animal. **3 a** sleeve of a garment. **b** arm support of a chair etc. **c** thing branching from a main stem (*an arm of the sea*). **d** control, means of reaching (*arm of the law*). □ **arm in arm** with arms linked. **at arm's length** at a distance. **with open arms** cordially. □ **armful** *n.* (*pl.* **-s**). [Old English]
arm[2] –*n.* **1** (usu. in *pl.*) weapon. **2** (in *pl.*) military profession. **3** branch of the military (e.g. infantry, cavalry). **4** (in *pl.*) heraldic devices (*coat of arms*). –*v.* **1** supply, or equip oneself, with weapons etc., esp. in preparation for war. **2** make (a bomb etc.) ready. □ **take up arms** go to war. **under arms** equipped for war. **up in arms** (usu. foll. by *against*, *about*) actively resisting, highly indignant. [Latin *arma* arms]
armada /ɑː'mɑːdə/ *n.* fleet of warships, esp. (**Armada**) that sent by Spain against England in 1588. [Spanish from Romanic]
armadillo /,ɑːmə'dɪləʊ/ *n.* (*pl.* **-s**) S. American mammal with a plated body and large claws. [Spanish *armado* armed man]
Armageddon /,ɑːmə'ged(ə)n/ *n.* huge battle or struggle, esp. marking the end of the world. [Rev. 16:16]
armament /'ɑːməmənt/ *n.* **1** (often in *pl.*) military equipment. **2** equipping for war. **3** force equipped. [Latin: related to ARM[2]]
armature /'ɑːmətʃə(r)/ *n.* **1** rotating coil or coils of a dynamo or electric motor. **2** iron bar placed across the poles of a horseshoe magnet to preserve its power. **3** metal framework on which a sculpture is moulded. [Latin *armatura*, = armour]
armband *n.* band worn around the upper arm to hold up a shirtsleeve, or as identification, or to aid swimming.
armchair *n.* **1** chair with arm supports. **2** (*attrib.*) theoretical rather than active (*armchair critic*).
armhole *n.* each of two holes for arms in a garment.
armistice /'ɑːmɪstɪs/ *n.* truce, esp. permanent. [Latin *arma* arms, *sisto* make stand]
Armistice Day *n.* anniversary of the armistice of 11 Nov. 1918.
armlet *n.* ornamental band worn round the arm.
armor etc. *US* var. of ARMOUR etc.
armorial /ɑː'mɔːrɪəl/ *adj.* of heraldry or coats of arms. [related to ARMOUR]
armour /'ɑːmə(r)/ –*n.* **1** protective usu. metal covering formerly worn in fighting. **2 a** (in full **armour-plate**) protective metal covering for an armed vehicle, ship, etc. **b** armed vehicles collectively. **3** protective covering or shell of an animal or plant. **4** heraldic devices. –*v.* (usu. as **armoured** *adj.*) provide with protective covering, and often guns (*armoured car*; *armoured train*). [Latin *armatura*: related to ARM[2]]
armourer *n.* **1** maker of arms or armour. **2** official in charge of arms.
armoury *n.* (*pl.* **-ies**) arsenal.
armpit *n.* hollow under the arm at the shoulder.
armrest *n.* = ARM[1] 3b.
arms race *n.* competitive accumulation of weapons by nations.
arm-wrestling *n.* trial of strength in which each party tries to force the other's arm down.
army *n.* (*pl.* **-ies**) **1** organized armed land force. **2** (prec. by *the*) the military profession. **3** (often foll. by *of*) very large number (*army of locusts*). **4** organized civilian body (*Salvation Army*). [French: related to ARM[2]]
arnica /'ɑːnɪkə/ *n.* **1** plant of the daisy family with yellow flowers. **2** medicine prepared from this. [origin unknown]
aroma /ə'rəʊmə/ *n.* **1** esp. pleasing smell, often of food. **2** subtle pervasive quality. [Greek, = spice]
aromatherapy *n.* use of aromatic plant extracts and oils in massage. □ **aromatherapist** *n.*
aromatic /,ærə'mætɪk/ –*adj.* **1** fragrant, spicy. **2** of organic compounds having an unsaturated ring, esp. containing a benzene ring. –*n.* aromatic substance. [Latin: related to AROMA]
arose *past* of ARISE.
around /ə'raʊnd/ –*adv.* **1** on every side; all round; round about. **2** *colloq.* **a** in existence; available. **b** near at hand. **3** here and there (*shop around*). –*prep.* **1** on or along the circuit of. **2** on every side of. **3** here and there in or near (*chairs

around the room). **4 a** round (*church around the corner*). **b** at a time near to (*came around four o'clock*). □ **have been around** *colloq.* be widely experienced.

arouse /ə'raʊz/ *v.* (**-sing**) **1** induce (esp. an emotion). **2** awake from sleep. **3** stir into activity. **4** stimulate sexually. □ **arousal** *n.* [*a-* intensive prefix]

arpeggio /ɑː'pedʒɪəʊ/ *n.* (*pl.* **-s**) *Mus.* notes of a chord played in succession. [Italian *arpa* harp]

arrack /'ærək/ *n.* (also **arak**) alcoholic spirit, esp. made from coco sap or rice. [Arabic]

arraign /ə'reɪn/ *v.* **1** indict, accuse. **2** find fault with; call into question (an action or statement). □ **arraignment** *n.* [Latin *ratio* reason]

arrange /ə'reɪndʒ/ *v.* (**-ging**) **1** put into order; classify. **2** plan or provide for; take measures (*arranged a meeting*; *arrange to see him*; *arranged for a taxi*). **3** agree (*arranged it with her*). **4** *Mus.* adapt (a composition) for a particular manner of performance. [French: related to RANGE]

arrangement *n.* **1** arranging or being arranged. **2** manner of this. **3** something arranged. **4** (in *pl.*) plans, measures (*made my own arrangements*). **5** *Mus.* composition adapted for performance in a particular way.

arrant /'ærənt/ *adj. literary* downright, utter (*arrant liar*). [var. of ERRANT, originally in *arrant* (= outlawed, roving) *thief* etc.]

arras /'ærəs/ *n. hist.* rich tapestry or wall-hanging. [*Arras* in France]

array /ə'reɪ/ —*n.* **1** imposing or well-ordered series or display. **2** ordered arrangement, esp. of troops (*battle array*). —*v.* **1** deck, adorn. **2** set in order; marshal (forces). [Latin *ad-*, READY]

arrears /ə'rɪəz/ *n.pl.* amount (esp. of work, rent, etc.) still outstanding or uncompleted. □ **in arrears** behind, esp. in payment. [medieval Latin *adretro* behindhand]

arrest /ə'rest/ —*v.* **1** lawfully seize (a suspect etc.). **2** stop or check the progress of. **3** attract (a person's attention). —*n.* **1** arresting or being arrested. **2** stoppage (*cardiac arrest*). [Latin *resto* remain]

arrester *n.* device for slowing an aircraft after landing.

arrière-pensée /,ærjerpɑ̃'seɪ/ *n.* **1** secret motive. **2** mental reservation. [French]

arris /'ærɪs/ *n. Archit.* sharp edge at the junction of two surfaces. [French *areste*, = ARÊTE]

arrival /ə'raɪv(ə)l/ *n.* **1** arriving; appearance on the scene. **2** person or thing that has arrived.

arrive /ə'raɪv/ *v.* (**-ving**) **1** (often foll. by *at*, *in*) reach a destination. **2** (foll. by *at*) reach (a conclusion etc.). **3** *colloq.* become successful. **4** *colloq.* (of a child) be born. **5** (of a time) come. [Latin *ripa* shore]

arriviste /,æriː'viːst/ *n.* ambitious or ruthless person. [French: related to ARRIVE]

arrogant /'ærəgənt/ *adj.* aggressively assertive or presumptuous. □ **arrogance** *n.* **arrogantly** *adv.* [related to ARROGATE]

arrogate /'ærə,geɪt/ *v.* (**-ting**) **1** (often foll. by *to* oneself) claim (power etc.) without right. **2** (often foll. by *to*) attribute unjustly (to a person). □ **arrogation** /-'geɪʃ(ə)n/ *n.* [Latin *rogo* ask]

arrow /'ærəʊ/ *n.* **1** pointed slender missile shot from a bow. **2** representation of this, esp. indicating direction. [Old English]

arrowhead *n.* **1** pointed tip of an arrow. **2** water-plant with arrow-shaped leaves.

arrowroot *n.* **1** nutritious starch. **2** plant yielding this.

arse /ɑːs/ *n.* (*US* **ass** /æs/) *coarse slang* buttocks. [Old English]

arsehole *n.* (*US* **asshole**) *coarse slang* **1** anus. **2** *offens.* contemptible person.

arsenal /'ɑːsən(ə)l/ *n.* **1** store, esp. of weapons. **2** place for the storage and manufacture of weapons and ammunition. [Arabic, = workshop]

arsenic —*n.* /'ɑːsənɪk/ **1** non-scientific name for arsenic trioxide, a highly poisonous white powder used in weed-killers etc. **2** *Chem.* brittle semi-metallic element. —*adj.* /ɑː'senɪk/ of or containing arsenic. [French, ultimately from Persian *zar* gold]

arson /'ɑːs(ə)n/ *n.* crime of deliberately setting fire to property. □ **arsonist** *n.* [Latin *ardeo ars-* burn]

art *n.* **1 a** human creative skill or its application. **b** work showing this. **2 a** (in *pl.*; prec. by *the*) branches of creative activity concerned with the production of imaginative designs, sounds, or ideas, e.g. painting, music, writing. **b** any one of these. **3** creative activity resulting in visual representation (*good at music but not art*). **4** human skill as opposed to nature (*art and nature combined*). **5** (often foll. by *of*) **a** skill, knack. **b** cunning; trick, stratagem. **6** (in *pl.*; usu. prec. by *the*) supposedly creative subjects (esp. languages, literature, and history) as opposed to scientific, technical, or vocational subjects. [Latin *ars art-*]

art deco /ɑ:t 'dekəʊ/ *n.* decorative art style of 1910–30, with geometric motifs and strong colours.

artefact /'ɑ:tɪ,fækt/ *n.* (also **artifact**) man-made object, esp. a tool or vessel as an archaeological item. [Latin *arte* by art, *facio* make]

arterial /ɑ:'tɪərɪəl/ *adj.* **1** of or like an artery. **2** (esp. of a road) main, important. [French: related to ARTERY]

arteriosclerosis /ɑ:,tɪərɪəʊsklɪə'rəʊsɪs/ *n.* loss of elasticity and thickening of artery walls, esp. in old age. [from ARTERY, SCLEROSIS]

artery /'ɑ:tərɪ/ *n.* (*pl.* **-ies**) **1** any of the blood-vessels carrying blood from the heart. **2** main road or railway line. [Greek, probably from *airō* raise]

artesian well /ɑ:'ti:zɪən/ *n.* well in which water rises to the surface by natural pressure through a vertically drilled hole. [*Artois*, old French province]

artful *adj.* crafty, deceitful. □ **artfully** *adv.*

arthritis /ɑ:'θraɪtɪs/ *n.* inflammation of a joint or joints. □ **arthritic** /-'θrɪtɪk/ *adj.* & *n.* [Greek *arthron* joint]

arthropod /'ɑ:θrə,pɒd/ *n.* invertebrate with a segmented body and jointed limbs, e.g. an insect, spider, or crustacean. [Greek *arthron* joint, *pous pod-* foot]

artichoke /'ɑ:tɪ,tʃəʊk/ *n.* **1** plant allied to the thistle. **2** (in full **globe artichoke**) its partly edible flower-head (see also JERUSALEM ARTICHOKE). [Italian from Arabic]

article /'ɑ:tɪk(ə)l/ *–n.* **1** item or thing. **2** non-fictional journalistic essay. **3** clause or item in an agreement or contract. **4** definite or indefinite article. *–v.* (**-ling**) employ under contract as a trainee. [Latin *articulus* from *artus* joint]

articled clerk *n.* trainee solicitor.

articular /ɑ:'tɪkjʊlə(r)/ *adj.* of a joint or joints. [Latin: related to ARTICLE]

articulate *–adj.* /ɑ:'tɪkjʊlət/ **1** fluent and clear in speech. **2** (of sound or speech) having clearly distinguishable parts. **3** having joints. *–v.* /ɑ:'tɪkjʊ,leɪt/ (**-ting**) **1 a** pronounce distinctly. **b** speak or express clearly. **2** (usu. in *passive*) connect by joints. **3** mark with apparent joints. **4** (often foll. by *with*) form a joint. □ **articulately** *adv.*

articulated lorry *n.* one with sections connected by a flexible joint.

articulation /ɑ:,tɪkjʊ'leɪʃ(ə)n/ *n.* **1 a** speaking or being spoken. **b** articulate utterance; speech. **2 a** act or mode of jointing. **b** joint. [Latin: related to ARTICULATE]

artifact var. of ARTEFACT.

artifice /'ɑ:tɪfɪs/ *n.* **1** trick or clever device. **2** cunning. **3** skill, ingenuity. [Latin *ars art-* art, *facio* make]

artificer /ɑ:'tɪfɪsə(r)/ *n.* **1** craftsman. **2** skilled military mechanic.

artificial /,ɑ:tɪ'fɪʃ(ə)l/ *adj.* **1** not natural (*artificial lake*). **2** imitating nature (*artificial flowers*). **3** affected, insincere. □ **artificiality** /-ʃɪ'ælɪtɪ/ *n.* **artificially** *adv.* [Latin: related to ARTIFICE]

artificial insemination *n.* non-sexual injection of semen into the uterus.

artificial intelligence *n.* use of computers for tasks normally regarded as needing human intelligence.

artificial respiration *n.* manual or mechanical stimulation of breathing.

artillery /ɑ:'tɪlərɪ/ *n.* (*pl.* **-ies**) **1** heavy guns used in land warfare. **2** branch of the army using these. □ **artilleryman** *n.* [French *artiller* equip]

artisan /,ɑ:tɪ'zæn/ *n.* skilled manual worker or craftsman. [Latin *artio* instruct in the arts]

artist /'ɑ:tɪst/ *n.* **1** practitioner of any of the arts, esp. painting. **2** artiste. **3** person using skill or taste. □ **artistry** *n.* [French *artiste* from Italian]

artiste /ɑ:'ti:st/ *n.* professional performer, esp. a singer or dancer.

artistic /ɑ:'tɪstɪk/ *adj.* **1** having natural skill in art. **2** skilfully or tastefully done. **3** of art or artists. □ **artistically** *adv.*

artless *adj.* **1** guileless, ingenuous. **2** natural. **3** clumsy. □ **artlessly** *adv.*

art nouveau /,ɑ: nu:'vəʊ/ *n.* art style of the late 19th century, with flowing lines.

artwork *n.* **1** illustrative material in printed matter. **2** works of art collectively (*exhibition of children's artwork*).

arty *adj.* (**-ier, -iest**) *colloq.* pretentiously or affectedly artistic.

arum /'eərəm/ *n.* plant with arrow-shaped leaves. [Greek *aron*]

-ary *suffix* forming adjectives (*contrary*; *primary*). [French *-aire*, Latin *-ari(u)s*]

Aryan /'eərɪən/ *–n.* **1** speaker of any of the languages of the Indo-European family. **2** *improperly* (in Nazi ideology) non-Jewish Caucasian. *–adj.* of Aryans. [Sanskrit]

As *symb.* arsenic.

as[1] /æz, əz/ *–adv.* & *conj.* (*adv.* as antecedent in main sentence; *conj.* in relative clause expressed or implied) to the extent to which . . . is or does etc. (*am as tall as he*; *am as tall as he is*; (*colloq.*) *am as tall as him*; *as recently as last week*). *–conj.* (with relative clause expressed or implied) **1** (with antecedent *so*) expressing result or purpose (*came early so as to meet us*). **2** (with antecedent adverb omitted) although (*good as it is* = although it is good). **3** (without antecedent adverb) **a** in the manner in which (*do as you like*; *rose as one man*). **b** in the capacity or form of (*I speak as your friend*; *Olivier as Hamlet*). **c** while (*arrived as I was eating*). **d** since, seeing that (*as you are here, we can talk*). **e** for instance (*cathedral cities, as York*). –*rel. pron.* (with verb of relative clause expressed or implied) **1** that, who, which (*I had the same trouble as you*; *he is a writer, as is his wife*; *such countries as France*). **2** (with a sentence as antecedent) a fact that (*he lost, as you know*). □ **as for** with regard to (*as for you, I think you are wrong*). **as from** on and after (a specified date). **as if** (or **though**) as would be the case if (*acts as if he were in charge*). **as it were** in a way; to some extent (*he is, as it were, infatuated*). **as long as** see LONG[1]. **as much** see MUCH. **as of 1** = *as from*. **2** as at (a specified time). **as per** see PER. **as regards** see REGARD. **as soon as** see SOON. **as such** see SUCH. **as though** see *as if*. **as to** with regard to. **as well 1** in addition. **2** advisable, desirable, reasonably. **as well as** in addition to. **as yet** until now or up to a particular time (*have received no news as yet*). [Old English, = ALSO]

as[2] /æs/ *n.* (*pl.* **asses**) Roman copper coin. [Latin]

asafoetida /,æsə'fetɪdə/ *n.* (*US* **asafetida**) resinous pungent plant gum used in cooking and formerly in medicine. [Persian *azā* mastic: related to FETID]

a.s.a.p. *abbr.* as soon as possible.

asbestos /æs'bestɒs/ *n.* **1** fibrous silicate mineral. **2** this as a heat-resistant or insulating material. [Greek, = unquenchable]

ascend /ə'send/ *v.* **1** move or slope upwards, rise. **2** climb; go up. □ **ascend the throne** become king or queen. [Latin *scando* climb]

ascendancy *n.* (often foll. by *over*) dominant power or control.

ascendant –*adj.* **1** rising. **2** *Astron.* rising towards the zenith. **3** *Astrol.* just above the eastern horizon. **4** predominant. –*n. Astrol.* ascendant point of the sun's apparent path. □ **in the ascendant** gaining or having power or authority.

ascension /ə'senʃ(ə)n/ *n.* **1** ascent. **2** (**Ascension**) ascent of Christ into heaven.

ascent /ə'sent/ *n.* **1** ascending, rising, or progressing. **2** upward slope or path etc.

ascertain /,æsə'teɪn/ *v.* find out for certain. □ **ascertainment** *n.* [French: related to CERTAIN]

ascetic /ə'setɪk/ –*adj.* severely abstinent; self-denying. –*n.* ascetic, esp. religious, person. □ **asceticism** /-tɪ,sɪz(ə)m/ *n.* [Greek *askeō* exercise]

ASCII /'æskɪ/ *abbr. Computing* American Standard Code for Information Interchange.

ascorbic acid /ə'skɔ:bɪk/ *n.* vitamin C, which prevents scurvy. [from A-, SCORBUTIC]

ascribe /ə'skraɪb/ *v.* (**-bing**) (usu. foll. by *to*) **1** attribute (*ascribes his health to exercise*). **2** regard as belonging. □ **ascription** /ə'skrɪpʃ(ə)n/ *n.* [Latin *scribo* write]

asepsis /eɪ'sepsɪs/ *n.* **1** absence of sepsis or harmful micro-organisms. **2** method of achieving asepsis in surgery. □ **aseptic** *adj.*

asexual /eɪ'sekʃʊəl/ *adj.* **1** without sex, sexual organs, or sexuality. **2** (of reproduction) not involving the fusion of gametes. □ **asexually** *adv.*

ash[1] *n.* **1** (often in *pl.*) powdery residue left after burning. **2** (*pl.*) human remains after cremation. **3** (**the Ashes**) *Cricket* trophy competed for by Australia and England. [Old English]

ash[2] *n.* **1** tree with silver-grey bark. **2** its hard, pale wood. [Old English]

ashamed /ə'ʃeɪmd/ *adj.* (usu. *predic.*) **1** embarrassed by shame (*ashamed of myself*). **2** (foll. by *to* + infin.) hesitant, reluctant out of shame (*am ashamed to say*). [Old English *a-* intensive prefix]

ashcan *n. US* dustbin.

ashen *adj.* like ashes, esp. grey or pale.

Ashkenazi /,æʃkə'nɑ:zɪ/ *n.* (*pl.* **-zim**) East European Jew. [Hebrew]

ashlar /'æʃlə(r)/ *n.* **1** large square-cut stone used in building; masonry made of these. **2** thin slabs of masonry used for facing walls. [Latin *axis* board]

ashore /ə'ʃɔ:(r)/ *adv.* towards or on the shore or land.

ashram /'æʃrəm/ *n.* place of religious retreat for Hindus. [Sanskrit]

ashtray *n.* small receptacle for cigarette ash, stubs, etc.

ashy *adj.* (**-ier, -iest**) **1** = ASHEN. **2** covered with ashes.

Asian /'eɪʃ(ə)n/ –*n.* **1** native of Asia. **2** person of Asian descent. –*adj.* of Asia. [Latin from Greek]

Asiatic /,eɪʃɪ'ætɪk/ –*n. offens.* Asian. –*adj.* Asian. [Latin from Greek]

aside /ə'saɪd/ –*adv.* to or on one side; away, apart. –*n.* words spoken aside, esp. confidentially to the audience by an actor.

asinine /'æsɪ,naɪn/ *adj.* like an ass, esp. stupid or stubborn. □ **asininity** /-'nɪnɪtɪ/ *n.* [Latin *asinus* ass]

ask /ɑ:sk/ *v.* **1** call for an answer to or about (*ask her about it*; *ask him his name*). **2** seek to obtain from someone (*ask a favour of*). **3** (usu. foll. by *out, in*, or *over*, or *to* (a function etc.)) invite (*must ask them over*; *asked her to dinner*). **4** (foll. by *for*) seek to obtain, meet, or be directed to (*ask for help*; *asking for you*; *ask for the bar*). □ **ask after** inquire about (esp. a person). **ask for it** *slang* invite trouble. [Old English]

askance /ə'skæns/ *adv.* sideways or squinting. □ **look askance at** regard suspiciously. [origin unknown]

askew /ə'skju:/ *–adv.* awry, crookedly. *–predic. adj.* oblique; awry.

aslant /ə'slɑ:nt/ *–adv.* obliquely or at a slant. *–prep.* obliquely across.

asleep /ə'sli:p/ *predic. adj.* & *adv.* **1 a** in or into a state of sleep. **b** inactive, inattentive. **2** (of a limb etc.) numb. **3** *euphem.* dead.

asp *n.* small venomous snake of North Africa or Southern Europe. [Greek *aspis*]

asparagus /ə'spærəgəs/ *n.* **1** plant of the lily family. **2** edible shoots of this. [Latin from Greek]

aspect /'æspekt/ *n.* **1** viewpoint, feature, etc. to be considered (*one aspect of the problem*). **2** appearance or look (*cheerful aspect*). **3** side of a building or location facing a particular direction (*southern aspect*). [Latin *adspicio* look at]

aspen /'æspən/ *n.* poplar with very tremulous leaves. [Old English: originally adj.]

asperity /ə'sperɪtɪ/ *n.* (*pl.* **-ies**) **1** sharpness of temper or tone. **2** roughness; rough excrescence. [Latin *asper* rough]

aspersion /ə'spɜ:ʃ(ə)n/ *n.* □ **cast aspersions on** attack the reputation of. [Latin *aspergo* besprinkle]

asphalt /'æsfælt/ *–n.* **1** dark bituminous pitch. **2** mixture of this with sand, gravel, etc., for surfacing roads etc. *–v.* surface with asphalt. [Latin from Greek]

asphodel /'æsfə,del/ *n.* **1** plant of the lily family. **2** *poet.* immortal flower growing in Elysium. [Latin from Greek]

asphyxia /æs'fɪksɪə/ *n.* lack of oxygen in the blood, causing unconsciousness or death; suffocation. □ **asphyxiant** *adj.* & *n.* [Greek *a-* not, *sphuxis* pulse]

asphyxiate /æs'fɪksɪ,eɪt/ *v.* (**-ting**) suffocate. □ **asphyxiation** /-'eɪʃ(ə)n/ *n.*

aspic /'æspɪk/ *n.* savoury jelly used esp. to contain game, eggs, etc. [French, = ASP, suggested by the colours of the jelly]

aspidistra /,æspɪ'dɪstrə/ *n.* house-plant with broad tapering leaves. [Greek *aspis* shield]

aspirant /'æspɪrənt/ *–adj.* aspiring. *–n.* person who aspires. [Latin: related to ASPIRE]

aspirate /'æspərət/ *–adj.* pronounced with an exhalation of breath; blended with the sound of *h*. *–n.* sound of *h*; consonant pronounced in this way. *–v.* /-,reɪt/ (**-ting**) **1** pronounce with breath or with initial *h*. **2** draw (fluid) by suction from a cavity etc.

aspiration /,æspɪ'reɪʃ(ə)n/ *n.* **1** ambition or desire. **2** drawing breath or *Phonet.* aspirating.

aspirator /'æspɪ,reɪtə(r)/ *n.* apparatus for aspirating fluid. [Latin: related to ASPIRE]

aspire /ə'spaɪə(r)/ *v.* (**-ring**) (usu. foll. by *to* or *after*, or *to* + infin.) have ambition or a strong desire. [Latin *aspiro* breathe upon]

aspirin /'æsprɪn/ *n.* (*pl.* same or **-s**) **1** white powder, acetylsalicylic acid, used to reduce pain and fever. **2** tablet of this. [German]

ass[1] *n.* **1 a** four-legged long-eared mammal related to the horse. **b** donkey. **2** stupid person. [Old English from Latin]

ass[2] *US* var. of ARSE.

assagai var. of ASSEGAI.

assail /ə'seɪl/ *v.* **1** attack physically or verbally. **2** tackle (a task) resolutely. □ **assailant** *n.* [Latin *salio* leap]

assassin /ə'sæsɪn/ *n.* killer, esp. of a political or religious leader. [Arabic, = hashish-eater]

assassinate /ə'sæsɪ,neɪt/ *v.* (**-ting**) kill for political or religious motives. □ **assassination** /-'neɪʃ(ə)n/ *n.*

assault /ə'sɔ:lt/ *–n.* **1** violent physical or verbal attack. **2** *Law* threat or display of violence against a person. *–v.* make an assault on. □ **assault and battery** *Law* threatening act resulting in physical harm to a person. [Latin: related to ASSAIL]

assay /ə'seɪ/ *–n.* testing of a metal or ore to determine its ingredients and quality. *–v.* make an assay of (a metal or ore). [French, var. of *essai* ESSAY]

assegai /'æsɪ,gaɪ/ *n.* (also **assagai**) light iron-tipped S. African spear. [Arabic, = the spear]

assemblage /ə'semblɪdʒ/ *n.* **1** assembling. **2** assembled group.

assemble /ə'semb(ə)l/ *v.* (**-ling**) **1** gather together; collect. **2** esp. *Mech.* fit together (components, a whole). [Latin *ad* to, *simul* together]

assembler /ə'semblə(r)/ *n.* **1** person who assembles a machine etc. **2** *Computing* **a** program for converting instructions written in low-level symbolic code into machine code. **b** the low-level symbolic code itself.

assembly /ə'semblɪ/ *n.* (*pl.* **-ies**) **1** assembling. **2** assembled group, esp. as a deliberative body. **3** assembling of components.

assembly line *n.* machinery arranged so that a product can be progressively assembled.

assent /ə'sent/ *–v.* (usu. foll. by *to*) **1** express agreement. **2** consent. *–n.* consent or approval, esp. official. □ **assenter** *n.* [Latin *sentio* think]

assert /ə'sɜ:t/ *v.* **1** declare, state clearly. **2** *refl.* insist on one's rights. **3** enforce a claim to (*assert one's rights*). [Latin *assero -sert-*]

assertion /ə'sɜ:ʃ(ə)n/ *n.* declaration, forthright statement.

assertive /ə'sɜ:tɪv/ *adj.* tending to assert oneself; forthright, positive. □ **assertively** *adv.* **assertiveness** *n.*

assess /ə'ses/ *v.* **1** estimate the size or quality of. **2** estimate the value of (property etc.) for taxation. □ **assessment** *n.* [Latin *assideo -sess-* sit by]

assessor *n.* **1** person who assesses (esp. for tax or insurance). **2** legal adviser on technical questions.

asset /'æset/ *n.* **1** useful or valuable person or thing. **2** (usu. in *pl.*) property and possessions, esp. that can be set against debts etc. [French *asez* from Latin *ad satis* to enough]

asset-stripping *n.* the taking over of a company and selling off of its assets to make a profit.

asseverate /ə'sevə,reɪt/ *v.* (**-ting**) declare solemnly. □ **asseveration** /-'reɪʃ(ə)n/ *n.* [Latin *severus* serious]

asshole *US* var. of ARSEHOLE.

assiduous /ə'sɪdjʊəs/ *adj.* **1** persevering, hard-working. **2** attending closely. □ **assiduity** /,æsɪ'dju:ɪtɪ/ *n.* **assiduously** *adv.* [Latin: related to ASSESS]

assign /ə'saɪn/ *–v.* **1** (usu. foll. by *to*) **a** allot as a share or responsibility. **b** appoint to a position, task, etc. **2** fix (a time, place, etc.). **3** (foll. by *to*) ascribe to (a reason, date, etc.) (*assigned the manuscript to 1832*). **4** (foll. by *to*) *Law* transfer formally (esp. property) to (another). *–n.* assignee. □ **assigner** *n.* **assignor** *n. Law.* [Latin *assigno* mark out]

assignation /,æsɪg'neɪʃ(ə)n/ *n.* **1** appointment to meet, esp. by lovers in secret. **2** assigning or being assigned.

assignee /,æsaɪ'ni:/ *n. Law* person to whom a right or property is assigned.

assignment /ə'saɪnmənt/ *n.* **1** task or mission. **2** assigning or being assigned. **3** legal transfer.

assimilate /ə'sɪmɪ,leɪt/ *v.* (**-ting**) **1** absorb or be absorbed, either physically or mentally. **2** (usu. foll. by *to, with*) make like; cause to resemble. □ **assimilable** *adj.* **assimilation** /-'leɪʃ(ə)n/ *n.* **assimilative** /-lətɪv/ *adj.* **assimilator** *n.* [Latin *similis* like]

assist /ə'sɪst/ *v.* (often foll. by *in* + verbal noun) help. □ **assistance** *n.* [Latin *assisto* stand by]

assistant *n.* **1** (often *attrib.*) person who helps, esp. as a subordinate. **2** = SHOP ASSISTANT.

assizes /ə'saɪzɪz/ *n.pl. hist.* court periodically administering the civil and criminal law. [French: related to ASSESS]

■ **Usage** In 1972 the civil jurisdiction of assizes in England and Wales was transferred to the High Court and the criminal jurisdiction to the Crown Court.

Assoc. *abbr.* Association.

associate *–v.* /ə'səʊʃɪ,eɪt/ (**-ting**) **1** connect mentally (*associate holly with Christmas*). **2** join or combine, esp. for a common purpose. **3** *refl.* declare oneself or be in agreement. **4** (usu. foll. by *with*) meet frequently or deal. *–n.* /ə'səʊʃɪət/ **1** partner, colleague. **2** friend, companion. **3** subordinate member of a society etc. *–adj.* /ə'səʊʃɪət/ **1** joined or allied. **2** of lower status (*associate member*). □ **associative** /-ʃɪətɪv/ *adj.* [Latin *socius* allied]

association /ə,səʊsɪ'eɪʃ(ə)n/ *n.* **1** group organized for a joint purpose; society. **2** associating or being associated. **3** companionship. **4** mental connection of ideas. [medieval Latin: related to ASSOCIATE]

Association Football *n.* football played with a round ball which may not be handled except by the goalkeepers.

assonance /'æsənəns/ *n.* partial resemblance of sound between two syllables e.g. *sonnet, porridge,* and *killed, cold, culled.* □ **assonant** *adj.* [Latin *sonus* sound]

assort /ə'sɔ:t/ *v.* **1** classify or arrange in sorts. **2** (usu. foll. by *with*) suit or harmonize with. [French: related to SORT]

assorted *adj.* **1** of various sorts, mixed. **2** classified. **3** matched (*ill-assorted pair*).

assortment *n.* diverse group or mixture.

assuage /ə'sweɪdʒ/ *v.* (**-ging**) **1** calm or soothe. **2** appease (an appetite). □ **assuagement** *n.* [Latin *suavis* sweet]

assume /ə'sju:m/ *v.* (**-ming**) **1** (usu. foll. by *that*) take to be true. **2** simulate (ignorance etc.). **3** undertake (an office etc.). **4** take or put on (an aspect, attribute, etc.) (*assumed immense importance*). [Latin *sumo* take]

assuming *adj.* arrogant, presumptuous.

assumption /əˈsʌmpʃ(ə)n/ *n.* **1** assuming. **2** thing assumed. **3** (**Assumption**) reception of the Virgin Mary bodily into heaven.

assurance /əˈʃʊərəns/ *n.* **1** emphatic declaration; guarantee. **2** insurance, esp. life insurance. **3** certainty. **4** self-confidence; assertiveness.

assure /əˈʃʊə(r)/ *v.* (**-ring**) **1** (often foll. by *of*) **a** convince. **b** tell (a person) confidently (*assured him all was well*). **2** ensure; guarantee (a result etc.). **3** insure (esp. a life). **4** (as **assured** *adj.*) **a** guaranteed. **b** self-confident. [Latin *securus* safe]

assuredly /əˈʃʊərɪdlɪ/ *adv.* certainly.

AST *abbr.* Atlantic Standard Time.

astatine /ˈæstəˌtiːn/ *n.* radioactive element, the heaviest of the halogens. [Greek *astatos* unstable]

aster *n.* plant with bright daisy-like flowers. [Greek, = star]

asterisk /ˈæstərɪsk/ —*n.* symbol (*) used to mark words or to indicate omission etc. —*v.* mark with an asterisk. [Greek, = little star]

astern /əˈstɜːn/ *adv.* (often foll. by *of*) **1** in or to the rear of a ship or aircraft. **2** backwards.

asteroid /ˈæstəˌrɔɪd/ *n.* **1** any of the minor planets orbiting the sun, mainly between the orbits of Mars and Jupiter. **2** starfish. [Greek: related to ASTER]

asthma /ˈæsmə/ *n.* respiratory condition marked by wheezing. [Greek *azō* breathe hard]

asthmatic /æsˈmætɪk/ —*adj.* of or suffering from asthma. —*n.* asthmatic person.

astigmatism /əˈstɪgməˌtɪz(ə)m/ *n.* eye or lens defect resulting in distorted images. □ **astigmatic** /ˌæstɪgˈmætɪk/ *adj.* [from A-, STIGMA]

astir /əˈstɜː(r)/ *predic. adj.* & *adv.* **1** in motion. **2** out of bed.

astonish /əˈstɒnɪʃ/ *v.* surprise greatly, amaze. □ **astonishment** *n.* [Latin *ex-* forth, *tono* thunder]

astound /əˈstaʊnd/ *v.* astonish greatly.

astraddle /əˈstræd(ə)l/ *adv.* astride.

astrakhan /ˌæstrəˈkæn/ *n.* **1** dark curly fleece of young Astrakhan lambs. **2** cloth imitating this. [*Astrakhan* in Russia]

astral /ˈæstr(ə)l/ *adj.* of the stars; starry. [Latin *astrum* star]

astray /əˈstreɪ/ *adv.* & *predic.adj.* out of the right way, erring. □ **go astray** be lost or mislaid. [Latin *extra* away, *vagor* wander]

astride /əˈstraɪd/ —*adv.* **1** (often foll. by *of*) with a leg on each side. **2** with legs apart. —*prep.* astride of; extending across.

astringent /əˈstrɪndʒ(ə)nt/ —*adj.* **1** checking bleeding by contracting body tissues. **2** severe, austere. —*n.* astringent substance. □ **astringency** *n.* [Latin *astringo* draw tight]

astrolabe /ˈæstrəˌleɪb/ *n.* instrument formerly used to measure the altitude of stars etc. [Greek, = star-taking]

astrology /əˈstrɒlədʒɪ/ *n.* study of supposed planetary influence on human affairs. □ **astrologer** *n.* **astrological** /ˌæstrəˈlɒdʒɪk(ə)l/ *adj.* **astrologist** *n.* [Greek *astron* star]

astronaut /ˈæstrəˌnɔːt/ *n.* crew member of a spacecraft. [Greek *astron* star, *nautēs* sailor]

astronautics /ˌæstrəˈnɔːtɪks/ *n.pl.* (treated as *sing.*) science of space travel. □ **astronautical** *adj.*

astronomical /ˌæstrəˈnɒmɪk(ə)l/ *adj.* (also **astronomic**) **1** of astronomy. **2** vast, gigantic. □ **astronomically** *adv.*

astronomy /əˈstrɒnəmɪ/ *n.* the scientific study of celestial bodies. □ **astronomer** *n.* [Greek *astron* star, *nemō* arrange]

astrophysics /ˌæstrəʊˈfɪzɪks/ *n.pl.* (treated as *sing.*) the study of the physics and chemistry of celestial bodies. □ **astrophysical** *adj.* **astrophysicist** /-sɪst/ *n.* [Greek *astron* star]

astute /əˈstjuːt/ *adj.* shrewd. □ **astutely** *adv.* **astuteness** *n.* [Latin *astus* craft]

asunder /əˈsʌndə(r)/ *adv. literary* apart.

asylum /əˈsaɪləm/ *n.* **1** sanctuary; protection, esp. for fugitives from the law (*seek asylum*). **2** *hist.* institution for the mentally ill or destitute. [Greek *a-* not, *sulon* right of seizure]

asymmetry /æˈsɪmɪtrɪ/ *n.* lack of symmetry. □ **asymmetric** /-ˈmetrɪk/ *adj.* **asymmetrical** /-ˈmetrɪk(ə)l/ *adj.* [Greek]

At *symb.* astatine.

at /æt, ət/ *prep.* **1** expressing position (*wait at the corner*; *at school*). **2** expressing a point in time (*at dawn*). **3** expressing a point in a scale (*at his best*). **4** expressing engagement in an activity etc. (*at war*). **5** expressing a value or rate (*sell at £10 each*). **6 a** with or with reference to (*annoyed at losing*; *came at a run*). **b** in response to (*starts at a touch*). **7** expressing motion or aim towards (*aim at the target*; *laughed at us*). □ **at all** see ALL. **at hand** see HAND. **at home** see HOME. **at it** engaged in an activity; working hard. **at once** see

ONCE. **at that 1** moreover (*a good one at that*). **2** then (*at that he left*). **at times** see TIME. [Old English]

atavism /'ætə,vɪz(ə)m/ *n.* **1** reappearance of a remote ancestral characteristic, throwback. **2** reversion to an earlier type. □ **atavistic** /-'vɪstɪk/ *adj.* [Latin *atavus* ancestor]

ataxia /ə'tæksɪə/ *n. Med.* imperfect control of bodily movements. [Greek *a-* without, *taxis* order]

ate *past* of EAT.

-ate[1] *suffix* forming nouns denoting status, function, or office (*doctorate*; *consulate*). [Latin]

-ate[2] *suffix* forming adjectives with the sense 'having, full of' (*foliate*; *passionate*). [Latin participial ending *-atus*]

atelier /ə'telɪ,eɪ/ *n.* workshop or artist's studio. [French]

atheism /'eɪθɪ,ɪz(ə)m/ *n.* belief that there is no God. □ **atheist** *n.* **atheistic** /-'ɪstɪk/ *adj.* [Greek *a-* not, *theos* god]

atherosclerosis /,æθərəʊsklə'rəʊsɪs/ *n.* degeneration of the arteries caused by a build-up of fatty deposits. [Greek *athērē* groats]

athirst /ə'θɜ:st/ *predic. adj. poet.* **1** (usu. foll. by *for*) eager. **2** thirsty.

athlete /'æθli:t/ *n.* person who engages in athletics, exercise, etc. [Greek *athlon* prize]

athlete's foot *n.* fungal foot condition.

athletic /æθ'letɪk/ *adj.* **1** of athletes or athletics. **2** physically strong or agile. □ **athletically** *adv.* **athleticism** /-,sɪz(ə)m/ *n.* [Latin: related to ATHLETE]

athletics *n.pl.* (usu. treated as *sing.*) physical exercises, esp. track and field events.

at-home *n.* social reception in a person's home.

-ation *suffix* **1** forming nouns denoting an action or an instance of it (*flirtation*; *hesitation*). **2** forming nouns denoting a result or product of action (*plantation*; *starvation*). [Latin *-atio*]

Atlantic /ət'læntɪk/ *adj.* of or adjoining the ocean between Europe and Africa to the east and America to the west. [Greek: related to ATLAS]

atlas /'ætləs/ *n.* book of maps or charts. [Greek *Atlas*, the Titan who held up the universe]

atmosphere /'ætməsfɪə(r)/ *n.* **1 a** gases enveloping the earth, any other planet, etc. **b** air in a room etc., esp. if fetid. **2** pervading tone or mood of a place, situation, or work of art. **3** unit of pressure equal to mean atmospheric pressure at sea level, 101,325 pascals. □ **atmospheric** /-'ferɪk/ *adj.* [Greek *atmos* vapour, SPHERE]

atmospherics /,ætməs'ferɪks/ *n.pl.* **1** electrical atmospheric disturbance, esp. caused by lightning. **2** interference with telecommunications caused by this.

atoll /'ætɒl/ *n.* ring-shaped coral reef enclosing a lagoon. [Maldive]

atom /'ætəm/ *n.* **1 a** smallest particle of a chemical element that can take part in a chemical reaction. **b** this as a source of nuclear energy. **2** minute portion or thing (*atom of pity*). [Greek *atomos* indivisible]

atom bomb *n.* bomb in which energy is released by nuclear fission.

atomic /ə'tɒmɪk/ *adj.* **1** of or using atomic energy or atomic bombs. **2** of atoms.

atomic bomb *n.* = ATOM BOMB.

atomic energy *n.* nuclear energy.

atomic mass *n.* mass of an atom measured in atomic mass units.

atomic mass unit *n.* unit of mass used to express atomic and molecular weights, equal to one-twelfth of the mass of an atom of carbon-12.

atomic number *n.* number of protons in the nucleus of an atom.

atomic theory *n.* theory that all matter consists of atoms.

atomic weight *n.* = RELATIVE ATOMIC MASS.

atomize /'ætə'maɪz/ *v.* (also **-ise**) (**-zing** or **-sing**) reduce to atoms or fine particles.

atomizer *n.* (also **-iser**) = AEROSOL 1.

atonal /eɪ'təʊn(ə)l/ *adj. Mus.* not written in any key or mode. □ **atonality** /-'nælɪtɪ/ *n.*

atone /ə'təʊn/ *v.* (**-ning**) (usu. foll. by *for*) make amends (for a wrong). [from ATONEMENT]

atonement *n.* **1** atoning. **2** (**the Atonement**) expiation by Christ of mankind's sins. [*at one* + -MENT]

atrium /'eɪtrɪəm/ *n.* (*pl.* **-s** or **atria**) **1 a** central court of an ancient Roman house. **b** (usu. skylit) central court rising through several storeys. **2** each of the two upper cavities of the heart. [Latin]

atrocious /ə'trəʊʃəs/ *adj.* **1** very bad or unpleasant (*atrocious manners*). **2** wicked (*atrocious cruelty*). □ **atrociously** *adv.* [Latin *atrox* cruel]

atrocity /ə'trɒsɪtɪ/ *n.* (*pl.* **-ies**) **1** wicked or cruel act. **2** extreme wickedness. [Latin: related to ATROCIOUS]

atrophy /'ætrəfɪ/ —*n.* wasting away, esp. through disuse; emaciation. —*v.* (**-ies, -ied**) suffer atrophy or cause atrophy in. [Greek *a-* without, *trophē* food]

atropine /'ætrə,pi:n/ *n.* poisonous alkaloid in deadly nightshade. [Greek *Atropos*, the Fate who cut the thread of life]

attach /ə'tætʃ/ *v.* **1** fasten, affix, join. **2** (in *passive*; foll. by *to*) be very fond of. **3** attribute or be attributable; assign (*can't attach a name to it*; *no blame attaches to us*). **4** accompany; form part of (*no conditions are attached*). **5** *refl.* (usu. foll. by *to*) take part in; join (*attached himself to the team*). **6** seize by legal authority. [French from Germanic]

attaché /ə'tæʃeɪ/ *n.* specialist member of an ambassador's staff.

attaché case *n.* small rectangular document case.

attachment *n.* **1** thing attached, esp. for a purpose. **2** affection, devotion. **3** attaching or being attached. **4** legal seizure. **5** temporary position in an organization.

attack /ə'tæk/ *–v.* **1** try to hurt or defeat using force. **2** criticize adversely. **3** act harmfully upon (*rust attacks metal*). **4** vigorously apply oneself to. **5** *Sport* try to gain ground or score (against). *–n.* **1** act of attacking. **2** offensive operation. **3** sudden onset of an illness. □ **attacker** *n.* [French from Italian]

attain /ə'teɪn/ *v.* **1** reach, gain, accomplish (a goal etc.). **2** (foll. by *to*) arrive at by effort or development. [Latin *attingo* reach]

attainment *n.* **1** (often in *pl.*) accomplishment or achievement. **2** attaining.

attar /'ætɑ:(r)/ *n.* perfume made from rose-petals. [Persian]

attempt /ə'tempt/ *–v.* **1** (often foll. by *to* + infin.) try to do or achieve (*attempted to explain*). **2** try to conquer (a mountain etc.). *–n.* (often foll. by *at*, *on*, or *to* + infin.) attempting; endeavour (*attempt at winning*; *attempt on his life*). [Latin *tempto* try]

attend /ə'tend/ *v.* **1 a** be present (at) (*attended the meeting*). **b** go regularly to (*attends church*). **2** escort. **3 a** (often foll. by *to*) turn or apply one's mind. **b** (foll. by *to*) deal with (*attend to the matter*). [Latin *tendo* stretch]

attendance *n.* **1** attending or being present. **2** number present (*high attendance*).

attendant *–n.* person escorting or providing a service (*cloakroom attendant*). *–adj.* **1** accompanying (*attendant costs*). **2** (often foll. by *on*) waiting (*attendant on the queen*).

attendee /,æten'di:/ *n.* person who attends (a meeting etc.).

attention /ə'tenʃ(ə)n/ *n.* **1** act or faculty of applying one's mind; notice (*attention wandered*; *attract his attention*). **2** consideration, care. **3** (in *pl.*) **a** courtesies. **b** sexual advances. **4** erect esp. military attitude of readiness.

attentive /ə'tentɪv/ *adj.* **1** concentrating; paying attention. **2** assiduously polite. □ **attentively** *adv.* **attentiveness** *n.*

attenuate /ə'tenjʊ,eɪt/ *v.* (**-ting**) **1** make thin. **2** reduce in force, value, etc. □ **attenuation** /-'eɪʃ(ə)n/ *n.* [Latin *tenuis* thin]

attest /ə'test/ *v.* **1** certify the validity of. **2** (foll. by *to*) bear witness to. □ **attestation** /,æte'steɪʃ(ə)n/ *n.* [Latin *testis* witness]

Attic /'ætɪk/ *–adj.* of ancient Athens or Attica, or the form of Greek used there. *–n.* Greek as used by the ancient Athenians. [Greek *Attikos*]

attic /'ætɪk/ *n.* space or room at the top of a house, usu. under the roof. [from Attic, with ref. to an architectural feature]

attire /ə'taɪə(r)/ *formal* *–n.* clothes, esp. formal. *–v.* (**-ring**) (usu. as **attired** *adj.*) dress, esp. formally. [French *à tire* in order]

attitude /'ætɪ,tju:d/ *n.* **1** opinion or way of thinking; behaviour reflecting this (*don't like his attitude*). **2** bodily posture; pose. **3** position of an aircraft etc. relative to given points. [Latin *aptus* fitted]

attitudinize /,ætɪ'tju:dɪ,naɪz/ *v.* (also **-ise**) (**-zing** or **-sing**) adopt (esp. affected) attitudes; pose.

attorney /ə'tɜ:nɪ/ *n.* (*pl.* **-s**) **1** person, esp. a lawyer, appointed to act for another in business or legal matters. **2** *US* qualified lawyer. [French *atorner* assign]

Attorney-General *n.* (*pl.* **Attorneys-General**) chief legal officer in some countries.

attract /ə'trækt/ *v.* **1** (also *absol.*) (of a magnet etc.) draw to itself or oneself. **2** arouse interest or admiration in. [Latin *traho* draw]

attraction /ə'trækʃ(ə)n/ *n.* **1 a** attracting or being attracted. **b** attractive quality (*can't see the attraction in it*). **c** person or thing that attracts. **2** *Physics* tendency of bodies to attract each other.

attractive /ə'træktɪv/ *adj.* **1** attracting (esp. interest or admiration). **2** aesthetically pleasing; good-looking. □ **attractively** *adv.*

attribute *–v.* /ə'trɪbju:t/ (**-ting**) (usu. foll. by *to*) **1** regard as belonging to, written or said by, etc. (*a poem attributed to Milton*). **2** ascribe to (a cause) (*delays attributed to snow*). *–n.* /'ætrɪ,bju:t/ **1** esp. characteristic quality ascribed to a person or thing. **2** object symbolizing or appropriate to a person, office, or status. □ **attributable** /ə'trɪbjʊtəb(ə)l/ *adj.* **attribution** /,ætrɪ'bju:ʃ(ə)n/ *n.* [Latin *tribuo* allot]

attributive /ə'trɪbjʊtɪv/ *adj. Gram.* (of an adjective or noun) preceding the word described, as *old* in *the old dog*.

attrition /ə'trɪʃ(ə)n/ *n.* **1** gradual wearing down (*war of attrition*). **2** abrasion, friction. [Latin *tero trit-* rub]

attune /ə'tju:n/ *v.* (**-ning**) **1** (usu. foll. by *to*) adjust (to a situation etc.). **2** *Mus.* tune. [related to TUNE]

atypical /eɪ'tɪpɪk(ə)l/ *adj.* not typical. □ **atypically** *adv.*

Au *symb.* gold. [Latin *aurum*]

aubergine /'əʊbə,ʒi:n/ *n.* plant with white or purple egg-shaped fruit used as a vegetable; eggplant. [French, ultimately from Sanskrit]

aubrietia /ɔ:'bri:ʃə/ *n.* (also **aubretia**) dwarf perennial rock-plant with purple or pink flowers. [*Aubriet*, name of an artist]

auburn /'ɔ:bən/ *adj.* reddish-brown (usu. of hair). [originally = yellowish white: from Latin *albus* white]

auction /'ɔ:kʃ(ə)n/ *—n.* sale in which articles are sold to the highest bidder. *—v.* sell by auction. [Latin *augeo auct-* increase]

auction bridge *n.* game in which players bid for the right to name trumps.

auctioneer /,ɔ:kʃə'nɪə(r)/ *n.* person who conducts auctions, esp. for a living.

audacious /ɔ:'deɪʃəs/ *adj.* **1** daring, bold. **2** impudent. □ **audacity** /ɔ:'dæsɪtɪ/ *n.* [Latin *audax* bold]

audible /'ɔ:dɪb(ə)l/ *adj.* able to be heard. □ **audibility** /-'bɪlɪtɪ/ *n.* **audibly** *adv.* [Latin *audio* hear]

audience /'ɔ:dɪəns/ *n.* **1 a** assembled listeners or spectators, esp. at a play, concert, etc. **b** people addressed by a film, book, etc. **2** formal interview with a superior. [Latin: related to AUDIBLE]

audio /'ɔ:dɪəʊ/ *n.* (usu. *attrib.*) sound or its reproduction. [Latin *audio* hear]

audio- *comb. form* hearing or sound.

audio frequency *n.* frequency able to be perceived by the human ear.

audiotape *n.* (also **audio tape**) **1 a** magnetic tape for recording sound. **b** a length of this. **2** a sound recording on tape.

audio typist *n.* person who types from a tape-recording.

audiovisual /,ɔ:dɪəʊ'vɪʒʊəl/ *adj.* (of teaching methods etc.) using both sight and sound.

audit /'ɔ:dɪt/ *—n.* official scrutiny of accounts. *—v.* (**-t-**) conduct an audit of.

audition /ɔ:'dɪʃ(ə)n/ *—n.* test of a performer's suitability or ability. *—v.* assess or be assessed at an audition. [Latin *audio* hear]

auditor *n.* person who audits accounts. [French from Latin]

auditorium /,ɔ:dɪ'tɔ:rɪəm/ *n.* (*pl.* **-s**) part of a theatre etc. for the audience. [Latin]

auditory /'ɔ:dɪtərɪ/ *adj.* of hearing.

au fait /əʊ 'feɪ/ *predic. adj.* (usu. foll. by *with*) conversant (*au fait with the rules*). [French]

Aug. *abbr.* August.

Augean /ɔ:'dʒi:ən/ *adj.* filthy. [Greek *Augeas*, a mythical king: his filthy stables were cleaned by Hercules diverting a river through them]

auger /'ɔ:gə(r)/ *n.* tool with a screw point for boring in wood. [Old English]

aught /ɔ:t/ *n. archaic* anything. [Old English]

augment /ɔ:g'ment/ *v.* make or become greater; increase. □ **augmentation** /-'teɪʃ(ə)n/ *n.* [Latin: related to AUCTION]

augmentative /ɔ:g'mentətɪv/ *adj.* augmenting.

au gratin /əʊ 'grætæ̃/ *adj.* cooked with a crust of breadcrumbs or melted cheese. [French]

augur /'ɔ:gə(r)/ *—v.* portend, serve as an omen (*augur well* or *ill*). *—n. hist.* Roman religious official interpreting natural phenomena in order to pronounce on proposed actions. [Latin]

augury /'ɔ:gjʊrɪ/ *n.* (*pl.* **-ies**) **1** omen. **2** interpretation of omens.

August /'ɔ:gəst/ *n.* eighth month of the year. [Latin *Augustus*, first Roman emperor]

august /ɔ:'gʌst/ *adj.* venerable, imposing. [Latin]

Augustan /ɔ:'gʌst(ə)n/ *adj.* **1** of the reign of Augustus, esp. as a flourishing literary period. **2** (of literature) refined and classical in style. [Latin: see AUGUST]

auk /ɔ:k/ *n.* black and white sea bird with short wings, e.g. the guillemot, puffin, etc. [Old Norse]

auld lang syne /,ɔ:ld læŋ 'saɪn/ *n.* times long past. [Scots, = old long since]

aunt /ɑ:nt/ *n.* **1** sister of one's father or mother. **2** uncle's wife. **3** *colloq.* (form of address by a child to) parent's female friend. [Latin *amita*]

auntie /'ɑ:ntɪ/ *n.* (also **aunty**) (*pl.* **-ies**) *colloq.* = AUNT.

Aunt Sally *n.* **1** game in which sticks or balls are thrown at a wooden dummy. **2** target of general abuse.

au pair /əʊ 'peə(r)/ *n.* young foreigner, esp. a woman, helping with housework etc. in exchange for board and lodging. [French]

aura /'ɔ:rə/ *n.* (*pl.* **-s**) **1** distinctive atmosphere. **2** subtle emanation. [Greek, = breeze]

aural /'ɔ:r(ə)l/ *adj.* of the ear or hearing. □ **aurally** *adv.* [Latin *auris* ear]

aureate /ˈɔːrɪət/ *adj. literary* **1** golden. **2** resplendent. [Latin *aurum* gold]

aureole /ˈɔːrɪ,əʊl/ *n.* (also **aureola** /ɔːˈriːələ/) **1** halo or circle of light, esp. in a religious painting. **2** corona round the sun or moon. [Latin, = golden (crown)]

au revoir /,əʊ rəˈvwɑː(r)/ *int.* & *n.* goodbye (until we meet again). [French]

auricle /ˈɔːrɪk(ə)l/ *n.* **1** each atrium of the heart. **2** external ear of animals. □ **auricular** /-ˈrɪk-/ *adj.* [related to AURICULA]

auricula /ɔːˈrɪkjʊlə/ *n.* (*pl.* **-s**) primula with ear-shaped leaves. [Latin, diminutive of *auris* ear]

auriferous /ɔːˈrɪfərəs/ *adj.* yielding gold. [Latin *aurifer* from *aurum* gold]

aurochs /ˈɔːrɒks/ *n.* (*pl.* same) extinct wild ox. [German]

aurora /ɔːˈrɔːrə/ *n.* (*pl.* **-s** or **aurorae** /-riː/) luminous phenomenon, usu. of streamers of light in the night sky above the northern (**aurora borealis** /,bɒrɪˈeɪlɪs/) or southern (**aurora australis** /ɔːˈstreɪlɪs/) magnetic pole. [Latin, = dawn, goddess of dawn]

auscultation /,ɔːskəlˈteɪʃ(ə)n/ *n.* listening, esp. to sounds from the heart, lungs, etc., for purposes of diagnosis. [Latin *ausculto* listen]

auspice /ˈɔːspɪs/ *n.* **1** (in *pl.*) patronage (esp. *under the auspices of*). **2** omen, premonition. [originally 'observation of bird-flight': Latin *avis* bird]

auspicious /ɔːˈspɪʃəs/ *adj.* promising well; favourable.

Aussie /ˈɒzɪ/ *slang* —*n.* **1** Australian. **2** Australia. —*adj.* Australian. [abbreviation]

austere /ɒˈstɪə(r)/ *adj.* (**-terer, -terest**) **1** severely simple. **2** morally strict. **3** stern, grim. [Greek *austēros*]

austerity /ɒˈsterɪtɪ/ *n.* (*pl.* **-ies**) being austere; hardship.

austral /ˈɔːstr(ə)l/ *adj.* **1** southern. **2** (**Austral**) of Australia or Australasia. [Latin *auster* south]

Australasian /,ɒstrəˈleɪʒ(ə)n/ *adj.* of Australasia, including Australia and the islands of the SW Pacific.

Australian /ɒˈstreɪlɪən/ —*n.* **1** native or national of Australia. **2** person of Australian descent. —*adj.* of Australia.

autarchy /ˈɔːtɑːkɪ/ *n.* absolute rule; despotism. [Greek *autos* self, *arkhē* rule]

autarky /ˈɔːtɑːkɪ/ *n.* self-sufficiency, esp. economic. [Greek *autos* self, *arkeō* suffice]

authentic /ɔːˈθentɪk/ *adj.* **1** of undisputed origin; genuine. **2** reliable, trustworthy. □ **authentically** *adv.* **authenticity** /-ˈtɪsɪtɪ/ *n.* [Greek *authentikos*]

authenticate /ɔːˈθentɪ,keɪt/ *v.* (**-ting**) establish as true, genuine, or valid. □ **authentication** /-ˈkeɪʃ(ə)n/ *n.*

author /ˈɔːθə(r)/ *n.* (*fem.* **authoress** /ˈɔːθrɪs/) **1** writer, esp. of books. **2** originator of an idea, event, etc. [Latin *auctor*]

authoritarian /ɔː,θɒrɪˈteərɪən/ —*adj.* favouring or enforcing strict obedience to authority. —*n.* authoritarian person.

authoritative /ɔːˈθɒrɪtətɪv/ *adj.* **1** reliable, esp. having authority. **2** official.

authority /ɔːˈθɒrɪtɪ/ *n.* (*pl.* **-ies**) **1 a** power or right to enforce obedience. **b** (often foll. by *for*, or *to* + infin.) delegated power. **2** (esp. in *pl.*) body having authority. **3** influence based on recognized knowledge or expertise. **4** expert. [Latin *auctoritas*]

authorize /ˈɔːθə,raɪz/ *v.* (also **-ise**) (**-zing** or **-sing**) **1** officially approve, sanction. **2** (foll. by *to* + infin.) give authority to (a person to do a thing). □ **authorization** /-ˈzeɪʃ(ə)n/ *n.*

Authorized Version *n.* English translation of the Bible made in 1611.

authorship *n.* **1** origin of a book etc. **2** profession of an author.

autism /ˈɔːtɪz(ə)m/ *n.* condition characterized by self-absorption and social withdrawal. □ **autistic** /ɔːˈtɪstɪk/ *adj.* [related to AUTO-]

auto /ˈɔːtəʊ/ *n.* (*pl.* **-s**) *US colloq.* car. [abbreviation of AUTOMOBILE]

auto- *comb. form* **1** self. **2** one's own. **3** of or by oneself or itself. [Greek *autos*]

autobahn /ˈɔːtəʊ,bɑːn/ *n.* (*pl.* **-s**) German, Austrian, or Swiss motorway. [German]

autobiography /,ɔːtəʊbaɪˈɒgrəfɪ/ *n.* (*pl.* **-ies**) **1** written account of one's own life. **2** this as a literary genre. □ **autobiographer** *n.* **autobiographical** /-,baɪəˈgræfɪk(ə)l/ *adj.*

autoclave /ˈɔːtə,kleɪv/ *n.* sterilizer using high-pressure steam. [Latin *clavus* nail or *clavis* key]

autocracy /ɔːˈtɒkrəsɪ/ *n.* (*pl.* **-ies**) **1** rule by an autocrat. **2** dictatorship. [Greek *kratos* power]

autocrat /ˈɔːtə,kræt/ *n.* **1** absolute ruler. **2** dictatorial person. □ **autocratic** /-ˈkrætɪk/ *adj.* **autocratically** /-ˈkrætɪkəlɪ/ *adv.*

autocross /ˈɔːtəʊ,krɒs/ *n.* motor racing across country or on unmade roads.

Autocue /ˈɔːtəʊ,kjuː/ *n. propr.* screen etc. from which a speaker reads a television script.

auto-da-fé /,ɔːtəʊdɑːˈfeɪ/ *n.* (*pl.* ***autos-da-fé*** /,ɔːtəʊz-/) **1** *hist.* ceremonial judgement of heretics by the Spanish Inquisition. **2** public burning of heretics. [Portuguese, = act of the faith]

autograph /'ɔ:tə,grɑ:f/ *–n.* signature, esp. that of a celebrity. *–v.* sign or write on in one's own hand. [Greek *graphō* write]

autoimmune /,ɔ:təʊɪ'mju:n/ *adj.* (of a disease) caused by antibodies produced against substances naturally present in the body.

automat /'ɔ:tə,mæt/ *n. US* **1** slot-machine. **2** cafeteria dispensing food and drink from slot-machines. [French: related to AUTOMATON]

automate /'ɔ:tə,meɪt/ *v.* (**-ting**) convert to or operate by automation.

automatic /,ɔ:tə'mætɪk/ *–adj.* **1** working by itself, without direct human intervention. **2 a** done spontaneously (*automatic reaction*). **b** following inevitably (*automatic penalty*). **3** (of a firearm) able to be loaded and fired continuously. **4** (of a vehicle or its transmission) using gears that change automatically. *–n.* **1** automatic machine, firearm, or tool. **2** vehicle with automatic transmission. □ **automatically** *adv.* [related to AUTOMATON]

automatic pilot *n.* device for keeping an aircraft or ship on a set course.

automation /,ɔ:tə'meɪʃ(ə)n/ *n.* use or introduction of automatic methods or equipment in place of manual labour.

automatism /ɔ:'tɒmə,tɪz(ə)m/ *n.* **1** involuntary action. **2** unthinking routine. [French: related to AUTOMATON]

automaton /ɔ:'tɒmət(ə)n/ *n.* (*pl.* **-mata** or **-s**) **1** machine controlled automatically; robot. **2** person acting like a robot. [Greek, = acting of itself]

automobile /'ɔ:təmə,bi:l/ *n. US* motor car. [French]

automotive /,ɔ:tə'məʊtɪv/ *adj.* of motor vehicles.

autonomous /ɔ:'tɒnəməs/ *adj.* **1** having self-government. **2** acting or free to act independently. [Greek *nomos* law]

autonomy *n.* **1** self-government. **2** personal freedom.

autopilot *n.* = AUTOMATIC PILOT.

autopsy /'ɔ:tɒpsɪ/ *n.* (*pl.* **-ies**) post-mortem. [Greek *autoptēs* eye-witness]

autoroute /'ɔ:təʊ,ru:t/ *n.* French motorway. [French]

autostrada /'ɔ:təʊ,strɑ:də/ *n.* (*pl.* *-s* or ***-strade*** /-deɪ/) Italian motorway. [Italian]

auto-suggestion /,ɔ:təʊsə'dʒestʃ(ə)n/ *n.* hypnotic or subconscious suggestion made to oneself.

autumn /'ɔ:təm/ *n.* **1** (often *attrib.*) season between summer and winter. **2** time of incipient decline. □ **autumnal** /ɔ:'tʌmn(ə)l/ *adj.* [Latin *autumnus*]

autumn equinox *n.* (also **autumnal equinox**) equinox about 22 Sept.

auxiliary /ɔ:g'zɪljərɪ/ *–adj.* **1** subsidiary, additional. **2** giving help. *–n.* (*pl.* **-ies**) **1** auxiliary person or thing. **2** (in *pl.*) foreign or allied troops in the service of a nation at war. **3** verb used to form tenses or moods of other verbs (e.g. *have* in *I have seen*). [Latin *auxilium* help]

auxin /'ɔ:ksɪn/ *n.* plant hormone that regulates growth.

AV *abbr.* Authorized Version.

avail /ə'veɪl/ *–v.* **1** help; be of use. **2** *refl.* (foll. by *of*) make use of, profit by. *–n.* use, profit (*of no avail*). [Latin *valeo* be strong]

available *adj.* **1** at one's disposal, obtainable. **2 a** (of a person) free, not committed. **b** able to be contacted. □ **availability** /-'bɪlɪtɪ/ *n.*

avalanche /'ævə,lɑ:ntʃ/ *n.* **1** rapidly sliding mass of snow and ice on a mountain. **2** sudden abundance (*avalanche of work*). [French]

avant-garde /,ævɑ̃'gɑ:d/ *–n.* pioneers or (esp. artistic) innovators. *–adj.* new; pioneering. [French, = vanguard]

avarice /'ævərɪs/ *n.* extreme greed for wealth. □ **avaricious** /-'rɪʃəs/ *adj.* [Latin *avarus* greedy]

avatar /'ævə,tɑ:(r)/ *n.* (in Hindu mythology) descent of a deity etc. to earth in bodily form. [Sanskrit, = descent]

Ave /'ɑ:veɪ/ *n.* (in full **Ave Maria**) prayer to the Virgin Mary (Luke 1:28). [Latin]

Ave. *abbr.* Avenue.

avenge /ə'vendʒ/ *v.* (**-ging**) **1** inflict retribution on behalf of. **2** take vengeance for (an injury). □ **be avenged** avenge oneself. [Latin *vindico*]

avenue /'ævə,nju:/ *n.* **1 a** broad esp. tree-lined road or street. **b** tree-lined path etc. **2** approach (*explored every avenue*). [French *avenir* come to]

aver /ə'vɜ:(r)/ *v.* (**-rr-**) *formal* assert, affirm. □ **averment** *n.* [Latin *verus* true]

average /'ævərɪdʒ/ *–n.* **1** usual amount, extent, or rate. **2** amount obtained by adding two or more numbers and dividing by how many there are. **3** (with ref. to speed etc.) ratio obtained by subtracting the inital from the final value of each element of the ratio (*average of 50 miles per hour*). **4** *Law* damage to or loss of a ship or cargo. *–adj.* **1 a** usual, ordinary. **b** mediocre. **2** constituting an average (*the average age is 72*). *–v.* (**-ging**) **1** amount on average to. **2** do on average. **3** estimate the average of. □ **average out (at)** result in an average (of). **law of averages** principle that if one of two extremes occurs the other will also. **on** (or **on an**) **average** as an average rate or estimate. [Arabic, = damaged goods]

averse /ə'vɜ:s/ *predic. adj.* (usu. foll. by *to*) opposed, disinclined. [Latin *verto vers-* turn]

aversion /ə'vɜ:ʃ(ə)n/ *n.* **1** (usu. foll. by *to, for*) dislike or unwillingness. **2** object of this.

avert /ə'vɜ:t/ *v.* (often foll. by *from*) **1** turn away (one's eyes or thoughts). **2** prevent or ward off (esp. danger).

Avesta /ə'vestə/ *n.* (usu. prec. by *the*) sacred writings of Zoroastrianism (cf. ZEND). [Persian]

aviary /'eɪvɪərɪ/ *n.* (*pl.* **-ies**) large cage or building for keeping birds. [Latin *avis* bird]

aviation /ˌeɪvɪ'eɪʃ(ə)n/ *n.* science or practice of flying aircraft. [Latin: related to AVIARY]

aviator /'eɪvɪˌeɪtə(r)/ *n.* person who flies aircraft.

avid /'ævɪd/ *adj.* eager, greedy. □ **avidity** /ə'vɪdɪtɪ/ *n.* **avidly** *adv.* [Latin *aveo* crave]

avionics /ˌeɪvɪ'ɒnɪks/ *n.pl.* (usu. treated as *sing.*) electronics as applied to aviation. [from AVIATION, ELECTRONICS]

avocado /ˌævə'kɑ:dəʊ/ *n.* (*pl.* **-s**) **1** (in full **avocado pear**) dark green edible pear-shaped fruit with yellowish-green creamy flesh. **2** tree bearing it. [Spanish from Aztec]

avocet /'ævəˌset/ *n.* long-legged wading bird with an upward-curved bill. [French from Italian]

avoid /ə'vɔɪd/ *v.* **1** keep away or refrain from. **2** escape; evade. **3** *Law* quash, annul. □ **avoidable** *adj.* **avoidance** *n.* [French]

avoirdupois /ˌævədə'pɔɪz/ *n.* (in full **avoirdupois weight**) system of weights based on a pound of 16 ounces or 7,000 grains. [French, = goods of weight]

avow /ə'vaʊ/ *v. formal* declare, confess. □ **avowal** *n.* **avowedly** /ə'vaʊɪdlɪ/ *adv.* [Latin *voco* call]

avuncular /ə'vʌŋkjʊlə(r)/ *adj.* like or of an uncle, esp. in manner. [Latin *avunculus* uncle]

await /ə'weɪt/ *v.* **1** wait for. **2** be in store for. [French: related to WAIT]

awake /ə'weɪk/ *–v.* (**-king**; *past* **awoke**; *past part.* **awoken**) **1** cease to sleep or arouse from sleep. **2** (often foll. by *to*) become or make alert, aware, or active. *–predic. adj.* **1** not asleep. **2** (often foll. by *to*) alert, aware. [Old English: related to A[2]]

awaken *v.* = AWAKE *v.*

■ **Usage** *Awake* and *awaken* are interchangeable but *awaken* is much rarer than *awake* as an intransitive verb.

---

award /ə'wɔ:d/ *–v.* give or order to be given as a payment or prize. *–n.* **1** thing or amount awarded. **2** judicial decision. [French]

aware /ə'weə(r)/ *predic. adj.* **1** (often foll. by *of* or *that*) conscious; having knowledge. **2** well-informed. □ **awareness** *n.* [Old English]

■ **Usage** *Aware* is also found used attributively in sense 2, as in '*a very aware person*', but this should be avoided in formal contexts.

---

awash /ə'wɒʃ/ *predic. adj.* **1** level with the surface of, and just covered by, water. **2** (foll. by *with*) overflowing, abounding.

away /ə'weɪ/ *–adv.* **1** to or at a distance from the place, person, or thing in question (*go, give, look, away*; *5 miles away*). **2** into non-existence (*explain, fade, away*). **3** constantly, persistently (*work away*). **4** without delay (*ask away*). *–attrib. adj. Sport* not played on one's own ground (*away match*). *–n. Sport* away match or win. [Old English: related to A[2], WAY]

awe /ɔ:/ *–n.* reverential fear or wonder. *–v.* (**awing**) inspire with awe. [Old Norse]

aweigh /ə'weɪ/ *predic. adj.* (of an anchor) clear of the bottom.

awe-inspiring *adj.* awesome; magnificent.

awesome /'ɔ:səm/ *adj.* inspiring awe; dreaded.

awful /'ɔ:fʊl/ *adj.* **1** *colloq.* very bad or unpleasant (*has awful writing*; *awful weather*). **2** (*attrib.*) as an intensifier (*awful lot of money*). **3** *poet.* inspiring awe.

awfully *adv.* **1** badly; unpleasantly (*played awfully*). **2** *colloq.* very (*awfully pleased*).

awhile /ə'waɪl/ *adv.* for a short time. [*a while*]

awkward /'ɔ:kwəd/ *adj.* **1** difficult to use or deal with. **2** clumsy, ungainly. **3 a** embarrassed. **b** embarrassing. [obsolete *awk* perverse]

awl *n.* small tool for piercing holes, esp. in leather. [Old English]

awn *n.* bristly head of a sheath of barley and other grasses. [Old Norse]

awning *n.* sheet of canvas etc. stretched on a frame as a shelter against the sun or rain. [origin uncertain]

awoke *past* of AWAKE.

awoken *past part.* of AWAKE.

AWOL /'eɪwɒl/ *abbr. colloq.* absent without leave.

awry /ə'raɪ/ *–adv.* **1** crookedly, askew. **2** amiss, wrong. *–predic. adj.* crooked; unsound.

axe /æks/ (*US* **ax**) *–n.* **1** chopping-tool with a handle and heavy blade. **2** (**the axe**) dismissal (of employees); abandonment of a project etc. *–v.* (**axing**) cut (esp. costs or staff) drastically; abandon (a project). □ **an axe to grind** private ends to serve. [Old English]

axial /'æksɪəl/ *adj.* of, forming, or placed round an axis.

axil /'æksɪl/ *n.* upper angle between a leaf and stem. [Latin *axilla* armpit]

axiom /'æksɪəm/ *n.* **1** established or accepted principle. **2** self-evident truth. □ **axiomatic** /-'mætɪk/ *adj.* [Greek *axios* worthy]

axis /'æksɪs/ *n.* (*pl.* **axes** /-si:z/) **1 a** imaginary line about which a body rotates. **b** line which divides a regular figure symmetrically. **2** fixed reference line for the measurement of coordinates etc. **3** (**the Axis**) alliance of Germany, Italy, and later Japan, in the war of 1939–45. [Latin, = axle]

axle /'æks(ə)l/ *n.* spindle on which a wheel is fixed or turns. [Old Norse]

axolotl /'æksə,lɒt(ə)l/ *n.* newtlike salamander, which in natural conditions retains its larval form of life. [Nahuatl, = water-servant]

ayatollah /,aɪə'tɒlə/ *n.* Shiite religious leader in Iran. [Persian from Arabic, = token of God]

aye /aɪ/ *–adv. archaic* or *dial.* yes. *–n.* affirmative answer or vote. [probably from *I*, expressing assent]

azalea /ə'zeɪlɪə/ *n.* a kind of rhododendron. [Greek *azaleos* dry]

azimuth /'æzɪməθ/ *n.* angular distance from a north or south point of the horizon to the intersection with the horizon of a vertical circle passing through a given celestial body. □ **azimuthal** /-'mʌθ(ə)l/ *adj.* [French from Arabic]

AZT *abbr.* drug intended for use against the Aids virus. [from the chemical name]

Aztec /'æztek/ *–n.* **1** member of the native Mexican people overthrown by the Spanish in 1519. **2** language of this people. *–adj.* of the Aztecs or their language. [Nahuatl, = men of the north]

azure /'æʒə(r)/ *–n.* **1** deep sky-blue colour. **2** *poet.* clear sky. *–adj.* deep sky-blue. [Arabic]

B

B[1] /biː/ *n.* (*pl.* **Bs** or **B's**) **1** (also **b**) second letter of the alphabet. **2** *Mus.* seventh note of the diatonic scale of C major. **3** second hypothetical person or example. **4** second highest category (of roads, academic marks, etc.). **5** (usu. **b**) *Algebra* second known quantity.

B[2] *symb.* boron.

B[3] *abbr.* (also **B.**) black (pencil-lead).

b. *abbr.* **1** born. **2** *Cricket* **a** bowled by. **b** bye.

BA *abbr.* **1** Bachelor of Arts. **2** British Airways.

Ba *symb.* barium.

baa /bɑː/ *–v.* (**baas**, **baaed** or **baa'd**) bleat. *–n.* sheep's cry. [imitative]

babble /'bæb(ə)l/ *–v.* (**-ling**) **1 a** talk, chatter, or say incoherently or excessively. **b** (of a stream etc.) murmur. **2** repeat or divulge foolishly. *–n.* **1** babbling. **2** murmur of voices, water, etc. [imitative]

babe *n.* **1** *literary* baby. **2** innocent or helpless person. **3** *US slang* young woman. [as BABY]

babel /'beɪb(ə)l/ *n.* **1** confused noise, esp. of voices. **2** scene of confusion. [Hebrew, = Babylon (Gen. 11)]

baboon /bə'buːn/ *n.* large long-nosed African and Arabian monkey. [French or medieval Latin]

baby /'beɪbɪ/ *–n.* (*pl.* **-ies**) **1** very young child. **2** childish person. **3** youngest member of a family etc. **4** (often *attrib.*) **a** very young animal. **b** small specimen. **5** *slang* sweetheart. **6** one's special concern etc. *–v.* (**-ies**, **-ied**) treat like a baby; pamper. □ **babyhood** *n.* **babyish** *adj.* [imitative of child's *ba ba*]

baby boom *n. colloq.* temporary increase in the birth rate.

Baby Buggy *n. propr.* a kind of child's pushchair.

baby carriage *n. US* pram.

baby grand *n.* small grand piano.

Babygro *n.* (*pl.* **-s**) *propr.* stretchy all-in-one baby suit.

babysit *v.* (**-tt-**; *past* and *past part.* **-sat**) look after a child while its parents are out. □ **babysitter** *n.*

baccalaureate /ˌbækə'lɔːrɪət/ *n.* final secondary school examination in France and many international schools. [medieval Latin *baccalaureus* bachelor]

baccarat /'bækəˌrɑː/ *n.* gambling card-game. [French]

bacchanal /'bækən(ə)l/ *–n.* **1** drunken revelry or reveller. **2** priest or follower of Bacchus. *–adj.* **1** of or like Bacchus. **2** drunkenly riotous. [Latin *Bacchus* from Greek, god of wine]

Bacchanalia /ˌbækə'neɪlɪə/ *n.pl.* **1** Roman festival of Bacchus. **2** (**bacchanalia**) drunken revelry.

bacchant /'bækənt/ *–n.* (*fem.* **bacchante** /bə'kæntɪ/) **1** priest or follower of Bacchus. **2** drunken reveller. *–adj.* **1** of or like Bacchus or his rites. **2** drunkenly riotous, roistering.

Bacchic /'bækɪk/ *adj.* = BACCHANAL *adj.*

baccy /'bækɪ/ *n.* (*pl.* **-ies**) *colloq.* tobacco. [abbreviation]

bachelor /'bætʃələ(r)/ *n.* **1** unmarried man. **2** person with a university first degree. □ **bachelorhood** *n.* [related to BACCALAUREATE]

bachelor girl *n.* independent young single woman.

bacillus /bə'sɪləs/ *n.* (*pl.* **bacilli** /-laɪ/) rod-shaped bacterium, esp. one causing disease. □ **bacillary** *adj.* [Latin, diminutive of *baculus* stick]

back *–n.* **1 a** rear surface of the human body from shoulder to hip. **b** upper surface of an animal's body. **c** spine (*broke his back*). **d** keel of a ship. **2** backlike surface (*back of the head, chair, shirt*). **3** reverse or more distant part (*back of the room*; *sat in the back*; *write it on the back*). **4** defensive player in football etc. *–adv.* **1** to the rear (*go back a bit*; *looked back*). **2** in or into a previous state, place, or time (*came back*; *put it back*; *back in June*). **3** at a distance (*stand back*). **4** in return (*pay back*). **5** in check (*hold him back*). *–v.* **1 a** give moral or financial support to. **b** bet on (a horse etc.). **2** (often foll. by *up*) move backwards. **3 a** put or serve as a back, background, or support to. **b** *Mus.* accompany. **4** lie at the back of (*beach backed by cliffs*). **5** (of the wind) move anticlockwise. *–adj.* **1** situated to the rear; remote, subsidiary (*back teeth*). **2** past; not current (*back pay*; *back issue*). **3** reversed (*back flow*). □ **back and forth** to and fro. **back down** withdraw from confrontation. **the back of beyond** very remote place. **back off 1** draw back, retreat. **2** = *back down*. **back on to** have its back adjoining (*backs on to a* *field*). **back out** (often foll. by *of*) withdraw from a commitment. **back-pedal** reverse one's action or opinion. **back to back** with backs adjacent and facing each other (*stood back to back*). **back up 1** give (esp. moral) support to. **2** *Computing* make a backup of (data, a disk, etc.). **get** (or **put**) **a person's back up** annoy a person. **get off a person's back** stop troubling a person. **turn one's back on** abandon; ignore. □ **backer** *n.* (in sense 1 of *v.*). **backless** *adj.* [Old English]

backache *n.* ache in the back.

back-bencher *n.* MP not holding a senior office.

backbiting *n.* malicious talk.

back-boiler *n.* boiler behind a domestic fire.

backbone *n.* **1** spine. **2** chief support. **3** firmness of character.

back-breaking *adj.* (esp. of manual work) extremely hard.

back-burner *n.* **on the back-burner** receiving little attention.

backchat *n. colloq.* verbal insolence.

backcloth *n.* **1** painted cloth at the back of a stage. **2** background to a scene or situation.

backcomb *v.* comb (the hair) towards the scalp to give it fullness.

back-crawl *n.* = BACKSTROKE.

backdate *v.* **(-ting) 1** make retrospectively valid. **2** put an earlier date to than the actual one.

back door *n.* secret or ingenious means.

backdrop *n.* = BACKCLOTH.

backfire *v.* **(-ring) 1** (of an engine or vehicle) ignite or explode too early in the cylinder or exhaust. **2** (of a plan etc.) rebound adversely on its originator.

back-formation *n.* **1** formation of a word from its seeming derivative (e.g. *laze* from *lazy*). **2** word so formed.

backgammon *n.* board-game with pieces moved according to throws of the dice. [from BACK + obsolete form of GAME[1]]

background *n.* **1** part of a scene or picture furthest from the observer. **2** (often *attrib.*) inconspicuous position (*kept in the background*; *background music*). **3** person's education, social circumstances, etc. **4** explanatory or contributory information or events.

backhand *–attrib. adj.* (of a stroke) made with the hand across one's body. *–n.* such a stroke.

backhanded /bæk'hændɪd/ *adj.* **1** made with the back of the hand. **2** indirect; ambiguous (*backhanded compliment*).

backhander *n.* **1 a** backhand stroke. **b** backhanded blow. **2** *slang* bribe.

backing *n.* **1 a** support, esp. financial or moral. **b** material used for a thing's back or support. **2** musical accompaniment, esp. to a pop singer.

backing track *n.* recorded musical accompaniment.

backlash *n.* **1** violent, usu. hostile, reaction. **2** sudden recoil in a mechanism.

backlist *n.* publisher's list of books still in print.

backlog *n.* arrears of work.

back number *n.* **1** out-of-date issue of a periodical. **2** *slang* out-of-date person or thing.

backpack *–n.* rucksack. *–v.* travel or hike with this. □ **backpacker** *n.*

back passage *n. colloq.* rectum.

backrest *n.* support for the back.

back room *n.* (often, with hyphen, *attrib.*) place where secret work is done.

back seat *n.* less prominent or important position.

back-seat driver *n.* person eager to advise without taking responsibility.

backside *n. colloq.* buttocks.

back slang *n.* slang using words spelt backwards (e.g. *yob*).

backslide *v.* **(-ding**; *past* **-slid**; *past part.* **-slid** or **-slidden)** return to bad habits etc.

backspace *v.* **(-cing)** move a typewriter carriage etc. back one or more spaces.

backspin *n.* backward spin making a ball bounce erratically.

backstage *adv.* & *adj.* behind the scenes.

backstairs *–n.pl.* rear or side stairs of a building. *–attrib. adj.* (also **backstair**) underhand; secret.

backstitch *n.* sewing with each stitch starting behind the end of the previous one.

back-stop *n.* **1** *Cricket* etc. **a** position directly behind the wicket-keeper. **b** fielder in this position. **2** last resort.

backstreet *–n.* side-street, alley. *–attrib. adj.* illicit; illegal (*backstreet abortion*).

backstroke *n.* swimming stroke done on the back.

back-to-back *adj.* (of houses) with a party wall at the rear.

back to front *adj.* **1** with back and front reversed. **2** in disorder.

back-to-nature *attrib. adj.* seeking a simpler way of life.

backtrack *v.* **1** retrace one's steps. **2** reverse one's policy or opinion.

backup *n.* (often *attrib.*) **1** support; reserve (*back-up team*). **2** *Computing* **a** making of spare copies of data for safety. **b** copy so made.

backward /ˈbækwəd/ *–adv.* = BACKWARDS. *–adj.* **1** towards the rear or starting-point (*backward look*). **2** reversed (*backward roll*). **3** slow to develop or progress. **4** hesitant, shy.

backwards *adv.* **1** away from one's front (*lean backwards*). **2 a** with the back foremost (*walk backwards*). **b** in reverse of the usual way (*count backwards*). **3 a** into a worse state. **b** into the past. **c** (of motion) back towards the starting-point (*roll backwards*). □ **backwards and forwards** to and fro. **bend** (or **fall** or **lean**) **over backwards** *colloq.* make every effort, esp. to be fair or helpful.

backwash *n.* **1** receding waves made by a ship etc. **2** repercussions.

backwater *n.* **1** peaceful, secluded, or dull place. **2** stagnant water fed from a stream.

backwoods *n.pl.* **1** remote uncleared forest land. **2** remote region. □ **backwoodsman** *n.*

backyard *n.* yard behind a house etc.

bacon /ˈbeɪkən/ *n.* cured meat from the back or sides of a pig. [French from Germanic]

bacteriology /ˌbæktɪərɪˈɒlədʒɪ/ *n.* the study of bacteria.

bacterium /bækˈtɪərɪəm/ *n.* (*pl.* **-ria**) unicellular micro-organism lacking an organized nucleus, esp. of a kind causing disease. □ **bacterial** *adj.* [Greek, = little stick]

■ **Usage** A common mistake is the use of the plural form *bacteria* as the singular. This should be avoided.

bad *–adj.* **(worse, worst) 1** inadequate, defective (*bad work, light*). **2** unpleasant (*bad weather*). **3** harmful (*is bad for you*). **4** (of food) decayed. **5** *colloq.* ill, injured (*feeling bad today*; *a bad leg*). **6** *colloq.* regretful, guilty (*feels bad about it*). **7** serious, severe (*a bad headache, mistake*). **8 a** morally unacceptable (*bad man*; *bad language*). **b** naughty. **9** not valid (*a bad cheque*). **10** (**badder, baddest**) esp. *US slang* excellent. *–n.* ill fortune; ruin. *–adv. US colloq.* badly. □ **not** (or **not so**) **bad** *colloq.* fairly good. **too bad** *colloq.* regrettable. [Old English]

bad blood *n.* ill feeling.

bad books see BOOK.

bad breath *n.* unpleasant-smelling breath.

bad debt *n.* debt that is not recoverable.

baddy *n.* (*pl.* **-ies**) *colloq.* villain in a story, film, etc.

bade see BID.

bad egg see EGG[1].

bad faith *n.* intent to deceive.

badge *n.* **1** small flat emblem worn to signify office, membership, etc., or as decoration. **2** thing that reveals a condition or quality. [origin unknown]

badger *–n.* nocturnal burrowing mammal with a black and white striped head. *–v.* pester, harass. [origin uncertain]

badinage /ˈbædɪˌnɑːʒ/ *n.* playful ridicule. [French]

bad lot *n.* person of bad character.

badly *adv.* **(worse, worst) 1** in a bad manner. **2** *colloq.* very much (*wants it badly*). **3** severely (*badly defeated*).

badminton /ˈbædmɪnt(ə)n/ *n.* game with rackets and a shuttlecock. [*Badminton* in S. England]

bad-mouth *v.* esp. *US slang* abuse verbally, put down.

bad news *n. colloq.* unpleasant or troublesome person or thing.

bad-tempered *adj.* irritable.

baffle /ˈbæf(ə)l/ *–v.* **(-ling) 1** perplex. **2** frustrate, hinder. *–n.* device that checks flow esp. of fluid or sound waves. □ **bafflement** *n.* [origin uncertain]

BAFTA /ˈbæftə/ *abbr.* British Association of Film and Television Arts.

bag *–n.* **1** soft open-topped receptacle. **2 a** piece of luggage. **b** woman's handbag. **3** (in *pl.*; usu. foll. by *of*) *colloq.* large amount (*bags of time*). **4** *slang derog.* woman. **5** animal's sac. **6** amount of game shot by one person. **7** (usu. in *pl.*) baggy skin under the eyes. **8** *slang* particular interest (*folk music is not my bag*). *–v.* **(-gg-) 1** *colloq.* **a** secure (*bagged the best seat*). **b** (often in phr. **bags I**) *colloq.* claim as being the first (*bags I go next*). **2** put in a bag. **3** (cause to) hang loosely; bulge. □ **in the bag** *colloq.* achieved, secured. □ **bagful** *n.* (*pl.* **-s**). [origin unknown]

bagatelle /ˌbægəˈtel/ *n.* **1** game in which small balls are struck into holes on a board. **2** mere trifle. **3** short piece of esp. piano music. [French from Italian]

bagel /ˈbeɪg(ə)l/ *n.* ring-shaped bread roll. [Yiddish]

baggage /ˈbægɪdʒ/ *n.* **1** luggage. **2** portable army equipment. **3** *joc.* or *derog.* girl or woman. **4** mental encumbrances. [French]

baggy *adj.* **(-ier, -iest)** hanging loosely. □ **baggily** *adv.* **bagginess** *n.*

bagpipe *n.* (usu. in *pl.*) musical instrument consisting of a windbag connected to reeded pipes.

baguette /bæˈget/ *n.* long thin French loaf. [French]

bah *int.* expressing contempt or disbelief. [French]

Baha'i /bəˈhaɪ/ *n.* (*pl.* **-s**) member of a monotheistic religion emphasizing religious unity and world peace. [Persian *bahá* splendour]

bail[1] *–n.* **1** money etc. pledged against the temporary release of an untried prisoner. **2** person(s) giving this. *–v.* (usu. foll. by *out*) **1** release or secure the release of (a prisoner) on payment of bail. **2** release from a difficulty; rescue. □ **on bail** released after payment of bail. [Latin *bajulus* carrier]

bail[2] *n.* **1** *Cricket* either of two crosspieces bridging the stumps. **2** bar holding the paper against a typewriter platen. **3** bar separating horses in an open stable. [French]

bail[3] *v.* (also **bale**) **1** (usu. foll. by *out*) scoop water out of (a boat etc.). **2** scoop (water etc.) out. □ **bail out** var. of *bale out* 1 (see BALE[1]). [French]

bailey /ˈbeɪlɪ/ *n.* (*pl.* **-s**) **1** outer wall of a castle. **2** court enclosed by it. [French: related to BAIL[2]]

Bailey bridge /ˈbeɪlɪ/ *n.* prefabricated military bridge for rapid assembly. [Sir D. *Bailey*, name of its designer]

bailiff /ˈbeɪlɪf/ *n.* **1** sheriff's officer who executes writs and carries out distraints. **2** landlord's agent or steward. [French: related to BAIL[1]]

bailiwick /ˈbeɪlɪwɪk/ *n.* **1** *Law* district of a bailiff. **2** *joc.* person's particular interest. [as BAILIFF, obsolete *wick* district]

bain-marie /ˌbænməˈriː/ *n.* (*pl.* **bains-marie** pronunc. same) pan of hot water holding a pan containing sauce etc. for slow heating. [French, translation of medieval Latin *balneum Mariae* bath of Maria (a supposed alchemist)]

bairn *n. Scot.* & *N.Engl.* child. [Old English: related to BEAR[1]]

bait *–n.* **1** food used to entice prey. **2** allurement. *–v.* **1** harass, torment, or annoy (a person or chained animal). **2** put bait on (a hook, trap, etc.). [Old Norse]

baize *n.* usu. green woollen felted material, used for coverings. [French pl. *baies* chestnut-coloured]

bake *v.* (**-king**) **1** cook or become cooked by dry heat, esp. in an oven. **2** *colloq.* (usu. as **be baking**) (of weather, a person, etc.) be very hot. **3** harden by heat. [Old English]

baked beans *n.pl.* baked haricot beans, usu. tinned in tomato sauce.

Bakelite /ˈbeɪkəˌlaɪt/ *n. propr.* plastic made from formaldehyde and phenol, used formerly for buttons, plates, etc. [German from *Baekeland*, name of its inventor]

baker *n.* person who bakes and sells bread, cakes, etc., esp. for a living.

Baker day *n. colloq.* day set aside for in-service training of teachers. [*Baker*, name of the Education Secretary responsible for introducing them]

baker's dozen *n.* thirteen.

bakery *n.* (*pl.* **-ies**) place where bread and cakes are made or sold.

Bakewell tart /ˈbeɪkwel/ *n.* open pastry case lined with jam and filled with almond paste. [*Bakewell* in Derbyshire]

baking-powder *n.* mixture of sodium bicarbonate, cream of tartar, etc., as a raising agent.

baking-soda *n.* sodium bicarbonate.

baklava /ˈbækləvə/ *n.* rich sweetmeat of flaky pastry, honey, and nuts. [Turkish]

baksheesh /ˈbækʃiːʃ/ *n.* gratuity, tip. [Persian]

Balaclava /ˌbæləˈklɑːvə/ *n.* (in full **Balaclava helmet**) usu. woollen covering for the whole head and neck, except for the face. [*Balaclava* in the Crimea, the site of a battle in 1854]

balalaika /ˌbæləˈlaɪkə/ *n.* guitar-like stringed instrument with a triangular body. [Russian]

balance /ˈbæləns/ *–n.* **1 a** even distribution of weight or amount. **b** stability of body or mind. **2** apparatus for weighing, esp. one with a central pivot, beam, and two scales. **3 a** counteracting weight or force. **b** (in full **balance-wheel**) regulating device in a clock etc. **4** decisive weight or amount (*balance of opinion*). **5 a** agreement or difference between credits and debits in an account. **b** amount still owing or outstanding (*will pay the balance*). **c** amount left over. **6 a** *Art* harmony and proportion. **b** *Mus.* relative volume of sources of sound. **7** (**the Balance**) zodiacal sign or constellation Libra. *–v.* (**-cing**) **1** bring into, keep, or be in equilibrium (*balanced a book on her head*; *balanced on one leg*). **2** (often foll. by *with*, *against*) offset or compare (one thing) with another (*balance the pros and cons*). **3** counteract, equal, or neutralize the weight or importance of. **4** (usu. as **balanced** *adj.*) make well-proportioned and harmonious (*balanced diet*; *balanced opinion*). **5 a** compare and esp. equalize the debits and credits of (an account). **b** (of an account) have credits and debits equal. □ **in the balance** uncertain; at a critical stage. **on balance** all things considered. [Latin *bilanx* scales]

balance of payments *n.* difference in value between payments into and out of a country.

balance of power *n.* **1** situation of roughly equal power among the chief

States of the world. **2** power held by a small group when larger groups are of equal strength.

balance of trade *n.* difference in value between imports and exports.

balance sheet *n.* statement giving the balance of an account.

balcony /ˈbælkənɪ/ *n.* (*pl.* **-ies**) **1** usu. balustraded platform on the outside of a building with access from an upper floor. **2** upper tier of seats in a theatre etc. □ **balconied** *adj.* [Italian]

bald /bɔːld/ *adj.* **1** lacking some or all hair on the scalp. **2** lacking the usual hair, feathers, leaves, etc. **3** *colloq.* with a worn surface (*bald tyre*). **4** plain, direct (*bald statement, style*). □ **balding** *adj.* (in senses 1–3). **baldly** *adv.* (in sense 4). **baldness** *n.* [Old English]

balderdash /ˈbɔːldəˌdæʃ/ *n.* nonsense. [origin unknown]

bale[1] *–n.* tightly bound bundle of merchandise or hay. *–v.* (**-ling**) make up into bales. □ **bale out 1** (also **bail out**) (of an airman) make an emergency parachute descent. **2** var. of BAIL[1] *v.* 2. [Dutch: related to BALL[1]]

bale[2] var. of BAIL[3].

baleen /bəˈliːn/ *n.* whalebone. [Latin *balaena* whale]

baleful /ˈbeɪlfʊl/ *adj.* **1** menacing in look, manner, etc. **2** malignant, destructive. □ **balefully** *adv.* [archaic *bale* evil]

balk var. of BAULK.

Balkan /ˈbɔːlkən/ *adj.* **1** of the region of SE Europe bounded by the Adriatic, Aegean, and Black Sea. **2** of its peoples or countries. [Turkish]

ball[1] /bɔːl/ *–n.* **1** sphere, esp. for use in a game. **2 a** ball-shaped object; material in the shape of a ball (*ball of snow, wool*). **b** rounded part of the body (*ball of the foot*). **3** cannon-ball. **4** single delivery or pass of a ball in cricket, baseball, football, etc. **5** (in *pl.*) *coarse slang* **a** testicles. **b** (usu. as *int.*) nonsense. **c** = BALLS-UP. **d** courage, 'guts'. *–v.* form into a ball. □ **balls up** *coarse slang* bungle; make a mess of. **on the ball** *colloq.* alert. [Old Norse]

ball[2] /bɔːl/ *n.* **1** formal social gathering for dancing. **2** *slang* enjoyable time (esp. *have a ball*). [Greek *ballō* throw]

ballad /ˈbæləd/ *n.* **1** poem or song narrating a popular story. **2** slow sentimental song. [Provençal: related to BALL[2]]

balladry *n.* ballad poetry.

ball-and-socket joint *n.* joint in which a rounded end lies in a concave socket.

ballast /ˈbæləst/ *–n.* **1** heavy material stabilizing a ship, the car of a balloon, etc. **2** coarse stone etc. as the bed of a railway track or road. **3** mixture of coarse and fine aggregate for making concrete. *–v.* provide with ballast. [Low German or Scandinavian]

ball-bearing *n.* **1** bearing in which the two halves are separated by a ring of small balls. **2** one of these balls.

ballboy *n.* (*fem.* **ballgirl**) (in tennis) boy or girl who retrieves balls.

ballcock *n.* floating ball on a hinged arm controlling the water level in a cistern.

ballerina /ˌbæləˈriːnə/ *n.* female ballet-dancer. [Italian: related to BALL[2]]

ballet /ˈbæleɪ/ *n.* **1** dramatic or representational style of dancing to music. **2** particular piece or performance of ballet. □ **balletic** /bəˈletɪk/ *adj.* [French: related to BALL[2]]

ballet-dancer *n.* dancer of ballet.

ball game *n.* **1 a** game played with a ball. **b** *US* baseball game. **2** esp. *US colloq.* affair; matter (*a whole new ball game*).

ballista /bəˈlɪstə/ *n.* (*pl.* **-stae** /-stiː/) (in ancient warfare) catapult for hurling large stones etc. [Latin from Greek *ballō* throw]

ballistic /bəˈlɪstɪk/ *adj.* of projectiles.

ballistic missile *n.* missile that is powered and guided but falls by gravity.

ballistics *n.pl.* (usu. treated as *sing.*) science of projectiles and firearms.

ballocking var. of BOLLOCKING.

ballocks var. of BOLLOCKS.

balloon /bəˈluːn/ *–n.* **1** small inflatable rubber toy or decoration. **2** large usu. round inflatable flying bag, often carrying a basket for passengers. **3** *colloq.* balloon shape enclosing dialogue etc. in a comic strip or cartoon. *–v.* **1** (cause to) swell out like a balloon. **2** travel by balloon. □ **balloonist** *n.* [French or Italian, = large ball]

ballot /ˈbælət/ *–n.* **1** occasion or system of voting, in writing and usu. secret. **2** total of such votes. **3** paper etc. used in voting. *–v.* (**-t-**) **1** (usu. foll. by *for*) **a** hold a ballot; give a vote. **b** draw lots for precedence etc. **2** take a ballot of (*balloted the members*). [Italian *ballotta*: related to BALLOON]

ballot-box *n.* sealed box for completed ballot-papers.

ballot-paper *n.* = BALLOT *n.* 3.

ballpark *n. US* **1** baseball ground. **2** *colloq.* sphere of activity, etc. **3** (*attrib.*) *colloq.* approximate. □ **in the right ballpark** *colloq.* approximately correct.

ball-point *n.* (in full **ball-point pen**) pen with a tiny ball as its writing point.

ballroom *n.* large room for dancing.

ballroom dancing *n.* formal social dancing.

balls-up *n. coarse slang* bungle, mess.

bally /'bælɪ/ *adj. & adv. slang* mild form of *bloody* (see BLOODY *adj.* 3). [alteration of BLOODY]

ballyhoo /,bælɪ'hu:/ *n.* **1** loud noise or fuss. **2** noisy publicity. [origin unknown]

balm /bɑ:m/ *n.* **1** aromatic ointment. **2** fragrant oil or resin exuded from certain trees and plants. **3** thing that heals or soothes. **4** aromatic herb. [Latin: related to BALSAM]

balmy /'bɑ:mɪ/ *adj.* (**-ier, -iest**) **1** mild and fragrant; soothing. **2** *slang* = BARMY. □ **balmily** *adv.* **balminess** *n.*

baloney var. of BOLONEY.

balsa /'bɔ:lsə/ *n.* **1** (in full **balsa-wood**) tough lightweight wood used for making models etc. **2** tropical American tree yielding it. [Spanish, = raft]

balsam /'bɔ:lsəm/ *n.* **1** resin exuded from various trees and shrubs. **2** ointment, esp. containing oil or turpentine. **3** tree or shrub yielding balsam. **4** any of several flowering plants. □ **balsamic** /-'sæmɪk/ *adj.* [Latin *balsamum*]

baluster /'bæləstə(r)/ *n.* short post or pillar supporting a rail. [Greek *balaustion* wild-pomegranate flower]

■ **Usage** *Baluster* is often confused with *banister*. A baluster is usually part of a balustrade whereas a banister supports a stair handrail.

balustrade /,bælə'streɪd/ *n.* railing supported by balusters, esp. on a balcony.

bamboo /bæm'bu:/ *n.* **1** tropical giant woody grass. **2** its stem, used for canes, furniture, etc. [Dutch from Malay]

bamboo-shoot *n.* young shoot of bamboo, eaten as a vegetable.

bamboozle /bæm'bu:z(ə)l/ *v.* (**-ling**) *colloq.* cheat; mystify. □ **bamboozlement** *n.* [origin unknown]

ban —*v.* (**-nn-**) forbid, prohibit, esp. formally. —*n.* formal prohibition (*ban on smoking*). [Old English, = summon]

banal /bə'nɑ:l/ *adj.* trite, commonplace. □ **banality** /-'nælɪtɪ/ *n.* (*pl.* **-ies**). **banally** *adv.* [French, related to BAN: originally = compulsory, hence = common]

banana /bə'nɑ:nə/ *n.* **1** long curved soft fruit with a yellow skin. **2** treelike plant bearing it. □ **go bananas** *slang* go mad. [Portuguese or Spanish, from an African name]

banana republic *n. derog.* small State, esp. in Central America, dependent on foreign capital.

band —*n.* **1** flat, thin strip or loop of paper, metal, cloth, etc., put round something esp. to hold or decorate it. **2 a** strip of material on a garment. **b** stripe. **3** group of esp. non-classical musicians. **4** organized group of criminals etc. **5** range of frequencies, wavelengths, or values. **6** belt connecting wheels or pulleys. —*v.* **1** (usu. foll. by *together*) unite. **2** put a band on. **3** mark with stripes. [Old Norse (related to BIND) and French]

bandage /'bændɪdʒ/ —*n.* strip of material used to bind a wound etc. —*v.* (**-ging**) bind with a bandage. [French: related to BAND]

bandanna /bæn'dænə/ *n.* large patterned handkerchief or neckerchief. [Portuguese from Hindi]

b. & b. *abbr.* bed and breakfast.

bandbox *n.* hatbox.

bandeau /'bændəʊ/ *n.* (*pl.* **-x** /-dəʊz/) narrow headband. [French]

banderole /,bændə'rəʊl/ *n.* **1** long narrow flag with a cleft end. **2** ribbon-like inscribed scroll. [Italian: related to BANNER]

bandicoot /'bændɪ,ku:t/ *n.* **1** catlike Australian marsupial. **2** (in full **bandicoot rat**) destructive rat in India. [Telugu, = pig-rat]

bandit /'bændɪt/ *n.* robber or outlaw, esp. one attacking travellers etc. □ **banditry** *n.* [Italian]

bandmaster *n.* conductor of a band.

bandog /'bændɒg/ *n.* fighting-dog bred for its strength and ferocity. [from BAND, DOG]

bandolier /,bændə'lɪə(r)/ *n.* (also **bandoleer**) shoulder belt with loops or pockets for cartridges. [Dutch or French]

band-saw *n.* mechanical saw with a blade formed by an endless toothed band.

bandsman *n.* player in a band.

bandstand *n.* outdoor platform for musicians.

bandwagon *n.* □ **climb** (or **jump**) **on the bandwagon** join a popular or successful cause etc.

bandwidth *n.* range of frequencies within a given band.

bandy[1] *adj.* (**-ier, -iest**) **1** (of the legs) curved so as to be wide apart at the knees. **2** (also **bandy-legged**) having bandy legs. [perhaps from obsolete *bandy* curved stick]

bandy[2] *v.* (**-ies, -ied**) **1** (often foll. by *about*) **a** pass (a story, rumour, etc.) to and fro. **b** discuss disparagingly (*bandied her name about*). **2** (often foll. by *with*) exchange (blows, insults, etc.). [perhaps from French]

bane *n.* **1** cause of ruin or trouble. **2** *poet.* ruin. **3** *archaic* (except in *comb.*) poison (*ratsbane*). □ **baneful** *adj.* [Old English]

bang *–n.* **1** loud short sound. **2** sharp blow. **3** *coarse slang* act of sexual intercourse. **4** *US* fringe cut straight across the forehead. *–v.* **1** strike or shut noisily (*banged the door shut*). **2** (cause to) make a bang. **3** *coarse slang* have sexual intercourse (with). *–adv.* **1** with a bang. **2** *colloq.* exactly (*bang in the middle*). □ **bang on** *colloq.* exactly right. **go bang 1** shut noisily. **2** explode. **3** (as **bang goes** etc.) *colloq.* be suddenly lost (*bang go my hopes*). [imitative]

banger *n.* **1** *slang* sausage. **2** *slang* noisy old car. **3** firework designed to go bang.

bangle /ˈbæŋg(ə)l/ *n.* rigid bracelet or anklet. [Hindi *bangri*]

banian var. of BANYAN.

banish /ˈbænɪʃ/ *v.* **1** condemn to exile. **2** dismiss (esp. from one's mind). □ **banishment** *n.* [Germanic: related to BAN]

banister /ˈbænɪstə(r)/ *n.* (also **bannister**) (usu. in *pl.*) uprights and handrail beside a staircase. [corruption of BALUSTER]

■ **Usage** See note at *baluster*.

---

banjo /ˈbændʒəʊ/ *n.* (*pl.* **-s** or **-es**) guitar-like stringed instrument with a circular body. □ **banjoist** *n.* [US southern corruption of *bandore* from Greek *pandoura* lute]

bank[1] *–n.* **1** sloping ground beside a river. **2** raised area, esp. in the sea; slope. **3** mass of cloud, fog, snow, etc. *–v.* **1** (often foll. by *up*) heap or rise into banks. **2** pack (a fire) tightly for slow burning. **3 a** (of a vehicle, aircraft, etc.) round a curve with one side higher than the other. **b** cause to do this. [Old Norse: related to BENCH]

bank[2] *–n.* **1** establishment for depositing, withdrawing, and borrowing money. **2** kitty in some gambling games. **3** storage place (*blood bank*). *–v.* **1** deposit (money etc.) in a bank. **2** (often foll. by *at*, *with*) keep money (at a bank). □ **bank on** *colloq.* rely on (*I'm banking on you*). [French *banque* or Italian *banca*: related to BANK[1]]

bank[3] *n.* row of similar objects, e.g. lights, switches, oars. [French *banc* from Germanic: related to BANK[1]]

bankable *adj.* certain to bring profit.

bank card *n.* (also **banker's card**) = CHEQUE CARD.

banker *n.* **1** owner or manager of a bank. **2** keeper of the bank in some gambling games.

bank holiday *n.* public holiday, when banks are officially closed.

banking *n.* business of running a bank.

banknote *n.* piece of paper money.

bankrupt /ˈbæŋkrʌpt/ *–adj.* **1** legally declared insolvent. **2** (often foll. by *of*) exhausted or drained (of emotion etc.). *–n.* insolvent person, esp. one whose assets are used to repay creditors. *–v.* make bankrupt. □ **bankruptcy** *n.* (*pl.* **-ies**). [Italian *banca rotta* broken bench: related to BANK[2]]

banksia /ˈbæŋksɪə/ *n.* Australian evergreen flowering shrub. [*Banks*, name of a naturalist]

banner *n.* **1** large sign bearing a slogan or design, esp. in a demonstration or procession; flag. **2** slogan, esp. political. [Latin *bandum* standard]

banner headline *n.* large, esp. front-page, newspaper headline.

bannister var. of BANISTER.

bannock /ˈbænək/ *n. Scot. & N.Engl.* round flat loaf, usu. unleavened. [Old English]

banns *n.pl.* notice announcing an intended marriage, read out in a parish church. [pl. of BAN]

banquet /ˈbæŋkwɪt/ *–n.* sumptuous, esp. formal, feast or dinner. *–v.* (**-t-**) attend, or entertain with, a banquet; feast. [French diminutive of *banc* bench]

banquette /bæŋˈket/ *n.* upholstered bench, esp. in a restaurant or bar. [French from Italian]

banshee /ˈbænʃiː/ *n. Ir. & Scot.* wailing female spirit warning of death in a house. [Irish, = fairy woman]

bantam /ˈbæntəm/ *n.* **1** a kind of small domestic fowl. **2** small but aggressive person. [apparently from *Bāntān* in Java]

bantamweight *n.* **1** weight in certain sports between flyweight and featherweight, in amateur boxing 51–4 kg. **2** sportsman of this weight.

banter *–n.* good-humoured teasing. *–v.* **1** tease. **2** exchange banter. [origin unknown]

Bantu /bænˈtuː/ *–n.* (*pl.* same or **-s**) **1** often *offens.* member of a large group of central and southern African Blacks. **2** group of languages spoken by them. *–adj.* of these peoples or languages. [Bantu, = people]

Bantustan /ˌbæntuːˈstɑːn/ *n. S.Afr.* often *offens.* = HOMELAND 2.

banyan /ˈbænjən/ *n.* (also **banian**) Indian fig tree with self-rooting branches. [Portuguese from Sanskrit, = trader]

baobab /ˈbeɪəʊˌbæb/ *n.* African tree with a massive trunk and large pulpy fruit. [probably African dial.]

bap *n.* soft flattish bread roll. [origin unknown]

baptism /'bæptɪz(ə)m/ *n.* symbolic admission to the Christian Church, with water and usu. name-giving. □ **baptismal** /-'tɪzm(ə)l/ *adj.* [Greek *baptizō* baptize]

baptism of fire *n.* **1** initiation into battle. **2** painful initiation into an activity.

baptist *n.* **1** person who baptizes, esp. John the Baptist. **2** (**Baptist**) Christian advocating baptism by total immersion.

baptistery /'bæptɪstərɪ/ *n.* (*pl.* **-ies**) **1 a** part of a church used for baptism. **b** *hist.* separate building used for baptism. **2** (in a Baptist chapel) receptacle used for immersion.

baptize /bæp'taɪz/ *v.* (also **-ise**) (**-zing** or **-sing**) **1** administer baptism to. **2** give a name or nickname to.

bar[1] *–n.* **1** long piece of rigid material, esp. used to confine or obstruct. **2 a** something of similar form (*bar of soap*; *bar of chocolate*). **b** band of colour or light. **c** heating element of an electric fire. **d** metal strip below the clasp of a medal, awarded as an extra distinction. **e** *Heraldry* narrow horizontal stripe across a shield. **3 a** counter for serving alcohol etc. on. **b** room or building containing it. **c** small shop or stall serving refreshments (*snack bar*). **d** counter for a special service (*heel bar*). **4 a** barrier. **b** restriction (*colour bar*; *bar to promotion*). **5** prisoner's enclosure in a lawcourt. **6** any of the sections into which a piece of music is divided by vertical lines. **7** (**the Bar**) *Law* **a** barristers collectively. **b** profession of barrister. *–v.* (**-rr-**) **1 a** fasten with a bar or bars. **b** (usu. foll. by *in*, *out*) shut or keep in or out. **2** obstruct, prevent. **3** (usu. foll. by *from*) prohibit, exclude. **4** mark with stripes. *–prep.* except. □ **be called to the Bar** be admitted as barrister. **behind bars** in prison. [French]

bar[2] *n.* esp. *Meteorol.* unit of pressure, 10^5 newtons per square metre, approx. one atmosphere. [Greek *baros* weight]

barathea /,bærə'θi:ə/ *n.* fine wool cloth. [origin unknown]

barb *–n.* **1** secondary backward-facing projection from an arrow, fish-hook, etc. **2** hurtful remark. **3** fleshy filament at the mouth of some fish. *–v.* **1** fit with a barb. **2** (as **barbed** *adj.*) (of a remark etc.) deliberately hurtful. [Latin *barba* beard]

barbarian /bɑ:'beərɪən/ *–n.* **1** uncultured or brutish person. **2** member of a primitive tribe etc. *–adj.* **1** rough and uncultured. **2** uncivilized. [Greek *barbaros* foreign]

barbaric /bɑ:'bærɪk/ *adj.* **1** uncultured; brutal, cruel. **2** primitive.

barbarism /'bɑ:bə,rɪz(ə)m/ *n.* **1** barbaric state or act. **2** non-standard word or expression.

barbarity /bɑ:'bærɪtɪ/ *n.* (*pl.* **-ies**) **1** savage cruelty. **2** brutal act.

barbarous /'bɑ:bərəs/ *adj.* = BARBARIC 1.

barbecue /'bɑ:bɪ,kju:/ *–n.* **1 a** meal cooked over charcoal etc. out of doors. **b** party for this. **2** grill etc. used for this. *–v.* (**-ues**, **-ued**, **-uing**) cook on a barbecue. [Spanish from Haitian]

barbed wire *n.* wire with interwoven sharp spikes, used in fences and barriers.

barbel /'bɑ:b(ə)l/ *n.* **1** freshwater fish with barbs. **2** = BARB *n.* 3. [Latin: related to BARB]

barbell *n.* iron bar with removable weights at each end, used for weightlifting.

barber *n.* person who cuts men's hair etc. by profession. [medieval Latin *barba* beard]

barberry /'bɑ:bərɪ/ *n.* (*pl.* **-ies**) **1** shrub with yellow flowers and red berries. **2** its berry. [French *berberis*]

barber-shop *n. colloq.* close harmony singing for four male voices.

barber's pole *n.* pole with spiral red and white stripes as a barber's sign.

barbican /'bɑ:bɪkən/ *n.* outer defence, esp. a double tower above a gate or drawbridge. [French]

barbie /'bɑ:bɪ/ *n. Austral. slang* barbecue. [abbreviation]

bar billiards *n.pl.* form of billiards with holes in the table.

barbiturate /bɑ:'bɪtjʊrət/ *n.* soporific or sedative drug from barbituric acid. [German, from the name *Barbara*]

barbituric acid /,bɑ:bɪ'tjʊərɪk/ *n.* organic acid from which barbiturates are derived.

Barbour /'bɑ:bə(r)/ *n. propr.* type of green waxed jacket. [*Barbour*, name of a draper]

barcarole /,bɑ:kə'rəʊl/ *n.* **1** gondoliers' song. **2** music imitating this. [Italian *barca* boat]

bar-code *n.* machine-readable striped code on packaging etc.

bard *n.* **1** *poet.* poet. **2 a** *hist.* Celtic minstrel. **b** prizewinner at an Eisteddfod. □ **bardic** *adj.* [Celtic]

bare *–adj.* **1** unclothed or uncovered. **2** leafless; unfurnished; empty. **3** plain, unadorned (*the bare truth*; *bare facts*). **4** (*attrib.*) scanty, just sufficient (*a bare majority*; *bare necessities*). *–v.* (**-ring**) uncover, reveal (*bared his teeth*; *bared his soul*). [Old English]

bareback *adj.* & *adv.* without a saddle.

barefaced *adj.* shameless, impudent.

barefoot *adj.* & *adv.* (also **barefooted** /-'fʊtɪd/) wearing nothing on the feet.

bareheaded /beə'hedɪd/ *adj.* & *adv.* wearing nothing on the head.

barely *adv.* **1** scarcely (*barely escaped*). **2** scantily (*barely furnished*).

bargain /'bɑːgɪn/ *–n.* **1 a** agreement on the terms of a sale etc. **b** this from the buyer's viewpoint (*a bad bargain*). **2** cheap thing. *–v.* (often foll. by *with, for*) discuss the terms of a sale etc. (*bargained with me; bargain for the table*). □ **bargain for** (or *colloq.* **on**) be prepared for; expect. **bargain on** rely on. **into the bargain** moreover. [French from Germanic]

barge *–n.* **1** long flat-bottomed cargo boat on a canal or river. **2** long ornamental pleasure boat. *–v.* **(-ging) 1** (foll. by *in, into*) **a** intrude rudely or awkwardly (*barged in on him*). **b** collide with (*barged into her*). **2** (often foll. by *around*) move clumsily about. [French: related to BARQUE]

bargeboard *n.* board fixed to the gable-end of a roof to hide the ends of the roof timbers. [perhaps from medieval Latin *bargus* gallows]

bargee /bɑː'dʒiː/ *n.* person sailing a barge.

bargepole *n.* □ **would not touch with a bargepole** refuse to be associated or concerned with.

baritone /'bærɪ,təʊn/ *n.* **1 a** second-lowest adult male singing voice. **b** singer with this voice. **2** instrument pitched second-lowest in its family. [Greek *barus* heavy, *tonos* tone]

barium /'beərɪəm/ *n.* white soft metallic element. [from BARYTA]

barium meal *n.* mixture swallowed to reveal the abdomen in X-rays.

bark[1] *–n.* **1** sharp explosive cry of a dog, fox, etc. **2** sound like this. *–v.* **1** (of a dog etc.) give a bark. **2** speak or utter sharply or brusquely. **3** *colloq.* cough harshly. □ **bark up the wrong tree** make false assumptions. [Old English]

bark[2] *–n.* tough outer skin of tree-trunks, branches, etc. *–v.* **1** graze (one's shin etc.). **2** strip bark from. [Scandinavian]

barker *n.* tout at an auction, sideshow, etc. [from BARK[1]]

barley /'bɑːlɪ/ *n.* **1** cereal used as food and in spirits. **2** (also **barleycorn**) its grain. [Old English]

barley sugar *n.* sweet made from sugar, usu. in twisted sticks.

barley water *n.* drink made from a boiled barley mixture.

barm *n.* froth on fermenting malt liquor. [Old English]

barmaid *n.* woman serving in a pub etc.

barman *n.* man serving in a pub etc.

bar mitzvah /bɑː 'mɪtsvə/ *n.* **1** religious initiation ceremony of a Jewish boy at 13. **2** boy undergoing this. [Hebrew, = son of the commandment]

barmy /'bɑːmɪ/ *adj.* **(-ier, -iest)** *slang* crazy, stupid. [from BARM: earlier, = frothy]

barn *n.* large farm building for storing grain etc. [Old English, = barley house]

barnacle /'bɑːnək(ə)l/ *n.* **1** marine crustacean clinging to rocks, ships' bottoms, etc. **2** tenacious attendant or follower. [French or medieval Latin]

barnacle goose *n.* Arctic goose.

barn dance *n.* **1** informal gathering for country dancing. **2** a kind of country dance.

barney /'bɑːnɪ/ *n.* (*pl.* **-s**) *colloq.* noisy quarrel. [perhaps dial.]

barn-owl *n.* a kind of owl frequenting barns.

barnstorm *v.* tour rural areas as an actor or political campaigner. □ **barnstormer** *n.*

barnyard *n.* area around a barn.

barograph /'bærə,grɑːf/ *n.* barometer equipped to record its readings. [Greek *baros* weight]

barometer /bə'rɒmɪtə(r)/ *n.* **1** instrument measuring atmospheric pressure, used in meteorology. **2** anything which reflects change. □ **barometric** /,bærə'metrɪk/ *adj.* [related to BAROGRAPH]

baron /'bærən/ *n.* **1** member of the lowest order of the British or foreign nobility. **2** powerful businessman, entrepreneur, etc. **3** *hist.* person holding lands from the sovereign. □ **baronial** /bə'rəʊnɪəl/ *adj.* [medieval Latin, = man]

baroness /'bærənɪs/ *n.* **1** woman holding the rank of baron. **2** baron's wife or widow.

baronet /'bærənɪt/ *n.* member of the lowest hereditary titled British order. □ **baronetcy** *n.* (*pl.* **-ies**).

baron of beef *n.* double sirloin.

barony /'bærənɪ/ *n.* (*pl.* **-ies**) domain or rank of a baron.

baroque /bə'rɒk/ *–adj.* **1** highly ornate and extravagant in style, esp. of European art etc. of the 17th and 18th c. **2** of this period. *–n.* baroque style or art. [Portuguese, originally = misshapen pearl]

bar person *n.* barmaid or barman.

barque /bɑːk/ *n.* **1** sailing-ship with the rear mast fore-and-aft rigged and other masts square-rigged. **2** *poet.* boat. [Provençal from Latin *barca*]

barrack[1] /'bærək/ *–n.* (usu. in *pl.*, often treated as *sing.*) **1** housing for soldiers. **2**

large bleak building. —*v.* lodge (soldiers etc.) in barracks. [Italian or Spanish]

barrack[2] /'bærək/ *v.* **1** shout or jeer at (players, a speaker, etc.). **2** (foll. by *for*) cheer for, encourage (a team etc.). [perhaps from Australian slang *borak* banter]

barracouta /ˌbærə'ku:tə/ *n.* (*pl.* same or **-s**) long slender fish of southern oceans. [var. of BARRACUDA]

barracuda /ˌbærə'ku:də/ *n.* (*pl.* same or **-s**) large tropical marine fish. [Spanish]

barrage /'bærɑ:ʒ/ *n.* **1** concentrated artillery bombardment. **2** rapid succession of questions or criticisms. **3** artificial barrier in a river etc. [French *barrer* BAR[1]]

barrage balloon *n.* large anchored balloon used as a defence against low-flying aircraft.

barratry /'bærətrɪ/ *n.* fraud or gross negligence by a ship's master or crew. [French *barat* deceit]

barre /bɑ:(r)/ *n.* horizontal bar at waist level, used in dance exercises. [French]

barré /'bæreɪ/ *n.* method of playing a chord on the guitar etc. with a finger laid across the strings at a particular fret. [French *barrer* bar]

barrel /'bær(ə)l/ —*n.* **1** cylindrical usu. convex container. **2** its contents. **3** measure of capacity (30 to 40 gallons). **4** cylindrical tube forming part of an object, e.g. a gun or a pen. —*v.* (**-ll-**; *US* **-l-**) put into a barrel or barrels. □ **over a barrel** *colloq.* helpless, at a person's mercy. [French]

barrel-organ *n.* mechanical musical instrument with a rotating pin-studded cylinder.

barren /'bærən/ *adj.* (**-er, -est**) **1 a** unable to bear young. **b** (of land, a tree, etc.) unproductive. **2** unprofitable, dull. □ **barrenness** *n.* [French]

barricade /ˌbærɪ'keɪd/ —*n.* barrier, esp. improvised. —*v.* (**-ding**) block or defend with this. [French *barrique* cask]

barrier /'bærɪə(r)/ *n.* **1** fence etc. that bars advance or access. **2** obstacle (*class barriers*). [Romanic: related to BAR[1]]

barrier cream *n.* protective skin cream.

barrier reef *n.* coral reef separated from the shore by a channel.

barring /'bɑ:rɪŋ/ *prep.* except, not including.

barrister /'bærɪstə(r)/ *n.* advocate entitled to practise in the higher courts. [from BAR[1]: cf. MINISTER]

barrow[1] /'bærəʊ/ *n.* **1** two-wheeled handcart. **2** = WHEELBARROW. [Old English: related to BEAR[1]]

barrow[2] /'bærəʊ/ *n.* ancient grave-mound. [Old English]

bar sinister *n.* = BEND SINISTER.

bartender *n.* person serving in a pub etc.

barter —*v.* **1** trade in goods without using money. **2** exchange (goods). —*n.* trade by bartering. [perhaps from French]

baryon /'bærɪˌɒn/ *n.* heavy elementary particle (i.e. a nucleon or a hyperon). [Greek *barus* heavy]

baryta /bə'raɪtə/ *n.* barium oxide or hydroxide. [from BARYTES]

barytes /bə'raɪti:z/ *n.* mineral form of barium sulphate. [Greek *barus* heavy]

basal /'beɪs(ə)l/ *adj.* of, at, or forming a base.

basalt /'bæsɔ:lt/ *n.* a dark volcanic rock. □ **basaltic** /bə'sɔ:ltɪk/ *adj.* [Latin *basaltes* from Greek]

base[1] —*n.* **1 a** part supporting from beneath or serving as a foundation. **b** notional support or foundation (*power base*). **2** principle or starting-point. **3** esp. *Mil.* headquarters. **4** main or important ingredient. **5** number in terms of which other numbers or logarithms are expressed. **6** substance capable of combining with an acid to form a salt. **7** *Baseball* etc. each of the four stations on a pitch. —*v.* (**-sing**) **1** (usu. foll. by *on*, *upon*) found or establish (a theory, hope, etc.). **2** station (*troops based in Malta*). [Greek *basis* stepping]

base[2] *adj.* **1** cowardly, despicable. **2** menial. **3** alloyed (*base coin*). **4** (of a metal) low in value. [Latin *bassus*]

baseball *n.* **1** game played esp. in the US with a circuit of four bases which batsmen must complete. **2** ball used in this.

baseless *adj.* unfounded, groundless.

baseline *n.* **1** line used as a base or starting-point. **2** line marking each end of a tennis-court.

basement /'beɪsmənt/ *n.* floor of a building below ground level.

base rate *n.* interest rate set by the Bank of England, used as the basis for other banks' rates.

bases *pl.* of BASE[1], BASIS.

bash —*v.* **1 a** strike bluntly or heavily. **b** (often foll. by *up*) *colloq.* attack violently. **c** (often foll. by *down*, *in*, etc.) damage or break by striking forcibly. **2** (foll. by *into*) collide with. —*n.* **1** heavy blow. **2** *slang* attempt. [imitative]

bashful /'bæʃfʊl/ *adj.* shy, diffident. □ **bashfully** *adv.* [as ABASHED]

BASIC /'beɪsɪk/ *n.* computer programming language using familiar English words. [*B*eginner's *A*ll-purpose *S*ymbolic *I*nstruction *C*ode]

basic /'beɪsɪk/ *–adj.* **1** serving as a base; fundamental. **2 a** simplest or lowest in level (*basic pay, needs*). **b** vulgar (*basic humour*). *–n.* (usu. in *pl.*) fundamental facts or principles. □ **basically** *adv.*

basic slag *n.* fertilizer containing phosphates formed as a by-product in steel manufacture.

basil /'bæz(ə)l/ *n.* aromatic herb used as flavouring. [Greek *basilikos* royal]

basilica /bə'zɪlɪkə/ *n.* **1** ancient Roman hall with an apse and colonnades, used as a lawcourt etc. **2** similar building as a Christian church. [Greek *basilikē (stoa)* royal (portico)]

basilisk /'bæzɪlɪsk/ *n.* **1** mythical reptile with lethal breath and glance. **2** small American crested lizard. [Greek, diminutive of *basileus* king]

basin /'beɪs(ə)n/ *n.* **1** round open vessel for holding liquids or preparing food in. **2** = WASH-BASIN. **3** hollow depression. **4** sheltered mooring area. **5** round valley. **6** area drained by a river. □ **basinful** *n.* (*pl.* **-s**). [medieval Latin *ba(s)cinus*]

basis /'beɪsɪs/ *n.* (*pl.* **bases** /-si:z/) **1** foundation or support. **2** main principle or ingredient (*on a friendly basis*). **3** starting-point for a discussion etc. [Greek: related to BASE[1]]

bask /bɑ:sk/ *v.* **1** relax in warmth and light. **2** (foll. by *in*) revel in (*basking in glory*). [Old Norse: related to BATHE]

basket /'bɑ:skɪt/ *n.* **1** container made of interwoven cane, reed, wire, etc. **2** amount held by this. **3** the goal in basketball, or a goal scored. **4** *Econ.* group or range (of currencies). [French]

basketball *n.* **1** game in which goals are scored by putting the ball through high nets. **2** ball used in this.

basketry *n.* **1** art of weaving cane etc. **2** work so produced.

basket weave *n.* weave like wickerwork.

basketwork *n.* = BASKETRY.

basking shark *n.* large shark which lies near the surface of the sea.

Basque /bæsk/ *–n.* **1** member of a people of the Western Pyrenees. **2** their language. *–adj.* of the Basques or their language. [Latin *Vasco*]

bas-relief /'bæsrɪ,li:f/ *n.* sculpture or carving with figures projecting slightly from the background. [French and Italian]

bass[1] /beɪs/ *–n.* **1 a** lowest adult male singing voice. **b** singer with this voice. **2** instrument pitched lowest in its family. **3** *colloq.* bass guitar or double-bass. **4** low-frequency output of a radio, record-player, etc. *–adj.* **1** lowest in musical pitch. **2** deep-sounding. □ **bassist** *n.* (in sense 3). [from BASE[2] altered after Italian *basso*]

bass[2] /bæs/ *n.* (*pl.* same or **-es**) **1** common perch. **2** other spiny-finned fish of the perch family. [Old English]

bass clef *n.* clef placing F below middle C on the second highest line of the staff.

basset /'bæsɪt/ *n.* (in full **basset-hound**) sturdy hunting-dog with a long body and short legs. [French diminutive of *bas* low]

bass guitar *n.* electric guitar tuned as a double-bass.

bassinet /,bæsɪ'net/ *n.* child's wicker cradle, usu. hooded. [French diminutive of *bassin* BASIN]

basso /'bæsəʊ/ *n.* (*pl.* **-s**) singer with a bass voice. [Italian, = BASS[1]]

bassoon /bə'su:n/ *n.* bass instrument of the oboe family. □ **bassoonist** *n.* [Italian: related to BASS[1]]

bast /bæst/ *n.* fibre from the inner bark of a tree (esp. the lime). [Old English]

bastard /'bɑ:stəd/ often *offens.* *–n.* **1** person born of an unmarried mother. **2** *slang* **a** unpleasant or despicable person. **b** person of a specified kind (*poor, lucky, bastard*). **3** *slang* difficult or awkward thing. *–attrib. adj.* **1** illegitimate by birth. **2** unauthorized, counterfeit, hybrid. □ **bastardy** *n.* (in sense 1 of *n.*). [French from medieval Latin]

bastardize *v.* (also **-ise**) (**-zing** or **-sing**) **1** corrupt, debase. **2** declare (a person) illegitimate.

baste[1] /beɪst/ *v.* (**-ting**) **1** moisten (meat) with fat etc. during cooking. **2** beat, thrash. [origin unknown]

baste[2] /beɪst/ *v.* (**-ting**) sew with large loose stitches, tack. [French from Germanic]

bastinado /,bæstɪ'neɪdəʊ/ *–n.* beating with a stick on the soles of the feet. *–v.* (**-es**, **-ed**) punish in this way. [Spanish *baston* stick]

bastion /'bæstɪən/ *n.* **1** projecting part of a fortification. **2** thing regarded as protecting (*bastion of freedom*). [Italian *bastire* build]

bat[1] *–n.* **1** implement with a handle, used for hitting balls in games. **2** turn with this. **3** batsman. *–v.* (**-tt-**) **1** hit with or as with a bat. **2** take a turn at batting. □ **off one's own bat** unprompted, unaided. [Old English from French]

bat[2] *n.* mouselike nocturnal flying mammal. [Scandinavian]

bat[3] *v.* (**-tt-**) □ **not** (or **never**) **bat an eyelid** *colloq.* show no reaction or emotion. [var. of obsolete *bate* flutter]

batch *–n.* **1** group of things or persons considered or dealt with together; instalment. **2** loaves produced at one baking. **3** *Computing* group of records processed as one unit. –*v.* arrange or deal with in batches. [related to BAKE]

bated /ˈbeɪtɪd/ *adj.* □ **with bated breath** very anxiously. [as ABATE]

bath /bɑːθ/ –*n.* (*pl.* **-s** /bɑːðz/) **1 a** (usu. plumbed-in) container for sitting in and washing the body. **b** its contents. **2** act of washing in it (*have a bath*). **3** (usu. in *pl.*) public building with baths or a swimming-pool. **4 a** vessel containing liquid for immersing something, e.g. a film for developing. **b** its contents. –*v.* **1** wash (esp. a baby) in a bath. **2** take a bath. [Old English]

Bath bun /bɑːθ/ *n.* round spiced bun with currants, often iced. [*Bath* in S. England]

Bath chair /bɑːθ/ *n.* wheelchair for invalids.

bath cube *n.* cube of soluble substance for scenting or softening bath-water.

bathe /beɪð/ –*v.* (**-thing**) **1** immerse oneself in water, esp. to swim or wash oneself. **2** immerse in, wash, or treat with liquid. **3** (of sunlight etc.) envelop. –*n.* swim. [Old English]

bathhouse *n.* public building with baths.

bathing-costume *n.* (also **bathing-suit**) garment worn for swimming.

bathos /ˈbeɪθɒs/ *n.* lapse in mood from the sublime to the absurd or trivial; anticlimax. □ **bathetic** /bəˈθetɪk/ *adj.* **bathotic** /bəˈθɒtɪk/ *adj.* [Greek, = depth]

bathrobe *n.* esp. *US* dressing-gown, esp. of towelling.

bathroom *n.* **1** room with a bath, wash-basin, etc. **2** *US* room with a lavatory.

bath salts *n.pl.* soluble powder or crystals for scenting or softening bath-water.

bathyscaphe /ˈbæθɪˌskæf/ *n.* manned vessel for deep-sea diving. [Greek *bathus* deep, *skaphos* ship]

bathysphere /ˈbæθɪˌsfɪə(r)/ *n.* vessel for deep-sea observation. [Greek *bathus* deep, SPHERE]

batik /bəˈtiːk/ *n.* **1** method of dyeing textiles by applying wax to parts to be left uncoloured. **2** cloth so treated. [Javanese, = painted]

batiste /bæˈtiːst/ *n.* fine linen or cotton cloth. [French from *Baptiste*, name of the first maker]

batman /ˈbætmən/ *n.* army officer's servant. [*bat* pack-saddle, from French]

baton /ˈbæt(ə)n/ *n.* **1** thin stick for conducting an orchestra etc. **2** short stick passed on in a relay race. **3** stick carried by a drum major. **4** staff of office. [French from Latin]

baton round *n.* rubber or plastic bullet.

batrachian /bəˈtreɪkɪən/ –*n.* amphibian that discards its gills and tail, esp. a frog or toad. –*adj.* of batrachians. [Greek *batrakhos* frog]

bats *predic. adj. slang* crazy. [originally pl. of BAT[2]]

batsman *n.* person who bats, esp. in cricket.

battalion /bəˈtæljən/ *n.* **1** army unit usu. of 300–1000 men. **2** large group with a common aim. [Italian *battaglia* BATTLE]

batten[1] /ˈbæt(ə)n/ –*n.* **1 a** long flat strip of squared timber. **b** horizontal strip of wood to which laths, tiles, etc., are fastened. **2** strip for securing tarpaulin over a ship's hatchway. –*v.* strengthen or (often foll. by *down*) fasten with battens. [French: related to BATTER]

batten[2] /ˈbæt(ə)n/ *v.* (foll. by *on*) thrive at the expense of (another). [Old Norse]

Battenberg /ˈbæt(ə)nˌbɜːg/ *n.* oblong sponge cake, usu. of two colours and covered with marzipan. [*Battenberg* in Germany]

batter[1] *v.* **1 a** strike hard and repeatedly. **b** (often foll. by *against*, *at*, etc.) pound insistently (*batter at the door*). **2** (often in *passive*) **a** subject to long-term violence (*battered baby, wife*). **b** criticize severely. □ **batterer** *n.* [French *battre* beat: related to BATTLE]

batter[2] *n.* mixture of flour, egg, and milk or water, used for pancakes etc. [French: related to BATTER[1]]

battered *adj.* coated in batter and deep-fried.

battering-ram *n. hist.* beam used in breaching fortifications.

battery /ˈbætərɪ/ *n.* (*pl.* **-ies**) **1** usu. portable container of an electrically charged cell or cells as a source of current. **2** (often *attrib.*) series of cages for the intensive breeding and rearing of poultry or cattle. **3** set of similar units of equipment; series, sequence. **4** emplacement for heavy guns. **5** *Law* unlawful physical violence against a person. [Latin: related to BATTLE]

battle /ˈbæt(ə)l/ –*n.* **1** prolonged fight between armed forces. **2** difficult struggle; contest (*battle for supremacy*; *battle of wits*). –*v.* (**-ling**) engage in battle; fight. □ **half the battle** key to the success of an undertaking. [Latin *battuo* beat]

battleaxe *n.* **1** large axe used in ancient warfare. **2** *colloq.* formidable older woman.

battlebus *n. colloq.* bus used by a politician during an election campaign as a mobile centre of operations.

battle-cruiser *n. hist.* warship of higher speed and lighter armour than a battleship.

battle-cry *n.* cry or slogan used in a battle or contest.

battledore /'bæt(ə)l,dɔ:(r)/ *n. hist.* **1** (in full **battledore and shuttlecock**) game played with a shuttlecock and rackets. **2** racket used in this. [perhaps from Provençal *batedor* beater]

battledress *n.* everyday uniform of a soldier.

battlefield *n.* (also **battleground**) scene of a battle.

battlement /'bæt(ə)lmənt/ *n.* (usu. in *pl.*) recessed parapet along the top of a wall, as part of a fortification. [French *batailler* fortify]

battle royal *n.* **1** battle of many combatants; free fight. **2** heated argument.

battleship *n.* heavily armoured warship.

batty /'bætɪ/ *adj.* (**-ier, -iest**) *slang* crazy. [from BAT[2]]

batwing *attrib. adj.* (esp. of a sleeve) shaped like a bat's wing.

bauble /'bɔ:b(ə)l/ *n.* showy worthless trinket or toy. [French *ba(u)bel* toy]

baulk /bɔ:lk, bɔ:k/ (also **balk**) *–v.* **1** (often foll. by *at*) jib, hesitate. **2 a** thwart, hinder. **b** disappoint. **3** miss, let slip (a chance etc.). *–n.* **1** hindrance; stumbling-block. **2** roughly-squared timber beam. [Old English]

bauxite /'bɔ:ksaɪt/ *n.* claylike mineral, the chief source of aluminium. [French from *Les Baux* in S. France]

bawdy /'bɔ:dɪ/ *–adj.* (**-ier, -iest**) humorously indecent. *–n.* such talk or writing. [*bawd* brothel-keeper from French *baudetrot*]

bawdy-house *n.* brothel.

bawl /bɔ:l/ *v.* **1** speak or shout noisily. **2** weep loudly. □ **bawl out** *colloq.* reprimand angrily. [imitative]

bay[1] *n.* broad curving inlet of the sea. [Spanish *bahia*]

bay[2] *n.* **1** laurel with deep green leaves. **2** (in *pl.*) bay wreath, for a victor or poet. [Latin *baca* berry]

bay[3] *n.* **1** recess; alcove in a wall. **2** compartment (*bomb bay*). **3** area specially allocated (*loading bay*). [French *baer* gape]

bay[4] *–adj.* (esp. of a horse) dark reddish-brown. *–n.* bay horse. [Latin *badius*]

bay[5] *–v.* bark or howl loudly and plaintively. *–n.* sound of this, esp. of hounds in close pursuit. □ **at bay** cornered, unable to escape. **keep at bay** hold off (a pursuer). [French *bayer* to bark]

bayberry *n.* (*pl.* **-ies**) fragrant N. American tree.

bay-leaf *n.* leaf of the bay-tree, used for flavouring.

bayonet /'beɪə,net/ *–n.* **1** stabbing blade attachable to the muzzle of a rifle. **2** electrical fitting pushed into a socket and twisted. *–v.* (**-t-**) stab with bayonet. [French, perhaps from *Bayonne* in SW France]

bay rum *n.* perfume distilled orig. from bayberry leaves in rum.

bay window *n.* window projecting outwards from a wall.

bazaar /bə'zɑ:(r)/ *n.* **1** oriental market. **2** fund-raising sale of goods, esp. for charity. [Persian]

bazooka /bə'zu:kə/ *n.* anti-tank rocket-launcher. [origin unknown]

BB *abbr.* double-black (pencil-lead).

BBC *abbr.* British Broadcasting Corporation.

BC *abbr.* British Columbia.

BC *abbr.* before Christ.

BCG *abbr.* Bacillus Calmette-Guérin, an anti-tuberculosis vaccine.

BD *abbr.* Bachelor of Divinity.

bdellium /'delɪəm/ *n.* **1** tree yielding resin. **2** this used in perfumes. [Latin from Greek]

Be *symb.* beryllium.

be /bi:, bɪ/ *v.* (*sing. present* **am**; **are** /ɑ:(r)/; **is** /ɪz/; *past* **was** /wɒz/; **were** /wɜ:(r)/; *pres. part.* **being**; *past part.* **been**) **1** exist, live (*I think, therefore I am*; *there is no God*). **2 a** occur; take place (*dinner is at eight*). **b** occupy a position (*he is in the garden*). **3** remain, continue (*let it be*). **4** linking subject and predicate, expressing: **a** identity (*she is the person*). **b** condition (*he is ill today*). **c** state or quality (*he is kind*; *they are my friends*). **d** opinion (*I am against hanging*). **e** total (*two and two are four*). **f** cost or significance (*it is £5 to enter*; *it is nothing to me*). **5** *v.aux.* **a** with a past participle to form the passive (*it was done*; *it is said*). **b** with a present participle to form continuous tenses (*we are coming*; *it is being cleaned*). **c** with an infinitive to express duty or commitment, intention, possibility, destiny, or hypothesis (*I am to tell you*; *we are to wait here*; *he is to come at four*; *it was not to be found*; *they were never to meet again*; *if I were to die*). □ **be at** occupy oneself with (*what is he at?*; *mice have been at the food*). **be off** *colloq.* go away; leave. **-to-be** of the future (in *comb.*: *bride-to-be*). [Old English]

be- *prefix* forming verbs: **1** (from transitive verbs) **a** all over; all round (*beset*). **b** thoroughly, excessively (*begrudge*). **2** (from intransitive verbs) expressing transitive action (*bemoan*; *bestride*). **3**

(from adjectives and nouns) expressing transitive action (*becalm*). **4** (from nouns) **a** affect with (*befog*). **b** treat as (*befriend*). **c** (forming adjectives in *-ed*) having (*bejewelled*). [Old English, = BY]

beach –*n.* pebbly or sandy shore esp. of the sea. –*v.* run or haul (a boat etc.) on to a beach. [origin unknown]

beachcomber *n.* person who searches beaches for articles of value.

beachhead *n.* fortified position set up on a beach by landing forces.

beacon /'bi:kən/ *n.* **1** fire or light set up high as a warning etc. **2** visible warning or guiding device (e.g. a lighthouse, navigation buoy, etc.). **3** radio transmitter whose signal helps fix the position of a ship or aircraft. **4** = BELISHA BEACON. [Old English]

bead –*n.* **1 a** small usu. rounded piece of glass, stone, etc., for threading to make necklaces etc., or sewing on to fabric, etc. **b** (in *pl.*) bead necklace; rosary. **2** drop of liquid. **3** small knob in the foresight of a gun. **4** inner edge of a pneumatic tyre gripping the rim of the wheel. –*v.* adorn with or as with beads. □ **draw a bead on** take aim at. [Old English, = prayer]

beading *n.* **1** moulding or carving like a series of beads. **2** bead of a tyre.

beadle /'bi:d(ə)l/ *n.* **1** ceremonial officer of a church, college, etc. **2** *Scot.* church officer serving the minister. **3** *hist.* minor parish disciplinary officer. [French from Germanic]

beady *adj.* (**-ier, -iest**) (esp. of the eyes) small, round, and bright.

beady-eyed *adj.* **1** with beady eyes. **2** observant.

beagle /'bi:g(ə)l/ *n.* small short-haired hound used for hunting hares. [French]

beak[1] *n.* **1 a** bird's horny projecting jaws. **b** similar jaw of a turtle etc. **2** *slang* hooked nose. **3** *hist.* pointed prow of a warship. **4** spout. [French from Celtic]

beak[2] *n. slang* **1** magistrate. **2** schoolmaster. [probably thieves' cant]

beaker *n.* **1** tall cup or tumbler. **2** lipped glass vessel for scientific experiments. [Old Norse]

be-all and end-all *n. colloq.* (often foll. by *of*) whole being, essence.

beam –*n.* **1** long sturdy piece of squared timber or metal used in house-building etc. **2** ray or shaft of light or radiation. **3** bright look or smile. **4** series of radio or radar signals as a guide to a ship or aircraft. **5** crossbar of a balance. **6 a** ship's breadth. **b** width of a person's hips. **7** (in *pl.*) horizontal cross-timbers of a ship. –*v.* **1** emit or direct (light, approval, etc.). **2 a** shine. **b** look or smile radiantly. □ **off** (or **off the**) **beam** *colloq.* mistaken. **on the beam** *colloq.* on the right track. [Old English, = tree]

bean *n.* **1 a** climber with edible seeds in long pods. **b** seed or pod of this. **2** similar seed of coffee etc. □ **full of beans** *colloq.* lively, exuberant. **not a bean** *slang* no money. [Old English]

beanbag *n.* **1** small bag filled with dried beans and used as a ball. **2** large bag filled usu. with polystyrene pieces and used as a chair.

beanfeast *n.* **1** *colloq.* celebration. **2** workers' annual dinner.

beano /'bi:nəʊ/ *n.* (*pl.* **-s**) *slang* celebration, party.

beanpole *n.* **1** support for bean plants. **2** *colloq.* tall thin person.

bean sprout *n.* sprout of a bean seed as food.

beanstalk *n.* stem of a bean plant.

bear[1] /beə(r)/ *v.* (*past* **bore**; *past part.* **borne** or **born**) **1** carry, bring, or take (esp. visibly) (*bear gifts*). **2** show; have, esp. characteristically (*bear marks of violence*; *bears no relation to the case*; *bore no name*). **3 a** produce, yield (fruit etc.). **b** give birth to (*has borne a son*; *was born last week*). **4 a** sustain (a weight, responsibility, cost, etc.). **b** endure (an ordeal, difficulty, etc.). **5** (usu. with *neg.* or *interrog.*) **a** tolerate (*can't bear him*). **b** admit of (*does not bear thinking about*). **6** carry mentally (*bear a grudge*). **7** veer in a given direction (*bear left*). **8** bring or provide (something needed) (*bear him company*). □ **bear down** press downwards. **bear down on** approach inexorably. **bear on** (or **upon**) be relevant to. **bear out** support or confirm as evidence. **bear up** not despair. **bear with** tolerate patiently. **bear witness** testify. [Old English]

bear[2] /beə(r)/ *n.* **1** any of several large heavy mammals with thick fur. **2** rough or uncouth person. **3** person who sells shares hoping to repurchase them more cheaply. □ **the Great Bear, the Little Bear** constellations near the North Pole. [Old English]

bearable *adj.* endurable.

beard –*n.* **1** facial hair on the chin etc. **2** similar tuft etc. on an animal. –*v.* oppose; defy. □ **bearded** *adj.* **beardless** *adj.* [Old English]

bearer *n.* **1** person or thing that bears, carries, or brings. **2** person who presents a cheque etc.

beargarden *n.* rowdy scene.

bear-hug *n.* tight embrace.

bearing *n.* **1** bodily attitude; general manner. **2** (foll. by *on, upon*) relevance to (*has no bearing on it*). **3** endurability (*beyond bearing*). **4** part of a machine supporting a rotating part. **5 a** direction or position relative to a fixed point. **b** (in *pl.*) knowledge of one's relative position. **6** *Heraldry* device or charge.

Béarnaise sauce /ˌbeɪəˈneɪz/ *n.* rich thick sauce with egg-yolks. [French]

bearskin *n.* **1** skin of a bear, esp. as a wrap etc. **2** tall furry hat worn by some regiments.

beast *n.* **1** animal, esp. a wild mammal. **2 a** brutal person. **b** *colloq.* objectionable person or thing. **3** (prec. by *the*) the animal nature in man. [Latin *bestia*]

beastly –*adj.* (**-ier, -iest**) **1** *colloq.* objectionable, unpleasant. **2** like a beast; brutal. –*adv. colloq.* very, extremely.

beast of prey *n.* animal which hunts animals for food.

beat –*v.* (*past* **beat**; *past part.* **beaten**) **1 a** strike persistently and violently. **b** strike (a carpet, drum, etc.) repeatedly. **2** (foll. by *against, at, on*, etc.) pound or knock repeatedly. **3 a** overcome; surpass. **b** be too hard for; perplex. **4** (often foll. by *up*) whisk (eggs etc.) vigorously. **5** (often foll. by *out*) shape (metal etc.) by blows. **6** (of the heart etc.) pulsate rhythmically. **7** (often foll. by *out*) **a** indicate (a tempo etc.) by tapping etc. **b** sound (a signal etc.) by striking a drum etc. (*beat a tattoo*). **8** move or cause (wings) to move up and down. **9** make (a path etc.) by trampling. **10** strike (bushes etc.) to rouse game. –*n.* **1 a** main accent in music or verse. **b** rhythm indicated by a conductor (*watch the beat*). **c** (in popular music) strong rhythm. **2 a** stroke or blow or measured sequence of strokes. **b** throbbing movement or sound. **3 a** police officer's route or area. **b** person's habitual round. –*predic. adj. slang* exhausted, tired out. □ **beat about the bush** not come to the point. **beat the bounds** mark parish boundaries by striking certain points with rods. **beat down 1** cause (a seller) to lower the price by bargaining. **2** strike to the ground (*beat the door down*). **3** (of the sun, rain, etc.) shine or fall relentlessly. **beat it** *slang* go away. **beat off** drive back (an attack etc.). **beat a person to it** arrive or do something before another person. **beat up** beat, esp. with punches and kicks. □ **beatable** *adj.* [Old English]

beater *n.* **1** implement for beating (esp. eggs). **2** person who rouses game at a shoot.

beatific /ˌbiːəˈtɪfɪk/ *adj.* **1** *colloq.* blissful (*beatific smile*). **2** making blessed. [Latin *beatus* blessed]

beatify /biːˈætɪˌfaɪ/ *v.* (**-ies, -ied**) **1** *RC Ch.* declare (a person) to be 'blessed', often as a step towards canonization. **2** make happy. □ **beatification** /-fɪˈkeɪʃ(ə)n/ *n.*

beatitude /biːˈætɪˌtjuːd/ *n.* **1** perfect bliss or happiness. **2** (in *pl.*) blessings in Matt. 5:3–11.

beatnik /ˈbiːtnɪk/ *n.* member of a movement of socially unconventional young people in the 1950s.

beat-up *adj. colloq.* dilapidated.

beau /bəʊ/ *n.* (*pl.* **-x** /bəʊz/) **1** esp. *US* admirer; boyfriend. **2** fop; dandy. [Latin *bellus* pretty]

Beaufort scale /ˈbəʊfət/ *n.* scale of wind speed ranging from 0 (calm) to 12 (hurricane). [Sir F. *Beaufort*, name of an admiral]

Beaujolais /ˈbəʊʒəˌleɪ/ *n.* red or white wine from the Beaujolais district of France.

Beaujolais Nouveau /nuːˈvəʊ/ *n.* Beaujolais wine sold in the first year of a vintage.

beauteous /ˈbjuːtɪəs/ *adj. poet.* beautiful.

beautician /bjuːˈtɪʃ(ə)n/ *n.* specialist in beauty treatment.

beautiful /ˈbjuːtɪˌfʊl/ *adj.* **1** having beauty, pleasing to the eye, ear, or mind etc. (*beautiful voice*). **2** pleasant, enjoyable (*had a beautiful time*). **3** excellent (*beautiful specimen*). □ **beautifully** *adv.*

beautify /ˈbjuːtɪˌfaɪ/ *v.* (**-ies, -ied**) make beautiful; adorn. □ **beautification** /-fɪˈkeɪʃ(ə)n/ *n.*

beauty /ˈbjuːtɪ/ *n.* (*pl.* **-ies**) **1** combination of shape, colour, sound, etc., that pleases the senses. **2** *colloq.* **a** excellent specimen (*what a beauty!*). **b** attractive feature; advantage (*that's the beauty of it!*). **3** beautiful woman. [Latin *bellus* pretty]

beauty parlour *n.* (also **beauty salon**) establishment for cosmetic treatment.

beauty queen *n.* woman judged most beautiful in a contest.

beauty spot *n.* **1** place of scenic beauty. **2** small facial mark considered to enhance the appearance.

beaux *pl.* of BEAU.

beaver[1] –*n.* **1 a** amphibious broad-tailed rodent which cuts down trees with its teeth and dams rivers. **b** its fur. **c** hat of this. **2** (**Beaver**) boy aged six or seven affiliated to the Scouts. –*v. colloq.* (usu. foll. by *away*) work hard. [Old English]

beaver[2] *n. hist.* lower face-guard of a helmet. [French, = bib]

beaver lamb *n.* lamb's wool made to look like beaver fur.

bebop /'bi:bɒp/ *n.* type of 1940s jazz with complex harmony and rhythms. [imitative]
becalm /bɪ'kɑ:m/ *v.* (usu. in *passive*) deprive (a ship) of wind.
became *past* of BECOME.
because /bɪ'kɒz/ *conj.* for the reason that; since. □ **because of** on account of; by reason of. [from BY, CAUSE]
béchamel /'beʃə,mel/ *n.* a kind of thick white sauce. [Marquis de *Béchamel*, name of a courtier]
beck[1] *n. N.Engl.* brook; mountain stream. [Old Norse]
beck[2] *n.* □ **at a person's beck and call** subject to a person's constant orders. [from BECKON]
beckon /'bekən/ *v.* **1** (often foll. by *to*) summon by gesture. **2** entice. [Old English]
become /bɪ'kʌm/ *v.* (**-ming**; *past* **became**; *past part.* **become**) **1** begin to be (*became king, famous*). **2** (often as **becoming** *adj.*) **a** look well on; suit (*blue becomes him*). **b** befit (*it ill becomes you to complain*; *becoming modesty*). □ **become of** happen to (*what became of her?*). [Old English: related to BE-]
becquerel /'bekə,rel/ *n.* SI unit of radioactivity. [*Becquerel*, name of a physicist]
bed –*n.* **1** piece of furniture for sleeping on. **2** any place used for sleep or rest. **3** garden plot, esp. for flowers. **4 a** bottom of the sea, a river, etc. **b** foundations of a road or railway. **5** stratum, layer. –*v.* (**-dd-**) **1** (usu. foll. by *down*) put or go to bed. **2** *colloq.* have sexual intercourse with. **3** (usu. foll. by *out*) plant in a garden bed. **4** cover up or fix firmly. **5** arrange as a layer. □ **bed of roses** life of ease. **go to bed 1** retire to sleep. **2** (often foll. by *with*) have sexual intercourse. [Old English]
B.Ed. *abbr.* Bachelor of Education.
bed and board *n.* lodging and food.
bed and breakfast *n.* **1** room and breakfast in a hotel etc. **2** establishment providing this.
bedaub /bɪ'dɔ:b/ *v.* smear or daub.
bedazzle /bɪ'dæz(ə)l/ *v.* (**-ling**) **1** dazzle. **2** confuse.
bedbug *n.* wingless parasite infesting beds etc.
bedclothes *n.pl.* sheets, blankets, etc., for a bed.
bedding *n.* **1** mattress and bedclothes. **2** litter for animals. **3** geological strata.
bedding plant *n.* plant set in a garden bed etc. when about to bloom.
bedeck /bɪ'dek/ *v.* adorn.
bedevil /bɪ'dev(ə)l/ *v.* (**-ll-**; *US* **-l-**) **1** trouble, vex. **2** confound, confuse. **3** torment, abuse. □ **bedevilment** *n.*
bedew /bɪ'dju:/ *v.* cover or sprinkle (as) with dew.
bedfellow *n.* **1** person who shares a bed. **2** associate.
bedizen /bɪ'daɪz(ə)n/ *v. poet.* deck out gaudily. [from BE-, obsolete *dizen* deck out]
bedjacket *n.* jacket worn when sitting up in bed.
bedlam /'bedləm/ *n.* uproar and confusion. [St Mary of *Bethlehem*, name of a hospital in London]
bedlinen *n.* sheets and pillowcases.
Bedlington terrier /'bedlɪŋt(ə)n/ *n.* terrier with a narrow head, long legs, and curly grey hair. [*Bedlington* in Northumberland]
Bedouin /'bedʊɪn/ *n.* (also **Beduin**) (*pl.* same) **1** nomadic desert Arab. **2** wandering person. [Arabic, = dwellers in the desert]
bedpan *n.* portable toilet for use in bed.
bedpost *n.* each of four upright supports of a bed.
bedraggled /bɪ'dræg(ə)ld/ *adj.* dishevelled, untidy.
bedrest *n.* recuperation in bed.
bedridden *adj.* confined to bed by infirmity.
bedrock *n.* **1** solid rock underlying alluvial deposits etc. **2** basic principles.
bedroom *n.* room for sleeping in.
bedside *n.* space beside esp. a patient's bed. □ **bedside manner** (esp. doctor's) way with patients.
bedsit *n.* (also **bedsitter**, **bedsitting room**) combined bedroom and sitting-room with cooking facilities.
bedsock *n.* warm sock worn in bed.
bedsore *n.* sore developed by lying in bed.
bedspread *n.* cover for a bed.
bedstead *n.* framework of a bed.
bedstraw *n.* small herbaceous plant.
bedtime *n.* (often *attrib.*) usual time for going to bed (*bedtime drink*).
Beduin var. of BEDOUIN.
bedwetting *n.* involuntary urination when asleep in bed.
bee *n.* **1** four-winged stinging insect, collecting nectar and pollen and producing wax and honey. **2** (usu. **busy bee**) busy person. **3** esp. *US* meeting for work or amusement. □ **a bee in one's bonnet** obsession. [Old English]
Beeb *n.* (prec. by *the*) *colloq.* BBC. [abbreviation]
beech *n.* **1** large smooth grey tree with glossy leaves. **2** its hard wood. [Old English]
beechmast *n.* (*pl.* same) fruit of the beech.
beef –*n.* **1** flesh of the ox, bull, or cow for eating. **2** *colloq.* male muscle. **3** (*pl.*

beeves or *US* **-s**) ox etc. bred for beef. **4** (*pl.* **-s**) *slang* complaint. *–v. slang* complain. □ **beef up** *slang* strengthen, reinforce. [Latin *bos bovis* ox]

beefburger *n.* = HAMBURGER.

beefcake *n.* esp. *US slang* male muscle.

beefeater *n.* **1** warder at the Tower of London. **2** Yeoman of the Guard.

beefsteak *n.* thick slice of beef for grilling or frying.

beef tea *n.* stewed beef extract for invalids.

beef tomato *n.* large tomato.

beefy *adj.* (**-ier, -iest**) **1** like beef. **2** solid, muscular. □ **beefiness** *n.*

beehive *n.* **1** artificial shelter for a colony of bees. **2** busy place.

bee-keeper *n.* keeper of bees. □ **bee-keeping** *n.*

beeline *n.* □ **make a beeline for** go directly to.

Beelzebub /biːˈelzɪˌbʌb/ *n.* the Devil. [Hebrew, = lord of the flies]

been *past part.* of BE.

beep *–n.* sound of a (esp. motor-car) horn. *–v.* emit a beep. [imitative]

beer *n.* **1** alcoholic drink made from fermented malt etc., flavoured with hops. **2** glass of this. □ **beer and skittles** amusement. [Old English]

beer-cellar *n.* cellar for storing, or selling and drinking, beer.

beer garden *n.* (also **beer hall**) garden (or large room) where beer is sold and drunk.

beer-mat *n.* small mat for a beer-glass.

beery *adj.* (**-ier, -iest**) **1** affected by beer-drinking. **2** like beer.

beeswax *–n.* **1** wax secreted by bees to make honeycombs. **2** this used to polish wood. *–v.* polish with beeswax.

beeswing *n.* filmy second crust on old port.

beet *n.* plant with an edible root (see BEETROOT, SUGAR BEET). [Old English]

beetle[1] /ˈbiːt(ə)l/ *–n.* insect, esp. black, with hard protective outer wings. *–v.* (**-ling**) *colloq.* (foll. by *about, off, etc.*) hurry, scurry. [Old English: related to BITE]

beetle[2] /ˈbiːt(ə)l/ *–adj.* projecting, shaggy, scowling (*beetle brows*). *–v.* (usu. as **beetling** *adj.*) (of brows, cliffs, etc.) project, overhang. [origin unknown]

beetle[3] /ˈbiːt(ə)l/ *n.* heavy tool for ramming, crushing, etc. [Old English: related to BEAT]

beetle-browed *adj.* with shaggy, projecting, or scowling eyebrows.

beetroot *n.* beet with an edible dark-red root.

beeves *pl.* of BEEF.

befall /bɪˈfɔːl/ *v.* (*past* **befell**; *past part.* **befallen**) *poet.* happen; happen to. [Old English: related to BE-]

befit /bɪˈfɪt/ *v.* (**-tt-**) be appropriate for; be incumbent on.

befog /bɪˈfɒg/ *v.* (**-gg-**) **1** confuse, obscure. **2** envelop in fog.

before /bɪˈfɔː(r)/ *–conj.* **1** earlier than the time when (*see me before you go*). **2** rather than that (*would die before he stole*). *–prep.* **1** earlier than (*before noon*). **2 a** in front of, ahead of (*before her in the queue*). **b** in the face of (*recoil before the attack*). **c** awaiting (*the future before them*). **3** rather than (*death before dishonour*). **4 a** in the presence of (*appear before the judge*). **b** for the attention of (*put before the committee*). *–adv.* **1** previously; already (*happened long before; done it before*). **2** ahead (*go before*). **3** on the front (*hit before and behind*). [Old English: related to BY, FORE]

Before Christ *adv.* (of a date) reckoned backwards from the birth of Christ.

beforehand *adv.* in advance, in readiness.

befriend /bɪˈfrend/ *v.* act as a friend to; help.

befuddle /bɪˈfʌd(ə)l/ *v.* (**-ling**) **1** make drunk. **2** confuse.

beg *v.* (**-gg-**) **1 a** (usu. foll. by *for*) ask for (food, money, etc.). **b** live by begging. **2** ask earnestly, humbly, or formally (for) (*begged for mercy; beg your indulgence; beg leave*). **3** (of a dog etc.) sit up with the front paws raised expectantly. **4** (foll. by *to* + infin.) take leave (*I beg to differ*). □ **beg off 1** ask to be excused from something. **2** get (a person) excused a penalty etc. **beg pardon** see PARDON. **beg the question** assume the truth of a proposition needing proof. **go begging** be unwanted or refused. [related to BID]

■ **Usage** The expression *beg the question* is often used incorrectly to mean (1) to avoid giving a straight answer, or (2) to invite the obvious question (that . . .). These uses should be avoided.

---

began *past* of BEGIN.

beget /bɪˈget/ *v.* (**-tt-**; *past* **begot**; *archaic* **begat**; *past part.* **begotten**) *literary* **1** father, procreate. **2** cause (*anger begets violence*). [Old English: related to BE-]

beggar /ˈbegə(r)/ *–n.* **1** person who lives by begging. **2** *colloq.* person (*cheeky beggar*). *–v.* **1** make poor. **2** be too extraordinary for (*beggar description*). □ **beggarly** *adj.*

begin /bɪˈgɪn/ *v.* (**-nn-**; *past* **began**; *past part.* **begun**) **1** perform the first part of; start (*begin work; begin crying; begin to understand*). **2** come into being (*war *began in 1939; Wales begins here*). **3** (usu. foll. by *to* + infin.) start at a certain time (*soon began to feel ill*). **4** be begun (*meeting began at 7*). **5 a** start speaking (*'No,' he began*). **b** take the first step, be the first (*shall I begin?*). **6** (usu. with *neg.*) *colloq.* show any likelihood (*can't begin to compete*). [Old English]

beginner *n.* trainee, learner. □ **beginner's luck** supposed good luck of a beginner.

beginning *n.* **1** time or place at which anything begins. **2** source or origin. **3** first part.

begone /bɪ'gɒn/ *int. poet.* go away at once! [*be gone*]

begonia /bɪ'gəʊnɪə/ *n.* garden plant with bright flowers and leaves. [*Bégon*, name of a patron of science]

begot *past* of BEGET.

begotten *past part.* of BEGET.

begrudge /bɪ'grʌdʒ/ *v.* (**-ging**) **1** resent; be dissatisfied at. **2** envy (a person) the possession of. □ **begrudgingly** *adv.*

beguile /bɪ'gaɪl/ *v.* (**-ling**) **1** charm; amuse. **2** wilfully divert, seduce. **3** (usu. foll. by *of*, *out of*, or *into* + verbal noun) delude; cheat. □ **beguilement** *n.* **beguiling** *adj.*

beguine /bɪ'gi:n/ *n.* popular dance of W. Indian origin. [French *béguin* infatuation]

begum /'beɪgəm/ *n.* in India or Pakistan: **1** Muslim woman of high rank. **2** (**Begum**) title of a married Muslim woman. [Turkish *bīgam* princess]

begun *past part.* of BEGIN.

behalf /bɪ'hɑ:f/ *n.* □ **on behalf of** (or **on a person's behalf**) in the interests of; as representative of. [earlier *bihalve* on the part of]

behave /bɪ'heɪv/ *v.* (**-ving**) **1** act or react (in a specified way) (*behaved well*). **2** (often *refl.*) conduct oneself properly (*behave yourself!*). **3** (of a machine etc.) work well (or in a specified way). [from BE-, HAVE]

behaviour /bɪ'heɪvjə(r)/ *n.* (*US* **behavior**) way of behaving or acting. □ **behavioural** *adj.*

behavioural science *n.* (*US* **behavioral**) the study of human behaviour.

behaviourism *n.* (*US* **behaviorism**) *Psychol.* theory that human behaviour is determined by conditioned response to stimuli, and that psychological disorders are best treated by altering behaviour. □ **behaviourist** *n.*

behead /bɪ'hed/ *v.* cut the head off (a person), esp. as execution. [Old English: related to BE-]

beheld *past* and *past part.* of BEHOLD.

behemoth /bɪ'hi:mɒθ/ *n.* huge creature or thing. [Hebrew (Job 40:15)]

behest /bɪ'hest/ *n. literary* command; request. [Old English]

behind /bɪ'haɪnd/ *–prep.* **1 a** in or to the rear of. **b** on the far side of (*behind the bush*). **c** hidden or implied by (*something behind that remark*). **2 a** in the past in relation to (*trouble is behind me now*). **b** late regarding (*behind schedule*). **3** inferior to; weaker than (*behind the others in maths*). **4 a** in support of (*she's behind us*). **b** responsible for (*the man behind the project*). **5** in the tracks of; following. *–adv.* **1 a** in or to the rear; further back (*the street behind*; *glance behind*). **b** on the further side (*wall with a field behind*). **2** remaining after others' departure (*stay behind*). **3** (usu. foll. by *with*) **a** in arrears. **b** late in finishing a task etc. (*getting behind*). **4** in a weak position; backward (*behind in Latin*). **5** following (*dog running behind*). *–n. colloq.* buttocks. □ **behind the scenes** see SCENE. **behind time** late. **behind the times** old-fashioned, antiquated. [Old English]

behindhand *adv.* & *predic.adj.* **1** (usu. foll. by *with*, *in*) late; in arrears. **2** out of date.

behold /bɪ'həʊld/ *v.* (*past & past part.* **beheld**) (esp. in *imper.*) *literary* see, observe. [Old English: related to BE-]

beholden *predic. adj.* (usu. foll. by *to*) under obligation.

behove /bɪ'həʊv/ *v.* (**-ving**) *formal* **1** be incumbent on. **2** befit (*it ill behoves him to protest*). [Old English: related to BE, HEAVE]

beige /beɪʒ/ *–n.* pale sandy fawn colour. *–adj.* of this colour. [French]

being /'bi:ɪŋ/ *n.* **1** existence. **2** nature or essence (of a person etc.) (*his whole being revolted*). **3** person or creature.

bejewelled /bɪ'dʒu:əld/ *adj.* (*US* **bejeweled**) adorned with jewels.

bel *n.* unit of relative power level, esp. of sound, corresponding to an intensity ratio of 10 to 1 (cf. DECIBEL). [*Bell*, name of the inventor of the telephone]

belabour /bɪ'leɪbə(r)/ *v.* (*US* **belabor**) **1** attack physically or verbally. **2** labour (a subject).

belated /bɪ'leɪtɪd/ *adj.* late or too late. □ **belatedly** *adv.*

belay /bɪ'leɪ/ *–v.* secure (a rope) by winding it round a peg etc. *–n.* act of belaying. [Dutch *beleggen*]

bel canto /bel 'kæntəʊ/ *n.* lyrical rich-toned style of operatic singing. [Italian, = fine song]

belch *–v.* **1** emit wind noisily through the mouth. **2** (of a chimney, gun, etc.)

send (smoke etc.) out or up. *–n.* act of belching. [Old English]

beleaguer /bɪ'li:gə(r)/ *v.* **1** besiege. **2** vex; harass. [Dutch *leger* camp]

belfry /'belfrɪ/ *n.* (*pl.* **-ies**) **1** bell tower. **2** space for bells in a church tower. [Germanic, probably = peace-protector]

Belial /'bi:lɪəl/ *n.* the Devil. [Hebrew, = worthless]

belie /bɪ'laɪ/ *v.* (**belying**) **1** give a false impression of (*its appearance belies its age*). **2** fail to fulfil or justify. [Old English: related to BE-]

belief /bɪ'li:f/ *n.* **1** firm opinion; acceptance (*that is my belief*). **2** religious conviction (*belief in the afterlife; has no belief*). **3** (usu. foll. by *in*) trust or confidence. [related to BELIEVE]

believe /bɪ'li:v/ *v.* (**-ving**) **1** accept as true or as conveying the truth (*I believe it; don't believe him*). **2** think, suppose. **3** (foll. by *in*) **a** have faith in the existence of (*believes in God*). **b** have confidence in (*believes in homoeopathy*). **c** have trust in as a policy (*believes in telling the truth*). **4** have (esp. religious) faith. □ **believable** *adj.* **believer** *n.* [Old English]

Belisha beacon /bə'li:ʃə/ *n.* flashing orange ball on a striped post, marking some pedestrian crossings. [Hore-*Belisha*, who introduced it]

belittle /bɪ'lɪt(ə)l/ *v.* (**-ling**) disparage, make appear insignificant. □ **belittlement** *n.*

bell[1] *–n.* **1** hollow esp. cup-shaped usu. metal object sounding a note when struck. **2** such a sound, esp. as a signal. **3** thing resembling a bell, esp. in sound or shape. *–v.* provide with a bell. □ **give a person a bell** *colloq.* telephone a person. [Old English]

bell[2] *–n.* cry of a stag. *–v.* make this cry. [Old English]

belladonna /,belə'dɒnə/ *n.* **1** deadly nightshade. **2** drug from this. [Italian, = fair lady]

bell-bottom *n.* **1** marked flare below the knee (of a trouser-leg). **2** (in *pl.*) trousers with these. □ **bell-bottomed** *adj.*

bellboy *n.* esp. *US* page in a hotel or club.

belle /bel/ *n.* beautiful or most beautiful woman. [French, feminine of BEAU]

belles-lettres /bel 'letr/ *n.pl.* (also treated as *sing.*) literary writings or studies. [French, = fine letters]

bellicose /'belɪ,kəʊs/ *adj.* eager to fight; warlike. [Latin *bellum* war]

belligerence /bɪ'lɪdʒərəns/ *n.* **1** aggressive behaviour. **2** status of a belligerent.

belligerent *–adj.* **1** engaged in war or conflict. **2** pugnacious. *–n.* belligerent nation or person. [Latin *belligero* wage war]

bell-jar *n.* bell-shaped glass cover or container for use in a laboratory.

bellow /'beləʊ/ *–v.* **1** emit a deep loud roar. **2** utter loudly. *–n.* loud roar. [origin uncertain]

bellows *n.pl.* (also treated as *sing.*) **1** device for driving air into or through something. **2** expandable part, e.g. of a camera. [related to BELLY]

bell-pull *n.* handle etc. sounding a bell when pulled.

bell-push *n.* button operating an electric bell.

bell-ringer *n.* ringer of church bells or handbells. □ **bell-ringing** *n.*

bell-wether *n.* **1** leading sheep of a flock. **2** ringleader.

belly /'belɪ/ *–n.* (*pl.* **-ies**) **1** trunk below the chest, containing the stomach and bowels. **2** stomach. **3** front of the body from waist to groin. **4** underside of an animal. **5** cavity or bulging part. *–v.* (**-ies, -ied**) (often foll. by *out*) swell; bulge. [Old English, = bag]

bellyache *–n. colloq.* stomach pain. *–v.* (**-ching**) *slang* complain noisily or persistently.

belly button *n. colloq.* navel.

belly-dance *n.* oriental dance performed by a woman, with voluptuous belly movements. □ **belly-dancer** *n.* **belly-dancing** *n.*

bellyflop *n. colloq.* dive with the belly landing flat on the water.

bellyful *n.* (*pl.* **-s**) **1** enough to eat. **2** *colloq.* more than one can tolerate.

belly-laugh *n.* loud unrestrained laugh.

belong /bɪ'lɒŋ/ *v.* **1** (foll. by *to*) **a** be the property of. **b** be correctly assigned to. **c** be a member of. **2** fit socially (*just doesn't belong*). **3** (foll. by *in, under*) be correctly placed or classified. [from BE-, obsolete *long* belong]

belongings *n.pl.* possessions or luggage.

Belorussian var. of BYELORUSSIAN.

beloved /bɪ'lʌvɪd/ *–adj.* much loved. *–n.* much loved person.

below /bɪ'ləʊ/ *–prep.* **1** lower in position, amount, status, etc., than. **2** beneath the surface of (*head below water*). **3** unworthy of. *–adv.* **1** at or to a lower point or level. **2** downstream. **3** further forward on a page or in a book (*as noted below*). [*be-* BY, LOW[1]]

belt *–n.* **1** strip of leather etc. worn esp. round the waist. **2** circular band in machinery; conveyor belt. **3** distinct strip of colour etc. **4** region or extent (*cotton belt*). **5** *slang* heavy blow. *–v.* **1** put a belt round; fasten with a belt. **2** *slang* hit hard. **3** *slang* rush, hurry. □ **below the belt** unfair(ly). **belt out** *slang* sing or play (music) loudly or vigorously. **belt up 1** *slang* be quiet. **2** *colloq.* put on a seat belt. **tighten one's belt** economize. **under one's belt** securely acquired. [Old English]

beluga /bɪ'lu:gə/ *n.* (*pl.* same or **-s**) **1 a** large kind of sturgeon. **b** caviare from it. **2** white whale. [Russian]

belvedere /'belvɪ,dɪə(r)/ *n.* raised summer-house or gallery for viewing scenery. [Italian, = beautiful view]

BEM *abbr.* British Empire Medal.

bemoan /bɪ'məʊn/ *v.* lament; complain about.

bemuse /bɪ'mju:z/ *v.* (**-sing**) puzzle, bewilder. [from BE-, MUSE[1]]

bench *n.* **1** long seat of wood or stone. **2** strong work-table. **3** (prec. by *the*) **a** judge's seat. **b** lawcourt. **c** judges and magistrates collectively. [Old English]

bencher *n.* senior member of an Inn of Court.

benchmark *n.* **1** surveyor's mark cut in a wall etc. as a reference point in measuring altitudes. **2** standard or point of reference.

bend[1] *–v.* (*past* and *past part.* **bent** except in *bended knee*) **1 a** force or adapt (something straight) into a curve or angle. **b** (of an object) be so altered. **2** (often foll. by *down*, *over*, etc.) curve, incline, or stoop (*road bends*; *bent down to pick it up*; *bent her head*). **3** interpret or modify (a rule) to suit oneself. **4** turn (one's steps, eyes, or energies) in a new direction. **5** be flexible, submit, or force to submit. *–n.* **1** curve; departure from a straight course. **2** bent part. **3** (in *pl.*; prec. by *the*) *colloq.* decompression sickness. □ **bend over backwards** see BACKWARDS. **round the bend** *colloq.* crazy, insane. □ **bendy** *adj.* (**-ier**, **-iest**). [Old English]

bend[2] *n.* **1** any of various knots. **2** *Heraldry* diagonal stripe from top right to bottom left of a shield (from its wearer's position). [Old English: related to BIND]

bender *n. slang* wild drinking-spree. [from BEND[1]]

bend sinister *n. Heraldry* diagonal stripe from top left to bottom right of a shield, as a sign of bastardy.

beneath /bɪ'ni:θ/ *–prep.* **1** unworthy of (*beneath him to reply*). **2** below, under. *–adv.* below, underneath. [Old English: related to BE-, NETHER]

Benedictine /,benɪ'dɪktɪn/ *–n.* **1** monk or nun of an order following the rule of St Benedict. **2** /-,ti:n/ *propr.* liqueur orig. made by Benedictines. *–adj.* of St Benedict or the Benedictines. [Latin *Benedictus* Benedict]

benediction /,benɪ'dɪkʃ(ə)n/ *n.* blessing, esp. at the end of a religious service or as a special Roman Catholic service. □ **benedictory** *adj.* [Latin *benedico* bless]

benefaction /,benɪ'fækʃ(ə)n/ *n.* **1** donation, gift. **2** giving or doing good. [Latin: related to BENEFIT]

benefactor /'benɪ,fæktə(r)/ *n.* (*fem.* **benefactress**) person giving (esp. financial) support.

benefice /'benɪfɪs/ *n.* a living from a church office.

beneficent /bɪ'nefɪs(ə)nt/ *adj.* doing good; generous, kind. □ **beneficence** *n.*

beneficial /,benɪ'fɪʃ(ə)l/ *adj.* advantageous; having benefits. □ **beneficially** *adv.*

beneficiary /,benɪ'fɪʃərɪ/ *n.* (*pl.* **-ies**) **1** person who benefits, esp. from a will. **2** holder of a benefice.

benefit /'benɪfɪt/ *–n.* **1** favourable or helpful factor etc. **2** (often in *pl.*) insurance or social security payment (*sickness benefit*). **3** public performance or game in aid of a charitable cause etc. *–v.* (**-t-**; *US* **-tt-**) **1** help; bring advantage to. **2** (often foll. by *from*, *by*) receive benefit. □ **the benefit of the doubt** concession that a person is innocent, correct, etc., although doubt exists. [Latin *benefactum* from *bene* well, *facio* do]

benevolent /bɪ'nevələnt/ *adj.* **1** well-wishing; actively friendly and helpful. **2** charitable (*benevolent fund*). □ **benevolence** *n.* [French from Latin *bene volens* well wishing]

Bengali /beŋ'gɔ:lɪ/ *–n.* (*pl.* **-s**) **1** native of Bengal. **2** language of Bengal. *–adj.* of Bengal, its people, or language. [*Bengal*, former Indian province]

benighted /bɪ'naɪtɪd/ *adj.* intellectually or morally ignorant.

benign /bɪ'naɪn/ *adj.* **1** gentle, mild, kindly. **2** fortunate, salutary. **3** (of a tumour etc.) not malignant. □ **benignly** *adv.* [Latin *benignus*]

benignant /bɪ'nɪgnənt/ *adj.* **1** kindly, esp. to inferiors. **2** salutary, beneficial. □ **benignancy** *n.*

benignity /bɪ'nɪgnɪtɪ/ *n.* (*pl.* **-ies**) kindliness.

bent[1] *past* and *past part.* of BEND[1] *v.* *–adj.* **1** curved, angular. **2** *slang* dishonest, illicit. **3** *slang* sexually deviant. **4** (foll. by *on*) set on doing or having. *–n.* **1** inclination or bias. **2** (foll. by *for*) talent.

bent[2] *n.* **1** reedy grass with stiff stems. **2** stiff flower stalk of a grass. [Old English]

bentwood *n.* wood artificially shaped for making furniture.

benumb /bɪ'nʌm/ *v.* **1** make numb; deaden. **2** paralyse (the mind or feelings).

benzene /'benzi:n/ *n.* colourless liquid found in coal tar and used as a volatile solvent etc. [as BENZOIN]

benzine /'benzi:n/ *n.* mixture of liquid hydrocarbons obtained from petroleum.

benzoin /'benzəʊɪn/ *n.* fragrant gum resin from an E. Asian tree. □ **benzoic** /-'zəʊɪk/ *adj.* [ultimately from Arabic *lubān jāwī* incense of Java]

benzol /'benzɒl/ *n.* benzene, esp. unrefined.

bequeath /bɪ'kwi:ð/ *v.* **1** leave to a person in a will. **2** hand down to posterity. [from BE-: related to QUOTH]

bequest /bɪ'kwest/ *n.* **1** bequeathing; bestowal by will. **2** thing bequeathed. [from BE-, obsolete *quiste* saying]

berate /bɪ'reɪt/ *v.* (**-ting**) scold, rebuke.

Berber /'bɜ:bə(r)/ *–n.* **1** member of the indigenous mainly Muslim Caucasian peoples of N. Africa. **2** language of these peoples. *–adj.* of the Berbers or their language. [Arabic]

berceuse /beə'sɜ:z/ *n.* (*pl.* *-s* pronunc. same) **1** lullaby. **2** instrumental piece in this style. [French]

bereave /bɪ'ri:v/ *v.* (**-ving**) (esp. as **bereaved** *adj.*) (often foll. by *of*) deprive of a relation, friend, etc., esp. by death. □ **bereavement** *n.* [Old English, from *reave* deprive of]

bereft /bɪ'reft/ *adj.* (foll. by *of*) deprived (*bereft of hope*). [past part. of BEREAVE]

beret /'bereɪ/ *n.* round flattish brimless cap of felt etc. [French: related to BIRETTA]

berg *n.* = ICEBERG. [abbreviation]

bergamot[1] /'bɜ:gə,mɒt/ *n.* **1** perfume from the fruit of a dwarf orange tree. **2** aromatic herb. [*Bergamo* in Italy]

bergamot[2] /'bɜ:gə,mɒt/ *n.* variety of fine pear. [Turkish, = prince's pear]

beriberi /,berɪ'berɪ/ *n.* nervous disease caused by a deficiency of vitamin B_1. [Sinhalese]

berk *n.* (also **burk**) *slang* stupid person. [*Berkshire Hunt*, rhyming slang for *cunt*]

berkelium /bɜ:'ki:lɪəm/ *n.* artificial radioactive metallic element. [*Berkeley* in US, where first made]

Bermuda shorts /bə'mju:də/ *n.pl.* (also **Bermudas**) close-fitting knee-length shorts. [*Bermuda* in the W. Atlantic]

berry *n.* (*pl.* **-ies**) **1** any small roundish juicy stoneless fruit. **2** *Bot.* fruit with its seeds enclosed in pulp (e.g. a banana or tomato). [Old English]

berserk /bə'zɜ:k/ *adj.* (esp. in **go berserk**) wild, frenzied; in a rage. [originally = Norse warrior: Icelandic, = bear-coat]

berth *–n.* **1** bunk on a ship, train, etc. **2** place for a ship to moor or be at anchor. **3** adequate room for a ship. **4** *colloq.* job. *–v.* **1** moor (a ship); be moored. **2** provide a sleeping place for. □ **give a wide berth to** stay away from. [probably from a special use of BEAR[1]]

beryl /'berɪl/ *n.* **1** transparent precious stone, esp. pale green. **2** mineral species including this, emerald, and aquamarine. [Greek *bērullos*]

beryllium /bə'rɪlɪəm/ *n.* hard white metallic element.

beseech /bɪ'si:tʃ/ *v.* (*past* and *past part.* **besought** /-'sɔ:t/ or **beseeched**) **1** (foll. by *for*, or *to* + infin.) entreat. **2** ask earnestly for. [from BE-, SEEK]

beset /bɪ'set/ *v.* (**-tt-**; *past* and *past part.* **beset**) **1** attack or harass persistently (*beset by worries*). **2** surround (a person etc.). [Old English: related to BE-]

beside /bɪ'saɪd/ *prep.* **1** at the side of; near. **2** compared with. **3** irrelevant to (*beside the point*). □ **beside oneself** frantic with worry etc. [Old English: related to BY, SIDE]

besides *–prep.* in addition to; apart from. *–adv.* also; moreover.

besiege /bɪ'si:dʒ/ *v.* (**-ging**) **1** lay siege to. **2** crowd round eagerly. **3** harass with requests.

besmirch /bɪ'smɜ:tʃ/ *v.* **1** soil, discolour. **2** dishonour.

besom /'bi:z(ə)m/ *n.* broom of twigs. [Old English]

besotted /bɪ'sɒtɪd/ *adj.* **1** infatuated. **2** intoxicated, stupefied.

besought *past* and *past part.* of BESEECH.

bespangle /bɪ'spæŋg(ə)l/ *v.* (**-ling**) adorn with spangles.

bespatter /bɪ'spætə(r)/ *v.* **1** spatter all over. **2** slander, defame.

bespeak /bɪ'spi:k/ *v.* (*past* **bespoke**; *past part.* **bespoken** or as *adj.* **bespoke**) **1** engage in advance. **2** order (goods). **3** suggest; be evidence of.

bespectacled /bɪ'spektək(ə)ld/ *adj.* wearing spectacles.

bespoke *past* and *past part.* of BESPEAK. *adj.* **1** made to order. **2** making goods to order.

bespoken *past part.* of BESPEAK.

best *–adj.* (*superl.* of GOOD) of the most excellent or desirable kind. *–adv.* (*superl.* of WELL[1]). **1** in the best manner. **2** to the greatest degree. **3** most usefully (*is best ignored*). *–n.* **1** that which is best (*the best is yet to come*). **2** chief merit or advantage (*brings out the best in him*). **3** (foll. by *of*) winning majority of (games played etc.) (*the best of five*). **4** one's best clothes. *–v. colloq.* defeat, outwit, outbid, etc. □ **at best** on the most optimistic view. **best bib and tucker** one's best clothes. **the best part of** most of. **do one's best** do all one can. **get the best of** defeat, outwit. **had best** would find it wisest to. **make the best of** derive what limited advantage one can from. [Old English]

bestial /'bestɪəl/ *adj.* **1** brutish, cruel. **2** of or like a beast. [Latin: related to BEAST]

bestiality /ˌbestɪ'ælɪtɪ/ *n.* **1** bestial behaviour. **2** sexual intercourse between a person and an animal.

bestiary /'bestɪərɪ/ *n.* (*pl.* **-ies**) medieval treatise on beasts. [medieval Latin: related to BEAST]

bestir /bɪ'stɜ:(r)/ *v.refl.* (**-rr-**) exert or rouse oneself.

best man *n.* bridegroom's chief attendant.

bestow /bɪ'stəʊ/ *v.* (foll. by *on, upon*) confer (a gift, right, etc.). □ **bestowal** *n.* [Old English *stow* a place]

bestrew /bɪ'stru:/ *v.* (*past part.* **bestrewed** or **bestrewn**) **1** (usu. foll. by *with*) strew (a surface). **2** lie scattered over.

bestride /bɪ'straɪd/ *v.* (**-ding**; *past* **-strode**; *past part.* **-stridden**) **1** sit astride on. **2** stand astride over.

best seller *n.* **1** book etc. sold in large numbers. **2** author of such a book.

bet *–v.* (**-tt-**; *past* and *past part.* **bet** or **betted**) **1** risk a sum of money on the result of a race, contest, etc. (*I don't bet on horses*). **2** risk (a sum) thus, or risk (a sum) against (a person) (*bet £10 on a horse*; *he bet me £10 I'd lose*). **3** *colloq.* feel sure. *–n.* **1** act of betting (*make a bet*). **2** money etc. staked (*put a bet on*). **3** *colloq.* opinion (*that's my bet*). **4** *colloq.* choice or possibility (*she's our best bet*). □ **you bet** *colloq.* you may be sure. [origin uncertain]

beta /'bi:tə/ *n.* **1** second letter of the Greek alphabet (*Β, β*). **2** second-class mark for a piece of work etc. [Latin from Greek]

beta-blocker *n.* drug preventing unwanted stimulation of the heart, used to treat angina and high blood pressure.

betake /bɪ'teɪk/ *v.refl.* (**-king**; *past* **betook**; *past part.* **betaken**) (foll. by *to*) go to (a place or person).

beta particle *n.* fast-moving electron emitted by the radioactive decay of substances.

betatron /'bi:təˌtrɒn/ *n.* apparatus for accelerating electrons in a circular path. [from BETA, ELECTRON]

betel /'bi:t(ə)l/ *n.* leaf chewed in the East with the betel-nut. [Portuguese from Malayalam]

betel-nut *n.* seed of a tropical palm.

bête noire /beɪt 'nwɑ:(r)/ *n.* (*pl.* ***bêtes noires*** pronunc. same) person or thing one hates or fears. [French, literally = 'black beast']

bethink /bɪ'θɪŋk/ *v.refl.* (*past* and *past part.* **bethought**) *formal* **1** reflect; stop to think. **2** be reminded by reflection. [Old English: related to BE-]

betide /bɪ'taɪd/ *v.* □ **woe betide a person** used as a warning (*woe betide us if we fail*). [from BE-, *tide* befall]

betimes /bɪ'taɪmz/ *adv. literary* early, in good time. [related to BY]

betoken /bɪ'təʊkən/ *v.* be a sign of; indicate. [Old English: related to BE-]

betony /'betənɪ/ *n.* (*pl.* **-ies**) purple-flowered plant. [French from Latin]

betook *past* of BETAKE.

betray /bɪ'treɪ/ *v.* **1** be disloyal or treacherous to (a friend, one's country, a person's trust, etc.). **2** reveal involuntarily or treacherously; be evidence of. **3** lead astray. □ **betrayal** *n.* [from BE-, obsolete *tray* from Latin *trado* hand over]

betroth /bɪ'trəʊð/ *v.* (usu. as **betrothed** *adj.*) engage to marry. □ **betrothal** *n.* [from BE-, TRUTH]

better *–adj.* (*compar.* of GOOD). **1** of a more excellent or desirable kind. **2** partly or fully recovered from illness (*feeling better*). *–adv.* (*compar.* of WELL[1]). **1** in a better manner. **2** to a greater degree. **3** more usefully (*is better forgotten*). *–n.* **1** better thing etc. (*the better of the two*). **2** (in *pl.*) one's superiors. *–v.* **1** improve on; surpass. **2** improve. **3** *refl.* improve one's position etc. □ **the better part of** most of. **get the better of** defeat, outwit. **had better** would find it prudent to. [Old English]

better half *n. colloq.* one's spouse.

betterment *n.* improvement.

better off *adj.* **1** in a better situation. **2** having more money.

betting *n.* **1** gambling by risking money on an unpredictable outcome. **2** odds offered on this. □ **what's the betting?** *colloq.* it is likely or to be expected (*what's the betting he'll be late?*).

betting-shop *n.* bookmaker's shop or office.

between /bɪ'twi:n/ *–prep.* **1 a** at a point in the area bounded by two or more other points in space, time, etc. (*between London and Dover*; *between now and Friday*). **b** along the extent of such an area (*no shops between here and the centre*; *numbers between 10 and 20*). **2** separating (*difference between right and wrong*). **3 a** shared by (*£5 between us*). **b** by joint action (*agreement between us*). **4** to and from (*runs between London and Oxford*). **5** taking one of (*choose between them*). *–adv.* (also **in between**) at a point or in the area bounded by two or more other points (*not fat or thin but in between*). □ **between ourselves** (or **you and me**) in confidence. [Old English: related to BY, TWO]

betwixt /bɪ'twɪkst/ *adv.* □ **betwixt and between** *colloq.* neither one thing nor the other. [Old English]

bevel /'bev(ə)l/ *–n.* **1** slope from the horizontal or vertical in carpentry etc.; sloping surface or edge. **2** tool for marking angles. *–v.* (**-ll-**; *US* **-l-**) **1** reduce (a square edge) to a sloping edge. **2** slope at an angle. [French]

bevel gear *n.* gear working another at an angle to it.

beverage /'bevərɪdʒ/ *n. formal* drink. [Latin *bibo* drink]

bevy /'bevɪ/ *n.* (*pl.* **-ies**) company (of quails, larks, women, etc.). [origin unknown]

bewail /bɪ'weɪl/ *v.* lament; wail over.

beware /bɪ'weə(r)/ *v.* (only in *imper.* or *infin.*; often foll. by *of*) be cautious (of) (*beware of the dog*; *beware the Ides of March*). [from BE, *ware* cautious]

bewilder /bɪ'wɪldə(r)/ *v.* perplex, confuse. □ **bewildering** *adj.* **bewilderment** *n.* [from BE-, obsolete *wilder* lose one's way]

bewitch /bɪ'wɪtʃ/ *v.* **1** enchant. **2** cast a spell on.

beyond /bɪ'jɒnd/ *–prep.* **1** at or to the further side of. **2** outside the scope or understanding of (*beyond repair*; *it is beyond me*). **3** more than. *–adv.* **1** at or to the further side. **2** further on. *–n.* (prec. by *the*) the unknown after death. [Old English: related to BY, YON]

bezel /'bez(ə)l/ *n.* **1** sloped edge of a chisel. **2** oblique faces of a cut gem. **3** groove holding a watch-glass or gem. [French]

bezique /bɪ'zi:k/ *n.* card-game for two. [French]

b.f. *abbr.* **1** *colloq.* bloody fool. **2** brought forward.

bhang /bæŋ/ *n.* Indian hemp used as a narcotic. [Portuguese from Sanskrit]

b.h.p. *abbr.* brake horsepower.

Bi *symb.* bismuth.

bi- *comb. form* forming nouns, adjectives, and verbs, meaning: **1** division into two (*biplane*; *bisect*). **2 a** occurring twice in every one or once in every two (*bi-weekly*). **b** lasting for two (*biennial*). **3** *Chem.* substance having a double proportion of what is indicated by the simple word (*bicarbonate*). **4** *Bot.* & *Zool.* having divided parts which are themselves similarly divided (*bipinnate*). [Latin]

biannual /baɪ'ænjʊəl/ *adj.* occurring etc. twice a year.

bias /'baɪəs/ *–n.* **1** (often foll. by *towards*, *against*) predisposition or prejudice. **2** *Statistics* distortion of a statistical result due to a neglected factor. **3** edge cut obliquely across the weave of a fabric. **4** *Sport* **a** irregular shape given to a bowl. **b** oblique course this causes it to run. *–v.* (**-s-** or **-ss-**) **1** (esp. as **biased** *adj.*) influence (usu. unfairly); prejudice. **2** give a bias to. □ **on the bias** obliquely, diagonally. [French]

bias binding *n.* strip of fabric cut obliquely and used to bind edges.

biathlon /baɪ'æθlən/ *n.* athletic contest in skiing and shooting or cycling and running. [from BI-, after PENTATHLON]

bib *n.* **1** piece of cloth etc. fastened round a child's neck while eating. **2** top front part of an apron, dungarees, etc. [origin uncertain]

bib-cock *n.* tap with a bent nozzle. [perhaps from BIB]

Bible /'baɪb(ə)l/ *n.* **1 a** (prec. by *the*) Christian scriptures of Old and New Testaments. **b** (**bible**) copy of these. **2** (**bible**) *colloq.* authoritative book. □ **biblical** /'bɪblɪk(ə)l/ *adj.* [Greek *biblia* books]

Bible-bashing *n.* (also **Bible-thumping**) *slang* aggressive fundamentalist preaching. □ **Bible-basher** *n.* (also **-thumper**).

bibliography /,bɪblɪ'ɒgrəfɪ/ *n.* (*pl.* **-ies**) **1** list of books on a specific subject, by a particular author, etc.; book containing this. **2** the study of books, their authorship, editions, etc. □ **bibliographer** *n.* **bibliographical** /-ə'græfɪk(ə)l/ *adj.* [Greek: related to BIBLE]

bibliophile /'bɪblɪə,faɪl/ *n.* lover or collector of books.

bibulous /'bɪbjʊləs/ *adj.* tending to drink alcohol. [Latin *bibo* drink]

bicameral /baɪ'kæmər(ə)l/ *adj.* (of a legislative body) having two chambers. [from BI-, Latin *camera* chamber]

bicarb /'baɪkɑ:b/ *n. colloq.* = BICARBONATE 2. [abbreviation]

bicarbonate /baɪ'kɑ:bənɪt/ *n.* **1** any acid salt of carbonic acid. **2** (in full **bicarbonate of soda**) sodium bicarbonate used as an antacid or in baking-powder.

bicentenary /ˌbaɪsen'tiːnərɪ/ *n.* (*pl.* **-ies**) **1** two-hundredth anniversary. **2** celebration of this.

bicentennial /ˌbaɪsen'tenɪəl/ esp. *US* *–n.* bicentenary. *–adj.* occurring every two hundred years.

biceps /'baɪseps/ *n.* (*pl.* same) muscle with two heads or attachments, esp. that bending the elbow. [Latin *caput* head]

bicker *v.* argue pettily. [origin unknown]

bicuspid /baɪ'kʌspɪd/ *–adj.* having two cusps. *–n.* the premolar tooth in humans. [from BI-, CUSP]

bicycle /'baɪsɪk(ə)l/ *–n.* pedal-driven two-wheeled vehicle. *–v.* (**-ling**) ride a bicycle. [Greek *kuklos* wheel]

bid *–v.* (**-dd-**; *past* **bid**, *archaic* **bade** /beɪd, bæd/; *past part.* **bid**, *archaic* **bidden**) **1** (*past* and *past part.* **bid**) **a** (esp. at an auction) make an offer (of) (*bid for the vase*; *bid £20*). **b** offer a service for a stated price. **2** *literary* command; invite (*bid the soldiers shoot*; *bade her start*). **3** *literary* utter (a greeting or farewell) to (*I bade him welcome*). **4** (*past* and *past part.* **bid**) *Cards* state before play how many tricks one intends to make. *–n.* **1** act of bidding. **2** amount bid. **3** *colloq.* attempt; effort (*bid for power*). □ **bidder** *n.* [Old English]

biddable *adj.* obedient.

bidding *n.* **1** command, request, or invitation. **2** bids at an auction or in a card-game.

biddy *n.* (*pl.* **-ies**) *slang* woman (esp. *old biddy*). [a form of the name *Bridget*]

bide *v.* (**-ding**) □ **bide one's time** wait for a good opportunity. [Old English]

bidet /'biːdeɪ/ *n.* low basin for sitting on to wash the genital area. [French, = pony]

biennial /baɪ'enɪəl/ *–adj.* lasting, or recurring every, two years. *–n.* plant that grows from seed one year and flowers and dies the following. [Latin *annus* year]

bier *n.* movable frame on which a coffin or corpse rests. [Old English]

biff *slang* *–n.* sharp blow. *–v.* strike (a person). [imitative]

bifid /'baɪfɪd/ *adj.* divided by a deep cleft into two parts. [Latin *findo* cleave]

bifocal /baɪ'fəʊk(ə)l/ *–adj.* having two focuses, esp. of a lens with a part for distant and a part for near vision. *–n.* (in *pl.*) bifocal spectacles.

bifurcate /'baɪfəˌkeɪt/ *–v.* (**-ting**) fork. *–adj.* forked; branched. □ **bifurcation** /-'keɪʃ(ə)n/ *n.* [Latin *furca* fork]

big *–adj.* (**bigger**, **biggest**) **1 a** of considerable size, amount, intensity, etc. **b** of a large or the largest size (*big toe*). **2** important (*my big day*). **3** adult, elder (*big sister*). **4** *colloq.* **a** boastful (*big words*). **b** often *iron.* generous (*big of him*). **c** ambitious (*big ideas*). **5** (usu. foll. by *with*) advanced in pregnancy (*big with child*). *–adv. colloq.* impressively or grandly (*think big*). □ **in a big way** *colloq.* with great enthusiasm, display, etc. □ **biggish** *adj.* [origin unknown]

bigamy /'bɪgəmɪ/ *n.* (*pl.* **-ies**) crime of marrying while still married to another person. □ **bigamist** *n.* **bigamous** *adj.* [Greek *gamos* marriage]

Big Apple *n. US slang* New York City.

big bang theory *n.* theory that the universe began with the explosion of dense matter.

Big Brother *n.* supposedly benevolent watchful dictator.

big end *n.* (in a vehicle) end of the connecting-rod, encircling the crankpin.

big-head *n. colloq.* conceited person. □ **big-headed** *adj.*

big-hearted *adj.* generous.

bight /baɪt/ *n.* **1** bay, inlet, etc. **2** loop of rope. [Old English]

big money *n.* large amounts of money.

big noise *n.* (also **big shot**) *colloq.* = BIGWIG.

bigot /'bɪgət/ *n.* obstinate believer who is intolerant of others. □ **bigoted** *adj.* **bigotry** *n.* [French]

big stick *n. colloq.* display of force.

big time *n.* (prec. by *the*) *slang* success, esp. in show business. □ **big-timer** *n.*

big top *n.* main tent in a circus.

big wheel *n.* Ferris wheel.

bigwig *n. colloq.* important person.

bijou /'biːʒuː/ *–n.* (*pl.* **-x** pronunc. same) jewel; trinket. *–attrib. adj.* (**bijou**) small and elegant. [French]

bike *colloq.* *–n.* bicycle or motor cycle. *–v.* (**-king**) ride a bike. □ **biker** *n.* [abbreviation]

bikini /bɪ'kiːnɪ/ *n.* (*pl.* **-s**) two-piece swimsuit for women. [*Bikini*, Pacific atoll]

bilateral /baɪ'lætər(ə)l/ *adj.* **1** of, on, or with two sides. **2** affecting or between two parties, countries, etc. □ **bilaterally** *adv.*

bilberry /'bɪlbərɪ/ *n.* (*pl.* **-ies**) **1** hardy N. European shrub of heaths and mountains. **2** its small dark-blue edible berry. [Scandinavian]

bile *n.* **1** bitter digestive fluid secreted by the liver. **2** bad temper; peevish anger. [Latin *bilis*]

bilge /bɪldʒ/ *n.* **1 a** the almost flat part of a ship's bottom. **b** (in full **bilge-water**) filthy water that collects there. **2** *slang* nonsense. [probably var. of BULGE]

bilharzia /bɪl'hɑ:tsɪə/ *n.* chronic tropical disease caused by a parasitic flatworm. [*Bilharz*, name of a physician]

biliary /'bɪlɪərɪ/ *adj.* of the bile. [French: related to BILE]

bilingual /baɪ'lɪŋgw(ə)l/ *–adj.* **1** able to speak two languages. **2** spoken or written in two languages. *–n.* bilingual person. □ **bilingualism** *n.* [Latin *lingua* tongue]

bilious /'bɪlɪəs/ *adj.* **1** affected by a disorder of the bile. **2** bad-tempered. [Latin: related to BILE]

bilk *v. slang* **1** cheat. **2** elude. **3** avoid paying (a creditor or debt). [origin uncertain]

Bill *n.* = OLD BILL. [diminutive of *William*]

bill[1] *–n.* **1** statement of charges for goods or services. **2** draft of a proposed law. **3** poster, placard. **4** programme of entertainment. **5** *US* banknote. *–v.* **1** send a statement of charges to. **2** put in the programme; announce. **3** (foll. by *as*) advertise as. [medieval Latin *bulla* seal]

bill[2] *–n.* **1** bird's beak. **2** narrow promontory. *–v.* (of doves etc.) stroke bills. □ **bill and coo** exchange caresses. [Old English]

bill[3] *n.* **1** *hist.* weapon with a hooked blade. **2** = BILLHOOK. [Old English]

billabong /'bɪlə,bɒŋ/ *n.* (in Australia) backwater of a river. [Aboriginal]

billboard *n.* large outdoor advertising hoarding.

billet[1] /'bɪlɪt/ *–n.* **1 a** place where troops etc. are lodged. **b** order to provide this. **2** *colloq.* job. *–v.* (**-t-**) (usu. foll. by *on, in, at*) quarter (soldiers etc.). [Anglo-French diminutive of BILL[1]]

billet[2] /'bɪlɪt/ *n.* **1** thick piece of firewood. **2** small metal bar. [French diminutive of *bille* tree-trunk]

billet-doux /,bɪlɪ'du:/ *n.* (*pl.* **billets-doux** /-'du:z/) often *joc.* love-letter. [French, = sweet note]

billhook *n.* pruning tool with a hooked blade.

billiards /'bɪljədz/ *n.* **1** game played on a table, with three balls struck with cues. **2** (**billiard**) (in *comb.*) used in billiards (*billiard-ball*). [French: related to BILLET[2]]

billion /'bɪljən/ *adj.* & *n.* (*pl.* same or (in sense 3) **-s**) **1** a thousand million (10^9). **2** (now less often) a million million (10^{12}). **3** (in *pl.*) *colloq.* a very large number (*billions of years*). □ **billionth** *adj.* & *n.* [French]

billionaire /,bɪljə'neə(r)/ *n.* person who has over a billion pounds, dollars, etc. [after MILLIONAIRE]

bill of exchange *n.* written order to pay a sum of money on a given date to the drawer or to a named payee.

bill of fare *n.* menu.

bill of lading *n.* detailed list of a ship's cargo.

billow /'bɪləʊ/ *–n.* **1** wave. **2** any large mass. *–v.* rise, fill, or surge in billows. □ **billowy** *adj.* [Old Norse]

billposter *n.* (also **billsticker**) person who pastes up advertisements on hoardings.

billy[1] /'bɪlɪ/ *n.* (*pl.* **-ies**) (in full **billycan**) *Austral.* tin or enamel outdoor cooking-pot. [perhaps from Aboriginal *billa* water]

billy[2] /'bɪlɪ/ *n.* (*pl.* **-ies**) (in full **billy-goat**) male goat. [from the name *Billy*]

bimbo /'bɪmbəʊ/ *n.* (*pl.* **-s** or **-es**) *slang* usu. *derog.* attractive but unintelligent young woman. [Italian, = little child]

bimetallic /,baɪmɪ'tælɪk/ *adj.* using or made of two metals. [French]

bin *n.* large receptacle for rubbish or storage. [Old English]

binary /'baɪnərɪ/ *–adj.* **1** of two parts, dual. **2** of the binary system. *–n.* (*pl.* **-ies**) **1** something having two parts. **2** binary number. [Latin *bini* two together]

binary star *n.* system of two stars orbiting each other.

binary system *n.* system using the digits 0 and 1 to code information, esp. in computing.

binaural /baɪ'nɔ:r(ə)l/ *adj.* **1** of or used with both ears. **2** (of sound) recorded using two microphones and usu. transmitted separately to the two ears. [from BI-, AURAL]

bind /baɪnd/ *–v.* (*past* and *past part.* **bound**) **1** tie or fasten tightly. **2** restrain forcibly. **3** (cause to) cohere. **4** compel; impose a duty on. **5 a** edge with braid etc. **b** fasten (the pages of a book) in a cover. **6** constipate. **7** ratify (a bargain, agreement, etc.). **8** (often foll. by *up*) bandage. *–n. colloq.* nuisance; restriction. □ **bind over** *Law* order (a person) to do something, esp. keep the peace. [Old English]

binder *n.* **1** cover for loose papers etc. **2** substance that binds things together. **3** *hist.* reaping-machine that binds grain into sheaves. **4** bookbinder.

bindery *n.* (*pl.* **-ies**) bookbinder's workshop.

binding *–n.* thing that binds, esp. the covers, glue, etc., of a book. *–adj.* obligatory.

bindweed *n.* **1** convolvulus. **2** honeysuckle or other climber.

bine *n.* **1** twisting stem of a climbing plant, esp. the hop. **2** flexible shoot. [dial. form of BIND]

bin end *n.* one of the last bottles from a bin of wine, usu. sold at a reduced price.

binge /bɪndʒ/ *slang* —*n.* bout of excessive eating, drinking, etc.; spree. —*v.* (**-ging**) indulge in a binge. [probably dial., = soak]

bingo /ˈbɪŋgəʊ/ *n.* gambling game in which each player has a card with numbers to be marked off as they are called. [origin uncertain]

bin-liner *n.* bag for lining a rubbish bin.

binman *n. colloq.* dustman.

binnacle /ˈbɪnək(ə)l/ *n.* case for a ship's compass. [Latin *habitaculum* dwelling]

binocular /baɪˈnɒkjʊlə(r)/ *adj.* for both eyes. [Latin *bini* two together, *oculus* eye]

binoculars /bɪˈnɒkjʊləz/ *n.pl.* instrument with a lens for each eye, for viewing distant objects.

binomial /baɪˈnəʊmɪəl/ —*n.* algebraic expression of the sum or the difference of two terms. —*adj.* of two terms. [Greek *nomos* part]

binomial theorem *n.* formula for finding any power of a binomial.

bint *n. slang*, usu. *offens.* girl or woman. [Arabic]

bio- *comb. form* **1** life (*biography*). **2** biological; of living things. [Greek *bios* life]

biochemistry /ˌbaɪəʊˈkemɪstrɪ/ *n.* the study of the chemistry of living organisms. □ **biochemical** *adj.* **biochemist** *n.*

biodegradable /ˌbaɪəʊdɪˈgreɪdəb(ə)l/ *adj.* capable of being decomposed by bacteria or other living organisms.

bioengineering /ˌbaɪəʊˌendʒɪˈnɪərɪŋ/ *n.* **1** the application of engineering techniques to biological processes. **2** the use of artificial tissues, organs, etc. to replace parts of the body, e.g. artificial limbs, pacemakers, etc.

biogenesis /ˌbaɪəʊˈdʒenɪsɪs/ *n.* **1** hypothesis that a living organism arises only from a similar living organism. **2** synthesis of substances by living organisms.

biography /baɪˈɒgrəfɪ/ *n.* (*pl.* **-ies**) **1** account of a person's life, written usu. by another. **2** these as a literary genre. □ **biographer** *n.* **biographical** /ˌbaɪəˈgræfɪk(ə)l/ *adj.* [French: related to BIO-]

biological /ˌbaɪəˈlɒdʒɪk(ə)l/ *adj.* of biology or living organisms. □ **biologically** *adv.*

biological clock *n.* innate mechanism controlling an organism's rhythmic physiological activities.

biological warfare *n.* use of toxins or micro-organisms against an enemy.

biology /baɪˈɒlədʒɪ/ *n.* the study of living organisms. □ **biologist** *n.* [German: related to BIO-]

bionic /baɪˈɒnɪk/ *adj.* having electronically operated body parts or the resulting superhuman powers. [from BIO- after ELECTRONIC]

bionics *n.pl.* (treated as *sing.*) the study of mechanical systems that function like living organisms.

biophysics /ˌbaɪəʊˈfɪzɪks/ *n.pl.* (treated as *sing.*) science of the application of the laws of physics to biological phenomena. □ **biophysical** *adj.* **biophysicist** *n.*

biopsy /ˈbaɪɒpsɪ/ *n.* (*pl.* **-ies**) examination of severed tissue for diagnosis. [Greek *bios* life, *opsis* sight]

biorhythm /ˈbaɪəʊˌrɪð(ə)m/ *n.* any recurring biological cycle thought to affect one's physical or mental state.

biosphere /ˈbaɪəʊˌsfɪə(r)/ *n.* regions of the earth's crust and atmosphere occupied by living things. [German: related to BIO-]

biosynthesis /ˌbaɪəʊˈsɪnθɪsɪs/ *n.* production of organic molecules by living organisms. □ **biosynthetic** /-ˈθetɪk/ *adj.*

biotechnology /ˌbaɪəʊtekˈnɒlədʒɪ/ *n.* branch of technology exploiting biological processes, esp. using micro-organisms, in industry, medicine, etc.

biotin /ˈbaɪətɪn/ *n.* vitamin of the B complex, found in egg-yolk, liver, and yeast. [Greek *bios* life]

bipartisan /ˌbaɪpɑːtɪˈzæn/ *adj.* of or involving two parties.

bipartite /baɪˈpɑːtaɪt/ *adj.* **1** of two parts. **2** shared by or involving two parties. [Latin *bipartio* divide in two]

biped /ˈbaɪped/ —*n.* two-footed animal. —*adj.* two-footed. □ **bipedal** /-ˈpiːd(ə)l/ *adj.* [Latin *bipes -edis*]

biplane /ˈbaɪpleɪn/ *n.* aeroplane with two sets of wings, one above the other.

bipolar /baɪˈpəʊlə(r)/ *adj.* having two poles or extremities.

birch —*n.* **1** tree with pale hard wood and thin peeling bark, bearing catkins. **2** bundle of birch twigs used for flogging. —*v.* beat with a birch. [Old English]

bird *n.* **1** two-legged feathered winged vertebrate, egg-laying and usu. able to fly. **2** *slang* young woman. **3** *slang* person. **4** *slang* prison; prison sentence. □ **a bird in the hand** something secured or certain. **the birds and the bees** *euphem.* sexual activity and reproduction. **birds of a feather** similar people. **for the birds** *colloq.* trivial, uninteresting. **get the bird** *slang* be rejected, esp. by an audience. [Old English]

bird-bath *n.* basin with water for birds to bathe in.

birdbrain *n. colloq.* stupid or flighty person. □ **birdbrained** *adj.*

birdcage *n.* cage for birds.

birdie *n.* **1** *colloq.* little bird. **2** *Golf* hole played in one under par.

birdlime *n.* sticky substance spread to trap birds.

bird-nesting *n.* hunting for birds' eggs.

bird of paradise *n.* bird, the male of which has brilliant plumage.

bird of passage *n.* **1** migrant. **2** habitual traveller.

bird of prey *n.* bird which hunts animals for food.

birdseed *n.* blend of seeds for caged birds.

bird's-eye view *n.* detached view from above.

birdsong *n.* musical cry of birds.

bird table *n.* platform on which food for birds is placed.

bird-watcher *n.* person who observes wild birds as a hobby. □ **bird-watching** *n.*

biretta /bɪ'retə/ *n.* square usu. black cap worn by Roman Catholic priests. [Latin *birrus* cape]

Biro /'baɪərəʊ/ *n.* (*pl.* **-s**) *propr.* a kind of ball-point pen. [*Biró*, name of its inventor]

birth *n.* **1** emergence of a baby or young from its mother's body. **2** beginning (*birth of civilization*). **3 a** ancestry (*of noble birth*). **b** high or noble birth; inherited position. □ **give birth to 1** produce (young). **2** be the cause of. [Old Norse]

birth certificate *n.* official document detailing a person's birth.

birth control *n.* contraception.

birthday *n.* **1** day on which one was born. **2** anniversary of this.

birthing pool *n.* large bath for giving birth in.

birthmark *n.* unusual coloured mark on one's body at or from birth.

birthplace *n.* place where one was born.

birth rate *n.* number of live births per thousand of population per year.

birthright *n.* inherited, esp. property, rights.

birthstone *n.* gem popularly associated with the month of one's birth.

biscuit /'bɪskɪt/ *n.* **1** flat thin unleavened cake, usu. crisp and sweet. **2** fired unglazed pottery. **3** light brown colour. [Latin *bis* twice, *coquo* cook]

bisect /baɪ'sekt/ *v.* divide into two (strictly, equal) parts. □ **bisection** *n.* **bisector** *n.* [from BI-, Latin *seco sect-* cut]

bisexual /baɪ'sekʃʊəl/ *–adj.* **1** feeling or involving sexual attraction to people of both sexes. **2** hermaphrodite. *–n.* bisexual person. □ **bisexuality** /-'ælɪtɪ/ *n.*

bishop /'bɪʃəp/ *n.* **1** senior clergyman in charge of a diocese. **2** mitre-shaped chess piece. [Greek *episkopos* overseer]

bishopric /'bɪʃəprɪk/ *n.* office or diocese of a bishop.

bismuth /'bɪzməθ/ *n.* **1** reddish-white metallic element used in alloys etc. **2** compound of it used medicinally. [German]

bison /'baɪs(ə)n/ *n.* (*pl.* same) wild hump-backed ox of Europe or N. America. [Latin from Germanic]

bisque[1] /bɪsk/ *n.* rich soup, esp. of lobster. [French]

bisque[2] /bɪsk/ *n.* advantage of one free point or stroke in certain games. [French]

bisque[3] /bɪsk/ *n.* = BISCUIT 2.

bistre /'bɪstə(r)/ *n.* (*US* **bister**) brownish pigment from wood soot. [French]

bistro /'bi:strəʊ/ *n.* (*pl.* **-s**) small informal restaurant. [French]

bit[1] *n.* **1** small piece or quantity. **2** (prec. by *a*) fair amount (*sold quite a bit*). **3** often *colloq.* short or small time, distance, or amount (*wait a bit*; *move up a bit*; *a bit tired*; *a bit of an idiot*). □ **bit by bit** gradually. **do one's bit** *colloq.* make a useful contribution. [Old English]

bit[2] *past* of BITE.

bit[3] *n.* **1** metal mouthpiece of a bridle. **2** tool or piece for boring or drilling. **3** cutting or gripping part of a plane, pincers, etc. [Old English]

bit[4] *n. Computing* unit of information expressed as a choice between two possibilities. [*binary digit*]

bitch *–n.* **1** female dog or other canine animal. **2** *slang offens.* spiteful woman. **3** *slang* unpleasant or difficult thing. *–v.* **1** speak scathingly or spitefully. **2** complain. [Old English]

bitchy *adj.* (**-ier**, **-iest**) *slang* spiteful. □ **bitchily** *adv.* **bitchiness** *n.*

bite *–v.* (**-ting**; *past* **bit**; *past part.* **bitten**) **1** cut or puncture with the teeth. **2** (foll. by *off*, *away*, etc.) detach thus. **3** (of an insect etc.) sting. **4** (of a wheel etc.) grip, penetrate. **5** accept bait or an inducement. **6** be harsh in effect, esp. intentionally. **7** (in *passive*) **a** swindle. **b** (foll. by *by*, *with*, etc.) be infected by (enthusiasm etc.). **8** *colloq.* worry, perturb. **9** cause smarting pain (*biting wind*). **10** be sharp or effective (*biting wit*). **11** (foll. by *at*) snap at. *–n.* **1** act of biting. **2** wound etc. made by biting. **3 a** mouthful of food. **b** snack. **4** taking of bait by a fish. **5** pungency (esp. of flavour). **6** incisiveness, sharpness. **7** position of the teeth when the jaws are closed. □ **bite the dust** *slang* die. **bite a person's head off** *colloq.* respond angrily. **bite one's lip** repress emotion etc. [Old English]

bit on the side *n. slang* sexual relationship involving infidelity.

bit part *n.* minor role.

bitter *–adj.* **1** having a sharp pungent taste; not sweet. **2** causing, showing, or feeling mental pain or resentment (*bitter memories*). **3 a** harsh; virulent (*bitter animosity*). **b** piercingly cold. *–n.* **1** beer flavoured with hops and tasting slightly bitter. **2** (in *pl.*) liquor flavoured esp. with wormwood, used in cocktails. □ **to the bitter end** to the very end in spite of difficulties. □ **bitterly** *adv.* **bitterness** *n.* [Old English]

bittern /ˈbɪt(ə)n/ *n.* wading bird of the heron family. [French *butor* from Latin *butio*]

bitter-sweet *–adj.* sweet with a bitter aftertaste. *–n.* **1** such sweetness. **2** = WOODY NIGHTSHADE.

bitty *adj.* (**-ier**, **-iest**) made up of bits; scrappy.

bitumen /ˈbɪtjʊmɪn/ *n.* tarlike mixture of hydrocarbons derived from petroleum. [Latin]

bituminous /bɪˈtjuːmɪnəs/ *adj.* of or like bitumen.

bituminous coal *n.* coal burning with a smoky flame.

bivalve /ˈbaɪvælv/ *–n.* aquatic mollusc with a hinged double shell, e.g. the oyster and mussel. *–adj.* with such a shell.

bivouac /ˈbɪvʊˌæk/ *–n.* temporary open encampment without tents. *–v.* (**-ck-**) make, or camp in, a bivouac. [French, probably from German]

biz *n. colloq.* business. [abbreviation]

bizarre /bɪˈzɑː(r)/ *adj.* strange; eccentric; grotesque. [French]

Bk *symb.* berkelium.

BL *abbr.* British Library.

blab *v.* (**-bb-**) **1** talk foolishly or indiscreetly. **2** reveal (a secret etc.); confess. [imitative]

blabber *–n.* (also **blabbermouth**) person who blabs. *–v.* (often foll. by *on*) talk foolishly or inconsequentially.

black *–adj.* **1** reflecting no light, colourless from lack of light (like coal or soot); completely dark. **2** (**Black**) of the human group with dark-coloured skin, esp. African. **3** (of the sky etc.) heavily overcast. **4** angry; gloomy (*black look, mood*). **5** implying disgrace etc. (*in his black books*). **6** wicked, sinister, deadly. **7** portending trouble (*things look black*). **8** comic but sinister (*black comedy*). **9** (of tea or coffee) without milk. **10** (of industrial labour or its products) boycotted, esp. by a trade union, in a strike etc. *–n.* **1** black colour or pigment. **2** black clothes or material (*dressed in black*). **3 a** (in a game) black piece, ball, etc. **b** player of this. **4** credit side of an account (*in the black*). **5** (**Black**) member of a dark-skinned race, esp. an African. *–v.* **1** make black (*blacked his boots*). **2** declare (goods etc.) 'black'. □ **black out 1** effect a blackout on. **2** undergo a blackout. [Old English]

black and blue *adj.* bruised.

black and white *–n.* writing or printing (*in black and white*). *–adj.* **1** (of a film etc.) monochrome. **2** consisting of extremes only, oversimplified.

black art *n.* = BLACK MAGIC.

blackball *v.* reject (a candidate) in a ballot.

black beetle *n.* the common cockroach.

black belt *n.* **1** highest grade of proficiency in judo, karate, etc. **2** holder of this grade, entitled to wear a black belt.

blackberry *n.* (*pl.* **-ies**) black fleshy edible fruit of the bramble.

blackbird *n.* common thrush of which the male is black with an orange beak.

blackboard *n.* board with a smooth dark surface for writing on with chalk.

black box *n.* flight-recorder.

blackcap *n.* small warbler, the male of which has a black-topped head.

Black Country *n.* (prec. by *the*) industrial area of the Midlands.

blackcurrant *n.* **1** cultivated flowering shrub. **2** its small dark edible berry.

Black Death *n.* (prec. by *the*) 14th-c. plague in Europe.

black economy *n.* unofficial and untaxed trade.

blacken *v.* **1** make or become black or dark. **2** defame, slander.

black eye *n.* bruised skin around the eye.

black flag *n.* flag of piracy.

blackfly *n.* **1** dark coloured thrips or aphid. **2** these collectively.

Black Forest gateau *n.* chocolate sponge with cherries and whipped cream.

Black Friar *n.* Dominican friar.

blackguard /ˈblægɑːd/ *n.* villain, scoundrel. □ **blackguardly** *adj.* [originally = menial]

blackhead *n.* black-topped pimple on the skin.

black hole *n.* region of space from which matter and radiation cannot escape.
black ice *n.* thin hard transparent ice on a road etc.
blacking *n.* black polish, esp. for shoes.
blackjack *n.* = PONTOON[1].
blacklead *n.* graphite.
blackleg –*n. derog.* person refusing to join a strike etc. –*v.* (**-gg-**) act as a blackleg.
blacklist –*n.* list of people in disfavour etc. –*v.* put on a blacklist.
black magic *n.* magic supposed to invoke evil spirits.
blackmail –*n.* **1 a** extortion of payment in return for silence. **b** payment so extorted. **2** use of threats or moral pressure. –*v.* **1** (try to) extort money etc. from by blackmail. **2** threaten, coerce. □ **blackmailer** *n.* [obsolete *mail* rent]
Black Maria /mə'raɪə/ *n. slang* police van.
black mark *n.* mark of discredit.
black market *n.* illicit trade in rationed, prohibited, or scarce commodities. □ **black marketeer** *n.*
Black Mass *n.* travesty of the Mass, in worship of Satan.
blackout *n.* **1** temporary loss of consciousness or memory. **2** loss of electric power, radio reception, etc. **3** compulsory darkness as a precaution against air raids. **4** temporary suppression of news. **5** sudden darkening of a theatre stage.
black pepper *n.* pepper made by grinding the whole dried pepper berry including the outer husk.
Black Power *n.* movement for Black rights and political power.
black pudding *n.* sausage of pork, dried pig's blood, suet, etc.
Black Rod *n.* principal usher of the House of Lords etc.
black sheep *n. colloq.* member of a family, group, etc. regarded as a disgrace or failure.
blackshirt *n. hist.* member of a Fascist organization.
blacksmith *n.* smith who works in iron.
black spot *n.* **1** place of danger or trouble. **2** plant disease producing black spots.
black tea *n.* tea that is fully fermented before drying.
blackthorn *n.* thorny shrub bearing white blossom and sloes.
black tie *n.* **1** black bow-tie worn with a dinner jacket. **2** *colloq.* man's formal evening dress.
black velvet *n.* mixture of stout and champagne.
Black Watch *n.* (prec. by *the*) Royal Highland Regiment.
black widow *n.* venomous spider of which the female devours the male.
bladder *n.* **1 a** sac in some animals, esp. that holding urine. **b** this adapted for various uses. **2** inflated blister in seaweed etc. [Old English]
bladderwrack *n.* brown seaweed with air bladders.
blade *n.* **1** cutting part of a knife etc. **2** flattened part of an oar, propeller, etc. **3 a** flat narrow leaf of grass etc. **b** broad thin part of a leaf. **4** flat bone, e.g. in the shoulder. [Old English]
blame –*v.* (**-ming**) **1** assign fault or responsibility to. **2** (foll. by *on*) fix responsibility for (an error etc.) on (*blamed it on his brother*). –*n.* **1** responsibility for an error etc. **2** blaming or attributing of responsibility (*got all the blame*). □ **be to blame** be responsible; deserve censure. □ **blameable** *adj.* **blameless** *adj.* **blameworthy** *adj.* [French: related to BLASPHEME]
blanch /blɑːntʃ/ *v.* **1** make or become white or pale. **2 a** peel (almonds etc.) by scalding. **b** immerse (vegetables etc.) briefly in boiling water. **3** whiten (a plant) by depriving it of light. [French: related to BLANK]
blancmange /blə'mɒndʒ/ *n.* sweet opaque jelly of flavoured cornflour and milk. [French, = white food]
bland *adj.* **1 a** mild, not irritating. **b** tasteless; insipid. **2** gentle in manner; suave. □ **blandly** *adv.* **blandness** *n.* [Latin *blandus* smooth]
blandish /'blændɪʃ/ *v.* flatter; coax. □ **blandishment** *n.* (usu. in *pl.*). [Latin: related to BLAND]
blank –*adj.* **1 a** (of paper) not written or printed on. **b** (of a document) with spaces left for a signature or details. **2 a** empty (*blank space*). **b** unrelieved (*blank wall*). **3 a** without interest, result, or expression (*blank face*). **b** having (temporarily) no knowledge etc. (*mind went blank*). **4** complete (*a blank refusal*; *blank despair*). –*n.* **1 a** unfilled space, esp. in a document. **b** document having blank spaces. **2** (in full **blank cartridge**) cartridge containing gunpowder but no bullet. **3** dash written instead of a word or letter. –*v.* (usu. foll. by *off*, *out*) screen, obscure. □ **draw a blank** get no response; fail. □ **blankly** *adv.* **blankness** *n.* [French *blanc* white, from Germanic]
blank cheque *n.* **1** cheque left for the payee to fill in. **2** *colloq.* unlimited freedom of action.
blanket /'blæŋkɪt/ –*n.* **1** large esp. woollen sheet used as a bed-covering etc. **2**

thick covering mass or layer. *–attrib. adj.* covering everything; inclusive. *–v.* (-t-) 1 cover. 2 stifle, suppress. [French: related to BLANK]

blanket bath *n.* body wash given to a bedridden patient.

blanket stitch *n.* stitch used to finish the edges of a blanket etc.

blank verse *n.* unrhymed verse, esp. iambic pentameters.

blare *–v.* **(-ring)** 1 sound or utter loudly. 2 make the sound of a trumpet. *–n.* blaring sound. [Low German or Dutch, imitative]

blarney /'blɑ:nɪ/ *–n.* cajoling talk; flattery. *–v.* **(-eys, -eyed)** flatter, cajole. [*Blarney*, castle near Cork]

blasé /'blɑ:zeɪ/ *adj.* bored or indifferent through over-familiarity. [French]

blaspheme /blæs'fi:m/ *v.* **(-ming)** 1 use religious names irreverently; treat a religious or sacred subject irreverently. 2 talk irreverently about; use blasphemy against. [Greek *blasphēmeō*]

blasphemy /'blæsfəmɪ/ *n.* (*pl.* **-ies**) 1 irreverent talk or treatment of a religious or sacred thing. 2 instance of this. □ **blasphemous** *adj.*

blast /blɑ:st/ *–n.* 1 strong gust of air. 2 a explosion. b destructive wave of air from this. 3 loud note from a wind instrument, car horn, etc. 4 *colloq.* severe reprimand. *–v.* 1 blow up with explosives. 2 wither, blight (*blasted oak*; *blasted her hopes*). 3 (cause to) make a loud noise. *–int.* expressing annoyance. □ **at full blast** *colloq.* at maximum volume, speed, etc. **blast off** take off from a launching site. [Old English]

blasted *colloq. –attrib. adj.* damned; annoying. *–adv.* damned; extremely.

blast-furnace *n.* smelting furnace into which hot air is driven.

blast-off *n.* launching of a rocket etc.

blatant /'bleɪt(ə)nt/ *adj.* 1 flagrant, unashamed. 2 loudly obtrusive. □ **blatantly** *adv.* [coined by Spenser]

blather /'blæðə(r)/ *–n.* (also **blether**) foolish talk. *–v.* talk foolishly. [Old Norse]

blaze[1] *–n.* 1 bright flame or fire. 2 violent outburst (of passion etc.). 3 brilliant display (*blaze of scarlet, of glory*). *–v.* **(-zing)** 1 burn or shine brightly or fiercely. 2 be consumed with anger, excitement, etc. □ **blaze away** (often foll. by *at*) 1 shoot continuously. 2 work vigorously. [Old English, = torch]

blaze[2] *–n.* 1 white mark on an animal's face. 2 mark cut on a tree, esp. to show a route. *–v.* **(-zing)** mark (a tree or a path) with blazes. □ **blaze a trail** show the way for others. [origin uncertain]

blazer *n.* jacket without matching trousers, esp. lightweight and often part of a uniform. [from BLAZE[1]]

blazon /'bleɪz(ə)n/ *–v.* 1 proclaim (esp. *blazon abroad*). 2 *Heraldry* describe or paint (arms). *–n. Heraldry* shield or coat of arms. □ **blazonment** *n.* **blazonry** *n.* [French, originally = shield]

bleach *–v.* whiten in sunlight or by a chemical process. *–n.* bleaching substance or process. [Old English]

bleak *adj.* 1 exposed, windswept. 2 dreary, grim. [Old Norse]

bleary /'blɪərɪ/ *adj.* **(-ier, -iest)** 1 dim; blurred. 2 indistinct. [Low German]

bleary-eyed *adj.* having dim sight.

bleat *–v.* 1 (of a sheep, goat, or calf) make a wavering cry. 2 (often foll. by *out*) speak or say plaintively. *–n.* bleating cry. [Old English]

bleed *–v.* (*past* and *past part.* **bled**) 1 emit blood. 2 draw blood from surgically. 3 *colloq.* extort money from. 4 (often foll. by *for*) suffer wounds or violent death. 5 a emit sap. b (of dye) come out in water. 6 empty (a system) of excess air or fluid. *–n.* act of bleeding. □ **one's heart bleeds** usu. *iron.* one is very sorrowful. [Old English]

bleeder *n. coarse slang* unpleasant or contemptible person.

bleeding *adj.* & *adv. coarse slang* expressing annoyance or antipathy.

bleep *–n.* intermittent high-pitched electronic sound. *–v.* 1 make a bleep. 2 summon with a bleeper. [imitative]

bleeper *n.* small electronic device bleeping to contact the carrier.

blemish /'blemɪʃ/ *–n.* flaw, defect, or stain. *–v.* spoil, mark, or stain. [French]

blench *v.* flinch, quail. [Old English]

blend *–v.* 1 mix together as required. 2 become one. 3 (often foll. by *with*, *in*) mingle; mix thoroughly. 4 (esp. of colours) merge imperceptibly; harmonize. *–n.* mixture. [Old Norse]

blender *n.* machine for liquidizing, chopping, or puréeing food.

blenny /'blenɪ/ *n.* (*pl.* **-ies**) small spiny-finned scaleless marine fish. [Greek *blennos* mucus]

bless *v.* (*past* and *past part.* **blessed**, *poet.* **blest**) 1 ask God to look favourably on, esp. by making the sign of the cross over. 2 consecrate (food etc.). 3 glorify (God). 4 attribute one's good luck to (stars etc.); thank. 5 (usu. in *passive*) make happy or successful (*blessed with children*). □ **bless me** (or **my soul**) exclamation of surprise etc. **bless you!** exclamation of endearment, gratitude, etc., or to a person who has just sneezed. [Old English]

blessed /'blesɪd, blest/ *adj.* (also *poet.* **blest**) **1** holy. **2** *euphem.* cursed (*blessed nuisance!*). **3** *RC Ch.* beatified. □ **blessedness** *n.*

blessing *n.* **1** invocation of (esp. divine) favour. **2** grace said at a meal. **3** benefit.

blether var. of BLATHER.

blew *past* of BLOW[1].

blight /blaɪt/ *–n.* **1** plant disease caused by insects etc. **2** such an insect etc. **3** harmful or destructive force. **4** ugly urban area. *–v.* **1** affect with blight. **2** harm, destroy. **3** spoil. [origin unknown]

blighter *n. colloq.* contemptible or annoying person.

Blighty /'blaɪtɪ/ *n. Mil. slang* England; home. [Hindustani, = foreign]

blimey /'blaɪmɪ/ *int. coarse slang* expression of surprise, contempt, etc. [(*God*) *blind me!*]

blimp *n.* **1** (also **(Colonel) Blimp**) reactionary person. **2** small non-rigid airship. **3** soundproof cover for a cine-camera. [origin uncertain]

blind /blaɪnd/ *–adj.* **1** lacking the power of sight. **2 a** without adequate foresight, discernment, or information (*blind effort*). **b** (often foll. by *to*) unwilling or unable to appreciate a factor etc. (*blind to argument*). **3** not governed by purpose or reason (*blind forces*). **4** reckless (*blind hitting*). **5 a** concealed (*blind ditch*). **b** closed at one end. **6** (of flying) using instruments only. **7** *Cookery* (of a flan case etc.) baked without a filling. *–v.* **1** deprive of sight. **2** rob of judgement; deceive; overawe. **3** *slang* go recklessly. *–n.* **1** screen for a window; awning. **2** thing used to hide the truth. **3** obstruction to sight or light. *–adv.* blindly. □ **blindly** *adv.* **blindness** *n.* [Old English]

blind alley *n.* **1** alley closed at one end. **2** futile course.

blind date *n. colloq.* date between two people who have not previously met.

blind drunk *adj. colloq.* extremely drunk.

blindfold *–v.* cover the eyes of (a person) with a tied cloth etc. *–n.* cloth etc. so used. *–adj.* & *adv.* **1** with eyes covered. **2** without due care. [originally *blindfelled* = struck blind]

blind man's buff *n.* game in which a blindfold player tries to catch others.

blind spot *n.* **1** point on the retina insensitive to light. **2** area where vision or understanding is lacking.

blindworm *n.* = SLOW-WORM.

blink *–v.* **1** shut and open the eyes quickly. **2** (often foll. by *back*) prevent (tears) by blinking. **3** shine unsteadily, flicker. *–n.* **1** act of blinking. **2** momentary gleam or glimpse. □ **blink at 1** look at while blinking. **2** ignore, shirk. **on the blink** *slang* not working properly; out of order. [Dutch, var. of BLENCH]

blinker *–n.* **1** (usu. in *pl.*) each of two screens on a bridle preventing lateral vision. **2** device that blinks. *–v.* **1** obscure with blinkers. **2** (as **blinkered** *adj.*) having narrow and prejudiced views.

blinking *adj.* & *adv. slang* expressing annoyance etc. (*it's blinking stupid*).

blip *–n.* **1** minor deviation or error. **2** quick popping sound. **3** small image on a radar screen. *–v.* (**-pp-**) make a blip. [imitative]

bliss *n.* **1** perfect joy. **2** being in heaven. □ **blissful** *adj.* **blissfully** *adv.* [Old English]

blister *–n.* **1** small bubble on the skin filled with watery fluid and caused by heat or friction. **2** similar swelling on plastic, wood, etc. *–v.* **1** come up in blisters. **2** raise a blister on. **3** attack sharply. [origin uncertain]

blithe /blaɪð/ *adj.* **1** cheerful, happy. **2** careless, casual. □ **blithely** *adv.* [Old English]

blithering /'blɪðərɪŋ/ *attrib. adj. colloq.* hopeless; contemptible (esp. in *blithering idiot*). [*blither*, var. of BLATHER]

BLitt. *abbr.* Bachelor of Letters. [Latin *Baccalaureus Litterarum*]

blitz /blɪts/ *colloq. –n.* **1 a** intensive or sudden (esp. aerial) attack. **b** intensive period of work etc. (*must have a blitz on this room*). **2** (**the Blitz**) German air raids on London in 1940. *–v.* inflict a blitz on. [abbreviation of BLITZKRIEG]

blitzkrieg /'blɪtskri:g/ *n.* intense military campaign intended to bring about a swift victory. [German, = lightning war]

blizzard /'blɪzəd/ *n.* severe snowstorm. [origin unknown]

bloat *v.* **1** inflate, swell. **2** (as **bloated** *adj.*) inflated with pride, wealth, or food. **3** cure (a herring) by salting and smoking lightly. [Old Norse]

bloater *n.* bloated herring.

blob *n.* small drop or spot. [imitative]

bloc *n.* group of governments etc. sharing a common purpose. [French: related to BLOCK]

block *–n.* **1** solid piece of hard material, esp. stone or wood. **2** this as a base for chopping etc., as a stand, or for mounting a horse from. **3 a** large building, esp. when subdivided. **b** group of buildings between streets. **4** obstruction. **5** two or more pulleys mounted in a case. **6** piece of wood or metal engraved for printing. **7** *slang* head. **8** (often *attrib.*) number of things as a unit, e.g. shares, theatre seats (*block booking*). **9** sheets of paper glued along one edge. *–v.* **1 a** (often foll. by *up*) obstruct. **b** impede. **2** restrict the use of. **3** *Cricket* stop (a ball) with a bat defensively. □ **block in 1** sketch roughly; plan. **2** confine. **block out 1** shut out (light, noise, a memory, view, etc.). **2** sketch roughly; plan. **block up** confine; enclose. [Low German or Dutch]

blockade /blɒ'keɪd/ *–n.* surrounding or blocking of a place by an enemy to prevent entry and exit. *–v.* (**-ding**) subject to a blockade.

blockage *n.* obstruction.

block and tackle *n.* system of pulleys and ropes, esp. for lifting.

blockbuster *n. slang* **1** thing of great power, esp. a very successful film, book, etc. **2** highly destructive bomb.

block capitals *n.pl.* (also **block letters**) letters printed without serifs, or written with each letter separate and in capitals.

blockhead *n.* stupid person.

blockhouse *n.* **1** reinforced concrete shelter. **2** *hist.* small fort of timber.

block vote *n.* vote proportional in power to the number of people a delegate represents.

bloke *n. slang* man, fellow. [Shelta]

blond (of a woman usu. **blonde**) *–adj.* (of a person, hair, or complexion) light-coloured, fair. *–n.* blond person. [Latin *blondus* yellow]

blood /blʌd/ *–n.* **1** usu. red fluid circulating in the arteries and veins of animals. **2** bloodshed, esp. killing. **3** passion, temperament. **4** race, descent, parentage (*of the same blood*). **5** relationship; relations (*blood is thicker than water*). **6** dandy. *–v.* **1** give (a hound) a first taste of blood. **2** initiate (a person). □ **in one's blood** inherent in one's character. [Old English]

blood bank *n.* store of blood for transfusion.

blood bath *n.* massacre.

blood count *n.* number of corpuscles in a specific amount of blood.

blood-curdling *adj.* horrifying.

blood donor *n.* person giving blood for transfusion.

blood group *n.* any of the types of human blood.

blood-heat *n.* normal human temperature, about 37 °C or 98.4 °F.

bloodhound *n.* large keen-scented dog used in tracking.

bloodless *adj.* **1** without blood or bloodshed. **2** unemotional. **3** pale. **4** feeble.

blood-letting *n.* surgical removal of blood.

blood-money *n.* **1** money paid as compensation for a death. **2** money paid to a killer.

blood orange *n.* red-fleshed orange.

blood-poisoning *n.* diseased condition caused by micro-organisms in the blood.

blood pressure *n.* pressure of the blood in the arteries etc., measured for diagnosis.

blood relation *n.* (also **blood relative**) relative by birth.

bloodshed *n.* killing.

bloodshot *adj.* (of an eyeball) inflamed.

blood sport *n.* sport involving the killing or wounding of animals.

bloodstain *n.* stain caused by blood. □ **bloodstained** *adj.*

bloodstream *n.* blood in circulation.

bloodsucker *n.* **1** leech. **2** extortioner. □ **bloodsucking** *adj.*

blood sugar *n.* amount of glucose in the blood.

blood test *n.* examination of blood, esp. for diagnosis.

bloodthirsty *adj.* (**-ier, -iest**) eager for bloodshed.

blood-vessel *n.* vein, artery, or capillary carrying blood.

bloody *–adj.* (**-ier, -iest**) **1** of, like, running with, or smeared with blood. **2 a** involving bloodshed. **b** bloodthirsty, cruel. **3** *coarse slang* expressing annoyance or antipathy, or as an intensifier (*bloody fool*; *a bloody sight better*). **4** red. *–adv. coarse slang* as an intensifier (*bloody awful*). *–v.* (**-ies, -ied**) stain with blood.

Bloody Mary *n.* mixture of vodka and tomato juice.

bloody-minded *adj. colloq.* deliberately uncooperative.

bloom *–n.* **1 a** flower, esp. cultivated. **b** state of flowering (*in bloom*). **2** one's prime (*in full bloom*). **3 a** healthy glow of the complexion. **b** fine powder on fresh fruit and leaves. *–v.* **1** bear flowers; be in flower. **2** be in one's prime; flourish. [Old Norse]

bloomer[1] *n.* **1** *slang* blunder. **2** plant that blooms in a specified way.

bloomer[2] *n.* long loaf with diagonal marks. [origin uncertain]

bloomers *n.pl.* **1** women's long loose knickers. **2** *hist.* women's loose knee-length trousers. [Mrs A. *Bloomer*, name of the originator]

blooming *–adj.* **1** flourishing; healthy. **2** *slang* an intensifier (*blooming miracle*). *–adv. slang* an intensifier (*blooming difficult*).

blossom /'blɒsəm/ –*n.* **1** flower or mass of flowers, esp. of a fruit-tree. **2** promising stage (*blossom of youth*). –*v.* **1** open into flower. **2** mature, thrive. [Old English]

blot –*n.* **1** spot or stain of ink etc. **2** disgraceful act or quality. **3** blemish. –*v.* (**-tt-**) **1** make a blot on, stain. **2** dry with blotting-paper. □ **blot one's copybook** damage one's reputation. **blot out 1** obliterate. **2** obscure (a view, sound, etc.). [probably Scandinavian]

blotch –*n.* **1** discoloured or inflamed patch on the skin. **2** irregular patch of colour. –*v.* cover with blotches. □ **blotchy** *adj.* (**-ier, -iest**). [obsolete *plotch*, BLOT]

blotter *n.* pad of blotting-paper.

blotting-paper *n.* absorbent paper for drying wet ink.

blotto /'blɒtəʊ/ *adj. slang* very drunk. [origin uncertain]

blouse /blaʊz/ –*n.* **1** woman's garment like a shirt. **2** upper part of a military uniform. –*v.* (**-sing**) make (a bodice etc.) full like a blouse. [French]

blouson /'blu:zɒn/ *n.* short blouse-shaped jacket. [French]

blow[1] /bləʊ/ –*v.* (*past* **blew**; *past part.* **blown**) **1** direct a current of air (at) esp. from the mouth. **2** drive or be driven by blowing (*blew the door open*). **3** (esp. of the wind) move rapidly. **4** expel by breathing (*blew smoke*). **5** sound or be sounded by blowing. **6** (*past part.* **blowed**) *slang* (esp. in *imper.*) curse, confound (*I'm blowed if I know*; *blow it!*). **7** clear (the nose) by blowing. **8** puff, pant. **9** *slang* depart suddenly (from). **10** shatter etc. by an explosion. **11** make or shape (glass or a bubble) by blowing. **12** **a** melt from overloading (*the fuse has blown*). **b** break or burst suddenly. **13** (of a whale) eject air and water. **14** break into with explosives. **15** *slang* **a** squander (*blew £20*). **b** bungle (an opportunity etc.). **c** reveal (a secret etc.). –*n.* **1** act of blowing. **2** **a** gust of wind or air. **b** exposure to fresh air. □ **be blowed if one will** *colloq.* be unwilling to. **blow a gasket** *slang* lose one's temper. **blow hot and cold** *colloq.* vacillate. **blow in 1** break inwards by an explosion. **2** *colloq.* arrive unexpectedly. **blow a person's mind** *slang* cause to have hallucinations etc.; astound. **blow off 1** escape or allow (steam etc.) to escape forcibly. **2** *slang* break wind noisily. **blow out 1** extinguish by blowing. **2** send outwards by an explosion. **blow over** (of trouble etc.) fade away. **blow one's top** *colloq.* explode in rage. **blow up 1** explode. **2** *colloq.* rebuke strongly. **3** inflate (a tyre etc.). **4** *colloq.* **a** enlarge (a photograph). **b** exaggerate. **5** *colloq.* arise, happen. **6** *colloq.* lose one's temper. [Old English]

blow[2] /bləʊ/ *n.* **1** hard stroke with a hand or weapon. **2** sudden shock or misfortune. [origin unknown]

blow-by-blow *attrib. adj.* (of a narrative etc.) detailed.

blow-dry –*v.* arrange (the hair) while drying it. –*n.* act of doing this.

blower *n.* **1** device for blowing. **2** *colloq.* telephone.

blowfly *n.* bluebottle.

blow-hole *n.* **1** nostril of a whale. **2** hole (esp. in ice) for breathing or fishing through. **3** vent for air, smoke, etc.

blow-job *n. coarse slang* instance of fellatio or cunnilingus.

blowlamp *n.* device with a very hot flame for burning off paint, plumbing, etc.

blown *past part.* of BLOW[1].

blow-out *n. colloq.* **1** burst tyre. **2** melted fuse. **3** huge meal.

blowpipe *n.* **1** tube for blowing air through, esp. to intensify a flame or to blow glass. **2** tube for propelling poisoned darts etc. by blowing.

blowtorch *n. US* = BLOWLAMP.

blow-up *n.* **1** *colloq.* enlargement (of a photograph etc.). **2** explosion.

blowy /'bləʊɪ/ *adj.* (**-ier, -iest**) windy.

blowzy /'blaʊzɪ/ *adj.* (**-ier, -iest**) **1** coarse-looking; red-faced. **2** slovenly. [obsolete *blowze* beggar's wench]

blub *v.* (**-bb-**) *slang* sob. [shortening of BLUBBER]

blubber –*n.* whale fat. –*v.* **1** sob loudly. **2** sob out (words). –*adj.* swollen, thick. [probably imitative]

bludgeon /'blʌdʒ(ə)n/ –*n.* heavy club. –*v.* **1** beat with this. **2** coerce. [origin unknown]

blue /blu:/ –*adj.* (**bluer, bluest**) **1** having the colour of a clear sky. **2** sad, depressed. **3** pornographic (*a blue film*). **4** politically conservative. –*n.* **1** blue colour or pigment. **2** blue clothes or material (*dressed in blue*). **3** person who represents a university in a sport, esp. Oxford or Cambridge. **4** Conservative party supporter. –*v.* (**blues, blued, bluing** or **blueing**) **1** make blue. **2** *slang* squander. □ **once in a blue moon** very rarely. **out of the blue** unexpectedly. [French from Germanic]

blue baby *n.* baby with a blue complexion due to a congenital heart defect.

bluebell *n.* woodland plant with bell-shaped blue flowers.

blueberry *n.* (*pl.* **-ies**) small blue-black edible fruit of various plants.

blue blood *n.* noble birth.

Blue Book *n.* report issued by Parliament or the Privy Council.

bluebottle *n.* large buzzing fly; blowfly.

blue cheese *n.* cheese with veins of blue mould.

blue-collar *attrib. adj.* (of a worker or work) manual; industrial.

blue-eyed boy *n. colloq.* favourite.

blue funk *n. colloq.* terror or panic.

bluegrass *n.* a kind of instrumental country-and-western music.

blue-pencil *v.* censor or cut (a manuscript, film, etc.).

Blue Peter *n.* blue flag with a white square flown by a ship about to leave port.

blueprint *n.* **1** photographic print of plans in white on a blue background. **2** detailed plan.

blue rinse *n.* bluish dye for grey hair.

blues *n.pl.* **1** (prec. by *the*) bout of depression. **2 a** (prec. by *the*; often treated as *sing.*) melancholic music of Black American origin, usu. in a twelve-bar sequence. **b** (*pl.* same) (as *sing.*) piece of such music (*played a blues*).

bluestocking *n.* usu. *derog.* intellectual or literary woman. [18th-c. Blue Stocking Society]

blue tit *n.* common tit with a blue crest.

blue whale *n.* rorqual, the largest known living mammal.

bluff[1] *–v.* pretend strength, confidence, etc. *–n.* act of bluffing. □ **call a person's bluff** challenge a person to prove a claim. [Dutch *bluffen* brag]

bluff[2] *–adj.* **1** blunt, frank, hearty. **2** vertical or steep and broad in front. *–n.* steep cliff or headland. [origin unknown]

bluish /ˈbluːɪʃ/ *adj.* fairly blue.

blunder *–n.* serious or foolish mistake. *–v.* **1** make a blunder. **2** move clumsily; stumble. [probably Scandinavian]

blunderbuss /ˈblʌndəˌbʌs/ *n. hist.* short large-bored gun. [Dutch *donderbus* thunder gun]

blunt *–adj.* **1** not sharp or pointed. **2** direct, outspoken. *–v.* make blunt or less sharp. □ **bluntly** *adv.* (in sense 2 of *adj.*). **bluntness** *n.* [probably Scandinavian]

blur *–v.* (**-rr-**) make or become unclear or less distinct; smear. *–n.* blurred object, sound, memory, etc. [perhaps related to BLEARY]

blurb *n.* promotional description, esp. of a book. [coined by G. Burgess 1907]

blurt *v.* (usu. foll. by *out*) utter abruptly, thoughtlessly, or tactlessly. [imitative]

blush *–v.* **1 a** become pink in the face from embarrassment or shame. **b** (of the face) redden thus. **2** feel embarrassed or ashamed. **3** redden. *–n.* **1** act of blushing. **2** pink tinge. [Old English]

blusher *n.* rouge.

bluster *–v.* **1** behave pompously or boisterously. **2** (of the wind etc.) blow fiercely. *–n.* bombastic talk; empty threats. □ **blustery** *adj.* [imitative]

BM *abbr.* **1** British Museum. **2** Bachelor of Medicine.

BMA *abbr.* British Medical Association.

B.Mus. *abbr.* Bachelor of Music.

BMX /ˌbiːemˈeks/ *n.* **1** organized bicycle-racing on a dirt-track. **2** bicycle used for this. [abbreviation of *b*icycle *m*oto*cross*]

BO *abbr. colloq.* body odour.

boa /ˈbəʊə/ *n.* **1** large snake which kills by crushing and suffocating. **2** long stole of feathers or fur. [Latin]

boa constrictor *n.* species of boa.

boar *n.* **1** male wild pig. **2** uncastrated male pig. [Old English]

board *–n.* **1 a** flat thin piece of sawn timber, usu. long and narrow. **b** material resembling this, of compressed fibres. **c** thin slab of wood etc. **d** thick stiff card used in bookbinding. **2** provision of regular meals, usu. with accommodation, for payment. **3** directors of a company; official administrative body. **4** (in *pl.*) stage of a theatre. **5** side of a ship. *–v.* **1** go on board (a ship, train, etc.). **2** receive, or provide with, meals and usu. lodging. **3** (usu. foll. by *up*) cover with boards; seal or close. □ **go by the board** be neglected or discarded. **on board** on or on to a ship, aircraft, oil rig, etc. **take on board** consider, take notice of; accept. [Old English]

boarder *n.* **1** person who boards, esp. at a boarding-school. **2** person who boards a ship, esp. an enemy.

board-game *n.* game played on a board.

boarding-house *n.* unlicensed establishment providing board and lodging, esp. to holiday-makers.

boarding-school *n.* school in which pupils live in term-time.

boardroom *n.* room in which a board of directors etc. meets regularly.

boast *–v.* **1** declare one's virtues, wealth, etc. with excessive pride. **2** own or have with pride (*hotel boasts a ballroom*). *–n.* **1** act of boasting. **2** thing one is proud of. [Anglo-French]

boastful *adj.* given to boasting. □ **boastfully** *adv.*

boat *–n.* **1** small vessel propelled on water by an engine, oars, or sails. **2** any ship. **3** long low jug for sauce etc. *–v.* go in a boat, esp. for pleasure. □ **in the same boat** having the same problems. [Old English]

boater *n.* flat-topped straw hat with a brim.

boat-hook *n.* long hooked pole for moving boats.
boat-house *n.* waterside shed for housing boats.
boating *n.* rowing or sailing as recreation.
boatman *n.* person who hires out boats or provides transport by boat.
boat people *n.pl.* refugees travelling by sea.
boatswain /'bəʊs(ə)n/ *n.* (also **bosun, bo'sun**) ship's officer in charge of equipment and crew.
boat-train *n.* train scheduled to meet or go on a boat.
bob[1] —*v.* (**-bb-**) **1** move quickly up and down. **2** (usu. foll. by *back, up*) bounce or emerge buoyantly or suddenly. **3** cut (the hair) in a bob. **4** curtsy. —*n.* **1** jerking or bouncing movement, esp. upward. **2** hairstyle with the hair hanging evenly above the shoulders. **3** weight on a pendulum etc. **4** horse's docked tail. **5** curtsy. [imitative]
bob[2] *n.* (*pl.* same) *hist. slang* shilling (now = 5 pence). [origin unknown]
bob[3] *n.* □ **bob's your uncle** *slang* expression of completion or success. [pet form of *Robert*]
bobbin /'bɒbɪn/ *n.* spool or reel for thread etc. [French]
bobble /'bɒb(ə)l/ *n.* small woolly ball on a hat etc. [diminutive of BOB[1]]
bobby *n.* (*pl.* **-ies**) *colloq.* police officer. [Sir *Robert* Peel, 19th-c. statesman]
bob-sled *n. US* = BOB-SLEIGH.
bob-sleigh —*n.* mechanically-steered and -braked racing sledge. —*v.* race in a bob-sleigh.
bobtail *n.* **1** docked tail. **2** horse or dog with this.
Boche /bɒʃ/ *n. slang derog.* German, esp. a soldier. [French]
bod *n. colloq.* person. [shortening of BODY]
bode *v.* (**-ding**) be a sign of, portend. □ **bode well** (or **ill**) be a good (or bad) sign. [Old English]
bodega /bəʊ'di:gə/ *n.* cellar or shop selling wine. [Spanish]
bodge var. of BOTCH.
bodice /'bɒdɪs/ *n.* **1** part of a woman's dress above the waist. **2** woman's vest-like undergarment. [originally *pair of bodies*]
bodily /'bɒdɪlɪ/ —*adj.* of the body. —*adv.* **1** as a whole body (*threw them bodily*). **2** in the flesh, in person.
bodkin /'bɒdkɪn/ *n.* blunt thick needle for drawing tape etc. through a hem. [origin uncertain]
body /'bɒdɪ/ *n.* (*pl.* **-ies**) **1** whole physical structure, including the bones, flesh, and organs, of a person or an animal, whether dead or alive. **2** = TRUNK 2. **3** main or central part; bulk or majority (*body of opinion*). **4 a** group regarded as a unit. **b** (usu. foll. by *of*) collection (*body of facts*). **5** quantity (*body of water*). **6** piece of matter (*heavenly body*). **7** *colloq.* person. **8** full or substantial quality of flavour, tone, etc. □ **in a body** all together. [Old English]
body-blow *n.* severe setback.
body-building *n.* exercises to enlarge and strengthen the muscles.
bodyguard *n.* person or group escorting and protecting another.
body language *n.* communication through gestures and poses.
body odour *n.* smell of the human body, esp. when unpleasant.
body politic *n.* nation or State as a corporate body.
body shop *n.* workshop where bodywork is repaired.
body stocking *n.* woman's undergarment covering the torso.
bodysuit *n.* close-fitting all-in-one garment for women, worn esp. for sport.
bodywork *n.* outer shell of a vehicle.
Boer /bɔ:(r)/ —*n.* South African of Dutch descent. —*adj.* of the Boers. [Dutch, = farmer]
boffin /'bɒfɪn/ *n. colloq.* research scientist. [origin unknown]
bog —*n.* **1 a** wet spongy ground. **b** stretch of this. **2** *slang* lavatory. —*v.* (**-gg-**) (foll. by *down*; usu. in *passive*) impede (*bogged down by snow*). □ **boggy** *adj.* (**-ier, -iest**). [Irish or Gaelic *bogach*]
bogey[1] /'bəʊgɪ/ *n.* (*pl.* **-eys**) *Golf* **1** score of one stroke more than par at any hole. **2** (formerly) par. [perhaps from *Bogey*, as an imaginary player]
bogey[2] /'bəʊgɪ/ *n.* (also **bogy**) (*pl.* **-eys** or **-ies**) **1** evil or mischievous spirit; devil. **2** awkward thing. **3** *slang* piece of dried nasal mucus. [originally (*Old*) *Bogey* the Devil]
bogeyman *n.* (also **bogyman**) frightening person etc.
boggle /'bɒg(ə)l/ *v.* (**-ling**) *colloq.* be startled or baffled (esp. *the mind boggles*). [probably dial. *boggle* BOGEY[2]]
bogie /'bəʊgɪ/ *n.* wheeled undercarriage below a locomotive etc. [origin unknown]
bogus /'bəʊgəs/ *adj.* sham, spurious. [origin unknown]
bogy var. of BOGEY[2].
bogyman var. of BOGEYMAN.
Bohemian /bəʊ'hi:mɪən/ —*n.* **1** native of Bohemia, a Czech. **2** (also **bohemian**) socially unconventional person, esp. an artist or writer. —*adj.* **1** of Bohemia or its people. **2** (also **bohemian**) socially

unconventional. □ **bohemianism** *n.* [*Bohemia*, part of Czechoslovakia]

boil[1] *–v.* **1 a** (of a liquid) start to bubble up and turn into vapour on reaching a certain temperature. **b** (of a vessel) contain boiling liquid (*kettle is boiling*). **2 a** bring to boiling-point. **b** cook in boiling liquid. **c** subject to boiling water, e.g. to clean. **3 a** move or seethe like boiling water. **b** be very angry. *–n.* act or process of boiling; boiling-point (*on the boil*; *bring to the boil*). □ **boil down 1** reduce in volume by boiling. **2** reduce to essentials. **3** (foll. by *to*) amount to. **boil over 1** spill over in boiling. **2** lose one's temper. [Latin *bullio* to bubble]

boil[2] *n.* inflamed pus-filled swelling under the skin. [Old English]

boiler *n.* **1** apparatus for heating a hot-water supply. **2** tank for heating water or turning it to steam. **3** tub for boiling laundry etc. **4** fowl etc. for boiling.

boiler-room *n.* room with a boiler and other heating equipment, esp. in a basement.

boiler suit *n.* protective outer garment of trousers and jacket in one.

boiling *adj. colloq.* very hot.

boiling-point *n.* **1** temperature at which a liquid begins to boil. **2** great excitement.

boisterous /'bɔɪstərəs/ *adj.* **1** noisily exuberant, rough. **2** (of the sea etc.) stormy. [origin unknown]

bold /bəʊld/ *adj.* **1** confidently assertive; adventurous, brave. **2** impudent. **3** vivid (*bold colours*). □ **make** (or **be**) **so bold as to** presume to; venture to. □ **boldly** *adv.* **boldness** *n.* [Old English]

bole *n.* trunk of a tree. [Old Norse]

bolero *n.* (*pl.* **-s**) **1** /bə'leərəʊ/ Spanish dance, or the music for it, in triple time. **2** /'bɒlərəʊ/ woman's short open jacket. [Spanish]

boll /bəʊl/ *n.* round seed-vessel of cotton, flax, etc. [Dutch]

bollard /'bɒlɑ:d/ *n.* **1** short post in the road, esp. on a traffic island. **2** short post on a quay or ship for securing a rope. [perhaps related to BOLE]

bollocking /'bɒləkɪŋ/ *n.* (also **ballocking**) *coarse slang* severe reprimand.

bollocks /'bɒləks/ *n.* (also **ballocks**) *coarse slang* **1** (usu. as *int.*) nonsense. **2** testicles. [Old English: related to BALL[1]]

boloney /bə'ləʊnɪ/ *n.* (also **baloney**) *slang* nonsense. [origin uncertain]

Bolshevik /'bɒlʃəvɪk/ *–n.* **1** *hist.* member of the radical faction of the Russian Social Democratic Party becoming the Communist Party in 1918. **2** Russian Communist. **3** any revolutionary socialist. *–adj.* **1** of the Bolsheviks. **2** Communist. □ **Bolshevism** *n.* **Bolshevist** *n.* [Russian, = member of the majority]

Bolshie /'bɒlʃɪ/ (also **Bolshy**) *slang –adj.* (usu. **bolshie**) **1** uncooperative; bad-tempered. **2** left-wing. *–n.* (*pl.* **-ies**) Bolshevik. [abbreviation]

bolster /'bəʊlstə(r)/ *–n.* long cylindrical pillow. *–v.* (usu. foll. by *up*) encourage, support, prop up. [Old English]

bolt[1] /bəʊlt/ *–n.* **1** sliding bar and socket used to fasten a door etc. **2** large metal pin with a thread, usu. used with a nut, to hold things together. **3** discharge of lightning. **4** act of bolting. *–v.* **1** fasten with a bolt. **2** (foll. by *in*, *out*) keep (a person etc.) in or out by bolting a door. **3** fasten together with bolts. **4 a** dash off suddenly. **b** (of a horse) suddenly gallop out of control. **5** gulp down (food) unchewed. **6** (of a plant) run to seed. *–adv.* (usu. in **bolt upright**) rigidly, stiffly. □ **bolt from the blue** complete surprise. [Old English]

bolt[2] /bəʊlt/ *v.* (also **boult**) sift (flour etc.). [French]

bolt-hole *n.* means of escape.

bomb /bɒm/ *–n.* **1** container filled with explosive, incendiary material, etc., designed to explode and cause damage. **2** (prec. by *the*) the atomic or hydrogen bomb. **3** *slang* large sum of money (*cost a bomb*). *–v.* **1** attack with bombs; drop bombs on. **2** (usu. foll. by *along*, *off*) *colloq.* go very quickly. □ **like a bomb** *colloq.* **1** very successfully. **2** very fast. [Greek *bombos* hum]

bombard /bɒm'bɑ:d/ *v.* **1** attack with heavy guns or bombs etc. **2** (often foll. by *with*) question or abuse persistently. **3** *Physics* direct a stream of high-speed particles at. □ **bombardment** *n.* [Latin: related to BOMB]

bombardier /,bɒmbə'dɪə(r)/ *n.* **1** non-commissioned officer in the artillery. **2** *US* crew member in an aircraft who aims and releases bombs.

bombast /'bɒmbæst/ *n.* pompous language; hyperbole. □ **bombastic** /-'bæstɪk/ *adj.* [earlier *bombace* cotton wool]

Bombay duck /bɒm'beɪ/ *n.* dried fish as a relish, esp. with curry. [corruption of *bombil*, native name of fish]

bombazine /'bɒmbə,zi:n/ *n.* twilled worsted dress-material. [Greek *bombux* silk]

bomber /'bɒmə(r)/ *n.* **1** aircraft equipped to drop bombs. **2** person using bombs, esp. illegally.

bomber jacket *n.* jacket gathered at the waist and cuffs.

bombshell *n.* **1** overwhelming surprise or disappointment. **2** artillery bomb. **3** *slang* very attractive woman.

bomb-site *n.* area where bombs have caused destruction.

bona fide /ˌbəʊnə ˈfaɪdɪ/ *–adj.* genuine; sincere. *–adv.* genuinely; sincerely. [Latin]

bonanza /bəˈnænzə/ *n.* **1** source of wealth or prosperity. **2** large output (esp. of a mine). [Spanish, = fair weather]

bon-bon *n.* sweet. [French *bon* good]

bond *–n.* **1** thing or force that unites or (usu. in *pl.*) restrains. **2** binding agreement. **3** *Commerce* certificate issued by a government or a company promising to repay borrowed money at a fixed rate of interest. **4** adhesiveness. **5** *Law* deed binding a person to make payment to another. **6** *Chem.* linkage between atoms in a molecule. *–v.* **1** hold or tie together. **2** connect or reinforce with a bond. **3** place (goods) in bond. □ **in bond** stored by Customs until duty is paid. [var. of BAND]

bondage *n.* **1** slavery. **2** subjection to constraint etc. **3** sexual practices involving constraint. [Anglo-Latin: related to BONDSMAN]

bonded *adj.* **1** stored in or for storing in bond (*bonded whisky, warehouse*). **2** (of a debt) secured by bonds.

bond paper *n.* high-quality writing-paper.

bondsman *n.* serf, slave. [Old English *bonda* husbandman]

bone *–n.* **1** any piece of hard tissue making up the skeleton in vertebrates. **2** (in *pl.*) **a** skeleton, esp. as remains. **b** body. **3** material of bones or similar material, e.g. ivory. **4** thing made of bone. **5** (in *pl.*) essentials (*the bones of an agreement*). **6** strip of stiffening in a corset etc. *–v.* (**-ning**) **1** remove the bones from. **2** stiffen with bone etc. □ **bone up** (often foll. by *on*) *colloq.* study intensively. **have a bone to pick** (usu. foll. by *with*) have cause for dispute (with a person). **make no bones about 1** be frank about. **2** not hesitate or scruple. □ **boneless** *adj.* [Old English]

bone china *n.* fine china made of clay mixed with bone ash.

bone-dry *adj.* completely dry.

bone-idle *adj.* utterly idle.

bone-marrow *n.* = MARROW 2.

bone-meal *n.* crushed bones, esp. as a fertilizer.

bone of contention *n.* source of dispute.

boneshaker *n.* decrepit or uncomfortable old vehicle.

bonfire *n.* large open-air fire, esp. for burning rubbish. [from BONE (because bones were once used), FIRE]

bongo /ˈbɒŋgəʊ/ *n.* (*pl.* **-s** or **-es**) either of a pair of small drums usu. held between the knees and played with the fingers. [American Spanish]

bonhomie /ˌbɒnɒˈmiː/ *n.* good-natured friendliness. [French]

bonk *–v.* **1** bang, bump. **2** *coarse slang* have sexual intercourse (with). *–n.* instance of bonking (*bonk on the head*). [imitative]

bonkers /ˈbɒŋkəz/ *predic. adj. slang* crazy. [origin unknown]

bon mot /bɔ̃ ˈməʊ/ *n.* (*pl.* ***bons mots*** /-məʊz/) witty saying. [French]

bonnet /ˈbɒnɪt/ *n.* **1 a** hat tied under the chin, worn esp. by babies. **b** Scotsman's floppy beret. **2** hinged cover over a vehicle's engine. [French]

bonny /ˈbɒnɪ/ *adj.* (**-ier, -iest**) esp. *Scot.* & *N.Engl.* **1 a** physically attractive. **b** healthy-looking. **2** good, pleasant. [perhaps from French *bon* good]

bonsai /ˈbɒnsaɪ/ *n.* (*pl.* same) **1** dwarfed tree or shrub. **2** art of growing these. [Japanese]

bonus /ˈbəʊnəs/ *n.* extra benefit or payment. [Latin, = good]

bon vivant /ˌbɔ̃ viːˈvɑ̃/ *n.* (*pl.* ***bon*** or ***bons vivants*** pronunc. same) person fond of good food and drink. [French]

bon voyage /ˌbɔ̃ vwɑːˈjɑːʒ/ *int.* expression of good wishes to a departing traveller. [French]

bony /ˈbəʊnɪ/ *adj.* (**-ier, -iest**) **1** thin with prominent bones. **2** having many bones. **3** of or like bone. □ **boniness** *n.*

boo *–int.* **1** expression of disapproval etc. **2** sound intended to surprise. *–n.* utterance of *boo*, esp. to a performer etc. *–v.* (**boos, booed**) **1** utter boos. **2** jeer at by booing. [imitative]

boob[1] *colloq. –n.* **1** silly mistake. **2** foolish person. *–v.* make a silly mistake. [shortening of BOOBY]

boob[2] *n. slang* woman's breast. [origin uncertain]

booby *n.* (*pl.* **-ies**) stupid or childish person. [Spanish *bobo*]

booby prize *n.* prize given for coming last.

booby trap *n.* **1** practical joke in the form of a trap. **2** disguised explosive device triggered by the unknowing victim.

boodle /ˈbuːd(ə)l/ *n. slang* money, esp. gained or used dishonestly. [Dutch *boedel* possessions]

boogie /ˈbuːgɪ/ *v.* (**-ies, -ied, -ieing**) *slang* dance to pop music.

boogie-woogie /ˌbu:gɪ'wu:gɪ/ *n.* style of playing blues or jazz on the piano. [origin unknown]

book /bʊk/ *–n.* **1 a** written or printed work with pages bound along one side. **b** work intended for publication. **2** bound blank sheets for notes, records, etc. **3** bound set of tickets, stamps, matches, etc. **4** (in *pl.*) set of records or accounts. **5** main division of a large literary work. **6** telephone directory. **7** *colloq.* magazine. **8** libretto, script, etc. **9** record of bets. *–v.* **1 a** (also *absol.*) reserve (a seat etc.) in advance. **b** engage (an entertainer etc.). **2 a** take the personal details of (an offender or rulebreaker). **b** enter in a book or list. □ **book in** register at a hotel etc. **book up 1** buy tickets in advance. **2** (as **booked up**) with all places reserved. **bring to book** call to account. **go by the book** proceed by the rules. **in a person's good** (or **bad**) **books** in (or out of) favour with a person. [Old English]

bookbinder *n.* person who binds books for a living. □ **bookbinding** *n.*

bookcase *n.* cabinet of shelves for books.

book club *n.* society in which selected books are available cheaply.

book end *n.* prop used to keep books upright.

bookie /'bʊkɪ/ *n. colloq.* = BOOKMAKER. [abbreviation]

booking *n.* reservation or engagement.

booking-hall *n.* (also **booking-office**) ticket office at a railway station etc.

bookish *adj.* **1** studious; fond of reading. **2** having knowledge mainly from books.

bookkeeper *n.* person who keeps accounts, esp. for a living. □ **bookkeeping** *n.*

booklet /'bʊklɪt/ *n.* small book usu. with a paper cover.

bookmaker *n.* professional taker of bets. □ **bookmaking** *n.*

bookmark *n.* thing used to mark a reader's place.

book-plate *n.* decorative personalized label stuck in a book.

bookseller *n.* dealer in books.

bookshop *n.* shop selling books.

bookstall *n.* stand selling books, newspapers, etc.

book token *n.* voucher exchangeable for books.

bookworm *n.* **1** *colloq.* devoted reader. **2** larva feeding on the paper and glue in books.

Boolean /'bu:lɪən/ *adj.* denoting a system of algebraic notation to represent logical propositions. [*Boole*, name of a mathematician]

Boolean logic *n.* use of 'and', 'or', and 'not' in retrieving information from a database.

boom[1] *–n.* deep resonant sound. *–v.* make or speak with a boom. [imitative]

boom[2] *–n.* period of economic prosperity or activity. *–v.* be suddenly prosperous. [perhaps from BOOM[1]]

boom[3] *n.* **1** pivoted spar to which a sail is attached. **2** long pole carrying a microphone, camera, etc. **3** barrier across a harbour etc. [Dutch, = BEAM]

boomerang /'bu:məˌræŋ/ *–n.* **1** flat V-shaped hardwood missile used esp. by Australian Aboriginals, able to return to its thrower. **2** plan that recoils on its originator. *–v.* (of a plan etc.) backfire. [Aboriginal]

boon[1] *n.* advantage; blessing. [Old Norse]

boon[2] *adj.* intimate, favourite (usu. *boon companion*). [French *bon* from Latin *bonus* good]

boor *n.* ill-mannered person. □ **boorish** *adj.* [Low German or Dutch]

boost *colloq.* *–v.* **1** promote or encourage. **2** increase, assist. **3** push from below. *–n.* act or result of boosting. [origin unknown]

booster *n.* **1** device for increasing power or voltage. **2** auxiliary engine or rocket for initial speed. **3** dose, injection, etc. renewing the effect of an earlier one.

boot[1] *–n.* **1** outer foot-covering reaching above the ankle. **2** luggage compartment of a car. **3** *colloq.* **a** firm kick. **b** (prec. by *the*) dismissal (*got the boot*). *–v.* **1** kick. **2** (often foll. by *out*) eject forcefully. **3** (usu. foll. by *up*) make (a computer) ready. □ **put the boot in 1** kick brutally. **2** harm a person. [Old Norse]

boot[2] *n.* □ **to boot** as well, in addition. [Old English]

bootblack *n. US* person who polishes boots and shoes.

bootee /bu:'ti:/ *n.* baby's soft shoe.

booth /bu:ð/ *n.* **1** small temporary structure used esp. as a market stall. **2** enclosure for telephoning, voting, etc. **3** cubicle in a restaurant etc. [Old Norse]

bootleg *–adj.* (esp. of alcohol) smuggled, illicit. *–v.* (**-gg-**) illicitly make or deal in (alcohol etc.). □ **bootlegger** *n.*

bootlicker *n. colloq.* toady.

boots *n.* hotel servant who cleans shoes etc.

bootstrap *n.* loop used to pull a boot on. □ **pull oneself up by one's bootstraps** better oneself.

booty /'bu:tɪ/ *n.* **1** loot, spoil. **2** *colloq.* prize or gain. [German]

booze *colloq.* –*n.* alcoholic drink. –*v.* (**-zing**) drink alcohol, esp. to excess. □ **boozy** *adj.* (**-ier**, **-iest**). [Dutch]

boozer *n. colloq.* **1** habitual drinker. **2** public house.

booze-up *n. slang* drinking bout.

bop[1] *colloq.* –*n.* **1 a** spell of dancing, esp. to pop music. **b** social occasion for this. **2** = BEBOP. –*v.* (**-pp-**) dance, esp. to pop music. □ **bopper** *n.* [abbreviation]

bop[2] *colloq.* –*v.* (**-pp-**) hit or punch, esp. lightly. –*n.* esp. light blow or hit. [imitative]

boracic /bəˈræsɪk/ *adj.* of borax.

boracic acid *n.* = BORIC ACID.

borage /ˈbɒrɪdʒ/ *n.* plant with leaves used as flavouring. [French ultimately from Arabic]

borax /ˈbɔːræks/ *n.* salt used in making glass and china, and as an antiseptic. [French ultimately from Persian]

Bordeaux /bɔːˈdəʊ/ *n.* (*pl.* same /-ˈdəʊz/) wine (esp. red) from the Bordeaux district in SW France.

border –*n.* **1** edge or boundary, or the part near it. **2 a** line or region separating two countries. **b** (**the Border**) boundary between Scotland and England (usu. **the Borders**), or N. Ireland and the Irish Republic. **3** esp. ornamental strip round an edge. **4** long narrow flower-bed (*herbaceous border*). –*v.* **1** be a border to. **2** provide with a border. **3** (usu. foll. by *on*, *upon*) **a** adjoin; come close to being. **b** resemble. [French from Germanic: related to BOARD]

Border collie *n.* sheepdog of the North Country.

borderer *n.* person living near a border.

borderland *n.* **1** district near a border. **2** condition between two extremes. **3** area for debate.

borderline –*n.* **1** line dividing two conditions. **2** line marking a boundary. –*adj.* **1** on the borderline. **2** barely acceptable.

Border terrier *n.* small rough-haired terrier.

bore[1] –*v.* (**-ring**) **1** make (a hole), esp. with a revolving tool. **2** make a hole in, hollow out. –*n.* **1** hollow of a firearm barrel or of a cylinder in an internal-combustion engine. **2** diameter of this. **3** deep hole made esp. to find water. [Old English]

bore[2] –*n.* tiresome or dull person or thing. –*v.* (**-ring**) weary by tedious talk or dullness. □ **bored** *adj.* **boring** *adj.* [origin unknown]

bore[3] *n.* high tidal wave in an estuary. [Scandinavian]

bore[4] *past* of BEAR[1].

boredom *n.* state of being bored. [from BORE[2]]

boric acid /ˈbɔːrɪk/ *n.* acid derived from borax, used as an antiseptic.

born *adj.* **1** existing as a result of birth. **2 a** of natural ability or quality (*a born leader*). **b** (usu. foll. by *to* + infin.) destined (*born lucky*; *born to be king*). **3** (in *comb.*) of a certain status by birth (*French-born*; *well-born*). [past part. of BEAR[1]]

born-again *attrib. adj.* converted (esp. to fundamentalist Christianity).

borne /bɔːn/ *past part.* of BEAR[1]. –*adj.* (in *comb.*) carried by (*airborne*).

boron /ˈbɔːrɒn/ *n.* non-metallic usu. crystalline element. [from BORAX, after *carbon*]

borough /ˈbʌrə/ *n.* **1** administrative division of London or New York City. **2** town with a corporation and privileges granted by royal charter. [Old English]

borrow /ˈbɒrəʊ/ *v.* **1 a** acquire temporarily, promising or intending to return. **b** obtain money thus. **2** use (another's idea, invention, etc.); plagiarize. □ **borrower** *n.* [Old English]

Borstal /ˈbɔːst(ə)l/ *n. hist.* residential institution for youth custody. [*Borstal* in Kent]

■ **Usage** This term has now been replaced by *detention centre* and *youth custody centre*.

---

bortsch /bɔːtʃ/ *n.* Russian soup of beetroot, cabbage, etc. [Russian]

borzoi /ˈbɔːzɔɪ/ *n.* large silky-coated dog. [Russian, = swift]

bosh *n.* & *int. slang* nonsense. [Turkish, = empty]

bosom /ˈbʊz(ə)m/ *n.* **1 a** person's (esp. woman's) breast. **b** *colloq.* each of a woman's breasts. **c** enclosure formed by the breast and arms. **2** emotional centre (*bosom of one's family*). [Old English]

bosom friend *n.* intimate friend.

boss[1] *colloq.* –*n.* employer, manager, or supervisor. –*v.* (usu. foll. by *about*, *around*) give orders to; order about. [Dutch *baas*]

boss[2] *n.* **1** round knob, stud, etc., esp. on the centre of a shield. **2** *Archit.* ornamental carving etc. at the junction of the ribs in a vault. [French]

bossa nova /ˌbɒsə ˈnəʊvə/ *n.* **1** dance like the samba. **2** music for this. [Portuguese, = new flair]

boss-eyed *adj. colloq.* **1** cross-eyed; blind in one eye. **2** crooked. [*boss* = bad shot, origin unknown]

bossy *adj.* (**-ier**, **-iest**) *colloq.* domineering. □ **bossiness** *n.*

bosun (also **bo'sun**) var. of BOATSWAIN.

botany /'bɒtənɪ/ *n.* the study of plants. □ **botanic** /bə'tænɪk/ *adj.* **botanical** /bə'tænɪk(ə)l/ *adj.* **botanist** *n.* [Greek *botanē* plant]

botch (also **bodge**) *–v.* **1** bungle; do badly. **2** patch clumsily. *–n.* bungled or spoilt work. [origin unknown]

both /bəʊθ/ *–adj. & pron.* the two, not only one (*both boys*; *both the boys*; *both of the boys*; *I like both*). *–adv.* with equal truth in two cases (*is both hot and dry*). [Old Norse]

bother /'bɒðə(r)/ *–v.* **1** trouble; worry, disturb. **2** (often foll. by *about*, *with*, or *to* + infin.) take the time or trouble (*didn't bother to tell me*; *shan't bother with dessert*). *–n.* **1 a** person or thing that bothers. **b** minor nuisance. **2** trouble, worry. *–int.* expressing irritation. [Irish *bodhraim* deafen]

botheration /,bɒðə'reɪʃ(ə)n/ *n. & int. colloq.* = BOTHER *n., int.*

bothersome /'bɒðəsəm/ *adj.* causing bother.

bottle /'bɒt(ə)l/ *–n.* **1** container, esp. glass or plastic, for storing liquid. **2** amount filling it. **3** baby's feeding-bottle. **4** = HOT-WATER BOTTLE. **5** metal cylinder for liquefied gas. **6** *slang* courage. *–v.* **(-ling) 1** put into, or preserve in, bottles or jars. **2** (foll. by *up*) conceal or restrain (esp. a feeling). □ **hit the bottle** *slang* drink heavily. [medieval Latin: related to BUTT[4]]

bottle bank *n.* place for depositing bottles for recycling.

bottle-feed *v.* feed (a baby) from a bottle as opposed to the breast.

bottle green *adj. & n.* (as adj. often hyphenated) dark green.

bottleneck *n.* **1** narrow congested area, esp. on a road. **2** impeding thing.

bottlenose dolphin *n.* dolphin with a bottle-shaped snout.

bottle party *n.* party to which guests bring bottles of drink.

bottom /'bɒtəm/ *–n.* **1 a** lowest point or part. **b** base. **c** underneath part. **d** furthest or inmost part. **2** *colloq.* **a** buttocks. **b** seat of a chair etc. **3 a** less honourable end of a table, class, etc. **b** person occupying this (*he's bottom of the class*). **4** ground below water. **5** basis or origin. **6** essential character. *–adj.* lowest, last. *–v.* **1** put a bottom to (a chair etc.). **2** find the extent of. **3** touch the bottom or lowest point (of). □ **at bottom** basically. **be at the bottom of** have caused. **bottom out** reach the lowest level. **get to the bottom of** fully investigate and explain. [Old English]

bottom drawer *n.* linen etc. stored by a woman for marriage.

bottomless *adj.* **1** without a bottom. **2** inexhaustible.

bottom line *n. colloq.* underlying truth; ultimate, esp. financial, criterion.

botulism /'bɒtjʊ,lɪz(ə)m/ *n.* poisoning caused by a bacillus in badly preserved food. [Latin *botulus* sausage]

bouclé /'bu:kleɪ/ *n.* **1** looped or curled yarn (esp. wool). **2** fabric made of this. [French, = curled]

boudoir /'bu:dwɑ:(r)/ *n.* woman's private room. [French *bouder* sulk]

bouffant /'bu:fɑ̃/ *adj.* (of a dress, hair, etc.) puffed out. [French]

bougainvillaea /,bu:gən'vɪlɪə/ *n.* tropical plant with large coloured bracts. [*Bougainville*, name of a navigator]

bough /baʊ/ *n.* main branch of a tree. [Old English]

bought *past* and *past part.* of BUY.

bouillon /'bu:jɒn/ *n.* clear broth. [French *bouillir* to boil]

boulder /'bəʊldə(r)/ *n.* large smooth rock. [Scandinavian]

boule /bu:l/ *n.* (also **boules** pronunc. same) French form of bowls played on rough ground. [French]

boulevard /'bu:lə,vɑ:d/ *n.* **1** broad tree-lined avenue. **2** esp. *US* broad main road. [French from German]

boult var. of BOLT[2].

bounce *–v.* **(-cing) 1** (cause to) rebound. **2** *slang* (of a cheque) be returned by a bank when there are no funds to meet it. **3** (foll. by *about*, *up*, *in*, *out*, etc.) jump, move, or rush boisterously. *–n.* **1 a** rebound. **b** power of rebounding. **2** *colloq.* **a** swagger, self-confidence. **b** liveliness. □ **bounce back** recover well after a setback. □ **bouncy** *adj.* **(-ier, -iest)**. [imitative]

bouncer *n.* **1** *slang* doorman ejecting troublemakers from a dancehall, club, etc. **2** = BUMPER 3.

bouncing *adj.* (esp. of a baby) big and healthy.

bound[1] *–v.* **1** spring, leap. **2** (of a ball etc.) bounce. *–n.* **1** springy leap. **2** bounce. [French *bondir* from Latin *bombus* hum]

bound[2] *–n.* (usu. in *pl.*) **1** limitation; restriction. **2** border, boundary. *–v.* **1** limit. **2** be the boundary of. □ **out of bounds** outside a permitted area. [French from medieval Latin]

bound[3] *adj.* **1** (usu. foll. by *for*) starting or having started (*bound for stardom*). **2** (in *comb.*) in a specified direction (*northbound*). [Old Norse, = ready]

bound[4] *past* and *past part.* of BIND. □ **bound to** certain to (*he's bound to come*). **bound up with** closely associated with.

boundary /'baʊndərɪ/ *n.* (*pl.* **-ies**) **1** line marking the limits of an area etc. **2** *Cricket* hit crossing the limits of the field, scoring 4 or 6 runs. [related to BOUND[2]]
bounden duty /'baʊnd(ə)n/ *n. formal* solemn responsibility. [archaic past part. of BIND]
bounder *n. colloq.* cad.
boundless *adj.* unlimited.
bounteous /'baʊntɪəs/ *adj. poet.* = BOUNTIFUL. [French: related to BOUNTY]
bountiful /'baʊntɪ,fʊl/ *adj.* **1** generous. **2** ample.
bounty /'baʊntɪ/ *n.* (*pl.* **-ies**) **1** generosity. **2** reward, esp. from the State. **3** gift. [French from Latin *bonus* good]
bouquet /bu:'keɪ/ *n.* **1** bunch of flowers, esp. professionally arranged. **2** scent of wine etc. **3** compliment. [French *bois* wood]
bouquet garni /,bu:'keɪ 'gɑ:nɪ/ *n.* (*pl.* **bouquets garnis** /,bu:keɪz 'gɑ:nɪ/) bunch or bag of herbs for seasoning.
bourbon /'bɜ:bən/ *n. US* whisky from maize and rye. [*Bourbon* County, Kentucky]
bourgeois /'bʊəʒwɑ:/ often *derog.* –*adj.* **1 a** conventionally middle-class. **b** materialistic. **2** capitalist. –*n.* (*pl.* same) bourgeois person. [French]
bourgeoisie /,bʊəʒwɑ:'zi:/ *n.* **1** capitalist class. **2** middle class. [French]
bourn /bɔ:n/ *n.* small stream. [var. of BURN[2]]
bourse /bʊəs/ *n.* **1** (**Bourse**) Paris Stock Exchange. **2** money-market. [French: related to PURSE]
bout *n.* **1** (often foll. by *of*) **a** spell (of work or activity). **b** attack (*bout of flu*). **2** wrestling- or boxing-match. [obsolete *bought* bending]
boutique /bu:'ti:k/ *n.* small shop selling esp. fashionable clothes. [French]
bouzouki /bu:'zu:kɪ/ *n.* (*pl.* **-s**) Greek form of mandolin. [modern Greek]
bovine /'bəʊvaɪn/ *adj.* **1** of cattle. **2** stupid, dull. [Latin *bos* ox]
bovine spongiform encephalopathy see BSE.
bow[1] /bəʊ/ –*n.* **1 a** slip-knot with a double loop. **b** ribbon etc. so tied. **2** curved piece of wood etc. with a string stretched across its ends, for shooting arrows. **3** rod with horsehair stretched along its length, for playing the violin etc. **4** shallow curve or bend; thing of this form. –*v.* (also *absol.*) use a bow on (a violin etc.). [Old English]
bow[2] /baʊ/ –*v.* **1** incline the head or body, esp. in greeting or acknowledgement. **2** submit (*bowed to the inevitable*). **3** cause (the head etc.) to incline. –*n.* act of bowing. □ **bow and scrape** toady. **bow down 1** bend or kneel esp. in submission or reverence. **2** make stoop; crush (*bowed down by care*). **bow out 1** exit (esp. formally). **2** withdraw; retire. **take a bow** acknowledge applause. [Old English]
bow[3] /baʊ/ *n.* **1** (often in *pl.*) front end of a boat. **2** rower nearest this. [Low German or Dutch: related to BOUGH]
bowdlerize /'baʊdlə,raɪz/ *v.* (also **-ise**) (**-zing** or **-sing**) expurgate (a book etc.). □ **bowdlerization** /-'zeɪʃ(ə)n/ *n.* [*Bowdler*, name of an expurgator of Shakespeare]
bowel /'baʊəl/ *n.* **1** (often in *pl.*) = INTESTINE. **2** (in *pl.*) innermost parts. [Latin *botulus* sausage]
bower /'baʊə(r)/ *n.* **1** arbour; summerhouse. **2** *poet.* inner room. [Old English, = dwelling]
bowerbird *n.* Australasian bird, the male of which constructs elaborate runs.
bowie /'bəʊɪ/ *n.* (in full **bowie knife**) a kind of long hunting-knife. [*Bowie*, name of an American soldier]
bowl[1] /bəʊl/ *n.* **1 a** usu. round deep basin for food or liquid. **b** contents of a bowl. **2** hollow part of a tobacco-pipe, spoon, etc. □ **bowlful** *n.* (*pl.* **-s**). [Old English]
bowl[2] /bəʊl/ –*n.* **1** hard heavy ball, made with a bias to run in a curve. **2** (in *pl.*; usu. treated as *sing.*) game played with these on grass. **3** spell or turn of bowling in cricket. –*v.* **1 a** roll (a ball etc.). **b** play bowls. **2** (also *absol.*) *Cricket* etc. **a** deliver (a ball, over, etc.). **b** (often foll. by *out*) dismiss (a batsman) by knocking down the wicket with a ball. **3** (often foll. by *along*) go along rapidly. □ **bowl out** *Cricket* etc. dismiss (a batsman or a side). **bowl over 1** knock down. **2** *colloq.* impress greatly, overwhelm. [Latin *bulla* bubble]
bow-legs *n.pl.* bandy legs. □ **bow-legged** *adj.*
bowler[1] *n.* **1** *Cricket* etc. player who bowls. **2** bowls-player.
bowler[2] *n.* (in full **bowler hat**) man's hard round felt hat. [*Bowler*, name of a hatter]
bowline /'bəʊlɪn/ *n.* **1** rope from a ship's bow keeping the sail taut against the wind. **2** knot forming a non-slipping loop at the end of a rope.
bowling *n.* the game of skittles, tenpin bowling, or bowls.
bowling-alley *n.* **1** long enclosure for skittles or tenpin bowling. **2** building with these.
bowling-green *n.* lawn for playing bowls.
bowman *n.* archer.

bowsprit /ˈbəʊsprɪt/ *n.* spar running forward from a ship's bow.

bowstring *n.* string of an archer's bow.

bow-tie *n.* necktie in the form of a bow.

bow-window *n.* curved bay window.

bow-wow –*int.* /baʊˈwaʊ/ imitation of a dog's bark. –*n.* /ˈbaʊwaʊ/ *colloq.* dog. [imitative]

box[1] –*n.* **1** container, usu. flat-sided and firm. **2** amount contained in a box. **3** compartment, e.g. in a theatre or lawcourt. **4** receptacle or kiosk for a special purpose (often in *comb.*: *money box*; *telephone box*). **5** facility at a newspaper office for receiving replies to an advertisement. **6** (prec. by *the*) *colloq.* television. **7** enclosed area or space. **8** area of print enclosed by a border. **9** light shield for the genitals in cricket etc. **10** (prec. by *the*) *Football colloq.* penalty area. –*v.* **1** put in or provide with a box. **2** (foll. by *in*, *up*) confine. [Latin *buxis*: related to BOX[3]]

box[2] –*v.* **1 a** take part in boxing. **b** fight (an opponent) at boxing. **2** slap (esp. a person's ears). –*n.* hard slap, esp. on the ears. [origin unknown]

box[3] *n.* **1** small evergreen tree with dark green leaves. **2** its fine hard wood. [Latin *buxus*, Greek *puxos*]

Box and Cox *n.* two people sharing accommodation etc. in shifts. [names of characters in a play (1847)]

box camera *n.* simple box-shaped camera.

boxer *n.* **1** person who boxes, esp. as a sport. **2** medium-size short-haired dog with a puglike face.

boxer shorts *n.pl.* men's loose underpants like shorts.

box girder *n.* hollow girder square in cross-section.

boxing *n.* fighting with the fists, esp. as a sport.

Boxing Day *n.* first weekday after Christmas. [from BOX[1], from the custom of giving Christmas-boxes]

boxing glove *n.* each of a pair of heavily padded gloves worn in boxing.

box junction *n.* road area marked with a yellow grid, which a vehicle should enter only if its exit is clear.

box number *n.* number for replies to a private advertisement in a newspaper.

box office *n.* ticket-office at a theatre etc.

box pleat *n.* arrangement of parallel pleats folding in alternate directions.

boxroom *n.* small room for storing boxes, cases, etc.

box spring *n.* each of a set of vertical springs in a frame, e.g. in a mattress.

boxwood *n.* = BOX[3] 2.

boxy *adj.* **(-ier, -iest)** cramped.

boy –*n.* **1** male child, son. **2** young man. **3** male servant etc. –*int.* expressing pleasure, surprise, etc. □ **boyhood** *n.* **boyish** *adj.* [origin uncertain]

boycott /ˈbɔɪkɒt/ –*v.* **1** refuse to have social or commercial relations with (a person, country, etc.). **2** refuse to handle (goods). –*n.* such a refusal. [Capt. *Boycott*, so treated from 1880]

boyfriend *n.* person's regular male companion or lover.

boyo /ˈbɔɪəʊ/ *n.* (*pl.* **-s**) *Welsh & Ir. colloq.* (esp. as a form of address) boy, mate.

boy scout *n.* = SCOUT *n.* 4.

BP *abbr.* **1** boiling-point. **2** blood pressure. **3** before the present (era). **4** British Petroleum. **5** British Pharmacopoeia.

Bq *abbr.* becquerel.

BR *abbr.* British Rail.

Br *symb.* bromine.

bra /brɑː/ *n.* undergarment worn by women to support the breasts. [abbreviation]

brace –*n.* **1** device that clamps or fastens tightly. **2** timber etc. strengthening a framework. **3** (in *pl.*) straps supporting trousers from the shoulders. **4** wire device for straightening the teeth. **5** (*pl.* same) pair (esp. of game). **6** rope for trimming a sail. **7** connecting mark { or } in printing. –*v.* **(-cing) 1** make steady by supporting. **2** fasten tightly to make firm. **3** (esp. as **bracing** *adj.*) invigorate, refresh. **4** (often *refl.*) prepare for a difficulty, shock, etc. [Latin *bracchia* arms]

brace and bit *n.* revolving tool for boring, with a D-shaped central handle.

bracelet /ˈbreɪslɪt/ *n.* **1** ornamental band or chain worn on the wrist or arm. **2** *slang* handcuff.

brachiosaurus /ˌbreɪkɪəˈsɔːrəs, ˌbræk-/ *n.* (*pl.* **-ruses**) plant-eating dinosaur with forelegs longer than its hind legs. [Latin from Greek *brakhīōn* arm, *sauros* lizard]

bracken /ˈbrækən/ *n.* **1** large coarse fern. **2** mass of these. [Old Norse]

bracket /ˈbrækɪt/ –*n.* **1** (esp. angled) support projecting from a vertical surface. **2** shelf fixed to a wall with this. **3** each of a pair of marks () [] {} enclosing words or figures. **4** group or classification (*income bracket*). –*v.* **(-t-) 1** enclose in brackets. **2** group or classify together. [Latin *bracae* breeches]

brackish /ˈbrækɪʃ/ *adj.* (of water etc.) slightly salty. [Low German or Dutch]

bract *n.* leaf-like and often brightly coloured part of a plant, growing before the flower. [Latin *bractea* thin sheet]

brad *n.* thin flat nail with a head on only one side. [Old Norse]

bradawl /ˈbrædɔːl/ *n.* small pointed tool for boring holes by hand.

brae /breɪ/ *n. Scot.* hillside. [Old Norse]

brag –*v.* (**-gg-**) talk boastfully. –*n.* **1** card-game like poker. **2** boastful statement or talk. [origin unknown]

braggart /ˈbrægət/ –*n.* boastful person. –*adj.* boastful.

Brahma /ˈbrɑːmə/ *n.* **1** Hindu Creator. **2** supreme divine Hindu reality. [Sanskrit, = creator]

Brahman /ˈbrɑːmən/ *n.* (also **brahman**) (*pl.* **-s**) **1** (also **Brahmin**) member of the highest or priestly Hindu caste. **2** = BRAHMA 2. □ **Brahmanic** /-ˈmænɪk/ *adj.* **Brahmanism** *n.*

braid –*n.* **1** woven band as edging or trimming. **2** *US* plait of hair. –*v.* **1** *US* plait. **2** trim with braid. □ **braiding** *n.* [Old English]

Braille /breɪl/ –*n.* system of writing and printing for the blind, with patterns of raised dots. –*v.* (**-ling**) print or transcribe in Braille. [*Braille*, name of its inventor]

brain –*n.* **1** organ of soft nervous tissue in the skull of vertebrates, the centre of sensation and of intellectual and nervous activity. **2 a** *colloq.* intelligent person. **b** (often in *pl.*) intelligence. **3** (usu. in *pl.*; prec. by *the*) *colloq.* cleverest person in a group; mastermind. **4** electronic device functioning like a brain. –*v.* **1** dash out the brains of. **2** *colloq.* strike hard on the head. □ **on the brain** *colloq.* obsessively in one's thoughts. [Old English]

brainchild *n. colloq.* person's clever idea or invention.

brain death *n.* irreversible brain damage causing the end of independent respiration, regarded as indicative of death. □ **brain-dead** *adj.*

brain drain *n. colloq.* loss of skilled personnel by emigration.

brainless *adj.* foolish.

brainpower *n.* mental ability or intelligence.

brainstorm *n.* **1** sudden mental disturbance. **2** *colloq.* mental lapse. **3** *US* brainwave. **4** pooling of spontaneous ideas about a problem etc. □ **brainstorming** *n.* (in sense 4).

brains trust *n.* group of experts answering questions, usu. publicly and impromptu.

brainwash *v.* implant ideas or esp. ideology into (a person) by repetition etc. □ **brainwashing** *n.*

brainwave *n.* **1** (usu. in *pl.*) electrical impulse in the brain. **2** *colloq.* sudden bright idea.

brainy *adj.* (**-ier, -iest**) intellectually clever.

braise /breɪz/ *v.* (**-sing**) stew slowly with a little liquid in a closed container. [French *braise* live coals]

brake[1] –*n.* **1** (often in *pl.*) device for stopping or slowing a wheel, vehicle, etc. **2** thing that impedes. –*v.* (**-king**) **1** apply a brake. **2** slow or stop with a brake. [probably obsolete *brake* = curb]

brake[2] *n.* large estate car. [var. of BREAK]

brake[3] –*n.* **1** toothed instrument for crushing flax and hemp. **2** (in full **brake harrow**) heavy harrow. –*v.* (**-king**) crush (flax or hemp). [Low German or Dutch: related to BREAK]

brake[4] *n.* thicket or clump of brushwood. [Old English]

brake drum *n.* cylinder attached to a wheel, on which the brake shoes press to brake.

brake horsepower *n.* power of an engine measured by the force needed to brake it.

brake lining *n.* strip of fabric increasing the friction of a brake shoe.

brake shoe *n.* long curved block which presses on a brake drum to brake.

bramble /ˈbræmb(ə)l/ *n.* wild thorny shrub, esp. the blackberry. □ **brambly** *adj.* [Old English]

brambling /ˈbræmblɪŋ/ *n.* the speckled finch. [German: related to BRAMBLE]

bran *n.* grain husks separated from flour. [French]

branch /brɑːntʃ/ –*n.* **1** limb of a tree or bough. **2** lateral extension or subdivision, esp. of a river, road, or railway. **3** subdivision of a family, knowledge, etc. **4** local office etc. of a large business. –*v.* (often foll. by *off*) **1** diverge. **2** divide into branches. □ **branch out** extend one's field of interest. [Latin *branca* paw]

brand –*n.* **1 a** particular make of goods. **b** identifying trade mark, label, etc. **2** (usu. foll. by *of*) characteristic kind (*brand of humour*). **3** identifying mark burned esp. on livestock. **4** iron used for this. **5** piece of burning or charred wood. **6** stigma; mark of disgrace. **7** *poet.* torch. –*v.* **1** mark with a hot iron. **2** stigmatize (*branded him a liar*). **3** impress unforgettably. **4** assign a trademark etc. to. [Old English]

brandish /ˈbrændɪʃ/ *v.* wave or flourish as a threat or display. [French from Germanic]

brand-new *adj.* completely new.

brandy /ˈbrændɪ/ *n.* (*pl.* **-ies**) strong alcoholic spirit distilled from wine or fermented fruit juice. [Dutch *brandewijn*]

brandy butter *n.* mixture of brandy, butter, and sugar.

brandy-snap *n.* crisp rolled gingerbread wafer usu. filled with cream.

bran-tub *n.* lucky dip with prizes hidden in bran.

brash *adj.* vulgarly self-assertive; impudent. □ **brashly** *adv.* **brashness** *n.* [dial.]

brass /brɑːs/ –*n.* **1** yellow alloy of copper and zinc. **2** brass objects collectively. **3** brass wind instruments. **4** *slang* money. **5** brass memorial tablet. **6** *colloq.* effrontery. –*adj.* made of brass. □ **brassed off** *slang* fed up. [Old English]

brass band *n.* band of brass instruments.

brasserie /ˈbræsərɪ/ *n.* restaurant, orig. one serving beer with food. [French *brasser* brew]

brassica /ˈbræsɪkə/ *n.* plant of the cabbage family. [Latin, = cabbage]

brassière /ˈbræzɪə(r)/ *n.* = BRA. [French]

brass monkey *n. coarse slang* used in various phrases to indicate extreme cold.

brass-rubbing *n.* **1** practice of taking impressions by rubbing heelball etc. over paper laid on engraved brasses. **2** impression obtained by this.

brass tacks *n.pl. slang* essential details.

brassy /ˈbrɑːsɪ/ *adj.* (**-ier, -iest**) **1** of or like brass. **2** impudent. **3** vulgarly showy. **4** loud and blaring.

brat *n.* usu. *derog.* child, esp. an ill-behaved one. [origin unknown]

bravado /brəˈvɑːdəʊ/ *n.* show of boldness. [Spanish]

brave –*adj.* **1** able or ready to face and endure danger, disgrace, or pain. **2** *formal* splendid, spectacular. –*n.* American Indian warrior. –*v.* (**-ving**) face bravely or defiantly. □ **bravely** *adv.* **braveness** *n.* **bravery** *n.* [ultimately Latin *barbarus* barbarian]

bravo /brɑːˈvəʊ/ –*int.* expressing approval. –*n.* (*pl.* **-s**) cry of 'bravo'. [French from Italian]

bravura /brəˈvjʊərə/ *n.* **1** brilliance of execution. **2** (often *attrib.*) passage of (esp. vocal) music requiring brilliant technique. [Italian]

brawl –*n.* noisy quarrel or fight. –*v.* **1** engage in a brawl. **2** (of a stream) run noisily. [Provençal]

brawn *n.* **1** muscular strength. **2** muscle; lean flesh. **3** jellied meat made from a pig's head. □ **brawny** *adj.* (**-ier, -iest**). [French from Germanic]

bray –*n.* **1** cry of a donkey. **2** harsh sound like this. –*v.* **1** make a bray. **2** utter harshly. [French *braire*]

braze *v.* (**-zing**) solder with an alloy of brass and zinc. [French *braser*]

brazen /ˈbreɪz(ə)n/ –*adj.* **1** shameless; insolent. **2** of or like brass. **3** harsh in sound. –*v.* (foll. by *out*) face or undergo defiantly (*brazen it out*). □ **brazenly** *adv.* [Old English]

brazier[1] /ˈbreɪzɪə(r)/ *n.* metal pan or stand holding burning coals etc. [French: related to BRAISE]

brazier[2] /ˈbreɪzɪə(r)/ *n.* worker in brass. [probably from BRASS]

Brazil /brəˈzɪl/ *n.* **1** tall S. American tree. **2** (in full **Brazil nut**) its large three-sided nut. [*Brazil* in S. America]

breach –*n.* **1** (often foll. by *of*) breaking or non-observation of a law, contract, etc. **2** breaking of relations; quarrel. **3** opening, gap. –*v.* **1** break through; make a gap in. **2** break (a law, contract, etc.). □ **step into the breach** help in a crisis, esp. as a replacement. [Germanic: related to BREAK]

breach of promise *n.* breaking of a promise, esp. to marry.

breach of the peace *n.* crime of causing a public disturbance.

bread /bred/ –*n.* **1** baked dough of flour and water, usu. leavened with yeast. **2** necessary food. **3** *slang* money. –*v.* coat with breadcrumbs for cooking. [Old English]

bread and butter –*n.* one's livelihood. –*attrib. adj.* (**bread-and-butter**) done or produced to earn a basic living.

breadboard *n.* **1** board for cutting bread on. **2** board for making an experimental model of an electric circuit.

breadcrumb *n.* small fragment of bread, esp. (in *pl.*) for use in cooking.

breadfruit *n.* **1** fruit which resembles new bread when roasted. **2** tropical evergreen tree bearing it.

breadline *n.* subsistence level (esp. *on the breadline*).

bread sauce *n.* white sauce thickened with breadcrumbs.

breadth /bredθ/ *n.* **1** distance or measurement from side to side of a thing. **2** freedom from prejudice or intolerance. [Old English: related to BROAD]

breadwinner *n.* person who works to support a family.

break /breɪk/ –*v.* (*past* **broke**; *past part.* **broken** /ˈbrəʊkən/) **1 a** separate into pieces under a blow or strain; shatter. **b** make or become inoperative. **c** break a bone in or dislocate (part of the body). **2 a** interrupt (*broke our journey*). **b** have an interval (*broke for tea*). **3** fail to keep (a law, promise, etc.). **4 a** make or become subdued or weak; (cause to) yield; destroy. **b** weaken the effect of (a fall, blow, etc.). **c** = *break in* 3c. **5** surpass (a record). **6** (foll. by *with*) end a friendship with (a person etc.). **7 a** be no longer subject to (a habit). **b** (foll. by *of*) free (a person) from a habit (*broke them of their addiction*). **8** reveal or be revealed (*broke the news*; *story broke*). **9 a** (of fine weather) change suddenly. **b** (of waves) curl over and foam. **c** (of the day) dawn. **d** (of clouds) move apart. **e** (of a storm) begin violently. **10** *Electr.* disconnect (a circuit). **11 a** (of the voice) change with emotion. **b** (of a boy's voice) change at puberty. **12 a** (often foll. by *up*) divide (a set etc.). **b** change (a banknote etc.) for coins. **13** ruin financially (see also BROKE *adj.*). **14** penetrate (e.g. a safe) by force. **15** decipher (a code). **16** make (a way, path, etc.) by force. **17** burst forth (*sun broke through*). **18 a** (of troops) disperse in confusion. **b** rupture (ranks). **19 a** (usu. foll. by *free*, *loose*, *out*, etc.) escape by a sudden effort. **b** escape or emerge from (prison, bounds, cover, etc.). **20** *Tennis etc.* win a game against (an opponent's service). **21** (of boxers etc.) come out of a clinch. **22** *Billiards* etc. disperse the balls at the start of a game. *–n.* **1 a** act or instance of breaking. **b** point of breaking; gap. **2** interval, interruption; pause. **3** sudden dash (esp. to escape). **4** *colloq.* piece of luck; fair chance. **5** *Cricket* deflection of a bowled ball on bouncing. **6** *Billiards* etc. **a** series of points scored during one turn. **b** opening shot dispersing the balls. □ **break away** make or become free or separate. **break the back of** do the hardest or greatest part of. **break down 1 a** fail mechanically; cease to function. **b** (of human relationships etc.) fail, collapse. **c** fail in (esp. mental) health. **d** collapse in tears or emotion. **2 a** demolish, destroy. **b** suppress (resistance). **c** force to yield. **3** analyse into components. **break even** make neither profit nor loss. **break a person's heart** see HEART. **break the ice** begin to overcome formality or shyness. **break in 1** enter by force, esp. with criminal intent. **2** interrupt. **3 a** accustom to a habit etc. **b** wear etc. until comfortable. **c** tame (an animal); accustom (a horse) to a saddle etc. **break in on** disturb; interrupt. **break into 1** enter forcibly. **2 a** burst forth with (a song, laughter, etc.). **b** change pace for (a faster one) (*broke into a gallop*). **3** interrupt. **break off 1** detach by breaking. **2** bring to an end. **3** cease talking etc. **break open** open forcibly. **break out 1** escape by force, esp. from prison. **2** begin suddenly. **3** (foll. by *in*) become covered in (a rash etc.). **4** exclaim. **break up 1** break into small pieces. **2** disperse; disband. **3** end the school term. **4** (cause to) terminate a relationship; disband. **break wind** release gas from the anus. [Old English]

breakable *–adj.* easily broken. *–n.* (esp. in *pl.*) breakable thing.

breakage *n.* **1 a** broken thing. **b** damage caused by breaking. **2** act or instance of breaking.

breakaway *n.* (often *attrib.*) breaking away; secession (*breakaway group*).

break-dancing *n.* acrobatic style of street-dancing.

breakdown *n.* **1 a** mechanical failure. **b** loss of (esp. mental) health. **2** collapse (*breakdown of communication*). **3** analysis (of statistics etc.).

breaker *n.* **1** heavy breaking wave. **2** person or thing that breaks something, esp. disused machinery.

breakfast /'brekfəst/ *–n.* first meal of the day. *–v.* have breakfast.

break-in *n.* illegal forced entry, esp. with criminal intent.

breaking and entering *n.* (formerly) the illegal entering of a building with intent to commit a felony.

breaking-point *n.* point of greatest strain.

breakneck *attrib. adj.* (of speed) dangerously fast.

break-out *n.* forcible escape.

breakthrough *n.* **1** major advance or discovery. **2** act of breaking through an obstacle etc.

breakup *n.* **1** disintegration or collapse. **2** dispersal.

breakwater *n.* barrier breaking the force of waves.

bream *n.* (*pl.* same) **1** yellowish arch-backed freshwater fish. **2** (in full **sea bream**) similar marine fish. [French from Germanic]

breast /brest/ *–n.* **1 a** either of two milk-secreting organs on a woman's chest. **b** corresponding part of a man's body. **2 a** chest. **b** corresponding part of an animal. **3** part of a garment that covers the breast. **4** breast as a source of nourishment or emotion. *–v.* **1** contend with. **2** reach the top of (a hill). □ **make a clean breast of** confess fully. [Old English]

breastbone *n.* thin flat vertical bone in the chest between the ribs.

breast-feed *v.* feed (a baby) from the breast.

breastplate *n.* armour covering the breast.

breast-stroke *n.* swimming stroke made by extending both arms forward and sweeping them back.

breastwork *n.* low temporary defence or parapet.

breath /breθ/ *n.* **1 a** air drawn into or expelled from the lungs. **b** one respiration of air. **c** breath as perceived by the senses. **2 a** slight movement of air. **b** whiff (of perfume etc.). **3** whisper, murmur (esp. of scandal). □ **catch one's breath 1** cease breathing momentarily in surprise etc. **2** rest to restore normal breathing. **hold one's breath** cease breathing temporarily. **out of breath** gasping for air, esp. after exercise. **take one's breath away** surprise, delight, etc. **under one's breath** in a whisper. [Old English]

Breathalyser /'breθə,laɪzə(r)/ *n.* (also **-lyzer**) *propr.* instrument for measuring alcohol levels in the breath exhaled into it. □ **breathalyse** *v.* (also **-lyze**) (**-sing** or **-zing**). [from BREATH, ANALYSE]

breathe /bri:ð/ *v.* (**-thing**) **1** draw air into and expel it from the lungs. **2** be or seem alive. **3 a** utter or sound (esp. quietly). **b** express (*breathed defiance*). **4** pause. **5** send out or take in (as if) with the breath (*breathed new life into them; breathed whisky*). **6** (of wine etc.) be exposed to the air. □ **breathe again** (or **freely**) feel relief.

breather /'bri:ðə(r)/ *n.* **1** *colloq.* brief pause for rest. **2** brief period in the fresh air.

breathing-space *n.* time to recover; pause.

breathless *adj.* **1** panting, out of breath. **2** holding the breath. **3** still, windless. □ **breathlessly** *adv.*

breathtaking *adj.* astounding; awe-inspiring. □ **breathtakingly** *adv.*

breath test *n.* test with a Breathalyser.

bred *past* and *past part.* of BREED.

breech *n.* back part of a rifle or gun barrel. [Old English]

breech birth *n.* (also **breech delivery**) delivery of a baby with the buttocks or feet foremost.

breeches /'brɪtʃɪz/ *n.pl.* short trousers, esp. fastened below the knee.

breeches buoy *n.* lifebuoy with canvas breeches for the user's legs.

breed *–v.* (*past* and *past part.* **bred**) **1** (of animals) produce young. **2** propagate; raise (animals). **3** yield; result in. **4** arise; spread. **5** bring up; train. **6** create (fissile material) by nuclear reaction. *–n.* **1** stock of similar animals or plants within a species, usu. developed by deliberate selection. **2** race; lineage. **3** sort, kind. □ **breeder** *n.* [Old English]

breeder reactor *n.* nuclear reactor creating surplus fissile material.

breeding *n.* **1** raising of offspring; propagation. **2** social behaviour; ancestry.

breeze[1] *–n.* **1** gentle wind. **2** *colloq.* quarrel. **3** esp. *US colloq.* easy task. *–v.* (**-zing**) (foll. by *in, out, along*, etc.) *colloq.* saunter casually. [probably Spanish and Portuguese *briza*]

breeze[2] *n.* small cinders. [French: related to BRAISE]

breeze-block *n.* lightweight building block, esp. of breeze mixed with sand and cement.

breezy *adj.* (**-ier, -iest**) **1** slightly windy. **2** *colloq.* cheerful, light-hearted, casual.

Bren *n.* (in full **Bren gun**) lightweight quick-firing machine-gun. [*Br*no in Czechoslovakia, *En*field in England]

brent *n.* (in full **brent-goose**) small migratory goose. [origin unknown]

brethren see BROTHER.

Breton /'bret(ə)n/ *–n.* **1** native of Brittany. **2** Celtic language of Brittany. *–adj.* of Brittany, its people, or language. [French, = BRITON]

breve *n.* **1** *Mus.* note twice the length of a semibreve. **2** mark (˘) indicating a short or unstressed vowel. [var. of BRIEF]

breviary /'bri:vɪərɪ/ *n.* (*pl.* **-ies**) book containing the Roman Catholic daily office. [Latin: related to BRIEF]

brevity /'brevɪtɪ/ *n.* **1** economy of expression; conciseness. **2** shortness (of time etc.). [Anglo-French: related to BRIEF]

brew /bru:/ *–v.* **1 a** make (beer etc.) by infusion, boiling, and fermentation. **b** make (tea etc.) by infusion. **2** undergo these processes. **3** gather force; threaten (*storm is brewing*). **4** concoct (a plan etc.). *–n.* **1** liquid or amount brewed; concoction. **2** process of brewing. □ **brew up** make tea. □ **brewer** *n.* [Old English]

brewery /'bru:ərɪ/ *n.* (*pl.* **-ies**) factory for brewing beer etc.

brew-up *n.* instance of making tea.

briar[1] var. of BRIER[1].

briar[2] var. of BRIER[2].

bribe *–v.* (**-bing**) (often foll. by *to* + infin.) persuade to act improperly in one's favour by a gift of money etc. *–n.* money or services offered in bribing. □ **bribery** *n.* [French *briber* beg]

bric-à-brac /'brɪkə,bræk/ *n.* (also **bric-a-brac**) cheap ornaments, trinkets, etc. [French]

brick *–n.* **1 a** small usu. rectangular block of fired or sun-dried clay used in building. **b** material of this. **2** child's toy block. **3** brick-shaped thing. **4** *slang* generous or loyal person. *–v.* (foll. by *in, up*) close or block with brickwork. *–adj.* **1** built of brick (*brick wall*). **2** (also **brick-red**) dull red. [Low German or Dutch]

brickbat *n.* **1** piece of brick, esp. as a missile. **2** insult.

brickie *n. slang* bricklayer.

bricklayer *n.* person who builds with bricks, esp. for a living. □ **bricklaying** *n.*

brickwork *n.* building or work in brick.

brickyard *n.* place where bricks are made.

bridal /'braɪd(ə)l/ *adj.* of a bride or wedding. [Old English]

bride *n.* woman on her wedding day and during the period just before and after it. [Old English]

bridegroom *n.* man on his wedding day and during the period just before and after it. [Old English]

bridesmaid *n.* girl or unmarried woman attending a bride at her wedding.

bridge[1] –*n.* **1 a** structure providing a way across a river, road, railway, etc. **b** thing joining or connecting. **2** operational superstructure on a ship. **3** upper bony part of the nose. **4** piece of wood on a violin etc. over which the strings are stretched. **5** = BRIDGEWORK. –*v.* (**-ging**) **1** be or make a bridge over. **2** reduce (a gap, deficiency, etc.). [Old English]

bridge[2] *n.* card-game derived from whist. [origin unknown]

bridgehead *n.* fortified position held on the enemy's side of a river etc.

bridge roll *n.* small soft bread roll.

bridgework *n. Dentistry* dental structure covering a gap, joined to the teeth on either side.

bridging loan *n.* loan to cover the interval between buying a house etc. and selling another.

bridle /'braɪd(ə)l/ –*n.* **1** headgear for controlling a horse, including reins and bit. **2** restraining thing. –*v.* (**-ling**) **1** put a bridle on. **2** curb, restrain. **3** (often foll. by *up*) express anger, offence, etc., esp. by throwing up the head and drawing in the chin. [Old English]

bridle-path *n.* (also **bridle-way**) rough path for riders or walkers.

Brie /bri:/ *n.* a kind of soft cheese. [*Brie* in N. France]

brief –*adj.* **1** of short duration. **2** concise; abrupt, brusque. **3** scanty (*brief skirt*). –*n.* **1** (in *pl.*) short pants. **2 a** summary of a case drawn up for counsel. **b** piece of work for a barrister. **3** instructions for a task. **4** papal letter on discipline. –*v.* **1** instruct (a barrister) by brief. **2** inform or instruct in advance. □ **hold a brief for** argue in favour of. **in brief** to sum up. □ **briefly** *adv.* **briefness** *n.* [Latin *brevis* short]

briefcase *n.* flat document case.

brier[1] /'braɪə(r)/ *n.* (also **briar**) wild rose or other prickly bush. [Old English]

brier[2] /'braɪə(r)/ *n.* (also **briar**) **1** white heath of S. Europe. **2** tobacco pipe made from its root. [French *bruyère*]

Brig. *abbr.* Brigadier.

brig[1] *n.* two-masted square-rigged ship. [abbreviation of BRIGANTINE]

brig[2] *n. Scot. & N.Engl.* bridge. [var. of BRIDGE[1]]

brigade /brɪ'geɪd/ *n.* **1** military unit, usu. three battalions, as part of a division. **2** group organized for a special purpose. [Italian *briga* strife]

brigadier /,brɪgə'dɪə(r)/ *n.* **1** officer commanding a brigade. **2** staff officer of similar standing.

brigand /'brɪgənd/ *n.* member of a robber band; bandit. □ **brigandage** *n.* [Italian *brigante*: related to BRIGADE]

brigantine /'brɪgən,ti:n/ *n.* two-masted ship with a square-rigged foremast and a fore-and-aft rigged mainmast. [French or Italian: related to BRIGAND]

bright /braɪt/ –*adj.* **1** emitting or reflecting much light; shining. **2** intense, vivid. **3** clever. **4** cheerful. –*adv.* esp. *poet.* brightly. □ **brightly** *adv.* **brightness** *n.* [Old English]

brighten *v.* make or become brighter.

Bright's disease /braɪts/ *n.* kidney disease. [*Bright*, name of a physician]

brill[1] *n.* (*pl.* same) European flat-fish. [origin unknown]

brill[2] *adj. colloq.* = BRILLIANT *adj.* 4. [abbreviation]

brilliant /'brɪlɪənt/ –*adj.* **1** very bright; sparkling. **2** outstandingly talented. **3** showy. **4** *colloq.* excellent. –*n.* diamond of the finest cut with many facets. □ **brilliance** *n.* **brilliantly** *adv.* [French *briller* shine, from Italian]

brilliantine /'brɪljən,ti:n/ *n.* dressing for making the hair glossy. [French: related to BRILLIANT]

brim –*n.* **1** edge or lip of a vessel. **2** projecting edge of a hat. –*v.* (**-mm-**) fill or be full to the brim. □ **brim over** overflow. [origin unknown]

brim-full *adj.* (also **brimful**) filled to the brim.

brimstone *n. archaic* sulphur. [from BURN[1], STONE]

brindled /'brɪnd(ə)ld/ *adj.* (esp. of domestic animals) brown or tawny with streaks of another colour. [Scandinavian]

brine *n.* **1** water saturated or strongly impregnated with salt. **2** sea water. [Old English]

bring *v.* (*past* and *past part.* **brought** /brɔ:t/) **1** come carrying; lead, accompany; convey. **2** cause or result in (*war brings misery*). **3** be sold for; produce as income. **4 a** prefer (a charge). **b** initiate (legal action). **5** cause to become or to reach a state (*brings me alive*; *cannot bring myself to agree*). **6** adduce (evidence, an argument, etc.). □ **bring about** cause to happen. **bring back** call to mind. **bring down 1** cause to fall. **2** lower (a price). **bring forth 1** give birth to. **2** cause. **bring forward 1** move to an earlier time. **2** transfer from the previous page or account. **3** draw attention to. **bring home to** cause to realize fully. **bring the house down** receive rapturous applause. **bring in 1** introduce. **2** yield as income or profit. **bring off** achieve successfully. **bring on** cause to happen, appear, or make progress. **bring out 1** emphasize; make evident. **2** publish. **bring over** convert to one's own side. **bring round 1** restore to consciousness. **2** persuade. **bring through** aid (a person) through adversity, esp. illness. **bring to** restore to consciousness (*brought him to*). **bring up 1** rear (a child). **2** vomit. **3** call attention to. **4** (*absol.*) stop suddenly. [Old English]

bring-and-buy sale *n.* charity sale at which people bring items for sale and buy those brought by others.

brink *n.* **1** extreme edge of land before a precipice, river, etc. **2** furthest point before danger, discovery, etc. □ **on the brink of** about to experience or suffer; in imminent danger of. [Old Norse]

brinkmanship *n.* pursuit (esp. habitual) of danger etc. to the brink of catastrophe.

briny –*adj.* (**-ier, -iest**) of brine or the sea; salty. –*n.* (prec. by *the*) *slang* the sea.

briquette /brɪ'ket/ *n.* block of compressed coal-dust as fuel. [French diminutive: related to BRICK]

brisk –*adj.* **1** quick, lively, keen (*brisk pace, trade*). **2** enlivening (*brisk wind*). –*v.* (often foll. by *up*) make or grow brisk. □ **briskly** *adv.* **briskness** *n.* [probably French BRUSQUE]

brisket /'brɪskɪt/ *n.* animal's breast, esp. as a joint of meat. [French]

brisling /'brɪzlɪŋ/ *n.* small herring or sprat. [Norwegian and Danish]

bristle /'brɪs(ə)l/ –*n.* short stiff hair, esp. one on an animal's back, used in brushes. –*v.* (**-ling**) **1 a** (of hair) stand upright. **b** make (hair) do this. **2** show irritation. **3** (usu. foll. by *with*) be covered or abundant (in). □ **bristly** *adj.* (**-ier, -iest**). [Old English]

Brit *n. colloq.* British person. [abbreviation]

Britannia /brɪ'tænjə/ *n.* personification of Britain, esp. as a helmeted woman with shield and trident. [Latin]

Britannia metal *n.* silvery alloy of tin, antimony, and copper.

Britannic /brɪ'tænɪk/ *adj.* (esp. in **His** or **Her Britannic Majesty**) of Britain.

Briticism /'brɪtɪ,sɪz(ə)m/ *n.* idiom used only in Britain. [after *Gallicism*]

British /'brɪtɪʃ/ –*adj.* of Great Britain, the British Commonwealth, or their people. –*n.* (prec. by *the*; treated as *pl.*) the British people. [Old English]

British English *n.* English as used in Great Britain.

British Legion *n.* = ROYAL BRITISH LEGION.

British summer time *n.* = SUMMER TIME.

British thermal unit *n.* amount of heat needed to raise 1 lb of water through one degree Fahrenheit, equivalent to 1.055×10^3 joules.

Briton /'brɪt(ə)n/ *n.* **1** inhabitant of S. Britain before the Roman conquest. **2** native or inhabitant of Great Britain. [Latin *Britto -onis*]

brittle /'brɪt(ə)l/ *adj.* hard and fragile; apt to break. □ **brittlely** *adv.* (also **brittly**). [Old English]

brittle-bone disease *n.* = OSTEOPOROSIS.

broach –*v.* **1** raise for discussion. **2** pierce (a cask) to draw liquor. **3** open and start using. –*n.* **1** bit for boring. **2** roasting-spit. [Latin *broccus* projecting]

broad /brɔ:d/ –*adj.* **1** large in extent from one side to the other; wide. **2** in breadth (*two metres broad*). **3** extensive (*broad acres*). **4** full and clear (*broad daylight*). **5** explicit (*broad hint*). **6** general (*broad intentions, facts*). **7** tolerant, liberal (*broad view*). **8** coarse (*broad humour*). **9** markedly regional (*broad Scots*). –*n.* **1** broad part (*broad of the back*). **2** *US slang* woman. **3** (**the Broads**) large areas of water in E. Anglia, formed where rivers widen. □ **broadly** *adv.* **broadness** *n.* [Old English]

broad bean *n.* **1** bean with large edible flat seeds. **2** one such seed.

broadcast –*v.* (*past* and *past part.* **broadcast**) **1** transmit by radio or television. **2** take part in such a transmission. **3** scatter (seed etc.). **4** disseminate (information) widely. –*n.* radio or television programme or transmission. □ **broadcaster** *n.* **broadcasting** *n.*

broadcloth *n.* fine cloth of wool, cotton, or silk.

broaden *v.* make or become broader.

broad gauge *n.* railway track with a wider than standard gauge.

broad-leaved *adj.* (of a tree) deciduous and hard-timbered.

broadloom *adj.* (esp. of carpet) woven in broad widths.

broad-minded *adj.* tolerant, liberal.

broadsheet *n.* **1** large-sized newspaper. **2** large sheet of paper printed on one side only.

broadside *n.* **1** vigorous verbal attack. **2** simultaneous firing of all guns from one side of a ship. **3** side of a ship above the water between the bow and quarter. □ **broadside on** sideways on.

broadsword *n.* broad-bladed sword, for cutting rather than thrusting.

brocade /brəˈkeɪd/ *–n.* rich fabric woven with a raised pattern. *–v.* **(-ding)** weave in this way. [Italian *brocco* twisted thread]

broccoli /ˈbrɒkəlɪ/ *n.* brassica with greenish flower-heads. [Italian]

brochure /ˈbrəʊʃə(r)/ *n.* pamphlet or booklet, esp. with descriptive information. [French *brocher* stitch]

broderie anglaise /ˌbrəʊdərɪ ɑ̃ˈgleɪz/ *n.* open embroidery on white linen etc. [French, = English embroidery]

brogue[1] /brəʊg/ *n.* **1** strong outdoor shoe with ornamental perforations. **2** rough shoe of untanned leather. [Gaelic and Irish *brōg* from Old Norse]

brogue[2] /brəʊg/ *n.* marked accent, esp. Irish. [perhaps related to BROGUE[1]]

broil *v.* esp. *US* **1** grill (meat). **2** make or become very hot, esp. from the sun. [French *bruler* burn]

broiler *n.* young chicken for broiling or roasting.

broke *past* of BREAK. *predic. adj. colloq.* having no money.

broken *past part.* of BREAK. *adj.* **1** having been broken. **2** reduced to despair; beaten. **3** (of language) badly spoken, esp. by a foreigner. **4** interrupted (*broken sleep*).

broken-down *adj.* **1** worn out by age, use, etc. **2** not functioning.

broken-hearted *adj.* overwhelmed with grief.

broken home *n.* family disrupted by divorce or separation.

broker *n.* **1** agent; middleman. **2** member of the Stock Exchange dealing in stocks and shares. **3** official appointed to sell or appraise distrained goods. □ **broking** *n.* [Anglo-French]

■ **Usage** In sense 2, brokers have officially been called *broker-dealers* in the UK since Oct. 1986, and entitled to act as agents and principals in share dealings.

brokerage *n.* broker's fee or commission.

brolly /ˈbrɒlɪ/ *n.* (*pl.* **-ies**) *colloq.* umbrella. [abbreviation]

bromide /ˈbrəʊmaɪd/ *n.* **1** any binary compound of bromine. **2** trite remark. **3** reproduction or proof on paper coated with silver bromide emulsion.

bromine /ˈbrəʊmiːn/ *n.* poisonous liquid element with a choking smell. [Greek *brōmos* stink]

bronchial /ˈbrɒŋkɪəl/ *adj.* of the bronchi (see BRONCHUS) or of the smaller tubes into which they divide.

bronchitis /brɒŋˈkaɪtɪs/ *n.* inflammation of the mucous membrane in the bronchial tubes.

bronchus /ˈbrɒŋkəs/ *n.* (*pl.* **-chi** /-kaɪ/) either of the two main divisions of the windpipe. [Latin from Greek]

bronco /ˈbrɒŋkəʊ/ *n.* (*pl.* **-s**) wild or half-tamed horse of the western US. [Spanish, = rough]

brontosaurus /ˌbrɒntəˈsɔːrəs/ *n.* (*pl.* **-ruses**) large plant-eating dinosaur with a long whiplike tail. [Greek *brontē* thunder, *sauros* lizard]

bronze *–n.* **1** alloy of copper and tin. **2** its brownish colour. **3** thing of bronze, esp. a sculpture. *–adj.* made of or coloured like bronze. *–v.* **(-zing)** make or become brown; tan. [French from Italian]

Bronze Age *n. Archaeol.* period when weapons and tools were usu. made of bronze.

bronze medal *n.* medal, usu. awarded as third prize.

brooch /brəʊtʃ/ *n.* ornamental hinged pin. [French *broche*: related to BROACH]

brood *–n.* **1** young of esp. a bird born or hatched at one time. **2** *colloq.* children in a family. *–v.* **1** worry or ponder (esp. resentfully). **2** (of a bird) sit on eggs to hatch them. [Old English]

broody *adj.* **(-ier, -iest) 1** (of a hen) wanting to brood. **2** sullenly thoughtful. **3** *colloq.* (of a woman) wanting pregnancy.

brook[1] /brʊk/ *n.* small stream. [Old English]

brook[2] /brʊk/ *v.* (usu. with *neg.*) *literary* tolerate, allow. [Old English]

broom *n.* **1** long-handled brush for sweeping. **2** shrub with bright yellow flowers. [Old English]

broomstick *n.* handle of a broom.

Bros. *abbr.* Brothers (esp. in the name of a firm).

broth *n.* thin soup of meat or fish stock. [Old English]

brothel /ˈbrɒθ(ə)l/ *n.* premises for prostitution. [originally = worthless fellow, from Old English]

brother /ˈbrʌðə(r)/ *n.* **1** man or boy in relation to his siblings. **2** close male friend or associate. **3** (*pl.* also **brethren** /ˈbreðrɪn/) **a** member of a male religious order, esp. a monk. **b** fellow Christian etc. **4** fellow human being. □ **brotherly** *adj.* [Old English]

brother german see GERMAN.

brotherhood *n.* **1** relationship between brothers. **2** association of people with a common interest. **3** community of feeling between human beings.

brother-in-law *n.* (*pl.* **brothers-in-law**) **1** one's wife's or husband's brother. **2** one's sister's or sister-in-law's husband.

brought *past* and *past part.* of BRING.

brouhaha /'bru:hɑ:,hɑ:/ *n.* commotion; sensation. [French]

brow /braʊ/ *n.* **1** forehead. **2** eyebrow. **3** summit of a hill etc. **4** edge of a cliff etc. [Old English]

browbeat *v.* (*past* **-beat**; *past part.* **-beaten**) intimidate, bully.

brown /braʊn/ *–adj.* **1** having the colour of dark wood or rich soil. **2** dark-skinned or suntanned. **3** (of bread) made from wholemeal or wheatmeal flour. *–n.* **1** brown colour or pigment. **2** brown clothes or material. *–v.* make or become brown. □ **browned off** *colloq.* fed up, disheartened. □ **brownish** *adj.* [Old English]

brown bear *n.* large N. American brown bear.

brown coal *n.* = LIGNITE.

Brownie *n.* **1** (in full **Brownie Guide**) junior Guide. **2** (**brownie**) small square of chocolate cake with nuts. **3** (**brownie**) benevolent elf.

Brownie point *n. colloq.* notional mark awarded for good conduct etc.

browning *n.* additive to colour gravy.

brown owl *n.* **1** any of various owls, esp. the tawny owl. **2** (**Brown Owl**) adult leader of Brownie Guides.

brown rice *n.* unpolished rice.

brown sugar *n.* unrefined or partially refined sugar.

browse /braʊz/ *–v.* (**-sing**) **1** read desultorily or look over goods for sale. **2** (often foll. by *on*) feed on leaves, twigs, etc. *–n.* **1** twigs, shoots, etc. as fodder. **2** act of browsing. [French *brost* bud]

brucellosis /,bru:sɪ'ləʊsɪs/ *n.* bacterial disease, esp. of cattle. [Sir D. *Bruce*, name of a physician]

bruise /bru:z/ *–n.* **1** discolouration of the skin caused esp. by a blow. **2** similar damage on a fruit etc. *–v.* (**-sing**) **1 a** inflict a bruise on. **b** hurt mentally. **2** be susceptible to bruising. [originally = crush, from Old English]

bruiser *n. colloq.* **1** large tough-looking person. **2** professional boxer.

bruit /bru:t/ *v.* (often foll. by *abroad*, *about*) spread (a report or rumour). [French, = noise]

brunch *n.* combined breakfast and lunch. [portmanteau word]

brunette /bru:'net/ *n.* woman with dark brown hair. [French diminutive]

brunt *n.* chief impact of an attack, task, etc. (esp. *bear the brunt of*). [origin unknown]

brush *–n.* **1** implement with bristles, hair, wire, etc. set into a block, for cleaning, painting, arranging the hair, etc. **2** act of brushing. **3** (usu. foll. by *with*) short esp. unpleasant encounter. **4** fox's bushy tail. **5** piece of carbon or metal as an electrical contact esp. with a moving part. **6** = BRUSHWOOD 2. *–v.* **1** sweep, scrub, treat, or tidy with a brush. **2** remove or apply with a brush. **3** graze in passing. □ **brush aside** dismiss curtly or lightly. **brush off** dismiss abruptly. **brush up 1** clean up or smarten. **2** revise (a subject). [French]

brush-off *n.* abrupt dismissal.

brush-up *n.* act of brushing up.

brushwood *n.* **1** undergrowth, thicket. **2** cut or broken twigs etc.

brushwork *n.* **1** use of the brush in painting. **2** painter's style in this.

brusque /brʊsk/ *adj.* abrupt or offhand. □ **brusquely** *adv.* **brusqueness** *n.* [Italian *brusco* sour]

Brussels sprout /'brʌs(ə)lz/ *n.* **1** brassica with small cabbage-like buds on a stem. **2** such a bud. [*Brussels* in Belgium]

brutal /'bru:t(ə)l/ *adj.* **1** savagely cruel. **2** harsh, merciless. □ **brutality** /-'tælɪtɪ/ *n.* (*pl.* **-ies**). **brutally** *adv.* [French: related to BRUTE]

brutalize /'bru:tə,laɪz/ *v.* (also **-ise**) (**-zing** or **-sing**) **1** make brutal. **2** treat brutally.

brute /bru:t/ *–n.* **1 a** brutal or violent person. **b** *colloq.* unpleasant person or difficult thing. **2** animal. *–attrib. adj.* **1** unthinking (*brute force*). **2** cruel; stupid; sensual. □ **brutish** *adj.* **brutishly** *adv.* **brutishness** *n.* [Latin *brutus* stupid]

bryony /'braɪənɪ/ *n.* (*pl.* **-ies**) climbing plant with red berries. [Latin from Greek]

BS *abbr.* **1** Bachelor of Surgery. **2** British Standard(s).

B.Sc. *abbr.* Bachelor of Science.

BSE *abbr.* bovine spongiform encephalopathy, a usu. fatal cattle disease.

BSI *abbr.* British Standards Institution.

BST *abbr.* **1** British Summer Time. **2** bovine somatotrophin, a growth hormone added to cattle-feed to boost milk production.

BT *abbr.* British Telecom.

Bt. *abbr.* Baronet.

B.th.u. *abbr.* (also **B.t.u.**, **BTU**, **B.Th.U.**) British thermal unit(s).

bubble /ˈbʌb(ə)l/ –*n.* **1 a** thin sphere of liquid enclosing air etc. **b** air-filled cavity in a liquid or solidified liquid. **2** transparent domed canopy. **3** visionary or unrealistic project. –*v.* (**-ling**) **1** rise in or send up bubbles. **2** make the sound of boiling. □ **bubble over** (often foll. by *with*) be exuberant. [imitative]

bubble and squeak *n.* cooked cabbage etc. fried with cooked potatoes.

bubble bath *n.* foaming preparation for adding to bath water.

bubble car *n.* small domed car.

bubble gum *n.* chewing-gum that can be blown into bubbles.

bubbly –*adj.* (**-ier**, **-iest**) **1** having or like bubbles. **2** exuberant. –*n. colloq.* champagne.

bubo /ˈbju:bəʊ/ *n.* (*pl.* **-es**) inflamed swelling in the armpit or groin. [Greek *boubōn* groin]

bubonic plague /bju:ˈbɒnɪk/ *n.* contagious disease with buboes.

buccaneer /ˌbʌkəˈnɪə(r)/ *n.* **1** pirate. **2** unscrupulous adventurer. □ **buccaneering** *n.* & *adj.* [French]

buck[1] –*n.* **1** male deer, hare, rabbit, etc. **2** *archaic* dandy. **3** (*attrib.*) *slang* male. –*v.* **1** (of a horse) jump upwards with its back arched. **2** (usu. foll. by *off*) throw (a rider) in this way. **3** (usu. foll. by *up*) *colloq.* **a** cheer up. **b** hurry up; make an effort. [Old English]

buck[2] *n. US slang* dollar. [origin unknown]

buck[3] *n. slang* (in poker) article placed before the next dealer. □ **pass the buck** *colloq.* shift responsibility (to another). [origin unknown]

bucket /ˈbʌkɪt/ –*n.* **1 a** round open container with a handle, for carrying or drawing water etc. **b** amount contained in this. **2** (in *pl.*) *colloq.* large quantities, esp. of rain or tears. **3** scoop in a water-wheel, dredger, etc. –*v.* (**-t-**) *colloq.* **1** (often foll. by *down*) (esp. of rain) pour heavily. **2** (often foll. by *along*) move or drive fast or bumpily. [Anglo-French]

bucket seat *n.* seat with a rounded back for one person, esp. in a car.

bucket-shop *n.* **1** unregistered broking agency. **2** *colloq.* travel agency specializing in cheap air tickets.

buckle /ˈbʌk(ə)l/ –*n.* clasp with a hinged pin for securing a belt, strap, etc. –*v.* (**-ling**) **1** (often foll. by *up*, *on*, etc.) fasten with a buckle. **2** (often foll. by *up*) (cause to) crumple under pressure. □ **buckle down** make a determined effort. [Latin *buccula* cheek-strap]

buckler *n. hist.* small round shield.

buckram /ˈbʌkrəm/ *n.* coarse linen etc. stiffened with paste etc. [French *boquerant*]

Buck's Fizz *n.* cocktail of champagne and orange juice. [*Buck's* Club in London]

buckshee /bʌkˈʃi:/ *adj.* & *adv. slang* free of charge. [corruption of BAKSHEESH]

buckshot *n.* coarse lead shot.

buckskin *n.* **1** leather from a buck's skin. **2** thick smooth cotton or woollen cloth.

buckthorn *n.* thorny shrub with berries formerly used as a purgative.

buck-tooth *n.* upper projecting tooth.

buckwheat *n.* seed of a plant related to rhubarb, used to make flour, or as an alternative to rice. [Dutch, = beech-wheat]

bucolic /bju:ˈkɒlɪk/ –*adj.* of shepherds; rustic, pastoral. –*n.* (usu. in *pl.*) pastoral poem or poetry. [Greek *boukolos* herdsman]

bud –*n.* **1 a** knoblike shoot from which a stem, leaf, or flower develops. **b** flower or leaf not fully open. **2** asexual outgrowth from an organism separating to form a new individual. –*v.* (**-dd-**) **1** form buds. **2** begin to grow or develop (*budding artist*). **3** graft a bud of (a plant) on to another. [origin unknown]

Buddha /ˈbʊdə/ *n.* **1** title of the Indian philosopher Gautama (5th c. BC) and his successors. **2** sculpture etc. of Buddha. [Sanskrit, = enlightened]

Buddhism /ˈbʊdɪz(ə)m/ *n.* Asian religion or philosophy founded by Gautama Buddha. □ **Buddhist** *n.* & *adj.*

buddleia /ˈbʌdlɪə/ *n.* shrub with fragrant flowers attractive to butterflies. [*Buddle*, name of a botanist]

buddy /ˈbʌdɪ/ *n.* (*pl.* **-ies**) esp. *US colloq.* friend or mate. [perhaps from BROTHER]

budge *v.* (**-ging**) (usu. with *neg.*) **1** move slightly. **2** (cause to) change an opinion. □ **budge up** (or **over**) make room for another person by moving. [French *bouger*]

budgerigar /ˈbʌdʒərɪˌgɑ:(r)/ *n.* small parrot, often kept as a cage-bird. [Aboriginal]

budget /ˈbʌdʒɪt/ –*n.* **1** amount of money needed or available. **2 a** (**the Budget**) Government's annual estimate or plan of revenue and expenditure. **b** similar estimate by a company etc. **3** (*attrib.*) inexpensive. –*v.* (**-t-**) (often foll. by *for*) allow or arrange for in a budget. □ **budgetary** *adj.* [Latin *bulga* bag]

budgie /ˈbʌdʒɪ/ *n. colloq.* = BUDGERIGAR. [abbreviation]

buff –*adj.* of a yellowish beige colour (*buff envelope*). –*n.* **1** this colour. **2** (in *comb.*) *colloq.* enthusiast (*railway buff*). **3** velvety dull-yellow ox-leather. –*v.* **1**

polish (metal etc.). **2** make (leather) velvety. □ **in the buff** *colloq.* naked. [originally = buffalo, from French *buffle*]

buffalo /ˈbʌfə,ləʊ/ *n.* (*pl.* same or -es) **1** wild ox of Africa or Asia. **2** N. American bison. [Greek *boubalos* ox]

buffer[1] *n.* **1** thing that deadens impact, esp. a device on a train or at the end of a track. **2** substance that maintains the constant acidity of a solution. **3** *Computing* temporary memory area or queue for data. [imitative]

buffer[2] *n. slang* silly or incompetent old man. [perhaps from BUFFER[1]]

buffer State *n.* small State between two larger ones, regarded as reducing friction.

buffet[1] /ˈbʊfeɪ/ *n.* **1** room or counter where refreshments are sold. **2** self-service meal of several dishes set out at once. **3** also /ˈbʌfɪt/ sideboard or recessed cupboard. [French, = stool]

buffet[2] /ˈbʌfɪt/ –*v.* (-t-) **1** strike repeatedly. **2** contend with (waves etc.). –*n.* **1** blow, esp. of the hand. **2** shock. [French diminutive of *bufe* blow]

buffet car *n.* railway coach serving refreshments.

buffoon /bəˈfuːn/ *n.* clownish or stupid person. □ **buffoonery** *n.* [Latin *buffo* clown]

bug –*n.* **1 a** any of various insects with mouthparts modified for piercing and sucking. **b** esp. *US* small insect. **2** *slang* virus; infection. **3** *slang* concealed microphone. **4** *slang* error in a computer program or system etc. **5** *slang* obsession, enthusiasm, etc. –*v.* (**-gg-**) **1** *slang* conceal a microphone in. **2** *slang* annoy. [origin unknown]

bugbear *n.* **1** cause of annoyance. **2** object of baseless fear. [*bug* = bogey]

bugger *coarse slang* (except in sense 2 of *n.* and 3 of *v.*) –*n.* **1 a** unpleasant or awkward person or thing. **b** person of a specified kind (*clever bugger!*). **2** person who commits buggery. –*v.* **1** as an exclamation of annoyance (*bugger it!*). **2** (often foll. by *up*) **a** ruin; spoil. **b** exhaust. **3** commit buggery with. –*int.* expressing annoyance. □ **bugger-all** nothing. **bugger about** (or **around**) (often foll. by *with*) mess about. **bugger off** (often in *imper.*) go away. [Latin *Bulgarus* Bulgarian heretic]

buggery *n.* **1** anal intercourse. **2** = BESTIALITY 2.

buggy /ˈbʌgɪ/ *n.* (*pl.* -ies) **1** small, sturdy, esp. open, motor vehicle. **2** lightweight pushchair. **3** light, horse-drawn vehicle for one or two people. [origin unknown]

bugle /ˈbjuːg(ə)l/ –*n.* brass military instrument like a small trumpet. –*v.* (**-ling**) **1** sound a bugle. **2** sound (a call etc.) on a bugle. □ **bugler** *n.* [Latin *buculus* young bull]

bugloss /ˈbjuːglɒs/ *n.* plant with bright blue tubular flowers, related to borage. [French *buglosse* from Greek, = ox-tongued]

build /bɪld/ –*v.* (*past* and *past part.* **built** /bɪlt/) **1** construct or cause to be constructed. **2 a** (often foll. by *up*) establish or develop (*built the business up*). **b** (often foll. by *on*) base (hopes, theories, etc.). **3** (as **built** *adj.*) of specified build (*sturdily built*). –*n.* **1** physical proportions (*slim build*). **2** style of construction; make. □ **build in** incorporate. **build on** add (an extension etc.). **build up 1** increase in size or strength. **2** praise; boost. **3** gradually become established. [Old English]

builder *n.* person who builds, esp. a building contractor.

building *n.* **1** permanent fixed structure e.g. a house, factory, or stable. **2** constructing of these.

building society *n.* public finance company paying interest to investors and lending capital for mortgages etc.

build-up *n.* **1** favourable advance publicity. **2** gradual approach to a climax. **3** accumulation or increase.

built *past* and *past part.* of BUILD.

built-in *adj.* integral.

built-up *adj.* **1** (of a locality) densely developed. **2** increased in height etc. by addition. **3** made of prefabricated parts.

bulb *n.* **1 a** globular base of the stem of some plants, sending roots downwards and leaves upwards. **b** plant grown from this, e.g. a daffodil. **2** = LIGHT-BULB. **3** object or part shaped like a bulb. [Latin *bulbus* from Greek, = onion]

bulbous *adj.* bulb-shaped; fat or bulging.

bulge –*n.* **1** irregular swelling. **2** *colloq.* temporary increase (*baby bulge*). –*v.* (**-ging**) swell outwards. □ **bulgy** *adj.* [Latin *bulga* bag]

bulimia /bjuːˈlɪmɪə/ *n.* (in full **bulimia nervosa**) disorder in which overeating alternates with self-induced vomiting, fasting, etc. [Greek *bous* ox, *limos* hunger]

bulk –*n.* **1 a** size; magnitude (esp. large). **b** large mass, body, etc. **c** large quantity. **2** (treated as *pl.* & usu. prec. by *the*) greater part or number (*the bulk of the applicants are women*). **3** roughage. –*v.* **1** seem (in size or importance) (*bulks large*). **2** make (a book etc.) thicker etc. □ **in bulk** in large quantities. [Old Norse]

bulk buying *n.* buying in quantity at a discount.

bulkhead *n.* upright partition in a ship, aircraft, etc.
bulky *adj.* (**-ier, -iest**) awkwardly large. □ **bulkiness** *n.*
bull[1] /bʊl/ *n.* **1 a** uncastrated male bovine animal. **b** male of the whale, elephant, etc. **2** (**the Bull**) zodiacal sign or constellation Taurus. **3** bull's-eye of a target. **4** person who buys shares hoping to sell them at a profit. □ **take the bull by the horns** face danger or a challenge boldly. □ **bullish** *adj.* [Old Norse]
bull[2] /bʊl/ *n.* papal edict. [Latin *bulla* seal]
bull[3] /bʊl/ *n.* **1** *slang* **a** nonsense. **b** unnecessary routine tasks. **2** absurdly illogical statement. [origin unknown]
bulldog *n.* **1** short-haired heavy-jowled sturdy dog. **2** tenacious and courageous person.
bulldog clip *n.* strong sprung clip for papers.
bulldoze *v.* (**-zing**) **1** clear with a bulldozer. **2** *colloq.* **a** intimidate. **b** make (one's way) forcibly.
bulldozer *n.* powerful tractor with a broad vertical blade at the front for clearing ground.
bullet /'bʊlɪt/ *n.* small pointed missile fired from a rifle, revolver, etc. [French diminutive of *boule* ball]
bulletin /'bʊlɪtɪn/ *n.* **1** short official news report. **2** society's regular list of information etc. [Italian diminutive: related to BULL[2]]
bulletproof *adj.* designed to protect from bullets.
bullfight *n.* public baiting, and usu. killing, of bulls. □ **bullfighter** *n.* **bullfighting** *n.*
bullfinch *n.* pink and black finch.
bullfrog *n.* large N. American frog with a booming croak.
bull-headed *adj.* obstinate, blundering.
bullion /'bʊlɪən/ *n.* gold or silver in bulk before coining, or valued by weight. [French: related to BOIL[1]]
bullock /'bʊlək/ *n.* castrated male of domestic cattle. [Old English diminutive of BULL[1]]
bullring *n.* arena for bullfights.
bull's-eye *n.* **1** centre of a target. **2** hard minty sweet. **3** hemispherical ship's window. **4** small circular window. **5 a** hemispherical lens. **b** lantern with this. **6** boss of glass in a blown glass sheet.
bullshit *coarse slang* –*n.* (often as *int.*) nonsense; pretended knowledge. –*v.* (**-tt-**) talk nonsense or as if one has specialist knowledge (to). □ **bullshitter** *n.* [from BULL[3]]
bull-terrier *n.* cross between a bulldog and a terrier.
bully[1] /'bʊlɪ/ –*n.* (*pl.* **-ies**) person coercing others by fear. –*v.* (**-ies, -ied**) persecute or oppress by force or threats. –*int.* (foll. by *for*) often *iron.* expressing approval (*bully for you*). [Dutch]
bully[2] /'bʊlɪ/ (in full **bully off**) –*n.* (*pl.* **-ies**) start of play in hockey in which two opponents strike each other's sticks three times and then go for the ball. –*v.* (**-ies, -ied**) start play in this way. [origin unknown]
bully[3] /'bʊlɪ/ *n.* (in full **bully beef**) corned beef. [French: related to BOIL[1]]
bulrush /'bʊlrʌʃ/ *n.* **1** a kind of tall rush. **2** *Bibl.* papyrus. [perhaps from BULL[1] = coarse + RUSH[2]]
bulwark /'bʊlwək/ *n.* **1** defensive wall, esp. of earth. **2** protecting person or thing. **3** (usu. in *pl.*) ship's side above deck. [Low German or Dutch]
bum[1] *n. slang* buttocks. [origin uncertain]
bum[2] *US slang* –*n.* loafer or tramp; dissolute person. –*v.* (**-mm-**) **1** (often foll. by *around*) loaf or wander around. **2** cadge. –*attrib. adj.* of poor quality. [German *Bummler* loafer]
bum-bag *n. slang* small pouch worn on a belt round the waist or hips.
bumble /'bʌmb(ə)l/ *v.* (**-ling**) **1** (foll. by *on*) speak in a rambling way. **2** (often as **bumbling** *adj.*) be inept; blunder. [from BOOM[1]]
bumble-bee *n.* large bee with a loud hum.
bumf *n. colloq.* usu. *derog.* papers, documents. [abbreviation of *bum-fodder* = toilet-paper]
bump –*n.* **1** dull-sounding blow or collision. **2** swelling or dent so caused. **3** uneven patch on a road etc. **4** prominence on the skull thought to indicate a mental faculty. –*v.* **1 a** hit or come against with a bump. **b** (often foll. by *against, into*) collide. **2** (often foll. by *against, on*) hurt or damage by striking (*bumped my head, the car*). **3** (usu. foll. by *along*) move along with jolts. –*adv.* with a bump; suddenly; violently. □ **bump into** *colloq.* meet by chance. **bump off** *slang* murder. **bump up** *colloq.* increase (prices etc.). □ **bumpy** *adj.* (**-ier, -iest**). [imitative]
bumper *n.* **1** horizontal bar at the front or back of a motor vehicle, reducing damage in a collision. **2** (usu. *attrib.*) unusually large or fine example (*bumper crop*). **3** *Cricket* ball rising high after pitching. **4** brim-full glass.
bumper car *n.* = DODGEM.
bumpkin /'bʌmpkɪn/ *n.* rustic or socially inept person. [Dutch]

bumptious /ˈbʌmpʃəs/ *adj.* offensively self-assertive or conceited. [from BUMP, after *fractious*]

bun *n.* **1** small sweet bread roll or cake, often with dried fruit. **2** hair coiled and pinned to the head. [origin unknown]

bunch –*n.* **1** things gathered together. **2** collection; lot (*best of the bunch*). **3** *colloq.* group; gang. –*v.* **1** make into a bunch; gather into close folds. **2** form into a group or crowd. [origin unknown]

bundle /ˈbʌnd(ə)l/ –*n.* **1** things tied or fastened together. **2** set of nerve fibres etc. **3** *slang* large amount of money. –*v.* (**-ling**) **1** (usu. foll. by *up*) tie or make into a bundle. **2** (usu. foll. by *into*) throw or move carelessly. **3** (usu. foll. by *out*, *off*, *away*, etc.) send away hurriedly. □ **be a bundle of nerves** (or **fun** etc.) be extremely nervous (or amusing etc.). **go a bundle on** *slang* admire; like. [Low German or Dutch]

bun fight *n. slang* tea party.

bung –*n.* stopper, esp. for a cask. –*v.* **1** stop with a bung. **2** *slang* throw. □ **bunged up** blocked up. [Dutch]

bungalow /ˈbʌŋɡə,ləʊ/ *n.* one-storeyed house. [Gujarati, = of Bengal]

bungee /ˈbʌndʒɪ/ *n.* (in full **bungee cord, rope**) elasticated cord or rope used for securing baggage or in bungee jumping.

bungee jumping *n.* sport of jumping from a height while secured by a bungee from the ankles or a harness.

bungle /ˈbʌŋɡ(ə)l/ –*v.* (**-ling**) **1** mismanage or fail at (a task). **2** work badly or clumsily. –*n.* bungled attempt or work. [imitative]

bunion /ˈbʌnjən/ *n.* swelling on the foot, esp. on the big toe. [French]

bunk[1] *n.* shelflike bed against a wall, esp. in a ship. [origin unknown]

bunk[2] *slang* –*v.* (often foll. by *off*) play truant (from). –*n.* (in **do a bunk**) leave or abscond hurriedly. [origin unknown]

bunk[3] *n. slang* nonsense, humbug. [shortening of BUNKUM]

bunk-bed *n.* each of two or more tiered beds forming a unit.

bunker *n.* **1** container for fuel. **2** reinforced underground shelter. **3** sandy hollow in a golf-course. [origin unknown]

bunkum /ˈbʌŋkəm/ *n.* nonsense, humbug. [*Buncombe* in US]

bunny /ˈbʌnɪ/ *n.* (*pl.* **-ies**) **1** child's name for a rabbit. **2** (in full **bunny girl**) club hostess, waitress, etc., wearing rabbit ears and tail. [dial. *bun* rabbit]

Bunsen burner /ˈbʌns(ə)n/ *n.* small adjustable gas burner used in a laboratory. [*Bunsen*, name of a chemist]

bunting[1] /ˈbʌntɪŋ/ *n.* small bird related to the finches. [origin unknown]

bunting[2] /ˈbʌntɪŋ/ *n.* **1** flags and other decorations. **2** loosely-woven fabric for these. [origin unknown]

buoy /bɔɪ/ –*n.* **1** anchored float as a navigation mark etc. **2** lifebuoy. –*v.* **1** (usu. foll. by *up*) **a** keep afloat. **b** encourage, uplift. **2** (often foll. by *out*) mark with a buoy. [Dutch, perhaps from Latin *boia* collar]

buoyant /ˈbɔɪənt/ *adj.* **1** able or apt to keep afloat. **2** resilient; exuberant. □ **buoyancy** *n.* [French or Spanish: related to BUOY]

BUPA /ˈbu:pə/ *abbr.* British United Provident Association, a private health insurance organization.

bur *n.* **1 a** prickly clinging seed-case or flower-head. **b** any plant having these. **2** clinging person. **3** var. of BURR *n.* 2. [Scandinavian]

burble /ˈbɜ:b(ə)l/ *v.* **1** talk ramblingly. **2** make a bubbling sound. [imitative]

burbot /ˈbɜ:bət/ *n.* (*pl.* same) eel-like freshwater fish. [French]

burden /ˈbɜ:d(ə)n/ –*n.* **1** load, esp. a heavy one. **2** oppressive duty, expense, emotion, etc. **3** bearing of loads (*beast of burden*). **4 a** refrain of a song. **b** chief theme of a speech, book, etc. –*v.* load with a burden; oppress. □ **burdensome** *adj.* [Old English: related to BIRTH]

burden of proof *n.* obligation to prove one's case.

burdock /ˈbɜ:dɒk/ *n.* plant with prickly flowers and docklike leaves. [from BUR, DOCK[3]]

bureau /ˈbjʊərəʊ/ *n.* (*pl.* **-x** or **-s** /-z/) **1 a** desk with drawers and usu. an angled hinged top. **b** *US* chest of drawers. **2 a** office or department for specific business. **b** government department. [French, originally = baize]

bureaucracy /bjʊəˈrɒkrəsɪ/ *n.* (*pl.* **-ies**) **1 a** government by central administration. **b** State etc. so governed. **2** government officials, esp. regarded as oppressive and inflexible. **3** conduct typical of these.

bureaucrat /ˈbjʊərə,kræt/ *n.* **1** official in a bureaucracy. **2** inflexible administrator. □ **bureaucratic** /-ˈkrætɪk/ *adj.* **bureaucratically** /-ˈkrætɪkəlɪ/ *adv.*

burette /bjʊəˈret/ *n.* (*US* **buret**) graduated glass tube with an end-tap for measuring liquid in chemical analysis. [French]

burgeon /ˈbɜ:dʒ(ə)n/ *v. literary* grow rapidly; flourish. [Latin *burra* wool]

burger /ˈbɜ:ɡə(r)/ *n. colloq.* hamburger. [abbreviation]

burgher /ˈbɜ:ɡə(r)/ *n.* citizen of a Continental town. [German or Dutch]

burglar /ˈbɜːglə(r)/ *n.* person who commits burglary. [Anglo-French]

burglary *n.* (*pl.* **-ies**) **1** illegal entry with intent to commit theft, do bodily harm, or do damage. **2** instance of this.

■ **Usage** Before 1968 in English law, burglary was a crime under statute and common law; since 1968 it has been a statutory crime only; cf. HOUSEBREAKING.

burgle /ˈbɜːg(ə)l/ *v.* (**-ling**) commit burglary (on).

burgomaster /ˈbɜːgə,mɑːstə(r)/ *n.* mayor of a Dutch or Flemish town. [Dutch]

burgundy /ˈbɜːgəndɪ/ *n.* (*pl.* **-ies**) **1** (also **Burgundy**) **a** red or white wine from Burgundy in E. France. **b** *hist.* similar wine from elsewhere. **2** dark red colour of this.

burial /ˈberɪəl/ *n.* **1 a** burying of a corpse. **b** funeral. **2** *Archaeol.* grave or its remains.

burin /ˈbjʊərɪn/ *n.* **1** tool for engraving copper or wood. **2** *Archaeol.* chisel-pointed flint tool. [French]

burk var. of BERK.

burlesque /bɜːˈlesk/ *–n.* **1 a** comic imitation, parody. **b** this as a genre. **2** *US* variety show, esp. with striptease. *–adj.* of or using burlesque. *–v.* (**-ques, -qued, -quing**) parody. [Italian *burla* mockery]

burly /ˈbɜːlɪ/ *adj.* (**-ier, -iest**) large and sturdy. [Old English]

burn[1] *–v.* (*past* and *past part.* **burnt** or **burned**) **1** (cause to) be consumed or destroyed by fire. **2** blaze or glow with fire. **3** (cause to) be injured or damaged by fire, heat, radiation, acid, etc. **4** use or be used as fuel etc. **5** char in cooking. **6** produce (a hole, mark, etc.) by fire or heat. **7 a** heat (clay, chalk, etc.). **b** harden (bricks) by fire. **8** colour, tan, or parch with heat or light. **9** (be) put to death by fire. **10** cauterize, brand. **11** make, be, or feel hot, esp. painfully. **12** (often foll. by *with*) (cause to) feel great emotion or passion (*burn with shame*). **13** *slang* drive fast. *–n.* mark or injury caused by burning. □ **burn one's boats** (or **bridges**) commit oneself irrevocably. **burn the candle at both ends** work etc. excessively. **burn down** destroy or be destroyed by burning. **burn one's fingers** suffer for meddling or rashness. **burn the midnight oil** read or work late. **burn out 1** be reduced to nothing by burning. **2** (cause to) fail by burning. **3** (usu. *refl.*) suffer exhaustion. **burn up 1** get rid of by fire. **2** begin to blaze. [Old English]

burn[2] *n. Scot.* brook. [Old English]

burner *n.* part of a gas cooker, lamp, etc. that emits the flame.

burning *adj.* **1** ardent, intense. **2** hotly discussed, vital, urgent.

burning-glass *n.* lens for concentrating the sun's rays to produce a flame.

burnish /ˈbɜːnɪʃ/ *v.* polish by rubbing. [French *brunir* from *brun* brown]

burnous /bɜːˈnuːs/ *n.* Arab or Moorish hooded cloak. [Arabic from Greek]

burn-out *n.* exhaustion. □ **burnt-out** *adj.*

burnt see BURN[1].

burnt ochre *n.* (also **burnt sienna** or **umber**) pigment darkened by burning.

burnt offering *n.* offering burnt on an altar as a sacrifice.

burp *colloq. –v.* **1** belch. **2** make (a baby) belch. *–n.* belch. [imitative]

burr *–n.* **1 a** whirring sound. **b** rough sounding of the letter *r*. **2** (also **bur**) **a** rough edge on metal or paper. **b** surgeon's or dentist's small drill. **3** var. of BUR 1, 2. *–v.* make a burr. [imitative]

burrow /ˈbʌrəʊ/ *–n.* hole or tunnel dug by a rabbit etc. as a dwelling or shelter. *–v.* **1** make a burrow. **2** make (a hole, one's way, etc.) (as) by digging. **3** (foll. by *into*) investigate, search. [apparently var. of BOROUGH]

bursar /ˈbɜːsə(r)/ *n.* **1** treasurer, esp. of a college. **2** holder of a bursary. [medieval Latin *bursarius* from *bursa* purse]

bursary *n.* (*pl.* **-ies**) grant, esp. a scholarship. [medieval Latin: related to BURSAR]

burst *–v.* (*past* and *past part.* **burst**) **1** (cause to) break violently apart; open forcibly from within. **2 a** (usu. foll. by *in, out*) make one's way suddenly or by force. **b** break away from or through (*river burst its banks*). **3** be full to overflowing. **4** appear or come suddenly (*burst into flame*). **5** (foll. by *into*) suddenly begin to shed (tears) or utter. **6** seem about to burst from effort, excitement, etc. *–n.* **1** act of bursting. **2** sudden issue or outbreak (*burst of flame*; *burst of applause*). **3** sudden effort, spurt. □ **burst out 1** suddenly begin (*burst out laughing*). **2** exclaim. [Old English]

burton /ˈbɜːt(ə)n/ *n.* □ **go for a burton** *slang* be lost, destroyed, or killed. [origin uncertain]

bury /ˈberɪ/ *v.* (**-ies, -ied**) **1** place (a corpse) in the earth, a tomb, or the sea. **2** lose by death (*buried two sons*). **3 a** put or hide under ground. **b** cover up; conceal. **4** consign to obscurity; forget. **5** (*refl.* or *passive*) involve deeply (*buried in a book*). □ **bury the hatchet** cease to quarrel. [Old English]

bus –*n.* (*pl.* **buses** or *US* **busses**) **1** large esp. public passenger vehicle, usu. travelling a fixed route. **2** *colloq.* car, aeroplane, etc. –*v.* (**buses** or **busses**, **bussed**, **bussing**) **1** go by bus. **2** *US* transport by bus, esp. to aid racial integration. [abbreviation of OMNIBUS]

busby /ˈbʌzbɪ/ *n.* (*pl.* **-ies**) tall fur hat worn by hussars etc. [origin unknown]

bush[1] /bʊʃ/ *n.* **1** shrub or clump of shrubs. **2** thing like a bush, esp. a clump of hair. **3** (esp. in Australia and Africa) uncultivated area; woodland or forest. [Old English and Old Norse]

bush[2] /bʊʃ/ –*n.* **1** metal lining for a hole enclosing a revolving shaft etc. **2** sleeve giving electrical insulation. –*v.* fit with a bush. [Dutch *busse* box]

bush-baby *n.* (*pl.* **-ies**) small African lemur.

bushed *adj. colloq.* tired out.

bushel /ˈbʊʃ(ə)l/ *n.* measure of capacity for corn, fruit, etc. (8 gallons or 36.4 litres). [French]

bushfire *n.* forest or scrub fire often spreading widely.

bushman *n.* **1** traveller or dweller in the Australian bush. **2** (**Bushman**) member or language of a S.African aboriginal people.

bush telegraph *n.* rapid spreading of information, rumour, etc.

bushy *adj.* (**-ier, -iest**) **1** growing thickly like a bush. **2** having many bushes.

business /ˈbɪznɪs/ *n.* **1** one's regular occupation or profession. **2** one's own concern. **3** task or duty. **4** serious work or activity. **5** (difficult or unpleasant) matter or affair. **6** thing(s) needing attention or discussion. **7** buying and selling; trade. **8** commercial firm. □ **mind one's own business** not meddle. [Old English: related to BUSY]

businesslike *adj.* efficient, systematic.

businessman *n.* (*fem.* **businesswoman**) man or woman engaged in trade or commerce.

business park *n.* area designed for commerce and light industry.

business person *n.* businessman or businesswoman.

busk *v.* perform esp. music in the street etc. for tips. □ **busker** *n.* [obsolete *busk* peddle]

bus lane *n.* part of a road mainly for use by buses.

busman *n.* bus driver.

busman's holiday *n.* holiday spent in an activity similar to one's regular work.

bus shelter *n.* shelter beside a bus-stop.

bus station *n.* centre where buses depart and arrive.

bus-stop *n.* **1** regular stopping-place of a bus. **2** sign marking this.

bust[1] *n.* **1** human chest, esp. of a woman; bosom. **2** sculpture of a person's head, shoulders, and chest. □ **busty** *adj.* (**-ier, -iest**). [French from Italian]

bust[2] –*v.* (*past* and *past part.* **busted** or **bust**) *colloq.* **1** break, burst. **2 a** raid, search. **b** arrest. –*adj.* (also **busted**) **1** broken, burst. **2** bankrupt. □ **bust up 1** collapse. **2** (esp. of a married couple) separate. [var. of BURST]

bustard /ˈbʌstəd/ *n.* large land bird that can run very fast. [Latin *avis tarda* slow bird ('slow' unexplained)]

buster *n.* esp. *US slang* mate; fellow. [from BUST[2]]

bustier /ˈbʌstɪˌeɪ/ *n.* strapless close-fitting bodice. [French]

bustle[1] /ˈbʌs(ə)l/ –*v.* (**-ling**) **1** (often foll. by *about*) (cause to) move busily and energetically. **2** (as **bustling** *adj.*) active, lively. –*n.* excited or energetic activity. [perhaps from obsolete *busk* prepare]

bustle[2] /ˈbʌs(ə)l/ *n. hist.* padding worn under a skirt to puff it out behind. [origin unknown]

bust-up *n.* **1** quarrel. **2** collapse.

busy /ˈbɪzɪ/ –*adj.* (**-ier, -iest**) **1** occupied or engaged in work etc. **2** full of activity; fussy (*busy evening, street; busy design*). **3** esp. *US* (of a telephone line) engaged. –*v.* (**-ies, -ied**) (often *refl.*) keep busy; occupy. □ **busily** *adv.* [Old English]

busybody *n.* (*pl.* **-ies**) meddlesome person.

busy Lizzie *n.* plant with abundant esp. red, pink, or white flowers.

but /bʌt, bət/ –*conj.* **1 a** nevertheless, however (*tried but failed*). **b** on the other hand; on the contrary (*I am old but you are young*). **2** except, otherwise than (*cannot choose but do it; what could we do but run?*). **3** without the result that (*it never rains but it pours*). –*prep.* except; apart from; other than (*all cried but me; nothing but trouble*). –*adv.* **1** only; no more than; only just (*we can but try; is but a child; had but arrived*). **2** in emphatic repetition; definitely (*would see nobody, but nobody*). –*rel. pron.* who not; that not (*not a man but feels pity*). –*n.* objection (*ifs and buts*). □ **but for** without the help or hindrance etc. of (*but for you I'd be rich*). **but one** (or **two** etc.) excluding one (or two etc.) from the number (*next door but one; last but one*). **but then** however (*I won, but then I am older*). [Old English]

butane /'bju:teɪn/ *n.* gaseous alkane hydrocarbon, used in liquefied form as fuel. [from BUTYL]

butch /bʊtʃ/ *adj. slang* masculine; tough-looking. [origin uncertain]

butcher /'bʊtʃə(r)/ *–n.* **1 a** person who deals in meat. **b** slaughterer. **2** brutal murderer. *–v.* **1** slaughter or cut up (an animal) for food. **2** kill wantonly or cruelly. **3** *colloq.* ruin through incompetence. [French *boc* BUCK[1]]

butchery *n.* (*pl.* **-ies**) **1** needless or cruel slaughter (of people). **2** butcher's trade.

butler *n.* principal manservant of a household. [French *bouteille* bottle]

butt[1] *–v.* **1** push or strike with the head or horns. **2** (cause to) meet edge to edge. *–n.* **1** push with the head. **2** join of two edges. □ **butt in** interrupt, meddle. [French from Germanic]

butt[2] *n.* **1** (often foll. by *of*) object of ridicule etc. **2 a** mound behind a target. **b** (in *pl.*) shooting-range. [French *but* goal]

butt[3] *n.* **1** thicker end, esp. of a tool or weapon. **2** stub of a cigarette etc. **3** esp. *US slang* buttocks. [Dutch]

butt[4] *n.* cask. [Latin *buttis*]

butter *–n.* **1** solidified churned cream, used as a spread and in cooking. **2** substance of similar texture (*peanut butter*). *–v.* spread, cook, or serve with butter. □ **butter up** *colloq.* flatter. [Greek *bouturon*]

butter-bean *n.* **1** flat, dried, white lima bean. **2** yellow-podded bean.

butter-cream *n.* mixture of butter, icing sugar, etc., as a filling etc. for a cake.

buttercup *n.* wild plant with yellow cup-shaped flowers.

butterfat *n.* essential fats of pure butter.

butter-fingers *n. colloq.* person prone to drop things.

butterfly *n.* (*pl.* **-flies**) **1** insect with four usu. brightly coloured wings. **2** (in *pl.*) *colloq.* nervous sensation in the stomach.

butterfly nut *n.* a kind of wing-nut.

butterfly stroke *n.* stroke in swimming, with arms raised and lifted forwards together.

butter-icing *n.* = BUTTER-CREAM.

buttermilk *n.* liquid left after churning butter.

butter muslin *n.* thin loosely-woven cloth, orig. for wrapping butter.

butterscotch *n.* brittle toffee made from butter, brown sugar, etc.

buttery[1] /'bʌtərɪ/ *n.* (*pl.* **-ies**) food store, esp. in a college; snack-bar etc. [related to BUTT[4]]

buttery[2] *adj.* like or containing butter.

buttock /'bʌtək/ *n.* **1** each of the two fleshy protuberances at the rear of the human trunk. **2** corresponding part of an animal. [*butt* ridge]

button /'bʌt(ə)n/ *–n.* **1** small disc etc. sewn to a garment as a fastener or worn as an ornament. **2** small round knob etc. pressed to operate electronic equipment. *–v.* = *button up* 1. □ **button up 1** fasten with buttons. **2** *colloq.* complete satisfactorily. **3** *colloq.* be silent. [French from Germanic]

buttonhole *–n.* **1** slit in cloth for a button. **2** flower etc. worn in a lapel buttonhole. *–v.* (**-ling**) *colloq.* accost and detain (a reluctant listener).

button mushroom *n.* young unopened mushroom.

buttress /'bʌtrɪs/ *–n.* **1** projecting support built against a wall. **2** source of help etc. *–v.* (often foll. by *up*) **1** support with a buttress. **2** support by argument etc. (*buttressed by facts*). [related to BUTT[1]]

butty /'bʌtɪ/ *n.* (*pl.* **-ies**) *N.Engl.* sandwich. [from BUTTER]

butyl /'bju:taɪl/ *n.* the univalent alkyl radical C_4H_9. [Latin *butyrum* BUTTER]

buxom /'bʌksəm/ *adj.* (esp. of a woman) plump and rosy; busty. [earlier = *pliant*: related to BOW[2]]

buy /baɪ/ *–v.* (**buys, buying**; *past* and *past part.* **bought** /bɔ:t/) **1 a** obtain for money etc. **b** serve to obtain (*money can't buy happiness*; *the best that money can buy*). **2 a** procure by bribery etc. **b** bribe. **3** get by sacrifice etc. **4** *slang* believe in, accept. **5** be a buyer for a store etc. *–n. colloq.* purchase. □ **buy in** buy a stock of. **buy into** pay for a share in (an enterprise). **buy off** pay to get rid of. **buy oneself out** obtain one's release (esp. from the armed services) by payment. **buy out** pay (a person) for ownership, an interest, etc. **buy up 1** buy as much as possible of. **2** absorb (a firm etc.) by purchase. [Old English]

buyer *n.* **1** person employed to purchase stock for a large store etc. **2** purchaser, customer.

buyer's market *n.* (also **buyers' market**) trading conditions favourable to buyers.

buyout *n.* purchase of a controlling share in a company etc.

buzz *–n.* **1** hum of a bee etc. **2** sound of a buzzer. **3 a** low murmur as of conversation. **b** stir; hurried activity (*buzz of excitement*). **4** *slang* telephone call. **5** *slang* thrill. *–v.* **1** hum. **2 a** summon with a buzzer. **b** *slang* telephone. **3 a** (often foll. by *about*) move busily. **b** (of a place) appear busy or full of excitement. □ **buzz off** *slang* go or hurry away. [imitative]

buzzard /'bʌzəd/ *n.* large bird of the hawk family. [Latin *buteo* falcon]

buzzer *n.* electrical buzzing device as a signal.

buzz-word *n. colloq.* fashionable technical or specialist word; catchword.

by /baɪ/ *–prep.* **1** near, beside (*sit by me*; *path by the river*). **2** through the agency or means of (*by proxy*; *poem by Donne*; *by bus*; *by cheating*; *divide by two*; *killed by robbers*). **3** not later than (*by next week*). **4 a** past, beyond (*drove by the church*). **b** through; via (*went by Paris*). **5** during (*by day*; *by daylight*). **6** to the extent of (*missed by a foot*; *better by far*). **7** according to; using as a standard or unit (*judge by appearances*; *paid by the hour*). **8** with the succession of (*worse by the minute*; *day by day*). **9** concerning; in respect of (*did our duty by them*; *Smith by name*). **10** used in mild oaths (*by God*). **11** expressing dimensions of an area etc. (*three feet by two*). **12** avoiding, ignoring (*passed us by*). **13** inclining to (*north by north-west*). *–adv.* **1** near (*sat by*). **2** aside; in reserve (*put £5 by*). **3** past (*marched by*). *–n.* (*pl.* **byes**) = BYE[1]. □ **by and by** before long; eventually. **by and large** on the whole. **by the by** (or **bye**) incidentally. **by oneself 1 a** unaided. **b** unprompted. **2** alone. [Old English]

by- *prefix* subordinate, incidental (*by-effect*; *byroad*).

bye[1] /baɪ/ *n.* **1** *Cricket* run scored from a ball that passes the batsman without being hit. **2** status of an unpaired competitor in a sport, who proceeds to the next round by default. [from BY as a noun]

bye[2] /baɪ/ *int.* (also **bye-bye**) *colloq.* = GOODBYE. [abbreviation]

by-election *n.* election to fill a vacancy arising between general elections.

Byelorussian /,byelə'rʌʃ(ə)n/ (also **Belorussian** /,belə-/) *–n.* native or language of Byelorussia in eastern Europe. *–adj.* of Byelorussia, its people, or language. [Russian from *belyĭ* white, *Russiya* Russia]

bygone *–adj.* past, antiquated. *–n.* (in phr. **let bygones be bygones**) forgive and forget past quarrels.

by-law *n.* regulation made by a local authority or corporation. [obsolete *by* town]

byline *n.* **1** line naming the writer of a newspaper article etc. **2** secondary line of work. **3** goal-line or touch-line.

bypass *–n.* **1** main road passing round a town or its centre. **2 a** secondary channel or pipe etc. used in emergencies. **b** alternative passage for the circulation of blood through the heart. *–v.* avoid, go round (a town, difficulty, etc.).

byplay *n.* secondary action, esp. in a play.

by-product *n.* **1** incidental product made in the manufacture of something else. **2** secondary result.

byre /'baɪə(r)/ *n.* cowshed. [Old English]

byroad *n.* minor road.

byssinosis /,bɪsɪ'nəʊsɪs/ *n.* lung disease caused by textile fibre dust. [Greek *bussinos* made of linen]

bystander *n.* person present but not taking part; onlooker.

byte /baɪt/ *n. Computing* group of eight binary digits, often representing one character. [origin uncertain]

byway *n.* **1** byroad or secluded path. **2** minor activity.

byword *n.* **1** person or thing as a notable example (*is a byword for luxury*). **2** familiar saying.

Byzantine /bɪ'zæntaɪn, baɪ-/ *–adj.* **1** of Byzantium or the E. Roman Empire. **2** of its highly decorated style of architecture. **3** (of a political situation etc.) complex, inflexible, or underhand. *–n.* citizen of Byzantium or the E. Roman Empire. □ **Byzantinism** *n.* **Byzantinist** *n.* [Latin *Byzantium*, now Istanbul]

C

C[1] /si:/ *n.* (*pl.* **Cs** or **C's**) **1** (also **c**) third letter of the alphabet. **2** *Mus.* first note of the diatonic scale of C major. **3** third hypothetical person or example. **4** third highest category etc. **5** *Algebra* (usu. **c**) third known quantity. **6** (as a roman numeral) 100. **7** (also ©) copyright.

C[2] *symb.* carbon.

C[3] *abbr.* (also **C.**) **1** Celsius, centigrade. **2** coulomb(s), capacitance.

c. *abbr.* **1** century. **2** cent(s).

c. *abbr. circa.*

Ca *symb.* calcium.

ca. *abbr. circa.*

CAA *abbr.* Civil Aviation Authority.

cab *n.* **1** taxi. **2** driver's compartment in a lorry, train, or crane etc. [abbreviation of CABRIOLET]

cabal /kə'bæl/ *n.* **1** secret intrigue. **2** political clique. [French from Latin]

cabaret /'kæbə,reɪ/ *n.* entertainment in a nightclub or restaurant. [French, = tavern]

cabbage /'kæbɪdʒ/ *n.* **1** vegetable with a round head and green or purple leaves. **2** = VEGETABLE 2. [French *caboche* head]

cabbage white *n.* butterfly whose caterpillars feed on cabbage leaves.

cabby *n.* (also **cabbie**) (*pl.* **-ies**) *colloq.* taxi-driver.

caber /'keɪbə(r)/ *n.* trimmed tree-trunk tossed as a sport in the Scottish Highlands. [Gaelic]

cabin /'kæbɪn/ *n.* **1** small shelter or house, esp. of wood. **2** room or compartment in an aircraft or ship for passengers or crew. **3** driver's cab. [French from Latin]

cabin-boy *n.* boy steward on a ship.

cabin cruiser *n.* large motor boat with accommodation.

cabinet /'kæbɪnɪt/ *n.* **1 a** cupboard or case for storing or displaying things. **b** casing of a radio, television, etc. **2** (**Cabinet**) committee of senior ministers in a government. [diminutive of CABIN]

cabinet-maker *n.* skilled joiner.

cable /'keɪb(ə)l/ —*n.* **1** encased group of insulated wires for transmitting electricity etc. **2** thick rope of wire or hemp. **3** cablegram. **4** (in full **cable stitch**) knitting stitch resembling twisted rope. —*v.* (**-ling**) transmit (a message) or inform (a person) by cablegram. [Latin *caplum* halter, from Arabic]

cable-car *n.* small cabin suspended on a looped cable, for carrying passengers up and down a mountain etc.

cablegram *n.* telegraph message sent by undersea cable.

cable television *n.* television transmission by cable to subscribers.

cabman *n.* driver of a cab.

caboodle /kə'bu:d(ə)l/ *n.* □ **the whole caboodle** *slang* the whole lot. [origin uncertain]

caboose /kə'bu:s/ *n.* **1** kitchen on a ship's deck. **2** *US* guard's van on a train etc. [Dutch]

cabriole /'kæbrɪ,əʊl/ *n.* a kind of esp. 18th-c. curved table or chair leg. [French: related to CAPRIOLE]

cabriolet /,kæbrɪəʊ'leɪ/ *n.* **1** car with a folding top. **2** light two-wheeled one-horse carriage with a hood. [French: related to CAPRIOLE]

cacao /kə'kaʊ/ *n.* (*pl.* **-s**) **1** seed from which cocoa and chocolate are made. **2** tree bearing these. [Spanish from Nahuatl]

cache /kæʃ/ —*n.* **1** hiding-place for treasure, stores, guns, etc. **2** things so hidden. —*v.* (**-ching**) put in a cache. [French *cacher* hide]

cachet /'kæʃeɪ/ *n.* **1** prestige. **2** distinguishing mark or seal. **3** flat capsule of medicine. [French *cacher* press]

cachou /'kæʃu:/ *n.* lozenge to sweeten the breath. [Portuguese *cachu* from Malay *kāchu*]

cack-handed /kæk'hændɪd/ *adj. colloq.* **1** clumsy. **2** left-handed. [dial. *cack* excrement]

cackle /'kæk(ə)l/ —*n.* **1** clucking of a hen etc. **2** raucous laugh. **3** noisy chatter. —*v.* (**-ling**) **1** emit a cackle. **2** chatter noisily. [imitative]

cacophony /kə'kɒfənɪ/ *n.* (*pl.* **-ies**) harsh discordant sound. □ **cacophonous** *adj.* [Greek *kakos* bad, *phōnē* sound]

cactus /'kæktəs/ *n.* (*pl.* **-ti** /-taɪ/ or **cactuses**) plant with a thick fleshy stem and usu. spines but no leaves. [Latin from Greek]

CAD *abbr.* computer-aided design.

cad *n.* man who behaves dishonourably. □ **caddish** *adj.* [abbreviation of CADDIE]

cadaver /kə'dævə(r)/ *n.* esp. *Med.* corpse. [Latin *cado* fall]

cadaverous /kə'dævərəs/ *adj.* corpse-like; very pale and thin.

caddie /ˈkædɪ/ (also **caddy**) *–n.* (*pl.* **-ies**) person who carries a golfer's clubs during play. *–v.* (**-ies, -ied, caddying**) act as a caddie. [French CADET]

caddis-fly /ˈkædɪs/ *n.* small nocturnal insect living near water. [origin unknown]

caddis-worm /ˈkædɪs/ *n.* (also **caddis**) larva of the caddis-fly. [origin unknown]

caddy[1] /ˈkædɪ/ *n.* (*pl.* **-ies**) small container for tea. [Malay]

caddy[2] var. of CADDIE.

cadence /ˈkeɪd(ə)ns/ *n.* **1** rhythm; the measure or beat of a sound or movement. **2** fall in pitch of the voice. **3** tonal inflection. **4** close of a musical phrase. [Latin *cado* fall]

cadenza /kəˈdenzə/ *n.* virtuoso passage for a soloist. [Italian: related to CADENCE]

cadet /kəˈdet/ *n.* young trainee for the armed services or police force. □ **cadetship** *n.* [French, ultimately from Latin *caput* head]

cadge *v.* (**-ging**) *colloq.* get or seek by begging. [origin unknown]

cadi /ˈkɑːdɪ/ *n.* (*pl.* **-s**) judge in a Muslim country. [Arabic]

cadmium /ˈkædmɪəm/ *n.* soft bluish-white metallic element. [Greek *kadmia* Cadmean (earth)]

cadre /ˈkɑːdə(r)/ *n.* **1** basic unit, esp. of servicemen. **2** group of esp. Communist activists. [French from Latin *quadrus* square]

caecum /ˈsiːkəm/ *n.* (*US* **cecum**) (*pl.* **-ca**) blind-ended pouch at the junction of the small and large intestines. [Latin *caecus* blind]

Caenozoic var. of CENOZOIC.

Caerphilly /keəˈfɪlɪ/ *n.* a kind of mild white cheese. [*Caerphilly* in Wales]

Caesar /ˈsiːzə(r)/ *n.* **1** title of Roman emperors. **2** autocrat. [Latin (C. Julius) *Caesar*]

Caesarean /sɪˈzeərɪən/ (*US* **Cesarean, Cesarian**) *–adj.* (of birth) effected by Caesarean section. *–n.* Caesarean section. [from CAESAR: Julius Caesar was supposedly born this way]

Caesarean section *n.* delivery of a child by cutting through the mother's abdomen.

caesium /ˈsiːzɪəm/ *n.* (*US* **cesium**) soft silver-white element. [Latin *caesius* blue-grey]

caesura /sɪˈzjʊərə/ *n.* (*pl.* **-s**) pause in a line of verse. □ **caesural** *adj.* [Latin *caedo* cut]

café /ˈkæfeɪ/ *n.* small coffee-house or restaurant. [French]

cafeteria /ˌkæfɪˈtɪərɪə/ *n.* self-service restaurant. [American Spanish, = coffee-shop]

caffeine /ˈkæfiːn/ *n.* alkaloid stimulant in tea-leaves and coffee beans. [French *café* coffee]

caftan /ˈkæftæn/ *n.* (also **kaftan**) **1** long tunic worn by men in the Near East. **2** long loose dress or shirt. [Turkish]

cage *–n.* **1** structure of bars or wires, esp. for confining animals or birds. **2** similar open framework, esp. a lift in mine etc. *–v.* (**-ging**) place or keep in a cage. [Latin *cavea*]

cagey /ˈkeɪdʒɪ/ *adj.* (also **cagy**) (**-ier, -iest**) *colloq.* cautious and non-committal. □ **cagily** *adv.* **caginess** *n.* [origin unknown]

cagoule /kəˈguːl/ *n.* thin hooded wind-proof jacket. [French]

cahoots /kəˈhuːts/ *n.pl.* □ **in cahoots** *slang* in collusion. [origin uncertain]

caiman var. of CAYMAN.

Cain *n.* □ **raise Cain** *colloq.* = *raise the roof*. [*Cain*, eldest son of Adam (Gen. 4)]

Cainozoic var. of CENOZOIC.

cairn *n.* **1** mound of stones as a monument or landmark. **2** (in full **cairn terrier**) small shaggy short-legged terrier. [Gaelic]

cairngorm /ˈkeəngɔːm/ *n.* semiprecious form of quartz. [*Cairngorms*, in Scotland]

caisson /ˈkeɪs(ə)n/ *n.* watertight chamber for underwater construction work. [Italian *cassone*]

cajole /kəˈdʒəʊl/ *v.* (**-ling**) persuade by flattery, deceit, etc. □ **cajolery** *n.* [French]

cake *–n.* **1** mixture of flour, butter, eggs, sugar, etc., baked in the oven and often iced and decorated. **2** other food in a flat round shape (*fish cake*). **3** flattish compact mass (*cake of soap*). *–v.* (**-king**) **1** form into a compact mass. **2** (usu. foll. by *with*) cover (with a hard or sticky mass). □ **have one's cake and eat it** *colloq.* enjoy both of two mutually exclusive alternatives. **a piece of cake** *colloq.* something easily achieved. **sell** (or **go**) **like hot cakes** *colloq.* be sold (or go) quickly; be popular. [Old Norse]

cakewalk *n.* **1** obsolete American Black dance. **2** *colloq.* easy task. **3** fairground entertainment consisting of a promenade moved by machinery.

Cal *abbr.* large calorie(s).

cal *abbr.* small calorie(s).

calabash /ˈkæləˌbæʃ/ *n.* **1** gourd-bearing tree of tropical America. **2** such a gourd, esp. as a vessel for water, etc. [French from Spanish]

calabrese /ˌkæləˈbriːs/ *n.* variety of broccoli. [Italian, = Calabrian]

calamine /'kælə,maɪn/ *n.* powdered form of zinc carbonate and ferric oxide used as a skin lotion. [French from Latin]

calamity /kə'læmɪtɪ/ *n.* (*pl.* **-ies**) disaster, great misfortune. □ **calamitous** *adj.* [French from Latin]

calcareous /kæl'keərɪəs/ *adj.* of or containing calcium carbonate. [related to CALX]

calceolaria /,kælsɪə'leərɪə/ *n.* plant with slipper-shaped flowers. [Latin *calceus* shoe]

calces *pl.* of CALX.

calciferol /kæl'sɪfə,rɒl/ *n.* vitamin (D_2) promoting calcium deposition in the bones. [related to CALX]

calciferous /kæl'sɪfərəs/ *adj.* yielding calcium salts, esp. calcium carbonate.

calcify /'kælsɪ,faɪ/ *v.* (**-ies, -ied**) **1** harden by the depositing of calcium salts. **2** convert or be converted to calcium carbonate. □ **calcification** /-fɪ'keɪʃ(ə)n/ *n.*

calcine /'kælsaɪn/ *v.* (**-ning**) decompose or be decomposed by strong heat. □ **calcination** /-'neɪʃ(ə)n/ *n.* [French or medieval Latin: related to CALX]

calcite /'kælsaɪt/ *n.* natural crystalline calcium carbonate. [Latin: related to CALX]

calcium /'kælsɪəm/ *n.* soft grey metallic element occurring in limestone, marble, chalk, etc. [related to CALX]

calcium carbide *n.* greyish solid used in the production of acetylene.

calcium carbonate *n.* white insoluble solid occurring as chalk, marble, etc.

calcium hydroxide *n.* white crystalline powder used in the manufacture of mortar.

calcium oxide *n.* white crystalline solid from which many calcium compounds are manufactured.

calculate /'kælkjʊ,leɪt/ *v.* (**-ting**) **1** ascertain or forecast esp. by mathematics or reckoning. **2** plan deliberately. **3** (foll. by *on*) rely on; reckon on. □ **calculable** *adj.* [Latin: related to CALCULUS]

calculated *adj.* **1** (of an action) done deliberately or with foreknowledge. **2** (foll. by *to* + infin.) designed or suitable; intended.

calculating *adj.* scheming, mercenary.

calculation /,kælkjʊ'leɪʃ(ə)n/ *n.* act, process, or result of calculating. [Latin: related to CALCULUS]

calculator *n.* device (esp. a small electronic one) for making mathematical calculations.

calculus /'kælkjʊləs/ *n.* (*pl.* **-luses** or **-li** /-,laɪ/) **1** particular method of mathematical calculation or reasoning. **2** stone or mineral mass in the body. [Latin, = small stone (used on an abacus)]

caldron var. of CAULDRON.

Caledonian /,kælɪ'dəʊnɪən/ *literary* —*adj.* of Scotland. —*n.* Scotsman. [Latin *Caledonia* N. Britain]

calendar /'kælɪndə(r)/ —*n.* **1** system fixing the year's beginning, length, and subdivision. **2** chart etc. showing such subdivisions. **3** timetable of dates, events, etc. —*v.* enter in a calendar; register (documents). [Latin: related to CALENDS]

calendar year *n.* period from 1 Jan. to 31 Dec. inclusive.

calender /'kælɪndə(r)/ —*n.* machine in which cloth, paper, etc. is rolled to glaze or smooth it. —*v.* press in a calender. [French]

calends /'kælendz/ *n.pl.* (also **kalends**) first of the month in the ancient Roman calendar. [Latin *Kalendae*]

calendula /kə'lendjʊlə/ *n.* plant with large yellow or orange flowers, esp. the marigold. [Latin diminutive of *Kalendae*]

calf[1] /kɑ:f/ *n.* (*pl.* **calves** /kɑ:vz/) **1** young cow or bull. **2** young of other animals, e.g. the elephant, deer, and whale. **3** calfskin. [Old English]

calf[2] /kɑ:f/ *n.* (*pl.* **calves**) fleshy hind part of the human leg below the knee. [Old Norse]

calf-love *n.* romantic adolescent love.

calfskin *n.* calf-leather.

calibrate /'kælɪ,breɪt/ *v.* (**-ting**) **1** mark (a gauge) with a scale of readings. **2** correlate the readings of (an instrument or system of measurement) with a standard. **3** determine the calibre of (a gun). □ **calibration** /-'breɪʃ(ə)n/ *n.*

calibre /'kælɪbə(r)/ *n.* (*US* **caliber**) **1 a** internal diameter of a gun or tube. **b** diameter of a bullet or shell. **2** strength or quality of character; ability, importance. [French from Italian from Arabic, = mould]

calices *pl.* of CALIX.

calico /'kælɪ,kəʊ/ —*n.* (*pl.* **-es** or *US* **-s**) **1** cotton cloth, esp. plain white or unbleached. **2** *US* printed cotton fabric. —*adj.* **1** of calico. **2** *US* multicoloured. [*Calicut* in India]

californium /,kælɪ'fɔ:nɪəm/ *n.* artificial radioactive metallic element. [*California* in US, where first made]

caliper var. of CALLIPER.

caliph /'keɪlɪf/ *n.* esp. *hist.* chief Muslim civil and religious ruler. □ **caliphate** *n.* [Arabic, = successor (of Muhammad)]

calisthenics var. of CALLISTHENICS.

calix var. of CALYX.

calk *US* var. of CAULK.

call /kɔ:l/ *–v.* **1 a** (often foll. by *out*) cry, shout; speak loudly. **b** (of a bird etc.) emit its characteristic sound. **2** communicate with by telephone or radio. **3** summon. **4** (often foll. by *at, in, on*) pay a brief visit. **5** order to take place (*called a meeting*). **6** name; describe as. **7** regard as (*I call that silly*). **8** rouse from sleep. **9** (foll. by *for*) demand. **10** (foll. by *on, upon*) appeal to (*called on us to be quiet*). **11** name (a suit) in bidding at cards. **12** guess the outcome of tossing a coin etc. *–n.* **1** shout, cry. **2 a** characteristic cry of a bird etc. **b** instrument for imitating it. **3** brief visit. **4 a** act of telephoning. **b** telephone conversation. **5 a** invitation, summons. **b** vocation. **6** need, occasion (*no call for rudeness*). **7** demand (*a call on one's time*). **8** signal on a bugle etc. **9** option of buying stock at a fixed price at a given date. **10** *Cards* **a** player's right or turn to make a bid. **b** bid made. □ **call in 1** withdraw from circulation. **2** seek the advice or services of. **call off 1** cancel (an arrangement). **2** order (an attacker or pursuer) to desist. **call out 1** summon to action. **2** order (workers) to strike. **call the shots** (or **tune**) *colloq.* be in control; take the initiative. **call up 1** telephone. **2** recall. **3** summon to military service. **on call** ready or available if required. [Old English from Old Norse]

call-box *n.* telephone box.

caller *n.* person who calls, esp. one who pays a visit or makes a telephone call.

call-girl *n.* prostitute accepting appointments by telephone.

calligraphy /kə'lɪgrəfɪ/ *n.* **1** handwriting, esp. when fine. **2** art of this. □ **calligrapher** *n.* **calligraphic** /-'græfɪk/ *adj.* **calligraphist** *n.* [Greek *kallos* beauty]

calling *n.* **1** profession or occupation. **2** vocation.

calliper /'kælɪpə(r)/ *n.* (also **caliper**) **1** (in *pl.*) compasses for measuring diameters. **2** metal splint to support the leg. [var. of CALIBRE]

callisthenics /,kælɪs'θenɪks/ *n.pl.* (also **calisthenics**) exercises for fitness and grace. □ **callisthenic** *adj.* [Greek *kallos* beauty, *sthenos* strength]

callosity /kə'lɒsɪtɪ/ *n.* (*pl.* **-ies**) area of hard thick skin. [Latin: related to CALLOUS]

callous /'kæləs/ *adj.* **1** unfeeling, insensitive. **2** (also **calloused**) (of skin) hardened. □ **callously** *adv.* **callousness** *n.* [Latin: related to CALLUS]

callow /'kæləʊ/ *adj.* inexperienced, immature. [Old English, = bald]

call-up *n.* summons to do military service.

callus /'kæləs/ *n.* (*pl.* **calluses**) **1** area of hard thick skin or tissue. **2** hard tissue formed round bone ends after a fracture. [Latin]

calm /kɑ:m/ *–adj.* **1** tranquil, quiet, windless. **2** serene; not agitated. *–n.* calm condition or period. *–v.* (often foll. by *down*) make or become calm. □ **calmly** *adv.* **calmness** *n.* [Greek *kauma* heat]

calomel /'kælə,mel/ *n.* compound of mercury used as a cathartic. [Greek *kalos* beautiful, *melas* black]

Calor gas /'kælə/ *n. propr.* liquefied butane gas stored under pressure in containers for domestic use. [Latin *calor* heat]

caloric /'kælərɪk/ *adj.* of heat or calories.

calorie /'kælərɪ/ *n.* (*pl.* **-ies**) unit of quantity of heat, the amount needed to raise the temperature of one gram (**small calorie**) or one kilogram (**large calorie**) of water by 1 °C. [Latin *calor* heat]

calorific /,kælə'rɪfɪk/ *adj.* producing heat.

calorimeter /,kælə'rɪmɪtə(r)/ *n.* instrument for measuring quantity of heat.

calumniate /kə'lʌmnɪ,eɪt/ *v.* (**-ting**) slander. [Latin]

calumny /'kæləmnɪ/ *n.* (*pl.* **-ies**) slander; malicious representation. □ **calumnious** /kə'lʌmnɪəs/ *adj.* [Latin]

calvados /'kælvə,dɒs/ *n.* apple brandy. [*Calvados* in France]

calve /kɑ:v/ *v.* (**-ving**) give birth to a calf. [Old English: related to CALF[1]]

calves *pl.* of CALF[1], CALF[2].

Calvinism /'kælvɪ,nɪz(ə)m/ *n.* theology of Calvin or his followers, stressing predestination and divine grace. □ **Calvinist** *n.* & *adj.* **Calvinistic** /-'nɪstɪk/ *adj.* [*Calvin*, name of a theologian]

calx *n.* (*pl.* **calces** /'kælsi:z/) powdery substance formed when an ore or mineral has been heated. [Latin *calx calc-* lime]

calypso /kə'lɪpsəʊ/ *n.* (*pl.* **-s**) W. Indian song with improvised usu. topical words and a syncopated rhythm. [origin unknown]

calyx /'keɪlɪks/ *n.* (*pl.* **calyces** /-lɪ,si:z/ or **-es**) (also **calix**) **1** sepals forming the protective case of a flower in bud. **2** cuplike cavity or structure. [Greek, = husk]

cam *n.* projection on a wheel etc., shaped to convert circular into reciprocal or variable motion. [Dutch *kam* comb]

camaraderie /,kæmə'rɑ:dərɪ/ *n.* friendly comradeship. [French]

camber *–n.* convex surface of a road, deck, etc. *–v.* build with a camber. [Latin *camurus* curved]

Cambrian /'kæmbrɪən/ *–adj.* **1** Welsh. **2** *Geol.* of the first period in the Palaeozoic era. *–n.* this period. [Welsh: related to CYMRIC]

cambric /'keɪmbrɪk/ *n.* fine linen or cotton fabric. [*Cambrai* in France]

Cambridge blue /'keɪmbrɪdʒ/ *adj.* & *n.* (as adj. often hyphenated) pale blue. [*Cambridge* in England]

camcorder /'kæm,kɔ:də(r)/ *n.* combined video camera and sound recorder. [from CAMERA, RECORDER]

came *past* of COME.

camel /'kæm(ə)l/ *n.* **1** long-legged ruminant with one hump (**Arabian camel**) or two humps (**Bactrian camel**). **2** fawn colour. [Greek]

camel-hair *n.* fine soft hair used in artists' brushes or for fabric.

camellia /kə'mi:lɪə/ *n.* evergreen shrub with shiny leaves and showy flowers. [*Camellus*, name of a botanist]

Camembert /'kæməm,beə(r)/ *n.* a kind of soft creamy pungent cheese. [*Camembert* in France]

cameo /'kæmɪəʊ/ *n.* (*pl.* **-s**) **1** small piece of hard stone carved in relief with a background of a different colour. **2 a** short descriptive literary sketch or acted scene. **b** small character part in a play or film, usu. brief and played by a distinguished actor. [French and medieval Latin]

camera /'kæmrə/ *n.* **1** apparatus for taking photographs or moving film. **2** equipment for converting images into electrical signals. □ **in camera** *Law* in private. [Latin: related to CHAMBER]

cameraman *n.* person who operates a camera professionally, esp. in film-making or television.

camiknickers /'kæmɪ,nɪkəz/ *n.pl.* women's knickers and vest combined. [from CAMISOLE, KNICKERS]

camisole /'kæmɪ,səʊl/ *n.* women's lightweight vest. [Italian or Spanish: related to CHEMISE]

camomile /'kæmə,maɪl/ *n.* (also **chamomile**) aromatic plant with daisy-like flowers used esp. to make tea. [Greek, = earth-apple]

camouflage /'kæmə,flɑ:ʒ/ *–n.* **1 a** disguising of soldiers, tanks, etc. so that they blend into the background. **b** such a disguise. **2** the natural blending colouring of an animal. **3** misleading or evasive behaviour etc. *–v.* (**-ging**) hide by camouflage. [French *camoufler* disguise]

camp[1] *–n.* **1** place where troops are lodged or trained. **2** temporary accommodation of huts, tents, etc., for detainees, holiday-makers, etc. **3** ancient fortified site. **4** party supporters etc. regarded collectively. *–v.* set up or spend time in a camp. [Latin *campus* level ground]

camp[2] *colloq.* *–adj.* **1** affected, effeminate, theatrically exaggerated. **2** homosexual. *–n.* camp manner or style. *–v.* behave or do in a camp way. □ **camp it up** overact; behave affectedly. □ **campy** *adj.* (**-ier**, **-iest**). [origin uncertain]

campaign /kæm'peɪn/ *–n.* **1** organized course of action, esp. to gain publicity. **2** military operations towards a particular objective. *–v.* take part in a campaign. □ **campaigner** *n.* [Latin: related to CAMP[1]]

campanile /,kæmpə'ni:lɪ/ *n.* bell-tower (usu. free-standing), esp. in Italy. [Italian *campana* 'bell', from Latin]

campanology /,kæmpə'nɒlədʒɪ/ *n.* **1** the study of bells. **2** bell-ringing. □ **campanologist** *n.* [Latin *campana* bell]

campanula /kæm'pænjʊlə/ *n.* plant with bell-shaped usu. blue, purple, or white flowers. [diminutive: related to CAMPANOLOGY]

camp-bed *n.* portable folding bed.

camper *n.* **1** person who camps. **2** large motor vehicle with beds etc.

camp-follower *n.* **1** civilian worker in a military camp. **2** disciple or adherent.

camphor /'kæmfə(r)/ *n.* pungent white crystalline substance used in making celluloid, medicine, and mothballs. [French ultimately from Sanskrit]

camphorate *v.* (**-ting**) impregnate or treat with camphor.

campion /'kæmpɪən/ *n.* wild plant with usu. pink or white notched flowers. [origin uncertain]

campsite *n.* place for camping.

campus /'kæmpəs/ *n.* (*pl.* **-es**) **1** grounds of a university or college. **2** esp. *US* a university. [Latin, = field]

CAMRA /'kæmrə/ *abbr.* Campaign for Real Ale.

camshaft *n.* shaft with one or more cams.

can[1] /kæn, kən/ *v.aux.* (*3rd sing. present* **can**; *past* **could** /kʊd/) **1 a** be able to; know how to. **b** be potentially capable of (*these storms can last for hours*). **2** be permitted to. [Old English, = know]

can[2] *–n.* **1** metal vessel for liquid. **2** sealed tin container for the preservation of food or drink. **3** (in *pl.*) *slang* headphones. **4** (prec. by *the*) *slang* **a** prison. **b** *US* lavatory. *–v.* (**-nn-**) put or preserve in a can. □ **in the can** *colloq.* completed, ready. [Old English]

Canada goose /'kænədə/ *n.* wild N. American goose with a brownish-grey body and white neck and breast.

canaille /kə'nɑ:ɪ/ *n.* rabble; populace. [French from Italian]

canal /kə'næl/ *n.* **1** artificial inland waterway. **2** tubular duct in a plant or animal. [Latin *canalis*]

canalize /'kænə,laɪz/ *v.* (also **-ise**) (**-zing** or **-sing**) **1** provide with or convert into a canal or canals. **2** channel. □ **canalization** /-'zeɪʃ(ə)n/ *n.* [French: related to CANAL]

canapé /'kænə,peɪ/ *n.* small piece of bread or pastry with a savoury topping. [French]

canard /'kænɑ:d/ *n.* unfounded rumour or story. [French, = duck]

canary /kə'neərɪ/ *n.* (*pl.* **-ies**) small songbird with yellow feathers. [*Canary* Islands]

canasta /kə'næstə/ *n.* card-game using two packs and resembling rummy. [Spanish, = basket]

cancan /'kænkæn/ *n.* lively stage-dance with high kicking. [French]

cancel /'kæns(ə)l/ *v.* (**-ll-**; *US* **-l-**) **1** revoke or discontinue (an arrangement). **2** delete (writing etc.). **3** mark (a ticket, stamp, etc.) to invalidate it. **4** annul; make void. **5** (often foll. by *out*) neutralize or counterbalance. **6** *Math.* strike out (an equal factor) on each side of an equation etc. □ **cancellation** /-'leɪʃ(ə)n/ *n.* [Latin: related to CHANCEL]

cancer *n.* **1 a** malignant tumour of body cells. **b** disease caused by this. **2** evil influence or corruption. **3** (**Cancer**) **a** constellation and fourth sign of the zodiac (the Crab). **b** person born when the sun is in this sign. □ **cancerous** *adj.* **cancroid** *adj.* [Latin, = crab]

candela /kæn'di:lə/ *n.* SI unit of luminous intensity. [Latin, = candle]

candelabrum /,kændɪ'lɑ:brəm/ *n.* (also **-bra**) (*pl.* **-bra**, *US* **-brums**, **-bras**) large branched candlestick or lamp-holder. [Latin: related to CANDELA]

■ **Usage** The form *candelabra* is, strictly speaking, the plural. However, *candelabra* (singular) and *candelabras* (plural) are often found in informal use.

candid /'kændɪd/ *adj.* **1** frank; open. **2** (of a photograph) taken informally, usu. without subject's knowledge. □ **candidly** *adv.* **candidness** *n.* [Latin *candidus* white]

candida /'kændɪdə/ *n.* fungus causing thrush. [Latin *candidus*: related to CANDID]

candidate /'kændɪdət/ *n.* **1** person nominated for or seeking office, an award, etc. **2** person or thing likely to gain some distinction or position. **3** person entered for an examination. □ **candidacy** *n.* **candidature** *n.* [Latin, = white-robed]

candle /'kænd(ə)l/ *n.* cylinder or block of wax or tallow with a central wick which gives light when burning. □ **cannot hold a candle to** is much inferior to. **not worth the candle** not justifying cost or trouble. [Latin *candela*]

candlelight *n.* light from candles. □ **candlelit** *adj.*

Candlemas /'kænd(ə)l,mæs/ *n.* feast of the Purification of the Virgin Mary (2 Feb.). [Old English: related to MASS[2]]

candlepower *n.* unit of luminous intensity.

candlestick *n.* holder for one or more candles.

candlewick *n.* **1** thick soft cotton yarn. **2** tufted material from this.

candour /'kændə(r)/ *n.* (*US* **candor**) frankness; openness. [Latin *candor*]

C. & W. *abbr.* country-and-western (music).

candy /'kændɪ/ —*n.* (*pl.* **-ies**) **1** (in full **sugar-candy**) sugar crystallized by repeated boiling and slow evaporation. **2** *US* sweets; a sweet. —*v.* (**-ies**, **-ied**) (usu. as **candied** *adj.*) preserve (fruit etc.) in candy. [French from Arabic]

candyfloss *n.* fluffy mass of spun sugar round a stick.

candystripe *n.* alternate stripes of white and a colour. □ **candystriped** *adj.*

candytuft /'kændɪ,tʌft/ *n.* plant with white, pink, or purple flowers in tufts. [*Candia* Crete, TUFT]

cane —*n.* **1 a** hollow jointed stem of giant reeds or grasses. **b** solid stem of slender palms. **2** = SUGAR CANE. **3** cane used for wickerwork etc. **4** cane used as a walking-stick, plant support, for punishment, etc. —*v.* (**-ning**) **1** beat with a cane. **2** weave cane into (a chair etc.). [Greek *kanna* reed]

cane-sugar *n.* sugar from sugar-cane.

canine /'keɪnaɪn/ —*adj.* of a dog or dogs. —*n.* **1** dog. **2** (in full **canine tooth**) pointed tooth between incisors and premolars. [Latin *canis* dog]

canister /'kænɪstə(r)/ *n.* **1** small container for tea etc. **2** cylinder of shot, tear-gas, etc., exploding on impact. [Greek *kanastron* wicker basket]

canker —*n.* **1** destructive disease of trees and plants. **2** ulcerous ear disease of animals. **3** corrupting influence. —*v.* **1** infect with canker. **2** corrupt. **3** (as **cankered** *adj.*) soured, malignant. □

cankerous *adj.* [Latin: related to CANCER]

canna /'kænə/ *n.* tropical plant with bright flowers and ornamental leaves. [Latin: related to CANE]

cannabis /'kænəbɪs/ *n.* **1** hemp plant. **2** parts of it used as a narcotic. [Latin from Greek]

canned *adj.* **1** pre-recorded (*canned music*). **2** sold in a can (*canned beer*). **3** *slang* drunk.

cannelloni /ˌkænə'ləʊnɪ/ *n.pl.* tubes of pasta stuffed with a savoury mixture. [Italian]

cannery /'kænərɪ/ *n.* (*pl.* **-ies**) canning-factory.

cannibal /'kænɪb(ə)l/ *n.* person or animal that eats its own species. □ **cannibalism** *n.* **cannibalistic** /-bə'lɪstɪk/ *adj.* [Spanish from Carib]

cannibalize /'kænɪbəˌlaɪz/ *v.* (also **-ise**) (**-zing** or **-sing**) use (a machine etc.) as a source of spare parts. □ **cannibalization** /-'zeɪʃ(ə)n/ *n.*

cannon /'kænən/ **–***n.* **1** *hist.* (*pl.* usu. same) large heavy esp. mounted gun. **2** *Billiards* hitting of two balls successively by the player's ball. **–***v.* (usu. foll. by *against*, *into*) collide. [Italian: related to CANE]

cannonade /ˌkænə'neɪd/ **–***n.* period of continuous heavy gunfire. **–***v.* (**-ding**) bombard with a cannonade. [Italian: related to CANNON]

cannon-ball *n. hist.* large ball fired by a cannon.

cannon-fodder *n.* soldiers regarded as expendable.

cannot /'kænɒt/ *v.aux.* can not.

canny /'kænɪ/ *adj.* (**-ier**, **-iest**) **1** shrewd, worldly-wise; thrifty. **2** *Scot.* & *N.Engl.* pleasant, agreeable. □ **cannily** *adv.* **canniness** *n.* [from CAN[1]]

canoe /kə'nu:/ **–***n.* small narrow boat with pointed ends, usu. paddled. **–***v.* (**-noes**, **-noed**, **-noeing**) travel in a canoe. □ **canoeist** *n.* [Spanish and Haitian]

canon /'kænən/ *n.* **1 a** general law, rule, principle, or criterion. **b** church decree or law. **2** (*fem.* **canoness**) member of a cathedral chapter. **3** body of (esp. sacred) writings accepted as genuine. **4** the part of the Roman Catholic Mass containing the words of consecration. **5** *Mus.* piece with different parts taking up the same theme successively. [Greek *kanōn* rule]

cañon var. of CANYON.

canonical /kə'nɒnɪk(ə)l/ **–***adj.* (also **canonic**) **1 a** according to canon law. **b** included in the canon of Scripture. **2** authoritative, accepted. **3** of a cathedral chapter or a member of it. **–***n.* (in *pl.*) canonical dress of clergy. [medieval Latin: related to CANON]

canonist /'kænənɪst/ *n.* expert in canon law.

canonize /'kænəˌnaɪz/ *v.* (also **-ise**) (**-zing** or **-sing**) **1 a** declare officially to be a saint, usu. with a ceremony. **b** regard as a saint. **2** admit to the canon of Scripture. **3** sanction by Church authority. □ **canonization** /-'zeɪʃ(ə)n/ *n.* [medieval Latin: related to CANON]

canon law *n.* ecclesiastical law.

canoodle /kə'nu:d(ə)l/ *v.* (**-ling**) *colloq.* kiss and cuddle. [origin unknown]

canopy /'kænəpɪ/ **–***n.* (*pl.* **-ies**) **1 a** covering suspended over a throne, bed, etc. **b** sky. **c** overhanging shelter. **2** *Archit.* rooflike projection over a niche etc. **3** expanding part of a parachute. **–***v.* (**-ies**, **-ied**) supply or be a canopy to. [Greek, = mosquito-net]

canst *archaic 2nd person sing.* of CAN[1].

cant[1] **–***n.* **1** insincere pious or moral talk. **2** language peculiar to a class, profession, etc.; jargon. **–***v.* use cant. [probably from Latin: related to CHANT]

cant[2] **–***n.* **1** slanting surface, bevel. **2** oblique push or jerk. **3** tilted position. **–***v.* push or pitch out of level; tilt. [Low German or Dutch, = edge]

can't /kɑ:nt/ *contr.* can not.

Cantab. /'kæntæb/ *abbr.* of Cambridge University. [Latin *Cantabrigiensis*]

cantabile /kæn'tɑ:bɪˌleɪ/ *Mus.* **–***adv.* & *adj.* in smooth flowing style. **–***n.* cantabile passage or movement. [Italian, = singable]

Cantabrigian /ˌkæntə'brɪdʒɪən/ **–***adj.* of Cambridge or its university. **–***n.* person from Cambridge or its university. [*Cantabrigia*, Latinized name of Cambridge]

cantaloup /'kæntəˌlu:p/ *n.* (also **cantaloupe**) small round ribbed melon. [*Cantaluppi* near Rome, where it was first grown in Europe]

cantankerous /kæn'tæŋkərəs/ *adj.* bad-tempered, quarrelsome. □ **cantankerously** *adv.* **cantankerousness** *n.* [origin uncertain]

cantata /kæn'tɑ:tə/ *n. Mus.* composition with vocal solos and usu. choral and orchestral accompaniment. [Italian: related to CHANT]

canteen /kæn'ti:n/ *n.* **1 a** restaurant for employees in an office, factory, etc. **b** shop for provisions in a barracks or camp. **2** case of cutlery. **3** soldier's or camper's water-flask. [Italian, = cellar]

canter **–***n.* horse's pace between a trot and a gallop. **–***v.* go or make go at a canter. [*Canterbury gallop* of medieval pilgrims]

canticle /'kæntɪk(ə)l/ *n.* song or chant with a biblical text. [Latin *canticum* CHANT]

cantilever /'kæntɪ,li:və(r)/ *n.* **1** bracket or beam etc. projecting from a wall to support a balcony etc. **2** beam or girder fixed at one end only. □ **cantilevered** *adj.* [origin unknown]

cantilever bridge *n.* bridge made of cantilevers projecting from piers and connected by girders.

canto /'kæntəʊ/ *n.* (*pl.* **-s**) division of a long poem. [Latin *cantus*: related to CHANT]

canton *–n.* /'kæntɒn/ subdivision of a country, esp. of Switzerland. *–v.* /kæn'tu:n/ put (troops) into quarters. [French, = corner: related to CANT[2]]

cantonment /kæn'tu:nmənt/ *n.* **1** lodging assigned to troops. **2** *hist.* permanent military station in India. [French: related to CANTON]

cantor /'kæntɔ:(r)/ *n.* **1** church choir leader. **2** precentor in a synagogue. [Latin, = singer]

canvas /'kænvəs/ *–n.* **1** strong coarse cloth used for sails and tents etc. and for oil-painting. **2** a painting on canvas, esp. in oils. *–v.* (**-ss-**; *US* **-s-**) cover with canvas. □ **under canvas 1** in tents. **2** with sails spread. [Latin: related to CANNABIS]

canvass /'kænvəs/ *–v.* **1** solicit votes, esp. from a constituency electorate. **2 a** ascertain the opinions of. **b** seek custom from. **3** propose (an idea or plan etc.). *–n.* canvassing, esp. of electors. □ **canvasser** *n.* [originally = toss in sheet, from CANVAS]

canyon /'kænjən/ *n.* (also **cañon**) deep gorge. [Spanish *cañón* tube]

CAP *abbr.* Common Agricultural Policy (of the EC).

cap *–n.* **1 a** soft brimless hat, usu. with a peak. **b** head-covering worn in a particular profession. **c** cap as a sign of membership of a sports team. **d** mortarboard. **2 a** cover like a cap (*kneecap*). **b** top for a bottle, jar, pen, camera lens, etc. **3** = DUTCH CAP. **4** = PERCUSSION CAP. **5** dental crown. *–v.* (**-pp-**) **1 a** put a cap on. **b** cover the top or end of. **c** set a limit to (*charge-capping*). **2** award a sports cap to. **3** form the top of. **4** surpass, excel. □ **cap in hand** humbly. **if the cap fits** (of a remark) if it applies to you, so be it. **to cap it all** after everything else. [Latin *cappa*]

capability /,keɪpə'bɪlɪtɪ/ *n.* (*pl.* **-ies**) **1** ability, power. **2** undeveloped or unused faculty.

capable /'keɪpəb(ə)l/ *adj.* **1** competent, able, gifted. **2** (foll. by *of*) **a** having the ability, fitness, etc. for. **b** admitting of (explanation, improvement, etc.). □ **capably** *adv.* [Latin *capio* hold]

capacious /kə'peɪʃəs/ *adj.* roomy. □ **capaciousness** *n.* [Latin *capax*: related to CAPABLE]

capacitance /kə'pæsɪt(ə)ns/ *n.* **1** ability to store electric charge. **2** ratio of change in the electric charge in a system to the corresponding change in its potential.

capacitor /kə'pæsɪtə(r)/ *n.* device able to store electric charge.

capacity /kə'pæsɪtɪ/ *n.* (*pl.* **-ies**) **1 a** power to contain, receive, experience, or produce (*capacity for heat, pain*, etc.). **b** maximum amount that can be contained or produced etc. **c** (*attrib.*) fully occupying the available space etc. (*capacity crowd*). **2** mental power. **3** position or function. **4** legal competence. □ **to capacity** fully. [Latin: related to CAPACIOUS]

caparison /kə'pærɪs(ə)n/ *literary –n.* **1** (usu. in *pl.*) horse's trappings. **2** equipment, finery. *–v.* adorn. [Spanish, = saddle-cloth]

cape[1] *n.* **1** sleeveless cloak. **2** this worn over or as part of a longer cloak or coat. [Latin *cappa* CAP]

cape[2] *n.* **1** headland, promontory. **2** (**the Cape**) the Cape of Good Hope. [Latin *caput* head]

caper[1] *–v.* jump or run playfully. *–n.* **1** playful leap. **2 a** prank. **b** *slang* illicit activity. □ **cut a caper** frolic. [abbreviation of CAPRIOLE]

caper[2] *n.* **1** bramble-like shrub. **2** (in *pl.*) its pickled buds used esp. in a sauce. [Greek *kapparis*]

capercaillie /,kæpə'keɪlɪ/ *n.* (also **capercailzie** /-lzɪ/) large European grouse. [Gaelic, = horse of the forest]

capillarity /,kæpɪ'lærɪtɪ/ *n.* the rise or depression of a liquid in a narrow tube. [French: related to CAPILLARY]

capillary /kə'pɪlərɪ/ *–attrib. adj.* **1** of or like a hair, esp. (of a tube) of very small diameter. **2** of the branching blood-vessels connecting arteries and veins. *–n.* (*pl.* **-ies**) **1** capillary tube. **2** capillary blood vessel. [Latin *capillus* hair]

capillary action *n.* = CAPILLARITY.

capital /'kæpɪt(ə)l/ *–n.* **1** chief town or city of a country or region. **2 a** money etc. with which a company starts in business. **b** accumulated wealth. **3** capitalists collectively. **4** capital letter. **5** head of a column or pillar. *–adj.* **1 a** principal, most important. **b** *colloq.* excellent. **2 a** involving punishment by death. **b** (of an error etc.) vitally harmful, fatal. **3** (of letters of the alphabet) large in size, used to begin sentences and names etc. □ **make capital out of**

use to one's advantage. [Latin *caput -itis* head]

capital gain *n.* profit from the sale of investments or property.

capital goods *n.pl.* machinery, plant, etc.

capitalism *n.* economic and political system dependent on private capital and profit-making.

capitalist *–n.* **1** person investing or possessing capital. **2** advocate of capitalism. *–adj.* of or favouring capitalism. □ **capitalistic** /-ˈlɪstɪk/ *adj.*

capitalize *v.* (also **-ise**) (**-zing** or **-sing**) **1** (foll. by *on*) use to one's advantage. **2** convert into or provide with capital. **3 a** write (a letter of the alphabet) as a capital. **b** begin (a word) with a capital letter. □ **capitalization** /-ˈzeɪʃ(ə)n/ *n.* [French: related to CAPITAL]

capital levy *n.* tax on wealth or property.

capital sum *n.* lump sum, esp. payable to an insured person.

capital transfer tax *n. hist.* tax levied on the transfer of capital by gift or bequest etc.

■ **Usage** This tax was replaced in 1986 by *inheritance tax*.

capitation /ˌkæpɪˈteɪʃ(ə)n/ *n.* tax or fee paid per person. [Latin: related to CAPITAL]

capitular /kəˈpɪtjʊlə(r)/ *adj.* of a cathedral chapter. [Latin *capitulum* CHAPTER]

capitulate /kəˈpɪtjʊˌleɪt/ *v.* (**-ting**) surrender. □ **capitulation** /-ˈleɪʃ(ə)n/ *n.* [medieval Latin, = put under headings]

capo /ˈkæpəʊ/ *n.* (*pl.* **-s**) device fitted across the strings of a guitar etc. to raise their pitch equally. [Italian *capo tasto* head stop]

capon /ˈkeɪpən/ *n.* castrated cock fattened for eating. [Latin *capo*]

cappuccino /ˌkæpʊˈtʃiːnəʊ/ *n.* (*pl.* **-s**) frothy milky coffee. [Italian, = CAPUCHIN]

caprice /kəˈpriːs/ *n.* **1 a** whim. **b** tendency to this. **2** lively or fanciful work of art, music, etc. [Italian *capriccio* sudden start]

capricious /kəˈprɪʃəs/ *adj.* subject to whims; unpredictable. □ **capriciously** *adv.* **capriciousness** *n.*

Capricorn /ˈkæprɪˌkɔːn/ *n.* **1** constellation and tenth sign of the zodiac (the Goat). **2** person born when the sun is in this sign. [Latin *caper -pri* goat, *cornu* horn]

capriole /ˈkæprɪˌəʊl/ *–n.* leap, caper, esp. of a trained horse. *–v.* (**-ling**) perform this. [Italian: related to CAPRICORN]

capsicum /ˈkæpsɪkəm/ *n.* **1** plant with edible fruits, esp. any of several varieties of pepper. **2** red, green, or yellow fruit of these. [Latin *capsa* case]

capsize /kæpˈsaɪz/ *v.* (**-zing**) (of a boat etc.) be overturned; overturn. [Spanish *capuzar* sink]

capstan /ˈkæpst(ə)n/ *n.* **1** thick revolving cylinder for winding a cable etc. **2** revolving spindle carrying the spool on a tape recorder. [Provençal]

capstan lathe *n.* lathe with a revolving tool-holder.

capsule /ˈkæpsjuːl/ *n.* **1** small edible soluble case enclosing medicine. **2** detachable compartment of a spacecraft or nose of a rocket. **3** enclosing membrane in the body. **4** dry fruit that releases its seeds when ripe. **5** (*attrib.*) concise; condensed. □ **capsular** *adj.* [Latin *capsa* case]

capsulize *v.* (also **-ise**) (**-zing** or **-sing**) put (information etc.) in compact form.

Capt. *abbr.* Captain.

captain /ˈkæptɪn/ *–n.* **1 a** chief, leader. **b** leader of a team. **2 a** commander of a ship. **b** pilot of a civil aircraft. **3** army officer next above lieutenant. *–v.* be captain of; lead. □ **captaincy** *n.* (*pl.* **-ies**). [Latin *caput* head]

caption /ˈkæpʃ(ə)n/ *–n.* **1** wording appended to an illustration, cartoon, etc. **2** wording on a cinema or television screen. **3** heading of a chapter, article, etc. *–v.* provide with a caption. [Latin *capio* take]

captious /ˈkæpʃəs/ *adj.* fault-finding. [Latin: related to CAPTION]

captivate /ˈkæptɪˌveɪt/ *v.* (**-ting**) fascinate; charm. □ **captivation** /-ˈveɪʃ(ə)n/ *n.* [Latin: related to CAPTIVE]

captive /ˈkæptɪv/ *–n.* confined or imprisoned person or animal. *–adj.* **1** taken prisoner; restrained. **2** unable to escape (*captive audience*). □ **captivity** /-ˈtɪvɪtɪ/ *n.* [Latin *capio capt-* take]

captor /ˈkæptə(r)/ *n.* person who captures. [Latin: related to CAPTIVE]

capture /ˈkæptʃə(r)/ *–v.* (**-ring**) **1 a** take prisoner; seize. **b** obtain by force or trickery. **2** portray; record on film etc. **3** absorb (a subatomic particle). **4** record (data) for use in a computer. *–n.* **1** act of capturing. **2** thing or person captured. [Latin: related to CAPTIVE]

Capuchin /ˈkæpjʊtʃɪn/ *n.* **1** Franciscan friar. **2** (**capuchin**) **a** monkey with cowl-like head hair. **b** pigeon with a cowl-like head and neck. [Italian *cappuccio* cowl]

capybara /ˌkæpɪˈbɑːrə/ *n.* large semi-aquatic S. American rodent. [Tupi]

car *n.* **1** (in full **motor car**) motor vehicle for a driver and small number of passengers. **2** (in *comb.*) road vehicle or railway carriage esp. of a specified kind (*tramcar*; *dining-car*). **3** *US* any railway carriage or van. **4** passenger compartment of a lift, balloon, etc. [French from Latin]

caracul var. of KARAKUL.

carafe /kə'ræf/ *n.* glass container for water or wine. [French from Arabic]

caramel /'kærə,mel/ *n.* **1 a** burnt sugar or syrup as a flavouring or colouring. **b** a kind of soft toffee. **2** light-brown colour. □ **caramelize** /-mə,laɪz/ *v.* (also **-ise**) (**-zing** or **-sing**). [French from Spanish]

carapace /'kærə,peɪs/ *n.* upper shell of a tortoise or crustacean. [French from Spanish]

carat /'kærət/ *n.* **1** unit of weight for precious stones (200 mg). **2** measure of purity of gold (pure gold = 24 carats). [French ultimately from Greek *keras* horn]

caravan /'kærə,væn/ –*n.* **1** vehicle equipped for living in and usu. towed by a car. **2** people travelling together, esp. across a desert. –*v.* (**-nn-**) travel or live in a caravan. □ **caravanner** *n.* [French from Persian]

caravanserai /,kærə'vænsə,raɪ/ *n.* Eastern inn with a central court. [Persian, = caravan place]

caravel /'kærə,vel/ *n.* (also **carvel** /'kɑ:v(ə)l/) *hist.* small light fast ship. [Greek *karabos*, literally 'horned beetle']

caraway /'kærə,weɪ/ *n.* plant with tiny white flowers. [Spanish from Arabic]

caraway seed *n.* fruit of the caraway as flavouring and a source of oil.

carb *n. colloq.* carburettor. [abbreviation]

carbide /'kɑ:baɪd/ *n.* **1** binary compound of carbon. **2** = CALCIUM CARBIDE.

carbine /'kɑ:baɪn/ *n.* short rifle orig. for cavalry use. [French]

carbohydrate /,kɑ:bə'haɪdreɪt/ *n.* energy-producing organic compound of carbon, hydrogen, and oxygen (e.g. starch, sugar).

carbolic /kɑ:'bɒlɪk/ *n.* (in full **carbolic acid**) phenol. [from CARBON]

carbolic soap *n.* soap containing carbolic.

car bomb *n.* terrorist bomb placed in or under a parked car.

carbon /'kɑ:bən/ *n.* **1** non-metallic element occurring naturally as diamond, graphite, and charcoal, and in all organic compounds. **2 a** = CARBON COPY. **b** = CARBON PAPER. **3** rod of carbon in an arc lamp. [Latin *carbo* charcoal]

carbonaceous /,kɑ:bə'neɪʃəs/ *adj.* **1** consisting of or containing carbon. **2** of or like coal or charcoal.

carbonate /'kɑ:bə,neɪt/ –*n. Chem.* salt of carbonic acid. –*v.* (**-ting**) fill with carbon dioxide. [French: related to CARBON]

carbon copy *n.* **1** copy made with carbon paper. **2** exact copy.

carbon dating *n.* determination of the age of an organic object from the ratio of isotopes, which changes as carbon-14 decays.

carbon dioxide *n.* gas occurring naturally in the atmosphere and formed by respiration.

carbon fibre *n.* thin strong crystalline filament of carbon used as a strengthening material.

carbon-14 *n.* radioisotope of mass 14, used in carbon dating.

carbonic /kɑ:'bɒnɪk/ *adj.* containing carbon.

carbonic acid *n.* weak acid formed from carbon dioxide in water.

carboniferous /,kɑ:bə'nɪfərəs/ –*adj.* **1** producing coal. **2** (**Carboniferous**) of the fifth period in the Palaeozoic era, with extensive formation of coal. –*n.* (**Carboniferous**) this period.

carbonize /'kɑ:bə,naɪz/ *v.* (also **-ise**) (**-zing** or **-sing**) **1** convert into carbon. **2** reduce to charcoal or coke. **3** coat with carbon. □ **carbonization** /-'zeɪʃ(ə)n/ *n.*

carbon monoxide *n.* toxic gas formed by the incomplete burning of carbon.

carbon paper *n.* thin carbon-coated paper used for making copies.

carbon tetrachloride *n.* colourless liquid used as a solvent.

carbon-12 *n.* stable isotope of carbon, used as a standard.

car-boot sale *n.* sale of goods from (tables stocked from) the boots of cars.

carborundum /,kɑ:bə'rʌndəm/ *n.* compound of carbon and silicon used esp. as an abrasive. [from CARBON, CORUNDUM]

carboy /'kɑ:bɔɪ/ *n.* large globular glass bottle usu. in a frame. [Persian]

carbuncle /'kɑ:bʌŋk(ə)l/ *n.* **1** severe skin abscess. **2** bright-red gem. [Latin: related to CARBON]

carburettor /,kɑ:bə'retə(r)/ *n.* (*US* **carburetor**) apparatus in an internal-combustion engine for mixing petrol and air to make an explosive mixture.

carcass /'kɑ:kəs/ *n.* (also **carcase**) **1** dead body of an animal, esp. as meat. **2** bones of a cooked bird. **3** *colloq.* human body; corpse. **4** framework. **5** worthless remains. [French]

carcinogen /kɑ:'sɪnədʒ(ə)n/ *n.* substance producing cancer. □ **carcinogenic** /-'dʒenɪk/ *adj.* [related to CARCINOMA]

carcinoma /ˌkɑːsɪˈnəʊmə/ *n.* (*pl.* **-s** or **-mata**) cancerous tumour. [Greek *karkinos* crab]

card[1] *n.* **1** thick stiff paper or thin pasteboard. **2 a** piece of this for writing or printing on, esp. to send greetings, to identify a person, or to record information. **b** small rectangular piece of plastic used for identity etc. **3 a** = PLAYING-CARD. **b** (in *pl.*) card-playing. **4** (in *pl.*) *colloq.* tax and national insurance documents etc., held by an employer. **5** programme of events at a race-meeting etc. **6** *colloq.* odd or amusing person. □ **card up one's sleeve** plan or secret weapon in reserve. **get one's cards** *colloq.* be dismissed from one's employment. **on the cards** possible or likely. **put** (or **lay**) **one's cards on the table** reveal one's resources, intentions, etc. [Greek *khartēs* papyrus-leaf]

card[2] *–n.* wire brush etc. for raising a nap on cloth etc. *–v.* brush or comb with a card. [Latin *caro* card (v.)]

cardamom /ˈkɑːdəməm/ *n.* seeds of an aromatic SE Asian plant used as a spice. [Latin from Greek]

cardboard *n.* pasteboard or stiff paper, esp. for making boxes.

cardboard city *n.* area where homeless people make shelters from cardboard boxes etc.

card-carrying *adj.* registered as a member (esp. of a political party or trade union).

card-game *n.* game using playing-cards.

cardiac /ˈkɑːdɪˌæk/ *adj.* of the heart. [Greek *kardia* heart]

cardigan /ˈkɑːdɪgən/ *n.* knitted jacket. [Earl of *Cardigan*]

cardinal /ˈkɑːdɪn(ə)l/ *–adj.* **1** chief, fundamental. **2** deep scarlet. *–n.* **1** (as a title **Cardinal**) leading Roman Catholic dignitary, one of the college electing the Pope. **2** small scarlet American songbird. [Latin *cardo -din-* hinge]

cardinal number *n.* number denoting quantity (1, 2, 3, etc.), as opposed to an ordinal number.

cardinal points *n.pl.* four main points of the compass (N., S., E., W.).

cardinal virtues *n.pl.* justice, prudence, temperance, and fortitude.

card index *n.* index with a card for each entry.

cardiogram /ˈkɑːdɪəʊˌgræm/ *n.* record of heart movements. [Greek *kardia* heart]

cardiograph /ˈkɑːdɪəʊˌgrɑːf/ *n.* instrument recording heart movements. □ **cardiographer** /-ˈɒgrəfə(r)/ *n.* **cardiography** /-ˈɒgrəfɪ/ *n.*

cardiology /ˌkɑːdɪˈɒlədʒɪ/ *n.* branch of medicine concerned with the heart. □ **cardiologist** *n.*

cardiovascular /ˌkɑːdɪəʊˈvæskjʊlə(r)/ *adj.* of the heart and blood-vessels.

cardoon /kɑːˈduːn/ *n.* thistle-like plant with leaves used as a vegetable. [French from Latin]

cardphone *n.* public telephone operated by a machine-readable card instead of money.

card-sharp *n.* (also **card-sharper**) swindler at card-games.

card-table *n.* (esp. folding) table for card-playing.

card vote *n.* = BLOCK VOTE.

care *–n.* **1** worry, anxiety. **2** cause of this. **3** serious attention; caution. **4 a** protection, looking after, charge. **b** = CHILD CARE. **5** thing to be done or seen to. *–v.* (**-ring**) **1** (usu. foll. by *about*, *for*, *whether*) feel concern or interest. **2** (usu. foll. by *for*) like, be fond of (*don't care for jazz*). **3** (foll. by *to* + infin.) wish or be willing (*would you care to try?*). □ **care for** provide for; look after. **care of** at the address of. **in care** (of a child) in local authority care. **not care a damn** etc. = *not give a damn* etc. (see GIVE). **take care 1** be careful. **2** (foll. by *to* + infin.) not fail or neglect. **take care of 1** look after. **2** deal with, dispose of. [Old English, = sorrow]

careen /kəˈriːn/ *v.* **1** turn (a ship) on one side for repair etc. **2** tilt, lean over. **3** swerve about. [Latin *carina* keel]

■ **Usage** Sense 3 of *careen* is influenced by the verb *career*.

career /kəˈrɪə(r)/ *–n.* **1** one's professional etc. progress through life. **2** profession or occupation, esp. as offering advancement. **3** (*attrib.*) **a** pursuing or wishing to pursue a career (*career woman*). **b** working permanently in a specified profession (*career diplomat*). **4** swift course (*in full career*). *–v.* **1** move or swerve about wildly. **2** go swiftly. [Latin: related to CAR]

careerist *n.* person predominantly concerned with personal advancement.

carefree *adj.* light-hearted; joyous.

careful *adj.* **1** painstaking, thorough. **2** cautious. **3** taking care; not neglecting (*careful to remind them*). □ **carefully** *adv.* **carefulness** *n.*

careless *adj.* **1** lacking care or attention. **2** unthinking, insensitive. **3** light-hearted. **4** (foll. by *of*) not concerned about. □ **carelessly** *adv.* **carelessness** *n.*

carer *n.* person who cares for a sick or elderly person, esp. a relative at home.

caress /kə'res/ –*v.* touch or stroke gently or lovingly. –*n.* loving or gentle touch. [Latin *carus* dear]
caret /'kærət/ *n.* mark (‸, ⁁) indicating a proposed insertion in printing or writing. [Latin, = is lacking]
caretaker *n.* **1** person employed to look after a house, building, etc. **2** (*attrib.*) exercising temporary authority (*caretaker government*).
careworn *adj.* showing the effects of prolonged worry.
cargo /'kɑ:gəʊ/ *n.* (*pl.* **-es** or **-s**) goods carried on a ship or aircraft. [Spanish: related to CHARGE]
Carib /'kærɪb/ –*n.* **1** aboriginal inhabitant of the southern W. Indies or adjacent coasts. **2** their language. –*adj.* of the Caribs. [Spanish from Haitian]
Caribbean /ˌkærə'bi:ən/ *adj.* of the Caribs or the W. Indies generally.
caribou /'kærɪˌbu:/ *n.* (*pl.* same) N. American reindeer. [French from American Indian]
caricature /'kærɪkətʃʊə(r)/ –*n.* **1** grotesque usu. comically exaggerated representation esp. of a person. **2** ridiculously poor imitation or version. –*v.* (**-ring**) make or give a caricature of. □ **caricaturist** *n.* [Italian *caricare* exaggerate]
caries /'keəri:z/ *n.* (*pl.* same) decay of a tooth or bone. [Latin]
carillon /kə'rɪljən/ *n.* **1** set of bells sounded either from a keyboard or mechanically. **2** tune played on bells. [French]
caring *adj.* **1** kind, humane. **2** (*attrib.*) concerned with looking after people (*caring professions*).
carioca /ˌkærɪ'əʊkə/ *n.* **1** Brazilian dance like the samba. **2** music for this. [Portuguese]
Carmelite /'kɑ:məˌlaɪt/ –*n.* **1** friar of the order of Our Lady of Carmel. **2** nun of a similar order. –*adj.* of the Carmelites. [Mt. *Carmel* in Palestine, where the order was founded]
carminative /'kɑ:mɪnətɪv/ –*adj.* relieving flatulence. –*n.* carminative drug. [Latin *carmino* heal by CHARM]
carmine /'kɑ:maɪn/ –*adj.* of vivid crimson colour. –*n.* **1** this colour. **2** carmine pigment made from cochineal. [probably from Latin *carmesinum* CRIMSON]
carnage /'kɑ:nɪdʒ/ *n.* great slaughter, esp. in battle. [Latin: related to CARNAL]
carnal /'kɑ:n(ə)l/ *adj.* **1** of the body or flesh; worldly. **2** sensual, sexual. □ **carnality** /-'nælɪtɪ/ *n.* [Latin *caro carn-* flesh]
carnation /kɑ:'neɪʃ(ə)n/ –*n.* **1** clove-scented pink. **2** rosy-pink colour. –*adj.* rosy-pink. [Italian: related to CARNAL because of the flesh-colour]
carnelian var. of CORNELIAN.
carnet /'kɑ:neɪ/ *n.* permit to drive across a frontier, use a camp-site, etc. [French, = notebook]
carnival /'kɑ:nɪv(ə)l/ *n.* **1 a** annual festivities including a parade through the streets in fancy dress. **b** festival preceding Lent. **2** merrymaking. **3** *US* funfair or circus. [Latin *carnem levo* put away meat]
carnivore /'kɑ:nɪˌvɔ:(r)/ *n.* carnivorous animal or plant, esp. a mammal of the order including cats, dogs, and bears.
carnivorous /kɑ:'nɪvərəs/ *adj.* (of an animal or plant) feeding on flesh. [Latin: related to CARNAL, *voro* devour]
carob /'kærəb/ *n.* seed pod of a Mediterranean tree used as a chocolate substitute. [Arabic *ḵarrūba*]
carol /'kær(ə)l/ –*n.* joyous song, esp. a Christmas hymn. –*v.* (**-ll-**; *US* **-l-**) **1** sing carols. **2** sing joyfully. [French]
Carolingian /ˌkærə'lɪndʒɪən/ –*adj.* of the Frankish dynasty founded by Charlemagne. –*n.* member of this dynasty. [Latin *Carolus* Charles]
carotene /'kærəˌti:n/ *n.* orange-coloured pigment found in carrots, tomatoes, etc., acting as a source of vitamin A. [Latin: related to CARROT]
carotid /kə'rɒtɪd/ –*n.* each of the two main arteries carrying blood to the head and neck. –*adj.* of these arteries. [Latin from Greek]
carouse /kə'raʊz/ –*v.* (**-sing**) have a lively drinking-party. –*n.* such a party. □ **carousal** *n.* **carouser** *n.* [German *gar aus* (drink) right out]
carousel /ˌkærə'sel/ *n.* **1** *US* merry-go-round. **2** rotating luggage delivery system at an airport etc. [French from Italian]
carp[1] *n.* (*pl.* same) freshwater fish often bred for food. [Provençal or Latin]
carp[2] *v.* find fault; complain pettily. □ **carper** *n.* [Old Norse, = brag]
carpal /'kɑ:p(ə)l/ –*adj.* of the bones in the wrist. –*n.* wrist-bone. [from CARPUS]
car park *n.* area for parking cars.
carpel /'kɑ:p(ə)l/ *n.* female reproductive organ of a flower. [Greek *karpos* fruit]
carpenter /'kɑ:pɪntə(r)/ –*n.* person skilled in woodwork. –*v.* **1** make or construct in wood. **2** construct; fit together. □ **carpentry** *n.* [Latin *carpentum* wagon]
carpet /'kɑ:pɪt/ –*n.* **1 a** thick fabric for covering floor or stairs. **b** piece of this. **2** thing resembling this etc. (*carpet of snow*). –*v.* (**-t-**) **1** cover with or as with carpet. **2** *colloq.* reprimand. □ **on the carpet** *colloq.* **1** being reprimanded. **2**

under consideration. **sweep under the carpet** conceal (a problem or difficulty). [Latin *carpo* pluck]

carpet-bag *n.* travelling-bag, orig. made of carpet-like material.

carpet-bagger *n. colloq.* **1** esp. *US* political candidate etc. without local connections. **2** unscrupulous opportunist.

carpeting *n.* **1** material for carpets. **2** carpets collectively.

carpet slipper *n.* soft slipper.

carpet-sweeper *n.* household implement for sweeping carpets.

car phone *n.* radio-telephone for use in a car etc.

carport *n.* roofed open-sided shelter for a car.

carpus /'kɑ:pəs/ *n.* (*pl.* **-pi** /-paɪ/) small bones forming the wrist in humans and similar parts in other mammals. [Latin from Greek]

carrageen /'kærə,gi:n/ *n.* (also **carragheen**) edible red seaweed. [origin uncertain]

carrel /'kær(ə)l/ *n.* small cubicle for a reader in a library. [French from medieval Latin]

carriage /'kærɪdʒ/ *n.* **1** railway passenger vehicle. **2** wheeled horse-drawn passenger vehicle. **3 a** conveying of goods. **b** cost of this. **4** carrying part of a machine (e.g. a typewriter). **5** gun-carriage. **6** bearing, deportment. [French: related to CARRY]

carriage clock *n.* portable clock with a handle.

carriageway *n.* the part of a road intended for vehicles.

carrier /'kærɪə(r)/ *n.* **1** person or thing that carries. **2** transport or freight company. **3** = CARRIER BAG. **4** framework on a bicycle for luggage or a passenger. **5** person or animal that may transmit disease etc. without suffering from it. **6** = AIRCRAFT-CARRIER.

carrier bag *n.* plastic or paper bag with handles.

carrier pigeon *n.* pigeon trained to carry messages.

carrier wave *n.* high-frequency electromagnetic wave modulated in amplitude or frequency to convey a signal.

carrion /'kærɪən/ *n.* **1** dead putrefying flesh. **2** something vile or filthy. [Latin *caro* flesh]

carrion crow *n.* crow feeding on carrion.

carrot /'kærət/ *n.* **1 a** plant with a tapering orange-coloured root. **b** this as a vegetable. **2** incentive. □ **carroty** *adj.* [Greek *karōton*]

carry /'kærɪ/ *–v.* (**-ies, -ied**) **1** support or hold up, esp. while moving. **2** convey with one or have on one's person. **3** conduct or transmit (*pipe carries water*). **4** (often foll. by *to*) take (a process etc.) to a specified point; continue; prolong (*carry into effect*; *carry a joke too far*). **5** involve, imply (*carries 6% interest*). **6** *Math.* transfer (a figure) to a column of higher value. **7** hold in a specified way (*carry oneself erect*). **8 a** (of a newspaper etc.) publish. **b** (of a radio or television station) broadcast. **9** keep a regular stock of. **10 a** (of sound) be audible at a distance. **b** (of a missile or gun etc.) travel or propel to a specified distance. **11 a** win victory or acceptance for (a proposal etc.). **b** win acceptance from (*carried the audience with her*). **c** win, capture (a prize, fortress, etc.). **12 a** endure the weight of; support. **b** be the driving force in (*you carry the department*). **13** be pregnant with. *–n.* (*pl.* **-ies**) **1** act of carrying. **2** *Golf* distance a ball travels before reaching the ground. □ **carry away 1** remove. **2** inspire. **3** deprive of self-control (*got carried away*). **carry the can** *colloq.* bear the responsibility or blame. **carry the day** be victorious or successful. **carry forward** transfer to a new page or account. **carry it off** do well under difficulties. **carry off 1** take away, esp. by force. **2** win (a prize). **3** (esp. of a disease) kill. **carry on 1** continue. **2** engage in (conversation or business). **3** *colloq.* behave strangely or excitedly. **4** (often foll. by *with*) *colloq.* flirt or have a love affair. **carry out** put (an idea etc.) into practice. **carry over 1** = *carry forward*. **2** postpone. **carry through 1** complete successfully. **2** bring safely out of difficulties. **carry weight** be influential or important. [Anglo-French: related to CAR]

carry-cot *n.* portable cot for a baby.

carryings-on *n.pl.* = CARRY-ON.

carry-on *n. slang* **1** fuss, excitement. **2** questionable behaviour. **3** flirtation or love affair.

carry-out *attrib. adj.* & *n.* esp. *Scot.* & *US* = TAKE-AWAY.

carsick *adj.* nauseous from car travel. □ **carsickness** *n.*

cart *–n.* **1** open usu. horse-drawn vehicle for carrying loads. **2** light vehicle for pulling by hand. *–v.* **1** convey in a cart. **2** *slang* carry or convey with effort. □ **put the cart before the horse** reverse the proper order or procedure. [Old English and Old Norse]

carte blanche /kɑ:t 'blɑ̃ʃ/ *n.* full discretionary power. [French, = blank paper]

cartel /kɑ:'tel/ *n.* union of suppliers etc. to control prices. [Italian diminutive: related to CARD[1]]

Cartesian /kɑ:'ti:zɪən/ *–adj.* of Descartes or his philosophy. *–n.* follower of Descartes. [Latin *Cartesius* Descartes]

Cartesian coordinates *n.pl.* system for locating a point by reference to its distance from axes intersecting at right angles.

cart-horse *n.* thickset horse.

Carthusian /kɑ:'θju:zɪən/ *–n.* monk of a contemplative order founded by St Bruno. *–adj.* of this order. [Latin: related to CHARTREUSE]

cartilage /'kɑ:tɪlɪdʒ/ *n.* firm flexible connective tissue, mainly replaced by bone in adulthood. □ **cartilaginous** /-'lædʒɪnəs/ *adj.* [French from Latin]

cartography /kɑ:'tɒgrəfɪ/ *n.* map-drawing. □ **cartographer** *n.* **cartographic** /-tə'græfɪk/ *adj.* [French *carte* map]

carton /'kɑ:t(ə)n/ *n.* light esp. cardboard box or container. [French: related to CARTOON]

cartoon /kɑ:'tu:n/ *n.* **1** humorous, esp. topical, drawing in a newspaper etc. **2** sequence of drawings telling a story. **3** animated sequence of these on film. **4** full-size preliminary design for a tapestry etc. □ **cartoonist** *n.* [Italian: related to CARD[1]]

cartouche /kɑ:'tu:ʃ/ *n.* **1** scroll-like ornamentation. **2** oval ring enclosing the name and title of a pharaoh. [French: related to CARTOON]

cartridge /'kɑ:trɪdʒ/ *n.* **1** case containing an explosive charge or bullet for firearms or blasting. **2** sealed container of film etc. **3** component carrying the stylus on a record-player. **4** ink-container for insertion in a pen. [French: related to CARTOON]

cartridge-belt *n.* belt with pockets or loops for cartridges.

cartridge paper *n.* thick paper for drawing etc.

cartwheel *n.* **1** wheel of a cart. **2** circular sideways handspring with arms and legs extended.

cart-wright *n.* maker of carts.

carve *v.* (**-ving**) **1** produce or shape by cutting. **2 a** cut patterns etc. in. **b** (foll. by *into*) form a pattern etc. from (*carved it into a bust*). **3** (also *absol.*) cut (meat etc.) into slices. □ **carve out 1** take from a larger whole. **2** establish (a career etc.) purposefully. **carve up 1** subdivide. **2** drive aggressively into the path of (another vehicle). [Old English]

carvel var. of CARAVEL.

carvel-built *adj.* (of a boat) made with planks flush, not overlapping.

carver *n.* **1** person who carves. **2** carving knife. **3** chair with arms, for a person carving.

carvery *n.* (*pl.* **-ies**) buffet or restaurant with joints displayed for carving.

carve-up *n. slang* sharing-out, esp. of spoils.

carving *n.* carved object, esp. as a work of art.

carving knife *n.* knife for carving meat.

Casanova /,kæsə'nəʊvə/ *n.* notorious womanizer. [Italian adventurer]

cascade /kæs'keɪd/ *–n.* **1** small waterfall, esp. one of series. **2** thing falling or arranged like a cascade. *–v.* (**-ding**) fall in or like a cascade. [Latin: related to CASE[1]]

cascara /kæs'kɑ:rə/ *n.* bark of a Californian buckthorn, used as a laxative. [Spanish]

case[1] *n.* **1** instance of something occurring. **2** hypothetical or actual situation. **3 a** person's illness, circumstances, etc., as regarded by a doctor, social worker, etc. **b** such a person. **4** matter under esp. police investigation. **5** suit at law. **6 a** sum of the arguments on one side, esp. in a lawsuit. **b** set of arguments (*have a good case*). **c** valid set of arguments (*have no case*). **7** *Gram.* **a** relation of a word to other words in a sentence. **b** form of a noun, adjective, or pronoun expressing this. **8** *colloq.* comical person. □ **in any case** whatever the truth is; whatever may happen. **in case 1** in the event that; if. **2** lest; in provision against a possibility (*took it in case*). **in case of** in the event of. **is** (or **is not**) **the case** is (or is not) so. [Latin *casus* from *cado* fall]

case[2] *–n.* **1** container or enclosing covering. **2** this with its contents. **3** protective outer covering. **4** item of luggage, esp. a suitcase. *–v.* (**-sing**) **1** enclose in a case. **2** (foll. by *with*) surround. **3** *slang* reconnoitre (a house etc.) before burgling it. [Latin *capsa* box]

case-harden *v.* **1** harden the surface of (esp. iron by carbonizing). **2** make callous.

case history *n.* record of a person's life or medical history for use in professional treatment.

casein /'keɪsɪɪn/ *n.* the main protein in milk and cheese. [Latin *caseus* cheese]

case-law *n.* law as established by the outcome of former cases.

casemate /'keɪsmeɪt/ *n.* **1** embrasured room in a fortress wall. **2** armoured enclosure for guns on a warship. [French and Italian]

casement /'keɪsmənt/ *n.* window or part of a window hinged to open like a door. [Anglo-Latin: related to CASE[2]]

casework *n.* social work concerned with studying a person's family and background. □ **caseworker** *n.*

cash –*n.* **1** money in coins or notes. **2** (also **cash down**) full payment at the time of purchase. **3** *colloq.* wealth. –*v.* give or obtain cash for (a note, cheque, etc.). □ **cash in 1** obtain cash for. **2** *colloq.* (usu. foll. by *on*) profit (from); take advantage (of). **cash up** count and check the day's takings. [Latin: related to CASE[2]]

cash and carry *n.* **1** system (esp. in wholesaling) of cash payment for goods taken away by the purchaser. **2** store where this operates.

cash-book *n.* book for recording receipts and cash payments.

cashcard *n.* plastic card for withdrawing money from a cash dispenser.

cash crop *n.* crop produced for sale.

cash desk *n.* counter etc. where payment is made in a shop.

cash dispenser *n.* automatic machine for the withdrawal of cash, esp. with a cashcard.

cashew /'kæʃu:/ *n.* **1** evergreen tree bearing kidney-shaped nuts. **2** this edible nut. [Portuguese from Tupi]

cash flow *n.* movement of money into and out of a business.

cashier[1] /kæ'ʃɪə(r)/ *n.* person dealing with cash transactions in a shop, bank, etc.

cashier[2] /kæ'ʃɪə(r)/ *v.* dismiss from service, esp. with disgrace. [French: related to QUASH]

cashmere /'kæʃmɪə(r)/ *n.* **1** fine soft wool, esp. that of a Kashmir goat. **2** material made from this. [*Kashmir* in Asia]

cash on delivery *n.* payment for goods when they are delivered.

cashpoint *n.* = CASH DISPENSER.

cash register *n.* till recording sales, totalling receipts, etc.

casing /'keɪsɪŋ/ *n.* protective or enclosing cover or material.

casino /kə'si:nəʊ/ *n.* (*pl.* **-s**) public room or building for gambling. [Italian diminutive of *casa* house]

cask /kɑ:sk/ *n.* **1** barrel, esp. for alcohol. **2** its contents. [French *casque* or Spanish *casco* helmet]

casket /'kɑ:skɪt/ *n.* **1** small often ornamental box for jewels etc. **2** *US* coffin. [Latin: related to CASE[2]]

cassata /kə'sɑ:tə/ *n.* ice-cream containing fruit and nuts. [Italian]

cassava /kə'sɑ:və/ *n.* **1** plant with starchy roots. **2** starch or flour from these, used e.g. in tapioca. [Taino]

casserole /'kæsə,rəʊl/ –*n.* **1** covered dish for cooking food in the oven. **2** food cooked in this. –*v.* (**-ling**) cook in a casserole. [Greek *kuathion* little cup]

cassette /kə'set/ *n.* sealed case containing magnetic tape, film etc., ready for insertion in a tape recorder, camera, etc. [French diminutive: related to CASE[2]]

cassia /'kæsɪə/ *n.* **1** tree from the leaves of which senna is extracted. **2** cinnamon-like bark of this used as a spice. [Greek *kasia* from Hebrew]

cassis /kæ'si:s/ *n.* blackcurrant flavouring for drinks etc. [French]

cassock /'kæsək/ *n.* long usu. black or red clerical garment. □ **cassocked** *adj.* [French from Italian]

cassoulet /'kæsʊ,leɪ/ *n.* ragout of meat and beans. [French]

cassowary /'kæsə,weərɪ/ *n.* (*pl.* **-ies**) large flightless Australasian bird. [Malay]

cast /kɑ:st/ –*v.* (*past* and *past part.* **cast**) **1** throw, esp. deliberately or forcefully. **2** (often foll. by *on, over*) **a** direct or cause (one's eyes, a glance, light, a shadow, a spell, etc.) to fall. **b** express (doubts, aspersions, etc.). **3** throw out (a fishing-line etc.) into the water. **4** let down (an anchor etc.). **5 a** throw off, get rid of. **b** shed or lose (horns, skin, a horseshoe, etc.). **6** register (a vote). **7 a** shape (molten metal etc.) in a mould. **b** make thus. **8 a** (usu. foll. by *as*) assign (an actor) to a role. **b** allocate roles in (a play etc.). **9** (foll. by *in, into*) arrange (facts etc.) in a specified form. **10** reckon, add up (accounts or figures). **11** calculate (a horoscope). –*n.* **1** throwing of a missile, dice, line, net, etc. **2 a** object made in a mould. **b** moulded mass of solidified material, esp. plaster for a broken limb. **3** actors in a play etc. **4** form, type, or quality. **5** tinge or shade of colour. **6** slight squint. **7** worm-cast. □ **cast about** (or **around**) search. **cast adrift** leave to drift. **cast aside** abandon. **cast loose** detach (oneself). **cast lots** see LOT. **cast off 1** abandon. **2** finish a piece of knitting. **3** set a ship free from a quay etc. **cast on** make the first row of a piece of knitting. **cast up 1** deposit on the shore. **2** add up (figures etc.). [Old Norse]

castanet /,kæstə'net/ *n.* (usu. in *pl.*) each of a pair of hand-held pieces of wood etc., clicked together as an accompaniment, esp. by Spanish dancers. [Latin: related to CHESTNUT]

castaway /'kɑ:stə,weɪ/ –*n.* shipwrecked person. –*adj.* shipwrecked.

caste /kɑ:st/ *n.* **1** any of the Hindu hereditary classes whose members have no social contact with other classes. **2** exclusive social class or system of classes. □ **lose caste** descend in social order. [Spanish and Portuguese: related to CHASTE]

casteism /'kɑ:stɪz(ə)m/ *n.* caste system.

castellated /'kæstə,leɪtɪd/ *adj.* **1** having battlements. **2** castle-like. □ **castellation** /-'leɪʃ(ə)n/ *n.* [medieval Latin: related to CASTLE]

caster var. of CASTOR.

castigate /'kæstɪ,geɪt/ *v.* **(-ting)** rebuke or punish severely. □ **castigation** /-'geɪʃ(ə)n/ *n.* **castigator** *n.* [Latin *castus* pure]

casting /'kɑ:stɪŋ/ *n.* cast, esp. of molten metal.

casting vote *n.* deciding vote when the votes on two sides are equal. [from an obsolete sense of *cast*, = turn the scale]

cast iron *n.* hard alloy of iron, carbon, and silicon cast in a mould.

cast-iron *adj.* **1** of cast iron. **2** very strong; rigid; unchallengeable.

castle /'kɑ:s(ə)l/ *–n.* **1** large fortified building with towers and battlements. **2** *Chess* = ROOK[2]. *–v.* **(-ling)** *Chess* move a rook next to the king and the king to the other side of the rook. □ **castles in the air** day-dream; impractical scheme. [Latin *castellum*]

cast-off *–adj.* abandoned, discarded. *–n.* cast-off thing, esp. a garment.

castor /'kɑ:stə(r)/ *n.* (also **caster**) **1** small swivelled wheel on the leg or underside of a piece of furniture. **2** small perforated container for sprinkling sugar, flour, etc. [from CAST]

castor oil /'kɑ:stə(r)/ *n.* oil from the seeds of a tropical plant, used as a purgative and lubricant. [origin uncertain]

castor sugar *n.* finely granulated white sugar.

castrate /kæ'streɪt/ *v.* **(-ting) 1** remove the testicles of; geld. **2** deprive of vigour. □ **castration** *n.* [Latin *castro*]

castrato /kæ'strɑ:təʊ/ *n.* (*pl.* **-ti** /-tɪ/) *hist.* castrated male soprano or alto singer. [Italian: related to CASTRATE]

casual /'kæʒʊəl/ *–adj.* **1** accidental; chance. **2** not regular or permanent (*casual work*). **3 a** unconcerned. **b** careless; unthinking. **4** (of clothes) informal. *–n.* **1** casual worker. **2** (usu. in *pl.*) casual clothes or shoes. □ **casually** *adv.* **casualness** *n.* [French and Latin: related to CASE[1]]

casualty /'kæʒʊəltɪ/ *n.* (*pl.* **-ies**) **1** person killed or injured in a war or accident. **2** thing lost or destroyed. **3** = CASUALTY DEPARTMENT. **4** accident, mishap. [medieval Latin: related to CASUAL]

casualty department *n.* part of a hospital where casualties are dealt with.

casuist /'kæʒʊɪst/ *n.* **1** person who uses clever but false reasoning in matters of conscience etc. **2** sophist, quibbler. □ **casuistic** /-'ɪstɪk/ *adj.* **casuistry** *n.* [Latin: related to CASE[1]]

cat *n.* **1** small soft-furred four-legged domesticated animal. **2** wild animal of the same family, e.g. lion, tiger. **3** *colloq.* malicious or spiteful woman. **4** = CAT-O'-NINE-TAILS. □ **the cat's whiskers** *colloq.* excellent person or thing. **let the cat out of the bag** reveal a secret. **like a cat on hot bricks** very agitated. **put** (or **set**) **the cat among the pigeons** cause trouble. **rain cats and dogs** rain hard. [Latin *cattus*]

cata- *prefix* **1** down. **2** wrongly. [Greek]

catabolism /kə'tæbə,lɪz(ə)m/ *n.* breakdown of complex molecules in living organisms to release energy; destructive metabolism. □ **catabolic** /,kætə'bɒlɪk/ *adj.* [Greek *katabolē* throwing down]

catachresis /,kætə'kri:sɪs/ *n.* (*pl.* **-chreses** /-si:z/) incorrect use of words. □ **catachrestic** /-'kri:stɪk/ *adj.* [Greek *khraomai* use]

cataclysm /'kætə,klɪz(ə)m/ *n.* **1 a** violent upheaval or disaster. **b** great change. **2** great flood. □ **cataclysmic** /-'klɪzmɪk/ *adj.* [Greek *kluzō* wash]

catacomb /'kætə,ku:m/ *n.* (often in *pl.*) underground cemetery, esp. Roman. [French from Latin]

catafalque /'kætə,fælk/ *n.* decorated bier, used esp. in State funerals or for lying in state. [French from Italian]

Catalan /'kætə,læn/ *–n.* native or language of Catalonia in Spain. *–adj.* of Catalonia. [French from Spanish]

catalepsy /'kætə,lepsɪ/ *n.* trance or seizure with unconsciousness and rigidity of the body. □ **cataleptic** /-'leptɪk/ *adj.* & *n.* [Greek *lēpsis* seizure]

catalogue /'kætə,lɒg/ (*US* **catalog**) *–n.* **1** complete alphabetical or otherwise ordered list of items, often with a description of each. **2** extensive list (*catalogue of disasters*). *–v.* **(-logues, -logued, -loguing;** *US* **-logs, -loged, -loging) 1** make a catalogue of. **2** enter in a catalogue. [Greek *legō* choose]

catalpa /kə'tælpə/ *n.* tree with long pods and showy flowers. [N. American Indian]

catalyse /'kætə,laɪz/ *v.* (*US* **-yze**) **(-sing** or **-zing)** produce (a reaction) by catalysis.

catalysis /kə'tælɪsɪs/ *n.* (*pl.* **-lyses** /-,si:z/) acceleration of a chemical reaction by a catalyst. [Greek *luō* set free]

catalyst /ˈkætəlɪst/ *n.* **1** substance that does not itself change, but speeds up a chemical reaction. **2** person or thing that precipitates change.

catalytic /ˌkætəˈlɪtɪk/ *adj.* of or involving catalysis.

catalytic converter *n.* device incorporated in a vehicle's exhaust system, with a catalyst for converting pollutant gases into harmless products.

catamaran /ˌkætəməˈræn/ *n.* **1** boat with parallel twin hulls. **2** raft of yoked logs or boats. [Tamil]

catamite /ˈkætəˌmaɪt/ *n.* passive partner (esp. a boy) in homosexual practices. [Latin, = Ganymede]

cat-and-dog *adj.* (of a relationship etc.) quarrelsome.

catapult /ˈkætəˌpʌlt/ *–n.* **1** forked stick etc. with elastic for shooting stones. **2** *Mil hist.* machine for hurling large stones etc. **3** device for launching a glider etc. *–v.* **1 a** hurl from or launch with a catapult. **b** fling forcibly. **2** leap or be hurled forcibly. [Latin from Greek]

cataract /ˈkætəˌrækt/ *n.* **1 a** large waterfall. **b** downpour; rush of water. **2** eye condition in which the lens becomes progressively opaque. [Greek *katarrhaktēs*, = down-rushing]

catarrh /kəˈtɑː(r)/ *n.* **1** inflammation of the mucous membrane of the nose, air-passages, etc. **2** mucus caused by this. □ **catarrhal** *adj.* [Greek *rheō* flow]

catastrophe /kəˈtæstrəfɪ/ *n.* **1** great and usu. sudden disaster. **2** denouement of a drama. □ **catastrophic** /-ˈstrɒfɪk/ *adj.* **catastrophically** /-ˈstrɒfɪkəlɪ/ *adv.* [Greek *strephō* turn]

catatonia /ˌkætəˈtəʊnɪə/ *n.* **1** schizophrenia with intervals of catalepsy and sometimes violence. **2** catalepsy. □ **catatonic** /-ˈtɒnɪk/ *adj.* & *n.* [Greek: related to CATA-, TONE]

cat burglar *n.* burglar who enters by climbing to an upper storey.

catcall *–n.* shrill whistle of disapproval. *–v.* make a catcall.

catch *–v.* (*past* and *past part.* **caught** /kɔːt/) **1** capture in a trap, one's hands, etc. **2** detect or surprise (esp. a guilty person). **3 a** intercept and hold (a moving thing) in the hands etc. **b** *Cricket* dismiss (a batsman) by catching the ball before it reaches the ground. **4 a** contract (a disease) from an infected person. **b** acquire (a quality etc.) from another. **5 a** reach in time and board (a train, bus, etc.). **b** be in time to see etc. (a person or thing about to leave or finish). **6** apprehend with the senses or mind (esp. a thing occurring quickly or briefly). **7** (of an artist etc.) reproduce faithfully. **8 a** (cause to) become fixed, entangled, or checked. **b** (often foll. by *on*) hit, deal a blow to (*caught his elbow on the table*). **9** draw the attention of; captivate (*caught his eye*; *caught her fancy*). **10** begin to burn. **11** reach or overtake (a person etc. ahead). **12** (foll. by *at*) try to grasp. *–n.* **1 a** act of catching. **b** *Cricket* etc. chance or act of catching the ball. **2 a** amount of a thing caught, esp. of fish. **b** thing or person caught or worth catching, esp. in marriage. **3 a** question, trick, etc., intended to deceive, incriminate, etc. **b** unexpected or hidden difficulty or disadvantage. **4** device for fastening a door or window etc. **5** *Mus.* round, esp. with words arranged to produce a humorous effect. □ **catch fire** see FIRE. **catch hold of** grasp, seize. **catch it** *slang* be punished. **catch on** *colloq.* **1** become popular. **2** understand what is meant. **catch out 1** detect in a mistake etc. **2** take unawares. **3** = sense 3b of *v.* **catch up 1 a** (often foll. by *with*) reach a person etc. ahead (*caught us up*; *caught up with us*). **b** (often foll. by *with, on*) make up arrears. **2** pick up hurriedly. **3** (often in *passive*) **a** involve; entangle (*caught up in crime*). **b** fasten up (*hair caught up in a ribbon*). [Latin *capto* try to catch]

catch-all *n.* (often *attrib.*) thing designed to be all-inclusive.

catch-as-catch-can *n.* wrestling with few holds barred.

catching *adj.* (of a disease, practice, etc.) infectious.

catchline *n.* short line of type, esp. at the head of copy or as a running headline.

catchment *n.* collection of rainfall.

catchment area *n.* **1** area served by a school, hospital, etc. **2** area from which rainfall flows into a river etc.

catchpenny *attrib. adj.* intended merely to sell quickly; superficially attractive.

catch-phrase *n.* phrase in frequent use.

catch-22 *n.* (often *attrib.*) *colloq.* unresolvable situation containing conflicting or mutually dependent conditions.

catchweight *–adj.* unrestricted as regards weight. *–n.* unrestricted weight category in sports.

catchword *n.* **1** phrase, word, or slogan in frequent current use. **2** word so placed as to draw attention.

catchy *adj.* (**-ier, -iest**) (of a tune) easy to remember, attractive.

cat door var. of CAT FLAP.

catechism /ˈkætɪˌkɪz(ə)m/ *n.* **1 a** principles of a religion in the form of questions and answers. **b** book containing this. **2** series of questions. [Church Latin: related to CATECHIZE]

catechist /ˈkætɪkɪst/ *n.* religious teacher, esp. one using a catechism.

catechize /ˈkætɪˌkaɪz/ *v.* (also **-ise**) (**-zing** or **-sing**) instruct by using a catechism. [Greek *katēkheō* cause to hear]

catechumen /ˌkætɪˈkjuːmən/ *n.* Christian convert under instruction before baptism. [Church Latin *catechumenus*]

categorical /ˌkætɪˈɡɒrɪk(ə)l/ *adj.* unconditional, absolute; explicit. □ **categorically** *adv.* [related to CATEGORY]

categorize /ˈkætɪɡəˌraɪz/ *v.* (also **-ise**) (**-zing** or **-sing**) place in a category. □ **categorization** /-ˈzeɪʃ(ə)n/ *n.*

category /ˈkætɪɡərɪ/ *n.* (*pl.* **-ies**) class or division (of things, ideas, etc.). [Greek, = statement]

cater /ˈkeɪtə(r)/ *v.* **1** supply food. **2** (foll. by *for*) provide what is needed or desired (*caters for all tastes*). **3** (foll. by *to*) pander to (esp. low tastes). [Anglo-French *acatour* buyer, from Latin *capto*: related to CATCH]

caterer *n.* professional supplier of food for social events.

caterpillar /ˈkætəˌpɪlə(r)/ *n.* **1** larva of a butterfly or moth. **2** (**Caterpillar**) **a** (in full **Caterpillar track** or **tread**) *propr.* steel band passing round the wheels of a tractor etc. for travel on rough ground. **b** vehicle with these. [Anglo-French, = hairy cat]

caterwaul /ˈkætəˌwɔːl/ *—v.* make the shrill howl of a cat. *—n.* this noise. [from CAT, *-waul* imitative]

catfish *n.* (*pl.* same) freshwater fish with whisker-like barbels round the mouth.

cat flap *n.* (also **cat door**) small swinging flap in an outer door, for a cat to pass in and out.

catgut *n.* material used for the strings of musical instruments and surgical sutures, made of intestines of the sheep, horse, etc. (but not cat).

catharsis /kəˈθɑːsɪs/ *n.* (*pl.* **catharses** /-siːz/) **1** emotional release in drama or art. **2** *Psychol.* freeing and elimination of repressed emotion. **3** emptying of the bowels. [Greek *katharos* clean]

cathartic /kəˈθɑːtɪk/ *—adj.* **1** effecting catharsis. **2** laxative. *—n.* laxative.

cathedral /kəˈθiːdr(ə)l/ *n.* principal church of a diocese. [Greek *kathedra* seat]

Catherine wheel /ˈkæθrɪn/ *n.* flat coiled firework spinning when lit. [St *Catherine*, who was martyred on a spiked wheel]

catheter /ˈkæθɪtə(r)/ *n.* tube inserted into a body cavity for introducing or removing fluid. [Greek *kathiēmi* send down]

cathode /ˈkæθəʊd/ *n. Electr.* **1** negative electrode in an electrolytic cell. **2** positive terminal of a battery etc. [Greek *kathodos* way down]

cathode ray *n.* beam of electrons from the cathode of a vacuum tube.

cathode-ray tube *n.* vacuum tube in which cathode rays produce a luminous image on a fluorescent screen.

catholic /ˈkæθlɪk/ *—adj.* **1** all-embracing; of wide sympathies or interests. **2** of interest or use to all; universal. **3** (**Catholic**) **a** Roman Catholic. **b** including all Christians, or all of the Western Church. *—n.* (**Catholic**) Roman Catholic. □ **Catholicism** /kəˈθɒlɪˌsɪz(ə)m/ *n.* **catholicity** /ˌkæθəˈlɪsɪtɪ/ *n.* [Greek *holos* whole]

cation /ˈkætˌaɪən/ *n.* positively charged ion. □ **cationic** /-ˈɒnɪk/ *adj.* [from CATA-, ION]

catkin *n.* small spike of usu. hanging flowers on a willow, hazel, etc. [Dutch, = kitten]

catlick *n. colloq.* perfunctory wash.

catmint *n.* pungent plant attractive to cats.

catnap *—n.* short sleep. *—v.* (**-pp-**) have a catnap.

catnip *n.* = CATMINT. [from CAT, dial. *nip* catmint]

cat-o'-nine-tails *n. hist.* whip with nine knotted lashes.

cat's cradle *n.* child's game of forming patterns from a loop of string.

Cat's-eye *n. propr.* reflector stud set into a road.

cat's-eye *n.* precious stone.

cat's-paw *n.* **1** person used as a tool by another. **2** slight breeze.

catsuit *n.* close-fitting garment with trouser legs, covering the whole body.

catsup *US* var. of KETCHUP.

cattery *n.* (*pl.* **-ies**) place where cats are boarded or bred.

cattle /ˈkæt(ə)l/ *n.pl.* large ruminant animals with horns and cloven hoofs, esp. bred for milk or meat. [Anglo-French *catel*: related to CAPITAL]

cattle-grid *n.* grid over a ditch, allowing people and vehicles but not livestock to pass over.

catty *adj.* (**-ier, -iest**) spiteful. □ **cattily** *adv.* **cattiness** *n.*

catwalk *n.* narrow footway or platform.

Caucasian /kɔːˈkeɪʒ(ə)n/ *—adj.* **1** of the White or light-skinned race. **2** of the Caucasus. *—n.* Caucasian person. [*Caucasus* in Georgia]

Caucasoid /ˈkɔːkəˌsɔɪd/ *adj.* of Caucasians.

caucus /'kɔ:kəs/ *n.* (*pl.* **-es**) **1** *US* meeting of party members, esp. in the Senate etc., to decide policy. **2** often *derog.* **a** meeting of a group within a larger organization or party. **b** such a group. [perhaps from Algonquian]
caudal /'kɔ:d(ə)l/ *adj.* **1** of or like a tail. **2** of the posterior part of the body. [Latin *cauda* tail]
caudate /'kɔ:deɪt/ *adj.* tailed.
caught *past* and *past part.* of CATCH.
caul /kɔ:l/ *n.* **1** membrane enclosing a foetus. **2** part of this occasionally found on a child's head at birth. [French]
cauldron /'kɔ:ldrən/ *n.* (also **caldron**) large deep vessel used for boiling. [Latin *caldarium* hot bath]
cauliflower /'kɒlɪˌflaʊə(r)/ *n.* cabbage with a large white flower-head. [French *chou fleuri* flowered cabbage]
cauliflower ear *n.* ear thickened by repeated blows.
caulk /kɔ:k/ *v.* (also **calk**) **1** stop up (the seams of a boat etc.). **2** make (esp. a boat) watertight. [Latin *calco* tread]
causal /'kɔ:z(ə)l/ *adj.* **1** of or forming a cause. **2** relating to cause and effect. □ **causally** *adv.*
causality /kɔ:'zælɪtɪ/ *n.* **1** relation of cause and effect. **2** principle that everything has a cause.
causation /kɔ:'zeɪʃ(ə)n/ *n.* **1** act of causing. **2** = CAUSALITY.
causative /'kɔ:zətɪv/ *adj.* acting as or expressing a cause.
cause /kɔ:z/ *–n.* **1 a** thing that produces an effect. **b** person or thing that occasions or produces something. **c** reason or motive. **2** adequate reason (*show cause*). **3** principle, belief, or purpose. **4 a** matter to be settled at law. **b** case offered at law (*plead a cause*). *–v.* (**-sing**) be the cause of, produce, make happen. [Latin *causa*]
cause célèbre /ˌkɔ:z se'lebr/ *n.* (*pl.* ***causes célèbres*** pronunc. same) lawsuit that attracts much interest. [French]
causerie /'kəʊzərɪ/ *n.* (*pl.* *-s* pronunc. same) informal article or talk. [French]
causeway /'kɔ:zweɪ/ *n.* **1** raised road across low ground or water. **2** raised path by a road. [Anglo-French *caucée* from Latin CALX]
caustic /'kɔ:stɪk/ *–adj.* **1** corrosive; burning. **2** sarcastic, biting. *–n.* caustic substance. □ **caustically** *adv.* **causticity** /-'tɪsɪtɪ/ *n.* [Greek *kaiō* burn]
caustic soda *n.* sodium hydroxide.
cauterize /'kɔ:təˌraɪz/ *v.* (also **-ise**) (**-zing** or **-sing**) burn (tissue), esp. to stop bleeding. [French: related to CAUSTIC]
caution /'kɔ:ʃ(ə)n/ *–n.* **1** attention to safety; prudence, carefulness. **2 a** *Law* warning, esp. a formal one. **b** warning and reprimand. **3** *colloq.* amusing or surprising person or thing. *–v.* **1** warn or admonish. **2** issue a caution to. [Latin *caveo* take heed]
cautionary *adj.* giving or serving as a warning.
cautious *adj.* having or showing caution. □ **cautiously** *adv.* **cautiousness** *n.*
cavalcade /ˌkævəl'keɪd/ *n.* procession or assembly of riders, vehicles, etc. [Italian: related to CHEVALIER]
cavalier /ˌkævə'lɪə(r)/ *–n.* **1** *hist.* (**Cavalier**) supporter of Charles I in the Civil War. **2** courtly gentleman. **3** *archaic* horseman. *–adj.* offhand, supercilious, curt. [related to CAVALCADE]
cavalry /'kævəlrɪ/ *n.* (*pl.* **-ies**) (usu. treated as *pl.*) soldiers on horseback or in armoured vehicles. [related to CAVALCADE]
cave *–n.* large hollow in the side of a cliff, hill, etc., or underground. *–v.* (**-ving**) explore caves. □ **cave in 1** (cause to) subside or collapse. **2** yield, give up. [Latin *cavus* hollow]
caveat /'kævɪˌæt/ *n.* **1** warning, proviso. **2** *Law* process in court to suspend proceedings. [Latin, = let him beware]
caveat emptor /'emptɔ:(r)/ *n.* principle that the buyer alone is responsible if dissatisfied. [Latin, = let the buyer beware]
caveman *n.* **1** prehistoric person living in caves. **2** crude person.
cavern /'kæv(ə)n/ *n.* cave, esp. a large or dark one. □ **cavernous** *adj.* [Latin *caverna*: related to CAVE]
caviare /'kævɪˌɑ:(r)/ *n.* (*US* **caviar**) pickled roe of sturgeon or other large fish. [Italian from Turkish]
cavil /'kævɪl/ *–v.* (**-ll-**, *US* **-l-**) (usu. foll. by *at, about*) make petty objections; carp. *–n.* trivial objection. [Latin *cavillor*]
cavity /'kævɪtɪ/ *n.* (*pl.* **-ies**) **1** hollow within a solid body. **2** decayed part of a tooth. [Latin: related to CAVE]
cavity wall *n.* double wall with a space between.
cavort /kə'vɔ:t/ *v.* caper excitedly. [origin uncertain]
cavy /'keɪvɪ/ *n.* (*pl.* **-ies**) small S. American rodent, esp. the guinea pig. [Latin from Galibi]
caw *–n.* harsh cry of a rook, crow, etc. *–v.* utter this cry. [imitative]
cayenne /keɪ'en/ *n.* (in full **cayenne pepper**) powdered red pepper. [Tupi]
cayman /'keɪmən/ *n.* (also **caiman**) (*pl.* **-s**) S. American alligator-like reptile. [Spanish and Portuguese from Carib]

CB *abbr.* **1** citizens' band. **2** Companion of the Order of the Bath.

CBE *abbr.* Commander of the Order of the British Empire.

CBI *abbr.* Confederation of British Industry.

cc *abbr.* (also **c.c.**) **1** cubic centimetre(s). **2** copy or copies (to).

CD *abbr.* **1** compact disc. **2** Civil Defence. **3** *Corps Diplomatique.*

Cd *symb.* cadmium.

cd *abbr.* candela.

CD-ROM /ˌsiːdiːˈrɒm/ *abbr.* compact disc read-only memory (for the retrieval of text or data on a VDU screen).

CDT *abbr.* craft, design, and technology.

CD-video /ˌsiːdiːˈvɪdɪəʊ/ *n.* (*pl.* **-s**) **1** system of simultaneously reproducing high-quality sound and video pictures from a compact disc. **2** such a compact disc.

Ce *symb.* cerium.

cease *formal* —*v.* (**-sing**) stop; bring or come to an end. —*n.* (in **without cease**) unending. [Latin *cesso*]

cease-fire *n.* **1** period of truce. **2** order to stop firing.

ceaseless *adj.* without end. □ **ceaselessly** *adv.*

cecum *US* var. of CAECUM.

cedar /ˈsiːdə(r)/ *n.* **1** spreading evergreen conifer. **2** its hard fragrant wood. [Greek *kedros*]

cede *v.* (**-ding**) *formal* give up one's rights to or possession of. [Latin *cedo cess-* yield]

cedilla /sɪˈdɪlə/ *n.* **1** mark written under *c*, esp. in French, to show it is sibilant (as in *façade*). **2** similar mark under *s* in Turkish etc. [Spanish diminutive of *zeda* Z]

Ceefax /ˈsiːfæks/ *n. propr.* teletext service provided by the BBC. [representing a pronunciation of *see*ing + *fac*simile]

ceilidh /ˈkeɪlɪ/ *n.* informal gathering for music, dancing, etc. [Gaelic]

ceiling /ˈsiːlɪŋ/ *n.* **1** upper interior surface of a room or other compartment. **2** upper limit. **3** maximum altitude a given aircraft can reach. [origin uncertain]

celandine /ˈselənˌdaɪn/ *n.* yellow-flowered plant. [Greek *khelidōn* a swallow]

celebrant /ˈselɪbrənt/ *n.* person who performs a rite, esp. the priest at the Eucharist.

celebrate /ˈselɪˌbreɪt/ *v.* (**-ting**) **1** mark with or engage in festivities. **2** perform (a rite or ceremony). **3** praise publicly. □ **celebration** /-ˈbreɪʃ(ə)n/ *n.* **celebrator** *n.* **celebratory** /-ˈbreɪtərɪ/ *adj.* [Latin *celeber* renowned]

celebrity /sɪˈlebrɪtɪ/ *n.* (*pl.* **-ies**) **1** well-known person. **2** fame. [Latin: related to CELEBRATE]

celeriac /sɪˈlerɪˌæk/ *n.* variety of celery. [from CELERY]

celerity /sɪˈlerɪtɪ/ *n. archaic* or *literary* swiftness. [Latin *celer* swift]

celery /ˈselərɪ/ *n.* plant with crisp long whitish leaf-stalks used as a vegetable. [Greek *selinon* parsley]

celesta /sɪˈlestə/ *n.* small keyboard instrument with steel plates struck to give a bell-like sound. [French: related to CELESTIAL]

celestial /sɪˈlestɪəl/ *adj.* **1** of the sky or heavenly bodies. **2** heavenly; divinely good; sublime. [Latin *caelum* sky]

celestial equator *n.* the great circle of the sky in the plane perpendicular to the earth's axis.

celestial sphere *n.* imaginary sphere, of any radius, of which the observer is the centre and in which celestial bodies are represented as lying.

celibate /ˈselɪbət/ —*adj.* **1** unmarried or committed to sexual abstention, esp. for religious reasons. **2** having no sexual relations. —*n.* celibate person. □ **celibacy** *n.* [Latin *caelebs* unmarried]

cell *n.* **1** small room, esp. in a prison or monastery. **2** small compartment, e.g. in a honeycomb. **3** small, active, esp. subversive, political group. **4 a** smallest structural and functional unit of living matter, consisting of cytoplasm and a nucleus enclosed in a membrane. **b** enclosed cavity in an organism etc. **5** vessel containing electrodes for current-generation or electrolysis. [Latin *cella*]

cellar /ˈselə(r)/ —*n.* **1** storage room below ground level in a house. **2** stock of wine in a cellar. —*v.* store in a cellar. [Latin *cellarium*: related to CELL]

cello /ˈtʃeləʊ/ *n.* (*pl.* **-s**) bass instrument of the violin family, held between the legs of the seated player. □ **cellist** *n.* [abbreviation of VIOLONCELLO]

Cellophane /ˈseləˌfeɪn/ *n. propr.* thin transparent viscose wrapping material. [from CELLULOSE: cf. DIAPHANOUS]

cellphone *n.* small portable radio-telephone.

cellular /ˈseljʊlə(r)/ *adj.* consisting of cells, of open texture; porous. □ **cellularity** /-ˈlærɪtɪ/ *n.* [French: related to CELL]

cellular radio *n.* system of mobile radio-telephone transmission with an area divided into 'cells', each served by a small transmitter.

cellulite /ˈseljʊˌlaɪt/ *n.* lumpy fat, esp. on the hips and thighs of women. [French: related to CELL]

celluloid /ˈseljʊˌlɔɪd/ *n.* **1** plastic made from camphor and cellulose nitrate. **2** cinema film.

cellulose /'seljʊ,ləʊs/ *n.* **1** carbohydrate forming plant-cell walls, used in textile fibres. **2** (in general use) paint or lacquer consisting of esp. cellulose acetate or nitrate in solution. [Latin: related to CELL]

Celsius /'selsɪəs/ *adj.* of a scale of temperature on which water freezes at 0° and boils at 100°. [name of an astronomer]

■ **Usage** See note at *centigrade*.

Celt /kelt/ *n.* (also **Kelt**) member of an ethnic group, including the inhabitants of Ireland, Wales, Scotland, Cornwall, and Brittany. [Latin from Greek]

Celtic /'keltɪk/ *–adj.* of the Celts. *–n.* group of Celtic languages, including Gaelic and Irish, Welsh, Cornish, and Breton.

cement /sɪ'ment/ *–n.* **1** powdery substance of calcined lime and clay, mixed with water to form mortar or used in concrete. **2** similar substance. **3** uniting factor or principle. **4** substance used in filling teeth, doing hip replacements, etc. *–v.* **1 a** unite with or as with cement. **b** establish or strengthen (a friendship etc.). **2** apply cement to. **3** line or cover with cement. □ **cementation** /,si:men'teɪʃ(ə)n/ *n.* [Latin *caedo* cut]

cemetery /'semɪtrɪ/ *n.* (*pl.* **-ies**) burial ground, esp. one not in a churchyard. [Greek *koimaō* put to sleep]

cenobite *US* var. of COENOBITE.

cenotaph /'senə,tɑ:f/ *n.* tomblike monument to a person whose body is elsewhere. [Greek *kenos* empty, *taphos* tomb]

Cenozoic /,si:nə'zəʊɪk/ (also **Cainozoic** /,kaɪnə-/, **Caenozoic** /,si:n-/) *–adj.* of the most recent geological era, marked by the evolution and development of mammals etc. *–n.* this era. [Greek *kainos* new, *zōion* animal]

censer /'sensə(r)/ *n.* vessel for burning incense. [Anglo-French: related to INCENSE[1]]

censor /'sensə(r)/ *–n.* official authorized to suppress or expurgate books, films, news, etc., on grounds of obscenity, threat to security, etc. *–v.* **1** act as a censor of. **2** make deletions or changes in. □ **censorial** /-'sɔ:rɪəl/ *adj.* **censorship** *n.* [Latin *censeo* assess]

■ **Usage** As a verb, *censor* is often confused with *censure*.

censorious /sen'sɔ:rɪəs/ *adj.* severely critical. □ **censoriously** *adv.*

censure /'senʃə(r)/ *–v.* (**-ring**) criticize harshly; reprove. *–n.* hostile criticism; disapproval. [Latin: related to CENSOR]

■ **Usage** As a verb, *censure* is often confused with *censor*.

census /'sensəs/ *n.* (*pl.* **-suses**) official count of population etc. [Latin: related to CENSOR]

cent *n.* **1 a** one-hundredth of a dollar or other decimal currency unit. **b** coin of this value. **2** *colloq.* very small amount. [Latin *centum* 100]

centaur /'sentɔ:(r)/ *n.* creature in Greek mythology with the upper half of a man and the lower half of a horse. [Latin from Greek]

centenarian /,sentɪ'neərɪən/ *–n.* person a hundred or more years old. *–adj.* a hundred or more years old.

centenary /sen'ti:nərɪ/ *–n.* (*pl.* **-ies**) **1** hundredth anniversary. **2** celebration of this. *–adj.* **1** of a centenary. **2** occurring every hundred years. [Latin *centeni* 100 each]

centennial /sen'tenɪəl/ *–adj.* **1** lasting for a hundred years. **2** occurring every hundred years. *–n.* *US* = CENTENARY *n.* [Latin *centum* 100: cf. BIENNIAL]

center *US* var. of CENTRE.

centerboard *US* var. of CENTREBOARD.

centerfold *US* var. of CENTREFOLD.

centesimal /sen'tesɪm(ə)l/ *adj.* reckoning or reckoned by hundredths. [Latin *centum* 100]

centi- *comb. form* **1** one-hundredth. **2** hundred. [Latin *centum* 100]

centigrade /'sentɪ,greɪd/ *adj.* **1** = CELSIUS. **2** having a scale of a hundred degrees. [Latin *gradus* step]

■ **Usage** In sense 1, *Celsius* is usually preferred in technical contexts.

centigram /'sentɪ,græm/ *n.* (also **centigramme**) metric unit of mass, equal to 0.01 gram.

centilitre /'sentɪ,li:tə(r)/ *n.* (*US* **centiliter**) 0.01 litre.

centime /'sɑ̃ti:m/ *n.* **1** one-hundredth of a franc. **2** coin of this value. [Latin *centum* 100]

centimetre /'sentɪ,mi:tə(r)/ *n.* (*US* **centimeter**) 0.01 metre.

centipede /'sentɪ,pi:d/ *n.* arthropod with a segmented wormlike body and many legs. [Latin *pes ped-* foot]

central /'sentr(ə)l/ *adj.* **1** of, at, or forming the centre. **2** from the centre. **3** chief, essential, most important. □ **centrality** /-'trælɪtɪ/ *n.* **centrally** *adv.*

central bank *n.* national bank issuing currency etc.

central heating *n.* method of heating a building by pipes, radiators, etc., fed from a central source.

centralism *n.* system that centralizes (esp. administration). □ **centralist** *n.*

centralize *v.* (also **-ise**) (**-zing** or **-sing**) **1** concentrate (esp. administration) at a single centre. **2** subject (a State) to this system. □ **centralization** /-'zeɪʃ(ə)n/ *n.*

central nervous system *n.* brain and spinal cord.

central processor *n.* (also **central processing unit**) principal operating part of a computer.

centre /'sentə(r)/ (*US* **center**) *–n.* **1** middle point. **2** pivot or axis of rotation. **3 a** place or buildings forming a central point or a main area for an activity (*shopping centre*; *town centre*). **b** (with a preceding word) equipment for a number of connected functions (*music centre*). **4** point of concentration or dispersion; nucleus, source. **5** political party or group holding moderate opinions. **6** filling in chocolate etc. **7** *Sport* **a** middle player in a line in some field games. **b** kick or hit from the side to the centre of a pitch. **8** (*attrib.*) of or at the centre. *–v.* (**-ring**) **1** (foll. by *in*, *on*, *round*) have as its main centre. **2** place in the centre. **3** (foll. by *in* etc.) concentrate. [Greek *kentron* sharp point]

■ **Usage** The use of the verb in sense 1 with *round* is common and used by good writers, but is still considered incorrect by some people.

centre back *n.* *Sport* middle player or position in a half-back line.

centreboard *n.* (*US* **centerboard**) board lowered through a boat's keel to prevent leeway.

centrefold *n.* (*US* **centerfold**) centre spread of a magazine etc., esp. with nude photographs.

centre forward *n.* *Sport* middle player or position in a forward line.

centre half *n.* = CENTRE BACK.

centre of gravity *n.* (also **centre of mass**) point at which the weight of a body may be considered to act.

centre-piece *n.* **1** ornament for the middle of a table. **2** principal item.

centre spread *n.* two facing middle pages of a newspaper etc.

centric *adj.* **1** at or near the centre. **2** from a centre. □ **centrical** *adj.* **centrically** *adv.*

centrifugal /,sentrɪ'fju:g(ə)l/ *adj.* moving or tending to move from a centre. □ **centrifugally** *adv.* [from CENTRE, Latin *fugio* flee]

centrifugal force *n.* apparent force that acts outwards on a body moving about a centre.

centrifuge /'sentrɪ,fju:dʒ/ *n.* rapidly rotating machine designed to separate liquids from solids etc.

centripetal /sen'trɪpɪt(ə)l/ *adj.* moving or tending to move towards a centre. □ **centripetally** *adv.* [Latin *peto* seek]

centripetal force *n.* force acting on a body causing it to move towards a centre.

centrist *n.* *Polit.* often *derog.* person holding moderate views. □ **centrism** *n.*

centurion /sen'tjʊərɪən/ *n.* commander of a century in the ancient Roman army. [Latin: related to CENTURY]

century /'sentʃərɪ/ *n.* (*pl.* **-ies**) **1 a** 100 years. **b** any century reckoned from the birth of Christ (*twentieth century* = 1901–2000; *fifth century* BC = 500–401 BC). **2** score etc. of 100 esp. by one batsman in cricket. **3** company in the ancient Roman army, orig. of 100 men. [Latin *centuria*: related to CENT]

■ **Usage** Strictly speaking, since the first century ran from the year 1–100, the first year of a given century should be that ending in 01. However, in popular use this has been moved back a year, and so the twenty-first century will commonly be regarded as running from 2000–2099.

cephalic /sə'fælɪk/ *adj.* of or in the head. [Greek *kephalē* head]

cephalopod /'sefələ,pɒd/ *n.* mollusc with a distinct tentacled head, e.g. the octopus. [from CEPHALIC, Greek *pous pod-* foot]

ceramic /sɪ'ræmɪk/ *–adj.* **1** made of (esp.) baked clay. **2** of ceramics. *–n.* ceramic article or product. [Greek *keramos* pottery]

ceramics *n.pl.* **1** ceramic products collectively. **2** (usu. treated as *sing.*) art of making ceramic articles.

cereal /'sɪərɪəl/ *–n.* **1 a** grain used for food. **b** wheat, maize, rye, etc. producing this. **2** breakfast food made from a cereal. *–adj.* of edible grain. [Latin *Ceres* goddess of agriculture]

cerebellum /,serɪ'beləm/ *n.* (*pl.* **-s** or **-bella**) part of the brain at the back of the skull. [Latin diminutive of CEREBRUM]

cerebral /'serɪbr(ə)l/ *adj.* **1** of the brain. **2** intellectual; unemotional. [related to CEREBRUM]

cerebral palsy *n.* paralysis resulting from brain damage before or at birth, involving spasm of the muscles and involuntary movements.

cerebration /,serɪ'breɪʃ(ə)n/ *n.* working of the brain.

cerebrospinal /,serɪbrəʊ'spaɪn(ə)l/ *adj.* of the brain and spine.

cerebrum /'serɪbrəm/ *n.* (*pl.* **-bra**) principal part of the brain in vertebrates, at the front of the skull. [Latin]

ceremonial /ˌserɪˈməʊnɪəl/ *–adj.* of or with ceremony; formal. *–n.* system of rites or ceremonies. □ **ceremonially** *adv.*

ceremonious /ˌserɪˈməʊnɪəs/ *adj.* fond of or characterized by ceremony; formal. □ **ceremoniously** *adv.*

ceremony /ˈserɪmənɪ/ *n.* (*pl.* **-ies**) **1** formal procedure, esp. at a public event or anniversary. **2** formalities, esp. ritualistic. **3** excessively polite behaviour. □ **stand on ceremony** insist on formality. [Latin *caerimonia* worship]

cerise /səˈriːz/ *n.* light clear red. [French: related to CHERRY]

cerium /ˈsɪərɪəm/ *n.* silvery metallic element of the lanthanide series. [*Ceres*, name of an asteroid]

CERN /sɜːn/ *abbr.* European Organization for Nuclear Research. [French *Conseil Européen pour la Recherche Nucléaire*, former title]

cert *n.* (esp. **dead cert**) *slang* a certainty. [abbreviation]

cert. *abbr.* **1** certificate. **2** certified.

certain /ˈsɜːt(ə)n/ *–adj.* **1 a** confident, convinced. **b** indisputable (*it is certain that he is guilty*). **2** (often foll. by *to* + infin.) sure; destined (*it is certain to rain*; *certain to win*). **3** unerring, reliable. **4** that need not be specified or may not be known to the reader or hearer (*of a certain age*; *a certain John Smith*). **5** some but not much (*a certain reluctance*). *–pron.* (as *pl.*) some but not all (*certain of them knew*). □ **for certain** without doubt. [Latin *certus*]

certainly *adv.* **1** undoubtedly. **2** (in answer) yes; by all means.

certainty *n.* (*pl.* **-ies**) **1 a** undoubted fact. **b** indubitable prospect. **2** absolute conviction. **3** reliable thing or person.

Cert. Ed. *abbr.* Certificate in Education.

certifiable /ˈsɜːtɪˌfaɪəb(ə)l/ *adj.* **1** able or needing to be certified. **2** *colloq.* insane.

certificate /səˈtɪfɪkət/ *–n.* formal document attesting a fact, esp. birth, marriage, or death, a medical condition, or a qualification. *–v.* /-ˌkeɪt/ (**-ting**) (esp. as **certificated** *adj.*) provide with, license, or attest by a certificate. □ **certification** /ˌsɜːtɪfɪˈkeɪʃ(ə)n/ *n.* [Latin: related to CERTIFY]

Certificate of Secondary Education *n. hist.* secondary-school leaving examination in England, Wales, and Northern Ireland.

■ **Usage** This examination was replaced in 1988 by the *General Certificate of Secondary Education* (GCSE).

certified cheque *n.* cheque guaranteed by a bank.

certify /ˈsɜːtɪˌfaɪ/ *v.* (**-ies, -ied**) **1** attest; attest to, esp. formally. **2** declare by certificate. **3** officially declare insane. [Latin *certus*]

certitude /ˈsɜːtɪˌtjuːd/ *n.* feeling of certainty. [Latin: related to CERTAIN]

cerulean /səˈruːlɪən/ *adj.* & *n. literary* deep sky-blue. [Latin *caeruleus*]

cervical /səˈvaɪk(ə)l, ˈsɜːvɪk(ə)l/ *adj.* of the neck or the cervix (*cervical vertebrae*). [related to CERVIX]

cervical screening *n.* mass routine examination for cervical cancer.

cervical smear *n.* specimen from the neck of the womb for examination.

cervix /ˈsɜːvɪks/ *n.* (*pl.* **cervices** /-ˌsiːz/) **1** necklike structure, esp. the neck of the womb. **2** the neck. [Latin]

Cesarean (also **Cesarian**) *US* var. of CAESAREAN.

cesium *US* var. of CAESIUM.

cessation /seˈseɪʃ(ə)n/ *n.* ceasing or pause. [Latin: related to CEASE]

cession /ˈseʃ(ə)n/ *n.* **1** ceding. **2** territory etc. ceded. [Latin: related to CEDE]

cesspit /ˈsespɪt/ *n.* (also **cesspool**) covered pit for the temporary storage of liquid waste or sewage. [origin uncertain]

cetacean /sɪˈteɪʃ(ə)n/ *–n.* marine mammal of the order Cetacea, e.g. the whale. *–adj.* of cetaceans. [Greek *kētos* whale]

cetane /ˈsiːteɪn/ *n.* liquid hydrocarbon used in standardizing ratings of diesel fuel. [from SPERMACETI]

Cf *symb.* californium.

cf. *abbr.* compare. [Latin *confer*]

CFC *abbr.* chlorofluorocarbon, a usu. gaseous compound of carbon, hydrogen, chlorine, and fluorine, used in refrigerants, aerosol propellants, etc., and thought to harm the ozone layer.

CFE *abbr.* College of Further Education.

cg *abbr.* centigram(s).

CH *abbr.* Companion of Honour.

Chablis /ˈʃæbliː/ *n.* (*pl.* same /-liːz/) very dry white wine from Chablis in E. France.

cha-cha /ˈtʃɑːtʃɑː/ *n.* (also **cha-cha-cha** /ˌtʃɑːtʃɑːˈtʃɑː/) **1** Latin American dance. **2** music for this. [American Spanish]

chaconne /ʃəˈkɒn/ *n.* **1** musical variations over a ground bass. **2** dance performed to this. [French from Spanish]

chafe *–v.* (**-fing**) **1** make or become sore or damaged by rubbing. **2** make or become annoyed; fret. **3** rub (esp. the skin to restore warmth or sensation). *–n.* sore caused by rubbing. [Latin *calefacio* make warm]

chaff /tʃɑːf/ *–n.* **1** separated husks of corn etc. **2** chopped hay or straw. **3** light-hearted teasing. **4** worthless things. *–v.* tease, banter. [Old English]

chaffinch /'tʃæfɪntʃ/ *n.* a common European finch. [Old English: related to CHAFF, FINCH]

chafing-dish *n.* vessel in which food is cooked or kept warm at table.

chagrin /'ʃægrɪn/ *–n.* acute annoyance or disappointment. *–v.* affect with chagrin. [French]

chain *–n.* **1 a** connected flexible series of esp. metal links. **b** thing resembling this. **2** (in *pl.*) fetters; restraining force. **3** sequence, series, or set. **4** group of associated hotels, shops, etc. **5** badge of office in the form of a chain worn round the neck. **6** unit of length (66 ft). *–v.* (often foll. by *up*) secure or confine with a chain. [Latin *catena*]

chain-gang *n. hist.* team of convicts chained together to work out of doors.

chain-mail *n.* armour made of interlaced rings.

chain reaction *n.* **1** chemical or nuclear reaction forming products which initiate further reactions. **2** series of events, each caused by the previous one.

chain-saw *n.* motor-driven saw with teeth on an endless chain.

chain-smoke *v.* smoke continually, esp. by lighting the next cigarette etc. from the previous one. □ **chain--smoker** *n.*

chain store *n.* one of a series of similar shops owned by one firm.

chair *–n.* **1** seat for one person usu. with a back. **2** professorship. **3 a** chairperson. **b** seat or office of a chairperson. **4** *US* = ELECTRIC CHAIR. *–v.* **1** preside over (a meeting). **2** carry (a person) aloft in triumph. □ **take the chair** preside over a meeting. [Greek *kathedra*]

chair-lift *n.* series of chairs on a looped cable, for carrying passengers up and down a mountain etc.

chairman *n.* (*fem.* also **chairwoman**) **1** person chosen to preside over a meeting. **2** permanent president of a committee, board of directors, etc.

chairperson *n.* chairman or chairwoman.

chaise /ʃeɪz/ *n.* esp. *hist.* horse-drawn usu. open carriage for one or two persons. [French]

chaise longue /ʃeɪz 'lɒŋ/ *n.* (*pl.* ***chaise longues*** or ***chaises longues*** pronunc. same) sofa with only one arm rest. [French, = long chair]

chalcedony /kæl'sedənɪ/ *n.* (*pl.* **-ies**) type of quartz with many varieties, e.g. onyx. [Latin from Greek]

chalet /'ʃæleɪ/ *n.* **1** Swiss mountain hut or cottage with overhanging eaves. **2** house in a similar style. **3** small cabin in a holiday camp etc. [Swiss French]

chalice /'tʃælɪs/ *n.* **1** goblet. **2** Eucharistic cup. [Latin CALIX]

chalk /tʃɔ:k/ *–n.* **1** white soft limestone. **2 a** similar substance, sometimes coloured, for writing or drawing. **b** piece of this. *–v.* **1** rub, mark, draw, or write with chalk. **2** (foll. by *up*) **a** write or record with chalk. **b** register or gain (success etc.). □ **by a long chalk** by far. □ **chalky** *adj.* (**-ier**, **-iest**). **chalkiness** *n.* [Latin CALX]

challenge /'tʃælɪndʒ/ *–n.* **1** summons to take part in a contest etc. or to prove or justify something. **2** demanding or difficult task. **3** objection made to a jury member. **4** call to respond. *–v.* (**-ging**) **1** issue a challenge to. **2** dispute, deny. **3** (as **challenging** *adj.*) stimulatingly difficult. **4** object to (a jury member, evidence, etc.). □ **challenger** *n.* [Latin *calumnia* calumny]

chalybeate /kə'lɪbɪət/ *adj.* (of water etc.) impregnated with iron salts. [Latin *chalybs* steel, from Greek]

chamber /'tʃeɪmbə(r)/ *n.* **1 a** hall used by a legislative or judicial body. **b** body that meets in it, esp. any of the houses of a parliament. **2** (in *pl.*) **a** rooms used by a barrister or barristers, esp. in Inns of Court. **b** judge's room for hearing cases not needing to be taken in court. **3** *archaic* room, esp. a bedroom. **4** *Mus.* (*attrib.*) of or for a small group of instruments. **5** cavity or compartment in the body, machinery, etc. (esp. the part of a gun-bore that contains the charge). [Greek *kamara* vault]

chamberlain /'tʃeɪmbəlɪn/ *n.* **1** officer managing a royal or noble household. **2** treasurer of a corporation etc. [Germanic: related to CHAMBER]

chambermaid *n.* woman who cleans hotel bedrooms.

Chamber of Commerce *n.* association to promote local commercial interests.

chamber-pot *n.* receptacle for urine etc., used in the bedroom.

chameleon /kə'mi:lɪən/ *n.* **1** small lizard able to change colour for camouflage. **2** variable or inconstant person. [Greek, = ground-lion]

chamfer /'tʃæmfə(r)/ *–v.* bevel symmetrically (a right-angled edge or corner). *–n.* bevelled surface at an edge or corner. [French *chant* edge, *fraint* broken]

chamois *n.* (*pl.* same /-wɑ:z/, /-mɪz/) **1** /'ʃæmwɑ:/ agile European and Asian mountain antelope. **2** /'ʃæmɪ/ (in full **chamois leather**) **a** soft leather from sheep, goats, deer, etc. **b** piece of this. [French]

chamomile var. of CAMOMILE.

champ[1] *–v.* munch or chew noisily. *–n.* chewing noise. □ **champ at the bit** be restlessly impatient. [imitative]

champ[2] *n. slang* champion. [abbreviation]

champagne /ʃæm'peɪn/ *n.* **1 a** white sparkling wine from Champagne. **b** similar wine from elsewhere. **2** pale cream colour. [*Champagne*, former province in E. France]

■ **Usage** The use of this word in sense 1b is, strictly speaking, incorrect.

champers /'ʃæmpəz/ *n. slang* champagne.

champion /'tʃæmpɪən/ *–n.* **1** (often *attrib.*) person or thing that has defeated or surpassed all rivals. **2** person who fights or argues for a cause or another person. *–v.* support the cause of, defend. *–adj. colloq.* splendid. *–adv. colloq.* splendidly. [medieval Latin *campio* fighter]

championship *n.* **1** (often in *pl.*) contest to decide the champion in a sport etc. **2** position of champion.

chance /tʃɑ:ns/ *–n.* **1** possibility. **2** (often in *pl.*) probability. **3** unplanned occurrence. **4** opportunity. **5** fortune; luck. **6** (often **Chance**) course of events regarded as a power; fate. *–attrib. adj.* fortuitous, accidental. *–v.* **(-cing) 1** *colloq.* risk. **2** happen (*I chanced to find it*). □ **by any chance** perhaps. **by chance** fortuitously. **chance one's arm** try though unlikely to succeed. **chance on** (or **upon**) happen to find, meet, etc. **game of chance** one decided by luck, not skill. **on the off chance** just in case (the unlikely occurs). **stand a chance** have a prospect of success etc. **take a chance** (or **chances**) risk failure; behave riskily. **take a** (or **one's**) **chance on** (or **with**) risk the consequences of. [Latin *cado* fall]

chancel /'tʃɑ:ns(ə)l/ *n.* part of a church near the altar. [Latin *cancelli* grating]

chancellery /'tʃɑ:nsələrɪ/ *n.* (*pl.* **-ies**) **1** chancellor's department, staff, or residence. **2** *US* office attached to an embassy or consulate.

chancellor /'tʃɑ:nsələ(r)/ *n.* **1** State or legal official. **2** head of government in some European countries. **3** non-resident honorary head of a university. [Latin *cancellarius* secretary]

Chancellor of the Exchequer *n.* UK finance minister.

chancery /'tʃɑ:nsərɪ/ *n.* (*pl.* **-ies**) **1** (**Chancery**) Lord Chancellor's division of the High Court of Justice. **2** records office. **3** chancellery. [contraction of CHANCELLERY]

chancy /'tʃɑ:nsɪ/ *adj.* **(-ier, -iest)** uncertain; risky. □ **chancily** *adv.*

chandelier /,ʃændə'lɪə(r)/ *n.* ornamental branched hanging support for lighting. [French: related to CANDLE]

chandler /'tʃɑ:ndlə(r)/ *n.* dealer in candles, oil, soap, paint, etc. [French: related to CANDLE]

change /tʃeɪndʒ/ *–n.* **1 a** making or becoming different. **b** alteration or modification. **2 a** money exchanged for money in larger units or a different currency. **b** money returned as the balance of that given in payment. **3** new experience; variety (*need a change*). **4** substitution of one thing for another (*change of scene*). **5** (in full **change of life**) *colloq.* menopause. **6** (usu. in *pl.*) one of the different orders in which bells can be rung. *–v.* **(-ging) 1** undergo, show, or subject to change; make or become different. **2 a** take or use another instead of; go from one to another (*change one's socks*; *changed trains*). **b** (usu. foll. by *for*) give up or get rid of in exchange (*changed the car for a van*). **3** give or get money in exchange for. **4** put fresh clothes or coverings on. **5** (often foll. by *with*) give and receive, exchange. **6** change trains etc. **7** (of the moon) arrive at a fresh phase. □ **change down** engage a lower gear. **change gear** engage a different gear. **change hands 1** pass to a different owner. **2** substitute one hand for the other. **change one's mind** adopt a different opinion or plan. **change over** change from one system or situation to another. **change one's tune 1** voice a different opinion from before. **2** become more respectful. **change up** engage a higher gear. **get no change out of** *slang* **1** get no information or help from. **2** fail to outwit (in business etc.). □ **changeful** *adj.* **changeless** *adj.* [Latin *cambio* barter]

changeable *adj.* **1** inconstant. **2** that can change or be changed.

changeling *n.* child believed to be substituted for another.

change of clothes *n.* second outfit in reserve.

change of heart *n.* conversion to a different view.

change-over *n.* change from one system to another.

channel /'tʃæn(ə)l/ *–n.* **1 a** piece of water wider than a strait, joining esp. two seas. **b** (**the Channel**) the English Channel. **2** medium of communication; agency. **3** band of frequencies used in radio and television transmission, esp.

by a particular station. **4** course in which anything moves. **5 a** hollow bed of water. **b** navigable part of a waterway. **6** passage for liquid. **7** lengthwise strip on recording tape etc. —*v.* (**-ll-**; *US* **-l-**) **1** guide, direct. **2** form channel(s) in. [Latin: related to CANAL]

chant /tʃɑːnt/ —*n.* **1** spoken singsong phrase. **2 a** simple tune used for singing unmetrical words, e.g. psalms. **b** song, esp. monotonous or repetitive. —*v.* **1** talk or repeat monotonously. **2** sing or intone (a psalm etc.). [Latin *canto* from *cano* sing]

chanter *n.* melody-pipe of bagpipes.

chanticleer /ˌtʃæntɪˈklɪə(r)/ *n.* name given to a domestic cock in stories. [French: related to CHANT, CLEAR]

chantry /ˈtʃɑːntrɪ/ *n.* (*pl.* **-ies**) **1** endowment for the singing of masses. **2** priests, chapel, etc., so endowed. [French: related to CHANT]

chaos /ˈkeɪɒs/ *n.* **1** utter confusion. **2** formless matter supposed to have existed before the creation of the universe. □ **chaotic** /keɪˈɒtɪk/ *adj.* **chaotically** /-ˈɒtɪkəlɪ/ *adv.* [Latin from Greek]

chap[1] *n. colloq.* man, boy, fellow. [abbreviation of CHAPMAN]

chap[2] —*v.* (**-pp-**) **1** (esp. of the skin) develop cracks or soreness. **2** (of the wind, cold, etc.) cause to chap. —*n.* (usu. in *pl.*) crack in the skin etc. [origin uncertain]

chaparral /ˌʃæpəˈræl/ *n. US* dense tangled brushwood. [Spanish]

chapatti /tʃəˈpɑːtɪ/ *n.* (also **chapati, chupatty**) (*pl.* **chapat(t)is** or **chupatties**) flat thin cake of unleavened bread. [Hindi]

chapel /ˈtʃæp(ə)l/ *n.* **1 a** place for private Christian worship in a Cathedral or large church, with its own altar. **b** this attached to a private house etc. **2 a** place of worship for Nonconformists. **b** chapel service. **3** members or branch of a printers' trade union at a place of work. [medieval Latin *cappa* cloak: the first chapel was a sanctuary in which St Martin's cloak (*cappella*) was preserved]

chaperon /ˈʃæpəˌrəʊn/ —*n.* person, esp. an older woman, ensuring propriety by accompanying a young unmarried woman on social occasions. —*v.* act as chaperon to. □ **chaperonage** /-rənɪdʒ/ *n.* [French from *chape* cope: related to CAPE[1]]

chaplain /ˈtʃæplɪn/ *n.* member of the clergy attached to a private chapel, institution, ship, regiment, etc. □ **chaplaincy** *n.* (*pl.* **-ies**). [Latin: related to CHAPEL]

chaplet /ˈtʃæplɪt/ *n.* **1** garland or circlet for the head. **2** short string of beads; rosary. [Latin: related to CAP]

chapman /ˈtʃæpmən/ *n. hist.* pedlar. [Old English: related to CHEAP, MAN]

chappie *n. colloq.* = CHAP[1].

chapter *n.* **1** main division of a book. **2** period of time (in a person's life etc.). **3 a** canons of a cathedral or members of a religious community. **b** meeting of these. [Latin diminutive of *caput* head]

chapter and verse *n.* exact reference or details.

chapter of accidents *n.* series of misfortunes.

char[1] *v.* (**-rr-**) **1** make or become black by burning; scorch. **2** burn to charcoal. [from CHARCOAL]

char[2] *colloq.* —*n.* = CHARWOMAN. —*v.* (**-rr-**) work as a charwoman. [Old English, = turn]

char[3] *n. slang* tea. [Chinese *cha*]

char[4] *n.* (*pl.* same) a kind of small trout. [origin unknown]

charabanc /ˈʃærəˌbæŋ/ *n. hist.* early form of motor coach. [French *char à bancs* seated carriage]

character /ˈkærɪktə(r)/ *n.* **1** collective qualities or characteristics that distinguish a person or thing. **2 a** moral strength. **b** reputation, esp. good reputation. **3 a** person in a novel, play, etc. **b** part played by an actor; role. **4** *colloq.* person, esp. an eccentric one. **5** printed or written letter, symbol, etc. **6** written description of a person's qualities. **7** characteristic (esp. of a biological species). □ **in** (or **out of**) **character** consistent (or inconsistent) with a person's character. □ **characterless** *adj.* [Greek *kharaktēr*]

characteristic /ˌkærɪktəˈrɪstɪk/ —*adj.* typical, distinctive. —*n.* characteristic feature or quality. □ **characteristically** *adv.*

characterize /ˈkærɪktəˌraɪz/ *v.* (also **-ise**) (**-zing** or **-sing**) **1 a** describe the character of. **b** (foll. by *as*) describe as. **2** be characteristic of. **3** impart character to. □ **characterization** /-ˈzeɪʃ(ə)n/ *n.*

charade /ʃəˈrɑːd/ *n.* **1** (usu. in *pl.*, treated as *sing.*) game of guessing a word from acted clues. **2** absurd pretence. [Provençal *charra* chatter]

charcoal /ˈtʃɑːkəʊl/ *n.* **1 a** form of carbon consisting of black residue from partially burnt wood etc. **b** piece of this for drawing. **c** a drawing in charcoal. **2** (in full **charcoal grey**) dark grey. [origin unknown]

charge —*v.* (**-ging**) **1 a** ask (an amount) as a price. **b** ask (a person) for an amount as a price. **2 a** (foll. by *to, up to*) debit the cost of to (a person or account).

b debit (a person or account). **3 a** (often foll. by *with*) accuse (of an offence). **b** (foll. by *that* + clause) make an accusation that. **4** (foll. by *to* + infin.) instruct or urge. **5** (foll. by *with*) entrust with. **6** make a rushing attack (on). **7** (often foll. by *up*) **a** give an electric charge to. **b** store energy in (a battery). **8** (often foll. by *with*) load or fill (a vessel, gun, etc.) to the full or proper extent. **9** (usu. as **charged** *adj.*) **a** (foll. by *with*) saturated with. **b** (usu. foll. by *with*) pervaded (with strong feelings etc.). *–n.* **1 a** price asked for services or goods. **b** financial liability or commitment. **2** accusation. **3 a** task, duty, commission. **b** care, custody. **c** person or thing entrusted. **4 a** impetuous rush or attack, esp. in battle. **b** signal for this. **5** appropriate amount of material to be put into a receptacle, mechanism, etc. at one time, esp. of explosive for a gun. **6 a** property of matter causing electrical phenomena. **b** quantity of this carried by the body. **c** energy stored chemically for conversion into electricity. **7** exhortation; directions, orders. **8** heraldic device or bearing. □ **in charge** having command. **take charge** (often foll. by *of*) assume control. □ **chargeable** *adj.* [Latin *carrus* CAR]

charge-capping *n.* imposition of an upper limit on the community charge leviable by a local authority.

charge card *n.* = CREDIT CARD.

chargé d'affaires /ˌʃɑːʒeɪ dæ'feə(r)/ *n.* (*pl.* **chargés** pronunc. same) **1** ambassador's deputy. **2** envoy to a minor country. [French]

charger *n.* **1** cavalry horse. **2** apparatus for charging a battery.

chariot /'tʃærɪət/ *n. hist.* two-wheeled vehicle drawn by horses, used in ancient warfare and racing. [French: related to CAR]

charioteer /ˌtʃærɪə'tɪə(r)/ *n.* chariot-driver.

charisma /kə'rɪzmə/ *n.* **1** power to inspire or attract others; exceptional charm. **2** divinely conferred power or talent. □ **charismatic** /-'mætɪk/ *adj.* [Greek *kharis* grace]

charitable /'tʃærɪtəb(ə)l/ *adj.* **1** generous in giving to those in need. **2** of or relating to a charity or charities. **3** generous in judging others. □ **charitably** *adv.*

charity /'tʃærɪtɪ/ *n.* (*pl.* **-ies**) **1** giving voluntarily to those in need. **2** organization set up to help those in need or for the common good. **3 a** kindness, benevolence. **b** tolerance in judging others. **c** love of fellow men. [Latin *caritas* from *carus* dear]

charlady *n.* = CHARWOMAN.

charlatan /'ʃɑːlət(ə)n/ *n.* person falsely claiming knowledge or skill. □ **charlatanism** *n.* [Italian, = babbler]

charleston /'tʃɑːlst(ə)n/ *n.* (also **Charleston**) lively dance of the 1920s with side-kicks from the knee. [*Charleston* in S. Carolina]

charlotte /'ʃɑːlɒt/ *n.* pudding of stewed fruit covered with bread etc. [French]

charm *–n.* **1** power or quality of delighting, arousing admiration, or influencing; fascination, attractiveness. **2** trinket on a bracelet etc. **3** object, act, or word(s) supposedly having magic power. *–v.* **1** delight, captivate. **2** influence or protect as if by magic (*a charmed life*). **3** obtain or gain by charm (*charmed his way into the BBC*). □ **charmer** *n.* [Latin *carmen* song]

charming *adj.* delightful. □ **charmingly** *adv.*

charnel-house /'tʃɑːn(ə)lˌhaʊs/ *n.* repository of corpses or bones. [Latin: related to CARNAL]

chart *–n.* **1** geographical map or plan, esp. for navigation. **2** sheet of information in the form of a table, graph, or diagram. **3** (usu. in *pl.*) *colloq.* listing of the currently best-selling pop records. *–v.* make a chart of, map. [Latin *charta*: related to CARD[1]]

charter *–n.* **1 a** document granting rights, issued esp. by a sovereign or legislature. **b** written constitution or description of an organization's functions etc. **2** contract to hire an aircraft, ship, etc., for a special purpose. *–v.* **1** grant a charter to. **2** hire (an aircraft, ship, etc.). [Latin *chartula*: related to CHART]

chartered *attrib. adj.* (of an accountant, engineer, librarian, etc.) qualified as a member of a professional body that has a royal charter.

charter flight *n.* flight by chartered aircraft.

Chartism *n. hist.* UK Parliamentary reform movement of 1837–48. □ **Chartist** *n.* [from CHARTER: name taken from 'People's Charter']

chartreuse /ʃɑː'trɜːz/ *n.* pale green or yellow brandy-based liqueur. [*Chartreuse*, monastery in S. France]

charwoman *n.* woman employed as a cleaner in a house.

chary /'tʃeərɪ/ *adj.* (**-ier**, **-iest**) **1** cautious, wary. **2** sparing; ungenerous. [Old English: related to CARE]

Charybdis see SCYLLA AND CHARYBDIS.

chase[1] *–v.* (**-sing**) **1** run after; pursue. **2** (foll. by *from*, *out of*, *to*, etc.) force to run away or flee. **3 a** (foll. by *after*) hurry in pursuit of. **b** (foll. by *round* etc.) *colloq.*

act or move about hurriedly. **4** (usu. foll. by *up*) *colloq.* pursue (a thing overdue). **5** *colloq.* **a** try to attain. **b** court persistently. **–***n.* **1** pursuit. **2** unenclosed hunting-land. **3** (prec. by *the*) hunting, esp. as a sport. [Latin *capto*: related to CATCH]

chase[2] *v.* (**-sing**) emboss or engrave (metal). [French: related to CASE[2]]

chaser *n.* **1** horse for steeplechasing. **2** *colloq.* drink taken after another of a different kind.

chasm /'kæz(ə)m/ *n.* **1** deep cleft or opening in the earth, rock, etc. **2** wide difference of feeling, interests, etc. [Latin from Greek]

chassis /'ʃæsɪ/ *n.* (*pl.* same /-sɪz/) **1** base-frame of a motor vehicle, carriage, etc. **2** frame to carry radio etc. components. [Latin: related to CASE[2]]

chaste /tʃeɪst/ *adj.* **1** abstaining from extramarital, or from all, sexual intercourse. **2** pure, virtuous. **3** simple, unadorned. □ **chastely** *adv.* **chasteness** *n.* [Latin *castus*]

chasten /'tʃeɪs(ə)n/ *v.* **1** (esp. as **chastening, chastened** *adjs.*) subdue, restrain. **2** discipline, punish.

chastise /tʃæs'taɪz/ *v.* (**-sing**) **1** rebuke severely. **2** punish, esp. by beating. □ **chastisement** *n.*

chastity /'tʃæstɪtɪ/ *n.* being chaste.

chasuble /'tʃæzjʊb(ə)l/ *n.* loose sleeveless usu. ornate outer vestment worn by a celebrant at Mass or the Eucharist. [Latin *casubla*]

chat **–***v.* (**-tt-**) talk in a light familiar way. **–***n.* **1** pleasant informal talk. **2** any of various songbirds. □ **chat up** *colloq.* chat to, esp. flirtatiously or with an ulterior motive. [shortening of CHATTER]

château /'ʃætəʊ/ *n.* (*pl.* **-x** /-təʊz/) large French country house or castle. [French: related to CASTLE]

chatelaine /'ʃætə,leɪn/ *n.* **1** mistress of a large house. **2** *hist.* set of short chains attached to a woman's belt, for carrying keys etc. [medieval Latin *castellanus*: related to CASTLE]

chatline *n.* telephone service which sets up a social conference call among youngsters.

chat show *n.* television or radio broadcast in which celebrities are interviewed informally.

chattel /'tʃæt(ə)l/ *n.* (usu. in *pl.*) movable possession. [French: related to CATTLE]

chatter **–***v.* **1** talk quickly, incessantly, trivially, or indiscreetly. **2** (of a bird, monkey, etc.) emit short quick sounds. **3** (of teeth) click repeatedly together. **–***n.* chattering talk or sounds. [imitative]

chatterbox *n.* talkative person.

chatty *adj.* (**-ier**, **-iest**) **1** fond of chatting. **2** resembling chat. □ **chattily** *adv.* **chattiness** *n.*

chauffeur /'ʃəʊfə(r)/ **–***n.* (*fem.* **chauffeuse** /-'fɜːz/) person employed to drive a car. **–***v.* drive (a car or person) as a chauffeur. [French, = stoker]

chauvinism /'ʃəʊvɪ,nɪz(ə)m/ *n.* **1** exaggerated or aggressive patriotism. **2** excessive or prejudiced support or loyalty for one's cause or group. [*Chauvin*, name of a character in a French play 1831]

chauvinist *n.* **1** person exhibiting chauvinism. **2** (in full **male chauvinist**) man who shows prejudice against women. □ **chauvinistic** /-'nɪstɪk/ *adj.* **chauvinistically** /-'nɪstɪkəlɪ/ *adv.*

cheap **–***adj.* **1** low in price; worth more than its cost. **2** charging low prices; offering good value. **3** of poor quality; inferior. **4** costing little effort and hence of little worth. **–***adv.* cheaply. □ **on the cheap** cheaply. □ **cheaply** *adv.* **cheapness** *n.* [Old English, = price, bargain]

cheapen *v.* make or become cheap; depreciate, degrade.

cheapjack **–***n.* seller of inferior goods at low prices. **–***adj.* inferior, shoddy.

cheapskate *n.* esp. *US colloq.* stingy person.

cheat **–***v.* **1 a** (often foll. by *into*, *out of*) deceive or trick. **b** (foll. by *of*) deprive of. **2** gain an unfair advantage by deception or breaking rules. **–***n.* **1** person who cheats. **2** trick, deception. □ **cheat on** *colloq.* be sexually unfaithful to. [from ESCHEAT]

check **–***v.* **1 a** examine the accuracy or quality of. **b** make sure, verify. **2 a** stop or slow the motion of; curb. **b** *colloq.* rebuke. **3** *Chess* directly threaten (the opposing king). **4** *US* agree on comparison. **5** *US* mark with a tick etc. **6** *US* deposit (luggage etc.). **–***n.* **1** means or act of testing or ensuring accuracy, quality, etc. **2 a** stopping or slowing of motion. **b** rebuff or rebuke. **c** person or thing that restrains. **3 a** pattern of small squares. **b** fabric so patterned. **c** (*attrib.*) so patterned. **4** (*also as int.*) *Chess* exposure of a king to direct attack. **5** *US* restaurant bill. **6** *US* = CHEQUE. **7** esp. *US* token of identification for left luggage etc. **8** *US Cards* counter used in games. **9** temporary loss of the scent in hunting. □ **check in 1** arrive or register at a hotel, airport, etc. **2** record the arrival of. **check into** register one's arrival at (a hotel etc.). **check off** mark on a list etc. as having been examined. **check on** examine, verify, keep watch on. **check out 1** (often foll. by *of*) leave a hotel etc. with due formalities. **2** esp. *US* investigate. **check up** make sure, verify. **check up on** = *check on*. [Persian, = king]

checked *adj.* having a check pattern.

checker[1] *n.* person etc. that examines, esp. in a factory etc.

checker[2] *n.* **1** var. of CHEQUER. **2** *US* **a** (in *pl.*, usu. treated as *sing.*) draughts. **b** piece used in this game.

check-in *n.* act or place of checking in.

checkmate –*n.* (also as *int.*) *Chess* check from which a king cannot escape. –*v.* (**-ting**) **1** *Chess* put into checkmate. **2** frustrate. [French: related to CHECK, Persian *māt* is dead]

checkout *n.* **1** act of checking out. **2** paydesk in a supermarket etc.

checkpoint *n.* place, esp. a barrier or entrance, where documents, vehicles, etc., are inspected.

check-up *n.* thorough (esp. medical) examination.

Cheddar /'tʃedə(r)/ *n.* a kind of firm smooth cheese. [*Cheddar* in Somerset]

cheek –*n.* **1 a** side of the face below the eye. **b** side-wall of the mouth. **2 a** impertinence; cool confidence. **b** impertinent speech. **3** *slang* buttock. –*v.* be impertinent to. □ **cheek by jowl** close together; intimate. [Old English]

cheek-bone *n.* bone below the eye.

cheeky *adj.* (**-ier**, **-iest**) impertinent. □ **cheekily** *adv.* **cheekiness** *n.*

cheep –*n.* weak shrill cry of a young bird. –*v.* make such a cry. [imitative]

cheer –*n.* **1** shout of encouragement or applause. **2** mood, disposition (*full of good cheer*). **3** (in *pl.*; as *int.*) *colloq.* **a** expressing good wishes on parting or before drinking. **b** expressing gratitude. –*v.* **1 a** applaud with shouts. **b** (usu. foll. by *on*) urge with shouts. **2** shout for joy. **3** gladden; comfort. □ **cheer up** make or become less depressed. [Latin *cara* face, from Greek]

cheerful *adj.* **1** in good spirits, noticeably happy. **2** bright, pleasant. □ **cheerfully** *adv.* **cheerfulness** *n.*

cheerio /ˌtʃɪrɪ'əʊ/ *int. colloq.* expressing good wishes on parting.

cheer-leader *n.* person who leads cheers of applause etc.

cheerless *adj.* gloomy, dreary.

cheery *adj.* (**-ier**, **-iest**) cheerful. □ **cheerily** *adv.* **cheeriness** *n.*

cheese /tʃi:z/ *n.* **1 a** food made from curds of milk. **b** cake of this with rind. **2** conserve with the consistency of soft cheese. □ **cheesy** *adj.* [Latin *caseus*]

cheeseburger *n.* hamburger with cheese in or on it.

cheesecake *n.* **1** tart filled with sweetened curds etc. **2** *slang* portrayal of women in a sexually stimulating manner.

cheesecloth *n.* thin loosely-woven cloth.

cheesed /tʃi:zd/ *adj. slang* (often foll. by *off*) bored, fed up. [origin unknown]

cheese-paring *adj.* stingy.

cheese plant *n.* climbing plant with holes in its leaves.

cheetah /'tʃi:tə/ *n.* swift-running spotted leopard-like feline. [Hindi]

chef /ʃef/ *n.* (usu. male) cook, esp. the chief cook in a restaurant. [French]

Chelsea bun /'tʃelsɪ/ *n.* currant bun in the form of a flat spiral. [*Chelsea* in London]

Chelsea pensioner *n.* inmate of the Chelsea Royal Hospital for old or disabled soldiers.

chemical /'kemɪk(ə)l/ –*adj.* of, made by, or employing chemistry or chemicals. –*n.* substance obtained or used in chemistry. □ **chemically** *adv.* [French or medieval Latin: related to ALCHEMY]

chemical engineering *n.* creation and operation of industrial chemical plants. □ **chemical engineer** *n.*

chemical warfare *n.* warfare using poison gas and other chemicals.

chemise /ʃə'mi:z/ *n. hist.* woman's loose-fitting undergarment or dress. [Latin *camisia* shirt]

chemist /'kemɪst/ *n.* **1** dealer in medicinal drugs etc. **2** expert in chemistry. [French: related to ALCHEMY]

chemistry /'kemɪstrɪ/ *n.* (*pl.* **-ies**) **1** branch of science dealing with the elements and the compounds they form and the reactions they undergo. **2** chemical composition and properties of a substance. **3** *colloq.* sexual attraction.

chemotherapy /ˌki:məʊ'θerəpɪ/ *n.* treatment of disease, esp. cancer, by chemical substances.

chenille /ʃə'ni:l/ *n.* **1** tufty velvety cord or yarn. **2** fabric of this. [French, = caterpillar, from Latin *canicula* little dog]

cheque /tʃek/ *n.* **1** written order to a bank to pay the stated sum from the drawer's account. **2** printed form on which this is written. [from CHECK]

cheque-book *n.* book of forms for writing cheques.

cheque card *n.* card issued by a bank to guarantee the honouring of cheques up to a stated amount.

chequer /'tʃekə(r)/ –*n.* **1** (often in *pl.*) pattern of squares often alternately coloured. **2** var. of CHECKER[2] 2. –*v.* **1** mark with chequers. **2** variegate; break the uniformity of. **3** (as **chequered** *adj.*) with varied fortunes (*chequered career*). [from EXCHEQUER]

cherish /ˈtʃerɪʃ/ *v.* **1** protect or tend lovingly. **2** hold dear, cling to (hopes, feelings, etc.). [French *cher* dear, from Latin *carus*]

cheroot /ʃəˈruːt/ *n.* cigar with both ends open. [French from Tamil]

cherry /ˈtʃerɪ/ –*n.* (*pl.* **-ies**) **1 a** small soft round stone-fruit. **b** tree bearing this or grown for its ornamental flowers. **c** its wood. **2** light red colour. –*adj.* of light red colour. [Greek *kerasos*]

cherub /ˈtʃerəb/ *n.* **1** (*pl.* **-im**) angelic being of the second order of the celestial hierarchy. **2 a** representation of a winged child or its head. **b** beautiful or innocent child. □ **cherubic** /tʃɪˈruːbɪk/ *adj.* [ultimately from Hebrew]

chervil /ˈtʃɜːvɪl/ *n.* herb used for flavouring. [Greek *khairephullon*]

Cheshire /ˈtʃeʃə(r)/ *n.* a kind of firm crumbly cheese. □ **like a Cheshire cat** with a broad fixed grin. [*Cheshire* in England]

chess *n.* game for two with 16 men each, played on a chessboard. [French: related to CHECK]

chessboard *n.* chequered board of 64 squares on which chess and draughts are played.

chessman *n.* any of the 32 pieces and pawns with which chess is played.

chest *n.* **1** large strong box. **2 a** part of the body enclosed by the ribs. **b** front surface of the body from the neck to the bottom of the ribs. **3** small cabinet for medicines etc. □ **get a thing off one's chest** *colloq.* disclose a secret etc. to relieve one's anxiety about it. [Latin *cista*]

chesterfield /ˈtʃestəˌfiːld/ *n.* sofa with arms and back of the same height and curved outwards at the top. [Earl of *Chesterfield*]

chestnut /ˈtʃesnʌt/ –*n.* **1 a** glossy hard brown edible nut. **b** tree bearing it. **2** = HORSE CHESTNUT. **3** wood of any chestnut. **4** horse of a reddish-brown colour. **5** *colloq.* stale joke etc. **6** reddish-brown. –*adj.* reddish-brown. [Greek *kastanea* nut]

chest of drawers *n.* piece of furniture consisting of a set of drawers in a frame.

chesty *adj.* (**-ier**, **-iest**) *colloq.* inclined to or symptomatic of chest disease. □ **chestily** *adv.* **chestiness** *n.*

cheval-glass /ʃəˈvæl/ *n.* tall mirror swung on an upright frame. [Latin *caballus* horse]

chevalier /ˌʃevəˈlɪə(r)/ *n.* member of certain orders of knighthood, or of the French Legion of Honour etc. [medieval Latin *caballarius* horseman]

chevron /ˈʃevrən/ *n.* V-shaped line or stripe. [Latin *caper* goat]

chew –*v.* work (food etc.) between the teeth. –*n.* **1** act of chewing. **2** chewy sweet. □ **chew on 1** work continuously between the teeth. **2** think about. **chew over 1** discuss, talk over. **2** think about. [Old English]

chewing-gum *n.* flavoured gum for chewing.

chewy *adj.* (**-ier**, **-iest**) **1** needing much chewing. **2** suitable for chewing. □ **chewiness** *n.*

chez /ʃeɪ/ *prep.* at the home of. [Latin *casa* cottage]

chi /kaɪ/ *n.* twenty-second letter of the Greek alphabet (X, χ). [Greek]

Chianti /kɪˈæntɪ/ *n.* (*pl.* **-s**) red wine from the Chianti area in Italy.

chiaroscuro /kɪˌɑːrəˈskʊərəʊ/ *n.* **1** treatment of light and shade in drawing and painting. **2** use of contrast in literature etc. [Italian, = clear dark]

chic /ʃiːk/ –*adj.* (**chic-er**, **chic-est**) stylish, elegant. –*n.* stylishness, elegance. [French]

chicane /ʃɪˈkeɪn/ –*n.* **1** artificial barrier or obstacle on a motor racecourse. **2** chicanery. –*v.* (**-ning**) *archaic* **1** use chicanery. **2** (usu. foll. by *into*, *out of*, etc.) cheat (a person). [French]

chicanery /ʃɪˈkeɪnərɪ/ *n.* (*pl.* **-ies**) **1** clever but misleading talk. **2** trickery, deception. [French]

chick *n.* **1** young bird. **2** *slang* young woman. [Old English: related to CHICKEN]

chicken /ˈtʃɪkɪn/ –*n.* **1 a** domestic fowl. **b** its flesh as food. **2** young bird of a domestic fowl. **3** youthful person (*is no chicken*). –*adj. colloq.* cowardly. –*v.* (foll. by *out*) *colloq.* withdraw through cowardice. [Old English]

chicken-feed *n.* **1** food for poultry. **2** *colloq.* trivial amount, esp. of money.

chickenpox *n.* infectious disease, esp. of children, with a rash of small blisters.

chicken-wire *n.* light wire netting with a hexagonal mesh.

chick-pea *n.* yellow pea-like seed used as a vegetable. [Latin *cicer*]

chickweed *n.* small weed with tiny white flowers.

chicle /ˈtʃɪk(ə)l/ *n.* milky juice of a tropical tree, used in chewing-gum. [Spanish from Nahuatl]

chicory /ˈtʃɪkərɪ/ *n.* (*pl.* **-ies**) **1** plant with leaves used in salads. **2** its root, roasted and ground and used with or instead of coffee. **3** esp. *US* = ENDIVE. [Greek *kikhorion*]

chide *v.* (*past* **chided** or **chid**; *past part.* **chided** or **chidden**) *archaic* scold, rebuke. [Old English]

chief –*n.* **1 a** leader or ruler. **b** head of a tribe, clan, etc. **2** head of a department;

highest official. *–adj.* **1** first in position, importance, influence, etc. **2** prominent, leading. [Latin *caput* head]

Chief Constable *n.* head of the police force of a county etc.

chiefly *adv.* above all; mainly but not exclusively.

Chief of Staff *n.* senior staff officer of a service or command.

chieftain /ˈtʃiːft(ə)n/ *n.* leader of a tribe, clan, etc. □ **chieftaincy** *n.* (*pl.* **-ies**). [Latin: related to CHIEF]

chiffchaff /ˈtʃɪftʃæf/ *n.* small European warbler. [imitative]

chiffon /ˈʃɪfɒn/ *n.* light diaphanous fabric of silk, nylon, etc. [French *chiffe* rag]

chignon /ˈʃiːnjɔ̃/ *n.* coil of hair at the back of a woman's head. [French]

chihuahua /tʃɪˈwɑːwə/ *n.* dog of a very small smooth-haired breed. [*Chihuahua* in Mexico]

chilblain /ˈtʃɪlbleɪn/ *n.* painful itching swelling on a hand, foot, etc., caused by exposure to cold. [from CHILL, *blain* inflamed sore, blister]

child /tʃaɪld/ *n.* (*pl.* **children** /ˈtʃɪldrən/) **1 a** young human being below the age of puberty. **b** unborn or newborn human being. **2** one's son or daughter. **3** (foll. by *of*) descendant, follower, or product of. **4** childish person. □ **childless** *adj.* [Old English]

child abuse *n.* maltreatment of a child, esp. by physical violence or sexual molestation.

child benefit *n.* regular payment by the State to the parents of a child up to a certain age.

childbirth *n.* giving birth to a child.

child care *n.* the care of children, esp. by a local authority.

childhood *n.* state or period of being a child.

childish *adj.* **1** of, like, or proper to a child. **2** immature, silly. □ **childishly** *adv.* **childishness** *n.*

childlike *adj.* having the good qualities of a child, such as innocence, frankness, etc.

child-minder *n.* person looking after children for payment.

child's play *n.* easy task.

chili var. of CHILLI.

chill *–n.* **1 a** unpleasant cold sensation; lowered body temperature. **b** feverish cold. **2** unpleasant coldness (of air, water, etc.). **3** depressing influence. **4** coldness of manner. *–v.* **1** make or become cold. **2** depress; horrify. **3** preserve (food or drink) by cooling. *–adj. literary* chilly. [Old English]

chilli /ˈtʃɪlɪ/ *n.* (also **chili**) (*pl.* **-es**) hot-tasting dried red capsicum pod. [Spanish from Aztec]

chilli con carne /kɒn ˈkɑːnɪ/ *n.* dish of chilli-flavoured mince and beans.

chilly *adj.* (**-ier, -iest**) **1** somewhat cold. **2** sensitive to the cold. **3** unfriendly; unemotional.

Chiltern Hundreds /ˈtʃɪlt(ə)n/ *n.pl.* Crown manor, whose administration is a nominal office for which an MP applies as a way of resigning from the House of Commons. [*Chiltern* Hills in S. England]

chime *–n.* **1** set of attuned bells. **2** sounds made by this. *–v.* (**-ming**) **1** (of bells) ring. **2** show (the time) by chiming. **3** (usu. foll. by *together, with*) be in agreement. □ **chime in 1** interject a remark. **2** join in harmoniously. **3** (foll. by *with*) agree with. [Old English: related to CYMBAL]

chimera /kaɪˈmɪərə/ *n.* **1** (in Greek mythology) monster with a lion's head, goat's body, and serpent's tail. **2** bogey. **3** wild or fantastic conception. □ **chimerical** /-ˈmerɪk(ə)l/ *adj.* [Latin from Greek]

chimney /ˈtʃɪmnɪ/ *n.* (*pl.* **-s**) **1** channel conducting smoke etc. up and away from a fire, engine, etc. **2** part of this above a roof. **3** glass tube protecting the flame of a lamp. **4** narrow vertical crack in a rock-face. [Latin *caminus* oven, from Greek]

chimney-breast *n.* projecting wall surrounding a chimney.

chimney-pot *n.* earthenware or metal pipe at the top of a chimney.

chimney-stack *n.* number of chimneys grouped in one structure.

chimney-sweep *n.* person who removes soot from inside chimneys.

chimp *n. colloq.* = CHIMPANZEE. [abbreviation]

chimpanzee /ˌtʃɪmpənˈziː/ *n.* small African manlike ape. [French from Kongo]

chin *n.* front of the lower jaw. □ **keep one's chin up** *colloq.* remain cheerful. **take on the chin** suffer a severe blow from; endure courageously. [Old English]

china /ˈtʃaɪnə/ *–n.* **1** fine white or translucent ceramic ware, porcelain, etc. **2** things made of this. *–adj.* made of china. [*China* in Asia]

china clay *n.* kaolin.

Chinaman /ˈtʃaɪnəmən/ *n.* **1** *archaic* or *derog.* (now usu. *offens.*) native of China. **2** *Cricket* ball bowled by a left-handed bowler that spins from off to leg.

chinchilla /tʃɪnˈtʃɪlə/ *n.* **1 a** small S. American rodent. **b** its soft grey fur. **2** breed of cat or rabbit. [Spanish *chinche* bug]

chine –*n.* **1 a** backbone. **b** joint of meat containing all or part of this. **2** ridge. –*v.* (**-ning**) cut (meat) through the backbone. [Latin *spina* SPINE]

Chinese /tʃaɪ'ni:z/ –*adj.* of China. –*n.* **1** Chinese language. **2** (*pl.* same) **a** native or national of China. **b** person of Chinese descent.

Chinese lantern *n.* **1** collapsible paper lantern. **2** plant with an orange-red papery calyx.

Chinese leaf *n.* lettuce-like cabbage.

Chink *n. slang offens.* a Chinese. [abbreviation]

chink[1] *n.* narrow opening; slit. [related to *chine* narrow ravine]

chink[2] –*v.* (cause to) make a sound like glasses or coins striking together. –*n.* this sound. [imitative]

chinless *adj. colloq.* weak or feeble in character.

chinless wonder *n.* ineffectual esp. upper-class person.

chinoiserie /ʃɪn'wɑ:zərɪ/ *n.* **1** imitation of Chinese motifs in painting and in decorating furniture. **2** object(s) in this style. [French]

chintz /tʃɪnts/ *n.* printed multicoloured usu. glazed cotton fabric. [Hindi from Sanskrit]

chintzy *adj.* (**-ier, -iest**) **1** like chintz. **2** gaudy, cheap. **3** characteristic of décor associated with chintz soft furnishings.

chin-wag *slang* –*n.* talk or chat. –*v.* (**-gg-**) chat.

chip –*n.* **1** small piece removed by chopping etc. **2** place or mark where a piece has been broken off. **3 a** strip of potato, usu. deep-fried. **b** *US* potato crisp. **4** counter used in some games to represent money. **5** = MICROCHIP. –*v.* (**-pp-**) **1** (often foll. by *off, away*) cut or break (a piece) from a hard material. **2** (often foll. by *at, away at*) cut pieces off (a hard material) to alter its shape etc. **3** be apt to break at the edge. **4** (usu. as **chipped** *adj.*) make (potatoes) into chips. □ **chip in** *colloq.* **1** interrupt. **2** contribute (money etc.). **a chip off the old block** child resembling its parent, esp. in character. **a chip on one's shoulder** *colloq.* inclination to feel resentful or aggrieved. **when the chips are down** *colloq.* when it comes to the point. [Old English]

chipboard *n.* board made from compressed wood chips.

chipmunk /'tʃɪpmʌŋk/ *n.* striped N. American ground squirrel. [Algonquian]

chipolata /,tʃɪpə'lɑ:tə/ *n.* small thin sausage. [French from Italian]

Chippendale /'tʃɪpən,deɪl/ *adj.* (of furniture) of an elegantly ornate 18th-c. style. [name of a cabinet-maker]

chiro- *comb. form* hand. [Greek *kheir*]

chiromancy /'kaɪərəʊ,mænsɪ/ *n.* palmistry. [Greek *mantis* seer]

chiropody /kɪ'rɒpədɪ/ *n.* treatment of the feet and their ailments. □ **chiropodist** *n.* [Greek *pous podos* foot]

chiropractic /,kaɪərəʊ'præktɪk/ *n.* treatment of disease by manipulation of esp. the spinal column. □ **chiropractor** /'kaɪərəʊ-/ *n.* [Greek *prattō* do]

chirp –*v.* **1** (of small birds, grasshoppers, etc.) utter a short sharp note. **2** speak or utter merrily. –*n.* chirping sound. [imitative]

chirpy *adj. colloq.* (**-ier, -iest**) cheerful, lively. □ **chirpily** *adv.* **chirpiness** *n.*

chirrup /'tʃɪrəp/ –*v.* (**-p-**) chirp, esp. repeatedly. –*n.* chirruping sound. [imitative]

chisel /'tʃɪz(ə)l/ –*n.* hand tool with a squared bevelled blade for shaping wood, stone, or metal. –*v.* **1** (**-ll-**; *US* **-l-**) cut or shape with a chisel. **2** (as **chiselled** *adj.*) (of facial features) clear-cut, fine. **3** *slang* cheat. [Latin *caedo* cut]

chit[1] *n.* **1** *derog.* or *joc.* young small woman (esp. *a chit of a girl*). **2** young child. [originally = whelp, cub]

chit[2] *n.* **1** note of requisition, of a sum owed, etc. **2** note or memorandum. [Hindi from Sanskrit]

chit-chat *n. colloq.* light conversation; gossip. [reduplication of CHAT]

chivalrous /'ʃɪvəlrəs/ *adj.* **1** gallant, honourable. **2** of or showing chivalry. □ **chivalrously** *adv.* [Latin: related to CHEVALIER]

chivalry /'ʃɪvəlrɪ/ *n.* **1** medieval knightly system with its religious, moral, and social code. **2** honour, courtesy, and readiness to help the weak. □ **chivalric** *adj.*

chive *n.* small plant with long onion-flavoured leaves. [Latin *cepa* onion]

chivvy /'tʃɪvɪ/ *v.* (**-ies, -ied**) urge persistently, nag. [probably from ballad of *Chevy Chase*]

chloral /'klɔ:r(ə)l/ *n.* **1** colourless liquid aldehyde used in making DDT. **2** (in full **chloral hydrate**) *Pharm.* crystalline solid made from this and used as a sedative. [French: related to CHLORINE, ALCOHOL]

chloride /'klɔ:raɪd/ *n.* **1** compound of chlorine and another element or group. **2** bleaching agent containing this.

chlorinate /'klɔ:rɪ,neɪt/ *v.* (**-ting**) impregnate or treat with chlorine. □ **chlorination** /-'neɪʃ(ə)n/ *n.*

chlorine /'klɔ:ri:n/ *n.* poisonous gaseous element used for purifying water etc. [Greek *khlōros* green]

chlorofluorocarbon see CFC.

chloroform /'klɒrə,fɔ:m/ *–n.* colourless volatile liquid formerly used as a general anaesthetic. *–v.* render unconscious with this. [from CHLORINE, FORMIC ACID]

chlorophyll /'klɒrəfɪl/ *n.* green pigment found in most plants. [Greek *khlōros* green, *phullon* leaf]

choc *n. colloq.* chocolate. [abbreviation]

choc-ice *n.* bar of ice-cream covered with chocolate.

chock *–n.* block or wedge to check the motion of a wheel etc. *–v.* make fast with chocks. [French]

chock-a-block *predic. adj.* (often foll. by *with*) crammed together or full.

chock-full *predic. adj.* (often foll. by *of*) crammed full.

chocolate /'tʃɒklət/ *–n.* **1 a** food preparation in the form of a paste or solid block made from ground cacao seeds and usu. sweetened. **b** sweet made of or coated with this. **c** drink containing this. **2** deep brown. *–adj.* **1** made from chocolate. **2** deep brown. [Aztec *chocolatl*]

choice *–n.* **1 a** act of choosing. **b** thing or person chosen. **2** range from which to choose. **3** power or opportunity to choose. *–adj.* of superior quality. [Germanic: related to CHOOSE]

choir /'kwaɪə(r)/ *n.* **1** regular group of singers, esp. in a church. **2** part of a cathedral or large church between the altar and nave. [Latin: related to CHORUS]

choirboy *n.* (*fem.* **choirgirl**) boy singer in a church choir.

choke *–v.* (**-king**) **1** stop the breathing of (a person or animal), esp. by constricting the windpipe or (of gas, smoke, etc.) by being unbreathable. **2** suffer a stoppage of breath. **3** make or become speechless from emotion. **4** retard the growth of or kill (esp. plants) by depriving of light etc. **5** (often foll. by *back*) suppress (feelings) with difficulty. **6** block or clog (a passage, tube, etc.). **7** (as **choked** *adj.*) *colloq.* disgusted, disappointed. *–n.* **1** valve in a carburettor controlling the intake of air. **2** device for smoothing the variations of an alternating current. □ **choke up** block (a channel etc.). [Old English]

choker *n.* close-fitting necklace.

cholecalciferol /,kɒlɪkæl'sɪfə,rɒl/ *n.* a vitamin (D_3) produced by the action of sunlight on a steroid in the skin. [from CHOLER, CALCIFEROL]

choler /'kɒlə(r)/ *n.* **1** *hist.* one of the four humours, bile. **2** *poet.* or *archaic* anger, irascibility. [Greek *kholē* bile]

cholera /'kɒlərə/ *n.* infectious often fatal bacterial disease of the small intestine. [related to CHOLER]

choleric /'kɒlərɪk/ *adj.* irascible, angry.

cholesterol /kə'lestə,rɒl/ *n.* sterol found in most body tissues, including the blood where high concentrations promote arteriosclerosis. [from CHOLER, Greek *stereos* stiff]

chomp *v.* = CHAMP[1]. [imitative]

choose /tʃu:z/ *v.* (**-sing**; *past* **chose** /tʃəʊz/; *past part.* **chosen**) **1** select out of a greater number. **2** (usu. foll. by *between*, *from*) take or select one or another. **3** (usu. foll. by *to* + infin.) decide, be determined. **4** select as (*was chosen leader*). □ **nothing** (or **little**) **to choose between them** they are very similar. [Old English]

choosy /'tʃu:zɪ/ *adj.* (**-ier, -iest**) *colloq.* fastidious. □ **choosiness** *n.*

chop[1] *–v.* (**-pp-**) **1** (usu. foll. by *off*, *down*, etc.) cut or fell by the blow of an axe etc. **2** (often foll. by *up*) cut into small pieces. **3** strike (esp. a ball) with a short heavy edgewise blow. *–n.* **1** cutting blow. **2** thick slice of meat (esp. pork or lamb) usu. including a rib. **3** short chopping stroke in cricket etc. **4** (prec. by *the*) *slang* **a** = SACK[1] *n.* 2. **b** killing or being killed. [related to CHAP[2]]

chop[2] *n.* (usu. in *pl.*) jaw. [origin unknown]

chop[3] *v.* (**-pp-**) □ **chop and change** vacillate; change direction frequently. **chop logic** argue pedantically. [perhaps related to CHEAP]

chopper *n.* **1 a** short axe with a large blade. **b** butcher's cleaver. **2** *colloq.* helicopter. **3** *colloq.* type of bicycle or motor cycle with high handlebars.

choppy *adj.* (**-ier, -iest**) (of the sea etc.) fairly rough. □ **choppily** *adv.* **choppiness** *n.* [from CHOP[1]]

chopstick *n.* each of a pair of sticks held in one hand as eating utensils by the Chinese, Japanese, etc. [pidgin English from Chinese, = nimble ones]

chopsuey /tʃɒp'su:ɪ/ *n.* (*pl.* **-s**) Chinese-style dish of meat fried with vegetables and rice. [Chinese, = mixed bits]

choral /'kɔ:r(ə)l/ *adj.* of, for, or sung by a choir or chorus. [medieval Latin: related to CHORUS]

chorale /kɒ'rɑ:l/ *n.* **1** simple stately hymn tune; harmonized form of this. **2** esp. *US* choir. [German: related to CHORAL]

chord[1] /kɔ:d/ *n.* group of notes sounded together. [originally *cord* from ACCORD]

chord[2] /kɔ:d/ *n.* **1** straight line joining the ends of an arc or curve. **2** *poet.* string of a harp etc. □ **strike a chord** elicit sympathy. [var. of CORD]
chordate /'kɔ:deɪt/ *–n.* animal having a cartilaginous skeletal rod at some stage of its development. *–adj.* of chordates. [Latin *chorda* CHORD[2] after *Vertebrata* etc.]
chore *n.* tedious or routine task, esp. domestic. [from CHAR[2]]
choreograph /'kɒrɪə,grɑ:f/ *v.* compose choreography for (a ballet etc.). □ **choreographer** /-ɪ'ɒgrəfə(r)/ *n.*
choreography /,kɒrɪ'ɒgrəfɪ/ *n.* design or arrangement of a ballet etc. □ **choreographic** /-ə'græfɪk/ *adj.* [Greek *khoreia* dance]
chorister /'kɒrɪstə(r)/ *n.* member of a choir, esp. a choirboy. [French: related to CHOIR]
chortle /'tʃɔ:t(ə)l/ *–n.* gleeful chuckle. *–v.* (**-ling**) utter or express with a chortle. [probably from CHUCKLE, SNORT]
chorus /'kɔ:rəs/ *–n.* (*pl.* **-es**) **1** group of singers; choir. **2** music composed for a choir. **3** refrain or main part of a song. **4** simultaneous utterance. **5** group of singers and dancers performing together. **6** *Gk Antiq.* **a** group of performers who comment on the action in a Greek play. **b** utterance made by it. **7** character speaking the prologue in a play. *–v.* (**-s-**) speak or utter simultaneously. [Latin from Greek]
chose *past* of CHOOSE.
chosen *past part.* of CHOOSE.
chough /tʃʌf/ *n.* bird with glossy blue-black plumage and red legs. [imitative]
choux pastry /ʃu:/ *n.* very light pastry enriched with eggs. [French]
chow *n.* **1** *slang* food. **2** dog of a Chinese breed with long woolly hair. [Chinese *chow-chow*]
chow mein /tʃaʊ 'meɪn/ *n.* Chinese-style dish of fried noodles with shredded meat or shrimps etc. and vegetables. [Chinese *chao mian* fried flour]
Christ /kraɪst/ *–n.* **1** title, also now treated as a name, given to Jesus. **2** Messiah as prophesied in the Old Testament. *–int. slang* expressing surprise, anger, etc. [Greek, = anointed]
christen /'krɪs(ə)n/ *v.* **1** baptize as a sign of admission to the Christian Church. **2** give a name to. **3** *colloq.* use for the first time. □ **christening** *n.* [Latin: related to CHRISTIAN]
Christendom /'krɪsəndəm/ *n.* Christians worldwide.
Christian /'krɪstʃ(ə)n/ *–adj.* **1** of Christ's teaching. **2** believing in or following the religion of Christ. **3** showing the associated qualities. **4** *colloq.* kind. *–n.* adherent of Christianity. [Latin *Christianus* of CHRIST]
Christian era *n.* era reckoned from Christ's birth.
Christianity /,krɪstɪ'ænɪtɪ/ *n.* **1** Christian religion. **2** being a Christian; Christian quality or character.
Christian name *n.* forename, esp. as given at baptism.
Christian Science *n.* a doctrine proclaiming the power of healing by prayer alone. □ **Christian Scientist** *n.*
Christmas /'krɪsməs/ *n.* **1** (also **Christmas Day**) annual festival of Christ's birth, celebrated on 25 Dec. **2** period around this. □ **Christmassy** *adj.* [Old English: related to CHRIST, MASS[2]]
Christmas-box *n.* present or gratuity given at Christmas.
Christmas Eve *n.* 24 Dec.
Christmas pudding *n.* rich boiled pudding of flour, suet, dried fruit, etc.
Christmas rose *n.* white-flowered winter-blooming hellebore.
Christmas tree *n.* evergreen tree or imitation of this set up and decorated at Christmas.
chromatic /krə'mætɪk/ *adj.* **1** of colour; in colours. **2** *Mus.* **a** of or having notes not belonging to a particular diatonic scale. **b** (of a scale) ascending or descending by semitones. □ **chromatically** *adv.* [Greek *khrōma -mat-* colour]
chromatin /'krəʊmətɪn/ *n.* chromosome material in a cell nucleus which stains with basic dyes. [Greek: related to CHROME]
chromatography /,krəʊmə'tɒgrəfɪ/ *n.* separation of the components of a mixture by slow passage through or over material which adsorbs them differently. [Greek: related to CHROME]
chrome /krəʊm/ *n.* **1** chromium, esp. as plating. **2** (in full **chrome yellow**) yellow pigment got from a certain compound of chromium. [Greek *khrōma* colour]
chromite /'krəʊmaɪt/ *n.* mineral of chromium and iron oxides.
chromium /'krəʊmɪəm/ *n.* metallic element used as a shiny decorative or protective coating.
chromium plate *n.* protective coating of chromium.
chromosome /'krəʊmə,səʊm/ *n.* threadlike structure, usu. found in the cell nucleus of animals and plants, carrying genes. [Greek: related to CHROME, *sōma* body]
chronic /'krɒnɪk/ *adj.* **1** (esp. of an illness) long-lasting. **2** having a chronic complaint. **3** *colloq.* very bad; intense, severe. **4** *colloq.* habitual, inveterate (*a*

chronic liar). □ **chronically** *adv.* [Greek *khronos* time]

■ **Usage** The use of *chronic* in sense 3 is very informal, and its use in sense 4 is considered incorrect by some people.

---

chronicle /ˈkrɒnɪk(ə)l/ *–n.* register of events in order of occurrence. *–v.* (**-ling**) record (events) thus. [Greek *khronika*: related to CHRONIC]

chronological /ˌkrɒnəˈlɒdʒɪk(ə)l/ *adj.* **1** according to order of occurrence. **2** of chronology. □ **chronologically** *adv.*

chronology /krəˈnɒlədʒɪ/ *n.* (*pl.* **-ies**) **1** science of determining dates. **2 a** arrangement of events etc. in order of occurrence. **b** table or document displaying this. [Greek *khronos* time, -LOGY]

chronometer /krəˈnɒmɪtə(r)/ *n.* time-measuring instrument, esp. one used in navigation. [from CHRONOLOGY, -METER]

chrysalis /ˈkrɪsəlɪs/ *n.* (*pl.* **-lises**) **1** pupa of a butterfly or moth. **2** case enclosing it. [Greek *khrusos* gold]

chrysanthemum /krɪˈsænθəməm/ *n.* garden plant of the daisy family blooming in autumn. [Greek, = gold flower]

chrysoberyl /ˈkrɪsəˌberɪl/ *n.* yellowish-green gem. [Greek *khrusos* gold, BERYL]

chrysolite /ˈkrɪsəˌlaɪt/ *n.* precious variety of olivine. [Greek *khrusos* gold, *lithos* stone]

chrysoprase /ˈkrɪsəˌpreɪz/ *n.* apple-green variety of chalcedony. [Greek *khrusos* gold, *prason* leek]

chub *n.* (*pl.* same) thick-bodied river fish. [origin unknown]

Chubb *n.* (in full **Chubb lock**) *propr.* lock with a device for fixing the bolt immovably should someone try to pick it. [*Chubb*, name of a locksmith]

chubby *adj.* (**-ier, -iest**) plump and rounded. [from CHUB]

chuck[1] *–v.* **1** *colloq.* fling or throw carelessly or casually. **2** (often foll. by *in, up*) *colloq.* give up; reject. **3** touch playfully, esp. under the chin. *–n.* **1** playful touch under the chin. **2** toss. □ **the chuck** *slang* dismissal; rejection. **chuck out** *colloq.* **1** expel (a person) from a gathering etc. **2** get rid of, discard. [perhaps from French *chuquer* knock]

chuck[2] *–n.* **1** cut of beef from neck to ribs. **2** device for holding a workpiece or bit. *–v.* fix to a chuck. [var. of CHOCK]

chuckle /ˈtʃʌk(ə)l/ *–v.* (**-ling**) laugh quietly or inwardly. *–n.* quiet or suppressed laugh. [*chuck* cluck]

chuff *v.* (of an engine etc.) work with a regular sharp puffing sound. [imitative]

chuffed *adj. slang* delighted. [dial. *chuff*]

chug *–v.* (**-gg-**) **1** emit a regular muffled explosive sound, as of an engine running slowly. **2** move with this sound. *–n.* chugging sound. [imitative]

chukka boot *n.* ankle-high leather boot.

chukker *n.* (also **chukka**) period of play in polo. [Sanskrit *cakra* wheel]

chum *n. colloq.* close friend. □ **chum up** (**-mm-**) (often foll. by *with*) become a close friend (of). □ **chummy** *adj.* (**-ier, -iest**). **chummily** *adv.* **chumminess** *n.* [abbreviation of *chamber-fellow*]

chump *n.* **1** *colloq.* foolish person. **2** thick end of a loin of lamb or mutton (*chump chop*). **3** short thick block of wood. □ **off one's chump** *slang* crazy. [blend of CHUNK, LUMP[1]]

chunk *n.* **1** thick piece cut or broken off. **2** substantial amount. [var. of CHUCK[2]]

chunky *adj.* (**-ier, -iest**) **1** consisting of or resembling chunks; thick, substantial. **2** small and sturdy. □ **chunkiness** *n.*

chunter *v. colloq.* mutter, grumble. [probably imitative]

chupatty var. of CHAPATTI.

church *n.* **1** building for public Christian worship. **2** public worship (*met after church*). **3** (**Church**) **a** body of all Christians. **b** clergy or clerical profession. **c** organized Christian society (*the early Church*). [Greek *kuriakon* Lord's (house)]

churchgoer *n.* person attending church regularly.

churchman *n.* member of the clergy or of a Church.

Church of England *n.* English Protestant Church.

churchwarden *n.* either of two elected lay representatives of an Anglican parish.

churchyard *n.* enclosed ground around a church used for burials.

churl *n.* **1** ill-bred person. **2** *archaic* peasant. [Old English, = man]

churlish *adj.* surly; mean. □ **churlishly** *adv.* **churlishness** *n.* [from CHURL]

churn *–n.* **1** large milk-can. **2** butter-making machine. *–v.* **1** agitate (milk or cream) in a churn. **2** produce (butter) in a churn. **3** (usu. foll. by *up*) upset, agitate. □ **churn out** produce in large quantities. [Old English]

chute[1] /ʃuːt/ *n.* sloping channel or slide for sending things to a lower level. [Latin *cado* fall]

chute[2] /ʃuːt/ *n. colloq.* parachute. [abbreviation]

chutney /ˈtʃʌtnɪ/ *n.* (*pl.* **-s**) pungent condiment of fruits, vinegar, spices, etc. [Hindi]

chutzpah /ˈxʊtspə/ *n. slang* shameless audacity. [Yiddish]
chyle /kaɪl/ *n.* milky fluid of food materials formed in the intestine after digestion. [Greek *khulos* juice]
chyme /kaɪm/ *n.* acid pulp formed from partly-digested food. [Greek *khumos* juice]
CIA *abbr.* (in the US) Central Intelligence Agency.
ciao /tʃaʊ/ *int. colloq.* **1** goodbye. **2** hello. [Italian]
cicada /sɪˈkɑːdə/ *n.* large transparent-winged insect making a rhythmic chirping sound. [Latin]
cicatrice /ˈsɪkətrɪs/ *n.* scar left by a wound. [Latin]
cicely /ˈsɪsəlɪ/ *n.* (*pl.* **-ies**) flowering plant related to parsley and chervil. [Greek *seselis*]
cicerone /ˌtʃɪtʃəˈrəʊnɪ/ *n.* (*pl.* **-roni** pronunc. same) person who guides sight-seers. [Latin *Cicero*, name of a Roman statesman]
CID *abbr.* Criminal Investigation Department.
-cide *suffix* **1** person or substance that kills (*regicide*; *insecticide*). **2** killing of (*infanticide*). [Latin *caedo* kill]
cider *n.* drink of fermented apple juice. [Hebrew, = strong drink]
cigar /sɪˈgɑː(r)/ *n.* tight roll of tobacco-leaves for smoking. [French or Spanish]
cigarette /ˌsɪgəˈret/ *n.* finely-cut tobacco rolled in paper for smoking. [French diminutive]
cilium /ˈsɪlɪəm/ *n.* (*pl.* **cilia**) **1** minute hairlike structure on the surface of many animal cells. **2** eyelash. □ **ciliary** *adj.* **ciliate** *adj.* [Latin, = eyelash]
cinch *n. colloq.* **1** sure thing; certainty. **2** easy task. [Spanish *cincha* saddle-girth]
cinchona /sɪŋˈkəʊnə/ *n.* **1 a** S. American evergreen tree or shrub. **b** its bark, containing quinine. **2** drug from this. [Countess of *Chinchón*]
cincture /ˈsɪŋktʃə(r)/ *n. literary* girdle, belt, or border. [Latin *cingo* gird]
cinder *n.* **1** residue of coal or wood etc. after burning. **2** (in *pl.*) ashes. [Old English *sinder* = slag]
Cinderella /ˌsɪndəˈrelə/ *n.* person or thing of unrecognized or disregarded merit or beauty. [name of a girl in a fairy tale]
cine- *comb. form* cinematographic (*cine-camera*). [abbreviation]
cinema /ˈsɪnəmə/ *n.* **1** theatre where films are shown. **2 a** films collectively. **b** art or industry of producing films. □ **cinematic** /-ˈmætɪk/ *adj.* [French: related to KINEMATIC]
cinematography /ˌsɪnɪməˈtɒgrəfɪ/ *n.* art of making films. □ **cinematographer** *n.* **cinematographic** /-ˌmætəˈgræfɪk/ *adj.*
cineraria /ˌsɪnəˈreərɪə/ *n.* composite plant with bright flowers and ash-coloured down on its leaves. [Latin *cinis -ner-* ashes]
cinnabar /ˈsɪnəˌbɑː(r)/ *n.* **1** bright red mercuric sulphide. **2** vermilion. **3** moth with reddish-marked wings. [Latin from Greek]
cinnamon /ˈsɪnəmən/ *n.* **1** aromatic spice from the bark of a SE Asian tree. **2** this tree. **3** yellowish-brown. [Greek *kinnamon*]
cinque /sɪŋk/ *n.* the five on dice. [Latin *quinque* five]
cinquefoil /ˈsɪŋkfɔɪl/ *n.* **1** plant with compound leaves of five leaflets. **2** *Archit.* five-cusped ornament in a circle or arch. [Latin: related to CINQUE, *folium* leaf]
Cinque Ports *n.pl.* group of (orig. five) ports in SE England with ancient privileges. [Latin *quinque portus* five ports]
cipher /ˈsaɪfə(r)/ (also **cypher**) —*n.* **1 a** secret or disguised writing. **b** thing so written. **c** key to it. **2** arithmetical symbol (0) used to occupy a vacant place in decimal etc. numeration. **3** person or thing of no importance. —*v.* write in cipher. [Arabic *ṣifr*]
circa /ˈsɜːkə/ *prep.* (preceding a date) about. [Latin]
circadian /sɜːˈkeɪdɪən/ *adj. Physiol.* occurring about once per day. [from CIRCA, Latin *dies* day]
circle /ˈsɜːk(ə)l/ —*n.* **1** round plane figure whose circumference is everywhere equidistant from its centre. **2** circular or roundish enclosure or structure. **3** curved upper tier of seats in a theatre etc. **4** circular route. **5** persons grouped round a centre of interest. **6** set or restricted group (*literary circles*). —*v.* (**-ling**) **1** (often foll. by *round*, *about*) move in a circle. **2 a** revolve round. **b** form a circle round. □ **come full circle** return to the starting-point. [Latin diminutive: related to CIRCUS]
circlet /ˈsɜːklɪt/ *n.* **1** small circle. **2** circular band, esp. as an ornament.
circuit /ˈsɜːkɪt/ *n.* **1** line or course enclosing an area; the distance round. **2 a** path of an electric current. **b** apparatus through which a current passes. **3 a** judge's itinerary through a district to hold courts. **b** such a district. **c** lawyers following a circuit. **4** chain of theatres, cinemas, etc. under a single management. **5** motor-racing track. **6** itinerary or specific sphere of operation (*election circuit*; *cabaret circuit*). **7** sequence of

sporting events or athletic exercises. [Latin: related to CIRCUM-, *eo it-* go]

circuit-breaker *n.* automatic device for interrupting an electric circuit.

circuitous /sɜːˈkjuːɪtəs/ *adj.* **1** indirect. **2** going a long way round.

circuitry /ˈsɜːkɪtrɪ/ *n.* (*pl.* **-ies**) **1** system of electric circuits. **2** equipment forming this.

circular /ˈsɜːkjʊlə(r)/ *–adj.* **1 a** having the form of a circle. **b** moving (roughly) in a circle, finishing at the starting-point (*circular walk*). **2** (of reasoning) using the point it is trying to prove as evidence for its conclusion, hence invalid. **3** (of a letter etc.) distributed to a number of people. *–n.* circular letter, leaflet, etc. □ **circularity** /-ˈlærɪtɪ/ *n.* [Latin: related to CIRCLE]

circularize *v.* (also **-ise**) (**-zing** or **-sing**) distribute circulars to.

circular saw *n.* power saw with a rapidly rotating toothed disc.

circulate /ˈsɜːkjʊˌleɪt/ *v.* (**-ting**) **1** be in circulation; spread. **2 a** put into circulation. **b** send circulars to. **3** move about among guests etc. [Latin: related to CIRCLE]

circulation /ˌsɜːkjʊˈleɪʃ(ə)n/ *n.* **1** movement to and fro, or from and back to a starting-point, esp. that of the blood from and to the heart. **2 a** transmission or distribution. **b** number of copies sold. □ **in** (or **out of**) **circulation** active (or not active) socially.

circulatory /ˌsɜːkjʊˈleɪtərɪ/ *adj.* of circulation, esp. of the blood.

circum- *comb. form* round, about. [Latin]

circumcise /ˈsɜːkəmˌsaɪz/ *v.* (**-sing**) cut off the foreskin or clitoris of. □ **circumcision** /-ˈsɪʒ(ə)n/ *n.* [Latin *caedo* cut]

circumference /səˈkʌmfərəns/ *n.* **1** enclosing boundary, esp. of a circle. **2** distance round. □ **circumferential** /-ˈrenʃ(ə)l/ *adj.* [Latin *fero* carry]

circumflex /ˈsɜːkəmˌfleks/ *n.* (in full **circumflex accent**) mark (^) placed over a vowel to show contraction, length, etc. [Latin: related to FLEX[1]]

circumlocution /ˌsɜːkəmləˈkjuːʃ(ə)n/ *n.* **1 a** roundabout expression. **b** evasive talk. **2** verbosity. □ **circumlocutory** /-ˈlɒkjʊtərɪ/ *adj.*

circumnavigate /ˌsɜːkəmˈnævɪˌgeɪt/ *v.* (**-ting**) sail round (esp. the world). □ **circumnavigation** /-ˈgeɪʃ(ə)n/ *n.*

circumscribe /ˈsɜːkəmˌskraɪb/ *v.* (**-bing**) **1** (of a line etc.) enclose or outline. **2** lay down the limits of; confine, restrict. **3** *Geom.* draw (a figure) round another, touching it at points but not cutting it. □ **circumscription** /-ˈskrɪpʃ(ə)n/ *n.* [Latin *scribo* write]

circumspect /ˈsɜːkəmˌspekt/ *adj.* cautious; taking everything into account. □ **circumspection** /-ˈspekʃ(ə)n/ *n.* **circumspectly** *adv.* [Latin *specio spect-* look]

circumstance /ˈsɜːkəmst(ə)ns/ *n.* **1** fact, occurrence, or condition, esp. (in *pl.*) connected with or influencing an event; (bad) luck (*victim of circumstance(s)*). **2** (in *pl.*) one's financial condition. **3** ceremony, fuss. □ **in** (or **under**) **the circumstances** the state of affairs being what it is. **in** (or **under**) **no circumstances** not at all; never. □ **circumstanced** *adj.* [Latin *sto* stand]

circumstantial /ˌsɜːkəmˈstænʃ(ə)l/ *adj.* **1** giving full details (*circumstantial account*). **2** (of evidence etc.) indicating a conclusion by inference from known facts hard to explain otherwise. □ **circumstantiality** /-ʃɪˈælɪtɪ/ *n.*

circumvent /ˌsɜːkəmˈvent/ *v.* **1** evade, find a way round. **2** baffle, outwit. □ **circumvention** *n.* [Latin *venio vent-* come]

circus /ˈsɜːkəs/ *n.* (*pl.* **-es**) **1** travelling show of performing acrobats, clowns, animals, etc. **2** *colloq.* **a** scene of lively action. **b** group of people in a common activity, esp. sport. **3** open space in a town, where several streets converge. **4** *Rom. Antiq.* arena for sports and games. [Latin, = ring]

cirrhosis /sɪˈrəʊsɪs/ *n.* chronic liver disease, as a result of alcoholism etc. [Greek *kirrhos* tawny]

cirrus /ˈsɪrəs/ *n.* (*pl.* **cirri** /-raɪ/) **1** white wispy cloud at high altitude. **2** tendril or appendage of a plant or animal. [Latin, = curl]

CIS *abbr.* Commonwealth of Independent States.

cisalpine /sɪsˈælpaɪn/ *adj.* on the south side of the Alps. [Latin *cis-* on this side of]

cissy var. of SISSY.

Cistercian /sɪˈstɜːʃ(ə)n/ *–n.* monk or nun of the order founded as a stricter branch of the Benedictines. *–adj.* of the Cistercians. [French *Cîteaux* in France]

cistern /ˈsɪst(ə)n/ *n.* **1** tank for storing water. **2** underground reservoir. [Latin *cista* box, from Greek]

cistus /ˈsɪstəs/ *n.* shrub with large white or red flowers. [Latin from Greek]

citadel /ˈsɪtəd(ə)l/ *n.* fortress, usu. on high ground, protecting or dominating a city. [French *citadelle*]

citation /saɪˈteɪʃ(ə)n/ *n.* **1** citing; passage cited. **2** *Mil.* mention in dispatches. **3** description of reasons for an award.

cite *v.* (**-ting**) **1** mention as an example etc. **2** quote (a book etc.) in support. **3** *Mil.* mention in dispatches. **4** summon to appear in court. [Latin *cieo* set in motion]

citizen /'sɪtɪz(ə)n/ *n.* **1** member of a State, either native or naturalized. **2** inhabitant of a city. **3** *US* civilian. □ **citizenry** *n.* **citizenship** *n.* [Anglo-French: related to CITY]

citizen's band *n.* system of local intercommunication by individuals on special radio frequencies.

citrate /'sɪtreɪt/ *n.* a salt of citric acid.

citric /'sɪtrɪk/ *adj.* derived from citrus fruit.

citric acid *n.* sharp-tasting acid in citrus fruits.

citron /'sɪtrən/ *n.* **1** tree with large lemon-like fruits. **2** this fruit. [French from Latin CITRUS]

citronella /,sɪtrə'nelə/ *n.* **1** a fragrant oil. **2** grass from S. Asia yielding it.

citrus /'sɪtrəs/ *n.* (*pl.* **-es**) **1** tree of a group including the lemon, orange, and grapefruit. **2** (in full **citrus fruit**) fruit of such a tree. [Latin]

city /'sɪtɪ/ *n.* (*pl.* **-ies**) **1** large town, strictly one created by charter and containing a cathedral. **2** (**the City**) **a** part of London governed by the Lord Mayor and Corporation. **b** business part of this. **c** commercial circles. [Latin *civitas*: related to CIVIC]

city-state *n.* esp. *hist.* city that with its surrounding territory forms an independent State.

civet /'sɪvɪt/ *n.* **1** (in full **civet-cat**) cat-like animal of Central Africa. **2** strong musky perfume obtained from it. [French ultimately from Arabic]

civic /'sɪvɪk/ *adj.* **1** of a city. **2** of citizens or citizenship. □ **civically** *adv.* [Latin *civis* citizen]

civic centre *n.* **1** area where municipal offices etc. are situated. **2** the offices themselves.

civics *n.pl.* (usu. treated as *sing.*) the study of the rights and duties of citizenship.

civil /'sɪv(ə)l/ *adj.* **1** of or belonging to citizens. **2** of ordinary citizens; non-military. **3** polite, obliging, not rude. **4** *Law* concerning private rights and not criminal offences. **5** (of the length of a day, year, etc.) fixed by custom or law, not natural or astronomical. □ **civilly** *adv.* [Latin *civilis*: related to CIVIC]

civil defence *n.* organizing of civilians for protection during wartime attacks.

civil disobedience *n.* refusal to comply with certain laws as a peaceful protest.

civil engineer *n.* one who designs or maintains roads, bridges, dams, etc.

civilian /sɪ'vɪlɪən/ –*n.* person not in the armed services or police force. –*adj.* of or for civilians.

civility /sɪ'vɪlɪtɪ/ *n.* (*pl.* **-ies**) **1** politeness. **2** act of politeness. [Latin: related to CIVIL]

civilization /,sɪvɪlaɪ'zeɪʃ(ə)n/ *n.* (also **-isation**) **1** advanced stage or system of social development. **2** peoples of the world that are regarded as having this. **3** a people or nation (esp. of the past) regarded as an element of social evolution (*Inca civilization*).

civilize /'sɪvɪ,laɪz/ *v.* (also **-ise**) (**-zing** or **-sing**) **1** bring out of a barbarous or primitive stage of society. **2** enlighten; refine and educate. [French: related to CIVIL]

civil liberty *n.* (often in *pl.*) freedom of action subject to the law.

civil list *n.* annual allowance voted by Parliament for the royal family's household expenses.

civil marriage *n.* one solemnized without religious ceremony.

civil rights *n.pl.* rights of citizens to freedom and equality.

civil servant *n.* member of the civil service.

civil service *n.* branches of State administration, excluding military and judicial branches and elected politicians.

civil war *n.* war between citizens of the same country.

civvies *n.pl. slang* civilian clothes. [abbreviation]

Civvy Street *n. slang* civilian life. [abbreviation]

Cl *symb.* chlorine.

cl *abbr.* centilitre(s).

clack –*v.* **1** make a sharp sound as of boards struck together. **2** chatter. –*n.* clacking noise or talk. [imitative]

clad *adj.* **1** clothed. **2** provided with cladding. [past part. of CLOTHE]

cladding *n.* covering or coating on a structure or material etc.

cladistics /klə'dɪstɪks/ *n.pl.* (usu. treated as *sing.*) *Biol.* method of classifying animals and plants on the basis of shared characteristics. [Greek *klados* branch]

claim –*v.* **1** state, declare, assert. **2** demand as one's due or property. **3** represent oneself as having or achieving (*claim victory*). **4** (foll. by *to* + infin.) profess. **5** have as an achievement or consequence (*fire claimed two victims*). **6** (of a thing) deserve (attention etc.). –*n.* **1** demand or request for a thing considered one's due (*lay claim to*; *put in a claim*). **2** (foll. by *to, on*) right or title to a thing. **3** assertion. **4** thing claimed. [Latin *clamo* call out]

claimant *n.* person making a claim, esp. in a lawsuit, or claiming State benefit.

clairvoyance /kleə'vɔɪəns/ *n.* supposed faculty of perceiving the future or things beyond normal sensory perception. □ **clairvoyant** *n.* & *adj.* [French: related to CLEAR, *voir* see]

clam –*n.* edible bivalve mollusc. –*v.* (**-mm-**) (foll. by *up*) *colloq.* refuse to talk. [related to CLAMP[1]]

clamber –*v.* climb laboriously using hands and feet. –*n.* difficult climb. [from CLIMB]

clammy /'klæmɪ/ *adj.* (**-ier, -iest**) unpleasantly damp and sticky. □ **clammily** *adv.* **clamminess** *n.* [*clam* to daub]

clamour /'klæmə(r)/ (*US* **clamor**) –*n.* **1** loud or vehement shouting or noise. **2** protest, demand. –*v.* **1** make a clamour. **2** utter with a clamour. □ **clamorous** *adj.* [Latin: related to CLAIM]

clamp[1] –*n.* **1** device, esp. a brace or band of iron etc., for strengthening or holding things together. **2** device for immobilizing an illegally parked vehicle. –*v.* **1** strengthen or fasten with a clamp; fix firmly. **2** immobilize (a vehicle) with a clamp. □ **clamp down** (usu. foll. by *on*) become stricter (about); suppress. [Low German or Dutch]

clamp[2] *n.* potatoes etc. stored under straw or earth. [Dutch: related to CLUMP]

clamp-down *n.* sudden policy of suppression.

clan *n.* **1** group of people with a common ancestor, esp. in the Scottish Highlands. **2** large family as a social group. **3** group with a strong common interest. [Gaelic]

clandestine /klæn'destɪn/ *adj.* surreptitious, secret. [Latin]

clang –*n.* loud resonant metallic sound. –*v.* (cause to) make a clang. [imitative: cf. Latin *clango* resound]

clanger *n. slang* mistake, blunder.

clangour /'klæŋgə(r)/ *n.* (*US* **clangor**) prolonged clanging. □ **clangorous** *adj.*

clank –*n.* sound as of metal on metal. –*v.* (cause to) make a clank. [imitative]

clannish *adj.* often *derog.* (of a family or group) associating closely with each other; inward-looking.

clansman *n.* (*fem.* **clanswoman**) member or fellow-member of a clan.

clap[1] –*v.* (**-pp-**) **1 a** strike the palms of one's hands together, esp. repeatedly as applause. **b** strike (the hands) together in this way. **2** applaud thus. **3** put or place quickly or with determination (*clapped him in prison*; *clap a tax on whisky*). **4** (foll. by *on*) give a friendly slap (*clapped him on the back*). –*n.* **1** act of clapping, esp. as applause. **2** explosive sound, esp. of thunder. **3** slap, pat. □ **clap eyes on** *colloq.* see. [Old English]

clap[2] *n. coarse slang* venereal disease, esp. gonorrhoea. [French]

clapped out *adj. slang* worn out; exhausted.

clapper *n.* tongue or striker of a bell. □ **like the clappers** *slang* very fast or hard.

clapperboard *n.* device in film-making of hinged boards struck together to synchronize the starting of picture and sound machinery.

claptrap *n.* insincere or pretentious talk, nonsense.

claque /klæk/ *n.* group of people hired to applaud. [French]

claret /'klærət/ *n.* **1** red wine, esp. from Bordeaux. **2** purplish-red. [French: related to CLARIFY]

clarify /'klærɪ,faɪ/ *v.* (**-ies, -ied**) **1** make or become clearer. **2 a** free (liquid etc.) from impurities. **b** make transparent. □ **clarification** /-fɪ'keɪʃ(ə)n/ *n.* [Latin: related to CLEAR]

clarinet /,klærɪ'net/ *n.* woodwind instrument with a single reed. □ **clarinettist** *n.* (*US* **clarinetist**). [French diminutive of *clarine*, a kind of bell]

clarion /'klærɪən/ *n.* **1** clear rousing sound. **2** *hist.* shrill war-trumpet. [Latin: related to CLEAR]

clarity /'klærɪtɪ/ *n.* clearness.

clash –*n.* **1 a** loud jarring sound as of metal objects struck together. **b** collision. **2 a** conflict. **b** discord of colours etc. –*v.* **1** (cause to) make a clashing sound. **2** collide; coincide awkwardly. **3** (often foll. by *with*) **a** come into conflict or be at variance. **b** (of colours) be discordant. [imitative]

clasp /klɑ:sp/ –*n.* **1** device with interlocking parts for fastening. **2 a** embrace. **b** grasp, handshake. **3** bar on a medal-ribbon. –*v.* **1** fasten with or as with a clasp. **2 a** grasp, hold closely. **b** embrace. [Old English]

clasp-knife *n.* folding knife, usu. with a catch to hold the blade open.

class /klɑ:s/ –*n.* **1** any set of persons or things grouped together, or graded or differentiated from others esp. by quality (*first class*; *economy class*). **2** division or order of society (*upper class*). **3** *colloq.* distinction, high quality. **4 a** group of students taught together. **b** occasion when they meet. **c** their course of instruction. **5** division of candidates by merit in an examination. **6** *Biol.* next grouping of organisms below a division or phylum. –*v.* assign to a class or category. □ **in a class of** (or **on**) **its** (or **one's**) **own** unequalled. □ **classless** *adj.* [Latin *classis* assembly]

class-conscious *adj.* aware of social divisions or one's place in them. □ **class-consciousness** *n.*

classic /'klæsɪk/ *–adj.* **1** first-class; of lasting value and importance. **2** very typical (*a classic case*). **3 a** of ancient Greek and Latin literature, art, etc. **b** (of style) simple, harmonious. **4** famous because long-established. *–n.* **1** classic writer, artist, work, or example. **2** (in *pl.*) ancient Greek and Latin. [Latin *classicus*: related to CLASS]

classical *adj.* **1 a** of ancient Greek or Roman literature or art. **b** (of a language) having the form used by ancient standard authors. **2** (of music) serious or conventional, or of the period from *c.*1750–1800. **3** restrained in style. □ **classicality** /-'kælɪtɪ/ *n.* **classically** *adv.*

classicism /'klæsɪ,sɪz(ə)m/ *n.* **1** following of a classic style. **2** classical scholarship. **3** ancient Greek or Latin idiom. □ **classicist** *n.*

classify /'klæsɪ,faɪ/ *v.* **(-ies, -ied) 1 a** arrange in classes or categories. **b** assign to a class or category. **2** designate as officially secret or not for general disclosure. □ **classifiable** *adj.* **classification** /-fɪ'keɪʃ(ə)n/ *n.* **classificatory** /-'keɪtərɪ/ *adj.* [French: related to CLASS]

classmate *n.* person in the same class at school.

classroom *n.* room where a class of students is taught.

classy *adj.* **(-ier, -iest)** *colloq.* superior, stylish. □ **classily** *adv.* **classiness** *n.*

clatter *–n.* sound as of hard objects struck together. *–v.* (cause to) make a clatter. [Old English]

clause /klɔ:z/ *n.* **1** *Gram.* part of a sentence, including a subject and predicate. **2** single statement in a treaty, law, contract, etc. □ **clausal** *adj.* [Latin *clausula*: related to CLOSE]

Clause 28 *n.* clause in the Local Government Bill (and later Act) banning local authorities from promoting homosexuality.

claustrophobia /,klɔ:strə'fəʊbɪə/ *n.* abnormal fear of confined places. □ **claustrophobic** *adj.* [Latin *claustrum* CLOISTER, -PHOBIA]

clavichord /'klævɪ,kɔ:d/ *n.* small keyboard instrument with a very soft tone. [medieval Latin: related to CLAVICLE]

clavicle /'klævɪk(ə)l/ *n.* collar-bone. [Latin *clavis* key]

claw *–n.* **1 a** pointed nail on an animal's foot. **b** foot armed with claws. **2** pincers of a shellfish. **3** device for grappling, holding, etc. *–v.* scratch, maul, or pull with claws or fingernails. [Old English]

claw back *v.* regain laboriously or gradually.

claw-hammer *n.* hammer with one side of the head forked for extracting nails.

clay *n.* **1** stiff sticky earth, used for making bricks, pottery, etc. **2** *poet.* substance of the human body. □ **clayey** *adj.* [Old English]

claymore /'kleɪmɔ:(r)/ *n. hist.* Scottish two-edged broadsword. [Gaelic, = great sword]

clay pigeon *n.* breakable disc thrown up from a trap as a target for shooting.

clean *–adj.* **1** free from dirt or impurities, unsoiled. **2** clear; unused; pristine (*clean air; clean page*). **3** not obscene or indecent. **4** attentive to personal hygiene and cleanliness. **5** complete, clear-cut. **6** showing no record of crime, disease, etc. **7** fair (*a clean fight*). **8** streamlined; well-formed. **9** adroit, skilful. **10** (of a nuclear weapon) producing relatively little fallout. *–adv.* **1** completely, outright, simply. **2** in a clean manner. *–v.* make or become clean. *–n.* act or process of cleaning. □ **clean out 1** clean thoroughly. **2** *slang* empty or deprive (esp. of money). **clean up 1 a** clear away (a mess). **b** (also *absol.*) put (things) tidy. **c** make (oneself) clean. **2** restore order or morality to. **3** *slang* acquire as or make a profit. **come clean** *colloq.* confess fully. **make a clean breast of** see BREAST. [Old English]

clean bill of health *n.* declaration that there is no disease or defect.

clean-cut *adj.* **1** sharply outlined or defined. **2** (of a person) clean and tidy.

cleaner *n.* **1** person employed to clean rooms etc. **2** establishment for cleaning clothes etc. **3** device or substance for cleaning. □ **take a person to the cleaners** *slang* **1** defraud or rob a person. **2** criticize severely.

cleanly[1] *adv.* in a clean way.

cleanly[2] /'klenlɪ/ *adj.* **(-ier, -iest)** habitually clean; with clean habits. □ **cleanliness** *n.*

cleanse /klenz/ *v.* **(-sing)** make clean or pure. □ **cleanser** *n.*

clean-shaven *adj.* without beard or moustache.

clean sheet *n.* (also **clean slate**) freedom from commitments or imputations; removal of these from one's record.

clean-up *n.* act of cleaning up.

clear *–adj.* **1** free from dirt or contamination. **2** (of weather, the sky, etc.) not dull. **3** transparent. **4 a** easily perceived; distinct; evident (*a clear voice; it is clear that*). **b** easily understood. **5** discerning readily and accurately (*clear mind*). **6** confident, convinced. **7** (of a conscience) free from guilt. **8** (of a road etc.) unobstructed. **9 a** net, without deduction. **b** complete (*three clear days*). **10** (often foll. by *of*) free, unhampered; unencumbered. –*adv.* **1** clearly. **2** completely (*got clear away*). **3** apart, out of contact (*keep clear*). –*v.* **1** make or become clear. **2** (often foll. by *of*) make or become free from obstruction etc. **3** (often foll. by *of*) show (a person) to be innocent. **4** approve (a person etc.) for a special duty, access, etc. **5** pass over or by, safely or without touching. **6** make (an amount of money) as a net gain or to balance expenses. **7** pass (a cheque) through a clearing-house. **8** pass through (customs etc.). **9** disappear (*mist cleared*). □ **clear the air** remove suspicion, tension, etc. **clear away 1** remove (esp. dishes etc.). **2** disappear. **clear the decks** prepare for action. **clear off** *colloq.* go away. **clear out 1** empty, tidy by emptying. **2** remove. **3** *colloq.* go away. **clear up 1** tidy up. **2** solve. **3** (of weather) become fine. **4** disappear (*acne has cleared up*). **clear a thing with** get approval or authorization for it from (a person). **in the clear** free from suspicion or difficulty. □ **clearly** *adj.* **clearness** *n.* [Latin *clarus*]

clearance *n.* **1** removal of obstructions etc. **2** space allowed for the passing of two objects or parts in machinery etc. **3** special authorization. **4 a** clearing by customs. **b** certificate showing this. **5** clearing of cheques. **6** clearing out.

clear-cut *adj.* sharply defined.

clear-headed *adj.* thinking clearly, sensible.

clearing *n.* open area in a forest.

clearing bank *n.* bank which is a member of a clearing-house.

clearing-house *n.* **1** bankers' establishment where cheques and bills are exchanged, only the balances being paid in cash. **2** agency for collecting and distributing information etc.

clear-out *n.* tidying by emptying and sorting.

clear-sighted *adj.* seeing, thinking, or understanding clearly.

clear-up *n.* **1** tidying up. **2** (usu. *attrib.*) solving of crimes (*clear-up rates*).

clearway *n.* main road (other than a motorway) on which vehicles may not normally stop.

cleat *n.* **1** piece of metal, wood, etc., bolted on for fastening ropes to, or to strengthen woodwork etc. **2** projecting piece on a spar, gangway, etc. to prevent slipping. [Old English]

cleavage /ˈkliːvɪdʒ/ *n.* **1** hollow between a woman's breasts. **2** division, splitting. **3** line along which rocks etc. split.

cleave[1] *v.* (**-ving**; *past* **clove** or **cleft** or **cleaved**; *past part.* **cloven** or **cleft** or **cleaved**) *literary* **1** chop or break apart; split, esp. along the grain or line of cleavage. **2** make one's way through (air or water). [Old English]

cleave[2] *v.* (**-ving**) (foll. by *to*) *literary* stick fast; adhere. [Old English]

cleaver *n.* butcher's heavy chopping tool.

clef *n. Mus.* symbol indicating the pitch of notes on a staff. [Latin *clavis* key]

cleft[1] *adj.* split, partly divided. [past part. of CLEAVE[1]]

cleft[2] *n.* split, fissure. [Old English: related to CLEAVE[1]]

cleft lip *n.* congenital cleft in the upper lip.

cleft palate *n.* congenital split in the roof of the mouth.

clematis /ˈklemətɪs/ *n.* climbing plant with white, pink, or purple flowers. [Greek]

clement /ˈklemənt/ *adj.* **1** (of weather) mild. **2** merciful. □ **clemency** *n.* [Latin *clemens*]

clementine /ˈkleməntiːn/ *n.* small tangerine-like citrus fruit. [French]

clench –*v.* **1** close (the teeth, fingers, etc.) tightly. **2** grasp firmly. –*n.* clenching action; clenched state. [Old English]

clerestory /ˈklɪəˌstɔːrɪ/ *n.* (*pl.* **-ies**) upper row of windows in a cathedral or large church, above the level of the aisle roofs. [*clear storey*]

clergy /ˈklɜːdʒɪ/ *n.* (*pl.* **-ies**) (usu. treated as *pl.*) those ordained for religious duties. [French (related to CLERIC) and Church Latin]

clergyman *n.* member of the clergy.

cleric /ˈklerɪk/ *n.* member of the clergy. [Greek *klērikos* from *klēros* lot, heritage]

clerical *adj.* **1** of clergy or clergymen. **2** of or done by clerks.

clerical collar *n.* stiff upright white collar fastening at the back.

clerihew /ˈklerɪˌhjuː/ *n.* short comic biographical verse in two rhyming couplets. [E. *Clerihew* Bentley, name of its inventor]

clerk /klɑːk/ –*n.* **1** person employed to keep records etc. **2** secretary or agent of a local council, court, etc. **3** lay officer of a church. –*v.* work as clerk. [Old English and French: related to CLERIC]

clever /ˈklevə(r)/ *adj.* (**-er, -est**) **1** skilful, talented; quick to understand and learn. **2** adroit, dexterous. **3** ingenious. □ **cleverly** *adv.* **cleverness** *n.* [Old English]

cliché /'kli:ʃeɪ/ *n.* **1** hackneyed phrase or opinion. **2** metal casting of a stereotype or electrotype. □ **clichéd** *adj.* (also **cliché'd**). [French]

click *–n.* slight sharp sound. *–v.* **1** (cause to) make a click. **2** *colloq.* **a** become clear or understood. **b** be popular. **c** (foll. by *with*) strike up a rapport. [imitative]

client /'klaɪənt/ *n.* **1** person using the services of a lawyer, architect, or other professional person. **2** customer. [Latin *cliens*]

clientele /ˌkli:ɒn'tel/ *n.* **1** clients collectively. **2** customers. [French and Latin: related to CLIENT]

cliff *n.* steep rock-face, esp. on a coast. [Old English]

cliff-hanger *n.* story etc. with a strong element of suspense.

climacteric /klaɪ'mæktərɪk/ *n.* period of life when fertility and sexual activity are in decline. [Greek: related to CLIMAX]

climate /'klaɪmɪt/ *n.* **1** prevailing weather conditions of an area. **2** region with particular weather conditions. **3** prevailing trend of opinion or feeling. □ **climatic** /-'mætɪk/ *adj.* **climatically** /-'mætɪkəlɪ/ *adv.* [Greek *klima*]

climax /'klaɪmæks/ *–n.* **1** event or point of greatest intensity or interest; culmination. **2** orgasm. *–v. colloq.* reach or bring to a climax. □ **climactic** /-'mæktɪk/ *adj.* [Greek, = ladder]

climb /klaɪm/ *–v.* **1** (often foll. by *up*) ascend, mount, go or come up. **2** grow up a wall etc. by clinging or twining. **3** progress, esp. in social rank. *–n.* **1** ascent by climbing. **2** hill etc. climbed or to be climbed. □ **climb down 1** descend, esp. using hands. **2** withdraw from a stance taken up in an argument etc. □ **climber** *n.* [Old English]

climb-down *n.* withdrawal from a stance taken up.

climbing-frame *n.* structure of joined bars etc. for children to climb on.

clime *n. literary* **1** region. **2** climate. [Latin: related to CLIMATE]

clinch *–v.* **1** confirm or settle (an argument, bargain, etc.) conclusively. **2** (of boxers etc.) become too closely engaged. **3** secure (a nail or rivet) by driving the point sideways when through. *–n.* **1 a** clinching action. **b** clinched state. **2** *colloq.* embrace. [var. of CLENCH]

clincher *n. colloq.* point or remark that settles an argument etc.

cling *v.* (*past* and *past part.* **clung**) **1** (often foll. by *to*) adhere. **2** (foll. by *to*) be unwilling to give up; be emotionally dependent on (a habit, idea, friend, etc.). **3** (often foll. by *to*) maintain grasp; keep hold; resist separation. □ **clingy** *adj.* (**-ier**, **-iest**). [Old English]

cling film *n.* thin transparent plastic covering for food.

clinic /'klɪnɪk/ *n.* **1** private or specialized hospital. **2** place or occasion for giving medical treatment or specialist advice. **3** gathering at a hospital bedside for medical teaching. [Greek *klinē* bed]

clinical *adj.* **1** of or for the treatment of patients. **2** dispassionate, coolly detached. **3** (of a room, building, etc.) bare, functional. □ **clinically** *adv.* [Greek: related to CLINIC]

clinical death *n.* death judged by professional observation of a person's condition.

clink[1] *–n.* sharp ringing sound. *–v.* (cause to) make a clink. [Dutch: imitative]

clink[2] *n. slang* prison. [origin unknown]

clinker *n.* **1** mass of slag or lava. **2** stony residue from burnt coal. [Dutch: related to CLINK[1]]

clinker-built *adj.* (of a boat) having external planks overlapping downwards and secured with clinched nails. [*clink*, Northern English var. of CLINCH]

clip[1] *–n.* **1** device for holding things together or for attaching something. **2** piece of jewellery fastened by a clip. **3** set of attached cartridges for a firearm. *–v.* (**-pp-**) fix with a clip. [Old English]

clip[2] *–v.* (**-pp-**) **1** cut (hair, wool, etc.) short with shears or scissors. **2** trim or remove the hair or wool of. **3** *colloq.* hit smartly. **4 a** omit (a letter etc.) from a word. **b** omit letters or syllables of (words uttered). **5** punch a hole in (a ticket) to show it has been used. **6** cut from a newspaper etc. **7** *slang* swindle, rob. *–n.* **1** act of clipping. **2** *colloq.* smart blow. **3** sequence from a motion picture. **4** yield of wool etc. **5** *colloq.* speed, esp. rapid. [Old Norse]

clipboard *n.* small board with a spring clip for holding papers etc.

clip-joint *n. slang* club etc. charging exorbitant prices.

clip-on *adj.* attached by a clip.

clipper *n.* **1** (usu. in *pl.*) instrument for clipping hair etc. **2** *hist.* fast sailing-ship.

clipping *n.* piece clipped, esp. from a newspaper.

clique /kli:k/ *n.* small exclusive group of people. □ **cliquey** *adj.* (**cliquier**, **cliquiest**). **cliquish** *adj.* [French]

clitoris /'klɪtərɪs/ *n.* small erectile part of the female genitals at the upper end of the vulva. □ **clitoral** *adj.* [Latin from Greek]

Cllr. *abbr.* Councillor.

cloak –*n.* **1** outdoor usu. long and sleeveless over-garment. **2** covering (*cloak of snow*). –*v.* **1** cover with a cloak. **2** conceal, disguise. □ **under the cloak of** using as pretext. [ultimately from medieval Latin *clocca* bell]

cloak-and-dagger *adj.* involving intrigue and espionage.

cloakroom *n.* **1** room where outdoor clothes or luggage may be left. **2** *euphem.* lavatory.

clobber[1] *v. slang* **1** hit; beat up. **2** defeat. **3** criticize severely. [origin unknown]

clobber[2] *n. slang* clothing, belongings. [origin unknown]

cloche /klɒʃ/ *n.* **1** small translucent cover for protecting outdoor plants. **2** (in full **cloche hat**) woman's close-fitting bell-shaped hat. [French, = bell, medieval Latin *clocca*]

clock[1] –*n.* **1** instrument for measuring and showing time. **2 a** measuring device resembling this. **b** *colloq.* speedometer, taximeter, or stopwatch. **3** *slang* person's face. **4** seed-head of the dandelion. –*v.* **1** *colloq.* **a** (often foll. by *up*) attain or register (a stated time, distance, or speed). **b** time (a race) with a stopwatch. **2** *slang* hit. □ **clock in** (or **on**) register one's arrival at work. **clock off** (or **out**) register one's departure from work. **round the clock** all day and (usu.) night. [medieval Latin *clocca* bell]

clock[2] *n.* ornamental pattern on the side of a stocking or sock near the ankle. [origin unknown]

clockwise *adj.* & *adv.* in a curve corresponding in direction to that of the hands of a clock.

clockwork *n.* **1** mechanism like that of a clock, with a spring and gears. **2** (*attrib.*) driven by clockwork. □ **like clockwork** smoothly, regularly, automatically.

clod *n.* lump of earth, clay, etc. [var. of CLOT]

cloddish *adj.* loutish, foolish, clumsy.

clodhopper *n.* (usu. in *pl.*) *colloq.* large heavy shoe.

clog –*n.* shoe with a thick wooden sole. –*v.* (**-gg-**) **1** (often foll. by *up*) obstruct or become obstructed; choke. **2** impede. [origin unknown]

cloister –*n.* **1** covered walk round a quadrangle, esp. in a college or ecclesiastical building. **2** monastic life or seclusion. –*v.* seclude. □ **cloistered** *adj.* **cloistral** *adj.* [Latin *claustrum*: related to CLOSE[2]]

clomp var. of CLUMP *v.* 2.

clone –*n.* **1 a** group of organisms produced asexually from one stock or ancestor. **b** one such organism. **2** *colloq.* person or thing regarded as identical to another. –*v.* (**-ning**) propagate as a clone. □ **clonal** *adj.* [Greek *klōn* twig]

clonk –*n.* abrupt heavy sound of impact. –*v.* **1** make this sound. **2** *colloq.* hit. [imitative]

close[1] /kləʊs/ –*adj.* **1** (often foll. by *to*) situated at a short distance or interval. **2 a** having a strong or immediate relation or connection (*close friend*). **b** in intimate friendship or association. **c** corresponding almost exactly (*close resemblance*). **3** in or almost in contact (*close combat*). **4** dense, compact, with no or only slight intervals. **5** (of a contest etc.) in which competitors are almost equal. **6** leaving no gaps or weaknesses, rigorous (*close reasoning*). **7** concentrated, searching. **8** (of air etc.) stuffy, humid. **9** closed, shut. **10** limited to certain persons etc. (*close corporation*). **11** hidden, secret; secretive. **12** niggardly. –*adv.* at only a short distance or interval. –*n.* **1** street closed at one end. **2** precinct of a cathedral. □ **at close quarters** very close together. □ **closely** *adv.* **closeness** *n.* [Latin *clausus* from *claudo* shut]

close[2] /kləʊz/ –*v.* (**-sing**) **1 a** shut. **b** block up. **2** bring or come to an end. **3** end the day's business. **4** bring or come closer or into contact. **5** make (an electric circuit etc.) continuous. –*n.* conclusion, end. □ **close down** (of a shop etc.) discontinue business. **close in 1** enclose. **2** come nearer. **3** (of days) get successively shorter. **close up 1** (often foll. by *to*) move closer. **2** shut. **3** block up. **4** (of an aperture) grow smaller. [Latin: related to CLOSE[1]]

closed book *n.* subject one does not understand.

closed-circuit *adj.* (of television) transmitted by wires to a restricted set of receivers.

closed shop *n.* business etc. where employees must belong to a specified trade union.

close harmony *n.* harmony in which the notes of a chord are close together.

close-knit *adj.* tightly interlocked; closely united in friendship.

close season *n.* season when the killing of game etc. is illegal.

close shave *n.* (also **close thing**) *colloq.* narrow escape.

closet /'klɒzɪt/ –*n.* **1** small room. **2** cupboard. **3** = WATER-CLOSET. **4** (*attrib.*) secret (*closet homosexual*). –*v.* (**-t-**) shut away, esp. in private conference or study. [French diminutive: related to CLOSE[2]]

close-up *n.* photograph etc. taken at close range.

closure /'kləʊʒə(r)/ *n.* **1** closing. **2** closed state. **3** procedure for ending a debate and taking a vote. [Latin: related to CLOSE[2]]

clot *–n.* **1** thick mass of coagulated liquid etc., esp. of blood. **2** *colloq.* foolish person. *–v.* (**-tt-**) form into clots. [Old English]

cloth *n.* **1** woven or felted material. **2** piece of this, esp. for a particular purpose; tablecloth, dishcloth, etc. **3** fabric for clothes. **4 a** status, esp. of the clergy, as shown by clothes. **b** (prec. by *the*) the clergy. [Old English]

clothe /kləʊð/ *v.* (**-thing**; *past* and *past part.* **clothed** or *formal* **clad**) **1** put clothes on; provide with clothes. **2** cover as with clothes. [Old English]

clothes /kləʊðz/ *n.pl.* **1** garments worn to cover the body and limbs. **2** bedclothes. [Old English]

clothes-horse *n.* frame for airing washed clothes.

clothes-line *n.* rope etc. on which clothes are hung to dry.

clothes-peg *n.* clip etc. for securing clothes to a clothes-line.

clothier /'kləʊðɪə(r)/ *n.* seller of men's clothes.

clothing /'kləʊðɪŋ/ *n.* clothes collectively.

clotted cream *n.* thick cream obtained by slow scalding.

cloud *–n.* **1** visible mass of condensed watery vapour floating high above the ground. **2** mass of smoke or dust. **3** (foll. by *of*) mass of insects etc. moving together. **4** state of gloom, trouble, or suspicion. *–v.* **1** cover or darken with clouds or gloom or trouble. **2** (often foll. by *over*, *up*) become overcast or gloomy. **3** make unclear. □ **on cloud nine** *colloq.* extremely happy. **under a cloud** out of favour, under suspicion. **with one's head in the clouds** day-dreaming. □ **cloudless** *adj.* [Old English]

cloudburst *n.* sudden violent rainstorm.

cloud chamber *n.* device containing vapour for tracking the paths of charged particles, X-rays, and gamma rays.

cloud-cuckoo-land *n.* fanciful or ideal place. [translation of Greek *Nephelokokkugia* in Aristophanes' *Birds*]

cloudy *adj.* (**-ier, -iest**) **1** (of the sky, weather) covered with clouds, overcast. **2** not transparent; unclear. □ **cloudily** *adv.* **cloudiness** *n.*

clout *–n.* **1** heavy blow. **2** *colloq.* influence, power of effective action. **3** *dial.* piece of cloth or clothing. *–v.* hit hard. [Old English]

clove[1] *n.* dried bud of a tropical plant used as a spice. [Latin *clavus* nail (from its shape)]

clove[2] *n.* small segment of a compound bulb, esp. of garlic. [Old English: related to CLEAVE[1]]

clove[3] *past* of CLEAVE[1].

clove hitch *n.* knot by which a rope is secured to a spar etc. [*clove*, old past part. of CLEAVE[1]]

cloven /'kləʊv(ə)n/ *adj.* split, partly divided. [past part. of CLEAVE[1]]

cloven hoof *n.* (also **cloven foot**) divided hoof, esp. of oxen, sheep, or goats, or of the Devil.

clover *n.* trefoil fodder plant. □ **in clover** in ease and luxury. [Old English]

clown *–n.* **1** comic entertainer, esp. in a circus. **2** foolish or playful person. *–v.* (often foll. by *about*, *around*) behave like a clown. [origin uncertain]

cloy *v.* satiate or sicken with sweetness, richness, etc. [obsolete *acloy* from Anglo-French: related to ENCLAVE]

club *–n.* **1** heavy stick with a thick end, esp. as a weapon. **2** stick with a head used in golf. **3** association of persons meeting periodically for a shared activity. **4** organization or premises offering members social amenities, meals, temporary residence, etc. **5 a** playing-card of the suit denoted by a black trefoil. **b** (in *pl.*) this suit. **6** commercial organization offering subscribers special deals (*book club*). *–v.* (**-bb-**) **1** beat with or as with a club. **2** (foll. by *together*, *with*) combine, esp. to raise a sum of money for a purpose. [Old Norse]

clubbable *adj.* sociable; fit for club membership.

club class *n.* class of fare on an aircraft etc. designed for business travellers.

club-foot *n.* congenitally deformed foot.

clubhouse *n.* premises of a (usu. sporting) club.

clubland *n.* area where there are many nightclubs.

club-root *n.* disease of cabbages etc. with swelling at the base of the stem.

club sandwich *n.* sandwich with two layers of filling between three slices of toast or bread.

cluck *–n.* guttural cry like that of a hen. *–v.* emit cluck(s). [imitative]

clue /klu:/ *–n.* **1** fact or idea that serves as a guide, or suggests a line of inquiry, in a problem or investigation. **2** piece of evidence etc. in the detection of a crime. **3** verbal formula as a hint to what is to be inserted in a crossword. *–v.* (**clues, clued, cluing** or **clueing**) provide a clue to. □ **clue in** (or **up**) *slang* inform.

not have a clue *colloq.* be ignorant or incompetent. [var. of Old English *clew*]

clueless *adj. colloq.* ignorant, stupid.

clump –*n.* (foll. by *of*) cluster or mass, esp. of trees. –*v.* **1 a** form a clump. **b** heap or plant together. **2** (also **clomp**) walk with a heavy tread. [Low German or Dutch]

clumsy /'klʌmzɪ/ *adj.* (**-ier, -iest**) **1** awkward in movement or shape; ungainly. **2** difficult to handle or use. **3** tactless. □ **clumsily** *adv.* **clumsiness** *n.* [obsolete *clumse* be numb with cold]

clung *past* and *past part.* of CLING.

clunk –*n.* dull sound as of thick pieces of metal meeting. –*v.* make such a sound. [imitative]

cluster –*n.* close group or bunch of similar people or things growing or occurring together. –*v.* **1** bring into, come into, or be in cluster(s). **2** (foll. by *round, around*) gather. [Old English]

clutch[1] –*v.* **1** seize eagerly; grasp tightly. **2** (foll. by *at*) try desperately to seize. –*n.* **1** tight grasp. **2** (in *pl.*) grasping hands; cruel or relentless grasp or control. **3 a** (in a vehicle) device for connecting and disconnecting the engine and the transmission. **b** pedal operating this. [Old English]

clutch[2] *n.* **1** set of eggs for hatching. **2** brood of chickens. [Old Norse, = hatch]

clutch bag *n.* slim flat handbag without handles.

clutter –*n.* **1** crowded and untidy collection of things. **2** untidy state. –*v.* (often foll. by *up, with*) crowd untidily, fill with clutter. [related to CLOT]

Cm *symb.* curium.

cm *abbr.* centimetre(s).

CMG *abbr.* Companion (of the Order) of St Michael and St George.

CND *abbr.* Campaign for Nuclear Disarmament.

CO *abbr.* Commanding Officer.

Co *symb.* cobalt.

Co. *abbr.* **1** company. **2** county.

co- *prefix* added to: **1** nouns, with the sense 'joint, mutual, common' (*co-author*; *coequality*). **2** adjectives and adverbs, with the sense 'jointly, mutually' (*coequal*). **3** verbs, with the sense 'together with another or others' (*cooperate*). [var. of COM-]

c/o *abbr.* care of.

coach –*n.* **1** single-decker bus, usu. comfortably equipped for long journeys. **2** railway carriage. **3** closed horse-drawn carriage. **4 a** instructor or trainer in a sport. **b** private tutor. –*v.* train or teach as a coach. [French from Magyar]

coachload *n.* group of tourists etc. taken by coach.

coachman *n.* driver of a horse-drawn carriage.

coachwork *n.* bodywork of a road or rail vehicle.

coagulate /kəʊ'ægjʊ,leɪt/ *v.* (**-ting**) **1** change from a fluid to a semisolid. **2** clot, curdle. □ **coagulant** *n.* **coagulation** /-'leɪʃ(ə)n/ *n.* [Latin *coagulum* rennet]

coal *n.* **1** hard black rock, mainly carbonized plant matter, found underground and used as a fuel. **2** piece of this, esp. one that is burning. □ **coals to Newcastle** something brought to a place where it is already plentiful. **haul** (or **call**) **over the coals** reprimand. [Old English]

coalesce /,kəʊə'les/ *v.* (**-cing**) come together and form a whole. □ **coalescence** *n.* **coalescent** *adj.* [Latin *alo* nourish]

coalface *n.* exposed working surface of coal in a mine.

coalfield *n.* extensive area yielding coal.

coal gas *n.* mixed gases formerly extracted from coal and used for lighting and heating.

coalition /,kəʊə'lɪʃ(ə)n/ *n.* **1** temporary alliance, esp. of political parties. **2** fusion into one whole. [medieval Latin: related to COALESCE]

coalman *n.* man who carries or delivers coal.

coalmine *n.* mine in which coal is dug. □ **coalminer** *n.*

coal-scuttle *n.* container for coal for a domestic fire.

coal tar *n.* thick black oily liquid distilled from coal and used as a source of benzene.

coal-tit *n.* small greyish bird with a black head.

coaming /'kəʊmɪŋ/ *n.* raised border round a ship's hatches etc. to keep out water. [origin unknown]

coarse *adj.* **1** rough or loose in texture; made of large particles. **2** lacking refinement; crude, obscene. □ **coarsely** *adv.* **coarseness** *n.* [origin unknown]

coarse fish *n.* freshwater fish other than salmon and trout.

coarsen *v.* make or become coarse.

coast –*n.* border of land near the sea; seashore. –*v.* **1** ride or move, usu. downhill, without the use of power. **2** make progress without much effort. **3** sail along the coast. □ **the coast is clear** there is no danger of being observed or caught. □ **coastal** *adj.* [Latin *costa* side]

coaster *n.* **1** ship that travels along the coast. **2** small tray or mat for a bottle or glass.

coastguard *n.* **1** member of a group of people employed to keep watch on coasts to save life, prevent smuggling, etc. **2** such a group.

coastline *n.* line of the seashore, esp. with regard to its shape.

coat *–n.* **1** outer garment with sleeves, usu. extending below the hips; overcoat or jacket. **2** animal's fur or hair. **3** covering of paint etc. laid on a surface at one time. *–v.* **1** (usu. foll. by *with*, *in*) cover with a coat or layer. **2** (of paint etc.) form a covering to. [French from Germanic]

coat-hanger see HANGER 2.

coating *n.* **1** layer of paint etc. **2** material for coats.

coat of arms *n.* heraldic bearings or shield of a person, family, or corporation.

coat of mail *n.* jacket covered with mail.

coat-tail *n.* each of the flaps formed by the back of a tailcoat.

coax *v.* **1** persuade gradually or by flattery. **2** (foll. by *out of*) obtain (a thing from a person) thus. **3** manipulate (a thing) carefully or slowly. [obsolete *cokes* a fool]

coaxial /kəʊ'æksɪəl/ *adj.* **1** having a common axis. **2** *Electr.* (of a cable or line) transmitting by means of two concentric conductors separated by an insulator.

cob *n.* **1** roundish lump. **2** domed loaf. **3** = CORN-COB. **4** large hazelnut. **5** sturdy riding-horse with short legs. **6** male swan. [origin unknown]

cobalt /'kəʊbɔ:lt/ *n.* **1** silvery-white metallic element. **2 a** pigment made from this. **b** its deep-blue colour. [German, probably = *Kobold* demon in mines]

cobber *n. Austral.* & *NZ colloq.* companion, friend. [origin uncertain]

cobble[1] /'kɒb(ə)l/ *–n.* (in full **cobblestone**) small rounded stone used for paving. *–v.* (**-ling**) pave with cobbles. [from COB]

cobble[2] /'kɒb(ə)l/ *v.* (**-ling**) **1** mend or patch up (esp. shoes). **2** (often foll. by *together*) join or assemble roughly. [from COBBLER]

cobbler *n.* **1** person who mends shoes professionally. **2** stewed fruit or meat topped with scones. **3** (in *pl.*) *slang* nonsense. [origin unknown]

COBOL /'kəʊbɒl/ *n.* computer language for use in commerce. [*co*mmon *b*usiness *o*riented *l*anguage]

cobra /'kəʊbrə/ *n.* venomous hooded snake of Africa and Asia. [Latin *colubra* snake]

cobweb *n.* **1** fine network spun by a spider from liquid it secretes. **2** thread of this. □ **cobwebby** *adj.* [obsolete *coppe* spider]

coca /'kəʊkə/ *n.* **1** S. American shrub. **2** its dried leaves, chewed as a stimulant. [Spanish from Quechua]

cocaine /kəʊ'keɪn/ *n.* drug from coca, used as a local anaesthetic and as a stimulant.

coccyx /'kɒksɪks/ *n.* (*pl.* **coccyges** /-,dʒi:z/) small triangular bone at the base of the spinal column. [Greek, = cuckoo (from shape of its bill)]

cochineal /,kɒtʃɪ'ni:l/ *n.* **1** scarlet dye used esp. for colouring food. **2** insects whose dried bodies yield this. [Latin *coccinus* scarlet, from Greek]

cock[1] *–n.* **1** male bird, esp. of the domestic fowl. **2** *slang* (as a form of address) friend; fellow. **3** *coarse slang* penis. **4** *slang* nonsense. **5 a** firing lever in a gun, raised to be released by the trigger. **b** cocked position of this. **6** tap or valve controlling flow. *–v.* **1** raise or make upright or erect. **2** turn or move (the eye or ear) attentively or knowingly. **3** set aslant; turn up the brim of (a hat). **4** raise the cock of (a gun). □ **at half cock** only partly ready. **cock a snook** see SNOOK[1]. **cock up** *slang* bungle; make a mess of. [Old English and French]

cock[2] *n.* conical heap of hay or straw. [perhaps from Scandinavian]

cockade /kɒ'keɪd/ *n.* rosette etc. worn in the hat as a badge. [French: related to COCK[1]]

cock-a-doodle-doo *n.* cock's crow.

cock-a-hoop *adj.* exultant.

cock-a-leekie *n.* Scottish soup of boiling fowl and leeks.

cock-and-bull story *n.* absurd or incredible account.

cockatoo /,kɒkə'tu:/ *n.* crested parrot. [Dutch from Malay]

cockchafer /'kɒk,tʃeɪfə(r)/ *n.* large pale-brown beetle. [from COCK[1]]

cock crow *n.* dawn.

cocker *n.* (in full **cocker spaniel**) small spaniel with a silky coat. [related to COCK[1]]

cockerel /'kɒkər(ə)l/ *n.* young cock. [diminutive of COCK[1]]

cock-eyed *adj. colloq.* **1** crooked, askew. **2** absurd, not practical. [from COCK[1]]

cock-fight *n.* fight between cocks as sport.

cockle /'kɒk(ə)l/ *n.* **1 a** edible bivalve shellfish. **b** its shell. **2** (in full **cockle-shell**) small shallow boat. **3** pucker or wrinkle in paper, glass, etc. □ **warm the cockles of one's heart** make one contented. [French *coquille* from Greek: related to CONCH]

cockney /ˈkɒknɪ/ *–n.* (*pl.* **-s**) **1** native of London, esp. of the East End. **2** dialect or accent used there. *–adj.* of cockneys or their dialect. [*cokeney* 'cock's egg']

cockpit *n.* **1 a** compartment for the pilot (and crew) of an aircraft or spacecraft. **b** driver's seat in a racing car. **c** space for the helmsman in some yachts. **2** arena of war or other conflict. **3** place for cock-fights.

cockroach /ˈkɒkrəʊtʃ/ *n.* flat dark-brown beetle-like insect infesting kitchens, bathrooms, etc. [Spanish *cucaracha*]

cockscomb /ˈkɒkskəʊm/ *n.* crest of a cock.

cocksure /ˌkɒkˈʃɔː(r)/ *adj.* arrogantly confident. [from COCK[1]]

cocktail /ˈkɒkteɪl/ *n.* **1** drink made of various spirits, fruit juices, etc. **2** appetizer containing shellfish or fruit. **3** any hybrid mixture. [origin unknown]

cocktail dress *n.* short evening dress worn at a drinks party.

cocktail stick *n.* small pointed stick for serving an olive, cherry, etc.

cock-up *n. slang* muddle or mistake.

cocky *adj.* (**-ier**, **-iest**) *colloq.* conceited, arrogant. □ **cockily** *adv.* **cockiness** *n.* [from COCK[1]]

coco /ˈkəʊkəʊ/ *n.* (*pl.* **-s**) coconut palm. [Portuguese and Spanish, = grimace]

cocoa /ˈkəʊkəʊ/ *n.* **1** powder made from crushed cacao seeds, often with other ingredients. **2** drink made from this. [altered from CACAO]

cocoa bean *n.* cacao seed.

cocoa butter *n.* fatty substance obtained from the cocoa bean.

coconut /ˈkəʊkəˌnʌt/ *n.* large brown seed of the coco, with a hard shell and edible white lining enclosing milky juice.

coconut matting *n.* matting made of fibre from coconut husks.

coconut shy *n.* fairground sideshow where balls are thrown to dislodge coconuts.

cocoon /kəˈkuːn/ *–n.* **1** silky case spun by insect larvae for protection as pupae. **2** protective covering. *–v.* wrap or coat in a cocoon. [Provençal *coca* shell]

cocotte /kəˈkɒt/ *n.* small fireproof dish for cooking and serving an individual portion. [French]

COD *abbr.* cash (*US* collect) on delivery.

cod[1] *n.* (*pl.* same) large sea fish. [origin unknown]

cod[2] *slang –n.* **1** parody. **2** hoax. *–v.* (**-dd-**) **1** perform a hoax. **2** parody. [origin unknown]

cod[3] *n. slang* nonsense. [abbreviation of CODSWALLOP]

coda /ˈkəʊdə/ *n.* **1** *Mus.* final additional passage of a piece or movement. **2** concluding section of a ballet. [Latin *cauda* tail]

coddle /ˈkɒd(ə)l/ *v.* (**-ling**) **1** treat as an invalid; protect attentively; pamper. **2** cook (an egg) in water below boiling point. □ **coddler** *n.* [a dialect form of *caudle* invalids' gruel]

code *–n.* **1** system of words, letters, symbols, etc., used to represent others for secrecy or brevity. **2** system of prearranged signals used to ensure secrecy in transmitting messages. **3** *Computing* piece of program text. **4** systematic set of laws etc. **5** prevailing standard of moral behaviour. *–v.* (**-ding**) put into code. [Latin CODEX]

codeine /ˈkəʊdiːn/ *n.* alkaloid derived from morphine, used to relieve pain. [Greek *kōdeia* poppy-head]

codependency /ˌkəʊdɪˈpendənsɪ/ *n.* addiction to a supportive role in a relationship. □ **codependent** *adj.* & *n.*

codex /ˈkəʊdeks/ *n.* (*pl.* **codices** /-dɪˌsiːz/) **1** ancient manuscript text in book form. **2** collection of descriptions of drugs etc. [Latin, = tablet, book]

codfish *n.* (*pl.* same) = COD[1].

codger /ˈkɒdʒə(r)/ *n.* (usu. in **old codger**) *colloq.* person, esp. a strange one. [origin uncertain]

codicil /ˈkəʊdɪsɪl/ *n.* addition to a will. [Latin diminutive of CODEX]

codify /ˈkəʊdɪˌfaɪ/ *v.* (**-ies**, **-ied**) arrange (laws etc.) systematically into a code. □ **codification** /-fɪˈkeɪʃ(ə)n/ *n.* **codifier** *n.*

codling[1] /ˈkɒdlɪŋ/ *n.* (also **codlin**) **1** a kind of cooking apple. **2** moth whose larva feeds on apples. [Anglo-French *quer de lion* lion-heart]

codling[2] *n.* small codfish.

cod-liver oil *n.* oil from cod livers, rich in vitamins D and A.

codpiece *n. hist.* bag or flap at the front of a man's breeches. [*cod* scrotum]

codswallop /ˈkɒdzˌwɒləp/ *n. slang* nonsense. [origin unknown]

coed /kəʊˈed/ *colloq. –n.* **1** school for both sexes. **2** esp. *US* female pupil of a coed school. *–adj.* coeducational. [abbreviation]

coeducation /ˌkəʊedjuːˈkeɪʃ(ə)n/ *n.* education of pupils of both sexes together. □ **coeducational** *adj.*

coefficient /ˌkəʊɪˈfɪʃ(ə)nt/ *n.* **1** *Math.* quantity placed before and multiplying an algebraic expression. **2** *Physics* multiplier or factor by which a property is measured (*coefficient of expansion*). [related to CO-, EFFICIENT]

coelacanth /ˈsiːləˌkænθ/ *n.* large sea fish formerly thought to be extinct. [Greek *koilos* hollow, *akantha* spine]

coelenterate /siː'lentəˌreɪt/ *n.* marine animal with a simple tube-shaped or cup-shaped body, e.g. jellyfish, corals, and sea anemones. [Greek *koilos* hollow, *enteron* intestine]

coeliac disease /'siːlɪˌæk/ *n.* disease of the small intestine, brought on by contact with dietary gluten. [Latin *coeliacus* from Greek *koilia* belly]

coenobite /'siːnəˌbaɪt/ *n.* (*US* **cenobite**) member of a monastic community. [Greek *koinos bios* common life]

coequal /kəʊ'iːkw(ə)l/ *adj.* & *n. archaic* or *literary* equal.

coerce /kəʊ'ɜːs/ *v.* (**-cing**) persuade or restrain by force. □ **coercible** *adj.* **coercion** /-'ɜːʃ(ə)n/ *n.* **coercive** *adj.* [Latin *coerceo* restrain]

coeval /kəʊ'iːv(ə)l/ *formal* –*adj.* of the same age; existing at the same time; contemporary. –*n.* coeval person or thing. □ **coevally** *adv.* [Latin *aevum* age]

coexist /ˌkəʊɪg'zɪst/ *v.* (often foll. by *with*) **1** exist together. **2** (esp. of nations) exist in mutual tolerance of each other's ideologies etc. □ **coexistence** *n.* **coexistent** *adj.*

coextensive /ˌkəʊɪk'stensɪv/ *adj.* extending over the same space or time.

C. of E. *abbr.* Church of England.

coffee /'kɒfɪ/ *n.* **1 a** drink made from roasted and ground beanlike seeds of a tropical shrub. **b** cup of this. **2 a** the shrub. **b** its seeds. **3** pale brown. [Turkish from Arabic]

coffee bar *n.* bar or café serving coffee and light refreshments from a counter.

coffee-mill *n.* small machine for grinding roasted coffee beans.

coffee morning *n.* morning gathering, esp. for charity, at which coffee is served.

coffee shop *n.* small informal restaurant, esp. in a hotel or department store.

coffee-table *n.* small low table.

coffee-table book *n.* large lavishly illustrated book.

coffer *n.* **1** large strong box for valuables. **2** (in *pl.*) treasury, funds. **3** sunken panel in a ceiling etc. [Latin *cophinus* basket]

coffer-dam *n.* watertight enclosure pumped dry to permit work below the waterline, e.g. building bridges etc. or repairing a ship.

coffin /'kɒfɪn/ *n.* box in which a corpse is buried or cremated. [Latin: related to COFFER]

cog *n.* **1** each of a series of projections on the edge of a wheel or bar transferring motion by engaging with another series. **2** unimportant member of an organization etc. [probably Scandinavian]

cogent /'kəʊdʒ(ə)nt/ *adj.* (of an argument etc.) convincing, compelling. □ **cogency** *n.* **cogently** *adv.* [Latin *cogo* drive]

cogitate /'kɒdʒɪˌteɪt/ *v.* (**-ting**) ponder, meditate. □ **cogitation** /-'teɪʃ(ə)n/ *n.* **cogitative** /-tətɪv/ *adj.* [Latin *cogito*]

cognac /'kɒnjæk/ *n.* high-quality brandy, properly that distilled in Cognac in W. France.

cognate /'kɒgneɪt/ –*adj.* **1** related to or descended from a common ancestor. **2** (of a word) having the same linguistic family or derivation. –*n.* **1** relative. **2** cognate word. [Latin *cognatus*]

cognate object *n. Gram.* object related in origin and sense to its verb (as in *live a good life*).

cognition /kɒg'nɪʃ(ə)n/ *n.* **1** knowing, perceiving, or conceiving as an act or faculty distinct from emotion and volition. **2** result of this. □ **cognitional** *adj.* **cognitive** /'kɒg-/ *adj.* [Latin *cognitio*: related to COGNIZANCE]

cognizance /'kɒgnɪz(ə)ns/ *n. formal* **1** knowledge or awareness; perception. **2** sphere of observation or concern. **3** *Heraldry* distinctive device or mark. [Latin *cognosco* get to know]

cognizant *adj.* (foll. by *of*) *formal* having knowledge or being aware of.

cognomen /kɒg'nəʊmen/ *n.* **1** nickname. **2** ancient Roman's third or fourth name designating a branch of a family, as in Marcus Tullius *Cicero*, or as an epithet, as in P. Cornelius Scipio *Africanus*. [Latin]

cognoscente /ˌkɒnjə'ʃentɪ/ *n.* (*pl.* **-ti** /-tɪ/) connoisseur. [Italian]

cog-wheel *n.* wheel with cogs.

cohabit /kəʊ'hæbɪt/ *v.* (**-t-**) (esp. of an unmarried couple) live together as husband and wife. □ **cohabitation** /-'teɪʃ(ə)n/ *n.* **cohabitee** /-'tiː/ *n.* [Latin *habito* dwell]

cohere /kəʊ'hɪə(r)/ *v.* (**-ring**) **1** (of parts or a whole) stick together, remain united. **2** (of reasoning etc.) be logical or consistent. [Latin *haereo haes-* stick]

coherent *adj.* **1** intelligible and articulate. **2** (of an argument etc.) consistent; easily followed. **3** cohering. **4** *Physics* (of waves) having a constant phase relationship. □ **coherence** *n.* **coherently** *adv.*

cohesion /kəʊ'hiːʒ(ə)n/ *n.* **1 a** sticking together. **b** tendency to cohere. **2** *Chem.* force with which molecules cohere. □ **cohesive** *adj.*

cohort /'kəʊhɔːt/ *n.* **1** ancient Roman military unit, one-tenth of a legion. **2** band of warriors. **3 a** persons banded together. **b** group of persons with a common statistical characteristic. [Latin]

coif *n. hist.* close-fitting cap. [Latin *cofia* helmet]

coiff /kwɑːf/ *v.* (usu. as **coiffed** *adj.*) dress or arrange (the hair). [French *coiffer*]

coiffeur /kwɑːˈfɜː(r)/ *n.* (*fem.* ***coiffeuse*** /-ˈfɜːz/) hairdresser. [French]

coiffure /kwɑːˈfjʊə(r)/ *n.* hairstyle. [French]

coil –*v.* **1** arrange or be arranged in spirals or concentric rings. **2** move sinuously. –*n.* **1** coiled arrangement. **2** coiled length of rope etc. **3** single turn of something coiled. **4** flexible loop as a contraceptive device in the womb. **5** coiled wire for the passage of an electric current and acting as an inductor. [Latin: related to COLLECT[1]]

coin –*n.* **1** stamped disc of metal as official money. **2** (*collect.*) metal money. –*v.* **1** make (coins) by stamping. **2** make (metal) into coins. **3** invent (esp. a new word or phrase). □ **coin money** make much money quickly. [Latin *cuneus* wedge]

coinage *n.* **1** coining. **2 a** coins. **b** system of coins in use. **3** invention, esp. of a word.

coin-box *n.* **1** telephone operated by inserting coins. **2** receptacle for these.

coincide /ˌkəʊɪnˈsaɪd/ *v.* (**-ding**) **1** occur at the same time. **2** occupy the same portion of space. **3** (often foll. by *with*) agree or be identical. [Latin: related to INCIDENT]

coincidence /kəʊˈɪnsɪd(ə)ns/ *n.* **1** coinciding. **2** remarkable concurrence of events etc. apparently by chance. □ **coincident** *adj.*

coincidental /kəʊˌɪnsɪˈdent(ə)l/ *adj.* in the nature of or resulting from a coincidence. □ **coincidentally** *adv.*

coir /ˈkɔɪə(r)/ *n.* coconut fibre used for ropes, matting, etc. [Malayalam *kāyar* cord]

coition /kəʊˈɪʃ(ə)n/ *n.* = COITUS. [Latin *coitio* from *eo* go]

coitus /ˈkəʊɪtəs/ *n.* sexual intercourse. □ **coital** *adj.* [Latin: related to COITION]

coitus interruptus /ˌɪntəˈrʌptəs/ *n.* sexual intercourse with withdrawal of the penis before ejaculation.

coke[1] –*n.* solid substance left after gases have been extracted from coal. –*v.* (**-king**) convert (coal) into coke. [dial. *colk* core]

coke[2] *n. slang* cocaine. [abbreviation]

Col. *abbr.* Colonel.

col *n.* depression in a chain of mountains. [Latin *collum* neck]

col. *abbr.* column.

col- see COM-.

cola /ˈkəʊlə/ *n.* (also **kola**) **1** W. African tree bearing seeds containing caffeine. **2** carbonated drink usu. flavoured with these. [West African]

colander /ˈkʌləndə(r)/ *n.* perforated vessel used to strain off liquid in cookery. [Latin *colo* strain]

cold /kəʊld/ –*adj.* **1** of or at a low temperature. **2** not heated; cooled after heat. **3** feeling cold. **4** lacking ardour, friendliness, or affection. **5 a** depressing, uninteresting. **b** (of colour) suggestive of cold. **6 a** dead. **b** *colloq.* unconscious. **7** (of a scent in hunting) grown faint. **8** (in games) far from finding what is sought. –*n.* **1 a** prevalence of low temperature. **b** cold weather or environment. **2** infection of the nose or throat with sneezing, catarrh, etc. –*adv.* unrehearsed. □ **in cold blood** without emotion, deliberately. **out in the cold** ignored, neglected. **throw** (or **pour**) **cold water on** be discouraging about. □ **coldly** *adv.* **coldness** *n.* [Old English]

cold-blooded *adj.* **1** having a body temperature varying with that of the environment. **2** callous; deliberately cruel. □ **cold-bloodedly** *adv.* **cold-bloodedness** *n.*

cold call –*n.* marketing call on a person who has previously not shown interest in the product. –*v.* visit or telephone (a person) in this way.

cold chisel *n.* chisel for cutting metal, stone, or brick.

cold comfort *n.* poor consolation.

cold cream *n.* ointment for cleansing and softening the skin.

cold feet *n.pl. colloq.* loss of nerve.

cold frame *n.* unheated glass-topped frame for growing small plants.

cold fusion *n.* nuclear fusion at room temperature, esp. as a possible energy source.

cold-hearted *adj.* lacking sympathy or kindness. □ **cold-heartedly** *adv.* **cold-heartedness** *n.*

cold shoulder –*n.* (prec. by *the*) intentional unfriendliness. –*v.* (**cold-shoulder**) be deliberately unfriendly towards.

cold sore *n.* inflammation and blisters in and around the mouth, caused by a virus infection.

cold storage *n.* **1** storage in a refrigerator. **2** temporary putting aside (of an idea etc.), postponement.

cold sweat *n.* sweating induced by fear or illness.

cold table *n.* selection of dishes of cold food.

cold turkey *n. slang* abrupt withdrawal from addictive drugs.

cold war *n.* hostility between nations without actual fighting.

cole *n.* (usu. in *comb.*) cabbage. [Latin *caulis*]

coleopteron /ˌkɒlɪ'ɒptəˌrɒn/ *n.* insect with front wings serving as sheaths, e.g. the beetle and weevil. □ **coleopterous** *adj.* [Greek *koleon* sheath, *pteron* wing]

coleslaw /'kəʊlslɔ:/ *n.* dressed salad of sliced raw cabbage etc. [from COLE, Dutch *sla* salad]

coleus /'kəʊlɪəs/ *n.* plant with variegated leaves. [Greek *koleon* sheath]

coley /'kəʊlɪ/ *n.* (*pl.* **-s**) any of several fish used as food, e.g. the rock-salmon. [origin uncertain]

colic /'kɒlɪk/ *n.* severe spasmodic abdominal pain. □ **colicky** *adj.* [Latin: related to COLON[2]]

colitis /kə'laɪtɪs/ *n.* inflammation of the lining of the colon.

collaborate /kə'læbəˌreɪt/ *v.* (**-ting**) (often foll. by *with*) **1** work together. **2** cooperate with an enemy. □ **collaboration** /-'reɪʃ(ə)n/ *n.* **collaborative** /-rətɪv/ *adj.* **collaborator** *n.* [Latin: related to LABOUR]

collage /'kɒlɑ:ʒ/ *n.* form or work of art in which various materials are fixed to a backing. [French, = gluing]

collagen /'kɒlədʒ(ə)n/ *n.* protein found in animal connective tissue, yielding gelatin on boiling. [Greek *kolla* glue]

collapse /kə'læps/ *—n.* **1** falling down or in of a structure; folding up; giving way. **2** sudden failure of a plan etc. **3** physical or mental breakdown; exhaustion *—v.* (**-sing**) **1** (cause to) undergo collapse. **2** *colloq.* lie or sit down and relax, esp. after prolonged effort. **3** fold up. □ **collapsible** *adj.* [Latin *labor laps-* slip]

collar /'kɒlə(r)/ *—n.* **1** neckband, upright or turned over. **2** band of leather etc. round an animal's neck. **3** band or ring or pipe in machinery. **4** piece of meat rolled up and tied. *—v.* **1** capture, seize. **2** *colloq.* accost. **3** *slang* appropriate. [Latin *collum* neck]

collar-bone *n.* bone joining the breastbone and shoulder-blade.

collate /kə'leɪt/ *v.* (**-ting**) **1** assemble and arrange systematically. **2** compare (texts, statements, etc.). □ **collator** *n.* [Latin: related to CONFER]

collateral /kə'lætər(ə)l/ *—n.* **1** security pledged as a guarantee for the repayment of a loan. **2** person having the same ancestor as another but by a different line. *—adj.* **1** descended from the same ancestor but by a different line. **2** side by side; parallel. **3 a** additional but subordinate. **b** contributory. **c** connected but aside from the main subject, course, etc. □ **collaterally** *adv.* [Latin: related to LATERAL]

collation /kə'leɪʃ(ə)n/ *n.* **1** collating. **2** thing collated. **3** light meal. [Latin: related to CONFER]

colleague /'kɒli:g/ *n.* fellow worker, esp. in a profession or business. [Latin *collega*]

collect[1] /kə'lekt/ *—v.* **1** bring or come together; assemble, accumulate. **2** systematically seek and acquire, esp. as a hobby. **3** obtain (contributions etc.) from a number of people. **4** call for; fetch. **5 a** *refl.* regain control of oneself. **b** concentrate (one's thoughts etc.). **c** (as **collected** *adj.*) not perturbed or distracted. *—adj.* & *adv. US* (of a telephone call, parcel, etc.) to be paid for by the receiver. [Latin *lego lect-* pick]

collect[2] /'kɒlekt/ *n.* short prayer of the Anglican or Roman Catholic Church. [Latin *collecta*: related to COLLECT[1]]

collectable /kə'lektɪb(ə)l/ (also **collectible**) *—adj.* worth collecting. *—n.* item sought by collectors.

collection /kə'lekʃ(ə)n/ *n.* **1** collecting or being collected. **2** things collected, esp. systematically. **3** money collected, esp. at a meeting or church service.

collective /kə'lektɪv/ *—adj.* of, by, or relating to a group or society as a whole; joint; shared. *—n.* **1** cooperative enterprise; its members. **2** = COLLECTIVE NOUN. □ **collectively** *adv.*

collective bargaining *n.* negotiation of wages etc. by an organized body of employees.

collective farm *n.* jointly-operated esp. State-owned amalgamation of several smallholdings.

collective noun *n.* singular noun denoting a collection or number of individuals (e.g. *assembly*, *family*, *troop*).

collective ownership *n.* ownership of land etc., by all for the benefit of all.

collectivism *n.* theory and practice of collective ownership of land and the means of production. □ **collectivist** *n.* & *adj.*

collectivize *v.* (also **-ise**) (**-zing** or **-sing**) organize on the basis of collective ownership. □ **collectivization** /-'zeɪʃ(ə)n/ *n.*

collector *n.* **1** person who collects things of interest. **2** person who collects money etc. due.

collector's item *n.* (also **collector's piece**) thing of interest to collectors.

colleen /kɒ'li:n/ *n. Ir.* girl. [Irish *cailín*]

college /'kɒlɪdʒ/ *n.* **1** establishment for further, higher, or professional education. **2** college premises (*lived in college*). **3** students and teachers in a college. **4** school. **5** organized body of persons with shared functions and privileges. [Latin: related to COLLEAGUE]

collegiate /kə'li:dʒɪət/ *adj.* **1** of, or constituted as, a college; corporate. **2** (of a university) consisting of different colleges.

collegiate church *n.* church endowed for a chapter of canons but without a bishop's see.

collide /kə'laɪd/ *v.* (**-ding**) (often foll. by *with*) come into collision or conflict. [Latin *collido -lis-* clash]

collie /'kɒlɪ/ *n.* sheepdog of an orig. Scottish breed. [perhaps from *coll* COAL]

collier /'kɒlɪə(r)/ *n.* **1** coalminer. **2 a** coal ship. **b** member of its crew. [from COAL]

colliery *n.* (*pl.* **-ies**) coalmine and its buildings.

collision /kə'lɪʒ(ə)n/ *n.* **1** violent impact of a moving body with another or with a fixed object. **2** clashing of interests etc. [Latin: related to COLLIDE]

collocate /'kɒlə,keɪt/ *v.* (**-ting**) juxtapose (a word etc.) with another. □ **collocation** /-'keɪʃ(ə)n/ *n.* [Latin: related to LOCUS]

colloid /'kɒlɔɪd/ *n.* **1** substance consisting of ultramicroscopic particles. **2** mixture of such particles dispersed in another substance. □ **colloidal** /kə'lɔɪd(ə)l/ *adj.* [Greek *kolla* glue]

colloquial /kə'ləʊkwɪəl/ *adj.* of ordinary or familiar conversation, informal. □ **colloquially** *adv.* [Latin: related to COLLOQUY]

colloquialism *n.* **1** colloquial word or phrase. **2** use of these.

colloquium /kə'ləʊkwɪəm/ *n.* (*pl.* **-s** or **-quia**) academic conference or seminar. [Latin: related to COLLOQUY]

colloquy /'kɒləkwɪ/ *n.* (*pl.* **-quies**) *literary* conversation, talk. [Latin *loquor* speak]

collude /kə'lu:d/ *v.* (**-ding**) conspire together. □ **collusion** *n.* **collusive** *adj.* [Latin *ludo lus-* play]

collywobbles /'kɒlɪ,wɒb(ə)lz/ *n.pl. colloq.* **1** rumbling or pain in the stomach. **2** apprehensive feeling. [from COLIC, WOBBLE]

cologne /kə'ləʊn/ *n.* eau-de-Cologne or similar toilet water. [abbreviation]

colon[1] /'kəʊlən/ *n.* punctuation mark (:), used esp. to mark illustration or antithesis. [Greek, = clause]

colon[2] /'kəʊlən/ *n.* lower and greater part of the large intestine. [Latin from Greek]

colonel /'kɜ:n(ə)l/ *n.* army officer in command of a regiment, ranking next below brigadier. □ **colonelcy** *n.* (*pl.* **-ies**). [Italian *colonnello*: related to COLUMN]

colonial /kə'ləʊnɪəl/ —*adj.* **1** of a colony or colonies. **2** of colonialism. —*n.* inhabitant of a colony.

colonialism *n.* **1** policy of acquiring or maintaining colonies. **2** *derog.* exploitation of colonies. □ **colonialist** *n.* & *adj.*

colonist /'kɒlənɪst/ *n.* settler in or inhabitant of a colony.

colonize /'kɒlə,naɪz/ *v.* (also **-ise**) (**-zing** or **-sing**) **1** establish a colony in. **2** join a colony. □ **colonization** /-'zeɪʃ(ə)n/ *n.*

colonnade /,kɒlə'neɪd/ *n.* row of columns, esp. supporting an entablature or roof. □ **colonnaded** *adj.* [French: related to COLUMN]

colony /'kɒlənɪ/ *n.* (*pl.* **-ies**) **1 a** settlement or settlers in a new country, fully or partly subject to the mother country. **b** their territory. **2 a** people of one nationality, occupation, etc., esp. forming a community in a city. **b** separate or segregated group (*nudist colony*). **3** group of animals, plants, etc., living close together. [Latin *colonia* farm]

colophon /'kɒlə,f(ə)n/ *n.* **1** publisher's imprint, esp. on the title-page. **2** tailpiece in a manuscript or book, giving the writer's or printer's name, date, etc. [Greek, = summit]

color etc. *US* var. of COLOUR etc.

Colorado beetle /,kɒlə'rɑ:dəʊ/ *n.* yellow and black beetle, with larva destructive to the potato plant. [*Colorado* in US]

coloration /,kʌlə'reɪʃ(ə)n/ *n.* (also **colouration**) **1** appearance as regards colour. **2** act or mode of colouring. [Latin: related to COLOUR]

coloratura /,kɒlərə'tʊərə/ *n.* **1** elaborate ornamentation of a vocal melody. **2** soprano skilled in this. [Italian: related to COLOUR]

colossal /kə'lɒs(ə)l/ *adj.* **1** huge. **2** *colloq.* splendid. □ **colossally** *adv.* [related to COLOSSUS]

colossus /kə'lɒsəs/ *n.* (*pl.* **-ssi** /-saɪ/ or **-ssuses**) **1** statue much bigger than life size. **2** gigantic or remarkable person etc. **3** imperial power personified. [Latin from Greek]

colostomy /kə'lɒstəmɪ/ *n.* (*pl.* **-ies**) operation on the colon to make an opening in the abdominal wall to provide an artificial anus. [from COLON[2]]

colour /'kʌlə(r)/ (*US* **color**) —*n.* **1** sensation produced on the eye by rays of light when resolved as by a prism into different wavelengths. **2** one, or any mixture, of the constituents into which light can be separated as in a spectrum or rainbow, sometimes including (loosely) black and white. **3** colouring substance, esp. paint. **4** use of all colours in photography etc. **5 a** pigmentation of the skin, esp. when dark. **b** this as ground for discrimination. **6** ruddiness of complexion. **7** (in *pl.*) appearance or aspect (*saw them in their true colours*). **8** (in *pl.*) **a** coloured ribbon or uniform etc. worn to signify membership of a school, club, team, etc. **b** flag of a regiment or ship. **9** quality, mood, or variety in music, literature, etc. **10** show of reason; pretext (*lend colour to*; *under colour of*). *–v.* **1** apply colour to, esp. by painting, dyeing, etc. **2** influence. **3** misrepresent, exaggerate. **4** take on colour; blush. □ **show one's true colours** reveal one's true character or intentions. [Latin *color*]

colouration var. of COLORATION.

colour bar *n.* racial discrimination against non-White people.

colour-blind *adj.* unable to distinguish certain colours. □ **colour-blindness** *n.*

colour code *–n.* use of colours as a means of identification. *–v.* **(colour-code)** identify by means of a colour code.

coloured (*US* **colored**) *–adj.* **1** having colour. **2** **(Coloured) a** wholly or partly of non-White descent. **b** *S.Afr.* of mixed White and non-White descent. *–n.* **1** **(Coloured) a** Coloured person. **b** *S.Afr.* person of mixed descent speaking Afrikaans or English as his or her mother tongue. **2** (in *pl.*) coloured clothing etc. for washing.

colourful *adj.* (*US* **color-**) **1** full of colour; bright. **2** full of interest; vivid. □ **colourfully** *adv.*

colouring *n.* (*US* **color-**) **1** appearance as regards colour, esp. facial complexion. **2** use or application of colour. **3** substance giving colour.

colourless *adj.* (*US* **color-**) **1** without colour. **2** lacking character or interest.

colour scheme *n.* arrangement of colours, esp. in interior design.

colour-sergeant *n.* senior sergeant of an infantry company.

colour supplement *n.* magazine with colour printing, as a supplement to a newspaper.

colposcopy /kɒl'pɒskəpɪ/ *n.* examination of the vagina and neck of the womb. □ **colposcope** /'kɒlpə,skəʊp/ *n.* [Greek *kolpos* womb]

colt /kəʊlt/ *n.* **1** young male horse. **2** *Sport* inexperienced player. □ **coltish** *adj.* [Old English]

colter *US* var. of COULTER.

coltsfoot *n.* (*pl.* **-s**) wild plant with large leaves and yellow flowers.

columbine /'kɒləm,baɪn/ *n.* garden plant with purple-blue flowers like a cluster of doves. [Latin *columba* dove]

column /'kɒləm/ *n.* **1** pillar, usu. of circular section and with a base and capital. **2** column-shaped object. **3** vertical cylindrical mass of liquid or vapour. **4** vertical division of a printed page. **5** part of a newspaper etc. regularly devoted to a particular subject. **6** vertical row of figures in accounts etc. **7** narrow-fronted arrangement of troops or armoured vehicles in successive lines. □ **columnar** /kə'lʌmnə(r)/ *adj.* **columned** *adj.* [French and Latin]

columnist /'kɒləmnɪst/ *n.* journalist contributing regularly to a newspaper etc.

com- *prefix* (also **co-**, **col-**, **con-**, **cor-**) with, together, jointly, altogether. [Latin *com-*, *cum* with]

■ **Usage** *Com-* is used before *b*, *m*, *p*, and occasionally before vowels and *f*; *co-* esp. before vowels, *h*, and *gn*; *col-* before *l*, *cor-* before *r*, and *con-* before other consonants.

coma /'kəʊmə/ *n.* (*pl.* **-s**) prolonged deep unconsciousness. [Latin from Greek]

comatose /'kəʊmə,təʊs/ *adj.* **1** in a coma. **2** drowsy, sleepy.

comb /kəʊm/ *–n.* **1 a** toothed strip of rigid material for tidying the hair. **b** similar curved decorative strip worn in the hair. **2** thing like a comb, esp. a device for tidying and straightening wool etc. **3** red fleshy crest of a fowl, esp. a cock. **4** honeycomb. *–v.* **1** draw a comb through (the hair). **2** dress (wool etc.) with a comb. **3** *colloq.* search (a place) thoroughly. □ **comb out 1** arrange (the hair) loosely by combing. **2** remove with a comb. **3** search out and get rid of. [Old English]

combat /'kɒmbæt/ *–n.* fight, struggle, contest. *–v.* **(-t-) 1** engage in combat (with). **2** oppose; strive against. [Latin: related to BATTLE]

combatant /'kɒmbət(ə)nt/ *–n.* person engaged in fighting. *–adj.* **1** fighting. **2** for fighting.

combative /'kɒmbətɪv/ *adj.* pugnacious.

combe var. of COOMB.

combination /,kɒmbɪ'neɪʃ(ə)n/ *n.* **1** combining or being combined. **2** combined set of things or people. **3** sequence of numbers or letters used to open a combination lock. **4** motor cycle with a side-car attached. **5** (in *pl.*) single undergarment for the body and legs. [Latin: related to COMBINE]

combination lock *n.* lock that can be opened only by a specific sequence of movements.

combine *–v.* /kəm'baɪn/ **(-ning) 1** join together; unite for a common purpose. **2**

possess (qualities usually distinct) together. **3** form or cause to form a chemical compound. **4** /'kɒmbaɪn/ harvest with a combine harvester. –*n.* /'kɒmbaɪn/ combination of esp. commercial interests. [Latin *bini* a pair]

combine harvester *n.* machine that reaps and threshes in one operation.

combings /'kəʊmɪŋz/ *n.pl.* hairs combed off.

combining form *n.* linguistic element used in combination with another to form a word (e.g. *Anglo-* = English).

combo /'kɒmbəʊ/ *n.* (*pl.* **-s**) *slang* small jazz or dance band. [abbreviation of COMBINATION]

combustible /kəm'bʌstɪb(ə)l/ –*adj.* capable of or used for burning. –*n.* combustible substance. □ **combustibility** /-'bɪlɪtɪ/ *n.* [Latin *comburo -bust-* burn up]

combustion /kəm'bʌstʃ(ə)n/ *n.* **1** burning. **2** development of light and heat from the chemical combination of a substance with oxygen.

come /kʌm/ –*v.* (**-ming**; *past* **came**; *past part.* **come**) **1** move, be brought towards, or reach a place. **2** reach a specified situation or result (*came to no harm*). **3** reach or extend to a specified point. **4** traverse or accomplish (with compl.: *have come a long way*). **5** occur, happen; (of time) arrive in due course (*how did you come to break your leg?*; *the day soon came*). **6** take or occupy a specified position in space or time (*Nero came after Claudius*). **7** become perceptible or known (*it will come to me*). **8** be available (*comes in three sizes*). **9** become (*come loose*). **10** (foll. by *from, of*) **a** be descended from. **b** be the result of (*that comes of complaining*). **11** *colloq.* play the part of; behave like (*don't come the bully with me*). **12** *slang* have an orgasm. **13** (in *subjunctive*) *colloq.* when a specified time is reached (*come next month*). **14** (as *int.*) expressing mild protest or encouragement (*come, it cannot be that bad*). –*n. slang* semen ejaculated. □ **come about** happen. **come across 1** meet or find by chance. **2** *colloq.* be effective or understood; give a specified impression. **come again** *colloq.* **1** make a further effort. **2** (as *imper.*) what did you say? **come along 1** make progress. **2** (as *imper.*) hurry up. **come apart** disintegrate. **come at 1** attack. **2** reach, get access to. **come away 1** become detached. **2** (foll. by *with*) be left with (an impression etc.). **come back 1** return. **2** recur to one's memory. **3** (often foll. by *in*) become fashionable or popular again. **come between 1** interfere with the relationship of. **2** separate. **come by 1** call on a visit. **2** obtain. **come clean** see CLEAN. **come down 1** lose position or wealth. **2** be handed down by tradition. **3** be reduced. **4** (foll. by *against, in favour of*) reach a decision. **5** (foll. by *to*) amount basically. **6** (foll. by *on*) criticize harshly; rebuke, punish. **7** (foll. by *with*) begin to suffer from (a disease). **come for 1** come to collect. **2** attack. **come forward 1** advance. **2** offer oneself for a task, post, etc. **come in 1** enter. **2** take a specified position in a race etc. (*came in third*). **3** become fashionable or seasonable. **4** (with compl.) prove to be (useful etc.) (*came in handy*). **5** have a part to play (*where do I come in?*). **6** be received (*news has just come in*). **7** begin speaking, interrupt. **8** return to base (*come in, number 9*). **9** (foll. by *for*) receive. **come into 1** enter, be brought into (collision, prominence, etc.). **2** receive, esp. as an heir. **come of age** see AGE. **come off 1** (of an action) succeed, occur. **2** fare (badly, well, etc.). **3** be detached or detachable (from). **come off it** *colloq.* expression of disbelief, disapproval, etc. **come on 1** advance. **2** make progress. **3** begin (*came on to rain*). **4** appear on a stage, field of play, etc. **5** (as *imper.*) expressing encouragement or disbelief. **6** = *come upon*. **come out 1** emerge; become known. **2** be published. **3 a** declare oneself. **b** openly declare that one is a homosexual. **4** go on strike. **5** (of a photograph or thing photographed) be produced satisfactorily and clearly. **6** attain a specified result in an examination etc. **7** (of a stain etc.) be removed. **8** make one's début in society etc. **9** (foll. by *in*) become covered with (*came out in spots*). **10** be solved. **11** (foll. by *with*) declare openly; disclose. **come over 1 a** come from some distance to visit etc. **b** come nearer. **2** change sides or one's opinion. **3 a** (of a feeling etc.) overtake or affect (a person). **b** *colloq.* feel suddenly (*came over faint*). **4** appear or sound in a specified way. **come round 1** pay an informal visit. **2** recover consciousness. **3** be converted to another person's opinion. **4** (of a date) recur. **come through** survive. **come to 1** recover consciousness. **2** amount to. **come to one's senses** see SENSE. **come up 1** arise; present itself; be mentioned or discussed. **2** attain position or wealth. **3** (often foll. by *to*) approach. **4** (foll. by *to*) match (a standard etc.). **5** (foll. by *with*) produce (an idea etc.). **6** (of a plant etc.) spring up out of the ground. **7** become brighter (e.g. with polishing). **come up against** be faced with or opposed by. **come upon**

1 meet or find by chance. **2** attack by surprise. [Old English]

comeback *n.* **1** return to a previous (esp. successful) state. **2** *slang* retaliation or retort.

Comecon /'kɒmɪ,kɒn/ *n.* economic association of Socialist countries in E. Europe. [abbreviation of *Council for Mutual Economic Assistance*]

comedian /kə'mi:dɪən/ *n.* **1** humorous entertainer. **2** comedy actor. **3** *slang* buffoon. [French]

comedienne /kə,mi:dɪ'en/ *n.* female comedian. [French feminine]

comedown *n.* **1** loss of status. **2** disappointment.

comedy /'kɒmədɪ/ *n.* (*pl.* **-ies**) **1 a** play, film, etc., of amusing character, usu. with a happy ending. **b** such works as a dramatic genre. **2** humour; amusing aspects. □ **comedic** /kə'mi:dɪk/ *adj.* [Greek: related to COMIC]

comedy of manners *n.* satirical play portraying the social behaviour of the upper classes.

come-hither *attrib. adj. colloq.* flirtatious, inviting.

comely /'kʌmlɪ/ *adj.* (**-ier, -iest**) *literary* handsome, good-looking. □ **comeliness** *n.* [Old English]

come-on *n. slang* enticement.

comer /'kʌmə(r)/ *n.* person who comes as an applicant etc. (*offered it to the first comer*).

comestibles /kə'mestɪb(ə)lz/ *n.pl. formal* or *joc.* food. [French from Latin]

comet /'kɒmɪt/ *n.* hazy object moving in a path about the sun, usu. with a nucleus of ice surrounded by gas and with a tail pointing away from the sun. [Greek *komētes*]

comeuppance /kʌm'ʌpəns/ *n. colloq.* deserved punishment. [*come up*, -ANCE]

comfit /'kʌmfɪt/ *n. archaic* sweet consisting of a nut etc. in sugar. [Latin: related to CONFECTION]

comfort /'kʌmfət/ *–n.* **1 a** state of physical well-being. **b** (usu. in *pl.*) things that make life easy or pleasant. **2** relief of suffering or grief, consolation. **3** person or thing giving consolation. *–v.* soothe in grief; console. [Latin *fortis* strong]

comfortable *adj.* **1** giving ease. **2** free from discomfort; at ease. **3** having an easy conscience. **4 a** having an adequate standard of living; free from financial worry. **b** sufficient (*comfortable income*). **5 a** with a wide margin (*comfortable win*). **b** appreciable (*comfortable margin*). □ **comfortably** *adv.*

comforter *n.* **1** person who comforts. **2** baby's dummy. **3** *archaic* woollen scarf.

comfortless *adj.* **1** dreary, cheerless. **2** without comfort.

comfort station *n. US euphem.* public lavatory.

comfrey /'kʌmfrɪ/ *n.* (*pl.* **-s**) tall bell-flowered plant growing in damp, shady places. [French from Latin]

comfy /'kʌmfɪ/ *adj.* (**-ier, -iest**) *colloq.* comfortable. [abbreviation]

comic /'kɒmɪk/ *–adj.* **1** of or like comedy. **2** funny. *–n.* **1** comedian. **2** periodical in the form of comic strips. □ **comical** *adj.* **comically** *adv.* [Greek *kōmos* revel]

comic strip *n.* sequence of drawings telling a story.

coming /'kʌmɪŋ/ *–attrib. adj.* **1** approaching, next (*the coming week*). **2** of potential importance (*coming man*). *–n.* arrival.

comity /'kɒmɪtɪ/ *n.* (*pl.* **-ies**) *formal* **1** courtesy, friendship. **2 a** association of nations etc. **b** (in full **comity of nations**) mutual recognition by nations of the laws and customs of others. [Latin *comis* courteous]

comma /'kɒmə/ *n.* punctuation mark (,) indicating a pause or break between parts of a sentence etc. [Greek, = clause]

command /kə'mɑ:nd/ *–v.* **1** (often foll. by *to* + infin., or *that* + clause) give a formal order or instruction to. **2** (also *absol.*) have authority or control over. **3** have at one's disposal or within reach (a skill, resources, etc.). **4** deserve and get (sympathy, respect, etc.). **5** dominate (a strategic position) from a superior height; look down over. *–n.* **1** order, instruction. **2** mastery, control, possession. **3** exercise or tenure of authority, esp. naval or military. **4 a** body of troops etc. **b** district under a commander. [Latin: related to MANDATE]

commandant /'kɒmən,dænt/ *n.* commanding officer, esp. of a military academy. [French or Italian or Spanish: related to COMMAND]

commandeer /,kɒmən'dɪə(r)/ *v.* **1** seize (esp. goods) for military use. **2** take arbitrary possession of. [Afrikaans *kommanderen*]

commander /kə'mɑ:ndə(r)/ *n.* **1** person who commands, esp. a naval officer next below captain. **2** (in full **knight commander**) member of a higher class in some orders of knighthood.

commander-in-chief *n.* (*pl.* **commanders-in-chief**) supreme commander, esp. of a nation's forces.

commanding *adj.* **1** exalted, impressive. **2** (of a position) giving a wide view. **3** (of an advantage etc.) substantial (*commanding lead*).

commandment *n.* divine command.

command module *n.* control compartment in a spacecraft.

commando /kə'mɑ:ndəʊ/ *n.* (*pl.* **-s**) **1** unit of shock troops. **2** member of this. [Portuguese: related to COMMAND]

Command Paper *n.* paper laid before Parliament by royal command.

command performance *n.* theatrical or film performance given at royal request.

commemorate /kə'memə,reɪt/ *v.* (**-ting**) **1** preserve in memory by a celebration or ceremony. **2** be a memorial of. □ **commemoration** /-'reɪʃ(ə)n/ *n.* **commemorative** /-rətɪv/ *adj.* [Latin: related to MEMORY]

commence /kə'mens/ *v.* (**-cing**) *formal* begin. [Latin: related to COM-, INITIATE]

commencement *n. formal* beginning.

commend /kə'mend/ *v.* **1** praise. **2** entrust, commit. **3** recommend. □ **commendation** /,kɒmen'deɪʃ(ə)n/ *n.* [Latin: related to MANDATE]

commendable *adj.* praiseworthy. □ **commendably** *adv.*

commensurable /kə'menʃərəb(ə)l/ *adj.* **1** (often foll. by *with, to*) measurable by the same standard. **2** (foll. by *to*) proportionate to. **3** *Math.* (of numbers) in a ratio equal to the ratio of integers. □ **commensurability** /-'bɪlɪtɪ/ *n.* [Latin: related to MEASURE]

commensurate /kə'menʃərət/ *adj.* **1** (usu. foll. by *with*) coextensive. **2** (often foll. by *to, with*) proportionate.

comment /'kɒment/ *—n.* **1** brief critical or explanatory remark or note; opinion. **2** commenting; criticism (*aroused much comment; his art is a comment on society*). *—v.* (often foll. by *on* or *that*) make (esp. critical) remarks. □ **no comment** *colloq.* I decline to answer your question. [Latin]

commentary /'kɒməntərɪ/ *n.* (*pl.* **-ies**) **1** descriptive spoken esp. broadcast account of an event or performance as it happens. **2** set of explanatory notes on a text etc. [Latin]

commentate /'kɒmən,teɪt/ *v.* (**-ting**) act as a commentator.

commentator *n.* **1** person who provides a commentary. **2** person who comments on current events. [Latin]

commerce /'kɒmɜ:s/ *n.* financial transactions, esp. buying and selling; trading. [Latin: related to MERCER]

commercial /kə'mɜ:ʃ(ə)l/ *—adj.* **1** of or engaged in commerce. **2** having financial profit as its primary aim. **3** (of chemicals) for industrial use. *—n.* television or radio advertisement. □ **commercially** *adv.*

commercial broadcasting *n.* broadcasting financed by advertising.

commercialism *n.* **1** commercial practices. **2** emphasis on financial profit.

commercialize *v.* (also **-ise**) (**-zing** or **-sing**) **1** exploit or spoil for profit. **2** make commercial. □ **commercialization** /-'zeɪʃ(ə)n/ *n.*

commercial traveller *n.* firm's representative visiting shops etc. to get orders.

Commie /'kɒmɪ/ *n. slang derog.* Communist. [abbreviation]

commination /,kɒmɪ'neɪʃ(ə)n/ *n. literary* threatening of divine vengeance. □ **comminatory** /'kɒmɪnətərɪ/ *adj.* [Latin: related to MENACE]

commingle /kə'mɪŋg(ə)l/ *v.* (**-ling**) *literary* mingle together.

comminute /'kɒmɪ,nju:t/ *v.* (**-ting**) **1** reduce to small fragments. **2** divide (property) into small portions. □ **comminution** /-'nju:ʃ(ə)n/ *n.* [Latin: related to MINUTE[2]]

comminuted fracture *n.* fracture producing multiple bone splinters.

commiserate /kə'mɪzə,reɪt/ *v.* (**-ting**) (usu. foll. by *with*) express or feel sympathy. □ **commiseration** /-'reɪʃ(ə)n/ *n.* [Latin: related to MISER]

commissar /'kɒmɪ,sɑ:(r)/ *n. hist.* **1** official of the Soviet Communist Party responsible for political education and organization. **2** head of a government department in the USSR. [Latin: related to COMMIT]

commissariat /,kɒmɪ'seərɪət/ *n.* **1** esp. *Mil.* **a** department for the supply of food etc. **b** food supplied. **2** *hist.* government department of the USSR. [related to COMMISSARY]

commissary /'kɒmɪsərɪ/ *n.* (*pl.* **-ies**) **1** deputy, delegate. **2** *US Mil.* store for supplies of food etc. [Latin: related to COMMIT]

commission /kə'mɪʃ(ə)n/ *—n.* **1 a** authority to perform a task etc. **b** person(s) entrusted with such authority. **c** task etc. given to such person(s). **2** order for something to be produced specially. **3 a** warrant conferring the rank of officer in the armed forces. **b** rank so conferred. **4** pay or percentage paid to an agent. **5** act of committing (a crime etc.). *—v.* **1** empower by commission. **2 a** give (an artist etc.) a commission for a piece of work. **b** order (a work) to be written etc. **3 a** give (an officer) command of a ship. **b** prepare (a ship) for active service. **4** bring (a machine etc.) into operation. □ **in** (or **out of**) **commission** ready (or not ready) for service. [Latin: related to COMMIT]

commission-agent *n.* bookmaker.

commissionaire /kəˌmɪʃə'neə(r)/ *n.* uniformed door-attendant. [French: related to COMMISSIONER]

commissioner *n.* **1** person appointed by a commission to perform a specific task, e.g. the head of the London police etc. **2** member of a government commission. **3** representative of government in a district, department, etc. [medieval Latin: related to COMMISSION]

Commissioner for Oaths *n.* solicitor authorized to administer an oath in an affidavit etc.

commit /kə'mɪt/ *v.* (**-tt-**) **1** do or make (a crime, blunder, etc.). **2** (usu. foll. by *to*) entrust or consign for safe keeping or treatment. **3** send (a person) to prison. **4** pledge or bind (esp. oneself) to a certain course or policy. **5** (as **committed** *adj.*) (often foll. by *to*) **a** dedicated. **b** obliged. □ **commit to memory** memorize. **commit to paper** write down. [Latin *committo -miss-*]

commitment *n.* **1** engagement or obligation. **2** committing or being committed. **3** dedication; committing oneself.

committal *n.* act of committing, esp. to prison.

committee /kə'mɪtɪ/ *n.* **1** body of persons appointed for a special function by (and usu. out of) a larger body. **2** (**Committee**) House of Commons sitting as a committee. [from COMMIT, -EE]

committee stage *n.* third of five stages of a bill's progress through Parliament.

commode /kə'məʊd/ *n.* **1** chamber-pot in a chair with a cover. **2** chest of drawers. [Latin *commodus* convenient]

commodious /kə'məʊdɪəs/ *adj.* roomy.

commodity /kə'mɒdɪtɪ/ *n.* (*pl.* **-ies**) article of trade, esp. a raw material or product as opposed to a service. [Latin: related to COMMODE]

commodore /'kɒməˌdɔ:(r)/ *n.* **1** naval officer above captain and below rear-admiral. **2** commander of a squadron or other division of a fleet. **3** president of a yacht-club. [French: related to COMMANDER]

common /'kɒmən/ *–adj.* (**-er, -est**) **1 a** occurring often. **b** ordinary; without special rank or position. **2 a** shared by, coming from, more than one (*common knowledge*). **b** belonging to the whole community; public. **3** *derog.* low-class; vulgar; inferior. **4** of the most familiar type (*common cold*). **5** *Math.* belonging to two or more quantities (*common denominator*). **6** *Gram.* (of gender) referring to individuals of either sex. *–n.* **1** piece of open public land. **2** *slang* = COMMON SENSE. □ **in common 1** in joint use; shared. **2** of joint interest. **in common with** in the same way as. [Latin *communis*]

commonality /ˌkɒmə'nælɪtɪ/ *n.* (*pl.* **-ies**) **1** sharing of an attribute. **2** common occurrence. **3** = COMMONALTY. [var. of COMMONALTY]

commonalty /'kɒmənəltɪ/ *n.* (*pl.* **-ies**) **1** the common people. **2** the general body (esp. of mankind). [medieval Latin: related to COMMON]

commoner *n.* **1** one of the common people (below the rank of peer). **2** university student without a scholarship. [medieval Latin: related to COMMON]

common ground *n.* point or argument accepted by both sides in a dispute.

common law *n.* unwritten law based on custom and precedent.

common-law husband *n.* (also **common-law wife**) partner recognized by common law without formal marriage.

commonly *adv.* usually, frequently; ordinarily.

Common Market *n.* European Community.

common noun *n. Gram.* name denoting a class of objects or a concept, not a particular individual.

common or garden *adj. colloq.* ordinary.

commonplace *–adj.* lacking originality; trite; ordinary. *–n.* **1** event, topic, etc. that is ordinary or usual. **2** trite remark. [translation of Latin *locus communis*]

common-room *n.* room for the social use of students or teachers at a college etc.

commons *n.pl.* **1** (**the Commons**) = HOUSE OF COMMONS. **2** the common people.

common sense *n.* sound practical sense.

commonsensical /ˌkɒmən'sensɪk(ə)l/ *adj.* having or marked by common sense.

common time *n. Mus.* four crotchets in a bar.

commonwealth *n.* **1** independent State or community, esp. a democratic republic. **2** (**the Commonwealth**) **a** association of the UK with States that were previously part of the British Empire. **b** republican government of Britain 1649–60. **3** federation of States.

commotion /kə'məʊʃ(ə)n/ *n.* confused and noisy disturbance, uproar. [Latin: related to COM-]

communal /'kɒmjʊn(ə)l/ *adj.* **1** shared between members of a group or community; for common use. **2** (of conflict etc.) between esp. ethnic or religious communities. □ **communally** *adv.* [Latin: related to COMMUNE[1]]

commune[1] /ˈkɒmju:n/ *n.* **1** group of people sharing accommodation, goods, etc. **2** small district of local government in France etc. [medieval Latin: related to COMMON]

commune[2] /kəˈmju:n/ *v.* (**-ning**) (usu. foll. by *with*) **1** speak intimately. **2** feel in close touch (with nature etc.). [French: related to COMMON]

communicable /kəˈmju:nɪkəb(ə)l/ *adj.* (esp. of a disease) able to be passed on. [Latin: related to COMMUNICATE]

communicant /kəˈmju:nɪkənt/ *n.* **1** person who receives Holy Communion. **2** person who imparts information. [related to COMMUNICATE]

communicate /kəˈmju:nɪˌkeɪt/ *v.* (**-ting**) **1** impart, transmit (news, heat, motion, feelings, disease, ideas, etc.). **2** succeed in conveying information. **3** (often foll. by *with*) relate socially; have dealings. **4** be connected (*they have communicating rooms*). □ **communicator** *n.* **communicatory** *adj.* [Latin: related to COMMON]

communication /kəˌmju:nɪˈkeɪʃ(ə)n/ *n.* **1 a** communicating or being communicated. **b** information etc. communicated. **c** letter, message, etc. **2** connection or means of access. **3** social dealings. **4** (in *pl.*) science and practice of transmitting information.

communication cord *n.* cord or chain pulled to stop a train in an emergency.

communication(s) satellite *n.* artificial satellite used to relay telephone circuits or broadcast programmes.

communicative /kəˈmju:nɪkətɪv/ *adj.* ready to talk and impart information.

communion /kəˈmju:nɪən/ *n.* **1** sharing, esp. of thoughts etc.; fellowship. **2** participation; sharing in common (*communion of interests*). **3** (**Communion** or **Holy Communion**) Eucharist. **4** body or group within the Christian faith (*the Methodist communion*). [Latin: related to COMMON]

communiqué /kəˈmju:nɪˌkeɪ/ *n.* official communication, esp. a news report. [French, = communicated]

communism /ˈkɒmjʊˌnɪz(ə)m/ *n.* **1 a** social system in which most property is publicly owned and each person works for the common benefit. **b** political theory advocating this. **2** (usu. **Communism**) the form of socialist society established in Cuba, China, etc., and previously, the USSR. [French: related to COMMON]

communist /ˈkɒmjʊnɪst/ *–n.* **1** person advocating communism. **2** (usu. **Communist**) supporter of Communism or member of a Communist Party. *–adj.* **1** of or relating to communism. **2** (usu. **Communist**) of Communists or a Communist Party. □ **communistic** /-ˈnɪstɪk/ *adj.*

Communist Party *n.* political party advocating communism or Communism.

community /kəˈmju:nɪtɪ/ *n.* (*pl.* **-ies**) **1** body of people living in one place, district, or country. **2** body of people having religion, ethnic origin, profession, etc., in common. **3** fellowship (*community of interest*). **4** commune. **5** joint ownership or liability. [Latin: related to COMMON]

community centre *n.* place providing social facilities for a neighbourhood.

community charge *n.* tax levied locally on every adult.

■ **Usage** The *community charge*, or *poll tax*, replaced household rates in 1989-90 and is itself to be replaced by a *council tax* in 1993.

---

community home *n.* centre housing young offenders and other juveniles.

community service *n.* unpaid work in the community, esp. by an offender.

community singing *n.* singing by a large group, esp. of old popular songs or hymns.

community spirit *n.* feeling of belonging to a community, expressed in mutual support etc.

commute /kəˈmju:t/ *v.* (**-ting**) **1** travel some distance to and from work. **2** (usu. foll. by *to*) change (a punishment) to one less severe. **3** (often foll. by *into, for*) change (one kind of payment or obligation) for another. **4** exchange. □ **commutable** *adj.* **commutation** /ˌkɒmjʊˈteɪʃ(ə)n/ *n.* [Latin *muto* change]

commuter *n.* person who commutes to and from work.

compact[1] *–adj.* /kəmˈpækt/ **1** closely or neatly packed together. **2** small and economically designed. **3** concise. **4** (of a person) small but well-proportioned. *–v.* /kəmˈpækt/ make compact. *–n.* /ˈkɒmpækt/ (in full **powder compact**) small flat case for face-powder. □ **compactly** *adv.* **compactness** *n.* [Latin *pango* fasten]

compact[2] /ˈkɒmpækt/ *n.* agreement, contract. [Latin: related to PACT]

compact disc /ˈkɒmpækt/ *n.* disc on which information or sound is recorded digitally and reproduced by reflection of laser light.

companion /kəm'pænjən/ *n.* **1 a** person who accompanies or associates with another. **b** (foll. by *in*, *of*) partner, sharer. **c** person employed to live with and assist another. **2** handbook or reference book. **3** thing that matches another. **4** (**Companion**) member of some orders of knighthood. [Latin *panis* bread]

companionable *adj.* sociable, friendly. □ **companionably** *adv.*

companionship *n.* friendship; being together.

companion-way *n.* staircase from a ship's deck to the saloon or cabins.

company /'kʌmpənɪ/ *n.* (*pl.* **-ies**) **1 a** number of people assembled. **b** guest(s). **2** person's associate(s). **3 a** commercial business. **b** partners in this. **4** actors etc. working together. **5** subdivision of an infantry battalion. **6** body of people combined for a common purpose (*the ship's company*). **7** being with another or others. □ **in company with** together with. **keep a person company** remain with a person to be sociable. **part company** (often foll. by *with*) cease to associate; separate; disagree. [French: related to COMPANION]

comparable /'kɒmpərəb(ə)l/ *adj.* (often foll. by *with*, *to*) able or fit to be compared. □ **comparability** /-'bɪlɪtɪ/ *n.* **comparably** *adv.* [Latin: related to COMPARE]

■ **Usage** Use of *comparable* with *to* and *with* corresponds to the senses of *compare*: *to* is more common.

comparative /kəm'pærətɪv/ *–adj.* **1** perceptible or estimated by comparison; relative (*in comparative comfort*). **2** of or involving comparison (*a comparative study*). **3** *Gram.* (of an adjective or adverb) expressing a higher degree of a quality (e.g. *braver*, *more quickly*). *–n.* *Gram.* comparative expression or word. □ **comparatively** *adv.* [Latin: related to COMPARE]

compare /kəm'peə(r)/ *–v.* (**-ring**) **1** (usu. foll. by *to*) express similarities in; liken. **2** (often foll. by *to*, *with*) estimate the similarity of. **3** (often foll. by *with*) bear comparison. **4** *Gram.* form comparative and superlative degrees of (an adjective or adverb). *–n.* *literary* comparison (*beyond compare*). □ **compare notes** exchange ideas or opinions. [Latin *compar* equal]

■ **Usage** In current use, *to* and *with* are generally interchangeable, but *with* often implies a greater element of formal analysis.

comparison /kəm'pærɪs(ə)n/ *n.* **1** comparing. **2** illustration or example of similarity. **3** capacity for being likened (*there's no comparison*). **4** (in full **degrees of comparison**) *Gram.* positive, comparative, and superlative forms of adjectives and adverbs. □ **bear** (or **stand**) **comparison** (often foll. by *with*) be able to be compared favourably. **beyond comparison 1** totally different in quality. **2** greatly superior; excellent.

compartment /kəm'pɑːtmənt/ *n.* **1** space within a larger space, separated by partitions. **2** watertight division of a ship. **3** area of activity etc. kept apart from others in a person's mind. [Latin: related to PART]

compartmental /ˌkɒmpɑːt'ment(ə)l/ *adj.* of or divided into compartments or categories.

compartmentalize *v.* (also **-ise**) (**-zing** or **-sing**) divide into compartments or categories.

compass /'kʌmpəs/ *n.* **1** instrument showing the direction of magnetic north and bearings from it. **2** (usu. in *pl.*) instrument for taking measurements and describing circles, with two arms connected at one end by a hinge. **3** circumference or boundary. **4** area, extent; scope; range. [Latin *passus* pace]

compassion /kəm'pæʃ(ə)n/ *n.* pity inclining one to help or be merciful. [Church Latin: related to PASSION]

compassionate /kəm'pæʃənət/ *adj.* showing compassion, sympathetic. □ **compassionately** *adv.*

compassionate leave *n.* leave granted on grounds of bereavement etc.

compatible /kəm'pætəb(ə)l/ *adj.* **1 a** able to coexist; well-suited. **b** (often foll. by *with*) consistent. **2** (of equipment etc.) able to be used in combination. □ **compatibility** /-'bɪlɪtɪ/ *n.* [medieval Latin: related to PASSION]

compatriot /kəm'pætrɪət/ *n.* fellow-countryman. [Latin *compatriota*]

compel /kəm'pel/ *v.* (**-ll-**) **1** force, constrain. **2** arouse irresistibly (*compels admiration*). **3** (as **compelling** *adj.*) rousing strong interest, conviction, or admiration. □ **compellingly** *adv.* [Latin *pello puls-* drive]

compendious /kəm'pendɪəs/ *adj.* comprehensive but brief. [Latin: related to COMPENDIUM]

compendium /kəm'pendɪəm/ *n.* (*pl.* **-s** or **-dia**) **1** concise summary or abridgement. **2** collection of table-games etc. [Latin]

compensate /'kɒmpenˌseɪt/ *v.* (**-ting**) **1 a** (often foll. by *for*) recompense (a person). **b** recompense (loss, damage, etc.). **2** (usu. foll. by *for* a thing) make amends. **3** counterbalance. **4** offset disability or frustration by development in another direction. □ **compensatory** /-'seɪtərɪ/ *adj.* [Latin *pendo pens-* weigh]

compensation /ˌkɒmpen'seɪʃ(ə)n/ *n.* **1** compensating or being compensated. **2** money etc. given as recompense.

compère /'kɒmpeə(r)/ *—n.* person who introduces a variety show etc. *—v.* (**-ring**) act as compère (to). [French, = godfather]

compete /kəm'pi:t/ *v.* (**-ting**) **1** take part in a contest etc. **2** (often foll. by *with*, *against* a person, *for* a thing) strive. [Latin *peto* seek]

competence /'kɒmpɪt(ə)ns/ *n.* (also **competency**) **1** ability; being competent. **2** income large enough to live on. **3** legal capacity.

competent *adj.* **1** adequately qualified or capable. **2** effective. □ **competently** *adv.* [Latin: related to COMPETE]

competition /ˌkɒmpə'tɪʃ(ə)n/ *n.* **1** (often foll. by *for*) competing. **2** event in which people compete. **3** the other people or trade competing; opposition. [Latin: related to COMPETE]

competitive /kəm'petɪtɪv/ *adj.* **1** of or involving competition. **2** (of prices etc.) comparing favourably with those of rivals. **3** having a strong urge to win. □ **competitiveness** *n.*

competitor /kəm'petɪtə(r)/ *n.* person who competes; rival, esp. in business.

compile /kəm'paɪl/ *v.* (**-ling**) **1 a** collect and arrange (material) into a list, book, etc. **b** produce (a book etc.) thus. **2** *Computing* translate (a programming language) into machine code. □ **compilation** /ˌkɒmpɪ'leɪʃ(ə)n/ *n.* [Latin *compilo* plunder]

compiler *n.* **1** person who compiles. **2** *Computing* program for translating a programming language into machine code.

complacent /kəm'pleɪs(ə)nt/ *adj.* smugly self-satisfied or contented. □ **complacence** *n.* **complacency** *n.* **complacently** *adv.* [Latin *placeo* please]

■ **Usage** *Complacent* is often confused with *complaisant*.

---

complain /kəm'pleɪn/ *v.* **1** express dissatisfaction. **2** (foll. by *of*) **a** say that one is suffering from (an ailment). **b** state a grievance concerning. **3** creak under strain. [Latin *plango* lament]

complainant *n.* plaintiff in certain lawsuits.

complaint *n.* **1** complaining. **2** grievance, cause of dissatisfaction. **3** ailment. **4** formal accusation.

complaisant /kəm'pleɪz(ə)nt/ *adj. formal* **1** deferential. **2** willing to please; acquiescent. □ **complaisance** *n.* [French: related to COMPLACENT]

■ **Usage** *Complaisant* is often confused with *complacent*.

---

complement *—n.* /'kɒmplɪmənt/ **1** thing that completes; counterpart. **2** full number needed. **3** word(s) added to a verb to complete the predicate of a sentence. **4** amount by which an angle is less than 90°. *—v.* /'kɒmplɪˌment/ **1** complete. **2** form a complement to. [Latin *compleo* fill up]

complementary /ˌkɒmplɪ'mentərɪ/ *adj.* **1** completing; forming a complement. **2** (of two or more things) complementing each other.

complementary medicine *n.* alternative medicine.

complete /kəm'pli:t/ *—adj.* **1** having all its parts; entire. **2** finished. **3** total, in every way. *—v.* (**-ting**) **1** finish. **2** make complete. **3** fill in (a form etc.). **4** conclude the sale or purchase of property. □ **complete with** having (as an important feature) (*comes complete with instructions*). □ **completely** *adv.* **completeness** *n.* **completion** *n.* [Latin: related to COMPLEMENT]

complex /'kɒmpleks/ *—n.* **1** building, series of rooms, etc., made up of related parts (*shopping complex*). **2** *Psychol.* group of usu. repressed feelings or thoughts which cause abnormal behaviour or mental states. **3** preoccupation; feeling of inadequacy. *—adj.* **1** complicated. **2** consisting of related parts; composite. □ **complexity** /kəm'pleksɪtɪ/ *n.* (*pl.* **-ies**). [Latin *complexus*]

complexion /kəm'plekʃ(ə)n/ *n.* **1** natural colour, texture, and appearance of the skin, esp. of the face. **2** aspect, character (*puts a different complexion on the matter*). [Latin: related to COMPLEX]

compliance /kəm'plaɪəns/ *n.* **1** obedience to a request, command, etc. **2** capacity to yield. □ **in compliance with** according to.

compliant *adj.* obedient; yielding. □ **compliantly** *adv.*

complicate /'kɒmplɪˌkeɪt/ *v.* (**-ting**) **1** make difficult or complex. **2** (as **complicated** *adj.*) complex; intricate. [Latin *plico* to fold]

complication /ˌkɒmplɪ'keɪʃ(ə)n/ *n.* **1 a** involved or confused condition or state. **b** complicating circumstance; difficulty. **2** (often in *pl.*) disease or condition aggravating or arising out of a previous one. [Latin: related to COMPLICATE]

complicity /kəmˈplɪsɪtɪ/ *n.* partnership in wrongdoing. [French: related to COMPLEX]

compliment *–n.* /ˈkɒmplɪmənt/ **1 a** polite expression of praise. **b** act implying praise. **2** (in *pl.*) **a** formal greetings accompanying a present etc. **b** praise. *–v.* /ˈkɒmplɪˌment/ (often foll. by *on*) congratulate; praise. [Latin: related to COMPLEMENT]

complimentary /ˌkɒmplɪˈmentərɪ/ *adj.* **1** expressing a compliment. **2** given free of charge.

compline /ˈkɒmplɪn/ *n.* **1** last of the canonical hours of prayer. **2** service during this. [Latin: related to COMPLY]

comply /kəmˈplaɪ/ *v.* **(-ies, -ied)** (often foll. by *with*) act in accordance (with a request or command). [Latin *compleo* fill up]

component /kəmˈpəʊnənt/ *–n.* part of a larger whole. *–adj.* being part of a larger whole. [Latin: related to COMPOUND[1]]

comport /kəmˈpɔːt/ *v.refl. literary* conduct oneself; behave. □ **comport with** suit, befit. □ **comportment** *n.* [Latin *porto* carry]

compose /kəmˈpəʊz/ *v.* **(-sing) 1** create in music or writing. **2** constitute; make up. **3** arrange artistically, neatly, or for a specified purpose. **4 a** (often *refl.*) calm; settle. **b** (as **composed** *adj.*) calm, self-possessed. **5** *Printing* **a** set up (type). **b** arrange (an article etc.) in type. □ **composed of** made up of, consisting of. □ **composedly** /-zɪdlɪ/ *adv.* [French: related to POSE]

■ **Usage** See note at *comprise*.

composer *n.* person who composes (esp. music).

composite /ˈkɒmpəzɪt/ *–adj.* **1** made up of parts. **2** of mixed Ionic and Corinthian style. **3** (of a plant) having a head of many flowers forming one bloom. *–n.* composite thing or plant. [Latin: related to COMPOSE]

composition /ˌkɒmpəˈzɪʃ(ə)n/ *n.* **1 a** act or method of putting together; composing. **b** thing composed, esp. music. **2** constitution of a substance. **3** school essay. **4** arrangement of the parts of a picture etc. **5** compound artificial substance. □ **compositional** *adj.*

compositor /kəmˈpɒzɪtə(r)/ *n.* person who sets up type for printing. [Latin: related to COMPOSE]

compos mentis /ˌkɒmpɒs ˈmentɪs/ *adj.* sane. [Latin]

compost /ˈkɒmpɒst/ *–n.* **1** mixture of decayed organic matter. **2** loam soil with fertilizer for growing plants. *–v.* **1** treat with compost. **2** make into compost. [Latin: related to COMPOSE]

composure /kəmˈpəʊʒə(r)/ *n.* tranquil manner. [from COMPOSE]

compote /ˈkɒmpəʊt/ *n.* fruit preserved or cooked in syrup. [French: related to COMPOSE]

compound[1] /ˈkɒmpaʊnd/ *–n.* **1** mixture of two or more things. **2** word made up of two or more existing words. **3** substance formed from two or more elements chemically united in fixed proportions. *–adj.* **1** made up of two or more ingredients or parts. **2** combined; collective. *–v.* /kəmˈpaʊnd/ **1** mix or combine (ingredients or elements). **2** increase or complicate (difficulties etc.). **3** make up (a composite whole). **4** settle (a matter) by mutual agreement. **5** *Law* condone or conceal (a liability or offence) for personal gain. **6** (usu. foll. by *with*) *Law* come to terms with a person. [Latin *compono -pos-* put together]

compound[2] /ˈkɒmpaʊnd/ *n.* **1** enclosure or fenced-in space. **2** enclosure, esp. in India, China, etc., in which a factory or house stands. [Malay *kampong*]

compound fracture *n.* fracture complicated by a wound.

compound interest *n.* interest payable on capital and its accumulated interest.

comprehend /ˌkɒmprɪˈhend/ *v.* **1** grasp mentally; understand. **2** include. [Latin *comprehendo* seize]

comprehensible *adj.* that can be understood. [Latin: related to COMPREHEND]

comprehension *n.* **1 a** understanding. **b** text set as a test of understanding. **2** inclusion.

comprehensive *–adj.* **1** including all or nearly all, inclusive. **2** (of motor insurance) providing protection against most risks. *–n.* (in full **comprehensive school**) secondary school for children of all abilities. □ **comprehensively** *adv.* **comprehensiveness** *n.*

compress *–v.* /kəmˈpres/ **1** squeeze together. **2** bring into a smaller space or shorter time. *–n.* /ˈkɒmpres/ pad of lint etc. pressed on to part of the body to relieve inflammation, stop bleeding, etc. □ **compressible** /kəmˈpresɪb(ə)l/ *adj.* [Latin: related to PRESS[1]]

compression /kəmˈpreʃ(ə)n/ *n.* **1** compressing. **2** reduction in volume of the fuel mixture in an internal-combustion engine before ignition.

compressor /kəmˈpresə(r)/ *n.* machine for compressing air or other gases.

comprise /kəmˈpraɪz/ *v.* **(-sing) 1** include. **2** consist of. **3** make up, compose. [French: related to COMPREHEND]

■ **Usage** The use of this word in sense 3 is considered incorrect and *compose* is generally preferred.

compromise /'kɒmprə,maɪz/ *–n.* **1** settlement of a dispute by mutual concession. **2** (often foll. by *between*) intermediate state between conflicting opinions, actions, etc. *–v.* (**-sing**) **1 a** settle a dispute by mutual concession. **b** modify one's opinions, demands, etc. **2** bring into disrepute or danger by indiscretion. [Latin: related to PROMISE]

comptroller /kən'trəʊlə(r)/ *n.* controller (used in the title of some financial officers). [var. of CONTROLLER]

compulsion /kəm'pʌlʃ(ə)n/ *n.* **1** compelling or being compelled; obligation. **2** irresistible urge. [Latin: related to COMPEL]

compulsive /kəm'pʌlsɪv/ *adj.* **1** compelling. **2** resulting or acting (as if) from compulsion (*compulsive gambler*). **3** irresistible (*compulsive entertainment*). □ **compulsively** *adv.* [medieval Latin: related to COMPEL]

compulsory /kəm'pʌlsərɪ/ *adj.* **1** required by law or a rule. **2** essential. □ **compulsorily** *adv.*

compulsory purchase *n.* enforced sale of land or property to a local authority etc.

compunction /kəm'pʌŋkʃ(ə)n/ *n.* **1** pricking of conscience. **2** slight regret; scruple. [Church Latin: related to POINT]

compute /kəm'pju:t/ *v.* (**-ting**) **1** reckon or calculate. **2** use a computer. □ **computation** /,kɒmpju:'teɪʃ(ə)n/ *n.* [Latin *puto* reckon]

computer *n.* electronic device for storing and processing data, making calculations, or controlling machinery.

computerize *v.* (also **-ise**) (**-zing** or **-sing**) **1** equip with a computer. **2** store, perform, or produce by computer. □ **computerization** /-'zeɪʃ(ə)n/ *n.*

computer-literate *adj.* able to use computers.

computer science *n.* the study of the principles and use of computers.

computer virus *n.* self-replicating code maliciously introduced into a computer program and intended to corrupt the system or destroy data.

comrade /'kɒmreɪd/ *n.* **1** associate or companion in some activity. **2** fellow socialist or Communist. □ **comradely** *adj.* **comradeship** *n.* [Spanish: related to CHAMBER]

con[1] *slang –n.* confidence trick. *–v.* (**-nn-**) swindle; deceive. [abbreviation]

con[2] *–n.* (usu. in *pl.*) reason against. *–prep.* & *adv.* against (cf. PRO[2]). [Latin *contra* against]

con[3] *n. slang* convict. [abbreviation]

con[4] *v.* (*US* **conn**) (**-nn-**) direct the steering of (a ship). [originally *cond* from French: related to CONDUCT]

con- see COM-.

concatenation /,kɒnkætɪ'neɪʃ(ə)n/ *n.* series of linked things or events. [Latin *catena* chain]

concave /'kɒnkeɪv/ *adj.* curved like the interior of a circle or sphere. □ **concavity** /-'kævɪtɪ/ *n.* [Latin: related to CAVE]

conceal /kən'si:l/ *v.* **1** keep secret. **2** hide. □ **concealment** *n.* [Latin *celo* hide]

concede /kən'si:d/ *v.* (**-ding**) **1** admit to be true. **2** admit defeat in. **3** grant (a right, privilege, etc.). [Latin: related to CEDE]

conceit /kən'si:t/ *n.* **1** personal vanity; pride. **2** *literary* **a** far-fetched comparison. **b** fanciful notion. [from CONCEIVE]

conceited *adj.* vain. □ **conceitedly** *adv.*

conceivable /kən'si:vəb(ə)l/ *adj.* capable of being grasped or imagined. □ **conceivably** *adv.*

conceive /kən'si:v/ *v.* (**-ving**) **1** become pregnant (with). **2 a** (often foll. by *of*) imagine, think. **b** (usu. in *passive*) formulate (a belief, plan, etc.). [Latin *concipio -cept-*]

concentrate /'kɒnsən,treɪt/ *–v.* (**-ting**) **1** (often foll. by *on*) focus one's attention or thought. **2** bring together to one point. **3** increase the strength of (a liquid etc.) by removing water etc. **4** (as **concentrated** *adj.*) intense, strong. *–n.* concentrated substance. [Latin: related to CENTRE]

concentration /,kɒnsən'treɪʃ(ə)n/ *n.* **1** concentrating or being concentrated. **2** mental attention. **3** something concentrated. **4** weight of a substance in a given amount of material.

concentration camp *n.* camp where political prisoners etc. are detained.

concentric /kən'sentrɪk/ *adj.* having a common centre. □ **concentrically** *adv.* [French or medieval Latin: related to CENTRE]

concept /'kɒnsept/ *n.* general notion; abstract idea. [Latin: related to CONCEIVE]

conception /kən'sepʃ(ə)n/ *n.* **1** conceiving or being conceived. **2** idea, plan. **3** understanding (*has no conception*). □ **conceptional** *adj.* [French from Latin: related to CONCEPT]

conceptual /kən'septʃʊəl/ *adj.* of mental conceptions or concepts. □ **conceptually** *adv.*

conceptualize *v.* (also **-ise**) (**-zing** or **-sing**) form a concept or idea of. □ **conceptualization** /-ˈzeɪʃ(ə)n/ *n.*

concern /kənˈsɜːn/ *—v.* **1 a** be relevant or important to. **b** relate to; be about. **2** (*refl.*; often foll. by *with*, *about*, *in*) interest or involve oneself. **3** worry, affect. *—n.* **1** anxiety, worry. **2 a** matter of interest or importance to one. **b** interest, connection (*has a concern in politics*). **3** business, firm. **4** *colloq.* complicated thing, contrivance. [Latin *cerno* sift]

concerned *adj.* **1** involved, interested. **2** troubled, anxious. □ **be concerned** (often foll. by *in*) take part. □ **concernedly** /-ɪdlɪ/ *adv.* **concernedness** /-ɪdnɪs/ *n.*

concerning *prep.* about, regarding.

concert /ˈkɒnsət/ *n.* **1** musical performance of usu. several separate compositions. **2** agreement. **3** combination of voices or sounds. [Italian: related to CONCERTO]

concerted /kənˈsɜːtɪd/ *adj.* **1** jointly arranged or planned. **2** *Mus.* arranged in parts for voices or instruments.

concertina /ˌkɒnsəˈtiːnə/ *—n.* musical instrument like an accordion but smaller. *—v.* (**-nas**, **-naed** /-nəd/ or **-na'd**, **-naing**) compress or collapse in folds like those of a concertina.

concerto /kənˈtʃeətəʊ/ *n.* (*pl.* **-s** or **-ti** /-tɪ/) composition for solo instrument(s) and orchestra. [Italian]

concert pitch *n.* pitch internationally agreed whereby the A above middle C = 440 Hz.

concession /kənˈseʃ(ə)n/ *n.* **1 a** conceding. **b** thing conceded. **2** reduction in price for a certain category of persons. **3 a** right to use land etc. **b** right to sell goods in a particular territory. □ **concessionary** *adj.* [Latin: related to CONCEDE]

concessive /kənˈsesɪv/ *adj. Gram.* (of a preposition or conjunction) introducing a phrase or clause which contrasts with the main clause (e.g. *in spite of*, *although*). [Latin: related to CONCEDE]

conch /kɒntʃ/ *n.* **1** thick heavy spiral shell of various marine gastropod molluscs. **2** any such gastropod. [Latin *concha*]

conchology /kɒŋˈkɒlədʒɪ/ *n.* the study of shells. [from CONCH]

concierge /ˌkɔ̃sɪˈeəʒ/ *n.* (esp. in France) door-keeper or porter of a block of flats etc. [French]

conciliate /kənˈsɪlɪˌeɪt/ *v.* (**-ting**) **1** make calm and amenable; pacify; gain the goodwill of. **2** reconcile. □ **conciliation** /-ˈeɪʃ(ə)n/ *n.* **conciliator** *n.* **conciliatory** /-ˈsɪlɪətərɪ/ *adj.* [Latin: related to COUNCIL]

concise /kənˈsaɪs/ *adj.* brief but comprehensive in expression. □ **concisely** *adv.* **conciseness** *n.* **concision** /-ˈsɪʒ(ə)n/ *n.* [Latin *caedo* cut]

conclave /ˈkɒnkleɪv/ *n.* **1** private meeting. **2** *RC Ch.* **a** assembly of cardinals for the election of a pope. **b** meeting-place for this. [Latin *clavis* key]

conclude /kənˈkluːd/ *v.* (**-ding**) **1** bring or come to an end. **2** (often foll. by *from* or *that*) infer. **3** settle (a treaty etc.). [Latin *concludo*: related to CLOSE[1]]

conclusion /kənˈkluːʒ(ə)n/ *n.* **1** ending, end. **2** judgement reached by reasoning. **3** summing-up. **4** settling (of peace etc.). **5** *Logic* proposition reached from given premisses. □ **in conclusion** lastly, to conclude. [Latin: related to CONCLUDE]

conclusive /kənˈkluːsɪv/ *adj.* decisive, convincing. □ **conclusively** *adv.* [Latin: related to CONCLUDE]

concoct /kənˈkɒkt/ *v.* **1** make by mixing ingredients. **2** invent (a story, lie, etc.). □ **concoction** /-ˈkɒkʃ(ə)n/ *n.* [Latin *coquo coct-* cook]

concomitant /kənˈkɒmɪt(ə)nt/ *—adj.* (often foll. by *with*) accompanying; occurring together. *—n.* accompanying thing. □ **concomitance** *n.* [Latin *comes comit-* companion]

concord /ˈkɒŋkɔːd/ *n.* agreement, harmony. □ **concordant** /kənˈkɔːd(ə)nt/ *adj.* [Latin *cor cord-* heart]

concordance /kənˈkɔːd(ə)ns/ *n.* **1** agreement. **2** alphabetical index of words used in a book or by an author. [medieval Latin: related to CONCORD]

concordat /kənˈkɔːdæt/ *n.* agreement, esp. between the Church and a State. [Latin: related to CONCORD]

concourse /ˈkɒŋkɔːs/ *n.* **1** crowd, gathering. **2** large open area in a railway station etc. [Latin: related to CONCUR]

concrete /ˈkɒŋkriːt/ *—adj.* **1 a** existing in a material form; real. **b** specific, definite (*concrete evidence*; *a concrete proposal*). **2** *Gram.* (of a noun) denoting a material object as opposed to a quality, state, etc. *—n.* (often *attrib.*) mixture of gravel, sand, cement, and water, used for building. *—v.* (**-ting**) cover with or embed in concrete. [Latin *cresco cret-* grow]

concretion /kənˈkriːʃ(ə)n/ *n.* **1** hard solid mass. **2** forming of this by coalescence. [Latin: related to CONCRETE]

concubine /ˈkɒŋkjʊˌbaɪn/ *n.* **1** *literary* or *joc.* mistress. **2** (among polygamous peoples) secondary wife. □ **concubinage** /kənˈkjuːbɪnɪdʒ/ *n.* [Latin *cubo* lie]

concupiscence /kən'kju:pɪs(ə)ns/ *n. formal* lust. □ **concupiscent** *adj.* [Latin *cupio* desire]
concur /kən'kɜ:(r)/ *v.* (**-rr-**) **1** (often foll. by *with*) have the same opinion. **2** coincide. [Latin *curro* run]
concurrent /kən'kʌrənt/ *adj.* **1** (often foll. by *with*) existing or in operation at the same time or together. **2** (of three or more lines) meeting at or tending towards one point. **3** agreeing, harmonious. □ **concurrence** *n.* **concurrently** *adv.*
concuss /kən'kʌs/ *v.* subject to concussion. [Latin *quatio* shake]
concussion /kən'kʌʃ(ə)n/ *n.* **1** temporary unconsciousness or incapacity due to a blow to the head, a fall, etc. **2** violent shaking.
condemn /kən'dem/ *v.* **1** express utter disapproval of. **2 a** find guilty; convict. **b** (usu. foll. by *to*) sentence to (a punishment). **3** pronounce (a building etc.) unfit for use. **4** (usu. foll. by *to*) doom or assign (to something unpleasant). □ **condemnation** /ˌkɒndem'neɪʃ(ə)n/ *n.* **condemnatory** /-'demnətərɪ/ *adj.* [Latin: related to DAMN]
condensation /ˌkɒnden'seɪʃ(ə)n/ *n.* **1** condensing or being condensed. **2** condensed liquid (esp. water on a cold surface). **3** abridgement. [Latin: related to CONDENSE]
condense /kən'dens/ *v.* (**-sing**) **1** make denser or more concentrated. **2** express in fewer words. **3** reduce or be reduced from a gas or vapour to a liquid. [Latin: related to DENSE]
condensed milk *n.* milk thickened by evaporation and sweetened.
condenser *n.* **1** apparatus or vessel for condensing vapour. **2** *Electr.* = CAPACITOR. **3** lens or system of lenses for concentrating light.
condescend /ˌkɒndɪ'send/ *v.* **1** be gracious enough (to do a thing) esp. while showing one's sense of dignity or superiority (*condescended to attend*). **2** (foll. by *to*) pretend to be on equal terms with (an inferior). **3** (as **condescending** *adj.*) patronizing. □ **condescendingly** *adv.* **condescension** /-'senʃ(ə)n/ *n.* [Latin: related to DESCEND]
condign /kən'daɪn/ *adj.* (of a punishment etc.) severe and well-deserved. [Latin *dignus* worthy]
condiment /'kɒndɪmənt/ *n.* seasoning or relish for food. [Latin *condio* pickle]
condition /kən'dɪʃ(ə)n/ *–n.* **1** stipulation; thing upon the fulfilment of which something else depends. **2 a** state of being or fitness of a person or thing. **b** ailment, abnormality (*heart condition*). **3** (in *pl.*) circumstances, esp. those affecting the functioning or existence of something (*good working conditions*). *–v.* **1 a** bring into a good or desired state. **b** make fit (esp. dogs or horses). **2** teach or accustom. **3 a** impose conditions on. **b** be essential to. □ **in** (or **out of**) **condition** in good (or bad) condition. **on condition that** with the stipulation that. [Latin *dico* say]
conditional *adj.* **1** (often foll. by *on*) dependent; not absolute; containing a condition. **2** *Gram.* (of a clause, mood, etc.) expressing a condition. □ **conditionally** *adv.* [Latin: related to CONDITION]
conditioned reflex *n.* reflex response to a non-natural stimulus, established by training.
conditioner *n.* agent that conditions, esp. the hair.
condole /kən'dəʊl/ *v.* (**-ling**) (foll. by *with*) express sympathy with (a person) over a loss etc. [Latin *condoleo* grieve with another]

■ **Usage** *Condole* is often confused with *console*[1].

condolence *n.* (often in *pl.*) expression of sympathy.
condom /'kɒndɒm/ *n.* contraceptive sheath worn by men. [origin unknown]
condominium /ˌkɒndə'mɪnɪəm/ *n.* **1** joint rule or sovereignty. **2** *US* building containing individually owned flats. [Latin *dominium* lordship]
condone /kən'dəʊn/ *v.* (**-ning**) forgive or overlook (an offence or wrongdoing). [Latin *dono* give]
condor /'kɒndɔ:(r)/ *n.* large S. American vulture. [Spanish from Quechua]
conduce /kən'dju:s/ *v.* (**-cing**) (foll. by *to*) contribute to (a result). [Latin: related to CONDUCT]
conducive *adj.* (often foll. by *to*) contributing or helping (towards something).
conduct *–n.* /'kɒndʌkt/ **1** behaviour. **2** activity or manner of directing or managing a business, war, etc. *–v.* /kən'dʌkt/ **1** lead or guide. **2** direct or manage (a business etc.). **3** (also *absol.*) be the conductor of (an orchestra etc.). **4** transmit (heat, electricity, etc.) by conduction. **5** *refl.* behave. [Latin *duco duct-* lead]
conductance /kən'dʌkt(ə)ns/ *n.* power of a specified material to conduct electricity.
conduction /kən'dʌkʃ(ə)n/ *n.* transmission of heat, electricity, etc. through a substance. [Latin: related to CONDUCT]
conductive /kən'dʌktɪv/ *adj.* transmitting (esp. heat, electricity, etc.). □ **conductivity** /ˌkɒndʌk'tɪvɪtɪ/ *n.*

conductor *n.* **1** person who directs an orchestra etc. **2** (*fem.* **conductress**) person who collects fares in a bus etc. **3** thing that conducts heat or electricity. [Latin: related to CONDUCT]

conduit /'kɒndɪt, -djʊɪt/ *n.* **1** channel or pipe conveying liquids. **2** tube or trough protecting insulated electric wires. [medieval Latin: related to CONDUCT]

cone *n.* **1** solid figure with a circular (or other curved) plane base, tapering to a point. **2** thing of similar shape. **3** dry fruit of a conifer. **4** ice-cream cornet. [Latin from Greek]

coney var. of CONY.

confab /'kɒnfæb/ *colloq.* –*n.* = CONFABULATION (see CONFABULATE). –*v.* (**-bb-**) = CONFABULATE. [abbreviation]

confabulate /kən'fæbjʊˌleɪt/ *v.* (**-ting**) converse, chat. □ **confabulation** /-'leɪʃ(ə)n/ *n.* [Latin: related to FABLE]

confection /kən'fekʃ(ə)n/ *n.* dish or delicacy made with sweet ingredients. [Latin *conficio* prepare]

confectioner *n.* maker or retailer of confectionery.

confectionery *n.* confections, esp. sweets.

confederacy /kən'fedərəsɪ/ *n.* (*pl.* **-ies**) league or alliance, esp. of confederate States. [French: related to CONFEDERATE]

confederate /kən'fedərət/ –*adj.* esp. *Polit.* allied. –*n.* **1** ally, esp. (in a bad sense) accomplice. **2** (**Confederate**) supporter of the Confederate States. –*v.* /-ˌreɪt/ (**-ting**) (often foll. by *with*) bring or come into alliance. [Latin: related to FEDERAL]

Confederate States *n.pl.* States which seceded from the US in 1860–1.

confederation /kənˌfedə'reɪʃ(ə)n/ *n.* **1** union or alliance, esp. of States. **2** confederating or being confederated.

confer /kən'fɜ:(r)/ *v.* (**-rr-**) **1** (often foll. by *on, upon*) grant or bestow. **2** (often foll. by *with*) converse, consult. □ **conferrable** *adj.* [Latin *confero collat-* bring together]

conference /'kɒnfərəns/ *n.* **1** consultation. **2** meeting for discussion. [French or medieval Latin: related to CONFER]

conferment /kən'fɜ:mənt/ *n.* conferring of a degree, honour, etc.

confess /kən'fes/ *v.* **1 a** (also *absol.*) acknowledge or admit (a fault, crime, etc.). **b** (foll. by *to*) admit to. **2** admit reluctantly. **3 a** (also *absol.*) declare (one's sins) to a priest. **b** (of a priest) hear the confession of. [Latin *confiteor -fess-*]

confessedly /kən'fesɪdlɪ/ *adv.* by one's own or general admission.

confession /kən'feʃ(ə)n/ *n.* **1 a** act of confessing. **b** thing confessed. **2** (in full **confession of faith**) declaration of one's beliefs or principles.

confessional –*n.* enclosed stall in a church in which the priest hears confessions. –*adj.* of confession.

confessor *n.* priest who hears confessions and gives spiritual counsel.

confetti /kən'fetɪ/ *n.* small pieces of coloured paper thrown by wedding guests at the bride and groom. [Italian]

confidant /'kɒnfɪˌdænt/ *n.* (*fem.* **confidante** pronunc. same) person trusted with knowledge of one's private affairs. [related to CONFIDE]

confide /kən'faɪd/ *v.* (**-ding**) **1** (foll. by *in*) talk confidentially to. **2** (usu. foll. by *to*) tell (a secret etc.) in confidence. **3** (foll. by *to*) entrust (an object of care, a task, etc.) to. [Latin *confido* trust]

confidence /'kɒnfɪd(ə)ns/ *n.* **1** firm trust. **2 a** feeling of reliance or certainty. **b** sense of self-reliance; boldness. **3** something told as a secret. □ **in confidence** as a secret. **in a person's confidence** trusted with a person's secrets. **take into one's confidence** confide in. [Latin: related to CONFIDE]

confidence trick *n.* swindle in which the victim is persuaded to trust the swindler. □ **confidence trickster** *n.*

confident *adj.* feeling or showing confidence; bold. □ **confidently** *adv.* [Italian: related to CONFIDE]

confidential /ˌkɒnfɪ'denʃ(ə)l/ *adj.* **1** spoken or written in confidence. **2** entrusted with secrets (*confidential secretary*). **3** confiding. □ **confidentiality** /-ʃɪ'ælɪtɪ/ *n.* **confidentially** *adv.*

configuration /kənˌfɪgjʊ'reɪʃ(ə)n/ *n.* **1** arrangement in a particular form. **2** form or figure resulting from this. **3** *Computing* hardware and its arrangement of connections etc. □ **configure** *v.* (**-ring**). [Latin: related to FIGURE]

confine –*v.* /kən'faɪn/ (**-ning**) **1** keep or restrict (within certain limits). **2** imprison. –*n.* /'kɒnfaɪn/ (usu. in *pl.*) limit, boundary. [Latin *finis* limit]

confinement *n.* **1** confining or being confined. **2** time of childbirth.

confirm /kən'fɜ:m/ *v.* **1** provide support for the truth or correctness of. **2** (foll. by *in*) encourage (a person) in (an opinion etc.). **3** establish more firmly (power, possession, etc.). **4** make formally valid. **5** administer the religious rite of confirmation to. [Latin: related to FIRM[1]]

confirmation /ˌkɒnfə'meɪʃ(ə)n/ *n.* **1** confirming or being confirmed. **2** rite confirming a baptized person as a member of the Christian Church.

confirmed *adj.* firmly settled in some habit or condition (*confirmed bachelor*).

confiscate /'kɒnfɪ,skeɪt/ *v.* (**-ting**) take or seize by authority. □ **confiscation** /-'skeɪʃ(ə)n/ *n.* [Latin: related to FISCAL]

conflagration /,kɒnflə'greɪʃ(ə)n/ *n.* great and destructive fire. [Latin: related to FLAGRANT]

conflate /kən'fleɪt/ *v.* (**-ting**) blend or fuse together (esp. two variant texts into one). □ **conflation** /-'fleɪʃ(ə)n/ *n.* [Latin *flo flat-* blow]

conflict *–n.* /'kɒnflɪkt/ **1 a** state of opposition. **b** fight, struggle. **2** (often foll. by *of*) clashing of opposed interests etc. *–v.* /kən'flɪkt/ clash; be incompatible. [Latin *fligo flict-* strike]

confluence /'kɒnfluəns/ *n.* **1** place where two rivers meet. **2 a** coming together. **b** crowd of people. [Latin *fluo* flow]

confluent *–adj.* flowing together, uniting. *–n.* stream joining another.

conform /kən'fɔ:m/ *v.* **1** comply with rules or general custom. **2** (foll. by *to, with*) comply with; be in accordance with. **3** (often foll. by *to*) be or make suitable. [Latin: related to FORM]

conformable *adj.* **1** (often foll. by *to*) similar. **2** (often foll. by *with*) consistent. **3** (often foll. by *to*) adaptable.

conformation /,kɒnfɔ:'meɪʃ(ə)n/ *n.* way a thing is formed; shape.

conformist /kən'fɔ:mɪst/ *–n.* person who conforms to an established practice. *–adj.* conforming, conventional. □ **conformism** *n.*

conformity *n.* **1** accordance with established practice. **2** agreement, suitability.

confound /kən'faʊnd/ *–v.* **1** perplex, baffle. **2** confuse (in one's mind). **3** *archaic* defeat, overthrow. *–int.* expressing annoyance (*confound you!*). [Latin *confundo -fus-* mix up]

confounded *attrib. adj. colloq.* damned.

confront /kən'frʌnt/ *v.* **1 a** face in hostility or defiance. **b** face up to and deal with. **2** (of a difficulty etc.) present itself to. **3** (foll. by *with*) bring (a person) face to face with (an accusation etc.). **4** meet or stand facing. □ **confrontation** /,kɒnfrʌn'teɪʃ(ə)n/ *n.* **confrontational** /,kɒnfrʌn'teɪʃən(ə)l/ *adj.* [French from medieval Latin]

Confucian /kən'fju:ʃ(ə)n/ *adj.* of Confucius or his philosophy. □ **Confucianism** *n.* [*Confucius*, name of a Chinese philosopher]

confuse /kən'fju:z/ *v.* (**-sing**) **1** perplex, bewilder. **2** mix up in the mind; mistake (one for another). **3** make indistinct (*confuse the issue*). **4** (often as **confused** *adj.*) throw into disorder. □ **confusedly** /-zɪdlɪ/ *adv.* **confusing** *adj.* [related to CONFOUND]

confusion *n.* confusing or being confused.

confute /kən'fju:t/ *v.* (**-ting**) prove (a person or argument) to be in error. □ **confutation** /,kɒnfju:'teɪʃ(ə)n/ *n.* [Latin]

conga /'kɒŋgə/ *–n.* **1** Latin-American dance, with a line of dancers one behind the other. **2** tall narrow drum beaten with the hands. *–v.* (**congas, congaed** /-gəd/ or **conga'd, congaing** /-gəɪŋ/) perform the conga. [Spanish *conga* (feminine), = of the Congo]

congeal /kən'dʒi:l/ *v.* **1** make or become semi-solid by cooling. **2** (of blood etc.) coagulate. □ **congelation** /,kɒndʒɪ'leɪʃ(ə)n/ *n.* [French from Latin *gelo* freeze]

congenial /kən'dʒi:nɪəl/ *adj.* **1** (often foll. by *with, to*) pleasant because like-minded. **2** (often foll. by *to*) suited or agreeable. □ **congeniality** /-'ælɪtɪ/ *n.* **congenially** *adv.* [from COM-, GENIAL]

congenital /kən'dʒenɪt(ə)l/ *adj.* **1** (esp. of disease) existing from birth. **2** as such from birth (*congenital liar*). □ **congenitally** *adv.* [Latin: related to COM-]

conger /'kɒŋgə(r)/ *n.* (in full **conger eel**) large marine eel. [Greek *goggros*]

congeries /kən'dʒɪəri:z/ *n.* (*pl.* same) disorderly collection; mass, heap. [Latin *congero* heap together]

■ **Usage** The form *congery*, formed under the misapprehension that *congeries* is plural only, is incorrect.

congest /kən'dʒest/ *v.* (esp. as **congested** *adj.*) affect with congestion. [Latin *congero -gest-* heap together]

congestion /kən'dʒestʃ(ə)n/ *n.* abnormal accumulation or obstruction, esp. of traffic etc. or of blood or mucus in part of the body.

conglomerate /kən'glɒmərət/ *–adj.* gathered into a rounded mass. *–n.* **1** heterogeneous mass. **2** group or corporation of merged firms. *–v.* /kən'glɒmə,reɪt/ (**-ting**) collect into a coherent mass. □ **conglomeration** /kən,glɒmə'reɪʃ(ə)n/ *n.* [Latin *glomus -eris* ball]

congratulate /kən'grætʃʊ,leɪt/ *v.* (**-ting**) (often foll. by *on*) **1** express pleasure at the happiness, good fortune, or excellence of (a person). **2** *refl.* think oneself fortunate or clever. □ **congratulatory** /-lətərɪ/ *adj.* [Latin *gratus* pleasing]

congratulation /kən,grætʃʊ'leɪʃ(ə)n/ *n.* **1** congratulating. **2** (usu. in *pl.*) expression of this.

congregate /'kɒŋgrɪ,geɪt/ *v.* (**-ting**) collect or gather into a crowd. [Latin *grex greg-* flock]

congregation /,kɒŋgrɪ'geɪʃ(ə)n/ *n.* **1** gathering of people, esp. for religious worship. **2** body of persons regularly attending a particular church etc. [Latin: related to CONGREGATE]

congregational *adj.* **1** of a congregation. **2** (**Congregational**) of or adhering to Congregationalism.

Congregationalism *n.* system whereby individual churches are largely self-governing. □ **Congregationalist** *n.*

congress /'kɒŋgres/ *n.* **1** formal meeting of delegates for discussion. **2** (**Congress**) national legislative body, esp. of the US. □ **congressional** /kən'greʃən(ə)l/ *adj.* [Latin *gradior gress-* walk]

congressman *n.* (*fem.* **congresswoman**) member of the US Congress.

congruent /'kɒŋgrʊənt/ *adj.* **1** (often foll. by *with*) suitable, agreeing. **2** *Geom.* (of figures) coinciding exactly when superimposed. □ **congruence** *n.* **congruency** *n.* [Latin *congruo* agree]

congruous /'kɒŋgrʊəs/ *adj.* suitable, agreeing; fitting. □ **congruity** /kən'gru:ɪtɪ/ *n.* [Latin: related to CONGRUENT]

conic /'kɒnɪk/ *adj.* of a cone. [Greek: related to CONE]

conical *adj.* cone-shaped.

conifer /'kɒnɪfə(r)/ *n.* tree usu. bearing cones. □ **coniferous** /kə'nɪfərəs/ *adj.* [Latin: related to CONE]

conjectural /kən'dʒektʃər(ə)l/ *adj.* based on conjecture.

conjecture /kən'dʒektʃə(r)/ —*n.* **1** formation of an opinion on incomplete information; guessing. **2** guess. —*v.* (**-ring**) guess. [Latin *conjectura* from *jacio* throw]

conjoin /kən'dʒɔɪn/ *v. formal* join, combine.

conjoint /kən'dʒɔɪnt/ *adj. formal* associated, conjoined.

conjugal /'kɒndʒʊg(ə)l/ *adj.* of marriage or the relationship of husband and wife. [Latin *conjux* consort]

conjugate —*v.* /'kɒndʒu,geɪt/ (**-ting**) **1** *Gram.* list the different forms of (a verb). **2 a** unite. **b** become fused. —*adj.* /'kɒndʒʊgət/ **1** joined together, paired. **2** fused. [Latin *jugum* yoke]

conjugation /,kɒndʒʊ'geɪʃ(ə)n/ *n.* *Gram.* system of verbal inflection.

conjunct /kən'dʒʌŋkt/ *adj.* joined together; combined; associated. [Latin from *juntus* joined]

conjunction /kən'dʒʌŋkʃ(ə)n/ *n.* **1** joining; connection. **2** *Gram.* word used to connect clauses or sentences or words in the same clause (e.g. *and*, *but*, *if*). **3** combination (of events or circumstances). **4** apparent proximity to each other of two bodies in the solar system.

conjunctiva /,kɒndʒʌŋk'taɪvə/ *n.* (*pl.* **-s**) mucous membrane covering the front of the eye and the lining inside the eyelids.

conjunctive /kən'dʒʌŋktɪv/ *adj.* **1** serving to join. **2** *Gram.* of the nature of a conjunction.

conjunctivitis /kən,dʒʌŋktɪ'vaɪtɪs/ *n.* inflammation of the conjunctiva.

conjure /'kʌndʒə(r)/ *v.* (**-ring**) **1** perform tricks which are seemingly magical, esp. by movements of the hands. **2** summon (a spirit or demon) to appear. **3** /kən'dʒʊə(r)/ *formal* appeal solemnly to. □ **conjure up 1** produce as if by magic. **2** evoke. [Latin *juro* swear]

conjuror *n.* (also **conjurer**) performer of conjuring tricks.

conk[1] *v.* (usu. foll. by *out*) *colloq.* **1** (of a machine etc.) break down. **2** (of a person) become exhausted and give up; fall asleep; faint; die. [origin unknown]

conk[2] *slang* —*n.* **1** nose or head. **2** punch on the nose or head. —*v.* hit on the nose or head. [perhaps = CONCH]

conker *n.* **1** fruit of the horse chestnut. **2** (in *pl.*) children's game played with conkers on strings. [dial. *conker* snail-shell]

con man *n.* confidence trickster.

conn *US* var. of CON[4].

connect /kə'nekt/ *v.* **1** (often foll. by *to*, *with*) join (two things, or one thing with another). **2** be joined or joinable. **3** (often foll. by *with*) associate mentally or practically. **4** (foll. by *with*) (of a train etc.) be timed to arrive with another, so passengers can transfer. **5** put into communication by telephone. **6 a** (usu. in *passive*; foll. by *with*) associate with others in relationships etc. **b** be meaningful or relevant. **7** *colloq.* hit or strike effectively. [Latin *necto nex-* bind]

connecting-rod *n.* rod between the piston and crankpin etc. in an internal combustion engine.

connection /kə'nekʃ(ə)n/ *n.* (also **connexion**) **1** connecting or being connected. **2** point at which two things are connected. **3** link, esp. by telephone. **4** connecting train etc. **5** (often in *pl.*) relative or associate, esp. one with influence. **6** relation of ideas. □ **in connection with** with reference to.

connective *adj.* connecting, esp. of body tissue connecting, separating, etc., organs etc.

connector *n.* thing that connects.
conning tower *n.* **1** superstructure of a submarine containing the periscope. **2** armoured wheel-house of a warship. [from CON[4]]
connive /kə'naɪv/ *v.* (**-ving**) **1** (foll. by *at*) disregard or tacitly consent to (a wrongdoing). **2** (usu. foll. by *with*) conspire. □ **connivance** *n.* [Latin *conniveo* shut the eyes]
connoisseur /,kɒnə'sɜ:(r)/ *n.* (often foll. by *of*, *in*) expert judge in matters of taste. [French *connaître* know]
connote /kə'nəʊt/ *v.* (**-ting**) **1** (of a word etc.) imply in addition to the literal or primary meaning. **2** mean, signify. □ **connotation** /,kɒnə'teɪʃ(ə)n/ *n.* **connotative** /'kɒnə,teɪtɪv/ *adj.* [medieval Latin: related to NOTE]
connubial /kə'nju:bɪəl/ *adj.* of marriage or the relationship of husband and wife. [Latin *nubo* marry]
conquer /'kɒŋkə(r)/ *v.* **1 a** overcome and control militarily. **b** be victorious. **2** overcome by effort. □ **conqueror** *n.* [Latin *conquiro* win]
conquest /'kɒŋkwest/ *n.* **1** conquering or being conquered. **2 a** conquered territory. **b** something won. **3** person whose affection has been won.
consanguineous /,kɒnsæŋ'gwɪnɪəs/ *adj.* descended from the same ancestor; akin. □ **consanguinity** *n.* [Latin *sanguis* blood]
conscience /'kɒnʃ(ə)ns/ *n.* moral sense of right and wrong, esp. as affecting behaviour. □ **in all conscience** *colloq.* by any reasonable standard. **on one's conscience** causing one feelings of guilt. **prisoner of conscience** person imprisoned by the State for his or her political or religious views. [Latin: related to SCIENCE]
conscience money *n.* sum paid to relieve one's conscience, esp. regarding a payment previously evaded.
conscience-stricken *adj.* (also **conscience-struck**) made uneasy by a bad conscience.
conscientious /,kɒnʃɪ'enʃəs/ *adj.* diligent and scrupulous. □ **conscientiously** *adv.* **conscientiousness** *n.* [medieval Latin: related to CONSCIENCE]
conscientious objector *n.* person who for reasons of conscience objects to military service etc.
conscious /'kɒnʃəs/ **—***adj.* **1** awake and aware of one's surroundings and identity. **2** (usu. foll. by *of* or *that*) aware, knowing. **3** (of actions, emotions, etc.) realized or recognized by the doer; intentional. **4** (in *comb.*) aware of; concerned with (*fashion-conscious*). **—***n.* (prec. by *the*) the conscious mind. □ **consciously** *adv.* **consciousness** *n.* [Latin *scio* know]
conscript **—***v.* /kən'skrɪpt/ summon for compulsory State (esp. military) service. **—***n.* /'kɒnskrɪpt/ conscripted person. □ **conscription** /kən'skrɪpʃ(ə)n/ *n.* [Latin *scribo* write]
consecrate /'kɒnsɪ,kreɪt/ *v.* (**-ting**) **1** make or declare sacred; dedicate formally to religious or divine purpose. **2** (foll. by *to*) devote to (a purpose). □ **consecration** /-'kreɪʃ(ə)n/ *n.* [Latin: related to SACRED]
consecutive /kən'sekjʊtɪv/ *adj.* **1 a** following continuously. **b** in an unbroken or logical order. **2** *Gram.* expressing a consequence. □ **consecutively** *adv.* [Latin *sequor secut-* follow]
consensus /kən'sensəs/ *n.* (often foll. by *of*; often *attrib.*) general agreement or opinion. [Latin: related to CONSENT]
consent /kən'sent/ **—***v.* (often foll. by *to*) express willingness, give permission, agree. **—***n.* voluntary agreement, permission. [Latin *sentio* feel]
consequence /'kɒnsɪkwəns/ *n.* **1** result or effect of what has gone before. **2** importance. □ **in consequence** as a result. **take the consequences** accept the results of one's choice or action. [Latin: related to CONSECUTIVE]
consequent *adj.* **1** (often foll. by *on*, *upon*) following as a result or consequence. **2** logically consistent.
consequential /,kɒnsɪ'kwenʃ(ə)l/ *adj.* **1** consequent; resulting indirectly. **2** important.
consequently *adv.* & *conj.* as a result; therefore.
conservancy /kən'sɜ:vənsɪ/ *n.* (*pl.* **-ies**) **1** body controlling a port, river, etc., or preserving the environment. **2** official environmental conservation. [Latin: related to CONSERVE]
conservation /,kɒnsə'veɪʃ(ə)n/ *n.* preservation, esp. of the natural environment. [Latin: related to CONSERVE]
conservationist *n.* supporter of environmental conservation.
conservation of energy *n.* principle that the total quantity of energy in any system that is not subject to external action remains constant.
conservative /kən'sɜ:vətɪv/ **—***adj.* **1 a** averse to rapid change. **b** (of views, taste, etc.) moderate, avoiding extremes. **2** (of an estimate etc.) purposely low. **3** (usu. **Conservative**) of Conservatives or the Conservative Party. **4** tending to conserve. **—***n.* **1** conservative person. **2** (usu. **Conservative**) supporter or member of the Conservative Party. □ **conservatism** *n.* [Latin: related to CONSERVE]

Conservative Party *n.* political party promoting free enterprise and private ownership.

conservatoire /kən'sɜ:və,twɑ:(r)/ *n.* (usu. European) school of music or other arts. [French from Italian]

conservatory /kən'sɜ:vətərɪ/ *n.* (*pl.* **-ies**) **1** greenhouse for tender plants, esp. attached to a house. **2** esp. *US* = CONSERVATOIRE. [Latin and Italian: related to CONSERVE]

conserve *–v.* /kən'sɜ:v/ (**-ving**) keep from harm or damage, esp. for later use. *–n.* /'kɒnsɜ:v/ fresh fruit jam. [Latin *servo* keep]

consider /kən'sɪdə(r)/ *v.* **1** contemplate mentally, esp. in order to reach a conclusion. **2** examine the merits of. **3** look attentively at. **4** take into account; show consideration or regard for. **5** (foll. by *that*) have the opinion. **6** regard as. **7** (as **considered** *adj.*) formed after careful thought (*a considered opinion*). □ **all things considered** taking everything into account. [French from Latin]

considerable *adj.* **1** much; a lot of (*considerable pain*). **2** notable, important. □ **considerably** *adv.*

considerate /kən'sɪdərət/ *adj.* thoughtful towards others; careful not to cause hurt or inconvenience. □ **considerately** *adv.* [Latin: related to CONSIDER]

consideration /kən,sɪdə'reɪʃ(ə)n/ *n.* **1** careful thought. **2** thoughtfulness for others; being considerate. **3** fact or thing taken into account. **4** compensation; payment or reward. □ **in consideration of** in return for; on account of. **take into consideration** make allowance for. **under consideration** being considered.

considering *–prep.* & *conj.* in view of; taking into consideration. *–adv. colloq.* taking everything into account (*not so bad, considering*).

consign /kən'saɪn/ *v.* (often foll. by *to*) **1** hand over; deliver. **2** assign; commit. **3** transmit or send (goods). □ **consignee** /,kɒnsaɪ'ni:/ *n.* **consignor** *n.* [Latin: related to SIGN]

consignment *n.* **1** consigning or being consigned. **2** goods consigned.

consist /kən'sɪst/ *v.* **1** (foll. by *of*) be composed; have as ingredients. **2** (foll. by *in*, *of*) have its essential features as specified. [Latin *sisto* stop]

consistency /kən'sɪstənsɪ/ *n.* (*pl.* **-ies**) **1** degree of density, firmness, or viscosity, esp. of thick liquids. **2** being consistent. [Latin: related to CONSIST]

consistent *adj.* **1** (usu. foll. by *with*) compatible or in harmony. **2** (of a person) constant to the same principles. □ **consistently** *adv.* [Latin: related to CONSIST]

consistory /kən'sɪstərɪ/ *n.* (*pl.* **-ies**) *RC Ch.* council of cardinals (with or without the pope). [Latin: related to CONSIST]

consolation /,kɒnsə'leɪʃ(ə)n/ *n.* **1** consoling or being consoled. **2** consoling thing or person. □ **consolatory** /kən'sɒlətərɪ/ *adj.*

consolation prize *n.* prize given to a competitor who just fails to win a main prize.

console[1] /kən'səʊl/ *v.* (**-ling**) comfort, esp. in grief or disappointment. [Latin: related to SOLACE]

■ **Usage** *Console* is often confused with *condole*, which is different in that it is always followed by *with*.

console[2] /'kɒnsəʊl/ *n.* **1** panel for switches, controls, etc. **2** cabinet for a television etc. **3** cabinet with the keyboards and stops of an organ. **4** bracket supporting a shelf etc. [French]

consolidate /kən'sɒlɪ,deɪt/ *v.* (**-ting**) **1** make or become strong or secure. **2** combine (territories, companies, debts, etc.) into one whole. □ **consolidation** /-'deɪʃ(ə)n/ *n.* **consolidator** *n.* [Latin: related to SOLID]

consommé /kən'sɒmeɪ/ *n.* clear soup from meat stock. [French]

consonance /'kɒnsənəns/ *n.* agreement, harmony. [Latin *sono* SOUND[1]]

consonant *–n.* **1** speech sound in which the breath is at least partly obstructed, and which forms a syllable by combining with a vowel. **2** letter(s) representing this. *–adj.* (foll. by *with*, *to*) consistent; in agreement or harmony. □ **consonantal** /-'nænt(ə)l/ *adj.*

consort[1] *–n.* /'kɒnsɔ:t/ wife or husband, esp. of royalty. *–v.* /kən'sɔ:t/ **1** (usu. foll. by *with*, *together*) keep company. **2** harmonize. [Latin: related to SORT]

consort[2] /'kɒnsɔ:t/ *n. Mus.* small group of players, singers, or instruments. [var. of CONCERT]

consortium /kən'sɔ:tɪəm/ *n.* (*pl.* **-tia** or **-s**) association, esp. of several business companies. [Latin: related to CONSORT[1]]

conspicuous /kən'spɪkjʊəs/ *adj.* **1** clearly visible; attracting notice. **2** noteworthy. □ **conspicuously** *adv.* [Latin *specio* look]

conspiracy /kən'spɪrəsɪ/ *n.* (*pl.* **-ies**) **1** secret plan to commit a crime; plot. **2** conspiring. [Latin: related to CONSPIRE]

conspiracy of silence *n.* agreement to say nothing.

conspirator /kən'spɪrətə(r)/ *n.* person who takes part in a conspiracy. □ **conspiratorial** /-'tɔ:rɪəl/ *adj.*

conspire /kən'spaɪə(r)/ *v.* (**-ring**) **1** combine secretly for an unlawful or harmful act. **2** (of events) seem to be working together. [Latin *spiro* breathe]

constable /'kʌnstəb(ə)l/ *n.* **1** (also **police constable**) police officer of the lowest rank. **2** governor of a royal castle. [Latin *comes stabuli* count of the stable]

constabulary /kən'stæbjʊlərɪ/ *n.* (*pl.* **-ies**) police force. [medieval Latin: related to CONSTABLE]

constancy /'kɒnstənsɪ/ *n.* being unchanging and dependable; faithfulness. [Latin: related to CONSTANT]

constant *–adj.* **1** continuous (*constant attention*). **2** occurring frequently (*constant complaints*). **3** unchanging, faithful, dependable. *–n.* **1** anything that does not vary. **2** *Math.* & *Physics* quantity or number that remains the same. □ **constantly** *adv.* [Latin *sto* stand]

constellation /,kɒnstə'leɪʃ(ə)n/ *n.* **1** group of fixed stars. **2** group of associated persons etc. [Latin *stella* star]

consternation /,kɒnstə'neɪʃ(ə)n/ *n.* anxiety, dismay. [Latin *sterno* throw down]

constipate /'kɒnstɪ,peɪt/ *v.* (**-ting**) (esp. as **constipated** *adj.*) affect with constipation. [Latin *stipo* cram]

constipation /,kɒnstɪ'peɪʃ(ə)n/ *n.* difficulty in emptying the bowels.

constituency /kən'stɪtjʊənsɪ/ *n.* (*pl.* **-ies**) **1** body of voters who elect a representative. **2** area so represented.

constituent /kən'stɪtjʊənt/ *–adj.* **1** composing or helping to make a whole. **2** able to make or change a constitution (*constituent assembly*). **3** electing. *–n.* **1** member of a constituency. **2** component part. [Latin: related to CONSTITUTE]

constitute /'kɒnstɪ,tju:t/ *v.* (**-ting**) **1** be the components or essence of; compose. **2 a** amount to (*this constitutes a warning*). **b** formally establish (*constitutes a precedent*). **3** give legal or constitutional form to. [Latin *constituo* establish]

constitution /,kɒnstɪ'tju:ʃ(ə)n/ *n.* **1** act or method of constituting; composition. **2** body of fundamental principles by which a State or other body is governed. **3** person's inherent state of health, strength, etc. [Latin: related to CONSTITUTE]

constitutional *–adj.* **1** of or in line with the constitution. **2** inherent (*constitutional weakness*). *–n.* walk taken regularly as healthy exercise. □ **constitutionality** /-'nælɪtɪ/ *n.* **constitutionally** *adv.*

constitutive /'kɒnstɪ,tju:tɪv/ *adj.* **1** able to form or appoint. **2** component. **3** essential.

constrain /kən'streɪn/ *v.* **1** compel. **2 a** confine forcibly; imprison. **b** restrict severely. **3** (as **constrained** *adj.*) forced, embarrassed. [Latin *stringo strict-* tie]

constraint /kən'streɪnt/ *n.* **1** constraining or being constrained. **2** restriction. **3** self-control.

constrict /kən'strɪkt/ *v.* make narrow or tight; compress. □ **constriction** *n.* **constrictive** *adj.* [Latin: related to CONSTRAIN]

constrictor *n.* **1** snake that kills by compressing. **2** muscle that contracts an organ or part of the body.

construct *–v.* /kən'strʌkt/ **1** make by fitting parts together; build, form. **2** *Geom.* delineate (a figure). *–n.* /'kɒnstrʌkt/ thing constructed, esp. by the mind. □ **constructor** /kən'strʌktə(r)/ *n.* [Latin *struo struct-* build]

construction /kən'strʌkʃ(ə)n/ *n.* **1** constructing or being constructed. **2** thing constructed. **3** interpretation or explanation. **4** syntactical arrangement of words. □ **constructional** *adj.*

constructive /kən'strʌktɪv/ *adj.* **1 a** tending to form a basis for ideas. **b** helpful, positive. **2** derived by inference. □ **constructively** *adv.*

construe /kən'stru:/ *v.* (**-strues, -strued, -struing**) **1** interpret. **2** (often foll. by *with*) combine (words) grammatically. **3** analyse the syntax of (a sentence). **4** translate literally. [Latin: related to CONSTRUCT]

consubstantial /,kɒnsəb'stænʃ(ə)l/ *adj. Theol.* of one substance. [Church Latin: related to SUBSTANCE]

consubstantiation /,kɒnsəb,stænʃɪ'eɪʃ(ə)n/ *n. Theol.* presence of Christ's body and blood together with the bread and wine in the Eucharist.

consul /'kɒns(ə)l/ *n.* **1** official appointed by a State to protect its citizens and interests in a foreign city. **2** *hist.* either of two chief magistrates in ancient Rome. □ **consular** /-sjʊlə(r)/ *adj.* **consulship** *n.* [Latin]

consulate /'kɒnsjʊlət/ *n.* **1** official building of a consul. **2** position of consul.

consult /kən'sʌlt/ *v.* **1** seek information or advice from. **2** (often foll. by *with*) refer to a person for advice etc. **3** take into account (feelings, interests, etc.). □ **consultative** *adj.* [Latin *consulo consult-* take counsel]

consultancy *n.* (*pl.* **-ies**) practice or position of a consultant.

consultant *n.* **1** person providing professional advice etc. **2** senior medical specialist in a hospital.

consultation /ˌkɒnsəl'teɪʃ(ə)n/ *n.* **1** meeting arranged to consult. **2** act or process of consulting.

consume /kən'sju:m/ *v.* (**-ming**) **1** eat or drink. **2** destroy. **3** preoccupy, possess (*consumed with rage*). **4** use up. □ **consumable** *adj.* & *n.* [Latin *consumo -sumpt-*]

consumer *n.* **1** person who consumes, esp. one who uses a product. **2** purchaser of goods or services.

consumer durable *n.* durable household product (e.g. a radio or washing-machine).

consumer goods *n.pl.* goods for consumers, not for producing other goods.

consumerism *n.* **1** protection of consumers' interests. **2** (often *derog.*) continual increase in the consumption of goods. □ **consumerist** *adj.*

consummate –*v.* /'kɒnsəˌmeɪt/ (**-ting**) **1** complete; make perfect. **2** complete (a marriage) by sexual intercourse. –*adj.* /kən'sʌmɪt/ complete, perfect; fully skilled. □ **consummation** /-'meɪʃ(ə)n/ *n.* [Latin *summus* utmost]

consumption /kən'sʌmpʃ(ə)n/ *n.* **1** consuming or being consumed. **2** amount consumed. **3** use by a particular group (*a film unsuitable for children's consumption*). **4** *archaic* tuberculosis of the lungs. **5** purchase and use of goods etc. [French: related to CONSUME]

consumptive /kən'sʌmptɪv/ *archaic* –*adj.* suffering or tending to suffer from consumption. –*n.* consumptive person. [medieval Latin: related to CONSUMPTION]

cont. *abbr.* **1** contents. **2** continued.

contact /'kɒntækt/ –*n.* **1** state or condition of touching, meeting, or communicating. **2** person who is or may be communicated with for information, assistance, etc. **3** connection for the passage of an electric current. **4** person likely to carry a contagious disease through being near an infected person. –*v.* **1** get in touch with (a person). **2** begin correspondence or personal dealings with. [Latin *tango tact-* touch]

contact lens *n.* small lens placed directly on the eyeball to correct vision.

contact print photographic print made by placing a negative directly on to printing paper and exposing it to light.

contagion /kən'teɪdʒ(ə)n/ *n.* **1 a** spreading of disease by bodily contact. **b** contagious disease. **2** moral corruption. [related to CONTACT]

contagious /kən'teɪdʒəs/ *adj.* **1 a** (of a person) likely to transmit a disease by contact. **b** (of a disease) transmitted in this way. **2** (of emotions etc.) likely to spread (*contagious enthusiasm*).

contain /kən'teɪn/ *v.* **1** hold or be capable of holding within itself; include, comprise. **2** (of measures) be equal to (*a gallon contains eight pints*). **3** prevent from moving or extending. **4** control or restrain (feelings etc.). **5** (of a number) be divisible by (a factor) without a remainder. [Latin *teneo* hold]

container *n.* **1** box, jar, etc., for holding things. **2** large metal box for transporting goods.

containerize *v.* (also **-ise**) (**-zing** or **-sing**) pack in or transport by container. □ **containerization** /-'zeɪʃ(ə)n/ *n.*

containment *n.* action or policy of preventing the expansion of a hostile country or influence.

contaminate /kən'tæmɪˌneɪt/ *v.* (**-ting**) **1** pollute, esp. with radioactivity. **2** infect. □ **contaminant** *n.* **contamination** /-'neɪʃ(ə)n/ *n.* **contaminator** *n.* [Latin *tamen-* related to *tango* touch]

contemplate /'kɒntəmˌpleɪt/ *v.* (**-ting**) **1** survey visually or mentally. **2** regard (an event) as possible. **3** intend (*he is not contemplating retiring*). **4** meditate. □ **contemplation** /-'pleɪʃ(ə)n/ *n.* [Latin]

contemplative /kən'templətɪv/ –*adj.* of or given to (esp. religious) contemplation; thoughtful. –*n.* person devoted to religious contemplation. [Latin: related to CONTEMPLATE]

contemporaneous /kənˌtempə'reɪnɪəs/ *adj.* (usu. foll. by *with*) existing or occurring at the same time. □ **contemporaneity** /-'ni:ɪtɪ/ *n.* [Latin: related to COM-, *tempus* time]

contemporary /kən'tempərərɪ/ –*adj.* **1** living or occurring at the same time. **2** of approximately the same age. **3** modern in style or design. –*n.* (*pl.* **-ies**) contemporary person or thing. [medieval Latin: related to CONTEMPORANEOUS]

contempt /kən'tempt/ *n.* **1** feeling that a person or thing deserves scorn or extreme reproach. **2** condition of being held in contempt. **3** (in full **contempt of court**) disobedience to or disrespect for a court of law. [Latin *temno tempt-* despise]

contemptible *adj.* deserving contempt. □ **contemptibly** *adv.*

contemptuous *adj.* (often foll. by *of*) feeling or showing contempt. □ **contemptuously** *adv.*

contend /kən'tend/ *v.* **1** (usu. foll. by *with*) fight, argue. **2** compete. **3** assert, maintain. □ **contender** *n.* [Latin: related to TEND[1]]

content[1] /kən'tent/ –*predic. adj.* **1** satisfied; adequately happy. **2** (foll. by *to* + infin.) willing. –*v.* make content;

satisfy. **–*n.*** contented state; satisfaction. □ **to one's heart's content** as much as one wishes. [Latin: related to CONTAIN]

content[2] /'kɒntent/ *n.* **1** (usu. in *pl.*) what is contained, esp. in a vessel, book, or house. **2** amount (of a constituent) contained (*high fat content*). **3** substance (of a speech etc.) as distinct from form. **4** capacity or volume. [medieval Latin: related to CONTAIN]

contented /kən'tentɪd/ *adj.* showing or feeling content; happy, satisfied. □ **contentedly** *adv.* **contentedness** *n.*

contention /kən'tenʃ(ə)n/ *n.* **1** dispute or argument; rivalry. **2** point contended for in an argument. [Latin: related to CONTEND]

contentious /kən'tenʃəs/ *adj.* **1** quarrelsome. **2** likely to cause an argument.

contentment *n.* satisfied state; tranquil happiness.

contest **–*n.*** /'kɒntest/ **1** contending; strife. **2** a competition. **–*v.*** /kən'test/ **1** dispute (a decision etc.). **2** contend or compete for; compete in (an election). [Latin *testis* witness]

contestant /kən'test(ə)nt/ *n.* person taking part in a contest.

context /'kɒntekst/ *n.* **1** parts that surround a word or passage and clarify its meaning. **2** relevant circumstances. □ **in** (or **out of**) **context** with (or without) the surrounding words or circumstances. □ **contextual** /kən'tekstjʊəl/ *adj.* **contextualize** /kən'tekstjʊə,laɪz/ *v.* (also **-ise**) (**-zing** or **-sing**). [Latin: related to TEXT]

contiguous /kən'tɪgjʊəs/ *adj.* (usu. foll. by *with, to*) touching; in contact. □ **contiguity** /,kɒntɪ'gju:ɪtɪ/ *n.* [Latin: related to CONTACT]

continent[1] /'kɒntɪnənt/ *n.* **1** any of the main continuous expanses of land (Europe, Asia, Africa, N. and S. America, Australia, Antarctica). **2** (**the Continent**) mainland of Europe as distinct from the British Isles. [Latin: related to CONTAIN]

continent[2] /'kɒntɪnənt/ *adj.* **1** able to control one's bowels and bladder. **2** exercising self-restraint, esp. sexually. □ **continence** *n.* [Latin: related to CONTAIN]

continental /,kɒntɪ'nent(ə)l/ *adj.* **1** of or characteristic of a continent. **2** (**Continental**) of or characteristic of mainland Europe.

continental breakfast *n.* light breakfast of coffee, rolls, etc.

continental quilt *n.* duvet.

continental shelf *n.* area of shallow seabed bordering a continent.

contingency /kən'tɪndʒənsɪ/ *n.* (*pl.* **-ies**) **1** event that may or may not occur. **2** something dependent on another uncertain event. [Latin: related to CONTINGENT]

contingent **–*adj.*** **1** (usu. foll. by *on, upon*) conditional, dependent (on an uncertain event or circumstance). **2 a** that may or may not occur. **b** fortuitous. **–*n.*** **1** body (of troops, ships, etc.) forming part of a larger group. **2** group of people sharing an interest, origin, etc. (*the Oxford contingent*). [Latin: related to CONTACT]

continual /kən'tɪnjʊəl/ *adj.* constantly or frequently recurring; always happening. □ **continually** *adv.* [French: related to CONTINUE]

■ **Usage** *Continual* is often confused with *continuous*. *Continual* is used of something that happens very frequently (e.g. *there were continual interruptions*), while *continuous* is used of something that happens without a pause (e.g. *continuous rain all day*).

continuance /kən'tɪnjʊəns/ *n.* **1** continuing in existence or operation. **2** duration.

continuation /kən,tɪnjʊ'eɪʃ(ə)n/ *n.* **1** continuing or being continued. **2** part that continues something else.

continue /kən'tɪnju:/ *v.* (**-ues, -ued, -uing**) **1** maintain, not stop (an action etc.) (*continued to read, reading*). **2** (also *absol.*) resume or prolong (a narrative, journey, etc.). **3** be a sequel to. **4** remain, stay (*will continue as manager; weather continued fine*). [Latin: related to CONTAIN]

continuity /,kɒntɪ'nju:ɪtɪ/ *n.* (*pl.* **-ies**) **1** state of being continuous. **2** a logical sequence. **3** detailed scenario of a film or broadcast. **4** linking of broadcast items.

continuo /kən'tɪnjʊəʊ/ *n.* (*pl.* **-s**) *Mus.* accompaniment providing a bass line, played usu. on a keyboard instrument. [Italian]

continuous /kən'tɪnjʊəs/ *adj.* uninterrupted, connected throughout in space or time. □ **continuously** *adv.* [Latin: related to CONTAIN]

■ **Usage** See note at *continual*.

continuous assessment *n.* evaluation of a pupil's progress throughout a course of study.

continuum /kən'tɪnjʊəm/ *n.* (*pl.* **-nua**) thing having a continuous structure. [Latin: related to CONTINUOUS]

contort /kən'tɔ:t/ *v.* twist or force out of its normal shape. □ **contortion** *n.* [Latin *torqueo tort-* twist]

contortionist /kən'tɔːʃənɪst/ *n.* entertainer who adopts contorted postures.

contour /'kɒntʊə(r)/ *–n.* **1** outline. **2** (in full **contour line**) line on a map joining points of equal altitude. *–v.* mark with contour lines. [Italian *contornare* draw in outline]

contra /'kɒntrə/ *n.* (*pl.* **-s**) member of a counter-revolutionary force in Nicaragua. [abbreviation of Spanish *contrarevolucionario* counter-revolutionary]

contra- *comb. form* against, opposite. [Latin]

contraband /'kɒntrə,bænd/ *–n.* **1** smuggled goods. **2** smuggling; illegal trade. *–adj.* forbidden to be imported or exported. [Spanish from Italian]

contraception /,kɒntrə'sepʃ(ə)n/ *n.* prevention of pregnancy; use of contraceptives. [from CONTRA-, CONCEPTION]

contraceptive /,kɒntrə'septɪv/ *–adj.* preventing pregnancy. *–n.* contraceptive device or drug.

contract *–n.* /'kɒntrækt/ **1** written or spoken agreement, esp. one enforceable by law. **2** document recording this. *–v.* /kən'trækt/ **1** make or become smaller. **2 a** (usu. foll. by *with*) make a contract. **b** (often foll. by *out*) arrange (work) to be done by contract. **3** become affected by (a disease). **4** enter into (marriage). **5** incur (a debt etc.). **6** draw together (the muscles, brow, etc.), or be drawn together. □ **contract in** (or **out**) choose to enter (or withdraw from or not enter) a scheme or commitment. [Latin *contractus*: related to TRACT[1]]

contractable *adj.* (of a disease) that can be contracted.

contract bridge *n.* bridge in which only tricks bid and won count towards the game.

contractible *adj.* that can be shrunk or drawn together.

contractile /kən'træktaɪl/ *adj.* capable of or producing contraction. □ **contractility** /,kɒntræk'tɪlɪtɪ/ *n.*

contraction /kən'trækʃ(ə)n/ *n.* **1** contracting or being contracted. **2** *Med.* shortening of the uterine muscles during childbirth. **3** shrinking, diminution. **4** shortened form of a word or words (e.g. *he's*).

contractor /kən'træktə(r)/ *n.* person who makes a contract, esp. to conduct building operations.

contractual /kən'træktʃʊəl/ *adj.* of or in the nature of a contract. □ **contractually** *adv.*

contradict /,kɒntrə'dɪkt/ *v.* **1** deny (a statement). **2** deny a statement made by (a person). **3** be in opposition to or in conflict with. □ **contradiction** *n.* **contradictory** *adj.* [Latin *dico dict-* say]

contradistinction /,kɒntrədɪ'stɪŋkʃ(ə)n/ *n.* distinction made by contrasting.

contraflow /'kɒntrə,fləʊ/ *n.* transfer of traffic from its usual half of the road to the other half by borrowing one or more of the other half's lanes.

contralto /kən'træltəʊ/ *n.* (*pl.* **-s**) **1** lowest female singing-voice. **2** singer with this voice. [Italian: related to CONTRA-, ALTO]

contraption /kən'træpʃ(ə)n/ *n.* machine or device, esp. a strange or cumbersome one. [origin unknown]

contrapuntal /,kɒntrə'pʌnt(ə)l/ *adj.* *Mus.* of or in counterpoint. □ **contrapuntally** *adv.* [Italian]

contrariwise /kən'treərɪ,waɪz/ *adv.* **1** on the other hand. **2** in the opposite way. **3** perversely.

contrary /'kɒntrərɪ/ *–adj.* **1** (usu. foll. by *to*) opposed in nature or tendency. **2** /kən'treərɪ/ perverse, self-willed. **3** (of a wind) unfavourable, impeding. **4** opposite in position or direction. *–n.* (prec. by *the*) the opposite. *–adv.* (foll. by *to*) in opposition or contrast (*contrary to expectations*). □ **on the contrary** expressing denial of what has just been implied or stated. **to the contrary** to the opposite effect. □ **contrariness** /kən'treərɪnɪs/ *n.* [Latin: related to CONTRA-]

contrast *–n.* /'kɒntrɑːst/ **1 a** juxtaposition or comparison showing differences. **b** difference so revealed. **2** (often foll. by *to*) thing or person having different qualities. **3** degree of difference between the tones in a television picture or photograph. *–v.* /kən'trɑːst/ (often foll. by *with*) **1** set together so as to reveal a contrast. **2** have or show a contrast. [Italian from Latin *sto* stand]

contravene /,kɒntrə'viːn/ *v.* (**-ning**) **1** infringe (a law etc.). **2** (of things) conflict with. □ **contravention** /-'venʃ(ə)n/ *n.* [Latin *venio* come]

contretemps /'kɒntrə,tɔ̃/ *n.* (*pl.* same /-,tɔ̃z/) **1** unfortunate occurrence. **2** unexpected mishap. [French]

contribute /kən'trɪbjuːt, 'kɒntrɪ,bjuːt/ *v.* (**-ting**) (often foll. by *to*) **1** give (time, money, etc.) towards a common purpose. **2** help to bring about a result etc. **3** (also *absol.*) supply (an article etc.) for publication with others. □ **contributor** /kən'trɪb-/ *n.* [Latin: related to TRIBUTE]

■ **Usage** The second pronunciation, stressed on the first syllable, is considered incorrect by some people.

contribution /,kɒntrɪ'bjuːʃ(ə)n/ *n.* **1** act of contributing. **2** thing contributed.

contributory /kən'trɪbjʊtərɪ/ *adj.* **1** that contributes. **2** using contributions.

contrite /'kɒntraɪt/ *adj.* penitent, feeling great guilt. □ **contritely** *adv.* **contrition** /kən'trɪʃ(ə)n/ *n.* [Latin: related to TRITE]

contrivance /kən'traɪv(ə)ns/ *n.* **1** something contrived, esp. a plan or mechanical device. **2** act of contriving.

contrive /kən'traɪv/ *v.* (**-ving**) **1** devise; plan or make resourcefully or with skill. **2** (often foll. by *to* + infin.) manage. [French from Latin]

contrived *adj.* artificial, forced.

control /kən'trəʊl/ *–n.* **1** power of directing. **2** power of restraining, esp. self-restraint. **3** means of restraint. **4** (usu. in *pl.*) means of regulating. **5** (usu. in *pl.*) switches and other devices by which a machine is controlled. **6** place where something is controlled or verified. **7** standard of comparison for checking the results of an experiment. *–v.* (**-ll-**) **1** have control of, regulate. **2** hold in check. **3** check, verify. □ **in control** (often foll. by *of*) directing an activity. **out of control** no longer manageable. **under control** being controlled; in order. □ **controllable** *adj.* [medieval Latin, = keep copy of accounts: related to CONTRA-, ROLL]

controller *n.* **1** person or thing that controls. **2** person in charge of expenditure.

control tower *n.* tall building at an airport etc. from which air traffic is controlled.

controversial /,kɒntrə'vɜ:ʃ(ə)l/ *adj.* causing or subject to controversy. [Latin: related to CONTROVERT]

controversy /'kɒntrə,vɜ:sɪ, kən'trɒvəsɪ/ *n.* (*pl.* **-ies**) prolonged argument or dispute. [Latin: related to CONTROVERT]

■ **Usage** The second pronunciation, stressed on the second syllable, is considered incorrect by some people.

controvert /'kɒntrə,vɜ:t/ *v.* dispute, deny. [Latin *verto vers-* turn]

contumacious /,kɒntju:'meɪʃəs/ *adj.* stubbornly or wilfully disobedient. □ **contumacy** /'kɒntjʊməsɪ/ *n.* (*pl.* **-ies**). [Latin *tumeo* swell]

contumely /'kɒntju:mlɪ/ *n.* **1** insolent language or treatment. **2** disgrace. [Latin: related to CONTUMACIOUS]

contuse /kən'tju:z/ *v.* (**-sing**) bruise. □ **contusion** *n.* [Latin *tundo tus-* thump]

conundrum /kə'nʌndrəm/ *n.* **1** riddle, esp. one with a pun in its answer. **2** hard question. [origin unknown]

conurbation /,kɒnɜ:'beɪʃ(ə)n/ *n.* extended urban area, esp. consisting of several towns and merging suburbs. [Latin *urbs* city]

convalesce /,kɒnvə'les/ *v.* (**-cing**) recover health after illness. [Latin *valeo* be well]

convalescent *–adj.* recovering from an illness. *–n.* convalescent person. □ **convalescence** *n.*

convection /kən'vekʃ(ə)n/ *n.* heat transfer by upward movement of a heated and less dense medium. [Latin *veho vect-* carry]

convector /kən'vektə(r)/ *n.* heating appliance that circulates warm air by convection.

convene /kən'vi:n/ *v.* (**-ning**) **1** summon or arrange (a meeting etc.). **2** assemble. [Latin *venio vent-* come]

convener *n.* (also **convenor**) **1** person who convenes a meeting. **2** senior trade union official at a workplace.

convenience /kən'vi:nɪəns/ *n.* **1** state of being convenient; suitability. **2** useful thing. **3** advantage. **4** lavatory, esp. a public one. □ **at one's convenience** at a time or place that suits one. [Latin: related to CONVENE]

convenience food *n.* food requiring little preparation.

convenient *adj.* **1 a** serving one's comfort or interests. **b** suitable. **c** free of trouble or difficulty. **2** available or occurring at a suitable time or place. **3** well situated (*convenient for the shops*). □ **conveniently** *adv.*

convent /'kɒnv(ə)nt/ *n.* **1** religious community, esp. of nuns, under vows. **2** premises occupied by this. [Latin: related to CONVENE]

conventicle /kən'ventɪk(ə)l/ *n.* esp. *hist.* secret or unlawful religious meeting, esp. of dissenters. [Latin: related to CONVENE]

convention /kən'venʃ(ə)n/ *n.* **1 a** general agreement on social behaviour etc. by implicit majority consent. **b** a custom or customary practice. **2** conference of people with a common interest. **3** a formal agreement, esp. between States. [Latin: related to CONVENE]

conventional *adj.* **1** depending on or according with convention. **2** (of a person) bound by social conventions. **3** usual; of agreed significance. **4** not spontaneous or sincere or original. **5** (of weapons etc.) non-nuclear. □ **conventionalism** *n.* **conventionality** /-'nælɪtɪ/ *n.* (*pl.* **-ies**). **conventionally** *adv.*

converge /kən'vɜ:dʒ/ *v.* (**-ging**) **1** come together or towards the same point. **2** (foll. by *on, upon*) approach from different directions. □ **convergence** *n.* **convergent** *adj.* [Latin *vergo* incline]

conversant /kən'vɜ:s(ə)nt/ *adj.* (foll. by *with*) well acquainted with. [French: related to CONVERSE[1]]

conversation /,kɒnvə'seɪʃ(ə)n/ *n.* **1** informal spoken communication. **2** instance of this. [Latin: related to CONVERSE[1]]

conversational *adj.* **1** of or in conversation. **2** colloquial. □ **conversationally** *adv.*

conversationalist *n.* person good at or fond of conversation.

converse[1] /kən'vɜ:s/ *v.* **(-sing)** (often foll. by *with*) talk. [Latin: related to CONVERT]

converse[2] /'kɒnvɜ:s/ *—adj.* opposite, contrary, reversed. *—n.* something, esp. a statement or proposition, that is opposite or reversed. □ **conversely** *adv.* [Latin: related to CONVERT]

conversion /kən'vɜ:ʃ(ə)n/ *n.* **1** converting or being converted. **2** converted building or part of this. [Latin: related to CONVERT]

convert *—v.* /kən'vɜ:t/ **1** (usu. foll. by *into*) change in form or function. **2** cause (a person) to change belief etc. **3** change (moneys etc.) into others of a different kind. **4** make structural alterations in (a building) for a new purpose. **5** (also *absol.*) *Rugby* score extra points from (a try) by a successful kick at the goal. *—n.* /'kɒnvɜ:t/ (often foll. by *to*) person converted to a different belief etc. [Latin *verto vers-* turn]

convertible /kən'vɜ:tɪb(ə)l/ *—adj.* able to be converted. *—n.* car with a folding or detachable roof. □ **convertibility** /-'bɪlɪtɪ/ *n.* [Latin: related to CONVERT]

convex /'kɒnveks/ *adj.* curved like the exterior of a circle or sphere. □ **convexity** /-'veksɪtɪ/ *n.* [Latin]

convey /kən'veɪ/ *v.* **1** transport or carry (goods, passengers, etc.). **2** communicate (an idea, meaning, etc.). **3** transfer the title to (a property). **4** transmit (sound etc.). □ **conveyable** *adj.* [Latin *via* way]

conveyance *n.* **1** conveying or being conveyed. **2** means of transport; vehicle. **3** *Law* **a** transfer of property. **b** document effecting this. □ **conveyancer** *n.* (in sense 3). **conveyancing** *n.* (in sense 3).

conveyor *n.* (also **conveyer**) person or thing that conveys.

conveyor belt *n.* endless moving belt for conveying articles, esp. in a factory.

convict *—v.* /kən'vɪkt/ **1** (often foll. by *of*) prove to be guilty (of a crime etc.). **2** declare guilty by a legal process. *—n.* /'kɒnvɪkt/ chiefly *hist.* person serving a prison sentence. [Latin *vinco vict-* conquer]

conviction /kən'vɪkʃ(ə)n/ *n.* **1** convicting or being convicted. **2** **a** being convinced. **b** firm belief. [Latin: related to CONVICT]

convince /kən'vɪns/ *v.* **(-cing)** firmly persuade. □ **convincible** *adj.* **convincing** *adj.* **convincingly** *adv.* [Latin: related to CONVICT]

convivial /kən'vɪvɪəl/ *adj.* fond of good company; sociable and lively. □ **conviviality** /-'ælɪtɪ/ *n.* [Latin *vivo* live]

convocation /,kɒnvə'keɪʃ(ə)n/ *n.* **1** convoking or being convoked. **2** large formal gathering. [Latin: related to CONVOKE]

convoke /kən'vəʊk/ *v.* **(-king)** *formal* call together; summon to assemble. [Latin *voco* call]

convoluted /'kɒnvə,lu:tɪd/ *adj.* **1** coiled, twisted. **2** complex. [Latin *volvo volut-* roll]

convolution /,kɒnvə'lu:ʃ(ə)n/ *n.* **1** coiling. **2** coil or twist. **3** complexity. **4** sinuous fold in the surface of the brain.

convolvulus /kən'vɒlvjʊləs/ *n.* (*pl.* **-luses**) twining plant, esp. bindweed. [Latin]

convoy /'kɒnvɔɪ/ *—n.* group of ships, vehicles, etc., travelling together or under escort. *—v.* escort, esp. with armed force. □ **in convoy** as a group. [French: related to CONVEY]

convulse /kən'vʌls/ *v.* **(-sing) 1** (usu. in *passive*) affect with convulsions. **2** cause to laugh uncontrollably. □ **convulsive** *adj.* **convulsively** *adv.* [Latin *vello vuls-* pull]

convulsion /kən'vʌlʃ(ə)n/ *n.* **1** (usu. in *pl.*) violent irregular motion of the limbs or body caused by involuntary contraction of muscles. **2** violent disturbance. **3** (in *pl.*) uncontrollable laughter.

cony /'kəʊnɪ/ *n.* (also **coney**) rabbit fur. [Latin *cuniculus*]

coo *—n.* soft murmuring sound as of a dove. *—v.* **(coos, cooed) 1** emit a coo. **2** talk or say in a soft or amorous voice. *—int. slang* expressing surprise or disbelief. [imitative]

cooee /'ku:i:/ *n.* & *int. colloq.* call used to attract attention. [imitative]

cook /kʊk/ *—v.* **1** prepare (food) by heating it. **2** (of food) undergo cooking. **3** *colloq.* falsify (accounts etc.). **4** (as **be cooking**) *colloq.* be happening or about to happen. *—n.* person who cooks, esp. professionally or in a specified way (*a good cook*). □ **cook up** *colloq.* concoct (a story, excuse, etc.). [Latin *coquus*]

cookbook *n. US* cookery book.

cook-chill *attrib. adj.* (of food, meals, etc.) sold in pre-cooked and refrigerated form.

cooker *n.* **1** appliance or vessel for cooking food. **2** fruit (esp. an apple) suitable for cooking.

cookery *n.* art or practice of cooking.

cookery book *n.* book containing recipes.

cookie /ˈkʊkɪ/ *n. US* **1** sweet biscuit. **2** *colloq.* person (*a tough cookie*). [Dutch *koekje*]

cool *–adj.* **1** of or at a fairly low temperature, fairly cold. **2** suggesting or achieving coolness. **3** calm, unexcited. **4** lacking enthusiasm. **5** unfriendly (*a cool reception*). **6** calmly audacious. **7** (prec. by *a*) *colloq.* at least (*cost a cool thousand*). **8** *slang* esp. *US* marvellous. *–n.* **1** coolness. **2** cool air or place. **3** *slang* calmness, composure. *–v.* (often foll. by *down, off*) make or become cool. □ **cool it** *slang* relax, calm down. □ **coolly** /ˈku:llɪ/ *adv.* **coolness** *n.* [Old English]

coolant *n.* cooling agent, esp. fluid.

cool-bag *n.* (also **cool-box**) insulated container for keeping food cool.

cooler *n.* **1** vessel in which a thing is cooled. **2** *US* refrigerator. **3** *slang* prison cell.

coolie /ˈku:lɪ/ *n.* unskilled native labourer in Eastern countries. [perhaps from *Kulī*, tribe in India]

cooling-off period *n.* interval to allow for a change of mind.

cooling tower *n.* tall structure for cooling hot water before reuse, esp. in industry.

coomb /ku:m/ *n.* (also **combe**) **1** valley on the side of a hill. **2** short valley running up from the coast. [Old English]

coon *n.* **1** *US* racoon. **2** *slang offens.* Black. [abbreviation]

coop *–n.* cage for keeping poultry. *–v.* (often foll. by *up, in*) confine (a person). [Latin *cupa* cask]

co-op /ˈkəʊɒp/ *n. colloq.* cooperative society or shop. [abbreviation]

cooper *n.* maker or repairer of casks, barrels, etc. [Low German or Dutch: related to COOP]

cooperate /kəʊˈɒpəˌreɪt/ *v.* (also **co-operate**) (**-ting**) **1** (often foll. by *with*) work or act together. **2** be helpful and do as one is asked. □ **cooperation** /-ˈreɪʃ(ə)n/ *n.* [related to CO-]

cooperative /kəʊˈɒpərətɪv/ (also **co-operative**) *–adj.* **1** willing to cooperate. **2** of or characterized by cooperation. **3** (of a business) owned and run jointly by its members, with profits shared. *–n.* cooperative farm, society, or business.

co-opt /kəʊˈɒpt/ *v.* appoint to membership of a body by invitation of the existing members. □ **co-option** *n.* **co-optive** *adj.* [Latin *coopto* from *opto* choose]

coordinate (also **co-ordinate**) *–v.* /kəʊˈɔ:dɪˌneɪt/ (**-ting**) **1** cause (parts, movements, etc.) to function together efficiently. **2** work or act together effectively. *–adj.* /kəʊˈɔ:dɪnət/ equal in rank or importance. *–n.* /kəʊˈɔ:dɪnət/ **1** *Math.* each of a system of values used to fix the position of a point, line, or plane. **2** (in *pl.*) matching items of clothing. □ **coordination** /-ˈneɪʃ(ə)n/ *n.* **coordinator** /-ˌneɪtə(r)/ *n.* [Latin *ordino*: related to ORDER]

coot *n.* **1** black aquatic bird with a white horny plate on its forehead. **2** *colloq.* stupid person. [probably Low German]

cop *slang –n.* **1** police officer. **2** capture or arrest (*it's a fair cop*). *–v.* (**-pp-**) **1** catch or arrest (an offender). **2** receive, suffer. **3** take, seize. □ **cop it** get into trouble; be punished. **cop out 1** withdraw; give up. **2** go back on a promise. **not much cop** of little value or use. [French *caper* seize]

copal /ˈkəʊp(ə)l/ *n.* resin of a tropical tree, used for varnish. [Spanish from Aztec]

copartner /kəʊˈpɑ:tnə(r)/ *n.* partner or associate. □ **copartnership** *n.*

cope[1] *v.* (**-ping**) (often foll. by *with*) deal effectively or contend; manage. [French: related to COUP]

cope[2] *–n.* priest's long cloaklike vestment. *–v.* (**-ping**) cover with a cope or coping. [Latin *cappa* CAP]

copeck /ˈkəʊpek/ *n.* (also **kopek, kopeck**) Russian coin worth one-hundredth of a rouble. [Russian *kopeĭka*]

Copernican system /kəˈpɜ:nɪkən/ *n.* theory that the planets (including the earth) move round the sun. [*Copernicus*, name of an astronomer]

copier /ˈkɒpɪə(r)/ *n.* machine that copies (esp. documents).

copilot /ˈkəʊˌpaɪlət/ *n.* second pilot in an aircraft.

coping *n.* top (usu. sloping) course of masonry in a wall. [from COPE[2]]

coping saw *n.* D-shaped saw for cutting curves in wood. [from COPE[1]]

coping-stone *n.* stone used in coping.

copious /ˈkəʊpɪəs/ *adj.* **1** abundant. **2** producing much. □ **copiously** *adv.* [Latin *copia* plenty]

cop-out *n.* cowardly evasion.

copper[1] *–n.* **1** malleable red-brown metallic element. **2** bronze coin. **3** large metal vessel for boiling esp. laundry. *–adj.* made of or coloured like copper. *–v.* cover with copper. [Latin *cuprum*]

copper[2] *n. slang* police officer. [from COP]

copper beech *n.* variety of beech with copper-coloured leaves.

copper-bottomed *adj.* **1** having a bottom sheathed with copper. **2** genuine or reliable.

copperhead *n.* venomous N. American or Australian snake.

copperplate *n.* **1 a** polished copper plate for engraving or etching. **b** print made from this. **2** ornate style of handwriting.

coppice /'kɒpɪs/ *n.* area of undergrowth and small trees. [medieval Latin: related to COUP]

copra /'kɒprə/ *n.* dried coconut-kernels. [Portuguese from Malayalam]

copse *n.* = COPPICE. [shortened form]

Copt *n.* **1** native Egyptian in the Hellenistic and Roman periods. **2** native Christian of the independent Egyptian Church. [French from Arabic]

Coptic *–n.* language of the Copts. *–adj.* of the Copts.

copula /'kɒpjʊlə/ *n.* (*pl.* **-s**) connecting word, esp. part of the verb *be* connecting subject and predicate. [Latin]

copulate /'kɒpjʊ,leɪt/ *v.* (**-ting**) (often foll. by *with*) (esp. of animals) have sexual intercourse. □ **copulation** /-'leɪʃ(ə)n/ *n.*

copy /'kɒpɪ/ *–n.* (*pl.* **-ies**) **1** thing made to imitate another. **2** single specimen of a publication or issue. **3** material to be printed, esp. regarded as good etc. reading matter (*the crisis will make exciting copy*). *–v.* (**-ies, -ied**) **1** make a copy of. **2** imitate, do the same as. [Latin *copia* transcript]

copybook *n.* **1** book containing models of handwriting for learners to imitate. **2** (*attrib.*) **a** tritely conventional. **b** exemplary.

copycat *n. colloq.* person who copies another, esp. slavishly.

copyist *n.* person who makes (esp. written) copies.

copyright *–n.* exclusive legal right to print, publish, perform, film, or record material. *–adj.* protected by copyright. *–v.* secure copyright for (material).

copy-typist *n.* typist who types from documents rather than dictation.

copywriter *n.* person who writes or prepares advertising copy for publication.

coq au vin /,kɒk əʊ 'væ̃/ *n.* casserole of chicken pieces in wine. [French]

coquette /kɒ'ket/ *n.* woman who flirts. □ **coquetry** /'kɒkɪtrɪ/ *n.* (*pl.* **-ies**). **coquettish** *adj.* [French diminutive: related to COCK[1]]

cor *int. slang* expressing surprise etc. [corruption of *God*]

cor- see COM-.

coracle /'kɒrək(ə)l/ *n.* small boat of wickerwork covered with watertight material. [Welsh]

coral /'kɒr(ə)l/ *–n.* hard red, pink, or white calcareous substance secreted by marine polyps for support and habitation. *–adj.* **1** red or pink, like coral. **2** made of coral. [Greek *korallion*]

coral island *n.* (also **coral reef**) island (or reef) formed by the growth of coral.

coralline /'kɒrə,laɪn/ *–n.* seaweed with a hard jointed stem. *–adj.* of or like coral. [French and Italian: related to CORAL]

cor anglais /kɔ:r 'ɒŋgleɪ/ *n.* (*pl.* **cors anglais** /kɔ:z/) alto woodwind instrument of the oboe family. [French]

corbel /'kɔ:b(ə)l/ *n.* projection of stone, timber, etc., jutting out from a wall to support a weight. □ **corbelled** *adj.* [Latin *corvus* crow]

cord *–n.* **1 a** flexible material like thick string, made from twisted strands. **b** piece of this. **2** similar structure in the body. **3 a** ribbed fabric, esp. corduroy. **b** (in *pl.*) corduroy trousers. **4** electric flex. *–v.* **1** fasten or bind with cord. **2** (as **corded** *adj.*) (of cloth) ribbed. [Greek *khordē* string]

cordial /'kɔ:dɪəl/ *–adj.* **1** heartfelt. **2** friendly. *–n.* fruit-flavoured drink. □ **cordiality** /-'ælɪtɪ/ *n.* **cordially** *adv.* [Latin *cor cord-* heart]

cordite /'kɔ:daɪt/ *n.* smokeless explosive. [from CORD, because of its appearance]

cordless *adj.* (of a hand-held electrical device) usable without a power cable because working from an internal source of energy or battery.

cordon /'kɔ:d(ə)n/ *–n.* **1** line or circle of police, soldiers, guards, etc., esp. preventing access. **2** ornamental cord or braid. **3** fruit-tree trained to grow as a single stem. *–v.* (often foll. by *off*) enclose or separate with a cordon of police etc. [Italian and French: related to CORD]

cordon bleu /,kɔ:dɒn 'blɜ:/ *Cookery* *–adj.* of the highest class. *–n.* cook of this class. [French]

cordon sanitaire /,kɔ:dɔ̃ ,sænɪ'teə(r)/ *n.* **1** guarded line between infected and uninfected districts. **2** measure designed to prevent the spread of undesirable influences.

corduroy /'kɔ:də,rɔɪ/ *n.* **1** thick cotton fabric with velvety ribs. **2** (in *pl.*) corduroy trousers. [*cord* = ribbed fabric]

core *–n.* **1** horny central part of certain fruits, containing the seeds. **2** central or most important part of anything (also *attrib.*: *core curriculum*). **3** inner central region of the earth. **4** part of a nuclear reactor containing fissile material. **5** *hist.* structural unit in a computer, storing one bit of data (see BIT[4]). **6** inner strand of an electric cable. **7** piece of soft iron forming the centre of an electromagnet or induction coil. –*v.* (**-ring**) remove the core from. □ **corer** *n.* [origin unknown]

co-respondent /ˌkəʊrɪˈspɒnd(ə)nt/ *n.* person cited in a divorce case as having committed adultery with the respondent.

corgi /ˈkɔːɡɪ/ *n.* (*pl.* **-s**) dog of a short-legged breed with a foxlike head. [Welsh]

coriander /ˌkɒrɪˈændə(r)/ *n.* **1** aromatic plant. **2** its seeds used for flavouring. [Greek *koriannon*]

Corinthian /kəˈrɪnθɪən/ *adj.* **1** of ancient Corinth in southern Greece. **2** *Archit.* of the order characterized by ornate decoration and acanthus leaves. [Latin from Greek]

cork –*n.* **1** buoyant light-brown bark of a S. European oak. **2** bottle-stopper of cork etc. **3** float of cork. **4** (*attrib.*) made of cork. –*v.* (often foll. by *up*) **1** stop or confine. **2** restrain (feelings etc.). [Spanish *alcorque*]

corkage *n.* charge made by a restaurant etc. for serving a customer's own wine etc.

corked *adj.* **1** stopped with a cork. **2** (of wine) spoilt by a decayed cork.

corker *n. slang* excellent person or thing.

corkscrew –*n.* **1** spiral device for extracting corks from bottles. **2** (often *attrib.*) thing with a spiral shape. –*v.* move spirally; twist.

corm *n.* underground swollen stem base of some plants. [Greek *kormos* lopped tree-trunk]

cormorant /ˈkɔːmərənt/ *n.* diving sea bird with black plumage. [Latin *corvus marinus* sea-raven]

corn[1] *n.* **1 a** cereal before or after harvesting, esp. the chief crop of a region. **b** grain or seed of a cereal plant. **2** *colloq.* something corny or trite. [Old English]

corn[2] *n.* small tender area of horny skin, esp. on the toe. [Latin *cornu* horn]

corn-cob *n.* cylindrical centre of a maize ear on which the grains grow.

corncrake *n.* rail inhabiting grassland and nesting on the ground.

corn dolly *n.* figure of plaited straw.

cornea /ˈkɔːnɪə/ *n.* transparent circular part of the front of the eyeball. □ **corneal** *adj.* [medieval Latin: related to CORN[2]]

corned *adj.* (esp. of beef) preserved in salt or brine. [from CORN[1]]

cornelian /kɔːˈniːlɪən/ *n.* (also **carnelian** /kɑː-/) dull red variety of chalcedony. [French]

corner –*n.* **1** place where converging sides or edges meet. **2** projecting angle, esp. where two streets meet. **3** internal space or recess formed by the meeting of two sides, esp. of a room. **4** difficult position, esp. one with no escape. **5** secluded place. **6** region or quarter, esp. a remote one. **7** action or result of buying or controlling the whole stock of a commodity. **8** *Boxing* & *Wrestling* corner of the ring where a contestant rests between rounds. **9** *Football* & *Hockey* free kick or hit from the corner of a pitch. –*v.* **1** force into a difficult or inescapable position. **2** establish a corner in (a commodity). **3** (esp. of or in a vehicle) go round a corner. [Latin: related to CORN[2]]

cornerstone *n.* **1 a** stone in the projecting angle of a wall. **b** foundation-stone. **2** indispensable part or basis.

cornet /ˈkɔːnɪt/ *n.* **1** brass instrument resembling a trumpet but shorter and wider. **2** conical wafer for holding ice-cream. □ **cornetist** /kɔːˈnetɪst/ *n.* (also **cornettist**). [Latin *cornu*: related to CORN[2]]

cornflake *n.* **1** (in *pl.*) breakfast cereal of toasted maize flakes. **2** flake of this cereal.

cornflour *n.* fine-ground flour, esp. of maize or rice.

cornflower *n.* plant with deep-blue flowers originally growing among corn.

cornice /ˈkɔːnɪs/ *n.* ornamental moulding, esp. round a room just below the ceiling or as the topmost part of an entablature. [French from Italian]

Cornish /ˈkɔːnɪʃ/ –*adj.* of Cornwall. –*n.* Celtic language of Cornwall.

Cornish pasty *n.* pastry envelope containing meat and vegetables.

corn on the cob *n.* maize cooked and eaten from the corn-cob.

cornucopia /ˌkɔːnjʊˈkəʊpɪə/ *n.* **1** horn overflowing with flowers, fruit, and corn, as a symbol of plenty. **2** abundant supply. [Latin: related to CORN[2], COPIOUS]

corny *adj.* (**-ier, -iest**) *colloq.* **1** banal. **2** feebly humorous. **3** sentimental. □ **cornily** *adv.* **corniness** *n.* [from CORN[1]]

corolla /kəˈrɒlə/ *n.* whorl of petals forming the inner envelope of a flower. [Latin diminutive of CORONA]

corollary /kəˈrɒlərɪ/ *n.* (*pl.* **-ies**) **1** proposition that follows from one already proved. **2** (often foll. by *of*) natural consequence. [Latin, = gratuity: related to COROLLA]

corona /kə'rəʊnə/ *n.* (*pl.* **-nae** /-ni:/) **1 a** halo round the sun or moon. **b** gaseous envelope of the sun, seen as an area of light around the moon during a total solar eclipse. **2** *Anat.* crownlike structure. **3** crownlike outgrowth from the inner side of a corolla. **4** glow around an electric conductor. □ **coronal** *adj.* [Latin, = crown]

coronary /'kɒrənərɪ/ *–adj. Anat.* resembling or encircling like a crown. *–n.* (*pl.* **-ies**) = CORONARY THROMBOSIS. [Latin: related to CORONA]

coronary artery *n.* artery supplying blood to the heart.

coronary thrombosis *n.* blockage caused by a blood clot in a coronary artery.

coronation /,kɒrə'neɪʃ(ə)n/ *n.* ceremony of crowning a sovereign or consort. [medieval Latin: related to CORONA]

coroner /'kɒrənə(r)/ *n.* official holding inquests on deaths thought to be violent or accidental. [Anglo-French: related to CROWN]

coronet /'kɒrənɪt/ *n.* **1** small crown. **2** circlet of precious materials, esp. as a headdress. [French diminutive: related to CROWN]

corpora *pl.* of CORPUS.

corporal[1] /'kɔ:pr(ə)l/ *n.* non-commissioned army or air-force officer ranking next below sergeant. [French from Italian]

corporal[2] /'kɔ:pər(ə)l/ *adj.* of the human body. □ **corporality** /-'rælɪtɪ/ *n.* [Latin *corpus* body]

corporal punishment *n.* physical punishment.

corporate /'kɔ:pərət/ *adj.* **1** forming a corporation. **2** of, belonging to, or united in a group. [Latin: related to CORPORAL[2]]

corporation /,kɔ:pə'reɪʃ(ə)n/ *n.* **1** group of people authorized to act as an individual, esp. in business. **2** municipal authorities of a borough, town, or city. **3** *joc.* large stomach.

corporative /'kɔ:pərətɪv/ *adj.* **1** of a corporation. **2** governed by or organized in corporations.

corporeal /kɔ:'pɔ:rɪəl/ *adj.* bodily, physical, material. □ **corporeality** /-'ælɪtɪ/ *n.* **corporeally** *adv.* [Latin: related to CORPORAL[2]]

corps /kɔ:(r)/ *n.* (*pl.* **corps** /kɔ:z/) **1 a** body of troops with special duties (*intelligence corps*). **b** main subdivision of an army in the field. **2** body of people engaged in a special activity (*diplomatic corps*). [French: related to CORPSE]

corps de ballet /,kɔ: də 'bæleɪ/ *n.* group of ensemble dancers in a ballet. [French]

corpse *n.* dead body. [Latin: related to CORPUS]

corpulent /'kɔ:pjʊlənt/ *adj.* physically bulky; fat. □ **corpulence** *n.* [Latin: related to CORPUS]

corpus /'kɔ:pəs/ *n.* (*pl.* **-pora**) body or collection of writings, texts, etc. [Latin, = body]

corpuscle /'kɔ:pʌs(ə)l/ *n.* minute body or cell in an organism, esp. (in *pl.*) the red or white cells in the blood of vertebrates. □ **corpuscular** /-'pʌskjʊlə(r)/ *adj.* [Latin diminutive of CORPUS]

corral /kə'rɑ:l/ *–n.* **1** *US* pen for cattle, horses, etc. **2** enclosure for capturing wild animals. *–v.* (**-ll-**) put or keep in a corral. [Spanish and Portuguese: related to KRAAL]

correct /kə'rekt/ *–adj.* **1** true, accurate. **2** proper, in accordance with taste or a standard. *–v.* **1** set right; amend. **2** mark errors in. **3** substitute a right thing for (a wrong one). **4 a** admonish (a person). **b** punish (a person or fault). **5** counteract (a harmful quality). **6** adjust (an instrument etc.). □ **correctly** *adv.* **correctness** *n.* **corrector** *n.* [Latin *rego rect-* guide]

correction /kə'rekʃ(ə)n/ *n.* **1** correcting or being corrected. **2** thing substituted for what is wrong. **3** *archaic* punishment. □ **correctional** *adj.* [Latin: related to CORRECT]

correctitude /kə'rektɪ,tju:d/ *n.* consciously correct behaviour. [from CORRECT, RECTITUDE]

corrective *–adj.* serving to correct or counteract something harmful. *–n.* corrective measure or thing. [Latin: related to CORRECT]

correlate /'kɒrə,leɪt/ *–v.* (**-ting**) (usu. foll. by *with*, *to*) have or bring into a mutual relation or dependence. *–n.* each of two related or complementary things. □ **correlation** /-'leɪʃ(ə)n/ *n.* [medieval Latin *correlatio*]

correlative /kə'relətɪv/ *–adj.* **1** (often foll. by *with*, *to*) having a mutual relation. **2** (of words) corresponding to each other and used together (as *neither* and *nor*). *–n.* correlative word or thing.

correspond /,kɒrɪ'spɒnd/ *v.* **1 a** (usu. foll. by *to*) be similar or equivalent. **b** (usu. foll. by *with*, *to*) be in agreement, not contradict. **2** (usu. foll. by *with*) exchange letters. □ **correspondingly** *adv.* [French from medieval Latin]

correspondence *n.* **1** agreement or similarity. **2 a** exchange of letters. **b** letters.

correspondence course *n.* course of study conducted by post.

correspondent *n.* **1** person who writes letters. **2** person employed to write or report for a newspaper or for broadcasting etc.

corridor /'kɒrɪ,dɔ:(r)/ *n.* **1** passage giving access into rooms. **2** passage in a train giving access into compartments. **3** strip of territory of one State passing through that of another. **4** route which an aircraft must follow, esp. over a foreign country. [French from Italian]

corridors of power *n.pl.* places where covert influence is said to be exerted in government.

corrigendum /,kɒrɪ'gendəm/ *n.* (*pl.* **-da**) error to be corrected. [Latin *corrigo*: related to CORRECT]

corrigible /'kɒrɪdʒɪb(ə)l/ *adj.* **1** able to be corrected. **2** submissive. □ **corrigibly** *adv.* [medieval Latin: related to CORRIGENDUM]

corroborate /kə'rɒbə,reɪt/ *v.* (**-ting**) confirm or give support to (a statement or belief etc.). □ **corroboration** /-'reɪʃ(ə)n/ *n.* **corroborative** /-rətɪv/ *adj.* **corroborator** *n.* [Latin *robur* strength]

corrode /kə'rəʊd/ *v.* (**-ding**) **1 a** wear away, esp. by chemical action. **b** decay. **2** destroy gradually. [Latin *rodo ros-* gnaw]

corrosion /kə'rəʊʒ(ə)n/ *n.* **1** corroding or being corroded. **2** corroded area. □ **corrosive** *adj.* & *n.*

corrugate /'kɒrə,geɪt/ *v.* (**-ting**) (esp. as **corrugated** *adj.*) form into alternate ridges and grooves, esp. to strengthen (*corrugated iron*). □ **corrugation** /-'geɪʃ(ə)n/ *n.* [Latin *ruga* wrinkle]

corrupt /kə'rʌpt/ *–adj.* **1** dishonest, esp. using bribery. **2** immoral; wicked. **3** (of a text etc.) made unreliable by errors or alterations. *–v.* make or become corrupt. □ **corruptible** *adj.* **corruptibility** /-'bɪlɪtɪ/ *n.* **corruption** *n.* **corruptive** *adj.* **corruptly** *adv.* **corruptness** *n.* [Latin *rumpo rupt-* break]

corsage /kɔ:'sɑ:ʒ/ *n.* small bouquet worn by women. [French: related to CORPSE]

corsair /'kɔ:seə(r)/ *n.* **1** pirate ship. **2** pirate. [French: related to COURSE]

corselette /'kɔ:slɪt/ *n.* combined corset and bra. [French *corslet* armour covering trunk]

corset /'kɔ:sɪt/ *n.* closely-fitting undergarment worn to shape the body or to support it after injury. □ **corsetry** *n.* [French diminutive: related to CORPSE]

cortège /kɔ:'teɪʒ/ *n.* procession, esp. for a funeral. [French]

cortex /'kɔ:teks/ *n.* (*pl.* **-tices** /-tɪ,si:z/) outer part of an organ, esp. of the brain or kidneys. □ **cortical** /-tɪk(ə)l/ *adj.* [Latin, = bark]

cortisone /'kɔ:tɪ,zəʊn/ *n.* hormone used esp. in treating inflammation and allergy. [abbreviation of chemical name]

corundum /kə'rʌndəm/ *n.* extremely hard crystallized alumina, used esp. as an abrasive. [Tamil from Sanskrit]

coruscate /'kɒrə,skeɪt/ *v.* (**-ting**) sparkle. □ **coruscation** /-'skeɪʃ(ə)n/ *n.* [Latin]

corvette /kɔ:'vet/ *n.* **1** small naval escort-vessel. **2** *hist.* warship with one tier of guns. [French from Dutch]

corymb /'kɒrɪmb/ *n.* flat-topped cluster of flowers with the outer flower-stalks proportionally longer. [Latin from Greek]

cos[1] *n.* lettuce with crisp narrow leaves. [*Kōs*, Greek island]

cos[2] /kɒz/ *abbr.* cosine.

cos[3] /kəz/ *conj. colloq.* because. [abbreviation]

cosec /'kəʊsek/ *abbr.* cosecant.

cosecant /kəʊ'si:kənt/ *n. Math.* ratio of the hypotenuse (in a right-angled triangle) to the side opposite an acute angle.

cosh[1] *colloq.* *–n.* heavy blunt weapon. *–v.* hit with a cosh. [origin unknown]

cosh[2] /kɒʃ, kɒs'eɪtʃ/ *abbr.* hyperbolic cosine.

co-signatory /kəʊ'sɪgnətərɪ/ *n.* (*pl.* **-ies**) person or State signing a treaty etc. jointly with others.

cosine /'kəʊsaɪn/ *n.* ratio of the side adjacent to an acute angle (in a right-angled triangle) to the hypotenuse.

cosmetic /kɒz'metɪk/ *–adj.* **1** beautifying, enhancing. **2** superficially improving or beneficial. **3** (of surgery or a prosthesis) imitating, restoring, or enhancing normal appearance. *–n.* cosmetic preparation, esp. for the face. □ **cosmetically** *adv.* [Greek, = ornament]

cosmic /'kɒzmɪk/ *adj.* **1** of the cosmos or its scale; universal (*of cosmic significance*). **2** of or for space travel.

cosmic rays *n.pl.* high-energy radiations from space etc.

cosmogony /kɒz'mɒgənɪ/ *n.* (*pl.* **-ies**) **1** origin of the universe. **2** theory about this. [Greek *-gonia* -begetting]

cosmology /kɒz'mɒlədʒɪ/ *n.* science or theory of the universe. □ **cosmological** /-mə'lɒdʒɪk(ə)l/ *adj.* **cosmologist** *n.* [from COSMOS, -LOGY]

cosmonaut /'kɒzmə,nɔ:t/ *n.* Soviet astronaut. [from COSMOS, Greek *nautēs* sailor]

cosmopolitan /ˌkɒzməˈpɒlɪt(ə)n/ *–adj.* **1** of, from, or knowing many parts of the world. **2** free from national limitations or prejudices. *–n.* cosmopolitan person. □ **cosmopolitanism** *n.* [Greek *politēs* citizen]

cosmos /ˈkɒzmɒs/ *n.* the universe as a well-ordered whole. [Greek]

Cossack /ˈkɒsæk/ *n.* member of a people of southern Russia. [Turki *quzzāq*]

cosset /ˈkɒsɪt/ *v.* (**-t-**) pamper. [dialect *cosset* = pet lamb, probably from Old English, = cottager]

cost *–v.* (*past* and *past part.* **cost**) **1** be obtainable for (a sum of money); have as a price. **2** involve as a loss or sacrifice (*it cost him his life*). **3** (*past* and *past part.* **costed**) fix or estimate the cost of. *–n.* **1** what a thing costs; price. **2** loss or sacrifice. **3** (in *pl.*) legal expenses. □ **at all costs** (or **at any cost**) whatever the cost or risk may be. [Latin *consto* stand at a price]

costal /ˈkɒst(ə)l/ *adj.* of the ribs. [Latin *costa* rib]

cost-effective *adj.* effective in relation to its cost.

costermonger /ˈkɒstəˌmʌŋgə(r)/ *n.* person who sells produce from a barrow. [*costard* large apple: related to COSTAL]

costing *n.* estimation of cost(s).

costive /ˈkɒstɪv/ *adj.* constipated. [Latin: related to CONSTIPATE]

costly *adj.* (**-ier, -iest**) costing much; expensive. □ **costliness** *n.*

cost of living *n.* level of prices esp. of basic necessities.

cost price *n.* price paid for a thing by one who later sells it.

costume /ˈkɒstjuːm/ *–n.* **1** style of dress, esp. of a particular place or time. **2** set of clothes. **3** clothing for a particular activity (*swimming-costume*). **4** actor's clothes for a part. *–v.* (**-ming**) provide with a costume. [Latin: related to CUSTOM]

costume jewellery *n.* artificial jewellery.

costumier /kɒˈstjuːmɪə(r)/ *n.* person who makes or deals in costumes. [French: related to COSTUME]

cosy /ˈkəʊzɪ/ (*US* **cozy**) *–adj.* (**-ier, -iest**) comfortable and warm; snug. *–n.* (*pl.* **-ies**) cover to keep a teapot etc. hot. □ **cosily** *adv.* **cosiness** *n.* [origin unknown]

cot[1] *n.* **1** small bed with high sides for a baby. **2** small light bed. [Hindi]

cot[2] *n.* **1** small shelter; cote. **2** *poet.* cottage. [Old English]

cot[3] *abbr.* cotangent.

cotangent /kəʊˈtændʒ(ə)nt/ *n.* ratio of the side adjacent to an acute angle (in a right-angled triangle) to the opposite side.

cot-death *n.* unexplained death of a sleeping baby.

cote *n.* shelter for animals or birds. [Old English]

coterie /ˈkəʊtərɪ/ *n.* exclusive group of people sharing interests. [French]

cotoneaster /kəˌtəʊnɪˈæstə(r)/ *n.* shrub bearing usu. bright red berries. [Latin *cotoneum* QUINCE]

cottage /ˈkɒtɪdʒ/ *n.* small simple house, esp. in the country. [Anglo-French: related to COT[2]]

cottage cheese *n.* soft white lumpy cheese made from skimmed milk curds.

cottage industry *n.* business activity carried on at home.

cottage pie *n.* dish of minced meat topped with mashed potato.

cottager *n.* person who lives in a cottage.

cotter *n.* **1** bolt or wedge for securing parts of machinery etc. **2** (in full **cotter pin**) split pin that can be opened after passing through a hole. [origin unknown]

cotton /ˈkɒt(ə)n/ *n.* **1** soft white fibrous substance covering the seeds of certain plants. **2** such a plant. **3** thread or cloth from this. □ **cotton on** (often foll. by *to*) *colloq.* begin to understand. [French from Arabic]

cotton wool *n.* fluffy wadding of a kind orig. made from raw cotton.

cotyledon /ˌkɒtɪˈliːd(ə)n/ *n.* embryonic leaf in seed-bearing plants. [Greek *kotulē* cup]

couch[1] *–n.* **1** upholstered piece of furniture for several people; sofa. **2** long padded seat with a headrest at one end. *–v.* **1** (foll. by *in*) express in (certain terms). **2** *archaic* (of an animal) lie, esp. in its lair. [Latin *colloco* lay in place]

couch[2] /kaʊtʃ, kuːtʃ/ *n.* (in full **couch grass**) a grass with long creeping roots. [var. of QUITCH]

couchette /kuːˈʃet/ *n.* **1** railway carriage with seats convertible into sleeping-berths. **2** berth in this. [French, = little bed]

couch potato *n. US slang* person who likes lazing at home.

cougar /ˈkuːgə(r)/ *n. US* puma. [French from Guarani]

cough /kɒf/ *–v.* **1** expel air etc. from the lungs with a sudden sharp sound. **2** (of an engine etc.) make a similar sound. **3** *slang* confess. *–n.* **1** act of coughing. **2** condition of respiratory organs causing coughing. □ **cough up 1** eject with coughs. **2** *slang* bring out or give (money or information) reluctantly. [imitative, related to Dutch *kuchen*]

cough mixture *n.* liquid medicine to relieve a cough.
could *past* of CAN[1]. *v. colloq.* feel inclined to (*I could murder him*).
couldn't /'kʊd(ə)nt/ *contr.* could not.
coulomb /'ku:lɒm/ *n.* SI unit of electric charge. [*Coulomb*, name of a physicist]
coulter /'kəʊltə(r)/ *n.* (*US* **colter**) vertical blade in front of a ploughshare. [Latin *culter* knife]
council /'kaʊns(ə)l/ *n.* **1 a** advisory, deliberative, or administrative body. **b** meeting of such a body. **2 a** local administrative body of a parish, district, town, etc. **b** (*attrib.*) provided by a local council (*council flat*). [Latin *concilium*]
councillor *n.* member of a (esp. local) council.
council tax *n.* proposed new local tax based on the value of a property and the number of people living in it, to replace the community charge.
counsel /'kaʊns(ə)l/ *–n.* **1** advice, esp. formally given. **2** consultation for advice. **3** (*pl.* same) legal adviser, esp. a barrister; body of these. *–v.* (**-ll-**; *US* **-l-**) **1** advise (a person). **2** give esp. professional advice to (a person) on personal problems. **3** recommend (a course of action). □ **keep one's own counsel** not confide in others. **take counsel** (usu. foll. by *with*) consult. □ **counselling** *n.* [Latin *consilium*]
counsellor *n.* (*US* **counselor**) **1** adviser. **2** person giving professional guidance on personal problems. **3** *US* barrister.
counsel of perfection *n.* ideal but impracticable advice.
count[1] *–v.* **1** determine the total number of, esp. by assigning successive numbers. **2** repeat numbers in ascending order. **3** (often foll. by *in*) include or be included in one's reckoning or plan. **4** consider or regard to be (lucky etc.). **5** (often foll. by *for*) have value; matter (*my opinion counts for little*). *–n.* **1 a** counting or being counted. **b** total of reckoning. **2** *Law* each charge in an indictment. □ **count against** be reckoned to the disadvantage of. **count one's blessings** be grateful for what one has. **count on** (or **upon**) rely on; expect. **count out 1** count while taking from a stock. **2** complete a count of ten seconds over (a fallen boxer etc.). **3** *colloq.* exclude, disregard. **4** *Polit.* procure the adjournment of (the House of Commons) when fewer than 40 members are present. **count up** find the sum of. **keep count** take note of how many there have been etc. **lose count** forget the number etc. counted. **out for the count 1** defeated. **2** unconscious; asleep. [Latin: related to COMPUTE]
count[2] *n.* foreign noble corresponding to an earl. [Latin *comes* companion]
countable *adj.* **1** that can be counted. **2** *Gram.* (of a noun) that can form a plural or be used with the indefinite article.
countdown *n.* **1** act of counting backwards to zero, esp. at the launching of a rocket etc. **2** period immediately before an event.
countenance /'kaʊntɪnəns/ *–n.* **1** the face or facial expression. **2** composure. **3** moral support. *–v.* (**-cing**) support, approve. [French: related to CONTAIN]
counter[1] *n.* **1** long flat-topped fitment in a shop etc., across which business is conducted. **2 a** small disc for playing or scoring in board-games etc. **b** token representing a coin. **3** apparatus for counting. □ **under the counter** surreptitiously, esp. illegally. [related to COUNT[1]]
counter[2] *–v.* **1 a** oppose, contradict. **b** meet by countermove. **2** *Boxing* give a return blow while parrying. *–adv.* in the opposite direction or manner. *–adj.* opposite. *–n.* parry; countermove. [related to COUNTER-]
counter- *comb. form* denoting: **1** retaliation, opposition, or rivalry (*counter-threat*). **2** opposite direction (*counter-clockwise*). **3** correspondence (*counterpart*; *countersign*). [Latin *contra* against]
counteract /,kaʊntə'rækt/ *v.* hinder or neutralize by contrary action. □ **counteraction** *n.* **counteractive** *adj.*
counter-attack *–n.* attack in reply to a preceding attack. *–v.* attack in reply.
counterbalance *–n.* weight or influence balancing another. *–v.* (**-cing**) act as a counterbalance to.
counter-clockwise /,kaʊntə 'klɒkwaɪz/ *adv.* & *adj. US* = ANTI-CLOCKWISE.
counter-espionage /,kaʊntər 'espɪə,nɑ:ʒ/ *n.* action taken against enemy spying.
counterfeit /'kaʊntəfɪt/ *–adj.* made in imitation; not genuine; forged. *–n.* a forgery or imitation. *–v.* imitate fraudulently; forge. [French]
counterfoil *n.* part of a cheque, receipt, etc., retained by the payer as a record. [from FOIL[2]]
counter-intelligence /,kaʊntərɪn 'telɪdʒ(ə)ns/ *n.* = COUNTER-ESPIONAGE.
countermand /,kaʊntə'mɑ:nd/ *–v.* **1** revoke (a command). **2** recall by a contrary order. *–n.* order revoking a previous one. [Latin: related to MANDATE]
countermeasure *n.* action taken to counteract a danger, threat, etc.

countermove *n.* move or action in opposition to another.

counterpane /'kaʊntə,peɪn/ *n.* bedspread. [medieval Latin *culcita puncta* quilted mattress]

counterpart *n.* **1** person or thing like another or forming the complement or equivalent to another. **2** duplicate.

counterpoint /'kaʊntə,pɔɪnt/ *n.* **1 a** art or practice of combining melodies according to fixed rules. **b** melody combined with another. **2** contrasting argument, plot, literary theme, etc. [medieval Latin *contrapunctum* marked opposite]

counterpoise /'kaʊntə,pɔɪz/ *–n.* **1** counterbalance. **2** state of equilibrium. *–v.* (**-sing**) counterbalance. [Latin *pensum* weight]

counter-productive /,kaʊntəprə'dʌktɪv/ *adj.* having the opposite of the desired effect.

counter-revolution /,kaʊntə,revə'lu:ʃ(ə)n/ *n.* revolution opposing a former one or reversing its results.

countersign *–v.* add a signature to (a document already signed by another). *–n.* **1** password spoken to a person on guard. **2** mark used for identification etc. [Italian: related to SIGN]

countersink *v.* (*past* and *past part.* **-sunk**) **1** shape (the rim of a hole) so that a screw or bolt can be inserted flush with the surface. **2** sink (a screw etc.) in such a hole.

counter-tenor *n.* **1** male alto singing-voice. **2** singer with this voice. [Italian: related to CONTRA-]

countervail /'kaʊntə,veɪl/ *v. literary* **1** counterbalance. **2** (often foll. by *against*) oppose, usu. successfully. [Latin *valeo* have worth]

counterweight *n.* counterbalancing weight.

countess /'kaʊntɪs/ *n.* **1** wife or widow of a count or earl. **2** woman holding the rank of count or earl. [Latin *comitissa*: related to COUNT²]

countless *adj.* too many to be counted.

countrified /'kʌntrɪ,faɪd/ *adj.* rustic in manner or appearance.

country /'kʌntrɪ/ *n.* (*pl.* **-ies**) **1** territory of a nation; State. **2** (often *attrib.*) rural districts as opposed to towns or the capital. **3** land of a person's birth or citizenship. **4** region with regard to its aspect, associations, etc. (*mountainous country*; *Hardy country*). **5** national population, esp. as voters. [medieval Latin *contrata* (*terra*) (land) lying opposite]

country-and-western *n.* type of folk music originated by Whites in the southern US.

country club *n.* sporting and social club in a rural setting.

country dance *n.* traditional dance, esp. English, usu. with couples facing each other in lines.

countryman *n.* (*fem.* **countrywoman**) **1** person living in a rural area. **2** (also **fellow-countryman**) person of one's own country.

country music *n.* = COUNTRY-AND-WESTERN.

countryside *n.* rural areas.

country-wide *adj.* & *adv.* extending throughout a nation.

county /'kaʊntɪ/ *–n.* (*pl.* **-ies**) **1** territorial division in some countries, forming the chief unit of local administration. **2** *US* political and administrative division of a State. *–adj.* of or like the gentry. [Latin *comitatus*: related to COUNT²]

county council *n.* elected governing body of an administrative county.

county court *n.* judicial court for civil cases.

county town *n.* administrative capital of a county.

coup /ku:/ *n.* (*pl.* **-s** /ku:z/) **1** successful stroke or move. **2** = COUP D'ÉTAT. [medieval Latin *colpus* blow]

coup de grâce /,ku: də 'grɑ:s/ *n.* finishing stroke. [French]

coup d'état /,ku: deɪ'tɑ:/ *n.* (*pl.* ***coups d'état*** pronunc. same) violent or illegal seizure of power. [French]

coupé /'ku:peɪ/ *n.* (*US* **coupe** /ku:p/) car with a hard roof, two doors, and usu. a sloping rear. [French *couper* cut]

couple /'kʌp(ə)l/ *–n.* **1 a** two (*a couple of girls*). **b** about two (*a couple of hours*). **2 a** two people who are married to, or in a sexual relationship with, each other. **b** pair of partners in a dance etc. *–v.* (**-ling**) **1** link together. **2** associate in thought or speech. **3** copulate. [Latin COPULA]

couplet /'kʌplɪt/ *n.* two successive lines of verse, usu. rhyming and of the same length. [French diminutive: related to COUPLE]

coupling /'kʌplɪŋ/ *n.* **1** link connecting railway carriages etc. **2** device for connecting parts of machinery.

coupon /'ku:pɒn/ *n.* **1** form etc. as an application for a purchase etc. **2** entry form for a football pool or other competition. **3** discount voucher given with a purchase. [French *couper* cut]

courage /'kʌrɪdʒ/ *n.* ability to disregard fear; bravery. □ **courage of one's convictions** courage to act on one's beliefs. [Latin *cor* heart]

courageous /kə'reɪdʒəs/ *adj.* brave. □ **courageously** *adv.*

courgette /kʊə'ʒet/ *n.* small vegetable marrow. [French]

courier /'kʊrɪə(r)/ *n.* **1** person employed to guide and assist tourists. **2** special messenger. [Latin *curro curs-* run]

course /kɔ:s/ *–n.* **1** onward movement or progression. **2** direction taken (*changed course*). **3** stretch of land or water for races; golf-course. **4** series of lessons etc. in a particular subject. **5** each successive part of a meal. **6** sequence of medical treatment etc. **7** line of conduct. **8** continuous horizontal layer of masonry, brick, etc. **9** channel in which water flows. *–v.* (**-sing**) **1** (esp. of liquid) run, esp. fast. **2** (also *absol.*) use hounds to hunt (esp. hares). □ **in course of** in the process of. **in the course of** during. **of course** naturally; as is or was to be expected; admittedly. [Latin *cursus*: related to COURIER]

courser *n. poet.* swift horse.

court /kɔ:t/ *–n.* **1** (in full **court of law**) **a** judicial body hearing legal cases. **b** = COURTROOM. **2** quadrangular area for games (*tennis-court*; *squash-court*). **3 a** yard surrounded by houses with entry from the street. **b** = COURTYARD. **4 a** the residence, retinue, and courtiers of a sovereign. **b** sovereign and councillors, constituting the ruling power. **c** assembly held by a sovereign; State reception. **5** attention paid to a person whose favour etc. is sought (*paid court to her*). *–v.* **1 a** try to win affection or favour of. **b** pay amorous attention to. **2** seek to win (applause, fame, etc.). **3** invite (misfortune) by one's actions. □ **go to court** take legal action. **out of court 1** without reaching trial. **2** not worthy of consideration. [Latin: related to COHORT]

court-card *n.* playing-card that is a king, queen, or jack.

courteous /'kɜ:tɪəs/ *adj.* polite, considerate. □ **courteously** *adv.* **courteousness** *n.* [French: related to COURT]

courtesan /ˌkɔ:tɪ'zæn/ *n.* prostitute, esp. one with wealthy or upper-class clients. [Italian: related to COURT]

courtesy /'kɜ:təsɪ/ *n.* (*pl.* **-ies**) courteous behaviour or act. □ **by courtesy of** with the formal permission of. [French: related to COURTEOUS]

courtesy light *n.* light in a car switched on by opening a door.

court-house *n.* **1** building in which a judicial court is held. **2** *US* building containing the administrative offices of a county.

courtier /'kɔ:tɪə(r)/ *n.* person who attends a sovereign's court. [Anglo-French: related to COURT]

courtly *adj.* (**-ier**, **-iest**) dignified, refined. □ **courtliness** *n.*

court martial /ˌkɔ:t 'mɑ:ʃ(ə)l/ *–n.* (*pl.* **courts martial**) judicial court trying members of the armed services. *–v.* (**court-martial**) (**-ll-**; *US* **-l-**) try by this.

court order *n.* direction issued by a court or judge.

courtroom *n.* room in which a court of law meets.

courtship *n.* **1** courting, wooing. **2** courting behaviour of animals, birds, etc.

court shoe *n.* woman's light, usu. high-heeled, shoe with a low-cut upper.

courtyard *n.* area enclosed by walls or buildings.

couscous /'ku:sku:s/ *n.* N. African dish of crushed wheat or coarse flour steamed over broth, often with meat or fruit added. [French from Arabic]

cousin /'kʌz(ə)n/ *n.* **1** (also **first cousin**) child of one's uncle or aunt. **2** person of a kindred race or nation. [Latin *consobrinus*]

■ **Usage** There is often some confusion as to the difference between *cousin*, *first cousin*, *second cousin*, *first cousin once removed*, etc. For definitions see *cousin*, *second cousin* and *remove v.* 5.

couture /ku:'tjʊə(r)/ *n.* design and manufacture of fashionable clothes. [French]

couturier /ku:'tjʊərɪˌeɪ/ *n.* fashion designer.

cove[1] *–n.* **1** small bay or creek. **2** sheltered recess. **3** moulding, esp. at the junction of a wall and a ceiling. *–v.* (**-ving**) **1** provide (a room etc.) with a cove. **2** slope (the sides of a fireplace) inwards. [Old English]

cove[2] *n. slang* fellow, chap. [cant: origin unknown]

coven /'kʌv(ə)n/ *n.* assembly of witches. [related to CONVENT]

covenant /'kʌvənənt/ *–n.* **1** agreement; contract. **2** *Law* sealed contract, esp. a deed of covenant. **3** (**Covenant**) *Bibl.* agreement between God and the Israelites. *–v.* agree, esp. by legal covenant. [French: related to CONVENE]

Coventry /'kɒvəntrɪ/ *n.* □ **send a person to Coventry** refuse to associate with or speak to a person. [*Coventry* in England]

cover /'kʌvə(r)/ *–v.* **1** (often foll. by *with*) protect or conceal with a cloth, lid, etc. **2 a** extend over; occupy the whole surface of. **b** (often foll. by *with*) strew thickly or thoroughly. **c** lie over. **3 a** protect; clothe. **b** (as **covered** *adj.*) wearing a hat; having a roof. **4** include; comprise; deal with. **5** travel (a specified distance). **6** describe as a reporter. **7** be enough to defray (*£20 should cover it*). **8 a** *refl.* take measures to protect oneself. **b** (*absol.*; foll. by *for*) stand in for. **9 a** aim a gun etc. at. **b** (of a fortress, guns, etc.) command (territory). **c** protect (an exposed person etc.) by being able to return fire. **10 a** esp. *Cricket* stand behind (another player) to stop any missed balls. **b** mark (an opposing player). **11** (of a stallion etc.) copulate with. **–*n*. 1** thing that covers, esp.: **a** lid. **b** book's binding. **c** either board of this. **d** envelope or wrapping (*under separate cover*). **2** shelter. **3 a** pretence; screen. **b** pretended identity. **c** *Mil.* supporting force protecting an advance party from attack. **4 a** funds, esp. obtainable from insurance to meet a liability or secure against loss. **b** insurance protection (*third-party cover*). **5** person acting as a substitute. **6** place-setting at table. **7** *Cricket* = COVER-POINT. □ **cover up** completely cover or conceal. **take cover** find shelter. [Latin *cooperio*]

coverage *n*. **1** area or amount covered. **2** amount of publicity received by an event etc.

coverall *n*. esp. *US* **1** thing that covers entirely. **2** (usu. in *pl.*) full-length protective garment.

cover charge *n*. service charge per head in a restaurant, nightclub, etc.

cover girl *n*. female model appearing on magazine covers etc.

covering letter *n*. (also **covering note**) explanatory letter sent with an enclosure.

coverlet /'kʌvəlɪt/ *n*. bedspread. [Anglo-French: related to COVER, *lit* bed]

cover note *n*. temporary certificate of insurance.

cover-point *n*. *Cricket* **1** fielding position covering point. **2** fielder at this position.

cover story *n*. news story in a magazine that is advertised etc. on the front cover.

covert /'kʌvət/ **–*adj*.** secret or disguised (*covert glance*). **–*n*. 1** shelter, esp. a thicket hiding game. **2** feather covering the base of a bird's flight-feather. □ **covertly** *adv*. [French: related to COVER]

cover-up *n*. concealment of facts.

covet /'kʌvɪt/ *v*. (**-t-**) desire greatly (esp. a thing belonging to another person). [French: related to CUPID]

covetous /'kʌvɪtəs/ *adj*. (usu. foll. by *of*) coveting; grasping. □ **covetously** *adv*.

covey /'kʌvɪ/ *n*. (*pl*. **-s**) **1** brood of partridges. **2** small group of people. [Latin *cubo* lie]

cow[1] *n*. **1** fully grown female of any esp. domestic bovine animal, used as a source of milk and beef. **2** female of other large animals, esp. the elephant, whale, and seal. **3** *derog. slang* woman. [Old English]

cow[2] *v*. intimidate or dispirit. [Old Norse]

coward /'kaʊəd/ *n*. person who is easily frightened. [Latin *cauda* tail]

cowardice /'kaʊədɪs/ *n*. lack of bravery.

cowardly *adj*. **1** of or like a coward; lacking courage. **2** (of an action) done against one who cannot retaliate.

cowbell *n*. bell worn round a cow's neck.

cowboy *n*. **1** (*fem*. **cowgirl**) person who tends cattle, esp. in the western US. **2** *colloq*. unscrupulous or incompetent person in business.

cower *v*. crouch or shrink back in fear or distress. [Low German]

cowherd *n*. person who tends cattle.

cowhide *n*. **1** cow's hide. **2** leather or whip made from this.

cowl *n*. **1** monk's cloak. **2** hood-shaped covering of a chimney or ventilating shaft. [Latin *cucullus*]

cow-lick *n*. projecting lock of hair.

cowling *n*. removable cover of a vehicle or aircraft engine.

co-worker /kəʊ'wɜ:kə(r)/ *n*. person who works with another.

cow-parsley *n*. hedgerow plant with lacelike umbels of flowers.

cow-pat *n*. flat round piece of cow-dung.

cowpox *n*. disease of cows, whose virus was formerly used in smallpox vaccination.

cowrie /'kaʊrɪ/ *n*. **1** tropical mollusc with a bright shell. **2** its shell as money in parts of Africa and S. Asia. [Urdu and Hindi]

co-write /kəʊ'raɪt/ *v*. write with another person. □ **co-writer** *n*.

cowslip /'kaʊslɪp/ *n*. primula with small yellow flowers. [obsolete *slyppe* dung]

cox **–*n*.** coxswain, esp. of a racing-boat. **–*v*.** act as cox (of). [abbreviation]

coxcomb /'kɒkskəʊm/ *n*. ostentatiously conceited man. □ **coxcombry** *n*. (*pl*. **-ies**). [= *cock's comb*]

coxswain /'kɒks(ə)n/ **–*n*. 1** person who steers, esp. a rowing-boat. **2** senior petty officer in a small ship. **–*v*.** act as coxswain (of). [*cock* ship's boat, SWAIN]

coy *adj*. **1** affectedly shy. **2** irritatingly reticent. □ **coyly** *adv*. **coyness** *n*. [French: related to QUIET]

coyote /kɔɪ'əʊtɪ/ *n*. (*pl*. same or **-s**) N. American wolflike wild dog. [Mexican Spanish]

coypu /'kɔɪpu:/ *n*. (*pl*. **-s**) aquatic beaver-like rodent native to S. America. [Araucan]

cozen /ˈkʌz(ə)n/ *v. literary* **1** cheat, defraud. **2** beguile. **3** act deceitfully. □ **cozenage** *n.* [cant]

cozy *US* var. of COSY.

c.p. *abbr.* candlepower.

Cpl. *abbr.* Corporal.

cps *abbr.* (also **c.p.s.**) **1** *Computing* characters per second. **2** *Sci.* cycles per second.

CPU *abbr. Computing* central processing unit.

Cr *symb.* chromium.

crab[1] *n.* **1 a** ten-footed crustacean, with the first pair of legs as pincers. **b** crab as food. **2** (**Crab**) sign or constellation Cancer. **3** (in full **crab-louse**) (often in *pl.*) parasitic louse transmitted sexually to esp. pubic hair. **4** machine for hoisting heavy weights. □ **catch a crab** *Rowing* jam an oar or miss the water. □ **crablike** *adj.* [Old English]

crab[2] *n.* **1** (in full **crab-apple**) small sour apple. **2** (in full **crab tree** or **crab-apple tree**) tree (esp. uncultivated) bearing this. **3** sour person. [origin unknown]

crab[3] *v.* (**-bb-**) *colloq.* **1** criticize; grumble. **2** spoil. [Low German *krabben*]

crabbed /ˈkræbɪd/ *adj.* **1** = CRABBY. **2** (of handwriting) ill-formed; illegible. [from CRAB[2]]

crabby *adj.* (**-ier, -iest**) irritable, morose. □ **crabbily** *adv.* **crabbiness** *n.*

crabwise *adv.* & *attrib.adj.* sideways or backwards.

crack –*n.* **1 a** sharp explosive noise. **b** sudden harshness or change in vocal pitch. **2** sharp blow. **3 a** narrow opening; break or split. **b** chink. **4** *colloq.* joke or malicious remark. **5** *colloq.* attempt. **6** *slang* crystalline form of cocaine broken into small pieces. –*v.* **1** break without separating the parts. **2** make or cause to make a sharp explosive sound. **3** break with a sharp sound. **4** give way or cause to give way (under torture etc.). **5** (of the voice) change pitch sharply; break. **6** *colloq.* find the solution to. **7** tell (a joke etc.). **8** *colloq.* hit sharply. **9** (as **cracked** *adj.*) crazy. **10** break (wheat) into coarse pieces. –*attrib. adj. colloq.* excellent; first-rate (*crack shot*). □ **crack a bottle** open a bottle, esp. of wine, and drink it. **crack down on** *colloq.* take severe measures against. **crack of dawn** daybreak. **crack up** *colloq.* **1** collapse under strain. **2** praise. **get cracking** *colloq.* begin promptly and vigorously. [Old English]

crack-brained *adj.* crazy.

crack-down *n. colloq.* severe measures (esp. against law-breakers).

cracker *n.* **1** paper cylinder pulled apart, esp. at Christmas, with a sharp noise and releasing a hat, joke, etc. **2** loud firework. **3** (usu. in *pl.*) instrument for cracking. **4** thin dry savoury biscuit. **5** *slang* attractive or admirable person. **6** *US* biscuit.

crackers *predic. adj. slang* crazy.

cracking *slang* –*adj.* **1** excellent. **2** (*attrib.*) fast and exciting. –*adv.* outstandingly.

crackle /ˈkræk(ə)l/ –*v.* (**-ling**) make repeated slight cracking sound (*radio crackled; fire was crackling*). –*n.* such a sound. □ **crackly** *adj.* [from CRACK]

crackling /ˈkræklɪŋ/ *n.* crisp skin of roast pork.

cracknel /ˈkrækn(ə)l/ *n.* light crisp biscuit. [Dutch: related to CRACK]

crackpot *slang* –*n.* eccentric person. –*adj.* mad, unworkable.

crack-up *n. colloq.* mental breakdown.

-cracy *comb. form* denoting a particular form of government etc. (*bureaucracy*). [Latin *-cratia*]

cradle /ˈkreɪd(ə)l/ –*n.* **1 a** baby's bed or cot, esp. on rockers. **b** place in which something begins, esp. civilization (*cradle of democracy*). **2** supporting framework or structure. –*v.* (**-ling**) **1** contain or shelter as in a cradle. **2** place in a cradle. [Old English]

cradle-snatcher *n. slang* admirer or lover of a much younger person.

cradle-song *n.* lullaby.

craft /krɑːft/ –*n.* **1** special skill or technique. **2** occupation needing this. **3** (*pl.* **craft**) **a** boat or vessel. **b** aircraft or spacecraft. **4** cunning or deceit. –*v.* make in a skilful way. [Old English]

craftsman *n.* (*fem.* **craftswoman**) **1** skilled worker. **2** person who practises a craft. □ **craftsmanship** *n.*

crafty *adj.* (**-ier, -iest**) cunning, artful, wily. □ **craftily** *adv.* **craftiness** *n.*

crag *n.* steep or rugged rock. [Celtic]

craggy *adj.* (**-ier, -iest**) (of facial features, landscape, etc.) rugged; rough-textured. □ **cragginess** *n.*

crake *n.* bird of the rail family, esp. the corncrake. [Old Norse, imitative of cry]

cram *v.* (**-mm-**) **1 a** fill to bursting; stuff. **b** (foll. by *in, into*; also *absol.*) force (a thing) in or into. **2** prepare intensively for an examination. **3** (often foll. by *with*) feed to excess. [Old English]

crammer *n.* person or institution that crams pupils for examinations.

cramp –*n.* **1** painful involuntary muscular contraction. **2** (also **cramp-iron**) metal bar with bent ends for holding masonry etc. together. –*v.* **1** affect with cramp. **2** (often foll. by *up*) confine narrowly. **3** restrict. **4** fasten with a cramp. □ **cramp a person's style** prevent a person from acting freely or naturally. [Low German or Dutch]

cramped *adj.* **1** (of a space) too small. **2** (of handwriting) small and with the letters close together.

crampon /'kræmpɒn/ *n.* (*US* **crampoon** /-'pu:n/) (usu. in *pl.*) spiked iron plate fixed to a boot for climbing on ice. [French: related to CRAMP]

cranberry /'krænbərɪ/ *n.* (*pl.* **-ies**) **1** shrub with small red acid berries. **2** this berry used in cookery. [German *Kranbeere* crane-berry]

crane –*n.* **1** machine with a long projecting arm for moving heavy objects. **2** tall wading bird with long legs, neck, and bill. –*v.* (**-ning**) (also *absol.*) stretch out (one's neck) in order to see something. [Old English]

crane-fly *n.* two-winged long-legged fly: also called DADDY-LONG-LEGS.

cranesbill *n.* wild geranium.

cranium /'kreɪnɪəm/ *n.* (*pl.* **-s** or **-nia**) **1** skull. **2** part of the skeleton enclosing the brain. □ **cranial** *adj.* **craniology** /-'ɒlədʒɪ/ *n.* [medieval Latin from Greek]

crank –*n.* **1** part of an axle or shaft bent at right angles for converting reciprocal into circular motion or vice versa. **2** eccentric person. –*v.* cause to move by means of a crank. □ **crank up** start (a car engine) with a crank. [Old English]

crankcase *n.* case enclosing a crankshaft.

crankpin *n.* pin by which a connecting-rod is attached to a crank.

crankshaft *n.* shaft driven by a crank.

cranky *adj.* (**-ier, -iest**) **1** *colloq.* eccentric. **2** working badly; shaky. **3** esp. *US* crotchety. □ **crankily** *adv.* **crankiness** *n.*

cranny /'krænɪ/ *n.* (*pl.* **-ies**) chink, crevice. □ **crannied** *adj.* [French]

crap *coarse slang* –*n.* **1** (often as *int.* or *attrib.*) nonsense, rubbish. **2** faeces. –*v.* (**-pp-**) defecate. □ **crappy** *adj.* (**-ier, -iest**). [Dutch]

crape *n.* crêpe, usu. of black silk, formerly used for mourning. [from CRÊPE]

craps *n.pl.* (also **crap game**) *US* gambling dice game. [origin uncertain]

crapulent /'kræpjʊlənt/ *adj.* suffering the effects of drunkenness. □ **crapulence** *n.* **crapulous** *adj.* [Latin *crapula* inebriation]

crash[1] –*v.* **1** (cause to) make a loud smashing noise. **2** throw, drive, move, or fall with a loud smash. **3** (often foll. by *into*) collide or fall, or cause (a vehicle etc.) to collide or fall, violently; overturn at high speed. **4** collapse financially. **5** *colloq.* gatecrash. **6** *Computing* (of a machine or system) fail suddenly. **7** *colloq.* pass (a red traffic-light etc.). **8** (often foll. by *out*) *slang* sleep, esp. on a floor etc. –*n.* **1** loud and sudden smashing noise. **2** violent collision or fall, esp. of a vehicle. **3** ruin, esp. financial. **4** *Computing* sudden failure of a machine or system. **5** (*attrib.*) done rapidly or urgently (*crash course in first aid*). –*adv.* with a crash (*go crash*). [imitative]

crash[2] *n.* coarse plain fabric of linen, cotton, etc. [Russian]

crash barrier *n.* barrier at the side or centre of a road etc.

crash-dive –*v.* **1 a** (of a submarine or its pilot) dive hastily in an emergency. **b** (of an aircraft or airman) dive and crash. **2** cause to crash-dive. –*n.* such a dive.

crash-helmet *n.* helmet worn esp. by motor cyclists.

crashing *adj. colloq.* overwhelming (*crashing bore*).

crash-land *v.* land or cause (an aircraft etc.) to land hurriedly with a crash. □ **crash landing** *n.*

crass *adj.* gross; grossly stupid. □ **crassly** *adv.* **crassness** *n.* [Latin *crassus* thick]

-crat *comb. form* member or supporter of a type of government etc.

crate –*n.* **1** slatted wooden case etc. for conveying esp. fragile goods. **2** *slang* old aircraft or other vehicle. –*v.* (**-ting**) pack in a crate. [perhaps from Dutch]

crater –*n.* **1** mouth of a volcano. **2** bowl-shaped cavity, esp. that made by a shell or bomb. **3** hollow on the surface of a planet or moon, caused by impact. –*v.* form a crater in. [Greek, = mixing-bowl]

-cratic *comb. form* (also **-cratical**) denoting a type of government etc. (*autocratic*). □ **-cratically** *comb. form* forming adverbs. [forming adverbs]

cravat /krə'væt/ *n.* man's scarf worn inside an open-necked shirt. [Serbo-Croatian, = Croat]

crave *v.* (**-ving**) (often foll. by *for*) long or beg for. [Old English]

craven /'kreɪv(ə)n/ *adj.* cowardly, abject. [probably French *cravanté* defeated]

craving *n.* strong desire or longing.

craw *n.* crop of a bird or insect. □ **stick in one's craw** be unacceptable. [Low German or Dutch]

crawfish /'krɔ:fɪʃ/ *n.* (*pl.* same) large marine spiny lobster. [var. of CRAYFISH]

crawl –*v.* **1** move slowly, esp. on hands and knees or with the body close to the ground etc. **2** walk or move slowly. **3** *colloq.* behave obsequiously. **4** (often foll. by *with*) be or appear to be covered or filled with crawling or moving things or people. **5** (esp. of the skin) creep. –*n.* **1**

crawling. **2** slow rate of movement. **3** high-speed overarm swimming stroke. [origin unknown]

crayfish /ˈkreɪfɪʃ/ *n.* (*pl.* same) **1** small lobster-like freshwater crustacean. **2** crawfish. [French *crevice*]

crayon /ˈkreɪən/ –*n.* stick or pencil of coloured chalk, wax, etc. –*v.* draw with crayons. [French *craie* chalk]

craze –*v.* (**-zing**) **1** (usu. as **crazed** *adj.*) make insane (*crazed with grief*). **2** produce fine surface cracks on (pottery glaze etc.); develop such cracks. –*n.* **1** usu. temporary enthusiasm (*craze for skateboarding*). **2** object of this. [perhaps from Old Norse]

crazy *adj.* (**-ier, -iest**) **1** *colloq.* insane or mad; foolish. **2** (usu. foll. by *about*) *colloq.* extremely enthusiastic. **3** (*attrib.*) (of paving etc.) made up of irregular pieces. □ **crazily** *adv.* **craziness** *n.*

creak –*n.* harsh scraping or squeaking sound. –*v.* **1** make a creak. **2 a** move stiffly or with a creaking noise. **b** be poorly constructed (*plot creaks*). [imitative]

creaky *adj.* (**-ier, -iest**) **1** liable to creak. **2 a** stiff or frail. **b** decrepit, outmoded. □ **creakiness** *n.*

cream –*n.* **1** fatty part of milk. **2** its yellowish-white colour. **3** creamlike cosmetic etc. **4** food or drink like or containing cream. **5** (usu. prec. by *the*) best part of something. –*v.* **1** take cream from (milk). **2** make creamy. **3** treat (the skin etc.) with cosmetic cream. **4** form a cream or scum. –*adj.* pale yellowish white. □ **cream off** take (esp. the best part) from a whole. [Latin *cramum* and Church Latin *chrisma* oil for anointing]

cream cheese *n.* soft rich cheese made from cream and unskimmed milk.

creamer *n.* **1** cream-substitute for adding to coffee. **2** jug for cream.

creamery *n.* (*pl.* **-ies**) **1** factory producing butter and cheese. **2** dairy.

cream of tartar *n.* purified tartar, used in medicine, baking powder, etc.

cream soda *n.* carbonated vanilla-flavoured soft drink.

cream tea *n.* afternoon tea with scones, jam, and cream.

creamy *adj.* (**-ier, -iest**) **1** like cream. **2** rich in cream. □ **creamily** *adv.* **creaminess** *n.*

crease –*n.* **1** line caused by folding or crushing. **2** *Cricket* line marking the position of a bowler or batsman. –*v.* (**-sing**) **1** make creases in. **2** develop creases. **3** *slang* (often foll. by *up*) make or become incapable through laughter. [from CREST]

create /kriːˈeɪt/ *v.* (**-ting**) **1** bring into existence; cause. **2** originate (*actor creates a part*). **3** invest with rank (*created him a lord*). **4** *slang* make a fuss. [Latin *creo*]

creation /kriːˈeɪʃ(ə)n/ *n.* **1** creating or being created. **2 a** (usu. **the Creation**) God's creating of the universe. **b** (usu. **Creation**) all created things, the universe. **3** product of the imagination, art, fashion, etc.

creative *adj.* **1** inventive, imaginative. **2** able to create. □ **creatively** *adv.* **creativeness** *n.* **creativity** /-ˈtɪvɪtɪ/ *n.*

creative accounting *n.* exploitation of loopholes in financial legislation to gain maximum advantage or present figures in a misleadingly favourable light.

creator *n.* **1** person who creates. **2** (as **the Creator**) God.

creature /ˈkriːtʃə(r)/ *n.* **1** any living being, esp. an animal. **2** person of a specified kind (*poor creature*). **3** subservient person. □ **creaturely** *adj.* [French from Latin: related to CREATE]

creature comforts *n.pl.* good food, warmth, etc.

crèche /kreʃ/ *n.* day nursery. [French]

credence /ˈkriːd(ə)ns/ *n.* belief. □ **give credence to** believe. [medieval Latin: related to CREDO]

credential /krɪˈdenʃ(ə)l/ *n.* (usu. in *pl.*) **1** certificates, references, etc., attesting to a person's education, character, etc. **2** letter(s) of introduction. [medieval Latin: related to CREDENCE]

credibility /ˌkredɪˈbɪlɪtɪ/ *n.* **1** being credible. **2** reputation, status.

credibility gap *n.* apparent difference between what is said and what is true.

credible /ˈkredɪb(ə)l/ *adj.* believable or worthy of belief. [Latin: related to CREDO]

■ **Usage** *Credible* is sometimes confused with *credulous*.

---

credit /ˈkredɪt/ –*n.* **1** source of honour, pride, etc. (*is a credit to the school*). **2** acknowledgement of merit. **3** good reputation. **4** belief or trust. **5 a** person's financial standing, esp. as regards money in the bank etc. **b** power to obtain goods etc. before payment. **6** (usu. in *pl.*) acknowledgement of a contributor's services to a film etc. **7** grade above pass in an examination. **8** reputation for solvency and honesty in business. **9 a** entry in an account of a sum paid into it. **b** sum entered. **c** side of an account recording such entries. **10** educational course counting towards a degree. –*v.* (**-t-**) **1** believe (*cannot credit it*). **2** (usu. foll. by *to, with*) enter on the credit side of an account. □ **credit a person with** ascribe (a good quality) to a person. **do credit to** (or **do a person credit**) enhance the reputation of. **on credit** with an arrangement to pay later. **to one's credit** in one's favour. [Italian or Latin: related to CREDO]

creditable *adj.* bringing credit or honour. □ **creditably** *adv.*

credit card *n.* plastic card from a bank etc. authorizing the purchase of goods on credit.

credit note *n.* note with a specific monetary value given by a shop etc. for goods returned.

creditor *n.* person to whom a debt is owing. [Latin: related to CREDIT]

credit rating *n.* estimate of a person's suitability for commercial credit.

creditworthy *adj.* considered suitable to receive commercial credit. □ **creditworthiness** *n.*

credo /'kri:dəʊ, 'kreɪ-/ *n.* (*pl.* **-s**) creed. [Latin, = I believe]

credulous /'kredjʊləs/ *adj.* too ready to believe; gullible. □ **credulity** /krɪ'dju:lɪtɪ/ *n.* **credulously** *adv.* [Latin: related to CREDO]

■ **Usage** *Credulous* is sometimes confused with *credible*.

creed *n.* **1** set of principles or beliefs. **2** system of religious belief. **3** (often **the Creed**) formal summary of Christian doctrine. [Latin: related to CREDO]

creek *n.* **1 a** inlet on a sea-coast. **b** short arm of a river. **2** esp. *US*, *Austral.*, & *NZ* tributary of a river; stream. □ **up the creek** *slang* **1** in difficulties. **2** crazy. [Old Norse and Dutch]

creel *n.* fisherman's large wicker basket. [origin unknown]

creep –*v.* (*past* and *past part.* **crept**) **1** move with the body prone and close to the ground. **2** move stealthily or timidly. **3** advance very gradually (*a feeling crept over her*). **4** *colloq.* act obsequiously in the hope of advancement. **5** (of a plant) grow along the ground or up a wall etc. **6** (as **creeping** *adj.*) developing slowly and steadily. **7** (of flesh) shiver or shudder from fear, horror, etc. –*n.* **1** act or spell of creeping. **2** (in *pl.*; prec. by *the*) *colloq.* feeling of revulsion or fear. **3** *slang* unpleasant person. **4** (of metals etc.) gradual change of shape under stress. [Old English]

creeper *n.* **1** climbing or creeping plant. **2** bird that climbs, esp. the treecreeper. **3** *slang* soft-soled shoe.

creepy *adj.* (**-ier, -iest**) *colloq.* feeling or causing horror or fear. □ **creepily** *adv.* **creepiness** *n.*

creepy-crawly /,kri:pɪ'krɔ:lɪ/ *n.* (*pl.* **-ies**) *colloq.* small crawling insect etc.

cremate /krɪ'meɪt/ *v.* (**-ting**) burn (a corpse etc.) to ashes. □ **cremation** *n.* [Latin *cremo* burn]

crematorium /,kremə'tɔ:rɪəm/ *n.* (*pl.* **-ria** or **-s**) place where corpses are cremated.

crème /krem/ *n.* **1** = CREAM *n.* 4. **2** liqueur (*crème de cassis*). [French, = cream]

crème brûlée /bru:'leɪ/ *n.* baked cream or custard pudding coated with caramel.

crème caramel *n.* custard coated with caramel.

crème de cassis /də kæ'si:s/ *n.* blackcurrant liqueur.

crème de la crème /də lɑ: 'krem/ *n.* best part; élite.

crème de menthe /də 'mɑ̃t, 'mɒnt/ *n.* peppermint liqueur.

crenellate /'krenə,leɪt/ *v.* (*US* **crenelate**) (**-ting**) provide (a tower etc.) with battlements. □ **crenellation** /-'leɪʃ(ə)n/ *n.* [French *crenel* embrasure]

Creole /'kri:əʊl/ –*n.* **1 a** descendant of European settlers in the W. Indies or Central or S. America. **b** White descendant of French settlers in the southern US. **c** person of mixed European and Black descent. **2** language formed from a European language and another (esp. African) language. –*adj.* **1** of Creoles. **2** (usu. **creole**) of Creole origin etc. (*creole cooking*). [French from Spanish]

creosote /'kri:ə,səʊt/ –*n.* **1** dark-brown oil distilled from coal tar, used as a wood-preservative. **2** oily fluid distilled from wood tar, used as an antiseptic. –*v.* (**-ting**) treat with creosote. [Greek *kreas* flesh, *sōtēr* preserver, because of its antiseptic properties]

crêpe /kreɪp/ *n.* **1** fine gauzy wrinkled fabric. **2** thin pancake with a savoury or sweet filling. **3** hard-wearing wrinkled sheet rubber used for the soles of shoes etc. □ **crêpey** *adj.* **crêpy** *adj.* [Latin: related to CRISP]

crêpe de Chine /də 'ʃi:n/ *n.* fine silk crêpe.

crêpe paper *n.* thin crinkled paper.

crêpe Suzette /su:'zet/ *n.* small dessert pancake flamed in alcohol.

crept *past* and *past part.* of CREEP.

crepuscular /krɪ'pʌskjʊlə(r)/ *adj.* **1 a** of twilight. **b** dim. **2** *Zool.* appearing or active in twilight. [Latin *crepusculum* twilight]

Cres. *abbr.* Crescent.

cresc. *abbr.* (also **cres.**) *Mus.* = CRESCENDO.

crescendo /krɪ'ʃendəʊ/ –*n.* (*pl.* **-s**) **1** *Mus.* gradual increase in loudness. **2**

progress towards a climax. *–adv. & adj.* increasing in loudness. [Italian: related to CRESCENT]

■ **Usage** *Crescendo* is sometimes wrongly used to mean the climax itself rather than progress towards it.

crescent /ˈkrez(ə)nt/ *–n.* **1** curved sickle shape as of the waxing or waning moon. **2** thing of this shape, esp. a street forming an arc. *–adj.* crescent-shaped. [Latin *cresco* grow]

cress *n.* any of various plants with pungent edible leaves. [Old English]

crest *–n.* **1 a** comb or tuft etc. on a bird's or animal's head. **b** plume etc. on a helmet etc. **2** top of a mountain, wave, roof, etc. **3** *Heraldry* **a** device above a coat of arms. **b** such a device on writing-paper etc. *–v.* **1** reach the crest of. **2** provide with a crest or serve as a crest to. **3** (of a wave) form a crest. □ **crested** *adj.* [Latin *crista*]

crestfallen *adj.* dejected, dispirited.

cretaceous /krɪˈteɪʃəs/ *–adj.* **1** of or like chalk. **2** **(Cretaceous)** *Geol.* of the last period of the Mesozoic era, with deposits of chalk. *–n.* **(Cretaceous)** *Geol.* this era or system. [Latin *creta* chalk]

cretin /ˈkretɪn/ *n.* **1** deformed and mentally retarded person, esp. as the result of thyroid deficiency. **2** *colloq.* stupid person. □ **cretinism** *n.* **cretinous** *adj.* [French *crétin*: related to CHRISTIAN]

cretonne /kreˈtɒn/ *n.* (often *attrib.*) heavy cotton upholstery fabric, usu. with a floral pattern. [*Creton* in Normandy]

crevasse /krəˈvæs/ *n.* deep open crack, esp. in a glacier. [Latin *crepo* crack]

crevice /ˈkrevɪs/ *n.* narrow opening or fissure, esp. in rock etc. [French: related to CREVASSE]

crew[1] /kruː/ *–n.* (often treated as *pl.*) **1 a** people manning a ship, aircraft, train, etc. **b** these as distinct from the captain or officers. **c** people working together; team. **2** *colloq.* gang. *–v.* **1** supply or act as a crew or crew member for. **2** act as a crew. [Latin *cresco* increase]

crew[2] *past* of CROW[2].

crew cut *n.* close-cropped hairstyle.

crewel /ˈkruːəl/ *n.* thin worsted yarn for tapestry and embroidery. [origin unknown]

crewel-work *n.* design in crewel.

crew neck *n.* round close-fitting neckline.

crib *–n.* **1 a** baby's small bed or cot. **b** model of the Nativity with a manger. **2** rack for animal fodder. **3** *colloq.* **a** translation of a text used by students. **b** plagiarized work etc. **4** *colloq.* **a** cribbage. **b** set of cards given to the dealer at cribbage. *–v.* **(-bb-)** (also *absol.*) **1** *colloq.* copy unfairly. **2** confine in a small space. **3** *colloq.* pilfer. [Old English]

cribbage /ˈkrɪbɪdʒ/ *n.* card-game for up to four players. [origin unknown]

crick *–n.* sudden painful stiffness, esp. in the neck. *–v.* cause this in. [origin unknown]

cricket[1] /ˈkrɪkɪt/ *n.* team game played on a grass pitch, with bowling at a wicket defended by a batting player of the other team. □ **not cricket** *colloq.* unfair behaviour. □ **cricketer** *n.* [origin uncertain]

cricket[2] /ˈkrɪkɪt/ *n.* grasshopper-like chirping insect. [French, imitative]

cri de cœur /ˌkriː də ˈkɜː(r)/ *n.* (*pl.* ***cris de cœur*** pronunc. same) passionate appeal, protest, etc. [French, = cry from the heart]

cried *past* and *past part.* of CRY.

crier /ˈkraɪə(r)/ *n.* (also **cryer**) **1** person who cries. **2** official making public announcements in a lawcourt or street. [related to CRY]

crikey /ˈkraɪkɪ/ *int. slang* expression of astonishment. [from CHRIST]

crime *n.* **1 a** offence punishable by law. **b** illegal acts (*resorted to crime*). **2** evil act (*crime against humanity*). **3** *colloq.* shameful act. [Latin *crimen*]

criminal /ˈkrɪmɪn(ə)l/ *–n.* person guilty of a crime. *–adj.* **1** of, involving, or concerning crime. **2** guilty of crime. **3** *Law* of or concerning criminal offences (*criminal code; criminal lawyer*). **4** *colloq.* scandalous, deplorable. □ **criminality** /-ˈnælɪtɪ/ *n.* **criminally** *adv.* [Latin: related to CRIME]

criminology /ˌkrɪmɪˈnɒlədʒɪ/ *n.* the study of crime. □ **criminologist** *n.*

crimp *–v.* **1** press into small folds; corrugate. **2** make waves in (hair). *–n.* crimped thing or form. [Low German or Dutch]

Crimplene /ˈkrɪmpliːn/ *n. propr.* synthetic crease-resistant fabric.

crimson /ˈkrɪmz(ə)n/ *–adj.* of a rich deep red. *–n.* this colour. [ultimately from Arabic: related to KERMES]

cringe /krɪndʒ/ *v.* **(-ging)** **1** shrink in fear; cower. **2** (often foll. by *to*) behave obsequiously. [related to CRANK]

crinkle /ˈkrɪŋk(ə)l/ *–n.* wrinkle or crease. *–v.* **(-ling)** form crinkles (in). □ **crinkly** *adj.* [related to CRINGE]

crinkle-cut *adj.* (of vegetables) with wavy edges.

crinoline /ˈkrɪnəlɪn/ *n.* **1** *hist.* stiffened or hooped petticoat. **2** stiff fabric of horsehair etc. used for linings, hats, etc. [French from Latin *crinis* hair, *linum* thread]

cripple /ˈkrɪp(ə)l/ *–n.* permanently lame person. *–v.* **(-ling) 1** make a cripple of; lame. **2** disable, weaken, or damage seriously (*crippled by strikes*). [Old English]

crisis /ˈkraɪsɪs/ *n.* (*pl.* **crises** /-siːz/) **1** time of danger or great difficulty. **2** decisive moment; turning-point. [Greek, = decision]

crisp *–adj.* **1** hard but brittle. **2 a** (of air) bracing. **b** (of style or manner) lively, brisk and decisive. **c** (of features etc.) neat, clear-cut. **d** (of paper) stiff and crackling. **e** (of hair) closely curling. *–n.* (in full **potato crisp**) potato sliced thinly, fried, and sold in packets. *–v.* make or become crisp. □ **crisply** *adv.* **crispness** *n.* [Latin *crispus* curled]

crispbread *n.* **1** thin crisp biscuit of crushed rye etc. **2** these collectively (*packet of crispbread*).

crispy *adj.* **(-ier, -iest)** crisp. □ **crispiness** *n.*

criss-cross *–n.* pattern of crossing lines. *–adj.* crossing; in cross lines. *–adv.* crosswise; at cross purposes. *–v.* **1 a** intersect repeatedly. **b** move crosswise. **2** mark or make with a criss-cross pattern. [*Christ's cross*]

criterion /kraɪˈtɪərɪən/ *n.* (*pl.* **-ria**) principle or standard of judgement. [Greek, = means of judging]

■ **Usage** The plural form of *criterion*, *criteria*, is often used incorrectly as the singular. In the singular *criterion* should always be used.

critic /ˈkrɪtɪk/ *n.* **1** person who criticizes. **2** person who reviews literary, artistic, etc. works. [Latin *criticus* from Greek *kritēs* judge]

critical *adj.* **1 a** fault-finding, censorious. **b** expressing or involving criticism. **2** skilful at or engaged in criticism. **3** providing textual criticism (*critical edition of Milton*). **4 a** of or at a crisis; dangerous, risky (*in a critical condition*). **b** decisive, crucial (*at the critical moment*). **5 a** *Math. & Physics* marking a transition from one state etc. to another (*critical angle*). **b** (of a nuclear reactor) maintaining a self-sustaining chain reaction. □ **critically** *adv.* **criticalness** *n.*

critical path *n.* sequence of stages determining the minimum time needed for an operation.

criticism /ˈkrɪtɪˌsɪz(ə)m/ *n.* **1 a** fault-finding; censure. **b** critical remark etc. **2 a** work of a critic. **b** analytical article, essay, etc.

criticize /ˈkrɪtɪˌsaɪz/ *v.* (also **-ise**) **(-zing** or **-sing)** (also *absol.*) **1** find fault with; censure. **2** discuss critically.

critique /krɪˈtiːk/ *n.* critical analysis. [French: related to CRITIC]

croak *–n.* deep hoarse sound, esp. of a frog. *–v.* **1** utter or speak with a croak. **2** *slang* die. [imitative]

croaky *adj.* **(-ier, -iest)** croaking; hoarse. □ **croakily** *adv.* **croakiness** *n.*

Croat /ˈkrəʊæt/ (also **Croatian** /krəʊˈeɪʃ(ə)n/) *–n.* **1 a** native of Croatia in SE Europe. **b** person of Croatian descent. **2** Slavonic dialect of the Croats. *–adj.* of the Croats or their dialect. [Serbo-Croatian *Hrvat*]

crochet /ˈkrəʊʃeɪ/ *–n.* needlework in which yarn is hooked to make a lacy patterned fabric. *–v.* **(crocheted** /-ʃeɪd/; **crocheting** /-ʃeɪɪŋ/) (also *absol.*) make using crochet. [French: related to CROTCHET]

crock[1] *n. colloq.* old or worn-out person or vehicle. [originally Scots]

crock[2] *n.* **1** earthenware pot or jar. **2** broken piece of this. [Old English]

crockery *n.* earthenware or china dishes, plates, etc. [related to CROCK[2]]

crocodile /ˈkrɒkəˌdaɪl/ *n.* **1 a** large tropical amphibious reptile with thick scaly skin, a long tail, and long jaws. **b** (often *attrib.*) its skin. **2** *colloq.* line of schoolchildren etc. walking in pairs. [Greek *krokodilos*]

crocodile tears *n.pl.* insincere grief.

crocus /ˈkrəʊkəs/ *n.* (*pl.* **-cuses**) small plant with white, yellow, or purple flowers, growing from a corm. [Latin from Greek]

Croesus /ˈkriːsəs/ *n.* person of great wealth. [name of a king of ancient Lydia]

croft *–n.* **1** enclosed piece of (usu. arable) land. **2** small rented farm in Scotland or N. England. *–v.* farm a croft; live as a crofter. [Old English]

crofter *n.* person who farms a croft.

Crohn's disease /krəʊnz/ *n.* chronic inflammatory disease of the alimentary tract. [E. *Crohn*, name of a US pathologist]

croissant /ˈkrwʌsɑ̃/ *n.* crescent-shaped breakfast roll. [French: related to CRESCENT]

cromlech /ˈkrɒmlek/ *n.* **1** dolmen. **2** prehistoric stone circle. [Welsh]

crone *n.* withered old woman. [Dutch *croonje* carcass]

crony /ˈkrəʊnɪ/ *n.* (*pl.* **-ies**) friend, companion. [Greek *khronios* long-lasting]

crook /krʊk/ *–n.* **1** hooked staff of a shepherd or bishop. **2 a** bend, curve, or hook. **b** hooked or curved thing. **3** *colloq.* rogue; swindler; criminal. *–v.* bend, curve. [Old Norse]

crooked /ˈkrʊkɪd/ *adj.* **(-er, -est) 1** not straight or level; bent. **2** *colloq.* not straightforward; dishonest, criminal. □ **crookedly** *adv.* **crookedness** *n.*

croon –*v.* sing, hum, or say in a low sentimental voice. –*n.* such singing etc. □ **crooner** *n.* [Low German or Dutch]

crop –*n.* **1 a** produce of cultivated plants, esp. cereals. **b** season's yield. **2** group, yield, etc., of one time or place (*a new crop of students*). **3** handle of a whip. **4 a** very short haircut. **b** cropping of hair. **5** pouch in a bird's gullet where food is prepared for digestion. –*v.* (**-pp-**) **1 a** cut off. **b** bite off. **2** cut (hair etc.) short. **3** (foll. by *with*) sow or plant (land) with a crop. **4** (of land) bear a crop. □ **crop up** occur unexpectedly. [Old English]

crop circle *n.* circle of crops that has been inexplicably flattened.

crop-eared *adj.* with the ears (esp. of animals) or hair cut short.

cropper *n.* crop-producing plant of a specified quality. □ **come a cropper** *slang* fall heavily; fail badly.

croquet /'krəʊkeɪ/ –*n.* **1** lawn game in which wooden balls are driven through hoops with mallets. **2** act of croqueting a ball. –*v.* (**croqueted** /-keɪd/; **croqueting** /-keɪɪŋ/) drive away (an opponent's ball) by placing and then striking one's own against it. [perhaps a dial. form of French *crochet* hook]

croquette /krə'ket/ *n.* ball of breaded and fried mashed potato etc. [French *croquer* crunch]

crosier /'krəʊzɪə(r)/ *n.* (also **crozier**) bishop's ceremonial hooked staff. [French *croisier* cross-bearer and *crossier* crook-bearer]

cross –*n.* **1** upright post with a transverse bar, as used in antiquity for crucifixion. **2 a** (**the Cross**) cross on which Christ was crucified. **b** representation of this as an emblem of Christianity. **c** = SIGN OF THE CROSS. **3** staff surmounted by a cross, carried in a religious procession. **4** thing or mark like a cross, esp. two short intersecting lines (+ or x). **5** cross-shaped military etc. decoration. **6 a** hybrid. **b** crossing of breeds etc. **7** (foll. by *between*) mixture of two things. **8** crosswise movement, pass in football, etc. **9** trial or affliction. –*v.* **1** (often foll. by *over*) go across. **2** intersect; (cause to) be across (*roads cross*; *cross one's legs*). **3 a** draw line(s) across. **b** mark (a cheque) with two parallel lines to indicate that it cannot be cashed. **4** (foll. by *off, out, through*) cancel etc. by drawing lines across. **5** (often *refl.*) make the sign of the cross on or over. **6 a** pass in opposite or different directions. **b** (of letters etc.) be sent at the same time. **c** (of telephone lines) be connected to an unwanted conversation. **7 a** cause to interbreed. **b** cross-fertilize (plants). **8** oppose or thwart (*crossed in love*). –*adj.* **1** (often foll. by *with*) peevish, angry. **2** (usu. *attrib.*) transverse; reaching from side to side. **3** (usu. *attrib.*) intersecting. **4** (usu. *attrib.*) contrary, opposed, reciprocal. □ **at cross purposes** misunderstanding; conflicting. **cross one's fingers** (or **keep one's fingers crossed**) **1** put one finger across another to ward off bad luck. **2** trust in good luck. **cross one's heart** make a solemn pledge, esp. by crossing one's front. **cross one's mind** occur to one, esp. transiently. **cross swords** (often foll. by *with*) argue or dispute. **cross wires** (or **get one's wires crossed**) **1** become wrongly connected by telephone. **2** have a misunderstanding. **on the cross** diagonally. □ **crossly** *adv.* **crossness** *n.* [Latin *crux*]

crossbar *n.* horizontal bar, esp. that on a man's bicycle.

cross-bench *n.* seat in the House of Lords for non-party members. □ **cross-bencher** *n.*

crossbill *n.* finch with a bill with crossed mandibles for opening pine cones.

crossbones see SKULL AND CROSSBONES.

crossbow *n.* bow fixed on a wooden stock, with a groove for an arrow.

cross-breed –*n.* **1** hybrid breed of animals or plants. **2** individual hybrid. –*v.* produce by crossing.

cross-check –*v.* check by alternative method(s). –*n.* such a check.

cross-country –*adj.* & *adv.* **1** across open country. **2** not keeping to main roads. –*n.* (*pl.* **-ies**) cross-country race.

cross-cut –*adj.* cut across the main grain. –*n.* diagonal cut, path, etc.

cross-cut saw *n.* saw for cross-cutting.

cross-dressing *n.* practice of dressing in the clothes of the opposite sex. □ **cross-dress** *v.*

crosse *n.* lacrosse stick. [French]

cross-examine *v.* question (esp. an opposing witness in a lawcourt). □ **cross-examination** *n.*

cross-eyed /'krɒsaɪd/ *adj.* having one or both eyes turned inwards.

cross-fertilize *v.* (also **-ise**) **1** fertilize (an animal or plant) from one of a different species. **2** interchange ideas etc. □ **cross-fertilization** *n.*

crossfire *n.* **1** firing in two crossing directions simultaneously. **2 a** attack or criticism from all sides. **b** combative exchange of views etc.

cross-grain *n.* grain in timber, running across the regular grain.

cross-grained *adj.* **1** having a cross-grain. **2** perverse, intractable.
cross-hatch *v.* shade with crossing parallel lines.
crossing *n.* **1** place where things (esp. roads) cross. **2** place for crossing a street etc. **3** journey across water.
cross-legged /krɒs'legd/ *adj.* (sitting) with legs folded one across the other.
crossover –*n.* **1** point or place of crossing. **2** process of crossing over, esp. from one style or genre to another. –*attrib. adj.* that crosses over, esp. from one style or genre to another.
crosspatch *n. colloq.* bad-tempered person.
crosspiece *n.* transverse beam etc.
cross-ply *adj.* (of a tyre) having fabric layers with crosswise cords.
cross-question *v.* = CROSS-EXAMINE.
cross-refer *v.* (-rr-) refer from one part of a book etc. to another.
cross-reference –*n.* reference from one part of a book etc. to another. –*v.* provide with cross-references.
crossroad *n.* (usu. in *pl.*) intersection of two or more roads. □ **at the crossroads** at the critical point.
cross-section *n.* **1 a** a cutting across a solid. **b** plane surface so produced. **c** drawing etc. of this. **2** representative sample. □ **cross-sectional** *adj.*
cross-stitch *n.* cross-shaped stitch.
crosstalk *n.* **1** unwanted signals between communication channels. **2** witty repartee.
crossways *adv.* = CROSSWISE.
crosswind *n.* wind blowing across one's path etc.
crosswise *adj.* & *adv.* **1** in the form of a cross; intersecting. **2** diagonal or diagonally.
crossword *n.* (also **crossword puzzle**) printed grid of squares and blanks for vertical and horizontal words to be filled in from clues.
crotch *n.* fork, esp. between legs (of a person, trousers, etc.). [related to CROOK]
crotchet /'krɒtʃɪt/ *n. Mus.* note equal to a quarter of a semibreve and usu. one beat. [French diminutive of *croc*: related to CROOK]
crotchety *adj.* peevish, irritable.
crouch –*v.* lower the body with limbs close to the chest; be in this position. –*n.* crouching; crouching position. [Old Norse: related to CROOK]
croup[1] /kru:p/ *n.* childhood inflammation of the larynx etc., with a hard cough. [imitative]
croup[2] /kru:p/ *n.* rump, esp. of a horse. [French: related to CROP]
croupier /'kru:pɪə(r)/ *n.* person running a gaming-table, raking in and paying out money etc. [French: related to CROUP[2]]
croûton /'kru:tɒn/ *n.* small cube of fried or toasted bread served with soup etc. [French: related to CRUST]
crow[1] /krəʊ/ *n.* **1** large black bird with a powerful black beak. **2** similar bird, e.g. the raven, rook, and jackdaw. □ **as the crow flies** in a straight line. [Old English]
crow[2] /krəʊ/ –*v.* **1** (*past* **crowed** or **crew** /kru:/) (of a cock) utter a loud cry. **2** (of a baby) utter happy cries. **3** (usu. foll. by *over*) gloat; show glee. –*n.* cry of a cock or baby. [Old English]
crowbar *n.* iron bar with a flattened end, used as a lever.
crowd –*n.* **1** large gathering of people. **2** spectators; audience. **3** *colloq.* particular set of people. **4** (prec. by *the*) majority. –*v.* **1 a** (cause to) come together in a crowd. **b** force one's way (*crowded into the cinema*). **2 a** (foll. by *into*) force or compress into a confined space. **b** (often foll. by *with*; usu. in *passive*) fill or make full of. **3** *colloq.* come aggressively close to. □ **crowd out** exclude by crowding. □ **crowdedness** *n.* [Old English]
crown –*n.* **1** monarch's jewelled head-dress. **2** (**the Crown**) **a** monarch as head of State. **b** power or authority of the monarchy. **3 a** wreath for the head as an emblem of victory. **b** award or distinction, esp. in sport. **4** crown-shaped ornament etc. **5** top part of the head, a hat, etc. **6 a** highest or central part (*crown of the road*). **b** thing that completes or forms a summit. **7 a** part of a tooth visible outside the gum. **b** artificial replacement for this. **8** former British coin worth five shillings. –*v.* **1** put a crown on (a person or head). **2** invest with a royal crown or authority. **3** be a crown to; rest on top of. **4 a** (often as **crowning** *adj.*) (cause to) be the reward, summit, or finishing touch to (*crowning glory*). **b** bring to a happy outcome. **5** fit a crown to (a tooth). **6** *slang* hit on the head. **7** promote (a piece in draughts) to king. [Latin *corona*]
Crown Colony *n.* British colony controlled by the Crown.
Crown Court *n.* court of criminal jurisdiction in England and Wales.
Crown Derby *n.* porcelain made at Derby and often marked with a crown.
crown glass *n.* glass without lead or iron used formerly in windows, now as optical glass of low refractive index.
crown jewels *n.pl.* sovereign's State regalia etc.
Crown prince *n.* male heir to a throne.

Crown princess *n.* **1** wife of a Crown prince. **2** female heir to a throne.
crown wheel *n.* wheel with teeth at right angles to its plane.
crow's-foot *n.* wrinkle near the eye.
crow's-nest *n.* shelter at a sailing-ship's masthead for a lookout man.
crozier var. of CROSIER.
CRT *abbr.* cathode-ray tube.
cru /kru:/ *n.* **1** French vineyard or wine region. **2** grade of wine. [French *crû* grown]
cruces *pl.* of CRUX.
crucial /'kru:ʃ(ə)l/ *adj.* **1** decisive, critical. **2** very important. □ **crucially** *adv.* [Latin *crux crucis* cross]

■ **Usage** The use of *crucial* in sense 2 should be restricted to informal contexts.

crucible /'kru:sɪb(ə)l/ *n.* **1** melting-pot for metals etc. **2** severe test. [medieval Latin: related to CRUCIAL]
cruciferous /kru:'sɪfərəs/ *adj.* having flowers with four petals arranged in a cross. [Latin: related to CRUCIAL]
crucifix /'kru:sɪfɪks/ *n.* model of a cross with the figure of Christ on it. [Latin *cruci fixus* fixed to a cross]
crucifixion /,kru:sɪ'fɪkʃ(ə)n/ *n.* **1** crucifying or being crucified. **2 (Crucifixion)** crucifixion of Christ. [Church Latin: related to CRUCIFIX]
cruciform /'kru:sɪ,fɔ:m/ *adj.* cross-shaped. [Latin *crux crucis* cross]
crucify /'kru:sɪ,faɪ/ *v.* **(-ies, -ied) 1** put to death by fastening to a cross. **2** persecute, torment. **3** *slang* defeat thoroughly; humiliate. [French: related to CRUCIFIX]
crud *n. slang* **1** deposit of grease etc. **2** unpleasant person. □ **cruddy** *adj.* **(-ier, -iest).** [var. of CURD]
crude /kru:d/ *–adj.* **1 a** in the natural state; not refined. **b** unpolished; lacking finish. **2 a** rude, blunt. **b** offensive, indecent. **3** inexact. *–n.* natural mineral oil. □ **crudely** *adv.* **crudeness** *n.* **crudity** *n.* [Latin *crudus* raw]
crudités /,kru:dɪ'teɪ/ *n.pl.* hors d'œuvre of mixed raw vegetables. [French]
cruel /'kru:əl/ *adj.* **(crueller, cruellest** or **crueler, cruelest) 1** causing pain or suffering, esp. deliberately. **2** harsh, severe (*a cruel blow*). □ **cruelly** *adv.* **cruelness** *n.* **cruelty** *n.* (*pl.* **-ies**). [Latin: related to CRUDE]
cruet /'kru:ɪt/ *n.* **1** set of small salt, pepper, etc. containers for use at table. **2** such a container. [Anglo-French diminutive: related to CROCK[2]]
cruise /kru:z/ *–v.* **(-sing) 1 a** travel by sea for pleasure, calling at ports. **b** sail about. **2** travel at a relaxed or economical speed. **3** achieve an objective, esp. win a race etc. with ease. **4** *slang* search for a sexual (esp. homosexual) partner in bars, streets, etc. *–n.* cruising voyage. [Dutch: related to CROSS]
cruise missile *n.* one able to fly low and guide itself.
cruiser *n.* **1** high-speed warship. **2** = CABIN CRUISER.
cruiserweight *n.* = LIGHT HEAVY-WEIGHT.
crumb /krʌm/ *–n.* **1 a** small fragment, esp. of bread. **b** small particle (*crumb of comfort*). **2** bread without crusts. **3** *slang* objectionable person. *–v.* cover with or break into breadcrumbs. [Old English]
crumble /'krʌmb(ə)l/ *–v.* **(-ling) 1** break or fall into small fragments. **2** (of power etc.) gradually disintegrate. *–n.* dish of stewed fruit with a crumbly topping.
crumbly *adj.* **(-ier, -iest)** consisting of, or apt to fall into, crumbs or fragments. □ **crumbliness** *n.*
crumbs /krʌmz/ *int. slang* expressing dismay or surprise. [euphemism for CHRIST]
crumby /'krʌmɪ/ *adj.* **(-ier, -iest) 1** like or covered in crumbs. **2** = CRUMMY.
crumhorn var. of KRUMMHORN.
crummy *adj.* **(-ier, -iest)** *slang* dirty, squalid; inferior, worthless. □ **crumminess** *n.* [var. of CRUMBY]
crumpet /'krʌmpɪt/ *n.* **1** soft flat yeasty cake toasted and buttered. **2** *joc.* or *offens.* sexually attractive woman or women. [origin uncertain]
crumple /'krʌmp(ə)l/ *–v.* **(-ling)** (often foll. by *up*) **1** crush or become crushed into creases or wrinkles. **2** collapse, give way. *–n.* crease or wrinkle. [obsolete *crump* curl up]
crunch *–v.* **1 a** crush noisily with the teeth. **b** grind under foot, wheels, etc. **2** (often foll. by *up, through*) make a crunching sound. *–n.* **1** crunching; crunching sound. **2** *colloq.* decisive event or moment. [imitative]
crunchy *adj.* **(-ier, -iest)** hard and crisp. □ **crunchiness** *n.*
crupper *n.* **1** strap looped under a horse's tail to hold the harness back. **2** hindquarters of a horse. [French: related to CROUP[2]]
crusade /kru:'seɪd/ *–n.* **1** *hist.* any of several medieval military expeditions made by Europeans to recover the Holy Land from the Muslims. **2** vigorous campaign for a cause. *–v.* **(-ding)** engage in a crusade. □ **crusader** *n.* [French: related to CROSS]
cruse /kru:z/ *n. archaic* earthenware pot. [Old English]

crush –*v.* **1** compress with force or violence, so as to break, bruise, etc. **2** reduce to powder by pressure. **3** crease or crumple. **4** defeat or subdue completely. –*n.* **1** act of crushing. **2** crowded mass of people. **3** drink from the juice of crushed fruit. **4** (usu. foll. by *on*) *colloq.* infatuation. [French]

crust –*n.* **1 a** hard outer part of bread. **b** hard dry scrap of bread. **c** *slang* livelihood. **2** pastry covering of a pie. **3** hard casing over a soft thing. **4** outer portion of the earth. **5** deposit, esp. from wine on a bottle. –*v.* cover or become covered with or form into a crust. [Latin *crusta* rind, shell]

crustacean /krʌ'steɪʃ(ə)n/ –*n.* esp. aquatic arthropod with a hard shell, e.g. the crab, lobster, and shrimp. –*adj.* of crustaceans.

crusty *adj.* **(-ier, -iest) 1** having a crisp crust. **2** irritable, curt. □ **crustily** *adv.* **crustiness** *n.*

crutch *n.* **1** usu. T-shaped support for a lame person fitting under the armpit. **2** support, prop. **3** crotch. [Old English]

crux *n.* (*pl.* **cruxes** or **cruces** /'kru:si:z/) decisive point at issue. [Latin, = cross]

cruzado /kru:'zɑ:dəʊ/ *n.* (*pl.* **-s**) chief monetary unit of Brazil. [Portuguese]

cruzeiro /kru:'zeərəʊ/ *n.* (*pl.* **-s**) one-thousandth of a cruzado. [Portuguese]

cry /kraɪ/ –*v.* **(cries, cried) 1** (often foll. by *out*) make a loud or shrill sound, esp. to express pain, grief, etc., or to appeal for help. **2** shed tears; weep. **3** (often foll. by *out*) say or exclaim loudly or excitedly. **4** (foll. by *for*) appeal, demand, or show a need for. **5** (of an animal, esp. a bird) make a loud call. –*n.* (*pl.* **cries**) **1** loud shout or scream of grief, pain, etc. **2** spell of weeping. **3** loud excited utterance. **4** urgent appeal. **5 a** public demand or opinion. **b** rallying call. **6** call of an animal. □ **cry down** disparage. **cry off** withdraw from an undertaking. **cry out for** need as an obvious requirement or solution. **cry wolf** see WOLF. [Latin *quirito*]

cry-baby *n.* person who weeps frequently.

cryer var. of CRIER.

crying *attrib. adj.* (of injustice etc.) flagrant, demanding redress.

cryogenics /,kraɪəʊ'dʒenɪks/ *n.* branch of physics dealing with very low temperatures. □ **cryogenic** *adj.* [Greek *kruos* frost, *-genēs* born]

crypt /krɪpt/ *n.* vault, esp. beneath a church, used usu. as a burial-place. [Latin *crypta* from Greek *kruptos* hidden]

cryptic *adj.* obscure in meaning; secret, mysterious. □ **cryptically** *adv.*

cryptogam /'krɪptə,gæm/ *n.* plant with no true flowers or seeds, e.g. ferns, mosses, and fungi. □ **cryptogamous** /-'tɒgəməs/ *adj.* [as CRYPT, Greek *gamos* marriage]

cryptogram /'krɪptə,græm/ *n.* text written in cipher. [related to CRYPT]

cryptography /krɪp'tɒgrəfɪ/ *n.* art of writing or solving ciphers. □ **cryptographer** *n.* **cryptographic** /-tə'græfɪk/ *adj.*

crystal /'krɪst(ə)l/ –*n.* **1 a** transparent colourless mineral, esp. rock crystal. **b** piece of this. **2 a** highly transparent glass; flint glass. **b** articles of this. **3** crystalline piece of semiconductor. **4** aggregation of molecules with a definite internal structure and the external form of a solid enclosed by symmetrically arranged plane faces. –*adj.* (usu. *attrib.*) made of, like, or clear as crystal. [Greek *krustallos*]

crystal ball *n.* glass globe used in crystal-gazing.

crystal-gazing *n.* supposed foretelling of the future by gazing into a crystal ball.

crystalline /'krɪstə,laɪn/ *adj.* **1** of, like, or clear as crystal. **2** having the structure and form of a crystal. □ **crystallinity** /-'lɪnɪtɪ/ *n.*

crystallize /'krɪstə,laɪz/ *v.* (also **-ise**) (**-zing** or **-sing**) **1** form into crystals. **2** (often foll. by *out*) (of ideas or plans) make or become definite. **3** make or become coated or impregnated with sugar (*crystallized fruit*). □ **crystallization** /-'zeɪʃ(ə)n/ *n.*

crystallography /,krɪstə'lɒgrəfɪ/ *n.* science of crystal formation and structure. □ **crystallographer** *n.*

crystalloid /'krɪstə,lɔɪd/ *n.* substance that in solution is able to pass through a semipermeable membrane.

Cs *symb.* caesium.

c/s *abbr.* cycles per second.

CSE *abbr. hist.* Certificate of Secondary Education.

■ **Usage** The CSE examination was replaced in 1988 by GCSE.

CS gas *n.* tear-gas used to control riots etc. [Corson and Stoughton, names of chemists]

CTC *abbr.* City Technology College.

Cu *symb.* copper. [Latin *cuprum*]

cu. *abbr.* cubic.

cub –*n.* **1** young of a fox, bear, lion, etc. **2** (**Cub**) (in full **Cub Scout**) junior Scout. **3** *colloq.* young newspaper reporter. –*v.* (**-bb-**) (also *absol.*) give birth to (cubs). [origin unknown]

cubby-hole /'kʌbɪ,həʊl/ *n.* **1** very small room. **2** snug space. [Low German]

cube –*n*. **1** solid contained by six equal squares. **2** cube-shaped block. **3** product of a number multiplied by its square. –*v*. (**-bing**) **1** find the cube of (a number). **2** cut (food etc.) into small cubes. [Latin from Greek]

cube root *n*. number which produces a given number when cubed.

cubic *adj*. **1** cube-shaped. **2** of three dimensions. **3** involving the cube (and no higher power) of a number (*cubic equation*).

cubical *adj*. cube-shaped.

cubicle /'kju:bɪk(ə)l/ *n*. **1** small screened space. **2** small separate sleeping-compartment. [Latin *cubo* lie]

cubic metre etc. *n*. volume of a cube whose edge is one metre etc.

cubism /'kju:bɪz(ə)m/ *n*. style in art, esp. painting, in which objects are represented geometrically. □ **cubist** *n*. & *adj*.

cubit /'kju:bɪt/ *n*. ancient measure of length, approximating to the length of a forearm. [Latin *cubitum* elbow]

cuboid /'kju:bɔɪd/ –*adj*. cube-shaped; like a cube. –*n*. *Geom*. rectangular parallelepiped.

cuckold /'kʌkəʊld/ –*n*. husband of an adulteress. –*v*. make a cuckold of. □ **cuckoldry** *n*. [French]

cuckoo /'kʊku:/ –*n*. bird having a characteristic cry, and laying its eggs in the nests of small birds. –*predic. adj. slang* crazy. [French, imitative]

cuckoo clock *n*. clock with the figure of a cuckoo emerging to make a call on the hour.

cuckoo-pint *n*. wild arum.

cuckoo-spit *n*. froth exuded by insect larvae on leaves, stems, etc.

cucumber /'kju:kʌmbə(r)/ *n*. **1** long green fleshy fruit, used in salads. **2** climbing plant yielding this. [French from Latin]

cud *n*. half-digested food returned to the mouth of ruminants for further chewing. [Old English]

cuddle /'kʌd(ə)l/ –*v*. (**-ling**) **1** hug, fondle. **2** nestle together, lie close and snug. –*n*. prolonged and fond hug. □ **cuddlesome** *adj*. [origin uncertain]

cuddly *adj*. (**-ier, -iest**) **1** (of a person, toy, etc.) soft and yielding. **2** given to cuddling.

cudgel /'kʌdʒ(ə)l/ –*n*. short thick stick used as a weapon. –*v*. (**-ll-**; *US* **-l-**) beat with a cudgel. [Old English]

cue[1] –*n*. **1 a** last words of an actor's speech as a signal to another to enter or speak. **b** similar signal to a musician etc. **2 a** stimulus to perception etc. **b** signal for action. **c** hint on appropriate behaviour. **3** cueing audio equipment (see sense 2 of *v*.). –*v*. (**cues, cued, cueing** or **cuing**) **1** give a cue to. **2** put (audio equipment) in readiness to play a particular section. □ **cue in 1** insert a cue for. **2** give information to. **on cue** at the correct moment. [origin unknown]

cue[2] *Billiards* etc. –*n*. long rod for striking a ball. –*v*. (**cues, cued, cueing** or **cuing**) strike (a ball) with or use a cue. [var. of QUEUE]

cue-ball *n*. ball to be struck with a cue.

cuff[1] *n*. **1** end part of a sleeve. **2** *US* trouser turn-up. **3** (in *pl*.) *colloq*. handcuffs. □ **off the cuff** *colloq*. without preparation, extempore. [origin unknown]

cuff[2] –*v*. strike with an open hand. –*n*. such a blow. [perhaps imitative]

cuff-link *n*. two joined studs etc. for fastening a cuff.

Cufic var. of KUFIC.

cuirass /kwɪ'ræs/ *n*. armour breast-plate and back-plate fastened together. [Latin *corium* leather]

cuisine /kwɪ'zi:n/ *n*. style or method of cooking. [French]

cul-de-sac /'kʌldə,sæk/ *n*. (*pl*. **culs-de-sac** pronunc. same, or **cul-de-sacs**) **1** road etc. with a dead end. **2** futile course. [French, = sack-bottom]

-cule *suffix* forming (orig. diminutive) nouns (*molecule*). [Latin *-culus*]

culinary /'kʌlɪnərɪ/ *adj*. of or for cooking. [Latin *culina* kitchen]

cull –*v*. **1** select or gather (*knowledge culled from books*). **2** gather (flowers etc.). **3 a** select (animals), esp. for killing. **b** reduce the population of (an animal) by selective slaughter. –*n*. **1** culling or being culled. **2** animal(s) culled. [French: related to COLLECT[1]]

culminate /'kʌlmɪ,neɪt/ *v*. (**-ting**) (usu. foll. by *in*) reach its highest or final point (*culminate in war*). □ **culmination** /-'neɪʃ(ə)n/ *n*. [Latin *culmen* top]

culottes /kju:'lɒts/ *n.pl*. women's trousers cut like a skirt. [French, = knee-breeches]

culpable /'kʌlpəb(ə)l/ *adj*. deserving blame. □ **culpability** /-'bɪlɪtɪ/ *n*. [Latin *culpo* blame]

culprit /'kʌlprɪt/ *n*. guilty person. [perhaps from Anglo-French *culpable*: see CULPABLE]

cult *n*. **1** religious system, sect, etc., esp. ritualistic. **2 a** devotion to a person or thing (*cult of aestheticism*). **b** fashion. **c** (*attrib*.) fashionable (*cult film*). [Latin: related to CULTIVATE]

cultivar /'kʌltɪ,vɑ:(r)/ *n*. plant variety produced by cultivation. [from CULTIVATE, VARIETY]

cultivate /'kʌltɪ,veɪt/ *v*. (**-ting**) **1** prepare and use (soil etc.) for crops or gardening. **2 a** raise (crops). **b** culture (bacteria etc.). **3 a** (often as **cultivated** *adj.*) improve (the mind, manners, etc.). **b** nurture (a person, friendship, etc.). □ **cultivable** *adj.* **cultivation** /-'veɪʃ(ə)n/ *n.* [Latin *colo cult-* till, worship]

cultivator *n.* **1** mechanical implement for breaking up the ground etc. **2** person or thing that cultivates.

cultural /'kʌltʃər(ə)l/ *adj.* of or relating to intellectual or artistic matters, or to a specific culture. □ **culturally** *adv.*

culture /'kʌltʃə(r)/ *–n.* **1 a** intellectual and artistic achievement or expression (*city lacking in culture*). **b** refined appreciation of the arts etc. (*person of culture*). **2** customs, achievements, etc. of a particular civilization or group (*Chinese culture*). **3** improvement by mental or physical training. **4** cultivation of plants; rearing of bees etc. **5** quantity of micro-organisms and nutrient material supporting their growth. *–v.* (**-ring**) maintain (bacteria etc.) in suitable growth conditions. [Latin: related to CULTIVATE]

cultured *adj.* having refined taste etc.

cultured pearl *n.* pearl formed by an oyster after the insertion of a foreign body into its shell.

culture shock *n.* disorientation felt by a person subjected to an unfamiliar way of life.

culture vulture *n. colloq.* person eager for cultural pursuits.

culvert /'kʌlvət/ *n.* underground channel carrying water under a road etc. [origin unknown]

cum *prep.* (usu. in *comb.*) with, combined with, also used as (*bedroom-cum-study*). [Latin]

cumbersome /'kʌmbəsəm/ *adj.* (also **cumbrous** /'kʌmbrəs/) inconveniently bulky etc.; unwieldy. [*cumber* hinder]

cumin /'kʌmɪn/ *n.* (also **cummin**) **1** plant with aromatic seeds. **2** these as flavouring. [Greek *kuminon*]

cummerbund /'kʌmə,bʌnd/ *n.* waist sash. [Hindustani and Persian]

cumquat var. of KUMQUAT.

cumulative /'kju:mjʊlətɪv/ *adj.* **1** increasing or increased progressively in amount, force, etc. (*cumulative evidence*). **2** formed by successive additions (*learning is a cumulative process*). □ **cumulatively** *adv.*

cumulus /'kju:mjʊləs/ *n.* (*pl.* **-li** /-,laɪ/) cloud formation of rounded masses heaped up on a flat base. [Latin, = heap]

cuneiform /'kju:nɪ,fɔ:m/ *–adj.* **1** wedge-shaped. **2** of or using wedge-shaped writing. *–n.* cuneiform writing. [Latin *cuneus* wedge]

cunnilingus /,kʌnɪ'lɪŋgəs/ *n.* oral stimulation of the female genitals. [Latin *cunnus* vulva, *lingo* lick]

cunning *–adj.* (**-er, -est**) **1** deceitful, clever, or crafty. **2** ingenious (*cunning device*). **3** *US* attractive, quaint. *–n.* **1** craftiness; deception. **2** skill, ingenuity. □ **cunningly** *adv.* [Old Norse: related to CAN[1]]

cunt *n. coarse slang* **1** female genitals. **2** *offens.* unpleasant person. [origin uncertain]

cup *–n.* **1** small bowl-shaped container for drinking from. **2 a** its contents. **b** = CUPFUL. **3** cup-shaped thing. **4** flavoured wine, cider, etc., usu. chilled. **5** cup-shaped trophy as a prize. **6** one's fate or fortune (*a bitter cup*). *–v.* (**-pp-**) **1** form (esp. the hands) into the shape of a cup. **2** take or hold as in a cup. □ **one's cup of tea** *colloq.* what interests or suits one. [medieval Latin *cuppa*]

cupboard /'kʌbəd/ *n.* recess or piece of furniture with a door and (usu.) shelves.

cupboard love *n.* false affection for gain.

Cup Final *n.* final match in a (esp. football) competition.

cupful *n.* (*pl.* **-s**) **1** amount held by a cup, esp. *US* a half-pint or 8-ounce measure. **2** full cup.

■ **Usage** A *cupful* is a measure, and so *three cupfuls* is a quantity regarded in terms of a cup; *three cups full* denotes the actual cups, as in *brought us three cups full of water*.

Cupid /'kju:pɪd/ *n.* **1** Roman god of love, represented as a naked winged boy archer. **2** (also **cupid**) representation of Cupid. [Latin *cupio* desire]

cupidity /kju:'pɪdɪtɪ/ *n.* greed; avarice. [Latin: related to CUPID]

Cupid's bow *n.* upper lip etc. shaped like an archery bow.

cupola /'kju:pələ/ *n.* **1** dome forming or adorning a roof. **2** revolving dome protecting mounted guns. **3** furnace for melting metals. □ **cupolaed** /-ləd/ *adj.* [Italian from Latin *cupa* cask]

cuppa /'kʌpə/ *n. colloq.* **1** cup of. **2** cup of tea. [corruption]

cupreous /'kju:prɪəs/ *adj.* of or like copper. [Latin: related to COPPER[1]]

cupric /'kju:prɪk/ *adj.* of copper.

cupro-nickel /,kju:prəʊ'nɪk(ə)l/ *n.* alloy of copper and nickel.

cup-tie *n.* match in a competition for a cup.

cur *n.* **1** mangy ill-tempered dog. **2** contemptible person. [perhaps from Old Norse *kurr* grumbling]

curable /'kjʊərəb(ə)l/ *adj.* able to be cured. □ **curability** /-'bɪlɪtɪ/ *n.*

curaçao /'kjʊərə,səʊ/ *n.* (*pl.* **-s**) orange-flavoured liqueur. [*Curaçao*, Caribbean island]

curacy /'kjʊərəsɪ/ *n.* (*pl.* **-ies**) curate's office or tenure of it.

curare /kjʊə'rɑ:rɪ/ *n.* extract of various plants, used by American Indians to poison arrows. [Carib]

curate /'kjʊərət/ *n.* assistant to a parish priest. [medieval Latin *curatus*: related to CURE]

curate's egg *n.* thing that is good in parts.

curative /'kjʊərətɪv/ *–adj.* tending or able to cure. *–n.* curative agent. [medieval Latin: related to CURATE]

curator /kjʊə'reɪtə(r)/ *n.* keeper or custodian of a museum etc. □ **curatorship** *n.* [Anglo-Latin: related to CURE]

curb *–n.* **1** check, restraint. **2** strap etc. passing under a horse's lower jaw, used as a check. **3** enclosing border, e.g. the frame round a well or a fender round a hearth. **4** = KERB. *–v.* **1** restrain. **2** put a curb on (a horse). [French: related to CURVE]

curd *n.* (often in *pl.*) coagulated acidic milk product made into cheese or eaten as food. [origin unknown]

curd cheese *n.* soft smooth cheese made from skimmed milk curds.

curdle /'kɜ:d(ə)l/ *v.* (**-ling**) form into curds; congeal. □ **make one's blood curdle** horrify one. [from CURD]

cure *–v.* (**-ring**) **1** (often foll. by *of*) restore to health; relieve (*cured of pleurisy*). **2** eliminate (disease, evil, etc.). **3** preserve (meat, fruit, etc.) by salting, drying, etc. **4** vulcanize (rubber); harden (plastic etc.). *–n.* **1** restoration to health. **2** thing effecting a cure. **3** course of treatment. **4** curacy. [Latin *cura* care]

curé /'kjʊəreɪ/ *n.* parish priest in France etc. [French]

cure-all *n.* panacea.

curette /kjʊə'ret/ *–n.* surgeon's small scraping-instrument. *–v.* (**-tting**) clean or scrape with this. □ **curettage** /-'retɪdʒ/ *n.* [French: related to CURE]

curfew /'kɜ:fju:/ *n.* **1** signal or time after which people must remain indoors. **2** *hist.* signal for extinction of fires at a fixed hour. [French: related to COVER, Latin FOCUS]

Curia /'kjʊərɪə/ *n.* (also **curia**) papal court; government departments of the Vatican. [Latin]

curie /'kjʊərɪ/ *n.* unit of radioactivity. [P. *Curie*, name of a scientist]

curio /'kjʊərɪəʊ/ *n.* (*pl.* **-s**) rare or unusual object. [abbreviation of CURIOSITY]

curiosity /,kjʊərɪ'ɒsɪtɪ/ *n.* (*pl.* **-ies**) **1** eager desire to know; inquisitiveness. **2** strange, rare, etc. object. [Latin: related to CURIOUS]

curious /'kjʊərɪəs/ *adj.* **1** eager to learn; inquisitive. **2** strange, surprising, odd. □ **curiously** *adv.* [Latin: related to CURE]

curium /'kjʊərɪəm/ *n.* artificial radioactive metallic element. [M. and P. *Curie*, name of scientists]

curl *–v.* **1** (often foll. by *up*) bend or coil into a spiral. **2** move in a spiral form. **3 a** (of the upper lip) be raised contemptuously. **b** cause (the lip) to do this. **4** play curling. *–n.* **1** lock of curled hair. **2** anything spiral or curved inwards. **3 a** curling movement. **b** being curled. □ **curl one's lip** express scorn. **curl up 1** lie or sit with the knees drawn up. **2** *colloq.* writhe in embarrassment etc. [Dutch]

curler *n.* pin or roller etc. for curling the hair.

curlew /'kɜ:lju:/ *n.* wading bird, usu. with a long slender bill. [French]

curlicue /'kɜ:lɪ,kju:/ *n.* decorative curl or twist. [from CURLY, CUE[2] or Q[1]]

curling *n.* game resembling bowls, played on ice with round flat stones.

curly *adj.* (**-ier, -iest**) **1** having or arranged in curls. **2** moving in curves. □ **curliness** *n.*

curly kale *n.* = KALE.

curmudgeon /kə'mʌdʒ(ə)n/ *n.* bad-tempered person. □ **curmudgeonly** *adj.* [origin unknown]

currant /'kʌrənt/ *n.* **1** small seedless dried grape. **2 a** any of various shrubs producing red, white, or black berries. **b** such a berry. [Anglo-French from *Corinth* in Greece]

currency /'kʌrənsɪ/ *n.* (*pl.* **-ies**) **1 a** money in use in a country. **b** other commodity used as money. **2** being current; prevalence (e.g. of words or ideas).

current *–adj.* **1** belonging to the present; happening now (*current events*). **2** (of money, opinion, rumour, etc.) in general circulation or use. *–n.* **1** body of moving water, air, etc., esp. passing through still water etc. **2 a** ordered movement of electrically charged particles. **b** quantity representing the intensity of this. **3** (usu. foll. by *of*) general tendency or course (of events, opinions, etc.). □ **currentness** *n.* [Latin *curro curs-* run]

current account *n.* instantly accessible bank account.

currently *adv.* at the present time; now.

curriculum /kə'rɪkjʊləm/ *n.* (*pl.* **-la**) subjects included in a course of study. [Latin, = course]

curriculum vitae /'vi:taɪ/ *n.* brief account of one's education, career, etc.
curry[1] /'kʌrɪ/ *–n.* (*pl.* **-ies**) meat, vegetables, etc., cooked in a spicy sauce, usu. served with rice. *–v.* (**-ies, -ied**) prepare or flavour with a curry sauce. [Tamil]
curry[2] /'kʌrɪ/ *v.* (**-ies, -ied**) **1** groom (a horse) with a curry-comb. **2** treat (tanned leather) to improve it. □ **curry favour** ingratiate oneself. [Germanic: related to READY]
curry-comb *n.* serrated device for grooming horses.
curry-powder *n.* mixture of turmeric, cumin, etc. for making curry.
curse *–n.* **1** solemn invocation of divine wrath on a person or thing. **2** supposed resulting evil. **3** violent or profane exclamation or oath. **4** thing causing evil or harm. **5** (prec. by *the*) *colloq.* menstruation. *–v.* (**-sing**) **1 a** utter a curse against. **b** (in *imper.*) may God curse. **2** (usu. in *passive*; foll. by *with*) afflict with. **3** swear profanely. [Old English]
cursed /'kɜ:sɪd/ *attrib. adj.* damned.
cursive /'kɜ:sɪv/ *–adj.* (of writing) with joined characters. *–n.* cursive writing. [medieval Latin, = running: related to CURRENT]
cursor /'kɜ:sə(r)/ *n.* **1** *Math.* etc. transparent slide with a hairline, forming part of a slide-rule. **2** *Computing* indicator on a VDU screen identifying esp. the position that the program will operate on with the next keystroke. [Latin, = runner: related to CURSIVE]
cursory /'kɜ:sərɪ/ *adj.* hasty, hurried. □ **cursorily** *adv.* **cursoriness** *n.* [Latin: related to CURSOR]
curt *adj.* noticeably or rudely brief. □ **curtly** *adv.* **curtness** *n.* [Latin *curtus* short]
curtail /kɜ:'teɪl/ *v.* cut short; reduce. □ **curtailment** *n.* [corruption of obsolete adj. *curtal*: related to CURT]
curtain /'kɜ:t(ə)n/ *–n.* **1** piece of cloth etc. hung as a screen, esp. at a window. **2 a** rise or fall of a stage curtain between acts or scenes. **b** = CURTAIN-CALL. **3** partition or cover. **4** (in *pl.*) *slang* the end. *–v.* **1** provide or cover with curtain(s). **2** (foll. by *off*) shut off with curtain(s). [Latin *cortina*]
curtain-call *n.* audience's applause summoning actors to take a bow.
curtain-raiser *n.* **1** short play before the main performance. **2** preliminary event.
curtilage /'kɜ:tɪlɪdʒ/ *n.* esp. *Law* area attached to a house and forming one enclosure with it. [French: related to COURT]
curtsy /'kɜ:tsɪ/ (also **curtsey**) *–n.* (*pl.* **-ies** or **-eys**) bending of the knees and lowering of the body made by a girl or woman in acknowledgement of applause or as a respectful greeting etc. *–v.* (**-ies, -ied** or **-eys, -eyed**) make a curtsy. [var. of COURTESY]
curvaceous /kɜ:'veɪʃəs/ *adj. colloq.* (esp. of a woman) having a shapely figure.
curvature /'kɜ:vətʃə(r)/ *n.* **1** curving. **2** curved form. **3** deviation of a curve or curved surface from a plane. [French from Latin: related to CURVE]
curve *–n.* **1** line or surface of which no part is straight or flat. **2** curved form or thing. **3** curved line on a graph. *–v.* (**-ving**) bend or shape to form a curve. □ **curved** *adj.* [Latin *curvus* curved]
curvet /kɜ:'vet/ *–n.* horse's frisky leap. *–v.* (**-tt-** or **-t-**) perform a curvet. [Italian diminutive: related to CURVE]
curvilinear /,kɜ:vɪ'lɪnɪə(r)/ *adj.* contained by or consisting of curved lines. □ **curvilinearly** *adv.* [from CURVE after *rectilinear*]
curvy *adj.* (**-ier, -iest**) **1** having many curves. **2** (of a woman's figure) shapely. □ **curviness** *n.*
cushion /'kʊʃ(ə)n/ *–n.* **1** bag stuffed with soft material, for sitting or leaning on etc. **2** protection against shock; measure to soften a blow. **3** padded rim of a billiard-table etc. **4** air supporting a hovercraft etc. *–v.* **1** provide or protect with cushion(s). **2** mitigate the adverse effects of. [Latin *culcita* mattress]
cushy /'kʊʃɪ/ *adj.* (**-ier, -iest**) *colloq.* (of a job etc.) easy and pleasant. [Hindi *ḳhūsh* pleasant]
cusp *n.* point at which two curves meet, e.g. the horn of a crescent moon etc. [Latin *cuspis -id-* point, apex]
cuss *colloq. –n.* **1** curse. **2** usu. *derog.* person; creature. *–v.* curse. [var. of CURSE]
cussed /'kʌsɪd/ *adj. colloq.* stubborn. □ **cussedness** *n.*
custard /'kʌstəd/ *n.* pudding or sweet sauce of eggs or flavoured cornflour and milk. [obsolete *crustade*: related to CRUST]
custodian /kʌ'stəʊdɪən/ *n.* guardian or keeper. □ **custodianship** *n.*
custody /'kʌstədɪ/ *n.* **1** guardianship; protective care. **2** imprisonment. □ **take into custody** arrest. □ **custodial** /kʌ'stəʊdɪəl/ *adj.* [Latin *custos -od-* guard]
custom /'kʌstəm/ *n.* **1 a** usual behaviour. **b** particular established way of behaving. **2** *Law* established usage having the force of law. **3** regular business dealings or customers. **4** (in *pl.*; also treated as *sing.*) **a** duty on imports and

exports. **b** official department administering this. **c** area at a port, frontier, etc., dealing with customs etc. [Latin *consuetudo*]

customary *adj.* in accordance with custom, usual. □ **customarily** *adv.* **customariness** *n.* [medieval Latin: related to CUSTOM]

custom-built *adj.* (also **custom-made**) made to order.

customer *n.* **1** person who buys goods or services from a shop or business. **2** *colloq.* person of a specified kind (*awkward customer*). [Anglo-French: related to CUSTOM]

custom-house *n.* customs office at a port or frontier etc.

customize *v.* (also **-ise**) (**-zing** or **-sing**) make or modify to order; personalize.

cut –*v.* (**-tt-**; *past* and *past part.* **cut**) **1** (also *absol.*) penetrate or wound with a sharp-edged instrument. **2** (often foll. by *into*) divide or be divided with a knife etc. **3** trim or detach by cutting. **4** (foll. by *loose*, *open*, etc.) loosen etc. by cutting. **5** (esp. as **cutting** *adj.*) wound (*cutting remark*). **6** (often foll. by *down*) reduce (wages, time, etc.) or cease (services etc.). **7 a** make (a coat, gem, key, record, etc.) by cutting. **b** make (a path, tunnel, etc.) by removing material. **8** perform, make (*cut a caper*; *cut a sorry figure*). **9** (also *absol.*) cross, intersect. **10** (foll. by *across*, *through*, etc.) traverse, esp. as a shorter way (*cut across the grass*). **11 a** deliberately ignore (a person one knows). **b** renounce (a connection). **12** esp. *US* deliberately miss (a class etc.). **13** *Cards* **a** divide (a pack) into two parts. **b** do this to select a dealer etc. **14 a** edit (film or tape). **b** (often in *imper.*) stop filming or recording. **c** (foll. by *to*) go quickly to (another shot). **15** switch off (an engine etc.). **16** chop (a ball). –*n.* **1** cutting. **2** division or wound made by cutting. **3** stroke with a knife, sword, whip, etc. **4 a** reduction (in wages etc.). **b** cessation (of power supply etc.). **5** removal of lines etc. from a play, film, etc. **6** wounding remark or act. **7** style of hair, garment, etc. achieved by cutting. **8** particular piece of butchered meat. **9** *colloq.* commission; share of profits. **10** stroke made by cutting. **11** deliberate ignoring of a person. **12** = WOODCUT. □ **a cut above** *colloq.* noticeably superior to. **be cut out** (foll. by *for*, or *to* + infin.) be suited. **cut across 1** transcend (normal limitations etc.). **2** see sense 10 of *v.* **cut and run** *slang* run away. **cut back 1** reduce (expenditure etc.). **2** prune (a tree etc.). **cut both ways 1** serve both sides of an argument etc. **2** (of an action) have both good and bad effects. **cut a corner** go across it. **cut corners** do a task etc. perfunctorily or incompletely, esp. to save time. **cut a dash** make a brilliant show. **cut a person dead** deliberately ignore (a person one knows). **cut down 1 a** bring or throw down by cutting. **b** kill by sword or disease. **2** see sense 6 of *v.* **3** reduce the length of (*cut down trousers to make shorts*). **4** (often foll. by *on*) reduce consumption (*cut down on beer*). **cut a person down to size** *colloq.* deflate a person's pretensions. **cut in 1** interrupt. **2** pull in too closely in front of another vehicle. **cut it fine** allow very little margin of time etc. **cut it out** (usu. in *imper.*) *slang* stop doing that. **cut one's losses** abandon an unprofitable scheme. **cut no ice** *slang* have no influence. **cut off 1** remove by cutting. **2 a** (often in *passive*) bring to an abrupt end or (esp. early) death. **b** intercept, interrupt. **c** disconnect (a person on the telephone). **3 a** prevent from travelling. **b** (as **cut off** *adj.*) isolated or remote. **4** disinherit. **cut out 1** remove from inside by cutting. **2** make by cutting from a larger whole. **3** omit. **4** *colloq.* stop doing or using (something) (*cut out chocolate*). **5** (cause to) cease functioning (*engine cut out*). **6** outdo or supplant (a rival). **cut short** interrupt; terminate. **cut one's teeth on** acquire experience from. **cut a tooth** have it appear through the gum. **cut up 1** cut into pieces. **2** (usu. in *passive*) distress greatly. **cut up rough** *slang* show anger or resentment. [Old English]

cut and dried *adj.* **1** completely decided; inflexible. **2** (of opinions etc.) ready-made, lacking freshness.

cut and thrust *n.* lively argument etc.

cutaneous /kju:'teɪnɪəs/ *adj.* of the skin. [Latin: related to CUTICLE]

cutaway *attrib. adj.* (of a diagram etc.) with parts of the exterior left out to reveal the interior.

cut-back *n.* cutting back, esp. a reduction in expenditure.

cute *adj. colloq.* **1** esp. *US* attractive, quaint. **2** clever, ingenious. □ **cutely** *adv.* **cuteness** *n.* [shortening of ACUTE]

cut glass *n.* (often hypenated when *attrib.*) glass with patterns cut on it.

cuticle /'kju:tɪk(ə)l/ *n.* dead skin at the base of a fingernail or toenail. [Latin diminutive of *cutis* skin]

cutis /'kju:tɪs/ *n.* true skin, beneath the epidermis. [Latin]

cutlass /'kʌtləs/ *n. hist.* short sword with a slightly curved blade. [Latin *cultellus*: related to CUTLER]

cutler *n.* person who makes or deals in knives etc. [Latin *cultellus* diminutive: related to COULTER]

cutlery /ˈkʌtlərɪ/ *n.* knives, forks, and spoons for use at table. [Anglo-French: related to CUTLER]

cutlet /ˈkʌtlɪt/ *n.* **1** neck-chop of mutton or lamb. **2** small piece of veal etc. for frying. **3** flat cake of minced meat or nuts and breadcrumbs etc. [French diminutive from Latin *costa* rib]

cut-off *n.* **1** (often *attrib.*) point at which something is cut off. **2** device for stopping a flow.

cut-out *n.* **1** figure cut out of paper etc. **2** device for automatic disconnection, the release of exhaust gases, etc.

cut-price *adj.* (also **cut-rate**) at a reduced price.

cutter *n.* **1 a** person or thing that cuts. **b** (in *pl.*) cutting tool. **2 a** small fast sailing-ship. **b** small boat carried by a large ship.

cutthroat –*n.* **1** murderer. **2** (in full **cutthroat razor**) razor with a long unguarded blade set in a handle. –*adj.* **1** (of competition) ruthless and intense. **2** (of a card-game) three-handed.

cutting –*n.* **1** piece cut from a newspaper etc. **2** piece cut from a plant for propagation. **3** excavated channel in a hillside etc. for a railway or road. –*adj.* see CUT *v.* 5. □ **cuttingly** *adv.*

cuttlefish /ˈkʌt(ə)lfɪʃ/ *n.* (*pl.* same or **-es**) mollusc with ten arms and ejecting a black fluid when threatened. [Old English]

cutwater *n.* **1** forward edge of a ship's prow. **2** wedge-shaped projection from the pier of a bridge.

cuvée /kju:ˈveɪ/ *n.* blend or batch of wine. [French, = vatful]

c.v. *abbr.* (also **CV**) curriculum vitae.

cwm /ku:m/ *n.* (in Wales) = COOMB. [Welsh]

cwt *abbr.* hundredweight.

-cy *suffix* denoting state, condition, or status (*idiocy*; *captaincy*). [Latin *-cia*, Greek *-kia*]

cyanic acid /saɪˈænɪk/ *n.* unstable colourless pungent acid gas. [Greek *kuanos* a blue mineral]

cyanide /ˈsaɪəˌnaɪd/ *n.* highly poisonous substance used in the extraction of gold and silver.

cyanogen /saɪˈænədʒ(ə)n/ *n.* highly poisonous gas used in fertilizers.

cyanosis /ˌsaɪəˈnəʊsɪs/ *n.* bluish skin due to oxygen-deficient blood.

cybernetics /ˌsaɪbəˈnetɪks/ *n.pl.* (usu. treated as *sing.*) science of communications and control systems in machines and living things. □ **cybernetic** *adj.* [Greek *kubernētēs* steersman]

cyberpunk /ˈsaɪbəˌpʌŋk/ *n.* science fiction writing combining high-tech plots with unconventional or nihilistic social values. [from CYBERNETICS, PUNK]

cycad /ˈsaɪkæd/ *n.* palmlike plant often growing to a great height. [Greek *koix* Egyptian palm]

cyclamate /ˈsaɪkləˌmeɪt/ *n.* former artificial sweetener. [chemical name]

cyclamen /ˈsɪkləmən/ *n.* **1** plant with pink, red, or white flowers with backward-turned petals. **2** cyclamen red or pink. [Latin from Greek]

cycle /ˈsaɪk(ə)l/ –*n.* **1 a** recurrent round or period (of events, phenomena, etc.). **b** time needed for this. **2 a** *Physics* etc. recurrent series of operations or states. **b** *Electr.* = HERTZ. **3** series of related songs, poems, etc. **4** bicycle, tricycle, etc. –*v.* (**-ling**) **1** ride a bicycle etc. **2** move in cycles. [Greek *kuklos* circle]

cycle-track *n.* (also **cycle-way**) path or road for bicycles.

cyclic /ˈsaɪklɪk/ *adj.* (also **cyclical** /ˈsɪklɪk(ə)l/) **1 a** recurring in cycles. **b** belonging to a chronological cycle. **2** with constituent atoms forming a ring. □ **cyclically** *adv.*

cyclist /ˈsaɪklɪst/ *n.* rider of a bicycle.

cyclo- *comb. form* circle, cycle, or cyclic.

cyclone /ˈsaɪkləʊn/ *n.* **1** winds rotating inwards to an area of low barometric pressure; depression. **2** violent hurricane of limited diameter. □ **cyclonic** /-ˈklɒnɪk/ *adj.* [Greek *kuklōma* wheel]

cyclotron /ˈsaɪkləˌtrɒn/ *n.* apparatus for the acceleration of charged atomic and subatomic particles revolving in a magnetic field.

cygnet /ˈsɪgnɪt/ *n.* young swan. [Latin *cygnus* swan from Greek]

cylinder /ˈsɪlɪndə(r)/ *n.* **1** uniform solid or hollow body with straight sides and a circular section. **2** thing of this shape, e.g. a container for liquefied gas, a piston-chamber in an engine. □ **cylindrical** /-ˈlɪndrɪk(ə)l/ *adj.* [Latin *cylindrus* from Greek]

cymbal /ˈsɪmb(ə)l/ *n.* concave disc, struck usu. with another to make a ringing sound. □ **cymbalist** *n.* [Latin from Greek]

cyme /saɪm/ *n.* flower cluster with a single terminal flower that develops first. □ **cymose** *adj.* [Greek *kuma* wave]

Cymric /ˈkɪmrɪk/ *adj.* Welsh. [Welsh *Cymru* Wales]

cynic /ˈsɪnɪk/ *n.* **1** person with a pessimistic view of human nature. **2** (**Cynic**) one of a school of ancient Greek philosophers showing contempt for ease and pleasure. □ **cynical** *adj.* **cynically** *adv.* **cynicism** /-ˌsɪz(ə)m/ *n.* [Greek *kuōn* dog]

cynosure /'saɪnə,zjʊə(r)/ *n.* centre of attraction or admiration. [Greek, = dog's tail (name for Ursa Minor)]

cypher var. of CIPHER.

cypress /'saɪprəs/ *n.* conifer with hard wood and dark foliage. [Greek *kuparissos*]

Cypriot /'sɪprɪət/ (also **Cypriote** /-əʊt/) *–n.* native or national of Cyprus. *–adj.* of Cyprus. [*Cyprus* in E. Mediterranean]

Cyrillic /sɪ'rɪlɪk/ *–adj.* of the alphabet used by the Slavonic peoples of the Orthodox Church, now used esp. for Russian and Bulgarian. *–n.* this alphabet. [St *Cyril*, d. 869]

cyst /sɪst/ *n.* sac formed in the body, containing liquid matter. [Greek *kustis* bladder]

cystic *adj.* **1** of the bladder. **2** like a cyst.

cystic fibrosis *n.* hereditary disease usu. with respiratory infections.

cystitis /sɪ'staɪtɪs/ *n.* inflammation of the bladder usu. causing frequent painful urination.

-cyte *comb. form* mature cell (*leucocyte*). [Greek *kutos* vessel]

cytology /saɪ'tɒlədʒɪ/ *n.* the study of cells. □ **cytological** /-tə'lɒdʒɪk(ə)l/ *adj.* **cytologist** *n.* [Greek *kutos* vessel]

cytoplasm /'saɪtəʊ,plæz(ə)m/ *n.* protoplasmic content of a cell apart from its nucleus. □ **cytoplasmic** /-'plæzmɪk/ *adj.*

czar var. of TSAR

Czech /tʃek/ *–n.* **1** native or national of Czechoslovakia. **2** one of the two official languages of Czechoslovakia. *–adj.* of Czechoslovakia, its people, or language. [Bohemian *Čech*]

Czechoslovak /,tʃekə'sləʊvæk/ (also **Czechoslovakian** /-slə'vækɪən/) *–n.* native or national of Czechoslovakia. *–adj.* of Czechoslovakia. [from CZECH, SLOVAK]

D

D[1] /diː/ *n.* (also **d**) (*pl.* **Ds** or **D's**) **1** fourth letter of the alphabet. **2** *Mus.* second note of the diatonic scale of C major. **3** (as a roman numeral) 500. **4** = DEE. **5** fourth highest class or category (of academic marks etc.).

D[2] *symb.* deuterium.

d. *abbr.* **1** died. **2** departs. **3** daughter. **4** *hist.* (pre-decimal) penny. [sense 4 from Latin DENARIUS]

'd *v. colloq.* (usu. after pronouns) had, would (*I'd*; *he'd*). [abbreviation]

dab[1] —*v.* (**-bb-**) **1** (often foll. by *at*) repeatedly press briefly and lightly with a cloth etc. (*dabbed at her eyes*). **2** press (a cloth etc.) thus. **3** (foll. by *on*) apply by dabbing. **4** (often foll. by *at*) aim a feeble blow; strike lightly. —*n.* **1** dabbing. **2** small amount thus applied (*dab of paint*). **3** light blow. **4** (in *pl.*) *slang* fingerprints. [imitative]

dab[2] *n.* (*pl.* same) a kind of marine flatfish. [origin unknown]

dabble /'dæb(ə)l/ *v.* (**-ling**) **1** (usu. foll. by *in*, *at*) engage (in an activity etc.) superficially. **2** move the feet, hands, etc. in esp. shallow liquid. **3** wet partly; stain, splash. □ **dabbler** *n.* [from DAB[1]]

dabchick /'dæbtʃɪk/ *n.* = LITTLE GREBE. [Old English]

dab hand *n.* (usu. foll. by *at*) *colloq.* expert. [*dab* adept, origin unknown]

da capo /dɑː 'kɑːpəʊ/ *adv. Mus.* repeat from the beginning. [Italian]

dace *n.* (*pl.* same) small freshwater fish related to the carp. [French *dars*: related to DART]

dacha /'dætʃə/ *n.* Russian country cottage. [Russian]

dachshund /'dækshʊnd/ *n.* dog of a short-legged long-bodied breed. [German, = badger-dog]

dactyl /'dæktɪl/ *n.* metrical foot consisting of one long syllable followed by two short syllables (–∪∪). □ **dactylic** /-'tɪlɪk/ *adj.* [Greek, = finger]

dad *n. colloq.* father. [imitative of a child's *da da*]

Dada /'dɑːdɑː/ *n.* early 20th-c. artistic and literary movement repudiating conventions. □ **Dadaism** /-də,ɪz(ə)m/ *n.* **Dadaist** /-dəɪst/ *n.* & *adj.* **Dadaistic** /-'ɪstɪk/ *adj.* [French *dada* hobby-horse]

daddy /'dædɪ/ *n.* (*pl.* **-ies**) *colloq.* father. [from DAD]

daddy-long-legs *n.* (*pl.* same) crane-fly.

dado /'deɪdəʊ/ *n.* (*pl.* **-s**) **1** lower, differently decorated, part of an interior wall. **2** plinth of a column. **3** cube of a pedestal between the base and the cornice. [Italian: related to DIE[2]]

daemon var. of DEMON 4.

daff *n. colloq.* = DAFFODIL. [abbreviation]

daffodil /'dæfədɪl/ *n.* spring bulb with a yellow trumpet-shaped flower. [related to ASPHODEL]

daft /dɑːft/ *adj. colloq.* silly, foolish, crazy. [Old English, = meek]

dagger *n.* **1** short pointed knife used as a weapon. **2** *Printing* = OBELUS. □ **at daggers drawn** in bitter enmity. **look daggers at** glare angrily at. [origin uncertain]

dago /'deɪgəʊ/ *n.* (*pl.* **-s**) *slang offens.* foreigner, esp. a Spaniard, Portuguese, or Italian. [Spanish *Diego* = James]

daguerreotype /də'gerəʊ,taɪp/ *n.* early photograph using a silvered plate and mercury vapour. [*Daguerre*, name of its inventor]

dahlia /'deɪlɪə/ *n.* large-flowered showy garden plant. [*Dahl*, name of a botanist]

Dáil /dɔɪl/ *n.* (in full **Dáil Éireann** /'eɪrən/) lower house of parliament in the Republic of Ireland. [Irish, = assembly (of Ireland)]

daily /'deɪlɪ/ —*adj.* done, produced, or occurring every day or every weekday. —*adv.* **1** every day. **2** constantly. —*n.* (*pl.* **-ies**) *colloq.* **1** daily newspaper. **2** cleaning woman.

daily bread *n.* necessary food; livelihood.

dainty /'deɪntɪ/ —*adj.* (**-ier**, **-iest**) **1** delicately pretty. **2** delicate or small. **3** (of food) choice. **4** fastidious; discriminating. —*n.* (*pl.* **-ies**) choice delicacy. □ **daintily** *adv.* **daintiness** *n.* [Latin *dignitas* DIGNITY]

daiquiri /'daɪkərɪ/ *n.* (*pl.* **-s**) cocktail of rum, lime juice, etc. [*Daiquiri* in Cuba]

dairy /'deərɪ/ *n.* (*pl.* **-ies**) **1** place for processing, distributing, or selling milk and its products. **2** (*attrib.*) of, containing, or used for, dairy products (and sometimes eggs) (*dairy cow*). [Old English]

dairying *n.* dairy farming and distribution.

dairymaid *n.* woman employed in a dairy.

dairyman *n.* dealer in dairy products.

dais /'deɪɪs/ *n.* low platform, usu. at the upper end of a hall. [Latin DISCUS disc, (later) table]

daisy /'deɪzɪ/ *n.* (*pl.* **-ies**) **1** small wild plant with white-petalled flowers. **2** plant with similar flowers. [Old English, = *day's eye*]

daisy wheel *n.* spoked disc bearing printing characters, used in word processors and typewriters.

dal var. of DHAL.

Dalai Lama /ˌdælaɪ 'lɑːmə/ *n.* spiritual head of Tibetan Buddhism. [Mongolian *dalai* ocean]

dale *n.* valley. [Old English]

dally /'dælɪ/ *v.* (**-ies, -ied**) **1** delay; waste time. **2** (often foll. by *with*) flirt, trifle. □ **dalliance** *n.* [French]

Dalmatian /dæl'meɪʃ(ə)n/ *n.* large white spotted short-haired dog. [*Dalmatia* in Croatia]

dal segno /dæl 'seɪnjəʊ/ *adv. Mus.* repeat from the point marked by a sign. [Italian, = from the sign]

dam[1] *–n.* **1** barrier across river etc., forming a reservoir or preventing flooding. **2** barrier made by beaver. *–v.* (**-mm-**) **1** provide or confine with a dam. **2** (often foll. by *up*) block up; obstruct. [Low German or Dutch]

dam[2] *n.* mother, esp. of a four-footed animal. [var. of DAME]

damage /'dæmɪdʒ/ *–n.* **1** harm or injury. **2** (in *pl.*) *Law* financial compensation for loss or injury. **3** (prec. by *the*) *slang* cost. *–v.* (**-ging**) inflict damage on. [Latin *damnum*]

damascene /'dæməˌsiːn/ *–v.* (**-ning**) decorate (metal) by etching or inlaying esp. with gold or silver. *–n.* design or article produced in this way. *–adj.* of this process. [*Damascus* in Syria]

damask /'dæməsk/ *–n.* reversible figured woven fabric, esp. white table linen. *–adj.* **1** made of damask. **2** velvety pink. *–v.* weave with figured designs. [as DAMASCENE]

damask rose *n.* old sweet-scented rose used to make attar.

dame *n.* **1** (**Dame**) **a** title given to a woman holding any of several orders of chivalry. **b** woman holding this title. **2** comic middle-aged female pantomime character, usu. played by a man. **3** *US slang* woman. [Latin *domina* lady]

dame-school *n. hist.* primary school kept by an elderly woman.

dammit /'dæmɪt/ *int. colloq.* damn it.

damn /dæm/ *–v.* **1** (often *absol.* or as *int.* of anger or annoyance, = *may God damn*) curse (a person or thing). **2** doom to hell; cause the damnation of. **3** condemn, censure (*review damning the book*). **4 a** (often as **damning** *adj.*) (of circumstance, evidence, etc.) show or prove to be guilty. **b** be the ruin of. *–n.* **1** uttered curse. **2** *slang* negligible amount. *–adj. & adv. colloq.* = DAMNED. □ **damn all** *slang* nothing at all. **damn well** *colloq.* (for emphasis) simply (*damn well do as I say*). **damn with faint praise** commend feebly, and so imply disapproval. **I'm** (or **I'll be**) **damned if** *colloq.* I certainly do not, will not, etc. **not give a damn** see GIVE. **well I'm** (or **I'll be**) **damned** *colloq.* exclamation of surprise etc. [Latin *damnum* loss]

damnable /'dæmnəb(ə)l/ *adj.* hateful, annoying. □ **damnably** *adv.*

damnation /dæm'neɪʃ(ə)n/ *–n.* eternal punishment in hell. *–int.* expressing anger.

damned /dæmd/ *colloq.* *–attrib. adj.* damnable. *–adv.* extremely (*damned hot*). □ **damned well** = *damn well.* **do one's damnedest** do one's utmost.

damp *–adj.* slightly wet. *–n.* slight diffused or condensed moisture, esp. when unwelcome. *–v.* **1** make damp; moisten. **2** (often foll. by *down*) **a** temper; mute (*damps my enthusiasm*). **b** make (a fire) burn less strongly by reducing the flow of air to it. **3** reduce or stop the vibration of (esp. strings of a musical instrument). □ **damply** *adv.* **dampness** *n.* [Low German]

damp course *n.* (also **damp-proof course**) layer of waterproof material in a wall near the ground, to prevent rising damp.

dampen *v.* **1** make or become damp. **2** (often foll. by *down*) = DAMP *v.* 2a.

damper *n.* **1** discouraging person or thing. **2** device that reduces shock or noise. **3** metal plate in a flue to control the draught. **4** *Mus.* pad silencing a piano string. □ **put a damper on** take the vigour or enjoyment out of.

damp squib *n.* unsuccessful attempt to impress etc.

damsel /'dæmz(ə)l/ *n. archaic* or *literary* young unmarried woman. [French diminutive: related to DAME]

damselfly *n.* insect like a dragonfly but with wings folded when resting.

damson /'dæmz(ə)n/ *n.* **1** (in full **damson plum**) small dark-purple plum. **2** dark-purple colour. [Latin: related to DAMASCENE]

dan *n.* **1** grade of proficiency in judo. **2** holder of such a grade. [Japanese]

dance /dɑːns/ *–v.* (**-cing**) **1** move rhythmically, usu. to music. **2** skip or jump about. **3** perform (a specified dance, role, etc.). **4** bob up and down. **5** dandle (a child). *–n.* **1 a** dancing as an art form. **b** style or form of this. **2** social gathering for dancing. **3** single round or turn of a dance. **4** music for dancing to. **5** lively motion. □ **dance attendance on** serve obsequiously. **lead a person a dance** (or **merry dance**) cause a person much trouble. □ **danceable** *adj.* **dancer** *n.* [French]

dancehall *n.* public hall for dancing.

d. and c. *n.* dilatation (of the cervix) and curettage (of the uterus).

dandelion /ˈdændɪˌlaɪən/ *n.* wild plant with jagged leaves, a yellow flower, and a fluffy seed-head. [French *dent-de-lion*, = lion's tooth]

dander *n. colloq.* temper, indignation. □ **get one's dander up** become angry. [origin uncertain]

dandify /ˈdændɪˌfaɪ/ *v.* (**-ies, -ied**) make a dandy.

dandle /ˈdænd(ə)l/ *v.* (**-ling**) bounce (a child) on one's knees etc. [origin unknown]

dandruff /ˈdændrʌf/ *n.* **1** flakes of dead skin in the hair. **2** this as a condition. [origin uncertain]

dandy /ˈdændɪ/ *–n.* (*pl.* **-ies**) **1** man greatly devoted to style and fashion. **2** *colloq.* excellent thing. *–adj.* (**-ier, -iest**) esp. *US colloq.* splendid. [perhaps from the name *Andrew*]

dandy-brush *n.* brush for grooming a horse.

Dane *n.* **1** native or national of Denmark. **2** *hist.* Viking invader of England in the 9th–11th c. [Old Norse]

danger /ˈdeɪndʒə(r)/ *n.* **1** liability or exposure to harm. **2** thing that causes or may cause harm. □ **in danger of** likely to incur or to suffer from. [earlier = 'power', from Latin *dominus* lord]

danger list *n.* list of those dangerously ill.

danger money *n.* extra payment for dangerous work.

dangerous *adj.* involving or causing danger. □ **dangerously** *adv.*

dangle /ˈdæŋg(ə)l/ *v.* (**-ling**) **1** be loosely suspended and able to sway. **2** hold or carry thus. **3** hold out (hope, temptation, etc.) enticingly. [imitative]

Danish /ˈdeɪnɪʃ/ *–adj.* of Denmark or the Danes. *–n.* **1** Danish language. **2** (prec. by *the*; treated as *pl.*) the Danish people. [Latin: related to DANE]

Danish blue *n.* white blue-veined cheese.

Danish pastry *n.* yeast cake topped with icing, fruit, nuts, etc.

dank *adj.* disagreeably damp and cold. □ **dankly** *adv.* **dankness** *n.* [probably Scandinavian]

daphne /ˈdæfnɪ/ *n.* any of various flowering shrubs. [Greek]

dapper *adj.* **1** neat and precise, esp. in dress. **2** sprightly. [Low German or Dutch *dapper* strong]

dapple /ˈdæp(ə)l/ *–v.* (**-ling**) mark or become marked with spots of colour or shade. *–n.* dappled effect. [origin unknown]

dapple-grey *adj.* (of an animal's coat) grey or white with darker spots.

dapple grey *n.* dapple-grey horse.

Darby and Joan /ˌdɑːbɪ ənd ˈdʒəʊn/ *n.* devoted old married couple. [names of a couple in an 18th-c. poem]

Darby and Joan club *n.* club for pensioners.

dare *–v.* (**-ring**; *3rd sing. present* usu. **dare** before an expressed or implied infinitive without *to*) **1** (foll. by infin. with or without *to*) have the courage or impudence (to) (*dare he do it?*; *if they dare to come*; *how dare you?*). **2** (usu. foll. by *to* + infin.) defy or challenge (*I dare you to own up*). *–n.* **1** act of daring. **2** challenge, esp. to prove courage. □ **I dare say 1** (often foll. by *that*) it is probable. **2** probably; I grant that much. [Old English]

daredevil *–n.* recklessly daring person. *–adj.* recklessly daring. □ **daredevilry** *n.*

daring *–n.* adventurous courage. *–adj.* adventurous, bold; prepared to take risks. □ **daringly** *adv.*

dariole /ˈdærɪˌəʊl/ *n.* dish cooked and served in a small mould. [French]

dark *–adj.* **1** with little or no light. **2** of deep or sombre colour. **3** (of a person) with dark colouring. **4** gloomy, dismal. **5** evil, sinister. **6** sullen, angry. **7** secret, mysterious. **8** ignorant, unenlightened. *–n.* **1** absence of light. **2** lack of knowledge. **3** dark area or colour, esp. in painting. □ **after dark** after nightfall. **the Dark Ages** (or **Age**) **1** period of European history from the 5th–10th c. **2** period of supposed unenlightenment. **in the dark 1** lacking information. **2** with no light. □ **darkish** *adj.* **darkly** *adv.* **darkness** *n.* [Old English]

darken *v.* make or become dark or darker. □ **never darken a person's door** keep away permanently. □ **darkener** *n.*

dark glasses *n.pl.* spectacles with dark-tinted lenses.

dark horse *n.* little-known person who is unexpectedly successful.

darkie var. of DARKY.

darkroom *n.* darkened room for photographic work.

darky *n.* (also **darkie**) (*pl.* -ies) *slang offens.* Black person.

darling /'dɑ:lɪŋ/ *–n.* **1** beloved, lovable, or endearing person or thing. **2** favourite. *–adj.* **1** beloved, lovable. **2** *colloq.* charming or pretty. [Old English: related to DEAR]

darn[1] *–v.* mend (cloth etc.) by filling a hole with stitching. *–n.* darned area. [origin uncertain]

darn[2] *v., int., adj., & adv. colloq.* = DAMN (in imprecatory senses). [corruption]

darned *adj. & adv. colloq.* = DAMNED.

darnel /'dɑ:n(ə)l/ *n.* grass growing in cereal crops. [origin unknown]

darner *n.* needle for darning.

darning *n.* **1** act of darning. **2** things to be darned.

dart *–n.* **1** small pointed missile. **2** (in *pl.*; usu. treated as *sing.*) indoor game of throwing darts at a dartboard to score points. **3** sudden rapid movement. **4** dartlike structure, e.g. an insect's sting. **5** tapering tuck in a garment. *–v.* (often foll. by *out, in, past,* etc.) move, send, or go suddenly or rapidly. [French from Germanic]

dartboard *n.* circular target in darts.

Darwinian /dɑ:'wɪnɪən/ *–adj.* of Darwin's theory of evolution. *–n.* adherent of this. □ **Darwinism** /'dɑ:-/ *n.* **Darwinist** /'dɑ:-/ *n.* [*Darwin*, name of a naturalist]

dash *–v.* **1** rush. **2** strike or fling forcefully, esp. so as to shatter (*dashed it to the ground*). **3** frustrate, dispirit (*dashed their hopes*). **4** *colloq.* (esp. **dash it** or **dash it all**) = DAMN *v.* 1. *–n.* **1** rush or onset; sudden advance. **2** horizontal stroke (–) in writing or printing to mark a pause etc. **3** impetuous vigour; capacity for or appearance of this. **4** *US* sprinting-race. **5** longer signal of two in Morse code (cf. DOT *n.* 2). **6** slight admixture, esp. of a liquid. **7** = DASHBOARD. □ **dash off** write or draw hurriedly. [imitative]

dashboard *n.* instrument panel of a vehicle or aircraft.

dashing *adj.* **1** spirited, lively. **2** showy. □ **dashingly** *adv.* **dashingness** *n.*

dastardly /'dæstədlɪ/ *adj.* cowardly, despicable. □ **dastardliness** *n.* [origin uncertain]

DAT /dæt/ *abbr.* digital audio tape.

data /'deɪtə/ *n.pl.* (also treated as *sing.*, although the singular form is strictly *datum*) **1** known facts used for inference or in reckoning. **2** quantities or characters operated on by a computer etc. [Latin *data* from *do* give]

■ **Usage** (1) In scientific, philosophical, and general use, this word is usually considered to denote a number of items and is thus treated as plural with *datum* as the singular. (2) In computing and allied subjects (and sometimes in general use), it is treated as a mass (or collective) noun and used with words like *this, that,* and *much,* with singular verbs, e.g. *useful data has been collected.* Some people consider use (2) to be incorrect but it is more common than use (1). However, *data* is not a singular countable noun and cannot be preceded by *a, every, each, either,* or *neither,* or be given a plural form *datas.*

data bank *n.* store or source of data.

database *n.* structured set of data held in a computer.

datable /'deɪtəb(ə)l/ *adj.* (often foll. by *to*) capable of being dated.

data capture *n.* entering of data into a computer.

data processing *n.* series of operations on data, esp. by a computer. □ **data processor** *n.*

date[1] *–n.* **1** day of the month, esp. as a number. **2** particular, esp. historical, day or year. **3** day, month, and year of writing etc., at the head of a document etc. **4** period to which a work of art etc. belongs. **5** time when an event takes place. **6** *colloq.* **a** appointment, esp. social with a person of the opposite sex. **b** *US* person to be met at this. *–v.* (**-ting**) **1** mark with a date. **2 a** assign a date to (an object, event, etc.). **b** (foll. by *to*) assign to a particular time, period, etc. **3** (often foll. by *from, back to,* etc.) have its origins at a particular time. **4** appear or expose as old-fashioned (*design that does not date; that hat dates you*). **5** *US colloq.* **a** make a date with. **b** go out together as sexual partners. □ **out of date** (*attrib.* **out-of-date**) old-fashioned, obsolete. **to date** until now. **up to date** (*attrib.* **up-to-date**) modern; fashionable; current. [French: related to DATA]

date[2] *n.* **1** dark oval single-stoned fruit. **2** (in full **date-palm**) tree bearing it. [Greek: related to DACTYL, from the shape of the leaf]

date-line *n.* **1** north–south line partly along the meridian 180° from Greenwich, to the east of which the date is a day earlier than to the west. **2** date and place of writing at the head of a newspaper article etc.

date-stamp *–n.* adjustable rubber stamp etc. used to record a date. *–v.* mark with a date-stamp.

dative /'deɪtɪv/ *Gram. –n.* case expressing the indirect object or recipient.

–adj. of or in this case. [Latin: related to DATA]

datum /'deɪtəm/ see DATA.

daub /dɔ:b/ *–v.* **1** spread (paint etc.) crudely or roughly. **2** coat or smear (a surface) with paint etc. **3** paint crudely or unskilfully. *–n.* **1** paint etc. daubed on a surface. **2** plaster, clay, etc., esp. coating laths or wattles to form a wall. **3** crude painting. [Latin: related to DE-, ALB]

daughter /'dɔ:tə(r)/ *n.* **1** girl or woman in relation to her parent(s). **2** female descendant. **3** (foll. by *of*) female member of a family etc. **4** (foll. by *of*) female descendant or inheritor of a quality etc. □ **daughterly** *adj.* [Old English]

daughter-in-law *n.* (*pl.* **daughters-in-law**) son's wife.

daunt /dɔ:nt/ *v.* discourage, intimidate. □ **daunting** *adj.* [Latin *domito* from *domo* tame]

dauntless *adj.* intrepid, persevering.

dauphin /'dɔ:fɪn/ *n. hist.* eldest son of the King of France. [French from Latin *delphinus* DOLPHIN, as a family name]

Davenport /'dævən,pɔ:t/ *n.* **1** small writing-desk with a sloping top. **2** *US* large sofa. [name of the maker]

davit /'dævɪt/ *n.* small crane on board ship, esp. for moving or holding a lifeboat. [French diminutive of *David*]

Davy /'deɪvɪ/ *n.* (*pl.* **-ies**) (in full **Davy lamp**) miner's safety lamp. [name of its inventor]

Davy Jones /,deɪvɪ 'dʒəʊnz/ *n. slang* (in full **Davy Jones's locker**) bottom of the sea, esp. as the sailors' graveyard. [origin unknown]

daw *n.* = JACKDAW. [Old English]

dawdle /'dɔ:d(ə)l/ *v.* (**-ling**) **1** walk slowly and idly. **2** waste time; procrastinate. [origin unknown]

dawn *–n.* **1** daybreak. **2** beginning or birth of something. *–v.* **1** (of a day) begin; grow light. **2** (often foll. by *on*, *upon*) begin to become obvious (to). [Old English]

dawn chorus *n.* bird-song at daybreak.

day *n.* **1** time between sunrise and sunset. **2 a** 24 hours as a unit of time. **b** corresponding period on other planets (*Martian day*). **3** daylight (*clear as day*). **4** time during which work is normally done (*eight-hour day*). **5 a** (also *pl.*) historical period (*in those days*). **b** (prec. by *the*) present time (*issues of the day*). **6** prime of a person's life (*have had my day*; *in my day*). **7** a future time (*will do it one day*). **8** date of a specific festival or event etc. (*graduation day*; *Christmas day*). **9** battle or contest (*win the day*). □ **all in a day's work** part of the normal routine. **at the end of the day** when all is said and done. **call it a day** end a period of activity. **day after day** without respite. **day and night** all the time. **day by day** gradually. **day in, day out** routinely, constantly. **not one's day** day when things go badly (for a person). **one of these days** soon. **one of those days** day when things go badly. **that will be the day** *colloq.* that will never happen. [Old English]

day-bed *n.* bed for daytime rest.

day-boy *n.* (also **day-girl**) non-boarding pupil, esp. at a boarding school.

daybreak *n.* first light in the morning.

day care *n.* care of young children, the elderly, the handicapped, etc. during the working day.

day centre *n.* place for care of the elderly or handicapped during the day.

day-dream *–n.* pleasant fantasy or reverie. *–v.* indulge in this. □ **day-dreamer** *n.*

daylight *n.* **1** light of day. **2** dawn. **3** visible gap, e.g. between boats in a race. **4** (usu. in *pl.*) *slang* life or consciousness (*scared the daylights out of me*; *beat the living daylights out of them*).

daylight robbery *n. colloq.* blatantly excessive charge.

daylight saving *n.* longer summer evening daylight, achieved by putting clocks forward.

day nursery *n.* nursery for children of working parents.

day off *n.* day's holiday.

day of reckoning *n.* time when something must be atoned for or avenged.

day release *n.* part-time education for employees.

day return *n.* reduced fare or ticket for a return journey in one day.

day-room *n.* room, esp. in an institution, used during the day.

day-school *n.* school for pupils living at home.

daytime *n.* part of the day when there is natural light.

day-to-day *adj.* mundane, routine.

day-trip *n.* trip completed in one day. □ **day-tripper** *n.*

daze *–v.* (**-zing**) stupefy, bewilder. *–n.* state of bewilderment. [Old Norse]

dazzle /'dæz(ə)l/ *–v.* (**-ling**) **1** blind or confuse temporarily with a sudden bright light. **2** impress or overpower with knowledge, ability, etc. *–n.* bright confusing light. □ **dazzling** *adj.* **dazzlingly** *adv.* [from DAZE]

dB *abbr.* decibel(s).

DBS *abbr.* **1** direct-broadcast satellite. **2** direct broadcasting by satellite.

DC *abbr.* **1** (also **dc**) direct current. **2** District of Columbia. **3** da capo.

DD *abbr.* Doctor of Divinity.

D-Day /'di:deɪ/ *n.* **1** day (6 June 1944) on which Allied forces invaded N. France. **2** important or decisive day. [*D* for *day*]
DDT *abbr.* colourless chlorinated hydrocarbon used as insecticide. [from the chemical name]
de- *prefix* **1** forming verbs and their derivatives: **a** down, away (*descend*; *deduct*). **b** completely (*denude*). **2** added to verbs and their derivatives to form verbs and nouns implying removal or reversal (*de-ice*; *decentralization*). [Latin]
deacon /'di:kən/ *n.* (*fem.* (in senses 2 and 3) **deaconess** /-'nes/) **1** (in Episcopal churches) minister below bishop and priest. **2** (in Nonconformist churches) lay officer. **3** (in the early Church) minister of charity. [Greek *diakonos* servant]
deactivate /di:'æktɪ,veɪt/ *v.* **(-ting)** make inactive or less reactive.
dead /ded/ *–adj.* **1** no longer alive. **2** *colloq.* extremely tired or unwell. **3** numb (*fingers are dead*). **4** (foll. by *to*) insensitive to. **5** no longer effective or in use; extinct. **6** (of a match, coal, etc.) extinguished. **7** inanimate. **8 a** lacking force or vigour. **b** (of sound) not resonant. **9** quiet; lacking activity (*dead season*). **10** (of a microphone, telephone, etc.) not transmitting sounds. **11** (of a ball in a game) out of play. **12** abrupt, complete (*come to a dead stop*; *a dead calm*; *dead certainty*). *–adv.* **1** absolutely, completely (*dead on target*; *dead tired*). **2** *colloq.* very, extremely (*dead easy*). *–n.* time of silence or inactivity (*dead of night*). □ **as dead as the** (or **a**) **dodo** entirely obsolete. **dead to the world** *colloq.* fast asleep; unconscious. [Old English]
dead beat *adj. colloq.* exhausted.
dead-beat *n. colloq.* derelict, tramp.
dead duck *n. slang* unsuccessful or useless person or thing.
deaden *v.* **1** deprive of or lose vitality, force, brightness, sound, feeling, etc. **2** (foll. by *to*) make insensitive.
dead end *n.* **1** closed end of road, passage, etc. **2** (often, with hyphen, *attrib.*) hopeless situation, job, etc.
deadhead *–n.* **1** faded flower-head. **2** non-paying passenger or spectator. **3** useless person. *–v.* remove deadheads from (a plant).
dead heat *n.* **1** race in which competitors tie. **2** result of such a race.
dead language *n.* language no longer spoken, e.g. Latin.
dead letter *n.* law or practice no longer observed or recognized.
deadline *n.* time-limit.
deadlock *–n.* **1** state of unresolved conflict. **2** lock requiring a key to open or close it. *–v.* bring or come to a standstill.
dead loss *n. colloq.* useless person or thing.
deadly *–adj.* **(-ier, -iest) 1** causing or able to cause fatal injury or serious damage. **2** intense, extreme (*deadly dullness*). **3** (of aim etc.) true; effective. **4** deathlike (*deadly pale*). **5** *colloq.* dreary, dull. *–adv.* **1** like death; as if dead (*deadly faint*). **2** extremely (*deadly serious*).
deadly nightshade *n.* poisonous plant with purple-black berries.
dead man's handle *n.* (also **dead man's pedal**) device on an electric train disconnecting the power supply if released.
dead march *n.* funeral march.
dead on *adj.* exactly right.
deadpan *adj.* & *adv.* lacking expression or emotion.
dead reckoning *n.* calculation of a ship's position from the log, compass, etc., when visibility is bad.
dead set *n.* determined attack. □ **be dead set against** strongly oppose. **be dead set on** be determined to do or get.
dead shot *n.* person who shoots extremely accurately.
dead weight *n.* (also **dead-weight**) **1 a** inert mass. **b** heavy burden. **2** debt not covered by assets. **3** total weight carried on a ship.
dead wood *n. colloq.* useless person(s) or thing(s).
deaf /def/ *adj.* **1** wholly or partly unable to hear. **2** (foll. by *to*) refusing to listen or comply. □ **turn a deaf ear** (usu. foll. by *to*) be unresponsive. □ **deafness** *n.* [Old English]
deaf-aid *n.* hearing-aid.
deaf-and-dumb alphabet *n.* (also **deaf-and-dumb language**) = SIGN LANGUAGE.

■ **Usage** *Sign language* is the preferred term in official use.

deafen *v.* (often as **deafening** *adj.*) overpower with noise or make deaf by noise, esp. temporarily. □ **deafeningly** *adv.*
deaf mute *n.* deaf and dumb person.
deal[1] *–v.* (*past* and *past part.* **dealt** /delt/) **1** (foll. by *with*) **a** take measures to resolve, placate, etc. **b** do business with; associate with. **c** discuss or treat (a subject). **2** (often foll. by *by*, *with*) behave in specified way (*dealt honourably by them*). **3** (foll. by *in*) sell (*deals in insurance*). **4** (often foll. by *out*, *round*) distribute to several people etc. **5** (also *absol.*) distribute (cards) to players. **6** administer (*was dealt a blow*). **7** assign, esp. providentially (*were dealt much

happiness). —*n.* **1** (usu. **a good** or **great deal**) *colloq.* **a** large amount (*good deal of trouble*). **b** considerably (*great deal better*). **2** *colloq.* business arrangement; transaction. **3** specified treatment (*a rough deal*). **4 a** dealing of cards. **b** player's turn to do this. [Old English]

deal[2] *n.* **1** fir or pine timber, esp. as boards of a standard size. **2** board of this. [Low German]

dealer *n.* **1** trader in (esp. retail) goods (*car-dealer*; *dealer in tobacco*). **2** player dealing at cards. **3** jobber on the Stock Exchange.

■ **Usage** In sense 3, this name has been merged with *broker* since Oct. 1986 (see BROKER 2, JOBBER 2).

dealings *n.pl.* contacts, conduct, or transactions.

dealt *past* and *past part.* of DEAL[1].

dean[1] *n.* **1 a** head of the chapter of a cathedral or collegiate church. **b** (usu. **rural dean**) clergyman supervising parochial clergy. **2 a** college or university official with disciplinary and advisory functions. **b** head of a university faculty or department or of a medical school. [Latin *decanus*]

dean[2] var. of DENE.

deanery *n.* (*pl.* **-ies**) **1** dean's house or position. **2** parishes presided over by a rural dean.

dear —*adj.* **1 a** beloved or much esteemed. **b** as a merely polite or ironic form (*my dear man*). **2** as a formula of address, esp. beginning a letter (*Dear Sir*). **3** (often foll. by *to*) precious; cherished. **4** (usu. in *superl.*) earnest (*my dearest wish*). **5 a** expensive. **b** having high prices. —*n.* (esp. as a form of address) dear person. —*adv.* at great cost (*will pay dear*). —*int.* expressing surprise, dismay, pity, etc. (*dear me!*; *oh dear!*). □ **for dear life** desperately. □ **dearly** *adv.* [Old English]

dearie *n.* my dear. □ **dearie me!** *int.* expressing surprise, dismay, etc.

dearth /dɜ:θ/ *n.* scarcity, lack.

death /deθ/ *n.* **1** irreversible ending of life; dying or being killed. **2** instance of this. **3** destruction; ending (*death of our hopes*). **4** being dead (*eyes closed in death*). **5** (usu. **Death**) personification of death, esp. as a skeleton. **6** lack of spiritual life. □ **at death's door** close to death. **be the death of 1** cause the death of. **2** be annoying or harmful to. **catch one's death** *colloq.* catch a serious chill etc. **do to death 1** kill. **2** overdo. **fate worse than death** *colloq.* very unpleasant experience. **put to death** kill or cause to be killed. **to death** to the utmost, extremely (*bored to death*). □ **deathlike** *adj.* [Old English]

deathbed *n.* bed where a person dies.

deathblow *n.* **1** blow etc. causing death. **2** event etc. that destroys or ends something.

death certificate *n.* official statement of a person's death.

death duty *n. hist.* property tax levied after death.

■ **Usage** This term was replaced in 1975 by *capital transfer tax* and in 1986 by *inheritance tax.*

deathly —*adj.* (**-ier, -iest**) suggestive of death (*deathly silence*). —*adv.* in a deathly way (*deathly pale*).

death-mask *n.* cast taken of a dead person's face.

death penalty *n.* punishment by death.

death rate *n.* number of deaths per thousand of population per year.

death-rattle *n.* gurgling in the throat sometimes heard at death.

death row *n. US* part of a prison for those sentenced to death.

death squad *n.* armed paramilitary group.

death-trap *n. colloq.* dangerous building, vehicle, etc.

death-warrant *n.* **1** order of execution. **2** anything that causes the end of an established practice etc.

death-watch *n.* (in full **death-watch beetle**) small beetle which makes a ticking sound, said to portend death.

death-wish *n. Psychol.* alleged usu. unconscious desire for death.

deb *n. colloq.* débutante. [abbreviation]

débâcle /deɪ'bɑ:k(ə)l/ *n.* (*US* **debacle**) **1 a** utter defeat or failure. **b** sudden collapse. **2** confused rush or rout. [French]

debag /di:'bæg/ *v.* (**-gg-**) *slang* remove the trousers of (a person), esp. as a joke.

debar /dɪ'bɑ:(r)/ *v.* (**-rr-**) (foll. by *from*) exclude; prohibit (*debarred from the club*). □ **debarment** *n.* [French: related to BAR[1]]

debark /di:'bɑ:k/ *v.* land from a ship. □ **debarkation** /-'keɪʃ(ə)n/ *n.* [French *débarquer*]

debase /dɪ'beɪs/ *v.* (**-sing**) **1** lower in quality, value, or character. **2** depreciate (a coin) by alloying etc. □ **debasement** *n.* [from DE-, (A)BASE]

debatable /dɪ'beɪtəb(ə)l/ *adj.* questionable; disputable. [related to DEBATE]

debate /dɪ'beɪt/ —*v.* (**-ting**) **1** (also *absol.*) discuss or dispute, esp. formally. **2** consider aspects of (a question); ponder. —*n.* **1** formal discussion on a particular matter. **2** discussion (*open to debate*). [French: related to BATTLE]

debauch /dɪˈbɔːtʃ/ *–v.* **1** (as **debauched** *adj.*) dissolute. **2** corrupt, deprave. **3** debase (taste or judgement). *–n.* bout of sensual indulgence. [French]

debauchee /ˌdɪbɔːˈtʃiː/ *n.* debauched person.

debauchery *n.* excessive sensual indulgence.

debenture /dɪˈbentʃə(r)/ *n.* acknowledgement of indebtedness, esp. a company bond providing for payment of interest at fixed intervals. [Latin *debentur* are owed]

debilitate /dɪˈbɪlɪˌteɪt/ *v.* (**-ting**) enfeeble, enervate. □ **debilitation** /-ˈteɪʃ(ə)n/ *n.* [Latin *debilis* weak]

debility /dɪˈbɪlɪtɪ/ *n.* feebleness, esp. of health.

debit /ˈdebɪt/ *–n.* **1** entry in an account recording a sum owed. **2** sum recorded. **3** total of such sums. **4** debit side of an account. *–v.* (**-t-**) **1** (foll. by *against, to*) enter on the debit side of an account (*debit £50 to my account*). **2** (foll. by *with*) charge (a person) with a debt (*debited me with £500*). [Latin *debitum* DEBT]

debonair /ˌdebəˈneə(r)/ *adj.* **1** cheerful, self-assured. **2** pleasant-mannered. [French]

debouch /dɪˈbaʊtʃ/ *v.* **1** (of troops or a stream) come out into open ground. **2** (often foll. by *into*) (of a river, road, etc.) merge into a larger body or area. □ **debouchment** *n.* [French *bouche* mouth]

debrief /diːˈbriːf/ *v. colloq.* question (a diplomat, pilot, etc.) about a completed mission or undertaking. □ **debriefing** *n.*

debris /ˈdebriː/ *n.* **1** scattered fragments, esp. of wreckage. **2** accumulation of loose rock etc. [French *briser* break]

debt /det/ *n.* **1** money etc. owed (*debt of gratitude*). **2** state of owing (*in debt; get into debt*). □ **in a person's debt** under obligation to a person. [Latin *debeo debit-* owe]

debt of honour *n.* debt not legally recoverable, esp. a sum lost in gambling.

debtor *n.* person owing money etc.

debug /diːˈbʌg/ *v.* (**-gg-**) *colloq.* **1** remove concealed microphones from (a room etc.). **2** remove defects from (a computer program etc.). **3** = DELOUSE.

debunk /diːˈbʌŋk/ *v. colloq.* expose (a person, claim, etc.) as spurious or false. □ **debunker** *n.*

début /ˈdeɪbjuː/ *n.* (*US* **debut**) first public appearance (as a performer etc.). [French]

débutante /ˈdebjuːˌtɑːnt/ *n.* (*US* **debutante**) (usu. wealthy) young woman making her social début.

Dec. *abbr.* December.

deca- *comb. form* ten. [Greek *deka* ten]

decade /ˈdekeɪd, dɪˈkeɪd/ *n.* **1** period of ten years. **2** series or group of ten. [Greek: related to DECA-]

■ **Usage** The second pronunciation given, with the stress on the second syllable, is considered incorrect by some people, even though it is much used in broadcasting.

decadence /ˈdekəd(ə)ns/ *n.* **1** moral or cultural decline. **2** immoral behaviour. □ **decadent** *adj.* & *n.* **decadently** *adv.* [Latin: related to DECAY]

decaffeinated /diːˈkæfɪˌneɪtɪd/ *adj.* with caffeine removed or reduced.

decagon /ˈdekəgən/ *n.* plane figure with ten sides and angles. □ **decagonal** /dɪˈkægən(ə)l/ *adj.* [Greek: related to DECA-, *-gōnos* -angled]

decahedron /ˌdekəˈhiːdrən/ *n.* solid figure with ten faces. □ **decahedral** *adj.* [after POLYHEDRON]

decalitre /ˈdekəˌliːtə(r)/ *n.* (*US* **-liter**) metric unit of capacity, equal to 10 litres.

Decalogue /ˈdekəˌlɒg/ *n.* Ten Commandments. [Greek: related to DECA-, *logos* word, reason]

decametre /ˈdekəˌmiːtə(r)/ *n.* (*US* **-meter**) metric unit of length, equal to 10 metres.

decamp /dɪˈkæmp/ *v.* **1** depart suddenly; abscond. **2** break up or leave camp. □ **decampment** *n.* [French: related to CAMP[1]]

decanal /dɪˈkeɪn(ə)l/ *adj.* **1** of a dean. **2** of the south side of a choir (where the dean sits). [Latin: related to DEAN[1]]

decant /dɪˈkænt/ *v.* **1** gradually pour off (esp. wine), esp. leaving the sediment behind. **2** transfer as if by pouring. [Greek *kanthos* lip of jug]

decanter *n.* stoppered glass container for decanted wine or spirit.

decapitate /dɪˈkæpɪˌteɪt/ *v.* (**-ting**) behead. □ **decapitation** /-ˈteɪʃ(ə)n/ *n.* [Latin: related to CAPITAL]

decapod /ˈdekəˌpɒd/ *n.* **1** crustacean with ten limbs for walking, e.g. the shrimp. **2** ten-tentacled mollusc, e.g. the squid. [Greek: related to DECA-, *pous pod-* foot]

decarbonize /diːˈkɑːbəˌnaɪz/ *v.* (also **-ise**) (**-zing** or **-sing**) remove the carbon etc. from (an internal-combustion engine etc.). □ **decarbonization** /-ˈzeɪʃ(ə)n/ *n.*

decathlon /dɪˈkæθlən/ *n.* athletic contest of ten events for all competitors.

□ **decathlete** /-li:t/ *n.* [from DECA-, Greek *athlon* contest]

decay /dɪ'keɪ/ *–v.* **1** (cause to) rot or decompose. **2** decline or cause to decline in quality, power, etc. **3** (usu. foll. by *to*) (of a substance) undergo change by radioactivity. *–n.* **1** rotten state; wasting away. **2** decline in health, quality, etc. **3** radioactive change. [Latin *cado* fall]

decease /dɪ'si:s/ *formal* esp. *Law –n.* death. *–v.* (**-sing**) die. [Latin *cedo* go]

deceased *formal –adj.* dead. *–n.* (usu. prec. by *the*) person who has died, esp. recently.

deceit /dɪ'si:t/ *n.* **1** deception, esp. by concealing the truth. **2** dishonest trick. [Latin *capio* take]

deceitful *adj.* using deceit. □ **deceitfully** *adv.* **deceitfulness** *n.*

deceive /dɪ'si:v/ *v.* (**-ving**) **1** make (a person) believe what is false; purposely mislead. **2** be unfaithful to, esp. sexually. **3** use deceit. □ **deceive oneself** persist in a mistaken belief. □ **deceiver** *n.*

decelerate /di:'seləˌreɪt/ *v.* (**-ting**) (cause to) reduce speed. □ **deceleration** /-'reɪʃ(ə)n/ *n.* [from DE-, ACCELERATE]

December /dɪ'sembə(r)/ *n.* twelfth month of the year. [Latin *decem* ten, originally 10th month of Roman year]

decency /'di:sənsɪ/ *n.* (*pl.* **-ies**) **1** correct, honourable, or modest behaviour. **2** (in *pl.*) proprieties; manners. [Latin: related to DECENT]

decennial /dɪ'senɪəl/ *adj.* lasting, recurring every, ten years. [Latin *decem* ten, *annus* year]

decent /'di:s(ə)nt/ *adj.* **1 a** conforming with standards of decency. **b** avoiding obscenity. **2** respectable. **3** acceptable, good enough. **4** kind, obliging. □ **decently** *adv.* [Latin *decet* is fitting]

decentralize /di:'sentrəˌlaɪz/ *v.* (also **-ise**) (**-zing** or **-sing**) **1** transfer (power etc.) from central to local authority. **2** reorganize to give greater local autonomy. □ **decentralization** /-'zeɪʃ(ə)n/ *n.*

deception /dɪ'sepʃ(ə)n/ *n.* **1** deceiving or being deceived. **2** thing that deceives. [Latin: related to DECEIVE]

deceptive /dɪ'septɪv/ *adj.* likely to deceive; misleading. □ **deceptively** *adv.* **deceptiveness** *n.*

deci- *comb. form* one-tenth. [Latin *decimus* tenth]

decibel /'desɪˌbel/ *n.* unit used in the comparison of sound levels or power levels of electrical signals.

decide /dɪ'saɪd/ *v.* (**-ding**) **1** (usu. foll. by *to*, *that*, or *on*, *about*) resolve after consideration (*decided to stay*; *decided quickly*; *weather decided me*; *decided on a blue hat*). **2** resolve or settle (an issue etc.). **3** (usu. foll. by *between*, *for*, *against*, *in favour of*, or *that*) give a judgement. □ **decidable** *adj.* [Latin *caedo* cut]

decided *adj.* **1** (usu. *attrib.*) definite, unquestionable (*decided tilt*). **2** positive, wilful, resolute.

decidedly *adv.* undoubtedly, undeniably.

decider *n.* **1** game, race, etc., as a tie-break. **2** person or thing that decides.

deciduous /dɪ'sɪdjʊəs/ *adj.* **1** (of a tree) shedding leaves annually. **2** (of leaves, horns, teeth, etc.) shed periodically. [Latin *cado* fall]

decigram /'desɪˌgræm/ *n.* (also **decigramme**) metric unit of mass, equal to 0.1 gram.

decilitre /'desɪˌli:tə(r)/ *n.* (*US* **-liter**) metric unit of capacity, equal to 0.1 litre.

decimal /'desɪm(ə)l/ *–adj.* **1** (of a system of numbers, weights, measures, etc.) based on the number ten. **2** of tenths or ten; reckoning or proceeding by tens. *–n.* decimal fraction. [Latin *decem* ten]

decimal fraction *n.* fraction expressed in tenths, hundredths, etc., esp. by units to the right of the decimal point (e.g. 0.61).

decimalize *v.* (also **-ise**) (**-zing** or **-sing**) **1** express as a decimal. **2** convert to a decimal system (esp. of coinage). □ **decimalization** /-'zeɪʃ(ə)n/ *n.*

decimal point *n.* dot placed before the fraction in a decimal fraction.

decimate /'desɪˌmeɪt/ *v.* (**-ting**) **1** destroy a large proportion of. **2** orig. *Rom. Hist.* kill or remove one in every ten of. □ **decimation** /-'meɪʃ(ə)n/ *n.*

■ **Usage** Sense 1 is now the usual sense, but it is considered inappropriate by some people. This word should not be used to mean 'defeat utterly'.

decimetre /'desɪˌmi:tə(r)/ *n.* (*US* **-meter**) metric unit of length, equal to 0.1 metre.

decipher /dɪ'saɪfə(r)/ *v.* **1** convert (coded information) into intelligible language. **2** determine the meaning of (unclear handwriting etc.). □ **decipherable** *adj.*

decision /dɪ'sɪʒ(ə)n/ *n.* **1** act or process of deciding. **2** resolution made after consideration (*made my decision*). **3** (often foll. by *of*) **a** settlement of a question. **b** formal judgement. **4** resoluteness. [Latin: related to DECIDE]

decisive /dɪ'saɪsɪv/ *adj.* **1** conclusive, settling an issue. **2** quick to decide. □ **decisively** *adv.* **decisiveness** *n.* [medieval Latin: related to DECIDE]

deck –*n.* **1 a** platform in a ship serving as a floor. **b** the accommodation on a particular deck of a ship. **2** floor or compartment of a bus etc. **3** section for playing discs or tapes etc. in a sound system. **4** esp. *US* pack of cards. **5** *slang* ground. –*v.* **1** (often foll. by *out*) decorate. **2** provide with or cover as a deck. □ **below deck(s)** in or into the space below the main deck. [Dutch, = cover]

deck-chair *n.* folding garden chair of wood and canvas.

-decker *comb. form* having a specified number of decks or layers (*double-decker*).

deck-hand *n.* cleaner on a ship's deck.

declaim /dɪˈkleɪm/ *v.* **1** speak or say as if addressing an audience. **2** (foll. by *against*) protest forcefully. □ **declamation** /ˌdekləˈmeɪʃ(ə)n/ *n.* **declamatory** /dɪˈklæmətərɪ/ *adj.* [Latin: related to CLAIM]

declaration /ˌdekləˈreɪʃ(ə)n/ *n.* **1** declaring. **2** formal, emphatic, or deliberate statement. [Latin: related to DECLARE]

declare /dɪˈkleə(r)/ *v.* (**-ring**) **1** announce openly or formally (*declare war*). **2** pronounce (*declared it invalid*). **3** (usu. foll. by *that*) assert emphatically. **4** acknowledge possession of (dutiable goods, income, etc.). **5** (as **declared** *adj.*) admitting to be such (*declared atheist*). **6** (also *absol.*) *Cricket* close (an innings) voluntarily before the team is out. **7** (also *absol.*) *Cards* name (the trump suit). □ **declare oneself** reveal one's intentions or identity. □ **declarative** /-ˈklærətɪv/ *adj.* **declaratory** /-ˈklærətərɪ/ *adj.* **declarer** *n.* [Latin *clarus* clear]

declassify /di:ˈklæsɪˌfaɪ/ *v.* (**-ies, -ied**) declare (information etc.) to be no longer secret. □ **declassification** /-fɪˈkeɪʃ(ə)n/ *n.*

declension /dɪˈklenʃ(ə)n/ *n.* **1** *Gram.* **a** variation of the form of a noun, pronoun, or adjective to show its grammatical case etc. **b** class of nouns with the same inflections. **2** deterioration, declining. [Latin: related to DECLINE]

declination /ˌdeklɪˈneɪʃ(ə)n/ *n.* **1** downward bend or turn. **2** angular distance of a star etc. north or south of the celestial equator. **3** deviation of a compass needle from true north. □ **declinational** *adj.* [Latin: related to DECLINE]

decline /dɪˈklaɪn/ –*v.* (**-ning**) **1** deteriorate; lose strength or vigour; decrease. **2** (also *absol.*) politely refuse (an invitation, challenge, etc.). **3** slope or bend downwards, droop. **4** *Gram.* state the forms of (a noun, pronoun, or adjective). –*n.* **1** gradual loss of vigour or excellence. **2** deterioration. [Latin *clino* bend]

declining years *n.pl.* old age.

declivity /dɪˈklɪvɪtɪ/ *n.* (*pl.* **-ies**) downward slope. [Latin *clivus* slope]

declutch /di:ˈklʌtʃ/ *v.* disengage the clutch of a motor vehicle.

decoction /dɪˈkɒkʃ(ə)n/ *n.* **1** boiling down to extract an essence. **2** the resulting liquid. [Latin *coquo* boil]

decode /di:ˈkəʊd/ *v.* (**-ding**) decipher. □ **decoder** *n.*

decoke *colloq.* –*v.* /di:ˈkəʊk/ (**-king**) decarbonize. –*n.* /ˈdi:kəʊk/ process of this.

décolletage /ˌdeɪkɒlˈtɑ:ʒ/ *n.* low neckline of a woman's dress etc. [French *collet* collar]

décolleté /deɪˈkɒlteɪ/ *adj.* (also ***décolletée***) (of a dress, woman, etc.) having or wearing a low neckline.

decompose /ˌdi:kəmˈpəʊz/ *v.* (**-sing**) **1** rot. **2** separate (a substance, light, etc.) into its elements. □ **decomposition** /ˌdi:kɒmpəˈzɪʃ(ə)n/ *n.*

decompress /ˌdi:kəmˈpres/ *v.* subject to decompression.

decompression /ˌdi:kəmˈpreʃ(ə)n/ *n.* **1** release from compression. **2** gradual reduction of high pressure on a deep-sea diver etc.

decompression chamber *n.* enclosed space for decompression.

decompression sickness *n.* condition caused by the sudden lowering of air pressure.

decongestant /ˌdi:kənˈdʒest(ə)nt/ *n.* medicine etc. that relieves nasal congestion.

decontaminate /ˌdi:kənˈtæmɪˌneɪt/ *v.* (**-ting**) remove contamination from. □ **decontamination** /-ˈneɪʃ(ə)n/ *n.*

décor /ˈdeɪkɔ:(r)/ *n.* furnishing and decoration of a room, stage set, etc. [French: related to DECORATE]

decorate /ˈdekəˌreɪt/ *v.* (**-ting**) **1** beautify, adorn. **2** paint, wallpaper, etc. (a room or building). **3** give a medal or award to. [Latin *decus -oris* beauty]

Decorated style *n. Archit.* highly ornamented late English Gothic style (14th c.).

decoration /ˌdekəˈreɪʃ(ə)n/ *n.* **1** decorating. **2** thing that decorates. **3** medal etc. worn as an honour. **4** (in *pl.*) flags, tinsel, etc., put up on a festive occasion.

decorative /ˈdekərətɪv/ *adj.* pleasing in appearance. □ **decoratively** *adv.*

decorator *n.* person who decorates for a living.

decorous /ˈdekərəs/ *adj.* having or showing decorum. □ **decorously** *adv.* **decorousness** *n.* [Latin *decorus* seemly]

decorum /dɪ'kɔ:rəm/ *n.* polite dignified behaviour. [as DECOROUS]
decoy –*n.* /'di:kɔɪ/ person or thing used as a lure; bait, enticement. –*v.* /dɪ'kɔɪ/ lure, esp. using a decoy. [Dutch]
decrease –*v.* /dɪ'kri:s/ (**-sing**) make or become smaller or fewer. –*n.* /'di:kri:s/ **1** decreasing. **2** amount of this. □ **decreasingly** *adv.* [Latin: related to DE-, *cresco* grow]
decree /dɪ'kri:/ –*n.* **1** official legal order. **2** legal judgement or decision, esp. in divorce cases. –*v.* (**-ees, -eed, -eeing**) ordain by decree. [Latin *decretum* from *cerno* sift]
decree absolute *n.* final order for completion of a divorce.
decree nisi /'naɪsaɪ/ *n.* provisional order for divorce, made absolute after a fixed period. [Latin *nisi* unless]
decrepit /dɪ'krepɪt/ *adj.* **1** weakened by age or infirmity. **2** dilapidated. □ **decrepitude** *n.* [Latin *crepo* creak]
decrescendo /ˌdi:krɪ'ʃendəʊ/ *adv., adj.,* & *n.* (*pl.* **-s**) = DIMINUENDO. [Italian: related to DECREASE]
decretal /dɪ'kri:t(ə)l/ *n.* papal decree. [Latin: related to DECREE]
decriminalize /di:'krɪmɪnəˌlaɪz/ *v.* (also **-ise**) (**-zing** or **-sing**) cease to treat as criminal. □ **decriminalization** /-'zeɪʃ(ə)n/ *n.*
decry /dɪ'kraɪ/ *v.* (**-ies, -ied**) disparage, belittle.
dedicate /'dedɪˌkeɪt/ *v.* (**-ting**) (often foll. by *to*) **1** devote (esp. oneself) to a special task or purpose. **2** address (a book etc.) to a friend, patron, etc. **3** devote (a building etc.) to a deity, saint, etc. **4** (as **dedicated** *adj.*) **a** (of a person) single-mindedly loyal to an aim, vocation, etc. **b** (of equipment, esp. a computer) designed for or assigned to a specific task. □ **dedicator** *n.* **dedicatory** *adj.* [Latin *dico* declare]
dedication /ˌdedɪ'keɪʃ(ə)n/ *n.* **1** dedicating or being dedicated. **2** words with which a book etc. is dedicated. [Latin: related to DEDICATE]
deduce /dɪ'dju:s/ *v.* (**-cing**) (often foll. by *from*) infer logically. □ **deducible** *adj.* [Latin *duco duct-* lead]
deduct /dɪ'dʌkt/ *v.* (often foll. by *from*) subtract, take away, or withhold (an amount, portion, etc.). [related to DEDUCE]
deductible *adj.* that may be deducted, esp. from tax or taxable income.
deduction /dɪ'dʌkʃ(ə)n/ *n.* **1 a** deducting. **b** amount deducted. **2 a** inferring of particular instances from a general law or principle. **b** conclusion deduced. [Latin: related to DEDUCE]
deductive *adj.* of or reasoning by deduction. □ **deductively** *adv.* [medieval Latin: related to DEDUCE]
dee *n.* **1** letter D. **2** thing shaped like this. [name of the letter D]
deed *n.* **1** thing done intentionally or consciously. **2** brave, skilful, or conspicuous act. **3** action (*kind in word and deed*). **4** legal document used esp. for transferring ownership of property. [Old English: related to DO[1]]
deed-box *n.* strong box for deeds etc.
deed of covenant *n.* agreement to pay a regular sum, esp. to charity.
deed poll *n.* deed made by one party only, esp. to change one's name.
deem *v. formal* consider, judge (*deem it my duty*). [Old English]
deemster *n.* judge in the Isle of Man. [from DEEM]
deep –*adj.* **1** extending far down or in (*deep water*; *deep wound*; *deep shelf*). **2** (*predic.*) **a** to or at a specified depth (*water 6 feet deep*). **b** in a specified number of ranks (*soldiers drawn up six deep*). **3** situated or coming from far down, back, or in (*deep in his pockets*; *deep sigh*). **4** low-pitched, full-toned (*deep voice*). **5** intense, extreme (*deep sleep*; *deep colour*; *deep interest*). **6** (*predic.*) fully absorbed or overwhelmed (*deep in a book*; *deep in debt*). **7** profound; difficult to understand (*too deep for me*). –*n.* **1** (prec. by *the*) *poet.* sea, esp. when deep. **2** abyss, pit, cavity. **3** (prec. by *the*) *Cricket* position of a fielder distant from the batsman. **4** deep state (*deep of the night*). –*adv.* deeply; far down or in (*dig deep*). □ **go off the deep end** *colloq.* give way to anger or emotion. **in deep water** in trouble or difficulty. □ **deeply** *adv.* [Old English]
deep breathing *n.* breathing with long breaths, esp. as exercise.
deepen *v.* make or become deep or deeper.
deep-freeze –*n.* cabinet for freezing and keeping food for long periods. –*v.* freeze or store in a deep-freeze.
deep-fry *v.* immerse in boiling fat to cook.
deep-laid *adj.* (of a scheme) secret and elaborate.
deep-rooted *adj.* (also **deep-seated**) firmly established, profound.
deer *n.* (*pl.* same) four-hoofed grazing animal, the male of which usu. has antlers. [Old English]
deerskin *n.* (often *attrib.*) leather from a deer's skin.
deerstalker *n.* soft cloth peaked cap with ear-flaps.

de-escalate /di:'eskə,leɪt/ *v.* make or become less intense. □ **de-escalation** /-'leɪʃ(ə)n/ *n.*

def *adj. slang* excellent. [perhaps from DEFINITE or DEFINITIVE]

deface /dɪ'feɪs/ *v.* (**-cing**) disfigure. □ **defacement** *n.* [French: related to FACE]

de facto /deɪ 'fæktəʊ/ *–adv.* in fact (whether by right or not). *–adj.* existing or so in fact (*a de facto ruler*). [Latin]

defalcate /'di:fæl,keɪt/ *v.* (**-ting**) *formal* misappropriate, esp. money. □ **defalcator** *n.* [Latin *defalcare* lop, from *falx* sickle]

defalcation /,di:fæl'keɪʃ(ə)n/ *n. formal* **1 a** misappropriation of money. **b** amount misappropriated. **2** shortcoming.

defame /dɪ'feɪm/ *v.* (**-ming**) libel; slander; speak ill of. □ **defamation** /,defə'meɪʃ(ə)n/ *n.* **defamatory** /dɪ 'fæmətərɪ/ *adj.* [Latin *fama* report]

default /dɪ'fɔ:lt/ *–n.* **1** failure to appear, pay, or act as one should. **2** preselected option adopted by a computer program when no alternative is specified. *–v.* fail to fulfil (esp. a legal) obligation. □ **by default** because of lack of an alternative or opposition. **in default of** because of the absence of. □ **defaulter** *n.* [French: related to FAIL]

defeat /dɪ'fi:t/ *–v.* **1** overcome in battle, a contest, etc. **2** frustrate, baffle. **3** reject (a motion etc.) by voting. *–n.* defeating or being defeated. [Latin: related to DIS-, FACT]

defeatism *n.* excessive readiness to accept defeat. □ **defeatist** *n.* & *adj.*

defecate /'defɪ,keɪt/ *v.* (**-ting**) evacuate the bowels. □ **defecation** /-'keɪʃ(ə)n/ *n.* [Latin *faex faecis* dregs]

defect *–n.* /'di:fekt/ fault, imperfection, shortcoming. *–v.* /dɪ'fekt/ leave one's country or cause for another. □ **defection** *n.* **defector** *n.* [Latin *deficio -fect-* fail]

defective /dɪ'fektɪv/ *adj.* having defect(s); imperfect. □ **defectiveness** *n.* [Latin: related to DEFECT]

defence /dɪ'fens/ *n.* (*US* **defense**) **1** defending, protection. **2** means of this. **3** (in *pl.*) fortifications. **4** justification, vindication. **5** defendant's case or counsel in a lawsuit. **6** defending play or players. □ **defenceless** *adj.* **defencelessly** *adv.* **defencelessness** *n.* [related to DEFEND]

defence mechanism *n.* **1** body's resistance to disease. **2** usu. unconscious mental process to avoid anxiety.

defend /dɪ'fend/ *v.* (also *absol.*) **1** (often foll. by *against*, *from*) resist an attack made on; protect. **2** uphold by argument. **3** conduct a defence (of) in a lawsuit. **4** compete to retain (a title etc.) in a contest. □ **defender** *n.* [Latin *defendo -fens-*]

defendant *n.* person etc. sued or accused in a lawcourt. [French: related to DEFEND]

defense *US* var. of DEFENCE.

defensible /dɪ'fensɪb(ə)l/ *adj.* **1** justifiable; supportable by argument. **2** able to be defended militarily. □ **defensibility** /-'bɪlɪtɪ/ *n.* **defensibly** *adv.* [Latin: related to DEFEND]

defensive *adj.* **1** done or intended for defence. **2** over-reacting to criticism. □ **on the defensive 1** expecting criticism. **2** *Mil* ready to defend. □ **defensively** *adv.* **defensiveness** *n.* [medieval Latin: related to DEFEND]

defer[1] /dɪ'fɜ:(r)/ *v.* (**-rr-**) postpone. □ **deferment** *n.* **deferral** *n.* [originally the same as DIFFER]

defer[2] /dɪ'fɜ:(r)/ *v.* (**-rr-**) (foll. by *to*) yield or make concessions to. [Latin *defero* carry away]

deference /'defərəns/ *n.* **1** courteous regard, respect. **2** compliance with another's wishes. □ **in deference to** out of respect for.

deferential /,defə'renʃ(ə)l/ *adj.* respectful. □ **deferentially** *adv.*

deferred payment *n.* payment by instalments.

defiance /dɪ'faɪəns/ *n.* open disobedience; bold resistance. [French: related to DEFY]

defiant *adj.* showing defiance; disobedient. □ **defiantly** *adv.*

deficiency /dɪ'fɪʃənsɪ/ *n.* (*pl.* **-ies**) **1** being deficient. **2** (usu. foll. by *of*) lack or shortage. **3** thing lacking. **4** deficit, esp. financial.

deficiency disease *n.* disease caused by the lack of an essential element of diet.

deficient *adj.* (often foll. by *in*) incomplete or insufficient in quantity, quality, etc. [Latin: related to DEFECT]

deficit /'defɪsɪt/ *n.* **1** amount by which a thing (esp. money) is too small. **2** excess of liabilities over assets. [French from Latin: related to DEFECT]

defile[1] /dɪ'faɪl/ *v.* (**-ling**) **1** make dirty; pollute. **2** desecrate, profane. □ **defilement** *n.* [earlier *defoul*, from French *defouler* trample down]

defile[2] /dɪ'faɪl/ *–n.* narrow gorge or pass. *–v.* (**-ling**) march in file. [French: related to FILE[1]]

define /dɪ'faɪn/ *v.* (**-ning**) **1** give the meaning of (a word etc.). **2** describe or explain the scope of (*define one's position*). **3** outline clearly (*well-defined image*). **4** mark out the boundary of. □ **definable** *adj.* [Latin *finis* end]

definite /'definit/ *adj.* **1** certain, sure. **2** clearly defined; not vague; precise. □ **definitely** *adv.* [Latin: related to DEFINE]

■ **Usage** See note at *definitive*.

definite article *n.* the word (*the* in English) preceding a noun and implying a specific instance.

definition /,defi'nɪʃ(ə)n/ *n.* **1 a** defining. **b** statement of the meaning of a word etc. **2** distinctness in outline, esp. of a photographic image. [Latin: related to DEFINE]

definitive /dɪ'fɪnɪtɪv/ *adj.* **1** (of an answer, verdict, etc.) decisive, unconditional, final. **2** (of a book etc.) most authoritative.

■ **Usage** In sense 1, this word is often confused with *definite*, which does not imply authority and conclusiveness. A *definite no* is a firm refusal, while a *definitive no* is an authoritative judgement or decision that something is not the case.

deflate /dɪ'fleɪt/ *v.* **(-ting) 1** empty (a tyre, balloon, etc.) of air, gas, etc.; be so emptied. **2** (cause to) lose confidence or conceit. **3 a** subject (a currency or economy) to deflation. **b** pursue this as a policy. [from DE-, INFLATE]

deflation /dɪ'fleɪʃ(ə)n/ *n.* **1** deflating or being deflated. **2** reduction of money in circulation, intended to combat inflation. □ **deflationary** *adj.*

deflect /dɪ'flekt/ *v.* **1** bend or turn aside from a course or purpose. **2** (often foll. by *from*) (cause to) deviate. □ **deflection** *n.* (also **deflexion**). **deflector** *n.* [Latin *flecto* bend]

deflower /dɪ'flaʊə(r)/ *v. literary* **1** deprive of virginity. **2** ravage, spoil. [Latin: related to FLOWER]

defoliate /di:'fəʊlɪ,eɪt/ *v.* **(-ting)** destroy the leaves of (trees or plants). □ **defoliant** *n.* **defoliation** /-'eɪʃ(ə)n/ *n.* [Latin: related to FOIL²]

deforest /di:'fɒrɪst/ *v.* clear of forests or trees. □ **deforestation** /-'steɪʃ(ə)n/ *n.*

deform /dɪ'fɔ:m/ *v.* make ugly or misshapen, disfigure. □ **deformation** /,di:fɔ:'meɪʃ(ə)n/ *n.* [Latin: related to FORM]

deformed *adj.* (of a person or limb) misshapen.

deformity *n.* (*pl.* **-ies**) **1** being deformed. **2** malformation, esp. of a body or limb.

defraud /dɪ'frɔ:d/ *v.* (often foll. by *of*) cheat by fraud. [Latin: related to FRAUD]

defray /dɪ'freɪ/ *v.* provide money for (a cost or expense). □ **defrayal** *n.* **defrayment** *n.* [medieval Latin *fredum* fine]

defrock /di:'frɒk/ *v.* deprive (esp. a priest) of office. [French: related to DE-, FROCK]

defrost /di:'frɒst/ *v.* **1** remove frost or ice from (a refrigerator, windscreen, etc.). **2** unfreeze (frozen food). **3** become unfrozen.

deft *adj.* neat; dexterous; adroit. □ **deftly** *adv.* **deftness** *n.* [var. of DAFT = 'meek']

defunct /dɪ'fʌŋkt/ *adj.* **1** no longer existing or used. **2** dead or extinct. □ **defunctness** *n.* [Latin *fungor* perform]

defuse /di:'fju:z/ *v.* **(-sing) 1** remove the fuse from (a bomb etc.). **2** reduce tension etc. in (a crisis, difficulty, etc.).

defy /dɪ'faɪ/ *v.* **(-ies, -ied) 1** resist openly; refuse to obey. **2** (of a thing) present insuperable obstacles to (*defies solution*). **3** (foll. by *to* + infin.) challenge (a person) to do or prove something. [Latin *fides* faith]

degenerate —*adj.* /dɪ'dʒenərət/ **1** having lost its usual or good qualities; immoral, degraded. **2** *Biol.* having changed to a lower type. —*n.* /dɪ'dʒenərət/ degenerate person or animal. —*v.* /dɪ'dʒenə,reɪt/ **(-ting)** become degenerate. □ **degeneracy** *n.* [Latin *genus* race]

degeneration /dɪ,dʒenə'reɪʃ(ə)n/ *n.* **1** becoming degenerate. **2** *Med.* morbid deterioration of body tissue etc. [Latin: related to DEGENERATE]

degrade /dɪ'greɪd/ *v.* **(-ding) 1** humiliate, dishonour. **2** reduce to a lower rank. **3** *Chem.* reduce to a simpler molecular structure. □ **degradation** /,degrə'deɪʃ(ə)n/ *n.* **degrading** *adj.* [Latin: related to GRADE]

degree /dɪ'gri:/ *n.* **1** stage in a scale, series, or process. **2** stage in intensity or amount (*in some degree*). **3** unit of measurement of an angle or arc. **4** unit in a scale of temperature, hardness, etc. **5** extent of burns. **6** academic rank conferred by a polytechnic, university, etc. **7** grade of crime (*first-degree murder*). **8** step in direct genealogical descent. **9** social rank. □ **by degrees** gradually. [Latin *gradus* step]

degrees of comparison see COMPARISON 4.

dehisce /di:'hɪs/ *v.* **(-cing)** (esp. of a pod, cut, etc.) gape or burst open. □ **dehiscence** *n.* **dehiscent** *adj.* [Latin *hio* gape]

dehumanize /di:'hju:mə,naɪz/ *v.* (also **-ise**) **(-zing** or **-sing) 1** take human qualities away from. **2** make impersonal. □ **dehumanization** /-'zeɪʃ(ə)n/ *n.*

dehydrate /,di:haɪ'dreɪt/ *v.* **(-ting) 1** remove water from (esp. foods). **2** make or become dry, esp. too dry. □ **dehydration** /-ˈdreɪʃ(ə)n/ *n.* [Greek *hudōr* water]

de-ice /diːˈaɪs/ *v.* **1** remove ice from. **2** prevent the formation of ice on. □ **de-icer** *n.*

deify /ˈdiːɪ,faɪ, ˈdeɪɪ-/ *v.* (**-ies, -ied**) make a god or idol of. □ **deification** /-fɪˈkeɪʃ(ə)n/ *n.* [Latin *deus* god]

deign /deɪn/ *v.* (foll. by *to* + infin.) think fit, condescend. [Latin *dignus* worthy]

deinstitutionalize /,diːɪnstɪˈtjuːʃənə,laɪz/ *v.* (also **-ise**) (**-zing** or **-sing**) (usu. as **deinstitutionalized** *adj.*) remove from an institution or help recover from the effects of institutional life. □ **deinstitutionalization** /-ˈzeɪʃ(ə)n/ *n.*

deism /ˈdiːɪz(ə)m, ˈdeɪ-/ *n.* reasoned belief in the existence of a god. □ **deist** *n.* **deistic** /-ˈɪstɪk/ *adj.* [Latin *deus* god]

deity /ˈdiːɪtɪ, ˈdeɪɪ-/ *n.* (*pl.* **-ies**) **1** god or goddess. **2** divine status or nature. **3** (**the Deity**) God. [French from Church Latin]

déjà vu /,deɪʒɑː ˈvuː/ *n.* **1** feeling of having already experienced the present situation. **2** something tediously familiar. [French, = already seen]

deject /dɪˈdʒekt/ *v.* (usu. as **dejected** *adj.*) make sad; depress. □ **dejectedly** *adv.* **dejection** *n.* [Latin *jacio* throw]

de jure /deɪ ˈjʊərɪ/ *–adj.* rightful. *–adv.* rightfully; by right. [Latin]

dekko /ˈdekəʊ/ *n.* (*pl.* **-s**) *slang* look, glance. [Hindi]

delay /dɪˈleɪ/ *–v.* **1** postpone; defer. **2** make or be late; loiter. *–n.* **1** delaying or being delayed. **2** time lost by this. **3** hindrance. [French]

delayed-action *attrib. adj.* (esp. of a bomb, camera, etc.) operating after a set interval.

delectable /dɪˈlektəb(ə)l/ *adj.* esp. *literary* delightful, delicious. □ **delectably** *adv.* [Latin: related to DELIGHT]

delectation /,diːlekˈteɪʃ(ə)n/ *n. literary* pleasure, enjoyment.

delegate *–n.* /ˈdelɪgət/ **1** elected representative sent to a conference. **2** member of a committee or delegation. *–v.* /ˈdelɪ,geɪt/ (**-ting**) **1** (often foll. by *to*) **a** commit (power etc.) to an agent or deputy. **b** entrust (a task) to another. **2** send or authorize (a person) as a representative. [Latin: related to LEGATE]

delegation /,delɪˈgeɪʃ(ə)n/ *n.* **1** group representing others. **2** delegating or being delegated.

delete /dɪˈliːt/ *v.* (**-ting**) remove (a letter, word, etc.), esp. by striking out. □ **deletion** *n.* [Latin *deleo*]

deleterious /,delɪˈtɪərɪəs/ *adj.* harmful. [Latin from Greek]

delft *n.* (also **delftware**) glazed, usu. blue and white, earthenware. [*Delft* in Holland]

deli /ˈdelɪ/ *n.* (*pl.* **-s**) *colloq.* delicatessen shop. [abbreviation]

deliberate *–adj.* /dɪˈlɪbərət/ **1 a** intentional. **b** considered; careful. **2** (of movement, thought, etc.) unhurried; cautious. *–v.* /dɪˈlɪbə,reɪt/ (**-ting**) **1** think carefully; consider. **2** discuss (*jury deliberated*). □ **deliberately** /dɪˈlɪbərətlɪ/ *adv.* [Latin *libra* balance]

deliberation /,dɪlɪbəˈreɪʃ(ə)n/ *n.* **1** careful consideration; discussion. **2** careful slowness.

deliberative /dɪˈlɪbərətɪv/ *adj.* (esp. of an assembly etc.) of or for deliberation or debate.

delicacy /ˈdelɪkəsɪ/ *n.* (*pl.* **-ies**) **1** being delicate (in all senses). **2** a choice food. [from DELICATE]

delicate /ˈdelɪkət/ *adj.* **1 a** fine in texture, quality, etc.; slender, slight. **b** (of a colour, flavour, etc.) subtle, hard to discern. **2** susceptible; weak, tender. **3 a** requiring tact; tricky (*delicate situation*). **b** (of an instrument) highly sensitive. **4** deft (*delicate touch*). **5** modest. **6** (esp. of actions) considerate. □ **delicately** *adv.* [Latin]

delicatessen /,delɪkəˈtes(ə)n/ *n.* **1** shop selling esp. exotic cooked meats, cheeses, etc. **2** (often *attrib.*) such foods. [French: related to DELICATE]

delicious /dɪˈlɪʃəs/ *adj.* highly enjoyable, esp. to taste or smell. □ **deliciously** *adv.* [Latin *deliciae* delights]

delight /dɪˈlaɪt/ *–v.* **1** (often as **delighted** *adj.*) please greatly (*her singing delighted us*; *delighted to help*). **2** (foll. by *in*) take great pleasure in (*delights in surprising everyone*). *–n.* **1** great pleasure. **2** thing that delights. □ **delighted** *adj.* **delightful** *adj.* **delightfully** *adv.* [Latin *delecto*]

delimit /dɪˈlɪmɪt/ *v.* (**-t-**) fix the limits or boundary of. □ **delimitation** /-ˈteɪʃ(ə)n/ *n.* [Latin: related to LIMIT]

delineate /dɪˈlɪnɪ,eɪt/ *v.* (**-ting**) portray by drawing etc. or in words. □ **delineation** /-ˈeɪʃ(ə)n/ *n.* [Latin: related to LINE[1]]

delinquent /dɪˈlɪŋkwənt/ *–n.* offender (*juvenile delinquent*). *–adj.* **1** guilty of a minor crime or misdeed. **2** failing in one's duty. □ **delinquency** *n.* [Latin *delinquo* offend]

deliquesce /,delɪˈkwes/ *v.* (**-cing**) **1** become liquid, melt. **2** dissolve in water absorbed from the air. □ **deliquescence** *n.* **deliquescent** *adj.* [Latin: related to LIQUID]

delirious /dɪˈlɪrɪəs/ *adj.* **1** affected with delirium. **2** wildly excited, ecstatic. □ **deliriously** *adv.*

delirium /dɪˈlɪrɪəm/ *n.* **1** disorder involving incoherent speech, hallucinations, etc., caused by intoxication, fever, etc. **2** great excitement, ecstasy. [Latin *lira* ridge between furrows]

delirium tremens /ˈtri:menz/ *n.* psychosis of chronic alcoholism involving tremors and hallucinations.

deliver /dɪˈlɪvə(r)/ *v.* **1 a** distribute (letters, goods, etc.) to their destination(s). **b** (often foll. by *to*) hand over. **2** (often foll. by *from*) save, rescue, or set free. **3 a** give birth to (*delivered a girl*). **b** assist at the birth of or in giving birth (*delivered six babies*). **4** utter (an opinion, speech, etc.). **5** (often foll. by *up*, *over*) abandon; resign (*delivered his soul up*). **6** launch or aim (a blow etc.). □ **be delivered of** give birth to. **deliver the goods** *colloq.* carry out an undertaking. [Latin *liber* free]

deliverance *n.* rescuing or being rescued.

delivery *n.* (*pl.* **-ies**) **1** delivering or being delivered. **2** regular distribution of letters etc. (*two deliveries a day*). **3** thing delivered. **4** childbirth. **5** deliverance. **6** style of throwing a ball, delivering a speech, etc. [Anglo-French: related to DELIVER]

dell *n.* small usu. wooded valley. [Old English]

delouse /di:ˈlaʊs/ *v.* (**-sing**) rid of lice.

Delphic /ˈdelfɪk/ *adj.* (also **Delphian** /-fɪən/) **1** obscure, ambiguous, or enigmatic. **2** of the ancient Greek oracle at Delphi.

delphinium /delˈfɪnɪəm/ *n.* (*pl.* **-s**) garden plant with tall spikes of usu. blue flowers. [Greek: related to DOLPHIN]

delta /ˈdeltə/ *n.* **1** triangular area of earth, alluvium etc. at the mouth of a river, formed by its diverging outlets. **2 a** fourth letter of the Greek alphabet (Δ, δ). **b** fourth-class mark for work etc. [Greek]

delta wing *n.* triangular swept-back wing of an aircraft.

delude /dɪˈlu:d/ *v.* (**-ding**) deceive, mislead. [Latin *ludo* mock]

deluge /ˈdelju:dʒ/ *—n.* **1** great flood. **2** (**the Deluge**) biblical Flood (Gen. 6-8). **3** overwhelming rush. **4** heavy fall of rain. *—v.* (**-ging**) flood or inundate (*deluged with complaints*). [Latin *diluvium*]

delusion /dɪˈlu:ʒ(ə)n/ *n.* **1** false belief, hope, etc. **2** hallucination. □ **delusive** *adj.* **delusory** *adj.* [related to DELUDE]

de luxe /də ˈlʌks/ *adj.* luxurious; superior; sumptuous. [French, = of luxury]

delve *v.* (**-ving**) **1** (often foll. by *in*, *into*) search or research energetically or deeply (*delved into his pocket, his family history*). **2** *poet.* dig. [Old English]

demagnetize /di:ˈmægnɪˌtaɪz/ *v.* (also **-ise**) (**-zing** or **-sing**) remove the magnetic properties of. □ **demagnetization** /-ˈzeɪʃ(ə)n/ *n.*

demagogue /ˈdeməˌgɒg/ *n.* (*US* **-gog**) political agitator appealing to mob instincts. □ **demagogic** /-ˈgɒgɪk/ *adj.* **demagogy** *n.* [Greek, = leader of the people]

demand /dɪˈmɑ:nd/ *—n.* **1** insistent and peremptory request. **2** desire for a commodity (*no demand for fur coats*). **3** urgent claim (*makes demands on her*). *—v.* **1** (often foll. by *of*, *from*, or *to* + infin., or *that* + clause) ask for insistently (*demanded to know*). **2** require (*task demanding skill*). **3** insist on being told (*demanded her age*). **4** (as **demanding** *adj.*) requiring skill, effort, attention, etc. (*demanding job; demanding child*). □ **in demand** sought after. **on demand** as soon as requested (*payable on demand*). [French from Latin: related to MANDATE]

demand feeding *n.* feeding a baby when it cries.

demarcation /ˌdi:mɑ:ˈkeɪʃ(ə)n/ *n.* **1** marking of a boundary or limits. **2** trade-union practice of restricting a specific job to one union. □ **demarcate** /ˈdi:-/ *v.* (**-ting**). [Spanish *marcar* MARK[1]]

dematerialize /ˌdi:məˈtɪərɪəˌlaɪz/ *v.* (also **-ise**) (**-zing** or **-sing**) make or become non-material; vanish. □ **dematerialization** /-ˈzeɪʃ(ə)n/ *n.*

demean /dɪˈmi:n/ *v.* (usu. *refl.*) lower the dignity of (*would not demean myself*). [from MEAN[2]]

demeanour /dɪˈmi:nə(r)/ *n.* (*US* **demeanor**) outward behaviour or bearing. [Latin *minor* threaten]

demented /dɪˈmentɪd/ *adj.* mad. □ **dementedly** *adv.* [Latin *mens* mind]

dementia /dɪˈmenʃə/ *n.* chronic insanity. [Latin: related to DEMENTED]

dementia praecox /ˈpri:kɒks/ *n.* *formal* schizophrenia.

demerara /ˌdeməˈreərə/ *n.* light-brown cane sugar. [*Demerara* in Guyana]

demerger /di:ˈmɜ:dʒə(r)/ *n.* dissolution of a commercial merger. □ **demerge** *v.* (**-ging**).

demerit /di:ˈmerɪt/ *n.* fault; blemish.

demesne /dɪˈmi:n, -ˈmeɪn/ *n.* **1 a** territory; domain. **b** land attached to a mansion etc. **c** landed property. **2** (usu. foll. by *of*) region or sphere. **3** *Law hist.* possession (of real property) as one's own. [Latin *dominicus* from *dominus* lord]

demi- *prefix* half; partly. [Latin *dimidius* half]

demigod /'demɪ,gɒd/ *n.* **1 a** partly divine being. **b** child of a god or goddess and a mortal. **2** *colloq.* godlike person.

demijohn /'demɪ,dʒɒn/ *n.* large bottle usu. in a wicker cover. [French]

demilitarize /di:'mɪlɪtə,raɪz/ *v.* (also **-ise**) (**-zing** or **-sing**) remove an army etc. from (a frontier, zone, etc.). □ **demilitarization** /-'zeɪʃ(ə)n/ *n.*

demi-monde /'demɪ,mɒnd/ *n.* **1** class of women considered to be of doubtful morality. **2** any semi-respectable group. [French, = half-world]

demise /dɪ'maɪz/ *—n.* **1** death; termination. **2** *Law* transfer of an estate, title, etc. by demising. *—v.* (**-sing**) *Law* transfer (an estate, title, etc.) by will, lease, or death. [Anglo-French: related to DISMISS]

demisemiquaver /,demɪ'semɪ,kweɪvə(r)/ *n. Mus.* note equal to half a semiquaver.

demist /di:'mɪst/ *v.* clear mist from (a windscreen etc.). □ **demister** *n.*

demo /'deməʊ/ *n.* (*pl.* **-s**) *colloq.* = DEMONSTRATION 2, 3. [abbreviation]

demob /di:'mɒb/ *colloq. —v.* (**-bb-**) demobilize. *—n.* demobilization. [abbreviation]

demobilize /di:'məʊbɪ,laɪz/ *v.* (also **-ise**) (**-zing** or **-sing**) disband (troops, ships, etc.). □ **demobilization** /-'zeɪʃ(ə)n/ *n.*

democracy /dɪ'mɒkrəsɪ/ *n.* (*pl.* **-ies**) **1 a** government by the whole population, usu. through elected representatives. **b** State so governed. **2** classless and tolerant society. [Greek *dēmokratia* rule of the people]

democrat /'demə,kræt/ *n.* **1** advocate of democracy. **2** (**Democrat**) (in the US) member of the Democratic Party.

democratic /,demə'krætɪk/ *adj.* **1** of, like, practising, or being a democracy. **2** favouring social equality. □ **democratically** *adv.*

Democratic Party *n.* more liberal of the two main US political parties.

democratize /dɪ'mɒkrə,taɪz/ *v.* (also **-ise**) (**-zing** or **-sing**) make democratic. □ **democratization** /-'zeɪʃ(ə)n/ *n.*

demodulate /di:'mɒdjʊ,leɪt/ *v.* (**-ting**) extract (a modulating signal) from its carrier. □ **demodulation** /,di:mɒdjʊ'leɪʃ(ə)n/ *n.*

demography /dɪ'mɒgrəfɪ/ *n.* the study of the statistics of births, deaths, disease, etc. □ **demographic** /,demə'græfɪk/ *adj.* **demographically** /,demə'græfɪkəlɪ/ *adv.* [Greek *dēmos* the people, -GRAPHY]

demolish /dɪ'mɒlɪʃ/ *v.* **1 a** pull down (a building). **b** destroy. **2** overthrow (an institution). **3** refute (an argument, theory, etc.). **4** *joc.* eat up voraciously. □ **demolition** /,demə'lɪʃ(ə)n/ *n.* [Latin *moles* mass]

demon /'di:mən/ *n.* **1 a** evil spirit or devil. **b** personification of evil passion. **2** (often *attrib.*) forceful or skilful performer (*demon player*). **3** cruel person. **4** (also **daemon**) supernatural being in ancient Greece. □ **demonic** /dɪ'mɒnɪk/ *adj.* [Greek *daimōn* deity]

demonetize /di:'mʌnɪ,taɪz/ *v.* (also **-ise**) (**-zing** or **-sing**) withdraw (a coin etc.) from use. □ **demonetization** /-'zeɪʃ(ə)n/ *n.* [French: related to DE-, MONEY]

demoniac /dɪ'məʊnɪ,æk/ *—adj.* **1** fiercely energetic or frenzied. **2** supposedly possessed by an evil spirit. **3** of or like demons. *—n.* demoniac person. □ **demoniacal** /,di:mə'naɪək(ə)l/ *adj.* **demoniacally** /,di:mə'naɪəkəlɪ/ *adv.* [Church Latin: related to DEMON]

demonism /'di:mə,nɪz(ə)m/ *n.* belief in demons.

demonolatry /,di:mə'nɒlətrɪ/ *n.* worship of demons. [from DEMON, Greek *latreuō* worship]

demonology /,di:mə'nɒlədʒɪ/ *n.* the study of demons etc.

demonstrable /'demɒnstrəb(ə)l, dɪ'mɒnstrəb(ə)l/ *adj.* able to be shown or proved. □ **demonstrably** *adv.*

demonstrate /'demən,streɪt/ *v.* (**-ting**) **1** show (feelings etc.). **2** describe and explain by experiment, practical use, etc. **3** logically prove or be proof of the truth or existence of. **4** take part in a public demonstration. **5** act as a demonstrator. [Latin *monstro* show]

demonstration /,demən'streɪʃ(ə)n/ *n.* **1** (foll. by *of*) show of feeling etc. **2** (esp. political) public meeting, march, etc. **3** the exhibiting etc. of specimens or experiments in esp. scientific teaching. **4** proof by logic, argument, etc. **5** *Mil.* display of military force.

demonstrative /dɪ'mɒnstrətɪv/ *adj.* **1** showing feelings readily; affectionate. **2** (usu. foll. by *of*) logically conclusive; giving proof (*demonstrative of their skill*). **3** *Gram.* (of an adjective or pronoun) indicating the person or thing referred to (e.g. *this*, *that*, *those*). □ **demonstratively** *adv.* **demonstrativeness** *n.*

demonstrator *n.* **1** person who demonstrates politically. **2** person who demonstrates machines etc. to prospective customers. **3** person who teaches by esp. scientific demonstration.

demoralize /dɪ'mɒrə,laɪz/ *v.* (also **-ise**) (**-zing** or **-sing**) destroy the morale of; dishearten. □ **demoralization** /-'zeɪʃ(ə)n/ *n.* [French]

demote /dɪ'məʊt/ *v.* (**-ting**) reduce to a lower rank or class. □ **demotion** /-'məʊʃ(ə)n/ *n.* [from DE-, PROMOTE]

demotic /dɪ'mɒtɪk/ *–n.* **1** colloquial form of a language. **2** simplified form of ancient Egyptian writing (cf. HIERATIC). *–adj.* **1** (esp. of language) colloquial or vulgar. **2** of ancient Egyptian or modern Greek demotic. [Greek *dēmos* the people]

demotivate /,di:'məʊtɪ,veɪt/ *v.* (**-ting**) (also *absol.*) cause to lose motivation or incentive. □ **demotivation** /-'veɪʃ(ə)n/ *n.*

demur /dɪ'mɜ:(r)/ *–v.* (**-rr-**) **1** (often foll. by *to, at*) raise objections. **2** *Law* put in a demurrer. *–n.* (usu. in *neg.*) objection; objecting (*agreed without demur*). [Latin *moror* delay]

demure /dɪ'mjʊə(r)/ *adj.* (**demurer, demurest**) **1** quiet, reserved; modest. **2** coy. □ **demurely** *adv.* **demureness** *n.* [French: related to DEMUR]

demurrer /dɪ'mʌrə(r)/ *n. Law* objection raised or exception taken.

demystify /di:'mɪstɪ,faɪ/ *v.* (**-ies, -ied**) remove the mystery from; clarify. □ **demystification** /-fɪ'keɪʃ(ə)n/ *n.*

den *n.* **1** wild animal's lair. **2** place of crime or vice (*opium den*). **3** small private room. [Old English]

denarius /dɪ'neərɪəs/ *n.* (*pl.* **denarii** /-rɪ,aɪ/) ancient Roman silver coin. [Latin *deni* by tens]

denary /'di:nərɪ/ *adj.* of ten; decimal.

denationalize /di:'næʃənə,laɪz/ *v.* (also **-ise**) (**-zing** or **-sing**) transfer (an industry etc.) from public to private ownership. □ **denationalization** /-'zeɪʃ(ə)n/ *n.*

denature /di:'neɪtʃə(r)/ *v.* (**-ring**) **1** change the properties of (a protein etc.) by heat, acidity, etc. **2** make (alcohol) undrinkable. [French]

dendrochronology /,dendrəʊkrə'nɒlədʒɪ/ *n.* **1** dating of trees by their annual growth rings. **2** the study of these. [Greek *dendron* tree]

dendrology /den'drɒlədʒɪ/ *n.* the study of trees. □ **dendrological** /-drə'lɒdʒɪk(ə)l/ *adj.* **dendrologist** *n.* [Greek *dendron* tree]

dene *n.* (also **dean**) narrow wooded valley. [Old English]

dengue /'deŋgɪ/ *n.* infectious tropical viral fever. [W. Indian Spanish from Swahili]

deniable /dɪ'naɪəb(ə)l/ *adj.* that may be denied.

denial /dɪ'naɪəl/ *n.* **1** denying the truth or existence of a thing. **2** refusal of a request or wish. **3** disavowal of a leader etc.

denier /'denjə(r)/ *n.* unit of weight measuring the fineness of silk, nylon, etc. [originally the name of a small coin, from Latin DENARIUS]

denigrate /'denɪ,greɪt/ *v.* (**-ting**) blacken the reputation of. □ **denigration** /-'greɪʃ(ə)n/ *n.* **denigrator** *n.* **denigratory** /-'greɪtərɪ/ *adj.* [Latin *niger* black]

denim /'denɪm/ *n.* **1** (often *attrib.*) hard-wearing usu. blue cotton twill used for jeans, overalls, etc. **2** (in *pl.*) *colloq.* jeans etc. made of this. [French *de* of, *Nîmes* in France]

denizen /'denɪz(ə)n/ *n.* **1** (usu. foll. by *of*) inhabitant or occupant. **2** foreigner having certain rights in an adopted country. **3** naturalized foreign word, animal, or plant. [Latin *de intus* from within]

denominate /dɪ'nɒmɪ,neɪt/ *v.* (**-ting**) give a name to, call, describe as. [Latin: related to NOMINATE]

denomination /dɪ,nɒmɪ'neɪʃ(ə)n/ *n.* **1** Church or religious sect. **2** class of measurement or money. **3** name, esp. a characteristic or class name. □ **denominational** *adj.* [Latin: related to DENOMINATE]

denominator /dɪ'nɒmɪ,neɪtə(r)/ *n.* number below the line in a vulgar fraction; divisor. [Latin *nomen* name]

denote /dɪ'nəʊt/ *v.* (**-ting**) **1** (often foll. by *that*) be a sign of; indicate; mean. **2** stand as a name for; signify. □ **denotation** /,di:nə'teɪʃ(ə)n/ *n.* [Latin: related to NOTE]

denouement /deɪ'nu:mɑ̃/ *n.* (also **dénouement**) **1** final unravelling of a plot or complicated situation. **2** final scene in a play, novel, etc. [French, from Latin *nodus* knot]

denounce /dɪ'naʊns/ *v.* (**-cing**) **1** accuse publicly; condemn. **2** inform against. **3** announce withdrawal from (an armistice, treaty, etc.). □ **denouncement** *n.* [Latin *nuntius* messenger]

de novo /di: 'nəʊvəʊ/ *adv.* starting again; anew. [Latin]

dense *adj.* **1** closely compacted; crowded together; thick. **2** *colloq.* stupid. □ **densely** *adv.* **denseness** *n.* [Latin *densus*]

density *n.* (*pl.* **-ies**) **1** denseness of thing(s) or a substance. **2** *Physics* degree of consistency measured by the quantity of mass per unit volume. **3** opacity of a photographic image.

dent *–n.* **1** slight hollow as made by a blow or pressure. **2** noticeable adverse effect (*dent in our funds*). –*v*. **1** mark with a dent. **2** adversely affect. [from INDENT]

dental /'dent(ə)l/ *adj*. **1** of the teeth or dentistry. **2** (of a consonant) produced with the tongue-tip against the upper front teeth (as *th*) or the ridge of the teeth (as *n, s, t*). [Latin *dens dent-* tooth]

dental floss *n*. thread used to clean between the teeth.

dental surgeon *n*. dentist.

dentate /'denteɪt/ *adj*. *Bot*. & *Zool*. toothed; with toothlike notches.

dentifrice /'dentɪfrɪs/ *n*. toothpaste or tooth powder. [Latin: related to DENTAL, *frico* rub]

dentine /'denti:n/ *n*. (*US* **dentin** /-tɪn/) hard dense tissue forming the bulk of a tooth.

dentist /'dentɪst/ *n*. person qualified to treat, extract, etc., teeth. □ **dentistry** *n*.

dentition /den'tɪʃ(ə)n/ *n*. **1** type, number, and arrangement of teeth in a species etc. **2** teething.

denture /'dentʃə(r)/ *n*. removable artificial tooth or teeth.

denuclearize /di:'nju:klɪə,raɪz/ *v*. (also **-ise**) (**-zing** or **-sing**) remove nuclear weapons from (a country etc.). □ **denuclearization** /-'zeɪʃ(ə)n/ *n*.

denude /dɪ'nju:d/ *v*. (**-ding**) **1** make naked or bare. **2** (foll. by *of*) strip of (covering, property, etc.). □ **denudation** /,di:nju:'deɪʃ(ə)n/ *n*. [Latin *nudus* naked]

denunciation /dɪ,nʌnsɪ'eɪʃ(ə)n/ *n*. denouncing; public condemnation. [Latin: related to DENOUNCE]

deny /dɪ'naɪ/ *v*. (**-ies, -ied**) **1** declare untrue or non-existent. **2** repudiate or disclaim. **3** (often foll. by *to*) withhold (a thing) from (*denied him the satisfaction; denied it to me*). □ **deny oneself** be abstinent. [Latin: related to NEGATE]

deodar /'di:ə,dɑ:(r)/ *n*. Himalayan cedar. [Sanskrit, = divine tree]

deodorant /di:'əʊdərənt/ *n*. (often *attrib*.) substance applied to the body or sprayed into the air to conceal smells. [related to ODOUR]

deodorize /di:'əʊdə,raɪz/ *v*. (also **-ise**) (**-zing** or **-sing**) remove or destroy the smell of. □ **deodorization** /-'zeɪʃ(ə)n/ *n*.

deoxyribonucleic acid /,di:ɒksɪ ,raɪbəʊnju'kleɪɪk/ see DNA. [from DE-, OXYGEN, RIBONUCLEIC ACID]

dep. *abbr*. **1** departs. **2** deputy.

depart /dɪ'pɑ:t/ *v*. **1 a** (often foll. by *from*) go away; leave. **b** (usu. foll. by *for*) start; set out. **2** (usu. foll. by *from*) deviate (*departs from good taste*). **3** esp. *formal* or *literary* leave by death; die (*departed this life*). [Latin *dispertio* divide]

departed –*adj*. bygone. –*n*. (prec. by *the*) *euphem*. dead person or people.

department *n*. **1** separate part of a complex whole, esp.: **a** a branch of administration (*Housing Department*). **b** a division of a school, college, etc., by subject (*physics department*). **c** a section of a large store (*hardware department*). **2** *colloq*. area of special expertise. **3** administrative district, esp. in France. [French: related to DEPART]

departmental /,di:pɑ:t'ment(ə)l/ *adj*. of a department. □ **departmentally** *adv*.

department store *n*. large shop with many departments.

departure /dɪ'pɑ:tʃə(r)/ *n*. **1** departing. **2** (often foll. by *from*) deviation (from the truth, a standard, etc.). **3** (often *attrib*.) departing of a train, aircraft, etc. (*departure lounge*). **4** new course of action or thought (*driving is rather a departure for him*).

depend /dɪ'pend/ *v*. **1** (often foll. by *on, upon*) be controlled or determined by (*it depends on luck*). **2** (foll. by *on, upon*) **a** need (*depends on his car*). **b** rely on (*I'm depending on good weather*). [Latin *pendeo* hang]

dependable *adj*. reliable. □ **dependability** /-'bɪlɪtɪ/ *n*. **dependableness** *n*. **dependably** *adv*.

dependant *n*. (*US* **dependent**) person supported, esp. financially, by another. [French: related to DEPEND]

dependence *n*. **1** depending or being dependent, esp. financially. **2** reliance; trust.

dependency *n*. (*pl*. **-ies**) country or province controlled by another.

dependent –*adj*. **1** (usu. foll. by *on, upon*) depending, conditional. **2** unable to do without (esp. a drug). **3** maintained at another's cost. **4** (of a clause etc.) subordinate to a sentence or word. –*n*. *US* var. of DEPENDANT.

depict /dɪ'pɪkt/ *v*. **1** represent in drawing or painting etc. **2** portray in words; describe. □ **depicter** *n*. (also **-tor**). **depiction** *n*. [Latin: related to PICTURE]

depilate /'depɪ,leɪt/ *v*. (**-ting**) remove hair from. □ **depilation** /-'leɪʃ(ə)n/ *n*. [Latin *pilus* hair]

depilatory /dɪ'pɪlətərɪ/ –*adj*. removing unwanted hair. –*n*. (*pl*. **-ies**) depilatory substance.

deplete /dɪ'pli:t/ *v*. (**-ting**) (esp. in *passive*) reduce in numbers, force, or quantity; exhaust. □ **depletion** *n*. [Latin *pleo* fill]

deplorable /dɪ'plɔ:rəb(ə)l/ *adj*. exceedingly bad. □ **deplorably** *adv*.

deplore /dɪ'plɔ:(r)/ *v.* **(-ring) 1** regret deeply. **2** find exceedingly bad. [Latin *ploro* wail]

deploy /dɪ'plɔɪ/ *v.* **1** spread out (troops) into a line ready for action. **2** use (arguments, forces, etc.) effectively. □ **deployment** *n.* [Latin *plico* fold]

depoliticize /ˌdi:pə'lɪtɪˌsaɪz/ *v.* (also **-ise**) **(-zing** or **-sing)** make non-political. □ **depoliticization** /-'zeɪʃ(ə)n/ *n.*

deponent /dɪ'pəʊnənt/ *—adj.* (of esp. a Latin or Greek verb) passive in form but active in meaning. *—n.* **1** deponent verb. **2** person making deposition under oath. [Latin *depono* put down, lay aside]

depopulate /di:'pɒpjʊˌleɪt/ *v.* **(-ting)** reduce the population of. □ **depopulation** /-'leɪʃ(ə)n/ *n.*

deport /dɪ'pɔ:t/ *v.* **1** remove forcibly or exile to another country; banish. **2** *refl.* behave (in a specified manner) (*deported himself well*). □ **deportation** /ˌdi:pɔ:'teɪʃ(ə)n/ *n.* (in sense 1). [Latin *porto* carry]

deportee /ˌdi:pɔ:'ti:/ *n.* deported person.

deportment /dɪ'pɔ:tmənt/ *n.* bearing, demeanour. [French: related to DEPORT]

depose /dɪ'pəʊz/ *v.* **(-sing) 1** remove from office, esp. dethrone. **2** *Law* (usu. foll. by *to*, or *that* + clause) testify, esp. on oath. [French from Latin: related to DEPOSIT]

deposit /dɪ'pɒzɪt/ *—n.* **1 a** money in a bank account. **b** anything stored for safe keeping. **2 a** payment made as a pledge for a contract or as an initial part payment for a thing bought. **b** returnable sum paid on the hire of an item. **3 a** natural layer of sand, rock, coal, etc. **b** layer of accumulated matter on a surface. *—v.* **(-t-) 1 a** put or lay down (*deposited the book on the shelf*). **b** (of water etc.) leave (matter etc.) lying. **2 a** store or entrust for keeping. **b** pay (a sum of money) into a bank account. **3** pay (a sum) as part of a larger sum or as a pledge for a contract. [Latin *pono posit-* put]

deposit account *n.* bank account that pays interest but is not usu. immediately accessible.

depositary *n.* (*pl.* **-ies**) person to whom a thing is entrusted. [Latin: related to DEPOSIT]

deposition /ˌdepə'zɪʃ(ə)n/ *n.* **1** deposing, esp. dethronement. **2** sworn evidence; giving of this. **3 (the Deposition)** taking down of Christ from the Cross. **4** depositing or being deposited. [Latin: related to DEPOSIT]

depositor *n.* person who deposits money, property, etc.

depository *n.* (*pl.* **-ies**) **1 a** storehouse. **b** store (of wisdom, knowledge, etc.). **2** = DEPOSITARY. [Latin: related to DEPOSIT]

depot /'depəʊ/ *n.* **1 a** storehouse, esp. for military supplies. **b** headquarters of a regiment. **2 a** place where vehicles, e.g. buses, are kept. **b** *US* railway or bus station. [French: related to DEPOSIT]

deprave /dɪ'preɪv/ *v.* **(-ving)** corrupt, esp. morally. [Latin *pravus* crooked]

depravity /dɪ'prævɪtɪ/ *n.* (*pl.* **-ies**) moral corruption; wickedness.

deprecate /'deprɪˌkeɪt/ *v.* **(-ting)** express disapproval of; deplore. □ **deprecation** /-'keɪʃ(ə)n/ *n.* **deprecatory** /-'keɪtərɪ/ *adj.* [Latin: related to PRAY]

■ **Usage** *Deprecate* is often confused with *depreciate.*

depreciate /dɪ'pri:ʃɪˌeɪt/ *v.* **(-ting) 1** diminish in value. **2** belittle. □ **depreciatory** /dɪ'pri:ʃɪətərɪ/ *adj.* [Latin: related to PRICE]

■ **Usage** *Depreciate* is often confused with *deprecate.*

depreciation /dɪˌpri:ʃɪ'eɪʃ(ə)n/ *n.* **1 a** decline in value, esp. due to wear and tear. **b** allowance made for this. **2** belittlement.

depredation /ˌdeprɪ'deɪʃ(ə)n/ *n.* (usu. in *pl.*) despoiling, ravaging. [Latin: related to PREY]

depress /dɪ'pres/ *v.* **1** make dispirited or sad. **2** push down; lower. **3** reduce the activity of (esp. trade). **4** (as **depressed** *adj.*) **a** miserable. **b** suffering from depression. □ **depressing** *adj.* **depressingly** *adv.* [Latin: related to PRESS[1]]

depressant *—adj.* reducing activity, esp. of a body function. *—n.* depressant substance.

depressed area *n.* area of economic depression.

depression /dɪ'preʃ(ə)n/ *n.* **1** extreme melancholy, often with a reduction in vitality and physical symptoms. **2** *Econ.* long period of slump. **3** lowering of atmospheric pressure; winds etc. caused by this. **4** hollow on a surface. **5** pressing down.

depressive *—adj.* **1** tending to depress (*depressive drug, influence*). **2** of or tending towards depression (*depressive illness*; *depressive father*). *—n.* person suffering from depression.

deprivation /ˌdeprɪ'veɪʃ(ə)n/ *n.* depriving or being deprived (*suffered many deprivations*).

deprive /dɪ'praɪv/ *v.* **(-ving) 1** (usu. foll. by *of*) prevent from having or enjoying. **2** (as **deprived** *adj.*) lacking what is needed for well-being; underprivileged.

□ **deprival** *n.* [Latin: related to PRIVATION]

Dept. *abbr.* Department.

depth *n.* **1 a** deepness. **b** measurement from the top down, from the surface inwards, or from front to back. **2** difficulty; abstruseness. **3 a** wisdom. **b** intensity of emotion etc. **4** intensity of colour, darkness, etc. **5** (usu. in *pl.*) **a** deep water or place; abyss. **b** low, depressed state. **c** lowest, central, or inmost part (*depths of the country*; *depth of winter*). □ **in depth** thoroughly. **out of one's depth 1** in water over one's head. **2** engaged in a task etc. too difficult for one. [related to DEEP]

depth-charge *n.* bomb exploding under water.

deputation /ˌdepjʊˈteɪʃ(ə)n/ *n.* delegation. [Latin: related to DEPUTE]

depute /dɪˈpju:t/ *v.* (**-ting**) (often foll. by *to*) **1** delegate (a task, authority, etc.). **2** authorize as representative. [Latin *puto* think]

deputize /ˈdepjʊˌtaɪz/ *v.* (also **-ise**) (**-zing** or **-sing**) (usu. foll. by *for*) act as deputy.

deputy /ˈdepjʊtɪ/ *n.* (*pl.* **-ies**) **1** person appointed to act for another (also *attrib.*: *deputy manager*). **2** parliamentary representative in some countries. [var. of DEPUTE]

derail /di:ˈreɪl/ *v.* (usu. in *passive*) cause (a train etc.) to leave the rails. □ **derailment** *n.* [French: related to RAIL[1]]

derange /dɪˈreɪndʒ/ *v.* (**-ging**) **1** make insane. **2** disorder, disturb. □ **derangement** *n.* [French: related to RANK[1]]

Derby /ˈdɑ:bɪ/ *n.* (*pl.* **-ies**) **1 a** annual flat horse-race at Epsom. **b** similar race elsewhere. **2** important sporting contest. **3** (**derby**) *US* bowler hat. [Earl of *Derby*]

derecognize /di:ˈrekəgˌnaɪz/ *v.* (also **-ise**) (**-zing** or **-sing**) cease to recognize the status of (esp. a trade union). □ **derecognition** /-ˈnɪʃ(ə)n/ *n.*

deregulate /di:ˈregjʊˌleɪt/ *v.* (**-ting**) remove regulations from. □ **deregulation** /-ˈleɪʃ(ə)n/ *n.*

derelict /ˈderəlɪkt/ **–adj. 1** (esp. of a property) dilapidated. **2** abandoned, ownerless. **–n. 1** vagrant. **2** abandoned property. [Latin: related to RELINQUISH]

dereliction /ˌderəˈlɪkʃ(ə)n/ *n.* **1** (usu. foll. by *of*) neglect; failure to carry out obligations. **2** abandoning or being abandoned.

derestrict /ˌdi:rɪˈstrɪkt/ *v.* remove restrictions (esp. speed limits) from. □ **derestriction** *n.*

deride /dɪˈraɪd/ *v.* (**-ding**) mock. □ **derision** /-ˈrɪʒ(ə)n/ *n.* [Latin *rideo* laugh]

de rigueur /də rɪˈgɜ:(r)/ *predic. adj.* required by fashion or etiquette (*drugs were de rigueur*). [French]

derisive /dɪˈraɪsɪv/ *adj.* = DERISORY. □ **derisively** *adv.* [from DERIDE]

derisory /dɪˈraɪsərɪ/ *adj.* **1** scoffing, ironical (*derisory cheers*). **2** ridiculously small (*derisory offer*).

derivation /ˌderɪˈveɪʃ(ə)n/ *n.* **1** deriving or being derived. **2 a** formation of a word from another or from a root. **b** tracing of the origin of a word. **c** statement of this.

derivative /dɪˈrɪvətɪv/ *–adj.* derived; not original (*his music is derivative*). *–n.* **1** derived word or thing. **2** *Math.* quantity measuring the rate of change of another.

derive /dɪˈraɪv/ *v.* (**-ving**) **1** (usu. foll. by *from*) get or trace from a source (*derived satisfaction from work*). **2** (foll. by *from*) arise from, originate in (*happiness derives from many things*). **3** (usu. foll. by *from*) show or state the origin or formation of (a word etc.). [Latin *rivus* stream]

dermatitis /ˌdɜ:məˈtaɪtɪs/ *n.* inflammation of the skin. [Greek *derma* skin, -ITIS]

dermatology /ˌdɜ:məˈtɒlədʒɪ/ *n.* the study of skin diseases. □ **dermatological** /-təˈlɒdʒɪk(ə)l/ *adj.* **dermatologist** *n.* [from DERMATITIS, -LOGY]

dermis /ˈdɜ:mɪs/ *n.* **1** (in general use) the skin. **2** layer of living tissue below the epidermis. [from EPIDERMIS]

derogate /ˈderəˌgeɪt/ *v.* (**-ting**) (foll. by *from*) *formal* detract from (merit, right, etc.). □ **derogation** /-ˈgeɪʃ(ə)n/ *n.* [Latin *rogo* ask]

derogatory /dɪˈrɒgətərɪ/ *adj.* disparaging; insulting (*derogatory remark*). □ **derogatorily** *adv.*

derrick /ˈderɪk/ *n.* **1** crane for heavy weights, with a movable pivoted arm. **2** framework over an oil well etc., holding the drilling machinery. [*Derrick*, name of a hangman]

derring-do /ˌderɪŋˈdu:/ *n. literary joc.* heroic courage or actions. [*daring to do*]

derris /ˈderɪs/ *n.* **1** tropical climbing plant. **2** insecticide made from its root. [Latin from Greek]

derv *n.* diesel oil for road vehicles. [*d*iesel-*e*ngined *r*oad-*v*ehicle]

dervish /ˈdɜ:vɪʃ/ *n.* member of a Muslim fraternity vowed to poverty and austerity. [Turkish from Persian, = poor]

DES *abbr.* Department of Education and Science.

desalinate /di:ˈsælɪˌneɪt/ *v.* (**-ting**) remove the salt from (esp. sea water). □ **desalination** /-ˈneɪʃ(ə)n/ *n.* [from SALINE]

descale /di:ˈskeɪl/ *v.* (**-ling**) remove scale from.

descant *–n.* /'deskænt/ **1** harmonizing treble melody above the basic melody, esp. of a hymn tune. **2** *poet.* melody; song. *–v.* /dɪs'kænt/ (foll. by *on, upon*) talk prosily, esp. in praise of. [Latin *cantus* song: related to CHANT]

descend /dɪ'send/ *v.* **1** go or come down. **2** sink, fall. **3** slope downwards. **4** (usu. foll. by *on*) make a sudden attack or visit. **5** (of property etc.) be passed on by inheritance. **6 a** sink in rank, quality, etc. **b** (foll. by *to*) stoop to (an unworthy act). □ **be descended from** have as an ancestor. □ **descendent** *adj.* [Latin *scando* climb]

descendant *n.* person or thing descended from another. [French: related to DESCEND]

descent /dɪ'sent/ *n.* **1** act or way of descending. **2** downward slope. **3** lineage, family origin. **4** decline; fall. **5** sudden attack.

describe /dɪ'skraɪb/ *v.* **(-bing) 1 a** state the characteristics, appearance, etc. of. **b** (foll. by *as*) assert to be; call (*described him as a liar*). **2 a** draw (esp. a geometrical figure). **b** move in (a specified way, esp. a curve) (*described a parabola through the air*). [Latin *scribo* write]

description /dɪ'skrɪpʃ(ə)n/ *n.* **1 a** describing or being described. **b** representation, esp. in words. **2** sort, kind (*no food of any description*). [Latin: related to DESCRIBE]

descriptive /dɪ'skrɪptɪv/ *adj.* describing, esp. vividly. [Latin: related to DESCRIBE]

descry /dɪ'skraɪ/ *v.* **(-ies, -ied)** *literary* catch sight of; discern. [French: related to CRY]

desecrate /'desɪ,kreɪt/ *v.* **(-ting)** violate (a sacred place etc.) with violence, profanity, etc. □ **desecration** /-'kreɪʃ(ə)n/ *n.* **desecrator** *n.* [from DE-, CONSECRATE]

desegregate /di:'segrɪ,geɪt/ *v.* **(-ting)** abolish racial segregation in. □ **desegregation** /-'geɪʃ(ə)n/ *n.*

deselect /,di:sɪ'lekt/ *v.* reject (a selected candidate, esp. a sitting MP) in favour of another. □ **deselection** *n.*

desensitize /di:'sensɪ,taɪz/ *v.* (also **-ise**) **(-zing** or **-sing)** reduce or destroy the sensitivity of. □ **desensitization** /-'zeɪʃ(ə)n/ *n.*

desert[1] /dɪ'zɜ:t/ *v.* **1** leave without intending to return. **2** (esp. as **deserted** *adj.*) forsake, abandon. **3** run away (esp. from military service). □ **deserter** *n.* (in sense 3). **desertion** *n.* [Latin *desero -sert-* leave]

desert[2] /'dezət/ *–n.* dry barren, esp. sandy, tract. *–adj.* uninhabited, desolate, barren. [Latin *desertus*: related to DESERT[1]]

desert[3] /dɪ'zɜ:t/ *n.* **1** (in *pl.*) deserved reward or punishment (*got his deserts*). **2** being worthy of reward or punishment. [French: related to DESERVE]

desert boot *n.* suede etc. ankle-high boot.

desertification /dɪ,zɜ:tɪfɪ'keɪʃ(ə)n/ *n.* making or becoming a desert.

desert island *n.* (usu. tropical) uninhabited island.

deserve /dɪ'zɜ:v/ *v.* **(-ving)** (often foll. by *to* + infin.) be worthy of (a reward, punishment, etc.) (*deserves a prize*). □ **deservedly** /-vɪdlɪ/ *adv.* [Latin *servio* serve]

deserving *adj.* (often foll. by *of*) worthy (esp. of help, praise, etc.).

déshabillé /,dezæ'bi:eɪ/ *n.* (also ***déshabille*** /,deɪzæ'bi:l/, **dishabille** /,dɪsə'bi:l/) state of partial undress. [French, = undressed]

desiccate /'desɪ,keɪt/ *v.* **(-ting)** remove moisture from (esp. food) (*desiccated coconut*). □ **desiccation** /-'keɪʃ(ə)n/ *n.* [Latin *siccus* dry]

desideratum /dɪ,zɪdə'rɑ:təm/ *n.* (*pl.* **-ta**) something lacking but desirable. [Latin: related to DESIRE]

design /dɪ'zaɪn/ *–n.* **1 a** preliminary plan or sketch for making something. **b** art of producing these. **2** lines or shapes forming a pattern or decoration. **3** plan, purpose, or intention. **4 a** arrangement or layout of a product. **b** established version of a product. *–v.* **1** produce a design for (a building, machine, etc.). **2** intend or plan (*designed for beginners*). **3** be a designer. □ **by design** on purpose. **have designs on** plan to appropriate, seduce, etc. [Latin *signum* mark]

designate *–v.* /'dezɪg,neɪt/ **(-ting) 1** (often foll. by *as*) appoint to an office or function. **2** specify (*designated times*). **3** (often foll. by *as*) describe as; style. **4** serve as the name or symbol of. *–adj.* /'dezɪgnət/ (after the noun) appointed to office but not yet installed. [Latin: related to DESIGN]

designation /,dezɪg'neɪʃ(ə)n/ *n.* **1** name, description, or title. **2** designating.

designedly /dɪ'zaɪnɪdlɪ/ *adv.* on purpose.

designer *n.* **1** person who designs e.g. clothing, machines, theatre sets; draughtsman. **2** (*attrib.*) bearing the label of a famous designer; prestigious.

designer drug *n.* synthetic analogue of an illegal drug.

designing *adj.* crafty, scheming.

desirable /dɪ'zaɪərəb(ə)l/ *adj.* **1** worth having or doing. **2** sexually attractive. □ **desirability** /-'bɪlɪtɪ/ *n.* **desirableness** *n.* **desirably** *adv.*

desire /dɪ'zaɪə(r)/ *–n.* **1 a** unsatisfied longing or wish. **b** expression of this; request. **2** sexual appetite. **3** something desired (*achieved his heart's desire*). *–v.* **(-ring) 1** (often foll. by *to* + infin., or *that* + clause) long for; wish. **2** request (*desires a rest*). [Latin *desidero* long for]

desirous *predic. adj.* **1** (usu. foll. by *of*) desiring, wanting (*desirous of stardom*). **2** wanting; hoping (*desirous to do the right thing*).

desist /dɪ'zɪst/ *v.* (often foll. by *from*) abstain; cease. [Latin *desisto*]

desk *n.* **1** piece of furniture with a surface for writing on, and often drawers. **2** counter in a hotel, bank, etc. **3** specialized section of a newspaper office (*sports desk*). **4** unit of two orchestral players sharing a stand. [Latin: related to DISCUS]

desktop *n.* **1** working surface of a desk. **2** (*attrib.*) (esp. of a microcomputer) for use on an ordinary desk.

desktop publishing *n.* printing with a desktop computer and high-quality printer.

desolate *–adj.* /'desələt/ **1** left alone; solitary. **2** uninhabited, ruined, dreary (*desolate moor*). **3** forlorn; wretched. *–v.* /'desə,leɪt/ **(-ting) 1** depopulate, devastate; lay waste. **2** (esp. as **desolated** *adj.*) make wretched. □ **desolately** /-lətlɪ/ *adv.* **desolateness** /-lətnɪs/ *n.* [Latin *solus* alone]

desolation /,desə'leɪʃ(ə)n/ *n.* **1** desolating or being desolated. **2** loneliness, grief, etc., esp. caused by desertion. **3** neglected, ruined, or empty state.

despair /dɪ'speə(r)/ *–n.* **1** complete loss or absence of hope. **2** cause of this. *–v.* (often foll. by *of*) lose or be without hope (*despaired of ever winning*). [Latin *spero* hope]

despatch var. of DISPATCH.

desperado /,despə'rɑ:dəʊ/ *n.* (*pl.* **-es** or *US* **-s**) desperate or reckless criminal etc. [as DESPERATE]

desperate /'despərət/ *adj.* **1** reckless from despair; violent and lawless. **2 a** extremely dangerous, serious, or bad (*desperate situation*). **b** staking all on a small chance (*desperate remedy*). **3** (usu. foll. by *for*) needing or desiring very much (*desperate for recognition*). □ **desperately** *adv.* **desperateness** *n.* **desperation** /-'reɪʃ(ə)n/ *n.* [Latin: related to DESPAIR]

despicable /'despɪkəb(ə)l, dɪ'spɪk-/ *adj.* vile; contemptible, esp. morally. □ **despicably** *adv.* [Latin *specio spect-* look at]

despise /dɪ'spaɪz/ *v.* **(-sing)** regard as inferior, worthless, or contemptible. [Latin: related to DESPICABLE]

despite /dɪ'spaɪt/ *prep.* in spite of. [Latin: related to DESPICABLE]

despoil /dɪ'spɔɪl/ *v. literary* (often foll. by *of*) plunder; rob; deprive. □ **despoliation** /dɪ,spəʊlɪ'eɪʃ(ə)n/ *n.* [Latin: related to SPOIL]

despondent /dɪ'spɒnd(ə)nt/ *adj.* in low spirits, dejected. □ **despondence** *n.* **despondency** *n.* **despondently** *adv.* [Latin: related to SPONSOR]

despot /'despɒt/ *n.* **1** absolute ruler. **2** tyrant. □ **despotic** /-'spɒtɪk/ *adj.* **despotically** /-'spɒtɪkəlɪ/ *adv.* [Greek *despotēs* master]

despotism /'despə,tɪz(ə)m/ *n.* **1** rule by a despot; tyranny. **2** country ruled by a despot.

des res /dez 'rez/ *n. slang* desirable residence. [abbreviation]

dessert /dɪ'zɜ:t/ *n.* **1** sweet course of a meal. **2** fruit, nuts, etc., served at the end of a meal. [French: related to DIS-, SERVE]

dessertspoon *n.* **1** medium-sized spoon for dessert. **2** amount held by this. □ **dessertspoonful** *n.* (*pl.* **-s**).

destabilize /di:'steɪbɪ,laɪz/ *v.* (also **-ise**) **(-zing** or **-sing) 1** make unstable. **2** subvert (esp. a foreign government). □ **destabilization** /-'zeɪʃ(ə)n/ *n.*

destination /,destɪ'neɪʃ(ə)n/ *n.* place a person or thing is bound for. [Latin: related to DESTINE]

destine /'destɪn/ *v.* **(-ning)** (often foll. by *to, for*, or *to* + infin.) appoint; preordain; intend (*destined him for the navy*). □ **be destined to** be fated or preordained to. [French from Latin]

destiny /'destɪnɪ/ *n.* (*pl.* **-ies**) **1 a** fate. **b** this regarded as a power. **2** particular person's fate etc. [French from Latin]

destitute /'destɪ,tju:t/ *adj.* **1** without food, shelter, etc. **2** (usu. foll. by *of*) lacking (*destitute of friends*). □ **destitution** /-'tju:ʃ(ə)n/ *n.* [Latin]

destroy /dɪ'strɔɪ/ *v.* **1** pull or break down; demolish. **2** kill (esp. an animal). **3** make useless; spoil. **4** ruin, esp. financially. **5** defeat. [Latin *struo struct-* build]

destroyer *n.* **1** person or thing that destroys. **2** fast armed warship escorting other ships.

destruct /dɪ'strʌkt/ *US* esp. *Astronaut.* *–v.* destroy (one's own rocket etc.) or be destroyed deliberately, esp. for safety. *–n.* destructing.

destructible *adj.* able to be destroyed. [Latin: related to DESTROY]

destruction /dɪ'strʌkʃ(ə)n/ *n.* **1** destroying or being destroyed. **2** cause of this. [Latin: related to DESTROY]

destructive *adj.* **1** (often foll. by *to, of*) destroying or tending to destroy. **2** negatively critical. □ **destructively** *adv.* **destructiveness** *n.*

desuetude /dɪ'sju:ɪ,tju:d/ *n. formal* state of disuse (*fell into desuetude*). [Latin *suesco* be accustomed]

desultory /'dezəltərɪ/ *adj.* **1** constantly turning from one subject to another. **2** disconnected; unmethodical. □ **desultorily** *adv.* [Latin *desultorius* superficial]

detach /dɪ'tætʃ/ *v.* **1** (often foll. by *from*) unfasten or disengage and remove. **2** send (troops etc.) on a separate mission. **3** (as **detached** *adj.*) **a** impartial; unemotional. **b** (esp. of a house) standing separate. □ **detachable** *adj.* [French: related to ATTACH]

detachment *n.* **1 a** aloofness; indifference. **b** impartiality. **2** detaching or being detached. **3** troops etc. detached for a specific purpose. [French: related to DETACH]

detail /'di:teɪl/ —*n.* **1** small particular; item. **2 a** these collectively (*eye for detail*). **b** treatment of them (*detail was unconvincing*). **3 a** minor decoration on a building etc. **b** small part of a picture etc. shown alone. **4** small military detachment. —*v.* **1** give particulars of. **2** relate circumstantially. **3** assign for special duty. **4** (as **detailed** *adj.*) **a** (of a picture, story, etc.) containing many details. **b** itemized (*detailed list*). □ **in detail** item by item, minutely. [French: related to TAIL²]

detain /dɪ'teɪn/ *v.* **1** keep waiting; delay. **2** keep in custody, lock up. □ **detainment** *n.* [Latin *teneo* hold]

detainee /,di:teɪ'ni:/ *n.* person kept in custody, esp. for political reasons.

detect /dɪ'tekt/ *v.* **1** discover or perceive (*detected a note of sarcasm*). **2** (often foll. by *in*) discover (a criminal); solve (a crime). □ **detectable** *adj.* **detector** *n.* [Latin *tego tect-* cover]

detection /dɪ'tekʃ(ə)n/ *n.* **1** detecting or being detected. **2** work of a detective.

detective /dɪ'tektɪv/ *n.* person, esp. a police officer, investigating crimes.

détente /deɪ'tɑ̃t/ *n.* easing of strained, esp. international, relations. [French, = relaxation]

detention /dɪ'tenʃ(ə)n/ *n.* **1** detaining or being detained. **2** being kept late in school as a punishment. [Latin: related to DETAIN]

detention centre *n.* short-term prison for young offenders.

deter /dɪ'tɜ:(r)/ *v.* (**-rr-**) (often foll. by *from*) discourage or prevent, esp. through fear. □ **determent** *n.* [Latin *terreo* frighten]

detergent /dɪ'tɜ:dʒ(ə)nt/ —*n.* synthetic cleansing agent used with water. —*adj.* cleansing. [Latin *tergeo* wipe]

deteriorate /dɪ'tɪərɪə,reɪt/ *v.* (**-ting**) become worse. □ **deterioration** /-'reɪʃ(ə)n/ *n.* [Latin *deterior* worse]

determinant /dɪ'tɜ:mɪnənt/ —*adj.* determining. —*n.* **1** determining factor etc. **2** quantity obtained by the addition of products of the elements of a square matrix according to a given rule. [Latin: related to DETERMINE]

determinate /dɪ'tɜ:mɪnət/ *adj.* limited, of definite scope or nature.

determination /dɪ,tɜ:mɪ'neɪʃ(ə)n/ *n.* **1** firmness of purpose; resoluteness. **2** process of deciding or determining.

determine /dɪ'tɜ:mɪn/ *v.* (**-ning**) **1** find out or establish precisely. **2** decide or settle; resolve. **3** be the decisive factor in regard to (*demand determines supply*). □ **be determined** be resolved. [Latin *terminus* boundary]

determined *adj.* showing determination; resolute, unflinching. □ **determinedly** *adv.*

determinism *n.* doctrine that human actions, events, etc. are determined by causes external to the will. □ **determinist** *n.* & *adj.* **deterministic** /-'nɪstɪk/ *adj.* **deterministically** /-'nɪstɪkəlɪ/ *adv.*

deterrent /dɪ'terənt/ —*adj.* deterring. —*n.* deterrent thing or factor (esp. nuclear weapons). □ **deterrence** *n.*

detest /dɪ'test/ *v.* hate violently, loathe. □ **detestation** /,di:te'steɪʃ(ə)n/ *n.* [Latin *detestor* from *testis* witness]

detestable *adj.* intensely disliked; hateful.

dethrone /di:'θrəʊn/ *v.* (**-ning**) remove from a throne, depose. □ **dethronement** *n.*

detonate /'detə,neɪt/ *v.* (**-ting**) set off (an explosive charge); be set off. □ **detonation** /-'neɪʃ(ə)n/ *n.* [Latin *tono* thunder]

detonator *n.* device for detonating explosives.

detour /'di:tʊə(r)/ *n.* divergence from a usual route; roundabout course. [French: related to TURN]

detoxify /di:'tɒksɪ,faɪ/ *v.* (**-ies, -ied**) remove poison or harmful substances from. □ **detoxification** /-fɪ'keɪʃ(ə)n/ *n.* [Latin *toxicum* poison]

detract /dɪ'trækt/ *v.* (foll. by *from*) take away (a part); diminish; make seem less valuable or important. [Latin *traho tract-* draw]

detractor *n.* person who criticizes unfairly. □ **detraction** *n.*

detriment /ˈdetrɪmənt/ *n.* **1** harm, damage. **2** cause of this. □ **detrimental** /-ˈment(ə)l/ *adj.* [Latin: related to TRITE]

detritus /dɪˈtraɪtəs/ *n.* gravel, sand, etc. produced by erosion; debris. [Latin: related to DETRIMENT]

de trop /də ˈtrəʊ/ *predic. adj.* not wanted, in the way. [French, = excessive]

deuce[1] /dju:s/ *n.* **1** two on dice or playing-cards. **2** *Tennis* score of 40 all. [Latin *duo duos* two]

deuce[2] /dju:s/ *n.* the Devil, esp. as an exclamation of surprise or annoyance (*who the deuce are you?*). [Low German *duus* two (being the worst throw at dice)]

deus ex machina /ˌdeɪʊs eks ˈmækɪnə/ *n.* unlikely agent resolving a seemingly hopeless situation, esp. in a play or novel. [Latin, = god from the machinery, i.e. in a theatre]

deuterium /dju:ˈtɪərɪəm/ *n.* stable isotope of hydrogen with a mass about double that of the usual isotope. [Greek *deuteros* second]

Deutschmark /ˈdɔɪtʃmɑ:k/ *n.* (also **Deutsche Mark** /ˈdɔɪtʃə mɑ:k/) chief monetary unit of Germany. [German: related to MARK[2]]

devalue /di:ˈvælju:/ *v.* (**-ues, -ued, -uing**) **1** reduce the value of. **2** reduce the value of (a currency) in relation to others or to gold. □ **devaluation** /-ˈeɪʃ(ə)n/ *n.*

devastate /ˈdevəˌsteɪt/ *v.* (**-ting**) **1** lay waste; cause great destruction to. **2** (often in *passive*) overwhelm with shock or grief. □ **devastation** /-ˈsteɪʃ(ə)n/ *n.* [Latin *vasto* lay waste]

devastating *adj.* crushingly effective; overwhelming. □ **devastatingly** *adv.*

develop /dɪˈveləp/ *v.* (**-p-**) **1 a** make or become bigger, fuller, more elaborate, etc. **b** bring or come to an active, visible, or mature state. **2** begin to exhibit or suffer from (*developed a rattle*). **3 a** build on (land). **b** convert (land) to new use. **4** treat (photographic film etc.) to make the image visible. □ **developer** *n.* [French]

developing country *n.* poor or primitive country.

development *n.* **1** developing or being developed. **2 a** stage of growth or advancement. **b** thing that has developed; new event or circumstance etc. (*latest developments*). **3** full-grown state. **4** developed land; group of buildings. □ **developmental** /-ˈment(ə)l/ *adj.*

development area *n.* area where new industries are encouraged by the State.

deviant /ˈdi:vɪənt/ *–adj.* deviating from what is normal, esp. sexually. *–n.* deviant person or thing. □ **deviance** *n.* **deviancy** *n.*

deviate /ˈdi:vɪˌeɪt/ *v.* (**-ting**) (often foll. by *from*) turn aside or diverge (from a course of action, rule, etc.). □ **deviation** /-ˈeɪʃ(ə)n/ *n.* [Latin *via* way]

device /dɪˈvaɪs/ *n.* **1** thing made or adapted for a special purpose. **2** plan, scheme, or trick. **3** design, esp. heraldic. □ **leave a person to his or her own devices** leave a person to do as he or she wishes. [French: related to DEVISE]

devil /ˈdev(ə)l/ *–n.* **1** (usu. **the Devil**) (in Christian and Jewish belief) supreme spirit of evil; Satan. **2 a** evil spirit; demon. **b** personified evil. **3 a** wicked person. **b** mischievously clever person. **4** *colloq.* person of a specified kind (*lucky devil*). **5** fighting spirit, mischievousness (*devil is in him tonight*). **6** *colloq.* awkward thing. **7** (**the devil** or **the Devil**) *colloq.* used as an exclamation of surprise or annoyance (*who the devil are you?*). **8** literary hack. **9** junior legal counsel. *–v.* (**-ll-**; *US* **-l-**) **1** cook (food) with hot seasoning. **2** act as devil for an author or barrister. **3** *US* harass, worry. □ **between the devil and the deep blue sea** in a dilemma. **a devil of** *colloq.* considerable, difficult, or remarkable. **devil's own** *colloq.* very difficult or unusual (*the devil's own job*). **the devil to pay** trouble to be expected. **speak (or talk) of the devil** said when person appears just after being mentioned. [Greek *diabolos* accuser, slanderer]

devilish *–adj.* **1** of or like a devil; wicked. **2** mischievous. *–adv. colloq.* very. □ **devilishly** *adv.*

devil-may-care *adj.* cheerful and reckless.

devilment *n.* mischief, wild spirits.

devilry *n.* (*pl.* **-ies**) **1** wickedness; reckless mischief. **2** black magic.

devil's advocate *n.* person who argues against a proposition to test it.

devils-on-horseback *n.pl.* savoury of prunes or plums wrapped in bacon.

devious /ˈdi:vɪəs/ *adj.* **1** not straightforward, underhand. **2** winding, circuitous. □ **deviously** *adv.* **deviousness** *n.* [Latin *via* way]

devise /dɪˈvaɪz/ *v.* (**-sing**) **1** carefully plan or invent. **2** *Law* leave (real estate) by will. [Latin: related to DIVIDE]

devoid /dɪˈvɔɪd/ *predic. adj.* (foll. by *of*) lacking or free from. [French: related to VOID]

devolution /ˌdiːvəˈluːʃ(ə)n/ *n.* delegation of power, esp. to local or regional administration. □ **devolutionist** *n.* & *adj.* [Latin: related to DEVOLVE]

devolve /dɪˈvɒlv/ *v.* **(-ving) 1** (foll. by *on*, *upon*, etc.) pass (work or duties) or be passed to (a deputy etc.). **2** (foll. by *on*, *to*, *upon*) (of property etc.) descend to. □ **devolvement** *n.* [Latin *volvo volut-* roll]

Devonian /dɪˈvəʊnɪən/ *–adj.* of the fourth period of the Palaeozoic era. *–n.* this period. [*Devon* in England]

devote /dɪˈvəʊt/ *v.* **(-ting)** (often *refl.*; foll. by *to*) apply or give over to (a particular activity etc.). [Latin *voveo vot-* vow]

devoted *adj.* loving; loyal. □ **devotedly** *adv.*

devotee /ˌdevəˈtiː/ *n.* **1** (usu. foll. by *of*) zealous enthusiast or supporter. **2** pious person.

devotion /dɪˈvəʊʃ(ə)n/ *n.* **1** (usu. foll. by *to*) great love or loyalty. **2 a** religious worship. **b** (in *pl.*) prayers. □ **devotional** *adj.* [Latin: related to DEVOTE]

devour /dɪˈvaʊə(r)/ *v.* **1** eat voraciously. **2** (of fire etc.) engulf, destroy. **3** take in eagerly (*devoured the play*). **4** preoccupy (*devoured by fear*). [Latin *voro* swallow]

devout /dɪˈvaʊt/ *adj.* earnestly religious or sincere. □ **devoutly** *adv.* **devoutness** *n.* [Latin: related to DEVOTE]

dew *n.* **1** condensed water vapour forming on cool surfaces at night. **2** similar glistening moisture. □ **dewy** *adj.* **(-ier, -iest).** [Old English]

dewberry *n.* (*pl.* **-ies**) bluish fruit like the blackberry.

dew-claw *n.* rudimentary inner toe on some dogs.

dewdrop *n.* drop of dew.

Dewey system /ˈdjuːɪ/ *n.* decimal system of library classification. [*Dewey*, name of a librarian]

dewlap *n.* loose fold of skin hanging from the throat of cattle, dogs, etc. [from DEW, LAP[1]]

dew point *n.* temperature at which dew forms.

dexter *adj.* esp. *Heraldry* on or of the right-hand side (observer's left) of a shield etc. [Latin, = on the right]

dexterity /dekˈsterɪtɪ/ *n.* **1** skill in using one's hands. **2** mental adroitness. [Latin: related to DEXTER]

dexterous /ˈdekstrəs/ *adj.* (also **dextrous**) having or showing dexterity. □ **dexterously** *adv.* **dexterousness** *n.*

dextrin /ˈdekstrɪn/ *n.* soluble gummy substance used as a thickening agent, adhesive, etc. [Latin *dextra* on or to the right]

dextrose /ˈdekstrəʊs/ *n.* form of glucose. [Latin *dextra* on or to the right]

DFC *abbr.* Distinguished Flying Cross.

DFM *abbr.* Distinguished Flying Medal.

dhal /dɑːl/ *n.* (also **dal**) **1** a kind of split pulse common in India. **2** dish made with this. [Hindi]

dharma /ˈdɑːmə/ *n. Ind.* **1** social custom; correct behaviour. **2** the Buddhist truth. **3** the Hindu moral law. [Sanskrit, = decree, custom]

dhoti /ˈdəʊtɪ/ *n.* (*pl.* **-s**) loincloth worn by male Hindus. [Hindi]

di-[1] *comb. form* two-, double. [Greek *dis* twice]

di-[2] *prefix* = DIS-.

di-[3] *prefix* form of DIA- before a vowel.

dia. *abbr.* diameter.

dia- *prefix* (also **di-** before a vowel) **1** through (*diaphanous*). **2** apart (*diacritical*). **3** across (*diameter*). [Greek *dia* through]

diabetes /ˌdaɪəˈbiːtiːz/ *n.* disease in which sugar and starch are not properly absorbed by the body. [Latin from Greek]

diabetic /ˌdaɪəˈbetɪk/ *–adj.* **1** of or having diabetes. **2** for diabetics. *–n.* person suffering from diabetes.

diabolical /ˌdaɪəˈbɒlɪk(ə)l/ *adj.* (also **diabolic**) **1** of the Devil. **2** devilish; inhumanly cruel or wicked. **3** extremely bad, clever, or annoying. □ **diabolically** *adv.* [Latin: related to DEVIL]

diabolism /daɪˈæbəˌlɪz(ə)m/ *n.* **1** worship of the Devil. **2** sorcery. [Greek: related to DEVIL]

diachronic /ˌdaɪəˈkrɒnɪk/ *adj.* of a thing's historical development. □ **diachronically** *adv.* [Greek *khronos* time]

diaconal /daɪˈækən(ə)l/ *adj.* of a deacon. [Church Latin: related to DEACON]

diaconate /daɪˈækənət/ *n.* **1** position of deacon. **2** body of deacons.

diacritic /ˌdaɪəˈkrɪtɪk/ *n.* sign (e.g. an accent or cedilla) indicating different sounds or values of a letter. [Greek: related to CRITIC]

diacritical *–adj.* distinguishing, distinctive. *–n.* (in full **diacritical mark** or **sign**) = DIACRITIC.

diadem /ˈdaɪəˌdem/ *n.* **1** crown or headband as a sign of sovereignty. **2** sovereignty. **3** crowning distinction. [Greek *deō* bind]

diaeresis /daɪˈɪərəsɪs/ *n.* (*pl.* **diaereses** /-ˌsiːz/) (*US* **dieresis**) mark (as in *naïve*) over a vowel to indicate that it is sounded separately. [Greek, = separation]

diagnose /ˈdaɪəgˌnəʊz/ *v.* **(-sing)** make a diagnosis of (a disease, fault, etc.).

diagnosis /ˌdaɪəg'nəʊsɪs/ *n.* (*pl.* **diagnoses** /-ˌsiːz/) **1 a** identification of a disease from its symptoms. **b** formal statement of this. **2** identification of the cause of a mechanical fault etc. [Greek *gignōskō* recognize]

diagnostic /ˌdaɪəg'nɒstɪk/ *–adj.* of or assisting diagnosis. *–n.* symptom. □ **diagnostically** *adv.* **diagnostician** /-nɒ'stɪʃ(ə)n/ *n.* [Greek: related to DIAGNOSIS]

diagnostics *n.* **1** (treated as *pl.*) *Computing* programs etc. used to identify faults in hardware or software. **2** (treated as *sing.*) science of diagnosing disease.

diagonal /daɪ'ægən(ə)l/ *–adj.* **1** crossing a straight-sided figure from corner to corner. **2** slanting, oblique. *–n.* straight line joining two opposite corners. □ **diagonally** *adv.* [Greek *gōnia* angle]

diagram /'daɪəˌgræm/ *n.* outline drawing, plan, or graphic representation of a machine, structure, process, etc. □ **diagrammatic** /-grə'mætɪk/ *adj.* **diagrammatically** /-grə'mætɪkəlɪ/ *adv.* [Greek: related to -GRAM]

dial /'daɪ(ə)l/ *–n.* **1** plate with a scale for measuring weight, volume, etc., indicated by a pointer. **2** movable numbered disc on a telephone for making connection. **3** face of a clock or watch, marking the hours etc. **4 a** plate or disc etc. on a radio or television for selecting a wavelength or channel. **b** similar device on other equipment. *–v.* (**-ll-**; *US* **-l-**) **1** (*also absol.*) select (a telephone number) with a dial. **2** measure, indicate, or regulate with a dial. [medieval Latin *diale* from *dies* day]

dialect /'daɪəˌlekt/ *n.* **1** regional form of speech. **2** variety of language with non-standard vocabulary, pronunciation, or grammar. □ **dialectal** /-'lekt(ə)l/ *adj.* [Greek *legō* speak]

dialectic /ˌdaɪə'lektɪk/ *n.* **1** art of investigating the truth by discussion and logical argument. **2** process whereby contradictions merge to form a higher truth. **3** any situation or discussion involving the juxtaposition or conflict of opposites. [Greek: related to DIALECT]

dialectical *adj.* of dialectic. □ **dialectically** *adv.*

dialectical materialism *n.* Marxist theory that political and historical events are due to the conflict of social forces arising from economic conditions.

dialectics *n.* (treated as *sing.* or *pl.*) = DIALECTIC *n.* 1.

dialling tone *n.* sound indicating that a telephone caller may dial.

dialogue /'daɪəˌlɒg/ *n.* (*US* **dialog**) **1 a** conversation. **b** this in written form. **2** discussion between people with different opinions. [Greek *legō* speak]

dialysis /daɪ'ælɪsɪs/ *n.* (*pl.* **dialyses** /-ˌsiːz/) **1** separation of particles in a liquid by differences in their ability to pass through a membrane into another liquid. **2** purification of the blood by this technique. [Greek *luō* set free]

diamanté /dɪə'mɑ̃teɪ/ *adj.* decorated with synthetic diamonds or another sparkling substance. [French *diamant* diamond]

diameter /daɪ'æmɪtə(r)/ *n.* **1** straight line passing through the centre of a circle or sphere to its edges; length of this. **2** transverse measurement; width, thickness. **3** unit of linear magnifying power. [Greek: related to -METER]

diametrical /ˌdaɪə'metrɪk(ə)l/ *adj.* (also **diametric**) **1** of or along a diameter. **2** (of opposites etc.) absolute. □ **diametrically** *adv.* [Greek: related to DIAMETER]

diamond /'daɪəmənd/ *n.* **1** very hard transparent precious stone of pure crystallized carbon. **2** rhombus. **3 a** playing-card of the suit denoted by a red rhombus. **b** (in *pl.*) this suit. [Greek: related to ADAMANT]

diamond wedding *n.* 60th (or 75th) wedding anniversary.

dianthus /daɪ'ænθəs/ *n.* flowering plant of the genus including the carnation. [Greek, = flower of Zeus]

diapason /ˌdaɪə'peɪz(ə)n/ *n.* **1** compass of a voice or musical instrument. **2** fixed standard of musical pitch. **3** either of two main organ-stops. [Greek, = through all (notes)]

diaper /'daɪəpə(r)/ *n.* *US* baby's nappy. [Greek *aspros* white]

diaphanous /daɪ'æfənəs/ *adj.* (of fabric etc.) light, delicate, and almost transparent. [Greek *phainō* show]

diaphragm /'daɪəˌfræm/ *n.* **1** muscular partition between the thorax and abdomen in mammals. **2** = DUTCH CAP. **3 a** *Photog.* plate or disc pierced with a circular hole to cut off marginal beams of light. **b** vibrating disc in a microphone, telephone, loudspeaker, etc. **4** device for varying the lens aperture in a camera etc. **5** thin sheet as a partition etc. [Greek *phragma* fence]

diapositive /ˌdaɪə'pɒzɪtɪv/ *n.* positive photographic slide or transparency.

diarist /'daɪərɪst/ *n.* person who keeps a diary.

diarrhoea /ˌdaɪə'rɪə/ *n.* (esp. *US* **diarrhea**) condition of excessively frequent and loose bowel movements. [Greek *rheō* flow]

diary /'daɪərɪ/ *n.* (*pl.* **-ies**) **1** daily record of events or thoughts. **2** book for this or for noting future engagements. [Latin *dies* day]
Diaspora /daɪ'æspərə/ *n.* **1** the dispersion of the Jews after their exile in 538 BC. **2** the dispersed Jews. [Greek]
diastase /'daɪə,steɪz/ *n.* enzyme converting starch to sugar. [Greek *diastasis* separation]
diatom /'daɪətəm/ *n.* one-cell alga found as plankton and forming fossil deposits. [Greek, = cut in half]
diatomic /,daɪə'tɒmɪk/ *adj.* consisting of two atoms.
diatonic /,daɪə'tɒnɪk/ *adj. Mus.* (of a scale, interval, etc.) involving only notes belonging to the prevailing key. [Greek: related to TONIC]
diatribe /'daɪə,traɪb/ *n.* forceful verbal attack or criticism; invective. [Greek *tribō* rub]
diazepam /daɪ'æzɪ,pæm/ *n.* a tranquillizing drug. [benzo*diazepine* + *am*]
dibble /'dɪb(ə)l/ *–n.* (also **dibber** /'dɪbə(r)/) hand tool for making holes for planting. *–v.* (**-ling**) sow, plant, or prepare (soil) with a dibble. [origin uncertain]
dice *–n.pl.* **1 a** small cubes with faces bearing 1–6 spots, used in games or gambling. **b** (treated as *sing.*) one of these cubes (see DIE[2]). **2** game played with dice. *–v.* (**-cing**) **1** take great risks, gamble (*dicing with death*). **2** cut into small cubes. [pl. of DIE[2]]

■ Usage See note at DIE[2].

dicey *adj.* (**dicier, diciest**) *slang* risky, unreliable.
dichotomy /daɪ'kɒtəmɪ/ *n.* (*pl.* **-ies**) division into two, esp. a sharply defined one. [Greek *dikho-* apart: related to TOME]

■ Usage The use of *dichotomy* to mean *dilemma* or *ambivalence* is considered incorrect in standard English.

dichromatic /,daɪkrəʊ'mætɪk/ *adj.* **1** two-coloured. **2** having vision sensitive to only two of the three primary colours.
dick[1] *n.* **1** *colloq.* (in certain set phrases) person (*clever dick*). **2** *coarse slang* penis. [*Dick*, pet form of *Richard*]
dick[2] *n. slang* detective. [perhaps an abbreviation]
dickens /'dɪkɪnz/ *n.* (usu. prec. by *how, what, why*, etc., *the*) *colloq.* (esp. in exclamations) deuce; the Devil (*what the dickens is it*). [probably the name *Dickens*]
Dickensian /dɪ'kenzɪən/ *adj.* **1** of the 19th-c. novelist Dickens or his work. **2** resembling situations in Dickens's work, esp. poverty.
dickhead *n. coarse slang* idiot. [from DICK[1]]
dicky /'dɪkɪ/ *–n.* (*pl.* **-ies**) *colloq.* false shirt-front. *–adj.* (**-ier, -iest**) *slang* unsound; unhealthy. [*Dicky*, pet form of *Richard*]
dicky-bird *n.* **1** child's word for a little bird. **2** word (*didn't say a dicky-bird*).
dicky bow *n. colloq.* bow-tie.
dicotyledon /,daɪkɒtɪ'li:d(ə)n/ *n.* flowering plant having two cotyledons. □ **dicotyledonous** *adj.*
dicta *pl.* of DICTUM.
Dictaphone /'dɪktə,fəʊn/ *n. propr.* machine for recording and playing back dictated words. [from DICTATE, PHONE]
dictate *–v.* /dɪk'teɪt/ (**-ting**) **1** say or read aloud (material to be written down or recorded). **2** state or order authoritatively or peremptorily. *–n.* /'dɪk-/ (usu. in *pl.*) authoritative instruction or requirement (*dictates of conscience, fashion*). □ **dictation** /dɪk'teɪʃ(ə)n/ *n.* [Latin *dicto* from *dico* say]
dictator /dɪk'teɪtə(r)/ *n.* **1** usu. unelected omnipotent ruler. **2** omnipotent person in any sphere. **3** domineering person. □ **dictatorship** *n.* [Latin: related to DICTATE]
dictatorial /,dɪktə'tɔ:rɪəl/ *adj.* **1** of or like a dictator. **2** overbearing. □ **dictatorially** *adv.* [Latin: related to DICTATOR]
diction /'dɪkʃ(ə)n/ *n.* manner of enunciation in speaking or singing. [Latin *dictio* from *dico dict-* say]
dictionary /'dɪkʃənərɪ/ *n.* (*pl.* **-ies**) **1** book listing (usu. alphabetically) and explaining the words of a language or giving corresponding words in another language. **2** reference book explaining the terms of a particular subject. [medieval Latin: related to DICTION]
dictum /'dɪktəm/ *n.* (*pl.* **dicta** or **-s**) **1** formal expression of opinion. **2** a saying. [Latin, neuter past part. of *dico* say]
did *past* of DO[1].
didactic /daɪ'dæktɪk/ *adj.* **1** meant to instruct. **2** (of a person) tediously pedantic. □ **didactically** *adv.* **didacticism** /-tɪ,sɪz(ə)m/ *n.* [Greek *didaskō* teach]
diddle /'dɪd(ə)l/ *v.* (**-ling**) *colloq.* swindle. [probably from *Diddler*, name of a character in a 19th-c. play]
diddums /'dɪdəmz/ *int.* often *iron.* expressing commiseration. [= *did 'em*, i.e. did they (tease you etc.)?]
didgeridoo /,dɪdʒərɪ'du:/ *n.* long tubular Australian Aboriginal musical instrument. [imitative]
didn't /'dɪd(ə)nt/ *contr.* did not.

die[1] /daɪ/ *v.* (**dies, died, dying** /ˈdaɪɪŋ/) **1** cease to live; expire, lose vital force. **2 a** come to an end, fade away (*his interest died*). **b** cease to function. **c** (of a flame) go out. **3** (foll. by *on*) die or cease to function while in the presence or charge of (a person). **4** (usu. foll. by *of, from, with*) be exhausted or tormented (*nearly died of boredom*). □ **be dying** (foll. by *for*, or *to* + infin.) wish for longingly or intently (*was dying for a drink*). **die away** fade to the point of extinction. **die back** (of a plant) decay from the tip towards the root. **die down** become fainter or weaker. **die hard** die reluctantly (*old habits die hard*). **die off** die one after another. **die out** become extinct, cease to exist. [Old Norse]

die[2] /daɪ/ *n.* **1** = DICE 1b. **2** (*pl.* **dies**) **a** engraved device for stamping coins, medals, etc. **b** device for stamping, cutting, or moulding material. □ **the die is cast** an irrevocable step has been taken. [Latin *datum* from *do* give]

■ **Usage** *Dice*, rather than *die*, is now the standard singular as well as plural form in the games sense (*one dice, two dice*).

die-casting *n.* process or product of casting from metal moulds.

die-hard *n.* conservative or stubborn person.

dielectric /ˌdaɪɪˈlektrɪk/ *–adj.* not conducting electricity. *–n.* dielectric substance.

dieresis *US* var. of DIAERESIS.

diesel /ˈdiːz(ə)l/ *n.* **1** (in full **diesel engine**) internal-combustion engine in which heat produced by the compression of air in the cylinder ignites the fuel. **2** vehicle driven by a diesel engine. **3** fuel for a diesel engine. [*Diesel*, name of an engineer]

diesel-electric *adj.* (of a locomotive etc.) driven by an electric current from a diesel-engined generator.

diesel oil *n.* heavy petroleum fraction used in diesel engines.

die-sinker *n.* engraver of dies.

die-stamping *n.* embossing paper etc. with die.

diet[1] /ˈdaɪət/ *–n.* **1** range of foods habitually eaten by a person or animal. **2** limited range of food to which a person is restricted. **3** thing regularly offered (*diet of half-truths*). *–v.* (**-t-**) restrict oneself to a special diet, esp. to slim. □ **dietary** *adj.* **dieter** *n.* [Greek *diaita* way of life]

diet[2] /ˈdaɪət/ *n.* **1** legislative assembly in certain countries. **2** *hist.* congress. [Latin *dieta*]

dietetic /ˌdaɪəˈtetɪk/ *adj.* of diet and nutrition. [Greek: related to DIET[1]]

dietetics *n.pl.* (usu. treated as *sing.*) the study of diet and nutrition.

dietitian /ˌdaɪəˈtɪʃ(ə)n/ *n.* (also **dietician**) expert in dietetics.

dif- *prefix* = DIS-.

differ *v.* **1** (often foll. by *from*) be unlike or distinguishable. **2** (often foll. by *with*) disagree. [Latin *differo, dilat-* bring apart]

difference /ˈdɪfrəns/ *n.* **1** being different or unlike. **2** degree of this. **3** way in which things differ. **4 a** quantity by which amounts differ. **b** remainder left after subtraction. **5** disagreement, dispute. □ **make a** (or **all the, no,** etc.) **difference** have a significant (or a very significant, or no) effect. **with a difference** having a new or unusual feature.

different *adj.* **1** (often foll. by *from* or *to*) unlike, of another nature. **2** distinct, separate. **3** unusual. □ **differently** *adv.*

■ **Usage** In sense 1, *different from* is more widely acceptable than *different to*, which is common in less formal use.

differential /ˌdɪfəˈrenʃ(ə)l/ *–adj.* **1** of, exhibiting, or depending on a difference. **2** *Math.* relating to infinitesimal differences. **3** constituting or relating to a specific difference. *–n.* **1** difference between things of the same kind. **2** difference in wages between industries or categories of employees in the same industry. **3** difference between rates of interest etc.

differential calculus *n.* method of calculating rates of change, maximum or minimum values, etc.

differential gear *n.* gear enabling a vehicle's rear wheels to revolve at different speeds on corners.

differentiate /ˌdɪfəˈrenʃɪˌeɪt/ *v.* (**-ting**) **1** constitute a difference between or in. **2** recognize as different; distinguish. **3** become different during development. **4** *Math.* calculate the derivative of. □ **differentiation** /-ˈeɪʃ(ə)n/ *n.*

difficult /ˈdɪfɪkəlt/ *adj.* **1 a** needing much effort or skill. **b** troublesome, perplexing. **2** (of a person) demanding. **3** problematic.

difficulty *n.* (*pl.* **-ies**) **1** being difficult. **2 a** difficult thing; problem, hindrance. **b** (often in *pl.*) distress, esp. financial (*in difficulties*). [Latin *difficultas*: related to FACULTY]

diffident /ˈdɪfɪd(ə)nt/ *adj.* shy, lacking self-confidence; excessively reticent. □ **diffidence** *n.* **diffidently** *adv.* [Latin *diffido* distrust]

diffract /dɪ'frækt/ *v.* break up (a beam of light) into a series of dark and light bands or coloured spectra, or (a beam of radiation or particles) into a series of high and low intensities. □ **diffraction** *n.* **diffractive** *adj.* [Latin *diffringo*: related to FRACTION]

diffuse –*adj.* /dɪ'fju:s/ **1** spread out, not concentrated. **2** not concise, wordy. –*v.* /dɪ'fju:z/ (**-sing**) **1** disperse or spread widely. **2** intermingle by diffusion. □ **diffusible** /dɪ'fju:zɪb(ə)l/ *adj.* **diffusive** /dɪ'fju:sɪv/ *adj.* [Latin: related to FOUND[3]]

diffusion /dɪ'fju:ʒ(ə)n/ *n.* **1** diffusing or being diffused. **2** interpenetration of substances by natural movement of their particles. [Latin: related to DIFFUSE]

dig –*v.* (**-gg-**; *past* and *past part.* **dug**) **1** (also *absol.*) break up and remove or turn over (ground etc.). **2** (foll. by *up*) break up the soil of (fallow land). **3** make (a hole, tunnel, etc.) by digging. **4** (often foll. by *up*, *out*) **a** obtain by digging. **b** (foll. by *up*, *out*) find or discover. **c** (foll. by *into*) search for information in (a book etc.). **5** (also *absol.*) excavate (an archaeological site). **6** *slang* like; understand. **7** (foll. by *in*, *into*) thrust (a sharp object); prod or nudge. **8** (foll. by *into*, *through*, *under*) make one's way by digging. –*n.* **1** piece of digging. **2** thrust or poke. **3** *colloq.* pointed remark. **4** archaeological excavation. **5** (in *pl.*) *colloq.* lodgings. □ **dig one's heels in** be obstinate. **dig in** *colloq.* begin eating. **dig oneself in 1** prepare a defensive trench or pit. **2** establish one's position. [Old English]

digest –*v.* /daɪ'dʒest/ **1** assimilate (food) in the stomach and bowels. **2** understand and assimilate mentally. **3** summarize. –*n.* /'daɪdʒest/ **1** periodical synopsis of current literature or news. **2** methodical summary, esp. of laws. □ **digestible** *adj.* [Latin *digero -gest-*]

digestion /daɪ'dʒestʃ(ə)n/ *n.* **1** process of digesting. **2** capacity to digest food.

digestive –*adj.* of or aiding digestion. –*n.* **1** substance aiding digestion. **2** (in full **digestive biscuit**) wholemeal biscuit.

digger *n.* **1** person or machine that digs, esp. a mechanical excavator. **2** *colloq.* Australian or New Zealander.

digit /'dɪdʒɪt/ *n.* **1** any numeral from 0 to 9. **2** finger or toe. [Latin, = finger, toe]

digital *adj.* **1** of digits. **2** (of a clock, watch, etc.) giving a reading by displayed digits. **3** (of a computer) operating on data represented by a series of digits. **4** (of a recording) with sound-information represented by digits for more reliable transmission. □ **digitally** *adv.* [Latin: related to DIGIT]

digital audio tape *n.* magnetic tape on which sound is recorded digitally.

digitalis /ˌdɪdʒɪ'teɪlɪs/ *n.* drug prepared from foxgloves, used to stimulate the heart. [related to DIGIT, from the form of the flowers]

digitize *v.* (also **-ise**) (**-zing** or **-sing**) convert (data etc.) into digital form, esp. for a computer. □ **digitization** /-'zeɪʃ(ə)n/ *n.*

dignified /'dɪgnɪˌfaɪd/ *adj.* having or showing dignity.

dignify /'dɪgnɪˌfaɪ/ *v.* (**-ies, -ied**) **1** confer dignity on; ennoble. **2** give a fine name to. [Latin *dignus* worthy]

dignitary /'dɪgnɪtərɪ/ *n.* (*pl.* **-ies**) person of high rank or office. [from DIGNITY]

dignity /'dɪgnɪtɪ/ *n.* (*pl.* **-ies**) **1** composed and serious manner. **2** worthiness, nobleness (*dignity of work*). **3** high rank or position. □ **beneath one's dignity** not worthy enough for one. **stand on one's dignity** insist on being treated with respect. [Latin *dignus* worthy]

digraph /'daɪgrɑ:f/ *n.* two letters representing one sound, e.g. *ph*, *ey* as in *ph*one, *key*. [from DI-[1], -GRAPH]

■ **Usage** *Digraph* is sometimes confused with *ligature*, which means two or more letters joined together.

digress /daɪ'gres/ *v.* depart from the main subject in speech or writing. □ **digression** *n.* [Latin *digredior -gress-*]

digs see DIG *n.* 5.

dike[1] var. of DYKE[1].

dike[2] var. of DYKE[2].

diktat /'dɪktæt/ *n.* categorical statement or decree. [German, = DICTATE]

dilapidated /dɪ'læpɪˌdeɪtɪd/ *adj.* in disrepair or ruin. □ **dilapidation** /-'deɪʃ(ə)n/ *n.* [Latin: related to DI-[2], *lapis* stone]

dilatation /ˌdaɪlə'teɪʃ(ə)n/ *n.* **1** dilating of the cervix, e.g. for surgical curettage. **2** dilation. [from DILATE]

dilate /daɪ'leɪt/ *v.* (**-ting**) **1** make or become wider or larger. **2** speak or write at length. □ **dilation** *n.* [Latin *latus* wide]

dilatory /'dɪlətərɪ/ *adj.* given to or causing delay. [Latin *dilatorius*: related to DIFFER]

dildo /'dɪldəʊ/ *n.* (*pl.* **-s**) artificial erect penis for sexual stimulation. [origin unknown]

dilemma /daɪ'lemə/ *n.* **1** situation in which a difficult choice has to be made. **2** difficult situation, predicament. [Greek *lēmma* premiss]

■ **Usage** The use of *dilemma* in sense 2 is considered incorrect by some people.

dilettante /ˌdɪlɪ'tæntɪ/ *n.* (*pl.* **dilettanti** /-tɪ/ or **-s**) dabbler in a subject. □ **dilettantism** *n.* [Italian *dilettare* DELIGHT]

diligent /'dɪlɪdʒ(ə)nt/ *adj.* **1** hard-working. **2** showing care and effort. □ **diligence** *n.* **diligently** *adv.* [French from Latin *diligo* love]

dill *n.* herb with aromatic leaves and seeds. [Old English]

dilly-dally /ˌdɪlɪ'dælɪ/ *v.* (**-ies, -ied**) *colloq.* **1** dawdle. **2** vacillate. [reduplication of DALLY]

dilute /daɪ'lju:t/ —*v.* (**-ting**) **1** reduce the strength of (a fluid) by adding water etc. **2** weaken or reduce in effect. —*adj.* diluted. □ **dilution** *n.* [Latin *diluo -lut-* wash away]

diluvial /daɪ'lu:vɪəl/ *adj.* of a flood, esp. of the Flood in Genesis. [Latin: related to DELUGE]

dim —*adj.* (**dimmer, dimmest**) **1 a** faintly luminous or visible; not bright. **b** indistinct. **2** not clearly perceived or remembered. **3** *colloq.* stupid. **4** (of the eyes) not seeing clearly. —*v.* (**-mm-**) make or become dim. □ **take a dim view of** *colloq.* disapprove of. □ **dimly** *adv.* **dimness** *n.* [Old English]

dime *n. US* ten-cent coin. [Latin *decima* tenth (part)]

dimension /daɪ'menʃ(ə)n/ —*n.* **1** measurable extent, as length, breadth, depth, etc. **2** (in *pl.*) size (*of huge dimensions*). **3** aspect, facet (*gained a new dimension*). —*v.* (usu. as **dimensioned** *adj.*) mark dimensions on (a diagram etc.). □ **dimensional** *adj.* [Latin *metior mens-* measure]

diminish /dɪ'mɪnɪʃ/ *v.* **1** make or become smaller or less. **2** (often as **diminished** *adj.*) lessen the reputation of (a person); humiliate. □ **law of diminishing returns** fact that expenditure etc. beyond a certain point ceases to produce a proportionate yield. [Latin: related to MINUTE[1]]

diminuendo /dɪˌmɪnjʊ'endəʊ/ *Mus.* —*n.* (*pl* **-s**) gradual decrease in loudness. —*adv.* & *adj.* decreasing in loudness. [Italian: related to DIMINISH]

diminution /ˌdɪmɪ'nju:ʃ(ə)n/ *n.* **1** diminishing or being diminished. **2** decrease. [Latin: related to DIMINISH]

diminutive /dɪ'mɪnjʊtɪv/ —*adj.* **1** tiny. **2** (of a word or suffix) implying smallness or affection. —*n.* diminutive word or suffix.

dimmer *n.* **1** (in full **dimmer switch**) device for varying the brightness of an electric light. **2** *US* **a** (in *pl.*) small parking lights on a vehicle. **b** headlight on low beam.

dimple /'dɪmp(ə)l/ —*n.* small hollow, esp. in the cheek or chin. —*v.* (**-ling**) form dimples (in). □ **dimply** *adj.* [probably Old English]

dim-wit *n. colloq.* stupid person. □ **dim-witted** *adj.*

DIN *n.* any of a series of German technical standards designating electrical connections, film speeds, and paper sizes. [German, from *D*eutsche *I*ndustrie-*N*orm]

din —*n.* prolonged loud confused noise. —*v.* (**-nn-**) (foll. by *into*) force (information) into a person by constant repetition; make a din. [Old English]

dinar /'di:nɑ:(r)/ *n.* chief monetary unit of Yugoslavia and several countries of the Middle East and N. Africa. [Arabic and Persian from Latin DENARIUS]

dine *v.* (**-ning**) **1 a** eat dinner. **b** (foll. by *on, upon*) eat for dinner. **2** (esp. in phr. **wine and dine**) entertain with food. □ **dine out** dine away from home. [French *diner* as DIS-, Latin *jejunus* fasting]

diner *n.* **1** person who dines. **2** dining-car. **3** *US* small restaurant. **4** small dining-room.

dinette /daɪ'net/ *n.* small room or alcove for eating meals.

ding —*v.* make a ringing sound. —*n.* ringing sound. [imitative]

dingbat /'dɪŋbæt/ *n. slang US* & *Austral.* stupid or eccentric person. [perhaps from *ding* to beat + BAT[1]]

ding-dong /'dɪŋdɒŋ/ *n.* **1** sound of two chimes, esp. as a doorbell. **2** *colloq.* heated argument or fight. [imitative]

dinghy /'dɪŋɪ/ *n.* (*pl.* **-ies**) **1** small boat carried by a ship. **2** small pleasure-boat. **3** small inflatable rubber boat. [Hindi]

dingle /'dɪŋg(ə)l/ *n.* deep wooded valley or dell. [origin unknown]

dingo /'dɪŋgəʊ/ *n.* (*pl.* **-es**) wild Australian dog. [Aboriginal]

dingy /'dɪndʒɪ/ *adj.* (**-ier, -iest**) dirty-looking, drab. □ **dingily** *adv.* **dinginess** *n.* [origin uncertain]

dining-car *n.* restaurant on a train.

dining-room *n.* room in which meals are eaten.

dinkum /'dɪŋkəm/ *adj. Austral.* & *NZ colloq.* genuine, honest, true. □ **dinkum oil** the honest truth. **fair dinkum 1** fair play. **2** genuine(ly), honest(ly), true, truly. [origin unknown]

dinky /'dɪŋkɪ/ *adj.* (**-ier, -iest**) *colloq.* neat and attractive; small, dainty. [Scots *dink*]

dinner *n.* **1** main meal of the day, either at midday or in the evening. **2** (in full **dinner-party**) formal evening meal,

esp. with guests. [French: related to DINE]
dinner-dance *n.* formal dinner followed by dancing.
dinner-jacket *n.* man's short usu. black formal jacket for evening wear.
dinner lady *n.* woman who supervises school dinners.
dinner service *n.* set of matching crockery for dinner.
dinosaur /'daɪnə,sɔː(r)/ *n.* **1** extinct, often enormous, reptile of the Mesozoic era. **2** unwieldy or unchanging system or organization. [Greek *deinos* terrible, SAURIAN]
dint –*n.* dent. –*v.* mark with dints. □ **by dint of** by force or means of. [Old English and Old Norse]
diocese /'daɪəsɪs/ *n.* district under the pastoral care of a bishop. □ **diocesan** /daɪ'ɒsɪs(ə)n/ *adj.* [Greek *dioikēsis* administration]
diode /'daɪəʊd/ *n.* **1** semiconductor allowing the flow of current in one direction only and having two terminals. **2** thermionic valve having two electrodes. [from DI-[1], ELECTRODE]
Dionysian /,daɪə'nɪzɪən/ *adj.* wildly sensual; unrestrained. [Greek *Dionusos* god of wine]
dioptre /daɪ'ɒptə(r)/ *n.* (*US* **diopter**) unit of refractive power of a lens. [Greek: related to DIA-, *opsis* sight]
diorama /,daɪə'rɑːmə/ *n.* **1** scenic painting lit to simulate sunrise etc. **2** small scene with three-dimensional figures, viewed through a window etc. **3** small-scale model or film-set. [from DIA-, Greek *horaō* see]
dioxide /daɪ'ɒksaɪd/ *n.* oxide containing two atoms of oxygen (*carbon dioxide*).
dip –*v.* (**-pp-**) **1** put or lower briefly into liquid etc.; immerse. **2 a** go below a surface or level. **b** (of income, activity, etc.) decline slightly, esp. briefly. **3** slope or extend downwards (*road dips*). **4** go under water and emerge quickly. **5** (foll. by *into*) look cursorily into (a book, subject, etc.). **6 a** (foll. by *into*) put a hand, ladle, etc., into (a container) to take something out. **b** use part of (one's resources) (*dipped into our savings*). **7** lower or be lowered, esp. in salute. **8** lower the beam of (headlights) to reduce dazzle. **9** colour (a fabric) by immersing it in dye. **10** wash (sheep) in disinfectant. –*n.* **1** dipping or being dipped. **2** liquid for dipping. **3** brief bathe in the sea etc. **4** downward slope or hollow in a road, skyline, etc. **5** sauce into which food is dipped. **6** candle made by dipping the wick in tallow. [Old English]
Dip. Ed. *abbr.* Diploma in Education.
diphtheria /dɪf'θɪərɪə, dɪp-/ *n.* acute infectious bacterial disease with inflammation of a mucous membrane esp. of the throat. [Greek *diphthera* skin, hide]

■ **Usage** The second pronunciation is considered incorrect by some people.

diphthong /'dɪfθɒŋ/ *n.* two written or spoken vowels pronounced in one syllable (as in *coin*, *loud*, *toy*). [Greek *phthoggos* voice]
diplodocus /dɪ'plɒdəkəs/ *n.* (*pl.* **-cuses**) giant plant-eating dinosaur with a long neck and tail. [Greek *diplous* double, *dokos* wooden beam]
diploma /dɪ'pləʊmə/ *n.* **1** certificate of qualification awarded by a college etc. **2** document conferring an honour or privilege. [Greek, = folded paper, from *diplous* double]
diplomacy /dɪ'pləʊməsɪ/ *n.* **1 a** management of international relations. **b** skill in this. **2** tact. [French: related to DIPLOMATIC]
diplomat /'dɪplə,mæt/ *n.* **1** member of a diplomatic service. **2** tactful person.
diplomatic /,dɪplə'mætɪk/ *adj.* **1** of or involved in diplomacy. **2** tactful. □ **diplomatically** *adv.* [French: related to DIPLOMA]
diplomatic bag *n.* container for dispatching official mail etc. to or from an embassy, usu. exempt from customs inspection.
diplomatic immunity *n.* exemption of diplomatic staff abroad from arrest, taxation, etc.
diplomatic service *n.* branch of the civil service concerned with the representation of a country abroad.
diplomatist /dɪ'pləʊmətɪst/ *n.* diplomat.
dipole /'daɪpəʊl/ *n.* **1** two equal and oppositely charged or magnetized poles separated by a distance. **2** molecule in which a concentration of positive charges is separated from a concentration of negative charges. **3** aerial consisting of a horizontal metal rod with a connecting wire at its core.
dipper *n.* **1** diving bird, esp. the water ouzel. **2** ladle.
dippy /'dɪpɪ/ *adj.* (**-ier, -iest**) *slang* crazy, silly. [origin uncertain]
dipso /'dɪpsəʊ/ *n.* (*pl.* **-s**) *colloq.* alcoholic. [abbreviation]
dipsomania /,dɪpsə'meɪnɪə/ *n.* alcoholism. □ **dipsomaniac** /-nɪ,æk/ *n.* [Greek *dipsa* thirst]
dipstick *n.* rod for measuring the depth of esp. oil in a vehicle's engine.
dip-switch *n.* switch for dipping a vehicle's headlights.

dipterous /'dɪptərəs/ *adj.* (of an insect) having two wings. [Greek *pteron* wing]

diptych /'dɪptɪk/ *n.* painting, esp. an altarpiece, on two hinged panels closing like a book. [Greek, = pair of writing-tablets, from *ptukhē* fold]

dire *adj.* **1 a** calamitous, dreadful. **b** ominous. **c** (*predic.*) *colloq.* very bad. **2** urgent (*in dire need*). [Latin]

direct /daɪ'rekt, dɪ-/ *–adj.* **1** extending or moving in a straight line or by the shortest route; not crooked or circuitous. **2** straightforward; frank. **3** with nothing or no-one in between; personal (*direct line*). **4** (of descent) lineal, not collateral. **5** complete, greatest possible (*the direct opposite*). *–adv.* **1** in a direct way or manner (*dealt with them direct*). **2** by the direct route (*sent direct to London*). *–v.* **1** control; govern or guide (*duty directs me*). **2** (foll. by *to* + infin., or *that* + clause) order (a person) to. **3** (foll. by *to*) **a** address (a letter etc.). **b** tell or show (a person) the way to (a place). **4** (foll. by *at, to, towards*) point, aim, or turn (a blow, attention, or remark). **5** (also *absol.*) supervise the performing, staging, etc., of (a film, play, etc.). □ **directness** *n.* [Latin *dirigo* from *rego rect-* guide]

direct access *n.* facility of retrieving data immediately from any part of a computer file.

direct current *n.* electric current flowing in one direction only.

direct debit *n.* regular debiting of a bank account at the request of the payee.

direct-grant school *n.* school funded by the Government and not a local authority.

direction /daɪ'rekʃ(ə)n, dɪ-/ *n.* **1** directing; supervision. **2** (usu. in *pl.*) order or instruction. **3** line along which, or point to or from which, a person or thing moves or looks. **4** tendency or scope of a theme, subject, etc.

directional *adj.* **1** of or indicating direction. **2** sending or receiving radio or sound waves in one particular direction.

directive /daɪ'rektɪv, dɪ-/ *–n.* order from an authority. *–adj.* serving to direct.

directly *–adv.* **1 a** at once; without delay, immediately (*directly after lunch*). **b** presently, shortly. **2** exactly (*directly opposite*). **3** in a direct manner. *–conj. colloq.* as soon as (*will tell you directly they come*).

direct object *n.* primary object of the action of a transitive verb.

director *n.* **1** person who directs or controls, esp. a member of the board of a company. **2** person who directs a film, play, etc. □ **directorial** /-'tɔ:rɪəl/ *adj.* **directorship** *n.*

directorate /daɪ'rektərət, dɪ-/ *n.* **1** board of directors. **2** office of director.

director-general *n.* chief executive of a large organization.

directory /daɪ'rektərɪ, dɪ-/ *n.* (*pl.* **-ies**) book with a list of telephone subscribers, inhabitants of a district, or members of a profession etc. [Latin: related to DIRECT]

directory enquiries *n.pl.* telephone service providing a subscriber's number on request.

directress *n.* woman director.

direct speech *n.* words actually spoken, not reported.

direct tax *n.* tax that one pays directly to the Government, esp. on income.

dirge *n.* **1** lament for the dead. **2** any dreary piece of music. [Latin imperative *dirige* = direct, used in the Office for the Dead]

dirham /'dɪəræm/ *n.* principal monetary unit of Morocco and the United Arab Emirates. [Arabic]

dirigible /'dɪrɪdʒɪb(ə)l/ *–adj.* capable of being guided. *–n.* dirigible balloon or airship. [related to DIRECT]

dirk *n.* short dagger. [origin unknown]

dirndl /'dɜ:nd(ə)l/ *n.* **1** dress with a close-fitting bodice and full skirt. **2** full skirt of this kind. [German]

dirt *n.* **1** unclean matter that soils. **2 a** earth, soil. **b** earth, cinders, etc., used to make the surface for a road etc. (usu. *attrib.*: *dirt track*). **3** foul or malicious words or talk. **4** excrement. □ **treat like dirt** treat with contempt. [Old Norse *drit* excrement]

dirt cheap *adj.* & *adv. colloq.* extremely cheap.

dirty *–adj.* (**-ier, -iest**) **1** soiled, unclean. **2** causing dirtiness (*dirty job*). **3** sordid, lewd, obscene. **4** unpleasant, dishonourable, unfair (*dirty trick*). **5** (of weather) rough, squally. **6** (of colour) muddied, dingy. *–adv. slang* **1** very (*a dirty great diamond*). **2** in a dirty manner (*talk dirty; act dirty*) (esp. in senses 3 and 4 of *adj.*). *–v.* (**-ies, -ied**) make or become dirty. □ **do the dirty on** *colloq.* play a mean trick on. □ **dirtily** *adv.* **dirtiness** *n.*

dirty look *n. colloq.* look of disapproval or disgust.

dirty old man *n. colloq.* lecherous man.

dirty weekend *n. colloq.* weekend spent with a lover.

dirty word *n.* **1** offensive or indecent word. **2** word for something disapproved of (*profit is a dirty word*).

dirty work *n.* dishonourable or illegal activity.

dis- *prefix* forming nouns, adjectives, and verbs implying: **1** negation or direct opposite (*dishonest*; *discourteous*). **2** reversal (*disengage*; *disorientate*). **3** removal of a thing or quality (*dismember*; *disable*). **4** separation (*distinguish*). **5** completeness or intensification (*disgruntled*). **6** expulsion from (*disbar*). [French *des-* or Latin *dis-*]

disability /ˌdɪsəˈbɪlɪtɪ/ *n.* (*pl.* **-ies**) **1** permanent physical or mental incapacity. **2** lack of some capacity etc., preventing action.

disable /dɪsˈeɪb(ə)l/ *v.* (**-ling**) **1** deprive of an ability or function. **2** (often as **disabled** *adj.*) physically incapacitate. □ **disablement** *n.*

disabuse /ˌdɪsəˈbju:z/ *v.* (**-sing**) (usu. foll. by *of*) free from a mistaken idea; disillusion.

disadvantage /ˌdɪsədˈvɑ:ntɪdʒ/ —*n.* **1** unfavourable circumstance or condition. **2** damage; loss. —*v.* (**-ging**) cause disadvantage to. □ **at a disadvantage** in an unfavourable position or aspect. □ **disadvantageous** /-ˌædvənˈteɪdʒəs/ *adj.*

disadvantaged *adj.* lacking normal opportunities through poverty, disability, etc.

disaffected /ˌdɪsəˈfektɪd/ *adj.* discontented (esp. politically); no longer loyal. □ **disaffection** *n.*

disagree /ˌdɪsəˈgri:/ *v.* (**-ees, -eed, -eeing**) (often foll. by *with*) **1** hold a different opinion. **2** (of factors) not correspond. **3** upset (*onions disagree with me*). □ **disagreement** *n.*

disagreeable *adj.* **1** unpleasant. **2** bad-tempered. □ **disagreeably** *adv.*

disallow /ˌdɪsəˈlaʊ/ *v.* refuse to allow or accept; prohibit.

disappear /ˌdɪsəˈpɪə(r)/ *v.* **1** cease to be visible. **2** cease to exist or be in circulation or use. **3** (of a person) go missing. □ **disappearance** *n.*

disappoint /ˌdɪsəˈpɔɪnt/ *v.* **1** fail to fulfil the desire or expectation of. **2** frustrate (a hope etc.). □ **disappointed** *adj.* **disappointing** *adj.*

disappointment *n.* **1** person or thing that disappoints. **2** being disappointed.

disapprobation /dɪsˌæprəˈbeɪʃ(ə)n/ *n. formal* disapproval.

disapprove /ˌdɪsəˈpru:v/ *v.* (**-ving**) (usu. foll. by *of*) have or express an unfavourable opinion. □ **disapproval** *n.*

disarm /dɪsˈɑ:m/ *v.* **1** take weapons etc. away from. **2** reduce or give up one's own weapons. **3** defuse (a bomb etc.). **4** make less angry, hostile, etc; charm, win over. □ **disarming** *adj.* (esp. in sense 4). **disarmingly** *adv.*

disarmament /dɪsˈɑ:məmənt/ *n.* reduction by a State of its armaments.

disarrange /ˌdɪsəˈreɪndʒ/ *v.* (**-ging**) bring into disorder. □ **disarrangement** *n.*

disarray /ˌdɪsəˈreɪ/ —*n.* disorder. —*v.* throw into disorder.

disassociate /ˌdɪsəˈsəʊʃɪˌeɪt/ *v.* (**-ting**) = DISSOCIATE. □ **disassociation** /-ˈeɪʃ(ə)n/ *n.*

disaster /dɪˈzɑ:stə(r)/ *n.* **1** great or sudden misfortune; catastrophe. **2** *colloq.* complete failure. □ **disastrous** *adj.* **disastrously** *adv.* [Latin *astrum* star]

disavow /ˌdɪsəˈvaʊ/ *v.* disclaim knowledge of or responsibility for. □ **disavowal** *n.*

disband /dɪsˈbænd/ *v.* break up; disperse. □ **disbandment** *n.*

disbar /dɪsˈbɑ:(r)/ *v.* (**-rr-**) deprive (a barrister) of the right to practise. □ **disbarment** *n.*

disbelieve /ˌdɪsbɪˈli:v/ *v.* (**-ving**) be unable or unwilling to believe; be sceptical. □ **disbelief** *n.* **disbelievingly** *adv.*

disburse /dɪsˈbɜ:s/ *v.* (**-sing**) pay out (money). □ **disbursal** *n.* **disbursement** *n.* [French: related to DIS-, BOURSE]

disc *n.* (also **disk** esp. *US* and in sense 4a) **1 a** flat thin circular object. **b** round flat or apparently flat surface or mark. **2** layer of cartilage between vertebrae. **3** gramophone record. **4 a** (usu. **disk**; in full **magnetic disk**) flat circular computer storage device. **b** (in full **optical disc**) disc for data recorded and read by laser. [Latin DISCUS]

discard /dɪsˈkɑ:d/ *v.* **1** reject as unwanted. **2** remove or put aside. [from DIS-, CARD[1]]

disc brake *n.* brake employing the friction of pads against a disc.

discern /dɪˈsɜ:n/ *v.* **1** perceive clearly with the mind or senses. **2** make out with effort. □ **discernible** *adj.* [Latin *cerno cret-* separate]

discerning *adj.* having good judgement. □ **discerningly** *adv.* **discernment** *n.*

discharge —*v.* /dɪsˈtʃɑ:dʒ/ (**-ging**) **1** release (a prisoner); allow (a patient, jury) to leave. **2** dismiss from office or employment. **3** fire (a gun etc.). **4** throw; eject. **5** emit, pour out (pus etc.). **6** (foll. by *into*) (of a river etc.) flow into (esp. the sea). **7 a** carry out (a duty or obligation). **b** relieve oneself of (a debt etc.). **c** relieve (a bankrupt) of residual liability. **8** *Law* cancel (an order of court). **9** release an electrical charge from. **10 a** relieve (a ship etc.) of cargo. **b** unload (cargo). **–***n.* /ˈdɪstʃɑːdʒ/ **1** discharging or being discharged. **2** certificate of release, dismissal, etc. **3** matter discharged; pus etc. **4** release of an electric charge, esp. with a spark.

disciple /dɪˈsaɪp(ə)l/ *n.* follower of a leader, teacher, etc., esp. of Christ. [Latin *disco* learn]

disciplinarian /ˌdɪsɪplɪˈneərɪən/ *n.* enforcer of or believer in firm discipline.

disciplinary /ˈdɪsɪplɪnərɪ/ *adj.* of or enforcing discipline.

discipline /ˈdɪsɪplɪn/ **–***n.* **1 a** control or order exercised over people or animals, e.g. over members of an organization. **b** system of rules for this. **2** training or way of life aimed at self-control and conformity. **3** branch of learning. **4** punishment. **–***v.* **(-ning) 1** punish. **2** control by training in obedience. [Latin *disciplina* from *disco* learn]

disc jockey *n.* presenter of recorded pop music.

disclaim /dɪsˈkleɪm/ *v.* **1** deny or disown. **2** renounce legal claim to.

disclaimer *n.* renunciation; statement disclaiming something.

disclose /dɪsˈkləʊz/ *v.* **(-sing)** make known; expose. □ **disclosure** *n.*

disco /ˈdɪskəʊ/ *colloq.* **–***n.* (*pl.* **-s**) = DISCOTHÈQUE. **–***v.* **(-es, -ed)** dance to disco music. [abbreviation]

discolour /dɪsˈkʌlə(r)/ *v.* (*US* **discolor**) cause to change from its normal colour; stain; tarnish. □ **discoloration** /-ˈreɪʃ(ə)n/ *n.*

discomfit /dɪsˈkʌmfɪt/ *v.* **(-t-)** disconcert, baffle, frustrate. □ **discomfiture** *n.* [French: related to DIS-, CONFECTION]

■ **Usage** *Discomfit* is sometimes confused with *discomfort.*

discomfort /dɪsˈkʌmfət/ **–***n.* **1** lack of comfort; slight pain or unease. **2** cause of this. **–***v.* make uncomfortable.

■ **Usage** As a verb, *discomfort* is sometimes confused with *discomfit.*

discompose /ˌdɪskəmˈpəʊz/ *v.* **(-sing)** disturb the composure of. □ **discomposure** *n.*

disco music *n.* popular dance music with a heavy bass rhythm.

disconcert /ˌdɪskənˈsɜːt/ *v.* disturb the composure of; fluster.

disconnect /ˌdɪskəˈnekt/ *v.* **1** break the connection of. **2** put out of action by disconnecting the parts. □ **disconnection** *n.*

disconnected *adj.* incoherent and illogical.

disconsolate /dɪsˈkɒnsələt/ *adj.* forlorn, unhappy, disappointed. □ **disconsolately** *adv.* [Latin: related to DIS-, SOLACE]

discontent /ˌdɪskənˈtent/ **–***n.* lack of contentment; dissatisfaction, grievance. **–***v.* (esp. as **discontented** *adj.*) make dissatisfied. □ **discontentment** *n.*

discontinue /ˌdɪskənˈtɪnjuː/ *v.* **(-ues, -ued, -uing) 1** come or bring to an end (*a discontinued line*). **2** give up, cease from (doing something). □ **discontinuance** *n.*

discontinuous *adj.* lacking continuity; intermittent. □ **discontinuity** /-ˌkɒntɪˈnjuːɪtɪ/ *n.*

discord /ˈdɪskɔːd/ *n.* **1** disagreement; strife. **2** harsh noise; clashing sounds. **3** lack of harmony in a chord. [Latin: related to DIS-, *cor cord-* heart]

discordant /dɪˈskɔːd(ə)nt/ *adj.* **1** disagreeing. **2** not in harmony; dissonant. □ **discordance** *n.* **discordantly** *adv.*

discothèque /ˈdɪskəˌtek/ *n.* **1** nightclub etc. for dancing to pop records. **2** professional lighting and sound equipment used for this. **3** party with such equipment. [French, = record-library]

discount **–***n.* /ˈdɪskaʊnt/ amount deducted from a full or normal price, esp. for prompt or advance payment. **–***v.* /dɪsˈkaʊnt/ **1** disregard as unreliable or unimportant. **2** deduct an amount from (a price etc.). **3** give or get the present worth of (an investment certificate which has yet to mature). □ **at a discount** below the usual price or true value.

discountenance /dɪˈskaʊntɪnəns/ *v.* **(-cing) 1** disconcert. **2** refuse to approve of.

discourage /dɪˈskʌrɪdʒ/ *v.* **(-ging) 1** deprive of courage or confidence. **2** dissuade, deter. **3** show disapproval of. □ **discouragement** *n.*

discourse **–***n.* /ˈdɪskɔːs/ **1** conversation. **2** lengthy treatment of a subject. **3** lecture, speech. **–***v.* /dɪˈskɔːs/ **(-sing) 1** converse. **2** speak or write at length on a subject. [Latin *curro curs-* run]

discourteous /dɪsˈkɜːtɪəs/ *adj.* lacking courtesy. □ **discourteously** *adv.* **discourtesy** *n.* (*pl.* **-ies**).

discover /dɪˈskʌvə(r)/ *v.* **1 a** find out or become aware of, by intention or chance. **b** be first to find or find out (*who discovered America?*). **2** find and promote as a new performer. □ **discoverer** *n.* [Latin *discooperio*: related to DIS-, COVER]

discovery *n.* (*pl.* **-ies**) **1** discovering or being discovered. **2** person or thing discovered.

discredit /dɪs'kredɪt/ *–n.* **1** harm to reputation. **2** person or thing causing this. **3** lack of credibility. *–v.* **(-t-) 1** harm the good reputation of. **2** cause to be disbelieved. **3** refuse to believe.

discreditable *adj.* bringing discredit; shameful. □ **discreditably** *adv.*

discreet /dɪ'skri:t/ *adj.* **(-er, -est) 1 a** circumspect. **b** tactful; judicious, prudent. **2** unobtrusive. □ **discreetly** *adv.* **discreetness** *n.* [Latin: related to DISCERN]

discrepancy /dɪs'krepənsɪ/ *n.* (*pl.* **-ies**) difference; inconsistency. □ **discrepant** *adj.* [Latin *discrepo* be discordant]

discrete /dɪ'skri:t/ *adj.* individually distinct; separate, discontinuous. □ **discreteness** *n.* [Latin: related to DISCERN]

discretion /dɪ'skreʃ(ə)n/ *n.* **1** being discreet. **2** prudence; good judgement. **3** freedom or authority to act according to one's judgement. □ **use one's discretion** act according to one's own judgement. □ **discretionary** *adj.* [Latin: related to DISCERN]

discriminate /dɪ'skrɪmɪˌneɪt/ *v.* **(-ting) 1** (often foll. by *between*) make or see a distinction. **2** (usu. foll. by *against* or *in favour of*) treat unfavourably or favourably, esp. on the basis of race, gender, etc. □ **discriminatory** /-nətərɪ/ *adj.* [Latin *discrimino*: related to DISCERN]

discriminating *adj.* showing good judgement or taste.

discrimination /dɪˌskrɪmɪ'neɪʃ(ə)n/ *n.* **1** unfavourable treatment based on racial, sexual, etc. prejudice. **2** good taste or judgement.

discursive /dɪ'skɜ:sɪv/ *adj.* tending to digress, rambling. [Latin *curro curs-* run]

discus /'dɪskəs/ *n.* (*pl.* **-cuses**) heavy thick-centred disc thrown in athletic events. [Latin from Greek]

discuss /dɪ'skʌs/ *v.* **1** talk about (*discussed their holidays*). **2** talk or write about (a subject) in detail. □ **discussion** *n.* [Latin *discutio -cuss-* disperse]

disdain /dɪs'deɪn/ *–n.* scorn, contempt. *–v.* **1** regard with disdain. **2** refrain or refuse out of disdain. □ **disdainful** *adj.* **disdainfully** *adv.* [Latin: related to DE-, DEIGN]

disease /dɪ'zi:z/ *n.* **1** unhealthy condition of the body or mind, plants, society, etc. **2** particular kind of disease. □ **diseased** *adj.* [French: related to DIS-, EASE]

disembark /ˌdɪsɪm'bɑ:k/ *v.* put or go ashore; get off an aircraft, bus, etc. □ **disembarkation** /-'keɪʃ(ə)n/ *n.*

disembarrass /ˌdɪsɪm'bærəs/ *v.* **1** (usu. foll. by *of*) relieve (of a load etc.). **2** free from embarrassment. □ **disembarrassment** *n.*

disembodied /ˌdɪsɪm'bɒdɪd/ *adj.* **1** (of the soul etc.) freed from the body or concrete form. **2** lacking a body. □ **disembodiment** *n.*

disembowel /ˌdɪsɪm'baʊəl/ *v.* **(-ll-;** *US* **-l-)** remove the bowels or entrails of. □ **disembowelment** *n.*

disenchant /ˌdɪsɪn'tʃɑ:nt/ *v.* disillusion. □ **disenchantment** *n.*

disencumber /ˌdɪsɪn'kʌmbə(r)/ *v.* free from encumbrance.

disenfranchise /ˌdɪsɪn'fræntʃaɪz/ *v.* (also **disfranchise**) **(-sing) 1** deprive of the right to vote or to be represented. **2** deprive of rights as a citizen or of a franchise held. □ **disenfranchisement** *n.*

disengage /ˌdɪsɪn'geɪdʒ/ *v.* **(-ging) 1** detach, loosen, release. **2** remove (troops) from battle etc. **3** become detached. **4** (as **disengaged** *adj.*) **a** at leisure. **b** uncommitted. □ **disengagement** *n.*

disentangle /ˌdɪsɪn'tæŋg(ə)l/ *v.* **(-ling)** free or become free of tangles or complications. □ **disentanglement** *n.*

disestablish /ˌdɪsɪ'stæblɪʃ/ *v.* **1** deprive (a Church) of State support. **2** terminate the establishment of. □ **disestablishment** *n.*

disfavour /dɪs'feɪvə(r)/ (*US* **disfavor**) *–n.* **1** disapproval or dislike. **2** being disliked. *–v.* regard or treat with disfavour.

disfigure /dɪs'fɪgə(r)/ *v.* **(-ring)** spoil the appearance of. □ **disfigurement** *n.*

disfranchise var. of DISENFRANCHISE.

disgorge /dɪs'gɔ:dʒ/ *v.* **(-ging) 1** eject from the throat. **2** pour forth. □ **disgorgement** *n.*

disgrace /dɪs'greɪs/ *–n.* **1** shame; ignominy. **2** shameful or very bad person or thing (*bus service is a disgrace*). *–v.* **(-cing) 1** bring shame or discredit on. **2** dismiss from a position of honour or favour. □ **in disgrace** out of favour. [Latin: related to DIS-, GRACE]

disgraceful *adj.* shameful; causing disgrace. □ **disgracefully** *adv.*

disgruntled /dɪs'grʌnt(ə)ld/ *adj.* discontented; sulky. □ **disgruntlement** *n.* [from DIS-, GRUNT]

disguise /dɪs'gaɪz/ *–v.* **(-sing) 1** conceal the identity of; make unrecognizable. **2** conceal (*disguised my anger*). *–n.* **1 a** costume, make-up, etc., used to disguise. **b** action, manner, etc., used to deceive. **2** disguised state. **3** practice of disguising. [French: related to DIS-]

disgust /dɪs'gʌst/ *–n.* strong aversion; repugnance. *–v.* cause disgust in. □ **disgusting** *adj.* **disgustingly** *adv.* [French or Italian: related to DIS-, GUSTO]

dish –*n.* **1 a** shallow flat-bottomed container for food. **b** its contents. **c** particular kind of food or food prepared to a particular recipe (*meat dish*). **2** (in *pl.*) crockery, pans, etc. after a meal (*wash the dishes*). **3 a** dish-shaped object or cavity. **b** = SATELLITE DISH. **4** *colloq.* sexually attractive person. –*v.* **1** *colloq.* outmanoeuvre, frustrate. **2** make dish-shaped. □ **dish out** *colloq.* distribute. **dish up 1** put (food) in dishes for serving. **2** *colloq.* present as a fact or argument. [Old English from Latin DISCUS]

dishabille var. of DÉSHABILLÉ.

disharmony /dɪsˈhɑːmənɪ/ *n.* lack of harmony; discord. □ **disharmonious** /-ˈməʊnɪəs/ *adj.*

dishcloth *n.* cloth for washing dishes.

dishearten /dɪsˈhɑːt(ə)n/ *v.* cause to lose courage, hope, or confidence. □ **disheartenment** *n.*

dishevelled /dɪˈʃev(ə)ld/ *adj.* (*US* **disheveled**) untidy; ruffled. □ **dishevelment** *n.* [from DIS-, *chevel* 'hair', from Latin *capillus*]

dishonest /dɪsˈɒnɪst/ *adj.* fraudulent or insincere. □ **dishonestly** *adv.* **dishonesty** *n.*

dishonour /dɪsˈɒnə(r)/ (*US* **dishonor**) –*n.* **1** loss of honour or respect; disgrace. **2** thing causing this. –*v.* **1** disgrace (*dishonoured his name*). **2** refuse to accept or pay (a cheque etc.).

dishonourable *adj.* (*US* **dishonorable**) **1** causing disgrace; ignominious. **2** unprincipled. □ **dishonourably** *adv.*

dishwasher *n.* machine or person that washes dishes.

dishy *adj.* (**-ier, -iest**) *colloq.* sexually attractive.

disillusion /ˌdɪsɪˈluːʒ(ə)n/ –*v.* free from an illusion or mistaken belief. –*n.* disillusioned state. □ **disillusionment** *n.*

disincentive /ˌdɪsɪnˈsentɪv/ *n.* thing discouraging action, effort, etc.

disincline /ˌdɪsɪnˈklaɪn/ *v.* (**-ning**) make unwilling or reluctant. □ **disinclination** /-klɪˈneɪʃ(ə)n/ *n.*

disinfect /ˌdɪsɪnˈfekt/ *v.* cleanse of infection, esp. with disinfectant. □ **disinfection** *n.*

disinfectant –*n.* substance that destroys germs etc. –*adj.* disinfecting.

disinformation /ˌdɪsɪnfəˈmeɪʃ(ə)n/ *n.* false information, propaganda.

disingenuous /ˌdɪsɪnˈdʒenjʊəs/ *adj.* insincere, not candid. □ **disingenuously** *adv.*

disinherit /ˌdɪsɪnˈherɪt/ *v.* (**-t-**) reject as one's heir; deprive of the right of inheritance. □ **disinheritance** *n.*

disintegrate /dɪsˈɪntɪˌgreɪt/ *v.* (**-ting**) **1** separate into component parts or fragments, break up. **2** *colloq.* break down, esp. mentally. **3** (of an atomic nucleus) emit particles or divide into smaller nuclei. □ **disintegration** /-ˈgreɪʃ(ə)n/ *n.*

disinter /ˌdɪsɪnˈtɜː(r)/ *v.* (**-rr-**) dig up (esp. a corpse). □ **disinterment** *n.*

disinterested /dɪsˈɪntrɪstɪd/ *adj.* **1** impartial. **2** uninterested. □ **disinterest** *n.* **disinterestedly** *adv.*

■ **Usage** Use of *disinterested* in sense 2 is common in informal use, but is widely regarded as incorrect. The use of the noun *disinterest* to mean 'lack of interest' is also objected to but it is rarely used in any other sense and the alternative *uninterest* is rare.

disinvest /ˌdɪsɪnˈvest/ *v.* reduce or dispose of one's investment. □ **disinvestment** *n.*

disjoint /dɪsˈdʒɔɪnt/ *v.* **1** take apart at the joints. **2** (as **disjointed** *adj.*) incoherent; disconnected. **3** disturb the working of; disrupt.

disjunction /dɪsˈdʒʌŋkʃ(ə)n/ *n.* separation.

disjunctive /dɪsˈdʒʌŋktɪv/ *adj.* **1** involving separation. **2** (of a conjunction) expressing an alternative, e.g. *or* in *is it wet or dry?*

disk var. of DISC (esp. *US* & *Computing*).

disk drive *n. Computing* mechanism for rotating a disk and reading or writing data from or to it.

diskette /dɪˈsket/ *n. Computing* = FLOPPY *n.*

dislike /dɪsˈlaɪk/ –*v.* (**-king**) have an aversion to; not like. –*n.* **1** feeling of repugnance or not liking. **2** object of this.

dislocate /ˈdɪsləˌkeɪt/ *v.* (**-ting**) **1** disturb the normal connection of (esp. a joint in the body). **2** disrupt. □ **dislocation** /-ˈkeɪʃ(ə)n/ *n.*

dislodge /dɪsˈlɒdʒ/ *v.* (**-ging**) disturb or move. □ **dislodgement** *n.*

disloyal /dɪsˈlɔɪəl/ *adj.* not loyal; unfaithful. □ **disloyally** *adv.* **disloyalty** *n.*

dismal /ˈdɪzm(ə)l/ *adj.* **1** gloomy; miserable. **2** dreary; sombre. **3** *colloq.* feeble, inept (*dismal attempt*). □ **dismally** *adv.* [medieval Latin *dies mali* unlucky days]

dismantle /dɪsˈmænt(ə)l/ *v.* (**-ling**) **1** take to pieces; pull down. **2** deprive of defences or equipment.

dismay /dɪsˈmeɪ/ –*n.* intense disappointment or despair. –*v.* fill with dismay. [French from Germanic: related to DIS-, MAY]

dismember /dɪsˈmembə(r)/ *v.* **1** remove the limbs from. **2** partition or divide up. □ **dismemberment** *n.*

dismiss /dɪs'mɪs/ *v.* **1** send away, esp. from one's presence; disperse. **2** terminate the employment of, esp. dishonourably; sack. **3** put from one's mind or emotions. **4** consider not worth talking or thinking about; treat summarily. **5** *Law* refuse further hearing to (a case). **6** *Cricket* put (a batsman or side) out (usu. for a stated score). □ **dismissal** *n.* [Latin *mitto miss-* send]

dismissive *adj.* dismissing rudely or casually; disdainful. □ **dismissively** *adv.* **dismissiveness** *n.*

dismount /dɪs'maʊnt/ *v.* **1 a** alight from a horse, bicycle, etc. **b** unseat. **2** remove (a thing) from its mounting.

disobedient /,dɪsə'bi:dɪənt/ *adj.* disobeying; rebellious. □ **disobedience** *n.* **disobediently** *adv.*

disobey /,dɪsə'beɪ/ *v.* refuse or fail to obey.

disoblige /,dɪsə'blaɪdʒ/ *v.* **(-ging)** refuse to help or cooperate with (a person).

disorder /dɪs'ɔ:də(r)/ *n.* **1** lack of order; confusion. **2** public disturbance; riot. **3** ailment or disease. □ **disordered** *adj.*

disorderly *adj.* **1** untidy; confused. **2** riotous, unruly. □ **disorderliness** *n.*

disorganize /dɪs'ɔ:gə,naɪz/ *v.* (also **-ise**) **(-zing** or **-sing) 1** throw into confusion or disorder. **2** (as **disorganized** *adj.*) badly organized; untidy. □ **disorganization** /-'zeɪʃ(ə)n/ *n.*

disorient /dɪs'ɔ:rɪənt/ *v.* = DISORIENTATE.

disorientate /dɪs'ɔ:rɪən,teɪt/ *v.* (also **disorient**) **(-ting)** confuse (a person), esp. as to his or her bearings. □ **disorientation** /-'teɪʃ(ə)n/ *n.*

disown /dɪs'əʊn/ *v.* deny or give up any connection with; repudiate.

disparage /dɪ'spærɪdʒ/ *v.* **(-ging) 1** criticize; belittle. **2** bring discredit on. □ **disparagement** *n.* [French: related to DIS-, *parage* rank]

disparate /'dɪspərət/ *adj.* essentially different; not comparable. □ **disparateness** *n.* [Latin *disparo* separate]

disparity /dɪ'spærɪtɪ/ *n.* (*pl.* **-ies**) inequality; difference; incongruity.

dispassionate /dɪ'spæʃ(ə)nət/ *adj.* free from emotion; impartial. □ **dispassionately** *adv.*

dispatch /dɪ'spætʃ/ (also **despatch**) *–v.* **1** send off to a destination or for a purpose. **2** perform (a task etc.) promptly; finish off. **3** kill, execute. **4** *colloq.* eat quickly. *–n.* **1** dispatching or being dispatched. **2 a** official written message, esp. military. **b** news report to a newspaper etc. **3** promptness, efficiency. [Italian *dispacciare* or Spanish *despachar*]

dispatch-box *n.* case for esp. parliamentary documents.

dispatch-rider *n.* messenger on a motor cycle.

dispel /dɪ'spel/ *v.* **(-ll-)** drive away; scatter (fears etc.). [Latin *pello* drive]

dispensable /dɪ'spensəb(ə)l/ *adj.* that can be dispensed with.

dispensary /dɪ'spensərɪ/ *n.* (*pl.* **-ies**) place where medicines are dispensed.

dispensation /,dɪspen'seɪʃ(ə)n/ *n.* **1** dispensing or distributing. **2** exemption from penalty, rule, etc. **3** ordering or management of the world by Providence.

dispense /dɪ'spens/ *v.* **(-sing) 1** distribute; deal out. **2** administer. **3** make up and give out (medicine etc.). **4** (foll. by *with*) do without; make unnecessary. [French from Latin *pendo pens-* weigh]

dispenser *n.* **1** person or thing that dispenses e.g. medicine, good advice. **2** automatic machine dispensing a specific amount.

disperse /dɪ'spɜ:s/ *v.* **(-sing) 1** go, send, drive, or scatter widely or in different directions. **2** send to or station at different points. **3** disseminate. **4** *Chem.* distribute (small particles) in a medium. **5** divide (white light) into its coloured constituents. □ **dispersal** *n.* **dispersive** *adj.* [Latin: related to DIS-, SPARSE]

dispersion /dɪ'spɜ:ʃ(ə)n/ *n.* **1** dispersing or being dispersed. **2 (the Dispersion)** = DIASPORA.

dispirit /dɪ'spɪrɪt/ *v.* (esp. as **dispiriting, dispirited** *adjs.*) make despondent, deject.

displace /dɪs'pleɪs/ *v.* **(-cing) 1** move from its place. **2** remove from office. **3** take the place of; oust.

displaced person *n.* refugee in war etc., or from persecution.

displacement *n.* **1** displacing or being displaced. **2** amount of fluid displaced by an object floating or immersed in it.

display /dɪ'spleɪ/ *–v.* **1** exhibit; show. **2** reveal; betray. *–n.* **1** displaying. **2 a** exhibition or show. **b** thing(s) displayed. **3** ostentation. **4** mating rituals of some birds etc. **5** what is shown on a visual display unit etc. [Latin *plico* fold]

displease /dɪs'pli:z/ *v.* **(-sing)** make upset or angry; annoy. □ **displeasure** /-'pleʒ(ə)r/ *n.*

disport /dɪ'spɔ:t/ *v.* (often *refl.*) play, frolic, enjoy oneself. [Anglo-French *porter* carry, from Latin]

disposable /dɪ'spəʊzəb(ə)l/ *–adj.* **1** intended to be used once and discarded. **2** able to be disposed of. *–n.* disposable article.

disposable income *n.* income after tax and other fixed payments.

disposal /dɪ'spəʊz(ə)l/ *n.* disposing of, e.g. waste. □ **at one's disposal** available.

■ **Usage** *Disposal* is the noun corresponding to the verb *dispose of* (get rid of, deal with, etc.). *Disposition* is the noun from *dispose* (arrange, incline).

dispose /dɪ'spəʊz/ *v.* **(-sing) 1** (usu. foll. by *to*, or *to* + infin.) **a** make willing; incline (*was disposed to agree*). **b** tend (*wheel was disposed to buckle*). **2** arrange suitably. **3** (as **disposed** *adj.*) have a specified inclination (*ill-disposed*; *well-disposed*). **4** determine events (*man proposes, God disposes*). □ **dispose of 1 a** deal with. **b** get rid of. **c** finish. **d** kill. **2** sell. **3** prove (an argument etc.) incorrect. [French: related to POSE]

disposition /ˌdɪspə'zɪʃ(ə)n/ *n.* **1** natural tendency; temperament. **2 a** ordering; arrangement (of parts etc.). **b** arrangement. **3** (usu. in *pl.*) preparations; plans.

■ **Usage** See note at *disposal*.

dispossess /ˌdɪspə'zes/ *v.* **1** (usu. foll. by *of*) (esp. as **dispossessed** *adj.*) deprive (a person) of. **2** dislodge; oust. □ **dispossession** /-'zeʃ(ə)n/ *n.*

disproof /dɪs'pruːf/ *n.* refutation.

disproportion /ˌdɪsprə'pɔːʃ(ə)n/ *n.* lack of proportion; being out of proportion. □ **disproportional** *adj.* **disproportionally** *adv.*

disproportionate *adj.* **1** out of proportion. **2** relatively too large or small etc. □ **disproportionately** *adv.*

disprove /dɪs'pruːv/ *v.* **(-ving)** prove false.

disputable /dɪ'spjuːtəb(ə)l/ *adj.* open to question; uncertain. □ **disputably** *adv.*

disputant /dɪ'spjuːt(ə)nt/ *n.* person in a dispute.

disputation /ˌdɪspjuː'teɪʃ(ə)n/ *n.* **1** debate, esp. formal. **2** argument; controversy.

disputatious *adj.* argumentative.

dispute /dɪ'spjuːt/ —*v.* **(-ting) 1** debate, argue. **2** discuss, esp. heatedly; quarrel. **3** question the truth or validity of (a statement etc.). **4** contend for (*disputed territory*). **5** resist, oppose. —*n.* **1** controversy; debate. **2** quarrel. **3** disagreement leading to industrial action. □ **in dispute 1** being argued about. **2** (of a workforce) involved in industrial action. [Latin *puto* reckon]

disqualify /dɪs'kwɒlɪˌfaɪ/ *v.* **(-ies, -ied) 1** debar from a competition or pronounce ineligible as a winner. **2** make or pronounce ineligible, unsuitable, or unqualified (*disqualified from driving*). □ **disqualification** /-fɪ'keɪʃ(ə)n/ *n.*

disquiet /dɪs'kwaɪət/ —*v.* make anxious. —*n.* anxiety; uneasiness.

disquietude *n.* disquiet.

disquisition /ˌdɪskwɪ'zɪʃ(ə)n/ *n.* discursive treatise or discourse. [Latin *quaero quaesit-* seek]

disregard /ˌdɪsrɪ'gɑːd/ —*v.* **1** ignore. **2** treat as unimportant. —*n.* indifference; neglect.

disrepair /ˌdɪsrɪ'peə(r)/ *n.* poor condition due to lack of repairs.

disreputable /dɪs'repjʊtəb(ə)l/ *adj.* **1** of bad reputation. **2** not respectable in character or appearance. □ **disreputably** *adv.*

disrepute /ˌdɪsrɪ'pjuːt/ *n.* lack of good reputation; discredit.

disrespect /ˌdɪsrɪ'spekt/ *n.* lack of respect; discourtesy. □ **disrespectful** *adj.* **disrespectfully** *adv.*

disrobe /dɪs'rəʊb/ *v.* **(-bing)** *literary* undress.

disrupt /dɪs'rʌpt/ *v.* **1** interrupt the continuity of; bring disorder to. **2** break apart. □ **disruption** *n.* **disruptive** *adj.* **disruptively** *adv.* [Latin: related to RUPTURE]

dissatisfy /dɪ'sætɪsˌfaɪ/ *v.* **(-ies, -ied)** make discontented; fail to satisfy. □ **dissatisfaction** /-'fækʃ(ə)n/ *n.*

dissect /dɪ'sekt/ *v.* **1** cut into pieces, esp. for examination or post mortem. **2** analyse or criticize in detail. □ **dissection** *n.* [Latin: related to SECTION]

dissemble /dɪ'semb(ə)l/ *v.* **(-ling) 1** be hypocritical or insincere. **2** disguise or conceal (a feeling, intention, etc.). [Latin *simulo* SIMULATE]

disseminate /dɪ'semɪˌneɪt/ *v.* **(-ting)** scatter about, spread (esp. ideas) widely. □ **dissemination** /-'neɪʃ(ə)n/ *n.* [Latin: related to DIS-, SEMEN]

dissension /dɪ'senʃ(ə)n/ *n.* angry disagreement. [Latin: related to DISSENT]

dissent /dɪ'sent/ —*v.* (often foll. by *from*) **1** disagree, esp. openly. **2** differ, esp. from the established or official opinion. —*n.* **1** such difference. **2** expression of this. [Latin: related to DIS-, *sentio* feel]

dissenter *n.* **1** person who dissents. **2** (**Dissenter**) Protestant dissenting from the Church of England.

dissentient /dɪ'senʃ(ə)nt/ —*adj.* disagreeing with the established or official view. —*n.* person who dissents.

dissertation /ˌdɪsə'teɪʃ(ə)n/ *n.* detailed discourse, esp. one submitted towards an academic degree. [Latin *disserto* discuss]

disservice /dɪs'sɜːvɪs/ *n.* harmful action, harm.

dissident /'dɪsɪd(ə)nt/ —*adj.* disagreeing, esp. with the established government, system, etc. —*n.* dissident person.

□ **dissidence** *n.* [Latin: related to DIS-, *sedeo* sit]

dissimilar /dɪ'sɪmɪlə(r)/ *adj.* unlike, not similar. □ **dissimilarity** /-'lærɪtɪ/ *n.* (*pl.* -ies).

dissimulate /dɪ'sɪmjʊ,leɪt/ *v.* (**-ting**) dissemble. □ **dissimulation** /-'leɪʃ(ə)n/ *n.* [Latin: related to DISSEMBLE]

dissipate /'dɪsɪ,peɪt/ *v.* (**-ting**) **1** disperse, disappear, dispel. **2** squander. **3** (as **dissipated** *adj.*) dissolute. [Latin *dissipare -pat-*]

dissipation /,dɪsɪ'peɪʃ(ə)n/ *n.* **1** dissolute way of life. **2** dissipating or being dissipated.

dissociate /dɪ'səʊʃɪ,eɪt/ *v.* (**-ting**) **1** disconnect or separate. **2** become disconnected. □ **dissociate oneself from** declare oneself unconnected with. □ **dissociation** /-'eɪʃ(ə)n/ *n.* **dissociative** /-ətɪv/ *adj.* [Latin: related to DIS-, ASSOCIATE]

dissoluble /dɪ'sɒljʊb(ə)l/ *adj.* that can be disintegrated, loosened, or disconnected.

dissolute /'dɪsə,lu:t/ *adj.* lax in morals; licentious. [Latin: related to DISSOLVE]

dissolution /,dɪsə'lu:ʃ(ə)n/ *n.* **1** dissolving or being dissolved, esp. of a partnership or of parliament for a new election. **2** breaking up, abolition (of an institution). **3** death.

dissolve /dɪ'zɒlv/ *v.* (**-ving**) **1** make or become liquid, esp. by immersion or dispersion in a liquid. **2** (cause to) disappear gradually. **3** dismiss (an assembly, esp. Parliament). **4** annul or put an end to (a partnership, marriage, etc.). **5** (often foll. by *into*) be overcome (by tears, laughter, etc.). [Latin: related to DIS-, *solvo solut-* loosen]

dissonant /'dɪsənənt/ *adj.* **1** harsh-toned; unharmonious. **2** incongruous. □ **dissonance** *n.* [Latin: related to DIS-, *sono* SOUND[1]]

dissuade /dɪ'sweɪd/ *v.* (**-ding**) (often foll. by *from*) discourage (a person); persuade against. □ **dissuasion** /-'sweɪʒ(ə)n/ *n.* **dissuasive** *adj.* [Latin: related to DIS-, *suadeo* advise]

dissyllable var. of DISYLLABLE.

distaff /'dɪstɑ:f/ *n.* cleft stick holding wool or flax for spinning by hand. [Old English]

distaff side *n.* female branch of a family.

distance /'dɪst(ə)ns/ —*n.* **1** being far off; remoteness. **2** space between two points. **3** distant point or place. **4** aloofness; reserve. **5** remoter field of vision (*in the distance*). **6** interval of time. —*v.* (**-cing**) (often *refl.*) **1** place or cause to seem far off; be aloof. **2** leave far behind in a race etc. □ **at a distance** far off. **keep one's distance** remain aloof. [Latin: related to DIS-, *sto* stand]

distant *adj.* **1** far away; at a specified distance (*three miles distant*). **2** remote in time, relationship, etc. (*distant prospect*; *distant relation*). **3** aloof. **4** abstracted (*distant stare*). **5** faint (*distant memory*). □ **distantly** *adv.*

distaste /dɪs'teɪst/ *n.* (usu. foll. by *for*) dislike; aversion. □ **distasteful** *adj.* **distastefully** *adv.* **distastefulness** *n.*

distemper[1] /dɪ'stempə(r)/ *hist* —*n.* paint using glue or size as a base, for use on walls. —*v.* paint with this. [Latin, = soak: see DISTEMPER[2]]

distemper[2] /dɪ'stempə(r)/ *n.* disease of esp. dogs, with coughing and weakness. [Latin: related to DIS-, *tempero* mingle]

distend /dɪ'stend/ *v.* swell out by pressure from within (*distended stomach*). □ **distensible** /-'stensɪb(ə)l/ *adj.* **distension** /-'stenʃ(ə)n/ *n.* [Latin: related to TEND[1]]

distich /'dɪstɪk/ *n.* verse couplet. [Greek *stikhos* line]

distil /dɪ'stɪl/ *v.* (*US* **distill**) (**-ll-**) **1** purify or extract the essence from (a substance) by vaporizing and condensing it and collecting the resulting liquid. **2** extract the essential meaning of (an idea etc.). **3** make (whisky, essence, etc.) by distilling raw materials. **4** fall or cause to fall in drops. □ **distillation** /-'leɪʃ(ə)n/ *n.* [Latin: related to DE-, *stillo* drip]

distiller *n.* person who distils, esp. alcoholic liquor.

distillery *n.* (*pl.* -ies) place where alcoholic liquor is distilled.

distinct /dɪ'stɪŋkt/ *adj.* **1** (often foll. by *from*) not identical; separate; different. **2** clearly perceptible. **3** unmistakable, decided (*distinct advantage*). □ **distinctly** *adv.* [Latin: related to DISTINGUISH]

distinction /dɪ'stɪŋkʃ(ə)n/ *n.* **1** discriminating or distinguishing. **2** difference between two things. **3** thing that differentiates or distinguishes. **4** special consideration or honour (*treat with distinction*). **5** excellence (*person of distinction*). **6** title or mark of honour. [Latin: related to DISTINGUISH]

distinctive *adj.* distinguishing, characteristic. □ **distinctively** *adv.* **distinctiveness** *n.*

distingué /dɪ'stæŋgeɪ/ *adj.* distinguished in appearance, manner, etc. [French]

distinguish /dɪ'stɪŋgwɪʃ/ *v.* **1** (often foll. by *from, between*) differentiate; see or draw distinctions. **2** be a mark or property of; characterize. **3** discover by listening, looking, etc. **4** (usu. *refl.*; often foll. by *by*) make prominent (*distinguished himself by winning*). □ **distinguishable** *adj.* [Latin: related to DIS-, *stinguo stinct-* extinguish]

distinguished *adj.* **1** eminent; famous. **2** dignified.

distort /dɪ'stɔ:t/ *v.* **1** pull or twist out of shape. **2** misrepresent (facts etc.). **3** transmit (sound etc.) inaccurately. □ **distortion** *n.* [Latin *torqueo tort-* twist]

distract /dɪ'strækt/ *v.* **1** (often foll. by *from*) draw away the attention of. **2** bewilder, perplex. **3** (as **distracted** *adj.*) confused, mad, or angry. **4** amuse, esp. to divert from pain etc. [Latin: related to DIS-, *traho tract-* draw]

distraction /dɪ'strækʃ(ə)n/ *n.* **1 a** distracting or being distracted. **b** thing that distracts. **2** relaxation; amusement. **3** confusion; frenzy, madness.

distrain /dɪ'streɪn/ *v.* (usu. foll. by *upon*) impose distraint (on a person, goods, etc.). [Latin: related to DIS-, *stringo strict-* draw tight]

distraint /dɪ'streɪnt/ *n.* seizure of goods to enforce payment.

distrait /dɪ'streɪ/ *adj.* (*fem.* ***distraite*** /-'streɪt/) inattentive; distraught. [French: related to DISTRACT]

distraught /dɪ'strɔ:t/ *adj.* distracted with worry, fear, etc.; extremely agitated. [related to DISTRAIT]

distress /dɪ'stres/ *—n.* **1** anguish or suffering caused by pain, sorrow, worry, etc. **2** poverty. **3** *Law* = DISTRAINT. *—v.* cause distress to, make unhappy. □ **in distress** suffering or in danger. □ **distressful** *adj.* [Romanic: related to DISTRAIN]

distressed *adj.* **1** suffering from distress. **2** impoverished. **3** (of furniture, clothing, etc.) aged, torn, etc. artificially.

distressed area *n.* region of high unemployment and poverty.

distribute /dɪ'strɪbju:t, 'dɪs-/ *v.* (**-ting**) **1** give shares of; deal out. **2** scatter; put at different points. **3** arrange; classify. [Latin *tribuo -but-* assign]

■ **Usage** The second pronunciation given, with the stress on the first syllable, is considered incorrect by some people.

distribution /ˌdɪstrɪ'bju:ʃ(ə)n/ *n.* **1** distributing or being distributed. **2 a** commercial dispersal of goods etc. **b** extent to which different classes etc. share in a nation's total wealth etc.

distributive /dɪ'strɪbjʊtɪv/ *—adj.* **1** of or produced by distribution. **2** *Logic* & *Gram.* referring to each individual of a class, not to the class collectively (e.g. *each*, *either*). *—n. Gram.* distributive word.

distributor *n.* **1** person or thing that distributes, esp. goods. **2** device in an internal-combustion engine for passing current to each spark-plug in turn.

district /'dɪstrɪkt/ *n.* **1** (often *attrib.*) area regarded as a geographical or administrative unit (*the Peak District*; *postal district*; *wine-growing district*). **2** administrative division of a county etc. [Latin: related to DISTRAIN]

district attorney *n.* (in the US) prosecuting officer of a district.

district nurse *n.* nurse who makes home visits in an area.

distrust /dɪs'trʌst/ *—n.* lack of trust; suspicion. *—v.* have no trust in. □ **distrustful** *adj.* **distrustfully** *adv.*

disturb /dɪ'stɜ:b/ *v.* **1** break the rest, calm, or quiet of. **2** agitate; worry. **3** move from a settled position (*disturbed my papers*). **4** (as **disturbed** *adj.*) emotionally or mentally unstable. [Latin: related to DIS-, *turba* tumult]

disturbance *n.* **1** disturbing or being disturbed. **2** tumult; uproar; agitation.

disunion /dɪs'ju:nɪən/ *n.* lack of union; separation; dissension.

disunite /ˌdɪsju:'naɪt/ *v.* (**-ting**) **1** remove the unity from. **2** separate. □ **disunity** /-'ju:nɪtɪ/ *n.*

disuse *—n.* /dɪs'ju:s/ disused state. *—v.* /-'ju:z/ (**-sing**) cease to use.

disyllable /daɪ'sɪləb(ə)l/ *n.* (also **dissyllable**) *Prosody* word or metrical foot of two syllables. □ **disyllabic** /-'læbɪk/ *adj.*

ditch *—n.* long narrow excavation esp. for drainage or as a boundary. *—v.* **1** make or repair ditches (*hedging and ditching*). **2** *slang* abandon; discard. □ **dull as ditch-water** extremely dull. [Old English]

dither /'dɪðə(r)/ *—v.* **1** hesitate; be indecisive. **2** tremble; quiver. *—n. colloq.* state of agitation or hesitation. □ **ditherer** *n.* **dithery** *adj.* [var. of *didder* DODDER[1]]

dithyramb /'dɪθɪˌræm/ *n.* **1** wild choral hymn in ancient Greece. **2** passionate or inflated poem etc. □ **dithyrambic** /-'ræmbɪk/ *adj.* [Latin from Greek]

ditto /'dɪtəʊ/ *n.* (*pl.* **-s**) **1** (in accounts, inventories, etc.) the aforesaid, the same. **2** *colloq.* (used to avoid repetition) the same (*came late today and ditto yesterday*). [Latin DICTUM]

■ **Usage** In sense 1, the word *ditto* is often replaced by 〃 under the word or sum to be repeated.

ditto marks *n.pl.* inverted commas etc. representing 'ditto'.

ditty /'dɪtɪ/ *n.* (*pl.* **-ies**) short simple song. [Latin: related to DICTATE]

diuretic /ˌdaɪjʊ'retɪk/ *–adj.* causing increased output of urine. *–n.* diuretic drug. [Greek: related to DIA-, *oureō* urinate]

diurnal /daɪ'ɜ:n(ə)l/ *adj.* **1** of the day or daytime. **2** daily. **3** occupying one day. □ **diurnally** *adv.* [Latin *diurnalis* from *dies* day]

diva /'di:və/ *n.* (*pl.* **-s**) great woman opera singer; prima donna. [Italian from Latin, = goddess]

divalent /daɪ'veɪlənt/ *adj. Chem.* having a valency of two.

divan /dɪ'væn/ *n.* low couch or bed without a back or ends. [ultimately from Persian *dīvān* bench]

dive *–v.* (**-ving**) **1** plunge head first into water. **2 a** (of an aircraft, person, etc.) plunge steeply downwards. **b** (of a submarine) submerge; go deeper. **3** (foll. by *into*) *colloq.* **a** put one's hand into (a pocket, handbag, etc.). **b** become enthusiastic about (a subject, meal, etc.). **4** move suddenly (*dived into a shop*). *–n.* **1** act of diving; plunge. **2** steep descent or fall. **3** *colloq.* disreputable nightclub, bar, etc. [Old English]

dive-bomb *v.* bomb (a target) from a diving aircraft. □ **dive-bomber** *n.*

diver *n.* **1** person who dives, esp. working under water. **2** diving bird.

diverge /daɪ'vɜ:dʒ/ *v.* (**-ging**) **1 a** spread out from a central point, become dispersed. **b** take different courses (*their interests diverged*). **2 a** (often foll. by *from*) depart from a set course. **b** (of opinions etc.) differ. □ **divergence** *n.* **divergent** *adj.* [Latin: related to DI-[2], *vergo* incline]

divers /'daɪvɜ:z/ *adj. archaic* various; several. [Latin: related to DIVERSE]

diverse /daɪ'vɜ:s/ *adj.* varied. [Latin: related to DI-[2], *verto vers-* turn]

diversify /daɪ'vɜ:sɪˌfaɪ/ *v.* (**-ies, -ied**) **1** make diverse; vary. **2** spread (investment) over several enterprises. **3** (often foll. by *into*) expand one's range of products. □ **diversification** /-fɪ'keɪʃ(ə)n/ *n.*

diversion /daɪ'vɜ:ʃ(ə)n/ *n.* **1** diverting or being diverted. **2 a** diverting of attention. **b** stratagem for this. **3** recreation, pastime. **4** alternative route when a road is temporarily closed. □ **diversionary** *adj.*

diversity /daɪ'vɜ:sɪtɪ/ *n.* variety.

divert /daɪ'vɜ:t/ *v.* **1 a** turn aside; deflect. **b** distract (attention). **2** (often as **diverting** *adj.*) entertain; amuse. [Latin: related to DIVERSE]

divest /daɪ'vest/ *v.* (usu. foll. by *of*) **1** unclothe; strip. **2** deprive, rid. [Latin: related to VEST]

divide /dɪ'vaɪd/ *–v.* (**-ding**) **1** (often foll. by *in, into*) separate into parts; break up; split. **2** (often foll. by *out*) distribute; deal; share. **3 a** separate (one thing) from another. **b** classify into parts or groups. **4** cause to disagree. **5 a** find how many times (a number) contains or is contained in another (*divide 20 by 4; divide 4 into 20*). **b** (of a number) be contained in (a number) without remainder (*4 divides into 20*). **6** *Parl.* vote (by members entering either of two lobbies) (*the House divided*). *–n.* **1** dividing line. **2** watershed. [Latin *divido -vis-*]

dividend /'dɪvɪˌdend/ *n.* **1** share of profits paid to shareholders or to winners in a football pool etc. **2** number to be divided. **3** benefit from an action. [Anglo-French: related to DIVIDE]

divider *n.* **1** screen etc. dividing a room. **2** (in *pl.*) measuring-compasses.

divination /ˌdɪvɪ'neɪʃ(ə)n/ *n.* supposed supernatural insight into the future etc. [Latin: related to DIVINE]

divine /dɪ'vaɪn/ *–adj.* (**diviner, divinest**) **1 a** of, from, or like God or a god. **b** sacred. **2** *colloq.* excellent; delightful. *–v.* (**-ning**) **1** discover by intuition or guessing. **2** foresee. **3** practise divination. *–n.* theologian or clergyman. □ **divinely** *adv.* [Latin *divinus*]

diviner *n.* person who practises divination.

diving-bell *n.* open-bottomed enclosure, supplied with air, for descent into deep water.

diving-board *n.* elevated board for diving from.

diving-suit *n.* watertight suit, usu. with helmet and air-supply, for work under water.

divining-rod *n.* = DOWSING-ROD.

divinity /dɪ'vɪnɪtɪ/ *n.* (*pl.* **-ies**) **1** being divine. **2** god; godhead. **3** theology.

divisible /dɪ'vɪzɪb(ə)l/ *adj.* capable of being divided. □ **divisibility** /-'bɪlɪtɪ/ *n.*

division /dɪ'vɪʒ(ə)n/ *n.* **1** dividing or being divided. **2** dividing one number by another. **3** disagreement (*division of opinion*). **4** *Parl.* separation of members for counting votes. **5** one of two or more parts into which a thing is divided. **6** unit of administration, esp. a group of army brigades, regiments, or teams in a sporting league. □ **divisional** *adj.*

division sign *n.* sign (÷) indicating that one quantity is to be divided by another.

divisive /dɪ'vaɪsɪv/ *adj.* causing disagreement. □ **divisively** *adv.* **divisiveness** *n.* [Latin: related to DIVIDE]

divisor /dɪ'vaɪzə(r)/ *n.* number by which another is divided.

divorce /dɪ'vɔːs/ *–n.* **1** legal dissolution of a marriage. **2** separation (*divorce between thought and feeling*). *–v.* (**-cing**) **1 a** (usu. as **divorced** *adj.*) (often foll. by *from*) legally dissolve the marriage of. **b** separate by divorce. **c** end one's marriage with. **2** separate (*divorced from reality*). [Latin: related to DIVERSE]

divorcee /ˌdɪvɔː'siː/ *n.* divorced person.

divot /'dɪvət/ *n.* piece of turf cut out by a blow, esp. by the head of a golf club. [origin unknown]

divulge /daɪ'vʌldʒ/ *v.* (**-ging**) disclose, reveal (a secret etc.). □ **divulgence** *n.* [Latin *divulgo* publish]

divvy /'dɪvɪ/ *colloq.* *–n.* (*pl.* **-ies**) dividend. *–v.* (**-ies, -ied**) (often foll. by *up*) share out. [abbreviation]

Diwali /diː'wɑːlɪ/ *n.* Hindu and Jainist festival with illuminations, held between September and November. [Sanskrit *dīpa* lamp]

Dixie /'dɪksɪ/ *n.* southern States of the US. [origin uncertain]

dixie /'dɪksɪ/ *n.* large iron cooking-pot used by campers etc. [Hindustani from Persian]

Dixieland *n.* **1** = DIXIE. **2** traditional kind of jazz.

DIY *abbr.* do-it-yourself.

dizzy /'dɪzɪ/ *–adj.* (**-ier, -iest**) **1 a** giddy. **b** feeling confused. **2** causing giddiness. *–v.* (**-ies, -ied**) **1** make dizzy. **2** bewilder. □ **dizzily** *adv.* **dizziness** *n.* [Old English]

DJ *abbr.* **1** dinner-jacket. **2** disc jockey.

djellaba /'dʒeləbə/ *n.* (also **jellaba**) loose hooded cloak (as) worn by Arab men. [Arabic]

dl *abbr.* decilitre(s).

D-layer *n.* lowest layer of the ionosphere. [*D* arbitrary]

D.Litt. *abbr.* Doctor of Letters. [Latin *Doctor Litterarum*]

DM *abbr.* Deutschmark.

dm *abbr.* decimetre(s).

D.Mus. *abbr.* Doctor of Music.

DNA *abbr.* deoxyribonucleic acid, esp. carrying genetic information in chromosomes.

D-notice *n.* government notice to news editors not to publish certain items for security reasons. [*d*efence, NOTICE]

do[1] /duː/ *–v.* (*3 sing. pres.* **does** /dʌz/; *past* **did**; *past part.* **done** /dʌn/; *pres. part.* **doing**) **1** perform, carry out, achieve, complete (work etc.) (*did his homework*; *a lot to do*). **2** produce, make, provide (*doing a painting*; *we do lunches*). **3** grant; impart (*does you good*; *do me a favour*). **4** act, behave, proceed (*do as I do*; *would do well to wait*). **5** work at (*do carpentry*; *do chemistry*). **6** be suitable or acceptable; satisfy (*will never do*; *will do me nicely*). **7** deal with; attend to (*do one's hair*). **8** fare; get on (*did badly in the test*). **9** solve; work out (*did the sum*). **10 a** traverse (a certain distance) (*did 50 miles today*). **b** travel at a specified speed (*was doing eighty*). **11** *colloq.* act or behave like; play the part of. **12** produce (a play, opera, etc.) (*will do Shakespeare*). **13 a** *colloq.* finish (*I've done in the garden*). **b** (as **done** *adj.*) be finished (*day is done*). **14** cook, esp. completely (*do it in the oven*; *potatoes aren't done*). **15** be in progress (*what's doing?*). **16** *colloq.* visit (*did the museums*). **17** *colloq.* **a** (often as **done** *adj.*) exhaust; tire out. **b** defeat, kill, ruin. **18** (foll. by *into*) translate or transform. **19** *colloq.* cater for (*they do one very well here*). **20** *slang* **a** rob (*did a big bank*). **b** swindle. **21** *slang* prosecute, convict (*done for shoplifting*). **22** *slang* undergo (a term of imprisonment). **23** *slang* take (an illegal drug). *–v.aux.* **1** in questions and negative statements or commands (*do you understand?*; *I don't smoke*; *don't be silly*). **2** *ellipt.* or in place of a verb (*you know her better than I do*; *I wanted to go and I did*; *tell me, do!*). **3** for emphasis (*I do want to*; *do tell me*; *they did go*). **4** in inversion (*rarely does it happen*). *–n.* (*pl.* **dos** or **do's**) *colloq.* elaborate party, operation, etc. □ **be done with** see DONE. **be nothing to do with 1** be no business of. **2** be unconnected with. **be to do with** be concerned or connected with. **do away with** *colloq.* **1** get rid of; abolish. **2** kill. **do down** *colloq.* **1** cheat, swindle. **2** overcome. **do for 1** be satisfactory or sufficient for. **2** *colloq.* (esp. as **done for** *adj.*) destroy, ruin, kill. **3** *colloq.* act as cleaner etc. for. **do in 1** *slang* **a** kill. **b** ruin. **2** *colloq.* exhaust, tire out. **do justice to** see JUSTICE. **do nothing for** (or **to**) *colloq.* not flatter or enhance. **do or die** persist recklessly. **do out** *colloq.* clean or redecorate (a room). **do a person out of** *colloq.* cheat of. **do over 1** *slang* attack; beat up. **2** *colloq.* redecorate, refurbish. **do proud** see PROUD. **dos and don'ts** rules of behaviour. **do something for** (or **to**) *colloq.* enhance the appearance or quality of. **do up 1** fasten. **2** *colloq.* **a** refurbish, renovate. **b** adorn, dress up. **do with** (prec. by *could*) would be glad of; would profit by (*could do with a rest*). **do without** manage without; forgo. **to do with** in connection with, related to (*what has that to do

with it?; *something to do with the weather*). □ **doable** *adj*. [Old English]
do[2] var. of DOH.
do. *abbr*. ditto.
Dobermann pinscher /ˌdəʊbəmən ˈpɪnʃə(r)/ *n*. large dog of a smooth-coated German breed with a docked tail. [German]
doc *n*. *colloq*. doctor. [abbreviation]
docile /ˈdəʊsaɪl/ *adj*. submissive, easily managed. □ **docilely** *adv*. **docility** /-ˈsɪlɪtɪ/ *n*. [Latin *doceo* teach]
dock[1] *–n*. **1** enclosed harbour for the loading, unloading, and repair of ships. **2** (in *pl*.) docks with wharves and offices. *–v*. **1** bring or come into dock. **2 a** join (spacecraft) together in space. **b** be joined thus. [Dutch *docke*]
dock[2] *n*. enclosure in a criminal court for the accused. [Flemish *dok* cage]
dock[3] *n*. weed with broad leaves. [Old English]
dock[4] *v*. **1** cut short (an animal's tail). **2** take away part of, reduce (wages, supplies, etc.). [Old English]
docker *n*. person employed to load and unload ships.
docket /ˈdɒkɪt/ *–n*. document or label listing goods delivered, jobs done, contents of a package, etc. *–v*. (**-t-**) label with, or enter on, a docket. [origin unknown]
dockland *n*. district near docks or former docks.
dockyard *n*. area with docks and equipment for building and repairing ships.
doctor /ˈdɒktə(r)/ *–n*. **1** qualified practitioner of medicine; physician. **2** person who holds a doctorate. *–v*. *colloq*. **1** treat medically. **2** castrate or spay. **3** patch up (machinery etc.). **4** adulterate. **5** tamper with, falsify. □ **what the doctor ordered** *colloq*. something welcome. [Latin *doceo* teach]
doctoral *adj*. of or for the degree of doctor.
doctorate /ˈdɒktərət/ *n*. highest university degree in any faculty, often honorary.
doctrinaire /ˌdɒktrɪˈneə(r)/ *adj*. applying theory or doctrine dogmatically. [French: related to DOCTRINE]
doctrine /ˈdɒktrɪn/ *n*. **1** what is taught; body of instruction. **2 a** principle of religious or political etc. belief. **b** set of such principles. □ **doctrinal** /-ˈtraɪn(ə)l/ *adj*. [Latin: related to DOCTOR]
docudrama /ˈdɒkjʊˌdrɑːmə/ *n*. television drama based on real events. [from DOCUMENTARY, DRAMA]
document *–n*. /ˈdɒkjʊmənt/ thing providing a record or evidence of events, agreement, ownership, identification, etc. *–v*. /ˈdɒkjʊˌment/ **1** prove by or support with documents. **2** record in a document. [Latin: related to DOCTOR]
documentary /ˌdɒkjʊˈmentərɪ/ *–adj*. **1** consisting of documents (*documentary evidence*). **2** providing a factual record or report. *–n*. (*pl*. **-ies**) documentary film etc.
documentation /ˌdɒkjʊmenˈteɪʃ(ə)n/ *n*. **1** collection and classification of information. **2** material so collected. **3** *Computing* etc. material explaining a system.
dodder[1] *v*. tremble or totter, esp. from age. □ **dodderer** *n*. **doddery** *adj*. [obsolete dial. *dadder*]
dodder[2] *n*. threadlike climbing parasitic plant. [origin uncertain]
doddle /ˈdɒd(ə)l/ *n*. *colloq*. easy task. [perhaps from *doddle* = TODDLE]
dodecagon /dəʊˈdekəgən/ *n*. plane figure with twelve sides. [Greek *dōdeka* twelve, *-gōnos* angled]
dodecahedron /ˌdəʊdekəˈhiːdrən/ *n*. solid figure with twelve faces. [from DODECAGON, Greek *hedra* base]
dodge *–v*. (**-ging**) **1** (often foll. by *about*, *behind*, *round*) move quickly to elude a pursuer, blow, etc. **2** evade by cunning or trickery. *–n*. **1** quick movement to avoid something. **2** clever trick or expedient. □ **dodger** *n*. [origin unknown]
dodgem /ˈdɒdʒəm/ *n*. small electrically-driven car in an enclosure at a funfair, bumped into others for fun. [from DODGE, 'EM]
dodgy *adj*. (**-ier, -iest**) *colloq*. unreliable, risky.
dodo /ˈdəʊdəʊ/ *n*. (*pl*. **-s**) large extinct bird of Mauritius etc. [Portuguese *doudo* simpleton]
DoE *abbr*. Department of the Environment.
doe *n*. (*pl*. same or **-s**) female fallow deer, reindeer, hare, or rabbit. [Old English]
doer /ˈduːə(r)/ *n*. **1** person who does something. **2** person who acts rather than theorizing.
does see DO[1].
doesn't /ˈdʌz(ə)nt/ *contr*. does not.
doff *v*. remove (a hat or clothes). [from *do off*]
dog *–n*. **1** four-legged flesh-eating animal akin to the fox and wolf, and of many breeds. **2** male of this, or of the fox or wolf. **3** *colloq*. **a** despicable person. **b** person of a specified kind (*lucky dog*). **4** mechanical device for gripping. **5** (in *pl*.; prec. by *the*) *colloq*. greyhound-racing. *–v*. (**-gg-**) follow closely; pursue, track. □ **go to the dogs** *slang* deteriorate, be ruined. **like a dog's dinner** *colloq*. smartly or flashily (dressed etc.). **not a**

dog's chance no chance at all. [Old English]
dogcart *n.* two-wheeled driving-cart with cross seats back to back.
dog-collar *n.* **1** collar for a dog. **2** *colloq.* clerical collar.
dog days *n.pl.* hottest period of the year.
doge /dəʊdʒ/ *n. hist.* chief magistrate of Venice or Genoa. [Italian from Latin *dux* leader]
dog-eared *adj.* (of a book etc.) with bent or worn corners.
dog-eat-dog *adj. colloq.* ruthlessly competitive.
dog-end *n. slang* cigarette-end.
dogfight *n.* **1** close combat between fighter aircraft. **2** rough fight.
dogfish *n.* (*pl.* same or **-es**) a kind of small shark.
dogged /'dɒgɪd/ *adj.* tenacious; grimly persistent. □ **doggedly** *adv.* **doggedness** *n.*
doggerel /'dɒgər(ə)l/ *n.* poor or trivial verse. [apparently from DOG]
doggo /'dɒgəʊ/ *adv.* □ **lie doggo** *slang* lie motionless or hidden.
doggy *–adj.* **1** of or like a dog. **2** devoted to dogs. *–n.* (also **doggie**) (*pl.* **-ies**) pet name for a dog.
doggy bag *n.* bag for leftovers given to a customer in a restaurant etc.
doggy-paddle *n.* (also **dog-paddle**) elementary swimming stroke like that of a dog.
doghouse *n. US* & *Austral.* dog's kennel. □ **in the doghouse** *slang* in disgrace.
dog in the manger *n.* person who stops others using a thing for which he or she has no use.
dogma /'dɒgmə/ *n.* **1** principle, tenet, or system of these, esp. of a Church or political party. **2** arrogant declaration of opinion. [Greek, = opinion]
dogmatic /dɒg'mætɪk/ *adj.* asserting or imposing personal opinions; intolerantly authoritative; arrogant. □ **dogmatically** *adv.*
dogmatism /'dɒgmə,tɪz(ə)m/ *n.* tendency to be dogmatic. □ **dogmatist** *n.*
dogmatize /'dɒgmə,taɪz/ *v.* (also **-ise**) (**-zing** or **-sing**) **1** speak dogmatically. **2** express (a principle etc.) as dogma.
do-gooder *n.* well-meaning but unrealistic or patronizing philanthropist or reformer.
dog-paddle var. of DOGGY-PADDLE.
dog-rose *n.* wild hedge-rose.
dogsbody *n.* (*pl.* **-ies**) *colloq.* drudge.
dog's breakfast *n.* (also **dog's dinner**) *colloq.* mess.
dog's life *n.* life of misery etc.
dog-star *n.* chief star of the constellation Canis Major or Minor, esp. Sirius.
dog-tired *adj.* tired out.
dog-tooth *n.* V-shaped pattern or moulding; chevron.
dogtrot *n.* gentle easy trot.
dogwatch *n.* either of two short watches on a ship (4–6 or 6–8 p.m.).
dogwood *n.* shrub with dark-red branches, greenish-white flowers, and purple berries.
DoH *abbr.* Department of Health.
doh /dəʊ/ *n.* (also **do**) *Mus.* first note of a major scale. [Italian *do*]
doily /'dɔɪlɪ/ *n.* (also **doyley**) (*pl.* **-ies** or **-eys**) small lacey usu. paper mat used on a plate for cakes etc. [*Doiley*, name of a draper]
doing /'du:ɪŋ/ *pres. part.* of DO[1]. *n.* **1 a** action (*famous for his doings*). **b** effort (*takes a lot of doing*). **2** (in *pl.*) *slang* unspecified things (*have we got all the doings?*).
doing-over *n. slang* attack, beating-up.
do-it-yourself *–adj.* (of work) done or to be done by a householder etc. *–n.* such work.
Dolby /'dɒlbɪ/ *n. propr.* electronic noise-reduction system used esp. in tape-recording to reduce hiss. [name of its inventor]
doldrums /'dɒldrəmz/ *n.pl.* (usu. prec. by *the*) **1** low spirits. **2** period of inactivity. **3** equatorial ocean region with little or no wind. [perhaps after *dull*, *tantrum*]
dole *–n.* **1** (usu. prec. by *the*) *colloq.* unemployment benefit. **2 a** charitable distribution. **b** thing given sparingly or reluctantly. *–v.* (**-ling**) (usu. foll. by *out*) distribute sparingly. □ **on the dole** *colloq.* receiving unemployment benefit. [Old English]
doleful /'dəʊlfʊl/ *adj.* **1** mournful, sad. **2** dreary, dismal. □ **dolefully** *adv.* **dolefulness** *n.* [Latin *doleo* grieve]
doll *–n.* **1** small model of esp. a baby or child as a child's toy. **2** *colloq.* **a** pretty but silly young woman. **b** attractive woman. **3** ventriloquist's dummy. *–v.* (foll. by *up*) *colloq.* dress smartly. [pet form of *Dorothy*]
dollar /'dɒlə(r)/ *n.* chief monetary unit in the US, Australia, etc. [Low German *daler* from German *Taler*]
dollop /'dɒləp/ *–n.* shapeless lump of food etc. *–v.* (**-p-**) (usu. foll. by *out*) serve in dollops. [perhaps from Scandinavian]
dolly *n.* (*pl.* **-ies**) **1** child's name for a doll. **2** movable platform for a cine-camera etc. **3** easy catch in cricket.
dolly-bird *n. colloq.* attractive and stylish young woman.

dolma /'dɒlmə/ *n.* (*pl.* **-s** or **dolmades** /-'mɑ:ðez/) E. European delicacy of spiced rice or meat etc. wrapped in vine or cabbage leaves. [modern Greek]

dolman sleeve /'dɒlmən/ *n.* loose sleeve cut in one piece with a bodice. [Turkish]

dolmen /'dɒlmən/ *n.* megalithic tomb with a large flat stone laid on upright ones. [French]

dolomite /'dɒlə,maɪt/ *n.* mineral or rock of calcium magnesium carbonate. [de *Dolomieu*, name of a French geologist]

dolour /'dɒlə(r)/ *n.* (*US* **dolor**) *literary* sorrow, distress. □ **dolorous** *adj.* [Latin *dolor* pain]

dolphin /'dɒlfɪn/ *n.* large porpoise-like sea mammal with a slender pointed snout. [Greek *delphis -in-*]

dolphinarium /,dɒlfɪ'neərɪəm/ *n.* (*pl.* **-s**) public aquarium for dolphins.

dolt /dəʊlt/ *n.* stupid person. □ **doltish** *adj.* [apparently related to obsolete *dol* = DULL]

Dom *n.* title of some Roman Catholic dignitaries, and Benedictine and Carthusian monks. [Latin *dominus* master]

-dom *suffix* forming nouns denoting: **1** condition (*freedom*). **2** rank, domain (*earldom*; *kingdom*). **3** class of people (or associated attitudes etc.) regarded collectively (*officialdom*). [Old English]

domain /də'meɪn/ *n.* **1** area under one rule; realm. **2** estate etc. under one control. **3** sphere of control or influence. [French: related to DEMESNE]

dome *–n.* **1** rounded (usu. hemispherical) vault forming a roof. **2** dome-shaped thing. *–v.* (**-ming**) (usu. as **domed** *adj.*) cover with or shape as a dome. [Latin *domus* house]

domestic /də'mestɪk/ *–adj.* **1** of the home, household, or family affairs. **2** of one's own country. **3** (of an animal) tamed, not wild. **4** fond of home life. *–n.* household servant. □ **domestically** *adv.* [Latin *domus* home]

domesticate /də'mestɪ,keɪt/ *v.* (**-ting**) **1** tame (an animal) to live with humans. **2** accustom to housework etc. □ **domestication** /-'keɪʃ(ə)n/ *n.* [medieval Latin: related to DOMESTIC]

domesticity /,dɒmə'stɪsɪtɪ/ *n.* **1** being domestic. **2** domestic or home life.

domestic science *n.* = HOME ECONOMICS.

domicile /'dɒmɪ,saɪl/ *–n.* **1** dwelling-place. **2** *Law* **a** place of permanent residence. **b** residing. *–v.* (**-ling**) (usu. as **domiciled** *adj.*) (usu. foll. by *at, in*) settle in a place. [Latin *domus* home]

domiciliary /,dɒmɪ'sɪlɪərɪ/ *adj. formal* (esp. of a doctor's etc. visit) to, at, or of a person's home. [medieval Latin: related to DOMICILE]

dominant /'dɒmɪnənt/ *–adj.* **1** dominating, prevailing. **2** (of an inherited characteristic) appearing in offspring even when the opposite characteristic is also inherited. *–n. Mus.* fifth note of the diatonic scale of any key. □ **dominance** *n.* **dominantly** *adv.*

dominate /'dɒmɪ,neɪt/ *v.* (**-ting**) **1** command, control. **2** be the most influential or obvious. **3** (of a high place) overlook. □ **domination** /-'neɪʃ(ə)n/ *n.* [Latin *dominor* from *dominus* lord]

domineer /,dɒmɪ'nɪə(r)/ *v.* (often as **domineering** *adj.*) behave arrogantly or tyrannically. [French: related to DOMINATE]

Dominican /də'mɪnɪkən/ *–adj.* of St Dominic or his order. *–n.* Dominican friar or nun. [Latin *Dominicus* Dominic]

dominion /də'mɪnjən/ *n.* **1** sovereignty, control. **2** realm; domain. **3** *hist.* self-governing territory of the British Commonwealth. [Latin *dominus* lord]

domino /'dɒmɪ,nəʊ/ *n.* (*pl.* **-es**) **1** any of 28 small oblong pieces marked with 0–6 pips in each half. **2** (in *pl.*) game played with these. **3** loose cloak with a mask. [French, probably as DOMINION]

domino effect *n.* (also **domino theory**) effect whereby (or theory that) one event precipitates others in causal sequence.

don[1] *n.* **1** university teacher, esp. a senior member of a college at Oxford or Cambridge. **2** (**Don**) Spanish title prefixed to a forename. [Spanish from Latin *dominus* lord]

don[2] *v.* (**-nn-**) put on (clothing). [= *do on*]

donate /dəʊ'neɪt/ *v.* (**-ting**) give (money etc.), esp. to charity. [from DONATION]

donation /dəʊ'neɪʃ(ə)n/ *n.* **1** donating or being donated. **2** thing, esp. money, donated. [Latin *donum* gift]

done /dʌn/ *adj.* **1** completed. **2** cooked. **3** *colloq.* socially acceptable (*the done thing*). **4** (often with *in*) *colloq.* tired out. **5** (esp. as *int.* in reply to an offer etc.) accepted. □ **be done with** have or be finished with. **done for** *colloq.* in serious trouble. **have done with** be rid of; finish dealing with. [past part. of DO[1]]

doner kebab /'dɒnə, 'dəʊ-/ *n.* spiced lamb cooked on a spit and served in slices, often with pitta bread. [Turkish: related to KEBAB]

donjon /'dɒndʒ(ə)n/ *n.* great tower or innermost keep of a castle. [archaic spelling of DUNGEON]

Don Juan /'dʒu:ən, 'wɑ:n/ *n.* seducer of women.

donkey /ˈdɒŋkɪ/ *n.* (*pl.* **-s**) **1** domestic ass. **2** *colloq.* stupid person. [perhaps from *Duncan*: cf. NEDDY]

donkey jacket *n.* thick weatherproof workman's jacket or fashion garment.

donkey's years *n.pl. colloq.* very long time.

donkey-work *n.* laborious part of a job.

Donna /ˈdɒnə/ *n.* title of an Italian, Spanish, or Portuguese lady. [Latin *domina* mistress]

donnish *adj.* like a college don; pedantic.

donor /ˈdəʊnə(r)/ *n.* **1** person who donates (e.g. to charity). **2** person who provides blood, semen, or an organ or tissue for medical use.

donor card *n.* official card authorizing the use of organs, carried by a donor.

don't /dəʊnt/ *–contr.* do not. *–n.* prohibition (*dos and don'ts*).

donut *US* var. of DOUGHNUT.

doodle /ˈdu:d(ə)l/ *–v.* (**-ling**) scribble or draw, esp. absent-mindedly. *–n.* such a scribble or drawing. [originally = foolish person]

doom *–n.* **1 a** grim fate or destiny. **b** death or ruin. **2** condemnation. *–v.* **1** (usu. foll. by *to*) condemn or destine. **2** (esp. as **doomed** *adj.*) consign to misfortune or destruction. [Old English, = STATUTE]

doomsday *n.* day of the Last Judgement. □ **till doomsday** for ever.

door *n.* **1 a** esp. hinged barrier for closing and opening the entrance to a building, room, cupboard, etc. **b** this as representing a house etc. (*lives two doors away*). **2 a** entrance or exit; doorway. **b** means of access. □ **close** (or **open**) **the door to** exclude (or create) an opportunity for. [Old English]

doorbell *n.* bell on a door rung by visitors to signal arrival.

door-keeper *n.* = DOORMAN.

doorknob *n.* knob turned to open a door.

doorman *n.* person on duty at the door to a large building.

doormat *n.* **1** mat at an entrance, for wiping shoes. **2** *colloq.* submissive person.

doorpost *n.* upright of a door-frame, on which the door is hung.

doorstep *–n.* **1** step or area in front of the outer door of a house etc. **2** *slang* thick slice of bread. *–v. colloq.* **1** go from door to door canvassing, selling, etc. **2** call upon or wait on the doorstep for (a person) in order to interview etc. □ **on one's doorstep** very near.

doorstop *n.* device for keeping a door open or to prevent it from striking the wall.

door-to-door *adj.* (of selling etc.) done at each house in turn.

doorway *n.* opening filled by a door.

dope *–n.* **1 a** *slang* narcotic. **b** drug etc. given to a horse, athlete, etc., to improve performance. **2** thick liquid used as a lubricant etc. **3** varnish. **4** *slang* stupid person. **5** *slang* information. *–v.* (**-ping**) **1** give or add a drug to. **2** apply dope to. [Dutch, = sauce]

dopey *adj.* (also **dopy**) (**dopier, dopiest**) *colloq.* **1** half asleep or stupefied as if by a drug. **2** stupid. □ **dopily** *adv.* **dopiness** *n.*

doppelgänger /ˈdɒp(ə)l,geŋə(r)/ *n.* apparition of a living person. [German, = double-goer]

Doppler effect *n.* increase (or decrease) in the frequency of sound, light, etc. waves caused by moving nearer to (or further from) the source. [*Doppler*, name of a physicist]

dorado /dəˈrɑ:dəʊ/ *n.* (*pl.* same or **-s**) sea-fish showing brilliant colours when dying out of water. [Spanish, = gilded]

Doric /ˈdɒrɪk/ *–adj.* **1** *Archit.* of the oldest and simplest of the Greek orders. **2** (of a dialect) broad, rustic. *–n.* rustic English or esp. Scots. [from *Dōris* in Greece]

dormant /ˈdɔ:mənt/ *adj.* **1** lying inactive; sleeping. **2** temporarily inactive. **3** (of plants) alive but not growing. □ **dormancy** *n.* [Latin *dormio* sleep]

dormer *n.* (in full **dormer window**) projecting upright window in a sloping roof. [French: related to DORMANT]

dormitory /ˈdɔ:mɪtərɪ/ *n.* (*pl.* **-ies**) **1** sleeping-room with several beds, esp. in a school or institution. **2** (in full **dormitory town** etc.) small commuter town or suburb. [Latin: related to DORMER]

Dormobile /ˈdɔ:mə,bi:l/ *n. propr.* motor caravan. [from DORMITORY, AUTOMOBILE]

dormouse /ˈdɔ:maʊs/ *n.* (*pl.* **-mice**) small mouselike hibernating rodent. [origin unknown]

dorsal /ˈdɔ:s(ə)l/ *adj.* of or on the back (*dorsal fin*). [Latin *dorsum* back]

dory /ˈdɔ:rɪ/ *n.* (*pl.* same or **-ies**) any of various edible marine fish, esp. the John Dory. [French *dorée* = gilded]

dosage /ˈdəʊsɪdʒ/ *n.* **1** size of a dose. **2** giving of a dose.

dose *–n.* **1** single portion of medicine. **2** experience of something (*dose of flu, laughter*). **3** amount of radiation received. **4** *slang* venereal infection. *–v.* (**-sing**) treat with or give doses of medicine to. [Greek *dosis* gift]

do-se-do /ˌdəʊsɪ'dəʊ/ *n.* (also **do-si-do**) (*pl.* **-s**) figure in which two dancers pass round each other back to back. [French *dos-à-dos,* = back to back]

dosh *n. slang* money. [origin unknown]

doss *v.* **1** (often foll. by *down*) *slang* sleep roughly or in a doss-house. **2** (often foll. by *about, around*) spend time idly. [probably originally = 'seat-back cover': from Latin *dorsum* back]

dosser *n. slang* **1** person who dosses. **2** = DOSS-HOUSE.

doss-house *n.* cheap lodging-house for vagrants.

dossier /'dɒsɪˌeɪ, -sɪə(r)/ *n.* file containing information about a person, event, etc. [French]

DoT *abbr.* Department of Transport.

dot –*n.* **1 a** small spot or mark. **b** this as part of *i* or *j*, or as a decimal point etc. **2** shorter signal of the two in Morse code. –*v.* (**-tt-**) **1 a** mark with dot(s). **b** place a dot over (a letter). **2** (often foll. by *about*) scatter like dots. **3** partly cover as with dots (*sea dotted with ships*). **4** *slang* hit. □ **dot the i's and cross the t's** *colloq.* **1** be minutely accurate. **2** add the final touches to a task etc. **on the dot** exactly on time. **the year dot** *colloq.* far in the past. [Old English]

dotage *n.* feeble-minded senility (*in his dotage*).

dotard /'dəʊtəd/ *n.* senile person.

dote *v.* (**-ting**) (foll. by *on*) be excessively fond of. □ **dotingly** *adv.* [origin uncertain]

dotted line *n.* line of dots on a document etc., esp. for writing a signature on.

dotterel /'dɒtər(ə)l/ *n.* small migrant plover. [from DOTE]

dottle /'dɒt(ə)l/ *n.* remnant of unburnt tobacco in a pipe. [from DOT]

dotty *adj.* (**-ier, -iest**) *colloq.* **1** crazy; eccentric. **2**(foll. by *about*) infatuated with. □ **dottiness** *n.*

double /'dʌb(ə)l/ –*adj.* **1** consisting of two parts or things; twofold. **2** twice as much or many (*double thickness*). **3** having twice the usual size, quantity, strength, etc. (*double bed*). **4 a** being double in part. **b** (of a flower) with two or more circles of petals. **5** ambiguous, deceitful (*double meaning*; *a double life*). –*adv.* **1** at or to twice the amount etc. (*counts double*). **2** two together (*sleep double*). –*n.* **1** double quantity (of spirits etc.) or thing; twice as much or many. **2** counterpart; person who looks exactly like another. **3** (in *pl.*) game between two pairs of players. **4** pair of victories. **5** bet in which winnings and stake from the first bet are transferred to the second. **6** doubling of an opponent's bid in bridge. **7** hit on the narrow ring between the two outer circles in darts. –*v.* (**-ling**) **1** make or become double; increase twofold; multiply by two. **2** amount to twice as much as. **3** fold or bend over on itself; become folded. **4 a** act (two parts) in the same play etc. **b** (often foll. by *for*) be understudy etc. **5** (usu. foll. by *as*) play a twofold role. **6** turn sharply in flight or pursuit. **7** sail round (a headland). **8** make a call in bridge increasing the value of the points to be won or lost on (an opponent's bid). □ **at the double** running, hurrying. **bent double** stooping. **double back** turn back in the opposite direction. **double or quits** gamble to decide whether a player's loss or debt be doubled or cancelled. **double up 1** (cause to) bend or curl up with pain or laughter. **2** share or assign to a shared room, quarters, etc. □ **doubly** *adv.* [Latin *duplus*]

double act *n.* comedy act by a duo.

double agent *n.* spy working for rival countries.

double-barrelled *adj.* **1** (of a gun) having two barrels. **2** (of a surname) hyphenated.

double-bass *n.* largest instrument of the violin family.

double bluff *n.* genuine action or statement disguised as a bluff.

double-book *v.* reserve (the same seat, room, etc.) for two people at once.

double-breasted *adj.* (of a coat etc.) overlapping across the body.

double-check *v.* verify twice.

double chin *n.* chin with a fold of loose flesh below it.

double cream *n.* thick cream with a high fat-content.

double-cross –*v.* deceive or betray (a supposed ally). –*n.* act of doing this. □ **double-crosser** *n.*

double-dealing –*n.* deceit, esp. in business. –*adj.* practising deceit.

double-decker *n.* **1** bus having an upper and lower deck. **2** *colloq.* sandwich with two layers of filling.

double Dutch *n. colloq.* gibberish.

double eagle *n.* figure of a two-headed eagle.

double-edged *adj.* **1** presenting both a danger and an advantage. **2** (of a knife etc.) having two cutting-edges.

double entendre /ˌdu:b(ə)l ɑ:n'tɑ:ndrə/ *n.* ambiguous phrase open to usu. indecent interpretation. [obsolete French]

double entry *n.* system of bookkeeping with entries debited in one account and credited in another.

double feature *n.* cinema programme with two full-length films.

double figures *n.pl.* numbers from 10 to 99.

double glazing *n.* two layers of glass in a window.

double helix *n.* pair of parallel helices with a common axis, esp. in the structure of a DNA molecule.

double-jointed *adj.* having joints that allow unusual bending.

double negative *n. Gram.* negative statement containing two negative elements (e.g. *he didn't say nothing*).

■ **Usage** The double negative is considered incorrect in standard English.

double-park *v.* (also *absol.*) park (a vehicle) alongside one already parked at the roadside.

double pneumonia *n.* pneumonia affecting both lungs.

double-quick *adj. & adv. colloq.* very quick or quickly.

double standard *n.* rule or principle not impartially applied.

doublet /ˈdʌblɪt/ *n.* **1** *hist.* man's short close-fitting jacket. **2** one of a pair of similar things. [French: related to DOUBLE]

double take *n.* delayed reaction to a situation etc.

double-talk *n.* (usu. deliberately) ambiguous or misleading speech.

double-think *n.* capacity to accept contrary opinions at the same time.

double time *n.* wages paid at twice the normal rate.

doubloon /dʌˈbluːn/ *n. hist.* Spanish gold coin. [French or Spanish: related to DOUBLE]

doubt /daʊt/ —*n.* **1** uncertainty; undecided state of mind. **2** cynicism. **3** uncertain state. **4** lack of full proof or clear indication. —*v.* **1** feel uncertain or undecided about. **2** hesitate to believe. **3** call in question. □ **in doubt** open to question. **no doubt** certainly; probably; admittedly. **without doubt** (or **a doubt**) certainly. [Latin *dubito* hesitate]

doubtful *adj.* **1** feeling doubt. **2** causing doubt. **3** unreliable. □ **doubtfully** *adv.* **doubtfulness** *n.*

doubtless *adv.* certainly; probably.

douche /duːʃ/ —*n.* **1** jet of liquid applied to part of the body for cleansing or medicinal purposes. **2** device for producing such a jet. —*v.* (**-ching**) **1** treat with a douche. **2** use a douche. [Latin: related to DUCT]

dough /dəʊ/ *n.* **1** thick mixture of flour etc. and liquid for baking. **2** *slang* money. □ **doughy** *adj.* (**-ier, -iest**). [Old English]

doughnut *n.* (*US* **donut**) small fried cake of sweetened dough.

doughnutting *n.* the clustering of politicians round a speaker during a televised debate to make him or her appear well supported.

doughty /ˈdaʊtɪ/ *adj.* (**-ier, -iest**) *archaic* valiant. □ **doughtily** *adv.* **doughtiness** *n.* [Old English]

dour /dʊə(r)/ *adj.* severe, stern, obstinate. [probably Gaelic *dúr* dull, obstinate]

douse *v.* (also **dowse**) (**-sing**) **1 a** throw water over. **b** plunge into water. **2** extinguish (a light). [origin uncertain]

dove /dʌv/ *n.* **1** bird with short legs, a small head, and a large breast. **2** gentle or innocent person. **3** advocate of peace or peaceful policies. [Old Norse]

dovecote /ˈdʌvkɒt/ *n.* (also **dovecot**) shelter with nesting-holes for domesticated pigeons.

dovetail —*n.* mortise and tenon joint shaped like a dove's spread tail. —*v.* **1** join with dovetails. **2** fit together; combine neatly.

dowager /ˈdaʊədʒə(r)/ *n.* **1** widow with a title or property from her late husband (*dowager duchess*). **2** *colloq.* dignified elderly woman. [French: related to DOWER]

dowdy /ˈdaʊdɪ/ *adj.* (**-ier, -iest**) **1** (of clothes) unattractively dull. **2** dressed dowdily. □ **dowdily** *adv.* **dowdiness** *n.* [origin unknown]

dowel /ˈdaʊəl/ —*n.* cylindrical peg for holding structural components together. —*v.* (**-ll-**; *US* **-l-**) fasten with a dowel. [Low German]

dowelling *n.* rods for cutting into dowels.

dower *n.* **1** widow's share for life of a husband's estate. **2** *archaic* dowry. [Latin *dos* dowry]

dower house *n.* smaller house near a big one, as part of a widow's dower.

Dow-Jones index /daʊˈdʒəʊnz/ *n.* (also **Dow-Jones average**) a figure indicating the relative price of shares on the New York Stock Exchange. [*Dow* and *Jones*, names of American economists]

down[1] —*adv.* **1** into or towards a lower place, esp. to the ground (*fall down*). **2** in a lower place or position (*blinds were down*). **3** to or in a place regarded as lower, esp.: **a** southwards. **b** away from a major city or a university. **4 a** in or into a low or weaker position or condition (*hit a man when he's down*; *down with a cold*). **b** losing by (*three goals down*; *£5 down*). **c** (of a computer system) out of action. **5** from an earlier to a later time (*down to 1600*). **6** to a finer or thinner consistency or smaller amount or size (*grind down*; *water down*; *boil down*). **7** cheaper (*bread is down*; *shares are down*). **8** into a more settled state (*calm down*). **9** in writing or recorded form (*copy it down*; *down on tape*; *down to speak next*). **10** paid or dealt with as a deposit or part (*£5 down, £20 to pay*; *three down, six to go*). **11** with the current or wind. **12** (of a crossword clue or answer) read vertically (*five down*). *–prep.* **1** downwards along, through, or into. **2** from the top to the bottom of. **3** along (*walk down the road*). **4** at or in a lower part of (*lives down the road*). *–attrib. adj.* **1** directed downwards (*a down draught*). **2** from a capital or centre (*down train*; *down platform*). *–v. colloq.* **1** knock or bring down. **2** swallow. *–n.* **1** act of putting down. **2** reverse of fortune (*ups and downs*). **3** *colloq.* period of depression. □ **be down to 1** be the responsibility of. **2** have nothing left but (*down to my last penny*). **3** be attributable to. **down on one's luck** *colloq.* temporarily unfortunate. **down to the ground** *colloq.* completely. **down tools** *colloq.* cease work, go on strike. **down with** expressing rejection of a specified person or thing. **have a down on** *colloq.* show hostility towards. [earlier *adown*: related to DOWN[3]]

down[2] *n.* **1 a** first covering of young birds. **b** bird's under-plumage. **c** fine soft feathers or hairs. **2** fluffy substance. [Old Norse]

down[3] *n.* **1** open rolling land. **2** (in *pl.*) chalk uplands, esp. in S. England. [Old English]

down-and-out *n.* destitute person. □ **down and out** *predic. adj.*

downbeat *–n. Mus.* accented beat, usu. the first of the bar. *–adj.* **1** pessimistic, gloomy. **2** relaxed.

downcast *adj.* **1** dejected. **2** (of eyes) looking downwards.

downer *n. slang* **1** depressant or tranquillizing drug. **2** depressing person or experience; failure. **3** = DOWNTURN.

downfall *n.* **1** fall from prosperity or power. **2** cause of this.

downgrade *v.* **(-ding)** reduce in rank or status.

downhearted *adj.* dejected. □ **downheartedly** *adv.* **downheartedness** *n.*

downhill *–adv.* in a descending direction. *–adj.* sloping down, declining. *–n.* **1** downhill race in skiing. **2** downward slope. □ **go downhill** *colloq.* deteriorate.

down in the mouth *adj.* looking unhappy.

downland *n.* = DOWN[3].

down-market *adj.* & *adv. colloq.* of or to the cheaper sector of the market.

down payment *n.* partial initial payment.

downpipe *n.* pipe to carry rainwater from a roof.

downpour *n.* heavy fall of rain.

downright *–adj.* **1** plain, straightforward. **2** utter (*downright nonsense*). *–adv.* thoroughly (*downright rude*).

Down's syndrome *n.* congenital disorder with mental retardation and physical abnormalities. [*Down*, name of a physician]

downstage *adj.* & *adv.* nearer the front of a theatre stage.

downstairs *–adv.* **1** down the stairs. **2** to or on a lower floor. *–attrib. adj.* situated downstairs. *–n.* lower floor.

downstream *adv.* & *adj.* in the direction in which a stream etc. flows.

down-to-earth *adj.* practical, realistic.

downtown esp. *US –attrib. adj.* of the lower or more central part of a town or city. *–n.* downtown area. *–adv.* in or into the downtown area.

downtrodden *adj.* oppressed; badly treated.

downturn *n.* decline, esp. in economic activity.

down under *adv. colloq.* in the antipodes, esp. Australia.

downward /'daʊnwəd/ *–adv.* (also **downwards**) towards what is lower, inferior, less important, or later. *–adj.* moving or extending downwards.

downwind *adj.* & *adv.* in the direction in which the wind is blowing.

downy *adj.* **(-ier, -iest) 1** of, like, or covered with down. **2** soft and fluffy.

dowry /'daʊərɪ/ *n.* (*pl.* **-ies**) property or money brought by a bride to her husband. [Anglo-French, = French *douaire* DOWER]

dowse[1] /daʊs/ *v.* **(-sing)** search for underground water or minerals by holding a stick or rod which dips abruptly when over the right spot. □ **dowser** *n.* [origin unknown]

dowse[2] var. of DOUSE.

dowsing-rod *n.* rod for dowsing.

doxology /dɒk'sɒlədʒɪ/ *n.* (*pl.* **-ies**) liturgical hymn etc. of praise to God. □ **doxological** /-sə'lɒdʒɪk(ə)l/ *adj.* [Greek *doxa* glory]

doyen /'dɔɪən/ *n.* (*fem.* **doyenne** /dɔɪ'en/) senior member of a group. [French: related to DEAN[1]]

doyley var. of DOILY.

doz. *abbr.* dozen.

doze *–v.* **(-zing)** sleep lightly; be half asleep. *–n.* short light sleep. □ **doze off** fall lightly asleep. [origin unknown]

dozen /'dʌz(ə)n/ *n.* **1** (prec. by *a* or a number) (*pl.* **dozen**) twelve (*a dozen eggs; two dozen eggs*). **2** set of twelve (*sold in dozens*). **3** (in *pl.*; usu. foll. by *of*) *colloq.* very many (*dozens of errors*). □ **talk nineteen to the dozen** talk incessantly. [Latin *duodecim* twelve]

dozy *adj.* (**-ier, -iest**) **1** drowsy. **2** *colloq.* stupid or lazy.

D.Phil. *abbr.* Doctor of Philosophy.

DPP *abbr.* Director of Public Prosecutions.

Dr *abbr.* Doctor.

drab *adj.* (**drabber, drabbest**) **1** dull, uninteresting. **2** of a dull brownish colour. □ **drably** *adv.* **drabness** *n.* [obsolete *drap* cloth]

drachm /dræm/ *n.* weight formerly used by apothecaries, = $^1/_8$ ounce. [Latin from Greek]

drachma /'drækmə/ *n.* (*pl.* **-s**) **1** chief monetary unit of Greece. **2** silver coin of ancient Greece. [Greek *drakhmē*]

Draconian /drə'kəʊnɪən/ *adj.* (of laws) very harsh, cruel. [*Drakōn*, name of an Athenian lawgiver]

draft /drɑ:ft/ *–n.* **1** preliminary written version of a speech, document, etc., or outline of a scheme. **2 a** written order for payment of money by a bank. **b** drawing of money by this. **3 a** detachment from a larger group. **b** selection of this. **4** *US* conscription. **5** *US* = DRAUGHT. *–v.* **1** prepare a draft of (a document, scheme, etc.). **2** select for a special duty or purpose. **3** *US* conscript. [phonetic spelling of DRAUGHT]

draftsman *n.* **1** person who drafts documents. **2** = DRAUGHTSMAN 1. [phonetic spelling of DRAUGHTSMAN]

drafty *US* var. of DRAUGHTY.

drag *–v.* (**-gg-**) **1** pull along with effort. **2 a** trail or allow to trail along the ground. **b** (often foll. by *on*) (of time, a meeting, etc.) go or pass slowly or tediously. **3 a** use a grapnel. **b** search the bottom of (a river etc.) with grapnels, nets, etc. **4** (often foll. by *to*) *colloq.* take (an esp. unwilling person) with one. **5** (foll. by *on, at*) draw on (a cigarette etc.). *–n.* **1 a** obstruction to progress. **b** retarding force or motion. **2** *colloq.* boring or tiresome person, duty, etc. **3 a** lure before hounds as a substitute for a fox. **b** hunt using this. **4** apparatus for dredging. **5** = DRAG-NET. **6** *slang* inhalation. **7** *slang* women's clothes worn by men. □ **drag one's feet** be deliberately slow or reluctant to act. **drag in** introduce (an irrelevant subject). **drag out** protract. **drag up** *colloq.* introduce or revive (an unwelcome subject). [Old English or Old Norse]

draggle /'dræg(ə)l/ *v.* (**-ling**) **1** make dirty, wet, or limp by trailing. **2** hang trailing. [from DRAG]

drag-net *n.* **1** net drawn through a river or across the ground to trap fish or game. **2** systematic hunt for criminals etc.

dragon /'drægən/ *n.* **1** mythical usu. winged monster like a reptile, able to breathe fire. **2** fierce woman. [Greek, = serpent]

dragonfly *n.* large insect with a long body and two pairs of transparent wings.

dragoon /drə'gu:n/ *–n.* **1** cavalryman. **2** fierce fellow. *–v.* (foll. by *into*) coerce or bully into. [French *dragon*: related to DRAGON]

drag queen *n. slang derog.* male homosexual transvestite.

drain *–v.* **1** draw off liquid from. **2** draw off (liquid). **3** flow or trickle away. **4** dry or become dry as liquid flows away. **5** exhaust of strength or resources. **6 a** drink to the dregs. **b** empty (a glass etc.) by drinking the contents. *–n.* **1 a** channel, conduit, or pipe carrying off liquid, sewage, etc. **b** tube for drawing off discharge etc. **2** constant outflow or expenditure. □ **down the drain** *colloq.* lost, wasted. [Old English: related to DRY]

drainage *n.* **1** draining. **2** system of drains. **3** what is drained off.

draining-board *n.* sloping grooved surface beside a sink for draining washed dishes.

drainpipe *n.* **1** pipe for carrying off water etc. **2** (*attrib.*) (of trousers) very narrow. **3** (in *pl.*) very narrow trousers.

drake *n.* male duck. [origin uncertain]

Dralon /'dreɪlɒn/ *n. propr.* **1** synthetic acrylic fibre. **2** fabric made from this. [invented word, after NYLON]

dram *n.* **1** small drink of spirits, esp. whisky. **2** = DRACHM. [Latin *drama*: related to DRACHM]

drama /'drɑ:mə/ *n.* **1** play for stage or broadcasting. **2** art of writing, acting, or presenting plays. **3** dramatic event or quality (*the drama of the situation*). [Latin from Greek *draō* do]

dramatic /drə'mætɪk/ *adj.* **1** of drama. **2** sudden and exciting or unexpected. **3** vividly striking. **4** (of a gesture etc.) theatrical. □ **dramatically** *adv.* [Greek: related to DRAMA]

dramatics *n.pl.* (often treated as *sing.*) **1** performance of plays. **2** exaggerated behaviour.

dramatis personae /ˌdræmətɪs pɜ:'səʊnaɪ/ *n.pl.* **1** characters in a play. **2** list of these. [Latin, = persons of the drama]

dramatist /ˈdræmətɪst/ *n.* writer of dramas.

dramatize /ˈdræmə,taɪz/ *v.* (also **-ise**) (**-zing** or **-sing**) **1** turn (a novel etc.) into a play. **2** make a dramatic scene of. **3** behave dramatically. □ **dramatization** /-ˈzeɪʃ(ə)n/ *n.*

drank *past* of DRINK.

drape *–v.* (**-ping**) **1** hang or cover loosely, adorn with cloth etc. **2** arrange (hangings etc.) esp. in folds. *–n.* (in *pl.*) *US* curtains. [Latin *drappus* cloth]

draper *n.* dealer in textile fabrics.

drapery *n.* (*pl.* **-ies**) **1** clothing or hangings arranged in folds. **2** draper's trade or fabrics.

drastic /ˈdræstɪk/ *adj.* far-reaching in effect; severe. □ **drastically** *adv.* [Greek *drastikos*: related to DRAMA]

drat *colloq.* *–v.* (**-tt-**) (usu. as *int.*) curse (*drat the thing!*). *–int.* expressing anger or annoyance. □ **dratted** *adj.* [(*Go*)*d rot*]

draught /drɑːft/ *n.* (*US* **draft**) **1** current of air in a room or chimney etc. **2** pulling, traction. **3** depth of water needed to float a ship. **4** drawing of liquor from a cask etc. **5 a** single act of drinking or inhaling. **b** amount drunk thus. **6** (in *pl.*) game for two with 12 pieces each on a draughtboard. **7 a** drawing in of a fishing-net. **b** fish so caught. □ **feel the draught** *colloq.* suffer from esp. financial hardship. [related to DRAW]

draught beer *n.* beer from the cask, not bottled or canned.

draughtboard *n.* = CHESSBOARD.

draught-horse *n.* horse for heavy work.

draughtsman *n.* **1** person who makes drawings, plans, or sketches. **2** piece in draughts. □ **draughtsmanship** *n.*

draughty *adj.* (*US* **drafty**) (**-ier**, **-iest**) (of a room etc.) letting in sharp currents of air. □ **draughtiness** *n.*

draw *–v.* (*past* **drew** /druː/; *past part.* **drawn**) **1** pull or cause to move towards or after one. **2** pull (a thing) up, over, or across. **3** pull (curtains etc.) open or shut. **4** take (a person) aside. **5** attract; bring; take in (*drew a deep breath*; *felt drawn to her*; *drew my attention*; *drew a crowd*). **6** (foll. by *at, on*) inhale from (a cigarette, pipe, etc.). **7** (also *absol.*) take out; remove (a tooth, gun, cork, card, etc.). **8** obtain or take from a source (*draw a salary*; *draw inspiration*; *drew £100 out*). **9 a** (also *absol.*) make (a line or mark). **b** produce (a picture) thus. **c** represent (something) thus. **10** (also *absol.*) finish (a contest or game) with equal scores. **11** proceed (*drew near the bridge*; *draw to a close*; *drew level*). **12** infer (a conclusion). **13 a** elicit, evoke (*draw criticism*). **b** induce (a person) to reveal facts etc. **14** haul up (water) from a well. **15** bring out (liquid from a tap etc. or blood from a wound). **16** extract a liquid essence from. **17** (of a chimney etc.) promote or allow a draught. **18** (of tea) infuse. **19 a** obtain by lot (*drew the winner*). **b** *absol.* draw lots. **20** (foll. by *on*) call on (a person or a person's skill etc.). **21** write out or compose (a cheque, document, etc.). **22** formulate or perceive (a comparison or distinction). **23** disembowel. **24** search (cover) for game. **25** drag (a badger or fox) from a hole. *–n.* **1** act of drawing. **2** person or thing attracting custom, attention, etc. **3** drawing of lots, raffle. **4** drawn game. **5** inhalation of smoke etc. □ **draw back** withdraw from an undertaking. **draw a bead on** see BEAD. **draw a blank** see BLANK. **draw in 1** (of days) become shorter. **2** persuade to join. **3** (of a train) arrive at a station. **draw in one's horns** become less assertive or ambitious. **draw the line at** set a limit of tolerance etc. at. **draw lots** see LOT. **draw on 1** approach, come near. **2** lead to. **3** allure. **4** put (gloves, boots, etc.) on. **draw out 1** prolong. **2** elicit. **3** induce to talk. **4** (of days) become longer. **5** (of a train) leave a station. **draw up 1** draft (a document etc.). **2** bring into order. **3** come to a halt. **4** make (oneself) erect. **quick on the draw** quick to react. [Old English]

drawback *n.* disadvantage.

drawbridge *n.* hinged retractable bridge, esp. over a moat.

drawer /ˈdrɔːə(r)/ *n.* **1** person or thing that draws, esp. a cheque etc. **2** also /drɔː(r)/ lidless boxlike storage compartment, sliding in and out of a table etc. (*chest of drawers*). **3** (in *pl.*) knickers, underpants.

drawing *n.* **1** art of representing by line with a pencil etc. **2** picture etc. made thus.

drawing-board *n.* board on which paper is fixed for drawing on.

drawing-pin *n.* flat-headed pin for fastening paper etc. to a surface.

drawing-room *n.* **1** room in a private house for sitting or entertaining in. **2** (*attrib.*) restrained, polite (*drawing-room manners*). [earlier *withdrawing-room*]

drawl *–v.* speak with drawn-out vowel sounds. *–n.* drawling utterance or way of speaking. [Low German or Dutch]

drawn *adj.* looking strained and tense.

drawstring *n.* string or cord threaded through a waistband, bag opening, etc.

dray *n.* low cart without sides for heavy loads, esp. beer-barrels. [related to DRAW]

dread /dred/ *–v.* fear greatly, esp. in advance. *–n.* great fear or apprehension. *–adj.* **1** dreaded. **2** *archaic* awe-inspiring, dreadful. [Old English]

dreadful *adj.* **1** terrible. **2** *colloq.* very annoying, very bad. □ **dreadfully** *adv.*

dreadlocks *n.pl.* Rastafarian hairstyle with hair hanging in tight braids on all sides.

dream *–n.* **1** series of scenes or feelings in the mind of a sleeping person. **2** day-dream or fantasy. **3** ideal, aspiration. **4** beautiful or ideal person or thing. *–v.* (*past* and *past part.* **dreamt** /dremt/ or **dreamed**) **1** experience a dream. **2** imagine as in a dream. **3** (with *neg.*) consider possible (*never dreamt that he would come*; *would not dream of it*). **4** (foll. by *away*) waste (time). **5** be inactive or unpractical. □ **dream up** imagine, invent. **like a dream** *colloq.* easily, effortlessly. □ **dreamer** *n.* [Old English]

dreamboat *n. colloq.* sexually attractive or ideal person.

dreamland *n.* ideal or imaginary land.

dreamy *adj.* (**-ier, -iest**) **1** given to day-dreaming or fantasy. **2** dreamlike; vague. **3** *colloq.* delightful. □ **dreamily** *adv.* **dreaminess** *n.*

dreary *adj.* (**-ier, -iest**) dismal, dull, gloomy. □ **drearily** *adv.* **dreariness** *n.* [Old English]

dredge[1] *–n.* apparatus used to scoop up oysters etc., or to clear mud etc., from a river or sea bed. *–v.* (**-ging**) **1** (often foll. by *up*) **a** bring up or clear (mud etc.) with a dredge. **b** bring up (something forgotten) (*dredged it all up*). **2** clean with or use a dredge. [origin uncertain]

dredge[2] *v.* (**-ging**) sprinkle with flour, sugar, etc. [earlier = sweetmeat, from French]

dredger[1] *n.* **1** boat with a dredge. **2** dredge.

dredger[2] *n.* container with a perforated lid, for sprinkling flour, sugar, etc.

dregs *n.pl.* **1** sediment; grounds, lees. **2** = SCUM *n.* 2 (*dregs of humanity*). [Old Norse]

drench *–v.* **1** wet thoroughly. **2** force (an animal) to take medicine. *–n.* dose of medicine for an animal. [Old English]

dress *–v.* **1 a** (also *absol.*) put clothes on. **b** have and wear clothes (*dresses well*). **2** put on evening dress. **3** arrange or adorn (hair, a shop window, etc.). **4** treat (a wound) esp. with a dressing. **5 a** prepare (poultry, crab, etc.) for cooking or eating. **b** add dressing to (a salad etc.). **6** apply manure to. **7** finish the surface of (fabric, leather, stone, etc.). **8** correct the alignment of (troops). **9** make (an artificial fly) for fishing. *–n.* **1** woman's garment of a bodice and skirt. **2** clothing, esp. a whole outfit. **3** formal or ceremonial costume. **4** external covering; outward form. □ **dress down** *colloq.* **1** reprimand or scold. **2** dress informally. **dress up 1** put on special clothes. **2** make (a thing) more attractive or interesting. [French *dresser*, ultimately related to DIRECT]

dressage /'dresɑːʒ/ *n.* training of a horse in obedience and deportment; display of this. [French]

dress circle *n.* first gallery in a theatre.

dress coat *n.* man's swallow-tailed evening coat.

dresser[1] *n.* kitchen sideboard with shelves for plates etc. [French *dresser* prepare]

dresser[2] *n.* **1** person who helps to dress actors or actresses. **2** surgeon's assistant in operations. **3** person who dresses in a specified way (*snappy dresser*).

dressing *n.* **1** putting one's clothes on. **2 a** sauce, esp. of oil and vinegar etc., for salads (*French dressing*). **b** sauce or stuffing etc. for food. **3** bandage, ointment, etc., for a wound. **4** size or stiffening used to coat fabrics. **5** compost etc. spread over land.

dressing-down *n. colloq.* scolding.

dressing-gown *n.* loose robe worn when one is not fully dressed.

dressing-room *n.* room for changing one's clothes, esp. in a theatre, or attached to a bedroom.

dressing-table *n.* table with a flat top, mirror, and drawers, used while applying make-up etc.

dressmaker *n.* person who makes women's clothes, esp. for a living. □ **dressmaking** *n.*

dress rehearsal *n.* final rehearsal in full costume.

dress-shield *n.* waterproof material in the armpit of a dress to protect it from sweat.

dress-shirt *n.* man's shirt worn with evening dress, usu. white with concealed buttons or studs.

dressy *adj.* (**-ier, -iest**) *colloq.* (of clothes or a person) smart, elaborate, elegant. □ **dressiness** *n.*

drew *past* of DRAW.

drey /dreɪ/ *n.* squirrel's nest. [origin unknown]

dribble /'drɪb(ə)l/ *–v.* (**-ling**) **1** allow saliva to flow from the mouth. **2** flow or allow to flow in drops. **3** (also *absol.*) esp. *Football* & *Hockey* move (the ball) forward with slight touches of the feet or stick. –*n.* **1** act of dribbling. **2** dribbling flow. [obsolete *drib* = DRIP]

driblet /ˈdrɪblɪt/ *n.* small quantity.

dribs and drabs *n.pl. colloq.* small scattered amounts.

dried *past* and *past part.* of DRY.

drier[1] *compar.* of DRY.

drier[2] /ˈdraɪə(r)/ *n.* (also **dryer**) device for drying hair, laundry, etc.

driest *superl.* of DRY.

drift –*n.* **1 a** slow movement or variation. **b** this caused by a current. **2** intention, meaning, etc. of what is said etc. **3** mass of snow etc. heaped up by the wind. **4** esp. *derog.* state of inaction. **5** slow deviation of a ship, aircraft, etc., from its course. **6** fragments of rock heaped up (*glacial drift*). **7** *S.Afr.* ford. –*v.* **1** be carried by or as if by a current of air or water. **2** progress casually or aimlessly (*drifted into teaching*). **3** pile or be piled into drifts. **4** (of a current) carry, cause to drift. [Old Norse and Germanic *trift* movement of cattle]

drifter *n.* **1** aimless person. **2** boat used for drift-net fishing.

drift-net *n.* net for sea fishing, allowed to drift.

driftwood *n.* wood floating on moving water or washed ashore.

drill[1] –*n.* **1** tool or machine for boring holes, sinking wells, etc. **2** instruction in military exercises. **3** routine procedure in an emergency (*fire-drill*). **4** thorough training, esp. by repetition. **5** *colloq.* recognized procedure (*what's the drill?*). –*v.* **1 a** make a hole in or through with a drill. **b** make (a hole) with a drill. **2** train or be trained by drill. [Dutch]

drill[2] –*n.* **1** machine for making furrows, sowing, and covering seed. **2** small furrow. **3** row of seeds sown by a drill. –*v.* plant in drills. [origin unknown]

drill[3] *n.* coarse twilled cotton or linen fabric. [Latin *trilix* having three threads]

drill[4] *n.* W. African baboon related to the mandrill. [probably native]

drily *adv.* (also **dryly**) in a dry manner.

drink –*v.* (*past* **drank**; *past part.* **drunk**) **1 a** (also *absol.*) swallow (liquid). **b** swallow the contents of (a vessel). **2** take alcohol, esp. to excess. **3** (of a plant, sponge, etc.) absorb (moisture). **4** bring (oneself etc.) to a specified condition by drinking. **5** wish (a person good health etc.) by drinking (*drank his health*). –*n.* **1 a** liquid for drinking. **b** draught or specified amount of this. **2 a** alcoholic liquor. **b** portion, glass, etc. of this. **c** excessive use of alcohol (*took to drink*). **3** (**the drink**) *colloq.* the sea. □ **drink in** listen eagerly to. **drink to** toast; wish success to. **drink up** (also *absol.*) drink all or the remainder of. □ **drinkable** *adj.* **drinker** *n.* [Old English]

drink-driver *n.* person who drives with excess alcohol in the blood. □ **drink-driving** *n.*

drip –*v.* (**-pp-**) **1** fall or let fall in drops. **2** (often foll. by *with*) be so wet as to shed drops. –*n.* **1 a** liquid falling in drops (*steady drip of rain*). **b** drop of liquid. **c** sound of dripping. **2** *colloq.* dull or ineffectual person. **3** = DRIP-FEED. □ **be dripping with** be full of or covered with. [Danish: cf. DROP]

drip-dry –*v.* dry or leave to dry crease-free when hung up. –*adj.* able to be drip-dried.

drip-feed –*v.* feed intravenously in drops. –*n.* **1** feeding thus. **2** apparatus for doing this.

dripping *n.* fat melted from roasted meat.

drippy *adj.* (**-ier**, **-iest**) *slang* ineffectual; sloppily sentimental.

drive –*v.* (**-ving**; *past* **drove**; *past part.* **driven** /ˈdrɪv(ə)n/) **1** urge forward, esp. forcibly. **2 a** compel (*was driven to complain*). **b** force into a specified state (*drove him mad*). **c** (often *refl.*) urge to overwork. **3 a** operate and direct (a vehicle, locomotive, etc.). **b** convey or be conveyed in a vehicle. **c** be competent to drive (a vehicle) (*does he drive?*). **d** travel in a private vehicle. **4** (of wind etc.) carry along, propel, esp. rapidly (*driven snow*; *driving rain*). **5 a** (often foll. by *into*) force (a stake, nail, etc.) into place by blows. **b** bore (a tunnel etc.). **6** effect or conclude forcibly (*drove a hard bargain*; *drove his point home*). **7** (of power) operate (machinery). **8** (usu. foll. by *at*) work hard; dash, rush. **9** hit (a ball) forcibly. –*n.* **1** journey or excursion in a vehicle. **2 a** (esp. scenic) street or road. **b** private road through a garden to a house. **3 a** motivation and energy. **b** inner urge (*sex-drive*). **4** forcible stroke of a bat etc. **5** organized effort (*membership drive*). **6 a** transmission of power to machinery, wheels, etc. **b** position of the steering-wheel in a vehicle (*left-hand drive*). **c** *Computing* = DISK DRIVE. **7** organized whist, bingo, etc. competition. □ **drive at** seek, intend, or mean (*what is he driving at?*). [Old English]

drive-in –*attrib. adj.* (of a bank, cinema, etc.) used while sitting in one's car. –*n.* such a bank, cinema, etc.

drivel /ˈdrɪv(ə)l/ –*n.* silly talk; nonsense. –*v.* (**-ll-**; *US* **-l-**, **-ling**) **1** talk drivel. **2** run at the mouth or nose. [Old English]

driven *past part.* of DRIVE.

drive-on *adj.* (of a ship) on to which vehicles may be driven.
driver *n.* **1** person who drives a vehicle. **2** golf-club for driving from a tee.
driveway *n.* = DRIVE *n.* 2b.
driving-licence *n.* licence permitting one to drive a vehicle.
driving test *n.* official test of competence to drive.
driving-wheel *n.* wheel communicating motive power in machinery.
drizzle /'drɪz(ə)l/ –*n.* very fine rain. –*v.* (**-ling**) (of rain) fall in very fine drops. □ **drizzly** *adj.* [Old English]
droll /drəʊl/ *adj.* quaintly amusing; strange, odd. □ **drollery** *n.* (*pl.* **-ies**). **drolly** /'drəʊllɪ/ *adv.* [French]
dromedary /'drɒmɪdərɪ/ *n.* (*pl.* **-ies**) one-humped (esp. Arabian) camel bred for riding. [Greek *dromas -ados* runner]
drone –*n.* **1** non-working male of the honey-bee. **2** idler. **3** deep humming sound. **4** monotonous speaking tone. **5** bass-pipe of bagpipes or its continuous note. –*v.* (**-ning**) **1** make a deep humming sound. **2** speak or utter monotonously. [Old English]
drool *v.* **1** slobber, dribble. **2** (often foll. by *over*) admire extravagantly. [from DRIVEL]
droop –*v.* **1** bend or hang down, esp. from weariness; flag. **2** (of the eyes) look downwards. –*n.* **1** drooping attitude. **2** loss of spirit. □ **droopy** *adj.* [Old Norse: related to DROP]
drop –*n.* **1 a** globule of liquid that hangs, falls, or adheres to a surface. **b** very small amount of liquid (*just a drop left*). **c** glass etc. of alcohol. **2 a** abrupt fall or slope. **b** amount of this (*drop of fifteen feet*). **c** act of dropping. **d** fall in prices, temperature, etc. **e** deterioration (*drop in status*). **3** drop-shaped thing, esp. a pendant or sweet. **4** curtain or scenery let down on to a stage. **5** (in *pl.*) liquid medicine used in drops (*eye drops*). **6** minute quantity. **7** *slang* hiding-place for stolen goods etc. **8** *slang* bribe. –*v.* (**-pp-**) **1** fall or let fall in drops, shed (tears, blood). **2** fall or allow to fall; let go. **3 a** sink down from exhaustion or injury. **b** die. **c** fall naturally (*drop asleep*; *drop into the habit*). **4 a** (cause to) cease or lapse; abandon. **b** *colloq.* cease to associate with or discuss. **5** set down (a passenger etc.) (*drop me here*). **6** utter or be uttered casually (*dropped a hint*). **7** send casually (*drop a line*). **8 a** fall or allow to fall in direction, amount, condition, degree, pitch, etc. (*voice dropped*; *wind dropped*; *we dropped the price*). **b** (of a person) jump down lightly; let oneself fall. **c** allow (trousers etc.) to fall to the ground. **9** omit (a letter) in speech (*drop one's h's*). **10** (as **dropped** *adj.*) in a lower position than usual (*dropped handlebars*; *dropped waist*). **11** give birth to (esp. a lamb). **12** lose (a game, point, etc.). **13** deliver by parachute etc. **14** *Football* send (a ball), or score (a goal), by a drop-kick. **15** *colloq.* dismiss or omit (*dropped from the team*). □ **at the drop of a hat** promptly, instantly. **drop back** (or **behind**) fall back; get left behind. **drop a brick** *colloq.* make an indiscreet or embarrassing remark. **drop a curtsy** curtsy. **drop in** (or **by**) *colloq.* visit casually. **drop off 1** fall asleep. **2** drop (a passenger). **drop out** *colloq.* cease to participate. □ **droplet** *n.* [Old English]
drop-curtain *n.* painted curtain lowered on to a stage.
drop-kick *n.* *Football* kick as the ball touches the ground having been dropped.
drop-out *n.* *colloq.* person who has dropped out of conventional society, a course of study, etc.
dropper *n.* device for releasing liquid in drops.
droppings *n.pl.* **1** dung of animals or birds. **2** thing that falls or has fallen in drops.
drop scone *n.* scone made by dropping a spoonful of mixture into the pan etc.
drop-shot *n.* (in tennis) shot dropping abruptly over the net.
dropsy /'drɒpsɪ/ *n.* = OEDEMA. □ **dropsical** *adj.* [earlier *hydropsy* from Greek *hudrōps*: related to HYDRO-]
drosophila /drə'sɒfɪlə/ *n.* fruit fly used in genetic research. [Greek, = dew-loving]
dross *n.* **1** rubbish. **2 a** scum from melted metals. **b** impurities. [Old English]
drought /draʊt/ *n.* prolonged absence of rain. [Old English]
drove[1] *past* of DRIVE.
drove[2] *n.* **1 a** moving crowd. **b** (in *pl.*) *colloq.* great number (*people arrived in droves*). **2** herd or flock driven or moving together. [Old English: related to DRIVE]
drover *n.* herder of cattle.
drown *v.* **1** kill or die by submersion in liquid. **2** submerge; flood; drench. **3** deaden (grief etc.) by drinking. **4** (often foll. by *out*) overpower (sound) with louder sound. [probably Old English]
drowse /draʊz/ *v.* (**-sing**) be lightly asleep. [from DROWSY]
drowsy /'draʊzɪ/ *adj.* (**-ier, -iest**) very sleepy, almost asleep. □ **drowsily** *adv.* **drowsiness** *n.* [probably Old English]
drub *v.* (**-bb-**) **1** beat, thrash. **2** defeat thoroughly. □ **drubbing** *n.* [Arabic *ḍaraba* beat]

drudge *–n.* person who does dull, laborious, or menial work. *–v.* (**-ging**) work laboriously, toil. □ **drudgery** *n.* [origin uncertain]

drug *–n.* **1** medicinal substance. **2** (esp. addictive) narcotic, hallucinogen, or stimulant. *–v.* (**-gg-**) **1** add a drug to (food or drink). **2 a** give a drug to. **b** stupefy. [French]

drugget /'drʌɡɪt/ *n.* coarse woven fabric used for floor coverings etc. [French]

druggist *n.* pharmacist. [related to DRUG]

drugstore *n. US* combined chemist's shop and café.

Druid /'dru:ɪd/ *n.* **1** priest of an ancient Celtic religion. **2** member of a modern Druidic order, esp. the Gorsedd. □ **Druidic** /-'ɪdɪk/ *adj.* **Druidism** *n.* [Latin from Celtic]

drum *–n.* **1** hollow esp. cylindrical percussion instrument covered at the end(s) with plastic etc. **2** (often in *pl.*) percussion section of an orchestra etc. **3** sound made by a drum. **4** thing resembling a drum, esp. a container, etc. **5** segment of a pillar. **6** eardrum. *–v.* (**-mm-**) **1** play a drum. **2** beat or tap continuously with the fingers etc. **3** (of a bird or insect) make a loud noise with the wings. □ **drum into** drive (a lesson or facts) into (a person) by persistence. **drum out** dismiss with ignominy. **drum up** summon or get by vigorous effort (*drum up support*). [Low German]

drumbeat *n.* stroke or sound of a stroke on a drum.

drum brake *n.* brake in which brake shoes on a vehicle press against the brake drum on a wheel.

drumhead *n.* part of a drum that is hit.

drum kit *n.* set of drums in a band etc.

drum machine *n.* electronic device that simulates percussion.

drum major *n.* leader of a marching band.

drum majorette *n.* female baton-twirling member of a parading group.

drummer *n.* player of drums.

drumstick *n.* **1** stick for beating drums. **2** lower leg of a dressed fowl.

drunk *–adj.* **1** lacking control from drinking alcohol. **2** (often foll. by *with*) overcome with joy, success, power, etc. *–n.* person who is drunk, esp. habitually. [past part. of DRINK]

drunkard /'drʌŋkəd/ *n.* person who is habitually drunk.

drunken *adj.* (usu. *attrib.*) **1** = DRUNK 1. **2** caused by or involving drunkenness (*drunken brawl*). **3** often drunk. □ **drunkenly** *adv.* **drunkenness** *n.*

drupe /dru:p/ *n.* fleshy stone-fruit, e.g. the olive and plum. [Latin from Greek]

dry /draɪ/ *–adj.* (**drier**; **driest**) **1** free from moisture, esp.: **a** with moisture having evaporated, drained away, etc. (*clothes are not dry yet*). **b** (of eyes) free from tears. **c** (of a climate etc.) with insufficient rain; not rainy (*dry spell*). **d** (of a river, well, etc.) dried up. **e** using or producing no moisture (*dry shampoo*; *dry cough*). **f** (of a shave) with an electric razor. **2** (of wine) not sweet (*dry sherry*). **3 a** plain, unelaborated (*dry facts*). **b** uninteresting (*dry book*). **4** (of a sense of humour) subtle, ironic, understated. **5** prohibiting the sale of alcohol (*a dry State*). **6** (of bread) without butter etc. **7** (of provisions etc.) solid, not liquid. **8** impassive. **9** (of a cow) not yielding milk. **10** *colloq.* thirsty (*feel dry*). *–v.* (**dries, dried**) **1** make or become dry. **2** (usu. as **dried** *adj.*) preserve (food etc.) by removing moisture. **3** (often foll. by *up*) *colloq.* forget one's lines. *–n.* (*pl.* **dries**) **1** act of drying. **2** dry ginger ale. **3** dry place (*come into the dry*). □ **dry out 1** make or become fully dry. **2** treat or be treated for alcoholism. **dry up 1** make or become utterly dry. **2** dry dishes. **3** *colloq.* (esp. in *imper.*) cease talking. **4** become unproductive. **5** (of supplies) run out. □ **dryness** *n.* [Old English]

dryad /'draɪæd/ *n.* wood nymph. [Greek *drus* tree]

dry battery *n.* (also **dry cell**) electric battery or cell in which electrolyte is absorbed in a solid.

dry-clean *v.* clean (clothes etc.) with solvents without water. □ **dry-cleaner** *n.*

dry dock *n.* dock that can be pumped dry for building or repairing ships.

dryer var. of DRIER[2].

dry-fly *attrib. adj.* (of fishing) with a floating artificial fly.

dry ice *n.* solid carbon dioxide used as a refrigerant.

dry land *n.* land as distinct from sea etc.

dryly var. of DRILY.

dry measure *n.* measure for dry goods.

dry rot *n.* decayed state of unventilated wood; fungi causing this.

dry run *n. colloq.* rehearsal.

dry-shod *adj.* & *adv.* without wetting one's shoes.

drystone *attrib. adj.* (of a wall etc.) built without mortar.

DSC *abbr.* Distinguished Service Cross.

D.Sc. *abbr.* Doctor of Science.

DSM *abbr.* Distinguished Service Medal.

DSO *abbr.* Distinguished Service Order.

DSS *abbr.* Department of Social Security (formerly DHSS).

DT *abbr.* (also **DT's** /di:'ti:z/) delirium tremens.

DTI *abbr.* Department of Trade and Industry.

dual /'dju:əl/ *–adj.* **1** in two parts; twofold. **2** double (*dual ownership*). *–n. Gram.* dual number or form. □ **duality** /-'ælɪtɪ/ *n.* [Latin *duo* two]

dual carriageway *n.* road with a dividing strip between traffic flowing in opposite directions.

dual control *n.* two linked sets of controls, enabling operation by either of two persons.

dub[1] *v.* (**-bb-**) **1** make (a person) a knight by touching his shoulders with a sword. **2** give (a person) a name, nickname, etc. **3** smear (leather) with grease. [French]

dub[2] *v.* (**-bb-**) **1** provide (a film etc.) with an, esp. translated, alternative soundtrack. **2** add (sound effects or music) to a film or broadcast. **3** transfer or make a copy of (recorded sound or images). [abbreviation of DOUBLE]

dubbin /'dʌbɪn/ *n.* (also **dubbing**) thick grease for softening and waterproofing leather. [see DUB[1] 3]

dubiety /dju:'baɪətɪ/ *n. literary* doubt. [Latin: related to DUBIOUS]

dubious /'dju:bɪəs/ *adj.* **1** hesitating, doubtful. **2** questionable; suspicious. **3** unreliable. □ **dubiously** *adv.* **dubiousness** *n.* [Latin *dubium* doubt]

ducal /'dju:k(ə)l/ *adj.* of or like a duke. [French: related to DUKE]

ducat /'dʌkət/ *n.* gold coin, formerly current in most of Europe. [medieval Latin *ducatus* DUCHY]

duchess /'dʌtʃɪs/ *n.* **1** duke's wife or widow. **2** woman holding the rank of duke. [medieval Latin *ducissa*: related to DUKE]

duchesse potatoes /du:'ʃes/ *n.pl.* mashed potatoes mixed with egg, baked or fried, and served as small cakes or used as piping. [French]

duchy /'dʌtʃɪ/ *n.* (*pl.* **-ies**) territory of a duke or duchess; royal dukedom of Cornwall or Lancaster. [medieval Latin *ducatus*: related to DUKE]

duck[1] *–n.* (*pl.* same or **-s**) **1 a** swimming-bird, esp. the domesticated form of the mallard or wild duck. **b** female of this. **c** its flesh as food. **2** score of 0 in cricket. **3** (also **ducks**) *colloq.* (esp. as a form of address) dear. *–v.* **1** bob down, esp. to avoid being seen or hit. **2 a** dip one's head briefly under water. **b** plunge (a person) briefly in water. **3** *colloq.* dodge (a task etc.). □ **like water off a duck's back** *colloq.* producing no effect. [Old English]

duck[2] *n.* **1** strong linen or cotton fabric. **2** (in *pl.*) trousers made of this. [Dutch]

duckbill *n.* (also **duck-billed platypus**) = PLATYPUS.

duckboard *n.* (usu. in *pl.*) path of wooden slats over muddy ground, in a trench, etc.

duckling *n.* young duck.

ducks and drakes *n.pl.* (usu. treated as *sing.*) game of making a flat stone skim the surface of water. □ **play ducks and drakes with** *colloq.* squander.

duckweed *n.* any of various plants growing on the surface of still water.

ducky *n.* (*pl.* **-ies**) *colloq.* (esp. as a form of address) dear.

duct *–n.* channel or tube for conveying a fluid, cable, bodily secretions, etc. (*tear ducts*). *–v.* convey through a duct. [Latin *ductus* from *duco duct-* lead]

ductile /'dʌktaɪl/ *adj.* **1** (of metal) capable of being drawn into wire; pliable. **2** easily moulded. **3** docile. □ **ductility** /-'tɪlɪtɪ/ *n.* [Latin: related to DUCT]

ductless gland *n.* gland secreting directly into the bloodstream.

dud *slang –n.* **1** useless or broken thing. **2** counterfeit article. **3** (in *pl.*) clothes, rags. *–adj.* useless, defective. [origin unknown]

dude /du:d/ *n. slang* **1** fellow. **2** *US* dandy. **3** *US* city-dweller staying on a ranch. [German dial. *dude* fool]

dudgeon /'dʌdʒ(ə)n/ *n.* resentment, indignation. □ **in high dudgeon** very angry. [origin unknown]

due *–adj.* **1** owing or payable. **2** (often foll. by *to*) merited; appropriate. **3** (foll. by *to*) that ought to be given or ascribed to (a person, cause, etc.) (*received the applause due to a hero*; *difficulty due to ignorance*). **4** (often foll. by *to* + infin.) expected or under an obligation at a certain time (*due to speak tonight*; *train due at 7.30*). **5** suitable, right, proper (*in due time*). *–n.* **1** what one owes or is owed (*give him his due*). **2** (usu. in *pl.*) fee or amount payable. *–adv.* (of a compass point) exactly, directly (*went due east*). □ **due to** because of (*he was late due to an accident*). [French from Latin *debeo* owe]

■ **Usage** The use of *due to* to mean 'because of' as in the example given is regarded as unacceptable by some people and could be avoided by substituting *his lateness was due to an accident*. Alternatively, *owing to* could be used.

duel /'dju:əl/ *–n.* **1** armed contest between two people, usu. to the death. **2** two-sided contest. *–v.* (**-ll-**; *US* **-l-**) fight a duel. □ **duellist** *n.* [Latin *duellum* war]

duenna /dju:'enə/ *n.* older woman acting as a chaperon to girls, esp. in Spain. [Spanish from Latin *domina* DON[1]]

duet /dju:'et/ *n.* musical composition for two performers. □ **duettist** *n.* [Latin *duo* two]

duff[1] *–n.* boiled pudding. *–adj. slang* worthless, counterfeit, useless. [var. of DOUGH]

duff[2] *v.* □ **duff up** *slang* beat; thrash. [perhaps from DUFFER]

duffer *n. colloq.* inefficient or stupid person; dunce. [origin uncertain]

duffle /'dʌf(ə)l/ *n.* (also **duffel**) heavy woollen cloth. [*Duffel* in Belgium]

duffle bag *n.* cylindrical canvas bag closed by a drawstring.

duffle-coat *n.* hooded overcoat of duffle, fastened with toggles.

dug[1] *past* and *past part.* of DIG.

dug[2] *n.* udder, teat. [origin unknown]

dugong /'du:gɒŋ/ *n.* (*pl.* same or **-s**) Asian sea-mammal. [Malay]

dugout *n.* **1 a** roofed shelter, esp. for troops in trenches. **b** underground shelter. **2** canoe made from a tree-trunk.

duke /dju:k/ *n.* **1** person holding the highest hereditary title of the nobility. **2** sovereign prince ruling a duchy or small State. □ **dukedom** *n.* [Latin *dux* leader]

dulcet /'dʌlsɪt/ *adj.* sweet-sounding. [Latin *dulcis* sweet]

dulcimer /'dʌlsɪmə(r)/ *n.* metal stringed instrument struck with two handheld hammers. [Latin: related to DULCET, *melos* song]

dull *–adj.* **1** tedious; not interesting. **2** (of the weather) overcast. **3** (of colour, light, sound, etc.) not bright, vivid, or clear. **4** (of a pain) indistinct; not acute (*a dull ache*). **5** slow-witted; stupid. **6** (of a knife-edge etc.) blunt. **7 a** (of trade etc.) sluggish, slow. **b** listless; depressed. **8** (of the ears, eyes, etc.) lacking keenness. *–v.* make or become dull. □ **dullness** *n.* **dully** /'dʌllɪ/ *adv.* [Low German or Dutch]

dullard /'dʌləd/ *n.* stupid person.

duly /'dju:lɪ/ *adv.* **1** in due time or manner. **2** rightly, properly.

dumb /dʌm/ *adj.* **1 a** unable to speak. **b** (of an animal) naturally dumb. **2** silenced by surprise, shyness, etc. **3** taciturn, reticent (*dumb insolence*). **4** suffered or done in silence (*dumb agony*). **5** *colloq.* stupid; ignorant. **6** disenfranchised; inarticulate (*dumb masses*). **7** (of a computer terminal etc.) able to transmit or receive but unable to process data. **8** giving no sound. [Old English]

dumb-bell *n.* **1** short bar with a weight at each end, for muscle-building etc. **2** *slang* stupid person, esp. a woman.

dumbfound /dʌm'faʊnd/ *v.* nonplus, make speechless with surprise. [from DUMB, CONFOUND]

dumbo /'dʊmbəʊ/ *n.* (*pl.* **-s**) *slang* stupid person. [from DUMB, -O]

dumb show *n.* gestures; mime.

dumbstruck *adj.* speechless with surprise.

dumb waiter *n.* small hand-operated lift for conveying food from kitchen to dining-room.

dumdum /'dʌmdʌm/ *n.* (in full **dumdum bullet**) soft-nosed bullet that expands on impact. [*Dum-Dum* in India]

dummy /'dʌmɪ/ *–n.* (*pl.* **-ies**) **1** model of a human figure, esp. as used to display clothes or by a ventriloquist or as a target. **2** (often *attrib.*) imitation object used to replace a real or normal one. **3** baby's rubber or plastic teat. **4** *colloq.* stupid person. **5** figurehead. **6** imaginary player in bridge etc., whose cards are exposed and played by a partner. *–attrib. adj.* sham; imitation. *–v.* (**-ies, -ied**) make a pretended pass or swerve in football etc. [from DUMB]

dummy run *n.* trial attempt; rehearsal.

dump *–n.* **1** place or heap for depositing rubbish. **2** *colloq.* unpleasant or dreary place. **3** temporary store of ammunition etc. *–v.* **1** put down firmly or clumsily. **2** deposit as rubbish. **3** *colloq.* abandon or get rid of. **4** sell (excess goods) to a foreign market at a low price. **5** copy (the contents of a computer memory etc.) as a diagnostic aid or for security. □ **dump on** esp. *US* criticize or abuse; get the better of. [origin uncertain]

dumpling /'dʌmplɪŋ/ *n.* **1** ball of dough boiled in stew or containing apple etc. **2** small fat person. [*dump* small round object]

dumps *n.pl.* (usu. in **down in the dumps**) *colloq.* low spirits. [Low German or Dutch: related to DAMP]

dump truck *n.* truck that tilts or opens at the back for unloading.

dumpy *adj.* (**-ier, -iest**) short and stout. □ **dumpily** *adv.* **dumpiness** *n.* [related to DUMPLING]

dun *–adj.* greyish-brown. *–n.* **1** dun colour. **2** dun horse. [Old English]

dunce *n.* person slow at learning; dullard. [*Duns* Scotus, name of a philosopher]

dunce's cap *n.* paper cone worn by a dunce.

dunderhead /'dʌndə,hed/ *n.* stupid person. [origin unknown]

dune *n.* drift of sand etc. formed by the wind. [Dutch: related to DOWN[3]]

dung –*n.* excrement of animals; manure. –*v.* apply dung to (land). [Old English]
dungaree /ˌdʌŋgəˈriː/ *n.* **1** coarse cotton cloth. **2** (in *pl.*) overalls or trousers of this. [Hindi]
dung-beetle *n.* beetle whose larvae develop in dung.
dungeon /ˈdʌndʒ(ə)n/ *n.* underground prison cell. [earlier *donjon* keep of a castle; ultimately from Latin *dominus* lord]
dunghill *n.* heap of dung or refuse.
dunk *v.* **1** dip (food) into liquid before eating. **2** immerse. [German *tunken* dip]
dunlin /ˈdʌnlɪn/ *n.* red-backed sandpiper. [probably from DUN]
dunnock /ˈdʌnək/ *n.* hedge sparrow. [apparently from DUN]
duo /ˈdjuːəʊ/ *n.* (*pl.* **-s**) **1** pair of performers. **2** duet. [Italian from Latin, = two]
duodecimal /ˌdjuːəʊˈdesɪm(ə)l/ *adj.* **1** of twelfths or twelve. **2** in or by twelves. [Latin *duodecim* twelve]
duodenum /ˌdjuːəʊˈdiːnəm/ *n.* (*pl.* **-s**) first part of the small intestine immediately below the stomach. □ **duodenal** *adj.* [medieval Latin: related to DUODECIMAL]
duologue /ˈdjuːəˌlɒg/ *n.* dialogue between two people. [from DUO, MONOLOGUE]
dupe –*n.* victim of deception. –*v.* (**-ping**) deceive, trick. [French]
duple /ˈdjuːp(ə)l/ *adj.* of two parts. [Latin *duplus*]
duple time *n. Mus.* rhythm with two beats to the bar.
duplex /ˈdjuːpleks/ –*n.* (often *attrib.*) esp. *US* **1** flat on two floors. **2** house subdivided for two families; semidetached house. –*adj.* **1** of two parts. **2** *Computing* (of a circuit) allowing simultaneous two-way transmission of signals. [Latin, = double]
duplicate –*adj.* /ˈdjuːplɪkət/ **1** identical. **2 a** having two identical parts. **b** doubled. **3** (of card-games) with the same hands played by different players. –*n.* /ˈdjuːplɪkət/ **1** identical thing, esp. a copy. **2** copy of a letter etc. –*v.* /ˈdjuːplɪˌkeɪt/ (**-ting**) **1** multiply by two; double. **2** make or be an exact copy of. **3** repeat (an action etc.), esp. unnecessarily. □ **in duplicate** in two exact copies. □ **duplication** /-ˈkeɪʃ(ə)n/ *n.* [Latin: related to DUPLEX]
duplicator *n.* machine for making multiple copies of a text etc.
duplicity /djuːˈplɪsɪtɪ/ *n.* double-dealing; deceitfulness. □ **duplicitous** *adj.* [Latin: related to DUPLEX]
durable /ˈdjʊərəb(ə)l/ –*adj.* **1** lasting; hard-wearing. **2** (of goods) with a relatively long useful life. –*n.* (in *pl.*) durable goods. □ **durability** /-ˈbɪlɪtɪ/ *n.* [Latin *durus* hard]
dura mater /ˌdjʊərə ˈmeɪtə(r)/ *n.* tough outermost membrane enveloping the brain and spinal cord. [medieval Latin = hard mother, translation of Arabic]
duration /djʊəˈreɪʃ(ə)n/ *n.* **1** time taken by an event. **2** specified length of time (*duration of a minute*). □ **for the duration 1** until the end of an event. **2** for a very long time. [medieval Latin: related to DURABLE]
duress /djʊəˈres/ *n.* **1** compulsion, esp. illegal use of threats or violence (*under duress*). **2** imprisonment. [Latin *durus* hard]
Durex /ˈdjʊəreks/ *n. propr.* condom. [origin uncertain]
during /ˈdjʊərɪŋ/ *prep.* throughout or at some point in. [Latin: related to DURABLE]
dusk *n.* darker stage of twilight. [Old English]
dusky *adj.* (**-ier, -iest**) **1** shadowy; dim. **2** dark-coloured; dark-skinned. □ **duskily** *adv.* **duskiness** *n.*
dust –*n.* **1** finely powdered earth or other material etc. (*pollen dust*). **2** dead person's remains. **3** confusion, turmoil. –*v.* **1** wipe the dust from (furniture etc.). **2 a** sprinkle with powder, sugar, etc. **b** sprinkle (sugar, powder, etc.). □ **dust down 1** dust the clothes of. **2** *colloq.* reprimand. **3** = *dust off*. **dust off 1** remove the dust from. **2** use again after a long period. **when the dust settles** when things quieten down. [Old English]
dustbin *n.* container for household refuse.
dust bowl *n.* desert made by drought or erosion.
dustcart *n.* vehicle collecting household refuse.
dust cover *n.* **1** = DUST-SHEET. **2** = DUST-JACKET.
duster *n.* cloth for dusting furniture etc.
dust-jacket *n.* paper cover on a hardback book.
dustman *n.* person employed to collect household refuse.
dustpan *n.* pan into which dust is brushed from the floor.
dust-sheet *n.* protective cloth over furniture.
dust-up *n. colloq.* fight, disturbance.
dusty *adj.* (**-ier, -iest**) **1** full of or covered with dust. **2** (of a colour) dull or muted. □ **not so dusty** *slang* fairly good. □ **dustily** *adv.* **dustiness** *n.*
dusty answer *n. colloq.* curt refusal.

Dutch *–adj.* of the Netherlands or its people or language. *–n.* **1** the Dutch language. **2** (prec. by *the*; treated as *pl.*) the people of the Netherlands. □ **go Dutch** share expenses on an outing etc. [Dutch]

dutch *n. slang* wife. [abbreviation of DUCHESS]

Dutch auction *n.* one in which the price is progressively reduced.

Dutch barn *n.* roof for hay etc., set on poles.

Dutch cap *n.* dome-shaped contraceptive device fitting over the cervix.

Dutch courage *n.* courage induced by alcohol.

Dutch elm disease *n.* fungus disease of elms.

Dutchman *n.* (*fem.* **Dutchwoman**) person of Dutch birth or nationality.

Dutch oven *n.* **1** metal box with the open side facing a fire. **2** covered cooking-pot for braising etc.

Dutch treat *n.* party, outing, etc., at which people pay for themselves.

Dutch uncle *n.* kind but firm adviser.

duteous /'dju:tɪəs/ *adj. literary* dutiful. □ **duteously** *adv.*

dutiable /'dju:tɪəb(ə)l/ *adj.* requiring the payment of duty.

dutiful /'dju:tɪ,fʊl/ *adj.* doing one's duty; obedient. □ **dutifully** *adv.*

duty /'dju:tɪ/ *n.* (*pl.* **-ies**) **1 a** moral or legal obligation; responsibility. **b** binding force of what is right. **2** tax on certain goods, imports, etc. **3** job or function arising from a business or office (*playground duty*). **4** deference; respect due to a superior. □ **do duty for** serve as or pass for (something else). **on** (or **off**) **duty** working (or not working). [Anglo-French: related to DUE]

duty-bound *adj.* obliged by duty.

duty-free *adj.* (of goods) on which duty is not payable.

duty-free shop *n.* shop at an airport etc. selling duty-free goods.

duvet /'du:veɪ/ *n.* thick soft quilt used instead of sheets and blankets. [French]

dwarf /dwɔ:f/ *–n.* (*pl.* -s or **dwarves** /dwɔ:vz/) **1** person, animal, or plant much below normal size. **2** small mythological being with magical powers. **3** small usu. dense star. *–v.* **1** stunt in growth. **2** make seem small. □ **dwarfish** *adj.* [Old English]

■ **Usage** In sense 1, with regard to people, the term *person of restricted growth* is now often preferred.

dwell *v.* (*past* and *past part.* **dwelt** or **dwelled**) live, reside. □ **dwell on** (or **upon**) think, write, or speak at length on. □ **dweller** *n.* [Old English, = lead astray]

dwelling *n.* house, residence.

dwindle /'dwɪnd(ə)l/ *v.* (**-ling**) **1** become gradually less or smaller. **2** lose importance. [Old English]

Dy *symb.* dysprosium.

dye /daɪ/ *–n.* **1** substance used to change the colour of hair, fabric, etc. **2** colour so produced. *–v.* (**dyeing**, **dyed**) **1** colour with dye. **2** dye a specified colour (*dyed it yellow*). □ **dyer** *n.* [Old English]

dyed-in-the-wool *adj.* (usu. *attrib.*) out and out; unchangeable.

dying /'daɪɪŋ/ *attrib. adj.* of, or at the time of, death (*dying words*).

dyke[1] /daɪk/ (also **dike**) *–n.* **1** embankment built to prevent flooding. **2** low wall of turf or stone. *–v.* (**-king**) provide or protect with dyke(s). [related to DITCH]

dyke[2] /daɪk/ *n.* (also **dike**) *slang* lesbian. [origin unknown]

dynamic /daɪ'næmɪk/ *adj.* **1** energetic; active. **2** *Physics* **a** of motive force. **b** of force in actual operation. **3** of dynamics. □ **dynamically** *adv.* [Greek *dunamis* power]

dynamics *n.pl.* **1** (usu. treated as *sing.*) **a** mathematical study of motion and the forces causing it. **b** branch of any science concerned with forces or changes. **2** motive forces in any sphere. **3** *Mus.* variation in loudness.

dynamism /'daɪnə,mɪz(ə)m/ *n.* energy; dynamic power.

dynamite /'daɪnə,maɪt/ *–n.* **1** high explosive mixture containing nitroglycerine. **2** potentially dangerous person etc. *–v.* (**-ting**) charge or blow up with dynamite.

dynamo /'daɪnə,məʊ/ *n.* (*pl.* -s) **1** machine converting mechanical into electrical energy, esp. by rotating coils of copper wire in a magnetic field. **2** *colloq.* energetic person. [abbreviation of *dynamo-electric machine*]

dynamometer /,daɪnə'mɒmɪtə(r)/ *n.* instrument measuring energy expended. [Greek: related to DYNAMIC]

dynast /'dɪnəst/ *n.* **1** ruler. **2** member of a dynasty. [Latin from Greek]

dynasty /'dɪnəstɪ/ *n.* (*pl.* **-ies**) **1** line of hereditary rulers. **2** succession of leaders in any field. □ **dynastic** /-'næstɪk/ *adj.* [Latin from Greek]

dyne /daɪn/ *n. Physics* force required to give a mass of one gram an acceleration of one centimetre per second per second. [Greek *dunamis* force]

dys- *prefix* bad, difficult. [Greek]

dysentery /'dɪsəntrɪ/ *n.* inflammation of the intestines, causing severe diarrhoea. [Greek *entera* bowels]
dysfunction /dɪs'fʌŋkʃ(ə)n/ *n.* abnormality or impairment of functioning.
dyslexia /dɪs'leksɪə/ *n.* abnormal difficulty in reading and spelling. □ **dyslectic** /-'lektɪk/ *adj.* & *n.* **dyslexic** *adj.* & *n.* [Greek *lexis* speech]
dysmenorrhoea /,dɪsmenə'rɪə/ *n.* painful or difficult menstruation.
dyspepsia /dɪs'pepsɪə/ *n.* indigestion. □ **dyspeptic** *adj.* & *n.* [Greek *peptos* digested]
dysphasia /,dɪs'feɪzɪə/ *n.* lack of coordination in speech, owing to brain damage. [Greek *dusphatos* hard to utter]
dysprosium /dɪs'prəʊzɪəm/ *n.* metallic element of the lanthanide series. [Greek *dusprositos* hard to get at]
dystrophy /'dɪstrəfɪ/ *n.* defective nutrition. [Greek *-trophē* nourishment]

E

E[1] /iː/ *n.* (also **e**) (*pl.* **Es** or **E's**) **1** fifth letter of the alphabet. **2** *Mus.* third note of the diatonic scale of C major.

E[2] *abbr.* (also **E.**) **1** east, eastern. **2** see E-NUMBER.

e- *prefix* see EX-[1] before some consonants.

each *–adj.* every one of two or more persons or things, regarded separately (*five in each class*). *–pron.* each person or thing (*each of us*). [Old English]

each other *pron.* one another.

each way *adj.* (of a bet) backing a horse etc. to win or to come second or third.

eager /ˈiːgə(r)/ *adj.* keen, enthusiastic (*eager to learn*; *eager for news*). □ **eagerly** *adv.* **eagerness** *n.* [Latin *acer* keen]

eager beaver *n. colloq.* very diligent person.

eagle /ˈiːg(ə)l/ *n.* **1 a** large bird of prey with keen vision and powerful flight. **b** this as a symbol, esp. of the US. **2** score of two strokes under par at any hole in golf. [Latin *aquila*]

eagle eye *n.* keen sight, watchfulness. □ **eagle-eyed** *adj.*

eaglet /ˈiːglɪt/ *n.* young eagle.

E. & O. E. *abbr.* errors and omissions excepted.

ear[1] *n.* **1** organ of hearing, esp. its external part. **2** faculty for discriminating sounds (*an ear for music*). **3** attention, esp. sympathetic (*give ear to*; *have a person's ear*). □ **all ears** listening attentively. **have** (or **keep**) **an ear to the ground** be alert to rumours or trends. **up to one's ears** (often foll. by *in*) *colloq.* deeply involved or occupied. [Old English]

ear[2] *n.* seed-bearing head of a cereal plant. [Old English]

earache *n.* pain in the inner ear.

eardrum *n.* membrane of the middle ear.

earful *n.* (*pl.* **-s**) *colloq.* **1** prolonged amount of talking. **2** strong reprimand.

earl /ɜːl/ *n.* British nobleman ranking between marquis and viscount. □ **earldom** *n.* [Old English]

Earl Marshal *n.* president of the College of Heralds, with ceremonial duties.

early /ˈɜːlɪ/ *–adj. & adv.* (**-ier, -iest**) **1** before the due, usual, or expected time. **2 a** not far on in the day or night, or in time (*early evening*; *at the earliest opportunity*). **b** prompt (*early payment appreciated*). **3** not far on in a period, development, or process of evolution; being the first stage (*Early English architecture*; *early spring*). **4** forward in flowering, ripening, etc. (*early peaches*). *–n.* (*pl.* **-ies**) (usu. in *pl.*) early fruit or vegetable. □ **earliness** *n.* [Old English: related to ERE]

early bird *n. colloq.* person who arrives, gets up, etc. early.

early days *n.pl.* too soon to expect results etc.

early on *adv.* at an early stage.

earmark *–v.* set aside for a special purpose. *–n.* identifying mark.

earn /ɜːn/ *v.* **1** bring in as income or interest. **2** be entitled to or obtain as the reward for work or merit. □ **earner** *n.* [Old English]

earnest /ˈɜːnɪst/ *adj.* intensely serious. □ **in earnest** serious, seriously, with determination. □ **earnestly** *adv.* **earnestness** *n.* [Old English]

earnings *n.pl.* money earned.

earphone *n.* device applied to the ear to receive a radio etc. communication.

earpiece *n.* part of a telephone etc. applied to the ear.

ear-piercing *–adj.* shrill. *–n.* piercing of the ears for wearing earrings.

earplug *n.* piece of wax etc. placed in the ear to protect against water, noise, etc.

earring *n.* jewellery worn on the ear.

earshot *n.* hearing-range (*within earshot*).

ear-splitting *adj.* excessively loud.

earth /ɜːθ/ *–n.* **1 a** (also **Earth**) the planet on which we live. **b** land and sea, as distinct from sky. **2 a** the ground (*fell to earth*). **b** soil, mould. **3** *Relig.* this world, as distinct from heaven or hell. **4** connection to the earth as the completion of an electrical circuit. **5** hole of a fox etc. **6** (prec. by *the*) *colloq.* huge sum; everything (*cost the earth*; *want the earth*). *–v.* **1** cover (plant-roots) with earth. **2** connect (an electrical circuit) to the earth. □ **come back** (or **down**) **to earth** return to realities. **gone to earth** in hiding. **on earth** *colloq.* existing anywhere; emphatically (*the happiest man on earth*; *looked like nothing on earth*; *what on earth have you done?*). **run to earth** find after a long search. □ **earthward** *adj. & adv.* **earthwards** *adv.* [Old English]

earthbound *adj.* **1** attached to the earth or earthly things. **2** moving towards the earth.

earthen *adj.* made of earth or baked clay.

earthenware *n.* pottery made of fired clay.

earthling *n.* inhabitant of the earth, esp. in science fiction.

earthly *adj.* **1** of the earth or human life on it; terrestrial. **2** (usu. with *neg.*) *colloq.* remotely possible (*is no earthly use*; *there wasn't an earthly reason*). □ **not an earthly** *colloq.* no chance or idea whatever.

earth mother *n.* sensual and maternal woman.

earthquake *n.* convulsion of the earth's surface as a result of faults in strata or volcanic action.

earth sciences *n.pl.* those concerned with the earth or part of it.

earth-shattering *adj. colloq.* traumatic, devastating. □ **earth-shatteringly** *adv.*

earthwork *n.* artificial bank of earth in fortification or road-building etc.

earthworm *n.* common worm living in the ground.

earthy *adj.* (**-ier, -iest**) **1** of or like earth or soil. **2** coarse, crude (*earthy humour*). □ **earthiness** *n.*

ear-trumpet *n.* trumpet-shaped device formerly used as a hearing-aid.

earwig *n.* small insect with pincers at its rear end. [from EAR[1], because they were once thought to enter the head through the ear]

ease /iːz/ *–n.* **1** facility, effortlessness. **2 a** freedom from pain or trouble. **b** freedom from constraint. *–v.* (**-sing**) **1** relieve from pain or anxiety. **2** (often foll. by *off*, *up*) **a** become less burdensome or severe. **b** begin to take it easy. **c** slow down; moderate one's behaviour etc. **3 a** relax; slacken; make a less tight fit. **b** move or be moved carefully into place (*eased it into position*). □ **at ease 1** free from anxiety or constraint. **2** *Mil.* in a relaxed attitude, with the feet apart. [Latin: related to ADJACENT]

easel /'iːz(ə)l/ *n.* stand for an artist's work, a blackboard, etc. [Dutch *ezel* ass]

easement *n.* legal right of way or similar right over another's land. [French: related to EASE]

easily /'iːzɪlɪ/ *adv.* **1** without difficulty. **2** by far (*easily the best*). **3** very probably (*it could easily snow*).

east *–n.* **1 a** point of the horizon where the sun rises at the equinoxes. **b** compass point corresponding to this. **c** direction in which this lies. **2** (usu. **the East**) **a** countries to the east of Europe. **b** States of eastern Europe. **3** eastern part of a country, town, etc. *–adj.* **1** towards, at, near, or facing the east. **2** from the east (*east wind*). *–adv.* **1** towards, at, or near the east. **2** (foll. by *of*) further east than. □ **to the east** (often foll. by *of*) in an easterly direction. [Old English]

eastbound *adj.* travelling or leading eastwards.

East End *n.* part of London east of the City. □ **East Ender** *n.*

Easter *n.* festival (held on a variable Sunday in March or April) commemorating Christ's resurrection. [Old English]

Easter egg *n.* artificial usu. chocolate egg given at Easter.

easterly *–adj.* & *adv.* **1** in an eastern position or direction. **2** (of a wind) from the east. *–n.* (*pl.* **-ies**) such a wind.

eastern *adj.* of or in the east. □ **easternmost** *adj.*

Eastern Church *n.* Orthodox Church.

easterner *n.* native or inhabitant of the east.

east-north-east *n.* point or direction midway between east and north-east.

east-south-east *n.* point or direction midway between east and south-east.

eastward *–adj.* & *adv.* (also **eastwards**) towards the east. *–n.* eastward direction or region.

easy /'iːzɪ/ *–adj.* (**-ier, -iest**) **1** not difficult; not requiring great effort. **2** free from pain, trouble, or anxiety. **3** free from constraint; relaxed and pleasant. **4** compliant. *–adv.* with ease; in an effortless or relaxed manner. *–int.* go or move carefully. □ **easy on the eye** (or **ear** etc.) *colloq.* pleasant to look at (or listen to etc.). **go easy** (foll. by *with*, *on*) be sparing or cautious. **I'm easy** *colloq.* I have no preference. **take it easy 1** proceed gently. **2** relax; work less. □ **easiness** *n.* [French: related to EASE]

easy chair *n.* large comfortable armchair.

easygoing *adj.* placid and tolerant.

Easy Street *n. colloq.* affluence.

eat *–v.* (*past* **ate** /et, eɪt/; *past part.* **eaten**) **1 a** take into the mouth, chew, and swallow (food). **b** consume food; take a meal. **c** devour (*eaten by a lion*). **2** (foll. by *away*, *at*, *into*) **a** destroy gradually, esp. by corrosion, disease, etc. **b** begin to consume or diminish (resources etc.). **3** *colloq.* trouble, vex (*what's eating you?*). *–n.* (in *pl.*) *colloq.* food. □ **eat one's heart out** suffer from excessive longing or envy. **eat out** have a meal away from home, esp. in a restaurant. **eat up 1** eat completely. **2** use or deal with rapidly or wastefully (*eats* *up petrol; eats up the miles*). **3** preoccupy (*eaten up with envy*). **eat one's words** retract them abjectly. [Old English]

eatable –*adj.* fit to be eaten. –*n.* (usu. in *pl.*) food.

eater *n.* **1** person who eats (*a big eater*). **2** eating apple etc.

eating apple etc. *n.* apple etc. suitable for eating raw.

eau-de-Cologne /ˌəʊdəkəˈləʊn/ *n.* toilet water orig. from Cologne. [French, = water of Cologne]

eaves /iːvz/ *n.pl.* underside of a projecting roof. [Old English]

eavesdrop *v.* (**-pp-**) listen to a private conversation. □ **eavesdropper** *n.*

ebb –*n.* movement of the tide out to sea. –*v.* (often foll. by *away*) **1** flow out to sea; recede. **2** decline (*life was ebbing away*). [Old English]

ebonite /ˈebəˌnaɪt/ *n.* vulcanite. [from EBONY]

ebony /ˈebənɪ/ –*n.* heavy hard dark wood of a tropical tree. –*adj.* **1** made of ebony. **2** black like ebony. [Greek *ebenos* ebony tree]

ebullient /ɪˈbʌlɪənt/ *adj.* exuberant. □ **ebullience** *n.* **ebulliency** *n.* **ebulliently** *adv.* [Latin: related to BOIL[1]]

EC *abbr.* **1** East Central. **2** European Community.

eccentric /ɪkˈsentrɪk/ –*adj.* **1** odd or capricious in behaviour or appearance. **2** (also **excentric**) **a** not placed, not having its axis placed, centrally. **b** (often foll. by *to*) (of a circle) not concentric (to another). **c** (of an orbit) not circular. –*n.* **1** eccentric person. **2** disc at the end of a shaft for changing rotatory into backward-and-forward motion. □ **eccentrically** *adv.* **eccentricity** /ˌeksenˈtrɪsətɪ/ *n.* [Greek: related to CENTRE]

Eccles cake /ˈek(ə)lz/ *n.* round cake of pastry filled with currants etc. [*Eccles* in N. England]

ecclesiastic /ɪˌkliːzɪˈæstɪk/ –*n.* clergyman. –*adj.* = ECCLESIASTICAL. [Greek *ekklēsia* church]

ecclesiastical *adj.* of the Church or clergy.

ECG *abbr.* electrocardiogram.

echelon /ˈeʃəˌlɒn/ *n.* **1** level in an organization, in society, etc.; those occupying it (often in *pl.*: *upper echelons*). **2** wedge-shaped formation of troops, aircraft, etc. [French, = ladder, from Latin *scala*]

echidna /ɪˈkɪdnə/ *n.* Australian egg-laying spiny mammal. [Greek, = viper]

echinoderm /ɪˈkaɪnəˌdɜːm/ *n.* (usu. spiny) sea animal of the group including the starfish and sea urchin. [Greek *ekhinos* sea-urchin, *derma* skin]

echo /ˈekəʊ/ –*n.* (*pl.* **-es**) **1 a** repetition of a sound by the reflection of sound waves. **b** sound so produced. **2** reflected radio or radar beam. **3** close imitation or imitator. **4** circumstance or event reminiscent of an earlier one. –*v.* (**-es, -ed**) **1 a** (of a place) resound with an echo. **b** (of a sound) be repeated; resound. **2** repeat (a sound) thus. **3 a** repeat (another's words). **b** imitate the opinions etc. of. [Latin from Greek]

echo chamber *n.* enclosure with sound-reflecting walls.

echoic /eˈkəʊɪk/ *adj.* (of a word) onomatopoeic.

echolocation *n.* location of objects by reflected sound.

echo-sounder *n.* depth-sounding device using timed echoes.

echt /ext/ *adj.* genuine. [German]

éclair /ɪˈkleə(r)/ *n.* small elongated iced cake of choux pastry filled with cream. [French, = lightning]

eclampsia /ɪˈklæmpsɪə/ *n.* convulsive condition occurring esp. in pregnant women. [ultimately from Greek]

éclat /eɪˈklɑː/ *n.* **1** brilliant display. **2** social distinction; conspicuous success. [French]

eclectic /ɪˈklektɪk/ –*adj.* selecting ideas, style, etc., from various sources. –*n.* eclectic person or philosopher. □ **eclectically** *adv.* **eclecticism** /-ˌsɪz(ə)m/ *n.* [Greek *eklegō* pick out]

eclipse /ɪˈklɪps/ –*n.* **1** obscuring of light from one heavenly body by another. **2** loss of light, importance, or prominence. –*v.* (**-sing**) **1** (of a heavenly body) cause the eclipse of (another). **2** intercept (light). **3** outshine, surpass. [Greek *ekleipsis*]

ecliptic /ɪˈklɪptɪk/ *n.* sun's apparent path among the stars during the year.

eclogue /ˈeklɒg/ *n.* short pastoral poem. [Greek: related to ECLECTIC]

eco- *comb. form* ecology, ecological (*ecoclimate*).

ecology /ɪˈkɒlədʒɪ/ *n.* **1** the study of the relations of organisms to one another and to their surroundings. **2** the study of the interaction of people with their environment. □ **ecological** /ˌiːkəˈlɒdʒɪk(ə)l/ *adj.* **ecologically** /ˌiːkəˈlɒdʒɪkəlɪ/ *adv.* **ecologist** *n.* [Greek *oikos* house]

economic /ˌiːkəˈnɒmɪk/ *adj.* **1** of economics. **2** profitable (*not economic to run buses on a Sunday*). **3** connected with trade and industry (*economic geography*). □ **economically** *adv.* [Greek: related to ECONOMY]

economical *adj.* sparing; avoiding waste. □ **economically** *adv.*
economics *n.pl.* (as *sing.*) **1** science of the production and distribution of wealth. **2** application of this to a particular subject (*the economics of publishing*).
economist /ɪ'kɒnəmɪst/ *n.* expert on or student of economics.
economize /ɪ'kɒnə,maɪz/ *v.* (also **-ise**) (**-zing** or **-sing**) **1** be economical; make economies; reduce expenditure. **2** (foll. by *on*) use sparingly.
economy /ɪ'kɒnəmɪ/ *n.* (*pl.* **-ies**) **1 a** community's system of wealth creation. **b** particular kind of this (*a capitalist economy*). **c** administration or condition of this. **2 a** careful management of (esp. financial) resources; frugality. **b** instance of this (*made many economies*). **3** sparing or careful use (*economy of language*). [Greek *oikonomia* household management]
economy class *n.* cheapest class of air travel.
economy-size *adj.* (of goods) consisting of a larger quantity for a proportionally lower cost.
ecosystem /'i:kəʊ,sɪstəm/ *n.* biological community of interacting organisms and their physical environment.
ecstasy /'ekstəsɪ/ *n.* (*pl.* **-ies**) **1** overwhelming joy or rapture. **2** *slang* type of hallucinogenic drug. □ **ecstatic** /-'stætɪk/ *adj.* **ecstatically** /-'stætɪkəlɪ/ *adv.* [Greek *ekstasis* standing outside oneself]
ECT *abbr.* electroconvulsive therapy.
ecto- *comb. form* outside. [Greek *ektos*]
ectomorph /'ektəʊ,mɔ:f/ *n.* person with a lean body. [Greek *morphē* form]
-ectomy *comb. form* denoting the surgical removal of part of the body (*appendectomy*). [Greek *ektomē* excision]
ectoplasm /'ektəʊ,plæz(ə)m/ *n.* supposed viscous substance exuding from the body of a spiritualistic medium during a trance. [from ECTO-, PLASMA]
ecu /'eɪkju:, -ku:/ *n.* (also **Ecu**) (*pl.* **-s**) European Currency Unit. [abbreviation]
ecumenical /,i:kju:'menɪk(ə)l/ *adj.* **1** of or representing the whole Christian world. **2** seeking worldwide Christian unity. □ **ecumenically** *adv.* **ecumenism** /i:'kju:mə,nɪz(ə)m/ *n.* [Greek *oikoumenikos* of the inhabited earth]
eczema /'eksɪmə/ *n.* inflammation of the skin, with itching and discharge. [Latin from Greek]
ed. *abbr.* **1** edited by. **2** edition. **3** editor. **4** educated.
-ed[1] *suffix* forming adjectives: **1** from nouns, meaning 'having, wearing, etc.' (*talented*; *trousered*). **2** from phrases of adjective and noun (*good-humoured*). [Old English]
-ed[2] *suffix* forming: **1** past tense and past participle of weak verbs (*needed*). **2** participial adjectives (*escaped prisoner*). [Old English]
Edam /'i:dæm/ *n.* round Dutch cheese with a red rind. [*Edam* in Holland]
eddy /'edɪ/ —*n.* (*pl.* **-ies**) **1** circular movement of water causing a small whirlpool. **2** movement of wind, smoke, etc. resembling this. —*v.* (**-ies**, **-ied**) whirl round in eddies. [Old English *ed-* again, back]
edelweiss /'eɪd(ə)l,vaɪs/ *n.* Alpine plant with white bracts. [German, = noble-white]
edema *US* var. of OEDEMA.
Eden /'i:d(ə)n/ *n.* place or state of great happiness, with reference to the abode of Adam and Eve at the Creation. [Hebrew, originally = delight]
edentate /ɪ'denteɪt/ —*adj.* having no or few teeth. —*n.* such a mammal. [Latin *dens dent-* tooth]
edge —*n.* **1** boundary-line or margin of an area or surface. **2** narrow surface of a thin object. **3** meeting-line of surfaces. **4 a** sharpened side of a blade. **b** sharpness. **5** brink of a precipice. **6** edge-like thing, esp. the crest of a ridge. **7** effectiveness, incisiveness; excitement. —*v.* (**-ging**) **1** advance, esp. gradually or furtively. **2 a** provide with an edge or border. **b** form a border to. **3** sharpen (a tool etc.). □ **have the edge on** (or **over**) have a slight advantage over. **on edge** tense and irritable. **set a person's teeth on edge** (of taste or sound) cause an unpleasant nervous sensation. **take the edge off** make less intense. [Old English]
edgeways *adv.* (also **edgewise**) with edge uppermost or foremost. □ **get a word in edgeways** contribute to a conversation when the dominant speaker pauses.
edging *n.* thing forming an edge or border.
edgy *adj.* (**-ier**, **-iest**) irritable; anxious. □ **edgily** *adv.* **edginess** *n.*
edible /'edɪb(ə)l/ *adj.* fit to be eaten. □ **edibility** /-'bɪlɪtɪ/ *n.* [Latin *edo* eat]
edict /'i:dɪkt/ *n.* order proclaimed by authority. [Latin *edico* proclaim]
edifice /'edɪfɪs/ *n.* building, esp. an imposing one. [Latin *aedis* dwelling]
edify /'edɪ,faɪ/ *v.* (**-ies**, **-ied**) improve morally or intellectually. □ **edification** /-fɪ'keɪʃ(ə)n/ *n.* [Latin *aedifico* build]
edit /'edɪt/ *v.* (**-t-**) **1** assemble, prepare, or modify (written material for publication). **2** be editor of (a newspaper etc.). **3**

take extracts from and collate (a film etc.) to form a unified sequence. **4 a** prepare (data) for processing by a computer. **b** alter (a text entered in a word processor etc.). **5 a** reword in order to correct, or to alter the emphasis. **b** (foll. by *out*) remove (a part) from a text etc. [Latin *edo edit-* give out]

edition /ɪ'dɪʃ(ə)n/ *n.* **1** edited or published form of a book etc. **2** copies of a book, newspaper, etc. issued at one time. **3** instance of a regular broadcast. **4** person or thing similar to another (*a miniature edition of her mother*).

editor *n.* **1** person who edits. **2** person who directs the preparation of a newspaper or broadcast news programme or a particular section of one (*sports editor*). **3** person who selects or commissions material for publication. **4** computer program for entering and modifying textual data. □ **editorship** *n.*

editorial /,edɪ'tɔ:rɪəl/ *–adj.* **1** of editing or editors. **2** written or approved by an editor. *–n.* article giving a newspaper's views on a current topic. □ **editorially** *adv.*

EDP *abbr.* electronic data processing.

educate /'edjʊ,keɪt/ *v.* (**-ting**) **1** give intellectual, moral, and social instruction to. **2** provide education for. □ **educable** /-kəb(ə)l/ *adj.* **educability** /-kə'bɪlɪtɪ/ *n.* **educative** /-kətɪv/ *adj.* **educator** *n.* [Latin *educo -are* rear]

educated *adj.* **1** having had an (esp. good) education. **2** resulting from this (*educated accent*). **3** based on experience or study (*educated guess*).

education /,edjʊ'keɪʃ(ə)n/ *n.* **1** systematic instruction. **2** particular kind of or stage in education (*a classical education; further education*). **3** development of character or mental powers. □ **educational** *adj.* **educationally** *adv.*

educationist *n.* (also **educationalist**) expert in educational methods.

educe /ɪ'dju:s/ *v.* (**-cing**) *literary* bring out or develop from latency. □ **eduction** /ɪ'dʌkʃ(ə)n/ *n.* [Latin *educo -ere* draw out]

Edwardian /ed'wɔ:dɪən/ *–adj.* of or characteristic of the reign of Edward VII (1901–10). *–n.* person of this period.

-ee *suffix* forming nouns denoting: **1** person affected by the verbal action (*employee*; *payee*). **2** person concerned with or described as (*absentee*; *refugee*). **3** object of smaller size (*bootee*). [French *-é* in past part.]

EEC *abbr.* European Economic Community.

■ **Usage** EC is the more correct term.

EEG *abbr.* electroencephalogram.

eel *n.* snakelike fish. [Old English]

-eer *suffix* forming: **1** nouns meaning 'person concerned with' (*auctioneer*). **2** verbs meaning 'be concerned with' (*electioneer*). [French *-ier* from Latin *-arius*]

eerie /'ɪərɪ/ *adj.* (**eerier, eeriest**) gloomy and strange; weird (*eerie silence*). □ **eerily** *adv.* **eeriness** *n.* [Old English]

ef- see EX-[1].

efface /ɪ'feɪs/ *v.* (**-cing**) **1** rub or wipe out (a mark, recollection, etc.). **2** surpass, eclipse. **3** *refl.* (usu. as **self-effacing** *adj.*) treat oneself as unimportant. □ **effacement** *n.* [French: related to FACE]

effect /ɪ'fekt/ *–n.* **1** result or consequence of an action etc. **2** efficacy (*had little effect*). **3** impression produced on a spectator, hearer, etc. (*lights gave a pretty effect; said it just for effect*). **4** (in *pl.*) property. **5** (in *pl.*) lighting, sound, etc., giving realism to a play, film, etc. **6** physical phenomenon (*Doppler effect*; *greenhouse effect*). *–v.* bring about (a change, cure, etc.). □ **bring** (or **carry**) **into effect** accomplish. **give effect to** make operative. **in effect** for practical purposes. **take effect** become operative. **to the effect that** the gist being that. **to that effect** having that result or implication. **with effect from** coming into operation at (a stated time). [Latin: related to FACT]

■ **Usage** *Effect* should not be confused with *affect* which, as a verb, has more meanings and is more common, but which does not exist as a noun.

effective *adj.* **1** producing the intended result. **2** impressive, striking. **3** actual, existing. **4** operative. □ **effectively** *adv.* **effectiveness** *n.*

effectual /ɪ'fektʃʊəl/ *adj.* **1** producing the required effect. **2** valid. □ **effectually** *adv.*

effeminate /ɪ'femɪnət/ *adj.* (of a man) womanish in appearance or manner. □ **effeminacy** *n.* **effeminately** *adv.* [Latin *femina* woman]

effervesce /,efə'ves/ *v.* (**-cing**) **1** give off bubbles of gas. **2** be lively. □ **effervescence** *n.* **effervescent** *adj.* [Latin: related to FERVENT]

effete /ɪ'fi:t/ *adj.* feeble, languid; effeminate. □ **effeteness** *n.* [Latin]

efficacious /,efɪ'keɪʃəs/ *adj.* producing the desired effect. □ **efficacy** /'efɪkəsɪ/ *n.* [Latin *efficax*: related to EFFICIENT]

efficient /ɪ'fɪʃ(ə)nt/ *adj.* **1** productive with minimum waste or effort. **2** (of a person) capable; acting effectively. □ **efficiency** *n.* **efficiently** *adv.* [Latin *facio* make]

effigy /ˈefɪdʒɪ/ *n.* (*pl.* **-ies**) sculpture or model of a person. □ **burn in effigy** burn a model of a person. [Latin *effigies* from *fingo* fashion]
effloresce /ˌeflɔːˈres/ *v.* (**-cing**) **1** burst into flower. **2 a** (of a substance) turn to a fine powder on exposure to air. **b** (of salts) come to the surface and crystallize. **c** (of a surface) become covered with salt particles. □ **efflorescence** *n.* **efflorescent** *adj.* [Latin *flos flor-* FLOWER]
effluence /ˈefluəns/ *n.* **1** flowing out of light, electricity, etc. **2** that which flows out. [Latin *fluo flux-* flow]
effluent *–adj.* flowing out. *–n.* **1** sewage or industrial waste discharged into a river etc. **2** stream or lake flowing from a larger body of water.
effluvium /ɪˈfluːvɪəm/ *n.* (*pl.* **-via**) unpleasant or noxious outflow. [Latin: related to EFFLUENCE]
effort /ˈefət/ *n.* **1** use of physical or mental energy. **2** determined attempt. **3** force exerted. **4** *colloq.* something accomplished. [Latin *fortis* strong]
effortless *adj.* easily done, requiring no effort. □ **effortlessly** *adv.* **effortlessness** *n.*
effrontery /ɪˈfrʌntərɪ/ *n.* (*pl.* **-ies**) impudent audacity. [Latin *frons front-* forehead]
effulgent /ɪˈfʌldʒ(ə)nt/ *adj. literary* radiant. □ **effulgence** *n.* [Latin *fulgeo* shine]
effuse /ɪˈfjuːz/ *v.* (**-sing**) **1** pour forth (liquid, light, etc.). **2** give out (ideas etc.). [Latin *fundo fus-* pour]
effusion /ɪˈfjuːʒ(ə)n/ *n.* **1** outpouring. **2** *derog.* unrestrained flow of words. [Latin: related to EFFUSE]
effusive /ɪˈfjuːsɪv/ *adj.* gushing, demonstrative. □ **effusively** *adv.* **effusiveness** *n.*
EFL *abbr.* English as a foreign language.
eft *n.* newt. [Old English]
Efta /ˈeftə/ *n.* (also **EFTA**) European Free Trade Association. [abbreviation]
e.g. *abbr.* for example. [Latin *exempli gratia*]
egalitarian /ɪˌgælɪˈteərɪən/ *–adj.* of or advocating equal rights for all. *–n.* egalitarian person. □ **egalitarianism** *n.* [French *égal* EQUAL]
egg[1] *n.* **1 a** body produced by females of birds, insects, etc. and capable of developing into a new individual. **b** egg of the domestic hen, used for food. **2** *Biol.* ovum. **3** *colloq.* person or thing of a specified kind (*good egg*). □ **with egg on one's face** *colloq.* looking foolish. □ **eggy** *adj.* [Old Norse]
egg[2] *v.* (foll. by *on*) urge. [Old Norse: related to EDGE]
eggcup *n.* cup for holding a boiled egg.
egg-flip *n.* (also **egg-nog**) drink of alcoholic spirit with beaten egg, milk, etc.
egghead *n. colloq.* intellectual; expert.
eggplant *n.* = AUBERGINE.
eggshell *–n.* shell of an egg. *–adj.* **1** (of china) thin and fragile. **2** (of paint) with a slight gloss.
egg-white *n.* white part round the yolk of an egg.
eglantine /ˈeglənˌtaɪn/ *n.* sweet-brier. [Latin *acus* needle]
ego /ˈiːgəʊ/ *n.* (*pl.* **-s**) **1** the self; the part of the mind that reacts to reality and has a sense of individuality. **2** self-esteem; self-conceit. [Latin, = I]
egocentric /ˌiːgəʊˈsentrɪk/ *adj.* self-centred.
egoism /ˈiːgəʊˌɪz(ə)m/ *n.* **1** self-interest as the moral basis of behaviour. **2** systematic selfishness. **3** = EGOTISM. □ **egoist** *n.* **egoistic** /-ˈɪstɪk/ *adj.* **egoistical** /-ˈɪstɪk(ə)l/ *adj.* **egoistically** /-ˈɪstɪkəlɪ/ *adv.*

■ **Usage** The senses of *egoism* and *egotism* overlap, but *egoism* alone is used as a term in philosophy and psychology to mean self-interest (often contrasted with *altruism*).

egotism /ˈiːgəˌtɪz(ə)m/ *n.* **1** self-conceit. **2** selfishness. □ **egotist** *n.* **egotistic** /-ˈtɪstɪk/ *adj.* **egotistical** /-ˈtɪstɪk(ə)l/ *adj.* **egotistically** /-ˈtɪstɪkəlɪ/ *adv.*

■ **Usage** See note at *egoism*.

ego-trip *n. colloq.* activity to boost one's own self-esteem or self-conceit.
egregious /ɪˈgriːdʒəs/ *adj.* **1** extremely bad. **2** *archaic* remarkable. [Latin *grex greg-* flock]
egress /ˈiːgres/ *n. formal* **1** exit. **2** right of going out. [Latin *egredior -gress-* walk out]
egret /ˈiːgrɪt/ *n.* a kind of heron with long white feathers. [French *aigrette*]
Egyptian /ɪˈdʒɪpʃ(ə)n/ *–adj.* of Egypt. *–n.* **1** native or national of Egypt. **2** language of the ancient Egyptians.
Egyptology /ˌiːdʒɪpˈtɒlədʒɪ/ *n.* the study of the language, history, and culture of ancient Egypt. □ **Egyptologist** *n.*
eh /eɪ/ *int. colloq.* **1** expressing enquiry or surprise. **2** inviting assent. **3** asking for repetition or explanation. [instinctive exclamation]
eider /ˈaɪdə(r)/ *n.* any of various large northern ducks. [Icelandic]
eiderdown *n.* quilt stuffed with soft material, esp. down.
eight /eɪt/ *adj.* & *n.* **1** one more than seven. **2** symbol for this (8, viii, VIII). **3** size etc. denoted by eight. **4** eight-oared

rowing-boat or its crew. **5** eight o'clock. [Old English]

eighteen /eɪ'ti:n/ *adj.* & *n.* **1** one more than seventeen. **2** symbol for this (18, xviii, XVIII). **3** size etc. denoted by eighteen. **4** **(18)** (of films) suitable only for persons of 18 years and over. □ **eighteenth** *adj.* & *n.* [Old English]

eightfold *adj.* & *adv.* **1** eight times as much or as many. **2** consisting of eight parts.

eighth *adj.* & *n.* **1** next after seventh. **2** one of eight equal parts of a thing. □ **eighthly** *adv.*

eightsome *n.* (in full **eightsome reel**) lively Scottish dance for eight people.

eighty /'eɪtɪ/ *adj.* & *n.* (*pl.* **-ies**) **1** eight times ten. **2** symbol for this (80, lxxx, LXXX). **3** (in *pl.*) numbers from 80 to 89, esp. the years of a century or of a person's life. □ **eightieth** *adj.* & *n.* [Old English]

einsteinium /aɪn'staɪnɪəm/ *n.* artificial radioactive metallic element. [*Einstein*, name of a physicist]

eisteddfod /aɪ'stedfəd/ *n.* congress of Welsh poets and musicians; festival for musical competitions etc. [Welsh]

either /'aɪðə(r), 'i:ðə(r)/ *–adj.* & *pron.* **1** one or the other of two (*either of you can go; you may have either book*). **2** each of two (*houses on either side of the road*). *–adv.* & *conj.* **1** as one possibility (*is either right or wrong*). **2** as one choice or alternative; which way you will (*either come in or go out*). **3** (with *neg.*) **a** any more than the other (*if you do not go, I shall not either*). **b** moreover (*there is no time to lose, either*). [Old English]

ejaculate /ɪ'dʒækjʊ,leɪt/ *v.* **(-ting)** (also *absol.*) **1** exclaim. **2** emit (semen) in orgasm. □ **ejaculation** /-'leɪʃ(ə)n/ *n.* **ejaculatory** *adj.* [Latin *ejaculor* dart out]

eject /ɪ'dʒekt/ *v.* **1** expel, compel to leave. **2** (of a pilot etc.) cause oneself to be propelled from an aircraft as an emergency measure. **3** cause to be removed, drop out, or pop up automatically from a gun, cassette-player, etc. **4** dispossess (a tenant). **5** emit, send out. □ **ejection** *n.* [Latin *ejicio eject-* throw out]

ejector *n.* device for ejecting.

ejector seat *n.* device in an aircraft for the emergency ejection of a pilot etc.

eke *v.* **(eking)** □ **eke out 1** supplement (income etc.). **2** make (a living) or support (an existence) with difficulty. [Old English]

elaborate *–adj.* /ɪ'læbərət/ **1** minutely worked out. **2** complicated. *–v.* /ɪ'læbə,reɪt/ **(-ting)** work out or explain in detail. □ **elaborately** /-rətlɪ/ *adv.* **elaborateness** /-rətnɪs/ *n.* **elaboration** /-'reɪʃ(ə)n/ *n.* [Latin: related to LABOUR]

élan /eɪ'lɑ̃/ *n.* vivacity, dash. [French]

eland /'i:lənd/ *n.* (*pl.* same or **-s**) large African antelope. [Dutch]

elapse /ɪ'læps/ *v.* **(-sing)** (of time) pass by. [Latin *elabor elaps-* slip away]

elastic /ɪ'læstɪk/ *–adj.* **1** able to resume its normal bulk or shape after contraction, dilation, or distortion. **2** springy. **3** flexible, adaptable. *–n.* elastic cord or fabric, usu. woven with strips of rubber. □ **elastically** *adv.* **elasticity** /,i:læ'stɪsətɪ/ *n.* [Greek *elastikos* propulsive]

elasticated /ɪ'læstɪ,keɪtɪd/ *adj.* (of fabric) made elastic by weaving with rubber thread.

elastic band *n.* = RUBBER BAND.

elastomer /ɪ'læstəmə(r)/ *n.* natural or synthetic rubber or rubber-like plastic. [from ELASTIC, after *isomer*]

elate /ɪ'leɪt/ *v.* **(-ting)** (esp. as **elated** *adj.*) make delighted or proud. □ **elatedly** *adv.* **elation** *n.* [Latin *effero elat-* raise]

elbow /'elbəʊ/ *–n.* **1 a** joint between the forearm and the upper arm. **b** part of a sleeve covering the elbow. **2** elbow-shaped bend etc. *–v.* (foll. by *in*, *out*, *aside*, etc.) **1** jostle or thrust (a person or oneself). **2** make (one's way) thus. □ **give a person the elbow** *colloq.* dismiss or reject a person. [Old English: related to ELL, BOW[1]]

elbow-grease *n.* *colloq.* vigorous polishing; hard work.

elbow-room *n.* sufficient room to move or work in.

elder[1] *–attrib. adj.* (of persons, esp. when related) senior; of greater age. *–n.* **1** older of two persons (*is my elder by ten years*). **2** (in *pl.*) persons of greater age or venerable because of age. **3** official in the early Christian Church and some modern Churches. [Old English: related to OLD]

elder[2] *n.* tree with white flowers and dark berries. [Old English]

elderberry *n.* (*pl.* **-ies**) berry of the elder tree.

elderly *adj.* rather old; past middle age.

elder statesman *n.* influential experienced older person, esp. a politician.

eldest *adj.* first-born; oldest surviving.

eldorado /,eldə'rɑ:dəʊ/ *n.* (*pl.* **-s**) **1** imaginary land of great wealth. **2** place of abundance or opportunity. [Spanish *el dorado* the gilded]

elecampane /,elɪkæm'peɪn/ *n.* plant with bitter aromatic leaves and roots. [Latin *enula* this plant, *campana* of the fields]

elect /ɪ'lekt/ *–v.* **1** (usu. foll. by *to* + infin.) choose. **2** choose by voting. *–adj.* **1** chosen. **2** select, choice. **3** (after the noun) chosen but not yet in office (*president elect*). [Latin *eligo elect-* pick out]

election /ɪ'lekʃ(ə)n/ *n.* **1** electing or being elected. **2** occasion of this.

electioneer /ɪ,lekʃə'nɪə(r)/ *v.* take part in an election campaign.

elective /ɪ'lektɪv/ *adj.* **1** chosen by or derived from election. **2** (of a body) having the power to elect. **3** optional, not urgently necessary.

elector *n.* **1** person who has the right to vote in an election. **2** (**Elector**) *hist.* (in the Holy Roman Empire) any of the German princes entitled to elect the Emperor. □ **electoral** *adj.*

electorate /ɪ'lektərət/ *n.* **1** body of all electors. **2** *hist.* office or territories of a German Elector.

electric /ɪ'lektrɪk/ *–adj.* **1** of, worked by, or charged with electricity; producing or capable of generating electricity. **2** causing or charged with excitement. *–n.* (in *pl.*) *colloq.* electrical equipment. [Greek *ēlektron* amber]

electrical *adj.* of electricity. □ **electrically** *adv.*

electric blanket *n.* blanket heated by an internal electric element.

electric chair *n.* electrified chair used for capital punishment.

electric eel *n.* eel-like fish able to give an electric shock.

electric eye *n. colloq.* photoelectric cell operating a relay when a beam of light is broken.

electric fire *n.* electrically operated portable domestic heater.

electric guitar *n.* guitar with a solid body and built-in pick-up rather than a soundbox.

electrician /,ɪlek'trɪʃ(ə)n/ *n.* person who installs or maintains electrical equipment for a living.

electricity /,ɪlek'trɪsɪtɪ/ *n.* **1** form of energy occurring in elementary particles (electrons, protons, etc.) and hence in larger bodies containing them. **2** science of electricity. **3** supply of electricity. **4** excitement.

electric shock *n.* effect of a sudden discharge of electricity through the body of a person etc.

electrify /ɪ'lektrɪ,faɪ/ *v.* (**-ies, -ied**) **1** charge with electricity. **2** convert to the use of electric power. **3** cause sudden excitement (*news was electrifying*). □ **electrification** /-fɪ'keɪʃ(ə)n/ *n.*

electro- *comb. form* of, by, or caused by electricity.

electrocardiogram /ɪ,lektrəʊ'kɑːdɪə,græm/ *n.* record traced by an electrocardiograph. [German: related to ELECTRO-]

electrocardiograph /ɪ,lektrəʊ'kɑːdɪə,grɑːf/ *n.* instrument recording the electric currents generated by a heartbeat.

electroconvulsive /ɪ,lektrəʊkən'vʌlsɪv/ *adj.* (of therapy) using convulsive response to electric shocks.

electrocute /ɪ'lektrə,kjuːt/ *v.* (**-ting**) kill by electric shock. □ **electrocution** /-'kjuːʃ(ə)n/ *n.* [from ELECTRO-, after *execute*]

electrode /ɪ'lektrəʊd/ *n.* conductor through which electricity enters or leaves an electrolyte, gas, vacuum, etc. [from ELECTRIC, Greek *hodos* way]

electrodynamics /ɪ,lektrəʊdaɪ'næmɪks/ *n.pl.* (usu. treated as *sing.*) the study of electricity in motion. □ **electrodynamic** *adj.*

electroencephalogram /ɪ,lektrəʊɪn'sefələ,græm/ *n.* record traced by an electroencephalograph. [German: related to ELECTRO-]

electroencephalograph /ɪ,lektrəʊɪn'sefələ,grɑːf/ *n.* instrument that records the electrical activity of the brain.

electrolyse /ɪ'lektrə,laɪz/ *v.* (*US* **-yze**) (**-sing**, *US* **-zing**) subject to or treat by electrolysis.

electrolysis /,ɪlek'trɒlɪsɪs/ *n.* **1** chemical decomposition by electric action. **2** destruction of tumours, hair-roots, etc., by this process. □ **electrolytic** /ɪ,lektrəʊ'lɪtɪk/ *adj.*

electrolyte /ɪ'lektrə,laɪt/ *n.* **1** solution able to conduct electricity, esp. in an electric cell or battery. **2** substance that can dissolve to produce this.

electromagnet /ɪ,lektrəʊ'mægnɪt/ *n.* soft metal core made into a magnet by passing an electric current through a coil surrounding it.

electromagnetic /ɪ,lektrəʊmæg'netɪk/ *adj.* having both electrical and magnetic properties. □ **electromagnetically** *adv.*

electromagnetism /ɪ,lektrəʊ'mægnɪ,tɪz(ə)m/ *n.* **1** magnetic forces produced by electricity. **2** the study of these.

electromotive /ɪ,lektrəʊ'məʊtɪv/ *adj.* producing or tending to produce an electric current.

electromotive force *n.* force set up in an electric circuit by a difference in potential.

electron /ɪ'lektrɒn/ *n.* stable elementary particle with a charge of negative electricity, found in all atoms and acting as the primary carrier of electricity in solids.

electronic /ˌɪlekˈtrɒnɪk/ *adj.* **1 a** produced by or involving the flow of electrons. **b** of electrons or electronics. **2** (of music) produced by electronic means. □ **electronically** *adv.*

electronic mail *n.* the sending of messages by a computer system; such messages.

electronics *n.pl.* (treated as *sing.*) science of the movement of electrons in a vacuum, gas, semiconductor, etc., esp. in devices in which the flow is controlled and utilized.

electronic tagging *n.* the attaching of electronic markers to people or goods, enabling them to be tracked down.

electron lens *n.* device for focusing a stream of electrons by means of electric or magnetic fields.

electron microscope *n.* microscope with high magnification and resolution, using electron beams instead of light.

electronvolt /ɪˈlektrɒnˌvəʊlt/ *n.* a unit of energy, the amount gained by an electron when accelerated through a potential difference of one volt.

electroplate /ɪˈlektrəʊˌpleɪt/ *–v.* (**-ting**) coat with a thin layer of chromium, silver, etc., by electrolysis. *–n.* electroplated articles.

electroscope /ɪˈlektrəˌskəʊp/ *n.* instrument for detecting and measuring electricity, esp. as an indication of the ionization of air by radioactivity. □ **electroscopic** /-ˈskɒpɪk/ *adj.*

electro-shock /ɪˌlektrəʊˈʃɒk/ *attrib. adj.* (of therapy) by means of electric shocks.

electrostatics /ɪˌlektrəʊˈstætɪks/ *n.pl.* (treated as *sing.*) the study of electricity at rest.

electrotechnology /ɪˌlektrəʊtekˈnɒlədʒɪ/ *n.* science of the application of electricity in technology.

electrotherapy /ɪˌlektrəʊˈθerəpɪ/ *n.* treatment of diseases by use of electricity.

elegant /ˈelɪgənt/ *adj.* **1** tasteful, refined, graceful. **2** ingeniously simple. □ **elegance** *n.* **elegantly** *adv.* [Latin: related to ELECT]

elegiac /ˌelɪˈdʒaɪək/ *–adj.* **1** used for elegies. **2** mournful. *–n.* (in *pl.*) elegiac verses. □ **elegiacally** *adv.*

elegy /ˈelɪdʒɪ/ *n.* (*pl.* **-ies**) **1** sorrowful poem or song, esp. for the dead. **2** poem in elegiac metre. [Latin from Greek]

element /ˈelɪmənt/ *n.* **1** component part; contributing factor. **2** any of the substances that cannot be resolved by chemical means into simpler substances. **3 a** any of the four substances (earth, water, air, and fire) in ancient and medieval philosophy. **b** a being's natural abode or environment. **4** *Electr.* wire that heats up in an electric heater, kettle, etc. **5** (in *pl.*) atmospheric agencies, esp. wind and storm. **6** (in *pl.*) rudiments of learning or of an art etc. **7** (in *pl.*) bread and wine of the Eucharist. □ **in one's element** in one's preferred situation, doing what one does well and enjoys. [French from Latin]

elemental /ˌelɪˈment(ə)l/ *adj.* **1** of or like the elements or the forces of nature; powerful. **2** essential, basic.

elementary /ˌelɪˈmentərɪ/ *adj.* **1** dealing with the simplest facts of a subject. **2** unanalysable.

elementary particle *n. Physics* subatomic particle, esp. one not known to consist of simpler ones.

elephant /ˈelɪf(ə)nt/ *n.* (*pl.* same or **-s**) largest living land animal, with a trunk and ivory tusks. [Greek *elephas*]

elephantiasis /ˌelɪfənˈtaɪəsɪs/ *n.* skin disease causing gross enlargement of limbs etc.

elephantine /ˌelɪˈfæntaɪn/ *adj.* **1** of elephants. **2 a** huge. **b** clumsy.

elevate /ˈelɪˌveɪt/ *v.* (**-ting**) **1** raise, lift up. **2** exalt in rank etc. **3** (usu. as **elevated** *adj.*) raise morally or intellectually. [Latin *levo* lift]

elevation /ˌelɪˈveɪʃ(ə)n/ *n.* **1 a** elevating or being elevated. **b** angle with the horizontal. **c** height above sea level etc. **d** high position. **2** drawing or diagram showing one side of a building.

elevator *n.* **1** *US* lift. **2** movable part of a tailplane for changing an aircraft's altitude. **3** hoisting machine.

eleven /ɪˈlev(ə)n/ *adj.* & *n.* **1** one more than ten. **2** symbol for this (11, xi, XI). **3** size etc. denoted by eleven. **4** team of eleven players at cricket, football, etc. **5** eleven o'clock. [Old English]

elevenfold *adj.* & *adv.* **1** eleven times as much or as many. **2** consisting of eleven parts.

eleven-plus *n.* esp. *hist.* examination taken at age 11–12 to determine the type of secondary school a child would enter.

elevenses /ɪˈlevənzɪz/ *n. colloq.* light refreshment taken at about 11 a.m.

eleventh *adj.* & *n.* **1** next after tenth. **2** each of eleven equal parts of a thing. □ **eleventh hour** last possible moment.

elf *n.* (*pl.* **elves** /elvz/) mythological being, esp. one that is small and mischievous. □ **elfish** *adj.* **elvish** *adj.* [Old English]

elfin *adj.* of elves; elflike.

elicit /ɪˈlɪsɪt/ *v.* (**-t-**) draw out (facts, a response, etc.), esp. with difficulty. [Latin *elicio*]

elide /ɪ'laɪd/ *v.* (**-ding**) omit (a vowel or syllable) in pronunciation. [Latin *elido elis-* crush out]

eligible /'elɪdʒɪb(ə)l/ *adj.* **1** (often foll. by *for*) fit or entitled to be chosen (*eligible for a rebate*). **2** desirable or suitable, esp. for marriage. □ **eligibility** /-'bɪlɪtɪ/ *n.* [Latin: related to ELECT]

eliminate /ɪ'lɪmɪ,neɪt/ *v.* (**-ting**) **1** remove, get rid of. **2** exclude from consideration. **3** exclude from a further stage of a competition through defeat etc. □ **elimination** /-'neɪʃ(ə)n/ *n.* **eliminator** *n.* [Latin *limen limin-* threshold]

elision /ɪ'lɪʒ(ə)n/ *n.* omission of a vowel or syllable in pronunciation (e.g. in *we'll*). [Latin: related to ELIDE]

élite /ɪ'li:t/ *n.* **1** (prec. by *the*) the best (of a group). **2** select group or class. **3** a size of letters in typewriting (12 per inch). [French: related to ELECT]

élitism *n.* recourse to or advocacy of leadership or dominance by a select group. □ **élitist** *n.* & *adj.*

elixir /ɪ'lɪksɪə(r)/ *n.* **1 a** alchemist's preparation supposedly able to change metals into gold or (in full **elixir of life**) to prolong life indefinitely. **b** remedy for all ills. **2** aromatic medicinal drug. [Latin from Arabic]

Elizabethan /ɪ,lɪzə'bi:θ(ə)n/ *–adj.* of the time of Queen Elizabeth I or II. *–n.* person of this time.

elk *n.* (*pl.* same or **-s**) large deer of northern parts of Europe, N. America, and Asia. [Old English]

ell *n. hist.* measure = 45 in. [Old English, = forearm]

ellipse /ɪ'lɪps/ *n.* regular oval, resulting when a cone is cut obliquely by a plane. [Greek *elleipsis* deficit]

ellipsis /ɪ'lɪpsɪs/ *n.* (*pl.* **ellipses** /-si:z/) **1** omission of words needed to complete a construction or sense. **2** set of three dots etc. indicating omission.

ellipsoid /ɪ'lɪpsɔɪd/ *n.* solid of which all the plane sections through one axis are circles and all the other plane sections are ellipses.

elliptic /ɪ'lɪptɪk/ *adj.* (also **elliptical**) of or in the form of an ellipse. □ **elliptically** *adv.*

elm *n.* **1** tree with rough serrated leaves. **2** its wood. [Old English]

elocution /,elə'kju:ʃ(ə)n/ *n.* art of clear and expressive speech. [Latin *loquor* speak]

elongate /'i:lɒŋ,geɪt/ *v.* (**-ting**) lengthen, extend. □ **elongation** /-'geɪʃ(ə)n/ *n.* [Latin *longus* long]

elope /ɪ'ləʊp/ *v.* (**-ping**) run away to marry secretly. □ **elopement** *n.* [Anglo-French]

eloquence /'eləkwəns/ *n.* fluent and effective use of language. [Latin *loquor* speak]

eloquent *adj.* **1** having eloquence. **2** (often foll. by *of*) expressive. □ **eloquently** *adv.*

else *adv.* **1** (prec. by indefinite or interrog. pron.) besides (*someone else*; *nowhere else*; *who else?*). **2** instead (*what else could I say?*). **3** otherwise; if not (*run, (or) else you will be late*). □ **or else** see OR[1]. [Old English]

elsewhere *adv.* in or to some other place.

elucidate /ɪ'lu:sɪ,deɪt/ *v.* (**-ting**) throw light on; explain. □ **elucidation** /-'deɪʃ(ə)n/ *n.* **elucidatory** *adj.* [Latin: related to LUCID]

elude /ɪ'lu:d/ *v.* (**-ding**) **1** escape adroitly from (danger, pursuit, etc.). **2** avoid compliance with (a law etc.) or fulfilment of (an obligation). **3** baffle (a person or memory etc.). □ **elusion** /-ʒ(ə)n/ *n.* [Latin *ludo* play]

elusive /ɪ'lu:sɪv/ *adj.* **1** difficult to find or catch. **2** difficult to remember. **3** avoiding the point raised. □ **elusiveness** *n.*

elver /'elvə(r)/ *n.* young eel. [from EEL, FARE]

elves *pl.* of ELF.

elvish see ELF.

Elysium /ɪ'lɪzɪəm/ *n.* **1** (also **Elysian Fields**) (in Greek mythology) abode of the blessed after death. **2** place of ideal happiness. □ **Elysian** *adj.* [Latin from Greek]

em *n. Printing* unit of measurement equal to the width of an M. [name of the letter *M*]

em-[1,2] see EN-[1,2].

'em /əm/ *pron. colloq.* them.

emaciate /ɪ'meɪsɪ,eɪt/ *v.* (**-ting**) (esp. as **emaciated** *adj.*) make abnormally thin or feeble. □ **emaciation** /-'eɪʃ(ə)n/ *n.* [Latin *macies* leanness]

email /'i:meɪl/ *n.* (also **e-mail**) = ELECTRONIC MAIL.

emanate /'emə,neɪt/ *v.* (**-ting**) (usu. foll. by *from*) issue or originate (from a source). □ **emanation** /-'neɪʃ(ə)n/ *n.* [Latin *mano* flow]

emancipate /ɪ'mænsɪ,peɪt/ *v.* (**-ting**) **1** free from social or political restraint. **2** (usu. as **emancipated** *adj.*) free from the inhibitions of moral or social conventions. **3** free from slavery. □ **emancipation** /-'peɪʃ(ə)n/ *n.* **emancipatory** /-pətərɪ/ *adj.* [Latin, = free from possession, from *manus* hand, *capio* take]

emasculate *–v.* /ɪ'mæskjʊ,leɪt/ (**-ting**) **1** deprive of force or vigour. **2** castrate. *–adj.* /ɪ'mæskjʊlət/ **1** deprived of force. **2**

castrated. **3** effeminate. □ **emasculation** /-ˈleɪʃ(ə)n/ *n.* [Latin: related to MALE]

embalm /ɪmˈbɑːm/ *v.* **1** preserve (a corpse) from decay. **2** preserve from oblivion. **3** make fragrant. □ **embalmment** *n.* [French: related to BALM]

embankment /ɪmˈbæŋkmənt/ *n.* bank constructed to keep back water or carry a road, railway, etc.

embargo /ɪmˈbɑːɡəʊ/ *–n.* (*pl.* **-es**) **1** order forbidding foreign ships to enter, or any ships to leave, a country's ports. **2** official suspension of an activity. *–v.* (**-es**, **-ed**) place under embargo. [Spanish: related to BAR[1]]

embark /ɪmˈbɑːk/ *v.* **1** (often foll. by *for*) put or go on board a ship or aircraft (to a destination). **2** (foll. by *on*, *in*) begin (an enterprise). □ **embarkation** /ˌembɑːˈkeɪʃ(ə)n/ *n.* (in sense 1). [French: related to BARQUE]

embarrass /ɪmˈbærəs/ *v.* **1** make (a person) feel awkward or ashamed. **2** (as **embarrassed** *adj.*) encumbered with debts. **3** encumber. □ **embarrassment** *n.* [Italian *imbarrare* bar in]

embassy /ˈembəsɪ/ *n.* (*pl.* **-ies**) **1 a** residence or offices of an ambassador. **b** ambassador and staff. **2** deputation to a foreign government. [French: related to AMBASSADOR]

embattled /ɪmˈbæt(ə)ld/ *adj.* **1** prepared or arrayed for battle. **2** fortified with battlements. **3** under heavy attack or in trying circumstances.

embed /ɪmˈbed/ *v.* (also **imbed**) (**-dd-**) (esp. as **embedded** *adj.*) fix firmly in a surrounding mass.

embellish /ɪmˈbelɪʃ/ *v.* **1** beautify, adorn. **2** enhance with fictitious additions. □ **embellishment** *n.* [French *bel*, BEAU]

ember *n.* (usu. in *pl.*) small piece of glowing coal etc. in a dying fire. [Old English]

ember days *n.pl.* days of fasting and prayer in the Christian Church, associated with ordinations. [Old English]

embezzle /ɪmˈbez(ə)l/ *v.* (**-ling**) divert (money etc.) fraudulently to one's own use. □ **embezzlement** *n.* **embezzler** *n.* [Anglo-French]

embitter /ɪmˈbɪtə(r)/ *v.* arouse bitter feelings in. □ **embitterment** *n.*

emblazon /ɪmˈbleɪz(ə)n/ *v.* **1** portray or adorn conspicuously. **2** adorn (a heraldic shield). □ **emblazonment** *n.*

emblem /ˈembləm/ *n.* **1** symbol. **2** (foll. by *of*) type, embodiment (*the very emblem of courage*). **3** heraldic or representative device. □ **emblematic** /-ˈmætɪk/ *adj.* [Greek, = insertion]

embody /ɪmˈbɒdɪ/ *v.* (**-ies**, **-ied**) **1** make (an idea etc.) actual or discernible. **2** (of a thing) be a tangible expression of. **3** include, comprise. □ **embodiment** *n.*

embolden /ɪmˈbəʊld(ə)n/ *v.* make bold; encourage.

embolism /ˈembəˌlɪz(ə)m/ *n.* obstruction of an artery by a clot, air-bubble, etc. [Latin from Greek]

embolus /ˈembələs/ *n.* (*pl.* **-li** /-ˌlaɪ/) object causing an embolism.

emboss /ɪmˈbɒs/ *v.* carve or decorate with a design in relief. □ **embossment** *n.* [related to BOSS[2]]

embouchure /ˈɒmbʊˌʃʊə(r)/ *n.* way of applying the mouth to the mouthpiece of a musical instrument. [French: related to EN-[1], *bouche* mouth]

embrace /ɪmˈbreɪs/ *–v.* (**-cing**) **1 a** hold closely in the arms. **b** (*absol.*, of two people) embrace each other. **2** clasp, enclose. **3** accept eagerly (an offer etc.). **4** adopt (a cause, idea, etc.). **5** include, comprise. **6** take in with the eye or mind. *–n.* act of embracing, clasp. □ **embraceable** *adj.* [Latin: related to BRACE]

embrasure /ɪmˈbreɪʒə(r)/ *n.* **1** bevelling of a wall at the sides of a window etc. **2** opening in a parapet for a gun etc. □ **embrasured** *adj.* [French *embraser* splay]

embrocation /ˌembrəˈkeɪʃ(ə)n/ *n.* liquid for rubbing on the body to relieve muscular pain. [Greek *embrokhē* lotion]

embroider /ɪmˈbrɔɪdə(r)/ *v.* **1** decorate (cloth etc.) with needlework. **2** embellish (a narrative). □ **embroiderer** *n.* [Anglo-French from Germanic]

embroidery *n.* (*pl.* **-ies**) **1** art of embroidering. **2** embroidered work. **3** inessential ornament. **4** fictitious additions (to a story etc.).

embroil /ɪmˈbrɔɪl/ *v.* (often foll. by *with*) involve (a person etc.) in a conflict or difficulties. □ **embroilment** *n.* [French *brouiller* mix]

embryo /ˈembrɪəʊ/ *n.* (*pl.* **-s**) **1 a** unborn or unhatched offspring. **b** human offspring in the first eight weeks from conception. **2** rudimentary plant in a seed. **3** thing in a rudimentary stage. **4** (*attrib.*) undeveloped, immature. □ **in embryo** undeveloped. □ **embryonic** /ˌembrɪˈɒnɪk/ *adj.* [Greek *bruō* grow]

embryology /ˌembrɪˈɒlədʒɪ/ *n.* the study of embryos.

emend /ɪˈmend/ *v.* edit (a text etc.) to make corrections. □ **emendation** /ˌiːmenˈdeɪʃ(ə)n/ *n.* [Latin *menda* fault]

■ **Usage** See note at *amend*.

---

emerald /ˈemər(ə)ld/ *–n.* **1** bright-green gem. **2** colour of this. *–adj.* bright green. [Greek *smaragdos*]

emerald green *adj.* & *n.* (as adj. often hyphenated) bright green.

Emerald Isle *n.* Ireland.

emerge /ɪ'mɜːdʒ/ *v.* (**-ging**) **1** come up or out into view. **2** (of facts etc.) become known, be revealed. **3** become recognized or prominent. **4** (of a question, difficulty, etc.) become apparent. □ **emergence** *n.* **emergent** *adj.* [Latin: related to MERGE]

emergency /ɪ'mɜːdʒənsɪ/ *n.* (*pl.* **-ies**) **1** sudden state of danger etc., requiring immediate action. **2 a** condition requiring immediate treatment. **b** patient with this. **3** (*attrib.*) for use in an emergency. [medieval Latin: related to EMERGE]

emeritus /ɪ'merɪtəs/ *adj.* retired but retaining one's title as an honour (*emeritus professor*). [Latin *mereor* earn]

emery /'emərɪ/ *n.* coarse corundum for polishing metal etc. [Greek *smēris* polishing powder]

emery-board *n.* emery-coated nail-file.

emetic /ɪ'metɪk/ *–adj.* that causes vomiting. *–n.* emetic medicine. [Greek *emeō* vomit]

emf *abbr.* (also **e.m.f.**) electromotive force.

emigrant /'emɪgrənt/ *–n.* person who emigrates. *–adj.* emigrating.

emigrate /'emɪˌgreɪt/ *v.* (**-ting**) leave one's own country to settle in another. □ **emigration** /-'greɪʃ(ə)n/ *n.* [Latin: related to MIGRATE]

émigré /'emɪˌgreɪ/ *n.* emigrant, esp. a political exile. [French]

eminence /'emɪnəns/ *n.* **1** distinction; recognized superiority. **2** piece of rising ground. **3** title used in addressing or referring to a cardinal (*Your Eminence*; *His Eminence*). [Latin: related to EMINENT]

éminence grise /ˌeɪmɪnɑ̃s 'griːz/ *n.* (*pl.* ***éminences grises*** pronunc. same) person who exercises power or influence without holding office. [French, = grey cardinal (orig. of Richelieu's secretary)]

eminent /'emɪnənt/ *adj.* distinguished, notable, outstanding. □ **eminently** *adv.* [Latin *emineo* jut out]

emir /e'mɪə(r)/ *n.* (also **amir** /ə'mɪə(r)/) title of various Muslim rulers. [French: from Arabic *'amīr*]

emirate /'emɪrət/ *n.* rank, domain, or reign of an emir.

emissary /'emɪsərɪ/ *n.* (*pl.* **-ies**) person sent on a diplomatic mission. [Latin: related to EMIT]

emit /ɪ'mɪt/ *v.* (**-tt-**) give or send out (heat, light, a smell, sound, etc.); discharge. □ **emission** /-ʃ(ə)n/ *n.* [Latin *emitto emiss-*]

emollient /ɪ'mɒlɪənt/ *–adj.* that softens or soothes the skin, feelings, etc. *–n.* emollient substance. [Latin *mollis* soft]

emolument /ɪ'mɒljʊmənt/ *n.* fee from employment, salary. [Latin]

emote /ɪ'məʊt/ *v.* (**-ting**) show excessive emotion.

emotion /ɪ'məʊʃ(ə)n/ *n.* **1** strong instinctive feeling such as love or fear. **2** emotional intensity or sensibility (*spoke with emotion*). [French: related to MOTION]

emotional *adj.* **1** of or expressing emotions. **2** especially liable to emotion. **3** arousing emotion. □ **emotionalism** *n.* **emotionally** *adv.*

■ **Usage** See note at *emotive*.

emotive /ɪ'məʊtɪv/ *adj.* **1** arousing emotion. **2** of emotion. [Latin: related to MOTION]

■ **Usage** Although the senses of *emotive* and *emotional* overlap, *emotive* is more common in the sense 'arousing emotion', as in *an emotive issue*, and is not used at all in sense 2 of *emotional*.

empanel /ɪm'pæn(ə)l/ *v.* (also **impanel**) (**-ll-**; *US* **-l-**) enter (a jury) on a panel.

empathize /'empəˌθaɪz/ *v.* (also **-ise**) (**-zing** or **-sing**) (usu. foll. by *with*) exercise empathy.

empathy /'empəθɪ/ *n.* ability to identify with a person or object. □ **empathetic** /-'θetɪk/ *adj.* [as PATHOS]

emperor /'empərə(r)/ *n.* sovereign of an empire. [Latin *impero* command]

emperor penguin *n.* largest known penguin.

emphasis /'emfəsɪs/ *n.* (*pl.* **emphases** /-ˌsiːz/) **1** importance or prominence attached to a thing (*emphasis on economy*). **2** stress laid on a word or syllable to make the meaning clear or show importance. **3** vigour or intensity of expression, feeling, etc. [Latin from Greek]

emphasize /'emfəˌsaɪz/ *v.* (also **-ise**) (**-zing** or **-sing**) put emphasis on, stress.

emphatic /ɪm'fætɪk/ *adj.* **1** forcibly expressive. **2** of words: **a** bearing the stress. **b** used to give emphasis. □ **emphatically** *adv.*

emphysema /ˌemfɪ'siːmə/ *n.* disease of the lungs causing breathlessness. [Greek *emphusaō* puff up]

empire /'empaɪə(r)/ *n.* **1** large group of States or countries under a single authority. **2** supreme dominion. **3** large commercial organization etc. owned or directed by one person. **4** (**the Empire**) *hist.* the British Empire. [Latin *imperium* dominion]

empire-building *n.* purposeful accumulation of territory, authority, etc.

empirical /ɪm'pɪrɪk(ə)l/ *adj.* (also **empiric**) based on observation, experience, or experiment, not on theory. □ **empirically** *adv.* **empiricism** /-,sɪz(ə)m/ *n.* **empiricist** /-sɪst/ *n.* [Greek *empeiria* experience]

emplacement /ɪm'pleɪsmənt/ *n.* **1** putting in position. **2** platform for guns. [French: related to PLACE]

employ /ɪm'plɔɪ/ *–v.* **1** use the services of (a person) in return for payment. **2** use (a thing, time, energy, etc.) to good effect. **3** keep (a person) occupied. *–n.* (in phr. **in the employ of**) employed by. □ **employable** *adj.* **employer** *n.* [Latin *implicor* be involved]

employee /,emplɔɪ'i:, -'plɔɪi/ *n.* person employed for wages.

employment *n.* **1** employing or being employed. **2** person's trade or profession.

employment office *n.* (formerly **employment exchange**) State-run employment agency.

emporium /em'pɔ:rɪəm/ *n.* (*pl.* **-s** or **-ria**) **1** large shop or store. **2** centre of commerce, market. [Greek *emporos* merchant]

empower /ɪm'paʊə(r)/ *v.* give authority to.

empress /'emprɪs/ *n.* **1** wife or widow of an emperor. **2** woman emperor. [French: related to EMPEROR]

empty /'emptɪ/ *–adj.* (**-ier, -iest**) **1** containing nothing. **2** (of a house etc.) unoccupied or unfurnished. **3** (of a vehicle etc.) without passengers etc. **4 a** hollow, insincere (*empty threats*). **b** without purpose (*an empty existence*). **c** vacuous (*an empty head*). **5** *colloq.* hungry. *–v.* (**-ies, -ied**) **1** remove the contents of. **2** (often foll. by *into*) transfer (contents). **3** become empty. **4** (of a river) discharge itself. *–n.* (*pl.* **-ies**) *colloq.* empty bottle etc. □ **emptiness** *n.* [Old English]

empty-handed *adj.* (usu. *predic.*) **1** bringing or taking nothing. **2** having achieved nothing.

empty-headed *adj.* foolish; lacking sense.

empyrean /,empaɪ'ri:ən/ *–n.* the highest heaven, as the sphere of fire or abode of God. *–adj.* of the empyrean. □ **empyreal** *adj.* [Greek *pur* fire]

EMS *abbr.* European Monetary System.

EMU /,i:em'ju:, 'i:mju:/ *abbr.* economic and monetary union; European monetary union.

emu /'i:mju:/ *n.* (*pl.* **-s**) large flightless Australian bird. [Portuguese]

emulate /'emjʊ,leɪt/ *v.* (**-ting**) **1** try to equal or excel. **2** imitate. □ **emulation** /-'leɪʃ(ə)n/ *n.* **emulative** /-lətɪv/ *adj.* **emulator** *n.* [Latin *aemulus* rival]

emulsify /ɪ'mʌlsɪ,faɪ/ *v.* (**-ies, -ied**) convert into an emulsion. □ **emulsification** /-fɪ'keɪʃ(ə)n/ *n.* **emulsifier** *n.*

emulsion /ɪ'mʌlʃ(ə)n/ *n.* **1** fine dispersion of one liquid in another, esp. as paint, medicine, etc. **2** mixture of a silver compound in gelatin etc. for coating photographic plate or film. **3** emulsion paint. [Latin *mulgeo* milk]

emulsion paint *n.* water-thinned paint.

en *n. Printing* unit of measurement equal to half an em. [name of the letter *N*]

en-[1] *prefix* (also **em-** before *b, p*) forming verbs, = IN-[2]: **1** from nouns, meaning 'put into or on' (*engulf*; *entrust*; *embed*). **2** from nouns or adjectives, meaning 'bring into the condition of' (*enslave*); often with the suffix *-en* (*enlighten*). **3** from verbs: **a** in the sense 'in, into, on' (*enfold*). **b** as an intensifier (*entangle*). [French *en-*, Latin *in-*]

en-[2] *prefix* (also **em-** before *b, p*) in, inside (*energy*; *enthusiasm*). [Greek]

-en *suffix* forming verbs: **1** from adjectives, usu. meaning 'make or become so or more so' (*deepen*; *moisten*). **2** from nouns (*happen*; *strengthen*). [Old English]

enable /ɪ'neɪb(ə)l/ *v.* (**-ling**) **1** (foll. by *to* + infin.) give (a person etc.) the means or authority. **2** make possible. **3** esp. *Computing* make (a device) operational; switch on.

enact /ɪ'nækt/ *v.* **1 a** ordain, decree. **b** make (a bill etc.) law. **2** play (a part on stage or in life). □ **enactive** *adj.*

enactment *n.* **1** law enacted. **2** process of enacting.

enamel /ɪ'næm(ə)l/ *–n.* **1** glasslike opaque ornamental or preservative coating on metal etc. **2 a** smooth hard coating. **b** a kind of hard gloss paint. **c** cosmetic simulating this, esp. nail varnish. **3** hard coating of a tooth. **4** painting done in enamel. *–v.* (**-ll-**; *US* **-l-**) inlay, coat, or portray with enamel. [Anglo-French from Germanic]

enamour /ɪ'næmə(r)/ *v.* (*US* **enamor**) (usu. in *passive*; foll. by *of*) inspire with love or delight. [French *amour* love]

en bloc /ɑ̃ 'blɒk/ *adv.* in a block; all at the same time. [French]

encamp /ɪn'kæmp/ *v.* settle in a (esp. military) camp. □ **encampment** *n.*

encapsulate /ɪn'kæpsjʊ,leɪt/ *v.* (**-ting**) **1** enclose in or as in a capsule. **2** express briefly, summarize. □ **encapsulation** /-'leɪʃ(ə)n/ *n.* [related to CAPSULE]

encase /ɪn'keɪs/ *v.* (**-sing**) enclose in or as in a case. □ **encasement** *n.*

encaustic /ɪn'kɔ:stɪk/ *–adj.* (of painting etc.) using pigments mixed with hot wax, which are burned in as an inlay. *–n.* **1** art of encaustic painting. **2** product of this. [Greek: related to CAUSTIC]

-ence *suffix* forming nouns expressing: **1** a quality or state or an instance of this (*patience*; *an impertinence*). **2** an action (*reference*). [French *-ence*, Latin *-erie*]

encephalitis /en,sefə'laɪtɪs/ *n.* inflammation of the brain. [Greek *egkephalos* brain]

encephalogram /en'sefələʊ,græm/ *n.* = ELECTROENCEPHALOGRAM.

encephalograph /en'sefələʊ,grɑ:f/ *n.* = ELECTROENCEPHALOGRAPH.

enchant /ɪn'tʃɑ:nt/ *v.* **1** charm, delight. **2** bewitch. □ **enchantedly** *adv.* **enchanting** *adj.* **enchantingly** *adv.* **enchantment** *n.*

enchanter *n.* (*fem.* **enchantress**) person who enchants, esp. by using magic.

encircle /ɪn'sɜ:k(ə)l/ *v.* **(-ling) 1** surround. **2** form a circle round. □ **encirclement** *n.*

enclave /'enkleɪv/ *n.* territory of one State surrounded by that of another. [Latin *clavis* key]

enclose /ɪn'kləʊz/ *v.* **(-sing) 1 a** surround with a wall, fence, etc. **b** shut in. **2** put in a receptacle (esp. in an envelope with a letter). **3** (usu. as **enclosed** *adj.*) seclude (a religious community) from the outside world. [Latin: related to INCLUDE]

enclosure /ɪn'kləʊʒə(r)/ *n.* **1** act of enclosing. **2** enclosed space or area, esp. at a sporting event. **3** thing enclosed with a letter. [French: related to ENCLOSE]

encode /ɪn'kəʊd/ *v.* **(-ding)** put into code.

encomium /ɪn'kəʊmɪəm/ *n.* (*pl.* **-s**) formal or high-flown praise. [Greek *kōmos* revelry]

encompass /ɪn'kʌmpəs/ *v.* **1** contain; include. **2** surround.

encore /'ɒŋkɔ:(r)/ *–n.* **1** audience's demand for the repetition of an item, or for a further item. **2** such an item. *–v.* **(-ring) 1** call for the repetition of (an item). **2** call back (a performer) for this. *–int.* /also -'kɔ:(r)/ again, once more. [French, = once again]

encounter /ɪn'kaʊntə(r)/ *–v.* **1** meet unexpectedly. **2** meet as an adversary. *–n.* meeting by chance or in conflict. [Latin *contra* against]

encourage /ɪn'kʌrɪdʒ/ *v.* **(-ging) 1** give courage or confidence to. **2** urge. **3** promote. □ **encouragement** *n.* [French: related to EN-[1]]

encroach /ɪn'krəʊtʃ/ *v.* **1** (foll. by *on*, *upon*) intrude on another's territory etc. **2** advance gradually beyond due limits. □ **encroachment** *n.* [French *croc* CROOK]

encrust /ɪn'krʌst/ *v.* **1** cover with or form a crust. **2** coat with a hard casing or deposit, sometimes for decoration. [French: related to EN-[1]]

encumber /ɪn'kʌmbə(r)/ *v.* **1** be a burden to. **2** hamper. [French from Romanic]

encumbrance /ɪn'kʌmbrəns/ *n.* **1** burden. **2** impediment.

-ency *suffix* forming nouns denoting quality or state (*efficiency*; *fluency*; *presidency*). [Latin *-entia*]

encyclical /ɪn'sɪklɪk(ə)l/ *–adj.* for wide circulation. *–n.* papal encyclical letter. [Greek: related to CYCLE]

encyclopedia /ɪn,saɪklə'pi:dɪə/ *n.* (also **-paedia**) book, often in a number of volumes, giving information on many subjects, or on many aspects of one subject. [Greek *egkuklios* all-round, *paideia* education]

encyclopedic *adj.* (also **-paedic**) (of knowledge or information) comprehensive.

end *–n.* **1 a** extreme limit. **b** extremity (*to the ends of the earth*). **2** extreme part or surface of a thing (*strip of wood with a nail in one end*). **3 a** finish (*no end to his misery*). **b** latter part. **c** death, destruction (*met an untimely end*). **d** result. **4** goal (*will do anything to achieve his ends*). **5** remnant (*cigarette-end*). **6** (prec. by *the*) *colloq.* the limit of endurability. **7** half of a sports pitch etc. occupied by one team or player. **8** part with which a person is concerned (*no problem at my end*). *–v.* **1** bring or come to an end, finish. **2** (foll. by *in*) result in. □ **end it all** (or **end it**) *colloq.* commit suicide. **end on** with the end facing one, or adjoining the end of the next object. **end to end** with the end of one adjoining the end of the next in a series. **end up** reach a specified state or action eventually (*ended up a drunkard*; *ended up making a fortune*). **in the end** finally. **keep one's end up** do one's part despite difficulties. **make ends meet** live within one's income. **no end** *colloq.* to a great extent. **no end of** *colloq.* much or many of. **on end 1** upright (*hair stood on end*). **2** continuously (*for three weeks on end*). **put an end to** stop, abolish, destroy. [Old English]

endanger /ɪn'deɪndʒə(r)/ *v.* place in danger.

endangered species *n.* species in danger of extinction.

endear /ɪn'dɪə(r)/ *v.* (usu. foll. by *to*) make dear. □ **endearing** *adj.*

endearment *n.* **1** an expression of affection. **2** liking, affection.

endeavour /ɪn'devə(r)/ (*US* **endeavor**) –*v.* (foll. by *to* + infin.) try earnestly. –*n.* earnest attempt. [from EN-[1], French *devoir* owe]

endemic /en'demɪk/ *adj.* (often foll. by *to*) regularly or only found among a particular people or in a particular region. □ **endemically** *adv.* [Greek *en-* in, *dēmos* the people]

ending *n.* **1** end or final part, esp. of a story. **2** inflected final part of a word.

endive /'endaɪv/ *n.* curly-leaved plant used in salads. [Greek *entubon*]

endless *adj.* **1** infinite; without end. **2** continual (*endless complaints*). **3** *colloq.* innumerable. **4** (of a belt, chain, etc.) having the ends joined for continuous action over wheels etc. □ **endlessly** *adv.* [Old English: related to END]

endmost *adj.* nearest the end.

endo- *comb. form* internal. [Greek *endon* within]

endocrine /'endəʊ,kraɪn/ *adj.* (of a gland) secreting directly into the blood. [Greek *krinō* sift]

endogenous /en'dɒdʒɪnəs/ *adj.* growing or originating from within.

endometrium /,endəʊ'mi:trɪəm/ *n.* membrane lining the womb. [Greek *mētra* womb]

endomorph /'endəʊ,mɔ:f/ *n.* person with a soft round body. [Greek *morphē* form]

endorse /ɪn'dɔ:s/ *v.* (also **indorse**) (**-sing**) **1** approve. **2** sign or write on (a document), esp. sign the back of (a cheque). **3** enter details of a conviction for an offence on (a driving-licence). □ **endorsement** *n.* [Latin *dorsum* back]

endoscope /'endəʊ,skəʊp/ *n.* instrument for viewing internal parts of the body.

endow /ɪn'daʊ/ *v.* **1** bequeath or give a permanent income to (a person, institution, etc.). **2** (esp. as **endowed** *adj.*) provide with talent, ability, etc. [Anglo-French: related to DOWER]

endowment *n.* **1** endowing. **2** endowed income. **3** (*attrib.*) denoting forms of insurance with payment of a sum on a specified date, or on the death of the insured person if earlier.

endowment mortgage *n.* mortgage linked to endowment insurance.

endpaper *n.* either of the blank leaves of paper at the beginning and end of a book.

end-product *n.* final product of manufacture etc.

endue /ɪn'dju:/ *v.* (also **indue**) (**-dues, -dued, -duing**) (foll. by *with*) provide (a person) with (qualities etc.). [Latin *induo* put on clothes]

endurance /ɪn'djʊərəns/ *n.* **1** power of enduring. **2** ability to withstand prolonged strain. [French: related to ENDURE]

endure /ɪn'djʊə(r)/ *v.* (**-ring**) **1** undergo (a difficulty etc.). **2** tolerate. **3** last. □ **endurable** *adj.* [Latin *durus* hard]

endways *adv.* (also **endwise**) **1** with end uppermost or foremost. **2** end to end.

enema /'enɪmə/ *n.* **1** introduction of fluid etc. into the rectum, esp. to flush out its contents. **2** fluid etc. used for this. [Greek *hiēmi* send]

enemy /'enəmɪ/ *n.* (*pl.* **-ies**) **1** person actively hostile to another. **2 a** (often *attrib.*) hostile nation or army. **b** member of this. **3** adversary or opponent (*enemy of progress*). [Latin: related to IN-[2], *amicus* friend]

energetic /,enə'dʒetɪk/ *adj.* full of energy, vigorous. □ **energetically** *adv.* [Greek: related to ENERGY]

energize /'enə,dʒaɪz/ *v.* (also **-ise**) (**-zing** or **-sing**) **1** give energy to. **2** provide (a device) with energy for operation.

energy /'enədʒɪ/ *n.* (*pl.* **-ies**) **1** capacity for activity, force, vigour. **2** capacity of matter or radiation to do work. [Greek *ergon* work]

enervate /'enə,veɪt/ *v.* (**-ting**) deprive of vigour or vitality. □ **enervation** /-'veɪʃ(ə)n/ *n.* [Latin: related to NERVE]

en famille /,ɑ̃ fæ'mi:/ *adv.* in or with one's family. [French, = in family]

enfant terrible /,ɑ̃fɑ̃ te'ri:bl/ *n.* (*pl.* ***enfants terribles*** pronunc. same) indiscreet or unruly person. [French, = terrible child]

enfeeble /ɪn'fi:b(ə)l/ *v.* (**-ling**) make feeble. □ **enfeeblement** *n.*

enfilade /,enfɪ'leɪd/ –*n.* gunfire directed along a line from end to end. –*v.* (**-ding**) direct an enfilade at. [French: related to FILE[1]]

enfold /ɪn'fəʊld/ *v.* **1** (usu. foll. by *in*, *with*) wrap; envelop. **2** clasp, embrace.

enforce /ɪn'fɔ:s/ *v.* (**-cing**) **1** compel observance of (a law etc.). **2** (foll. by *on*) impose (an action or one's will, etc.) on. □ **enforceable** *adj.* **enforcement** *n.* **enforcer** *n.* [Latin: related to FORCE[1]]

enfranchise /ɪn'fræntʃaɪz/ *v.* (**-sing**) **1** give (a person) the right to vote. **2** give (a town) municipal rights, esp. representation in parliament. **3** *hist.* free (a slave etc.). □ **enfranchisement** /-ɪzmənt/ *n.* [French: related to FRANK]

engage /ɪn'geɪdʒ/ *v.* (**-ging**) **1** employ or hire (a person). **2 a** (usu. in *passive*) occupy (*are you engaged tomorrow?*). **b** hold fast (a person's attention). **3** (usu.

in *passive*) bind by a promise, esp. of marriage. **4** arrange beforehand to occupy (a room, seat, etc.). **5 a** interlock (parts of a gear etc.). **b** (of a gear etc.) become interlocked. **6 a** come into battle with. **b** bring (troops) into battle with. **c** come into battle with (an enemy etc.). **7** take part (*engage in politics*). **8** (foll. by *that* + clause or *to* + infin.) undertake. [French: related to GAGE[1]]

engaged *adj.* **1** pledged to marry. **2** (of a person) occupied, busy. **3** (of a telephone line, toilet, etc.) in use.

engagement *n.* **1** engaging or being engaged. **2** appointment with another person. **3** betrothal. **4** battle.

engaging *adj.* attractive, charming. □ **engagingly** *adv.*

engender /ɪn'dʒendə(r)/ *v.* give rise to; produce (a feeling etc.). [related to GENUS]

engine /'endʒɪn/ *n.* **1** mechanical contrivance of parts working together, esp. as a source of power (*steam engine*). **2 a** railway locomotive. **b** = FIRE-ENGINE. [Latin *ingenium* device]

engineer /ˌendʒɪ'nɪə(r)/ —*n.* **1** person skilled in a branch of engineering. **2** person who makes or is in charge of engines etc. (*ship's engineer*). **3** person who designs and constructs military works; soldier so trained. **4** contriver. —*v.* **1** contrive, bring about. **2** act as an engineer. **3** construct or manage as an engineer. [medieval Latin: related to ENGINE]

engineering *n.* application of science to the design, building, and use of machines etc. (*civil engineering*).

English /'ɪŋglɪʃ/ —*adj.* of England or its people or language. —*n.* **1** language of England, now used in the UK, US, and most Commonwealth countries. **2** (prec. by *the*; treated as *pl.*) the people of England. [Old English]

Englishman *n.* (*fem.* **Englishwoman**) person who is English by birth or descent.

engorged /ɪn'gɔːdʒd/ *adj.* **1** crammed full. **2** congested with fluid, esp. blood. [French: related to EN-[1], GORGE]

engraft /ɪn'grɑːft/ *v.* (also **ingraft**) **1** *Bot.* (usu. foll. by *into, on*) graft. **2** implant. **3** (usu. foll. by *into*) incorporate.

engrave /ɪn'greɪv/ *v.* (**-ving**) **1** (often foll. by *on*) carve (a text or design) on a hard surface. **2** inscribe (a surface) thus. **3** (often foll. by *on*) impress deeply (on a person's memory). □ **engraver** *n.* [from GRAVE[3]]

engraving *n.* print made from an engraved plate.

engross /ɪn'grəʊs/ *v.* **1** absorb the attention of; occupy fully. **2** write out in larger letters or in legal form. □ **engrossment** *n.* [Anglo-French: related to EN-[1]]

engulf /ɪn'gʌlf/ *v.* flow over and swamp; overwhelm. □ **engulfment** *n.*

enhance /ɪn'hɑːns/ *v.* (**-cing**) intensify (qualities, powers, etc.); improve (something already good). □ **enhancement** *n.* [Anglo-French from Latin *altus* high]

enigma /ɪ'nɪgmə/ *n.* **1** puzzling thing or person. **2** riddle or paradox. □ **enigmatic** /ˌenɪg'mætɪk/ *adj.* **enigmatically** /ˌenɪg'mætɪkəlɪ/ *adv.* [Latin from Greek]

enjoin /ɪn'dʒɔɪn/ *v.* **1** command or order. **2** (often foll. by *on*) impose (an action). **3** (usu. foll. by *from*) *Law* prohibit by injunction (from doing a thing). [Latin *injungo* attach]

enjoy /ɪn'dʒɔɪ/ *v.* **1** take pleasure in. **2** have the use or benefit of. **3** experience (*enjoy good health*). □ **enjoy oneself** experience pleasure. □ **enjoyment** *n.* [French]

enjoyable *adj.* pleasant. □ **enjoyably** *adv.*

enkephalin /en'kefəlɪn/ *n.* either of two morphine-like peptides in the brain thought to control levels of pain. [Greek *egkephalos* brain]

enkindle /ɪn'kɪnd(ə)l/ *v.* (**-ling**) cause to flare up, arouse.

enlarge /ɪn'lɑːdʒ/ *v.* (**-ging**) **1** make or become larger or wider. **2** (often foll. by *on, upon*) describe in greater detail. **3** reproduce a photograph on a larger scale. □ **enlargement** *n.* [French: related to LARGE]

enlarger *n.* apparatus for enlarging photographs.

enlighten /ɪn'laɪt(ə)n/ *v.* **1** (often foll. by *on*) inform (about a subject). **2** (as **enlightened** *adj.*) progressive.

enlightenment *n.* **1** enlightening or being enlightened. **2** (**the Enlightenment**) 18th-c. philosophy of reason and individualism.

enlist /ɪn'lɪst/ *v.* **1** enrol in the armed services. **2** secure as a means of help or support. □ **enlistment** *n.*

enliven /ɪn'laɪv(ə)n/ *v.* make lively or cheerful; brighten (a picture etc.); inspirit. □ **enlivenment** *n.*

en masse /ɑ̃ 'mæs/ *adv.* all together. [French]

enmesh /ɪn'meʃ/ *v.* entangle in or as in a net.

enmity /'enmɪtɪ/ *n.* (*pl.* **-ies**) **1** state of being an enemy. **2** hostility. [Romanic: related to ENEMY]

ennoble /ɪ'nəʊb(ə)l/ *v.* (**-ling**) **1** make noble. **2** make (a person) a noble. □

ennoblement *n.* [French: related to EN-[1]]

ennui /ɒ'nwi:/ *n.* mental weariness from idleness or lack of interest; boredom. [French: related to ANNOY]

enormity /ɪ'nɔ:mɪtɪ/ *n.* (*pl.* **-ies**) **1** monstrous wickedness; monstrous crime. **2** serious error. **3** great size. [Latin *enormitas*]

■ **Usage** Sense 3 is commonly found, but is regarded as incorrect by some people.

enormous /ɪ'nɔ:məs/ *adj.* extremely large. □ **enormously** *adv.* [Latin *enormis*: related to NORM]

enough /ɪ'nʌf/ *—adj.* as much or as many as required (*enough apples*). *—n.* sufficient amount or quantity (*we have enough*). *—adv.* **1** adequately (*warm enough*). **2** fairly (*sings well enough*). **3** quite (*you know well enough what I mean*). □ **have had enough of** want no more of; be satiated with or tired of. **sure enough** as expected. [Old English]

en passant /ˌɑ̃ pæ'sɑ̃/ *adv.* in passing; casually (*mentioned it en passant*). [French, = in passing]

enprint /'enprɪnt/ *n.* standard-sized photograph. [*en*larged *print*]

enquire /ɪn'kwaɪə(r)/ *v.* (**-ring**) **1** seek information; ask; ask a question. **2** = INQUIRE. **3** (foll. by *after*, *for*) ask about (a person, a person's health, etc.). □ **enquirer** *n.* [Latin *quaero quaesit-* seek]

enquiry *n.* (*pl.* **-ies**) **1** act of asking or seeking information. **2** = INQUIRY.

enrage /ɪn'reɪdʒ/ *v.* (**-ging**) make furious. [French: related to EN-[1]]

enrapture /ɪn'ræptʃə(r)/ *v.* (**-ring**) delight intensely.

enrich /ɪn'rɪtʃ/ *v.* **1** make rich or richer. **2** make more nutritive. **3** increase the strength, wealth, value, or contents of. □ **enrichment** *n.* [French: related to EN-[1]]

enrol /ɪn'rəʊl/ *v.* (*US* **enroll**) (**-ll-**) **1** enlist. **2 a** write the name of (a person) on a list. **b** incorporate as a member. **c** enrol oneself, esp. for a course of study. □ **enrolment** *n.* [French: related to EN-[1]]

en route /ɑ̃ 'ru:t/ *adv.* on the way. [French]

ensconce /ɪn'skɒns/ *v.* (**-cing**) (usu. *refl.* or in *passive*) establish or settle comfortably. [from *sconce* small fortification]

ensemble /ɒn'sɒmb(ə)l/ *n.* **1 a** thing viewed as the sum of its parts. **b** general effect of this. **2** set of clothes worn together. **3** group of performers working together. **4** *Mus.* concerted passage for an ensemble. [Latin *simul* at the same time]

enshrine /ɪn'ʃraɪn/ *v.* (**-ning**) **1** enclose in a shrine. **2** protect, make inviolable. □ **enshrinement** *n.*

enshroud /ɪn'ʃraʊd/ *v. literary* **1** cover with or as with a shroud. **2** obscure.

ensign /'ensaɪn, -s(ə)n/ *n.* **1** banner or flag, esp. the military or naval flag of a nation. **2** standard-bearer. **3 a** *hist.* lowest commissioned infantry officer. **b** *US* lowest commissioned naval officer. [French: related to INSIGNIA]

ensilage /'ensɪlɪdʒ/ *—n.* = SILAGE. *—v.* (**-ging**) preserve (fodder) by ensilage. [French: related to SILO]

enslave /ɪn'sleɪv/ *v.* (**-ving**) make (a person) a slave. □ **enslavement** *n.*

ensnare /ɪn'sneə(r)/ *v.* (**-ring**) catch in or as in a snare. □ **ensnarement** *n.*

ensue /ɪn'sju:/ *v.* (**-sues, -sued, -suing**) happen later or as a result. [Latin *sequor* follow]

en suite /ɑ̃ 'swi:t/ *—adv.* forming a single unit (*bedroom with bathroom en suite*). *—adj.* **1** forming a single unit (*en suite bathroom*). **2** with a bathroom attached (*seven en suite bedrooms*). [French, = in sequence]

ensure /ɪn'ʃʊə(r)/ *v.* (**-ring**) **1** make certain. **2** (usu. foll. by *against*) make safe (*ensure against risks*). □ **ensurer** *n.* [Anglo-French: related to ASSURE]

ENT *abbr.* ear, nose, and throat.

-ent *suffix* **1** forming adjectives denoting attribution of an action (*consequent*) or state (*existent*). **2** forming agent nouns (*president*). [Latin *-ent-* present participial stem of verbs]

entablature /ɪn'tæblətʃə(r)/ *n.* upper part of a classical building supported by columns including an architrave, frieze, and cornice. [Italian: related to TABLE]

entail /ɪn'teɪl/ *—v.* **1** necessitate or involve unavoidably (*entails much effort*). **2** *Law* bequeath (an estate) to a specified line of beneficiaries so that it cannot be sold or given away. *—n. Law* **1** entailed estate. **2** succession to such an estate. [related to TAIL[2]]

entangle /ɪn'tæŋg(ə)l/ *v.* (**-ling**) **1** catch or hold fast in a snare, tangle, etc. **2** involve in difficulties. **3** complicate. □ **entanglement** *n.*

entente /ɒn'tɒnt/ *n.* friendly understanding between States. [French]

entente cordiale *n.* entente, esp. between Britain and France from 1904.

enter *v.* **1** go or come in or into. **2** come on stage (also as a direction: *enter Macbeth*). **3** penetrate (*bullet entered his arm*). **4** write (name, details, etc.) in a list, book, etc. **5** register, record the name of as a competitor (*entered for the long jump*). **6 a** become a member of (a society or profession). **b** enrol in a school etc. **7** make known; present for consideration (*enter a protest*). **8** record formally (before a court of law etc.). **9** (foll. by *into*) **a** engage in (conversation etc.). **b** subscribe to; bind oneself by (an agreement, contract, etc.). **c** form part of (a calculation, plan, etc.). **d** sympathize with (feelings). **10** (foll. by *on*, *upon*) **a** begin; begin to deal with. **b** assume the functions of (an office) or possession of (property). [Latin *intra* within]

enteric /en'terɪk/ *adj.* of the intestines. □ **enteritis** /,entə'raɪtɪs/ *n.* [Greek *enteron* intestine]

enterprise /'entə,praɪz/ *n.* **1** undertaking, esp. a challenging one. **2** readiness to engage in such undertakings. **3** business firm or venture. [Latin *prehendo* grasp]

enterprising *adj.* showing enterprise; resourceful, energetic. □ **enterprisingly** *adv.*

entertain /,entə'teɪn/ *v.* **1** occupy agreeably. **2 a** receive as a guest. **b** receive guests. **3** cherish, consider (an idea etc.). [Latin *teneo* hold]

entertainer *n.* person who entertains, esp. professionally.

entertaining *adj.* amusing, diverting. □ **entertainingly** *adv.*

entertainment *n.* **1** entertaining or being entertained. **2** thing that entertains; performance.

enthral /ɪn'θrɔ:l/ *v.* (*US* **enthrall**) (**-ll-**) captivate, please greatly. □ **enthralment** *n.* [from EN-[1], THRALL]

enthrone /ɪn'θrəʊn/ *v.* (**-ning**) place on a throne, esp. ceremonially. □ **enthronement** *n.*

enthuse /ɪn'θju:z/ *v.* (**-sing**) *colloq.* be or make enthusiastic.

enthusiasm /ɪn'θju:zɪ,æz(ə)m/ *n.* **1** (often foll. by *for*, *about*) strong interest or admiration, great eagerness. **2** object of enthusiasm. [Greek *entheos* inspired by a god]

enthusiast *n.* person full of enthusiasm. [Church Latin: related to ENTHUSIASM]

enthusiastic /ɪn,θju:zɪ'æstɪk/ *adj.* having enthusiasm. □ **enthusiastically** *adv.*

entice /ɪn'taɪs/ *v.* (**-cing**) attract by the offer of pleasure or reward. □ **enticement** *n.* **enticing** *adj.* **enticingly** *adv.* [French *enticier* probably from Romanic]

entire /ɪn'taɪə(r)/ *adj.* **1** whole, complete. **2** unbroken. **3** unqualified, absolute. **4** in one piece; continuous. [Latin: related to INTEGER]

entirely *adv.* **1** wholly. **2** solely.

entirety /ɪn'taɪərətɪ/ *n.* (*pl.* **-ies**) **1** completeness. **2** (usu. foll. by *of*) sum total. □ **in its entirety** in its complete form.

entitle /ɪn'taɪt(ə)l/ *v.* (**-ling**) **1** (usu. foll. by *to*) give (a person) a just claim or right. **2** give a title to. □ **entitlement** *n.* [Latin: related to TITLE]

entity /'entɪtɪ/ *n.* (*pl.* **-ies**) **1** thing with distinct existence. **2** thing's existence in itself. [Latin *ens ent-* being]

entomb /ɪn'tu:m/ *v.* **1** place in a tomb. **2** serve as a tomb for. □ **entombment** *n.* [French: related to TOMB]

entomology /,entə'mɒlədʒɪ/ *n.* the study of insects. □ **entomological** /-mə'lɒdʒɪk(ə)l/ *adj.* **entomologist** *n.* [Greek *entomon* insect]

entourage /'ɒntʊə,rɑ:ʒ/ *n.* people attending an important person. [French]

entr'acte /'ɒntrækt/ *n.* **1** interval between acts of a play. **2** music or dance performed during this. [French]

entrails /'entreɪlz/ *n.pl.* **1** bowels, intestines. **2** innermost parts of a thing. [Latin *inter* among]

entrance[1] /'entrəns/ *n.* **1** place for entering. **2** going or coming in. **3** right of admission. **4** coming of an actor on stage. **5** (in full **entrance fee**) admission fee. [French: related to ENTER]

entrance[2] /ɪn'trɑ:ns/ *v.* (**-cing**) **1** enchant, delight. **2** put into a trance. □ **entrancement** *n.* **entrancing** *adj.* **entrancingly** *adv.*

entrant /'entrənt/ *n.* person who enters (an examination, profession, etc.). [French: related to ENTER]

entrap /ɪn'træp/ *v.* (**-pp-**) **1** catch in or as in a trap. **2** beguile. □ **entrapment** *n.* [related to EN-[1]]

entreat /ɪn'tri:t/ *v.* ask earnestly, beg. [related to EN-[1]]

entreaty *n.* (*pl.* **-ies**) earnest request.

entrecôte /'ɒntrə,kəʊt/ *n.* boned steak off the sirloin. [French, = between-rib]

entrée /'ɒntreɪ/ *n.* **1** dish served between the fish and meat courses. **2** *US* main dish. **3** right of admission. [French]

entrench /ɪn'trentʃ/ *v.* **1 a** establish firmly (in a position, office, etc.). **b** (as **entrenched** *adj.*) (of an attitude etc.) not easily modified. **2** surround with a trench as a fortification. □ **entrenchment** *n.*

entrepôt /'ɒntrə,pəʊ/ *n.* warehouse for goods in transit. [French]

entrepreneur /,ɒntrəprə'nɜ:(r)/ *n.* **1** person who undertakes a commercial venture. **2** contractor acting as an intermediary. □ **entrepreneurial** *adj.*

entrepreneurialism *n.* (also **entrepreneurism**). [French: related to ENTERPRISE]

entropy /'entrəpɪ/ *n.* **1** *Physics* measure of the disorganization or degradation of the universe, resulting in a decrease in available energy. **2** *Physics* measure of the unavailability of a system's thermal energy for conversion into mechanical work. [Greek: related to EN-[2], *tropē* transformation]

entrust /ɪn'trʌst/ *v.* (also **intrust**) **1** (foll. by *to*) give (a person or thing) into the care of a person. **2** (foll. by *with*) assign responsibility for (a person or thing) to (a person) (*entrusted him with my camera*).

entry /'entrɪ/ *n.* (*pl.* **-ies**) **1 a** going or coming in. **b** liberty to do this. **2** place of entrance; door, gate, etc. **3** passage between buildings. **4 a** item entered in a diary, list, etc. **b** recording of this. **5 a** person or thing competing in a race etc. **b** list of competitors. [Romanic: related to ENTER]

Entryphone *n. propr.* intercom at the entrance of a building or flat for callers to identify themselves.

entwine /ɪn'twaɪn/ *v.* (**-ning**) twine round, interweave.

E-number /'i:,nʌmbə(r)/ *n.* E plus a number, the EC designation for food additives.

enumerate /ɪ'nju:mə,reɪt/ *v.* (**-ting**) **1** specify (items). **2** count. □ **enumeration** /-'reɪʃ(ə)n/ *n.* **enumerative** /-rətɪv/ *adj.* [Latin: related to NUMBER]

enumerator *n.* person employed in census-taking.

enunciate /ɪ'nʌnsɪ,eɪt/ *v.* (**-ting**) **1** pronounce (words) clearly. **2** express in definite terms. □ **enunciation** /-'eɪʃ(ə)n/ *n.* [Latin *nuntio* announce]

enuresis /,enjʊə'ri:sɪs/ *n.* involuntary urination. [Greek *enoureō* urinate in]

envelop /ɪn'veləp/ *v.* (**-p-**) **1** wrap up or cover completely. **2** completely surround. □ **envelopment** *n.* [French]

envelope /'envə,ləʊp/ *n.* **1** folded paper container for a letter etc. **2** wrapper, covering. **3** gas container of a balloon or airship.

enviable /'envɪəb(ə)l/ *adj.* likely to excite envy, desirable. □ **enviably** *adv.*

envious /'envɪəs/ *adj.* feeling or showing envy. □ **enviously** *adv.* [Anglo-French: related to ENVY]

environment /ɪn'vaɪərənmənt/ *n.* **1** surroundings, esp. as affecting lives. **2** circumstances of living. **3** *Computing* overall structure within which a user, computer, or program operates. □ **environmental** /-'ment(ə)l/ *adj.* **environmentally** /-'mentəlɪ/ *adv.* [French *environ* surroundings]

environmentalist /ɪn,vaɪərən'mentəlɪst/ *n.* person concerned with the protection of the natural environment. □ **environmentalism** *n.*

environs /ɪn'vaɪərənz/ *n.pl.* district round a town etc.

envisage /ɪn'vɪzɪdʒ/ *v.* (**-ging**) **1** have a mental picture of (a thing not yet existing). **2** imagine as possible or desirable. [French: related to VISAGE]

envoy /'envɔɪ/ *n.* **1** messenger or representative. **2** (in full **envoy extraordinary**) diplomatic agent ranking below ambassador. [French *envoyer* send, from Latin *via* way]

envy /'envɪ/ *–n.* (*pl.* **-ies**) **1** discontent aroused by another's better fortune etc. **2** object of this feeling. *–v.* (**-ies, -ied**) feel envy of (a person etc.). [Latin *invidia*, from *video* see]

enwrap /ɪn'ræp/ *v.* (**-pp-**) (often foll. by *in*) *literary* wrap, enfold.

enzyme /'enzaɪm/ *n.* protein catalyst of a specific biochemical reaction. [Greek *enzumos* leavened]

Eocene /'i:əʊ,si:n/ *Geol.* *–adj.* of the second epoch of the Tertiary period. *–n.* this epoch. [Greek *ēōs* dawn, *kainos* new]

eolian harp *US* var. of AEOLIAN HARP.

eolithic /,i:ə'lɪθɪk/ *adj.* of the period preceding the palaeolithic age. [Greek *ēōs* dawn, *lithos* stone]

eon var. of AEON.

EP *abbr.* extended-play (gramophone record).

epaulette /'epə,let/ *n.* (*US* **epaulet**) ornamental shoulder-piece on a coat etc., esp. on a uniform. [French *épaule* shoulder]

épée /eɪ'peɪ/ *n.* sharp-pointed sword, used (with the end blunted) in fencing. [French: related to SPATHE]

ephedrine /'efədrɪn/ *n.* alkaloid drug used to relieve asthma, etc. [*Ephedra*, genus of plants yielding it]

ephemera /ɪ'femərə/ *n.pl.* things of only short-lived relevance. [Latin: related to EPHEMERAL]

ephemeral /ɪ'femər(ə)l/ *adj.* lasting or of use for only a short time; transitory. [Greek: related to EPI-, *hēmera* day]

epi- *prefix* **1** upon. **2** above. **3** in addition. [Greek]

epic /'epɪk/ *–n.* **1** long poem narrating the adventures or deeds of one or more heroic or legendary figures. **2** book or film based on an epic narrative. *–adj.* **1** of or like an epic. **2** grand, heroic. [Greek *epos* song]

epicene /'epɪ,si:n/ *–adj.* **1** of, for, denoting, or used by both sexes. **2** having characteristics of both sexes or of neither sex. *–n.* epicene person. [Greek *koinos* common]

epicentre /'epɪ,sentə(r)/ *n.* (*US* **epicenter**) **1** point at which an earthquake reaches the earth's surface. **2** central point of a difficulty. [Greek: related to CENTRE]

epicure /'epɪ,kjʊə(r)/ *n.* person with refined tastes, esp. in food and drink. □ **epicurism** *n.* [medieval Latin: related to EPICUREAN]

Epicurean /,epɪkjʊə'ri:ən/ *–n.* **1** disciple or student of the Greek philosopher Epicurus. **2** (**epicurean**) devotee of (esp. sensual) enjoyment. *–adj.* **1** of Epicurus or his ideas. **2** (**epicurean**) characteristic of an epicurean. □ **Epicureanism** *n.* [Latin from Greek]

epidemic /,epɪ'demɪk/ *–n.* widespread occurrence of a disease in a community at a particular time. *–adj.* in the nature of an epidemic. [Greek *epi* against, *dēmos* the people]

epidemiology /,epɪdi:mɪ'ɒlədʒɪ/ *n.* the study of epidemic diseases and their control. □ **epidemiologist** *n.*

epidermis /,epɪ'dɜ:mɪs/ *n.* outer layer of the skin. □ **epidermal** *adj.* [Greek *derma* skin]

epidiascope /,epɪ'daɪə,skəʊp/ *n.* optical projector capable of giving images of both opaque and transparent objects. [from EPI-, DIA-, -SCOPE]

epidural /,epɪ'djʊər(ə)l/ *–adj.* (of an anaesthetic) introduced into the space around the dura mater of the spinal cord. *–n.* epidural anaesthetic. [from EPI-, DURA MATER]

epiglottis /,epɪ'glɒtɪs/ *n.* flap of cartilage at the root of the tongue, depressed during swallowing to cover the windpipe. □ **epiglottal** *adj.* [Greek *glōtta* tongue]

epigram /'epɪ,græm/ *n.* **1** short poem with a witty ending. **2** pointed saying. □ **epigrammatic** /-grə'mætɪk/ *adj.* [Greek: related to -GRAM]

epigraph /'epɪ,grɑ:f/ *n.* inscription. [Greek: related to -GRAPH]

epilepsy /'epɪ,lepsɪ/ *n.* nervous disorder with convulsions and often loss of consciousness. [Greek *lambanō* take]

epileptic /,epɪ'leptɪk/ *–adj.* of epilepsy. *–n.* person with epilepsy. [French: related to EPILEPSY]

epilogue /'epɪ,lɒg/ *n.* **1** short piece ending a literary work. **2** speech addressed to the audience by an actor at the end of a play. [Greek *logos* speech]

epiphany /ɪ'pɪfənɪ/ *n.* (*pl.* **-ies**) **1** (**Epiphany**) **a** manifestation of Christ to the Magi. **b** festival of this on 6 January. **2** manifestation of a god or demigod. [Greek *phainō* show]

episcopacy /ɪ'pɪskəpəsɪ/ *n.* (*pl.* **-ies**) **1** government by bishops. **2** (prec. by *the*) the bishops.

episcopal /ɪ'pɪskəp(ə)l/ *adj.* **1** of a bishop or bishops. **2** (of a Church) governed by bishops. □ **episcopally** *adv.* [Church Latin: related to BISHOP]

episcopalian /ɪ,pɪskə'peɪlɪən/ *–adj.* **1** of episcopacy. **2** of an episcopal Church or (**Episcopalian**) the Episcopal Church. *–n.* **1** adherent of episcopacy. **2** (**Episcopalian**) member of the Episcopal Church. □ **episcopalianism** *n.*

episcopate /ɪ'pɪskəpət/ *n.* **1** the office or tenure of a bishop. **2** (prec. by *the*) the bishops collectively. [Church Latin: related to BISHOP]

episiotomy /ɪ,pɪzɪ'ɒtəmɪ/ *n.* (*pl.* **-ies**) surgical cut made at the vaginal opening during childbirth, to aid delivery. [Greek *epision* pubic region]

episode /'epɪ,səʊd/ *n.* **1** event or group of events as part of a sequence. **2** each of the parts of a serial story or broadcast. **3** incident or set of incidents in a narrative. [Greek *eisodos* entry]

episodic /,epɪ'sɒdɪk/ *adj.* **1** consisting of separate episodes. **2** irregular, sporadic. □ **episodically** *adv.*

epistemology /ɪ,pɪstɪ'mɒlədʒɪ/ *n.* philosophy of knowledge. □ **epistemological** /-mə'lɒdʒɪk(ə)l/ *adj.* [Greek *epistēmē* knowledge]

epistle /ɪ'pɪs(ə)l/ *n.* **1** *joc.* letter. **2** (**Epistle**) any of the apostles' letters in the New Testament. **3** poem etc. in the form of a letter. [Greek *epistolē* from *stellō* send]

epistolary /ɪ'pɪstələrɪ/ *adj.* of or in the form of a letter or letters. [Latin: related to EPISTLE]

epitaph /'epɪ,tɑ:f/ *n.* words written in memory of a dead person, esp. as a tomb inscription. [Greek *taphos* tomb]

epithelium /,epɪ'θi:lɪəm/ *n.* (*pl.* **-s** or **-lia** /-lɪə/) tissue forming the outer layer of the body and lining many hollow structures. □ **epithelial** *adj.* [Greek *thēlē* teat]

epithet /'epɪ,θet/ *n.* **1** adjective etc. expressing a quality or attribute. **2** this as a term of abuse. [Greek *tithēmi* place]

epitome /ɪ'pɪtəmɪ/ *n.* **1** person or thing embodying a quality etc. **2** thing representing another in miniature. [Greek *temnō* cut]

epitomize *v.* (also **-ise**) (**-zing** or **-sing**) make or be a perfect example of (a quality etc.).

EPNS *abbr.* electroplated nickel silver.

epoch /'i:pɒk/ *n.* **1** period of history etc. marked by notable events. **2** beginning of an era. **3** *Geol.* division of a period, corresponding to a set of strata. □ **epochal** /'epək(ə)l/ *adj.* [Greek, = pause]

epoch-making *adj.* remarkable; very important.

eponym /'epənɪm/ *n.* **1** word, place-name, etc., derived from a person's name. **2** person whose name is used in this way. □ **eponymous** /ɪ'pɒnɪməs/ *adj.* [Greek *onoma* name]

EPOS /'i:pɒs/ *abbr.* electronic point-of-sale (equipment recording stock, sales, etc. in shops).

epoxy /ɪ'pɒksɪ/ *adj.* relating to or derived from a compound with one oxygen atom and two carbon atoms bonded in a triangle. [from EPI-, OXYGEN]

epoxy resin *n.* synthetic thermosetting resin.

epsilon /'epsɪ,lɒn/ *n.* fifth letter of the Greek alphabet (*E*, *ϵ*). [Greek]

Epsom salts /'epsəm/ *n.* magnesium sulphate used as a purgative etc. [*Epsom* in S. England]

equable /'ekwəb(ə)l/ *adj.* **1** not varying. **2** moderate (*equable climate*). **3** (of a person) not easily disturbed. □ **equably** *adv.* [related to EQUAL]

equal /'i:kw(ə)l/ —*adj.* **1** (often foll. by *to*, *with*) the same in quantity, quality, size, degree, level, etc. **2** evenly balanced (*an equal contest*). **3** having the same rights or status (*human beings are essentially equal*). **4** uniform in application or effect. —*n.* person or thing equal to another, esp. in rank or quality. —*v.* (**-ll-**; *US* **-l-**) **1** be equal to. **2** achieve something that is equal to. □ **be equal to** have the ability or resources for. [Latin *aequalis*]

equality /ɪ'kwɒlɪtɪ/ *n.* being equal. [Latin: related to EQUAL]

equalize *v.* (also **-ise**) (**-zing** or **-sing**) **1** make or become equal. **2** reach one's opponent's score. □ **equalization** /-'zeɪʃ(ə)n/ *n.*

equalizer *n.* (also **-iser**) equalizing score or goal etc.

equally *adv.* **1** in an equal manner (*treated them equally*). **2** to an equal degree (*equally important*).

■ **Usage** In sense 2, construction with *as* (e.g. *equally as important*) is often found, but is considered incorrect by some people.

equal opportunity *n.* (often in *pl.*) opportunity to compete on equal terms, regardless of sex, race, etc.

equanimity /,ekwə'nɪmɪtɪ/ *n.* composure, evenness of temper, esp. in adversity. [Latin *aequus* even, *animus* mind]

equate /ɪ'kweɪt/ *v.* (**-ting**) **1** (usu. foll. by *to*, *with*) regard as equal or equivalent. **2** (foll. by *with*) be equal or equivalent to. □ **equatable** *adj.* [Latin *aequo aequat-*: related to EQUAL]

equation /ɪ'kweɪʒ(ə)n/ *n.* **1** equating or making equal; being equal. **2** statement that two mathematical expressions are equal (indicated by the sign =). **3** formula indicating a chemical reaction by means of symbols.

equator *n.* **1** imaginary line round the earth or other body, equidistant from the poles. **2** = CELESTIAL EQUATOR. [medieval Latin: related to EQUATE]

equatorial /,ekwə'tɔ:rɪəl/ *adj.* of or near the equator.

equerry /'ekwərɪ/ *n.* (*pl.* **-ies**) officer attending the British royal family. [French *esquierie* stable]

equestrian /ɪ'kwestrɪən/ —*adj.* **1** of horse-riding. **2** on horseback. —*n.* rider or performer on horseback. □ **equestrianism** *n.* [Latin *equestris* from *equus* horse]

equi- *comb. form* equal. [Latin: related to EQUAL]

equiangular /,i:kwɪ'æŋgjʊlə(r)/ *adj.* having equal angles.

equidistant /,i:kwɪ'dɪst(ə)nt/ *adj.* at equal distances.

equilateral /,i:kwɪ'lætər(ə)l/ *adj.* having all its sides equal in length.

equilibrium /,i:kwɪ'lɪbrɪəm/ *n.* (*pl.* **-ria** /-rɪə/ or **-s**) **1** state of physical balance. **2** state of composure. [Latin *libra* balance]

equine /'ekwaɪn/ *adj.* of or like a horse. [Latin *equus* horse]

equinoctial /,i:kwɪ'nɒkʃ(ə)l/ —*adj.* happening at or near the time of an equinox. —*n.* (in full **equinoctial line**) = CELESTIAL EQUATOR. [Latin: related to EQUINOX]

equinox /'i:kwɪ,nɒks/ *n.* time or date (twice each year) at which the sun crosses the celestial equator, when day and night are of equal length. [Latin *nox noctis* night]

equip /ɪ'kwɪp/ *v.* (**-pp-**) supply with what is needed. [Old Norse *skipa* to man a ship]

equipage /'ekwɪpɪdʒ/ *n.* **1** *archaic* **a** requisites. **b** outfit. **2** *hist.* carriage and horses with attendants. [French: related to EQUIP]

equipment *n.* **1** necessary articles, clothing, etc. **2** equipping or being equipped. [French: related to EQUIP]

equipoise /'ekwɪ,pɔɪz/ *n.* **1** equilibrium. **2** counterbalancing thing.

equitable /'ekwɪtəb(ə)l/ *adj.* **1** fair, just. **2** *Law* valid in equity as distinct from law. □ **equitably** *adv.* [French: related to EQUITY]

equitation /,ekwɪ'teɪʃ(ə)n/ *n.* horsemanship; horse-riding. [Latin *equito* ride a horse]

equity /'ekwɪtɪ/ *n.* (*pl.* **-ies**) **1** fairness. **2** principles of justice used to correct or supplement the law. **3 a** value of the shares issued by a company. **b** (in *pl.*) stocks and shares not bearing fixed interest. [Latin *aequitas*: related to EQUAL]

equivalent /ɪ'kwɪvələnt/ *–adj.* **1** (often foll. by *to*) equal in value, amount, importance, etc. **2** corresponding. **3** having the same meaning or result. *–n.* equivalent thing, amount, etc. □ **equivalence** *n.* [Latin: related to VALUE]

equivocal /ɪ'kwɪvək(ə)l/ *adj.* **1** of double or doubtful meaning. **2** of uncertain nature. **3** (of a person etc.) questionable. □ **equivocally** *adv.* [Latin *voco* call]

equivocate /ɪ'kwɪvə,keɪt/ *v.* (**-ting**) use ambiguity to conceal the truth. □ **equivocation** /-'keɪʃ(ə)n/ *n.* **equivocator** *n.* [Latin: related to EQUIVOCAL]

ER *abbr.* Queen Elizabeth. [Latin *Elizabetha Regina*]

Er *symb.* erbium.

er /ɜ:(r)/ *int.* expressing hesitation. [imitative]

-er[1] *suffix* forming nouns from nouns, adjectives, and verbs, denoting: **1** person, animal, or thing that does (*cobbler*; *poker*). **2** person or thing that is (*foreigner*; *four-wheeler*). **3** person concerned with (*hatter*; *geographer*). **4** person from (*villager*; *sixth-former*). [Old English]

-er[2] *suffix* forming the comparative of adjectives (*wider*) and adverbs (*faster*). [Old English]

-er[3] *suffix* used in a slang distortion of the word (*rugger*). [probably an extension of -ER[1]]

era /'ɪərə/ *n.* **1** system of chronology reckoning from a noteworthy event (*Christian era*). **2** large period, esp. regarded historically. **3** date at which an era begins. **4** major division of geological time. [Latin, = number (pl. of *aes* money)]

eradicate /ɪ'rædɪ,keɪt/ *v.* (**-ting**) root out; destroy completely. □ **eradicable** *adj.* **eradication** /-'keɪʃ(ə)n/ *n.* **eradicator** *n.* [Latin *radix -icis* root]

erase /ɪ'reɪz/ *v.* (**-sing**) **1** rub out; obliterate. **2** remove all traces of. **3** remove recorded material from (magnetic tape or disk). [Latin *rado ras-* scrape]

eraser *n.* thing that erases, esp. a piece of rubber etc. for removing pencil etc. marks.

erasure /ɪ'reɪʒə(r)/ *n.* **1** erasing. **2** erased word etc.

erbium /'ɜ:bɪəm/ *n.* metallic element of the lanthanide series. [*Ytterby* in Sweden]

ere /eə(r)/ *prep.* & *conj. poet.* or *archaic* before (of time) (*ere noon*; *ere they come*). [Old English]

erect /ɪ'rekt/ *–adj.* **1** upright, vertical. **2** (of the penis etc.) enlarged and rigid, esp. in sexual excitement. **3** (of hair) bristling. *–v.* **1** set up; build. **2** establish. □ **erection** *n.* **erectly** *adv.* **erectness** *n.* [Latin *erigere erect-* set up]

erectile *adj.* that can become erect (esp. of body tissue in sexual excitement). [French: related to ERECT]

erg *n.* unit of work or energy. [Greek *ergon* work]

ergo /'ɜ:gəʊ/ *adv.* therefore. [Latin]

ergonomics /,ɜ:gə'nɒmɪks/ *n.* the study of the relationship between people and their working environment. □ **ergonomic** *adj.* [Greek *ergon* work]

ergot /'ɜ:gət/ *n.* disease of rye etc. caused by a fungus. [French]

Erin /'ɪərɪn/ *n. poet.* Ireland. [Irish]

ERM *abbr.* Exchange Rate Mechanism.

ermine /'ɜ:mɪn/ *n.* (*pl.* same or **-s**) **1** stoat, esp. when white in winter. **2** its white fur, used to trim robes etc. [French]

Ernie /'ɜ:nɪ/ *n.* device for drawing prize-winning numbers of Premium Bonds. [*e*lectronic *r*andom *n*umber *i*ndicator *e*quipment]

erode /ɪ'rəʊd/ *v.* (**-ding**) wear away, destroy gradually. □ **erosion** *n.* **erosive** *adj.* [Latin *rodo ros-* gnaw]

erogenous /ɪ'rɒdʒɪnəs/ *adj.* (of a part of the body) particularly sensitive to sexual stimulation. [Greek (as EROTIC), -GENOUS]

erotic /ɪ'rɒtɪk/ *adj.* of or causing sexual love, esp. tending to arouse sexual desire or excitement. □ **erotically** *adv.* [Greek *erōs* sexual love]

erotica *n.pl.* erotic literature or art.

eroticism /ɪ'rɒtɪ,sɪz(ə)m/ *n.* **1** erotic character. **2** use of or response to erotic images or stimulation.

err /ɜ:(r)/ *v.* **1** be mistaken or incorrect. **2** do wrong; sin. [Latin *erro* stray]

errand /'erənd/ *n.* **1** short journey, esp. on another's behalf, to take a message, collect goods, etc. **2** object of such a journey. [Old English]

errand of mercy *n.* journey to relieve suffering etc.

errant /'erənt/ *adj.* **1** erring. **2** *literary* or *archaic* travelling in search of adventure (*knight errant*). □ **errantry** *n.* (in sense 2). [from ERR: sense 2 ultimately from Latin *iter* journey]

erratic /ɪ'rætɪk/ *adj.* **1** inconsistent in conduct, opinions, etc. **2** uncertain in movement. □ **erratically** *adv.* [Latin: related to ERR]

erratum /ɪ'rɑ:təm/ *n.* (*pl.* **errata**) error in printing or writing. [Latin: related to ERR]

erroneous /ɪ'rəʊnɪəs/ *adj.* incorrect. □ **erroneously** *adv.* [Latin: related to ERR]

error /'erə(r)/ *n.* **1** mistake. **2** condition of being morally wrong (*led into error*). **3** degree of inaccuracy in a calculation etc. (*2% error*). [Latin: related to ERR]

ersatz /'eəzæts/ *adj.* & *n.* substitute, imitation. [German]

Erse /ɜ:s/ *–adj.* Irish or Highland Gaelic. *–n.* the Gaelic language. [early Scots form of IRISH]

erstwhile /'ɜ:stwaɪl/ *–adj.* former, previous. *–adv. archaic* formerly. [related to ERE]

eructation /ˌi:rʌk'teɪʃ(ə)n/ *n. formal* belching. [Latin *ructo* belch]

erudite /'eru:ˌdaɪt/ *adj.* learned. □ **erudition** /-'dɪʃ(ə)n/ *n.* [Latin *eruditus* instructed: related to RUDE]

erupt /ɪ'rʌpt/ *v.* **1** break out suddenly or dramatically. **2** (of a volcano) eject lava etc. **3** (of a rash etc.) appear on the skin. □ **eruption** *n.* **eruptive** *adj.* [Latin *erumpo erupt-* break out]

-ery *suffix* (also **-ry**) forming nouns denoting: **1** class or kind (*greenery*; *machinery*; *citizenry*). **2** employment; state or condition (*dentistry*; *slavery*). **3** place of work or cultivation or breeding (*brewery*; *rookery*). **4** behaviour (*mimicry*). **5** often *derog.* all that has to do with (*popery*). [French *-erie*]

erysipelas /ˌerɪ'sɪpɪləs/ *n.* disease causing fever and a deep red inflammation of the skin. [Latin from Greek]

erythrocyte /ɪ'rɪθrəʊˌsaɪt/ *n.* red blood cell. [Greek *eruthros* red, -CYTE]

Es *symb.* einsteinium.

escalate /'eskəˌleɪt/ *v.* (**-ting**) **1** increase or develop (usu. rapidly) by stages. **2** make or become more intense. □ **escalation** /-'leɪʃ(ə)n/ *n.* [from ESCALATOR]

escalator *n.* moving staircase consisting of a circulating belt forming steps. [Latin *scala* ladder]

escalope /'eskəˌlɒp/ *n.* thin slice of boneless meat, esp. veal. [French, originally = shell]

escapade /'eskəˌpeɪd/ *n.* piece of reckless behaviour. [French from Provençal or Spanish: related to ESCAPE]

escape /ɪ'skeɪp/ *–v.* (**-ping**) **1** (often foll. by *from*) get free of restriction or control. **2** (of gas etc.) leak. **3** succeed in avoiding punishment etc. **4** get free of (a person, grasp, etc.). **5** avoid (a commitment, danger, etc.). **6** elude the notice or memory of (*nothing escapes you*; *name escaped me*). **7** (of words etc.) issue unawares from (a person etc.). *–n.* **1** act or instance of escaping. **2** means of escaping (often *attrib.*: *escape hatch*). **3** leakage of gas etc. **4** temporary relief from unpleasant reality. [Latin *cappa* cloak]

escape clause *n. Law* clause specifying conditions under which a contracting party is free from an obligation.

escapee /ɪskeɪ'pi:/ *n.* person who has escaped.

escapement *n.* part of a clock etc. that connects and regulates the motive power. [French: related to ESCAPE]

escape velocity *n.* minimum velocity needed to escape from the gravitational field of a body.

escapism *n.* pursuit of distraction and relief from reality. □ **escapist** *n.* & *adj.*

escapology /ˌeskə'pɒlədʒɪ/ *n.* techniques of escaping from confinement, esp. as entertainment. □ **escapologist** *n.*

escarpment /ɪ'skɑ:pmənt/ *n.* long steep slope at the edge of a plateau etc. [French from Italian: related to SCARP]

eschatology /ˌeskə'tɒlədʒɪ/ *n.* theology of death and final destiny. □ **eschatological** /-tə'lɒdʒɪk(ə)l/ *adj.* [Greek *eskhatos* last]

escheat /ɪs'tʃi:t/ *hist. –n.* **1** reversion of property to the State etc. in the absence of legal heirs. **2** property so affected. *–v.* **1** hand over (property) as an escheat. **2** confiscate. **3** revert by escheat. [Latin *cado* fall]

eschew /ɪs'tʃu:/ *v. formal* avoid; abstain from. □ **eschewal** *n.* [Germanic: related to SHY[1]]

escort *–n.* /'eskɔ:t/ **1** one or more persons, vehicles, etc., accompanying a person, vehicle, etc., for protection or as a mark of status. **2** person accompanying a person of the opposite sex socially. *–v.* /ɪ'skɔ:t/ act as an escort to. [French from Italian]

escritoire /ˌeskrɪ'twɑ:(r)/ *n.* writing-desk with drawers etc. [French from Latin *scriptorium* writing-room]

escudo /e'skju:dəʊ/ *n.* (*pl.* **-s**) chief monetary unit of Portugal. [Spanish and Portuguese from Latin *scutum* shield]

escutcheon /ɪ'skʌtʃ(ə)n/ *n.* shield or emblem bearing a coat of arms. [Latin *scutum* shield]

Eskimo /'eskɪ,məʊ/ *–n.* (*pl.* same or **-s**) **1** member of a people inhabiting N. Canada, Alaska, Greenland, and E. Siberia. **2** language of this people. *–adj.* of Eskimos or their language. [Algonquian]

■ **Usage** The Eskimos of N. America prefer the name *Inuit*.

ESN *abbr.* educationally subnormal.
esophagus *US* var. of OESOPHAGUS.
esoteric /,esəʊ'terɪk/ *adj.* intelligible only to those with special knowledge. □ **esoterically** *adv.* [Greek *esō* within]
ESP *abbr.* extrasensory perception.
espadrille /,espə'drɪl/ *n.* light canvas shoe with a plaited fibre sole. [Provençal: related to ESPARTO]
espalier /ɪ'spælɪə(r)/ *n.* **1** lattice-work along which the branches of a tree or shrub are trained. **2** tree or shrub so trained. [French from Italian]
esparto /e'spɑ:təʊ/ *n.* (*pl.* **-s**) (in full **esparto grass**) coarse grass of Spain and N. Africa, used to make good-quality paper etc. [Greek *sparton* rope]
especial /ɪ'speʃ(ə)l/ *adj.* notable. [Latin: related to SPECIAL]
especially *adv.* **1** in particular. **2** much more than in other cases. **3** particularly.
Esperanto /,espə'ræntəʊ/ *n.* an artificial language designed for universal use. [Latin *spero* hope]
espionage /'espɪə,nɑ:ʒ/ *n.* spying or use of spies. [French: related to SPY]
esplanade /,esplə'neɪd/ *n.* **1** long open level area for walking on, esp. beside the sea. **2** level space separating a fortress from a town. [Latin *planus* level]
espousal /ɪ'spaʊz(ə)l/ *n.* **1** (foll. by *of*) espousing of (a cause etc.). **2** *archaic* marriage, betrothal.
espouse /ɪ'spaʊz/ *v.* (**-sing**) **1** adopt or support (a cause, doctrine, etc.). **2** *archaic* **a** (usu. of a man) marry. **b** (usu. foll. by *to*) give (a woman) in marriage. [Latin *spondeo* betroth]
espresso /e'spresəʊ/ *n.* (also **expresso** /ek'spresəʊ/) (*pl.* **-s**) strong black coffee made under steam pressure. [Italian, = pressed out]
esprit /e'spri:/ *n.* sprightliness, wit. □ ***esprit de corps*** /də 'kɔ:(r)/ devotion to and pride in one's group. [French: related to SPIRIT]
espy /ɪ'spaɪ/ *v.* (**-ies, -ied**) catch sight of. [French: related to SPY]
Esq. *abbr.* Esquire.
-esque *suffix* forming adjectives meaning 'in the style of' or 'resembling' (*Kafkaesque*). [French from Latin *-iscus*]
esquire /ɪ'skwaɪə(r)/ *n.* **1** (usu. as abbr. **Esq.**) title added to a man's surname when no other title is used, esp. as a form of address for letters. **2** *archaic* = SQUIRE. [French from Latin *scutum* shield]
-ess *suffix* forming nouns denoting females (*actress*; *lioness*). [Greek *-issa*]
essay *–n.* /'eseɪ/ **1** short piece of writing on a given subject. **2** (often foll. by *at, in*) *formal* attempt. *–v.* /e'seɪ/ attempt. □ **essayist** *n.* [Latin *exigo* weigh: cf. ASSAY]
essence /'es(ə)ns/ *n.* **1** fundamental nature; inherent characteristics. **2 a** extract got by distillation etc. **b** perfume. □ **of the essence** indispensable. **in essence** fundamentally. [Latin *esse* be]
essential /ɪ'senʃ(ə)l/ *–adj.* **1** necessary; indispensable. **2** of or constituting the essence of a person or thing. *–n.* (esp. in *pl.*) basic or indispensable element or thing. □ **essentially** *adv.* [Latin: related to ESSENCE]
essential oil *n.* volatile oil derived from a plant etc. with its characteristic odour.
-est *suffix* forming the superlative of adjectives (*widest*; *nicest*; *happiest*) and adverbs (*soonest*). [Old English]
establish /ɪ'stæblɪʃ/ *v.* **1** set up (a business, system, etc.) on a permanent basis. **2** (foll. by *in*) settle (a person or oneself) in some capacity. **3** (esp. as **established** *adj.*) **a** achieve permanent acceptance for (a custom, belief, etc.). **b** place (a fact etc.) beyond dispute. [Latin *stabilio* make firm]
Established Church *n.* the Church recognized by the State.
establishment *n.* **1** establishing or being established. **2 a** business organization or public institution. **b** place of business. **c** residence. **3 a** staff of an organization. **b** household. **4** organized body permanently maintained. **5** Church system organized by law. **6** (**the Establishment**) social group with authority or influence and resisting change.
estate /ɪ'steɪt/ *n.* **1** property consisting of much land and usu. a large house. **2** modern residential or industrial area with an integrated design or purpose. **3** person's assets and liabilities, esp. at death. **4** property where rubber, tea, grapes, etc., are cultivated. **5** order or class forming (or regarded as) part of the body politic. **6** *archaic* or *literary* state or position in life (*the estate of holy matrimony*). □ **the Three Estates** Lords Spiritual (the heads of the Church), Lords Temporal (the peerage), and the Commons. [French *estat*, from Latin *sto stat-* stand]

estate agent *n.* person whose business is the sale or lease of buildings and land on behalf of others.
estate car *n.* car with a continuous area for rear passengers and luggage.
estate duty *n. hist.* death duty.

■ **Usage** Estate duty was replaced in 1975 by *capital transfer tax* and in 1986 by *inheritance tax.*

esteem /ɪ'sti:m/ *–v.* **1** (usu. in *passive*) have a high regard for. **2** *formal* consider (*esteemed it an honour*). *–n.* high regard; favour. [Latin: related to ESTIMATE]
ester /'estə(r)/ *n. Chem.* a compound produced by replacing the hydrogen of an acid by an organic radical. [German]
estimable /'estɪməb(ə)l/ *adj.* worthy of esteem; admirable. [Latin: related to ESTEEM]
estimate *–n.* /'estɪmət/ **1** approximate judgement, esp. of cost, value, size, etc. **2** statement of approximate charge for work to be undertaken. *–v.* /'estɪ,meɪt/ (**-ting**) (*also absol.*) **1** form an estimate or opinion of. **2** (foll. by *that*) make a rough calculation. **3** (often foll. by *at*) form an estimate; adjudge. □ **estimator** /-,meɪtə(r)/ *n.* [Latin *aestimo* fix the price of]
estimation /,estɪ'meɪʃ(ə)n/ *n.* **1** estimating. **2** judgement of worth. [Latin: related to ESTIMATE]
Estonian /e'stəʊnɪən/ *–n.* **1 a** native or national of Estonia in eastern Europe. **b** person of Estonian descent. **2** language of Estonia. *–adj.* of Estonia, its people, or language.
estrange /ɪ'streɪndʒ/ *v.* (**-ging**) **1** (usu. in *passive*; often foll. by *from*) alienate; make hostile or indifferent. **2** (as **estranged** *adj.*) (of a husband or wife) no longer living with his or her spouse. □ **estrangement** *n.* [Latin: related to STRANGE]
estrogen *US* var. of OESTROGEN.
estrus *US* var. of OESTRUS.
estuary /'estjʊərɪ/ *n.* (*pl.* **-ies**) wide tidal river mouth. [Latin *aestus* tide]
ETA *abbr.* estimated time of arrival.
eta /'i:tə/ *n.* seventh letter of the Greek alphabet (H, η). [Greek]
et al. /et 'æl/ *abbr.* and others. [Latin *et alii*]
etc. *abbr.* = ET CETERA.
et cetera /et 'setrə/ (also **etcetera**) *–adv.* **1** and the rest. **2** and so on. *–n.* (in *pl.*) the usual extras. [Latin]
etch *v.* **1 a** reproduce (a picture etc.) by engraving it on a metal plate with acid (esp. to print copies). **b** engrave (a plate) in this way. **2** practise this craft. **3** (foll. by *on, upon*) impress deeply (esp. on the mind). □ **etcher** *n.* [Dutch *etsen*]
etching *n.* **1** print made from an etched plate. **2** art of producing these plates.
eternal /ɪ'tɜ:n(ə)l/ *adj.* **1** existing always; without an end or (usu.) beginning. **2** unchanging. **3** *colloq.* constant; too frequent (*eternal nagging*). □ **eternally** *adv.* [Latin *aeternus*]
eternal triangle *n.* two people of one sex and one person of the other involved in a complex emotional relationship.
eternity /ɪ'tɜ:nɪtɪ/ *n.* (*pl.* **-ies**) **1** infinite (esp. future) time. **2** endless life after death. **3** being eternal. **4** *colloq.* (often prec. by *an*) a very long time. [Latin: related to ETERNAL]
eternity ring *n.* finger-ring esp. set with gems all round.
-eth var. of -TH.
ethanal /'eθə,næl/ *n.* = ACETALDEHYDE.
ethane /'i:θeɪn/ *n.* gaseous hydrocarbon of the alkane series. [from ETHER]
ether /'i:θə(r)/ *n.* **1** *Chem.* colourless volatile organic liquid used as an anaesthetic or solvent. **2** clear sky; upper regions of the air. **3** *hist.* **a** medium formerly assumed to permeate all space. **b** medium through which electromagnetic waves were formerly thought to be transmitted. [Greek *aithō* burn]
ethereal /ɪ'θɪərɪəl/ *adj.* **1** light, airy. **2** highly delicate, esp. in appearance. **3** heavenly. □ **ethereally** *adv.* [Greek: related to ETHER]
ethic /'eθɪk/ *–n.* set of moral principles (*the Quaker ethic*). *–adj.* = ETHICAL. [Greek: related to ETHOS]
ethical *adj.* **1** relating to morals, esp. as concerning human conduct. **2** morally correct. **3** (of a drug etc.) not advertised to the general public, and usu. available only on prescription. □ **ethically** *adv.*
ethics *n.pl.* (also treated as *sing.*) **1** moral philosophy. **2 a** moral principles. **b** set of these.
Ethiopian /,i:θɪ'əʊpɪən/ *–n.* **1** native or national of Ethiopia in NE Africa. **2** person of Ethiopian descent. *–adj.* of Ethiopia.
ethnic /'eθnɪk/ *adj.* **1 a** (of a social group) having a common national or cultural tradition. **b** (of music, clothing, etc.) inspired by or resembling those of an exotic people. **2** denoting origin by birth or descent rather than nationality (*ethnic Turks*). □ **ethnically** *adv.* [Greek *ethnos* nation]
ethnology /eθ'nɒlədʒɪ/ *n.* the comparative study of peoples. □ **ethnological** /-nə'lɒdʒɪk(ə)l/ *adj.* **ethnologist** *n.*
ethos /'i:θɒs/ *n.* characteristic spirit or attitudes of a community etc. [Greek *ēthos* character]

ethyl /'eθɪl/ *n.* (*attrib.*) a radical derived from ethane, present in alcohol and ether. [German: related to ETHER]

ethylene /'eθɪˌli:n/ *n.* a hydrocarbon of the alkene series.

etiolate /'i:tɪəˌleɪt/ *v.* (**-ting**) **1** make (a plant) pale by excluding light. **2** give a sickly colour to (a person). □ **etiolation** /-'leɪʃ(ə)n/ *n.* [Latin *stipula* straw]

etiology *US* var. of AETIOLOGY.

etiquette /'etɪˌket/ *n.* conventional rules of social behaviour or professional conduct. [French: related to TICKET]

Etruscan /ɪ'trʌskən/ *–adj.* of ancient Etruria in Italy. *–n.* **1** native of Etruria. **2** language of Etruria. [Latin *Etruscus*]

et seq. *abbr.* (also **et seqq.**) and the following (pages etc.). [Latin *et sequentia*]

-ette *suffix* forming nouns meaning: **1** small (*kitchenette*). **2** imitation or substitute (*flannelette*). **3** female (*usherette*). [French]

étude /'eɪtju:d/ *n.* = STUDY *n.* 6. [French, = study]

etymology /ˌetɪ'mɒlədʒɪ/ *n.* (*pl.* **-ies**) **1 a** derivation and development of a word in form and meaning. **b** account of these. **2** the study of word origins. □ **etymological** /-mə'lɒdʒɪk(ə)l/ *adj.* **etymologist** *n.* [Greek *etumos* true]

Eu *symb.* europium.

eu- *comb. form* well, easily. [Greek]

eucalyptus /ju:kə'lɪptəs/ *n.* (*pl.* **-tuses** or **-ti** /-taɪ/) (also **eucalypt** *pl.* **-s**) **1** tall evergreen Australasian tree. **2** its oil, used as an antiseptic etc. [from EU-, Greek *kaluptos* covered]

Eucharist /'ju:kərɪst/ *n.* **1** Christian sacrament in which consecrated bread and wine are consumed. **2** consecrated elements, esp. the bread. □ **Eucharistic** /-'rɪstɪk/ *adj.* [Greek, = thanksgiving]

eugenics /ju:'dʒenɪks/ *n.pl.* (also treated as *sing.*) improvement of the qualities of a race by control of inherited characteristics. □ **eugenic** *adj.* **eugenically** *adv.* [from EU-, Greek *gen-* produce]

eukaryote /ju:'kærɪˌəʊt/ *n.* organism consisting of a cell or cells in which the genetic material is contained within a distinct nucleus. □ **eukaryotic** /-'ɒtɪk/ *adj.* [from EU-, *karyo-* from Greek *karuon* kernel, *-ote* as in ZYGOTE]

eulogize /'ju:ləˌdʒaɪz/ *v.* (also **-ise**) (**-zing** or **-sing**) praise in speech or writing. □ **eulogistic** /-'dʒɪstɪk/ *adj.*

eulogy /'ju:lədʒɪ/ *n.* (*pl.* **-ies**) **1** speech or writing in praise of a person. **2** expression of praise. [Latin from Greek]

eunuch /'ju:nək/ *n.* castrated man, esp. one formerly employed at an oriental harem or court. [Greek, = bedchamber attendant]

euphemism /'ju:fɪˌmɪz(ə)m/ *n.* **1** mild or vague expression substituted for a harsher or more direct one (e.g. *pass over* for *die*). **2** use of such expressions. □ **euphemistic** /-'mɪstɪk/ *adj.* **euphemistically** /-'mɪstɪkəlɪ/ *adv.* [Greek *phēmē* speaking]

euphonium /ju:'fəʊnɪəm/ *n.* brass instrument of the tuba family. [related to EUPHONY]

euphony /'ju:fənɪ/ *n.* (*pl.* **-ies**) **1** pleasantness of sound, esp. of a word or phrase. **2** pleasant sound. □ **euphonious** /-'fəʊnɪəs/ *adj.* [Greek *phōnē* sound]

euphoria /ju:'fɔ:rɪə/ *n.* intense feeling of well-being and excitement. □ **euphoric** /-'fɒrɪk/ *adj.* [Greek *pherō* bear]

Eurasian /jʊə'reɪʒ(ə)n/ *–adj.* **1** of mixed European and Asian parentage. **2** of Europe and Asia. *–n.* Eurasian person.

eureka /jʊə'ri:kə/ *int.* I have found it! (announcing a discovery etc.). [Greek *heurēka*]

Euro- *comb. form* Europe, European. [abbreviation]

Eurodollar /'jʊərəʊˌdɒlə(r)/ *n.* dollar held in a bank outside the US.

European /jʊərə'pɪən/ *–adj.* **1** of or in Europe. **2** originating in, native to, or characteristic of Europe. *–n.* **1 a** native or inhabitant of Europe. **b** person descended from natives of Europe. **2** person favouring European integration. [Greek *Eurōpē* Europe]

europium /jʊ'rəʊpɪəm/ *n.* metallic element of the lanthanide series. [from the name *Europe*]

Eustachian tube /ju:'steɪʃ(ə)n/ *n.* tube from the pharynx to the cavity of the middle ear. [*Eustachio*, name of an anatomist]

euthanasia /ju:θə'neɪzɪə/ *n.* bringing about of a gentle death in the case of incurable and painful disease. [Greek *thanatos* death]

eV *abbr.* electronvolt.

evacuate /ɪ'vækjʊˌeɪt/ *v.* (**-ting**) **1 a** remove (people) from a place of danger. **b** empty (a place) in this way. **2** make empty. **3** (of troops) withdraw from (a place). **4** empty (the bowels etc.). □ **evacuation** /-'eɪʃ(ə)n/ *n.* [Latin *vacuus* empty]

evacuee /ɪˌvækju:'i:/ *n.* person evacuated.

evade /ɪ'veɪd/ *v.* (**-ding**) **1 a** escape from, avoid, esp. by guile or trickery. **b** avoid doing (one's duty etc.). **c** avoid answering (a question). **2** avoid paying (tax). [Latin *evado* escape]

evaluate /ɪ'væljʊ,eɪt/ *v.* **(-ting) 1** assess, appraise. **2** find or state the number or amount of. ▫ **evaluation** /-'eɪʃ(ə)n/ *n.* [French: related to VALUE]

evanesce /,evə'nes/ *v.* **(-cing)** *literary* fade from sight. [Latin *vanus* empty]

evanescent /,evə'nes(ə)nt/ *adj.* quickly fading. ▫ **evanescence** *n.*

evangelical /,i:væn'dʒelɪk(ə)l/ *–adj.* **1** of or according to the teaching of the gospel. **2** of the Protestant school maintaining the doctrine of salvation by faith. *–n.* member of this. ▫ **evangelicalism** *n.* **evangelically** *adv.* [Greek: related to EU-, ANGEL]

evangelism /ɪ'vændʒə,lɪz(ə)m/ *n.* preaching or spreading of the gospel.

evangelist *n.* **1** writer of one of the four Gospels. **2** preacher of the gospel. ▫ **evangelistic** /-'lɪstɪk/ *adj.*

evangelize *v.* (also **-ise**) **(-zing** or **-sing) 1** (also *absol.*) preach the gospel to. **2** convert to Christianity. ▫ **evangelization** /-'zeɪʃ(ə)n/ *n.*

evaporate /ɪ'væpə,reɪt/ *v.* **(-ting) 1** turn from solid or liquid into vapour. **2** (cause to) lose moisture as vapour. **3** (cause to) disappear. ▫ **evaporable** *adj.* **evaporation** /-'reɪʃ(ə)n/ *n.* [Latin: related to VAPOUR]

evaporated milk *n.* unsweetened milk concentrated by evaporation.

evasion /ɪ'veɪʒ(ə)n/ *n.* **1** evading. **2** evasive answer. [Latin: related to EVADE]

evasive /ɪ'veɪsɪv/ *adj.* **1** seeking to evade. **2** not direct in one's answers etc. ▫ **evasively** *adv.* **evasiveness** *n.*

eve *n.* **1** evening or day before a festival etc. (*Christmas Eve*; *eve of the funeral*). **2** time just before an event (*eve of the election*). **3** *archaic* evening. [= EVEN²]

even¹ /'i:v(ə)n/ *–adj.* **(evener, evenest) 1** level; smooth. **2 a** uniform in quality; constant. **b** equal in amount or value etc. **c** equally balanced. **3** (of a person's temper etc.) equable, calm. **4 a** (of a number) divisible by two without a remainder. **b** bearing such a number (*no parking on even dates*). **c** not involving fractions; exact (*in even dozens*). *–adv.* **1** inviting comparison of the assertion, negation, etc., with an implied one that is less strong or remarkable (*never even opened* [let alone read] *the letter*; *ran even faster* [not just as fast as before]). **2** introducing an extreme case (*even you must realize it*). *–v.* (often foll. by *up*) make or become even. ▫ **even now 1** now as well as before. **2** at this very moment. **even so** nevertheless. **even though** despite the fact that. **get** (or **be**) **even with** have one's revenge on. ▫ **evenly** *adv.* **evenness** *n.* [Old English]

even² /'i:v(ə)n/ *n. poet.* evening. [Old English]

even chance *n.* equal chance of success or failure.

even-handed *adj.* impartial.

evening /'i:vnɪŋ/ *n.* end part of the day, esp. from about 6 p.m. to bedtime. [Old English: related to EVEN²]

evening dress *n.* formal dress for evening wear.

evening primrose *n.* plant with pale-yellow flowers that open in the evening.

evening star *n.* planet, esp. Venus, conspicuous in the west after sunset.

even money *n.* betting odds offering the gambler the chance of winning the amount staked.

evens *n.pl.* = EVEN MONEY.

evensong *n.* service of evening prayer in the Church of England. [from EVEN²]

event /ɪ'vent/ *n.* **1** thing that happens. **2** fact of a thing's occurring. **3** item in a (esp. sports) programme. ▫ **at all events** (or **in any event**) whatever happens. **in the event** as it turns (or turned) out. **in the event of** if (a specified thing) happens. **in the event that** if it happens that. [Latin *venio vent-* come]

■ **Usage** The phrase *in the event that* is considered awkward by some people. It can usually be avoided by rephrasing, e.g. *in the event that it rains* can be replaced by *in the event of rain.*

eventful *adj.* marked by noteworthy events. ▫ **eventfully** *adv.*

eventide /'i:v(ə)n,taɪd/ *n. archaic* or *poet.* = EVENING. [related to EVEN²]

eventing *n.* participation in equestrian competitions, esp. dressage and show-jumping. [see EVENT 3]

eventual /ɪ'ventʃʊəl/ *adj.* occurring in due course, ultimate. ▫ **eventually** *adv.* [from EVENT]

eventuality /ɪ,ventʃʊ'ælɪtɪ/ *n.* (*pl.* **-ies**) possible event or outcome.

eventuate /ɪ'ventʃʊ,eɪt/ *v.* **(-ting)** (often foll. by *in*) result.

ever /'evə(r)/ *adv.* **1** at all times; always (*ever hopeful*; *ever after*). **2** at any time (*have you ever smoked?*; *nothing ever happens*). **3** (used for emphasis) in any way; at all (*how ever did you do it?*). **4** (in *comb.*) constantly (*ever-present*). **5** (foll. by *so, such*) *colloq.* very; very much (*ever so easy*; *thanks ever so*). ▫ **did you ever?** *colloq.* did you ever hear or see the like? **ever since** throughout the period since. [Old English]

■ **Usage** When *ever* is used with a question word for emphasis it is written separately (see sense 3). When used with a relative pronoun or adverb to give it indefinite or general force, *ever* is written as one word with the relative pronoun or adverb, e.g. *however it's done, it's difficult*.

evergreen –*adj.* retaining green leaves all year round. –*n.* evergreen plant.
everlasting –*adj.* **1** lasting for ever or for a long time. **2** (of flowers) keeping their shape and colour when dried. –*n.* **1** eternity. **2** everlasting flower.
evermore *adv.* for ever; always.
every /'evrɪ/ *adj.* **1** each single (*heard every word*). **2** each at a specified interval in a series (*comes every four days*). **3** all possible (*every prospect of success*). □ **every bit as** *colloq.* (in comparisons) quite as. **every now and again** (or **then**) from time to time. **every other** each second in a series (*every other day*). **every so often** occasionally. [Old English: related to EVER, EACH]
everybody *pron.* every person.
everyday *attrib. adj.* **1** occurring every day. **2** used on ordinary days. **3** commonplace.
Everyman *n.* ordinary or typical human being. [name of a character in a 15th-c. morality play]
everyone *pron.* everybody.
every one *n.* each one.
everything *pron.* **1** all things. **2** most important thing (*speed is everything*).
everywhere *adv.* **1** in every place. **2** *colloq.* in many places.
evict /ɪ'vɪkt/ *v.* expel (a tenant etc.) by legal process. □ **eviction** *n.* [Latin *evinco evict-* conquer]
evidence /'evɪd(ə)ns/ –*n.* **1** (often foll. by *for*, *of*) available facts, circumstances, etc. indicating whether or not a thing is true or valid. **2** *Law* **a** information tending to prove a fact or proposition. **b** statements or proofs admissible as testimony in a lawcourt. –*v.* (**-cing**) be evidence of. □ **in evidence** conspicuous. **Queen's** (or **King's** or **State's**) **evidence** *Law* evidence for the prosecution given by a participant in the crime at issue. [Latin *video* see]
evident *adj.* plain or obvious; manifest. [Latin: related to EVIDENCE]
evidential /,evɪ'denʃ(ə)l/ *adj.* of or providing evidence.
evidently *adv.* **1** seemingly; as it appears. **2** as shown by evidence.
evil /'i:v(ə)l/ –*adj.* **1** morally bad; wicked. **2** harmful. **3** disagreeable (*evil temper*). –*n.* **1** evil thing. **2** wickedness. □ **evilly** *adv.* [Old English]
evildoer *n.* sinner. □ **evildoing** *n.*
evil eye *n.* gaze that is superstitiously believed to cause harm.
evince /ɪ'vɪns/ *v.* (**-cing**) indicate, display (a quality, feeling, etc.). [Latin: related to EVICT]
eviscerate /ɪ'vɪsə,reɪt/ *v.* (**-ting**) disembowel. □ **evisceration** /-'reɪʃ(ə)n/ *n.* [Latin: related to VISCERA]
evocative /ɪ'vɒkətɪv/ *adj.* evoking (esp. feelings or memories). □ **evocatively** *adv.* **evocativeness** *n.*
evoke /ɪ'vəʊk/ *v.* (**-king**) inspire or draw forth (memories, a response, etc.). □ **evocation** /,evə'keɪʃ(ə)n/ *n.* [Latin *voco* call]
evolution /,i:və'lu:ʃ(ə)n/ *n.* **1** gradual development. **2** development of species from earlier forms, as an explanation of their origins. **3** unfolding of events etc. (*evolution of the plot*). **4** change in the disposition of troops or ships. □ **evolutionary** *adj.* [Latin: related to EVOLVE]
evolutionist *n.* person who regards evolution as explaining the origin of species.
evolve /ɪ'vɒlv/ *v.* (**-ving**) **1** develop gradually and naturally. **2** devise (a theory, plan, etc.). **3** unfold. **4** give off (gas, heat, etc.). [Latin *volvo volut-* roll]
ewe /ju:/ *n.* female sheep. [Old English]
ewer /'ju:ə(r)/ *n.* water-jug with a wide mouth. [Latin *aqua* water]
ex[1] *prep.* (of goods) sold from (*ex-works*). [Latin, = out of]
ex[2] *n. colloq.* former husband or wife. [see EX-[1] 2]
ex-[1] *prefix* (also before some consonants **e-**, **ef-** before *f*) **1** forming verbs meaning: **a** out, forth (*exclude*; *exit*). **b** upward (*extol*). **c** thoroughly (*excruciate*). **d** bring into a state (*exasperate*). **e** remove or free from (*expatriate*; *exonerate*). **2** forming nouns from titles of office, status, etc., meaning 'formerly' (*ex-president*; *ex-wife*). [Latin from *ex* out of]
ex-[2] *prefix* out (*exodus*). [Greek]
exacerbate /ek'sæsə,beɪt/ *v.* (**-ting**) **1** make (pain etc.) worse. **2** irritate (a person). □ **exacerbation** /-'beɪʃ(ə)n/ *n.* [Latin *acerbus* bitter]
exact /ɪg'zækt/ –*adj.* **1** accurate; correct in all details (*exact description*). **2** precise. –*v.* **1** demand and enforce payment of (money etc.). **2** demand; insist on; require. □ **exactness** *n.* [Latin *exigo exact-* require]
exacting *adj.* **1** making great demands. **2** requiring much effort.
exaction /ɪg'zækʃ(ə)n/ *n.* **1** exacting or being exacted. **2 a** illegal or exorbitant demand; extortion. **b** sum or thing exacted.

exactitude *n.* exactness, precision.
exactly *adv.* **1** precisely. **2** (said in reply) I quite agree.
exact science *n.* a science in which absolute precision is possible.
exaggerate /ɪg'zædʒəˌreɪt/ *v.* (**-ting**) **1** (also *absol.*) make (a thing) seem larger or greater etc. than it really is. **2** increase beyond normal or due proportions (*exaggerated politeness*). □ **exaggeration** /-'reɪʃ(ə)n/ *n.* [Latin *agger* heap]
exalt /ɪg'zɔ:lt/ *v.* **1** raise in rank or power etc. **2** praise highly. **3** (usu. as **exalted** *adj.*) make lofty or noble (*exalted aims*; *exalted style*). □ **exaltation** /ˌegzɔ:l'teɪʃ(ə)n/ *n.* [Latin *altus* high]
exam /ɪg'zæm/ *n.* = EXAMINATION 3.
examination /ɪgˌzæmɪ'neɪʃ(ə)n/ *n.* **1** examining or being examined. **2** detailed inspection. **3** test of proficiency or knowledge by questions. **4** formal questioning of a witness etc. in court.
examine /ɪg'zæmɪn/ *v.* (**-ning**) **1** inquire into the nature or condition etc. of. **2** look closely at. **3** test the proficiency of. **4** check the health of (a patient). **5** formally question in court. □ **examinee** /-'ni:/ *n.* **examiner** *n.* [Latin *examen* tongue of a balance]
example /ɪg'zɑ:mp(ə)l/ *n.* **1** thing characteristic of its kind or illustrating a general rule. **2** person, thing, or piece of conduct, in terms of its fitness to be imitated. **3** circumstance or treatment seen as a warning to others. **4** problem or exercise designed to illustrate a rule. □ **for example** by way of illustration. [Latin *exemplum*: related to EXEMPT]
exasperate /ɪg'zɑ:spəˌreɪt/ *v.* (**-ting**) irritate intensely. □ **exasperation** /-'reɪʃ(ə)n/ *n.* [Latin *asper* rough]
ex cathedra /ˌeks kə'θi:drə/ *adj.* & *adv.* with full authority (esp. of a papal pronouncement). [Latin, = from the chair]
excavate /'ekskəˌveɪt/ *v.* (**-ting**) **1 a** make (a hole or channel) by digging. **b** dig out material from (the ground). **2** reveal or extract by digging. **3** (also *absol.*) *Archaeol.* dig systematically to explore (a site). □ **excavation** /-'veɪʃ(ə)n/ *n.* **excavator** *n.* [Latin *excavo*: related to CAVE]
exceed /ɪk'si:d/ *v.* **1** (often foll. by *by* an amount) be more or greater than. **2** go beyond or do more than is warranted by (a set limit, esp. of one's authority, instructions, or rights). **3** surpass. [Latin *excedo -cess-* go beyond]
exceedingly /ɪk'si:dɪŋlɪ/ *adv.* extremely.
excel /ɪk'sel/ *v.* (**-ll-**) **1** surpass. **2** be pre-eminent. [Latin *excello* be eminent]
excellence /'eksələns/ *n.* outstanding merit or quality. [Latin: related to EXCEL]
Excellency *n.* (*pl.* **-ies**) (usu. prec. by *Your*, *His*, *Her*, *Their*) title used in addressing or referring to certain high officials.
excellent *adj.* extremely good.
excentric var. of ECCENTRIC (in technical senses).
except /ɪk'sept/ *—v.* exclude from a general statement, condition, etc. *—prep.* (often foll. by *for*) not including; other than (*all failed except him*; *is all right except that it is too long*). *—conj. archaic* unless (*except he be born again*). [Latin *excipio -cept-* take out]
excepting *prep.* = EXCEPT *prep.*

■ **Usage** *Excepting* should be used only after *not* and *always*; otherwise, *except* should be used.

exception /ɪk'sepʃ(ə)n/ *n.* **1** excepting or being excepted. **2** thing that has been or will be excepted. **3** instance that does not follow a rule. □ **take exception** (often foll. by *to*) object. **with the exception of** except.
exceptionable *adj.* open to objection.

■ **Usage** *Exceptionable* is sometimes confused with *exceptional*.

exceptional *adj.* **1** forming an exception; unusual. **2** outstanding. □ **exceptionally** *adv.*

■ **Usage** See note at *exceptionable*.

excerpt *—n.* /'eksɜ:pt/ short extract from a book, film, etc. *—v.* /ɪk'sɜ:pt/ (also *absol.*) take excerpts from. □ **excerption** /-'sɜ:pʃ(ə)n/ *n.* [Latin *carpo* pluck]
excess /ɪk'ses, 'ekses/ *—n.* **1** exceeding. **2** amount by which one thing exceeds another. **3 a** overstepping of accepted limits of moderation, esp. in eating or drinking. **b** (in *pl.*) immoderate behaviour. **4** part of an insurance claim to be paid by the insured. *—attrib. adj.* usu. /'ekses/ **1** that exceeds a limited or prescribed amount. **2** required as extra payment (*excess postage*). □ **in** (or **to**) **excess** exceeding the proper amount or degree. **in excess of** more than; exceeding. [Latin: related to EXCEED]
excess baggage *n.* (also **excess luggage**) baggage exceeding a weight allowance and liable to an extra charge.
excessive /ɪk'sesɪv/ *adj.* too much or too great. □ **excessively** *adv.*
exchange /ɪks'tʃeɪndʒ/ *—n.* **1** giving of one thing and receiving of another in its place. **2** giving of money for its equivalent in the money of the same or another

country. **3** centre where telephone connections are made. **4** place where merchants, bankers, etc. transact business. **5 a** office where information is given or a service provided. **b** employment office. **6** system of settling debts without the use of money, by bills of exchange. **7** short conversation. –*v.* (**-ging**) **1** (often foll. by *for*) give or receive (one thing) in place of another. **2** give and receive as equivalents. **3** (often foll. by *with*) make an exchange. □ **in exchange** (often foll. by *for*) as a thing exchanged (for). □ **exchangeable** *adj.* [French: related to CHANGE]

exchange rate *n.* value of one currency in terms of another.

exchequer /ɪks'tʃekə(r)/ *n.* **1** former government department in charge of national revenue. **2** royal or national treasury. **3** money of a private individual or group. [medieval Latin *scaccarium* chessboard]

■ **Usage** With reference to sense 1, the functions of this department in the UK now belong to the Treasury, although the name formally survives, esp. in the title *Chancellor of the Exchequer*.

excise[1] /'eksaɪz/ –*n.* **1** tax on goods produced or sold within the country of origin. **2** tax on certain licences. –*v.* (**-sing**) **1** charge excise on. **2** force (a person) to pay excise. [Dutch *excijs* from Romanic: related to Latin CENSUS tax]

excise[2] /ɪk'saɪz/ *v.* (**-sing**) **1** remove (a passage from a book etc.). **2** cut out (an organ etc.) by surgery. □ **excision** /ɪk'sɪʒ(ə)n/ *n.* [Latin *excido* cut out]

excitable /ɪk'saɪtəb(ə)l/ *adj.* easily excited. □ **excitability** /-'bɪlɪtɪ/ *n.* **excitably** *adv.*

excite /ɪk'saɪt/ *v.* (**-ting**) **1 a** rouse the emotions of (a person). **b** arouse (feelings etc.). **c** arouse sexually. **2** provoke (an action etc.). **3** stimulate (an organism, tissue, etc.) to activity. [Latin *cieo* stir up]

excitement *n.* **1** excited state of mind. **2** exciting thing.

exciting *adj.* arousing great interest or enthusiasm. □ **excitingly** *adv.*

exclaim /ɪk'skleɪm/ *v.* **1** cry out suddenly. **2** (foll. by *that*) utter by exclaiming. [Latin: related to CLAIM]

exclamation /,ekskləˈmeɪʃ(ə)n/ *n.* **1** exclaiming. **2** word(s) exclaimed. [Latin: related to EXCLAIM]

exclamation mark *n.* punctuation mark (!) indicating exclamation.

exclamatory /ɪk'sklæmətərɪ/ *adj.* of or serving as an exclamation.

exclude /ɪk'sklu:d/ *v.* (**-ding**) **1** keep out (a person or thing) from a place, group, privilege, etc. **2** remove from consideration (*no theory can be excluded*). **3** make impossible, preclude (*excluded all doubt*). □ **exclusion** *n.* [Latin *excludo -clus-* shut out]

exclusive /ɪk'sklu:sɪv/ –*adj.* **1** excluding other things. **2** (*predic.*; foll. by *of*) not including; except for. **3** tending to exclude others, esp. socially. **4** high-class. **5** not obtainable elsewhere or not published elsewhere. –*n.* article etc. published by only one newspaper etc. □ **exclusively** *adv.* **exclusiveness** *n.* **exclusivity** /-'sɪvɪtɪ/ *n.* [medieval Latin: related to EXCLUDE]

excommunicate –*v.* /,ekskə'mju:nɪ,keɪt/ (**-ting**) officially exclude (a person) from membership and esp. sacraments of the Church. –*adj.* /,ekskə'mju:nɪkət/ excommunicated. –*n.* /,ekskə'mju:nɪkət/ excommunicated person. □ **excommunication** /-'keɪʃ(ə)n/ *n.* [Latin: related to COMMON]

excoriate /eks'kɔ:rɪ,eɪt/ *v.* (**-ting**) **1 a** remove skin from (a person etc.) by abrasion. **b** strip off (skin). **2** censure severely. □ **excoriation** /-'eɪʃ(ə)n/ *n.* [Latin *corium* hide]

excrement /'ekskrɪmənt/ *n.* faeces. □ **excremental** /-'ment(ə)l/ *adj.* [Latin: related to EXCRETE]

excrescence /ɪk'skres(ə)ns/ *n.* **1** abnormal or morbid outgrowth on the body or a plant. **2** ugly addition. □ **excrescent** *adj.* [Latin *cresco* grow]

excreta /ɪk'skri:tə/ *n.pl.* faeces and urine. [Latin: related to EXCRETE]

excrete /ɪk'skri:t/ *v.* (**-ting**) (of an animal or plant) expel (waste matter). □ **excretion** *n.* **excretory** *adj.* [Latin *cerno cret-* sift]

excruciating /ɪk'skru:ʃɪ,eɪtɪŋ/ *adj.* causing acute mental or physical pain. □ **excruciatingly** *adv.* [Latin *crucio* torment]

exculpate /'ekskʌl,peɪt/ *v.* (**-ting**) *formal* (often foll. by *from*) free from blame; clear of a charge. □ **exculpation** /-'peɪʃ(ə)n/ *n.* **exculpatory** /-'kʌlpətərɪ/ *adj.* [Latin *culpa* blame]

excursion /ɪk'skɜ:ʃ(ə)n/ *n.* journey (usu. a day-trip) to a place and back, made for pleasure. [Latin *excurro* run out]

excursive /ɪk'skɜ:sɪv/ *adj. literary* digressive.

excuse –*v.* /ɪk'skju:z/ (**-sing**) **1** try to lessen the blame attaching to (a person, act, or fault). **2** (of a fact) serve as a reason to judge (a person or act) less severely. **3** (often foll. by *from*) release (a person) from a duty etc. **4** forgive (a fault or offence). **5** (foll. by *for*) forgive (a person) for (a fault). **6** *refl.* leave with apologies. *–n.* /ɪk'skju:s/ **1** reason put forward to mitigate or justify an offence. **2** apology (*made my excuses*). □ **be excused** be allowed to leave the room etc. or be absent. **excuse me** polite preface to an interruption etc., or to disagreeing. □ **excusable** /-'kju:zəb(ə)l/ *adj.* [Latin *causa* accusation]

ex-directory /,eksdə'rektərɪ/ *adj.* not listed in a telephone directory, at one's own request.

execrable /'eksɪkrəb(ə)l/ *adj.* abominable. [Latin: related to EXECRATE]

execrate /'eksɪ,kreɪt/ *v.* **(-ting) 1** express or feel abhorrence for. **2** (also *absol.*) curse (a person or thing). □ **execration** /-'kreɪʃ(ə)n/ *n.* [Latin *exsecror* curse: related to SACRED]

execute /'eksɪ,kju:t/ *v.* **(-ting) 1** carry out, perform (a plan, duty etc.). **2** carry out a design for (a product of art or skill). **3** carry out a death sentence on. **4** make (a legal instrument) valid by signing, sealing, etc. [Latin *sequor* follow]

execution /,eksɪ'kju:ʃ(ə)n/ *n.* **1** carrying out; performance. **2** technique or style of performance in the arts, esp. music. **3** carrying out of a death sentence. [Latin: related to EXECUTE]

executioner *n.* official who carries out a death sentence.

executive /ɪg'zekjʊtɪv/ *–n.* **1** person or body with managerial or administrative responsibility. **2** branch of a government etc. concerned with executing laws, agreements, etc. *–adj.* concerned with executing laws, agreements, etc., or with other administration or management. [medieval Latin: related to EXECUTE]

executor /ɪg'zekjʊtə(r)/ *n.* (*fem.* **executrix** /-trɪks/) person appointed by a testator to administer his or her will. □ **executorial** /-'tɔ:rɪəl/ *adj.*

exegesis /,eksɪ'dʒi:sɪs/ *n.* (*pl.* **exegeses** /-si:z/) critical explanation of a text, esp. of Scripture. □ **exegetic** /-'dʒetɪk/ *adj.* [Greek *hēgeomai* lead]

exemplar /ɪg'zemplə(r)/ *n.* **1** model. **2** typical or parallel instance. [Latin: related to EXAMPLE]

exemplary *adj.* **1** fit to be imitated; outstandingly good. **2** serving as a warning. **3** illustrative. [Latin: related to EXAMPLE]

exemplify /ɪg'zemplɪ,faɪ/ *v.* **(-ies, -ied) 1** illustrate by example. **2** be an example of. □ **exemplification** /-fɪ'keɪʃ(ə)n/ *n.*

exempt /ɪg'zempt/ *–adj.* (often foll. by *from*) free from an obligation or liability etc. imposed on others. *–v.* (foll. by *from*) make exempt. □ **exemption** *n.* [Latin *eximo -empt-* take out]

exercise /'eksə,saɪz/ *–n.* **1** activity requiring physical effort, done to sustain or improve health. **2** mental or spiritual activity, esp. as practice to develop a faculty. **3** task devised as exercise. **4 a** use or application of a mental faculty, right, etc. **b** practice of an ability, quality, etc. **5** (often in *pl.*) military drill or manoeuvres. *–v.* **(-sing) 1** use or apply (a faculty, right, etc.). **2** perform (a function). **3 a** take (esp. physical) exercise. **b** provide (an animal) with exercise. **4 a** tax the powers of. **b** perplex, worry. [Latin *exerceo* keep busy]

exert /ɪg'zɜ:t/ *v.* **1** bring to bear, use (a quality, force, influence, etc.). **2** *refl.* (often foll. by *for*, or *to* + infin.) use one's efforts or endeavours; strive. □ **exertion** *n.* [Latin *exsero exsert-* put forth]

exeunt /'eksɪ,ʌnt/ *v.* (as a stage direction) (actors) leave the stage. [Latin: related to EXIT]

exfoliate /eks'fəʊlɪ,eɪt/ *v.* **(-ting) 1** come off in scales or layers. **2** throw off layers of bark. □ **exfoliation** /-'eɪʃ(ə)n/ *n.* [Latin *folium* leaf]

ex gratia /eks 'greɪʃə/ *–adv.* as a favour; not from (esp. legal) obligation. *–attrib. adj.* granted on this basis. [Latin, = from favour]

exhale /eks'heɪl/ *v.* **(-ling) 1** breathe out. **2** give off or be given off in vapour. □ **exhalation** /,ekshə'leɪʃ(ə)n/ *n.* [French from Latin *halo* breathe]

exhaust /ɪg'zɔ:st/ *–v.* **1** consume or use up the whole of. **2** (often as **exhausted** *adj.* or **exhausting** *adj.*) tire out. **3** study or expound (a subject) completely. **4** (often foll. by *of*) empty (a vessel etc.) of its contents. *–n.* **1** waste gases etc. expelled from an engine after combustion. **2** (also **exhaust-pipe**) pipe or system by which these are expelled. **3** process of expulsion of these gases. □ **exhaustible** *adj.* [Latin *haurio haust-* drain]

exhaustion /ɪg'zɔ:stʃ(ə)n/ *n.* **1** exhausting or being exhausted. **2** total loss of strength.

exhaustive /ɪg'zɔ:stɪv/ *adj.* thorough, comprehensive. □ **exhaustively** *adv.* **exhaustiveness** *n.*

exhibit /ɪg'zɪbɪt/ *–v.* **(-t-) 1** show or reveal, esp. publicly. **2** display (a quality etc.). *–n.* item displayed, esp. in an exhibition or as evidence in a lawcourt. □ **exhibitor** *n.* [Latin *exhibeo -hibit-*]

exhibition /,eksɪ'bɪʃ(ə)n/ *n.* **1** display (esp. public) of works of art etc. **2** exhibiting or being exhibited. **3** scholarship, esp. from the funds of a school, college, etc.

exhibitioner *n.* student who has been awarded an exhibition.

exhibitionism *n.* **1** tendency towards attention-seeking behaviour. **2** *Psychol.* compulsion to display one's genitals in public. □ **exhibitionist** *n.*

exhilarate /ɪg'zɪlə,reɪt/ *v.* (often as **exhilarating** *adj.* or **exhilarated** *adj.*) enliven, gladden; raise the spirits of. □ **exhilaration** /-'reɪʃ(ə)n/ *n.* [Latin *hilaris* cheerful]

exhort /ɪg'zɔ:t/ *v.* (often foll. by *to* + infin.) urge strongly or earnestly. □ **exhortation** /,egzɔ:'teɪʃ(ə)n/ *n.* **exhortative** /-tətɪv/ *adj.* **exhortatory** /-tətərɪ/ *adj.* [Latin *exhortor* encourage]

exhume /eks'hju:m/ *v.* (**-ming**) dig up (esp. a buried corpse). □ **exhumation** /-'meɪʃ(ə)n/ *n.* [Latin *humus* ground]

exigency /'eksɪdʒənsɪ/ *n.* (*pl.* **-ies**) (also **exigence**) **1** urgent need or demand. **2** emergency. □ **exigent** *adj.* [Latin *exigo* EXACT]

exiguous /eg'zɪgjʊəs/ *adj.* scanty, small. □ **exiguity** /-'gju:ɪtɪ/ *n.* [Latin]

exile /'eksaɪl/ —*n.* **1** expulsion from one's native land or (**internal exile**) native town etc. **2** long absence abroad. **3** exiled person. —*v.* (**-ling**) send into exile. [French from Latin]

exist /ɪg'zɪst/ *v.* **1** have a place in objective reality. **2** (of circumstances etc.) occur; be found. **3** live with no pleasure. **4** continue in being. **5** live. [Latin *existo*]

existence *n.* **1** fact or manner of being or existing. **2** continuance in life or being. **3** all that exists. □ **existent** *adj.*

existential /,egzɪ'stenʃ(ə)l/ *adj.* **1** of or relating to existence. **2** *Philos.* concerned with existence, esp. with human existence as viewed by existentialism. □ **existentially** *adv.*

existentialism *n.* philosophical theory emphasizing the existence of the individual as a free and self-determining agent. □ **existentialist** *n.* & *adj.*

exit /'eksɪt/ —*n.* **1** passage or door by which to leave a room etc. **2** act or right of going out. **3** place where vehicles can leave a motorway etc. **4** actor's departure from the stage. —*v.* (**-t-**) **1** go out of a room etc. **2** leave the stage (also as a direction: *exit Macbeth*). [Latin *exeo exit-* go out]

exit poll *n.* poll of people leaving a polling-station, asking how they voted.

exo- *comb. form* external. [Greek *exō* outside]

exocrine /'eksəʊ,kraɪn/ *adj.* (of a gland) secreting through a duct. [Greek *krinō* sift]

exodus /'eksədəs/ *n.* **1** mass departure. **2** (**Exodus**) Biblical departure of the Israelites from Egypt. [Greek *hodos* way]

ex officio /,eks ə'fɪʃɪəʊ/ *adv.* & *attrib. adj.* by virtue of one's office. [Latin]

exonerate /ɪg'zɒnə,reɪt/ *v.* (**-ting**) (often foll. by *from*) free or declare free from blame etc. □ **exoneration** /-'reɪʃ(ə)n/ *n.* [Latin *onus oner-* burden]

exorbitant /ɪg'zɔ:bɪt(ə)nt/ *adj.* (of a price, demand, etc.) grossly excessive. [Latin: related to ORBIT]

exorcize /'eksɔ:,saɪz/ *v.* (also **-ise**) (**-zing** or **-sing**) **1** expel (a supposed evil spirit) by prayers etc. **2** (often foll. by *of*) free (a person or place) in this way. □ **exorcism** *n.* **exorcist** *n.* [Greek *horkos* oath]

exordium /ek'sɔ:dɪəm/ *n.* (*pl.* **-s** or **-dia**) introductory part, esp. of a discourse or treatise. [Latin *exordior* begin]

exotic /ɪg'zɒtɪk/ —*adj.* **1** introduced from a foreign country; not native. **2** strange or unusual. —*n.* exotic person or thing. □ **exotically** *adv.* [Greek *exō* outside]

exotica *n.pl.* strange or rare objects.

expand /ɪk'spænd/ *v.* **1** increase in size or importance. **2** (often foll. by *on*) give a fuller account. **3** become more genial. **4** set or write out in full. **5** spread out flat. □ **expandable** *adj.* [Latin *pando pans-* spread]

expanse /ɪk'spæns/ *n.* wide continuous area of land, space, etc.

expansible *adj.* that can be expanded.

expansion /ɪk'spænʃ(ə)n/ *n.* **1** expanding or being expanded. **2** enlargement of the scale or scope of a business.

expansionism *n.* advocacy of expansion, esp. of a State's territory. □ **expansionist** *n.* & *adj.*

expansive /ɪk'spænsɪv/ *adj.* **1** able or tending to expand. **2** extensive. **3** (of a person etc.) effusive, open. □ **expansively** *adv.* **expansiveness** *n.*

expat /eks'pæt/ *n.* & *adj. colloq.* expatriate. [abbreviation]

expatiate /ɪk'speɪʃɪ,eɪt/ *v.* (**-ting**) (usu. foll. by *on*, *upon*) speak or write at length. □ **expatiation** /-'eɪʃ(ə)n/ *n.* **expatiatory** /-ʃɪətərɪ/ *adj.* [Latin *spatium* SPACE]

expatriate —*adj.* /eks'pætrɪət/ **1** living abroad. **2** exiled. —*n.* /eks'pætrɪət, -'peɪtrɪət/ expatriate person. —*v.* /eks'pætrɪ,eɪt/ (**-ting**) **1** expel (a person) from his or her native country. **2** *refl.* renounce one's citizenship. □ **expatriation** /-'eɪʃ(ə)n/ *n.* [Latin *patria* native land]

expect /ɪk'spekt/ *v.* **1 a** regard as likely. **b** look for as appropriate or one's due (*I expect cooperation*). **2** *colloq.* think, suppose. □ **be expecting** *colloq.* be pregnant (with). [Latin *specto* look]

expectancy /ɪk'spektənsɪ/ *n.* (*pl.* **-ies**) **1** state of expectation. **2** prospect. **3** (foll. by *of*) prospective chance.

expectant /ɪk'spekt(ə)nt/ *adj.* **1** hopeful, expecting. **2** having an expectation. **3** pregnant. □ **expectantly** *adv.*

expectation /,ekspek'teɪʃ(ə)n/ *n.* **1** expecting or anticipation. **2** thing expected. **3** (foll. by *of*) probability of an event. **4** (in *pl.*) one's prospects of inheritance.

expectorant /ek'spektərənt/ *—adj.* causing expectoration. *—n.* expectorant medicine.

expectorate /ek'spektə,reɪt/ *v.* (**-ting**) (also *absol.*) cough or spit out (phlegm etc.). □ **expectoration** /-'reɪʃ(ə)n/ *n.* [Latin *pectus pector-* breast]

expedient /ɪk'spi:dɪənt/ *—adj.* advantageous; advisable on practical rather than moral grounds. *—n.* means of attaining an end; resource. □ **expedience** *n.* **expediency** *n.* [related to EXPEDITE]

expedite /'ekspɪ,daɪt/ *v.* (**-ting**) **1** assist the progress of. **2** accomplish (business) quickly. [Latin *expedio* from *pes ped-* foot]

expedition /,ekspɪ'dɪʃ(ə)n/ *n.* **1** journey or voyage for a particular purpose, esp. exploration. **2** people etc. undertaking this. **3** speed. [Latin: related to EXPEDITE]

expeditionary *adj.* of or used in an expedition.

expeditious /,ekspɪ'dɪʃəs/ *adj.* acting or done with speed and efficiency.

expel /ɪk'spel/ *v.* (**-ll-**) (often foll. by *from*) **1** deprive (a person) of membership etc. of a school, society, etc. **2** force out, eject. **3** order or force to leave a building etc. [Latin *pello puls-* drive]

expend /ɪk'spend/ *v.* spend or use up (money, time, etc.). [Latin *pendo pens-* weigh]

expendable *adj.* that may be sacrificed or dispensed with; not worth preserving or saving.

expenditure /ɪk'spendɪtʃə(r)/ *n.* **1** spending or using up. **2** thing (esp. money) expended.

expense /ɪk'spens/ *n.* **1** cost incurred; payment of money. **2** (usu. in *pl.*) **a** costs incurred in doing a job etc. **b** amount paid to reimburse this. **3** thing on which money is spent. □ **at the expense of** so as to cause loss or harm to; costing. [Latin *expensa*: related to EXPEND]

expense account *n.* list of an employee's expenses payable by the employer.

expensive *adj.* costing or charging much. □ **expensively** *adv.* **expensiveness** *n.*

experience /ɪk'spɪərɪəns/ *—n.* **1** observation of or practical acquaintance with facts or events. **2** knowledge or skill resulting from this. **3** event or activity participated in or observed (*a rare experience*). *—v.* (**-cing**) **1** have experience of; undergo. **2** feel. [Latin *experior -pert-* try]

experienced *adj.* **1** having had much experience. **2** skilled from experience (*experienced driver*).

experiential /ɪk,spɪərɪ'enʃ(ə)l/ *adj.* involving or based on experience. □ **experientially** *adv.*

experiment /ɪk'sperɪmənt/ *—n.* procedure adopted in the hope of success, or for testing a hypothesis etc., or to demonstrate a known fact. *—v.* /also -,ment/ (often foll. by *on, with*) make an experiment. □ **experimentation** /-men'teɪʃ(ə)n/ *n.* **experimenter** *n.* [Latin: related to EXPERIENCE]

experimental /ɪk,sperɪ'ment(ə)l/ *adj.* **1** based on or making use of experiment. **2** used in experiments. □ **experimentalism** *n.* **experimentally** *adv.*

expert /'ekspɜ:t/ *—adj.* **1** (often foll. by *at, in*) having special knowledge of or skill in a subject. **2** (*attrib.*) involving or resulting from this (*expert advice*). *—n.* (often foll. by *at, in*) person with special knowledge or skill. □ **expertly** *adv.* [Latin: related to EXPERIENCE]

expertise /,ekspɜ:'ti:z/ *n.* expert skill, knowledge, or judgement. [French]

expiate /'ekspɪ,eɪt/ *v.* (**-ting**) pay the penalty for or make amends for (wrongdoing). □ **expiable** /'ekspɪəb(ə)l/ *adj.* **expiation** /-'eɪʃ(ə)n/ *n.* **expiatory** *adj.* [Latin *expio*: related to PIOUS]

expire /ɪk'spaɪə(r)/ *v.* (**-ring**) **1** (of a period of time, validity, etc.) come to an end. **2** cease to be valid. **3** die. **4** (also *absol.*) breathe out (air etc.). □ **expiration** /,ekspɪ'reɪʃ(ə)n/ *n.* **expiratory** *adj.* (in sense 4). [Latin *spirare* breathe]

expiry *n.* end of validity or duration.

explain /ɪk'spleɪn/ *v.* **1 a** make clear or intelligible (also *absol.*: *let me explain*). **b** make known in detail. **2** (foll. by *that*) say by way of explanation. **3** account for (one's conduct etc.). □ **explain away** minimize the significance of by explanation. **explain oneself 1** make one's meaning clear. **2** give an account of one's motives or conduct. [Latin *explano* from *planus* flat]

explanation /,ekspla'neɪʃ(ə)n/ *n.* **1** explaining. **2** statement or circumstance that explains something.

explanatory /ɪk'splænətərɪ/ *adj.* serving or designed to explain.

expletive /ɪk'spli:tɪv/ *n.* swear-word or exclamation. [Latin *expleo* fill out]

explicable /ɪk'splɪkəb(ə)l/ *adj.* that can be explained.

explicate /'ekspli,keɪt/ *v.* **(-ting) 1** develop the meaning of (an idea etc.). **2** explain (esp. a literary text). □ **explication** /-'keɪʃ(ə)n/ *n.* [Latin *explico -plicat-* unfold]

explicit /ɪk'splɪsɪt/ *adj.* **1** expressly stated, not merely implied; stated in detail. **2** definite. **3** outspoken. □ **explicitly** *adv.* **explicitness** *n.* [Latin: related to EXPLICATE]

explode /ɪk'spləʊd/ *v.* **(-ding) 1 a** expand suddenly with a loud noise owing to a release of internal energy. **b** cause (a bomb etc.) to explode. **2** give vent suddenly to emotion, esp. anger. **3** (of a population etc.) increase suddenly or rapidly. **4** show (a theory etc.) to be false or baseless. **5** (as **exploded** *adj.*) (of a drawing etc.) showing the components of a mechanism somewhat separated but in the normal relative positions. [Latin *explodo -plos-* hiss off the stage]

exploit *–n.* /'eksplɔɪt/ daring feat. *–v.* /ɪk'splɔɪt/ **1** make use of (a resource etc.). **2** usu. *derog.* utilize or take advantage of (esp. a person) for one's own ends. □ **exploitation** /,eksplɔɪ'teɪʃ(ə)n/ *n.* **exploitative** /ɪk'splɔɪtətɪv/ *adj.* **exploiter** *n.* [Latin: related to EXPLICATE]

explore /ɪk'splɔ:(r)/ *v.* **(-ring) 1** travel through (a country etc.) to learn about it. **2** inquire into. **3** *Surgery* examine (a part of the body) in detail. □ **exploration** /,eksplə'reɪʃ(ə)n/ *n.* **exploratory** /-'splɒrətrɪ/ *adj.* **explorer** *n.* [Latin *exploro* search out]

explosion /ɪk'spləʊʒ(ə)n/ *n.* **1** exploding. **2** loud noise caused by this. **3** sudden outbreak of feeling. **4** rapid or sudden increase. [Latin: related to EXPLODE]

explosive /ɪk'spləʊsɪv/ *–adj.* **1** able, tending, or likely to explode. **2** likely to cause a violent outburst etc.; dangerously tense. *–n.* explosive substance. □ **explosiveness** *n.*

Expo /'ekspəʊ/ *n.* (also **expo**) (*pl.* **-s**) large international exhibition. [abbreviation of EXPOSITION 4]

exponent /ɪk'spəʊnənt/ *n.* **1** person who promotes an idea etc. **2** practitioner of an activity, profession, etc. **3** person who explains or interprets something. **4** type or representative. **5** raised symbol beside a numeral indicating how many of the number are to be multiplied together (e.g. 2^3 = 2 x 2 x 2). [Latin *expono* EXPOUND]

exponential /,ekspə'nenʃ(ə)l/ *adj.* **1** of or indicated by a mathematical exponent. **2** (of an increase etc.) more and more rapid.

export *–v.* /ɪk'spɔ:t, 'ek-/ sell or send (goods or services) to another country. *–n.* /'ekspɔ:t/ **1** exporting. **2 a** exported article or service. **b** (in *pl.*) amount exported. □ **exportation** /-'teɪʃ(ə)n/ *n.* **exporter** /ɪk'spɔ:tə(r)/ *n.* [Latin *porto* carry]

expose /ɪk'spəʊz/ *v.* **(-sing)** (esp. as **exposed** *adj.*) **1** leave uncovered or unprotected, esp. from the weather. **2** (foll. by *to*) **a** put at risk of. **b** subject to (an influence etc.). **3** *Photog.* subject (a film) to light, esp. by operation of a camera. **4** reveal the identity or fact of. **5** exhibit, display. □ **expose oneself** display one's body, esp. one's genitals, indecently in public. [Latin *pono* put]

exposé /ek'spəʊzeɪ/ *n.* **1** orderly statement of facts. **2** revelation of something discreditable. [French]

exposition /,ekspə'zɪʃ(ə)n/ *n.*, **1** explanatory account. **2** explanation or commentary. **3** *Mus.* part of a movement in which the principal themes are presented. **4** large public exhibition. [Latin: related to EXPOUND]

ex post facto /,eks pəʊst 'fæktəʊ/ *adj.* & *adv.* with retrospective action or force. [Latin, = in the light of subsequent events]

expostulate /ɪk'spɒstjʊ,leɪt/ *v.* **(-ting)** (often foll. by *with* a person) make a protest; remonstrate. □ **expostulation** /-'leɪʃ(ə)n/ *n.* **expostulatory** /-lətərɪ/ *adj.* [Latin: related to POSTULATE]

exposure /ɪk'spəʊʒə(r)/ *n.* (foll. by *to*) **1** exposing or being exposed. **2** physical condition resulting from being exposed to the elements. **3** *Photog.* **a** exposing a film etc. to the light. **b** duration of this. **c** section of film etc. affected by it.

expound /ɪk'spaʊnd/ *v.* **1** set out in detail. **2** explain or interpret. [Latin *pono posit-* place]

express /ɪk'spres/ *–v.* **1** represent or make known in words or by gestures, conduct, etc. **2** *refl.* communicate what one thinks, feels, or means. **3** esp. *Math.* represent by symbols. **4** squeeze out (liquid or air). **5** send by express service. *–adj.* **1** operating at high speed. **2** also /'ekspres/ definitely stated. **3** delivered by a specially fast service. *–adv.* **1** at high speed. **2** by express messenger or train. *–n.* **1** express train etc. **2** *US* service for the rapid transport of parcels etc. □ **expressible** *adj.* **expressly** *adv.* (in sense 2 of *adj.*). [Latin *exprimo -press-* squeeze out]

expression /ɪk'spreʃ(ə)n/ *n.* **1** expressing or being expressed. **2** word or phrase expressed. **3** person's facial appearance, indicating feeling. **4** conveying of feeling in music, speaking, dance, etc. **5** depiction of feeling etc. in art. **6** *Math.* collection of symbols expressing a quantity. □ **expressionless** *adj.* [French: related to EXPRESS]

expressionism *n.* style of painting, music, drama, etc., seeking to express emotion rather than the external world. □ **expressionist** *n.* & *adj.*

expressive /ɪk'spresɪv/ *adj.* **1** full of expression (*expressive look*). **2** (foll. by *of*) serving to express. □ **expressively** *adv.* **expressiveness** *n.*

expresso var. of ESPRESSO.

expressway *n. US* motorway.

expropriate /ɪk'sprəʊprɪ,eɪt/ *v.* (**-ting**) **1** take away (property) from its owner. **2** (foll. by *from*) dispossess. □ **expropriation** /-'eɪʃ(ə)n/ *n.* **expropriator** *n.* [Latin *proprium* property]

expulsion /ɪk'spʌlʃ(ə)n/ *n.* expelling or being expelled. □ **expulsive** *adj.* [Latin: related to EXPEL]

expunge /ɪk'spʌndʒ/ *v.* (**-ging**) erase, remove (objectionable matter) from a book etc. [Latin *expungo* prick out (for deletion)]

expurgate /'ekspə,geɪt/ *v.* (**-ting**) **1** remove objectionable matter from (a book etc.). **2** remove (such matter). □ **expurgation** /-'geɪʃ(ə)n/ *n.* **expurgator** *n.* [Latin: related to PURGE]

exquisite /'ekskwɪzɪt/ *adj.* **1** extremely beautiful or delicate. **2** keenly felt (*exquisite pleasure*). **3** highly sensitive (*exquisite taste*). □ **exquisitely** *adv.* [Latin *exquiro -quisit-* seek out]

ex-serviceman /eks'sɜ:vɪsmən/ *n.* man formerly a member of the armed forces.

ex-servicewoman /eks'sɜ:vɪs,wʊmən/ *n.* woman formerly a member of the armed forces.

extant /ek'stænt/ *adj.* still existing. [Latin *ex(s)to* exist]

extemporaneous /ɪk,stempə'reɪnɪəs/ *adj.* spoken or done without preparation. □ **extemporaneously** *adv.* [from EXTEMPORE]

extemporary /ɪk'stempərərɪ/ *adj.* = EXTEMPORANEOUS. □ **extemporarily** *adv.*

extempore /ɪk'stempərɪ/ *adj.* & *adv.* without preparation. [Latin]

extemporize /ɪk'stempə,raɪz/ *v.* (also **-ise**) (**-zing** or **-sing**) improvise. □ **extemporization** /-'zeɪʃ(ə)n/ *n.*

extend /ɪk'stend/ *v.* **1** lengthen or make larger in space or time. **2** stretch or lay out at full length. **3** (foll. by *to, over*) reach or be or make continuous over a specified area. **4** (foll. by *to*) have a specified scope (*permit does not extend to camping*). **5** offer or accord (an invitation, hospitality, kindness, etc.). **6** (usu. *refl.* or in *passive*) tax the powers of (an athlete, horse, etc.). □ **extendible** *adj.* (also **extensible**). [Latin *extendo -tens-*: related to TEND[1]]

extended family *n.* family including relatives living near.

extended-play *adj.* (of a gramophone record) playing for somewhat longer than most singles.

extension /ɪk'stenʃ(ə)n/ *n.* **1** extending or being extended. **2** part enlarging or added on to a main building etc. **3** additional part. **4 a** subsidiary telephone on the same line as the main one. **b** its number. **5** additional period of time. **6** extramural instruction by a university or college.

extensive /ɪk'stensɪv/ *adj.* **1** covering a large area. **2** far-reaching. □ **extensively** *adv.* **extensiveness** *n.* [Latin: related to EXTEND]

extent /ɪk'stent/ *n.* **1** space over which a thing extends. **2** range, scope, degree. [Anglo-French: related to EXTEND]

extenuate /ɪk'stenjʊ,eɪt/ *v.* (often as **extenuating** *adj.*) make (guilt or an offence) seem less serious by reference to another factor. □ **extenuation** /-'eɪʃ(ə)n/ *n.* [Latin *tenuis* thin]

exterior /ɪk'stɪərɪə(r)/ *–adj.* **1** of or on the outer side. **2** coming from outside. *–n.* **1** outward aspect or surface of a building etc. **2** outward demeanour. **3** outdoor scene in filming. [Latin]

exterminate /ɪk'stɜ:mɪ,neɪt/ *v.* (**-ting**) destroy utterly (esp. a living thing). □ **extermination** /-'neɪʃ(ə)n/ *n.* **exterminator** *n.* [Latin: related to TERMINAL]

external /ɪk'stɜ:n(ə)l/ *–adj.* **1 a** of or on the outside or visible part. **b** coming from the outside or an outside source. **2** relating to a country's foreign affairs. **3** outside the conscious subject (*the external world*). **4** (of medicine etc.) for use on the outside of the body. **5** for students taking the examinations of a university without attending it. *–n.* (in *pl.*) **1** outward features or aspect. **2** external circumstances. **3** inessentials. □ **externality** /,ekstɜ:'nælɪtɪ/ *n.* **externally** *adv.* [Latin *externus* outer]

externalize *v.* (also **-ise**) (**-zing** or **-sing**) give or attribute external existence to. □ **externalization** /-'zeɪʃ(ə)n/ *n.*

extinct /ɪk'stɪŋkt/ *adj.* **1** that has died out. **2 a** no longer burning. **b** (of a volcano) that no longer erupts. **3** obsolete. [Latin *ex(s)tinguo -stinct-* quench]

extinction /ɪk'stɪŋkʃ(ə)n/ *n.* **1** making or becoming extinct. **2** extinguishing or being extinguished. **3** total destruction or annihilation.

extinguish /ɪk'stɪŋgwɪʃ/ *v.* **1** cause (a flame, light, etc.) to die out. **2** destroy. **3** terminate. **4** wipe out (a debt). □ **extinguishable** *adj.*

extinguisher *n.* = FIRE EXTINGUISHER.

extirpate /'ekstə,peɪt/ *v.* (**-ting**) root out; destroy completely. □ **extirpation** /-'peɪʃ(ə)n/ *n.* [Latin *ex(s)tirpo* from *stirps* stem of tree]

extol /ɪk'stəʊl/ *v.* (**-ll-**) praise enthusiastically. [Latin *tollo* raise]

extort /ɪk'stɔ:t/ *v.* obtain by coercion. [Latin *torqueo tort-* twist]

extortion /ɪk'stɔ:ʃ(ə)n/ *n.* **1** act of extorting, esp. money. **2** illegal exaction. □ **extortioner** *n.* **extortionist** *n.*

extortionate /ɪk'stɔ:ʃənət/ *adj.* (of a price etc.) exorbitant. □ **extortionately** *adv.*

extra /'ekstrə/ *–adj.* additional; more than usual or necessary or expected. *–adv.* **1** more than usually. **2** additionally (*was charged extra*). *–n.* **1** extra thing. **2** thing for which an extra charge is made. **3** person engaged temporarily for a minor part in a film. **4** special issue of a newspaper etc. **5** *Cricket* run scored other than from a hit with the bat. [probably from EXTRAORDINARY]

extra- *comb. form* **1** outside, beyond. **2** beyond the scope of. [Latin *extra* outside]

extra cover *n. Cricket* **1** fielding position on a line between cover-point and mid-off, but beyond these. **2** fielder at this position.

extract *–v.* /ɪk'strækt/ **1** remove or take out, esp. by effort or force. **2** obtain (money, an admission, etc.) against a person's will. **3** obtain (a natural resource) from the earth. **4** select or reproduce for quotation or performance. **5** obtain (juice etc.) by pressure, distillation, etc. **6** derive (pleasure etc.). **7** find (the root of a number). *–n.* /'ekstrækt/ **1** short passage from a book etc. **2** preparation containing a concentrated constituent of a substance (*malt extract*). [Latin *traho tract-* draw]

extraction /ɪk'strækʃ(ə)n/ *n.* **1** extracting or being extracted. **2** removal of a tooth. **3** lineage, descent (*of Indian extraction*). [Latin: related to EXTRACT]

extractive /ɪk'stræktɪv/ *adj.* of or involving extraction.

extractor /ɪk'stræktə(r)/ *n.* **1** person or machine that extracts. **2** (*attrib.*) (of a device) that extracts bad air etc.

extracurricular /,ekstrəkə'rɪkjʊlə(r)/ *adj.* not part of the normal curriculum.

extraditable /'ekstrə,daɪtəb(ə)l/ *adj.* **1** liable to extradition. **2** (of a crime) warranting extradition.

extradite /'ekstrə,daɪt/ *v.* (**-ting**) hand over (a person accused or convicted of a crime) to the foreign State etc. in which the crime was committed. □ **extradition** /-'dɪʃ(ə)n/ *n.* [French: related to TRADITION]

extramarital /,ekstrə'mærɪt(ə)l/ *adj.* (esp. of sexual relations) occurring outside marriage.

extramural /,ekstrə'mjʊər(ə)l/ *adj.* additional to normal teaching or studies, esp. for non-resident students.

extraneous /ɪk'streɪnɪəs/ *adj.* **1** of external origin. **2** (often foll. by *to*) **a** separate from the object to which it is attached etc. **b** irrelevant, unrelated. [Latin *extraneus*]

extraordinary /ɪk'strɔ:dɪnərɪ/ *adj.* **1** unusual or remarkable. **2** unusually great. **3** (of a meeting, official, etc.) additional; specially employed. □ **extraordinarily** *adv.* [Latin]

extrapolate /ɪk'stræpə,leɪt/ *v.* (**-ting**) (also *absol.*) calculate approximately from known data etc. (others which lie outside the range of those known). □ **extrapolation** /-'leɪʃ(ə)n/ *n.* [from EXTRA-, INTERPOLATE]

extrasensory /,ekstrə'sensərɪ/ *adj.* derived by means other than the known senses, e.g. by telepathy.

extraterrestrial /,ekstrətɪ'restrɪəl/ *–adj.* outside the earth or its atmosphere. *–n.* (in science fiction) being from outer space.

extravagant /ɪk'strævəgənt/ *adj.* **1** spending money excessively. **2** excessive; absurd. **3** costing much. □ **extravagance** *n.* **extravagantly** *adv.* [Latin *vagor* wander]

extravaganza /ɪk,strævə'gænzə/ *n.* **1** spectacular theatrical or television production. **2** fanciful literary, musical, or dramatic composition. [Italian]

extreme /ɪk'stri:m/ *–adj.* **1** of a high, or the highest, degree (*extreme danger*). **2** severe (*extreme measures*). **3** outermost. **4** on the far left or right of a political party. **5** utmost; last. *–n.* **1** (often in *pl.*) either of two things as remote or as different as possible. **2** thing at either end. **3** highest degree. **4** *Math.* first or last term of a ratio or series. □ **go to extremes** take an extreme course of action. **in the extreme** to an extreme degree. □ **extremely** *adv.* [French from Latin]

extreme unction *n.* last rites in the Roman Catholic and Orthodox Churches.

extremist *n.* (also *attrib.*) person with extreme views. □ **extremism** *n.*

extremity /ɪk'stremɪtɪ/ *n.* (*pl.* **-ies**) **1** extreme point; very end. **2** (in *pl.*) the hands and feet. **3** condition of extreme adversity. [Latin: related to EXTREME]

extricate /'ekstrɪˌkeɪt/ *v.* (**-ting**) (often foll. by *from*) free or disentangle from a difficulty etc. □ **extricable** *adj.* **extrication** /-'keɪʃ(ə)n/ *n.* [Latin *tricae* perplexities]

extrinsic /ek'strɪnsɪk/ *adj.* **1** not inherent or intrinsic. **2** (often foll. by *to*) extraneous; not belonging. □ **extrinsically** *adv.* [Latin *extrinsecus* outwardly]

extrovert /'ekstrəˌvɜ:t/ **–***n.* **1** outgoing person. **2** person mainly concerned with external things. **–***adj.* typical of or with the nature of an extrovert. □ **extroversion** /-'vɜ:ʃ(ə)n/ *n.* **extroverted** *adj.* [Latin *verto* turn]

extrude /ɪk'stru:d/ *v.* (**-ding**) **1** (foll. by *from*) thrust or force out. **2** shape metal, plastics, etc. by forcing them through a die. □ **extrusion** *n.* **extrusive** *adj.* [Latin *extrudo -trus-* thrust out]

exuberant /ɪg'zju:bərənt/ *adj.* **1** lively, high-spirited. **2** (of a plant etc.) prolific. **3** (of feelings etc.) abounding. □ **exuberance** *n.* **exuberantly** *adv.* [Latin *uber* fertile]

exude /ɪg'zju:d/ *v.* (**-ding**) **1** ooze out. **2** emit (a smell). **3** display (an emotion etc.) freely. □ **exudation** /-'deɪʃ(ə)n/ *n.* [Latin *sudo* sweat]

exult /ɪg'zʌlt/ *v.* be joyful. □ **exultation** /-'teɪʃ(ə)n/ *n.* **exultant** *adj.* **exultantly** *adv.* [Latin *ex(s)ulto* from *salio salt-* leap]

-ey var. of -Y².

eye /aɪ/ **–***n.* **1** organ of sight. **2** eye characterized by the colour of the iris (*has blue eyes*). **3** region round the eye (*eyes swollen from weeping*). **4** (in *sing.* or *pl.*) sight. **5** particular visual ability (*a straight eye*). **6** thing like an eye, esp.: **a** a spot on a peacock's tail. **b** a leaf bud of a potato. **7** calm region at the centre of a hurricane etc. **8** hole of a needle. **–***v.* (**eyes, eyed, eyeing** or **eying**) (often foll. by *up*) watch or observe closely, esp. admiringly or with suspicion. □ **all eyes** watching intently. **an eye for an eye** retaliation in kind. **have an eye for** be discerning about. **have one's eye on** wish or plan to procure. **have eyes for** be interested in; wish to acquire. **keep an eye on 1** watch. **2** look after. **keep an eye open** (or **out**) (often foll. by *for*) watch carefully. **keep one's eyes open** (or **peeled** or **skinned**) watch out; be on the alert. **make eyes** (or **sheep's eyes**) (foll. by *at*) look amorously or flirtatiously at. **one in the eye** (foll. by *for*) disappointment or setback. **see eye to eye** (often foll. by *with*) agree. **set eyes on** see. **up to the** (or **one's**) **eyes in** deeply engaged or involved in. **with one's eyes shut** (or **closed**) with little effort. **with an eye to** with a view to. [Old English]

eyeball **–***n.* ball of the eye within the lids and socket. **–***v.* *US slang* look or stare (at).

eyeball to eyeball *adv. colloq.* confronting closely.

eyebath *n.* small vessel for applying lotion etc. to the eye.

eyebright *n.* plant used as a remedy for weak eyes.

eyebrow *n.* line of hair on the ridge above the eye-socket. □ **raise one's eyebrows** show surprise, disbelief, or disapproval.

eye-catching *adj. colloq.* striking.

eyeful *n.* (*pl.* **-s**) *colloq.* **1** (esp. in phr. **get an eyeful (of)**) good look; as much as the eye can take in. **2** visually striking person or thing. **3** thing thrown or blown into the eye.

eyeglass *n.* lens to assist defective sight.

eyehole *n.* hole to look through.

eyelash *n.* each of the hairs growing on the edges of the eyelids.

eyelet /'aɪlɪt/ *n.* **1** small hole for string or rope etc. to pass through. **2** metal ring strengthening this. [French *oillet* from Latin *oculus*]

eyelid *n.* either of the folds of skin closing to cover the eye.

eye-liner *n.* cosmetic applied as a line round the eye.

eye-opener *n. colloq.* enlightening experience; unexpected revelation.

eyepiece *n.* lens or lenses to which the eye is applied at the end of an optical instrument.

eye-shade *n.* device to protect the eyes, esp. from strong light.

eye-shadow *n.* coloured cosmetic applied to the eyelids.

eyesight *n.* faculty or power of seeing.

eyesore *n.* ugly thing.

eye strain *n.* fatigue of the eye muscles.

eye-tooth *n.* canine tooth in the upper jaw just under the eye.

eyewash *n.* **1** lotion for the eyes. **2** *slang* nonsense; insincere talk.

eyewitness *n.* person who saw a thing happen and can tell of it.

eyrie /'ɪərɪ/ *n.* **1** nest of a bird of prey, esp. an eagle, built high up. **2** house etc. perched high up. [French *aire* lair, from Latin *agrum* piece of ground]

F

F[1] /ef/ *n.* (also **f**) (*pl.* **Fs** or **F's**) **1** sixth letter of the alphabet. **2** *Mus.* fourth note of the diatonic scale of C major.

F[2] *abbr.* (also **F.**) **1** Fahrenheit. **2** farad(s). **3** fine (pencil-lead).

F[3] *symb.* fluorine.

f *abbr.* (also **f.**) **1** female. **2** feminine. **3** following page etc. **4** *Mus.* forte. **5** folio. **6** focal length.

FA *abbr.* Football Association.

fa var. of FAH.

fab *adj. colloq.* fabulous, marvellous. [abbreviation]

fable /'feɪb(ə)l/ *n.* **1 a** fictional, esp. supernatural, story. **b** moral tale, esp. with animals as characters. **2** legendary tales collectively (*in fable*). **3 a** lie. **b** thing only supposed to exist. [Latin *fabula* discourse]

fabled *adj.* celebrated; legendary.

fabric /'fæbrɪk/ *n.* **1** woven material; cloth. **2** walls, floor, and roof of a building. **3** essential structure. [Latin *faber* metal-worker]

fabricate /'fæbrɪ,keɪt/ *v.* (**-ting**) **1** construct, esp. from components. **2** invent (a story etc.). **3** forge (a document). □ **fabrication** /-'keɪʃ(ə)n/ *n.* **fabricator** *n.* [Latin: related to FABRIC]

fabulous /'fæbjʊləs/ *adj.* **1** incredible. **2** *colloq.* marvellous. **3** legendary. □ **fabulously** *adv.* [Latin: related to FABLE]

façade /fə'sɑːd/ *n.* **1** face or front of a building. **2** outward appearance, esp. a deceptive one. [French: related to FACE]

face —*n.* **1** front of the head from forehead to chin. **2** facial expression. **3** coolness, effrontery. **4** surface, esp.: **a** the side of a mountain etc. (*north face*). **b** = COALFACE. **c** *Geom.* each surface of a solid. **d** the façade of a building. **e** the dial of a clock etc. **5** functional side of a tool etc. **6** = TYPEFACE. **7** aspect (*unacceptable face of capitalism*). —*v.* (**-cing**) **1** look or be positioned towards or in a certain direction. **2** be opposite. **3** meet resolutely. **4** confront (*faces us with a problem*). **5 a** coat the surface of (a thing). **b** put a facing on (a garment). □ **face the music** *colloq.* take unpleasant consequences without flinching. **face up to** accept bravely. **have the face** be shameless enough. **in face** (or **the face**) **of** despite. **lose face** be humiliated. **on the face of it** apparently. **put a bold** (or **brave**) **face on it** accept difficulty etc. cheerfully. **save face** avoid humiliation. **set one's face against** oppose stubbornly. **to a person's face** openly in a person's presence. [Latin *facies*]

face-cloth *n.* cloth for washing one's face.

face-flannel *n.* = FACE-CLOTH.

faceless *adj.* **1** without identity; characterless. **2** purposely not identifiable.

face-lift *n.* **1** (also **face-lifting**) cosmetic surgery to remove wrinkles etc. **2** improvement to appearance, efficiency, etc.

face-pack *n.* skin preparation for the face.

facer *n. colloq.* sudden difficulty.

facet /'fæsɪt/ *n.* **1** aspect. **2** side of a cut gem etc. [French: related to FACT]

facetious /fə'siːʃəs/ *adj.* intending or intended to be amusing, esp. inappropriately. □ **facetiously** *adv.* [Latin *facetia* jest]

face to face *adv.* & *adj.* (also **face-to-face** when *attrib.*) (often foll. by *with*) facing; confronting each other.

face value *n.* **1** nominal value of money. **2** superficial appearance or implication.

facia var. of FASCIA.

facial /'feɪʃ(ə)l/ —*adj.* of or for the face. —*n.* beauty treatment for the face. □ **facially** *adv.*

facile /'fæsaɪl/ *adj.* usu. *derog.* **1** easily achieved but of little value. **2** glib, fluent. [Latin *facio* do]

facilitate /fə'sɪlɪ,teɪt/ *v.* (**-ting**) ease (a process etc.). □ **facilitation** /-'teɪʃ(ə)n/ *n.* [Italian: related to FACILE]

facility /fə'sɪlɪtɪ/ *n.* (*pl.* **-ies**) **1** ease; absence of difficulty. **2** fluency, dexterity. **3** (esp. in *pl.*) opportunity or equipment for doing something. [Latin: related to FACILE]

facing *n.* **1** layer of material covering part of a garment etc. for contrast or strength. **2** outer covering on a wall etc.

facsimile /fæk'sɪmɪlɪ/ *n.* exact copy, esp. of writing, printing, a picture, etc. [Latin, = make like]

fact *n.* **1** thing that is known to exist or to be true. **2** (usu. in *pl.*) item of verified information. **3** truth, reality. **4** thing assumed as the basis for argument. □ **before** (or **after**) **the fact** before (or after) the committing of a crime. **in** (or **in point of**) **fact 1** in reality. **2** in short. [Latin *factum* from *facio* do]

faction /ˈfækʃ(ə)n/ *n.* small organized dissentient group within a larger one, esp. in politics. □ **factional** *adj.* [Latin: related to FACT]

-faction *comb. form* forming nouns of action from verbs in *-fy* (*satisfaction*). [Latin *-factio*]

factious /ˈfækʃəs/ *adj.* of, characterized by, or inclined to faction. [Latin: related to FACTION]

factitious /fækˈtɪʃəs/ *adj.* **1** specially contrived. **2** artificial. [Latin: related to FACT]

fact of life *n.* something that must be accepted.

factor /ˈfæktə(r)/ *n.* **1** circumstance etc. contributing to a result. **2** whole number etc. that when multiplied with another produces a given number. **3 a** business agent. **b** *Scot.* land-agent, steward. **c** agent, deputy. [Latin: related to FACT]

factorial /fækˈtɔːrɪəl/ *–n.* product of a number and all the whole numbers below it. *–adj.* of a factor or factorial.

factorize *v.* (also **-ise**) (**-zing** or **-sing**) resolve into factors. □ **factorization** /-ˈzeɪʃ(ə)n/ *n.*

factory /ˈfæktərɪ/ *n.* (*pl.* **-ies**) building(s) in which goods are manufactured. [ultimately from Latin *factorium*]

factory farm *n.* farm using intensive or industrial methods of livestock rearing. □ **factory farming** *n.*

factotum /fækˈtəʊtəm/ *n.* (*pl.* **-s**) employee who does all kinds of work. [medieval Latin: related to FACT, TOTAL]

facts and figures *n.pl.* precise details.

factsheet *n.* information leaflet, esp. accompanying a television programme.

facts of life *n.pl.* (prec. by *the*) information about sexual functions and practices.

factual /ˈfæktʃʊəl/ *adj.* based on or concerned with fact. □ **factually** *adv.*

faculty /ˈfækəltɪ/ *n.* (*pl.* **-ies**) **1** aptitude for a particular activity. **2** inherent mental or physical power. **3 a** group of related university departments. **b** *US* teaching staff of a university or college. **4** authorization, esp. by a Church authority. [Latin: related to FACILE]

fad *n.* **1** craze. **2** peculiar notion. □ **faddish** *adj.* [probably from *fiddle-faddle*]

faddy *adj.* (**-ier, -iest**) having petty likes and dislikes. □ **faddiness** *n.*

fade *–v.* (**-ding**) **1** lose or cause to lose colour, light, or sound; slowly diminish. **2** lose freshness or strength. **3** (foll. by *in, out*) *Cinematog.* etc. cause (a picture or sound) to appear or disappear, increase or decrease, gradually. *–n.* action of fading. □ **fade away 1** *colloq.* languish, grow thin. **2** die away; disappear. [French *fade* dull]

faeces /ˈfiːsiːz/ *n.pl.* (*US* **feces**) waste matter discharged from the bowels. □ **faecal** /ˈfiːk(ə)l/ *adj.* [Latin]

faff *v. colloq.* (often foll. by *about, around*) fuss, dither. [imitative]

fag[1] *–n.* **1** *colloq.* tedious task. **2** *slang* cigarette. **3** (at public schools) junior boy who runs errands for a senior. *–v.* (**-gg-**) **1** (often foll. by *out*) *colloq.* exhaust. **2** (at public schools) act as a fag. [origin unknown]

fag[2] *n. US slang offens.* male homosexual. [abbreviation of FAGGOT]

fag-end *n. slang* cigarette-end.

faggot /ˈfægət/ *n.* (*US* **fagot**) **1** ball of seasoned chopped liver etc., baked or fried. **2** bundle of sticks etc. **3** *slang offens.* **a** unpleasant woman. **b** *US* male homosexual. [French from Italian]

fah /fɑː/ *n.* (also **fa**) *Mus.* fourth note of a major scale. [Latin *famuli*: see GAMUT]

Fahrenheit /ˈfærən,haɪt/ *adj.* of a scale of temperature on which water freezes at 32° and boils at 212°. [*Fahrenheit*, name of a physicist]

faience /ˈfaɪɑ̃s/ *n.* decorated and glazed earthenware and porcelain. [French from *Faenza* in Italy]

fail *–v.* **1** not succeed (*failed to qualify*). **2** be or judge to be unsuccessful in (an examination etc.). **3** be unable; neglect (*failed to appear*). **4** disappoint. **5** be absent or insufficient. **6** become weaker; cease functioning (*health is failing; engine failed*). **7** become bankrupt. *–n.* failure in an examination. □ **without fail** for certain, whatever happens. [Latin *fallo* deceive]

failed *adj.* unsuccessful (*failed actor*).

failing *–n.* fault, weakness. *–prep.* in default of.

fail-safe *adj.* reverting to a safe condition when faulty etc.

failure /ˈfeɪljə(r)/ *n.* **1** lack of success; failing. **2** unsuccessful person or thing. **3** non-performance. **4** breaking down or ceasing to function (*heart failure*). **5** running short of supply etc. [Anglo-French: related to FAIL]

fain *archaic –predic. adj.* (foll. by *to* + infin.) willing or obliged to. *–adv.* gladly (esp. *would fain*). [Old English]

faint *–adj.* **1** indistinct, pale, dim. **2** weak or giddy. **3** slight. **4** feeble; timid. **5** (also **feint**) (of paper) with inconspicuous ruled lines. *–v.* **1** lose consciousness. **2** become faint. *–n.* act or state of fainting. □ **faintly** *adv.* **faintness** *n.* [French: related to FEIGN]

faint-hearted *adj.* cowardly, timid.

fair[1] *–adj.* **1** just, equitable; in accordance with the rules. **2** blond; light or pale. **3 a** moderate in quality or amount. **b** satisfactory. **4** (of weather) fine; (of the wind) favourable. **5** clean, clear (*fair copy*). **6** *archaic* beautiful. *–adv.* **1** in a just manner. **2** exactly, completely. □ **in a fair way to** likely to. □ **fairness** *n.* [Old English]

fair[2] *n.* **1** stalls, amusements, etc., for public entertainment. **2** periodic market, often with entertainments. **3** exhibition, esp. commercial. [Latin *feriae* holiday]

fair and square *adv.* exactly; straightforwardly.

fair dinkum see DINKUM.

fair dos *n.pl.* (esp. as *int.*) *colloq.* fair shares; fair treatment.

fair game *n.* legitimate target or object.

fairground *n.* outdoor area where a fair is held.

fairing *n.* streamlining structure added to a ship, aircraft, vehicle, etc.

Fair Isle *n.* (also *attrib.*) multicoloured knitwear design characteristic of Fair Isle. [*Fair Isle* in the Shetlands]

fairly *adv.* **1** in a fair manner. **2** moderately (*fairly good*). **3** quite, rather (*fairly narrow*).

fair play *n.* just treatment or behaviour.

fair sex *n.* (prec. by *the*) women.

fairway *n.* **1** navigable channel. **2** part of a golf-course between a tee and its green, kept free of rough grass.

fair-weather friend *n.* unreliable friend or ally.

fairy *n.* (*pl.* **-ies**) **1** (often *attrib.*) small winged legendary being. **2** *slang offens.* male homosexual. [French: related to FAY, -ERY]

fairy cake *n.* small iced sponge cake.

fairy godmother *n.* benefactress.

fairyland *n.* **1** home of fairies. **2** enchanted region.

fairy lights *n.pl.* small decorative coloured lights.

fairy ring *n.* ring of darker grass caused by fungi.

fairy story *n.* (also **fairy tale**) **1** tale about fairies. **2** incredible story; lie.

fait accompli /ˌfeɪt əˈkɒmpli:/ *n.* thing that has been done and is not capable of alteration. [French]

faith *n.* **1** complete trust or confidence. **2** firm, esp. religious, belief. **3** religion or creed (*Christian faith*). **4** loyalty, trustworthiness. [Latin *fides*]

faithful *adj.* **1** showing faith. **2** (often foll. by *to*) loyal, trustworthy. **3** accurate (*faithful account*). **4** (**the Faithful**) the believers in a religion. □ **faithfulness** *n.*

faithfully *adv.* in a faithful manner. □ **Yours faithfully** formula for ending a formal letter when it begins 'Dear Sir' or 'Dear Madam'.

faithless *adj.* **1** false, unreliable, disloyal. **2** without religious faith.

fake *–n.* false or counterfeit thing or person. *–adj.* counterfeit; not genuine. *–v.* (**-king**) **1** make a fake or imitation of (*faked my signature*). **2** feign (a feeling, illness, etc.). [German *fegen* sweep]

fakir /ˈfeɪkɪə(r)/ *n.* Muslim or (rarely) Hindu religious beggar or ascetic. [Arabic, = poor man]

falcon /ˈfɔ:lkən/ *n.* small hawk sometimes trained to hunt. [Latin *falco*]

falconry *n.* breeding and training of hawks.

fall /fɔ:l/ *–v.* (*past* **fell**; *past part.* **fallen**) **1** go or come down freely; descend. **2** (often foll. by *over*) come suddenly to the ground from loss of balance etc. **3 a** hang or slope down. **b** (foll. by *into*) (of a river etc.) discharge into. **4 a** sink lower; decline, esp. in power, status, etc. **b** subside. **5** occur (*falls on a Monday*). **6** (of the face) show dismay or disappointment. **7** yield to temptation. **8** take or have a particular direction or place (*his eye fell on me*; *accent falls on the first syllable*). **9 a** find a place; be naturally divisible. **b** (foll. by *under*, *within*) be classed among. **10** come by chance or duty (*it fell to me to answer*). **11 a** pass into a specified condition (*fell ill*). **b** become (*fall asleep*). **12** be defeated or captured. **13** die. **14** (foll. by *on*, *upon*) **a** attack. **b** meet with. **c** embrace or embark on avidly. **15** (foll. by *to* + verbal noun) begin (*fell to wondering*). *–n.* **1** act of falling. **2** that which falls or has fallen, e.g. snow. **3** recorded amount of rainfall etc. **4** overthrow (*fall of Rome*). **5 a** succumbing to temptation. **b** (**the Fall**) Adam's sin and its results. **6** (also **Fall**) *US* autumn. **7** (esp. in *pl.*) waterfall etc. **8** wrestling-bout; throw in wrestling. □ **fall about** *colloq.* be helpless with laughter. **fall away 1** (of a surface) incline abruptly. **2** become few or thin; gradually vanish. **3** desert. **fall back** retreat. **fall back on** have recourse to in difficulty. **fall behind 1** be outstripped; lag. **2** be in arrears. **fall down** (often foll. by *on*) *colloq.* fail. **fall for** *colloq.* be captivated or deceived by. **fall foul of** come into conflict with. **fall in 1** take one's place in military formation. **2** collapse inwards. **fall in with 1** meet by chance. **2** agree with. **3** coincide with. **fall off 1** become detached. **2** decrease, deteriorate. **fall out 1** quarrel. **2** (of the hair, teeth, etc.) become detached. **3** *Mil.* come out of formation. **4** result; occur. **fall over backwards**

see BACKWARDS. **fall over oneself** *colloq.* **1** be eager. **2** stumble through haste, confusion, etc. **fall short** be deficient. **fall short of** fail to reach or obtain. **fall through** fail; miscarry. **fall to** begin, e.g. eating or working. [Old English]

fallacy /'fæləsɪ/ *n.* (*pl.* **-ies**) **1** mistaken belief. **2** faulty reasoning; misleading argument. □ **fallacious** /fə'leɪʃəs/ *adj.* [Latin *fallo* deceive]

fall guy *n. slang* easy victim; scapegoat.

fallible /'fælɪb(ə)l/ *adj.* capable of making mistakes. □ **fallibility** /-'bɪlɪtɪ/ *n.* **fallibly** *adv.* [medieval Latin: related to FALLACY]

falling star *n.* meteor.

Fallopian tube /fə'ləʊpɪən/ *n.* either of two tubes along which ova travel from the ovaries to the womb. [*Fallopius*, name of an anatomist]

fallout *n.* radioactive nuclear debris.

fallow /'fæləʊ/ *–adj.* **1** (of land) ploughed but left unsown. **2** uncultivated. *–n.* fallow land. [Old English]

fallow deer *n.* small deer with a white-spotted reddish-brown summer coat. [Old English *fallow* pale brownish or reddish yellow]

false /fɔ:ls/ *adj.* **1** wrong, incorrect. **2** spurious, artificial. **3** improperly so called (*false acacia*). **4** deceptive. **5** (foll. by *to*) deceitful, treacherous, or unfaithful. □ **falsely** *adv.* **falseness** *n.* [Latin *falsus*: related to FAIL]

false alarm *n.* alarm given needlessly.

falsehood *n.* **1** untrue thing. **2 a** act of lying. **b** lie.

false pretences *n.pl.* misrepresentations made with intent to deceive (esp. *under false pretences*).

falsetto /fɔ:l'setəʊ/ *n.* male singing voice above the normal range. [Italian diminutive: related to FALSE]

falsies /'fɔ:lsɪz/ *n.pl. colloq.* pads worn to make the breasts seem larger.

falsify /'fɔ:lsɪ,faɪ/ *v.* (**-ies, -ied**) **1** fraudulently alter. **2** misrepresent. □ **falsification** /-fɪ'keɪʃ(ə)n/ *n.* [French or medieval Latin: related to FALSE]

falsity *n.* being false.

falter /'fɔ:ltə(r)/ *v.* **1** stumble; go unsteadily. **2** lose courage. **3** speak hesitatingly. [origin uncertain]

fame *n.* **1** renown; being famous. **2** *archaic* reputation. [Latin *fama*]

famed *adj.* (foll. by *for*) famous; much spoken of.

familial /fə'mɪlɪəl/ *adj.* of a family or its members.

familiar /fə'mɪlɪə(r)/ *–adj.* **1 a** (often foll. by *to*) well known. **b** often met (with). **2** (foll. by *with*) knowing a thing well. **3** (often foll. by *with*) well acquainted (with a person). **4** informal, esp. presumptuously so. *–n.* **1** close friend. **2** (in full **familiar spirit**) supposed attendant of a witch etc. □ **familiarity** /-'ærɪtɪ/ *n.* **familiarly** *adv.* [Latin: related to FAMILY]

familiarize *v.* (also **-ise**) (**-zing** or **-sing**) (usu. foll. by *with*) make (a person or oneself) conversant or well acquainted. □ **familiarization** /-'zeɪʃ(ə)n/ *n.*

family /'fæmɪlɪ/ *n.* (*pl.* **-ies**) **1** set of relations, esp. parents and children. **2 a** members of a household. **b** person's children. **3** all the descendants of a common ancestor. **4** group of similar objects, people, etc. **5** group of related genera of animals or plants. □ **in the family way** *colloq.* pregnant. [Latin *familia*]

family allowance *n.* former name for CHILD BENEFIT.

family credit *n.* regular State payment to a family with an income below a certain level.

family man *n.* man who has a wife and children, esp. one fond of family life.

family name *n.* surname.

family planning *n.* birth control.

family tree *n.* genealogical chart.

famine /'fæmɪn/ *n.* extreme scarcity, esp. of food. [Latin *fames* hunger]

famish /'fæmɪʃ/ *v.* (usu. in *passive*) make or become extremely hungry. □ **be famished** (or **famishing**) *colloq.* be very hungry. [Romanic: related to FAMINE]

famous /'feɪməs/ *adj.* **1** (often foll. by *for*) celebrated; well-known. **2** *colloq.* excellent. □ **famously** *adv.* [Latin: related to FAME]

fan[1] *–n.* **1** apparatus, usu. with rotating blades, for ventilation etc. **2** folding semicircular device waved to cool oneself. **3** thing spread out like a fan (*fan tracery*). *–v.* (**-nn-**) **1** blow air on, with or as with a fan. **2** (of a breeze) blow gently on. **3** (usu. foll. by *out*) spread out like a fan. [Latin *vannus* winnowing-basket]

fan[2] *n.* devotee of a particular activity, performer, etc. (*film fan*). [abbreviation of FANATIC]

fanatic /fə'nætɪk/ *–n.* person obsessively devoted to a belief, activity, etc. *–adj.* excessively enthusiastic. □ **fanatical** *adj.* **fanatically** *adv.* **fanaticism** /-tɪ,sɪz(ə)m/ *n.* [Latin *fanum* temple]

fan belt *n.* belt driving a fan to cool the radiator in a vehicle.

fancier /'fænsɪə(r)/ *n.* connoisseur (*dog-fancier*).

fanciful /'fænsɪ,fʊl/ *adj.* **1** imaginary. **2** indulging in fancies. □ **fancifully** *adv.*

fan club *n.* club of devotees.

fancy /'fænsɪ/ *–n.* (*pl.* **-ies**) **1** inclination. **2** whim. **3** supposition. **4 a** faculty of imagination. **b** mental image. *–adj.* (**-ier, -iest**) **1** ornamental. **2** extravagant. *–v.* (**-ies, -ied**) **1** (foll. by *that*) be inclined to suppose. **2** *colloq.* feel a desire for (*fancy a drink?*). **3** *colloq.* find sexually attractive. **4** *colloq.* value (oneself, one's ability, etc.) unduly highly. **5** (in *imper.*) exclamation of surprise. **6** imagine. □ **take a fancy to** become (esp. inexplicably) fond of. **take a person's fancy** suddenly attract or please. □ **fanciable** *adj.* (in sense 3 of *v.*). **fancily** *adv.* **fanciness** *n.* [contraction of FANTASY]

fancy dress *n.* costume for masquerading at a party.

fancy-free *adj.* without (esp. emotional) commitments.

fancy man *n. slang derog.* **1** woman's lover. **2** pimp.

fancy woman *n. slang derog.* mistress.

fandango /fæn'dæŋgəʊ/ *n.* (*pl.* **-es** or **-s**) **1** lively Spanish dance for two. **2** music for this. [Spanish]

fanfare /'fænfeə(r)/ *n.* short showy or ceremonious sounding of trumpets etc. [French]

fang *n.* **1** canine tooth, esp. of a dog or wolf. **2** tooth of a venomous snake. **3** root of a tooth or its prong. [Old English]

fan-jet *n.* = TURBOFAN.

fanlight *n.* small, orig. semicircular, window over a door or another window.

fan mail *n.* letters from fans.

fanny *n.* (*pl.* **-ies**) **1** *coarse slang* the female genitals. **2** *US slang* the buttocks. [origin unknown]

fantail *n.* pigeon with a broad tail.

fantasia /fæn'teɪzɪə/ *n.* free or improvisatory musical or other composition, or one based on familiar tunes. [Italian: related to FANTASY]

fantasize /'fæntəˌsaɪz/ *v.* (also **-ise**) (**-zing** or **-sing**) **1** day-dream. **2** imagine; create a fantasy about.

fantastic /fæn'tæstɪk/ *adj.* **1** *colloq.* excellent, extraordinary. **2** extravagantly fanciful. **3** grotesque, quaint. □ **fantastically** *adv.* [Greek: related to FANTASY]

fantasy /'fæntəsɪ/ *n.* (*pl.* **-ies**) **1** imagination, esp. when unrelated to reality (*lives in the realm of fantasy*). **2** mental image, day-dream. **3** fantastic invention or composition. [Greek *phantasia* appearance]

far (**further, furthest** or **farther, farthest**) *–adv.* **1** at, to, or by a great distance (*far away*; *far off*; *far out*). **2** a long way (off) in space or time (*are you travelling far?*). **3** to a great extent or degree; by much (*far better*; *far too early*). *–adj.* **1** remote; distant (*far country*). **2** more distant (*far end of the hall*). **3** extreme (*far left*). □ **as far as 1** right up to (a place). **2** to the extent that. **by far** by a great amount. **a far cry** a long way. **far from** very different from being; almost the opposite of (*far from being fat*). **go far 1** achieve much. **2** contribute greatly. **go too far** overstep the limit (of propriety etc.). **so far 1** to such an extent; to this point. **2** until now. **so** (or **in so**) **far as** (or **that**) to the extent that. **so far so good** satisfactory up to now. [Old English]

farad /'færəd/ *n.* SI unit of capacitance. [*Faraday*, name of a physicist]

far and away *adv.* by a very large amount.

far and wide *adv.* over a large area.

far-away *adj.* **1** remote. **2** (of a look or voice) dreamy, distant.

farce *n.* **1 a** low comedy with a ludicrously improbable plot. **b** this branch of drama. **2** absurdly futile proceedings; pretence. □ **farcical** *adj.* [Latin *farcio* to stuff, used metaphorically of interludes etc.]

fare *–n.* **1 a** price of a journey on public transport. **b** fare-paying passenger. **2** range of food. *–v.* (**-ring**) progress; get on (*how did you fare?*). [Old English]

Far East *n.* (prec. by *the*) China, Japan, and other countries of E. Asia.

fare-stage *n.* **1** section of bus etc. route for which a fixed fare is charged. **2** stop marking this.

farewell /feə'wel/ *–int.* goodbye. *–n.* leave-taking.

far-fetched *adj.* unconvincing, incredible.

far-flung *adj.* **1** widely scattered. **2** remote.

far gone *adj. colloq.* very ill, drunk, etc.

farina /fə'ri:nə/ *n.* **1** flour or meal of cereal, nuts, or starchy roots. **2** starch. □ **farinaceous** /ˌfærɪ'neɪʃəs/ *adj.* [Latin]

farm *–n.* **1** land and its buildings under one management for growing crops, rearing animals, etc. **2** such land etc. for a specified purpose (*trout-farm*). **3** = FARMHOUSE. *–v.* **1 a** use (land) for growing crops, rearing animals, etc. **b** be a farmer; work on a farm. **2** breed (fish etc.) commercially. **3** (often foll. by *out*) delegate or subcontract (work) to others. □ **farming** *n.* [French *ferme* from Latin *firma* fixed payment]

farmer *n.* owner or manager of a farm.

farm-hand *n.* worker on a farm.

farmhouse *n.* house attached to a farm.

farmstead /'fɑ:msted/ *n.* farm and its buildings.

farmyard *n.* yard attached to a farmhouse.

far-off *adj.* remote.
far-out *adj.* **1** distant. **2** *slang* avant-garde, unconventional. **3** *slang* excellent.
farrago /fə'rɑ:gəʊ/ *n.* (*pl.* **-s** or *US* **-es**) medley, hotchpotch. [Latin, = mixed fodder, from *far* corn]
far-reaching *adj.* widely influential or applicable.
farrier /'færɪə(r)/ *n.* smith who shoes horses. [Latin *ferrum* iron, horseshoe]
farrow /'færəʊ/ *–n.* **1** litter of pigs. **2** birth of a litter. *–v.* (also *absol.*) (of a sow) produce (pigs). [Old English]
far-seeing *adj.* showing foresight; wise.
Farsi /'fɑ:sɪ/ *n.* modern Persian language. [Persian]
far-sighted *adj.* **1** having foresight, prudent. **2** esp. *US* = LONG-SIGHTED.
fart *coarse slang –v.* **1** emit wind from the anus. **2** (foll. by *about, around*) behave foolishly. *–n.* **1** an emission of wind from the anus. **2** unpleasant or foolish person. [Old English]
farther var. of FURTHER (esp. of physical distance).
farthest var. of FURTHEST (esp. of physical distance).
farthing /'fɑ:ðɪŋ/ *n. hist.* coin and monetary unit worth a quarter of an old penny. [Old English: related to FOURTH]

■ **Usage** The farthing was withdrawn from circulation in 1961.

farthingale /'fɑ:ðɪŋ,geɪl/ *n. hist.* hooped petticoat. [Spanish *verdugo* rod]
fasces /'fæsi:z/ *n.pl.* **1** *Rom. Hist.* bundle of rods with a projecting axe-blade, as a magistrate's symbol of power. **2** emblems of authority. [Latin, pl. of *fascis* bundle]
fascia /'feɪʃə/ *n.* (also **facia**) (*pl.* **-s**) **1 a** instrument panel of a vehicle. **b** similar panel etc. for operating machinery. **2** strip with a name etc. over a shop-front. **3 a** long flat surface in classical architecture. **b** flat surface, usu. of wood, covering the ends of rafters. **4** stripe or band. [Latin, = band, door-frame]
fascicle /'fæsɪk(ə)l/ *n.* section of a book that is published in instalments. [Latin diminutive: related to FASCES]
fascinate /'fæsɪ,neɪt/ *v.* **(-ting) 1** capture the interest of; attract. **2** paralyse (a victim) with fear. □ **fascination** /-'neɪʃ(ə)n/ *n.* [Latin *fascinum* spell]
Fascism /'fæʃɪz(ə)m/ *n.* **1** extreme totalitarian right-wing nationalist movement in Italy (1922–43). **2** (also **fascism**) any similar movement. □ **Fascist** *n.* & *adj.* (also **fascist**). **Fascistic** /-'ʃɪstɪk/ *adj.* (also **fascistic**). [Italian *fascio* bundle, organized group]
fashion /'fæʃ(ə)n/ *–n.* **1** current popular custom or style, esp. in dress. **2** manner of doing something. *–v.* (often foll. by *into*) make or form. □ **after** (or **in**) **a fashion** to some extent, barely acceptably. **in** (or **out of**) **fashion** fashionable (or not fashionable). [Latin *factio*: related to FACT]
fashionable *adj.* **1** following or suited to current fashion. **2** of or favoured by high society. □ **fashionableness** *n.* **fashionably** *adv.*
fast[1] /fɑ:st/ *–adj.* **1** rapid, quick-moving. **2** capable of or intended for high speed (*fast car; fast road*). **3** (of a clock etc.) ahead of the correct time. **4** (of a pitch etc.) causing the ball to bounce quickly. **5** firm; firmly fixed or attached (*fast knot; fast friendship*). **6** (of a colour) not fading. **7** pleasure seeking, dissolute. **8** (of photographic film etc.) needing only short exposure. *–adv.* **1** quickly; in quick succession. **2** firmly, tightly (*stand fast*). **3** soundly, completely (*fast asleep*). □ **pull a fast one** *colloq.* perpetrate deceit. [Old English]
fast[2] /fɑ:st/ *–v.* abstain from food, or certain food, for a time. *–n.* act or period of fasting. [Old English]
fastback *n.* **1** car with a sloping rear. **2** such a rear.
fast breeder *n.* (also **fast breeder reactor**) reactor using fast neutrons.
fasten /'fɑ:s(ə)n/ *v.* **1** make or become fixed or secure. **2** (foll. by *in, up*) lock securely; shut in. **3** (foll. by *on, upon*) direct (a look, thoughts, etc.) towards. **4** (foll. by *on, upon*) **a** take hold of. **b** single out. **5** (foll. by *off*) fix with a knot or stitches. □ **fastener** *n.* [Old English: related to FAST[1]]
fastening /'fɑ:snɪŋ/ *n.* device that fastens something; fastener.
fast food *n.* restaurant food that is quickly produced and served.
fastidious /fæ'stɪdɪəs/ *adj.* **1** excessively discriminatory; fussy. **2** easily disgusted; squeamish. □ **fastidiously** *adv.* **fastidiousness** *n.* [Latin *fastidium* loathing]
fastness /'fɑ:stnɪs/ *n.* stronghold. [Old English]
fast neutron *n.* neutron with high kinetic energy.
fast worker *n. colloq.* person who rapidly makes esp. sexual advances.
fat *–n.* **1** natural oily or greasy substance found esp. in animal bodies. **2** part of meat etc. containing this. *–adj.* **(fatter, fattest) 1** corpulent; plump. **2** containing much fat. **3** fertile. **4 a** thick (*fat book*). **b** substantial (*fat cheque*). **5** *colloq. iron.* very little; not much (*a fat chance; a fat lot*). *–v.* **(-tt-)** make or

become fat. □ **the fat is in the fire** trouble is imminent. **kill the fatted calf** celebrate, esp. at a prodigal's return (Luke 15). **live off** (or **on**) **the fat of the land** live luxuriously. □ **fatless** *adj.* **fatness** *n.* **fattish** *adj.* [Old English]

fatal /'feɪt(ə)l/ *adj.* **1** causing or ending in death (*fatal accident*). **2** (often foll. by *to*) ruinous (*fatal mistake*). **3** fateful. □ **fatally** *adv.* [Latin: related to FATE]

fatalism *n.* **1** belief in predetermination. **2** submissive acceptance. □ **fatalist** *n.* **fatalistic** /-'lɪstɪk/ *adj.* **fatalistically** /-'lɪstɪkəlɪ/ *adv.*

fatality /fə'tælətɪ/ *n.* (*pl.* **-ies**) **1** death by accident or in war etc. **2** fatal influence. **3** predestined liability to disaster.

fate *—n.* **1** supposed power predetermining events. **2 a** the future so determined. **b** individual's destiny or fortune. **3** death, destruction. *—v.* **(-ting) 1** (usu. in *passive*) preordain (*fated to win*). **2** (as **fated** *adj.*) doomed. □ **fate worse than death** see DEATH. [Italian and Latin *fatum*]

fateful *adj.* **1** important, decisive. **2** controlled by fate. □ **fatefully** *adv.*

fat-head *n. colloq.* stupid person.

fat-headed *adj.* stupid.

father /'fɑːðə(r)/ *—n.* **1** male parent. **2** (usu. in *pl.*) forefather. **3** originator, early leader. **4** (**Fathers** or **Fathers of the Church**) early Christian theologians. **5** (also **Father**) (often as a title or form of address) priest. **6** (**the Father**) (in Christian belief) first person of the Trinity. **7** (**Father**) venerable person, esp. as a title in personifications (*Father Time*). **8** (usu. in *pl.*) elders (*city fathers*). *—v.* **1** beget. **2** originate (a scheme etc.). □ **fatherhood** *n.* **fatherless** *adj.* [Old English]

father-figure *n.* older man respected and trusted like a father.

father-in-law *n.* (*pl.* **fathers-in-law**) father of one's husband or wife.

fatherland *n.* one's native country.

fatherly *adj.* like or of a father.

Father's Day *n.* day on which cards and presents are given to fathers.

fathom /'fæð(ə)m/ *—n.* (*pl.* often **fathom** when prec. by a number) measure of six feet, esp. in depth soundings. *—v.* **1** comprehend. **2** measure the depth of (water). □ **fathomable** *adj.* [Old English]

fathomless *adj.* too deep to fathom.

fatigue /fə'tiːg/ *—n.* **1** extreme tiredness. **2** weakness in metals etc. caused by repeated stress. **3 a** non-military army duty. **b** (in *pl.*) clothing worn for this. *—v.* **(-gues, -gued, -guing)** cause fatigue in. [Latin *fatigo* exhaust]

fatstock *n.* livestock fattened for slaughter.

fatten *v.* make or become fat.

fatty *adj.* **(-ier, -iest)** like or containing fat.

fatty acid *n.* organic compound consisting of a hydrocarbon chain and a terminal carboxyl group.

fatuous /'fætjʊəs/ *adj.* vacantly silly; purposeless, idiotic. □ **fatuity** /fə'tjuːɪtɪ/ *n.* (*pl.* **-ies**). **fatuously** *adv.* **fatuousness** *n.* [Latin *fatuus*]

fatwa /'fætwɑː/ *n.* legal decision or ruling by an Islamic religious leader. [Arabic]

faucet /'fɔːsɪt/ *n.* esp. *US* tap. [French *fausset* vent-peg]

fault /fɔːlt/ *—n.* **1** defect or imperfection of character, structure, appearance, etc. **2** responsibility for wrongdoing, error, etc. (*your own fault*). **3** break in an electric circuit. **4** transgression, offence. **5 a** *Tennis* etc. incorrect service. **b** (in showjumping) penalty for error. **6** break in rock strata. *—v.* **1** find fault with; blame. **2** *Geol.* **a** break the continuity of (strata). **b** show a fault. □ **at fault** guilty; to blame. **find fault** (often foll. by *with*) criticize; complain. **to a fault** excessively (*generous to a fault*). [Latin *fallo* deceive]

fault-finder *n.* complaining person.

fault-finding *n.* continual criticism.

faultless *adj.* perfect. □ **faultlessly** *adv.*

faulty *adj.* **(-ier, -iest)** having faults; imperfect. □ **faultily** *adv.* **faultiness** *n.*

faun /fɔːn/ *n.* Latin rural deity with goat's horns, legs, and tail. [Latin *Faunus*]

fauna /'fɔːnə/ *n.* (*pl.* **-s** or **-nae** /-niː/) animal life of a region or period. [Latin *Fauna*, name of a rural goddess]

faux pas /fəʊ 'pɑː/ *n.* (*pl.* same /'pɑːz/) tactless mistake; blunder. [French, = false step]

favour /'feɪvə(r)/ (*US* **favor**) *—n.* **1** kind act (*did it as a favour*). **2** approval, goodwill; friendly regard (*gained their favour*). **3** partiality. **4** badge, ribbon, etc., as an emblem of support. *—v.* **1** regard or treat with favour or partiality. **2** support, promote, prefer. **3** be to the advantage of; facilitate. **4** tend to confirm (an idea etc.). **5** (foll. by *with*) oblige. **6** (as **favoured** *adj.*) having special advantages. □ **in favour 1** approved of. **2** (foll. by *of*) **a** in support of. **b** to the advantage of. **out of favour** disapproved of. [Latin *faveo* be kind to]

favourable /'feɪvərəb(ə)l/ *adj.* (*US* **favorable**) **1** well-disposed; propitious;

approving. **2** promising, auspicious. **3** helpful, suitable. □ **favourably** *adv.*

favourite /'feɪvərɪt/ (*US* **favorite**) –*adj.* preferred to all others (*favourite book*). –*n.* **1** favourite person or thing. **2** *Sport* competitor thought most likely to win. [Italian: related to FAVOUR]

favouritism *n.* (*US* **favoritism**) unfair favouring of one person etc. at the expense of another.

fawn[1] –*n.* **1** deer in its first year. **2** light yellowish brown. –*adj.* fawn-coloured. –*v.* (also *absol.*) give birth to (a fawn). [Latin: related to FOETUS]

fawn[2] *v.* **1** (often foll. by *on, upon*) behave servilely, cringe. **2** (of esp. a dog) show extreme affection. [Old English]

fax –*n.* **1** transmission of an exact copy of a document etc. electronically. **2** copy produced by this. –*v.* transmit in this way. [abbreviation of FACSIMILE]

fay *n. literary* fairy. [Latin *fata* pl., = goddesses of destiny]

faze *v.* (**-zing**) (often as **fazed** *adj.*) *colloq.* disconcert, disorientate. [origin unknown]

FBA *abbr.* Fellow of the British Academy.

FBI *abbr.* Federal Bureau of Investigation.

FC *abbr.* Football Club.

FCO *abbr.* Foreign and Commonwealth Office.

FE *abbr.* further education.

Fe *symb.* iron. [Latin *ferrum*]

fealty /'fi:əltɪ/ *n.* (*pl.* **-ies**) **1** *hist.* fidelity to a feudal lord. **2** allegiance. [Latin: related to FIDELITY]

fear –*n.* **1 a** panic or distress caused by a sense of impending danger, pain, etc. **b** cause of this. **c** state of alarm (*in fear*). **2** (often foll. by *of*) dread, awe (towards) (*fear of heights*). **3** danger (*little fear of failure*). –*v.* **1** feel fear about or towards. **2** (foll. by *for*) feel anxiety about (*feared for my life*). **3** (often foll. by *that*) foresee or expect with unease, fear, or regret (*fear the worst*; *I fear that you are wrong*). **4** (foll. by verbal noun) shrink from (*feared meeting his ex-wife*). **5** revere (esp. God). □ **for fear of** (or **that**) to avoid the risk of (or that). **no fear** *colloq.* certainly not! [Old English]

fearful *adj.* **1** (usu. foll. by *of* or *that*) afraid. **2** terrible, awful. **3** *colloq.* extreme, esp. unpleasant (*fearful row*). □ **fearfully** *adv.* **fearfulness** *n.*

fearless *adj.* (often foll. by *of*) not afraid, brave. □ **fearlessly** *adv.* **fearlessness** *n.*

fearsome *adj.* frightening. □ **fearsomely** *adv.*

feasible /'fi:zɪb(ə)l/ *adj.* practicable, possible. □ **feasibility** /-'bɪlɪtɪ/ *n.* **feasibly** *adv.* [Latin *facio* do]

■ **Usage** *Feasible* should not be used to mean 'possible' or 'probable' in the sense 'likely'. 'Possible' or 'probable' should be used instead.

feast –*n.* **1** large or sumptuous meal. **2** sensual or mental pleasure. **3** religious festival. **4** annual village festival. –*v.* **1** (often foll. by *on*) partake of a feast; eat and drink sumptuously. **2** regale. □ **feast one's eyes on** look with pleasure at. [Latin *festus* joy]

feat *n.* remarkable act or achievement. [Latin: related to FACT]

feather /'feðə(r)/ –*n.* **1** one of the structures forming a bird's plumage, with a horny stem and fine strands. **2** (*collect.*) **a** plumage. **b** game-birds. –*v.* **1** cover or line with feathers. **2** turn (an oar) edgeways through the air. □ **feather in one's cap** a personal achievement. **feather one's nest** enrich oneself. **in fine** (or **high**) **feather** *colloq.* in good spirits. □ **feathery** *adj.* [Old English]

feather bed *n.* bed with a feather-stuffed mattress.

feather-bed *v.* (**-dd-**) cushion, esp. financially.

feather-brain *n.* (also **feather-head**) silly or absent-minded person. □ **feather-brained** *adj.* (also **feather-headed**).

feathering *n.* **1** bird's plumage. **2** feathers of an arrow. **3** feather-like structure or marking.

featherweight *n.* **1 a** weight in certain sports between bantamweight and lightweight, in amateur boxing 54–7kg. **b** sportsman of this weight. **2** very light person or thing. **3** (usu. *attrib.*) unimportant thing.

feature /'fi:tʃə(r)/ –*n.* **1** distinctive or characteristic part of a thing. **2** (usu. in *pl.*) part of the face. **3** (esp. specialized) article in a newspaper etc. **4** (in full **feature film**) main film in a cinema programme. –*v.* (**-ring**) **1** make a special display of; emphasize. **2** have as or be a central participant or topic in a film, broadcast, etc. □ **featureless** *adj.* [Latin *factura* formation: related to FACT]

Feb. *abbr.* February.

febrifuge /'febrɪ,fju:dʒ/ *n.* medicine or treatment for fever. [Latin *febris* fever]

febrile /'fi:braɪl/ *adj.* of fever; feverish. [Latin *febris* fever]

February /'febrʊərɪ/ *n.* (*pl.* **-ies**) second month of the year. [Latin *februa* purification feast]

fecal *US* var. of FAECAL.

feces *US* var. of FAECES.

feckless /'feklɪs/ *adj.* **1** feeble, ineffective. **2** unthinking, irresponsible. [Scots *feck* from *effeck* var. of EFFECT]

fecund /'fekənd/ *adj.* **1** prolific, fertile. **2** fertilizing. □ **fecundity** /fɪ'kʌndətɪ/ *n.* [Latin]

fecundate /'fekən,deɪt/ *v.* **(-ting) 1** make fruitful. **2** fertilize. □ **fecundation** /-'deɪʃ(ə)n/ *n.*

fed *past* and *past part.* of FEED. □ **fed up** (often foll. by *with*) discontented or bored.

federal /'fedər(ə)l/ *adj.* **1** of a system of government in which self-governing States unite for certain functions etc. **2** of such a federation (*federal laws*). **3** of or favouring centralized government. **4** (**Federal**) *US* of the Northern States in the Civil War. **5** comprising an association of largely independent units. □ **federalism** *n.* **federalist** *n.* **federalize** *v.* (also **-ise**) (**-zing** or **-sing**). **federalization** /-'zeɪʃ(ə)n/ *n.* **federally** *adv.* [Latin *foedus* covenant]

federal reserve *n.* (in the US) reserve cash available to banks.

federate –*v.* /'fedə,reɪt/ **(-ting)** unite on a federal basis. –*adj.* /'fedərət/ federally organized. □ **federative** /'fedərətɪv/ *adj.*

federation /,fedə'reɪʃ(ə)n/ *n.* **1** federal group. **2** act of federating. [Latin: related to FEDERAL]

fee *n.* **1** payment made for professional advice or services etc. **2 a** charge for a privilege, examination, admission to a society, etc. (*enrolment fee*). **b** money paid for the transfer to another employer of a footballer etc. **3** (in *pl.*) regular payments (esp. to a school). **4** *Law* inherited estate, unlimited (**fee simple**) or limited (**fee tail**) as to category of heir. [medieval Latin *feudum*]

feeble /'fi:b(ə)l/ *adj.* (**feebler, feeblest**) **1** weak, infirm. **2** lacking strength, energy, or effectiveness. □ **feebly** *adv.* [Latin *flebilis* lamentable]

feeble-minded *adj.* mentally deficient.

feed –*v.* (*past* and *past part.* **fed**) **1 a** supply with food. **b** put food into the mouth of. **2** give as food, esp. to animals. **3** (usu. foll. by *on*) (esp. of animals, or *colloq.* of people) eat. **4** (often foll. by *on*) nourish or be nourished by; benefit from. **5 a** keep (a fire, machine, etc.) supplied with fuel etc. **b** (foll. by *into*) supply (material) to a machine etc. **c** (often foll. by *into*) (of a river etc.) flow into a lake etc. **d** keep (a meter) supplied with coins to ensure continuity. **6** *slang* supply (an actor etc.) with cues. **7** *Sport* send passes to (a player). **8** gratify (vanity etc.). **9** provide (advice, information, etc.) to. –*n.* **1** food, esp. for animals or infants. **2** feeding; giving of food. **3** *colloq.* meal. **4 a** raw material for a machine etc. **b** provision of or device for this. □ **feed back** produce feedback. **feed up** fatten. [Old English]

feedback *n.* **1** public response to an event, experiment, etc. **2** *Electronics* **a** return of a fraction of an output signal to the input. **b** signal so returned.

feeder *n.* **1** person or thing that feeds, esp. in specified manner. **2** baby's feeding-bottle. **3** bib. **4** tributary stream. **5** branch road, railway line, etc. linking with a main system. **6** main carrying electricity to a distribution point. **7** feeding apparatus in a machine.

feel –*v.* (*past* and *past part.* **felt**) **1 a** examine or search by touch. **b** (*absol.*) have the sensation of touch (*unable to feel*). **2** perceive or ascertain by touch (*feel the warmth*). **3** experience, exhibit, or be affected by (an emotion, conviction, etc.) (*felt strongly about it; felt the rebuke*). **4** (foll. by *that*) have an impression (*I feel that I am right*). **5** consider, think (*I feel it useful*). **6** seem (*air feels chilly*). **7** be consciously; consider oneself (*I feel happy*). **8** (foll. by *for, with*) have sympathy or pity. **9** (often foll. by *up*) *slang* fondle sexually. –*n.* **1** feeling; testing by touch. **2** sensation characterizing a material, situation, etc. **3** sense of touch. □ **feel like** have a wish or inclination for. **feel up to** be ready to face or deal with. **feel one's way** proceed cautiously. **get the feel of** become accustomed to using. [Old English]

feeler *n.* **1** organ in certain animals for touching or searching for food. **2** tentative proposal (*put out feelers*).

feeling –*n.* **1 a** capacity to feel; sense of touch (*lost all feeling*). **b** physical sensation. **2 a** (often foll. by *of*) emotional reaction (*feeling of despair*). **b** (in *pl.*) emotional susceptibilities (*hurt my feelings*). **3** particular sensitivity (*feeling for literature*). **4 a** opinion or notion (*had a feeling she would*). **b** general sentiment. **5** sympathy or compassion. **6** emotional sensibility or intensity (*played with feeling*). –*adj.* sensitive, sympathetic; heartfelt. □ **feelingly** *adv.*

feet *pl.* of FOOT.

feign /feɪn/ *v.* simulate; pretend (*feign madness*). [Latin *fingo fict-* mould, contrive]

feint /feɪnt/ –*n.* **1** sham attack or diversionary blow. **2** pretence. –*v.* make a feint. –*adj.* = FAINT *adj.* 5. [French: related to FEIGN]

feldspar /ˈfeldspɑː(r)/ *n.* (also **felspar**) common aluminium silicate of potassium, sodium, or calcium. □ **feldspathic** /-ˈspæθɪk/ *adj.* [German *Feld* field, *Spat(h)* SPAR[3]]

felicitate /fəˈlɪsɪˌteɪt/ *v.* (**-ting**) *formal* congratulate. □ **felicitation** /-ˈteɪʃ(ə)n/ *n.* (usu. in *pl.*). [Latin *felix* happy]

felicitous /fəˈlɪsɪtəs/ *adj. formal* apt; pleasantly ingenious; well-chosen.

felicity /fəˈlɪsɪtɪ/ *n.* (*pl.* **-ies**) *formal* **1** intense happiness. **2 a** capacity for apt expression. **b** well-chosen phrase. [Latin *felix* happy]

feline /ˈfiːlaɪn/ *–adj.* **1** of the cat family. **2** catlike. *–n.* animal of the cat family. □ **felinity** /fɪˈlɪnɪtɪ/ *n.* [Latin *feles* cat]

fell[1] *past* of FALL *v.*

fell[2] *v.* **1** cut down (esp. a tree). **2** strike or knock down. **3** stitch down (the edge of a seam). [Old English]

fell[3] *n. N.Engl.* **1** hill. **2** stretch of hills or moorland. [Old Norse]

fell[4] *adj. poet.* or *rhet.* ruthless, destructive. □ **at** (or **in**) **one fell swoop** in a single (orig. deadly) action. [French: related to FELON]

fell[5] *n.* animal's hide or skin with its hair. [Old English]

fellatio /fɪˈleɪʃɪəʊ/ *n.* oral stimulation of the penis. [Latin *fello* suck]

feller *n.* = FELLOW 1.

felloe /ˈfeləʊ/ *n.* (also **felly** /ˈfelɪ/) (*pl.* **-s** or **-ies**) outer circle (or a section of it) of a wheel. [Old English]

fellow /ˈfeləʊ/ *n.* **1** *colloq.* man or boy (*poor fellow!*). **2** (usu. in *pl.*) person in a group etc.; comrade (*separated from their fellows*). **3** counterpart; one of a pair. **4** equal; peer. **5 a** incorporated senior member of a college. **b** elected graduate paid to do research. **6** member of a learned society. **7** (*attrib.*) of the same group etc. (*fellow-countryman*). [Old English from Old Norse]

fellow-feeling *n.* sympathy.

fellowship *n.* **1** friendly association with others, companionship. **2** body of associates. **3** status or income of a fellow of a college or society.

fellow-traveller *n.* **1** person who travels with another. **2** sympathizer with the Communist Party.

felly var. of FELLOE.

felon /ˈfelən/ *n.* person who has committed a felony. [medieval Latin *fello*]

felony *n.* (*pl.* **-ies**) serious, usu. violent, crime. □ **felonious** /fɪˈləʊnɪəs/ *adj.*

felspar var. of FELDSPAR.

felt[1] *–n.* cloth of matted and pressed fibres of wool etc. *–v.* **1** make into felt; mat. **2** cover with felt. **3** become matted. [Old English]

felt[2] *past* and *past part.* of FEEL.

felt-tipped pen *n.* (also **felt-tip pen**) pen with a fibre point.

felucca /fɪˈlʌkə/ *n.* small Mediterranean coasting vessel with oars and/or sails. [Arabic *fulk*]

female /ˈfiːmeɪl/ *–adj.* **1** of the sex that can give birth or produce eggs. **2** (of plants) fruit-bearing. **3** of women or female animals or plants. **4** (of a screw, socket, etc.) hollow to receive an inserted part. *–n.* female person, animal, or plant. [Latin diminutive of *femina* woman, assimilated to *male*]

feminine /ˈfemɪnɪn/ *–adj.* **1** of women. **2** having womanly qualities. **3** of or denoting the female gender. *–n.* feminine gender or word. □ **femininity** /-ˈnɪnɪtɪ/ *n.* [Latin: related to FEMALE]

feminism /ˈfemɪˌnɪz(ə)m/ *n.* advocacy of women's rights and sexual equality. □ **feminist** *n.* & *adj.*

femme fatale /ˌfæm fæˈtɑːl/ *n.* (*pl.* ***femmes fatales*** pronunc. same) dangerously seductive woman. [French]

femur /ˈfiːmə(r)/ *n.* (*pl.* **-s** or **femora** /ˈfemərə/) thigh-bone. □ **femoral** /ˈfemər(ə)l/ *adj.* [Latin]

fen *n.* **1** low marshy land. **2** (**the Fens**) low-lying areas in Cambridgeshire etc. [Old English]

fence *–n.* **1** barrier, railing, etc., enclosing a field, garden, etc. **2** large upright jump for horses. **3** *slang* receiver of stolen goods. **4** guard or guide in machinery. *–v.* (**-cing**) **1** surround with or as with a fence. **2** (foll. by *in*, *off*, *up*) enclose, separate, or seal, with or as with a fence. **3** practise fencing with a sword. **4** be evasive. **5** *slang* deal in (stolen goods). □ **fencer** *n.* [from DEFENCE]

fencing *n.* **1** set of, or material for, fences. **2** sword-fighting, esp. as a sport.

fend *v.* **1** (foll. by *for*) look after (esp. oneself). **2** (usu. foll. by *off*) ward off. [from DEFEND]

fender *n.* **1** low frame bordering a fireplace. **2** *Naut.* padding protecting a ship against impact. **3** *US* vehicle's bumper.

fennel /ˈfen(ə)l/ *n.* yellow-flowered fragrant herb used for flavouring. [Latin *fenum* hay]

fenugreek /ˈfenjuːˌgriːk/ *n.* leguminous plant with aromatic seeds used for flavouring. [Latin, = Greek hay]

feral /ˈfer(ə)l/ *adj.* **1** wild; uncultivated. **2** (of an animal) escaped and living wild. **3** brutal. [Latin *ferus* wild]

ferial /ˈfɪərɪəl/ *adj. Eccl.* (of a day) not a festival or fast. [Latin *feria* FAIR[2]]

ferment *–n.* /ˈfɜːment/ **1** excitement, unrest. **2 a** fermentation. **b** fermenting-agent. *–v.* /fəˈment/ **1** undergo or subject to fermentation. **2** excite; stir up. [Latin *fermentum*: related to FERVENT]

fermentation /ˌfɜːmenˈteɪʃ(ə)n/ *n.* **1** breakdown of a substance by yeasts and bacteria etc., esp. of sugar in making alcohol. **2** agitation, excitement. □ **fermentative** /-ˈmentətɪv/ *adj.* [Latin: related to FERMENT]

fermium /ˈfɜːmɪəm/ *n.* transuranic artificial radioactive metallic element. [*Fermi*, name of a physicist]

fern *n.* (*pl.* same or **-s**) flowerless plant usu. having feathery fronds. □ **ferny** *adj.* [Old English]

ferocious /fəˈrəʊʃəs/ *adj.* fierce, savage. □ **ferociously** *adv.* **ferocity** /fəˈrɒsɪtɪ/ *n.* [Latin *ferox*]

-ferous *comb. form* (usu. **-iferous**) forming adjectives with the sense 'bearing', 'having' (*odoriferous*). [Latin *fero* bear]

ferrel var. of FERRULE.

ferret /ˈferɪt/ —*n.* small polecat used in catching rabbits, rats, etc. —*v.* **1** hunt with ferrets. **2** (often foll. by *out*, *about*, etc.) rummage; search out (secrets, criminals, etc.). [Latin *fur* thief]

ferric /ˈferɪk/ *adj.* **1** of iron. **2** containing iron in a trivalent form. [Latin *ferrum* iron]

Ferris wheel /ˈferɪs/ *n.* tall revolving vertical wheel with passenger cars in fairgrounds etc. [*Ferris*, name of its inventor]

ferro- *comb. form* **1** iron. **2** (of alloys) containing iron. [related to FERRIC]

ferroconcrete /ˌferəʊˈkɒŋkriːt/ —*n.* reinforced concrete. —*adj.* made of this.

ferrous /ˈferəs/ *adj.* **1** containing iron. **2** containing iron in a divalent form.

ferrule /ˈferuːl/ *n.* (also **ferrel** /ˈfer(ə)l/) **1** ring or cap on the lower end of a stick, umbrella, etc. **2** band strengthening or forming a joint. [Latin *viriae* bracelet]

ferry —*n.* (*pl.* **-ies**) **1** boat or aircraft etc. for esp. regular transport, esp. across water. **2** place or service of ferrying. —*v.* (**-ies, -ied**) **1** convey or go in a ferry. **2** (of a boat etc.) cross water regularly. **3** transport, esp. regularly, from place to place. □ **ferryman** *n.* [Old Norse]

fertile /ˈfɜːtaɪl/ *adj.* **1 a** (of soil) abundantly productive. **b** fruitful. **2 a** (of a seed, egg, etc.) capable of growth. **b** (of animals and plants) able to reproduce. **3** (of the mind) inventive. **4** (of nuclear material) able to become fissile by the capture of neutrons. □ **fertility** /-ˈtɪlɪtɪ/ *n.* [French from Latin]

fertilize /ˈfɜːtɪˌlaɪz/ *v.* (also **-ise**) (**-zing** or **-sing**) **1** make (soil etc.) fertile. **2** cause (an egg, female animal, etc.) to develop by mating etc. □ **fertilization** /-ˈzeɪʃ(ə)n/ *n.*

fertilizer *n.* (also **-iser**) substance added to soil to make it more fertile.

fervent /ˈfɜːv(ə)nt/ *adj.* ardent, intense (*fervent admirer*). □ **fervency** *n.* **fervently** *adv.* [Latin *ferveo* boil]

fervid /ˈfɜːvɪd/ *adj.* ardent, intense. □ **fervidly** *adv.* [Latin: related to FERVENT]

fervour /ˈfɜːvə(r)/ *n.* (*US* **fervor**) passion, zeal. [Latin: related to FERVENT]

fescue /ˈfeskjuː/ *n.* a pasture and fodder grass. [Latin *festuca* stalk, straw]

festal /ˈfest(ə)l/ *adj.* **1** joyous, merry. **2** of a feast or festival. □ **festally** *adv.* [Latin: related to FEAST]

fester *v.* **1** make or become septic. **2** cause continuing anger or bitterness. **3** rot, stagnate. [Latin FISTULA]

festival /ˈfestɪv(ə)l/ *n.* **1** day or period of celebration. **2** series of cultural events in a town etc. (*Bath Festival*). [French: related to FESTIVE]

festive /ˈfestɪv/ *adj.* **1** of or characteristic of a festival. **2** joyous. □ **festively** *adv.* **festiveness** *n.* [Latin: related to FEAST]

festivity /feˈstɪvɪtɪ/ *n.* (*pl.* **-ies**) **1** gaiety, rejoicing. **2** (in *pl.*) celebration; party.

festoon /feˈstuːn/ —*n.* curved hanging chain of flowers, leaves, ribbons, etc. —*v.* (often foll. by *with*) adorn with or form into festoons; decorate elaborately. [Italian: related to FESTIVE]

Festschrift /ˈfestʃrɪft/ *n.* (also **festschrift**) (*pl.* **-en** or **-s**) collection of writings published in honour of a scholar. [German, = festival-writing]

feta /ˈfetə/ *n.* soft white esp. ewe's-milk cheese made esp. in Greece. [Greek *pheta*]

fetal *US* var. of FOETAL.

fetch —*v.* **1** go for and bring back (*fetch a doctor*). **2** be sold for (a price) (*fetched £10*). **3** cause (blood, tears, etc.) to flow. **4** draw (breath), heave (a sigh). **5** *colloq.* give (a blow etc.) (*fetched him a slap*). —*n.* **1** act of fetching. **2** dodge, trick. □ **fetch and carry** do menial tasks. **fetch up** *colloq.* **1** arrive, come to rest. **2** vomit. [Old English]

fetching *adj.* attractive. □ **fetchingly** *adv.*

fête /feɪt/ —*n.* **1** outdoor fund-raising event with stalls and amusements etc. **2** festival. **3** saint's day. —*v.* (**-ting**) honour or entertain lavishly. [French: related to FEAST]

fetid /ˈfetɪd/ *adj.* (also **foetid**) stinking. [Latin *feteo* stink]

fetish /ˈfetɪʃ/ *n.* **1** *Psychol.* abnormal object of sexual desire. **2 a** object worshipped by primitive peoples. **b** obsessional cause (*makes a fetish of punctuality*). □ **fetishism** *n.* **fetishist** *n.* **fetishistic** /-'ʃɪstɪk/ *adj.* [Portuguese *feitiço* charm]

fetlock /'fetlɒk/ *n.* back of a horse's leg above the hoof with a tuft of hair. [ultimately related to FOOT]

fetter –*n.* **1** shackle for the ankles. **2** (in *pl.*) captivity. **3** restraint. –*v.* **1** put into fetters. **2** restrict. [Old English]

fettle /'fet(ə)l/ *n.* condition or trim (*in fine fettle*). [Old English]

fetus *US* var. of FOETUS.

feu /fju:/ *Scot.* –*n.* **1** perpetual lease at a fixed rent. **2** land so held. –*v.* (**feus, feued, feuing**) grant (land) on feu. [French: related to FEE]

feud[1] /fju:d/ –*n.* prolonged hostility, esp. between families, tribes, etc. –*v.* conduct a feud. [Germanic: related to FOE]

feud[2] /fju:d/ *n.* = FIEF. [medieval Latin *feudum* FEE]

feudal /'fju:d(ə)l/ *adj.* **1** of, like, or according to the feudal system. **2** reactionary (*feudal attitude*). □ **feudalism** *n.* **feudalistic** /-'lɪstɪk/ *adj.*

feudal system *n.* medieval system of land tenure with allegiance and service due to the landowner.

fever –*n.* **1 a** abnormally high temperature, often with delirium etc. **b** disease characterized by this (*scarlet fever*). **2** nervous excitement; agitation. –*v.* (esp. as **fevered** *adj.*) affect with fever or excitement. [Latin *febris*]

feverfew /'fi:və,fju:/ *n.* aromatic bushy plant, used formerly to reduce fever, now to cure migraine. [Latin *febrifuga*: related to FEVER, *fugo* drive away]

feverish *adj.* **1** having symptoms of fever. **2** excited, restless. □ **feverishly** *adv.* **feverishness** *n.*

fever pitch *n.* state of extreme excitement.

few –*adj.* not many (*few doctors smoke*). –*n.* (as *pl.*) **1** (prec. by *a*) some but not many (*a few of his friends were there*). **2** not many (*few are chosen*). **3** (prec. by *the*) **a** the minority. **b** the elect. □ **a good few** *colloq.* fairly large number (of). **no fewer than** as many as (a specified number). **not a few** a considerable number. [Old English]

few and far between *predic. adj.* scarce.

fey /feɪ/ *adj.* **1 a** strange, other-worldly; whimsical. **b** clairvoyant. **2** *Scot.* fated to die soon. [Old English, = doomed to die]

fez *n.* (*pl.* **fezzes**) man's flat-topped conical red cap worn by some Muslims. [Turkish]

ff *abbr. Mus.* fortissimo.

ff. *abbr.* following pages etc.

fiancé /fɪ'ɒnseɪ/ *n.* (*fem.* **fiancée** pronunc. same) person one is engaged to. [French]

fiasco /fɪ'æskəʊ/ *n.* (*pl.* **-s**) ludicrous or humiliating failure or breakdown. [Italian, = bottle]

fiat /'faɪæt/ *n.* **1** authorization. **2** decree. [Latin, = let it be done]

fib –*n.* trivial lie. –*v.* (**-bb-**) tell a fib. □ **fibber** *n.* [perhaps from *fible-fable*, a reduplication of FABLE]

fibre /'faɪbə(r)/ *n.* (*US* **fiber**) **1** thread or filament forming tissue or textile. **2** piece of threadlike glass. **3** substance formed of fibres, or able to be spun, woven, etc. **4** structure; character (*moral fibre*). **5** roughage. [French from Latin *fibra*]

fibreboard *n.* (*US* **fiberboard**) board of compressed wood or other plant fibres.

fibreglass *n.* (*US* **fiberglass**) **1** fabric made from woven glass fibres. **2** plastic reinforced by glass fibres.

fibre optics *n.pl.* optics using thin glass fibres, usu. for the transmission of modulated light to carry signals.

fibril /'faɪbrɪl/ *n.* small fibre. [diminutive of FIBRE]

fibroid /'faɪbrɔɪd/ –*adj.* of, like, or containing fibrous tissue or fibres. –*n.* benign fibrous tumour growing in the womb.

fibrosis /faɪ'brəʊsɪs/ *n.* thickening and scarring of connective tissue. [from FIBRE, -OSIS]

fibrositis /,faɪbrə'saɪtɪs/ *n.* rheumatic inflammation of fibrous tissue. [from FIBRE, -ITIS]

fibrous /'faɪbrəs/ *adj.* of or like fibres.

fibula /'fɪbjʊlə/ *n.* (*pl.* **fibulae** /-,li:/ or **-s**) small outer bone between the knee and the ankle. □ **fibular** *adj.* [Latin, = brooch]

-fic *suffix* (usu. as **-ific**) forming adjectives meaning 'producing', 'making' (*prolific*; *pacific*). [Latin *facio* make]

-fication *suffix* (usu. as **-ification**) forming nouns of action from verbs in *-fy* (*purification*; *simplification*).

fiche /fi:ʃ/ *n.* (*pl.* same or **-s**) microfiche. [abbreviation]

fickle /'fɪk(ə)l/ *adj.* inconstant, changeable, disloyal. □ **fickleness** *n.* **fickly** *adv.* [Old English]

fiction /'fɪkʃ(ə)n/ *n.* **1** non-factual literature, esp. novels. **2** invented idea, thing, etc. **3** generally accepted falsehood (*polite fiction*). □ **fictional** *adj.* **fictionalize** *v.* (also **-ise**) (**-zing** or **-sing**). [Latin: related to FEIGN]

fictitious /fɪk'tɪʃəs/ *adj.* imaginary, unreal; not genuine.

fiddle /'fɪd(ə)l/ –*n.* **1** *colloq.* or *derog.* stringed instrument played with a bow, esp. a violin. **2** *colloq.* cheat or fraud. **3**

fiddly task. –*v.* (**-ling**) **1 a** (often foll. by *with, at*) play restlessly. **b** (often foll. by *about*) move aimlessly; waste time. **c** (usu. foll. by *with*) adjust, tinker; tamper. **2** *slang* **a** cheat, swindle. **b** falsify. **c** get by cheating. **3** play (a tune) on the fiddle. □ **as fit as a fiddle** in very good health. **play second** (or **first**) **fiddle** take a subordinate (or leading) role. [Old English]

fiddle-faddle –*n.* trivial matters. –*v.* (**-ling**) fuss, trifle. –*int.* nonsense! [reduplication of FIDDLE]

fiddler *n.* **1** fiddle-player. **2** *slang* swindler, cheat. **3** small N. American crab.

fiddlesticks *int.* nonsense.

fiddling *adj.* **1** petty, trivial. **2** *colloq.* = FIDDLY.

fiddly *adj.* (**-ier, -iest**) *colloq.* awkward or tiresome to do or use.

fidelity /fɪ'delɪtɪ/ *n.* **1** faithfulness, loyalty. **2** strict accuracy. **3** precision in sound reproduction (*high fidelity*). [Latin *fides* faith]

fidget /'fɪdʒɪt/ –*v.* (**-t-**) **1** move or act restlessly or nervously. **2** be or make uneasy. –*n.* **1** person who fidgets. **2** (usu. in *pl.*) restless movements or mood. □ **fidgety** *adj.* [obsolete or dial. *fidge* twitch]

fiduciary /fɪ'dju:ʃərɪ/ –*adj.* **1 a** of a trust, trustee, or trusteeship. **b** held or given in trust. **2** (of paper currency) dependent on public confidence or securities. –*n.* (*pl.* **-ies**) trustee. [Latin *fiducia* trust]

fie /faɪ/ *int. archaic* expressing disgust, shame, etc. [French from Latin]

fief *n.* **1** land held under the feudal system or in fee. **2** person's sphere of operation. [French: related to FEE]

field –*n.* **1** area of esp. cultivated enclosed land. **2** area rich in some natural product (*gas field*). **3** land for a game etc. (*football field*). **4** participants in a contest, race, or sport, or all except those specified. **5** *Cricket* **a** the side fielding. **b** fielder. **6** expanse of ice, snow, sea, sky, etc. **7 a** battlefield. **b** (*attrib.*) (of artillery etc.) light and mobile. **8** area of activity or study (*in his own field*). **9** *Physics* **a** region in which a force is effective (*gravitational field*). **b** force exerted in this. **10** range of perception (*field of view*). **11** (*attrib.*) **a** (of an animal or plant) wild (*field mouse*). **b** in the natural environment, not in a laboratory etc. (*field test*). **12 a** background of a picture, coin, flag, etc. **b** *Heraldry* surface of an escutcheon. **13** *Computing* part of a record representing an item of data. –*v.* **1 a** act as a fieldsman in cricket etc. **b** stop and return (the ball) in cricket etc. **2** select to play in a game. **3** deal with (questions, an argument, etc.). □ **hold the field** not be superseded. **play the field** *colloq.* date many partners. **take the field** begin a campaign. [Old English]

field-day *n.* **1** exciting or successful time. **2** military exercise or review.

fielder *n.* = FIELDSMAN.

field events *n.pl.* athletic events other than races.

fieldfare *n.* thrush with grey plumage.

field-glasses *n.pl.* outdoor binoculars.

field marshal *n.* army officer of the highest rank.

field mouse *n.* small long-tailed rodent.

field officer *n.* army officer of field rank.

field of honour *n.* battlefield.

field rank *n.* army rank above captain and below general.

fieldsman *n. Cricket, Baseball,* etc. member (other than the bowler or pitcher) of the fielding side.

field sports *n.pl.* outdoor sports, esp. hunting, shooting, and fishing.

field telegraph *n.* movable military telegraph.

fieldwork *n.* **1** practical surveying, science, sociology, etc. conducted in the natural environment. **2** temporary fortification. □ **fieldworker** *n.*

fiend *n.* **1** evil spirit, demon. **2 a** wicked or cruel person. **b** mischievous or annoying person. **3** *slang* devotee (*fitness fiend*). **4** difficult or unpleasant thing. □ **fiendish** *adj.* **fiendishly** *adv.* [Old English]

fierce *adj.* (**fiercer, fiercest**) **1** violently aggressive or frightening. **2** eager, intense. **3** unpleasantly strong or intense (*fierce heat*). □ **fiercely** *adv.* **fierceness** *n.* [Latin *ferus* savage]

fiery /'faɪərɪ/ *adj.* (**-ier, -iest**) **1** consisting of or flaming with fire. **2** bright red. **3** hot; burning. **4 a** flashing, ardent (*fiery eyes*). **b** pugnacious; spirited (*fiery temper*). □ **fierily** *adv.* **fieriness** *n.*

fiesta /fɪ'estə/ *n.* holiday, festivity, or religious festival. [Spanish]

FIFA /'fi:fə/ *abbr.* International Football Federation. [French *Fédération Internationale de Football Association*]

fife *n.* small shrill flute used in military music. □ **fifer** *n.* [German *Pfeife* PIPE or French *fifre*]

fifteen /fɪf'ti:n/ *adj.* & *n.* **1** one more than fourteen. **2** symbol for this (15, xv, XV). **3** size etc. denoted by fifteen. **4** team of fifteen players, esp. in Rugby. **5** (**15**) (of a film) for persons of 15 and over. □ **fifteenth** *adj.* & *n.* [Old English: related to FIVE, -TEEN]

fifth *adj.* & *n.* **1** next after fourth. **2** any of five equal parts of a thing. **3** *Mus.* interval or chord spanning five consecutive notes in a diatonic scale (e.g. C to G). □ **fifthly** *adv.* [Old English: related to FIVE]

fifth column *n.* traitorous group within a country at war etc. □ **fifth-columnist** *n.*

fifty /'fɪftɪ/ *adj.* & *n.* (*pl.* **-ies**) **1** five times ten. **2** symbol for this (50, l, L). **3** (in *pl.*) numbers from 50 to 59, esp. the years of a century or of a person's life. □ **fiftieth** *adj.* & *n.* [Old English]

fifty-fifty *–adj.* equal. *–adv.* equally, half and half.

fig[1] *n.* **1** soft pulpy fruit with many seeds. **2** (in full **fig-tree**) tree bearing figs. □ **not care** (or **give**) **a fig** not care at all. [Latin *ficus*]

fig[2] *n.* **1** dress or equipment (*in full fig*). **2** condition or form (*in good fig*). [obsolete *feague*: related to FAKE]

fig. *abbr.* figure.

fight /faɪt/ *–v.* (*past* and *past part.* **fought** /fɔ:t/) **1** (often foll. by *against*, *with*) contend or contend with in war, battle, single combat, etc. **2** engage in (a battle, duel, etc.). **3** contend (an election); maintain (a lawsuit, cause, etc.) against an opponent. **4** strive to achieve something or to overcome (disease, fire, etc.). **5** make (one's way) by fighting. *–n.* **1 a** combat. **b** boxing-match. **c** battle. **2** conflict, struggle, or effort. **3** power or inclination to fight (*no fight left*). □ **fight back 1** counter-attack. **2** suppress (tears etc.). **fight for 1** fight on behalf of. **2** fight to secure. **fight a losing battle** struggle without hope of success. **fight off** repel with effort. **fight out** (usu. **fight it out**) settle by fighting. **fight shy of** avoid. **put up a fight** offer resistance. [Old English]

fighter *n.* **1** person or animal that fights. **2** fast military aircraft designed for attacking other aircraft.

fighting chance *n.* slight chance of success if an effort is made.

fighting fit *n.* fit and ready; at the peak of fitness.

fig-leaf *n.* **1** leaf of a fig-tree. **2** concealing device, esp. for the genitals (Gen. 3:7).

figment /'fɪgmənt/ *n.* invented or imaginary thing. [Latin: related to FEIGN]

figuration /,fɪgjʊ'reɪʃ(ə)n/ *n.* **1 a** act or mode of formation; form. **b** shape or outline. **2** ornamentation. [Latin: related to FIGURE]

figurative /'fɪgərətɪv/ *adj.* **1** metaphorical, not literal. **2** characterized by figures of speech. **3** of pictorial or sculptural representation. □ **figuratively** *adv.* [Latin: related to FIGURE]

figure /'fɪgə(r)/ *–n.* **1** external form or bodily shape. **2 a** silhouette, human form (*figure on the lawn*). **b** person of a specified kind or appearance (*public figure*; *cut a poor figure*). **3 a** human form in drawing, sculpture, etc. **b** image or likeness. **4** two- or three-dimensional space enclosed by lines or surface(s), e.g. a triangle or sphere. **5 a** numerical symbol or number, esp. 0–9. **b** amount; estimated value (*cannot put a figure on it*). **c** (in *pl.*) arithmetical calculations. **6** diagram or illustration. **7** decorative pattern. **8** movement or sequence in a set dance etc. **9** *Mus.* succession of notes from which longer passages are developed. **10** (in full **figure of speech**) metaphor, hyperbole, etc. *–v.* (**-ring**) **1** appear or be mentioned, esp. prominently. **2** represent pictorially. **3** imagine; picture mentally. **4** embellish with a pattern etc. (*figured satin*). **5** calculate; do arithmetic. **6** symbolize. **7** esp. *US* **a** understand, consider. **b** *colloq.* make sense; be likely (*that figures*). □ **figure on** *US* count on, expect. **figure out** work out by arithmetic or logic. [Latin *figura*: related to FEIGN]

figured bass *n. Mus.* = CONTINUO.

figurehead *n.* **1** nominal leader. **2** wooden bust or figure at a ship's prow.

figure-skating *n.* skating in prescribed patterns. □ **figure-skater** *n.*

figurine /,fɪgjə'ri:n/ *n.* statuette. [Italian: related to FIGURE]

filament /'fɪləmənt/ *n.* **1** threadlike body or fibre. **2** conducting wire or thread in an electric bulb etc. □ **filamentous** /-'mentəs/ *adj.* [Latin *filum* thread]

filbert /'fɪlbət/ *n.* **1** the cultivated hazel with edible nuts. **2** this nut. [Anglo-French, because ripe about St *Philibert's* day]

filch *v.* pilfer, steal. [origin unknown]

file[1] *–n.* **1** folder, box, etc., for holding loose papers. **2** papers kept in this. **3** *Computing* collection of (usu. related) data stored under one name. **4** line of people or things one behind another. *–v.* (**-ling**) **1** place (papers) in a file or among (esp. public) records. **2** submit (a petition for divorce, a patent application, etc.). **3** (of a reporter) send (copy) to a newspaper. **4** walk in a line. [Latin *filum* thread]

file[2] *–n.* tool with a roughened surface for smoothing or shaping wood, metal, fingernails, etc. *–v.* (**-ling**) smooth or shape with a file. [Old English]

filial /'fɪlɪəl/ *adj.* of or due from a son or daughter. □ **filially** *adv.* [Latin *filius, -a* son, daughter]

filibuster /'fɪlɪ,bʌstə(r)/ *–n.* **1** obstruction of progress in a legislative assembly, esp. by prolonged speaking. **2** esp. *US* person who engages in this. *–v.* act as a filibuster (against). □ **filibusterer** *n.* [Dutch: related to FREEBOOTER]

filigree /'fɪlɪ,gri:/ *n.* **1** fine ornamental work in gold etc. wire. **2** similar delicate work. □ **filigreed** *adj.* [Latin *filum* thread, *granum* seed]

filing *n.* (usu. in *pl.*) particle rubbed off by a file.

filing cabinet *n.* cabinet with drawers for storing files.

Filipino /,fɪlɪ'pi:nəʊ/ *–n.* (*pl.* **-s**) native or national of the Philippines. *–adj.* of the Philippines or Filipinos. [Spanish, = Philippine]

fill *–v.* **1** (often foll. by *with*) make or become full. **2** occupy completely; spread over or through. **3** block up (a cavity in a tooth); drill and put a filling into (a decayed tooth). **4** appoint a person to hold or (of a person) hold (a post). **5** hold (an office). **6** carry out or supply (an order, commission, etc.). **7** occupy (vacant time). **8** (of a sail) be distended by wind. **9** (usu. as **filling** *adj.*) (esp. of food) satisfy, satiate. *–n.* **1** as much as one wants or can bear (*eat your fill*). **2** enough to fill something. □ **fill the bill** be suitable or adequate. **fill in 1** complete (a form, document, etc.). **2 a** complete (a drawing etc.) within an outline. **b** fill (an outline) in this way. **3** fill (a hole etc.) completely. **4** (often foll. by *for*) act as a substitute. **5** occupy oneself during (spare time). **6** *colloq.* inform (a person) more fully. **7** *slang* thrash, beat. **fill out 1** enlarge to the required size. **2** become enlarged or plump. **3** *US* fill in (a document etc.). **fill up 1** make or become completely full. **2** fill in (a document etc.). **3** fill the petrol tank of (a car etc.). [Old English]

filler *n.* **1** material used to fill a cavity or increase bulk. **2** small item filling space in a newspaper etc.

fillet /'fɪlɪt/ *–n.* **1 a** boneless piece of meat or fish. **b** (in full **fillet steak**) undercut of a sirloin. **2** ribbon etc. binding the hair. **3** thin narrow strip or ridge. **4** narrow flat band between mouldings. *–v.* (**-t-**) **1** remove bones from (fish or meat) or divide into fillets. **2** bind or provide with fillet(s). [Latin *filum* thread]

filling *n.* material that fills a tooth, sandwich, pie, etc.

filling-station *n.* garage selling petrol etc.

fillip /'fɪlɪp/ *–n.* **1** stimulus, incentive. **2** flick with a finger or thumb. *–v.* (**-p-**) **1** stimulate. **2** flick. [imitative]

filly /'fɪlɪ/ *n.* (*pl.* **-ies**) **1** young female horse. **2** *colloq.* girl or young woman. [Old Norse]

film *–n.* **1** thin coating or covering layer. **2** strip or sheet of plastic etc. coated with light-sensitive emulsion for exposure in a camera. **3 a** story, episode, etc., on film, with the illusion of movement. **b** (in *pl.*) the cinema industry. **4** slight veil or haze etc. **5** dimness or morbid growth affecting the eyes. *–v.* **1** make a photographic film of (a scene, story, etc.). **2** cover or become covered with or as with a film. [Old English]

film-goer *n.* person who frequents the cinema.

filmsetting *n.* typesetting by projecting characters on to photographic film. □ **film-set** *v.* **film-setter** *n.*

film star *n.* celebrated film actor or actress.

film-strip *n.* series of transparencies in a strip for projection.

filmy *adj.* (**-ier, -iest**) **1** thin and translucent. **2** covered with or as with a film.

Filofax /'faɪləʊ,fæks/ *n. propr.* a type of loose-leaf personal organizer. [from FILE[1], FACT]

filo pastry /'fi:ləʊ, 'faɪ-/ *n.* (also **phyllo pastry**) leaved pastry like strudel pastry. [Greek *phullon* leaf]

filter *–n.* **1** porous device for removing impurities etc. from a liquid or gas passed through it. **2** = FILTER TIP. **3** screen or attachment for absorbing or modifying light, X-rays, etc. **4** device for suppressing unwanted electrical or sound waves. **5** arrangement for filtering traffic. *–v.* **1** (cause to) pass through a filter. **2** (foll. by *through, into,* etc.) make way gradually. **3** (foll. by *out*) (cause to) leak. **4** allow (traffic) or (of traffic) be allowed to pass to the left or right at a junction. [Germanic: related to FELT[1]]

filter-paper *n.* porous paper for filtering.

filter tip *n.* **1** filter on a cigarette removing some impurities. **2** cigarette with this. □ **filter-tipped** *adj.*

filth *n.* **1** repugnant or extreme dirt. **2** obscenity. [Old English: related to FOUL]

filthy *–adj.* (**-ier, -iest**) **1** extremely or disgustingly dirty. **2** obscene. **3** *colloq.* (of weather) very unpleasant. *–adv.* **1** filthily (*filthy dirty*). **2** *colloq.* extremely (*filthy rich*). □ **filthily** *adv.* **filthiness** *n.*

filthy lucre *n.* **1** dishonourable gain. **2** *joc.* money.

filtrate /'fɪltreɪt/ *–v.* (**-ting**) filter. *–n.* filtered liquid. □ **filtration** /-'treɪʃ(ə)n/ *n.* [related to FILTER]

fin *n.* **1** (usu. thin) flat external organ of esp. fish, for propelling, steering, etc. (*dorsal fin*). **2** similar stabilizing projection on an aircraft, car, etc. **3** underwater swimmer's flipper. □ **finned** *adj.* [Old English]

finagle /fɪ'neɪg(ə)l/ *v.* (**-ling**) *colloq.* act or obtain dishonestly. □ **finagler** *n.* [dial. *fainaigue* cheat]

final /'faɪn(ə)l/ *–adj.* **1** situated at the end, coming last. **2** conclusive, decisive. *–n.* **1** last or deciding heat or game in sports etc. (*Cup Final*). **2** last daily edition of a newspaper. **3** (usu. in *pl.*) examinations at the end of a degree course. □ **finally** *adv.* [Latin *finis* end]

final cause *n. Philos.* ultimate purpose.

final clause *n. Gram.* clause expressing purpose.

finale /fɪ'nɑːlɪ/ *n.* last movement or section of a piece of music or drama etc. [Italian: related to FINAL]

finalist *n.* competitor in the final of a competition etc.

finality /faɪ'nælɪtɪ/ *n.* (*pl.* **-ies**) **1** fact of being final. **2** final act etc. [Latin: related to FINAL]

finalize *v.* (also **-ise**) (**-zing** or **-sing**) put into final form; complete. □ **finalization** /-'zeɪʃ(ə)n/ *n.*

final solution *n.* Nazi policy (1941–5) of exterminating European Jews.

finance /'faɪnæns/ *–n.* **1** management of (esp. public) money. **2** monetary support for an enterprise. **3** (in *pl.*) money resources of a State, company, or person. *–v.* (**-cing**) provide capital for. [French: related to FINE[2]]

finance company *n.* (also **finance house**) company providing money, esp. for hire-purchase transactions.

financial /faɪ'nænʃ(ə)l/ *adj.* of finance. □ **financially** *adv.*

financial year *n.* year as reckoned for taxing or accounting, esp. from 6 April.

financier /faɪ'nænsɪə(r)/ *n.* capitalist; entrepreneur. [French: related to FINANCE]

finch *n.* small seed-eating bird, esp. a crossbill, canary, or chaffinch. [Old English]

find /faɪnd/ *–v.* (*past* and *past part.* **found**) **1 a** discover or get by chance or effort (*found a key*). **b** become aware of. **2 a** obtain, succeed in obtaining; receive (*idea found acceptance*). **b** summon up (*found courage*). **3** seek out and provide or supply (*will find you a book*; *finds his own meals*). **4** discover by study etc. (*find the answer*). **5 a** perceive or experience (*find no sense in it*). **b** (often in *passive*) discover to be present (*not found in Shakespeare*). **c** discover from experience (*finds England too cold*). **6** *Law* (of a jury, judge, etc.) decide and declare (*found him guilty*). **7** reach by a natural process (*water finds its own level*). *–n.* **1** discovery of treasure etc. **2** valued thing or person newly discovered. □ **all found** (of wages) with board and lodging provided free. **find fault** see FAULT. **find favour** prove acceptable. **find one's feet 1** become able to walk. **2** develop independence. **find oneself 1** discover that one is (*found herself agreeing*). **2** discover one's vocation. **find out 1** discover or detect (a wrongdoer etc.). **2** (often foll. by *about*) get information. **3** discover (*find out where we are*). **4** (often foll. by *about*) discover the truth, a fact, etc. (*he never found out*). [Old English]

finder *n.* **1** person who finds. **2** small telescope attached to a large one to locate an object. **3** viewfinder.

finding *n.* (often in *pl.*) conclusion reached by an inquiry etc.

fine[1] *–adj.* **1 a** of high quality; excellent (*fine painting*). **b** good, satisfactory (*that will be fine*). **2 a** pure, refined. **b** (of gold or silver) containing a specified proportion of pure metal. **3** imposing, dignified (*fine buildings*). **4** in good health (*I'm fine*). **5** (of weather etc.) bright and clear. **6 a** thin; sharp. **b** in small particles. **c** worked in slender thread. **7** euphemistic; flattering (*fine words*). **8** ornate, showy. **9** fastidious, affectedly refined. *–adv.* **1** finely. **2** *colloq.* very well (*suits me fine*). *–v.* (**-ning**) **1** (often foll. by *away*, *down*, *off*) make or become finer, thinner, more tapering, or less coarse. **2** (often foll. by *down*) make or become clear (esp. of beer etc.). □ **not to put too fine a point on it** to speak bluntly. □ **finely** *adv.* **fineness** *n.* [French *fin* from Latin *finio* FINISH]

fine[2] *–n.* money to be paid as a penalty. *–v.* (**-ning**) punish by a fine (*fined him £5*). □ **in fine** in short. [French *fin* settlement of a dispute, from Latin *finis* end]

fine arts *n.pl.* poetry, music, and the visual arts, esp. painting, sculpture, and architecture.

finery /'faɪnərɪ/ *n.* showy dress or decoration. [from FINE[1]]

fines herbes /fiːnz 'eəb/ *n.pl.* mixed herbs used in cooking. [French, = fine herbs]

fine-spun *adj.* **1** delicate. **2** (of theory etc.) too subtle, unpractical.

finesse /fɪ'nes/ *–n.* **1** refinement. **2** subtle manipulation. **3** artfulness; tact. **4** *Cards* attempt to win a trick with a card that is not the highest held. –*v.* (**-ssing**) **1** use or achieve by finesse. **2** *Cards* **a** make a finesse. **b** play (a card) as a finesse. [French: related to FINE[1]]

fine-tooth comb *n.* comb with close-set teeth. □ **go over with a fine-tooth comb** check or search thoroughly.

fine-tune *v.* make small adjustments to (a mechanism etc.).

finger /'fɪŋgə(r)/ –*n.* **1** any of the terminal projections of the hand (usu. excluding the thumb). **2** part of a glove etc. for a finger. **3** finger-like object or structure (*fish finger*). **4** *colloq.* small measure of liquor. –*v.* touch, feel, or turn about with the fingers. □ **get** (or **pull**) **one's finger out** *slang* start to act. **lay a finger on** touch, however slightly. **put one's finger on** locate or identify exactly. □ **fingerless** *adj.* [Old English]

finger-board *n.* part of the neck of a stringed instrument on which the fingers press to vary the pitch.

finger-bowl *n.* (also **finger-glass**) small bowl for rinsing the fingers during a meal.

finger-dry *v.* dry and style (the hair) by running one's fingers through it.

fingering *n.* **1** technique etc. of using the fingers, esp. in playing music. **2** indication of this in a musical score.

finger-mark *n.* mark left by a finger.

fingernail *n.* nail of each finger.

finger-plate *n.* plate fixed to a door to prevent finger-marks.

fingerprint –*n.* impression of a finger-tip on a surface, used in detecting crime. –*v.* record the fingerprints of.

finger-stall *n.* protective cover for an injured finger.

fingertip *n.* tip of a finger. □ **have at one's fingertips** be thoroughly familiar with (a subject etc.).

finial /'fɪnɪəl/ *n.* ornamental top or end of a roof, gable, etc. [Anglo-French: related to FINE[1]]

finicky /'fɪnɪkɪ/ *adj.* (also **finical**, **finicking**) **1** over-particular, fastidious. **2** detailed; fiddly. □ **finickiness** *n.* [perhaps from FINE[1]]

finis /'fɪnɪs/ *n.* end, esp. of a book. [Latin]

finish /'fɪnɪʃ/ –*v.* **1 a** (often foll. by *off*) bring or come to an end or the end of; complete; cease. **b** (usu. foll. by *off*) *colloq.* kill; vanquish. **c** (often foll. by *off*, *up*) consume or complete consuming (food or drink). **2** treat the surface of (cloth, woodwork, etc.). –*n.* **1 a** end, last stage, completion. **b** point at which a race etc. ends. **2** method, material, etc. used for surface treatment of wood, cloth, etc. (*mahogany finish*). □ **finish up** (often foll. by *in*, *by*) end (*finished up by crying*). **finish with** have no more to do with, complete using etc. [Latin *finis* end]

finishing-school *n.* private college preparing girls for fashionable society.

finishing touch *n.* (also **finishing touches**) final enhancing details.

finite /'faɪnaɪt/ *adj.* **1** limited; not infinite. **2** (of a part of a verb) having a specific number and person. [Latin: related to FINISH]

Finn *n.* native or national of Finland; person of Finnish descent. [Old English]

finnan /'fɪnən/ *n.* (in full **finnan haddock**) smoke-cured haddock. [*Findhorn, Findon,* in Scotland]

Finnic *adj.* of the group of peoples or languages related to the Finns or Finnish.

Finnish –*adj.* of the Finns or their language. –*n.* language of the Finns.

fino /'fiːnəʊ/ *n.* (*pl.* **-s**) light-coloured dry sherry. [Spanish, = fine]

fiord /fjɔːd/ *n.* (also **fjord**) long narrow sea inlet, as in Norway. [Norwegian]

fipple /'fɪp(ə)l/ *n.* plug at the mouth-end of a wind instrument. [origin unknown]

fipple flute *n.* flute played by blowing endwise, e.g. a recorder.

fir *n.* **1** (in full **fir-tree**) evergreen coniferous tree with needles growing singly on the stems. **2** its wood. □ **firry** *adj.* [Old Norse]

fir-cone *n.* fruit of the fir.

fire –*n.* **1 a** combustion of substances with oxygen, giving out light and heat. **b** flame or incandescence. **2** destructive burning (*forest fire*). **3 a** burning fuel in a grate, furnace, etc. **b** = ELECTRIC FIRE. **c** = GAS FIRE. **4** firing of guns. **5 a** fervour, spirit, vivacity. **b** poetic inspiration. **6** burning heat, fever. –*v.* (**-ring**) **1** (often foll. by *at*, *into*, *on*) **a** shoot (a gun, missile, etc.). **b** shoot a gun or missile etc. **2** produce (a broadside, salute, etc.) by shooting guns etc. **3** (of a gun etc.) be discharged. **4** explode or kindle (an explosive). **5** deliver or utter rapidly (*fired insults at us*). **6** *slang* dismiss (an employee). **7** set fire to intentionally. **8** catch fire. **9** (of esp. an internal-combustion engine) undergo ignition. **10** supply (a furnace, engine, etc.) with fuel. **11** stimulate; enthuse. **12** bake, dry, or cure (pottery, bricks, tea, tobacco, etc.). **13** become or cause to become heated, excited, red, or glowing. □ **catch fire** begin to burn. **fire away** *colloq.* begin; go ahead. **on fire 1** burning. **2** excited. **set fire to** (or **set on fire**) ignite, kindle. **set the world** (or **Thames**) **on fire** do something remarkable or sensational. **under fire 1** being shot at. **2** being rigorously criticized or questioned. [Old English]

fire-alarm *n.* device warning of fire.

fire and brimstone *n.* supposed torments of hell.

firearm *n.* (usu. in *pl.*) gun, pistol, or rifle.

fire-ball *n.* **1** large meteor. **2** ball of flame or lightning. **3** energetic person.

fire-bomb *n.* incendiary bomb.

firebox *n.* place where fuel is burned in a steam engine or boiler.

firebrand *n.* **1** piece of burning wood. **2** person causing trouble or unrest.

fire-break *n.* obstacle to the spread of fire in a forest etc., esp. an open space.

fire-brick *n.* fireproof brick in a grate.

fire brigade *n.* body of professional firefighters.

fireclay *n.* clay used to make fire-bricks.

firecracker *n. US* explosive firework.

firedamp *n.* miners' name for methane, which is explosive when mixed with air.

firedog *n.* andiron.

fire door *n.* fire-resistant door preventing the spread of fire.

fire-drill *n.* rehearsal of the procedures to be used in case of fire.

fire-eater *n.* **1** conjuror who appears to swallow fire. **2** quarrelsome person.

fire-engine *n.* vehicle carrying hoses, firefighters, etc.

fire-escape *n.* emergency staircase etc. for use in a fire.

fire extinguisher *n.* apparatus discharging foam etc. to extinguish a fire.

firefighter *n.* = FIREMAN 1.

firefly *n.* beetle emitting phosphorescent light, e.g. the glow-worm.

fire-guard *n.* protective screen placed in front of a fireplace.

fire-irons *n.pl.* tongs, poker, and shovel for a domestic fire.

firelight *n.* light from a fire in a fireplace.

fire-lighter *n.* inflammable material used to start a fire in a grate.

fireman *n.* **1** member of a fire brigade. **2** person who tends a steam engine or steamship furnace.

fireplace *n.* **1** place for a domestic fire, esp. a recess in a wall. **2** structure surrounding this.

fire-power *n.* destructive capacity of guns etc.

fire-practice *n.* fire-drill.

fireproof *–adj.* able to resist fire or great heat. *–v.* make fireproof.

fire-raiser *n.* arsonist. □ **fire-raising** *n.*

fire-screen *n.* **1** ornamental screen for a fireplace. **2** screen against the direct heat of a fire. **3** fire-guard.

fire-ship *n. hist.* ship set on fire and directed against an enemy's ships etc.

fireside *n.* **1** area round a fireplace. **2** home or home-life.

fire station *n.* headquarters of a fire brigade.

fire-storm *n.* high wind or storm following a fire caused by bombs.

fire-trap *n.* building without fire-escapes etc.

fire-watcher *n.* person keeping watch for fires, esp. those caused by bombs.

fire-water *n. colloq.* strong alcoholic liquor.

firewood *n.* wood as fuel.

firework *n.* **1** device that burns or explodes spectacularly when lit. **2** (in *pl.*) outburst of passion, esp. anger.

firing *n.* **1** discharge of guns. **2** fuel.

firing-line *n.* **1** front line in a battle. **2** centre of activity etc.

firing-squad *n.* **1** soldiers ordered to shoot a condemned person. **2** group firing the salute at a military funeral.

firm[1] *–adj.* **1 a** solid or compact. **b** fixed, stable, steady. **2 a** resolute, determined. **b** steadfast, constant (*firm belief*; *firm friend*). **3** (of an offer etc.) definite; not conditional. *–adv.* firmly (*stand firm*). *–v.* (often foll. by *up*) make or become firm, secure, compact, or solid. □ **firmly** *adv.* **firmness** *n.* [Latin *firmus*]

firm[2] *n.* business concern or its partners. [Latin *firma*: cf. FIRM[1]]

firmament /'fɜːməmənt/ *n. literary* the sky regarded as a vault or arch. [Latin: related to FIRM[1]]

firmware *n. Computing* permanent kind of software.

firry see FIR.

first *–adj.* **1** earliest in time or order (*took the first bus*). **2** foremost in rank or importance (*First Lord of the Treasury*). **3** most willing or likely (*the first to admit it*). **4** basic or evident (*first principles*). *–n.* **1** (prec. by *the*) person or thing first mentioned or occurring. **2** first occurrence of something notable. **3** place in the first class in an examination. **4** first gear. **5 a** first place in a race. **b** winner of this. *–adv.* **1** before any other person or thing (*first of all*; *first and foremost*). **2** before someone or something else (*get this done first*). **3** for the first time (*when did you first see her?*). **4** in preference; rather (*will see him damned first*). □ **at first** at the beginning. **at first hand** directly from the original source. **first past the post** (of an electoral system) selecting a candidate or party by simple majority. **from the first** from the beginning. **in the first place** as first consideration. [Old English]

first aid *n.* emergency medical treatment.

first-born *–adj.* eldest. *–n.* person's eldest child.

first class *–n.* **1** best group or category. **2** best accommodation in a train, ship, etc. **3** mail given priority. **4** highest division in an examination. *–adj.* & *adv.* (**first-class**) **1** of or by the first class. **2** excellent.

first cousin see COUSIN.

first-day cover *n.* envelope with stamps postmarked on their first day of issue.

first-degree *adj.* denoting non-serious surface burns.

first finger *n.* finger next to the thumb.

first floor *n.* (*US* **second floor**) floor above the ground floor.

first-foot *Scot.* *–n.* first person to cross a threshold in the New Year. *–v.* be a first-foot.

first-fruit *n.* (usu. in *pl.*) **1** first agricultural produce of a season, esp. as offered to God. **2** first results of work etc.

firsthand *adj.* & *adv.* from the original source; direct.

First Lady *n.* (in the US) wife of the President.

first light *n.* dawn.

firstly *adv.* in the first place, first (cf. FIRST *adv.*).

first mate *n.* (on a merchant ship) second in command.

first name *n.* personal or Christian name.

first night *n.* first public performance of a play etc.

first offender *n.* criminal without previous convictions.

first officer *n.* = FIRST MATE.

first person see PERSON.

first post *n.* (also **last post**) bugle-call as a signal to retire for the night.

first-rate *adj.* **1** excellent. **2** *colloq.* very well (*feeling first-rate*).

first thing *adv. colloq.* before anything else; very early.

firth *n.* (also **frith**) **1** narrow inlet of sea. **2** estuary. [Old Norse: related to FIORD]

fiscal /'fɪsk(ə)l/ *–adj.* of public revenue. *–n.* **1** legal official in some countries. **2** *Scot.* = PROCURATOR FISCAL. [Latin *fiscus* treasury]

fiscal year *n.* = FINANCIAL YEAR.

fish[1] *–n.* (*pl.* same or **-es**) **1** vertebrate cold-blooded animal with gills and fins living wholly in water. **2** any of various non-vertebrate animals living wholly in water, e.g. the cuttlefish, shellfish, and jellyfish. **3** fish as food. **4** *colloq.* person of a specified, usu. unpleasant, kind (*an odd fish*). **5** (**the Fish** or **Fishes**) sign or constellation Pisces. *–v.* **1** try to catch fish. **2** fish in (a certain river, pool, etc.). **3** (foll. by *for*) **a** search for. **b** seek indirectly (*fishing for compliments*). **4** (foll. by *up*, *out*, etc.) retrieve with effort. □ **drink like a fish** drink alcohol excessively. **fish out of water** person out of his or her element. **other fish to fry** other matters to attend to. [Old English]

fish[2] *n.* flat or curved plate of iron, wood, etc., used to strengthen a beam, joint, or mast. [French *ficher* fix, from Latin *figere* FIX]

fish-bowl *n.* (usu. round) glass bowl for pet fish.

fish cake *n.* breaded cake of fish and mashed potato, usu. fried.

fisher *n.* **1** animal that catches fish. **2** *archaic* fisherman. [Old English]

fisherman *n.* man who catches fish as a livelihood or for sport.

fishery *n.* (*pl.* **-ies**) **1** place where fish are caught or reared. **2** industry of fishing or breeding fish.

fish-eye lens *n.* very wide-angle lens with a highly-curved front.

fish farm *n.* place where fish are bred for food.

fish finger *n.* small oblong piece of fish in batter or breadcrumbs.

fish-hook *n.* barbed hook for catching fish.

fishing *n.* catching fish.

fishing-line *n.* thread with a baited hook etc. for catching fish.

fishing-rod *n.* tapering usu. jointed rod for fishing.

fish-kettle *n.* oval pan for boiling fish.

fish-knife *n.* knife for eating or serving fish.

fish-meal *n.* ground dried fish as fertilizer or animal feed.

fishmonger *n.* dealer in fish.

fishnet *n.* (often *attrib.*) open-meshed fabric (*fishnet stockings*).

fish-plate *n.* flat piece of iron etc. connecting railway rails or positioning masonry.

fish-slice *n.* flat slotted cooking utensil.

fishtail *n.* device etc. shaped like a fish's tail.

fishwife *n.* **1** coarse-mannered or noisy woman. **2** woman who sells fish.

fishy *adj.* (**-ier**, **-iest**) **1** of or like fish. **2** *slang* dubious, suspect. □ **fishily** *adv.* **fishiness** *n.*

fissile /'fɪsaɪl/ *adj.* **1** capable of undergoing nuclear fission. **2** tending to split. [Latin: related to FISSURE]

fission /'fɪʃ(ə)n/ *–n.* **1** splitting of a heavy atomic nucleus, with a release of energy. **2** cell division as a mode of reproduction. *–v.* (cause to) undergo fission. □ **fissionable** *adj.* [Latin: related to FISSURE]

fission bomb *n.* atomic bomb.

fissure /ˈfɪʃə(r)/ *–n.* crack or split, usu. long and narrow. *–v.* (**-ring**) split, crack. [Latin *findo fiss-* cleave]

fist *n.* tightly closed hand. □ **fistful** *n.* (*pl.* -s). [Old English]

fisticuffs /ˈfɪstɪˌkʌfs/ *n.pl.* fighting with the fists. [probably from obsolete *fisty* (from FIST), CUFF[2]]

fistula /ˈfɪstjʊlə/ *n.* (*pl.* **-s** or **-lae** /-ˌliː/) abnormal or artificial passage between an organ and the body surface or between two organs. □ **fistular** *adj.* **fistulous** *adj.* [Latin, = pipe]

fit[1] *–adj.* (**fitter, fittest**) **1 a** well suited. **b** qualified, competent, worthy. **c** in suitable condition, ready. **d** (foll. by *for*) good enough (*fit for a king*). **2** in good health or condition. **3** proper, becoming, right (*it is fit that*). *–v.* (**-tt-**) **1 a** (also *absol.*) be of the right shape and size for (*dress fits her*; *key doesn't fit*). **b** (often foll. by *in, into*) be correctly positioned (*that bit fits here*). **c** find room for (*fit another on here*). **2** make suitable or competent; adapt (*fitted for battle*). **3** (usu. foll. by *with*) supply. **4** fix in place (*fit a lock on the door*). **5** = *fit on*. **6** befit, become (*it fits the occasion*). *–n.* way in which a garment, component, etc., fits (*tight fit*). *–adv.* (foll. by *to* + infin.) *colloq.* so that; likely (*laughing fit to bust*). □ **fit the bill** = *fill the bill*. **fit in 1** (often foll. by *with*) be compatible; accommodate (*tried to fit in with their plans*). **2** find space or time for (*dentist fitted me in*). **fit on** try on (a garment). **fit out** (or **up**) (often foll. by *with*) equip. **see** (or **think**) **fit** (often foll. by *to* + infin.) decide or choose (a specified action). □ **fitly** *adv.* **fitness** *n.* [origin unknown]

fit[2] *n.* **1** sudden esp. epileptic seizure with unconsciousness or convulsions. **2** sudden brief bout or burst (*fit of giggles; fit of coughing*). □ **by** (or **in**) **fits and starts** spasmodically. **have a fit** *colloq.* be greatly surprised or outraged. **in fits** laughing uncontrollably. [Old English]

fitful *adj.* spasmodic or intermittent. □ **fitfully** *adv.*

fitment *n.* (usu. in *pl.*) fixed item of furniture.

fitted *adj.* **1** made to fit closely or exactly (*fitted carpet*). **2** provided with built-in fittings etc. (*fitted kitchen*). **3** built-in (*fitted cupboards*).

fitter *n.* **1** mechanic who fits together and adjusts machinery. **2** supervisor of the cutting, fitting, etc. of garments.

fitting *–n.* **1** trying-on of a garment etc. for adjustment before completion. **2** (in *pl.*) fixtures and fitments of a building. *–adj.* proper, becoming, right. □ **fittingly** *adv.*

five *adj.* & *n.* **1** one more than four. **2** symbol for this (5, v, V). **3** size etc. denoted by five. **4** set or team of five. **5** five o'clock (*is it five yet?*). **6** *Cricket* hit scoring five runs. [Old English]

fivefold *adj.* & *adv.* **1** five times as much or as many. **2** consisting of five parts.

five o'clock shadow *n.* beard-growth visible in the latter part of the day.

fiver *n. colloq.* five-pound note.

fives *n.* game in which a ball is hit with a gloved hand or bat against the walls of a court.

five-star *adj.* of the highest class.

fivestones *n.* jacks played with five pieces of metal etc. and usu. no ball.

fix *–v.* **1** make firm or stable; fasten, secure. **2** decide, settle, specify (a price, date, etc.). **3** mend, repair. **4** implant in the mind. **5 a** (foll. by *on, upon*) direct (the eyes etc.) steadily, set. **b** attract and hold (the attention, eyes, etc.). **c** (foll. by *with*) single out with one's look etc. **6** place definitely, establish. **7** determine the exact nature, position, etc., of; refer (a thing) to a definite place or time; identify, locate. **8 a** make (the eyes, features, etc.) rigid. **b** (of eyes, features, etc.) become rigid. **9** *US colloq.* prepare (food or drink). **10** congeal or become congealed. **11** *colloq.* punish, kill, deal with (a person). **12** *colloq.* **a** bribe or threaten into supporting. **b** gain a fraudulent result of (a race etc.). **13** *slang* inject a narcotic. **14** make (a colour, photographic image, etc.) fast or permanent. **15** (of a plant etc.) assimilate (nitrogen or carbon dioxide). *–n.* **1** *colloq.* dilemma, predicament. **2 a** finding one's position by bearings etc. **b** position found in this way. **3** *slang* dose of an addictive drug. □ **be fixed** (usu. foll. by *for*) *colloq.* be situated (regarding) (*how is he fixed for money?*). **fix on** (or **upon**) choose, decide on. **fix up 1** arrange, organize. **2** accommodate. **3** (often foll. by *with*) provide (a person) (*fixed me up with a job*). □ **fixable** *adj.* [Latin *figo fix-*]

fixate /fɪkˈseɪt/ *v.* (**-ting**) **1** direct one's gaze on. **2** *Psychol.* (usu. in *passive*; often foll. by *on, upon*) cause (a person) to become abnormally attached to a person or thing. [Latin: related to FIX]

fixation /fɪkˈseɪʃ(ə)n/ *n.* **1** state of being fixated. **2** obsession, monomania. **3** coagulation. **4** process of assimilating a gas to form a solid compound.

fixative /ˈfɪksətɪv/ *–adj.* tending to fix or secure. *–n.* fixative substance.

fixedly /ˈfɪksɪdlɪ/ *adv.* intently.

fixed star *n. Astron.* seemingly motionless star.

fixer *n.* **1** person or thing that fixes. **2** *Photog.* substance for fixing a photographic image etc. **3** *colloq.* person who makes esp. illicit deals.

fixings *n.pl. US* **1** apparatus or equipment. **2** trimmings for a dish, dress, etc.

fixity *n.* fixed state; stability; permanence.

fixture /ˈfɪkstʃə(r)/ *n.* **1 a** something fixed in position. **b** *colloq.* seemingly immovable person or thing (*seems to be a fixture*). **2 a** sporting event, esp. a match, race, etc. **b** date agreed for this. **3** (in *pl.*) articles attached to a house or land and regarded as legally part of it.

fizz –*v.* **1** make a hissing or spluttering sound. **2** (of a drink) effervesce. –*n.* **1** effervescence. **2** *colloq.* effervescent drink, esp. champagne. [imitative]

fizzle /ˈfɪz(ə)l/ –*v.* (**-ling**) make a feeble hiss. –*n.* such a sound. □ **fizzle out** end feebly. [imitative]

fizzy *adj.* (**-ier, -iest**) effervescent. □ **fizziness** *n.*

fjord var. of FIORD.

fl. *abbr.* **1** floruit. **2** fluid.

flab *n. colloq.* fat; flabbiness. [imitative, or from FLABBY]

flabbergast /ˈflæbə,gɑːst/ *v.* (esp. as **flabbergasted** *adj.*) *colloq.* astonish; dumbfound. [origin uncertain]

flabby /ˈflæbɪ/ *adj.* (**-ier, -iest**) **1** (of flesh etc.) limp; flaccid. **2** feeble. □ **flabbiness** *n.* [alteration of *flappy*: related to FLAP]

flaccid /ˈflæksɪd/ *adj.* limp, flabby, drooping. □ **flaccidity** /-ˈsɪdɪtɪ/ *n.* [Latin *flaccus* limp]

flag[1] –*n.* **1 a** usu. oblong or square piece of cloth, attachable by one edge to a pole or rope as a country's emblem or standard, a signal, etc. **b** small toy etc. resembling a flag. **2** adjustable strip of metal etc. indicating a taxi's availability for hire. –*v.* (**-gg-**) **1 a** grow tired; lag (*was soon flagging*). **b** hang down; droop. **2** mark out with or as if with a flag or flags. **3** (often foll. by *that*) inform or communicate by flag-signals. □ **flag down** signal to stop. [origin unknown]

flag[2] –*n.* (also **flagstone**) **1** flat usu. rectangular paving stone. **2** (in *pl.*) pavement of these. –*v.* (**-gg-**) pave with flags. [probably Scandinavian]

flag[3] *n.* plant with a bladed leaf (esp. the iris). [origin unknown]

flag-day *n.* fund-raising day for a charity, esp. with the sale of small paper flags etc. in the street.

flagellant /ˈflædʒələnt/ –*n.* person who scourges himself, herself, or others as a religious discipline or as a sexual stimulus. –*adj.* of flagellation. [Latin *flagellum* whip]

flagellate /ˈflædʒə,leɪt/ *v.* (**-ting**) scourge, flog. □ **flagellation** /-ˈleɪʃ(ə)n/ *n.*

flagellum /fləˈdʒeləm/ *n.* (*pl.* **-gella**) **1** long lashlike appendage on some microscopic organisms. **2** runner; creeping shoot. [Latin, = whip]

flageolet /,flædʒəˈlet/ *n.* small flute blown at the end. [French from Provençal]

flag of convenience *n.* foreign flag under which a ship is registered, usu. to avoid regulations or financial charges.

flag-officer *n.* admiral, vice admiral, or rear admiral, or the commodore of a yacht-club.

flag of truce *n.* white flag requesting a truce.

flagon /ˈflægən/ *n.* **1** large bottle, usu. holding a quart (1.13 litres), esp. of wine, cider, etc. **2** large vessel for wine etc., usu. with a handle, spout, and lid. [Latin *flasco* FLASK]

flag-pole *n.* = FLAGSTAFF.

flagrant /ˈfleɪgrənt/ *adj.* blatant; notorious; scandalous. □ **flagrancy** *n.* **flagrantly** *adv.* [Latin *flagro* blaze]

flagship *n.* **1** ship with an admiral on board. **2** leader in a category etc.; exemplar.

flagstaff *n.* pole on which a flag may be hoisted.

flagstone *n.* = FLAG[2].

flag-waving *n.* populist agitation, chauvinism.

flail –*n.* wooden staff with a short heavy stick swinging from it, used for threshing. –*v.* **1** wave or swing wildly. **2** beat with or as with a flail. [Latin *flagellum* whip]

flair *n.* **1** natural talent in a specific area (*flair for languages*). **2** style, finesse. [French *flairer* to smell]

flak *n.* **1** anti-aircraft fire. **2** adverse criticism; abuse. [German, *Fl*iegerabwehr*k*anone, 'aviator-defence-gun']

flake –*n.* **1** small thin light piece of snow etc. **2** thin broad piece peeled or split off. **3** dogfish etc. as food. –*v.* (**-king**) (often foll. by *away*, *off*) **1** take off or come away in flakes. **2** sprinkle with or fall in flakes. □ **flake out** *colloq.* fall asleep or drop from exhaustion; faint. [origin unknown]

flak jacket *n.* protective reinforced military jacket.

flaky *adj.* (**-ier, -iest**) **1** of, like, or in flakes. **2** esp. *US slang* crazy, eccentric.

flaky pastry *n.* crumblier version of puff pastry.

flambé /ˈflɒmbeɪ/ *adj.* (of food) covered with alcohol and set alight briefly (following a noun: *pancakes flambé*). [French: related to FLAME]

flamboyant /flæm'bɔɪənt/ *adj.* **1** ostentatious; showy. **2** floridly decorated or coloured. □ **flamboyance** *n.* **flamboyantly** *adv.* [French: related to FLAMBÉ]

flame –*n.* **1 a** ignited gas. **b** portion of this (*flame flickered; burst into flames*). **2 a** bright light or colouring. **b** brilliant orange-red colour. **3 a** strong passion, esp. love (*fan the flame*). **b** *colloq.* sweetheart. –*v.* (**-ming**) **1** (often foll. by *away, forth, out, up*) burn; blaze. **2** (often foll. by *out, up*) **a** (of passion) break out. **b** (of a person) become angry. **3** shine or glow like flame. [Latin *flamma*]

flamenco /flə'meŋkəʊ/ *n.* (*pl.* **-s**) **1** style of Spanish Gypsy guitar music with singing. **2** dance performed to this. [Spanish, = Flemish]

flame-thrower *n.* weapon for throwing a spray of flame.

flaming *adj.* **1** emitting flames. **2** very hot (*flaming June*). **3** *colloq.* **a** passionate (*flaming row*). **b** expressing annoyance (*that flaming dog*). **4** bright-coloured.

flamingo /flə'mɪŋgəʊ/ *n.* (*pl.* **-s** or **-es**) tall long-necked wading bird with mainly pink plumage. [Provençal: related to FLAME]

flammable /'flæməb(ə)l/ *adj.* inflammable. □ **flammability** /-'bɪlɪtɪ/ *n.* [Latin: related to FLAME]

■ **Usage** *Flammable* is often used because *inflammable* can be mistaken for a negative (the true negative being *non-flammable*).

flan *n.* **1** pastry case with a savoury or sweet filling. **2** sponge base with a sweet topping. [medieval Latin *flado -onis*]

flange /flændʒ/ *n.* projecting flat rim etc., for strengthening or attachment. [origin uncertain]

flank –*n.* **1** side of the body between ribs and hip. **2** side of a mountain, building, etc. **3** right or left side of an army etc. –*v.* (often in *passive*) be at or move along the side of (*road flanked by mountains*). [French from Germanic]

flannel /'flæn(ə)l/ –*n.* **1 a** woven woollen usu. napless fabric. **b** (in *pl.*) flannel garments, esp. trousers. **2** face-cloth, esp. towelling. **3** *slang* nonsense; flattery. –*v.* (**-ll-**; *US* **-l-**) **1** *slang* flatter. **2** wash with a flannel. [Welsh *gwlanen* from *gwlān* wool]

flannelette /,flænə'let/ *n.* napped cotton fabric like flannel.

flap –*v.* (**-pp-**) **1** move or be moved up and down; beat. **2** *colloq.* be agitated or panicky. **3** sway; flutter. **4** (usu. foll. by *away, off*) strike (flies etc.) with flat object; drive. **5** *colloq.* (of ears) listen intently. –*n.* **1** piece of cloth, wood, etc. attached by one side esp. to cover a gap, e.g. a pocket-cover, the folded part of an envelope, a table-leaf. **2** motion of a wing, arm, etc. **3** *colloq.* agitation; panic (*in a flap*). **4** aileron. **5** light blow with something flat. □ **flappy** *adj.* [probably imitative]

flapdoodle /flæp'du:d(ə)l/ *n. colloq.* nonsense. [origin unknown]

flapjack *n.* **1** sweet oatcake. **2** esp. *US* pancake.

flapper *n.* **1** person apt to panic. **2** *slang* (in the 1920s) young unconventional woman.

flare –*v.* (**-ring**) **1** widen gradually (*flared trousers*). **2** (cause to) blaze brightly and unsteadily. **3** burst out, esp. angrily. –*n.* **1 a** dazzling irregular flame or light. **b** sudden outburst of flame. **2** flame or bright light used as a signal or to illuminate a target etc. **3 a** gradual widening, esp. of a skirt or trousers. **b** (in *pl.*) wide-bottomed trousers. □ **flare up** burst into a sudden blaze, anger, activity, etc. [origin unknown]

flare-path *n.* line of lights on a runway to guide aircraft.

flare-up *n.* sudden outburst.

flash –*v.* **1** (cause to) emit a brief or sudden light; (cause to) gleam. **2** send or reflect like a sudden flame (*eyes flashed fire*). **3 a** burst suddenly into view or perception (*answer flashed upon me*). **b** move swiftly (*train flashed past*). **4 a** send (news etc.) by radio, telegraph, etc. **b** signal to (a person) with lights. **5** *colloq.* show ostentatiously (*flashed her ring*). **6** *slang* indecently expose oneself. –*n.* **1** sudden bright light or flame, e.g. of lightning. **2** an instant (*in a flash*). **3** sudden brief feeling, display of wit, etc. (*flash of hope*). **4** = NEWSFLASH. **5** *Photog.* = FLASHLIGHT 1. **6** *Mil.* coloured cloth patch on a uniform. **7** bright patch of colour. –*adj. colloq.* gaudy; showy; vulgar (*flash car*). [imitative]

flashback *n.* scene set in an earlier time than the main action.

flash bulb *n. Photog.* bulb for a flashlight.

flash-cube *n. Photog.* set of four flash bulbs in a cube, operated in turn.

flasher *n.* **1** *slang* man who indecently exposes himself. **2** automatic device for switching lights rapidly on and off.

flash-gun *n.* device operating a camera flashlight.

flashing *n.* (usu. metal) strip used to prevent water penetration at a roof joint etc. [dial.]

flash in the pan *n.* promising start followed by failure.

flash-lamp *n.* portable flashing electric lamp.

flashlight *n.* **1** light giving an intense flash, used for night or indoor photography. **2** *US* electric torch.

flashpoint *n.* **1** temperature at which vapour from oil etc. will ignite in air. **2** point at which anger etc. is expressed.

flashy *adj.* (**-ier, -iest**) showy; gaudy; cheaply attractive. □ **flashily** *adv.* **flashiness** *n.*

flask /flɑːsk/ *n.* **1** narrow-necked bulbous bottle for wine etc. or used in chemistry. **2** = HIP-FLASK. **3** = VACUUM FLASK. [Latin *flasca, flasco*: cf. FLAGON]

flat[1] *–adj.* (**flatter, flattest**) **1 a** horizontally level. **b** even; smooth; unbroken. **c** level and shallow (*flat cap*). **2** unqualified; downright (*flat refusal*). **3 a** dull; lifeless; monotonous (*in a flat tone*). **b** dejected. **4** (of a fizzy drink) having lost its effervescence. **5** (of an accumulator, battery, etc.) having exhausted its charge. **6** *Mus.* **a** below true or normal pitch (*violins are flat*). **b** (of a key) having a flat or flats in the signature. **c** (as **B, E,** etc. **flat**) semitone lower than B, E, etc. **7** (of a tyre) punctured; deflated. *–adv.* **1** at full length; spread out (*lay flat; flat against the wall*). **2** *colloq.* **a** completely, absolutely (*flat broke*). **b** exactly (*in five minutes flat*). **3** *Mus.* below the true or normal pitch (*sings flat*). *–n.* **1** flat part or thing (*flat of the hand*). **2** level ground, esp. a plain or swamp. **3** *Mus.* **a** note lowered a semitone below natural pitch. **b** sign (♭) indicating this. **4** (as **the flat**) flat racing or its season. **5** *Theatr.* flat scenery on a frame. **6** esp. *US colloq.* flat tyre. □ **flat out 1** at top speed. **2** using all one's strength etc. **that's flat** *colloq.* that is definite. □ **flatly** *adv.* **flatness** *n.* **flattish** *adj.* [Old Norse]

flat[2] *n.* set of rooms, usu. on one floor, as a residence. □ **flatlet** *n.* [obsolete *flet* floor, dwelling, from Germanic: related to FLAT[1]]

flat-fish *n.* sole, plaice, etc. with both eyes on one side of a flattened body.

flat foot *n.* foot with a flattened arch.

flat-footed *n.* **1** having flat feet. **2** *colloq.* **a** uninspired. **b** unprepared. **c** resolute.

flat-iron *n. hist.* domestic iron heated on a fire etc.

flatmate *n.* person sharing a flat.

flat race *n.* horse race without jumps, over level ground. □ **flat racing** *n.*

flat rate *n.* unvarying rate or charge.

flat spin *n.* **1** *Aeron.* a nearly horizontal spin. **2** *colloq.* state of panic.

flatten *v.* **1** make or become flat. **2** *colloq.* **a** humiliate. **b** knock down.

flatter *v.* **1** compliment unduly, esp. for gain or advantage. **2** (usu. *refl.*; usu. foll. by *that*) congratulate or delude (oneself etc.) (*he flatters himself that he can sing*). **3** (of colour, style, portrait, painter etc.) enhance the appearance of (*that blouse flatters you*). **4** cause to feel honoured. □ **flatterer** *n.* **flattering** *adj.* **flatteringly** *adv.* [French]

flattery *n.* exaggerated or insincere praise.

flatulent /ˈflætjʊlənt/ *adj.* **1 a** causing intestinal wind. **b** caused by or suffering from this. **2** (of speech etc.) inflated, pretentious. □ **flatulence** *n.* [Latin *flatus* blowing]

flatworm *n.* worm with a flattened body, e.g. flukes.

flaunt *v.* (often *refl.*) display proudly; show off; parade. [origin unknown]

■ **Usage** *Flaunt* is often confused with *flout* which means 'to disobey contemptuously'.

---

flautist /ˈflɔːtɪst/ *n.* flute-player. [Italian: related to FLUTE]

flavour /ˈfleɪvə(r)/ (*US* **flavor**) *–n.* **1** mingled sensation of smell and taste (*cheesy flavour*). **2** characteristic quality (*romantic flavour*). **3** (usu. foll. by *of*) slight admixture (*flavour of failure*). *–v.* give flavour to; season. □ **flavourless** *adj.* **flavoursome** *adj.* [French]

flavouring *n.* (*US* **flavoring**) substance used to flavour food or drink.

flavour of the month *n.* (also **flavour of the week**) temporary trend or fashion.

flaw[1] *–n.* **1** imperfection; blemish. **2** crack, chip, etc. **3** invalidating defect. *–v.* crack; damage; spoil. □ **flawless** *adj.* **flawlessly** *adv.* [Old Norse]

flaw[2] *n.* squall of wind. [Low German or Dutch]

flax *n.* **1** blue-flowered plant cultivated for its textile fibre and its seeds. **2** flax fibres. [Old English]

flaxen *adj.* **1** of flax. **2** (of hair) pale yellow.

flax-seed *n.* linseed.

flay *v.* **1** strip the skin or hide off, esp. by beating. **2** criticize severely. **3** peel off (skin, bark, peel, etc.). **4** extort money etc. from. [Old English]

flea *n.* small wingless jumping parasitic insect. □ **a flea in one's ear** sharp reproof. [Old English]

fleabag *n. slang* shabby or unattractive person or thing.

flea-bite *n.* **1** bite of a flea. **2** trivial injury or inconvenience.

flea-bitten *adj.* **1** bitten by or infested with fleas. **2** shabby.

flea market *n.* street market selling second-hand goods etc.

flea-pit *n.* dingy dirty cinema etc.

fleck –*n.* **1** small patch of colour or light. **2** particle, speck. –*v.* mark with flecks. [Old Norse, or Low German or Dutch]

flection *US* var. of FLEXION.

fled *past* and *past part.* of FLEE.

fledge *v.* (**-ging**) **1** provide or deck (an arrow etc.) with feathers. **2** bring up (a young bird) until it can fly. **3** (as **fledged** *adj.*) **a** able to fly. **b** independent; mature. [obsolete adj. *fledge* fit to fly]

fledgling /ˈfledʒlɪŋ/ *n.* (also **fledgeling**) **1** young bird. **2** inexperienced person.

flee *v.* (*past* and *past part.* **fled**) **1** (often foll. by *from*, *before*) **a** run away (from); leave abruptly (*fled the room*). **b** seek safety by fleeing. **2** vanish. [Old English]

fleece –*n.* **1 a** woolly coat of a sheep etc. **b** wool sheared from a sheep at one time. **2** thing resembling a fleece, esp. soft fabric for lining etc. –*v.* (**-cing**) **1** (often foll. by *of*) strip of money, valuables, etc.; swindle. **2** shear (sheep etc.). **3** cover as if with a fleece (*sky fleeced with clouds*). □ **fleecy** *adj.* (**-ier**, **-iest**). [Old English]

fleet –*n.* **1 a** warships under one commander-in-chief. **b** (prec. by *the*) nation's warships etc.; navy. **2** number of vehicles in one company etc. –*adj.* *poet. literary* swift, nimble. [Old English]

fleeting *adj.* transitory; brief. □ **fleetingly** *adv.*

Fleming /ˈflemɪŋ/ *n.* **1** native of medieval Flanders. **2** member of a Flemish-speaking people of N. and W. Belgium. [Old English]

Flemish /ˈflemɪʃ/ –*adj.* of Flanders. –*n.* language of the Flemings. [Dutch]

flesh *n.* **1 a** soft, esp. muscular, substance between the skin and bones of an animal or a human. **b** plumpness; fat. **2** the body, esp. as sinful. **3** pulpy substance of a fruit etc. **4 a** visible surface of the human body. **b** (also **flesh-colour**) yellowish pink colour. **5** animal or human life. □ **all flesh** all animate creation. **flesh out** make or become substantial. **in the flesh** in person. **one's own flesh and blood** near relatives. [Old English]

flesh and blood –*n.* **1** the body or its substance. **2** humankind. **3** human nature, esp. as fallible. –*adj.* real, not imaginary.

fleshly *adj.* (**-lier**, **-liest**) **1** bodily; sensual. **2** mortal. **3** worldly.

fleshpots *n.pl.* luxurious living.

flesh-wound *n.* superficial wound.

fleshy *adj.* (**-ier**, **-iest**) of flesh; plump, pulpy. □ **fleshiness** *n.*

fleur-de-lis /ˌflɜːdəˈliː/ *n.* (also **fleur-de-lys**) (*pl.* **fleurs-** pronunc. same) **1** iris flower. **2** *Heraldry* **a** lily of three petals. **b** former royal arms of France. [French, = flower of lily]

flew *past* of FLY[1].

flews *n.pl.* hanging lips of a bloodhound etc. [origin unknown]

flex[1] *v.* **1** bend (a joint, limb, etc.) or be bent. **2** move (a muscle) or (of a muscle) be moved to bend a joint. [Latin *flecto flex-* bend]

flex[2] *n.* flexible insulated electric cable. [abbreviation of FLEXIBLE]

flexible /ˈfleksɪb(ə)l/ *adj.* **1** capable of bending without breaking; pliable. **2** manageable. **3** adaptable; variable (*works flexible hours*). □ **flexibility** /-ˈbɪlɪtɪ/ *n.* **flexibly** *adv.* [Latin *flexibilis*: related to FLEX[1]]

flexion /ˈflekʃ(ə)n/ *n.* (*US* **flection**) **1** bending or being bent, esp. of a limb or joint. **2** bent part; curve. [Latin *flexio*: related to FLEX[1]]

flexitime /ˈfleksɪˌtaɪm/ *n.* system of flexible working hours. [from FLEXIBLE]

flibbertigibbet /ˌflɪbətɪˈdʒɪbɪt/ *n.* gossiping, frivolous, or restless person. [imitative]

flick –*n.* **1 a** light sharp blow with a whip etc. **b** sudden release of a bent digit, esp. to propel a small object. **2** sudden movement or jerk, esp. of the wrist in throwing etc. **3** *colloq.* **a** cinema film. **b** (in *pl.*; prec. by *the*) the cinema. –*v.* **1** (often foll. by *away*, *off*) strike or move with a flick (*flicked the ash off*). **2** give a flick with (a whip etc.). □ **flick through 1** turn over (cards, pages, etc.). **2 a** turn over the pages etc. of, by a rapid movement of the fingers. **b** glance through (a book etc.). [imitative]

flicker –*v.* **1** (of light or flame) shine or burn unsteadily. **2** flutter. **3** (of hope etc.) waver. –*n.* **1** flickering movement or light. **2** brief spell (of hope etc.). □ **flicker out** die away. [Old English]

flick-knife *n.* knife with a blade that springs out when a button is pressed.

flier var. of FLYER.

flight[1] /flaɪt/ *n.* **1 a** act or manner of flying. **b** movement or passage through the air. **2 a** journey through the air or in space. **b** timetabled airline journey. **3** flock of birds, insects, etc. **4** (usu. foll. by *of*) series, esp. of stairs. **5** imaginative excursion or sally (*flight of fancy*). **6** (usu. foll. by *of*) volley (*flight of arrows*). **7** tail of a dart. [Old English: related to FLY[1]]

flight[2] /flaɪt/ *n.* fleeing, hasty retreat. □ **put to flight** cause to flee. **take (or take to) flight** flee. [Old English]

flight bag *n.* small zipped shoulder bag for air travel.

flight-deck *n.* **1** deck of an aircraft-carrier. **2** control room of a large aircraft.

flightless *adj.* (of a bird etc.) unable to fly.

flight lieutenant *n.* RAF officer next below squadron leader.

flight path *n.* planned course of an aircraft etc.

flight-recorder *n.* device in an aircraft recording technical details of a flight.

flight sergeant *n.* RAF rank next above sergeant.

flighty *adj.* (**-ier, -iest**) (usu. of a girl) frivolous, fickle, changeable. □ **flightiness** *n.*

flimsy /'flɪmzɪ/ *adj.* (**-ier, -iest**) **1** insubstantial, rickety (*flimsy structure*). **2** (of an excuse etc.) unconvincing. **3** (of clothing) thin. □ **flimsily** *adv.* **flimsiness** *n.* [origin uncertain]

flinch *v.* draw back in fear etc.; wince. [French from Germanic]

fling –*v.* (*past* and *past part.* **flung**) **1** throw or hurl forcefully or hurriedly. **2** (foll. by *on, off*) put on or take off (clothes) carelessly or rapidly. **3** put or send suddenly or violently (*was flung into jail*). **4** rush, esp. angrily (*flung out of the room*). **5** (foll. by *away*) discard rashly. –*n.* **1** act of flinging; throw. **2** bout of wild behaviour. **3** whirling Scottish dance, esp. the Highland fling. [Old Norse]

flint *n.* **1 a** hard grey siliceous stone. **b** piece of this, esp. as a primitive tool or weapon. **2** piece of hard alloy used to give a spark. **3** anything hard and unyielding. □ **flinty** *adj.* (**-ier, -iest**). [Old English]

flintlock *n. hist.* old type of gun fired by a spark from a flint.

flip[1] –*v.* (**-pp-**) **1** flick or toss (a coin, pellet, etc.) so that it spins in the air. **2** turn (a small object) over; flick. **3** *slang* = *flip one's lid.* –*n.* **1** act of flipping. **2** *colloq.* short trip. –*adj. colloq.* glib; flippant. □ **flip one's lid** *slang* lose self-control; go mad. **flip through** = *flick through.* [probably from FILLIP]

flip[2] *n.* **1** = EGG-FLIP. **2** drink of heated beer and spirit. [perhaps from FLIP[1]]

flip chart *n.* large pad of paper on a stand.

flip-flop *n.* (usu. rubber) sandal with a thong between the toes. [imitative]

flippant /'flɪpənt/ *adj.* frivolous; disrespectful; offhand. □ **flippancy** *n.* **flippantly** *adv.* [from FLIP[1]]

flipper *n.* **1** broad flat limb of a turtle, penguin, etc., used in swimming. **2** similar rubber foot attachment for underwater swimming. **3** *slang* hand.

flipping *adj.* & *adv. slang* expressing annoyance, or as an intensifier.

flip side *n. colloq.* **1** reverse side of a gramophone record. **2** reverse or less important side of something.

flirt –*v.* **1** (usu. foll. by *with*) try to attract sexually but without serious intent. **2** (usu. foll. by *with*) superficially engage in; trifle. –*n.* person who flirts. □ **flirtation** /-'teɪʃ(ə)n/ *n.* **flirtatious** /-'teɪʃəs/ *adj.* **flirtatiously** /-'teɪʃəslɪ/ *adv.* **flirtatiousness** /-'teɪʃəsnɪs/ *n.* [imitative]

flit –*v.* (**-tt-**) **1** move lightly, softly, or rapidly. **2** make short flights. **3** *colloq.* disappear secretly to escape creditors etc. –*n.* act of flitting. [Old Norse: related to FLEET]

flitch *n.* side of bacon. [Old English]

flitter *v.* flit about; flutter. [from FLIT]

flitter-mouse *n.* = BAT[2].

float –*v.* **1 a** (cause to) rest or move on the surface of a liquid. **b** set (a stranded ship) afloat. **2** *colloq.* **a** move in a leisurely way. **b** (often foll. by *before*) hover before the eye or mind. **3** (often foll. by *in*) move or be suspended freely in a liquid or gas. **4 a** start or launch (a company, scheme, etc.). **b** offer (stock, shares, etc.) on the stock market. **5** *Commerce* cause or allow to have a fluctuating exchange rate. **6** circulate or cause (a rumour or idea) to circulate. –*n.* **1** thing that floats, esp.: **a** a raft. **b** a light object as an indicator of a fish biting or supporting a fishing-net. **c** a hollow structure enabling an aircraft to float on water. **d** a floating device on water, petrol, etc., controlling the level. **2** small esp. electrically-powered vehicle or cart (*milk float*). **3** decorated platform or tableau on a lorry in a procession etc. **4 a** supply of loose change in a shop, at a fête, etc. **b** petty cash. **5** *Theatr.* (in *sing.* or *pl.*) footlights. **6** tool for smoothing plaster. □ **floatable** *adj.* [Old English]

floatation var. of FLOTATION.

floating *adj.* not settled; variable (*floating population*).

floating dock *n.* floating structure usable as a dry dock.

floating kidney *n.* abnormally movable kidney.

floating rib *n.* lower rib not attached to the breastbone.

floating voter *n.* voter without fixed allegiance.

floaty *adj.* (esp. of fabric) light and airy. [from FLOAT]

flocculent /'flɒkjʊlənt/ *adj.* like or in tufts of wool etc.; downy. □ **flocculence** *n.* [related to FLOCK[2]]

flock[1] *–n.* **1** animals of one kind as a group or unit. **2** large crowd of people. **3** people in the care of a priest or teacher etc. *–v.* (usu. foll. by *to, in, out, together*) congregate; mass; troop. [Old English]

flock[2] *n.* **1** lock or tuft of wool, cotton, etc. **2** (also in *pl.*; often *attrib.*) wool-refuse etc. used for quilting and stuffing. [Latin *floccus*]

flock-paper *n.* (also **flock-wallpaper**) wallpaper with a raised flock pattern.

floe *n.* sheet of floating ice. [Norwegian]

flog *v.* (**-gg-**) **1 a** beat with a whip, stick, etc. **b** make work through violent effort (*flogged the engine*). **2** (often foll. by *off*) *slang* sell. □ **flog a dead horse** waste one's efforts. **flog to death** *colloq.* talk about or promote at tedious length. [origin unknown]

flood /flʌd/ *–n.* **1 a** overflowing or influx of water, esp. over land; inundation. **b** the water that overflows. **2** outpouring; torrent (*flood of tears*). **3** inflow of the tide (also in *comb.*: *flood-tide*). **4** *colloq.* floodlight. **5** (**the Flood**) the flood described in Genesis. *–v.* **1** overflow, cover, or be covered with or as if with a flood (*bathroom flooded*; *flooded with enquiries*). **2** irrigate. **3** deluge (a mine etc.) with water. **4** (often foll. by *in, through*) come in great quantities (*complaints flooded in*). **5** overfill (a carburettor) with petrol. **6** have a uterine haemorrhage. □ **flood out** drive out (of one's home etc.) with a flood. [Old English]

floodgate *n.* **1** gate for admitting or excluding water, esp. in a lock. **2** (usu. in *pl.*) last restraint against tears, rain, anger, etc.

floodlight *–n.* large powerful light (usu. one of several) to illuminate a building, sports ground, etc. *–v.* illuminate with floodlights. □ **floodlit** *adj.*

flood-tide *n.* exceptionally high tide caused esp. by the moon.

floor /flɔ:(r)/ *–n.* **1** lower supporting surface of a room. **2 a** bottom of the sea, a cave, etc. **b** any level area. **3** all the rooms etc. on one level of a building; storey. **4 a** (in a legislative assembly) place where members sit and speak. **b** right to speak next in a debate (*gave him the floor*). **5** minimum of prices, wages, etc. **6** *colloq.* ground. *–v.* **1** provide with a floor; pave. **2** knock or bring (a person) down. **3** *colloq.* confound, baffle. **4** *colloq.* overcome. **5** serve as the floor of (*lino floored the hall*). □ **from the floor** (of a speech etc.) given by a member of the audience. **take the floor 1** begin to dance. **2** speak in a debate. [Old English]

floorboard *n.* long wooden board used for flooring.

floorcloth *n.* cloth for washing the floor.

flooring *n.* material of which a floor is made.

floor manager *n.* stage-manager of a television production.

floor plan *n.* diagram of the rooms etc. on one storey.

floor show *n.* nightclub entertainment.

floozie /'flu:zɪ/ *n.* (also **floozy**) (*pl.* **-ies**) *colloq.* esp. disreputable girl or woman. [origin unknown]

flop *–v.* (**-pp-**) **1** sway about heavily or loosely. **2** (often foll. by *down, on, into*) fall or sit etc. awkwardly or suddenly. **3** *slang* fail; collapse (*play flopped*). **4** make a dull soft thud or splash. *–n.* **1** flopping movement or sound. **2** *slang* failure. *–adv.* with a flop. [var. of FLAP]

floppy *–adj.* (**-ier, -iest**) tending to flop; flaccid. *–n.* (*pl.* **-ies**) (in full **floppy disk**) *Computing* flexible disc for the storage of data. □ **floppiness** *n.*

flora /'flɔ:rə/ *n.* (*pl.* **-s** or **florae** /-ri:/) **1** plant life of a region or period. **2** list or book of these. [Latin *Flora*, name of the goddess of flowers]

floral *adj.* of, decorated with, or depicting flowers. □ **florally** *adv.* [Latin]

Florentine /'flɒrən,taɪn/ *–adj.* of Florence in Italy. *–n.* native or citizen of Florence. [Latin]

floret /'flɒrɪt/ *n.* **1** each of the small flowers making up a composite flower-head. **2** each stem of a head of cauliflower, broccoli, etc. **3** small flower. [Latin *flos* FLOWER]

floribunda /,flɒrɪ'bʌndə/ *n.* plant, esp. a rose, bearing dense clusters of flowers. [related to FLORET: cf. MORIBUND]

florid /'flɒrɪd/ *adj.* **1** ruddy (*florid complexion*). **2** elaborately ornate; showy. □ **floridly** *adv.* **floridness** *n.* [Latin: related to FLOWER]

florin /'flɒrɪn/ *n. hist.* **1** British two-shilling coin now worth 10 pence. **2** English or foreign gold or silver coin. [Italian *fiorino*: related to FLORIST]

florist /'flɒrɪst/ *n.* person who deals in or grows flowers. [Latin *flos* FLOWER]

floruit /'flɒrʊɪt/ *–v.* flourished; lived and worked (of a painter, writer, etc., whose exact dates are unknown). *–n.* period or date of working etc. [Latin, = he or she flourished]

floss *–n.* **1** rough silk of a silkworm's cocoon. **2** silk thread used in embroidery. **3** = DENTAL FLOSS. *–v.* (also *absol.*) clean (teeth) with dental floss. □ **flossy** *adj.* [French *floche*]

flotation /fləʊˈteɪʃ(ə)n/ *n.* (also **floatation**) launching or financing of a commercial enterprise etc. [from FLOAT]

flotilla /fləˈtɪlə/ *n.* **1** small fleet. **2** fleet of small ships. [Spanish]

flotsam /ˈflɒtsəm/ *n.* wreckage found floating. [Anglo-French: related to FLOAT]

flotsam and jetsam *n.* **1** odds and ends. **2** vagrants.

flounce[1] *–v.* (**-cing**) (often foll. by *away, about, off, out*) go or move angrily or impatiently (*flounced out in a huff*). *–n.* flouncing movement. [origin unknown]

flounce[2] *–n.* frill on a dress, skirt, etc. *–v.* (**-cing**) trim with flounces. [alteration of *frounce* pleat, from French]

flounder[1] *–v.* **1** struggle helplessly as if wading in mud. **2** do a task clumsily. *–n.* act of floundering. [imitative]

flounder[2] *n.* (*pl.* same) **1** edible European flat-fish. **2** N. American flat-fish. [Anglo-French, probably Scandinavian]

flour *–n.* **1** meal or powder from ground wheat etc. **2** any fine powder. *–v.* sprinkle with flour. □ **floury** *adj.* (**-ier, -iest**). **flouriness** *n.* [different spelling of FLOWER 'best part']

flourish /ˈflʌrɪʃ/ *–v.* **1 a** grow vigorously; thrive. **b** prosper. **c** be in one's prime. **2** wave, brandish. *–n.* **1** showy gesture. **2** ornamental curve in handwriting. **3** *Mus.* ornate passage or fanfare. [Latin *floreo* from *flos* FLOWER]

flout *–v.* disobey (the law etc.) contemptuously; mock; insult. *–n.* flouting speech or act. [Dutch *fluiten* whistle: related to FLUTE]

■ **Usage** *Flout* is often confused with *flaunt* which means 'to display proudly, show off'.

flow /fləʊ/ *–v.* **1** glide along as a stream. **2** (of liquid, blood, etc.) gush out; be spilt. **3** (of blood, money, electric current, etc.) circulate. **4** move smoothly or steadily. **5** (of a garment, hair, etc.) hang gracefully. **6** (often foll. by *from*) be caused by. **7** (esp. of the tide) be in flood. **8** (of wine) be plentiful. **9** (foll. by *with*) *archaic* be plentifully supplied with (*flowing with milk and honey*). *–n.* **1 a** flowing movement or mass. **b** flowing liquid (*stop the flow*). **c** outpouring; stream (*flow of complaints*). **2** rise of a tide or river (*ebb and flow*). [Old English]

flow chart *n.* (also **flow diagram** or **flow sheet**) diagram of the movement or action in a complex activity.

flower /ˈflaʊə(r)/ *–n.* **1** part of a plant from which the fruit or seed is developed. **2** blossom, esp. used for decoration. **3** plant cultivated for its flowers. *–v.* **1** bloom or cause (a plant) to bloom; blossom. **2** reach a peak. □ **the flower of** the best of. **in flower** blooming. □ **flowered** *adj.* [Latin *flos flor-*]

flower-bed *n.* garden bed for flowers.

flower-head *n.* = HEAD *n.* 3 c.

flower people *n.* hippies with flowers as symbols of peace and love.

flowerpot *n.* pot for growing a plant in.

flower power *n.* peace and love, esp. as a political idea.

flowers of sulphur *n.* fine powder produced when sulphur evaporates and condenses.

flowery *adj.* **1** florally decorated. **2** (of style, speech, etc.) high-flown; ornate. **3** full of flowers. □ **floweriness** *n.*

flowing *adj.* **1** (of style etc.) fluent; easy. **2** (of a line, curve, etc.) smoothly continuous. **3** (of hair etc.) unconfined. □ **flowingly** *adv.*

flown *past part.* of FLY[1].

flu /flu:/ *n. colloq.* influenza. [abbreviation]

fluctuate /ˈflʌktʃʊˌeɪt/ *v.* (**-ting**) vary irregularly; rise and fall. □ **fluctuation** /-ˈeɪʃ(ə)n/ *n.* [Latin *fluctus* wave]

flue /flu:/ *n.* **1** smoke-duct in a chimney. **2** channel for conveying heat. [origin unknown]

fluent /ˈflu:ənt/ *adj.* **1** (of speech, style, etc.) flowing, natural. **2** verbally facile, esp. in a foreign language (*fluent in German*). □ **fluency** *n.* **fluently** *adv.* [Latin *fluo* flow]

fluff *–n.* **1** soft fur, feathers, or fabric particles etc. **2** *slang* mistake in a performance etc. *–v.* **1** (often foll. by *up*) shake into or become a soft mass. **2** *colloq.* make a fluff; bungle. □ **bit of fluff** *slang offens.* attractive woman. □ **fluffy** *adj.* (**-ier, -iest**). **fluffiness** *n.* [probably dial. alteration of *flue* fluff]

flugelhorn /ˈflu:g(ə)lˌhɔ:n/ *n.* valved brass wind instrument like a cornet. [German *Flügel* wing, *Horn* horn]

fluid /ˈflu:ɪd/ *–n.* **1** substance, esp. a gas or liquid, whose shape is determined by its confines. **2** fluid part or secretion. *–adj.* **1** able to flow and alter shape freely. **2** constantly changing (*situation is fluid*). □ **fluidity** /-ˈɪdɪtɪ/ *n.* **fluidly** *adv.* **fluidness** *n.* [Latin: related to FLUENT]

fluid ounce *n.* one-twentieth, or *US* one-sixteenth, of a pint.

fluke[1] /flu:k/ *–n.* lucky accident (*won by a fluke*). *–v.* (**-king**) achieve by a fluke. □ **fluky** *adj.* (**-ier, -iest**). [origin uncertain]

fluke[2] /flu:k/ *n.* **1** parasitic flatworm, e.g. the liver fluke. **2** flat-fish, esp. a flounder. [Old English]

fluke[3] /flu:k/ *n.* **1** broad triangular plate on an anchor arm. **2** lobe of a whale's tail. [perhaps from FLUKE[2]]

flummery /'flʌmərɪ/ *n.* (*pl.* **-ies**) **1** flattery; nonsense. **2** sweet dish made with beaten eggs, sugar, etc. [Welsh *llymru*]

flummox /'flʌməks/ *v. colloq.* bewilder, disconcert. [origin unknown]

flung *past* and *past part.* of FLING.

flunk *v. US colloq.* fail (esp. an exam). [origin unknown]

flunkey /'flʌŋkɪ/ *n.* (also **flunky**) (*pl.* **-eys** or **-ies**) usu. *derog.* **1** liveried footman. **2** toady; snob. **3** *US* cook, waiter, etc. [origin uncertain]

fluoresce /flʊə'res/ *v.* (**-scing**) be or become fluorescent. [from FLUORESCENT]

fluorescence *n.* **1** light radiation from certain substances. **2** property of absorbing invisible light and emitting visible light. [from FLUORSPAR, after *opalescence*]

fluorescent *adj.* of, having, or showing fluorescence.

fluorescent lamp *n.* (also **fluorescent bulb**) esp. tubular lamp or bulb radiating largely by fluorescence.

fluoridate /'flʊərɪ,deɪt/ *v.* (**-ting**) add fluoride to (drinking-water etc.), esp. to prevent tooth decay. □ **fluoridation** /-'deɪʃ(ə)n/ *n.*

fluoride /'flʊəraɪd/ *n.* binary compound of fluorine.

fluorinate /'flʊərɪ,neɪt/ *v.* (**-ting**) **1** = FLUORIDATE. **2** introduce fluorine into (a compound). □ **fluorination** /-'neɪʃ(ə)n/ *n.*

fluorine /'flʊəri:n/ *n.* poisonous pale-yellow gaseous element. [French: related to FLUORSPAR]

fluorite /'flʊəraɪt/ *n.* mineral form of calcium fluoride. [Italian: related to FLUORSPAR]

fluorocarbon /,flʊərəʊ'kɑ:bən/ *n.* compound of a hydrocarbon with fluorine atoms.

fluorspar /'flʊəspɑ:(r)/ *n.* = FLUORITE. [*fluor* a mineral used as flux, from Latin *fluo* flow]

flurry /'flʌrɪ/ *–n.* (*pl.* **-ies**) **1** gust or squall (of snow, rain, etc.). **2** sudden burst of activity, excitement, etc.; commotion. *–v.* (**-ies, -ied**) confuse; agitate. [imitative]

flush[1] *–v.* **1** blush, redden, glow warmly (*he flushed with embarrassment*). **2** (usu. as **flushed** *adj.*) cause to glow or blush (often foll. by *with*: *he was flushed with pride*). **3 a** cleanse (a drain, lavatory, etc.) by a flow of water. **b** (often foll. by *away, down*) dispose of in this way. **4** rush out, spurt. *–n.* **1** blush or glow. **2 a** rush of water. **b** cleansing of a drain, lavatory, etc. thus. **3** rush of esp. elation or triumph. **4** freshness; vigour. **5 a** (also **hot flush**) sudden feeling of heat during menopause. **b** feverish redness or temperature etc. *–adj.* **1** level, in the same plane. **2** *colloq.* having plenty of money. [perhaps = FLUSH[3]]

flush[2] *n.* hand of cards all of one suit, esp. in poker. [Latin *fluxus* FLUX]

flush[3] *v.* **1** cause (esp. a game-bird) to fly up. **2** (of a bird) fly up and away. □ **flush out 1** reveal. **2** drive out. [imitative]

fluster *–v.* **1** make or become nervous or confused (*he flusters easily*). **2** bustle. *–n.* confused or agitated state. [origin unknown]

flute /flu:t/ *–n.* **1 a** high-pitched woodwind instrument held sideways. **b** any similar wind instrument. **2** ornamental vertical groove in a column. *–v.* (**-ting**) **1** play, or play (a tune etc.) on, the flute. **2** speak or sing etc. in a high voice. **3** make grooves in. □ **fluting** *n.* **fluty** *adj.* (in sense 1a of *n.*). [French]

flutter *–v.* **1** flap (the wings) in flying or trying to fly. **2** fall quiveringly (*fluttered to the ground*). **3** wave or flap quickly. **4** move about restlessly. **5** (of a pulse etc.) beat feebly or irregularly. *–n.* **1** act of fluttering. **2** tremulous excitement (*caused a flutter*). **3** *slang* small bet, esp. on a horse. **4** abnormally rapid heartbeat. **5** rapid variation of pitch, esp. of recorded sound. [Old English]

fluvial /'flu:vɪəl/ *adj.* of or found in rivers. [Latin *fluvius* river]

flux *n.* **1** process of flowing or flowing out. **2** discharge. **3** continuous change (*state of flux*). **4** substance mixed with a metal etc. to aid fusion. [Latin *fluxus* from *fluo flux-* flow]

fly[1] /flaɪ/ *–v.* (**flies**; *past* **flew** /flu:/; *past part.* **flown** /fləʊn/) **1 a** (of an aircraft, bird, etc.) move through the air or space under control, esp. with wings. **b** travel through the air or space. **2** control the flight of or transport in (esp. an aircraft). **3 a** cause to fly or remain aloft. **b** (of a flag, hair, etc.) wave or flutter. **4** pass, move, or rise quickly. **5 a** flee; flee from. **b** *colloq.* depart hastily. **6** be driven, forced, or scattered (*sent me flying*). **7** (foll. by *at, upon*) **a** hasten or spring violently. **b** attack or criticize fiercely. *–n.* (*pl.* **-ies**) **1** (usu. in *pl.*) **a** concealing flap, esp. over a trouser-fastening. **b** this fastening. **2** flap at a tent entrance. **3** (in *pl.*) space above a stage where scenery and lighting are suspended. **4** act of flying. □ **fly high** be ambitious; prosper. **fly in the face of** disregard or disobey. **fly a kite** test opinion. **fly off the handle** *colloq.* lose one's temper. [Old English]

fly[2] /flaɪ/ *n.* (*pl.* **flies**) **1** insect with two usu. transparent wings. **2** other winged insect, e.g. a firefly. **3** disease of plants or animals caused by flies. **4** (esp. artificial) fly as bait in fishing. □ **like flies** in large numbers (usu. of people dying etc.). **no flies on** (**him** etc.) *colloq.* (he is) very astute. [Old English]

fly[3] /flaɪ/ *adj. slang* knowing, clever, alert. [origin unknown]

fly-away *adj.* (of hair) fine and difficult to control.

fly-blown *adj.* tainted, esp. by flies.

fly-by-night –*adj.* unreliable. –*n.* unreliable person.

flycatcher *n.* bird catching insects during short flights from a chosen perch.

flyer *n.* (also **flier**) *colloq.* **1** airman or airwoman. **2** thing that flies in a specified way (*poor flyer*). **3** fast-moving animal or vehicle. **4** ambitious or outstanding person. **5** small handbill.

fly-fish *v.* fish with a fly.

fly-half *n. Rugby* stand-off half.

flying –*adj.* **1** fluttering, waving, or hanging loose. **2** hasty, brief (*flying visit*). **3** designed for rapid movement. **4** (of an animal) leaping with winglike membranes etc. –*n.* flight, esp. in an aircraft. □ **with flying colours** with distinction.

flying boat *n.* boatlike seaplane.

flying buttress *n.* (usu. arched) buttress running from the upper part of a wall to an outer support and transmitting the thrust of the roof or vault.

flying doctor *n.* doctor who uses an aircraft to visit patients.

flying fish *n.* tropical fish with winglike fins for gliding through the air.

flying fox *n.* fruit-eating bat with a foxlike head.

flying officer *n.* RAF rank next below flight lieutenant.

flying picket *n.* mobile industrial strike picket.

flying saucer *n.* supposed alien spaceship.

flying squad *n.* rapidly mobile police detachment etc.

flying start *n.* **1** start (of a race etc.) in which the starting-point is crossed at full speed. **2** vigorous start (of an enterprise etc.).

fly in the ointment *n.* minor irritation or setback.

flyleaf *n.* blank leaf at the beginning or end of a book.

fly on the wall *n.* unnoticed observer.

flyover *n.* bridge carrying one road or railway over another.

fly-paper *n.* sticky treated paper for catching flies.

fly-past *n.* ceremonial flight of aircraft.

fly-post *v.* fix (posters etc.) illegally on walls etc. [FLY[1]]

flysheet *n.* **1** canvas cover over a tent for extra protection. **2** short tract or circular. [FLY[1]]

fly-tip *v.* illegally dump (waste). □ **fly-tipper** *n.* [FLY[1]]

fly-trap *n.* plant that catches flies.

flyweight *n.* **1** weight in certain sports between light flyweight and bantamweight, in amateur boxing 48–51 kg. **2** sportsman of this weight.

flywheel *n.* heavy wheel on a revolving shaft to regulate machinery or accumulate power.

FM *abbr.* **1** field marshal. **2** frequency modulation.

Fm *symb.* fermium.

f-number /ˈef,nʌmbə(r)/ *n.* ratio of the focal length to the effective diameter of a camera lens. [from *f*ocal]

FO *abbr.* Flying Officer.

foal –*n.* young of a horse or related animal. –*v.* give birth to (a foal). □ **in** (or **with**) **foal** (of a mare etc.) pregnant. [Old English]

foam –*n.* **1** mass of small bubbles formed on or in liquid by agitation, fermentation, etc. **2** froth of saliva or sweat. **3** substance resembling these, e.g. spongy rubber or plastic. –*v.* emit or run with foam; froth. □ **foam at the mouth** be very angry. □ **foamy** *adj.* (**-ier**, **-iest**). [Old English]

fob[1] *n.* **1** chain of a pocket-watch. **2** small pocket for a watch etc. **3** tab on a key-ring. [German]

fob[2] *v.* (**-bb-**) □ **fob off 1** (often foll. by *with* a thing) deceive into accepting something inferior. **2** (often foll. by *on* or *on to* a person) offload (an unwanted thing). [cf. obsolete *fop* dupe]

focal /ˈfəʊk(ə)l/ *adj.* of or at a focus. [Latin: related to FOCUS]

focal distance *n.* (also **focal length**) distance between the centre of a mirror or lens and its focus.

focal point *n.* **1** = FOCUS *n.* 1. **2** centre of interest or activity.

fo'c's'le var. of FORECASTLE.

focus /ˈfəʊkəs/ –*n.* (*pl.* **focuses** or **foci** /ˈfəʊsaɪ/) **1 a** point at which rays or waves meet after reflection or refraction. **b** point from which rays etc. appear to proceed. **2 a** point at which an object must be situated for a lens or mirror to give a well-defined image. **b** adjustment of the eye or a lens to give a clear image. **c** state of clear definition (*out of focus*). **3** = FOCAL POINT 2. –*v.* (**-s-** or **-ss-**) **1** bring into focus. **2** adjust the focus of (a lens or eye). **3** concentrate or be concentrated on. **4** converge or make converge to a focus. [Latin, = hearth]

fodder *n.* dried hay or straw etc. as animal food. [Old English]

FOE /*colloq.* fəʊ/ *abbr.* Friends of the Earth.

foe *n.* esp. *poet.* enemy. [Old English]

foetid var. of FETID.

foetus /ˈfi:təs/ *n.* (*US* **fetus**) (*pl.* **-tuses**) unborn mammalian offspring, esp. a human embryo of eight weeks or more. □ **foetal** *adj.* [Latin *fetus* offspring]

fog –*n.* **1** thick cloud of water droplets or smoke suspended at or near the earth's surface. **2** cloudiness on a photographic negative etc. **3** uncertain or confused position or state. –*v.* (**-gg-**) **1** cover or become covered with or as with fog. **2** perplex. [perhaps a back-formation from FOGGY]

fog-bank *n.* mass of fog at sea.

fog-bound *adj.* unable to travel because of fog.

fogey var. of FOGY.

foggy *adj.* (**-ier, -iest**) **1** full of fog. **2** of or like fog. **3** vague, indistinct. □ **not have the foggiest** *colloq.* have no idea at all. □ **fogginess** *n.* [perhaps from *fog* long grass]

foghorn *n.* **1** horn warning ships in fog. **2** *colloq.* loud penetrating voice.

fog-lamp *n.* powerful lamp for use in fog.

fogy /ˈfəʊgɪ/ *n.* (also **fogey**) (*pl.* **-ies** or **-eys**) dull old-fashioned person (esp. *old fogy*). [origin unknown]

foible /ˈfɔɪb(ə)l/ *n.* minor weakness or idiosyncrasy. [French: related to FEEBLE]

foil[1] *v.* frustrate, baffle, defeat. [perhaps from French *fouler* trample]

foil[2] *n.* **1** metal rolled into a very thin sheet. **2** person or thing setting off another to advantage. [Latin *folium* leaf]

foil[3] *n.* light blunt fencing sword. [origin unknown]

foist *v.* (foll. by *on*) force (a thing or oneself) on to an unwilling person. [Dutch *vuisten* take in the hand]

fold[1] /fəʊld/ –*v.* **1 a** bend or close (a flexible thing) over upon itself. **b** (foll. by *back*, *over*, *down*) bend part of (a thing) (*fold down the flap*). **2** become or be able to be folded. **3** (foll. by *away*, *up*) make compact by folding. **4** (often foll. by *up*) *colloq.* collapse, cease to function. **5** enfold (esp. *fold in the arms* or *to the breast*). **6** (foll. by *about*, *round*) clasp (the arms). **7** (foll. by *in*) mix (an ingredient with others) gently. –*n.* **1** folding. **2** line made by folding. **3** folded part. **4** hollow among hills. **5** curvature of geological strata. □ **fold one's arms** place or entwine them across the chest. **fold one's hands** clasp them. [Old English]

fold[2] /fəʊld/ –*n.* **1** = SHEEPFOLD. **2** religious group or congregation. –*v.* enclose (sheep) in a fold. [Old English]

-fold *suffix* forming adjectives and adverbs from cardinal numbers, meaning: **1** in an amount multiplied by (*repaid tenfold*). **2** with so many parts (*threefold blessing*). [originally = 'folded in so many layers']

folder *n.* folding cover or holder for loose papers.

foliaceous /ˌfəʊlɪˈeɪʃəs/ *adj.* **1** of or like leaves. **2** laminated. [Latin: related to FOIL[2]]

foliage /ˈfəʊlɪɪdʒ/ *n.* leaves, leafage. [French *feuillage* from *feuille* leaf]

foliar /ˈfəʊlɪə(r)/ *adj.* of leaves. [as FOLIATE]

foliar feed *n.* fertilizer supplied to the leaves of plants.

foliate –*adj.* /ˈfəʊlɪət/ **1** leaflike. **2** having leaves. –*v.* /ˈfəʊlɪˌeɪt/ (**-ting**) split or beat into thin layers. □ **foliation** /-ˈeɪʃ(ə)n/ *n.* [Latin *folium* leaf]

folio /ˈfəʊlɪəʊ/ –*n.* (*pl.* **-s**) **1** leaf of paper etc., esp. numbered only on the front. **2** sheet of paper folded once making two leaves of a book. **3** book of such sheets. –*adj.* (of a book) made of folios, of the largest size. □ **in folio** made of folios. [Latin, ablative of *folium* leaf]

folk /fəʊk/ *n.* (*pl.* same or **-s**) **1** (treated as *pl.*) people in general or of a specified class (*few folk about*; *townsfolk*). **2** (in *pl.*) (usu. **folks**) one's parents or relatives. **3** (treated as *sing.*) a people or nation. **4** (in full **folk-music**) (treated as *sing.*) *colloq.* traditional music or modern music in this style. **5** (*attrib.*) of popular origin (*folk art*). [Old English]

folk-dance *n.* dance of popular origin.

folklore *n.* traditional beliefs and stories of a people; the study of these.

folk-singer *n.* singer of folk-songs.

folk-song *n.* song of popular or traditional origin or style.

folksy /ˈfəʊksɪ/ *adj.* (**-ier, -iest**) **1** of or like folk art, culture, etc. **2** friendly, unpretentious. □ **folksiness** *n.*

folk-tale *n.* traditional story.

folkweave *n.* rough loosely woven fabric.

follicle /ˈfɒlɪk(ə)l/ *n.* small sac or vesicle in the body, esp. one containing a hair-root. □ **follicular** /fɒˈlɪkjʊlə(r)/ *adj.* [Latin diminutive of *follis* bellows]

follow /ˈfɒləʊ/ *v.* **1** (often foll. by *after*) go or come after (a person or thing ahead). **2** go along (a road etc.). **3** come after in order or time (*dessert followed*; *proceed as follows*). **4** take as a guide or leader. **5** conform to. **6** practise (a trade or profession). **7** undertake (a course of study etc.). **8** understand (a speaker, argument, etc.). **9** take an interest in (current affairs etc.). **10** (foll. by *with*) provide with a sequel or successor. **11** happen after something else; ensue. **12 a** be necessarily true as a consequence. **b** (foll. by *from*) result. □ **follow on 1** continue. **2** (of a cricket team) have to bat twice in succession. **follow out** carry out (instructions etc.). **follow suit 1** play a card of the suit led. **2** conform to another's actions. **follow through 1** continue to a conclusion. **2** continue the movement of a stroke after hitting the ball. **follow up** (foll. by *with*) **1** develop, supplement. **2** investigate further. [Old English]

follower *n.* **1** supporter or devotee. **2** person who follows.

following *–prep.* after in time; as a sequel to. *–n.* supporters or devotees. *–adj.* that follows or comes after. □ **the following 1** what follows. **2** now to be given or named (*answer the following*).

follow-on *n. Cricket* instance of following on.

follow-through *n.* action of following through.

follow-up *n.* subsequent or continued action.

folly *n.* (*pl.* **-ies**) **1** foolishness. **2** foolish act, behaviour, idea, etc. **3** fanciful ornamental building created for display. [French *folie* from *fol* mad, FOOL[1]]

foment /fə'ment/ *v.* instigate or stir up (trouble, discontent, etc.). □ **fomentation** /ˌfəʊmen'teɪʃ(ə)n/ *n.* [Latin *foveo* heat, cherish]

fond *adj.* **1** (foll. by *of*) liking. **2 a** affectionate. **b** doting. **3** (of beliefs etc.) foolishly optimistic or credulous. □ **fondly** *adv.* **fondness** *n.* [obsolete *fon* fool, be foolish]

fondant /'fɒnd(ə)nt/ *n.* soft sugary sweet. [French = melting: related to FUSE[1]]

fondle /'fɒnd(ə)l/ *v.* (**-ling**) caress. [related to FOND]

fondue /'fɒndju:/ *n.* dish of melted cheese. [French, = melted: related to FUSE[1]]

font[1] *n.* receptacle in a church for baptismal water. [Latin *fons font-* fountain]

font[2] var. of FOUNT[2].

fontanelle /ˌfɒntə'nel/ *n.* (*US* **fontanel**) membranous space in an infant's skull at the angles of the parietal bones. [Latin *fontanella* little FOUNTAIN]

food *n.* **1 a** substance taken in to maintain life and growth. **b** solid food (*food and drink*). **2** mental stimulus (*food for thought*). [Old English]

food additive *n.* substance added to food to colour or flavour it etc.

food-chain *n.* series of organisms each dependent on the next for food.

foodie /'fu:dɪ/ *n. colloq.* person who makes a cult of food; gourmet.

food poisoning *n.* illness due to bacteria etc. in food.

food processor *n.* machine for chopping and mixing food.

foodstuff *n.* substance used as food.

food value *n.* nourishing power of a food.

fool[1] *–n.* **1** rash, unwise, or stupid person. **2** *hist.* jester; clown. **3** dupe. *–v.* **1** deceive. **2** (foll. by *into* or *out of*) trick; cheat. **3** joke or tease. **4** (foll. by *about, around*) play or trifle. □ **act** (or **play**) **the fool** behave in a silly way. **be no** (or **nobody's**) **fool** be shrewd or prudent. **make a fool of** make (a person or oneself) look foolish; trick, deceive. [Latin *follis* bellows]

fool[2] *n.* dessert of fruit purée with cream or custard. [perhaps from FOOL[1]]

foolery *n.* foolish behaviour.

foolhardy *adj.* (**-ier, -iest**) rashly or foolishly bold; reckless. □ **foolhardily** *adv.* **foolhardiness** *n.*

foolish *adj.* lacking good sense or judgement; unwise. □ **foolishly** *adv.* **foolishness** *n.*

foolproof *adj.* (of a procedure, mechanism, etc.) incapable of misuse or mistake.

foolscap /'fu:lskæp/ *n.* large size of paper, about 330 x 200 (or 400) mm. [from a watermark of a *fool's cap*]

fool's paradise *n.* illusory happiness.

foot *–n.* (*pl.* **feet**) **1 a** part of the leg below the ankle. **b** part of a sock etc. covering this. **2 a** lowest part of a page, stairs, etc. **b** end of a bed where the feet rest. **c** part of a chair, appliance, etc. on which it rests. **3** step, pace, or tread (*fleet of foot*). **4** (*pl.* **feet** or **foot**) linear measure of 12 inches (30.48 cm). **5** metrical unit of verse forming part of a line. **6** *hist.* infantry. *–v.* **1** pay (a bill). **2** (usu. as **foot it**) go or traverse on foot. □ **feet of clay** fundamental weakness in a respected person. **have one's** (or **both**) **feet on the ground** be practical. **have one foot in the grave** be near death or very old. **my foot!** *int.* expressing strong contradiction. **on foot** walking. **put one's feet up** *colloq.* take a rest. **put one's foot down** *colloq.* **1** insist firmly. **2** accelerate a vehicle. **put one's foot in it** *colloq.* make a tactless blunder. **under one's feet** in the way. **under foot** on the ground. □ **footless** *adj.* [Old English]

footage *n.* **1** a length of TV or cinema film etc. **2** length in feet.

foot-and-mouth disease *n.* contagious viral disease of cattle etc.

football *n.* **1** large inflated ball of leather or plastic. **2** outdoor team game played with this. □ **footballer** *n.*

football pool *n.* (also **football pools** *pl.*) large-scale organized gambling on the results of football matches.

footbrake *n.* foot-operated brake on a vehicle.

footbridge *n.* bridge for pedestrians.

footfall *n.* sound of a footstep.

foot-fault *n.* (in tennis) placing of the foot over the baseline while serving.

foothill *n.* any of the low hills at the base of a mountain or range.

foothold *n.* **1** secure place for a foot when climbing etc. **2** secure initial position.

footing *n.* **1** foothold; secure position (*lost his footing*). **2** operational basis. **3** relative position or status (*on an equal footing*). **4** (often in *pl.*) foundations of a wall.

footle /'fu:t(ə)l/ *v.* (**-ling**) (usu. foll. by *about*) *colloq.* potter or fiddle about. [origin uncertain]

footlights *n.pl.* row of floor-level lights at the front of a stage.

footling /'fu:tlɪŋ/ *adj. colloq.* trivial, silly.

footloose *adj.* free to act as one pleases.

footman *n.* liveried servant.

footmark *n.* footprint.

footnote *n.* note printed at the foot of a page.

footpad *n. hist.* unmounted highwayman.

footpath *n.* path for pedestrians; pavement.

footplate *n.* platform for the crew in a locomotive.

footprint *n.* impression left by a foot or shoe.

footrest *n.* stool, rail, etc. for the feet.

Footsie /'fʊtsɪ/ *n.* = FT-SE. [respelling of FT-SE]

footsie /'fʊtsɪ/ *n. colloq.* amorous play with the feet.

footsore *adj.* with sore feet, esp. from walking.

footstep *n.* **1** step taken in walking. **2** sound of this. □ **follow in a person's footsteps** do as another did before.

footstool *n.* stool for resting the feet on when sitting.

footway *n.* path for pedestrians.

footwear *n.* shoes, socks, etc.

footwork *n.* use or agility of the feet in sports, dancing, etc.

fop *n.* dandy. □ **foppery** *n.* **foppish** *adj.* [perhaps from obsolete *fop* fool]

for /fə(r), fɔ:(r)/ *—prep.* **1** in the interest or to the benefit of; intended to go to (*did it all for my country*; *these flowers are for you*). **2** in defence, support, or favour of. **3** suitable or appropriate to (*a dance for beginners*; *not for me to say*). **4** in respect of or with reference to; regarding (*usual for ties to be worn*; *ready for bed*). **5** representing or in place of (*MP for Lincoln*; *here for my uncle*). **6** in exchange with, at the price of, corresponding to (*swapped it for a cake*; *give me £5 for it*; *bought it for £5*; *word for word*). **7** as a consequence of (*fined for speeding*; *decorated for bravery*; *here's £5 for your trouble*). **8 a** with a view to; in the hope or quest of; in order to get (*go for a walk*; *send for a doctor*; *did it for the money*). **b** on account of (*could not speak for laughing*). **9** to reach; towards (*left for Rome*). **10** so as to start promptly at (*meet at seven for eight*). **11** through or over (a distance or period); during (*walked for miles*). **12** as being (*for the last time*; *I for one refuse*). **13** in spite of; notwithstanding (*for all your fine words*). **14** considering or making due allowance in respect of (*good for a beginner*). *—conj.* because, since, seeing that. □ **be for it** *colloq.* be about to be punished etc. **for all (that)** in spite of, although. **for ever** for all time (cf. FOREVER). [Old English reduced form of FORE]

for- *prefix* forming verbs etc. meaning: **1** away, off (*forget*; *forgive*). **2** prohibition (*forbid*). **3** abstention or neglect (*forgo*; *forsake*). [Old English]

forage /'fɒrɪdʒ/ *—n.* **1** food for horses and cattle. **2** searching for food. *—v.* **1** search for food; rummage. **2** collect food from. **3** get by foraging. [Germanic: related to FODDER]

forage cap *n.* infantry undress cap.

forasmuch as /,fɒrəz'mʌtʃ/ *conj. archaic* because, since. [from *for as much*]

foray /'fɒreɪ/ *—n.* sudden attack; raid. *—v.* make a foray. [French: related to FODDER]

forbade (also **forbad**) *past* of FORBID.

forbear[1] /fɔ:'beə(r)/ *v.* (*past* **forbore**; *past part.* **forborne**) *formal* abstain or desist (from) (*could not forbear (from) speaking out*; *forbore to mention it*). [Old English: related to BEAR[1]]

forbear[2] var. of FOREBEAR.

forbearance *n.* patient self-control; tolerance.

forbid /fə'bɪd/ *v.* (**forbidding**; *past* **forbade** /-'bæd/ or **forbad**; *past part.* **forbidden**) **1** (foll. by *to* + infin.) order not (*I forbid you to go*). **2** refuse to allow (a thing, or a person to have a thing). **3** refuse a person entry to. □ **God forbid!** may it not happen! [Old English: related to BID]

forbidden degrees *n.pl.* (also **prohibited degrees**) family relationship too close for marriage to be permitted.
forbidden fruit *n.* something desired esp. because not allowed.
forbidding *adj.* stern, threatening. □ **forbiddingly** *adv.*
forbore *past* of FORBEAR[1].
forborne *past part.* of FORBEAR[1].
force[1] –*n.* **1** power; strength, impetus; intense effort. **2** coercion, compulsion. **3** **a** military strength. **b** organized body of soldiers, police, etc. **4** **a** moral, intellectual, or legal power, influence, or validity. **b** person etc. with such power (*force for good*). **5** effect; precise significance. **6** **a** influence tending to cause a change in the motion of a body. **b** intensity of this. –*v.* (**-cing**) **1** compel or coerce (a person) by force. **2** make a forcible entry into; break open by force. **3** drive or propel violently or against resistance. **4** make (a way) by force. **5** (foll. by *on, upon*) impose or press on (a person). **6** cause, produce, or attain by effort (*forced a smile*; *forced an entry*). **7** strain or increase to the utmost. **8** artificially hasten the growth of (a plant). **9** seek quick results from; accelerate (*force the pace*). □ **force a person's hand** make a person act prematurely or unwillingly. **force the issue** make an immediate decision necessary. **in force** **1** valid (*laws now in force*). **2** in great strength or numbers (*attacked in force*). [Latin *fortis* strong]
force[2] *n. N.Engl.* waterfall. [Old Norse]
forced labour *n.* compulsory labour, esp. in prison.
forced landing *n.* emergency landing of an aircraft.
forced march *n.* long and vigorous march, esp. by troops.
force-feed *v.* force (esp. a prisoner) to take food.
forceful *adj.* vigorous, powerful, impressive. □ **forcefully** *adv.* **forcefulness** *n.*
force majeure /ˌfɔːs mæˈʒɜː(r)/ *n.* **1** irresistible force. **2** unforeseeable circumstances excusing a person from the fulfilment of a contract. [French]
forcemeat *n.* minced seasoned meat for stuffing or garnish. [related to FARCE]
forceps /ˈfɔːseps/ *n.* (*pl.* same) surgical pincers. [Latin]
forcible *adj.* done by or involving force; forceful. □ **forcibly** *adv.* [French: related to FORCE[1]]
ford –*n.* shallow place where a river or stream may be crossed by wading, in a vehicle, etc. –*v.* cross (water) at a ford. □ **fordable** *adj.* [Old English]
fore –*adj.* situated in front. –*n.* front part; bow of a ship. –*int.* (in golf) warning to a person in the path of a ball. □ **to the fore** in or into a conspicuous position. [Old English]
fore- *prefix* forming: **1** verbs meaning: **a** in front (*foreshorten*). **b** beforehand (*forewarn*). **2** nouns meaning: **a** situated in front of (*forecourt*). **b** front part of (*forehead*). **c** of or near the bow of a ship (*forecastle*). **d** preceding (*forerunner*).
fore and aft –*adv.* at bow and stern; all over the ship. –*adj.* (**fore-and-aft**) (of a sail or rigging) lengthwise.
forearm[1] /ˈfɔːrɑːm/ *n.* the arm from the elbow to the wrist or fingertips.
forearm[2] /fɔːrˈɑːm/ *v.* arm beforehand, prepare.
forebear /ˈfɔːbeə(r)/ *n.* (also **forbear**) (usu. in *pl.*) ancestor. [from FORE, obsolete *beer*: related to BE]
forebode /fɔːˈbəʊd/ *v.* (**-ding**) **1** be an advance sign of, portend. **2** (often foll. by *that*) have a presentiment of (usu. evil).
foreboding *n.* expectation of trouble.
forecast –*v.* (*past* and *past part.* **-cast** or **-casted**) predict; estimate beforehand. –*n.* prediction, esp. of weather. □ **forecaster** *n.*
forecastle /ˈfəʊks(ə)l/ *n.* (also **fo'c's'le**) forward part of a ship, formerly the living quarters.
foreclose /fɔːˈkləʊz/ *v.* (**-sing**) **1** stop (a mortgage) from being redeemable. **2** repossess the mortgaged property of (a person) when a loan is not duly repaid. **3** exclude, prevent. □ **foreclosure** *n.* [Latin *foris* outside, CLOSE[2]]
forecourt *n.* **1** part of a filling-station with petrol pumps. **2** enclosed space in front of a building.
forefather *n.* (usu. in *pl.*) ancestor of a family or people.
forefinger *n.* finger next to the thumb.
forefoot *n.* front foot of an animal.
forefront *n.* **1** leading position. **2** foremost part.
forego var. of FORGO.
foregoing /fɔːˈgəʊɪŋ/ *adj.* preceding; previously mentioned.
foregone conclusion /ˈfɔːgɒn/ *n.* easily predictable result.
foreground *n.* **1** part of a view or picture nearest the observer. **2** most conspicuous position. [Dutch: related to FORE-, GROUND[1]]
forehand *n.* **1** (in tennis etc.) stroke played with the palm of the hand facing forward. **2** (*attrib.*) (also **forehanded**) of or made with a forehand.
forehead /ˈfɒrɪd/ *n.* the part of the face above the eyebrows.
foreign /ˈfɒrən/ *adj.* **1** of, from, in, or characteristic of, a country or language

other than one's own. **2** dealing with other countries (*foreign service*). **3** of another district, society, etc. **4** (often foll. by *to*) unfamiliar, alien. **5** coming from outside (*foreign body*). □ **foreignness** *n.* [Latin *foris* outside]

Foreign and Commonwealth Office *n.* British government department dealing with foreign affairs.

foreigner *n.* person born in or coming from another country.

foreign legion *n.* body of foreign volunteers in the (esp. French) army.

foreign minister *n.* (also **foreign secretary**) government minister in charge of foreign affairs.

Foreign Office *n. hist.* or *informal* = FOREIGN AND COMMONWEALTH OFFICE.

foreknow /fɔ:'nəʊ/ *v.* (*past* **-knew**, *past part.* **-known**) *literary* know beforehand. □ **foreknowledge** /fɔ:'nɒlɪdʒ/ *n.*

foreland *n.* cape, promontory.

foreleg *n.* front leg of an animal.

forelimb *n.* front limb of an animal.

forelock *n.* lock of hair just above the forehead. □ **touch one's forelock** defer to a person of higher social rank.

foreman *n.* **1** worker supervising others. **2** president and spokesman of a jury.

foremast *n.* mast nearest the bow of a ship.

foremost *–adj.* **1** most notable, best. **2** first, front. *–adv.* most importantly (*first and foremost*). [Old English]

forename *n.* first or Christian name.

forenoon *n.* morning.

forensic /fə'rensɪk/ *adj.* **1** of or used in courts of law (*forensic science*; *forensic medicine*). **2** of or involving forensic science (*sent for forensic examination*).□ **forensically** *adv.* [Latin *forensis*: related to FORUM]

■ **Usage** Use of *forensic* in sense 2 is common but considered an illogical extension of sense 1 by some people.

foreordain /,fɔ:rɔ:'deɪn/ *v.* destine beforehand.

forepaw *n.* front paw of an animal.

foreplay *n.* stimulation preceding sexual intercourse.

forerunner *n.* **1** predecessor. **2** herald.

foresail /'fɔ:seɪl/ *n.* principal sail on a foremast.

foresee /fɔ:'si:/ *v.* (*past* **-saw**; *past part.* **-seen**) see or be aware of beforehand. □ **foreseeable** *adj.*

foreshadow /fɔ:'ʃædəʊ/ *v.* be a warning or indication of (a future event).

foreshore *n.* shore between high- and low-water marks.

foreshorten /fɔ:'ʃɔ:t(ə)n/ *v.* show or portray (an object) with the apparent shortening due to visual perspective.

foresight *n.* **1** regard or provision for the future. **2** foreseeing. **3** front sight of a gun.

foreskin *n.* fold of skin covering the end of the penis.

forest *–n.* **1** (often *attrib.*) large area of trees and undergrowth. **2** trees in this. **3** large number or dense mass. *–v.* **1** plant with trees. **2** convert into a forest. [Latin *forestis*: related to FOREIGN]

forestall /fɔ:'stɔ:l/ *v.* **1** prevent by advance action. **2** deal with beforehand. [from FORE-, STALL[1]]

forester *n.* **1** person managing a forest or skilled in forestry. **2** dweller in a forest.

forestry *n.* science or management of forests.

foretaste *n.* small preliminary experience of something.

foretell /fɔ:'tel/ *v.* (*past* and *past part.* **-told**) **1** predict, prophesy. **2** indicate the approach of.

forethought *n.* **1** care or provision for the future. **2** deliberate intention.

forever /fə'revə(r)/ *adv.* continually, persistently (*is forever complaining*) (cf. *for ever*).

forewarn /fɔ:'wɔ:n/ *v.* warn beforehand.

forewoman *n.* **1** female worker with supervisory responsibilities. **2** president and spokeswoman of a jury.

foreword *n.* introductory remarks at the beginning of a book, often not by the author.

forfeit /'fɔ:fɪt/ *–n.* **1** penalty. **2** thing surrendered as a penalty. *–adj.* lost or surrendered as a penalty. *–v.* (**-t-**) lose the right to, surrender as a penalty. □ **forfeiture** *n.* [French *forfaire* transgress, from Latin *foris* outside, *facio* do]

forgather /fɔ:'gæðə(r)/ *v.* assemble; associate. [Dutch]

forgave *past* of FORGIVE.

forge[1] *–v.* (**-ging**) **1** make or write in fraudulent imitation. **2** shape (metal) by heating and hammering. *–n.* **1** furnace or workshop etc. for melting or refining metal. **2** blacksmith's workshop; smithy. □ **forger** *n.* [Latin *fabrica*: related to FABRIC]

forge[2] *v.* (**-ging**) move forward gradually or steadily. □ **forge ahead 1** take the lead. **2** progress rapidly. [perhaps an alteration of FORCE[1]]

forgery /'fɔ:dʒərɪ/ *n.* (*pl.* **-ies**) **1** act of forging. **2** forged document etc.

forget /fə'get/ *v.* (**forgetting**; *past* **forgot**; *past part.* **forgotten** or *US* **forgot**)

1 (often foll. by *about*) lose remembrance of; not remember. **2** neglect or overlook. **3** cease to think of. □ **forget oneself 1** act without dignity. **2** act selflessly. □ **forgettable** *adj.* [Old English]

forgetful *adj.* **1** apt to forget, absent-minded. **2** (often foll. by *of*) neglectful. □ **forgetfully** *adj.* **forgetfulness** *n.*

forget-me-not *n.* plant with small blue flowers.

forgive /fəˈgɪv/ *v.* (**-ving**; *past* **forgave**; *past part.* **forgiven**) **1** cease to feel angry or resentful towards; pardon. **2** remit (a debt). □ **forgivable** *adj.* [Old English]

forgiveness *n.* forgiving or being forgiven.

forgiving *adj.* inclined to forgive.

forgo /fɔːˈgəʊ/ *v.* (also **forego**) (**-goes**; *past* **-went**; *past part.* **-gone**) go without; relinquish. [Old English]

forgot *past* of FORGET.

forgotten *past part.* of FORGET.

fork –*n.* **1** pronged item of cutlery. **2** similar large tool used for digging, lifting, etc. **3** forked support for a bicycle wheel. **4 a** divergence of a branch, road, etc. into two parts. **b** place of this. **c** either part. –*v.* **1** form a fork or branch by separating into two parts. **2** take one road at a fork. **3** dig, lift, etc., with a fork. □ **fork out** *slang* pay, esp. reluctantly. [Latin *furca* pitchfork]

fork-lift truck *n.* vehicle with a fork for lifting and carrying loads.

forlorn /fəˈlɔːn/ *adj.* **1** sad and abandoned. **2** in a pitiful state. □ **forlornly** *adv.* [*lorn* = past part. of obsolete *leese* LOSE]

forlorn hope *n.* faint remaining hope or chance. [Dutch *verloren hoop* lost troop]

form –*n.* **1** shape; arrangement of parts; visible aspect. **2** person or animal as visible or tangible. **3** mode of existence or manifestation. **4** kind or variety (*a form of art*). **5** printed document with blank spaces for information to be inserted. **6** class in a school. **7** customary method. **8** set order of words. **9** etiquette or specified adherence to it (*good or bad form*). **10** (prec. by *the*) correct procedure (*knows the form*). **11 a** (of an athlete, horse, etc.) condition of health and training. **b** racing history of a horse etc. **12** state or disposition (*in great form*). **13** any of the spellings, inflections, etc. of a word. **14** arrangement and style in a literary or musical composition. **15** long low bench. **16** hare's lair. –*v.* **1** make or be made (*formed a straight line*; *puddles formed*). **2** make up or constitute. **3** develop or establish as a concept, institution, or practice (*form an idea*; *form a habit*). **4** (foll. by *into*) mould or organize to become (*formed ourselves into a circle*). **5** (often foll. by *up*) (of troops etc.) bring or move into formation. **6** train or instruct. □ **off form** not playing or performing well. **on form** playing or performing well. **out of form** not fit for racing etc. [Latin *forma*]

-form *comb. form* (usu. as **-iform**) forming adjectives meaning: **1** having the form of (*cruciform*). **2** having so many forms (*multiform*).

formal /ˈfɔːm(ə)l/ *adj.* **1** in accordance with rules, convention, or ceremony (*formal dress*; *formal occasion*). **2** precise or symmetrical (*formal garden*). **3** prim or stiff. **4** perfunctory, in form only. **5** drawn up etc. correctly; explicit (*formal agreement*). **6** of or concerned with (outward) form, not content or matter. □ **formally** *adv.* [Latin: related to FORM]

formaldehyde /fɔːˈmældɪˌhaɪd/ *n.* colourless pungent gas used as a disinfectant and preservative. [from FORMIC ACID, ALDEHYDE]

formalin /ˈfɔːməlɪn/ *n.* solution of formaldehyde in water.

formalism *n.* strict adherence to external form without regard to content, esp. in art. □ **formalist** *n.*

formality /fɔːˈmælɪtɪ/ *n.* (*pl.* **-ies**) **1 a** formal, esp. meaningless, act, regulation, or custom. **b** thing done simply to comply with a rule. **2** rigid observance of rules or convention.

formalize *v.* (also **-ise**) (**-zing** or **-sing**) **1** give definite (esp. legal) form to. **2** make formal. □ **formalization** /-ˈzeɪʃ(ə)n/ *n.*

format /ˈfɔːmæt/ –*n.* **1** shape and size (of a book, etc.). **2** style or manner of procedure etc. **3** *Computing* arrangement of data etc. –*v.* (**-tt-**) **1** arrange or put into a format. **2** *Computing* prepare (a storage medium) to receive data. [Latin *formatus* shaped: related to FORM]

formation /fɔːˈmeɪʃ(ə)n/ *n.* **1** forming. **2** thing formed. **3** particular arrangement (e.g. of troops). **4** rocks or strata with a common characteristic. [Latin: related to FORM]

formative /ˈfɔːmətɪv/ *adj.* serving to form or fashion; of formation (*formative years*).

forme *n.* *Printing* body of type secured in a chase ready for printing. [var. of FORM]

former *attrib. adj.* **1** of the past, earlier, previous (*in former times*). **2** (**the former**) (often *absol.*) the first or first-mentioned of two. [related to FOREMOST]

-former *comb. form* pupil in a specified form (*fourth-former*).
formerly *adv.* in former times.
Formica /fɔːˈmaɪkə/ *n. propr.* hard durable plastic laminate used for surfaces. [origin uncertain]
formic acid /ˈfɔːmɪk/ *n.* colourless irritant volatile acid contained in fluid emitted by ants; methanoic acid. [Latin *formica* ant]
formidable /ˈfɔːmɪdəb(ə)l, fɔːˈmɪd-/ *adj.* **1** inspiring dread, awe, or respect. **2** hard to overcome or deal with. □ **formidably** *adv.* [Latin *formido* fear]

■ **Usage** The second pronunciation given, with the stress on the second syllable, is common but considered incorrect by some people.

formless *adj.* without definite or regular form. □ **formlessness** *n.*
formula /ˈfɔːmjʊlə/ *n.* (*pl.* **-s** or (esp. in senses 1, 2) **-lae** /-ˌliː/) **1** chemical symbols showing the constituents of a substance. **2** mathematical rule expressed in symbols. **3 a** fixed form of esp. ceremonial or polite words. **b** words used to formulate a treaty etc. **4 a** list of ingredients. **b** *US* infant's food. **5** classification of a racing car, esp. by engine capacity. □ **formulaic** /-ˈleɪɪk/ *adj.* [Latin, diminutive of *forma* FORM]
formulary /ˈfɔːmjʊlərɪ/ *n.* (*pl.* **-ies**) **1** collection of esp. religious formulas or set forms. **2** *Pharm.* compendium of drug formulae. [French or medieval Latin: related to FORMULA]
formulate /ˈfɔːmjʊˌleɪt/ *v.* (**-ting**) **1** express in a formula. **2** express clearly and precisely. □ **formulation** /-ˈleɪʃ(ə)n/ *n.*
fornicate /ˈfɔːnɪˌkeɪt/ *v.* (**-ting**) *archaic* or *joc.* (of people not married to each other) have sexual intercourse. □ **fornication** /-ˈkeɪʃ(ə)n/ *n.* **fornicator** *n.* [Latin *fornix* brothel]
forsake /fɔːˈseɪk/ *v.* (**-king**; *past* **forsook** /-ˈsʊk/; *past part.* **forsaken**) *literary* **1** give up; renounce. **2** desert, abandon. [Old English]
forsooth /fəˈsuːθ/ *adv. archaic* or *joc.* truly; no doubt. [Old English: related to FOR, SOOTH]
forswear /fɔːˈsweə(r)/ *v.* (*past* **forswore**; *past part.* **forsworn**) **1** abjure; renounce. **2** (as **forsworn** *adj.*) perjured. □ **forswear oneself** perjure oneself. [Old English]
forsythia /fɔːˈsaɪθɪə/ *n.* shrub with bright yellow flowers in early spring. [*Forsyth*, name of a botanist]
fort *n.* fortified military building or position. [Latin *fortis* strong]
forte[1] /ˈfɔːteɪ/ *n.* person's strong point or speciality. [feminine of French FORT]
forte[2] /ˈfɔːteɪ/ *Mus.* —*adj.* loud. —*adv.* loudly. —*n.* loud playing or passage. [Italian: related to FORT]
forth *adv. archaic* except in set phrases **1** forward; into view (*bring forth*; *come forth*). **2** onwards in time (*from this time forth*). **3** forwards (*back and forth*). **4** out from a starting-point (*set forth*). □ **and so forth** see SO[1]. [Old English]
forthcoming /fɔːθˈkʌmɪŋ, ˈfɔːθ-/ *adj.* **1** coming or available soon. **2** produced when wanted. **3** (of a person) informative, responsive.
forthright *adj.* **1** outspoken; straightforward. **2** decisive. [Old English]
forthwith /fɔːθˈwɪθ/ *adv.* at once; without delay. [from FORTH]
fortification /ˌfɔːtɪfɪˈkeɪʃ(ə)n/ *n.* **1** act of fortifying. **2** (usu. in *pl.*) defensive works, walls, etc.
fortify /ˈfɔːtɪˌfaɪ/ *v.* (**-ies**, **-ied**) **1** provide with fortifications. **2** strengthen physically, mentally, or morally. **3** strengthen (wine) with alcohol. **4** increase the nutritive value of (food, esp. with vitamins). [Latin *fortis* strong]
fortissimo /fɔːˈtɪsɪˌməʊ/ *Mus.* —*adj.* very loud. —*adv.* very loudly. —*n.* (*pl.* **-s** or **-mi** /-ˌmiː/) very loud playing or passage. [Italian, superlative of FORTE[2]]
fortitude /ˈfɔːtɪˌtjuːd/ *n.* courage in pain or adversity. [Latin *fortis* strong]
fortnight *n.* two weeks. [Old English, = *fourteen nights*]
fortnightly —*adj.* done, produced, or occurring once a fortnight. —*adv.* every fortnight. —*n.* (*pl.* **-ies**) fortnightly magazine etc.
Fortran /ˈfɔːtræn/ *n.* (also **FORTRAN**) computer language used esp. for scientific calculations. [from *for*mula *tran*slation]
fortress /ˈfɔːtrɪs/ *n.* fortified building or town. [Latin *fortis* strong]
fortuitous /fɔːˈtjuːɪtəs/ *adj.* happening by esp. lucky chance; accidental. □ **fortuitously** *adv.* **fortuitousness** *n.* **fortuity** *n.* (*pl.* **-ies**). [Latin *forte* by chance]
fortunate /ˈfɔːtʃənət/ *adj.* **1** lucky. **2** auspicious. □ **fortunately** *adv.* [Latin *fortunatus*: related to FORTUNE]
fortune /ˈfɔːtʃ(ə)n, -tʃuːn/ *n.* **1 a** chance or luck in human affairs. **b** person's destiny. **2** (in *sing.* or *pl.*) luck that befalls a person or enterprise. **3** good luck. **4** prosperity. **5** *colloq.* great wealth. □ **make a** (or **one's**) **fortune** become very rich. [Latin *fortuna*]
fortune-teller *n.* person who claims to foretell one's destiny. □ **fortune-telling** *n.*
forty /ˈfɔːtɪ/ *adj.* & *n.* (*pl.* **-ies**) **1** four times ten. **2** symbol for this (40, xl, XL). **3** (in *pl.*) numbers from 40 to 49, esp. the

years of a century or of a person's life. □ **fortieth** *adj.* & *n.* [Old English: related to FOUR]

forty winks *n. colloq.* short sleep.

forum /'fɔ:rəm/ *n.* **1** place of or meeting for public discussion. **2** court or tribunal. **3** *hist.* public square in an ancient Roman city used for judicial and other business. [Latin]

forward /'fɔ:wəd/ *–adj.* **1** onward; towards the front. **2** lying in the direction in which one is moving. **3** precocious; bold; presumptuous. **4** relating to the future (*forward contract*). **5 a** approaching maturity or completion. **b** (of a plant etc.) early. *–n.* attacking player near the front in football, hockey, etc. *–adv.* **1** to the front; into prominence (*come forward; move forward*). **2** in advance; ahead (*sent them forward*). **3** onward so as to make progress (*no further forward*). **4** towards the future (*from this time forward*). **5** (also **forwards**) **a** towards the front in the direction one is facing. **b** in the normal direction of motion. **c** with continuous forward motion (*rushing forward*). *–v.* **1 a** send (a letter etc.) on to a further destination. **b** dispatch (goods etc.). **2** help to advance; promote. [Old English: related to FORTH, -WARD]

forwent *past* of FORGO.

fosse /fɒs/ *n.* long ditch or trench, esp. in a fortification. [Latin *fossa*]

fossil /'fɒs(ə)l/ *–n.* **1** remains or impression of a (usu. prehistoric) plant or animal hardened in rock. **2** *colloq.* antiquated or unchanging person or thing. *–attrib. adj.* of or like a fossil; antiquated. □ **fossilize** *v.* (also **-ise**) (**-zing** or **-sing**). **fossilization** /-'zeɪʃ(ə)n/ *n.* [Latin *fodio foss-* dig]

fossil fuel *n.* natural fuel extracted from the ground.

foster *–v.* **1 a** promote the growth or development of. **b** encourage or harbour (a feeling). **2 a** bring up (another's child). **b** (of a local authority etc.) assign (a child) to be fostered. **3** (of circumstances) be favourable to. *–attrib. adj.* **1** having a family connection by fostering (*foster-brother; foster-parent*). **2** concerned with fostering a child (*foster care; foster home*). [Old English: related to FOOD]

fought *past* and *past part.* of FIGHT.

foul *–adj.* **1** offensive; loathsome, stinking. **2** soiled, filthy. **3** *colloq.* disgusting, awful. **4 a** noxious (*foul air*). **b** clogged, choked. **5** obscenely abusive (*foul language*). **6** unfair; against the rules (*by fair means or foul*). **7** (of the weather) rough, stormy. **8** (of a rope etc.) entangled. *–n.* **1** *Sport* foul stroke or play. **2** collision, entanglement. *–adv.* unfairly. *–v.* **1** make or become foul. **2** (of an animal) foul with excrement. **3** *Sport* commit a foul against (a player). **4** (often foll. by *up*) **a** (cause to) become entangled or blocked. **b** bungle. **5** collide with. □ **foully** *adv.* **foulness** *n.* [Old English]

foul-mouthed *adj.* using obscene or offensive language.

foul play *n.* **1** unfair play in games. **2** treacherous or violent act, esp. murder.

foul-up *n.* muddle, bungle.

found[1] *past* and *past part.* of FIND.

found[2] *v.* **1** establish (an institution etc.); initiate, originate. **2** be the original builder of (a town etc.). **3** lay the base of (a building). **4** (foll. by *on, upon*) construct or base (a story, theory, rule, etc.) on. □ **founder** *n.* [Latin *fundus* bottom]

found[3] *v.* **1 a** melt and mould (metal). **b** fuse (materials for glass). **2** make by founding. □ **founder** *n.* [Latin *fundo fus-* pour]

foundation /faʊn'deɪʃ(ə)n/ *n.* **1 a** solid ground or base beneath a building. **b** (usu. in *pl.*) lowest part of a building, usu. below ground level. **2** material base. **3** basis, underlying principle. **4 a** establishing (esp. an endowed institution). **b** college, hospital, etc. so founded; its revenues. **5** (in full **foundation garment**) woman's supporting undergarment, e.g. a corset. [Latin: related to FOUND[2]]

foundation-stone *n.* **1** stone laid ceremonially at the founding of a building. **2** basis.

founder *v.* **1** (of a ship) fill with water and sink. **2** (of a plan etc.) fail. **3** (of a horse or its rider) stumble, fall lame, stick in mud etc. [related to FOUND[2]]

founding father *n.* American statesman at the time of the Revolution.

foundling /'faʊndlɪŋ/ *n.* abandoned infant of unknown parentage. [related to FIND]

foundry /'faʊndrɪ/ *n.* (*pl.* **-ies**) workshop for or business of casting metal.

fount[1] *n. poet.* spring or fountain; source. [back-formation from FOUNTAIN]

fount[2] /fɒnt/ *n.* (also **font**) set of printing-type of same face and size. [French: related to FOUND[3]]

fountain /'faʊntɪn/ *n.* **1 a** spouting jet or jets of water as an ornament or for drinking. **b** structure for this. **2** spring. **3** (often foll. by *of*) source. [Latin *fontana* from *fons font-* spring]

fountain-head *n.* source.

fountain-pen *n.* pen with a reservoir or cartridge for ink.

four /fɔ:(r)/ *adj.* & *n.* **1** one more than three. **2** symbol for this (4, iv, IV). **3** size etc. denoted by four. **4** team or crew of four; four-oared rowing-boat. **5** four o'clock. □ **on all fours** on hands and knees. [Old English]

fourfold *adj.* & *adv.* **1** four times as much or as many. **2** of four parts.

four-in-hand *n.* four-horse carriage with one driver.

four-letter word *n.* short obscene word.

four-poster *n.* bed with four posts supporting a canopy.

foursome *n.* **1** group of four people. **2** golf match between two pairs.

four-square *–adj.* **1** solidly based. **2** steady, resolute. *–adv.* steadily, resolutely.

four-stroke *adj.* (of an internal-combustion engine) having a cycle of four strokes of the piston with the cylinder firing once.

fourteen /fɔːˈtiːn/ *adj.* & *n.* **1** one more than thirteen. **2** symbol for this (14, xiv, XIV). **3** size etc. denoted by fourteen. □ **fourteenth** *adj.* & *n.* [Old English: related to FOUR, -TEEN]

fourth *adj.* & *n.* **1** next after third. **2** any of four equal parts of a thing. □ **fourthly** *adv.* [Old English: related to FOUR]

fourth estate *n.* the press.

four-wheel drive *n.* drive acting on all four wheels of a vehicle.

fowl *–n.* (*pl.* same or **-s**) **1** chicken kept for eggs and meat. **2** poultry as food. **3** *archaic* (except in *comb.*) bird (*guinea-fowl*; *wildfowl*). *–v.* catch or hunt wildfowl. [Old English]

fox *–n.* **1 a** wild canine animal with a bushy tail and red or grey fur. **b** its fur. **2** cunning person. *–v.* **1** deceive, baffle, trick. **2** (usu. as **foxed** *adj.*) discolour (leaves of a book etc.) with brownish marks. □ **foxlike** *adj.* [Old English]

foxglove *n.* tall plant with purple or white flowers like glove-fingers.

foxhole *n.* hole in the ground used as a shelter etc. in battle.

foxhound *n.* a kind of hound bred and trained to hunt foxes.

fox-hunting *n.* hunting foxes with hounds.

fox-terrier *n.* a kind of short-haired terrier.

foxtrot *–n.* **1** ballroom dance with slow and quick steps. **2** music for this. *–v.* (**-tt-**) perform this.

foxy *adj.* (**-ier, -iest**) **1** foxlike. **2** sly or cunning. **3** reddish-brown. □ **foxily** *adv.* **foxiness** *n.*

foyer /ˈfɔɪeɪ/ *n.* entrance-hall in a hotel, theatre, etc. [French, = hearth, home, from Latin FOCUS]

FPA *abbr.* Family Planning Association.

Fr *symb.* francium.

Fr. *abbr.* **1** Father. **2** French.

fr. *abbr.* franc(s).

fracas /ˈfrækɑː/ *n.* (*pl.* same /-kɑːz/) noisy disturbance or quarrel. [French from Italian]

fraction /ˈfrækʃ(ə)n/ *n.* **1** part of a whole number (e.g. $^1/_2$, 0.5). **2** small part, piece, or amount. **3** portion of a mixture obtained by distillation etc. □ **fractional** *adj.* **fractionally** *adv.* [Latin *frango fract-* break]

fractious /ˈfrækʃəs/ *adj.* irritable, peevish. [from FRACTION in obsolete sense 'brawling']

fracture /ˈfræktʃə(r)/ *–n.* breakage, esp. of a bone or cartilage. *–v.* (**-ring**) cause a fracture in; suffer fracture. [Latin: related to FRACTION]

fragile /ˈfrædʒaɪl/ *adj.* **1** easily broken; weak. **2** delicate; not strong. □ **fragility** /frəˈdʒɪlɪtɪ/ *n.* [Latin: related to FRACTURE]

fragment *–n.* /ˈfrægmənt/ **1** part broken off. **2** extant remains or unfinished portion (of a book etc.). *–v.* /frægˈment/ break or separate into fragments. □ **fragmental** /-ˈment(ə)l/ *adj.* **fragmentary** /ˈfrægməntərɪ/ *adj.* **fragmentation** /-ˈteɪʃ(ə)n/ *n.* [Latin: related to FRACTION]

fragrance /ˈfreɪgrəns/ *n.* **1** sweetness of smell. **2** sweet scent. [Latin *fragro* smell sweet]

fragrant *adj.* sweet-smelling.

frail *adj.* **1** fragile, delicate. **2** morally weak. □ **frailly** *adv.* **frailness** *n.* [Latin: related to FRAGILE]

frailty *n.* (*pl.* **-ies**) **1** frail quality. **2** weakness, foible.

frame *–n.* **1** case or border enclosing a picture, window, door, etc. **2** basic rigid supporting structure of a building, vehicle, etc. **3** (in *pl.*) structure of spectacles holding the lenses. **4** human or animal body, esp. as large or small. **5 a** established order or system (*the frame of society*). **b** construction, build, structure. **6** temporary state (esp. in **frame of mind**). **7** single complete image on a cinema film or transmitted in a series of lines by television. **8 a** triangular structure for positioning balls in snooker etc. **b** round of play in snooker etc. **9** boxlike structure of glass etc. for protecting plants. **10** *US slang* = FRAME-UP. *–v.* (**-ming**) **1 a** set in a frame. **b** serve as a frame for. **2** construct, put together, devise. **3** (foll. by *to, into*) adapt or fit. **4** *slang* concoct a false charge or evidence against; devise a plot against. **5** articulate (words). [Old English, = be helpful]

frame of reference *n.* **1** set of standards or principles governing behaviour, thought, etc. **2** system of geometrical axes for defining position.

frame-up *n. slang* conspiracy to convict an innocent person.

framework *n.* **1** essential supporting structure. **2** basic system.

franc *n.* unit of currency of France, Belgium, Switzerland, etc. [French: related to FRANK]

franchise /'fræntʃaɪz/ *–n.* **1** right to vote in State elections. **2** full membership of a corporation or State; citizenship. **3** authorization to sell a company's goods etc. in a particular area. **4** right or privilege granted to a person or corporation. *–v.* **(-sing)** grant a franchise to. [French *franc* FRANK]

Franciscan /fræn'sɪskən/ *–adj.* of St Francis or his order. *–n.* Franciscan friar or nun. [Latin *Franciscus* Francis]

francium /'fræŋkɪəm/ *n.* radioactive metallic element. [*France*, the discoverer's country]

Franco- *comb. form* French and (*Franco-German*). [Latin: related to FRANK]

franglais /'frɑ̃gleɪ/ *n.* corrupt version of French using many English words and idioms. [French *français* French, *anglais* English]

Frank *n.* member of the Germanic people that conquered Gaul in the 6th c. □ **Frankish** *adj.* [Old English]

frank *–adj.* **1** candid, outspoken. **2** undisguised. **3** open. *–v.* mark (a letter) to record the payment of postage. *–n.* franking signature or mark. □ **frankly** *adv.* **frankness** *n.* [Latin *francus* free: related to FRANK]

Frankenstein /'fræŋkən,staɪn/ *n.* (in full **Frankenstein's monster**) thing that becomes terrifying to its maker. [*Frankenstein*, name of a character in and title of a novel by Mary Shelley]

frankfurter /'fræŋk,fɜ:tə(r)/ *n.* seasoned smoked sausage. [German from *Frankfurt* in Germany]

frankincense /'fræŋkɪn,sens/ *n.* aromatic gum resin burnt as incense. [French: related to FRANK in obsolete sense 'high quality', INCENSE[1]]

frantic /'fræntɪk/ *adj.* **1** wildly excited; frenzied. **2** hurried, anxious; desperate, violent. **3** *colloq.* extreme. □ **frantically** *adv.* [Latin: related to FRENETIC]

frappé /'fræpeɪ/ *adj.* iced, cooled. [French]

fraternal /frə'tɜ:n(ə)l/ *adj.* **1** of brothers, brotherly; comradely. **2** (of twins) developed from separate ova and not necessarily similar. □ **fraternally** *adv.* [Latin *frater* brother]

fraternity /frə'tɜ:nɪtɪ/ *n.* (*pl.* **-ies**) **1** religious brotherhood. **2** group with common interests, or of the same professional class. **3** *US* male students' society. **4** brotherliness. [Latin: related to FRATERNAL]

fraternize /'frætə,naɪz/ *v.* (also **-ise**) (**-zing** or **-sing**) (often foll. by *with*) **1** associate; make friends. **2** enter into friendly relations with enemies etc. □ **fraternization** /-'zeɪʃ(ə)n/ *n.* [French and Latin: related to FRATERNAL]

fratricide /'frætrɪ,saɪd/ *n.* **1** killing of one's brother or sister. **2** person who does this. □ **fratricidal** /-'saɪd(ə)l/ *adj.* [Latin *frater* brother]

Frau /fraʊ/ *n.* (*pl.* **Frauen** /'fraʊən/) (often as a title) married or widowed German-speaking woman. [German]

fraud /frɔ:d/ *n.* **1** criminal deception. **2** dishonest artifice or trick. **3** person or thing that is not what it claims to be. [Latin *fraus fraud-*]

fraudulent /'frɔ:djʊlənt/ *adj.* of, involving, or guilty of fraud. □ **fraudulence** *n.* **fraudulently** *adv.* [Latin: related to FRAUD]

fraught /frɔ:t/ *adj.* **1** (foll. by *with*) filled or charged with (danger etc.). **2** *colloq.* distressing; tense. [Dutch *vracht* FREIGHT]

Fräulein /'frɔɪlaɪn/ *n.* (often as a title or form of address) unmarried German-speaking woman. [German]

fray[1] *v.* **1** wear through or become worn; esp. (of woven material) unravel at the edge. **2** (of nerves, temper, etc.) become strained. [Latin *frico* rub]

fray[2] *n.* **1** conflict, fight. **2** brawl. [related to AFFRAY]

frazzle /'fræz(ə)l/ *colloq.* *–n.* worn, exhausted, or shrivelled state (*burnt to a frazzle*). *–v.* (**-ling**) (usu. as **frazzled** *adj.*) wear out; exhaust. [origin uncertain]

freak *–n.* **1** (often *attrib.*) monstrosity; abnormal person or thing (*freak storm*). **2** *colloq.* **a** unconventional person. **b** fanatic of a specified kind (*health freak*). **c** drug addict. *–v.* (often foll. by *out*) *colloq.* **1** become or make very angry. **2** (cause to) undergo hallucinations etc., esp. as a result of drug abuse. **3** adopt an unconventional lifestyle. □ **freakish** *adj.* **freaky** *adj.* (**-ier, -iest**). [probably from dial.]

freckle /'frek(ə)l/ *–n.* small light brown spot on the skin. *–v.* (**-ling**) (usu. as **freckled** *adj.*) spot or be spotted with freckles. □ **freckly** *adj.* [Old Norse]

free *–adj.* (**freer** /'fri:ə(r)/; **freest** /'fri:ɪst/) **1** not a slave or under another's control; having personal rights and social and political liberty. **2** (of a State, its citizens, etc.) autonomous; democratic. **3 a** unrestricted; not confined or fixed. **b** not imprisoned. **c** released from duties etc. **d** independent (*free agent*). **4** (foll. by *of*, *from*) **a** exempt from (tax etc.). **b** not containing or subject to (*free of preservatives*; *free from disease*). **5** (foll. by *to* + infin.) permitted; at liberty to. **6** costing nothing. **7 a** clear of duties etc. (*am free tomorrow*). **b** not in use (*bathroom is free*). **8** spontaneous, unforced (*free offer*). **9** available to all. **10** lavish (*free with their money*). **11** frank, unreserved. **12** (of literary style) informal, unmetrical. **13** (of translation) not literal. **14** familiar, impudent. **15** (of stories etc.) slightly indecent. **16** *Chem.* not combined (*free oxygen*). **17** (of power or energy) disengaged, available. *–adv.* **1** freely. **2** without cost or payment. *–v.* (**frees, freed**) **1** make free; liberate. **2** (foll. by *of*, *from*) relieve from. **3** disentangle, clear. □ **for free** *colloq.* free of charge, gratis. **free on board** (or **rail**) without charge for delivery to a ship or railway wagon. □ **freely** *adv.* [Old English]

-free *comb. form* free of or from (*worry-free*; *duty-free*).

free and easy *adj.* informal, relaxed.

freebie /'fri:bɪ/ *n. colloq.* thing given free of charge.

freeboard *n.* part of a ship's side between the water-line and deck.

freebooter *n.* pirate. [Dutch *vrijbuiter*: related to FREE, BOOTY]

free-born *adj.* not born a slave.

Free Church *n.* Nonconformist Church.

freedman *n.* emancipated slave.

freedom /'fri:dəm/ *n.* **1** condition of being free or unrestricted. **2** personal or civic liberty. **3** liberty of action (*freedom to leave*). **4** (foll. by *from*) exemption from. **5** (foll. by *of*) **a** honorary membership or citizenship (*freedom of the city*). **b** unrestricted use of (a house etc.). [Old English]

freedom fighter *n.* terrorist or rebel claiming to fight for freedom.

free enterprise *n.* freedom of private business from State control.

free fall *n.* movement under the force of gravity only.

free fight *n.* general fight in which all present join.

freefone *n.* (also **Freefone, -phone**) system whereby certain telephone calls, esp. on business, can be made without cost to the caller.

free-for-all *n.* free fight, unrestricted discussion, etc.

free-form *attrib. adj.* of irregular shape or structure.

freehand *–adj.* (of a drawing etc.) done without special instruments. *–adv.* in a freehand manner.

free hand *n.* freedom to act at one's own discretion.

free-handed *adj.* generous.

freehold *–n.* **1** complete ownership of property for an unlimited period. **2** such land or property. *–adj.* owned thus. □ **freeholder** *n.*

free house *n.* public house not controlled by a brewery.

free kick *n.* kick granted in football as a minor penalty.

freelance *–n.* **1** (also **freelancer**) person, usu. self-employed, working for several employers on particular assignments. **2** (*attrib.*) (*freelance editor*). *–v.* (**-cing**) act as a freelance. *–adv.* as a freelance. [*free lance*, a medieval mercenary]

freeloader *n. slang* sponger. □ **freeload** /-'ləʊd/ *v.*

free love *n.* sexual freedom.

freeman *n.* **1** person who has the freedom of a city etc. **2** person who is not a slave or serf.

free market *n.* market governed by unrestricted competition.

Freemason /'fri:,meɪs(ə)n/ *n.* member of an international fraternity for mutual help, having elaborate secret rituals. □ **Freemasonry** *n.*

freephone var. of FREEFONE.

free port *n.* **1** port without customs duties. **2** port open to all traders.

freepost *n.* system of business post where postage is paid by the addressee.

free radical *n. Chem.* atom or group of atoms with one or more unpaired electrons.

free-range *adj.* **1** (of hens etc.) roaming freely; not kept in a battery. **2** (of eggs) produced by such hens.

freesia /'fri:zjə/ *n.* African bulb with fragrant flowers. [*Freese*, name of a physician]

free speech *n.* right of expression.

free spirit *n.* independent or uninhibited person.

free-spoken *adj.* forthright.

free-standing *adj.* not supported by another structure.

freestyle *n.* **1** swimming race in which any stroke may be used. **2** wrestling allowing almost any hold.

freethinker /fri:'θɪŋkə(r)/ *n.* person who rejects dogma or authority, esp. in religious belief. □ **freethinking** *n.* & *adj.*

free trade *n.* trade without import restrictions etc.

free vote *n.* parliamentary vote not subject to party discipline.

freeway *n. US* motorway.

free wheel *n.* driving wheel of a bicycle, able to revolve with the pedals at rest.

free-wheel *v.* **1** ride a bicycle with the pedals at rest. **2** act without constraint.

free will *n.* **1** power of acting independently of necessity or fate. **2** ability to act without coercion (*did it of my own free will*).

free world *n. hist.* non-Communist countries' collective name for themselves.

freeze –*v.* (**-zing**; *past* **froze**; *past part.* **frozen**) **1 a** turn into ice or another solid by cold. **b** make or become rigid from the cold. **2** be or feel very cold. **3** cover or become covered with ice. **4** (foll. by *to, together*) adhere by frost. **5** refrigerate (food) below freezing point. **6 a** make or become motionless through fear, surprise, etc. **b** (as **frozen** *adj*) devoid of emotion (*frozen smile*). **7** make (assets etc.) unrealizable. **8** fix (prices, wages, etc.) at a certain level. **9** stop (the movement in a film). –*n.* **1** period or state of frost. **2** fixing or stabilization of prices, wages, etc. **3** (in full **freeze-frame**) still film-shot. □ **freeze up** obstruct or be obstructed by ice. [Old English]

freeze-dry *v.* preserve (food) by freezing and then drying in a vacuum.

freezer *n.* refrigerated cabinet etc. for preserving frozen food at very low temperatures.

freeze-up *n.* period or state of extreme cold.

freezing point *n.* temperature at which a liquid, esp. water, freezes.

freight /freɪt/ –*n.* **1** transport of goods in containers or by water or air, or (*US*) by land. **2** goods transported; cargo, load. **3** charge for the transport of goods. –*v.* transport as or load with freight. [Low German or Dutch *vrecht*]

freighter *n.* **1** ship or aircraft for carrying freight. **2** *US* freight-wagon.

freightliner *n.* train carrying goods in containers.

French –*adj.* **1** of France, its people, or language. **2** having French characteristics. –*n.* **1** the French language. **2** (**the French**) (*pl.*) the people of France. **3** *colloq.* dry vermouth. □ **Frenchness** *n.* [Old English: related to FRANK]

French bean *n.* kidney or haricot bean as unripe sliced pods or ripe seeds.

French bread *n.* long crisp loaf.

French Canadian *n.* Canadian whose principal language is French.

French chalk *n.* a kind of talc used for marking cloth, as a dry lubricant, etc.

French dressing *n.* salad dressing of seasoned vinegar and oil.

French fried potatoes *n.pl.* (also **French fries**) *US* potato chips.

French horn *n.* coiled brass wind instrument with a wide bell.

Frenchify /ˈfrentʃɪˌfaɪ/ *v.* (**-ies**, **-ied**) (usu. as **Frenchified** *adj.*) *colloq.* make French in form, manners, etc.

French kiss *n.* open-mouthed kiss.

French leave *n.* absence without permission.

French letter *n. colloq.* condom.

Frenchman *n.* man of French birth or nationality.

French polish *n.* shellac polish for wood. □ **French-polish** *v.*

French window *n.* glazed door in an outside wall.

Frenchwoman *n.* woman of French birth or nationality.

frenetic /frəˈnetɪk/ *adj.* (also **phrenetic**) **1** frantic, frenzied. **2** fanatic. □ **frenetically** *adv.* [Greek *phrēn* mind]

frenzy /ˈfrenzɪ/ –*n.* (*pl.* **-ies**) wild or delirious excitement, agitation, or fury. –*v.* (**-ies**, **-ied**) (usu. as **frenzied** *adj.*) drive to frenzy. □ **frenziedly** *adv.* [medieval Latin: related to FRENETIC]

frequency /ˈfriːkwənsɪ/ *n.* (*pl.* **-ies**) **1** commonness of occurrence. **2** frequent occurrence. **3** rate of recurrence (of a vibration etc.); number of repetitions in a given time, esp. per second. [related to FREQUENT]

frequency modulation *n. Electronics* modulation by varying carrier-wave frequency.

frequent –*adj.* /ˈfriːkwənt/ **1** occurring often or in close succession. **2** habitual, constant. –*v.* /frɪˈkwent/ attend or go to habitually. □ **frequently** /ˈfriːkwəntlɪ/ *adv.* [Latin *frequens -ent-* crowded]

frequentative /frɪˈkwentətɪv/ *Gram.* –*adj.* (of a verb etc.) expressing frequent repetition or intensity. –*n.* frequentative verb etc.

fresco /ˈfreskəʊ/ *n.* (*pl.* **-s**) painting done in water-colour on a wall or ceiling before the plaster is dry. [Italian, = fresh]

fresh –*adj.* **1** newly made or obtained. **2 a** other, different; new (*start a fresh page*; *fresh ideas*). **b** additional (*fresh supplies*). **3** (foll. by *from*) lately arrived. **4** not stale, musty, or faded. **5** (of food) not preserved; newly caught, grown, etc. **6** not salty (*fresh water*). **7 a** pure, untainted, refreshing (*fresh air*). **b** bright and pure in colour (*fresh complexion*). **8** (of wind) brisk. **9** *colloq.* cheeky; amorously impudent. **10** inexperienced. –*adv.* newly, recently (esp. in *comb.*: *fresh-baked*). □ **freshly** *adv.* **freshness** *n.* [Old English *fersc* and French *freis*]

freshen *v.* **1** make or become fresh. **2** (foll. by *up*) **a** wash, tidy oneself, etc. **b** revive.
fresher *n. colloq.* first-year student at university or (*US*) high school.
freshet /ˈfreʃɪt/ *n.* **1** rush of fresh water flowing into the sea. **2** flood of a river.
freshman *n.* = FRESHER.
freshwater *attrib. adj.* (of fish etc.) not of the sea.
fret[1] *–v.* (**-tt-**) **1** be worried or distressed. **2** worry, vex. **3** wear or consume by gnawing or rubbing. *–n.* worry, vexation. [Old English: related to FOR, EAT]
fret[2] *–n.* ornamental pattern of straight lines joined usu. at right angles. *–v.* (**-tt-**) embellish with a fret or with carved or embossed work. [French *freter*]
fret[3] *n.* each of a series of bars or ridges on the finger-board of a guitar etc. to guide fingering. [origin unknown]
fretful *adj.* anxious, irritable. □ **fretfully** *adv.*
fretsaw *n.* narrow saw on a frame for cutting thin wood in patterns.
fretwork *n.* ornamental work in wood done with a fretsaw.
Freudian /ˈfrɔɪdɪən/ *–adj.* of Freud, his theories, or his method of psychoanalysis. *–n.* follower of Freud.
Freudian slip *n.* unintentional verbal error revealing subconscious feelings.
Fri. *abbr.* Friday.
friable /ˈfraɪəb(ə)l/ *adj.* easily crumbled. □ **friability** /-ˈbɪlɪtɪ/ *n.* [Latin *frio* crumble]
friar /ˈfraɪə(r)/ *n.* member of a male non-enclosed Roman Catholic order, e.g. Carmelites and Franciscans. [Latin *frater* brother]
friar's balsam *n.* tincture of benzoin etc. used esp. as an inhalant.
friary /ˈfraɪərɪ/ *n.* (*pl.* **-ies**) monastery for friars.
fricassee /ˈfrɪkəˌseɪ/ *–n.* pieces of meat served in a thick sauce. *–v.* (**fricassees, fricasseed**) make a fricassee of. [French]
fricative /ˈfrɪkətɪv/ *–adj.* (of a consonant) sounded by friction of the breath in a narrow opening. *–n.* such a consonant (e.g. *f*, *th*). [Latin *frico* rub]
friction /ˈfrɪkʃ(ə)n/ *n.* **1** rubbing of one object against another. **2** the resistance encountered in so moving. **3** clash of wills, opinions, etc. □ **frictional** *adj.* [Latin: related to FRICATIVE]
Friday /ˈfraɪdeɪ/ *–n.* day of the week following Thursday. *–adv. colloq.* **1** on Friday. **2** (**Fridays**) on Fridays; each Friday. [Old English]
fridge *n. colloq.* = REFRIGERATOR. [abbreviation]
fridge-freezer *n.* combined refrigerator and freezer.
friend /frend/ *n.* **1** person one likes and chooses to spend time with (usu. without sexual or family bonds). **2** sympathizer, helper. **3** ally or neutral person (*friend or foe?*). **4** person already mentioned (*our friend at the bank*). **5** regular supporter of an institution. **6** (**Friend**) Quaker. [Old English]
friendly *–adj.* (**-ier, -iest**) **1** outgoing, well-disposed, kindly. **2 a** (often foll. by *with*) on amicable terms. **b** not hostile. **3** (in *comb.*) not harming; helping (*ozone-friendly*; *user-friendly*). **4** = USER-FRIENDLY. *–n.* (*pl.* **-ies**) = FRIENDLY MATCH. *–adv.* in a friendly manner. □ **friendliness** *n.*
friendly match *n.* match played for enjoyment rather than competition.
Friendly Society *n.* insurance society insuring against illness etc.
friendship *n.* friendly relationship or feeling.
frier var. of FRYER.
Friesian /ˈfriːzɪən/ *n.* one of a breed of black and white dairy cattle orig. from Friesland. [var. of FRISIAN]
frieze /friːz/ *n.* **1** part of an entablature between the architrave and cornice. **2** horizontal band of sculpture filling this. **3** band of decoration, esp. at the top of a wall. [Latin *Phrygium* (*opus*) Phrygian (work)]
frig *v.* (**-gg-**) *coarse slang* **1** = FUCK *v.* **2** masturbate. [perhaps imitative]
frigate /ˈfrɪgɪt/ *n.* **1** naval escort-vessel. **2** *hist.* warship. [French from Italian]
fright /fraɪt/ *n.* **1 a** sudden or extreme fear. **b** instance of this (*gave me a fright*). **2** grotesque-looking person or thing. □ **take fright** become frightened. [Old English]
frighten *v.* **1** fill with fright (*the bang frightened me*; *frightened of dogs*). **2** (foll. by *away*, *off*, *out of*, *into*) drive by fright. □ **frightening** *adj.* **frighteningly** *adv.*
frightful *adj.* **1 a** dreadful, shocking. **b** ugly. **2** *colloq.* extremely bad. **3** *colloq.* extreme (*frightful rush*). □ **frightfully** *adv.*
frigid /ˈfrɪdʒɪd/ *adj.* **1** unfriendly, cold (*frigid stare*). **2** (of a woman) sexually unresponsive. **3** (esp. of a climate or air) cold. □ **frigidity** /-ˈdʒɪdɪtɪ/ *n.* [Latin *frigus* (n.) cold]
frill *–n.* **1** strip of gathered or pleated material as an ornamental edging. **2** (in *pl.*) unnecessary embellishments. *–v.* decorate with a frill. □ **frilly** *adj.* (**-ier, -iest**). [origin unknown]
fringe *–n.* **1** border of tassels or loose threads. **2** front hair hanging over the

forehead. **3** outer limit of an area, population, etc. (often *attrib.*: *fringe theatre*). **4** unimportant area or part. –*v.* (**-ging**) **1** adorn with a fringe. **2** serve as a fringe to. [Latin *fimbria*]

fringe benefit *n.* employee's benefit additional to salary.

frippery /ˈfrɪpərɪ/ *n.* (*pl.* **-ies**) **1** showy finery, esp. in dress. **2** empty display in speech, literary style, etc. **3** (usu. in *pl.*) knick-knacks. [French *friperie*]

Frisbee /ˈfrɪzbɪ/ *n. propr.* concave plastic disc for skimming through the air as an outdoor game. [perhaps from *Frisbie* bakery pie-tins]

Frisian /ˈfrɪzɪən/ –*adj.* of Friesland. –*n.* native or language of Friesland. [Latin *Frisii* (n. pl.) from Old Frisian *Frīsa*]

frisk –*v.* **1** leap or skip playfully. **2** *slang* search (a person) for a weapon etc. by feeling. –*n.* **1** playful leap or skip. **2** *slang* frisking of a person. [French *frisque* lively]

frisky *adj.* (**-ier, -iest**) lively, playful. □ **friskily** *adv.* **friskiness** *n.*

frisson /ˈfriːsɒn/ *n.* emotional thrill. [French]

frith var. of FIRTH.

fritillary /frɪˈtɪlərɪ/ *n.* (*pl.* **-ies**) **1** plant with bell-like flowers. **2** butterfly with red-brown wings chequered with black. [Latin *fritillus* dice-box]

fritter[1] *v.* (usu. foll. by *away*) waste (money, time, etc.) triflingly. [obsolete *fritter*(*s*) fragments]

fritter[2] *n.* fruit, meat, etc. coated in batter and fried. [French *friture* from Latin *frigo* FRY[1]]

frivolous /ˈfrɪvələs/ *adj.* **1** not serious, silly, shallow. **2** paltry, trifling. □ **frivolity** /-ˈvɒlɪtɪ/ *n.* (*pl.* **-ies**). **frivolously** *adv.* **frivolousness** *n.* [Latin]

frizz –*v.* form (hair) into tight curls. –*n.* frizzed hair or state. [French *friser*]

frizzle[1] /ˈfrɪz(ə)l/ *v.* (**-ling**) **1** fry or cook with a sizzling noise. **2** (often foll. by *up*) burn or shrivel. [obsolete *frizz*: related to FRY[1], with imitative ending]

frizzle[2] /ˈfrɪz(ə)l/ –*v.* (**-ling**) form into tight curls. –*n.* frizzled hair. [perhaps related to FRIZZ]

frizzy *adj.* (**-ier, -iest**) in tight curls.

fro *adv.* back (now only in *to and fro*: see TO). [Old Norse: related to FROM]

frock *n.* **1** woman's or girl's dress. **2** monk's or priest's gown. **3** smock. [French from Germanic]

frock-coat *n.* man's long-skirted coat.

frog[1] *n.* **1** small smooth tailless leaping amphibian. **2** (**Frog**) *slang offens.* Frenchman. □ **frog in one's throat** *colloq.* hoarseness. [Old English]

frog[2] *n.* horny substance in the sole of a horse's foot. [origin uncertain: perhaps a use of FROG[1]]

frog[3] *n.* ornamental coat-fastening of a button and loop. [origin unknown]

frogman *n.* person with a rubber suit, flippers, and an oxygen supply for underwater swimming.

frogmarch *v.* hustle forward with the arms pinned behind.

frog-spawn *n.* frog's eggs.

frolic /ˈfrɒlɪk/ –*v.* (**-ck-**) play about cheerfully. –*n.* **1** cheerful play. **2** prank. **3** merry party. [Dutch *vrolijk* (adj.) from *vro* glad]

frolicsome *adj.* merry, playful.

from /frəm, frɒm/ *prep.* expressing separation or origin, followed by: **1** person, place, time, etc., that is the starting-point (*dinner is served from 8*; *from start to finish*). **2** place, object, etc. at a specified distance etc. (*10 miles from Rome*; *far from sure*). **3 a** source (*gravel from a pit*; *quotations from Shaw*). **b** giver or sender (*not heard from her*). **4** thing or person avoided, deprived, etc. (*released him from prison*; *took his gun from him*). **5** reason, cause, motive (*died from fatigue*; *did it from jealousy*). **6** thing distinguished or unlike (*know black from white*). **7** lower limit (*from 10 to 20 boats*). **8** state changed for another (*from being poor he became rich*). **9** adverb or preposition of time or place (*from long ago*; *from abroad*; *from under the bed*). □ **from time to time** occasionally. [Old English]

fromage frais /ˌfrɒmɑːʒ ˈfreɪ/ *n.* smooth low-fat soft cheese.

frond *n.* leaflike part of a fern or palm. [Latin *frons frond-* leaf]

front /frʌnt/ –*n.* **1** side or part most prominent or important, or nearer the spectator or direction of motion (*front of the house*). **2 a** line of battle. **b** ground towards an enemy. **c** scene of actual fighting. **3 a** activity compared to a military front. **b** organized political group. **4** demeanour, bearing. **5** forward or conspicuous position. **6 a** bluff. **b** pretext. **7** person etc. as a cover for subversive or illegal activities. **8** promenade. **9** forward edge of advancing cold or warm air. **10** auditorium of a theatre. **11** breast of a garment (*spilt food down his front*). –*attrib. adj.* **1** of the front. **2** situated in front. –*v.* **1** (foll. by *on, to, towards, upon*) have the front facing or directed towards. **2** (foll. by *for*) *slang* act as a front or cover for. **3** provide with or have a front (*fronted with stone*). **4** lead (a band, organization, etc.). □ **in front** in an advanced or facing position.

in front of 1 ahead of, in advance of. **2** in the presence of. [Latin *frons front-* face]

frontage *n.* **1** front of a building. **2** land next to a street or water etc. **3** extent of a front. **4 a** the way a thing faces. **b** outlook.

frontal *adj.* **1** of or on the front (*frontal view*; *frontal attack*). **2** of the forehead (*frontal bone*).

front bench *n.* seats in Parliament occupied by leading members of the government and opposition.

front-bencher *n.* MP occupying the front bench.

frontier /'frʌntɪə(r)/ *n.* **1 a** border between two countries. **b** district on each side of this. **2** limits of attainment or knowledge in a subject. **3** *US* borders between settled and unsettled country. □ **frontiersman** *n.*

frontispiece /'frʌntɪs,pi:s/ *n.* illustration facing the title-page of a book. [Latin: related to FRONT, *specio* look]

front line *n.* foremost part of an army or group under attack.

front runner *n.* favourite in a race etc.

frost –*n.* **1 a** frozen dew or vapour. **b** consistent temperature below freezing point. **2** cold dispiriting atmosphere. –*v.* **1** (usu. foll. by *over*, *up*) become covered with frost. **2 a** cover with or as with frost. **b** injure (a plant etc.) with frost. **3** make (glass) non-transparent by roughening its surface. [Old English: related to FREEZE]

frostbite *n.* injury to body tissues due to freezing. □ **frostbitten** *adj.*

frosting *n.* icing.

frosty *adj.* (**-ier, -iest**) **1** cold with frost. **2** covered with or as with frost. **3** unfriendly in manner. □ **frostily** *adv.* **frostiness** *n.*

froth –*n.* **1** foam. **2** idle or amusing talk etc. –*v.* **1** emit or gather froth. **2** cause (beer etc.) to foam. □ **frothy** *adj.* (**-ier, -iest**). [Old Norse]

frown –*v.* **1** wrinkle one's brows, esp. in displeasure or concentration. **2** (foll. by *at*, *on*) disapprove of. –*n.* **1** act of frowning. **2** look of displeasure or concentration. [French]

frowsty /'fraʊstɪ/ *adj.* (**-ier, -iest**) fusty, stuffy. [var. of FROWZY]

frowzy /'fraʊzɪ/ *adj.* (also **frowsy**) (**-ier, -iest**) **1** fusty. **2** slatternly, dingy. [origin unknown]

froze *past* of FREEZE.

frozen *past part.* of FREEZE.

FRS *abbr.* Fellow of the Royal Society.

fructify /'frʌktɪ,faɪ/ *v.* (**-ies, -ied**) **1** bear fruit. **2** make fruitful. [Latin: related to FRUIT]

fructose /'frʌktəʊz/ *n.* sugar in honey, fruits, etc. [Latin: related to FRUIT]

frugal /'fru:g(ə)l/ *adj.* **1** sparing or thrifty, esp. as regards food. **2** meagre, cheap. □ **frugality** /-'gælɪtɪ/ *n.* **frugally** *adv.* [Latin]

fruit /fru:t/ –*n.* **1 a** seed-bearing part of a plant or tree; this as food. **b** these collectively. **2** (usu. in *pl.*) vegetables, grains, etc. as food (*fruits of the earth*). **3** (usu. in *pl.*) profits, rewards. –*v.* (cause to) bear fruit. [Latin *fructus* from *fruor* enjoy]

fruit cake *n.* cake containing dried fruit.

fruit cocktail *n.* diced fruit salad.

fruiterer *n.* dealer in fruit.

fruitful *adj.* **1** producing much fruit. **2** successful, profitable. □ **fruitfully** *adv.*

fruition /fru:'ɪʃ(ə)n/ *n.* **1** bearing of fruit. **2** realization of aims or hopes. [Latin: related to FRUIT]

fruit juice *n.* juice of fruit, esp. as a drink.

fruitless *adj.* **1** not bearing fruit. **2** useless, unsuccessful. □ **fruitlessly** *adv.*

fruit machine *n.* coin-operated gaming machine using symbols representing fruit.

fruit sugar *n.* fructose.

fruity *adj.* (**-ier, -iest**) **1 a** of fruit. **b** tasting or smelling like fruit. **2** (of a voice etc.) deep and rich. **3** *colloq.* slightly indecent or suggestive. □ **fruitily** *adv.* **fruitiness** *n.*

frump *n.* dowdy unattractive woman. □ **frumpish** *adj.* **frumpy** *adj.* (**-ier, -iest**). [perhaps dial. *frumple* wrinkle]

frustrate /frʌ'streɪt/ *v.* (**-ting**) **1** make (efforts) ineffective. **2** prevent (a person) from achieving a purpose. **3** (as **frustrated** *adj.*) **a** discontented because unable to achieve one's aims. **b** sexually unfulfilled. □ **frustrating** *adj.* **frustratingly** *adv.* **frustration** *n.* [Latin *frustra* in vain]

frustum /'frʌstəm/ *n.* (*pl.* **-ta** or **-s**) *Geom.* remaining part of a decapitated cone or pyramid. [Latin, = piece cut off]

fry[1] –*v.* (**fries, fried**) cook or be cooked in hot fat. –*n.* (*pl.* **fries**) **1** offal, usu. eaten fried (*lamb's fry*). **2** fried food, esp. meat. [Latin *frigo*]

fry[2] *n.pl.* young or newly hatched fishes. [Old Norse, = seed]

fryer *n.* (also **frier**) **1** person who fries. **2** vessel for frying esp. fish.

frying-pan *n.* shallow pan used in frying. □ **out of the frying-pan into the fire** from a bad situation to a worse one.

fry-up *n. colloq.* fried bacon, eggs, etc.

ft *abbr.* foot, feet.

FT-SE *abbr.* Financial Times Stock Exchange 100 share index (based on the share values of Britain's largest public companies).

fuchsia /'fju:ʃə/ *n.* shrub with drooping red, purple, or white flowers. [*Fuchs*, name of a botanist]

fuck *coarse slang* –*v.* **1** have sexual intercourse (with). **2** (often *absol.* or as *int.* expressing annoyance) curse (a person or thing). **3** (foll. by *about, around*) mess about; fool around. **4** (as **fucking** *adj., adv.*) expressing annoyance etc. –*n.* **1 a** act of sexual intercourse. **b** partner in this. **2** slightest amount (*don't give a fuck*). □ **fuck-all** nothing. **fuck off** go away. **fuck up 1** bungle. **2** disturb emotionally. □ **fucker** *n.* (often as a term of abuse). [origin unknown]

■ **Usage** Although widely used, *fuck* is still considered to be the most offensive word in the English language by many people. In discussions about bad language it is frequently referred to as *the 'f' word*.

fuck-up *n. coarse slang* bungle or muddle.

fuddle /'fʌd(ə)l/ –*v.* (**-ling**) confuse or stupefy, esp. with alcohol. –*n.* **1** confusion. **2** intoxication. [origin unknown]

fuddy-duddy /'fʌdɪ,dʌdɪ/ *slang* –*adj.* old-fashioned or quaintly fussy. –*n.* (*pl.* **-ies**) such a person. [origin unknown]

fudge –*n.* **1** soft toffee-like sweet made of milk, sugar, butter, etc. **2** piece of dishonesty or faking. –*v.* (**-ging**) make or do clumsily or dishonestly; fake (*fudge the results*). [origin uncertain]

fuehrer var. of FÜHRER.

fuel /'fju:əl/ –*n.* **1** material for burning or as a source of heat, power, or nuclear energy. **2** food as a source of energy. **3** thing that sustains or inflames passion etc. –*v.* (**-ll-**; *US* **-l-**) **1** supply with, take in, or get, fuel. **2** inflame (feeling etc.). [French from Latin]

fuel cell *n.* cell producing electricity direct from a chemical reaction.

fug *n. colloq.* close stuffy atmosphere. □ **fuggy** *adj.* [origin unknown]

fugitive /'fju:dʒɪtɪv/ –*n.* (often foll. by *from*) person who flees, e.g. from justice or an enemy. –*adj.* **1** fleeing. **2** transient, fleeting. [Latin *fugio* flee]

fugue /fju:g/ *n.* piece of music in which a short melody or phrase is introduced by one part and taken up and developed by others. □ **fugal** *adj.* [Latin *fuga* flight]

führer /'fjʊərə(r)/ *n.* (also **fuehrer**) tyrannical leader. [German]

-ful *comb. form* forming: **1** adjectives from **a** nouns, meaning full of or having qualities of (*beautiful*; *masterful*). **b** adjectives (*direful*). **c** verbs, meaning 'apt to' (*forgetful*). **2** nouns (*pl.* **-fuls**) meaning 'amount that fills' (*handful*; *spoonful*).

fulcrum /'fʌlkrəm/ *n.* (*pl.* **-s** or **-cra**) point on which a lever is supported. [Latin *fulcio* to prop]

fulfil /fʊl'fɪl/ *v.* (*US* **fulfill**) (**-ll-**) **1** carry out (a task, prophecy, promise, etc.). **2 a** satisfy (conditions, a desire, prayer, etc.). **b** (as **fulfilled** *adj.*) completely happy. **3** answer (a purpose). □ **fulfil oneself** realize one's potential. □ **fulfilment** *n.* [Old English: related to FULL[1], FILL]

full[1] /fʊl/ –*adj.* **1** holding all it can (*bucket is full*; *full of water*). **2** having eaten all one can or wants. **3** abundant, copious, satisfying (*a full life*; *full details*). **4** (foll. by *of*) having an abundance of (*full of vitality*). **5** (foll. by *of*) engrossed in (*full of himself*). **6** complete, perfect (*full membership*; *in full bloom*). **7** (of tone) deep and clear. **8** plump, rounded (*full figure*). **9** (of clothes) ample, hanging in folds. –*adv.* **1** very (*knows full well*). **2** quite, fully (*full six miles*). **3** exactly (*full on the nose*). □ **full up** *colloq.* completely full. **in full 1** without abridgement. **2** to or for the full amount. **in full view** entirely visible. **to the full** to the utmost extent. [Old English]

full[2] /fʊl/ *v.* clean and thicken (cloth). [from FULLER]

full back *n.* defensive player near the goal in football, hockey, etc.

full-blooded *adj.* **1** vigorous, hearty, sensual. **2** not hybrid.

full-blown *adj.* fully developed.

full board *n.* provision of bed and all meals at a hotel etc.

full-bodied *adj.* rich in quality, tone, etc.

fuller *n.* person who fulls cloth. □ **fuller's earth** type of clay used in fulling. [Latin *fullo*]

full-frontal *adj.* **1** (of a nude figure) fully exposed at the front. **2** explicit, unrestrained.

full house *n.* **1** maximum attendance at a theatre etc. **2** hand in poker with three of a kind and a pair.

full-length *adj.* **1** not shortened. **2** (of a mirror, portrait, etc.) showing the whole figure.

full moon *n.* **1** moon with its whole disc illuminated. **2** time of this.

fullness *n.* being full. □ **the fullness of time** the appropriate or destined time.

full-scale *adj.* not reduced in size, complete.

full stop *n.* **1** punctuation mark (.) at the end of a sentence or an abbreviation. **2** complete cessation.

full term *n.* completion of a normal pregnancy.

full-time –*adj.* for or during the whole of the working week (*full-time job*). –*adv.* on a full-time basis (*work full-time*).

full-timer *n.* person who does a full-time job.

fully *adv.* **1** completely, entirely (*am fully aware*). **2** at least (*fully 60*).

fully-fashioned *adj.* (of clothing) shaped to fit closely.

fulmar /'fʊlmə(r)/ *n.* Arctic sea bird related to the petrel. [Old Norse: related to FOUL, *mar* gull]

fulminant /'fʊlmɪnənt/ *adj.* **1** fulminating. **2** (of a disease etc.) developing suddenly. [Latin: related to FULMINATE]

fulminate /'fʊlmɪˌneɪt/ *v.* (**-ting**) **1** criticize loudly and forcefully. **2** explode violently; flash. □ **fulmination** /-'neɪʃ(ə)n/ *n.* [Latin *fulmen -min-* lightning]

fulsome /'fʊlsəm/ *adj.* excessive, cloying, insincere (*fulsome praise*). □ **fulsomely** *adv.* [from FULL[1]]

■ **Usage** The phrase *fulsome praise* is sometimes wrongly used to mean generous praise rather than excessive praise.

fumble /'fʌmb(ə)l/ –*v.* (**-ling**) **1** use the hands awkwardly, grope about. **2** handle clumsily or nervously (*fumbled the ball*). –*n.* act of fumbling. [Low German *fummeln*]

fume –*n.* (usu. in *pl.*) exuded gas, smoke, or vapour, esp. when harmful or unpleasant. –*v.* (**-ming**) **1** emit fumes or as fumes. **2** be very angry. **3** subject (oak, film, etc.) to fumes to darken. [Latin *fumus* smoke]

fumigate /'fju:mɪˌgeɪt/ *v.* (**-ting**) disinfect or purify with fumes. □ **fumigation** /-'geɪʃ(ə)n/ *n.* **fumigator** *n.* [Latin: related to FUME]

fun –*n.* **1** lively or playful amusement. **2** source of this. **3** mockery, ridicule (*figure of fun*). –*attrib. adj. colloq.* amusing, enjoyable (*a fun thing to do*). □ **for fun** (or **for the fun of it**) not for a serious purpose. **in fun** as a joke, not seriously. **make fun of** (or **poke fun at**) ridicule, tease. [obsolete *fun*, *fon*: related to FOND]

■ **Usage** The use of *fun* as an attributive adjective is common in informal use, but is considered incorrect by some people.

function /'fʌŋkʃ(ə)n/ –*n.* **1 a** proper or necessary role, activity, or purpose. **b** official or professional duty. **2** public or social occasion. **3** *Math.* quantity whose value depends on the varying values of others. –*v.* fulfil a function, operate. [Latin *fungor funct-* perform]

functional *adj.* **1** of or serving a function. **2** practical rather than attractive. **3** affecting the function of a bodily organ but not its structure. □ **functionally** *adv.*

functionalism *n.* belief that a thing's function should determine its design. □ **functionalist** *n.* & *adj.*

functionary *n.* (*pl.* **-ies**) official performing certain duties.

fund –*n.* **1** permanently available stock (*fund of knowledge*). **2** sum of money, esp. set apart for a purpose. **3** (in *pl.*) money resources. –*v.* **1** provide with money. **2** make (a debt) permanent at fixed interest. □ **in funds** *colloq.* having money to spend. [Latin *fundus* bottom]

fundamental /ˌfʌndə'ment(ə)l/ –*adj.* of or being a base or foundation; essential, primary. –*n.* **1** (usu. in *pl.*) fundamental principle. **2** *Mus.* fundamental note. □ **fundamentally** *adv.* [Latin: related to FOUND[2]]

fundamentalism *n.* strict adherence to traditional religious beliefs or doctrines. □ **fundamentalist** *n.* & *adj.*

fundamental note *n. Mus.* lowest note of a chord.

fundamental particle *n.* elementary particle.

fund-raiser *n.* person raising money for a cause, enterprise, etc. □ **fund-raising** *n.*

funeral /'fju:nər(ə)l/ –*n.* **1** ceremonial burial or cremation of a corpse. **2** *slang* one's (usu. unpleasant) concern (*that's your funeral*). –*attrib. adj.* of or used at funerals. [Latin *funus funer-*]

funeral director *n.* undertaker.

funeral parlour *n.* establishment where corpses are prepared for funerals.

funerary /'fju:nərərɪ/ *adj.* of or used at funerals.

funereal /fju:'nɪərɪəl/ *adj.* **1** of or appropriate to a funeral. **2** dismal, dark. □ **funereally** *adv.*

funfair *n.* fair with amusements and sideshows.

fungicide /'fʌndʒɪˌsaɪd/ *n.* substance that kills fungus. □ **fungicidal** /-'saɪd(ə)l/ *adj.*

fungoid /'fʌŋgɔɪd/ –*adj.* fungus-like. –*n.* fungoid plant.

fungus /'fʌŋgəs/ *n.* (*pl.* **-gi** /-gaɪ/ or **-guses**) **1** mushroom, toadstool, or allied plant, including moulds, feeding on organic matter. **2** *Med.* spongy morbid growth. □ **fungal** *adj.* **fungous** *adj.* [Latin]

funicular /fju:'nɪkjʊlə(r)/ –*adj.* (of a mountain railway) operating by cable with ascending and descending cars counterbalanced. **–*n.*** funicular railway. [Latin *funiculus* diminutive of *funis* rope]

funk[1] *slang* **–*n.*** **1** fear, panic. **2** coward. **–*v.*** **1** evade through fear. **2** be afraid (of). [origin uncertain]

funk[2] *n. slang* funky music. [origin uncertain]

funky *adj.* (**-ier, -iest**) *slang* (esp. of jazz or rock music) earthy, bluesy, with a heavy rhythm.

funnel /ˈfʌn(ə)l/ **–*n.*** **1** tube widening at the top, for pouring liquid etc. into a small opening. **2** metal chimney on a steam engine or steamship. **–*v.*** (**-ll-**; *US* **-l-**) guide or move through or as through a funnel. [Provençal *fonilh* from Latin *(in)fundibulum*]

funny /ˈfʌnɪ/ *adj.* (**-ier, -iest**) **1** amusing, comical. **2** strange, peculiar. **3** *colloq.* **a** slightly unwell. **b** eccentric. □ **funnily** *adv.* **funniness** *n.* [from FUN]

funny-bone *n.* part of the elbow over which a very sensitive nerve passes.

fun run *n. colloq.* uncompetitive sponsored run for charity.

fur **–*n.*** **1 a** short fine soft animal hair. **b** hide with fur on it, used esp. for clothing. **2** garment of or lined with fur. **3** (*collect.*) animals with fur. **4** fur-like coating on the tongue, in a kettle, etc. **–*v.*** (**-rr-**) **1** (esp. as **furred** *adj.*) line or trim with fur. **2** (often foll. by *up*) (of a kettle etc.) become coated with fur. □ **make the fur fly** *colloq.* cause a disturbance, stir up trouble. [French from Germanic]

furbelow /ˈfɜːbɪˌləʊ/ *n.* **1** (in *pl.*) showy ornaments. **2** *archaic* gathered strip or border of a skirt or petticoat. [French *falbala*]

furbish /ˈfɜːbɪʃ/ *v.* (often foll. by *up*) = REFURBISH. [French from Germanic]

furcate /ˈfɜːkeɪt/ **–*adj.*** forked, branched. **–*v.*** (**-ting**) fork, divide. □ **furcation** /fɜːˈkeɪʃ(ə)n/ *n.* [Latin: related to FORK]

furious /ˈfjʊərɪəs/ *adj.* **1** very angry. **2** raging, frantic. □ **furiously** *adv.* [Latin: related to FURY]

furl *v.* **1** roll up and secure (a sail etc.). **2** become furled. [French *ferler*]

furlong /ˈfɜːlɒŋ/ *n.* eighth of a mile. [Old English: related to FURROW, LONG[1]]

furlough /ˈfɜːləʊ/ **–*n.*** leave of absence, esp. military. **–*v.*** *US* **1** grant furlough to. **2** spend furlough. [Dutch: related to FOR-, LEAVE[1]]

furnace /ˈfɜːnɪs/ *n.* **1** enclosed structure for intense heating by fire, esp. of metals or water. **2** very hot place. [Latin *fornax* from *fornus* oven]

furnish /ˈfɜːnɪʃ/ *v.* **1** provide (a house, room, etc.) with furniture. **2** (often foll. by *with*) supply. [French from Germanic]

furnished *adj.* (of a house etc.) let with furniture.

furnisher *n.* **1** person who sells furniture. **2** person who furnishes.

furnishings *n.pl.* furniture and fitments in a house, room, etc.

furniture /ˈfɜːnɪtʃə(r)/ *n.* **1** movable equipment of a house, room, etc., e.g. tables, beds. **2** *Naut.* ship's equipment. **3** accessories, e.g. the handles and lock on a door. [French: related to FURNISH]

furore /fjʊəˈrɔːrɪ/ *n.* (*US* **furor** /ˈfjʊərɔː(r)/) **1** uproar; fury. **2** enthusiastic admiration. [Latin: related to FURY]

furrier /ˈfʌrɪə(r)/ *n.* dealer in or dresser of furs. [French]

furrow /ˈfʌrəʊ/ **–*n.*** **1** narrow trench made by a plough. **2** rut, groove, wrinkle. **3** ship's track. **–*v.*** **1** plough. **2** make furrows in. [Old English]

furry /ˈfɜːrɪ/ *adj.* (**-ier, -iest**) like or covered with fur.

further /ˈfɜːðə(r)/ **–*adv.*** (also **farther** /ˈfɑːðə(r)/) **1** more distant in space or time. **2** to a greater extent, more (*will enquire further*). **3** in addition (*I may add further*). **–*adj.*** (also **farther** /ˈfɑːðə(r)/) **1** more distant or advanced. **2** more, additional (*further details*). **–*v.*** promote or favour (a scheme etc.). [Old English: related to FORTH]

■ **Usage** The form *farther* is used esp. with reference to physical distance, although *further* is preferred by many people even in this sense.

furtherance *n.* furthering of a scheme etc.

further education *n.* education for those above school age.

furthermore /ˌfɜːðəˈmɔː(r)/ *adv.* in addition, besides.

furthest /ˈfɜːðɪst/ (also **farthest** /ˈfɑːðɪst/) **–*adj.*** most distant. **–*adv.*** to or at the greatest distance.

■ **Usage** The form *farthest* is used esp. with reference to physical distance, although *furthest* is preferred by many people even in this sense.

furtive /ˈfɜːtɪv/ *adj.* sly, stealthy. □ **furtively** *adv.* **furtiveness** *n.* [Latin *fur* thief]

fury /ˈfjʊərɪ/ *n.* (*pl.* **-ies**) **1 a** wild and passionate anger. **b** fit of rage. **2** violence of a storm, disease, etc. **3** (**Fury**) (usu. in *pl.*) (in Greek mythology) avenging goddess. **4** avenging spirit. **5** angry or malignant woman. □ **like fury** *colloq.* with great force or effort. [Latin *furia*]

furze *n.* = GORSE. □ **furzy** *adj.* [Old English]

fuse[1] /fjuːz/ *–v.* (**-sing**) **1** melt with intense heat. **2** blend into one whole by melting. **3** provide (an electric circuit) with a fuse. **4 a** (of an appliance) fail owing to the melting of a fuse. **b** cause to do this. *–n.* device with a strip or wire of easily melted metal placed in an electric circuit so as to interrupt an excessive current by melting. [Latin *fundo fus-* melt]

fuse[2] /fjuːz/ (also **fuze**) *–n.* **1** device of combustible matter for igniting a bomb or explosive charge. **2** component made of this in a shell, mine, etc. *–v.* (**-sing**) fit a fuse to. [Latin *fusus* spindle]

fuselage /ˈfjuːzəˌlɑːʒ/ *n.* body of an aeroplane. [French from *fuseau* spindle]

fusible /ˈfjuːzɪb(ə)l/ *adj.* that can be melted. □ **fusibility** /-ˈbɪlɪtɪ/ *n.* [Latin: related to FUSE[1]]

fusil /ˈfjuːzɪl/ *n. hist.* light musket. [Latin *focus* fire]

fusilier /ˌfjuːzɪˈlɪə(r)/ *n.* member of any of several British regiments formerly armed with fusils. [French: related to FUSIL]

fusillade /ˌfjuːzɪˈleɪd/ *n.* **1** period of continuous discharge of firearms. **2** sustained outburst of criticism etc.

fusion /ˈfjuːʒ(ə)n/ *n.* **1** fusing or melting. **2** blending. **3** coalition. **4** = NUCLEAR FUSION. [Latin: related to FUSE[1]]

fuss *–n.* **1** excited commotion, bustle. **2** excessive concern about a trivial thing. **3** sustained protest or dispute. *–v.* **1** behave with nervous concern. **2** agitate, worry. □ **make a fuss** complain vigorously. **make a fuss of** (or **over**) treat (a person or animal) affectionately. □ **fusser** *n.* [origin unknown]

fusspot *n. colloq.* person given to fussing.

fussy *adj.* (**-ier, -iest**) **1** inclined to fuss. **2** over-elaborate. **3** fastidious. □ **fussily** *adv.* **fussiness** *n.*

fustian /ˈfʌstɪən/ *–n.* **1** thick usu. dark twilled cotton cloth. **2** bombast. *–adj.* **1** made of fustian. **2** bombastic. **3** worthless. [French]

fusty /ˈfʌstɪ/ *adj.* (**-ier, -iest**) **1** musty, stuffy. **2** antiquated. □ **fustiness** *n.* [French *fust* cask, from Latin *fustis* cudgel]

futile /ˈfjuːtaɪl/ *adj.* **1** useless, ineffectual. **2** frivolous. □ **futility** /-ˈtɪlɪtɪ/ *n.* [Latin *futilis* leaky, futile]

futon /ˈfuːtɒn/ *n.* Japanese quilted mattress used as a bed; this sold with a low wooden frame, often convertible into a couch. [Japanese]

future /ˈfjuːtʃə(r)/ *–adj.* **1** about to happen, be, or become. **2 a** of time to come. **b** *Gram.* (of a tense) describing an event yet to happen. *–n.* **1** time to come. **2** future events. **3** future condition of a person, country, etc. **4** prospect of success etc. (*no future in it*). **5** *Gram.* future tense. **6** (in *pl.*) *Stock Exch.* goods etc. sold for future delivery. □ **in future** from now onwards. [Latin *futurus* future part. of *sum* be]

future perfect *n. Gram.* tense giving the sense 'will have done'.

futurism *n.* 20th-century artistic movement departing from traditional forms and celebrating technology and dynamism. □ **futurist** *n.* & *adj.*

futuristic /ˌfjuːtʃəˈrɪstɪk/ *adj.* **1** suitable for the future; ultra-modern. **2** of futurism.

futurity /fjuːˈtjʊərɪtɪ/ *n.* (*pl.* **-ies**) *literary* **1** future time. **2** (in *sing.* or *pl.*) future events.

futurology /ˌfjuːtʃəˈrɒlədʒɪ/ *n.* forecasting of the future, esp. from present trends.

fuze var. of FUSE[2].

fuzz *n.* **1** fluff. **2** fluffy or frizzed hair. **3** *slang* **a** (prec. by *the*) the police. **b** police officer. [probably Low German or Dutch]

fuzzy *adj.* (**-ier, -iest**) **1** like fuzz, fluffy. **2** blurred, indistinct. □ **fuzzily** *adv.* **fuzziness** *n.*

-fy *suffix* forming: **1** verbs from nouns, meaning: **a** make, produce (*pacify*). **b** make into (*deify*; *petrify*). **2** verbs from adjectives, meaning 'bring or come into a state' (*Frenchify*; *solidify*). **3** verbs in a causative sense (*horrify*; *stupefy*). [French *-fier* from Latin *facio* make]

G

G[1] /dʒi:/ *n.* (also **g**) (*pl.* **Gs** or **G's**) **1** seventh letter of the alphabet. **2** *Mus.* fifth note of the diatonic scale of C major.

G[2] *abbr.* (also **G.**) **1** gauss. **2** giga-. **3** gravitational constant.

g *abbr.* (also **g.**) **1** gram(s). **2 a** gravity. **b** acceleration due to gravity.

Ga *symb.* gallium.

gab *n. colloq.* talk, chatter. [var. of GOB[1]]

gabardine /ˌgæbə'di:n/ *n.* (also **gaberdine**) **1** twill-woven cloth, esp. of worsted. **2** raincoat etc. made of this. [French *gauvardine*]

gabble /'gæb(ə)l/ *–v.* (**-ling**) talk or utter unintelligibly or too fast. *–n.* fast unintelligible talk. [Dutch, imitative]

gaberdine var. of GABARDINE.

gable /'geɪb(ə)l/ *n.* **1** triangular upper part of a wall at the end of a ridged roof. **2** gable-topped wall. □ **gabled** *adj.* [Old Norse and French]

gad *v.* (**-dd-**) (foll. by *about*) go about idly or in search of pleasure. [obsolete *gadling* companion]

gadabout *n.* person who gads about.

gadfly *n.* **1** fly that bites cattle and horses. **2** irritating person. [obsolete *gad* spike]

gadget /'gædʒɪt/ *n.* small mechanical device or tool. □ **gadgetry** *n.* [origin unknown]

gadolinium /ˌgædə'lɪnɪəm/ *n.* metallic element of the lanthanide series. [*Gadolin*, name of a mineralogist]

gadwall /'gædwɔ:l/ *n.* brownish-grey freshwater duck. [origin unknown]

Gael /geɪl/ *n.* **1** Scottish Celt. **2** Gaelic-speaking Celt. [Gaelic *Gaidheal*]

Gaelic /'geɪlɪk, 'gæ-/ *–n.* Celtic language of Ireland and Scotland. *–adj.* of the Celts or the Celtic languages.

gaff[1] *–n.* **1 a** stick with an iron hook for landing large fish. **b** barbed fishing-spear. **2** spar to which the head of a fore-and-aft sail is bent. *–v.* seize (a fish) with a gaff. [Provençal *gaf* hook]

gaff[2] *n. slang* □ **blow the gaff** reveal a plot or secret. [origin unknown]

gaffe /gæf/ *n.* blunder; indiscreet act or remark. [French]

gaffer /'gæfə(r)/ *n.* **1** old fellow. **2** *colloq.* foreman, boss. **3** chief electrician in a film or television production unit. [probably from GODFATHER]

gag *–n.* **1** thing thrust into or tied across the mouth, esp. to prevent speaking or crying out. **2** joke or comic scene. **3** parliamentary closure. **4** thing restricting free speech. *–v.* (**-gg-**) **1** apply a gag to. **2** silence; deprive of free speech. **3** choke, retch. **4** make gags as a comedian etc. [origin uncertain]

gaga /'gɑ:gɑ:/ *adj. slang* **1** senile. **2** slightly crazy. [French]

gage[1] *n.* **1** pledge; thing deposited as security. **2** symbol of a challenge to fight, esp. a glove thrown down. [Germanic: related to WED, WAGE]

gage[2] *US* var. of GAUGE.

gaggle /'gæg(ə)l/ *n.* **1** flock of geese. **2** *colloq.* disorganized group of people. [imitative]

gaiety /'geɪətɪ/ *n.* (*US* **gayety**) **1** being gay; mirth. **2** merrymaking. **3** bright appearance. [French: related to GAY]

gaily *adv.* in a gay or careless manner (*gaily decorated*; *gaily announced their departure*).

gain *–v.* **1** obtain or win (*gain advantage*; *gain recognition*). **2** acquire as profits etc., earn. **3** (often foll. by *in*) get more of, improve (*gain momentum*; *gain in experience*). **4** benefit, profit. **5** (of a clock etc.) become fast; become fast by (a specified amount of time). **6** (often foll. by *on, upon*) come closer to a person or thing pursued. **7 a** reclaim (land from the sea). **b** win (a battle). **8** reach (a desired place). *–n.* **1** increase of wealth etc.; profit, improvement. **2** (in *pl.*) sums of money got by trade etc. **3** increase in amount. □ **gain ground 1** advance. **2** (foll. by *on*) catch up (a person pursued). [French from Germanic]

gainful *adj.* **1** (of employment) paid. **2** lucrative. □ **gainfully** *adv.*

gainsay /geɪn'seɪ/ *v.* deny, contradict. [Old Norse: related to AGAINST, SAY]

gait *n.* manner of walking or forward motion. [Old Norse]

gaiter *n.* covering of cloth, leather, etc., for the lower leg. [French *guêtre*]

gal *n. slang* girl. [representing a variant pronunciation]

gal. *abbr.* (also **gall.**) gallon(s).

gala /'gɑ:lə/ *n.* festive occasion or gathering (*swimming gala*). [ultimately from French *gale* rejoicing from Germanic]

galactic /gə'læktɪk/ *adj.* of a galaxy or galaxies.

galantine /'gælən,ti:n/ *n.* white meat boned, stuffed, spiced, etc., and served cold. [French from Latin]

galaxy /'gæləksɪ/ *n.* (*pl.* **-ies**) **1** independent system of stars, gas, dust, etc., in space. **2** (**the Galaxy**) Milky Way. **3** (foll. by *of*) brilliant company (*galaxy of talent*). [Greek *gala* milk]

gale *n.* **1** very strong wind or storm. **2** outburst, esp. of laughter. [origin unknown]

gall[1] /gɔ:l/ *n.* **1** *slang* impudence. **2** rancour. **3** bitterness. **4** bile. [Old Norse]

gall[2] /gɔ:l/ *—n.* **1** sore made by chafing. **2** mental soreness or its cause. **3** place rubbed bare. *—v.* **1** rub sore. **2** vex, humiliate. [Low German or Dutch *galle*]

gall[3] /gɔ:l/ *n.* growth produced by insects etc. on plants and trees, esp. on oak. [Latin *galla*]

gall. *abbr.* var. of GAL.

gallant *—adj.* /'gælənt/ **1** brave. **2** fine, stately. **3** /gə'lænt/ very attentive to women. *—n.* /gə'lænt/ ladies' man. □ **gallantly** /'gæləntlɪ/ *adv.* [French *galer* make merry]

gallantry /'gæləntrɪ/ *n.* (*pl.* **-ies**) **1** bravery. **2** devotion to women. **3** polite act or speech.

gall bladder *n.* organ storing bile.

galleon /'gælɪən/ *n. hist.* warship (usu. Spanish). [French or Spanish: related to GALLEY]

galleria /,gælə'ri:ə/ *n.* collection of small shops under a single roof. [Italian]

gallery /'gælərɪ/ *n.* (*pl.* **-ies**) **1** room or building for showing works of art. **2** balcony, esp. in a church, hall, etc. (*minstrels' gallery*). **3** highest balcony in a theatre. **4 a** covered walk partly open at the side; colonnade. **b** narrow passage in the thickness of a wall or on corbels, open towards the interior of the building. **5** long narrow room or passage (*shooting-gallery*). **6** horizontal underground passage in a mine etc. **7** group of spectators at a golf-match etc. □ **play to the gallery** seek to win approval by appealing to popular taste. [French *galerie*]

galley /'gælɪ/ *n.* (*pl.* **-s**) **1** *hist.* **a** long flat single-decked vessel usu. rowed by slaves or criminals. **b** ancient Greek or Roman warship. **2** ship's or aircraft's kitchen. **3** *Printing* (in full **galley proof**) proof in continuous form before division into pages. [Latin *galea*]

galley-slave *n.* drudge.

Gallic /'gælɪk/ *adj.* **1** French or typically French. **2** of Gaul or the Gauls. [Latin *Gallicus*]

Gallicism /'gælɪ,sɪz(ə)m/ *n.* French idiom. [related to GALLIC]

gallinaceous /,gælɪ'neɪʃəs/ *adj.* of the order including domestic poultry, pheasants, etc. [Latin *gallina* hen]

gallium /'gælɪəm/ *n.* soft bluish-white metallic element. [Latin *Gallia* France: so named patriotically by its discoverer Lecoq]

gallivant /'gælɪ,vænt/ *v. colloq.* gad about. [origin uncertain]

Gallo- *comb. form* French. [Latin]

gallon /'gælən/ *n.* **1** measure of capacity equal to eight pints (4.5 litres; for wine, or *US*, 3.8 litres). **2** (in *pl.*) *colloq.* large amount. [French]

gallop /'gæləp/ *—n.* **1** fastest pace of a horse etc., with all the feet off the ground together in each stride. **2** ride at this pace. *—v.* (**-p-**) **1 a** (of a horse etc. or its rider) go at a gallop. **b** make (a horse etc.) gallop. **2** read, talk, etc., fast. **3** progress rapidly (*galloping inflation*). [French: related to WALLOP]

gallows /'gæləʊz/ *n.pl.* (usu. treated as *sing.*) structure, usu. of two uprights and a crosspiece, for hanging criminals. [Old Norse]

gallstone *n.* small hard mass forming in the gall bladder.

Gallup poll /'gæləp/ *n.* = OPINION POLL. [*Gallup*, name of a statistician]

galore /gə'lɔ:(r)/ *adv.* in plenty (*whisky galore*). [Irish *go leor* enough]

galosh /gə'lɒʃ/ *n.* (also **golosh**) (usu. in *pl.*) overshoe, usu. of rubber. [French]

galumph /gə'lʌmf/ *v.* (esp. as **galumphing** *adj.*) *colloq.* move noisily or clumsily. [coined by Lewis Carroll, perhaps from GALLOP, TRIUMPH]

galvanic /gæl'vænɪk/ *adj.* **1 a** producing an electric current by chemical action. **b** (of electricity) produced by chemical action. **2 a** sudden and remarkable (*had a galvanic effect*). **b** stimulating; full of energy. □ **galvanically** *adv.*

galvanize /'gælvə,naɪz/ *v.* (also **-ise**) (**-zing** or **-sing**) **1** (often foll. by *into*) rouse forcefully, esp. by shock or excitement (*was galvanized into action*). **2** stimulate by or as by electricity. **3** coat (iron) with zinc to protect against rust. □ **galvanization** /-'zeɪʃ(ə)n/ *n.* [*Galvani*, name of a physiologist]

galvanometer /,gælvə'nɒmɪtə(r)/ *n.* instrument for detecting and measuring small electric currents. □ **galvanometric** /-nə'metrɪk/ *adj.*

gambit /'gæmbɪt/ *n.* **1** chess opening in which a player sacrifices a piece or pawn to secure an advantage. **2** opening move in a discussion etc. **3** trick or device. [Italian *gambetto* tripping up]

gamble /'gæmb(ə)l/ *—v.* (**-ling**) **1** play games of chance for money. **2 a** bet (a sum of money) in gambling. **b** (often foll. by *away*) lose by gambling. **3** risk much in the hope of great gain. **4** (foll. by *on*) act in the hope of. —*n.* **1** risky undertaking. **2** spell of gambling. □ **gambler** *n.*

gamboge /gæm'bəʊdʒ/ *n.* gum resin used as a yellow pigment and as a purgative. [*Cambodia* in SE Asia]

gambol /'gæmb(ə)l/ —*v.* (**-ll-**; *US* **-l-**) skip or jump about playfully. —*n.* frolic, caper. [French *gambade* leap, from Italian *gamba* leg]

game[1] —*n.* **1** form of play or sport, esp. a competitive one with rules. **2** portion of play forming a scoring unit, e.g. in bridge or tennis. **3** (in *pl.*) series of athletic etc. contests (*Olympic Games*). **4 a** piece of fun, jest (*didn't mean to upset you; it was only a game*). **b** (in *pl.*) dodges, tricks (*none of your games!*). **5** *colloq.* **a** scheme (*so that's your game*). **b** type of activity or business (*have been in the antiques game a long time*). **6 a** wild animals or birds hunted for sport or food. **b** their flesh as food. —*adj.* spirited; eager and willing (*are you game for a walk?*). —*v.* (**-ming**) gamble for money stakes. □ **the game is up** scheme is revealed or foiled. **on the game** *slang* working as a prostitute. □ **gamely** *adv.* [Old English]

game[2] *adj. colloq.* (of a leg, arm, etc.) crippled. [origin unknown]

gamecock *n.* cock bred and trained for cock-fighting.

gamekeeper *n.* person employed to breed and protect game.

gamelan /'gæmə,læn/ *n.* **1** SE Asian orchestra mainly of percussion instruments. **2** type of xylophone used in this. [Javanese]

gamesmanship *n.* art of winning games by gaining psychological advantage.

gamester /'geɪmstə(r)/ *n.* gambler.

gamete /'gæmi:t/ *n.* mature germ cell able to unite with another in sexual reproduction. □ **gametic** /gə'metɪk/ *adj.* [Greek, = wife]

gamin /'gæmɪn/ *n.* **1** street urchin. **2** impudent child. [French]

gamine /gæ'mi:n/ *n.* **1** girl gamin. **2** girl with mischievous charm. [French]

gamma /'gæmə/ *n.* **1** third letter of the Greek alphabet (Γ, γ). **2** third-class mark for a piece of work etc. [Greek]

gamma radiation *n.* (also **gamma rays**) electromagnetic radiation of shorter wavelength than X-rays.

gammon /'gæmən/ *n.* **1** bottom piece of a flitch of bacon including a hind leg. **2** ham of a pig cured like bacon. [French: related to JAMB]

gammy /'gæmɪ/ *adj.* (**-ier, -iest**) *slang* = GAME[2]. [dial. form of GAME[2]]

gamut /'gæmət/ *n.* entire range or scope. □ **run the gamut of** experience or perform the complete range of. [Latin *gamma ut*, words arbitrarily taken as names of notes]

gamy *adj.* (**-ier, -iest**) smelling or tasting like high game.

gander *n.* **1** male goose. **2** *slang* look, glance (*take a gander*). [Old English]

gang *n.* **1** band of persons associating for some (usu. antisocial or criminal) purpose. **2** set of workers, slaves, or prisoners. □ **gang up** *colloq.* **1** (often foll. by *with*) act together. **2** (foll. by *on*) combine against. [Old Norse]

ganger *n.* foreman of a gang of workers.

gangling /'gæŋglɪŋ/ *adj.* (of a person) loosely built; lanky. [frequentative of Old English *gang* go]

ganglion /'gæŋglɪən/ *n.* (*pl.* **-lia** or **-s**) structure containing an assemblage of nerve cells. □ **ganglionic** /-'ɒnɪk/ *adj.* [Greek]

gangly /'gæŋglɪ/ *adj.* (**-ier, -iest**) = GANGLING.

gangplank *n.* movable plank for boarding or disembarking from a ship etc.

gangrene /'gæŋgri:n/ *n.* death of body tissue, usu. resulting from obstructed circulation. □ **gangrenous** /-grɪnəs/ *adj.* [Greek *gaggraina*]

gangster *n.* member of a gang of violent criminals.

gangue /gæŋ/ *n.* valueless earth etc. in which ore is found. [German: related to GANG]

gangway *n.* **1** passage, esp. between rows of seats. **2 a** opening in a ship's bulwarks. **b** bridge from ship to shore.

gannet /'gænɪt/ *n.* **1** large diving sea bird. **2** *slang* greedy person. [Old English]

gantry /'gæntrɪ/ *n.* (*pl.* **-ies**) structure supporting a travelling crane, railway or road signals, rocket-launching equipment, etc. [probably *gawn*, a dial. form of GALLON, + TREE]

gaol var. of JAIL.

gaolbird var. of JAILBIRD.

gaolbreak var. of JAILBREAK.

gaoler var. of JAILER.

gap *n.* **1** empty space, interval; deficiency. **2** breach in a hedge, fence, etc. **3** wide divergence in views etc. □ **gappy** *adj.* [Old Norse]

gape —*v.* (**-ping**) **1 a** open one's mouth wide. **b** be or become wide open; split. **2** (foll. by *at*) stare at. —*n.* **1** open-mouthed stare; open mouth. **2** rent, opening. [Old Norse]

garage /'gærɑ:ʒ, -rɪdʒ/ *–n.* **1** building for housing a vehicle. **2** establishment selling petrol etc., or repairing and selling vehicles. *–v.* (**-ging**) put or keep in a garage. [French]

garb *–n.* clothing, esp. of a distinctive kind. *–v.* (usu. in *passive* or *refl.*) dress. [Germanic: related to GEAR]

garbage /'gɑ:bɪdʒ/ *n.* **1** esp. *US* refuse. **2** *colloq.* nonsense. [Anglo-French]

garble /'gɑ:b(ə)l/ *v.* (**-ling**) **1** (esp. as **garbled** *adj.*) unintentionally distort or confuse (facts, messages, etc.). **2** make a (usu. unfair) selection from (facts, statements, etc.). [Italian from Arabic]

garden /'gɑ:d(ə)n/ *–n.* **1** piece of ground for growing flowers, fruit, or vegetables, and as a place of recreation. **2** (esp. in *pl.*) grounds laid out for public enjoyment. **3** (*attrib.*) cultivated (*garden plants*). *–v.* cultivate or tend a garden. □ **gardening** *n.* [Germanic: related to YARD[2]]

garden centre *n.* place where plants and garden equipment are sold.

garden city *n.* town spaciously laid out with parks etc.

gardener *n.* person who gardens, esp. for a living.

gardenia /gɑ:'di:nɪə/ *n.* tree or shrub with large fragrant flowers. [*Garden*, name of a naturalist]

garden party *n.* party held on a lawn or in a garden.

garfish /'gɑ:fɪʃ/ *n.* (*pl.* same or **-es**) fish with a long spearlike snout. [Old English, = spear-fish]

gargantuan /gɑ:'gæntjʊən/ *adj.* gigantic. [from the name *Gargantua*, a giant in Rabelais]

gargle /'gɑ:g(ə)l/ *–v.* (**-ling**) wash (the throat) with a liquid kept in motion by breathing through it. *–n.* liquid for gargling. [French: related to GARGOYLE]

gargoyle /'gɑ:gɔɪl/ *n.* grotesque carved face or figure, esp. as a spout from the gutter of a building. [French, = throat]

garibaldi /,gærɪ'bɔ:ldɪ/ *n.* (*pl.* **-s**) biscuit containing a layer of currants. [*Garibaldi*, name of an Italian patriot]

garish /'geərɪʃ/ *adj.* obtrusively bright; showy; gaudy. □ **garishly** *adv.* **garishness** *n.* [obsolete *gaure* stare]

garland /'gɑ:lənd/ *–n.* wreath of flowers etc., worn on the head or hung as a decoration. *–v.* adorn or crown with a garland or garlands. [French]

garlic /'gɑ:lɪk/ *n.* plant of the onion family with a pungent bulb used in cookery. □ **garlicky** *adj.* [Old English, = spear-leek]

garment /'gɑ:mənt/ *n.* **1** article of dress. **2** outward covering. [French: related to GARNISH]

garner /'gɑ:nə(r)/ *–v.* **1** collect. **2** store. *–n. literary* storehouse or granary. [Latin: related to GRANARY]

garnet /'gɑ:nɪt/ *n.* glassy silicate mineral, esp. a red kind used as a gem. [medieval Latin *granatum* POMEGRANATE]

garnish /'gɑ:nɪʃ/ *–v.* decorate (esp. food). *–n.* decoration, esp. to food. [French *garnir* from Germanic]

garotte var. of GARROTTE.

garret /'gærɪt/ *n.* attic or room in a roof. [French, = watch-tower: related to GARRISON]

garrison /'gærɪs(ə)n/ *–n.* troops stationed in a town etc. to defend it. *–v.* (**-n-**) **1** provide with or occupy as a garrison. **2** place on garrison duty. [French *garir* defend, from Germanic]

garrotte /gə'rɒt/ (also **garotte**; *US* **garrote**) *–v.* (**-ting**) execute or kill by strangulation, esp. with a wire collar. *–n.* device used for this. [French or Spanish]

garrulous /'gærʊləs/ *adj.* talkative. □ **garrulity** /gə'ru:lɪtɪ/ *n.* **garrulousness** *n.* [Latin]

garter *n.* **1** band worn to keep a sock or stocking up. **2** (**the Garter**) **a** highest order of English knighthood. **b** badge or membership of this. [French]

garter stitch *n.* plain knitting stitch.

gas *–n.* (*pl.* **-es**) **1** any airlike substance (i.e. not solid or liquid) moving freely to fill any space available. **2** such a substance (esp. found naturally or extracted from coal) used as fuel (also *attrib.*: *gas cooker*; *gas industry*). **3** nitrous oxide or other gas as an anaesthetic. **4** poisonous gas used in war. **5** *US colloq.* petrol, gasoline. **6** *slang* idle talk; boasting. **7** *slang* enjoyable or amusing thing or person. *–v.* (**gases, gassed, gassing**) **1** expose to gas, esp. to kill. **2** *colloq.* talk idly or boastfully. [Dutch invented word based on Greek *khaos* = CHAOS]

gasbag *n. slang* idle talker.

gas chamber *n.* room filled with poisonous gas to kill people or animals.

gaseous /'gæsɪəs/ *adj.* of or like gas.

gas fire *n.* domestic heater burning gas.

gas-fired *adj.* using gas as fuel.

gash *–n.* long deep slash, cut, or wound. *–v.* make a gash in; cut. [French]

gasholder *n.* large receptacle for storing gas; gasometer.

gasify /'gæsɪ,faɪ/ *v.* (**-ies, -ied**) convert into gas. □ **gasification** /-fɪ'keɪʃ(ə)n/ *n.*

gasket /'gæskɪt/ *n.* sheet or ring of rubber etc., shaped to seal the junction of metal surfaces. [French *garcette*]

gaslight *n.* light from burning gas.

gasman *n.* man who installs or services gas appliances, or reads gas meters.

gas mask *n.* respirator as a protection against poison gas.

gasoline /'gæsə,li:n/ *n.* (also **gasolene**) *US* petrol.

gasometer /gæ'sɒmɪtə(r)/ *n.* large tank from which gas is distributed by pipes. [French *gazomètre*: related to GAS, -METER]

gasp /gɑ:sp/ *–v.* **1** catch one's breath with an open mouth as in exhaustion or astonishment. **2** utter with gasps. *–n.* convulsive catching of breath. [Old Norse]

gas ring *n.* hollow ring perforated with gas jets, for cooking etc.

gassy *adj.* **(-ier, -iest) 1 a** of or like gas. **b** full of gas. **2** *colloq.* verbose.

gasteropod var. of GASTROPOD.

gastric /'gæstrɪk/ *adj.* of the stomach. [French: related to GASTRO-]

gastric flu *n. colloq.* intestinal disorder of unknown cause.

gastric juice *n.* digestive fluid secreted by the stomach glands.

gastritis /gæ'straɪtɪs/ *n.* inflammation of the stomach.

gastro- *comb. form* stomach. [Greek *gastēr* stomach]

gastro-enteritis /,gæstrəʊ,entə'raɪtɪs/ *n.* inflammation of the stomach and intestines.

gastronome /'gæstrə,nəʊm/ *n.* gourmet. [Greek *gastēr* stomach, *nomos* law]

gastronomy /gæ'strɒnəmɪ/ *n.* science or art of good eating and drinking. □ **gastronomic** /,gæstrə'nɒmɪk/ *adj.* **gastronomical** /,gæstrə'nɒmɪk(ə)l/ *adj.* **gastronomically** /,gæstrə'nɒmɪkəlɪ/ *adv.*

gastropod /'gæstrə,pɒd/ *n.* (also **gasteropod**) mollusc that moves by means of a ventral muscular organ, e.g. a snail. [from GASTRO-, Greek *pous pod-* foot]

gasworks *n.* place where gas is manufactured for lighting and heating.

gate *–n.* **1** barrier, usu. hinged, used to close an opening made for entrance and exit through a wall, fence, etc. **2** such an opening. **3** means of entrance or exit. **4** numbered place of access to aircraft at an airport. **5** device regulating the passage of water in a lock etc. **6 a** number of people entering by payment at the gates of a sports ground etc. **b** amount of money taken thus. **7 a** electrical signal that causes or controls the passage of other signals. **b** electrical circuit with an output that depends on the combination of several inputs. *–v.* **(-ting)** confine to college or school as a punishment. □ **gated** *adj.* [Old English]

gateau /'gætəʊ/ *n.* (*pl.* **-s** or **-x** /-təʊz/) large rich cake filled with cream etc. [French]

gatecrasher *n.* uninvited guest at a party etc. □ **gatecrash** *v.*

gatehouse *n.* house standing by or over a gateway, esp. to a large house or park.

gateleg *n.* (in full **gateleg table**) table with folding flaps supported by legs swung open like a gate. □ **gatelegged** *adj.*

gatepost *n.* post at either side of a gate.

gateway *n.* **1** opening which can be closed with a gate. **2** means of access (*gateway to the South*; *gateway to success*).

gather /'gæðə(r)/ *–v.* **1** bring or come together; accumulate. **2** pick or collect as harvest. **3** infer or deduce. **4 a** increase (*gather speed*). **b** collect (*gather dust*). **5** summon up (energy etc.). **6** draw together in folds or wrinkles. **7** (often as **gathering** *adj.*) come to a head (*gathering storm*). **8** develop a purulent swelling. *–n.* fold or pleat. □ **gather up** bring together; pick up from the ground; draw into a small compass. [Old English]

gathering *n.* **1** assembly. **2** purulent swelling. **3** group of leaves taken together in bookbinding.

GATT /gæt/ *abbr.* General Agreement on Tariffs and Trade.

gauche /gəʊʃ/ *adj.* **1** socially awkward. **2** tactless. □ **gauchely** *adv.* **gaucheness** *n.* [French]

gaucherie /'gəʊʃə,ri:/ *n.* gauche manners or act. [French: related to GAUCHE]

gaucho /'gaʊtʃəʊ/ *n.* (*pl.* **-s**) cowboy from the S. American pampas. [Spanish from Quechua]

gaudy /'gɔ:dɪ/ *adj.* **(-ier, -iest)** tastelessly showy. □ **gaudily** *adv.* **gaudiness** *n.* [obsolete *gaud* ornament, from Latin *gaudeo* rejoice]

gauge /geɪdʒ/ (*US* **gage**: see also sense 6) *–n.* **1** standard measure, esp. of the capacity or contents of a barrel, fineness of a textile, diameter of a bullet, or thickness of sheet metal. **2** instrument for measuring pressure, width, length, thickness, etc. **3** distance between rails or opposite wheels. **4** capacity, extent. **5** criterion, test. **6** (usu. **gage**) *Naut.* position relative to the wind. *–v.* **(-ging) 1** measure exactly. **2** measure the capacity or content of. **3** estimate (a person, situation, etc.). [French]

Gaul /gɔ:l/ *n.* inhabitant of ancient Gaul. [French from Germanic]

Gaulish *–adj.* of the Gauls. *–n.* their language.

gaunt /gɔ:nt/ *adj.* **1** lean, haggard. **2** grim, desolate. □ **gauntness** *n.* [origin unknown]

gauntlet[1] /ˈgɔ:ntlɪt/ *n.* **1** stout glove with a long loose wrist. **2** *hist.* armoured glove. □ **pick up** (or **take up**) **the gauntlet** accept a challenge. **throw down the gauntlet** issue a challenge. [French diminutive of *gant* glove]

gauntlet[2] /ˈgɔ:ntlɪt/ *n.* □ **run the gauntlet 1** undergo harsh criticism. **2** pass between two rows of people and receive blows from them, as a punishment or ordeal. [Swedish *gatlopp* from *gata* lane, *lopp* course]

gauss /gaʊs/ *n.* (*pl.* same) unit of magnetic flux density. [*Gauss*, name of a mathematician]

gauze /gɔ:z/ *n.* **1** thin transparent fabric of silk, cotton, etc. **2** fine mesh of wire etc. □ **gauzy** *adj.* (**-ier**, **-iest**). [French from *Gaza* in Palestine]

gave *past* of GIVE.

gavel /ˈgæv(ə)l/ *n.* hammer used for calling attention by an auctioneer, chairman, or judge. [origin unknown]

gavotte /gəˈvɒt/ *n.* **1** old French dance. **2** music for this. [French from Provençal]

gawk *–v. colloq.* gawp. *–n.* awkward or bashful person. [obsolete *gaw* GAZE]

gawky *adj.* (**-ier**, **-iest**) awkward or ungainly. □ **gawkily** *adv.* **gawkiness** *n.*

gawp *v. colloq.* stare stupidly or obtrusively. [related to YELP]

gay *–adj.* **1** light-hearted, cheerful. **2** brightly coloured. **3** *colloq.* homosexual. **4** *colloq.* careless, thoughtless (*gay abandon*). *–n. colloq.* (esp. male) homosexual. □ **gayness** *n.* [French]

■ **Usage** Sense 3 is generally informal in tone, but is favoured by homosexual groups.

gayety *US* var. of GAIETY.

gaze *–v.* (**-zing**) (foll. by *at*, *into*, *on*, etc.) look fixedly. *–n.* intent look. [origin unknown]

gazebo /gəˈzi:bəʊ/ *n.* (*pl.* **-s**) summer-house, turret, etc., with a wide view. [perhaps a fanciful formation from GAZE]

gazelle /gəˈzel/ *n.* (*pl.* same or **-s**) small graceful antelope. [Arabic *ġazāl*]

gazette /gəˈzet/ *–n.* **1** newspaper (used in the title). **2** official publication with announcements etc. *–v.* (**-tting**) announce or name in an official gazette. [French from Italian]

gazetteer /ˌgæzɪˈtɪə(r)/ *n.* geographical index. [Italian: related to GAZETTE]

gazpacho /gæˈspætʃəʊ/ *n.* (*pl.* **-s**) cold Spanish soup. [Spanish]

gazump /gəˈzʌmp/ *v. colloq.* **1** raise the price of a property after accepting an offer from (a buyer). **2** swindle. [origin unknown]

gazunder /gəˈzʌndə(r)/ *v. colloq.* lower an offer made to (a seller) for a property just before the exchange of contracts. [from GAZUMP, UNDER]

GB *abbr.* Great Britain.

GBH *abbr.* grievous bodily harm.

GC *abbr.* George Cross.

GCE *abbr.* General Certificate of Education.

GCHQ *abbr.* Government Communications Headquarters.

GCSE *abbr.* General Certificate of Secondary Education.

Gd *symb.* gadolinium.

GDP *abbr.* gross domestic product.

GDR *abbr. hist.* German Democratic Republic.

Ge *symb.* germanium.

gear *–n.* **1** (often in *pl.*) **a** set of toothed wheels that work together, esp. those connecting the engine of a vehicle to the road wheels. **b** particular setting of these (*first gear*). **2** equipment, apparatus, or tackle. **3** *colloq.* clothing. *–v.* **1** (foll. by *to*) adjust or adapt to. **2** (often foll. by *up*) equip with gears. **3** (foll. by *up*) make ready or prepared. **4** put in gear. □ **in gear** with a gear engaged. **out of gear** with no gear engaged. [Old Norse]

gearbox *n.* **1** set of gears with its casing, esp. in a vehicle. **2** the casing itself.

gearing *n.* set or arrangement of gears.

gear lever *n.* (also **gear shift**) lever used to engage or change gear.

gearwheel *n.* toothed wheel in a set of gears.

gecko /ˈgekəʊ/ *n.* (*pl.* **-s**) tropical house-lizard. [Malay]

gee[1] *int.* (also **gee whiz** /wɪz/) esp. *US colloq.* expression of surprise etc. [perhaps an abbreviation of JESUS]

gee[2] *int.* (usu. foll. by *up*) command to a horse etc. to start or go faster. [origin unknown]

gee-gee /ˈdʒi:dʒi:/ *n. colloq.* (a child's word for) a horse.

geese *pl.* of GOOSE.

geezer /ˈgi:zə(r)/ *n. slang* person, esp. an old man. [dial. *guiser* mummer]

Geiger counter /ˈgaɪgə(r)/ *n.* device for detecting and measuring radioactivity. [*Geiger*, name of a physicist]

geisha /ˈgeɪʃə/ *n.* (*pl.* same or **-s**) Japanese woman trained to entertain men. [Japanese]

gel *–n.* **1** semi-solid jelly-like colloid. **2** jelly-like substance used for setting the hair. *–v.* (**-ll-**) **1** form a gel. **2** = JELL 2. [from GELATIN]

gelatin /'dʒelətɪn/ *n.* (also **gelatine** /-,ti:n/) transparent tasteless substance from skin, tendons, etc., used in cookery, photography, etc. □ **gelatinize** /dʒɪ'lætɪ,naɪz/ *v.* (also **-ise**) (**-zing** or **-sing**). [Italian: related to JELLY]

gelatinous /dʒɪ'lætɪnəs/ *adj.* of a jelly-like consistency.

geld /geld/ *v.* castrate. [Old Norse]

gelding *n.* gelded animal, esp. a horse.

gelignite /'dʒelɪg,naɪt/ *n.* explosive made from nitroglycerine. [from GELATIN, IGNEOUS]

gem *–n.* **1** precious stone, esp. cut and polished or engraved. **2** thing or person of great beauty or worth. *–v.* (**-mm-**) adorn with or as with gems. [Latin *gemma* bud, jewel]

geminate *–adj.* /'dʒemɪnət/ combined in pairs. *–v.* /'dʒemɪ,neɪt/ (**-ting**) **1** double, repeat. **2** arrange in pairs. □ **gemination**/-'neɪʃ(ə)n/*n.*[Latin: related to GEMINI]

Gemini /'dʒemɪ,naɪ/ *n.* (*pl.* **-s**) **1** constellation and third sign of the zodiac (the Twins). **2** person born when the sun is in this sign. [Latin, = twins]

gemma *n.* (*pl.* **gemmae** /-mi:/) small cellular body in plants such as mosses, that separates from the mother-plant and starts a new one. □ **gemmation** /-'meɪʃ(ə)n/ *n.* [Latin, see GEM]

gemstone *n.* precious stone used as a gem.

Gen. *abbr.* General.

gen *slang –n.* information. *–v.* (**-nn-**) (foll. by *up*) gain or give information. [probably *gen*eral information]

-gen *comb. form Chem.* that which produces (*hydrogen*; *antigen*). [Greek *-genes* born]

gendarme /'ʒɒndɑ:m/ *n.* (in French-speaking countries) police officer. [French *gens d'armes* men of arms]

gender *n.* **1 a** classification roughly corresponding to the two sexes and sexlessness. **b** class of noun according to this classification (see MASCULINE, FEMININE, NEUTER). **2** a person's sex. [Latin GENUS]

gene /dʒi:n/ *n.* unit in a chromosome determining heredity. [German]

genealogy /,dʒi:nɪ'ælədʒɪ/ *n.* (*pl.* **-ies**) **1** descent traced continuously from an ancestor, pedigree. **2** study of pedigrees. **3** organism's line of development from earlier forms. □ **genealogical** /-ə'lɒdʒɪk(ə)l/ *adj.* **genealogically** /-ə'lɒdʒɪkəlɪ/ *adv.* **genealogist** *n.* [Greek *genea* race]

genera *pl.* of GENUS.

general /'dʒenər(ə)l/ *–adj.* **1** including or affecting all or most parts or cases of things. **2** prevalent, usual (*the general feeling*). **3** not partial or particular or local. **4** not limited in application, true of all or nearly all cases (*as a general rule*). **5** not restricted or specialized (*general knowledge*; *general hospital*). **6** not detailed (*general idea*). **7** vague (*spoke only in general terms*). **8** chief, head; having overall authority (*general manager*; *Secretary-General*). *–n.* **1 a** army officer next below Field Marshal. **b** = *lieutenant general* (see LIEUTENANT COLONEL), MAJOR-GENERAL. **2** commander of an army. **3** strategist (*a great general*). **4** head of a religious order, e.g. of Jesuits etc. □ **in general 1** as a normal rule; usually. **2** for the most part. [Latin *generalis*]

general anaesthetic *n.* anaesthetic affecting the whole body, usu. with loss of consciousness.

General Certificate of Education *n.* examination set esp. for secondary-school pupils at advanced level (and, formerly, ordinary level) in England, Wales and Northern Ireland.

General Certificate of Secondary Education *n.* examination replacing and combining the GCE ordinary level and CSE examinations.

general election *n.* national parliamentary election.

generalissimo /,dʒenərə'lɪsɪ,məʊ/ *n.* (*pl.* **-s**) commander of a combined military and naval and air force, or of combined armies. [Italian superlative]

generality /,dʒenə'rælɪtɪ/ *n.* (*pl.* **-ies**) **1** general statement or rule. **2** general applicability. **3** lack of detail. **4** (foll. by *of*) main body or majority.

generalize /'dʒenərə,laɪz/ *v.* (also **-ise**) (**-zing** or **-sing**) **1 a** speak in general or indefinite terms. **b** form general notions. **2** reduce to a general statement. **3** infer (a rule etc.) from particular cases. **4** bring into general use. □ **generalization** /-'zeɪʃ(ə)n/ *n.*

generally *adv.* **1** usually; in most respects or cases (*generally get up early*; *was generally well-behaved*). **2** in a general sense; without regard to particulars or exceptions (*generally speaking*). **3** for the most part (*not generally known*).

general meeting *n.* meeting open to all the members of a society etc.

general practice *n.* work of a general practitioner.

general practitioner *n.* community doctor treating cases of all kinds in the first instance.

general staff *n.* staff assisting a military commander at headquarters.

general strike *n.* simultaneous strike of workers in all or most trades.

generate /ˈdʒenə,reɪt/ *v.* (**-ting**) bring into existence; produce. [Latin: related to GENUS]

generation /,dʒenəˈreɪʃ(ə)n/ *n.* **1** all the people born at about the same time. **2** single stage in a family history (*three generations were present in the photograph*). **3** stage in (esp. technological) development (*fourth-generation computers*). **4** average time in which children are ready to take the place of their parents (about 30 years). **5** production, esp. of electricity. **6** procreation. □ **first-** (or **second-, third-,** etc.) **generation** *attrib.* designating a person who emigrated to a place (or whose parents or grandparents etc. emigrated). [Latin: related to GENERATE]

generation gap *n.* differences of outlook between different generations.

generative /ˈdʒenərətɪv/ *adj.* **1** of procreation. **2** productive.

generator *n.* **1** machine for converting mechanical into electrical energy. **2** apparatus for producing gas, steam, etc.

generic /dʒɪˈnerɪk/ *adj.* **1** characteristic of or relating to a class; general, not specific or special. **2** *Biol.* characteristic of or belonging to a genus. □ **generically** *adv.* [Latin: related to GENUS]

generous /ˈdʒenərəs/ *adj.* **1** giving or given freely. **2** magnanimous, unprejudiced. **3** abundant, copious. □ **generosity** /-ˈrɒsɪtɪ/ *n.* **generously** *adv.* [Latin: related to GENUS]

genesis /ˈdʒenɪsɪs/ *n.* **1** origin; mode of formation. **2** (**Genesis**) first book of the Old Testament, with an account of the Creation. [Greek *gen-* be produced]

gene therapy *n.* introduction of normal genes into cells in place of defective or missing ones in order to correct genetic disorders.

genetic /dʒɪˈnetɪk/ *adj.* **1** of genetics or genes. **2** of or in origin. □ **genetically** *adv.* [from GENESIS]

genetic code *n.* arrangement of genetic information in chromosomes.

genetic engineering *n.* manipulation of DNA to modify hereditary features.

genetic fingerprinting *n.* (also **genetic profiling**) identifying individuals by DNA patterns.

genetics *n.pl.* (treated as *sing.*) the study of heredity and the variation of inherited characteristics. □ **geneticist** /-tɪsɪst/ *n.*

genial /ˈdʒiːnɪəl/ *adj.* **1** jovial, sociable, kindly. **2** (of the climate) mild and warm; conducive to growth. **3** cheering. □ **geniality** /-ˈælɪtɪ/ *n.* **genially** *adv.* [Latin: related to GENIUS]

genie /ˈdʒiːnɪ/ *n.* (*pl.* **genii** /-nɪ,aɪ/) (in Arabian tales) spirit or goblin with magical powers. [French *génie* GENIUS: cf. JINNEE]

genital /ˈdʒenɪt(ə)l/ *–adj.* of animal reproduction or the reproductive organs. *–n.* (in *pl.*) external reproductive organs. [Latin *gigno genit-* beget]

genitalia /,dʒenɪˈteɪlɪə/ *n.pl.* genitals. [Latin, neuter pl. of *genitalis*: see GENITAL]

genitive /ˈdʒenɪtɪv/ *Gram.* *–n.* case expressing possession or close association, corresponding to *of*, *from*, etc. *–adj.* of or in this case. [Latin: related to GENITAL]

genius /ˈdʒiːnɪəs/ *n.* (*pl.* **geniuses**) **1 a** exceptional intellectual or creative power or other natural ability or tendency. **b** person with this. **2** tutelary spirit of a person, place, etc. **3** person or spirit powerfully influencing a person for good or evil. **4** prevalent feeling or association etc. of a people or place. [Latin]

genocide /ˈdʒenə,saɪd/ *n.* deliberate extermination of a people or nation. □ **genocidal** /-ˈsaɪd(ə)l/ *adj.* [Greek *genos* race, -CIDE]

genome /ˈdʒiːnəʊm/ *n.* **1** the haploid set of chromosomes of an organism. **2** the genetic material of an organism.

-genous *comb. form* forming adjectives meaning 'produced' (*endogenous*).

genre /ˈʒɑ̃rə/ *n.* **1** kind or style of art etc. **2** painting of scenes from ordinary life. [French: related to GENDER]

gent *n. colloq.* **1** gentleman. **2** (**the Gents**) *colloq.* men's public lavatory. [shortening of GENTLEMAN]

genteel /dʒenˈtiːl/ *adj.* **1** affectedly refined or stylish. **2** upper-class. □ **genteelly** *adv.* [French *gentil*: related to GENTLE]

gentian /ˈdʒenʃ(ə)n/ *n.* mountain plant usu. with blue flowers. [Latin *gentiana* from *Gentius*, king of Illyria]

Gentile /ˈdʒentaɪl/ *–adj.* not Jewish; heathen. *–n.* person who is not Jewish. [Latin *gentilis* from *gens* family]

gentility /dʒenˈtɪlɪtɪ/ *n.* **1** social superiority. **2** genteel manners or behaviour. [French: related to GENTLE]

gentle /ˈdʒent(ə)l/ *adj.* (**gentler**, **gentlest**) **1** not rough or severe; mild, kind (*a gentle nature*). **2** moderate (*gentle breeze*). **3** (of birth, pursuits, etc.) honourable, of or fit for gentlefolk. **4** quiet; requiring patience (*gentle art*). □ **gentleness** *n.* **gently** *adv.* [Latin: related to GENTILE]

gentlefolk *n.pl.* people of good family.

gentleman *n.* **1** man (in polite or formal use). **2** chivalrous well-bred man. **3** man of good social position (*country gentleman*). **4** man of gentle birth attached to a royal household (*gentleman in waiting*). **5** (in *pl.*) (as a form of address) male audience or part of this.

gentlemanly *adj.* like or befitting a gentleman.

gentleman's agreement *n.* (also **gentlemen's agreement**) agreement binding in honour but not enforceable.

gentlewoman *n. archaic* woman of good birth or breeding.

gentrification /ˌdʒentrɪfɪ'keɪʃ(ə)n/ *n.* upgrading of a working-class urban area by the arrival of more affluent residents. □ **gentrify** /-ˌfaɪ/ *v.* (**-ies, -ied**).

gentry /'dʒentrɪ/ *n.pl.* **1** people next below the nobility. **2** *derog.* people (*these gentry*). [French: related to GENTLE]

genuflect /'dʒenju:ˌflekt/ *v.* bend one's knee, esp. in worship. □ **genuflection** /-'flekʃ(ə)n/ *n.* (also **genuflexion**). [Latin *genu* knee, *flecto* bend]

genuine /'dʒenju:ɪn/ *adj.* **1** really coming from its reputed source etc. **2** properly so called; not sham; sincere. □ **genuinely** *adv.* **genuineness** *n.* [Latin]

genus /'dʒi:nəs/ *n.* (*pl.* **genera** /'dʒenərə/) **1** taxonomic category of animals or plants with common structural characteristics, usu. containing several species. **2** (in logic) kind of things including subordinate kinds or species. **3** *colloq.* kind, class. [Latin *genus -eris*]

geo- *comb. form* earth. [Greek *gē*]

geocentric /ˌdʒi:əʊ'sentrɪk/ *adj.* **1** considered as viewed from the earth's centre. **2** having the earth as the centre. □ **geocentrically** *adv.*

geode /'dʒi:əʊd/ *n.* **1** cavity lined with crystals. **2** rock containing this. [Greek *geōdēs* earthy]

geodesic /ˌdʒi:əʊ'di:zɪk/ *adj.* (also **geodetic** /-'detɪk/) of geodesy.

geodesic line *n.* shortest possible line between two points on a curved surface.

geodesy /dʒi:'ɒdɪsɪ/ *n.* the study of the shape and area of the earth. [Greek *geōdaisia*]

geographical /ˌdʒi:ə'græfɪk(ə)l/ *adj.* (also **geographic**) of geography. □ **geographically** *adv.*

geographical mile *n.* distance of one minute of longitude or latitude at the equator (about 1.85 km).

geography /dʒɪ'ɒgrəfɪ/ *n.* **1** science of the earth's physical features, resources, climate, population, etc. **2** features or arrangement of an area, rooms, etc. □ **geographer** *n.* [Latin from Greek]

geology /dʒɪ'ɒlədʒɪ/ *n.* **1** science of the earth's crust, strata, origin of its rocks, etc. **2** geological features of a district. □ **geological** /ˌdʒi:ə'lɒdʒɪk(ə)l/ *adj.* **geologically** /ˌdʒi:ə'lɒdʒɪkəlɪ/ *adv.* **geologist** *n.*

geometric /ˌdʒi:ə'metrɪk/ *adj.* (also **geometrical**) **1** of geometry. **2** (of a design etc.) with regular lines and shapes. □ **geometrically** *adv.*

geometric progression *n.* progression with a constant ratio between successive quantities (as 1, 3, 9, 27).

geometry /dʒɪ'ɒmətrɪ/ *n.* science of the properties and relations of lines, surfaces, and solids. □ **geometrician** /-'trɪʃ(ə)n/ *n.* [from GEO-, -METRY]

geophysics /ˌdʒi:əʊ'fɪzɪks/ *n.pl.* (treated as *sing.*) physics of the earth.

Geordie /'dʒɔ:dɪ/ *n.* native of Tyneside. [name *George*]

George Cross /dʒɔ:dʒ/ *n.* decoration for bravery awarded esp. to civilians. [King *George* VI]

georgette /dʒɔ:'dʒet/ *n.* thin dress-material similar to crêpe. [*Georgette* de la Plante, name of a dressmaker]

Georgian[1] /'dʒɔ:dʒ(ə)n/ *adj.* of the time of Kings George I–IV or of George V and VI.

Georgian[2] /'dʒɔ:dʒ(ə)n/ –*adj.* of Georgia in eastern Europe or the US. –*n.* **1** native or language of Georgia in eastern Europe. **2** native of Georgia in the US.

geranium /dʒə'reɪnɪəm/ *n.* (*pl.* **-s**) **1** (in general use) cultivated pelargonium. **2** herb or shrub bearing fruit shaped like a crane's bill. [Greek *geranos* crane]

gerbil /'dʒɜ:bɪl/ *n.* (also **jerbil**) mouse-like desert rodent with long hind legs. [French: related to JERBOA]

geriatric /ˌdʒerɪ'ætrɪk/ –*adj.* **1** of old people. **2** *colloq.* old, outdated. –*n.* old person. [Greek *gēras* old age, *iatros* doctor]

geriatrics *n.pl.* (usu. treated as *sing.*) branch of medicine or social science dealing with the health and care of old people. □ **geriatrician** /-ə'trɪʃ(ə)n/ *n.*

germ *n.* **1** micro-organism, esp. one causing disease. **2** portion of an organism capable of developing into a new one; rudiment of an animal or plant in seed (*wheat germ*). **3** thing that may develop; elementary principle. □ **germy** *adj.* (**-ier, -iest**). [Latin *germen* sprout]

German /'dʒɜ:mən/ –*n.* **1 a** native or national of Germany. **b** person of German descent. **2** language of Germany. –*adj.* of Germany or its people or language. [Latin *Germanus*]

german /'dʒɜ:mən/ *adj.* (placed after *brother*, *sister*, or *cousin*) having both parents the same, or both grandparents the same on one side (*brother german*; *cousin german*). [Latin *germanus*]

germander /dʒɜ:'mændə(r)/ *n.* plant of the mint family. [Greek, = ground-oak]

germane /dʒɜ:'meɪn/ *adj.* (usu. foll. by *to*) relevant (to a subject). [var. of GERMAN]

Germanic /dʒɜ:'mænɪk/ *–adj.* **1** having German characteristics. **2** *hist.* of the Germans. **3** of the Scandinavians, Anglo-Saxons, or Germans. *–n.* **1** the branch of Indo-European languages which includes English, German, Dutch, and the Scandinavian languages. **2** the primitive language of Germanic peoples.

germanium /dʒɜ:'meɪnɪəm/ *n.* brittle greyish-white semi-metallic element. [related to GERMAN]

German measles *n.pl.* disease like mild measles; rubella.

Germano- *comb. form* German.

German shepherd *n.* (also **German shepherd dog**) = ALSATIAN.

German silver *n.* white alloy of nickel, zinc, and copper.

germicide /'dʒɜ:mɪ,saɪd/ *n.* substance that destroys germs. □ **germicidal** /-'saɪd(ə)l/ *adj.*

germinal /'dʒɜ:mɪn(ə)l/ *adj.* **1** of germs. **2** in the earliest stage of development. **3** productive of new ideas. □ **germinally** *adv.* [related to GERM]

germinate /'dʒɜ:mɪ,neɪt/ *v.* **(-ting) 1** sprout, bud, or develop. **2** cause to do this. □ **germination** /-'neɪʃ(ə)n/ *n.* **germinative** *adj.* [Latin: related to GERM]

germ warfare *n.* use of germs to spread disease in war.

gerontology /,dʒerɒn'tɒlədʒɪ/ *n.* the study of old age and the process of ageing. [Greek *gerōn geront-* old man]

gerrymander /'dʒerɪ,mændə(r)/ *–v.* manipulate the boundaries of (a constituency etc.) so as to give undue influence to some party or class. *–n.* this practice. [Governor *Gerry* of Massachusetts]

gerund /'dʒerənd/ *n.* verbal noun, in English ending in *-ing* (e.g. *do you mind my asking you?*). [Latin]

gesso /'dʒesəʊ/ *n.* (*pl.* **-es**) gypsum as used in painting or sculpture. [Italian: related to GYPSUM]

Gestapo /ge'stɑ:pəʊ/ *n. hist.* Nazi secret police. [German, from *Ge*heime *Sta*atspolizei]

gestation /dʒe'steɪʃ(ə)n/ *n.* **1 a** process of carrying or being carried in the uterus between conception and birth. **b** this period. **2** development of a plan, idea, etc. □ **gestate** *v.* **(-ting)**. [Latin *gesto* carry]

gesticulate /dʒe'stɪkjʊ,leɪt/ *v.* **(-ting) 1** use gestures instead of, or to reinforce, speech. **2** express thus. □ **gesticulation** /-'leɪʃ(ə)n/ *n.* [Latin: related to GESTURE]

gesture /'dʒestʃə(r)/ *–n.* **1** significant movement of a limb or the body. **2** use of such movements, esp. as a rhetorical device. **3** action to evoke a response or convey intention, usu. friendly. *–v.* **(-ring)** gesticulate. [Latin *gestura* from *gero* wield]

get /get/ *v.* (**getting**; *past* **got**; *past part.* **got** or *US* **gotten**) (and in *comb.*) **1** come into possession of; receive or earn (*get a job*; *got £200 a week*; *got first prize*). **2** fetch or procure (*get my book for me*; *got a new car*). **3** go to reach or catch (a bus, train, etc.). **4** prepare (a meal etc.). **5** (cause to) reach some state or become (*get rich*; *get married*; *get to be famous*; *got them ready*; *got him into trouble*). **6** obtain as a result of calculation. **7** contract (a disease etc.). **8** establish contact by telephone etc. with; receive (a broadcast signal). **9** experience or suffer; have inflicted on one; receive as one's lot or penalty (*got four years in prison*). **10 a** succeed in bringing, placing, etc. (*get it round the corner*; *get it on to the agenda*). **b** (cause to) succeed in coming or going (*will get you there somehow*; *got absolutely nowhere*; *got home*). **11** (prec. by *have*) **a** possess (*have not got a penny*). **b** (foll. by *to* + infin.) be bound or obliged (*have got to see you*). **12** (foll. by *to* + infin.) induce; prevail upon (*got them to help me*). **13** *colloq.* understand (a person or an argument) (*have you got that?*; *I get your point*; *do you get me?*). **14** *colloq.* harm, injure, kill, esp. in retaliation (*I'll get you for that*). **15** *colloq.* **a** annoy. **b** affect emotionally. **c** attract. **16** (foll. by *to* + infin.) develop an inclination (*am getting to like it*). **17** (foll. by verbal noun) begin (*get going*). **18** establish (an idea etc.) in one's mind. **19** *archaic* beget. □ **get about 1** travel extensively or fast; go from place to place. **2** begin walking etc. (esp. after illness). **get across 1** communicate (an idea etc.). **2** (of an idea etc.) be communicated. **get ahead** make progress (esp. in a career etc.). **get along** (or **on**) (foll. by *together*, *with*) live harmoniously. **get around** = *get about*. **get at 1** reach; get hold of. **2** *colloq.* imply. **3** *colloq.* nag, criticize. **get away 1** escape, start. **2** (as *int.*) *colloq.* expressing disbelief or scepticism. **3** (foll. by *with*) escape blame or punishment for. **get back at** *colloq.* retaliate against. **get by** *colloq.* manage, even if with difficulty. **get cracking** see CRACK. **get down 1** alight, descend (from a vehicle, ladder, etc.). **2** record in writing. **get a person down** depress or deject him or her. **get down to** begin working on. **get hold of** grasp, secure, acquire, obtain; make contact with (a person). **get in 1** arrive, obtain a place in a college etc. **2** be elected. **get it** *slang* be punished. **get off 1** *colloq.* (cause to) be acquitted; escape with little or no punishment. **2** start. **3** alight from (a bus etc.). **4** (foll. by *with*, *together*) *colloq.* form an amorous or sexual relationship, esp. quickly. **get on 1** make progress; manage. **2** enter (a bus etc.). **3** = *get along*. **4** (usu. as **getting on** adj.) grow old. **get on to** *colloq.* **1** make contact with. **2** understand; become aware of. **get out 1** leave or escape or help to do this. **2** manage to go outdoors. **3** alight from a vehicle. **4** transpire; become known. **get out of** avoid or escape (a duty etc.). **get over 1** recover from (an illness, upset, etc.). **2** overcome (a difficulty). **3** = *get across*. **get a thing over** (or **over with**) complete (a tedious task) quickly. **get one's own back** *colloq.* have one's revenge. **get rid of** see RID. **get round 1** coax or cajole (a person) esp. to secure a favour. **2** evade (a law etc.). **get round to** deal with (a task etc.) in due course. **get somewhere** make progress; be initially successful. **get there** *colloq.* **1** succeed. **2** understand what is meant. **get through 1** pass (an examination etc.). **2** finish or use up (esp. resources). **3** make contact by telephone. **4** (foll. by *to*) succeed in making (a person) understand. **get together** gather, assemble. **get up 1** rise from sitting etc., or from bed after sleeping etc. **2** (of wind etc.) begin to be strong. **3** prepare or organize. **4** produce or stimulate (*get up steam*). **5** (often *refl.*) dress or arrange elaborately; arrange the appearance of. **6** (foll. by *to*) *colloq.* indulge or be involved in (*always getting up to mischief*). [Old Norse]

get-at-able /get'ætəb(ə)l/ *adj. colloq.* accessible.

getaway *n.* escape, esp. after a crime.

get-out *n.* means of avoiding something.

get-together *n. colloq.* social gathering.

get-up *n. colloq.* style or arrangement of dress etc.

get-up-and-go *n. colloq.* energy, enthusiasm.

geyser /'gi:zə(r)/ *n.* **1** intermittent hot spring. **2** apparatus for heating water. [Icelandic *Geysir* from *geysa* to gush]

ghastly /'gɑ:stlɪ/ *adj.* (**-ier, -iest**) **1** horrible, frightful. **2** *colloq.* unpleasant. **3** deathlike, pallid. □ **ghastliness** *n.* [obsolete *gast* terrify]

ghee /gi:/ *n.* Indian clarified butter. [Hindi from Sanskrit]

gherkin /'gɜ:kɪn/ *n.* small pickled cucumber. [Dutch]

ghetto /'getəʊ/ *n.* (*pl.* **-s**) **1** part of a city occupied by a minority group. **2** *hist.* Jewish quarter in a city. **3** segregated group or area. [Italian]

ghetto-blaster *n. slang* large portable radio, esp. for playing loud pop music.

ghillie var. of GILLIE.

ghost /gəʊst/ *–n.* **1** supposed apparition of a dead person or animal; disembodied spirit. **2** shadow or semblance (*not a ghost of a chance*). **3** secondary image in a defective telescope or television picture. *–v.* (often foll. by *for*) act as ghost-writer of (a work). □ **ghostliness** *n.* **ghostly** *adj.* (**-ier, -iest**). [Old English]

ghosting *n.* appearance of a 'ghost' image in a television picture.

ghost town *n.* town with few or no remaining inhabitants.

ghost train *n.* (at a funfair) open-topped miniature railway in which the rider experiences ghoulish sights, sounds, etc.

ghost-writer *n.* person who writes on behalf of the credited author.

ghoul /gu:l/ *n.* **1** person morbidly interested in death etc. **2** evil spirit or phantom. **3** spirit in Muslim folklore preying on corpses. □ **ghoulish** *adj.* **ghoulishly** *adv.* [Arabic]

GHQ *abbr.* General Headquarters.

ghyll var. of GILL³.

GI /dʒi:'aɪ/ *n.* (often *attrib.*) soldier in the US army. [abbreviation of *g*overnment (or *g*eneral) *i*ssue]

giant /'dʒaɪənt/ *–n.* **1** (*fem.* **giantess**) imaginary or mythical being of human form but superhuman size. **2** person or thing of great size, ability, courage, etc. *–attrib. adj.* **1** gigantic. **2** of a very large kind. [Greek *gigas gigant-*]

gibber *v.* jabber inarticulately. [imitative]

gibberish *n.* unintelligible or meaningless speech; nonsense.

gibbet /'dʒɪbɪt/ *–n. hist.* **1 a** gallows. **b** post with an arm on which an executed criminal was hung. **2** (prec. by *the*) death by hanging. *–v.* (**-t-**) **1** put to death by hanging. **2** expose or hang up on a gibbet. [French *gibet*]

gibbon /'gɪbən/ *n.* long-armed SE Asian anthropoid ape. [French]

gibbous /'gɪbəs/ *adj.* **1** convex. **2** (of a moon or planet) having the bright part greater than a semicircle and less than a circle. **3** humpbacked. [Latin *gibbus* hump]

gibe (also **jibe**) *–v.* (**-bing**) (often foll. by *at*) jeer, mock. *–n.* jeering remark, taunt. [perhaps from French *giber* handle roughly]

giblets /ˈdʒɪblɪts/ *n.pl.* edible organs etc. of a bird, removed and usu. cooked separately. [French *gibelet* game stew]
giddy /ˈgɪdɪ/ *adj.* (**-ier**, **-iest**) **1** dizzy, tending to fall or stagger. **2 a** mentally intoxicated (*giddy with success*). **b** excitable, frivolous, flighty. **3** making dizzy (*giddy heights*). □ **giddily** *adv.* **giddiness** *n.* [Old English]
gift /gɪft/ *n.* **1** thing given; present. **2** natural ability or talent. **3** the power to give (*in his gift*). **4** giving. **5** *colloq.* easy task. [Old Norse: related to GIVE]
gifted *adj.* talented; intelligent.
gift of the gab *n. colloq.* eloquence, loquacity.
gift token *n.* (also **gift voucher**) voucher used as a gift and exchangeable for goods.
gift-wrap *v.* wrap attractively as a gift.
gig[1] /gɪg/ *n.* **1** light two-wheeled one-horse carriage. **2** light ship's boat for rowing or sailing. **3** rowing-boat esp. for racing. [probably imitative]
gig[2] /gɪg/ *colloq.* –*n.* engagement to play music etc., usu. for one night. –*v.* (**-gg-**) perform a gig. [origin unknown]
giga- *comb. form* one thousand million (10^9). [Greek: related to GIANT]
gigantic /dʒaɪˈgæntɪk/ *adj.* huge, giant-like. □ **gigantically** *adv.* [Latin: related to GIANT]
giggle /ˈgɪg(ə)l/ –*v.* (**-ling**) laugh in half-suppressed spasms. –*n.* **1** such a laugh. **2** *colloq.* amusing person or thing; joke (*did it for a giggle*). □ **giggly** *adj.* (**-ier**, **-iest**). [imitative]
gigolo /ˈdʒɪgəˌləʊ/ *n.* (*pl.* **-s**) young man paid by an older woman to be her escort or lover. [French]
gild[1] /gɪld/ *v.* (*past part.* **gilded** or as adj. in sense 1 **gilt**) **1** cover thinly with gold. **2** tinge with a golden colour. **3** give a false brilliance to. □ **gild the lily** try to improve what is already satisfactory. [Old English: related to GOLD]
gild[2] var. of GUILD.
gill[1] /gɪl/ *n.* (usu. in *pl.*) **1** respiratory organ in a fish etc. **2** vertical radial plate on the underside of a mushroom etc. **3** flesh below a person's jaws and ears. [Old Norse]
gill[2] /dʒɪl/ *n.* unit of liquid measure equal to $^1/_4$ pint. [French]
gill[3] /gɪl/ *n.* (also **ghyll**) **1** deep usu. wooded ravine. **2** narrow mountain torrent. [Old Norse]
gillie /ˈgɪlɪ/ *n.* (also **ghillie**) *Scot.* man or boy attending a person hunting or fishing. [Gaelic]
gillyflower /ˈdʒɪlɪˌflaʊə(r)/ *n.* clove-scented flower, e.g. a wallflower or the clove-scented pink. [French *gilofre*]
gilt[1] /gɪlt/ –*adj.* **1** thinly covered with gold. **2** gold-coloured. –*n.* **1** gilding. **2** gilt-edged security. [from GILD[1]]
gilt[2] /gɪlt/ *n.* young sow. [Old Norse]
gilt-edged *adj.* (of securities, stocks, etc.) having a high degree of reliability.
gimbals /ˈdʒɪmb(ə)lz/ *n.pl.* contrivance of rings and pivots for keeping instruments horizontal in ships, aircraft, etc. [var. of *gimmal* from French *gemel* double finger-ring]
gimcrack /ˈdʒɪmkræk/ –*adj.* showy but flimsy and worthless. –*n.* showy ornament; knick-knack. [origin unknown]
gimlet /ˈgɪmlɪt/ *n.* small tool with a screw-tip for boring holes. [French]
gimlet eye *n.* eye with a piercing glance.
gimmick /ˈgɪmɪk/ *n.* trick or device, esp. to attract attention or publicity. □ **gimmickry** *n.* **gimmicky** *adj.* [origin unknown]
gimp /gɪmp/ *n.* (also **gymp**) **1** twist of silk etc. with cord or wire running through it. **2** fishing-line of silk etc. bound with wire. [Dutch]
gin[1] *n.* spirit made from grain or malt and flavoured with juniper berries. [Dutch *geneva*: related to JUNIPER]
gin[2] –*n.* **1** snare, trap. **2** machine separating cotton from its seeds. **3** a kind of crane and windlass. –*v.* (**-nn-**) **1** treat (cotton) in a gin. **2** trap. [French: related to ENGINE]
ginger /ˈdʒɪndʒə(r)/ –*n.* **1 a** hot spicy root usu. powdered for use in cooking, or preserved in syrup, or candied. **b** plant having this root. **2** light reddish-yellow. **3** spirit, mettle. –*adj.* of a ginger colour. –*v.* **1** flavour with ginger. **2** (foll. by *up*) enliven. □ **gingery** *adj.* [Old English and French, ultimately from Sanskrit]
ginger ale *n.* ginger-flavoured non-alcoholic drink.
ginger beer *n.* mildly alcoholic or non-alcoholic cloudy drink made from fermented ginger and syrup.
gingerbread –*n.* ginger-flavoured treacle cake. –*attrib. adj.* gaudy, tawdry.
ginger group *n.* group urging a party or movement to stronger policy or action.
gingerly –*adv.* in a careful or cautious manner. –*adj.* showing great care or caution. [perhaps from French *gensor* delicate]
ginger-nut *n.* ginger-flavoured biscuit.
gingham /ˈgɪŋəm/ *n.* plain-woven cotton cloth, esp. striped or checked. [Dutch from Malay]

gingivitis /ˌdʒɪndʒɪˈvaɪtɪs/ *n.* inflammation of the gums. [Latin *gingiva* GUM[2], -ITIS]

ginkgo /ˈgɪŋkgəʊ/ *n.* (*pl.* **-s**) tree with fan-shaped leaves and yellow flowers. [Chinese, = silver apricot]

ginormous /dʒaɪˈnɔːməs/ *adj. slang* enormous. [from GIANT, ENORMOUS]

gin rummy *n.* a form of the card-game rummy.

ginseng /ˈdʒɪnseŋ/ *n.* **1** plant found in E. Asia and N. America. **2** root of this used as a medicinal tonic. [Chinese]

gippy tummy /ˈdʒɪpɪ/ *n. colloq.* diarrhoea affecting visitors to hot countries. [from EGYPTIAN]

Gipsy var. of GYPSY.

giraffe /dʒɪˈrɑːf/ *n.* (*pl.* same or **-s**) large four-legged African animal with a long neck and forelegs. [French, ultimately from Arabic]

gird /gɜːd/ *v.* (*past* and *past part.* **girded** or **girt**) **1** encircle, attach, or secure, with a belt or band. **2** enclose or encircle. **3** (foll. by *round*) place (a cord etc.) round. □ **gird** (or **gird up**) **one's loins** prepare for action. [Old English]

girder *n.* iron or steel beam or compound structure for bridge-building etc.

girdle[1] /ˈgɜːd(ə)l/ –*n.* **1** belt or cord worn round the waist. **2** corset. **3** thing that surrounds. **4** bony support for the limbs (*pelvic girdle*). –*v.* (**-ling**) surround with a girdle. [Old English]

girdle[2] /ˈgɜːd(ə)l/ *n. Scot.* & *N.Engl.* var. of GRIDDLE.

girl /gɜːl/ *n.* **1** female child, daughter. **2** *colloq.* young woman. **3** *colloq.* girlfriend. **4** female servant. □ **girlhood** *n.* **girlish** *adj.* **girly** *adj.* [origin uncertain]

girl Friday *n.* female helper or follower.

girlfriend *n.* **1** person's regular female companion or lover. **2** female friend.

girlie *adj. colloq.* (of a magazine etc.) depicting young women in erotic poses.

girl scout *n.* = SCOUT *n.* 4.

giro /ˈdʒaɪrəʊ/ –*n.* (*pl.* **-s**) **1** system of credit transfer between banks, Post Offices, etc. **2** cheque or payment by giro. –*v.* (**-es, -ed**) pay by giro. [German from Italian]

girt see GIRD.

girth /gɜːθ/ *n.* **1** distance round a thing. **2** band round the body of a horse to secure the saddle etc. [Old Norse: related to GIRD]

gismo /ˈgɪzməʊ/ *n.* (also **gizmo**) (*pl.* **-s**) *slang* gadget. [origin unknown]

gist *n.* substance or essence of a matter. [Latin *jaceo* LIE[1]]

git /gɪt/ *n. slang* silly or contemptible person. [*get* (n.), = fool]

gîte /ʒiːt/ *n.* furnished holiday house in the French countryside. [French]

give /gɪv/ –*v.* (**-ving**; *past* **gave**; *past part.* **given** /ˈgɪv(ə)n/) **1** transfer the possession of freely; hand over as a present; donate. **2 a** transfer temporarily; provide with (*gave him the dog to hold*; *gave her a new hip*). **b** administer (medicine). **c** deliver (a message). **3** (usu. foll. by *for*) make over in exchange or payment. **4 a** confer; grant (a benefit, honour, etc.). **b** accord; bestow (love, time, etc.). **c** pledge (*gave his word*). **5 a** perform (an action etc.) (*gave a jump*; *gave a performance*; *gave an interview*). **b** utter; declare (*gave a shriek*; *gave the batsman out*). **6** (in *passive*; foll. by *to*) be inclined to or fond of (*is given to boasting*; *is given to strong drink*). **7** yield to pressure; collapse. **8** yield as a product or result (*gives an average of 7*). **9 a** consign, put (*gave him into custody*). **b** sanction the marriage of (a daughter etc.). **10** devote; dedicate (*gave his life to the cause*). **11** present; offer; show; hold out (*gives no sign of life*; *gave her his arm*; *give me an example*). **12** impart; be a source of; cause (*gave me a cold*; *gave me trouble*; *gave much pain*). **13** concede (*I give you the benefit of the doubt*). **14** deliver (a judgement etc.) authoritatively. **15** provide (a party, meal, etc.) as host. **16** (in past part.) assume or grant or specify (*given the circumstances*; *in a given situation*; *given that we earn so little*). **17** (*absol.*) *colloq.* tell what one knows. –*n.* capacity to yield or comply; elasticity. □ **give and take 1** exchange of words, ideas, blows, etc. **2** ability to compromise. **give away 1** transfer as a gift. **2** hand over (a bride) to a bridegroom. **3** reveal (a secret etc.). **give the game** (or **show**) **away** reveal a secret or intention. **give in 1** yield; acknowledge defeat. **2** hand in (a document etc.) to an official etc. **give it to a person** *colloq.* scold or punish. **give me** I prefer (*give me Greece any day*). **give off** emit (fumes etc.). **give oneself up to 1** abandon oneself to (despair etc.). **2** addict oneself to. **give on to** (or **into**) (of a window, corridor, etc.) overlook or lead into. **give or take** *colloq.* accepting as a margin of error in estimating. **give out 1** announce; emit; distribute. **2** be exhausted. **3** run short. **give over 1** *colloq.* stop or desist. **2** hand over. **3** devote. **give rise to** cause. **give a person to understand** inform or assure. **give up 1** resign; surrender. **2** part with. **3** deliver (a wanted person etc.). **4** pronounce incurable or insoluble; renounce hope of. **5** renounce or cease (an activity). **give way** yield under pressure, give precedence. **not give a damn** (or **monkey's** or **toss** etc.) *colloq.* not care at all. □ **giver** *n.* [Old English]

give-away *n. colloq.* **1** unintentional revelation. **2** thing given as a gift or at a low price.

gizmo var. of GISMO.

gizzard /'gɪzəd/ *n.* **1** second part of a bird's stomach, for grinding food. **2** muscular stomach of some fish etc. [French]

glacé /'glæseɪ/ *adj.* **1** (of fruit, esp. cherries) preserved in sugar. **2** (of cloth etc.) smooth; polished. [French]

glacé icing *n.* icing made with icing sugar and water.

glacial /'gleɪʃ(ə)l/ *adj.* **1** of ice. **2** *Geol.* characterized or produced by ice. [Latin *glacies* ice]

glacial period *n.* period when an exceptionally large area was covered by ice.

glaciated /'gleɪsɪ,eɪtɪd/ *adj.* **1** marked or polished by the action of ice. **2** covered by glaciers or ice sheets. □ **glaciation** /-'eɪʃ(ə)n/ *n.* [*glaciate* freeze, from Latin: related to GLACIAL]

glacier /'glæsɪə(r)/ *n.* mass of land ice formed by the accumulation of snow on high ground. [French: related to GLACIAL]

glad *adj.* (**gladder, gladdest**) **1** (*predic.*) pleased. **2** expressing or causing pleasure (*glad cry*; *glad news*). **3** ready and willing (*am glad to help*). □ **be glad of** find useful. □ **gladly** *adv.* **gladness** *n.* [Old English]

gladden /'glæd(ə)n/ *v.* make or become glad.

glade *n.* open space in a forest. [origin unknown]

glad eye *n.* (prec. by *the*) *colloq.* amorous glance.

glad hand *n. colloq.* hearty welcome.

gladiator /'glædɪ,eɪtə(r)/ *n. hist.* trained fighter in ancient Roman shows. □ **gladiatorial** /-ɪə'tɔːrɪəl/ *adj.* [Latin *gladius* sword]

gladiolus /,glædɪ'əʊləs/ *n.* (*pl.* **-li** /-laɪ/) plant of the lily family with sword-shaped leaves and flower-spikes. [Latin, diminutive of *gladius* sword]

glad rags *n.pl. colloq.* best clothes.

gladsome *adj. poet.* cheerful, joyous.

Gladstone bag /'glædst(ə)n/ *n.* bag with two compartments joined by a hinge. [*Gladstone*, name of a statesman]

glair *n.* **1** white of egg. **2** adhesive preparation made from this. [French]

glam *adj. colloq.* glamorous. [abbreviation]

glamorize /'glæmə,raɪz/ *v.* (also **-ise**) (**-zing** or **-sing**) make glamorous or attractive.

glamour /'glæmə(r)/ *n.* (*US* **glamor**) **1** physical, esp. cosmetic, attractiveness. **2** alluring or exciting beauty or charm. □ **glamorous** *adj.* **glamorously** *adv.* [var. of GRAMMAR in obsolete sense 'magic']

glance /glɑːns/ —*v.* (**-cing**) **1** (often foll. by *down*, *up*, *over*, etc.) look briefly, direct one's eye. **2** strike at an angle and glide off an object (*glancing blow*; *ball glanced off his bat*). **3** (usu. foll. by *over*) refer briefly or indirectly to a subject or subjects. **4** (of light etc.) flash or dart. —*n.* **1** brief look. **2** flash or gleam. **3** glancing stroke in cricket. □ **at a glance** immediately upon looking. [origin uncertain]

gland *n.* **1** organ or similar structure secreting substances for use in the body or for ejection. **2** *Bot.* similar organ in a plant. [Latin *glandulae* pl.]

glanders /'glændəz/ *n.pl.* contagious disease of horses. [French *glandre*: related to GLAND]

glandular /'glændjʊlə(r)/ *adj.* of a gland or glands.

glandular fever *n.* infectious disease with swelling of the lymph glands.

glare —*v.* (**-ring**) **1** look fiercely or fixedly. **2** shine dazzlingly or oppressively. —*n.* **1 a** strong fierce light, esp. sunshine. **b** oppressive public attention (*glare of publicity*). **2** fierce or fixed look. **3** tawdry brilliance. [Low German or Dutch]

glaring *adj.* **1** obvious, conspicuous (*glaring error*). **2** shining oppressively. □ **glaringly** *adv.*

glasnost /'glæznɒst/ *n.* (in the former Soviet Union) policy of more open government and access to information. [Russian, = openness]

glass /glɑːs/ —*n.* **1 a** (often *attrib.*) hard, brittle, usu. transparent substance, made by fusing sand with soda and lime etc. **b** substance of similar properties. **2** glass objects collectively. **3 a** glass drinking vessel. **b** its contents. **4** mirror. **5** glazed frame for plants. **6** barometer. **7** covering of a watch-face. **8** lens. **9** (in *pl.*) **a** spectacles. **b** binoculars. —*v.* (usu. as **glassed** *adj.*) fit with glass. □ **glassful** *n.* (*pl.* **-s**). [Old English]

glass-blowing *n.* blowing semi-molten glass to make glassware.

glass fibre *n.* filaments of glass made into fabric or embedded in plastic as reinforcement.

glasshouse *n.* **1** greenhouse. **2** *slang* military prison.

glass-paper *n.* paper coated with glass particles, for smoothing and polishing.
glassware *n.* articles made of glass.
glass wool *n.* mass of fine glass fibres for packing and insulation.
glassy *adj.* (**-ier**, **-iest**) **1** like glass. **2** (of the eye, expression, etc.) abstracted; dull; fixed.
Glaswegian /glæz'wi:dʒ(ə)n/ *–adj.* of Glasgow. *–n.* native of Glasgow. [after *Norwegian*]
glaucoma /glɔ:'kəʊmə/ *n.* eye-condition with increased pressure in the eyeball and gradual loss of sight. □ **glaucomatous** *adj.* [Greek *glaukos* greyish blue]
glaze *–v.* (**-zing**) **1** fit (a window etc.) with glass or (a building) with windows. **2 a** cover (pottery etc.) with a glaze. **b** fix (paint) on pottery thus. **3** cover (pastry, cloth, etc.) with a glaze. **4** (often foll. by *over*) (of the eyes) become glassy. **5** give a glassy surface to. *–n.* **1** vitreous substance for glazing pottery. **2** smooth shiny coating on food etc. **3** thin coat of transparent paint to modify underlying tone. **4** surface formed by glazing. [from GLASS]
glazier /'gleɪzjə(r)/ *n.* person whose trade is glazing windows etc.
gleam *–n.* faint or brief light or show. *–v.* emit gleams, shine. [Old English]
glean *v.* **1** acquire (facts etc.) in small amounts. **2** gather (corn left by reapers). [French]
gleanings *n.pl.* things gleaned, esp. facts.
glebe *n.* piece of land as part of a clergyman's benefice and providing income. [Latin *gl(a)eba* clod, soil]
glee *n.* **1** mirth; delight. **2** part-song for three or more (esp. male) voices. [Old English]
gleeful *adj.* joyful. □ **gleefully** *adv.* **gleefulness** *n.*
glen *n.* narrow valley. [Gaelic]
glengarry /glen'gærɪ/ *n.* (*pl.* **-ies**) brimless Scottish hat cleft down the centre and with ribbons at the back. [*Glengarry* in Scotland]
glib *adj.* (**glibber**, **glibbest**) speaking or spoken quickly or fluently but without sincerity. □ **glibly** *adv.* **glibness** *n.* [obsolete *glibbery* slippery, perhaps imitative]
glide *–v.* (**-ding**) **1** move smoothly and continuously. **2** (of an aircraft or pilot) fly without engine-power. **3** pass gradually or imperceptibly. **4** go stealthily. **5** cause to glide. *–n.* gliding movement. [Old English]
glide path *n.* aircraft's line of descent to land.
glider *n.* light aircraft without an engine.
glimmer *–v.* shine faintly or intermittently. *–n.* **1** feeble or wavering light. **2** (also **glimmering**) (usu. foll. by *of*) small sign (of hope etc.). [probably Scandinavian]
glimpse *–n.* (often foll. by *of*, *at*) **1** brief view or look. **2** faint transient appearance (*glimpses of the truth*). *–v.* (**-sing**) have a brief view of (*glimpsed his face in the crowd*). [related to GLIMMER]
glint *–v.* flash, glitter. *–n.* flash, sparkle. [probably Scandinavian]
glissade /glɪ'sɑ:d/ *–n.* **1** controlled slide down a snow slope in mountaineering. **2** gliding step in ballet. *–v.* (**-ding**) perform a glissade. [French]
glissando /glɪ'sændəʊ/ *n.* (*pl.* **-di** /-dɪ/ or **-s**) *Mus.* continuous slide of adjacent notes. [French *glissant* sliding: related to GLISSADE]
glisten /'glɪs(ə)n/ *–v.* shine like a wet or polished surface. *–n.* glitter; sparkle. [Old English]
glitch *n. colloq.* sudden irregularity or malfunction (of equipment etc.). [origin unknown]
glitter *–v.* **1** shine with a bright reflected light; sparkle. **2** (usu. foll. by *with*) be showy or splendid. *–n.* **1** sparkle. **2** showiness. **3** tiny pieces of sparkling material as decoration etc. □ **glittery** *adj.* [Old Norse]
glitterati /ˌglɪtə'rɑ:tɪ/ *n.pl. slang* rich fashionable people. [from GLITTER, LITERATI]
glitz *n. slang* showy glamour. □ **glitzy** *adj.* (**-ier**, **-iest**). [from GLITTER, RITZY]
gloaming /'gləʊmɪŋ/ *n. Scot.* or *poet.* twilight. [Old English]
gloat *–v.* (often foll. by *over* etc.) look or consider with greed, malice, etc. *–n.* act of gloating. [origin unknown]
glob *n. colloq.* mass or lump of semiliquid substance, e.g. mud. [perhaps from BLOB, GOB^2]
global /'gləʊb(ə)l/ *adj.* **1** worldwide (*global conflict*). **2** all-embracing. □ **globally** *adv.* [French: related to GLOBE]
global warming *n.* increase in the temperature of the earth's atmosphere caused by the greenhouse effect.
globe *n.* **1 a** (prec. by *the*) the earth. **b** spherical representation of it with a map on the surface. **2** spherical object, e.g. a fish-bowl, lamp, etc. [Latin *globus*]
globe artichoke *n.* the partly edible head of the artichoke plant.
globe-trotter *n. colloq.* person who travels widely. □ **globe-trotting** *n.* & *attrib. adj.*
globular /'glɒbjʊlə(r)/ *adj.* **1** globe-shaped. **2** composed of globules.
globule /'glɒbju:l/ *n.* small globe or round particle or drop. [Latin *globulus*]

globulin /ˈglɒbjʊlɪn/ *n.* molecule-transporting protein in plant and animal tissues.
glockenspiel /ˈglɒkən,spiːl/ *n.* musical instrument with bells or metal bars or tubes struck by hammers. [German, = bell-play]
gloom *n.* **1** darkness; obscurity. **2** melancholy; despondency. [origin unknown]
gloomy *adj.* (**-ier**, **-iest**) **1** dark; unlit. **2** depressed or depressing. □ **gloomily** *adv.* **gloominess** *n.*
glorify /ˈglɔːrɪ,faɪ/ *v.* (**-ies**, **-ied**) **1** make glorious. **2** make seem better or more splendid than it is. **3** (as **glorified** *adj.*) invested with more attractiveness, importance, etc. than it has in reality (*glorified waitress*). **4** extol. □ **glorification** /-fɪˈkeɪʃ(ə)n/ *n.* [Latin: related to GLORY]
glorious /ˈglɔːrɪəs/ *adj.* **1** possessing or conferring glory; illustrious. **2** *colloq.* often *iron.* splendid, excellent (*glorious day*; *glorious muddle*). □ **gloriously** *adv.*
glory /ˈglɔːrɪ/ *–n.* (*pl.* **-ies**) **1** renown, fame; honour. **2** adoring praise. **3** resplendent majesty, beauty, etc. **4** thing that brings renown, distinction, or pride. **5** heavenly bliss and splendour. **6** *colloq.* state of exaltation, prosperity, etc. **7** halo of a saint etc. *–v.* (**-ies**, **-ied**) (often foll. by *in*) pride oneself. [Latin *gloria*]
glory-hole *n. colloq.* untidy room, cupboard, etc.
gloss[1] *–n.* **1** surface shine or lustre. **2** deceptively attractive appearance. **3** (in full **gloss paint**) paint giving a glossy finish. *–v.* make glossy. □ **gloss over** seek to conceal, esp. by mentioning only briefly. [origin unknown]
gloss[2] *–n.* **1** explanatory comment added to a text, e.g. in the margin. **2** interpretation or paraphrase. *–v.* add a gloss to (a text, word, etc.). [Latin *glossa* tongue]
glossary /ˈglɒsərɪ/ *n.* (*pl.* **-ies**) **1** list or dictionary of technical or special words. **2** collection of glosses. [Latin: related to GLOSS[2]]
glossy *–adj.* (**-ier**, **-iest**) **1** smooth and shiny (*glossy paper*). **2** printed on such paper. *–n.* (*pl.* **-ies**) *colloq.* glossy magazine or photograph. □ **glossily** *adv.* **glossiness** *n.*
glottal /ˈglɒt(ə)l/ *adj.* of the glottis.
glottal stop *n.* sound produced by the sudden opening or shutting of the glottis.
glottis /ˈglɒtɪs/ *n.* opening at the upper end of the windpipe and between the vocal cords. [Greek]
Gloucester /ˈglɒstə(r)/ *n.* (usu. **double Gloucester**, orig. a richer kind) cheese made in Gloucestershire. [*Gloucester* in England]
glove /glʌv/ *–n.* **1** hand-covering for protection, warmth, etc., usu. with separate fingers. **2** boxing glove. *–v.* (**-ving**) cover or provide with gloves. [Old English]
glove compartment *n.* recess for small articles in the dashboard of a car etc.
glove puppet *n.* small puppet fitted on the hand and worked by the fingers.
glover *n.* glove-maker.
glow /gləʊ/ *–v.* **1 a** emit light and heat without flame. **b** shine as if heated in this way. **2** (often foll. by *with*) **a** (of the body) be heated. **b** show or feel strong emotion (*glowed with pride*). **3** show a warm colour. **4** (as **glowing** *adj.*) expressing pride or satisfaction (*glowing report*). *–n.* **1** glowing state. **2** bright warm colour. **3** feeling of satisfaction or well-being. [Old English]
glower /ˈglaʊə(r)/ *–v.* **1** (often foll. by *at*) look angrily. **2** look dark or threatening. *–n.* glowering look. [origin uncertain]
glow-worm *n.* beetle whose wingless female emits light from the end of the abdomen.
gloxinia /glɒkˈsɪnɪə/ *n.* American tropical plant with large bell-shaped flowers. [*Gloxin*, name of a botanist]
glucose /ˈgluːkəʊz/ *n.* sugar found in the blood or in fruit juice etc., and as a constituent of starch, cellulose, etc. [Greek *gleukos* sweet wine]
glue /gluː/ *–n.* adhesive substance. *–v.* (**glues**, **glued**, **gluing** or **glueing**) **1** fasten or join with glue. **2** keep or put very close (*eye glued to the keyhole*). □ **gluey** /ˈgluːɪ/ *adj.* (**gluier**, **gluiest**). [Latin *glus*: related to GLUTEN]
glue ear *n.* blocking of the Eustachian tube, esp. in children.
glue-sniffing *n.* inhalation of fumes from adhesives as an intoxicant. □ **glue-sniffer** *n.*
glum *adj.* (**glummer**, **glummest**) dejected; sullen. □ **glumly** *adv.* **glumness** *n.* [var. of GLOOM]
glut *–v.* (**-tt-**) **1** feed (a person, one's stomach, etc.) or indulge (a desire etc.) to the full; satiate. **2** fill to excess. **3** overstock (a market). *–n.* **1** supply exceeding demand. **2** full indulgence; surfeit. [French *gloutir* swallow: related to GLUTTON]
glutamate /ˈgluːtə,meɪt/ *n.* salt or ester of glutamic acid, esp. a sodium salt used to enhance the flavour of food.

glutamic acid /glu:'tæmɪk/ *n.* amino acid normally found in proteins. [from GLUTEN, AMINE]

gluten /'glu:t(ə)n/ *n.* mixture of proteins present in cereal grains; sticky protein substance left when starch is washed out of flour. [Latin *gluten -tin-* glue]

glutinous /'glu:tɪnəs/ *adj.* sticky; like glue. [Latin: related to GLUTEN]

glutton /'glʌt(ə)n/ *n.* **1** greedy eater. **2** (often foll. by *for*) *colloq.* person insatiably eager (*glutton for work*). **3** voracious animal of the weasel family. □ **gluttonous** *adj.* **gluttonously** *adv.* [Latin *gluttio* SWALLOW[1]]

glutton for punishment *n.* person eager to take on hard or unpleasant tasks.

gluttony /'glʌtənɪ/ *n.* greed or excess in eating. [French: related to GLUTTON]

glycerine /'glɪsə,ri:n/ *n.* (also **glycerol**, *US* **glycerin** /-rɪn/) thick sweet colourless liquid used as medicine, ointment, etc., and in explosives. [Greek *glukeros* sweet]

glycerol /'glɪsə,rɒl/ *n.* = GLYCERINE.

glycogen /'glaɪkədʒ(ə)n/ *n.* polysaccharide serving as a store of carbohydrates, esp. in animal tissues.

glycolysis /glaɪ'kɒlɪsɪs/ *n.* breakdown of glucose by enzymes with the release of energy.

GM *abbr.* George Medal.

gm *abbr.* gram(s).

G-man /'dʒi:mæn/ *n. US colloq.* federal criminal-investigation officer. [from Government]

GMS *abbr.* grant maintained status.

GMT *abbr.* Greenwich Mean Time.

gnarled /nɑ:ld/ *adj.* (of a tree, hands, etc.) knobbly, twisted, rugged. [var. of *knarled*: related to KNURL]

gnash /næʃ/ *v.* **1** grind (the teeth). **2** (of the teeth) strike together. [Old Norse]

gnat /næt/ *n.* small two-winged biting fly. [Old English]

gnaw /nɔ:/ *v.* **1 a** (usu. foll. by *away* etc.) wear away by biting. **b** (often foll. by *at*, *into*) bite persistently. **2 a** corrode; wear away. **b** (of pain, fear, etc.) torment. [Old English]

gneiss /naɪs/ *n.* coarse-grained metamorphic rock of feldspar, quartz, and mica. [German]

gnome /nəʊm/ *n.* **1 a** dwarfish legendary spirit or goblin living underground. **b** figure of this as a garden ornament. **2** (esp. in *pl.*) *colloq.* person with sinister influence, esp. financial (*gnomes of Zurich*). □ **gnomish** *adj.* [French]

gnomic /'nəʊmɪk/ *adj.* of aphorisms; sententious. [Greek *gnōmē* opinion]

gnomon /'nəʊmɒn/ *n.* rod or pin etc. on a sundial, showing the time by its shadow. [Greek, = indicator]

gnostic /'nɒstɪk/ *—adj.* **1** of knowledge; having special mystical knowledge. **2** (**Gnostic**) concerning the Gnostics. *—n.* (**Gnostic**) (usu. in *pl.*) early Christian heretic claiming mystical knowledge. □ **Gnosticism** /-,sɪz(ə)m/ *n.* [Greek *gnōsis* knowledge]

GNP *abbr.* gross national product.

gnu /nu:/ *n.* (*pl.* same or **-s**) oxlike antelope. [Bushman *nqu*]

go[1] /gəʊ/ *—v.* (*3rd sing. present* **goes** /gəʊz/; *past* **went**; *past part.* **gone** /gɒn/) **1 a** start moving or be moving from one place or point in time to another; travel, proceed. **b** (foll. by *and* + verb) *colloq.* expressing annoyance (*you went and told him*). **2** (foll. by verbal noun) make a special trip for; participate in (*went skiing; goes running*). **3** lie or extend in a certain direction (*the road goes to London*). **4** leave; depart (*they had to go*). **5** move, act, work, etc. (*clock doesn't go*). **6 a** make a specified movement (*go like this with your foot*). **b** make a sound (often of a specified kind) (*gun went bang; door bell went*). **c** (of an animal) make (its characteristic cry) (*the cow went 'moo'*). **d** *colloq.* say (*so he goes to me 'Why didn't you like it?'*). **7** be in a specified state (*go hungry; went in fear of his life*). **8 a** pass into a specified condition (*gone bad; went to sleep*). **b** *colloq.* die. **c** proceed or escape in a specified condition (*poet went unrecognized*). **9** (of time or distance) pass, elapse; be traversed (*ten days to go before Easter; the last mile went quickly*). **10 a** (of a document, verse, song, etc.) have a specified content or wording (*the tune goes like this*). **b** be current or accepted (*so the story goes*). **c** be suitable; fit; match (*the shoes don't go with the hat; those pinks don't go*). **d** be regularly kept or put (*the forks go here*). **e** find room; fit (*this won't go into the cupboard*). **11 a** turn out, proceed; take a course or view (*things went well; Liverpool went Labour*). **b** be successful (*make the party go*). **12 a** be sold (*went for £1; went cheap*). **b** (of money) be spent. **13 a** be relinquished or abolished (*the car will have to go*). **b** fail, decline; give way, collapse (*his sight is going; the bulb has gone*). **14** be acceptable or permitted; be accepted without question (*anything goes; what I say goes*). **15** (often foll. by *by*, *with*, *on*, *upon*) be guided by; judge or act on or in harmony with (*have nothing to go on; a good rule to go by*). **16** attend regularly (*goes to school*). **17** (foll. by pres. part.) *colloq.*

proceed (often foolishly) to do (*went running to the police*; *don't go making him angry*). **18** act or proceed to a certain point (*will go so far and no further*; *went as high as £100*). **19** (of a number) be capable of being contained in another (*6 into 5 won't go*). **20** (usu. foll. by *to*) be allotted or awarded; pass (*first prize went to the girl*). **21** (foll. by *to*, *towards*) amount to; contribute to (*12 inches go to make a foot*; *this will go towards your holiday*). **22** (in *imper.*) begin motion (a starter's order in a race) (*ready, steady, go!*). **23** (usu. foll. by *by*, *under*) be known or called (*goes by the name of Droopy*). **24** *colloq.* proceed to (*go jump in the lake*). **25** (foll. by *for*) apply to (*that goes for me too*). *–n.* (*pl.* **goes**) **1** mettle; animation (*has a lot of go in her*). **2** vigorous activity (*it's all go*). **3** *colloq.* success (*made a go of it*). **4** *colloq.* attempt; turn (*I'll have a go*; *it's my go*). *–adj. colloq.* functioning properly (*all systems are go*). □ **go about 1** set to work at. **2** be socially active. **3** (foll. by pres. part.) make a habit of doing. **go ahead** proceed without hesitation. **go along with** agree to or with. **go back on** fail to keep (a promise etc.). **go begging** see BEG. **go down 1 a** (of an amount) become less through use (*coffee has gone down*). **b** subside (*the flood went down*). **c** decrease in price. **2 a** (of a ship) sink. **b** (of the sun) set. **c** (of a curtain) fall. **3** deteriorate; (of a computer system etc.) cease to function. **4** be recorded in writing. **5** be swallowed. **6** (often foll. by *with*) find acceptance. **7** *colloq.* leave university. **8** *colloq.* be sent to prison. **go down with** become ill with (a disease). **go far** see FAR. **go for 1** go to fetch. **2** pass or be accounted as (*went for nothing*). **3** prefer; choose. **4** *colloq.* strive to attain (*go for it!*). **5** *colloq.* attack (*the dog went for him*). **go halves** (often foll. by *with*) share equally. **go in 1** enter a room, house, etc. **2** (of the sun etc.) become obscured by cloud. **go in for 1** enter as a competitor. **2** take as one's style, pursuit, etc. **go into 1** enter (a profession, hospital, etc.). **2** investigate. **go a long way 1** (often foll.by *towards*) have a great effect. **2** (of food, money, etc.) last a long time, buy much. **3** = *go far*. **go off 1** explode. **2** leave the stage. **3** wear off. **4** (esp. of foodstuffs) deteriorate. **5** go to sleep. **6** *colloq.* begin to dislike (*I've gone off him*). **go off well** (or **badly** etc.) (of an enterprise etc.) be received or accomplished well (or badly etc.). **go on 1** (often foll. by pres. part.) continue, persevere (*decided to go on with it*; *went on trying*). **2** *colloq.* **a** talk at great length. **b** (foll. by *at*) nag. **3** (foll. by *to* + infin.) proceed (*went on to become a star*). **4** (also **go upon**) *colloq.* use as evidence. **go out 1** leave a room, house, etc. **2** be extinguished. **3** be broadcast. **4** (often foll. by *with*) be courting. **5** cease to be fashionable. **6** *colloq.* lose consciousness. **7** (usu. foll. by *to*) (of the heart etc.) expand with sympathy etc. towards (*my heart goes out to them*). **8** (of a tide) ebb to low tide. **go over 1** inspect the details of; rehearse; retouch. **2** (of a play etc.) be received, esp. favourably. **go round 1** spin, revolve. **2** (of food etc.) suffice for everybody. **3** (usu. foll. by *to*) visit informally. **go slow** work slowly, as a form of industrial action. **go through 1** be dealt with or completed. **2** discuss or scrutinize in detail. **3** perform. **4** undergo. **5** *colloq.* use up; spend (money etc.). **go through with** complete. **go to blazes** (or **hell**) *slang* exclamation of dismissal, contempt, etc. **go too far** see FAR. **go under** sink; fail; succumb. **go up 1** rise in price. **2** *colloq.* enter university. **3** be consumed (in flames etc.); explode. **go with 1** be harmonious with; match. **2** be courting. **go without** manage without; forgo (also *absol.*). **have a go at 1** attack. **2** attempt. **on the go** *colloq.* **1** in constant motion. **2** constantly working. [Old English: *went* is originally a past of WEND]

go[2] /gəʊ/ *n.* Japanese board-game. [Japanese]

goad *–v.* **1** urge on with a goad. **2** (usu. foll. by *on*, *into*) irritate; stimulate. *–n.* **1** spiked stick used for urging cattle forward. **2** anything that torments or incites. [Old English]

go-ahead *–n.* permission to proceed. *–adj.* enterprising.

goal *n.* **1** object of ambition or effort; destination. **2 a** structure into or through which the ball has to be sent to score in certain games. **b** point won. **3** point marking the finish of a race. [origin unknown]

goalie *n. colloq.* = GOALKEEPER.

goalkeeper *n.* player defending a goal.

goalpost *n.* either of the two upright posts of a goal.

goat *n.* **1** hardy domesticated mammal, with horns and (in the male) a beard. **2** lecherous man. **3** *colloq.* foolish person. **4** (**the Goat**) zodiacal sign or constellation Capricorn. □ **get a person's goat** *colloq.* irritate a person. [Old English]

goatee /gəʊ'ti:/ *n.* small pointed beard.

goatherd *n.* person who tends goats.

goatskin *n.* **1** skin of a goat. **2** garment or bottle made of goatskin.

gob[1] *n. slang* mouth. [origin unknown]

gob[2] *slang –n.* clot of slimy matter. *–v.* (**-bb-**) spit. [French *go(u)be* mouthful]

gobbet /'gɒbɪt/ *n.* **1** piece or lump of flesh, food, etc. **2** extract from a text, esp. one set for translation or comment. [French diminutive of *gobe* GOB[2]]

gobble[1] /'gɒb(ə)l/ *v.* (**-ling**) eat hurriedly and noisily. [from GOB[2]]

gobble[2] /'gɒb(ə)l/ *v.* (**-ling**) **1** (of a turkeycock) make a characteristic guttural sound. **2** make such a sound when speaking. [imitative]

gobbledegook /'gɒb(ə)ldɪ,gu:k/ *n.* (also **gobbledygook**) *colloq.* pompous or unintelligible jargon. [probably imitative of a turkeycock]

go-between *n.* intermediary.

goblet /'gɒblɪt/ *n.* drinking-vessel with a foot and stem. [French diminutive of *gobel* cup]

goblin /'gɒblɪn/ *n.* mischievous ugly dwarflike creature of folklore. [Anglo-French]

gobsmacked *adj. slang* flabbergasted.

gob-stopper *n.* large hard sweet.

goby /'gəʊbɪ/ *n.* (*pl.* **-ies**) small fish with ventral fins joined to form a disc or sucker. [Greek *kōbios* GUDGEON[1]]

go-cart *n.* var. of GO-KART.

god *n.* **1 a** (in many religions) superhuman being or spirit worshipped as having power over nature, human fortunes, etc. **b** image, idol, etc., symbolizing a god. **2** (**God**) (in Christian and other monotheistic religions) creator and ruler of the universe. **3** adored or greatly admired person. **4** (in *pl.*) *Theatr.* gallery. □ **God forbid** may it not happen! **God knows 1** it is beyond all knowledge. **2** I call God to witness that. **God willing** if Providence allows. [Old English]

godchild *n.* person in relation to his or her godparent.

god-daughter *n.* female godchild.

goddess /'gɒdɪs/ *n.* **1** female deity. **2** adored woman.

godfather *n.* **1** male godparent. **2** esp. *US* person directing an illegal organization, esp. the Mafia.

God-fearing *adj.* earnestly religious.

God-forsaken *adj.* dismal.

godhead *n.* (also **Godhead**) **1 a** state of being God or a god. **b** divine nature. **2** deity. **3** (**the Godhead**) God.

godless *adj.* **1** impious; wicked. **2** without a god. **3** not recognizing God. □ **godlessness** *n.*

godlike *adj.* resembling God or a god.

godly *adj.* (**-ier, -iest**) pious, devout. □ **godliness** *n.*

godmother *n.* female godparent.

godparent *n.* person who presents a child at baptism and responds on the child's behalf.

godsend *n.* unexpected but welcome event or acquisition.

godson *n.* male godchild.

Godspeed /gɒd'spi:d/ *int.* expression of good wishes to a person starting a journey.

goer *n.* **1** person or thing that goes (*slow goer*). **2** (often in *comb.*) person who attends, esp. regularly (*churchgoer*). **3** *colloq.* **a** lively or persevering person. **b** sexually promiscuous person.

go-getter *n. colloq.* aggressively enterprising person.

goggle /'gɒg(ə)l/ –*v.* (**-ling**) **1 a** (often foll. by *at*) look with wide-open eyes. **b** (of the eyes) be rolled about; protrude. **2** roll (the eyes). –*adj.* (usu. *attrib.*) (of the eyes) protuberant or rolling. –*n.* (in *pl.*) spectacles for protecting the eyes. [probably imitative]

goggle-box *n. colloq.* television set.

go-go *adj. colloq.* (of a dancer, music, etc.) in modern style; lively, erotic, and rhythmic.

going /'gəʊɪŋ/ –*n.* **1** act or process of going. **2 a** condition of the ground for walking, riding, etc. **b** progress affected by this. –*adj.* **1** in or into action (*set the clock going*). **2** existing, available (*there's cold beef going*). **3** current, prevalent (*the going rate*). □ **get going** start steadily talking, working, etc. **going on fifteen** etc. esp. *US* approaching one's fifteenth etc. birthday. **going on for** approaching (a time, age, etc.). **going strong** continuing vigorously. **going to** intending to; about to. **to be going on with** to start with; for the time being. **while the going is good** while conditions are favourable.

going concern *n.* thriving business.

going-over *n.* (*pl.* **goings-over**) **1** *colloq.* inspection or overhaul. **2** *slang* thrashing.

goings-on *n.pl.* (esp. morally suspect) behaviour.

goitre /'gɔɪtə(r)/ *n.* (*US* **goiter**) morbid enlargement of the thyroid gland. [Latin *guttur* throat]

go-kart *n.* (also **go-cart**) miniature racing car with a skeleton body.

gold /gəʊld/ –*n.* **1** precious yellow metallic element. **2** colour of gold. **3 a** coins or articles made of gold. **b** wealth. **4** something precious or beautiful. **5** = GOLD MEDAL. –*adj.* **1** made wholly or chiefly of gold. **2** coloured like gold. [Old English]

goldcrest *n.* tiny bird with a golden crest.

gold-digger *n. slang* woman who cultivates men to obtain money from them.

gold-dust *n.* gold in fine particles as often found naturally.

golden *adj.* **1 a** made or consisting of gold. **b** yielding gold. **2** coloured or shining like gold (*golden hair*). **3** precious; excellent.

golden age *n.* period of a nation's greatest prosperity, cultural merit, etc.

golden eagle *n.* large eagle with yellow-tipped head-feathers.

golden handshake *n. colloq.* payment given on redundancy or early retirement.

golden jubilee *n.* fiftieth anniversary.

golden mean *n.* the principle of moderation.

golden retriever *n.* retriever with a thick golden-coloured coat.

golden rod *n.* plant with a spike of yellow flowers.

golden rule *n.* basic principle of action, esp. 'do as you would be done by'.

golden wedding *n.* fiftieth anniversary of a wedding.

gold-field *n.* district in which gold occurs naturally.

goldfinch *n.* songbird with a yellow band across each wing.

goldfish *n.* (*pl.* same or **-es**) small reddish-golden Chinese carp.

gold foil *n.* gold beaten into a thin sheet.

gold leaf *n.* gold beaten into a very thin sheet.

gold medal *n.* medal of gold, usu. awarded as first prize.

gold-mine *n.* **1** place where gold is mined. **2** *colloq.* source of great wealth.

gold plate *n.* **1** vessels made of gold. **2** material plated with gold.

gold-plate *v.* plate with gold.

gold-rush *n.* rush to a newly-discovered gold-field.

goldsmith *n.* worker in gold.

gold standard *n.* system by which the value of a currency is defined in terms of gold.

golf –*n.* game in which a small hard ball is driven with clubs into a series of 18 or 9 holes with the fewest possible strokes. –*v.* play golf. □ **golfer** *n.* [origin unknown]

golf ball *n.* **1** ball used in golf. **2** *colloq.* small ball used in some electric typewriters to carry the type.

golf club *n.* **1** club used in golf. **2** association for playing golf. **3** premises of this.

golf-course *n.* (also **golf-links**) course on which golf is played.

golliwog /'gɒlɪ,wɒg/ *n.* black-faced soft doll with fuzzy hair. [origin uncertain]

golly[1] /'gɒlɪ/ *int.* expressing surprise. [euphemism for GOD]

golly[2] /'gɒlɪ/ *n.* (*pl.* **-ies**) *colloq.* = GOLLIWOG. [abbreviation]

golosh var. of GALOSH.

gonad /'gəʊnæd/ *n.* animal organ producing gametes, esp. the testis or ovary. [Greek *gonē* seed]

gondola /'gɒndələ/ *n.* **1** light flat-bottomed boat used on Venetian canals. **2** car suspended from an airship or balloon, or attached to a ski-lift. [Italian]

gondolier /,gɒndə'lɪə(r)/ *n.* oarsman on a gondola. [Italian: related to GONDOLA]

gone /gɒn/ *adj.* **1** (of time) past (*not until gone nine*). **2 a** lost; hopeless. **b** dead. **3** *colloq.* pregnant for a specified time (*already three months gone*). **4** *slang* completely enthralled or entranced, esp. by rhythmic music, drugs, etc. □ **be gone** depart; leave temporarily (cf. BEGONE). **gone on** *slang* infatuated with. [past part. of GO[1]]

goner /'gɒnə(r)/ *n. slang* person or thing that is doomed or irrevocably lost.

gong *n.* **1** metal disc with a turned rim, giving a resonant note when struck. **2** saucer-shaped bell. **3** *slang* medal. [Malay]

gonorrhoea /,gɒnə'rɪə/ *n.* (*US* **gonorrhea**) venereal disease with inflammatory discharge from the urethra or vagina. [Greek, = semen-flux]

goo *n. colloq.* **1** sticky or slimy substance. **2** sickly sentiment. [origin unknown]

good /gʊd/ –*adj.* (**better, best**) **1** having the right or desired qualities; adequate. **2 a** (of a person) efficient, competent (*good at French*; *good driver*). **b** effective, reliable (*good brakes*). **3 a** kind. **b** morally excellent; virtuous (*good deed*). **c** well-behaved (*good child*). **4** enjoyable, agreeable (*good party*; *good news*). **5** thorough, considerable (*a good wash*). **6 a** not less than (*waited a good hour*). **b** considerable in number, quality, etc. (*a good many people*). **7** beneficial (*milk is good for you*). **8 a** valid, sound (*good reason*). **b** financially sound (*his credit is good*). **9** in exclamations of surprise (*good heavens!*). **10** (sometimes patronizing) commendable, worthy (*good old George*; *my good man*). **11** in courteous greetings and farewells (*good morning*). –*n.* **1** (only in *sing.*) that which is good; what is beneficial or morally right (*only good can come of it*; *what good will it do?*). **2** (in *pl.*) **a** movable property or merchandise. **b** things to be transported. **c** (prec. by *the*) *colloq.* what one has undertaken to supply (esp. *deliver the goods*). –*adv. US colloq.* well (*doing pretty good*). □ **as good as** practically. **be (a certain amount) to the good** have as net profit or advantage. **for good (and all)** finally, permanently. **good for 1** beneficial to; having a good effect on. **2** able to perform. **3** able to be trusted to pay. **good riddance** see RIDDANCE. **have the goods on a person** *slang* have information about a person giving one an advantage over him or her. **in good faith** with honest or sincere intentions. **in good time 1** with no risk of being late. **2** (also **all in good time**) in due course but without haste. **to the good** having as profit or benefit. [Old English]

good book *n.* (prec. by *the*) the Bible.

goodbye /gʊd'baɪ/ (*US* **goodby**) *–int.* expressing good wishes on parting, ending a telephone conversation, etc. *–n.* (*pl.* **-byes** or *US* **-bys**) parting; farewell. [from *God be with you!*]

good faith *n.* sincerity of intention.

good-for-nothing *–adj.* worthless. *–n.* worthless person.

Good Friday *n.* Friday before Easter Sunday, commemorating the Crucifixion.

good-hearted *adj.* kindly, well-meaning.

good humour *n.* genial mood.

good-humoured *adj.* cheerful, amiable. □ **good-humouredly** *adv.*

goodie var. of GOODY *n.*

good job *n.* fortunate state of affairs.

good-looking *adj.* handsome.

goodly /'gʊdlɪ/ *adj.* (**-ier**, **-iest**) **1** handsome. **2** of imposing size etc. □ **goodliness** *n.* [Old English]

good nature *n.* friendly disposition.

good-natured *adj.* kind, patient; easy-going. □ **good-naturedly** *adv.*

goodness *–n.* **1** virtue; excellence. **2** kindness (*had the goodness to wait*). **3** what is beneficial in a thing. *–int.* (esp. as a substitution for 'God') expressing surprise, anger, etc. (*goodness me!*; *goodness knows*). [Old English]

good-tempered *adj.* having a good temper; not easily annoyed.

goodwill *n.* **1** kindly feeling. **2** established reputation of a business etc. as enhancing its value. **3** willingness to undertake unpaid duties.

good will *n.* intention that good will result (see also GOODWILL).

good works *n.pl.* charitable acts.

goody *–n.* (also **goodie**) (*pl.* **-ies**) **1** *colloq.* good or favoured person. **2** (usu. in *pl.*) something good or attractive, esp. to eat. *–int.* expressing childish delight.

goody-goody *colloq.* *–n.* (*pl.* **-ies**) smug or obtrusively virtuous person. *–adj.* obtrusively or smugly virtuous.

gooey *adj.* (**gooier**, **gooiest**) *slang* **1** viscous, sticky. **2** sentimental. [from GOO]

goof *slang* *–n.* **1** foolish or stupid person. **2** mistake. *–v.* **1** bungle. **2** blunder. [Latin *gufus* coarse]

goofy *adj.* (**-ier**, **-iest**) *slang* **1** stupid. **2** having protruding or crooked front teeth.

googly /'gu:glɪ/ *n.* (*pl.* **-ies**) *Cricket* ball bowled so as to bounce in an unexpected direction. [origin unknown]

goon *n.* *slang* **1** stupid person. **2** esp. *US* ruffian hired by racketeers etc. [origin uncertain]

goose *n.* (*pl.* **geese**) **1 a** large water-bird with webbed feet and a broad bill. **b** female of this (opp. GANDER 1). **c** flesh of a goose as food. **2** *colloq.* simpleton. [Old English]

gooseberry /'gʊzbərɪ/ *n.* (*pl.* **-ies**) **1** yellowish-green berry with juicy flesh. **2** thorny shrub bearing this. □ **play gooseberry** *colloq.* be an unwanted extra person. [origin uncertain]

goose-flesh *n.* (also **goose-pimples**; *US* **goose-bumps**) bristling state of the skin produced by cold, fright, etc.

goose-step *n.* military marching step in which the knees are kept stiff.

gopher /'gəʊfə(r)/ *n.* American burrowing rodent, ground-squirrel, or burrowing tortoise. [origin uncertain]

Gordian /'gɔ:dɪən/ *adj.* □ **cut the Gordian knot** solve a problem by force or by evasion. [*Gordius* king of Phrygia, who tied a knot later cut by Alexander the Great]

gore[1] *n.* blood shed and clotted. [Old English, = dirt]

gore[2] *v.* (**-ring**) pierce with a horn, tusk, etc. [origin unknown]

gore[3] *–n.* **1** wedge-shaped piece in a garment. **2** triangular or tapering piece in an umbrella etc. *–v.* (**-ring**) shape (a garment) with a gore. [Old English, = triangle of land]

gorge *–n.* **1** narrow opening between hills. **2** act of gorging. **3** contents of the stomach. *–v.* (**-ging**) **1** feed greedily. **2 a** (often *refl.*) satiate. **b** devour greedily. □ **one's gorge rises at** one is sickened by. [French, = throat]

gorgeous /'gɔ:dʒəs/ *adj.* **1** richly coloured, sumptuous. **2** *colloq.* very pleasant, splendid (*gorgeous weather*). **3** *colloq.* strikingly beautiful. □ **gorgeously** *adv.* [French]

gorgon /'gɔ:gən/ *n.* **1** (in Greek mythology) each of three snake-haired sisters (esp. Medusa) with the power to turn anyone who looked at them to stone. **2** frightening or repulsive woman. [Greek *gorgos* terrible]

Gorgonzola /ˌgɔ:gən'zəʊlə/ *n.* type of rich cheese with bluish-green veins. [*Gorgonzola* in Italy]

gorilla /gəˈrɪlə/ *n.* largest anthropoid ape, native to Africa. [Greek, perhaps from African = wild man]

gormless /ˈgɔ:mlɪs/ *adj. colloq.* foolish, lacking sense. □ **gormlessly** *adv.* [originally *gaumless* from dial. *gaum* understanding]

gorse *n.* spiny yellow-flowered shrub; furze. □ **gorsy** *adj.* [Old English]

Gorsedd /ˈgɔ:seð/ *n.* Druidic order, meeting before the eisteddfod. [Welsh, literally 'throne']

gory *adj.* (**-ier, -iest**) **1** involving bloodshed; bloodthirsty. **2** covered in gore. □ **gorily** *adv.* **goriness** *n.*

gosh *int.* expressing surprise. [euphemism for GOD]

goshawk /ˈgɒshɔ:k/ *n.* large short-winged hawk. [Old English: related to GOOSE, HAWK[1]]

gosling /ˈgɒzlɪŋ/ *n.* young goose. [Old Norse: related to GOOSE]

go-slow *n.* working slowly, as a form of industrial action.

gospel /ˈgɒsp(ə)l/ *n.* **1** teaching or revelation of Christ. **2** (**Gospel**) **a** record of Christ's life in the first four books of the New Testament. **b** each of these books. **c** portion from one of them read at a service. **3** (also **gospel truth**) thing regarded as absolutely true. **4** (in full **gospel music**) Black American religious singing. [Old English: related to GOOD, SPELL[1] = news]

gossamer /ˈgɒsəmə(r)/ —*n.* **1** filmy substance of small spiders' webs. **2** delicate filmy material. —*adj.* light and flimsy as gossamer. [origin uncertain]

gossip /ˈgɒsɪp/ —*n.* **1 a** unconstrained talk or writing, esp. about persons or social incidents. **b** idle talk. **2** person who indulges in gossip. —*v.* (**-p-**) talk or write gossip. □ **gossipy** *adj.* [Old English, originally 'godparent', hence 'familiar acquaintance']

gossip column *n.* section of a newspaper devoted to gossip about well-known people. □ **gossip columnist** *n.*

got *past* and *past part.* of GET.

Goth *n.* **1** member of a Germanic tribe that invaded the Roman Empire in the 3rd–5th c. **2** uncivilized or ignorant person. [Old English *Gota* and Greek *Gothoi*]

goth *n.* **1** style of rock music with an intense or droning blend of guitars, bass, and drums, often with apocalyptic or mystical lyrics. **2** performer or devotee of this music, or member of the subculture favouring black clothing and white-painted faces with black make-up.

Gothic —*adj.* **1** of the Goths. **2** in the style of architecture prevalent in W. Europe in the 12th–16th c., characterized by pointed arches. **3** (of a novel etc.) in a style popular in the 18th–19th c., with supernatural or horrifying events. **4** barbarous, uncouth. —*n.* **1** Gothic language. **2** Gothic architecture. [Latin: related to GOTH]

gotten *US past part.* of GET.

gouache /gʊˈɑ:ʃ/ *n.* **1** method of painting in opaque pigments ground in water and thickened with a gluelike substance. **2** these pigments. [French from Italian]

Gouda /ˈgaʊdə/ *n.* flat round usu. Dutch cheese. [*Gouda* in Holland]

gouge /gaʊdʒ/ —*n.* chisel with a concave blade. —*v.* (**-ging**) **1** cut with or as with a gouge. **2** (foll. by *out*) force out (esp. an eye with the thumb) with or as with a gouge. [Latin *gubia*]

goulash /ˈgu:læʃ/ *n.* highly-seasoned Hungarian stew of meat and vegetables. [Magyar *gulyás-hús*, = herdsman's meat]

gourd /gʊəd/ *n.* **1 a** fleshy usu. large fruit with a hard skin. **b** climbing or trailing plant of the cucumber family bearing this. **2** dried skin of the gourd-fruit, used as a drinking-vessel etc. [Latin *cucurbita*]

gourmand /ˈgʊəmənd/ *n.* **1** glutton. **2** gourmet. [French]

■ **Usage** The use of *gourmand* in sense 2 is considered incorrect by some people.

gourmandise /ˈgʊəmɑ̃ˌdi:z/ *n.* gluttony.

gourmet /ˈgʊəmeɪ/ *n.* connoisseur of good food. [French]

gout *n.* disease with inflammation of the smaller joints, esp. of the toe. □ **gouty** *adj.* [Latin *gutta* drop]

govern /ˈgʌv(ə)n/ *v.* **1** rule or control with authority; conduct the policy and affairs of. **2** influence or determine (a person or course of action). **3** be a standard or principle for. **4** check or control (esp. passions). **5** *Gram.* (esp. of a verb or preposition) have (a noun or pronoun or its case) depending on it. [Greek *kubernaō* steer]

governance /ˈgʌvənəns/ *n.* **1** act or manner of governing. **2** function of governing. [French: related to GOVERN]

governess /ˈgʌvənɪs/ *n.* woman employed to teach children in a private household.

government /ˈgʌvənmənt/ *n.* **1** act or manner of governing. **2** system by which a State is governed. **3 a** body of persons governing a State. **b** (usu. **Government**) particular ministry in office. **4** the State as an agent. □ **governmental** /-'ment(ə)l/ *adj.*

governor *n.* **1** ruler. **2 a** official governing a province, town, etc. **b** representative of the Crown in a colony. **3** executive head of each State of the US. **4** officer commanding a fortress etc. **5** head or member of the governing body of an institution. **6** official in charge of a prison. **7 a** *slang* one's employer. **b** *slang* one's father. **8** *Mech.* automatic regulator controlling the speed of an engine etc. □ **governorship** *n.*

Governor-General *n.* representative of the Crown in a Commonwealth country that regards the Queen as Head of State.

gown *n.* **1** loose flowing garment, esp. a woman's long dress. **2** official robe of an alderman, judge, cleric, academic, etc. **3** surgeon's overall. [Latin *gunna* fur]

goy *n.* (*pl.* **-im** or **-s**) Jewish name for a non-Jew. [Hebrew, = people]

GP *abbr.* general practitioner.

GPO *abbr.* General Post Office.

gr *abbr.* (also **gr.**) **1** gram(s). **2** grains. **3** gross.

grab *–v.* (**-bb-**) **1** seize suddenly. **2** take greedily or unfairly. **3** *slang* attract the attention of, impress. **4** (foll. by *at*) snatch at. **5** (of brakes) act harshly or jerkily. *–n.* **1** sudden clutch or attempt to seize. **2** mechanical device for clutching. [Low German or Dutch]

grace *–n.* **1** attractiveness, esp. in elegance of proportion or manner or movement. **2** courteous good will (*had the grace to apologize*). **3** attractive feature; accomplishment (*social graces*). **4 a** (in Christian belief) the unmerited favour of God. **b** state of receiving this. **5** goodwill, favour. **6** delay granted as a favour (*a year's grace*). **7** short thanksgiving before or after a meal. **8 (Grace)** (in Greek mythology) each of three beautiful sister goddesses, bestowers of beauty and charm. **9 (Grace)** (prec. by *His*, *Her*, *Your*) forms of description or address for a duke, duchess, or archbishop. *–v.* (**-cing**) (often foll. by *with*) add grace to; confer honour on (*graced us with his presence*). □ **with good** (or **bad**) **grace** as if willingly (or reluctantly). [Latin *gratia*]

graceful *adj.* having or showing grace or elegance. □ **gracefully** *adv.* **gracefulness** *n.*

graceless *adj.* lacking grace, elegance, or charm.

grace-note *n. Mus.* extra note as an embellishment.

gracious /'greɪʃəs/ *–adj.* **1** kind; indulgent and beneficent to inferiors. **2** (of God) merciful, benign. *–int.* expressing surprise. □ **graciously** *adv.* **graciousness** *n.* [Latin: related to GRACE]

gracious living *n.* elegant way of life.

gradate /grə'deɪt/ *v.* (**-ting**) **1** (cause to) pass gradually from one shade to another. **2** arrange in steps or grades of size etc.

gradation /grə'deɪʃ(ə)n/ *n.* (usu. in *pl.*) **1** stage of transition or advance. **2 a** certain degree in rank, intensity, etc. **b** arrangement in such degrees. □ **gradational** *adj.* [Latin: related to GRADE]

grade *–n.* **1 a** certain degree in rank, merit, proficiency, etc. **b** class of persons or things of the same grade. **2** mark indicating the quality of a student's work. **3** *US* class in school. **4** gradient, slope. *–v.* (**-ding**) **1** arrange in grades. **2** (foll. by *up*, *down*, *off*, *into*, etc.) pass gradually between grades, or into a grade. **3** give a grade to (a student). **4** reduce (a road etc.) to easy gradients. [Latin *gradus* step]

gradient /'greɪdɪənt/ *n.* **1** stretch of road, railway, etc., that slopes. **2** amount of such a slope. [probably from GRADE after *salient*]

gradual /'grædʒʊəl/ *adj.* **1** progressing by degrees. **2** not rapid, steep, or abrupt. □ **gradually** *adv.* [Latin: related to GRADE]

gradualism /'grædʒʊə,lɪz(ə)m/ *n.* policy of gradual reform.

graduate *–n.* /'grædʒʊət/ person holding an academic degree. *–v.* /'grædʒʊ,eɪt/ (**-ting**) **1** obtain an academic degree. **2** (foll. by *to*) move up to (a higher grade of activity etc.). **3** mark out in degrees or parts. **4** arrange in gradations; apportion (e.g. tax) according to a scale. □ **graduation** /,grædʒʊ'eɪʃ(ə)n/ *n.* [medieval Latin *graduor* take a degree: related to GRADE]

Graeco-Roman /,gri:kəʊ'rəʊmən/ *adj.* of the Greeks and Romans.

graffiti /grə'fi:tɪ/ *n.pl.* (*sing.* **graffito**) writing or drawing scribbled, scratched, or sprayed on a surface. [Italian *graffio* a scratch]

■ **Usage** The singular or collective use of the form *graffiti* is considered incorrect by some people, but it is frequently found, e.g. *graffiti has appeared*.

graft[1] /grɑ:ft/ *–n.* **1** *Bot.* **a** shoot or scion inserted into a slit of stock, from which it receives sap. **b** place where a graft is inserted. **2** *Surgery* piece of living tissue, organ, etc., transplanted surgically. **3** *slang* hard work. *–v.* **1** (often foll. by *into*, *on*, *together*, etc.) insert (a scion) as a graft. **2** transplant (living tissue). **3** (foll. by *in*, *on*) insert or fix (a thing) permanently to another. **4** *slang* work hard. [Greek *graphion* stylus]

graft[2] /grɑ:ft/ *colloq.* –*n.* **1** practices, esp. bribery, used to secure illicit gains in politics or business. **2** such gains. –*v.* seek or make such gains. [origin unknown]

Grail *n.* (in full **Holy Grail**) (in medieval legend) cup or platter used by Christ at the Last Supper. [medieval Latin *gradalis* dish]

grain –*n.* **1** fruit or seed of a cereal. **2** (*collect.*) wheat or any allied grass used as food; corn. **3** small hard particle of salt, sand, etc. **4** unit of weight, 0.0648 gram. **5** smallest possible quantity (*not a grain of truth in it*). **6** roughness of surface. **7** texture of skin, wood, stone, etc. **8 a** pattern of lines of fibre in wood or paper. **b** lamination in stone etc. –*v.* **1** paint in imitation of the grain of wood etc. **2** give a granular surface to. **3** form into grains. □ **against the grain** contrary to one's natural inclination or feeling. □ **grainy** *adj.* (**-ier, -iest**). [Latin *granum*]

gram *n.* (also **gramme**) metric unit of mass equal to one-thousandth of a kilogram. [Greek *gramma* small weight]

-gram *comb. form* forming nouns denoting a thing written or recorded (often in a certain way) (*anagram*; *epigram*; *telegram*). [Greek *gramma* thing written]

graminaceous /ˌgræmɪˈneɪʃəs/ *adj.* of or like grass. [Latin *gramen* grass]

graminivorous /ˌgræmɪˈnɪvərəs/ *adj.* feeding on grass, cereals, etc.

grammar /ˈgræmə(r)/ *n.* **1** the study or rules of a language's inflections or other means of showing the relation between words. **2** observance or application of the rules of grammar (*bad grammar*). **3** book on grammar. [Greek *gramma* letter]

grammarian /grəˈmeərɪən/ *n.* expert in grammar or linguistics.

grammar school *n.* esp. *hist.* selective State secondary school with a mainly academic curriculum.

grammatical /grəˈmætɪk(ə)l/ *adj.* of or conforming to the rules of grammar. □ **grammatically** *adv.*

gramme var. of GRAM.

gramophone /ˈgræməˌfəʊn/ *n.* = RECORD-PLAYER. [inversion of *phonogram*: as PHONO-, -GRAM]

gramophone record = RECORD *n.* 3.

grampus /ˈgræmpəs/ *n.* (*pl.* **-puses**) a kind of dolphin with a blunt snout. [Latin *crassus piscis* fat fish]

gran *n. colloq.* grandmother. [abbreviation]

granadilla /ˌgrænəˈdɪlə/ *n.* passion-fruit. [Spanish, diminutive of *granada* pomegranate]

granary /ˈgrænərɪ/ *n.* (*pl.* **-ies**) **1** storehouse for threshed grain. **2** region producing, and esp. exporting, much corn. [Latin: related to GRAIN]

grand –*adj.* **1** splendid, magnificent, imposing, dignified. **2** main; of chief importance. **3** (**Grand**) of the highest rank (*Grand Duke*). **4** *colloq.* excellent, enjoyable. **5** belonging to high society. **6** (in *comb.*) (in names of family relationships) denoting the second degree of ascent or descent (*granddaughter*). –*n.* **1** = GRAND PIANO. **2** (*pl.* same) (usu. in *pl.*) esp. *US slang* a thousand dollars or pounds. □ **grandly** *adv.* **grandness** *n.* [Latin *grandis* full-grown]

grandad *n.* (also **grand-dad**) *colloq.* **1** grandfather. **2** elderly man.

grandchild *n.* child of one's son or daughter.

granddaughter *n.* female grandchild.

grandee /grænˈdi:/ *n.* **1** Spanish or Portuguese nobleman of the highest rank. **2** person of high rank. [Spanish and Portuguese *grande*: related to GRAND]

grandeur /ˈgrændʒə(r)/ *n.* **1** majesty, splendour; dignity of appearance or bearing. **2** high rank, eminence. **3** nobility of character. [French: related to GRAND]

grandfather *n.* male grandparent.

grandfather clock *n.* clock in a tall wooden case, driven by weights.

grandiloquent /ˌgrænˈdɪləkwənt/ *adj.* pompous or inflated in language. □ **grandiloquence** *n.* [Latin: related to GRAND, *-loquus* from *loquor* speak]

grandiose /ˈgrændɪˌəʊs/ *adj.* **1** producing or meant to produce an imposing effect. **2** planned on an ambitious scale. □ **grandiosity** /-ˈɒsɪtɪ/ *n.* [Italian: related to GRAND]

grand jury *n.* esp. *US* jury selected to examine the validity of an accusation prior to trial.

grandma /ˈgrænmɑ:/ *n. colloq.* grandmother.

grand mal /grɑ̃ ˈmæl/ *n.* serious form of epilepsy with loss of consciousness. [French, = great sickness]

grand master *n.* chess-player of the highest class.

grandmother *n.* female grandparent.

Grand National *n.* steeplechase held annually at Aintree, Liverpool.

grand opera *n.* opera on a serious theme, or in which the entire libretto (including dialogue) is sung.

grandpa /ˈgrænpɑ:/ *n. colloq.* grandfather.

grandparent *n.* parent of one's father or mother.

grand piano *n.* large full-toned piano with horizontal strings.

Grand Prix /grɑ̃ 'pri:/ *n.* any of several important international motor or motor-cycle racing events. [French, = great or chief prize]

grandsire *n. archaic* grandfather.

grand slam *n.* **1** *Sport* winning of all of a group of matches etc. **2** *Bridge* winning of 13 tricks.

grandson *n.* male grandchild.

grandstand *n.* main stand for spectators at a racecourse etc.

grand total *n.* sum of other totals.

grand tour *n. hist.* cultural tour of Europe.

grange /greɪndʒ/ *n.* country house with farm-buildings. [Latin *granica*: related to GRAIN]

graniferous /grə'nɪfərəs/ *adj.* producing grain or a grainlike seed. [Latin: related to GRAIN]

granite /'grænɪt/ *n.* granular crystalline rock of quartz, mica, etc., used for building. [Italian *granito*: related to GRAIN]

granivorous /grə'nɪvərəs/ *adj.* feeding on grain. [Latin: related to GRAIN]

granny /'grænɪ/ *n.* (also **grannie**) (*pl.* **-ies**) *colloq.* grandmother. [diminutive of *grannam* from archaic *grandam*: related to GRAND, DAME]

granny flat *n.* part of a house made into self-contained accommodation for an elderly relative.

granny knot *n.* reef-knot crossed the wrong way and therefore insecure.

grant /grɑ:nt/ *–v.* **1 a** consent to fulfil (a request etc.). **b** allow (a person) to have (a thing). **2** give formally; transfer legally. **3** (often foll. by *that*) admit as true; concede. *–n.* **1** process of granting. **2** sum of money given by the State. **3** legal conveyance by written instrument. □ **take for granted 1** assume something to be true or valid. **2** cease to appreciate through familiarity. □ **grantor** /-'tɔ:(r)/ *n.* (esp. in sense 2 of *v.*). [French *gr(e)anter* var. of *creanter* from Latin *credo* entrust]

grant-maintained *adj.* (of a school) funded by central rather than local government.

granular /'grænjʊlə(r)/ *adj.* of or like grains or granules. □ **granularity** /-'lærɪtɪ/ *n.* [Latin: related to GRANULE]

granulate /'grænjʊ,leɪt/ *v.* (**-ting**) **1** form into grains. **2** roughen the surface of. □ **granulation** /-'leɪʃ(ə)n/ *n.*

granule /'grænju:l/ *n.* small grain. [Latin diminutive of *granum*: related to GRAIN]

grape *n.* berry (usu. green, purple, or black) growing in clusters on a vine, used as fruit and in making wine. [French, probably from *grappe* hook]

grapefruit *n.* (*pl.* same) large round usu. yellow citrus fruit.

grape hyacinth *n.* plant of the lily family with clusters of usu. blue flowers.

grapeshot *n. hist.* small balls used as charge in a cannon and scattering when fired.

grapevine *n.* **1** vine. **2** *colloq.* the means of transmission of a rumour.

graph /grɑ:f/ *–n.* diagram showing the relation between variable quantities, usu. of two variables, each measured along one of a pair of axes. *–v.* plot or trace on a graph. [abbreviation of *graphic formula*]

-graph *comb. form* forming nouns and verbs meaning: **1** thing written or drawn etc. in a specified way (*photograph*). **2** instrument that records (*seismograph*).

-grapher *comb. form* forming nouns denoting a person concerned with a subject (*geographer*; *radiographer*). [Greek *-graphō* write]

graphic /'græfɪk/ *adj.* **1** of or relating to the visual or descriptive arts, esp. writing and drawing. **2** vividly descriptive. □ **graphically** *adv.* [Greek *graphē* writing]

-graphic *comb. form* (also **-graphical**) forming adjectives corresponding to nouns in *-graphy*.

graphic arts *n.pl.* visual and technical arts involving design or the use of lettering.

graphic novel *n.* novel in comic-strip format.

graphics *n.pl.* (usu. treated as *sing.*) **1** products of the graphic arts. **2** use of diagrams in calculation and design.

graphite /'græfaɪt/ *n.* crystalline allotropic form of carbon used as a lubricant, in pencils, etc. □ **graphitic** /-'fɪtɪk/ *adj.* [German *Graphit* from Greek *graphō* write]

graphology /grə'fɒlədʒɪ/ *n.* the study of handwriting, esp. as a supposed guide to character. □ **graphologist** *n.* [Greek: related to GRAPHIC]

graph paper *n.* paper printed with a network of lines as a basis for drawing graphs.

-graphy *comb. form* forming nouns denoting: **1** descriptive science (*geography*). **2** technique of producing images (*photography*). **3** style or method of writing etc. (*calligraphy*).

grapnel /'græpn(ə)l/ *n.* **1** device with iron claws, for dragging or grasping. **2**

small anchor with several flukes. [French *grapon*: related to GRAPE]

grapple /ˈgræp(ə)l/ *–v.* (**-ling**) **1** (often foll. by *with*) fight in close combat. **2** (foll. by *with*) try to manage (a difficult problem etc.). **3 a** grip with the hands; come to close quarters with. **b** seize with or as with a grapnel. *–n.* **1 a** hold or grip in or as in wrestling. **b** contest at close quarters. **2** clutching-instrument; grapnel. [French *grapil*: related to GRAPNEL]

grappling-iron *n.* (also **grappling-hook**) = GRAPNEL.

grasp /grɑːsp/ *–v.* **1 a** clutch at; seize greedily. **b** hold firmly. **2** (foll. by *at*) try to seize; accept avidly. **3** understand or realize (a fact or meaning). *–n.* **1** firm hold; grip. **2** (foll. by *of*) **a** mastery (*a grasp of the situation*). **b** mental hold. □ **grasp the nettle** tackle a difficulty boldly. [earlier *grapse*: related to GROPE]

grasping *adj.* avaricious.

grass /grɑːs/ *–n.* **1 a** any of a group of wild plants with green blades that are eaten by ruminants. **b** plant of the family which includes cereals, reeds, and bamboos. **2** pasture land. **3** grass-covered ground, lawn. **4** grazing (*out to grass*). **5** *slang* marijuana. **6** *slang* informer. *–v.* **1** cover with turf. **2** *US* provide with pasture. **3** *slang* **a** betray, esp. to the police. **b** inform the police. □ **grassy** *adj.* (**-ier**, **-iest**). [Old English]

grasshopper *n.* jumping and chirping insect.

grassland *n.* large open area covered with grass, esp. used for grazing.

grass roots *n.pl.* **1** fundamental level or source. **2** ordinary people; rank and file of an organization, esp. a political party.

grass snake *n.* common harmless European snake.

grass widow *n.* (also **grass widower**) person whose husband (or wife) is away for a prolonged period.

grate[1] *v.* (**-ting**) **1** reduce to small particles by rubbing on a serrated surface. **2** (often foll. by *against*, *on*) rub with a harsh scraping sound. **3** utter in a harsh tone. **4** (often foll. by *on*) **a** sound harshly. **b** have an irritating effect. **5** grind (one's teeth). **6** creak. □ **grater** *n.* [French from Germanic]

grate[2] *n.* **1** fireplace or furnace. **2** metal frame confining fuel in this. [Latin *cratis* hurdle]

grateful /ˈgreɪtfʊl/ *adj.* **1** thankful; feeling or showing gratitude. **2** pleasant, acceptable. □ **gratefully** *adv.* [obsolete *grate* from Latin *gratus*]

gratify /ˈgrætɪˌfaɪ/ *v.* (**-ies**, **-ied**) **1 a** please, delight. **b** please by compliance. **2** yield to (a feeling or desire). □ **gratification** /-fɪˈkeɪʃ(ə)n/ *n.* [Latin: related to GRATEFUL]

grating *n.* **1** framework of parallel or crossed metal bars. **2** *Optics* set of parallel wires, lines ruled on glass, etc.

gratis /ˈgrɑːtɪs/ *adv.* & *adj.* free; without charge. [Latin]

gratitude /ˈgrætɪˌtjuːd/ *n.* being thankful; readiness to return kindness. [Latin: related to GRATEFUL]

gratuitous /grəˈtjuːɪtəs/ *adj.* **1** given or done free of charge. **2** uncalled-for; lacking good reason. □ **gratuitously** *adv.* **gratuitousness** *n.* [Latin, = spontaneous]

gratuity /grəˈtjuːɪtɪ/ *n.* (*pl.* **-ies**) = TIP[3] *n.* 1. [Latin: related to GRATEFUL]

grave[1] *n.* **1** trench dug in the ground for the burial of a corpse; mound or memorial stone placed over this. **2** (prec. by *the*) death. [Old English]

grave[2] *–adj.* **1 a** serious, weighty, important. **b** dignified, solemn, sombre. **2** extremely serious or threatening. *–n.* /grɑːv/ = GRAVE ACCENT. □ **gravely** *adv.* [Latin *gravis* heavy]

grave[3] *v.* (**-ving**; *past part.* **graven** or **graved**) **1** (foll. by *in*, *on*) fix indelibly (on one's memory). **2** *archaic* engrave, carve. [Old English]

grave accent /grɑːv/ *n.* a mark (`) placed over a vowel to denote pronunciation, length, etc.

gravedigger *n.* person who digs graves.

gravel /ˈgræv(ə)l/ *–n.* **1** mixture of coarse sand and small stones, used for paths etc. **2** *Med.* aggregations of crystals formed in the urinary tract. *–v.* (**-ll-**; *US* **-l-**) lay or strew with gravel. [French diminutive, perhaps of *grave* shore]

gravelly *adj.* **1** of or like gravel. **2** (of a voice) deep and rough-sounding.

graven *past part.* of GRAVE[3].

graven image *n.* idol.

Graves /grɑːv/ *n.* light usu. white wine from Graves in France.

gravestone *n.* stone (usu. inscribed) marking a grave.

graveyard *n.* burial ground.

gravid /ˈgrævɪd/ *adj.* pregnant. [Latin *gravidus*: related to GRAVE[2]]

gravimeter /grəˈvɪmɪtə(r)/ *n.* instrument measuring the difference in the force of gravity between two places. [Latin: related to GRAVE[2]]

gravimetry /grəˈvɪmɪtrɪ/ *n.* measurement of weight. □ **gravimetric** /ˌgrævɪˈmetrɪk/ *adj.*

gravitate /ˈgrævɪˌteɪt/ *v.* (**-ting**) **1** (foll. by *to*, *towards*) move or be attracted to. **2 a** move or tend by force of gravity towards. **b** sink by or as if by gravity. [related to GRAVE[2]]

gravitation /ˌgrævɪˈteɪʃ(ə)n/ *n. Physics* **1** force of attraction between any particle of matter in the universe and any other. **2** effect of this, esp. the falling of bodies to the earth. □ **gravitational** *adj.*

gravity /ˈgrævɪtɪ/ *n.* **1 a** force that attracts a body to the centre of the earth etc. **b** degree of intensity of this. **c** gravitational force. **2** property of having weight. **3 a** importance, seriousness. **b** solemnity. [Latin: related to GRAVE[2]]

gravy /ˈgreɪvɪ/ *n.* (*pl.* **-ies**) **1** juices exuding from meat during and after cooking. **2** sauce for food, made from these etc. [perhaps from a misreading of French *grané* from *grain* spice, GRAIN]

gravy-boat *n.* boat-shaped vessel for serving gravy.

gravy train *n. slang* source of easy financial benefit.

gray *US* var. of GREY.

grayling *n.* (*pl.* same) silver-grey freshwater fish. [from GREY, -LING]

graze[1] *v.* (**-zing**) **1** (of cattle, sheep, etc.) eat growing grass. **2 a** feed (cattle etc.) on growing grass. **b** feed on (grass). **3** pasture cattle. [Old English: related to GRASS]

graze[2] *–v.* (**-zing**) **1** rub or scrape (part of the body, esp. the skin). **2 a** touch lightly in passing. **b** (foll. by *against*, *along*, etc.) move with a light passing contact. *–n.* abrasion. [perhaps from GRAZE[1], as if 'take off the grass close to the ground']

grazier /ˈgreɪzɪə(r)/ *n.* **1** person who feeds cattle for market. **2** *Austral.* large-scale sheep-farmer etc. [from GRASS]

grazing *n.* grassland suitable for pasturage.

grease /gri:s/ *–n.* **1** oily or fatty matter, esp. as a lubricant. **2** melted fat of a dead animal. *–v.* (**-sing**) smear or lubricate with grease. □ **grease the palm of** *colloq.* bribe. [Latin *crassus* (adj.) fat]

greasepaint *n.* make-up used by actors.

greaseproof *adj.* impervious to grease.

greaser *n. slang* member of a gang of youths with long hair and motor cycles.

greasy *adj.* (**-ier, -iest**) **1 a** of or like grease. **b** smeared or covered with grease. **c** containing or having too much grease. **2 a** slippery. **b** (of a person or manner) unpleasantly unctuous. □ **greasily** *adv.* **greasiness** *n.*

great /greɪt/ *–adj.* **1 a** of a size, amount, extent, or intensity considerably above the normal or average (*a great hole*; *great fun*). **b** also with implied admiration, contempt, etc., esp. in exclamations (*you great idiot!*; *great stuff!*). **c** reinforcing other words denoting size, quantity, etc. (*great big hole*). **2** important, pre-eminent (*the great thing is not to get caught*). **3** grand, imposing (*great occasion*). **4** distinguished. **5** remarkable in ability, character, etc. (*great men*; *great thinker*). **6** (foll. by *at*, *on*) competent, well-informed. **7** fully deserving the name of; doing a thing extensively (*great reader*; *great believer in tolerance*). **8** (also **greater**) the larger of the name, species, etc. (*great auk*; *greater celandine*). **9** *colloq.* very enjoyable or satisfactory (*had a great time*). **10** (in *comb.*) (in names of family relationships) denoting one degree further removed upwards or downwards (*great-uncle*; *great-great-grandmother*). *–n.* **1** great or outstanding person or thing. **2** (in *pl.*) (**Greats**) *colloq.* (at Oxford University) honours course or final examinations in classics and philosophy. □ **greatness** *n.* [Old English]

Great Bear see BEAR[2].

great circle *n.* circle on the surface of a sphere whose plane passes through the sphere's centre.

greatcoat *n.* heavy overcoat.

Great Dane *n.* dog of a large short-haired breed.

great deal *n.* = DEAL[1] *n.* 1.

greatly *adv.* much; by a considerable amount (*greatly admired*; *greatly superior*).

great tit *n.* Eurasian songbird with black and white head markings.

Great War *n.* world war of 1914–18.

greave *n.* (usu. in *pl.*) armour for the shin. [French, = shin]

grebe *n.* a kind of diving bird. [French]

Grecian /ˈgri:ʃ(ə)n/ *adj.* (of architecture or facial outline) Greek. [Latin *Graecia* Greece]

Grecian nose *n.* straight nose that continues the line of the forehead without a dip.

greed *n.* excessive desire, esp. for food or wealth. [from GREEDY]

greedy *adj.* (**-ier, -iest**) **1** having or showing greed. **2** (foll. by *for*, or *to* + infin.) very eager. □ **greedily** *adv.* **greediness** *n.* [Old English]

Greek *–n.* **1 a** native or national of Greece. **b** person of Greek descent. **2** language of Greece. *–adj.* of Greece or its people or language; Hellenic. □ **Greek to me** *colloq.* incomprehensible to me. [Old English ultimately from Greek *Graikoi*]

Greek cross *n.* cross with four equal arms.

green *–adj.* **1** of the colour between blue and yellow in the spectrum; coloured like grass. **2** covered with leaves or grass. **3** (of fruit etc. or wood) unripe or unseasoned. **4** not dried, smoked, or tanned. **5** inexperienced, gullible. **6 a** (of the complexion) pale, sickly-hued. **b** jealous, envious. **7** young, flourishing. **8** not withered or worn out (*a green old age*). **9** (also **Green**) concerned with protection of the environment as a political principle; not harmful to the environment. –*n.* **1** green colour or pigment. **2** green clothes or material. **3 a** piece of public grassy land (*village green*). **b** grassy area used for a special purpose (*putting-green*). **4** (in *pl.*) green vegetables. **5** (also **Green**) supporter of an environmentalist group or party. –*v.* make or become green. □ **greenish** *adj.* **greenly** *adv.* **greenness** *n.* [Old English]

green belt *n.* area of open land round a city, designated for preservation.

green card *n.* international insurance document for motorists.

greenery *n.* green foliage or growing plants.

green-eyed *adj. colloq.* jealous.

greenfinch *n.* finch with green and yellow plumage.

green fingers *n. pl. colloq.* skill in growing plants.

greenfly *n.* **1** green aphid. **2** these collectively.

greengage *n.* roundish green variety of plum. [Sir W. *Gage*, name of a botanist]

greengrocer *n.* retailer of fruit and vegetables.

greengrocery *n.* (*pl.* **-ies**) **1** greengrocer's business. **2** goods sold by a greengrocer.

greenhorn *n.* inexperienced person; new recruit.

greenhouse *n.* light structure with the sides and roof mainly of glass, for rearing plants.

greenhouse effect *n.* trapping of the sun's warmth in the lower atmosphere of the earth, caused by an increase in carbon dioxide, methane, etc.

greenhouse gas *n.* any of the gases, esp. carbon dioxide and methane, that contribute to the greenhouse effect.

green light *n.* **1** signal to proceed on a road, railway, etc. **2** *colloq.* permission to proceed with a project.

Green Paper *n.* preliminary report of Government proposals, for discussion.

green pound *n.* exchange rate for the pound for payments for agricultural produce in the EC.

green revolution *n.* greatly increased crop production in underdeveloped countries.

green-room *n.* room in a theatre for actors and actresses who are off stage.

green-stick fracture *n.* bone-fracture, esp. in children, in which one side of the bone is broken and one only bent.

greenstuff *n.* vegetation; green vegetables.

greensward *n.* expanse of grassy turf.

green tea *n.* tea made from steam-dried leaves.

Greenwich Mean Time /'grenɪtʃ/ *n.* local time on the meridian of Greenwich, used as an international basis of time-reckoning.

greenwood *n.* a wood in summer.

greeny *adj.* greenish.

greet[1] *v.* **1** address politely or welcomingly on meeting or arrival. **2** receive or acknowledge in a specified way. **3** (of a sight, sound, etc.) become apparent to or noticed by. [Old English]

greet[2] *v. Scot.* weep. [Old English]

greeting *n.* **1** act or instance of welcoming etc. **2** words, gestures, etc., used to greet a person. **3** (often in *pl.*) expression of goodwill.

greetings card *n.* decorative card sent to convey greetings.

gregarious /grɪ'geərɪəs/ *adj.* **1** fond of company. **2** living in flocks or communities. □ **gregariousness** *n.* [Latin *grex gregis* flock]

Gregorian calendar /grɪ'gɔ:rɪən/ *n.* calendar introduced in 1582 by Pope Gregory XIII.

Gregorian chant /grɪ'gɔ:rɪən/ *n.* plainsong ritual music, named after Pope Gregory I.

gremlin /'gremlɪn/ *n. colloq.* imaginary mischievous sprite regarded as responsible for mechanical faults etc. [origin unknown]

grenade /grɪ'neɪd/ *n.* small bomb thrown by hand (**hand-grenade**) or shot from a rifle. [French: related to POMEGRANATE]

grenadier /,grenə'dɪə(r)/ *n.* **1** (**Grenadiers** or **Grenadier Guards**) first regiment of the royal household infantry. **2** *hist.* soldier armed with grenades.

grew *past* of GROW.

grey /greɪ/ (*US* **gray**) –*adj.* **1** of a colour intermediate between black and white. **2** dull, dismal. **3 a** (of hair) turning white with age etc. **b** having grey hair. **4** anonymous, unidentifiable. –*n.* **1 a** grey colour or pigment. **b** grey clothes or material (*dressed in grey*). **2** grey or white horse. –*v.* make or become grey. □ **greyish** *adj.* **greyness** *n.* [Old English]

grey area *n.* situation or topic not clearly defined.

Grey Friar *n.* Franciscan friar.

greyhound *n.* dog of a tall slender breed capable of high speed. [Old English, = bitch-hound]

greylag *n.* (in full **greylag goose**) European wild goose. [from GREY]

grey matter *n.* **1** the darker tissues of the brain and spinal cord. **2** *colloq.* intelligence.

grey squirrel *n.* American squirrel brought to Europe in the 19th c.

grid *n.* **1** grating. **2** system of numbered squares printed on a map and forming the basis of map references. **3** network of lines, electric-power connections, gas-supply lines, etc. **4** pattern of lines marking the starting-places on a motor-racing track. **5** perforated electrode controlling the flow of electrons in a thermionic valve etc. **6** arrangement of town streets in a rectangular pattern. [from GRIDIRON]

griddle /'grɪd(ə)l/ *n.* circular iron plate placed over a source of heat for baking etc. [Latin *cratis* hurdle]

gridiron /'grɪd,aɪən/ *n.* cooking utensil of metal bars for broiling or grilling. [related to GRIDDLE]

grief *n.* **1** intense sorrow. **2** cause of this. □ **come to grief** meet with disaster. [French: related to GRIEVE]

grievance /'gri:v(ə)ns/ *n.* real or fancied cause for complaint. [French: related to GRIEF]

grieve *v.* (**-ving**) **1** cause grief to. **2** suffer grief. [Latin: related to GRAVE[2]]

grievous *adj.* **1** (of pain etc.) severe. **2** causing grief. **3** injurious. **4** flagrant, heinous. □ **grievously** *adv.* [French: related to GRIEVE]

grievous bodily harm *n. Law* serious injury inflicted intentionally.

griffin /'grɪfɪn/ *n.* (also **gryphon** /-f(ə)n/) fabulous creature with an eagle's head and wings and a lion's body. [Latin *gryphus* from Greek]

griffon /'grɪf(ə)n/ *n.* **1** dog of a small terrier-like breed. **2** large vulture. **3** = GRIFFIN. [French, = GRIFFIN]

grill –*n.* **1 a** device on a cooker for radiating heat downwards. **b** = GRIDIRON. **2** food cooked on a grill. **3** (in full **grill room**) restaurant specializing in grilled food. –*v.* **1** cook or be cooked under a grill or on a gridiron. **2** subject or be subjected to extreme heat. **3** subject to severe questioning. [French: related to GRIDDLE]

grille *n.* (also **grill**) **1** grating or latticed screen, used as a partition etc. **2** metal grid protecting the radiator of a vehicle.

grilse *n.* (*pl.* same or **-s**) young salmon that has returned to fresh water from the sea for the first time. [origin unknown]

grim *adj.* (**grimmer**, **grimmest**) **1** of stern or forbidding appearance. **2** harsh, merciless. **3** ghastly, joyless (*has a grim truth in it*). **4** unpleasant, unattractive. □ **grimly** *adv.* **grimness** *n.* [Old English]

grimace /'grɪməs, -'meɪs/ –*n.* distortion of the face made in disgust etc. or to amuse. –*v.* (**-cing**) make a grimace. [French from Spanish]

grime –*n.* soot or dirt ingrained in a surface. –*v.* (**-ming**) blacken with grime; befoul. □ **griminess** *n.* **grimy** *adj.* (**-ier**, **-iest**). [Low German or Dutch]

grin –*v.* (**-nn-**) **1 a** smile broadly, showing the teeth. **b** make a forced, unrestrained, or stupid smile. **2** express by grinning. –*n.* act of grinning. □ **grin and bear it** take pain etc. stoically. [Old English]

grind /graɪnd/ –*v.* (*past* and *past part.* **ground**) **1** reduce to small particles or powder by crushing. **2 a** sharpen or smooth by friction. **b** rub or rub together gratingly. **3** (often foll. by *down*) oppress; harass with exactions. **4 a** (often foll. by *away*) work or study hard. **b** (foll. by *out*) produce with effort. –*n.* **1** act or instance of grinding. **2** *colloq.* hard dull work (*the daily grind*). **3** size of ground particles. □ **grind to a halt** stop laboriously. [Old English]

grinder *n.* **1** person or thing that grinds, esp. a machine. **2** molar tooth.

grindstone *n.* **1** thick revolving disc used for grinding, sharpening, and polishing. **2** a kind of stone used for this. □ **keep one's nose to the grindstone** work hard and continuously.

grip –*v.* (**-pp-**) **1 a** grasp tightly. **b** take a firm hold, esp. by friction. **2** compel the attention of. –*n.* **1 a** firm hold; tight grasp. **b** manner of grasping or holding. **2** power of holding attention. **3 a** intellectual mastery. **b** effective control of one's behaviour etc. (*lose one's grip*). **4 a** part of a machine that grips. **b** part by which a weapon etc. is held. **5** = HAIR-GRIP. **6** travelling bag. □ **come** (or **get**) **to grips with** approach purposefully; begin to deal with. [Old English]

gripe –*v.* (**-ping**) **1** *colloq.* complain. **2** affect with gastric pain. –*n.* **1** (usu. in *pl.*) colic. **2** *colloq.* complaint. **3** grip, clutch. [Old English]

Gripe Water *n. propr.* preparation to relieve colic in infants.

grisly /'grɪzlɪ/ *adj.* (**-ier**, **-iest**) causing horror, disgust, or fear. □ **grisliness** *n.* [Old English]

grist *n.* corn to grind. □ **grist to the** (or **a person's**) **mill** source of profit or advantage. [Old English: related to GRIND]

gristle /'grɪs(ə)l/ *n.* tough flexible animal tissue; cartilage. □ **gristly** /-slɪ/ *adj.* [Old English]

grit –*n.* **1** particles of stone or sand, esp. as irritating or hindering. **2** coarse sandstone. **3** *colloq.* pluck, endurance. –*v.* (**-tt-**) **1** spread grit on (icy roads etc.). **2** clench (the teeth). **3** make a grating sound. □ **gritter** *n.* **gritty** *adj.* (**-ier**, **-iest**). [Old English]

grits *n.pl.* **1** coarsely ground grain, esp. oatmeal. **2** oats that have been husked but not ground. [Old English]

grizzle /'grɪz(ə)l/ *v.* (**-ling**) *colloq.* **1** (esp. of a child) cry fretfully. **2** complain whiningly. □ **grizzly** *adj.* [origin unknown]

grizzled /'grɪz(ə)ld/ *adj.* **1** (of hair) grey or streaked with grey. **2** having grizzled hair. [*grizzle* grey from French *grisel*]

grizzly /'grɪzlɪ/ –*adj.* (**-ier**, **-iest**) grey, grey-haired. –*n.* (*pl.* **-ies**) (in full **grizzly bear**) large variety of brown bear, found in N. America.

groan –*v.* **1 a** make a deep sound expressing pain, grief, or disapproval. **b** utter with groans. **2** (usu. foll. by *under*, *beneath*, *with*) be loaded or oppressed. –*n.* sound made in groaning. [Old English]

groat *n. hist.* silver coin worth four old pence. [Low German or Dutch: related to GREAT]

groats *n.pl.* hulled or crushed grain, esp. oats. [Old English]

grocer *n.* dealer in food and household provisions. [Anglo-French *grosser* from Latin *grossus* GROSS]

grocery *n.* (*pl.* **-ies**) **1** grocer's trade or shop. **2** (in *pl.*) goods, esp. food, sold by a grocer.

grog *n.* drink of spirit (orig. rum) and water. [origin uncertain]

groggy *adj.* (**-ier**, **-iest**) incapable or unsteady. □ **groggily** *adv.* **grogginess** *n.*

groin[1] –*n.* **1** depression between the belly and the thigh. **2** *Archit.* **a** edge formed by intersecting vaults. **b** arch supporting a vault. –*v.* *Archit.* build with groins. [origin uncertain]

groin[2] *US* var. of GROYNE.

grommet /'grɒmɪt/ *n.* (also **grummet** /'grʌmɪt/) **1** metal, plastic, or rubber eyelet placed in a hole to protect or insulate a rope or cable etc. passed through it. **2** tube passed through the eardrum to make a communication with the middle ear. [French]

groom –*n.* **1** person employed to take care of horses. **2** = BRIDEGROOM. **3** *Mil.* any of certain officers of the Royal Household. –*v.* **1 a** curry or tend (a horse). **b** give a neat appearance to (a person etc.). **2** (of an ape etc.) clean and comb the fur of (its fellow). **3** prepare or train (a person) for a particular purpose or job. [origin unknown]

groove –*n.* **1** channel or elongated hollow, esp. one made to guide motion or receive a corresponding ridge. **2** spiral track cut in a gramophone record. –*v.* (**-ving**) **1** make a groove or grooves in. **2** *slang* enjoy oneself. [Dutch]

groovy *adj.* (**-ier**, **-iest**) **1** *slang* excellent. **2** of or like a groove.

grope –*v.* (**-ping**) **1** (usu. foll. by *for*) feel about or search blindly. **2** (foll. by *for*, *after*) search mentally. **3** feel (one's way) towards something. **4** *slang* fondle clumsily for sexual pleasure. –*n.* act of groping. [Old English]

grosgrain /'grəʊgreɪn/ *n.* corded fabric of silk etc. [French, = coarse grain: related to GROSS, GRAIN]

gros point /grəʊ 'pwæ̃/ *n.* cross-stitch embroidery on canvas. [French: related to GROSS, POINT]

gross /grəʊs/ –*adj.* **1** overfed, bloated. **2** (of a person, manners, or morals) coarse, unrefined, or indecent. **3** flagrant (*gross negligence*). **4** total; not net (*gross tonnage*). **5** (of the senses etc.) dull. –*v.* produce as gross profit. –*n.* (*pl.* same) amount equal to twelve dozen. □ **grossly** *adv.* **grossness** *n.* [Latin *grossus*]

gross domestic product *n.* total value of goods produced and services provided in a country in one year.

gross national product *n.* gross domestic product plus the total of net income from abroad.

grotesque /grəʊ'tesk/ –*adj.* **1** comically or repulsively distorted. **2** incongruous, absurd. –*n.* **1** decorative form interweaving human and animal features. **2** comically distorted figure or design. □ **grotesquely** *adv.* **grotesqueness** *n.* [Italian: related to GROTTO]

grotto /'grɒtəʊ/ *n.* (*pl.* **-es** or **-s**) **1** picturesque cave. **2** artificial ornamental cave. [Italian *grotta* from Greek *kruptē* CRYPT]

grotty /'grɒtɪ/ *adj.* (**-ier**, **-iest**) *slang* unpleasant, dirty, shabby, unattractive. [shortening of GROTESQUE]

grouch *colloq.* –*v.* grumble. –*n.* **1** discontented person. **2** fit of grumbling or the sulks. □ **grouchy** *adj.* (**-ier**, **-iest**). [related to GRUDGE]

ground[1] –*n.* **1 a** surface of the earth, esp. as contrasted with the air around it. **b** part of this specified in some way (*low ground*). **2 a** position, area, or distance on the earth's surface. **b** extent of a subject dealt with (*the book covers a lot* *of ground*). **3** (often in *pl.*) reason, justification. **4** area of a special kind or use (often in *comb.*: *cricket-ground*; *fishing-grounds*). **5** (in *pl.*) enclosed land attached to a house etc. **6** area or basis for agreement etc. (*common ground*). **7** (in painting etc.) the surface giving the predominant colour. **8** (in *pl.*) solid particles, esp. of coffee, forming a residue. **9** *US Electr.* = EARTH *n.* 4. **10** bottom of the sea. **11** floor of a room etc. **12** (in full **ground bass**) *Mus.* short theme in the bass constantly repeated with the upper parts of the music varied. **13** (*attrib.*) (of animals) living on or in the ground; (of plants) dwarfish or trailing. –*v.* **1** refuse authority for (a pilot or an aircraft) to fly. **2 a** run (a ship) aground; strand. **b** (of a ship) run aground. **3** (foll. by *in*) instruct thoroughly (in a subject). **4** (often as **grounded** *adj.*) (foll. by *on*) base (a principle, conclusion, etc.) on. **5** *US Electr.* = EARTH *v.* □ **break new** (or **fresh**) **ground** treat a subject previously not dealt with. **get off the ground** *colloq.* make a successful start. **give** (or **lose**) **ground** retreat, decline. **go to ground 1** (of a fox etc.) enter its earth etc. **2** (of a person) become inaccessible for a prolonged period. **hold one's ground** not retreat. **on the grounds of** because of. [Old English]

ground[2] *past* and *past part.* of GRIND.

ground control *n.* personnel directing the landing etc. of aircraft etc.

ground cover *n.* low-growing plants covering the surface of the earth.

ground elder *n.* garden weed spreading by means of underground stems.

ground floor *n.* floor of a building at ground level.

ground frost *n.* frost on the surface of the ground or in the top layer of soil.

ground glass *n.* **1** glass made non-transparent by grinding etc. **2** glass ground to a powder.

grounding *n.* basic training or instruction.

groundless *adj.* without motive or foundation.

groundnut *n.* = PEANUT 1, 2.

ground-plan *n.* **1** plan of a building at ground level. **2** general outline of a scheme.

ground-rent *n.* rent for land leased for building.

groundsel /'graʊns(ə)l/ *n.* wild plant with small yellow flowers, used as a food for cage-birds etc. [Old English]

groundsheet *n.* waterproof sheet for spreading on the ground.

groundsman *n.* person who maintains a sports ground.

ground speed *n.* aircraft's speed relative to the ground.

ground swell *n.* heavy sea caused by a distant or past storm or an earthquake.

groundwater *n.* water found in soil or in pores, crevices, etc., in rock.

groundwork *n.* preliminary or basic work.

group /gru:p/ –*n.* **1** number of persons or things located close together, or considered or classed together. **2** number of people working together etc. **3** number of commercial companies under common ownership. **4** ensemble playing popular music. **5** division of an air force etc. –*v.* **1** form or be formed into a group. **2** (often foll. by *with*) place in a group or groups. [Italian *gruppo*]

group captain *n.* RAF officer next below air commodore.

groupie *n. slang* ardent follower of touring pop groups, esp. a young woman seeking sexual relations with their members.

group therapy *n.* therapy in which people are brought together to assist one another psychologically.

grouse[1] *n.* (*pl.* same) **1** game-bird with a plump body and feathered legs. **2** its flesh as food. [origin uncertain]

grouse[2] *colloq.* –*v.* (**-sing**) grumble or complain. –*n.* complaint. [origin unknown]

grout –*n.* thin fluid mortar. –*v.* provide or fill with grout. [origin uncertain]

grove *n.* small wood or group of trees. [Old English]

grovel /'grɒv(ə)l/ *v.* (**-ll-**; *US* **-l-**) **1** behave obsequiously. **2** lie prone in abject humility. □ **grovelling** *adj.* [obsolete *grovelling* (adv.) from Old Norse *á grúfu* face down]

grow /grəʊ/ *v.* (*past* **grew**; *past part.* **grown**) **1** increase in size, height, quantity, degree, etc. **2** develop or exist as a living plant or natural product. **3 a** produce (plants etc.) by cultivation. **b** allow (a beard etc.) to develop. **4** become gradually (*grow rich*). **5** (foll. by *on*) become gradually more favoured by. **6** (in *passive*; foll. by *over* etc.) be covered with a growth. □ **grow out of 1** become too large to wear. **2** become too mature to retain (a habit etc.). **3** develop from. **grow up 1** advance to maturity. **2** (of a custom) arise. [Old English]

grower *n.* **1** (often in *comb.*) person growing produce (*fruit-grower*). **2** plant that grows in a specified way (*fast grower*).

growing pains *n.pl.* **1** early difficulties in the development of a project etc. **2** neuralgic pain in children's legs due to fatigue etc.

growl –*v.* **1 a** (often foll. by *at*) make a low guttural sound, usu. of anger. **b** murmur angrily. **2** rumble. **3** (often foll. by *out*) utter with a growl. –*n.* **1** growling sound. **2** angry murmur. **3** rumble. [probably imitative]

grown *past part.* of GROW.

grown-up –*adj.* adult. –*n.* adult person.

growth /grəʊθ/ *n.* **1** act or process of growing. **2** increase in size or value. **3** something that has grown or is growing. **4** *Med.* morbid formation.

growth industry *n.* industry that is developing rapidly.

groyne *n.* (*US* **groin**) timber, stone, or concrete wall built at right angles to the coast to check beach erosion. [dial. *groin* snout, from French]

grub –*n.* **1** larva of an insect. **2** *colloq.* food. –*v.* (**-bb-**) **1** dig superficially. **2** (foll. by *up*, *out*) **a** extract by digging. **b** extract (information etc.) by searching in books etc. **3** rummage. [Old English]

grubby *adj.* (**-ier, -iest**) **1** dirty. **2** of or infested with grubs. □ **grubbily** *adv.* **grubbiness** *n.*

grudge –*n.* persistent feeling of ill will or resentment. –*v.* (**-ging**) **1** be resentfully unwilling to give or allow. **2** (foll. by verbal noun or *to* + infin.) be reluctant to do. [French]

gruel /'gru:əl/ *n.* liquid food of oatmeal etc. boiled in milk or water. [French from Germanic]

gruelling *adj.* (*US* **grueling**) extremely demanding or tiring.

gruesome /'gru:səm/ *adj.* horrible, grisly, disgusting. □ **gruesomely** *adv.* [Scandinavian]

gruff *adj.* **1 a** (of a voice) low and harsh. **b** (of a person) having a gruff voice. **2** surly. □ **gruffly** *adv.* **gruffness** *n.* [Low German or Dutch *grof* coarse]

grumble /'grʌmb(ə)l/ –*v.* (**-ling**) **1** complain peevishly. **2** rumble. –*n.* **1** complaint. **2** rumble. □ **grumbler** *n.* [obsolete *grumme*]

grummet var. of GROMMET.

grumpy /'grʌmpɪ/ *adj.* (**-ier, -iest**) morosely irritable. □ **grumpily** *adv.* **grumpiness** *n.* [imitative]

grunt –*n.* **1** low guttural sound made by a pig. **2** similar sound. –*v.* **1** make a grunt. **2** make a similar sound, esp. to express discontent. **3** utter with a grunt. [Old English, imitative]

Gruyère /'gru:jeə(r)/ *n.* a firm pale cheese. [*Gruyère* in Switzerland]

gryphon var. of GRIFFIN.

G7 *attrib. adj.* designating the world's seven richest nations. [*G*roup of *Seven*]

G-string /'dʒi:strɪŋ/ *n.* **1** *Mus.* string sounding the note G. **2** narrow strip of cloth etc. covering only the genitals and attached to a string round the waist.

G-suit /'dʒi:su:t, -sju:t/ *n.* garment with inflatable pressurized pouches, worn by pilots and astronauts to enable them to withstand high acceleration. [*g* = gravity, SUIT]

GT *n.* high-performance saloon car. [Italian *gran turismo* great touring]

guano /'gwɑ:nəʊ/ *n.* (*pl.* **-s**) **1** excrement of sea birds, used as manure. **2** artificial manure, esp. that made from fish. [Spanish from Quechua]

guarantee /,gærən'ti:/ –*n.* **1 a** formal promise or assurance, esp. that something is of a specified quality and durability. **b** document giving such an undertaking. **2** = GUARANTY. **3** person making a guaranty or giving a security. –*v.* (**-tees, -teed**) **1 a** give or serve as a guarantee for. **b** provide with a guarantee. **2** give a promise or assurance. **3** (foll. by *to*) secure the possession of (a thing) for a person. [related to WARRANT]

guarantor /,gærən'tɔ:(r)/ *n.* person who gives a guarantee or guaranty.

guaranty /'gærəntɪ/ *n.* (*pl.* **-ies**) **1** written or other undertaking to answer for the payment of a debt or for the performance of an obligation by another person liable in the first instance. **2** thing serving as security.

guard /gɑ:d/ –*v.* **1** (often foll. by *from*, *against*) watch over and defend or protect. **2** keep watch by (a door etc.) to control entry or exit. **3** supervise (prisoners etc.) and prevent from escaping. **4** keep (thoughts or speech) in check. **5** (foll. by *against*) take precautions. –*n.* **1** state of vigilance. **2** person who protects or keeps watch. **3** soldiers etc. protecting a place or person; escort. **4** official in general charge of a train. **5** part of an army detached for some purpose (*advance guard*). **6** (in *pl.*) (usu. **Guards**) body of troops nominally employed to guard a monarch. **7** thing that protects (*fire-guard*). **8** *US* prison warder. **9** defensive posture or motion in boxing etc. □ **be on** (or **keep** or **stand**) **guard** keep watch. **off** (or **off one's**) **guard** unprepared for some surprise or difficulty. **on** (or **on one's**) **guard** prepared for all contingencies. [Germanic: related to WARD]

guarded *adj.* (of a remark etc.) cautious. □ **guardedly** *adv.*

guardhouse *n.* building used to accommodate a military guard or to detain prisoners.

guardian /'gɑ:dɪən/ *n.* **1** protector, keeper. **2** person having legal custody of another, esp. a minor. □ **guardianship** *n.* [French: related to WARD, WARDEN]

guardroom *n.* room serving the same purpose as a guardhouse.

guardsman *n.* soldier belonging to a body of guards or regiment of Guards.

guava /'gwɑ:və/ *n.* **1** edible pale orange fruit with pink flesh. **2** tree bearing this. [Spanish]

gubernatorial /ˌgju:bənə'tɔ:rɪəl/ *adj.* esp. *US* of or relating to a governor. [Latin *gubernator* governor]

gudgeon[1] /'gʌdʒ(ə)n/ *n.* small freshwater fish often used as bait. [French *goujon* from Latin *gobio* GOBY]

gudgeon[2] /'gʌdʒ(ə)n/ *n.* **1** a kind of pivot. **2** tubular part of a hinge. **3** socket for a rudder. **4** pin holding two blocks of stone etc. together. [French diminutive: related to GOUGE]

guelder rose /'geldə(r)/ *n.* shrub with round bunches of creamy-white flowers. [Dutch from *Gelderland* in the Netherlands]

Guernsey /'gɜ:nzɪ/ *n.* (*pl.* **-s**) **1** one of a breed of dairy cattle from Guernsey in the Channel Islands. **2** (**guernsey**) type of thick woollen sweater.

guerrilla /gə'rɪlə/ *n.* (also **guerilla**) member of a small independently acting (usu. political) group taking part in irregular fighting. [Spanish diminutive: related to WAR]

guess /ges/ —*v.* **1** (often *absol.*) estimate without calculation or measurement. **2** form a hypothesis or opinion about; conjecture; think likely. **3** conjecture or estimate correctly. **4** (foll. by *at*) make a conjecture about. —*n.* estimate, conjecture. □ **I guess** *colloq.* I think it likely; I suppose. [origin uncertain]

guesswork *n.* process of or results got by guessing.

guest /gest/ *n.* **1** person invited to visit another's house or to have a meal etc. at another's expense. **2** person lodging at a hotel etc. **3** outside performer invited to take part with a regular body of performers. [Old Norse]

guest-house *n.* private house offering paid accommodation.

guestimate /'gestɪmət/ *n.* (also **guesstimate**) *colloq.* estimate based on a mixture of guesswork and calculation. [from GUESS, ESTIMATE]

guff *n. slang* empty talk. [imitative]

guffaw /gʌ'fɔ:/ —*n.* boisterous laugh. —*v.* utter a guffaw. [imitative]

guidance /'gaɪd(ə)ns/ *n.* **1** advice or direction for solving a problem etc. **2** guiding or being guided.

guide /gaɪd/ —*n.* **1** person who leads or shows the way. **2** person who conducts tours. **3** adviser. **4** directing principle. **5** book with essential information on a subject, esp. = GUIDEBOOK. **6** thing marking a position or guiding the eye. **7** bar etc. directing the motion of something. **8** (**Guide**) member of a girls' organization similar to the Scouts. —*v.* (**-ding**) **1** act as guide to. **2** be the principle or motive of. [French from Germanic]

guidebook *n.* book of information about a place for tourists etc.

guided missile *n.* missile under remote control or directed by equipment within itself.

guide-dog *n.* dog trained to guide a blind person.

guideline *n.* principle directing action.

Guider *n.* adult leader of Guides.

guild /gɪld/ *n.* (also **gild**) **1** association of people for mutual aid or the pursuit of a common goal. **2** medieval association of craftsmen or merchants. [Low German or Dutch *gilde*]

guilder /'gɪldə(r)/ *n.* chief monetary unit of the Netherlands. [alteration of Dutch *gulden* golden]

guildhall *n.* meeting-place of a medieval guild; town hall.

guile /gaɪl/ *n.* cunning or sly behaviour; treachery, deceit. □ **guileful** *adj.* **guileless** *adj.* [French from Scandinavian]

guillemot /'gɪlɪˌmɒt/ *n.* fast-flying sea bird nesting on cliffs etc. [French]

guillotine /'gɪləˌti:n/ —*n.* **1** machine with a blade sliding vertically in grooves, used for beheading. **2** device for cutting paper etc. **3** method of preventing delay in the discussion of a legislative bill by fixing times at which various parts of it must be voted on. —*v.* (**-ning**) use a guillotine on. [*Guillotin*, name of a physician]

guilt /gɪlt/ *n.* **1** fact of having committed a specified or implied offence. **2** feeling of having done wrong. [Old English]

guiltless *adj.* (often foll. by *of* an offence) innocent.

guilty *adj.* (**-ier, -iest**) **1** culpable of or responsible for a wrong. **2** conscious of or affected by guilt. **3** causing a feeling of guilt (*a guilty secret*). **4** (often foll. by *of*) having committed a (specified) offence. □ **guiltily** *adv.* **guiltiness** *n.* [Old English: related to GUILT]

guinea /'gɪnɪ/ *n.* **1** *hist.* sum of 21 old shillings (£1.05). **2** *hist.* former British gold coin first coined for the African trade. [*Guinea* in W. Africa]

guinea-fowl *n.* African fowl with slate-coloured white-spotted plumage.

guinea-pig *n.* **1** domesticated S. American cavy. **2** person used in an experiment.

guipure /'gi:pjʊə(r)/ *n.* heavy lace of linen pieces joined by embroidery. [French]

guise /gaɪz/ *n.* **1** assumed appearance; pretence. **2** external appearance. [Germanic: related to WISE²]

guitar /gɪ'tɑ:(r)/ *n.* usu. six-stringed musical instrument played with the fingers or a plectrum. □ **guitarist** *n.* [Greek *kithara* harp]

Gujarati /ˌgʊdʒə'rɑ:tɪ/ (also **Gujerati**) *–n.* (*pl.* **-s**) **1** native of Gujarat. **2** language of Gujarat. *–adj.* of Gujarat, its people, or language. [*Gujarat*, State in India]

gulch *n. US* ravine, esp. one in which a torrent flows. [origin uncertain]

gulf *n.* **1** stretch of sea consisting of a deep inlet with a narrow mouth. **2** deep hollow; chasm. **3** wide difference of feelings, opinion, etc. [Greek *kolpos*]

Gulf Stream *n.* warm current flowing from the Gulf of Mexico to Newfoundland where it is deflected across the Atlantic Ocean.

gull¹ *n.* long-winged web-footed sea bird. [probably Welsh *gwylan*]

gull² *v.* dupe, fool. [perhaps from obsolete *gull* yellow from Old Norse]

gullet /'gʌlɪt/ *n.* food-passage extending from the mouth to the stomach. [Latin *gula* throat]

gullible *adj.* easily persuaded or deceived. □ **gullibility** /-'bɪlɪtɪ/ *n.* [from GULL²]

gully /'gʌlɪ/ *n.* (*pl.* **-ies**) **1** water-worn ravine. **2** gutter or drain. **3** *Cricket* fielding position between point and slips. [French *goulet*: related to GULLET]

gulp *–v.* **1** (often foll. by *down*) swallow hastily, greedily, or with effort. **2** swallow gaspingly or with difficulty; choke. **3** (foll. by *down*, *back*) suppress (esp. tears). *–n.* **1** act of gulping. **2** large mouthful of a drink. [Dutch *gulpen*, imitative]

gum¹ *–n.* **1 a** viscous secretion of some trees and shrubs. **b** adhesive substance made from this. **2** *US* chewing gum. **3** = GUMDROP. **4** = GUM ARABIC. **5** = GUM-TREE. *–v.* (**-mm-**) **1** (usu. foll. by *down*, *together*, etc.) fasten with gum. **2** apply gum to. □ **gum up** *colloq.* interfere with the smooth running of. [Greek *kommi* from Egyptian *kemai*]

gum² *n.* (usu. in *pl.*) firm flesh around the roots of the teeth. [Old English]

gum³ *n.* □ **by gum!** *colloq.* by God! [corruption of *God*]

gum arabic *n.* gum exuded by some kinds of acacia.

gumboil *n.* small abscess on the gum.

gumboot *n.* rubber boot.

gumdrop *n.* hard translucent sweet made with gelatin etc.

gummy¹ *adj.* (**-ier**, **-iest**) **1** sticky. **2** exuding gum.

gummy² *adj.* (**-ier**, **-iest**) toothless.

gumption /'gʌmpʃ(ə)n/ *n. colloq.* **1** resourcefulness, initiative. **2** common sense. [origin unknown]

gum-tree *n.* tree exuding gum, esp. a eucalyptus. □ **up a gum-tree** *colloq.* in great difficulties.

gun *–n.* **1** weapon consisting of a metal tube from which bullets or other missiles are propelled with great force, esp. by a contained explosion. **2** starting pistol. **3** device for discharging insecticide, grease etc., in the required direction. **4** member of a shooting-party. **5** *US* gunman. *–v.* (**-nn-**) **1 a** (usu. foll. by *down*) shoot (a person) with a gun. **b** shoot at with a gun. **2** go shooting. **3** (foll. by *for*) seek out determinedly to attack or rebuke. □ **go great guns** *colloq.* proceed vigorously or successfully. **stick to one's guns** *colloq.* maintain one's position under attack. [perhaps an abbreviation of the Scandinavian woman's name *Gunnhildr*, applied to cannon etc.]

gunboat *n.* small vessel with heavy guns.

gunboat diplomacy *n.* political negotiation backed by the threat of force.

gun-carriage *n.* wheeled support for a gun.

gun-cotton *n.* explosive made by steeping cotton in acids.

gun dog *n.* dog trained to retrieve game shot by sportsmen.

gunfight *n. US* fight with firearms. □ **gunfighter** *n.*

gunfire *n.* firing of a gun or guns.

gunge /gʌndʒ/ *colloq. –n.* sticky or viscous matter. *–v.* (**-ging**) (usu. foll. by *up*) clog with gunge. □ **gungy** *adj.* [origin uncertain]

gung-ho /gʌŋ'həʊ/ *adj.* zealous, arrogantly eager. [Chinese *gonghe* work together]

gunman *n.* man armed with a gun, esp. when committing a crime.

gun-metal *n.* **1** a dull bluish-grey colour. **2** alloy formerly used for guns.

gunnel var. of GUNWALE.

gunner *n.* **1** artillery soldier (esp. as an official term for a private). **2** *Naut.* warrant-officer in charge of a battery, magazine, etc. **3** member of an aircraft crew who operates a gun.

gunnery *n.* **1** construction and management of large guns. **2** firing of guns.

gunny /'gʌnɪ/ *n.* (*pl.* **-ies**) **1** coarse sacking, usu. of jute fibre. **2** sack made of this. [Hindi and Marathi]

gunpoint *n.* □ **at gunpoint** threatened with a gun or an ultimatum etc.

gunpowder *n.* explosive made of salt-petre, sulphur, and charcoal.

gunrunner *n.* person engaged in the illegal sale or importing of firearms. □ **gunrunning** *n.*

gunshot *n.* **1** shot fired from a gun. **2** range of a gun (*within gunshot*).

gunslinger *n.* esp. *US slang* gunman.

gunsmith *n.* maker and repairer of small firearms.

gunwale /'gʌn(ə)l/ *n.* (also **gunnel**) upper edge of the side of a boat or ship. [from GUN, WALE, because it was formerly used to support guns]

guppy /'gʌpɪ/ *n.* (*pl.* **-ies**) freshwater fish of the W. Indies and S. America frequently kept in aquariums. [*Guppy*, name of a clergyman]

gurgle /'gɜ:g(ə)l/ *–v.* (**-ling**) **1** make a bubbling sound as of water from a bottle. **2** utter with such a sound. *–n.* gurgling sound. [probably imitative]

Gurkha /'gɜ:kə/ *n.* **1** member of the dominant Hindu race in Nepal. **2** Nepalese soldier serving in the British army. [Sanskrit]

gurnard /'gɜ:nəd/ *n.* (*pl.* same or **-s**) marine fish with a large spiny head and finger-like pectoral rays. [French]

guru /'gʊru:/ *n.* (*pl.* **-s**) **1** Hindu spiritual teacher or head of a religious sect. **2** influential or revered teacher. [Hindi]

gush *–v.* **1** emit or flow in a sudden and copious stream. **2** speak or behave effusively. *–n.* **1** sudden or copious stream. **2** effusive manner. [probably imitative]

gusher *n.* **1** oil well from which oil flows without being pumped. **2** effusive person.

gusset /'gʌsɪt/ *n.* **1** piece let into a garment etc. to strengthen or enlarge it. **2** bracket strengthening an angle of a structure. [French]

gust *–n.* **1** sudden strong rush of wind. **2** burst of rain, smoke, emotion, etc. *–v.* blow in gusts. □ **gusty** *adj.* (**-ier**, **-iest**). [Old Norse]

gusto /'gʌstəʊ/ *n.* zest; enjoyment. [Latin *gustus* taste]

gut *–n.* **1** the intestine. **2** (in *pl.*) the bowel or entrails. **3** (in *pl.*) *colloq.* personal courage and determination; perseverance. **4** *slang* stomach, belly. **5** (in *pl.*) **a** contents. **b** essence. **6** **a** material for violin strings etc. **b** material for fishing-lines made from the silk-glands of silkworms. **7** (*attrib.*) **a** instinctive (*a gut reaction*). **b** fundamental (*a gut issue*). *–v.* (**-tt-**) **1** remove or destroy the internal fittings of (a house etc.). **2** remove the guts of (a fish). □ **hate a person's guts** *colloq.* dislike a person intensely. [Old English]

gutless *adj. colloq.* lacking courage or energy.

gutsy *adj.* (**-ier**, **-iest**) *colloq.* **1** courageous. **2** greedy.

gutta-percha /ˌgʌtə'pɜ:tʃə/ *n.* tough rubbery substance obtained from latex. [Malay]

gutted *adj. slang* utterly exhausted or fed-up.

gutter *–n.* **1** shallow trough below the eaves of a house, or a channel at the side of a street, to carry off rainwater. **2** (prec. by *the*) poor or degraded background or environment. **3** open conduit. **4** groove. *–v.* (of a candle) burn unsteadily and melt away rapidly. [Latin *gutta* drop]

guttering *n.* **1** gutters of a building etc. **2** material for gutters.

gutter press *n.* sensational newspapers.

guttersnipe *n.* street urchin.

guttural /'gʌtər(ə)l/ *–adj.* **1** throaty, harsh-sounding. **2** *Phonet.* (of a consonant) produced in the throat or by the back of the tongue and palate. **3** of the throat. *–n. Phonet.* guttural consonant (e.g. *k*, *g*). □ **gutturally** *adv.* [Latin *guttur* throat]

guv *n. slang* = GOVERNOR 7. [abbreviation]

guy[1] /gaɪ/ *–n.* **1** *colloq.* man; fellow. **2** effigy of Guy Fawkes burnt on 5 Nov. *–v.* ridicule. [*Guy* Fawkes, name of a conspirator]

guy[2] /gaɪ/ *–n.* rope or chain to secure a tent or steady a crane-load etc. *–v.* secure with a guy or guys. [probably Low German]

guzzle /'gʌz(ə)l/ *v.* (**-ling**) eat or drink greedily. [probably French *gosiller* from *gosier* throat]

gybe /dʒaɪb/ *v.* (*US* **jibe**) (**-bing**) **1** (of a fore-and-aft sail or boom) swing across. **2** cause (a sail) to do this. **3** (of a ship or its crew) change course so that this happens. [Dutch]

gym /dʒɪm/ *n. colloq.* **1** gymnasium. **2** gymnastics. [abbreviation]

gymkhana /dʒɪm'kɑ:nə/ *n.* horse-riding competition. [Hindustani *gendkhāna* ball-house, assimilated to GYMNASIUM]

gymnasium /dʒɪm'neɪzɪəm/ *n.* (*pl.* **-s** or **-sia**) room or building equipped for gymnastics. [Greek *gumnos* naked]

gymnast /'dʒɪmnæst/ *n.* person who does gymnastics, esp. an expert.

gymnastic /dʒɪm'næstɪk/ *adj.* of or involving gymnastics. □ **gymnastically** *adv.*

gymnastics *n.pl.* (also treated as *sing.*) **1** exercises performed in order to develop or display physical agility. **2** other forms of physical or mental agility.

gymnosperm /'dʒɪmnəʊ,spɜ:m/ *n.* any of a group of plants having seeds unprotected by an ovary, including conifers, cycads, and ginkgos. [Greek *gumnos* naked]

gymp var. of GIMP.

gymslip *n.* sleeveless tunic worn by schoolgirls.

gynae /'gaɪnɪ/ *n.* (also **gynie**) *colloq.* gynaecology. [abbreviation].

gynaecology /,gaɪnɪ'kɒlədʒɪ/ *n.* (*US* **gynecology**) science of the physiological functions and diseases of women. □ **gynaecological** /-kə'lɒdʒɪk(ə)l/ *adj.* **gynaecologist** *n.* [Greek *gunē gunaik-* woman, -LOGY]

gypsum /'dʒɪpsəm/ *n.* mineral used esp. to make plaster of Paris. [Greek *gupsos*]

Gypsy *n.* (also **Gipsy**) (*pl.* **-ies**) **1** member of a nomadic people of Europe and N. America, of Hindu origin with dark skin and hair. **2** (**gypsy**) person resembling or living like a Gypsy. [from EGYPTIAN]

gyrate /,dʒaɪə'reɪt/ *v.* (**-ting**) move in a circle or spiral; revolve, whirl. □ **gyration** /-'reɪʃ(ə)n/ *n.* **gyratory** *adj.* [Greek: related to GYRO-]

gyrfalcon /'dʒɜ:,fɔ:lkən/ *n.* large falcon of the northern hemisphere. [French from Old Norse]

gyro /'dʒaɪərəʊ/ *n.* (*pl.* **-s**) *colloq.* = GYROSCOPE. [abbreviation]

gyro- *comb. form* rotation. [Greek *guros* ring]

gyrocompass /'dʒaɪərəʊ,kʌmpəs/ *n.* compass giving true north and bearings from it by means of a gyroscope.

gyroscope /'dʒaɪərə,skəʊp/ *n.* rotating wheel whose axis is free to turn but maintains a fixed direction unless perturbed, esp. used for stabilization or with the compass in an aircraft, ship, etc.

H

H[1] /eɪtʃ/ *n.* (also **h**) (*pl.* **Hs** or **H's**) **1** eighth letter of the alphabet (see AITCH). **2** anything having the form of an H (esp. in *comb.*: *H-girder*).
H[2] *abbr.* (also **H.**) **1** (of a pencil-lead) hard. **2** (water) hydrant. **3** *slang* heroin. **4** henry(s).
H[3] *symb.* hydrogen.
h. *abbr.* (also **h**) **1** hecto-. **2** (also ***h***) height. **3** hot. **4** hour(s).
Ha *symb.* hahnium.
ha[1] /hɑː/ (also **hah**) *–int.* expressing surprise, derision, triumph, etc. (cf. HA HA). *–v.* in **hum and ha**: see HUM. [imitative]
ha[2] *abbr.* hectare(s).
habeas corpus /ˌheɪbɪəs ˈkɔːpəs/ *n.* writ requiring a person to be brought before a judge or into court, esp. to investigate the lawfulness of his or her detention. [Latin, = you must have the body]
haberdasher /ˈhæbəˌdæʃə(r)/ *n.* dealer in dress accessories and sewing-goods. □ **haberdashery** *n.* (*pl.* **-ies**). [probably Anglo-French]
habiliment /həˈbɪlɪmənt/ *n.* (usu. in *pl.*) *archaic* clothes. [French from *habiller* fit out]
habit /ˈhæbɪt/ *n.* **1** settled or regular tendency or practice (often foll. by *of* + verbal noun: *has a habit of ignoring me*). **2** practice that is hard to give up. **3** mental constitution or attitude. **4** dress, esp. of a religious order. [Latin *habeo habit-* have]
habitable *adj.* suitable for living in. □ **habitability** /-ˈbɪlɪtɪ/ *n.* [Latin *habito* inhabit]
habitat /ˈhæbɪˌtæt/ *n.* natural home of an animal or plant. [Latin, = it dwells]
habitation /ˌhæbɪˈteɪʃ(ə)n/ *n.* **1** inhabiting (*fit for habitation*). **2** house or home.
habit-forming *adj.* causing addiction.
habitual /həˈbɪtʃʊəl/ *adj.* **1** done constantly or as a habit. **2** regular, usual. **3** given to a (specified) habit (*habitual smoker*). □ **habitually** *adv.*
habituate /həˈbɪtʃʊˌeɪt/ *v.* (**-ting**) (often foll. by *to*) accustom. □ **habituation** /-ˈeɪʃ(ə)n/ *n.* [Latin: related to HABIT]
habitué /həˈbɪtʃʊˌeɪ/ *n.* habitual visitor or resident. [French]
háček /ˈhætʃek/ *n.* diacritic (ˇ) placed over a letter to modify its sound in some languages. [Czech, diminutive of *hák* hook]
hachures /hæˈʃjʊə(r)/ *n.pl.* parallel lines on a map indicating the degree of steepness of hills. [French: related to HATCH[3]]
hacienda /ˌhæsɪˈendə/ *n.* (in Spanish-speaking countries) estate with a dwelling-house. [Spanish, from Latin *facienda* things to be done]
hack[1] *–v.* **1** cut or chop roughly. **2** *Football etc.* kick the shin of (an opponent). **3** (often foll. by *at*) deliver cutting blows. **4** cut (one's way) through foliage etc. **5** *colloq.* gain unauthorized access to (data in a computer). **6** *slang* manage, cope with; tolerate. **7** (as **hacking** *adj.*) (of a cough) short, dry, and frequent. *–n.* **1** kick with the toe of a boot. **2** gash or wound, esp. from a kick. **3 a** mattock. **b** miner's pick. [Old English]
hack[2] *–n.* **1 a** = HACKNEY. **b** horse let out for hire. **2** person hired to do dull routine work, esp. writing. *–attrib. adj.* **1** used as a hack. **2** typical of a hack; commonplace (*hack work*). *–v.* ride on horseback on a road at an ordinary pace. [abbreviation of HACKNEY]
hacker *n.* **1** person or thing that hacks or cuts roughly. **2** *colloq.* **a** person whose hobby is computing or computer programming. **b** person who uses a computer to gain unauthorized access to a computer network.
hackle /ˈhæk(ə)l/ *n.* **1 a** (in *pl.*) erectile hairs on an animal's neck, rising when it is angry or alarmed. **b** feather(s) on the neck of a domestic cock etc. **2** steel comb for dressing flax. □ **make one's hackles rise** cause one to be angry or indignant. [Old English]
hackney /ˈhæknɪ/ *n.* (*pl.* **-s**) horse for ordinary riding. [*Hackney* in London]
hackney carriage *n.* taxi.
hackneyed /ˈhæknɪd/ *adj.* (of a phrase etc.) made trite by overuse.
hacksaw *n.* saw with a narrow blade set in a frame, for cutting metal.
had *past* and *past part.* of HAVE.
haddock /ˈhædək/ *n.* (*pl.* same) N. Atlantic marine fish used as food. [probably French]
Hades /ˈheɪdiːz/ *n.* (in Greek mythology) the underworld. [Greek, originally a name of Pluto]
hadj var. of HAJJ.
hadji var. of HAJJI.
hadn't /ˈhæd(ə)nt/ *contr.* had not.

haemal /'hi:m(ə)l/ *adj.* (*US* **hem-**) of the blood. [Greek *haima* blood]

haematite /'hi:mə,taɪt/ *n.* (*US* **hem-**) a ferric oxide ore. [Latin: related to HAEMAL]

haematology /,hi:mə'tɒlədʒɪ/ *n.* (*US* **hem-**) the study of the blood. □ **haematologist** *n.*

haemoglobin /,hi:mə'gləʊbɪn/ *n.* (*US* **hem-**) oxygen-carrying substance in the red blood cells of vertebrates. [from GLOBULIN]

haemophilia /,hi:mə'fɪlɪə/ *n.* (*US* **hem-**) hereditary failure of the blood to clot normally with the tendency to bleed severely from even a slight injury. [Greek *haima* blood, *philia* loving]

haemophiliac *n.* (*US* **hem-**) person with haemophilia.

haemorrhage /'hemərɪdʒ/ (*US* **hem-**) –*n.* **1** profuse loss of blood from a ruptured blood-vessel. **2** damaging loss, esp. of people or assets. –*v.* (**-ging**) suffer a haemorrhage. [Greek *haima* blood, *rhēgnumi* burst]

haemorrhoids /'hemə,rɔɪdz/ *n.pl.* (*US* **hem-**) swollen veins in the wall of the anus; piles. [Greek *haima* blood, *-rhoos* -flowing]

hafnium /'hæfnɪəm/ *n.* silvery lustrous metallic element. [Latin *Hafnia* Copenhagen]

haft /hɑ:ft/ *n.* handle of a dagger, knife, etc. [Old English]

hag *n.* **1** ugly old woman. **2** witch. [Old English]

haggard /'hægəd/ *adj.* looking exhausted and distraught. [French *hagard*]

haggis /'hægɪs/ *n.* Scottish dish of offal boiled in a sheep's stomach with suet, oatmeal, etc. [origin unknown]

haggle /'hæg(ə)l/ –*v.* (**-ling**) (often foll. by *about*, *over*) bargain persistently. –*n.* haggling. [Old Norse]

hagio- *comb. form* of saints. [Greek *hagios* holy]

hagiography /,hægɪ'ɒgrəfɪ/ *n.* writing about saints' lives. □ **hagiographer** *n.*

hagiology /,hægɪ'ɒlədʒɪ/ *n.* literature dealing with the lives and legends of saints.

hagridden *adj.* afflicted by nightmares or anxieties.

hah var. of HA[1].

ha ha /hɑ:'hɑ:/ *int.* representing laughter (*iron.* when spoken). [Old English]

ha-ha /'hɑ:hɑ:/ *n.* ditch with a wall in it, forming a boundary or fence without interrupting the view. [French]

hahnium /'hɑ:nɪəm/ *n.* artificially produced radioactive element. [*Hahn*, name of a chemist]

haiku /'haɪku:/ *n.* (*pl.* same) very short Japanese three-part poem of usu. 17 syllables. [Japanese]

hail[1] –*n.* **1** pellets of frozen rain. **2** (foll. by *of*) barrage or onslaught. –*v.* **1 a** (prec. by *it* as subject) hail falls. **b** come down forcefully. **2** pour down (blows, words, etc.). [Old English]

hail[2] –*v.* **1** signal to (a taxi etc.) to stop. **2** greet enthusiastically. **3** acclaim (*hailed him king*). **4** (foll. by *from*) originate or come (*hails from Leeds*). –*int. archaic* or *joc.* expressing greeting. –*n.* act of hailing. [Old Norse *heill*: related to WASSAIL]

hail-fellow-well-met *adj.* friendly, esp. too friendly towards strangers.

Hail Mary *n.* the Ave Maria (see AVE).

hailstone *n.* pellet of hail.

hailstorm *n.* period of heavy hail.

hair *n.* **1 a** any of the fine threadlike strands growing from the skin of mammals, esp. from the human head. **b** these collectively (*has long hair*). **2** thing resembling a hair. **3** elongated cell growing from a plant. **4** very small quantity or extent (also *attrib.*: *hair crack*). □ **get in a person's hair** *colloq.* annoy a person. **keep one's hair on** *colloq.* keep calm; not get angry. **let one's hair down** *colloq.* enjoy oneself by abandoning restraint. **make one's hair stand on end** *colloq.* horrify one. **not turn a hair** remain unmoved or unaffected. □ **hairless** *adj.* [Old English]

hairbrush *n.* brush for tidying the hair.

haircloth *n.* stiff cloth woven from hair.

haircut *n.* **1** act of cutting the hair (*needs a haircut*). **2** style in which the hair is cut.

hairdo *n.* (*pl.* **-s**) style of or act of styling the hair.

hairdresser *n.* **1** person who cuts and styles the hair, esp. for a living. **2** hairdresser's shop. □ **hairdressing** *n.*

hair-drier *n.* (also **hair-dryer**) device for drying the hair with warm air.

hairgrip *n.* flat hairpin with the ends close together.

hairline *n.* **1** edge of a person's hair, esp. on the forehead. **2** very narrow line, crack (usu. **hairline crack**), etc.

hairnet *n.* piece of netting for confining the hair.

hair of the dog *n.* further alcoholic drink taken to cure the effects of drink.

hairpiece *n.* quantity of hair augmenting a person's natural hair.

hairpin *n.* U-shaped pin for fastening the hair.

hairpin bend *n.* sharp U-shaped bend in a road.

hair-raising *adj.* terrifying.

hair's breadth *n.* a tiny amount or margin.

hair shirt *n.* shirt of haircloth, worn formerly by penitents and ascetics.

hair-slide *n.* clip for keeping the hair in place.

hair-splitting *adj.* & *n.* quibbling.

hairspray *n.* liquid sprayed on the hair to keep it in place.

hairspring *n.* fine spring regulating the balance-wheel in a watch.

hairstyle *n.* particular way of arranging the hair. □ **hairstylist** *n.*

hair-trigger *n.* trigger of a firearm set for release at the slightest pressure.

hairy *adj.* (**-ier, -iest**) **1** covered with hair. **2** *slang* frightening, dangerous. □ **hairiness** *n.*

hajj /hædʒ/ *n.* (also **hadj**) Islamic pilgrimage to Mecca. [Arabic]

hajji /ˈhædʒɪ/ *n.* (also **hadji**) (*pl.* **-s**) Muslim who has made the pilgrimage to Mecca. [Persian from Arabic]

haka /ˈhɑːkə/ *n. NZ* **1** Maori ceremonial war dance with chanting. **2** imitation of this by a sports team before a match. [Maori]

hake *n.* (*pl.* same) marine fish resembling the cod, used as food. [origin uncertain]

halal /hɑːˈlɑːl/ *n.* (also **hallal**) (often *attrib.*) meat from an animal killed according to Muslim law. [Arabic]

halberd /ˈhælbəd/ *n. hist.* combined spear and battleaxe. [French from German]

halcyon /ˈhælsɪən/ *adj.* calm, peaceful, happy (*halcyon days*). [Greek, = kingfisher, because it was reputed to calm the sea at midwinter]

hale *adj.* strong and healthy (esp. in **hale and hearty**). [var. of WHOLE]

half /hɑːf/ —*n.* (*pl.* **halves** /hɑːvz/) **1** either of two (esp. equal) parts into which a thing is divided. **2** *colloq.* half a pint, esp. of beer. **3** *Sport* either of two equal periods of play. **4** *colloq.* half-price fare or ticket, esp. for a child. **5** *colloq.* = HALF-BACK. —*adj.* **1** amounting to half (*half the men*). **2** forming a half (*a half share*). —*adv.* **1** (often in *comb.*) to the extent of half; partly (*half cooked*). **2** to some extent (esp. in idiomatic phrases: *half dead*; *am half convinced*). **3** (in reckoning time) by the amount of half (an hour etc.) (*half past two*). □ **at half cock** see COCK[1]. **by half** (prec. by *too* + adj.) excessively (*too clever by half*). **by halves** imperfectly or incompletely (*does nothing by halves*). **half a mind** see MIND. **half the time** see TIME. **not half 1** *slang* extremely, violently (*he didn't half swear*). **2** not nearly (*not half long enough*). **3** *colloq.* not at all (*not half bad*). [Old English]

■ **Usage** In sense 3 of the adverb, the word 'past' is often omitted in colloquial usage, e.g. *came at half two*. In some parts of Scotland and Ireland this means 'half past one'.

half-and-half *adj.* being half one thing and half another.

half-back *n. Sport* player between the forwards and full backs.

half-baked *adj. colloq.* **1** not thoroughly thought out; foolish. **2** (of enthusiasm etc.) only partly committed.

half board *n.* provision of bed, breakfast, and one main meal at a hotel etc.

half-breed *n. offens.* = HALF-CASTE.

half-brother *n.* brother with whom one has only one parent in common.

half-caste *n. offens.* person of mixed race.

half-crown *n.* (also **half a crown**) former coin and monetary unit worth 2s. 6d. ($12^1/_2$p).

half-cut *adj. slang* fairly drunk.

half-dozen *n.* (also **half a dozen**) *colloq.* six, or about six.

half-duplex *n. Computing* (of a circuit) allowing the two-way transmission of signals but not simultaneously.

half-hardy *adj.* (of a plant) able to grow in the open except in severe frost.

half-hearted *adj.* lacking enthusiasm. □ **half-heartedly** *adv.* **half-heartedness** *n.*

half hitch *n.* knot formed by passing the end of a rope round its standing part and then through the loop.

half holiday *n.* half a day as holiday.

half-hour *n.* **1** (also **half an hour**) period of 30 minutes. **2** point of time 30 minutes after any hour o'clock. □ **half-hourly** *adj.* & *adv.*

half-life *n.* time taken for radioactivity etc. to fall to half its original value.

half-light *n.* dim imperfect light.

half-mast *n.* position of a flag halfway down a mast, as a mark of respect for a deceased person.

half measures *n.pl.* unsatisfactory compromise or inadequate policy.

half moon *n.* **1** moon when only half its surface is illuminated. **2** time when this occurs. **3** semicircular object.

half nelson see NELSON.

halfpenny /ˈheɪpnɪ/ *n.* (*pl.* **-pennies** or **-pence** /ˈheɪpəns/) former coin worth half a penny.

■ **Usage** The halfpenny was withdrawn from circulation in 1984.

half-sister *n.* sister with whom one has only one parent in common.

half-term *n.* short holiday halfway through a school term.

half-timbered *adj.* having walls with a timber frame and a brick or plaster filling.

half-time *n.* **1** mid-point of a game or contest. **2** short break occurring at this time.

half-title *n.* title or short title of a book printed on the front of the leaf preceding the title-page.

halftone *n.* photographic illustration in which various tones of grey are produced from small and large black dots.

half-truth *n.* statement that (esp. deliberately) conveys only part of the truth.

half-volley *n.* (in ball games) playing of the ball as soon as it bounces off the ground.

halfway *–adv.* **1** at a point midway between two others (*halfway to Rome*). **2** to some extent, more or less (*is halfway acceptable*). *–adj.* situated halfway (*reached a halfway point*).

halfway house *n.* **1** compromise. **2** halfway point in a progression. **3** centre for rehabilitating ex-prisoners etc. **4** inn midway between two towns.

halfwit *n.* foolish or stupid person. □ **halfwitted** *adj.*

halibut /'hælɪbət/ *n.* (*pl.* same) large marine flat-fish used as food. [from HOLY (perhaps because eaten on holy days), *butt* flat-fish]

halitosis /,hælɪ'təʊsɪs/ *n.* = BAD BREATH. [Latin *halitus* breath]

hall /hɔ:l/ *n.* **1** area into which the front entrance of a house etc. opens. **2** large room or building for meetings, concerts, etc. **3** large country house or estate. **4** (in full **hall of residence**) residence for students. **5** (in a college etc.) dining-room. **6** premises of a guild (*Fishmongers' Hall*). **7** large public room in a palace etc. [Old English]

hallal var. of HALAL.

hallelujah var. of ALLELUIA.

halliard var. of HALYARD.

hallmark *–n.* **1** mark indicating the standard of gold, silver, and platinum. **2** distinctive feature. *–v.* stamp with a hallmark.

hallo var. of HELLO.

halloo /hə'lu:/ *int.* inciting dogs to the chase or calling attention. [perhaps from *hallow* pursue with shouts]

hallow /'hæləʊ/ *v.* **1** make holy, consecrate. **2** honour as holy. [Old English: related to HOLY]

Hallowe'en /,hæləʊ'i:n/ *n.* eve of All Saints' Day, 31 Oct.

hallucinate /hə'lu:sɪ,neɪt/ *v.* (**-ting**) experience hallucinations. □ **hallucinant** *adj.* & *n.* [Greek *alussō* be uneasy]

hallucination /hə,lu:sɪ'neɪʃ(ə)n/ *n.* illusion of seeing or hearing something not actually present. □ **hallucinatory** /hə'lu:sɪnətərɪ/ *adj.*

hallucinogen /hə'lu:sɪnədʒ(ə)n/ *n.* drug causing hallucinations. □ **hallucinogenic** /-'dʒenɪk/ *adj.*

hallway *n.* entrance-hall or corridor.

halm var. of HAULM.

halo /'heɪləʊ/ *–n.* (*pl.* **-es**) **1** disc or circle of light shown surrounding the head of a sacred person. **2** glory associated with an idealized person etc. **3** circle of white or coloured light round a luminous body, esp. the sun or moon. *–v.* (**-es, -ed**) surround with a halo. [Greek *halōs* threshing-floor, disc of the sun or moon]

halogen /'hælədʒ(ə)n/ *n.* any of the non-metallic elements (fluorine, chlorine, bromine, iodine, and astatine) which form a salt (e.g. sodium chloride) when combined with a metal. [Greek *hals halos* salt, -GEN]

halon /'heɪlɒn/ *n.* any of various gaseous compounds of carbon, bromine, and other halogens, used to extinguish fires. [related to HALOGEN]

halt[1] /hɔ:lt/ *–n.* **1** stop (usu. temporary) (*come to a halt*). **2** minor stopping-place on a local railway line. *–v.* stop; come or bring to a halt. □ **call a halt (to)** decide to stop. [German: related to HOLD]

halt[2] /hɔ:lt/ *–v.* (esp. as **halting** *adj.*) proceed hesitantly. *–adj. archaic* lame. □ **haltingly** *adv.* [Old English]

halter /'hɔ:ltə(r)/ *n.* **1** headstall and rope for leading or tying up a horse etc. **2 a** strap round the neck holding a dress etc. up and leaving the shoulders and back bare. **b** (also **halterneck**) dress etc. held by this. [Old English]

halva /'hælvə/ *n.* confection of sesame flour and honey etc. [Yiddish from Turkish *helva* from Arabic *ḥalwa*]

halve /hɑ:v/ *v.* (**-ving**) **1** divide into two halves or parts; share equally between two. **2** reduce by half. **3** *Golf* use the same number of strokes as one's opponent in (a hole or match).

halves *pl.* of HALF.

halyard /'hæljəd/ *n.* (also **halliard**) rope or tackle for raising or lowering a sail, yard, etc. [archaic *hale* drag forcibly]

ham *–n.* **1 a** upper part of a pig's leg salted and dried or smoked for food. **b** meat from this. **2** back of the thigh; thigh and buttock. **3** *colloq.* (often *attrib.*) inexpert or unsubtle actor or piece of acting. **4** *colloq.* operator of an amateur radio station. *–v.* (**-mm-**) (usu. in **ham it up**) *colloq.* overact. [Old English]

hamburger /'hæm,bɜ:gə(r)/ *n.* cake of minced beef, usu. eaten in a soft bread roll. [*Hamburg* in Germany]

ham-fisted *adj.* (also **ham-handed**) *colloq.* clumsy.

Hamitic /hə'mɪtɪk/ *–n.* group of African languages including ancient Egyptian and Berber. *–adj.* of this group. [from the name *Ham* (Gen. 10:6 ff.)]

hamlet /'hæmlɪt/ *n.* small village, esp. without a church. [French *hamelet* diminutive]

hammer *–n.* **1 a** tool with a heavy metal head at right angles to its handle, used for driving nails etc. **b** similar device, as for exploding the charge in a gun, striking the strings of a piano, etc. **2** auctioneer's mallet. **3** metal ball attached to a wire for throwing in an athletic contest. *–v.* **1 a** hit or beat with or as with a hammer. **b** strike loudly. **2 a** drive in (nails) with a hammer. **b** fasten or secure by hammering (*hammered the lid down*). **3** (usu. foll. by *in*) inculcate (ideas, knowledge, etc.) forcefully or repeatedly. **4** *colloq.* defeat utterly; beat up. **5** (foll. by *at*, *away at*) work hard or persistently at. □ **come under the hammer** be sold at auction. **hammer out 1** make flat or smooth by hammering. **2** work out details of (a plan etc.) laboriously. **3** play (a tune, esp. on the piano) loudly or clumsily. □ **hammering** *n.* (esp. in sense 4 of *v.*). [Old English]

hammer and sickle *n.* symbols of the industrial worker and peasant used as an emblem of the former USSR and international communism.

hammer and tongs *adv. colloq.* with great vigour and commotion.

hammerhead *n.* shark with a flattened head and with eyes in lateral extensions of it.

hammerlock *n. Wrestling* hold in which the arm is twisted and bent behind the back.

hammer-toe *n.* toe bent permanently downwards.

hammock /'hæmək/ *n.* bed of canvas or rope network suspended by cords at the ends. [Spanish from Carib]

hammy *adj.* (**-ier**, **-iest**) *colloq.* over-theatrical.

hamper[1] *n.* large basket, usu. with a hinged lid and containing food. [French *hanap* goblet]

hamper[2] *v.* prevent the free movement of; hinder. [origin unknown]

hamster *n.* mouselike rodent with a short tail and large cheek-pouches for storing food. [German]

hamstring *–n.* **1** each of five tendons at the back of the knee. **2** great tendon at the back of the hock in quadrupeds. *–v.* (*past* and *past part.* **-strung** or **-stringed**) **1** cripple by cutting the hamstrings of (a person or animal). **2** impair the activity or efficiency of.

hand *–n.* **1 a** end part of the human arm beyond the wrist. **b** (in other primates) end part of a forelimb. **2 a** (often in *pl.*) control, management, custody, disposal (*is in good hands*). **b** agency or influence (*suffered at their hands*). **c** share in an action; active support (*had a hand in it*; *give me a hand*). **3** thing like a hand, esp. the pointer of a clock. **4** right or left side or direction relative to a person or thing. **5 a** skill (*has a hand for making pastry*). **b** person skilful in some respect. **6** person who does or makes something, esp. distinctively (*picture by the same hand*). **7** person's writing or its style. **8** person etc. as a source (*at first hand*). **9** pledge of marriage. **10** manual worker, esp. at a factory or farm; member of a ship's crew. **11 a** playing-cards dealt to a player. **b** round of play. **12** *colloq.* burst of applause. **13** unit of measure of a horse's height, 4 inches (10.16 cm). **14** forehock of pork. **15** (*attrib.*) **a** operated by or held in the hand (*hand-drill*). **b** done by hand, not machine (*hand-knitted*). *–v.* **1** (foll. by *in*, *to*, *over*, etc.) deliver; transfer by hand or otherwise. **2** *colloq.* give away too readily (*handed them the advantage*). □ **all hands** entire crew or workforce. **at hand 1** close by. **2** about to happen. **by hand 1** by a person, not a machine. **2** delivered privately, not by post. **from hand to mouth** satisfying only one's immediate needs. **get** (or **have** or **keep**) **one's hand in** become (or be or remain) in practice. **hand down 1** pass ownership or use of to a later generation etc. **2 a** transmit (a decision) from a higher court etc. **b** *US* express (an opinion or verdict). **hand it to** *colloq.* award deserved praise to. **hand on** pass (a thing) to the next in a series. **hand out 1** serve, distribute. **2** award, allocate (*handed out stiff penalties*). **hand round** serve, distribute. **hands down** with no difficulty; completely. **in hand 1** receiving attention. **2** in reserve. **3** under control. **on hand** available. **on the one** (or **the other**) **hand** from one (or another) point of view. **out of hand 1** out of control. **2** peremptorily (*refused out of hand*). **put** (or **set**) **one's hand to** start work on; engage in. **to hand** within easy reach; available. **turn one's hand to** undertake (as a new activity). □ **-handed** *adj.* (in *comb.*). [Old English]

handbag *n.* small bag carried esp. by a woman.

handball *n.* **1** game with a ball thrown by hand among players or against a wall. **2** *Football* intentional touching of the ball, constituting a foul.

handbell *n.* small bell for ringing by hand, esp. one of a set.

handbill *n.* printed notice distributed by hand.

handbook *n.* short manual or guide-book.

handbrake *n.* brake operated by hand.

h. & c. *abbr.* hot and cold (water).

handcart *n.* small cart pushed or drawn by hand.

handclap *n.* clapping of the hands.

handcraft *–n.* = HANDICRAFT. *–v.* make by handicraft.

handcuff *–n.* each of a pair of linked metal rings for securing a prisoner's wrist(s). *–v.* put handcuffs on.

handful *n.* (*pl.* **-s**) **1** quantity that fills the hand. **2** small number or amount. **3** *colloq.* troublesome person or task.

hand-grenade see GRENADE.

handgun *n.* small firearm held in and fired with one hand.

handhold *n.* something for the hand to grip on (in climbing etc.).

handicap /'hændɪ,kæp/ *–n.* **1** physical or mental disability. **2** thing that makes progress or success difficult. **3 a** disadvantage imposed on a superior competitor to make chances more equal. **b** race etc. in which this is imposed. **4** number of strokes by which a golfer normally exceeds par for a course. *–v.* (**-pp-**) **1** impose a handicap on. **2** place at a disadvantage. [*hand i'* (= in) *cap* describing a kind of sporting lottery]

handicapped *adj.* suffering from a physical or mental disability.

handicraft /'hændɪ,krɑːft/ *n.* work requiring manual and artistic skill. [from earlier HANDCRAFT]

hand in glove *adj.* in collusion or association.

hand in hand *adv.* **1** in close association (*power and money go hand in hand*). **2** (**hand-in-hand**) holding hands.

handiwork /'hændɪ,wɜːk/ *n.* work done or a thing made by hand, or by a particular person. [Old English]

handkerchief /'hæŋkətʃɪf/ *n.* (*pl.* **-s** or **-chieves** /-,tʃiːvz/) square of cloth for wiping one's nose etc.

handle /'hænd(ə)l/ *–n.* **1** part by which a thing is held, carried, or controlled. **2** fact that may be taken advantage of (*gave a handle to his critics*). **3** *colloq.* personal title. *–v.* (**-ling**) **1** touch, feel, operate, or move with the hands. **2** manage, deal with (*can handle people*). **3** deal in (goods). **4** treat (a subject). [Old English: related to HAND]

handlebar *n.* (usu. in *pl.*) steering-bar of a bicycle etc.

handlebar moustache *n.* thick moustache with curved ends.

handler *n.* **1** person who handles or deals in something. **2** person who trains and looks after an animal (esp. a police dog).

handmade *adj.* made by hand (as opposed to machine).

handmaid *n.* (also **handmaiden**) *archaic* female servant.

hand-me-down *n.* article of clothing etc. passed on from another person.

hand-out *n.* **1** thing given free to a needy person. **2** statement given to the press etc.; notes given out in a class etc.

hand-over *n.* handing over.

hand-over-fist *adv. colloq.* with rapid progress.

hand-pick *v.* choose carefully or personally.

handrail *n.* narrow rail for holding as a support.

handsaw *n.* saw worked by one hand.

handset *n.* telephone mouthpiece and earpiece as one unit.

handshake *n.* clasping of a person's hand as a greeting etc.

hands off *–int.* warning not to touch or interfere with something. *–adj.* & *adv.* (also **hands-off**) not requiring the manual use of controls.

handsome /'hænsəm/ *adj.* (**handsomer, handsomest**) **1** (usu. of a man) good-looking. **2** (of an object) imposing, attractive. **3 a** generous, liberal (*handsome present*). **b** (of a price, fortune, etc.) considerable. □ **handsomely** *adv.*

hands on (also **hands-on**) *–adj.* & *adv.* of or requiring personal operation at a keyboard. *–attrib. adj.* practical rather than theoretical (*lacks hands-on experience*).

handspring *n.* gymnastic feat consisting of a handstand, somersaulting, and landing in a standing position.

handstand *n.* supporting oneself on one's hands with one's feet in the air.

hand-to-hand *adj.* (of fighting) at close quarters.

handwork *n.* work done with the hands. □ **handworked** *adj.*

handwriting *n.* **1** writing done with a pen, pencil, etc. **2** person's particular style of this. □ **handwritten** *adj.*

handy *adj.* (**-ier, -iest**) **1** convenient to handle or use; useful. **2** ready to hand. **3** clever with the hands. □ **handily** *adv.* **handiness** *n.*

handyman *n.* person able to do occasional repairs etc.; odd-job man.

hang –*v.* (*past* and *past part.* **hung** except in sense 7) **1 a** secure or cause to be supported from above, esp. with the lower part free. **b** (foll. by *up*, *on*, *on to*, etc.) attach by suspending from the top. **2** set up (a door etc.) on hinges. **3** place (a picture) on a wall or in an exhibition. **4** attach (wallpaper) to a wall. **5** (foll. by *on*) *colloq.* blame (a thing) on (a person) (*can't hang that on me*). **6** (foll. by *with*) decorate by suspending pictures etc. (*hall hung with tapestries*). **7** (*past* and *past part.* **hanged**) **a** suspend or be suspended by the neck with a noosed rope until dead, esp. as a form of capital punishment. **b** as a mild oath (*hang the expense*). **8** let droop (*hang one's head*). **9** suspend (meat or game) from a hook and leave until dry, tender, or high. **10** be or remain hung (in various senses). **11** remain static in the air. **12** (often foll. by *over*) be present or imminent, esp. oppressively or threateningly (*a hush hung over the room*). **13** (foll. by *on*) **a** be contingent or dependent on (*everything hangs on his reply*). **b** listen closely to (*hangs on my every word*). –*n.* way a thing hangs or falls. □ **get the hang of** *colloq.* understand the technique or meaning of. **hang about** (or **around**) **1 a** stand about or spend time aimlessly; not move away. **b** linger near (a person or place). **2** (often foll. by *with*) *colloq.* associate with. **hang back** show reluctance to act or move. **hang fire** be slow in taking action or in progressing. **hang heavily** (or **heavy**) (of time) seem to pass slowly. **hang in** *US colloq.* **1** persist, persevere. **2** linger. **hang on 1** (often foll. by *to*) continue to hold or grasp. **2** (foll. by *to*) retain; fail to give back. **3** *colloq.* **a** wait for a short time. **b** (in telephoning) not ring off during a pause in the conversation. **4** *colloq.* continue; persevere. **hang out 1** suspend from a window, clothes-line, etc. **2 a** protrude downwards (*shirt hanging out*). **b** (foll. by *of*) lean out of (a window etc.). **3** *slang* frequent or live in a place. **hang together 1** make sense. **2** remain associated. **hang up 1** hang from a hook etc. **2** (often foll. by *on*) end a telephone conversation by replacing the receiver (*he hung up on me*). **3** (usu. in *passive*, foll. by *on*) *slang* be a psychological problem or obsession for (*is hung up on her father*). **not care** (or **give**) **a hang** *colloq.* not care at all. [Old English]

hangar /ˈhæŋə(r)/ *n.* building for housing aircraft etc. [French]

hangdog *adj.* shamefaced.

hanger *n.* **1** person or thing that hangs. **2** (in full **coat-hanger**) shaped piece of wood etc. for hanging clothes on.

hanger-on *n.* (*pl.* **hangers-on**) follower or dependant, esp. an unwelcome one.

hang-glider *n.* glider with a fabric wing on a light frame, from which the operator is suspended. □ **hang-glide** *v.* **hang-gliding** *n.*

hanging *n.* **1** execution by suspending by the neck. **2** (usu. in *pl.*) draperies hung on a wall etc.

hangman *n.* **1** executioner who hangs condemned persons. **2** word-game for two players, with failed guesses recorded by drawing a representation of a gallows.

hangnail *n.* = AGNAIL.

hang-out *n. slang* place frequented by a person; haunt.

hangover *n.* **1** severe headache etc. from drinking too much alcohol. **2** survival from the past.

hang-up *n. slang* emotional problem or inhibition.

hank *n.* coil or skein of wool or thread etc. [Old Norse]

hanker *v.* (foll. by *for*, *after*, or *to* + infin.) long for; crave. □ **hankering** *n.* [from obsolete *hank*]

hanky /ˈhæŋkɪ/ *n.* (also **hankie**) (*pl.* **-ies**) *colloq.* handkerchief. [abbreviation]

hanky-panky /ˌhæŋkɪˈpæŋkɪ/ *n. slang* **1** naughtiness, esp. sexual. **2** double-dealing; trickery. [origin unknown]

Hanoverian /ˌhænəˈvɪərɪən/ *adj.* of British sovereigns from George I to Victoria. [*Hanover* in Germany]

Hansard /ˈhænsɑːd/ *n.* official verbatim record of debates in the British Parliament. [*Hansard*, name of its first printer]

Hansen's disease /ˈhæns(ə)nz/ *n.* leprosy. [*Hansen*, name of a physician]

hansom /ˈhænsəm/ *n.* (in full **hansom cab**) *hist.* two-wheeled horse-drawn cab. [*Hansom*, name of an architect]

Hanukkah /ˈhɑːnəkə/ *n.* Jewish festival of lights, commemorating the purification of the Temple in 165 BC. [Hebrew *ḥănukkāh* consecration]

haphazard /hæpˈhæzəd/ *adj.* done etc. by chance; random. □ **haphazardly** *adv.* [archaic *hap* chance, luck, from Old Norse *happ*]

hapless *adj.* unlucky.

haploid /ˈhæplɔɪd/ *adj.* (of an organism or cell) with a single set of chromosomes. [Greek *haplous* single, *eidos* form]

happen /ˈhæpən/ *v.* **1** occur (by chance or otherwise). **2** (foll. by *to* + infin.) have the (good or bad) fortune to (*I happened to meet her*). **3** (foll. by *to*) be the (esp. unwelcome) fate or experience of (*what happened to you?*). **4** (foll. by *on*) encounter or discover by chance. □ **as it happens** in fact; in reality. [related to HAPHAZARD]

happening *n.* **1** event. **2** improvised or spontaneous theatrical etc. performance.

happy /'hæpɪ/ *adj.* (**-ier, -iest**) **1** feeling or showing pleasure or contentment. **2 a** fortunate; characterized by happiness. **b** (of words, behaviour, etc.) apt, pleasing. □ **happily** *adv.* **happiness** *n.*

happy-go-lucky *adj.* cheerfully casual.

happy hour *n.* time of the day when goods, esp. drinks, are sold at reduced prices.

happy medium *n.* compromise; avoidance of extremes.

hara-kiri /ˌhærə'kɪrɪ/ *n.* ritual suicide by disembowelment with a sword, formerly practised by samurai to avoid dishonour. [Japanese *hara* belly, *kiri* cutting]

harangue /hə'ræŋ/ *–n.* lengthy and earnest speech. *–v.* (**-guing**) make a harangue to; lecture. [French *arenge* from medieval Latin]

harass /'hærəs, hə'ræs/ *v.* **1** trouble and annoy continually. **2** make repeated attacks on. □ **harassment** *n.* [French]

■ **Usage** The second pronunciation given, with the stress on the second syllable, is common, but is considered incorrect by some people.

harbinger /'hɑ:bɪndʒə(r)/ *n.* **1** person or thing that announces or signals the approach of another. **2** forerunner. [Germanic: related to HARBOUR]

harbour /'hɑ:bə(r)/ (*US* **harbor**) *–n.* **1** place of shelter for ships. **2** shelter; refuge. *–v.* **1** give shelter to (esp. a criminal). **2** keep in one's mind (esp. resentment etc.). [Old English, = army shelter]

harbour-master *n.* official in charge of a harbour.

hard *–adj.* **1** (of a substance etc.) firm and solid. **2 a** difficult to understand, explain, or accomplish. **b** (foll. by *to* + infin.) not easy to (*hard to please*). **3** difficult to bear (*a hard life*). **4** unfeeling; severely critical. **5** (of a season or the weather) severe. **6** unpleasant to the senses, harsh (*hard colours*). **7 a** strenuous, enthusiastic, intense (*a hard worker*). **b** severe, uncompromising (*a hard bargain*). **c** *Polit.* extreme; most radical (*the hard right*). **8 a** (of liquor) strongly alcoholic. **b** (of drugs) potent and addictive. **c** (of pornography) highly obscene. **9** (of water) containing mineral salts that make lathering difficult. **10** established; not disputable (*hard facts*). **11** (of currency, prices, etc.) high; not likely to fall in value. **12** (of a consonant) guttural (as *c* in *cat*, *g* in *go*). *–adv.* strenuously, intensely, copiously (*try hard*; *raining hard*). □ **be hard on 1** be difficult for. **2** be severe in one's treatment or criticism of. **3** be unpleasant to (the senses). **be hard put to it** (usu. foll. by *to* + infin.) find it difficult. **hard by** close by. **hard on** (or **upon**) close to in pursuit etc. □ **hardish** *adj.* **hardness** *n.* [Old English]

hard and fast *adj.* (of a rule or distinction) definite, unalterable, strict.

hardback *–adj.* bound in boards covered with cloth etc. *–n.* hardback book.

hardbitten *adj. colloq.* tough and cynical; hard-headed.

hardboard *n.* stiff board made of compressed and treated wood pulp.

hard-boiled *adj.* **1** (of an egg) boiled until the white and yolk are solid. **2** *colloq.* (of a person) tough, shrewd.

hard cash *n.* negotiable coins and banknotes.

hard copy *n.* material printed by a computer on paper.

hardcore *n.* solid material, esp. rubble, as road-foundation.

hard core *n.* **1** irreducible nucleus. **2** *colloq.* **a** the most committed members of a society etc. **b** conservative or reactionary minority (see also HARDCORE).

hard-core *adj.* **1** forming a nucleus. **2** blatant, uncompromising. **3** (of pornography) explicit, obscene.

hard disk *n. Computing* large-capacity rigid usu. magnetic storage disk.

hard-done-by *adj.* unfairly treated.

harden *v.* **1** make or become hard or harder. **2** become, or make (one's attitude etc.), less sympathetic. **3** (of prices etc.) cease to fall or fluctuate. □ **harden off** inure (a plant) to the cold by gradually increasing its exposure.

hardening of the arteries *n.* = ARTERIOSCLEROSIS.

hard-headed *adj.* practical; not sentimental. □ **hard-headedness** *n.*

hard-hearted *adj.* unfeeling. □ **hard-heartedness** *n.*

hardihood /'hɑ:dɪˌhʊd/ *n.* boldness, daring.

hard labour *n.* heavy manual work as a punishment, esp. in a prison.

hard line *n.* unyielding adherence to a policy. □ **hard-liner** *n.*

hard luck *n.* worse fortune than one deserves.

hardly *adv.* **1** scarcely; only just (*hardly knew me*). **2** only with difficulty (*can hardly see*). **3** surely not (*can hardly have realised*). □ **hardly any**

almost no; almost none. **hardly ever** very seldom.

hard-nosed *adj. colloq.* realistic, uncompromising.

hard of hearing *adj.* somewhat deaf.

hard-on *n. coarse slang* erection of the penis.

hard pad *n.* form of distemper in dogs etc.

hard palate *n.* front part of the palate.

hard-pressed *adj.* **1** closely pursued. **2** burdened with urgent business.

hard roe see ROE[1].

hard sell *n.* aggressive salesmanship.

hardship *n.* **1** severe suffering or privation. **2** circumstance causing this.

hard shoulder *n.* hard surface alongside a motorway for stopping on in an emergency.

hard tack *n. Naut.* ship's biscuit.

hardtop *n.* car with a rigid (usu. detachable) roof.

hard up *adj.* short of money.

hardware *n.* **1** tools and household articles of metal etc. **2** heavy machinery or armaments. **3** mechanical and electronic components of a computer etc.

hard-wearing *adj.* able to stand much wear.

hardwood *n.* wood from a deciduous broad-leaved tree.

hard-working *adj.* diligent.

hardy /'hɑ:dɪ/ *adj.* (**-ier, -iest**) **1** robust; capable of enduring difficult conditions. **2** (of a plant) able to grow in the open air all year. □ **hardiness** *n.* [French *hardi* made bold]

hardy annual *n.* annual plant that may be sown in the open.

hare *–n.* mammal like a large rabbit, with long ears, short tail, and long hind legs. *–v.* (**-ring**) run rapidly. [Old English]

harebell *n.* plant with pale-blue bell-shaped flowers.

hare-brained *adj.* rash, wild.

harelip *n.* often *offens.* = CLEFT LIP.

harem /'hɑ:ri:m/ *n.* **1** women of a Muslim household. **2** their quarters. [Arabic, = sanctuary]

haricot /'hærɪˌkəʊ/ *n.* (in full **haricot bean**) variety of French bean with small white seeds dried and used as a vegetable. [French]

hark *v.* (usu. in *imper.*) *archaic* listen attentively. □ **hark back** revert to earlier topic. [Old English]

harlequin /'hɑ:lɪkwɪn/ *–n.* (**Harlequin**) name of a mute character in pantomime, usu. masked and dressed in a diamond-patterned costume. *–attrib. adj.* in varied colours. [French]

harlequinade /ˌhɑ:lɪkwɪ'neɪd/ *n.* **1** part of a pantomime featuring Harlequin. **2** piece of buffoonery.

harlot /'hɑ:lət/ *n. archaic* prostitute. □ **harlotry** *n.* [French, = knave]

harm *–n.* hurt, damage. *–v.* cause harm to. □ **out of harm's way** in safety. [Old English]

harmful *adj.* causing or likely to cause harm. □ **harmfully** *adv.* **harmfulness** *n.*

harmless *adj.* **1** not able or likely to cause harm. **2** inoffensive. □ **harmlessly** *adv.* **harmlessness** *n.*

harmonic /hɑ:'mɒnɪk/ *–adj.* of or relating to harmony; harmonious. *–n. Mus.* overtone accompanying (and forming a note with) a fundamental at a fixed interval. □ **harmonically** *adv.*

harmonica *n.* small rectangular musical instrument played by blowing and sucking air through it.

harmonious /hɑ:'məʊnɪəs/ *adj.* **1** sweet-sounding; tuneful. **2** forming a pleasing or consistent whole. **3** free from disagreement or dissent. □ **harmoniously** *adv.*

harmonium /hɑ:'məʊnɪəm/ *n.* keyboard instrument in which the notes are produced by air driven through metal reeds by foot-operated bellows. [Latin: related to HARMONY]

harmonize /'hɑ:məˌnaɪz/ *v.* (also **-ise**) (**-zing** or **-sing**) **1** add notes to (a melody) to produce harmony. **2** bring into or be in harmony. **3** make or form a pleasing or consistent whole. □ **harmonization** /-'zeɪʃ(ə)n/ *n.*

harmony /'hɑ:mənɪ/ *n.* (*pl.* **-ies**) **1** combination of simultaneously sounded musical notes to produce chords and chord progressions, esp. as creating a pleasing effect. **2 a** apt or aesthetic arrangement of parts. **b** pleasing effect of this. **3** agreement, concord. □ **in harmony 1** in agreement. **2** (of singing etc.) producing chords; not discordant. [Greek *harmonia* joining]

harness /'hɑ:nɪs/ *–n.* **1** equipment of straps etc. by which a horse is fastened to a cart etc. and controlled. **2** similar arrangement for fastening a thing to a person's body. *–v.* **1 a** put a harness on. **b** (foll. by *to*) attach by harness to. **2** make use of (natural resources), esp. to produce energy. □ **in harness** in the routine of daily work. [French *harneis* military equipment]

harp *–n.* large upright stringed instrument plucked with the fingers. *–v.* (foll. by *on, on about*) talk repeatedly and tediously about. □ **harpist** *n.* [Old English]

harpoon /hɑ:'pu:n/ *–n.* barbed spear-like missile with a rope attached, for catching whales etc. *–v.* spear with a harpoon. [Greek *harpē* sickle]

harpsichord /'hɑ:psɪ,kɔ:d/ *n.* keyboard instrument with horizontal strings plucked mechanically. □ **harpsichordist** *n.* [Latin *harpa* harp, *chorda* string]

harpy /'hɑ:pɪ/ *n.* (*pl.* **-ies**) **1** mythological monster with a woman's head and body and a bird's wings and claws. **2** grasping unscrupulous person. [Greek *harpuiai* snatchers]

harridan /'hærɪd(ə)n/ *n.* bad-tempered old woman. [origin uncertain]

harrier /'hærɪə(r)/ *n.* **1** hound used for hunting hares. **2** group of cross-country runners. **3** hawklike bird of prey. [from HARE, HARRY]

harrow /'hærəʊ/ *–n.* heavy frame with iron teeth dragged over ploughed land to break up clods etc. *–v.* **1** draw a harrow over (land). **2** (usu. as **harrowing** *adj.*) distress greatly. [Old Norse *hervi*]

harry /'hærɪ/ *v.* (**-ies, -ied**) **1** ravage or despoil. **2** harass. [Old English]

harsh *adj.* **1** unpleasantly rough or sharp, esp. to the senses. **2** severe, cruel. □ **harshen** *v.* **harshly** *adv.* **harshness** *n.* [Low German]

hart *n.* (*pl.* same or **-s**) male of the (esp. red) deer, esp. after its 5th year. [Old English]

hartebeest /'hɑ:tɪ,bi:st/ *n.* large African antelope with curving horns. [Afrikaans]

harum-scarum /,heərəm'skeərəm/ *colloq. –adj.* wild and reckless. *–n.* such a person. [rhyming formation on HARE, SCARE]

harvest /'hɑ:vɪst/ *–n.* **1 a** process of gathering in crops etc. **b** season of this. **2** season's yield. **3** product of any action. *–v.* gather as harvest, reap. [Old English]

harvester *n.* **1** reaper. **2** reaping-machine, esp. with sheaf-binding.

harvest festival *n.* Christian thanksgiving service for the harvest.

harvest moon *n.* full moon nearest to the autumn equinox (22 or 23 Sept.).

harvest mouse *n.* small mouse nesting in the stalks of growing grain.

has *3rd sing. present* of HAVE.

has-been *n. colloq.* person or thing of declined importance.

hash[1] *–n.* **1** dish of cooked meat cut into small pieces and reheated. **2 a** mixture; jumble. **b** mess. **3** recycled material. *–v.* (often foll. by *up*) recycle (old material). □ **make a hash of** *colloq.* make a mess of; bungle. **settle a person's hash** *colloq.* deal with and subdue a person. [French *hacher* cut up]

hash[2] *n. colloq.* hashish. [abbreviation]

hashish /'hæʃɪʃ/ *n.* resinous product of hemp, smoked or chewed as a narcotic. [Arabic]

haslet /'hæzlɪt/ *n.* pieces of (esp. pig's) offal cooked together, usu. as a meat loaf. [French *hastelet*]

hasn't /'hæz(ə)nt/ *contr.* has not.

hasp /hɑ:sp/ *n.* hinged metal clasp fitting over a staple and secured by a padlock. [Old English]

hassle /'hæs(ə)l/ *colloq. –n.* trouble; problem; argument. *–v.* (**-ling**) harass, annoy. [originally a dial. word]

hassock /'hæsək/ *n.* thick firm cushion for kneeling on. [Old English]

haste /heɪst/ *–n.* urgency of movement or action; excessive hurry. *–v.* (**-ting**) *archaic* = HASTEN 1. □ **in haste** quickly, hurriedly. **make haste** hurry; be quick. [French from Germanic]

hasten /'heɪs(ə)n/ *v.* **1** make haste; hurry. **2** cause to occur or be ready or be done sooner.

hasty /'heɪstɪ/ *adj.* (**-ier, -iest**) **1** hurried; acting too quickly. **2** said, made, or done too quickly or too soon; rash. □ **hastily** *adv.* **hastiness** *n.*

hat *n.* **1** (esp. outdoor) covering for the head. **2** *colloq.* person's present capacity (*wearing his managerial hat*). □ **keep it under one's hat** *colloq.* keep it secret. **pass the hat round** collect contributions of money. **take one's hat off to** *colloq.* acknowledge admiration for. [Old English]

hatband *n.* band of ribbon etc. round a hat above the brim.

hatbox *n.* box to hold a hat, esp. for travelling.

hatch[1] *n.* **1** opening in a wall between a kitchen and dining-room for serving food. **2** opening or door in an aircraft etc. **3 a** = HATCHWAY. **b** cover for this. [Old English]

hatch[2] *–v.* **1 a** (often foll. by *out*) (of a young bird or fish etc.) emerge from the egg. **b** (of an egg) produce a young animal. **2** incubate (an egg). **3** (also foll. by *up*) devise (a plot etc.). *–n.* **1** act of hatching. **2** brood hatched. [earlier *hacche*, from Germanic]

hatch[3] *v.* mark with close parallel lines. □ **hatching** *n.* [French *hacher*: related to HASH[1]]

hatchback *n.* car with a sloping back hinged at the top to form a door.

hatchet /'hætʃɪt/ *n.* light short-handled axe. [French *hachette*]

hatchet man *n. colloq.* person hired to kill, dismiss, or otherwise harm another.

hatchway *n.* opening in a ship's deck for raising and lowering cargo.

hate –*v.* (**-ting**) **1** dislike intensely. **2** *colloq.* **a** dislike. **b** be reluctant (to do something) (*I hate to disturb you*; *I hate fighting*). –*n.* **1** hatred. **2** *colloq.* hated person or thing. [Old English]

hateful *adj.* arousing hatred.

hatpin *n.* long pin for securing a hat to the hair.

hatred /'heɪtrɪd/ *n.* extreme dislike or ill will.

hatstand *n.* stand with hooks for hanging hats etc. on.

hatter *n.* maker or seller of hats.

hat trick *n.* **1** *Cricket* taking of three wickets by the same bowler with three successive balls. **2** three consecutive successes etc.

haughty /'hɔ:tɪ/ *adj.* (**-ier, -iest**) arrogant and disdainful. □ **haughtily** *adv.* **haughtiness** *n.* [*haught, haut* from French, = high]

haul /hɔ:l/ –*v.* **1** pull or drag forcibly. **2** transport by lorry, cart, etc. **3** turn a ship's course. **4** *colloq.* (usu. foll. by *up*) bring for reprimand or trial. –*n.* **1** hauling. **2** amount gained or acquired. **3** distance to be traversed (*a short haul*). □ **haul over the coals** see COAL. [French *haler* from Old Norse *hala*]

haulage *n.* **1** commercial transport of goods. **2** charge for this.

haulier /'hɔ:lɪə(r)/ *n.* person or firm engaged in the transport of goods.

haulm /hɔ:m/ *n.* (also **halm**) **1** stalk or stem. **2** stalks or stems of peas, beans, etc., collectively. [Old English]

haunch /hɔ:ntʃ/ *n.* **1** fleshy part of the buttock with the thigh. **2** leg and loin of a deer etc. as food. [French from Germanic]

haunt /hɔ:nt/ –*v.* **1** (of a ghost) visit (a place) regularly. **2** frequent (a place). **3** linger in the mind of. –*n.* place frequented by a person or animal. [French from Germanic]

haunting *adj.* (of a memory, melody, etc.) tending to linger in the mind; poignant, evocative.

haute couture /,əʊt ku:'tjʊə(r)/ *n.* high fashion; leading fashion houses or their products. [French]

haute cuisine /,əʊt kwɪ'zi:n/ *n.* high-class cookery. [French]

hauteur /əʊ'tɜ:(r)/ *n.* haughtiness. [French]

have /hæv, həv/ –*v.* (**-ving**; *3rd sing. present* **has** /hæz/; *past* and *past part.* **had**) **1** as an auxiliary verb with *past part.* or *ellipt.*, to form the perfect, pluperfect, and future perfect tenses, and the conditional mood (*has, had, will have, seen*; *had I known, I would have gone*; *yes, I have*). **2** own or be able to use; be provided with (*has a car*; *had no time*). **3** hold in a certain relationship (*has a sister*; *had no equals*). **4** contain as a part or quality (*box has a lid*; *has big eyes*). **5 a** experience (*had a good time, a shock, a pain*). **b** be subjected to a specified state (*had my car stolen*; *book has a page missing*). **c** cause (a person or thing) to be in a particular state or take particular action (*had him sacked*; *had us worried*; *had my hair cut*; *had a copy made*; *had them to stay*). **6 a** engage in (an activity) (*have an argument, sex*). **b** hold (a meeting, party, etc.). **7** eat or drink (*had a beer*). **8** (usu. in *neg.*) accept or tolerate; permit to (*I won't have it*; *won't have you say that*). **9 a** feel (*have no doubt*; *has nothing against me*). **b** show (mercy, pity, etc.). **c** (foll. by *to* + infin.) show by action that one is influenced by (a feeling, quality, etc.) (*have the sense to stop*). **10 a** give birth to (offspring). **b** conceive mentally (an idea etc.). **11** receive, obtain (*had a letter from him*; *not a ticket to be had*). **12** be burdened with or committed to (*has a job to do*). **13 a** have obtained (a qualification) (*has six O levels*). **b** know (a language) (*has no Latin*). **14** *slang* **a** get the better of (*I had him there*). **b** (usu. in *passive*) cheat, deceive (*you were had*). **15** *coarse slang* have sexual intercourse with. –*n.* **1** (usu. in *pl.*) *colloq.* person with wealth or resources. **2** *slang* swindle. □ **had best** see BEST. **had better** see BETTER. **have got to** *colloq.* = *have to*. **have had it** *colloq.* **1** have missed one's chance. **2** have passed one's prime. **3** have been killed, defeated, etc. **have it 1** (foll. by *that*) maintain that. **2** win a decision in a vote etc. **3** *colloq.* have found the answer etc. **have it away** (or **off**) *coarse slang* have sexual intercourse. **have it in for** *colloq.* be hostile or ill-disposed towards. **have it out** (often foll. by *with*) *colloq.* attempt to settle a dispute by argument. **have on 1** wear (clothes). **2** have (an engagement). **3** *colloq.* tease, hoax. **have to** be obliged to, must. **have up** *colloq.* bring (a person) before a judge, interviewer, etc. [Old English]

haven /'heɪv(ə)n/ *n.* **1** refuge. **2** harbour, port. [Old English]

have-not *n.* (usu. in *pl.*) *colloq.* person lacking wealth or resources.

haven't /'hæv(ə)nt/ *contr.* have not.

haver /'heɪvə(r)/ *v.* **1** vacillate, hesitate. **2** *dial.* talk foolishly. [origin unknown]

haversack /ˈhævə,sæk/ *n.* stout canvas bag carried on the back or over the shoulder. [German *Habersack*, = oats-sack]

havoc /ˈhævək/ *n.* widespread destruction; great disorder. [French *havo(t)*]

haw[1] *n.* hawthorn berry. [Old English]

haw[2] see HUM.

hawfinch *n.* large finch with a thick beak for cracking seeds. [from HAW[1], FINCH]

hawk[1] *–n.* **1** bird of prey with a curved beak, rounded short wings, and a long tail. **2** *Polit.* person who advocates aggressive policies. *–v.* hunt with a hawk. □ **hawkish** *adj.* [Old English]

hawk[2] *v.* carry about or offer (goods) for sale. [back-formation from HAWKER]

hawk[3] *v.* **1** clear the throat noisily. **2** (foll. by *up*) bring (phlegm etc.) up from the throat. [imitative]

hawker *n.* person who travels about selling goods. [Low German or Dutch]

hawk-eyed *adj.* keen-sighted.

hawser /ˈhɔːzə(r)/ *n.* thick rope or cable for mooring or towing a ship. [French, *haucier* hoist, from Latin *altus* high]

hawthorn *n.* thorny shrub with small dark-red berries. [related to HAW[1]]

hay *n.* grass mown and dried for fodder. □ **make hay (while the sun shines)** seize opportunities. [Old English]

haycock *n.* conical heap of hay.

hay fever *n.* allergy with asthmatic symptoms etc., caused by pollen or dust.

haymaking *n.* mowing grass and spreading it to dry. □ **haymaker** *n.*

haystack *n.* (also **hayrick**) packed pile of hay with a pointed or ridged top.

haywire *adj. colloq.* badly disorganized, out of control.

hazard /ˈhæzəd/ *–n.* **1** danger or risk. **2** source of this. **3** *Golf* obstacle, e.g. a bunker. *–v.* **1** venture (*hazard a guess*). **2** risk. [Arabic *az-zahr* chance, luck]

hazardous *adj.* risky.

haze *n.* **1** thin atmospheric vapour. **2** mental obscurity or confusion. [back-formation from HAZY]

hazel /ˈheɪz(ə)l/ *n.* **1** hedgerow shrub bearing round brown edible nuts. **2** greenish-brown. [Old English]

hazelnut *n.* nut of the hazel.

hazy *adj.* (**-ier, -iest**) **1** misty. **2** vague, indistinct. **3** confused, uncertain. □ **hazily** *adv.* **haziness** *n.* [origin unknown]

HB *abbr.* (of pencil-lead) hard black.

H-bomb /ˈeɪtʃbɒm/ *n.* = HYDROGEN BOMB. [from H[3]]

HCF *abbr.* highest common factor.

HE *abbr.* **1** His or Her Excellency. **2** His Eminence. **3** high explosive.

He *symb.* helium.

he /hiː/ *–pron.* (*obj.* **him**; *poss.* **his**; *pl.* **they**) **1** the man, boy, or male animal previously named or in question. **2** person etc. of unspecified sex (*if anyone comes he will have to wait*; *he who hesitates*). *–n.* **1** male; man. **2** (in *comb.*) male (*he-goat*). [Old English]

head /hed/ *–n.* **1** upper part of the human body, or foremost or upper part of an animal's body, containing the brain, mouth, and sense-organs. **2 a** seat of intellect (*use your head*). **b** mental aptitude or tolerance (*a good head for business*; *no head for heights*). **3** thing like a head in form or position, esp.: **a** the operative part of a tool. **b** the top of a nail. **c** the leaves or flowers at the top of a stem. **d** foam on the top of a glass of beer etc. **4 a** person in charge, esp. the principal teacher of a school. **b** position of command. **5** front part of a queue etc. **6** upper end of a table or bed etc. **7** top or highest part of a page, stairs, etc. **8 a** individual person as a unit (*£10 per head*). **b** (*pl.* same) individual animal as a unit (*20 head*). **9 a** side of a coin bearing the image of a head. **b** (usu. in *pl.*) this as a choice when tossing a coin. **10 a** source of a river etc. **b** end of a lake at which a river enters it. **11** height or length of a head as a measure. **12** part of a machine in contact with or very close to what is being worked on, esp.: **a** the part of a tape recorder that touches the moving tape and converts signals. **b** the part of a record-player that holds the playing cartridge and stylus. **13** (usu. in phr. **come to a head**) climax, crisis. **14 a** confined body of water or steam in an engine etc. **b** pressure exerted by this. **15** promontory (esp. in place-names) (*Beachy Head*). **16** heading or headline. **17** fully developed top of a boil etc. **18** *colloq.* headache. **19** (*attrib.*) chief, principal. *–v.* **1** be at the head or front of. **2** be in charge of. **3** provide with a head or heading. **4** (often foll. by *for*) face, move, or direct in a specified direction (*is heading for trouble*). **5** hit (a ball etc.) with the head. □ **above** (or **over**) **one's head** beyond one's understanding. **come to a head** reach a crisis. **get it into one's head** (foll. by *that*) **1** adopt a mistaken idea. **2** form a definite plan. **give a person his** (or **her**) **head** allow a person to act freely. **go to one's head 1** make one slightly drunk. **2** make one conceited. **head off 1** get ahead of so as to intercept and turn aside. **2** forestall. **keep** (or **lose**) **one's head** remain (or fail to remain) calm. **off one's head** *slang* crazy. **off the top of one's head** *colloq.* impromptu. **on one's** (or **one's own**) **head** as one's own responsibility.

out of one's head *slang* crazy. **over one's head 1** beyond one's understanding. **2** without one's rightful knowledge or involvement, esp. of action taken by a subordinate consulting one's own superior. **3** with disregard for one's own (stronger) claim (*was promoted over my head*). **put heads together** consult together. **take it into one's head** (foll. by *that* + clause or *to* + infin.) decide, esp. impetuously. **turn a person's head** make a person conceited. [Old English]

headache *n.* **1** continuous pain in the head. **2** *colloq.* worrying problem. □ **headachy** *adj.*

headband *n.* band worn round the head as decoration or to confine the hair.

headbanger *n. slang* **1** person who shakes his or her head violently to the rhythm of music; fan of loud music. **2** crazy or eccentric person.

headboard *n.* upright panel at the head of a bed.

head-butt *–n.* thrust with the head into the chin or body of another person. *–v.* attack with a head-butt.

headcount *n.* **1** counting of individual people. **2** total number of people, esp. employees.

headdress *n.* covering for the head.

header *n.* **1** *Football* shot or pass made with the head. **2** *colloq.* headlong fall or dive. **3** brick etc. laid at right angles to the face of a wall. **4** (in full **header-tank**) tank of water etc. maintaining pressure in a plumbing system.

head first *adv.* **1** with the head foremost. **2** precipitately.

headgear *n.* hat or headdress.

head-hunting *n.* **1** collecting of the heads of dead enemies as trophies. **2** seeking of (esp. senior) staff by approaching people employed elsewhere. □ **head-hunt** *v.* **head-hunter** *n.*

heading *n.* **1 a** title at the head of a page or section of a book etc. **b** section of a subject of discourse etc. **2** horizontal passage made in preparation for building a tunnel, or in a mine.

head in the sand *n.* refusal to acknowledge danger or difficulty.

headlamp *n.* = HEADLIGHT.

headland *n.* promontory.

headlight *n.* **1** strong light at the front of a vehicle. **2** beam from this.

headline *n.* **1** heading at the top of an article or page, esp. in a newspaper. **2** (in *pl.*) summary of the most important items in a news bulletin.

headlock *n. Wrestling* hold with an arm round the opponent's head.

headlong *adv.* & *adj.* **1** with the head foremost. **2** in a rush.

headman *n.* chief man of a tribe etc.

headmaster *n.* (*fem.* **headmistress**) = HEAD TEACHER.

head-on *adj.* & *adv.* **1** with the front foremost (*head-on crash*). **2** in direct confrontation.

head over heels *–n.* turning over completely in forward motion as in a somersault etc. *–adv.* utterly (*head over heels in love*).

headphones *n.pl.* set of earphones fitting over the head, for listening to audio equipment etc.

headquarters *n.* (as *sing.* or *pl.*) administrative centre of an organization.

headrest *n.* support for the head, esp. on a seat.

headroom *n.* space or clearance above a vehicle, person's head, etc.

headscarf *n.* scarf worn round the head and tied under the chin.

headset *n.* headphones, often with a microphone attached.

headship *n.* position of head or chief, esp. in a school.

headshrinker *n. slang* psychiatrist.

headstall *n.* part of a halter or bridle fitting round a horse's head.

head start *n.* advantage granted or gained at an early stage.

headstone *n.* stone set up at the head of a grave.

headstrong *adj.* self-willed.

head teacher *n.* teacher in charge of a school.

headwaters *n.pl.* streams flowing from the sources of a river.

headway *n.* **1** progress. **2** ship's rate of progress. **3** headroom.

head wind *n.* wind blowing from directly in front.

headword *n.* word forming a heading.

heady *adj.* (**-ier, -iest**) **1** (of liquor) potent. **2** intoxicating, exciting. **3** impulsive, rash. **4** headachy. □ **headily** *adv.* **headiness** *n.*

heal *v.* **1** (often foll. by *up*) become sound or healthy again. **2** cause to heal. **3** put right (differences etc.). **4** alleviate (sorrow etc.). □ **healer** *n.* [Old English: related to WHOLE]

health /helθ/ *n.* **1** state of being well in body or mind. **2** person's mental or physical condition. **3** soundness, esp. financial or moral. [Old English: related to WHOLE]

health centre *n.* building containing various local medical services and doctors' practices.

health farm *n.* establishment offering improved health by a regime of dieting, exercise, etc.

health food *n.* natural food, thought to promote good health.

healthful *adj.* conducive to good health; beneficial.

health service *n.* public service providing medical care.

health visitor *n.* trained nurse who visits mothers and babies, or the sick or elderly, at home.

healthy *adj.* (**-ier, -iest**) **1** having, showing, or promoting good health. **2** indicative of (esp. moral or financial) health (*a healthy sign*). **3** substantial (*won by a healthy 40 seconds*). □ **healthily** *adv.* **healthiness** *n.*

heap –*n.* **1** disorderly pile. **2** (esp. in *pl.*) *colloq.* large number or amount. **3** *slang* dilapidated vehicle. –*v.* **1** (foll. by *up*, *together*, etc.) collect or be collected in a heap. **2** (foll. by *with*) load copiously with. **3** (foll. by *on*, *upon*) give or offer copiously (*heaped insults on them*). [Old English]

hear *v.* (*past* and *past part.* **heard** /hɜːd/) **1** (also *absol.*) perceive with the ear. **2** listen to (*heard them on the radio*). **3** listen judicially to (a case etc.). **4** be told or informed. **5** (foll. by *from*) be contacted by, esp. by letter or telephone. **6** be ready to obey (an order). **7** grant (a prayer). □ **have heard of** be aware of the existence of. **hear! hear!** *int.* expressing agreement. **hear a person out** listen to all a person says. **will not hear of** will not allow. □ **hearer** *n.* [Old English]

hearing *n.* **1** faculty of perceiving sounds. **2** range within which sounds may be heard (*within hearing*). **3** opportunity to state one's case (*a fair hearing*). **4** trial of a case before a court.

hearing-aid *n.* small device to amplify sound, worn by a partially deaf person.

hearken /ˈhɑːkən/ *v. archaic* (often foll. by *to*) listen. [Old English: related to HARK]

hearsay *n.* rumour, gossip.

hearse /hɜːs/ *n.* vehicle for conveying the coffin at a funeral. [French *herse* harrow, from Latin *hirpex* large rake]

heart /hɑːt/ *n.* **1** hollow muscular organ maintaining the circulation of blood by rhythmic contraction and dilation. **2** region of the heart; the breast. **3 a** centre of thought, feeling, and emotion (esp. love). **b** capacity for feeling emotion (*has no heart*). **4 a** courage or enthusiasm (*take heart*). **b** mood or feeling (*change of heart*). **5 a** central or innermost part of something. **b** essence (*heart of the matter*). **6** compact tender inner part of a lettuce etc. **7 a** heart-shaped thing. **b** conventional representation of a heart with two equal curves meeting at a point at the bottom and a cusp at the top. **8 a** playing-card of the suit denoted by a red figure of a heart. **b** (in *pl.*) this suit. □ **at heart 1** in one's inmost feelings. **2** basically. **break a person's heart** overwhelm a person with sorrow. **by heart** from memory. **give** (or **lose**) **one's heart** (often foll. by *to*) fall in love (with). **have the heart** (usu. with *neg.*; foll. by *to* + infin.) be insensitive or hard-hearted enough (*didn't have the heart to ask him*). **take to heart** be much affected by. **to one's heart's content** see CONTENT[1]. **with all one's heart** sincerely; with all goodwill. [Old English]

heartache *n.* mental anguish.

heart attack *n.* sudden occurrence of coronary thrombosis.

heartbeat *n.* pulsation of the heart.

heartbreak *n.* overwhelming distress. □ **heartbreaking** *adj.* **heartbroken** *adj.*

heartburn *n.* burning sensation in the chest from indigestion.

hearten *v.* make or become more cheerful. □ **heartening** *adj.*

heart failure *n.* failure of the heart to function properly, esp. as a cause of death.

heartfelt *adj.* sincere; deeply felt.

hearth /hɑːθ/ *n.* **1** floor of a fireplace. **2** the home. [Old English]

hearthrug *n.* rug laid before a fireplace.

heartily *adv.* **1** in a hearty manner. **2** very (*am heartily sick of it*).

heartland *n.* central part of an area.

heartless *adj.* unfeeling, pitiless. □ **heartlessly** *adv.*

heart-lung machine *n.* machine that temporarily takes over the functions of the heart and lungs.

heart-rending *adj.* very distressing.

heart-searching *n.* examination of one's own feelings and motives.

heartsick *adj.* despondent.

heartstrings *n.pl.* one's deepest feelings.

heartthrob *n. colloq.* person for whom one has (esp. immature) romantic feelings.

heart-to-heart –*attrib. adj.* (of a conversation etc.) candid, intimate. –*n.* candid or personal conversation.

heart-warming *adj.* emotionally rewarding or uplifting.

heartwood *n.* dense inner part of a tree-trunk, yielding the hardest timber.

hearty *adj.* (**-ier, -iest**) **1** strong, vigorous. **2** (of a meal or appetite) large. **3** warm, friendly. □ **heartiness** *n.*

heat –*n.* **1** condition of being hot. **2** *Physics* form of energy arising from the motion of bodies' molecules. **3** hot weather. **4** warmth of feeling; anger or excitement. **5** (foll. by *of*) most intense part or period of activity (*heat of battle*). **6** (usu. preliminary or trial) round in a race etc. **–***v*. **1** make or become hot or warm. **2** inflame. □ **on heat** (of mammals, esp. females) sexually receptive. [Old English]

heated *adj*. angry; impassioned. □ **heatedly** *adv*.

heater *n*. stove or other heating device.

heath *n*. **1** area of flattish uncultivated land with low shrubs. **2** plant growing on a heath, esp. heather. [Old English]

heathen /ˈhiːð(ə)n/ **–***n*. **1** person not belonging to a predominant religion, esp. not a Christian, Jew, or Muslim. **2** person regarded as lacking culture or moral principles. **–***adj*. **1** of heathens. **2** having no religion. [Old English]

heather /ˈheðə(r)/ *n*. any of various shrubs growing esp. on moors and heaths. [origin unknown]

Heath Robinson /hiːθ ˈrɒbɪns(ə)n/ *adj*. absurdly ingenious and impracticable. [name of a cartoonist]

heating *n*. **1** imparting or generation of heat. **2** equipment used to heat a building etc.

heatproof **–***adj*. able to resist great heat. **–***v*. make heatproof.

heat shield *n*. device to protect (esp. a spacecraft) from excessive heat.

heatwave *n*. period of unusually hot weather.

heave **–***v*. (**-ving**; *past* and *past part*. **heaved** or esp. *Naut*. **hove** /həʊv/) **1** lift or haul with great effort. **2** utter with effort (*heaved a sigh*). **3** *colloq*. throw. **4** rise and fall rhythmically or spasmodically. **5** *Naut*. haul by rope. **6** retch. **–***n*. heaving. □ **heave in sight** come into view. **heave to** esp. *Naut*. bring or be brought to a standstill. [Old English]

heaven /ˈhev(ə)n/ *n*. **1** place regarded in some religions as the abode of God and the angels, and of the blessed after death. **2** place or state of supreme bliss. **3** *colloq*. delightful thing. **4** (usu. **Heaven**) God, Providence (often as an exclamation or mild oath: *Heavens*). **5** (**the heavens**) esp. *poet*. the sky as seen from the earth, in which the sun, moon, and stars appear. □ **heavenward** *adv*. (also **heavenwards**). [Old English]

heavenly *adj*. **1** of heaven; divine. **2** of the heavens or sky. **3** *colloq*. very pleasing; wonderful.

heavenly bodies *n.pl*. the sun, stars, planets, etc.

heaven-sent *adj*. providential.

heavier-than-air *attrib. adj*. (of an aircraft) weighing more than the air it displaces.

heavy /ˈhevɪ/ **–***adj*. (**-ier, -iest**) **1** of great or unusually high weight; difficult to lift. **2** of great density (*heavy metal*). **3** abundant, considerable (*heavy crop*; *heavy traffic*). **4** severe, intense, extensive (*heavy fighting*; *a heavy sleep*). **5** doing a thing to excess (*heavy drinker*). **6** striking or falling with force; causing strong impact (*heavy blows*; *heavy rain*; *heavy sea*; *a heavy fall*). **7** (of machinery, artillery, etc.) very large of its kind; large in calibre etc. **8** needing much physical effort (*heavy work*). **9** carrying heavy weapons (*the heavy brigade*). **10** serious or sombre in tone or attitude; dull, tedious. **11 a** hard to digest. **b** hard to read or understand. **12** (of bread etc.) too dense from not having risen. **13** (of ground) difficult to traverse or work. **14** oppressive; hard to endure (*heavy demands*). **15 a** coarse, ungraceful (*heavy features*). **b** unwieldy. **–***n*. (*pl*. **-ies**) **1** *colloq*. large violent person; thug (esp. hired). **2** villainous or tragic role or actor. **3** (usu. in *pl*.) *colloq*. serious newspaper. **4** anything large or heavy of its kind, e.g. a vehicle. **–***adv*. heavily (esp. in *comb*.: *heavy-laden*). □ **heavy on** using a lot of (*heavy on petrol*). **make heavy weather of** see WEATHER. □ **heavily** *adv*. **heaviness** *n*. **heavyish** *adj*. [Old English]

heavy-duty *adj*. intended to withstand hard use.

heavy going *n*. slow or difficult progress.

heavy-handed *adj*. **1** clumsy. **2** overbearing, oppressive. □ **heavy-handedly** *adv*. **heavy-handedness** *n*.

heavy-hearted *adj*. sad, doleful.

heavy hydrogen *n*. = DEUTERIUM.

heavy industry *n*. industry producing metal, machinery, etc.

heavy metal *n*. **1** heavy guns. **2** metal of high density. **3** *colloq*. loud kind of rock music with a pounding rhythm.

heavy petting *n*. erotic fondling that stops short of intercourse.

heavy water *n*. water composed of deuterium and oxygen.

heavyweight *n*. **1 a** weight in certain sports, in amateur boxing over 81 kg. **b** sportsman of this weight. **2** person etc. of above average weight. **3** *colloq*. person of influence or importance.

hebdomadal /hebˈdɒməd(ə)l/ *adj*. *formal* weekly, esp. meeting weekly. [Greek *hepta* seven]

hebe /ˈhiːbɪ/ *n*. evergreen flowering shrub from New Zealand. [Greek goddess *Hēbē*]

Hebraic /hiːˈbreɪɪk/ *adj*. of Hebrew or the Hebrews.

Hebrew /'hi:bru:/ *–n.* **1** member of a Semitic people orig. centred in ancient Palestine. **2 a** their language. **b** modern form of this, used esp. in Israel. *–adj.* **1** of or in Hebrew. **2** of the Hebrews or the Jews. [Hebrew, = one from the other side of the river]

heck *int. colloq.* mild exclamation of surprise or dismay. [a form of HELL]

heckle /'hek(ə)l/ *–v.* (**-ling**) interrupt and harass (a public speaker). *–n.* act of heckling. □ **heckler** *n.* [var. of HACKLE]

hectare /'hekteə(r)/ *n.* metric unit of square measure, 100 ares (2.471 acres or 10,000 square metres). [French: related to HECTO-, ARE[2]]

hectic /'hektɪk/ *adj.* **1** busy and confused; excited. **2** feverish. □ **hectically** *adv.* [Greek *hektikos* habitual]

hecto- *comb. form* hundred. [Greek *hekaton*]

hectogram /'hektə,græm/ *n.* (also **hectogramme**) metric unit of mass equal to 100 grams.

hector /'hektə(r)/ *–v.* bully, intimidate. *–n.* bully. [from the name *Hector* in the *Iliad*]

he'd /hi:d/ *contr.* **1** he had. **2** he would.

hedge *–n.* **1** fence or boundary of dense bushes or shrubs. **2** protection against possible loss. *–v.* (**-ging**) **1** surround or bound with a hedge. **2** (foll. by *in*) enclose. **3 a** reduce one's risk of loss on (a bet or speculation) by compensating transactions on the other side. **b** avoid committing oneself. [Old English]

hedgehog *n.* small insect-eating mammal with a piglike snout and a coat of spines, rolling itself up into a ball when attacked.

hedge-hop *v.* fly at a very low altitude.

hedgerow *n.* row of bushes etc. forming a hedge.

hedge sparrow *n.* common grey and brown bird; the dunnock.

hedonism /'hi:də,nɪz(ə)m/ *n.* **1** belief in pleasure as mankind's proper aim. **2** behaviour based on this. □ **hedonist** *n.* **hedonistic** /-'nɪstɪk/ *adj.* [Greek *hēdonē* pleasure]

heebie-jeebies /,hi:bɪ'ji:bɪz/ *n.pl.* (prec. by *the*) *slang* nervous anxiety, tension. [origin unknown]

heed *–v.* attend to; take notice of. *–n.* careful attention. □ **heedful** *adj.* **heedless** *adj.* **heedlessly** *adv.* [Old English]

hee-haw /'hi:hɔ:/ *–n.* bray of a donkey. *–v.* make a braying sound. [imitative]

heel[1] *–n.* **1** back of the foot below the ankle. **2 a** part of a sock etc. covering this. **b** part of a shoe etc. supporting this. **3** thing like a heel in form or position. **4** crust end of a loaf of bread. **5** *colloq.* scoundrel. **6** (as *int.*) command to a dog to walk close to its owner's heel. *–v.* **1** fit or renew a heel on (a shoe etc.). **2** touch the ground with the heel as in dancing. **3** (foll. by *out*) *Rugby* pass the ball with the heel. □ **at heel 1** (of a dog) close behind. **2** (of a person etc.) under control. **at** (or **on**) **the heels of** following closely after (a person or event). **cool** (or **kick**) **one's heels** be kept waiting. **down at heel 1** (of a shoe) with the heel worn down. **2** (of a person) shabby. **take to one's heels** run away. **to heel 1** (of a dog) close behind. **2** (of a person etc.) under control. **turn on one's heel** turn sharply round. [Old English]

heel[2] *–v.* (often foll. by *over*) **1** (of a ship etc.) lean over. **2** cause (a ship etc.) to do this. *–n.* act or amount of heeling. [obsolete *heeld*, from Germanic]

heel[3] var. of HELE.

heelball *n.* **1** mixture of hard wax and lampblack used by shoemakers for polishing. **2** this or a similar mixture used in brass-rubbing.

hefty /'heftɪ/ *adj.* (**-ier, -iest**) **1** (of a person) big and strong. **2** (of a thing) large, heavy, powerful. □ **heftily** *adv.* **heftiness** *n.* [*heft* weight: related to HEAVE]

hegemony /hɪ'gemənɪ/ *n.* leadership, esp. by one State of a confederacy. [Greek *hēgemōn* leader]

Hegira /'hedʒɪrə/ *n.* (also **Hejira**) **1** Muhammad's flight from Mecca in AD 622. **2** Muslim era reckoned from this date. [Arabic *hijra* departure]

heifer /'hefə(r)/ *n.* young cow, esp. one that has not had more than one calf. [Old English]

height /haɪt/ *n.* **1** measurement from base to top or head to foot. **2** elevation above the ground or a recognized level. **3** considerable elevation (*situated at a height*). **4** high place or area. **5** top. **6 a** most intense part or period (*battle was at its height*). **b** extreme example (*the height of fashion*). [Old English]

heighten *v.* make or become higher or more intense.

heinous /'heɪnəs/ *adj.* utterly odious or wicked. [French *haïr* hate]

heir /eə(r)/ *n.* (*fem.* **heiress**) person entitled to property or rank as the legal successor of its former holder. [Latin *heres hered-*]

heir apparent *n.* heir whose claim cannot be set aside by the birth of another heir.

heirloom *n.* **1** piece of personal property that has been in a family for several generations. **2** piece of property as part of an inheritance.

heir presumptive *n.* heir whose claim may be set aside by the birth of another heir.
Hejira var. of HEGIRA.
held *past* and *past part.* of HOLD[1].
hele /hi:l/ *v.* (**-ling**) (also **heel**) (foll. by *in*) set (a plant) in the ground temporarily and cover its roots. [Old English]
helical /'hi:lɪk(ə)l/ *adj.* having the form of a helix.
helices *pl.* of HELIX.
helicopter /'helɪ,kɒptə(r)/ *n.* wingless aircraft obtaining lift and propulsion from horizontally revolving overhead blades. [Greek: related to HELIX, *pteron* wing]
helio- *comb. form* sun. [Greek *hēlios* sun]
heliocentric /,hi:lɪə'sentrɪk/ *adj.* **1** regarding the sun as centre. **2** considered as viewed from the sun's centre.
heliograph /'hi:lɪə,grɑ:f/ *–n.* **1** signalling apparatus reflecting sunlight in flashes. **2** message sent by means of this. *–v.* send (a message) by heliograph.
heliotrope /'hi:lɪə,trəʊp/ *n.* plant with fragrant purple flowers. [Greek: related to HELIO-, *trepō* turn]
heliport /'helɪ,pɔ:t/ *n.* place where helicopters take off and land.
helium /'hi:lɪəm/ *n.* light inert gaseous element used in airships and as a refrigerant. [related to HELIO-]
helix /'hi:lɪks/ *n.* (*pl.* **helices** /'hi:lɪ,si:z/) spiral curve (like a corkscrew) or coiled curve (like a watch spring). [Latin from Greek]
hell *–n.* **1** place regarded in some religions as the abode of the dead, or of devils and condemned sinners. **2** place or state of misery or wickedness. *–int.* expressing anger, surprise, etc. □ **the hell** (usu. prec. by *what*, *where*, *who*, etc.) expressing anger, disbelief, etc. (*who the hell is this?*; *the hell you are!*). **beat** etc. **the hell out of** *colloq.* beat etc. without restraint. **come hell or high water** no matter what the difficulties. **for the hell of it** *colloq.* just for fun. **get hell** *colloq.* be severely scolded or punished. **give a person hell** *colloq.* scold or punish a person. **a** (or **one**) **hell of a** *colloq.* outstanding example of (*a hell of a mess*; *one hell of a party*). **like hell** *colloq.* **1** not at all. **2** recklessly, exceedingly. [Old English]
he'll /hi:l/ *contr.* he will; he shall.
hell-bent *adj.* (foll. by *on*) recklessly determined.
hellebore /'helɪ,bɔ:(r)/ *n.* evergreen plant with usu. white, purple, or green flowers, e.g. the Christmas rose. [Greek (*h*)*elleborus*]
Hellene /'heli:n/ *n.* **1** native of modern Greece. **2** ancient Greek. □ **Hellenic** /he'lenɪk, -'li:nɪk/ *adj.* [Greek]
Hellenism /'helɪ,nɪz(ə)m/ *n.* (esp. ancient) Greek character or culture. □ **Hellenist** *n.*
Hellenistic /,helɪ'nɪstɪk/ *adj.* of Greek history, language, and culture of the late 4th to the late 1st c. BC.
hell-fire *n.* fire(s) regarded as existing in hell.
hell for leather *adv.* at full speed.
hell-hole *n.* oppressive or unbearable place.
hellish *–adj.* **1** of or like hell. **2** *colloq.* extremely difficult or unpleasant. *–adv. colloq.* extremely (*hellish expensive*). □ **hellishly** *adv.*
hello /hə'ləʊ/ (also **hallo**, **hullo**) *–int.* expression of informal greeting, or of surprise, or to call attention. *–n.* (*pl.* **-s**) cry of 'hello'. [var. of earlier *hollo*]
Hell's Angel *n.* member of a gang of male motor-cycle enthusiasts notorious for outrageous and violent behaviour.
helm *n.* tiller or wheel for controlling a ship's rudder. □ **at the helm** in control; at the head of an organization etc. [Old English]
helmet /'helmɪt/ *n.* protective head-covering worn by a policeman, motor cyclist, etc. [French from Germanic]
helmsman /'helmzmən/ *n.* person who steers a ship.
helot /'helət/ *n.* serf, esp. (**Helot**) of a class in ancient Sparta. [Latin from Greek]
help *–v.* **1** provide with the means towards what is needed or sought (*helped me with my work*; *helped me* (*to*) *pay my debts*; *helped him on with his coat*). **2** (often *absol.*) be of use or service to (*does that help?*). **3** contribute to alleviating (a pain or difficulty). **4** prevent or remedy (*it can't be helped*). **5** (usu. with *neg.*) **a** refrain from (*can't help it*; *could not help laughing*). **b** *refl.* refrain from acting (*couldn't help himself*). **6** (often foll. by *to*) serve (a person with food). *–n.* **1** helping or being helped (*need your help*; *came to our help*). **2** person or thing that helps. **3** *colloq.* domestic assistant or assistance. **4** remedy or escape (*there is no help for it*). □ **help oneself** (often foll. by *to*) **1** serve oneself (with food etc.). **2** take without permission. **help a person out** give a person help, esp. in difficulty. □ **helper** *n.* [Old English]
helpful *adj.* giving help; useful. □ **helpfully** *adv.* **helpfulness** *n.*
helping *n.* portion of food at a meal.

helpless *adj.* **1** lacking help or protection; defenceless. **2** unable to act without help. □ **helplessly** *adv.* **helplessness** *n.*
helpline *n.* telephone service providing help with problems.
helpmate *n.* helpful companion or partner.
helter-skelter /ˌheltəˈskeltə(r)/ *–adv.* & *adj.* in disorderly haste. *–n.* (at a fairground) external spiral slide round a tower. [imitative]
hem[1] *–n.* border of cloth where the edge is turned under and sewn down. *–v.* (**-mm-**) turn down and sew in the edge of (cloth etc.). □ **hem in** confine; restrict the movement of. [Old English]
hem[2] *–int.* calling attention or expressing hesitation by a slight cough. *–n.* utterance of this. *–v.* (**-mm-**) say *hem*; hesitate in speech. □ **hem and haw** = *hum and haw* (see HUM[1]). [imitative]
hemal etc. *US* var. of HAEMAL etc.
he-man *n.* masterful or virile man.
hemi- *comb. form* half. [Greek, = Latin *semi-*]
hemipterous /heˈmɪptərəs/ *adj.* of the insect order including aphids, bugs, and cicadas, with piercing or sucking mouthparts. [Greek *pteron* wing]
hemisphere /ˈhemɪˌsfɪə(r)/ *n.* **1** half a sphere. **2** half of the earth, esp. as divided by the equator (into *northern* and *southern hemisphere*) or by a line passing through the poles (into *eastern* and *western hemisphere*). □ **hemispherical** /-ˈsferɪk(ə)l/ *adj.* [Greek: related to HEMI-, SPHERE]
hemline *n.* lower edge of a skirt etc.
hemlock /ˈhemlɒk/ *n.* **1** poisonous plant with fernlike leaves and small white flowers. **2** poison made from this. [Old English]
hemp *n.* **1** (in full **Indian hemp**) Asian herbaceous plant. **2** its fibre used to make rope and stout fabrics. **3** narcotic drug made from the hemp plant. [Old English]
hempen *adj.* made of hemp.
hemstitch *–n.* decorative stitch. *–v.* hem with this stitch.
hen *n.* female bird, esp. of a domestic fowl. [Old English]
henbane *n.* poisonous hairy plant with an unpleasant smell.
hence *adv.* **1** from this time (*two years hence*). **2** for this reason (*hence we seem to be wrong*). **3** *archaic* from here. [Old English]
henceforth *adv.* (also **henceforward**) from this time onwards.
henchman *n.* usu. *derog.* trusted supporter. [Old English *hengst* horse, MAN]
henge /hendʒ/ *n.* prehistoric monument consisting of a circle of stone or wood uprights. [*Stonehenge* in S. England]
henna /ˈhenə/ *–n.* **1** tropical shrub. **2** reddish dye made from it and used to colour hair. *–v.* (**hennaed**, **hennaing**) dye with henna. [Arabic]
hen-party *n. colloq.* social gathering of women only.
henpeck *v.* (usu. in *passive*) (of a wife) constantly nag (her husband).
henry /ˈhenrɪ/ *n.* (*pl.* **-s** or **-ies**) *Electr.* SI unit of inductance. [*Henry*, name of a physicist]
hep var. of HIP[4].
hepatic /hɪˈpætɪk/ *adj.* of the liver. [Greek *hēpar -atos* liver]
hepatitis /ˌhepəˈtaɪtɪs/ *n.* inflammation of the liver. [related to HEPATIC]
hepta- *comb. form* seven. [Greek]
heptagon /ˈheptəgən/ *n.* plane figure with seven sides and angles. □ **heptagonal** /-ˈtægən(ə)l/ *adj.* [Greek: related to HEPTA-, *-gōnos* angled]
her *–pron.* **1** *objective case* of SHE (*I like her*). **2** *colloq.* she (*it's her all right*; *am older than her*). *–poss. pron.* (*attrib.*) of or belonging to her or herself (*her house*; *her own business*). [Old English dative and genitive of SHE]
herald /ˈher(ə)ld/ *–n.* **1** official messenger bringing news. **2** forerunner, harbinger. **3 a** *hist.* officer responsible for State ceremonial and etiquette. **b** official concerned with pedigrees and coats of arms. *–v.* proclaim the approach of; usher in. □ **heraldic** /-ˈrældɪk/ *adj.* [French from Germanic]
heraldry *n.* **1** art or knowledge of a herald. **2** coats of arms.
herb *n.* **1** any non-woody seed-bearing plant. **2** plant with leaves, seeds, or flowers used for flavouring, food, medicine, scent, etc. □ **herby** *adj.* (**-ier**, **-iest**). [Latin *herba*]
herbaceous /hɜːˈbeɪʃəs/ *adj.* of or like herbs.
herbaceous border *n.* garden border containing esp. perennial flowering plants.
herbage *n.* vegetation collectively, esp. as pasture.
herbal *–adj.* of herbs in medicinal and culinary use. *–n.* book describing the medicinal and culinary uses of herbs.
herbalist *n.* **1** dealer in medicinal herbs. **2** writer on herbs.
herbarium /hɜːˈbeərɪəm/ *n.* (*pl.* **-ria**) **1** systematically arranged collection of dried plants. **2** book, room, etc. for these.
herbicide /ˈhɜːbɪˌsaɪd/ *n.* poison used to destroy unwanted vegetation.

herbivore /'hɜ:bɪ,vɔ:(r)/ *n.* animal that feeds on plants. □ **herbivorous** /-'bɪvərəs/ *adj.* [Latin *voro* devour]

Herculean /,hɜ:kjʊ'li:ən/ *adj.* having or requiring great strength or effort. [from the name *Hercules*, Latin alteration of Greek *Hēraklēs*]

herd –*n.* **1** a number of animals, esp. cattle, feeding or travelling or kept together. **2** (prec. by *the*) *derog.* large number of people; mob (*tends to follow the herd*). –*v.* **1** (cause to) go in a herd (*herded together for warmth; herded the cattle into the field*). **2** look after (sheep, cattle, etc.). [Old English]

herd instinct *n.* (prec. by *the*) tendency to think and act as a crowd.

herdsman *n.* man who owns or tends a herd.

here –*adv.* **1** in or at or to this place or position (*come here; sit here*). **2** indicating a person's presence or a thing offered (*my son here will show you; here is your coat*). **3** at this point in the argument, situation, etc. (*here I have a question*). –*n.* this place (*get out of here; lives near here; fill it up to here*). –*int.* **1** calling attention: short for *come here*, *look here*, etc. (*here, where are you going with that?*). **2** indicating one's presence in a roll-call: short for *I am here*. □ **here goes!** *colloq.* expression indicating the start of a bold act. **here's to** I drink to the health of. **here we are** *colloq.* said on arrival at one's destination. **here we go again** *colloq.* the same, usu. undesirable, events are recurring. **here you are** said on handing something to somebody. **neither here nor there** of no importance. [Old English]

hereabouts /,hɪərə'baʊts/ *adv.* (also **hereabout**) near this place.

hereafter /hɪər'ɑ:ftə(r)/ –*adv.* from now on; in the future. –*n.* **1** the future. **2** life after death.

here and now *adv.* at this very moment; immediately.

here and there *adv.* in various places.

hereby /hɪə'baɪ/ *adv.* by this means; as a result of this.

hereditable /hɪ'redɪtəb(ə)l/ *adj.* that can be inherited. [Latin: related to HEIR]

hereditary /hɪ'redɪtərɪ/ *adj.* **1** (of a disease, instinct, etc.) able to be passed down genetically from one generation to another. **2 a** descending by inheritance. **b** holding a position by inheritance. [Latin: related to HEIR]

heredity /hɪ'redɪtɪ/ *n.* **1 a** passing on of physical or mental characteristics genetically. **b** these characteristics. **2** genetic constitution.

Hereford /'herɪfəd/ *n.* animal of a breed of red and white beef cattle. [*Hereford* in England]

herein /hɪə'rɪn/ *adv. formal* in this matter, book, etc.

hereinafter /,hɪərɪn'ɑ:ftə(r)/ *adv.* esp. *Law formal* **1** from this point on. **2** in a later part of this document etc.

hereof /hɪər'ɒv/ *adv. formal* of this.

heresy /'herəsɪ/ *n.* (*pl.* **-ies**) **1** esp. *RC Ch.* religious belief or practice contrary to orthodox doctrine. **2** opinion contrary to what is normally accepted or maintained. [Greek *hairesis* choice]

heretic /'herətɪk/ *n.* **1** person believing in or practising religious heresy. **2** holder of an unorthodox opinion. □ **heretical** /hɪ'retɪk(ə)l/ *adj.*

hereto /hɪə'tu:/ *adv. formal* to this matter.

heretofore /,hɪətʊ'fɔ:(r)/ *adv. formal* before this time.

hereupon /,hɪərə'pɒn/ *adv.* after this; in consequence of this.

herewith /hɪə'wɪð/ *adv.* with this (esp. of an enclosure in a letter etc.).

heritable /'herɪtəb(ə)l/ *adj.* **1** *Law* capable of being inherited or of inheriting. **2** *Biol.* genetically transmissible from parent to offspring. [French: related to HEIR]

heritage /'herɪtɪdʒ/ *n.* **1** what is or may be inherited. **2** inherited circumstances, benefits, etc. **3** a nation's historic buildings, monuments, countryside, etc., esp. when regarded as worthy of preservation.

hermaphrodite /hɜ:'mæfrə,daɪt/ –*n.* person, animal, or plant having both male and female reproductive organs. –*adj.* combining both sexes. □ **hermaphroditic** /-'dɪtɪk/ *adj.* [from *Hermaphroditus*, son of *Hermes* and *Aphrodite* who became joined in one body to a nymph]

hermetic /hɜ:'metɪk/ *adj.* with an airtight closure. □ **hermetically** *adv.* [from the Greek god *Hermes*, regarded as the founder of alchemy]

hermit /'hɜ:mɪt/ *n.* person (esp. an early Christian) living in solitude and austerity. □ **hermitic** /-'mɪtɪk/ *adj.* [Greek *erēmos* solitary]

hermitage *n.* **1** hermit's dwelling. **2** secluded dwelling.

hermit-crab *n.* crab that lives in a mollusc's cast-off shell.

hernia /'hɜ:nɪə/ *n.* protrusion of part of an organ through the wall of the body cavity containing it. [Latin]

hero /'hɪərəʊ/ *n.* (*pl.* **-es**) **1** person noted or admired for nobility, courage, outstanding achievements, etc. **2** chief male character in a play, story, etc. [Greek *hērōs*]

heroic /hɪ'rəʊɪk/ *–adj.* of, fit for, or like a hero; very brave. *–n.* (in *pl.*) **1** high-flown language or sentiments. **2** unduly bold behaviour. □ **heroically** *adv.*

heroin /'herəʊɪn/ *n.* addictive analgesic drug derived from morphine, often used as a narcotic. [German: related to HERO, from the effect on the user's self-esteem]

heroine /'herəʊɪn/ *n.* **1** woman noted or admired for nobility, courage, outstanding achievements, etc. **2** chief female character in a play, story, etc. [Greek: related to HERO]

heroism /'herəʊ,ɪz(ə)m/ *n.* heroic conduct or qualities. [French *héroïsme*: related to HERO]

heron /'herən/ *n.* long-legged wading bird with a long S-shaped neck. [French from Germanic]

hero-worship *–n.* idealization of an admired person. *–v.* idolize.

herpes /'hɜ:pi:z/ *n.* virus disease causing skin blisters. [Greek *herpō* creep]

Herr /heə(r)/ *n.* (*pl.* **Herren** /'herən/) **1** title of a German man; Mr. **2** German man. [German]

herring /'herɪŋ/ *n.* (*pl.* same or **-s**) N. Atlantic fish used as food. [Old English]

herring-bone *n.* stitch or weave consisting of a series of small 'V' shapes making a zigzag pattern.

herring-gull *n.* large gull with dark wing-tips.

hers /hɜ:z/ *poss. pron.* the one or ones belonging to or associated with her (*it is hers*; *hers are over there*). □ **of hers** of or belonging to her (*friend of hers*).

herself /hə'self/ *pron.* **1 a** *emphat. form* of SHE or HER (*she herself will do it*). **b** *refl. form* of HER (*she has hurt herself*). **2** in her normal state of body or mind (*does not feel quite herself today*). □ **be herself** see ONESELF. **by herself** see *by oneself*. [Old English: related to HER, SELF]

hertz *n.* (*pl.* same) SI unit of frequency, equal to one cycle per second. [*Hertz*, name of a physicist]

he's /hi:z, hɪz/ *contr.* **1** he is. **2** he has.

hesitant /'hezɪt(ə)nt/ *adj.* hesitating; irresolute. □ **hesitance** *n.* **hesitancy** *n.* **hesitantly** *adv.*

hesitate /'hezɪ,teɪt/ *v.* (**-ting**) **1** show or feel indecision or uncertainty; pause in doubt (*hesitated over her choice*). **2** be reluctant (*I hesitate to say so*). □ **hesitation** /-'teɪʃ(ə)n/ *n.* [Latin *haereo haes-* stick fast]

hessian /'hesɪən/ *n.* strong coarse sacking made of hemp or jute. [*Hesse* in Germany]

hetero- *comb. form* other, different. [Greek *heteros* other]

heterodox /'hetərəʊ,dɒks/ *adj.* not orthodox. □ **heterodoxy** *n.* [from HETERO-, Greek *doxa* opinion]

heterodyne /'hetərəʊ,daɪn/ *adj. Radio* relating to the production of a lower frequency from the combination of two almost equal high frequencies. [from HETERO-, Greek *dunamis* force]

heterogeneous /,hetərəʊ'dʒi:nɪəs/ *adj.* **1** diverse in character. **2** varied in content. □ **heterogeneity** /-dʒɪ'ni:ɪtɪ/ *n.* [Latin from Greek *genos* kind]

heteromorphic /,hetərəʊ'mɔ:fɪk/ *adj.* (also **heteromorphous** /-'mɔ:fəs/) *Biol.* of dissimilar forms. □ **heteromorphism** *n.*

heterosexual /,hetərəʊ'sekʃʊəl/ *–adj.* feeling or involving sexual attraction to the opposite sex. *–n.* heterosexual person. □ **heterosexuality** /-'ælɪtɪ/ *n.*

het up *predic. adj. colloq.* overwrought. [*het*, a dial. word = heated]

heuristic /hjʊə'rɪstɪk/ *adj.* **1** allowing or assisting to discover. **2** proceeding to a solution by trial and error. [Greek *heuriskō* find]

hew *v.* (*past part.* **hewn** /hju:n/ or **hewed**) **1** chop or cut with an axe, sword, etc. **2** cut into shape. [Old English]

hex *–v.* **1** practise witchcraft. **2** bewitch. *–n.* magic spell. [German]

hexa- *comb. form* six. [Greek]

hexadecimal /,heksə'desɪm(ə)l/ *adj.* esp. *Computing* of a system of numerical notation that has 16 (the figures 0 to 9 and the letters A to F) rather than 10 as a base.

hexagon /'heksəgən/ *n.* plane figure with six sides and angles. □ **hexagonal** /-'sægən(ə)l/ *adj.* [Greek: related to HEXA-, *-gōnos* angled]

hexagram /'heksə,græm/ *n.* figure formed by two intersecting equilateral triangles.

hexameter /hek'sæmɪtə(r)/ *n.* line of verse with six metrical feet.

hey /heɪ/ *int.* calling attention or expressing joy, surprise, inquiry, etc. [imitative]

heyday /'heɪdeɪ/ *n.* time of greatest success or prosperity. [Low German]

hey presto! *int.* conjuror's phrase on completing a trick.

Hezbollah /,hezbə'lɑ:/ *n.* (also **Hiz-**) /,hɪz-/ extreme Shiite Muslim group, active esp. in Lebanon. [Arabic *hizbullah* Party of God]

HF *abbr.* high frequency.

Hf *symb.* hafnium.

Hg *symb.* mercury. [Latin *hydrargyrum*]

hg *abbr.* hectogram(s).

HGV *abbr.* heavy goods vehicle.

HH *abbr.* **1** Her or His Highness. **2** His Holiness. **3** (of pencil-lead) double-hard.

hi /haɪ/ *int.* calling attention or as a greeting.

hiatus /haɪ'eɪtəs/ *n.* (*pl.* **-tuses**) **1** break or gap in a series or sequence. **2** break between two vowels coming together but not in the same syllable, as in *though oft the ear*. [Latin *hio* gape]

hibernate /'haɪbə,neɪt/ *v.* (**-ting**) (of an animal) spend the winter in a dormant state. □ **hibernation** /-'neɪʃ(ə)n/ *n.* [Latin *hibernus* wintry]

Hibernian /haɪ'bɜ:nɪən/ *archaic poet.* –*adj.* of Ireland. –*n.* native of Ireland. [Latin *Hibernia* Ireland]

hibiscus /hɪ'bɪskəs/ *n.* (*pl.* **-cuses**) cultivated shrub with large bright-coloured flowers. [Greek *hibiskos* marsh mallow]

hiccup /'hɪkʌp/ (also **hiccough**) –*n.* **1** involuntary spasm of the diaphragm causing a characteristic sound 'hic'. **2** temporary or minor stoppage or difficulty. –*v.* (**-p-**) make a hiccup. [imitative]

hick *n.* (often *attrib.*) esp. *US colloq.* country bumpkin, provincial. [familiar form of *Richard*]

hickory /'hɪkərɪ/ *n.* (*pl.* **-ies**) **1** N. American tree yielding wood and nutlike edible fruits. **2** the tough heavy wood of this. [Virginian *pohickery*]

hid *past* of HIDE[1].

hidden *past part.* of HIDE[1].

hidden agenda *n.* secret motivation behind a policy, statement, etc.; ulterior motive.

hide[1] –*v.* (**-ding**; *past* **hid**; *past part.* **hidden**) **1** put or keep out of sight. **2** conceal oneself. **3** (usu. foll. by *from*) keep (a fact) secret. **4** conceal. –*n.* camouflaged shelter used for observing wildlife. □ **hider** *n.* [Old English]

hide[2] *n.* **1** animal's skin, esp. when tanned or dressed. **2** *colloq.* the human skin, esp. the backside. [Old English]

hide-and-seek *n.* game in which players hide and another searches for them.

hideaway *n.* hiding-place or place of retreat.

hidebound *adj.* **1** narrow-minded. **2** constricted by tradition.

hideous /'hɪdɪəs/ *adj.* **1** very ugly, revolting. **2** *colloq.* unpleasant. □ **hideosity** /-'ɒsɪtɪ/ *n.* (*pl.* **-ies**). **hideously** *adv.* [Anglo-French *hidous*]

hide-out *n. colloq.* hiding-place.

hiding[1] *n. colloq.* a thrashing. □ **on a hiding to nothing** with no chance of succeeding. [from HIDE[2]]

hiding[2] *n.* **1** act of hiding. **2** state of remaining hidden (*go into hiding*). [from HIDE[1]]

hiding-place *n.* place of concealment.

hierarchy /'haɪə,rɑ:kɪ/ *n.* (*pl.* **-ies**) system of grades of status or authority ranked one above the other. □ **hierarchical** /-'rɑ:kɪk(ə)l/ *adj.* [Greek *hieros* sacred, *arkhō* rule]

hieratic /,haɪə'rætɪk/ *adj.* **1** of priests. **2** of the ancient Egyptian hieroglyphic writing as used by priests. [Greek *hiereus* priest]

hieroglyph /'haɪərəglɪf/ *n.* picture representing a word, syllable, or sound, as used in ancient Egyptian etc. [Greek *hieros* sacred, *gluphō* carve]

hieroglyphic /,haɪərə'glɪfɪk/ –*adj.* of or written in hieroglyphs. –*n.* (in *pl.*) hieroglyphs; hieroglyphic writing.

hi-fi /'haɪfaɪ/ *colloq.* –*adj.* of high fidelity. –*n.* (*pl.* **-s**) set of high-fidelity equipment. [abbreviation]

higgledy-piggledy /,hɪgəldɪ'pɪgəldɪ/ *adv.* & *adj.* in confusion or disorder. [origin uncertain]

high /haɪ/ –*adj.* **1 a** of great vertical extent (*high building*). **b** (*predic.*; often in *comb.*) of a specified height (*one inch high*; *waist-high*). **2 a** far above ground or sea level etc. (*high altitude*). **b** inland, esp. when raised (*High Asia*). **3** extending above the normal level (*jersey with a high neck*). **4 a** of exalted quality (*high minds*). **b** lavish; superior (*high living*; *high fashion*). **5** of exalted rank (*high society*; *is high in the Government*). **6 a** great; intense; extreme; powerful (*high praise*; *high temperature*). **b** greater than normal (*high prices*). **c** extreme or very traditional in religious or political opinion (*high Tory*). **7** performed at, to, or from a considerable height (*high diving*; *high flying*). **8** (often foll. by *on*) *colloq.* intoxicated by alcohol or esp. drugs. **9** (of a sound etc.) of high frequency; shrill. **10** (of a period, age, time, etc.) at its peak (*high noon*; *high summer*; *High Renaissance*). **11 a** (of meat etc.) beginning to go bad; off. **b** (of game) well-hung and slightly decomposed. –*n.* **1** high, or the highest, level or figure. **2** area of high pressure; anticyclone. **3** *slang* euphoric state, esp. drug-induced (*am on a high*). –*adv.* **1** far up; aloft (*flew the flag high*). **2** in or to a high degree. **3** at a high price. **4** (of a sound) at or to a high pitch. □ **high opinion of** favourable opinion of. **on high** in or to heaven or a high place. **on one's high horse** *colloq.* acting arrogantly. [Old English]

high altar *n.* chief altar in a church.

high and dry *adj.* stranded; aground.

high and low *adv.* everywhere (*searched high and low*).

high and mighty *adj. colloq.* arrogant.

highball *n. US* drink of spirits and soda etc., served with ice in a tall glass.
highbrow *colloq.* –*adj.* intellectual; cultural. –*n.* intellectual or cultured person.
high chair *n.* infant's chair with long legs and a tray for meals.
High Church *n.* section of the Church of England emphasizing ritual, priestly authority, and sacraments.
high-class *adj.* of high quality.
high colour *n.* flushed complexion.
high command *n.* army commander-in-chief and associated staff.
High Commission *n.* embassy from one Commonwealth country to another. □ **High Commissioner** *n.*
High Court *n.* (also in England **High Court of Justice**) supreme court of justice for civil cases.
high day *n.* festal day.
higher animal *n.* (also **higher plant**) animal or plant evolved to a high degree.
higher education *n.* education at university etc.
high explosive *n.* extremely explosive substance used in shells, bombs, etc.
highfalutin /,haɪfə'lu:tɪn/ *adj.* (also **highfaluting** /-tɪŋ/) *colloq.* pompous, pretentious. [origin unknown]
high fidelity *n.* high-quality sound reproduction with little distortion.
high-flown *adj.* (of language etc.) extravagant, bombastic.
high-flyer *n.* (also **high-flier**) **1** ambitious person. **2** person or thing of great potential. □ **high-flying** *adj.*
high frequency *n.* frequency, esp. in radio, of 3–30 megahertz.
high gear *n.* gear such that the driven end of a transmission revolves faster than the driving end.
high-handed *adj.* disregarding others' feelings; overbearing. □ **high-handedly** *adv.* **high-handedness** *n.*
high heels *n.pl.* women's shoes with high heels.
high jinks *n.pl.* boisterous fun.
high jump *n.* **1** athletic event consisting of jumping over a high bar. **2** *colloq.* drastic punishment (*he's for the high jump*).
highland –*n.* (usu. in *pl.*) **1** area of high land. **2** (**the Highlands**) mountainous part of Scotland. –*adj.* of or in a highland or the Highlands. □ **highlander** *n.* (also **Highlander**). [Old English, = promontory: related to HIGH]
Highland cattle *n.* cattle of a shaggy-haired breed with long curved horns.
Highland fling see FLING *n.* 3.
high-level *adj.* **1** (of negotiations etc.) conducted by high-ranking people. **2** *Computing* (of a programming language) not machine-dependent and usu. at a level of abstraction close to natural language.
highlight –*n.* **1** moment or detail of vivid interest; outstanding feature. **2** (in a painting etc.) bright area. **3** (usu. in *pl.*) light streak in the hair produced by bleaching. –*v.* **1** bring into prominence; draw attention to. **2** mark with a highlighter.
highlighter *n.* marker pen for emphasizing a printed word etc. by overlaying it with colour.
highly /'haɪlɪ/ *adv.* **1** in a high degree (*highly amusing*; *commend it highly*). **2** favourably (*think highly of him*).
highly-strung *adj.* very sensitive or nervous.
high-minded *adj.* having high moral principles. □ **high-mindedly** *adv.* **high-mindedness** *n.*
highness *n.* **1** state of being high (*highness of taxation*). **2** (**Highness**) title used when addressing or referring to a prince or princess (*Her Highness*; *Your Royal Highness*).
high-octane *adj.* (of fuel used in internal-combustion engines) not detonating readily during the power stroke.
high-pitched *adj.* **1** (of a sound) high. **2** (of a roof) steep.
high point *n.* the maximum or best state reached.
high-powered *adj.* **1** having great power or energy. **2** important or influential.
high pressure *n.* **1** high degree of activity or exertion. **2** atmospheric condition with the pressure above average.
high priest *n.* (*fem.* **high priestess**) **1** chief priest, esp. Jewish. **2** head of a cult.
high-ranking *adj.* of high rank, senior.
high-rise –*attrib. adj.* (of a building) having many storeys. –*n.* such a building.
high-risk *attrib. adj.* involving or exposed to danger (*high-risk sports*).
high road *n.* main road.
high school *n.* **1** grammar school. **2** *US* & *Scot.* secondary school.
high sea *n.* (also **high seas**) open seas not under any country's jurisdiction.
high season *n.* busiest period at a resort etc.
high-speed *attrib. adj.* operating at great speed.
high-spirited *adj.* vivacious; cheerful; lively.
high spot *n.* important place or feature.
high street *n.* principal shopping street of a town.

high table *n.* dining-table for the most important guests or members.
high tea *n.* evening meal usu. consisting of a cooked dish, bread and butter, tea, etc.
high-tech *adj.* **1** employing, requiring, or involved in high technology. **2** imitating styles more usual in industry etc.
high technology *n.* advanced technological development, esp. in electronics.
high tension *n.* = HIGH VOLTAGE.
high tide *n.* time or level of the tide at its peak.
high time *n.* time that is overdue (*it is high time they arrived*).
high treason *n.* = TREASON.
high-up *n. colloq.* person of high rank.
high voltage *n.* electrical potential large enough to injure or damage.
high water *n.* = HIGH TIDE.
high-water mark *n.* level reached at high water.
highway *n.* **1 a** public road. **b** main route. **2** direct course of action (*on the highway to success*).
Highway Code *n.* official booklet of guidance for road-users.
highwayman *n. hist.* robber of travellers etc., usu. mounted.
high wire *n.* high tightrope.
hijack /'haɪdʒæk/ *–v.* **1** seize control of (a vehicle etc.), esp. to force it to a different destination. **2** seize (goods) in transit. **3** take control of (talks etc.) by force or subterfuge. *–n.* a hijacking. □ **hijacker** *n.* [origin unknown]
hike *–n.* **1** long walk, esp. in the country for pleasure. **2** rise in prices etc. *–v.* (**-king**) **1** go for a hike. **2** walk laboriously. **3** (usu. foll. by *up*) hitch up (clothing etc.); become hitched up. **4** (usu. foll. by *up*) raise (prices etc.). □ **hiker** *n.* [origin unknown]
hilarious /hɪ'leərɪəs/ *adj.* **1** exceedingly funny. **2** boisterously merry. □ **hilariously** *adv.* **hilarity** /-'lærɪtɪ/ *n.* [Greek *hilaros* cheerful]
hill *n.* **1** naturally raised area of land, lower than a mountain. **2** (often in *comb.*) heap, mound (*anthill*). **3** sloping piece of road. □ **over the hill** *colloq.* past the prime of life. [Old English]
hill-billy *n. US colloq.*, often *derog.* person from a remote rural area in a southern State.
hillock /'hɪlək/ *n.* small hill, mound.
hillside *n.* sloping side of a hill.
hilltop *n.* top of a hill.
hillwalking *n.* hiking in hilly country. □ **hillwalker** *n.*
hilly *adj.* (**-ier, -iest**) having many hills. □ **hilliness** *n.*
hilt *n.* handle of a sword, dagger, etc. □ **up to the hilt** completely. [Old English]
him *pron.* **1** *objective case* of HE (*I saw him*). **2** *colloq.* he (*it's him again*; *taller than him*). [Old English, dative of HE]
himself /hɪm'self/ *pron.* **1 a** *emphat. form* of HE or HIM (*he himself will do it*). **b** *refl. form* of HIM (*he has hurt himself*). **2** in his normal state of body or mind (*does not feel quite himself today*). □ **be himself** see ONESELF. **by himself** see *by oneself*. [Old English: related to HIM, SELF]
hind[1] /haɪnd/ *adj.* at the back (*hind leg*). [Old English *hindan* from behind]
hind[2] /haɪnd/ *n.* female (esp. red) deer, esp. in and after the third year. [Old English]
hinder[1] /'hɪndə(r)/ *v.* impede; delay. [Old English]
hinder[2] /'haɪndə(r)/ *adj.* rear, hind (*the hinder part*). [Old English]
Hindi /'hɪndɪ/ *n.* **1** group of spoken dialects of N. India. **2** literary form of Hindustani, an official language of India. [Urdu *Hind* India]
hindmost *adj.* furthest behind.
hindquarters *n.pl.* hind legs and rump of a quadruped.
hindrance /'hɪndrəns/ *n.* **1** hindering; being hindered. **2** thing that hinders.
hindsight *n.* wisdom after the event.
Hindu /'hɪndu:/ *–n.* (*pl.* **-s**) follower of Hinduism. *–adj.* of Hindus or Hinduism. [Urdu *Hind* India]
Hinduism /'hɪndu:ˌɪz(ə)m/ *n.* main religious and social system of India, including the belief in reincarnation, several gods, and a caste system.
Hindustani /ˌhɪndʊ'stɑ:nɪ/ *n.* language based on Hindi, used as a lingua franca in much of India. [from HINDU, *stān* country]
hinge /hɪndʒ/ *–n.* **1** movable joint on which a door, lid, etc., turns or swings. **2** principle on which all depends. *–v.* (**-ging**) **1** (foll. by *on*) depend (on a principle, an event, etc.). **2** attach or be attached by a hinge. [related to HANG]
hinny /'hɪnɪ/ *n.* (*pl.* **-ies**) offspring of a female donkey and a male horse. [Greek *hinnos*]
hint *–n.* **1** slight or indirect indication or suggestion. **2** small piece of practical information. **3** very small trace; suggestion (*a hint of perfume*). *–v.* suggest slightly or indirectly. □ **hint at** give a hint of; refer indirectly to. **take a hint** heed a hint. [obsolete *hent* grasp]
hinterland /'hɪntəˌlænd/ *n.* **1** district beyond a coast or river's banks. **2** area served by a port or other centre. [German]
hip[1] *n.* projection of the pelvis and the upper part of the thigh-bone. [Old English]

hip[2] *n.* fruit of a rose, esp. wild. [Old English]

hip[3] *int.* introducing a united cheer (*hip, hip, hooray*). [origin unknown]

hip[4] *adj.* (also **hep**) (**-pper, -ppest**) *slang* trendy, stylish. [origin unknown]

hip-bath *n.* portable bath in which one sits immersed to the hips.

hip-bone *n.* bone forming the hip.

hip-flask *n.* small flask for spirits.

hip hop *n.* (also **hip-hop**) subculture combining rap music, graffiti art, and break-dancing. [from HIP[4]]

hippie /'hɪpɪ/ *n.* (also **hippy**) (*pl.* **-ies**) *colloq.* (esp. in the 1960s) person rejecting convention, typically with long hair, jeans, beads, etc., and taking hallucinogenic drugs. [from HIP[4]]

hippo /'hɪpəʊ/ *n.* (*pl.* **-s**) *colloq.* hippopotamus. [abbreviation]

hip-pocket *n.* trouser-pocket just behind the hip.

Hippocratic oath /,hɪpə'krætɪk/ *n.* statement of ethics of the medical profession. [*Hippocrates*, name of a Greek physician]

hippodrome /'hɪpə,drəʊm/ *n.* **1** music-hall or dancehall. **2** (in classical antiquity) course for chariot races etc. [Greek *hippos* horse, *dromos* race]

hippopotamus /,hɪpə'pɒtəməs/ *n.* (*pl.* **-muses** or **-mi** /-,maɪ/) large African mammal with short legs and thick skin, living by rivers, lakes, etc. [Greek *hippos* horse, *potamos* river]

hippy[1] var. of HIPPIE.

hippy[2] *adj.* having large hips.

hipster[1] *–attrib. adj.* (of a garment) hanging from the hips rather than the waist. *–n.* (in *pl.*) such trousers.

hipster[2] *n. slang* hip person.

hire *–v.* (**-ring**) **1** purchase the temporary use of (a thing) (*hired a van*). **2** esp. *US* employ (a person). *–n.* **1** hiring or being hired. **2** payment for this. □ **for** (or **on**) **hire** ready to be hired. **hire out** grant the temporary use of (a thing) for payment. □ **hireable** *adj.* **hirer** *n.* [Old English]

hireling *n.* usu. *derog.* person who works (only) for money.

hire purchase *n.* system of purchase by paying in instalments.

hirsute /'hɜ:sju:t/ *adj.* hairy. [Latin]

his /hɪz/ *poss. pron.* **1** (*attrib.*) of or belonging to him or himself (*his house*; *his own business*). **2** the one or ones belonging to or associated with him (*it is his*; *his are over there*). □ **of his** of or belonging to him (*friend of his*). [Old English, genitive of HE]

Hispanic /hɪ'spænɪk/ *–adj.* **1** of Spain or Spain and Portugal. **2** of Spain and other Spanish-speaking countries. *–n.* Spanish-speaking person living in the US. [Latin *Hispania* Spain]

hiss *–v.* **1** make a sharp sibilant sound, as of the letter *s*. **2** express disapproval of by hisses. **3** whisper urgently or angrily. *–n.* **1** sharp sibilant sound as of the letter *s*. **2** *Electronics* interference at audio frequencies. [imitative]

histamine /'hɪstə,mi:n/ *n.* chemical compound in body tissues etc., associated with allergic reactions. [from HISTOLOGY, AMINE]

histogram /'hɪstə,græm/ *n.* statistical diagram of rectangles with areas proportional to the value of a number of variables. [Greek *histos* mast]

histology /hɪ'stɒlədʒɪ/ *n.* the study of tissue structure. [Greek *histos* web]

historian /hɪ'stɔ:rɪən/ *n.* **1** writer of history. **2** person learned in history.

historic /hɪ'stɒrɪk/ *adj.* **1** famous or important in history or potentially so (*historic moment*). **2** *Gram.* (of a tense) used to narrate past events.

historical *adj.* **1** of or concerning history (*historical evidence*). **2** (of the study of a subject) showing its development over a period. **3** factual, not fictional or legendary. **4** belonging to the past, not the present. **5** (of a novel etc.) dealing with historical events. □ **historically** *adv.*

historicism /hɪ'stɒrɪ,sɪz(ə)m/ *n.* **1** theory that social and cultural phenomena are determined by history. **2** belief that historical events are governed by laws.

historicity /,hɪstə'rɪsɪtɪ/ *n.* historical truth or authenticity.

historiography /hɪ,stɔ:rɪ'ɒgrəfɪ/ *n.* **1** the writing of history. **2** the study of this. □ **historiographer** *n.*

history /'hɪstərɪ/ *n.* (*pl.* **-ies**) **1** continuous record of (esp. public) events. **2 a** the study of past events, esp. human affairs. **b** total accumulation of past events, esp. relating to human affairs or a particular nation, person, thing, etc. **3** eventful past (*this house has a history*). **4** (foll. by *of*) past record (*had a history of illness*). **5 a** systematic or critical account of or research into past events etc. **b** similar record or account of natural phenomena. **6** historical play. □ **make history** do something memorable. [Greek *historia* inquiry]

histrionic /,hɪstrɪ'ɒnɪk/ *–adj.* (of behaviour) theatrical, dramatic. *–n.* (in *pl.*) insincere and dramatic behaviour designed to impress. [Latin *histrio* actor]

hit *–v.* (**-tt-**; *past* and *past part.* **hit**) **1 a** strike with a blow or missile. **b** (of a moving body) strike with force (*the* *plane hit the ground*). **c** reach (a target etc.) with a directed missile (*hit the wicket*). **2** cause to suffer; affect adversely. **3** (often foll. by *at, against*) direct a blow. **4** (often foll. by *against, on*) knock (a part of the body) (*hit his head*). **5** achieve, reach (*hit the right tone*; *can't hit the high notes*). **6** *colloq.* **a** encounter (*hit a snag*). **b** arrive at (*hit town*). **c** indulge heavily in, esp. liquor etc. (*hit the bottle*). **7** esp. *US slang* rob or kill. **8** occur forcefully to (*it only hit him later*). **9** **a** propel (a ball etc.) with a bat etc. to score runs or points. **b** score in this way (*hit a six*). –*n.* **1** **a** blow, stroke. **b** collision. **2** shot etc. that hits its target. **3** *colloq.* popular success. □ **hit back** retaliate. **hit below the belt** **1** esp. *Boxing* give a foul blow. **2** treat or behave unfairly. **hit the hay** (or **sack**) *colloq.* go to bed. **hit it off** (often foll. by *with, together*) *colloq.* get on well (with a person). **hit the nail on the head** state the truth exactly. **hit on** (or **upon**) find by chance. **hit out** deal vigorous physical or verbal blows. **hit the road** *slang* depart. **hit the roof** see ROOF. [Old English from Old Norse]

hit-and-run *attrib. adj.* **1** (of a driver, raider, etc.) causing damage or injury and leaving the scene immediately. **2** (of an accident, attack, etc.) perpetrated by such a person or people.

hitch –*v.* **1** fasten or be fastened with a loop, hook, etc.; tether. **2** move (a thing) slightly or with a jerk. **3** *colloq.* **a** = HITCHHIKE. **b** obtain (a lift) by hitchhiking. –*n.* **1** temporary obstacle or snag. **2** abrupt pull or push. **3** noose or knot of various kinds. **4** *colloq.* free ride in a vehicle. □ **get hitched** *colloq.* marry. **hitch up** lift (esp. clothing) with a jerk. [origin uncertain]

hitchhike *v.* (**-king**) travel by seeking free lifts in passing vehicles. □ **hitchhiker** *n.*

hi-tech /'haɪtek/ *adj.* = HIGH-TECH. [abbreviation]

hither /'hɪðə(r)/ *adv. formal* to or towards this place. [Old English]

hither and thither *adv.* to and fro.

hitherto /ˌhɪðə'tu:/ *adv.* until this time, up to now.

hit list *n. slang* list of prospective victims.

hit man *n. slang* hired assassin.

hit-or-miss *adj.* liable to error, random.

hit parade *n. colloq.* list of the current best-selling pop records.

Hittite /'hɪtaɪt/ –*n.* member or language of an ancient people of Asia Minor and Syria. –*adj.* of the Hittites. [Hebrew]

HIV *abbr.* human immunodeficiency virus, either of two viruses causing Aids.

hive *n.* beehive. □ **hive off** (**-ving**) separate from a larger group. [Old English]

hives *n.pl.* skin-eruption, esp. nettle-rash. [origin unknown]

Hizbollah var. of HEZBOLLAH.

HM *abbr.* Her (or His) Majesty('s).

HMG *abbr.* Her (or His) Majesty's Government.

HMI *abbr.* Her (or His) Majesty's Inspector (of Schools).

HMS *abbr.* Her (or His) Majesty's Ship.

HMSO *abbr.* Her (or His) Majesty's Stationery Office.

HNC *abbr.* Higher National Certificate.

HND *abbr.* Higher National Diploma.

Ho *symb.* holmium.

ho /həʊ/ *int.* expressing triumph, derision, etc., or calling attention. [natural exclamation]

hoard –*n.* stock or store (esp. of money or food). –*v.* amass and store. □ **hoarder** *n.* [Old English]

hoarding *n.* **1** large, usu. wooden, structure used to carry advertisements etc. **2** temporary fence round a building site etc. [obsolete *hoard* from French: *hourd*]

hoar-frost *n.* frozen water vapour on vegetation etc. [Old English]

hoarse *adj.* **1** (of the voice) rough and deep; husky, croaking. **2** having such a voice. □ **hoarsely** *adv.* **hoarseness** *n.* [Old Norse]

hoary *adj.* (**-ier**, **-iest**) **1** **a** (of hair) grey or white with age. **b** having such hair; aged. **2** old and trite (*hoary joke*). [Old English]

hoax –*n.* humorous or malicious deception. –*v.* deceive (a person) with a hoax. [probably a shortening of *hocus* in HOCUS-POCUS]

hob *n.* **1** flat heating surface with hotplates or burners, on a cooker or as a separate unit. **2** flat metal shelf at the side of a fireplace for heating a pan etc. [perhaps var. of HUB]

hobble /'hɒb(ə)l/ –*v.* (**-ling**) **1** walk lamely; limp. **2** tie together the legs of (a horse etc.) to prevent it from straying. –*n.* **1** uneven or infirm gait. **2** rope etc. for hobbling a horse etc. [probably Low German]

hobby /'hɒbɪ/ *n.* (*pl.* **-ies**) leisure-time activity pursued for pleasure. [from the name *Robin*]

hobby-horse *n.* **1** child's toy consisting of a stick with a horse's head. **2** favourite subject or idea.

hobgoblin /'hɒbˌgɒblɪn/ *n.* mischievous imp; bogy. [from HOBBY, GOBLIN]

hobnail *n.* heavy-headed nail for boot-soles. [from HOB]

hobnob /'hɒbnɒb/ *v.* (**-bb-**) (usu. foll. by *with*) mix socially or informally. [*hab nab* have or not have]

hobo /'həʊbəʊ/ *n.* (*pl.* **-es** or **-s**) *US* wandering worker; tramp. [origin unknown]

Hobson's choice /'hɒbs(ə)nz/ *n.* choice of taking the thing offered or nothing. [*Hobson*, name of a carrier who let out horses thus]

hock[1] *n.* joint of a quadruped's hind leg between the knee and the fetlock. [Old English]

hock[2] *n.* German white wine from the Rhineland. [*Hochheim* in Germany]

hock[3] *v.* esp. *US colloq.* pawn; pledge. □ **in hock 1** in pawn. **2** in debt. **3** in prison. [Dutch]

hockey /'hɒkɪ/ *n.* team game with hooked sticks and a small hard ball. [origin unknown]

hocus-pocus /ˌhəʊkəs'pəʊkəs/ *n.* deception; trickery. [sham Latin]

hod *n.* **1** V-shaped trough on a pole used for carrying bricks etc. **2** portable receptacle for coal. [French *hotte* pannier]

hodgepodge var. of HOTCHPOTCH.

Hodgkin's disease /'hɒdʒkɪnz/ *n.* malignant disease of lymphatic tissues, usu. characterized by enlargement of the lymph nodes. [*Hodgkin*, name of a physician]

hoe *–n.* long-handled tool with a blade, used for weeding etc. *–v.* (**hoes, hoed, hoeing**) weed (crops); loosen (earth); dig up with a hoe. [French from Germanic]

hog *–n.* **1** castrated male pig. **2** *colloq.* greedy person. *–v.* (**-gg-**) *colloq.* take greedily; hoard selfishly; monopolize. □ **go the whole hog** *colloq.* do something completely or thoroughly. □ **hoggish** *adj.* [Old English]

hogmanay /'hɒgməˌneɪ/ *n. Scot.* New Year's Eve. [probably French]

hogshead *n.* **1** large cask. **2** liquid or dry measure (about 50 gallons). [from HOG: the reason for the name is unknown]

hogwash *n. colloq.* nonsense, rubbish.

ho-ho /həʊ'həʊ/ *int.* **1** representing a deep jolly laugh. **2** expressing surprise, triumph, or derision. [reduplication of HO]

hoick *v. colloq.* (often foll. by *out*) lift or pull, esp. with a jerk. [perhaps var. of HIKE]

hoi polloi /ˌhɔɪ pə'lɔɪ/ *n.* the masses; the common people. [Greek, = the many]

■ **Usage** This phrase is often preceded by *the*, which is, strictly speaking, unnecessary, since *hoi* means 'the'.

hoist *–v.* **1** raise or haul up. **2** raise by means of ropes and pulleys etc. *–n.* **1** act of hoisting, lift. **2** apparatus for hoisting. □ **hoist with one's own petard** caught by one's own trick etc. [earlier *hoise*, probably from Low German]

hoity-toity /ˌhɔɪtɪ'tɔɪtɪ/ *adj.* haughty. [obsolete *hoit* romp]

hokum /'həʊkəm/ *n.* esp. *US slang* **1** sentimental, sensational, or unreal material in a film or play etc. **2** bunkum; rubbish. [origin unknown]

hold[1] /həʊld/ *–v.* (*past* and *past part.* **held**) **1 a** keep fast; grasp (esp. in the hands or arms). **b** (also *refl.*) keep or sustain (a thing, oneself, one's head, etc.) in a particular position. **c** grip so as to control (*hold the reins*). **2** have the capacity for, contain (*holds two pints*). **3** possess, gain, or have, esp.: **a** be the owner or tenant of (land, property, stocks, etc.). **b** gain or have gained (a qualification, record, etc.). **c** have the position of (a job or office). **d** keep possession of (a place etc.), esp. against attack. **4** remain unbroken; not give way (*roof held under the storm*). **5** celebrate or conduct (a meeting, festival, conversation, etc.). **6 a** keep (a person etc.) in a place or condition (*held him in suspense*). **b** detain, esp. in custody. **7 a** engross (*book held him for hours*). **b** dominate (*held the stage*). **8** (foll. by *to*) keep (a person etc.) to (a promise etc.). **9** (of weather) continue fine. **10** think, believe; assert (*held it to be plain*; *held that the earth was flat*). **11** regard with a specified feeling (*held him in contempt*). **12** cease; restrain (*hold your fire*). **13** keep or reserve (*please hold our seats*). **14** be able to drink (alcohol) without effect (*can't hold his drink*). **15** (of a court etc.) lay down; decide. **16** *Mus.* sustain (a note). **17** = *hold the line*. *–n.* **1** (foll. by *on, over*) influence or power over (*has a strange hold over me*). **2** manner of holding in wrestling etc. **3** grasp (*take hold of him*). **4** (often in *comb.*) thing to hold by (*seized the handhold*). □ **hold (a thing) against (a person)** resent or regard it as discreditable to (a person). **hold back 1** impede the progress of; restrain. **2** keep for oneself. **3** (often foll. by *from*) hesitate; refrain. **hold one's breath** see BREATH. **hold down 1** repress. **2** *colloq.* be competent enough to keep (one's job etc.). **hold the fort 1** act as a temporary substitute. **2** cope in an emergency. **hold forth** speak at length or tediously. **hold one's ground** see GROUND[1]. **hold hands** grasp one another by the hand as a sign of affection or for support or guidance. **hold it** cease action or movement. **hold**

the line not ring off (in a telephone connection). **hold one's nose** compress the nostrils to avoid a bad smell. **hold off 1** delay, not begin. **2** keep one's distance. **hold on 1** keep one's grasp on something. **2** wait a moment. **3** = *hold the line.* **hold out 1** stretch forth (a hand etc.). **2** offer (an inducement etc.). **3** maintain resistance. **4** persist or last. **hold out for** continue to demand. **hold out on** *colloq.* refuse something to (a person). **hold over** postpone. **hold one's own** maintain one's position; not be beaten. **hold one's tongue** *colloq.* remain silent. **hold to ransom 1** keep prisoner until a ransom is paid. **2** demand concessions from by threats. **hold up 1** support, sustain. **2** exhibit, display. **3** hinder, obstruct. **4** stop and rob by force. **hold water** (of reasoning) be sound, bear examination. **hold with** (usu. with *neg.*) *colloq.* approve of. **on hold** holding the telephone line. **take hold** (of a custom or habit) become established. **with no holds barred** with no restrictions of method. □ **holder** *n.* [Old English]

hold[2] /həʊld/ *n.* cavity in the lower part of a ship or aircraft for cargo. [Old English: related to HOLLOW]

holdall *n.* large soft travelling bag.

holding *n.* **1** tenure of land. **2** stocks, property, etc. held.

holding company *n.* company created to hold the shares of other companies, which it then controls.

hold-up *n.* **1** stoppage or delay. **2** robbery by force.

hole *n.* **1 a** empty space in a solid body. **b** opening in or through something. **2** animal's burrow. **3** (in games) cavity or receptacle for a ball. **4** *colloq.* small or dingy place. **5** *colloq.* awkward situation. **6** *Golf* **a** point scored by a player who gets the ball from tee to hole with the fewest strokes. **b** terrain or distance from tee to hole. □ **hole up** *US colloq.* hide oneself. **make a hole in** use a large amount of. □ **holey** *adj.* [Old English]

hole-and-corner *adj.* secret; underhand.

hole in the heart *n. colloq.* congenital defect in the heart membrane.

holiday /'hɒlɪ,deɪ/ *–n.* **1** (often in *pl.*) extended period of recreation, esp. spent away from home or travelling; break from work. **2** day of festivity or recreation when no work is done, esp. a religious festival etc. *–v.* spend a holiday. [Old English: related to HOLY, DAY]

holiday camp *n.* place for holiday-makers with facilities on site.

holiday-maker *n.* person on holiday.

holier-than-thou *adj. colloq.* self-righteous.

holiness /'həʊlɪnɪs/ *n.* **1** being holy or sacred. **2** (**Holiness**) title used when addressing or referring to the Pope. [Old English: related to HOLY]

holism /'həʊlɪz(ə)m/ *n.* (also **wholism**) **1** *Philos.* theory that certain wholes are greater than the sum of their parts. **2** *Med.* treating of the whole person rather than the symptoms of a disease. □ **holistic** /-'lɪstɪk/ *adj.* [Greek *holos* whole]

hollandaise sauce /,hɒlən'deɪz/ *n.* creamy sauce of melted butter, egg-yolks, vinegar, etc. [French]

holler *v.* & *n. US colloq.* shout. [French *holà* hello!]

hollow /'hɒləʊ/ *–adj.* **1 a** having a cavity; not solid. **b** sunken (*hollow cheeks*). **2** (of a sound) echoing. **3** empty; hungry. **4** meaningless (*hollow victory*). **5** insincere (*hollow laugh*). *–n.* **1** hollow place; hole. **2** valley; basin. *–v.* (often foll. by *out*) make hollow; excavate. *–adv. colloq.* completely (*beaten hollow*). □ **hollowly** *adv.* **hollowness** *n.* [Old English]

holly /'hɒlɪ/ *n.* (*pl.* **-ies**) evergreen shrub with prickly leaves and red berries. [Old English]

hollyhock /'hɒlɪ,hɒk/ *n.* tall plant with showy flowers. [from HOLY, obsolete *hock* mallow]

holm /həʊm/ *n.* (in full **holm-oak**) evergreen oak with holly-like young leaves. [dial. *holm* holly]

holmium /'həʊlmɪəm/ *n.* metallic element of the lanthanide series. [Latin *Holmia* Stockholm]

holocaust /'hɒlə,kɔ:st/ *n.* **1** large-scale destruction, esp. by fire or nuclear war. **2** (**the Holocaust**) mass murder of the Jews by the Nazis 1939–45. [Greek *holos* whole, *kaustos* burnt]

hologram /'hɒlə,græm/ *n.* photographic pattern that gives a three-dimensional image when illuminated by coherent light. [Greek *holos* whole, -GRAM]

holograph /'hɒlə,grɑ:f/ *–adj.* wholly written by hand by the person named as the author. *–n.* holograph document. [Greek *holos* whole, -GRAPH]

holography /hə'lɒgrəfɪ/ *n.* the study or production of holograms.

hols /hɒlz/ *n.pl. colloq.* holidays. [abbreviation]

holster /'həʊlstə(r)/ *n.* leather case for a pistol or revolver, worn on a belt etc. [Dutch]

holy /'həʊlɪ/ *adj.* (**-ier, -iest**) **1** morally and spiritually excellent or perfect, and to be revered. **2** belonging to or devoted to God. **3** consecrated, sacred. [Old English: related to WHOLE]

Holy Communion see COMMUNION.

Holy Ghost *n.* = HOLY SPIRIT.

Holy Grail see GRAIL.

Holy Land *n.* area between the River Jordan and the Mediterranean Sea.

holy of holies *n.* **1** sacred inner chamber of the Jewish temple. **2** thing regarded as most sacred.

holy orders *n.pl.* the status of a bishop, priest, or deacon.

Holy Roman Empire *n. hist.* Western part of the Roman Empire as revived by Charlemagne in 800 AD.

Holy See *n.* papacy or papal court.

Holy Spirit *n.* Third Person of the Trinity, God as spiritually acting.

Holy Week *n.* week before Easter.

Holy Writ *n.* holy writings, esp. the Bible.

homage /'hɒmɪdʒ/ *n.* tribute, expression of reverence (*pay homage to*). [Latin *homo* man]

Homburg /'hɒmbɜːg/ *n.* man's felt hat with a narrow curled brim and a lengthwise dent in the crown. [*Homburg* in Germany]

home –*n.* **1 a** place where one lives; fixed residence. **b** dwelling-house. **2** family circumstances (*comes from a good home*). **3** native land. **4** institution caring for people or animals. **5** place where a thing originates, is kept, or is native or most common. **6 a** finishing-point in a race. **b** (in games) place where one is safe; goal. **7** *Sport* home match or win. –*attrib. adj.* **1 a** of or connected with one's home. **b** carried on, done, or made, at home. **2** in one's own country (*home industries*; *the home market*). **3** *Sport* played on one's own ground etc. (*home match*). –*adv.* **1** to, at, or in one's home or country (*go home*; *is he home yet?*). **2** to the point aimed at; completely (*drove the nail home*). –*v.* (**-ming**) **1** (esp. of a trained pigeon) return home. **2** (often foll. by *on*, *in on*) (of a vessel, missile, etc.) be guided towards a destination or target. **3** send or guide homewards. □ **at home 1** in one's house or native land. **2** at ease (*make yourself at home*). **3** (usu. foll. by *in*, *on*, *with*) familiar or well informed. **4** available to callers. [Old English]

home and dry *predic. adj.* having achieved one's aim.

home-brew *n.* beer or other alcoholic drink brewed at home.

home-coming *n.* arrival at home.

Home Counties *n.pl.* the counties closest to London.

home economics *n.pl.* the study of household management.

home farm *n.* principal farm on an estate, providing produce for the owner.

home-grown *adj.* grown or produced at home.

Home Guard *n. hist.* British citizen army organized for defence in 1940.

home help *n.* person helping with housework etc., esp. one provided by a local authority.

homeland *n.* **1** one's native land. **2** any of several partially self-governing areas in S. Africa reserved for Black South Africans (the official name for a Bantustan).

homeless *adj.* lacking a home. □ **homelessness** *n.*

homely *adj.* (**-ier, -iest**) **1** simple, plain, unpretentious. **2** *US* (of facial appearance) plain, unattractive. **3** comfortable, cosy. □ **homeliness** *n.*

home-made *adj.* made at home.

Home Office *n.* British government department dealing with law and order, immigration, etc., in England and Wales.

homeopathy *US* var. of HOMOEOPATHY.

Homeric /həʊ'merɪk/ *adj.* **1** of, or in the style of, Homer. **2** of Bronze Age Greece as described in Homer's poems.

home rule *n.* government of a country or region by its own citizens.

Home Secretary *n.* Secretary of State in charge of the Home Office.

homesick *adj.* depressed by absence from home. □ **homesickness** *n.*

homespun –*adj.* **1** made of yarn spun at home. **2** plain, simple. –*n.* homespun cloth.

homestead /'həʊmsted/ *n.* house, esp. a farmhouse, and outbuildings.

home truth *n.* basic but unwelcome information about oneself.

homeward /'həʊmwəd/ –*adv.* (also **homewards**) towards home. –*adj.* going towards home.

homework *n.* **1** work to be done at home, esp. by a school pupil. **2** preparatory work or study.

homey *adj.* (also **homy**) (**-mier, -miest**) suggesting home; cosy.

homicide /'hɒmɪˌsaɪd/ *n.* **1** killing of a human being by another. **2** person who kills a human being. □ **homicidal** /-'saɪd(ə)l/ *adj.* [Latin *homo* man]

homily /'hɒmɪlɪ/ *n.* (*pl.* **-ies**) **1** sermon. **2** tedious moralizing discourse. □ **homiletic** /-'letɪk/ *adj.* [Greek *homilia*]

homing *attrib. adj.* **1** (of a pigeon) trained to fly home. **2** (of a device) for guiding to a target etc.

hominid /'hɒmɪnɪd/ –*adj.* of the primate family including humans and their fossil ancestors. –*n.* member of this family. [Latin *homo homin-* man]

hominoid /'hɒmɪ,nɔɪd/ *–adj.* like a human. *–n.* animal resembling a human.

homo /'həʊməʊ/ *n.* (*pl.* **-s**) *colloq. offens.* homosexual. [abbreviation]

homo- *comb. form* same. [Greek *homos* same]

homoeopathy /,həʊmɪ'ɒpəθɪ/ *n.* (*US* **homeopathy**) treatment of disease by minute doses of drugs that in a healthy person would produce symptoms of the disease. □ **homoeopath** /'həʊmɪəʊ,pæθ/ *n.* **homoeopathic** /-'pæθɪk/ *adj.* [Greek *homoios* like: related to PATHOS]

homogeneous /,həʊməʊ'dʒi:nɪəs/ *adj.* **1** of the same kind. **2** consisting of parts all of the same kind; uniform. □ **homogeneity** /-dʒɪ'ni:ɪtɪ/ *n.* **homogeneously** *adv.* [from HOMO-, Greek *genos* kind]

■ **Usage** *Homogeneous* is often confused with *homogenous* which is a term in biology meaning 'similar owing to common descent'.

homogenize /hə'mɒdʒɪ,naɪz/ *v.* (also **-ise**) (**-zing** or **-sing**) **1** make homogeneous. **2** treat (milk) so that the fat droplets are emulsified and the cream does not separate.

homograph /'hɒmə,grɑ:f/ *n.* word spelt like another but of different meaning or origin (e.g. POLE[1], POLE[2]).

homologous /hə'mɒləgəs/ *adj.* **1 a** having the same relation, relative position, etc. **b** corresponding. **2** *Biol.* (of organs etc.) similar in position and structure but not necessarily in function. [from HOMO-, Greek *logos* ratio]

homology /hə'mɒlədʒɪ/ *n.* homologous state or relation; correspondence.

homonym /'hɒmənɪm/ *n.* **1** word spelt or pronounced like another but of different meaning; homograph or homophone. **2** namesake. [from HOMO-, *onoma* name]

homophobia /,həʊmə'fəʊbɪə/ *n.* hatred or fear of homosexuals. □ **homophobe** /'həʊm-/ *n.* **homophobic** *adj.*

homophone /'hɒmə,fəʊn/ *n.* word pronounced like another but of different meaning or origin (e.g. *pair*, *pear*). [from HOMO-, Greek *phōnē* sound]

Homo sapiens /,həʊməʊ 'sæpɪenz/ *n.* modern humans regarded as a species. [Latin, = wise man]

homosexual /,həʊməʊ'sekʃʊəl/ *–adj.* feeling or involving sexual attraction only to people of the same sex. *–n.* homosexual person. □ **homosexuality** /-'ælɪtɪ/ *n.* [from HOMO-, SEXUAL]

homy var. of HOMEY.

Hon. *abbr.* **1** Honorary. **2** Honourable.

hone *–n.* whetstone, esp. for razors. *–v.* (**-ning**) sharpen on or as on a hone. [Old English]

honest /'ɒnɪst/ *–adj.* **1** fair and just; not cheating or stealing. **2** free of deceit and untruthfulness; sincere. **3** fairly earned (*an honest living*). **4** blameless but undistinguished. *–adv. colloq.* genuinely, really. [Latin *honestus*]

honestly *adv.* **1** in an honest way. **2** really (*I don't honestly know*).

honesty *n.* **1** being honest. **2** truthfulness. **3** plant with purple or white flowers and flat round semi-transparent seed-pods.

honey /'hʌnɪ/ *n.* (*pl.* **-s**) **1** sweet sticky yellowish fluid made by bees from nectar. **2** colour of this. **3 a** sweetness. **b** sweet thing. **4** esp. *US* (usu. as a form of address) darling. [Old English]

honey-bee *n.* common hive-bee.

honeycomb *–n.* **1** bees' wax structure of hexagonal cells for honey and eggs. **2** pattern arranged hexagonally. *–v.* **1** fill with cavities or tunnels, undermine. **2** mark with a honeycomb pattern. [Old English]

honeydew *n.* **1** sweet sticky substance excreted by aphids on leaves and stems. **2** variety of melon.

honeyed *adj.* (of words, flattery, etc.) sweet, sweet-sounding.

honeymoon *–n.* **1** holiday taken by a newly married couple. **2** initial period of enthusiasm or goodwill. *–v.* spend a honeymoon. □ **honeymooner** *n.*

honeysuckle *n.* climbing shrub with fragrant yellow or pink flowers.

honk *–n.* **1** sound of a car horn. **2** cry of a wild goose. *–v.* (cause to) make a honk. [imitative]

honky-tonk /'hɒŋkɪ,tɒŋk/ *n. colloq.* **1** ragtime piano music. **2** cheap or disreputable nightclub etc. [origin unknown]

honor *US* var. of HONOUR.

honorable *US* var. of HONOURABLE.

honorarium /,ɒnə'reərɪəm/ *n.* (*pl.* **-s** or **-ria**) fee, esp. a voluntary payment for professional services rendered without the normal fee. [Latin: related to HONOUR]

honorary /'ɒnərərɪ/ *adj.* **1** conferred as an honour (*honorary degree*). **2** (of an office or its holder) unpaid.

honorific /,ɒnə'rɪfɪk/ *adj.* **1** conferring honour. **2** implying respect.

honour /'ɒnə(r)/ (*US* **honor**) *–n.* **1** high respect, public regard. **2** adherence to what is right or an accepted standard of conduct. **3** nobleness of mind, magnanimity (*honour among thieves*). **4** thing conferred as a distinction, esp. an official award for bravery or achievement. **5** privilege, special right (*had the* *honour of being invited*). **6 a** exalted position. **b** (**Honour**) (prec. by *your*, *his*, etc.) title of a circuit judge etc. **7** (foll. by *to*) person or thing that brings honour (*an honour to her profession*). **8 a** chastity (of a woman). **b** reputation for this. **9** (in *pl.*) specialized degree course or special distinction in an examination. **10** (in card-games) the four or five highest-ranking cards. **11** *Golf* the right of driving off first. *–v.* **1** respect highly. **2** confer honour on. **3** accept or pay (a bill or cheque) when due. □ **do the honours** perform the duties of a host to guests etc. **in honour of** as a celebration of. **on one's honour** (usu. foll. by *to* + infin.) under a moral obligation. [Latin *honor* repute]

honourable *adj.* (*US* **honorable**) **1** deserving, bringing, or showing honour. **2** (**Honourable**) title indicating distinction, given to certain high officials, the children of certain ranks of the nobility, and (in the House of Commons) to MPs. □ **honourably** *adv.*

hooch *n. US colloq.* alcoholic liquor, esp. inferior or illicit whisky. [Alaskan]

hood[1] /hʊd/ *–n.* **1 a** covering for the head and neck, esp. as part of a garment. **b** separate hoodlike garment. **2** folding top of a car etc. **3** *US* bonnet of a car etc. **4** protective cover. *–v.* cover with or as with a hood. [Old English]

hood[2] /hʊd/ *n. US slang* gangster, gunman. [abbreviation of HOODLUM]

-hood *suffix* forming nouns: **1** of condition or state (*childhood; falsehood*). **2** designating a group (*sisterhood*; *neighbourhood*). [Old English]

hooded *adj.* **1** having a hood. **2** (of an animal) having a hoodlike part (*hooded crow*).

hoodlum /ˈhu:dləm/ *n.* **1** street hooligan, young thug. **2** gangster. [origin unknown]

hoodoo /ˈhu:du:/ *n.* esp. *US* **1 a** bad luck. **b** thing or person that brings this. **2** voodoo. [alteration of VOODOO]

hoodwink *v.* deceive, delude. [from HOOD[1]: originally = 'blindfold']

hoof /hu:f/ *n.* (*pl.* **-s** or **hooves** /-vz/) horny part of the foot of a horse etc. □ **hoof it** *slang* go on foot. [Old English]

hoo-ha /ˈhu:hɑ:/ *n. slang* commotion. [origin unknown]

hook /hʊk/ *–n.* **1 a** bent or curved piece of metal etc. for catching hold or for hanging things on. **b** (in full **fish-hook**) bent piece of wire for catching fish. **2** curved cutting instrument (*reaping-hook*). **3** bend in a river, curved strip of land, etc. **4 a** hooking stroke. **b** *Boxing* short swinging blow. *–v.* **1** grasp or secure with hook(s). **2** catch with or as with a hook. **3** *slang* steal. **4** (in sports) send (the ball) in a curve or deviating path. **5** *Rugby* secure (the ball) and pass it backward with the foot in the scrum. □ **by hook or by crook** by one means or another. **off the hook 1** *colloq.* out of difficulty or trouble. **2** (of a telephone receiver) not on its rest. [Old English]

hookah /ˈhʊkə/ *n.* oriental tobacco-pipe with a long tube passing through water for cooling the smoke as it is drawn through. [Urdu from Arabic, = casket]

hook and eye *n.* small metal hook and loop as a fastener on a garment.

hooked *adj.* **1** hook-shaped. **2** (often foll. by *on*) *slang* addicted or captivated.

hooker *n.* **1** *Rugby* player in the front row of the scrum who tries to hook the ball. **2** *slang* prostitute.

hookey /ˈhʊkɪ/ *n. US* □ **play hookey** *slang* play truant. [origin unknown]

hook, line, and sinker *adv.* entirely.

hook-up *n.* connection, esp. of broadcasting equipment.

hookworm *n.* worm with hooklike mouthparts, infesting humans and animals.

hooligan /ˈhu:lɪgən/ *n.* young ruffian. □ **hooliganism** *n.* [origin unknown]

hoop *–n.* **1** circular band of metal, wood, etc., esp. as part of a framework. **2** ring bowled along by a child, or for circus performers to jump through. **3** arch through which balls are hit in croquet. *–v.* bind or encircle with hoop(s). □ **be put** (or **go**) **through the hoop** (or **hoops**) undergo rigorous testing. [Old English]

hoop-la *n.* fairground game with rings thrown to encircle a prize.

hoopoe /ˈhu:pu:/ *n.* salmon-pink bird with black and white wings and a large erectile crest. [Latin *upupa* (imitative of its cry)]

hooray /hʊˈreɪ/ *int.* = HURRAH.

Hooray Henry /ˈhu:reɪ/ *n. slang* loud upper-class young man.

hoot *–n.* **1** owl's cry. **2** sound made by a car's horn etc. **3** shout expressing scorn or disapproval. **4** *colloq.* **a** laughter. **b** cause of this. **5** (also **two hoots**) *slang* anything at all, in the slightest degree (*don't care a hoot*; *doesn't matter two hoots*). *–v.* **1** utter or make hoot(s). **2** greet or drive away with scornful hoots. **3** sound (a car horn etc.). [imitative]

hooter *n.* **1** thing that hoots, esp. a car's horn or a siren. **2** *slang* nose.

Hoover *–n. propr.* vacuum cleaner. *–v.* (**hoover**) **1** (also *absol.*) clean with a vacuum cleaner. **2** (foll. by *up*) **a** suck up with a vacuum cleaner. **b** clean a room etc. with a vacuum cleaner. [name of the manufacturer]

hooves *pl.* of HOOF.

hop[1] –*v.* (**-pp-**) **1** (of a bird, frog, etc.) spring with two or all feet at once. **2** (of a person) jump on one foot. **3** move or go quickly (*hopped over the fence*). **4** cross (a ditch etc.) by hopping. –*n.* **1** hopping movement. **2** *colloq.* informal dance. **3** short journey, esp. a flight. □ **hop in** (or **out**) *colloq.* get into (or out of) a car etc. **hop it** *slang* go away. **on the hop** *colloq.* unprepared (*caught on the hop*). [Old English]

hop[2] *n.* **1** climbing plant bearing cones. **2** (in *pl.*) its ripe cones, used to flavour beer. [Low German or Dutch]

hope –*n.* **1** expectation and desire for a thing. **2** person or thing giving cause for hope. **3** what is hoped for. –*v.* (**-ping**) **1** feel hope. **2** expect and desire. **3** feel fairly confident. □ **hope against hope** cling to a mere possibility. [Old English]

hopeful –*adj.* **1** feeling hope. **2** causing or inspiring hope. **3** likely to succeed, promising. –*n.* person likely to succeed.

hopefully *adv.* **1** in a hopeful manner. **2** it is to be hoped (*hopefully, we will succeed*).

■ **Usage** The use of *hopefully* in sense 2 is common, but is considered incorrect by some people.

hopeless *adj.* **1** feeling no hope. **2** admitting no hope (*hopeless case*). **3** incompetent. □ **hopelessly** *adv.* **hopelessness** *n.*

hopper[1] *n.* **1** container tapering downward to an opening for discharging its contents. **2** hopping insect.

hopper[2] *n.* hop-picker.

hopping mad *predic. adj. colloq.* very angry.

hopscotch *n.* children's game of hopping over squares marked on the ground to retrieve a stone etc. [from HOP[1], SCOTCH]

horde *n.* usu. *derog.* large group, gang. [Turkish *ordū* camp]

horehound /'hɔ:haʊnd/ *n.* herbaceous plant yielding a bitter aromatic juice used against coughs etc. [Old English, = hoary herb]

horizon /hə'raɪz(ə)n/ *n.* **1** line at which the earth and sky appear to meet. **2** limit of mental perception, experience, interest, etc. □ **on the horizon** (of an event) just imminent or becoming apparent. [Greek *horizō* bound]

horizontal /,hɒrɪ'zɒnt(ə)l/ –*adj.* **1** parallel to the plane of the horizon, at right angles to the vertical. **2** of or concerned with the same work, status, etc. (*it was a horizontal move rather than promotion*). –*n.* horizontal line, plane, etc. □ **horizontality** /-'tælɪtɪ/ *n.* **horizontally** *adv.*

hormone /'hɔ:məʊn/ *n.* **1** regulatory substance produced in an organism and transported in tissue fluids to stimulate cells or tissues into action. **2** similar synthetic substance. □ **hormonal** /-'məʊn(ə)l/ *adj.* [Greek *hormaō* impel]

hormone replacement therapy *n.* treatment to relieve menopausal symptoms by boosting a woman's oestrogen levels.

horn *n.* **1 a** hard outgrowth, often curved and pointed, on the head of esp. hoofed animals. **b** each of two branched appendages on the head of (esp. male) deer. **c** hornlike projection on animals, e.g. a snail's tentacle. **2** substance of which horns are made. **3** *Mus.* **a** = FRENCH HORN. **b** wind instrument played by lip vibration, orig. made of horn, now usu. of brass. **4** instrument sounding a warning. **5** receptacle or instrument made of horn. **6** horn-shaped projection. **7** extremity of the moon or other crescent. **8** arm of a river etc. □ **horn in** *slang* intrude, interfere. □ **horned** *adj.* **hornist** *n.* (in sense 3 of *n.*). [Old English]

hornbeam *n.* tree with a hard tough wood.

hornbill *n.* bird with a hornlike excrescence on its large curved bill.

hornblende /'hɔ:nblend/ *n.* dark-brown, black, or green mineral occurring in many rocks. [German]

hornet /'hɔ:nɪt/ *n.* large wasp capable of inflicting a serious sting. [Low German or Dutch]

horn of plenty *n.* a cornucopia.

hornpipe *n.* **1** lively dance (esp. associated with sailors). **2** music for this.

horn-rimmed *adj.* (esp. of spectacles) having rims made of horn or a similar substance.

horny *adj.* (**-ier, -iest**) **1** of or like horn. **2** hard like horn. **3** *slang* sexually excited. □ **horniness** *n.*

horology /hə'rɒlədʒɪ/ *n.* art of measuring time or making clocks, watches, etc. □ **horological** /,hɒrə'lɒdʒɪk(ə)l/ *adj.* [Greek *hōra* time]

horoscope /'hɒrə,skəʊp/ *n.* **1** forecast of a person's future from a diagram showing the relative positions of the stars and planets at his or her birth. **2** such a diagram. [Greek *hōra* time, *skopos* observer]

horrendous /hə'rendəs/ *adj.* horrifying. □ **horrendously** *adv.* [Latin: related to HORRIBLE]

horrible /ˈhɒrɪb(ə)l/ *adj.* **1** causing or likely to cause horror. **2** *colloq.* unpleasant. □ **horribly** *adv.* [Latin *horreo* bristle, shudder at]

horrid /ˈhɒrɪd/ *adj.* **1** horrible, revolting. **2** *colloq.* unpleasant (*horrid weather*).

horrific /həˈrɪfɪk/ *adj.* horrifying. □ **horrifically** *adv.*

horrify /ˈhɒrɪˌfaɪ/ *v.* (**-ies, -ied**) arouse horror in; shock. □ **horrifying** *adj.*

horror /ˈhɒrə(r)/ *–n.* **1** painful feeling of loathing and fear. **2 a** (often foll. by *of*) intense dislike. **b** (often foll. by *at*) *colloq.* intense dismay. **3 a** person or thing causing horror. **b** *colloq.* bad or mischievous person etc. **4** (in *pl.*; prec. by *the*) fit of depression, nervousness, etc. *–attrib. adj.* (of films etc.) designed to interest by arousing feelings of horror.

hors d'œuvre /ɔːˈdɜːvr/ *n.* food served as an appetizer at the start of a meal. [French, = outside the work]

horse *–n.* **1 a** large four-legged mammal with flowing mane and tail, used for riding and to carry and pull loads. **b** adult male horse; stallion or gelding. **c** (*collect.*; as *sing.*) cavalry. **2** vaulting-block. **3** supporting frame (*clothes-horse*). *–v.* (**-sing**) (foll. by *around*) fool about. □ **from the horse's mouth** *colloq.* (of information etc.) from the original or an authoritative source. [Old English]

horseback *n.* □ **on horseback** mounted on a horse.

horsebox *n.* closed vehicle for transporting horse(s).

horse-brass *n.* brass ornament orig. for a horse's harness.

horse chestnut *n.* **1** large tree with upright conical clusters of flowers. **2** dark brown fruit of this.

horse-drawn *adj.* (of a vehicle) pulled by a horse or horses.

horseflesh *n.* **1** flesh of a horse, esp. as food. **2** horses collectively.

horsefly *n.* any of various biting insects troublesome esp. to horses.

Horse Guards *n.pl.* cavalry brigade of the household troops.

horsehair *n.* hair from the mane or tail of a horse, used for padding etc.

horseman *n.* **1** rider on horseback. **2** skilled rider. □ **horsemanship** *n.*

horseplay *n.* boisterous play.

horsepower *n.* (*pl.* same) imperial unit of power (about 750 watts), esp. for measuring the power of an engine.

horse-race *n.* race between horses with riders. □ **horse-racing** *n.*

horseradish *n.* plant with a pungent root used to make a sauce.

horse sense *n. colloq.* plain common sense.

horseshoe *n.* **1** U-shaped iron shoe for a horse. **2** thing of this shape.

horsetail *n.* **1** horse's tail. **2** plant resembling it.

horsewhip *–n.* whip for driving horses. *–v.* (**-pp-**) beat with a horsewhip.

horsewoman *n.* **1** woman who rides on horseback. **2** skilled woman rider.

horsy *adj.* (**-ier, -iest**) **1** of or like a horse. **2** concerned with or devoted to horses.

horticulture /ˈhɔːtɪˌkʌltʃə(r)/ *n.* art of garden cultivation. □ **horticultural** /-ˈkʌltʃər(ə)l/ *adj.* **horticulturist** /-ˈkʌltʃərɪst/ *n.* [Latin *hortus* garden, CULTURE]

hosanna /həʊˈzænə/ *n.* & *int.* shout of adoration (Matt. 21:9, 15, etc.). [Hebrew]

hose /həʊz/ *–n.* **1** (also **hose-pipe**) flexible tube for conveying water. **2 a** (*collect.*; as *pl.*) stockings and socks. **b** *hist.* breeches (*doublet and hose*). *–v.* (**-sing**) (often foll. by *down*) water, spray, or drench with a hose. [Old English]

hosier /ˈhəʊzɪə(r)/ *n.* dealer in hosiery.

hosiery *n.* stockings and socks.

hospice /ˈhɒspɪs/ *n.* **1** home for people who are ill (esp. terminally) or destitute. **2** lodging for travellers, esp. one kept by a religious order. [Latin: related to HOST[2]]

hospitable /hɒˈspɪtəb(ə)l/ *adj.* giving hospitality. □ **hospitably** *adv.* [Latin *hospito* entertain: related to HOST[2]]

hospital /ˈhɒspɪt(ə)l/ *n.* **1** institution providing medical and surgical treatment and nursing care for ill and injured people. **2** *hist.* hospice. [Latin: related to HOST[2]]

hospitality /ˌhɒspɪˈtælɪtɪ/ *n.* friendly and generous reception and entertainment of guests or strangers.

hospitalize /ˈhɒspɪtəˌlaɪz/ *v.* (also **-ise**) (**-zing** or **-sing**) send or admit (a patient) to hospital. □ **hospitalization** /-ˈzeɪʃ(ə)n/ *n.*

host[1] /həʊst/ *n.* (usu. foll. by *of*) large number of people or things. [Latin *hostis* enemy, army]

host[2] /həʊst/ *–n.* **1** person who receives or entertains another as a guest. **2** compère. **3** *Biol.* animal or plant having a parasite. **4** recipient of a transplanted organ etc. **5** landlord of an inn. *–v.* be host to (a person) or of (an event). [Latin *hospes hospitis* host, guest]

host[3] /həʊst/ *n.* (usu. prec. by *the*; often **Host**) bread consecrated in the Eucharist. [Latin *hostia* victim]

hostage /ˈhɒstɪdʒ/ *n.* person seized or held as security for the fulfilment of a condition. [Latin *obses obsidis* hostage]

hostel /ˈhɒst(ə)l/ *n.* **1** house of residence or lodging for students, nurses, etc. **2** = YOUTH HOSTEL. [medieval Latin: related to HOSPITAL]

hostelling *n.* (*US* **hosteling**) practice of staying in youth hostels. □ **hosteller** *n.*

hostelry *n.* (*pl.* **-ies**) *archaic* inn.

hostess /ˈhəʊstɪs/ *n.* **1** woman who receives or entertains a guest. **2** woman employed to entertain customers at a nightclub etc. **3** stewardess on an aircraft etc. [related to HOST[2]]

hostile /ˈhɒstaɪl/ *adj.* **1** of an enemy. **2** (often foll. by *to*) unfriendly, opposed. □ **hostilely** *adv.* [Latin: related to HOST[1]]

hostility /hɒˈstɪlɪtɪ/ *n.* (*pl.* **-ies**) **1** being hostile, enmity. **2** state of warfare. **3** (in *pl.*) acts of warfare.

hot –*adj.* (**hotter, hottest**) **1** having a high temperature. **2** causing a sensation of heat (*hot flush*). **3** (of pepper, spices, etc.) pungent. **4** (of a person) feeling heat. **5 a** ardent, passionate, excited. **b** (often foll. by *for, on*) eager, keen (*in hot pursuit*). **c** angry or upset. **6** (of news etc.) fresh, recent. **7** *Hunting* (of the scent) fresh, recent. **8 a** (of a player, competitor, or feat) very skilful, formidable. **b** (foll. by *on*) knowledgeable about. **9** (esp. of jazz) strongly rhythmical. **10** *slang* (of stolen goods) difficult to dispose of because identifiable. **11** *slang* radioactive. –*v.* (**-tt-**) (usu. foll. by *up*) *colloq.* **1** make or become hot. **2** make or become more active, exciting, or dangerous. □ **have the hots for** *slang* be sexually attracted to. **hot under the collar** angry, resentful, embarrassed. **like hot cakes** see CAKE. **make it** (or **things**) **hot for a person** persecute a person. □ **hotly** *adv.* **hotness** *n.* **hottish** *adj.* [Old English]

hot air *n. slang* empty or boastful talk.

hot-air balloon *n.* balloon containing air heated by burners below it, causing it to rise.

hotbed *n.* **1** (foll. by *of*) environment conducive to (vice, intrigue, etc.). **2** bed of earth heated by fermenting manure.

hot-blooded *adj.* ardent, passionate.

hotchpotch /ˈhɒtʃpɒtʃ/ *n.* (also **hodgepodge** /ˈhɒdʒpɒdʒ/) confused mixture or jumble, esp. of ideas. [French *hochepot* shake pot]

hot cross bun *n.* bun marked with a cross and traditionally eaten on Good Friday.

hot dog *n. colloq.* hot sausage in a soft roll.

hotel /həʊˈtel/ *n.* (usu. licensed) establishment providing accommodation and meals for payment. [French: related to HOSTEL]

hotelier /həʊˈtelɪə(r)/ *n.* hotel-keeper.

hot flush see FLUSH[1].

hotfoot –*adv.* in eager haste. –*v.* hurry eagerly (esp. *hotfoot it*).

hot gospeller *n. colloq.* eager preacher of the gospel.

hothead *n.* impetuous person. □ **hotheaded** *adj.* **hotheadedness** *n.*

hothouse *n.* **1** heated (mainly glass) building for rearing tender plants. **2** environment conducive to the rapid growth or development of something.

hot line *n.* direct exclusive telephone etc. line, esp. for emergencies.

hot money *n.* capital frequently transferred.

hotplate *n.* heated metal plate etc. (or a set of these) for cooking food or keeping it hot.

hotpot *n.* casserole of meat and vegetables topped with potato.

hot potato *n. colloq.* contentious matter.

hot rod *n.* vehicle modified to have extra power and speed.

hot seat *n. slang* **1** position of difficult responsibility. **2** electric chair.

hot spot *n.* **1** small region that is relatively hot. **2** lively or dangerous place.

hot stuff *n. colloq.* **1** formidably capable or important person or thing. **2** sexually attractive person. **3** erotic book, film, etc.

hot-tempered *adj.* impulsively angry.

Hottentot /ˈhɒtən,tɒt/ *n.* **1** member of a SW African Negroid people. **2** their language. [Afrikaans]

hot water *n. colloq.* difficulty or trouble.

hot-water bottle *n.* (usu. rubber) container filled with hot water to warm a bed.

houmous var. of HUMMUS.

hound –*n.* **1** dog used in hunting. **2** *colloq.* despicable man. –*v.* harass or pursue. [Old English]

hour /aʊə(r)/ *n.* **1** twenty-fourth part of a day and night, 60 minutes. **2** time of day, point in time (*a late hour; what is the hour?*). **3** (in *pl.* with preceding numerals in form 18.00, 20.30, etc.) this number of hours and minutes past midnight on the 24-hour clock (*will assemble at 20.00 hours*). **4 a** period for a specific purpose (*lunch hour; keep regular hours*). **b** (in *pl.*) fixed working or open period (*office hours; opening hours*). **5** short period of time (*an idle hour*). **6** present time (*question of the hour*). **7** time for action etc. (*the hour has come*). **8** expressing distance by travelling time (*we are an hour from London*). **9** *RC Ch.* prayers to be said at one of seven fixed times of day (*book of *hours*). **10** (prec. by *the*) each time o'clock of a whole number of hours (*buses leave on the hour*; *on the half hour*; *at a quarter past the hour*). □ **after hours** after closing-time. [Greek *hōra*]

hourglass *n.* two vertically connected glass bulbs containing sand taking an hour to pass from upper to lower bulb.

houri /'hʊərɪ/ *n.* (*pl.* **-s**) beautiful young woman of the Muslim Paradise. [Persian from Arabic, = dark-eyed]

hourly *–adj.* **1** done or occurring every hour. **2** frequent. **3** per hour (*hourly wage*). *–adv.* **1** every hour. **2** frequently.

house *–n.* /haʊs/ (*pl.* /'haʊzɪz/) **1** building for human habitation. **2** building for a special purpose or for animals or goods (*opera-house*; *summer-house*; *hen-house*). **3 a** religious community. **b** its buildings. **4 a** body of pupils living in the same building at a boarding-school. **b** such a building. **c** division of a day-school for games, competitions, etc. **5** royal family or dynasty (*House of York*). **6 a** firm or institution. **b** its premises. **7 a** legislative or deliberative assembly. **b** building for this. **8** audience or performance in a theatre etc. **9** *Astrol.* twelfth part of the heavens. *–v.* /haʊz/ (**-sing**) **1** provide with a house or other accommodation. **2** store (goods etc.). **3** enclose or encase (a part or fitting). **4** fix in a socket, mortise, etc. □ **keep house** provide for or manage a household. **like a house on fire 1** vigorously, fast. **2** successfully, excellently. **on the house** free. **put** (or **set**) **one's house in order** make necessary reforms. [Old English]

house-agent *n.* agent for the sale and letting of houses.

house arrest *n.* detention in one's own house, not in prison.

houseboat *n.* boat equipped for living in.

housebound *adj.* confined to one's house through illness etc.

housebreaking /'haʊs,breɪkɪŋ/ *n.* act of breaking into a building, esp. in daytime, to commit a crime. □ **house-breaker** *n.*

■ **Usage** In 1968 housebreaking was replaced as a statutory crime in English law by *burglary*.

housecoat *n.* woman's informal indoor coat or gown.

housefly *n.* common fly often entering houses.

household *n.* **1** occupants of a house as a unit. **2** house and its affairs.

householder *n.* **1** person who owns or rents a house. **2** head of a household.

household troops *n.pl.* troops nominally guarding the sovereign.

household word *n.* (also **household name**) **1** familiar name or saying. **2** familiar person or thing.

house-hunting *n.* seeking a house to buy or rent.

house-husband *n.* man who does a wife's traditional household duties.

housekeeper *n.* person, esp. a woman, employed to manage a household.

housekeeping *n.* **1** management of household affairs. **2** money allowed for this. **3** operations of maintenance, record-keeping, etc., in an organization.

house lights *n.pl.* lights in a theatre auditorium.

housemaid *n.* female servant in a house.

housemaid's knee *n.* inflammation of the kneecap.

houseman *n.* resident junior doctor at a hospital etc.

house-martin *n.* black and white bird nesting on house walls etc.

housemaster *n.* (*fem.* **housemistress**) teacher in charge of a house, esp. at a boarding-school.

house music *n.* style of pop music, typically using drum machines and synthesized bass lines with sparse repetitive vocals and a fast beat.

house of cards *n.* insecure scheme etc.

House of Commons *n.* elected chamber of Parliament.

House of Keys *n.* (in the Isle of Man) elected chamber of the Tynwald.

House of Lords *n.* chamber of Parliament that is mainly hereditary.

house party *n.* group of guests staying at a country house etc.

house-plant *n.* plant grown indoors.

house-proud *adj.* attentive to the care and appearance of the home.

houseroom *n.* space or accommodation in one's house. □ **not give houseroom to** not have in any circumstances.

housetop *n.* roof of a house. □ **shout** etc. **from the housetops** announce publicly.

house-trained *adj.* **1** (of animals) trained to be clean in the house. **2** *colloq.* well-mannered.

house-warming *n.* party celebrating a move to a new home.

housewife *n.* **1** woman who manages a household and usu. does not have a full-time paid job. **2** /'hʌzɪf/ case for needles, thread, etc. □ **housewifely** *adj.* [from HOUSE, WIFE = woman]

housework *n.* regular housekeeping work, e.g. cleaning and cooking.

housey-housey /ˌhaʊzɪ'haʊzɪ/ *n.* (also **housie-housie**) *slang* gambling form of lotto.

housing /'haʊzɪŋ/ *n.* **1 a** dwelling-houses collectively. **b** provision of these. **2** shelter, lodging. **3** rigid casing for machinery etc. **4** hole or niche cut in one piece of wood for another to fit into.

housing estate *n.* residential area planned as a unit.

hove *past* of HEAVE.

hovel /'hɒv(ə)l/ *n.* small miserable dwelling. [origin unknown]

hover /'hɒvə(r)/ *–v.* **1** (of a bird etc.) remain in one place in the air. **2** (often foll. by *about, round*) wait close at hand, linger. *–n.* **1** hovering. **2** state of suspense. [obsolete *hove* hover]

hovercraft *n.* (*pl.* same) vehicle travelling on a cushion of air provided by a downward blast.

hoverport *n.* terminal for hovercraft.

how *–interrog. adv.* **1** by what means, in what way (*how do you do it?*; *tell me how you do it*; *how could you?*). **2** in what condition, esp. of health (*how are you?*; *how do things stand?*). **3 a** to what extent (*how far is it?*; *how would you like to take my place?*; *how we laughed!*). **b** to what extent good or well, what ... like (*how was the film?*; *how did they play?*). *–rel. adv.* in whatever way, as (*do it how you can*). *–conj. colloq.* that (*told us how he'd been in India*). □ **how about** *colloq.* would you like (*how about a quick swim?*). **how do you do?** a formal greeting. **how many** what number. **how much 1** what amount. **2** what price. **how's that? 1** what is your opinion or explanation of that? **2** *Cricket* (said to an umpire) is the batsman out or not? [Old English]

howbeit /haʊ'bi:ɪt/ *adv. archaic* nevertheless.

howdah /'haʊdə/ *n.* (usu. canopied) seat for riding on an elephant or camel. [Urdu *hawda*]

however /haʊ'evə(r)/ *adv.* **1 a** in whatever way (*do it however you want*). **b** to whatever extent (*must go however inconvenient*). **2** nevertheless.

howitzer /'haʊɪtsə(r)/ *n.* short gun for the high-angle firing of shells. [Czech *houfnice* catapult]

howl *–n.* **1** long loud doleful cry of a dog etc. **2** prolonged wailing noise. **3** loud cry of pain, rage, derision, or laughter. *–v.* **1** make a howl. **2** weep loudly. **3** utter with a howl. □ **howl down** prevent (a speaker) from being heard by howls of derision. [imitative]

howler *n. colloq.* glaring mistake.

howsoever /ˌhaʊsəʊ'evə(r)/ *adv. formal* **1** in whatsoever way. **2** to whatsoever extent.

hoy *int.* used to call attention. [natural cry]

hoyden /'hɔɪd(ə)n/ *n.* boisterous girl. [Dutch *heiden*: related to HEATHEN]

h.p. *abbr.* (also **hp**) **1** horsepower. **2** hire purchase.

HQ *abbr.* headquarters.

hr. *abbr.* hour.

HRH *abbr.* Her or His Royal Highness.

hrs. *abbr.* hours.

HRT *abbr.* hormone replacement therapy.

HT *abbr.* high tension.

hub *n.* **1** central part of a wheel, rotating on or with the axle. **2** centre of interest, activity, etc. [origin uncertain]

hubble-bubble /'hʌb(ə)l,bʌb(ə)l/ *n.* **1** simple hookah. **2** bubbling sound. **3** confused talk. [imitative]

hubbub /'hʌbʌb/ *n.* **1** confused noise of talking. **2** disturbance. [perhaps of Irish origin]

hubby /'hʌbɪ/ *n.* (*pl.* **-ies**) *colloq.* husband. [abbreviation]

hubris /'hju:brɪs/ *n.* arrogant pride or presumption. □ **hubristic** /-'brɪstɪk/ *adj.* [Greek]

huckleberry /'hʌkəlbərɪ/ *n.* **1** low-growing N. American shrub. **2** blue or black fruit of this. [probably an alteration of *hurtleberry*, WHORTLEBERRY]

huckster *–n.* aggressive salesman; hawker. *–v.* **1** haggle. **2** hawk (goods). [Low German]

huddle /'hʌd(ə)l/ *–v.* (**-ling**) **1** (often foll. by *up*) crowd together; nestle closely. **2** (often foll. by *up*) curl one's body into a small space. **3** heap together in a muddle. *–n.* **1** confused or crowded mass. **2** *colloq.* close or secret conference (esp. in **go into a huddle**). [perhaps from Low German]

hue *n.* **1** colour, tint. **2** variety or shade of colour. [Old English]

hue and cry *n.* loud outcry. [French *huer* shout]

huff *–n. colloq.* fit of petty annoyance. *–v.* **1** blow air, steam, etc. **2** (esp. **huff and puff**) bluster self-importantly but ineffectually. **3** *Draughts* remove (an opponent's piece) as a forfeit. □ **in a huff** *colloq.* annoyed and offended. [imitative of blowing]

huffy *adj.* (**-ier, -iest**) *colloq.* **1** apt to take offence. **2** offended. □ **huffily** *adv.* **huffiness** *n.*

hug *–v.* (**-gg-**) **1** squeeze tightly in one's arms, esp. with affection. **2** (of a bear) squeeze (a person) between its forelegs. **3** keep close to; fit tightly around. *–n.* **1** strong clasp with the arms. **2** squeezing grip in wrestling. [probably Scandinavian]

huge *adj.* **1** extremely large; enormous. **2** (of an abstract thing) very great. □ **hugeness** *n.* [French *ahuge*]

hugely *adv.* **1** extremely (*hugely successful*). **2** very much (*enjoyed it hugely*).

hugger-mugger /'hʌgə,mʌgə(r)/ *–adj.* & *adv.* **1** in secret. **2** confused; in confusion. *–n.* **1** secrecy. **2** confusion. [origin uncertain]

Huguenot /'hju:gə,nəʊ/ *n. hist.* French Protestant. [French]

huh /hə/ *int.* expressing disgust, surprise, etc. [imitative]

hula /'hu:lə/ *n.* (also **hula-hula**) Polynesian dance performed by women, with flowing arm movements. [Hawaiian]

hula hoop *n.* large hoop spun round the body.

hulk *n.* **1** body of a dismantled ship. **2** *colloq.* large clumsy-looking person or thing. [Old English]

hulking *adj. colloq.* bulky; clumsy.

hull[1] *n.* body of a ship, airship, etc. [perhaps related to HOLD[2]]

hull[2] *–n.* outer covering of a fruit, esp. the pod of peas and beans, the husk of grain, or the green calyx of a strawberry. *–v.* remove the hulls from (fruit etc.). [Old English]

hullabaloo /,hʌləbə'lu:/ *n.* uproar. [reduplication of *hallo*, *hullo*, etc.]

hullo var. of HELLO.

hum *–v.* (**-mm-**) **1** make a low steady continuous sound like a bee. **2** sing with closed lips. **3** utter a slight inarticulate sound. **4** *colloq.* be active (*really made things hum*). **5** *colloq.* smell unpleasantly. *–n.* **1** humming sound. **2** *colloq.* bad smell. □ **hum and haw** (or **ha**) hesitate; be indecisive. [imitative]

human /'hju:mən/ *–adj.* **1** of or belonging to the species *Homo sapiens*. **2** consisting of human beings (*the human race*). **3** of or characteristic of humankind, esp. as being weak, fallible, etc. (*is only human*). **4** showing warmth, sympathy, etc. (*is very human*). *–n.* human being. [Latin *humanus*]

human being *n.* man, woman, or child.

human chain *n.* line of people formed for passing things along, as a protest, etc.

humane /hju:'meɪn/ *adj.* **1** benevolent, compassionate. **2** inflicting the minimum of pain. **3** (of learning) tending to civilize. □ **humanely** *adv.* **humaneness** *n.*

humane killer *n.* instrument for the painless slaughter of animals.

humanism /'hju:mə,nɪz(ə)m/ *n.* **1** non-religious philosophy based on liberal human values. **2** (often **Humanism**) literary culture, esp. that of the Renaissance. □ **humanist** *n.* **humanistic** /-'nɪstɪk/ *adj.*

humanitarian /hju:,mænɪ'teərɪən/ *–n.* person who seeks to promote human welfare. *–adj.* of humanitarians. □ **humanitarianism** *n.*

humanity /hju:'mænɪtɪ/ *n.* (*pl.* **-ies**) **1 a** the human race. **b** human beings collectively. **c** being human. **2** humaneness, benevolence. **3** (in *pl.*) subjects concerned with human culture, e.g. language, literature, and history.

humanize *v.* (also **-ise**) (**-zing** or **-sing**) make human or humane. □ **humanization** /-'zeɪʃ(ə)n/ *n.* [French: related to HUMAN]

humankind *n.* human beings collectively.

humanly *adv.* **1** by human means (*if it is humanly possible*). **2** in a human manner.

human nature *n.* general characteristics and feelings of mankind.

human rights *n.pl.* rights held to be common to all.

human shield *n.* person(s) placed in the line of fire in order to discourage attack.

humble /'hʌmb(ə)l/ *–adj.* **1** having or showing low self-esteem. **2** of low social or political rank. **3** modest in size, pretensions, etc. *–v.* (**-ling**) **1** make humble; abase. **2** lower the rank or status of. □ **eat humble pie** apologize humbly; accept humiliation. □ **humbleness** *n.* **humbly** *adv.* [Latin *humilis*: related to HUMUS]

humbug /'hʌmbʌg/ *–n.* **1** lying or deception; hypocrisy. **2** impostor. **3** hard boiled striped peppermint sweet. *–v.* (**-gg-**) **1** be or behave like an impostor. **2** deceive, hoax. [origin unknown]

humdinger /'hʌm,dɪŋə(r)/ *n. slang* excellent or remarkable person or thing. [origin unknown]

humdrum /'hʌmdrʌm/ *adj.* commonplace, dull, monotonous. [a reduplication of HUM]

humerus /'hju:mərəs/ *n.* (*pl.* **-ri** /-,raɪ/) the bone of the upper arm. □ **humeral** *adj.* [Latin, = shoulder]

humid /'hju:mɪd/ *adj.* (of the air or climate) warm and damp. [Latin *humidus*]

humidifier /hju:'mɪdɪ,faɪə(r)/ *n.* device for keeping the atmosphere moist in a room etc.

humidify /hju:'mɪdɪ,faɪ/ *v.* (**-ies**, **-ied**) make (air etc.) humid.

humidity /hju:'mɪdɪtɪ/ *n.* (*pl.* **-ies**) **1** dampness. **2** degree of moisture, esp. in the atmosphere.

humiliate /hju:'mɪlɪˌeɪt/ *v.* (**-ting**) injure the dignity or self-respect of. □ **humiliating** *adj.* **humiliation** /-'eɪʃ(ə)n/ *n.* [Latin: related to HUMBLE]

humility /hju:'mɪlɪtɪ/ *n.* **1** humbleness, meekness. **2** humble condition. [French: related to HUMILIATE]

hummingbird *n.* small tropical bird that makes a humming sound with its wings when it hovers.

hummock /'hʌmək/ *n.* hillock or hump. [origin unknown]

hummus /'hʊmʊs/ *n.* (also **houmous**) dip or appetizer made from ground chick-peas, sesame oil, lemon, and garlic. [Turkish]

humor *US* var. of HUMOUR.

humoresque /ˌhju:mə'resk/ *n.* short lively piece of music. [German *Humoreske*]

humorist *n.* humorous writer, talker, or actor.

humorous /'hju:mərəs/ *adj.* showing humour or a sense of humour. □ **humorously** *adv.*

humour /'hju:mə(r)/ (*US* **humor**) *–n.* **1 a** quality of being amusing or comic. **b** the expression of humour in literature, speech, etc. **2** (in full **sense of humour**) ability to perceive or express humour. **3** state of mind; inclination (*bad humour*). **4** (in full **cardinal humour**) *hist.* each of the four fluids (blood, phlegm, choler, melancholy), thought to determine a person's physical and mental qualities. *–v.* gratify or indulge (a person or taste etc.). □ **out of humour** displeased. □ **humourless** *adj.* [Latin *humor* moisture]

hump *–n.* **1** rounded protuberance on a camel's back, or as an abnormality on a person's back. **2** rounded raised mass of earth etc. **3** critical point in an undertaking. **4** (prec. by *the*) *slang* fit of depression or vexation (*gave me the hump*). *–v.* **1** (often foll. by *about*) *colloq.* lift or carry (heavy objects etc.) with difficulty. **2** make hump-shaped. [probably Low German or Dutch]

humpback *n.* **1 a** deformed back with a hump. **b** person with this. **2** whale with a dorsal fin forming a hump. □ **humpbacked** *adj.*

humpback bridge *n.* small bridge with a steep ascent and descent.

humph /həmf/ *int.* & *n.* inarticulate sound of doubt or dissatisfaction. [imitative]

humus /'hju:məs/ *n.* organic constituent of soil formed by the decomposition of vegetation. [Latin, = soil]

Hun *n.* **1** *offens.* German (esp. in military contexts). **2** member of a warlike Asiatic nomadic people who ravaged Europe in the 4th–5th c. **3** vandal. □ **Hunnish** *adj.* [Old English]

hunch *–v.* bend or arch into a hump. *–n.* **1** intuitive feeling or idea. **2** hump. [origin unknown]

hunchback *n.* = HUMPBACK 1. □ **hunchbacked** *adj.*

hundred /'hʌndrəd/ *adj.* & *n.* (*pl.* **hundreds** or (in sense 1) **hundred**) (in *sing.*, prec. by *a* or *one*) **1** ten times ten. **2** symbol for this (100, c, C). **3** (in *sing.* or *pl.*) *colloq.* a large number. **4** (in *pl.*) the years of a specified century (*the seventeen hundreds*). **5** *hist.* subdivision of a county or shire, having its own court. □ **hundredfold** *adj.* & *adv.* **hundredth** *adj.* & *n.* [Old English]

hundreds and thousands *n.pl.* tiny coloured sweets for decorating cakes etc.

hundredweight *n.* (*pl.* same or **-s**) **1** unit of weight equal to 112 lb, or *US* equal to 100 lb. **2** unit of weight equal to 50 kg.

hung *past* and *past part.* of HANG.

Hungarian /hʌŋ'geərɪən/ *–n.* **1 a** native or national of Hungary. **b** person of Hungarian descent. **2** language of Hungary. *–adj.* of Hungary or its people or language. [medieval Latin]

hunger /'hʌŋgə(r)/ *–n.* **1 a** lack of food. **b** feeling of discomfort or exhaustion caused by this. **2** (often foll. by *for*, *after*) strong desire. *–v.* **1** (often foll. by *for*, *after*) crave or desire. **2** feel hunger. [Old English]

hunger strike *n.* refusal of food as a protest.

hung-over *adj. colloq.* suffering from a hangover.

hung parliament *n.* parliament in which no party has a clear majority.

hungry /'hʌŋgrɪ/ *adj.* (**-ier**, **-iest**) **1** feeling or showing hunger; needing food. **2** inducing hunger (*hungry work*). **3** craving (*hungry for news*). □ **hungrily** *adv.* [Old English]

hunk *n.* **1** large piece cut off (*hunk of bread*). **2** *colloq.* sexually attractive man. □ **hunky** *adj.* (**-ier**, **-iest**). [probably Dutch]

hunky-dory /ˌhʌŋkɪ'dɔ:rɪ/ *adj.* esp. *US colloq.* excellent. [origin unknown]

hunt *–v.* **1** (also *absol.*) **a** pursue and kill (wild animals, esp. foxes, or game) for sport or food. **b** use (a horse or hounds) for hunting. **c** (of an animal) chase (its prey). **2** (foll. by *after*, *for*) seek, search. **3** (of an engine etc.) run alternately too fast and too slow. **4** scour (a district) for game. **5** (as **hunted** *adj.*) (of a look etc.)

terrified as if being hunted. *–n.* **1** practice or instance of hunting. **2 a** association of people hunting with hounds. **b** area for hunting. □ **hunt down** pursue and capture. □ **hunting** *n.* [Old English]

hunter *n.* **1 a** (*fem.* **huntress**) person or animal that hunts. **b** horse used in hunting. **2** person who seeks something. **3** pocket-watch with a hinged cover protecting the glass.

hunter's moon *n.* next full moon after the harvest moon.

huntsman *n.* **1** hunter. **2** hunt official in charge of hounds.

hurdle /'hɜ:d(ə)l/ *–n.* **1 a** each of a series of light frames to be cleared by athletes in a race. **b** (in *pl.*) hurdle-race. **2** obstacle or difficulty. **3** portable rectangular frame used as a temporary fence etc. *–v.* (**-ling**) **1** run in a hurdle-race. **2** fence off etc. with hurdles. [Old English]

hurdler /'hɜ:dlə(r)/ *n.* **1** athlete who runs in hurdle-races. **2** maker of hurdles.

hurdy-gurdy /,hɜ:dɪ'gɜ:dɪ/ *n.* (*pl.* **-ies**) **1** droning musical instrument played by turning a handle. **2** *colloq.* barrel-organ. [imitative]

hurl *–v.* **1** throw with great force. **2** utter (abuse etc.) vehemently. *–n.* forceful throw. [imitative]

hurley /'hɜ:lɪ/ *n.* **1** (also **hurling**) Irish game resembling hockey. **2** stick used in this.

hurly-burly /,hɜ:lɪ'bɜ:lɪ/ *n.* boisterous activity; commotion. [a reduplication of HURL]

hurrah /hʊ'rɑ:/ *int.* & *n.* (also **hurray** /hʊ'reɪ/) exclamation of joy or approval. [earlier *huzza*, origin uncertain]

hurricane /'hʌrɪkən/ *n.* **1** storm with a violent wind, esp. a W. Indian cyclone. **2** *Meteorol.* wind of 65 knots (75 m.p.h.) or more, force 12 on the Beaufort scale. [Spanish and Portuguese from Carib]

hurricane-lamp *n.* oil-lamp designed to resist a high wind.

hurry /'hʌrɪ/ *–n.* **1** great or eager haste. **2** (with *neg.* or *interrog.*) need for haste (*there is no hurry*; *what's the hurry?*). *–v.* (**-ies, -ied**) **1** move or act hastily. **2** cause to hurry. **3** (as **hurried** *adj.*) hasty; done rapidly. □ **hurry along** (or **up**) (cause to) make haste. **in a hurry 1** hurrying. **2** *colloq.* easily or readily (*you will not beat that in a hurry*). □ **hurriedly** *adv.* [imitative]

hurt *–v.* (*past* and *past part.* **hurt**) **1** (also *absol.*) cause pain or injury to. **2** cause mental pain or distress to. **3** suffer pain (*my arm hurts*). *–n.* **1** injury. **2** harm, wrong. [French *hurter* knock]

hurtful *adj.* causing (esp. mental) hurt. □ **hurtfully** *adv.*

hurtle /'hɜ:t(ə)l/ *v.* (**-ling**) **1** move or hurl rapidly or noisily. **2** come with a crash. [from HURT in the obsolete sense 'strike hard']

husband /'hʌzbənd/ *–n.* married man, esp. in relation to his wife. *–v.* use (resources) economically; eke out. [Old English, = house-dweller]

husbandry *n.* **1** farming. **2** management of resources.

hush *–v.* make or become silent or quiet. *–int.* calling for silence. *–n.* expectant stillness or silence. □ **hush up** suppress public mention of (an affair). [*husht*, an obsolete exclamation, taken as a past part.]

hush-hush *adj. colloq.* highly secret, confidential.

hush money *n. slang* money paid to ensure discretion.

husk *–n.* **1** dry outer covering of some fruits or seeds. **2** worthless outside part of a thing. *–v.* remove husk(s) from. [probably Low German]

husky[1] /'hʌskɪ/ *adj.* (**-ier, -iest**) **1** (of a person or voice) dry in the throat; hoarse. **2** of or full of husks. **3** dry as a husk. **4** tough, strong, hefty. □ **huskily** *adv.* **huskiness** *n.*

husky[2] /'hʌskɪ/ *n.* (*pl.* **-ies**) dog of a powerful breed used in the Arctic for pulling sledges. [perhaps from corruption of ESKIMO]

huss *n.* dogfish as food. [origin unknown]

hussar /hʊ'zɑ:(r)/ *n.* soldier of a light cavalry regiment. [Magyar *huszár*]

hussy /'hʌsɪ/ *n.* (*pl.* **-ies**) *derog.* impudent or promiscuous girl or woman. [contraction of HOUSEWIFE]

hustings /'hʌstɪŋz/ *n.* election campaign or proceedings. [Old English, = house of assembly, from Old Norse]

hustle /'hʌs(ə)l/ *–v.* (**-ling**) **1** jostle, bustle. **2** (foll. by *into*, *out of*, etc.) force, coerce, or hurry (*hustled them out of the room*; *was hustled into agreeing*). **3** *slang* **a** solicit business. **b** engage in prostitution. **4** *slang* obtain by energetic activity. *–n.* act or instance of hustling. □ **hustler** *n.* [Dutch]

hut *n.* small simple or crude house or shelter. [French *hutte* from Germanic]

hutch *n.* box or cage for rabbits etc. [French *huche*]

hyacinth /'haɪəsɪnθ/ *n.* **1** bulbous plant with racemes of bell-shaped (esp. purplish-blue) fragrant flowers. **2** purplish-blue. [Greek *huakinthos*]

hyaena var. of HYENA.

hybrid /'haɪbrɪd/ *–n.* **1** offspring of two plants or animals of different species or varieties. **2** thing composed of diverse elements, e.g. a word with parts taken from different languages. *–adj.* **1** bred as a hybrid. **2** heterogeneous. □ **hybridism** *n.* [Latin]

hybridize *v.* (also **-ise**) (**-zing** or **-sing**) **1** subject (a species etc.) to crossbreeding. **2 a** produce hybrids. **b** (of an animal or plant) interbreed. □ **hybridization** /-'zeɪʃ(ə)n/ *n.*

hydra /'haɪdrə/ *n.* **1** freshwater polyp with a tubular body and tentacles. **2** something hard to destroy. [Greek, a mythical snake with many heads that grew again when cut off]

hydrangea /haɪ'dreɪndʒə/ *n.* shrub with globular clusters of white, pink, or blue flowers. [Greek *hudōr* water, *aggos* vessel]

hydrant /'haɪdrənt/ *n.* outlet (esp. in a street) with a nozzle for a hose, for drawing water from the main. [as HYDRO-]

hydrate /'haɪdreɪt/ *–n.* compound in which water is chemically combined with another compound or an element. *–v.* (**-ting**) **1** combine chemically with water. **2** cause to absorb water. □ **hydration** /-'dreɪʃ(ə)n/ *n.* [French: related to HYDRO-]

hydraulic /haɪ'drɔ:lɪk/ *adj.* **1** (of water, oil, etc.) conveyed through pipes or channels. **2** (of a mechanism etc.) operated by liquid moving in this way (*hydraulic brakes*). □ **hydraulically** *adv.* [Greek *hudōr* water, *aulos* pipe]

hydraulics *n.pl.* (usu. treated as *sing.*) science of the conveyance of liquids through pipes etc., esp. as motive power.

hydride /'haɪdraɪd/ *n.* compound of hydrogen with an element.

hydro /'haɪdrəʊ/ *n.* (*pl.* **-s**) *colloq.* **1** hotel or clinic etc., orig. providing hydropathic treatment. **2** hydroelectric power plant. [abbreviation]

hydro- *comb. form* **1** having to do with water (*hydroelectric*). **2** combined with hydrogen (*hydrochloric*). [Greek *hudro-* from *hudōr* water]

hydrocarbon /ˌhaɪdrəʊ'kɑ:bən/ *n.* compound of hydrogen and carbon.

hydrocephalus /ˌhaɪdrə'sefələs/ *n.* accumulated fluid in the brain, esp. in young children. □ **hydrocephalic** /-sɪ'fælɪk/ *adj.* [Greek *kephalē* head]

hydrochloric acid /ˌhaɪdrə'klɒrɪk/ *n.* solution of the colourless gas hydrogen chloride in water.

hydrocyanic acid /ˌhaɪdrəsaɪ'ænɪk/ *n.* highly poisonous liquid smelling of bitter almonds; prussic acid.

hydrodynamics /ˌhaɪdrəʊdaɪ'næmɪks/ *n.pl.* (usu. treated as *sing.*) science of forces acting on or exerted by fluids (esp. liquids). □ **hydrodynamic** *adj.*

hydroelectric /ˌhaɪdrəʊɪ'lektrɪk/ *adj.* **1** generating electricity by water-power. **2** (of electricity) so generated. □ **hydroelectricity** /-'trɪsɪtɪ/ *n.*

hydrofoil /'haɪdrəˌfɔɪl/ *n.* **1** boat equipped with planes for lifting its hull out of the water to increase its speed. **2** such a plane.

hydrogen /'haɪdrədʒ(ə)n/ *n.* tasteless odourless gas, the lightest element, occurring in water and all organic compounds. □ **hydrogenous** /-'drɒdʒɪnəs/ *adj.* [French: related to HYDRO-, -GEN]

hydrogenate /haɪ'drɒdʒɪˌneɪt/ *v.* (**-ting**) charge with or cause to combine with hydrogen. □ **hydrogenation** /-'neɪʃ(ə)n/ *n.*

hydrogen bomb *n.* immensely powerful bomb utilizing the explosive fusion of hydrogen nuclei.

hydrogen peroxide *n.* viscous unstable liquid with strong oxidizing properties.

hydrogen sulphide *n.* poisonous unpleasant-smelling gas formed by rotting animal matter.

hydrography /haɪ'drɒgrəfɪ/ *n.* science of surveying and charting seas, lakes, rivers, etc. □ **hydrographer** *n.* **hydrographic** /-drə'græfɪk/ *adj.*

hydrology /haɪ'drɒlədʒɪ/ *n.* science of the properties of water, esp. of its movement in relation to land. □ **hydrologist** *n.*

hydrolyse /'haɪdrəˌlaɪz/ *v.* (*US* **-lyze**) (**-sing** or **-zing**) decompose by hydrolysis.

hydrolysis /haɪ'drɒlɪsɪs/ *n.* chemical reaction of a substance with water, usu. resulting in decomposition. [Greek *lusis* dissolving]

hydrometer /haɪ'drɒmɪtə(r)/ *n.* instrument for measuring the density of liquids.

hydropathy /haɪ'drɒpəθɪ/ *n.* (medically unorthodox) treatment of disease by water. □ **hydropathic** /ˌhaɪdrə'pæθɪk/ *adj.* [related to PATHOS]

hydrophilic /ˌhaɪdrə'fɪlɪk/ *adj.* **1** having an affinity for water. **2** wettable by water. [Greek *philos* loving]

hydrophobia /ˌhaɪdrə'fəʊbɪə/ *n.* **1** aversion to water, esp. as a symptom of rabies in humans. **2** rabies, esp. in humans. □ **hydrophobic** *adj.*

hydroplane /'haɪdrəˌpleɪn/ *n.* **1** light fast motor boat that skims over water. **2** finlike attachment enabling a submarine to rise and descend.

hydroponics /ˌhaɪdrə'pɒnɪks/ *n.* growing plants without soil, in sand, gravel, or liquid, with added nutrients. [Greek *ponos* labour]

hydrosphere /ˈhaɪdrəˌsfɪə(r)/ *n.* waters of the earth's surface.

hydrostatic /ˌhaɪdrəˈstætɪk/ *adj.* of the equilibrium of liquids and the pressure exerted by liquid at rest. [related to STATIC]

hydrostatics *n.pl.* (usu. treated as *sing.*) mechanics of the hydrostatic properties of liquids.

hydrotherapy /ˌhaɪdrəˈθerəpɪ/ *n.* use of water, esp. swimming, in the treatment of arthritis, paralysis, etc.

hydrous /ˈhaɪdrəs/ *adj.* containing water. [related to HYDRO-]

hydroxide /haɪˈdrɒksaɪd/ *n.* compound containing oxygen and hydrogen as either a hydroxide ion or a hydroxyl group.

hydroxyl /haɪˈdrɒksɪl/ *n.* (*attrib.*) univalent group containing hydrogen and oxygen.

hyena /haɪˈiːnə/ *n.* (also **hyaena**) dog-like flesh-eating mammal. [Latin from Greek]

hygiene /ˈhaɪdʒiːn/ *n.* **1** conditions or practices, esp. cleanliness, conducive to maintaining health. **2** science of maintaining health. □ **hygienic** /-ˈdʒiːnɪk/ *adj.* **hygienically** /-ˈdʒiːnɪkəlɪ/ *adv.* **hygienist** *n.* [Greek *hugiēs* healthy]

hygrometer /haɪˈgrɒmɪtə(r)/ *n.* instrument for measuring the humidity of the air or a gas. [Greek *hugros* wet]

hygroscope /ˈhaɪgrəˌskəʊp/ *n.* instrument indicating but not measuring the humidity of the air.

hygroscopic /ˌhaɪgrəˈskɒpɪk/ *adj.* **1** of the hygroscope. **2** (of a substance) tending to absorb moisture from the air.

hymen /ˈhaɪmen/ *n.* membrane at the opening of the vagina, usu. broken at the first occurrence of sexual intercourse. [Greek *humēn* membrane]

hymenopterous /ˌhaɪməˈnɒptərəs/ *adj.* of an order of insects having four transparent wings, including bees, wasps, and ants. [Greek, = membrane-winged]

hymn /hɪm/ —*n.* **1** song of esp. Christian praise. **2** crusading theme (*hymn of freedom*). —*v.* praise or celebrate in hymns. [Greek *humnos*]

hymnal /ˈhɪmn(ə)l/ *n.* book of hymns. [medieval Latin: related to HYMN]

hymnology /hɪmˈnɒlədʒɪ/ *n.* (*pl.* **-ies**) **1** the composition or study of hymns. **2** hymns collectively. □ **hymnologist** *n.*

hyoscine /ˈhaɪəˌsiːn/ *n.* poisonous alkaloid found in plants of the nightshade family, used to prevent motion sickness etc. [Greek *huoskuamos* henbane from *hus huos* pig, *kuamos* bean]

hype /haɪp/ *slang* —*n.* extravagant or intensive promotion of a product etc. —*v.* (**-ping**) promote with hype. [origin unknown]

hyped up *adj. slang* nervously excited or stimulated. [shortening of HYPODERMIC]

hyper /ˈhaɪpə(r)/ *adj. slang* hyperactive, highly-strung. [abbreviation of HYPERACTIVE]

hyper- *prefix* meaning: **1** over, beyond, above (*hypersonic*). **2** too (*hypersensitive*). [Greek *huper* over]

hyperactive /ˌhaɪpəˈræktɪv/ *adj.* (of a person) abnormally active.

hyperbola /haɪˈpɜːbələ/ *n.* (*pl.* **-s** or **-lae** /-ˌliː/) plane curve produced when a cone is cut by a plane that makes a larger angle with the base than the side of the cone makes. □ **hyperbolic** /ˌhaɪpəˈbɒlɪk/ *adj.* [Greek *hyperbolē*, = excess: related to HYPER-, *ballō* throw]

hyperbole /haɪˈpɜːbəlɪ/ *n.* exaggeration, esp. for effect. □ **hyperbolical** /-ˈbɒlɪk(ə)l/ *adj.*

hyperbolic function *n.* function related to a rectangular hyperbola, e.g. a hyperbolic cosine or sine.

hypercritical /ˌhaɪpəˈkrɪtɪk(ə)l/ *adj.* excessively critical. □ **hypercritically** *adv.*

hyperglycaemia /ˌhaɪpəglaɪˈsiːmɪə/ *n.* (*US* **hyperglycemia**) excess of glucose in the bloodstream. [from HYPER-, Greek *glukus* sweet, *haima* blood]

hypermarket /ˈhaɪpəˌmɑːkɪt/ *n.* very large supermarket.

hypermedia /ˈhaɪpəˌmiːdɪə/ *n.* provision of several media (e.g. audio, video, and graphics) on one computer system, with cross-references from one to another (often *attrib.*: *hypermedia database*).

hypersensitive /ˌhaɪpəˈsensɪtɪv/ *adj.* excessively sensitive. □ **hypersensitivity** /-ˈtɪvɪtɪ/ *n.*

hypersonic /ˌhaɪpəˈsɒnɪk/ *adj.* **1** of speeds of more than five times that of sound. **2** of sound-frequencies above about a thousand million hertz.

hypertension /ˌhaɪpəˈtenʃ(ə)n/ *n.* **1** abnormally high blood pressure. **2** great emotional tension.

hypertext /ˈhaɪpəˌtekst/ *n.* provision of several texts on one computer system, with cross-references from one to another.

hyperthermia /ˌhaɪpəˈθɜːmɪə/ *n.* abnormally high body-temperature. [from HYPER-, Greek *thermē* heat]

hyperthyroidism /ˌhaɪpəˈθaɪrɔɪˌdɪz(ə)m/ *n.* overactivity of the thyroid gland, resulting in an increased rate of metabolism.

hyperventilation /ˌhaɪpəˌventɪˈleɪʃ(ə)n/ *n.* abnormally rapid breathing. □ **hyperventilate** *v.* (**-ting**).

hyphen /ˈhaɪf(ə)n/ *–n.* sign (-) used to join words semantically or syntactically (e.g. *fruit-tree*, *pick-me-up*, *rock-forming*), to indicate the division of a word at the end of a line, or to indicate a missing or implied element (as in *man- and womankind*). *–v.* = HYPHENATE. [Greek *huphen* together]

hyphenate /ˈhaɪfəˌneɪt/ *v.* (**-ting**) **1** write (a compound word) with a hyphen. **2** join (words) with a hyphen. □ **hyphenation** /-ˈneɪʃ(ə)n/ *n.*

hypnosis /hɪpˈnəʊsɪs/ *n.* **1** state like sleep in which the subject acts only on external suggestion. **2** artificially produced sleep. [Greek *hupnos* sleep]

hypnotherapy /ˌhɪpnəʊˈθerəpɪ/ *n.* treatment of mental disorders by hypnosis.

hypnotic /hɪpˈnɒtɪk/ *–adj.* **1** of or producing hypnosis. **2** inducing sleep. *–n.* hypnotic drug or influence. □ **hypnotically** *adv.* [Greek: related to HYPNOSIS]

hypnotism /ˈhɪpnəˌtɪz(ə)m/ *n.* the study or practice of hypnosis. □ **hypnotist** *n.*

hypnotize /ˈhɪpnəˌtaɪz/ *v.* (also **-ise**) (**-zing** or **-sing**) **1** produce hypnosis in. **2** fascinate; capture the mind of.

hypo[1] /ˈhaɪpəʊ/ *n.* sodium thiosulphate (incorrectly called hyposulphite) used as a photographic fixer. [abbreviation]

hypo[2] /ˈhaɪpəʊ/ *n.* (*pl.* **-s**) *slang* = HYPODERMIC *n.* [abbreviation]

hypo- *prefix* **1** under (*hypodermic*). **2** below normal (*hypotension*). **3** slightly. [Greek *hupo* under]

hypocaust /ˈhaɪpəˌkɔːst/ *n.* space for underfloor hot-air heating in ancient Roman houses. [from HYPO-, *kaustos* burnt]

hypochondria /ˌhaɪpəˈkɒndrɪə/ *n.* abnormal and ill-founded anxiety about one's health. [Latin from Greek, = soft parts of the body below the ribs, where melancholy was thought to arise]

hypochondriac /ˌhaɪpəˈkɒndrɪˌæk/ *–n.* person given to hypochondria. *–adj.* of or affected by hypochondria.

hypocrisy /hɪˈpɒkrɪsɪ/ *n.* (*pl.* **-ies**) **1** false claim to virtue; insincerity, pretence. **2** instance of this. [Greek, = acting, feigning]

hypocrite /ˈhɪpəkrɪt/ *n.* person given to hypocrisy. □ **hypocritical** /-ˈkrɪtɪk(ə)l/ *adj.* **hypocritically** /-ˈkrɪtɪkəlɪ/ *adv.*

hypodermic /ˌhaɪpəˈdɜːmɪk/ *–adj.* **1** of the area beneath the skin. **2 a** injected beneath the skin. **b** (of a syringe, etc.) used to do this. *–n.* hypodermic injection or syringe. [from HYPO-, Greek *derma* skin]

hypotension /ˌhaɪpəʊˈtenʃ(ə)n/ *n.* abnormally low blood pressure.

hypotenuse /haɪˈpɒtəˌnjuːz/ *n.* side opposite the right angle of a right-angled triangle. [Greek, = subtending line]

hypothalamus /ˌhaɪpəˈθæləməs/ *n.* (*pl.* **-mi** /-ˌmaɪ/) region of the brain controlling body-temperature, thirst, hunger, etc. □ **hypothalamic** *adj.* [Latin: related to HYPO-, Greek *thalamos* inner room]

hypothermia /ˌhaɪpəʊˈθɜːmɪə/ *n.* abnormally low body-temperature. [from HYPO-, Greek *thermē* heat]

hypothesis /haɪˈpɒθɪsɪs/ *n.* (*pl.* **-theses** /-ˌsiːz/) proposition or supposition made as the basis for reasoning or investigation. [Greek, = foundation]

hypothesize /haɪˈpɒθɪˌsaɪz/ *v.* (also **-ise**) (**-zing** or **-sing**) form or assume a hypothesis.

hypothetical /ˌhaɪpəˈθetɪk(ə)l/ *adj.* **1** of, based on, or serving as a hypothesis. **2** supposed; not necessarily true. □ **hypothetically** *adv.*

hypothyroidism /ˌhaɪpəʊˈθaɪrɔɪˌdɪz(ə)m/ *n.* subnormal activity of the thyroid gland, resulting in cretinism. □ **hypothyroid** *n.* & *adj.*

hypoventilation /ˌhaɪpəʊˌventɪˈleɪʃ(ə)n/ *n.* abnormally slow breathing.

hyssop /ˈhɪsəp/ *n.* **1** small bushy aromatic herb, formerly used medicinally. **2** *Bibl.* plant whose twigs were used for sprinkling in Jewish rites. [ultimately from Greek *hyssōpos*, of Semitic origin]

hysterectomy /ˌhɪstəˈrektəmɪ/ *n.* (*pl.* **-ies**) surgical removal of the womb. [Greek *hustera* womb, -ECTOMY]

hysteresis /ˌhɪstəˈriːsɪs/ *n.* phenomenon whereby changes in an effect lag behind changes in its cause. [Greek *husteros* coming after]

hysteria /hɪˈstɪərɪə/ *n.* **1** wild uncontrollable emotion or excitement. **2** functional disturbance of the nervous system, of psychoneurotic origin. [Greek *hustera* womb]

hysteric /hɪˈsterɪk/ *n.* **1** (in *pl.*) **a** fit of hysteria. **b** *colloq.* overwhelming laughter (*we were in hysterics*). **2** hysterical person.

hysterical *adj.* **1** of or affected with hysteria. **2** uncontrollably emotional. **3** *colloq.* extremely funny. □ **hysterically** *adv.*

Hz *abbr.* hertz.

I

I[1] /aɪ/ *n.* (also **i**) (*pl.* **Is** or **I's**) **1** ninth letter of the alphabet. **2** (as a roman numeral) 1.

I[2] /aɪ/ *pron.* (*obj.* **me**; *poss.* **my, mine**; *pl.* **we**) used by a speaker or writer to refer to himself or herself. [Old English]

I[3] *symb.* iodine.

I[4] *abbr.* (also **I.**) **1** Island(s). **2** Isle(s).

-ial var. of -AL.

iambic /aɪ'æmbɪk/ *Prosody –adj.* of or using iambuses. *–n.* (usu. in *pl.*) iambic verse.

iambus /aɪ'æmbəs/ *n.* (*pl.* **-buses** or **-bi** /-baɪ/) metrical foot consisting of one short followed by one long syllable (◡ –). [Greek, = lampoon]

-ian var. of -AN.

IBA *abbr.* Independent Broadcasting Authority.

Iberian /aɪ'bɪərɪən/ *–adj.* of Iberia, the peninsula comprising Spain and Portugal; of Spain and Portugal. *–n.* native or language of Iberia. [Latin *Iberia*]

ibex /'aɪbeks/ *n.* (*pl.* **-es**) wild mountain goat with thick curved ridged horns. [Latin]

ibid. /'ɪbɪd/ *abbr.* in the same book or passage etc. [Latin *ibidem* in the same place]

-ibility *suffix* forming nouns from, or corresponding to, adjectives in *-ible*.

ibis /'aɪbɪs/ *n.* (*pl.* **-es**) wading bird with a curved bill, long neck, and long legs. [Greek, from Egyptian]

-ible *suffix* forming adjectives meaning 'that may or may be' (*forcible*; *possible*). [Latin]

-ibly *suffix* forming adverbs corresponding to adjectives in *-ible*.

Ibo /'i:bəʊ/ *n.* (also **Igbo**) (*pl.* same or **-s**) **1** member of a Black people of SE Nigeria. **2** their language. [native name]

-ic *suffix* **1** forming adjectives (*Arabic*; *classic*; *public*) and nouns (*critic*; *epic*; *mechanic*; *music*). **2** combined in higher valence or degree of oxidation (*ferric*; *sulphuric*). [Latin *-icus*, Greek *-ikos*]

-ical *suffix* forming adjectives corresponding to nouns or adjectives in *-ic* or *-y* (*classical*; *historical*).

ice *–n.* **1 a** frozen water. **b** sheet of this on water. **2** ice-cream or water-ice (*ate an ice*). *–v.* (**icing**) **1** mix with or cool in ice (*iced drinks*). **2** (often foll. by *over*, *up*) **a** cover or become covered with ice. **b** freeze. **3** cover (a cake etc.) with icing. □ **on ice 1** performed by skaters. **2** *colloq.* in reserve. **on thin ice** in a risky situation. [Old English]

ice age *n.* glacial period.

ice-axe *n.* cutting tool used by mountaineers.

iceberg *n.* large floating mass of ice. □ **the tip of the iceberg** small perceptible part of something very large or complex. [Dutch]

iceberg lettuce *n.* crisp type of round lettuce.

ice blue *adj.* & *n.* (as adj. often hyphenated) very pale blue.

icebox *n.* **1** compartment in a refrigerator for making or storing ice. **2** *US* refrigerator.

ice-breaker *n.* **1** ship designed to break through ice. **2** joke, incident, etc. that breaks the ice.

ice bucket *n.* bucket holding ice, used to chill wine.

icecap *n.* permanent covering of ice, esp. in polar regions.

ice-cream *n.* sweet creamy frozen food, usu. flavoured.

ice-cube *n.* small block of ice for drinks etc.

ice-field *n.* expanse of ice, esp. in polar regions.

ice hockey *n.* form of hockey played on ice.

Icelander /'aɪsləndə(r)/ *n.* **1** native or national of Iceland. **2** person of Icelandic descent.

Icelandic /aɪs'lændɪk/ *–adj.* of Iceland. *–n.* language of Iceland.

ice lolly *n.* (also **iced lolly**) flavoured ice on a stick.

ice-pack *n.* **1** = PACK ICE. **2** ice applied to the body for medical purposes.

ice-pick *n.* tool with a spike for splitting up ice.

ice-plant *n.* plant with speckled leaves.

ice-rink *n.* = RINK *n.* 1.

ice-skate *–n.* boot with a blade beneath, for skating on ice. *–v.* skate on ice. □ **ice-skater** *n.*

ichneumon /ɪk'nju:mən/ *n.* **1** (in full **ichneumon fly**) small wasp depositing eggs in or on the larva of another as food for its own larva. **2** mongoose noted for destroying crocodile eggs. [Greek from *ikhnos* footstep]

ichthyology /,ɪkθɪ'ɒlədʒɪ/ *n.* the study of fishes. □ **ichthyological** /-ə'lɒdʒɪk(ə)l/ *adj.* **ichthyologist** *n.* [Greek *ikhthus* fish]

ichthyosaurus /ˌɪkθɪəˈsɔːrəs/ *n.* (also **ichthyosaur** /ˈɪkθɪəˌsɔːr/) (*pl.* **-sauruses** or **-saurs**) extinct marine reptile with four flippers and usu. a large tail. [Greek *ikhthus* fish, *sauros* lizard]

-ician *suffix* forming nouns denoting persons skilled in subjects having nouns usu. ending in *-ic* or *-ics* (*magician*; *politician*). [French *-icien*]

icicle /ˈaɪsɪk(ə)l/ *n.* hanging tapering piece of ice, formed from dripping water. [from ICE, obsolete *ickle* icicle]

icing *n.* **1** coating of sugar etc. on a cake or biscuit. **2** formation of ice on a ship or aircraft. □ **icing on the cake** inessential though attractive addition or enhancement.

icing sugar *n.* finely powdered sugar.

icon /ˈaɪkɒn/ *n.* (also **ikon**) **1** painting of Christ etc., esp. in the Eastern Church. **2** image or statue. **3** symbol on a VDU screen of a program, option, or window, esp. for selection. □ **iconic** /aɪˈkɒnɪk/ *adj.* [Greek *eikōn* image]

iconoclast /aɪˈkɒnəˌklæst/ *n.* **1** person who attacks cherished beliefs. **2** *hist.* person destroying religious images. □ **iconoclasm** *n.* **iconoclastic** /-ˈklæstɪk/ *adj.* [Greek: related to ICON, *klaō* break]

iconography /ˌaɪkəˈnɒgrəfɪ/ *n.* **1** the illustration of a subject by drawings or figures. **2** the study of portraits, esp. of an individual, or of artistic images or symbols. [Greek: related to ICON]

iconostasis /ˌaɪkəˈnɒstəsɪs/ *n.* (*pl.* **-stases** /-ˌsiːz/) (in the Eastern Church) screen bearing icons. [Greek: related to ICON]

icosahedron /ˌaɪkəsəˈhiːdrən/ *n.* solid figure with twenty faces. [Greek *eikosi* twenty, *hedra* base]

-ics *suffix* (treated as *sing.* or *pl.*) forming nouns denoting arts, sciences, etc. (*athletics*; *politics*).

icy /ˈaɪsɪ/ *adj.* (**-ier**, **-iest**) **1** very cold. **2** covered with or abounding in ice. **3** (of a tone or manner) unfriendly, hostile. □ **icily** *adv.* **iciness** *n.*

ID *abbr.* identification, identity (*ID card*).

id *n.* person's inherited unconscious psychological impulses. [Latin, = that]

I'd /aɪd/ *contr.* **1** I had. **2** I should; I would.

-ide *suffix Chem.* forming nouns denoting binary compounds of an element (*sodium chloride*; *lead sulphide*; *calcium carbide*). [extended from OXIDE]

idea /aɪˈdɪə/ *n.* **1** plan etc. formed by mental effort (*an idea for a book*). **2 a** mental impression or concept. **b** vague belief or fancy (*had an idea you were married*). **3** intention or purpose (*the idea is to make money*). **4** archetype or pattern. **5** ambition or aspiration (*have ideas*; *put ideas into a person's head*). □ **have no idea** *colloq.* **1** not know at all. **2** be completely incompetent. **not one's idea of** *colloq.* not what one regards as (*not my idea of a holiday*). [Greek, = form, kind]

ideal /aɪˈdiːəl/ —*adj.* **1** answering to one's highest conception; perfect. **2** existing only in idea; visionary. —*n.* perfect type, thing, concept, principle, etc., esp. as a standard to emulate. [French: related to IDEA]

idealism *n.* **1** forming or pursuing ideals, esp. unrealistically. **2** representation of things in ideal form. **3** system of thought in which objects are held to be in some way dependent on the mind. □ **idealist** *n.* **idealistic** /-ˈlɪstɪk/ *adj.* **idealistically** /-ˈlɪstɪkəlɪ/ *adv.*

idealize *v.* (also **-ise**) (**-zing** or **-sing**) regard or represent as ideal or perfect. □ **idealization** /-ˈzeɪʃ(ə)n/ *n.*

ideally *adv.* **1** in ideal circumstances. **2** according to an ideal.

idée fixe /ˌiːdeɪ ˈfiːks/ *n.* (*pl.* ***idées fixes*** pronunc. same) dominating idea; obsession. [French, = fixed idea]

identical /aɪˈdentɪk(ə)l/ *adj.* **1** (often foll. by *with*) (of different things) absolutely alike. **2** one and the same. **3** (of twins) developed from a single ovum. □ **identically** *adv.* [Latin *identicus*: related to IDENTITY]

identification /aɪˌdentɪfɪˈkeɪʃ(ə)n/ *n.* **1** identifying. **2** means of identifying (also *attrib.*: *identification card*).

identification parade *n.* group of people from whom a suspect is to be identified.

identify /aɪˈdentɪˌfaɪ/ *v.* (**-ies**, **-ied**) **1** establish the identity of; recognize. **2** select or discover (*identify the best solution*). **3** (also *refl.*; foll. by *with*) associate inseparably or very closely (with a party, policy, etc.). **4** (often foll. by *with*) treat as identical. **5** (foll. by *with*) put oneself in the place of (another person). □ **identifiable** *adj.* [medieval Latin *identifico*: related to IDENTITY]

Identikit /aɪˈdentɪkɪt/ *n.* (often *attrib.*) *propr.* picture of esp. a wanted suspect assembled from standard components using witnesses' descriptions. [from IDENTITY, KIT]

identity /aɪˈdentɪtɪ/ *n.* (*pl.* **-ies**) **1 a** condition of being a specified person or thing. **b** individuality, personality (*felt he had lost his identity*). **2** identification or the result of it (*mistaken identity*; *identity card*). **3** absolute sameness (*identity of interests*). **4** *Algebra* **a** equality of two expressions for all values of the quantities. **b** equation expressing this. [Latin *identitas* from *idem* same]

ideogram /ˈɪdɪəˌɡræm/ *n.* character symbolizing a thing without indicating the sounds in its name (e.g. a numeral, Chinese characters). [Greek *idea* form, -GRAM]

ideograph /ˈɪdɪəˌɡrɑːf/ *n.* = IDEOGRAM. □ **ideographic** /-ˈɡræfɪk/ *adj.* **ideography** /ˌɪdɪˈɒɡrəfɪ/ *n.*

ideologue /ˈaɪdɪəˌlɒɡ/ *n.* often *derog.* adherent of an ideology. [French: related to IDEA]

ideology /ˌaɪdɪˈɒlədʒɪ/ *n.* (*pl.* **-ies**) **1** ideas at the basis of an economic or political theory (*Marxist ideology*). **2** characteristic thinking of a class etc. (*bourgeois ideology*). □ **ideological** /-əˈlɒdʒɪk(ə)l/ *adj.* **ideologically** /-əˈlɒdʒɪkəlɪ/ *adv.* **ideologist** *n.* [French: related to IDEA, -LOGY]

ides /aɪdz/ *n.pl.* day of the ancient Roman month (the 15th day of March, May, July, and October, the 13th of other months). [Latin *idus*]

idiocy /ˈɪdɪəsɪ/ *n.* (*pl.* **-ies**) **1** foolishness; foolish act. **2** extreme mental imbecility.

idiom /ˈɪdɪəm/ *n.* **1** phrase etc. established by usage and not immediately comprehensible from the words used (e.g. *over the moon*, *see the light*). **2** form of expression peculiar to a language etc. **3** language of a people or country. **4** characteristic mode of expression in art etc. [Greek *idios* own]

idiomatic /ˌɪdɪəˈmætɪk/ *adj.* **1** relating or conforming to idiom. **2** characteristic of a particular language. □ **idiomatically** *adv.*

idiosyncrasy /ˌɪdɪəʊˈsɪŋkrəsɪ/ *n.* (*pl.* **-ies**) attitude, behaviour, or opinion peculiar to a person; anything highly individual or eccentric. □ **idiosyncratic** /-ˈkrætɪk/ *adj.* **idiosyncratically** /-ˈkrætɪkəlɪ/ *adv.* [Greek *idios* private, *sun* with, *krasis* mixture]

idiot /ˈɪdɪət/ *n.* **1** stupid person. **2** mentally deficient person incapable of rational conduct. □ **idiotic** /-ˈɒtɪk/ *adj.* **idiotically** /-ˈɒtɪkəlɪ/ *adv.* [Greek *idiotēs*, = private citizen, ignorant person]

idle /ˈaɪd(ə)l/ —*adj.* (**idler, idlest**) **1** lazy, indolent. **2** not in use; not working. **3** (of time etc.) unoccupied. **4** purposeless; groundless (*idle rumour*). **5** useless, ineffective (*idle protest*). —*v.* (**-ling**) **1** be idle. **2** run (an engine) or (of an engine) be run slowly without doing any work. **3** (foll. by *away*) pass (time etc.) in idleness. □ **idleness** *n.* **idler** *n.* **idly** *adv.* [Old English]

idol /ˈaɪd(ə)l/ *n.* **1** image of a deity etc. as an object of worship. **2** object of excessive or supreme adulation. [Greek *eidōlon* image, phantom]

idolater /aɪˈdɒlətə(r)/ *n.* **1** worshipper of idols. **2** devoted admirer. □ **idolatrous** *adj.* **idolatry** *n.* [related to IDOL, Greek *latreuō* worship]

idolize *v.* (also **-ise**) (**-zing** or **-sing**) **1** venerate or love excessively. **2** make an idol of. □ **idolization** /-ˈzeɪʃ(ə)n/ *n.*

idyll /ˈɪdɪl/ *n.* **1** short description, esp. in verse, of a peaceful or romantic, esp. rural, scene or incident. **2** such a scene or incident. [Greek *eidullion*]

idyllic /ɪˈdɪlɪk/ *adj.* **1** blissfully peaceful and happy. **2** of or like an idyll. □ **idyllically** *adv.*

i.e. *abbr.* that is to say. [Latin *id est*]

-ie see -Y^2.

if —*conj.* **1** introducing a conditional clause: **a** on the condition or supposition that; in the event that (*if he comes I will tell him*; *if you are tired we can rest*). **b** (with past tense) implying that the condition is not fulfilled (*if I knew I would say*). **2** even though (*I'll finish it, if it takes me all day*). **3** whenever (*if I am not sure I ask*). **4** whether (*see if you can find it*). **5** expressing a wish, surprise, or request (*if I could just try!*; *if it isn't my old hat!*; *if you wouldn't mind?*). —*n.* condition, supposition (*too many ifs about it*). □ **if only 1** even if for no other reason than (*I'll come if only to see her*). **2** (often *ellipt.*) expression of regret; I wish that (*if only I had thought of it*). [Old English]

iffy *adj.* (**-ier, -iest**) *colloq.* uncertain; dubious.

Igbo var. of IBO.

igloo /ˈɪɡluː/ *n.* Eskimo dome-shaped dwelling, esp. of snow. [Eskimo, = house]

igneous /ˈɪɡnɪəs/ *adj.* **1** of fire; fiery. **2** (esp. of rocks) volcanic. [Latin *ignis* fire]

ignite /ɪɡˈnaɪt/ *v.* (**-ting**) **1** set fire to. **2** catch fire. **3** provoke or excite (feelings etc.). [Latin *ignio ignit-* set on fire]

ignition /ɪɡˈnɪʃ(ə)n/ *n.* **1** mechanism for, or the action of, starting combustion in an internal-combustion engine. **2** igniting or being ignited.

ignoble /ɪɡˈnəʊb(ə)l/ *adj.* (**-bler, -blest**) **1** dishonourable. **2** of low birth, position, or reputation. □ **ignobly** *adv.* [Latin: related to IN-^{1}, NOBLE]

ignominious /ˌɪɡnəˈmɪnɪəs/ *adj.* shameful, humiliating. □ **ignominiously** *adv.* [Latin: related to IGNOMINY]

ignominy /'ɪgnəmɪnɪ/ *n.* dishonour, infamy. [Latin: related to IN-[1], Latin *(g)nomen* name]

ignoramus /,ɪgnə'reɪməs/ *n.* (*pl.* **-muses**) ignorant person. [Latin, = we do not know: related to IGNORE]

ignorance /'ɪgnərəns/ *n.* lack of knowledge. [French from Latin: related to IGNORE]

ignorant *adj.* **1** (often foll. by *of, in*) lacking knowledge (esp. of a fact or subject). **2** *colloq.* uncouth. □ **ignorantly** *adv.*

ignore /ɪg'nɔ:(r)/ *v.* **(-ring)** refuse to take notice of; intentionally disregard. [Latin *ignoro* not know]

iguana /ɪg'wɑ:nə/ *n.* large American, W. Indian, or Pacific lizard with a dorsal crest. [Spanish from Carib *iwana*]

iguanodon /ɪ'gwɑ:nə,dɒn/ *n.* large plant-eating dinosaur with small fore-limbs. [from IGUANA, which it resembles, after *mastodon* etc.]

ikebana /,ɪkɪ'bɑ:nə/ *n.* art of Japanese flower arrangement. [Japanese, = living flowers]

ikon var. of ICON.

il- *prefix* assim. form of IN-[1], IN-[2] before *l.*

ileum /'ɪlɪəm/ *n.* (*pl.* **ilea**) third and last portion of the small intestine. [Latin *ilium*]

ilex /'aɪleks/ *n.* (*pl.* **-es**) **1** tree or shrub of the genus including the common holly. **2** holm-oak. [Latin]

iliac /'ɪlɪ,æk/ *adj.* of the lower body (*iliac artery*). [Latin *ilia* flanks]

ilk *n.* **1** *colloq.*, usu. *derog.* sort, family, class, etc. **2** (in **of that ilk**) *Scot.* of the ancestral estate with the same name as the family (*Guthrie of that ilk*). [Old English]

ill *–adj.* (*attrib.* except in sense 1) **1** (usu. *predic.*) not in good health; unwell. **2** wretched, unfavourable (*ill fortune*; *ill luck*). **3** harmful (*ill effects*). **4** hostile, unkind (*ill feeling*). **5** faulty, unskilful (*ill management*). **6** (of manners or conduct) improper. *–adv.* **1** badly, wrongly, imperfectly (*ill-matched*; *ill-provided*). **2** scarcely (*can ill afford it*). **3** unfavourably (*spoke ill of them*). *–n.* **1** injury, harm. **2** evil. □ **ill at ease** embarrassed, uneasy. [Old Norse]

I'll /aɪl/ *contr.* I shall; I will.

ill-advised *adj.* foolish; imprudent.

ill-assorted *adj.* badly matched; mixed.

ill-bred *adj.* badly brought up; rude.

ill-defined *adj.* not clearly defined.

ill-disposed *adj.* **1** (often foll. by *towards*) unfavourably disposed. **2** malevolent.

illegal /ɪ'li:g(ə)l/ *adj.* **1** not legal. **2** criminal. □ **illegality** /-'gælɪtɪ/ *n.* (*pl.* **-ies**). **illegally** *adv.*

illegible /ɪ'ledʒɪb(ə)l/ *adj.* not legible. □ **illegibility** /-'bɪlɪtɪ/ *n.* **illegibly** *adv.*

illegitimate /,ɪlɪ'dʒɪtɪmət/ *adj.* **1** born of parents not married to each other. **2** unlawful. **3** improper. **4** wrongly inferred. □ **illegitimacy** *n.* **illegitimately** *adv.*

ill-fated *adj.* destined to or bringing bad fortune.

ill-favoured *adj.* unattractive.

ill-founded *adj.* (of an idea etc.) baseless.

ill-gotten *adj.* gained unlawfully or wickedly.

ill health *n.* poor physical or mental condition.

ill humour *n.* irritability.

illiberal /ɪ'lɪbər(ə)l/ *adj.* **1** intolerant, narrow-minded. **2** without liberal culture; vulgar. **3** stingy; mean. □ **illiberality** /-'rælɪtɪ/ *n.* **illiberally** *adv.*

illicit /ɪ'lɪsɪt/ *adj.* unlawful, forbidden. □ **illicitly** *adv.*

illiterate /ɪ'lɪtərət/ *–adj.* **1** unable to read. **2** uneducated. *–n.* illiterate person. □ **illiteracy** *n.* **illiterately** *adv.*

ill-mannered *adj.* having bad manners; rude.

ill-natured *adj.* churlish, unkind.

illness *n.* **1** disease. **2** being ill.

illogical /ɪ'lɒdʒɪk(ə)l/ *adj.* devoid of or contrary to logic. □ **illogicality** /-'kælɪtɪ/ *n.* (*pl.* **-ies**). **illogically** *adv.*

ill-omened *adj.* doomed.

ill-tempered *adj.* morose, irritable.

ill-timed *adj.* done or occurring at an inappropriate time.

ill-treat *v.* treat badly; abuse.

illuminate /ɪ'lu:mɪ,neɪt/ *v.* **(-ting) 1** light up; make bright. **2** decorate (buildings etc.) with lights. **3** decorate (a manuscript etc.) with gold, colour, etc. **4** help to explain (a subject etc.). **5** enlighten spiritually or intellectually. **6** shed lustre on. □ **illuminating** *adj.* **illumination** /-'neɪʃ(ə)n/ *n.* **illuminative** *adj.* [Latin *lumen* light]

illumine /ɪ'lju:mɪn/ *v.* **(-ning)** *literary* **1** light up; make bright. **2** enlighten.

ill-use *v.* = ILL-TREAT.

illusion /ɪ'lu:ʒ(ə)n/ *n.* **1** false impression or belief. **2** state of being deceived by appearances. **3** figment of the imagination. □ **be under the illusion** (foll. by *that*) believe mistakenly. □ **illusive** *adj.* **illusory** *adj.* [Latin *illudo* mock]

illusionist *n.* conjuror.

illustrate /'ɪlə,streɪt/ *v.* **(-ting) 1 a** provide (a book etc.) with pictures. **b** elucidate by drawings, pictures, examples, etc. **2** serve as an example of. □ **illustrator** *n.* [Latin *lustro* light up]

illustration /ˌɪləˈstreɪʃ(ə)n/ *n.* **1** drawing or picture in a book, magazine, etc. **2** explanatory example. **3** illustrating.

illustrative /ˈɪləstrətɪv/ *adj.* (often foll. by *of*) explanatory; exemplary.

illustrious /ɪˈlʌstrɪəs/ *adj.* distinguished, renowned. [Latin *illustris*: related to ILLUSTRATE]

ill will *n.* bad feeling; animosity.

im- *prefix* assim. form of IN-[1], IN-[2] before *b*, *m*, or *p*.

I'm /aɪm/ *contr.* I am.

image /ˈɪmɪdʒ/ *–n.* **1** representation of an object, e.g. a statue. **2** reputation or persona of a person, company, etc. **3** appearance as seen in a mirror or through a lens. **4** mental picture or idea. **5** simile or metaphor. *–v.* (**-ging**) **1** make an image of; portray. **2** reflect, mirror. **3** describe or imagine vividly. □ **be the image of** be or look exactly like. [Latin *imago imagin-*]

imagery *n.* **1** figurative illustration, esp. in literature. **2** images; statuary, carving. **3** mental images collectively.

imaginary /ɪˈmædʒɪnərɪ/ *adj.* **1** existing only in the imagination. **2** *Math.* being the square root of a negative quantity. [Latin: related to IMAGE]

imagination /ɪˌmædʒɪˈneɪʃ(ə)n/ *n.* **1** mental faculty of forming images or concepts of objects or situations not existent or not directly experienced. **2** mental creativity or resourcefulness.

imaginative /ɪˈmædʒɪnətɪv/ *adj.* having or showing imagination. □ **imaginatively** *adv.* **imaginativeness** *n.*

imagine /ɪˈmædʒɪn/ *v.* (**-ning**) **1 a** form a mental image or concept of. **b** picture to oneself. **2** think of as probable (*can't imagine he'd be so stupid*). **3** guess (*can't imagine what he is doing*). **4** suppose (*I imagine you'll need help*). □ **imaginable** *adj.* [Latin *imaginor*]

imago /ɪˈmeɪgəʊ/ *n.* (*pl.* **-s** or **imagines** /ɪˈmædʒɪˌni:z/) fully developed stage of an insect, e.g. a butterfly. [Latin: see IMAGE]

imam /ɪˈmɑ:m/ *n.* **1** leader of prayers in a mosque. **2** title of various Muslim leaders. [Arabic]

imbalance /ɪmˈbæləns/ *n.* **1** lack of balance. **2** disproportion.

imbecile /ˈɪmbɪˌsi:l/ *–n.* **1** *colloq.* stupid person. **2** person with a mental age of about five. *–adj.* mentally weak; stupid, idiotic. □ **imbecilic** /-ˈsɪlɪk/ *adj.* **imbecility** /-ˈsɪlɪtɪ/ *n.* (*pl.* **-ies**). [French from Latin]

imbed var. of EMBED.

imbibe /ɪmˈbaɪb/ *v.* (**-bing**) **1** drink (esp. alcohol). **2 a** assimilate (ideas etc.). **b** absorb (moisture etc.). **3** inhale (air etc.). [Latin *bibo* drink]

imbroglio /ɪmˈbrəʊlɪəʊ/ *n.* (*pl.* **-s**) **1** confused or complicated situation. **2** confused heap. [Italian: related to IN-[2], BROIL]

imbue /ɪmˈbju:/ *v.* (**-bues, -bued, -buing**) (often foll. by *with*) **1** inspire or permeate (with feelings, opinions, or qualities). **2** saturate. **3** dye. [Latin *imbuo*]

IMF *abbr.* International Monetary Fund.

imitate /ˈɪmɪˌteɪt/ *v.* (**-ting**) **1** follow the example of; copy. **2** mimic. **3** make a copy of. **4** be like. □ **imitable** *adj.* **imitator** *n.* [Latin *imitor -tat-*]

imitation /ˌɪmɪˈteɪʃ(ə)n/ *n.* **1** imitating or being imitated. **2** copy. **3** counterfeit (often *attrib.*: *imitation leather*).

imitative /ˈɪmɪtətɪv/ *adj.* **1** (often foll. by *of*) imitating; following a model or example. **2** (of a word) reproducing a natural sound (e.g. *fizz*), or otherwise suggestive (e.g. *blob*).

immaculate /ɪˈmækjʊlət/ *adj.* **1** perfectly clean and tidy. **2** perfect (*immaculate timing*). **3** innocent, faultless. □ **immaculately** *adv.* **immaculateness** *n.* [Latin: related to IN-[1], *macula* spot]

Immaculate Conception *n. RC Ch.* doctrine that the Virgin Mary was without original sin from conception.

immanent /ˈɪmənənt/ *adj.* **1** (often foll. by *in*) naturally present, inherent. **2** (of God) omnipresent. □ **immanence** *n.* [Latin: related to IN-[2], *maneo* remain]

immaterial /ˌɪməˈtɪərɪəl/ *adj.* **1** unimportant; irrelevant. **2** not material; incorporeal. □ **immateriality** /-ˈælɪtɪ/ *n.*

immature /ˌɪməˈtjʊə(r)/ *adj.* **1** not mature. **2** undeveloped, esp. emotionally. **3** unripe. □ **immaturely** *adv.* **immaturity** *n.*

immeasurable /ɪˈmeʒərəb(ə)l/ *adj.* not measurable; immense. □ **immeasurably** *adv.*

immediate /ɪˈmi:dɪət/ *adj.* **1** occurring or done at once (*immediate reply*). **2** nearest, next; direct (*immediate vicinity*; *immediate future*; *immediate cause of death*). **3** most pressing or urgent (*our immediate concern*). □ **immediacy** *n.* **immediateness** *n.* [Latin: related to IN-[1], MEDIATE]

immediately *–adv.* **1** without pause or delay. **2** without intermediary. *–conj.* as soon as.

immemorial /ˌɪmɪˈmɔ:rɪəl/ *adj.* ancient beyond memory or record (*from time immemorial*).

immense /ɪˈmens/ *adj.* **1** extremely large; huge. **2** considerable (*immense* *difference*). □ **immenseness** *n.* **immensity** *n.* [Latin *metior mens-* measure]

immensely *adv.* **1** *colloq.* very much (*enjoyed myself immensely*). **2** to an immense degree (*immensely rich*).

immerse /ɪ'mɜ:s/ *v.* (**-sing**) **1 a** (often foll. by *in*) dip, plunge. **b** submerge (a person). **2** (often *refl.* or in *passive*; often foll. by *in*) absorb or involve deeply. **3** (often foll. by *in*) bury, embed. [Latin *mergo mers-* dip]

immersion /ɪ'mɜ:ʃ(ə)n/ *n.* **1** immersing or being immersed. **2** baptism by total bodily immersion. **3** mental absorption.

immersion heater *n.* electric device immersed in a liquid to heat it, esp. in a hot-water tank.

immigrant /'ɪmɪgrənt/ **–***n.* person who immigrates. **–***adj.* **1** immigrating. **2** of immigrants.

immigrate /'ɪmɪ,greɪt/ *v.* come into a country and settle. □ **immigration** /-'greɪʃ(ə)n/ *n.* [related to IN-[2], MIGRATE]

imminent /'ɪmɪnənt/ *adj.* impending; about to happen (*war is imminent*). □ **imminence** *n.* **imminently** *adv.* [Latin *immineo* be impending]

immiscible /ɪ'mɪsɪb(ə)l/ *adj.* (often foll. by *with*) not able to be mixed. □ **immiscibility** /-'bɪlɪtɪ/ *n.*

immobile /ɪ'məʊbaɪl/ *adj.* **1** not moving. **2** unable to move or be moved. □ **immobility** /-'bɪlɪtɪ/ *n.*

immobilize /ɪ'məʊbɪ,laɪz/ *v.* (also **-ise**) (**-zing** or **-sing**) **1** make or keep immobile. **2** keep (a limb or patient) still for healing purposes. □ **immobilization** /-'zeɪʃ(ə)n/ *n.*

immoderate /ɪ'mɒdərət/ *adj.* excessive; lacking moderation. □ **immoderately** *adv.*

immodest /ɪ'mɒdɪst/ *adj.* **1** lacking modesty; conceited. **2** shameless, indecent. □ **immodestly** *adv.* **immodesty** *n.*

immolate /'ɪmə,leɪt/ *v.* (**-ting**) kill or offer as a sacrifice. □ **immolation** /-'leɪʃ(ə)n/ *n.* [Latin, = sprinkle with meal]

immoral /ɪ'mɒr(ə)l/ *adj.* **1** not conforming to accepted morality; morally wrong. **2** sexually promiscuous or deviant. □ **immorality** /,ɪmə'rælɪtɪ/ *n.* (*pl.* **-ies**). **immorally** *adv.*

immortal /ɪ'mɔ:t(ə)l/ **–***adj.* **1 a** living for ever; not mortal. **b** divine. **2** unfading. **3** famous for all time. **–***n.* **1 a** immortal being. **b** (in *pl.*) gods of antiquity. **2** person, esp. an author, remembered long after death. □ **immortality** /,ɪmɔ:'tælɪtɪ/ *n.* **immortalize** *v.* (also **-ise**) (**-zing** or **-sing**). **immortally** *adv.*

immovable /ɪ'mu:vəb(ə)l/ *adj.* (also **immoveable**) **1** not able to be moved. **2** steadfast, unyielding. **3** emotionless. **4** not subject to change (*immovable law*). **5** motionless. **6** (of property) consisting of land, houses, etc. □ **immovability** /-'bɪlɪtɪ/ *n.* **immovably** *adv.*

immune /ɪ'mju:n/ *adj.* **1 a** (often foll. by *against*, *from*, *to*) protected against infection through inoculation etc. **b** relating to immunity (*immune system*). **2** (foll. by *from*, *to*) exempt from or proof against a charge, duty, criticism, etc. [Latin *immunis* exempt]

immunity *n.* (*pl.* **-ies**) **1** ability of an organism to resist infection by means of antibodies and white blood cells. **2** (often foll. by *from*) freedom or exemption.

immunize /'ɪmju:,naɪz/ *v.* (also **-ise**) (**-zing** or **-sing**) make immune, usu. by inoculation. □ **immunization** /-'zeɪʃ(ə)n/ *n.*

immunodeficiency /,ɪmju:,nəʊdɪ'fɪʃənsɪ/ *n.* reduction in normal immune defences.

immunoglobulin /,ɪmju:nəʊ'glɒbjʊlɪn/ *n.* any of a group of related proteins functioning as antibodies.

immunology /,ɪmju:'nɒlədʒɪ/ *n.* the study of immunity. □ **immunological** /-nə'lɒdʒɪk(ə)l/ *adj.* **immunologist** *n.*

immunotherapy /,ɪmju:nəʊ'θerəpɪ/ *n.* prevention or treatment of disease with substances that stimulate the immune response.

immure /ɪ'mjʊə(r)/ *v.* (**-ring**) **1** imprison. **2** *refl.* shut oneself away. [Latin *murus* wall]

immutable /ɪ'mju:təb(ə)l/ *adj.* unchangeable. □ **immutability** /-'bɪlɪtɪ/ *n.* **immutably** *adv.*

imp *n.* **1** mischievous child. **2** small devil or sprite. [Old English, = young shoot]

impact **–***n.* /'ɪmpækt/ **1** effect of sudden forcible contact between two solid bodies etc.; collision. **2** strong effect or impression. **–***v.* /ɪm'pækt/ **1** press or fix firmly. **2** (as **impacted** *adj.*) (of a tooth) wedged between another tooth and the jaw. **3** (often foll. by *on*) have an impact on. □ **impaction** /ɪm'pækʃ(ə)n/ *n.* [Latin: related to IMPINGE]

impair /ɪm'peə(r)/ *v.* damage, weaken. □ **impairment** *n.* [Latin, = make worse, from *pejor*]

impala /ɪm'pɑ:lə/ *n.* (*pl.* same or **-s**) small African antelope. [Zulu]

impale /ɪm'peɪl/ *v.* (**-ling**) transfix or pierce with a sharp stake etc. □ **impalement** *n.* [Latin *palus* PALE[2]]

impalpable /ɪm'pælpəb(ə)l/ *adj.* **1** not easily grasped by the mind; intangible. **2** imperceptible to the touch. **3** (of powder) very fine. □ **impalpability** /-'bɪlɪtɪ/ *n.* **impalpably** *adv.*

impanel var. of EMPANEL.

impart /ɪm'pɑ:t/ *v.* (often foll. by *to*) **1** communicate (news etc.). **2** give a share of (a thing). [Latin: related to PART]

impartial /ɪm'pɑ:ʃ(ə)l/ *adj.* treating all alike; unprejudiced, fair. □ **impartiality** /-ʃɪ'ælɪtɪ/ *n.* **impartially** *adv.*

impassable /ɪm'pɑ:səb(ə)l/ *adj.* not able to be traversed. □ **impassability** /-'bɪlɪtɪ/ *n.* **impassableness** *n.* **impassably** *adv.*

impasse /'æmpæs, 'ɪm-/ *n.* deadlock. [French: related to PASS[1]]

impassible /ɪm'pæsɪb(ə)l/ *adj.* **1** impassive. **2** incapable of feeling, emotion, or injury. □ **impassibility** /-'bɪlɪtɪ/ *n.* **impassibly** *adv.* [Latin *patior pass-* suffer]

impassioned /ɪm'pæʃ(ə)nd/ *adj.* filled with passion; ardent. [Italian *impassionato*: related to PASSION]

impassive /ɪm'pæsɪv/ *adj.* incapable of or not showing emotion; serene. □ **impassively** *adv.* **impassiveness** *n.* **impassivity** /-'sɪvɪtɪ/ *n.*

impasto /ɪm'pæstəʊ/ *n. Art* technique of laying on paint thickly. [Italian]

impatiens /ɪm'peɪʃɪ,enz/ *n.* any of several plants including the busy Lizzie. [Latin: related to IMPATIENT]

impatient /ɪm'peɪʃ(ə)nt/ *adj.* **1** lacking, or showing a lack of, patience or tolerance. **2** restlessly eager. **3** (foll. by *of*) intolerant of. □ **impatience** *n.* **impatiently** *adv.*

impeach /ɪm'pi:tʃ/ *v.* **1** charge with a crime against the State, esp. treason. **2** *US* charge (a public official) with misconduct. **3** call in question, disparage. □ **impeachable** *adj.* **impeachment** *n.* [French *empecher* from Latin *pedica* fetter]

impeccable /ɪm'pekəb(ə)l/ *adj.* faultless, exemplary. □ **impeccability** /-'bɪlɪtɪ/ *n.* **impeccably** *adv.* [related to IN-[1], Latin *pecco* sin]

impecunious /,ɪmpɪ'kju:nɪəs/ *adj.* having little or no money. □ **impecuniosity** /-'ɒsɪtɪ/ *n.* **impecuniousness** *n.* [related to PECUNIARY]

impedance /ɪm'pi:d(ə)ns/ *n.* total effective resistance of an electric circuit etc. to an alternating current. [from IMPEDE]

■ **Usage** *Impedance* is sometimes confused with *impediment*, which means 'a hindrance' or 'a speech defect'.

impede /ɪm'pi:d/ *v.* **(-ding)** obstruct; hinder. [Latin *impedio* from *pes ped-* foot]

impediment /ɪm'pedɪmənt/ *n.* **1** hindrance or obstruction. **2** speech defect, e.g. a stammer. [Latin: related to IMPEDE]

■ **Usage** See note at *impedance.*

impedimenta /ɪm,pedɪ'mentə/ *n.pl.* **1** encumbrances. **2** baggage, esp. of an army.

impel /ɪm'pel/ *v.* **(-ll-) 1** drive, force, or urge. **2** propel. [Latin *pello* drive]

impend /ɪm'pend/ *v.* (often foll. by *over*) **1** (of a danger, event, etc.) be threatening or imminent. **2** hang. □ **impending** *adj.* [Latin *pendeo* hang]

impenetrable /ɪm'penɪtrəb(ə)l/ *adj.* **1** not able to be penetrated. **2** inscrutable. **3** inaccessible to ideas, influences, etc. □ **impenetrability** /-'bɪlɪtɪ/ *n.* **impenetrableness** *n.* **impenetrably** *adv.*

impenitent /ɪm'penɪt(ə)nt/ *adj.* not sorry, unrepentant. □ **impenitence** *n.*

imperative /ɪm'perətɪv/ *–adj.* **1** urgent; obligatory. **2** commanding, peremptory. **3** *Gram.* (of a mood) expressing a command (e.g. *come here!*). *–n.* **1** *Gram.* imperative mood. **2** command. **3** essential or urgent thing. [Latin *impero* command]

imperceptible /,ɪmpə'septɪb(ə)l/ *adj.* **1** not perceptible. **2** very slight, gradual, or subtle. □ **imperceptibility** /-'bɪlɪtɪ/ *n.* **imperceptibly** *adv.*

imperfect /ɪm'pɜ:fɪkt/ *–adj.* **1** not perfect; faulty, incomplete. **2** *Gram.* (of a tense) denoting action in progress but not completed (e.g. *they were singing*). *–n.* imperfect tense. □ **imperfectly** *adv.*

imperfection /,ɪmpə'fekʃ(ə)n/ *n.* **1** state of being imperfect. **2** fault, blemish.

imperial /ɪm'pɪərɪəl/ *adj.* **1** of or characteristic of an empire or similar sovereign State. **2 a** of an emperor. **b** majestic, august; authoritative. **3** (of non-metric weights and measures) statutory in the UK, esp. formerly (*imperial gallon*). □ **imperially** *adv.* [Latin *imperium* dominion]

imperialism *n.* **1** imperial rule or system. **2** usu. *derog.* policy of dominating other nations by acquiring dependencies etc. □ **imperialist** *n.* & *adj.* **imperialistic** /-'lɪstɪk/ *adj.*

imperil /ɪm'perɪl/ *v.* **(-ll-;** *US* **-l-)** endanger.

imperious /ɪm'pɪərɪəs/ *adj.* overbearing, domineering. □ **imperiously** *adv.* **imperiousness** *n.*

imperishable /ɪm'perɪʃəb(ə)l/ *adj.* not able to perish, indestructible.

impermanent /ɪm'pɜ:mənənt/ *adj.* not permanent. □ **impermanence** *n.* **impermanency** *n.*

impermeable /ɪm'pɜ:mɪəb(ə)l/ *adj.* not permeable, not allowing fluids to pass through. □ **impermeability** /-'bɪlɪtɪ/ *n.*

impermissible /,ɪmpə'mɪsɪb(ə)l/ *adj.* not allowable.

impersonal /ɪm'pɜ:sən(ə)l/ *adj.* **1** without personal reference; objective, impartial. **2** without human attributes; cold, unfeeling. **3** *Gram.* **a** (of a verb) used esp. with *it* as a subject (e.g. *it is snowing*). **b** (of a pronoun) = INDEFINITE. □ **impersonality** /-'nælɪtɪ/ *n.* **impersonally** *adv.*

impersonate /ɪm'pɜ:sə,neɪt/ *v.* **(-ting) 1** pretend to be (another person), esp. as entertainment or fraud. **2** act (a character). □ **impersonation** /-'neɪʃ(ə)n/ *n.* **impersonator** *n.* [from IN-[2], Latin PERSONA]

impertinent /ɪm'pɜ:tɪnənt/ *adj.* **1** insolent, disrespectful. **2** esp. *Law* irrelevant. □ **impertinence** *n.* **impertinently** *adv.*

imperturbable /,ɪmpə'tɜ:bəb(ə)l/ *adj.* not excitable; calm. □ **imperturbability** /-'bɪlɪtɪ/ *n.* **imperturbably** *adv.*

impervious /ɪm'pɜ:vɪəs/ *adj.* (usu. foll. by *to*) **1** impermeable. **2** not responsive (to argument etc.).

impetigo /,ɪmpɪ'taɪgəʊ/ *n.* contagious skin infection forming pimples and sores. [Latin *impeto* assail]

impetuous /ɪm'petjʊəs/ *adj.* **1** acting or done rashly or with sudden energy. **2** moving forcefully or rapidly. □ **impetuosity** /-'ɒsɪtɪ/ *n.* **impetuously** *adv.* **impetuousness** *n.* [Latin: related to IMPETUS]

impetus /'ɪmpɪtəs/ *n.* **1** force with which a body moves. **2** driving force or impulse. [Latin *impeto* assail]

impiety /ɪm'paɪətɪ/ *n.* (*pl.* **-ies**) **1** lack of piety or reverence. **2** act etc. showing this.

impinge /ɪm'pɪndʒ/ *v.* **(-ging)** (usu. foll. by *on, upon*) **1** make an impact or effect. **2** encroach. □ **impingement** *n.* [Latin *pango pact-* fix]

impious /'ɪmpɪəs/ *adj.* **1** not pious. **2** wicked, profane.

impish *adj.* of or like an imp; mischievous. □ **impishly** *adv.* **impishness** *n.*

implacable /ɪm'plækəb(ə)l/ *adj.* unable to be appeased. □ **implacability** /-'bɪlɪtɪ/ *n.* **implacably** *adv.*

implant –*v.* /ɪm'plɑ:nt/ **1** (often foll. by *in*) insert or fix. **2** (often foll. by *in*) instil (an idea etc.) in a person's mind. **3** plant. **4 a** insert (tissue etc.) in a living body. **b** (in *passive*) (of a fertilized ovum) become attached to the wall of the womb. –*n.* /'ɪmplɑ:nt/ thing implanted, esp. a piece of tissue. □ **implantation** /-'teɪʃ(ə)n/ *n.* [Latin: related to PLANT]

implausible /ɪm'plɔ:zɪb(ə)l/ *adj.* not plausible. □ **implausibility** /-'bɪlɪtɪ/ *n.* **implausibly** *adv.*

implement –*n.* /'ɪmplɪmənt/ tool, instrument, utensil. –*v.* /'ɪmplɪ,ment/ put (a decision, plan, contract, etc.) into effect. □ **implementation** /,ɪmplɪmen'teɪʃ(ə)n/ *n.* [Latin *impleo* fulfil]

implicate /'ɪmplɪ,keɪt/ *v.* **(-ting) 1** (often foll. by *in*) show (a person) to be involved (in a crime etc.). **2** imply. [Latin *plico* fold]

implication /,ɪmplɪ'keɪʃ(ə)n/ *n.* **1** thing implied. **2** implicating or implying.

implicit /ɪm'plɪsɪt/ *adj.* **1** implied though not plainly expressed. **2** absolute, unquestioning (*implicit belief*). □ **implicitly** *adv.* [Latin: related to IMPLICATE]

implode /ɪm'pləʊd/ *v.* **(-ding)** (cause to) burst inwards. □ **implosion** /ɪm'pləʊʒ(ə)n/ *n.* [from IN-[2]: cf. EXPLODE]

implore /ɪm'plɔ:(r)/ *v.* **(-ring) 1** (often foll. by *to* + infin.) entreat (a person). **2** beg earnestly for. [Latin *ploro* weep]

imply /ɪm'plaɪ/ *v.* **(-ies, -ied) 1** (often foll. by *that*) strongly suggest or insinuate without directly stating (*what are you implying?*). **2** signify, esp. as a consequence (*silence implies guilt*). [Latin: related to IMPLICATE]

impolite /,ɪmpə'laɪt/ *adj.* **(impolitest)** ill-mannered, uncivil, rude. □ **impolitely** *adv.* **impoliteness** *n.*

impolitic /ɪm'pɒlɪtɪk/ *adj.* inexpedient, unwise. □ **impoliticly** *adv.*

imponderable /ɪm'pɒndərəb(ə)l/ –*adj.* **1** not able to be estimated. **2** very light; weightless. –*n.* (usu. in *pl.*) imponderable thing. □ **imponderability** /-'bɪlɪtɪ/ *n.* **imponderably** *adv.*

import –*v.* /ɪm'pɔ:t, 'ɪm-/ **1** bring in (esp. foreign goods or services) to a country. **2** imply, indicate, signify. –*n.* /'ɪmpɔ:t/ **1** (esp. in *pl.*) imported article or service. **2** importing. **3** what is implied; meaning. **4** importance. □ **importation** /,ɪmpɔ:'teɪʃ(ə)n/ *n.* **importer** /ɪm'pɔ:tə(r)/ *n.* [Latin *importo* carry in]

important /ɪm'pɔ:t(ə)nt/ *adj.* **1** (often foll. by *to*) of great effect or consequence; momentous. **2** (of a person) having high rank or authority. **3** pompous. □ **importance** *n.* **importantly** *adv.* [Latin *importo* carry in, signify]

importunate /ɪm'pɔ:tjʊnət/ *adj.* making persistent or pressing requests. □ **importunity** /,ɪmpɔ:'tju:nɪtɪ/ *n.* [Latin *importunus* inconvenient]

importune /,ɪmpə'tju:n/ *v.* **(-ning) 1** pester (a person) with requests. **2** solicit as a prostitute.

impose /ɪm'pəʊz/ *v.* **(-sing) 1** (often foll. by *on, upon*) lay (a tax, duty, charge, or obligation) on. **2** enforce compliance with. **3** also *refl.* (foll. by *on*, *upon*, or *absol.*) take advantage of (*will not impose on you any longer*). **4** (often foll. by *on*, *upon*) inflict (a thing) on. [Latin *impono*]

imposing *adj.* impressive, formidable, esp. in appearance.

imposition /ˌɪmpəˈzɪʃ(ə)n/ *n.* **1** imposing or being imposed. **2** unfair demand or burden. **3** tax, duty.

impossible /ɪmˈpɒsɪb(ə)l/ *adj.* **1** not possible. **2** *colloq.* not easy, convenient, or believable. **3** *colloq.* (esp. of a person) outrageous, intolerable. □ **impossibility** /-ˈbɪlɪtɪ/ *n.* (*pl.* **-ies**). **impossibly** *adv.*

impost[1] /ˈɪmpəʊst/ *n.* tax, duty, or tribute. [Latin *impono impost-* impose]

impost[2] /ˈɪmpəʊst/ *n.* upper course of a pillar, carrying an arch.

impostor /ɪmˈpɒstə(r)/ *n.* (also **imposter**) **1** person who assumes a false character or pretends to be someone else. **2** swindler.

imposture /ɪmˈpɒstʃə(r)/ *n.* fraudulent deception.

impotent /ˈɪmpət(ə)nt/ *adj.* **1** powerless, ineffective. **2** (of a male) unable to achieve an erection or orgasm. □ **impotence** *n.*

impound /ɪmˈpaʊnd/ *v.* **1** confiscate. **2** take legal possession of. **3** shut up (animals) in a pound.

impoverish /ɪmˈpɒvərɪʃ/ *v.* make poor. □ **impoverishment** *n.* [French: related to POVERTY]

impracticable /ɪmˈpræktɪkəb(ə)l/ *adj.* not practicable. □ **impracticability** /-ˈbɪlɪtɪ/ *n.* **impracticably** *adv.*

impractical /ɪmˈpræktɪk(ə)l/ *adj.* **1** not practical. **2** esp. *US* not practicable. □ **impracticality** /-ˈkælɪtɪ/ *n.*

imprecation /ˌɪmprɪˈkeɪʃ(ə)n/ *n. formal* oath, curse. [Latin *precor* pray]

imprecise /ˌɪmprɪˈsaɪs/ *adj.* not precise. □ **imprecisely** *adv.* **impreciseness** *n.* **imprecision** /-ˈsɪʒ(ə)n/ *n.*

impregnable /ɪmˈpregnəb(ə)l/ *adj.* strong enough to be secure against attack. □ **impregnability** /-ˈbɪlɪtɪ/ *n.* **impregnably** *adv.* [French: related to IN-[1], Latin *prehendo* take]

impregnate /ˈɪmpregˌneɪt/ *v.* (**-ting**) **1** (often foll. by *with*) fill or saturate. **2** (often foll. by *with*) imbue (with feelings etc.). **3 a** make (a female) pregnant. **b** fertilize (an ovum). □ **impregnation** /-ˈneɪʃ(ə)n/ *n.* [Latin: related to PREGNANT]

impresario /ˌɪmprɪˈsɑːrɪəʊ/ *n.* (*pl.* **-s**) organizer of public entertainment, esp. a theatrical etc. manager. [Italian]

impress *–v.* /ɪmˈpres/ **1** (often foll. by *with*) **a** affect or influence deeply. **b** affect (a person) favourably (*was most impressed*). **2** (often foll. by *on*) emphasize (an idea etc.) (*must impress on you the need to be prompt*). **3 a** (often foll. by *on*) imprint or make (a mark). **b** mark (a thing) with a stamp, seal, etc. *–n.* /ˈɪmpres/ **1** mark made by a seal, stamp, etc. **2** characteristic mark or quality. □ **impressible** /ɪmˈpresɪb(ə)l/ *adj.* [French: related to PRESS[1]]

impression /ɪmˈpreʃ(ə)n/ *n.* **1** effect (esp. on the mind or feelings). **2** notion or belief (esp. vague or mistaken). **3** imitation of a person or sound, esp. as entertainment. **4 a** impressing. **b** mark impressed. **5** unaltered reprint from standing type or plates. **6** number of copies of a book etc. issued at one time. **7** print taken from a wood or copper engraving.

impressionable *adj.* easily influenced. □ **impressionability** /-ˈbɪlɪtɪ/ *n.* **impressionably** *adv.*

impressionism /ɪmˈpreʃəˌnɪz(ə)m/ *n.* **1** style or movement in art concerned with conveying the effect of natural light on objects. **2** style of music or writing seeking to convey esp. fleeting feelings or experience. □ **impressionist** *n.* **impressionistic** /-ˈnɪstɪk/ *adj.*

impressive /ɪmˈpresɪv/ *adj.* arousing respect, approval, or admiration. □ **impressively** *adv.* **impressiveness** *n.*

imprimatur /ˌɪmprɪˈmɑːtə(r)/ *n.* **1** *RC Ch.* licence to print (a religious book etc.). **2** official approval. [Latin, = let it be printed]

■ **Usage** *Imprimatur* is sometimes confused with sense 2 of *imprint*.

imprint *–v.* /ɪmˈprɪnt/ **1** (often foll. by *on*) impress firmly, esp. on the mind. **2 a** (often foll. by *on*) make a stamp or impression of (a figure etc.) on a thing. **b** make an impression on (a thing) with a stamp etc. *–n.* /ˈɪmprɪnt/ **1** impression, stamp. **2** printer's or publisher's name etc. printed in a book.

■ **Usage** See note at *imprimatur*.

imprison /ɪmˈprɪz(ə)n/ *v.* **1** put in prison. **2** confine. □ **imprisonment** *n.*

improbable /ɪmˈprɒbəb(ə)l/ *adj.* **1** unlikely. **2** difficult to believe. □ **improbability** /-ˈbɪlɪtɪ/ *n.* **improbably** *adv.*

improbity /ɪmˈprəʊbɪtɪ/ *n.* (*pl.* **-ies**) **1** wickedness; dishonesty. **2** wicked or dishonest act.

impromptu /ɪmˈprɒmptjuː/ *–adj.* & *adv.* extempore, unrehearsed. *–n.* (*pl.* **-s**) **1** extempore performance or speech.

2 short, usu. solo, instrumental composition, often improvisatory in style. [French from Latin *in promptu* in readiness]

improper /ɪm'prɒpə(r)/ *adj.* **1** unseemly; indecent. **2** inaccurate, wrong. □ **improperly** *adv.*

improper fraction *n.* fraction in which the numerator is greater than or equal to the denominator.

impropriety /,ɪmprə'praɪətɪ/ *n.* (*pl.* **-ies**) **1** lack of propriety; indecency. **2** instance of this. **3** incorrectness.

improve /ɪm'pru:v/ *v.* (**-ving**) **1 a** make or become better. **b** (foll. by *on*, *upon*) produce something better than. **2** (as **improving** *adj.*) giving moral benefit (*improving literature*). □ **improvable** *adj.* **improvement** *n.* [Anglo-French *emprower* from French *prou* profit]

improvident /ɪm'prɒvɪd(ə)nt/ *adj.* **1** lacking foresight. **2** profligate; wasteful. **3** incautious. □ **improvidence** *n.* **improvidently** *adv.*

improvise /'ɪmprə,vaɪz/ *v.* (**-sing**) (also *absol.*) **1** compose or perform (music, verse, etc.) extempore. **2** provide or construct from materials not intended for the purpose. □ **improvisation** /-'zeɪʃ(ə)n/ *n.* **improvisational** /-'zeɪʃən(ə)l/ *adj.* **improvisatory** /-'zeɪtərɪ/ *adj.* [Latin *improvisus* unforeseen]

imprudent /ɪm'pru:d(ə)nt/ *adj.* unwise, indiscreet. □ **imprudence** *n.* **imprudently** *adv.*

impudent /'ɪmpjʊd(ə)nt/ *adj.* impertinent. □ **impudence** *n.* **impudently** *adv.* [Latin *pudeo* be ashamed]

impugn /ɪm'pju:n/ *v.* challenge or call in question. □ **impugnment** *n.* [Latin *pugno* fight]

impulse /'ɪmpʌls/ *n.* **1** sudden urge (*felt an impulse to laugh*). **2** tendency to follow such urges (*man of impulse*). **3** impelling; a push. **4** impetus. **5** *Physics* **a** large temporary force producing a change of momentum (e.g. a hammer-blow). **b** change of momentum so produced. **6** wave of excitation in a nerve. [Latin: related to PULSE[1]]

impulse buying *n.* purchasing goods on impulse.

impulsion /ɪm'pʌlʃ(ə)n/ *n.* **1** impelling. **2** mental impulse. **3** impetus.

impulsive /ɪm'pʌlsɪv/ *adj.* **1** tending to act on impulse. **2** done on impulse. **3** tending to impel. □ **impulsively** *adv.* **impulsiveness** *n.*

impunity /ɪm'pju:nɪtɪ/ *n.* exemption from punishment, bad consequences, etc. □ **with impunity** without punishment etc. [Latin *poena* penalty]

impure /ɪm'pjʊə(r)/ *adj.* **1** adulterated. **2** dirty. **3** unchaste.

impurity *n.* (*pl.* **-ies**) **1** being impure. **2** impure thing or part.

impute /ɪm'pju:t/ *v.* (**-ting**) (foll. by *to*) attribute (a fault etc.) to. □ **imputation** /-'teɪʃ(ə)n/ *n.* [Latin *puto* reckon]

In *symb.* indium.

in *–prep.* **1** expressing inclusion or position within limits of space, time, circumstance, etc. (*in England*; *in bed*; *in 1989*; *in the rain*). **2 a** within (a certain time) (*finished it in two hours*). **b** after (a certain time) (*will be leaving in an hour*). **3** with respect to (*blind in one eye*; *good in parts*). **4** as a proportionate part of (*one in three failed*; *gradient of one in six*). **5** with the form or arrangement of (*packed in tens*; *falling in folds*). **6** as a member of (*in the army*). **7** involved with (*is in banking*). **8** as the content of (*there is something in what you say*). **9** within the ability of (*does he have it in him?*). **10** having the condition of; affected by (*in bad health*; *in danger*). **11** having as a purpose (*in search of*; *in reply to*). **12** by means of or using as material (*drawn in pencil*; *modelled in bronze*). **13 a** using as the language of expression (*written in French*). **b** (of music) having as its key (*symphony in C*). **14** (of a word) having as a beginning or ending (*words in un-*). **15** wearing (*in blue*; *in a suit*). **16** with the identity of (*found a friend in Mary*). **17** (of an animal) pregnant with (*in calf*). **18** into (with a verb of motion or change: *put it in the box*; *cut it in two*). **19** introducing an indirect object after a verb (*believe in*; *engage in*; *share in*). **20** forming adverbial phrases (*in any case*; *in reality*; *in short*). *–adv.* expressing position within limits, or motion to such a position: **1** into a room, house, etc. (*come in*). **2** at home, in one's office, etc. (*is not in*). **3** so as to be enclosed (*locked in*). **4** in a publication (*is the advert in?*). **5** in or to the inward side (*rub it in*). **6 a** in fashion or season (*long skirts are in*). **b** elected or in office (*Democrat got in*). **7** favourable (*their luck was in*). **8** *Cricket* (of a player or side) batting. **9** (of transport) at the platform etc. (*the train is in*). **10** (of a season, order, etc.) having arrived or been received. **11** (of a fire) continuing to burn. **12** denoting effective action (*join in*). **13** (of the tide) at the highest point. **14** (in *comb.*) *colloq.* denoting prolonged concerted action, esp. by large numbers (*sit-in*; *teach-in*). *–adj.* **1** internal; living in; inside (*in-patient*). **2** fashionable (*the in thing to do*). **3** confined to a small group (*in-joke*). □ **in all** see ALL. **in between** see

BETWEEN *adv.* **in for 1** about to undergo or get. **2** competing in or for. **in on** sharing in; privy to. **ins and outs** (often foll. by *of*) all the details. **in so far as** see FAR. **in that** because; in so far as. **in with** on good terms with; favourably placed for (*in with a chance*). [Old English]

in. *abbr.* inch(es).

in-[1] *prefix* (also **il-**, **im-**, **ir-**) added to: **1** adjectives, meaning 'not' (*inedible*; *insane*). **2** nouns, meaning 'without, lacking' (*inaction*). [Latin]

in-[2] *prefix* (also **il-**, **im-**, **ir-**) in, on, into, towards, within (*induce*; *influx*; *insight*; *intrude*). [from IN, or from Latin *in* (prep.)]

inability /ˌɪnəˈbɪlɪtɪ/ *n.* **1** being unable. **2** lack of power or means.

in absentia /ˌɪn æbˈsentɪə/ *adv.* in (his, her, or their) absence. [Latin]

inaccessible /ˌɪnəkˈsesɪb(ə)l/ *adj.* **1** not accessible. **2** (of a person) unapproachable. □ **inaccessibility** /-ˈbɪlɪtɪ/ *n.*

inaccurate /ɪnˈækjʊrət/ *adj.* not accurate. □ **inaccuracy** *n.* (*pl.* **-ies**). **inaccurately** *adv.*

inaction /ɪnˈækʃ(ə)n/ *n.* lack of action.

inactive /ɪnˈæktɪv/ *adj.* **1** not active. **2** not operating. **3** indolent. □ **inactivity** /-ˈtɪvɪtɪ/ *n.*

inadequate /ɪnˈædɪkwət/ *adj.* **1** not adequate; insufficient. **2** (of a person) incompetent; weak. □ **inadequacy** *n.* (*pl.* **-ies**). **inadequately** *adv.*

inadmissible /ˌɪnədˈmɪsɪb(ə)l/ *adj.* that cannot be admitted or allowed. □ **inadmissibility** /-ˈbɪlɪtɪ/ *n.* **inadmissibly** *adv.*

inadvertent /ˌɪnədˈvɜːt(ə)nt/ *adj.* **1** unintentional. **2** negligent, inattentive. □ **inadvertence** *n.* **inadvertently** *adv.* [from IN-[1] + obsolete *advertent* 'attentive']

inadvisable /ˌɪnədˈvaɪzəb(ə)l/ *adj.* not advisable. □ **inadvisability** /-ˈbɪlɪtɪ/ *n.*

inalienable /ɪnˈeɪlɪənəb(ə)l/ *adj.* that cannot be transferred to another or taken away (*inalienable rights*).

inamorato /ɪnˌæməˈrɑːtəʊ/ *n.* (*fem.* **inamorata**) (*pl.* **-s**) *literary* lover. [Italian *inamorato*: related to IN-[2], Latin *amor* love]

inane /ɪˈneɪn/ *adj.* **1** silly, senseless. **2** empty, void. □ **inanely** *adv.* **inanity** /-ˈænɪtɪ/ *n.* (*pl.* **-ies**). [Latin *inanis*]

inanimate /ɪnˈænɪmət/ *adj.* **1** not endowed with, or deprived of, animal life (*an inanimate object*). **2** spiritless, dull.

inapplicable /ˌɪnəˈplɪkəb(ə)l/ *adj.* (often foll. by *to*) not applicable or relevant. □ **inapplicability** /-ˈbɪlɪtɪ/ *n.*

inapposite /ɪnˈæpəzɪt/ *adj.* not apposite.

inappropriate /ˌɪnəˈprəʊprɪət/ *adj.* not appropriate. □ **inappropriately** *adv.* **inappropriateness** *n.*

inapt /ɪnˈæpt/ *adj.* **1** not apt or suitable. **2** unskilful. □ **inaptitude** *n.*

inarticulate /ˌɪnɑːˈtɪkjʊlət/ *adj.* **1** unable to express oneself clearly. **2** (of speech) not articulate; indistinct. **3** dumb. **4** esp. *Anat.* not jointed. □ **inarticulately** *adv.*

inasmuch /ˌɪnəzˈmʌtʃ/ *adv.* (foll. by *as*) **1** since, because. **2** to the extent that. [from *in as much*]

inattentive /ˌɪnəˈtentɪv/ *adj.* **1** not paying attention. **2** neglecting to show courtesy. □ **inattention** *n.* **inattentively** *adv.*

inaudible /ɪnˈɔːdɪb(ə)l/ *adj.* unable to be heard. □ **inaudibly** *adv.*

inaugural /ɪˈnɔːgjʊr(ə)l/ *–adj.* of or for an inauguration. *–n.* inaugural speech, lecture, etc. [French from Latin *auguro* take omens: related to AUGUR]

inaugurate /ɪˈnɔːgjʊˌreɪt/ *v.* (**-ting**) **1** admit formally to office. **2** begin (an undertaking) or initiate the public use of (a building etc.), with a ceremony. **3** begin, introduce. □ **inauguration** /-ˈreɪʃ(ə)n/ *n.* **inaugurator** *n.*

inauspicious /ˌɪnɔːˈspɪʃəs/ *adj.* **1** ill-omened, not favourable. **2** unlucky. □ **inauspiciously** *adv.* **inauspiciousness** *n.*

in-between *attrib. adj. colloq.* intermediate.

inboard /ˈɪnbɔːd/ *–adv.* within the sides or towards the centre of a ship, aircraft, or vehicle. *–adj.* situated inboard.

inborn /ˈɪnbɔːn, ɪnˈbɔːn/ *adj.* existing from birth; natural, innate.

inbred /ɪnˈbred/ *adj.* **1** inborn. **2** produced by inbreeding.

inbreeding /ɪnˈbriːdɪŋ/ *n.* breeding from closely related animals or persons. □ **inbreed** *v.* (*past* and *past part.* **-bred**).

inbuilt /ɪnˈbɪlt/ *adj.* built-in.

Inc. *abbr. US* Incorporated.

Inca /ˈɪŋkə/ *n.* member of a people of Peru before the Spanish conquest. [Quechua, = lord]

incalculable /ɪnˈkælkjʊləb(ə)l/ *adj.* **1** too great for calculation. **2** not calculable beforehand. **3** uncertain, unpredictable. □ **incalculability** /-ˈbɪlɪtɪ/ *n.* **incalculably** *adv.*

incandesce /ˌɪnkænˈdes/ *v.* (**-cing**) (cause to) glow with heat.

incandescent *adj.* **1** glowing with heat. **2** shining. **3** (of artificial light) produced by a glowing filament etc. □ **incandescence** *n.* [Latin *candeo* be white]

incantation /ˌɪnkæn'teɪʃ(ə)n/ *n.* magical formula; spell, charm. □ **incantational** *adj.* [Latin *canto* sing]
incapable /ɪn'keɪpəb(ə)l/ *adj.* **1 a** not capable. **b** too honest, kind, etc., to do something (*incapable of hurting anyone*). **2** not capable of rational conduct (*drunk and incapable*). □ **incapability** /-'bɪlɪtɪ/ *n.* **incapably** *adv.*
incapacitate /ˌɪnkə'pæsɪˌteɪt/ *v.* (**-ting**) make incapable or unfit.
incapacity /ˌɪnkə'pæsɪtɪ/ *n.* **1** inability; lack of power. **2** legal disqualification.
incarcerate /ɪn'kɑːsəˌreɪt/ *v.* (**-ting**) imprison. □ **incarceration** /-'reɪʃ(ə)n/ *n.* [medieval Latin *carcer* prison]
incarnate *–adj.* /ɪn'kɑːnət/ embodied in flesh, esp. in human form (*is the devil incarnate*). *–v.* /'ɪnkɑːˌneɪt, -'kɑːneɪt/ (**-ting**) **1** embody in flesh. **2** put (an idea etc.) into concrete form. **3** be the living embodiment of (a quality). [Latin *incarnor* be made flesh: related to CARNAGE]
incarnation /ˌɪnkɑː'neɪʃ(ə)n/ *n.* **1 a** embodiment in (esp. human) flesh. **b** (**the Incarnation**) the embodiment of God in Christ. **2** (often foll. by *of*) living type (of a quality etc.).
incautious /ɪn'kɔːʃəs/ *adj.* heedless, rash. □ **incautiously** *adv.*
incendiary /ɪn'sendɪərɪ/ *–adj.* **1** (of a bomb) designed to cause fires. **2 a** of arson. **b** guilty of arson. **3** inflammatory. *–n.* (*pl.* **-ies**) **1** incendiary bomb. **2** arsonist. □ **incendiarism** *n.* [Latin *incendo -cens-* set fire to]
incense[1] /'ɪnsens/ *n.* **1** gum or spice producing a sweet smell when burned. **2** smoke of this, esp. in religious ceremonial. [Church Latin *incensum*]
incense[2] /ɪn'sens/ *v.* (**-sing**) make angry. [Latin: related to INCENDIARY]
incentive /ɪn'sentɪv/ *–n.* **1** motive or incitement. **2** payment or concession encouraging effort in work. *–attrib. adj.* serving to motivate or incite (*incentive scheme*). [Latin *incentivus* that sets the tune]
inception /ɪn'sepʃ(ə)n/ *n.* beginning. [Latin *incipio -cept-* begin]
inceptive /ɪn'septɪv/ *adj.* **1 a** beginning. **b** initial. **2** (of a verb) denoting the beginning of an action.
incessant /ɪn'ses(ə)nt/ *adj.* unceasing, continual, repeated. □ **incessantly** *adv.* [Latin *cesso* cease]
incest /'ɪnsest/ *n.* sexual intercourse between persons too closely related to marry. [Latin *castus* chaste]
incestuous /ɪn'sestjʊəs/ *adj.* **1** of or guilty of incest. **2** having relationships restricted to a particular group or organization. □ **incestuously** *adv.*
inch *–n.* **1** linear measure of $^1/_{12}$ of a foot (2.54 cm). **2** (as a unit of rainfall) 1 inch depth of water. **3** (as a unit of map-scale) so many inches representing 1 mile. **4** small amount (usu. with *neg.*: *would not yield an inch*). *–v.* move gradually. □ **every inch** entirely (*looked every inch a queen*). **within an inch of** almost to the point of. [Old English from Latin *uncia* OUNCE[1]]
inchoate /ɪn'kəʊeɪt/ *adj.* **1** just begun. **2** undeveloped. □ **inchoation** /-'eɪʃ(ə)n/ *n.* [Latin *inchoo, incoho* begin]

■ **Usage** *Inchoate* is sometimes used incorrectly to mean 'chaotic' or 'incoherent'.

incidence /'ɪnsɪd(ə)ns/ *n.* **1** (often foll. by *of*) range, scope, extent, or rate of occurrence or influence (of disease, tax, etc.). **2** falling of a line, ray, particles, etc., on a surface. **3** coming into contact with a thing. [Latin *cado* fall]
incident *–n.* **1** occurrence, esp. a minor one. **2** public disturbance (*the march took place without incident*). **3** clash of armed forces (*frontier incident*). **4** distinct piece of action in a play, film, etc. *–adj.* **1** (often foll. by *to*) apt to occur; naturally attaching. **2** (often foll. by *on, upon*) (of light etc.) falling. [Latin *cado* fall]
incidental /ˌɪnsɪ'dent(ə)l/ *–adj.* (often foll. by *to*) **1** small and relatively unimportant, minor; supplementary. **2** not essential. *–n.* (usu. in *pl.*) minor detail, expense, event, etc.
incidentally *adv.* **1** by the way. **2** in an incidental way.
incidental music *n.* background music in a film, broadcast, etc.
incinerate /ɪn'sɪnəˌreɪt/ *v.* (**-ting**) burn to ashes. □ **incineration** /-'reɪʃ(ə)n/ *n.* [medieval Latin *cinis ciner-* ashes]
incinerator *n.* furnace or device for incineration.
incipient /ɪn'sɪpɪənt/ *adj.* **1** beginning. **2** in an early stage. [Latin *incipio* begin]
incise /ɪn'saɪz/ *v.* (**-sing**) **1** make a cut in. **2** engrave. [Latin *caedo* cut]
incision /ɪn'sɪʒ(ə)n/ *n.* **1** cutting, esp. by a surgeon. **2** cut made in this way.
incisive /ɪn'saɪsɪv/ *adj.* **1** sharp. **2** clear and effective.
incisor /ɪn'saɪzə(r)/ *n.* cutting-tooth, esp. at the front of the mouth.
incite /ɪn'saɪt/ *v.* (**-ting**) (often foll. by *to*) urge or stir up. □ **incitement** *n.* [Latin *cito* rouse]
incivility /ˌɪnsɪ'vɪlɪtɪ/ *n.* (*pl.* **-ies**) **1** rudeness. **2** impolite act.
inclement /ɪn'klemənt/ *adj.* (of the weather) severe or stormy. □ **inclemency** *n.*

inclination /ˌɪnklɪˈneɪʃ(ə)n/ *n.* **1** disposition or propensity. **2** liking, affection. **3** slope, slant. **4** angle between lines. **5** dip of a magnetic needle. **6** slow nod of the head. [Latin: related to INCLINE]

incline *–v.* /ɪnˈklaɪn/ (**-ning**) **1** (usu. in *passive*) **a** dispose or influence (*am inclined to think so*; *does not incline me to agree*; *don't feel inclined*). **b** have a specified tendency (*the door is inclined to bang*). **2 a** be disposed (*I incline to think so*). **b** (often foll. by *to*, *towards*) tend. **3** (cause to) lean, usu. from the vertical; slope. **4** bend forward or downward. *–n.* /ˈɪnklaɪn/ slope. □ **incline one's ear** listen favourably. [Latin *clino* bend]

inclined plane *n.* sloping plane used e.g. to reduce work in raising a load.

include /ɪnˈkluːd/ *v.* (**-ding**) **1** comprise or reckon in as part of a whole. **2** (as **including** *prep.*) if we include (*six, including me*). **3** put in a certain category etc. □ **inclusion** /-ʒ(ə)n/ *n.* [Latin *includo -clus-* enclose, from *claudo* shut]

inclusive /ɪnˈkluːsɪv/ *adj.* **1** (often foll. by *of*) including. **2** including the limits stated (*pages 7 to 26 inclusive*). **3** including all or much (*inclusive terms*). □ **inclusively** *adv.* **inclusiveness** *n.*

incognito /ˌɪnkɒɡˈniːtəʊ/ *–predic. adj.* & *adv.* with one's name or identity kept secret. *–n.* (*pl.* **-s**) **1** person who is incognito. **2** pretended identity. [Italian, = unknown: related to IN-[1], COGNITION]

incognizant /ɪnˈkɒɡnɪz(ə)nt/ *adj. formal* unaware. □ **incognizance** *n.*

incoherent /ˌɪnkəʊˈhɪərənt/ *adj.* **1** unintelligible. **2** lacking logic or consistency; not clear. □ **incoherence** *n.* **incoherently** *adv.*

incombustible /ˌɪnkəmˈbʌstɪb(ə)l/ *adj.* that cannot be burnt.

income /ˈɪnkʌm/ *n.* money received, esp. periodically or in a year, from one's work, investments, etc. [from IN, COME]

income tax *n.* tax levied on income.

incoming *–adj.* **1** coming in (*incoming telephone calls*). **2** succeeding another (*incoming tenant*). *–n.* (usu. in *pl.*) revenue, income.

incommensurable /ˌɪnkəˈmenʃərəb(ə)l/ *adj.* (often foll. by *with*) **1** not commensurable. **2** having no common factor, integral or fractional. □ **incommensurability** /-ˈbɪlɪtɪ/ *n.*

incommensurate /ˌɪnkəˈmenʃərət/ *adj.* **1** (often foll. by *with*, *to*) out of proportion; inadequate. **2** = INCOMMENSURABLE.

incommode /ˌɪnkəˈməʊd/ *v.* (**-ding**) *formal* **1** inconvenience. **2** trouble, annoy.

incommodious *adj. formal* too small for comfort; inconvenient.

incommunicable /ˌɪnkəˈmjuːnɪkəb(ə)l/ *adj.* that cannot be communicated.

incommunicado /ˌɪnkəˌmjuːnɪˈkɑːdəʊ/ *adj.* **1** without means of communication. **2** (of a prisoner) in solitary confinement. [Spanish *incomunicado*]

incommunicative /ˌɪnkəˈmjuːnɪkətɪv/ *adj.* uncommunicative.

incomparable /ɪnˈkɒmpərəb(ə)l/ *adj.* without an equal; matchless. □ **incomparability** /-ˈbɪlɪtɪ/ *n.* **incomparably** *adv.*

incompatible /ˌɪnkəmˈpætɪb(ə)l/ *adj.* not compatible. □ **incompatibility** /-ˈbɪlɪtɪ/ *n.*

incompetent /ɪnˈkɒmpɪt(ə)nt/ *–adj.* lacking the necessary skill. *–n.* incompetent person. □ **incompetence** *n.*

incomplete /ˌɪnkəmˈpliːt/ *adj.* not complete.

incomprehensible /ɪnˌkɒmprɪˈhensɪb(ə)l/ *adj.* that cannot be understood.

incomprehension /ɪnˌkɒmprɪˈhenʃ(ə)n/ *n.* failure to understand.

inconceivable /ˌɪnkənˈsiːvəb(ə)l/ *adj.* **1** that cannot be imagined. **2** *colloq.* most unlikely. □ **inconceivably** *adv.*

inconclusive /ˌɪnkənˈkluːsɪv/ *adj.* (of an argument, evidence, or action) not decisive or convincing.

incongruous /ɪnˈkɒŋɡrʊəs/ *adj.* **1** out of place; absurd. **2** (often foll. by *with*) out of keeping. □ **incongruity** /-ˈɡruːɪtɪ/ *n.* (*pl.* **-ies**). **incongruously** *adv.*

inconsequent /ɪnˈkɒnsɪkwənt/ *adj.* **1** irrelevant. **2** lacking logical sequence. **3** disconnected. □ **inconsequence** *n.*

inconsequential /ɪnˌkɒnsɪˈkwenʃ(ə)l/ *adj.* **1** unimportant. **2** = INCONSEQUENT. □ **inconsequentially** *adv.*

inconsiderable /ˌɪnkənˈsɪdərəb(ə)l/ *adj.* **1** of small size, value, etc. **2** not worth considering. □ **inconsiderably** *adv.*

inconsiderate /ˌɪnkənˈsɪdərət/ *adj.* (of a person or action) lacking regard for others; thoughtless. □ **inconsiderately** *adv.* **inconsiderateness** *n.*

inconsistent /ˌɪnkənˈsɪst(ə)nt/ *adj.* not consistent. □ **inconsistency** *n.* (*pl.* **-ies**). **inconsistently** *adv.*

inconsolable /ˌɪnkənˈsəʊləb(ə)l/ *adj.* (of a person, grief, etc.) that cannot be consoled. □ **inconsolably** *adv.*

inconspicuous /ˌɪnkənˈspɪkjʊəs/ *adj.* not conspicuous; not easily noticed. □ **inconspicuously** *adv.* **inconspicuousness** *n.*

inconstant /ɪnˈkɒnst(ə)nt/ *adj.* **1** fickle, changeable. **2** variable, not fixed. □ **inconstancy** *n.* (*pl.* **-ies**).

incontestable /ˌɪnkən'testəb(ə)l/ *adj.* that cannot be disputed. □ **incontestably** *adv.*

incontinent /ɪn'kɒntɪnənt/ *adj.* **1** unable to control the bowels or bladder. **2** lacking self-restraint (esp. in sexual matters). □ **incontinence** *n.*

incontrovertible /ˌɪnkɒntrə'vɜ:tɪb(ə)l/ *adj.* indisputable, undeniable. □ **incontrovertibly** *adv.*

inconvenience /ˌɪnkən'vi:nɪəns/ *–n.* **1** lack of ease or comfort; trouble. **2** cause or instance of this. *–v.* (**-cing**) cause inconvenience to.

inconvenient *adj.* causing trouble, difficulty, or discomfort; awkward. □ **inconveniently** *adv.*

incorporate *–v.* /ɪn'kɔ:pəˌreɪt/ (**-ting**) **1** include as a part or ingredient (*incorporated all the latest features*). **2** (often foll. by *in*, *with*) unite (in one body). **3** admit as a member of a company etc. **4** (esp. as **incorporated** *adj.*) form into a legal corporation. *–adj.* /ɪn'kɔ:pərət/ incorporated. □ **incorporation** /-'reɪʃ(ə)n/ *n.* [Latin *corpus* body]

incorporeal /ˌɪnkɔ:'pɔ:rɪəl/ *adj.* without physical or material existence. □ **incorporeally** *adv.* **incorporeity** /-pə'ri:ɪtɪ/ *n.*

incorrect /ˌɪnkə'rekt/ *adj.* **1** not correct or true. **2** improper, unsuitable. □ **incorrectly** *adv.*

incorrigible /ɪn'kɒrɪdʒɪb(ə)l/ *adj.* (of a person or habit) that cannot be corrected or improved. □ **incorrigibility** /-'bɪlɪtɪ/ *n.* **incorrigibly** *adv.*

incorruptible /ˌɪnkə'rʌptɪb(ə)l/ *adj.* **1** that cannot be corrupted, esp. by bribery. **2** that cannot decay. □ **incorruptibility** /-'bɪlɪtɪ/ *n.* **incorruptibly** *adv.*

increase *–v.* /ɪŋ'kri:s/ (**-sing**) make or become greater or more numerous. *–n.* /'ɪnkri:s/ **1** growth, enlargement. **2** (of people, animals, or plants) multiplication. **3** amount or extent of an increase. □ **on the increase** increasing. [Latin *cresco* grow]

increasingly /ɪŋ'kri:sɪŋlɪ/ *adv.* more and more.

incredible /ɪn'kredɪb(ə)l/ *adj.* **1** that cannot be believed. **2** *colloq.* amazing, extremely good. □ **incredibility** /-'bɪlɪtɪ/ *n.* **incredibly** *adv.*

incredulous /ɪn'kredjʊləs/ *adj.* unwilling to believe; showing disbelief. □ **incredulity** /ˌɪnkrɪ'dju:lɪtɪ/ *n.* **incredulously** *adv.*

increment /'ɪŋkrɪmənt/ *n.* increase or added amount, esp. on a fixed salary scale. □ **incremental** /-'ment(ə)l/ *adj.* [Latin *cresco* grow]

incriminate /ɪn'krɪmɪˌneɪt/ *v.* (**-ting**) **1** make (a person) appear to be guilty. **2** charge with a crime. □ **incrimination** /-'neɪʃ(ə)n/ *n.* **incriminatory** *adj.* [Latin: related to CRIME]

incrustation /ˌɪnkrʌ'steɪʃ(ə)n/ *n.* **1** encrusting. **2** crust or hard coating. **3** deposit on a surface. [Latin: related to CRUST]

incubate /'ɪŋkjʊˌbeɪt/ *v.* (**-ting**) **1** hatch (eggs) by sitting on them or by artificial heat. **2** cause (micro-organisms) to develop. **3** develop slowly. [Latin *cubo* lie]

incubation /ˌɪŋkjʊ'beɪʃ(ə)n/ *n.* **1** incubating. **2** period between infection and the appearance of the first symptoms.

incubator *n.* apparatus providing artificial warmth for hatching eggs, rearing premature babies, or developing micro-organisms.

incubus /'ɪŋkjʊbəs/ *n.* (*pl.* **-buses** or **-bi** /-ˌbaɪ/) **1** demon formerly believed to have sexual intercourse with sleeping women. **2** nightmare. **3** oppressive person or thing. [Latin: as INCUBATE]

inculcate /'ɪnkʌlˌkeɪt/ *v.* (**-ting**) (often foll. by *upon*, *in*) urge or impress (a habit or idea) persistently. □ **inculcation** /-'keɪʃ(ə)n/ *n.* [Latin *calco* tread]

incumbency /ɪn'kʌmbənsɪ/ *n.* (*pl.* **-ies**) office or tenure of an incumbent.

incumbent *–adj.* **1** resting as a duty (*it is incumbent on you to do it*). **2** (often foll. by *on*) lying, pressing. **3** currently holding office (*the incumbent president*). *–n.* holder of an office or post, esp. a benefice. [Latin *incumbo* lie upon]

incunabulum /ˌɪnkju:'næbjʊləm/ *n.* (*pl.* **-la**) **1** early printed book, esp. from before 1501. **2** (in *pl.*) early stages of a thing. [Latin, (in *pl.*) = swaddling-clothes]

incur /ɪn'kɜ:(r)/ *v.* (**-rr-**) bring on oneself (danger, blame, loss, etc.). [Latin *curro* run]

incurable /ɪn'kjʊərəb(ə)l/ *–adj.* that cannot be cured. *–n.* incurable person. □ **incurability** /-'bɪlɪtɪ/ *n.* **incurably** *adv.*

incurious /ɪn'kjʊərɪəs/ *adj.* lacking curiosity.

incursion /ɪn'kɜ:ʃ(ə)n/ *n.* invasion or attack, esp. sudden or brief. □ **incursive** *adj.* [Latin: related to INCUR]

incurve /ɪn'kɜ:v/ *v.* (**-ving**) **1** bend into a curve. **2** (as **incurved** *adj.*) curved inwards. □ **incurvation** /-'veɪʃ(ə)n/ *n.*

indebted /ɪn'detɪd/ *adj.* (usu. foll. by *to*) owing gratitude or money. □ **indebtedness** *n.* [French *endetté*: related to DEBT]

indecent /ɪn'di:s(ə)nt/ *adj.* **1** offending against decency. **2** unbecoming; unsuitable (*indecent haste*). □ **indecency** *n.* (*pl.* **-ies**). **indecently** *adv.*

indecent assault *n.* sexual attack not involving rape.

indecent exposure *n.* exposing one's genitals in public.

indecipherable /ˌɪndɪˈsaɪfərəb(ə)l/ *adj.* that cannot be deciphered.

indecision /ˌɪndɪˈsɪʒ(ə)n/ *n.* inability to decide; hesitation.

indecisive /ˌɪndɪˈsaɪsɪv/ *adj.* **1** (of a person) not decisive; hesitating. **2** not conclusive (*an indecisive battle*). □ **indecisively** *adv.* **indecisiveness** *n.*

indeclinable /ˌɪndɪˈklaɪnəb(ə)l/ *adj. Gram.* that cannot be declined; having no inflections.

indecorous /ɪnˈdekərəs/ *adj.* **1** improper, undignified. **2** in bad taste. □ **indecorously** *adv.*

indeed /ɪnˈdiːd/ *–adv.* **1** in truth; really. **2** admittedly. *–int.* expressing irony, incredulity, etc.

indefatigable /ˌɪndɪˈfætɪgəb(ə)l/ *adj.* unwearying, unremitting. □ **indefatigably** *adv.*

indefeasible /ˌɪndɪˈfiːzɪb(ə)l/ *adj. literary* (esp. of a claim, rights, etc.) that cannot be forfeited or annulled. □ **indefeasibly** *adv.*

indefensible /ˌɪndɪˈfensɪb(ə)l/ *adj.* that cannot be defended or justified. □ **indefensibility** /-ˈbɪlɪtɪ/ *n.* **indefensibly** *adv.*

indefinable /ˌɪndɪˈfaɪnəb(ə)l/ *adj.* that cannot be defined; mysterious. □ **indefinably** *adv.*

indefinite /ɪnˈdefɪnɪt/ *adj.* **1** vague, undefined. **2** unlimited. **3** (of adjectives, adverbs, and pronouns) not determining the person etc. referred to (e.g. *some, someone, anyhow*).

indefinite article *n.* word (e.g. *a*, *an* in English) preceding a noun and implying 'any of several'.

indefinitely *adv.* **1** for an unlimited time (*was postponed indefinitely*). **2** in an indefinite manner.

indelible /ɪnˈdelɪb(ə)l/ *adj.* that cannot be rubbed out or removed. □ **indelibly** *adv.* [Latin *deleo* efface]

indelicate /ɪnˈdelɪkət/ *adj.* **1** coarse, unrefined. **2** tactless. □ **indelicacy** *n.* (*pl.* **-ies**). **indelicately** *adv.*

indemnify /ɪnˈdemnɪˌfaɪ/ *v.* (**-ies, -ied**) **1** (often foll. by *from, against*) secure (a person) in respect of harm, a loss, etc. **2** (often foll. by *for*) exempt from a penalty. **3** compensate. □ **indemnification** /-fɪˈkeɪʃ(ə)n/ *n.* [Latin *indemnis* free from loss]

indemnity /ɪnˈdemnɪtɪ/ *n.* (*pl.* **-ies**) **1 a** compensation for damage. **b** sum exacted by a victor in war. **2** security against loss. **3** exemption from penalties.

indent *–v.* /ɪnˈdent/ **1** make or impress marks, notches, dents, etc. in. **2** start (a line of print or writing) further from the margin than others. **3** draw up (a legal document) in duplicate. **4 a** (often foll. by *on*, *upon* a person, *for* a thing) make a requisition. **b** order (goods) by requisition. *–n.* /ˈɪndent/ **1 a** order (esp. from abroad) for goods. **b** official requisition for stores. **2** indented line. **3** indentation. **4** indenture. [Latin *dens dentis* tooth]

indentation /ˌɪndenˈteɪʃ(ə)n/ *n.* **1** indenting or being indented. **2** notch.

indention /ɪnˈdenʃ(ə)n/ *n.* **1** indenting, esp. in printing. **2** notch.

indenture /ɪnˈdentʃə(r)/ *–n.* **1** (usu. in *pl.*) sealed agreement or contract. **2** formal list, certificate, etc. *–v.* (**-ring**) *hist.* bind by indentures, esp. as an apprentice. [Anglo-French: related to INDENT]

independent /ˌɪndɪˈpend(ə)nt/ *–adj.* **1 a** (often foll. by *of*) not depending on authority or control. **b** self-governing. **2 a** not depending on another person for one's opinions or livelihood. **b** (of income or resources) making it unnecessary to earn one's living. **3** unwilling to be under an obligation to others. **4** acting independently of any political party. **5** not depending on something else for its validity etc. (*independent proof*). **6** (of broadcasting, a school, etc.) not supported by public funds. *–n.* person who is politically independent. □ **independence** *n.* **independently** *adv.*

in-depth *adj.* thorough.

indescribable /ˌɪndɪˈskraɪbəb(ə)l/ *adj.* **1** too good or bad etc. to be described. **2** that cannot be described. □ **indescribably** *adv.*

indestructible /ˌɪndɪˈstrʌktɪb(ə)l/ *adj.* that cannot be destroyed. □ **indestructibility** /-ˈbɪlɪtɪ/ *n.* **indestructibly** *adv.*

indeterminable /ˌɪndɪˈtɜːmɪnəb(ə)l/ *adj.* that cannot be ascertained or settled. □ **indeterminably** *adv.*

indeterminate /ˌɪndɪˈtɜːmɪnət/ *adj.* **1** not fixed in extent, character, etc. **2** left doubtful; vague. **3** *Math.* of no fixed value. □ **indeterminacy** *n.*

indeterminate vowel *n.* vowel /ə/ heard in '*a* mom*e*nt *a*go'.

index /ˈɪndeks/ *–n.* (*pl.* **indexes** or **indices** /ˈɪndɪˌsiːz/) **1** alphabetical list of subjects etc. with references, usu. at the end of a book. **2** = CARD INDEX. **3** measure of prices or wages compared with a previous month, year, etc. (*retail price index*). **4** *Math.* exponent of a number. **5** pointer, sign, or indicator. *–v.* **1** provide (a book etc.) with an index. **2** enter in an index. **3** relate (wages etc.) to a price index. □ **indexation** /-ˈseɪʃ(ə)n/ *n.* (in sense 3 of *v.*). [Latin]

index finger *n.* forefinger.
index-linked *adj.* related to the value of a price index.
Indiaman /ˈɪndɪəmən/ *n.* (*pl.* **-men**) *hist.* ship engaged in trade with India or the East Indies.
Indian /ˈɪndɪən/ *–n.* **1 a** native or national of India. **b** person of Indian descent. **2** (in full **American Indian**) **a** original inhabitant of America. **b** any of the languages of the American Indians. *–adj.* **1** of India or the subcontinent comprising India, Pakistan, and Bangladesh. **2** of the original peoples of America.
Indian corn *n.* maize.
Indian elephant *n.* the elephant of India, smaller than the African elephant.
Indian file *n.* = SINGLE FILE.
Indian hemp see HEMP 1.
Indian ink *n.* **1** black pigment. **2** ink made from this.
Indian summer *n.* **1** dry warm weather in late autumn. **2** late tranquil period of life.
indiarubber *n.* rubber for erasing pencil marks etc.
indicate /ˈɪndɪˌkeɪt/ *v.* (**-ting**) (often foll. by *that*) **1** point out; make known. **2** be a sign of; show the presence of. **3** call for; require (*stronger measures are indicated*). **4** state briefly. **5** give as a reading or measurement. **6** point by hand; use a vehicle's indicator (*failed to indicate*). □ **indication** /-ˈkeɪʃ(ə)n/ *n.* [Latin *dico* make known]
indicative /ɪnˈdɪkətɪv/ *–adj.* **1** (foll. by *of*) suggestive; serving as an indication. **2** *Gram.* (of a mood) stating a fact. *–n. Gram.* **1** indicative mood. **2** verb in this mood.
indicator *n.* **1** flashing light on a vehicle showing the direction in which it is about to turn. **2** person or thing that indicates. **3** device indicating the condition of a machine etc. **4** recording instrument. **5** board giving information, esp. times of trains etc.
indicatory /ɪnˈdɪkətərɪ/ *adj.* (often foll. by *of*) indicative.
indices *pl.* of INDEX.
indict /ɪnˈdaɪt/ *v.* accuse formally by legal process. [Anglo-French: related to IN-², DICTATE]
indictable *adj.* **1** (of an offence) making the doer liable to be charged with a crime. **2** (of a person) so liable.
indictment *n.* **1 a** indicting, accusation. **b** document containing this. **2** thing that serves to condemn or censure (*an indictment of society*).
indie /ˈɪndɪ/ *colloq. –adj.* (of a pop group or record label) independent, not belonging to one of the major companies. *–n.* such a group or label. [abbreviation of INDEPENDENT]
indifference /ɪnˈdɪfrəns/ *n.* **1** lack of interest or attention. **2** unimportance.
indifferent *adj.* **1** (foll. by *to*) showing indifference; unconcerned. **2** neither good nor bad. **3** of poor quality or ability. □ **indifferently** *adv.*
indigenous /ɪnˈdɪdʒɪnəs/ *adj.* (often foll. by *to*) native or belonging naturally to a place. [Latin: from a root *gen-* be born]
indigent /ˈɪndɪdʒ(ə)nt/ *adj. formal* needy, poor. □ **indigence** *n.* [Latin *egeo* need]
indigestible /ˌɪndɪˈdʒestɪb(ə)l/ *adj.* **1** difficult or impossible to digest. **2** too complex to read or understand. □ **indigestibility** /-ˈbɪlɪtɪ/ *n.*
indigestion /ˌɪndɪˈdʒestʃ(ə)n/ *n.* **1** difficulty in digesting food. **2** pain caused by this.
indignant /ɪnˈdɪgnənt/ *adj.* feeling or showing indignation. □ **indignantly** *adv.* [Latin *dignus* worthy]
indignation /ˌɪndɪgˈneɪʃ(ə)n/ *n.* anger at supposed injustice etc.
indignity /ɪnˈdɪgnɪtɪ/ *n.* (*pl.* **-ies**) **1** humiliating treatment or quality. **2** insult.
indigo /ˈɪndɪˌgəʊ/ *n.* (*pl.* **-s**) **1** colour between blue and violet in the spectrum. **2** dye of this colour. [Greek *indikon* Indian dye]
indirect /ˌɪndaɪˈrekt/ *adj.* **1** not going straight to the point. **2** (of a route etc.) not straight. **3 a** not directly sought (*indirect result*). **b** not primary (*indirect cause*). □ **indirectly** *adv.*
indirect object *n. Gram.* person or thing affected by a verbal action but not primarily acted on (e.g. *him* in *give him the book*).
indirect question *n. Gram.* question in indirect speech.
indirect speech *n.* = REPORTED SPEECH.
indirect tax *n.* tax on goods and services, not on income or profits.
indiscernible /ˌɪndɪˈsɜːnɪb(ə)l/ *adj.* that cannot be discerned.
indiscipline /ɪnˈdɪsɪplɪn/ *n.* lack of discipline.
indiscreet /ˌɪndɪˈskriːt/ *adj.* **1** not discreet. **2** injudicious, unwary. □ **indiscreetly** *adv.*
indiscretion /ˌɪndɪˈskreʃ(ə)n/ *n.* indiscreet conduct or action.
indiscriminate /ˌɪndɪˈskrɪmɪnət/ *adj.* making no distinctions; done or acting at random (*indiscriminate shooting*). □ **indiscriminately** *adv.*

indispensable /ˌɪndɪ'spensəb(ə)l/ *adj.* that cannot be dispensed with; necessary. □ **indispensability** /-'bɪlɪtɪ/ *n.* **indispensably** *adv.*

indisposed /ˌɪndɪ'spəʊzd/ *adj.* **1** slightly unwell. **2** averse or unwilling. □ **indisposition** /-spə'zɪʃ(ə)n/ *n.*

indisputable /ˌɪndɪ'spju:təb(ə)l/ *adj.* that cannot be disputed. □ **indisputably** *adv.*

indissoluble /ˌɪndɪ'sɒljʊb(ə)l/ *adj.* **1** that cannot be dissolved or broken up. **2** firm and lasting. □ **indissolubly** *adv.*

indistinct /ˌɪndɪ'stɪŋkt/ *adj.* **1** not distinct. **2** confused, obscure. □ **indistinctly** *adv.*

indistinguishable /ˌɪndɪ'stɪŋgwɪʃəb(ə)l/ *adj.* (often foll. by *from*) not distinguishable.

indite /ɪn'daɪt/ *v.* **(-ting)** *formal* or *joc.* **1** put (a speech etc.) into words. **2** write (a letter etc.). [French: related to INDICT]

indium /'ɪndɪəm/ *n.* soft silvery-white metallic element occurring in zinc ores. [Latin *indicum* INDIGO]

individual /ˌɪndɪ'vɪdʒʊəl/ *–adj.* **1** of, for, or characteristic of, a single person etc. **2 a** single (*individual words*). **b** particular; not general. **3** having a distinct character. **4** designed for use by one person. *–n.* **1** single member of a class. **2** single human being. **3** *colloq.* person (*a tiresome individual*). **4** distinctive person. [medieval Latin: related to DIVIDE]

individualism *n.* **1** social theory favouring free action by individuals. **2** being independent or different. □ **individualist** *n.* **individualistic** /-'lɪstɪk/ *adj.*

individuality /ˌɪndɪvɪdʒʊ'ælɪtɪ/ *n.* **1** individual character, esp. when strongly marked. **2** separate existence.

individualize *v.* (also **-ise**) **(-zing** or **-sing) 1** give an individual character to. **2** (esp. as **individualized** *adj.*) personalize (*individualized notepaper*).

individually *adv.* **1** one by one. **2** personally. **3** distinctively.

indivisible /ˌɪndɪ'vɪzɪb(ə)l/ *adj.* not divisible.

Indo- *comb. form* Indian; Indian and.

indoctrinate /ɪn'dɒktrɪˌneɪt/ *v.* **(-ting)** teach to accept a particular belief uncritically. □ **indoctrination** /-'neɪʃ(ə)n/ *n.*

Indo-European /ˌɪndəʊjʊərə'pɪən/ *–adj.* **1** of the family of languages spoken over most of Europe and Asia as far as N. India. **2** of the hypothetical parent language of this family. *–n.* **1** Indo-European family of languages. **2** hypothetical parent language of these.

indolent /'ɪndələnt/ *adj.* lazy; averse to exertion. □ **indolence** *n.* **indolently** *adv.* [Latin *doleo* suffer pain]

indomitable /ɪn'dɒmɪtəb(ə)l/ *adj.* **1** unconquerable. **2** unyielding. □ **indomitably** *adv.* [Latin: related to IN-[1], *domito* tame]

indoor *adj.* of, done, or for use in a building or under cover.

indoors /ɪn'dɔ:z/ *adv.* into or in a building.

indorse var. of ENDORSE.

indrawn /'ɪndrɔ:n/ *adj.* (of breath etc.) drawn in.

indubitable /ɪn'dju:bɪtəb(ə)l/ *adj.* that cannot be doubted. □ **indubitably** *adv.* [Latin *dubito* doubt]

induce /ɪn'dju:s/ *v.* **(-cing) 1** prevail on; persuade. **2** bring about. **3 a** bring on (labour) artificially. **b** bring on labour in (a mother). **c** speed up the birth of (a baby). **4** produce (a current) by induction. **5** infer; deduce. □ **inducible** *adj.* [Latin *duco duct-* lead]

inducement *n.* attractive offer; incentive; bribe.

induct /ɪn'dʌkt/ *v.* (often foll. by *to, into*) **1** introduce into office, install (into a benefice etc.). **2** *archaic* lead (to a seat, into a room, etc.); install. [related to INDUCE]

inductance *n.* property of an electric circuit generating an electromotive force by virtue of the current flowing through it.

induction /ɪn'dʌkʃ(ə)n/ *n.* **1** act of inducting or inducing. **2** act of bringing on (esp. labour) by artificial means. **3** inference of a general law from particular instances. **4** (often *attrib.*) formal introduction to a new job etc. (*induction course*). **5** *Electr.* **a** production of an electric or magnetic state by the proximity (without contact) of an electrified or magnetized body. **b** production of an electric current by a change of magnetic field. **6** drawing of the fuel mixture into the cylinders of an internal-combustion engine.

inductive /ɪn'dʌktɪv/ *adj.* **1** (of reasoning etc.) based on induction. **2** of electric or magnetic induction.

inductor *n.* component (in an electric circuit) having inductance.

indue var. of ENDUE.

indulge /ɪn'dʌldʒ/ *v.* **(-ging) 1** (often foll. by *in*) take pleasure freely. **2** yield freely to (a desire etc.). **3** (also *refl.*) gratify the wishes of. **4** *colloq.* take alcoholic liquor. [Latin *indulgeo* give free rein to]

indulgence *n.* **1** indulging or being indulgent. **2** thing indulged in. **3** *RC Ch.* remission of punishment still due after absolution. **4** privilege granted.

indulgent *adj.* **1** lenient; ready to overlook faults etc. **2** indulging. □ **indulgently** *adv.*

industrial /ɪn'dʌstrɪəl/ *adj.* **1** of, engaged in, or for use in or serving the needs of industries. **2** (of a nation etc.) having developed industries. □ **industrially** *adv.*

industrial action *n.* strike or other disruptive action by workers as a protest.

industrial estate *n.* area of land zoned for factories etc.

industrialism *n.* system in which manufacturing industries are prevalent.

industrialist *n.* owner or manager in industry.

industrialize *v.* (also **-ise**) (**-zing** or **-sing**) make (a nation etc.) industrial. □ **industrialization** /-'zeɪʃ(ə)n/ *n.*

industrial relations *n.pl.* relations between management and workers.

industrious /ɪn'dʌstrɪəs/ *adj.* hard-working. □ **industriously** *adv.*

industry /'ɪndəstrɪ/ *n.* (*pl.* **-ies**) **1 a** branch of production or manufacture; commercial enterprise. **b** these collectively. **2** concerted activity (*a hive of industry*). **3** diligence. [Latin *industria*]

-ine *suffix* **1** forming adjectives, meaning 'belonging to, of the nature of' (*Alpine*; *asinine*). **2** forming feminine nouns (*heroine*). [Latin *-inus*]

inebriate —*v.* /ɪ'ni:brɪ,eɪt/ (**-ting**) **1** make drunk. **2** excite. —*adj.* /ɪ'ni:brɪət/ drunken. —*n.* /ɪ'ni:brɪət/ drunkard. □ **inebriation** /-'eɪʃ(ə)n/ *n.* **inebriety** /-'braɪətɪ/ *n.* [Latin *ebrius* drunk]

inedible /ɪn'edɪb(ə)l/ *adj.* not suitable for eating.

ineducable /ɪn'edjʊkəb(ə)l/ *adj.* incapable of being educated.

ineffable /ɪn'efəb(ə)l/ *adj.* **1** too great for description in words. **2** that must not be uttered. □ **ineffability** /-'bɪlɪtɪ/ *n.* **ineffably** *adv.* [Latin *effor* speak out]

ineffective /,ɪnɪ'fektɪv/ *adj.* not achieving the desired effect or results. □ **ineffectively** *adv.* **ineffectiveness** *n.*

ineffectual /,ɪnɪ'fektʃʊəl/ *adj.* ineffective, feeble. □ **ineffectually** *adv.* **ineffectualness** *n.*

inefficient /,ɪnɪ'fɪʃ(ə)nt/ *adj.* **1** not efficient or fully capable. **2** (of a machine etc.) wasteful. □ **inefficiency** *n.* **inefficiently** *adv.*

inelegant /ɪn'elɪgənt/ *adj.* **1** ungraceful. **2** unrefined. □ **inelegance** *n.* **inelegantly** *adv.*

ineligible /ɪn'elɪdʒɪb(ə)l/ *adj.* not eligible or qualified. □ **ineligibility** /-'bɪlɪtɪ/ *n.*

ineluctable /,ɪnɪ'lʌktəb(ə)l/ *adj.* inescapable, unavoidable. [Latin *luctor* strive]

inept /ɪ'nept/ *adj.* **1** unskilful. **2** absurd, silly. **3** out of place. □ **ineptitude** *n.* **ineptly** *adv.* [Latin: related to APT]

inequable /ɪn'ekwəb(ə)l/ *adj.* **1** unfair. **2** not uniform.

inequality /,ɪnɪ'kwɒlɪtɪ/ *n.* (*pl.* **-ies**) **1** lack of equality. **2** variability. **3** unevenness.

inequitable /ɪn'ekwɪtəb(ə)l/ *adj.* unfair, unjust.

inequity /ɪn'ekwɪtɪ/ *n.* (*pl.* **-ies**) unfairness, injustice.

ineradicable /,ɪnɪ'rædɪkəb(ə)l/ *adj.* that cannot be rooted out.

inert /ɪ'nɜ:t/ *adj.* **1** without inherent power of action, motion, or resistance. **2** not reacting chemically with other substances (*inert gas*). **3** sluggish, slow; lifeless. [Latin *iners -ert-*: related to ART]

inertia /ɪ'nɜ:ʃə/ *n.* **1** *Physics* property of matter by which it continues in its existing state of rest or motion unless an external force is applied. **2 a** inertness, lethargy. **b** tendency to remain unchanged (*inertia of the system*). □ **inertial** *adj.* [Latin: related to INERT]

inertia reel *n.* reel allowing a seat-belt to unwind freely but locking on impact etc.

inertia selling *n.* sending of unsolicited goods in the hope of making a sale.

inescapable /,ɪnɪ'skeɪpəb(ə)l/ *adj.* that cannot be escaped or avoided.

inessential /,ɪnɪ'senʃ(ə)l/ —*adj.* not necessary; dispensable. —*n.* inessential thing.

inestimable /ɪn'estɪməb(ə)l/ *adj.* too great, precious, etc., to be estimated. □ **inestimably** *adv.*

inevitable /ɪn'evɪtəb(ə)l/ —*adj.* **1** unavoidable; sure to happen. **2** *colloq.* tiresomely familiar. —*n.* (prec. by *the*) inevitable fact, event, etc. □ **inevitability** /-'bɪlɪtɪ/ *n.* **inevitably** *adv.* [Latin *evito* avoid]

inexact /,ɪnɪg'zækt/ *adj.* not exact. □ **inexactitude** *n.* **inexactly** *adv.*

inexcusable /,ɪnɪk'skju:zəb(ə)l/ *adj.* that cannot be excused or justified. □ **inexcusably** *adv.*

inexhaustible /,ɪnɪg'zɔ:stɪb(ə)l/ *adj.* that cannot be used up, endless.

inexorable /ɪn'eksərəb(ə)l/ *adj.* relentless; unstoppable. □ **inexorably** *adv.* [Latin *exoro* entreat]

inexpedient /,ɪnɪk'spi:dɪənt/ *adj.* not expedient.

inexpensive /,ɪnɪk'spensɪv/ *adj.* not expensive.

inexperience /ˌɪnɪk'spɪərɪəns/ *n.* lack of experience, knowledge, or skill. □ **inexperienced** *adj.*

inexpert /ɪn'ekspɜ:t/ *adj.* unskilful; lacking expertise.

inexpiable /ɪn'ekspɪəb(ə)l/ *adj.* that cannot be expiated or appeased.

inexplicable /ˌɪnɪk'splɪkəb(ə)l/ *adj.* that cannot be explained. □ **inexplicably** *adv.*

inexpressible /ˌɪnɪk'spresɪb(ə)l/ *adj.* that cannot be expressed. □ **inexpressibly** *adv.*

inextinguishable /ˌɪnɪk'stɪŋgwɪʃəb(ə)l/ *adj.* that cannot be extinguished or destroyed.

in extremis /ˌɪn ɪk'stri:mɪs/ *adj.* **1** at the point of death. **2** in great difficulties; in an emergency. [Latin]

inextricable /ˌɪnɪk'strɪkəb(ə)l/ *adj.* **1** inescapable. **2** that cannot be separated, loosened, or solved. □ **inextricably** *adv.*

INF *abbr.* intermediate-range nuclear forces.

infallible /ɪn'fælɪb(ə)l/ *adj.* **1** incapable of error. **2** unfailing; sure to succeed. **3** (of the Pope) incapable of doctrinal error. □ **infallibility** /-'bɪlɪtɪ/ *n.* **infallibly** *adv.*

infamous /'ɪnfəməs/ *adj.* notoriously bad. □ **infamously** *adv.* **infamy** /'ɪnfəmɪ/ *n.* (*pl.* **-ies**).

infant /'ɪnf(ə)nt/ *n.* **1 a** child during the earliest period of its life. **b** schoolchild below the age of seven years. **2** (esp. *attrib.*) thing in an early stage of its development. **3** *Law* person under 18. □ **infancy** *n.* [Latin *infans* unable to speak]

infanta /ɪn'fæntə/ *n. hist.* daughter of a Spanish or Portuguese king. [Spanish and Portuguese: related to INFANT]

infanticide /ɪn'fæntɪˌsaɪd/ *n.* **1** killing of an infant, esp. soon after birth. **2** person who kills an infant.

infantile /'ɪnfənˌtaɪl/ *adj.* **1** of or like infants. **2** childish, immature. □ **infantilism** /ɪn'fæntɪˌlɪz(ə)m/ *n.*

infantile paralysis *n.* poliomyelitis.

infantry /'ɪnfəntrɪ/ *n.* (*pl.* **-ies**) body of foot-soldiers; foot-soldiers collectively. [Italian *infante* youth, foot-soldier]

infantryman *n.* soldier of an infantry regiment.

infarct /'ɪnfɑ:kt/ *n.* small area of dead tissue caused by an inadequate blood supply. □ **infarction** /ɪn'fɑ:kʃ(ə)n/ *n.* [Latin *farcio farct-* stuff]

infatuate /ɪn'fætjʊˌeɪt/ *v.* (**-ting**) (usu. as **infatuated** *adj.*) **1** inspire with intense usu. transitory fondness or admiration. **2** affect with extreme folly. □ **infatuation** /-'eɪʃ(ə)n/ *n.* [Latin: related to FATUOUS]

infect /ɪn'fekt/ *v.* **1** affect or contaminate with a germ, virus, or disease. **2** imbue, taint. [Latin *inficio -fect-* taint]

infection /ɪn'fekʃ(ə)n/ *n.* **1 a** infecting or being infected. **b** instance of this; disease. **2** communication of disease, esp. by air, water, etc.

infectious *adj.* **1** infecting. **2** (of a disease) transmissible by infection. **3** (of emotions etc.) quickly affecting or spreading to others. □ **infectiously** *adv.* **infectiousness** *n.*

infelicity /ˌɪnfɪ'lɪsɪtɪ/ *n.* (*pl.* **-ies**) **1** inapt expression etc. **2** unhappiness. □ **infelicitous** *adj.*

infer /ɪn'fɜ:(r)/ *v.* (**-rr-**) **1** deduce or conclude. **2** imply. □ **inferable** *adj.* [Latin *fero* bring]

■ **Usage** The use of *infer* in sense 2 is considered incorrect by some people.

inference /'ɪnfərəns/ *n.* **1** act of inferring. **2** thing inferred. □ **inferential** /-'renʃ(ə)l/ *adj.*

inferior /ɪn'fɪərɪə(r)/ —*adj.* **1** (often foll. by *to*) lower in rank, quality, etc. **2** of poor quality. **3** situated below. **4** written or printed below the line. —*n.* person inferior to another, esp. in rank. [Latin, comparative of *inferus*]

inferiority /ɪnˌfɪərɪ'ɒrɪtɪ/ *n.* being inferior.

inferiority complex *n.* feeling of inadequacy, sometimes marked by compensating aggressive behaviour.

infernal /ɪn'fɜ:n(ə)l/ *adj.* **1** of hell; hellish. **2** *colloq.* detestable, tiresome. □ **infernally** *adv.* [Latin *infernus* low]

inferno /ɪn'fɜ:nəʊ/ *n.* (*pl.* **-s**) **1** raging fire. **2** scene of horror or distress. **3** hell. [Italian: related to INFERNAL]

infertile /ɪn'fɜ:taɪl/ *adj.* **1** not fertile. **2** unable to have offspring. □ **infertility** /-fə'tɪlɪtɪ/ *n.*

infest /ɪn'fest/ *v.* (esp. of vermin) overrun (a place). □ **infestation** /-'steɪʃ(ə)n/ *n.* [Latin *infestus* hostile]

infidel /'ɪnfɪd(ə)l/ —*n.* unbeliever in esp. the supposed true religion. —*adj.* **1** of infidels. **2** unbelieving. [Latin *fides* faith]

infidelity /ˌɪnfɪ'delɪtɪ/ *n.* (*pl.* **-ies**) unfaithfulness, esp. adultery. [Latin: related to INFIDEL]

infield *n. Cricket* the part of the ground near the wicket.

infighting *n.* **1** conflict or competitiveness between colleagues. **2** boxing within arm's length.

infill /'ɪnfɪl/ —*n.* **1** material used to fill a hole, gap, etc. **2** filling gaps (esp. in a row of buildings). —*v.* fill in (a cavity etc.).

infilling *n.* = INFILL *n.*

infiltrate /ˈɪnfɪlˌtreɪt/ *v.* **(-ting) 1 a** enter (a territory, political party, etc.) gradually and imperceptibly. **b** cause to do this. **2** permeate by filtration. **3** (often foll. by *into, through*) introduce (fluid) by filtration. □ **infiltration** /-ˈtreɪʃ(ə)n/ *n.* **infiltrator** *n.* [from IN-², FILTRATE]

infinite /ˈɪnfɪnɪt/ *–adj.* **1** boundless, endless. **2** very great or many. *–n.* **1 (the Infinite)** God. **2 (the infinite)** infinite space. □ **infinitely** *adv.* [Latin: related to IN-¹, FINITE]

infinitesimal /ˌɪnfɪnɪˈtesɪm(ə)l/ *–adj.* infinitely or very small. *–n.* infinitesimal amount. □ **infinitesimally** *adv.*

infinitive /ɪnˈfɪnɪtɪv/ *–n.* form of a verb expressing the verbal notion without a particular subject, tense, etc. (e.g. *see* in *we came to see, let him see*). *–adj.* having this form.

infinitude /ɪnˈfɪnɪˌtjuːd/ *n. literary* = INFINITY 1, 2.

infinity /ɪnˈfɪnɪtɪ/ *n.* (*pl.* **-ies**) **1** being infinite; boundlessness. **2** infinite number or extent. **3** infinite distance (*gaze into infinity*). **4** *Math.* infinite quantity.

infirm /ɪnˈfɜːm/ *adj.* physically weak, esp. through age.

infirmary *n.* (*pl.* **-ies**) **1** hospital. **2** sick-quarters in a school etc.

infirmity *n.* (*pl.* **-ies**) **1** being infirm. **2** particular physical weakness.

infix /ɪnˈfɪks/ *v.* fasten or fix in.

in flagrante delicto /ˌɪn fləˌgræntɪ dɪˈlɪktəʊ/ *adv.* in the very act of committing an offence. [Latin, = in blazing crime]

inflame /ɪnˈfleɪm/ *v.* **(-ming) 1** provoke to strong feeling, esp. anger. **2** cause inflammation in; make hot. **3** aggravate. **4** catch or set on fire. **5** light up with or as with flames.

inflammable /ɪnˈflæməb(ə)l/ *adj.* easily set on fire or excited. □ **inflammability** /-ˈbɪlɪtɪ/ *n.*

■ **Usage** Where there is a danger of *inflammable* being understood to mean the opposite, i.e. 'not easily set on fire', *flammable* can be used to avoid confusion.

inflammation /ˌɪnfləˈmeɪʃ(ə)n/ *n.* **1** inflaming. **2** bodily condition with heat, swelling, redness, and usu. pain.

inflammatory /ɪnˈflæmətərɪ/ *adj.* **1** tending to cause anger etc. **2** of inflammation.

inflatable /ɪnˈfleɪtəb(ə)l/ *–adj.* that can be inflated. *–n.* inflatable object.

inflate /ɪnˈfleɪt/ *v.* **(-ting) 1** distend with air or gas. **2** (usu. foll. by *with*; usu. in *passive*) puff up (with pride etc.). **3 a** cause inflation of (the currency). **b** raise (prices) artificially. **4** (as **inflated** *adj.*) (esp. of language, opinions, etc.) bombastic, overblown, exaggerated. [Latin *inflo -flat-*]

inflation /ɪnˈfleɪʃ(ə)n/ *n.* **1** inflating. **2** *Econ.* **a** general increase in prices. **b** increase in the supply of money regarded as causing this. □ **inflationary** *adj.*

inflect /ɪnˈflekt/ *v.* **1** change the pitch of (the voice). **2 a** change the form of (a word) to express grammatical relation. **b** undergo such a change. **3** bend, curve. □ **inflective** *adj.* [Latin *flecto flex-* bend]

inflection /ɪnˈflekʃ(ə)n/ *n.* (also **inflexion**) **1** inflecting or being inflected. **2 a** inflected word. **b** suffix etc. used to inflect. **3** modulation of the voice. □ **inflectional** *adj.* [Latin: related to INFLECT]

inflexible /ɪnˈfleksɪb(ə)l/ *adj.* **1** unbendable. **2** unbending. □ **inflexibility** /-ˈbɪlɪtɪ/ *n.* **inflexibly** *adv.*

inflexion var. of INFLECTION.

inflict /ɪnˈflɪkt/ *v.* (usu. foll. by *on*) **1** deal (a blow etc.). **2** often *joc.* impose (suffering, oneself, etc.) on (*shall not inflict myself on you any longer*). □ **infliction** *n.* **inflictor** *n.* [Latin *fligo flict-* strike]

inflight *attrib. adj.* occurring or provided during a flight.

inflorescence /ˌɪnfləˈres(ə)ns/ *n.* **1 a** complete flower-head of a plant. **b** arrangement of this. **2** flowering. [Latin: related to IN-², FLOURISH]

inflow *n.* **1** flowing in. **2** something that flows in.

influence /ˈɪnflʊəns/ *–n.* **1** (usu. foll. by *on*) effect a person or thing has on another. **2** (usu. foll. by *over, with*) moral ascendancy or power. **3** thing or person exercising this. *–v.* **(-cing)** exert influence on; affect. □ **under the influence** *colloq.* drunk. [Latin *influo* flow in]

influential /ˌɪnflʊˈenʃ(ə)l/ *adj.* having great influence. □ **influentially** *adv.*

influenza /ˌɪnflʊˈenzə/ *n.* virus infection causing fever, aches, and catarrh. [Italian: related to INFLUENCE]

influx /ˈɪnflʌks/ *n.* flowing in, esp. of people or things into a place. [Latin: related to FLUX]

info /ˈɪnfəʊ/ *n. colloq.* information. [abbreviation]

inform /ɪnˈfɔːm/ *v.* **1** tell (*informed them of their rights*). **2** (usu. foll. by *against, on*) give incriminating information about a person to the authorities. [Latin: related to FORM]

informal /ɪnˈfɔːm(ə)l/ *adj.* **1** without formality. **2** not formal. □ **informality** /-ˈmælɪtɪ/ *n.* (*pl.* **-ies**). **informally** *adv.*

informant *n.* giver of information.

information /ˌɪnfəˈmeɪʃ(ə)n/ *n.* **1 a** something told; knowledge. **b** items of knowledge; news. **2** charge or complaint lodged with a court etc.

information retrieval *n.* the tracing of information stored in books, computers, etc.

information technology *n.* the study or use of processes (esp. computers, telecommunications, etc.) for storing, retrieving, and sending information.

informative /ɪnˈfɔːmətɪv/ *adj.* giving information; instructive.

informed *adj.* **1** knowing the facts. **2** having some knowledge.

informer *n.* person who informs, esp. against others.

infra /ˈɪnfrə/ *adv.* below, further on (in a book etc.). [Latin, = below]

infra- *comb. form* below.

infraction /ɪnˈfrækʃ(ə)n/ *n.* infringement. [Latin: related to INFRINGE]

infra dig /ˌɪnfrə ˈdɪg/ *predic. adj. colloq.* beneath one's dignity. [Latin *infra dignitatem*]

infrared /ˌɪnfrəˈred/ *adj.* of or using rays with a wavelength just longer than the red end of the visible spectrum.

infrastructure /ˈɪnfrəˌstrʌktʃə(r)/ *n.* **1 a** basic structural foundations of a society or enterprise. **b** roads, bridges, sewers, etc., regarded as a country's economic foundation. **2** permanent installations as a basis for military etc. operations.

infrequent /ɪnˈfriːkwənt/ *adj.* not frequent. □ **infrequently** *adv.*

infringe /ɪnˈfrɪndʒ/ *v.* **(-ging) 1** break or violate (a law, another's rights, etc.). **2** (usu. foll. by *on*) encroach; trespass. □ **infringement** *n.* [Latin *frango fract-* break]

infuriate /ɪnˈfjʊərɪˌeɪt/ *v.* make furious; irritate greatly. □ **infuriating** *adj.* **infuriatingly** *adv.* [medieval Latin: related to FURY]

infuse /ɪnˈfjuːz/ *v.* **(-sing) 1** (usu. foll. by *with*) fill (with a quality). **2** steep (tea leaves etc.) in liquid to extract the content; be steeped thus. **3** (usu. foll. by *into*) instil (life etc.). [Latin *infundo -fus-*: related to FOUND[3]]

infusible /ɪnˈfjuːzɪb(ə)l/ *adj.* that cannot be melted. □ **infusibility** /-ˈbɪlɪtɪ/ *n.*

infusion /ɪnˈfjuːʒ(ə)n/ *n.* **1 a** infusing. **b** liquid extract obtained thus. **2** infused element.

-ing[1] *suffix* forming nouns from verbs denoting: **1** verbal action or its result (*asking*). **2** material associated with a process etc. (*piping*; *washing*). **3** occupation or event (*banking*; *wedding*). [Old English]

-ing[2] *suffix* **1** forming the present participle of verbs (*asking*; *fighting*), often as adjectives (*charming*; *strapping*). **2** forming adjectives from nouns (*hulking*) and verbs (*balding*). [Old English]

ingenious /ɪnˈdʒiːnɪəs/ *adj.* **1** clever at inventing, organizing, etc. **2** cleverly contrived. □ **ingeniously** *adv.* [Latin *ingenium* cleverness]

■ **Usage** *Ingenious* is sometimes confused with *ingenuous*.

ingénue /ˌæʒeɪˈnjuː/ *n.* **1** unsophisticated young woman. **2** such a part in a play. [French: related to INGENUOUS]

ingenuity /ˌɪndʒɪˈnjuːɪtɪ/ *n.* inventiveness, cleverness.

ingenuous /ɪnˈdʒenjʊəs/ *adj.* **1** artless. **2** frank. □ **ingenuously** *adv.* [Latin *ingenuus* free-born, frank]

■ **Usage** *Ingenuous* is sometimes confused with *ingenious*.

ingest /ɪnˈdʒest/ *v.* **1** take in (food etc.). **2** absorb (knowledge etc.). □ **ingestion** /ɪnˈdʒestʃ(ə)n/ *n.* [Latin *gero* carry]

inglenook /ˈɪŋg(ə)lˌnʊk/ *n.* space within the opening on either side of a large fireplace. [perhaps Gaelic *aingeal* fire, light]

inglorious /ɪnˈglɔːrɪəs/ *adj.* **1** shameful. **2** not famous.

ingoing *adj.* going in.

ingot /ˈɪŋgət/ *n.* (usu. oblong) piece of cast metal, esp. gold. [origin uncertain]

ingraft var. of ENGRAFT.

ingrained /ɪnˈgreɪnd/ *adj.* **1** deeply rooted; inveterate. **2** (of dirt etc.) deeply embedded.

ingratiate /ɪnˈgreɪʃɪˌeɪt/ *v.refl.* **(-ting)** (usu. foll. by *with*) bring oneself into favour. □ **ingratiating** *adj.* **ingratiatingly** *adv.* [Latin *in gratiam* into favour]

ingratitude /ɪnˈgrætɪˌtjuːd/ *n.* lack of due gratitude.

ingredient /ɪnˈgriːdɪənt/ *n.* component part in a mixture. [Latin *ingredior* enter into]

ingress /ˈɪngres/ *n.* act or right of going in. [Latin *ingressus*: related to INGREDIENT]

ingrowing *adj.* (esp. of a toenail) growing into the flesh. □ **ingrown** *adj.*

inguinal /ˈɪŋgwɪn(ə)l/ *adj.* of the groin. [Latin *inguen* groin]

inhabit /ɪnˈhæbɪt/ *v.* **(-t-)** dwell in; occupy. □ **inhabitable** *adj.* [Latin: related to HABIT]

inhabitant *n.* person etc. who inhabits a place.

inhalant /ɪn'heɪlənt/ *n.* medicinal substance for inhaling.

inhale /ɪn'heɪl/ *v.* **(-ling)** (often *absol.*) breathe in (air, gas, smoke, etc.). □ **inhalation** /-hə'leɪʃ(ə)n/ *n.* [Latin *halo* breathe]

inhaler *n.* device for administering an inhalant, esp. to relieve asthma.

inhere /ɪn'hɪə(r)/ *v.* **(-ring)** be inherent. [Latin *haereo haes-* stick]

inherent /ɪn'herənt/ *adj.* (often foll. by *in*) existing in something as an essential or permanent attribute. □ **inherence** *n.* **inherently** *adv.*

inherit /ɪn'herɪt/ *v.* **(-t-) 1** receive (property, rank, title, etc.) by legal succession. **2** derive (a characteristic) from one's ancestors. **3** derive (a situation etc.) from a predecessor. □ **inheritable** *adj.* **inheritor** *n.* [Latin *heres* heir]

inheritance *n.* **1** thing that is inherited. **2** inheriting.

inheritance tax *n.* tax levied on property acquired by gift or inheritance.

■ **Usage** This tax was introduced in 1986 to replace *capital transfer tax*.

inhibit /ɪn'hɪbɪt/ *v.* **(-t-) 1** hinder, restrain, or prevent (action or progress). **2** (as **inhibited** *adj.*) suffering from inhibition. **3** (usu. foll. by *from* + verbal noun) prohibit (a person etc.). □ **inhibitory** *adj.* [Latin *inhibeo -hibit-* hinder]

inhibition /ˌɪnhɪ'bɪʃ(ə)n/ *n.* **1** *Psychol.* restraint on the direct expression of an instinct. **2** *colloq.* emotional resistance to a thought, action, etc. **3** inhibiting or being inhibited.

inhospitable /ˌɪnhɒ'spɪtəb(ə)l/ *adj.* **1** not hospitable. **2** (of a region etc.) not affording shelter, favourable conditions, etc. □ **inhospitably** *adv.*

in-house *adj.* & *adv.* within an institution, company, etc.

inhuman /ɪn'hju:mən/ *adj.* brutal; unfeeling; barbarous. □ **inhumanity** *n.* (*pl.* **-ies**). **inhumanly** *adv.*

inhumane /ˌɪnhju:'meɪn/ *adj.* = INHUMAN. □ **inhumanely** *adv.*

inimical /ɪ'nɪmɪk(ə)l/ *adj.* **1** hostile. **2** harmful. □ **inimically** *adv.* [Latin *inimicus* enemy]

inimitable /ɪ'nɪmɪtəb(ə)l/ *adj.* impossible to imitate. □ **inimitably** *adv.*

iniquity /ɪ'nɪkwɪtɪ/ *n.* (*pl.* **-ies**) **1** wickedness. **2** gross injustice. □ **iniquitous** *adj.* [French from Latin *aequus* just]

initial /ɪ'nɪʃ(ə)l/ —*adj.* of or at the beginning. —*n.* initial letter, esp. (in *pl.*) those of a person's names. —*v.* **(-ll-;** *US* **-l-)** mark or sign with one's initials. □ **initially** *adv.* [Latin *initium* beginning]

initial letter *n.* first letter of a word.

initiate —*v.* /ɪ'nɪʃɪˌeɪt/ **(-ting) 1** begin; set going; originate. **2 a** admit (a person) into a society, office, etc., esp. with a ritual. **b** instruct (a person) in a subject. —*n.* /ɪ'nɪʃɪət/ (esp. newly) initiated person. □ **initiation** /-'eɪʃ(ə)n/ *n.* **initiator** *n.* **initiatory** /ɪ'nɪʃjətərɪ/ *adj.* [Latin *initium* beginning]

initiative /ɪ'nɪʃətɪv/ *n.* **1** ability to initiate things; enterprise (*lacks initiative*). **2** first step. **3** (prec. by *the*) power or right to begin. □ **have the initiative** esp. *Mil.* be able to control the enemy's movements. [French: related to INITIATE]

inject /ɪn'dʒekt/ *v.* **1 a** (usu. foll. by *into*) drive (a solution, medicine, etc.) by or as if by a syringe. **b** (usu. foll. by *with*) fill (a cavity etc.) by injecting. **c** administer medicine etc. to (a person) by injection. **2** place (a quality, money, etc.) into something. □ **injection** *n.* **injector** *n.* [Latin *injicere -ject-* from *jacio* throw]

injudicious /ˌɪndʒu:'dɪʃəs/ *adj.* unwise; ill-judged.

injunction /ɪn'dʒʌŋkʃ(ə)n/ *n.* **1** authoritative order. **2** judicial order restraining a person or body from an act, or compelling redress to an injured party. [Latin: related to ENJOIN]

injure /'ɪndʒə(r)/ *v.* **(-ring) 1** harm or damage. **2** do wrong to. [back-formation from INJURY]

injured *adj.* **1** harmed or hurt. **2** offended.

injurious /ɪn'dʒʊərɪəs/ *adj.* **1** hurtful. **2** (of language) insulting. **3** wrongful.

injury /'ɪndʒərɪ/ *n.* (*pl.* **-ies**) **1** physical harm or damage. **2** offence to feelings etc. **3** esp. *Law* wrongful action or treatment. [Latin *injuria*]

injury time *n.* extra playing-time at a football etc. match to compensate for time lost in dealing with injuries.

injustice /ɪn'dʒʌstɪs/ *n.* **1** lack of fairness. **2** unjust act. □ **do a person an injustice** judge a person unfairly. [French from Latin: related to IN-[1]]

ink —*n.* **1** coloured fluid or paste used for writing, printing, etc. **2** black liquid ejected by a cuttlefish etc. —*v.* **1** (usu. foll. by *in*, *over*, etc.) mark with ink. **2** cover (type etc.) with ink. [Greek *egkauston* purple ink used by Roman emperors]

inkling /'ɪŋklɪŋ/ *n.* (often foll. by *of*) slight knowledge or suspicion; hint. [origin unknown]

inkstand *n.* stand for one or more ink bottles.

ink-well *n.* pot for ink, usu. housed in a hole in a desk.

inky *adj.* **(-ier, -iest)** of, as black as, or stained with ink. □ **inkiness** *n.*

inland –*adj.* /ˈɪnlənd/ **1** in the interior of a country. **2** carried on within a country. –*adv.* /ɪnˈlænd/ in or towards the interior of a country.

Inland Revenue *n.* government department assessing and collecting taxes.

in-law *n.* (often in *pl.*) relative by marriage.

inlay –*v.* /ɪnˈleɪ/ (*past* and *past part.* **inlaid** /ɪnˈleɪd/) **1** embed (a thing in another) so that the surfaces are even. **2** decorate (a thing with inlaid work). –*n.* /ˈɪnleɪ/ **1** inlaid work. **2** material inlaid. **3** filling shaped to fit a tooth-cavity. [from IN-[2], LAY[1]]

inlet /ˈɪnlət/ *n.* **1** small arm of the sea, a lake, or a river. **2** piece inserted. **3** way of entry. [from IN, LET[1]]

in loco parentis /ɪn ˌləʊkəʊ pəˈrentɪs/ *adv.* (acting) for or instead of a parent. [Latin]

inmate *n.* occupant of a hospital, prison, institution, etc. [probably from INN, MATE[1]]

in memoriam /ˌɪn mɪˈmɔːrɪˌæm/ *prep.* in memory of (a dead person). [Latin]

inmost *adj.* most inward. [Old English]

inn *n.* **1** pub, sometimes with accommodation. **2** *hist.* house providing accommodation, esp. for travellers. [Old English: related to IN]

innards /ˈɪnədz/ *n.pl. colloq.* entrails. [special pronunciation of INWARD]

innate /ɪˈneɪt/ *adj.* inborn; natural. □ **innately** *adv.* [Latin *natus* born]

inner –*adj.* (usu. *attrib.*) **1** inside; interior. **2** (of thoughts, feelings, etc.) deeper. –*n. Archery* **1** division of the target next to the bull's-eye. **2** shot striking this. □ **innermost** *adj.* [Old English, comparative of IN]

inner city *n.* central area of a city, esp. regarded as having particular problems (also (with hyphen) *attrib.*: *inner-city housing*).

inner man *n.* (also **inner woman**) **1** soul or mind. **2** *joc.* stomach.

inner tube *n.* separate inflatable tube inside a pneumatic tyre.

innings *n.* (*pl.* same) **1** esp. *Cricket* part of a game during which a side is batting. **2** period during which a government, party, person, etc. is in office or can achieve something. [obsolete *in* (verb) = go in]

innkeeper *n.* person who keeps an inn.

innocent /ˈɪnəs(ə)nt/ –*adj.* **1** free from moral wrong. **2** (usu. foll. by *of*) not guilty (of a crime etc.). **3** simple; guileless. **4** harmless. –*n.* innocent person, esp. a young child. □ **innocence** *n.* **innocently** *adv.* [Latin *noceo* hurt]

innocuous /ɪˈnɒkjʊəs/ *adj.* harmless. [Latin *innocuus*: related to INNOCENT]

Inn of Court *n.* each of the four legal societies admitting people to the English bar.

innovate /ˈɪnəˌveɪt/ *v.* (**-ting**) bring in new methods, ideas, etc.; make changes. □ **innovation** /-ˈveɪʃ(ə)n/ *n.* **innovative** *adj.* **innovator** *n.* **innovatory** *adj.* [Latin *novus* new]

innuendo /ˌɪnjʊˈendəʊ/ *n.* (*pl.* **-es** or **-s**) allusive remark or hint, usu. disparaging or with a double meaning. [Latin, = by nodding at: related to IN-[2], *nuo* nod]

Innuit var. of INUIT.

innumerable /ɪˈnjuːmərəb(ə)l/ *adj.* too many to be counted. □ **innumerably** *adv.*

innumerate /ɪˈnjuːmərət/ *adj.* having no knowledge of basic mathematics. □ **innumeracy** *n.*

inoculate /ɪˈnɒkjʊˌleɪt/ *v.* (**-ting**) treat (a person or animal) with vaccine or serum to promote immunity against a disease. □ **inoculation** /-ˈleɪʃ(ə)n/ *n.* [Latin *oculus* eye, bud]

inoffensive /ˌɪnəˈfensɪv/ *adj.* not objectionable; harmless.

inoperable /ɪnˈɒpərəb(ə)l/ *adj. Surgery* that cannot successfully be operated on.

inoperative /ɪnˈɒpərətɪv/ *adj.* not working or taking effect.

inopportune /ɪnˈɒpəˌtjuːn/ *adj.* not appropriate, esp. not timely.

inordinate /ɪnˈɔːdɪnət/ *adj.* excessive. □ **inordinately** *adv.* [Latin: related to ORDAIN]

inorganic /ˌɪnɔːˈgænɪk/ *adj.* **1** *Chem.* (of a compound) not organic, usu. of mineral origin. **2** without organized physical structure. **3** extraneous.

in-patient *n.* patient who lives in hospital while under treatment.

input /ˈɪnpʊt/ –*n.* **1** what is put in or taken in. **2** place where energy, information, etc., enters a system. **3** action of putting in or feeding in. **4** contribution of information etc. –*v.* (**inputting**; *past* and *past part.* **input** or **inputted**) (often foll. by *into*) **1** put in. **2** supply (data, programs, etc., to a computer etc.).

inquest /ˈɪŋkwest/ *n.* **1** *Law* inquiry by a coroner's court into the cause of a death. **2** *colloq.* discussion analysing the outcome of a game, election, etc. [Romanic: related to INQUIRE]

inquietude /ɪnˈkwaɪɪˌtjuːd/ *n.* uneasiness. [Latin: related to QUIET]

inquire /ɪnˈkwaɪə(r)/ *v.* (**-ring**) **1** seek information formally; make a formal investigation. **2** = ENQUIRE. [Latin *quaero quisit-* seek]

inquiry *n.* (*pl.* **-ies**) **1** investigation, esp. an official one. **2** = ENQUIRY.

inquisition /ˌɪnkwɪˈzɪʃ(ə)n/ *n.* **1** intensive search or investigation. **2** judicial or official inquiry. **3 (the Inquisition)** *RC Ch. hist.* ecclesiastical tribunal for the violent suppression of heresy, esp. in Spain. □ **inquisitional** *adj.* [Latin: related to INQUIRE]

inquisitive /ɪnˈkwɪzɪtɪv/ *adj.* **1** unduly curious; prying. **2** seeking knowledge. □ **inquisitively** *adv.* **inquisitiveness** *n.*

inquisitor /ɪnˈkwɪzɪtə(r)/ *n.* **1** official investigator. **2** *hist.* officer of the Inquisition.

inquisitorial /ɪnˌkwɪzɪˈtɔːrɪəl/ *adj.* **1** of or like an inquisitor. **2** prying. □ **inquisitorially** *adv.*

inquorate /ɪnˈkwɔːreɪt/ *adj.* not constituting a quorum.

in re /ɪn ˈriː/ *prep.* = RE[1]. [Latin]

INRI *abbr.* Jesus of Nazareth, King of the Jews. [Latin *Iesus Nazarenus Rex Iudaeorum*]

inroad *n.* **1** (often in *pl.*) encroachment; using up of resources etc. **2** hostile attack.

inrush *n.* rapid influx.

insalubrious /ˌɪnsəˈluːbrɪəs/ *adj.* (of a climate or place) unhealthy.

insane *adj.* **1** mad. **2** *colloq.* extremely foolish. □ **insanely** *adv.* **insanity** /-ˈsænɪtɪ/ *n.* (*pl.* **-ies**).

insanitary /ɪnˈsænɪtərɪ/ *adj.* not sanitary; dirty.

insatiable /ɪnˈseɪʃəb(ə)l/ *adj.* **1** unable to be satisfied. **2** extremely greedy. □ **insatiability** /-ˈbɪlɪtɪ/ *n.* **insatiably** *adv.*

insatiate /ɪnˈseɪʃɪət/ *adj.* never satisfied.

inscribe /ɪnˈskraɪb/ *v.* (**-bing**) **1 a** (usu. foll. by *in, on*) write or carve (words etc.) on a surface, page, etc. **b** (usu. foll. by *with*) mark (a surface) with characters. **2** (usu. foll. by *to*) write an informal dedication in or on (a book etc.). **3** enter the name of (a person) on a list or in a book. **4** *Geom.* draw (a figure) within another so that points of it lie on the boundary of the other. [Latin *scribo* write]

inscription /ɪnˈskrɪpʃ(ə)n/ *n.* **1** words inscribed. **2** inscribing. □ **inscriptional** *adj.* [Latin: related to INSCRIBE]

inscrutable /ɪnˈskruːtəb(ə)l/ *adj.* mysterious, impenetrable. □ **inscrutability** /-ˈbɪlɪtɪ/ *n.* **inscrutably** *adv.* [Latin *scrutor* search]

insect /ˈɪnsekt/ *n.* small invertebrate of a class characteristically having a head, thorax, abdomen, two antennae, three pairs of thoracic legs, and usu. one or two pairs of thoracic wings. [Latin: related to SECTION]

insecticide /ɪnˈsektɪˌsaɪd/ *n.* substance for killing insects.

insectivore /ɪnˈsektɪˌvɔː(r)/ *n.* **1** animal that feeds on insects. **2** plant which captures and absorbs insects. □ **insectivorous** /-ˈtɪvərəs/ *adj.* [from INSECT, Latin *voro* devour]

insecure /ˌɪnsɪˈkjʊə(r)/ *adj.* **1 a** unsafe; not firm. **b** (of a surface etc.) liable to give way. **2** uncertain; lacking confidence. □ **insecurity** /-ˈkjʊrɪtɪ/ *n.*

inseminate /ɪnˈsemɪˌneɪt/ *v.* (**-ting**) **1** introduce semen into. **2** sow (seed etc.). □ **insemination** /-ˈneɪʃ(ə)n/ *n.* [Latin: related to SEMEN]

insensate /ɪnˈsenseɪt/ *adj.* **1** without physical sensation. **2** without sensibility. **3** stupid. [Latin: related to SENSE]

insensible /ɪnˈsensɪb(ə)l/ *adj.* **1** unconscious. **2** (usu. foll. by *of, to*) unaware (*insensible of her needs*). **3** callous. **4** too small or gradual to be perceived. □ **insensibility** /-ˈbɪlɪtɪ/ *n.* **insensibly** *adv.*

insensitive /ɪnˈsensɪtɪv/ *adj.* (often foll. by *to*) **1** unfeeling; boorish; crass. **2** not sensitive to physical stimuli. □ **insensitively** *adv.* **insensitiveness** *n.* **insensitivity** /-ˈtɪvɪtɪ/ *n.*

insentient /ɪnˈsenʃ(ə)nt/ *adj.* not sentient; inanimate.

inseparable /ɪnˈsepərəb(ə)l/ *adj.* (esp. of friends) unable or unwilling to be separated. □ **inseparability** /-ˈbɪlɪtɪ/ *n.* **inseparably** *adv.*

insert –*v.* /ɪnˈsɜːt/ place or put (a thing) into another. –*n.* /ˈɪnsɜːt/ something (esp. pages) inserted. [Latin *sero sert-* join]

insertion /ɪnˈsɜːʃ(ə)n/ *n.* **1** inserting. **2** thing inserted.

in-service *attrib. adj.* (of training) for those actively engaged in the profession or activity concerned.

inset –*n.* /ˈɪnset/ **1 a** extra section inserted in a book etc. **b** small map etc. within the border of a larger one. **2** piece let into a dress etc. –*v.* /ɪnˈset/ (**insetting**; *past* and *past part.* **inset** or **insetted**) **1** put in as an inset. **2** decorate with an inset.

inshore *adv.* & *adj.* at sea but close to the shore.

inside –*n.* /ɪnˈsaɪd/ **1 a** inner side. **b** inner part; interior. **2** side away from the road. **3** (usu. in *pl.*) *colloq.* stomach and bowels. –*adj.* /ˈɪnsaɪd/ **1** situated on or in the inside. **2** *Football & Hockey* nearer to the centre of the field. –*adv.* /ɪnˈsaɪd/ **1** on, in, or to the inside. **2** *slang* in prison. –*prep.* /ɪnˈsaɪd/ **1** on the inner side of; within. **2** in less than (*inside an hour*). □ **inside out 1** with the inner surface turned outwards. **2** thoroughly (*knew his subject inside out*).

inside information *n.* information not normally accessible to outsiders.

inside job *n. colloq.* crime committed by a person living or working on the premises burgled etc.

insider /ɪn'saɪdə(r)/ *n.* **1** person who is within an organization etc. **2** person privy to a secret.

insider dealing *n. Stock Exch.* illegal practice of trading to one's own advantage through having access to confidential information.

insidious /ɪn'sɪdɪəs/ *adj.* **1** proceeding inconspicuously but harmfully. **2** crafty. □ **insidiously** *adv.* **insidiousness** *n.* [Latin *insidiae* ambush]

insight *n.* (usu. foll. by *into*) **1** capacity of understanding hidden truths etc. **2** instance of this.

insignia /ɪn'sɪgnɪə/ *n.* (treated as *sing.* or *pl.*) badge. [Latin *signum* sign]

insignificant /,ɪnsɪg'nɪfɪkənt/ *adj.* **1** unimportant. **2** meaningless. □ **insignificance** *n.*

insincere /,ɪnsɪn'sɪə(r)/ *adj.* not sincere. □ **insincerely** *adv.* **insincerity** /-'serɪtɪ/ *n.* (*pl.* **-ies**).

insinuate /ɪn'sɪnjʊ,eɪt/ *v.* (**-ting**) **1** hint obliquely, esp. unpleasantly. **2** (often *refl.*; usu. foll. by *into*) **a** introduce (a person etc.) into favour etc., by subtle manipulation. **b** introduce (a thing, oneself, etc.) deviously into a place. □ **insinuation** /-'eɪʃ(ə)n/ *n.* [Latin *sinuo* curve]

insipid /ɪn'sɪpɪd/ *adj.* **1** lacking vigour or character; dull. **2** tasteless. □ **insipidity** /-'pɪdɪtɪ/ *n.* **insipidly** *adv.* [Latin *sapio* have savour]

insist /ɪn'sɪst/ *v.* (usu. foll. by *on* or *that*; also *absol.*) maintain or demand assertively (*insisted on my going*; *insisted that he was innocent*). [Latin *sisto* stand]

insistent *adj.* **1** (often foll. by *on*) insisting. **2** forcing itself on the attention. □ **insistence** *n.* **insistently** *adv.*

in situ /ɪn 'sɪtju:/ *adv.* in its proper or original place. [Latin]

insobriety /,ɪnsə'braɪətɪ/ *n.* intemperance, esp. in drinking.

insofar /,ɪnsəʊ'fɑ:(r)/ *adv.* = *in so far* (see FAR).

insole *n.* fixed or removable inner sole of a boot or shoe.

insolent /'ɪnsələnt/ *adj.* impertinently insulting. □ **insolence** *n.* **insolently** *adv.* [Latin *soleo* be accustomed]

insoluble /ɪn'sɒljʊb(ə)l/ *adj.* **1** incapable of being solved. **2** incapable of being dissolved. □ **insolubility** /-'bɪlɪtɪ/ *n.* **insolubly** *adv.*

insolvent /ɪn'sɒlv(ə)nt/ *–adj.* unable to pay one's debts; bankrupt. *–n.* insolvent person. □ **insolvency** *n.*

insomnia /ɪn'sɒmnɪə/ *n.* sleeplessness, esp. habitual. [Latin *somnus* sleep]

insomniac /ɪn'sɒmnɪ,æk/ *n.* person suffering from insomnia.

insomuch /,ɪnsəʊ'mʌtʃ/ *adv.* **1** (foll. by *that*) to such an extent. **2** (foll. by *as*) inasmuch. [originally *in so much*]

insouciant /ɪn'su:sɪənt/ *adj.* carefree; unconcerned. □ **insouciance** *n.* [French *souci* care]

inspect /ɪn'spekt/ *v.* **1** look closely at. **2** examine officially. □ **inspection** *n.* [Latin *specio spect-* look]

inspector *n.* **1** person who inspects. **2** official employed to supervise. **3** police officer next above sergeant in rank. □ **inspectorate** *n.*

inspector of taxes *n.* Inland Revenue official responsible for assessing taxes.

inspiration /,ɪnspə'reɪʃ(ə)n/ *n.* **1 a** creative force or influence. **b** person etc. stimulating creativity etc. **c** divine influence, esp. on the writing of Scripture etc. **2** sudden brilliant idea. □ **inspirational** *adj.*

inspire /ɪn'spaɪə(r)/ *v.* (**-ring**) **1** stimulate (a person) to esp. creative activity. **2 a** (usu. foll. by *with*) animate (a person) with a feeling. **b** create (a feeling) in a person (*inspires confidence*). **3** prompt; give rise to (*a poem inspired by love*). **4** (as **inspired** *adj.*) characterized by inspiration. □ **inspiring** *adj.* [Latin *spiro* breathe]

inspirit /ɪn'spɪrɪt/ *v.* (**-t-**) **1** put life into; animate. **2** encourage.

inst. *abbr.* = INSTANT *adj.* 4 (*the 6th inst.*).

instability /,ɪnstə'bɪlɪtɪ/ *n.* **1** lack of stability. **2** unpredictability in behaviour etc.

install /ɪn'stɔ:l/ *v.* (also **instal**) (**-ll-**) **1** place (equipment etc.) in position ready for use. **2** place (a person) in an office or rank with ceremony. **3** establish (oneself, a person, etc.). □ **installation** /,ɪnstə'leɪʃ(ə)n/ *n.* [Latin: related to STALL[1]]

instalment *n.* (*US* **installment**) **1** any of several usu. equal payments for something. **2** any of several parts, esp. of a broadcast or published story. [Anglo-French *estaler* fix]

instance /'ɪnst(ə)ns/ *–n.* **1** example or illustration of. **2** particular case (*that's not true in this instance*). *–v.* (**-cing**) cite as an instance. □ **for instance** as an example. **in the first** (or **second** etc.) **instance** in the first (or second etc.) place; at the first (or second etc.) stage (of a proceeding). [French from Latin *instantia* contrary example]

instant –*adj.* **1** occurring immediately. **2** (of food etc.) processed for quick preparation. **3** urgent; pressing. **4** *Commerce* of the current month (*the 6th instant*). –*n.* **1** precise moment (*come here this instant*). **2** short space of time (*in an instant*). [Latin *insto* be urgent]

instantaneous /ˌɪnstənˈteɪnɪəs/ *adj.* occurring or done in an instant. □ **instantaneously** *adv.*

instantly *adv.* immediately; at once.

instead /ɪnˈsted/ *adv.* **1** (foll. by *of*) in place of. **2** as an alternative.

instep *n.* **1** inner arch of the foot between the toes and the ankle. **2** part of a shoe etc. over or under this. [ultimately from IN-², STEP]

instigate /ˈɪnstɪˌgeɪt/ *v.* (**-ting**) **1** bring about by incitement or persuasion. **2** urge on, incite. □ **instigation** /-ˈgeɪʃ(ə)n/ *n.* **instigator** *n.* [Latin *stigo* prick]

instil /ɪnˈstɪl/ *v.* (*US* **instill**) (**-ll-**) (often foll. by *into*) **1** introduce (a feeling, idea, etc.) into a person's mind etc. gradually. **2** put (a liquid) into something in drops. □ **instillation** /-ˈleɪʃ(ə)n/ *n.* **instilment** *n.* [Latin *stillo* drop]

instinct –*n.* /ˈɪnstɪŋkt/ **1 a** innate pattern of behaviour, esp. in animals. **b** innate impulse. **2** intuition. –*predic. adj.* /ɪnˈstɪŋkt/ (foll. by *with*) imbued, filled (with life, beauty, etc.). □ **instinctive** /-ˈstɪŋktɪv/ *adj.* **instinctively** /-ˈstɪŋktɪvlɪ/ *adv.* **instinctual** /-ˈstɪŋktjʊəl/ *adj.* [Latin *stinguo* prick]

institute /ˈɪnstɪˌtjuːt/ –*n.* **1** society or organization for the promotion of science, education, etc. **2** its premises. –*v.* (**-ting**) **1** establish; found. **2** initiate (an inquiry etc.). **3** (usu. foll. by *to*, *into*) appoint (a person) as a cleric in a church etc. [Latin *statuo* set up]

institution /ˌɪnstɪˈtjuːʃ(ə)n/ *n.* **1** organization or society founded for a particular purpose. **2** established law, practice, or custom. **3** *colloq.* (of a person etc.) familiar object. **4** instituting or being instituted.

institutional *adj.* **1** of or like an institution. **2** typical of institutions. □ **institutionally** *adv.*

institutionalize *v.* (also **-ise**) (**-zing** or **-sing**) **1** (as **institutionalized** *adj.*) made dependent after a long period in an institution. **2** place or keep (a person) in an institution. **3** make institutional.

instruct /ɪnˈstrʌkt/ *v.* **1** teach (a person) a subject etc.; train. **2** (usu. foll. by *to* + infin.) direct; command. **3** *Law* **a** employ (a lawyer). **b** inform. □ **instructor** *n.* [Latin *instruo -struct-* build, teach]

instruction /ɪnˈstrʌkʃ(ə)n/ *n.* **1** (often in *pl.*) **a** order. **b** direction (as to how a thing works etc.). **2** teaching (*course of instruction*). □ **instructional** *adj.*

instructive /ɪnˈstrʌktɪv/ *adj.* tending to instruct; enlightening.

instrument /ˈɪnstrəmənt/ *n.* **1** tool or implement, esp. for delicate or scientific work. **2** (in full **musical instrument**) device for producing musical sounds. **3 a** thing used in performing an action. **b** person made use of. **4** measuring-device, esp. in an aeroplane. **5** formal, esp. legal, document. [Latin *instrumentum*: related to INSTRUCT]

instrumental /ˌɪnstrəˈment(ə)l/ *adj.* **1** serving as an instrument or means. **2** (of music) performed on instruments. **3** of, or arising from, an instrument (*instrumental error*).

instrumentalist *n.* performer on a musical instrument.

instrumentality /ˌɪnstrəmenˈtælɪtɪ/ *n.* agency or means.

instrumentation /ˌɪnstrəmenˈteɪʃ(ə)n/ *n.* **1 a** provision or use of instruments. **b** instruments collectively. **2 a** arrangement of music for instruments. **b** the particular instruments used in a piece.

insubordinate /ˌɪnsəˈbɔːdɪnət/ *adj.* disobedient; rebellious. □ **insubordination** /-ˈneɪʃ(ə)n/ *n.*

insubstantial /ˌɪnsəbˈstænʃ(ə)l/ *adj.* **1** lacking solidity or substance. **2** not real.

insufferable /ɪnˈsʌfərəb(ə)l/ *adj.* **1** intolerable. **2** unbearably conceited etc. □ **insufferably** *adv.*

insufficient /ˌɪnsəˈfɪʃ(ə)nt/ *adj.* not sufficient; inadequate. □ **insufficiency** *n.* **insufficiently** *adv.*

insular /ˈɪnsjʊlə(r)/ *adj.* **1 a** of or like an island. **b** separated or remote. **2** narrow-minded. □ **insularity** /-ˈlærɪtɪ/ *n.* [Latin *insula* island]

insulate /ˈɪnsjʊˌleɪt/ *v.* (**-ting**) **1** prevent the passage of electricity, heat, or sound from (a thing, room, etc.) by interposing non-conductors. **2** isolate. □ **insulation** /-ˈleɪʃ(ə)n/ *n.* **insulator** *n.* [Latin *insula* island]

insulin /ˈɪnsjʊlɪn/ *n.* hormone regulating the amount of glucose in the blood, the lack of which causes diabetes. [Latin *insula* island]

insult –*v.* /ɪnˈsʌlt/ **1** speak to or treat with scornful abuse. **2** offend the self-respect or modesty of. –*n.* /ˈɪnsʌlt/ insulting remark or action. □ **insulting** *adj.* **insultingly** *adv.* [Latin *insulto* leap on, assail]

insuperable /ɪnˈsuːpərəb(ə)l/ *adj.* **1** (of a barrier) impossible to surmount. **2** (of a difficulty etc.) impossible to overcome.

□ **insuperability** /-'bɪlɪtɪ/ *n.* **insuperably** *adv.* [Latin *supero* overcome]
insupportable /ˌɪnsə'pɔ:təb(ə)l/ *adj.* **1** unable to be endured. **2** unjustifiable.
insurance /ɪn'ʃʊərəns/ *n.* **1** insuring. **2** **a** sum paid for this. **b** sum paid out as compensation for theft, damage, etc. [French: related to ENSURE]
insure /ɪn'ʃʊə(r)/ *v.* (**-ring**) (often foll. by *against*; also *absol.*) secure compensation in the event of loss or damage to (property, life, a person, etc.) by advance regular payments. [var. of ENSURE]
insured *n.* (usu. prec. by *the*) person etc. covered by insurance.
insurer *n.* person or company selling insurance policies.
insurgent /ɪn'sɜ:dʒ(ə)nt/ *–adj.* in active revolt. *–n.* rebel. □ **insurgence** *n.* [Latin *surgo surrect-* rise]
insurmountable /ˌɪnsə'maʊntəb(ə)l/ *adj.* unable to be surmounted or overcome.
insurrection /ˌɪnsə'rekʃ(ə)n/ *n.* rebellion. □ **insurrectionist** *n.* [Latin: related to INSURGENT]
insusceptible /ˌɪnsə'septɪb(ə)l/ *adj.* not susceptible.
intact /ɪn'tækt/ *adj.* **1** undamaged; entire. **2** untouched. □ **intactness** *n.* [Latin *tango tact-* touch]
intaglio /ɪn'tɑ:lɪəʊ/ *n.* (*pl.* **-s**) **1** gem with an incised design. **2** engraved design. [Italian: related to IN-[2], TAIL[2]]
intake *n.* **1** action of taking in. **2** **a** number (of people etc.), or amount, taken in or received. **b** such people etc. (*this year's intake*). **3** place where water is taken into a pipe, or fuel or air enters an engine etc.
intangible /ɪn'tændʒɪb(ə)l/ *–adj.* **1** unable to be touched. **2** unable to be grasped mentally. *–n.* thing that cannot be precisely assessed or defined. □ **intangibility** /-'bɪlɪtɪ/ *n.* **intangibly** *adv.* [Latin: related to INTACT]
integer /'ɪntɪdʒə(r)/ *n.* whole number. [Latin, = untouched, whole]
integral /'ɪntɪgr(ə)l/ *–adj.* also /-'teg-/ **1** **a** of or necessary to a whole. **b** forming a whole. **c** complete. **2** of or denoted by an integer. *–n. Math.* quantity of which a given function is the derivative. □ **integrally** *adv.* [Latin: related to INTEGER]

■ **Usage** The alternative pronunciation given for the adjective, stressed on the second syllable, is considered incorrect by some people.

integral calculus *n.* mathematics concerned with finding integrals, their properties and application, etc.
integrate /'ɪntɪˌgreɪt/ *v.* (**-ting**) **1** **a** combine (parts) into a whole. **b** complete by the addition of parts. **2** bring or come into equal membership of society, a school, etc. **3** desegregate, esp. racially (a school etc.). **4** *Math.* calculate the integral of. □ **integration** /-'greɪʃ(ə)n/ *n.*
integrated circuit *n. Electronics* small chip etc. of material replacing several separate components in a conventional electronic circuit.
integrity /ɪn'tegrɪtɪ/ *n.* **1** moral excellence; honesty. **2** wholeness; soundness. [Latin: related to INTEGER]
integument /ɪn'tegjʊmənt/ *n.* natural outer covering, as a skin, husk, rind, etc. [Latin *tego* cover]
intellect /'ɪntəˌlekt/ *n.* **1** **a** faculty of reasoning, knowing, and thinking. **b** understanding. **2** clever or knowledgeable person. [Latin: related to INTELLIGENT]
intellectual /ˌɪntə'lektʃʊəl/ *–adj.* **1** of or appealing to the intellect. **2** possessing a highly developed intellect. **3** requiring the intellect. *–n.* intellectual person. □ **intellectuality** /-'ælɪtɪ/ *n.* **intellectualize** *v.* (also **-ise**) (**-zing** or **-sing**). **intellectually** *adv.*
intelligence /ɪn'telɪdʒ(ə)ns/ *n.* **1** **a** intellect; understanding. **b** quickness of understanding. **2** **a** the collecting of information, esp. of military or political value. **b** information so collected. **c** people employed in this.
intelligence quotient *n.* number denoting the ratio of a person's intelligence to the average.
intelligent *adj.* **1** having or showing intelligence, esp. of a high level. **2** clever. □ **intelligently** *adv.* [Latin *intelligo -lect-* understand]
intelligentsia /ˌɪntelɪ'dʒentsɪə/ *n.* class of intellectuals regarded as possessing culture and political initiative. [Russian *intelligentsiya*]
intelligible /ɪn'telɪdʒɪb(ə)l/ *adj.* able to be understood. □ **intelligibility** /-'bɪlɪtɪ/ *n.* **intelligibly** *adv.*
intemperate /ɪn'tempərət/ *adj.* **1** immoderate. **2** **a** given to excessive drinking of alcohol. **b** excessively indulgent in one's appetites. □ **intemperance** *n.*
intend /ɪn'tend/ *v.* **1** have as one's purpose (*we intend to go*; *we intend going*). **2** (usu. foll. by *for*, *as*) design or destine (a person or a thing) (*I intend him to go*; *I intend it as a warning*). [Latin *tendo* stretch]
intended *–adj.* done on purpose. *–n. colloq.* one's fiancé or fiancée.
intense /ɪn'tens/ *adj.* (**intenser**, **intensest**) **1** existing in a high degree; violent;

forceful; extreme (*intense joy*; *intense cold*). **2** very emotional. □ **intensely** *adv.* **intenseness** *n.* [Latin *intensus* stretched]

■ **Usage** *Intense* is sometimes confused with *intensive*, and wrongly used to describe a course of study etc.

intensifier /ɪn'tensɪˌfaɪə(r)/ *n.* **1** thing that makes something more intense. **2** word or prefix used to give force or emphasis, e.g. *thundering* in *a thundering nuisance*.

intensify *v.* **(-ies, -ied)** make or become intense or more intense. □ **intensification** /-fɪ'keɪʃ(ə)n/ *n.*

intensity *n.* (*pl.* **-ies**) **1** intenseness. **2** amount of some quality, e.g. force, brightness, etc.

intensive *adj.* **1** thorough, vigorous; directed to a single point, area, or subject (*intensive study*; *intensive bombardment*). **2** of or relating to intensity. **3** serving to increase production in relation to costs (*intensive farming*). **4** (usu. in *comb.*) *Econ.* making much use of (*labour-intensive*). **5** (of an adjective, adverb, etc.) expressing intensity, e.g. *really* in *my feet are really cold*. □ **intensively** *adv.* **intensiveness** *n.*

■ **Usage** See note at *intense*.

intensive care *n.* **1** constant monitoring etc. of a seriously ill patient. **2** part of a hospital devoted to this.

intent /ɪn'tent/ —*n.* intention; purpose (*with intent to defraud*). —*adj.* **1** (usu. foll. by *on*) **a** resolved, determined. **b** attentively occupied. **2** (esp. of a look) earnest; eager. □ **to all intents and purposes** practically; virtually. □ **intently** *adv.* **intentness** *n.* [Latin *intentus*]

intention /ɪn'tenʃ(ə)n/ *n.* **1** thing intended; aim, purpose. **2** intending (*done without intention*).

intentional *adj.* done on purpose. □ **intentionally** *adv.*

inter /ɪn'tɜ:(r)/ *v.* **(-rr-)** bury (a corpse etc.). [Latin *terra* earth]

inter- *comb. form* **1** between, among (*intercontinental*). **2** mutually, reciprocally (*interbreed*). [Latin *inter* between, among]

interact /ˌɪntər'ækt/ *v.* act on each other. □ **interaction** *n.*

interactive *adj.* **1** reciprocally active. **2** (of a computer or other electronic device) allowing a two-way flow of information between it and a user. □ **interactively** *adv.*

inter alia /ˌɪntər 'eɪlɪə/ *adv.* among other things. [Latin]

interbreed /ˌɪntə'bri:d/ *v.* (*past* and *past part.* **-bred**) **1** (cause to) breed with members of a different race or species to produce a hybrid. **2** breed within one family etc.

intercalary /ɪn'tɜ:kələrɪ/ *attrib. adj.* **1 a** (of a day or a month) inserted in the calendar to harmonize it with the solar year. **b** (of a year) having such an addition. **2** interpolated. [Latin *calo* proclaim]

intercede /ˌɪntə'si:d/ *v.* **(-ding)** (usu. foll. by *with*) intervene on behalf of another; plead. [Latin: related to CEDE]

intercept /ˌɪntə'sept/ *v.* **1** seize, catch, or stop (a person or thing) going from one place to another. **2** (usu. foll. by *from*) cut off (light etc.). □ **interception** *n.* **interceptive** *adj.* **interceptor** *n.* [Latin *intercipio -cept-* from *capio* take]

intercession /ˌɪntə'seʃ(ə)n/ *n.* interceding. □ **intercessor** *n.* [Latin: related to INTERCEDE]

interchange —*v.* /ˌɪntə'tʃeɪndʒ/ **(-ging)** **1** (of two people) exchange (things) with each other. **2** put each of (two things) in the other's place; alternate. —*n.* /'ɪntəˌtʃeɪndʒ/ **1** (often foll. by *of*) exchange between two people etc. **2** alternation. **3** road junction where traffic streams do not cross.

interchangeable *adj.* that can be interchanged, esp. without affecting the way a thing works. □ **interchangeably** *adv.*

inter-city /ˌɪntə'sɪtɪ/ *adj.* existing or travelling between cities.

intercom /'ɪntəˌkɒm/ *n. colloq.* **1** system of intercommunication by radio or telephone. **2** instrument used in this. [abbreviation]

intercommunicate /ˌɪntəkə'mju:nɪˌkeɪt/ *v.* **(-ting)** **1** communicate reciprocally. **2** (of rooms etc.) open into each other. □ **intercommunication** /-'keɪʃ(ə)n/ *n.*

intercommunion /ˌɪntəkə'mju:nɪən/ *n.* **1** mutual communion. **2** mutual action or relationship, esp. between Christian denominations.

interconnect /ˌɪntəkə'nekt/ *v.* connect with each other. □ **interconnection** /-'nekʃ(ə)n/ *n.*

intercontinental /ˌɪntəˌkɒntɪ'nent(ə)l/ *adj.* connecting or travelling between continents.

intercourse /'ɪntəˌkɔ:s/ *n.* **1** communication or dealings between individuals, nations, etc. **2** = SEXUAL INTERCOURSE. [Latin: related to COURSE]

interdenominational /ˌɪntədɪˌnɒmɪ'neɪʃən(ə)l/ *adj.* concerning more than one (religious) denomination.

interdepartmental /ˌɪntəˌdiːpɑːt'ment(ə)l/ *adj.* concerning more than one department.

interdependent /ˌɪntədɪ'pend(ə)nt/ *adj.* dependent on each other. □ **interdependence** *n.*

interdict –*n.* /'ɪntədɪkt/ **1** authoritative prohibition. **2** *RC Ch.* sentence debarring a person, or esp. a place, from ecclesiastical functions and privileges. –*v.* /ˌɪntə'dɪkt/ **1** prohibit (an action). **2** forbid the use of. **3** (usu. foll. by *from* + verbal noun) restrain (a person). **4** (usu. foll. by *to*) forbid (a thing) to a person. □ **interdiction** /-'dɪkʃ(ə)n/ *n.* **interdictory** /-'dɪktərɪ/ *adj.* [Latin *dico* say]

interdisciplinary /ˌɪntəˌdɪsɪ'plɪnərɪ/ *adj.* of or between more than one branch of learning.

interest /'ɪntrəst/ –*n.* **1 a** concern; curiosity (*have no interest in fishing*). **b** quality exciting curiosity etc. (*this book lacks interest*). **2** subject, hobby, etc., in which one is concerned. **3** advantage or profit (*it is in my interest to go*). **4** money paid for the use of money lent. **5 a** thing in which one has a stake or concern (*business interests*). **b** financial stake (in an undertaking etc.). **c** legal concern, title, or right (in property). **6 a** party or group with a common interest (*the brewing interest*). **b** principle or cause with which this is concerned. –*v.* **1** excite the curiosity or attention of. **2** (usu. foll. by *in*) cause (a person) to take a personal interest. **3** (as **interested** *adj.*) having a private interest; not impartial or disinterested. [Latin, = it matters]

interesting *adj.* causing curiosity; holding the attention. □ **interestingly** *adv.*

interface –*n.* **1** surface forming a boundary between two regions. **2** means or place of interaction between two systems etc.; interaction (*the interface between psychology and education*). **3** esp. *Computing* apparatus for connecting two pieces of equipment so that they can be operated jointly. –*v.* (**-cing**) (often foll. by *with*) **1** connect with (another piece of equipment etc.) by an interface. **2** interact.

■ **Usage** The use of the noun and verb in sense 2 is deplored by some people.

interfacing *n.* stiffish material between two layers of fabric in collars etc.

interfere /ˌɪntə'fɪə(r)/ *v.* (**-ring**) **1** (usu. foll. by *with*) **a** (of a person) meddle; obstruct a process etc. **b** (of a thing) be a hindrance. **2** (usu. foll. by *in*) intervene, esp. without invitation or necessity. **3** (foll. by *with*) *euphem.* molest or assault sexually. **4** (of light or other waves) combine so as to cause interference. [Latin *ferio* strike]

interference *n.* **1** act of interfering. **2** fading or disturbance of received radio signals. **3** *Physics* combination of two or more wave motions to form a resultant wave in which the displacement is reinforced or cancelled.

interferon /ˌɪntə'fɪərɒn/ *n.* any of various proteins inhibiting the development of a virus in a cell etc.

interfuse /ˌɪntə'fjuːz/ *v.* (**-sing**) **1 a** (usu. foll. by *with*) mix (a thing) with; intersperse. **b** blend (things). **2** (of two things) blend with each other. □ **interfusion** /-'fjuːʒ(ə)n/ *n.* [Latin: related to FUSE[1]]

intergalactic /ˌɪntəgə'læktɪk/ *adj.* of or situated between galaxies.

interim /'ɪntərɪm/ –*n.* intervening time. –*adj.* provisional, temporary. [Latin, = in the interim]

interior /ɪn'tɪərɪə(r)/ –*adj.* **1** inner. **2** inland. **3** internal; domestic. **4** (usu. foll. by *to*) situated further in or within. **5** existing in the mind. **6** coming from inside. –*n.* **1** interior part; inside. **2** interior part of a region. **3** home affairs of a country (*Minister of the Interior*). **4** representation of the inside of a room etc. [Latin]

interior decoration *n.* decoration of the interior of a building etc. □ **interior decorator** *n.*

interior design *n.* design of the interior of a building. □ **interior designer** *n.*

interject /ˌɪntə'dʒekt/ *v.* **1** utter (words) abruptly or parenthetically. **2** interrupt. [Latin *jacio* throw]

interjection /ˌɪntə'dʒekʃ(ə)n/ *n.* exclamation, esp. as a part of speech (e.g. *ah!*, *dear me!*).

interlace /ˌɪntə'leɪs/ *v.* (**-cing**) **1** bind intricately together; interweave. **2** cross each other intricately. □ **interlacement** *n.*

interlard /ˌɪntə'lɑːd/ *v.* (usu. foll. by *with*) mix (writing or speech) with unusual words or phrases. [French]

interleave /ˌɪntə'liːv/ *v.* (**-ving**) insert (usu. blank) leaves between the leaves of (a book etc.).

interline /ˌɪntə'laɪn/ *v.* (**-ning**) put an extra layer of material between the fabric of (a garment) and its lining.

interlink /ˌɪntə'lɪŋk/ *v.* link or be linked together.

interlock /ˌɪntə'lɒk/ –*v.* **1** engage with each other by overlapping. **2** lock or clasp within each other. –*n.* **1** machine-knitted fabric with fine stitches. **2** mechanism for preventing a set of operations from being performed in any but the prescribed sequence.

interlocutor /ˌɪntəˈlɒkjʊtə(r)/ *n. formal* person who takes part in a conversation. [Latin *loquor* speak]

interlocutory *adj. formal* **1** of dialogue. **2** (of a decree etc.) given provisionally in a legal action.

interloper /ˈɪntəˌləʊpə(r)/ *n.* **1** intruder. **2** person who interferes in others' affairs, esp. for profit. [after *landloper* vagabond, from Dutch *loopen* run]

interlude /ˈɪntəˌlu:d/ *n.* **1 a** pause between the acts of a play. **b** something performed during this pause. **2** contrasting event, time, etc. in the middle of something (*comic interlude*). **3** piece of music played between other pieces etc. [medieval Latin *ludus* play]

intermarry /ˌɪntəˈmærɪ/ *v.* (**-ies, -ied**) (foll. by *with*) (of races, castes, families, etc.) become connected by marriage. □ **intermarriage** /-rɪdʒ/ *n.*

intermediary /ˌɪntəˈmi:dɪərɪ/ *–n.* (*pl.* **-ies**) intermediate person or thing, esp. a mediator. *–adj.* acting as mediator; intermediate.

intermediate /ˌɪntəˈmi:dɪət/ *–adj.* coming between two things in time, place, order, character, etc. *–n.* **1** intermediate thing. **2** chemical compound formed by one reaction and then used in another. [Latin *intermedius*]

interment /ɪnˈtɜ:mənt/ *n.* burial.

■ **Usage** *Interment* is sometimes confused with *internment*, which means 'confinement'.

---

intermezzo /ˌɪntəˈmetsəʊ/ *n.* (*pl.* **-mezzi** /-tsɪ/ or **-s**) **1 a** short connecting instrumental movement in a musical work. **b** similar independent piece. **2** short light dramatic or other performance inserted between the acts of a play. [Italian]

interminable /ɪnˈtɜ:mɪnəb(ə)l/ *adj.* **1** endless. **2** tediously long. □ **interminably** *adv.*

intermingle /ˌɪntəˈmɪŋg(ə)l/ *v.* (**-ling**) mix together; mingle.

intermission /ˌɪntəˈmɪʃ(ə)n/ *n.* **1** pause or cessation. **2** interval in a cinema etc. [Latin: related to INTERMITTENT]

intermittent /ˌɪntəˈmɪt(ə)nt/ *adj.* occurring at intervals; not continuous. □ **intermittently** *adv.* [Latin *mitto miss-* let go]

intermix /ˌɪntəˈmɪks/ *v.* mix together.

intern *–n.* /ˈɪntɜ:n/ (also **interne**) esp. *US* = HOUSEMAN. *–v.* /ɪnˈtɜ:n/ oblige (a prisoner, alien, etc.) to reside within prescribed limits. □ **internment** *n.* [French: related to INTERNAL]

■ **Usage** *Internment* is sometimes confused with *interment*, which means 'burial'.

---

internal /ɪnˈtɜ:n(ə)l/ *adj.* **1** of or situated in the inside or invisible part. **2** of the inside of the body (*internal injuries*). **3** of a nation's domestic affairs. **4** (of a student) attending a university etc. as well as taking its examinations. **5** used or applying within an organization. **6 a** intrinsic. **b** of the mind or soul. □ **internality** /-ˈnælɪtɪ/ *n.* **internally** *adv.* [medieval Latin *internus* internal]

internal-combustion engine *n.* engine with its motive power generated by the explosion of gases or vapour with air in a cylinder.

internal evidence *n.* evidence derived from the contents of the thing discussed.

internalize *v.* (also **-ise**) (**-zing** or **-sing**) *Psychol.* make (attitudes, behaviour, etc.) part of one's nature by learning or unconscious assimilation. □ **internalization** /-ˈzeɪʃ(ə)n/ *n.*

international /ˌɪntəˈnæʃən(ə)l/ *–adj.* **1** existing or carried on between nations. **2** agreed on or used by all or many nations. *–n.* **1 a** contest, esp. in sport, between teams representing different countries. **b** member of such a team. **2** (**International**) any of four successive associations for socialist or Communist action. □ **internationality** /-ˈnælɪtɪ/ *n.* **internationally** *adv.*

internationalism *n.* advocacy of a community of interests among nations. □ **internationalist** *n.*

internationalize *v.* (also **-ise**) (**-zing** or **-sing**) **1** make international. **2** bring under the protection or control of two or more nations.

interne var. of INTERN *n.*

internecine /ˌɪntəˈni:saɪn/ *adj.* mutually destructive. [Latin *internecinus* deadly]

internee /ˌɪntɜ:ˈni:/ *n.* person interned.

interpenetrate /ˌɪntəˈpenɪˌtreɪt/ *v.* (**-ting**) **1** penetrate each other. **2** pervade. □ **interpenetration** /-ˈtreɪʃ(ə)n/ *n.*

interpersonal /ˌɪntəˈpɜ:sən(ə)l/ *adj.* between persons, social (*interpersonal skills*).

interplanetary /ˌɪntəˈplænɪtərɪ/ *adj.* **1** between planets. **2** of travel between planets.

interplay /ˈɪntəˌpleɪ/ *n.* reciprocal action.

Interpol /ˈɪntəˌpɒl/ *n.* International Criminal Police Organization. [abbreviation]

interpolate /ɪn'tɜ:pə,leɪt/ *v.* (**-ting**) **1 a** insert (words) in a book etc., esp. misleadingly. **b** make such insertions in (a book etc.). **2** interject (a remark) in a conversation. **3** estimate (values) between known ones in the same range. □ **interpolation** /-'leɪʃ(ə)n/ *n.* **interpolator** *n.* [Latin *interpolo* furbish]

interpose /,ɪntə'pəʊz/ *v.* (**-sing**) **1** (often foll. by *between*) insert (a thing) between others. **2** say (words) as an interruption; interrupt. **3** exercise or advance (a veto or objection) so as to interfere. **4** (foll. by *between*) intervene (between parties). □ **interposition** /-pə'zɪʃ(ə)n/ *n.* [Latin *pono* put]

interpret /ɪn'tɜ:prɪt/ *v.* (**-t-**) **1** explain the meaning of (words, a dream, etc.). **2** make out or bring out the meaning of (creative work). **3** act as an interpreter. **4** explain or understand (behaviour etc.) in a specified manner. □ **interpretation** /-'teɪʃ(ə)n/ *n.* **interpretative** *adj.* **interpretive** *adj.* [Latin *interpres -pretis* explainer]

interpreter *n.* person who interprets, esp. one who translates foreign speech orally.

interracial /,ɪntə'reɪʃ(ə)l/ *adj.* between or affecting different races.

interregnum /,ɪntə'regnəm/ *n.* (*pl.* **-s**) **1** interval when the normal government or leadership is suspended, esp. between successive reigns or regimes. **2** interval, pause. [Latin *regnum* reign]

interrelate /,ɪntərɪ'leɪt/ *v.* (**-ting**) **1** relate (two or more things) to each other. **2** (of two or more things) relate to each other. □ **interrelation** *n.* **interrelationship** *n.*

interrogate /ɪn'terə,geɪt/ *v.* (**-ting**) question (a person), esp. closely or formally. □ **interrogation** /-'geɪʃ(ə)n/ *n.* **interrogator** *n.* [Latin *rogo* ask]

interrogative /,ɪntə'rɒgətɪv/ *–adj.* of, like, or used in a question. *–n.* interrogative word (e.g. *what?*).

interrogatory /,ɪntə'rɒgətərɪ/ *–adj.* questioning (*interrogatory tone*). *–n.* (*pl.* **-ies**) formal set of questions.

interrupt /,ɪntə'rʌpt/ *v.* **1** break the continuous progress of (an action, speech, person speaking, etc.). **2** obstruct (a person's view etc.). □ **interruption** *n.* [Latin: related to RUPTURE]

interrupter *n.* (also **interruptor**) **1** person or thing that interrupts. **2** device for interrupting, esp. an electric circuit.

intersect /,ɪntə'sekt/ *v.* **1** divide (a thing) by crossing it. **2** (of lines, roads, etc.) cross each other. [Latin: related to SECTION]

intersection /,ɪntə'sekʃ(ə)n/ *n.* **1** intersecting. **2** place where two roads intersect. **3** point or line common to lines or planes that intersect.

intersperse /,ɪntə'spɜ:s/ *v.* (**-sing**) **1** (often foll. by *between, among*) scatter. **2** (foll. by *with*) vary (a thing) by scattering other things among it. □ **interspersion** *n.* [Latin: related to SPARSE]

interstate /'ɪntə,steɪt/ *adj.* existing or carried on between States, esp. those of the US.

interstellar /,ɪntə'stelə(r)/ *adj.* between stars.

interstice /ɪn'tɜ:stɪs/ *n.* **1** intervening space. **2** chink or crevice. [Latin *interstitium* from *sisto* stand]

interstitial /,ɪntə'stɪʃ(ə)l/ *adj.* of, forming, or occupying interstices. □ **interstitially** *adv.*

intertwine /,ɪntə'twaɪn/ *v.* (**-ning**) (often foll. by *with*) entwine (together).

interval /'ɪntəv(ə)l/ *n.* **1** intervening time or space. **2** pause or break, esp. between the parts of a performance. **3** difference in pitch between two sounds. □ **at intervals** here and there; now and then. [Latin *intervallum* space between ramparts]

intervene /,ɪntə'vi:n/ *v.* (**-ning**) **1** occur in time between events. **2** interfere; prevent or modify events. **3** be situated between things. **4** come in as an extraneous factor. [Latin *venio vent-* come]

intervention /,ɪntə'venʃ(ə)n/ *n.* **1** intervening. **2** interference, esp. by a State. **3** mediation.

interventionist *n.* person who favours intervention.

interview /'ɪntə,vju:/ *–n.* **1** oral examination of an applicant. **2** conversation with a reporter, for a broadcast or publication. **3** meeting face to face, esp. for consultation. *–v.* hold an interview with. □ **interviewee** /-vju:'i:/ *n.* **interviewer** *n.* [French *entrevue*: related to INTER-, *vue* sight]

interwar /,ɪntə'wɔ:(r)/ *attrib. adj.* existing in the period between two wars.

interweave /,ɪntə'wi:v/ *v.* (**-ving**; *past* **-wove**; *past part.* **-woven**) **1** weave together. **2** blend intimately.

intestate /ɪn'testeɪt/ *–adj.* not having made a will before death. *–n.* person who has died intestate. □ **intestacy** /-təsɪ/ *n.* [Latin: related to TESTAMENT]

intestine /ɪn'testɪn/ *n.* (in *sing.* or *pl.*) lower part of the alimentary canal. □ **intestinal** *adj.* [Latin *intus* within]

intifada /,ɪntɪ'fɑ:də/ *n.* Arab uprising. [Arabic]

intimacy /'ɪntɪməsɪ/ *n.* (*pl.* **-ies**) **1** state of being intimate. **2** intimate remark or act; sexual intercourse.

intimate[1] /'ɪntɪmət/ *–adj.* **1** closely acquainted; familiar (*intimate friend*). **2** private and personal. **3** (usu. foll. by *with*) having sexual relations. **4** (of knowledge) detailed, thorough. **5** (of a relationship between things) close. *–n.* close friend. □ **intimately** *adv.* [Latin *intimus* inmost]

intimate[2] /'ɪntɪ,meɪt/ *v.* (**-ting**) **1** (often foll. by *that*) state or make known. **2** imply, hint. □ **intimation** /-'meɪʃ(ə)n/ *n.* [Latin *intimo* announce: related to INTIMATE[1]]

intimidate /ɪn'tɪmɪ,deɪt/ *v.* (**-ting**) frighten or overawe, esp. to subdue or influence. □ **intimidation** /-'deɪʃ(ə)n/ *n.* [medieval Latin: related to TIMID]

into /'ɪntʊ, 'ɪntə/ *prep.* **1** expressing motion or direction to a point on or within (*walked into a tree*; *ran into the house*). **2** expressing direction of attention etc. (*will look into it*). **3** expressing a change of state (*turned into a dragon*; *separated into groups*). **4** after the beginning of (*five minutes into the game*). **5** *colloq.* interested in. [Old English: related to IN, TO]

intolerable /ɪn'tɒlərəb(ə)l/ *adj.* that cannot be endured. □ **intolerably** *adv.*

intolerant /ɪn'tɒlərənt/ *adj.* not tolerant, esp. of others' beliefs or behaviour. □ **intolerance** *n.*

intonation /,ɪntə'neɪʃ(ə)n/ *n.* **1** modulation of the voice; accent. **2** intoning. **3** accuracy of musical pitch. [medieval Latin: related to INTONE]

intone /ɪn'təʊn/ *v.* (**-ning**) **1** recite (prayers etc.) with prolonged sounds, esp. in a monotone. **2** utter with a particular tone. [medieval Latin: related to IN-[2]]

in toto /ɪn 'təʊtəʊ/ *adv.* completely. [Latin]

intoxicant /ɪn'tɒksɪkənt/ *–adj.* intoxicating. *–n.* intoxicating substance.

intoxicate /ɪn'tɒksɪ,keɪt/ *v.* (**-ting**) **1** make drunk. **2** excite or elate beyond self-control. □ **intoxication** /-'keɪʃ(ə)n/ *n.* [medieval Latin: related to TOXIC]

intra- *prefix* on the inside, within. [Latin *intra* inside]

intractable /ɪn'træktəb(ə)l/ *adj.* **1** hard to control or deal with. **2** difficult, stubborn. □ **intractability** /-'bɪlɪtɪ/ *n.* **intractably** *adv.*

intramural /,ɪntrə'mjʊər(ə)l/ *adj.* **1** situated or done within the walls of an institution etc. **2** forming part of normal university etc. studies. □ **intramurally** *adv.* [Latin *murus* wall]

intramuscular /,ɪntrə'mʌskjʊlə(r)/ *adj.* in or into muscle tissue.

intransigent /ɪn'trænsɪdʒ(ə)nt/ *–adj.* uncompromising, stubborn. *–n.* intransigent person. □ **intransigence** *n.* [Spanish *los intransigentes* extremists]

intransitive /ɪn'trænsɪtɪv/ *adj.* (of a verb) not taking a direct object.

intrauterine /,ɪntrə'ju:tə,raɪn/ *adj.* within the womb.

intravenous /,ɪntrə'vi:nəs/ *adj.* in or into a vein or veins. □ **intravenously** *adv.*

in-tray *n.* tray for incoming documents.

intrepid /ɪn'trepɪd/ *adj.* fearless; very brave. □ **intrepidity** /-trɪ'pɪdɪtɪ/ *n.* **intrepidly** *adv.* [Latin *trepidus* alarmed]

intricate /'ɪntrɪkət/ *adj.* very complicated; perplexingly detailed. □ **intricacy** /-kəsɪ/ *n.* (*pl.* **-ies**). **intricately** *adv.* [Latin: related to IN-[2], *tricae* tricks]

intrigue *–v.* /ɪn'tri:g/ (**-gues**, **-gued**, **-guing**) **1** (foll. by *with*) **a** carry on an underhand plot. **b** use secret influence. **2** arouse the curiosity of. *–n.* /'ɪntri:g/ **1** underhand plot or plotting. **2** secret arrangement (*amorous intrigues*). □ **intriguing** *adj.* esp. in sense 2 of *v.* **intriguingly** *adv.* [French from Italian *intrigo*]

intrinsic /ɪn'trɪnzɪk/ *adj.* inherent, essential (*intrinsic value*). □ **intrinsically** *adv.* [Latin *intrinsecus* inwardly]

intro /'ɪntrəʊ/ *n.* (*pl.* **-s**) *colloq.* introduction. [abbreviation]

intro- *comb. form* into. [Latin]

introduce /,ɪntrə'dju:s/ *v.* (**-cing**) **1** (foll. by *to*) make (a person or oneself) known by name to another, esp. formally. **2** announce or present to an audience. **3** bring (a custom etc.) into use. **4** bring (legislation) before Parliament etc. **5** (foll. by *to*) initiate (a person) in a subject. **6** insert. **7** bring in; usher in; bring forward. **8** occur just before the start of. **9** put on sale for the first time. □ **introducible** *adj.* [Latin *duco* lead]

introduction /,ɪntrə'dʌkʃ(ə)n/ *n.* **1** introducing or being introduced. **2** formal presentation of one person to another. **3** explanatory section at the beginning of a book etc. **4** introductory treatise. **5** thing introduced.

introductory /,ɪntrə'dʌktərɪ/ *adj.* serving as an introduction; preliminary.

introit /'ɪntrɔɪt/ *n.* psalm or antiphon sung or said as the priest approaches the altar for the Eucharist. [Latin *introitus* entrance]

introspection /,ɪntrə'spekʃ(ə)n/ *n.* examination of one's own thoughts. □ **introspective** *adj.* [Latin *specio spect-* look]

introvert /'ɪntrə,vɜ:t/ *–n.* **1** person predominantly concerned with his or her own thoughts. **2** shy thoughtful person.

–adj. (also **introverted**) characteristic of an introvert. □ **introversion** /-'vɜ:ʃ(ə)n/ *n.*

intrude /ɪn'tru:d/ *v.* (**-ding**) (foll. by *on*, *upon*, *into*) **1** come uninvited or unwanted. **2** force on a person. [Latin *trudo trus-* thrust]

intruder *n.* person who intrudes, esp. a trespasser.

intrusion /ɪn'tru:ʒ(ə)n/ *n.* **1** intruding. **2** influx of molten rock between existing strata etc. □ **intrusive** *adj.*

intrust var. of ENTRUST.

intuition /,ɪntju:'ɪʃ(ə)n/ *n.* immediate insight or understanding without conscious reasoning. □ **intuit** /ɪn'tju:ɪt/ *v.* **intuitional** *adj.* [Latin *tueor tuit-* look]

intuitive /ɪn'tju:ɪtɪv/ *adj.* of, possessing, or perceived by intuition. □ **intuitively** *adv.* **intuitiveness** *n.* [medieval Latin: related to INTUITION]

Inuit /'ɪnʊɪt/ *n.* (also **Innuit**) (*pl.* same or **-s**) N. American Eskimo. [Eskimo *inuit* people]

inundate /'ɪnʌn,deɪt/ *v.* (**-ting**) (often foll. by *with*) **1** flood. **2** overwhelm. □ **inundation** /-'deɪʃ(ə)n/ *n.* [Latin *unda* wave]

inure /ɪ'njʊə(r)/ *v.* (**-ring**) **1** (often in *passive*; foll. by *to*) accustom (a person) to an esp. unpleasant thing. **2** *Law* take effect. □ **inurement** *n.* [Anglo-French: related to IN, *eure* work, from Latin *opera*]

invade /ɪn'veɪd/ *v.* (**-ding**) (often *absol.*) **1** enter (a country etc.) under arms to control or subdue it. **2** swarm into. **3** (of a disease) attack. **4** encroach upon (a person's rights, esp. privacy). □ **invader** *n.* [Latin *vado vas-* go]

invalid[1] /'ɪnvəlɪd, -,li:d/ *–n.* person enfeebled or disabled by illness or injury. *–attrib. adj.* **1** of or for invalids. **2** sick, disabled. *–v.* (**-d-**) **1** (often foll. by *out* etc.) remove (an invalid) from active service. **2** (usu. in *passive*) disable (a person) by illness. □ **invalidism** *n.* **invalidity** /,ɪnvə'lɪdɪtɪ/ *n.* [Latin: related to IN-[1]]

invalid[2] /ɪn'vælɪd/ *adj.* not valid. □ **invalidity** /,ɪnvə'lɪdɪtɪ/ *n.*

invalidate /ɪn'vælɪ,deɪt/ *v.* (**-ting**) make (a claim etc.) invalid. □ **invalidation** /-'deɪʃ(ə)n/ *n.*

invaluable /ɪn'væljʊəb(ə)l/ *adj.* above valuation; very valuable. □ **invaluably** *adv.*

invariable /ɪn'veərɪəb(ə)l/ *adj.* **1** unchangeable. **2** always the same. **3** *Math.* constant. □ **invariably** *adv.*

invasion /ɪn'veɪʒ(ə)n/ *n.* invading or being invaded.

invasive /ɪn'veɪsɪv/ *adj.* **1** (of weeds, cancer cells, etc.) tending to spread. **2** (of surgery) involving large incisions etc. **3** tending to encroach.

invective /ɪn'vektɪv/ *n.* strong verbal attack. [Latin: related to INVEIGH]

inveigh /ɪn'veɪ/ *v.* (foll. by *against*) speak or write with strong hostility. [Latin *invehor -vect-* assail]

inveigle /ɪn'veɪg(ə)l/ *v.* (**-ling**) (foll. by *into*, or *to* + infin.) entice; persuade by guile. □ **inveiglement** *n.* [Anglo-French from French *aveugler* to blind]

invent /ɪn'vent/ *v.* **1** create by thought, originate (a method, device, etc.). **2** concoct (a false story etc.). □ **inventor** *n.* [Latin *invenio -vent-* find]

invention /ɪn'venʃ(ə)n/ *n.* **1** inventing or being invented. **2** thing invented. **3** fictitious story. **4** inventiveness.

inventive *adj.* able to invent; imaginative. □ **inventively** *adv.* **inventiveness** *n.*

inventory /'ɪnvəntərɪ/ *–n.* (*pl.* **-ies**) **1** complete list of goods etc. **2** goods listed in this. *–v.* (**-ies, -ied**) **1** make an inventory of. **2** enter (goods) in an inventory. [medieval Latin: related to INVENT]

inverse /ɪn'vɜ:s/ *–adj.* inverted in position, order, or relation. *–n.* **1** inverted state. **2** (often foll. by *of*) the direct opposite. [Latin: related to INVERT]

inverse proportion *n.* (also **inverse ratio**) relation between two quantities such that one increases in proportion as the other decreases.

inversion /ɪn'vɜ:ʃ(ə)n/ *n.* **1** turning upside down. **2** reversal of a normal order, position, or relation.

invert /ɪn'vɜ:t/ *v.* **1** turn upside down. **2** reverse the position, order, or relation of. [Latin *verto vers-* turn]

invertebrate /ɪn'vɜ:tɪbrət/ *–adj.* (of an animal) not having a backbone. *–n.* invertebrate animal.

inverted commas *n.pl.* = QUOTATION MARKS.

invest /ɪn'vest/ *v.* **1 a** (often foll. by *in*) apply or use (money), esp. for profit. **b** (foll. by *in*) put money for profit into (stocks etc.). **2** (often foll. by *in*) devote (time etc.) to an enterprise. **3** (foll. by *in*) *colloq.* buy (something useful). **4 a** (foll. by *with*) provide or credit (a person etc. with qualities) (*invested her with magical importance*; *invested his tone with irony*). **b** (foll. by *in*) attribute or entrust (qualities or feelings) to (a person etc.) (*power invested in the doctor*). **5** (often foll. by *with, in*) clothe with the insignia of office; install in an office. □ **investor** *n.* [Latin *vestis* clothing]

investigate /ɪn'vestɪ,geɪt/ *v.* (**-ting**) **1** inquire into; examine. **2** make a systematic inquiry. □ **investigation** /-'geɪʃ(ə)n/ *n.* **investigative** /-gətɪv/

adj. **investigator** *n.* **investigatory** /-gətərɪ/ *adj.* [Latin *vestigo* track]

investiture /ɪn'vestɪ,tʃə(r)/ *n.* formal investing of a person with honours or rank. [medieval Latin: related to INVEST]

investment *n.* **1** investing. **2** money invested. **3** property etc. in which money is invested.

investment trust *n.* trust that buys and sells shares in selected companies to make a profit for its members.

inveterate /ɪn'vetərət/ *adj.* **1** (of a person) confirmed in a habit etc. **2** (of a habit etc.) long-established. □ **inveteracy** *n.* [Latin *vetus* old]

invidious /ɪn'vɪdɪəs/ *adj.* likely to cause resentment or anger (*invidious position*; *invidious task*). [Latin *invidiosus*: related to ENVY]

invigilate /ɪn'vɪdʒɪ,leɪt/ *v.* (**-ting**) supervise people taking an exam. □ **invigilation** /-'leɪʃ(ə)n/ *n.* **invigilator** *n.* [Latin: related to VIGIL]

invigorate /ɪn'vɪgə,reɪt/ *v.* (**-ting**) give vigour or strength to. □ **invigorating** *adj.* [medieval Latin: related to VIGOUR]

invincible /ɪn'vɪnsɪb(ə)l/ *adj.* unconquerable. □ **invincibility** /-'bɪlɪtɪ/ *n.* **invincibly** *adv.* [Latin *vinco* conquer]

inviolable /ɪn'vaɪələb(ə)l/ *adj.* not to be violated or dishonoured. □ **inviolability** /-'bɪlɪtɪ/ *n.* **inviolably** *adv.*

inviolate /ɪn'vaɪələt/ *adj.* **1** not violated. **2** safe (from violation or harm). □ **inviolacy** *n.*

invisible /ɪn'vɪzɪb(ə)l/ *adj.* not visible to the eye. □ **invisibility** /-'bɪlɪtɪ/ *n.* **invisibly** *adv.*

invisible exports *n.pl.* (also **invisible imports** etc.) intangible commodities, esp. services, involving payment between countries.

invitation /,ɪnvɪ'teɪʃ(ə)n/ *n.* **1** inviting or being invited. **2** letter or card etc. used to invite.

invite –*v.* /ɪn'vaɪt/ (**-ting**) **1** (often foll. by *to*, or *to* + infin.) ask (a person) courteously to come, or to do something. **2** make a formal courteous request for. **3** tend to call forth unintentionally. **4 a** attract. **b** be attractive. –*n.* /'ɪnvaɪt/ *colloq.* invitation. [Latin *invito*]

inviting *adj.* **1** attractive. **2** tempting. □ **invitingly** *adv.*

in vitro /ɪn 'vi:trəʊ/ *adv.* (of biological processes) taking place in a test-tube or other laboratory environment. [Latin, = in glass]

invocation /,ɪnvə'keɪʃ(ə)n/ *n.* **1** invoking or being invoked, esp. in prayer. **2** summoning of supernatural beings, e.g. the Muses, for inspiration. **3** *Eccl.* the words 'In the name of the Father' etc. used to preface a sermon etc. □ **invocatory** /ɪn'vɒkətərɪ/ *adj.* [Latin: related to INVOKE]

invoice /'ɪnvɔɪs/ –*n.* bill for usu. itemized goods or services. –*v.* (**-cing**) **1** send an invoice to. **2** make an invoice of. [earlier *invoyes* pl. of *invoy*: related to ENVOY]

invoke /ɪn'vəʊk/ *v.* (**-king**) **1** call on (a deity etc.) in prayer or as a witness. **2** appeal to (the law, a person's authority, etc.). **3** summon (a spirit) by charms etc. **4** ask earnestly for (vengeance etc.). [Latin *voco* call]

involuntary /ɪn'vɒləntərɪ/ *adj.* **1** done without exercising the will; unintentional. **2** (of a muscle) not under the control of the will. □ **involuntarily** *adv.* **involuntariness** *n.*

involute /'ɪnvə,lu:t/ *adj.* **1** involved, intricate. **2** curled spirally. [Latin: related to INVOLVE]

involuted *adj.* complicated, abstruse.

involution /,ɪnvə'lu:ʃ(ə)n/ *n.* **1** involving. **2** intricacy. **3** curling inwards. **4** part that curls inwards.

involve /ɪn'vɒlv/ *v.* (**-ving**) **1** (often foll. by *in*) cause (a person or thing) to share the experience or effect (of a situation, activity, etc.). **2** imply, entail, make necessary. **3** (often foll. by *in*) implicate (a person) in a charge, crime, etc. **4** include or affect in its operations. **5** (as **involved** *adj.*) **a** (often foll. by *in*) concerned. **b** complicated in thought or form. **c** amorously associated. □ **involvement** *n.* [Latin *volvo* roll]

invulnerable /ɪn'vʌlnərəb(ə)l/ *adj.* that cannot be wounded, damaged, or hurt, physically or mentally. □ **invulnerability** /-'bɪlɪtɪ/ *n.* **invulnerably** *adv.*

inward /'ɪnwəd/ –*adj.* **1** directed towards the inside; going in. **2** situated within. **3** mental, spiritual. –*adv.* (also **inwards**) **1** towards the inside. **2** in the mind or soul. [Old English: related to IN, -WARD]

inwardly *adv.* **1** on the inside. **2** in the mind or soul. **3** not aloud.

inwrought /ɪn'rɔ:t/ *adj.* **1** (often foll. by *with*) (of a fabric) decorated (with a pattern). **2** (often foll. by *in*, *on*) (of a pattern) wrought (in or on a fabric).

iodide /'aɪə,daɪd/ *n.* any compound of iodine with another element or group.

iodine /'aɪə,di:n/ *n.* **1** black crystalline element forming a violet vapour. **2** solution of this as an antiseptic. [French *iode* from Greek *iōdēs* violet-like]

IOM *abbr.* Isle of Man.

ion /'aɪən/ *n.* atom or group of atoms that has lost one or more electrons (= CATION), or gained one or more electrons (= ANION). [Greek, = going]

-ion *suffix* (usu. as **-sion, -tion, -xion**) forming nouns denoting: **1** verbal action (*excision*). **2** instance of this (*a suggestion*). **3** resulting state or product (*vexation*; *concoction*). [Latin *-io*]

Ionic /aɪ'ɒnɪk/ *adj.* of the order of Greek architecture characterized by a column with scroll-shapes on either side of the capital. [from *Ionia* in Greek Asia Minor]

ionic /aɪ'ɒnɪk/ *adj.* of or using ions. □ **ionically** *adv.*

ionize *v.* (also **-ise**) (**-zing** or **-sing**) convert or be converted into an ion or ions. □ **ionization** /-'zeɪʃ(ə)n/ *n.*

ionizer *n.* device producing ions to improve the quality of the air.

ionosphere /aɪ'ɒnə,sfɪə(r)/ *n.* ionized region of the atmosphere above the stratosphere, reflecting radio waves. □ **ionospheric** /-'sferɪk/ *adj.*

iota /aɪ'əʊtə/ *n.* **1** ninth letter of the Greek alphabet (*Ι, ι*). **2** (usu. with *neg.*) a jot. [Greek *iōta*]

IOU /,aɪəʊ'ju:/ *n.* signed document acknowledging a debt. [from *I owe you*]

IOW *abbr.* Isle of Wight.

IPA *abbr.* International Phonetic Alphabet.

ipecacuanha /,ɪpɪ,kækjʊ'ɑ:nə/ *n.* root of a S. American shrub, used as an emetic and purgative. [Portuguese from S. American Indian, = emetic creeper]

ipso facto /,ɪpsəʊ 'fæktəʊ/ *adv.* by that very fact. [Latin]

IQ *abbr.* intelligence quotient.

ir- *prefix* assim. form of IN-[1], IN-[2] before *r*.

IRA *abbr.* Irish Republican Army.

Iranian /ɪ'reɪnɪən/ *–adj.* **1** of Iran (formerly Persia). **2** of the group of languages including Persian. *–n.* **1** native or national of Iran. **2** person of Iranian descent.

Iraqi /ɪ'rɑ:kɪ/ *–adj.* of Iraq. *–n.* (*pl.* **-s**) **1** **a** native or national of Iraq. **b** person of Iraqi descent. **2** the form of Arabic spoken in Iraq.

irascible /ɪ'ræsɪb(ə)l/ *adj.* irritable; hot-tempered. □ **irascibility** /-'bɪlɪtɪ/ *n.* **irascibly** *adv.* [Latin *irascor* grow angry, from *ira* anger]

irate /aɪ'reɪt/ *adj.* angry, enraged. □ **irately** *adv.* **irateness** *n.* [Latin *iratus* from *ira* anger]

ire /'aɪə(r)/ *n. literary* anger. [Latin *ira*]

iridaceous /,ɪrɪ'deɪʃəs/ *adj.* of the iris family of plants.

iridescent /,ɪrɪ'des(ə)nt/ *adj.* **1** showing rainbow-like luminous colours. **2** changing colour with position. □ **iridescence** *n.*

iridium /ɪ'rɪdɪəm/ *n.* hard white metallic element of the platinum group.

iris /'aɪərɪs/ *n.* **1** circular coloured membrane behind the cornea of the eye, with a circular opening (pupil) in the centre. **2** plant of a family with bulbs or tuberous roots, sword-shaped leaves, and showy flowers. **3** adjustable diaphragm for regulating the size of a central hole, esp. for the admission of light to a lens. [Greek *iris iridos* rainbow]

Irish /'aɪərɪʃ/ *–adj.* of Ireland or its people. *–n.* **1** Celtic language of Ireland. **2** (prec. by *the*; treated as *pl.*) the people of Ireland. [Old English]

Irish bull *n.* = BULL[3].

Irish coffee *n.* coffee with a dash of whiskey and a little sugar, topped with cream.

Irishman *n.* man who is Irish by birth or descent.

Irish stew *n.* stew of mutton, potato, and onion.

Irishwoman *n.* woman who is Irish by birth or descent.

irk *v.* irritate, bore, annoy. [origin unknown]

irksome *adj.* annoying, tiresome. □ **irksomely** *adv.*

iron /'aɪən/ *–n.* **1** grey metallic element used for tools and constructions and found in some foods, e.g. spinach. **2** this as a symbol of strength or firmness (*man of iron*; *iron will*). **3** tool made of iron. **4** implement with a flat base which is heated to smooth clothes etc. **5** golf club with an iron or steel sloping face. **6** (usu. in *pl.*) fetter. **7** (usu. in *pl.*) stirrup. **8** (often in *pl.*) iron support for a malformed leg. *–adj.* **1** made of iron. **2** very robust. **3** unyielding, merciless. *–v.* smooth (clothes etc.) with an iron. □ **iron out** remove (difficulties etc.). [Old English]

Iron Age *n.* period when iron replaced bronze in the making of tools and weapons.

ironclad *–adj.* **1** clad or protected with iron. **2** impregnable. *–n. hist.* warship protected by iron plates.

Iron Cross *n.* German military decoration.

Iron Curtain *n. hist.* former notional barrier to the passage of people and information between the Soviet bloc and the West.

ironic /aɪ'rɒnɪk/ *adj.* (also **ironical**) using or displaying irony. □ **ironically** *adv.*

ironing *n.* clothes etc. for ironing or just ironed.

ironing-board *n.* narrow folding table on which clothes etc. are ironed.

iron in the fire *n.* undertaking, opportunity (usu. in *pl.*: *too many irons in the fire*).

iron lung *n.* rigid case fitted over a patient's body for administering prolonged artificial respiration.

ironmaster *n.* manufacturer of iron.

ironmonger *n.* dealer in hardware etc. □ **ironmongery** *n.* (*pl.* **-ies**).

iron rations *n.pl.* small emergency supply of food.

ironstone *n.* **1** rock containing much iron. **2** a kind of hard white pottery.

ironware *n.* articles made of iron.

ironwork *n.* **1** articles made of iron. **2** work in iron.

ironworks *n.* (as *sing.* or *pl.*) factory where iron is smelted or iron goods are made.

irony /'aɪrənɪ/ *n.* (*pl.* **-ies**) **1** expression of meaning, often humorous or sarcastic, using language of a different or opposite tendency. **2** apparent perversity of an event or circumstance in reversing human intentions. **3** *Theatr.* use of language with one meaning for a privileged audience and another for those addressed or concerned. [Greek *eirōneia* pretended ignorance]

irradiate /ɪ'reɪdɪ,eɪt/ *v.* (**-ting**) **1** subject to radiation. **2** shine upon; light up. **3** throw light on (a subject). □ **irradiation** /ɪ,reɪdɪ'eɪʃ(ə)n/ *n.* [Latin *irradio* shine on, from *radius* ray]

irrational /ɪ'ræʃən(ə)l/ *adj.* **1** illogical; unreasonable. **2** not endowed with reason. **3** *Math.* not commensurate with the natural numbers. □ **irrationality** /-'nælɪtɪ/ *n.* **irrationally** *adv.*

irreconcilable /ɪ'rekən,saɪləb(ə)l/ *adj.* **1** implacably hostile. **2** (of ideas etc.) incompatible. □ **irreconcilability** /-'bɪlɪtɪ/ *n.* **irreconcilably** *adv.*

irrecoverable /,ɪrɪ'kʌvərəb(ə)l/ *adj.* not able to be recovered or remedied. □ **irrecoverably** *adv.*

irredeemable /,ɪrɪ'di:məb(ə)l/ *adj.* **1** not able to be redeemed. **2** hopeless. □ **irredeemably** *adv.*

irredentist /,ɪrɪ'dentɪst/ *n.* person advocating the restoration to his or her country of any territory formerly belonging to it. □ **irredentism** *n.* [Italian *irredenta* unredeemed]

irreducible /,ɪrɪ'dju:sɪb(ə)l/ *adj.* not able to be reduced or simplified. □ **irreducibility** /-'bɪlɪtɪ/ *n.* **irreducibly** *adv.*

irrefutable /,ɪrɪ'fju:təb(ə)l/ *adj.* that cannot be refuted. □ **irrefutably** *adv.*

irregular /ɪ'regjʊlə(r)/ —*adj.* **1** not regular; unsymmetrical, uneven; varying in form. **2** not occurring at regular intervals. **3** contrary to a rule, principle, or custom; abnormal. **4** (of troops) not belonging to the regular army. **5** (of a verb, noun, etc.) not inflected according to the usual rules. **6** disorderly. —*n.* (in *pl.*) irregular troops. □ **irregularity** /-'lærɪtɪ/ *n.* (*pl.* **-ies**). **irregularly** *adv.*

irrelevant /ɪ'relɪv(ə)nt/ *adj.* (often foll. by *to*) not relevant. □ **irrelevance** *n.* **irrelevancy** *n.* (*pl.* **-ies**).

irreligious /,ɪrɪ'lɪdʒəs/ *adj.* lacking or hostile to religion; irreverent.

irremediable /,ɪrɪ'mi:dɪəb(ə)l/ *adj.* that cannot be remedied. □ **irremediably** *adv.*

irremovable /,ɪrɪ'mu:vəb(ə)l/ *adj.* that cannot be removed. □ **irremovably** *adv.*

irreparable /ɪ'repərəb(ə)l/ *adj.* (of an injury, loss, etc.) that cannot be rectified or made good. □ **irreparably** *adv.*

irreplaceable /,ɪrɪ'pleɪsəb(ə)l/ *adj.* that cannot be replaced.

irrepressible /,ɪrɪ'presɪb(ə)l/ *adj.* that cannot be repressed or restrained. □ **irrepressibly** *adv.*

irreproachable /,ɪrɪ'prəʊtʃəb(ə)l/ *adj.* faultless, blameless. □ **irreproachably** *adv.*

irresistible /,ɪrɪ'zɪstɪb(ə)l/ *adj.* too strong, delightful, or convincing to be resisted. □ **irresistibly** *adv.*

irresolute /ɪ'rezə,lu:t/ *adj.* **1** hesitant. **2** lacking in resoluteness. □ **irresolutely** *adv.* **irresoluteness** *n.* **irresolution** /-'lu:ʃ(ə)n, -'lju:ʃ(ə)n/ *n.*

irrespective /,ɪrɪ'spektɪv/ *adj.* (foll. by *of*) not taking into account; regardless of.

irresponsible /,ɪrɪ'spɒnsɪb(ə)l/ *adj.* **1** acting or done without due sense of responsibility. **2** not responsible for one's conduct. □ **irresponsibility** /-'bɪlɪtɪ/ *n.* **irresponsibly** *adv.*

irretrievable /,ɪrɪ'tri:vəb(ə)l/ *adj.* that cannot be retrieved or restored. □ **irretrievably** *adv.*

irreverent /ɪ'revərənt/ *adj.* lacking reverence. □ **irreverence** *n.* **irreverently** *adv.*

irreversible /,ɪrɪ'vɜ:sɪb(ə)l/ *adj.* not reversible or alterable. □ **irreversibly** *adv.*

irrevocable /ɪ'revəkəb(ə)l/ *adj.* **1** unalterable. **2** gone beyond recall. □ **irrevocably** *adv.*

irrigate /'ɪrɪ,geɪt/ *v.* (**-ting**) **1 a** water (land) by means of channels etc. **b** (of a stream etc.) supply (land) with water. **2** supply (a wound etc.) with a constant flow of liquid. □ **irrigable** *adj.* **irrigation** /-'geɪʃ(ə)n/ *n.* **irrigator** *n.* [Latin *rigo* moisten]

irritable /'ɪrɪtəb(ə)l/ *adj.* **1** easily annoyed. **2** (of an organ etc.) very sensitive to contact. □ **irritability** /-'bɪlɪtɪ/ *n.* **irritably** *adv.* [Latin: related to IRRITATE]

irritant /'ɪrɪt(ə)nt/ *–adj.* causing irritation. *–n.* irritant substance.

irritate /'ɪrɪ,teɪt/ *v.* **(-ting) 1** excite to anger; annoy. **2** stimulate discomfort in (a part of the body). **3** *Biol.* stimulate (an organ) to action. □ **irritating** *adj.* **irritation** /-'teɪʃ(ə)n/ *n.* **irritative** *adj.* [Latin *irrito*]

irrupt /ɪ'rʌpt/ *v.* (foll. by *into*) enter forcibly or violently. □ **irruption** *n.* [Latin: related to RUPTURE]

is *3rd sing. present* of BE.

ISBN *abbr.* international standard book number.

-ise var. of -IZE.

■ **Usage** See note at *-ize*.

-ish *suffix* forming adjectives: **1** from nouns, meaning: **a** having the qualities of (*boyish*). **b** of the nationality of (*Danish*). **2** from adjectives, meaning 'somewhat' (*thickish*). **3** *colloq.* denoting an approximate age or time of day (*fortyish*; *six-thirtyish*). [Old English]

isinglass /'aɪzɪŋ,glɑ:s/ *n.* **1** gelatin obtained from fish, esp. sturgeon, and used in making jellies, glue, etc. **2** mica. [Dutch *huisenblas* sturgeon's bladder]

Islam /'ɪzlɑ:m/ *n.* **1** the religion of the Muslims, proclaimed by Muhammad. **2** the Muslim world. □ **Islamic** /ɪz'læmɪk/ *adj.* [Arabic, = submission (to God)]

island /'aɪlənd/ *n.* **1** piece of land surrounded by water. **2** = TRAFFIC ISLAND. **3** detached or isolated thing. [Old English *īgland*; first syllable influenced by ISLE]

islander *n.* native or inhabitant of an island.

isle /aɪl/ *n. poet.* (and in place-names) island, esp. a small one. [French *ile* from Latin *insula*]

islet /'aɪlɪt/ *n.* **1** small island. **2** *Anat.* structurally distinct portion of tissue. [French diminutive of ISLE]

ism /'ɪz(ə)m/ *n. colloq.* usu. *derog.* any distinctive doctrine or practice. [from -ISM]

-ism *suffix* forming nouns, esp. denoting: **1** action or its result (*baptism*; *organism*). **2** system or principle (*Conservatism*; *jingoism*). **3** state or quality (*heroism*; *barbarism*). **4** basis of prejudice or discrimination (*racism*; *sexism*). **5** peculiarity in language (*Americanism*). [Greek *-ismos*]

isn't /'ɪz(ə)nt/ *contr.* is not.

iso- *comb. form* equal. [Greek *isos* equal]

isobar /'aɪsəʊ,bɑ:(r)/ *n.* line on a map connecting places with the same atmospheric pressure. □ **isobaric** /-'bærɪk/ *adj.* [Greek *baros* weight]

isochronous /aɪ'sɒkrənəs/ *adj.* **1** occurring at the same time. **2** occupying equal time.

isolate /'aɪsə,leɪt/ *v.* **(-ting) 1 a** place apart or alone. **b** place (a contagious or infectious patient etc.) in quarantine. **2** separate (a substance) from a mixture. **3** insulate (electrical apparatus), esp. by a physical gap; disconnect. □ **isolation** /-'leɪʃ(ə)n/ *n.* [Latin *insulatus* made into an island]

isolationism *n.* policy of holding aloof from the affairs of other countries or groups. □ **isolationist** *n.*

isomer /'aɪsəmə(r)/ *n.* one of two or more compounds with the same molecular formula but a different arrangement of atoms. □ **isomeric** /-'merɪk/ *adj.* **isomerism** /aɪ'sɒmə,rɪz(ə)m/ *n.* [Greek ISO-, *meros* share]

isometric /,aɪsəʊ'metrɪk/ *adj.* **1** of equal measure. **2** (of muscle action) developing tension while the muscle is prevented from contracting. **3** (of a drawing etc.) with the plane of projection at equal angles to the three principal axes of the object shown. [Greek *isometria* equality of measure]

isomorphic /,aɪsəʊ'mɔ:fɪk/ *adj.* (also **isomorphous** /-fəs/) exactly corresponding in form and relations. [from ISO-, Greek *morphē* form]

isosceles /aɪ'sɒsɪ,li:z/ *adj.* (of a triangle) having two sides equal. [from ISO-, Greek *skelos* leg]

isotherm /'aɪsəʊ,θɜ:m/ *n.* line on a map connecting places with the same temperature. □ **isothermal** /-'θɜ:m(ə)l/ *adj.* [from ISO-, Greek *thermē* heat]

isotope /'aɪsə,təʊp/ *n.* one of two or more forms of an element differing from each other in relative atomic mass, and in nuclear but not chemical properties. □ **isotopic** /-'tɒpɪk/ *adj.* [from ISO-, Greek *topos* place]

isotropic /,aɪsəʊ'trɒpɪk/ *adj.* having the same physical properties in all directions. □ **isotropy** /aɪ'sɒtrəpɪ/ *n.* [from ISO-, Greek *tropos* turn]

Israeli /ɪz'reɪlɪ/ *–adj.* of the modern State of Israel. *–n.* (*pl.* **-s**) **1** native or national of Israel. **2** person of Israeli descent. [Hebrew]

Israelite /'ɪzrɪə,laɪt/ *n. hist.* native of ancient Israel; Jew. [Hebrew]

issue /'ɪʃu:/ *–n.* **1 a** act of giving out or circulating shares, notes, stamps, etc. **b** quantity of coins, copies of a newspaper, etc., circulated at one time. **c** each of a regular series of a magazine etc. (*the May issue*). **2 a** outgoing, outflow. **b** way out, outlet, esp. the place of the emergence of a stream etc. **3** point in question; important subject of debate or litigation. **4** result; outcome. **5** *Law* children, progeny (*without male issue*). *–v.* (**issues, issued, issuing**) **1** *literary* go or come out. **2 a** send forth; publish; put into circulation. **b** supply, esp. officially or authoritatively (foll. by *to, with*: *issued passports to them*; *issued them with passports*). **3 a** (often foll. by *from*) be derived or result. **b** (foll. by *in*) end, result. **4** (foll. by *from*) emerge from a condition. □ **at issue** under discussion; in dispute. **join** (or **take**) **issue** (foll. by *with* a person etc., *about, on, over* a subject) disagree or argue. [Latin *exitus*: related to EXIT]

-ist *suffix* forming personal nouns denoting: **1** adherent of a system etc. in *-ism*: (*Marxist*; *fatalist*). **2** person pursuing, using, or concerned with something as an interest or profession (*balloonist*; *tobacconist*). **3** person who does something expressed by a verb in *-ize* (*plagiarist*). **4** person who subscribes to a prejudice or practises discrimination (*racist*; *sexist*). [Greek *-istēs*]

isthmus /ˈɪsməs/ *n.* (*pl.* **-es**) narrow piece of land connecting two larger bodies of land. [Greek *isthmos*]

IT *abbr.* information technology.

it *pron.* (*poss.* **its**; *pl.* **they**) **1** thing (or occasionally an animal or child) previously named or in question (*took a stone and threw it*). **2** person in question (*Who is it? It is I*). **3** as the subject of an impersonal verb (*it is raining*; *it is winter*; *it is two miles to Bath*). **4** as a substitute for a deferred subject or object (*it is silly to talk like that*; *I take it that you agree*). **5** as a substitute for a vague object (*brazen it out*). **6** as the antecedent to a relative word or clause (*it was an owl that I heard*). **7** exactly what is needed. **8** extreme limit of achievement. **9** *colloq.* **a** sexual intercourse. **b** sex appeal. **10** (in children's games) player who has to perform a required feat. □ **that's it** *colloq.* that is: **1** what is required. **2** the difficulty. **3** the end, enough. [Old English]

Italian /ɪˈtæljən/ *–n.* **1 a** native or national of Italy. **b** person of Italian descent. **2** Romance language of Italy. *–adj.* of or relating to Italy.

Italianate /ɪˈtæljəˌneɪt/ *adj.* of Italian style or appearance.

Italian vermouth *n.* sweet kind of vermouth.

italic /ɪˈtælɪk/ *–adj.* **1 a** of the sloping kind of letters now used esp. for emphasis and in foreign words. **b** (of handwriting) compact and pointed like early Italian handwriting. **2** (**Italic**) of ancient Italy. *–n.* **1** letter in italic type. **2** this type. [Latin *italicus*: related to ITALIAN]

italicize /ɪˈtælɪˌsaɪz/ *v.* (also **-ise**) (**-zing** or **-sing**) print in italics.

itch *–n.* **1** irritation in the skin. **2** impatient desire. **3** (prec. by *the*) (in general use) scabies. *–v.* **1** feel an irritation in the skin. **2** feel a desire to do something (*itching to tell you*). [Old English]

itching palm *n.* avarice.

itchy *adj.* (**-ier, -iest**) having or causing an itch. □ **have itchy feet** *colloq.* **1** be restless. **2** have a strong urge to travel. □ **itchiness** *n.*

it'd /ˈɪtəd/ *contr. colloq.* **1** it had. **2** it would.

-ite *suffix* forming nouns meaning 'a person or thing connected with' (*Israelite*; *Trotskyite*; *graphite*; *dynamite*). [Greek *-itēs*]

item /ˈaɪtəm/ *n.* **1** any of a number of enumerated things. **2** separate or distinct piece of news etc. [Latin, = in like manner]

itemize *v.* (also **-ise**) (**-zing** or **-sing**) state item by item. □ **itemization** /-ˈzeɪʃ(ə)n/ *n.*

iterate /ˈɪtəˌreɪt/ *v.* (**-ting**) repeat; state repeatedly. □ **iteration** /-ˈreɪʃ(ə)n/ *n.* **iterative** /-rətɪv/ *adj.* [Latin *iterum* again]

-itic *suffix* forming adjectives and nouns corresponding to nouns in *-ite*, *-itis*, etc. (*Semitic*; *arthritic*). [Latin *-iticus*, Greek *-itikos*]

itinerant /aɪˈtɪnərənt/ *–adj.* travelling from place to place. *–n.* itinerant person. [Latin *iter itiner-* journey]

itinerary /aɪˈtɪnərərɪ/ *n.* (*pl.* **-ies**) **1** detailed route. **2** record of travel. **3** guidebook.

-itis *suffix* forming nouns, esp.: **1** names of inflammatory diseases (*appendicitis*). **2** *colloq.* with ref. to conditions compared to diseases (*electionitis*). [Greek]

it'll /ˈɪt(ə)l/ *contr. colloq.* it will; it shall.

its *poss. pron.* of it; of itself.

it's *contr.* **1** it is. **2** it has.

itself /ɪtˈself/ *pron.* emphatic and refl. form of IT. □ **be itself** see ONESELF. **by itself** see *by oneself*. **in itself** viewed in its essential qualities (*not in itself a bad thing*). [Old English: related to IT, SELF]

ITV *abbr.* Independent Television.

-ity *suffix* forming nouns denoting: **1** quality or condition (*humility*; *purity*). **2** instance of this (*monstrosity*). [Latin *-itas*]

IUD *abbr.* intrauterine (contraceptive) device.

I've /aɪv/ *contr.* I have.

-ive *suffix* forming adjectives meaning 'tending to', and corresponding nouns (*suggestive*; *corrosive*; *palliative*). [Latin *-ivus*]

IVF *abbr. in vitro* fertilization.

ivory /ˈaɪvərɪ/ *n.* (*pl.* **-ies**) **1** hard substance of the tusks of an elephant etc. **2** creamy-white colour of this. **3** (usu. in *pl.*) **a** article made of ivory. **b** *slang* thing made of or resembling ivory, esp. a piano key or a tooth. [Latin *ebur*]

ivory tower *n.* seclusion or withdrawal from the harsh realities of life (often *attrib.*: *ivory tower professors*).

ivy /ˈaɪvɪ/ *n.* (*pl.* **-ies**) climbing evergreen shrub with shiny five-angled leaves. [Old English]

-ize *suffix* (also **-ise**) forming verbs, meaning: **1** make or become such (*Americanize*; *realize*). **2** treat in such a way (*monopolize*; *pasteurize*). **3 a** follow a special practice (*economize*). **b** have a specified feeling (*sympathize*). □ **-ization** /-ˈzeɪʃ(ə)n/ *suffix* forming nouns. [Greek *-izō*]

■ **Usage** The form *-ize* has been in use in English since the 16th c.; it is widely used in American English, but is not an Americanism. The alternative spelling *-ise* (reflecting a French influence) is in common use, esp. in British English, and is obligatory in certain cases: (*a*) where it forms part of a larger word-element, such as *-mise* (= sending) in *compromise*, and *-prise* (= taking) in *surprise*; and (*b*) in verbs corresponding to nouns with *-s-* in the stem, such as *advertise* and *televise*.

J

J[1] /dʒeɪ/ *n.* (also **j**) (*pl.* **Js** or **J's**) tenth letter of the alphabet.

J[2] *abbr.* (also **J.**) joule(s).

jab –*v.* (**-bb-**) **1 a** poke roughly. **b** stab. **2** (foll. by *into*) thrust (a thing) hard or abruptly. –*n.* **1** abrupt blow, thrust, or stab. **2** *colloq.* hypodermic injection. [var. of *job* = prod]

jabber –*v.* **1** chatter volubly. **2** utter (words) in this way. –*n.* chatter; gabble. [imitative]

jabot /'ʒæbəʊ/ *n.* ornamental frill etc. on the front of a shirt or blouse. [French]

jacaranda /,dʒækə'rændə/ *n.* tropical American tree with trumpet-shaped blue flowers or hard scented wood. [Tupi]

jacinth /'dʒæsɪnθ/ *n.* reddish-orange zircon used as a gem. [Latin: related to HYACINTH]

jack –*n.* **1** device for raising heavy objects, esp. vehicles. **2** court-card with a picture of a soldier, page, etc. **3** ship's flag, esp. showing nationality. **4** device using a single-pronged plug to connect an electrical circuit. **5** small white target ball in bowls. **6 a** = JACKSTONE. **b** (in *pl.*) game of jackstones. **7** (**Jack**) familiar form of *John*, esp. typifying the common man, male animal, etc. (*I'm all right, Jack*). –*v.* (usu. foll. by *up*) **1** raise with or as with a jack (in sense 1). **2** *colloq.* raise (e.g. prices). □ **every man jack** every person. **jack in** *slang* abandon (an attempt etc.). [familiar form of the name *John*]

jackal /'dʒæk(ə)l/ *n.* **1** African or Asian wild animal of the dog family, scavenging in packs for food. **2** *colloq.* menial. [Persian]

jackanapes /'dʒækə,neɪps/ *n. archaic* rascal. [earlier *Jack Napes*, supposed to refer to the Duke of Suffolk]

jackass *n.* **1** male ass. **2** stupid person.

jackboot *n.* **1** military boot reaching above the knee. **2** this as a militaristic or fascist symbol.

jackdaw *n.* grey-headed bird of the crow family.

jacket /'dʒækɪt/ *n.* **1 a** short coat with sleeves. **b** protective or supporting garment (*life-jacket*). **2** casing or covering round a boiler etc. **3** = DUST-JACKET. **4** skin of a potato. **5** animal's coat. [French]

jacket potato *n.* potato baked in its skin.

Jack Frost *n.* frost personified.

jack-in-the-box *n.* toy figure that springs out of a box.

jackknife –*n.* **1** large clasp-knife. **2** dive in which the body is bent and then straightened. –*v.* (**-fing**) (of an articulated vehicle) fold against itself in an accident.

jack of all trades *n.* multi-skilled person.

jack-o'-lantern *n.* **1** will-o'-the wisp. **2** pumpkin lantern.

jack plane *n.* medium-sized joinery plane.

jack plug *n.* plug for use with a jack (see JACK *n.* 4).

jackpot *n.* large prize, esp. accumulated in a game, lottery, etc. □ **hit the jackpot** *colloq.* **1** win a large prize. **2** have remarkable luck or success.

jackrabbit *n. US* large prairie hare.

Jack Russell /dʒæk 'rʌs(ə)l/ *n.* short-legged breed of terrier.

jackstone *n.* **1** metal etc. piece used in tossing-games. **2** (in *pl.*) game with a ball and jackstones.

Jack tar *n.* sailor.

Jacobean /,dʒækə'bi:ən/ –*adj.* **1** of the reign of James I. **2** (of furniture) heavy and dark in style. –*n.* Jacobean person. [Latin *Jacobus* James]

Jacobite /'dʒækə,baɪt/ *n. hist.* supporter of James II after his flight, or of the Stuarts.

Jacquard /'dʒækɑ:d/ *n.* **1** apparatus with perforated cards, for weaving figured fabrics. **2** (in full **Jacquard loom**) loom with this. **3** fabric or article so made. [name of its inventor]

Jacuzzi /dʒə'ku:zɪ/ *n.* (*pl.* **-s**) *propr.* large bath with massaging underwater jets of water. [name of its inventor and manufacturers]

jade[1] *n.* **1** hard usu. green stone used for ornaments etc. **2** green colour of jade. [Spanish *ijada* from Latin *ilia* flanks (named as a cure for colic)]

jade[2] *n.* **1** inferior or worn-out horse. **2** *derog.* disreputable woman. [origin unknown]

jaded *adj.* tired out; surfeited.

j'adoube /ʒɑ:'du:b/ *int. Chess* declaration of the intention to adjust a piece without moving it. [French, = I adjust]

jag[1] –*n.* sharp projection of rock etc. –*v.* (**-gg-**) **1** cut or tear unevenly. **2** make indentations in. [imitative]

jag[2] *n. slang* **1** drinking bout. **2** period of indulgence in an activity, emotion, etc. [originally dial., = load]

jagged /'dʒægɪd/ *adj.* **1** unevenly cut or torn. **2** deeply indented. □ **jaggedly** *adv.* **jaggedness** *n.*

jaguar /'dʒægjʊə(r)/ *n.* large American flesh-eating spotted animal of the cat family. [Tupi]

jail (also **gaol**) *–n.* **1** place for the detention of prisoners. **2** confinement in a jail. *–v.* put in jail. [French *jaiole*, ultimately from Latin *cavea* cage]

jailbird *n.* (also **gaolbird**) prisoner or habitual criminal.

jailbreak *n.* (also **gaolbreak**) escape from jail.

jailer *n.* (also **gaoler**) person in charge of a jail or prisoners.

Jain /dʒaɪn/ *–n.* adherent of an Indian religion resembling Buddhism. *–adj.* of this religion. □ **Jainism** *n.* **Jainist** *n.* & *adj.* [Hindi]

jalap /'dʒæləp/ *n.* purgative drug from the tuberous roots of a Mexican climbing plant. [Spanish *Xalapa*, name of a Mexican city, from Aztec]

jalopy /dʒə'lɒpɪ/ *n.* (*pl.* **-ies**) *colloq.* dilapidated old vehicle. [origin unknown]

jalousie /'ʒælʊˌzi:/ *n.* slatted blind or shutter to keep out rain etc. and control light. [French: related to JEALOUSY]

jam[1] *–v.* (**-mm-**) **1 a** (usu. foll. by *into*, *together*, etc.) squeeze, cram, or wedge into a space. **b** become wedged. **2** cause (machinery etc.) to become wedged or (of machinery etc.) become wedged and unworkable. **3 a** block (a passage, road, etc.) by crowding etc. **b** (foll. by *in*) obstruct the exit of (*was jammed in*). **4** (usu. foll. by *on*) apply (brakes etc.) forcefully or abruptly. **5** make (a radio transmission) unintelligible by interference. **6** *colloq.* (in jazz etc.) improvise with other musicians. *–n.* **1** squeeze, crush. **2** crowded mass (*traffic jam*). **3** *colloq.* predicament. **4** stoppage (of a machine etc.) due to jamming. **5** (in full **jam session**) *colloq.* (in jazz etc.) improvised ensemble playing. [imitative]

jam[2] *n.* **1** conserve of boiled fruit and sugar. **2** *colloq.* easy or pleasant thing (*money for jam*). □ **jam tomorrow** promise of future treats etc. that never materialize. [perhaps from JAM[1]]

jamb /dʒæm/ *n.* side post or side face of a doorway, window, or fireplace. [French *jambe* leg, from Latin]

jamboree /ˌdʒæmbə'ri:/ *n.* **1** celebration. **2** large rally of Scouts. [origin unknown]

jamjar *n.* glass jar for jam.

jammy *adj.* (**-ier**, **-iest**) **1** covered with jam. **2** *colloq.* **a** lucky. **b** profitable.

jam-packed *adj. colloq.* full to capacity.

Jan. *abbr.* January.

jangle /'dʒæŋg(ə)l/ *–v.* (**-ling**) **1** (cause to) make a (esp. harsh) metallic sound. **2** irritate (the nerves etc.) by discord etc. *–n.* harsh metallic sound. [French]

janitor /'dʒænɪtə(r)/ *n.* **1** doorkeeper. **2** caretaker. [Latin *janua* door]

January /'dʒænjʊərɪ/ *n.* (*pl.* **-ies**) first month of the year. [Latin *Janus*, guardian god of doors]

Jap *n.* & *adj. colloq.* often *offens.* = JAPANESE. [abbreviation]

japan /dʒə'pæn/ *–n.* hard usu. black varnish, orig. from Japan. *–v.* (**-nn-**) **1** varnish with japan. **2** make black and glossy. [*Japan* in E. Asia]

Japanese /ˌdʒæpə'ni:z/ *–n.* (*pl.* same) **1 a** native or national of Japan. **b** person of Japanese descent. **2** language of Japan. *–adj.* of Japan, its people, or its language.

jape *–n.* practical joke. *–v.* (**-ping**) play a joke. [origin unknown]

japonica /dʒə'pɒnɪkə/ *n.* flowering shrub with bright red flowers and round edible fruits. [Latinized name for *Japanese*]

jar[1] *n.* **1 a** container, usu. of glass and cylindrical. **b** contents of this. **2** *colloq.* glass of beer. [French from Arabic]

jar[2] *–v.* (**-rr-**) **1** (often foll. by *on*) (of sound, manner, etc.) sound discordant, grate (on the nerves etc.). **2 a** (often foll. by *against*, *on*) (cause to) strike (esp. part of the body) with vibration or shock (*jarred his neck*). **b** vibrate with shock etc. **3** (often foll. by *with*) be at variance or in conflict. *–n.* **1** jarring sound or sensation. **2** physical shock or jolt. [imitative]

jar[3] *n.* □ **on the jar** ajar. [obsolete *char* turn: see AJAR, CHAR[2]]

jardinière /ˌʒɑ:dɪ'njeə(r)/ *n.* **1** ornamental pot or stand for plants. **2** dish of mixed vegetables. [French]

jargon /'dʒɑ:gən/ *n.* **1** words or expressions used by a particular group or profession (*medical jargon*). **2** debased or pretentious language. [French]

jasmine /'dʒæzmɪn/ *n.* ornamental shrub with white or yellow flowers. [French from Arabic from Persian]

jasper /'dʒæspə(r)/ *n.* opaque quartz, usu. red, yellow, or brown. [French from Latin from Greek *iaspis*]

jaundice /'dʒɔ:ndɪs/ *–n.* **1** yellowing of the skin etc. caused by liver disease, bile disorder, etc. **2** disordered (esp. mental) vision. **3** envy. *–v.* (**-cing**) **1** affect with jaundice. **2** (esp. as **jaundiced** *adj.*) affect (a person) with envy, resentment, etc. [French *jaune* yellow]

jaunt /dʒɔ:nt/ *–n.* short pleasure trip. *–v.* take a jaunt. [origin unknown]

jaunting car *n.* light horse-drawn vehicle formerly used in Ireland.

jaunty /ˈdʒɔ:ntɪ/ *adj.* (**-ier, -iest**) **1** cheerful and self-confident. **2** sprightly. □ **jauntily** *adv.* **jauntiness** *n.* [French: related to GENTLE]

Javanese /,dʒɑ:vəˈni:z/ *–n.* (*pl.* same) **1 a** native of Java. **b** person of Javanese descent. **2** language of Java. *–adj.* (also **Javan**) of Java, its people, or its language. [*Java* in Indonesia]

javelin /ˈdʒævəlɪn/ *n.* light spear thrown in sport or, formerly, as a weapon. [French]

jaw *–n.* **1 a** upper or lower bony structure in vertebrates containing the teeth. **b** corresponding parts of certain invertebrates. **2 a** (in *pl.*) the mouth with its bones and teeth. **b** narrow mouth of a valley, channel, etc. **c** gripping parts of a tool etc. **d** grip (*jaws of death*). **3** *colloq.* tedious talk (*hold your jaw*). *–v. colloq.* speak, esp. at tedious length. [French]

jawbone *n.* lower jaw in most mammals.

jaw-breaker *n. colloq.* long or hard word.

jay *n.* noisy European bird of the crow family with vivid plumage. [Latin *gaius*, *gaia*, perhaps from the name *Gaius*: cf. *jackdaw*, *robin*]

jaywalk *v.* cross a road carelessly or dangerously. □ **jaywalker** *n.*

jazz *–n.* **1** rhythmic syncopated esp. improvised music of Black US origin. **2** *slang* pretentious talk or behaviour (*all that jazz*). *–v.* play or dance to jazz. □ **jazz up** brighten or enliven. □ **jazzer** *n.* [origin uncertain]

jazzman *n.* jazz-player.

jazzy *adj.* (**-ier, -iest**) **1** of or like jazz. **2** vivid, showy.

JCB *n. propr.* mechanical excavator with a shovel and a digging arm. [*J. C. Bamford*, name of the makers]

JCR *abbr.* Junior Common (or Combination) Room.

jealous /ˈdʒeləs/ *adj.* **1** resentful of rivalry in love. **2** (often foll. by *of*) envious (of a person etc.). **3** (often foll. by *of*) fiercely protective (of rights etc.). **4** (of God) intolerant of disloyalty. **5** (of inquiry, supervision, etc.) vigilant. □ **jealously** *adv.* [medieval Latin *zelosus*: related to ZEAL]

jealousy *n.* (*pl.* **-ies**) **1** jealous state or feeling. **2** instance of this. [French: related to JEALOUS]

jeans /dʒi:nz/ *n.pl.* casual esp. denim trousers. [earlier *geane fustian*, = material from Genoa]

Jeep *n. propr.* small sturdy esp. military vehicle with four-wheel drive. [originally US, from the initials of *general purposes*]

jeepers /ˈdʒi:pəz/ *int. US slang* expressing surprise etc. [corruption of *Jesus*]

jeer *–v.* (often foll. by *at*) scoff derisively; deride. *–n.* taunt. □ **jeeringly** *adv.* [origin unknown]

jehad var. of JIHAD.

Jehovah /dʒəˈhəʊvə/ *n.* Hebrew name of God in the Old Testament. [Hebrew *yahveh*]

Jehovah's Witness *n.* member of a millenarian Christian sect rejecting the supremacy of the State and religious institutions over personal conscience, faith, etc.

jejune /dʒɪˈdʒu:n/ *adj.* **1** intellectually unsatisfying; shallow, meagre, scanty, dry. **2** puerile. **3** (of land) barren. [Latin *jejunus*]

jejunum /dʒɪˈdʒu:nəm/ *n.* small intestine between the duodenum and ileum. [Latin: related to JEJUNE]

Jekyll and Hyde /,dʒekɪl ənd ˈhaɪd/ *n.* person having opposing good and evil personalities. [names of a character in a story by R. L. Stevenson]

jell *v. colloq.* **1** set as jelly. **2** (of ideas etc.) take a definite form; cohere. [back-formation from JELLY]

jellaba var. of DJELLABA.

jellify /ˈdʒelɪ,faɪ/ *v.* (**-ies, -ied**) turn into jelly; make or become like jelly. □ **jellification** /-fɪˈkeɪʃ(ə)n/ *n.*

jelly /ˈdʒelɪ/ *–n.* (*pl.* **-ies**) **1 a** (usu. fruit-flavoured) translucent dessert set with gelatin. **b** similar preparation as a jam, condiment, or sweet (*redcurrant jelly*). **c** similar preparation from meat, bones, etc., and gelatin (*marrowbone jelly*). **2** any similar substance. **3** *slang* gelignite. *–v.* (**-ies, -ied**) (cause to) set as or in a jelly, congeal (*jellied eels*). □ **jelly-like** *adj.* [French *gelée* from Latin *gelo* freeze]

jelly baby *n.* jelly-like baby-shaped sweet.

jellyfish *n.* (*pl.* same or **-es**) marine animal with a jelly-like body and stinging tentacles.

jemmy /ˈdʒemɪ/ *n.* (*pl.* **-ies**) burglar's short crowbar. [from the name *James*]

jenny /ˈdʒenɪ/ *n.* (*pl.* **-ies**) **1** *hist.* = SPINNING-JENNY. **2** female donkey. [from the name *Janet*]

jenny-wren *n.* female wren.

jeopardize /ˈdʒepə,daɪz/ *v.* (also **-ise**) (**-zing** or **-sing**) endanger.

jeopardy /ˈdʒepədɪ/ *n.* danger, esp. severe. [obsolete French *iu parti* divided play]

jerbil var. of GERBIL.

jerboa /dʒɜː'bəʊə/ *n.* small jumping desert rodent. [Arabic]

jeremiad /,dʒerɪ'maɪæd/ *n.* doleful complaint or lamentation. [Church Latin: related to JEREMIAH]

Jeremiah /,dʒerɪ'maɪə/ *n.* dismal prophet, denouncer of the times. [Lamentations of *Jeremiah*, in the Old Testament]

jerk[1] –*n.* **1** sharp sudden pull, twist, twitch, start, etc. **2** spasmodic muscular twitch. **3** (in *pl.*) *colloq.* exercises (*physical jerks*). **4** *slang* fool. –*v.* move, pull, thrust, twist, throw, etc., with a jerk. □ **jerk off** *coarse slang* masturbate. [imitative]

jerk[2] *v.* cure (beef) by cutting it in long slices and drying it in the sun. [Quechua *echarqui* dried fish in strips]

jerkin /'dʒɜːkɪn/ *n.* **1** sleeveless jacket. **2** *hist.* man's close-fitting, esp. leather, jacket. [origin unknown]

jerky *adj.* **(-ier, -iest) 1** moving suddenly or abruptly. **2** spasmodic. □ **jerkily** *adv.* **jerkiness** *n.*

jeroboam /,dʒerə'bəʊəm/ *n.* wine bottle of 4–12 times the ordinary size. [*Jeroboam* in the Old Testament]

Jerry /'dʒerɪ/ *n.* (*pl.* **-ies**) *slang* **1** German (esp. soldier). **2** Germans collectively. [probably an alteration of *German*]

jerry /'dʒerɪ/ *n.* (*pl.* **-ies**) *slang* chamber-pot. [probably an abbreviation of JEROBOAM]

jerry-builder *n.* incompetent builder using cheap materials. □ **jerry-building** *n.* **jerry-built** *adj.* [origin uncertain]

jerrycan *n.* (also **jerrican**) a kind of (orig. German) petrol- or water-can. [from JERRY]

jersey /'dʒɜːzɪ/ *n.* (*pl.* **-s**) **1 a** knitted usu. woollen pullover. **b** plain-knitted (orig. woollen) fabric. **2** (**Jersey**) light brown dairy cow from Jersey. [*Jersey* in the Channel Islands]

Jerusalem artichoke /dʒə'ruːsələm/ *n.* **1** a kind of sunflower with edible tubers. **2** this as a vegetable. [corruption of Italian *girasole* sunflower]

jest –*n.* **1** joke; fun. **2 a** raillery, banter. **b** object of derision. –*v.* joke; fool about. □ **in jest** in fun. [Latin *gesta* exploits]

jester *n. hist.* professional clown at a medieval court etc.

Jesuit /'dʒezjʊɪt/ *n.* member of the Society of Jesus, a Roman Catholic order. [Latin *Jesus*, founder of the Christian religion]

Jesuitical /,dʒezjʊ'ɪtɪk(ə)l/ *adj.* **1** of the Jesuits. **2** often *offens.* equivocating, casuistic.

Jesus /'dʒiːzəs/ *int. colloq.* exclamation of surprise, dismay, etc. [name of the founder of the Christian religion]

jet[1] –*n.* **1** stream of water, gas, flame, etc., shot esp. from a small opening. **2** spout or nozzle for this purpose. **3** jet engine or jet plane. –*v.* **(-tt-) 1** spurt out in jets. **2** *colloq.* send or travel by jet plane. [French *jeter* throw from Latin *jacto*]

jet[2] *n.* (often *attrib.*) hard black lignite often carved and highly polished. [French *jaiet* from *Gagai* in Asia Minor]

jet black *adj.* & *n.* (as adj. often hyphenated) deep glossy black.

jet engine *n.* engine using jet propulsion, esp. of an aircraft.

jet lag *n.* exhaustion etc. felt after a long flight across time zones.

jet plane *n.* plane with a jet engine.

jet-propelled *adj.* **1** having jet propulsion. **2** very fast.

jet propulsion *n.* propulsion by the backward ejection of a high-speed jet of gas etc.

jetsam /'dʒetsəm/ *n.* objects washed ashore, esp. jettisoned from a ship. [contraction of JETTISON]

jet set *n.* wealthy people who travel widely, esp. for pleasure. □ **jet-setter** *n.* **jet-setting** *n.* & *attrib. adj.*

jettison /'dʒetɪs(ə)n/ –*v.* **1 a** throw (esp. heavy material) overboard to lighten a ship etc. **b** drop (goods) from an aircraft. **2** abandon; get rid of. –*n.* jettisoning. [Anglo-French *getteson*: related to JET[1]]

jetty /'dʒetɪ/ *n.* (*pl.* **-ies**) **1** pier or breakwater to protect or defend a harbour, coast, etc. **2** landing-pier. [French *jetee*: related to JET[1]]

Jew /dʒuː/ *n.* **1** person of Hebrew descent or whose religion is Judaism. **2** *slang offens.* miserly person. [Greek *ioudaios*]

■ **Usage** The stereotype conveyed in sense 2 is deeply offensive. It arose from historical associations of Jews as moneylenders in medieval England.

---

jewel /'dʒuːəl/ –*n.* **1 a** precious stone. **b** this used in watchmaking. **2** jewelled personal ornament. **3** precious person or thing. –*v.* (**-ll-**; *US* **-l-**) (esp. as **jewelled** *adj.*) adorn or set with jewels. [French]

jeweller *n.* (*US* **jeweler**) maker of or dealer in jewels or jewellery.

jewellery /'dʒuːəlrɪ/ *n.* (also **jewelry**) rings, brooches, necklaces, etc., regarded collectively.

Jewess /'dʒuːes/ *n.* often *offens.* woman or girl of Hebrew descent or whose religion is Judaism.

Jewish *adj.* **1** of Jews. **2** of Judaism. □ **Jewishness** *n.*

Jewry /'dʒʊərɪ/ *n.* Jews collectively.

jew's harp *n.* small musical instrument held between the teeth.

Jezebel /'dʒezə,bel/ *n.* shameless or immoral woman. [*Jezebel* in the Old Testament]

jib[1] *n.* **1** triangular staysail. **2** projecting arm of a crane. [origin unknown]

jib[2] *v.* (**-bb-**) **1** (esp. of a horse) stop and refuse to go on. **2** (foll. by *at*) show aversion to. □ **jibber** *n.* [origin unknown]

jibe[1] var. of GIBE.

jibe[2] *US* var. of GYBE.

jiff *n.* (also **jiffy**, *pl.* **-ies**) *colloq.* short time; moment (*in a jiffy*). [origin unknown]

Jiffy bag /'dʒɪfɪ/ *n. propr.* padded envelope.

jig *–n.* **1 a** lively leaping dance. **b** music for this. **2** device that holds a piece of work and guides the tools operating on it. *–v.* (**-gg-**) **1** dance a jig. **2** (often foll. by *about*) move quickly and jerkily up and down; fidget. **3** work on or equip with a jig or jigs. [origin unknown]

jigger /'dʒɪgə(r)/ *n.* **1** *Billiards colloq.* cue-rest. **2 a** measure of spirits etc. **b** small glass holding this. [partly from JIG]

jiggered /'dʒɪgəd/ *adj. colloq.* (as a mild oath) confounded (*I'll be jiggered*). [euphemism]

jiggery-pokery /,dʒɪgərɪ'pəʊkərɪ/ *n. colloq.* trickery; swindling. [origin uncertain]

jiggle /'dʒɪg(ə)l/ *–v.* (**-ling**) (often foll. by *about* etc.) shake or jerk lightly; fidget. *–n.* light shake. [from JIG]

jigsaw *n.* **1 a** (in full **jigsaw puzzle**) picture on board or wood etc. cut into irregular interlocking pieces to be reassembled as a pastime. **b** problem consisting of various pieces of information. **2** mechanical fretsaw with a fine blade.

jihad /dʒɪ'hæd/ *n.* (also **jehad**) Muslim holy war against unbelievers. [Arabic *jihād*]

jilt *v.* abruptly reject or abandon (esp. a lover). [origin unknown]

Jim Crow /'krəʊ/ *n. US colloq.* **1** segregation of Blacks. **2** *offens.* a Black. [nickname]

jim-jams /'dʒɪmdʒæmz/ *n.pl.* **1** *slang* = DELIRIUM TREMENS. **2** *colloq.* nervousness; depression. [fanciful reduplication]

jingle /'dʒɪŋg(ə)l/ *–n.* **1** mixed ringing or clinking noise. **2 a** repetition of sounds in a phrase etc. **b** short catchy verse or song in advertising etc. *–v.* (**-ling**) **1** (cause to) make a jingling sound. **2** (of writing) be full of alliteration, rhymes, etc. [imitative]

jingo /'dʒɪŋgəʊ/ *n.* (*pl.* **-es**) supporter of war; blustering patriot. □ **by jingo!** mild oath. □ **jingoism** *n.* **jingoist** *n.* **jingoistic** /-'ɪstɪk/ *adj.* [conjuror's word]

jink *–v.* **1** move elusively; dodge. **2** elude by dodging. *–n.* dodging or eluding. [originally Scots: imitative]

jinnee /dʒɪ'ni:/ *n.* (also **jinn**, **djinn** /dʒɪn/) (*pl.* **jinn** or **djinn**) (in Muslim mythology) spirit in human or animal form having power over people. [Arabic]

jinx *colloq.* *–n.* person or thing that seems to cause bad luck. *–v.* (esp. as **jinxed** *adj.*) subject to bad luck. [perhaps var. of *jynx* wryneck, charm]

jitter *colloq.* *–n.* (**the jitters**) extreme nervousness. *–v.* be nervous; act nervously. □ **jittery** *adj.* **jitteriness** *n.* [origin unknown]

jitterbug *–n.* **1** nervous person. **2** *hist.* fast popular dance. *–v.* (**-gg-**) *hist.* dance the jitterbug.

jiu-jitsu var. of JU-JITSU.

jive *–n.* **1** lively dance popular esp. in the 1950s. **2** music for this. *–v.* (**-ving**) dance to or play jive music. □ **jiver** *n.* [origin uncertain]

Jnr. *abbr.* Junior.

job *–n.* **1** piece of work to be done; task. **2** position in, or piece of, paid employment. **3** *colloq.* difficult task (*had a job to find it*). **4** *slang* crime, esp. a robbery. **5** state of affairs etc. (*bad job*). *–v.* (**-bb-**) **1** do jobs; do piece-work. **2** deal in stocks; buy and sell (stocks or goods). **3** deal corruptly with (a matter). □ **just the job** *colloq.* exactly what is wanted. **make a job** (or **good job**) **of** do well. **on the job** *colloq.* **1** at work. **2** engaged in sexual intercourse. **out of a job** unemployed. [origin unknown]

jobber *n.* **1** person who jobs. **2** *hist.* principal or wholesaler on the Stock Exchange.

■ **Usage** Up to Oct. 1986 jobbers were permitted to deal only with brokers, not directly with the public. From Oct. 1986 the name ceased to be in official use (see BROKER 2).

jobbery *n.* corrupt dealing.

jobbing *attrib. adj.* freelance; pieceworking (*jobbing gardener*).

jobcentre *n.* local government office advertising available jobs.

job-hunt *v. colloq.* seek employment.

jobless *adj.* unemployed. □ **joblessness** *n.*

job lot *n.* mixed lot bought at auction etc.

Job's comforter /dʒəʊbz/ *n.* person who intends to comfort but increases distress. [*Job* in the Old Testament]

jobs for the boys *n.pl. colloq.* appointments for members of one's own group etc.

job-sharing *n.* sharing of a full-time job by two or more people. □ **job-share** *n.* & *v.*

jobsheet *n.* sheet for recording details of jobs done.

Jock *n. slang* Scotsman. [Scots form of the name *Jack*]

jockey /'dʒɒkɪ/ —*n.* (*pl.* **-s**) rider in horse-races, esp. professional. —*v.* (**-eys, -eyed**) **1** trick, cheat, or outwit. **2** (foll. by *away*, *out*, *into*, etc.) manoeuvre (a person). □ **jockey for position** manoeuvre for advantage. [diminutive of JOCK]

jockstrap *n.* support or protection for the male genitals, worn esp. in sport. [slang *jock* genitals]

jocose /dʒə'kəʊs/ *adj.* playful; jocular. □ **jocosely** *adv.* **jocosity** /-'kɒsɪtɪ/ *n.* (*pl.* **-ies**). [Latin *jocus* jest]

jocular /'dʒɒkjʊlə(r)/ *adj.* **1** fond of joking. **2** humorous. □ **jocularity** /-'lærɪtɪ/ *n.* (*pl.* **-ies**). **jocularly** *adv.*

jocund /'dʒɒkənd/ *adj. literary* merry, cheerful. □ **jocundity** /dʒə'kʌndɪtɪ/ *n.* (*pl.* **-ies**). **jocundly** *adv.* [French from Latin *jucundus* pleasant]

jodhpurs /'dʒɒdpəz/ *n.pl.* riding breeches tight below the knee. [*Jodhpur* in India]

Joe Bloggs *n. colloq.* hypothetical average man.

jog —*v.* (**-gg-**) **1** run slowly, esp. as exercise. **2** push or jerk, esp. unsteadily. **3** nudge, esp. to alert. **4** stimulate (the memory). **5** (often foll. by *on*, *along*) trudge; proceed ploddingly (*must jog on somehow*). **6** (of a horse) trot. —*n.* **1** spell of jogging; slow walk or trot. **2** push, jerk, or nudge. [probably imitative]

jogger *n.* person who jogs, esp. for exercise.

joggle /'dʒɒg(ə)l/ —*v.* (**-ling**) move in jerks. —*n.* slight shake.

jogtrot *n.* slow regular trot.

john /dʒɒn/ *n. US slang* lavatory. [from the name *John*]

John Bull /dʒɒn/ *n.* England or the typical Englishman. [name of a character in an 18th-c. satire]

John Dory /dʒɒn 'dɔ:rɪ/ *n.* (*pl.* same or **-ies**) edible marine fish. [see DORY]

johnny /'dʒɒnɪ/ *n.* (*pl.* **-ies**) **1** *slang* condom. **2** *colloq.* fellow; man. [diminutive of *John*]

johnny-come-lately *n. colloq.* newcomer; upstart.

joie de vivre /ˌʒwɑ: də 'vi:vrə/ *n.* exuberance; high spirits. [French, = joy of living]

join —*v.* **1** (often foll. by *to*, *together*) put together; fasten, unite (with one or several things or people). **2** connect (points) by a line etc. **3** become a member of (a club, organization, etc.). **4 a** take one's place with (a person, group, etc.). **b** (foll. by *in*, *for*, etc.) take part with (others) in an activity etc. (*joined them in prayer*). **5** (often foll. by *with*, *to*) come together; be united. **6** (of a river etc.) be or become connected or continuous with. —*n.* point, line, or surface at which things are joined. □ **join battle** begin fighting. **join forces** combine efforts. **join hands 1** clasp hands. **2** combine in an action etc. **join in** (also *absol.*) take part in (an activity). **join up 1** enlist for military service. **2** (often foll. by *with*) unite, connect. [Latin *jungo junct-*]

joiner *n.* **1** maker of finished wood fittings. **2** *colloq.* person who joins an organization or who readily joins societies etc. □ **joinery** *n.* (in sense 1).

joint —*n.* **1** place at which two or more things or parts of a structure are joined; device for joining these. **2** point at which two bones fit together. **3** division of an animal carcass as meat. **4** *slang* restaurant, bar, etc. **5** *slang* marijuana cigarette. **6** *Geol.* crack in rock. —*adj.* **1** held, done by, or belonging to, two or more persons etc. (*joint mortgage*; *joint action*). **2** sharing with another (*joint author*; *joint favourite*). —*v.* **1** connect by joint(s). **2** divide at a joint or into joints. □ **out of joint 1** (of a bone) dislocated. **2** out of order. □ **jointly** *adv.* [French: related to JOIN]

joint stock *n.* capital held jointly; common fund.

joint-stock company *n.* company formed on the basis of a joint stock.

jointure /'dʒɔɪntʃə(r)/ —*n.* estate settled on a wife by her husband for use after his death. —*v.* provide with a jointure. [Latin: related to JOIN]

joist *n.* supporting beam in a floor, ceiling, etc. [French *giste* from Latin *jaceo* lie]

jojoba /həʊ'həʊbə/ *n.* plant with seeds yielding an oily extract used in cosmetics etc. [Mexican Spanish]

joke —*n.* **1** thing said or done to cause laughter; witticism. **2** ridiculous person or thing. —*v.* (**-king**) make jokes; tease (*only joking*). □ **no joke** *colloq.* serious matter. □ **jokingly** *adv.* **joky** *adj.* (also **jokey**). **jokily** *adv.* **jokiness** *n.* [probably Latin *jocus* jest]

joker *n.* **1** person who jokes. **2** *slang* person. **3** playing-card used in some games.

jollify /ˈdʒɒlɪ,faɪ/ *v.* (**-ies**, **-ied**) make merry. □ **jollification** /-fɪˈkeɪʃ(ə)n/ *n.*

jollity /ˈdʒɒlɪtɪ/ *n.* (*pl.* **-ies**) merry-making; festivity. [French *joliveté*: related to JOLLY[1]]

jolly[1] /ˈdʒɒlɪ/ *–adj.* (**-ier**, **-iest**) **1** cheerful; merry. **2** festive, jovial. **3** *colloq.* pleasant, delightful. *–adv. colloq.* very. *–v.* (**-ies**, **-ied**) (usu. foll. by *along*) *colloq.* coax or humour in a friendly way. *–n.* (*pl.* **-ies**) *colloq.* party or celebration. □ **jollily** *adv.* **jolliness** *n.* [French *jolif* gay, pretty: perhaps related to YULE]

jolly[2] /ˈdʒɒlɪ/ *n.* (*pl.* **-ies**) (in full **jolly boat**) clinker-built ship's boat smaller than a cutter. [origin unknown: perhaps related to YAWL]

Jolly Roger *n.* pirates' black flag, usu. with skull and crossbones.

jolt /dʒəʊlt/ *–v.* **1** disturb or shake (esp. in a moving vehicle) with a jerk. **2** shock; perturb. **3** move along jerkily. *–n.* **1** jerk. **2** surprise or shock. □ **jolty** *adj.* (**-ier**, **-iest**). [origin unknown]

Jonah /ˈdʒəʊnə/ *n.* person who seems to bring bad luck. [*Jonah* in the Old Testament]

jonquil /ˈdʒɒŋkwɪl/ *n.* narcissus with small fragrant yellow or white flowers. [ultimately from Latin *juncus* rush plant]

josh *slang –v.* **1** tease, banter. **2** indulge in ridicule. *–n.* good-natured or teasing joke. [origin unknown]

joss *n.* Chinese idol. [ultimately from Latin *deus* god]

joss-stick *n.* incense-stick for burning.

jostle /ˈdʒɒs(ə)l/ *–v.* (**-ling**) **1** (often foll. by *away*, *from*, *against*, etc.) push against; elbow, esp. roughly or in a crowd. **2** (foll. by *with*) struggle roughly. *–n.* jostling. [from JOUST]

jot *–v.* (**-tt-**) (usu. foll. by *down*) write briefly or hastily. *–n.* very small amount (*not one jot*). [Greek IOTA]

jotter *n.* small pad or notebook.

jotting *n.* (usu. in *pl.*) jotted note.

joule /dʒu:l/ *n.* SI unit of work or energy. [*Joule*, name of a physicist]

journal /ˈdʒɜ:n(ə)l/ *n.* **1** newspaper or periodical. **2** daily record of events; diary. **3** book in which transactions and accounts are entered. **4** part of a shaft or axle that rests on bearings. [Latin *diurnalis* DIURNAL]

journalese /,dʒɜ:nəˈli:z/ *n.* hackneyed writing characteristic of newspapers.

journalism *n.* profession of writing for or editing newspapers etc.

journalist *n.* person writing for or editing newspapers etc. □ **journalistic** /-ˈlɪstɪk/ *adj.*

journey /ˈdʒɜ:nɪ/ *–n.* (*pl.* **-s**) **1** act of going from one place to another, esp. at a long distance. **2** time taken for this (*a day's journey*). *–v.* (**-s**, **-ed**) make a journey. [French *jornee* day, day's work or travel, from Latin *diurnus* daily]

journeyman *n.* **1** qualified mechanic or artisan who works for another. **2** *derog.* reliable but not outstanding worker.

joust /dʒaʊst/ *hist. –n.* combat between two knights on horseback with lances. *–v.* engage in a joust. □ **jouster** *n.* [French *jouste* from Latin *juxta* near]

Jove *n.* (in Roman mythology) Jupiter. □ **by Jove!** exclamation of surprise etc. [Latin *Jupiter Jov-*]

jovial /ˈdʒəʊvɪəl/ *adj.* merry, convivial, hearty. □ **joviality** /-ˈælɪtɪ/ *n.* **jovially** *adv.* [Latin *jovialis*: related to JOVE]

jowl[1] *n.* **1** jaw or jawbone. **2** cheek (*cheek by jowl*). [Old English]

jowl[2] *n.* loose hanging skin on the throat or neck. □ **jowly** *adj.* [Old English]

joy *n.* **1** (often foll. by *at*, *in*) pleasure; extreme gladness. **2** thing causing joy. **3** *colloq.* satisfaction, success (*got no joy*). □ **joyful** *adj.* **joyfully** *adv.* **joyfulness** *n.* **joyless** *adj.* **joyous** *adj.* **joyously** *adv.* [French *joie* from Latin *gaudium*]

joyride *colloq. –n.* pleasure ride in esp. a stolen car. *–v.* (**-ding**; *past* **-rode**; *past part* **-ridden**) go for a joyride. □ **joyrider** *n.*

joystick *n.* **1** *colloq.* control column of an aircraft. **2** lever controlling movement of an image on a VDU screen etc.

JP *abbr.* Justice of the Peace.

Jr. *abbr.* Junior.

jubilant /ˈdʒu:bɪlənt/ *adj.* exultant, rejoicing. □ **jubilance** *n.* **jubilantly** *adv.* **jubilation** /-ˈleɪʃ(ə)n/ *n.* [Latin *jubilo* shout]

jubilee /ˈdʒu:bɪ,li:/ *n.* **1** anniversary, esp. the 25th or 50th. **2** time of rejoicing. [Hebrew, ultimately, = ram's-horn trumpet]

Judaic /dʒu:ˈdeɪɪk/ *adj.* of or characteristic of the Jews. [Greek: related to JEW]

Judaism /ˈdʒu:deɪ,ɪz(ə)m/ *n.* religion of the Jews.

Judas /ˈdʒu:dəs/ *n.* traitor. [*Judas* Iscariot who betrayed Christ]

judder *–v.* shake noisily or violently. *–n.* juddering. [imitative: cf. *shudder*]

judge /dʒʌdʒ/ *–n.* **1** public official appointed to hear and try legal cases. **2** person appointed to decide in a contest, dispute, etc. **3 a** person who decides a question. **b** person regarded as having judgement of a specified type (*am no *judge; good judge of art*). –*v.* (**-ging**) **1** form an opinion or judgement (about); estimate, appraise. **2** act as a judge (of). **3** **a** try (a case) at law. **b** pronounce sentence on. **4** (often foll. by *to* + infin. or *that* + clause) conclude, consider. [Latin *judex judic-*]

judgement *n.* (also **judgment**) **1** critical faculty; discernment (*error of judgement*). **2** good sense. **3** opinion or estimate (*in my judgement*). **4** sentence of a court of justice. **5** often *joc.* deserved misfortune. □ **against one's better judgement** contrary to what one really feels to be advisable.

judgemental /dʒʌdʒ'ment(ə)l/ *adj.* (also **judgmental**) **1** of or by way of judgement. **2** condemning, critical. □ **judgementally** *adv.*

Judgement Day *n.* (in Judaism, Christianity, and Islam) day on which mankind will be judged by God.

judicature /'dʒu:dɪkətʃə(r)/ *n.* **1** administration of justice. **2** judge's position. **3** judges collectively. [medieval Latin *judico* judge]

judicial /dʒu:'dɪʃ(ə)l/ *adj.* **1** of, done by, or proper to a court of law. **2** having the function of judgement (*judicial assembly*). **3** of or proper to a judge. **4** impartial. □ **judicially** *adv.* [Latin *judicium* judgement]

judiciary /dʒu:'dɪʃərɪ/ *n.* (*pl.* **-ies**) judges of a State collectively.

judicious /dʒu:'dɪʃəs/ *adj.* sensible, prudent. □ **judiciously** *adv.*

judo /'dʒu:dəʊ/ *n.* sport derived from ju-jitsu. [Japanese, = gentle way]

jug –*n.* **1** deep vessel for liquids, with a handle and a lip for pouring. **2** contents of this. **3** *slang* prison. –*v.* (**-gg-**) (usu. as **jugged** *adj.*) stew or boil (esp. hare) in a casserole etc. □ **jugful** *n.* (*pl.* **-s**). [origin uncertain]

juggernaut /'dʒʌgə,nɔ:t/ *n.* **1** large heavy lorry etc. **2** overwhelming force or object. [Hindi *Jagannath*, = lord of the world]

juggle /'dʒʌg(ə)l/ –*v.* (**-ling**) **1** **a** (often foll. by *with*) keep several objects in the air at once by throwing and catching. **b** perform such feats with (balls etc.). **2** deal with (several activities) at once. **3** (often foll. by *with*) misrepresent or rearrange (facts) adroitly. –*n.* **1** juggling. **2** fraud. □ **juggler** *n.* [French from Latin *jocus* jest]

Jugoslav var. of YUGOSLAV.

jugular /'dʒʌgjʊlə(r)/ –*adj.* of the neck or throat. –*n.* = JUGULAR VEIN. [Latin *jugulum* collar-bone]

jugular vein *n.* any of several large veins in the neck carrying blood from the head.

juice /dʒu:s/ *n.* **1** liquid part of vegetables or fruits. **2** animal fluid, esp. a secretion (*gastric juice*). **3** *colloq.* petrol; electricity. [French from Latin]

juicy *adj.* (**-ier, -iest**) **1** full of juice; succulent. **2** *colloq.* interesting; racy, scandalous. **3** *colloq.* profitable. □ **juicily** *adv.* **juiciness** *n.*

ju-jitsu /dʒu:'dʒɪtsu:/ *n.* (also **jiu-jitsu, ju-jutsu**) Japanese system of unarmed combat and physical training. [Japanese *jūjutsu* gentle skill]

ju-ju /'dʒu:dʒu:/ *n.* **1** charm or fetish of some W. African peoples. **2** supernatural power attributed to this. [perhaps French *joujou* toy]

jujube /'dʒu:dʒu:b/ *n.* small flavoured jelly-like lozenge. [Greek *zizuphon*]

ju-jutsu var. of JU-JITSU.

jukebox /'dʒu:kbɒks/ *n.* coin-operated record-playing machine. [Black *juke* disorderly]

Jul. *abbr.* July.

julep /'dʒu:lep/ *n.* **1** **a** sweet drink, esp. as a vehicle for medicine. **b** medicated drink as a mild stimulant etc. **2** *US* iced and flavoured spirits and water (*mint julep*). [Persian *gulāb* rose-water]

Julian /'dʒu:lɪən/ *adj.* of Julius Caesar. [Latin *Julius*]

Julian calendar *n.* calendar introduced by Julius Caesar, with a year of 365 days, every fourth year having 366.

julienne /,dʒu:lɪ'en/ –*n.* vegetables cut into short thin strips. –*adj.* cut into thin strips. [French from name *Jules* or *Julien*]

Juliet cap /'dʒu:lɪət/ *n.* small net skull-cap worn by brides etc. [*Juliet* in Shakespeare's *Romeo & Juliet*]

July /dʒu:'laɪ/ *n.* (*pl.* **Julys**) seventh month of the year. [Latin *Julius* Caesar]

jumble /'dʒʌmb(ə)l/ –*v.* (**-ling**) (often foll. by *up*) confuse; mix up; muddle. –*n.* **1** confused state or heap; muddle. **2** articles in a jumble sale. [probably imitative]

jumble sale *n.* sale of second-hand articles, esp. for charity.

jumbo /'dʒʌmbəʊ/ *n.* (*pl.* **-s**) *colloq.* **1** (*often attrib.*) large animal (esp. an elephant), person, or thing (*jumbo packet*). **2** (in full **jumbo jet**) large airliner for several hundred passengers. [probably from MUMBO-JUMBO]

■ **Usage** In sense 2, *jumbo* is usu. applied specifically to the Boeing 747.

jump –*v.* **1** rise off the ground etc. by sudden muscular effort in the legs. **2** (often foll. by *up, from, in, out,* etc.) move suddenly or hastily (*jumped into the car*). **3** jerk or twitch from shock or excitement etc. **4** **a** change, esp. advance in status or rise, rapidly (*prices jumped*). **b** cause to do this. **5** (often foll. by *about*) change the subject etc. rapidly. **6** pass over (an obstacle etc.) by jumping. **7** skip (a passage in a book etc.). **8** cause (a horse etc.) to jump. **9** (foll. by *to*, *at*) reach (a conclusion) hastily. **10** (of a train) leave (the rails). **11** pass (a red traffic-light etc.). **12** get on or off (a train etc.) quickly, esp. illegally or dangerously. **13** attack (a person) unexpectedly. *–n.* **1** act of jumping. **2** sudden jerk caused by shock or excitement. **3** abrupt rise in amount, value, status, etc. **4** obstacle to be jumped. **5 a** sudden transition. **b** gap in a series, logical sequence, etc. □ **jump at** accept eagerly. **jump bail** fail to appear for trial having been released on bail. **jump down a person's throat** *colloq.* reprimand or contradict a person fiercely. **jump the gun** *colloq.* begin prematurely. **jump on** *colloq.* attack or criticize severely. **jump out of one's skin** *colloq.* be extremely startled. **jump the queue** take unfair precedence. **jump ship** (of a seaman) desert. **jump to it** *colloq.* act promptly. **one jump ahead** one stage further on than a rival etc. [imitative]

jumped-up *adj. colloq.* upstart.

jumper[1] *n.* **1** knitted pullover. **2** loose outer jacket worn by sailors. **3** *US* pinafore dress. [probably *jump* short coat]

jumper[2] *n.* **1** person or animal that jumps. **2** short wire used to make or break an electrical circuit.

jumping bean *n.* seed of a Mexican plant that jumps with the movement of a larva inside.

jump-jet *n.* vertical take-off jet aircraft.

jump-lead *n.* cable for conveying current from the battery of one vehicle to that of another.

jump-off *n.* deciding round in show-jumping.

jump-start *–v.* start (a vehicle) by pushing it or with jump-leads. *–n.* act of jump-starting.

jump suit *n.* one-piece garment for the whole body.

jumpy *adj.* (**-ier, -iest**) **1** nervous; easily startled. **2** making sudden movements. □ **jumpiness** *n.*

Jun. *abbr.* **1** June. **2** Junior.

junction /'dʒʌŋkʃ(ə)n/ *n.* **1** joint; joining-point. **2** place where railway lines or roads meet. **3** joining. [Latin: related to JOIN]

junction box *n.* box containing a junction of electric cables etc.

juncture /'dʒʌŋktʃə(r)/ *n.* **1** critical convergence of events; point of time (*at this juncture*). **2** joining-point. **3** joining.

June *n.* sixth month of the year. [Latin *Junius* from *Juno*, name of a goddess]

Jungian /'jʊŋɪən/ *–adj.* of the Swiss psychologist Carl Jung or his theories. *–n.* supporter of Jung or of his theories.

jungle /'dʒʌŋg(ə)l/ *n.* **1 a** land overgrown with tangled vegetation, esp. in the tropics. **b** an area of this. **2** wild tangled mass. **3** place of bewildering complexity, confusion, or struggle. □ **law of the jungle** state of ruthless competition. □ **jungly** *adj.* [Hindi from Sanskrit]

junior /'dʒu:nɪə(r)/ *–adj.* **1** (often foll. by *to*) inferior in age, standing, or position. **2** the younger (esp. appended to the name of a son for distinction from his father). **3** of the lower or lowest position (*junior partner*). **4** (of a school) for younger pupils, usu. aged 7–11. *–n.* **1** junior person. **2** person at the lowest level (in an office etc.). [Latin, comparative of *juvenis* young]

junior common room *n.* (also **junior combination room**) **1** common-room for undergraduates in a college. **2** undergraduates of a college collectively.

juniper /'dʒu:nɪpə(r)/ *n.* evergreen shrub or tree with prickly leaves and dark-purple berry-like cones. [Latin *juniperus*]

junk[1] *–n.* **1** discarded articles; rubbish. **2** anything regarded as of little value. **3** *slang* narcotic drug, esp. heroin. *–v.* discard as junk. [origin unknown]

junk[2] *n.* flat-bottomed sailing-vessel in the China seas. [Javanese *djong*]

junk bond *n.* bond bearing high interest but deemed to be a risky investment.

junket /'dʒʌŋkɪt/ *–n.* **1** pleasure outing. **2** official's tour at public expense. **3** sweetened and flavoured milk curds. **4** feast. *–v.* (**-t-**) feast, picnic. [French *jonquette* rush-basket (used for junket 3 and 4), from Latin *juncus* rush]

junk food *n.* food, such as sweets and crisps, with low nutritional value.

junkie *n. slang* drug addict.

junk mail *n.* unsolicited advertising matter sent by post.

junk shop *n.* second-hand or cheap antiques shop.

junta /'dʒʌntə/ *n.* (usu. military) clique taking power in a *coup d'état*. [Spanish: related to JOIN]

jural /'dʒʊər(ə)l/ *adj.* **1** of law. **2** of rights and obligations. [Latin *jus jur-* law, right]

Jurassic /dʒʊə'ræsɪk/ *Geol. –adj.* of the second period of the Mesozoic era. *–n.* this era or system. [French from *Jura* mountains]

juridical /dʒʊə'rɪdɪk(ə)l/ *adj.* **1** of judicial proceedings. **2** relating to the law. [Latin *jus jur-* law, *dico* say]

jurisdiction /ˌdʒʊərɪs'dɪkʃ(ə)n/ *n.* **1** (often foll. by *over*, *of*) administration of justice. **2 a** legal or other authority. **b** extent of this; territory it extends over. □ **jurisdictional** *adj.*

jurisprudence /ˌdʒʊərɪs'pru:d(ə)ns/ *n.* science or philosophy of law. □ **jurisprudential** /-'denʃ(ə)l/ *adj.*

jurist /'dʒʊərɪst/ *n.* expert in law. □ **juristic** /-'rɪstɪk/ *adj.*

juror /'dʒʊərə(r)/ *n.* **1** member of a jury. **2** person taking an oath.

jury /'dʒʊərɪ/ *n.* (*pl.* **-ies**) **1** body of usu. twelve people giving a verdict in a court of justice. **2** body of people awarding prizes in a competition.

jury-box *n.* enclosure for the jury in a lawcourt.

jury-rigged /'dʒʊərɪrɪgd/ *adj. Naut.* having temporary makeshift rigging. [origin uncertain]

just *–adj.* **1** morally right or fair. **2** (of treatment etc.) deserved (*just reward*). **3** well-grounded; justified (*just anger*). **4** right in amount etc.; proper. *–adv.* **1** exactly (*just what I need*). **2** a little time ago; very recently (*has just seen them*). **3** *colloq.* simply, merely (*just good friends*; *just doesn't make sense*). **4** barely; no more than (*just managed it*). **5** *colloq.* positively; indeed (*just splendid*; *won't I just tell him!*). **6** quite (*not just yet*). □ **just about** *colloq.* almost exactly; almost completely. **just in case** as a precaution. **just now 1** at this moment. **2** a little time ago. **just the same** = *all the same*. **just so 1** exactly arranged (*everything just so*). **2** it is exactly as you say. □ **justly** *adv.* **justness** *n.* [Latin *justus* from *jus* right]

justice /'dʒʌstɪs/ *n.* **1** justness, fairness. **2** authority exercised in the maintenance of right. **3** judicial proceedings (*brought to justice*; *Court of Justice*). **4** magistrate; judge. □ **do justice to 1** treat fairly. **2** appreciate properly. **do oneself justice** perform at one's best. **with justice** reasonably. [Latin *justitia*]

Justice of the Peace *n.* unpaid lay magistrate appointed to hear minor cases.

justifiable /'dʒʌstɪˌfaɪəb(ə)l/ *adj.* able to be justified. □ **justifiably** *adv.*

justify /'dʒʌstɪˌfaɪ/ *v.* (**-ies**, **-ied**) **1** show the justice or correctness of (a person, act, assertion, etc.). **2** (esp. in *passive*) cite or constitute adequate grounds for (conduct, a claim, etc.); vindicate. **3** (as **justified** *adj.*) just, right (*justified in assuming*). **4** *Printing* adjust (a line of type) to give even margins. □ **justification** /-fɪ'keɪʃ(ə)n/ *n.* **justificatory** /-fɪˌkeɪtərɪ/ *adj.*

jut *–v.* (**-tt-**) (often foll. by *out*, *forth*) protrude, project. *–n.* projection. [var. of JET[1]]

jute *n.* **1** fibre from the bark of an E. Indian plant, used esp. for sacking, mats, etc. **2** plant yielding this. [Bengali]

juvenile /'dʒu:vəˌnaɪl/ *–adj.* **1 a** youthful. **b** of or for young people. **2** often *derog.* immature (*juvenile behaviour*). *–n.* **1** young person. **2** actor playing a juvenile part. [Latin *juvenis* young]

juvenile court *n.* court for children under 17.

juvenile delinquency *n.* offences committed by people below the age of legal responsibility. □ **juvenile delinquent** *n.*

juvenilia /ˌdʒu:və'nɪlɪə/ *n.pl.* author's or artist's youthful works.

juxtapose /ˌdʒʌkstə'pəʊz/ *v.* (**-sing**) **1** place (things) side by side. **2** (foll. by *to*, *with*) place (a thing) beside another. □ **juxtaposition** /-pə'zɪʃ(ə)n/ *n.* **juxtapositional** /-pə'zɪʃən(ə)l/ *adj.* [Latin *juxta* next, *pono* put]

K

K[1] /keɪ/ *n.* (also **k**) (*pl.* **Ks** or **K's**) eleventh letter of the alphabet.

K[2] *abbr.* (also **K.**) **1** kelvin(s). **2** King, King's. **3** Köchel (catalogue of Mozart's works). **4** (also **k**) (prec. by a numeral) **a** *Computing* unit of 1,024 (i.e. 2^{10}) bytes or bits, or loosely 1,000. **b** 1,000. [sense 4 as abbreviation of KILO-]

K[3] *symb.* potassium. [Latin *Kalium*]

k *abbr.* **1** kilo-. **2** knot(s).

Kaffir /'kæfə(r)/ *n.* **1** *hist.* member or language of a S. African people of the Bantu family. **2** *S.Afr. offens.* any Black African. [Arabic, = infidel]

Kafkaesque /,kæfkə'resk/ *adj.* impenetrably oppressive or nightmarish, as in the fiction of Franz Kafka.

kaftan var. of CAFTAN.

kaiser /'kaɪzə(r)/ *n. hist.* emperor, esp. of Germany, Austria, or the Holy Roman Empire. [Latin CAESAR]

kalashnikov /kə'læʃnɪ,kɒf/ *n.* type of Soviet rifle or sub-machine-gun. [Russian]

kale *n.* variety of cabbage, esp. with wrinkled leaves and no heart. [northern English var. of COLE]

kaleidoscope /kə'laɪdə,skəʊp/ *n.* **1** tube containing mirrors and pieces of coloured glass etc. producing changing reflected patterns when shaken. **2** constantly changing pattern, group, etc. □ **kaleidoscopic** /-'skɒpɪk/ *adj.* [Greek *kalos* beautiful, *eidos* form, -SCOPE]

kalends var. of CALENDS.

kaleyard *n. Scot.* kitchen garden.

kamikaze /,kæmɪ'kɑ:zɪ/ *–n. hist.* **1** explosive-laden Japanese aircraft deliberately crashed on a ship etc. during the war of 1939–45. **2** pilot of this. *–attrib. adj.* **1** of a kamikaze. **2** reckless, esp. suicidal. [Japanese, = divine wind]

kangaroo /,kæŋgə'ru:/ *n.* (*pl.* **-s**) Australian marsupial with strong hind legs for jumping. [Aboriginal]

kangaroo court *n.* illegal court, e.g. held by strikers or mutineers.

kaolin /'keɪəlɪn/ *n.* fine soft white clay used esp. for porcelain and in medicines. [Chinese *kao-ling* high hill]

kapok /'keɪpɒk/ *n.* fine fibrous cotton-like substance from a tropical tree, used for padding. [Malay]

kappa /'kæpə/ *n.* tenth letter of the Greek alphabet (*Κ*, *κ*). [Greek]

kaput /kə'pʊt/ *predic. adj. slang* broken, ruined. [German]

karabiner /,kærə'bi:nə(r)/ *n.* coupling link used by mountaineers. [German, literally 'carbine']

karakul /'kærə,kʊl/ *n.* (also **caracul**) **1** Asian sheep with a dark curled fleece when young. **2** fur of or like this. [Russian]

karaoke /,kærɪ'əʊkɪ/ *n.* entertainment in nightclubs etc. with customers singing to a backing track. [Japanese, = empty orchestra]

karate /kə'rɑ:tɪ/ *n.* Japanese system of unarmed combat using the hands and feet as weapons. [Japanese, = empty hand]

karma /'kɑ:mə/ *n. Buddhism & Hinduism* person's actions in previous lives, believed to decide his or her fate in future existences. [Sanskrit, = action, fate]

kauri /kaʊ'rɪ/ *n.* (*pl.* **-s**) coniferous New Zealand tree yielding timber and resin. [Maori]

kayak /'kaɪæk/ *n.* **1** Eskimo one-man canoe of wood and sealskins. **2** small covered canoe. [Eskimo]

kazoo /kə'zu:/ *n.* toy musical instrument into which the player sings or hums. [origin uncertain]

KBE *abbr.* Knight Commander of the Order of the British Empire.

KC *abbr.* King's Counsel.

kc/s *abbr.* kilocycles per second.

kea /'ki:ə, 'keɪə/ *n.* New Zealand parrot with brownish-green and red plumage. [Maori, imitative]

kebab /kɪ'bæb/ *n.* pieces of meat, vegetables, etc. cooked on a skewer (cf. DONER KEBAB, SHISH KEBAB). [Urdu from Arabic]

kedge *–v.* (**-ging**) **1** move (a ship) with a hawser attached to a small anchor. **2** (of a ship) move in this way. *–n.* (in full **kedge-anchor**) small anchor for this purpose. [origin uncertain]

kedgeree /,kedʒə'ri:/ *n.* dish of fish, rice, hard-boiled eggs, etc. [Hindi]

keel *–n.* main lengthwise member of the base of a ship etc. *–v.* **1** (often foll. by *over*) (cause to) fall down or over. **2** turn keel upwards. □ **on an even keel** steady; balanced. [Old Norse]

keelhaul *v.* **1** drag (a person) under the keel of a ship as a punishment. **2** scold or rebuke severely.

keelson /ˈkiːls(ə)n/ *n.* (also **kelson** /ˈkels(ə)n/) line of timber fixing a ship's floor-timbers to its keel. [origin uncertain]

keen[1] *adj.* **1** enthusiastic, eager. **2** (foll. by *on*) enthusiastic about, fond of. **3** (of the senses) sharp. **4** intellectually acute. **5** (of a knife etc.) sharp. **6** (of a sound, light, etc.) penetrating, vivid. **7** (of a wind etc.) piercingly cold. **8** (of a pain etc.) acute. **9** (of a price) competitive. □ **keenly** *adv.* **keenness** *n.* [Old English]

keen[2] *–n.* Irish wailing funeral song. *–v.* (often foll. by *over, for*) wail mournfully, esp. at a funeral. [Irish *caoine* from *caoinim* wail]

keep *–v.* (*past* and *past part.* **kept**) **1** have continuous charge of; retain possession of. **2** (foll. by *for*) retain or reserve for (a future time) (*kept it for later*). **3** retain or remain in a specified condition, position, place, etc. (*keep cool*; *keep out*; *keep them happy*; *knives are kept here*). **4** (foll. by *from*) restrain, hold back. **5** detain (*what kept you?*). **6** observe, honour, or respect (a law, custom, commitment, secret, etc.) (*keep one's word*; *keep the sabbath*). **7** own and look after (animals). **8 a** clothe, feed, maintain, etc. (a person, oneself, etc.). **b** (foll. by *in*) maintain (a person) with a supply of. **9** carry on; manage (a business etc.). **10** maintain (a diary, house, accounts, etc.) regularly and in proper order. **11** normally have on sale (*do you keep buttons?*). **12** guard or protect (a person or place). **13** preserve (*keep order*). **14** (foll. by verbal noun) continue; repeat habitually (*keeps telling me*). **15** continue to follow (a way or course). **16 a** (esp. of food) remain in good condition. **b** (of news etc.) not suffer from delay in telling. **17** (often foll. by *to*) remain in (one's bed, room, etc.). **18** maintain (a person) as one's mistress etc. (*kept woman*). *–n.* **1** maintenance, food, etc. (*hardly earn your keep*). **2** *hist.* tower, esp. the central stronghold of a castle. □ **for keeps** *colloq.* permanently, indefinitely. **how are you keeping?** how are you? **keep at** (cause to) persist with. **keep away** (often foll. by *from*) avoid, prevent from being near. **keep back 1** remain or keep at a distance. **2** retard the progress of. **3** conceal. **4** withhold (*kept back £50*). **keep down 1** hold in subjection. **2** keep low in amount. **3** stay hidden. **4** not vomit (food eaten). **keep one's hair on** see HAIR. **keep one's hand in** see HAND. **keep in with** remain on good terms with. **keep off 1** (cause to) stay away from. **2** ward off. **3** abstain from. **4** avoid (a subject) (*let's keep off religion*). **keep on 1** continue; do continually (*kept on laughing*). **2** continue to employ. **3** (foll. by *at*) nag. **keep out 1** keep or remain outside. **2** exclude. **keep to 1** adhere to (a course, promise, etc.). **2** confine oneself to. **keep to oneself 1** avoid contact with others. **2** keep secret. **keep track of** see TRACK. **keep under** repress. **keep up 1** maintain (progress, morale, etc.). **2** keep in repair etc. **3** carry on (a correspondence etc.). **4** prevent from going to bed. **5** (often foll. by *with*) not fall behind. **keep up with the Joneses** compete socially with one's neighbours. [Old English]

keeper *n.* **1** person who looks after or is in charge of animals, people, or a thing. **2** custodian of a museum, forest, etc. **3 a** = WICKET-KEEPER. **b** = GOALKEEPER. **4 a** sleeper in a pierced ear. **b** ring that keeps another on the finger.

keep-fit *n.* regular physical exercises.

keeping *n.* **1** custody, charge (*in safe keeping*). **2** agreement, harmony (esp. *in* or *out of keeping* (*with*)).

keepsake *n.* souvenir, esp. of a person.

keg *n.* small barrel. [Old Norse]

keg beer *n.* beer kept in a metal keg under pressure.

kelp *n.* **1** large brown seaweed suitable for manure. **2** its calcined ashes, formerly a source of sodium, potassium, etc. [origin unknown]

kelpie /ˈkelpɪ/ *n. Scot.* **1** malevolent water-spirit, usu. in the form of a horse. **2** Australian sheepdog. [origin unknown]

kelson var. of KEELSON.

Kelt var. of CELT.

kelt *n.* salmon or sea trout after spawning. [origin unknown]

kelter var. of KILTER.

kelvin /ˈkelvɪn/ *n.* SI unit of thermodynamic temperature. [*Kelvin*, name of a physicist]

Kelvin scale *n.* scale of temperature with zero at absolute zero.

ken *–n.* range of knowledge or sight (*beyond my ken*). *–v.* (**-nn-**; *past* and *past part.* **kenned** or **kent**) *Scot.* & *N.Engl.* **1** recognize at sight. **2** know. [Old English, = make known: related to CAN[1]]

kendo /ˈkendəʊ/ *n.* Japanese fencing with two-handed bamboo swords. [Japanese, = sword-way]

kennel /ˈken(ə)l/ *–n.* **1** small shelter for a dog. **2** (in *pl.*) breeding or boarding place for dogs. *–v.* (**-ll-**; *US* **-l-**) put into or keep in a kennel. [French *chenil* from Latin *canis* dog]

kent *past* and *past part.* of KEN.

Kenyan /ˈkenjən/ *–adj.* of Kenya in E. Africa. *–n.* **1** native or national of Kenya. **2** person of Kenyan descent.

kepi /ˈkeɪpɪ/ *n.* (*pl.* **-s**) French military cap with a horizontal peak. [French *képi*]

kept *past* and *past part.* of KEEP.

keratin /ˈkerətɪn/ *n.* fibrous protein in hair, feathers, hooves, claws, horns, etc. [Greek *keras kerat-* horn]

kerb *n.* stone edging to a pavement or raised path. [var. of CURB]

kerb-crawling *n. colloq.* driving slowly in order to engage a prostitute.

kerb drill *n.* precautions before crossing a road.

kerbstone *n.* stone forming part of a kerb.

kerchief /ˈkɜːtʃɪf/ *n.* **1** headscarf, neckerchief. **2** *poet.* handkerchief. [Anglo-French *courchef*: related to COVER, CHIEF]

kerfuffle /kəˈfʌf(ə)l/ *n. colloq.* fuss, commotion. [originally Scots]

kermes /ˈkɜːmɪz/ *n.* **1** female of an insect with a berry-like appearance. **2** (in full **kermes oak**) evergreen oak on which this feeds. **3** red dye made from these insects dried. [Arabic]

kernel /ˈkɜːn(ə)l/ *n.* **1** (usu. soft) edible centre within the hard shell of a nut, fruit stone, seed, etc. **2** whole seed of a cereal. **3** essence of anything. [Old English: related to CORN[1]]

kerosene /ˈkerəˌsiːn/ *n.* (also **kerosine**) esp. *US* fuel oil for use in jet engines, boilers, etc.; paraffin oil. [Greek *kēros* wax]

kestrel /ˈkestr(ə)l/ *n.* small hovering falcon. [origin uncertain]

ketch *n.* small two-masted sailing-boat. [probably from CATCH]

ketchup /ˈketʃʌp/ *n.* (*US* **catsup** /ˈkætsəp/) spicy esp. tomato sauce used as a condiment. [Chinese]

ketone /ˈkiːtəʊn/ *n.* any of a class of organic compounds including propanone (acetone). [German *Keton*, alteration of *Aketon* ACETONE]

kettle /ˈket(ə)l/ *n.* vessel for boiling water in. □ **a different kettle of fish** a different matter altogether. **a fine** (or **pretty**) **kettle of fish** *iron.* an awkward state of affairs. [Old Norse]

kettledrum *n.* large bowl-shaped drum.

key[1] /kiː/ —*n.* (*pl.* **-s**) **1** (usu. metal) instrument for moving the bolt of a lock. **2** similar implement for operating a switch. **3** instrument for grasping screws, nuts, etc., or for winding a clock etc. **4** (often in *pl.*) finger-operated button or lever on a typewriter, piano, computer terminal, etc. **5** means of advance, access, etc. (*key to success*). **6** (*attrib.*) essential (*key element*). **7 a** solution or explanation. **b** word or system for solving a cipher or code. **c** explanatory list of symbols used in a map, table, etc. **8** *Mus.* system of notes related to each other and based on a particular note (*key of C major*). **9** tone or style of thought or expression. **10** piece of wood or metal inserted between and securing others. **11** coat of wall plaster between the laths securing other coats. **12** roughness of a surface helping the adhesion of plaster etc. **13** winged fruit of the sycamore etc. **14** device for making or breaking an electric circuit. —*v.* (**keys, keyed**) **1** (foll. by *in, on*, etc.) fasten with a pin, wedge, bolt, etc. **2** (often foll. by *in*) enter (data) by means of a keyboard. **3** roughen (a surface) to help the adhesion of plaster etc. **4** (foll. by *to*) align or link (one thing to another). □ **keyed up** tense, nervous, excited. [Old English]

key[2] /kiː/ *n.* low-lying island or reef, esp. in the W. Indies. [Spanish *cayo*]

keyboard —*n.* **1** set of keys on a typewriter, computer, piano, etc. **2** electronic musical instrument with keys arranged as on a piano. —*v.* enter (data) by means of a keyboard. □ **keyboarder** *n.* (in sense 1 of *n.*). **keyboardist** *n.* (in sense 2 of *n.*).

keyhole *n.* hole in a door etc. for a key.

keyhole surgery *n. colloq.* minimally invasive surgery carried out through a very small incision.

Keynesian /ˈkeɪnzɪən/ *adj.* of the economic theories of J. M. Keynes, esp. regarding State intervention in the economy.

keynote *n.* **1** (esp. *attrib.*) prevailing tone or idea, esp. in a speech, conference, etc. **2** *Mus.* note on which a key is based.

keypad *n.* miniature keyboard etc. for a portable electronic device, telephone, etc.

keypunch —*n.* device for recording data by means of punched holes or notches on cards or paper tape. —*v.* record (data) thus.

key-ring *n.* ring for keeping keys on.

key signature *n. Mus.* any of several combinations of sharps or flats indicating the key of a composition.

keystone *n.* **1** central principle of a system, policy, etc. **2** central locking stone in an arch.

keystroke *n.* single depression of a key on a keyboard, esp. as a measure of work.

keyword *n.* **1** key to a cipher etc. **2 a** word of great significance. **b** significant word used in indexing.

KG *abbr.* Knight of the Order of the Garter.

kg *abbr.* kilogram(s).

KGB *n.* State security police of the former USSR. [Russian abbreviation, = committee of State security]
khaki /'kɑ:kɪ/ *–adj.* dull brownish-yellow. *–n.* (*pl.* **-s**) **1** khaki fabric or uniform. **2** dull brownish-yellow colour. [Urdu, = dusty]
khan /kɑ:n/ *n.* title of rulers and officials in Central Asia, Afghanistan, etc. □ **khanate** *n.* [Turki, = lord]
kHz *abbr.* kilohertz.
kibbutz /kɪ'bʊts/ *n.* (*pl.* **kibbutzim** /-'tsi:m/) communal esp. farming settlement in Israel. [Hebrew, = gathering]
kibosh /'kaɪbɒʃ/ *n. slang* nonsense. □ **put the kibosh on** put an end to. [origin unknown]
kick *–v.* **1** strike, strike out, or propel forcibly, with the foot or hoof. **2** (often foll. by *at*, *against*) protest at; rebel against. **3** *slang* give up (a habit). **4** (often foll. by *out* etc.) expel or dismiss forcibly. **5** *refl.* be annoyed with oneself. **6** *Football* score (a goal) by a kick. *–n.* **1** kicking action or blow. **2** *colloq.* **a** sharp stimulant effect, esp. of alcohol. **b** (often in *pl.*) thrill (*did it for kicks*). **3** strength, resilience (*no kick left*). **4** *colloq.* specified temporary interest (*on a jogging kick*). **5** recoil of a gun when fired. □ **kick about** (or **around**) *colloq.* **1 a** drift idly from place to place. **b** be unused or unwanted. **2 a** treat roughly. **b** discuss unsystematically. **kick the bucket** *slang* die. **kick one's heels** see HEEL. **kick off 1 a** *Football* start or resume a match. **b** *colloq.* begin. **2** remove (shoes etc.) by kicking. **kick over the traces** see TRACE[2]. **kick up** (or **kick up a fuss, dust**, etc.) *colloq.* create a disturbance; object. **kick a person upstairs** dispose of a person by promotion etc. [origin unknown]
kickback *n. colloq.* **1** recoil. **2** (usu. illegal) payment for help or favours, esp. in business.
kick-off *n. Football* start or resumption of a match.
kickstand *n.* rod for supporting a bicycle or motor cycle when stationary.
kick-start *–n.* (also **kick-starter**) device to start the engine of a motor cycle etc. by the downward thrust of a pedal. *–v.* start (a motor cycle etc.) in this way.
kid[1] *–n.* **1** young goat. **2** leather from this. **3** *colloq.* child. *–v.* (**-dd-**) (of a goat) give birth. □ **handle with kid gloves** treat carefully. [Old Norse]
kid[2] *v.* (also *refl.*) (**-dd-**) *colloq.* deceive, trick, tease (*don't kid yourself*; *only kidding*). □ **no kidding** *slang* that is the truth. [origin uncertain]
kiddie /'kɪdɪ/ *n.* (also **kiddy**) (*pl.* **-ies**) *slang* = KID[1] *n.* 3.
kiddo /'kɪdəʊ/ *n.* (*pl.* **-s**) *slang* = KID[1] *n.* 3.
kidnap *v.* (**-pp-**; *US* **-p-**) **1** abduct (a person etc.), esp. to obtain a ransom. **2** steal (a child). □ **kidnapper** *n.* [from KID[1], *nap* = NAB]
kidney /'kɪdnɪ/ *n.* (*pl.* **-s**) **1** either of two organs in the abdominal cavity of vertebrates which remove nitrogenous wastes from the blood and excrete urine. **2** animal's kidney as food. [origin unknown]
kidney bean *n.* red-skinned dried bean.
kidney machine *n.* machine able to take over the function of a damaged kidney.
kidney-shaped *adj.* having one side concave and the other convex.
kill *–v.* **1** (also *absol.*) deprive of life or vitality; cause death or the death of. **2** destroy (feelings etc.). **3** *refl. colloq.* **a** overexert oneself (*don't kill yourself trying*). **b** laugh heartily. **4** *colloq.* overwhelm with amusement. **5** switch off (a light, engine, etc.). **6** *Computing colloq.* delete. **7** *colloq.* cause pain or discomfort to (*my feet are killing me*). **8** pass (time, or a specified period) usu. while waiting (*an hour to kill before the interview*). **9** defeat (a bill in Parliament). **10 a** *Tennis* etc. hit (the ball) so that it cannot be returned. **b** stop (the ball) dead. **11** make ineffective (taste, sound, pain, etc.) (*carpet killed the sound*). *–n.* **1** act of killing (esp. in hunting). **2** animal(s) killed, esp. by a hunter. **3** *colloq.* destruction or disablement of an enemy aircraft etc. □ **dressed to kill** dressed showily or alluringly. **in at the kill** present at a successful conclusion. **kill off 1** destroy completely. **2** (of an author) bring about the death of (a fictional character). **kill or cure** (usu. *attrib.*) (of a remedy etc.) drastic, extreme. **kill two birds with one stone** achieve two aims at once. **kill with kindness** spoil with overindulgence. [perhaps related to QUELL]
killer *n.* **1 a** person, animal, or thing that kills. **b** murderer. **2** *colloq.* **a** impressive, formidable, or excellent thing. **b** hilarious joke.
killer instinct *n.* **1** innate tendency to kill. **2** ruthless streak.
killer whale *n.* dolphin with a prominent dorsal fin.
killing *–n.* **1 a** causing of death. **b** instance of this. **2** *colloq.* great (esp. financial) success (*make a killing*). *–adj. colloq.* **1** very funny. **2** exhausting.
killjoy *n.* gloomy or censorious person, esp. at a party etc.
kiln *n.* furnace or oven for burning, baking, or drying, esp. for calcining

lime or firing pottery etc. [Old English from Latin *culina* kitchen]

kilo /'ki:ləʊ/ *n.* (*pl.* **-s**) kilogram. [French, abbreviation]

kilo- *comb. form* 1,000 (esp. in metric units). [Greek *khilioi*]

kilobyte /'kɪlə,baɪt/ *n. Computing* 1,024 (i.e. 2^{10}) bytes as a measure of memory size etc.

kilocalorie /'kɪlə,kælərɪ/ *n.* = *large calorie* (see CALORIE).

kilocycle /'kɪlə,saɪk(ə)l/ *n. hist.* kilohertz.

kilogram /'kɪlə,græm/ *n.* (also **-gramme**) SI unit of mass, approx. 2.205 lb.

kilohertz /'kɪlə,hɜ:ts/ *n.* 1,000 hertz, 1,000 cycles per second.

kilojoule /'kɪlə,dʒu:l/ *n.* 1,000 joules, esp. as a measure of the energy value of foods.

kilolitre /'kɪlə,li:tə(r)/ *n.* (*US* **-liter**) 1,000 litres (220 imperial gallons).

kilometre /'kɪlə,mi:tə(r), kɪ'lɒmɪtə(r)/ *n.* (*US* **-meter**) 1,000 metres (approx. 0.62 miles). □ **kilometric** /,kɪlə'metrɪk/ *adj.*

■ **Usage** The second pronunciation given, with the stress on the second syllable, is considered incorrect by some people.

kiloton /'kɪlə,tʌn/ *n.* (also **kilotonne**) unit of explosive power equivalent to 1,000 tons of TNT.

kilovolt /'kɪlə,vɒlt/ *n.* 1,000 volts.

kilowatt /'kɪlə,wɒt/ *n.* 1,000 watts.

kilowatt-hour *n.* electrical energy equivalent to a power consumption of 1,000 watts for one hour.

kilt *–n.* pleated knee-length usu. tartan skirt, traditionally worn by Highland men. *–v.* **1** tuck up (the skirts) round the body. **2** (esp. as **kilted** *adj.*) gather in vertical pleats. [Scandinavian]

kilter /'kɪltə(r)/ *n.* (also **kelter** /'kel-/) good working order (esp. *out of kilter*). [origin unknown]

kimono /kɪ'məʊnəʊ/ *n.* (*pl.* **-s**) **1** long sashed Japanese robe. **2** similar dressing-gown. [Japanese]

kin *–n.* one's relatives or family. *–predic. adj.* related. [Old English]

-kin *suffix* forming diminutive nouns (*catkin*; *manikin*). [Dutch]

kind /kaɪnd/ *–n.* **1** race, species, or natural group of animals, plants, etc. (*human kind*). **2** class, type, sort, variety. **3** natural way, fashion, etc. (*true to kind*). *–adj.* (often foll. by *to*) friendly, generous, or benevolent. □ **in kind 1** in the same form, likewise (*was insulted and replied in kind*). **2** (of payment) in goods or labour, not money. **3** character, quality (*differ in degree but not in kind*). **kind of** *colloq.* to some extent (*I kind of expected it*). **a kind of** loosely resembling (*he's a kind of doctor*). [Old English]

■ **Usage** In sense 2 of the noun, *these kinds of* is usually preferred to *these kind of*.

kindergarten /'kɪndə,gɑ:t(ə)n/ *n.* class or school for very young children. [German, = children's garden]

kind-hearted *adj.* of a kind disposition. □ **kind-heartedly** *adv.* **kind-heartedness** *n.*

kindle /'kɪnd(ə)l/ *v.* (**-ling**) **1** light, catch, or set on fire. **2** arouse or inspire. **3** become aroused or animated. [Old Norse]

kindling *n.* small sticks etc. for lighting fires.

kindly[1] *adv.* **1** in a kind manner (*spoke kindly*). **2** often *iron.* please (*kindly go away*). □ **look kindly upon** regard sympathetically. **take kindly to** be pleased by; like.

kindly[2] *adj.* (**-ier, -iest**) **1** kind, kind-hearted. **2** (of a climate etc.) pleasant, mild. □ **kindlily** *adv.* **kindliness** *n.*

kindness *n.* **1** being kind. **2** kind act.

kindred /'kɪndrɪd/ *–adj.* related, allied, or similar. *–n.* **1** one's relations collectively. **2** blood relationship. **3** resemblance in character. [Old English, = kinship]

kindred spirit *n.* person like or in sympathy with oneself.

kinematics /,kɪnɪ'mætɪks/ *n.pl.* (usu. treated as *sing.*) branch of mechanics concerned with the motion of objects without reference to cause. □ **kinematic** *adj.* [Greek *kinēma -matos* motion]

kinetic /kɪ'netɪk/ *adj.* of or due to motion. □ **kinetically** *adv.* [Greek *kineo* move]

kinetic art *n.* sculpture etc. designed to move.

kinetic energy *n.* energy of motion.

kinetics *n.pl.* **1** = DYNAMICS 1a. **2** (usu. treated as *sing.*) branch of physical chemistry measuring and studying the rates of chemical reactions.

king *n.* **1** (as a title usu. **King**) male sovereign, esp. a hereditary ruler. **2** preeminent person or thing (*oil king*). **3** (*attrib.*) large (or the largest) kind of plant, animal, etc. (*king penguin*). **4** *Chess* piece which must be checkmated for a win. **5** crowned piece in draughts. **6** court-card depicting a king. **7** (**the King**) national anthem when the sovereign is male. □ **kingly** *adj.* **kingship** *n.* [Old English]

King Charles spaniel *n.* small black and tan spaniel.
kingcup *n.* marsh marigold.
kingdom *n.* **1** territory or State ruled by a king or queen. **2** spiritual reign or sphere of God. **3** domain. **4** division of the natural world (*plant kingdom*). **5** specified sphere (*kingdom of the heart*). [Old English]
kingdom come *n. colloq.* the next world.
kingfisher *n.* small bird with brightly coloured plumage, diving for fish etc.
King of Arms *n.* a chief herald.
king of beasts *n.* lion.
king of birds *n.* eagle.
kingpin *n.* **1** main, large, or vertical bolt, esp. as a pivot. **2** essential person or thing.
king-post *n.* upright post from the tie-beam of a roof to the apex of a truss.
King's Counsel *n.* = QUEEN'S COUNSEL.
King's English *n.* = QUEEN'S ENGLISH.
King's evidence see EVIDENCE.
King's Guide *n.* = QUEEN'S GUIDE.
King's highway *n.* = QUEEN'S HIGHWAY.
king-size *adj.* (also **-sized**) very large.
King's Proctor *n.* = QUEEN'S PROCTOR.
King's Scout *n.* = QUEEN'S SCOUT.
kink –*n.* **1 a** twist or bend in wire etc. **b** tight wave in hair. **2** mental twist or quirk, esp. when perverse. –*v.* (cause to) form a kink. [Low German or Dutch]
kinky *adj.* (**-ier, -iest**) **1** *colloq.* **a** sexually perverted or unconventional. **b** (of clothing etc.) bizarre and sexually provocative. **2** having kinks. □ **kinkily** *adv.* **kinkiness** *n.*
kinsfolk *n.pl.* one's blood relations.
kinship *n.* **1** blood relationship. **2** likeness; sympathy.
kinsman *n.* (*fem.* **kinswoman**) **1** blood relation. **2** relation by marriage.

■ **Usage** Use of *kinsman* in sense 2 is considered incorrect by some people.

kiosk /'ki:ɒsk/ *n.* **1** light open-fronted booth selling food, newspapers, tickets, etc. **2** telephone box. [Turkish from Persian]
kip *slang* –*n.* **1** sleep; nap. **2** bed or cheap lodgings. –*v.* (**-pp-**) (often foll. by *down*) sleep. [cf. Danish *kippe* mean hut]
kipper /'kɪpə(r)/ –*n.* fish, esp. a herring, split, salted, dried, and usu. smoked. –*v.* cure (a herring etc.) thus. [origin uncertain]
kir /kɜ:(r)/ *n.* dry white wine with *crème de cassis.*
kirby-grip /'kɜ:bɪgrɪp/ *n.* (also **Kirbigrip** *propr.*) type of sprung hairgrip. [*Kirby*, name of the manufacturer]
kirk *n. Scot.* & *N.Engl.* **1** church. **2** (**the Kirk** or **the Kirk of Scotland**) Church of Scotland. [Old Norse *kirkja* = CHURCH]
Kirk-session *n.* lowest court in the Church of Scotland.
kirsch /kɪəʃ/ *n.* brandy distilled from cherries. [German, = cherry]
kismet /'kɪzmet/ *n.* destiny, fate. [Turkish from Arabic]
kiss –*v.* **1** touch with the lips, esp. as a sign of love, affection, greeting, or reverence. **2** (of two people) touch each others' lips in this way. **3** lightly touch. –*n.* **1** touch with the lips. **2** light touch. □ **kiss and tell** recount one's sexual exploits. **kiss a person's arse** *coarse slang* toady to. **kiss the dust** submit abjectly. [Old English]
kiss-curl *n.* small curl of hair on the forehead, nape, etc.
kisser *n.* **1** person who kisses. **2** *slang* mouth; face.
kiss of death *n.* apparent good luck etc. which causes ruin.
kiss of life *n.* mouth-to-mouth resuscitation.
kissogram /'kɪsə,græm/ *n.* (also **Kissagram** *propr.*) novelty telegram or greeting delivered with a kiss.
kit –*n.* **1** articles, equipment, etc. for a specific purpose (*first-aid kit*). **2** specialized, esp. sports, clothing or uniform (*football kit*). **3** set of parts needed to assemble furniture, a model, etc. –*v.* (**-tt-**) (often foll. by *out, up*) equip with kit. [Dutch]
kitbag *n.* large usu. cylindrical bag used for a soldier's or traveller's kit.
kitchen /'kɪtʃɪn/ *n.* **1** place where food is prepared and cooked. **2** kitchen fitments (*half-price kitchens*). [Latin *coquina*]
kitchenette /,kɪtʃɪ'net/ *n.* small kitchen or cooking area.
kitchen garden *n.* garden with vegetables, fruit, herbs, etc.
kitchenware *n.* cooking utensils.
kite *n.* **1** light framework with a thin covering flown on a string in the wind. **2** soaring bird of prey. [Old English]
Kitemark *n.* official kite-shaped mark on goods approved by the British Standards Institution.
kith *n.* □ **kith and kin** friends and relations. [Old English, originally 'knowledge': related to CAN[1]]
kitsch /kɪtʃ/ *n.* (often *attrib.*) vulgar, pretentious, or worthless art. □ **kitschy** *adj.* (**-ier, -iest**). [German]
kitten /'kɪt(ə)n/ –*n.* young cat, ferret, etc. –*v.* (of a cat etc.) give birth (to). □

have kittens *colloq.* be very upset or anxious. [Anglo-French diminutive of *chat* CAT]

kittenish *adj.* playful, lively, or flirtatious.

kittiwake /'kɪtɪ,weɪk/ *n.* a kind of small seagull. [imitative of its cry]

kitty[1] /'kɪtɪ/ *n.* (*pl.* **-ies**) **1** fund of money for communal use. **2** pool in some card-games. [origin unknown]

kitty[2] /'kɪtɪ/ *n.* (*pl.* **-ies**) childish name for a kitten or cat.

kiwi /'ki:wi:/ *n.* (*pl.* **-s**) **1** flightless long-billed New Zealand bird. **2** (**Kiwi**) *colloq.* New Zealander. [Maori]

kiwi fruit *n.* green-fleshed fruit of a climbing plant.

kJ *abbr.* kilojoule(s).

kl *abbr.* kilolitre(s).

Klaxon /'klæks(ə)n/ *n. propr.* horn or warning hooter. [name of the manufacturer]

Kleenex /'kli:neks/ *n.* (*pl.* same or **-es**) *propr.* disposable paper handkerchief.

kleptomania /,kleptə'meɪnɪə/ *n.* obsessive apparently motiveless urge to steal. □ **kleptomaniac** /-nɪ,æk/ *n.* & *adj.* [Greek *kleptēs* thief]

km *abbr.* kilometre(s).

knack *n.* **1** acquired faculty or trick of doing a thing. **2** habit (*a knack of offending people*). [origin unknown]

knacker –*n.* buyer of useless horses etc. for slaughter, or of old houses, ships, etc. for the materials. –*v. slang* (esp. as **knackered** *adj.*) exhaust, wear out. [origin unknown]

knapsack /'næpsæk/ *n.* soldier's or hiker's usu. canvas bag carried on the back. [German *knappen* bite, SACK[1]]

knapweed /'næpwi:d/ *n.* plant with thistle-like purple flowers. [from *knop* ornamental knob or tuft]

knave *n.* **1** rogue, scoundrel. **2** = JACK *n.* 2. □ **knavery** *n.* (*pl.* **-ies**). **knavish** *adj.* [Old English, originally = boy, servant]

knead *v.* **1 a** work into a dough, paste, etc. by pummelling. **b** make (bread, pottery, etc.) thus. **2** massage (muscles etc.) as if kneading. [Old English]

knee –*n.* **1 a** (often *attrib.*) joint between the thigh and the lower leg in humans. **b** corresponding joint in other animals. **c** area around this. **d** lap (*sat on his knee*). **2** part of a garment covering the knee. –*v.* (**knees, kneed, kneeing**) **1** touch or strike with the knee (*kneed him in the groin*). **2** *colloq.* make (trousers) bulge at the knee. □ **bring a person** (or **thing**) **to his** (or **her** or **its**) **knees** reduce to submission or a state of weakness. [Old English]

knee-bend *n.* bending of the knee, esp. as a physical exercise.

knee-breeches *n.pl.* close-fitting trousers to the knee or just below.

kneecap –*n.* **1** convex bone in front of the knee. **2** protective covering for the knee. –*v.* (**-pp-**) *slang* (of a terrorist) shoot (a person) in the knee or leg as a punishment.

knee-deep *adj.* **1** (usu. foll. by *in*) **a** immersed up to the knees. **b** deeply involved. **2** so deep as to reach the knees.

knee-high *adj.* so high as to reach the knees.

knee-jerk *n.* **1** sudden involuntary kick caused by a blow on the tendon just below the knee. **2** (*attrib.*) predictable, automatic, stereotyped.

kneel *v.* (*past* and *past part.* **knelt** /nelt/ or esp. *US* **kneeled**) fall or rest on the knees or a knee. [Old English: related to KNEE]

knee-length *adj.* reaching the knees.

kneeler *n.* **1** cushion for kneeling on. **2** person who kneels.

knees-up *n. colloq.* lively party or gathering.

knell –*n.* **1** sound of a bell, esp. for a death or funeral. **2** announcement, event, etc., regarded as an ill omen. –*v.* **1** ring a knell. **2** proclaim by or as by a knell. [Old English]

knelt *past* and *past part.* of KNEEL.

knew *past* of KNOW.

knickerbocker /'nɪkə,bɒkə(r)/ *n.* (in *pl.*) loose-fitting breeches gathered at the knee or calf. [the pseudonym of W. Irving, author of *History of New York*]

Knickerbocker Glory *n.* ice-cream served with fruit etc. in a tall glass.

knickers *n.pl.* woman's or girl's undergarment for the lower torso. [abbreviation of KNICKERBOCKER]

knick-knack /'nɪknæk/ *n.* (also **nick-nack**) trinket or small dainty ornament etc. [from KNACK in the obsolete sense 'trinket']

knife –*n.* (*pl.* **knives**) **1** metal blade for cutting or as a weapon, with usu. one long sharp edge fixed in a handle. **2** cutting-blade in a machine. **3** (as **the knife**) surgical operation. –*v.* (**-fing**) cut or stab with a knife. □ **at knife-point** threatened with a knife or an ultimatum etc. **get** (or **have got**) **one's knife into** treat maliciously, persecute. [Old English]

knife-edge *n.* **1** edge of a knife. **2** position of extreme danger or uncertainty.

knife-pleat *n.* narrow flat usu. overlapping pleat on a skirt etc.

knight /naɪt/ –*n.* **1** man awarded a non-hereditary title (*Sir*) by a sovereign. **2** *hist.* **a** man, usu. noble, raised to honourable military rank after service as a page and squire. **b** military follower, attendant, or lady's champion in a war or tournament. **3** man devoted to a cause, woman, etc. **4** *Chess* piece usu. shaped like a horse's head. **–***v.* confer a knighthood on. □ **knighthood** *n.* **knightly** *adj. poet.* [Old English, originally = boy]

knight commander see COMMANDER.

knight errant *n.* **1** medieval knight in search of chivalrous adventures. **2** chivalrous or quixotic man. □ **knight-errantry** *n.*

knit *v.* (**-tt-**; *past* and *past part.* **knitted** or (esp. in senses 2–4) **knit**) **1** (also *absol.*) **a** make (a garment etc.) by interlocking loops of esp. wool with knitting-needles or a knitting-machine. **b** make (a plain stitch) in knitting (*knit one, purl one*). **2** momentarily wrinkle (the forehead) or (of the forehead) become momentarily wrinkled. **3** (often foll. by *together*) make or become close or compact. **4** (often foll. by *together*) (of a broken bone) become joined; heal. □ **knit up** make or repair by knitting. □ **knitter** *n.* [Old English]

knitting *n.* work being knitted.

knitting-machine *n.* machine for knitting.

knitting-needle *n.* thin pointed rod used esp. in pairs for knitting by hand.

knitwear *n.* knitted garments.

knives *pl.* of KNIFE.

knob *n.* **1** rounded protuberance, esp. at the end or on the surface of a thing, e.g. the handle of a door, drawer, a radio control, etc. **2** small piece (of butter etc.). □ **with knobs on** *slang* that and more (*same to you with knobs on*). □ **knobby** *adj.* **knoblike** *adj.* [Low German *knobbe* knot, knob]

knobbly /'nɒblɪ/ *adj.* (**-ier, -iest**) hard and lumpy. [*knobble*, diminutive of KNOB]

knock **–***v.* **1 a** strike with an audible sharp blow. **b** (often foll. by *at*) strike (a door etc.) to gain admittance. **2** make (a hole etc.) by knocking. **3** (usu. foll. by *in, out, off*, etc.) drive (a thing, person, etc.) by striking (*knocked the ball into the hole; knocked those ideas out of him*). **4** *slang* criticize. **5 a** (of an engine) make a thumping or rattling noise. **b** = PINK[3]. **6** *coarse slang offens.* = *knock off* 6. **–***n.* **1** act or sound of knocking. **2** knocking sound in esp. an engine. □ **knock about** (or **around**) *colloq.* **1** strike repeatedly; treat roughly. **2 a** wander aimlessly or adventurously. **b** be present, esp. by chance (*a cup knocking about somewhere*). **c** (usu. foll. by *with*) be associated socially. **knock back 1** *slang* eat or drink, esp. quickly. **2** *slang* disconcert. **knock down 1** strike (esp. a person) to the ground. **2** demolish. **3** (usu. foll. by *to*) (at an auction) sell (an article) to a bidder by a knock with a hammer. **4** *colloq.* lower the price of (an article). **5** *US slang* steal. **knock off 1** strike off with a blow. **2** *colloq.* finish (work) (*knocked off at 5.30; knocked off work early*). **3** *colloq.* produce (a work of art etc.) or do (a task) rapidly. **4** (often foll. by *from*) deduct (a sum) from a price etc. **5** *slang* steal. **6** *coarse slang offens.* have sexual intercourse with (a woman). **7** *slang* kill. **knock on the head** *colloq.* put an end to (a scheme etc.). **knock on** (or **knock**) **wood** *US* = *touch wood.* **knock out 1** make unconscious by a blow on the head. **2** defeat (a boxer) by knocking him or her down for a count of 10. **3** defeat, esp. in a knockout competition. **4** *slang* astonish. **5** (often *refl.*) *colloq.* exhaust. **knock sideways** *colloq.* astonish, shock. **knock spots off** defeat easily. **knock together** assemble hastily or roughly. **knock up 1** make hastily. **2** waken by a knock at the door. **3** esp. *US slang* make pregnant. **4** practise tennis etc. before formal play begins. [Old English]

knockabout *attrib. adj.* **1** (of comedy) boisterous; slapstick. **2** (of clothes) hard-wearing.

knock-down *attrib. adj.* **1** overwhelming. **2** (of a price) very low. **3** (of a price at auction) reserve. **4** (of furniture etc.) easily dismantled and reassembled.

knocker *n.* **1** hinged esp. metal instrument on a door for knocking with. **2** (in *pl.*) *coarse slang* woman's breasts.

knocking-shop *n. slang* brothel.

knock knees *n.pl.* abnormal curvature of the legs inwards at the knee. □ **knock-kneed** *adj.*

knock-on effect *n.* secondary, indirect, or cumulative effect.

knockout *n.* **1** act of making unconscious by a blow. **2** (usu. *attrib.*) *Boxing* etc. such a blow. **3** competition in which the loser in each round is eliminated (also *attrib.*: *knockout round*). **4** *colloq.* outstanding or irresistible person or thing.

knock-up *n.* practice at tennis etc.

knoll /nəʊl/ *n.* hillock, mound. [Old English]

knot[1] **–***n.* **1 a** intertwining of rope, string, hair, etc., so as to fasten. **b** set method of this (*reef knot*). **c** knotted ribbon etc. as an ornament. **d** tangle in hair, knitting, etc. **2** unit of a ship's or aircraft's speed, equivalent to one nautical mile per hour. **3** (usu. foll. by *of*) cluster (*knot of journalists*). **4** bond, esp. of marriage. **5** hard lump of organic tissue. **6 a** hard mass in a tree-trunk where a branch grows out. **b** round cross-grained piece in timber marking this. **7** central point in a problem etc. —*v.* (**-tt-**) **1** tie in a knot. **2** entangle. **3** unite closely. □ **at a rate of knots** *colloq.* very fast. **tie in knots** *colloq.* baffle or confuse completely. [Old English]

knot[2] *n.* small sandpiper. [origin unknown]

knotgrass *n.* wild plant with creeping stems and small pink flowers.

knot-hole *n.* hole in timber where a knot has fallen out.

knotty *adj.* (**-ier, -iest**) **1** full of knots. **2** puzzling (*knotty problem*).

know /nəʊ/ *v.* (*past* **knew**; *past part.* **known** /nəʊn/) **1** (often foll. by *that, how, what,* etc.) **a** have in the mind; have learnt; be able to recall (*knows a lot about cars*). **b** (also *absol.*) be aware of (a fact) (*I think he knows*). **c** have a good command of (*knew German*; *knows his tables*). **2** be acquainted or friendly with. **3 a** (often foll. by *to* + infin.) recognize; identify (*I knew him at once*; *knew them to be rogues*). **b** (foll. by *from*) be able to distinguish (*did not know him from Adam*). **4** be subject to (*joy knew no bounds*). **5** have personal experience of (fear etc.). **6** (as **known** *adj.*) **a** publicly acknowledged (*known fact*). **b** *Math.* (of a quantity etc.) having a value that can be stated. **7** have understanding or knowledge. □ **in the know** *colloq.* knowing inside information. **know of** be aware of; have heard of (*not that I know of*). **know one's own mind** be decisive, not vacillate. **know what's what** have knowledge of the world, life, etc. **you know** *colloq.* **1** implying something generally known etc. (*you know, the pub on the corner*). **2** expression used as a gap-filler in conversation. **you never know** it is possible. □ **knowable** *adj.* [Old English]

know-all *n. colloq.* person who claims or seems to know everything.

know-how *n.* practical knowledge; natural skill.

knowing *adj.* **1** suggesting that one has inside information (*a knowing look*). **2** showing knowledge; shrewd.

knowingly *adv.* **1** consciously; intentionally (*wouldn't knowingly hurt him*). **2** in a knowing manner (*smiled knowingly*).

knowledge /ˈnɒlɪdʒ/ *n.* **1 a** (usu. foll. by *of*) awareness or familiarity (of or with a person or thing) (*have no knowledge of that*). **b** person's range of information. **2 a** (usu. foll. by *of*) understanding of a subject etc. (*good knowledge of Greek*). **b** sum of what is known (*every branch of knowledge*). □ **to my knowledge** as far as I know.

knowledgeable *adj.* (also **knowledgable**) well-informed; intelligent. □ **knowledgeability** /-ˈbɪlɪtɪ/ *n.* **knowledgeably** *adv.*

known *past part.* of KNOW.

knuckle /ˈnʌk(ə)l/ —*n.* **1** bone at a finger-joint, esp. that connecting the finger to the hand. **2 a** knee- or ankle-joint of a quadruped. **b** this as a joint of meat, esp. of bacon or pork. —*v.* (**-ling**) strike, press, or rub with the knuckles. □ **knuckle down** (often foll. by *to*) **1** apply oneself seriously (to a task etc.). **2** (also **knuckle under**) give in; submit. [Low German or Dutch diminutive of *knoke* bone]

knuckleduster *n.* metal guard worn over the knuckles in fighting, esp. in order to inflict greater damage.

knuckle sandwich *n. slang* punch in the mouth.

knurl *n.* small projecting knob, ridge, etc. [Low German or Dutch]

KO *abbr.* knockout.

koala /kəʊˈɑːlə/ *n.* (in full **koala bear**) small Australian bearlike marsupial with thick grey fur. [Aboriginal]

kohl /kəʊl/ *n.* black powder used as eye make-up, esp. in Eastern countries. [Arabic]

kohlrabi /kəʊlˈrɑːbɪ/ *n.* (*pl.* **-bies**) cabbage with an edible turnip-like swollen stem. [German, from Italian *cavolo rapa*]

kola var. of COLA.

kolkhoz /kʌlˈxɔːz/ *n.* collective farm in the former USSR. [Russian]

koodoo var. of KUDU.

kook *n. US slang* crazy or eccentric person. □ **kooky** *adj.* (**-ier, -iest**). [probably from CUCKOO]

kookaburra /ˈkʊkəˌbʌrə/ *n.* Australian kingfisher with a strange laughing cry. [Aboriginal]

kopek (also **kopeck**) var. of COPECK.

koppie /ˈkɒpɪ/ *n.* (also **kopje**) *S.Afr.* small hill. [Afrikaans *koppie* little head]

Koran /kɔːˈrɑːn/ *n.* Islamic sacred book. [Arabic, = recitation]

Korean /kəˈrɪən/ —*n.* **1** native or national of N. or S. Korea. **2** language of Korea. —*adj.* of Korea, its people, or language.

kosher /ˈkəʊʃə(r)/ —*adj.* **1** (of food or a food-shop) fulfilling the requirements of Jewish law. **2** *colloq.* correct, genuine, legitimate. —*n.* kosher food or shop. [Hebrew, = proper]

kowtow /kaʊˈtaʊ/ —*n. hist.* Chinese custom of kneeling with the forehead touching the ground, esp. in submission. —*v.* **1** (usu. foll. by *to*) act obsequiously. **2** *hist.* perform the kowtow. [Chinese, = knock the head]

k.p.h. *abbr.* kilometres per hour.

Kr *symb.* krypton.

kraal /krɑ:l/ *n. S.Afr.* **1** village of huts enclosed by a fence. **2** enclosure for cattle or sheep. [Afrikaans from Portuguese *curral*, of Hottentot origin]

Kraut /kraʊt/ *n. slang offens.* German. [shortening of SAUERKRAUT]

kremlin /'kremlɪn/ *n.* **1** (**the Kremlin**) **a** citadel in Moscow. **b** Russian Government housed within it. **2** citadel within a Russian town. [Russian]

krill *n.* tiny planktonic crustaceans. [Norwegian *kril* tiny fish]

krona /'krəʊnə/ *n.* **1** (*pl.* **kronor**) chief monetary unit of Sweden. **2** (*pl.* **kronur**) chief monetary unit of Iceland. [Swedish and Icelandic, = CROWN]

krone /'krəʊnə/ *n.* (*pl.* **kroner**) chief monetary unit of Denmark and Norway. [Danish and Norwegian, = CROWN]

krugerrand /'kru:gə,rɑ:nt/ *n.* S. African gold coin. [*Kruger*, name of a S. African statesman]

krummhorn /'krʌmhɔ:n/ *n.* (also **crumhorn**) medieval wind instrument. [German]

krypton /'krɪptɒn/ *n.* inert gaseous element used in fluorescent lamps etc. [Greek *kruptō* hide]

Kt. *abbr.* Knight.

kt. *abbr.* knot.

Ku *symb.* kurchatovium.

kudos /'kju:dɒs/ *n. colloq.* glory; renown. [Greek]

kudu /'ku:du:/ *n.* (also **koodoo**) (*pl.* same or **-s**) African antelope with white stripes and corkscrew-shaped ridged horns. [Xhosa]

Kufic /'kju:fɪk/ (also **Cufic**) —*n.* early angular form of the Arabic alphabet used esp. in decorative inscriptions. —*adj.* of or in this script. [from *Kufa*, city in Iraq]

Ku Klux Klan /,ku:klʌks'klæn/ *n.* secret White racist society in the southern US. [origin uncertain]

kümmel /'kʊm(ə)l/ *n.* sweet liqueur flavoured with caraway and cumin seeds. [German: related to CUMIN]

kumquat /'kʌmkwɒt/ *n.* (also **cumquat**) **1** small orange-like fruit. **2** shrub or small tree yielding this. [Chinese *kin kü* gold orange]

kung fu /kʌŋ 'fu:/ *n.* Chinese form of karate. [Chinese]

kurchatovium /,kɜ:tʃə'təʊvɪəm/ *n.* = RUTHERFORDIUM. [*Kurchatov*, name of a Russian physicist]

kV *abbr.* kilovolt(s).

kW *abbr.* kilowatt(s).

kWh *abbr.* kilowatt-hour(s).

kyle /kaɪl/ *n.* (in Scotland) narrow channel, strait. [Gaelic *caol* strait]

L

L[1] /el/ *n.* (also **l**) (*pl.* **Ls** or **L's**) **1** twelfth letter of the alphabet. **2** (as a roman numeral) 50.

L[2] *abbr.* (also **L.**) **1** learner driver. **2** Lake.

l *abbr.* (also **l.**) **1** left. **2** line. **3** litre(s).

£ *abbr.* pound(s) (money). [Latin *libra*]

LA *abbr.* Los Angeles.

La *symb.* lanthanum.

la var. of LAH.

Lab. *abbr.* Labour.

lab *n. colloq.* laboratory. [abbreviation]

label /'leɪb(ə)l/ *–n.* **1** piece of paper etc. attached to an object to give information about it. **2** short classifying phrase applied to a person etc. **3** logo, title, or trademark of a company. *–v.* (**-ll-**; *US* **-l-**) **1** attach a label to. **2** (usu. foll. by *as*) assign to a category. **3** replace (an atom) by an atom of a usu. radioactive isotope as a means of identification. [French]

labial /'leɪbɪəl/ *–adj.* **1 a** of the lips. **b** of, like, or serving as a lip. **2** (of a sound) requiring partial or complete closure of the lips. *–n.* labial sound (e.g. *p*, *m*, *v*). [Latin *labia* lips]

labium /'leɪbɪəm/ *n.* (*pl.* **labia**) (usu. in *pl.*) each fold of skin of the two pairs enclosing the vulva. [Latin, = lip]

labor etc. *US & Austral.* var. of LABOUR etc.

laboratory /lə'bɒrətərɪ/ *n.* (*pl.* **-ies**) room, building, or establishment for scientific experiments, research, chemical manufacture, etc. [Latin: related to LABORIOUS]

laborious /lə'bɔ:rɪəs/ *adj.* **1** needing hard work or toil. **2** (esp. of literary style) showing signs of toil. □ **laboriously** *adv.* [Latin: related to LABOUR]

labour /'leɪbə(r)/ (*US & Austral.* **labor**) *–n.* **1** physical or mental work; exertion. **2 a** workers, esp. manual, considered as a political and economic force. **b** (**Labour**) Labour Party. **3** process of childbirth. **4** particular task. *–v.* **1** work hard; exert oneself. **2 a** elaborate needlessly (*don't labour the point*). **b** (as **laboured** *adj.*) done with great effort; not spontaneous. **3** (often foll. by *under*) suffer under (a delusion etc.). **4** proceed with trouble or difficulty. [French from Latin *labor*, *-oris*]

labour camp *n.* prison camp enforcing a regime of hard labour.

Labour Day *n.* May 1 (or in the US and Canada the first Monday in September), celebrated in honour of working people.

labourer *n.* (*US* **laborer**) person doing unskilled, usu. manual, work for wages.

Labour Exchange *n. colloq.* or *hist.* employment exchange.

Labour Party *n.* political party formed to represent the interests of working people.

labour-saving *adj.* designed to reduce or eliminate work.

Labrador /'læbrə,dɔ:(r)/ *n.* retriever of a breed with a black or golden coat. [*Labrador* in Canada]

laburnum /lə'bɜ:nəm/ *n.* tree with drooping golden flowers yielding poisonous seeds. [Latin]

labyrinth /'læbərɪnθ/ *n.* **1** complicated network of passages etc. **2** intricate or tangled arrangement. **3** the complex structure of the inner ear. □ **labyrinthine** /-'rɪnθaɪn/ *adj.* [Latin from Greek]

lac *n.* resinous substance secreted as a protective covering by a SE Asian insect. [Hindustani]

lace *–n.* **1** fine open fabric or trimming, made by weaving thread in patterns. **2** cord etc. passed through holes or hooks for fastening shoes etc. *–v.* (**-cing**) **1** (usu. foll. by *up*) fasten or tighten with a lace or laces. **2** add spirits to (a drink). **3** (often foll. by *through*) pass (a shoelace etc.) through. [Latin *laqueus* noose]

lacerate /'læsə,reɪt/ *v.* (**-ting**) **1** mangle or tear (esp. flesh etc.). **2** cause pain to (the feelings etc.). □ **laceration** /-'reɪʃ(ə)n/ *n.* [Latin *lacer* torn]

lace-up *–n.* shoe fastened with a lace. *–attrib. adj.* (of a shoe etc.) fastened by a lace or laces.

lachrymal /'lækrɪm(ə)l/ *adj.* (also **lacrimal**) of or for tears (*lacrimal duct*). [Latin *lacrima* tear]

lachrymose /'lækrɪ,məʊs/ *adj. formal* given to weeping; tearful.

lack *–n.* (usu. foll. by *of*) want, deficiency. *–v.* be without or deficient in. [Low German or Dutch]

lackadaisical /,lækə'deɪzɪk(ə)l/ *adj.* unenthusiastic; listless; idle. □ **lackadaisically** *adv.* [from archaic *lackaday*]

lackey /'lækɪ/ *n.* (*pl.* **-s**) **1** servile follower; toady. **2** footman, manservant. [Catalan *alacay*]

lacking *adj.* absent or deficient (*money was lacking*; *is lacking in determination*).

lacklustre *adj.* (*US* **lackluster**) **1** lacking in vitality etc. **2** dull.

laconic /lə'kɒnɪk/ *adj.* terse, using few words. □ **laconically** *adv.* [Greek *Lakōn* Spartan]

lacquer /'lækə(r)/ –*n.* **1** varnish made of shellac or a synthetic substance. **2** substance sprayed on the hair to keep it in place. –*v.* coat with lacquer. [French *lacre* LAC]

lacrimal var. of LACHRYMAL.

lacrosse /lə'krɒs/ *n.* game like hockey, but with the ball carried in a crosse. [French *la* the, CROSSE]

lactate[1] /læk'teɪt/ *v.* (**-ting**) (of mammals) secrete milk. [as LACTATION]

lactate[2] /'lækteɪt/ *n.* salt or ester of lactic acid.

lactation /læk'teɪʃ(ə)n/ *n.* **1** secretion of milk. **2** suckling. [Latin: related to LACTIC]

lacteal /'læktɪəl/ –*adj.* **1** of milk. **2** conveying chyle etc. –*n.* (in *pl.*) *Anat.* vessels which absorb fats. [Latin *lacteus*: related to LACTIC]

lactic /'læktɪk/ *adj.* of milk. [Latin *lac lactis* milk]

lactic acid *n.* acid formed esp. in sour milk.

lactose /'læktəʊs/ *n.* sugar that occurs in milk.

lacuna /lə'kju:nə/ *n.* (*pl.* **lacunae** /-ni:/ or **-s**) **1** gap. **2** missing portion etc., esp. in an ancient MS etc. [Latin: related to LAKE[1]]

lacy /'leɪsɪ/ *adj.* (**-ier, -iest**) of or resembling lace fabric.

lad *n.* **1** boy, youth. **2** *colloq.* man. [origin unknown]

ladder –*n.* **1** set of horizontal bars fixed between two uprights and used for climbing up or down. **2** vertical strip of unravelled stitching in a stocking etc. **3** hierarchical structure, esp. as a means of career advancement. –*v.* **1** cause a ladder in (a stocking etc.). **2** develop a ladder. [Old English]

ladder-back *n.* upright chair with a back resembling a ladder.

lade *v.* (**-ding**; *past part.* **laden**) **1 a** load (a ship). **b** ship (goods). **2** (as **laden** *adj.*) (usu. foll. by *with*) loaded, burdened. [Old English]

la-di-da /,lɑ:dɪ'dɑ:/ *adj. colloq.* pretentious or snobbish, esp. in manner or speech. [imitative]

ladies' man *n.* (also **lady's man**) man fond of female company.

ladle /'leɪd(ə)l/ –*n.* deep long-handled spoon used for serving liquids. –*v.* (**-ling**) (often foll. by *out*) transfer (liquid) with a ladle. [Old English]

lady /'leɪdɪ/ *n.* (*pl.* **-ies**) **1 a** woman regarded as being of superior social status or as having refined manners. **b** (**Lady**) title of peeresses, female relatives of peers, the wives and widows of knights, etc. **2** (often *attrib.*) woman; female (*ask that lady*; *lady butcher*). **3** *colloq.* wife, girlfriend. **4** ruling woman (*lady of the house*). **5** (**the Ladies** or **Ladies'**) women's public lavatory. [Old English, = loaf-kneader]

ladybird *n.* small beetle, usu. red with black spots.

Lady chapel *n.* chapel dedicated to the Virgin Mary.

Lady Day *n.* Feast of the Annunciation, 25 Mar.

lady-in-waiting *n.* lady attending a queen or princess.

lady-killer *n.* habitual seducer of women.

ladylike *adj.* like or befitting a lady.

ladyship *n.* □ **her** (or **your**) **ladyship** respectful form of reference or address to a Lady.

lady's man var. of LADIES' MAN.

lady's slipper *n.* plant of the orchid family with a slipper-shaped lip on its flowers.

lag[1] –*v.* (**-gg-**) fall behind; not keep pace. –*n.* delay. [origin uncertain]

lag[2] –*v.* (**-gg-**) enclose in heat-insulating material. –*n.* insulating cover. [Old Norse]

lag[3] *n. slang* habitual convict. [origin unknown]

lager /'lɑ:gə(r)/ *n.* a kind of light effervescent beer. [German, = store]

lager lout *n. colloq.* youth behaving badly as a result of heavy drinking.

laggard /'lægəd/ *n.* person who lags behind.

lagging *n.* material used to lag a boiler etc. against loss of heat.

lagoon /lə'gu:n/ *n.* stretch of salt water separated from the sea by a sandbank, reef, etc. [Latin LACUNA pool]

lah /lɑ:/ *n.* (also **la**) *Mus.* sixth note of a major scale. [Latin *labii*, word arbitrarily taken]

laid *past* and *past part.* of LAY[1].

laid-back *adj.* relaxed; easygoing.

laid paper *n.* paper with the surface marked in fine ribs.

laid up *adj.* **1** confined to bed or the house. **2** (of a ship) out of service.

lain *past part.* of LIE[1].

lair *n.* **1** wild animal's resting-place. **2** person's hiding-place. [Old English]

laird *n. Scot.* landed proprietor. [from LORD]

laissez-faire /,leseɪ'feə(r)/ *n.* (also ***laisser-faire***) policy of non-interference. [French, = let act]

laity /'leɪɪtɪ/ *n.* lay people, as distinct from the clergy. [from LAY[2]]

lake[1] *n.* large body of water surrounded by land. [Latin *lacus*]
lake[2] *n.* **1** reddish pigment orig. made from lac. **2** pigment obtained by combining an organic colouring matter with a metallic oxide, hydroxide, or salt. [var. of LAC]
Lake District *n.* (also **the Lakes**) region of lakes in Cumbria.
lakh /læk/ *n. Ind.* (usu. foll. by *of*) hundred thousand (rupees etc.). [Hindustani *lākh*]
lam *v.* (**-mm-**) *slang* thrash; hit. [perhaps Scandinavian]
lama /'lɑ:mə/ *n.* Tibetan or Mongolian Buddhist monk. [Tibetan]
lamasery /'lɑ:məsərɪ/ *n.* (*pl.* **-ies**) monastery of lamas. [French]
lamb /læm/ *–n.* **1** young sheep. **2** its flesh as food. **3** mild, gentle, or kind person. *–v.* give birth to lambs. □ **The Lamb** (or **Lamb of God**) name for Christ. [Old English]
lambada /læm'bɑ:də/ *n.* fast erotic Brazilian dance in which couples dance with their stomachs touching each other. [Portuguese, = a beating]
lambaste /læm'beɪst/ *v.* (**-ting**) (also **lambast** /-'bæst/) *colloq.* thrash, beat. [from LAM, BASTE[1]]
lambda /'læmdə/ *n.* eleventh letter of the Greek alphabet (Λ, λ). [Greek]
lambent /'læmbənt/ *adj.* **1** (of a flame or a light) playing on a surface. **2** (of the eyes, sky, wit, etc.) lightly brilliant. □ **lambency** *n.* [Latin *lambo* lick]
lambswool *n.* soft fine wool from a young sheep.
lame *–adj.* **1** disabled in the foot or leg. **2 a** (of an excuse etc.) unconvincing; feeble. **b** (of verse etc.) halting. *–v.* (**-ming**) make lame; disable. □ **lamely** *adv.* **lameness** *n.* [Old English]
lamé /'lɑ:meɪ/ *n.* fabric with gold or silver threads interwoven. [French]
lame duck *n.* helpless person or firm.
lament /lə'ment/ *–n.* **1** passionate expression of grief. **2** song etc. of mourning etc. *–v.* (also *absol.*) **1** express or feel grief for or about. **2** (as **lamented** *adj.*) used to refer to a recently dead person. □ **lament for** (or **over**) mourn or regret. [Latin *lamentor*]
lamentable /'læməntəb(ə)l/ *adj.* deplorable, regrettable. □ **lamentably** *adv.*
lamentation /ˌlæmən'teɪʃ(ə)n/ *n.* **1** lamenting. **2** lament.
lamina /'læmɪnə/ *n.* (*pl.* **-nae** /-ˌni:/) thin plate or scale. □ **laminar** *adj.* [Latin]
laminate *–v.* /'læmɪˌneɪt/ (**-ting**) **1** beat or roll into thin plates. **2** overlay with metal plates, a plastic layer, etc. **3** split into layers. *–n.* /'læmɪnət/ laminated structure, esp. of layers fixed together. *–adj.* /'læmɪnət/ in the form of thin plates. □ **lamination** /-'neɪʃ(ə)n/ *n.*
Lammas /'læməs/ *n.* (in full **Lammas Day**) first day of August, formerly kept as harvest festival. [Old English: related to LOAF[1], MASS[2]]
lamp *n.* **1** device for producing a steady light, esp.: **a** an electric bulb, and usu. its holder. **b** an oil-lamp. **c** a gas-jet and mantle. **2** device producing esp. ultraviolet or infrared radiation. [Greek *lampas* torch]
lampblack *n.* pigment made from soot.
lamplight *n.* light from a lamp.
lamplighter *n. hist.* person who lit street lamps.
lampoon /læm'pu:n/ *–n.* satirical attack on a person etc. *–v.* satirize. □ **lampoonist** *n.* [French *lampon*]
lamppost *n.* tall post supporting a street-light.
lamprey /'læmprɪ/ *n.* (*pl.* **-s**) eel-like aquatic animal with a sucker mouth. [Latin *lampreda*]
lampshade *n.* translucent cover for a lamp.
Lancastrian /læŋ'kæstrɪən/ *–n.* **1** native of Lancashire or Lancaster. **2** *hist.* member or supporter of the House of Lancaster in the Wars of the Roses. *–adj.* of or concerning Lancashire or Lancaster, or the House of Lancaster. [*Lancaster* in Lancashire]
lance /lɑ:ns/ *–n.* long spear, esp. one used by a horseman. *–v.* (**-cing**) **1** prick or cut open with a lancet. **2** pierce with a lance. [French from Latin]
lance-corporal *n.* lowest rank of NCO in the Army.
lanceolate /'lɑ:nsɪələt/ *adj.* shaped like a lance-head, tapering at each end.
lancer *n.* **1** *hist.* soldier of a cavalry regiment armed with lances. **2** (in *pl.*) **a** quadrille. **b** music for this.
lancet /'lɑ:nsɪt/ *n.* small broad two-edged surgical knife with a sharp point.
lancet arch *n.* (also **lancet light** or **window**) narrow arch or window with a pointed head.
land *–n.* **1** solid part of the earth's surface. **2 a** expanse of country; ground, soil. **b** this in relation to its use, quality, etc., or as a basis for agriculture. **3** country, nation, State. **4 a** landed property. **b** (in *pl.*) estates. *–v.* **1 a** set or go ashore. **b** (often foll. by *at*) disembark. **2** bring (an aircraft) to the ground or another surface. **3** alight on the ground etc. **4** bring (a fish) to land. **5** (also *refl.*; often foll. by *up*) *colloq.* bring to, reach, or find oneself in a certain situation or place. **6** *colloq.* **a** deal (a person etc. a blow etc.). **b** (foll. by *with*) present (a

person) with (a problem, job, etc.). **7** *colloq.* win or obtain (a prize, job, etc.). □ **how the land lies** what is the state of affairs. **land on one's feet** attain a good position, job, etc., by luck. □ **landless** *adj.* [Old English]

land-agent *n.* **1** steward of an estate. **2** agent for the sale of estates.

landau /'lændɔː/ *n.* four-wheeled enclosed carriage with a divided top. [*Landau* in Germany]

landed *adj.* **1** owning land. **2** consisting of land.

landfall *n.* approach to land, esp. after a sea or air journey.

landfill *n.* **1** waste material etc. used to landscape or reclaim land. **2** process of disposing of rubbish in this way.

land-girl *n.* woman doing farm work, esp. in wartime.

landing *n.* **1** platform at the top of or part way up a flight of stairs. **2** coming to land. **3** place where ships etc. land.

landing-craft *n.* craft designed for putting troops and equipment ashore.

landing-gear *n.* undercarriage of an aircraft.

landing-stage *n.* platform for disembarking goods and passengers.

landlady *n.* **1** woman who owns and lets land or premises. **2** woman who keeps a public house, boarding-house, etc.

land line *n.* means of telecommunication over land.

landlocked *adj.* almost or entirely enclosed by land.

landlord *n.* **1** man who owns and lets land or premises. **2** man who keeps a public house, boarding-house, etc.

landlubber *n.* person unfamiliar with the sea.

landmark *n.* **1** conspicuous object in a district, landscape, etc. **2** prominent and critical event etc.

land mass *n.* large area of land.

land-mine *n.* explosive mine laid in or on the ground.

landowner *n.* owner of (esp. much) land. □ **landowning** *adj.* & *n.*

landscape /'lændskeɪp/ —*n.* **1** scenery as seen in a broad view. **2** (often *attrib.*) picture representing this; this genre of painting. —*v.* (**-ping**) improve (a piece of land) by landscape gardening. [Dutch *landscap*]

landscape gardening *n.* laying out of grounds to resemble natural scenery.

landslide *n.* **1** sliding down of a mass of land from a mountain, cliff, etc. **2** overwhelming victory in an election.

landslip *n.* = LANDSLIDE 1.

lane *n.* **1** narrow road. **2** division of a road for a stream of traffic. **3** strip of track etc. for a competitor in a race. **4** path regularly followed by a ship, aircraft, etc. **5** gangway between crowds of people. [Old English]

language /'læŋgwɪdʒ/ *n.* **1** use of words in an agreed way as a method of human communication. **2** system of words of a particular community or country etc. **3 a** faculty of speech. **b** style of expression; use of words, etc. (*poetic language*). **4** system of symbols and rules for writing computer programs. **5** any method of communication. **6** professional or specialized vocabulary. [Latin *lingua* tongue]

language laboratory *n.* room equipped with tape recorders etc. for learning a foreign language.

languid /'læŋgwɪd/ *adj.* lacking vigour; idle; inert. □ **languidly** *adv.* [related to LANGUISH]

languish /'læŋgwɪʃ/ *v.* lose or lack vitality. □ **languish for** droop or pine for. **languish under** suffer under (depression, confinement, etc.). [Latin *langueo*]

languor /'læŋgə(r)/ *n.* **1** lack of energy; idleness. **2** soft or tender mood or effect. **3** oppressive stillness. □ **languorous** *adj.*

lank *adj.* **1** (of hair, grass, etc.) long and limp. **2** thin and tall. [Old English]

lanky *adj.* (**-ier, -iest**) ungracefully thin and long or tall. □ **lankiness** *n.*

lanolin /'lænəlɪn/ *n.* fat found on sheep's wool and used in cosmetics etc. [Latin *lana* wool, *oleum* OIL]

lantern /'lænt(ə)n/ *n.* **1** lamp with a transparent case protecting a flame etc. **2** raised structure on a dome, room, etc., glazed to admit light. **3** light-chamber of a lighthouse. [Greek *lamptēr* torch]

lantern jaws *n.pl.* long thin jaws and chin.

lanthanide /'lænθə,naɪd/ *n.* any element of the lanthanide series. [German: related to LANTHANUM]

lanthanide series *n. Chem.* series of 15 metallic elements from lanthanum to lutetium in the periodic table, having similar chemical properties.

lanthanum /'lænθənəm/ *n.* metallic element, first of the lanthanide series. [Greek *lanthanō* escape notice]

lanyard /'lænjəd/ *n.* **1** cord worn round the neck or the shoulder, to which a knife etc. may be attached. **2** *Naut.* short rope or line used for securing, tightening, etc. [French *laniere*, assimilated to YARD[1]]

Laodicean /,leɪəʊdɪ'siːən/ half-hearted, esp. in religion or politics. [*Laodicea* in Asia Minor (Rev. 3:16)]

lap[1] *n.* **1** front of the body from the waist to the knees of a sitting person. **2** clothing covering this. □ **in the lap of the gods** beyond human control. **in the lap of luxury** in extremely luxurious surroundings. [Old English]

lap[2] *–n.* **1 a** one circuit of a racetrack etc. **b** section of a journey etc. **2 a** amount of overlapping. **b** overlapping part. **3** single turn of thread etc. round a reel etc. *–v.* (**-pp-**) **1** lead or overtake (a competitor in a race) by one or more laps. **2** (often foll. by *about, round*) fold or wrap (a garment etc.) round. **3** (usu. foll. by *in*) enfold in wraps etc. **4** (as **lapped** *adj.*) (usu. foll. by *in*) enfolded caressingly. **5** cause to overlap. [probably from LAP[1]]

lap[3] *–v.* (**-pp-**) **1 a** (esp. of an animal) drink with the tongue. **b** (usu. foll. by *up, down*) consume (liquid) greedily. **c** (usu. foll. by *up*) consume (gossip, praise, etc.) greedily. **2** (of waves etc.) ripple; make a lapping sound against (the shore). *–n.* **1 a** act of lapping. **b** amount of liquid taken up. **2** sound of wavelets. [Old English]

lap-dog *n.* small pet dog.

lapel /ləˈpel/ *n.* part of either side of a coat-front etc., folded back against itself. [from LAP[1]]

lapidary /ˈlæpɪdərɪ/ *–adj.* **1** concerned with stone or stones. **2** engraved upon stone. **3** concise, well-expressed, epigrammatic. *–n.* (*pl.* **-ies**) cutter, polisher, or engraver, of gems. [Latin *lapis lapid-* stone]

lapis lazuli /ˌlæpɪs ˈlæzjʊlɪ/ *n.* **1** blue mineral used as a gemstone. **2** bright blue pigment. **3** its colour. [related to LAPIDARY, AZURE]

Laplander /ˈlæpˌlændə(r)/ *n.* native or inhabitant of Lapland; Lapp. [as LAPP]

lap of honour *n.* ceremonial circuit of a racetrack etc. by a winner.

Lapp *n.* **1** member of a Mongol people of N. Scandinavia and NW Russia. **2** their language. [Swedish]

lappet /ˈlæpɪt/ *n.* **1** small flap or fold of a garment etc. **2** hanging piece of flesh. [from LAP[1]]

lapse *–n.* **1** slight error; slip of memory etc. **2** weak or careless decline into an inferior state. **3** (foll. by *of*) passage of time. *–v.* (**-sing**) **1** fail to maintain a position or standard. **2** (foll. by *into*) fall back into an inferior or previous state. **3** (of a right or privilege etc.) become invalid through disuse, failure to renew, etc. **4** (as **lapsed** *adj.*) that has lapsed. [Latin *lapsus* from *labor laps-* slip]

laptop *n.* (often *attrib.*) portable microcomputer suitable for use while travelling.

lapwing /ˈlæpwɪŋ/ *n.* plover with a shrill cry. [Old English: related to LEAP, WINK: from its mode of flight]

larboard /ˈlɑːbəd/ *n.* & *adj. archaic* = PORT[3]. [originally *ladboard*, perhaps 'side on which cargo was taken in': related to LADE]

larceny /ˈlɑːsənɪ/ *n.* (*pl.* **-ies**) theft of personal property. □ **larcenous** *adj.* [Anglo-French from Latin *latrocinium*]

■ **Usage** In 1968 *larceny* was replaced as a statutory crime in English law by *theft*.

larch *n.* **1** deciduous coniferous tree with bright foliage. **2** its wood. [Latin *larix -icis*]

lard *–n.* pig fat used in cooking etc. *–v.* **1** insert strips of fat or bacon in (meat etc.) before cooking. **2** (foll. by *with*) garnish (talk etc.) with strange terms. [French = bacon, from Latin *lardum*]

larder *n.* room or large cupboard for storing food.

lardy *adj.* like lard.

lardy-cake *n.* cake made with lard, currants, etc.

large *adj.* **1** of relatively great size or extent. **2** of the larger kind (*large intestine*). **3** comprehensive. **4** pursuing an activity on a large scale (*large farmer*). □ **at large 1** at liberty. **2** as a body or whole. **3** at full length, with all details. □ **largeness** *n.* **largish** *adj.* [Latin *largus* copious]

large as life *adj. colloq.* in person, esp. prominently.

largely *adv.* to a great extent (*largely my own fault*).

large-scale *adj.* made or occurring on a large scale.

largesse /lɑːˈʒes/ *n.* (also **largess**) money or gifts freely given. [Latin *largus*: related to LARGE]

largo /ˈlɑːɡəʊ/ *Mus. –adv.* & *adj.* in a slow tempo and dignified style. *–n.* (*pl.* **-s**) largo passage or movement. [Italian, = broad]

lariat /ˈlærɪət/ *n.* **1** lasso. **2** tethering-rope. [Spanish *la reata*]

lark[1] *n.* small bird with a tuneful song, esp. the skylark. [Old English]

lark[2] *colloq. –n.* **1** frolic; amusing incident. **2** type of activity (*fed up with this digging lark*). *–v.* (foll. by *about*) play tricks. [origin uncertain]

larkspur *n.* plant with a spur-shaped calyx.

larva /ˈlɑːvə/ *n.* (*pl.* **-vae** /-viː/) stage of an insect's development between egg and pupa. □ **larval** *adj.* [Latin, = ghost]

laryngeal /lə'rɪndʒɪəl/ *adj.* of the larynx.

laryngitis /,lærɪn'dʒaɪtɪs/ *n.* inflammation of the larynx.

larynx /'lærɪŋks/ *n.* (*pl.* **larynges** /lə'rɪndʒi:z/ or **-xes**) hollow organ in the throat holding the vocal cords. [Latin from Greek]

lasagne /lə'sænjə/ *n.* pasta in the form of sheets. [Italian pl., from Latin *lasanum* cooking-pot]

lascivious /lə'sɪvɪəs/ *adj.* **1** lustful. **2** inciting to lust. □ **lasciviously** *adv.* [Latin]

laser /'leɪzə(r)/ *n.* device that generates an intense beam of coherent light, or other electromagnetic radiation, in one direction. [*l*ight *a*mplification by *s*timulated *e*mission of *r*adiation]

lash *–v.* **1** make a sudden whiplike movement. **2** beat with a whip etc. **3** (often foll. by *against*, *down*, etc.) (of rain etc.) beat, strike. **4** criticize harshly. **5** rouse, incite. **6** (foll. by *down*, *together*, etc.) fasten with a cord etc. *–n.* **1** sharp blow made by a whip etc. **2** flexible end of a whip. **3** eyelash. □ **lash out 1** speak or hit out angrily. **2** *colloq.* spend money extravagantly. [imitative]

lashings *n.pl. colloq.* (foll. by *of*) plenty.

lass *n.* esp. *Scot.* & *N.Engl.* or *poet.* girl. [Old Norse]

Lassa fever /'læsə/ *n.* acute febrile viral disease of tropical Africa. [*Lassa* in Nigeria]

lassitude /'læsɪ,tju:d/ *n.* **1** languor. **2** disinclination to exert oneself. [Latin *lassus* tired]

lasso /læ'su:/ *–n.* (*pl.* **-s** or **-es**) rope with a noose at one end, esp. for catching cattle. *–v.* (**-es**, **-ed**) catch with a lasso. [Spanish *lazo*: related to LACE]

last[1] /lɑ:st/ *–adj.* **1** after all others; coming at or belonging to the end. **2** most recent; next before a specified time (*last Christmas*). **3** only remaining (*last chance*). **4** (prec. by *the*) least likely or suitable (*the last person I'd want*). **5** lowest in rank (*last place*). *–adv.* **1** after all others (esp. in *comb.*: *last-mentioned*). **2** on the most recent occasion (*when did you last see him?*). **3** lastly. *–n.* **1** person or thing that is last, last-mentioned, most recent, etc. **2** (prec. by *the*) last mention or sight etc. (*shall never hear the last of it*). **3** last performance of certain acts (*breathed his last*). **4** (prec. by *the*) the end; death (*fighting to the last*). □ **at last** (or **long last**) in the end; after much delay. [Old English, = latest]

last[2] /lɑ:st/ *v.* **1** remain unexhausted or alive for a specified or considerable time (*food to last a week*). **2** continue for a specified time (*match lasts an hour*). □ **last out** be strong enough or sufficient for the whole of a given period. [Old English]

last[3] /lɑ:st/ *n.* shoemaker's model for shaping a shoe etc. □ **stick to one's last** not meddle in what one does not understand. [Old English]

last-ditch *attrib. adj.* (of an attempt etc.) final, desperate.

lasting *adj.* permanent; durable.

lastly *adv.* finally; in the last place.

last minute *n.* (also **last moment**) the time just before an important event (often (with hyphen) *attrib.*: *last-minute panic*).

last name *n.* surname.

last post *n.* bugle-call at military funerals or as a signal to retire for the night.

last rites *n.pl.* rites for a person about to die.

last straw *n.* (prec. by *the*) slight addition to a burden that makes it finally unbearable.

last trump *n.* (prec. by *the*) trumpet-blast to wake the dead on Judgement Day.

last word *n.* (prec. by *the*) **1** final or definitive statement. **2** (often foll. by *in*) latest fashion.

lat. *abbr.* latitude.

latch *–n.* **1** bar with a catch and lever as a fastening for a gate etc. **2** spring-lock preventing a door from being opened from the outside without a key. *–v.* fasten with a latch. □ **latch on** (often foll. by *to*) *colloq.* **1** attach oneself (to). **2** understand. **on the latch** fastened by the latch (sense 1) only. [Old English]

latchkey *n.* (*pl.* **-s**) key of an outer door.

late *–adj.* **1** after the due or usual time; occurring or done after the proper time. **2 a** far on in the day or night or in a specified period. **b** far on in development. **3** flowering or ripening towards the end of the season. **4** no longer alive; no longer having the specified status, former (*my late husband*; *the late prime minister*). **5** of recent date. *–adv.* **1** after the due or usual time. **2** far on in time. **3** at or till a late hour. **4** at a late stage of development. **5** formerly but not now (*late of the Scillies*). □ **late in the day** *colloq.* at a late stage in the proceedings. □ **lateness** *n.* [Old English]

latecomer *n.* person who arrives late.

lateen /lə'ti:n/ *adj.* (of a ship) rigged with a lateen sail. [French *voile latine* Latin sail]

lateen sail *n.* triangular sail on a long yard at an angle of 45° to the mast.

lately *adv.* not long ago; recently. [Old English: related to LATE]

latent /'leɪt(ə)nt/ *adj.* existing but not developed or manifest; concealed, dormant. □ **latency** *n.* [Latin *lateo* be hidden]
latent heat *n. Physics* heat required to convert a solid into a liquid or vapour, or a liquid into a vapour, without change of temperature.
lateral /'lætər(ə)l/ *–adj.* **1** of, at, towards, or from the side or sides. **2** descended from the sibling of a person in direct line. *–n.* lateral shoot or branch. □ **laterally** *adv.* [Latin *latus later-* side]
lateral thinking *n.* method of solving problems other than by using conventional logic.
latex /'leɪteks/ *n.* (*pl.* **-xes**) **1** milky fluid of esp. the rubber tree. **2** synthetic product resembling this. [Latin, = liquid]
lath /lɑ:θ/ *n.* (*pl.* **laths** /lɑ:θs, lɑ:ðz/) thin flat strip of wood. [Old English]
lathe /leɪð/ *n.* machine for shaping wood, metal, etc., by rotating the article against cutting tools. [origin uncertain]
lather /'lɑ:ðə(r)/ *–n.* **1** froth produced by agitating soap etc. and water. **2** frothy sweat. **3** state of agitation. *–v.* **1** (of soap etc.) form a lather. **2** cover with lather. **3** *colloq.* thrash. [Old English]
Latin /'lætɪn/ *–n.* language of ancient Rome and its empire. *–adj.* **1** of or in Latin. **2** of the countries or peoples using languages descended from Latin. **3** of the Roman Catholic Church. [Latin *Latium* district around Rome]
Latin America *n.* parts of Central and S. America where Spanish or Portuguese is the main language.
Latinate /'lætɪ,neɪt/ *adj.* having the character of Latin.
Latinize /'lætɪ,naɪz/ *v.* (also **-ise**) (**-zing** or **-sing**) give a Latin form to. □ **Latinization** /-'zeɪʃ(ə)n/ *n.*
latish *adj.* & *adv.* fairly late.
latitude /'lætɪ,tju:d/ *n.* **1 a** angular distance on a meridian north or south of the equator. **b** (usu. in *pl.*) regions or climes. **2** tolerated variety of action or opinion. □ **latitudinal** /-'tju:dɪn(ə)l/ *adj.* [Latin *latus* broad]
latitudinarian /,lætɪ,tju:dɪ'neərɪən/ *–adj.* liberal, esp. in religion. *–n.* latitudinarian person.
latrine /lə'tri:n/ *n.* communal lavatory, esp. in a camp. [Latin *latrina*]
latter *adj.* **1 a** second-mentioned of two, or last-mentioned of three or more. **b** (prec. by *the*; usu. *absol.*) the second- or last-mentioned person or thing. **2** nearer the end (*latter part of the year*). **3** recent. **4** of the end of a period, the world, etc. [Old English, = later]

■ **Usage** The use of *latter* to mean 'last mentioned of three or more' is considered incorrect by some people.

latter-day *attrib. adj.* modern, contemporary.
Latter-day Saints *n.pl.* Mormons' name for themselves.
latterly *adv.* **1** recently. **2** in the latter part of life or a period.
lattice /'lætɪs/ *n.* **1** structure of crossed laths or bars with spaces between, used as a screen, fence, etc. **2** regular periodic arrangement of atoms, ions, or molecules. □ **latticed** *adj.* [French *lattis* from *latte* LATH]
lattice window *n.* window with small panes set in diagonally crossing strips of lead.
Latvian /'lætvɪən/ *–n.* **1 a** native or national of Latvia in eastern Europe. **b** person of Latvian descent. **2** language of Latvia. *–adj.* of Latvia, its people, or language.
laud /lɔ:d/ *–v.* praise or extol. *–n.* **1** praise; hymn of praise. **2** (in *pl.*) the first morning prayer of the Roman Catholic Church. [Latin *laus laud-*]
laudable *adj.* commendable. □ **laudability** /-'bɪlɪtɪ/ *n.* **laudably** *adv.*

■ **Usage** *Laudable* is sometimes confused with *laudatory*.

laudanum /'lɔ:dnəm/ *n.* solution prepared from opium. [perhaps from medieval Latin]
laudatory /'lɔ:dətərɪ/ *adj.* praising.

■ **Usage** *Laudatory* is sometimes confused with *laudable*.

laugh /lɑ:f/ *–v.* **1** make the sounds and movements usual in expressing lively amusement, scorn, etc. **2** express by laughing. **3** (foll. by *at*) ridicule, make fun of. *–n.* **1** sound, act, or manner of laughing. **2** *colloq.* comical thing. □ **laugh off** get rid of (embarrassment or humiliation) by joking. **laugh up one's sleeve** laugh secretly. [Old English]
laughable *adj.* ludicrous; amusing. □ **laughably** *adv.*
laughing *n.* laughter. □ **no laughing matter** serious matter. □ **laughingly** *adv.*
laughing-gas *n.* nitrous oxide as an anaesthetic.
laughing jackass *n.* = KOOKABURRA.
laughing stock *n.* person or thing open to general ridicule.
laughter /'lɑ:ftə(r)/ *n.* act or sound of laughing. [Old English]
launch[1] /lɔ:ntʃ/ *–v.* **1** set (a vessel) afloat. **2** hurl or send forth (a weapon, rocket, etc.). **3** start or set in motion (an

enterprise, person, etc.). **4** formally introduce (a new product) with publicity etc. **5** (foll. by *out*, *into*, etc.) **a** make a start on (an enterprise etc.). **b** burst into (strong language etc.). —*n.* act of launching. [Anglo-Norman *launcher*: related to LANCE]

launch[2] /lɔ:ntʃ/ *n.* **1** large motor boat. **2** man-of-war's largest boat. [Spanish *lancha*]

launcher *n.* structure to hold a rocket during launching.

launch pad *n.* (also **launching pad**) platform with a supporting structure, for launching rockets from.

launder /'lɔ:ndə(r)/ *v.* **1** wash and iron (clothes etc.). **2** *colloq.* transfer (funds) to conceal their origin. [French: related to LAVE]

launderette /lɔ:n'dret/ *n.* (also **laundrette**) establishment with coin-operated washing-machines and driers for public use.

laundress /'lɔ:ndrɪs/ *n.* woman who launders, esp. professionally.

laundry /'lɔ:ndrɪ/ *n.* (*pl.* **-ies**) **1 a** place for washing clothes etc. **b** firm washing clothes etc. commercially. **2** clothes or linen for laundering or newly laundered.

laureate /'lɒrɪət, 'lɔ:-/ —*adj.* wreathed with laurel as a mark of honour. —*n.* = POET LAUREATE. □ **laureateship** *n.* [related to LAUREL]

laurel /'lɒr(ə)l/ *n.* **1** = BAY². **2** (in *sing.* or *pl.*) wreath of bay-leaves as an emblem of victory or poetic merit. **3** any of various plants with dark-green glossy leaves. □ **look to one's laurels** beware of losing one's pre-eminence. **rest on one's laurels** see REST. [Latin *laurus* bay]

lav *n. colloq.* lavatory. [abbreviation]

lava /'lɑ:və/ *n.* matter flowing from a volcano and solidifying as it cools. [Latin *lavo* wash]

lavatorial /ˌlævə'tɔ:rɪəl/ *adj.* of or like lavatories; (esp. of humour) relating to excretion.

lavatory /'lævətərɪ/ *n.* (*pl.* **-ies**) **1** receptacle for urine and faeces, usu. with a means of disposal. **2** room or compartment containing this. [Latin: related to LAVA]

lavatory paper *n.* = TOILET PAPER.

lave *v.* (**-ving**) *literary* **1** wash, bathe. **2** (of water) wash against; flow along. [Latin *lavo* wash]

lavender /'lævɪndə(r)/ *n.* **1 a** evergreen shrub with purple aromatic flowers. **b** its flowers and stalks dried and used to scent linen etc. **2** pale mauve colour. [Latin *lavandula*]

lavender-water *n.* light perfume made with distilled lavender.

laver /'leɪvə(r), 'lɑ:-/ *n.* edible seaweed. [Latin]

lavish /'lævɪʃ/ —*adj.* **1** giving or producing in large quantities; profuse. **2** generous. —*v.* (often foll. by *on*) bestow or spend (money, effort, praise, etc.) abundantly. □ **lavishly** *adv.* [French *lavasse* deluge: related to LAVE]

law *n.* **1 a** rule enacted or customary in a community and recognized as commanding or forbidding certain actions. **b** body of such rules. **2** controlling influence of laws; respect for laws. **3** laws collectively as a social system or subject of study. **4** binding force (*her word is law*). **5** (prec. by *the*) **a** the legal profession. **b** *colloq.* the police. **6** (in *pl.*) jurisprudence. **7 a** the judicial remedy. **b** the lawcourts as providing this (*go to law*). **8** rule of action or procedure. **9** regularity in natural occurrences (*laws of nature*; *law of gravity*). **10** divine commandments. □ **be a law unto oneself** do what one considers right; disregard custom. **lay down the law** be dogmatic or authoritarian. **take the law into one's own hands** redress a grievance by one's own means, esp. by force. [Old English from Old Norse, = thing laid down]

law-abiding *adj.* obedient to the laws.

lawbreaker *n.* person who breaks the law. □ **lawbreaking** *n.* & *adj.*

lawcourt *n.* court of law.

lawful *adj.* conforming with or recognized by law; not illegal. □ **lawfully** *adv.* **lawfulness** *n.*

lawgiver *n.* person who formulates laws; legislator.

lawless *adj.* **1** having no laws or law enforcement. **2** disregarding laws. □ **lawlessness** *n.*

Law Lord *n.* member of the House of Lords qualified to perform its legal work.

lawmaker *n.* legislator.

lawn[1] *n.* piece of closely-mown grass in a garden etc. [French *launde* glade]

lawn[2] *n.* fine linen or cotton. [probably from *Laon* in France]

lawnmower *n.* machine for cutting lawns.

lawn tennis *n.* tennis played with a soft ball on outdoor grass or a hard court.

lawrencium /lə'rensɪəm/ *n.* artificially made transuranic metallic element. [*Lawrence*, name of a physicist]

lawsuit *n.* bringing of a dispute, claim, etc. before a lawcourt.

lawyer /'lɔ:jə(r)/ *n.* legal practitioner, esp. a solicitor.

lax *adj.* **1** lacking care or precision. **2** not strict. □ **laxity** *n.* **laxly** *adv.* **laxness** *n.* [Latin *laxus* loose]

laxative /ˈlæksətɪv/ *–adj.* facilitating evacuation of the bowels. *–n.* laxative medicine. [Latin: related to LAX]

lay[1] *–v.* (*past* and *past part.* **laid**) **1** place on a surface, esp. horizontally or in the proper or specified place. **2** put or bring into the required position or state (*lay carpet*). **3** make by laying (*lay foundations*). **4** (often *absol.*) (of a hen bird) produce (an egg). **5** cause to subside or lie flat. **6** (usu. foll. by *on*); attribute or impute (blame etc.). **7** prepare or make ready (a plan or trap). **8** prepare (a table) for a meal. **9** arrange the material for (a fire). **10** put down as a wager; stake. **11** (foll. by *with*) coat or strew (a surface). **12** *slang offens.* have sexual intercourse with (esp. a woman). *–n.* **1** way, position, or direction in which something lies. **2** *slang offens.* partner (esp. female) in, or act of, sexual intercourse. □ **lay about one** hit out on all sides. **lay aside 1** put to one side. **2** cease to consider. **lay at the door of** impute to. **lay bare** expose, reveal. **lay claim to** claim as one's own. **lay down 1** put on a flat surface. **2** give up (an office). **3** formulate (a rule). **4** store (wine) for maturing. **5** sacrifice (one's life). **lay (one's) hands on** obtain, locate. **lay hands on** seize or attack. **lay hold of** seize. **lay in** provide oneself with a stock of. **lay into** *colloq.* punish or scold harshly. **lay it on thick** (or **with a trowel**) *colloq.* flatter or exaggerate grossly. **lay low** overthrow or humble. **lay off 1** discharge (unneeded workers) temporarily; make redundant. **2** *colloq.* desist. **lay on 1** provide. **2** impose. **3** inflict (blows). **4** spread on (paint etc.). **lay open 1** break the skin of. **2** (foll. by *to*) expose (to criticism etc.). **lay out 1** spread out, expose to view. **2** prepare (a corpse) for burial. **3** *colloq.* knock unconscious. **4** arrange (grounds etc.) according to a design. **5** expend (money). **lay to rest** bury in a grave. **lay up** store, save. **lay waste** ravage, destroy. [Old English]

■ **Usage** The intransitive use of *lay*, meaning *lie*, as in *she was laying on the floor*, is incorrect in standard English.

lay[2] *adj.* **1 a** non-clerical. **b** not ordained into the clergy. **2 a** not professionally qualified. **b** of or done by such persons. [Greek *laos* people]

lay[3] *n.* **1** short poem meant to be sung. **2** song. [French]

lay[4] *past* of LIE[1].

layabout *n.* habitual loafer or idler.

lay-by *n.* (*pl.* **-bys**) area at the side of a road where vehicles may stop.

layer *–n.* **1** thickness of matter, esp. one of several, covering a surface. **2** person or thing that lays. **3** hen that lays eggs. **4** shoot fastened down to take root while attached to the parent plant. *–v.* **1** arrange in layers. **2** cut (hair) in layers. **3** propagate (a plant) by a layer.

layette /leɪˈet/ *n.* set of clothing etc. for a newborn child. [French from Dutch]

lay figure *n.* **1** jointed figure of a human body used by artists for arranging drapery on etc. **2** unrealistic character in a novel etc. [Dutch *led* joint]

layman *n.* (*fem.* **laywoman**) **1** non-ordained member of a Church. **2** person without professional or specialized knowledge.

lay-off *n.* temporary discharge of workers; a redundancy.

layout *n.* **1** way in which land, a building, printed matter, etc., is arranged or set out. **2** something arranged in a particular way; display.

lay reader *n.* lay person licensed to conduct some religious services.

laze *–v.* (**-zing**) **1** spend time idly. **2** (foll. by *away*) pass (time) idly. *–n.* spell of lazing. [back-formation from LAZY]

lazy *adj.* (**-ier, -iest**) **1** disinclined to work, doing little work. **2** of or inducing idleness. □ **lazily** *adv.* **laziness** *n.* [perhaps from Low German]

lazybones *n.* (*pl.* same) *colloq.* lazy person.

lb *abbr.* pound(s) (weight). [Latin *libra*]

LBC *abbr.* London Broadcasting Company.

l.b.w. *abbr.* leg before wicket.

l.c. *abbr.* **1** = LOC. CIT. **2** lower case.

LCD *abbr.* **1** liquid crystal display. **2** lowest (or least) common denominator.

LCM *abbr.* lowest (or least) common multiple.

L/Cpl *abbr.* Lance-Corporal.

Ld. *abbr.* Lord.

LEA *abbr.* Local Education Authority.

lea *n. poet.* meadow, field. [Old English]

leach *v.* **1** make (a liquid) percolate through some material. **2** subject (bark, ore, ash, or soil) to the action of percolating fluid. **3** (foll. by *away, out*) remove (soluble matter) or be removed in this way. [Old English]

lead[1] /liːd/ *–v.* (*past* and *past part.* **led**) **1** cause to go with one, esp. by guiding or going in front. **2 a** direct the actions or opinions of. **b** (often foll. by *to*, or *to* + infin.) guide by persuasion or example (*what led you to think that*). **3** (also *absol.*) provide access to; bring to a certain position (*gate leads you into a field; road leads to Lincoln*). **4** pass or go through (a life etc. of a specified kind). **5 a** have the first place in. **b** (*absol.*) go first; be ahead in a race etc. **c** (*absol.*) be pre-eminent in some field. **6** be in charge of (*leads a team*). **7** (also *absol.*) play (a card) or a card of (a particular suit) as first player in a round. **8** (foll. by *to*) result in. **9** (foll. by *with*) (of a newspaper or news broadcast) have as its main story (*led with the royal wedding*). **10** (foll. by *through*) make (a liquid, strip of material, etc.) pass through a certain course. *–n.* **1** guidance given by going in front; example. **2 a** leading place (*take the lead*). **b** amount by which a competitor is ahead of the others. **3** clue. **4** strap etc. for leading a dog etc. **5** conductor (usu. a wire) conveying electric current to an appliance. **6 a** chief part in a play etc. **b** person playing this. **c** (*attrib.*) chief performer or instrument of a specified type (*lead guitar*). **7** *Cards* **a** act or right of playing first. **b** card led. □ **lead by the nose** cajole into compliance. **lead off** begin. **lead on** entice dishonestly. **lead up the garden path** *colloq.* mislead. **lead up to** form a preparation for; direct conversation towards. [Old English]

lead[2] /led/ *–n.* **1** heavy bluish-grey soft metallic element. **2 a** graphite. **b** thin length of this in a pencil. **3** lump of lead used in sounding water. **4** (in *pl.*) **a** strips of lead covering a roof. **b** piece of lead-covered roof. **5** (in *pl.*) lead frames holding the glass of a lattice etc. **6** blank space between lines of print. *–v.* **1** cover, weight, or frame with lead. **2** space (printed matter) with leads. [Old English]

leaden /'led(ə)n/ *adj.* **1** of or like lead. **2** heavy or slow. **3** lead-coloured.

leader *n.* **1 a** person or thing that leads. **b** person followed by others. **2** principal player in a music group or of the first violins in an orchestra. **3** = LEADING ARTICLE. **4** shoot of a plant at the apex of a stem or of the main branch. □ **leadership** *n.*

lead-free *adj.* (of petrol) without added lead compounds.

lead-in *n.* introduction, opening, etc.

leading[1] *adj.* chief; most important.

leading[2] /'ledɪŋ/ *n. Printing* = LEAD[2] *n.* 6.

leading aircraftman *n.* rank above aircraftman in the RAF.

leading article *n.* newspaper article giving editorial opinion.

leading light *n.* prominent and influential person.

leading note *n. Mus.* seventh note of a diatonic scale.

leading question *n.* question prompting the answer wanted.

■ **Usage** *Leading question* does not mean a 'principal' or 'loaded' or 'searching' question.

lead pencil *n.* pencil of graphite in wood.

lead-poisoning *n.* poisoning by absorption of lead into the body.

leaf *–n.* (*pl.* **leaves**) **1** each of several flattened usu. green structures of a plant, growing usu. on the side of a stem. **2 a** foliage regarded collectively. **b** state of bearing leaves (*tree in leaf*). **3** single thickness of paper. **4** very thin sheet of metal etc. **5** hinged part, extra section, or flap of a table etc. *–v.* **1** put forth leaves. **2** (foll. by *through*) turn over the pages of (a book etc.). □ **leafage** *n.* **leafy** *adj.* (**-ier, -iest**). [Old English]

leaflet /'li:flɪt/ *–n.* **1** sheet of paper, pamphlet, etc. giving information. **2** young leaf. **3** *Bot.* division of a compound leaf. *–v.* (**-t-**) distribute leaflets (to).

leaf-mould *n.* soil or compost consisting chiefly of decayed leaves.

leaf-stalk *n.* stalk joining a leaf to a stem.

league[1] /li:g/ *–n.* **1** people, countries, groups, etc., combining for a particular purpose. **2** agreement to combine in this way. **3** group of sports clubs which compete for a championship. **4** class of contestants etc. *–v.* (**-gues, -gued, -guing**) (often foll. by *together*) join in a league. □ **in league** allied, conspiring. [Latin *ligo* bind]

league[2] /li:g/ *n. hist.* varying measure of distance, usu. about three miles. [Latin from Celtic]

league table *n.* list in ranked order of success etc.

leak *–n.* **1 a** hole through which matter passes accidentally in or out. **b** matter passing through thus. **c** act of passing through thus. **2 a** similar escape of electrical charge. **b** charge that escapes. **3** disclosure of secret information. *–v.* **1 a** pass through a leak. **b** lose or admit through a leak. **2** disclose (secret information). **3** (often foll. by *out*) become known. □ **have** (or **take**) **a leak** *slang* urinate. □ **leaky** *adj.* (**-ier, -iest**). [Low German or Dutch]

leakage *n.* action or result of leaking.

lean[1] *–v.* (*past* and *past part.* **leaned** or **leant** /lent/) **1** (often foll. by *across, back, over*, etc.) be or place in a sloping position; incline from the perpendicular. **2** (foll. by *against, on, upon*) (cause to) rest for support against etc. **3** (foll. by *on, upon*) rely on. **4** (foll. by *to, towards*) be inclined or partial to. –*n.* deviation from the perpendicular; inclination. □ **lean on** *colloq.* put pressure on (a person) to act in a certain way. **lean over backwards** see BACKWARDS. [Old English]

lean[2] –*adj.* **1** (of a person or animal) thin; having no superfluous fat. **2** (of meat) containing little fat. **3** meagre. –*n.* lean part of meat. □ **leanness** *n.* [Old English]

leaning *n.* tendency or partiality.

lean-to *n.* (*pl.* **-tos**) building with its roof leaning against a larger building or a wall.

lean years *n.pl.* years of scarcity.

leap –*v.* (*past* and *past part.* **leaped** or **leapt** /lept/) jump or spring forcefully. –*n.* forceful jump. □ **by leaps and bounds** with startlingly rapid progress. **leap in the dark** daring step or enterprise. [Old English]

leap-frog –*n.* game in which players vault with parted legs over others bending down. –*v.* (**-gg-**) **1** perform such a vault (over). **2** overtake alternately.

leap year *n.* year with 366 days (including 29th Feb. as an intercalary day).

learn /lɜ:n/ *v.* (*past* and *past part.* **learned** /lɜ:nt, lɜ:nd/ or **learnt**) **1** gain knowledge of or skill in. **2** commit to memory. **3** (foll. by *of*) be told about. **4** (foll. by *that, how*, etc.) become aware of. **5** receive instruction. **6** *archaic* or *dial.* teach. [Old English]

learned /'lɜ:nɪd/ *adj.* **1** having much knowledge acquired by study. **2** showing or requiring learning (*a learned work*). **3** (of a publication) academic.

learner *n.* **1** person who is learning a subject or skill. **2** (in full **learner driver**) person who is learning to drive and has not yet passed a driving test.

learning *n.* knowledge acquired by study.

lease –*n.* contract by which the owner of property allows another to use it for a specified time, usu. in return for payment. –*v.* (**-sing**) grant or take on lease. □ **new lease of** (*US* **on**) **life** improved prospect of living, or of use after repair. [Anglo-French *lesser* let, from Latin *laxo* loosen]

leasehold *n.* **1** holding of property by lease. **2** property held by lease. □ **leaseholder** *n.*

leash –*n.* strap for holding a dog etc.; lead. –*v.* **1** put a leash on. **2** restrain. □ **straining at the leash** eager to begin. [French *lesse*: related to LEASE]

least –*adj.* **1** smallest, slightest. **2** (of a species etc.) very small. –*n.* the least amount. –*adv.* in the least degree. □ **at least 1** at any rate. **2** (also **at the least**) not less than. **in the least** (or **the least**) (usu. with *neg.*) at all (*not in the least offended*). **to say the least** putting the case moderately. [Old English, superlative of LESS]

least common denominator *n.* = LOWEST COMMON DENOMINATOR.

least common multiple *n.* = LOWEST COMMON MULTIPLE.

leather /'leðə(r)/ –*n.* **1** material made from the skin of an animal by tanning etc. **2** piece of leather for polishing with. **3** leather part(s) of a thing. **4** *slang* cricket-ball or football. **5** (in *pl.*) leather clothes. –*v.* **1** beat, thrash. **2** cover with leather. **3** polish or wipe with a leather. [Old English]

leatherback *n.* large marine turtle with a leathery shell.

leather-bound *adj.* bound in leather.

leatherette /,leðə'ret/ *n.* imitation leather.

leather-jacket *n.* crane-fly grub with a tough skin.

leathery *adj.* **1** like leather. **2** tough.

leave[1] *v.* (**-ving**; *past* and *past part.* **left**) **1 a** go away from. **b** (often foll. by *for*) depart. **2** cause to or let remain; depart without taking. **3** (also *absol.*) cease to reside at or belong to or work for. **4** abandon; cease to live with (one's family etc.). **5** have remaining after one's death. **6** bequeath. **7** (foll. by *to* + infin.) allow (a person or thing) to do something independently. **8** (foll. by *to*) commit to another person etc. (*leave that to me*). **9 a** abstain from consuming or dealing with. **b** (in *passive*; often foll. by *over*) remain over. **10 a** deposit or entrust (a thing) to be attended to in one's absence (*left a message with his secretary*). **b** depute (a person) to perform a function in one's absence. **11** allow to remain or cause to be in a specified state or position (*left the door open*; *left me exhausted*). □ **leave alone** refrain from disturbing, not interfere with. **leave a person cold** not impress or excite a person. **leave off 1** come to or make an end. **2** discontinue. **leave out** omit; exclude. [Old English]

leave[2] *n.* **1** (often foll. by *to* + infin.) permission. **2 a** (in full **leave of absence**) permission to be absent from duty. **b** period for which this lasts. □ **on leave** legitimately absent from duty. **take one's leave (of)** bid farewell (to). **take leave of one's senses** go mad. [Old English]

leaved *adj.* having a leaf or leaves, esp. (in *comb.*) of a specified kind or number (*four-leaved clover*).

leaven /'lev(ə)n/ *–n.* **1** substance causing dough to ferment and rise. **2** pervasive transforming influence; admixture. *–v.* **1** ferment (dough) with leaven. **2** permeate and transform; modify with a tempering element. [Latin *levo* lift]

leaves *pl.* of LEAF.

leave-taking *n.* act of taking one's leave.

leavings *n.pl.* things left over.

Lebanese /,lebə'ni:z/ *–adj.* of Lebanon. *–n.* (*pl.* same) **1** native or national of Lebanon. **2** person of Lebanese descent.

lech *colloq. –v.* (often foll. by *after*) lust. *–n.* **1** lecherous man. **2** lust. [back-formation from LECHER]

lecher *n.* lecherous man. [French *lechier* live in debauchery]

lecherous *adj.* lustful, having excessive sexual desire. □ **lecherously** *adv.*

lechery *n.* excessive sexual desire.

lectern /'lekt(ə)n/ *n.* **1** stand for holding a book in a church etc. **2** similar stand for a lecturer etc. [Latin *lectrum* from *lego* read]

lecture /'lektʃə(r)/ *–n.* **1** talk giving specified information to a class etc. **2** long serious speech, esp. as a reprimand. *–v.* (**-ring**) **1** (often foll. by *on*) deliver lecture(s). **2** talk seriously or reprovingly to. □ **lectureship** *n.* [Latin: related to LECTERN]

lecturer *n.* person who lectures, esp. as a teacher in higher education.

LED *abbr.* light-emitting diode.

led *past* and *past part.* of LEAD[1].

lederhosen /'leɪdə,həʊz(ə)n/ *n.pl.* leather shorts as worn by some men in Bavaria etc. [German, = leather trousers]

ledge *n.* narrow horizontal or shelflike projection. [origin uncertain]

ledger *n.* main record of the accounts of a business. [Dutch]

lee *n.* **1** shelter given by a close object (*under the lee of*). **2** (in full **lee side**) side away from the wind. [Old English]

leech *n.* **1** bloodsucking worm formerly much used medically. **2** person who sponges on others. [Old English]

leek *n.* **1** plant of the onion family with flat leaves forming a cylindrical bulb, used as food. **2** this as a Welsh national emblem. [Old English]

leer *–v.* look slyly, lasciviously, or maliciously. *–n.* leering look. [perhaps from obsolete *leer* cheek]

leery *adj.* (**-ier, -iest**) *slang* **1** knowing, sly. **2** (foll. by *of*) wary.

lees /li:z/ *n.pl.* **1** sediment of wine etc. **2** dregs. [French]

leeward /'li:wəd, *Naut.* 'lu:əd/ *–adj.* & *adv.* on or towards the side sheltered from the wind. *–n.* leeward region or side.

leeway *n.* **1** allowable scope of action. **2** sideways drift of a ship to leeward of the desired course.

left[1] *–adj.* **1** on or towards the west side of the human body, or of any object, when facing north. **2** (also **Left**) *Polit.* of the Left. *–adv.* on or to the left side. *–n.* **1** left-hand part, region, or direction. **2** *Boxing* **a** left hand. **b** blow with this. **3** (often **Left**) group or section favouring socialism; socialists collectively. [Old English, originally = 'weak, worthless']

left[2] *past* and *past part.* of LEAVE[1].

left bank *n.* bank of a river on the left facing downstream.

left-hand *attrib. adj.* **1** on or towards the left side of a person or thing. **2** done with the left hand. **3** (of a screw) = LEFT-HANDED 4b.

left-handed *adj.* **1** naturally using the left hand for writing etc. **2** (of a tool etc.) for use by the left hand. **3** (of a blow) struck with the left hand. **4 a** turning to the left. **b** (of a screw) turned anticlockwise to tighten. **5** awkward, clumsy. **6 a** (of a compliment) ambiguous. **b** of doubtful sincerity. □ **left-handedly** *adv.* **left-handedness** *n.*

left-hander *n.* **1** left-handed person. **2** left-handed blow.

leftism *n.* socialist political principles. □ **leftist** *n.* & *adj.*

left luggage *n.* luggage deposited for later retrieval.

leftmost *adj.* furthest to the left.

leftover *–n.* (usu. in *pl.*) surplus item (esp. of food). *–attrib. adj.* remaining over, surplus.

leftward /'leftwəd/ *–adv.* (also **leftwards**) towards the left. *–adj.* going towards or facing the left.

left wing *–n.* **1** more socialist section of a political party or system. **2** left side of a football etc. team on the field. *–adj.* (**left-wing**) socialist, radical. □ **left-winger** *n.*

lefty /'leftɪ/ *n.* (*pl.* **-ies**) *colloq.* **1** *Polit.* often *derog.* left-winger. **2** left-handed person.

leg *n.* **1** each of the limbs on which a person or animal walks and stands. **2** leg of an animal or bird as food. **3** part of a garment covering a leg. **4** support of a chair, table, etc. **5** *Cricket* the half of the field (divided lengthways) in which the batsman's feet are placed. **6 a** section of a journey. **b** section of a relay race. **c** stage in a competition. □ **leg it** (**-gg-**) *colloq.* walk or run hard. **not have a leg to stand on** be unable to support one's argument by facts or sound reasons. **on one's last legs** near death or the end of usefulness etc. □ **legged** /legd, 'legɪd/ *adj.* (also in *comb.*). [Old Norse]

legacy /'legəsɪ/ *n.* (*pl.* **-ies**) **1** gift left in a will. **2** thing handed down by a predecessor. [Latin *lego* bequeath]

legal /'li:g(ə)l/ *adj.* **1** of or based on law; concerned with law. **2** appointed or required by law. **3** permitted by law. □ **legally** *adv.* [Latin *lex leg-* law]

legal aid *n.* State assistance for legal advice or action.

legalese /,li:gə'li:z/ *n. colloq.* technical language of legal documents.

legalistic /,li:gə'lɪstɪk/ *adj.* adhering excessively to a law or formula. □ **legalism** /'li:gə,lɪz(ə)m/ *n.* **legalist** /'li:gəlɪst/ *n.*

legality /lɪ'gælɪtɪ/ *n.* (*pl.* **-ies**) **1** lawfulness. **2** (in *pl.*) obligations imposed by law.

legalize /'li:gə,laɪz/ *v.* (also **-ise**) (**-zing** or **-sing**) **1** make lawful. **2** bring into harmony with the law. □ **legalization** /-'zeɪʃ(ə)n/ *n.*

legal separation see SEPARATION.

legal tender *n.* currency that cannot legally be refused in payment of a debt.

legate /'legət/ *n.* ambassador of the Pope. [Latin *lego* depute]

legatee /,legə'ti:/ *n.* recipient of a legacy. [Latin *lego* bequeath]

legation /lɪ'geɪʃ(ə)n/ *n.* **1** diplomatic minister and his or her staff. **2** this minister's official residence. [Latin: related to LEGATE]

legato /lɪ'gɑ:təʊ/ *Mus.* *–adv.* & *adj.* in a smooth flowing manner. *–n.* (*pl.* **-s**) **1** legato passage. **2** legato playing. [Italian, = bound, from *ligo* bind]

leg before *–adj.* & *adv.* (in full **leg before wicket**) *Cricket* (of a batsman) out because of stopping the ball, other than with the bat or hand, which would otherwise have hit the wicket. *–n.* such a dismissal.

leg-bye *n. Cricket* run scored from a ball that touches the batsman.

legend /'ledʒ(ə)nd/ *n.* **1 a** traditional story; myth. **b** these collectively. **2** *colloq.* famous or remarkable event or person. **3** inscription. **4** explanation on a map etc. of symbols used. [Latin *legenda* what is to be read]

legendary *adj.* **1** of, based on, or described in a legend. **2** *colloq.* remarkable.

legerdemain /,ledʒədə'meɪn/ *n.* **1** sleight of hand. **2** trickery, sophistry. [French, = light of hand]

leger line /'ledʒə(r)/ *n. Mus.* short line added for notes above or below the range of a staff. [var. of LEDGER]

legging *n.* (usu. in *pl.*) **1** close-fitting knitted trousers for women or children. **2** stout protective outer covering for the lower leg.

leggy *adj.* (**-ier**, **-iest**) **1** long-legged. **2** long-stemmed and weak. □ **legginess** *n.*

legible /'ledʒɪb(ə)l/ *adj.* clear enough to read; readable. □ **legibility** /-'bɪlɪtɪ/ *n.* **legibly** *adv.* [Latin *lego* read]

legion /'li:dʒ(ə)n/ *–n.* **1** division of 3,000–6,000 men in the ancient Roman army. **2** large organized body. *–predic. adj.* great in number (*his good works were legion*). [Latin *legio -onis*]

legionary *–adj.* of a legion or legions. *–n.* (*pl.* **-ies**) member of a legion.

legionnaire /,li:dʒə'neə(r)/ *n.* member of a legion. [French: related to LEGION]

legionnaires' disease *n.* form of bacterial pneumonia.

legislate /'ledʒɪs,leɪt/ *v.* (**-ting**) make laws. □ **legislator** *n.* [from LEGISLATION]

legislation /,ledʒɪs'leɪʃ(ə)n/ *n.* **1** law-making. **2** laws collectively. [Latin *lex legis* law, *latus* past part. of *fero* carry]

legislative /'ledʒɪslətɪv/ *adj.* of or empowered to make legislation.

legislature /'ledʒɪs,leɪtʃə(r), -lətʃə(r)/ *n.* legislative body of a State.

legit /lɪ'dʒɪt/ *adj. colloq.* legitimate (in sense 2). [abbreviation]

legitimate /lɪ'dʒɪtəmət/ *adj.* **1** (of a child) born of parents married to each other. **2** lawful, proper, regular. **3** logically acceptable. □ **legitimacy** *n.* **legitimately** *adv.* [Latin *legitimo* legitimize, from *lex legis* law]

legitimatize /lɪ'dʒɪtəmə,taɪz/ *v.* (also **-ise**) (**-zing** or **-sing**) legitimize.

legitimize /lɪ'dʒɪtə,maɪz/ *v.* (also **-ise**) (**-zing** or **-sing**) **1** make legitimate. **2** serve as a justification for. □ **legitimization** /-'zeɪʃ(ə)n/ *n.*

legless *adj.* **1** having no legs. **2** *slang* very drunk.

Lego /'legəʊ/ *n. propr.* toy consisting of interlocking plastic building blocks. [Danish *legetøj* toys]

leg-of-mutton sleeve *n.* sleeve which is full and loose on the upper arm but close-fitting on the forearm.

leg-pull *n. colloq.* hoax.

leg-room *n.* space for the legs of a seated person.

legume /'legju:m/ *n.* **1** leguminous plant. **2** edible part of a leguminous plant. [Latin *legumen -minis* from *lego* pick, because pickable by hand]

leguminous /lɪ'gju:mɪnəs/ *adj.* of the family of plants with seeds in pods (e.g. peas and beans).

leg up *n.* help given to mount a horse etc., or to overcome an obstacle or problem; boost.

leg warmer *n.* either of a pair of tubular knitted garments covering the leg from ankle to knee or thigh.

lei /'leɪiː/ *n.* Polynesian garland of flowers. [Hawaiian]

leisure /'leʒə(r)/ *n.* **1** free time. **2** enjoyment of free time. □ **at leisure 1** not occupied. **2** in an unhurried manner. **at one's leisure** when one has time. [Anglo-French *leisour* from Latin *licet* it is allowed]

leisure centre *n.* public building with sports facilities etc.

leisured *adj.* having ample leisure.

leisurely /'leʒəlɪ/ *–adj.* unhurried, relaxed. *–adv.* without hurry. □ **leisureliness** *n.*

leisurewear *n.* informal clothes, esp. sportswear.

leitmotif /'laɪtməʊˌtiːf/ *n.* (also **leitmotiv**) recurrent theme in a musical etc. composition representing a particular person, idea, etc. [German: related to LEAD[1], MOTIVE]

lemming /'lemɪŋ/ *n.* small Arctic rodent reputed to rush into the sea and drown during migration. [Norwegian]

lemon /'lemən/ *n.* **1 a** yellow oval citrus fruit with acidic juice. **b** tree bearing it. **2** pale yellow colour. **3** *colloq.* person or thing regarded as a failure. □ **lemony** *adj.* [Arabic *laimūn*]

lemonade /ˌlemə'neɪd/ *n.* **1** drink made from lemon juice. **2** synthetic substitute for this.

lemon balm *n.* bushy plant smelling and tasting of lemon.

lemon curd *n.* (also **lemon cheese**) creamy conserve made from lemons.

lemon geranium *n.* lemon-scented pelargonium.

lemon sole /'lemən/ *n.* (*pl.* same or **-s**) flat-fish of the plaice family. [French *limande*]

lemur /'liːmə(r)/ *n.* tree-dwelling primate of Madagascar. [Latin *lemures* ghosts]

lend *v.* (*past* and *past part.* **lent**) **1** (usu. foll. by *to*) grant (to a person) the use of (a thing) on the understanding that it or its equivalent shall be returned. **2** allow the use of (money) at interest. **3** bestow or contribute (*lends a certain charm*). □ **lend an ear** listen. **lend a hand** help. **lend itself to** (of a thing) be suitable for. □ **lender** *n.* [Old English: related to LOAN]

length *n.* **1** measurement or extent from end to end. **2** extent in or of time. **3** distance a thing extends. **4** length of a horse, boat, etc., as a measure of the lead in a race. **5** long stretch or extent. **6** degree of thoroughness in action (*went to great lengths*). **7** piece of a certain length (*length of cloth*). **8** *Prosody* quantity of a vowel or syllable. **9** *Cricket* **a** distance from the batsman at which the ball pitches. **b** proper amount of this. **10** length of a swimming-pool as a measure of distance swum. □ **at length 1** in detail. **2** after a long time. [Old English: related to LONG[1]]

lengthen *v.* make or become longer.

lengthways *adv.* in a direction parallel with a thing's length.

lengthwise *–adv.* lengthways. *–adj.* lying or moving lengthways.

lengthy *adj.* (**-ier, -iest**) of unusual or tedious length. □ **lengthily** *adv.* **lengthiness** *n.*

lenient /'liːnɪənt/ *adj.* merciful, not severe. □ **lenience** *n.* **leniency** *n.* **leniently** *adv.* [Latin *lenis* gentle]

lens /lenz/ *n.* **1** piece of a transparent substance with one or (usu.) both sides curved for concentrating or dispersing light-rays esp. in optical instruments. **2** combination of lenses used in photography. **3** transparent substance behind the iris of the eye. **4** = CONTACT LENS. [Latin *lens lent-* lentil (from the similarity of shape)]

Lent *n. Eccl.* period of fasting and penitence from Ash Wednesday to Holy Saturday. □ **Lenten** *adj.* [Old English, = spring]

lent *past* and *past part.* of LEND.

lentil /'lentɪl/ *n.* **1** pea-like plant. **2** its seed, esp. used as food. [Latin *lens*]

lento /'lentəʊ/ *Mus. –adj.* slow. *–adv.* slowly. [Italian]

Leo /'liːəʊ/ *n.* (*pl.* **-s**) **1** constellation and fifth sign of the zodiac (the Lion). **2** person born when the sun is in this sign. [Latin]

leonine /'liːəˌnaɪn/ *adj.* **1** like a lion. **2** of or relating to lions. [Latin: related to LEO]

leopard /'lepəd/ *n.* large African or Asian animal of the cat family with a black-spotted yellowish or all black coat, panther. [Greek *leōn* lion, *pardos* panther]

leotard /'liːəˌtɑːd/ *n.* close-fitting one-piece garment worn by dancers etc. [*Léotard*, name of a trapeze artist]

leper /'lepə(r)/ *n.* **1** person with leprosy. **2** person who is shunned. [Greek *lepros* scaly]

lepidopterous /ˌlepɪ'dɒptərəs/ *adj.* of the order of insects with four scale-covered wings, including butterflies and moths. □ **lepidopterist** *n.* [Greek *lepis -idos* scale, *pteron* wing]

leprechaun /'leprəˌkɔːn/ *n.* small mischievous sprite in Irish folklore. [Irish *lu* small, *corp* body]

leprosy /ˈleprəsɪ/ *n.* contagious disease that damages the skin and nerves. □ **leprous** *adj.* [related to LEPER]

lesbian /ˈlezbɪən/ *–n.* homosexual woman. *–adj.* of female homosexuality. □ **lesbianism** *n.* [*Lesbos*, name of an island in the Aegean Sea]

lese-majesty /liːz ˈmædʒɪstɪ/ *n.* **1** treason. **2** insult to a sovereign or ruler. **3** presumptuous conduct. [French *lèse-majesté* injured sovereignty]

lesion /ˈliːʒ(ə)n/ *n.* **1** damage. **2** injury. **3** morbid change in the functioning or texture of an organ etc. [Latin *laedo laes-* injure]

less *–adj.* **1** smaller in extent, degree, duration, number, etc. **2** of smaller quantity, not so much (*less meat*). **3** *colloq.* fewer (*less biscuits*). *–adv.* to a smaller extent, in a lower degree. *–n.* smaller amount, quantity, or number (*will take less; for less than £10*). *–prep.* minus (*made £1,000 less tax*). [Old English]

■ **Usage** The use of *less* to mean 'fewer', as in sense 3, is regarded as incorrect in standard English.

-less *suffix* forming adjectives and adverbs: **1** from nouns, meaning 'not having, without, free from' (*powerless*). **2** from verbs, meaning 'not accessible to, affected by, or performing the action of the verb' (*fathomless; ceaseless*). [Old English]

lessee /leˈsiː/ *n.* (often foll. by *of*) person holding a property by lease. [French: related to LEASE]

lessen /ˈles(ə)n/ *v.* make or become less, diminish.

lesser *adj.* (usu. *attrib.*) not so great as the other(s) (*lesser evil; lesser mortals*).

lesson /ˈles(ə)n/ *n.* **1** spell of teaching. **2** (in *pl.*; foll. by *in*) systematic instruction. **3** thing learnt by a pupil. **4** experience that serves to warn or encourage (*let that be a lesson*). **5** passage from the Bible read aloud during a church service. [French *leçon* from Latin *lego lect-*]

lessor /leˈsɔː(r)/ *n.* person who lets a property by lease. [Anglo-French: related to LEASE]

lest *conj. formal* **1** in order that not, for fear that (*lest he forget*). **2** that (*afraid lest we should be late*). [Old English: related to LESS]

■ **Usage** *Lest* is followed by *should* or the subjunctive (see examples above).

let[1] *–v.* (**-tt-**; *past* and *past part.* **let**) **1 a** allow to, not prevent or forbid. **b** cause to (*let me know*). **2** (foll. by *into*) allow to enter. **3** grant the use of (rooms, land, etc.) for rent or hire. **4** allow or cause (liquid or air) to escape (*let blood*). **5** *aux.* supplying the first and third persons of the imperative in exhortations (*let us pray*), commands (*let it be done at once; let there be light*), assumptions, etc. (*let AB equal CD*). *–n.* act of letting a house, room, etc. □ **let alone 1** not to mention, far less or more (*hasn't got a television, let alone a video*). **2** = *let be*. **let be** not interfere with, attend to, or do. **let down 1** lower. **2** fail to support or satisfy, disappoint. **3** lengthen (a garment). **4** deflate (a tyre). **let down gently** reject or disappoint without humiliating. **let drop** (or **fall**) drop (esp. a word or hint) intentionally or by accident. **let go 1** release. **2 a** (often foll. by *of*) lose one's hold. **b** lose hold of. **let oneself go 1** act spontaneously. **2** neglect one's appearance or habits. **let in 1** allow to enter (*let the dog in; let in a flood of light*). **2** (foll. by *for*) involve (a person, often oneself) in loss or difficulty. **3** (foll. by *on*) allow (a person) to share a secret, privileges, etc. **let loose** release, unchain. **let off 1 a** fire (a gun). **b** explode (a bomb). **2** allow or cause (steam etc.) to escape. **3 a** not punish or compel. **b** (foll. by *with*) punish lightly. **let off steam** release pent-up energy or feeling. **let on** *colloq.* **1** reveal a secret. **2** pretend. **let out 1** release. **2** reveal (a secret etc.). **3** make (a garment) looser. **4** put out to rent or to contract. **let rip 1** act without restraint. **2** speak violently. **let up** *colloq.* **1** become less intense or severe. **2** relax one's efforts. **to let** available for rent. [Old English]

let[2] *–n.* obstruction of a ball or player in tennis etc., requiring the ball to be served again. *–v.* (**-tt-**; *past* and *past part.* **letted** or **let**) *archaic* hinder, obstruct. □ **without let or hindrance** unimpeded. [Old English: related to LATE]

-let *suffix* forming nouns, usu. diminutive (*flatlet*) or denoting articles of ornament or dress (*anklet*). [French]

let-down *n.* disappointment.

lethal /ˈliːθ(ə)l/ *adj.* causing or sufficient to cause death. □ **lethally** *adv.* [Latin *letum* death]

lethargy /ˈleθədʒɪ/ *n.* **1** lack of energy. **2** morbid drowsiness. □ **lethargic** /lɪˈθɑːdʒɪk/ *adj.* **lethargically** /lɪˈθɑːdʒɪkəlɪ/ *adv.* [Greek *lēthargos* forgetful]

let-out *n. colloq.* opportunity to escape a commitment etc.

letter *–n.* **1** character representing one or more of the sounds used in speech. **2 a** written or printed message, usu. sent in an envelope by post. **b** (in *pl.*) addressed legal or formal document. **3** precise terms of a statement, the strict verbal interpretation (*letter of the law*). **4** (in *pl.*) **a** literature. **b** acquaintance with books, erudition. –*v.* **1** inscribe letters on. **2** classify with letters. □ **to the letter** with adherence to every detail. [French from Latin *littera*]

letter-bomb *n.* terrorist explosive device in the form of a postal packet.

letter-box *n.* box or slot into which letters are posted or delivered.

lettered *adj.* well-read or educated.

letterhead *n.* **1** printed heading on stationery. **2** stationery with this.

letter of credit *n.* letter from a bank authorizing the bearer to draw money from another bank.

letterpress *n.* **1** printed words of an illustrated book. **2** printing from raised type.

lettuce /'letɪs/ *n.* plant with crisp leaves used in salads. [Latin *lactuca* from *lac lact-* milk]

let-up *n. colloq.* **1** reduction in intensity. **2** relaxation of effort.

leuco- *comb. form* white. [Greek *leukos* white]

leucocyte /'lu:kə,saɪt/ *n.* white blood cell.

leukaemia /lu:'ki:mɪə/ *n.* (*US* **leukemia**) malignant disease in which the bone-marrow etc. produces too many leucocytes. [Greek *leukos* white, *haima* blood]

Levant /lɪ'vænt/ *n.* (prec. by *the*) *archaic* eastern Mediterranean countries. [French, = point of sunrise, from Latin *levo* lift]

Levantine /'levən,taɪn/ –*adj.* of or trading to the Levant. –*n.* native or inhabitant of the Levant.

levee /'levɪ/ *n. US* **1** embankment against river floods. **2** natural embankment built up by a river. **3** landing-place. [French *levée* past part. of *lever* raise: related to LEVY]

level /'lev(ə)l/ –*n.* **1** horizontal line or plane. **2** height or value reached; position on a real or imaginary scale (*eye level*; *sugar level*; *danger level*). **3** social, moral, or intellectual standard. **4** plane of rank or authority (*talks at Cabinet level*). **5** instrument giving a line parallel to the plane of the horizon. **6** level surface. **7** flat tract of land. –*adj.* **1** flat and even; not bumpy. **2** horizontal. **3** (often foll. by *with*) **a** on the same horizontal plane as something else. **b** having equality with something else. **4** even, uniform, equable, or well-balanced. –*v.* (**-ll-**; *US* **-l-**) **1** make level. **2** raze. **3** (also *absol.*) aim (a missile or gun). **4** (also *absol.*; foll. by *at, against*) direct (an accusation etc.). □ **do one's level best** *colloq.* do one's utmost. **find one's level** reach the right social, intellectual, etc. position. **level down** bring down to a standard. **level off** make or become level. **level out** make or become level. **level up** bring up to a standard. **on the level 1** honestly, without deception. **2** honest, truthful. **on a level with 1** in the same horizontal plane as. **2** equal with. [Latin diminutive of *libra* balance]

level crossing *n.* crossing of a railway and a road, or two railways, at the same level.

level-headed *adj.* sensible, mentally well-balanced. □ **level-headedness** *n.*

leveller *n.* (*US* **leveler**) **1** person who advocates the abolition of social distinctions. **2** person or thing that levels.

level pegging *n.* equality of scores etc.

lever /'li:və(r)/ –*n.* **1** bar resting on a pivot, used to prise. **2** bar pivoted about a fulcrum (fixed point) which can be acted upon by a force (effort) in order to move a load. **3** projecting handle moved to operate a mechanism. **4** means of exerting moral pressure. –*v.* **1** use a lever. **2** (often foll. by *away, out, up,* etc.) lift, move, etc. with a lever. [Latin *levo* raise]

leverage *n.* **1** action or power of a lever. **2** power to accomplish a purpose.

leveraged buyout /'levərɪdʒd/ *n.* buyout in which outside capital is used to enable the management to buy up the company.

■ **Usage** The pronunciation is American because the practice takes place mainly in the US.

leveret /'levərɪt/ *n.* young hare, esp. one in its first year. [Latin *lepus lepor-* hare]

leviathan /lɪ'vaɪəθ(ə)n/ *n.* **1** *Bibl.* sea-monster. **2** very large or powerful thing. [Latin from Hebrew]

Levis /'li:vaɪz/ *n.pl. propr.* type of (orig. blue) denim jeans or overalls reinforced with rivets. [*Levi* Strauss, name of the manufacturer]

levitate /'levɪ,teɪt/ *v.* (**-ting**) **1** rise and float in the air (esp. with reference to spiritualism). **2** cause to do this. □ **levitation** /-'teɪʃ(ə)n/ *n.* [Latin *levis* light, after GRAVITATE]

levity /'levɪtɪ/ *n.* lack of serious thought, frivolity. [Latin *levis* light]

levy /'levɪ/ –*v.* (**-ies, -ied**) **1** impose or collect compulsorily (payment etc.). **2** enrol (troops etc.). **3** wage (war). –*n.* (*pl.* **-ies**) **1 a** collecting of a contribution, tax, etc. **b** contribution etc. levied. **2 a** act of enrolling troops etc. **b** (in *pl.*) troops enrolled. [Latin *levo* raise]

lewd /lju:d/ *adj.* **1** lascivious. **2** obscene. [Old English, originally = lay, vulgar]

lexical /'leksɪk(ə)l/ *adj.* **1** of the words of a language. **2** of or as of a lexicon. [Greek *lexikos, lexikon*: see LEXICON]

lexicography /,leksɪ'kɒgrəfɪ/ *n.* compiling of dictionaries. □ **lexicographer** *n.* [from LEXICON, -GRAPHY]

lexicon /'leksɪkən/ *n.* **1** dictionary, esp. of Greek, Hebrew, Syriac, or Arabic. **2** vocabulary of a person etc. [Greek *lexis* word]

Leyden jar /'laɪd(ə)n/ *n.* early capacitor consisting of a glass jar with layers of metal foil on the outside and inside. [*Leyden* (now *Leiden*) in Holland]

LF *abbr.* low frequency.

Li *symb.* lithium.

liability /,laɪə'bɪlɪtɪ/ *n.* (*pl.* **-ies**) **1** being liable. **2** troublesome responsibility; handicap. **3** (in *pl.*) debts etc. for which one is liable.

liable /'laɪəb(ə)l/ *predic. adj.* **1** legally bound. **2** (foll. by *to*) subject to. **3** (foll. by *to* + infin.) under an obligation. **4** (foll. by *to*) exposed or open to (something undesirable). **5** (foll. by *to* + infin.) apt, likely (*it is liable to rain*). **6** (foll. by *for*) answerable. [French *lier* bind, from Latin *ligo*]

■ **Usage** Use of *liable* in sense 5, though common, is considered incorrect by some people.

liaise /lɪ'eɪz/ *v.* (**-sing**) (foll. by *with, between*) *colloq.* establish cooperation, act as a link. [back-formation from LIAISON]

liaison /lɪ'eɪzɒn/ *n.* **1** communication or cooperation. **2** illicit sexual relationship. [French *lier* bind: see LIABLE]

liana /lɪ'ɑ:nə/ *n.* climbing plant of tropical forests. [French]

liar /'laɪə(r)/ *n.* person who tells a lie or lies.

Lib. *abbr.* Liberal.

lib *n. colloq.* (in names of political movements) liberation. [abbreviation]

libation /laɪ'beɪʃ(ə)n/ *n.* **1** pouring out of a drink-offering to a god. **2** such a drink-offering. [Latin]

libel /'laɪb(ə)l/ *–n.* **1** *Law* **a** published false statement that is damaging to a person's reputation. **b** act of publishing this. **2** false and defamatory misrepresentation or statement. *–v.* (**-ll-**; *US* **-l-**) **1** defame by libellous statements. **2** *Law* publish a libel against. □ **libellous** *adj.* [Latin *libellus* diminutive of *liber* book]

liberal /'lɪbər(ə)l/ *–adj.* **1** abundant, ample. **2** giving freely, generous. **3** open-minded. **4** not strict or rigorous. **5** for the general broadening of the mind (*liberal studies*). **6 a** favouring moderate political and social reform. **b** (**Liberal**) of or characteristic of Liberals. *–n.* **1** person of liberal views. **2** (**Liberal**) supporter or member of a Liberal Party. □ **liberalism** *n.* **liberality** /-'rælɪtɪ/ *n.* **liberally** *adv.* [Latin *liber* free]

■ **Usage** In the UK the name *Liberal* was discontinued in official political use in 1988 when the party regrouped to form the *Social and Liberal Democratic Party*. In 1989 this name was officially replaced by *Liberal Democratic Party*.

Liberal Democrat *n.* member of the party formed from the Liberal Party and the Social Democratic Party.

■ **Usage** See note at *liberal*.

liberalize /'lɪbərə,laɪz/ *v.* (also **-ise**) (**-zing** or **-sing**) make or become more liberal or less strict. □ **liberalization** /-'zeɪʃ(ə)n/ *n.*

liberate /'lɪbə,reɪt/ *v.* (**-ting**) **1** (often foll. by *from*) set free. **2** free (a country etc.) from an oppressor or enemy. **3** (often as **liberated** *adj.*) free (a person) from rigid social conventions. □ **liberation** /-'reɪʃ(ə)n/ *n.* **liberator** *n.* [Latin *liberare liberat-* from *liber* free]

libertine /'lɪbə,ti:n/ *–n.* licentious person, rake. *–adj.* licentious. [Latin, = freedman, from *liber* free]

liberty /'lɪbətɪ/ *n.* (*pl.* **-ies**) **1** freedom from captivity etc. **2** right or power to do as one pleases. **3** (usu. in *pl.*) right or privilege granted by authority. □ **at liberty 1** free. **2** (foll. by *to* + infin.) permitted. **take liberties** (often foll. by *with*) behave in an unduly familiar manner. [Latin: related to LIBERAL]

libidinous /lɪ'bɪdɪnəs/ *adj.* lustful. [Latin: related to LIBIDO]

libido /lɪ'bi:,dəʊ/ *n.* (*pl.* **-s**) psychic drive or energy, esp. that associated with sexual desire. □ **libidinal** /lɪ'bɪdɪn(ə)l/ *adj.* [Latin, = lust]

Libra /'li:brə/ *n.* **1** constellation and seventh sign of the zodiac (the Scales). **2** person born when the sun is in this sign. [Latin, = pound weight]

librarian /laɪ'breərɪən/ *n.* person in charge of or assisting in a library. □ **librarianship** *n.*

library /'laɪbrərɪ/ *n.* (*pl.* **-ies**) **1** collection of books. **2** room or building where these are kept. **3 a** similar collection of films, records, computer routines, etc. **b** place where these are kept. **4** set of books issued in similar bindings. [Latin *liber* book]

libretto /lɪ'bretəʊ/ *n.* (*pl.* **-ti** /-tɪ/ or **-s**) text of an opera etc. □ **librettist** *n.* [Italian, = little book]

lice *pl.* of LOUSE.

licence /'laɪs(ə)ns/ *n.* (*US* **license**) **1** official permit to own or use something, do something, or carry on a trade. **2** permission. **3** liberty of action, esp. when excessive. **4** writer's or artist's deliberate deviation from fact, correct grammar, etc. (*poetic licence*). [Latin *licet* it is allowed]

license /'laɪs(ə)ns/ *v.* (**-sing**) **1** grant a licence to. **2** authorize the use of (premises) for a certain purpose.

licensee /,laɪsən'si:/ *n.* holder of a licence, esp. to sell alcoholic liquor.

licentiate /laɪ'senʃɪət/ *n.* holder of a certificate of professional competence. [medieval Latin: related to LICENCE]

licentious /laɪ'senʃəs/ *adj.* sexually promiscuous. [Latin: related to LICENCE]

lichee var. of LYCHEE.

lichen /'laɪkən, 'lɪtʃ(ə)n/ *n.* plant composed of a fungus and an alga in association, growing on and colouring rocks, tree-trunks, etc. [Greek *leikhēn*]

lich-gate *n.* (also **lych-gate**) roofed gateway to a churchyard where a coffin awaits the clergyman's arrival. [from *lich* = corpse]

licit /'lɪsɪt/ *adj. formal* permitted, lawful. [Latin: related to LICENCE]

lick *–v.* **1** pass the tongue over. **2** bring into a specified condition by licking (*licked it all up*; *licked it clean*). **3** (of a flame etc.) play lightly over. **4** *colloq.* defeat. **5** *colloq.* thrash. *–n.* **1** act of licking with the tongue. **2** *colloq.* fast pace (*at a lick*). **3** smart blow. □ **lick a person's boots** be servile. **lick into shape** make presentable or efficient. **lick one's lips** (or **chops**) look forward with relish. **lick one's wounds** be in retirement regaining strength etc. after defeat. [Old English]

lick and a promise *n. colloq.* hasty performance of a task, esp. washing oneself.

licorice var. of LIQUORICE.

lid *n.* **1** hinged or removable cover, esp. for a container. **2** = EYELID. □ **put the lid on** *colloq.* **1** be the culmination of. **2** put a stop to. □ **lidded** *adj.* (also in *comb.*). [Old English]

lido /'li:dəʊ, 'laɪ-/ *n.* (*pl.* **-s**) public open-air swimming-pool or bathing-beach. [*Lido*, name of a beach near Venice]

lie[1] /laɪ/ *–v.* (**lies**; **lying**; *past* **lay**; *past part.* **lain**) **1** be in or assume a horizontal position on a surface; be at rest on something. **2** (of a thing) rest flat on a surface. **3** remain undisturbed or undiscussed etc. (*let matters lie*). **4 a** be kept, remain, or be in a specified state or place (*lie hidden*; *lie in wait*; *books lay unread*). **b** (of abstract things) exist; be in a certain position or relation (*answer lies in education*). **5 a** be situated (*village lay to the east*). **b** be spread out to view. *–n.* way, direction, or position in which a thing lies. □ **lie down** assume a lying position; have a short rest. **lie down under** accept (an insult etc.) without protest. **lie in** stay in bed late in the morning. **lie low 1** keep quiet or unseen. **2** be discreet about one's intentions. **lie with** be the responsibility of (a person) (*decision lies with you*). **take lying down** (usu. with *neg.*) accept (an insult etc.) without protest. [Old English]

■ **Usage** The transitive use of *lie*, meaning *lay*, as in *lie her on the bed*, is incorrect in standard English.

lie[2] /laɪ/ *–n.* **1** intentionally false statement (*tell a lie*). **2** something that deceives. *–v.* (**lies, lied, lying**) **1** tell a lie or lies. **2** (of a thing) be deceptive. □ **give the lie to** show the falsity of (a supposition etc.). [Old English]

lied /li:d/ *n.* (*pl.* **lieder**) German song, esp. of the Romantic period. [German]

lie-detector *n.* instrument supposedly determining whether a person is lying, by testing for certain physiological changes.

lie-down *n.* short rest.

liege usu. *hist.* *–adj.* entitled to receive, or bound to give, feudal service or allegiance. *–n.* **1** (in full **liege lord**) feudal superior or sovereign. **2** (usu. in *pl.*) vassal, subject. [medieval Latin *laeticus*, probably from Germanic]

lie-in *n.* prolonged stay in bed in the morning.

lien /'li:ən/ *n. Law* right to hold another's property until a debt on it is paid. [Latin *ligo* bind]

lie of the land *n.* state of affairs.

lieu /lju:/ *n.* □ **in lieu 1** instead. **2** (foll. by *of*) in the place of. [Latin *locus* place]

Lieut. *abbr.* Lieutenant.

lieutenant /lef'tenənt/ *n.* **1 a** army officer next in rank below captain. **b** naval officer next in rank below lieutenant commander. **2** deputy. □ **lieutenancy** *n.* (*pl.* **-ies**). [French: related to LIEU place, TENANT holder]

lieutenant colonel *n.* (also **lieutenant commander** or **general**) officers ranking next below colonel, commander, or general.

life *n.* (*pl.* **lives**) **1** capacity for growth, functional activity, and continual change until death. **2** living things and their activity (*insect life*; *is there life on Mars?*). **3 a** period during which life lasts, or the period from birth to the present time or from the present time to death (*have done it all my life; will regret it all my life*). **b** duration of a thing's existence or ability to function. **4** **a** person's state of existence as a living individual (*sacrificed their lives*). **b** living person (*many lives were lost*). **5** **a** individual's actions or fortunes; manner of existence (*start a new life*). **b** particular aspect of this (*private life*). **6** business and pleasures of the world (*in Paris you really see life*). **7** energy, liveliness (*full of life*). **8** biography. **9** *colloq.* = LIFE SENTENCE. □ **for dear** (or **one's**) **life** as if or in order to escape death. **for life** for the rest of one's life. **not on your life** *colloq.* most certainly not. [Old English]

life assurance *n.* = LIFE INSURANCE.

lifebelt *n.* buoyant belt for keeping a person afloat.

lifeblood *n.* **1** blood, as being necessary to life. **2** vital factor or influence.

lifeboat *n.* **1** special boat for rescuing those in distress at sea. **2** ship's small boat for use in emergency.

lifebuoy *n.* buoyant support for keeping a person afloat.

life cycle *n.* series of changes in the life of an organism, including reproduction.

lifeguard *n.* expert swimmer employed to rescue bathers from drowning.

Life Guards *n.pl.* regiment of the royal household cavalry.

life insurance *n.* insurance for a sum to be paid on the death of the insured person.

life-jacket *n.* buoyant jacket for keeping a person afloat.

lifeless *adj.* **1** dead. **2** unconscious. **3** lacking movement or vitality. □ **lifelessly** *adv.* [Old English]

lifelike *adj.* closely resembling life or the person or thing represented.

lifeline *n.* **1** rope etc. used for life-saving. **2** sole means of communication or transport.

lifelong *adj.* lasting a lifetime.

life peer *n.* peer whose title lapses on death.

life-preserver *n.* **1** short stick with a heavily loaded end. **2** life-jacket etc.

lifer *n. slang* person serving a life sentence.

life sciences *n.pl.* biology and related subjects.

life sentence *n.* sentence of imprisonment for an indefinite period.

life-size *adj.* (also **-sized**) of the same size as the person or thing represented.

lifestyle *n.* way of life of a person or group.

life-support machine *n.* respirator.

lifetime *n.* duration of a person's life.

lift —*v.* **1** (often foll. by *up, off, out*, etc.) raise or remove to a higher position. **2** go up; be raised; yield to an upward force. **3** give an upward direction to (the eyes or face). **4** elevate to a higher plane of thought or feeling. **5** (of fog etc.) rise, disperse. **6** remove (a barrier or restriction). **7** transport (supplies, troops, etc.) by air. **8** *colloq.* **a** steal. **b** plagiarize (a passage of writing etc.). **9** dig up (esp. potatoes etc.). —*n.* **1** lifting or being lifted. **2** ride in another person's vehicle (*gave them a lift*). **3** **a** apparatus for raising and lowering persons or things to different floors of a building etc. **b** apparatus for carrying persons up or down a mountain etc. **4** **a** transport by air. **b** quantity of goods transported by air. **5** upward pressure which air exerts on an aerofoil. **6** supporting or elevating influence; feeling of elation. □ **lift down** pick up and bring to a lower position. [Old Norse: related to LOFT]

lift-off *n.* vertical take-off of a spacecraft or rocket.

ligament /ˈlɪgəmənt/ *n.* band of tough fibrous tissue linking bones. [Latin *ligo* bind]

ligature /ˈlɪgətʃə(r)/ —*n.* **1** tie or bandage. **2** *Mus.* slur, tie. **3** two or more letters joined, e.g. æ. **4** bond; thing that unites. —*v.* (**-ring**) bind or connect with a ligature. [Latin *ligo* bind]

■ **Usage** Sense 3 of this word is sometimes confused with *digraph*, which means 'two separate letters together representing one sound'.

light[1] /laɪt/ —*n.* **1** the natural agent (electromagnetic radiation) that stimulates sight and makes things visible. **2** the medium or condition of the space in which this is present (*just enough light to see*). **3** appearance of brightness (*saw a distant light*). **4** source of light, e.g. the sun, a lamp, fire, etc. **5** (often in *pl.*) traffic-light. **6** **a** flame or spark serving to ignite. **b** device producing this. **7** aspect in which a thing is regarded (*appeared in a new light*). **8** **a** mental illumination. **b** spiritual illumination by divine truth. **9** vivacity etc. in a person's face, esp. in the eyes. **10** eminent person (*leading light*). **11** bright parts of a picture etc. **12** window or opening in a wall to let light in. —*v.* (*past* **lit**; *past part.* **lit** or **lighted**) (*attrib.*) **1** set burning; begin to burn. **2** (often foll. by *up*) provide with light or lighting; make prominent by means of light. **3** show (a person) the way or surroundings with a light. **4** (usu. foll. by *up*) (of the face or eyes) brighten with animation, pleasure, etc. *–adj.* **1** well provided with light; not dark. **2** (of a colour) pale (*light blue*; *light-blue ribbon*). □ **bring** (or **come**) **to light** reveal or be revealed. **in a good** (or **bad**) **light** giving a favourable (or unfavourable) impression. **in the light of** taking account of. **light up 1** *colloq.* begin to smoke a cigarette etc. **2** = sense 2 of *v.* **3** = sense 4 of *v.* □ **lightish** *adj.* [Old English]

light[2] /laɪt/ *–adj.* **1** not heavy. **2 a** relatively low in weight, amount, density, intensity, etc. (*light arms, traffic, metal, rain*). **b** deficient in weight (*light coin*). **3 a** carrying or suitable for small loads (*light railway*). **b** (of a ship) unladen. **c** carrying only light arms, armaments, etc. **4** (of food) easy to digest. **5** (of entertainment, music, etc.) intended for amusement only; not profound. **6** (of sleep or a sleeper) easily disturbed. **7** easily borne or done (*light duties*). **8** nimble; quick-moving (*light step*; *light rhythm*). **9** (of a building etc.) graceful, elegant. **10 a** free from sorrow; cheerful (*light heart*). **b** giddy (*light in the head*). *–adv.* **1** in a light manner (*tread light*; *sleep light*). **2** with a minimum load (*travel light*). *–v.* (*past* and *past part.* **lit** or **lighted**) (foll. by *on, upon*) come upon or find by chance. □ **make light of** treat as unimportant. □ **lightish** *adj.* **lightly** *adv.* **lightness** *n.* [Old English]

light-bulb *n.* glass bulb containing an inert gas and a metal filament, providing light when an electric current is passed through it.

lighten[1] *v.* **1 a** make or become lighter in weight. **b** reduce the weight or load of. **2** bring relief to (the mind etc.). **3** mitigate (a penalty).

lighten[2] *v.* **1** shed light on. **2** make or grow bright.

lighter[1] *n.* device for lighting cigarettes etc.

lighter[2] *n.* boat, usu. flat-bottomed, for transferring goods from a ship to a wharf or another ship. [Dutch: related to LIGHT[2] in the sense 'unload']

lighter-than-air *attrib. adj.* (of an aircraft) weighing less than the air it displaces.

light-fingered *adj.* given to stealing.

light flyweight *n.* **1** amateur boxing weight up to 48 kg. **2** amateur boxer of this weight.

light-footed *adj.* nimble.

light-headed *adj.* giddy, delirious. □ **light-headedness** *n.*

light-hearted *adj.* **1** cheerful. **2** (unduly) casual. □ **light-heartedly** *adv.*

light heavyweight *n.* **1** weight in certain sports between middleweight and heavyweight, in amateur boxing 75–81 kg: also called CRUISERWEIGHT. **2** sportsman of this weight.

lighthouse *n.* tower etc. containing a beacon light to warn or guide ships at sea.

light industry *n.* manufacture of small or light articles.

lighting *n.* **1** equipment in a room or street etc. for producing light. **2** arrangement or effect of lights.

lighting-up time *n.* time after which vehicles must show the prescribed lights.

light meter *n.* instrument for measuring the intensity of the light, esp. to show the correct photographic exposure.

light middleweight *n.* **1** weight in amateur boxing of 67–71 kg. **2** amateur boxer of this weight.

lightning *–n.* flash of bright light produced by an electric discharge between clouds or between clouds and the ground. *–attrib. adj.* very quick. [from LIGHTEN[2]]

lightning-conductor *n.* (also **lightning-rod**) metal rod or wire fixed to an exposed part of a building or to a mast to divert lightning into the earth or sea.

lights *n.pl.* lungs of sheep, pigs, etc., used as a food esp. for pets. [from LIGHT[2]: cf. LUNG]

lightship *n.* moored or anchored ship with a beacon light.

lightweight *–adj.* **1** of below average weight. **2** of little importance or influence. *–n.* **1** lightweight person, animal, or thing. **2 a** weight in certain sports between featherweight and welterweight, in amateur boxing 57–60 kg. **b** sportsman of this weight.

light welterweight *n.* **1** weight in amateur boxing of 60–63.5 kg. **2** amateur boxer of this weight.

light-year *n.* distance light travels in one year, nearly 6 million million miles.

ligneous /ˈlɪgnɪəs/ *adj.* **1** (of a plant) woody. **2** of the nature of wood. [Latin *lignum* wood]

lignite /ˈlɪgnaɪt/ *n.* brown coal of woody texture.

lignum vitae /ˌlɪgnəm ˈvaɪtɪ/ *n.* a hard-wooded tree. [Latin, = wood of life]

likable var. of LIKEABLE.

like[1] *–adj.* (**more like, most like**) **1 a** having some or all of the qualities of another, each other, or an original. **b** resembling in some way, such as (*good writers like Dickens*). **2** characteristic of (*not like them to be late*). **3** in a suitable state or mood for (*felt like* working; *felt like a cup of tea*). –*prep.* in the manner of; to the same degree as (*drink like a fish*; *acted like an idiot*). –*adv.* **1** *slang* so to speak (*did a quick getaway, like*). **2** *colloq.* likely, probably (*as like as not*). –*conj. colloq.* **1** as (*cannot do it like you do*). **2** as if (*ate like they were starving*). –*n.* **1** counterpart; equal; similar person or thing. **2** (prec. by *the*) thing or things of the same kind (*will never do the like again*). □ **and the like** and similar things. **like anything** *colloq.* very much, vigorously. **the likes of** *colloq.* a person such as. **more like it** *colloq.* nearer what is required. **what is he** (or **it** etc.) **like?** what sort of person is he (or thing is it etc.)? [Old English]

■ **Usage** The use of *like* as a conjunction is considered incorrect by some people.

like[2] –*v.* (**-king**) **1** find agreeable or enjoyable. **2 a** choose to have; prefer (*like my tea weak*). **b** wish for or be inclined to (*would like a nap*; *should like to come*). –*n.* (in *pl.*) things one likes or prefers. [Old English]

-like *comb. form* forming adjectives from nouns, meaning 'similar to, characteristic of' (*doglike*; *shell-like*; *tortoise-like*).

■ **Usage** In formations not generally current the hyphen should be used. It may be omitted when the first element is of one syllable, unless it ends in *-l*.

likeable *adj.* (also **likable**) pleasant; easy to like. □ **likeably** *adv.*

likelihood /'laɪklɪˌhʊd/ *n.* probability. □ **in all likelihood** very probably.

likely /'laɪklɪ/ –*adj.* (**-ier, -iest**) **1** probable; such as may well happen or be true. **2** to be reasonably expected (*not likely to come now*). **3** promising; apparently suitable (*a likely spot*). –*adv.* probably. □ **not likely!** *colloq.* certainly not, I refuse. [Old Norse: related to LIKE[1]]

like-minded *adj.* having the same tastes, opinions, etc.

liken *v.* (foll. by *to*) point out the resemblance of (a person or thing to another). [from LIKE[1]]

likeness *n.* **1** (usu. foll. by *between*, *to*) resemblance. **2** (foll. by *of*) semblance or guise (*in the likeness of a ghost*). **3** portrait, representation.

likewise *adv.* **1** also, moreover. **2** similarly (*do likewise*).

liking *n.* **1** what one likes; one's taste (*is it to your liking?*). **2** (foll. by *for*) regard or fondness; taste or fancy.

lilac /'laɪlək/ –*n.* **1** shrub with fragrant pinkish-violet or white blossoms. **2** pale pinkish-violet colour. –*adj.* of this colour. [Persian]

liliaceous /ˌlɪlɪ'eɪʃəs/ *adj.* of the lily family. [related to LILY]

lilliputian /ˌlɪlɪ'pju:ʃ(ə)n/ –*n.* diminutive person or thing. –*adj.* diminutive. [*Lilliput* in Swift's *Gulliver's Travels*]

Lilo /'laɪləʊ/ *n.* (also **Li-Lo** *propr.*) (*pl.* **-s**) type of inflatable mattress. [from *lie low*]

lilt –*n.* **1** light springing rhythm. **2** tune with this. –*v.* (esp. as **lilting** *adj.*) speak etc. with a lilt; have a lilt. [origin unknown]

lily /'lɪlɪ/ *n.* (*pl.* **-ies**) **1** bulbous plant with large trumpet-shaped flowers on a tall stem. **2** heraldic fleur-de-lis. [Latin *lilium*]

lily-livered *adj.* cowardly.

lily of the valley *n.* plant with white bell-shaped fragrant flowers.

lily white *adj.* & *n.* (as adj. often hyphenated) pure white.

limb[1] /lɪm/ *n.* **1** arm, leg, or wing. **2** large branch of a tree. **3** branch of a cross. □ **out on a limb** isolated. [Old English]

limb[2] /lɪm/ *n.* specified edge of the sun, moon, etc. [Latin *limbus* hem, border]

limber[1] –*adj.* **1** lithe. **2** flexible. –*v.* (usu. foll. by *up*) **1** make (oneself or a part of the body etc.) supple. **2** warm up in preparation for athletic etc. activity. [origin uncertain]

limber[2] –*n.* detachable front part of a gun-carriage. –*v.* attach a limber to. [perhaps from Latin *limo -onis* shaft]

limbo[1] /'lɪmbəʊ/ *n.* (*pl.* **-s**) **1** (in some Christian beliefs) supposed abode of the souls of unbaptized infants, and of the just who died before Christ. **2** intermediate state or condition of awaiting a decision etc. [Latin *in limbo*: related to LIMB[2]]

limbo[2] /'lɪmbəʊ/ *n.* (*pl.* **-s**) W. Indian dance in which the dancer bends backwards to pass under a horizontal bar which is progressively lowered. [W. Indian word, perhaps = LIMBER[1]]

lime[1] –*n.* **1** (in full **quicklime**) white substance (calcium oxide) obtained by heating limestone. **2** (in full **slaked lime**) calcium hydroxide obtained by reacting quicklime with water, used as a fertilizer and in making mortar. –*v.* (**-ming**) treat with lime. □ **limy** *adj.* (**-ier, -iest**). [Old English]

lime[2] *n.* **1 a** fruit like a lemon but green, rounder, smaller, and more acid. **b** tree which produces this fruit. **2** (in full **lime-green**) yellowish-green colour. [French from Arabic]

lime[3] *n.* (in full **lime-tree**) tree with heart-shaped leaves and fragrant

creamy blossom. [alteration of *line* = Old English *lind* = LINDEN]
limekiln *n.* kiln for heating limestone.
limelight *n.* **1** intense white light used formerly in theatres. **2** (prec. by *the*) the glare of publicity.
limerick /'lɪmərɪk/ *n.* humorous five-line verse with a rhyme-scheme *aabba*. [origin uncertain]
limestone *n.* rock composed mainly of calcium carbonate.
Limey /'laɪmɪ/ *n.* (*pl.* **-s**) *US slang offens.* British person (orig. a sailor) or ship. [from LIME[2], because of the former enforced consumption of lime juice in the British Navy]
limit /'lɪmɪt/ *–n.* **1** point, line, or level beyond which something does not or may not extend or pass. **2** greatest or smallest amount permissible. *–v.* (**-t-**) **1** set or serve as a limit to. **2** (foll. by *to*) restrict. □ **be the limit** *colloq.* be intolerable. **within limits** with some degree of freedom. □ **limitless** *adj.* [Latin *limes limit-* boundary, frontier]
limitation /ˌlɪmɪ'teɪʃ(ə)n/ *n.* **1** limiting or being limited. **2** limit (of ability etc.) (often in *pl.*: *know one's limitations*). **3** limiting circumstance.
limited *adj.* **1** confined within limits. **2** not great in scope or talents. **3** restricted to a few examples (*limited edition*). **4** (after a company name) being a limited company.
limited company *n.* (also **limited liability company**) company whose owners are legally responsible only to a specified amount for its debts.
limn /lɪm/ *v. archaic* paint. [French *luminer* from Latin *lumino* ILLUMINATE]
limo /'lɪməʊ/ *n.* (*pl.* **-s**) *US colloq.* limousine. [abbreviation]
limousine /ˌlɪmʊ'ziːn/ *n.* large luxurious car. [French]
limp[1] *–v.* walk or proceed lamely or awkwardly. *–n.* lame walk. [perhaps from obsolete *limphalt*: related to HALT[2]]
limp[2] *adj.* **1** not stiff or firm. **2** without energy or will. □ **limply** *adv.* **limpness** *n.* [perhaps from LIMP[1]]
limpet /'lɪmpɪt/ *n.* marine gastropod with a conical shell, sticking tightly to rocks. [Old English]
limpet mine *n.* delayed-action mine attached to a ship's hull.
limpid /'lɪmpɪd/ *adj.* clear, transparent. □ **limpidity** /-'pɪdɪtɪ/ *n.* [Latin]
linage /'laɪnɪdʒ/ *n.* **1** number of lines in printed or written matter. **2** payment by the line.
linchpin *n.* **1** pin passed through an axle-end to keep a wheel in position. **2** person or thing vital to an organization etc. [Old English *lynis* = axle-tree]
linctus /'lɪŋktəs/ *n.* syrupy medicine, esp. a soothing cough mixture. [Latin *lingo* lick]
linden /'lɪnd(ə)n/ *n.* lime-tree. [Old English *lind(e)*]
line[1] *–n.* **1** continuous mark made on a surface. **2** similar mark, esp. a furrow or wrinkle. **3** use of lines in art. **4 a** straight or curved continuous extent of length without breadth. **b** track of a moving point. **5** contour or outline (*has a slimming line*). **6 a** curve connecting all points having a specified common property. **b** (**the Line**) the Equator. **7 a** limit or boundary. **b** mark limiting the area of play, the starting or finishing point in a race, etc. **8 a** row of persons or things. **b** direction as indicated by them. **c** *US* queue. **9 a** row of printed or written words. **b** portion of verse written in one line. **10** (in *pl.*) **a** piece of poetry. **b** words of an actor's part. **c** specified amount of text etc. to be written out as a school punishment. **11** short letter or note (*drop me a line*). **12** length of cord, rope, etc., usu. serving a specified purpose. **13 a** wire or cable for a telephone or telegraph. **b** connection by means of this. **14 a** single track of a railway. **b** one branch or route of a railway system, or the whole system under one management. **15 a** regular succession of buses, ships, aircraft, etc., plying between certain places. **b** company conducting this. **16** connected series of persons following one another in time (esp. several generations of a family); stock. **17** course or manner of procedure, conduct, thought, etc. (*along these lines*; *don't take that line*). **18** direction, course, or channel (*lines of communication*). **19** department of activity; branch of business. **20** type of product (*new line in hats*). **21 a** connected series of military fieldworks. **b** arrangement of soldiers or ships side by side. **22** each of the very narrow horizontal sections forming a television picture. **23** level of the base of most letters in printing and writing. *–v.* (**-ning**) **1** mark with lines. **2** cover with lines. **3** position or stand at intervals along (*crowds lined the route*). □ **all along the line** at every point. **bring into line** make conform. **come into line** conform. **get a line on** *colloq.* get information about. **in line for** likely to receive. **in** (or **out of**) **line with** in (or not in) accordance with. **lay** (or **put**) **it on the line** *colloq.* speak frankly. **line up 1** arrange or be arranged in a line or lines. **2** have ready. **out of line** not in alignment; inappropriate. [Latin *linea* from *linum* flax]

line[2] *v.* **(-ning) 1** cover the inside surface of (a garment, box, etc.) with a layer of usu. different material. **2** serve as a lining for. **3** *colloq.* fill, esp. plentifully. [obsolete *line* linen used for linings]

lineage /'lɪnɪɪdʒ/ *n.* lineal descent; ancestry. [Latin: related to LINE[1]]

lineal /'lɪnɪəl/ *adj.* **1** in the direct line of descent or ancestry. **2** linear. □ **lineally** *adv.*

lineament /'lɪnɪəmənt/ *n.* (usu. in *pl.*) distinctive feature or characteristic, esp. of the face. [Latin: related to LINE[1]]

linear /'lɪnɪə(r)/ *adj.* **1** of or in lines. **2** long and narrow and of uniform breadth. □ **linearity** /-'ærɪtɪ/ *n.* **linearly** *adv.*

Linear B *n.* form of Bronze Age writing found in Greece: an earlier undeciphered form **(Linear A)** also exists.

lineation /ˌlɪnɪ'eɪʃ(ə)n/ *n.* marking with or drawing of lines.

line-drawing *n.* drawing in which images are produced with lines.

linen /'lɪnɪn/ *–n.* **1** cloth woven from flax. **2** (*collect.*) articles made or orig. made of linen, as sheets, shirts, underwear, etc. *–adj.* made of linen. [Old English: related to Latin *linum* flax]

linen basket *n.* basket for dirty washing.

line of fire *n.* expected path of gunfire etc.

line of vision *n.* straight line along which an observer looks.

line-out *n.* (in Rugby) parallel lines of opposing forwards at right angles to the touchline for the throwing in of the ball.

line printer *n.* machine that prints output from a computer a line at a time.

liner[1] *n.* ship or aircraft etc. carrying passengers on a regular line.

liner[2] *n.* removable lining.

linesman *n.* umpire's or referee's assistant who decides whether a ball has fallen within the playing area or not.

line-up *n.* **1** line of people for inspection. **2** arrangement of persons in a team, band, etc.

ling[1] *n.* (*pl.* same) long slender marine fish. [probably Dutch]

ling[2] *n.* any of various heathers. [Old Norse]

-ling *suffix* **1** denoting a person or thing: **a** connected with (*hireling*). **b** having the property of being (*weakling*) or undergoing (*starveling*). **2** denoting a diminutive (*duckling*), often derogatory (*lordling*). [Old English]

linger /'lɪŋgə(r)/ *v.* **1** stay about. **2** (foll. by *over*, *on*, etc.) dally (*linger over dinner*; *lingered on the final note*). **3** (esp. of an illness) be protracted. **4** (often foll. by *on*) be slow in dying. [Old English *lengan*: related to LONG[1]]

lingerie /'læʒərɪ/ *n.* women's underwear and nightclothes. [French *linge* linen]

lingo /'lɪŋgəʊ/ *n.* (*pl.* **-s** or **-es**) *colloq.* **1** foreign language. **2** vocabulary of a special subject or group. [probably from Portuguese *lingoa* from Latin *lingua* tongue]

lingua franca /ˌlɪŋgwə 'fræŋkə/ *n.* (*pl.* **lingua francas**) **1** language used in common by speakers with different native languages. **2** system for mutual understanding. [Italian, = Frankish tongue]

lingual /'lɪŋgw(ə)l/ *adj.* **1** of or formed by the tongue. **2** of speech or languages. □ **lingually** *adv.* [Latin *lingua* tongue, language]

linguist /'lɪŋgwɪst/ *n.* person skilled in languages or linguistics.

linguistic /lɪŋ'gwɪstɪk/ *adj.* of language or the study of languages. □ **linguistically** *adv.*

linguistics *n.* the study of language and its structure.

liniment /'lɪnɪmənt/ *n.* embrocation. [Latin *linio* smear]

lining *n.* material which lines a surface etc.

link *–n.* **1** one loop or ring of a chain etc. **2 a** connecting part; one in a series. **b** state or means of connection. **3** cufflink. *–v.* **1** (foll. by *together*, *to*, *with*) connect or join (two things or one to another). **2** clasp or intertwine (hands or arms). **3** (foll. by *on*, *to*, *in to*) be joined; attach oneself to (a system, company, etc.). □ **link up** (foll. by *with*) connect or combine. [Old Norse]

linkage *n.* **1** linking or being linked, esp. the linking of quite different political issues in negotiations. **2** link or system of links.

linkman *n.* person providing continuity in a broadcast programme.

links *n.pl.* (treated as *sing.* or *pl.*) golf-course. [Old English, = rising ground]

link-up *n.* act or result of linking up.

Linnaean *adj.* of Linnaeus or his system of classifying plants and animals.

■ **Usage** This word is spelt *Linnean* in *Linnean Society*.

linnet /'lɪnɪt/ *n.* brown-grey finch. [French *linette* from *lin* flax, because it eats flax-seed]

lino /'laɪnəʊ/ *n.* (*pl.* **-s**) linoleum. [abbreviation]

linocut *n.* **1** design carved in relief on a block of linoleum. **2** print made from this.

linoleum /lɪ'nəʊlɪəm/ *n.* canvas-backed material thickly coated with a preparation of linseed oil and powdered cork etc., esp. as a floor covering. [Latin *linum* flax, *oleum* oil]
linseed /'lɪnsi:d/ *n.* seed of flax. [Old English: related to LINE[1]]
linseed oil *n.* oil extracted from linseed and used in paint and varnish.
linsey-woolsey /ˌlɪnzɪ'wʊlzɪ/ *n.* fabric of coarse wool woven on a cotton warp. [probably from *Lindsey* in Suffolk + WOOL]
lint *n.* **1** linen or cotton with a raised nap on one side, used for dressing wounds. **2** fluff. [perhaps from French *linette* from *lin* flax]
lintel /'lɪnt(ə)l/ *n.* horizontal timber, stone, etc., across the top of a door or window. [French: related to LIMIT]
lion /'laɪən/ *n.* **1** (*fem.* **lioness**) large tawny flesh-eating wild cat of Africa and S. Asia. **2** (**the Lion**) zodiacal sign or constellation Leo. **3** brave or celebrated person. [Latin *leo*]
lion-heart *n.* courageous person. □ **lion-hearted** *adj.*
lionize *v.* (also **-ise**) (**-zing** or **-sing**) treat as a celebrity.
lion's share *n.* largest or best part.
lip –*n.* **1** either of the two fleshy parts forming the edges of the mouth-opening. **2** edge of a cup, vessel, etc., esp. the part shaped for pouring from. **3** *colloq.* impudent talk. –*v.* (**-pp-**) **1** touch with the lips; apply the lips to. **2** touch lightly. □ **lipped** *adj.* (also in *comb.*). [Old English]
lipid /'lɪpɪd/ *n.* any of a group of fatlike substances that are insoluble in water but soluble in organic solvents, including fatty acids, oils, waxes, and steroids. [Greek *lipos* fat]
liposuction /'laɪpəʊˌsʌkʃ(ə)n/ *n.* technique in cosmetic surgery for removing excess fat from under the skin by suction.
lip-read *v.* understand (speech) from observing a speaker's lip-movements.
lip-service *n.* insincere expression of support etc.
lipstick *n.* stick of cosmetic for colouring the lips.
liquefy /'lɪkwɪˌfaɪ/ *v.* (**-ies, -ied**) make or become liquid. □ **liquefaction** /-'fækʃ(ə)n/ *n.* [Latin: related to LIQUID]
liqueur /lɪ'kjʊə(r)/ *n.* any of several strong sweet alcoholic spirits. [French]
liquid /'lɪkwɪd/ –*adj.* **1** having a consistency like that of water or oil, flowing freely but of constant volume. **2** having the qualities of water in appearance. **3** (of sounds) clear and pure. **4** (of assets) easily converted into cash. –*n.* **1** liquid substance. **2** *Phonet.* sound of *l* or *r*. [Latin *liqueo* be liquid]
liquidate /'lɪkwɪˌdeɪt/ *v.* (**-ting**) **1** wind up the affairs of (a firm) by ascertaining liabilities and apportioning assets. **2** pay off (a debt). **3** wipe out, kill. □ **liquidator** *n.* [medieval Latin: related to LIQUID]
liquidation /ˌlɪkwɪ'deɪʃ(ə)n/ *n.* liquidating, esp. of a firm. □ **go into liquidation** (of a firm etc.) be wound up and have its assets apportioned.
liquid crystal *n.* turbid liquid with some order in its molecular arrangement.
liquid crystal display *n.* visual display in electronic devices, in which the reflectivity of a matrix of liquid crystals changes as a signal is applied.
liquidity /lɪ'kwɪdɪtɪ/ *n.* (*pl.* **-ies**) **1** state of being liquid. **2** availability of liquid assets.
liquidize /'lɪkwɪˌdaɪz/ *v.* (also **-ise**) (**-zing** or **-sing**) reduce to a liquid state.
liquidizer *n.* (also **-iser**) machine for liquidizing foods.
liquor /'lɪkə(r)/ *n.* **1** alcoholic (esp. distilled) drink. **2** other liquid, esp. that produced in cooking. [Latin: related to LIQUID]
liquorice /'lɪkərɪs, -rɪʃ/ *n.* (also **licorice**) **1** black root extract used as a sweet and in medicine. **2** plant from which it is obtained. [Greek *glukus* sweet, *rhiza* root]
lira /'lɪərə/ *n.* (*pl.* **lire** pronunc. same or /-reɪ/) **1** chief monetary unit of Italy. **2** chief monetary unit of Turkey. [Latin *libra* pound]
lisle /laɪl/ *n.* fine cotton thread for stockings etc. [*Lille* in France]
lisp –*n.* speech defect in which *s* is pronounced like *th* in *thick* and *z* is pronounced like *th* in *this*. –*v.* speak or utter with a lisp. [Old English]
lissom /'lɪsəm/ *adj.* lithe, agile. [ultimately from LITHE]
list[1] –*n.* **1** number of items, names, etc., written or printed together as a record or aid to memory. **2** (in *pl.*) **a** palisades enclosing an area for a tournament. **b** scene of a contest. –*v.* **1** make a list of. **2** enter in a list. **3** (as **listed** *adj.*) **a** (of securities) approved for dealings on the Stock Exchange. **b** (of a building) of historical importance and officially protected. □ **enter the lists** issue or accept a challenge. [Old English]
list[2] –*v.* (of a ship etc.) lean over to one side. –*n.* process or instance of listing. [origin unknown]
listen /'lɪs(ə)n/ *v.* **1 a** make an effort to hear something. **b** attentively hear a person speaking. **2** (foll. by *to*) **a** give

attention with the ear. **b** take notice of; heed. **3** (also **listen out**) (often foll. by *for*) seek to hear by waiting alertly. □ **listen in 1** tap a telephonic communication. **2** use a radio receiving set. [Old English]

listener *n.* **1** person who listens. **2** person who listens to the radio.

listeria /lɪ'stɪərɪə/ *n.* any of several bacteria infecting humans and animals eating contaminated food. [*Lister*, name of a surgeon]

listless *adj.* lacking energy or enthusiasm. □ **listlessly** *adv.* **listlessness** *n.* [from obsolete *list* inclination]

list price *n.* price of something as shown in a published list.

lit *past* and *past part.* of LIGHT[1], LIGHT[2].

litany /'lɪtənɪ/ *n.* (*pl.* **-ies**) **1 a** series of supplications to God recited by a priest etc. with set responses by the congregation. **b** (**the Litany**) that in the Book of Common Prayer. **2** tedious recital (*litany of woes*). [Greek *litaneia* prayer]

litchi var. of LYCHEE.

liter *US* var. of LITRE.

literacy /'lɪtərəsɪ/ *n.* ability to read and write. [Latin *littera* letter]

literal /'lɪtər(ə)l/ *–adj.* **1** taking words in their basic sense without metaphor or allegory. **2** corresponding exactly to the original words (*literal translation*). **3** prosaic; matter-of-fact. **4** so called without exaggeration (*literal bankruptcy*). **5** of a letter or the letters of the alphabet. *–n.* misprint. □ **literally** *adv.* [Latin *littera* letter]

literalism *n.* insistence on a literal interpretation; adherence to the letter. □ **literalist** *n.*

literary /'lɪtərərɪ/ *adj.* **1** of or concerned with books or literature etc. **2** (of a word or idiom) used chiefly by writers; formal. □ **literariness** *n.* [Latin: related to LETTER]

literate /'lɪtərət/ *–adj.* able to read and write; educated. *–n.* literate person.

literati /,lɪtə'rɑ:tɪ/ *n.pl.* the class of learned people.

literature /'lɪtərətʃə(r)/ *n.* **1** written works, esp. those valued for form and style. **2** writings of a country or period or on a particular subject. **3** literary production. **4** *colloq.* printed matter, leaflets, etc.

lithe /laɪð/ *adj.* flexible, supple. [Old English]

lithium /'lɪθɪəm/ *n.* soft silver-white metallic element. [Greek *lithion* from *lithos* stone]

litho /'laɪθəʊ/ *colloq.* *–n.* = LITHOGRAPHY. *–v.* (**-oes, -oed**) lithograph. [abbreviation]

lithograph /'lɪθə,grɑ:f/ *–n.* lithographic print. *–v.* print by lithography. [Greek *lithos* stone]

lithography /lɪ'θɒgrəfɪ/ *n.* process of printing from a plate so treated that ink adheres only to the design to be printed. □ **lithographer** *n.* **lithographic** /,lɪθə'græfɪk/ *adj.* **lithographically** /,lɪθə'græfɪkəlɪ/ *adv.*

Lithuanian /,lɪθju:'eɪnɪən/ *–n.* **1 a** native or national of Lithuania in eastern Europe. **b** person of Lithuanian descent. **2** language of Lithuania. *–adj.* of Lithuania, its people, or language.

litigant /'lɪtɪgənt/ *–n.* party to a lawsuit. *–adj.* engaged in a lawsuit. [related to LITIGATE]

litigate /'lɪtɪ,geɪt/ *v.* (**-ting**) **1** go to law. **2** contest (a point) at law. □ **litigation** /-'geɪʃ(ə)n/ *n.* **litigator** *n.* [Latin *lis lit-* lawsuit]

litigious /lɪ'tɪdʒəs/ *adj.* **1** fond of litigation. **2** contentious. [Latin: related to LITIGATE]

litmus /'lɪtməs/ *n.* dye from lichens, turned red by acid and blue by alkali. [Old Norse, = dye-moss]

litmus paper *n.* paper stained with litmus, used to test for acids or alkalis.

litmus test *n.* *colloq.* real or ultimate test.

litotes /laɪ'təʊti:z/ *n.* (*pl.* same) ironic understatement, esp. using the negative (e.g. *I shan't be sorry* for *I shall be glad*). [Greek *litos* plain, meagre]

litre /'li:tə(r)/ *n.* (*US* **liter**) metric unit of capacity equal to 1 cubic decimetre (1.76 pints). [Greek *litra*]

Litt.D. *abbr.* Doctor of Letters. [Latin *Litterarum Doctor*]

litter *–n.* **1 a** refuse, esp. paper, discarded in a public place. **b** odds and ends lying about. **2** young animals brought forth at one birth. **3** vehicle containing a couch and carried on men's shoulders or by animals. **4** a kind of stretcher for the sick and wounded. **5** straw etc., as bedding for animals. **6** granulated material for use as an animal's, esp. a cat's, toilet indoors. *–v.* **1** make (a place) untidy with refuse. **2** give birth to (whelps etc.). **3 a** provide (a horse etc.) with litter as bedding. **b** spread straw etc. on (a stable-floor etc.). [Latin *lectus* bed]

litterbug *n.* *colloq.* person who drops litter in the street etc.

litter-lout *n.* *colloq.* = LITTERBUG.

little /'lɪt(ə)l/ *–adj.* (**littler, littlest; less** or **lesser, least**) **1** small in size, amount, degree, etc.; often used affectionately or condescendingly (*friendly little chap*; *silly little fool*). **2 a** short in stature. **b** of short distance or duration. **3** (prec. by *a*) a certain though small amount of (*give me a little butter*). **4** trivial (*questions every little thing*). **5** only a small amount (*had little sleep*). **6** operating on a small scale; humble, ordinary (*the little shopkeeper*; *the little man*). **7** smaller or the smallest of the name (*little hand of a clock*; *little auk*). **8** young or younger (*little boy*; *my little sister*). *–n.* **1** not much; only a small amount (*got little out of it*; *did what little I could*). **2** (usu. prec. by *a*) **a** a certain but no great amount (*knows a little of everything*). **b** short time or distance (*after a little*). *–adv.* (**less, least**) **1** to a small extent only (*little-known author*; *little more than speculation*). **2** not at all; hardly (*they little thought*). **3** (prec. by *a*) somewhat (*is a little deaf*). [Old English]

Little Bear see BEAR².

little by little *adv.* by degrees; gradually.

little end *n.* the smaller end of a connecting-rod, attached to the piston.

little grebe *n.* small water-bird of the grebe family.

little people *n.pl.* (prec. by *the*) fairies.

little woman *n.* (prec. by *the*) *colloq.* often *derog.* one's wife.

littoral /ˈlɪtər(ə)l/ *–adj.* of or on the shore. *–n.* region lying along a shore. [Latin *litus litor-* shore]

liturgy /ˈlɪtədʒɪ/ *n.* (*pl.* **-ies**) **1** prescribed form of public worship. **2** (**the Liturgy**) the Book of Common Prayer. □ **liturgical** /-ˈtɜːdʒɪk(ə)l/ *adj.* **liturgically** /-ˈtɜːdʒɪkəlɪ/ *adv.* [Greek *leitourgia* public worship]

livable var. of LIVEABLE.

live¹ /lɪv/ *v.* (**-ving**) **1** have life; be or remain alive. **2** have one's home (*lives up the road*). **3** (foll. by *on*) subsist or feed (*lives on fruit*). **4** (foll. by *on, off*) depend for subsistence (*lives off the State*; *lives on a pension*). **5** (foll. by *on, by*) sustain one's position (*live on their reputation*; *lives by his wits*). **6 a** spend or pass (*lived a full life*). **b** express in one's life (*lives his faith*). **7** conduct oneself, arrange one's habits, etc., in a specified way (*live quietly*). **8** (often foll. by *on*) (of a person or thing) survive; remain (*memory lived on*). **9** enjoy life to the full (*not really living*). □ **live and let live** condone others' failings so as to be similarly tolerated. **live down** cause (past guilt, a scandal, etc.) to be forgotten by blameless conduct thereafter. **live for** regard as one's life's purpose (*lives for her music*). **live in** (or **out**) reside on (or off) the premises of one's work. **live it up** *colloq.* live gaily and extravagantly. **live a lie** keep up a pretence. **live together** (esp. of a couple not married to each other) share a home and have a sexual relationship. **live up to** fulfil. **live with 1** share a home with. **2** tolerate. [Old English]

live² /laɪv/ *–adj.* **1** (*attrib.*) that is alive; living. **2** (of a broadcast, performance, etc.) heard or seen at the time of its performance or with an audience present. **3** of current interest or importance (*a live issue*). **4** glowing, burning (*live coals*). **5** (of a match, bomb, etc.) not yet kindled or exploded. **6** (of a wire etc.) charged with or carrying electricity. *–adv.* **1** in order to make a live broadcast (*going live now to the House of Commons*). **2** as a live performance etc. (*show went out live*). [from ALIVE]

liveable /ˈlɪvəb(ə)l/ *adj.* (also **livable**) **1** *colloq.* (usu. **liveable-in**) (of a house etc.) fit to live in. **2** (of a life) worth living. **3** *colloq.* (usu. **liveable-with**) (of a person) easy to live with.

lived-in *adj.* **1** (of a room etc.) showing signs of habitation. **2** *colloq.* (of a face) marked by experience.

live-in *attrib. adj.* (of a sexual partner, employee, etc.) cohabiting; resident.

livelihood /ˈlaɪvlɪˌhʊd/ *n.* means of living; job, income. [Old English: related to LIFE]

livelong /ˈlɪvlɒŋ/ *adj.* in its entire length (*the livelong day*). [from obsolete *lief*, assimilated to LIVE¹]

lively /ˈlaɪvlɪ/ *adj.* (**-ier, -iest**) **1** full of life; vigorous, energetic. **2** vivid (*lively imagination*). **3** cheerful. **4** *joc.* exciting, dangerous (*made things lively for him*). □ **liveliness** *n.* [Old English]

liven /ˈlaɪv(ə)n/ *v.* (often foll. by *up*) *colloq.* make or become lively, cheer up.

liver¹ /ˈlɪvə(r)/ *n.* **1** large glandular organ in the abdomen of vertebrates. **2** liver of some animals as food. [Old English]

liver² /ˈlɪvə(r)/ *n.* person who lives in a specified way (*a fast liver*).

liveried /ˈlɪvərɪd/ *adj.* wearing livery.

liverish /ˈlɪvərɪʃ/ *adj.* **1** suffering from a liver disorder. **2** peevish, glum.

Liverpudlian /ˌlɪvəˈpʌdlɪən/ *–n.* native of Liverpool. *–adj.* of Liverpool. [*Liverpool* in NW England]

liver sausage *n.* sausage of cooked liver etc.

liverwort *n.* small mosslike or leafless plant sometimes lobed like a liver.

livery /ˈlɪvərɪ/ *n.* (*pl.* **-ies**) **1** distinctive uniform of a member of a City Company or of a servant. **2** distinctive guise or marking (*birds in their winter livery*). **3** distinctive colour scheme in which a company's vehicles etc. are painted. □ **at livery** (of a horse) kept for the owner for a fixed charge. [Anglo-French *liveré*, past part. of *livrer* DELIVER]

livery stable *n.* stable where horses are kept at livery or let out for hire.

lives *pl.* of LIFE.

livestock *n.* (usu. treated as *pl.*) animals on a farm, kept for use or profit.

live wire *n.* spirited person.

livid /'lɪvɪd/ *adj.* **1** *colloq.* furious. **2** of a bluish leaden colour (*livid bruise*). [Latin]

living /'lɪvɪŋ/ *–n.* **1** being alive (*that's what living is all about*). **2** livelihood. **3** position held by a clergyman, providing an income. *–adj.* **1** contemporary; now alive. **2** (of a likeness) exact, lifelike. **3** (of a language) still in vernacular use. □ **within living memory** within the memory of people still alive.

living-room *n.* room for general day use.

living wage *n.* wage on which one can live without privation.

lizard /'lɪzəd/ *n.* reptile with usu. a long body and tail, four legs, and a rough or scaly hide. [Latin *lacertus*]

LJ *abbr.* (*pl.* **L JJ**) Lord Justice.

'll *v.* (usu. after pronouns) shall, will (*I'll; that'll*). [abbreviation]

llama /'lɑ:mə/ *n.* S. American ruminant kept as a beast of burden and for its soft woolly fleece. [Spanish from Quechua]

LL B *abbr.* Bachelor of Laws. [Latin *legum baccalaureus*]

LL D *abbr.* Doctor of Laws. [Latin *legum doctor*]

LL M *abbr.* Master of Laws. [Latin *legum magister*]

Lloyd's /lɔɪdz/ *n.* incorporated society of underwriters in London. [*Lloyd*, proprietor of the coffee-house where the society originally met]

Lloyd's List *n.* daily publication devoted to shipping news.

Lloyd's Register *n.* annual classified list of all ships.

ln *abbr.* natural logarithm.

lo /ləʊ/ *int. archaic* look. □ **lo and behold** *joc.* formula introducing mention of a surprising fact. [Old English]

loach *n.* (*pl.* same or **-es**) small freshwater fish. [French]

load *–n.* **1 a** what is carried or to be carried. **b** amount usu. or actually carried (often in *comb.*: *lorry-load of bricks*). **2** burden or commitment of work, responsibility, care, etc. **3** *colloq.* **a** (in *pl.*; often foll. by *of*) plenty, a lot (*loads of money, people*). **b** (**a load of**) a quantity (*a load of nonsense*). **4** amount of power carried by an electric circuit or supplied by a generating station. *–v.* **1 a** put a load on or aboard. **b** place (a load) aboard a ship, on a vehicle, etc. **2** (often foll. by *up*) (of a vehicle or person) take a load aboard. **3** (often foll. by *with*) burden, strain (*loaded with food*). **4** (also **load up**) (foll. by *with*) overburden, overwhelm (*loaded us with work, with abuse*). **5 a** put ammunition in (a gun), film in (a camera), a cassette in (a tape recorder), a program in (a computer), etc. **b** put (a film, cassette, etc.) into a device. **6** give a bias to. □ **get a load of** *slang* take note of. [Old English, = way]

loaded *adj.* **1** *slang* **a** rich. **b** drunk. **c** *US* drugged. **2** (of dice etc.) weighted. **3** (of a question or statement) carrying some hidden implication.

loader *n.* **1** loading-machine. **2** (in *comb.*) gun, machine, lorry, etc., loaded in a specified way (*breech-loader; front-loader*). □ **-loading** *adj.* (in *comb.*) (in sense 2).

load line *n.* = PLIMSOLL LINE.

loadstone var. of LODESTONE.

loaf[1] *n.* (*pl.* **loaves**) **1** unit of baked bread, usu. of a standard size or shape. **2** other food made in the shape of a loaf and cooked. **3** *slang* head as the seat of common sense. [Old English]

loaf[2] *v.* (often foll. by *about, around*) spend time idly; hang about. [back-formation from LOAFER]

loafer *n.* **1** idle person. **2** (**Loafer**) *propr.* flat soft-soled leather shoe. [origin uncertain]

loam *n.* rich soil of clay, sand, and humus. □ **loamy** *adj.* [Old English]

loan *–n.* **1** thing lent, esp. a sum of money. **2** lending or being lent. *–v.* lend (money, works of art, etc.). □ **on loan** being lent. [Old English]

loan shark *n. colloq.* person who lends money at exorbitant rates of interest.

loath *predic. adj.* (also **loth**) disinclined, reluctant (*loath to admit it*). [Old English]

loathe /ləʊð/ *v.* (**-thing**) detest, hate. □ **loathing** *n.* [Old English]

loathsome /'ləʊðsəm/ *adj.* arousing hatred or disgust; repulsive.

loaves *pl.* of LOAF[1].

lob *–v.* (**-bb-**) hit or throw (a ball etc.) slowly or in a high arc. *–n.* such a ball. [probably Low German or Dutch]

lobar /'ləʊbə(r)/ *adj.* of a lobe, esp. of the lung (*lobar pneumonia*).

lobate /'ləʊbeɪt/ *adj.* having a lobe or lobes.

lobby /'lɒbɪ/ *–n.* (*pl.* **-ies**) **1** porch, anteroom, entrance-hall, or corridor. **2 a** (in the House of Commons) large hall used esp. for interviews between MPs and the public. **b** (also **division lobby**) each of two corridors to which MPs retire to vote. **3 a** body of lobbyists (*anti-abortion lobby*). **b** organized rally of lobbying members of the public. **4** (prec. by *the*) group of journalists who receive unattributable briefings from the government (*lobby correspondent*). –*v.* (**-ies, -ied**) **1** solicit the support of (an influential person). **2** (of members of the public) inform in order to influence (legislators, an MP, etc.). **3** frequent a parliamentary lobby. [Latin *lobia* lodge]

lobbyist *n.* person who lobbies an MP etc., esp. professionally.

lobe *n.* **1** lower soft pendulous part of the outer ear. **2** similar part of other organs, esp. the brain, liver, and lung. □ **lobed** *adj.* [Greek *lobos* lobe, pod]

lobelia /lə'bi:lɪə/ *n.* plant with bright, esp. blue, flowers. [*Lobel*, name of a botanist]

lobotomy /lə'bɒtəmɪ/ *n.* (*pl.* **-ies**) incision into the frontal lobe of the brain, formerly used in some cases of mental disorder. [from LOBE]

lobscouse /'lɒbskaʊs/ *n.* sailor's dish of meat stewed with vegetables and ship's biscuit. [origin unknown]

lobster *n.* **1** marine crustacean with two pincer-like claws. **2** its flesh as food. [Latin *locusta* lobster, LOCUST]

lobster-pot *n.* basket for trapping lobsters.

lobworm *n.* large earthworm used as fishing-bait. [from LOB in obsolete sense 'pendulous object']

local /'ləʊk(ə)l/ –*adj.* **1** belonging to, existing in, or peculiar to a particular place (*local history*). **2** of the neighbourhood (*local paper*). **3** of or affecting a part and not the whole (*local anaesthetic*). **4** (of a telephone call) to a nearby place and charged at a lower rate. –*n.* **1** inhabitant of a particular place. **2** (often prec. by *the*) *colloq.* local public house. **3** local anaesthetic. □ **locally** *adv.* [Latin *locus* place]

local authority *n.* administrative body in local government.

local colour *n.* touches of detail in a story etc. designed to provide a realistic background.

locale /ləʊ'kɑ:l/ *n.* scene or locality of an event or occurrence. [French *local*]

local government *n.* system of administration of a county, district, parish, etc., by the elected representatives of those who live there.

locality /ləʊ'kælɪtɪ/ *n.* (*pl.* **-ies**) **1** district. **2** site or scene of a thing. **3** thing's position. [Latin: related to LOCAL]

localize /'ləʊkə,laɪz/ *v.* (also **-ise**) (**-zing** or **-sing**) **1** restrict or assign to a particular place. **2** invest with the characteristics of a particular place. **3** decentralize.

local time *n.* time in a particular place.

local train *n.* train stopping at all the stations on its route.

locate /ləʊ'keɪt/ *v.* (**-ting**) **1** discover the exact place of. **2** establish in a place; situate. **3** state the locality of. [Latin: related to LOCAL]

■ **Usage** In standard English, it is not acceptable to use *locate* to mean merely 'find' as in *can't locate my key*.

location /ləʊ'keɪʃ(ə)n/ *n.* **1** particular place. **2** locating. **3** natural, not studio, setting for a film etc. (*filmed on location*).

loc. cit. *abbr.* in the passage cited. [Latin *loco citato*]

loch /lɒk, lɒx/ *n. Scot.* lake or narrow inlet of the sea. [Gaelic]

loci *pl.* of LOCUS.

lock[1] –*n.* **1** mechanism for fastening a door etc., with a bolt that requires a key of a particular shape to work it. **2** confined section of a canal or river within sluice-gates, for moving boats from one level to another. **3 a** turning of a vehicle's front wheels. **b** (in full **full lock**) maximum extent of this. **4** interlocked or jammed state. **5** wrestling-hold that keeps an opponent's limb fixed. **6** (in full **lock forward**) player in the second row of a Rugby scrum. **7** mechanism for exploding the charge of a gun. –*v.* **1 a** fasten with a lock. **b** (foll. by *up*) shut (a house etc.) thus. **c** (of a door etc.) be lockable. **2 a** (foll. by *up*, *in*, *into*) enclose (a person or thing) by locking. **b** (foll. by *up*) *colloq.* imprison (a person). **3** (often foll. by *up*, *away*) store inaccessibly (*capital locked up in land*). **4** (foll. by *in*) hold fast (in sleep, an embrace, a struggle, etc.). **5** (usu. in *passive*) (of land, hills, etc.) enclose. **6** make or become rigidly fixed. **7** (cause to) jam or catch. □ **lock on to** (of a missile etc.) automatically find and then track (a target). **lock out 1** keep out by locking the door. **2** (of an employer) subject (employees) to a lockout. **under lock and key** locked up. □ **lockable** *adj.* [Old English]

lock[2] *n.* **1** portion of hair that hangs together. **2** (in *pl.*) the hair of the head (*golden locks*). [Old English]

locker *n.* (usu. lockable) cupboard or compartment, esp. for public use.

locket /'lɒkɪt/ *n.* small ornamental case for a portrait or lock of hair, worn on a chain round the neck. [French diminutive of *loc* latch, LOCK[1]]

lockjaw *n.* form of tetanus in which the jaws become rigidly closed.

lock-keeper *n.* person in charge of a river or canal lock.

lockout *n.* employer's exclusion of employees from the workplace until certain terms are agreed to.

locksmith *n.* maker and mender of locks.

lock, stock, and barrel *adv.* completely.

lock-up –*n.* **1** house or room for the temporary detention of prisoners. **2** premises that can be locked up, esp. a small shop. –*attrib. adj.* that can be locked up (*lock-up garage*).

loco[1] /'ləʊkəʊ/ *n.* (*pl.* **-s**) *colloq.* locomotive engine. [abbreviation]

loco[2] /'ləʊkəʊ/ *predic. adj. slang* crazy. [Spanish]

locomotion /,ləʊkə'məʊʃ(ə)n/ *n.* motion or the power of motion from place to place. [Latin LOCUS, MOTION]

locomotive /,ləʊkə'məʊtɪv/ –*n.* engine for pulling trains. –*adj.* of, having, or effecting locomotion.

locum tenens /,ləʊkəm 'ti:nenz, 'te-/ *n.* (*pl.* **locum tenentes** /tɪ'nenti:z/) (also *colloq.* **locum**) deputy acting esp. for a doctor or clergyman. [Latin, = (one) holding a place]

locus /'ləʊkəs/ *n.* (*pl.* **loci** /-saɪ/) **1** position or locality. **2** line or curve etc. formed by all the points satisfying certain conditions, or by the defined motion of a point, line, or surface. [Latin, = place]

locus classicus /,ləʊkəs 'klæsɪkəs/ *n.* (*pl.* ***loci classici*** /,ləʊsaɪ 'klæsɪ,saɪ/) best known or most authoritative passage on a subject. [Latin: related to LOCUS]

locust /'ləʊkəst/ *n.* African or Asian grasshopper migrating in swarms and consuming all vegetation. [Latin *locusta* locust, LOBSTER]

locution /lə'kju:ʃ(ə)n/ *n.* **1** word, phrase, or idiom. **2** style of speech. [Latin *loquor locut-* speak]

lode *n.* vein of metal ore. [var. of LOAD]

lodestar *n.* **1** star used as a guide in navigation, esp. the pole star. **2 a** guiding principle. **b** object of pursuit. [from LODE in obsolete sense 'way, journey']

lodestone *n.* (also **loadstone**) **1** magnetic oxide of iron. **2 a** piece of this used as a magnet. **b** thing that attracts.

lodge –*n.* **1** small house at the entrance to a park or grounds of a large house, occupied by a gatekeeper etc. **2** small house used in the sporting seasons (*hunting lodge*). **3** porter's room at the gate of a college, factory, etc. **4** members or meeting-place of a branch of a society such as the Freemasons. **5** beaver's or otter's lair. –*v.* (**-ging**) **1 a** reside or live, esp. as a lodger. **b** provide with temporary accommodation. **2** submit or present (a complaint etc.) for attention. **3** become fixed or caught; stick. **4** deposit (money etc.) for security. **5** (foll. by *in*, *with*) place (power etc.) in a person. [French *loge*: related to LEAF]

lodger *n.* person paying for accommodation in another's house.

lodging *n.* **1** temporary accommodation (*a lodging for the night*). **2** (in *pl.*) room or rooms rented for lodging in.

loess /'ləʊɪs/ *n.* deposit of fine wind-blown soil, esp. in the basins of large rivers. [Swiss German, = loose]

loft –*n.* **1** attic. **2** room over a stable. **3** gallery in a church or hall. **4** pigeon-house. **5** backward slope on the face of a golf-club. **6** lofting stroke. –*v.* send (a ball etc.) high up. [Old English, = air, upper room]

lofty *adj.* (**-ier, -iest**) **1** (of things) of imposing height. **2** haughty, aloof. **3** exalted, noble (*lofty ideals*). □ **loftily** *adv.* **loftiness** *n.*

log[1] –*n.* **1** unhewn piece of a felled tree; any large rough piece of wood, esp. cut for firewood. **2** *hist.* floating device for gauging a ship's speed. **3** record of events occurring during the voyage of a ship or aircraft. **4** any systematic record of deeds, experiences, etc. **5** = LOGBOOK. –*v.* (**-gg-**) **1 a** enter (a ship's speed, or other transport details) in a logbook. **b** enter (data etc.) in a regular record. **2** attain (a distance, speed, etc., thus recorded) (*had logged over 600 miles*). **3** cut into logs. □ **log in** = *log on*. **log on** (or **off**) open (or close) one's online access to a computer system. **sleep like a log** sleep soundly. [origin unknown]

log[2] *n.* logarithm. [abbreviation]

logan /'ləʊgən/ *n.* (in full **logan-stone**) poised heavy stone rocking at a touch. [= (dial.) *logging*, = rocking]

loganberry /'ləʊgənbərɪ/ *n.* (*pl.* **-ies**) dark red fruit, hybrid of a blackberry and a raspberry. [*Logan*, name of a horticulturalist]

logarithm /'lɒgə,rɪð(ə)m/ *n.* one of a series of arithmetic exponents tabulated to simplify computation by making it possible to use addition and subtraction instead of multiplication and division. □ **logarithmic** /-'rɪðmɪk/ *adj.* **logarithmically** /-'rɪðmɪkəlɪ/ *adv.* [Greek *logos* reckoning, *arithmos* number]

logbook *n.* **1** book containing a detailed record or log. **2** vehicle registration document.

log cabin *n.* hut built of logs.

logger *n. US* lumberjack.

loggerhead /'lɒgə,hed/ *n.* □ **at loggerheads** (often foll. by *with*) disagreeing or disputing. [probably dial. from *logger* wooden block]

loggia /'ləʊdʒə, 'lɒ-/ *n.* open-sided gallery or arcade. [Italian, = LODGE]

logging *n.* work of cutting and preparing forest timber.

logic /'lɒdʒɪk/ *n.* **1 a** science of reasoning. **b** particular system or method of reasoning. **2 a** chain of reasoning (regarded as sound or unsound). **b** use of or ability in argument. **3** inexorable force, compulsion, or consequence (*the logic of events*). **4 a** principles used in designing a computer etc. **b** circuits using this. □ **logician** /lə'dʒɪʃ(ə)n/ *n.* [related to -LOGIC]

-logic *comb. form* (also **-logical**) forming adjectives corresponding esp. to nouns in *-logy* (*pathological*; *zoological*). [Greek *-logikos*]

logical *adj.* **1** of or according to logic (*the logical conclusion*). **2** correctly reasoned. **3** defensible or explicable on the ground of consistency. **4** capable of correct reasoning. □ **logicality** /-'kælɪtɪ/ *n.* **logically** *adv.* [Greek *logos* word, reason]

-logist *comb. form* forming nouns meaning 'person skilled in *-logy*' (*geologist*).

logistics /lə'dʒɪstɪks/ *n.pl.* **1** organization of (orig. military) services and supplies. **2** organization of any complex operation. □ **logistic** *adj.* **logistical** *adj.* **logistically** *adv.* [French *loger* lodge]

log-jam *n.* deadlock.

logo /'ləʊgəʊ/ *n.* (*pl.* **-s**) emblem of an organization used in its display material etc. [abbreviation of *logotype* from Greek *logos* word]

-logy *comb. form* forming nouns denoting: **1** a subject of study (*biology*). **2** speech or discourse or a characteristic of this (*trilogy*; *tautology*; *phraseology*). [Greek *-logia* from *logos* word]

loin *n.* **1** (in *pl.*) side and back of the body between the ribs and the hip-bones. **2** joint of meat from this part of an animal. [French *loigne* from Latin *lumbus*]

loincloth *n.* cloth worn round the hips, esp. as a sole garment.

loiter *v.* **1** stand about idly; linger. **2** go slowly with frequent stops. □ **loiter with intent** linger in order to commit a felony. □ **loiterer** *n.* [Dutch]

loll *v.* **1** stand, sit, or recline in a lazy attitude. **2** hang loosely. [imitative]

lollipop /'lɒlɪˌpɒp/ *n.* hard sweet on a stick. [origin uncertain]

lollipop man *n.* (also **lollipop lady**) *colloq.* warden using a circular sign on a pole to stop traffic for children to cross the road.

lollop /'lɒləp/ *v.* (**-p-**) *colloq.* **1** flop about. **2** move in ungainly bounds. [probably from LOLL, TROLLOP]

lolly /'lɒlɪ/ *n.* (*pl.* **-ies**) **1** *colloq.* lollipop. **2** = ICE LOLLY. **3** *slang* money. [abbreviation]

Londoner /'lʌndənə(r)/ *n.* native or inhabitant of London.

London pride /'lʌnd(ə)n/ *n.* pink-flowered saxifrage.

lone *attrib. adj.* **1** solitary; without companions. **2** isolated. **3** unmarried, single (*lone parent*). [from ALONE]

lone hand *n.* **1** hand played or player playing against the rest at cards. **2** person or action without allies.

lonely /'ləʊnlɪ/ *adj.* (**-ier, -iest**) **1** without companions (*lonely existence*). **2** sad because of this. **3** unfrequented, isolated, uninhabited. □ **loneliness** *n.*

lonely hearts *n.pl.* people seeking friendship or marriage through a newspaper column, club, etc.

loner *n.* person or animal that prefers to be alone.

lonesome /'ləʊnsəm/ *adj.* esp. *US* **1** lonely. **2** making one feel forlorn (*a lonesome song*).

lone wolf *n.* loner.

long[1] *–adj.* (**longer** /'lɒŋgə(r)/; **longest** /'lɒŋgɪst/) **1** measuring much from end to end in space or time. **2** (following a measurement) in length or duration (*2 metres long*; *two months long*). **3 a** consisting of many items (*a long list*). **b** seemingly more than the stated amount; tedious (*ten long miles*). **4** of elongated shape. **5** lasting or reaching far back or forward in time (*long friendship*). **6** far-reaching; acting at a distance; involving a great interval or difference. **7** (of a vowel or syllable) having the greater of the two recognized durations. **8** (of odds or a chance) reflecting a low level of probability. **9** (of stocks) bought in large quantities in advance, with the expectation of a rise in price. **10** (foll. by *on*) *colloq.* well supplied with. *–n.* long interval or period (*will not take long*; *won't be long*). *–adv.* (**longer** /'lɒŋgə(r)/; **longest** /'lɒŋgɪst/) **1** by or for a long time (*long before*; *long ago*). **2** (following nouns of duration) throughout a specified time (*all day long*). **3** (in *compar.*) after an implied point of time (*shall not wait any longer*). □ **as** (or **so**) **long as** provided that. **before long** soon. **in the long run** (or **term**) eventually, ultimately. **the long and the short of it 1** all that need be said. **2** the eventual outcome. **not by a long shot** (or **chalk**) by no means. □ **longish** *adj.* [Old English]

long[2] *v.* (foll. by *for* or *to* + infin.) have a strong wish or desire for. [Old English, = seem LONG[1] to]

long. *abbr.* longitude.
longboat *n.* sailing-ship's largest boat.
longbow *n.* bow drawn by hand and shooting a long feathered arrow.
long-distance –*attrib. adj.* travelling or operating between distant places. –*adv.* between distant places (*phone long-distance*).
long division *n.* division of numbers with details of the calculations written down.
long-drawn *adj.* (also **long-drawn-out**) prolonged.
longeron /'lɒndʒərən/ *n.* longitudinal member of a plane's fuselage. [French]
longevity /lɒn'dʒevɪtɪ/ *n. formal* long life. [Latin *longus* long, *aevum* age]
long face *n.* dismal expression.
longhand *n.* ordinary handwriting.
long haul *n.* **1** transport over a long distance. **2** prolonged effort or task.
longing /'lɒŋɪŋ/ –*n.* intense desire. –*adj.* having or showing this. □ **longingly** *adv.*
long in the tooth *predic. adj. colloq.* old.
longitude /'lɒŋgɪ,tju:d, 'lɒndʒ-/ *n.* **1** angular distance east or west from a standard meridian such as Greenwich to the meridian of any place. **2** angular distance of a celestial body, esp. along the ecliptic. [Latin *longitudo* length, from *longus* long]
longitudinal /,lɒŋgɪ'tju:dɪn(ə)l, ,lɒndʒ-/ *adj.* **1** of or in length. **2** running lengthwise. **3** of longitude. □ **longitudinally** *adv.*
long johns *n.pl. colloq.* long underpants.
long jump *n.* athletic contest of jumping as far as possible along the ground in one leap.
long-life *adj.* (of milk etc.) treated to prolong its period of usability.
long-lived *adj.* having a long life; durable.
long-lost *attrib. adj.* that has been lost for a long time.
long-playing *adj.* (of a gramophone record) playing for about 20–30 minutes on each side.
long-range *adj.* **1** having a long range. **2** relating to a period of time far into the future (*long-range weather forecast*).
long-running *adj.* continuing for a long time (*a long-running musical*).
longshore *attrib. adj.* **1** existing on or frequenting the shore. **2** directed along the shore. [from *along shore*]
longshoreman *n. US* docker.
long shot *n.* **1** wild guess or venture. **2** bet at long odds.
long sight *n.* ability to see clearly only what is comparatively distant.
long-sighted *adj.* **1** having long sight. **2** far-sighted. □ **long-sightedness** *n.*
long-standing *adj.* that has long existed.
long-suffering *adj.* bearing provocation patiently.
long-term *adj.* of or for a long period of time (*long-term plans*).
long wave *n.* radio wave of frequency less than 300 kHz.
longways *adv.* (also **longwise**) = LENGTHWAYS.
long-winded *adj.* (of a speech or writing) tediously lengthy.
loo *n. colloq.* lavatory. [origin uncertain]
loofah /'lu:fə/ *n.* rough bath-sponge made from the dried pod of a type of gourd. [Arabic]
look /lʊk/ –*v.* **1 a** (often foll. by *at, down, up,* etc.) use one's sight; turn one's eyes in some direction. **b** turn one's eyes on; examine (*looked me in the eyes*; *looked us up and down*). **2 a** make a visual or mental search (*I'll look in the morning*). **b** (foll. by *at*) consider, examine (*must look at the facts*). **3** (foll. by *for*) search for, seek, be on the watch for. **4** inquire (*when one looks deeper*). **5** have a specified appearance; seem (*look a fool*; *future looks bleak*). **6** (foll. by *to*) **a** consider; be concerned about (*look to the future*). **b** rely on (*look to me for support*). **7** (foll. by *into*) investigate. **8** (foll. by *what, where, whether*, etc.) ascertain or observe by sight. **9** (of a thing) face some direction. **10** indicate (emotion etc.) by one's looks. **11** (foll. by *that*) take care; make sure. **12** (foll. by *to* + infin.) aim (*am looking to finish it soon*). –*n.* **1** act of looking; gaze, glance. **2** (in *sing.* or *pl.*) appearance of a face; expression. **3** appearance of a thing (*by the look of it*). **4** style, fashion (*this year's look*; *the wet look*). –*int.* (also **look here!**) calling attention, expressing a protest, etc. □ **look after** attend to; take care of. **look one's age** appear as old as one really is. **look back 1** (foll. by *on, to*) turn one's thoughts to (something past). **2** (usu. with *neg.*) cease to progress (*he's never looked back*). **look down on** (or **look down one's nose at**) regard with contempt or superiority. **look forward to** await (an expected event) eagerly or with specified feelings. **look in** make a short visit or call. **look on 1** (often foll. by *as*) regard. **2** be a spectator. **look oneself** appear well (esp. after illness etc.). **look out 1** direct one's sight or put one's head out of a window etc. **2** (often foll. by *for*) be vigilant or prepared. **3** (foll. by *on, over*, etc.) have or afford an outlook. **4** search for and produce. **5** (as *imper.*) warning of immediate danger

etc. **look over** inspect. **look smart** make haste. **look up 1** search for (esp. information in a book). **2** *colloq.* visit (a person). **3** improve in prospect. **look up to** respect or admire. **not like the look of** find alarming or suspicious. [Old English]

look-alike *n.* person or thing closely resembling another.

looker *n.* **1** person of a specified appearance (*good-looker*). **2** *colloq.* attractive woman.

looker-on *n.* (*pl.* **lookers-on**) spectator.

look-in *n. colloq.* chance of participation or success (*never gets a look-in*).

looking-glass *n.* mirror.

lookout *n.* **1** watch or looking out (*on the lookout*). **2 a** observation-post. **b** person etc. stationed to keep watch. **3** prospect (*it's a bad lookout*). **4** *colloq.* person's own concern (*that's your lookout*).

loom[1] *n.* apparatus for weaving. [Old English]

loom[2] *v.* **1** appear dimly, esp. as a vague and often threatening shape. **2** (of an event) be ominously close. [probably Low German or Dutch]

loon *n.* **1** a kind of diving bird. **2** *colloq.* crazy person (cf. LOONY). [Old Norse]

loony *slang* –*n.* (*pl.* **-ies**) lunatic. –*adj.* (**-ier, -iest**) crazy. □ **looniness** *n.* [abbreviation]

loony-bin *n. slang offens.* mental home or hospital.

loop –*n.* **1 a** figure produced by a curve, or a doubled thread etc., that crosses itself. **b** thing, path, etc., forming this figure. **2** similarly shaped attachment used as a fastening. **3** ring etc. as a handle etc. **4** contraceptive coil. **5** (in full **loop-line**) railway or telegraph line that diverges from a main line and joins it again. **6** skating or aerobatic manoeuvre describing a loop. **7** complete circuit for an electric current. **8** endless band of tape or film allowing continuous repetition. **9** sequence of computer operations repeated until some condition is satisfied. –*v.* **1** form or bend into a loop. **2** fasten with a loop or loops. **3** form a loop. **4** (also **loop the loop**) fly in a circle vertically. [origin unknown]

loophole *n.* **1** means of evading a rule etc. without infringing it. **2** narrow vertical slit in the wall of a fort etc.

loopy *adj.* (**-ier, -iest**) *slang* crazy, daft.

loose –*adj.* **1** not tightly held, fixed, etc. (*loose handle; loose stones*). **2** free from bonds or restraint. **3** not held together (*loose papers*). **4** not compact or dense (*loose soil*). **5** inexact (*loose translation*). **6** morally lax. **7** (of the tongue) indiscreet. **8** tending to diarrhoea. **9** (in *comb.*) loosely (*loose-fitting*). –*v.* (**-sing**) **1** free; untie or detach; release. **2** relax (*loosed my hold*). **3** discharge (a missile). □ **at a loose end** unoccupied. **on the loose 1** escaped from captivity. **2** enjoying oneself freely. □ **loosely** *adv.* **looseness** *n.* **loosish** *adj.* [Old Norse]

loose cover *n.* removable cover for an armchair etc.

loose-leaf *adj.* (of a notebook etc.) with pages that can be removed and replaced.

loosen *v.* make or become loose or looser. □ **loosen a person's tongue** make a person talk freely. **loosen up 1** relax. **2** limber up.

loot –*n.* **1** spoil, booty. **2** *slang* money. –*v.* **1** rob or steal, esp. after rioting etc. **2** plunder. □ **looter** *n.* [Hindi]

lop *v.* (**-pp-**) **1 a** (often foll. by *off, away*) cut or remove (a part or parts) from a whole, esp. branches from a tree. **b** remove branches from (a tree). **2** (often foll. by *off*) remove (items) as superfluous. [Old English]

lope –*v.* (**-ping**) run with a long bounding stride. –*n.* long bounding stride. [Old Norse: related to LEAP]

lop-eared *adj.* having drooping ears. [related to LOB]

lopsided *adj.* unevenly balanced. □ **lopsidedness** *n.* [related to LOB]

loquacious /lə'kweɪʃəs/ *adj.* talkative. □ **loquacity** /-'kwæsɪtɪ/ *n.* [Latin *loquor* speak]

loquat /'ləʊkwɒt/ *n.* **1** small yellow egg-shaped fruit. **2** tree bearing it. [Chinese]

lord –*n.* **1** master or ruler. **2** *hist.* feudal superior, esp. of a manor. **3** peer of the realm or person with the title *Lord*. **4** (**Lord**) (often prec. by *the*) God or Christ. **5** (**Lord**) **a** prefixed as the designation of a marquis, earl, viscount, or baron, or (to the Christian name) of the younger son of a duke or marquis. **b** (**the Lords**) = HOUSE OF LORDS. –*int.* (**Lord, good Lord**, etc.) expressing surprise, dismay, etc. □ **lord it over** domineer. [Old English, = bread-keeper: related to LOAF[1], WARD]

Lord Chamberlain *n.* official in charge of the Royal Household.

Lord Chancellor *n.* (also **Lord High Chancellor**) highest officer of the Crown, presiding in the House of Lords etc.

Lord Chief Justice *n.* president of the Queen's Bench Division.

Lord Lieutenant *n.* **1** chief executive authority and head of magistrates in each county. **2** *hist.* viceroy of Ireland.

lordly *adj.* (**-ier, -iest**) **1** haughty, imperious. **2** suitable for a lord. □ **lordliness** *n.*

Lord Mayor *n.* title of the mayor in some large cities.

Lord Privy Seal *n.* senior Cabinet minister without official duties.

lords and ladies *n.* wild arum.

Lord's Day *n.* Sunday.

lordship *n.* **1** (usu. **Lordship**) title used in addressing or referring to a man with the rank of Lord (*Your Lordship; His Lordship*). **2** (foll. by *over*) dominion, rule.

Lord's Prayer *n.* the Our Father.

Lords spiritual *n.pl.* bishops in the House of Lords.

Lord's Supper *n.* Eucharist.

Lords temporal *n.pl.* members of the House of Lords other than bishops.

lore *n.* body of traditions and knowledge on a subject or held by a particular group (*bird lore; Gypsy lore*). [Old English: related to LEARN]

lorgnette /lɔːˈnjet/ *n.* pair of eyeglasses or opera-glasses on a long handle. [French *lorgner* to squint]

lorn *adj. archaic* desolate, forlorn. [Old English, past part. of LOSE]

lorry /ˈlɒrɪ/ *n.* (*pl.* **-ies**) large vehicle for transporting goods etc. [origin uncertain]

lose /luːz/ *v.* (**-sing**; *past* and *past part.* **lost**) **1** be deprived of or cease to have, esp. by negligence. **2** be deprived of (a person) by death. **3** become unable to find, follow, or understand (*lose one's way*). **4** let or have pass from one's control or reach (*lost my chance; lost his composure*). **5** be defeated in (a game, lawsuit, battle, etc.). **6** get rid of (*lost our pursuers; lose weight*). **7** forfeit (a right to a thing). **8** spend (time, efforts, etc.) to no purpose. **9 a** suffer loss or detriment. **b** be worse off. **10** cause (a person) the loss of (*will lose you your job*). **11** (of a clock etc.) become slow; become slow by (a specified time). **12** (in *passive*) disappear, perish; be dead (*lost at sea; is a lost art*). □ **be lost** (or **lose oneself**) **in** be engrossed in. **be lost on** be wasted on, or not noticed or appreciated by. **be lost to** be no longer affected by or accessible to (*is lost to pity; is lost to the world*). **be lost without** be dependent on (*am lost without my diary*). **get lost** *slang* (usu. in *imper.*) go away. **lose face** see FACE. **lose out 1** (often foll. by *on*) *colloq.* be unsuccessful; not get a full chance or advantage (in). **2** (foll. by *to*) be beaten in competition or replaced by. [Old English]

loser *n.* **1** person or thing that loses, esp. a contest (*is a bad loser*). **2** *colloq.* person who regularly fails.

loss *n.* **1** losing or being lost. **2** thing or amount lost. **3** detriment resulting from losing. □ **at a loss** (sold etc.) for less than was paid for it. **be at a loss** be puzzled or uncertain. [probably back-formation from LOST]

loss-leader *n.* item sold at a loss to attract customers.

lost *past* and *past part.* of LOSE.

lost cause *n.* hopeless undertaking.

lot *n.* **1** *colloq.* (prec. by *a* or in *pl.*) **a** a large number or amount (*a lot of people; lots of milk*). **b** *colloq.* much (*a lot warmer; smiles a lot*). **2 a** each of a set of objects used to make a chance selection. **b** this method of deciding (*chosen by lot*). **3** share or responsibility resulting from it. **4** person's destiny, fortune, or condition. **5** (esp. *US*) plot; allotment of land (*parking lot*). **6** article or set of articles for sale at an auction etc. **7** group of associated persons or things. □ **cast** (or **draw**) **lots** decide by lots. **throw in one's lot with** decide to share the fortunes of. **the** (or **the whole**) **lot** the total number or quantity. **a whole lot** *colloq.* very much (*is a whole lot better*). [Old English]

■ **Usage** In sense 1a, *a lot of* is somewhat informal, but acceptable in serious writing, whereas *lots of* is not acceptable.

---

loth var. of LOATH.

Lothario /ləˈθɑːrɪəʊ, -ˈθeərɪəʊ/ *n.* (*pl.* **-s**) libertine. [name of a character in a play]

lotion /ˈləʊʃ(ə)n/ *n.* medicinal or cosmetic liquid preparation applied externally. [Latin *lavo lot-* wash]

lottery /ˈlɒtərɪ/ *n.* (*pl.* **-ies**) **1** means of raising money by selling numbered tickets and giving prizes to the holders of numbers drawn at random. **2** thing whose success is governed by chance. [Dutch: related to LOT]

lotto /ˈlɒtəʊ/ *n.* game of chance like bingo, but with numbers drawn by players instead of called. [Italian]

lotus /ˈləʊtəs/ *n.* **1** legendary plant inducing luxurious languor when eaten. **2** a kind of water lily etc., esp. used symbolically in Hinduism and Buddhism. [Greek *lōtos*]

lotus-eater *n.* person given to indolent enjoyment.

lotus position *n.* cross-legged position of meditation with the feet resting on the thighs.

loud *–adj.* **1** strongly audible, noisy. **2** (of colours etc.) gaudy, obtrusive. *–adv.* loudly. □ **out loud** aloud. □ **loudish**

adj. **loudly** *adv.* **loudness** *n.* [Old English]

loud hailer *n.* electronic device for amplifying the voice.

loudspeaker *n.* apparatus that converts electrical signals into sound.

lough /lɒk, lɒx/ *n. Ir.* lake, arm of the sea. [Irish: related to LOCH]

lounge –*v.* (**-ging**) **1** recline comfortably; loll. **2** stand or move about idly. –*n.* **1** place for lounging, esp.: **a** a sitting-room in a house. **b** a public room (e.g. in a hotel). **c** a place in an airport etc. with seats for waiting passengers. **2** spell of lounging. [origin uncertain]

lounge bar *n.* more comfortable bar in a pub etc.

lounge suit *n.* man's suit for ordinary day (esp. business) wear.

lour *v.* (also **lower**) **1** frown; look sullen. **2** (of the sky etc.) look dark and threatening. [origin unknown]

louse –*n.* **1** (*pl.* **lice**) parasitic insect. **2** (*pl.* **louses**) *slang* contemptible person. –*v.* (**-sing**) delouse. □ **louse up** *slang* make a mess of. [Old English]

lousy /'laʊzɪ/ *adj.* (**-ier**, **-iest**) **1** *colloq.* very bad; disgusting; ill (*feel lousy*). **2** (often foll. by *with*) *colloq.* well supplied, teeming. **3** infested with lice. □ **lousily** *adv.* **lousiness** *n.*

lout *n.* rough-mannered person. □ **loutish** *adj.* [origin uncertain]

louvre /'lu:və(r)/ *n.* (also **louver**) **1** each of a set of overlapping slats designed to admit air and some light and exclude rain. **2** domed structure on a roof with side openings for ventilation etc. □ **louvred** *adj.* [French *lover* skylight]

lovable /'lʌvəb(ə)l/ *adj.* (also **loveable**) inspiring love or affection.

lovage /'lʌvɪdʒ/ *n.* herb used for flavouring etc. [French *levesche* from Latin *ligusticum* Ligurian]

lovat /'lʌvət/ *n.* & *adj.* muted green. [*Lovat* in Scotland]

love /lʌv/ –*n.* **1** deep affection or fondness. **2** sexual passion. **3** sexual relations. **4 a** beloved one; sweetheart (often as a form of address). **b** *colloq.* form of address regardless of affection. **5** *colloq.* person of whom one is fond. **6** affectionate greetings (*give him my love*). **7** (in games) no score; nil. –*v.* (**-ving**) **1** feel love or a deep fondness for. **2** delight in; admire; greatly cherish. **3** *colloq.* like very much (*loves books*). **4** (foll. by verbal noun, or *to* + infin.) be inclined, esp. as a habit; greatly enjoy (*children love dressing up*; *loves to run*). □ **fall in love** (often foll. by *with*) suddenly begin to love. **for love** for pleasure not profit. **for the love of** for the sake of. **in love** (often foll. by *with*) enamoured (of). **make love** (often foll. by *to*) **1** have sexual intercourse (with). **2** *archaic* pay amorous attention (to). **not for love or money** *colloq.* not in any circumstances. [Old English]

loveable var. of LOVABLE.

love affair *n.* romantic or sexual relationship between two people.

love-bird *n.* parrot, esp. one seeming to show great affection for its mate.

love bite *n.* bruise made by a partner's biting etc. during lovemaking.

love-child *n.* child of unmarried parents.

love-hate relationship *n.* intense relationship involving ambivalent emotions.

love-in-a-mist *n.* blue-flowered cultivated plant.

loveless *adj.* unloving or unloved or both.

love-lies-bleeding *n.* cultivated plant with drooping spikes of purple-red blooms.

lovelorn *adj.* pining from unrequited love.

lovely –*adj.* (**-ier**, **-iest**) **1** *colloq.* pleasing, delightful. **2** beautiful. –*n.* (*pl.* **-ies**) *colloq.* pretty woman. □ **lovely and** *colloq.* delightfully (*lovely and warm*). □ **loveliness** *n.* [Old English]

lovemaking *n.* **1** sexual play, esp. intercourse. **2** *archaic* courtship.

love-nest *n. colloq.* secluded retreat for (esp. illicit) lovers.

lover *n.* **1** person in love with another. **2** person with whom another is having sexual relations. **3** (in *pl.*) unmarried couple in love or having sexual relations. **4** person who likes or enjoys a specified thing (*music lover*).

love-seat *n.* small sofa in the shape of an S, with two seats facing in opposite directions.

lovesick *adj.* languishing with love.

lovey-dovey /ˌlʌvɪ'dʌvɪ/ *adj. colloq.* fondly affectionate, sentimental.

loving –*adj.* feeling or showing love; affectionate. –*n.* affection; love. □ **lovingly** *adv.*

loving-cup *n.* two-handled drinking-cup.

low[1] /ləʊ/ –*adj.* **1** not high or tall (*low wall*). **2 a** not elevated in position (*low altitude*). **b** (of the sun) near the horizon. **3** of or in humble rank or position (*of low birth*). **4** of small or less than normal amount, extent, or intensity (*low temperature*; *low in calories*). **5** small or reduced in quantity (*stocks are low*). **6** coming below the normal level (*low neck*). **7** dejected; lacking vigour (*feeling low*). **8** (of a sound) not shrill or loud. **9** not exalted or sublime; commonplace. **10** unfavourable (*low opinion*). **11** abject, mean, vulgar (*low cunning*; *low slang*). **12** (of a geological period) earlier. *–n.* **1** low or the lowest level or number (*pound reached a new low*). **2** area of low pressure. *–adv.* **1** in or to a low position or state. **2** in a low tone (*speak low*). **3** (of a sound) at or to a low pitch. □ **lowish** *adj.* **lowness** *n.* [Old Norse]

low[2] /ləʊ/ *–n.* sound made by cattle; moo. *–v.* make this sound. [Old English]

low-born *adj.* of humble birth.

lowbrow *–adj.* not intellectual or cultured. *–n.* lowbrow person.

Low Church *n.* section of the Church of England attaching little importance to ritual, priestly authority, and the sacraments.

low-class *adj.* of low quality or social class.

low comedy *n.* comedy bordering on farce.

Low Countries *n.pl.* the Netherlands, Belgium, and Luxembourg.

low-down *–adj.* mean, dishonourable. *–n. colloq.* (prec. by *the*; usu. foll. by *on*) relevant information.

lower[1] *–adj.* (*compar.* of LOW[1]). **1** less high in position or status. **2** situated below another part (*lower lip*). **3 a** situated on less high land (*Lower Egypt*). **b** situated to the South (*Lower California*). **4** (of a mammal, plant, etc.) evolved to only a slight degree. *–adv.* in or to a lower position, status, etc. □ **lowermost** *adj.*

lower[2] *v.* **1** let or haul down. **2** make or become lower. **3** degrade.

lower[3] var. of LOUR.

lower case *n.* small letters.

lower class *n.* working class.

Lower House *n.* larger and usu. elected body in a legislature, esp. the House of Commons.

lowest *adj.* (*superl.* of LOW[1]) least high in position or status.

lowest common denominator *n.* **1** *Math.* lowest common multiple of the denominators of several fractions. **2** the worst or most vulgar common feature of members of a group.

lowest common multiple *n. Math.* least quantity that is a multiple of two or more given quantities.

low frequency *n.* frequency, esp. in radio, 30 to 300 kilohertz.

low gear *n.* gear such that the driven end of a transmission revolves slower than the driving end.

low-grade *adj.* of low quality.

low-key *adj.* lacking intensity, restrained.

lowland *–n.* (usu. in *pl.*) low-lying country. *–adj.* of or in lowland. □ **lowlander** *n.*

low-level *adj.* (of a computer language) close in form to machine code.

lowly *adj.* (**-ier, -iest**) humble; unpretentious. □ **lowliness** *n.*

low-lying *adj.* near to the ground or sea level.

low-pitched *adj.* **1** (of a sound) low. **2** (of a roof) having only a slight slope.

low pressure *n.* **1** low degree of activity or exertion. **2** atmospheric condition with the pressure below average.

low-rise *–adj.* (of a building) having few storeys. *–n.* such a building.

low season *n.* period of fewest visitors at a resort etc.

Low Sunday *n.* Sunday after Easter.

low tide *n.* (also **low water**) time or level of the tide at its ebb.

loyal /ˈlɔɪəl/ *adj.* **1** (often foll. by *to*) faithful. **2** steadfast in allegiance etc. □ **loyally** *adv.* **loyalty** *n.* (*pl.* **-ies**). [Latin: related to LEGAL]

loyalist *n.* **1** person who remains loyal to the legitimate sovereign etc. **2** (**Loyalist**) (esp. extremist) supporter of union between Great Britain and Northern Ireland. □ **loyalism** *n.*

loyal toast *n.* toast to the sovereign.

lozenge /ˈlɒzɪndʒ/ *n.* **1** rhombus. **2** small sweet or medicinal tablet to be dissolved in the mouth. **3** lozenge-shaped object. [French]

LP *abbr.* long-playing (record).

L-plate *n.* sign bearing the letter L, attached to a vehicle to show that it is being driven by a learner. [from PLATE]

LPO *abbr.* London Philharmonic Orchestra.

LSD *abbr.* lysergic acid diethylamide, a powerful hallucinogenic drug.

l.s.d. /ˌelesˈdiː/ *n.* (also **£.s.d.**) **1** *hist.* pounds, shillings, and pence (in former British currency). **2** money, riches. [Latin *librae, solidi, denarii*]

LSE *abbr.* London School of Economics.

LSO *abbr.* London Symphony Orchestra.

Lt. *abbr.* **1** Lieutenant. **2** light.

Ltd. *abbr.* Limited.

Lu *symb.* lutetium.

lubber *n.* clumsy fellow, lout. [origin uncertain]

lubricant /ˈluːbrɪkənt/ *n.* substance used to reduce friction.

lubricate /ˈluːbrɪˌkeɪt/ *v.* (**-ting**) **1** apply oil or grease etc. to. **2** make slippery. □ **lubrication** /-ˈkeɪʃ(ə)n/ *n.* **lubricator** *n.* [Latin *lubricus* slippery]

lubricious /luːˈbrɪʃəs/ *adj.* **1** slippery, evasive. **2** lewd. □ **lubricity** *n.* [Latin: related to LUBRICATE]

lucerne /lu:'sɜ:n/ *n.* = ALFALFA. [Provençal, = glow-worm, referring to its shiny seeds]

lucid /'lu:sɪd/ *adj.* **1** expressing or expressed clearly. **2** sane. □ **lucidity** /-'sɪdɪtɪ/ *n.* **lucidly** *adv.* **lucidness** *n.* [Latin *lux luc-* light]

Lucifer /'lu:sɪfə(r)/ *n.* Satan. [Latin: related to LUCID, *fero* bring]

luck *n.* **1** good or bad fortune. **2** circumstances of life (beneficial or not) brought by this. **3** good fortune; success due to chance (*in luck*; *out of luck*). □ **no such luck** *colloq.* unfortunately not. [Low German or Dutch]

luckless *adj.* unlucky; ending in failure.

lucky *adj.* **(-ier, -iest) 1** having or resulting from good luck. **2** bringing good luck (*lucky charm*). □ **luckily** *adv.*

lucky dip *n.* tub containing articles varying in value and chosen at random.

lucrative /'lu:krətɪv/ *adj.* profitable. □ **lucratively** *adv.* **lucrativeness** *n.* [Latin: related to LUCRE]

lucre /'lu:kə(r)/ *n. derog.* financial gain. [Latin *lucrum* gain]

Luddite /'lʌdaɪt/ *–n.* **1** person opposed to industrial progress or new technology. **2** *hist.* member of a band of English artisans who destroyed machinery (1811–16). *–adj.* of the Luddites. □ **Ludditism** *n.* [Ned *Lud*, name of a destroyer of machinery]

ludicrous /'lu:dɪkrəs/ *adj.* absurd, ridiculous, laughable. □ **ludicrously** *adv.* **ludicrousness** *n.* [Latin *ludicrum* stage play]

ludo /'lu:dəʊ/ *n.* simple board-game played with dice and counters. [Latin, = I play]

luff *v.* (also *absol.*) **1** steer (a ship) nearer the wind. **2** raise or lower (a crane's jib). [French, probably from Low German]

lug *–v.* **(-gg-) 1** drag or carry with effort. **2** pull hard. *–n.* **1** hard or rough pull. **2** *colloq.* ear. **3** projection on an object by which it may be carried, fixed in place, etc. [probably Scandinavian]

luggage /'lʌgɪdʒ/ *n.* suitcases, bags, etc., for a traveller's belongings. [from LUG]

lugger /'lʌgə(r)/ *n.* small ship with four-cornered sails. [from LUGSAIL]

lughole *n. slang* ear.

lugsail *n.* four-cornered sail on a yard. [probably from LUG]

lugubrious /lu:'gu:brɪəs/ *adj.* doleful. □ **lugubriously** *adv.* **lugubriousness** *n.* [Latin *lugeo* mourn]

lugworm *n.* large marine worm used as bait. [origin unknown]

lukewarm *adj.* **1** moderately warm; tepid. **2** unenthusiastic, indifferent. [Old English (now dial.) *luke* warm, WARM]

lull *–v.* **1** soothe or send to sleep. **2** (usu. foll. by *into*) deceive (a person) into undue confidence (*lulled into a false sense of security*). **3** allay (suspicions etc.), usu. by deception. **4** (of noise, a storm, etc.) abate or fall quiet. *–n.* temporary quiet period. [imitative]

lullaby /'lʌlə,baɪ/ *n.* (*pl.* **-ies**) soothing song to send a child to sleep. [related to LULL]

lumbago /lʌm'beɪgəʊ/ *n.* rheumatic pain in the muscles of the lower back. [Latin *lumbus* loin]

lumbar /'lʌmbə(r)/ *adj.* of the lower back area. [as LUMBAGO]

lumbar puncture *n.* withdrawal of spinal fluid from the lower back for diagnosis.

lumber /'lʌmbə(r)/ *–n.* **1** disused and cumbersome articles. **2** partly prepared timber. *–v.* **1** (usu. foll. by *with*) leave (a person etc.) with something unwanted or unpleasant. **2** (usu. foll. by *up*) obstruct, fill inconveniently. **3** cut and prepare forest timber. **4** move in a slow clumsy way. [origin uncertain]

lumberjack *n.* person who fells and transports lumber.

lumber-jacket *n.* jacket of the kind worn by lumberjacks.

lumber-room *n.* room where disused things are kept.

luminary /'lu:mɪnərɪ/ *n.* (*pl.* **-ies**) **1** *literary* natural light-giving body. **2** wise or inspiring person. **3** celebrated member of a group (*show-business luminaries*). [Latin *lumen lumin-* light]

luminescence /,lu:mɪ'nes(ə)ns/ *n.* emission of light without heat. □ **luminescent** *adj.*

luminous /'lu:mɪnəs/ *adj.* **1** shedding light. **2** phosphorescent, visible in darkness (*luminous paint*). □ **luminosity** /-'nɒsɪtɪ/ *n.*

lump[1] *–n.* **1** compact shapeless mass. **2** tumour; swelling, bruise. **3** heavy, dull, or ungainly person. **4** (prec. by *the*) *slang* casual workers in the building trade. *–v.* **1** (usu. foll. by *together* etc.) treat as all alike; put together in a lump. **2** (of sauce etc.) become lumpy. □ **lump in the throat** feeling of pressure there, caused by emotion. [Scandinavian]

lump[2] *v. colloq.* put up with ungraciously (*like it or lump it*). [imitative]

lumpectomy /lʌm'pektəmɪ/ *n.* (*pl.* **-ies**) surgical removal of a lump from the breast.

lumpish *adj.* **1** heavy and clumsy. **2** stupid, lethargic.

lump sugar *n.* sugar in cubes.

lump sum *n.* **1** sum covering a number of items. **2** money paid down at once.

lumpy *adj.* (**-ier, -iest**) full of or covered with lumps. □ **lumpily** *adv.* **lumpiness** *n.*

lunacy /'lu:nəsɪ/ *n.* (*pl.* **-ies**) **1** insanity. **2** mental unsoundness. **3** great folly. [Latin: related to LUNAR]

lunar /'lu:nə(r)/ *adj.* of, like, concerned with, or determined by the moon. [Latin *luna* moon]

lunar module *n.* small craft for travelling between the moon and a spacecraft in orbit around it.

lunar month *n.* **1** period of the moon's revolution, esp. the interval between new moons (about $29^1/_2$ days). **2** (in general use) four weeks.

lunate /'lu:neɪt/ *adj.* crescent-shaped.

lunatic /'lu:nətɪk/ *–n.* **1** insane person. **2** wildly foolish person. *–adj.* insane; extremely reckless or foolish. [related to LUNACY]

lunatic asylum *n. hist.* mental home or hospital.

lunatic fringe *n.* extreme or eccentric minority group.

lunation /lu:'neɪʃ(ə)n/ *n.* interval between new moons, about $29^1/_2$ days. [medieval Latin: related to LUNAR]

lunch *–n.* midday meal. *–v.* **1** take lunch. **2** entertain to lunch. [shortening of LUNCHEON]

luncheon /'lʌntʃ(ə)n/ *n. formal* lunch. [origin unknown]

luncheon meat *n.* tinned meat loaf of pork etc.

luncheon voucher *n.* voucher issued to employees and exchangeable for food at many restaurants and shops.

lung *n.* either of the pair of respiratory organs in humans and many other vertebrates. [Old English: related to LIGHT[2]]

lunge *–n.* **1** sudden movement forward. **2** the basic attacking move in fencing. **3** long rope on which a horse is held and made to circle round its trainer. *–v.* (**-ging**) (usu. foll. by *at, out*) deliver or make a lunge. [French *allonger* from *long* LONG[1]]

lupin /'lu:pɪn/ *n.* cultivated plant with long tapering spikes of flowers. [related to LUPINE]

lupine /'lu:paɪn/ *adj.* of or like wolves. [Latin *lupinus* from *lupus* wolf]

lupus /'lu:pəs/ *n.* autoimmune inflammatory skin disease. [Latin, = wolf]

lurch[1] *–n.* stagger; sudden unsteady movement or leaning. *–v.* stagger; move or progress unsteadily. [originally Naut., of uncertain origin]

lurch[2] *n.* □ **leave in the lurch** desert (a friend etc.) in difficulties. [obsolete French *lourche* a kind of backgammon]

lurcher /'lɜ:tʃə(r)/ *n.* crossbred dog, usu. a working dog crossed with a greyhound. [related to LURK]

lure *–v.* (**-ring**) **1** (usu. foll. by *away, into*) entice. **2** recall with a lure. *–n.* **1** thing used to entice. **2** (usu. foll. by *of*) enticing quality (of a pursuit etc.). **3** falconer's apparatus for recalling a hawk. [French from Germanic]

Lurex /'ljʊəreks/ *n. propr.* **1** type of yarn incorporating a glittering metallic thread. **2** fabric made from this.

lurid /'ljʊərɪd/ *adj.* **1** bright and glaring in colour. **2** sensational, shocking (*lurid details*). **3** ghastly, wan (*lurid complexion*). □ **luridly** *adv.* [Latin]

lurk *v.* **1** linger furtively. **2 a** lie in ambush. **b** (usu. foll. by *in, under, about,* etc.) hide, esp. for sinister purposes. **3** (as **lurking** *adj.*) dormant (*a lurking suspicion*). [perhaps from LOUR]

luscious /'lʌʃəs/ *adj.* **1** richly sweet in taste or smell. **2** (of style) over-rich. **3** voluptuously attractive. [perhaps related to DELICIOUS]

lush[1] *adj.* **1** (of vegetation) luxuriant and succulent. **2** luxurious. **3** *slang* excellent. [origin uncertain]

lush[2] *n. slang* alcoholic, drunkard. [origin uncertain]

lust *–n.* **1** strong sexual desire. **2** (usu. foll. by *for, of*) passionate desire for or enjoyment of (*lust for power*; *lust of battle*). **3** sensuous appetite regarded as sinful (*lusts of the flesh*). *–v.* (usu. foll. by *after, for*) have a strong or excessive (esp. sexual) desire. □ **lustful** *adj.* **lustfully** *adv.* [Old English]

lustre /'lʌstə(r)/ *n.* (*US* **luster**) **1** gloss, shining surface. **2** brilliance, splendour. **3** iridescent glaze on pottery and porcelain. □ **lustrous** *adj.* [Latin *lustro* illumine]

lusty *adj.* (**-ier, -iest**) **1** healthy and strong. **2** vigorous, lively. □ **lustily** *adv.* **lustiness** *n.* [from LUST]

lutanist var. of LUTENIST.

lute[1] /lu:t/ *n.* guitar-like instrument with a long neck and a pear-shaped body. [Arabic]

lute[2] /lu:t/ *–n.* clay or cement for making joints airtight etc. *–v.* (**-ting**) apply lute to. [Latin *lutum* mud]

lutenist /'lu:tənɪst/ *n.* (also **lutanist**) lute-player. [related to LUTE[1]]

lutetium /lu:'ti:ʃəm/ *n.* silvery metallic element, the heaviest of the lanthanide series. [*Lutetia*, ancient name of Paris]

Lutheran /'lu:θərən/ *–n.* **1** follower of Luther. **2** member of the Lutheran Church. *–adj.* of Luther, or the Protestant Reformation and the doctrines associated with him. □ **Lutheranism** *n.* [Martin *Luther*, religious reformer]

lux *n.* (*pl.* same) the SI unit of illumination. [Latin]
luxuriant /lʌg'zjʊərɪənt/ *adj.* **1** growing profusely. **2** exuberant. **3** florid. □ **luxuriance** *n.* **luxuriantly** *adv.* [Latin: related to LUXURY]

■ **Usage** *Luxuriant* is sometimes confused with *luxurious.*

luxuriate /lʌg'zjʊərɪ,eɪt/ *v.* (**-ting**) **1** (foll. by *in*) take self-indulgent delight in, enjoy as a luxury. **2** relax in comfort.
luxurious /lʌg'zjʊərɪəs/ *adj.* **1** supplied with luxuries. **2** extremely comfortable. **3** fond of luxury. □ **luxuriously** *adv.* [Latin: related to LUXURY]

■ **Usage** *Luxurious* is sometimes confused with *luxuriant.*

luxury /'lʌkʃərɪ/ *n.* (*pl.* **-ies**) **1** choice or costly surroundings, possessions, etc. **2** thing giving comfort or enjoyment but not essential. **3** (*attrib.*) comfortable and expensive (*luxury flat*). [Latin *luxus* abundance]
LV *abbr.* luncheon voucher.
Lw *symb.* lawrencium.
-ly[1] *suffix* forming adjectives, esp. from nouns, meaning: **1** having the qualities of (*princely*). **2** recurring at intervals of (*daily*). [Old English]
-ly[2] *suffix* forming adverbs from adjectives (*boldly*; *happily*). [Old English]
lychee /'laɪtʃɪ, 'lɪ-/ *n.* (also **litchi, lichee**) **1** sweet white juicy fruit in a brown skin. **2** tree, orig. from China, bearing this. [Chinese]
lych-gate var. of LICH-GATE.
Lycra /'laɪkrə/ *n. propr.* elastic polyurethane fabric used esp. for sportswear.
lye /laɪ/ *n.* **1** water made alkaline with wood ashes. **2** any alkaline solution for washing. [Old English]
lying *pres. part.* of LIE[1], LIE[2].
lymph /lɪmf/ *n.* **1** colourless fluid from the tissues of the body, containing white blood cells. **2** this fluid used as a vaccine. [Latin *lympha*]
lymphatic /lɪm'fætɪk/ *adj.* **1** of, secreting, or conveying lymph. **2** (of a person) pale, flabby, or sluggish.
lymphatic system *n.* network of vessels conveying lymph.
lymph gland *n.* (also **lymph node**) small mass of tissue in the lymphatic system.
lymphoma /lɪm'fəʊmə/ *n.* (*pl.* **-s** or **-mata**) tumour of the lymph nodes.
lynch /lɪntʃ/ *v.* (of a mob) put (a person) to death without a legal trial. □ **lynching** *n.* [originally *US*, after *Lynch*, 18th-c. Justice of the Peace in Virginia]
lynch law *n.* procedure followed when a person is lynched.
lynx /lɪŋks/ *n.* (*pl.* same or **-es**) wild cat with a short tail and spotted fur. [Greek *lugx*]
lynx-eyed *adj.* keen-sighted.
lyre /'laɪə(r)/ *n.* ancient U-shaped stringed instrument. [Greek *lura*]
lyre-bird *n.* Australian bird, the male of which has a lyre-shaped tail display.
lyric /'lɪrɪk/ *–adj.* **1** (of poetry) expressing the writer's emotions, usu. briefly and in stanzas. **2** (of a poet) writing in this manner. **3** meant or fit to be sung, songlike. *–n.* **1** lyric poem. **2** (in *pl.*) words of a song. [Latin: related to LYRE]
lyrical *adj.* **1** = LYRIC. **2** resembling, or using language appropriate to, lyric poetry. **3** *colloq.* highly enthusiastic (*wax lyrical about*). □ **lyrically** *adv.*
lyricism /'lɪrɪ,sɪz(ə)m/ *n.* quality of being lyric.
lyricist /'lɪrɪsɪst/ *n.* writer of (esp. popular) lyrics.
lysergic acid diethylamide /laɪ'sɜ:dʒɪk, ,daɪə'θaɪlə,maɪd/ *n.* = LSD. [from hydro*lys*is, *erg*ot, -IC]
-lysis *comb. form* forming nouns denoting disintegration or decomposition (*electrolysis*). [Greek *lusis* loosening]
-lyte *suffix* forming nouns denoting substances that can be decomposed (*electrolyte*). [Greek *lutos* loosened]

M

M[1] /em/ *n.* (*pl.* **Ms** or **M's**) **1** thirteenth letter of the alphabet. **2** (as a roman numeral) 1,000.

M[2] *abbr.* (also **M.**) **1** Master. **2** *Monsieur.* **3** motorway. **4** mega-.

m *abbr.* (also **m.**) **1** male. **2** masculine. **3** married. **4** mile(s). **5** metre(s). **6** million(s). **7** minute(s). **8** milli-.

MA *abbr.* Master of Arts.

ma /mɑː/ *n. colloq.* mother. [abbreviation of MAMA]

ma'am /mæm, mɑːm, məm/ *n.* madam (used esp. in addressing a royal lady). [contraction]

mac *n.* (also **mack**) *colloq.* mackintosh. [abbreviation]

macabre /mə'kɑːbr/ *adj.* grim, gruesome. [French]

macadam /mə'kædəm/ *n.* **1** broken stone as material for road-making. **2** = TARMACADAM. □ **macadamize** *v.* (also **-ise**) (**-zing** or **-sing**). [*McAdam*, name of a surveyor]

macadamia /ˌmækə'deɪmɪə/ *n.* edible seed of an Australian tree. [*Macadam*, name of a chemist]

macaque /mə'kæk/ *n.* a kind of monkey, e.g. the rhesus monkey and Barbary ape, with prominent cheek-pouches. [Portuguese, = monkey]

macaroni /ˌmækə'rəʊnɪ/ *n.* small pasta tubes. [Italian from Greek]

macaroon /ˌmækə'ruːn/ *n.* small almond cake or biscuit. [Italian: related to MACARONI]

macaw /mə'kɔː/ *n.* long-tailed brightly coloured American parrot. [Portuguese *macao*]

McCarthyism /mə'kɑːθɪˌɪz(ə)m/ *n. hist.* hunting out and sacking of Communists in the US. [*McCarthy*, name of a senator]

McCoy /mə'kɔɪ/ *n.* □ **the real McCoy** *colloq.* the real thing; the genuine article. [origin uncertain]

mace[1] *n.* **1** staff of office, esp. symbol of the Speaker's authority in the House of Commons. **2** person bearing this. [French from Romanic]

mace[2] *n.* dried outer covering of the nutmeg as a spice. [Latin *macir*]

macédoine /'mæsɪˌdwɑːn/ *n.* mixed vegetables or fruit, esp. diced or jellied. [French]

macerate /'mæsəˌreɪt/ *v.* (**-ting**) **1** soften by soaking. **2** waste away by fasting. □ **maceration** /-'reɪʃ(ə)n/ *n.* [Latin]

Mach /mɑːk/ *n.* (in full **Mach number**) ratio of the speed of a body to the speed of sound in the surrounding medium. [*Mach*, name of a physicist]

machete /mə'ʃetɪ/ *n.* broad heavy knife, esp. of Central America. [Spanish from Latin]

machiavellian /ˌmækɪə'velɪən/ *adj.* elaborately cunning; scheming, unscrupulous. □ **machiavellianism** *n.* [*Machiavelli*, name of a political writer]

machination /ˌmækɪ'neɪʃ(ə)n/ *n.* (usu. in *pl.*) plot, intrigue. □ **machinate** /'mæk-/ *v.* (**-ting**). [Latin: related to MACHINE]

machine /mə'ʃiːn/ –*n.* **1** apparatus for applying mechanical power, having several interrelated parts. **2** particular machine, esp. a vehicle or an electrical or electronic apparatus. **3** controlling system of an organization etc. (*party machine*). **4** person who acts mechanically. **5** (esp. in *comb.*) mechanical dispenser with slots for coins (*cigarette machine*). –*v.* (**-ning**) make or operate on with a machine. [Greek *mēkhanē*]

machine code *n.* (also **machine language**) computer language for a particular computer.

machine-gun –*n.* automatic gun giving continuous fire. –*v.* (**-nn-**) shoot at with a machine-gun.

machine-readable *adj.* in a form that a computer can process.

machinery *n.* (*pl.* **-ies**) **1** machines. **2** mechanism. **3** (usu. foll. by *of*) organized system. **4** (usu. foll. by *for*) means devised.

machine tool *n.* mechanically operated tool.

machinist *n.* **1** person who operates a machine, esp. a sewing-machine or a machine tool. **2** person who makes machinery.

machismo /mə'kɪzməʊ, -'tʃɪzməʊ/ *n.* being macho; masculine pride. [Spanish]

macho /'mætʃəʊ/ *adj.* aggressively masculine. [from MACHISMO]

Mach one *n.* (also **Mach two** etc.) the speed (or twice etc. the speed) of sound.

macintosh var. of MACKINTOSH.

mack var. of MAC.

mackerel /'mækr(ə)l/ *n.* (*pl.* same or **-s**) marine fish used as food. [Anglo-French]

mackerel sky *n.* sky dappled with rows of small white fleecy clouds.

mackintosh /'mækɪn,tɒʃ/ *n.* (also **macintosh**) **1** waterproof coat or cloak. **2** cloth waterproofed with rubber. [*Macintosh*, name of its inventor]

macramé /mə'krɑ:mɪ/ *n.* **1** art of knotting cord or string in patterns to make decorative articles. **2** work so made. [Arabic, = bedspread]

macro- *comb. form* **1** long. **2** large, large-scale. [Greek *makros* long]

macrobiotic /,mækrəʊbaɪ'ɒtɪk/ *–adj.* of a diet intended to prolong life, esp. consisting of wholefoods. *–n.* (in *pl.*; treated as *sing.*) theory of such a diet. [Greek *bios* life]

macrocarpa /,mækrəʊ'kɑ:pə/ *n.* evergreen tree, often cultivated for hedges or wind-breaks. [Greek MACRO-, *karpos* fruit]

macrocosm /'mækrəʊ,kɒz(ə)m/ *n.* **1** universe. **2** the whole of a complex structure. [from MACRO-, COSMOS]

macroeconomics /,mækrəʊ,i:kə'nɒmɪks/ *n.* the study of the economy as a whole. □ **macroeconomic** *adj.*

macron /'mækrɒn/ *n.* mark (¯) over a long or stressed vowel. [Greek, neuter of *makros* long]

macroscopic /,mækrəʊ'skɒpɪk/ *adj.* **1** visible to the naked eye. **2** regarded in terms of large units.

macula /'mækjʊlə/ *n.* (*pl.* **-lae** /-,li:/) dark, esp. permanent, spot in the skin. □ **maculation** /-'leɪʃ(ə)n/ *n.* [Latin, = spot, mesh]

mad *adj.* (**madder, maddest**) **1** insane; frenzied. **2** wildly foolish. **3** (often foll. by *about, on*) *colloq.* wildly excited or infatuated. **4** *colloq.* angry. **5** (of an animal) rabid. **6** wildly light-hearted. □ **like mad** *colloq.* with great energy or enthusiasm. □ **madness** *n.* [Old English]

madam /'mædəm/ *n.* **1** polite or respectful form of address or mode of reference to a woman. **2** *colloq.* conceited or precocious girl or young woman. **3** woman brothel-keeper. [related to MADAME]

Madame /mə'dɑ:m, 'mædəm/ *n.* **1** (*pl.* **Mesdames** /meɪ'dɑ:m, -'dæm/) Mrs or madam (used of or to a French-speaking woman). **2** (**madame**) = MADAM 1. [French *ma dame* my lady]

madcap *–adj.* wildly impulsive. *–n.* wildly impulsive person.

mad cow disease *n. colloq.* = BSE.

madden *v.* **1** make or become mad. **2** irritate. □ **maddening** *adj.* **maddeningly** *adv.*

madder /'mædə(r)/ *n.* **1** herbaceous plant with yellowish flowers. **2 a** red dye from its root. **b** its synthetic substitute. [Old English]

made *past* and *past part.* of MAKE. *–adj.* **1** built or formed (*well-made*). **2** successful (*self-made man*; *be made*). □ **have** (or **have got**) **it made** *colloq.* be sure of success. **made for** ideally suited to. **made of** consisting of. **made of money** *colloq.* very rich.

Madeira /mə'dɪərə/ *n.* **1** fortified white wine from Madeira. **2** (in full **Madeira cake**) a kind of sponge cake.

Mademoiselle /,mædəmwə'zel/ *n.* (*pl.* **Mesdemoiselles**) /,meɪdm-/ **1** Miss or madam (used of or to an unmarried French-speaking woman). **2** (**mademoiselle**) **a** young Frenchwoman. **b** French governess. [French *ma* my, *demoiselle* DAMSEL]

made to measure *adj.* tailor-made.

madhouse *n.* **1** *colloq.* scene of confused uproar. **2** *archaic* mental home or hospital.

madly *adv.* **1** in a mad manner. **2** *colloq.* **a** passionately. **b** extremely.

madman *n.* man who is mad.

Madonna /mə'dɒnə/ *n.* **1** (prec. by *the*) the Virgin Mary. **2** (**madonna**) picture or statue of her. [Italian, = my lady]

madrigal /'mædrɪg(ə)l/ *n.* part-song, usu. unaccompanied, for several voices. [Italian]

madwoman *n.* woman who is mad.

maelstrom /'meɪlstrəm/ *n.* **1** great whirlpool. **2** state of confusion. [Dutch]

maenad /'mi:næd/ *n.* **1** bacchante. **2** frenzied woman. □ **maenadic** /-'nædɪk/ *adj.* [Greek *mainomai* rave]

maestro /'maɪstrəʊ/ *n.* (*pl.* **maestri** /-strɪ/ or **-s**) **1** distinguished musician, esp. a conductor, composer, or teacher. **2** great performer in any sphere. [Italian]

Mae West /meɪ 'west/ *n. slang* inflatable life-jacket. [name of a film actress]

Mafia /'mæfɪə/ *n.* **1** organized body of criminals, orig. in Sicily, now also in Italy and the US. **2** (**mafia**) group regarded as exerting an intimidating and corrupt power. [Italian dial., = bragging]

Mafioso /,mæfɪ'əʊsəʊ/ *n.* (*pl.* **Mafiosi** /-sɪ/) member of the Mafia. [Italian: related to MAFIA]

mag *n. colloq.* = MAGAZINE 1. [abbreviation]

magazine /,mægə'zi:n/ *n.* **1** illustrated periodical publication containing articles, stories, etc. **2** chamber holding cartridges to be fed automatically to the breech of a gun. **3** similar device in a slide projector etc. **4** military store for arms etc. **5** store for explosives. [Arabic *makāzin*]

magenta /mə'dʒentə/ *–n.* **1** shade of crimson. **2** aniline crimson dye. *–adj.* of or coloured with magenta. [*Magenta* in N. Italy]

maggot /'mægət/ *n.* larva, esp. of the housefly or bluebottle. □ **maggoty** *adj.* [perhaps an alteration of *maddock*, from Old Norse]

magi *pl.* of MAGUS.

magic /'mædʒɪk/ –*n.* **1 a** supposed art of influencing or controlling events supernaturally. **b** witchcraft. **2** conjuring tricks. **3** inexplicable influence. **4** enchanting quality or phenomenon. –*adj.* **1** of magic. **2** producing surprising results. **3** *colloq.* wonderful, exciting. –*v.* (**-ck-**) change or create by or as if by magic. □ **like magic** very rapidly. **magic away** cause to disappear as if by magic. [Greek *magikos*: related to MAGUS]

magical *adj.* **1** of magic. **2** resembling, or produced as if by, magic. **3** wonderful, enchanting. □ **magically** *adv.*

magic eye *n.* photoelectric device used for detection, automatic control, etc.

magician /mə'dʒɪʃ(ə)n/ *n.* **1** person skilled in magic. **2** conjuror.

magic lantern *n.* primitive form of slide projector.

magisterial /,mædʒɪ'stɪərɪəl/ *adj.* **1** imperious. **2** authoritative. **3** of a magistrate. □ **magisterially** *adv.* [medieval Latin: related to MASTER]

magistracy /'mædʒɪstrəsɪ/ *n.* (*pl.* **-ies**) **1** magisterial office. **2** magistrates collectively.

magistrate /'mædʒɪ,streɪt/ *n.* **1** civil officer administering the law. **2** official conducting a court for minor cases and preliminary hearings. [Latin: related to MASTER]

magma /'mægmə/ *n.* (*pl.* **-s**) molten rock under the earth's crust, from which igneous rock is formed by cooling. [Greek *massō* knead]

Magna Carta /,mægnə 'kɑ:tə/ *n.* (also **Magna Charta**) charter of liberty obtained from King John in 1215. [medieval Latin, = great charter]

magnanimous /mæg'nænɪməs/ *adj.* nobly generous; not petty in feelings or conduct. □ **magnanimity** /,mægnə'nɪmɪtɪ/ *n.* **magnanimously** *adv.* [Latin *magnus* great, *animus* mind]

magnate /'mægneɪt/ *n.* wealthy and influential person, usu. in business. [Latin *magnus* great]

magnesia /mæg'ni:ʃə, -ʒə/ *n.* **1** magnesium oxide. **2** hydrated magnesium carbonate, used as an antacid and laxative. [*Magnesia* in Asia Minor]

magnesium /mæg'ni:zɪəm/ *n.* silvery metallic element.

magnet /'mægnɪt/ *n.* **1** piece of iron, steel, alloy, ore, etc., having the properties of attracting iron and of pointing approximately north and south when suspended. **2** lodestone. **3** person or thing that attracts. [Greek *magnēs -ētos* of Magnesia: related to MAGNESIA]

magnetic /mæg'netɪk/ *adj.* **1 a** having the properties of a magnet. **b** produced or acting by magnetism. **2** capable of being attracted by or acquiring the properties of a magnet. **3** strongly attractive (*magnetic personality*). □ **magnetically** *adv.*

magnetic field *n.* area of force around a magnet.

magnetic mine *n.* underwater mine detonated by the approach of a large mass of metal, e.g. a ship.

magnetic needle *n.* piece of magnetized steel used as an indicator on the dial of a compass etc.

magnetic north *n.* point indicated by the north end of a magnetic needle.

magnetic pole *n.* point near the north or south pole where a magnetic needle dips vertically.

magnetic storm *n.* disturbance of the earth's magnetic field by charged particles from the sun etc.

magnetic tape *n.* plastic strip coated with magnetic material for recording sound or pictures.

magnetism /'mægnɪ,tɪz(ə)m/ *n.* **1 a** magnetic phenomena and their science. **b** property of producing these. **2** attraction; personal charm.

magnetize /'mægnɪ,taɪz/ *v.* (also **-ise**) (**-zing** or **-sing**) **1** give magnetic properties to. **2** make into a magnet. **3** attract as a magnet does. □ **magnetizable** *adj.* **magnetization** /-'zeɪʃ(ə)n/ *n.*

magneto /mæg'ni:təʊ/ *n.* (*pl.* **-s**) electric generator using permanent magnets (esp. for the ignition of an internal-combustion engine). [abbreviation of *magneto-electric*]

Magnificat /mæg'nɪfɪ,kæt/ *n.* hymn of the Virgin Mary used as a canticle. [from its opening word]

magnification /,mægnɪfɪ'keɪʃ(ə)n/ *n.* **1** magnifying or being magnified. **2** degree of this.

magnificent /mæg'nɪfɪs(ə)nt/ *adj.* **1** splendid, stately. **2** *colloq.* fine, excellent. □ **magnificence** *n.* **magnificently** *adv.* [Latin *magnificus* from *magnus* great]

magnify /'mægnɪ,faɪ/ *v.* (**-ies, -ied**) **1** make (a thing) appear larger than it is, as with a lens. **2** exaggerate. **3** intensify. **4** *archaic* extol. □ **magnifiable** *adj.* **magnifier** *n.* [Latin: related to MAGNIFICENT]

magnifying glass *n.* lens used to magnify.

magnitude /'mægnɪ,tju:d/ *n.* **1** largeness. **2** size. **3** importance. **4 a** degree of brightness of a star. **b** class of stars arranged according to this (*of the third magnitude*). □ **of the first magnitude** very important. [Latin *magnus* great]

magnolia /mæg'nəʊlɪə/ *n.* **1** tree with dark-green foliage and waxy flowers. **2** creamy-pink colour. [*Magnol*, name of a botanist]

magnox /'mægnɒks/ *n.* magnesium-based alloy used to enclose uranium fuel elements in some nuclear reactors. [*mag*nesium *no ox*idation]

magnum /'mægnəm/ *n.* (*pl.* **-s**) wine bottle twice the normal size. [Latin, neuter of *magnus* great]

magnum opus /,mægnəm 'əʊpəs/ *n.* great work of art, literature, etc., esp. an artist's most important work. [Latin]

magpie /'mægpaɪ/ *n.* **1** a kind of crow with a long tail and black and white plumage. **2** chatterer. **3** indiscriminate collector. [from *Mag*, abbreviation of *Margaret*, PIE^2]

magus /'meɪgəs/ *n.* (*pl.* **magi** /'meɪdʒaɪ/) **1** priest of ancient Persia. **2** sorcerer. **3** (**the Magi**) the 'wise men' from the East (Matt. 2:1–12). [Persian *magus*]

Magyar /'mægjɑ:(r)/ **–***n.* **1** member of the chief ethnic group in Hungary. **2** their language. **–***adj.* of this people. [native name]

maharaja /,mɑ:hə'rɑ:dʒə/ *n.* (also **maharajah**) *hist.* title of some Indian princes. [Hindi, = great rajah]

maharanee /,mɑ:hə'rɑ:nɪ/ *n.* (also **maharani**) (*pl.* **-s**) *hist.* maharaja's wife or widow. [Hindi, = great ranee]

maharishi /,mɑ:hə'rɪʃɪ/ *n.* (*pl.* **-s**) great Hindu sage. [Hindi]

mahatma /mə'hætmə/ *n.* **1** (in India etc.) revered person. **2** one of a class of persons supposed by some Buddhists to have preternatural powers. [Sanskrit, = great soul]

mah-jong /mɑ:'dʒɒŋ/ *n.* (also **mah-jongg**) game played with 136 or 144 pieces called tiles. [Chinese dial. *ma-tsiang* sparrows]

mahlstick var. of MAULSTICK.

mahogany /mə'hɒgənɪ/ *n.* (*pl.* **-ies**) **1** reddish-brown tropical wood used for furniture. **2** its colour. [origin unknown]

mahonia /mə'həʊnɪə/ *n.* evergreen shrub with yellow bell-shaped flowers. [French or Spanish]

mahout /mə'haʊt/ *n.* (in India etc.) elephant-driver. [Hindi from Sanskrit]

maid *n.* **1** female servant. **2** *archaic* or *poet.* girl, young woman. [abbreviation of MAIDEN]

maiden /'meɪd(ə)n/ *n.* **1 a** *archaic* or *poet.* girl; young unmarried woman. **b** (*attrib.*) unmarried (*maiden aunt*). **2** = MAIDEN OVER. **3** (*attrib.*) (of a female animal) unmated. **4** (often *attrib.*) **a** horse that has never won a race. **b** race open only to such horses. **5** (*attrib.*) first (*maiden speech*; *maiden voyage*). □ **maidenhood** *n.* **maidenly** *adj.* [Old English]

maidenhair *n.* fern with hairlike stalks and delicate fronds.

maidenhead *n.* **1** virginity. **2** hymen.

maiden name *n.* woman's surname before marriage.

maiden over *n.* over in cricket in which no runs are scored.

maid of honour *n.* **1** unmarried lady attending a queen or princess. **2** esp. *US* principal bridesmaid.

maidservant *n.* female servant.

mail^1 **–***n.* **1 a** letters and parcels etc. carried by post. **b** postal system. **c** one complete delivery or collection of mail. **2** email. **3** vehicle carrying mail. **–***v.* send by post or email. [French *male* wallet]

mail^2 *n.* armour of metal rings or plates. [French *maille* from Latin *macula*]

mailbag *n.* large sack for carrying mail.

mailbox *n. US* letter-box.

mailing list *n.* list of people to whom advertising matter etc. is posted.

mail order *n.* purchase of goods by post.

mailshot *n.* advertising material sent to potential customers.

maim *v.* cripple, disable, mutilate. [French *mahaignier*]

main **–***adj.* **1** chief, principal. **2** exerted to the full (*by main force*). **–***n.* **1** principal duct etc. for water, sewage, etc. **2** (usu. in *pl.*; prec. by *the*) **a** central distribution network for electricity, gas, water, etc. **b** domestic electricity supply as distinct from batteries. **3** *poet.* high seas (*Spanish Main*). □ **in the main** mostly. [Old English]

main brace *n.* brace attached to the main yard.

main chance *n.* (prec. by *the*) one's own interests.

mainframe *n.* **1** central processing unit of a large computer. **2** (often *attrib.*) large computer system.

mainland *n.* large continuous extent of land, excluding neighbouring islands.

mainline *v.* (**-ning**) *slang* **1** take drugs intravenously. **2** inject (drugs) intravenously. □ **mainliner** *n.*

main line *n.* railway line linking large cities.

mainly /'meɪnlɪ/ *adv.* mostly; chiefly.

mainmast *n.* principal mast of a ship.

mainsail /'meɪnseɪl, -s(ə)l/ *n.* **1** (in a square-rigged vessel) lowest sail on the mainmast. **2** (in a fore-and-aft rigged vessel) sail set on the after part of the mainmast.

mainspring *n.* **1** principal spring of a watch, clock, etc. **2** chief motivating force; incentive.

mainstay *n.* **1** chief support. **2** stay from the maintop to the foot of the foremast.

mainstream *n.* **1** (often *attrib.*) ultimately prevailing trend in opinion, fashion, etc. **2** type of swing jazz, esp. with solo improvisation. **3** principal current of a river etc.

maintain /meɪn'teɪn/ *v.* **1** cause to continue; keep up (an activity etc.). **2** support by work, expenditure, etc. **3** assert as true. **4** preserve (a house, machine, etc.) in good repair. **5** provide means for. [Latin *manus* hand, *teneo* hold]

maintained school *n.* school supported from public funds, State school.

maintenance /'meɪntənəns/ *n.* **1** maintaining or being maintained. **2 a** provision of the means to support life. **b** alimony. [French: related to MAINTAIN]

maintop *n.* platform above the head of the lower mainmast.

maintopmast *n.* mast above the head of the lower mainmast.

main yard *n.* yard on which the mainsail is extended.

maiolica /mə'jɒlɪkə/ *n.* (also **majolica**) white tin-glazed earthenware decorated with metallic colours or enamelled. [Italian, from the former name of Majorca]

maisonette /,meɪzə'net/ *n.* **1** flat on more than one floor. **2** small house. [French *maisonnette* diminutive of *maison* house]

maize *n.* **1** cereal plant of N. America. **2** cobs or grain of this. [French or Spanish]

Maj. *abbr.* Major.

majestic /mə'dʒestɪk/ *adj.* stately and dignified; imposing. □ **majestically** *adv.*

majesty /'mædʒɪstɪ/ *n.* (*pl.* **-ies**) **1** stateliness, dignity, or authority, esp. of bearing, language, etc. **2 a** royal power. **b** (**Majesty**) (prec. by *His, Her, Your*) forms of description or address for a sovereign or a sovereign's wife or widow (*Your Majesty*; *Her Majesty the Queen Mother*). [Latin *majestas*: related to MAJOR]

majolica var. of MAIOLICA.

major /'meɪdʒə(r)/ *–adj.* **1** relatively great in size, intensity, scope, or importance. **2** (of surgery) serious. **3** *Mus.* **a** (of a scale) having intervals of a semitone above its third and seventh notes. **b** (of an interval) greater by a semitone than a minor interval (*major third*). **c** (of a key) based on a major scale. **4** of full legal age. *–n.* **1 a** army officer next below lieutenant-colonel. **b** officer in charge of a band section (*drum major*). **2** person of full legal age. **3** *US* **a** student's main subject or course. **b** student of this. *–v.* (foll. by *in*) *US* study or qualify in (a subject) as one's main subject. [Latin, comparative of *magnus* great]

major-domo /,meɪdʒə'dəʊməʊ/ *n.* (*pl.* **-s**) chief steward of a great household. [medieval Latin *major domus* highest official of the household]

majorette /,meɪdʒə'ret/ *n.* = DRUM MAJORETTE. [abbreviation]

major-general *n.* officer next below a lieutenant-general.

majority /mə'dʒɒrɪtɪ/ *n.* (*pl.* **-ies**) **1** (usu. foll. by *of*) greater number or part. **2 a** number of votes by which a candidate wins. **b** party etc. receiving the greater number of votes. **3** full legal age. **4** rank of major. [medieval Latin: related to MAJOR]

■ **Usage** In sense 1, *majority* is strictly used only with countable nouns, as in *the majority of people*, and not (e.g.) *the majority of the work*.

majority rule *n.* principle that the greater number should exercise the greater power.

make *–v.* (**-king**; *past* and *past part.* **made**) **1** construct; create; form from parts or other substances. **2** cause or compel (*made me do it*). **3 a** cause to exist; bring about (*made a noise*). **b** cause to become or seem (*made him angry*; *made a fool of me*; *made him a knight*). **4** compose; prepare; write (*made her will*; *made a film*). **5** constitute; amount to; be reckoned as (*2 and 2 make 4*). **6 a** undertake (*made a promise*; *make an effort*). **b** perform (an action etc.) (*made a face*; *made a bow*). **7** gain, acquire, procure (money, a living, a profit, etc.). **8** prepare (tea, coffee, a meal, etc.). **9 a** arrange (a bed) for use. **b** arrange and light materials for (a fire). **10 a** proceed (*made towards the river*). **b** (foll. by *to* + infin.) act as if with the intention to (*he made to go*). **11** *colloq.* **a** arrive at (a place) or in time for (a train etc.). **b** manage to attend; manage to attend on (a certain day) or at (a certain time) (*couldn't make the meeting last* *week; can make any day except Friday*). **c** achieve a place in (*made the first eleven*). **12** establish or enact (a distinction, rule, law, etc.). **13** consider to be; estimate as (*what do you make the total?*). **14** secure the success or advancement of (*his second novel made him; it made my day*). **15** accomplish (a distance, speed, score, etc.). **16 a** become by development (*made a great leader*). **b** serve as (*makes a useful seat*). **17** (usu. foll. by *out*) represent as (*makes him out a liar*). **18** form in the mind (*make a decision*). **19** (foll. by *it* + compl.) **a** determine, establish, or choose (*let's make it Tuesday*). **b** bring to (a chosen value etc.) (*make it a dozen*). *–n.* **1** type or brand of manufacture. **2** way a thing is made. □ **make away with 1** = *make off with.* **2** = *do away with.* **make believe** pretend. **make the best of** see BEST. **make a clean breast** see BREAST. **make a clean sweep** see SWEEP. **make a day** (or **night** etc.) **of it** devote a whole day (or night etc.) to an activity. **make do 1** manage with the inadequate means available. **2** (foll. by *with*) manage with (something) as an inferior substitute. **make for 1** tend to result in. **2** proceed towards (a place). **3** attack. **make good 1** repay, repair, or compensate for. **2** achieve (a purpose); be successful. **make the grade** succeed. **make it** *colloq.* **1** succeed in reaching, esp. in time. **2** succeed. **make it up 1** be reconciled. **2** remedy a deficit. **make it up to** remedy negligence, an injury, etc. to (a person). **make love** see LOVE. **make a meal of** see MEAL[1]. **make merry** see MERRY. **make money** acquire wealth. **make the most of** see MOST. **make much** (or **little**) **of** treat as important (or unimportant). **make a name for oneself** see NAME. **make no bones about** see BONE. **make nothing of 1** treat as trifling. **2** be unable to understand, use, or deal with. **make of 1** construct from. **2** conclude from or about (*can you make anything of it?*). **make off** depart hastily. **make off with** carry away; steal. **make or break** cause the success or ruin of. **make out 1** discern or understand. **2** assert; pretend. **3** *colloq.* progress; fare. **4** write out (a cheque etc.) or fill in (a form). **make over 1** transfer the possession of. **2** refashion. **make up 1** act to overcome (a deficiency). **2** complete (an amount etc.). **3** (foll. by *for*) compensate for. **4** be reconciled. **5** put together; prepare (*made up the medicine*). **6** concoct (a story). **7** apply cosmetics (to). **8** prepare (a bed) with fresh linen. **make up one's mind** decide. **make up to** curry favour with. **make water** urinate. **make way 1** (often foll. by *for*) allow room to pass. **2** (foll. by *for*) be superseded by. **make one's way** go; prosper. **on the make** *colloq.* intent on gain. [Old English]

make-believe *–n.* pretence. *–attrib. adj.* pretended.

maker *n.* **1** person who makes. **2** (**Maker**) God.

makeshift *–adj.* temporary. *–n.* temporary substitute or device.

make-up *n.* **1** cosmetics, as used generally or by actors. **2** character, temperament, etc. **3** composition (of a thing).

makeweight *n.* **1** small quantity added to make up the weight. **2** person or thing supplying a deficiency.

making *n.* (in *pl.*) **1** earnings; profit. **2** essential qualities or ingredients (*has the makings of a pilot*). □ **be the making of** ensure the success of. **in the making** in the course of being made or formed. [Old English: related to MAKE]

mal- *comb. form* **1 a** bad, badly (*malpractice; maltreat*). **b** faulty (*malfunction*). **2** not (*maladroit*). [French *mal* badly, from Latin *male*]

malachite /'mælə,kaɪt/ *n.* green mineral used for ornament. [Greek *molokhitis*]

maladjusted /,mælə'dʒʌstɪd/ *adj.* (of a person) unable to adapt to or cope with the demands of a social environment. □ **maladjustment** *n.*

maladminister /,mæləd'mɪnɪstə(r)/ *v.* manage badly or improperly. □ **maladministration** /-'streɪʃ(ə)n/ *n.*

maladroit /,mælə'drɔɪt/ *adj.* clumsy; bungling. [French: related to MAL-]

malady /'mælədɪ/ *n.* (*pl.* **-ies**) ailment, disease. [French *malade* sick]

malaise /mə'leɪz/ *n.* **1** general bodily discomfort or lassitude. **2** feeling of unease or demoralization. [French: related to EASE]

malapropism /'mæləprɒ,pɪz(ə)m/ *n.* comical misuse of a word in mistake for one sounding similar, e.g. *alligator* for *allegory*. [Mrs *Malaprop*, name of a character in Sheridan's *The Rivals*]

malaria /mə'leərɪə/ *n.* recurrent fever caused by a parasite transmitted by a mosquito bite. □ **malarial** *adj.* [Italian, = bad air]

malarkey /mə'lɑːkɪ/ *n. colloq.* humbug; nonsense. [origin unknown]

Malay /mə'leɪ/ *–n.* **1** member of a people predominating in Malaysia and Indonesia. **2** their language. *–adj.* of this people or language. □ **Malayan** *n.* & *adj.* [Malay *malāyu*]

malcontent /'mælkən,tent/ *–n.* discontented person. *–adj.* discontented. [French: related to MAL-]

male *–adj.* **1** of the sex that can beget offspring by fertilization. **2** of men or male animals, plants, etc.; masculine. **3** (of plants or flowers) containing stamens but no pistil. **4** (of parts of machinery etc.) designed to enter or fill the corresponding hollow part (*male screw*). *–n.* male person or animal. □ **maleness** *n.* [Latin *masculus* from *mas* a male]

male chauvinist *n.* = CHAUVINIST 2.

malediction /,mælı'dıkʃ(ə)n/ *n.* **1** curse. **2** utterance of a curse. □ **maledictory** *adj.* [Latin *maledictio*: related to MAL-]

malefactor /'mælı,fæktə(r)/ *n.* criminal; evil-doer. □ **malefaction** /-'fækʃ(ə)n/ *n.* [Latin *male* badly, *facio fact-* do]

male menopause *n. colloq.* crisis of potency, confidence, etc., supposed to afflict some men in middle life.

malevolent /mə'levələnt/ *adj.* wishing evil to others. □ **malevolence** *n.* **malevolently** *adv.* [Latin *volo* wish]

malfeasance /mæl'fi:z(ə)ns/ *n. formal* misconduct, esp. in an official capacity. [French: related to MAL-]

malformation /,mælfɔ:'meıʃ(ə)n/ *n.* faulty formation. □ **malformed** /-'fɔ:md/ *adj.*

malfunction /mæl'fʌŋkʃ(ə)n/ *–n.* failure to function normally. *–v.* fail to function normally.

malice /'mælıs/ *n.* **1** desire to harm or cause difficulty to others; ill-will. **2** *Law* harmful intent. [Latin *malus* bad]

malice aforethought *n. Law* intention to commit a crime, esp. murder.

malicious /mə'lıʃəs/ *adj.* given to or arising from malice. □ **maliciously** *adv.*

malign /mə'laın/ *–adj.* **1** (of a thing) injurious. **2** (of a disease) malignant. **3** malevolent. *–v.* speak ill of; slander. □ **malignity** /mə'lıgnıtı/ *n.* [Latin *malus* bad]

malignant /mə'lıgnənt/ *adj.* **1 a** (of a disease) very virulent or infectious. **b** (of a tumour) spreading or recurring; cancerous. **2** harmful; feeling or showing intense ill-will. □ **malignancy** *n.* **malignantly** *adv.* [Latin: related to MALIGN]

malinger /mə'lıŋgə(r)/ *v.* pretend to be ill, esp. to escape work. □ **malingerer** *n.* [French *malingre* sickly]

mall /mæl, mɔ:l/ *n.* **1** sheltered walk or promenade. **2** shopping precinct. [*The Mall*, street in London]

mallard /'mælɑ:d/ *n.* (*pl.* same) a kind of wild duck. [French]

malleable /'mælıəb(ə)l/ *adj.* **1** (of metal etc.) that can be shaped by hammering. **2** easily influenced; pliable. □ **malleability** /-'bılıtı/ *n.* **malleably** *adv.* [medieval Latin: related to MALLET]

mallet /'mælıt/ *n.* **1** hammer, usu. of wood. **2** implement for striking a croquet or polo ball. [Latin *malleus* hammer]

mallow /'mæləʊ/ *n.* plant with hairy stems and leaves and pink or purple flowers. [Latin *malva*]

malmsey /'mɑ:mzı/ *n.* a strong sweet wine. [Low German or Dutch from *Monemvasia* in Greece]

malnourished /mæl'nʌrıʃt/ *adj.* suffering from malnutrition. □ **malnourishment** *n.*

malnutrition /,mælnju:'trıʃ(ə)n/ *n.* condition resulting from the lack of foods necessary for health.

malodorous /mæl'əʊdərəs/ *adj.* evil-smelling.

malpractice /mæl'præktıs/ *n.* improper, negligent, or criminal professional conduct.

malt /mɔ:lt/ *–n.* **1** barley, or other grain, steeped, germinated, and dried, for brewing etc. **2** *colloq.* malt whisky; malt liquor. *–v.* convert (grain) into malt. □ **malty** *adj.* (**-ier**, **-iest**). [Old English]

malted milk *n.* drink made from dried milk and extract of malt.

Maltese /mɔ:l'ti:z/ *–n.* (*pl.* same) native or language of Malta. *–adj.* of Malta.

Maltese cross *n.* cross with the arms broadening outwards, often indented at the ends.

Malthusian /mæl'θju:zıən/ *adj.* of Malthus's doctrine that the population should be restricted so as to prevent an increase beyond its means of subsistence. □ **Malthusianism** *n.* [*Malthus*, name of a clergyman]

maltose /'mɔ:ltəʊz/ *n.* sugar made from starch by enzymes in malt, saliva, etc. [French: related to MALT]

maltreat /mæl'tri:t/ *v.* ill-treat. □ **maltreatment** *n.* [French: related to MAL-]

malt whisky *n.* whisky made solely from malted barley.

mama /mə'mɑ:/ *n.* (also **mamma**) *archaic* mother. [imitative of child's *ma, ma*]

mamba /'mæmbə/ *n.* venomous African snake. [Zulu *imamba*]

mambo /'mæmbəʊ/ *n.* (*pl.* **-s**) Latin American dance like the rumba. [American Spanish]

mamma var. of MAMA.

mammal /'mæm(ə)l/ *n.* warm-blooded vertebrate of the class secreting milk to feed its young. □ **mammalian** /-'meɪlɪən/ *adj.* & *n.* [Latin *mamma* breast]

mammary /'mæmərɪ/ *adj.* of the breasts.

mammogram /'mæmə,græm/ *n.* image obtained by mammography. [Latin *mamma* breast]

mammography /mæ'mɒgrəfɪ/ *n.* X-ray technique for screening the breasts for tumours etc.

Mammon /'mæmən/ *n.* wealth regarded as a god or evil influence. [Aramaic *māmōn*]

mammoth /'mæməθ/ *–n.* large extinct elephant with a hairy coat and curved tusks. *–adj.* huge. [Russian]

man *–n.* (*pl.* **men**) **1** adult human male. **2 a** human being; person. **b** the human race. **3 a** workman (*the manager spoke to the men*). **b** manservant, valet. **4** (usu. in *pl.*) soldiers, sailors, etc., esp. non-officers. **5** suitable or appropriate person; expert (*he is your man; the man for the job*). **6 a** husband (*man and wife*). **b** *colloq.* boyfriend, lover. **7** human being of a specified type or historical period (*Renaissance man*; *Peking man*). **8** piece in chess, draughts, etc. **9** *colloq.* as a form of address. **10** person pursued; opponent (*police caught their man*). *–v.* (**-nn-**) **1** supply with a person or people for work or defence. **2** work, service, or defend (*man the pumps*). **3** fill (a post). □ **as one man** in unison. **be one's own man** be independent. **to a man** without exception. □ **manlike** *adj.* [Old English]

man about town *n.* fashionable socializer.

manacle /'mænək(ə)l/ *–n.* (usu. in *pl.*) **1** fetter for the hand; handcuff. **2** restraint. *–v.* (**-ling**) fetter with manacles. [Latin *manus* hand]

manage /'mænɪdʒ/ *v.* (**-ging**) **1** organize; regulate; be in charge of. **2** succeed in achieving; contrive (*managed to come*; *managed a smile*; *managed to ruin the day*). **3** (often foll. by *with*) succeed with limited resources etc.; be able to cope. **4** succeed in controlling. **5** (often prec. by *can* etc.) **a** cope with (*couldn't manage another bite*). **b** be free to attend on or at (*can manage Monday*). **6** use or wield (a tool etc.). □ **manageable** *adj.* [Latin *manus* hand]

management *n.* **1** managing or being managed. **2 a** administration of business or public undertakings. **b** people engaged in this, esp. those controlling a workforce.

manager *n.* **1** person controlling or administering a business or part of a business. **2** person controlling the affairs, training, etc. of a person or team in sports, entertainment, etc. **3** person of a specified level of skill in household or financial affairs etc. (*a good manager*). □ **managerial** /,mænɪ'dʒɪərɪəl/ *adj.*

manageress /,mænɪdʒə'res/ *n.* woman manager, esp. of a shop, hotel, etc.

managing director *n.* director with executive control or authority.

mañana /mæn'jɑːnə/ *–adv.* tomorrow (esp. to indicate procrastination). *–n.* indefinite future. [Spanish]

man-at-arms *n.* (*pl.* **men-at-arms**) *archaic* soldier.

manatee /,mænə'tiː/ *n.* large aquatic plant-eating mammal. [Spanish from Carib]

Mancunian /mæŋ'kjuːnɪən/ *–n.* native of Manchester. *–adj.* of Manchester. [Latin *Mancunium*]

mandala /'mændələ/ *n.* circular figure as a religious symbol of the universe. [Sanskrit]

mandamus /mæn'deɪməs/ *n.* judicial writ issued as a command to an inferior court, or ordering a person to perform a public or statutory duty. [Latin, = we command]

mandarin /'mændərɪn/ *n.* **1** (**Mandarin**) official language of China. **2** *hist.* Chinese official. **3** powerful person, esp. a top civil servant. **4** (in full **mandarin orange**) = TANGERINE 1. [Hindi *mantrī*]

mandate /'mændeɪt/ *–n.* **1** official command or instruction. **2** authority given by electors to a government, trade union, etc. **3** authority to act for another. *–v.* (**-ting**) instruct (a delegate) how to act or vote. [Latin *mandatum*, past part. of *mando* command]

mandatory /'mændətərɪ/ *adj.* **1** compulsory. **2** of or conveying a command. □ **mandatorily** *adv.* [Latin: related to MANDATE]

mandible /'mændɪb(ə)l/ *n.* **1** jaw, esp. the lower jaw in mammals and fishes. **2** upper or lower part of a bird's beak. **3** either half of the crushing organ in the mouth-parts of an insect etc. [Latin *mando* chew]

mandolin /,mændə'lɪn/ *n.* a kind of lute with paired metal strings plucked with a plectrum. □ **mandolinist** *n.* [French from Italian]

mandrake /'mændreɪk/ *n.* poisonous narcotic plant with large yellow fruit. [Greek *mandragoras*]

mandrel /'mændr(ə)l/ *n.* **1** lathe-shaft to which work is fixed while being turned. **2** cylindrical rod round which metal or other material is forged or shaped. [origin unknown]

mandrill /'mændrɪl/ *n.* large W. African baboon. [probably from MAN, DRILL[4]]
mane *n.* **1** long hair on the neck of a horse, lion, etc. **2** *colloq.* person's long hair. [Old English]
manège /mæ'neɪʒ/ *n.* (also **manege**) **1** riding-school. **2** movements of a trained horse. **3** horsemanship. [Italian: related to MANAGE]
maneuver *US* var. of MANOEUVRE.
man Friday *n.* male helper or follower.
manful *adj.* brave; resolute. □ **manfully** *adv.*
manganese /'mæŋgə,ni:z/ *n.* **1** grey brittle metallic element. **2** black mineral oxide of this used in glass-making etc. [Italian: related to MAGNESIA]
mange /meɪndʒ/ *n.* skin disease in hairy and woolly animals. [French *mangeue* itch, from Latin *manduco* chew]
mangel-wurzel /'mæŋg(ə)l,wɜ:z(ə)l/ *n.* (also **mangold-** /'mæŋg(ə)ld-/) large beet used as cattle food. [German *Mangold* beet, *Wurzel* root]
manger /'meɪndʒə(r)/ *n.* box or trough for horses or cattle to feed from. [Latin: related to MANGE]
mange-tout /mɑ̃ʒ'tu:/ *n.* a kind of pea eaten in the pod. [French, = eat-all]
mangle[1] /'mæŋg(ə)l/ **–***n.* machine of two or more cylinders for squeezing water from and pressing wet clothes. **–***v.* (**-ling**) press (clothes etc.) in a mangle. [Dutch *mangel*]
mangle[2] /'mæŋg(ə)l/ *v.* (**-ling**) **1** hack or mutilate by blows. **2** spoil (a text etc.) by gross blunders. **3** cut roughly so as to disfigure. [Anglo-French *ma(ha)ngler*: probably related to MAIM]
mango /'mæŋgəʊ/ *n.* (*pl.* **-es** or **-s**) **1** tropical fruit with yellowish flesh. **2** tree bearing this. [Tamil *mānkāy*]
mangold-wurzel var. of MANGEL-WURZEL.
mangrove /'mæŋgrəʊv/ *n.* tropical tree or shrub growing in shore-mud with many tangled roots above ground. [origin unknown]
mangy /'meɪndʒɪ/ *adj.* (**-ier**, **-iest**) **1** having mange. **2** squalid; shabby.
manhandle *v.* (**-ling**) **1** *colloq.* handle (a person) roughly. **2** move by human effort.
manhole *n.* covered opening in a pavement, sewer, etc. for workmen to gain access.
manhood *n.* **1** state of being a man. **2 a** manliness; courage. **b** a man's sexual potency. **3** men of a country etc.
man-hour *n.* work done by one person in one hour.
manhunt *n.* organized search for a person, esp. a criminal.
mania /'meɪnɪə/ *n.* **1** mental illness marked by excitement and violence. **2** (often foll. by *for*) excessive enthusiasm; obsession. [Greek *mainomai* be mad]
-mania *comb. form* **1** denoting a special type of mental disorder (*megalomania*). **2** denoting enthusiasm or admiration (*Beatlemania*).
maniac /'meɪnɪ,æk/ **–***n.* **1** *colloq.* person behaving wildly (*too many maniacs on the road*). **2** *colloq.* obsessive enthusiast. **3** person suffering from mania. **–***adj.* of or behaving like a maniac. □ **maniacal** /mə'naɪək(ə)l/ *adj.* **maniacally** /mə'naɪəkəlɪ/ *adv.*
-maniac *comb. form* forming adjectives and nouns meaning 'affected with -mania' or 'a person affected with -mania' (*nymphomaniac*).
manic /'mænɪk/ *adj.* **1** of or affected by mania. **2** *colloq.* wildly excited; frenzied; excitable. □ **manically** *adv.*
manic-depressive **–***adj.* relating to a mental disorder with alternating periods of elation and depression. **–***n.* person with such a disorder.
manicure /'mænɪ,kjʊə(r)/ **–***n.* cosmetic treatment of the hands and fingernails. **–***v.* (**-ring**) give a manicure to (the hands or a person). □ **manicurist** *n.* [Latin *manus* hand, *cura* care]
manifest /'mænɪ,fest/ **–***adj.* clear or obvious to the eye or mind. **–***v.* **1** show (a quality or feeling) by one's acts etc. **2** show plainly to the eye or mind. **3** be evidence of; prove. **4** *refl.* (of a thing) reveal itself. **5** (of a ghost) appear. **–***n.* cargo or passenger list. □ **manifestation** /-'steɪʃ(ə)n/ *n.* **manifestly** *adv.* [Latin *manifestus*]
manifesto /,mænɪ'festəʊ/ *n.* (*pl.* **-s**) declaration of policies, esp. by a political party. [Italian: related to MANIFEST]
manifold /'mænɪ,fəʊld/ **–***adj.* **1** many and various. **2** having various forms, parts, applications, etc. **–***n.* **1** manifold thing. **2** pipe or chamber branching into several openings. [Old English: related to MANY, -FOLD]
manikin /'mænɪkɪn/ *n.* little man; dwarf. [Dutch]
Manila /mə'nɪlə/ *n.* **1** (in full **Manila hemp**) strong fibre of a kind of tree native to the Philippines. **2** (also **manila**) strong brown paper made from this. [*Manila* in the Philippines]
man in the street *n.* ordinary person.
manipulate /mə'nɪpjʊ,leɪt/ *v.* (**-ting**) **1** handle, esp. with skill. **2** manage (a person, situation, etc.) to one's own advantage, esp. unfairly. **3** move (part of a patient's body) by hand in order to increase flexion etc. **4** *Computing* edit or move (text, data, etc.). □ **manipulable**

/-ləb(ə)l/ *adj.* **manipulation** /-'leɪʃ(ə)n/ *n.* **manipulator** *n.* [Latin *manus* hand]

manipulative /mə'nɪpjʊlətɪv/ *adj.* tending to exploit a situation, person, etc., for one's own ends. □ **manipulatively** *adv.*

mankind *n.* **1** human species. **2** male people.

manky /'mæŋkɪ/ *adj.* (**-ier, -iest**) *colloq.* **1** bad, inferior, defective. **2** dirty. [obsolete *mank* defective]

manly *adj.* (**-ier, -iest**) **1** having qualities associated with a man (e.g. strength and courage). **2** befitting a man. □ **manliness** *n.*

man-made *adj.* (of textiles) artificial, synthetic.

manna /'mænə/ *n.* **1** substance miraculously supplied as food to the Israelites in the wilderness (Exod. 16). **2** unexpected benefit (esp. *manna from heaven*). [Old English ultimately from Hebrew]

manned *adj.* (of a spacecraft etc.) having a human crew.

mannequin /'mænɪkɪn/ *n.* **1** fashion model. **2** window dummy. [French, = MANIKIN]

manner /'mænə(r)/ *n.* **1** way a thing is done or happens. **2** (in *pl.*) **a** social behaviour (*good manners*). **b** polite behaviour (*has no manners*). **c** modes of life; social conditions. **3** outward bearing, way of speaking, etc. **4** style (*in the manner of Rembrandt*). **5** kind, sort (*not by any manner of means*). □ **in a manner of speaking** in a way; so to speak. **to the manner born** *colloq.* naturally at ease in a particular situation etc. [Latin *manus* hand]

mannered *adj.* **1** (in *comb.*) having specified manners (*ill-mannered*). **2** esp. *Art* full of mannerisms.

mannerism *n.* **1** habitual gesture or way of speaking etc. **2 a** stylistic trick in art etc. **b** excessive use of these. □ **mannerist** *n.*

mannerly *adj.* well-mannered, polite.

mannish *adj.* **1** (of a woman) masculine in appearance or manner. **2** characteristic of a man. □ **mannishly** *adv.*

manoeuvre /mə'nu:və(r)/ (*US* **maneuver**) *—n.* **1** planned and controlled movement of a vehicle or body of troops etc. **2** (in *pl.*) large-scale exercise of troops, ships, etc. **3** agile or skilful movement. **4** artful plan. *—v.* (**-ring**) **1** move (a thing, esp. a vehicle) carefully. **2** perform or cause (troops etc.) to perform manoeuvres. **3 a** (usu. foll. by *into, out of*, etc.) manipulate (a person, thing, etc.) by scheming or adroitness. **b** use artifice. □ **manoeuvrable** *adj.* **manoeuvrability** /-vrə'bɪlɪtɪ/ *n.* [medieval Latin *manu operor* work with the hand]

man of letters *n.* scholar or author.

man of the world see WORLD.

man-of-war *n.* (*pl.* **men-of-war**) warship.

manor /'mænə(r)/ *n.* **1** (also **manor-house**) large country house with lands. **2** *hist.* feudal lordship over lands. **3** *slang* district covered by a police station. □ **manorial** /mə'nɔ:rɪəl/ *adj.* [Latin *maneo* remain]

manpower *n.* number of people available for work, service, etc.

manqué /'mɒŋkeɪ/ *adj.* (placed after noun) that might have been but is not (*an actor manqué*). [French]

mansard /'mænsɑ:d/ *n.* roof with four sloping sides, each of which becomes steeper halfway down. [*Mansart*, name of an architect]

manse /mæns/ *n.* ecclesiastical residence, esp. a Scottish Presbyterian minister's house. [medieval Latin: related to MANOR]

manservant *n.* (*pl.* **menservants**) male servant.

mansion /'mænʃ(ə)n/ *n.* **1** large grand house. **2** (in *pl.*) large building divided into flats. [Latin: related to MANOR]

manslaughter *n.* unintentional but not accidental unlawful killing of a human being.

mantel /'mænt(ə)l/ *n.* mantelpiece or mantelshelf. [var. of MANTLE]

mantelpiece *n.* **1** structure of wood, marble, etc. above and around a fireplace. **2** = MANTELSHELF.

mantelshelf *n.* shelf above a fireplace.

mantilla /mæn'tɪlə/ *n.* lace scarf worn by Spanish women over the hair and shoulders. [Spanish: related to MANTLE]

mantis /'mæntɪs/ *n.* (*pl.* same or **mantises**) (in full **praying mantis**) predatory insect that holds its forelegs like hands folded in prayer. [Greek, = prophet]

mantissa /mæn'tɪsə/ *n.* part of a logarithm after the decimal point. [Latin, = makeweight]

mantle /'mænt(ə)l/ *—n.* **1** loose sleeveless cloak. **2** covering (*mantle of snow*). **3** fragile lacelike tube fixed round a gas-jet to give an incandescent light. **4** region between the crust and the core of the earth. *—v.* (**-ling**) clothe; conceal, envelop. [Latin *mantellum* cloak]

man to man *adv.* candidly.

mantra /'mæntrə/ *n.* **1** Hindu or Buddhist devotional incantation. **2** Vedic hymn. [Sanskrit, = instrument of thought]

mantrap *n.* trap for catching trespassers etc.

manual /'mænjʊəl/ *–adj.* **1** of or done with the hands (*manual labour*). **2 a** worked by hand, not automatically (*manual gear-change*). **b** (of a vehicle) worked by manual gear-change. *–n.* **1** reference book. **2** organ keyboard played with the hands, not the feet. **3** *colloq.* vehicle with manual transmission. □ **manually** *adv.* [Latin *manus* hand]

manufacture /,mænjʊ'fæktʃə(r)/ *–n.* **1** making of articles, esp. in a factory etc. **2** branch of industry (*woollen manufacture*). *–v.* (**-ring**) **1** make (articles), esp. on an industrial scale. **2** invent or fabricate (evidence, a story, etc.). □ **manufacturer** *n.* [Latin *manufactum* made by hand]

manure /mə'njʊə(r)/ *–n.* fertilizer, esp. dung. *–v.* (**-ring**) apply manure to (land etc.). [Anglo-French *mainoverer* MANOEUVRE]

manuscript /'mænjʊskrɪpt/ *–n.* **1** text written by hand. **2** author's handwritten or typed text. **3** handwritten form (*produced in manuscript*). *–adj.* written by hand. [medieval Latin *manuscriptus* written by hand]

Manx /mæŋks/ *–adj.* of the Isle of Man. *–n.* **1** former Celtic language of the Isle of Man. **2** (prec. by *the*; treated as *pl.*) Manx people. [Old Norse]

Manx cat *n.* tailless variety of cat.

many /'menɪ/ *–adj.* (**more**; **most**) great in number; numerous (*many people*). *–n.* (as *pl.*) **1** many people or things. **2** (prec. by *the*) the majority of people. □ **a good** (or **great**) **many** a large number. **many's the time** often. **many a time** many times. [Old English]

Maoism /'maʊɪz(ə)m/ *n.* Communist doctrines of Mao Zedong. □ **Maoist** *n.* & *adj.* [*Mao* Zedong, name of a Chinese statesman]

Maori /'maʊrɪ/ *–n.* (*pl.* same or **-s**) **1** member of the aboriginal people of New Zealand. **2** their language. *–adj.* of this people. [native name]

map *–n.* **1 a** flat representation of the earth's surface, or part of it. **b** diagram of a route etc. **2** similar representation of the stars, sky, moon, etc. **3** diagram showing the arrangement or components of a thing. *–v.* (**-pp-**) **1** represent on a map. **2** *Math.* associate each element of (a set) with one element of another set. □ **map out** plan in detail. [Latin *mappa* napkin]

maple /'meɪp(ə)l/ *n.* **1** any of various trees or shrubs grown for shade, ornament, wood, or sugar. **2** its wood. [Old English]

maple-leaf *n.* emblem of Canada.

maple sugar *n.* sugar produced by evaporating the sap of some kinds of maple.

maple syrup *n.* syrup made by evaporating maple sap or dissolving maple sugar.

maquette /mə'ket/ *n.* preliminary model or sketch. [Italian *macchia* spot]

Maquis /mæ'ki:, 'mæ-/ *n.* (*pl.* same) **1** French resistance movement during the German occupation (1940–45). **2** member of this. [French, = brushwood]

Mar. *abbr.* March.

mar *v.* (**-rr-**) spoil; disfigure. [Old English]

marabou /'mærə,bu:/ *n.* **1** large W. African stork. **2** its down as trimming etc. [French from Arabic]

maraca /mə'rækə/ *n.* clublike bean-filled gourd etc., shaken rhythmically in pairs in Latin American music. [Portuguese]

maraschino /,mærə'ski:nəʊ/ *n.* (*pl.* **-s**) sweet liqueur made from black cherries. [Italian]

maraschino cherry *n.* cherry preserved in maraschino and used in cocktails etc.

marathon /'mærəθ(ə)n/ *n.* **1** long-distance running race, usu. of 26 miles 385 yards (42.195 km). **2** long-lasting or difficult undertaking etc. [*Marathon* in Greece, scene of a decisive battle in 490 BC: a messenger supposedly ran with news of the outcome to Athens]

maraud /mə'rɔ:d/ *v.* **1** make a plundering raid (on). **2** pilfer systematically. □ **marauder** *n.* [French *maraud* rogue]

marble /'mɑ:b(ə)l/ *–n.* **1** crystalline limestone capable of taking a polish, used in sculpture and architecture. **2** (often *attrib.*) **a** anything of marble (*marble clock*). **b** anything like marble in hardness, coldness, etc. (*her features were marble*). **3 a** small, esp. glass, ball as a toy. **b** (in *pl.*; treated as *sing.*) game using these. **4** (in *pl.*) *slang* one's mental faculties (*he's lost his marbles*). **5** (in *pl.*) collection of sculptures (*Elgin Marbles*). *–v.* (**-ling**) **1** (esp. as **marbled** *adj.*) stain or colour (paper, soap, etc.) to look like variegated marble. **2** (as **marbled** *adj.*) (of meat) striped with fat and lean. [Latin *marmor* from Greek]

marble cake *n.* mottled cake of light and dark sponge.

marbling *n.* **1** colouring or marking like marble. **2** streaks of fat in lean meat.

marcasite /'mɑ:kə,saɪt/ *n.* **1** yellowish crystalline iron sulphide. **2** crystals of this used in jewellery. [Arabic *markashita*]

March *n.* third month of the year. [Latin *Martius* of Mars]

march[1] *–v.* **1** (cause to) walk in a military manner with a regular tread (*army marched past; marched him away*). **2 a** walk purposefully. **b** (often foll. by *on*) (of events etc.) continue unrelentingly (*time marches on*). **3** (foll. by *on*) advance towards (a military objective). *–n.* **1 a** act of marching. **b** uniform military step (*slow march*). **2** long difficult walk. **3** procession as a demonstration. **4** (usu. foll. by *of*) progress or continuity (*march of events*). **5 a** music to accompany a march. **b** similar musical piece. □ **marcher** *n.* [French *marcher*]

march[2] *–n. hist.* **1** (usu. in *pl.*) boundary, frontier (esp. between England and Scotland or Wales). **2** tract of land between two countries, esp. disputed. *–v.* (foll. by *upon, with*) (of a country, an estate, etc.) border on. [French *marche* from medieval Latin *marca*]

March hare *n.* hare exuberant in the breeding season (*mad as a March hare*).

marching orders *n.pl.* **1** order for troops to mobilize etc. **2** dismissal (*gave him his marching orders*).

marchioness /ˌmɑːʃəˈnes/ *n.* **1** wife or widow of a marquess. **2** woman holding the rank of marquess. [medieval Latin: related to MARCH[2]]

march past *–n.* marching of troops past a saluting-point at a review. *–v.* (of troops) carry out a march past.

Mardi Gras /ˌmɑːdɪ ˈɡrɑː/ *n.* **1 a** Shrove Tuesday in some Catholic countries. **b** merrymaking on this day. **2** last day of a carnival etc. [French, = fat Tuesday]

mare[1] *n.* female equine animal, esp. a horse. [Old English]

mare[2] /ˈmɑːreɪ/ *n.* (*pl.* **maria** /ˈmɑːrɪə/ or **-s**) **1** large dark flat area on the moon, once thought to be sea. **2** similar area on Mars. [Latin, = sea]

mare's nest *n.* illusory discovery.

mare's tail *n.* **1** tall slender marsh plant. **2** (in *pl.*) long straight streaks of cirrus cloud.

margarine /ˌmɑːdʒəˈriːn/ *n.* butter-substitute made from vegetable oils or animal fats with milk etc. [Greek *margaron* pearl]

marge *n. colloq.* margarine. [abbreviation]

margin /ˈmɑːdʒɪn/ *–n.* **1** edge or border of a surface. **2** blank border flanking print etc. **3** amount by which a thing exceeds, falls short, etc. (*won by a narrow margin*). **4** lower limit (*his effort fell below the margin*). *–v.* (**-n-**) provide with a margin or marginal notes. [Latin *margo -ginis*]

marginal *adj.* **1** of or written in a margin. **2 a** of or at the edge. **b** insignificant (*of merely marginal interest*). **3** (of a parliamentary seat etc.) held by a small majority. **4** close to the limit, esp. of profitability. **5** (of land) difficult to cultivate; unprofitable. **6** barely adequate. □ **marginally** *adv.* [medieval Latin: related to MARGIN]

marginal cost *n.* cost added by making one extra copy etc.

marginalia /ˌmɑːdʒɪˈneɪlɪə/ *n.pl.* marginal notes.

marginalize /ˈmɑːdʒɪnəˌlaɪz/ *v.* (also **-ise**) (**-zing** or **-sing**) make or treat as insignificant. □ **marginalization** /-ˈzeɪʃ(ə)n/ *n.*

margin of error *n.* allowance for miscalculation etc.

marguerite /ˌmɑːɡəˈriːt/ *n.* ox-eye daisy. [Latin *margarita* pearl]

maria *pl.* of MARE[2].

marigold /ˈmærɪˌɡəʊld/ *n.* plant with golden or bright yellow flowers. [*Mary* (probably the Virgin), *gold* (dial.) marigold]

marijuana /ˌmærɪˈhwɑːnə/ *n.* (also **marihuana**) dried leaves etc. of hemp, smoked in cigarettes as a drug. [American Spanish]

marimba /məˈrɪmbə/ *n.* **1** xylophone played by natives of Africa and Central America. **2** modern orchestral instrument derived from this. [Congolese]

marina /məˈriːnə/ *n.* harbour for pleasure-yachts etc. [Latin: related to MARINE]

marinade /ˌmærɪˈneɪd, ˈmæ-/ *–n.* **1** mixture of wine, vinegar, oil, spices, etc., for soaking meat, fish, etc. before cooking. **2** meat, fish, etc., so soaked. *–v.* (**-ding**) soak in a marinade. [Spanish *marinar* pickle in brine: related to MARINE]

marinate /ˈmærɪˌneɪt/ *v.* (**-ting**) = MARINADE. □ **marination** /-ˈneɪʃ(ə)n/ *n.* [French: related to MARINE]

marine /məˈriːn/ *–adj.* **1** of, found in, or produced by the sea. **2 a** of shipping or naval matters (*marine insurance*). **b** for use at sea. *–n.* **1** soldier trained to serve on land or sea. **2** country's shipping, fleet, or navy (*merchant marine*). [Latin *mare* sea]

mariner /ˈmærɪnə(r)/ *n.* seaman.

marionette /ˌmærɪəˈnet/ *n.* puppet worked by strings. [French: related to *Mary*]

marital /ˈmærɪt(ə)l/ *adj.* of marriage or marriage relations. [Latin *maritus* husband]

maritime /ˈmærɪˌtaɪm/ *adj.* **1** connected with the sea or seafaring (*maritime insurance*). **2** living or found near the sea. [Latin: related to MARINE]

marjoram /'mɑːdʒərəm/ *n.* aromatic herb used in cookery. [French from medieval Latin]

mark[1] *–n.* **1** spot, sign, stain, scar, etc., on a surface etc. **2** (esp. in *comb.*) **a** written or printed symbol (*question mark*). **b** number or letter denoting proficiency, conduct, etc. (*black mark*; *46 marks out of 50*). **3** (usu. foll. by *of*) sign of quality, character, feeling, etc. (*mark of respect*). **4 a** sign, seal, etc., of identification. **b** cross etc. made as a signature by an illiterate person. **5** lasting effect (*war left its mark*). **6 a** target etc. (*missed the mark*). **b** standard, norm (*his work falls below the mark*). **7** line etc. indicating a position. **8** (usu. **Mark**) (followed by a numeral) particular design, model, etc., of a car, aircraft, etc. (*Mark 2 Ford Granada*). **9** runner's starting-point in a race. *–v.* **1 a** make a mark on. **b** mark with initials, name, etc. to identify etc. **2** correct and assess (a student's work etc.). **3** attach a price to (*marked the doll at £5*). **4** notice or observe (*marked his agitation*). **5 a** characterize (*day was marked by storms*). **b** acknowledge, celebrate (*marked the occasion with a toast*). **6** name or indicate on a map etc. (*the pub isn't marked*). **7** keep close to (an opponent in sport) to hinder him or her. **8** (as **marked** *adj.*) have natural marks (*is marked with dark spots*). □ **beside** (or **off** or **wide of**) **the mark 1** irrelevant. **2** not accurate. **make one's mark** attain distinction; make an impression. **one's mark** *colloq.* opponent, object, etc., of one's own size etc. (*the little one's more my mark*). **mark down 1** reduce the price of (goods etc.). **2** make a written note of. **3** reduce the examination marks of. **mark off** separate by a boundary etc. **mark out 1** plan (a course of action etc.). **2** destine (*marked out for success*). **3** trace out (boundaries etc.). **mark time 1** march on the spot without moving forward. **2** act routinely while awaiting an opportunity to advance. **mark up 1** add a proportion to the price of (goods etc.) for profit. **2** mark or correct (text etc.). **off the mark 1** having made a start. **2** = *beside the mark*. **on the mark** ready to start. **on your mark** (or **marks**) get ready to start (esp. a race). **up to the mark** normal (esp. of health). [Old English]

mark[2] *n.* = DEUTSCHMARK. [German]

mark-down *n.* reduction in price.

marked *adj.* **1** having a visible mark. **2** clearly noticeable (*marked difference*). **3** (of playing-cards) marked on their backs to assist cheating. □ **markedly** /-kɪdlɪ/ *adv.*

marked man *n.* person singled out, esp. for attack.

marker *n.* **1** thing marking a position etc. **2** person or thing that marks. **3** broad-tipped felt-tipped pen. **4** scorer in a game.

market /'mɑːkɪt/ *–n.* **1** gathering of buyers and sellers of provisions, livestock, etc. **2** space for this. **3** (often foll. by *for*) demand for a commodity etc. (*no market for sheds*). **4** place or group providing such a demand. **5** conditions etc. for buying or selling; rate of purchase and sale (*market is sluggish*). **6** = STOCK MARKET. *–v.* **(-t-) 1** offer for sale, esp. by advertising etc. **2** *archaic* buy or sell goods in a market. □ **be in the market for** wish to buy. **be on the market** be offered for sale. **put on the market** offer for sale. □ **marketer** *n.* **marketing** *n.* [Latin *mercor* buy]

marketable *adj.* able or fit to be sold. □ **marketability** /-'bɪlɪtɪ/ *n.*

market-day *n.* day on which a market is regularly held.

marketeer /ˌmɑːkɪ'tɪə(r)/ *n.* **1** supporter of the EC and British membership of it. **2** marketer.

market garden *n.* farm where vegetables and fruit are grown for sale in markets.

market-place *n.* **1** open space for a market. **2** commercial world.

market price *n.* price in current dealings.

market research *n.* surveying of consumers' needs and preferences.

market town *n.* town where a market is held.

market value *n.* value if offered for sale.

marking *n.* (usu. in *pl.*) **1** identification mark. **2** colouring of an animal's fur etc.

marksman *n.* skilled shot, esp. with a pistol or rifle. □ **marksmanship** *n.*

mark-up *n.* **1** amount added to a price by the retailer for profit. **2** corrections in a text.

marl *–n.* soil of clay and lime, used as fertilizer. *–v.* apply marl to. □ **marly** *adj.* [medieval Latin *margila*]

marlin /'mɑːlɪn/ *n.* (*pl.* same or **-s**) *US* long-nosed marine fish. [from MARLINSPIKE]

marlinspike *n.* pointed iron tool used to separate strands of rope etc. [*marling* from Dutch *marlen* from *marren* bind]

marmalade /'mɑːməˌleɪd/ *n.* preserve of citrus fruit, usu. oranges. [Portuguese *marmelo* quince]

Marmite /'mɑːmaɪt/ *n. propr.* thick brown spread made from yeast and vegetable extract. [French, = cooking-pot]

marmoreal /mɑː'mɔːrɪəl/ *adj.* of or like marble. [Latin: related to MARBLE]

marmoset /'mɑːmə,zet/ *n.* small monkey with a long bushy tail. [French]

marmot /'mɑːmət/ *n.* heavy-set burrowing rodent with a short bushy tail. [Latin *mus* mouse, *mons* mountain]

marocain /'mærə,keɪn/ *n.* fabric of ribbed crêpe. [French, = Moroccan]

maroon[1] /mə'ruːn/ *adj.* & *n.* brownish-crimson. [French *marron* chestnut]

maroon[2] /mə'ruːn/ *v.* **1** leave (a person) isolated, esp. on an island. **2** (of weather etc.) cause (a person) to be forcibly detained. [French *marron* wild person, from Spanish *cimarrón*]

marque /mɑːk/ *n.* make of car, as distinct from a specific model (*the Jaguar marque*). [French, = MARK[1]]

marquee /mɑː'kiː/ *n.* large tent for social functions etc. [French *marquise*]

marquess /'mɑːkwɪs/ *n.* British nobleman ranking between duke and earl. [var. of MARQUIS]

marquetry /'mɑːkɪtrɪ/ *n.* inlaid work in wood, ivory, etc. [French: related to MARQUE]

marquis /'mɑːkwɪs/ *n.* (*pl.* **-quises**) foreign nobleman ranking between duke and count. [French: related to MARCH[2]]

marquise /mɑː'kiːz/ *n.* **1** wife or widow of a marquis. **2** woman holding the rank of marquis.

marram /'mærəm/ *n.* shore grass that binds sand. [Old Norse, = sea-haulm]

marriage /'mærɪdʒ/ *n.* **1** legal union of a man and a woman for cohabitation and often procreation. **2** act or ceremony marking this. **3** particular such union (*a happy marriage*). **4** intimate union, combination. [French *marier* MARRY]

marriageable *adj.* free, ready, or fit for marriage. □ **marriageability** /-'bɪlɪtɪ/ *n.*

marriage bureau *n.* company arranging introductions with a view to marriage.

marriage certificate *n.* certificate verifying a legal marriage.

marriage guidance *n.* counselling of people with marital problems.

marriage licence *n.* licence to marry.

marriage lines *n.pl.* marriage certificate.

marriage of convenience *n.* loveless marriage for gain.

marriage settlement *n.* legal property arrangement between spouses.

married –*adj.* **1** united in marriage. **2** of marriage (*married name*; *married life*). –*n.* (usu. in *pl.*) married person (*young marrieds*).

marron glacé /,mærɒn 'glæseɪ/ *n.* (*pl.* **marrons glacés** pronunc. same) chestnut preserved in syrup. [French]

marrow /'mærəʊ/ *n.* **1** large fleshy usu. striped gourd eaten as a vegetable. **2** soft fatty substance in the cavities of bones. **3** essential part. □ **to the marrow** right through. [Old English]

marrowbone *n.* bone containing edible marrow.

marrowfat *n.* a kind of large pea.

marry /'mærɪ/ *v.* (**-ies, -ied**) **1** take, join, or give in marriage. **2 a** enter into marriage. **b** (foll. by *into*) become a member of (a family) by marriage. **3 a** unite intimately, combine. **b** pair (socks etc.). □ **marry off** find a spouse for. **marry up** link, join. [Latin *maritus* husband]

Marsala /mɑː'sɑːlə/ *n.* a dark sweet fortified dessert wine. [*Marsala* in Sicily]

Marseillaise /,mɑːseɪ'jeɪz/ *n.* French national anthem. [French *Marseille* in France]

marsh *n.* (often *attrib.*) low watery land. □ **marshy** *adj.* (**-ier, -iest**). **marshiness** *n.* [Old English]

marshal /'mɑːʃ(ə)l/ –*n.* **1** (**Marshal**) high-ranking officer of state or in the armed forces (*Earl Marshal*; *Field Marshal*). **2** officer arranging ceremonies, controlling racecourses, crowds, etc. –*v.* (**-ll-**) **1** arrange (soldiers, one's thoughts, etc.) in due order. **2** conduct (a person) ceremoniously. [French *mareschal*]

marshalling yard *n.* yard for assembling goods trains etc.

Marshal of the Royal Air Force *n.* highest rank in the RAF.

marsh gas *n.* methane.

marshland *n.* land consisting of marshes.

marshmallow /mɑːʃ'mæləʊ/ *n.* soft sticky sweet made of sugar, albumen, gelatin, etc. [MARSH MALLOW]

marsh mallow *n.* shrubby herbaceous plant.

marsh marigold *n.* golden-flowered plant.

marsupial /mɑː'suːpɪəl/ –*n.* mammal giving birth to underdeveloped young subsequently carried in a pouch. –*adj.* of or like a marsupial. [Greek *marsupion* pouch]

mart *n.* **1** trade centre. **2** auction-room. **3** market. [Dutch: related to MARKET]

Martello /mɑː'teləʊ/ *n.* (*pl.* **-s**) (in full **Martello tower**) small circular coastal fort. [Cape *Mortella* in Corsica]

marten /'mɑːtɪn/ *n.* weasel-like carnivore with valuable fur. [Dutch from French]

martial /ˈmɑːʃ(ə)l/ *adj.* **1** of warfare. **2** warlike. [Latin *martialis* of Mars]

martial arts *n.pl.* oriental fighting sports such as judo and karate.

martial law *n.* military government with ordinary law suspended.

Martian /ˈmɑːʃ(ə)n/ *–adj.* of the planet Mars. *–n.* hypothetical inhabitant of Mars. [Latin]

martin /ˈmɑːtɪn/ *n.* a kind of swallow, esp. the house-martin and sand-martin. [probably St *Martin*, name of a 4th-c. bishop]

martinet /ˌmɑːtɪˈnet/ *n.* strict disciplinarian. [*Martinet*, name of a drill-master]

martingale /ˈmɑːtɪŋˌgeɪl/ *n.* strap(s) preventing a horse from rearing etc. [French, origin uncertain]

Martini /mɑːˈtiːnɪ/ *n.* (*pl.* **-s**) **1** *propr.* type of vermouth. **2** cocktail of gin and French vermouth. [*Martini* and Rossi, name of a firm selling vermouth]

Martinmas /ˈmɑːtɪnməs/ *n.* St Martin's day, 11 Nov. [from MASS[2]]

martyr /ˈmɑːtə(r)/ *–n.* **1 a** person killed for persisting in a belief. **b** person who suffers for a cause etc. **c** person who suffers or pretends to suffer to get pity etc. **2** (foll. by *to*) *colloq.* constant sufferer from (an ailment). *–v.* **1** put to death as a martyr. **2** torment. □ **martyrdom** *n.* [Greek *martur* witness]

marvel /ˈmɑːv(ə)l/ *–n.* **1** wonderful thing. **2** (foll. by *of*) wonderful example of (a quality). *–v.* (**-ll-**; *US* **-l-**) (foll. by *at* or *that*) feel surprise or wonder. [Latin *miror* wonder at]

marvellous /ˈmɑːvələs/ *adj.* (*US* **marvelous**) **1** astonishing. **2** excellent. □ **marvellously** *adv.* [French: related to MARVEL]

Marxism /ˈmɑːksɪz(ə)m/ *n.* political and economic theories of Marx, predicting the overthrow of capitalism and common ownership of the means of production in a classless society. □ **Marxist** *n.* & *adj.*

Marxism-Leninism *n.* Marxism as developed by Lenin. □ **Marxist-Leninist** *n.* & *adj.*

marzipan /ˈmɑːzɪˌpæn/ *–n.* paste of ground almonds, sugar, etc., used in confectionery. *–v.* (**-nn-**) cover with marzipan. [German from Italian]

mascara /mæˈskɑːrə/ *n.* cosmetic for darkening the eyelashes. [Italian, = mask]

mascot /ˈmæskɒt/ *n.* person, animal, or thing supposed to bring luck. [Provençal *masco* witch]

masculine /ˈmæskjʊlɪn/ *–adj.* **1** of men. **2** having manly qualities. **3** of or denoting the male gender. *–n.* masculine gender or word. □ **masculinity** /-ˈlɪnɪtɪ/ *n.* [Latin: related to MALE]

maser /ˈmeɪzə(r)/ *n.* device used to amplify or generate coherent electromagnetic radiation in the microwave range. [*m*icrowave *a*mplification by *s*timulated *e*mission of *r*adiation]

mash *–n.* **1** soft or confused mixture. **2** mixture of boiled grain, bran, etc., fed to horses etc. **3** *colloq.* mashed potatoes. **4** mixture of malt and hot water used in brewing. **5** soft pulp made by crushing, mixing with water, etc. *–v.* **1** crush (potatoes etc.) to a pulp. **2** *dial.* **a** infuse (tea). **b** (of tea) draw. □ **masher** *n.* [Old English]

mask /mɑːsk/ *–n.* **1** covering for all or part of the face as a disguise or for protection against infection etc. **2** respirator. **3** likeness of a person's face, esp. one from a mould (*death-mask*). **4** disguise, pretence (*throw off the mask*). *–v.* **1** cover with a mask. **2** conceal. **3** protect. [Arabic *maskara* buffoon]

masking tape *n.* adhesive tape used in decorating to protect areas where paint is not wanted.

masochism /ˈmæsəˌkɪz(ə)m/ *n.* **1** sexual perversion involving one's own pain or humiliation. **2** *colloq.* enjoyment of what appears to be painful or tiresome. □ **masochist** *n.* **masochistic** /-ˈkɪstɪk/ *adj.* **masochistically** /-ˈkɪstɪkəlɪ/ *adv.* [von Sacher-*Masoch*, name of a novelist]

mason /ˈmeɪs(ə)n/ *n.* **1** person who builds with stone. **2** (**Mason**) Freemason. [French]

Masonic /məˈsɒnɪk/ *adj.* of Freemasons.

masonry *n.* **1 a** stonework. **b** work of a mason. **2** (**Masonry**) Freemasonry.

masque /mɑːsk/ *n.* musical drama with mime, esp. in the 16th and 17th c. [var. of MASK]

masquerade /ˌmæskəˈreɪd/ *–n.* **1** false show, pretence. **2** masked ball. *–v.* (**-ding**) (often foll. by *as*) appear falsely or in disguise. [Spanish *máscara* mask]

mass[1] *–n.* **1** shapeless body of matter. **2** dense aggregation of objects (*mass of fibres*). **3** (in *sing.* or *pl.*; usu. foll. by *of*) large number or amount. **4** (usu. foll. by *of*) unbroken expanse (of colour etc.). **5** (prec. by *the*) **a** the majority. **b** (in *pl.*) ordinary people. **6** *Physics* quantity of matter a body contains. **7** (*attrib.*) on a large scale (*mass hysteria*; *mass audience*). *–v.* assemble into a mass or as one body. [Latin *massa* from Greek]

mass[2] *n.* (often **Mass**) **1** Eucharist, esp. in the Roman Catholic Church. **2** celebration of this. **3** liturgy used in this. **4** musical setting of parts of this. [Latin *missa* dismissal]

massacre /'mæsəkə(r)/ *–n.* **1** mass killing. **2** utter defeat or destruction. *–v.* (**-ring**) **1** kill (esp. many people) cruelly or violently. **2** *colloq.* defeat heavily. [French]

massage /'mæsɑːʒ/ *–n.* rubbing and kneading of the muscles and joints with the hands, to relieve stiffness, cure strains, stimulate, etc. *–v.* (**-ging**) **1** apply massage to. **2** manipulate (statistics etc.) to give an acceptable result. **3** flatter (a person's ego etc.). [French]

massage parlour *n.* **1** establishment providing massage. **2** *euphem.* brothel.

masseur /mæ'sɜː(r)/ *n.* (*fem.* **masseuse** /mæ'sɜːz/) person who gives massage for a living. [French: related to MASSAGE]

massif /'mæsiːf/ *n.* compact group of mountain heights. [French: related to MASSIVE]

massive /'mæsɪv/ *adj.* **1** large and heavy or solid. **2** (of the features, head, etc.) relatively large or solid. **3** exceptionally large or severe (*massive heart attack*). **4** substantial, impressive. □ **massively** *adv.* **massiveness** *n.* [Latin: related to MASS[1]]

mass media *n.pl.* = MEDIA 2.

mass noun *n. Gram.* noun that is not normally countable and cannot be used with the indefinite article (e.g. *bread*).

mass production *n.* mechanical production of large quantities of a standardized article. □ **mass-produce** *v.*

mast[1] /mɑːst/ *n.* **1** long upright post of timber etc. on a ship's keel to support sails. **2** post etc. for supporting a radio or television aerial. **3** flag-pole (*half-mast*). □ **before the mast** as an ordinary seaman. □ **masted** *adj.* (also in *comb.*). **master** *n.* (also in *comb.*). [Old English]

mast[2] /mɑːst/ *n.* fruit of the beech, oak, etc., esp. as food for pigs. [Old English]

mastectomy /mæs'tektəmɪ/ *n.* (*pl.* **-ies**) surgical removal of a breast. [Greek *mastos* breast]

master /'mɑːstə(r)/ *–n.* **1** person having control or ownership (*master of the house*; *dog obeyed his master*; *master of the hunt*). **2** captain of a merchant ship. **3** male teacher. **4** prevailing person. **5 a** skilled tradesman able to teach others (often *attrib.*: *master carpenter*). **b** skilled practitioner (*master of innuendo*). **6** holder of a usu. post-graduate university degree (*Master of Arts*). **7** revered teacher in philosophy etc. **8** great artist. **9** *Chess* etc. player at international level. **10** original copy of a film, recording, etc., from which others can be made. **11** (**Master**) title for a boy not old enough to be called *Mr.* **12** *archaic* employer. *–attrib. adj.* **1** commanding, superior (*master hand*). **2** main, principal (*master bedroom*). **3** controlling others (*master plan*). *–v.* **1** overcome, defeat. **2** gain full knowledge of or skill in. [Latin *magister*]

master-class *n.* class given by a famous musician etc.

masterful *adj.* **1** imperious, domineering. **2** masterly. □ **masterfully** *adv.*

■ **Usage** *Masterful* is normally used of a person, whereas *masterly* is used of achievements, abilities, etc.

master-key *n.* key that opens several different locks.

masterly *adj.* very skilful.

■ **Usage** See note at *masterful.*

mastermind *–n.* **1** person with an outstanding intellect. **2** person directing a scheme etc. *–v.* plan and direct (a scheme etc.).

Master of Ceremonies *n.* **1** person introducing speakers at a banquet or entertainers in a variety show. **2** person in charge of a ceremonial or social occasion.

Master of the Rolls *n.* judge who presides over the Court of Appeal.

masterpiece *n.* **1** outstanding piece of artistry or workmanship. **2** person's best work.

master-stroke *n.* skilful tactic etc.

master-switch *n.* switch controlling the supply of electricity etc. to an entire system.

mastery *n.* **1** control, dominance. **2** (often foll. by *of*) comprehensive knowledge or skill.

masthead *n.* **1** top of a ship's mast, esp. as a place of observation or punishment. **2** title of a newspaper etc. at the head of the front page or editorial page.

mastic /'mæstɪk/ *n.* **1** gum or resin from the mastic tree, used in making varnish. **2** (in full **mastic tree**) evergreen tree yielding this. **3** waterproof filler and sealant. [Greek *mastikhē*]

masticate /'mæstɪ,keɪt/ *v.* (**-ting**) grind or chew (food) with one's teeth. □ **mastication** /-'keɪʃ(ə)n/ *n.* **masticatory** *adj.* [Latin from Greek]

mastiff /'mæstɪf/ *n.* dog of a large strong breed with drooping ears. [Latin *mansuetus* tame]

mastitis /mæ'staɪtɪs/ *n.* inflammation of the breast or udder. [Greek *mastos* breast]

mastodon /'mæstə,dɒn/ *n.* (*pl.* same or **-s**) large extinct mammal resembling the elephant. [Greek *mastos* breast, *odous* tooth]

mastoid /'mæstɔɪd/ *–adj.* shaped like a breast. *–n.* **1** = MASTOID PROCESS. **2** (usu. in *pl.*) *colloq.* inflammation of the mastoid process. [Greek *mastos* breast]

mastoid process *n.* conical prominence on the temporal bone behind the ear.

masturbate /'mæstə,beɪt/ *v.* (**-ting**) (usu. *absol.*) sexually arouse (oneself or another) by manual stimulation of the genitals. □ **masturbation** /-'beɪʃ(ə)n/ *n.* [Latin]

mat[1] *–n.* **1** small piece of coarse material on a floor, esp. for wiping one's shoes on. **2** piece of cork, rubber, etc., to protect a surface from a hot dish etc. placed on it. **3** padded floor covering in gymnastics, wrestling, etc. *–v.* (**-tt-**) (esp. as **matted** *adj.*) entangle or become entangled in a thick mass (*matted hair*). □ **on the mat** *slang* being reprimanded. [Old English]

mat[2] var. of MATT.

matador /'mætə,dɔ:(r)/ *n.* bullfighter whose task is to kill the bull. [Spanish from *matar* kill: related to *mate* in CHECKMATE]

match[1] *–n.* **1** contest or game in which players or teams compete. **2 a** person as an equal contender (*meet one's match*). **b** person or thing exactly like or corresponding to another. **3** marriage. **4** person viewed as a marriage prospect. *–v.* **1** correspond (to); be like or alike; harmonize (with) (*his socks do not match*; *curtains match the wallpaper*). **2** equal. **3** (foll. by *against*, *with*) place in conflict or competition with. **4** find material etc. that matches (another) (*can you match this silk?*). **5** find a person or thing suitable for another. □ **match up** (often foll. by *with*) fit to form a whole; tally. **match up to** be as good as or equal to. [Old English]

match[2] *n.* **1** short thin piece of wood etc. with a combustible tip. **2** wick or cord etc. for firing a cannon etc. [French *mesche*]

matchboard *n.* tongued and grooved board fitting with similar boards.

matchbox *n.* box for holding matches.

matchless *adj.* incomparable.

matchmaker *n.* person who arranges marriages or schemes to bring couples together. □ **matchmaking** *n.*

match point *n.* *Tennis* etc. **1** position when one side needs only one more point to win the match. **2** this point.

matchstick *n.* stem of a match.

matchwood *n.* **1** wood suitable for matches. **2** minute splinters.

mate[1] *–n.* **1** friend or fellow worker. **2** *colloq.* form of address, esp. to another man. **3 a** each of a breeding pair, esp. of birds. **b** *colloq.* partner in marriage. **c** (in *comb.*) fellow member or joint occupant of (*team-mate*; *room-mate*). **4** officer on a merchant ship. **5** assistant to a skilled worker (*plumber's mate*). *–v.* (**-ting**) (often foll. by *with*) **1** come or bring together for breeding. **2** *Mech.* fit well. [Low German]

mate[2] *n.* & *v.* (**-ting**) *Chess* = CHECKMATE.

mater /'meɪtə(r)/ *n.* *slang* mother. [Latin]

■ **Usage** *Mater* is now only found in jocular or affected use.

---

material /mə'tɪərɪəl/ *–n.* **1** matter from which a thing is made. **2** cloth, fabric. **3** (in *pl.*) things needed for an activity (*building materials*). **4** person or thing of a specified kind or suitable for a purpose (*officer material*). **5** (in *sing.* or *pl.*) information etc. for a book etc. **6** (in *sing.* or *pl.*, often foll. by *of*) elements, constituent parts, or substance. *–adj.* **1** of matter; corporeal; not spiritual. **2** of bodily comfort etc. (*material well-being*). **3** (often foll. by *to*) important, significant, relevant. [Latin *materia* MATTER]

materialism *n.* **1** greater interest in material possessions and comfort than in spiritual values. **2** *Philos.* theory that nothing exists but matter. □ **materialist** *n.* **materialistic** /-'lɪstɪk/ *adj.* **materialistically** /-'lɪstɪkəlɪ/ *adv.*

materialize *v.* (also **-ise**) (**-zing** or **-sing**) **1** become actual fact; happen. **2** *colloq.* appear or be present. **3** represent in or assume bodily form. □ **materialization** /-'zeɪʃ(ə)n/ *n.*

materially *adv.* substantially, significantly.

matériel /mə,tɪərɪ'el/ *n.* means, esp. materials and equipment in warfare. [French]

maternal /mə'tɜ:n(ə)l/ *adj.* **1** of or like a mother; motherly. **2** related through the mother (*maternal uncle*). **3** of the mother in pregnancy and childbirth. □ **maternally** *adv.* [Latin *mater* mother]

maternity /mə'tɜ:nɪtɪ/ *n.* **1** motherhood. **2** motherliness. **3** (*attrib.*) for women during pregnancy and childbirth (*maternity leave*; *maternity dress*). [French from medieval Latin: related to MATERNAL]

matey (also **maty**) *–adj.* (**-tier**, **-tiest**) sociable; familiar, friendly. *–n.* (*pl.* **-s**) *colloq.* (as a form of address) mate. □ **mateyness** *n.* (also **matiness**). **matily** *adv.*

math *n. US colloq.* mathematics. [abbreviation]

mathematical /ˌmæθəˈmætɪk(ə)l/ *adj.* **1** of mathematics. **2** rigorously precise. □ **mathematically** *adv.*

mathematical tables *n.pl.* tables of logarithms and trigonometric values etc.

mathematics /ˌmæθəˈmætɪks/ *n.pl.* **1** (also treated as *sing.*) abstract science of number, quantity, and space. **2** (as *pl.*) use of this in calculation etc. □ **mathematician** /-məˈtɪʃ(ə)n/ *n.* [Greek *manthanō* learn]

maths *n. colloq.* mathematics. [abbreviation]

matinée /ˈmætɪˌneɪ/ *n.* (*US* **matinee**) afternoon performance in the theatre, cinema, etc. [French from *matin* morning: related to MATINS]

matinée coat *n.* (also **matinée jacket**) baby's short knitted coat.

matinée idol *n.* handsome actor.

matins /ˈmætɪnz/ *n.* (also **mattins**) (as *sing.* or *pl.*) morning prayer, esp. in the Church of England. [Latin *matutinus* of the morning]

matriarch /ˈmeɪtrɪˌɑːk/ *n.* female head of a family or tribe. □ **matriarchal** /-ˈɑːk(ə)l/ *adj.* [Latin *mater* mother]

matriarchy /ˈmeɪtrɪˌɑːkɪ/ *n.* (*pl.* **-ies**) female-dominated system of society, with descent through the female line.

matrices *pl.* of MATRIX.

matricide /ˈmeɪtrɪˌsaɪd, ˈmæ-/ *n.* **1** killing of one's mother. **2** person who does this. [Latin: related to MATER, -CIDE]

matriculate /məˈtrɪkjʊˌleɪt/ *v.* (**-ting**) enrol at a college or university. □ **matriculation** /-ˈleɪʃ(ə)n/ *n.* [medieval Latin: related to MATRIX]

matrimony /ˈmætrɪmənɪ/ *n.* rite or state of marriage. □ **matrimonial** /-ˈməʊnɪəl/ *adj.* [Latin *matrimonium*: related to MATER]

matrix /ˈmeɪtrɪks/ *n.* (*pl.* **matrices** /-ˌsiːz/ or **-es**) **1** mould in which a thing is cast or shaped. **2** place etc. in which a thing is developed. **3** rock in which gems, fossils, etc., are embedded. **4** *Math.* rectangular array of elements treated as a single element. [Latin, = womb]

matron /ˈmeɪtrən/ *n.* **1** woman in charge of nursing in a hospital. **2** married, esp. staid, woman. **3** woman nurse and housekeeper at a school etc. [Latin *matrona*: related to MATER]

■ **Usage** In sense 1, *senior nursing officer* is now the official term.

matronly *adj.* like a matron, esp. portly or staid.

matron of honour *n.* married woman attending the bride at a wedding.

matt (also **mat**) *–adj.* not shiny or glossy; dull. *–n.* (in full **matt paint**) paint giving a dull flat finish. [French: related to MATE[2]]

matter *–n.* **1** physical substance having mass and occupying space, as distinct from mind and spirit. **2** specified substance (*colouring matter*; *reading matter*). **3** (prec. by *the*; often foll. by *with*) (thing) amiss (*something the matter with him*). **4** content as distinct from style, form, etc. **5** (often foll. by *of*, *for*) situation etc. under consideration or as an occasion for (regret etc.) (*matter for concern*; *matter of discipline*). **6** pus or a similar substance discharged from the body. *–v.* (often foll. by *to*) be of importance; have significance. □ **as a matter of fact** in reality; actually. **for that matter 1** as far as that is concerned. **2** and indeed also. **a matter of** approximately; amounting to (*a matter of 40 years*). **no matter 1** (foll. by *when*, *how*, etc.) regardless of. **2** it is of no importance. [Latin *materia* timber, substance]

matter of course *n.* natural or expected thing.

matter-of-fact *adj.* **1** unimaginative, prosaic. **2** unemotional. □ **matter-of-factly** *adv.* **matter-of-factness** *n.*

matter of life and death *n.* matter of vital importance.

matting *n.* fabric for mats.

mattins var. of MATINS.

mattock /ˈmætək/ *n.* agricultural tool like a pickaxe, with an adze and a chisel edge. [Old English]

mattress /ˈmætrɪs/ *n.* stuffed, or air- or water-filled cushion the size of a bed. [Arabic *almaṭraḥ*]

maturate /ˈmætjʊˌreɪt/ *v.* (**-ting**) (of a boil etc.) come to maturation. [Latin: related to MATURE]

maturation /ˌmætjʊˈreɪʃ(ə)n/ *n.* **1** maturing or being matured. **2** formation of pus. [French or medieval Latin: related to MATURE]

mature /məˈtʃʊə(r)/ *–adj.* (**maturer, maturest**) **1 a** fully developed, adult. **b** sensible, wise. **2** ripe; seasoned. **3** (of thought etc.) careful, considered. **4** (of a bill, insurance policy, etc.) due, payable. *–v.* (**-ring**) **1** develop fully; ripen. **2** perfect (a plan etc.). **3** (of a bill, insurance policy, etc.) become due or payable. □ **maturely** *adv.* **matureness** *n.* **maturity** *n.* [Latin *maturus* timely]

mature student *n.* adult student.

matutinal /ˌmætjuːˈtaɪn(ə)l, məˈtjuːtɪn(ə)l/ *adj.* of the morning; early. [Latin: related to MATINS]

maty var. of MATEY.

maudlin /'mɔːdlɪn/ *adj.* weakly or tearfully sentimental, esp. from drunkenness. [French *Madeleine*, referring to pictures of Mary Magdalen weeping]

maul /mɔːl/ —*v.* **1** tear the flesh of; claw. **2** handle roughly. **3** damage by criticism. —*n.* **1** *Rugby* loose scrum. **2** brawl. **3** heavy hammer. [Latin *malleus* hammer]

maulstick /'mɔːlstɪk/ *n.* (also **mahlstick**) stick held to support the hand in painting. [Dutch *malen* paint]

maunder /'mɔːndə(r)/ *v.* **1** talk ramblingly. **2** move or act listlessly or idly. [origin unknown]

Maundy /'mɔːndɪ/ *n.* distribution of Maundy money. [French *mandé* from Latin *mandatum* command]

Maundy money *n.* specially minted silver coins distributed by the British sovereign on Maundy Thursday.

Maundy Thursday *n.* Thursday before Easter.

mausoleum /,mɔːsə'liːəm/ *n.* magnificent tomb. [from *Mausōlos*, king of Caria, whose tomb had this name]

mauve /məʊv/ —*adj.* pale purple. —*n.* this colour. □ **mauvish** *adj.* [Latin: related to MALLOW]

maverick /'mævərɪk/ *n.* **1** unorthodox or independent-minded person. **2** *US* unbranded calf or yearling. [*Maverick*, name of an owner of unbranded cattle]

maw *n.* **1** stomach of an animal or *colloq.* greedy person. **2** jaws or throat of a voracious animal. [Old English]

mawkish /'mɔːkɪʃ/ *adj.* feebly sentimental; sickly. □ **mawkishly** *adv.* **mawkishness** *n.* [obsolete *mawk* MAGGOT]

max. *abbr.* maximum.

maxi /'mæksɪ/ *n.* (*pl.* **-s**) *colloq.* maxi-coat, -skirt, etc. [abbreviation]

maxi- *comb. form* very large or long. [abbreviation of MAXIMUM; cf. MINI-]

maxilla /mæk'sɪlə/ *n.* (*pl.* **-llae** /-liː/) jaw or jawbone, esp. (in vertebrates) the upper jaw. □ **maxillary** *adj.* [Latin]

maxim /'mæksɪm/ *n.* general truth or rule of conduct briefly expressed. [French or medieval Latin: related to MAXIMUM]

maxima *pl.* of MAXIMUM.

maximal /'mæksɪm(ə)l/ *adj.* of or being a maximum.

maximize /'mæksɪ,maɪz/ *v.* (also **-ise**) (**-zing** or **-sing**) make as large or great as possible. □ **maximization** /-'zeɪʃ(ə)n/ *n.* [Latin: related to MAXIMUM]

■ **Usage** *Maximize* should not be used in standard English to mean 'to make as good as possible' or 'to make the most of'.

maximum /'mæksɪməm/ —*n.* (*pl.* **-ma**) highest possible amount, size, etc. —*adj.* greatest in amount, size, etc. [Latin *maximus* greatest]

May *n.* **1** fifth month of the year. **2** (**may**) hawthorn, esp. in blossom. [Latin *Maius* of the goddess Maia]

may *v.aux.* (*3rd sing. present* **may**; *past* **might** /maɪt/) **1** expressing: **a** (often foll. by *well* for emphasis) possibility (*it may be true*; *you may well lose your way*). **b** permission (*may I come in?*). **c** a wish (*may he live to regret it*). **d** uncertainty or irony (*who may you be?*; *who are you, may I ask?*). **2** in purpose clauses and after *wish*, *fear*, etc. (*hope he may succeed*). □ **be that as it may** (or **that is as may be**) it is possible (but) (*be that as it may, I still want to go*). **may as well** = *might as well* (see MIGHT[1]). [Old English]

■ **Usage** In sense 1b, both *can* and *may* are used to express permission; in more formal contexts *may* is preferred since *can* also denotes capability (*can I move?* = am I physically able to move?; *may I move?* = am I allowed to move?).

Maya /'mɑːjə/ *n.* **1** (*pl.* same or **-s**) member of an ancient Indian people of Central America. **2** their language. □ **Mayan** *adj.* & *n.* [native name]

maybe /'meɪbiː/ *adv.* perhaps. [from *it may be*]

May Day *n.* 1 May as a Spring festival or international holiday in honour of workers.

mayday /'meɪdeɪ/ *n.* international radio distress-signal. [representing pronunciation of French *m'aidez* help me]

mayflower *n.* any of various flowers that bloom in May.

mayfly *n.* a kind of insect living briefly in spring.

mayhem /'meɪhem/ *n.* destruction; havoc. [Anglo-French *mahem*: related to MAIM]

mayn't /'meɪənt/ *contr.* may not.

mayonnaise /,meɪə'neɪz/ *n.* **1** thick creamy dressing of egg-yolks, oil, vinegar, etc. **2** dish dressed with this (*egg mayonnaise*). [French]

mayor /meə(r)/ *n.* **1** head of the corporation of a city or borough. **2** head of a district council with the status of a borough. □ **mayoral** *adj.* [Latin: related to MAJOR]

mayoralty /'meərəltɪ/ *n.* (*pl.* **-ies**) **1** office of mayor. **2** period of this.

mayoress /'meərɪs/ *n.* **1** woman mayor. **2** wife or official consort of a mayor.

maypole *n.* decorated pole for dancing round on May Day.

May queen *n.* girl chosen to preside over May Day festivities.
maze *n.* **1** network of paths and hedges designed as a puzzle for those who enter it. **2** labyrinth. **3** confused network, mass, etc. [related to AMAZE]
mazurka /mə'zɜ:kə/ *n.* **1** lively Polish dance in triple time. **2** music for this. [French or German from Polish]
MB *abbr.* **1** Bachelor of Medicine. **2** *Computing* megabyte. [sense 1 from Latin *Medicinae Baccalaureus*]
MBA *abbr.* Master of Business Administration.
MBE *abbr.* Member of the Order of the British Empire.
MBO *abbr.* management buyout.
MC *abbr.* **1** Master of Ceremonies. **2** Military Cross. **3** Member of Congress.
MCC *abbr.* Marylebone Cricket Club.
McCarthyism, McCoy see at MACC-.
MD *abbr.* **1** Doctor of Medicine. **2** Managing Director. [sense 1 from Latin *Medicinae Doctor*]
Md *symb.* mendelevium.
ME *abbr.* myalgic encephalomyelitis, a condition with prolonged flu-like symptoms and depression.
me[1] /mi:/ *pron.* **1** *objective case* of I[2] (*he saw me*). **2** *colloq.* = I[2] (*it's me all right; is taller than me*). [Old English accusative and dative of I[2]]
me[2] /mi:/ *n.* (also **mi**) *Mus.* third note of a major scale. [Latin *mira*, word arbitrarily taken]
mea culpa /ˌmeɪə 'kʊlpə/ *–n.* acknowledgement of error. *–int.* expressing this. [Latin, = by my fault]
mead *n.* alcoholic drink of fermented honey and water. [Old English]
meadow /'medəʊ/ *n.* **1** piece of grassland, esp. one used for hay. **2** low marshy ground, esp. near a river. □ **meadowy** *adj.* [Old English]
meadowsweet *n.* fragrant meadow and marsh plant with creamy-white flowers.
meagre /'mi:gə(r)/ *adj.* (*US* **meager**) **1** scant in amount or quality. **2** lean, thin. [Anglo-French *megre* from Latin *macer*]
meal[1] *n.* **1** occasion when food is eaten. **2** the food eaten at a meal. □ **make a meal of** *colloq.* treat (a task etc.) too laboriously or fussily. [Old English]
meal[2] *n.* **1** grain or pulse ground to powder. **2** *Scot.* oatmeal. **3** *US* maize flour. [Old English]
meals on wheels *n.pl.* (usu. treated as *sing.*) regular voluntary esp. lunch deliveries to old people, invalids, etc.
meal-ticket *n. colloq.* person or thing that is a source of maintenance or income.
mealtime *n.* usual time of eating.
mealy *adj.* (**-ier, -iest**) **1** of, like, or containing meal. **2** (of a complexion) pale. □ **mealiness** *n.*
mealy-mouthed *adj.* afraid to speak plainly.
mean[1] *v.* (*past* and *past part.* **meant** /ment/) **1** have as one's purpose or intention (*meant no harm by it; I didn't mean to break it*). **2** design or destine for a purpose (*meant to be used*). **3** intend to convey or refer to (*I mean Richmond in Surrey*). **4** (often foll. by *that*) entail, involve, portend, signify (*this means war; means that he is dead*). **5** (of a word) have as its equivalent in the same or another language. **6** (foll. by *to*) be of specified importance to (*that means a lot to me*). □ **mean business** *colloq.* be in earnest. **mean it** not be joking or exaggerating. **mean well** have good intentions. [Old English]
mean[2] *adj.* **1** niggardly; not generous. **2** ignoble, small-minded. **3** (of capacity, understanding, etc.) inferior, poor. **4** shabby; inadequate (*mean hovel*). **5 a** malicious, ill-tempered. **b** *US* vicious or aggressive in behaviour. **6** *US colloq.* skilful, formidable (*a mean fighter*). □ **no mean** a very good (*no mean feat*). □ **meanly** *adv.* **meanness** *n.* [Old English]
mean[3] *–n.* **1** median point (*mean between modesty and pride*). **2 a** term midway between the first and last terms of an arithmetical etc. progression. **b** quotient of the sum of several quantities and their number; average. *–adj.* **1** (of a quantity) equally far from two extremes. **2** calculated as a mean. [Latin *medianus* MEDIAN]
meander /mɪ'ændə(r)/ *–v.* **1** wander at random. **2** (of a stream) wind about. *–n.* **1** (in *pl.*) sinuous windings of a river, path, etc. **2** circuitous journey. [Greek *Maiandros*, a winding river in ancient Phrygia]
meanie /'mi:nɪ/ *n.* (also **meany**) (*pl.* **-ies**) *colloq.* niggardly or small-minded person.
meaning *–n.* **1** what is meant. **2** significance. **3** importance. *–adj.* expressive, significant (*meaning glance*). □ **meaningly** *adv.*
meaningful *adj.* **1** full of meaning; significant. **2** *Logic* able to be interpreted. □ **meaningfully** *adv.* **meaningfulness** *n.*
meaningless *adj.* having no meaning or significance. □ **meaninglessly** *adv.* **meaninglessness** *n.*
means *n.pl.* **1** (often treated as *sing.*) action, agent, device, or method producing a result (*means of quick travel*). **2 a** money resources (*live beyond one's*

means). **b** wealth (*man of means*). □ **by all means** certainly. **by means of** by the agency etc. of. **by no means** certainly not. [from MEAN[3]]

mean sea level *n.* level halfway between high and low water.

means test –*n.* inquiry into income as a basis for eligibility for State benefit etc. –*v.* (**means-test**) subject to or base on a means test.

meant *past* and *past part.* of MEAN[1].

meantime –*adv.* = MEANWHILE. –*n.* intervening period (esp. *in the meantime*).

■ **Usage** As an adverb, *meantime* is less common than *meanwhile*.

meanwhile –*adv.* **1** in the intervening period of time. **2** at the same time. –*n.* intervening period (esp. *in the meanwhile*).

meany var. of MEANIE.

measles /'mi:z(ə)lz/ *n.pl.* (also treated as *sing.*) infectious viral disease marked by a red rash. [Low German *masele* or Dutch *masel*]

measly /'mi:zlɪ/ *adj.* (**-ier**, **-iest**) *colloq.* meagre, contemptible.

measure /'meʒə(r)/ –*n.* **1** size or quantity found by measuring. **2** system or unit of measuring (*liquid measure*; *20 measures of wheat*). **3** rod, tape, vessel, etc. for measuring. **4** (often foll. by *of*) degree, extent, or amount (*a measure of wit*). **5** factor determining evaluation etc. (*sales are the measure of popularity*). **6** (usu. in *pl.*) suitable action to achieve some end. **7** legislative bill, act, etc. **8** prescribed extent or quantity. **9** poetic metre. **10** mineral stratum (*coal measures*). –*v.* (**-ring**) **1** ascertain the extent or quantity of (a thing) by comparison with a known standard. **2** be of a specified size. **3** ascertain the size of (a person) for clothes. **4** estimate (a quality etc.) by some criterion. **5** (often foll. by *off*) mark (a line etc. of a given length). **6** (foll. by *out*) distribute in measured quantities. **7** (foll. by *with*, *against*) bring (oneself or one's strength etc.) into competition with. □ **beyond measure** excessively. **for good measure** as a finishing touch. **in some measure** partly. **measure up 1** take the measurements (of). **2** (often foll. by *to*) have the qualifications (for). □ **measurable** *adj.* [Latin *mensura* from *metior* measure]

measured *adj.* **1** rhythmical; regular (*measured tread*). **2** (of language) carefully considered.

measureless *adj.* not measurable; infinite.

measurement *n.* **1** measuring. **2** amount measured. **3** (in *pl.*) detailed dimensions.

meat *n.* **1** animal flesh as food. **2** (often foll. by *of*) substance; chief part. □ **meatless** *adj.* [Old English]

meatball *n.* small round ball of minced meat.

meat loaf *n.* minced meat etc. moulded and baked.

meat safe *n.* ventilated cupboard for storing meat.

meaty *adj.* (**-ier**, **-iest**) **1** full of meat; fleshy. **2** of or like meat. **3** substantial, full of interest, satisfying. □ **meatiness** *n.*

Mecca /'mekə/ *n.* place one aspires to visit. [*Mecca*, Muslim holy city in Arabia]

mechanic /mɪ'kænɪk/ *n.* person skilled in using or repairing machinery. [Latin: related to MACHINE]

mechanical *adj.* **1** of machines or mechanisms. **2** working or produced by machinery. **3** (of an action etc.) automatic; repetitive. **4** (of an agency, principle, etc.) belonging to mechanics. **5** of mechanics as a science. □ **mechanically** *adv.* [Latin: related to MECHANIC]

mechanical engineer *n.* person qualified in the design, construction, etc. of machines.

mechanics *n.pl.* (usu. treated as *sing.*) **1** branch of applied mathematics dealing with motion etc. **2** science of machinery. **3** routine technical aspects of a thing (*mechanics of local government*).

mechanism /'mekə,nɪz(ə)m/ *n.* **1** structure or parts of a machine. **2** system of parts working together. **3** process; method (*defence mechanism*; *no mechanism for complaints*). □ **mechanistic** /-'nɪstɪk/ *adj.* [Greek: related to MACHINE]

mechanize /'mekə,naɪz/ *v.* (also **-ise**) (**-zing** or **-sing**) **1** introduce machines in (a factory etc.). **2** make mechanical. **3** equip with tanks, armoured cars, etc. □ **mechanization** /-'zeɪʃ(ə)n/ *n.*

Med *n. colloq.* Mediterranean Sea. [abbreviation]

medal /'med(ə)l/ *n.* commemorative metal disc etc., esp. awarded for military or sporting prowess. [Latin: related to METAL]

medallion /mɪ'dæljən/ *n.* **1** large medal. **2** thing so shaped, e.g. a decorative panel etc. [Italian: related to MEDAL]

medallist /'medəlɪst/ *n.* (*US* **medalist**) winner of a (specified) medal (*gold medallist*).

meddle /'med(ə)l/ *v.* (**-ling**) (often foll. by *with*, *in*) interfere in others' concerns. □ **meddler** *n.* [Latin: related to MIX]

meddlesome *adj.* interfering.

media /'mi:dɪə/ *n.pl.* **1** *pl.* of MEDIUM. **2** (usu. prec. by *the*) mass communications (esp. newspapers and broadcasting) regarded collectively.

■ **Usage** *Media* is commonly used with a singular verb (e.g. *the media is biased*), but this is not generally accepted (cf. DATA).

mediaeval var. of MEDIEVAL.

medial /'mi:dɪəl/ *adj.* = MEDIAN. □ **medially** *adv.* [Latin *medius* middle]

median /'mi:dɪən/ —*adj.* situated in the middle. —*n.* **1** straight line drawn from any vertex of a triangle to the middle of the opposite side. **2** middle value of a series. [Latin: related to MEDIAL]

mediate /'mi:dɪ,eɪt/ *v.* (**-ting**) **1** (often foll. by *between*) intervene (between disputants) to settle a quarrel etc. **2** bring about (a result) thus. □ **mediation** /-'eɪʃ(ə)n/ *n.* **mediator** *n.* [Latin *medius* middle]

medic /'medɪk/ *n. colloq.* medical practitioner or student. [Latin *medicus* physician]

medical /'medɪk(ə)l/ —*adj.* of medicine in general or as distinct from surgery (*medical ward*). —*n. colloq.* medical examination. □ **medically** *adv.*

medical certificate *n.* certificate of fitness or unfitness for work etc.

medical examination *n.* examination to determine a person's physical fitness.

medical officer *n.* person in charge of the health services of a local authority etc.

medical practitioner *n.* physician or surgeon.

medicament /mɪ'dɪkəmənt/ *n.* = MEDICINE 2.

Medicare /'medɪ,keə(r)/ *n.* US federally funded health insurance scheme for the elderly. [from MEDICAL, CARE]

medicate /'medɪ,keɪt/ *v.* (**-ting**) **1** treat medically. **2** impregnate with medicine etc. □ **medicative** /-kətɪv/ *adj.* [Latin *medicare medicat-*]

medication /,medɪ'keɪʃ(ə)n/ *n.* **1** = MEDICINE 2. **2** treatment using drugs.

medicinal /mɪ'dɪsɪn(ə)l/ *adj.* (of a substance) healing. □ **medicinally** *adv.*

medicine /'meds(ə)n/ *n.* **1** science or practice of the diagnosis, treatment, and prevention of disease, esp. as distinct from surgery. **2** drug etc. for the treatment or prevention of disease, esp. taken by mouth. □ **take one's medicine** submit to something disagreeable. [Latin *medicina*]

medicine man *n.* tribal, esp. N. American Indian, witch-doctor.

medieval /,medɪ'i:v(ə)l/ *adj.* (also **mediaeval**) **1** of the Middle Ages. **2** *colloq.* old-fashioned. [Latin *medium aevum* middle age]

medieval history *n.* history of the 5th–15th c.

medieval Latin *n.* Latin of about AD 600–1500.

mediocre /,mi:dɪ'əʊkə(r)/ *adj.* **1** indifferent in quality. **2** second-rate. [Latin *mediocris*]

mediocrity /,mi:dɪ'ɒkrɪtɪ/ *n.* (*pl.* **-ies**) **1** being mediocre. **2** mediocre person.

meditate /'medɪ,teɪt/ *v.* (**-ting**) **1** (often foll. by *on*, *upon*) engage in (esp. religious) contemplation. **2** plan mentally. □ **meditation** /-'teɪʃ(ə)n/ *n.* **meditator** *n.* [Latin *meditor*]

meditative /'medɪtətɪv/ *adj.* **1** inclined to meditate. **2** indicative of meditation, thoughtful. □ **meditatively** *adv.* **meditativeness** *n.*

Mediterranean /,medɪtə'reɪnɪən/ *adj.* of the sea bordered by S. Europe, SW Asia, and N. Africa, or its surrounding region (*Mediterranean cookery*). [Latin *mediterraneus* inland]

medium /'mi:dɪəm/ —*n.* (*pl.* **media** or **-s**) **1** middle quality, degree, etc. between extremes (*find a happy medium*). **2** means of communication (*medium of television*). **3** substance, e.g. air, through which sense-impressions are conveyed. **4** physical environment etc. of a living organism. **5** means. **6** material or form used by an artist, composer, etc. **7** liquid (e.g. oil or gel) used for diluting paints. **8** (*pl.* **-s**) person claiming to communicate with the dead. —*adj.* **1** between two qualities, degrees, etc. **2** average (*of medium height*). [Latin *medius* middle]

medium-range *adj.* (of an aircraft, missile, etc.) able to travel a medium distance.

medium wave *n.* radio wave of frequency between 300 kHz and 3 MHz.

medlar /'medlə(r)/ *n.* **1** tree bearing small brown apple-like fruits, eaten when decayed. **2** such a fruit. [French *medler* from Greek *mespilē*]

medley /'medlɪ/ *n.* (*pl.* **-s**) **1** varied mixture. **2** collection of tunes etc. played as one piece. [French *medlee*]

medulla /mɪ'dʌlə/ *n.* **1** inner part of certain organs etc., e.g. the kidney. **2** soft internal tissue of plants. □ **medullary** *adj.* [Latin]

medulla oblongata /ˌɒblɒŋˈɡɑːtə/ *n.* hindmost part of the brain, formed from a continuation of the spinal cord.

medusa /mɪˈdjuːsə/ *n.* (*pl.* **medusae** /-siː/ or **-s**) jellyfish. [Greek *Medousa*, name of a Gorgon]

meek *adj.* humble and submissive or gentle. □ **meekly** *adv.* **meekness** *n.* [Old Norse]

meerkat /ˈmɪəkæt/ *n.* S. African mongoose. [Dutch, = sea-cat]

meerschaum /ˈmɪəʃəm/ *n.* **1** soft white clay-like substance. **2** tobacco-pipe with its bowl made from this. [German, = sea-foam]

meet[1] –*v.* (*past* and *past part.* **met**) **1** encounter (a person etc.) or (of two or more people) come together by accident or design; come face to face (with) (*met on the bridge*). **2** be present by design at the arrival of (a person, train, etc.). **3** come or seem to come together or into contact (with); join (*where the sea and the sky meet; jacket won't meet*). **4** make the acquaintance of (*delighted to meet you; all met at Oxford*). **5** come together for business, worship, etc. (*union met management*). **6 a** deal with or answer (a demand, objection, etc.) (*met the proposal with hostility*). **b** satisfy or conform with (*agreed to meet the new terms*). **7** pay (a bill etc.); honour (a cheque) (*meet the cost*). **8** (often foll. by *with*) experience, encounter, or receive (*met their death; met with hostility*). **9** confront in battle etc. –*n.* **1** assembly for a hunt. **2** assembly for sport, esp. athletics. □ **make ends meet** see END. **meet the case** be adequate. **meet the eye** be visible or evident. **meet a person half way** compromise with. **meet up** *colloq.* (often foll. by *with*) = sense 1 of *v.* **meet with 1** see sense 8 of *v.* **2** receive (a reaction) (*met with her approval*). **3** esp. *US* = sense 1 of *v.* [Old English]

meet[2] *adj. archaic* fitting, proper. [related to METE]

meeting *n.* **1** coming together. **2** assembly of esp. a society, committee, etc. **3** = RACE MEETING.

mega /ˈmeɡə/ *slang* –*adj.* **1** excellent. **2** enormous. –*adv.* extremely.

mega- *comb. form* **1** large. **2** one million (10^6) in the metric system of measurement. **3** *slang* extremely; very big (*mega-stupid; mega-project*). [Greek *megas* great]

megabuck /ˈmeɡəˌbʌk/ *n. US slang* million dollars.

megabyte /ˈmeɡəˌbaɪt/ *n. Computing* 1,048,576 (i.e. 2^{20}) bytes as a measure of data capacity, or loosely 1,000,000.

megadeath /ˈmeɡəˌdeθ/ *n.* death of one million people (in war).

megahertz /ˈmeɡəˌhɜːts/ *n.* (*pl.* same) one million hertz, esp. as a measure of radio frequency.

megalith /ˈmeɡəlɪθ/ *n.* large stone, esp. as a prehistoric monument or part of one. □ **megalithic** /-ˈlɪθɪk/ *adj.* [Greek *lithos* stone]

megalomania /ˌmeɡələˈmeɪnɪə/ *n.* **1** mental disorder producing delusions of grandeur. **2** passion for grandiose schemes. □ **megalomaniac** *adj.* & *n.* [Greek *megas* great, MANIA]

megalosaurus /ˌmeɡələˈsɔːrəs/ *n.* (*pl.* **-ruses**) large flesh-eating dinosaur with stout hind legs and small forelimbs. [Greek *megas* great, *sauros* lizard]

megaphone /ˈmeɡəˌfəʊn/ *n.* large funnel-shaped device for amplifying the voice. [Greek *megas* great, *phōnē* sound]

megastar /ˈmeɡəˌstɑː(r)/ *n. colloq.* very famous entertainer etc.

megaton /ˈmeɡəˌtʌn/ *n.* unit of explosive power equal to one million tons of TNT.

megavolt /ˈmeɡəˌvəʊlt/ *n.* one million volts, esp. as a unit of electromotive force.

megawatt /ˈmeɡəˌwɒt/ *n.* one million watts, esp. as a measure of electrical power.

megohm /ˈmeɡəʊm/ *n.* one million ohms.

meiosis /maɪˈəʊsɪs/ *n.* (*pl.* **meioses** /-siːz/) **1** cell division that results in gametes with half the normal chromosome number. **2** = LITOTES. [Greek *meiōn* less]

melamine /ˈmeləˌmiːn/ *n.* **1** white crystalline compound producing resins. **2** (in full **melamine resin**) plastic made from this and used esp. for laminated coatings. [from arbitrary *melam*, AMINE]

melancholia /ˌmelənˈkəʊlɪə/ *n.* depression and anxiety. [Latin: related to MELANCHOLY]

melancholy /ˈmelənkəlɪ/ –*n.* **1** pensive sadness. **2 a** mental depression. **b** tendency to this. –*adj.* sad; saddening, depressing; expressing sadness. □ **melancholic** /-ˈkɒlɪk/ *adj.* [Greek *melas* black, *kholē* bile]

mélange /meɪˈlɑ̃ʒ/ *n.* mixture, medley. [French *mêler* mix]

melanin /ˈmelənɪn/ *n.* dark pigment in the hair, skin, etc., causing tanning in sunlight. [Greek *melas* black]

melanoma /ˌmeləˈnəʊmə/ *n.* malignant skin tumour.

Melba toast /ˈmelbə/ *n.* very thin crisp toast. [*Melba*, name of a soprano]

meld *v.* merge, blend. [origin uncertain]

mêlée /'meleɪ/ *n.* (*US* **melee**) **1** confused fight, skirmish, or scuffle. **2** muddle. [French: related to MEDLEY]

mellifluous /mɪ'lɪflʊəs/ *adj.* (of a voice etc.) pleasing, musical, flowing. □ **mellifluously** *adv.* **mellifluousness** *n.* [Latin *mel* honey, *fluo* flow]

mellow /'meləʊ/ –*adj.* **1** (of sound, colour, light) soft and rich, free from harshness. **2** (of character) gentle; mature. **3** genial, jovial. **4** *euphem.* partly intoxicated. **5** (of fruit) soft, sweet, and juicy. **6** (of wine) well-matured, smooth. **7** (of earth) rich, loamy. –*v.* make or become mellow. □ **mellowly** *adv.* **mellowness** *n.* [origin unknown]

melodeon /mɪ'ləʊdɪən/ *n.* (also **melodion**) **1** small organ similar to the harmonium. **2** small German accordion. [from MELODY, HARMONIUM]

melodic /mɪ'lɒdɪk/ *adj.* of melody; melodious. □ **melodically** *adv.* [Greek: related to MELODY]

melodious /mɪ'ləʊdɪəs/ *adj.* **1** of, producing, or having melody. **2** sweet-sounding. □ **melodiously** *adv.* **melodiousness** *n.* [French: related to MELODY]

melodrama /'melə,drɑːmə/ *n.* **1** sensational play etc. appealing blatantly to the emotions. **2** this type of drama. **3** theatrical language, behaviour, etc. □ **melodramatic** /-drə'mætɪk/ *adj.* **melodramatically** /-drə'mætɪkəlɪ/ *adv.* [Greek *melos* music, DRAMA]

melody /'melədɪ/ *n.* (*pl.* **-ies**) **1** single notes arranged to make a distinctive recognizable pattern; tune. **2** principal part in harmonized music. **3** musical arrangement of words. **4** sweet music, tunefulness. [Greek *melos* song: related to ODE]

melon /'melən/ *n.* **1** sweet fleshy fruit of various climbing plants of the gourd family. **2** such a gourd. [Greek *mēlon* apple]

melt *v.* **1** become liquefied or change to liquid by the action of heat; dissolve. **2** (as **molten** *adj.*) (esp. of metals etc.) liquefied by heat (*molten lava*; *molten lead*). **3** (of food) be delicious, seeming to dissolve in the mouth. **4** soften, or (of a person, the heart, etc.) be softened, by pity, love, etc. (*a melting look*). **5** (usu. foll. by *into*) merge imperceptibly; change into (*night melted into dawn*). **6** (often foll. by *away*) (of a person) leave or disappear unobtrusively (*melted into the background*). □ **melt away** disappear by or as if by liquefaction. **melt down 1** melt (esp. metal) for reuse. **2** become liquid and lose structure. [Old English]

meltdown *n.* **1** melting of a structure, esp. the overheated core of a nuclear reactor. **2** disastrous event, esp. a rapid fall in share values.

melting point *n.* temperature at which a solid melts.

melting-pot *n.* place for mixing races, theories, etc.

member /'membə(r)/ *n.* **1** person etc. belonging to a society, team, group, etc. **2** (**Member**) person elected to certain assemblies etc. **3** part of a larger structure, e.g. of a group of figures or a mathematical set. **4 a** part or organ of the body, esp. a limb. **b** = PENIS. [Latin *membrum* limb]

membership *n.* **1** being a member. **2** number or body of members.

membrane /'membreɪn/ *n.* **1** pliable sheetlike tissue connecting or lining organs in plants and animals. **2** thin pliable sheet or skin. □ **membranous** /'membrənəs/ *adj.* [Latin *membrana* skin, parchment: related to MEMBER]

memento /mɪ'mentəʊ/ *n.* (*pl.* **-es** or **-s**) souvenir of a person or event. [Latin, imperative of *memini* remember]

memento mori /mɪ,mentəʊ 'mɔːrɪ/ *n.* skull etc. as a reminder of death. [Latin, = remember you must die]

memo /'meməʊ/ *n.* (*pl.* **-s**) *colloq.* memorandum. [abbreviation]

memoir /'memwɑː(r)/ *n.* **1** historical account etc. written from personal knowledge or special sources. **2** (in *pl.*) autobiography, esp. partial or dealing with specific events or people. **3** essay on a learned subject. [French *mémoire*: related to MEMORY]

memorabilia /,memərə'bɪlɪə/ *n.pl.* souvenirs of memorable events. [Latin: related to MEMORABLE]

memorable /'memərəb(ə)l/ *adj.* **1** worth remembering. **2** easily remembered. □ **memorably** *adv.* [Latin *memor* mindful]

memorandum /,memə'rændəm/ *n.* (*pl.* **-da** or **-s**) **1** note or record for future use. **2** informal written message, esp. in business, diplomacy, etc. [see MEMORABLE]

memorial /mɪ'mɔːrɪəl/ –*n.* object etc. established in memory of a person or event. –*attrib. adj.* commemorating (*memorial service*). [Latin: related to MEMORY]

memorize /'memə,raɪz/ *v.* (also **-ise**) (**-zing** or **-sing**) commit to memory.

memory /'memərɪ/ *n.* (*pl.* **-ies**) **1** faculty by which things are recalled to or kept in the mind. **2 a** this in an individual (*my memory is failing*). **b** store of things remembered (*deep in my memory*). **3** recollection; remembrance, esp. of a person etc.; person or thing remembered (*memory of better times*; *his mother's memory*). **4** storage capacity of a computer etc. **5** posthumous reputation (*his memory lives on*; *of blessed memory*). **6** length of remembered time of a specific person, group, etc. (*within living memory*). **7** remembering (*deed worthy of memory*). □ **from memory** as remembered (without checking). **in memory of** to keep alive the remembrance of. [Latin *memoria* from *memor* mindful]

memory lane *n.* (usu. prec. by *down, along*) *joc.* sentimental remembering.

memsahib /'memsɑ:b/ *n. Anglo-Ind. hist.* Indian name for a European married woman in India. [from MA'AM, SAHIB]

men *pl.* of MAN.

menace /'menɪs/ –*n.* **1** threat. **2** dangerous thing or person. **3** *joc.* pest, nuisance. –*v.* (**-cing**) threaten. □ **menacingly** *adv.* [Latin *minax* from *minor* threaten]

ménage /meɪ'nɑ:ʒ/ *n.* household. [Latin: related to MANOR]

ménage à trois /meɪˌnɑ:ʒ ɑ: 'trwɑ:/ *n.* (*pl.* ***ménages à trois***) household of three, usu. a married couple and a lover. [French, = household of three]

menagerie /mɪ'nædʒərɪ/ *n.* small zoo. [French: related to MÉNAGE]

mend –*v.* **1** restore to good condition; repair. **2** regain health. **3** improve (*mend matters*). –*n.* darn or repair in material etc. □ **mend one's ways** reform oneself. **on the mend** recovering, esp. in health. [Anglo-French: related to AMEND]

mendacious /men'deɪʃəs/ *adj.* lying, untruthful. □ **mendacity** /-'dæsɪtɪ/ *n.* (*pl.* **-ies**). [Latin *mendax*]

mendelevium /ˌmendə'li:vɪəm/ *n.* artificially made transuranic radioactive metallic element. [*Mendeleev*, name of a chemist]

Mendelian /men'di:lɪən/ *adj.* of Mendel's theory of heredity by genes. [*Mendel*, name of a botanist]

mendicant /'mendɪkənt/ –*adj.* **1** begging. **2** (of a friar) living solely on alms. –*n.* **1** beggar. **2** mendicant friar. [Latin *mendicus* beggar]

mending *n.* **1** action of repairing. **2** things, esp. clothes, to be mended.

menfolk *n.pl.* men, esp. the men of a family.

menhir /'menhɪə(r)/ *n.* usu. prehistoric monument of a tall upright stone. [Breton *men* stone, *hir* long]

menial /'mi:nɪəl/ –*adj.* (of esp. work) degrading, servile. –*n.* domestic servant. [Anglo-French *meinie* retinue]

meninges /mɪ'nɪndʒi:z/ *n.pl.* three membranes enclosing the brain and spinal cord. [Greek *mēnigx* membrane]

meningitis /ˌmenɪn'dʒaɪtɪs/ *n.* (esp. viral) infection and inflammation of the meninges.

meniscus /mɪ'nɪskəs/ *n.* (*pl.* **menisci** /-saɪ/) **1** curved upper surface of liquid in a tube. **2** lens convex on one side and concave on the other. [Greek *mēniskos* crescent, from *mēnē* moon]

menopause /'menəˌpɔ:z/ *n.* **1** ceasing of menstruation. **2** period in a woman's life (usu. 45–55) when this occurs. □ **menopausal** /-'pɔ:z(ə)l/ *adj.* [Greek *mēn* month, PAUSE]

menorah /mɪ'nɔ:rə/ *n.* seven-branched Jewish candelabrum. [Hebrew, = candlestick]

menses /'mensi:z/ *n.pl.* flow of menstrual blood etc. [Latin, pl. of *mensis* month]

mens rea /menz 'ri:ə/ *n. Law* criminal intent. [Latin, = guilty mind]

menstrual /'menstrʊəl/ *adj.* of menstruation. [Latin *menstruus* monthly]

menstrual cycle *n.* process of ovulation and menstruation.

menstruate /'menstrʊˌeɪt/ *v.* (**-ting**) undergo menstruation.

menstruation /ˌmenstrʊ'eɪʃ(ə)n/ *n.* process of discharging blood etc. from the uterus, usu. at monthly intervals from puberty to menopause.

mensuration /ˌmensjʊə'reɪʃ(ə)n/ *n.* **1** measuring. **2** measuring of lengths, areas, and volumes. [Latin: related to MEASURE]

menswear *n.* clothes for men.

-ment *suffix* **1** forming nouns expressing the means or result of verbal action (*abridgement*; *embankment*). **2** forming nouns from adjectives (*merriment*; *oddment*). [Latin *-mentum*]

mental /'ment(ə)l/ *adj.* **1** of, in, or done by the mind. **2** caring for mental patients. **3** *colloq.* insane. □ **mentally** *adv.* [Latin *mens ment-* mind]

mental age *n.* degree of mental development in terms of the average age at which such development is attained.

mental block *n.* inability due to subconscious mental factors.

mental deficiency *n.* abnormally low intelligence.

mentality /men'tælɪtɪ/ *n.* (*pl.* **-ies**) mental character or disposition; kind or degree of intelligence.

mental patient *n.* sufferer from mental illness.

mental reservation *n.* silent qualification made while seeming to agree.

menthol /'menθɒl/ *n.* mint-tasting organic alcohol found in oil of peppermint etc., used as a flavouring and to relieve local pain. [Latin: related to MINT[1]]

mentholated /'menθə,leɪtɪd/ *adj.* treated with or containing menthol.

mention /'menʃ(ə)n/ *–v.* **1** refer to briefly or by name. **2** reveal or disclose (*do not mention this to anyone*). **3** (usu. as **mention in dispatches**) award a minor military honour to in war. *–n.* **1** reference, esp. by name. **2** minor military or other honour. □ **don't mention it** polite reply to an apology or thanks. **not to mention** and also. [Latin *mentio*]

mentor /'mentɔ:(r)/ *n.* experienced and trusted adviser. [*Mentor* in Homer's *Odyssey*]

menu /'menju:/ *n.* **1** list of dishes available in a restaurant etc., or to be served at a meal. **2** *Computing* list of options displayed on a VDU. [Latin: related to MINUTE[2]]

MEP *abbr.* Member of the European Parliament.

Mephistophelean /,mefɪstə'fi:lɪən/ *adj.* fiendish. [*Mephistopheles*, evil spirit to whom Faust sold his soul in German legend]

mercantile /'mɜ:kən,taɪl/ *adj.* **1** of trade, trading. **2** commercial. [Latin: related to MERCHANT]

mercantile marine *n.* merchant shipping.

Mercator projection /mɜ:'keɪtə/ *n.* (also **Mercator's projection**) map of the world projected on to a cylinder so that all the parallels of latitude have the same length as the equator. [*Mercator*, name of a geographer]

mercenary /'mɜ:sɪnərɪ/ *–adj.* primarily concerned with or working for money etc. *–n.* (*pl.* **-ies**) hired soldier in foreign service. □ **mercenariness** *n.* [Latin from *merces* reward]

mercer *n.* dealer in textile fabrics. [Latin *merx merc-* goods]

mercerize /'mɜ:sə,raɪz/ *v.* (also **-ise**) (**-zing** or **-sing**) treat (cotton) with caustic alkali to strengthen and make lustrous. [*Mercer*, name of its alleged inventor]

merchandise /'mɜ:tʃən,daɪz/ *–n.* goods for sale. *–v.* (**-sing**) **1** trade, traffic (in). **2** advertise or promote (goods, an idea, or a person). [French: related to MERCHANT]

merchant /'mɜ:tʃ(ə)nt/ *n.* **1** wholesale trader, esp. with foreign countries. **2** esp. *US & Scot.* retail trader. **3** *colloq.* usu. *derog.* person devoted to a specified activity etc. (*speed merchant*). [Latin *mercor* trade (v.)]

merchantable *adj.* saleable, marketable.

merchant bank *n.* bank dealing in commercial loans and finance.

merchantman *n.* (*pl.* **-men**) merchant ship.

merchant navy *n.* nation's commercial shipping.

merchant ship *n.* ship carrying merchandise.

merciful /'mɜ:sɪ,fʊl/ *adj.* showing mercy. □ **mercifulness** *n.*

mercifully *adv.* **1** in a merciful manner. **2** fortunately (*mercifully, the sun came out*).

merciless /'mɜ:sɪləs/ *adj.* showing no mercy. □ **mercilessly** *adv.*

mercurial /mɜ:'kjʊərɪəl/ *adj.* **1** (of a person) volatile. **2** of or containing mercury. [Latin: related to MERCURY]

mercury /'mɜ:kjʊrɪ/ *n.* **1** silvery heavy liquid metallic element used in barometers, thermometers, etc. **2** (**Mercury**) planet nearest to the sun. □ **mercuric** /-'kjʊərɪk/ *adj.* **mercurous** *adj.* [Latin *Mercurius*, Roman messenger-god]

mercy /'mɜ:sɪ/ *–n.* (*pl.* **-ies**) **1** compassion or forbearance towards defeated enemies or offenders or as a quality. **2** act of mercy. **3** (*attrib.*) done out of compassion (*mercy killing*). **4** thing to be thankful for (*small mercies*). *–int.* expressing surprise or fear. □ **at the mercy of 1** in the power of. **2** liable to danger or harm from. **have mercy on** (or **upon**) show mercy to. [Latin *merces* reward, pity]

mere[1] /mɪə(r)/ *attrib. adj.* (**merest**) being solely or only what is specified (*a mere boy*; *no mere theory*). □ **merely** *adv.* [Latin *merus* unmixed]

mere[2] /mɪə(r)/ *n. dial.* or *poet.* lake. [Old English]

meretricious /,merə'trɪʃəs/ *adj.* showily but falsely attractive. [Latin *meretrix* prostitute]

merganser /mɜ:'gænsə(r)/ *n.* (*pl.* same or **-s**) a diving duck. [Latin *mergus* diver, *anser* goose]

merge *v.* (**-ging**) **1** (often foll. by *with*) **a** combine. **b** join or blend gradually. **2** (foll. by *in*) (cause to) lose character and identity in (something else). [Latin *mergo* dip]

merger *n.* combining, esp. of two commercial companies etc. into one.

meridian /mə'rɪdɪən/ *n.* **1 a** circle of constant longitude, passing through a given place and the terrestrial poles. **b** corresponding line on a map. **2** (often *attrib.*) prime; full splendour. [Latin *meridies* midday]

meridional *adj.* **1** of or in the south (esp. of Europe). **2** of a meridian.

meringue /mə'ræŋ/ *n.* **1** sugar, whipped egg-whites, etc., baked crisp. **2** small cake of this, esp. filled with whipped cream. [French]

merino /mə'ri:nəʊ/ *n.* (*pl.* **-s**) **1** (in full **merino sheep**) variety of sheep with long fine wool. **2** soft cashmere-like material, orig. of merino wool. **3** fine woollen yarn. [Spanish]

merit /'merɪt/ *—n.* **1** quality of deserving well. **2** excellence, worth. **3** (usu. in *pl.*) **a** thing that entitles one to reward or gratitude. **b** intrinsic rights and wrongs (*merits of a case*). *—v.* (**-t-**) deserve. [Latin *meritum* value, from *mereor* deserve]

meritocracy /,merɪ'tɒkrəsɪ/ *n.* (*pl.* **-ies**) **1** government by those selected for merit. **2** group selected in this way. **3** society governed thus.

meritorious /,merɪ'tɔ:rɪəs/ *adj.* praiseworthy.

merlin /'mɜ:lɪn/ *n.* small falcon. [Anglo-French]

mermaid *n.* legendary creature with a woman's head and trunk and a fish's tail. [from MERE[2] 'sea', MAID]

merry /'merɪ/ *adj.* (**-ier, -iest**) **1 a** joyous. **b** full of laughter or gaiety. **2** *colloq.* slightly drunk. □ **make merry** be festive. □ **merrily** *adv.* **merriment** *n.* **merriness** *n.* [Old English]

merry-go-round *n.* **1 a** fairground ride with revolving model horses or cars. **b** = ROUNDABOUT 2a. **2** cycle of bustling activity.

merrymaking *n.* festivity, fun. □ **merrymaker** *n.*

mésalliance /meɪ'zælɪ,ɑ̃s/ *n.* marriage with a social inferior. [French]

mescal /'meskæl/ *n.* peyote cactus. [Spanish from Nahuatl]

mescal buttons *n.pl.* disc-shaped dried tops from the mescal, esp. as an intoxicant.

mescaline /'meskə,lɪn/ *n.* (also **mescalin**) hallucinogenic alkaloid present in mescal buttons.

Mesdames *pl.* of MADAME.

Mesdemoiselles *pl.* of MADEMOISELLE.

mesembryanthemum /mɪ,zembrɪ'ænθɪməm/ *n.* S. African fleshy-leaved plant with bright daisy-like flowers that open fully in sunlight. [Greek, = noon flower]

mesh *—n.* **1** network fabric or structure. **2** each of the open spaces in a net or sieve etc. **3** (in *pl.*) **a** network. **b** snare. *—v.* **1** (often foll. by *with*) (of the teeth of a wheel) be engaged. **2** be harmonious. **3** catch in a net. □ **in mesh** (of the teeth of wheels) engaged. [Dutch]

mesmerize /'mezmə,raɪz/ *v.* (also **-ise**) (**-zing** or **-sing**) **1** hypnotize. **2** fascinate, spellbind. □ **mesmerism** *n.* **mesmerizingly** *adv.* [*Mesmer*, name of a physician]

meso- *comb. form* middle, intermediate. [Greek *mesos* middle]

mesolithic /,mezəʊ'lɪθɪk/ *adj.* of the part of the Stone Age between the palaeolithic and neolithic periods. [Greek *lithos* stone]

mesomorph /'mesəʊ,mɔ:f/ *n.* person with a compact muscular body. [Greek *morphē* form]

meson /'mi:zɒn/ *n.* elementary particle believed to help hold nucleons together in the atomic nucleus. [from MESO-]

mesosphere /'mesəʊ,sfɪə(r)/ *n.* region of the atmosphere from the top of the stratosphere to an altitude of about 80 km.

Mesozoic /,mesəʊ'zəʊɪk/ *—adj.* of the geological era marked by the development of dinosaurs, and the first mammals, birds, and flowering plants. *—n.* this era. [Greek *zōion* animal]

mess *—n.* **1** dirty or untidy state of things. **2** state of confusion, embarrassment, or trouble. **3** something spilt etc. **4** disagreeable concoction. **5 a** soldiers etc. dining together. **b** army dining-hall. **c** meal taken there. **6** domestic animal's excreta. **7** *archaic* portion of liquid or pulpy food. *—v.* **1** (often foll. by *up*) make a mess of; dirty; muddle. **2** *US* (foll. by *with*) interfere with. **3** take one's meals. **4** *colloq.* defecate. □ **make a mess of** bungle. **mess about** (or **around**) **1** potter; fiddle. **2** *colloq.* make things awkward or inconvenient for (a person). [Latin *missus* course of a meal: related to MESSAGE]

message /'mesɪdʒ/ *n.* **1** communication sent by one person to another. **2** exalted or spiritual communication. **3** (in *pl.*) *Scot., Ir., & N.Engl.* shopping. □ **get the message** *colloq.* understand (a hint etc.). [Latin *mitto miss-* send]

Messeigneurs *pl.* of MONSEIGNEUR.

messenger /'mesɪndʒə(r)/ *n.* person who carries a message.

Messiah /mɪ'saɪə/ *n.* **1 a** promised deliverer of the Jews. **b** Christ regarded as this. **2** liberator of an oppressed people. [Hebrew, = anointed]

Messianic /,mesɪ'ænɪk/ *adj.* **1** of the Messiah. **2** inspired by hope or belief in a Messiah. [French: related to MESSIAH]

Messieurs *pl.* of MONSIEUR.

mess kit *n.* soldier's cooking and eating utensils.

Messrs /'mesəz/ *pl.* of MR. [abbreviation of MESSIEURS]

messy *adj.* (**-ier, -iest**) **1** untidy or dirty. **2** causing or accompanied by a mess. **3**

difficult to deal with; awkward. □ **messily** *adv.* **messiness** *n.*

met[1] *past* and *past part.* of MEET[1].

met[2] *adj. colloq.* **1** meteorological. **2** metropolitan. **3** (**the Met**) **a** (in full **the Met Office**) Meteorological Office. **b** Metropolitan Police in London. [abbreviation]

meta- *comb. form* **1** denoting change of position or condition (*metabolism*). **2** denoting position: **a** behind, after, or beyond (*metaphysics*). **b** of a higher or second-order kind (*metalanguage*). [Greek *meta* with, after]

metabolism /mɪ'tæbə,lɪz(ə)m/ *n.* all the chemical processes in a living organism producing energy and growth. □ **metabolic** /,metə'bɒlɪk/ *adj.* [Greek *metabolē* change: related to META-, Greek *ballō* throw]

metabolite /mɪ'tæbə,laɪt/ *n.* substance formed in or necessary for metabolism.

metabolize /mɪ'tæbə,laɪz/ *v.* (also **-ise**) (**-zing** or **-sing**) process or be processed by metabolism.

metacarpus /,metə'kɑːpəs/ *n.* (*pl.* **-carpi** /-paɪ/) **1** part of the hand between the wrist and the fingers. **2** set of five bones in this. □ **metacarpal** *adj.* [related to META-, CARPUS]

metal /'met(ə)l/ —*n.* **1 a** any of a class of workable elements such as gold, silver, iron, or tin, usu. good conductors of heat and electricity and forming basic oxides. **b** alloy of any of these. **2** molten material for making glass. **3** (in *pl.*) rails of a railway line. **4** = ROAD-METAL. —*adj.* made of metal. —*v.* (**-ll-**; *US* **-l-**) **1** make or mend (a road) with road-metal. **2** cover or fit with metal. [Greek *metallon* mine]

metalanguage /'metə,læŋgwɪdʒ/ *n.* **1** form of language used to discuss language. **2** system of propositions about propositions.

metal detector *n.* electronic device for locating esp. buried metal.

metallic /mɪ'tælɪk/ *adj.* **1** of or like metal or metals (*metallic taste*). **2** sounding like struck metal. **3** shiny (*metallic blue*). □ **metallically** *adv.*

metalliferous /,metə'lɪfərəs/ *adj.* (of rocks) containing metal.

metallize /'metə,laɪz/ *v.* (also **-ise**; *US* **metalize**) (**-zing** or **-sing**) **1** render metallic. **2** coat with a thin layer of metal.

metallography /,metə'lɒgrəfɪ/ *n.* descriptive science of metals.

metalloid /'metə,lɔɪd/ *n.* element intermediate in properties between metals and non-metals, e.g. boron, silicon, and germanium.

metallurgy /mɪ'tælədʒɪ, 'metə,lɜːdʒɪ/ *n.* **1** science of metals and their application. **2** extraction and purification of metals. □ **metallurgic** /,metə'lɜːdʒɪk/ *adj.* **metallurgical** /,metə'lɜːdʒɪk(ə)l/ *adj.* **metallurgist** *n.* [Greek *metallon* METAL, *-ourgia* working]

metalwork *n.* **1** art of working in metal. **2** metal objects collectively. □ **metalworker** *n.*

metamorphic /,metə'mɔːfɪk/ *adj.* **1** of metamorphosis. **2** (of rock) transformed naturally, e.g. by heat or pressure. □ **metamorphism** *n.* [from META-, Greek *morphē* form]

metamorphose /,metə'mɔːfəʊz/ *v.* (**-sing**) (often foll. by *to*, *into*) change in form or nature.

metamorphosis /,metə'mɔːfəsɪs, -'fəʊsɪs/ *n.* (*pl.* **-phoses** /-,siːz/) **1** change of form, esp. from a pupa to an insect etc. **2** change of character, conditions, etc. [Greek *morphē* form]

metaphor /'metə,fɔː(r)/ *n.* **1** application of a name or description to something to which it is not literally applicable (e.g. *a glaring error*). **2** instance of this. □ **metaphoric** /-'fɒrɪk/ *adj.* **metaphorical** /-'fɒrɪk(ə)l/ *adj.* **metaphorically** /-'fɒrɪkəlɪ/ *adv.* [Latin from Greek]

metaphysic /,metə'fɪzɪk/ *n.* system of metaphysics.

metaphysical *adj.* **1** of metaphysics. **2** *colloq.* excessively abstract or theoretical. **3** (of esp. 17th-c. English poetry) subtle and complex in imagery.

metaphysics /,metə'fɪzɪks/ *n.pl.* (usu. treated as *sing.*) **1** branch of philosophy dealing with the nature of existence, truth, and knowledge. **2** *colloq.* abstract talk; mere theory. [Greek, as having followed physics in Aristotle's works]

metastasis /me'tæstəsɪs/ *n.* (*pl.* **-stases** /-,siːz/) transference of a bodily function, disease, etc., from one part or organ to another. [Greek, = removal]

metatarsus /,metə'tɑːsəs/ *n.* (*pl.* **-tarsi** /-saɪ/) **1** part of the foot between the ankle and the toes. **2** set of five bones in this. □ **metatarsal** *adj.* [related to META-, TARSUS]

mete *v.* (**-ting**) (usu. foll. by *out*) *literary* apportion or allot (punishment or reward). [Old English]

meteor /'miːtɪə(r)/ *n.* **1** small solid body from outer space that becomes incandescent when entering the earth's atmosphere. **2** streak of light from a meteor. [Greek *meteōros* lofty]

meteoric /,miːtɪ'ɒrɪk/ *adj.* **1** rapid; dazzling (*meteoric rise to fame*). **2** of meteors. □ **meteorically** *adv.*

meteorite /'miːtɪə,raɪt/ *n.* fallen meteor, or fragment of natural rock or metal from outer space.

meteoroid /'miːtɪə,rɔɪd/ *n.* small body that becomes visible as it passes through the earth's atmosphere as a meteor.

meteorology /,miːtɪə'rɒlədʒɪ/ *n.* the study of atmospheric phenomena, esp. for forecasting the weather. □ **meteorological** /-rə'lɒdʒɪk(ə)l/ *adj.* **meteorologist** *n.* [Greek *meteōrologia*: related to METEOR]

meter[1] /'miːtə(r)/ —*n.* **1** instrument that measures or records, esp. gas, electricity, etc. used, distance travelled, etc. **2** = PARKING-METER. —*v.* measure or record by meter. [from METE]

meter[2] *US* var. of METRE[1,2].

-meter *comb. form* **1** forming nouns denoting measuring instruments (*barometer*). **2** forming nouns denoting lines of poetry with a specified number of measures (*pentameter*). [Greek *metron* measure]

methadone /'meθə,dəʊn/ *n.* narcotic analgesic drug used esp. as a substitute for morphine or heroin. [6-di*meth*yl-*a*mino-4, 4-*d*iphenyl-3-heptan*one*]

methanal /'meθə,næl/ *n.* = FORMALDEHYDE. [from METHANE, ALDEHYDE]

methane /'miːθeɪn/ *n.* colourless odourless inflammable gaseous hydrocarbon, the main constituent of natural gas. [from METHYL]

methanoic acid /,meθə'nəʊɪk/ *n.* = FORMIC ACID. [related to METHANE]

methanol /'meθə,nɒl/ *n.* colourless volatile inflammable liquid, used as a solvent. [from METHANE, ALCOHOL]

methinks /mɪ'θɪŋks/ *v.* (*past* **methought** /mɪ'θɔːt/) *archaic* it seems to me. [Old English: related to ME[1], THINK]

method /'meθəd/ *n.* **1** way of doing something; systematic procedure. **2** orderliness; regular habits. □ **method in one's madness** sense in apparently foolish or strange behaviour. [Greek: related to META-, *hodos* way]

methodical /mɪ'θɒdɪk(ə)l/ *adj.* characterized by method or order. □ **methodically** *adv.*

Methodist /'meθədɪst/ —*n.* member of a Protestant denomination originating in the 18th-c. Wesleyan evangelistic movement. —*adj.* of Methodists or Methodism. □ **Methodism** *n.*

methodology /,meθə'dɒlədʒɪ/ *n.* (*pl.* **-ies**) **1** body of methods used in a particular activity. **2** science of method. □ **methodological** /-də'lɒdʒɪk(ə)l/ *adj.* **methodologically** /-də'lɒdʒɪkəlɪ/ *adv.*

methought *past* of METHINKS.

meths *n. colloq.* methylated spirit. [abbreviation]

methyl /'meθɪl, 'miːθaɪl/ *n.* univalent hydrocarbon radical CH_3, present in many organic compounds. [Greek *methu* wine, *hulē* wood]

methyl alcohol *n.* = METHANOL.

methylate /'meθɪ,leɪt/ *v.* (**-ting**) **1** mix or impregnate with methanol. **2** introduce a methyl group into (a molecule or compound).

methylated spirit *n.* (also **methylated spirits** *n.pl.*) alcohol treated to make it unfit for drinking and exempt from duty.

meticulous /mə'tɪkjʊləs/ *adj.* **1** giving great attention to detail. **2** very careful and precise. □ **meticulously** *adv.* **meticulousness** *n.* [Latin *metus* fear]

métier /'metjeɪ/ *n.* **1** one's trade, profession, or field of activity. **2** one's forte. [Latin: related to MINISTER]

metonymy /mɪ'tɒnɪmɪ/ *n.* substitution of the name of an attribute or adjunct for that of the thing meant (e.g. *Crown* for *king*, *the turf* for *horse-racing*). [Greek: related to META-, *onuma* name]

metre[1] /'miːtə(r)/ *n.* (*US* **meter**) metric unit and the base SI unit of linear measure, equal to about 39.4 inches. □ **metreage** /'miːtərɪdʒ/ *n.* [Greek *metron* measure]

metre[2] /'miːtə(r)/ *n.* (*US* **meter**) **1 a** poetic rhythm, esp. as determined by the number and length of feet in a line. **b** metrical group or measure. **2** basic rhythm of music. [related to METRE[1]]

metre-kilogram-second *n.* denoting a system of measure using the metre, kilogram, and second.

metric /'metrɪk/ *adj.* of or based on the metre. [French: related to METRE[1]]

-metric *comb. form* (also **-metrical**) forming adjectives corresponding to nouns in *-meter* and *-metry* (*thermometric*; *geometric*).

metrical *adj.* **1** of or composed in metre (*metrical psalms*). **2** of or involving measurement (*metrical geometry*). □ **metrically** *adv.* [Greek: related to METRE[2]]

metricate *v.* (**-ting**) convert to a metric system. □ **metrication** /-'keɪʃ(ə)n/ *n.*

metric system *n.* decimal measuring system with the metre, litre, and gram (or kilogram) as units of length, volume, and mass.

metric ton *n.* (also **metric tonne**) 1,000 kilograms (2205 lb).

metro /'metrəʊ/ *n.* (*pl.* **-s**) underground railway system, esp. in Paris. [French shortened from *métropolitain* metropolitan]

metronome /'metrə,nəʊm/ *n.* device ticking at a selected rate to mark time for musicians. [Greek *metron* measure, *nomos* law]

metropolis /mɪ'trɒpəlɪs/ *n.* chief city, capital. [Greek *mētēr* mother, *polis* city]

metropolitan /,metrə'pɒlɪt(ə)n/ *–adj.* **1** of a metropolis. **2** of or forming a mother country as distinct from its colonies etc. (*metropolitan France*). *–n.* **1** bishop having authority over the bishops of a province. **2** inhabitant of a metropolis.

-metry *comb. form* forming nouns denoting procedures and systems involving measurement (*geometry*).

mettle /'met(ə)l/ *n.* **1** quality or strength of character. **2** spirit, courage. □ **on one's mettle** keen to do one's best. □ **mettlesome** *adj.* [from METAL *n.*]

MeV *abbr.* mega-electronvolt(s).

mew[1] *–n.* characteristic cry of a cat, gull, etc. *–v.* utter this sound. [imitative]

mew[2] *n.* gull, esp. the common gull. [Old English]

mewl *v.* **1** whimper. **2** mew like a cat. [imitative]

mews /mju:z/ *n.* (treated as *sing.*) stabling round a yard etc., now used esp. for housing. [originally sing. *mew* 'cage for hawks': French from Latin *muto* change]

Mexican /'meksɪkən/ *–n.* **1** native or national of Mexico. **2** person of Mexican descent. *–adj.* of Mexico or its people. [Spanish]

mezzanine /'metsə,ni:n, 'mez-/ *n.* storey between two others (usu. between the ground and first floors). [Italian: related to MEDIAN]

mezzo /'metsəʊ/ *Mus.* *–adv.* half, moderately. *–n.* (in full **mezzo-soprano**) (*pl.* **-s**) **1** female singing-voice between soprano and contralto. **2** singer with this voice. [Latin *medius* middle]

mezzo forte *adj.* & *adv.* fairly loud(ly).

mezzo piano *adj.* & *adv.* fairly soft(ly).

mezzotint /'metsəʊtɪnt/ *n.* **1** method of printing or engraving in which a plate is roughened by scraping to produce tones and halftones. **2** print so produced. [Italian: related to MEZZO, TINT]

mf *abbr.* mezzo forte.

Mg *symb.* magnesium.

mg *abbr.* milligram(s).

Mgr. *abbr.* **1** Manager. **2** *Monseigneur*. **3** Monsignor.

MHz *abbr.* megahertz.

mi var. of ME[2].

mi. *abbr. US* mile(s).

miaow /mɪ'aʊ/ *–n.* characteristic cry of a cat. *–v.* make this cry. [imitative]

miasma /mɪ'æzmə/ *n.* (*pl.* **-mata** or **-s**) *archaic* infectious or noxious vapour. [Greek, = defilement]

mica /'maɪkə/ *n.* silicate mineral found as glittering scales in granite etc. or in crystals separable into thin transparent plates. [Latin, = crumb]

mice *pl.* of MOUSE.

Michaelmas /'mɪkəlməs/ *n.* feast of St Michael, 29 September. [related to MASS[2]]

Michaelmas daisy *n.* autumn-flowering aster.

mick *n. slang offens.* Irishman. [pet form of *Michael*]

mickey /'mɪkɪ/ *n.* (also **micky**) □ **take the mickey** (often foll. by *out of*) *slang* tease, mock, ridicule. [origin uncertain]

Mickey Finn /,mɪkɪ 'fɪn/ *n. slang* drugged drink intended to make the victim unconscious. [origin uncertain]

mickle /'mɪk(ə)l/ *n.* (also **muckle** /'mʌk(ə)l/) *archaic* or *Scot.* large amount. □ **many a little makes a mickle** (also *erroneously* **many a mickle makes a muckle**) small amounts accumulate. [Old Norse]

micky var. of MICKEY.

micro /'maɪkrəʊ/ *n.* (*pl.* **-s**) *colloq.* **1** = MICROCOMPUTER. **2** = MICROPROCESSOR.

micro- *comb. form* **1** small (*microchip*). **2** denoting a factor of one millionth (10^{-6}) (*microgram*). [Greek *mikros* small]

microbe /'maɪkrəʊb/ *n.* micro-organism (esp. a bacterium causing disease or fermentation). □ **microbial** /-'krəʊbɪəl/ *adj.* **microbic** /-'krəʊbɪk/ *adj.* [Greek *mikros* small, *bios* life]

microbiology /,maɪkrəʊbaɪ'ɒlədʒɪ/ *n.* the study of micro-organisms. □ **microbiologist** *n.*

microchip /'maɪkrəʊtʃɪp/ *n.* small piece of semiconductor (usu. silicon) used to carry integrated circuits.

microcircuit /'maɪkrəʊ,sɜ:kɪt/ *n.* integrated circuit on a microchip.

microclimate /'maɪkrəʊ,klaɪmɪt/ *n.* small localized climate, e.g. inside a greenhouse.

microcomputer /'maɪkrəʊkəm,pju:tə(r)/ *n.* small computer with a microprocessor as its central processor.

microcosm /'maɪkrə,kɒz(ə)m/ *n.* (often foll. by *of*) miniature representation, e.g. mankind or a community seen as a small-scale model of the universe; epitome. □ **microcosmic** /-'kɒzmɪk/ *adj.* [from MICRO-, COSMOS]

microdot /'maɪkrəʊ,dɒt/ *n.* microphotograph of a document etc. reduced to the size of a dot.

micro-electronics /,maɪkrəʊɪlek'trɒnɪks/ *n.* design, manufacture, and use of microchips and microcircuits.

microfiche /'maɪkrəʊ,fi:ʃ/ *n.* (*pl.* same or **-s**) small flat piece of film bearing microphotographs of documents etc. [from MICRO-, French *fiche* slip of paper]

microfilm /'maɪkrəʊfɪlm/ *–n.* length of film bearing microphotographs of documents etc. *–v.* photograph on microfilm.

microlight /'maɪkrəʊ,laɪt/ *n.* a kind of motorized hang-glider.

micromesh /'maɪkrəʊ,meʃ/ *n.* (often *attrib.*) fine-meshed material, esp. nylon.

micrometer /maɪ'krɒmɪtə(r)/ *n.* gauge for accurate small-scale measurement.

micron /'maɪkrɒn/ *n.* one-millionth of a metre. [Greek *mikros* small]

micro-organism /,maɪkrəʊ'ɔ:gə,nɪz(ə)m/ *n.* microscopic organism, e.g. bacteria, protozoa, and viruses.

microphone /'maɪkrə,fəʊn/ *n.* instrument for converting sound waves into electrical energy for reconversion into sound after transmission or recording. [from MICRO-, Greek *phōnē* sound]

microphotograph /,maɪkrəʊ'fəʊtə,grɑ:f/ *n.* photograph reduced to a very small size. [from MICRO-]

microprocessor /,maɪkrəʊ'prəʊsesə(r)/ *n.* integrated circuit containing all the functions of a computer's central processing unit.

microscope /'maɪkrə,skəʊp/ *n.* instrument with lenses for magnifying objects or details invisible to the naked eye. [from MICRO-, -SCOPE]

microscopic /,maɪkrə'skɒpɪk/ *adj.* **1** visible only with a microscope. **2** extremely small. **3** of or by means of a microscope. □ **microscopically** *adv.*

microscopy /maɪ'krɒskəpɪ/ *n.* use of microscopes.

microsecond /'maɪkrəʊ,sekənd/ *n.* one-millionth of a second.

microsurgery /'maɪkrəʊ,sɜ:dʒərɪ/ *n.* intricate surgery using microscopes.

microwave /'maɪkrəʊ,weɪv/ *–n.* **1** electromagnetic wave with a wavelength in the range 0.001–0.3m. **2** (in full **microwave oven**) oven using microwaves to cook or heat food quickly. *–v.* (**-ving**) cook in a microwave oven.

micturition /,mɪktjʊə'rɪʃ(ə)n/ *n.* *formal* urination. [Latin]

mid *attrib. adj.* (usu. in *comb.*) the middle of (*mid-air*; *mid-June*). [Old English]

midday *n.* (often *attrib.*) middle of the day; noon. [Old English: related to MID, DAY]

midden /'mɪd(ə)n/ *n.* **1** dunghill. **2** refuse heap. [Scandinavian: related to MUCK]

middle /'mɪd(ə)l/ *–attrib. adj.* **1** at an equal distance, time, or number from extremities; central. **2** intermediate in rank, quality, etc. **3** average (*of middle height*). *–n.* **1** (often foll. by *of*) middle point, position, or part. **2** waist. □ **in the middle of 1** in the process of. **2** during. [Old English]

middle age *n.* period between youth and old age. □ **middle-aged** *adj.*

Middle Ages *n.* (prec. by *the*) period of European history from *c.*1000 to 1453.

middle-age spread *n.* (also **middle-aged spread**) increased bodily girth at middle age.

middlebrow *colloq.* *–adj.* having or appealing to non-intellectual or conventional tastes. *–n.* middlebrow person.

middle C *n.* C near the middle of the piano keyboard, (in notation) the note between the treble and bass staves.

middle class *n.* social class between the upper and the lower, including professional and business workers. □ **middle-class** *adj.*

middle distance *n.* **1** (in a landscape) part between the foreground and the background. **2** *Athletics* race distance of esp. 400 or 800 metres.

middle ear *n.* cavity behind the eardrum.

Middle East *n.* (prec. by *the*) area covered by countries from Egypt to Iran inclusive. □ **Middle Eastern** *adj.*

Middle English *n.* English language from *c.*1150 to 1500.

middle game *n.* central phase of a chess game.

middleman *n.* **1** trader who handles a commodity between producer and consumer. **2** intermediary.

middle name *n.* **1** name between first name and surname. **2** *colloq.* person's most characteristic quality (*tact is my middle name*).

middle-of-the-road *adj.* **1** moderate; avoiding extremes. **2** of general appeal.

middle school *n.* school for children from about 9 to 13 years.

middle-sized *adj.* of medium size.

middleweight *n.* **1** weight in certain sports between welterweight and light heavyweight, in amateur boxing 71–5 kg. **2** sportsman of this weight.

middling *–adj.* moderately good. *–adv.* fairly, moderately.

midfield *n.* *Football* central part of the pitch, away from the goals. □ **midfielder** *n.*

midge *n.* gnatlike insect. [Old English]

midget /'mɪdʒɪt/ *n.* **1** extremely small person or thing. **2** (*attrib.*) very small.

MIDI /'mɪdɪ/ *n.* (also **midi**) an interface allowing electronic musical instruments, synthesizers, and computers to be interconnected and used simultaneously. [abbreviation of *m*usical *i*nstrument *d*igital *i*nterface]

midi system *n.* set of compact stacking components of hi-fi equipment.

midland *–n.* **1** (**the Midlands**) inland counties of central England. **2** middle part of a country. *–adj.* of or in the midland or Midlands.

mid-life *n.* middle age.

mid-life crisis *n.* crisis of self-confidence in early middle age.

midnight *n.* middle of the night; 12 o'clock at night. [Old English]

midnight blue *adj.* & *n.* (as adj. often hyphenated) very dark blue.

midnight sun *n.* sun visible at midnight during the summer in polar regions.

mid-off *n. Cricket* position of the fielder near the bowler on the off side.

mid-on *n. Cricket* position of the fielder near the bowler on the on side.

midriff *n.* front of the body just above the waist. [Old English, = mid-belly]

midshipman *n.* naval officer ranking next above a cadet.

midships *adv.* = AMIDSHIPS.

midst *–prep. poet.* amidst. *–n.* middle. □ **in the midst of** among; in the middle of. **in our** (or **your** or **their**) **midst** among us (or you or them). [related to MID]

midstream *–n.* middle of a stream etc. *–adv.* (also **in midstream**) in the middle of an action etc. (*abandoned the project midstream*).

midsummer *n.* period of or near the summer solstice, about 21 June. [Old English]

Midsummer Day *n.* (also **Midsummer's Day**) 24 June.

midsummer madness *n.* extreme folly.

midway *adv.* in or towards the middle of the distance between two points.

Midwest *n.* region of the US adjoining the northern Mississippi.

midwicket *n. Cricket* position of a fielder on the leg side opposite the middle of the pitch.

midwife *n.* person trained to assist at childbirth. □ **midwifery** /-ˌwɪfrɪ, -'wɪfərɪ/ *n.* [originally = with-woman]

midwinter *n.* period of or near the winter solstice, about 22 Dec. [Old English]

mien *n. literary* person's look or bearing. [probably obsolete *demean*]

miff *v. colloq.* (usu. as **miffed** *adj.*) offend. [origin uncertain]

M.I.5 *abbr.* UK department of Military Intelligence concerned with State security.

■ **Usage** This term is not in official use.

might[1] /maɪt/ *past* of MAY, used esp.: **1** in reported speech, expressing possibility (*said he might come*) or permission (*asked if I might leave*) (cf. MAY 1, 2). **2** (foll. by perfect infin.) expressing a possibility based on a condition not fulfilled (*if you'd looked you might have found it*). **3** (foll. by present infin. or perfect infin.) expressing complaint that an obligation or expectation is not or has not been fulfilled (*they might have asked*). **4** expressing a request (*you might call in at the butcher's*). **5** *colloq.* **a** = MAY 1 (*it might be true*). **b** (in tentative questions) = MAY 2 (*might I have the pleasure of this dance?*). **c** = MAY 1d (*who might you be?*). □ **might as well** expressing lukewarm acquiescence (*might as well try*).

might[2] /maɪt/ *n.* strength, power. □ **with might and main** with all one's power. [Old English: related to MAY]

might-have-been *n. colloq.* **1** past possibility that no longer applies. **2** person of unfulfilled promise.

mightn't /'maɪt(ə)nt/ *contr.* might not.

mighty *–adj.* (**-ier, -iest**) **1** powerful, strong. **2** massive, bulky. **3** *colloq.* great, considerable. *–adv. colloq.* very (*mighty difficult*). □ **mightily** *adv.* **mightiness** *n.* [Old English: related to MIGHT[2]]

mignonette /ˌmɪnjə'net/ *n.* plant with fragrant grey-green flowers. [French, diminutive of *mignon* small]

migraine /'mi:greɪn, 'maɪ-/ *n.* recurrent throbbing headache often with nausea and visual disturbance. [Greek *hēmikrania*: related to HEMI-, CRANIUM]

migrant /'maɪgrənt/ *–adj.* migrating. *–n.* migrant person or animal, esp. a bird.

migrate /maɪ'greɪt/ *v.* (**-ting**) **1** move from one place and settle in another, esp. abroad. **2** (of a bird or fish) change its habitation seasonally. **3** move under natural forces. □ **migration** /-'greɪʃ(ə)n/ *n.* **migrator** *n.* **migratory** /'maɪgrətərɪ/ *adj.* [Latin *migro*]

mikado /mɪ'kɑ:dəʊ/ *n.* (*pl.* **-s**) *hist.* emperor of Japan. [Japanese, = august door]

mike *n. colloq.* microphone. [abbreviation]

mil *n.* one-thousandth of an inch, as a unit of measure for the diameter of wire etc. [Latin *mille* thousand]

milady /mɪ'leɪdɪ/ *n.* (*pl.* **-ies**) (esp. as a form of address) English noblewoman. [French from *my Lady*]

milage var. of MILEAGE.

milch *adj.* giving milk. [Old English: related to MILK]

milch cow *n.* source of easy profit.

mild /maɪld/ *–adj.* **1** (esp. of a person) gentle and conciliatory. **2** not severe or harsh. **3** (of the weather) moderately warm. **4** (of flavour etc.) not sharp or strong. **5** tame, feeble; lacking vivacity. *–n.* dark mild draught beer (cf. BITTER). □ **mildish** *adj.* **mildness** *n.* [Old English]

mildew /ˈmɪldju:/ *–n.* **1** destructive growth of minute fungi on plants. **2** similar growth on damp paper, leather, etc. *–v.* taint or be tainted with mildew. □ **mildewy** *adj.* [Old English]

mildly *adv.* in a mild fashion. □ **to put it mildly** as an understatement.

mild-mannered *adj.* = MILD 1.

mild steel *n.* strong and tough steel not readily tempered.

mile *n.* **1** (also **statute mile**) unit of linear measure equal to 1,760 yards (approx. 1.6 kilometres). **2** (in *pl.*) *colloq.* great distance or amount (*miles better*). **3** race extending over a mile. [Latin *mille* thousand]

mileage *n.* (also **milage**) **1** number of miles travelled, esp. by a vehicle per unit of fuel. **2** *colloq.* profit, advantage.

miler *n. colloq.* person or horse specializing in races of one mile.

milestone *n.* **1** stone beside a road marking a distance in miles. **2** significant event or point in a life, history, project, etc.

milfoil /ˈmɪlfɔɪl/ *n.* common yarrow with small white flowers. [Latin: related to MILE, FOIL[2]]

milieu /mɪˈljɜ:, ˈmi:-/ *n.* (*pl.* **milieux** or **-s** /-ˈljɜ:z/) person's environment or social surroundings. [French]

militant /ˈmɪlɪt(ə)nt/ *–adj.* **1** combative; aggressively active in support of a cause. **2** engaged in warfare. *–n.* militant person. □ **militancy** *n.* **militantly** *adv.* [Latin: related to MILITATE]

militarism /ˈmɪlɪtəˌrɪz(ə)m/ *n.* **1** aggressively military policy etc. **2** military spirit. □ **militarist** *n.* **militaristic** /-ˈrɪstɪk/ *adj.*

militarize /ˈmɪlɪtəˌraɪz/ *v.* (also **-ise**) (**-zing** or **-sing**) **1** equip with military resources. **2** make military or warlike. **3** imbue with militarism. □ **militarization** /-ˈzeɪʃ(ə)n/ *n.*

military /ˈmɪlɪtərɪ/ *–adj.* of or characteristic of soldiers or armed forces. *–n.* (as *sing.* or *pl.*; prec. by *the*) the army. □ **militarily** *adv.* [Latin *miles milit-* soldier]

military honours *n.pl.* burial rites of a soldier, royalty, etc., performed by the military.

military police *n.* (as *pl.*) army police force disciplining soldiers.

militate /ˈmɪlɪˌteɪt/ *v.* (**-ting**) (usu. foll. by *against*) have force or effect; tell. [Latin: related to MILITARY]

■ **Usage** *Militate* is often confused with *mitigate*.

militia /mɪˈlɪʃə/ *n.* military force, esp. one conscripted in an emergency. □ **militiaman** *n.* [Latin, = military service]

milk *–n.* **1** opaque white fluid secreted by female mammals for the nourishment of their young. **2** milk of cows, goats, or sheep as food. **3** milklike juice of the coconut etc. *–v.* **1** draw milk from (a cow etc.). **2** exploit (a person or situation) to the utmost. [Old English]

milk and honey *n.* abundance; prosperity.

milk and water *n.* feeble or insipid writing, speech, etc.

milk chocolate *n.* chocolate made with milk.

milk float *n.* small usu. electric vehicle used in delivering milk.

milkmaid *n.* girl or woman who milks cows or works in a dairy.

milkman *n.* person who sells or delivers milk.

Milk of Magnesia *n. propr.* white suspension of magnesium hydroxide usu. in water, taken as an antacid or laxative.

milk-powder *n.* dehydrated milk.

milk pudding *n.* pudding, esp. of rice, baked with milk.

milk round *n.* **1** fixed route for milk delivery. **2** regular trip with calls at several places.

milk run *n.* routine expedition etc.

milk shake *n.* drink of whisked milk, flavouring, etc.

milksop *n.* weak or timid man or youth.

milk tooth *n.* temporary tooth in young mammals.

milky *adj.* (**-ier**, **-iest**) **1** of, like, or mixed with milk. **2** (of a gem or liquid) cloudy; not clear. □ **milkiness** *n.*

Milky Way *n.* luminous band of stars; the Galaxy.

mill *–n.* **1 a** building fitted with a mechanical device for grinding corn. **b** such a device. **2** device for grinding any solid to powder etc. (*pepper-mill*). **3 a** building fitted with machinery for manufacturing processes etc. (*cotton-mill*). **b** such machinery. *–v.* **1** grind (corn), produce (flour), or hull (seeds) in a mill. **2** (esp. as **milled** *adj.*) produce a ribbed edge on (a coin). **3** cut or shape (metal) with a rotating tool. **4** (often foll. by *about*, *around*) move aimlessly, esp. in a confused mass. □ **go** (or **put**) **through the mill** undergo (or cause to undergo) intensive work, pain, training, etc. [Latin *molo* grind]

millefeuille /mi:l'fɜ:j/ *n.* rich cake of puff pastry split and filled with jam, cream, etc. [French, = thousand-leaf]

millennium /mɪ'lenɪəm/ *n.* (*pl.* **-s** or **millennia**) **1** period of 1,000 years, esp. that of Christ's prophesied reign on earth (Rev. 20:1–5). **2** (esp. future) period of happiness and prosperity. □ **millennial** *adj.* [Latin *mille* thousand]

millepede var. of MILLIPEDE.

miller *n.* **1** proprietor or tenant of a mill, esp. a corn-mill. **2** person operating a milling machine. [related to MILL]

miller's thumb *n.* small spiny freshwater fish.

millesimal /mɪ'lesɪm(ə)l/ *–adj.* **1** thousandth. **2** of, belonging to, or dealing with, a thousandth or thousandths. *–n.* thousandth part. [Latin *mille* thousand]

millet /'mɪlɪt/ *n.* **1** cereal plant bearing small nutritious seeds. **2** seed of this. [Latin *milium*]

millet-grass *n.* tall woodland grass.

milli- *comb. form* thousand, esp. denoting a factor of one thousandth. [Latin *mille* thousand]

milliard /'mɪljəd/ *n.* one thousand million. [French *mille* thousand]

■ **Usage** *Milliard* is now largely superseded by *billion*.

millibar /'mɪlɪ,bɑ:(r)/ *n.* unit of atmospheric pressure equivalent to 100 pascals.

milligram /'mɪlɪ,græm/ *n.* (also **-gramme**) one-thousandth of a gram.

millilitre /'mɪlɪ,li:tə(r)/ *n.* (*US* **-liter**) one-thousandth of a litre (0.002 pint).

millimetre /'mɪlɪ,mi:tə(r)/ *n.* (*US* **-meter**) one-thousandth of a metre (0.039 in.).

milliner /'mɪlɪnə(r)/ *n.* person who makes or sells women's hats. □ **millinery** *n.* [*Milan* in Italy]

million /'mɪljən/ *n.* & *adj.* (*pl.* same or (in sense 2) **-s**) (in *sing.* prec. by *a* or *one*) **1** thousand thousand. **2** (in *pl.*) *colloq.* very large number. **3** million pounds or dollars. □ **millionth** *adj.* & *n.* [French, probably from Italian *mille* thousand]

millionaire /,mɪljə'neə(r)/ *n.* (*fem.* **millionairess**) person who has over a million pounds, dollars, etc. [French *millionnaire*: related to MILLION]

millipede /'mɪlɪ,pi:d/ *n.* (also **millepede**) small crawling invertebrate with a long segmented body with two pairs of legs on each segment. [Latin *mille* thousand, *pes ped-* foot]

millisecond /'mɪlɪ,sekənd/ *n.* one-thousandth of a second.

millpond *n.* pool of water retained by a dam for operating a mill-wheel. □ **like a millpond** (of water) very calm.

mill-race *n.* current of water that drives a mill-wheel.

millstone *n.* **1** each of two circular stones for grinding corn. **2** heavy burden or responsibility.

mill-wheel *n.* wheel used to drive a water-mill.

millworker *n.* factory worker.

millwright *n.* person who designs or builds mills.

milometer /maɪ'lɒmɪtə(r)/ *n.* instrument for measuring the number of miles travelled by a vehicle.

milord /mɪ'lɔ:d/ *n.* (esp. as a form of address) English nobleman. [French from *my lord*]

milt *n.* **1** spleen in mammals. **2** sperm-filled reproductive gland or the sperm of a male fish. [Old English]

mime *–n.* **1** acting without words, using only gestures. **2** performance using mime. **3** (also **mime artist**) mime actor. *–v.* (**-ming**) **1** (also *absol.*) convey by mime. **2** (often foll. by *to*) mouth words etc. in time with a soundtrack (*mime to a record*). [Greek *mimos*]

mimeograph /'mɪmɪə,grɑ:f/ *–n.* **1** machine which duplicates from a stencil. **2** copy so produced. *–v.* reproduce by this process. [Greek *mimeomai* imitate]

mimetic /mɪ'metɪk/ *adj.* of or practising imitation or mimicry. [Greek *mimētikos*: see MIMEOGRAPH]

mimic /'mɪmɪk/ *–v.* (**-ck-**) **1** imitate (a person, gesture, etc.) esp. to entertain or ridicule. **2** copy minutely or servilely. **3** resemble closely. *–n.* person skilled in imitation. □ **mimicry** *n.* [Greek *mimikos*: related to MIME]

mimosa /mɪ'məʊzə/ *n.* **1** shrub with globular usu. yellow flowers. **2** acacia plant with showy yellow flowers. [Latin: related to MIME]

Min. *abbr.* **1** Minister. **2** Ministry.

min. *abbr.* **1** minute(s). **2** minimum. **3** minim (fluid measure).

mina var. of MYNA.

minaret /,mɪnə'ret/ *n.* slender turret next to a mosque, from which the muezzin calls at hours of prayer. [French or Spanish from Turkish from Arabic]

minatory /'mɪnətərɪ/ *adj. formal* threatening, menacing. [Latin *minor* threaten]

mince *–v.* (**-cing**) **1** cut up or grind (esp. meat) finely. **2** (usu. as **mincing** *adj.*) speak or esp. walk effeminately or affectedly. *–n.* minced meat. □ **mince matters** (or **one's words**) (usu. with *neg.*) speak evasively or unduly mildly.

□ **mincer** *n.* [Latin *minutia* something small]

mincemeat *n.* mixture of currants, sugar, spices, suet, etc. □ **make mincemeat of** utterly defeat.

mince pie *n.* pie containing mincemeat.

mind /maɪnd/ –*n.* **1 a** seat of consciousness, thought, volition, and feeling. **b** attention, concentration (*mind keeps wandering*). **2** intellect. **3** memory (*can't call it to mind*). **4** opinion (*of the same mind*). **5** way of thinking or feeling (*the Victorian mind*). **6** focussed will (*put one's mind to it*). **7** sanity (*lose one's mind*). **8** person in regard to mental faculties (*a great mind*). –*v.* **1** object; be upset (*do you mind if I smoke?*; *minded terribly when she left*). **2** (often foll. by *out*) heed; take care (to) (*mind you come on time*; *mind the step*; *mind how you go*; *mind out!*). **3** look after (*mind the house*). **4** apply oneself to, concern oneself with (*mind my own business*). □ **be in two minds** be undecided. **do you mind!** *iron.* expression of annoyance. **have a good** (or **half a**) **mind to** feel inclined to (*I've a good mind to report you*). **have (it) in mind** intend. **in one's mind's eye** in one's imagination. **mind one's Ps & Qs** be careful in one's behaviour. **mind you** used to qualify a statement (*mind you, it wasn't easy*). **never mind 1** let alone; not to mention. **2** used to comfort or console. **3** (also **never you mind**) used to evade a question. **to my mind** in my opinion. [Old English]

mind-blowing *adj. slang* **1** mind-boggling; overwhelming. **2** (esp. of drugs etc.) inducing hallucinations.

mind-boggling *adj. colloq.* unbelievable, startling.

minded *adj.* **1** (in *comb.*) **a** inclined to think in some specified way, or with a specified interest (*mathematically minded*; *fair-minded*; *car-minded*). **b** having a specified kind of mind (*high-minded*). **2** (usu. foll. by *to* + infin.) disposed or inclined.

minder *n.* **1** (often in *comb.*) person employed to look after a person or thing (*child minder*). **2** *slang* bodyguard.

mindful *adj.* (often foll. by *of*) taking heed or care; giving thought (to). □ **mindfully** *adv.*

mindless *adj.* **1** lacking intelligence; brutish (*mindless violence*). **2** not requiring thought or skill (*mindless work*). **3** (usu. foll. by *of*) heedless of (advice etc.). □ **mindlessly** *adv.* **mindlessness** *n.*

mind-read *v.* discern the thoughts of (another person). □ **mind-reader** *n.*

mine[1] *poss. pron.* the one(s) of or belonging to me (*it is mine*; *mine are over there*). □ **of mine** of or belonging to me (*a friend of mine*). [Old English]

mine[2] –*n.* **1** excavation to extract metal, coal, salt, etc. **2** abundant source (of information etc.). **3** military explosive device placed in the ground or in the water. –*v.* **(-ning) 1** obtain (metal, coal, etc.) from a mine. **2** (also *absol.*, often foll. by *for*) dig in (the earth etc.) for ore etc. or to tunnel. **3** lay explosive mines under or in. □ **mining** *n.* [French]

minefield *n.* **1** area planted with explosive mines. **2** *colloq.* hazardous subject or situation.

minelayer *n.* ship or aircraft for laying explosive mines.

miner *n.* person who works in a mine. [French: related to MINE[2]]

mineral /ˈmɪnər(ə)l/ *n.* (often *attrib.*) **1** inorganic substance. **2** substance obtained by mining. **3** (often in *pl.*) artificial mineral water or similar carbonated drink. [French or medieval Latin: related to MINE[2]]

mineralize /ˈmɪnərəˌlaɪz/ *v.* (also **-ise**) (**-zing** or **-sing**) impregnate (water etc.) with a mineral substance.

mineralogy /ˌmɪnəˈrælədʒɪ/ *n.* the study of minerals. □ **mineralogical** /-rəˈlɒdʒɪk(ə)l/ *adj.* **mineralogist** *n.*

mineral water *n.* **1** natural water often containing dissolved salts. **2** artificial imitation of this, esp. soda water.

minestrone /ˌmɪnɪˈstrəʊnɪ/ *n.* soup containing vegetables and pasta, beans, or rice. [Italian]

minesweeper *n.* ship for clearing explosive mines from the sea.

mineworker *n.* miner.

Ming *n.* (often *attrib.*) Chinese porcelain made during the Ming dynasty (1368–1644).

mingle /ˈmɪŋg(ə)l/ *v.* (**-ling**) **1** mix, blend. **2** (often foll. by *with*) mix socially. [Old English]

mingy /ˈmɪndʒɪ/ *adj.* (**-ier, -iest**) *colloq.* mean, stingy. □ **mingily** *adv.* [probably from MEAN[2], STINGY]

mini /ˈmɪnɪ/ *n.* (*pl.* **-s**) **1** *colloq.* miniskirt. **2** (**Mini**) *propr.* make of small car. [abbreviation]

mini- *comb. form* miniature; small of its kind (*minibus*).

miniature /ˈmɪnɪtʃə(r)/ –*adj.* **1** much smaller than normal. **2** represented on a small scale. –*n.* **1** any miniature object. **2** detailed small-scale portrait. **3** this genre. □ **in miniature** on a small scale. □ **miniaturist** *n.* (in senses 2 and 3 of *n.*). [Latin *minium* red lead]

miniaturize *v.* (also **-ise**) (**-zing** or **-sing**) produce in a smaller version;

make small. □ **miniaturization** /-ˈzeɪʃ(ə)n/ *n.*

minibus /ˈmɪnɪ,bʌs/ *n.* small bus for about twelve passengers.

minicab /ˈmɪnɪ,kæb/ *n.* car used as a taxi, hireable only by telephone.

minicomputer /ˈmɪnɪkəm,pju:tə(r)/ *n.* computer of medium power.

minim /ˈmɪnɪm/ *n.* **1** *Mus.* note equal to two crotchets or half a semibreve. **2** one-sixtieth of a fluid drachm, about a drop. [Latin *minimus* least]

minima *pl.* of MINIMUM.

minimal /ˈmɪnɪm(ə)l/ *adj.* **1** very minute or slight. **2** being a minimum. □ **minimally** *adv.*

minimalism /ˈmɪnɪmə,lɪz(ə)m/ *n.* **1** *Art* use of simple or primary forms, often geometric and massive. **2** *Mus.* repetition of short phrases incorporating changes very gradually. □ **minimalist** *n.* & *adj.*

minimize /ˈmɪnɪ,maɪz/ *v.* (also **-ise**) (**-zing** or **-sing**) **1** reduce to, or estimate at, the smallest possible amount or degree. **2** estimate or represent at less than true value or importance. □ **minimization** /-ˈzeɪʃ(ə)n/ *n.*

minimum /ˈmɪnɪməm/ (*pl.* **minima**) *–n.* least possible or attainable amount (*reduced to a minimum*). *–adj.* that is a minimum. [Latin: related to MINIM]

minimum wage *n.* lowest wage permitted by law or agreement.

minion /ˈmɪnjən/ *n. derog.* servile subordinate. [French *mignon*]

minipill /ˈmɪnɪpɪl/ *n.* contraceptive pill containing a progestogen only (not oestrogen).

miniseries /ˈmɪnɪ,sɪərɪz/ *n.* (*pl.* same) short series of related television programmes.

miniskirt /ˈmɪnɪ,skɜ:t/ *n.* very short skirt.

minister /ˈmɪnɪstə(r)/ *–n.* **1** head of a government department. **2** clergyman, esp. in the Presbyterian and Nonconformist Churches. **3** diplomat, usu. ranking below an ambassador. *–v.* (usu. foll. by *to*) help, serve, look after (a person, cause, etc.). □ **ministerial** /-ˈstɪərɪəl/ *adj.* [Latin, = servant]

Minister of State *n.* government minister, esp. holding a rank below that of Head of Department.

Minister of the Crown *n. Parl.* member of the Cabinet.

Minister without Portfolio *n.* government minister not in charge of a specific department of State.

ministration /,mɪnɪˈstreɪʃ(ə)n/ *n.* **1** (usu. in *pl.*) help or service (*kind ministrations*). **2** ministering, esp. in religious matters. **3** (usu. foll. by *of*) supplying of help, justice, etc. □ **ministrant** /ˈmɪnɪstrənt/ *adj.* & *n.* [Latin: related to MINISTER]

ministry /ˈmɪnɪstrɪ/ *n.* (*pl.* **-ies**) **1 a** government department headed by a minister. **b** building for this. **2 a** (prec. by *the*) vocation, office, or profession of a religious minister. **b** period of tenure of this. **3** (prec. by *the*) body of ministers of a government or religion. **4** period of government under one prime minister. **5** ministering, ministration. [Latin: related to MINISTER]

mink *n.* (*pl.* same or **-s**) **1** small semi-aquatic stoatlike animal bred for its thick brown fur. **2** this fur. **3** coat of this. [Swedish]

minnow /ˈmɪnəʊ/ *n.* small freshwater carp. [Old English]

Minoan /mɪˈnəʊən/ *–adj.* of the Bronze Age civilization centred on Crete (*c.*3000–1100 BC). *–n.* person of this civilization. [*Minos*, legendary king of Crete]

minor /ˈmaɪnə(r)/ *–adj.* **1** lesser or comparatively small in size or importance (*minor poet*). **2** *Mus.* **a** (of a scale) having intervals of a semitone above its second, fifth, and seventh notes. **b** (of an interval) less by a semitone than a major interval. **c** (of a key) based on a minor scale. *–n.* **1** person under full legal age. **2** *US* student's subsidiary subject or course. *–v.* (foll. by *in*) *US* study (a subject) as a subsidiary. [Latin, = less]

minority /maɪˈnɒrɪtɪ/ *n.* (*pl.* **-ies**) **1** (often foll. by *of*) smaller number or part, esp. in politics. **2** state of having less than half the votes or support (*in the minority*). **3** small group of people differing from others in race, religion, language, etc. **4** (*attrib.*) of or done by the minority (*minority interests*). **5 a** being under full legal age. **b** period of this. [French or medieval Latin: related to MINOR]

minster *n.* **1** large or important church. **2** church of a monastery. [Old English: related to MONASTERY]

minstrel /ˈmɪnstr(ə)l/ *n.* **1** medieval singer or musician. **2** (usu. in *pl.*) entertainer with a blacked face singing ostensibly Black songs in a group. [related to MINISTER]

mint[1] *n.* **1** aromatic herb used in cooking. **2** peppermint. **3** peppermint sweet. □ **minty** *adj.* (**-ier, -iest**). [Latin *menta* from Greek]

mint[2] *–n.* **1** (esp. State) establishment where money is coined. **2** *colloq.* vast sum (*making a mint*). –*v.* **1** make (a coin) by stamping metal. **2** invent, coin (a word, phrase, etc.). □ **in mint condition** as new. [Latin *moneta*]

minuet /ˌmɪnjʊ'et/ –*n.* **1** slow stately dance for two in triple time. **2** music for this, often as a movement in a suite etc. –*v.* (**-t-**) dance a minuet. [French diminutive]

minus /'maɪnəs/ –*prep.* **1** with the subtraction of (*7 minus 4 equals 3*). **2** below zero (*minus 2°*). **3** *colloq.* lacking (*returned minus their dog*). –*adj.* **1** *Math.* negative. **2** *Electronics* having a negative charge. –*n.* **1** = MINUS SIGN. **2** *Math.* negative quantity. **3** *colloq.* disadvantage. [Latin, neuter of MINOR]

minuscule /'mɪnəˌskju:l/ *adj. colloq.* extremely small or unimportant. [Latin diminutive: related to MINUS]

minus sign *n.* the symbol –, indicating subtraction or a negative value.

minute[1] /'mɪnɪt/ –*n.* **1** sixtieth part of an hour. **2** distance covered in one minute (*ten minutes from the shops*). **3 a** moment (*expecting her any minute*). **b** (prec. by *the*) *colloq.* present time (*not here at the minute*). **c** (prec. by *the*, foll. by a clause) as soon as (*the minute you get back*). **4** sixtieth part of an angular degree. **5** (in *pl.*) summary of the proceedings of a meeting. **6** official memorandum authorizing or recommending a course of action. –*v.* (**-ting**) **1** record in minutes. **2** send the minutes of a meeting to. □ **up to the minute** completely up to date. [Latin *minuo* lessen]

minute[2] /maɪ'nju:t/ *adj.* (**-est**) **1** very small. **2** accurate, detailed. □ **minutely** *adv.* [Latin *minutus*: related to MINUTE[1]]

minute steak *n.* thin quickly-cooked slice of steak.

minutiae /maɪ'nju:ʃɪi:/ *n.pl.* very small, precise, or minor details. [Latin: related to MINUTE[1]]

minx /mɪŋks/ *n.* pert, sly, or playful girl. [origin unknown]

Miocene /'mɪəˌsi:n/ *Geol.* –*adj.* of the fourth epoch of the Tertiary period. –*n.* this epoch. [Greek *meiōn* less, *kainos* new]

miracle /'mɪrək(ə)l/ *n.* **1** extraordinary, supposedly supernatural, event. **2** remarkable occurrence or development (*economic miracle*). **3** (usu. foll. by *of*) remarkable specimen (*a miracle of ingenuity*). [Latin *mirus* wonderful]

miracle play *n.* medieval play on biblical themes.

miraculous /mɪ'rækjʊləs/ *adj.* **1** being a miracle. **2** supernatural. **3** remarkable, surprising. □ **miraculously** *adv.* [French or medieval Latin: related to MIRACLE]

mirage /'mɪrɑ:ʒ/ *n.* **1** optical illusion caused by atmospheric conditions, esp. the appearance of a pool of water in a desert etc. from the reflection of light. **2** illusory thing. [Latin *miro* look at]

MIRAS /'maɪræs/ *abbr.* mortgage interest relief at source.

mire –*n.* **1** area of swampy ground. **2** mud, dirt. –*v.* (**-ring**) **1** plunge or sink in a mire. **2** involve in difficulties. **3** bespatter; besmirch. □ **miry** *adj.* [Old Norse]

mirror /'mɪrə(r)/ –*n.* **1** polished surface, usu. of coated glass, reflecting an image. **2** anything reflecting or illuminating a state of affairs etc. –*v.* reflect in or as in a mirror. [Latin *miro* look at]

mirror image *n.* identical image or reflection with left and right reversed.

mirth *n.* merriment, laughter. □ **mirthful** *adj.* [Old English: related to MERRY]

mis-[1] *prefix* added to verbs and verbal derivatives: meaning 'amiss', 'badly', 'wrongly', 'unfavourably' (*mislead*; *misshapen*; *mistrust*). [Old English]

mis-[2] *prefix* occurring in some verbs, nouns, and adjectives meaning 'badly', 'wrongly', 'amiss', 'ill-', or having a negative force (*misadventure*; *mischief*). [Latin *minus*]

misadventure /ˌmɪsəd'ventʃə(r)/ *n.* **1** *Law* accident without crime or negligence (*death by misadventure*). **2** bad luck. **3** a misfortune.

misalliance /ˌmɪsə'laɪəns/ *n.* unsuitable alliance, esp. a marriage.

misanthrope /'mɪsənˌθrəʊp/ *n.* (also **misanthropist** /mɪ'sænθrəpɪst/) **1** person who hates mankind. **2** person who avoids human society. □ **misanthropic** /-'θrɒpɪk/ *adj.* **misanthropically** /-'θrɒpɪkəlɪ/ *adv.* [Greek *misos* hatred, *anthrōpos* man]

misanthropy /mɪ'sænθrəpɪ/ *n.* condition or habits of a misanthrope.

misapply /ˌmɪsə'plaɪ/ *v.* (**-ies, -ied**) apply (esp. funds) wrongly. □ **misapplication** /mɪsˌæplɪ'keɪʃ(ə)n/ *n.*

misapprehend /ˌmɪsæprɪ'hend/ *v.* misunderstand (words, a person). □ **misapprehension** /-'henʃ(ə)n/ *n.*

misappropriate /ˌmɪsə'prəʊprɪˌeɪt/ *v.* (**-ting**) take (another's money etc.) for one's own use; embezzle. □ **misappropriation** /-'eɪʃ(ə)n/ *n.*

misbegotten /ˌmɪsbɪ'gɒt(ə)n/ *adj.* **1** illegitimate, bastard. **2** contemptible, disreputable.

misbehave /ˌmɪsbɪ'heɪv/ *v.* & *refl.* (**-ving**) behave badly. □ **misbehaviour** *n.*

misc. *abbr.* miscellaneous.

miscalculate /ˌmɪs'kælkjʊˌleɪt/ *v.* (**-ting**) calculate wrongly. □ **miscalculation** /-'leɪʃ(ə)n/ *n.*

miscarriage /ˈmɪs,kærɪdʒ/ *n.* spontaneous premature expulsion of a foetus from the womb.

miscarriage of justice *n.* failure of the judicial system to attain justice.

miscarry /mɪsˈkærɪ/ *v.* (**-ies, -ied**) **1** (of a woman) have a miscarriage. **2** (of a plan etc.) fail.

miscast /mɪsˈkɑːst/ *v.* (*past* and *past part.* **-cast**) allot an unsuitable part to (an actor) or unsuitable actors to (a play etc.).

miscegenation /,mɪsɪdʒɪˈneɪʃ(ə)n/ *n.* interbreeding of races, esp. of Whites and non-Whites. [related to MIX, GENUS]

miscellaneous /,mɪsəˈleɪnɪəs/ *adj.* **1** of mixed composition or character. **2** (foll. by a plural noun) of various kinds. □ **miscellaneously** *adv.* [Latin *misceo* mix]

miscellany /mɪˈselənɪ/ *n.* (*pl.* **-ies**) **1** mixture, medley. **2** book containing various literary compositions. [Latin: related to MISCELLANEOUS]

mischance /mɪsˈtʃɑːns/ *n.* **1** bad luck. **2** instance of this. [French: related to MIS-[2]]

mischief /ˈmɪstʃɪf/ *n.* **1** troublesome, but not malicious, conduct, esp. of children (*get into mischief*). **2** playfulness; malice (*eyes full of mischief*). **3** harm, injury (*do someone a mischief*). □ **make mischief** create discord. [French: related to MIS-[2], *chever* happen]

mischievous /ˈmɪstʃɪvəs/ *adj.* **1** (of a person) disposed to mischief. **2** (of conduct) playful; malicious. **3** harmful. □ **mischievously** *adv.* **mischievousness** *n.*

miscible /ˈmɪsɪb(ə)l/ *adj.* capable of being mixed. □ **miscibility** /-ˈbɪlɪtɪ/ *n.* [medieval Latin: related to MIX]

misconceive /,mɪskənˈsiːv/ *v.* (**-ving**) **1** (often foll. by *of*) have a wrong idea or conception. **2** (as **misconceived** *adj.*) badly planned, organized, etc. □ **misconception** /-ˈsepʃ(ə)n/ *n.* [from MIS-[1]]

misconduct /mɪsˈkɒndʌkt/ *n.* improper or unprofessional behaviour.

misconstrue /,mɪskənˈstruː/ *v.* (**-strues, -strued, -struing**) interpret wrongly. □ **misconstruction** /-ˈstrʌkʃ(ə)n/ *n.*

miscopy /mɪsˈkɒpɪ/ *v.* (**-ies, -ied**) copy inaccurately.

miscount /mɪsˈkaʊnt/ —*v.* (also *absol.*) count inaccurately. —*n.* inaccurate count.

miscreant /ˈmɪskrɪənt/ *n.* vile wretch, villain. [French: related to MIS-[2], *creant* believer]

misdeed /mɪsˈdiːd/ *n.* evil deed, wrongdoing, crime. [Old English]

misdemeanour /,mɪsdɪˈmiːnə(r)/ *n.* (*US* **misdemeanor**) **1** misdeed. **2** *hist* indictable offence less serious than a felony. [from MIS-[1]]

misdiagnose /,mɪsˈdaɪəg,nəʊz/ *v.* (**-sing**) diagnose incorrectly. □ **misdiagnosis** /-ˈnəʊsɪs/ *n.*

misdial /mɪsˈdaɪəl/ *v.* (also *absol.*) (**-ll-**; *US* **-l-**) dial (a telephone number etc.) incorrectly.

misdirect /,mɪsdaɪˈrekt/ *v.* direct wrongly. □ **misdirection** *n.*

misdoing /mɪsˈduːɪŋ/ *n.* misdeed.

miser /ˈmaɪzə(r)/ *n.* **1** person who hoards wealth and lives miserably. **2** avaricious person. □ **miserly** *adj.* [Latin, = wretched]

miserable /ˈmɪzərəb(ə)l/ *adj.* **1** wretchedly unhappy or uncomfortable. **2** contemptible, mean. **3** causing wretchedness or discomfort (*miserable weather*). □ **miserableness** *n.* **miserably** *adv.* [Latin: related to MISER]

misericord /mɪˈzerɪ,kɔːd/ *n.* projection under a choir stall seat serving (when the seat is turned up) to support a person standing. [Latin *misericordia* pity]

misery /ˈmɪzərɪ/ *n.* (*pl.* **-ies**) **1** condition or feeling of wretchedness. **2** cause of this. **3** *colloq.* constantly depressed or discontented person. [Latin: related to MISER]

misfield /mɪsˈfiːld/ —*v.* (also *absol.*) (in cricket, baseball, etc.) field (the ball) badly. —*n.* instance of this. [from MIS-[1]]

misfire /mɪsˈfaɪə(r)/ —*v.* (**-ring**) **1** (of a gun, motor engine, etc.) fail to go off or start or function smoothly. **2** (of a plan etc.) fail to have the intended effect. —*n.* such failure.

misfit *n.* **1** person unsuited to an environment, occupation, etc. **2** garment etc. that does not fit.

misfortune /mɪsˈfɔːtʃ(ə)n/ *n.* **1** bad luck. **2** instance of this.

misgive /mɪsˈgɪv/ *v.* (**-ving**; *past* **-gave**; *past part.* **-given** /-ˈgɪv(ə)n/) (of a person's mind, heart, etc.) fill (a person) with suspicion or foreboding.

misgiving /mɪsˈgɪvɪŋ/ *n.* (usu. in *pl.*) feeling of mistrust or apprehension.

misgovern /mɪsˈgʌv(ə)n/ *v.* govern badly. □ **misgovernment** *n.*

misguided /mɪsˈgaɪdɪd/ *adj.* mistaken in thought or action. □ **misguidedly** *adv.* **misguidedness** *n.*

mishandle /mɪsˈhænd(ə)l/ *v.* (**-ling**) **1** deal with incorrectly or inefficiently. **2** handle roughly or rudely.

mishap /ˈmɪshæp/ *n.* unlucky accident.

mishear /mɪsˈhɪə(r)/ *v.* (*past* and *past part.* **-heard** /-ˈhɜːd/) hear incorrectly or imperfectly.

mishit –*v.* /mɪs'hɪt/ (**-tt-**; *past* and *past part.* **-hit**) hit (a ball etc.) badly. –*n.* /'mɪshɪt/ faulty or bad hit.

mishmash *n.* confused mixture. [reduplication of MASH]

misinform /ˌmɪsɪn'fɔ:m/ *v.* give wrong information to, mislead. □ **misinformation** /-fə'meɪʃ(ə)n/ *n.* [from MIS-¹]

misinterpret /ˌmɪsɪn'tɜ:prɪt/ *v.* (**-t-**) **1** interpret wrongly. **2** draw a wrong inference from. □ **misinterpretation** /-'teɪʃ(ə)n/ *n.*

M.I.6 *abbr.* UK department of Military Intelligence concerned with espionage.

■ **Usage** This term is not in official use.

misjudge /mɪs'dʒʌdʒ/ *v.* (**-ging**) (also *absol.*) **1** judge wrongly. **2** have a wrong opinion of. □ **misjudgement** *n.* (also **-judgment**).

miskey /mɪs'ki:/ *v.* (**-keys, -keyed**) key (data) wrongly.

mislay /mɪs'leɪ/ *v.* (*past* and *past part.* **-laid**) accidentally put (a thing) where it cannot readily be found.

mislead /mɪs'li:d/ *v.* (*past* and *past part.* **-led**) cause to infer what is not true; deceive. □ **misleading** *adj.* [Old English]

mismanage /mɪs'mænɪdʒ/ *v.* (**-ging**) manage badly or wrongly. □ **mismanagement** *n.* [from MIS-¹]

mismatch –*v.* /mɪs'mætʃ/ match unsuitably or incorrectly. –*n.* /'mɪsmætʃ/ bad match.

misnomer /mɪs'nəʊmə(r)/ *n.* **1** name or term used wrongly. **2** wrong use of a name or term. [Anglo-French: related to MIS-², *nommer* to name]

misogyny /mɪ'sɒdʒɪnɪ/ *n.* hatred of women. □ **misogynist** *n.* **misogynistic** /-'nɪstɪk/ *adj.* [Greek *misos* hatred, *gunē* woman]

misplace /mɪs'pleɪs/ *v.* (**-cing**) **1** put in the wrong place. **2** bestow (affections, confidence, etc.) on an inappropriate object. □ **misplacement** *n.*

misprint –*n.* /'mɪsprɪnt/ printing error. –*v.* /mɪs'prɪnt/ print wrongly.

misprision /mɪs'prɪʒ(ə)n/ *n. Law* **1** (in full **misprision of a felony** or **of treason**) deliberate concealment of one's knowledge of a crime or treason. **2** wrong action or omission. [Anglo-French: related to MIS-², *prendre* take]

mispronounce /ˌmɪsprə'naʊns/ *v.* (**-cing**) pronounce (a word etc.) wrongly. □ **mispronunciation** /-ˌnʌnsɪ'eɪʃ(ə)n/ *n.* [from MIS-¹]

misquote /mɪs'kwəʊt/ *v.* (**-ting**) quote inaccurately. □ **misquotation** /-'teɪʃ(ə)n/ *n.*

misread /mɪs'ri:d/ *v.* (*past* and *past part.* **-read** /-'red/) read or interpret wrongly.

misrepresent /ˌmɪsreprɪ'zent/ *v.* represent wrongly; give a false account or idea of. □ **misrepresentation** /-'teɪʃ(ə)n/ *n.*

misrule /mɪs'ru:l/ –*n.* bad government; disorder. –*v.* (**-ling**) govern badly.

miss¹ –*v.* **1** (also *absol.*) fail to hit, reach, find, catch, etc. (an object or goal). **2** fail to catch (a bus, train, etc.) or see (an event) or meet (a person). **3** fail to seize (an opportunity etc.) (*missed my chance*). **4** fail to hear or understand (*missed what you said*). **5 a** regret the loss or absence of (*did you miss me?*). **b** notice the loss or absence of (*won't be missed until evening*). **6** avoid (*go early to miss the traffic*). **7** (of an engine etc.) fail, misfire. –*n.* failure to hit, reach, attain, connect, etc. □ **be missing** not have (*am missing a page*) (see also MISSING). **give (a thing) a miss** *colloq.* not attend or partake of (*gave the party a miss*). **miss out 1** omit, leave out. **2** (usu. foll. by *on*) *colloq.* fail to get or experience. [Old English]

miss² *n.* **1** (**Miss**) **a** title of an unmarried woman or girl. **b** title of a beauty queen (*Miss World*). **2** title used to address a female schoolteacher, shop assistant, etc. **3** girl or unmarried woman. [from MISTRESS]

missal /'mɪs(ə)l/ *n. RC Ch.* **1** book containing the texts for the Mass throughout the year. **2** book of prayers. [Latin *missa* MASS²]

missel thrush var. of MISTLE THRUSH.

misshapen /mɪs'ʃeɪpən/ *adj.* ill-shaped, deformed, distorted. [from MIS-¹, *shapen* (archaic) = shaped]

missile /'mɪsaɪl/ *n.* **1** object or weapon suitable for throwing at a target or for discharge from a machine. **2** weapon directed by remote control or automatically. [Latin *mitto miss-* send]

missing *adj.* **1** not in its place; lost. **2** (of a person) not yet traced or confirmed as alive but not known to be dead. **3** not present.

missing link *n.* **1** thing lacking to complete a series. **2** hypothetical intermediate type, esp. between humans and apes.

mission /'mɪʃ(ə)n/ *n.* **1 a** task or goal assigned to a person or group. **b** journey undertaken as part of this. **c** person's vocation. **2** military or scientific operation or expedition. **3** body of persons sent to conduct negotiations or propagate a religious faith. **4** missionary post. [Latin: related to MISSILE]

missionary /ˈmɪʃənərɪ/ *–adj.* of or concerned with religious missions. *–n.* (*pl.* **-ies**) person doing missionary work. [Latin: related to MISSION]

missionary position *n. colloq.* position for sexual intercourse with the woman lying on her back and the man lying on top and facing her.

missis /ˈmɪsɪz/ *n.* (also **missus** /-səz/) *colloq.* or *joc.* **1** form of address to a woman. **2** wife. □ **the missis** my or your wife. [from MISTRESS]

missive /ˈmɪsɪv/ *n.* **1** *joc.* letter. **2** official letter. [Latin: related to MISSILE]

misspell /mɪsˈspel/ *v.* (*past* and *past part.* **-spelt** or **-spelled**) spell wrongly.

misspend /mɪsˈspend/ *v.* (*past* and *past part.* **-spent**) (esp. as **misspent** *adj.*) spend amiss or wastefully.

misstate /mɪsˈsteɪt/ *v.* (**-ting**) state wrongly or inaccurately. □ **misstatement** *n.*

missus var. of MISSIS.

mist *–n.* **1 a** diffuse cloud of minute water droplets near the ground. **b** condensed vapour obscuring glass etc. **2** dimness or blurring of the sight caused by tears etc. **3** cloud of particles resembling mist. *–v.* (usu. foll. by *up*, *over*) cover or become covered with mist or as with mist. [Old English]

mistake /mɪˈsteɪk/ *–n.* **1** incorrect idea or opinion; thing incorrectly done or thought. **2** error of judgement. *–v.* (**-king**; *past* **mistook** /-ˈstʊk/; *past part.* **mistaken**) **1** misunderstand the meaning of. **2** (foll. by *for*) wrongly take or identify (*mistook me for you*). **3** choose wrongly (*mistake one's vocation*). [Old Norse: related to MIS-[1], TAKE]

mistaken /mɪˈsteɪkən/ *adj.* **1** wrong in opinion or judgement. **2** based on or resulting from this (*mistaken loyalty*; *mistaken identity*). □ **mistakenly** *adv.*

mister /ˈmɪstə(r)/ *n. colloq.* or *joc.* form of address to a man. [from MASTER; cf. MR]

mistime /mɪsˈtaɪm/ *v.* (**-ming**) say or do at the wrong time. [related to MIS-[1]]

mistle thrush /ˈmɪs(ə)l/ *n.* (also **missel thrush**) large thrush with a spotted breast, feeding on mistletoe berries. [Old English]

mistletoe /ˈmɪs(ə)lˌtəʊ/ *n.* parasitic plant with white berries growing on apple and other trees. [Old English]

mistook *past* of MISTAKE.

mistral /ˈmɪstr(ə)l/ *n.* cold N. or NW wind in S. France. [Latin: related to MASTER]

mistreat /mɪsˈtriːt/ *v.* treat badly. □ **mistreatment** *n.*

mistress /ˈmɪstrɪs/ *n.* **1** female head of a household. **2 a** woman in authority. **b** female owner of a pet. **3** female teacher. **4** woman having an illicit sexual relationship with a (usu. married) man. [French *maistre* MASTER, -ESS]

mistrial /mɪsˈtraɪəl/ *n.* trial rendered invalid by error.

mistrust /mɪsˈtrʌst/ *–v.* **1** be suspicious of. **2** feel no confidence in. *–n.* **1** suspicion. **2** lack of confidence. □ **mistrustful** *adj.* **mistrustfully** *adv.*

misty *adj.* (**-ier**, **-iest**) **1** of or covered with mist. **2** dim in outline. **3** obscure, vague (*misty idea*). □ **mistily** *adv.* **mistiness** *n.* [Old English: related to MIST]

misunderstand /ˌmɪsʌndəˈstænd/ *v.* (*past* and *past part.* **-understood** /-ˈstʊd/) **1** understand incorrectly. **2** misinterpret the words or actions of (a person).

misunderstanding *n.* **1** failure to understand correctly. **2** slight disagreement or quarrel.

misusage /mɪsˈjuːsɪdʒ/ *n.* **1** wrong or improper usage. **2** ill-treatment.

misuse *–v.* /mɪsˈjuːz/ (**-sing**) **1** use wrongly; apply to the wrong purpose. **2** ill-treat. *–n.* /mɪsˈjuːs/ wrong or improper use or application.

MIT *abbr.* Massachusetts Institute of Technology.

mite[1] *n.* small arachnid, esp. of a kind found in cheese etc. [Old English]

mite[2] *n.* **1** any small monetary unit. **2** small object or person, esp. a child. **3** modest contribution. [probably the same as MITE[1]]

miter *US* var. of MITRE.

mitigate /ˈmɪtɪˌgeɪt/ *v.* (**-ting**) make less intense or severe. □ **mitigation** /-ˈgeɪʃ(ə)n/ *n.* [Latin *mitis* mild]

mitigating circumstances *n.pl.* circumstances permitting greater leniency.

mitosis /maɪˈtəʊsɪs/ *n. Biol.* type of cell division that results in two nuclei each having the same number and kind of chromosomes as the parent nucleus. □ **mitotic** /-ˈtɒtɪk/ *adj.* [Greek *mitos* thread]

mitre /ˈmaɪtə(r)/ (*US* **miter**) *–n.* **1** tall deeply-cleft headdress worn by bishops and abbots, esp. as a symbol of office. **2** joint of two pieces of wood etc. at an angle of 90°, such that the line of junction bisects this angle. *–v.* (**-ring**) **1** bestow a mitre on. **2** join with a mitre. [Greek *mitra* turban]

mitt *n.* **1** (also **mitten**) glove with only two compartments, one for the thumb and the other for all four fingers. **2** glove leaving the fingers and thumb-tip exposed. **3** *slang* hand or fist. **4** baseball glove. [Latin: related to MOIETY]

mix –*v.* **1** combine or put together (two or more substances or things) so that the constituents of each are diffused among those of the other(s). **2** prepare (a compound, cocktail, etc.) by combining the ingredients. **3** combine (activities etc.) (*mix business and pleasure*). **4 a** join, be mixed, or combine, esp. readily (*oil and water will not mix*). **b** be compatible. **c** be sociable (*must learn to mix*). **5 a** (foll. by *with*) (of a person) be harmonious or sociable with; have regular dealings with. **b** (foll. by *in*) participate in. **6** drink different kinds of (alcoholic liquor) in close succession. **7** combine (two or more sound signals) into one. –*n.* **1 a** mixing; mixture. **b** proportion of materials in a mixture. **2** ingredients prepared commercially for making a cake, concrete, etc. □ **be mixed up in** (or **with**) be involved in or with (esp. something undesirable). **mix it** *colloq.* start fighting. **mix up 1** mix thoroughly. **2** confuse. [back-formation from MIXED]

mixed /mɪkst/ *adj.* **1** of diverse qualities or elements. **2** containing persons from various backgrounds etc. **3** for persons of both sexes (*mixed school*). [Latin *misceo* mix]

mixed bag *n.* diverse assortment.

mixed blessing *n.* thing having advantages and disadvantages.

mixed doubles *n.pl. Tennis* doubles game with a man and a woman on each side.

mixed economy *n.* economic system combining private and State enterprise.

mixed farming *n.* farming of both crops and livestock.

mixed feelings *n.pl.* mixture of pleasure and dismay about something.

mixed grill *n.* dish of various grilled meats and vegetables etc.

mixed marriage *n.* marriage between persons of different race or religion.

mixed metaphor *n.* combination of inconsistent metaphors (e.g. *this tower of strength will forge ahead*).

mixed-up *adj. colloq.* mentally or emotionally confused; socially ill-adjusted.

mixer *n.* **1** machine for mixing foods etc. **2** person who manages socially in a specified way (*a good mixer*). **3** (usu. soft) drink to be mixed with another. **4** device that receives two or more separate signals from microphones etc. and combines them in a single output.

mixer tap *n.* tap through which both hot and cold water can be drawn together.

mixture /'mɪkstʃə(r)/ *n.* **1** process or result of mixing. **2** combination of ingredients, qualities, characteristics, etc. [Latin: related to MIXED]

mix-up *n.* confusion, misunderstanding.

mizen /'mɪz(ə)n/ *n.* (also **mizzen**) (in full **mizen-sail**) lowest fore-and-aft sail of a fully rigged ship's mizen-mast. [Italian: related to MEZZANINE]

mizen-mast *n.* (also **mizzen-mast**) mast next aft of the mainmast.

ml *abbr.* **1** millilitre(s). **2** mile(s).

M.Litt. *abbr.* Master of Letters. [Latin *Magister Litterarum*]

Mlle *abbr.* (*pl.* **-s**) Mademoiselle.

MM *abbr.* **1** Messieurs. **2** Military Medal.

mm *abbr.* millimetre(s).

Mme *abbr.* (*pl.* **-s**) Madame.

Mn *symb.* manganese.

mnemonic /nɪ'mɒnɪk/ –*adj.* of or designed to aid the memory. –*n.* mnemonic word, verse, etc. □ **mnemonically** *adv.* [Greek *mnēmōn* mindful]

MO *abbr.* **1** Medical Officer. **2** money order.

Mo *symb.* molybdenum.

mo /məʊ/ *n.* (*pl.* **-s**) *colloq.* moment. [abbreviation]

moa /'məʊə/ *n.* (*pl.* **-s**) extinct flightless New Zealand bird resembling the ostrich. [Maori]

moan –*n.* **1** long murmur expressing physical or mental suffering or pleasure. **2** low plaintive sound of wind etc. **3** *colloq.* complaint; grievance. –*v.* **1** make a moan or moans. **2** *colloq.* complain, grumble. **3** utter with moans. □ **moaner** *n.* [Old English]

moat *n.* defensive ditch round a castle etc., usu. filled with water. [French *mote* mound]

mob –*n.* **1** disorderly crowd; rabble. **2** (prec. by *the*) usu. *derog.* the populace. **3** *colloq.* gang; group. –*v.* (**-bb-**) crowd round in order to attack or admire. [Latin *mobile vulgus* excitable crowd]

mob-cap *n. hist.* woman's large indoor cap covering all the hair. [obsolete *mob*, originally = slut]

mobile /'məʊbaɪl/ –*adj.* **1** movable; able to move easily or get out and about. **2** (of the face etc.) readily changing its expression. **3** (of a shop etc.) accommodated in a vehicle so as to serve various places. **4** (of a person) able to change his or her social status. –*n.* decoration that may be hung so as to turn freely. □ **mobility** /mə'bɪlɪtɪ/ *n.* [Latin *moveo* move]

mobile home *n.* large caravan usu. permanently parked and used as a residence.

mobilize /ˈməʊbɪˌlaɪz/ *v.* (also **-ise**) (**-zing** or **-sing**) esp. *Mil.* make or become ready for service or action. □ **mobilization** /-ˈzeɪʃ(ə)n/ *n.*

Möbius strip /ˈmɜːbɪəs/ *n. Math.* one-sided surface formed by joining the ends of a narrow rectangle after twisting one end through 180°. [*Möbius*, name of a mathematician]

mobster *n. slang* gangster.

moccasin /ˈmɒkəsɪn/ *n.* soft flat-soled shoe orig. worn by N. American Indians. [American Indian]

mocha /ˈmɒkə/ *n.* **1** coffee of fine quality. **2** flavouring made with this. [*Mocha*, port on the Red Sea]

mock *–v.* **1** (often foll. by *at*) ridicule; scoff (at); act with scorn or contempt for. **2** mimic contemptuously. **3** defy or delude contemptuously. *–attrib. adj.* **1** sham, imitation. **2** as a trial run (*mock exam*). *–n.* (in *pl.*) *colloq.* mock examinations. □ **mockingly** *adv.* [French *moquer*]

mocker *n.* person who mocks. □ **put the mockers on** *slang* **1** bring bad luck to. **2** put a stop to.

mockery *n.* (*pl.* **-ies**) **1** derision, ridicule. **2** counterfeit or absurdly inadequate representation. **3** ludicrously or insultingly futile action etc.

mockingbird *n.* bird that mimics the notes of other birds.

mock orange *n.* white-flowered heavy-scented shrub.

mock turtle soup *n.* soup made from a calf's head etc. to resemble turtle soup.

mock-up *n.* experimental model or replica of a proposed structure etc.

MOD *abbr.* Ministry of Defence.

mod *colloq. –adj.* modern. *–n.* young person (esp. in the 1960s) of a group known for its smart modern dress. [abbreviation]

modal /ˈməʊd(ə)l/ *adj.* **1** of mode or form, not of substance. **2** *Gram.* **a** of the mood of a verb. **b** (of an auxiliary verb, e.g. *would*) used to express the mood of another verb. **3** *Mus.* denoting a style of music using a particular mode. [Latin: related to MODE]

mod cons *n.pl.* modern conveniences.

mode *n.* **1** way in which a thing is done. **2** prevailing fashion or custom. **3** *Mus.* any of several types of scale. [French and Latin *modus* measure]

model /ˈmɒd(ə)l/ *–n.* **1** representation in three dimensions of an existing person or thing or of a proposed structure, esp. on a smaller scale (often *attrib.*: *model train*). **2** simplified description of a system etc., to assist calculations and predictions. **3** figure in clay, wax, etc., to be reproduced in another material. **4** particular design or style, esp. of a car. **5 a** exemplary person or thing. **b** (*attrib.*) ideal, exemplary. **6** person employed to pose for an artist or photographer or to wear clothes etc. for display. **7** garment etc. by a well-known designer, or a copy of this. *–v.* (**-ll-**; *US* **-l-**) **1 a** fashion or shape (a figure) in clay, wax, etc. **b** (foll. by *after*, *on*, etc.) form (a thing in imitation of). **2 a** act or pose as a model. **b** (of a person acting as a model) display (a garment). [Latin: related to MODE]

modem /ˈməʊdem/ *n.* combined device for modulation and demodulation, e.g. between a computer and a telephone line. [portmanteau word]

moderate *–adj.* /ˈmɒdərət/ **1** avoiding extremes; temperate in conduct or expression. **2** fairly large or good. **3** (of the wind) of medium strength. **4** (of prices) fairly low. *–n.* /ˈmɒdərət/ person who holds moderate views, esp. in politics. *–v.* /ˈmɒdəˌreɪt/ (**-ting**) **1** make or become less violent, intense, rigorous, etc. **2** (also *absol.*) act as moderator of or to. □ **moderately** /-rətlɪ/ *adv.* **moderateness** /-rətnəs/ *n.* [Latin]

moderation /ˌmɒdəˈreɪʃ(ə)n/ *n.* **1** moderateness. **2** moderating. □ **in moderation** in a moderate manner or degree.

moderato /ˌmɒdəˈrɑːtəʊ/ *adj.* & *adv.* *Mus.* at a moderate pace. [Italian]

moderator *n.* **1** arbitrator, mediator. **2** presiding officer. **3** Presbyterian minister presiding over an ecclesiastical body. **4** *Physics* substance used in a nuclear reactor to retard neutrons.

modern /ˈmɒd(ə)n/ *–adj.* **1** of present and recent times. **2** in current fashion; not antiquated. *–n.* person living in modern times. □ **modernity** /-ˈdɜːnɪtɪ/ *n.* [Latin *modo* just now]

modern English *n.* English from about 1500 onwards.

modernism *n.* modern ideas or methods, esp. in art. □ **modernist** *n.* & *adj.*

modernize *v.* (also **-ise**) (**-zing** or **-sing**) **1** make modern; adapt to modern needs or habits. **2** adopt modern ways or views. □ **modernization** /-ˈzeɪʃ(ə)n/ *n.*

modest /ˈmɒdɪst/ *adj.* **1** having or expressing a humble or moderate estimate of one's own merits. **2** diffident, bashful. **3** decorous. **4** moderate or restrained in amount, extent, severity, etc. **5** unpretentious, not extravagant. □ **modestly** *adv.* **modesty** *n.* [French from Latin]

modicum /ˈmɒdɪkəm/ *n.* (foll. by *of*) small quantity. [Latin: related to MODE]

modification /ˌmɒdɪfɪˈkeɪʃ(ə)n/ *n.* **1** modifying or being modified. **2** change made. □ **modificatory** /'mɒd-/ *adj.* [Latin: related to MODIFY]

modify /'mɒdɪ,faɪ/ *v.* (**-ies, -ied**) **1** make less severe or extreme. **2** make partial changes in. **3** *Gram.* qualify or expand the sense of (a word etc.). [Latin: related to MODE]

modish /'məʊdɪʃ/ *adj.* fashionable. □ **modishly** *adv.*

modiste /mɒ'di:st/ *n.* milliner; dressmaker. [French: related to MODE]

modulate /'mɒdjʊ,leɪt/ *v.* (**-ting**) **1 a** regulate or adjust. **b** moderate. **2** adjust or vary the tone or pitch of (the speaking voice). **3** alter the amplitude or frequency of (a wave) by using a wave of a lower frequency to convey a signal. **4** *Mus.* (cause to) change from one key to another. □ **modulation** /-'leɪʃ(ə)n/ *n.* [Latin: related to MODULE]

module /'mɒdju:l/ *n.* **1** standardized part or independent unit in construction, esp. of furniture, a building, or an electronic system. **2** independent self-contained unit of a spacecraft. **3** unit or period of training or education. □ **modular** *adj.* [Latin: related to MODULUS]

modulus /'mɒdjʊləs/ *n.* (*pl.* **moduli** /-,laɪ/) *Math.* constant factor or ratio. [Latin, = measure: related to MODE]

modus operandi /,məʊdəs ,ɒpə'rændɪ/ *n.* (*pl.* ***modi operandi*** /,məʊdɪ/) method of working. [Latin, = way of operating]

modus vivendi /,məʊdəs vɪ'vendɪ/ *n.* (*pl.* ***modi vivendi*** /,məʊdɪ/) **1** way of living or coping. **2** arrangement between people who agree to differ. [Latin, = way of living]

mog *n.* (also **moggie**) *slang* cat. [originally a dial. word]

Mogadon /'mɒgə,dɒn/ *n. propr.* hypnotic drug used to treat insomnia.

mogul /'məʊg(ə)l/ *n.* **1** *colloq.* important or influential person. **2** (**Mogul**) *hist.* **a** Mongolian. **b** (often **the Great Mogul**) emperor of Delhi in the 16th–19th c. [Persian and Arabic: related to MONGOL]

mohair /'məʊheə(r)/ *n.* **1** hair of the angora goat. **2** yarn or fabric from this. [ultimately from Arabic, = choice]

Mohammedan var. of MUHAMMADAN.

Mohican /məʊ'hi:kən/ —*adj.* (of a hairstyle) with the head shaved except for a strip of hair from the middle of the forehead to the back of the neck, often worn in long spikes. —*n.* such a hairstyle. [*Mohicans*, N. American Indian people]

moiety /'mɔɪətɪ/ *n.* (*pl.* **-ies**) *Law* or *literary* **1** half. **2** each of the two parts of a thing. [Latin *medietas* from *medius* middle]

moire /mwɑ:(r)/ *n.* (in full **moire antique**) watered fabric, usu. silk. [French: related to MOHAIR]

moiré /'mwɑ:reɪ/ *adj.* **1** (of silk) watered. **2** (of metal) having a clouded appearance. [French: related to MOIRE]

moiré pattern *n.* pattern observed when one pattern of lines etc. is superimposed on another.

moist *adj.* slightly wet; damp. [French]

moisten /'mɔɪs(ə)n/ *v.* make or become moist.

moisture /'mɔɪstʃə(r)/ *n.* water or other liquid diffused in a small quantity as vapour, or within a solid, or condensed on a surface.

moisturize *v.* (also **-ise**) (**-zing** or **-sing**) make less dry (esp. the skin by use of a cosmetic). □ **moisturizer** *n.*

molar /'məʊlə(r)/ —*adj.* (usu. of a mammal's back teeth) serving to grind. —*n.* molar tooth. [Latin *mola* millstone]

molasses /mə'læsɪz/ *n.pl.* (treated as *sing.*) **1** uncrystallized syrup extracted from raw sugar. **2** *US* treacle. [Portuguese from Latin *mel* honey]

mold *US* var. of MOULD[1], MOULD[2], MOULD[3].

molder *US* var. of MOULDER.

molding *US* var. of MOULDING.

moldy *US* var. of MOULDY.

mole[1] *n.* **1** small burrowing mammal with dark velvety fur and very small eyes. **2** *slang* spy established in a position of trust in an organization. [Low German or Dutch]

mole[2] *n.* small permanent dark spot on the skin. [Old English]

mole[3] *n.* **1** massive structure serving as a pier, breakwater, or causeway. **2** artificial harbour. [Latin *moles* mass]

mole[4] *n. Chem.* the SI unit of amount of a substance equal to the quantity containing as many elementary units as there are atoms in 0.012 kg of carbon-12. [German *Mol* from *Molekül* MOLECULE]

molecular /mə'lekjʊlə(r)/ *adj.* of, relating to, or consisting of molecules. □ **molecularity** /-'lærɪtɪ/ *n.*

molecular weight *n.* = RELATIVE MOLECULAR MASS.

molecule /'mɒlɪ,kju:l/ *n.* **1** smallest fundamental unit (usu. a group of atoms) of a chemical compound that can take part in a chemical reaction. **2** (in general use) small particle. [Latin diminutive: related to MOLE[3]]

molehill *n.* small mound thrown up by a mole in burrowing. □ **make a mountain out of a molehill** overreact to a minor difficulty.

molest /mə'lest/ *v.* **1** annoy or pester (a person). **2** attack or interfere with (a person), esp. sexually. □ **molestation**

/-ˈsteɪʃ(ə)n/ *n.* **molester** *n.* [Latin *molestus* troublesome]

moll *n. slang* **1** gangster's female companion. **2** prostitute. [pet form of *Mary*]

mollify /ˈmɒlɪˌfaɪ/ *v.* **(-ies, -ied)** appease. □ **mollification** /-fɪˈkeɪʃ(ə)n/ *n.* [Latin *mollis* soft]

mollusc /ˈmɒləsk/ *n.* (*US* **mollusk**) invertebrate with a soft body and usu. a hard shell, e.g. snails and oysters. [Latin *molluscus* soft]

mollycoddle /ˈmɒlɪˌkɒd(ə)l/ *v.* **(-ling)** coddle, pamper. [related to MOLL, CODDLE]

Molotov cocktail /ˈmɒləˌtɒf/ *n.* crude incendiary device, usu. a bottle filled with inflammable liquid. [*Molotov*, name of a Russian statesman]

molt *US* var. of MOULT.

molten /ˈməʊlt(ə)n/ *adj.* melted, esp. made liquid by heat. [from MELT]

molto /ˈmɒltəʊ/ *adv. Mus.* very. [Latin *multus* much]

molybdenum /məˈlɪbdɪnəm/ *n.* silver-white metallic element added to steel to give strength and resistance to corrosion. [Greek *molubdos* lead]

mom *n. US colloq.* mother. [abbreviation of MOMMA]

moment /ˈməʊmənt/ *n.* **1** very brief portion of time. **2** an exact point of time (*I came the moment you called*). **3** importance (*of no great moment*). **4** product of a force and the distance from its line of action to a point. □ **at the moment** now. **in a moment** very soon. **man** (or **woman** etc.) **of the moment** the one of importance at the time in question. [Latin: related to MOMENTUM]

momentary *adj.* lasting only a moment; transitory. □ **momentarily** *adv.* [Latin: related to MOMENT]

moment of truth *n.* time of crisis or test.

momentous /məˈmentəs/ *adj.* very important. □ **momentously** *adv.* **momentousness** *n.*

momentum /məˈmentəm/ *n.* (*pl.* **momenta**) **1** quantity of motion of a moving body, the product of its mass and velocity. **2** impetus gained by movement. **3** strength or continuity derived from an initial effort. [Latin *moveo* move]

momma /ˈmɒmə/ *n. US colloq.* mother. [var. of MAMA]

mommy /ˈmɒmɪ/ *n.* (*pl.* **-ies**) esp. *US colloq.* = MUMMY[1].

Mon. *abbr.* Monday.

monad /ˈmɒnæd, ˈməʊ-/ *n.* **1** the number one; unit. **2** *Philos.* ultimate unit of being (e.g. a soul, an atom, a person, God). □ **monadic** /məˈnædɪk/ *adj.* [Greek *monas -ados* unit]

monarch /ˈmɒnək/ *n.* sovereign with the title of king, queen, emperor, empress, or equivalent. □ **monarchic** /məˈnɑːkɪk/ *adj.* **monarchical** /məˈnɑːkɪk(ə)l/ *adj.* [Greek: related to MONO-, *arkhō* rule]

monarchism *n.* the advocacy of monarchy. □ **monarchist** *n.* [French: related to MONARCH]

monarchy *n.* (*pl.* **-ies**) **1** form of government with a monarch at the head. **2** State with this. □ **monarchial** /məˈnɑːkɪəl/ *adj.* [Greek: related to MONARCH]

monastery /ˈmɒnəstrɪ/ *n.* (*pl.* **-ies**) residence of a community of monks. [Latin *monasterium* from Greek *monazō* alone]

monastic /məˈnæstɪk/ *adj.* of or like monasteries or monks, nuns, etc. □ **monastically** *adv.* **monasticism** /-ˌsɪz(ə)m/ *n.* [Greek: related to MONASTERY]

Monday /ˈmʌndeɪ/ —*n.* day of the week following Sunday. —*adv. colloq.* **1** on Monday. **2** (**Mondays**) on Mondays; each Monday. [Old English]

monetarism /ˈmʌnɪtəˌrɪz(ə)m/ *n.* control of the supply of money as the chief method of stabilizing the economy. □ **monetarist** *n.* & *adj.*

monetary /ˈmʌnɪtərɪ/ *adj.* **1** of the currency in use. **2** of or consisting of money. [Latin: related to MONEY]

money /ˈmʌnɪ/ *n.* **1** coins and banknotes as a medium of exchange. **2** (*pl.* **-eys** or **-ies**) (in *pl.*) sums of money. **3 a** wealth. **b** wealth as power (*money talks*). **c** rich person or family (*married into money*). □ **for my money** in my opinion; for my preference. **in the money** *colloq.* having or winning a lot of money. **money for jam** (or **old rope**) *colloq.* profit for little or no trouble. [Latin *moneta*]

moneybags *n.pl.* (treated as *sing.*) *colloq.* usu. *derog.* wealthy person.

moneyed /ˈmʌnɪd/ *adj.* wealthy.

money-grubber *n. colloq.* person greedily intent on amassing money. □ **money-grubbing** *n.* & *adj.*

moneylender *n.* person who lends money at interest.

moneymaker *n.* **1** person who earns much money. **2** thing, idea, etc., that produces much money. □ **money-making** *n.* & *adj.*

money market *n.* trade in short-term stocks, loans, etc.

money order *n.* order for payment of a specified sum, issued by a bank or Post Office.

money-spinner *n.* thing that brings in a profit.

money's worth see ONE'S MONEY'S WORTH.

monger /ˈmʌŋgə(r)/ *n.* (usu. in *comb.*) **1** dealer, trader (*fishmonger*). **2** usu. *derog.* promoter, spreader (*warmonger*; *scaremonger*). [Latin *mango* dealer]

Mongol /ˈmɒŋg(ə)l/ *–adj.* **1** of the Asian people of Mongolia. **2** resembling this people. **3** (**mongol**) often *offens.* suffering from Down's syndrome. *–n.* **1** Mongolian. **2** (**mongol**) often *offens.* person suffering from Down's syndrome. [native name: perhaps from *mong* brave]

Mongolian /mɒŋˈgəʊlɪən/ *–n.* **1** native or inhabitant of Mongolia. **2** language of Mongolia. *–adj.* of or relating to Mongolia or its people or language.

mongolism /ˈmɒŋgə,lɪz(ə)m/ *n.* = DOWN'S SYNDROME.

■ **Usage** The term *Down's syndrome* is now preferred.

Mongoloid /ˈmɒŋgə,lɔɪd/ *–adj.* **1** characteristic of the Mongolians, esp. in having a broad flat yellowish face. **2** (**mongoloid**) often *offens.* having the characteristic symptoms of Down's syndrome. *–n.* Mongoloid or mongoloid person.

mongoose /ˈmɒŋgu:s/ *n.* (*pl.* **-s**) small flesh-eating civet-like mammal. [Marathi]

mongrel /ˈmʌŋgr(ə)l/ *–n.* **1** dog of no definable type or breed. **2** other animal or plant resulting from the crossing of different breeds or types. *–adj.* of mixed origin, nature, or character. [related to MINGLE]

monies see MONEY 2.

monism /ˈmɒnɪz(ə)m/ *n.* **1** doctrine that only one ultimate principle or being exists. **2** theory denying the duality of matter and mind. □ **monist** *n.* **monistic** /-ˈnɪstɪk/ *adj.* [Greek *monos* single]

monitor /ˈmɒnɪtə(r)/ *–n.* **1** person or device for checking or warning. **2** school pupil with disciplinary or other special duties. **3 a** television receiver used in a studio to select or verify the picture being broadcast. **b** = VISUAL DISPLAY UNIT. **4** person who listens to and reports on foreign broadcasts etc. **5** detector of radioactive contamination. *–v.* **1** act as a monitor of. **2** maintain regular surveillance over. **3** regulate the strength of (a recorded or transmitted signal). [Latin *moneo* warn]

monitory /ˈmɒnɪtərɪ/ *adj. literary* giving or serving as a warning. [Latin *monitorius*: related to MONITOR]

monk /mʌŋk/ *n.* member of a religious community of men living under vows. □ **monkish** *adj.* [Greek *monakhos* from *monos* alone]

monkey /ˈmʌŋkɪ/ *–n.* (*pl.* **-eys**) **1** any of various primates, including marmosets, baboons etc., esp. a small long-tailed kind. **2** mischievous person, esp. a child. *–v.* (**-eys, -eyed**) **1** (often foll. by *with*) tamper or play mischievous tricks. **2** (foll. by *around*, *about*) fool around. [origin unknown]

monkey business *n. colloq.* mischief.

monkey-nut *n.* peanut.

monkey-puzzle *n.* tree with hanging prickly branches.

monkey tricks *n.pl. colloq.* mischief.

monkey wrench *n.* wrench with an adjustable jaw.

monkshood /ˈmʌŋkshʊd/ *n.* poisonous plant with hood-shaped flowers.

mono /ˈmɒnəʊ/ *colloq.* *–adj.* monophonic. *–n.* monophonic reproduction. [abbreviation]

mono- *comb. form* (usu. **mon-** before a vowel) one, alone, single. [Greek *monos* alone]

monochromatic /,mɒnəkrəˈmætɪk/ *adj.* **1** (of light or other radiation) of a single colour or wavelength. **2** containing only one colour. □ **monochromatically** *adv.*

monochrome /ˈmɒnə,krəʊm/ *–n.* photograph or picture done in one colour or different tones of this, or in black and white only. *–adj.* having or using only one colour or in black and white only. [from MONO-, Greek *khrōma* colour]

monocle /ˈmɒnək(ə)l/ *n.* single eyeglass. □ **monocled** *adj.* [Latin: related to MONO-, *oculus* eye]

monocotyledon /,mɒnə,kɒtɪˈli:d(ə)n/ *n.* flowering plant with one cotyledon. □ **monocotyledonous** *adj.*

monocular /məˈnɒkjʊlə(r)/ *adj.* with or for one eye. [related to MONOCLE]

monody /ˈmɒnədɪ/ *n.* (*pl.* **-ies**) **1** ode sung by a single actor in a Greek tragedy. **2** poem lamenting a person's death. □ **monodist** *n.* [Greek: related to MONO-, ODE]

monogamy /məˈnɒgəmɪ/ *n.* practice or state of being married to one person at a time. □ **monogamous** *adj.* [Greek *gamos* marriage]

monogram /ˈmɒnə,græm/ *n.* two or more letters, esp. a person's initials, interwoven as a device.

monograph /ˈmɒnə,grɑ:f/ *n.* treatise on a single subject.

monolingual /,mɒnəʊˈlɪŋgw(ə)l/ *adj.* speaking or using only one language.

monolith /ˈmɒnəlɪθ/ *n.* **1** single block of stone, esp. shaped into a pillar etc. **2** person or thing like a monolith in being massive, immovable, or solidly uniform. □ **monolithic** /-ˈlɪθɪk/ *adj.* [Greek *lithos* stone]

monologue /ˈmɒnəˌlɒg/ *n.* **1 a** scene in a drama in which a person speaks alone. **b** dramatic composition for one performer. **2** long speech by one person in a conversation etc. [French from Greek *monologos* speaking alone]

monomania /ˌmɒnəˈmeɪnɪə/ *n.* obsession by a single idea or interest. □ **monomaniac** *n.* & *adj.*

monophonic /ˌmɒnəˈfɒnɪk/ *adj.* (of sound-reproduction) using only one channel of transmission. [Greek *phōnē* sound]

monoplane /ˈmɒnəˌpleɪn/ *n.* aeroplane with one set of wings.

monopolist /məˈnɒpəlɪst/ *n.* person who has or advocates a monopoly. □ **monopolistic** /-ˈlɪstɪk/ *adj.*

monopolize /məˈnɒpəˌlaɪz/ *v.* (also **-ise**) (**-zing** or **-sing**) **1** obtain exclusive possession or control of (a trade or commodity etc.). **2** dominate or prevent others from sharing in (a conversation etc.). □ **monopolization** /-ˈzeɪʃ(ə)n/ *n.* **monopolizer** *n.*

monopoly /məˈnɒpəlɪ/ *n.* (*pl.* **-ies**) **1 a** exclusive possession or control of the trade in a commodity or service. **b** this conferred as a privilege by the State. **2** (foll. by *of*, *US on*) exclusive possession, control, or exercise. [Greek *pōleō* sell]

monorail /ˈmɒnəʊˌreɪl/ *n.* railway with a single-rail track.

monosodium glutamate /ˌmɒnəʊ ˈsəʊdɪəm ˈgluːtəˌmeɪt/ *n.* sodium salt of glutamic acid used to enhance the flavour of food. [Latin *gluten* glue]

monosyllable /ˈmɒnəˌsɪləb(ə)l/ *n.* word of one syllable. □ **monosyllabic** /-ˈlæbɪk/ *adj.*

monotheism /ˈmɒnəˌθiːɪz(ə)m/ *n.* doctrine that there is only one god. □ **monotheist** *n.* **monotheistic** /-ˈɪstɪk/ *adj.*

monotone /ˈmɒnəˌtəʊn/ *–n.* **1** sound or utterance continuing or repeated on one note without change of pitch. **2** sameness of style in writing. *–adj.* without change of pitch.

monotonous /məˈnɒtənəs/ *adj.* lacking in variety; tedious through sameness. □ **monotonously** *adv.* **monotony** *n.*

monovalent /ˌmɒnəˈveɪlənt/ *adj.* univalent.

monoxide /məˈnɒksaɪd/ *n.* oxide containing one oxygen atom.

Monseigneur /ˌmɒnsenˈjɜː(r)/ *n.* (*pl.* ***Messeigneurs*** /ˌmesenˈjɜː(r)/) title given to an eminent French person, esp. a prince, cardinal, archbishop, or bishop. [French *mon* my, SEIGNEUR]

Monsieur /məˈsjɜː(r)/ *n.* (*pl.* **Messieurs** /meˈsjɜː(r)/) title used of or to a French-speaking man, corresponding to Mr or sir. [French *mon* my, *sieur* lord]

Monsignor /mɒnˈsiːnjə(r)/ *n.* (*pl.* ***-nori*** /-ˈnjɔːrɪ/) title of various Roman Catholic priests and officials. [Italian: related to MONSEIGNEUR]

monsoon /mɒnˈsuːn/ *n.* **1** wind in S. Asia, esp. in the Indian Ocean. **2** rainy season accompanying the summer monsoon. [Arabic *mawsim*]

monster *n.* **1** imaginary creature, usu. large and frightening, made up of incongruous elements. **2** inhumanly cruel or wicked person. **3** misshapen animal or plant. **4** large, usu. ugly, animal or thing. **5** (*attrib.*) huge. [Latin *monstrum* from *moneo* warn]

monstrance *n. RC Ch.* vessel in which the host is exposed for veneration. [Latin *monstro* show]

monstrosity /mɒnˈstrɒsɪtɪ/ *n.* (*pl.* **-ies**) **1** huge or outrageous thing. **2** monstrousness. **3** = MONSTER 3. [Latin: related to MONSTROUS]

monstrous *adj.* **1** like a monster; abnormally formed. **2** huge. **3 a** outrageously wrong or absurd. **b** atrocious. □ **monstrously** *adv.* **monstrousness** *n.* [Latin: related to MONSTER]

montage /mɒnˈtɑːʒ/ *n.* **1** selection, cutting, and piecing together as a consecutive whole, of separate sections of cinema or television film. **2 a** composite whole made from juxtaposed photographs etc. **b** production of this. [French: related to MOUNT[1]]

month /mʌnθ/ *n.* **1** (in full **calendar month**) **a** each of twelve periods into which a year is divided. **b** period of time between the same dates in successive calendar months. **2** period of 28 days. [Old English]

monthly *–adj.* done, produced, or occurring once every month. *–adv.* every month. *–n.* (*pl.* **-ies**) monthly periodical.

month of Sundays *n. colloq.* very long period.

monument /ˈmɒnjʊmənt/ *n.* **1** anything enduring that serves to commemorate or celebrate, esp. a structure or building. **2** stone etc. placed over a grave or in a church etc. in memory of the dead. **3** ancient building or site etc. that has been preserved. **4** lasting reminder. [Latin *moneo* remind]

monumental /ˌmɒnjʊˈment(ə)l/ *adj.* **1 a** extremely great; stupendous (*monumental effort*). **b** (of a work of art etc.) massive and permanent. **2** of or serving as a monument. □ **monumentally** *adv.*

monumental mason *n.* maker of tombstones etc.

moo *–n.* (*pl.* **-s**) cry of cattle. *–v.* (**moos, mooed**) make this sound. [imitative]

mooch *v. colloq.* **1** (usu. foll. by *about, around*) wander aimlessly around. **2** esp. *US* cadge; steal. [probably from French *muchier* skulk]

mood[1] *n.* **1** state of mind or feeling. **2** fit of bad temper or depression. □ **in the mood** (usu. foll. by *for*, or *to* + infin.) inclined. [Old English]

mood[2] *n.* **1** *Gram.* form or set of forms of a verb indicating whether it expresses a fact, command, wish, etc. (*subjunctive mood*). **2** distinction of meaning expressed by different moods. [alteration of MODE]

moody *–adj.* (**-ier, -iest**) given to changes of mood; gloomy, sullen. *–n.* (*pl.* **-ies**) *colloq.* bad mood; tantrum. □ **moodily** *adv.* **moodiness** *n.* [related to MOOD[1]]

moon *–n.* **1 a** natural satellite of the earth, orbiting it monthly, illuminated by the sun and reflecting some light to the earth. **b** this regarded in terms of its waxing and waning in a particular month (*new moon*). **c** the moon when visible (*there is no moon tonight*). **2** satellite of any planet. **3** (prec. by *the*) *colloq.* something desirable but unattainable (*promised me the moon*). *–v.* **1** wander about aimlessly or listlessly. **2** *slang* expose one's buttocks. □ **many moons ago** a long time ago. **moon over** act dreamily thinking about (a loved one). **over the moon** *colloq.* extremely happy. □ **moonless** *adj.* [Old English]

moonbeam *n.* ray of moonlight.

moon boot *n.* thickly-padded boot for low temperatures.

moon-face *n.* round face.

Moonie /'mu:nɪ/ *n. colloq. offens.* member of the Unification Church. [Sun Myung *Moon*, name of its founder]

moonlight *–n.* **1** light of the moon. **2** (*attrib.*) lit by the moon. *–v.* (**-lighted**) *colloq.* have two paid occupations, esp. one by day and one by night. □ **moonlighter** *n.*

moonlight flit *n.* hurried departure by night, esp. to avoid paying a debt.

moonlit *adj.* lit by the moon.

moonscape *n.* **1** surface or landscape of the moon. **2** area resembling this; wasteland.

moonshine *n.* **1** foolish or unrealistic talk or ideas. **2** *slang* illicitly distilled or smuggled alcohol.

moonshot *n.* launching of a spacecraft to the moon.

moonstone *n.* feldspar of pearly appearance.

moonstruck *adj.* slightly mad.

moony *adj.* (**-ier, -iest**) listless; stupidly dreamy.

Moor /mʊə(r), mɔ:(r)/ *n.* member of a Muslim people of NW Africa. □ **Moorish** *adj.* [Greek *Mauros*]

moor[1] /mʊə(r), mɔ:(r)/ *n.* **1** open uncultivated upland, esp. when covered with heather. **2** tract of ground preserved for shooting. [Old English]

moor[2] /mʊə(r), mɔ:(r)/ *v.* attach (a boat etc.) to a fixed object. □ **moorage** *n.* [probably Low German]

moorhen *n.* small waterfowl.

mooring *n.* **1** (often in *pl.*) place where a boat etc. is moored. **2** (in *pl.*) set of permanent anchors and chains.

moorland *n.* extensive area of moor.

moose *n.* (*pl.* same) N. American deer; elk. [Narragansett]

moot *–adj.* debatable, undecided (*moot point*). *–v.* raise (a question) for discussion. *–n. hist.* assembly. [Old English]

mop *–n.* **1** bundle of yarn or cloth or a sponge on the end of a stick, for cleaning floors etc. **2** similarly-shaped implement for various purposes. **3** thick mass of hair. **4** mopping or being mopped (*gave it a mop*). *–v.* (**-pp-**) **1** wipe or clean with or as with a mop. **2 a** wipe tears or sweat etc. from (one's face etc.). **b** wipe away (tears etc.). □ **mop up 1** wipe up with or as with a mop. **2** *colloq.* absorb. **3** dispatch; make an end of. **4 a** complete the occupation of (a district etc.) by capturing or killing enemy troops left there. **b** capture or kill (stragglers). [origin uncertain]

mope *–v.* (**-ping**) **1** be depressed or listless. **2** wander about listlessly. *–n.* person who mopes. □ **mopy** *adj.* (**-ier, -iest**). [origin unknown]

moped /'məʊped/ *n.* two-wheeled low-powered motor vehicle with pedals. [Swedish: related to MOTOR, PEDAL]

moquette /mɒ'ket/ *n.* thick pile or looped material used for upholstery etc. [French]

moraine /mə'reɪn/ *n.* area of debris carried down and deposited by a glacier. [French]

moral /'mɒr(ə)l/ *–adj.* **1 a** concerned with goodness or badness of human character or behaviour, or with the distinction between right and wrong. **b** concerned with accepted rules and standards of human behaviour. **2 a** virtuous in general conduct. **b** capable of moral action. **3** (of rights or duties etc.) founded on moral not actual law. **4** associated with the psychological rather than the physical (*moral courage; moral support*). *–n.* **1** moral lesson of a fable, story, event, etc. **2** (in *pl.*) moral behaviour, e.g. in sexual conduct.

□ **morally** *adv.* [Latin *mos mor-* custom]

morale /mə'rɑ:l/ *n.* confidence, determination, etc. of a person or group. [French *moral*: related to MORAL]

moralist /'mɒrəlɪst/ *n.* **1** person who practises or teaches morality. **2** person who follows a natural system of ethics. □ **moralistic** /-'lɪstɪk/ *adj.*

morality /mə'rælɪtɪ/ *n.* (*pl.* **-ies**) **1** degree of conformity to moral principles. **2** right moral conduct. **3** science of morals. **4** particular system of morals (*commercial morality*).

morality play *n. hist.* drama with personified abstract qualities and including a moral lesson.

moralize /'mɒrə,laɪz/ *v.* (also **-ise**) (**-zing** or **-sing**) **1** (often foll. by *on*) indulge in moral reflection or talk. **2** make moral or more moral. □ **moralization** /-'zeɪʃ(ə)n/ *n.*

moral law *n.* the conditions to be satisfied by any right course of action.

moral philosophy *n.* branch of philosophy concerned with ethics.

moral victory *n.* defeat that has some of the satisfactory elements of victory.

morass /mə'ræs/ *n.* **1** entanglement; confusion. **2** *literary* bog. [French *marais* related to MARSH]

moratorium /,mɒrə'tɔ:rɪəm/ *n.* (*pl.* **-s** or **-ria**) **1** (often foll. by *on*) temporary prohibition or suspension (of an activity). **2 a** legal authorization to debtors to postpone payment. **b** period of this postponement. [Latin *moror* delay]

morbid *adj.* **1 a** (of the mind, ideas, etc.) unwholesome. **b** given to morbid feelings. **2** *colloq.* melancholy. **3** *Med.* of the nature of or indicative of disease. □ **morbidity** /-'bɪdɪtɪ/ *n.* **morbidly** *adv.* [Latin *morbus* disease]

mordant /'mɔ:d(ə)nt/ *–adj.* **1** (of sarcasm etc.) caustic, biting. **2** pungent, smarting. **3** corrosive or cleansing. **4** serving to fix dye. *–n.* mordant substance. [Latin *mordeo* bite]

more /mɔ:(r)/ *–adj.* greater in quantity or degree; additional (*more problems than last time*; *bring some more water*). *–n.* greater quantity, number, or amount (*more than three people*; *more to it than meets the eye*). *–adv.* **1** to a greater degree or extent. **2** forming the comparative of adjectives and adverbs, esp. those of more than one syllable (*more absurd*; *more easily*). □ **more and more** to an increasing degree. **more of** to a greater extent. **more or less** approximately; effectively; nearly. **what is more** as an additional point. [Old English]

moreish /'mɔ:rɪʃ/ *adj.* (also **morish**) *colloq.* (of food) causing a desire for more.

morello /mə'reləʊ/ *n.* (*pl.* **-s**) sour kind of dark cherry. [Italian, = blackish]

moreover /mɔ:'rəʊvə(r)/ *adv.* besides, in addition to what has been said.

mores /'mɔ:reɪz/ *n.pl.* customs or conventions of a community. [Latin, pl. of *mos* custom]

morganatic /,mɔ:gə'nætɪk/ *adj.* **1** (of a marriage) between a person of high rank and one of lower rank, the spouse and children having no claim to the possessions or title of the person of higher rank. **2** (of a spouse) married in this way. [Latin *morganaticus* from Germanic, = 'morning gift', from a husband to his wife on the morning after consummation of a marriage]

morgue /mɔ:g/ *n.* **1** mortuary. **2** (in a newspaper office) room or file of miscellaneous information. [French, originally the name of a Paris mortuary]

moribund /'mɒrɪ,bʌnd/ *adj.* **1** at the point of death. **2** lacking vitality. [Latin *morior* die]

morish var. of MOREISH.

Mormon /'mɔ:mən/ *n.* member of the Church of Jesus Christ of Latter-Day Saints. □ **Mormonism** *n.* [*Mormon*, name of the supposed author of the book on which Mormonism is founded]

morn *n. poet.* morning. [Old English]

mornay /'mɔ:neɪ/ *n.* cheese-flavoured white sauce. [origin uncertain]

morning *n.* **1** early part of the day, ending at noon or lunch-time (*this morning*; *during the morning*). **2** *attrib.* taken, occurring, or appearing during the morning (*morning coffee*). □ **in the morning** *colloq.* tomorrow morning. [from MORN]

morning after *n. colloq.* = HANGOVER 1.

morning-after pill *n.* contraceptive pill taken some hours after intercourse.

morning coat *n.* coat with tails, and with the front cut away.

morning dress *n.* man's morning coat and striped trousers.

morning glory *n.* twining plant with trumpet-shaped flowers.

morning sickness *n.* nausea felt in esp. early pregnancy.

morning star *n.* planet, usu. Venus, seen in the east before sunrise.

morocco /mə'rɒkəʊ/ *n.* (*pl.* **-s**) fine flexible leather of goatskin tanned with sumac. [*Morocco* in NW Africa]

moron /'mɔ:rɒn/ *n.* **1** *colloq.* very stupid person. **2** adult with a mental age of 8–12. □ **moronic** /mə'rɒnɪk/ *adj.* [Greek *mōros* foolish]

morose /mə'rəʊs/ *adj.* sullen, gloomy. □ **morosely** *adv.* **moroseness** *n.* [Latin *mos mor-* manner]

morpheme /'mɔ:fi:m/ *n. Linguistics* meaningful unit of a language that cannot be further divided (e.g. *in*, *come*, *-ing*, forming *incoming*). [Greek *morphē* form]

morphia /'mɔ:fiə/ *n.* (in general use) = MORPHINE.

morphine /'mɔ:fi:n/ *n.* narcotic drug from opium, used to relieve pain. [Latin *Morpheus* god of sleep]

morphology /mɔ:'fɒlədʒɪ/ *n.* the study of the forms of things, esp. of animals and plants and of words and their structure. □ **morphological** /-fə'lɒdʒɪk(ə)l/ *adj.* [Greek *morphē* form]

morris dance /'mɒrɪs/ *n.* traditional English dance in fancy costume, with ribbons and bells. □ **morris dancer** *n.* **morris dancing** *n.* [*morys*, var. of *Moorish*: related to MOOR]

morrow *n.* (usu. prec. by *the*) *literary* the following day. [related to MORN]

Morse /mɔ:s/ *–n.* (in full **Morse code**) code in which letters are represented by combinations of long and short light or sound signals. *–v.* (**-sing**) signal by Morse code. [*Morse*, name of an electrician]

morsel /'mɔ:s(ə)l/ *n.* mouthful; small piece (esp. of food). [Latin *morsus* bite]

mortal *–adj.* **1** subject to death. **2** causing death; fatal. **3** (of combat) fought to the death. **4** associated with death (*mortal agony*). **5** (of an enemy) implacable. **6** (of pain, fear, an affront, etc.) intense, very serious. **7** *colloq.* long and tedious (*for two mortal hours*). **8** *colloq.* conceivable, imaginable (*every mortal thing*; *of no mortal use*). *–n.* human being. □ **mortally** *adv.* [Latin *mors mort-* death]

mortality /mɔ:'tælɪtɪ/ *n.* (*pl.* **-ies**) **1** being subject to death. **2** loss of life on a large scale. **3 a** number of deaths in a given period etc. **b** (in full **mortality rate**) death rate.

mortal sin *n.* sin that deprives the soul of divine grace.

mortar /'mɔ:tə(r)/ *–n.* **1** mixture of lime or cement, sand, and water, for bonding bricks or stones. **2** short large-bore cannon for firing shells at high angles. **3** vessel in which ingredients are pounded with a pestle. *–v.* **1** plaster or join with mortar. **2** bombard with mortar shells. [Latin *mortarium*]

mortarboard *n.* **1** academic cap with a stiff flat square top. **2** flat board for holding mortar.

mortgage /'mɔ:gɪdʒ/ *–n.* **1 a** conveyance of property to a creditor as security for a debt (usu. one incurred by the purchase of the property). **b** deed effecting this. **2** sum of money lent by this. *–v.* (**-ging**) convey (a property) by mortgage. □ **mortgageable** *adj.* [French, = dead pledge: related to GAGE[1]]

mortgagee /,mɔ:gɪ'dʒi:/ *n.* creditor in a mortgage.

mortgager /'mɔ:gɪdʒə(r)/ *n.* (also **mortgagor** /-'dʒɔ:(r)/) debtor in a mortgage.

mortgage rate *n.* rate of interest charged by a mortgagee.

mortice var. of MORTISE.

mortician /mɔ:'tɪʃ(ə)n/ *n. US* undertaker. [Latin *mors mort-* death]

mortify /'mɔ:tɪ,faɪ/ *v.* (**-ies**, **-ied**) **1 a** cause (a person) to feel shamed, humiliated, or sorry. **b** wound (a person's feelings). **2** bring (the body, the flesh, the passions, etc.) into subjection by self-denial or discipline. **3** (of flesh) be affected by gangrene or necrosis. □ **mortification** /-fɪ'keɪʃ(ə)n/ *n.* **mortifying** *adj.* [Latin: related to MORTICIAN]

mortise /'mɔ:tɪs/ (also **mortice**) *–n.* hole in a framework designed to receive the end of another part, esp. a tenon. *–v.* (**-sing**) **1** join securely, esp. by mortise and tenon. **2** cut a mortise in. [French from Arabic]

mortise lock *n.* lock recessed in the frame of a door etc.

mortuary /'mɔ:tjʊərɪ/ *–n.* (*pl.* **-ies**) room or building in which dead bodies are kept until burial or cremation. *–attrib. adj.* of death or burial. [medieval Latin *mortuus* dead]

Mosaic /məʊ'zeɪɪk/ *adj.* of Moses. [French from *Moses* in the Old Testament]

mosaic /məʊ'zeɪɪk/ *n.* **1 a** picture or pattern produced by arranging small variously coloured pieces of glass or stone etc. **b** this as an art form. **2** diversified thing. **3** (*attrib.*) of or like a mosaic. [Greek: ultimately related to MUSE[2]]

Mosaic Law *n.* the laws attributed to Moses and listed in the Pentateuch.

moselle /məʊ'zel/ *n.* dry white wine from the Moselle valley in Germany.

mosey /'məʊzɪ/ *v.* (**-eys**, **-eyed**) (often foll. by *along*) *slang* go in a leisurely manner. [origin unknown]

Moslem var. of MUSLIM.

mosque /mɒsk/ *n.* Muslim place of worship. [Arabic *masgid*]

mosquito /mɒs'ki:təʊ/ *n.* (*pl.* **-es**) biting insect, esp. one of which the female punctures the skin with a long proboscis to suck blood. [Spanish and Portuguese, diminutive of *mosca* fly]

mosquito-net *n.* net to keep off mosquitoes.

moss *n.* **1** small flowerless plant growing in dense clusters in bogs, on the ground, trees, stones, etc. **2** *Scot.* & *N.Engl.* bog, esp. a peatbog. □ **mossy** *adj.* (**-ier, -iest**). [Old English]

most /məʊst/ *–adj.* **1** greatest in quantity or degree. **2** the majority of (*most people think so*). *–n.* **1** greatest quantity or number (*this is the most I can do*). **2** the majority (*most of them are missing*). *–adv.* **1** in the highest degree. **2** forming the superlative of adjectives and adverbs, esp. those of more than one syllable (*most absurd*; *most easily*). **3** *US colloq.* almost. □ **at most** no more or better than (*this is at most a makeshift*). **at the most 1** as the greatest amount. **2** not more than. **for the most part 1** mainly. **2** usually. **make the most of** employ to the best advantage. [Old English]

-most *suffix* forming superlative adjectives and adverbs from prepositions and other words indicating relative position (*foremost*; *uttermost*). [Old English]

mostly *adv.* **1** mainly. **2** usually.

Most Reverend *n.* title of archbishops.

MOT *abbr.* (in full **MOT test**) compulsory annual test of vehicles of more than a specified age. [*M*inistry *o*f *T*ransport]

mot /məʊ/ *n.* (*pl.* ***mots*** pronunc. same) = BON MOT. [French, = word]

mote *n.* speck of dust. [Old English]

motel /məʊ'tel/ *n.* roadside hotel for motorists. [from *mo*tor ho*tel*]

motet /məʊ'tet/ *n. Mus.* short religious choral work. [French: related to MOT]

moth *n.* **1** nocturnal insect like a butterfly but without clubbed antennae. **2** insect of this type breeding in cloth etc., on which its larva feeds. [Old English]

mothball *n.* ball of naphthalene etc. placed in stored clothes to deter moths. □ **in mothballs** stored unused for a considerable time.

moth-eaten *adj.* **1** damaged by moths. **2** time-worn.

mother /'mʌðə(r)/ *–n.* **1** female parent. **2** woman, quality, or condition etc. that gives rise to something else (*necessity is the mother of invention*). **3** (in full **Mother Superior**) head of a female religious community. *–v.* **1** treat as a mother does. **2** give birth to; be the mother or origin of. □ **motherhood** *n.* **motherless** *adj.* [Old English]

Mother Carey's chicken *n.* = STORM PETREL 1.

mother country *n.* country in relation to its colonies.

mother earth *n.* the earth as mother of its inhabitants.

Mothering Sunday *n.* = MOTHER'S DAY.

mother-in-law *n.* (*pl.* **mothers-in--law**) husband's or wife's mother.

motherland *n.* one's native country.

motherly *adj.* kind or tender like a mother. □ **motherliness** *n.*

mother-of-pearl *n.* smooth iridescent substance forming the inner layer of the shell of oysters etc.

Mother's Day *n.* day when mothers are honoured with presents, (in the UK) the fourth Sunday in Lent, (in the US) the second Sunday in May.

mother tongue *n.* native language.

mothproof *–adj.* (of clothes) treated so as to repel moths. *–v.* treat (clothes) in this way.

motif /məʊ'ti:f/ *n.* **1** theme that is repeated and developed in an artistic work. **2** decorative design or pattern. **3** ornament sewn separately on a garment. [French: related to MOTIVE]

motion /'məʊʃ(ə)n/ *–n.* **1** moving; changing position. **2** gesture. **3** formal proposal put to a committee, legislature, etc. **4** application to a court for an order. **5 a** an evacuation of the bowels. **b** (in *sing.* or *pl.*) faeces. *–v.* (often foll. by *to* + infin.) **1** direct (a person) by a gesture. **2** (often foll. by *to* a person) make a gesture directing (*motioned to me to leave*). □ **go through the motions** do something perfunctorily or superficially. **in motion** moving; not at rest. **put** (or **set**) **in motion** set going or working. □ **motionless** *adj.* [Latin: related to MOVE]

motion picture *n.* (esp. US) cinema film.

motivate /'məʊtɪ,veɪt/ *v.* (**-ting**) **1** supply a motive to; be the motive of. **2** cause (a person) to act in a particular way. **3** stimulate the interest of (a person in an activity). □ **motivation** /-'veɪʃ(ə)n/ *n.* **motivational** /-'veɪʃən(ə)l/ *adj.*

motive /'məʊtɪv/ *–n.* **1** what induces a person to act in a particular way. **2** = MOTIF. *–adj.* **1** tending to initiate movement. **2** concerned with movement. [Latin *motivus*: related to MOVE]

motive power *n.* moving or impelling power, esp. a source of energy used to drive machinery.

mot juste /məʊ'ʒu:st/ *n.* (*pl.* ***mots justes*** pronunc. same) most appropriate expression.

motley /'mɒtlɪ/ *–adj.* (**-lier, -liest**) **1** diversified in colour. **2** of varied character (*a motley crew*). *–n. hist.* jester's particoloured costume. [origin unknown]

moto-cross /'məʊtəʊ,krɒs/ *n.* cross-country racing on motor cycles. [from MOTOR, CROSS]

motor /'məʊtə(r)/ *–n.* **1** thing that imparts motion. **2** machine (esp. one using electricity or internal combustion) supplying motive power for a vehicle or other machine. **3** = CAR 1. **4** (*attrib.*) **a** giving, imparting, or producing motion. **b** driven by a motor (*motor-mower*). **c** of or for motor vehicles. **d** *Anat.* relating to muscular movement or the nerves activating it. *–v.* go or convey in a motor vehicle. [Latin: related to MOVE]

motor bike *n. colloq.* = MOTOR CYCLE.

motor boat *n.* motor-driven boat.

motorcade /'məʊtə,keɪd/ *n.* procession of motor vehicles. [from MOTOR, after *cavalcade*]

motor car *n.* = CAR 1.

motor cycle *n.* two-wheeled motor vehicle without pedal propulsion. □ **motor cyclist** *n.*

motorist *n.* driver of a car.

motorize *v.* (also **-ise**) (**-zing** or **-sing**) **1** equip with motor transport. **2** provide with a motor.

motorman *n.* driver of an underground train, tram, etc.

motor scooter see SCOOTER.

motor vehicle *n.* road vehicle powered by an internal-combustion engine.

motorway *n.* road for fast travel, with separate carriageways and limited access.

Motown /'məʊtaʊn/ *n. propr.* music with elements of rhythm and blues, associated with Detroit. [*Motor Town*, = Detroit in US]

mottle *v.* (**-ling**) (esp. as **mottled** *adj.*) mark with spots or smears of colour. [back-formation from MOTLEY]

motto /'mɒtəʊ/ *n.* (*pl.* **-es**) **1** maxim adopted as a rule of conduct. **2** phrase or sentence accompanying a coat of arms. **3** appropriate inscription. **4** joke, maxim, etc. in a paper cracker. [Italian: related to MOT]

mould[1] /məʊld/ (*US* **mold**) *–n.* **1** hollow container into which a substance is poured or pressed to harden into a required shape. **2 a** vessel for shaping puddings etc. **b** pudding etc. made in this way. **3** form or shape. **4** frame or template for producing mouldings. **5** character or type (*in heroic mould*). *–v.* **1** make (an object) in a required shape or from certain ingredients (*moulded out of clay*). **2** give shape to. **3** influence the development of. [French *modle* from Latin MODULUS]

mould[2] /məʊld/ *n.* (*US* **mold**) furry growth of fungi occurring esp. in moist warm conditions. [Old Norse]

mould[3] /məʊld/ *n.* (*US* **mold**) **1** loose earth. **2** upper soil of cultivated land, esp. when rich in organic matter. [Old English]

moulder *v.* (*US* **molder**) **1** decay to dust. **2** (foll. by *away*) rot or crumble. **3** deteriorate. [from MOULD[3]]

moulding *n.* (*US* **molding**) **1** ornamentally shaped outline of plaster etc. as an architectural feature, esp. in a cornice. **2** similar feature in woodwork etc.

mouldy *adj.* (*US* **moldy**) (**-ier, -iest**) **1** covered with mould. **2** stale; out of date. **3** *colloq.* dull, miserable. □ **mouldiness** *n.*

moult /məʊlt/ (*US* **molt**) *–v.* (also *absol.*) shed (feathers, hair, a shell etc.) in the process of renewing plumage, a coat, etc. *–n.* moulting. [Latin *muto* change]

mound *n.* **1** raised mass of earth, stones, etc. **2** heap or pile; large quantity. **3** hillock. [origin unknown]

mount[1] *–v.* **1** ascend; climb on to. **2 a** get up on (a horse etc.) to ride it. **b** set on horseback. **c** (as **mounted** *adj.*) serving on horseback (*mounted police*). **3 a** (often foll. by *up*) accumulate. **b** (of a feeling) increase. **4** (often foll. by *on, in*) set (an object) on a support or in a backing, frame, etc., esp. for viewing. **5** organize, arrange, set in motion (a play, exhibition, attack, guard, etc.). **6** (of a male animal) get on to (a female) to copulate. *–n.* **1** backing, etc. on which a picture etc. is set for display. **2** horse for riding. **3** setting for a gem etc. [Latin: related to MOUNT[2]]

mount[2] *n. archaic* (except before a name): mountain, hill (*Mount Everest*). [Latin *mons mont-*]

mountain /'maʊntɪn/ *n.* **1** large abrupt natural elevation of the ground. **2** large heap or pile; huge quantity. **3** large surplus stock (*butter mountain*). □ **make a mountain out of a molehill** see MOLEHILL. [Latin: related to MOUNT[2]]

mountain ash *n.* tree with scarlet berries; rowan.

mountain bike *n.* sturdy bike with many gears for riding over rough terrain.

mountaineer /,maʊntɪ'nɪə(r)/ *–n.* person who practises mountain-climbing. *–v.* climb mountains as a sport. □ **mountaineering** *n.*

mountain lion *n.* puma.

mountainous *adj.* **1** having many mountains. **2** huge.

mountain range *n.* continuous line of mountains.

mountain sickness *n.* sickness caused by thin air at great heights.
mountainside *n.* sloping side of a mountain.
mountebank /'maʊntɪ,bæŋk/ *n.* **1** swindler; charlatan. **2** *hist.* itinerant quack. [Italian, = mount on bench]
Mountie *n. colloq.* member of the Royal Canadian Mounted Police. [abbreviation]
mounting *n.* **1** = MOUNT[1] *n.* 1. **2** in senses of MOUNT[1] *v.*
mourn /mɔ:n/ *v.* (often foll. by *for, over*) feel or show deep sorrow or regret for (a dead person, a lost thing, a past event, etc.). [Old English]
mourner *n.* person who mourns, esp. at a funeral.
mournful *adj.* doleful, sad, expressing mourning. □ **mournfully** *adv.* **mournfulness** *n.*
mourning *n.* **1** expressing of sorrow for a dead person, esp. by wearing black clothes. **2** such clothes.
mouse –*n.* (*pl.* **mice**) **1** small rodent, esp. of a kind infesting houses. **2** timid or feeble person. **3** (*pl.* **-s**) *Computing* small hand-held device controlling the cursor on a VDU screen. –*v.* /also maʊz/ (**-sing**) (of a cat, owl, etc.) hunt mice. □ **mouser** *n.* [Old English]
mousetrap *n.* **1** trap for catching mice. **2** (often *attrib.*) *colloq.* poor quality cheese.
moussaka /mʊ'sɑ:kə/ *n.* (also **mousaka**) Greek dish of minced meat, aubergine, etc. [Greek or Turkish]
mousse /mu:s/ *n.* **1 a** dessert of whipped cream, eggs, etc., usu. flavoured with fruit or chocolate. **b** meat or fish purée made with whipped cream etc. **2** foamy substance applied to the hair to enable styling. [French, = froth]
moustache /mə'stɑ:ʃ/ *n.* (*US* **mustache**) hair left to grow on a man's upper lip. [Greek *mustax*]
mousy /'maʊsɪ/ *adj.* (**-ier**, **-iest**) **1** of or like a mouse. **2** (of a person) timid, feeble. **3** nondescript light brown.
mouth /maʊθ/ –*n.* (*pl.* **mouths** /maʊðz/) **1 a** external opening in the head, through which most animals take in food and emit communicative sounds. **b** (in humans and some animals) cavity behind it containing the means of biting and chewing and the vocal organs. **2** opening of a container, cave, trumpet, etc. **3** place where a river enters the sea. **4** an individual as needing sustenance (*an extra mouth to feed*). **5** *colloq.* **a** meaningless or ineffectual talk. **b** impudent talk; cheek. –*v.* /maʊð/ (**-thing**) **1** say or speak by moving the lips but with no sound. **2** utter or speak insincerely or without understanding (*mouthing platitudes*). □ **put words into a person's mouth** represent a person as having said something. **take the words out of a person's mouth** say what another was about to say. [Old English]
mouthful *n.* (*pl.* **-s**) **1** quantity of food etc. that fills the mouth. **2** small quantity. **3** *colloq.* long or complicated word or phrase.
mouth-organ *n.* = HARMONICA.
mouthpiece *n.* **1** part of a musical instrument, telephone, etc., placed next to the lips. **2** *colloq.* person who speaks for another or others.
mouth-to-mouth *adj.* (of resuscitation) in which a person breathes into a subject's lungs through the mouth.
mouthwash *n.* liquid antiseptic etc. for rinsing the mouth or gargling.
mouth-watering *adj.* (of food etc.) having a delicious smell or appearance.
movable /'mu:vəb(ə)l/ *adj.* (also **moveable**) **1** that can be moved. **2** variable in date from year to year (*movable feast*). [related to MOVE]
move /mu:v/ –*v.* (**-ving**) **1** (cause to) change position or posture. **2** put or keep in motion; rouse, stir. **3 a** take a turn in a board-game. **b** change the position of (a piece) in a board-game. **4** (often foll. by *about, away, off*, etc.) go or proceed. **5** take action, esp. promptly (*moved to reduce crime*). **6** make progress (*project is moving fast*). **7** (also *absol.*) change (one's home or place of work). **8** (foll. by *in*) be socially active in (a specified group etc.) (*moves in the best circles*). **9** affect (a person) with (usu. tender) emotion. **10** (foll. by *to*) provoke (a person to laughter etc.) (*was moved to tears*). **11** (foll. by *to*, or *to* + infin.) prompt or incline (a person to a feeling or action). **12** (cause to) change one's attitude (*nothing can move me on this issue*). **13 a** cause (the bowels) to be evacuated. **b** (of the bowels) be evacuated. **14** (often foll. by *that*) propose in a meeting, etc. **15** (foll. by *for*) make a formal request or application. **16** sell; be sold. –*n.* **1** act or process of moving. **2** change of house, premises, etc. **3** step taken to secure an object. **4 a** changing of the position of a piece in a board-game. **b** player's turn to do this. □ **get a move on** *colloq.* hurry up. **make a move** take action. **move along** (or **on**) advance, progress, esp. to avoid crowding etc. **move away** go to live in another area. **move heaven and earth** (foll. by *to* + infin.) make extraordinary efforts. **move in 1** take up residence in a new home. **2** get into a position of readiness or proximity (for an offensive action

etc.). **move in with** start to share accommodation with (an existing resident). **move out** leave one's home. **move over** (or **up**) adjust one's position to make room for another. **on the move** moving. [Latin *moveo*]

moveable var. of MOVABLE.

movement *n.* **1 a** moving or being moved. **b** instance of this (*watched his every movement*). **2** moving parts of a mechanism (esp. a clock or watch). **3 a** body of persons with a common object (*peace movement*). **b** campaign undertaken by them. **4** (in *pl.*) person's activities and whereabouts. **5** *Mus.* principal division of a longer musical work. **6** motion of the bowels. **7** rise or fall in price(s) on the stock market. **8** progress.

mover *n.* **1** person, animal, or thing that moves or dances, esp. in a specified way. **2** person who moves a proposition. **3** (also **prime mover**) originator.

movie *n.* esp. *US colloq.* cinema film.

moving *adj.* emotionally affecting. □ **movingly** *adv.*

moving staircase *n.* escalator.

mow /məʊ/ *v.* (*past part.* **mowed** or **mown**) **1** (also *absol.*) cut (grass, hay, etc.) with a scythe or machine. **2** cut down the produce of (a field) or the grass etc. of (a lawn) by mowing. □ **mow down** kill or destroy randomly or in great numbers. □ **mower** *n.* [Old English]

mozzarella /ˌmɒtsəˈrelə/ *n.* Italian curd cheese, orig. of buffalo milk. [Italian]

MP *abbr.* Member of Parliament.

mp *abbr.* mezzo piano.

m.p.g. *abbr.* miles per gallon.

m.p.h. *abbr.* miles per hour.

M.Phil. *abbr.* Master of Philosophy.

Mr /ˈmɪstə(r)/ *n.* (*pl.* **Messrs**) **1** title of a man without a higher title (*Mr Jones*). **2** title prefixed to a designation of office etc. (*Mr President*; *Mr Speaker*). [abbreviation of MISTER]

Mrs /ˈmɪsɪz/ *n.* (*pl.* same) title of a married woman without a higher title (*Mrs Jones*). [abbreviation of MISTRESS]

MS *abbr.* **1** (*pl.* **MSS** /emˈesɪz/) manuscript. **2** multiple sclerosis.

Ms /mɪz, məz/ *n.* title of a married or unmarried woman without a higher title. [combination of MRS, MISS[2]]

M.Sc. *abbr.* Master of Science.

MS-DOS /ˌemesˈdɒs/ *abbr. propr. Computing* Microsoft disk operating system.

Mt. *abbr.* Mount.

mu /mjuː/ *n.* **1** twelfth Greek letter (M, μ). **2** (μ, as a symbol) = MICRO- 2. [Greek]

much –*adj.* **1** existing or occurring in a great quantity (*much trouble*; *too much noise*). **2** (prec. by *as, how, that*, etc.) with relative sense (*I don't know how much money you want*). –*n.* **1** a great quantity (*much of that is true*). **2** (prec. by *as, how, that*, etc.) with relative sense (*we do not need that much*). **3** (usu. in *neg.*) noteworthy or outstanding example (*not much to look at*). –*adv.* **1** in a great degree (*much to my surprise*; *is much the same*; *I much regret it*; *much annoyed*; *much better*; *much the best*). **2** for a large part of one's time; often (*he is not here much*). □ **as much** so (*I thought as much*). **a bit much** *colloq.* excessive, immoderate. **much as** even though (*cannot come, much as I would like to*). **much of a muchness** very nearly the same. **not much of a** *colloq.* a rather poor. [from MICKLE]

mucilage /ˈmjuːsɪlɪdʒ/ *n.* **1** viscous substance obtained from plants. **2** adhesive gum. [Latin: related to MUCUS]

muck –*n.* **1** *colloq.* dirt or filth; anything disgusting. **2** farmyard manure. **3** *colloq.* mess. –*v.* **1** (usu. foll. by *up*) *colloq.* **a** bungle (a job). **b** make dirty or untidy. **2** (foll. by *out*) remove manure from. □ **make a muck of** *colloq.* bungle. **muck about** (or **around**) *colloq.* **1** potter or fool about. **2** (foll. by *with*) fool or interfere with. **muck in** (often foll. by *with*) *colloq.* share tasks etc. equally. [Scandinavian]

mucker *n. slang* friend, mate. [probably from *muck in*: related to MUCK]

muckle var. of MICKLE.

muckrake *v.* (**-king**) search out and reveal scandal. □ **muckraker** *n.* **muckraking** *n.*

muck-spreader *n.* machine for spreading dung. □ **muck-spreading** *n.*

mucky *adj.* (**-ier, -iest**) covered with muck, dirty.

mucous /ˈmjuːkəs/ *adj.* of or covered with mucus. □ **mucosity** /-ˈkɒsɪtɪ/ *n.* [Latin *mucosus*: related to MUCUS]

mucous membrane *n.* mucus-secreting tissue lining body cavities etc.

mucus /ˈmjuːkəs/ *n.* slimy substance secreted by a mucous membrane. [Latin]

mud *n.* soft wet earth. □ **fling** (or **sling** or **throw**) **mud** speak disparagingly or slanderously. **one's name is mud** one is in disgrace. [German]

muddle –*v.* (**-ling**) (often foll. by *up*) **1** bring into disorder. **2** bewilder, confuse. –*n.* **1** disorder. **2** confusion. □ **muddle along** (or **on**) progress in a haphazard way. **muddle through** succeed despite one's inefficiency. [perhaps Dutch, related to MUD]

muddle-headed *adj.* mentally disorganized, confused.

muddy –*adj.* (**-ier, -iest**) **1** like mud. **2** covered in or full of mud. **3** (of liquid, colour, or sound) not clear, impure. **4** vague, confused. **–*v.*** **(-ies, -ied)** make muddy. □ **muddiness** *n.*

mudflap *n.* flap hanging behind the wheel of a vehicle, to prevent splashes.

mud-flat *n.* stretch of muddy land uncovered at low tide.

mudguard *n.* curved strip over a bicycle wheel etc. to protect the rider from splashes.

mud pack *n.* cosmetic paste applied thickly to the face.

mud-slinger *n. colloq.* person given to making abusive or disparaging remarks. □ **mud-slinging** *n.*

muesli /'mu:zlɪ, 'mju:-/ *n.* breakfast food of crushed cereals, dried fruits, nuts, etc., eaten with milk. [Swiss German]

muezzin /mu:'ezɪn/ *n.* Muslim crier who proclaims the hours of prayer. [Arabic]

muff[1] *n.* covering, esp. of fur, for keeping the hands or ears warm. [Dutch *mof*]

muff[2] *v. colloq.* **1** bungle. **2** miss (a catch, ball, etc.). [origin unknown]

muffin /'mʌfɪn/ *n.* **1** light flat round spongy cake, eaten toasted and buttered. **2** *US* similar round cake made from batter or dough. [origin unknown]

muffle *v.* **(-ling) 1** (often foll. by *up*) wrap or cover for warmth, or to deaden sound. **2** (usu. as **muffled** *adj.*) stifle (an utterance). [perhaps French *moufle* thick glove, MUFF[1]]

muffler *n.* **1** wrap or scarf worn for warmth. **2** thing used to deaden sound. **3** *US* silencer of a vehicle.

mufti /'mʌftɪ/ *n.* civilian clothes (*in mufti*). [Arabic]

mug[1] **–*n.*** **1 a** drinking-vessel, usu. cylindrical with a handle and no saucer. **b** its contents. **2** *slang* gullible person. **3** *slang* face or mouth. **–*v.*** **(-gg-)** attack and rob, esp. in public. □ **a mug's game** *colloq.* foolish or unprofitable activity. □ **mugger** *n.* **mugful** *n.* (*pl.* **-s**). **mugging** *n.* [Scandinavian]

mug[2] *v.* **(-gg-)** (usu. foll. by *up*) *slang* learn (a subject) by concentrated study. [origin unknown]

muggins /'mʌgɪnz/ *n.* (*pl.* same or **mugginses**) *colloq.* gullible person (often meaning oneself: *so muggins had to pay*). [perhaps from the surname]

muggy *adj.* **(-ier, -iest)** (of weather etc.) oppressively humid. □ **mugginess** *n.* [Old Norse]

mug shot *n. slang* photograph of a face, esp. for police records.

Muhammadan /mə'hæməd(ə)n/ *n.* & *adj.* (also **Mohammedan**) = MUSLIM. [*Muhammad*, name of a prophet]

■ **Usage** The term *Muhammadan* is not used by Muslims, and is often regarded as offensive.

mujahidin /ˌmʊdʒəhə'di:n/ *n.pl.* (also **mujahedin, -deen**) guerrilla fighters in Islamic countries, esp. Muslim fundamentalists. [Persian and Arabic: related to JIHAD]

mulatto /mju:'lætəʊ/ *n.* (*pl.* **-s** or **-es**) person of mixed White and Black parentage. [Spanish *mulato* young mule]

mulberry /'mʌlbərɪ/ *n.* (*pl.* **-ies**) **1** tree bearing edible purple or white berries, and leaves used to feed silkworms. **2** its fruit. **3** dark-red or purple. [Latin *morum* mulberry, BERRY]

mulch –*n.* layer of wet straw, leaves, or plastic, etc., spread around or over a plant to enrich or insulate the soil. **–*v.*** treat with mulch. [Old English, = soft]

mule[1] *n.* **1** offspring of a male donkey and a female horse, or (in general use) of a female donkey and a male horse (cf. HINNY). **2** stupid or obstinate person. **3** (in full **spinning mule**) a kind of spinning-machine. [Latin *mulus*]

mule[2] *n.* backless slipper. [French]

muleteer /ˌmju:lə'tɪə(r)/ *n.* mule-driver. [French *muletier*: related to MULE[1]]

mulish *adj.* stubborn.

mull[1] *v.* (often foll. by *over*) ponder, consider. [probably Dutch]

mull[2] *v.* warm (wine or beer) with added sugar, spices, etc. [origin unknown]

mull[3] *n. Scot.* promontory. [origin uncertain]

mullah /'mʌlə/ *n.* Muslim learned in theology and sacred law. [ultimately Arabic *mawlā*]

mullet *n.* (*pl.* same) any of several kinds of marine fish valued for food. [Greek *mullos*]

mulligatawny /ˌmʌlɪgə'tɔ:nɪ/ *n.* highly seasoned soup orig. from India. [Tamil, = pepper-water]

mullion /'mʌljən/ *n.* vertical bar dividing the lights in a window. □ **mullioned** *adj.* [probably French *moinel* middle: related to MEAN[3]]

multi- *comb. form* many. [Latin *multus* much, many]

multi-access /ˌmʌltɪ'ækses/ *adj.* (of a computer system) allowing access to the central processor from several terminals simultaneously.

multicoloured /'mʌltɪˌkʌləd/ *adj.* of many colours.

multicultural /ˌmʌltɪ'kʌltʃər(ə)l/ *adj.* of several cultural groups. □ **multiculturalism** *n.*

multidirectional /ˌmʌltɪdaɪ'rekʃən(ə)l/ *adj.* of, involving, or operating in several directions.

multifarious /ˌmʌltɪ'feərɪəs/ *adj.* **1** many and various. **2** of great variety. □ **multifariousness** *n.* [Latin *multifarius*]

multiform /'mʌltɪˌfɔːm/ *adj.* **1** having many forms. **2** of many kinds.

multilateral /ˌmʌltɪ'lætər(ə)l/ *adj.* **1** (of an agreement etc.) in which three or more parties participate. **2** having many sides. □ **multilaterally** *adv.*

multilingual /ˌmʌltɪ'lɪŋgw(ə)l/ *adj.* in, speaking, or using several languages.

multimedia –*attrib. adj.* using more than one medium of communication. –*n.* = HYPERMEDIA.

multimillion /'mʌltɪˌmɪljən/ *attrib. adj.* costing or involving several million (pounds, dollars, etc.) (*multimillion dollar fraud*).

multimillionaire /ˌmʌltɪˌmɪljə'neə(r)/ *n.* person with a fortune of several millions.

multinational /ˌmʌltɪ'næʃən(ə)l/ –*adj.* **1** operating in several countries. **2** of several nationalities. –*n.* multinational company.

multiple /'mʌltɪp(ə)l/ –*adj.* **1** having several parts, elements, or components. **2** many and various. –*n.* number that contains another without a remainder (*56 is a multiple of 7*). [Latin *multiplus*: related to MULTIPLEX]

multiple-choice *adj.* (of an examination question) accompanied by several possible answers from which the correct one has to be chosen.

multiple sclerosis see SCLEROSIS.

multiplex /'mʌltɪˌpleks/ *adj.* manifold; of many elements. [Latin: related to MULTI-, *-plex -plicis* -fold]

multiplicand /ˌmʌltɪplɪ'kænd/ *n.* quantity to be multiplied by another.

multiplication /ˌmʌltɪplɪ'keɪʃ(ə)n/ *n.* multiplying.

multiplication sign *n.* sign (×) to indicate that one quantity is to be multiplied by another.

multiplication table *n.* list of multiples of a particular number, usu. from 1 to 12.

multiplicity /ˌmʌltɪ'plɪsɪtɪ/ *n.* (*pl.* **-ies**) **1** manifold variety. **2** (foll. by *of*) great number.

multiplier /'mʌltɪˌplaɪə(r)/ *n.* quantity by which a given number is multiplied.

multiply /'mʌltɪˌplaɪ/ *v.* (**-ies, -ied**) **1** (also *absol.*) obtain from (a number) another that is a specified number of times its value (*multiply 6 by 4 and you get 24*). **2** increase in number, esp. by procreation. **3** produce a large number of (instances etc.). **4 a** breed (animals). **b** propagate (plants). [Latin *multiplico*: related to MULTIPLEX]

multi-purpose /ˌmʌltɪ'pɜːpəs/ *attrib. adj.* having several purposes.

multiracial /ˌmʌltɪ'reɪʃ(ə)l/ *adj.* of several races.

multi-storey /ˌmʌltɪ'stɔːrɪ/ *attrib. adj.* having several storeys.

multitude *n.* **1** (often foll. by *of*) great number. **2** large gathering of people; crowd. **3** (**the multitude**) the common people. [French from Latin]

multitudinous /ˌmʌltɪ'tjuːdɪnəs/ *adj.* **1** very numerous. **2** consisting of many individuals. [Latin: related to MULTITUDE]

multi-user /ˌmʌltɪ'juːzə(r)/ *attrib. adj.* (of a computer system) having a number of simultaneous users.

mum[1] *n. colloq.* = MUMMY[1].

mum[2] *adj. colloq.* silent (*keep mum*). □ **mum's the word** say nothing. [imitative]

mumble –*v.* (**-ling**) speak or utter indistinctly. –*n.* indistinct utterance or sound. [related to MUM[2]]

mumbo-jumbo /ˌmʌmbəʊ'dʒʌmbəʊ/ *n.* (*pl.* **-s**) **1** meaningless or ignorant ritual. **2** meaningless or unnecessarily complicated language; nonsense. [*Mumbo Jumbo*, name of a supposed African idol]

mummer *n.* actor in a traditional mime. [French *momeur*: cf. MUM[2]]

mummery *n.* (*pl.* **-ies**) **1** ridiculous (esp. religious) ceremonial. **2** performance by mummers. [French *momerie*: related to MUMMER]

mummify /'mʌmɪˌfaɪ/ *v.* (**-ies, -ied**) preserve (a body) as a mummy. □ **mummification** /-fɪ'keɪʃ(ə)n/ *n.*

mummy[1] /'mʌmɪ/ *n.* (*pl.* **-ies**) *colloq.* mother. [imitative of a child's pronunciation]

mummy[2] /'mʌmɪ/ *n.* (*pl.* **-ies**) body of a human being or animal embalmed for burial, esp. in ancient Egypt. [Persian *mūm* wax]

mumps *n.pl.* (treated as *sing.*) infectious disease with swelling of the neck and face. [imitative of mouth-shape]

munch *v.* eat steadily with a marked action of the jaws. [imitative]

mundane /mʌn'deɪn/ *adj.* **1** dull, routine. **2** of this world. □ **mundanely** *adv.* **mundanity** /-'dænɪtɪ/ *n.* [Latin *mundus* world]

mung *n.* (in full **mung bean**) leguminous Indian plant used as food. [Hindi *mūng*]

municipal /mjuː'nɪsɪp(ə)l/ *adj.* of a municipality or its self-government. □ **municipalize** *v.* (also **-ise**) (**-zing** or

-sing). **municipally** *adv.* [Latin *municipium* free city]

municipality /mju:ˌnɪsɪ'pælɪtɪ/ *n.* (*pl.* **-ies**) **1** town or district having local self-government. **2** governing body of this area.

munificent /mju:'nɪfɪs(ə)nt/ *adj.* (of a giver or a gift) splendidly generous. □ **munificence** *n.* [Latin *munus* gift: related to -FIC]

muniment *n.* (usu. in *pl.*) document kept as evidence of rights or privileges etc. [Latin *munio* fortify]

munition /mju:'nɪʃ(ə)n/ *n.* (usu. in *pl.*) military weapons, ammunition etc. [Latin, = fortification: related to MUNIMENT]

muon /'mju:ɒn/ *n.* *Physics* unstable elementary particle like an electron, but with a much greater mass. [μ (MU), the symbol for it]

mural /'mjʊər(ə)l/ *–n.* painting executed directly on a wall. *–adj.* of, on, or like a wall. [Latin *murus* wall]

murder *–n.* **1** intentional unlawful killing of a human being by another. **2** *colloq.* unpleasant, troublesome, or dangerous state of affairs. *–v.* **1** kill (a human being) intentionally and unlawfully. **2** *colloq.* **a** utterly defeat. **b** spoil by a bad performance, mispronunciation, etc. □ **cry blue murder** *colloq.* make an extravagant outcry. **get away with murder** *colloq.* do whatever one wishes and escape punishment. □ **murderer** *n.* **murderess** *n.* [Old English]

murderous *adj.* **1** (of a person, weapon, action, etc.) capable of, intending, or involving murder or great harm. **2** *colloq.* extremely arduous or unpleasant.

murk *n.* darkness, poor visibility. [probably Scandinavian]

murky *adj.* (**-ier**, **-iest**) **1** dark, gloomy. **2** (of darkness, liquid, etc.) thick, dirty. **3** suspiciously obscure (*murky past*). □ **murkily** *adv.* **murkiness** *n.*

murmur /'mɜ:mə(r)/ *–n.* **1** subdued continuous sound, as made by waves, a brook, etc. **2** softly spoken or nearly inarticulate utterance. **3** subdued expression of discontent. *–v.* **1** make a murmur. **2** utter (words) in a low voice. **3** (usu. foll. by *at*, *against*) complain in low tones, grumble. [Latin]

Murphy's Law /'mɜ:fɪz/ *n. joc.* any of various maxims about the perverseness of things. [*Murphy*, Irish surname]

murrain /'mʌrɪn/ *n.* infectious disease of cattle. [Anglo-French *moryn*]

Mus.B. *abbr.* (also **Mus. Bac.**) Bachelor of Music. [Latin *Musicae Baccalaureus*]

Muscadet /'mʌskəˌdeɪ/ *n.* **1** a dry white wine from the Loire region of France. **2** variety of grape used for this. [*Muscadet* grape]

muscat /'mʌskət/ *n.* **1** sweet usu. fortified white wine made from musk-flavoured grapes. **2** this grape. [Provençal: related to MUSK]

muscatel /ˌmʌskə'tel/ *n.* **1** = MUSCAT. **2** raisin from a muscat grape.

muscle /'mʌs(ə)l/ *–n.* **1** fibrous tissue producing movement in or maintaining the position of an animal body. **2** part of an animal body that is composed of muscles. **3** strength, power. *–v.* (**-ling**) (foll. by *in*, *in on*) *colloq.* force oneself on others; intrude by forceful means. □ **not move a muscle** be completely motionless. [Latin diminutive of *mus* mouse]

muscle-bound *adj.* with muscles stiff and inflexible through excessive exercise.

muscle-man *n.* man with highly developed muscles.

Muscovite /'mʌskəˌvaɪt/ *–n.* native or citizen of Moscow. *–adj.* of Moscow. [from *Muscovy*, principality of Moscow]

Muscovy duck /'mʌskəvɪ/ *n.* crested duck with red markings on its head. [*Muscovy*, principality of Moscow]

muscular /'mʌskjʊlə(r)/ *adj.* **1** of or affecting the muscles. **2** having well-developed muscles. **3** robust. □ **muscularity** /-'lærɪtɪ/ *n.*

muscular Christianity *n.* Christian life of cheerful physical activity as described in the writings of Charles Kingsley.

muscular dystrophy *n.* hereditary progressive wasting of the muscles.

musculature /'mʌskjʊlətʃə(r)/ *n.* muscular system of a body or organ.

Mus.D. *abbr.* (also **Mus. Doc.**) Doctor of Music. [Latin *Musicae Doctor*]

muse[1] /mju:z/ *v.* (**-sing**) **1** (usu. foll. by *on*, *upon*) ponder, reflect. **2** say meditatively. [French]

muse[2] /mju:z/ *n.* **1** (in Greek and Roman mythology) any of the nine goddesses who inspire poetry, music, etc. **2** (usu. prec. by *the*) poet's inspiration. [Greek *Mousa*]

museum /mju:'zɪəm/ *n.* building used for storing and exhibiting objects of historical, scientific, or cultural interest. [Greek: related to MUSE[2]]

museum piece *n.* **1** specimen of art etc. fit for a museum. **2** *derog.* old-fashioned or quaint person or object.

mush *n.* **1** soft pulp. **2** feeble sentimentality. **3** *US* maize porridge. [apparently var. of MASH]

mushroom *–n.* **1** edible fungus with a stem and domed cap. **2** pinkish-brown colour of this. *–v.* appear or develop rapidly. [French *mousseron* from Latin]

mushroom cloud *n.* mushroom-shaped cloud from a nuclear explosion.

mushy *adj.* (**-ier, -iest**) **1** like mush; soft. **2** feebly sentimental. □ **mushiness** *n.*

music /'mju:zɪk/ *n.* **1** art of combining vocal or instrumental sounds in a harmonious or expressive way. **2** sounds so produced. **3** musical composition. **4** written or printed score of this. **5** pleasant natural sound. □ **music to one's ears** something one is pleased to hear. [Greek: related to MUSE[2]]

musical –*adj.* **1** of music. **2** (of sounds etc.) melodious, harmonious. **3** fond of, sensitive to, or skilled in music. **4** set to or accompanied by music. –*n.* musical film or play. □ **musicality** /-'kælɪtɪ/ *n.* **musically** *adv.*

musical box *n.* box containing a mechanism which plays a tune.

musical chairs *n.pl.* **1** party game in which the players compete in successive rounds for a decreasing number of chairs. **2** series of changes or political manoeuvring etc.

music centre *n.* equipment combining radio, record-player, tape recorder, etc.

music-hall *n.* **1** variety entertainment with singing, dancing, etc. **2** theatre for this.

musician /mju:'zɪʃ(ə)n/ *n.* person who plays a musical instrument, esp. professionally. □ **musicianly** *adj.* **musicianship** *n.* [French: related to MUSIC]

musicology /,mju:zɪ'kɒlədʒɪ/ *n.* the academic study of music. □ **musicologist** *n.* **musicological** /-kə'lɒdʒɪk(ə)l/ *adj.*

music stand *n.* support for sheet music.

music stool *n.* piano stool.

musk *n.* **1** substance secreted by the male musk deer and used in perfumes. **2** plant which orig. had a smell of musk. □ **musky** *adj.* (**-ier, -iest**). **muskiness** *n.* [Latin *muscus* from Persian]

musk deer *n.* small hornless Asian deer.

musket *n. hist.* infantryman's (esp. smooth-bored) light gun. [Italian *moschetto* crossbow bolt]

musketeer /,mʌskə'tɪə(r)/ *n. hist.* soldier armed with a musket.

musketry /'mʌskɪtrɪ/ *n.* **1** muskets; soldiers armed with muskets. **2** knowledge of handling small arms.

musk ox *n.* shaggy N. American ruminant with curved horns.

muskrat *n.* **1** large N. American aquatic rodent with a musky smell. **2** its fur.

musk-rose *n.* rambling rose smelling of musk.

Muslim /'mʊzlɪm, 'mʌ-/ (also **Moslem** /'mɒzləm/) –*n.* follower of the Islamic religion. –*adj.* of the Muslims or their religion. [Arabic: related to ISLAM]

muslin /'mʌzlɪn/ *n.* fine delicately woven cotton fabric. [Italian *Mussolo* Mosul in Iraq]

musquash /'mʌskwɒʃ/ *n.* = MUSKRAT. [Algonquian]

mussel /'mʌs(ə)l/ *n.* bivalve mollusc, esp. of the kind used for food. [Old English: related to MUSCLE]

must[1] –*v.aux.* (*present* **must**; *past* **had to** or in indirect speech **must**) (foll. by infin., or *absol.*) **1 a** be obliged to (*you must go to school*). **b** in ironic questions (*must you slam the door?*). **2** be certain to (*you must be her sister*). **3** ought to (*must see what can be done*). **4** expressing insistence (*must ask you to leave*). **5** (foll. by *not* + infin.) **a** not be permitted to, be forbidden to (*must not smoke*). **b** ought not; need not (*mustn't think he's angry; must not worry*). **c** expressing insistence that something should not be done (*they must not be told*). –*n. colloq.* thing that should not be missed (*this exhibition is a must*). □ **I must say** often *iron.* I cannot refrain from saying (*I must say he tries hard; a fine way to behave, I must say*). **must needs** see NEEDS. [Old English]

■ **Usage** In sense 1a, the negative (i.e. lack of obligation) is expressed by *not have to* or *need not*; *must not* denotes positive forbidding, as in *you must not smoke*.

must[2] *n.* grape juice before fermentation is complete. [Old English from Latin]

mustache *US* var. of MOUSTACHE.

mustang *n.* small wild horse of Mexico and California. [Spanish]

mustard /'mʌstəd/ *n.* **1 a** plant with slender pods and yellow flowers. **b** seeds of this crushed into a paste and used as a spicy condiment. **2** plant eaten at the seedling stage, often with cress. **3** brownish-yellow colour. [Romanic: related to MUST[2]]

mustard gas *n.* colourless oily liquid, whose vapour is a powerful irritant.

muster –*v.* **1** collect (orig. soldiers) for inspection, to check numbers, etc. **2** collect, gather together. **3** summon (courage etc.). –*n.* assembly of persons for inspection. □ **pass muster** be accepted as adequate. [Latin *monstro* show]

mustn't /'mʌs(ə)nt/ *contr.* must not.

musty *adj.* (**-ier, -iest**) **1** mouldy, stale. **2** dull, antiquated. □ **mustily** *adv.* **mustiness** *n.* [perhaps an alteration of *moisty*: related to MOIST]

mutable /'mju:təb(ə)l/ *adj. literary* liable to change. □ **mutability** /-'bɪlɪtɪ/ *n.* [Latin *muto* change]

mutagen /'mju:tədʒ(ə)n/ *n.* agent promoting genetic mutation. □ **mutagenic** /-'dʒenɪk/ *adj.* **mutagenesis** /-'dʒenɪsɪs/ *n.* [from MUTATION, -GEN]

mutant /'mju:t(ə)nt/ *–adj.* resulting from mutation. *–n.* mutant organism or gene.

mutate /mju:'teɪt/ *v.* (**-ting**) (cause to) undergo mutation.

mutation /mju:'teɪʃ(ə)n/ *n.* **1** change, alteration. **2** genetic change which, when transmitted to offspring, gives rise to heritable variations. **3** mutant. [Latin *muto* change]

mutatis mutandis /mu:,tɑ:tɪs mu:'tændɪs/ *adv.* (in comparing cases) making the necessary alterations. [Latin]

mute /mju:t/ *–adj.* **1** silent, refraining from or temporarily bereft of speech. **2** (of a person or animal) dumb. **3** not expressed in speech (*mute protest*). **4** (of a letter) not pronounced. *–n.* **1** dumb person. **2** device for damping the sound of a musical instrument. **3** unsounded consonant. *–v.* (**-ting**) **1** deaden or soften the sound of (esp. a musical instrument). **2 a** tone down, make less intense. **b** (as **muted** *adj.*) (of colours etc.) subdued. □ **mutely** *adv.* **muteness** *n.* [Latin *mutus*]

mute button *n.* device on a telephone to temporarily prevent the caller from hearing what is being said at the receiver's end, or on a television etc. to temporarily turn off the sound.

mute swan *n.* common white swan.

mutilate /'mju:tɪ,leɪt/ *v.* (**-ting**) **1 a** deprive (a person or animal) of a limb or organ. **b** destroy the use of (a limb or organ). **2** excise or damage part of (a book etc.). □ **mutilation** /-'leɪʃ(ə)n/ *n.* [Latin *mutilus* maimed]

mutineer /,mju:tɪ'nɪə(r)/ *n.* person who mutinies. [Romanic: related to MOVE]

mutinous /'mju:tɪnəs/ *adj.* rebellious; ready to mutiny. □ **mutinously** *adv.*

mutiny /'mju:tɪnɪ/ *–n.* (*pl.* **-ies**) open revolt, esp. by soldiers or sailors against their officers. *–v.* (**-ies, -ied**) (often foll. by *against*) revolt; engage in mutiny.

mutt *n.* **1** *slang* ignorant or stupid person. **2** *derog.* dog. [abbreviation of MUTTON-HEAD]

mutter *–v.* **1** (also *absol.*) utter (words) in a barely audible manner. **2** (often foll. by *against, at*) murmur or grumble. *–n.* **1** muttered words or sounds. **2** muttering. [related to MUTE]

mutton *n.* flesh of sheep as food. [medieval Latin *multo* sheep]

mutton dressed as lamb *n. colloq.* middle-aged or elderly woman dressed to appear younger.

mutton-head *n. colloq.* stupid person.

mutual /'mju:tʃʊəl/ *adj.* **1** (of feelings, actions, etc.) experienced or done by each of two or more parties to or towards the other(s) (*mutual affection*). **2** *colloq.* common to two or more persons (*a mutual friend*). **3** having the same (specified) relationship to each other (*mutual well-wishers*). □ **mutuality** /-'ælɪtɪ/ *n.* **mutually** *adv.* [Latin *mutuus* borrowed]

■ **Usage** The use of *mutual* in sense 2, although often found, is considered incorrect by some people, for whom *common* is preferable.

Muzak /'mju:zæk/ *n.* **1** *propr.* system of piped music used in public places. **2** (**muzak**) recorded light background music. [fanciful var. of MUSIC]

muzzle *–n.* **1** projecting part of an animal's face, including the nose and mouth. **2** guard, usu. of straps or wire, put over an animal's nose and mouth to stop it biting or feeding. **3** open end of a firearm. *–v.* (**-ling**) **1** put a muzzle on. **2** impose silence on. [medieval Latin *musum*]

muzzy *adj.* (**-ier, -iest**) **1** mentally hazy. **2** blurred, indistinct. □ **muzzily** *adv.* **muzziness** *n.* [origin unknown]

MW *abbr.* **1** megawatt(s). **2** medium wave.

my /maɪ/ *poss. pron.* (*attrib.*) **1** of or belonging to me. **2** affectionate, patronizing, etc. form of address (*my dear boy*). **3** in expressions of surprise (*my God!*; *oh my!*). **4** *colloq.* indicating a close relative etc. of the speaker (*my Johnny's ill again*). □ **my Lady** (or **Lord**) form of address to certain titled persons. [from MINE[1]]

myalgia /maɪ'ældʒə/ *n.* muscular pain. □ **myalgic** *adj.* [Greek *mus* muscle]

mycelium /maɪ'si:lɪəm/ (*pl.* **-lia**) microscopic threadlike parts of a fungus. [Greek *mukēs* mushroom]

Mycenaean /,maɪsɪ'ni:ən/ *–adj.* of the late Bronze Age civilization in Greece (*c.*1500–1100 BC), depicted in the Homeric poems. *–n.* person of this civilization. [Latin *Mycenaeus*]

mycology /maɪ'kɒlədʒɪ/ *n.* **1** the study of fungi. **2** fungi of a particular region. □ **mycologist** *n.* [Greek *mukēs* mushroom]

myna /'maɪnə/ *n.* (also **mynah, mina**) talking bird of the starling family. [Hindi]

myopia /maɪ'əʊpɪə/ *n.* **1** short-sightedness. **2** lack of imagination or insight. □

myopic /-ˈɒpɪk/ *adj.* **myopically** /-ˈɒpɪkəlɪ/ *adv.* [Greek *muō* shut, *ōps* eye]

myriad /ˈmɪrɪəd/ *literary* –*n.* an indefinitely great number. –*adj.* innumerable. [Greek *murioi* 10,000]

myrrh /mɜ:(r)/ *n.* gum resin used in perfume, medicine, incense, etc. [Latin *myrrha* from Greek]

myrtle /ˈmɜ:t(ə)l/ *n.* evergreen shrub with shiny leaves and white scented flowers. [Greek *murtos*]

myself /maɪˈself/ *pron.* **1** *emphat. form* of I[2] or ME[1] (*I saw it myself*). **2** *refl. form* of ME[1] (*I was angry with myself*). □ **be myself** see ONESELF. **I myself** I for my part (*I myself am doubtful*). [Old English: related to ME[1], SELF]

mysterious /mɪˈstɪərɪəs/ *adj.* full of or wrapped in mystery. □ **mysteriously** *adv.* [French: related to MYSTERY]

mystery /ˈmɪstərɪ/ *n.* (*pl.* **-ies**) **1** secret, hidden, or inexplicable matter. **2** secrecy or obscurity. **3** (*attrib.*) secret, undisclosed (*mystery guest*). **4** practice of making a secret of things (*engaged in mystery and intrigue*). **5** (in full **mystery story**) fictional work dealing with a puzzling event, esp. a murder. **6** a religious truth divinely revealed. **7** (in *pl.*) **a** secret religious rites of the ancient Greeks, Romans, etc. **b** *archaic* Eucharist. [Greek *mustērion*: related to MYSTIC]

mystery play *n.* miracle play.

mystery tour *n.* pleasure trip to an unspecified destination.

mystic /ˈmɪstɪk/ –*n.* person who seeks by contemplation etc. to achieve unity with the Deity, or who believes in the spiritual apprehension of truths that are beyond the understanding. –*adj.* = MYSTICAL. □ **mysticism** /-ˌsɪz(ə)m/ *n.* [Greek *mustēs* initiated person]

mystical *adj.* **1** of mystics or mysticism. **2** mysterious; occult; of hidden meaning. **3** spiritually allegorical or symbolic. □ **mystically** *adv.*

mystify /ˈmɪstɪˌfaɪ/ *v.* (**-ies, -ied**) **1** bewilder, confuse. **2** wrap in mystery. □ **mystification** /-fɪˈkeɪʃ(ə)n/ *n.* [French: related to MYSTIC or MYSTERY]

mystique /mɪˈsti:k/ *n.* atmosphere of mystery and veneration attending some activity, person, profession, etc. [French: related to MYSTIC]

myth /mɪθ/ *n.* **1** traditional story usu. involving supernatural or imaginary persons and embodying popular ideas on natural or social phenomena etc. **2** such narratives collectively. **3** widely held but false notion. **4** fictitious person, thing, or idea. **5** allegory (*Platonic myth*). □ **mythical** *adj.* **mythically** *adv.* [Greek *muthos*]

mythology /mɪˈθɒlədʒɪ/ *n.* (*pl.* **-ies**) **1** body of myths. **2** the study of myths. □ **mythological** /-θəˈlɒdʒɪk(ə)l/ *adj.* **mythologize** *v.* (also **-ise**) (**-zing** or **-sing**). [Greek: related to MYTH]

myxomatosis /ˌmɪksəməˈtəʊsɪs/ *n.* viral disease of rabbits. [Greek *muxa* mucus]

N

N[1] /en/ *n.* (also **n**) (*pl.* **Ns** or **N's**) **1** fourteenth letter of the alphabet. **2** (usu. **n**) indefinite number. □ **to the nth degree** to the utmost.

N[2] *abbr.* (also **N.**) **1** North; Northern. **2** New.

N[3] *symb.* nitrogen.

n *abbr.* (also **n.**) **1** name. **2** neuter.

Na *symb.* sodium. [Latin *natrium*]

NAAFI /'næfɪ/ *abbr.* **1** Navy, Army, and Air Force Institutes. **2** canteen for servicemen run by the NAAFI.

nab *v.* (**-bb-**) *slang* **1** arrest; catch in wrongdoing. **2** grab. [origin unknown]

nacho /'nætʃəʊ, 'nɑ:-/ *n.* (*pl.* **-s**) tortilla chip, usu. topped with melted cheese and spices etc. [origin uncertain]

nacre /'neɪkə(r)/ *n.* mother-of-pearl from any shelled mollusc. □ **nacreous** /'neɪkrɪəs/ *adj.* [French]

nadir /'neɪdɪə(r)/ *n.* **1** part of the celestial sphere directly below an observer. **2** lowest point; time of deep despair. [Arabic, = opposite]

naevus /'ni:vəs/ *n.* (*US* **nevus**) (*pl.* **naevi** /-vaɪ/) **1** raised red birthmark. **2** = MOLE[2]. [Latin]

naff *adj. slang* **1** unfashionable. **2** rubbishy. [origin unknown]

nag[1] *v.* (**-gg-**) **1 a** persistently criticize or scold. **b** (often foll. by *at*) find fault or urge, esp. persistently. **2** (of a pain) be persistent. [originally a dial. word]

nag[2] *n. colloq.* horse. [origin unknown]

naiad /'naɪæd/ *n.* water-nymph. [Latin from Greek]

nail –*n.* **1** small metal spike hammered in to join things together or as a peg or decoration. **2** horny covering on the upper surface of the tip of the human finger or toe. –*v.* **1** fasten with a nail or nails. **2** secure or get hold of (a person or thing). **3** keep (attention etc.) fixed. **4** expose or discover (a lie or liar). □ **nail down 1** bind (a person) to a promise etc. **2** define precisely. **3** fasten (a thing) with nails. **nail in a person's coffin** something thought to increase the risk of death. **on the nail** (esp. of payment) without delay. [Old English]

nail-file *n.* roughened metal or emery strip used for smoothing the nails.

nail polish *n.* (also **nail varnish**) varnish, usu. coloured, applied to the nails.

naïve /naɪ'i:v/ *adj.* (also **naive**) **1** innocent; unaffected. **2** foolishly credulous. **3** (of art) produced in a sophisticated society but lacking conventional expertise. □ **naïvely** *adv.* **naïvety** *n.* (also *naïveté*). [Latin *nativus* NATIVE]

naked /'neɪkɪd/ *adj.* **1** without clothes; nude. **2** without its usual covering. **3** undisguised (*the naked truth*). **4** (of a light, flame, sword, etc.) unprotected or unsheathed. □ **nakedly** *adv.* **nakedness** *n.* [Old English]

naked eye *n.* (prec. by *the*) unassisted vision, e.g. without a telescope etc.

namby-pamby /,næmbɪ'pæmbɪ/ –*adj.* insipidly pretty or sentimental; weak. –*n.* (*pl.* **-ies**) namby-pamby person. [fanciful formulation on the name of the writer *Ambrose* Philips]

name –*n.* **1** word by which an individual person, family, animal, place, or thing is spoken of etc. **2 a** (usu. abusive) term used of a person etc. (*called him names*). **b** word denoting an object or esp. a class of objects etc. (*what is the name of those flowers?*). **3** famous person. **4** reputation, esp. a good one. –*v.* (**-ming**) **1** give a name to. **2** state the name of. **3** mention; specify; cite. **4** nominate. □ **have to one's name** possess. **in the name of** as representing; by virtue of (*in the name of the law*). **in name only** not in reality. **make a name for oneself** become famous. □ **nameable** *adj.* [Old English]

name-day *n.* feast-day of the saint after whom a person is named.

name-dropping *n.* familiar mention of famous people as a form of boasting.

nameless *adj.* **1** having or showing no name. **2** unnamed (*our informant, who shall be nameless*). **3** too horrific to be named (*nameless vices*).

namely *adv.* that is to say; in other words.

name-plate *n.* plate or panel bearing the name of an occupant of a room etc.

namesake *n.* person or thing having the same name as another. [probably from *for the name's sake*]

nan *n.* (also **nana**, **nanna** /'nænə/) *colloq.* grandmother. [childish pronunciation]

nancy /'nænsɪ/ *n.* (*pl.* **-ies**) (in full **nancy boy**) *slang offens.* effeminate man, esp. a homosexual. [pet form of *Ann*]

nanny /'nænɪ/ *n.* (*pl.* **-ies**) **1** child's nurse. **2** *colloq.* grandmother. **3** (in full

nanny-goat) female goat. [related to NANCY]

nano- *comb. form* denoting a factor of 10^{-9} (*nanosecond*). [Greek *nanos* dwarf]

nap[1] *–v.* (**-pp-**) sleep lightly or briefly. *–n.* short sleep or doze, esp. by day. □ **catch a person napping** detect in negligence etc; catch off guard. [Old English]

nap[2] *n.* raised pile on textiles, esp. velvet. [Low German or Dutch]

nap[3] *–n.* **1** form of whist in which players declare the number of tricks they expect to take. **2** racing tip claimed to be almost a certainty. *–v.* (**-pp-**) name (a horse etc.) as a probable winner. □ **go nap 1** attempt to take all five tricks in nap. **2** risk everything. [*Napoleon*]

napalm /'neɪpɑːm/ *–n.* thick jellied hydrocarbon mixture used in bombs. *–v.* attack with napalm bombs. [from NAPHTHALENE, PALM[1]]

nape *n.* back of the neck. [origin unknown]

naphtha /'næfθə/ *n.* inflammable hydrocarbon distilled from coal etc. [Latin from Greek]

naphthalene /'næfθəˌliːn/ *n.* white crystalline substance produced by distilling coal tar.

napkin /'næpkɪn/ *n.* **1** piece of linen etc. for wiping the lips, fingers, etc. at meals. **2** baby's nappy. [French *nappe* from Latin *mappa* MAP]

nappy /'næpɪ/ *n.* (*pl.* **-ies**) piece of towelling etc. wrapped round a baby to absorb or retain urine and faeces. [from NAPKIN]

narcissism /'nɑːsɪˌsɪz(ə)m/ *n.* excessive or erotic interest in oneself. □ **narcissistic** /-'sɪstɪk/ *adj.* [*Narkissos*, name of a youth in Greek myth who fell in love with his reflection]

narcissus /nɑː'sɪsəs/ *n.* (*pl.* **-cissi** /-saɪ/) any of several flowering bulbs, including the daffodil. [Latin from Greek]

narcosis /nɑː'kəʊsɪs/ *n.* **1** state of insensibility. **2** induction of this. [Greek *narkē* numbness]

narcotic /nɑː'kɒtɪk/ *–adj.* **1** (of a substance) inducing drowsiness etc. **2** (of a drug) affecting the mind. *–n.* narcotic substance, drug, or influence. [Greek *narkōtikos*]

nark *slang –n.* police informer or decoy. *–v.* annoy. [Romany *nāk* nose]

narrate /nə'reɪt/ *v.* (**-ting**) **1** give a continuous story or account of. **2** provide a spoken accompaniment for (a film etc.). □ **narration** /nə'reɪʃ(ə)n/ *n.* **narrator** *n.* [Latin *narro*]

narrative /'nærətɪv/ *–n.* ordered account of connected events. *–adj.* of or by narration.

narrow /'nærəʊ/ *–adj.* (**-er, -est**) **1 a** of small width. **b** confined or confining (*within narrow bounds*). **2** of limited scope (*in the narrowest sense*). **3** with little margin (*narrow escape*). **4** precise; exact. **5** = NARROW-MINDED. *–n.* (usu. in *pl.*) narrow part of a strait, river, pass, street, etc. *–v.* become or make narrow; contract; lessen. □ **narrowly** *adv.* **narrowness** *n.* [Old English]

narrow boat *n.* canal boat.

narrow-minded *adj.* rigid or restricted in one's views, intolerant. □ **narrow-mindedness** *n.*

narwhal /'nɑːw(ə)l/ *n.* Arctic white whale, the male of which has a long tusk. [Dutch from Danish]

NASA /'næsə/ *abbr.* (in the US) National Aeronautics and Space Administration.

nasal /'neɪz(ə)l/ *–adj.* **1** of the nose. **2** (of a letter or a sound) pronounced with the breath passing through the nose, e.g. *m, n, ng*. **3** (of the voice or speech) having many nasal sounds. *–n.* nasal letter or sound. □ **nasalize** *v.* (also **-ise**) (**-zing** or **-sing**). **nasally** *adv.* [Latin *nasus* nose]

nascent /'næs(ə)nt/ *adj.* **1** in the act of being born. **2** just beginning to be; not yet mature. □ **nascency** /'næsənsɪ/ *n.* [Latin: related to NATAL]

nasturtium /nə'stɜːʃəm/ *n.* trailing plant with edible leaves and bright orange, yellow, or red flowers. [Latin]

nasty /'nɑːstɪ/ *–adj.* (**-ier, -iest**) **1** highly unpleasant. **2** difficult to negotiate. **3** (of a person or animal) ill-natured. *–n.* (*pl.* **-ies**) *colloq.* horror film, esp. one on video and depicting cruelty or killing. □ **nastily** *adv.* **nastiness** *n.* [origin unknown]

nasty piece of work *n. colloq.* unpleasant or contemptible person.

Nat. *abbr.* **1** National. **2** Nationalist. **3** Natural.

natal /'neɪt(ə)l/ *adj.* of or from one's birth. [Latin *natalis* from *nascor nat-* be born]

nation /'neɪʃ(ə)n/ *n.* community of people of mainly common descent, history, language, etc., forming a State or inhabiting a territory. [Latin: related to NATAL]

national /'næʃən(ə)l/ *–adj.* **1** of a, or the, nation. **2** characteristic of a particular nation. *–n.* **1** citizen of a specified country. **2** fellow-countryman. **3** (**the National**) = GRAND NATIONAL. □ **nationally** *adv.*

national anthem *n.* song adopted by a nation, intended to inspire patriotism.

national curriculum *n.* common programme of study for pupils in the maintained schools of England and Wales, with tests at specified ages.

national debt *n.* money owed by a State because of loans to it.

National Front *n.* UK political party with extreme reactionary views on immigration etc.

national grid *n.* **1** network of high-voltage electric power lines between major power stations. **2** metric system of geographical coordinates used in maps of the British Isles.

National Health *n.* (also **National Health Service**) system of national medical care paid for mainly by taxation.

National Insurance *n.* system of compulsory payments by employed persons (supplemented by employers) to provide State assistance in sickness etc.

nationalism *n.* **1** patriotic feeling, principles, etc. **2** policy of national independence. □ **nationalist** *n.* & *adj.* **nationalistic** /-'lɪstɪk/ *adj.*

nationality /ˌnæʃə'nælɪtɪ/ *n.* (*pl.* **-ies**) **1** status of belonging to a particular nation (*has British nationality*). **2** condition of being national; distinctive national qualities. **3** ethnic group forming a part of one or more political nations.

nationalize /'næʃənəˌlaɪz/ *v.* (also **-ise**) (**-zing** or **-sing**) **1** take (railways, industry, land, etc.) into State ownership. **2** make national. □ **nationalization** /-'zeɪʃ(ə)n/ *n.*

national park *n.* area of natural beauty protected by the State for the use of the public.

national service *n. hist.* conscripted peacetime military service.

nationwide *adj.* & *adv.* extending over the whole nation.

native /'neɪtɪv/ *–n.* **1 a** (usu. foll. by *of*) person born in a specified place. **b** local inhabitant. **2** often *offens.* member of a non-White indigenous people, as regarded by colonial settlers. **3** (usu. foll. by *of*) indigenous animal or plant. *–adj.* **1** inherent; innate. **2** of one's birth (*native country*). **3** (usu. foll. by *to*) belonging to a specified place. **4** (esp. of a non-European) indigenous; born in a place. **5** (of metal etc.) found in a pure or uncombined state. [Latin: related to NATAL]

nativity /nə'tɪvɪtɪ/ *n.* (*pl.* **-ies**) **1** (esp. **the Nativity**) **a** Christ's birth. **b** festival of Christ's birth. **2** birth. [Latin: related to NATIVE]

NATO /'neɪtəʊ/ *abbr.* (also **Nato**) North Atlantic Treaty Organization.

natter *colloq. –v.* chatter idly. *–n.* aimless chatter. [imitative, originally dial.]

natterjack /'nætəˌdʒæk/ *n.* a kind of small toad. [perhaps from NATTER]

natty /'nætɪ/ *adj.* (**-ier**, **-iest**) *colloq.* trim; smart. □ **nattily** *adv.* [cf. NEAT]

natural /'nætʃər(ə)l/ *–adj.* **1 a** existing in or caused by nature (*natural landscape*). **b** uncultivated (*in its natural state*). **2** in the course of nature (*died of natural causes*). **3** not surprising; to be expected (*natural for her to be upset*). **4** unaffected, spontaneous. **5** innate (*natural talent for music*). **6** not disguised or altered (as by make-up etc.). **7** likely or suited by its or their nature to be such (*natural enemies*; *natural leader*). **8** physically existing (*the natural world*). **9** illegitimate. **10** *Mus.* (of a note) not sharpened or flattened (*B natural*). *–n.* **1** *colloq.* (usu. foll. by *for*) person or thing naturally suitable, adept, etc. **2** *Mus.* **a** sign (♮) denoting a return to natural pitch. **b** natural note. □ **naturalness** *n.* [Latin: related to NATURE]

natural gas *n.* gas found in the earth's crust, not manufactured.

natural history *n.* the study of animals or plants.

naturalism *n.* **1** theory or practice in art and literature of realistic representation. **2 a** theory of the world that excludes the supernatural or spiritual. **b** moral or religious system based on this. □ **naturalistic** /-'lɪstɪk/ *adj.*

naturalist *n.* **1** person who studies natural history. **2** adherent of naturalism.

naturalize *v.* (also **-ise**) (**-zing** or **-sing**) **1** admit (a foreigner) to citizenship. **2** successfully introduce (an animal, plant, etc.) into another region. **3** adopt (a foreign word, custom, etc.). □ **naturalization** /-'zeɪʃ(ə)n/ *n.*

natural law *n.* **1** unchanging moral principles common to all human beings. **2** correct statement of an invariable sequence between specified conditions and a specified phenomenon.

naturally *adv.* **1** in a natural manner. **2** (qualifying a whole sentence) as might be expected; of course.

natural number *n.* whole number greater than 0.

natural resources *n.pl.* materials or conditions occurring in nature and capable of economic exploitation.

natural science *n.* **1** the study of the natural or physical world. **2** (in *pl.*) sciences used for this.

natural selection *n.* Darwinian theory of the survival and propagation of organisms best adapted to their environment.

nature /'neɪtʃə(r)/ *n.* **1** thing's or person's innate or essential qualities or character. **2** (often **Nature**) **a** physical power causing all material phenomena. **b** these phenomena. **3** kind or class (*things of this nature*). **4** inherent impulses determining character or action. □ **by nature** innately. **in** (or **by**) **the nature of things 1** inevitable. **2** inevitably. [Latin *natura*: related to NATAL]

natured *adj.* (in *comb.*) having a specified disposition (*good-natured*).

nature reserve *n.* tract of land managed so as to preserve its flora, fauna, physical features, etc.

nature trail *n.* signposted path through the countryside designed to draw attention to natural phenomena.

naturism *n.* nudism. □ **naturist** *n.*

naught /nɔːt/ *archaic* or *literary* –*n.* nothing, nought. –*adj.* (usu. *predic.*) worthless; useless. □ **come to naught** come to nothing, fail. **set at naught** despise. [Old English: related to NO[2], WIGHT]

naughty /ˈnɔːtɪ/ *adj.* (**-ier, -iest**) **1** (esp. of children) disobedient; badly behaved. **2** *colloq. joc.* indecent. □ **naughtily** *adv.* **naughtiness** *n.* [from NAUGHT]

nausea /ˈnɔːsɪə/ *n.* **1** inclination to vomit. **2** revulsion. [Greek *naus* ship]

nauseate /ˈnɔːsɪˌeɪt/ *v.* (**-ting**) affect with nausea. □ **nauseating** *adj.* **nauseatingly** *adv.*

nauseous /ˈnɔːsɪəs/ *adj.* **1** causing nausea. **2** inclined to vomit (*feel nauseous*). **3** disgusting; loathsome.

nautical /ˈnɔːtɪk(ə)l/ *adj.* of sailors or navigation. [Greek *nautēs* sailor]

nautical mile *n.* unit of approx. 2,025 yards (1,852 metres).

nautilus /ˈnɔːtɪləs/ *n.* (*pl.* **nautiluses** or **nautili** /-ˌlaɪ/) cephalopod mollusc with a spiral shell, esp. (**pearly nautilus**) one having a chambered shell. [Greek *nautilos*: related to NAUTICAL]

naval /ˈneɪv(ə)l/ *adj.* **1** of the or a navy. **2** of ships. [Latin *navis* ship]

nave[1] *n.* central part of a church, usu. from the west door to the chancel excluding the side aisles. [Latin *navis* ship]

nave[2] *n.* hub of a wheel. [Old English]

navel /ˈneɪv(ə)l/ *n.* depression in the centre of the belly marking the site of attachment of the umbilical cord. [Old English]

navel orange *n.* orange with a navel-like formation at the top.

navigable /ˈnævɪgəb(ə)l/ *adj.* **1** (of a river etc.) suitable for ships to pass through. **2** seaworthy. **3** steerable. □ **navigability** /-ˈbɪlɪtɪ/ *n.* [Latin: related to NAVIGATE]

navigate /ˈnævɪˌgeɪt/ *v.* (**-ting**) **1** manage or direct the course of (a ship or aircraft) using maps and instruments. **2 a** sail on (a sea, river, etc.). **b** fly through (the air). **3** (in a car etc.) assist the driver by map-reading etc. **4** sail a ship; sail in a ship. □ **navigator** *n.* [Latin *navigo* from *navis*]

navigation /ˌnævɪˈgeɪʃ(ə)n/ *n.* **1** act or process of navigating. **2** art or science of navigating. □ **navigational** *adj.*

navvy /ˈnævɪ/ –*n.* (*pl.* **-ies**) labourer employed in building or excavating roads, canals, etc. –*v.* (**-ies, -ied**) work as a navvy. [abbreviation of *navigator*]

navy /ˈneɪvɪ/ *n.* (*pl.* **-ies**) **1** (often **the Navy**) **a** whole body of a State's ships of war, including crews, maintenance systems, etc. **b** officers and men of a navy. **2** (in full **navy blue**) dark-blue colour as of naval uniforms. **3** *poet.* fleet of ships. [Romanic *navia* ship: related to NAVAL]

nay –*adv.* **1** or rather; and even; and more than that (*large, nay, huge*). **2** *archaic* = NO[2] *adv.* 1. –*n.* utterance of 'nay'; 'no' vote. [Old Norse, = not ever]

Nazarene /ˈnæzəˌriːn/ –*n.* **1 a** (prec. by *the*) Christ. **b** (esp. in Jewish or Muslim use) Christian. **2** native or inhabitant of Nazareth. –*adj.* of Nazareth. [Latin from Greek]

Nazi /ˈnɑːtsɪ/ –*n.* (*pl.* **-s**) *hist.* member of the German National Socialist party. –*adj.* of the Nazis or Nazism. □ **Nazism** *n.* [representing pronunciation of *Nati-* in German *Nationalsozialist*]

NB *abbr.* note well. [Latin *nota bene*]

Nb *symb.* niobium.

NCB *abbr. hist.* National Coal Board.

■ **Usage** Since 1987 the official name has been *British Coal.*

NCO *abbr.* non-commissioned officer.

NCP *abbr.* National Car Parks.

Nd *symb.* neodymium.

NE *abbr.* **1** north-east. **2** north-eastern.

Ne *symb.* neon.

Neanderthal /nɪˈændəˌtɑːl/ *adj.* of the type of human widely distributed in palaeolithic Europe, with a retreating forehead and massive brow-ridges. [region in W. Germany]

neap *n.* (in full **neap tide**) tide at the times of the month when there is least difference between high and low water. [Old English]

Neapolitan /nɪəˈpɒlɪt(ə)n/ –*n.* native or citizen of Naples. –*adj.* of Naples. [Greek *Neapolis* Naples]

near –*adv.* **1** (often foll. by *to*) to or at a short distance in space or time. **2** closely (*as near as one can guess*). –*prep.* **1** to or at a short distance from (in space, time, condition, or resemblance). **2** (in *comb.*) almost (*near-hysterical*). –*adj.* **1** close (to), not far (in place or time) (*my flat's very near; the man nearest you; in* *the near future*). **2 a** closely related. **b** intimate. **3** (of a part of a vehicle, animal, or road) on the left side. **4** close; narrow (*near escape*). **5** similar (to) (*is nearer the original*). **6** *colloq.* niggardly. –*v.* approach; draw near to. □ **come** (or **go**) **near** (foll. by verbal noun, or *to* + verbal noun) be on the point of, almost succeed in. **near at hand** within easy reach. **near the knuckle** *colloq.* verging on the indecent. □ **nearish** *adj.* **nearness** *n.* [Old Norse, originally = nigher: related to NIGH]

nearby –*adj.* near in position. –*adv.* close; not far away.

Near East *n.* (prec. by *the*) region comprising the countries of the eastern Mediterranean. □ **Near Eastern** *adj.*

nearly *adv.* **1** almost. **2** closely. □ **not nearly** nothing like.

near miss *n.* **1** bomb etc. falling close to the target. **2** narrowly avoided collision. **3** not quite successful attempt.

nearside *n.* (often *attrib.*) left side of a vehicle, animal, etc.

near-sighted *adj.* = SHORT-SIGHTED.

near thing *n.* narrow escape.

neat *adj.* **1** tidy and methodical. **2** elegantly simple. **3** brief, clear, and pointed. **4 a** cleverly executed. **b** dexterous. **5** (of esp. alcoholic liquor) undiluted. □ **neatly** *adv.* **neatness** *n.* [French *net* from Latin *nitidus* shining]

neaten *v.* make neat.

neath *prep. poet.* beneath. [from BENEATH]

nebula /ˈnebjʊlə/ *n.* (*pl.* **nebulae** /-ˌliː/) cloud of gas and dust seen in the night sky, sometimes glowing and sometimes appearing as a dark silhouette. □ **nebular** *adj.* [Latin, = mist]

nebulous *adj.* **1** cloudlike. **2** indistinct, vague. [Latin: related to NEBULA]

NEC *abbr.* National Executive Committee.

necessary /ˈnesəsərɪ/ –*adj.* **1** requiring to be done; requisite, essential. **2** determined, existing, or happening by natural laws etc., not by free will; inevitable. –*n.* (*pl.* **-ies**) (usu. in *pl.*) any of the basic requirements of life. □ **the necessary** *colloq.* **1** money. **2** an action etc. needed for a purpose. □ **necessarily** /ˈnes-, -ˈserɪlɪ/ *adv.* [Latin *necesse* needful]

necessitarian /nɪˌsesɪˈteərɪən/ –*n.* person who holds that all action is predetermined and free will is impossible. –*adj.* of such a person or theory. □ **necessitarianism** *n.*

necessitate /nɪˈsesɪˌteɪt/ *v.* (**-ting**) make necessary (esp. as a result) (*will necessitate some sacrifice*).

necessitous /nɪˈsesɪtəs/ *adj.* poor; needy.

necessity /nɪˈsesɪtɪ/ *n.* (*pl.* **-ies**) **1** indispensable thing. **2** pressure of circumstances. **3** imperative need. **4** want; poverty. **5** constraint or compulsion regarded as a natural law governing all human action. □ **of necessity** unavoidably.

neck –*n.* **1 a** part of the body connecting the head to the shoulders. **b** part of a garment round the neck. **2** something resembling a neck; narrow part of a cavity, vessel, or object such as a bottle or violin. **3** length of a horse's head and neck as a measure of its lead in a race. **4** flesh of an animal's neck as food. **5** *slang* impudence. –*v. colloq.* kiss and caress amorously. □ **get it in the neck** *colloq.* **1** be severely reprimanded or punished. **2** suffer a severe blow. **up to one's neck** (often foll. by *in*) *colloq.* very deeply involved; very busy. [Old English]

neck and neck *adj.* & *adv.* (running) level in a race etc.

neckband *n.* strip of material round the neck of a garment.

neckerchief *n.* square of cloth worn round the neck. [from KERCHIEF]

necklace /ˈnekləs/ *n.* **1** chain or string of beads, precious stones, etc., worn round the neck. **2** *S.Afr.* tyre soaked or filled with petrol, placed round a victim's neck, and set alight.

neckline *n.* edge or shape of a garment-opening at the neck.

necktie *n.* esp. *US* = TIE *n.* 2.

necro- *comb. form* corpse. [Greek *nekros* corpse]

necromancy /ˈnekrəʊˌmænsɪ/ *n.* **1** divination by supposed communication with the dead. **2** magic. □ **necromancer** *n.* [from NECRO-, *mantis* seer]

necrophilia /ˌnekrəˈfɪlɪə/ *n.* morbid and esp. sexual attraction to corpses.

necropolis /neˈkrɒpəlɪs/ *n.* ancient cemetery or burial place. [Greek: related to NECRO-, *polis* city]

necrosis /neˈkrəʊsɪs/ *n.* death of tissue. □ **necrotic** /-ˈkrɒtɪk/ *adj.* [Greek *nekroō* kill]

nectar /ˈnektə(r)/ *n.* **1** sugary substance produced by plants and made into honey by bees. **2** (in Greek and Roman mythology) the drink of the gods. **3** drink compared to this. □ **nectarous** *adj.* [Latin from Greek]

nectarine /ˈnektərɪn/ *n.* smooth-skinned variety of peach. [from NECTAR]

NEDC *abbr.* National Economic Development Council.

neddy /'nedɪ/ *n.* (*pl.* **-ies**) *colloq.* **1** donkey. **2** (**Neddy**) = NEDC. [pet form of *Edward*]

née /neɪ/ *adj.* (*US* **nee**) (used in adding a married woman's maiden name after her surname) born (*Mrs Ann Hall, née Brown*). [French, feminine past part. of *naître* be born]

need –*v.* **1** stand in want of; require. **2** (foll. by *to* + infin.; *3rd sing. present neg. or interrog.* **need** without *to*) be under the necessity or obligation (*needs to be done well; he need not come; need you ask?*). –*n.* **1** requirement (*my needs are few*). **2** circumstances requiring some course of action (*no need to worry; if need be*). **3** destitution; poverty. **4** crisis; emergency (*failed them in their need*). □ **have need of** require. **need not have** did not need to (but did). [Old English]

needful *adj.* requisite. □ **needfully** *adv.*

needle /'ni:d(ə)l/ –*n.* **1 a** very thin pointed rod of smooth steel etc. with a slit ('eye') for thread at the blunt end, used in sewing. **b** larger plastic, wooden, etc. slender rod without an eye, used in knitting etc. **2** pointer on a dial. **3** any of several small thin pointed instruments, esp.: **a** the end of a hypodermic syringe. **b** = STYLUS 1. **4 a** obelisk (*Cleopatra's Needle*). **b** pointed rock or peak. **5** leaf of a fir or pine tree. **6** (**the needle**) *slang* fit of bad temper or nervousness. –*v.* (**-ling**) *colloq.* irritate; provoke. [Old English]

needlecord *n.* fine-ribbed corduroy fabric.

needle-point *n.* **1** lace made with needles, not bobbins. **2** = GROS OF PETIT POINT.

needless *adj.* **1** unnecessary. **2** uncalled for. □ **needlessly** *adv.*

needlewoman *n.* **1** seamstress. **2** woman or girl with specified sewing skill.

needlework *n.* sewing or embroidery.

needs *adv. archaic* (usu. prec. or foll. by *must*) of necessity.

needy *adj.* (**-ier**, **-iest**) poor; destitute. □ **neediness** *n.*

ne'er /neə(r)/ *adv. poet.* = NEVER. [contraction]

ne'er-do-well –*n.* good-for-nothing person. –*adj.* good-for-nothing.

nefarious /nɪ'feərɪəs/ *adj.* wicked. [Latin *nefas* wrong *n.*]

neg. *abbr.* esp. *Photog.* negative.

negate /nɪ'geɪt/ *v.* (**-ting**) **1** nullify. **2** assert or imply the non-existence of. [Latin *nego* deny]

negation /nɪ'geɪʃ(ə)n/ *n.* **1** absence or opposite of something actual or positive. **2** act of denying. **3** negative statement. **4** negative or unreal thing.

negative /'negətɪv/ –*adj.* **1** expressing or implying denial, prohibition, or refusal (*negative answer*). **2** (of a person or attitude) lacking positive attributes. **3** marked by the absence of qualities (*negative reaction*). **4** of the opposite nature to a thing regarded as positive. **5** (of a quantity) less than zero, to be subtracted from others or from zero. **6** *Electr.* **a** of the kind of charge carried by electrons. **b** containing or producing such a charge. –*n.* **1** negative statement or word. **2** *Photog.* **a** image with black and white reversed or colours replaced by complementary ones, from which positive pictures are obtained. **b** developed film or plate bearing such an image. –*v.* (**-ving**) **1** refuse to accept or countenance; veto. **2** disprove. **3** contradict (a statement). **4** neutralize (an effect). □ **in the negative** with negative effect. □ **negatively** *adv.* **negativity** /-'tɪvɪtɪ/ *n.*

negativism *n.* negative attitude; extreme scepticism.

neglect /nɪ'glekt/ –*v.* **1** fail to care for or to do; be remiss about. **2** (foll. by *to* + infin.) fail; overlook the need to. **3** not pay attention to; disregard. –*n.* **1** negligence. **2** neglecting or being neglected. **3** (usu. foll. by *of*) disregard. □ **neglectful** *adj.* **neglectfully** *adv.* [Latin *neglego neglect-*]

negligée /'neglɪ,ʒeɪ/ *n.* (also **negligee**, **négligé**) woman's flimsy dressing-gown. [French, past part. of *négliger* NEGLECT]

negligence /'neglɪdʒ(ə)ns/ *n.* **1** lack of proper care and attention. **2** culpable carelessness. □ **negligent** *adj.* **negligently** *adv.* [Latin: related to NEGLECT]

negligible /'neglɪdʒɪb(ə)l/ *adj.* not worth considering; insignificant. □ **negligibly** *adv.* [French: related to NEGLECT]

negotiable /nɪ'gəʊʃəb(ə)l/ *adj.* **1** open to discussion. **2** able to be negotiated.

negotiate /nɪ'gəʊʃɪ,eɪt/ *v.* (**-ting**) **1** (usu. foll. by *with*) confer in order to reach an agreement. **2** arrange (an affair) or bring about (a result) by negotiating. **3** find a way over, through, etc. (an obstacle, difficulty, etc.). **4** convert (a cheque etc.) into money. □ **negotiation** /-ʃɪ'eɪʃ(ə)n/ *n.* **negotiator** *n.* [Latin *negotium* business]

Negress /'ni:grɪs/ *n.* female Negro.

■ **Usage** The term *Negress* is often considered offensive; *Black* is usually preferred.

Negritude /'negrɪ,tju:d/ *n.* **1** state of being Black. **2** affirmation of Black culture. [French]

Negro /'ni:grəʊ/ *–n.* (*pl.* **-es**) member of a dark-skinned race orig. native to Africa. *–adj.* **1** of Negroes. **2** (as **negro**) *Zool.* black or dark. [Latin *niger nigri* black]

■ **Usage** The term *Negro* is often considered offensive; *Black* is usually preferred.

Negroid /'ni:grɔɪd/ *–adj.* (of physical features etc.) characteristic of Black people. *–n.* Black.

neigh /neɪ/ *–n.* cry of a horse. *–v.* make a neigh. [Old English]

neighbour /'neɪbə(r)/ (*US* **neighbor**) *–n.* **1** person living next door to or near or nearest another. **2** fellow human being. **3** person or thing near or next to another. *–v.* border on; adjoin. [Old English: related to NIGH, BOOR]

neighbourhood *n.* (*US* **neighborhood**) **1** district; vicinity. **2** people of a district. □ **in the neighbourhood of** roughly; about.

neighbourhood watch *n.* organized local vigilance by householders to discourage crime.

neighbourly *adj.* (*US* **neighborly**) like a good neighbour; friendly; kind. □ **neighbourliness** *n.*

neither /'naɪðə(r), 'ni:ð-/ *–adj.* & *pron.* (foll. by sing. verb) not the one nor the other (of two things); not either (*neither of the accusations is true*; *neither of them knows*; *neither wish was granted*; *neither went to the fair*). *–adv.* **1** not either; not on the one hand (foll. by *nor*; introducing the first of two or more things in the negative: *neither knowing nor caring*; *neither the teachers nor the parents nor the children*). **2** also not (*if you do not, neither shall I*). *–conj. archaic* nor yet; nor (*I know not, neither can I guess*). [Old English: related to NO², WHETHER]

nelson /'nels(ə)n/ *n.* wrestling-hold in which one arm is passed under the opponent's arm from behind and the hand is applied to the neck (**half nelson**), or both arms and hands are applied (**full nelson**). [apparently from the name *Nelson*]

nematode /'nemə,təʊd/ *n.* worm with a slender unsegmented cylindrical shape. [Greek *nēma* thread]

nem. con. *abbr.* with no one dissenting. [Latin *nemine contradicente*]

nemesis /'neməsɪs/ *n.* (*pl.* **nemeses** /-,si:z/) **1** retributive justice. **2** downfall caused by this. [Greek, = retribution]

neo- *comb. form* **1** new, modern. **2** new form of. [Greek *neos* new]

neoclassicism /,ni:əʊ'klæsɪ,sɪz(ə)m/ *n.* revival of classical style or treatment in the arts. □ **neoclassical** /-k(ə)l/ *adj.*

neodymium /,ni:ə'dɪmɪəm/ *n.* metallic element of the lanthanide series. [from NEO-, Greek *didumos* twin]

neolithic /,ni:ə'lɪθɪk/ *adj.* of the later part of the Stone Age. [Greek *lithos* stone]

neologism /ni:'ɒlə,dʒɪz(ə)m/ *n.* **1** new word. **2** coining of new words. [Greek *logos* word]

neon /'ni:ɒn/ *n.* inert gaseous element giving an orange glow when electricity is passed through it. [Greek, = new]

neophyte /'ni:ə,faɪt/ *n.* **1** new convert. **2** *RC Ch.* novice of a religious order. **3** beginner. [Greek *phuton* plant]

nephew /'nefju:/ *n.* son of one's brother or sister or of one's spouse's brother or sister. [Latin *nepos*]

nephritic /nɪ'frɪtɪk/ *adj.* **1** of or in the kidneys. **2** of nephritis. [Greek *nephros* kidney]

nephritis /nɪ'fraɪtɪs/ *n.* inflammation of the kidneys.

ne plus ultra /,neɪ plʊs 'ʊltrɑ:/ *n.* **1** furthest attainable point. **2** acme, perfection. [Latin, = not further beyond]

nepotism /'nepə,tɪz(ə)m/ *n.* favouritism shown to relatives in conferring offices. [Italian *nepote* nephew]

neptunium /nep'tju:nɪəm/ *n.* transuranic metallic element produced when uranium atoms absorb bombarding neutrons. [*Neptune*, name of a planet]

nerd *n.* (also **nurd**) esp. *US slang* foolish, feeble, or uninteresting person. [origin uncertain]

nereid /'nɪərɪɪd/ *n.* sea-nymph. [Latin from Greek]

nerve *–n.* **1 a** fibre or bundle of fibres that transmits impulses of sensation or motion between the brain or spinal cord and other parts of the body. **b** material constituting these. **2 a** coolness in danger; bravery. **b** *colloq.* impudence. **3** (in *pl.*) nervousness; mental or physical stress. *–v.* (**-ving**) **1** (usu. *refl.*) brace (oneself) to face danger etc. **2** give strength, vigour, or courage to. □ **get on a person's nerves** irritate a person. [Latin *nervus* sinew, bowstring]

nerve cell *n.* cell transmitting impulses in nerve tissue.

nerve-centre *n.* **1** group of closely connected nerve-cells. **2** centre of control.

nerve gas *n.* poisonous gas affecting the nervous system.

nerveless *adj.* **1** lacking vigour. **2** (of style) diffuse.

nerve-racking *adj.* causing mental strain.

nervous *adj.* **1** easily upset, timid, highly strung. **2** anxious. **3** affecting the nerves. **4** (foll. by *of* + verbal noun) afraid (*am nervous of meeting them*). □ **nervously** *adv.* **nervousness** *n.*

nervous breakdown *n.* period of mental illness, usu. resulting from severe stress.

nervous system *n.* body's network of nerve cells.

nervy *adj.* (**-ier**, **-iest**) *colloq.* nervous; easily excited.

nescient /ˈnesɪənt/ *adj. literary* (foll. by *of*) lacking knowledge. □ **nescience** *n.* [Latin *ne-* not, *scio* know]

-ness *suffix* forming nouns from adjectives, expressing: **1** state or condition, or an instance of this (*happiness*; *a kindness*). **2** something in a certain state (*wilderness*). [Old English]

nest –*n.* **1** structure or place where a bird lays eggs and shelters its young. **2** any creature's breeding-place or lair. **3** snug retreat or shelter. **4** brood or swarm. **5** group or set of similar objects, often of different sizes and fitting one inside the other (*nest of tables*). –*v.* **1** use or build a nest. **2** take wild birds' nests or eggs. **3** (of objects) fit together or one inside another. [Old English]

nest egg *n.* sum of money saved for the future.

nestle /ˈnes(ə)l/ *v.* (**-ling**) **1** (often foll. by *down*, *in*, etc.) settle oneself comfortably. **2** press oneself against another in affection etc. **3** (foll. by *in*, *into*, etc.) push (a head or shoulder etc.) affectionately or snugly. **4** lie half hidden or embedded. [Old English]

nestling /ˈnestlɪŋ/ *n.* bird too young to leave its nest.

net[1] –*n.* **1** open-meshed fabric of cord, rope, etc. **2** piece of net used esp. to restrain, contain, or delimit, or to catch fish etc. **3** structure with a net used in various games. –*v.* (**-tt-**) **1 a** cover, confine, or catch with a net. **b** procure as with a net. **2** hit (a ball) into the net, esp. of a goal. [Old English]

net[2] (also **nett**) –*adj.* **1** (esp. of money) remaining after all necessary deductions. **2** (of a price) not reducible. **3** (of a weight) excluding that of the packaging etc. **4** (of an effect, result, etc.) ultimate, actual. –*v.* (**-tt-**) gain or yield (a sum) as net profit. [French: related to NEAT]

netball *n.* team game in which goals are scored by throwing a ball through a high horizontal ring from which a net hangs.

nether /ˈneðə(r)/ *adj. archaic* = LOWER[1]. [Old English]

nether regions *n.pl.* (also **nether world**) hell; the underworld.

net profit *n.* actual gain after working expenses have been paid.

nett var. of NET[2].

netting *n.* **1** netted fabric. **2** piece of this.

nettle /ˈnet(ə)l/ –*n.* **1** plant with jagged leaves covered with stinging hairs. **2** plant resembling this. –*v.* (**-ling**) irritate, provoke. [Old English]

nettle-rash *n.* skin eruption like nettle stings.

network –*n.* **1** arrangement of intersecting horizontal and vertical lines. **2** complex system of railways etc. **3** people connected by the exchange of information etc., professionally or socially. **4** system of connected electrical conductors. **5** group of broadcasting stations connected for the simultaneous broadcast of a programme. **6** chain of interconnected computers. –*v.* broadcast on a network.

neural /ˈnjʊər(ə)l/ *adj.* of a nerve or the central nervous system. [Greek *neuron* nerve]

neuralgia /njʊəˈrældʒə/ *n.* intense pain along a nerve, esp. in the head or face. □ **neuralgic** *adj.*

neuritis /njʊəˈraɪtɪs/ *n.* inflammation of a nerve or nerves.

neuro- *comb. form* nerve or nerves. [Greek *neuron* nerve]

neurology /njʊəˈrɒlədʒɪ/ *n.* the study of nerve systems. □ **neurological** /-rəˈlɒdʒɪk(ə)l/ *adj.* **neurologist** *n.*

neuron /ˈnjʊərɒn/ *n.* (also **neurone** /-rəʊn/) nerve cell.

neurosis /njʊəˈrəʊsɪs/ *n.* (*pl.* **neuroses** /-siːz/) irrational or disturbed behaviour pattern, associated with nervous distress.

neurosurgery /ˌnjʊərəʊˈsɜːdʒərɪ/ *n.* surgery on the nervous system, esp. the brain or spinal cord. □ **neurosurgeon** *n.* **neurosurgical** *adj.*

neurotic /njʊəˈrɒtɪk/ –*adj.* **1** caused by or relating to neurosis. **2** suffering from neurosis. **3** *colloq.* abnormally sensitive or obsessive. –*n.* neurotic person. □ **neurotically** *adv.*

neuter /ˈnjuːtə(r)/ –*adj.* **1** neither masculine nor feminine. **2** (of a plant) having neither pistils nor stamen. **3** (of an insect) sexually undeveloped. –*n.* **1** neuter gender or word. **2 a** non-fertile insect, esp. a worker bee or ant. **b** castrated animal. –*v.* castrate or spay. [Latin]

neutral /ˈnjuːtr(ə)l/ –*adj.* **1** not supporting either of two opposing sides, impartial. **2** belonging to a neutral State etc.

(*neutral ships*). **3** indistinct, vague, indeterminate. **4** (of a gear) in which the engine is disconnected from the driven parts. **5** (of colours) not strong or positive; grey or beige. **6** *Chem.* neither acid nor alkaline. **7** *Electr.* neither positive nor negative. **8** *Biol.* sexually undeveloped; asexual. –*n.* **1 a** neutral State or person. **b** citizen of a neutral State. **2** neutral gear. □ **neutrality** /-'trælɪtɪ/ *n.* [Latin *neutralis* of neuter gender]

neutralize *v.* (also **-ise**) (**-zing** or **-sing**) **1** make neutral. **2** make ineffective by an opposite force or effect. **3** exempt or exclude (a place) from the sphere of hostilities. □ **neutralization** /-'zeɪʃ(ə)n/ *n.*

neutrino /njuː'triːnəʊ/ *n.* (*pl.* **-s**) elementary particle with zero electric charge and probably zero mass. [Italian, diminutive of *neutro* neutral: related to NEUTER]

neutron /'njuːtrɒn/ *n.* elementary particle of about the same mass as a proton but without an electric charge. [from NEUTRAL]

neutron bomb *n.* bomb producing neutrons and little blast, destroying life but not property.

never /'nevə(r)/ *adv.* **1 a** at no time; on no occasion; not ever. **b** *colloq.* as an emphatic negative (*I never heard you come in*). **2** not at all (*never fear*). **3** *colloq.* (expressing surprise) surely not (*you never left the door open!*). □ **well I never!** expressing great surprise. [Old English, = not ever]

nevermore *adv.* at no future time.

never-never *n.* (often prec. by *the*) *colloq.* hire purchase.

nevertheless /ˌnevəðə'les/ *adv.* in spite of that; notwithstanding.

nevus *US* var. of NAEVUS.

new –*adj.* **1 a** of recent origin or arrival. **b** made, discovered, acquired, or experienced recently or now for the first time. **2** in original condition; not worn or used. **3 a** renewed; reformed (*new life*; *the new order*). **b** reinvigorated (*felt like a new person*). **4** different from a recent previous one (*has a new job*). **5** (often foll. by *to*) unfamiliar or strange (*all new to me*). **6** (usu. prec. by *the*) often *derog.* **a** later, modern. **b** newfangled. **c** given to new or modern ideas. **d** recently affected by social change (*the new rich*). **7** (often prec. by *the*) advanced in method or theory. **8** (in place-names) discovered or founded later than and named after (*New York*). –*adv.* (usu. in *comb.*) newly, recently (*new-found*; *new-baked*). □ **newish** *adj.* **newness** *n.* [Old English]

New Age *n.* set of beliefs replacing traditional Western culture, with alternative approaches to religion, medicine, the environment, etc.

new arrival *n. colloq.* newborn child.

newborn *adj.* recently born.

new broom *n.* new employee etc. eager to make changes.

newcomer *n.* **1** person who has recently arrived. **2** beginner in some activity.

newel /'njuːəl/ *n.* **1** supporting central post of winding stairs. **2** (also **newel post**) top or bottom supporting post of a stair-rail. [Latin *nodus* knot]

newfangled /njuː'fæŋg(ə)ld/ *adj. derog.* different from what one is used to; objectionably new. [= new taken]

newly *adv.* **1** recently. **2** afresh, anew.

newly-wed *n.* recently married person.

new mathematics *n.pl.* (also **new maths**) (also treated as *sing.*) system of elementary maths teaching with an emphasis on investigation and set theory.

new moon *n.* **1** moon when first seen as a crescent after conjunction with the sun. **2** time of its appearance.

new potatoes *n.pl.* earliest potatoes of a new crop.

news /njuːz/ *n.pl.* (usu. treated as *sing.*) **1** information about important or interesting recent events, esp. when published or broadcast. **2** (prec. by *the*) broadcast report of news. **3** newly received or noteworthy information. [from NEW]

newsagent *n.* seller of or shop selling newspapers etc.

newscast *n.* radio or television broadcast of news reports.

newscaster *n.* = NEWSREADER.

news conference *n.* press conference.

newsflash *n.* single item of important news, broadcast urgently and often interrupting other programmes.

newsletter *n.* informal printed report issued periodically to members of a club etc.

newspaper *n.* **1** printed publication of loose folded sheets containing news, advertisements, correspondence, etc. **2** paper forming this (*wrapped in newspaper*).

Newspeak *n.* ambiguous euphemistic language used esp. in political propaganda. [an artificial official language in Orwell's *Nineteen Eighty-Four*]

newsprint *n.* low-quality paper on which newspapers are printed.

newsreader *n.* person who reads out broadcast news bulletins.

newsreel *n.* short cinema film of recent events.

news room *n.* room in a newspaper or broadcasting office where news is processed.

news-sheet *n.* simple form of newspaper; newsletter.

news-stand *n.* stall for the sale of newspapers.

new star *n.* nova.

new style *n.* dating reckoned by the Gregorian Calendar.

news-vendor *n.* newspaper-seller.

newsworthy *adj.* topical; noteworthy as news.

newsy *adj.* **(-ier, -iest)** *colloq.* full of news.

newt *n.* small amphibian with a well-developed tail. [*ewt*, with *n* from *an*: var. of *evet* EFT]

New Testament *n.* part of the Bible concerned with the life and teachings of Christ and his earliest followers.

newton /'nju:t(ə)n/ *n.* SI unit of force that, acting on a mass of one kilogram, increases its velocity by one metre per second every second. [*Newton*, name of a scientist]

new town *n.* town planned and built all at once with government funds.

New Wave *n.* a style of rock music.

New World *n.* North and South America.

new year *n.* year just begun or about to begin; first few days of a year.

New Year's Day *n.* 1 January.

New Year's Eve *n.* 31 December.

next *–adj.* **1** (often foll. by *to*) being, positioned, or living nearest. **2** nearest in order of time; soonest encountered (*next Friday*; *ask the next person you see*). *–adv.* **1** (often foll. by *to*) in the nearest place or degree (*put it next to mine*). **2** on the first or soonest occasion (*when we next meet*). *–n.* next person or thing. *–prep. colloq.* next to. □ **next to** almost (*next to nothing left*). [Old English, superlative of NIGH]

next-best *adj.* the next in order of preference.

next door *adj.* & *adv.* (as adj. often hyphenated) in the next house or room.

next of kin *n.sing.* & *pl.* closest living relative(s).

next world *n.* (prec. by *the*) life after death.

nexus /'neksəs/ *n.* (*pl.* same) connected group or series. [Latin *necto nex-* bind]

NHS *abbr.* National Health Service.

NI *abbr.* **1** Northern Ireland. **2** National Insurance.

Ni *symb.* nickel.

niacin /'naɪəsɪn/ *n.* = NICOTINIC ACID. [shortening]

nib *n.* **1** pen-point. **2** (in *pl.*) shelled and crushed coffee or cocoa beans. [Low German or Dutch]

nibble /'nɪb(ə)l/ *–v.* **(-ling) 1** (foll. by *at*) **a** take small bites at. **b** take cautious interest in. **2** eat in small amounts. **3** bite at gently, cautiously, or playfully. *–n.* **1** act of nibbling. **2** very small amount of food. [Low German or Dutch]

nibs *n.* □ **his nibs** *joc. colloq.* mock title used with reference to an important or self-important person. [origin unknown]

nice *adj.* **1** pleasant, satisfactory. **2** (of a person) kind, good-natured. **3** *iron.* bad or awkward (*nice mess*). **4** fine or subtle (*nice distinction*). **5** fastidious; delicately sensitive. **6** (foll. by an adj., often with *and*) satisfactory in terms of the quality described (*a nice long time*; *nice and warm*). □ **nicely** *adv.* **niceness** *n.* **nicish** *adj.* (also **niceish**). [originally = foolish, from Latin *nescius* ignorant]

nicety /'naɪsɪtɪ/ *n.* (*pl.* **-ies**) **1** subtle distinction or detail. **2** precision. □ **to a nicety** with exactness.

niche /ni:ʃ/ *n.* **1** shallow recess, esp. in a wall. **2** comfortable or apt position in life or employment. **3** position from which an entrepreneur exploits a gap in the market; profitable corner of the market. [Latin *nidus* nest]

nick *–n.* **1** small cut or notch. **2** *slang* **a** prison. **b** police station. **3** *colloq.* condition (*in good nick*). *–v.* **1** make a nick or nicks in. **2** *slang* **a** steal. **b** arrest, catch. □ **in the nick of time** only just in time. [origin uncertain]

nickel /'nɪk(ə)l/ *n.* **1** silver-white metallic element, used esp. in magnetic alloys. **2** *colloq.* US five-cent coin. [German]

nickel silver *n.* = GERMAN SILVER.

nickel steel *n.* type of stainless steel with chromium and nickel.

nicker *n.* (*pl.* same) *slang* pound sterling. [origin unknown]

nick-nack var. of KNICK-KNACK.

nickname /'nɪkneɪm/ *–n.* familiar or humorous name given to a person or thing instead of or as well as the real name. *–v.* **(-ming) 1** give a nickname to. **2** call by a nickname. [earlier *eke-name*, with *n* from *an*: *eke* = addition, from Old English: related to EKE]

nicotine /'nɪkə,ti:n/ *n.* poisonous alkaloid present in tobacco. [French from *Nicot*, introducer of tobacco into France]

nicotinic acid /,nɪkə'tɪnɪk/ *n.* vitamin of the B complex.

nictitate /'nɪktɪ,teɪt/ *v.* **(-ting)** blink or wink. □ **nictitation** /-'teɪʃ(ə)n/ *n.* [Latin]

nictitating membrane *n.* transparent third eyelid in amphibians, birds, and some other animals.

niece *n.* daughter of one's brother or sister or of one's spouse's brother or sister. [Latin *neptis* granddaughter]

niff *n.* & *v. colloq.* smell, stink. □ **niffy** *adj.* (**-ier, -iest**). [originally a dial. word]

nifty /ˈnɪftɪ/ *adj.* (**-ier, -iest**) *colloq.* **1** clever, adroit. **2** smart, stylish. [origin uncertain]

niggard /ˈnɪgəd/ *n.* stingy person. [probably of Scandinavian origin]

niggardly *adj.* stingy. □ **niggardliness** *n.*

nigger *n. offens.* Black or dark-skinned person. [Spanish NEGRO]

niggle /ˈnɪg(ə)l/ *v.* (**-ling**) **1** be over-attentive to details. **2** find fault in a petty way. **3** *colloq.* irritate; nag pettily. □ **niggling** *adj.* [origin unknown]

nigh /naɪ/ *adv., prep.,* & *adj. archaic* or *dial.* near. [Old English]

night /naɪt/ *n.* **1** period of darkness between one day and the next; time from sunset to sunrise. **2** nightfall. **3** darkness of night. **4** night or evening appointed for some activity regarded in a certain way (*last night of the Proms*). [Old English]

nightbird *n.* person who is most active at night.

nightcap *n.* **1** *hist.* cap worn in bed. **2** hot or alcoholic drink taken at bedtime.

nightclub *n.* club providing refreshment and entertainment late at night.

nightdress *n.* woman's or child's loose garment worn in bed.

nightfall *n.* end of daylight.

nightgown *n.* = NIGHTDRESS.

nightie *n. colloq.* nightdress.

nightingale /ˈnaɪtɪŋ,geɪl/ *n.* small reddish-brown bird, of which the male sings melodiously, esp. at night. [Old English, = night-singer]

nightjar *n.* nocturnal bird with a characteristic harsh cry.

night-life *n.* entertainment available at night in a town.

night-light *n.* dim light kept burning in a bedroom at night.

night-long *adj.* & *adv.* throughout the night.

nightly –*adj.* **1** happening, done, or existing in the night. **2** recurring every night. –*adv.* every night.

nightmare *n.* **1** frightening dream. **2** *colloq.* frightening or unpleasant experience or situation. **3** haunting fear. □ **nightmarish** *adj.* [evil spirit (incubus) once thought to lie on and suffocate sleepers: Old English *mære* incubus]

night safe *n.* safe with access from the outer wall of a bank for the deposit of money etc. when the bank is closed.

night school *n.* institution providing classes in the evening.

nightshade *n.* any of various plants with poisonous berries. [Old English]

nightshirt *n.* long shirt worn in bed.

nightspot *n.* nightclub.

night-time *n.* time of darkness.

night-watchman *n.* **1** person employed to keep watch at night. **2** *Cricket* inferior batsman sent in near the close of a day's play.

nihilism /ˈnaɪɪ,lɪz(ə)m/ *n.* **1** rejection of all religious and moral principles. **2** belief that nothing really exists. □ **nihilist** *n.* **nihilistic** /-ˈlɪstɪk/ *adj.* [Latin *nihil* nothing]

-nik *suffix* forming nouns denoting a person associated with a specified thing or quality (*beatnik*). [Russian (as SPUTNIK) and Yiddish]

Nikkei index /ˈnɪkeɪ/ *n.* (also **Nikkei average**) a figure indicating the relative price of representative shares on the Tokyo Stock Exchange. [Japanese]

nil *n.* nothing; no number or amount (esp. as a score in games). [Latin]

nimble /ˈnɪmb(ə)l/ *adj.* (**-bler, -blest**) quick and light in movement or function; agile. □ **nimbly** *adv.* [Old English, = quick to seize]

nimbus /ˈnɪmbəs/ *n.* (*pl.* **nimbi** /-baɪ/ or **nimbuses**) **1** halo. **2** rain-cloud. [Latin, = cloud]

Nimby /ˈnɪmbɪ/ –*adj.* objecting to the siting of unpleasant developments in one's own locality. –*n.* (*pl.* **-ies**) person who so objects. [*n*ot *i*n *m*y *b*ack *y*ard]

nincompoop /ˈnɪŋkəm,puːp/ *n.* foolish person. [origin unknown]

nine *adj.* & *n.* **1** one more than eight. **2** symbol for this (9, ix, IX). **3** size etc. denoted by nine. [Old English]

nine days' wonder *n.* person or thing that is briefly famous.

ninefold *adj.* & *adv.* **1** nine times as much or as many. **2** consisting of nine parts.

ninepin *n.* **1** (in *pl.*; usu. treated as *sing.*) game in which nine pins are bowled at. **2** pin used in this game.

nineteen /naɪnˈtiːn/ *adj.* & *n.* **1** one more than eighteen. **2** symbol for this (19, xix, XIX). **3** size etc. denoted by nineteen. □ **talk nineteen to the dozen** see DOZEN. □ **nineteenth** *adj.* & *n.* [Old English]

ninety /ˈnaɪntɪ/ *adj.* & *n.* (*pl.* **-ies**) **1** product of nine and ten. **2** symbol for this (90, xc, XC). **3** (in *pl.*) numbers from 90 to 99, esp. the years of a century or of a person's life. □ **ninetieth** *adj.* & *n.* [Old English]

ninny /ˈnɪnɪ/ *n.* (*pl.* **-ies**) foolish person. [origin uncertain]

ninth /naɪnθ/ *adj.* & *n.* **1** next after eighth. **2** any of nine equal parts of a thing. □ **ninthly** *adv.*

niobium /naɪˈəʊbɪəm/ *n.* rare metallic element occurring naturally. [*Niobe* in Greek legend]

Nip *n. slang offens.* Japanese person. [abbreviation of *Nipponese* from Japanese *Nippon* Japan]

nip[1] –*v.* (**-pp-**) **1** pinch, squeeze, or bite sharply. **2** (often foll. by *off*) remove by pinching etc. **3** (of the cold etc.) cause pain or harm to. **4** (foll. by *in*, *out*, etc.) *colloq.* go nimbly or quickly. –*n.* **1 a** pinch, sharp squeeze. **b** bite. **2** biting cold. □ **nip in the bud** suppress or destroy (esp. an idea) at an early stage. [Low German or Dutch]

nip[2] *n.* small quantity of spirits. [from *nipperkin* small measure]

nipper *n.* **1** person or thing that nips. **2** claw of a crab etc. **3** *colloq.* young child. **4** (in *pl.*) any tool for gripping or cutting.

nipple /ˈnɪp(ə)l/ *n.* **1** small projection in which the mammary ducts of either sex of mammals terminate and from which in females milk is secreted for the young. **2** teat of a feeding-bottle. **3** device like a nipple in function. **4** nipple-like protuberance. [perhaps from *neb* tip]

nippy *adj.* (**-ier, -iest**) *colloq.* **1** quick, nimble. **2** chilly. [from NIP[1]]

nirvana /nɪəˈvɑːnə, nɜː-/ *n.* (in Buddhism) perfect bliss attained by the extinction of individuality. [Sanskrit, = extinction]

Nissen hut /ˈnɪs(ə)n/ *n.* tunnel-shaped hut of corrugated iron with a cement floor. [*Nissen*, name of an engineer]

nit *n.* **1** egg or young form of a louse or other parasitic insect. **2** *slang* stupid person. [Old English]

nit-picking *n.* & *adj. colloq.* fault-finding in a petty manner.

nitrate –*n.* /ˈnaɪtreɪt/ **1** any salt or ester of nitric acid. **2** potassium or sodium nitrate as a fertilizer. –*v.* /naɪˈtreɪt/ (**-ting**) treat, combine, or impregnate with nitric acid. □ **nitration** /-ˈtreɪʃ(ə)n/ *n.* [French: related to NITRE]

nitre /ˈnaɪtə(r)/ *n.* (*US* **niter**) saltpetre. [Greek *nitron*]

nitric /ˈnaɪtrɪk/ *adj.* of or containing nitrogen.

nitric acid *n.* colourless corrosive poisonous liquid.

nitride /ˈnaɪtraɪd/ *n.* binary compound of nitrogen. [from NITRE]

nitrify /ˈnaɪtrɪˌfaɪ/ *v.* (**-ies, -ied**) **1** impregnate with nitrogen. **2** convert into nitrites or nitrates. □ **nitrification** /-fɪˈkeɪʃ(ə)n/ *n.* [French: related to NITRE]

nitrite /ˈnaɪtraɪt/ *n.* any salt or ester of nitrous acid. [from NITRE]

nitro- *comb. form* of or containing nitric acid, nitre, or nitrogen. [Greek: related to NITRE]

nitrogen /ˈnaɪtrədʒ(ə)n/ *n.* gaseous element that forms four-fifths of the atmosphere. □ **nitrogenous** /-ˈtrɒdʒɪnəs/ *adj.* [French]

nitroglycerine /ˌnaɪtrəʊˈglɪsərɪn/ *n.* (*US* **nitroglycerin**) explosive yellow liquid made by reacting glycerol with a mixture of concentrated sulphuric and nitric acids.

nitrous oxide /ˈnaɪtrəs/ *n.* colourless gas used as an anaesthetic. [Latin: related to NITRE]

nitty-gritty /ˌnɪtɪˈgrɪtɪ/ *n. slang* realities or practical details of a matter. [origin uncertain]

nitwit *n. colloq.* stupid person. [perhaps from NIT, WIT]

NNE *abbr.* north-north-east.

NNW *abbr.* north-north-west.

No[1] *symb.* nobelium.

No[2] var. of NOH.

No. *abbr.* number. [Latin *numero*, ablative of *numerus* number]

no[1] /nəʊ/ *adj.* **1** not any (*there is no excuse*). **2** not a, quite other than (*is no fool*). **3** hardly any (*did it in no time*). **4** used elliptically in a notice etc., to forbid etc. the thing specified (*no parking*). □ **no way** *colloq.* **1** it is impossible. **2** I will not agree etc. **no wonder** see WONDER. [related to NONE]

no[2] /nəʊ/ –*adv.* **1** indicating that the answer to the question is negative, the statement etc. made or course of action intended or conclusion arrived at is not correct or satisfactory, the request or command will not be complied with, or the negative statement made is correct. **2** (foll. by *compar.*) by no amount; not at all (*no better than before*). –*n.* (*pl.* **noes**) **1** utterance of the word *no*. **2** denial or refusal. **3** 'no' vote. □ **no longer** not now or henceforth as formerly. **or no** or not (*pleasant or no, it is true*). [Old English]

nob[1] *n. slang* person of wealth or high social position. [origin unknown]

nob[2] *n. slang* head. [from KNOB]

no-ball *n. Cricket* unlawfully delivered ball.

nobble /ˈnɒb(ə)l/ *v.* (**-ling**) *slang* **1** try to influence (e.g. a judge), esp. unfairly. **2** tamper with (a racehorse) to prevent its winning. **3** steal. **4** seize, catch. [dial. *knobble* beat]

nobelium /nəʊˈbiːlɪəm/ *n.* artificially produced radioactive transuranic metallic element. [from *Nobel*: see NOBEL PRIZE]

Nobel prize /nəʊ'bel/ *n.* any of six international prizes awarded annually for physics, chemistry, physiology or medicine, literature, economics, and the promotion of peace. [from *Nobel*, Swedish chemist and engineer, who endowed them]

nobility /nəʊ'bɪlɪtɪ/ *n.* (*pl.* **-ies**) **1** nobleness of character, mind, birth, or rank. **2** class of nobles, highest social class.

noble /'nəʊb(ə)l/ *–adj.* (**nobler, noblest**) **1** belonging to the aristocracy. **2** of excellent character; magnanimous. **3** of imposing appearance. *–n.* nobleman, noblewoman. □ **nobleness** *n.* **nobly** *adv.* [Latin *(g)nobilis*]

noble gas *n.* any of a group of gaseous elements that almost never combine with other elements.

nobleman *n.* peer.

noblesse oblige /nəʊˌbles ɒ'bli:ʒ/ *n.* privilege entails responsibility. [French]

noblewoman *n.* peeress.

nobody /'nəʊbədɪ/ *–pron.* no person. *–n.* (*pl.* **-ies**) person of no importance.

no claim bonus *n.* (also **no claims bonus**) reduction of an insurance premium after an agreed period without a claim.

nocturnal /nɒk'tɜ:n(ə)l/ *adj.* of or in the night; done or active by night. [Latin *nox noct-* night]

nocturne /'nɒktɜ:n/ *n.* **1** *Mus.* short romantic composition, usu. for piano. **2** picture of a night scene. [French]

nod *–v.* (**-dd-**) **1** incline one's head slightly and briefly in assent, greeting, or command. **2** let one's head fall forward in drowsiness; be drowsy. **3** incline (one's head). **4** signify (assent etc.) by a nod. **5** (of flowers, plumes, etc.) bend downwards and sway. **6** make a mistake due to a momentary lack of alertness or attention. *–n.* nodding of the head. □ **nod off** *colloq.* fall asleep. [origin unknown]

noddle /'nɒd(ə)l/ *n. colloq.* head. [origin unknown]

noddy *n.* (*pl.* **-ies**) **1** simpleton. **2** tropical sea bird. [origin unknown]

node *n.* **1 a** part of a plant stem from which leaves emerge. **b** knob on a root or branch. **2** natural swelling. **3** either of two points at which a planet's orbit intersects the plane of the ecliptic or the celestial equator. **4** point of minimum disturbance in a standing wave system. **5** point at which a curve intersects itself. **6** component in a computer network. □ **nodal** *adj.* [Latin *nodus* knot]

nodule /'nɒdju:l/ *n.* **1** small rounded lump of anything. **2** small tumour, node, or ganglion, or a swelling on the root of a legume containing bacteria etc. □ **nodular** *adj.* [Latin diminutive: related to NODE]

Noel /nəʊ'el/ *n.* Christmas. [Latin: related to NATAL]

noggin /'nɒgɪn/ *n.* **1** small mug. **2** small measure, usu. $^{1}/_{4}$ pint, of spirits. **3** *slang* head. [origin unknown]

no go *adj.* (usu. hyphenated when *attrib.*) *colloq.* impossible, hopeless; forbidden (*tried to get him to agree, but it was clearly no go*; *no-go area*).

Noh /nəʊ/ *n.* (also **No**) traditional Japanese drama. [Japanese]

noise /nɔɪz/ *–n.* **1** sound, esp. a loud or unpleasant one. **2** series or confusion of loud sounds. **3** irregular fluctuations accompanying a transmitted signal. **4** (in *pl.*) conventional remarks, or speechlike sounds without actual words (*made sympathetic noises*). *–v.* (**-sing**) (usu. in *passive*) make public; spread abroad (a person's fame or a fact). [Latin NAUSEA]

noiseless *adj.* making little or no noise. □ **noiselessly** *adv.*

noisome /'nɔɪsəm/ *adj. literary* **1** harmful, noxious. **2** evil-smelling. [from ANNOY]

noisy *adj.* (**-ier, -iest**) **1** making much noise. **2** full of noise. □ **noisily** *adv.* **noisiness** *n.*

nomad /'nəʊmæd/ *n.* **1** member of a tribe roaming from place to place for pasture. **2** wanderer. □ **nomadic** /-'mædɪk/ *adj.* [Greek *nomas nomad-* from *nemō* to pasture]

no man's land *n.* **1** space between two opposing armies. **2** area not assigned to any owner.

nom de plume /ˌnɒm də 'plu:m/ *n.* (*pl.* ***noms de plume*** pronunc. same) writer's assumed name. [sham French, = pen-name]

nomen /'nəʊmen/ *n.* ancient Roman's second or family name, as in Marcus *Tullius* Cicero. [Latin, = name]

nomenclature /nəʊ'menklətʃə(r)/ *n.* **1** person's or community's system of names for things. **2** terminology of a science etc. [Latin *nomen* name, *calo* call]

nominal /'nɒmɪn(ə)l/ *adj.* **1** existing in name only; not real or actual. **2** (of a sum of money etc.) very small. **3** of or in names (*nominal and essential distinctions*). **4** of, as, or like a noun. □ **nominally** *adv.* [Latin *nomen* name]

nominalism *n.* doctrine that universals or general ideas are mere names. □

nominalist *n.* **nominalistic** /-'lɪstɪk/ *adj.*
nominal value *n.* face value.
nominate /'nɒmɪ,neɪt/ *v.* **(-ting) 1** propose (a candidate) for election. **2** appoint to an office. **3** name or appoint (a date or place). □ **nomination** /-'neɪʃ(ə)n/ *n.* **nominator** *n.* [Latin: related to NOMINAL]
nominative /'nɒmɪnətɪv/ *Gram.* –*n.* case expressing the subject of a verb. –*adj.* of or in this case.
nominee /,nɒmɪ'ni:/ *n.* person who is nominated.
non- *prefix* giving the negative sense of words with which it is combined. [Latin *non* not]

■ **Usage** The number of words that can be formed from the suffix *non-* is unlimited; consequently, only the most current and noteworthy can be given here.

nonagenarian /,nəʊnədʒɪ'neərɪən/ *n.* person from 90 to 99 years old. [Latin *nonageni* ninety each]
non-aggression /,nɒnə'greʃ(ə)n/ *n.* lack of or restraint from aggression (often *attrib.*: *non-aggression pact*).
nonagon /'nɒnəgən/ *n.* plane figure with nine sides and angles. [Latin *nonus* ninth, after HEXAGON]
non-alcoholic /,nɒnælkə'hɒlɪk/ *adj.* containing no alcohol.
non-aligned /,nɒnə'laɪnd/ *adj.* (of a State) not aligned with a major power. □ **non-alignment** *n.*
non-belligerent /,nɒnbə'lɪdʒərənt/ –*adj.* not engaged in hostilities. –*n.* non-belligerent State etc.
nonce *n.* □ **for the nonce** for the time being; for the present occasion. [from *for than anes* = for the one]
nonce-word *n.* word coined for one occasion.
nonchalant /'nɒnʃələnt/ *adj.* calm and casual. □ **nonchalance** *n.* **nonchalantly** *adv.* [French *chaloir* be concerned]
non-com /'nɒnkɒm/ *n. colloq.* non-commissioned officer. [abbreviation]
non-combatant /nɒn'kɒmbət(ə)nt/ *n.* person not fighting in a war, esp. a civilian, army chaplain, etc.
non-commissioned /,nɒnkə'mɪʃ(ə)nd/ *adj.* (of an officer) not holding a commission.
noncommittal /,nɒnkə'mɪt(ə)l/ *adj.* avoiding commitment to a definite opinion or course of action.
non compos mentis /,nɒn kɒmpɒs 'mentɪs/ *adj.* (also ***non compos***) not in one's right mind. [Latin, = not having control of one's mind]
non-conductor /,nɒnkən'dʌktə(r)/ *n.* substance that does not conduct heat or electricity.
nonconformist /,nɒnkən'fɔ:mɪst/ *n.* **1** person who does not conform to the doctrine or discipline of an established Church, esp. (**Nonconformist**) member of a (usu. Protestant) sect dissenting from the Anglican Church. **2** person who does not conform to a prevailing principle.
nonconformity /,nɒnkən'fɔ:mɪtɪ/ *n.* **1** nonconformists as a body, or their principles. **2** (usu. foll. by *to*) failure to conform. **3** lack of correspondence between things.
non-contributory /,nɒnkən'trɪbjʊtərɪ/ *adj.* not involving contributions.
non-cooperation /,nɒnkəʊ,ɒpə'reɪʃ(ə)n/ *n.* failure to cooperate.
nondescript /'nɒndɪskrɪpt/ –*adj.* lacking distinctive characteristics, not easily classified. –*n.* nondescript person or thing. [related to DESCRIBE]
non-drinker /nɒn'drɪŋkə(r)/ *n.* person who does not drink alcoholic liquor.
non-driver /nɒn'draɪvə(r)/ *n.* person who does not drive a motor vehicle.
none /nʌn/ –*pron.* **1** (foll. by *of*) **a** not any of (*none of this concerns me*; *none of them have found it*). **b** not any one of (*none of them has come*). **2 a** no persons (*none but fools believe it*). **b** no person (*none but a fool believes it*). **3** (usu. with the preceding noun implied) not any (*you have money and I have none*). –*adv.* (foll. by *the* + compar., or *so, too*) by no amount; not at all (*am none the wiser*). [Old English, = not one]

■ **Usage** In sense 1b, the verb following *none* can be singular or plural according to meaning.

nonentity /nɒ'nentɪtɪ/ *n.* (*pl.* **-ies**) **1** person or thing of no importance. **2 a** non-existence. **b** non-existent thing. [medieval Latin]
nones /nəʊnz/ *n.pl.* day of the ancient Roman month (the 7th day of March, May, July, and October, the 5th of other months). [Latin *nonus* ninth]
non-essential /,nɒnɪ'senʃ(ə)l/ –*adj.* not essential. –*n.* non-essential thing.
nonetheless /,nʌnðə'les/ *adv.* (also **none the less**) nevertheless.
non-event /,nɒnɪ'vent/ *n.* insignificant event, esp. contrary to hopes or expectations.
non-existent /,nɒnɪg'zɪst(ə)nt/ *adj.* not existing.
non-fattening /nɒn'fætənɪŋ/ *adj.* (of food) not containing many calories.

non-ferrous /nɒn'ferəs/ *adj.* (of a metal) other than iron or steel.

non-fiction /nɒn'fɪkʃ(ə)n/ *n.* literary work other than fiction.

non-flammable /nɒn'flæməb(ə)l/ *adj.* not inflammable.

non-interference /ˌnɒnɪntə'fɪərəns/ *n.* = NON-INTERVENTION.

non-intervention /ˌnɒnɪntə'venʃ(ə)n/ *n.* (esp. political) principle or practice of not becoming involved in others' affairs.

non-member *n.* person who is not a member.

non-nuclear /nɒn'nju:klɪə(r)/ *adj.* **1** not involving nuclei or nuclear energy. **2** (of a State etc.) not having nuclear weapons.

non-observance /ˌnɒnəb'zɜ:v(ə)ns/ *n.* failure to observe (an agreement, requirement, etc.).

non-operational /ˌnɒnɒpə'reɪʃən(ə)l/ *adj.* **1** that does not operate. **2** out of order.

nonpareil /ˌnɒnpə'reɪl, 'nɒnpər(ə)l/ *–adj.* unrivalled or unique. *–n.* such a person or thing. [French *pareil*]

non-partisan /ˌnɒnpɑ:tɪ'zæn/ *adj.* not partisan.

non-party /nɒn'pɑ:tɪ/ *adj.* independent of political parties.

non-payment /nɒn'peɪmənt/ *n.* failure to pay; lack of payment.

nonplus /nɒn'plʌs/ *v.* (**-ss-**) completely perplex. [Latin *non plus* not more]

non-profit-making /nɒn'prɒfɪt ˌmeɪkɪŋ/ *adj.* (of an enterprise) not conducted primarily to make a profit.

non-proliferation /ˌnɒnprəˌlɪfə'reɪʃ(ə)n/ *n.* prevention of an increase in something, esp. possession of nuclear weapons.

non-resident /nɒn'rezɪd(ə)nt/ *–adj.* **1** not residing in a particular place. **2** (of a post) not requiring the holder to reside at the place of work. *–n.* non-resident person. □ **non-residential** /-'denʃ(ə)l/ *adj.*

non-resistance /ˌnɒnrɪ'zɪst(ə)ns/ *n.* practice or principle of not resisting authority.

non-returnable /ˌnɒnrɪ'tɜ:nəb(ə)l/ *adj.* that is not to be returned.

non-sectarian /ˌnɒnsek'teərɪən/ *adj.* not sectarian.

nonsense /'nɒns(ə)ns/ *n.* **1** (often as *int.*) absurd or meaningless words or ideas. **2** foolish or extravagant conduct. □ **nonsensical** /-'sensɪk(ə)l/ *adj.* **nonsensically** /-'sensɪkəlɪ/ *adv.*

non sequitur /nɒn 'sekwɪtə(r)/ *n.* conclusion that does not logically follow from the premisses. [Latin, = it does not follow]

non-slip /nɒn'slɪp/ *adj.* **1** that does not slip. **2** that inhibits slipping.

non-smoker /nɒn'sməʊkə(r)/ *n.* **1** person who does not smoke. **2** train compartment etc. where smoking is forbidden. □ **non-smoking** *adj.*

non-specialist /nɒn'speʃəlɪst/ *n.* person who is not a specialist (in a particular subject).

non-specific /ˌnɒnspɪ'sɪfɪk/ *adj.* that cannot be specified.

non-standard /nɒn'stændəd/ *adj.* not standard.

non-starter /nɒn'stɑ:tə(r)/ *n. colloq.* person or scheme that is unlikely to succeed.

non-stick /nɒn'stɪk/ *adj.* that does not allow things to stick to it.

non-stop /nɒn'stɒp/ *–adj.* **1** (of a train etc.) not stopping at intermediate places. **2** done without a stop or intermission. *–adv.* without stopping.

non-swimmer /nɒn'swɪmə(r)/ *n.* person who cannot swim.

non-toxic /nɒn'tɒksɪk/ *adj.* not toxic.

non-transferable /ˌnɒntræns'fɜ:rəb(ə)l/ *adj.* that may not be transferred.

non-U /nɒn'ju:/ *adj. colloq.* not characteristic of the upper class. [from U²]

non-union /nɒn'ju:nɪən/ *adj.* **1** not belonging to a trade union. **2** not done or made by trade-union members.

non-verbal /nɒn'vɜ:b(ə)l/ *adj.* not involving words or speech.

non-violence /nɒn'vaɪələns/ *n.* avoidance of violence, esp. as a principle. □ **non-violent** *adj.*

non-voting /nɒn'vəʊtɪŋ/ *adj.* **1** not having or using a vote. **2** (of shares) not entitling the holder to vote.

non-White /nɒn'waɪt/ *–adj.* not White. *–n.* non-White person.

noodle¹ /'nu:d(ə)l/ *n.* strip or ring of pasta. [German]

noodle² /'nu:d(ə)l/ *n.* **1** simpleton. **2** *slang* head. [origin unknown]

nook /nʊk/ *n.* corner or recess; secluded place. [origin unknown]

noon *n.* twelve o'clock in the day, midday. [Latin *nona* (*hora*) ninth (hour): originally = 3 p.m.]

noonday *n.* midday.

no one *n.* no person; nobody.

noose *–n.* **1** loop with a running knot. **2** snare, bond. *–v.* (**-sing**) catch with or enclose in a noose. [French *no(u)s* from Latin *nodus* NODE]

nor *conj.* and not; and not either (*neither one thing nor the other*; *can neither read nor write*). [contraction of obsolete *nother*: related to NO², WHETHER]

nor' /nɔ:(r)/ *n., adj., & adv.* (esp. in compounds) = NORTH (*nor'wester*). [abbreviation]

Nordic /'nɔ:dɪk/ *–adj.* of the tall blond long-headed Germanic people of Scandinavia. *–n.* Nordic person. [French *nord* north]

Norfolk jacket /'nɔ:fək/ *n.* man's loose belted jacket with box pleats. [*Norfolk* in England]

norm *n.* **1** standard, pattern, or type. **2** standard amount of work etc. **3** customary behaviour etc. [Latin *norma* carpenter's square]

normal /'nɔ:m(ə)l/ *–adj.* **1** conforming to a standard; regular, usual, typical. **2** free from mental or emotional disorder. **3** *Geom.* (of a line) at right angles, perpendicular. *–n.* **1 a** normal value of a temperature etc. **b** usual state, level, etc. **2** line at right angles. □ **normalcy** *n.* esp. *US.* **normality** /-'mælɪtɪ/ *n.* [Latin *normalis*: related to NORM]

normal distribution *n.* function that represents the distribution of many random variables as a symmetrical bell-shaped graph.

normalize *v.* (also **-ise**) (**-zing** or **-sing**) **1** make or become normal. **2** cause to conform. □ **normalization** /-'zeɪʃ(ə)n/ *n.*

normally *adv.* **1** in a normal manner. **2** usually.

Norman /'nɔ:mən/ *–n.* **1** native or inhabitant of medieval Normandy. **2** descendant of the people of mixed Scandinavian and Frankish origin established there in the 10th c. **3** Norman French. **4** style of architecture found in Britain under the Normans. *–adj.* **1** of the Normans. **2** of the Norman style of architecture. [Old Norse, = NORTHMAN]

Norman Conquest *n.* conquest of England by William of Normandy in 1066.

Norman French *n.* French as spoken by the Normans or (after 1066) in English lawcourts.

normative /'nɔ:mətɪv/ *adj.* of or establishing a norm. [Latin: related to NORM]

Norn *n.* any of three goddesses of destiny in Scandinavian mythology. [Old Norse]

Norse *–n.* **1** Norwegian language. **2** Scandinavian language-group. *–adj.* of ancient Scandinavia, esp. Norway. □ **Norseman** *n.* [Dutch *noor(d)sch* northern]

north *–n.* **1 a** point of the horizon 90° anticlockwise from east. **b** compass point corresponding to this. **c** direction in which this lies. **2** (usu. **the North**) **a** part of a country or town lying to the north. **b** the industrialized nations. *–adj.* **1** towards, at, near, or facing the north. **2** from the north (*north wind*). *–adv.* **1** towards, at, or near the north. **2** (foll. by *of*) further north than. □ **to the north** (often foll. by *of*) in a northerly direction. [Old English]

North American *–adj.* of North America. *–n.* native or inhabitant of North America, esp. a citizen of the US or Canada.

northbound *adj.* travelling or leading northwards.

north country *n.* northern England.

North-East *n.* part of a country or town to the north-east.

north-east *–n.* **1** point of the horizon midway between north and east. **2** direction in which this lies. *–adj.* of, towards, or coming from the north-east. *–adv.* towards, at, or near the north-east.

northeaster /nɔ:θ'i:stə(r)/ *n.* north-east wind.

north-easterly *adj.* & *adv.* = NORTH-EAST.

north-eastern *adj.* on the north-east side.

northerly /'nɔ:ðəlɪ/ *–adj.* & *adv.* **1** in a northern position or direction. **2** (of wind) from the north. *–n.* (*pl.* **-ies**) such a wind.

northern /'nɔ:ð(ə)n/ *adj.* of or in the north. □ **northernmost** *adj.* [Old English]

northerner *n.* native or inhabitant of the north.

Northern hemisphere *n.* the half of the earth north of the equator.

northern lights *n.pl.* aurora borealis.

Northman *n.* native of Scandinavia, esp. Norway. [Old English]

north-north-east *n.* point or direction midway between north and north-east.

north-north-west *n.* point or direction midway between north and north-west.

North Pole *n.* northernmost point of the earth's axis of rotation.

North Star *n.* pole star.

northward /'nɔ:θwəd/ *–adj.* & *adv.* (also **northwards**) towards the north. *–n.* northward direction or region.

North-West *n.* part of a country or town to the north-west.

north-west *–n.* **1** point of the horizon midway between north and west. **2** direction in which this lies. *–adj.* of, towards, or coming from the north-west. *–adv.* towards, at, or near the north-west.

northwester /nɔ:θ'westə(r)/ *n.* north-west wind.

north-westerly *adj.* & *adv.* = NORTH-WEST.

north-western *adj.* on the north-west side.

Norwegian /nɔ:'wi:dʒ(ə)n/ *–n.* **1 a** native or national of Norway. **b** person of Norwegian descent. **2** language of Norway. *–adj.* of or relating to Norway. [medieval Latin *Norvegia* from Old Norse, = northway]

nor'wester /nɔ:'westə(r)/ *n.* north-wester.

Nos. *pl.* of No.

nose /nəʊz/ *–n.* **1** organ above the mouth of a human or animal, used for smelling and breathing. **2 a** sense of smell. **b** ability to detect a particular thing (*a nose for scandal*). **3** odour or perfume of wine etc. **4** front end or projecting part of a thing, e.g. of a car or aircraft. *–v.* **(-sing) 1** (usu. foll. by *about*, *around*, etc.) pry or search. **2** (often foll. by *out*) **a** perceive the smell of, discover by smell. **b** detect. **3** thrust one's nose against or into. **4** make one's way cautiously forward. □ **by a nose** by a very narrow margin. **get up a person's nose** *slang* annoy a person. **keep one's nose clean** *slang* stay out of trouble. **put a person's nose out of joint** *colloq.* annoy; make envious. **turn up one's nose** (usu. foll. by *at*) *colloq.* show disdain. **under a person's nose** *colloq.* right before a person. **with one's nose in the air** haughtily. [Old English]

nosebag *n.* bag containing fodder, hung on a horse's head.

noseband *n.* lower band of a bridle, passing over the horse's nose.

nosebleed *n.* bleeding from the nose.

nosedive *–n.* **1** steep downward plunge by an aeroplane. **2** sudden plunge or drop. *–v.* **(-ving)** make a nosedive.

nosegay *n.* small bunch of flowers.

nose-to-tail *adj.* & *adv.* (of vehicles) one close behind another.

nosh *slang* *–v.* eat. *–n.* **1** food or drink. **2** *US* snack. [Yiddish]

nosh-up *n. slang* large meal.

nostalgia /nɒ'stældʒə/ *n.* **1** (often foll. by *for*) yearning for a past period. **2** severe homesickness. □ **nostalgic** *adj.* **nostalgically** *adv.* [Greek *nostos* return home]

nostril /'nɒstr(ə)l/ *n.* either of the two openings in the nose. [Old English, = nose-hole]

nostrum /'nɒstrəm/ *n.* **1** quack remedy, patent medicine. **2** pet scheme, esp. for political or social reform. [Latin, = 'of our own make']

nosy *adj.* **(-ier, -iest)** *colloq.* inquisitive, prying. □ **nosily** *adv.* **nosiness** *n.*

Nosy Parker *n. colloq.* busybody.

not *adv.* expressing negation, esp.: **1** (also **n't** joined to a preceding verb) following an auxiliary verb or *be* or (in a question) the subject of such a verb (*I cannot say*; *she isn't there*; *am I not right?*). **2** used elliptically for a negative phrase etc. (*Is she coming? — I hope not*; *Do you want it? — Certainly not!*). □ **not at all** (in polite reply to thanks) there is no need for thanks. **not half** see HALF. **not quite 1** almost. **2** noticeably not (*not quite proper*). [contraction of NOUGHT]

■ **Usage** The use of *not* with verbs other than auxiliaries or *be* is now *archaic* except with participles and infinitives (*not knowing*, *I cannot say*; *we asked them not to come*).

notable /'nəʊtəb(ə)l/ *–adj.* worthy of note; remarkable, eminent. *–n.* eminent person. □ **notability** /-'bɪlɪtɪ/ *n.* **notably** *adv.* [Latin *noto* NOTE]

notary /'nəʊtərɪ/ *n.* (*pl.* **-ies**) (in full **notary public**) solicitor etc. who attests or certifies deeds etc. □ **notarial** /nəʊ'teərɪəl/ *adj.* [Latin *notarius* secretary]

notation /nəʊ'teɪʃ(ə)n/ *n.* **1** representation of numbers, quantities, the pitch and duration of musical notes, etc., by symbols. **2** any set of such symbols. [Latin: related to NOTE]

notch *–n.* V-shaped indentation on an edge or surface. *–v.* **1** make notches in. **2** (usu. foll. by *up*) record or score with or as with notches. [Anglo-French]

note *–n.* **1** brief written record as an aid to memory (often in *pl.*: *make notes*). **2** observation, usu. unwritten, of experiences etc. (*compare notes*). **3** short or informal letter. **4** formal diplomatic communication. **5** short annotation or additional explanation in a book etc. **6 a** = BANKNOTE. **b** written promise of payment. **7 a** notice, attention (*worthy of note*). **b** eminence (*person of note*). **8 a** single musical tone of definite pitch. **b** written sign representing its pitch and duration. **c** key of a piano etc. **9** quality or tone of speaking, expressing mood or attitude etc. (*note of optimism*). *–v.* **(-ting) 1** observe, notice; give attention to. **2** (often foll. by *down*) record as a thing to be remembered or observed. **3** (in *passive*; often foll. by *for*) be well known. □ **hit** (or **strike**) **the right note** speak or act in exactly the right manner. [Latin *nota* mark (n.), *noto* mark (v.)]

notebook *n.* small book for making notes in.

notecase *n.* wallet for holding banknotes.

notelet /ˈnəʊtlɪt/ *n.* small folded usu. decorated sheet of paper for an informal letter.

notepaper *n.* paper for writing letters.

noteworthy *adj.* worthy of attention; remarkable.

nothing /ˈnʌθɪŋ/ *–n.* **1** not anything (*nothing has been done*). **2** no thing (often foll. by compl.: *I see nothing that I want*). **3** person or thing of no importance. **4** non-existence; what does not exist. **5** no amount; nought. *–adv.* not at all, in no way. □ **be** (or **have**) **nothing to do with 1** have no connection with. **2** not be involved or associated with. **for nothing 1** at no cost. **2** to no purpose. **have nothing on 1** be naked. **2** have no engagements. **nothing doing** *colloq.* **1** no prospect of success or agreement. **2** I refuse. [Old English: related to NO[1], THING]

nothingness *n.* **1** non-existence. **2** worthlessness, triviality.

notice /ˈnəʊtɪs/ *–n.* **1** attention, observation (*escaped my notice*). **2** displayed sheet etc. bearing an announcement. **3 a** intimation or warning, esp. a formal one. **b** formal announcement or declaration of intention to end an agreement or leave employment at a specified time. **4** short published review of a new play, book, etc. *–v.* (**-cing**) (often foll. by *that*, *how*, etc.) perceive, observe. □ **at short** (or **a moment's**) **notice** with little warning. **take notice** (or **no notice**) show signs (or no signs) of interest. **take notice of 1** observe. **2** act upon. [Latin *notus* known]

noticeable *adj.* perceptible; noteworthy. □ **noticeably** *adv.*

notice-board *n.* board for displaying notices.

notifiable /ˈnəʊtɪˌfaɪəb(ə)l/ *adj.* (of a disease etc.) that must be notified to the health authorities.

notify /ˈnəʊtɪˌfaɪ/ *v.* (**-ies, -ied**) **1** (often foll. by *of* or *that*) inform or give formal notice to (a person). **2** make known. □ **notification** /-fɪˈkeɪʃ(ə)n/ *n.* [Latin *notus* known]

notion /ˈnəʊʃ(ə)n/ *n.* **1 a** concept or idea; conception. **b** opinion. **c** vague view or understanding. **2** inclination or intention. [Latin *notio*: related to NOTIFY]

notional *adj.* hypothetical, imaginary. □ **notionally** *adv.*

notorious /nəʊˈtɔːrɪəs/ *adj.* well-known, esp. unfavourably. □ **notoriety** /-təˈraɪətɪ/ *n.* **notoriously** *adv.* [Latin *notus* known]

notwithstanding /ˌnɒtwɪðˈstændɪŋ/ *–prep.* in spite of; without prevention by. *–adv.* nevertheless. [from NOT, WITHSTAND]

nougat /ˈnuːgɑː/ *n.* sweet made from sugar or honey, nuts, and egg-white. [French from Provençal]

nought /nɔːt/ *n.* **1** digit 0; cipher. **2** *poet.* or *archaic* nothing. [Old English: related to NOT, AUGHT]

noughts and crosses *n.pl.* pencil-and-paper game in which players seek to complete a row of three noughts or three crosses.

noun *n.* word used to name a person, place, or thing. [Latin *nomen* name]

nourish /ˈnʌrɪʃ/ *v.* **1** sustain with food. **2** foster or cherish (a feeling etc.). □ **nourishing** *adj.* [Latin *nutrio* to feed]

nourishment *n.* sustenance, food.

nous /naʊs/ *n.* **1** *colloq.* common sense; gumption. **2** *Philos.* mind, intellect. [Greek]

nouveau riche /ˌnuːvəʊ ˈriːʃ/ *n.* (*pl.* ***nouveaux riches*** pronunc. same) person who has recently acquired (usu. ostentatious) wealth. [French, = new rich]

nouvelle cuisine /ˌnuːvel kwɪˈziːn/ *n.* modern style of cookery avoiding heaviness and emphasizing presentation. [French, = new cookery]

Nov. *abbr.* November.

nova /ˈnəʊvə/ *n.* (*pl.* **novae** /-viː/ or **-s**) star showing a sudden burst of brightness and then subsiding. [Latin, = new]

novel[1] /ˈnɒv(ə)l/ *n.* fictitious prose story of book length. [Latin *novus* new]

novel[2] /ˈnɒv(ə)l/ *adj.* of a new kind or nature. [Latin *novus* new]

novelette /ˌnɒvəˈlet/ *n.* short novel.

novelist /ˈnɒvəlɪst/ *n.* writer of novels.

novella /nəˈvelə/ *n.* (*pl.* **-s**) short novel or narrative story. [Italian: related to NOVEL[1]]

novelty /ˈnɒvəltɪ/ *n.* (*pl.* **-ies**) **1** newness. **2** new or unusual thing or occurrence. **3** small toy or trinket. [related to NOVEL[2]]

November /nəʊˈvembə(r)/ *n.* eleventh month of the year. [Latin *novem* nine, originally the 9th month of the Roman year]

novena /nəˈviːnə/ *n.* *RC Ch.* devotion consisting of special prayers or services on nine successive days. [Latin *novem* nine]

novice /ˈnɒvɪs/ *n.* **1 a** probationary member of a religious order. **b** new convert. **2** beginner. [Latin *novicius*, from *novus* new]

noviciate /nəˈvɪʃɪət/ *n.* (also **novitiate**) **1** period of being a novice. **2** religious novice. **3** novices' quarters. [medieval Latin: related to NOVICE]

now *–adv.* **1** at the present or mentioned time. **2** immediately (*I must go now*). **3** by this time. **4** under the present circumstances (*I cannot now agree*). **5** on this further occasion (*what do you want now?*). **6** in the immediate past (*just now*). **7** (esp. in a narrative) then, next (*the police now arrived*). **8** (without reference to time, giving various tones to a sentence) surely, I insist, I wonder, etc. (*now what do you mean by that?*; *oh come now!*). *–conj.* (often foll. by *that*) as a consequence of the fact (*now that I am older*). *–n.* this time; the present. □ **for now** until a later time (*goodbye for now*). **now and again** (or **then**) from time to time; intermittently. [Old English]

nowadays /'naʊə,deɪz/ *–adv.* at the present time or age; in these times. *–n.* the present time.

nowhere /'nəʊweə(r)/ *–adv.* in or to no place. *–pron.* no place. □ **get nowhere** make no progress. **nowhere near** not nearly. [Old English]

no-win *attrib. adj.* of or designating a situation in which success is impossible.

nowt *n. colloq.* or *dial.* nothing. [from NOUGHT]

noxious /'nɒkʃəs/ *adj.* harmful, unwholesome. [Latin *noxa* harm]

nozzle /'nɒz(ə)l/ *n.* spout on a hose etc. from which a jet issues. [diminutive of NOSE]

Np *symb.* neptunium.

nr. *abbr.* near.

NS *abbr.* new style.

NSPCC *abbr.* National Society for the Prevention of Cruelty to Children.

NSW *abbr.* New South Wales.

NT *abbr.* **1** New Testament. **2** Northern Territory (of Australia). **3** National Trust.

n't see NOT.

nth see N[1].

nu /nju:/ *n.* thirteenth letter of the Greek alphabet (Ν, ν). [Greek]

nuance /'nju:ɑ̃s/ *n.* subtle shade of meaning, feeling, colour, etc. [Latin *nubes* cloud]

nub *n.* **1** point or gist (of a matter or story). **2** (also **nubble**) small lump, esp. of coal. □ **nubbly** *adj.* [related to KNOB]

nubile /'nju:baɪl/ *adj.* (of a woman) marriageable or sexually attractive. □ **nubility** /-'bɪlɪtɪ/ *n.* [Latin *nubo* become the wife of]

nuclear /'nju:klɪə(r)/ *adj.* **1** of, relating to, or constituting a nucleus. **2** using nuclear energy.

nuclear bomb *n.* bomb using the release of energy by nuclear fission or fusion or both.

nuclear energy *n.* energy obtained by nuclear fission or fusion.

nuclear family *n.* a couple and their child or children.

nuclear fission *n.* nuclear reaction in which a heavy nucleus splits spontaneously or on impact with another particle, with the release of energy.

nuclear fuel *n.* source of nuclear energy.

nuclear fusion *n.* nuclear reaction in which atomic nuclei of low atomic number fuse to form a heavier nucleus with the release of energy.

nuclear physics *n.pl.* (treated as *sing.*) physics of atomic nuclei.

nuclear power *n.* **1** power generated by a nuclear reactor. **2** country that has nuclear weapons.

nuclear reactor *n.* device in which a nuclear fission chain reaction is used to produce energy.

nuclear weapon *n.* weapon using the release of energy by nuclear fission or fusion or both.

nucleate /'nju:klɪ,eɪt/ *–adj.* having a nucleus. *–v.* (**-ting**) form or form into a nucleus. [Latin: related to NUCLEUS]

nucleic acid /nju:'kli:ɪk/ *n.* either of two complex organic molecules (DNA and RNA), present in all living cells.

nucleon /'nju:klɪ,ɒn/ *n.* proton or neutron.

nucleus /'nju:klɪəs/ *n.* (*pl.* **nuclei** /-lɪ,aɪ/) **1 a** central part or thing round which others are collected. **b** kernel of an aggregate or mass. **2** initial part meant to receive additions. **3** central core of an atom. **4** large dense part of a cell, containing the genetic material. [Latin, = kernel, diminutive of *nux nuc-* nut]

nude *–adj.* naked, bare, unclothed. *–n.* **1** painting, sculpture, etc. of a nude human figure. **2** nude person. □ **in the nude** naked. □ **nudity** *n.* [Latin *nudus*]

nudge *–v.* (**-ging**) **1** prod gently with the elbow to attract attention. **2** push gradually. *–n.* prod; gentle push. [origin unknown]

nudist /'nju:dɪst/ *n.* person who advocates or practises going unclothed. □ **nudism** *n.*

nugatory /'nju:gətərɪ/ *adj.* **1** futile, trifling. **2** inoperative; not valid. [Latin *nugae* jests]

nugget /'nʌgɪt/ *n.* **1** lump of gold etc., as found in the earth. **2** lump of anything. **3** something valuable. [apparently from dial. *nug* lump]

nuisance /'nju:s(ə)ns/ *n.* person, thing, or circumstance causing trouble or annoyance. [French, = hurt, from *nuire nuis-* injure, from Latin *noceo* to hurt]

nuke *colloq. –n.* nuclear weapon. *–v.* (**-king**) attack with nuclear weapons. [abbreviation]

null *adj.* **1** (esp. **null and void**) invalid. **2** non-existent. **3** without character or expression. □ **nullity** *n.* [Latin *nullus* none]

nullify /ˈnʌlɪˌfaɪ/ *v.* (**-ies, -ied**) neutralize, invalidate. □ **nullification** /-fɪˈkeɪʃ(ə)n/ *n.*

numb /nʌm/ *–adj.* (often foll. by *with*) deprived of feeling; paralysed. *–v.* **1** make numb. **2** stupefy, paralyse. □ **numbness** *n.* [obsolete *nome* past part. of *nim* take: related to NIMBLE]

number *–n.* **1 a** arithmetical value representing a particular quantity. **b** word, symbol, or figure representing this. **c** arithmetical value showing position in a series (*registration number*). **2** (often foll. by *of*) total count or aggregate (*the number of accidents has decreased*). **3** numerical reckoning (*the laws of number*). **4 a** (in *sing.* or *pl.*) quantity, amount (*a large number of people*; *only in small numbers*). **b** (**a number of**) several (of). **c** (in *pl.*) numerical preponderance (*force of numbers*). **5** person or thing having a place in a series, esp. a single issue of a magazine, an item in a programme, etc. **6** company, collection, group (*among our number*). **7** *Gram.* **a** classification of words by their singular or plural forms. **b** such a form. *–v.* **1** include (*I number you among my friends*). **2** assign a number or numbers to. **3** amount to (a specified number). **4** count. □ **one's days are numbered** one does not have long to live. **have a person's number** *colloq.* understand a person's real motives, character, etc. **one's number is up** *colloq.* one is doomed to die soon. **without number** innumerable. [Latin *numerus*]

■ **Usage** In sense 4b, *a number of* is normally used with a plural verb: *a number of problems remain.*

number crunching *n. colloq.* process of making complex calculations.

numberless *adj.* innumerable.

number one *–n. colloq.* oneself. *–adj.* most important (*the number one priority*).

number-plate *n.* plate on a vehicle showing its registration number.

numerable /ˈnjuːmərəb(ə)l/ *adj.* that can be counted. [Latin: related to NUMBER]

numeral /ˈnjuːmər(ə)l/ *–n.* symbol or group of symbols denoting a number. *–adj.* of or denoting a number. [Latin: related to NUMBER]

numerate /ˈnjuːmərət/ *adj.* acquainted with the basic principles of mathematics. □ **numeracy** *n.* [Latin *numerus* number, after *literate*]

numeration /ˌnjuːməˈreɪʃ(ə)n/ *n.* **1** method or process of numbering. **2** calculation. [Latin: related to NUMBER]

numerator /ˈnjuːməˌreɪtə(r)/ *n.* number above the line in a vulgar fraction showing how many of the parts indicated by the denominator are taken (e.g. 2 in $^2/_3$). [Latin: related to NUMBER]

numerical /njuːˈmerɪk(ə)l/ *adj.* of or relating to a number or numbers. □ **numerically** *adv.* [medieval Latin: related to NUMBER]

numerology /ˌnjuːməˈrɒlədʒɪ/ *n.* the study of the supposed occult significance of numbers.

numerous /ˈnjuːmərəs/ *adj.* **1** many. **2** consisting of many. [Latin: related to NUMBER]

numinous /ˈnjuːmɪnəs/ *adj.* **1** indicating the presence of a divinity. **2** spiritual, awe-inspiring. [Latin *numen* deity]

numismatic /ˌnjuːmɪzˈmætɪk/ *adj.* of or relating to coins or medals. [Greek *nomisma* coin]

numismatics *n.pl.* (usu. treated as *sing.*) the study of coins or medals. □ **numismatist** /-ˈmɪzmətɪst/ *n.*

numskull *n.* stupid person. [from NUMB]

nun *n.* member of a religious community of women living under certain vows. [Latin *nonna*]

nuncio /ˈnʌnsɪəʊ/ *n.* (*pl.* **-s**) papal ambassador. [Latin *nuntius* messenger]

nunnery *n.* (*pl.* **-ies**) religious house of nuns.

nuptial /ˈnʌpʃ(ə)l/ *–adj.* of marriage or weddings. *–n.* (usu. in *pl.*) wedding. [Latin *nubo nupt-* wed]

nurd var. of NERD.

nurse /nɜːs/ *–n.* **1** person trained to care for the sick or infirm or to provide medical advice and treat minor medical problems. **2** = NURSEMAID. *–v.* (**-sing**) **1 a** work as a nurse. **b** attend to (a sick person). **2** feed or be fed at the breast. **3** hold or treat carefully. **4 a** foster; promote the development of. **b** harbour (a grievance etc.). [Latin: related to NOURISH]

nurseling var. of NURSLING.

nursemaid *n.* woman in charge of a child or children.

nursery /ˈnɜːsərɪ/ *n.* (*pl.* **-ies**) **1 a** a room or place equipped for young children. **b** = DAY NURSERY. **2** place where plants are reared for sale. [probably Anglo-French: related to NURSE]

nurseryman *n.* owner of or worker in a plant nursery.

nursery rhyme *n.* simple traditional song or rhyme for children.

nursery school *n.* school for children between the ages of three and five.

nursery slopes *n.pl.* gentle slopes for novice skiers.
nursing home *n.* privately run hospital or home for invalids, old people, etc.
nursling /'nɜ:slɪŋ/ *n.* (also **nurseling**) infant that is being suckled.
nurture /'nɜ:tʃə(r)/ —*n.* **1** bringing up, fostering care. **2** nourishment. —*v.* (**-ring**) bring up; rear. [French: related to NOURISH]
nut *n.* **1 a** fruit consisting of a hard or tough shell around an edible kernel. **b** this kernel. **2** pod containing hard seeds. **3** small usu. hexagonal flat piece of metal etc. with a threaded hole through it for screwing on the end of a bolt to secure it. **4** *slang* person's head. **5** *slang* crazy or eccentric person. **6** small lump (of coal etc.). **7** (in *pl.*) *coarse slang* testicles.□ **do one's nut** *slang* be extremely angry. [Old English]
nutcase *n. slang* crazy person.
nutcracker *n.* (usu. in *pl.*) device for cracking nuts.
nuthatch *n.* small bird which climbs up and down tree-trunks.
nutmeg *n.* **1** hard aromatic seed used as a spice and in medicine. **2** E. Indian tree bearing this. [French *nois* nut, *mugue* MUSK]
nutria /'nju:trɪə/ *n.* coypu fur. [Spanish, = otter]
nutrient /'nju:trɪənt/ —*n.* substance that provides essential nourishment. —*adj.* serving as or providing nourishment. [Latin *nutrio* nourish]
nutriment /'nju:trɪmənt/ *n.* **1** nourishing food. **2** intellectual or artistic etc. nourishment.
nutrition /nju:'trɪʃ(ə)n/ *n.* food, nourishment. □ **nutritional** *adj.* **nutritionist** *n.*
nutritious /nju:'trɪʃəs/ *adj.* efficient as food.
nutritive /'nju:trɪtɪv/ *adj.* **1** of nutrition. **2** nutritious.
nuts *predic. adj. slang* crazy, mad.□ **be nuts about** (or **on**) *colloq.* be very fond of. [pl. of NUT]
nuts and bolts *n.pl. colloq.* practical details.
nutshell *n.* hard exterior covering of a nut. □ **in a nutshell** in a few words.
nutter *n. slang* crazy person.
nutty *adj.* (**-ier, -iest**) **1 a** full of nuts. **b** tasting like nuts. **2** *slang* crazy. □ **nuttiness** *n.*
nux vomica /nʌks 'vɒmɪkə/ *n.* **1** E. Indian tree. **2** seeds of this tree, containing strychnine. [Latin, = abscess nut]
nuzzle /'nʌz(ə)l/ *v.* (**-ling**) **1** prod or rub gently with the nose. **2** (foll. by *into*, *against*, *up to*) press the nose gently. **3** nestle; lie snug. [from NOSE]
NW *abbr.* **1** north-west. **2** north-western.
NY *abbr. US* New York.
nylon /'naɪlɒn/ *n.* **1** tough light elastic synthetic fibre. **2** nylon fabric. **3** (in *pl.*) stockings of nylon. [invented word]
nymph /nɪmf/ *n.* **1** mythological semi-divine spirit regarded as a maiden and associated with an aspect of nature, esp. rivers and woods. **2** *poet.* beautiful young woman. **3** immature form of some insects. [Greek *numphē* nymph, bride]
nympho /'nɪmfəʊ/ *n.* (*pl.* **-s**) *colloq.* nymphomaniac. [abbreviation]
nymphomania /,nɪmfə'meɪnɪə/ *n.* excessive sexual desire in a woman. □ **nymphomaniac** *n.* & *adj.* [from NYMPH, -MANIA]
NZ *abbr.* New Zealand.

O

O[1] /əʊ/ *n.* (also **o**) (*pl.* **Os** or **O's**) **1** fifteenth letter of the alphabet. **2** (**0**) nought, zero.
O[2] *abbr.* (also **O.**) Old.
O[3] *symb.* oxygen.
O[4] /əʊ/ *int.* **1** var. of OH. **2** prefixed to a name in the vocative (*O God*). [natural exclamation]
o' /ə/ *prep.* of, on (esp. in phrases: *o'clock*; *will-o'-the-wisp*). [abbreviation]
-o *suffix* forming usu. *slang* or *colloq.* variants or derivatives (*beano*; *wino*). [perhaps from OH]
-o- *suffix* terminal vowel of comb. forms (*neuro-*; *Franco-*). [originally Greek]

■ **Usage** This suffix is often elided before a vowel, as in *neuralgia*.

oaf *n.* (*pl.* **-s**) **1** awkward lout. **2** stupid person. □ **oafish** *adj.* **oafishly** *adv.* **oafishness** *n.* [Old Norse: related to ELF]
oak *n.* **1** acorn-bearing tree with lobed leaves. **2** its durable wood. **3** (*attrib.*) of oak. **4** (**the Oaks**) (treated as *sing.*) annual race at Epsom for fillies. [Old English]
oak-apple *n.* (also **oak-gall**) a kind of growth formed on oak trees by the larvae of certain wasps.
oakum /'əʊkəm/ *n.* loose fibre obtained by picking old rope to pieces and used esp. in caulking. [Old English, = off-comb]
OAP *abbr.* old-age pensioner.
oar /ɔ:(r)/ *n.* **1** pole with a blade used to propel a boat by leverage against the water. **2** rower. □ **put one's oar in** interfere. [Old English]
oarsman *n.* (*fem.* **oarswoman**) rower. □ **oarsmanship** *n.*
oasis /əʊ'eɪsɪs/ *n.* (*pl.* **oases** /-si:z/) **1** fertile place in a desert. **2** area or period of calm in the midst of turbulence. [Latin from Greek]
oast *n.* kiln for drying hops. [Old English]
oast-house *n.* building containing an oast.
oat *n.* **1 a** hardy cereal plant grown as food. **b** (in *pl.*) grain yielded by this. **2** oat plant or a variety of it. **3** (in *pl.*) *slang* sexual gratification. □ **off one's oats** *colloq.* not hungry. □ **oaten** *adj.* [Old English]
oatcake *n.* thin oatmeal biscuit.
oath *n.* (*pl.* **-s** /əʊðz/) **1** solemn declaration naming God etc. as witness. **2** profanity, curse. □ **on** (or **under**) **oath** having sworn a solemn oath. [Old English]
oatmeal *n.* **1** meal ground from oats. **2** greyish-fawn colour flecked with brown.
OAU *abbr.* Organization of African Unity.
OB *abbr.* outside broadcast.
ob. *abbr.* he or she died. [Latin *obiit*]
ob- *prefix* (also **oc-** before *c*, **of-** before *f*, **op-** before *p*) esp. in words from Latin, meaning: **1** exposure. **2** meeting or facing. **3** direction. **4** resistance. **5** hindrance or concealment. **6** finality or completeness. [Latin *ob* towards, against, in the way of]
obbligato /,ɒblɪ'gɑ:təʊ/ *n.* (*pl.* **-s**) *Mus.* accompaniment forming an integral part of a composition. [Italian, = obligatory]
obdurate /'ɒbdjʊrət/ *adj.* **1** stubborn. **2** hardened. □ **obduracy** *n.* [Latin *duro* harden]
OBE *abbr.* Officer of the Order of the British Empire.
obedient /əʊ'bi:dɪənt/ *adj.* **1** obeying or ready to obey. **2** submissive to another's will. □ **obedience** *n.* **obediently** *adv.* [Latin: related to OBEY]
obeisance /əʊ'beɪs(ə)ns/ *n.* **1** bow, curtsy, or other respectful gesture. **2** homage. □ **obeisant** *adj.* [French: related to OBEY]
obelisk /'ɒbəlɪsk/ *n.* tapering usu. four-sided stone pillar as a monument or landmark. [Greek diminutive: related to OBELUS]
obelus /'ɒbələs/ *n.* (*pl.* **obeli** /-,laɪ/) dagger-shaped reference mark (†). [Greek, = pointed pillar, SPIT[2]]
obese /əʊ'bi:s/ *adj.* very fat. □ **obesity** *n.* [Latin *edo* eat]
obey /əʊ'beɪ/ *v.* **1 a** carry out the command of. **b** carry out (a command). **2** do what one is told to do. **3** be actuated by (a force or impulse). [Latin *obedio* from *audio* hear]
obfuscate /'ɒbfʌ,skeɪt/ *v.* (**-ting**) **1** obscure or confuse (a mind, topic, etc.). **2** stupefy, bewilder. □ **obfuscation** /-'keɪʃ(ə)n/ *n.* [Latin *fuscus* dark]
obituary /ə'bɪtjʊərɪ/ *n.* (*pl.* **-ies**) **1** notice of a death or deaths. **2** account of the life of a deceased person. **3** (*attrib.*) of or

serving as an obituary. [Latin *obitus* death]

object –*n.* /ˈɒbdʒɪkt/ **1** material thing that can be seen or touched. **2** person or thing to which action or feeling is directed (*object of attention*). **3** thing sought or aimed at. **4** *Gram.* noun or its equivalent governed by an active transitive verb or by a preposition. **5** *Philos.* thing external to the thinking mind or subject. –*v.* /əbˈdʒekt/ (often foll. by *to, against*) **1** express opposition, disapproval, or reluctance. **2** protest. □ **no object** not forming an important or restricting factor (*money no object*). □ **objector** /əbˈdʒektə(r)/ *n.* [Latin *jacio ject-* throw]

objectify /əbˈdʒektɪˌfaɪ/ *v.* (**-ies, -ied**) present as an object; express in concrete form.

objection /əbˈdʒekʃ(ə)n/ *n.* **1** expression or feeling of opposition or disapproval. **2** objecting. **3** adverse reason or statement. [Latin: related to OBJECT]

objectionable *adj.* **1** unpleasant, offensive. **2** open to objection. □ **objectionably** *adv.*

objective /əbˈdʒektɪv/ –*adj.* **1** external to the mind; actually existing. **2** dealing with outward things or exhibiting facts uncoloured by feelings or opinions. **3** *Gram.* (of a case or word) in the form appropriate to the object. –*n.* **1** something sought or aimed at. **2** *Gram.* objective case. □ **objectively** *adv.* **objectivity** /ˌɒbdʒekˈtɪvɪtɪ/ *n.* [medieval Latin: related to OBJECT]

object-lesson *n.* striking practical example of some principle.

objet d'art /ˌɒbʒeɪ ˈdɑː/ *n.* (*pl.* ***objets d'art*** pronunc. same) small decorative object. [French, = object of art]

oblate /ˈɒbleɪt/ *adj. Geom.* (of a spheroid) flattened at the poles. [Latin: related to OB-; cf. PROLATE]

oblation /əʊˈbleɪʃ(ə)n/ *n.* thing offered to a divine being. [Latin: related to OFFER]

obligate /ˈɒblɪˌgeɪt/ *v.* (**-ting**) bind (a person) legally or morally (*was obligated to attend*). [Latin: related to OBLIGE]

obligation /ˌɒblɪˈgeɪʃ(ə)n/ *n.* **1** constraining power of a law, duty, contract, etc. **2** duty, task. **3** binding agreement. **4** indebtedness for a service or benefit (*be under an obligation*). [Latin: related to OBLIGE]

obligatory /əˈblɪgətərɪ/ *adj.* **1** binding. **2** compulsory. □ **obligatorily** *adv.* [Latin: related to OBLIGE]

oblige /əˈblaɪdʒ/ *v.* (**-ging**) **1** constrain, compel. **2** be binding on. **3** do (a person) a small favour, help. **4** (as **obliged** *adj.*) indebted, grateful. □ **much obliged** thank you. [Latin *obligo* bind]

obliging *adj.* accommodating, helpful. □ **obligingly** *adv.*

oblique /əˈbliːk/ –*adj.* **1** slanting; at an angle. **2** not going straight to the point; indirect. **3** *Gram.* (of a case) other than nominative or vocative. –*n.* oblique stroke (/). □ **obliquely** *adv.* **obliqueness** *n.* **obliquity** /əˈblɪkwɪtɪ/ *n.* [French from Latin]

obliterate /əˈblɪtəˌreɪt/ *v.* (**-ting**) blot out, destroy, leave no clear traces of. □ **obliteration** /-ˈreɪʃ(ə)n/ *n.* [Latin *oblitero* erase, from *litera* letter]

oblivion /əˈblɪvɪən/ *n.* state of having or being forgotten. [Latin *obliviscor* forget]

oblivious /əˈblɪvɪəs/ *adj.* unaware or unconscious. □ **obliviously** *adv.* **obliviousness** *n.*

oblong /ˈɒblɒŋ/ –*adj.* rectangular with adjacent sides unequal. –*n.* oblong figure or object. [Latin *oblongus* longish]

obloquy /ˈɒbləkwɪ/ *n.* **1** being generally ill spoken of. **2** abuse. [Latin *obloquium* contradiction, from *loquor* speak]

obnoxious /əbˈnɒkʃəs/ *adj.* offensive, objectionable. □ **obnoxiously** *adv.* **obnoxiousness** *n.* [Latin *noxa* injury]

oboe /ˈəʊbəʊ/ *n.* woodwind double-reed instrument with a piercing plaintive tone. □ **oboist** /ˈəʊbəʊɪst/ *n.* [French *hautbois* from *haut* high, *bois* wood]

obscene /əbˈsiːn/ *adj.* **1** offensively indecent. **2** *colloq.* highly offensive. **3** *Law* (of a publication) tending to deprave or corrupt. □ **obscenely** *adv.* **obscenity** /-ˈsenɪtɪ/ *n.* (pl. **-ies**). [Latin *obsc(a)enus* abominable]

obscurantism /ˌɒbskjʊəˈræntɪz(ə)m/ *n.* opposition to knowledge and enlightenment. □ **obscurantist** *n.* & *adj.* [Latin *obscurus* dark]

obscure /əbˈskjʊə(r)/ –*adj.* **1** not clearly expressed or easily understood. **2** unexplained. **3** dark. **4** indistinct. **5** hidden; unnoticed. **6** (of a person) undistinguished, hardly known. –*v.* (**-ring**) **1** make obscure or unintelligible. **2** conceal. □ **obscurity** *n.* [French from Latin]

obsequies /ˈɒbsɪkwɪz/ *n.pl.* funeral rites. [Latin *obsequiae*]

obsequious /əbˈsiːkwɪəs/ *adj.* servile, fawning. □ **obsequiously** *adv.* **obsequiousness** *n.* [Latin *obsequor* comply with]

observance /əbˈzɜːv(ə)ns/ *n.* **1** keeping or performing of a law, duty, etc. **2** rite or ceremony.

observant *adj.* **1** acute in taking notice. **2** attentive in observance. □ **observantly** *adv.*

observation /ˌɒbzəˈveɪʃ(ə)n/ *n.* **1** observing or being observed. **2** power of perception. **3** remark, comment. **4** thing observed by esp. scientific study. □ **observational** *adj.*

observatory /əbˈzɜːvətərɪ/ *n.* (*pl.* **-ies**) building for astronomical or other observation.

observe /əbˈzɜːv/ *v.* (**-ving**) **1** perceive, become aware of. **2** watch carefully. **3 a** follow or keep (rules etc.). **b** celebrate or perform (an occasion, rite, etc.). **4** remark. **5** take note of scientifically. □ **observable** *adj.* [Latin *servo* watch, keep]

observer *n.* **1** person who observes. **2** interested spectator. **3** person who attends a meeting etc. to note the proceedings but does not participate.

obsess /əbˈses/ *v.* fill the mind of (a person) continually; preoccupy. □ **obsessive** *adj.* & *n.* **obsessively** *adv.* **obsessiveness** *n.* [Latin *obsideo obsess-* besiege]

obsession /əbˈseʃ(ə)n/ *n.* **1** obsessing or being obsessed. **2** persistent idea dominating a person's mind. □ **obsessional** *adj.* **obsessionally** *adv.*

obsidian /əbˈsɪdɪən/ *n.* dark glassy rock formed from lava. [Latin from *Obsius*, discoverer of a similar stone]

obsolescent /ˌɒbsəˈles(ə)nt/ *adj.* becoming obsolete. □ **obsolescence** *n.* [Latin *soleo* be accustomed]

obsolete /ˈɒbsəˌliːt/ *adj.* no longer used, antiquated.

obstacle /ˈɒbstək(ə)l/ *n.* thing that obstructs progress. [Latin *obsto* stand in the way]

obstetrician /ˌɒbstəˈtrɪʃ(ə)n/ *n.* specialist in obstetrics.

obstetrics /əbˈstetrɪks/ *n.pl.* (treated as *sing.*) branch of medicine and surgery dealing with childbirth. □ **obstetric** *adj.* [Latin *obstetrix* midwife, from *obsto* be present]

obstinate /ˈɒbstɪnət/ *adj.* **1** stubborn, intractable. **2** firmly continuing in one's action or opinion despite advice. □ **obstinacy** *n.* **obstinately** *adv.* [Latin *obstino* persist]

obstreperous /əbˈstrepərəs/ *adj.* **1** turbulent, unruly. **2** noisy. □ **obstreperously** *adv.* **obstreperousness** *n.* [Latin *obstrepo* shout at]

obstruct /əbˈstrʌkt/ *v.* **1** block up; make hard or impossible to pass along or through. **2** prevent or retard the progress of. [Latin *obstruo obstruct-* block up]

obstruction /əbˈstrʌkʃ(ə)n/ *n.* **1** obstructing or being obstructed. **2** thing that obstructs, blockage. **3** *Sport* act of unlawfully obstructing another player.

obstructive *adj.* causing or intended to cause an obstruction. □ **obstructively** *adv.* **obstructiveness** *n.*

obtain /əbˈteɪn/ *v.* **1** acquire, secure; have granted to one, get. **2** be in vogue, prevail. □ **obtainable** *adj.* [Latin *teneo* hold]

obtrude /əbˈtruːd/ *v.* (**-ding**) **1** be or become obtrusive. **2** (often foll. by *on*, *upon*) thrust (oneself, a matter, etc.) importunately forward. □ **obtrusion** *n.* [Latin *obtrudo* thrust against]

obtrusive /əbˈtruːsɪv/ *adj.* **1** unpleasantly noticeable. **2** obtruding oneself. □ **obtrusively** *adv.* **obtrusiveness** *n.*

obtuse /əbˈtjuːs/ *adj.* **1** dull-witted. **2** (of an angle) between 90° and 180°. **3** of blunt form; not sharp-pointed or sharp-edged. □ **obtuseness** *n.* [Latin *obtundo obtus-* beat against, blunt]

obverse /ˈɒbvɜːs/ *n.* **1** counterpart, opposite. **2** side of a coin or medal etc. bearing the head or principal design. **3** front, proper, or top side of a thing. [Latin *obverto obvers-* turn towards]

obviate /ˈɒbvɪˌeɪt/ *v.* (**-ting**) get round or do away with (a need, inconvenience, etc.). [Latin *obvio* prevent]

obvious /ˈɒbvɪəs/ *adj.* easily seen, recognized, or understood. □ **obviously** *adv.* **obviousness** *n.* [Latin *ob viam* in the way]

OC *abbr.* Officer Commanding.

oc- see OB-.

ocarina /ˌɒkəˈriːnə/ *n.* small egg-shaped musical wind instrument. [Italian *oca* goose]

occasion /əˈkeɪʒ(ə)n/ —*n.* **1 a** special event or happening. **b** time of this. **2** reason, need. **3** suitable juncture, opportunity. **4** immediate but subordinate cause. —*v.* cause, esp. incidentally. □ **on occasion** now and then; when the need arises. [Latin *occido occas-* go down]

occasional *adj.* **1** happening irregularly and infrequently. **2** made or meant for, or acting on, a special occasion. □ **occasionally** *adv.*

occasional table *n.* small table for use as required.

Occident /ˈɒksɪd(ə)nt/ *n. poet.* or *rhet.* **1** (prec. by *the*) West. **2** western Europe. **3** Europe and America as distinct from the Orient. [Latin *occidens -entis* setting, sunset, west]

occidental /ˌɒksɪˈdent(ə)l/ —*adj.* **1** of the Occident. **2** western. —*n.* native of the Occident.

occiput /ˈɒksɪˌpʌt/ *n.* back of the head. □ **occipital** /-ˈsɪpɪt(ə)l/ *adj.* [Latin *caput* head]

occlude /əˈkluːd/ *v.* (**-ding**) **1** stop up or close. **2** *Chem.* absorb and retain (gases). **3** (as **occluded** *adj.*) *Meteorol.* (of a frontal system) formed when a cold front overtakes a warm front, raising warm air from ground level. □ **occlusion** *n.* [Latin *occludo occlus-* close up]

occult /ɒ'kʌlt, 'ɒ-/ *adj.* **1** involving the supernatural; mystical. **2** esoteric. □ **the occult** occult phenomena generally. [Latin *occulo occult-* hide]

occupant /'ɒkjʊpənt/ *n.* person who occupies, esp. lives in, a place etc. □ **occupancy** *n.* (*pl.* **-ies**). [Latin: related to OCCUPY]

occupation /ˌɒkjʊ'peɪʃ(ə)n/ *n.* **1** person's employment or profession. **2** pastime. **3** occupying or being occupied. **4** taking or holding of a country etc. by force.

occupational *adj.* **1** of or connected with one's occupation. **2** (of a disease, hazard, etc.) connected with one's occupation.

occupational therapy *n.* programme of mental or physical activity to assist recovery from disease or injury.

occupier /'ɒkjʊˌpaɪə(r)/ *n.* person living in a house etc. as its owner or tenant.

occupy /'ɒkjʊˌpaɪ/ *v.* (**-ies, -ied**) **1** live in; be the tenant of. **2** take up or fill (space, time, or a place). **3** hold (a position or office). **4** take military possession of. **5** place oneself in (a building etc.) forcibly or without authority as a protest. **6** keep busy or engaged. [Latin *occupo* seize]

occur /ə'kɜ:(r)/ *v.* (**-rr-**) **1** come into being as an event or process. **2** exist or be encountered in some place or conditions. **3** (foll. by *to*) come into the mind of. [Latin *occurro* befall]

occurrence /ə'kʌrəns/ *n.* **1** occurring. **2** incident or event.

ocean /'əʊʃ(ə)n/ *n.* **1** large expanse of sea, esp. each of the main areas called the Atlantic, Pacific, Indian, Arctic, and Antarctic Oceans. **2** (often in *pl.*) *colloq.* very large expanse or quantity. □ **oceanic** /ˌəʊʃɪ'ænɪk/ *adj.* [Greek *ōkeanos*]

ocean-going *adj.* (of a ship) able to cross oceans.

oceanography /ˌəʊʃə'nɒgrəfɪ/ *n.* the study of the oceans. □ **oceanographer** *n.*

ocelot /'ɒsɪˌlɒt/ *n.* leopard-like cat of S. and Central America. [French from Nahuatl]

ochre /'əʊkə(r)/ *n.* (*US* **ocher**) **1** earth used as yellow, brown, or red pigment. **2** pale brownish-yellow colour. □ **ochreous** /'əʊkrɪəs/ *adj.* [Greek *ōkhra*]

o'clock /ə'klɒk/ *adv.* of the clock (used to specify the hour) (*6 o'clock*).

Oct. *abbr.* October.

octa- *comb. form* (also **oct-** before a vowel) eight. [Latin *octo*, Greek *oktō* eight]

octagon /'ɒktəgən/ *n.* plane figure with eight sides and angles. □ **octagonal** /-'tægən(ə)l/ *adj.* [Greek: related to OCTA-, *-gōnos* -angled]

octahedron /ˌɒktə'hi:drən/ *n.* (*pl.* **-s**) solid figure contained by eight (esp. triangular) plane faces. □ **octahedral** *adj.* [Greek]

octane /'ɒkteɪn/ *n.* colourless inflammable hydrocarbon occurring in petrol. [from OCTA-]

octane number *n.* (also **octane rating**) figure indicating the antiknock properties of a fuel.

octave /'ɒktɪv/ *n.* **1** *Mus.* **a** interval between (and including) two notes, one having twice or half the frequency of vibration of the other. **b** eight notes occupying this interval. **c** each of the two notes at the extremes of this interval. **2** eight-line stanza. [Latin *octavus* eighth]

octavo /ɒk'teɪvəʊ/ *n.* (*pl.* **-s**) **1** size of a book or page given by folding a sheet of standard size three times to form eight leaves. **2** book or sheet of this size. [Latin: related to OCTAVE]

octet /ɒk'tet/ *n.* (also **octette**) **1 a** musical composition for eight performers. **b** the performers. **2** group of eight. [Italian or German: related to OCTA-]

octo- *comb. form* (also **oct-** before a vowel) eight. [see OCTA-]

October /ɒk'təʊbə(r)/ *n.* tenth month of the year. [Latin *octo* eight, originally the 8th month of the Roman year]

octogenarian /ˌɒktəʊdʒɪ'neərɪən/ *n.* person from 80 to 89 years old. [Latin *octogeni* 80 each]

octopus /'ɒktəpəs/ *n.* (*pl.* **-puses**) sea mollusc with eight suckered tentacles. [Greek: related to OCTO-, *pous* foot]

ocular /'ɒkjʊlə(r)/ *adj.* of, for, or by the eyes; visual. [Latin *oculus* eye]

oculist /'ɒkjʊlɪst/ *n.* specialist in the treatment of the eyes.

OD /əʊ'di:/ *slang* —*n.* drug overdose. —*v.* (**OD's, OD'd, OD'ing**) take an overdose. [abbreviation]

odd *adj.* **1** strange, remarkable, eccentric. **2** casual, occasional (*odd jobs; odd moments*). **3** not normally considered; unconnected (*in some odd corner; picks up odd bargains*). **4 a** (of numbers) not integrally divisible by two, e.g. 1, 3, 5. **b** bearing such a number (*no parking on odd dates*). **5** left over when the rest have been distributed or divided into pairs (*odd sock*). **6** detached from a set or series (*odd volumes*). **7** (appended to a number, sum, weight, etc.) somewhat more than (*forty odd*; *forty-odd people*). **8** by which a round number, given sum, etc., is exceeded (*we have 102 – do you want the odd 2?*). □ **oddly** *adv.* **oddness** *n.* [Old Norse *oddi* angle, point, third or odd number]

oddball *n. colloq.* eccentric person.

oddity /'ɒdɪtɪ/ *n.* (*pl.* **-ies**) **1** strange person, thing, or occurrence. **2** peculiar trait. **3** strangeness.

odd man out *n.* person or thing differing from the others in a group in some respect.

oddment *n.* **1** odd article; something left over. **2** (in *pl.*) miscellaneous articles.

odds *n.pl.* **1** ratio between the amounts staked by the parties to a bet, based on the expected probability either way. **2** balance of probability or advantage (*the odds are against it*; *the odds are in your favour*). **3** difference giving an advantage (*it makes no odds*). □ **at odds** (often foll. by *with*) in conflict or at variance. **over the odds** above the normal price etc. [apparently from ODD]

odds and ends *n.pl.* miscellaneous articles or remnants.

odds-on –*n.* state when success is more likely than failure. –*adj.* (of a chance) better than even; likely.

ode *n.* lyric poem of exalted style and tone. [Greek *ōidē* song]

odious /'əʊdɪəs/ *adj.* hateful, repulsive. □ **odiously** *adv.* **odiousness** *n.* [related to ODIUM]

odium /'əʊdɪəm/ *n.* widespread dislike or disapproval of a person or action. [Latin, = hatred]

odometer /əʊ'dɒmɪtə(r), ɒ-/ *n. US* = MILOMETER. [Greek *hodos* way]

odoriferous /ˌəʊdə'rɪfərəs/ *adj.* diffusing a (usu. agreeable) odour. [Latin: related to ODOUR]

odour /'əʊdə(r)/ *n.* (*US* **odor**) **1** smell or fragrance. **2** quality or trace (*an odour of intolerance*). **3** regard, repute (*in bad odour*). □ **odorous** *adj.* **odourless** *adj.* [Latin *odor*]

odyssey /'ɒdɪsɪ/ *n.* (*pl.* **-s**) long adventurous journey. [title of the Homeric epic poem on the adventures of Odysseus]

OECD *abbr.* Organization for Economic Cooperation and Development.

oedema /ɪ'di:mə/ *n.* (*US* **edema**) accumulation of excess fluid in body tissues, causing swelling. [Greek *oideō* swell]

Oedipus complex /'i:dɪpəs/ *n.* child's, esp. a boy's, subconscious sexual desire for the parent of the opposite sex. □ **Oedipal** *adj.* [Greek *Oidipous*, who unknowingly married his mother]

o'er /'əʊə(r)/ *adv.* & *prep. poet.* = OVER. [contraction]

oesophagus /i:'sɒfəgəs/ *n.* (*US* **esophagus**) (*pl.* **-gi** /-ˌdʒaɪ/ or **-guses**) passage from the mouth to the stomach; gullet. [Greek]

oestrogen /'i:strədʒ(ə)n/ *n.* (*US* **estrogen**) **1** sex hormone developing and maintaining female characteristics of the body. **2** this produced artificially for use in medicine. [Greek *oistros* frenzy, -GEN]

oestrus /'i:strəs/ *n.* (also **oestrum**, *US* **estrus**) recurring period of sexual receptivity in many female mammals. □ **oestrous** *adj.* [Greek *oistros* frenzy]

œuvre /'ɜ:vr/ *n.* works of a creative artist regarded collectively. [French, = work, from Latin *opera*]

of /ɒv, əv/ *prep.* expressing: **1** origin or cause (*paintings of Turner*; *died of cancer*). **2** material or substance (*house of cards*; *built of bricks*). **3** belonging or connection (*thing of the past*; *articles of clothing*; *head of the business*). **4** identity or close relation (*city of Rome*; *a pound of apples*; *a fool of a man*). **5** removal or separation (*north of the city*; *got rid of them*; *robbed us of £1000*). **6** reference or direction (*beware of the dog*; *suspected of lying*; *very good of you*; *short of money*). **7** objective relation (*love of music*; *in search of peace*). **8** partition, classification, or inclusion (*no more of that*; *part of the story*; *this sort of book*). **9** description, quality, or condition (*the hour of prayer*; *person of tact*; *girl of ten*; *on the point of leaving*). **10** *US* time in relation to the following hour (*a quarter of three*). □ **be of** possess, give rise to (*is of great interest*). **of an evening** (or **morning** etc.) *colloq.* **1** on most evenings (or mornings etc.). **2** at some time in the evenings (or mornings etc.). **of late** recently. **of old** formerly. [Old English]

of- see OB-.

Off. *abbr.* **1** Office. **2** Officer.

off –*adv.* **1** away; at or to a distance (*drove off*; *3 miles off*). **2** out of position; not on, touching, or attached; loose, separate, gone (*has come off*; *take your coat off*). **3** so as to be rid of (*sleep it off*). **4** so as to break continuity or continuance; discontinued, stopped (*turn off the radio*; *take a day off*; *the game is off*). **5** not available on a menu etc. (*chips are off*). **6** to the end; entirely; so as to be clear (*clear off*; *finish off*; *pay off*). **7** situated as regards money, supplies, etc. (*well off*). **8** off stage (*noises off*). **9** (of food etc.) beginning to decay. –*prep.* **1 a** from; away, down, or up from (*fell off the chair*; *took something off the price*). **b** not on (*off the pitch*). **2 a** temporarily relieved of or abstaining from (*off duty*).

b temporarily not attracted by (*off his food*). **c** not achieving (*off form*). **3** using as a source or means of support (*live off the land*). **4** leading from; not far from (*a street off the Strand*). **5** at a short distance to sea from (*sank off Cape Horn*). –*adj.* **1** far, further (*off side of the wall*). **2** (of a part of a vehicle, animal, or road) right (*the off front wheel*). **3** *Cricket* designating the half of the field (as divided lengthways through the pitch) to which the striker's feet are pointed. **4** *colloq.* **a** annoying, unfair (*that's really off*). **b** somewhat unwell (*feeling a bit off*). –*n.* **1** the off side in cricket. **2** start of a race. □ **off and on** intermittently; now and then. **off the cuff** see CUFF[1]. **off the peg** see PEG. [var. of OF]

■ **Usage** The use of *off of* for the preposition *off* (sense 1a), e.g. *picked it up off of the floor*, is non-standard and should be avoided.

offal /ˈɒf(ə)l/ *n.* **1** less valuable edible parts of a carcass, esp. the heart, liver, etc. **2** refuse, scraps. [Dutch *afval*: related to OFF, FALL]

offbeat –*adj.* **1** not coinciding with the beat. **2** eccentric, unconventional. –*n.* any of the unaccented beats in a bar.

off-centre *adj.* & *adv.* not quite centrally placed.

off chance *n.* (prec. by *the*) remote possibility.

off colour *predic. adj.* **1** unwell. **2** *US* somewhat indecent.

offcut *n.* remnant of timber, paper, etc., after cutting.

off-day *n. colloq.* day when one is not at one's best.

offence /əˈfens/ *n.* (*US* **offense**) **1** illegal act; transgression. **2** upsetting of feelings, insult; umbrage (*give offence*; *take offence*). **3** aggressive action. [related to OFFEND]

offend /əˈfend/ *v.* **1** cause offence to, upset. **2** displease, anger. **3** (often foll. by *against*) do wrong; transgress. □ **offender** *n.* **offending** *adj.* [Latin *offendo offens-* strike against, displease]

offense *US* var. of OFFENCE.

offensive /əˈfensɪv/ –*adj.* **1** causing offence; insulting. **2** disgusting. **3 a** aggressive, attacking. **b** (of a weapon) for attacking. –*n.* aggressive action, attitude, or campaign. □ **offensively** *adv.* **offensiveness** *n.*

offer –*v.* **1** present for acceptance, refusal, or consideration. **2** (foll. by *to* + infin.) express readiness or show intention. **3** provide; give an opportunity for. **4** make available for sale. **5** present to the attention. **6** present (a sacrifice etc.). **7** present itself; occur (*as opportunity offers*). **8** attempt (violence, resistance, etc.). –*n.* **1** expression of readiness to do or give if desired, or to buy or sell. **2** amount offered. **3** proposal (esp. of marriage). **4** bid. □ **on offer** for sale at a certain (esp. reduced) price. [Latin *offero oblat-*]

offering *n.* **1** contribution or gift, esp. of money. **2** thing offered as a sacrifice etc.

offertory /ˈɒfətərɪ/ *n.* (*pl.* **-ies**) **1** offering of the bread and wine at the Eucharist. **2** collection of money at a religious service. [Church Latin: related to OFFER]

offhand –*adj.* curt or casual in manner. –*adv.* without preparation or thought (*can't say offhand*). □ **offhanded** *adj.* **offhandedly** *adv.* **offhandedness** *n.*

office /ˈɒfɪs/ *n.* **1** room or building used as a place of business, esp. for clerical or administrative work. **2** room or area for a particular business (*ticket office*). **3** local centre of a large business (*our London office*). **4** position with duties attached to it. **5** tenure of an official position (*hold office*). **6** (**Office**) quarters, staff, or collective authority of a Government department etc. (*Foreign Office*). **7** duty, task, function. **8** (usu. in *pl.*) piece of kindness; service (esp. *through the good offices of*). **9** authorized form of worship. [Latin *officium* from *opus* work, *facio fic-* do]

officer /ˈɒfɪsə(r)/ *n.* **1** person holding a position of authority or trust, esp. one with a commission in the army, navy, air force, etc. **2** policeman or policewoman. **3** holder of a post in a society (e.g. the president or secretary).

official /əˈfɪʃ(ə)l/ –*adj.* **1** of an office or its tenure. **2** characteristic of officials and bureaucracy. **3** properly authorized. –*n.* person holding office or engaged in official duties. □ **officialdom** *n.* **officially** *adv.*

officialese /əˌfɪʃəˈliːz/ *n. derog.* language characteristic of official documents.

official secrets *n.pl.* confidential information involving national security.

officiate /əˈfɪʃɪˌeɪt/ *v.* (**-ting**) **1** act in an official capacity. **2** conduct a religious service. □ **officiation** /-ˈeɪʃ(ə)n/ *n.* **officiator** *n.*

officious /əˈfɪʃəs/ *adj.* **1** domineering. **2** intrusive in correcting etc. □ **officiously** *adv.* **officiousness** *n.*

offing *n.* more distant part of the sea in view. □ **in the offing** not far away; likely to appear or happen soon. [probably from OFF]

off-key *adj.* & *adv.* **1** out of tune. **2** not quite fitting.

off-licence *n.* **1** shop selling alcoholic drink. **2** licence for this.

offline *Computing* *–adj.* not online. *–adv.* with a delay between the production of data and its processing; not under direct computer control.

offload *v.* get rid of (esp. something unpleasant) by passing it to someone else.

off-peak *adj.* used or for use at times other than those of greatest demand.

off-piste *adj.* (of skiing) away from prepared ski runs.

offprint *n.* printed copy of an article etc. originally forming part of a larger publication.

offscreen *adj.* & *adv.* beyond the range of a film camera etc.; when not being filmed.

off-season *n.* time of the year when business etc. is slack.

offset *–n.* **1** side-shoot from a plant serving for propagation. **2** compensation, consideration or amount diminishing or neutralizing the effect of a contrary one. **3** sloping ledge in a wall etc. **4** bend in a pipe etc. to carry it past an obstacle. **5** (often *attrib.*) method of printing in which ink is transferred from a plate or stone to a rubber surface and from there to paper etc. (*offset litho*). *–v.* (**-setting**; *past* and *past part.* **-set**) **1** counterbalance, compensate. **2** print by the offset process.

offshoot *n.* **1** side-shoot or branch. **2** derivative.

offshore *adj.* **1** at sea some distance from the shore. **2** (of the wind) blowing seawards.

offside *–adj.* (of a player in a field game) in a position where he or she may not play the ball. *–n.* (often *attrib.*) right side of a vehicle, animal, etc.

offspring *n.* (*pl.* same) **1** person's child, children, or descendants. **2** animal's young or descendants. **3** result. [Old English: see OFF, SPRING]

off-stage *adj.* & *adv.* not on the stage; not visible to the audience.

off-street *adj.* (esp. of parking) other than on a street.

off-the-wall *adj.* esp. *US slang* crazy, absurd, outlandish.

off white *adj.* & *n.* (as adj. often hyphenated) white with a grey or yellowish tinge.

oft *adv. archaic* often. [Old English]

often /ˈɒf(ə)n/ *adv.* (**oftener**, **oftenest**) **1 a** frequently; many times. **b** at short intervals. **2** in many instances.

oft-times *adv.* often.

ogee /ˈəʊdʒiː/ *n.* S-shaped line or moulding. [apparently from OGIVE]

ogive /ˈəʊdʒaɪv/ *n.* **1** pointed arch. **2** diagonal rib of a vault. [French]

ogle /ˈəʊg(ə)l/ *–v.* (**-ling**) look amorously or lecherously (at). *–n.* amorous or lecherous look. [probably Low German or Dutch]

ogre /ˈəʊgə(r)/ *n.* (*fem.* **ogress** /-grɪs/) **1** man-eating giant in folklore. **2** terrifying person. □ **ogreish** /ˈəʊgərɪʃ/ *adj.* (also **ogrish**). [French]

oh /əʊ/ *int.* (also **O**) expressing surprise, pain, entreaty, etc. □ **oh** (or **o**) **for** I wish I had. [var. of O⁴]

ohm /əʊm/ *n.* SI unit of electrical resistance. [*Ohm*, name of a physicist]

OHMS *abbr.* on Her (or His) Majesty's Service.

oho /əʊˈhəʊ/ *int.* expressing surprise or exultation. [from O⁴, HO]

OHP *abbr.* overhead projector.

oi *int.* calling attention or expressing alarm etc. [var. of HOY]

-oid *suffix* forming adjectives and nouns, denoting form or resemblance (*asteroid*; *rhomboid*; *thyroid*). [Greek *eidos* form]

oil *–n.* **1** any of various viscous, usu. inflammable liquids insoluble in water (*cooking oil*; *drill for oil*). **2** petroleum. **3** (in *comb.*) using oil as fuel (*oil-heater*). **4 a** (usu. in *pl.*) = OIL-PAINT. **b** picture painted in oil-paints. *–v.* **1** apply oil to; lubricate. **2** impregnate or treat with oil (*oiled silk*). □ **oil the wheels** help make things go smoothly. [Latin *oleum* olive oil]

oilcake *n.* compressed linseed from which the oil has been extracted, used as fodder or manure.

oilcan *n.* can with a long nozzle for oiling machinery.

oilcloth *n.* fabric, esp. canvas, waterproofed with oil or another substance.

oil-colour var. of OIL-PAINT.

oiled *adj. slang* drunk.

oilfield *n.* area yielding mineral oil.

oil-fired *adj.* using oil as fuel.

oil of turpentine *n.* volatile pungent oil distilled from turpentine, used as a solvent in mixing paints and varnishes, and in medicine.

oil-paint *n.* (also **oil-colour**) paint made by mixing powdered pigment in oil. □ **oil-painting** *n.*

oil rig *n.* structure with equipment for drilling an oil well.

oilskin *n.* **1** cloth waterproofed with oil. **2 a** garment of this. **b** (in *pl.*) suit of this.

oil slick *n.* patch of oil, esp. on the sea.

oilstone *n.* fine-grained flat stone used with oil for sharpening flat tools, e.g. chisels, planes, etc.

oil well *n.* well from which mineral oil is drawn.

oily *adj.* (**-ier, -iest**) **1** of or like oil. **2** covered or soaked with oil. **3** (of a manner etc.) fawning, unctuous, ingratiating. □ **oiliness** *n.*

ointment /ˈɔɪntmənt/ *n.* smooth greasy healing or cosmetic preparation for the skin. [Latin *unguo* anoint]

OK /əʊˈkeɪ/ (also **okay**) *colloq.* —*adj.* (often as *int.*) all right; satisfactory. —*adv.* well, satisfactorily. —*n.* (*pl.* **OKs**) approval, sanction. —*v.* (**OK's, OK'd, OK'ing**) approve, sanction. [originally US: probably abbreviation of *orl* (or *oll*) *korrect*, jocular form of 'all correct']

okapi /əʊˈkɑːpɪ/ *n.* (*pl.* same or **-s**) African giraffe-like mammal but with a shorter neck and striped body. [Mbuba]

okay var. of OK.

okra /ˈəʊkrə/ *n.* tall, orig. African plant with long ridged seed-pods used for food. [West African native name]

-ol *suffix* in the names of alcohols or analogous compounds. [from ALCOHOL and Latin *oleum* oil]

old /əʊld/ *adj.* (**older, oldest**) **1 a** advanced in age; far on in the natural period of existence. **b** not young or near its beginning. **2** made long ago. **3** long in use. **4** worn, dilapidated, or shabby from the passage of time. **5** having the characteristics of age (*child has an old face*). **6** practised, inveterate (*old offender*). **7** belonging to the past; lingering on; former (*old times*; *old memories*; *our old house*). **8** dating from far back; long established or known; ancient, primeval (*an old family*; *old friends*; *old as the hills*). **9** (appended to a period of time) of age (*is four years old*; *four-year-old boy*; *a four-year-old*). **10** (of a language) as used in former or earliest times. **11** *colloq.* as a term of affection or casual reference (*good old Charlie*; *old thing*). □ **oldish** *adj.* **oldness** *n.* [Old English]

old age *n.* later part of normal life.

old-age pension *n.* = RETIREMENT PENSION. □ **old-age pensioner** *n.*

Old Bill *n. slang* **1** the police. **2** a policeman.

old boy *n.* **1** former male pupil of a school. **2** *colloq.* **a** elderly man. **b** (as a form of address) = OLD MAN.

old boy network *n. colloq.* preferment in employment, esp. of fellow ex-pupils of public schools.

old country *n.* (prec. by *the*) native country of colonists etc.

olden *attrib. adj. archaic* old; of old.

old-fashioned *adj.* showing or favouring the tastes of former times.

old girl *n.* **1** former female pupil of a school. **2** *colloq.* **a** elderly woman. **b** affectionate term of address to a girl or woman.

Old Glory *n. US* US national flag.

old gold *adj.* & *n.* (as adj. often hyphenated) dull brownish-gold colour.

old guard *n.* original, past, or conservative members of a group.

old hand *n.* person with much experience.

old hat *adj. colloq.* hackneyed.

oldie *n. colloq.* old person or thing.

old lady *n. colloq.* one's mother or wife.

old maid *n.* **1** *derog.* elderly unmarried woman. **2** prim and fussy person.

old man *n. colloq.* **1** one's husband or father. **2** affectionate form of address to a boy or man.

old man's beard *n.* wild clematis, with grey fluffy hairs round the seeds.

old master *n.* **1** great artist of former times, esp. of the 13th–17th c. in Europe. **2** painting by such a painter.

Old Nick *n. colloq.* the Devil.

Old Norse *n.* **1** Germanic language from which the Scandinavian languages are derived. **2** language of Norway and its colonies until the 14th c.

old school *n.* traditional attitudes or people having them.

old school tie *n.* excessive loyalty to traditional values and to former pupils of one's own, esp. public, school.

old soldier *n.* (also **old stager** or **timer**) experienced person.

old style *n.* dating reckoned by the Julian calendar.

Old Testament *n.* part of the Bible containing the scriptures of the Hebrews.

old-time *attrib. adj.* belonging to former times (*old-time dancing*).

old timer var. of OLD SOLDIER.

old wives' tale *n.* unscientific belief.

old woman *n. colloq.* **1** one's wife or mother. **2** fussy or timid man.

Old World *n.* Europe, Asia, and Africa.

old year *n.* year just ended or ending.

oleaginous /ˌəʊlɪˈædʒɪnəs/ *adj.* **1** like or producing oil. **2** oily. [Latin: related to OIL]

oleander /ˌəʊlɪˈændə(r)/ *n.* evergreen flowering Mediterranean shrub. [Latin]

O level *n. hist.* = ORDINARY LEVEL. [abbreviation]

olfactory /ɒlˈfæktərɪ/ *adj.* of the sense of smell (*olfactory nerves*). [Latin *oleo* smell, *facio* make]

oligarch /ˈɒlɪˌgɑːk/ *n.* member of an oligarchy. [Greek *oligoi* few]

oligarchy *n.* (*pl.* **-ies**) **1** government, or State governed, by a small group of people. **2** members of such a government. □ **oligarchic** /-ˈgɑːkɪk/ *adj.* **oligarchical** /-ˈgɑːkɪk(ə)l/ *adj.*

Oligocene /'ɒlɪgə,si:n/ *–adj.* of the third geological epoch of the Tertiary period. *–n.* this epoch. [Greek *oligos* little, *kainos* new]

olive /'ɒlɪv/ *–n.* **1** small oval hard-stoned fruit, green when unripe and bluish-black when ripe. **2** tree bearing this. **3** its wood. **4** olive green. *–adj.* **1** olive-green. **2** (of the complexion) yellowish-brown. [Latin *oliva* from Greek]

olive branch *n.* gesture of reconciliation or peace.

olive green *adj.* & *n.* (as adj. often hyphenated) dull yellowish green.

olive oil *n.* cooking-oil extracted from olives.

olivine /'ɒlɪ,vi:n/ *n.* mineral (usu. olive-green) composed of magnesium-iron silicate.

Olympiad /ə'lɪmpɪ,æd/ *n.* **1 a** period of four years between Olympic games, used by the ancient Greeks in dating events. **b** four-yearly celebration of the ancient Olympic Games. **2** celebration of the modern Olympic Games. **3** regular international contest in chess etc. [Greek *Olumpias Olumpiad-*: related to OLYMPIC]

Olympian /ə'lɪmpɪən/ *–adj.* **1 a** of Olympus. **b** celestial. **2** (of manners etc.) magnificent, condescending, superior. **3** = OLYMPIC. *–n.* **1** Greek god dwelling on Olympus. **2** person of superhuman ability or calm. **3** competitor in the Olympic games. [from Mt. *Olympus* in Greece, or as OLYMPIC]

Olympic /ə'lɪmpɪk/ *–adj.* of the Olympic games. *–n.pl.* (**the Olympics**) Olympic games. [Greek from *Olympia* in S. Greece]

Olympic games *n.pl.* **1** ancient Greek athletic festival held at Olympia every four years. **2** modern international revival of this.

OM *abbr.* Order of Merit.

ombudsman /'ɒmbʊdzmən/ *n.* (*pl.* **-men**) official appointed to investigate complaints against public authorities. [Swedish, = legal representative]

omega /'əʊmɪgə/ *n.* **1** last (24th) letter of the Greek alphabet (Ω, ω). **2** last of a series; final development. [Greek *ō mega* = great O]

omelette /'ɒmlɪt/ *n.* beaten eggs fried and often folded round a savoury filling. [French]

omen /'əʊmən/ *–n.* **1** event or object portending good or evil. **2** prophetic significance (*of good omen*). *–v.* (usu. in *passive*) portend. [Latin]

omicron /ə'maɪkrən, 'ɒmɪ-/ *n.* fifteenth letter of the Greek alphabet (Ο, ο). [Greek *o mikron* = small o]

ominous /'ɒmɪnəs/ *adj.* **1** threatening. **2** of evil omen; inauspicious. □ **ominously** *adv.* [Latin: related to OMEN]

omission /əʊ'mɪʃ(ə)n/ *n.* **1** omitting or being omitted. **2** thing omitted.

omit /əʊ'mɪt/ *v.* (**-tt-**) **1** leave out; not insert or include. **2** leave undone. **3** (foll. by verbal noun or *to* + infin.) fail or neglect. [Latin *omitto omiss-*]

omni- *comb. form* all. [Latin *omnis* all]

omnibus /'ɒmnɪbəs/ *–n.* **1** *formal* bus. **2** volume containing several literary works previously published separately. *–adj.* **1** serving several purposes at once. **2** comprising several items. [Latin, = for all]

omnipotent /ɒm'nɪpət(ə)nt/ *adj.* having great or absolute power. □ **omnipotence** *n.* [Latin: related to POTENT]

omnipresent /,ɒmnɪ'prez(ə)nt/ *adj.* present everywhere. □ **omnipresence** *n.* [Latin: related to PRESENT[1]]

omniscient /ɒm'nɪsɪənt/ *adj.* knowing everything or much. □ **omniscience** *n.* [Latin *scio* know]

omnivorous /ɒm'nɪvərəs/ *adj.* **1** feeding on both plant and animal material. **2** reading, observing, etc. everything that comes one's way. □ **omnivore** /'ɒmnɪ,vɔ:(r)/ *n.* **omnivorousness** *n.* [Latin *voro* devour]

on *–prep.* **1** (so as to be) supported by, attached to, covering, or enclosing (*sat on a chair*; *stuck on the wall*; *rings on her fingers*; *leaned on his elbow*). **2** carried with; about the person of (*have you a pen on you?*). **3** (of time) exactly at; during (*on 29 May*; *on the hour*; *on schedule*; *closed on Tuesday*). **4** immediately after or before (*I saw them on my return*). **5** as a result of (*on further examination*). **6** (so as to be) having membership etc. of or residence at or in (*is on the board of directors*; *lives on the continent*). **7** supported, succoured, or fuelled by (*lives on a grant*; *lives on sandwiches*; *runs on diesel*). **8** close to; just by (*house on the sea*; *lives on the main road*). **9** in the direction of; against. **10** so as to threaten; touching or striking (*advanced on Rome*; *pulled a knife on me*; *a punch on the nose*). **11** having as an axis or pivot (*turned on his heels*). **12** having as a basis or motive (*works on a ratchet*; *arrested on suspicion*). **13** having as a standard, confirmation, or guarantee (*had it on good authority*; *did it on purpose*; *I promise on my word*). **14** concerning or about (*writes on frogs*). **15** using or engaged with (*is on the pill*; *here on business*). **16** so as to affect (*walked out on her*). **17** at the expense of (*the drinks are on me*; *the joke is on him*). **18** added to (*disaster on disaster*; *ten pence on a pint of beer*). **19** in a specified manner or state (*on the cheap*; *on the run*). *–adv.* **1** (so as to be) covering or in contact (*put your boots on*). **2** in the appropriate direction; towards something (*look on*). **3** further forward; in an advanced position or state (*time is getting on*). **4** with continued movement or action (*play on*). **5** in operation or activity (*light is on*; *chase was on*). **6** due to take place as planned (*is the party still on?*). **7** *colloq.* **a** willing to participate or approve, make a bet, etc. (*you're on*). **b** practicable or acceptable (*that's just not on*). **8** being shown or performed (*a good film on tonight*). **9** on stage. **10** on duty. **11** forward (*head on*). *–adj. Cricket* designating the part of the field on the striker's side and in front of the wicket. *–n. Cricket* the on side. □ **be on about** *colloq.* discuss, esp. tediously. **be on at** *colloq.* nag or grumble at. **be on to** *colloq.* realize the significance or intentions of. **on and off** intermittently; now and then. **on and on** continually; at tedious length. **on time** punctual, punctually. **on to** to a position on. [Old English]

onanism /'əʊnə,nɪz(ə)m/ *n. literary* masturbation. [*Onan*, biblical person]

ONC *abbr.* Ordinary National Certificate.

once /wʌns/ *–adv.* **1** on one occasion only. **2** at some point or period in the past. **3** ever or at all (*if you once forget it*). **4** multiplied by one. *–conj.* as soon as. *–n.* one time or occasion (*just the once*). □ **all at once 1** suddenly. **2** all together. **at once 1** immediately. **2** simultaneously. **for once** on this (or that) occasion, even if at no other. **once again** (or **more**) another time. **once and for all** (or **once for all**) in a final manner, esp. after much hesitation. **once** (or **every once**) **in a while** from time to time. **once or twice** a few times. **once upon a time** at some unspecified time in the past. [originally genitive of ONE]

once-over *n. colloq.* rapid inspection.

oncogene /'ɒŋkə,dʒi:n/ *n.* gene which can transform a cell into a cancer cell. [Greek *ogkos* mass]

oncology /ɒŋ'kɒlədʒɪ/ *n.* the study of tumours. [Greek *ogkos* mass]

oncoming *adj.* approaching from the front.

OND *abbr.* Ordinary National Diploma.

one /wʌn/ *–adj.* **1** single and integral in number. **2** (with a noun implied) a single person or thing of the kind expressed or implied (*one of the best*; *a nasty one*). **3** particular but undefined, esp. as contrasted with another (*that is one view*; *one night last week*). **4** only such (*the one man who can do it*). **5** forming a unity (*one and undivided*). **6** identical; the same (*of one opinion*). *–n.* **1 a** lowest cardinal number. **b** thing numbered with it. **2** unity; a unit (*one is half of two*; *came in ones and twos*). **3** single thing, person, or example (often referring to a noun previously expressed or implied: *the big dog and the small one*). **4** *colloq.* drink (*a quick one*; *have one on me*). **5** story or joke (*the one about the parrot*). *–pron.* **1** person of a specified kind (*loved ones*; *like one possessed*). **2** any person, as representing people in general (*one is bound to lose in the end*). **3** I, me. □ **all one** (often foll. by *to*) a matter of indifference. **at one** in agreement. **one and all** everyone. **one by one** singly, successively. **one day 1** on an unspecified day. **2** at some unspecified future date. **one or two** *colloq.* a few. [Old English]

■ **Usage** The use of the pronoun *one* to mean 'I' or 'me' (e.g. *one would like to help*) is often regarded as an affectation.

one another *pron.* each the other or others (as a formula of reciprocity: *love one another*).

one-armed bandit *n. colloq.* fruit machine with a long handle.

one-horse *attrib. adj.* **1** using a single horse. **2** *colloq.* small, poorly equipped.

one-horse race *n.* contest in which one competitor is far superior to all the others.

one-liner *n.* short joke or remark in a play, comedy routine, etc.

one-man *attrib. adj.* involving or operated by only one man.

oneness *n.* **1** singleness. **2** uniqueness. **3** agreement. **4** sameness.

one-night stand *n.* **1** single performance of a play etc. in a place. **2** *colloq.* sexual liaison lasting only one night.

one-off *–attrib. adj.* made or done as the only one; not repeated. *–n.* one-off occurrence, achievement, etc.

onerous /'əʊnərəs/ *adj.* burdensome. □ **onerousness** *n.* [Latin: related to ONUS]

oneself *pron.* reflexive and emphatic form of *one* (*kill oneself*; *do it oneself*). □ **be oneself** act in one's normal unconstrained manner.

one-sided *adj.* unfair, partial. □ **one-sidedly** *adv.* **one-sidedness** *n.*

one's money's-worth *n.* good value for one's money.

one-time *attrib. adj.* former.

one-to-one *adj.* & *adv.* **1** involving or between only two people. **2** with one member of one group corresponding to one of another.

one-track mind *n.* mind preoccupied with one subject.

one-up *adj. colloq.* having a particular advantage. □ **one-upmanship** *n.*

one-way *adj.* allowing movement, travel, etc., in one direction only.

ongoing *adj.* **1** continuing. **2** in progress.

onion /ˈʌnjən/ *n.* vegetable with an edible bulb of a pungent smell and flavour. □ **oniony** *adj.* [Latin *unio -onis*]

online *Computing* –*adj.* directly connected, so that a computer immediately receives an input from or sends an output to a peripheral process etc.; carried out while so connected or under direct computer control. –*adv.* with the processing of data carried out simultaneously with its production; while connected to a computer; under direct computer control.

onlooker *n.* spectator. □ **onlooking** *adj.*

only /ˈəʊnlɪ/ –*adv.* **1** solely, merely, exclusively; and no one or nothing more besides (*needed six only*; *is only a child*). **2** no longer ago than (*saw them only yesterday*). **3** not until (*arrives only on Tuesday*). **4** with no better result than (*hurried home only to find her gone*). –*attrib. adj.* **1** existing alone of its or their kind (*their only son*). **2** best or alone worth considering (*the only place to eat*). –*conj. colloq.* except that; but (*I would go, only I feel ill*). [Old English: related to ONE]

■ **Usage** In informal English *only* is usually placed between the subject and verb regardless of what it refers to (e.g. *I only want to talk to you*); in more formal English it is often placed more exactly, esp. to avoid ambiguity (e.g. *I want to talk only to you*). In speech, intonation usually serves to clarify the sense.

only too *adv.* extremely.

o.n.o. *abbr.* or near offer.

onomatopoeia /ˌɒnəˌmætəˈpiːə/ *n.* formation of a word from a sound associated with what is named (e.g. *cuckoo*, *sizzle*). □ **onomatopoeic** *adj.* [Greek *onoma* name, *poieō* make]

onrush *n.* onward rush.

onscreen *adj. & adv.* within the range of a film camera etc; when being filmed.

onset *n.* **1** attack. **2** impetuous beginning.

onshore *adj.* **1** on the shore. **2** (of the wind) blowing landwards from the sea.

onside *adj.* (of a player in a field game) not offside.

onslaught /ˈɒnslɔːt/ *n.* fierce attack. [Dutch: related to ON, *slag* blow]

on-street *adj.* (esp. of parking) along a street.

onto *prep.* = *on to.*

■ **Usage** The form *onto* is still not fully accepted in the way that *into* is, although it is in wide use. It is however useful in distinguishing sense as between *we drove on to the beach* (i.e. in that direction) and *we drove onto the beach* (i.e. in contact with it).

ontology /ɒnˈtɒlədʒɪ/ *n.* branch of metaphysics dealing with the nature of being. □ **ontological** /-təˈlɒdʒɪk(ə)l/ *adj.* **ontologically** /-təˈlɒdʒɪkəlɪ/ *adv.* **ontologist** *n.* [Greek *ont-* being]

onus /ˈəʊnəs/ *n.* (*pl.* **onuses**) burden, duty, responsibility. [Latin]

onward /ˈɒnwəd/ –*adv.* (also **onwards**) **1** forward, advancing. **2** into the future (*from 1985 onwards*). –*adj.* forward, advancing.

onyx /ˈɒnɪks/ *n.* semiprecious variety of agate with coloured layers. [Greek *onux*]

oodles /ˈuːd(ə)lz/ *n.pl. colloq.* very great amount. [origin unknown]

ooh /uː/ *int.* expressing surprise, delight, pain, etc. [natural exclamation]

oolite /ˈəʊəˌlaɪt/ *n.* granular limestone. □ **oolitic** /-ˈlɪtɪk/ *adj.* [Greek *ōion* egg]

oompah /ˈʊmpɑː/ *n. colloq.* rhythmical sound of deep brass instruments. [imitative]

oomph /ʊmf/ *n. slang* **1** energy, enthusiasm. **2** attractiveness, esp. sex appeal. [origin uncertain]

oops /ʊps/ *int. colloq.* on making an obvious mistake. [natural exclamation]

ooze[1] –*v.* (**-zing**) **1** trickle or leak slowly out. **2** (of a substance) exude fluid. **3** (often foll. by *with*) exude (a feeling) freely (*oozed* (*with*) *charm*). –*n.* sluggish flow. □ **oozy** *adj.* [Old English]

ooze[2] *n.* wet mud. □ **oozy** *adj.* [Old English]

op *n. colloq.* operation. [abbreviation]

op. *abbr.* opus.

op- see OB-.

opacity /əʊˈpæsɪtɪ/ *n.* opaqueness. [Latin: related to OPAQUE]

opal /ˈəʊp(ə)l/ *n.* semiprecious stone usu. of a milky or bluish colour and sometimes showing changing colours. [Latin]

opalescent /ˌəʊpəˈles(ə)nt/ *adj.* iridescent. □ **opalescence** *n.*

opaline /ˈəʊpəˌlaɪn/ *adj.* opal-like, opalescent.

opaque /əʊˈpeɪk/ *adj.* (**opaquer**, **opaquest**) **1** not transmitting light. **2** impenetrable to sight. **3** unintelligible.

4 unintelligent, stupid. □ **opaquely** *adv.* **opaqueness** *n.* [Latin *opacus* shaded]

op art *n. colloq.* = OPTICAL ART. [abbreviation]

op. cit. *abbr.* in the work already quoted. [Latin *opere citato*]

OPEC /ˈəʊpek/ *abbr.* Organization of Petroleum Exporting Countries.

open /ˈəʊpən/ *–adj.* **1** not closed, locked, or blocked up; allowing access. **2** unenclosed, unconfined, unobstructed (*the open road*; *open views*). **3 a** uncovered, bare, exposed (*open drain*; *open wound*). **b** (of a goal etc.) unprotected, undefended. **4** undisguised, public, manifest (*open hostilities*). **5** expanded, unfolded, or spread out (*had the map open on the table*). **6** (of a fabric) not close; with gaps. **7 a** frank and communicative. **b** open-minded. **8 a** accessible to visitors or customers; ready for business. **b** (of a meeting) admitting all, not restricted to members etc. **9** (of a race, competition, scholarship, etc.) unrestricted as to who may compete. **10** (foll. by *to*) **a** willing to receive (*is open to offers*). **b** (of a choice, offer, or opportunity) available (*three courses open to us*). **c** vulnerable to, allowing of (*open to abuse*; *open to doubt*). **11** (of a return ticket) not restricted as to the day of travel. *–v.* **1** make or become open or more open. **2** (foll. by *into*, *on to*, etc.) (of a door, room, etc.) give access as specified (*opened on to a patio*). **3 a** start, establish, or set going (a business, activity, etc.) (*opened a new shop*; *opened fire*). **b** start (*conference opens today*). **4** (often foll. by *with*) start; begin speaking, writing, etc. (*show opens with a song*; *he opened with a joke*). **5** ceremonially declare (a building etc.) in use. *–n.* **1** (prec. by *the*) **a** open space, country, or air. **b** public notice; general attention (esp. *into the open*). **2** open championship or competition etc. □ **open a person's eyes** enlighten a person. **open out 1** unfold. **2** develop, expand. **3** become communicative. **open up 1** unlock (premises). **2** make accessible. **3** reveal; bring to notice. **4** accelerate. **5** begin shooting or sounding. □ **openness** *n.* [Old English]

open air *n.* outdoors. □ **open-air** *attrib. adj.*

open-and-shut *adj.* straightforward.

open book *n.* person who is easily understood.

opencast *adj.* (of a mine or mining) with removal of the surface layers and working from above, not from shafts.

Open College *n.* college offering training and vocational courses mainly by correspondence.

open day *n.* day when the public may visit a place normally closed to them.

open-door *attrib. adj.* open, accessible.

open-ended *adj.* having no predetermined limit.

opener *n.* **1** device for opening tins, bottles, etc. **2** *colloq.* first item on a programme etc.

open-handed *adj.* generous.

open-hearted *adj.* frank and kindly.

open-heart surgery *n.* surgery with the heart exposed and the blood made to bypass it.

open house *n.* hospitality for all visitors.

opening *–n.* **1** aperture or gap. **2** opportunity. **3** beginning; initial part. *–attrib. adj.* initial, first (*opening remarks*).

opening-time *n.* time at which public houses may legally open for custom.

open letter *n.* letter of protest etc. addressed to an individual and published in a newspaper etc.

openly *adv.* **1** frankly. **2** publicly.

open-minded *adj.* accessible to new ideas; unprejudiced.

open-mouthed *adj.* aghast with surprise.

open-plan *adj.* (of a house, office, etc.) having large undivided rooms.

open prison *n.* prison with few restraints on prisoners' movements.

open question *n.* matter on which different views are legitimate.

open sandwich *n.* sandwich without a top slice of bread.

open sea *n.* expanse of sea away from land.

open secret *n.* supposed secret known to many.

open society *n.* society with freedom of belief.

Open University *n.* university teaching mainly by broadcasting and correspondence, and open to those without academic qualifications.

open verdict *n.* verdict affirming that a crime has been committed but not specifying the criminal or (in case of violent death) the cause.

openwork *n.* pattern with intervening spaces in metal, leather, lace, etc.

opera[1] /ˈɒpərə/ *n.* **1 a** drama set to music for singers and instrumentalists. **b** this as a genre. **2** opera-house. [Italian from Latin, = labour, work]

opera[2] *pl.* of OPUS.

operable /ˈɒpərəb(ə)l/ *adj.* **1** that can be operated. **2** suitable for treatment by surgical operation. [Latin: related to OPERATE]

opera-glasses *n.pl.* small binoculars for use at the opera or theatre.

opera-house *n.* theatre for operas.

operate /'ɒpə,reɪt/ *v.* (**-ting**) **1** work, control. **2** be in action; function. **3 a** perform a surgical operation. **b** conduct a military etc. action. **c** be active in business etc. **4** bring about. [Latin *operor* work: related to OPUS]

operatic /,ɒpə'rætɪk/ *adj.* of or like an opera or opera singer (*an operatic voice*). □ **operatically** *adv.*

operatics /,ɒpə'rætɪks/ *n.pl.* production and performance of operas.

operating system *n.* basic software that enables the running of a computer program.

operating theatre *n.* room for surgical operations.

operation /,ɒpə'reɪʃ(ə)n/ *n.* **1** action, scope, or method of working or operating. **2** active process. **3** piece of work, esp. one in a series (*begin operations*). **4** act of surgery on a patient. **5** military manoeuvre. **6** financial transaction. **7** state of functioning (*in operation*). **8** subjection of a number etc. to a process affecting its value or form, e.g. multiplication. [Latin: related to OPERATE]

operational *adj.* **1** of or engaged in or used for operations. **2** able or ready to function. □ **operationally** *adv.*

operational research *n.* the application of scientific principles to business etc. management.

operations research *n.* = OPERATIONAL RESEARCH.

operative /'ɒpərətɪv/ *—adj.* **1** in operation; having effect. **2** having the main relevance (*'may' is the operative word*). **3** of or by surgery. *—n.* worker, esp. a skilled one. [Latin: related to OPERATE]

operator *n.* **1** person operating a machine etc., esp. connecting lines in a telephone exchange. **2** person engaging in business. **3** *colloq.* person acting in a specified way (*smooth operator*). **4** symbol or function denoting an operation in mathematics, computing, etc.

operculum /ə'pɜːkjʊləm/ *n.* (*pl.* **-cula**) **1** fish's gill-cover. **2** any of various other parts covering or closing an aperture in an animal or plant. [Latin *operio* cover (v.)]

operetta /,ɒpə'retə/ *n.* **1** light opera. **2** one-act or short opera. [Italian, diminutive of OPERA[1]]

ophidian /əʊ'fɪdɪən/ *—n.* member of a suborder of reptiles including snakes. *—adj.* **1** of this order. **2** snakelike. [Greek *ophis* snake]

ophthalmia /ɒf'θælmɪə/ *n.* inflammation of the eye. [Greek *ophthalmos* eye]

ophthalmic /ɒf'θælmɪk/ *adj.* of or relating to the eye and its diseases.

ophthalmic optician *n.* optician qualified to prescribe as well as dispense spectacles etc.

ophthalmology /,ɒfθæl'mɒlədʒɪ/ *n.* the study of the eye. □ **ophthalmologist** *n.*

ophthalmoscope /ɒf'θælmə,skəʊp/ *n.* instrument for examining the eye.

opiate /'əʊpɪət/ *—adj.* **1** containing opium. **2** narcotic, soporific. *—n.* **1** drug containing opium, usu. to ease pain or induce sleep. **2** soothing influence. [Latin: related to OPIUM]

opine /əʊ'paɪn/ *v.* (**-ning**) (often foll. by *that*) *literary* hold or express as an opinion. [Latin *opinor* believe]

opinion /ə'pɪnjən/ *n.* **1** unproven belief. **2** view held as probable. **3** what one thinks about something. **4** piece of professional advice (*a second opinion*). **5** estimation (*low opinion of*). [Latin: related to OPINE]

opinionated /ə'pɪnjə,neɪtɪd/ *adj.* dogmatic in one's opinions.

opinion poll *n.* assessment of public opinion by questioning a representative sample.

opium /'əʊpɪəm/ *n.* drug made from the juice of a certain poppy, used esp. as an analgesic and narcotic. [Latin from Greek *opion*]

opossum /ə'pɒsəm/ *n.* **1** tree-living American marsupial. **2** *Austral.* & *NZ* = POSSUM 2. [Virginian Indian]

opp. *abbr.* opposite.

opponent /ə'pəʊnənt/ *n.* person who opposes. [Latin *oppono opposit-* set against]

opportune /'ɒpə,tjuːn/ *adj.* **1** well-chosen or especially favourable (*opportune moment*). **2** (of an action or event) well-timed. [Latin *opportunus* (of the wind) driving towards the PORT[1]]

opportunism /,ɒpə'tjuːnɪz(ə)m, 'ɒp-/ *n.* adaptation of one's policy or judgement to circumstances or opportunity, esp. regardless of principle. □ **opportunist** *n.* **opportunistic** /-'nɪstɪk/ *adj.* **opportunistically** /-'nɪstɪkəlɪ/ *adv.*

opportunity /,ɒpə'tjuːnɪtɪ/ *n.* (*pl.* **-ies**) favourable chance or opening offered by circumstances.

opposable /ə'pəʊzəb(ə)l/ *adj. Zool.* (of the thumb in primates) capable of facing and touching the other digits on the same hand.

oppose /ə'pəʊz/ *v.* (**-sing**) **1** set oneself against; resist; argue or compete against. **2** (foll. by *to*) place in opposition or contrast. □ **as opposed to** in contrast with. □ **opposer** *n.* [Latin: related to OPPONENT]

opposite /ˈɒpəzɪt/ *–adj.* **1** facing, on the other side (*opposite page*; *the house opposite*). **2** (often foll. by *to*, *from*) contrary; diametrically different (*opposite opinion*). *–n.* opposite thing, person, or term. *–adv.* facing, on the other side (*lives opposite*). *–prep.* **1** facing (*sat opposite me*). **2** in a complementary role to (another actor etc.).

opposite number *n.* person holding an equivalent position in another group etc.

opposite sex *n.* (prec. by *the*) either sex in relation to the other.

opposition /ˌɒpəˈzɪʃ(ə)n/ *n.* **1** resistance, antagonism. **2** being hostile or in conflict or disagreement. **3** contrast, antithesis. **4 a** group or party of opponents or competitors. **b (the Opposition)** chief parliamentary party opposed to that in office. **5** act of placing opposite. **6** diametrically opposite position of two celestial bodies. [Latin: related to POSITION]

oppress /əˈpres/ *v.* **1** keep in subservience. **2** govern or treat cruelly. **3** weigh down (with cares or unhappiness). □ **oppression** *n.* **oppressor** *n.* [Latin: related to PRESS[1]]

oppressive *adj.* **1** oppressing. **2** (of weather) close and sultry. □ **oppressively** *adv.* **oppressiveness** *n.*

opprobrious /əˈprəʊbrɪəs/ *adj.* (of language) very scornful; abusive.

opprobrium /əˈprəʊbrɪəm/ *n.* **1** disgrace. **2** cause of this. [Latin, = infamy, reproach]

oppugn /əˈpjuːn/ *v. literary* controvert, call in question. [Latin *oppugno* fight against]

opt *v.* (usu. foll. by *for*) make a choice, decide. □ **opt out** (often foll. by *of*) choose not to participate (in). [Latin *opto* choose, wish]

optative /ˈɒptətɪv, ɒpˈteɪtɪv/ *Gram.* *–adj.* (esp. of a mood in Greek) expressing a wish. *–n.* optative mood or form. [Latin: related to OPT]

optic /ˈɒptɪk/ *adj.* of the eye or sight (*optic nerve*). [Greek *optos* seen]

optical *adj.* **1** of sight; visual. **2** of or according to optics. **3** aiding sight. □ **optically** *adv.*

optical art *n.* art using contrasting colours to create the illusion of movement.

optical disc see DISC.

optical fibre *n.* thin glass fibre through which light can be transmitted to carry signals.

optical illusion *n.* **1** image which deceives the eye. **2** mental misapprehension caused by this.

optician /ɒpˈtɪʃ(ə)n/ *n.* **1** maker, seller, or prescriber of spectacles and contact lenses etc. **2** person trained in the detection and correction of poor eyesight. [medieval Latin: related to OPTIC]

optics *n.pl.* (treated as *sing.*) science of light and vision.

optimal /ˈɒptɪm(ə)l/ *adj.* best or most favourable. [Latin *optimus* best]

optimism /ˈɒptɪˌmɪz(ə)m/ *n.* **1** inclination to hopefulness and confidence. **2** *Philos.* belief that this world is as good as it could be or that good must ultimately prevail over evil. □ **optimist** *n.* **optimistic** /-ˈmɪstɪk/ *adj.* **optimistically** /-ˈmɪstɪkəlɪ/ *adv.* [Latin *optimus* best]

optimize /ˈɒptɪˌmaɪz/ *v.* (also **-ise**) (**-zing** or **-sing**) make the best or most effective use of. □ **optimization** /-ˈzeɪʃ(ə)n/ *n.*

optimum /ˈɒptɪməm/ *–n.* (*pl.* **optima**) **1** most favourable conditions (for growth etc.). **2** best practical solution. *–adj.* = OPTIMAL. [Latin, neuter of *optimus* best]

option /ˈɒpʃ(ə)n/ *n.* **1 a** choosing; choice. **b** thing that is or may be chosen. **2** liberty to choose. **3** right to buy or sell at a specified price within a set time. □ **keep** (or **leave**) **one's options open** not commit oneself. [Latin: related to OPT]

optional *adj.* not obligatory. □ **optionally** *adv.*

optional extra *n.* item costing extra if one chooses to have it.

opulent /ˈɒpjʊlənt/ *adj.* **1** wealthy. **2** luxurious. **3** abundant. □ **opulence** *n.* [Latin *opes* wealth]

opus /ˈəʊpəs/ *n.* (*pl.* **opuses** or **opera** /ˈɒpərə/) **1** musical composition numbered as one of a composer's works (*Beethoven, opus 15*). **2** any artistic work (cf. MAGNUM OPUS). [Latin, = work]

or[1] *conj.* **1** introducing an alternative (*white or black*; *take it or leave it*; *whether or not*). **2** introducing an alternative name (*the lapwing or peewit*). **3** introducing an afterthought (*came in laughing – or was it crying?*). **4** = *or else* 1 (*run or you'll be late*). □ **or else 1** otherwise (*run, or else you will be late*). **2** *colloq.* expressing a warning or threat (*be good or else*). [Old English]

or[2] *n. Heraldry* gold. [Latin *aurum* gold]

-or *suffix* forming nouns denoting esp. an agent (*actor*; *escalator*) or condition (*error*; *horror*). [Latin]

oracle /ˈɒrək(ə)l/ *n.* **1 a** place at which divine advice or prophecy was sought in classical antiquity. **b** response given. **c** prophet or prophetess at an oracle. **2** person or thing regarded as a source of wisdom etc. **3 (Oracle)** *propr.* teletext service provided by Independent Television. □ **oracular** /ɒ'rækjʊlə(r)/ *adj.* [Latin *oraculum* from *oro* speak]

oral /'ɔ:r(ə)l/ *–adj.* **1** by word of mouth; spoken; not written (*oral examination*). **2** done or taken by the mouth (*oral sex*; *oral contraceptive*). *–n. colloq.* spoken examination. □ **orally** *adv.* [Latin *os oris* mouth]

orange /'ɒrɪndʒ/ *–n.* **1 a** roundish reddish-yellow juicy citrus fruit. **b** tree bearing this. **2** its colour. *–adj.* orange-coloured. [Arabic *nāranj*]

orangeade /,ɒrɪndʒ'eɪd/ *n.* orange-flavoured, usu. fizzy, drink.

Orangeman *n.* member of a political society formed in 1795 to support Protestantism in Ireland. [William of *Orange*]

orangery /'ɒrɪndʒərɪ/ *n.* (*pl.* **-ies**) place, esp. a building, where orange-trees are cultivated.

orang-utan /ɔ:,ræŋu:'tæn/ *n.* (also **orang-outang** /-u:'tæŋ/) large reddish-haired long-armed anthropoid ape of the E. Indies. [Malay, = wild man]

oration /ɔ:'reɪʃ(ə)n/ *n.* formal or ceremonial speech. [Latin *oratio* discourse, prayer, from *oro* speak, pray]

orator /'ɒrətə(r)/ *n.* **1** person making a formal speech. **2** eloquent public speaker. [Latin: related to ORATION]

oratorio /,ɒrə'tɔ:rɪəʊ/ *n.* (*pl.* **-s**) semi-dramatic work for orchestra and voices, esp. on a sacred theme. [Church Latin]

oratory /'ɒrətərɪ/ *n.* (*pl.* **-ies**) **1** art of or skill in public speaking. **2** small private chapel. □ **oratorical** /-'tɒrɪk(ə)l/ *adj.* [French and Latin *oro* speak, pray]

orb *n.* **1** globe surmounted by a cross as part of coronation regalia. **2** sphere, globe. **3** *poet.* celestial body. **4** *poet.* eye. [Latin *orbis* ring]

orbicular /ɔ:'bɪkjʊlə(r)/ *adj. formal* circular or spherical. [Latin *orbiculus* diminutive of *orbis* ring]

orbit /'ɔ:bɪt/ *–n.* **1 a** curved course of a planet, satellite, etc. **b** one complete passage around a body. **2** range or sphere of action. **3** eye socket. *–v.* (**-t-**) **1** move in orbit round. **2** put into orbit. □ **orbiter** *n.* [Latin *orbitus* circular]

orbital *adj.* **1** of an orbit or orbits. **2** (of a road) passing round the outside of a town.

orca /'ɔ:kə/ *n.* any of various cetaceans, esp. the killer whale. [Latin]

Orcadian /ɔ:'keɪdɪən/ *–adj.* of Orkney. *–n.* native of Orkney. [Latin *Orcades* Orkney Islands]

orch. *abbr.* **1** orchestrated by. **2** orchestra.

orchard /'ɔ:tʃəd/ *n.* piece of enclosed land with fruit-trees. [Latin *hortus* garden]

orchestra /'ɔ:kɪstrə/ *n.* **1** large group of instrumentalists combining strings, woodwinds, brass, and percussion. **2** (in full **orchestra pit**) part of a theatre etc. where the orchestra plays, usu. in front of the stage and on a lower level. □ **orchestral** /-'kestr(ə)l/ *adj.* [Greek, = area for the chorus in drama]

orchestrate /'ɔ:kɪ,streɪt/ *v.* (**-ting**) **1** arrange or compose for orchestral performance. **2** arrange (elements) to achieve a desired result. □ **orchestration** /-'streɪʃ(ə)n/ *n.*

orchid /'ɔ:kɪd/ *n.* any of various plants with brilliant flowers. [Greek *orkhis*, originally = testicle]

ordain /ɔ:'deɪn/ *v.* **1** confer holy orders on. **2** decree, order. [Latin *ordino*: related to ORDER]

ordeal /ɔ:'di:l/ *n.* **1** painful or horrific experience; severe trial. **2** *hist.* test of an accused person by subjection to severe pain, with survival taken as proof of innocence. [Old English]

order *–n.* **1 a** condition in which every part, unit, etc. is in its right place; tidiness. **b** specified sequence, succession, etc. (*alphabetical order*; *the order of events*). **2** authoritative command, direction, instruction, etc. **3** state of obedience to law, authority, etc. **4 a** direction to supply or pay something. **b** goods etc. to be supplied. **5** social class; its members (*the lower orders*). **6** kind; sort (*talents of a high order*). **7** constitution or nature of the world, society, etc. (*the moral order*; *the order of things*). **8** taxonomic rank below a class and above a family. **9** religious fraternity with a common rule of life. **10** grade of the Christian ministry. **11** any of the five classical styles of architecture (*Doric order*). **12 a** company of persons distinguished by a particular honour (*Order of the Garter*). **b** insignia worn by its members. **13** *Eccl.* the stated form of divine service (*the order of confirmation*). **14** system of rules or procedure (at meetings etc.) (*point of order*). *–v.* **1** command; bid; prescribe. **2** command or direct (a person) to a specified destination (*ordered them home*). **3** direct a waiter, tradesman, etc. to supply (*ordered dinner*; *ordered a new suit*). **4** (often as **ordered** *adj.*) put in order; regulate (*an ordered life*). **5** (of God, fate, etc.) ordain. □ **in** (or **out of**) **order 1** in the correct (or incorrect) sequence or position. **2** fit (or not fit) for use. **3** according (or not according) to the rules at a meeting etc. **in order that** with the intention; so that. **in order to** with the purpose of doing; with a view to. **of** (or **in**) **the order of** approximately. **on order** ordered but not yet received. **order about** command officiously. **to order** as specified by the customer. [Latin *ordo ordin-* row, command, etc.]

Order in Council *n.* executive order, often approved by Parliament but not debated.

orderly *–adj.* **1** methodically arranged or inclined, tidy. **2** well-behaved. *–n.* (*pl.* **-ies**) **1** male cleaner in a hospital. **2** soldier who carries orders for an officer etc. □ **orderliness** *n.*

orderly room *n.* room in a barracks used for company business.

order of the day *n.* **1** prevailing state of things. **2** principal action, procedure, or programme.

order-paper *n.* written or printed order of the day's proceedings, esp. in Parliament.

ordinal /'ɔ:dɪn(ə)l/ *n.* (in full **ordinal number**) number defining position in a series, e.g. 'first', 'second', 'third', etc. [Latin: related to ORDER]

ordinance /'ɔ:dɪnəns/ *n.* **1** decree. **2** religious rite. [Latin: related to ORDAIN]

ordinand /'ɔ:dɪnənd/ *n.* candidate for ordination. [Latin: related to ORDAIN]

ordinary /'ɔ:dɪnərɪ/ *–adj.* **1** normal, usual. **2** commonplace, unexceptional. *–n.* (*pl.* **-ies**) *RC Ch.* **1** parts of a service that do not vary from day to day. **2** rule or book laying down the order of service. □ **in the ordinary way** in normal circumstances. **out of the ordinary** unusual. □ **ordinarily** *adv.* **ordinariness** *n.* [Latin: related to ORDER]

ordinary level *n. hist.* lowest level of the GCE examination.

ordinary seaman *n.* sailor of the lowest rank.

ordinate /'ɔ:dɪnət/ *n. Math.* coordinate measured usu. vertically. [Latin: related to ORDAIN]

ordination /,ɔ:dɪ'neɪʃ(ə)n/ *n.* conferring of holy orders, ordaining.

ordnance /'ɔ:dnəns/ *n.* **1** artillery; military supplies. **2** government service dealing with these. [contraction of ORDINANCE]

Ordnance Survey *n.* official survey of the UK producing detailed maps.

Ordovician /,ɔ:də'vɪʃɪən/ *–adj.* of the second period in the Palaeozoic era. *–n.* this period. [Latin *Ordovices*, an ancient British tribe in N. Wales]

ordure /'ɔ:djʊə(r)/ *n.* dung. [Latin *horridus*: related to HORRID]

ore /ɔ:(r)/ *n.* solid rock or mineral from which metal or other valuable minerals may be extracted. [Old English]

oregano /,ɒrɪ'gɑ:nəʊ/ *n.* dried wild marjoram as seasoning. [Spanish, = ORIGAN]

organ /'ɔ:gən/ *n.* **1 a** musical instrument having pipes supplied with air from bellows and operated by keyboards and pedals. **b** instrument producing similar sounds electronically. **c** harmonium. **2 a** part of an animal or plant body serving a particular function (*vocal organs*; *digestive organs*). **b** esp. *joc.* penis. **3** medium of communication, esp. a newspaper representing a party or interest. [Greek *organon* tool]

organdie /'ɔ:gəndɪ/ *n.* fine translucent muslin, usu. stiffened. [French]

organ-grinder *n.* player of a barrel-organ.

organic /ɔ:'gænɪk/ *adj.* **1** of or affecting a bodily organ or organs. **2** (of a plant or animal) having organs or an organized physical structure. **3** produced without the use of artificial fertilizers, pesticides, etc. **4** (of a chemical compound etc.) containing carbon. **5 a** structural, inherent. **b** constitutional. **6** organized or systematic (*an organic whole*). □ **organically** *adv.* [Greek: related to ORGAN]

organic chemistry *n.* chemistry of carbon compounds.

organism /'ɔ:gə,nɪz(ə)m/ *n.* **1** individual plant or animal. **2** living being with interdependent parts. **3** system made up of interdependent parts. [French: related to ORGANIZE]

organist /'ɔ:gənɪst/ *n.* organ-player.

organization /,ɔ:gənaɪ'zeɪʃ(ə)n/ *n.* (also **-isation**) **1** organizing or being organized. **2** organized body, system, or society. □ **organizational** *adj.*

organize /'ɔ:gə,naɪz/ *v.* (also **-ise**) (**-zing** or **-sing**) **1 a** give an orderly structure to, systematize. **b** make arrangements for (a person or oneself). **2** initiate, arrange for. **3** (often *absol.*) **a** enlist (a person or group) in a trade union, political party, etc. **b** form (a trade union etc.). **4** (esp. as **organized** *adj.*) make organic; make into living tissue. □ **organizer** *n.* [Latin: related to ORGAN]

organ-loft *n.* gallery for an organ.

organza /ɔ:'gænzə/ *n.* thin stiff transparent silk or synthetic dress fabric. [origin uncertain]

orgasm /'ɔ:gæz(ə)m/ *–n.* climax of sexual excitement. *–v.* have a sexual orgasm. □ **orgasmic** /-'gæzmɪk/ *adj.* [Greek, = excitement]

orgy /'ɔ:dʒɪ/ *n.* (*pl.* **-ies**) **1** wild party with indiscriminate sexual activity. **2** excessive indulgence in an activity. □ **orgiastic** /-'æstɪk/ *adj.* [Greek *orgia* pl.]

oriel /'ɔ:rɪəl/ *n.* (in full **oriel window**) projecting window of an upper storey. [French]

orient /'ɔ:rɪənt/ *–n.* (**the Orient**) countries east of the Mediterranean, esp. E. Asia. *–v.* **1 a** place or determine the position of with the aid of a compass; find the bearings of. **b** (often foll. by *towards*) direct. **2** place (a building etc.) to face east. **3** turn eastward or in a specified direction. □ **orient oneself** determine how one stands in relation to one's surroundings. [Latin *oriens -entis* rising, sunrise, east]

oriental /,ɔ:rɪ'ent(ə)l/ (often **Oriental**) *–adj.* of the East, esp. E. Asia; of the Orient. *–n.* native of the Orient.

orientate /'ɔ:rɪən,teɪt/ *v.* (**-ting**) = ORIENT *v.* [apparently from ORIENT]

orientation /,ɔ:rɪən'teɪʃ(ə)n/ *n.* **1** orienting or being oriented. **2 a** relative position. **b** person's attitude or adjustment in relation to circumstances. **3** introduction to a subject or situation; briefing. □ **orientational** *adj.*

orienteering /,ɔ:rɪən'tɪərɪŋ/ *n.* competitive sport in which runners cross open country with a map, compass, etc. [Swedish]

orifice /'ɒrɪfɪs/ *n.* opening, esp. the mouth of a cavity. [Latin *os or-* mouth, *facio* make]

origami /,ɒrɪ'gɑ:mɪ/ *n.* art of folding paper into decorative shapes. [Japanese]

origan /'ɒrɪgən/ *n.* (also **origanum** /ə'rɪgənəm/) wild marjoram. [Latin from Greek]

origin /'ɒrɪdʒɪn/ *n.* **1** starting-point; source. **2** (often in *pl.*) ancestry, parentage. **3** *Math.* point from which coordinates are measured. [Latin *origo origin-* from *orior* rise]

original /ə'rɪdʒɪn(ə)l/ *–adj.* **1** existing from the beginning; earliest; innate. **2** inventive; creative; not derivative or imitative. **3** not copied or translated; by the artist etc. himself (*in the original Greek; has an original Rembrandt*). *–n.* original model, pattern, picture, etc. from which another is copied or translated. □ **originality** /-'nælɪtɪ/ *n.* **originally** *adv.*

original sin *n.* innate human sinfulness held to be a result of the Fall.

originate /ə'rɪdʒɪ,neɪt/ *v.* (**-ting**) **1** cause to begin; initiate. **2** have as an origin; begin. □ **origination** /-'neɪʃ(ə)n/ *n.* **originator** *n.*

oriole /'ɔ:rɪəʊl/ *n.* (in full **golden oriole**) bird with black and yellow plumage in the male. [Latin *aurum* gold]

ormolu /'ɔ:mə,lu:/ *n.* **1** (often *attrib.*) gilded bronze; gold-coloured alloy. **2** articles made of or decorated with ormolu. [French *or moulu* powdered gold]

ornament /'ɔ:nəmənt/ *–n.* **1 a** thing used to adorn or decorate. **b** quality or person bringing honour or distinction. **2** decoration, esp. on a building (*tower rich in ornament*). **3** musical embellishment. *–v.* /'ɔ:nə,ment/ adorn; beautify. □ **ornamental** /-'ment(ə)l/ *adj.* **ornamentation** /-men'teɪʃ(ə)n/ *n.* [Latin *orno* adorn]

ornate /ɔ:'neɪt/ *adj.* **1** elaborately adorned. **2** (of literary style) convoluted; flowery. □ **ornately** *adv.* **ornateness** *n.* [Latin: related to ORNAMENT]

ornithology /,ɔ:nɪ'θɒlədʒɪ/ *n.* the study of birds. □ **ornithological** /-θə'lɒdʒɪk(ə)l/ *adj.* **ornithologist** *n.* [Greek *ornis ornith-* bird]

orotund /'ɒrə,tʌnd/ *adj.* **1** (of the voice) full, round; imposing. **2** (of writing, style, etc.) pompous; pretentious. [Latin *ore rotundo* with rounded mouth]

orphan /'ɔ:f(ə)n/ *–n.* child whose parents are dead. *–v.* bereave (a child) of its parents. [Latin from Greek, = bereaved]

orphanage *n.* home for orphans.

orrery /'ɒrərɪ/ *n.* (*pl.* **-ies**) clockwork model of the solar system. [Earl of *Orrery*]

orris /'ɒrɪs/ *n.* **1** a kind of iris. **2** = ORRIS ROOT. [alteration of IRIS]

orris root *n.* fragrant iris root used in perfumery etc.

ortho- *comb. form* **1** straight. **2** right, correct. [Greek *orthos* straight]

orthodontics /,ɔ:θə'dɒntɪks/ *n.pl.* (treated as *sing.*) correction of irregularities in the teeth and jaws. □ **orthodontic** *adj.* **orthodontist** *n.* [Greek *odous odont-* tooth]

orthodox /'ɔ:θə,dɒks/ *adj.* **1** holding usual or accepted opinions, esp. on religion, morals, etc. **2** generally approved, conventional (*orthodox medicine*). **3** (also **Orthodox**) (of Judaism) strictly traditional. □ **orthodoxy** *n.* [Greek *doxa* opinion]

Orthodox Church *n.* Eastern Church with the Patriarch of Constantinople as its head, and including the national Churches of Russia, Romania, Greece, etc.

orthography /ɔ:'θɒgrəfɪ/ *n.* (*pl.* **-ies**) spelling (esp. with reference to its correctness). □ **orthographic** /-'græfɪk/ *adj.* [Greek *orthographia*]

orthopaedics /,ɔ:θə'pi:dɪks/ *n.pl.* (treated as *sing.*) (*US* **-pedics**) branch of medicine dealing with the correction of diseased, deformed, or injured bones or muscles. □ **orthopaedic** *adj.* **orthopaedist** *n.* [Greek *pais paid-* child]

ortolan /'ɔ:tələn/ *n.* European bunting, eaten as a delicacy. [Latin *hortus* garden]

-ory *suffix* **1** forming nouns denoting a place (*dormitory*; *refectory*). **2** forming adjectives and nouns relating to or involving a verbal action (*accessory*; *compulsory*). [Latin *-orius, -orium*]

OS *abbr.* **1** old style. **2** ordinary seaman. **3** Ordnance Survey. **4** outsize.

Os *symb.* osmium.

Oscar /'ɒskə(r)/ *n.* any of the statuettes awarded by the US Academy of Motion Picture Arts and Sciences for excellence in film acting, directing, etc. [man's name]

oscillate /'ɒsɪ,leɪt/ *v.* **(-ting) 1** (cause to) swing to and fro. **2** vacillate; vary between extremes. **3** (of an electric current) undergo high-frequency alternations. □ **oscillation** /-'leɪʃ(ə)n/ *n.* **oscillator** *n.* [Latin *oscillo* swing]

oscillo- *comb. form* oscillation, esp. of an electric current.

oscilloscope /ə'sɪlə,skəʊp/ *n.* device for viewing oscillations by a display on the screen of a cathode-ray tube.

-ose *suffix* forming adjectives denoting possession of a quality (*grandiose*; *verbose*). [Latin *-osus*]

osier /'əʊzɪə(r)/ *n.* **1** willow used in basketwork. **2** shoot of this. [French]

-osis *suffix* denoting a process or condition (*apotheosis*; *metamorphosis*), esp. a pathological state (*neurosis*; *thrombosis*). [Latin or Greek]

-osity *suffix* forming nouns from adjectives in *-ose* and *-ous* (*verbosity*; *curiosity*). [Latin *-ositas*]

osmium /'ɒzmɪəm/ *n.* heavy hard bluish-white metallic element. [Greek *osmē* smell]

osmosis /ɒz'məʊsɪs/ *n.* **1** passage of a solvent through a semi-permeable membrane into a more concentrated solution. **2** process by which something is acquired by absorption. □ **osmotic** /-'mɒtɪk/ *adj.* [Greek *ōsmos* push]

osprey /'ɒspreɪ/ *n.* (*pl.* **-s**) large bird of prey feeding on fish. [Latin *ossifraga* from *os* bone, *frango* break]

osseous /'ɒsɪəs/ *adj.* **1** of bone. **2** bony. [Latin *os oss-* bone]

ossicle /'ɒsɪk(ə)l/ *n.* small bone or piece of bonelike substance. [Latin diminutive: related to OSSEOUS]

ossify /'ɒsɪ,faɪ/ *v.* **(-ies, -ied) 1** turn into bone; harden. **2** make or become rigid, callous, or unprogressive. □ **ossification** /-fɪ'keɪʃ(ə)n/ *n.* [Latin: related to OSSEOUS]

ostensible /ɒ'stensɪb(ə)l/ *adj.* concealing the real; professed. □ **ostensibly** *adv.* [Latin *ostendo ostens-* show]

ostensive /ɒ'stensɪv/ *adj.* directly showing.

ostentation /,ɒsten'teɪʃ(ə)n/ *n.* **1** pretentious display of wealth etc. **2** showing off. □ **ostentatious** *adj.* **ostentatiously** *adv.*

osteo- *comb. form* bone. [Greek *osteon*]

osteoarthritis /,ɒstɪəʊɑ:'θraɪtɪs/ *n.* degenerative disease of joint cartilage. □ **osteoarthritic** /-'θrɪtɪk/ *adj.*

osteopathy /,ɒstɪ'ɒpəθɪ/ *n.* treatment of disease through the manipulation of bones. □ **osteopath** /'ɒstɪə,pæθ/ *n.*

osteoporosis /,ɒstɪəʊpə'rəʊsɪs/ *n.* condition of brittle bones caused esp. by hormonal changes or deficiency of calcium or vitamin D.

ostler /'ɒslə(r)/ *n. hist.* stableman at an inn. [related to HOSTEL]

ostracize /'ɒstrə,saɪz/ *v.* (also **-ise**) (**-zing** or **-sing**) exclude from society; refuse to associate with. □ **ostracism** /-,sɪz(ə)m/ *n.* [Greek (*ostrakon* potsherd, on which a vote was recorded in ancient Athens to expel a powerful or unpopular citizen)]

ostrich /'ɒstrɪtʃ/ *n.* **1** large African swift-running flightless bird. **2** person who refuses to acknowledge an awkward truth. [Latin *avis* bird, *struthio* (from Greek) ostrich]

OT *abbr.* Old Testament.

other /'ʌðə(r)/ *—adj.* **1** not the same as one or some already mentioned or implied; separate in identity or distinct in kind (*other people*; *use other means*). **2 a** further; additional (*a few other examples*). **b** second of two (*open your other eye*). **3** (prec. by *the*) only remaining (*must be in the other pocket*; *where are the other two?*). **4** (foll. by *than*) apart from. *—n.* or *pron.* other person or thing. (*some others have come*; *give me one other*; *where are the others?*). □ **other than 1** except (*never speaks to me other than to insult me*; *has no friends other than me*). **2** differently; not (*cannot do other than laugh*; *never appears other than happy*). [Old English]

■ **Usage** In sense 2 of *other than*, *otherwise* is standard except in less formal use.

other day *n.* (also **other night**) (prec. by *the*) a few days (or nights) ago.

other half *n. colloq.* one's wife or husband.

otherwise *—adv.* **1** or else; in different circumstances (*hurry, otherwise we'll be late*). **2** in other respects (*is otherwise very suitable*). **3** in a different way (*could not have acted otherwise*). **4** as an alternative (*otherwise known as Jack*). *–adj.* (*predic.*) different (*the matter is quite otherwise*). [Old English: related to WISE[2]]

■ **Usage** See note at *other*.

other woman *n.* (prec. by *the*) married man's mistress.

other-worldly *adj.* **1** of another world. **2** dreamily distracted from mundane life.

otiose /ˈəʊtɪəʊs/ *adj.* serving no practical purpose; not required. [Latin *otium* leisure]

OTT *abbr. colloq.* over-the-top.

otter /ˈɒtə(r)/ *n.* **1** aquatic fish-eating mammal with webbed feet and thick brown fur. **2** its fur. [Old English]

Ottoman /ˈɒtəmən/ *–adj.* **1** of the dynasty of Osman (or Othman) I or the empire ruled by his descendants. **2** Turkish. *–n.* (*pl.* **-s**) **1** Turk of the Ottoman period. **2** (**ottoman**) upholstered seat without back or arms, sometimes a box with a padded top. [French from Arabic]

OU *abbr.* **1** Open University. **2** Oxford University.

oubliette /ˌuːblɪˈet/ *n.* secret dungeon with a trapdoor entrance. [French *oublier* forget]

ouch *int.* expressing sharp or sudden pain. [imitative]

ought /ɔːt/ *v.aux.* (as present and past, the only form now in use) **1** expressing duty or rightness (*we ought to be thankful; it ought to have been done long ago*). **2** advisability (*you ought to see a dentist*). **3** probability (*it ought to rain soon*). □ **ought not** negative form of *ought* (*he ought not to have stolen it*). [Old English, past of OWE]

oughtn't /ˈɔːt(ə)nt/ *contr.* ought not.

Ouija /ˈwiːdʒə/ *n.* (in full **Ouija board**) *propr.* board marked with letters or signs and used with a movable pointer to try to obtain messages at a seance. [French *oui*, German *ja*, yes]

ounce *n.* **1** unit of weight, $^1/_{16}$ lb or approx. 28 g. **2** very small quantity. [Latin *uncia* twelfth part of a pound or a foot]

our *poss. pron.* **1** of or belonging to us or society (*our children's future*). **2** *colloq.* indicating a relative, friend, etc. of the speaker (*our Barry works there; our friend here*). [Old English]

Our Father *n.* prayer beginning with these words (Matt. 6:9-13).

Our Lady *n.* Virgin Mary.

Our Lord *n.* Christ.

ours *poss. pron.* the one or ones belonging to or associated with us (*it is ours; ours are best; a friend of ours*).

ourself *pron. archaic* = MYSELF as used by a sovereign etc.

ourselves *pron.* **1 a** *emphat. form* of WE or US (*we did it ourselves*). **b** *refl. form* of US (*we are pleased with ourselves*). **2** in our normal state of body or mind (*not quite ourselves today*). □ **be ourselves** see ONESELF. **by ourselves** see *by oneself*.

-ous *suffix* **1** forming adjectives meaning 'abounding in, characterized by, of the nature of' (*envious; glorious; mountainous; poisonous*). **2** *Chem.* denoting a state of lower valence than *-ic* (*ferrous; sulphurous*). [Anglo-French *-ous*, from Latin *asus*]

ousel var. of OUZEL.

oust *v.* drive out or expel, esp. by seizing the place of. [Latin *obsto* oppose]

out *–adv.* **1** away from or not in or at a place etc. (*keep him out; get out; tide is out*). **2** indicating: **a** dispersal away from a centre etc. (*share out*). **b** coming or bringing into the open (*call out; will look it out for you*). **c** need for attentiveness (*watch out; listen out*). **3** not in one's house, office, etc. (*tell them I'm out*). **4** to or at an end; completely (*tired out; die out; fight it out; my luck was out; typed it out*). **5** (of a fire, candle, etc.) not burning. **6** in error (*was 3% out*). **7** *colloq.* unconscious (*is out cold*). **8** (of a limb etc.) dislocated (*put his arm out*). **9** (of a political party etc.) not in office. **10** (of a jury) considering its verdict. **11** (of workers) on strike. **12** (of a secret) revealed. **13** (of a flower) open. **14** (of a book, record, etc.) published, on sale. **15** (of a star) visible after dark. **16** no longer in fashion (*turn-ups are out*). **17** (of a batsman etc.) dismissed from batting. **18** not worth considering (*that idea is out*). **19** (prec. by *superl.*) *colloq.* known to exist (*the best game out*). **20** (of a mark etc.) removed (*washed the stain out*). *–prep.* out of (*looked out the window*). *–n.* way of escape. *–v.* come or go out; emerge (*murder will out*). □ **out for** intent on, determined to get. **out of 1** from within. **2** not within. **3** from among. **4** beyond the range of (*out of reach*). **5** so as to be without, lacking (*was swindled out of his money; out of sugar*). **6** from (*get money out of him*). **7** because of (*asked out of curiosity*). **8** by the use of (*what did you make it out of?*). **out of bounds** see BOUND[2]. **out of date** see DATE[1]. **out of order** see ORDER. **out of pocket** see POCKET. **out of the question** see QUESTION. **out of sorts** see SORT. **out of this world** see WORLD. **out of the way** see WAY. **out to** determined to. [Old English]

■ **Usage** The use of *out* as a preposition, e.g. *he walked out the room*, is non-standard. *Out of* should be used.

out- *prefix* in senses **1** so as to surpass or exceed (*outdo*). **2** external, separate (*outline*). **3** out of; away from; outward (*outgrowth*).

outage /'aʊtɪdʒ/ *n.* period during which a power-supply etc. is not operating.

out and about *adj.* active outdoors (esp. after an illness etc.).

out and out –*adj.* thorough; complete. –*adv.* thoroughly.

outback *n.* remote inland areas of Australia.

outbalance /aʊt'bæləns/ *v.* (**-cing**) outweigh.

outbid /aʊt'bɪd/ *v.* (**-bidding**; *past* and *past part.* **-bid**) bid higher than.

outboard motor *n.* portable engine attached to the outside of a boat.

outbreak *n.* sudden eruption of anger, war, disease, fire, etc.

outbuilding *n.* shed, barn, etc. detached from a main building.

outburst *n.* **1** verbal explosion of anger etc. **2** bursting out (*outburst of steam*).

outcast –*n.* person rejected by family or society. –*adj.* rejected; homeless.

outclass /aʊt'klɑːs/ *v.* surpass in quality.

outcome *n.* result.

outcrop *n.* **1 a** emergence of a stratum etc. at a surface. **b** stratum etc. emerging. **2** noticeable manifestation.

outcry *n.* (*pl.* **-ies**) strong public protest.

outdated /aʊt'deɪtɪd/ *adj.* out of date; obsolete.

outdistance /aʊt'dɪst(ə)ns/ *v.* (**-cing**) leave (a competitor) behind completely.

outdo /aʊt'duː/ *v.* (**-doing**; *3rd sing. present* **-does**; *past* **-did**; *past part.* **-done**) exceed, excel, surpass.

outdoor *attrib. adj.* **1** done, existing, or used out of doors. **2** fond of the open air (*an outdoor type*).

outdoors /aʊt'dɔːz/ –*adv.* in or into the open air. –*n.* the open air.

outer *adj.* **1** outside; external (*pierced the outer layer*). **2** farther from the centre or the inside. □ **outermost** *adj.*

outer space *n.* universe beyond the earth's atmosphere.

outface /aʊt'feɪs/ *v.* (**-cing**) disconcert by staring or by a display of confidence.

outfall *n.* outlet of a river, drain, etc.

outfield *n.* outer part of a cricket or baseball pitch. □ **outfielder** *n.*

outfit *n.* **1** set of clothes or equipment. **2** *colloq.* group of people regarded as an organization.

outfitter *n.* supplier of clothing.

outflank /aʊt'flæŋk/ *v.* **1** extend beyond the flank of (an enemy). **2** outmanoeuvre, outwit.

outflow *n.* **1** outward flow. **2** amount that flows out.

outfox /aʊt'fɒks/ *v.* outwit.

outgoing –*adj.* **1** friendly. **2** retiring from office. **3** going out. –*n.* (in *pl.*) expenditure.

outgrow /aʊt'grəʊ/ *v.* (*past* **-grew**; *past part.* **-grown**) **1** grow too big for. **2** leave behind (a childish habit etc.). **3** grow faster or taller than.

outgrowth *n.* **1** offshoot. **2** natural product or development.

outhouse *n.* small building adjoining or apart from a house.

outing *n.* pleasure trip, excursion.

outlandish /aʊt'lændɪʃ/ *adj.* bizarre, strange. □ **outlandishly** *adv.* **outlandishness** *n.* [Old English, from *outland* foreign country]

outlast /aʊt'lɑːst/ *v.* last longer than.

outlaw –*n.* **1** fugitive from the law. **2** *hist.* person deprived of the protection of the law. –*v.* **1** declare (a person) an outlaw. **2** make illegal; proscribe.

outlay *n.* expenditure.

outlet *n.* **1** means of exit or escape. **2** means of expressing feelings. **3 a** market for goods. **b** shop (*retail outlet*).

outline –*n.* **1** rough draft. **2** summary. **3** sketch consisting of only contour lines. **4** (in *sing.* or *pl.*) **a** lines enclosing or indicating an object. **b** contour. **c** external boundary. **5** (in *pl.*) main features or principles. –*v.* (**-ning**) **1** draw or describe in outline. **2** mark the outline of.

outlive /aʊt'lɪv/ *v.* (**-ving**) **1** live longer than (a person). **2** live beyond (a period or date).

outlook *n.* **1** prospect, view. **2** mental attitude.

outlying *adj.* far from a centre; remote.

outmanoeuvre /ˌaʊtmə'nuːvə(r)/ *v.* (**-ring**) (*US* **-maneuver**) secure an advantage over by skilful manoeuvring.

outmatch /aʊt'mætʃ/ *v.* be more than a match for.

outmoded /aʊt'məʊdɪd/ *adj.* **1** outdated. **2** out of fashion.

outnumber /aʊt'nʌmbə(r)/ *v.* exceed in number.

out of doors *adj.* & *adv.* in or into the open air.

out of it *predic. adj.* **1** (of a person) not included; forlorn. **2** *colloq.* unconscious, dazed.

outpace /aʊt'peɪs/ *v.* (**-cing**) **1** go faster than. **2** outdo in a contest.

outpatient *n.* non-resident hospital patient.

outplacement *n.* assistance in finding a new job after redundancy.

outpost *n.* **1** detachment posted at a distance from an army. **2** distant branch or settlement (*outpost of empire*).

outpouring *n.* (usu. in *pl.*) copious expression of emotion.

output *–n.* **1** amount produced (by a machine, worker, etc.). **2** electrical power etc. delivered by an apparatus. **3** printout, results, etc. from a computer. **4** place where energy, information, etc. leaves a system. *–v.* (**-tt-**; *past* and *past part.* **-put** or **-putted**) (of a computer) supply (results etc.).

outrage *–n.* **1** extreme violation of others' rights, sentiments, etc. **2** gross offence or indignity. **3** fierce resentment. *–v.* (**-ging**) **1** subject to outrage. **2** commit an outrage against. **3** shock and anger. [French *outrer* exceed, from Latin *ultra* beyond]

outrageous /aʊt'reɪdʒəs/ *adj.* **1** immoderate. **2** shocking. **3** immoral, offensive. □ **outrageously** *adv.*

outrank /aʊt'ræŋk/ *v.* be superior in rank to.

outré /'u:treɪ/ *adj.* eccentric, unconventional. [French, past part. of *outrer*: see OUTRAGE]

outrider *n.* mounted guard or motor cyclist riding ahead of a procession etc.

outrigger *n.* **1** spar or framework projecting over the side of a ship, racing boat, or canoe to give stability. **2** boat fitted with this.

outright *–adv.* **1** altogether, entirely. **2** not gradually. **3** without reservation, openly. *–adj.* **1** downright, complete. **2** undisputed (*outright winner*).

outrun /aʊt'rʌn/ *v.* (**-nn-**; *past* **-ran**; *past part.* **-run**) **1** run faster or farther than. **2** go beyond (a point or limit).

outsell /aʊt'sel/ *v.* (*past* and *past part.* **-sold**) **1** sell more than. **2** be sold in greater quantities than.

outset *n.* □ **at** (or **from**) **the outset** from the beginning.

outshine /aʊt'ʃaɪn/ *v.* (**-ning**; *past* and *past part.* **-shone**) **1** shine brighter than. **2** surpass in excellence etc.

outside *–n.* /aʊt'saɪd, 'aʊtsaɪd/ **1** external side or surface; outer parts. **2** external appearance; outward aspect. **3** position on the outer side (*gate opens from the outside*). *–adj.* /'aʊtsaɪd/ **1 a** of, on, or nearer the outside; outer. **b** not in the main building (*outside toilet*). **2** not belonging to a particular group or organization (*outside help*). **3** (of a chance etc.) remote; very unlikely. **4** (of an estimate etc.) the greatest or highest possible (*the outside price*). **5** (of a player in football etc.) positioned nearest to the edge of the field (*outside left*). *–adv.* /aʊt'saɪd/ **1** on or to the outside. **2** in or to the open air. **3** not within, enclosed, or included. **4** *slang* not in prison. *–prep.* /aʊt'saɪd/ **1** not in; to or at the exterior of. **2** external to, not included in, beyond the limits of. □ **at the outside** (of an estimate etc.) at the most. **from the outside** from an objective or impartial standpoint.

outside broadcast *n.* one not made in a studio.

outside interest *n.* hobby etc. not connected with one's work.

outsider /aʊt'saɪdə(r)/ *n.* **1** non-member of some group, organization, profession, etc. **2** competitor thought to have little chance.

outside world *n.* society outside the confines of an institution etc.

outsize *adj.* unusually large.

outskirts *n.pl.* outer area of a town etc.

outsmart /aʊt'smɑ:t/ *v.* outwit, be cleverer than.

outspoken /aʊt'spəʊkən/ *adj.* saying openly what one thinks; frank. □ **outspokenly** *adv.* **outspokenness** *n.*

outspread /aʊt'spred/ *–adj.* spread out; expanded. *–v.* spread out; expand.

outstanding /aʊt'stændɪŋ/ *adj.* **1** conspicuous because of excellence. **2 a** (of a debt) not yet settled. **b** still to be dealt with (*work outstanding*). □ **outstandingly** *adv.*

outstation *n.* remote branch or outpost.

outstay /aʊt'steɪ/ *v.* stay longer than (one's welcome etc.).

outstretched /aʊt'stretʃt/ *adj.* stretched out.

outstrip /aʊt'strɪp/ *v.* (**-pp-**) **1** go faster than. **2** surpass, esp. competitively.

out-take *n.* film or tape sequence rejected in editing.

out-tray *n.* tray for outgoing documents etc.

outvote /aʊt'vəʊt/ *v.* (**-ting**) defeat by a majority of votes.

outward /'aʊtwəd/ *–adj.* **1** situated on or directed towards the outside. **2** going out. **3** bodily, external, apparent. *–adv.* (also **outwards**) in an outward direction; towards the outside. □ **outwardly** *adv.* [Old English: related to OUT-, -WARD]

outward bound *adj.* going away from home.

outwardness *n.* external existence; objectivity.

outwards var. of OUTWARD *adv.*

outweigh /aʊt'weɪ/ *v.* exceed in weight, value, importance, or influence.

outwit /aʊt'wɪt/ *v.* (**-tt-**) be too clever for; overcome by greater ingenuity.

outwork *n.* **1** advanced or detached part of a fortification. **2** work done off the premises of the firm etc. which supplies it. □ **outworker** *n.* (in sense 2).

ouzel /'u:z(ə)l/ *n.* (also **ousel**) **1** (in full **ring ouzel**) white-breasted thrush. **2** (in full **water ouzel**) diving bird; dipper. [Old English, = blackbird]

ouzo /'u:zəʊ/ *n.* (*pl.* **-s**) Greek aniseed-flavoured spirit. [Greek]

ova *pl.* of OVUM.

oval /'əʊv(ə)l/ *–adj.* **1** egg-shaped, ellipsoidal. **2** having the outline of an egg, elliptical. *–n.* **1** egg-shaped or elliptical closed curve. **2** thing with an oval outline. [Latin: related to OVUM]

ovary /'əʊvərɪ/ *n.* (*pl.* **-ies**) **1** each of the female reproductive organs in which ova are produced. **2** hollow base of the carpel of a flower. □ **ovarian** /ə'veərɪən/ *adj.*

ovation /əʊ'veɪʃ(ə)n/ *n.* enthusiastic reception, esp. applause. [Latin *ovo* exult]

oven /'ʌv(ə)n/ *n.* enclosed compartment for heating or cooking food etc. [Old English]

ovenproof *adj.* suitable for use in an oven; heat-resistant.

oven-ready *adj.* (of food) prepared before sale for immediate cooking in the oven.

ovenware *n.* dishes for cooking food in the oven.

over *–adv.* expressing movement, position, or state above or beyond something stated or implied: **1** outward and downward from a brink or from any erect position (*knocked me over*). **2** so as to cover or touch a whole surface (*paint it over*). **3** so as to produce a fold or reverse position (*bend it over; turn it over*). **4 a** across a street or other space (*cross over; came over from France*). **b** for a visit etc. (*invited them over*). **5** with transference or change from one hand, part, etc., to another (*went over to the enemy; swapped them over*). **6** with motion above something; so as to pass across something (*climb over; fly over; boil over*). **7** from beginning to end with repetition or detailed consideration (*think it over; did it six times over*). **8** in excess; in addition, besides (*left over*). **9** for or until a later time (*hold it over*). **10** at an end; settled; completely finished (*crisis is over; it's over between us; get it over with*). **11** (in full **over to you**) (as *int.*) (in radio conversations etc.) it is your turn to speak. **12** umpire's call to change ends in cricket. *–prep.* **1** above, in, or to a position higher than. **2** out and down from; down from the edge of (*fell over the cliff*). **3** so as to cover (*hat over his eyes*). **4** above and across; so as to clear; on or to the other side of (*flew over Scotland; bridge over the Avon; look over the wall*). **5** concerning; while occupied with (*laughed over it; fell asleep over a book*). **6 a** in superiority of; superior to; in charge of (*victory over them; reign over two kingdoms*). **b** in preference to. **7 a** throughout (*travelled over most of Africa*). **b** so as to deal with completely (*went over the plans*). **8 a** for or through the duration of (*stay over Monday night; over the years*). **b** during the course of (*did it over the weekend*). **9** beyond; more than (*bids of over £50; is he over 18?*). **10** transmitted by (*heard it over the radio*). **11** in comparison with (*gained 20% over last year*). **12** recovered from (*am over my cold; got over it in time*). *–n.* **1** sequence of six balls in cricket bowled from one end of the pitch. **2** play resulting from this. *–adj.* (see also OVER-). **1** upper, outer. **2** superior. **3** extra. □ **over again** once again, again from the beginning. **over against** in contrast with. **over all** taken as a whole. **over and above** in addition to; not to mention. **over and over** repeatedly. **over one's head** see HEAD. **over the hill** see HILL. **over the moon** see MOON. **over the way** (in a street etc.) facing or opposite. [Old English]

over- *prefix* **1** excessively. **2** upper, outer. **3** = OVER in various senses (*overshadow*). **4** completely (*overawe; overjoyed*).

over-abundance /,əʊvərə'bʌnd(ə)ns/ *n.* excessive quantity. □ **over-abundant** *adj.*

overact /,əʊvər'ækt/ *v.* act (a role) in an exaggerated manner.

over-active /,əʊvər'æktɪv/ *adj.* excessively active.

overall *–attrib. adj.* /'əʊvər,ɔ:l/ **1** total, inclusive of all (*overall cost*). **2** taking everything into account, general (*overall improvement*). *–adv.* /,əʊvər'ɔ:l/ **1** including everything (*cost £50 overall*). **2** on the whole, generally (*did well overall*). *–n.* /'əʊvər,ɔ:l/ **1** protective outer garment. **2** (in *pl.*) protective outer trousers or suit.

overambitious /,əʊvəræm'bɪʃəs/ *adj.* excessively ambitious.

over-anxious /,əʊvər'æŋkʃəs/ *adj.* excessively anxious.

overarm *adj.* & *adv.* with the hand above the shoulder (*bowl overarm; overarm service*).

overate *past* of OVEREAT.

overawe /,əʊvər'ɔ:/ *v.* (**-wing**) overcome with awe.

overbalance /,əʊvə'bæləns/ *v.* (**-cing**) **1** lose balance and fall. **2** cause to do this.

overbear /ˌəʊvəˈbeə(r)/ *v.* (*past* **-bore**; *past part.* **-borne**) **1** (as **overbearing** *adj.*) **a** domineering, bullying. **b** overpowering. **2** bear down by weight, force, or emotion. **3** repress by power or authority.

overbid –*v.* /ˌəʊvəˈbɪd/ (**-dd-**; *past* and *past part.* **-bid**) make a higher bid than. –*n.* /ˈəʊvəbɪd/ bid that is higher than another, or higher than is justified.

overblown /ˌəʊvəˈbləʊn/ *adj.* **1** inflated or pretentious. **2** (of a flower) past its prime.

overboard *adv.* from a ship into the water (*fall overboard*). □ **go overboard** *colloq.* **1** be highly enthusiastic. **2** behave immoderately.

overbook /ˌəʊvəˈbʊk/ *v.* (also *absol.*) make too many bookings for (an aircraft, hotel, etc.).

overbore *past* of OVERBEAR.

overborne *past part.* of OVERBEAR.

overburden /ˌəʊvəˈbɜːd(ə)n/ *v.* burden (a person, thing, etc.) to excess.

overcame *past* of OVERCOME.

overcast /ˈəʊvəˌkɑːst/ *adj.* **1** (of the sky) covered with cloud. **2** (in sewing) edged with stitching to prevent fraying.

overcautious /ˌəʊvəˈkɔːʃəs/ *adj.* excessively cautious.

overcharge /ˌəʊvəˈtʃɑːdʒ/ *v.* (**-ging**) **1** charge too high a price to (a person). **2** put too much charge into (a battery, gun, etc.). **3** put excessive detail into (a description, picture, etc.).

overcoat *n.* warm outdoor coat.

overcome /ˌəʊvəˈkʌm/ *v.* (**-ming**; *past* **-came**; *past part.* **-come**) **1** prevail over, master, be victorious. **2** (usu. as **overcome** *adj.*) **a** make faint (*overcome by smoke*). **b** (usu. foll. by *with, by*) make weak or helpless (*overcome with grief*).

overcompensate /ˌəʊvəˈkɒmpenˌseɪt/ *v.* (**-ting**) **1** (usu. foll. by *for*) compensate excessively. **2** strive exaggeratedly to make amends etc.

overconfident /ˌəʊvəˈkɒnfɪd(ə)nt/ *adj.* excessively confident.

overcook /ˌəʊvəˈkʊk/ *v.* cook too much or for too long.

overcrowd /ˌəʊvəˈkraʊd/ *v.* (usu. as **overcrowded** *adj.*) fill beyond what is usual or comfortable. □ **overcrowding** *n.*

overdevelop /ˌəʊvədɪˈveləp/ *v.* (**-p-**) **1** develop too much. **2** *Photog.* treat with developer for too long.

overdo /ˌəʊvəˈduː/ *v.* (**-doing**; *3rd sing. present* **-does**; *past* **-did**; *past part.* **-done**) **1** carry to excess, go too far. **2** (esp. as **overdone** *adj.*) overcook. □ **overdo it** (or **things**) *colloq.* exhaust oneself.

overdose *n.* excessive dose of a drug etc.

overdraft *n.* **1** overdrawing of a bank account. **2** amount by which an account is overdrawn.

overdraw /ˌəʊvəˈdrɔː/ *v.* (*past* **-drew**; *past part.* **-drawn**) **1** draw more from (a bank account) than the amount credited. **2** (as **overdrawn** *adj.*) having overdrawn one's account.

overdress /ˌəʊvəˈdres/ *v.* dress with too much formality.

overdrive *n.* **1** mechanism in a vehicle providing a gear above top gear for economy at high speeds. **2** state of high activity.

overdue /ˌəʊvəˈdjuː/ *adj.* past the due time for payment, arrival, return, etc.

overeager /ˌəʊvərˈiːgə(r)/ *adj.* excessively eager.

overeat /ˌəʊvərˈiːt/ *v.* (*past* **-ate**; *past part.* **-eaten**) eat too much.

overemphasize /ˌəʊvərˈemfəˌsaɪz/ *v.* (also **-ise**) (**-zing** or **-sing**) give too much emphasis to.

overenthusiasm /ˌəʊvərɪnˈθjuːzɪˌæz(ə)m/ *n.* excessive enthusiasm. □ **overenthusiastic** /-ˈæstɪk/ *adj.* **overenthusiastically** /-ˈæstɪkəlɪ/ *adv.*

overestimate –*v.* /ˌəʊvərˈestɪˌmeɪt/ (**-ting**) form too high an estimate of. –*n.* /ˌəʊvərˈestɪmət/ too high an estimate. □ **overestimation** /-ˈmeɪʃ(ə)n/ *n.*

overexcite /ˌəʊvərɪkˈsaɪt/ *v.* (**-ting**) excite excessively. □ **overexcitement** *n.*

overexert /ˌəʊvərɪgˈzɜːt/ *v.* exert too much. □ **overexertion** /-ɪgˈzɜːʃ(ə)n/ *n.*

overexpose /ˌəʊvərɪkˈspəʊz/ *v.* (**-sing**) **1** expose too much to the public. **2** expose (film) too long. □ **overexposure** *n.*

overfeed /ˌəʊvəˈfiːd/ *v.* (*past* and *past part.* **-fed**) feed excessively.

overfill /ˌəʊvəˈfɪl/ *v.* fill to excess or to overflowing.

overfish /ˌəʊvəˈfɪʃ/ *v.* deplete (a stream etc.) by too much fishing.

overflow –*v.* /ˌəʊvəˈfləʊ/ **1** flow over (the brim etc.). **2 a** (of a receptacle etc.) be so full that the contents overflow. **b** (of contents) overflow a container. **3** (of a crowd etc.) extend beyond the limits of (a room etc.). **4** flood (a surface or area). **5** (of kindness, a harvest, etc.) be very abundant. –*n.* /ˈəʊvəˌfləʊ/ **1** what overflows or is superfluous. **2** outlet for excess water etc.

overfly /ˌəʊvəˈflaɪ/ *v.* (**-flies**; *past* **-flew**; *past part.* **-flown**) fly over or beyond (a place or territory).

overfond /ˌəʊvəˈfɒnd/ *adj.* (often foll. by *of*) having too great an affection or liking for (*overfond of chocolate*; *overfond parent*).

overfull /ˌəʊvə'fʊl/ *adj.* filled excessively.
overground *adj.* **1** raised above the ground. **2** not underground.
overgrown /ˌəʊvə'grəʊn/ *adj.* **1** grown too big. **2** wild; covered with weeds etc. □ **overgrowth** *n.*
overhang –*v.* /ˌəʊvə'hæŋ/ (*past* and *past part.* **-hung**) project or hang over. –*n.* /'əʊvəˌhæŋ/ **1** overhanging. **2** overhanging part or amount.
overhaul –*v.* /ˌəʊvə'hɔːl/ **1** thoroughly examine the condition of and repair if necessary. **2** overtake. –*n.* /'əʊvəˌhɔːl/ thorough examination, with repairs if necessary.
overhead –*adv.* /ˌəʊvə'hed/ **1** above head height. **2** in the sky. –*adj.* /'əʊvəˌhed/ placed overhead. –*n.* /'əʊvəˌhed/ (in *pl.*) routine administrative and maintenance expenses of a business.
overhead projector *n.* projector for producing an enlarged image of a transparency.
overhear /ˌəʊvə'hɪə(r)/ *v.* (*past* and *past part.* **-heard**) (also *absol.*) hear unintentionally or as an eavesdropper.
overheat /ˌəʊvə'hiːt/ *v.* **1** make or become too hot. **2** cause inflation (in) by placing excessive pressure on resources at a time of expanding demand. **3** (as **overheated** *adj.*) overexcited.
overindulge /ˌəʊvərɪn'dʌldʒ/ *v.* (**-ging**) indulge to excess. □ **overindulgence** *n.* **overindulgent** *adj.*
overjoyed /ˌəʊvə'dʒɔɪd/ *adj.* filled with great joy.
overkill *n.* **1** excess of capacity to kill or destroy. **2** excess.
overland /'əʊvəˌlænd, -'lænd/ *adj.* & *adv.* **1** by land. **2** not by sea.
overlap –*v.* /ˌəʊvə'læp/ (**-pp-**) **1** (cause to) partly cover and extend beyond (*don't overlap them*). **2** (of two things) be placed so that one overlaps the other (*overlapping tiles*). **3** partly coincide. –*n.* /'əʊvəˌlæp/ **1** overlapping. **2** overlapping part or amount.
over-large /ˌəʊvə'lɑːdʒ/ *adj.* too large.
overlay –*v.* /ˌəʊvə'leɪ/ (*past* and *past part.* **-laid**) **1** lay over. **2** (foll. by *with*) cover (a thing) with (a coating etc.). –*n.* /'əʊvəˌleɪ/ thing laid over another.
overleaf /ˌəʊvə'liːf/ *adv.* on the other side of the leaf of a book.
overlie /ˌəʊvə'laɪ/ *v.* (**-lying**; *past* **-lay**; *past part.* **-lain**) **1** lie on top of. **2** smother (a child etc.) thus.
overload –*v.* /ˌəʊvə'ləʊd/ **1** load excessively (with baggage, work, etc.). **2** put too great a demand on (an electrical circuit etc.). –*n.* /'əʊvəˌləʊd/ excessive quantity or demand.
over-long /ˌəʊvə'lɒŋ/ *adj.* & *adv.* too long.
overlook /ˌəʊvə'lʊk/ *v.* **1** fail to notice; tolerate. **2** have a view of from above. **3** supervise.
overlord *n.* supreme lord.
overly *adv.* excessively; too.
overman /ˌəʊvə'mæn/ *v.* (**-nn-**) provide with too large a crew, staff, etc.
over-much /ˌəʊvə'mʌtʃ/ –*adv.* to too great an extent. –*adj.* excessive.
overnight /ˌəʊvə'naɪt/ –*adv.* **1** for a night. **2** during the night. **3** instantly, suddenly. –*adj.* **1** done or for use etc. overnight. **2** instant (*overnight success*).
over-particular /ˌəʊvəpə'tɪkjʊlə(r)/ *adj.* excessively particular or fussy.
overpass *n.* road or railway line that passes over another by means of a bridge.
overpay /ˌəʊvə'peɪ/ *v.* (*past* and *past part.* **-paid**) pay too highly or too much. □ **overpayment** *n.*
overplay /ˌəʊvə'pleɪ/ *v.* give undue importance to; overemphasize. □ **overplay one's hand** act on an unduly optimistic estimation of one's chances.
overpopulated /ˌəʊvə'pɒpjʊˌleɪtɪd/ *adj.* having too large a population. □ **overpopulation** /-'leɪʃ(ə)n/ *n.*
overpower /ˌəʊvə'paʊə(r)/ *v.* **1** subdue, conquer. **2** (esp. as **overpowering** *adj.*) be too intense or overwhelming for (*overpowering smell*). □ **overpoweringly** *adv.*
overprice /ˌəʊvə'praɪs/ *v.* (**-cing**) price too highly.
overprint –*v.* /ˌəʊvə'prɪnt/ print over (a surface already printed). –*n.* /'əʊvəprɪnt/ words etc. overprinted.
overproduce /ˌəʊvəprə'djuːs/ *v.* (**-cing**) **1** (often *absol.*) produce more of (a commodity) than is wanted. **2** produce (a play, recording, etc.) to an excessive degree. □ **overproduction** /-'dʌkʃ(ə)n/ *n.*
overprotective /ˌəʊvəprə'tektɪv/ *adj.* excessively protective.
overqualified /ˌəʊvə'kwɒlɪˌfaɪd/ *adj.* too highly qualified for a particular job etc.
overrate /ˌəʊvə'reɪt/ *v.* (**-ting**) **1** assess or value too highly. **2** (as **overrated** *adj.*) not as good as it is said to be.
overreach /ˌəʊvə'riːtʃ/ *v.* outwit, cheat. □ **overreach oneself** fail by attempting too much.
overreact /ˌəʊvərɪ'ækt/ *v.* respond more forcibly than is justified. □ **overreaction** *n.*
override –*v.* /ˌəʊvə'raɪd/ (**-ding**; *past* **-rode**; *past part.* **-ridden**) **1** (often as **overriding** *adj.*) have priority over

(*overriding consideration*). **2 a** intervene and make ineffective. **b** interrupt the action of (an automatic device), esp. to take manual control. *–n.* /ˈəʊvəˌraɪd/ **1** suspension of an automatic function. **2** device for this.

overrider *n.* each of a pair of projecting pieces on the bumper of a car.

overripe /ˌəʊvəˈraɪp/ *adj.* excessively ripe.

overrule /ˌəʊvəˈru:l/ *v.* (**-ling**) **1** set aside (a decision etc.) by superior authority. **2** reject a proposal of (a person) in this way.

overrun /ˌəʊvəˈrʌn/ *v.* (**-nn-**; *past* **-ran**; *past part.* **-run**) **1** swarm or spread over. **2** conquer (a territory) by force. **3** (usu. *absol.*) exceed (an allotted time).

overseas *–adv.* /ˌəʊvəˈsi:z/ across the sea; abroad. *–attrib. adj.* /ˈəʊvəˌsi:z/ of places across the sea; foreign.

oversee /ˌəʊvəˈsi:/ *v.* (**-sees**; *past* **-saw**; *past part.* **-seen**) officially supervise (workers etc.); superintend. □ **overseer** *n.*

over-sensitive /ˌəʊvəˈsensɪtɪv/ *adj.* excessively sensitive; easily hurt or quick to react. □ **over-sensitiveness** *n.* **over-sensitivity** /-ˈtɪvɪtɪ/ *n.*

oversew *v.* (*past part.* **-sewn** or **-sewed**) sew (two edges) with stitches passing over the join.

oversexed /ˌəʊvəˈsekst/ *adj.* having unusually strong sexual desires.

overshadow /ˌəʊvəˈʃædəʊ/ *v.* **1** appear much more prominent or important than. **2** cast into the shade.

overshoe *n.* outer protective shoe worn over an ordinary one.

overshoot /ˌəʊvəˈʃu:t/ *v.* (*past* and *past part.* **-shot**) **1** pass or send beyond (a target or limit). **2** fly beyond or taxi too far along (the runway) when landing or taking off. □ **overshoot the mark** go beyond what is intended or proper.

oversight *n.* **1** failure to do or notice something. **2** inadvertent mistake. **3** supervision.

oversimplify /ˌəʊvəˈsɪmplɪˌfaɪ/ *v.* (**-ies, -ied**) (also *absol.*) distort (a problem etc.) by stating it in too simple terms. □ **oversimplification** /-fɪˈkeɪʃ(ə)n/ *n.*

oversize *adj.* (also **-sized**) of greater than the usual size.

oversleep /ˌəʊvəˈsli:p/ *v.* (*past* and *past part.* **-slept**) sleep beyond the intended time of waking.

overspecialize /ˌəʊvəˈspeʃəˌlaɪz/ *v.* (also **-ise**) (**-zing** or **-sing**) concentrate too much on one aspect or area. □ **overspecialization** /-ˈzeɪʃ(ə)n/ *n.*

overspend /ˌəʊvəˈspend/ *v.* (*past* and *past part.* **-spent**) spend too much or beyond one's means.

overspill *n.* **1** what is spilt over or overflows. **2** surplus population moving to a new area.

overspread /ˌəʊvəˈspred/ *v.* (*past* and *past part.* **-spread**) **1** cover the surface of. **2** (as **overspread** *adj.*) (usu. foll. by *with*) covered.

overstate /ˌəʊvəˈsteɪt/ *v.* (**-ting**) **1** state too strongly. **2** exaggerate. □ **overstatement** *n.*

overstay /ˌəʊvəˈsteɪ/ *v.* stay longer than (one's welcome etc.).

oversteer *–n.* /ˈəʊvəˌstɪə(r)/ tendency of a vehicle to turn more sharply than was intended. *–v.* /ˌəʊvəˈstɪə(r)/ (of a vehicle) exhibit oversteer.

overstep /ˌəʊvəˈstep/ *v.* (**-pp-**) pass beyond (a permitted or acceptable limit). □ **overstep the mark** violate conventions of behaviour etc.

overstock /ˌəʊvəˈstɒk/ *v.* stock excessively.

overstrain /ˌəʊvəˈstreɪn/ *v.* strain too much.

overstretch /ˌəʊvəˈstretʃ/ *v.* **1** stretch too much. **2** (esp. as **overstretched** *adj.*) make excessive demands on (resources, a person, etc.).

overstrung *adj.* **1** /ˌəʊvəˈstrʌŋ/ (of a person, nerves, etc.) too highly strung. **2** /ˈəʊvəˌstrʌŋ/ (of a piano) with strings in sets crossing each other obliquely.

overstuffed /ˌəʊvəˈstʌft/ *adj.* **1** (of furniture) made soft and comfortable by thick upholstery. **2** stuffed too full.

oversubscribe /ˌəʊvəsəbˈskraɪb/ *v.* (**-bing**) (usu. as **oversubscribed** *adj.*) subscribe for more than the amount available of (shares, tickets, places, etc.).

overt /əʊˈvɜ:t/ *adj.* done openly; unconcealed. □ **overtly** *adv.* [French, past part. of *ouvrir* open]

overtake /ˌəʊvəˈteɪk/ *v.* (**-king**; *past* **-took**; *past part.* **-taken**) **1** (also *absol.*) catch up with and pass while travelling in the same direction. **2** (of misfortune etc.) come suddenly upon.

overtax /ˌəʊvəˈtæks/ *v.* **1** make excessive demands on. **2** tax too heavily.

over-the-top *adj. colloq.* excessive.

overthrow *–v.* /ˌəʊvəˈθrəʊ/ (*past* **-threw**; *past part.* **-thrown**) **1** remove forcibly from power. **2** conquer, overcome. *–n.* /ˈəʊvəˌθrəʊ/ defeat, downfall.

overtime *–n.* **1** time worked in addition to regular hours. **2** payment for this. *–adv.* in addition to regular hours.

overtone *n.* **1** *Mus.* any of the tones above the lowest in a harmonic series. **2** subtle extra quality or implication.

overture /ˈəʊvəˌtjʊə(r)/ *n.* **1** orchestral piece opening an opera etc. **2** composition in this style. **3** (usu. in *pl.*) **a** opening of negotiations. **b** formal proposal or offer. [French: related to OVERT]

overturn /ˌəʊvəˈtɜːn/ *v.* **1** (cause to) fall down or over. **2** reverse; overthrow.

overuse –*v.* /ˌəʊvəˈjuːz/ (**-sing**) use too much. –*n.* /ˌəʊvəˈjuːs/ excessive use.

overview *n.* general survey.

overweening /ˌəʊvəˈwiːnɪŋ/ *adj.* arrogant, presumptuous.

overweight –*adj.* /ˌəʊvəˈweɪt/ above an allowed or suitable weight. –*n.* /ˈəʊvəˌweɪt/ excess weight; preponderance.

overwhelm /ˌəʊvəˈwelm/ *v.* **1** overpower with emotion or a burden. **2** overcome by force of numbers. **3** bury or drown beneath a huge mass.

overwhelming *adj.* **1** too great to resist or overcome (*an overwhelming desire to laugh*). **2** by a great number (*the overwhelming majority*). □ **overwhelmingly** *adv.*

overwind /ˌəʊvəˈwaɪnd/ *v.* (*past* and *past part.* **-wound**) wind (a watch etc.) beyond the proper stopping point.

overwork /ˌəʊvəˈwɜːk/ –*v.* **1** (cause to) work too hard. **2** weary or exhaust with too much work. **3** (esp. as **overworked** *adj.*) make excessive use of (*an overworked phrase*). **4** (as **overworked** *adj.*) = OVERWROUGHT 2. –*n.* excessive work.

overwrought /ˌəʊvəˈrɔːt/ *adj.* **1** overexcited, nervous, distraught. **2** too elaborate.

ovi- *comb. form* egg, ovum. [from OVUM]

oviduct /ˈəʊvɪˌdʌkt/ *n.* tube through which an ovum passes from the ovary.

oviform /ˈəʊvɪˌfɔːm/ *adj.* egg-shaped.

ovine /ˈəʊvaɪn/ *adj.* of or like sheep. [Latin *ovis* sheep]

oviparous /əʊˈvɪpərəs/ *adj.* producing young from eggs hatching after leaving the body. [from OVUM, Latin *-parus* bearing]

ovoid /ˈəʊvɔɪd/ *adj.* (of a solid) egg-shaped. [related to OVUM]

ovulate /ˈɒvjʊˌleɪt/ *v.* (**-ting**) produce ova or ovules, or discharge them from the ovary. □ **ovulation** /-ˈleɪʃ(ə)n/ *n.* [related to OVUM]

ovule /ˈɒvjuːl/ *n.* structure that contains the germ cell in a female plant. [related to OVUM]

ovum /ˈəʊvəm/ *n.* (*pl.* **ova**) female egg-cell from which young develop after fertilization. [Latin, = egg]

ow *int.* expressing sudden pain. [natural exclamation]

owe /əʊ/ *v.* (**owing**) **1 a** be under obligation (to a person etc.) to pay or repay (money, gratitude, etc.). **b** (usu. foll. by *for*) be in debt. **2** have a duty to render (*owe allegiance*). **3** (usu. foll. by *to*) be indebted to a person or thing for (*we owe our success to the weather*). [Old English]

owing /ˈəʊɪŋ/ *predic. adj.* **1** owed; yet to be paid. **2** (foll. by *to*) **a** caused by. **b** (as *prep.*) because of.

■ **Usage** The use of *owing to* as a preposition meaning 'because of' is entirely acceptable (e.g. *couldn't come owing to the snow*), unlike this use of *due to*.

owl *n.* **1** nocturnal bird of prey with large eyes and a hooked beak. **2** solemn or wise-looking person. □ **owlish** *adj.* [Old English]

owlet *n.* small or young owl.

own /əʊn/ –*adj.* (prec. by possessive) **1 a** belonging to oneself or itself; not another's (*saw it with my own eyes*). **b** individual, peculiar, particular (*has its own charm*). **2** used to emphasize identity rather than possession (*cooks his own meals*). **3** (*absol.*) private property (*is it your own?*). –*v.* **1** have as property; possess. **2** admit as valid, true, etc. **3** acknowledge paternity, authorship, or possession of. □ **come into one's own 1** receive one's due. **2** achieve recognition. **get one's own back** get revenge. **hold one's own** maintain one's position. **of one's own** belonging to oneself. **on one's own 1** alone, independent. **2** independently, without help. **own up** (often foll. by *to*) confess frankly. □ **-owned** *adj.* (in *comb.*). [Old English]

own brand *n.* (often *attrib.*) goods manufactured specially for a retailer and bearing the retailer's name.

owner *n.* person who owns something. □ **ownership** *n.*

owner-occupier *n.* person who owns and occupies a house.

own goal *n.* **1** goal scored by mistake against the scorer's own side. **2** act etc. that has the unintended effect of harming one's own interests.

owt *n. colloq.* or *dial.* anything. [var. of AUGHT]

ox *n.* (*pl.* **oxen**) **1** large usu. horned ruminant used for draught, milk, and meat. **2** castrated male of a domesticated species of cattle. [Old English]

oxalic acid /ɒkˈsælɪk/ *n.* very poisonous and sour acid found in sorrel and rhubarb leaves. [Greek *oxalis* wood sorrel]

oxbow *n.* loop formed by a horseshoe bend in a river.

Oxbridge *n.* (also *attrib.*) Oxford and Cambridge universities regarded together, esp. in contrast to newer ones. [portmanteau word]

oxen *pl.* of ox.

ox-eye daisy *n.* daisy with white petals and a yellow centre.

Oxf. *abbr.* Oxford.

Oxfam *abbr.* Oxford Committee for Famine Relief.

Oxford blue /'ɒksfəd/ *adj.* & *n.* (as adj. often hyphenated) a dark blue, often with a purple tinge.

oxhide *n.* **1** hide of an ox. **2** leather from this.

oxidation /,ɒksɪ'deɪʃ(ə)n/ *n.* process of oxidizing. [French: related to OXIDE]

oxide /'ɒksaɪd/ *n.* binary compound of oxygen. [French: related to OXYGEN]

oxidize /'ɒksɪ,daɪz/ *v.* (also **-ise**) (**-zing** or **-sing**) **1** combine with oxygen. **2** make or become rusty. **3** coat (metal) with oxide. □ **oxidization** /-'zeɪʃ(ə)n/ *n.*

Oxon *abbr.* (esp. in degree titles) of Oxford University. [Latin *Oxoniensis*: related to OXONIAN]

Oxonian /ɒk'səʊnɪən/ *–adj.* of Oxford or Oxford University. *–n.* **1** member of Oxford University. **2** native or inhabitant of Oxford. [*Oxonia* Latinized name of *Ox(en)ford*]

oxtail *n.* tail of an ox, often used in making soup.

oxyacetylene /,ɒksɪə'setɪ,li:n/ *adj.* of or using a mixture of oxygen and acetylene, esp. in cutting or welding metals.

oxygen /'ɒksɪdʒ(ə)n/ *n.* tasteless odourless gaseous element essential to plant and animal life. [Greek *oxus* sharp, -GEN (because it was thought to be present in all acids)]

oxygenate /'ɒksɪdʒə,neɪt/ *v.* (**-ting**) supply, treat, or mix with oxygen; oxidize.

oxygen tent *n.* tentlike enclosure supplying a patient with air rich in oxygen.

oxymoron /,ɒksɪ'mɔ:rɒn/ *n.* figure of speech in which apparently contradictory terms appear in conjunction (e.g. *faith unfaithful kept him falsely true*). [Greek, = pointedly foolish, from *oxus* sharp, *mōros* dull]

oyez /əʊ'jes/ *int.* (also **oyes**) uttered, usu. three times, by a public crier or a court officer to command attention. [Anglo-French, = hear!, from Latin *audio*]

oyster *–n.* **1** bivalve mollusc, esp. an edible kind, sometimes producing a pearl. **2** symbol of all one desires (*the world is my oyster*). **3** oyster white. *–adj.* oyster-white. [Greek *ostreon*]

oyster-catcher *n.* wading sea bird.

oyster white *adj.* & *n.* (as adj. often hyphenated) greyish white.

oz *abbr.* ounce(s). [Italian *onza* ounce]

ozone /'əʊzəʊn/ *n.* **1** *Chem.* unstable form of oxygen with three atoms in a molecule, having a pungent odour. **2** *colloq.* **a** invigorating air at the seaside etc. **b** exhilarating influence. [Greek *ozō* smell (v.)]

ozone-friendly *adj.* not containing chemicals destructive to the ozone layer.

ozone layer *n.* layer of ozone in the stratosphere that absorbs most of the sun's ultraviolet radiation.

P

P[1] /piː/ *n.* (also **p**) (*pl.* **Ps** or **P's**) sixteenth letter of the alphabet.

P[2] *abbr.* (also **P.**) **1** (on road signs) parking. **2** *Chess* pawn. **3** (also Ⓟ) proprietary.

P[3] *symb.* phosphorus.

p *abbr.* (also **p.**) **1** penny, pence. **2** page. **3** piano (softly).

PA *abbr.* **1** personal assistant. **2** public address (system).

Pa *symb.* protactinium.

pa /pɑː/ *n. colloq.* father. [abbreviation of PAPA]

p.a. *abbr.* per annum.

pabulum /'pæbjʊləm/ *n.* food, esp. for the mind. [Latin]

pace[1] *—n.* **1 a** single step in walking or running. **b** distance covered in this. **2** speed in walking or running. **3** rate of movement or progression. **4** way of walking or running; gait (*ambling pace*). *—v.* **(-cing) 1 a** walk slowly and evenly (*pace up and down*). **b** (of a horse) amble. **2** traverse by pacing. **3** set the pace for (a rider, runner, etc.). **4** (foll. by *out*) measure by pacing. □ **keep pace** (often foll. by *with*) advance at an equal rate (to). **put a person** etc. **through his** (or **her**) **paces** test a person's qualities in action etc. **set the pace** determine the speed; lead. [French *pas* from Latin *passus*]

pace[2] /'pɑːtʃeɪ, 'peɪsɪ/ *prep.* (in stating a contrary opinion) with due respect to (the person named). [Latin, ablative of *pax* peace]

pace bowler *n. Cricket* fast bowler.

pacemaker *n.* **1** competitor who sets the pace in a race. **2** natural or artificial device for stimulating the heart muscle.

pace-setter *n.* **1** leader. **2** = PACEMAKER 1.

pachyderm /'pækɪˌdɜːm/ *n.* thick-skinned mammal, esp. an elephant or rhinoceros. □ **pachydermatous** /-'dɜːmətəs/ *adj.* [Greek *pakhus* thick, *derma* skin]

pacific /pə'sɪfɪk/ *—adj.* **1** peaceful; tranquil. **2** (**Pacific**) of or adjoining the Pacific. *—n.* (**the Pacific**) ocean between America to the east and Asia to the west. [Latin *pax pacis* peace]

pacifier /'pæsɪˌfaɪə(r)/ *n.* **1** person or thing that pacifies. **2** *US* baby's dummy.

pacifism /'pæsɪˌfɪz(ə)m/ *n.* belief that war and violence are morally unjustifiable. □ **pacifist** *n.* & *adj.*

pacify /'pæsɪˌfaɪ/ *v.* **(-ies, -ied) 1** appease (a person, anger, etc.). **2** bring (a country etc.) to a state of peace. □ **pacification** /-fɪ'keɪʃ(ə)n/ *n.* **pacificatory** /pə'sɪfɪkətərɪ/ *adj.*

pack[1] *—n.* **1 a** collection of things wrapped up or tied together for carrying. **b** = BACKPACK. **2** set of packaged items. **3** usu. *derog.* lot or set (*pack of lies*; *pack of thieves*). **4** set of playing-cards. **5** group of hounds, wild animals, etc. **6** organized group of Cub Scouts or Brownies. **7** *Rugby* team's forwards. **8** = FACE-PACK. **9** = PACK ICE. *—v.* **1** (often foll. by *up*) **a** fill (a suitcase, bag, etc.) with clothes etc. **b** put (things) in a bag or suitcase, esp. for travelling. **2** (often foll. by *in, into*) crowd or cram (*packed a lot into a few hours*; *packed in like sardines*). **3** (esp. in *passive*; often foll. by *with*) fill (*restaurant was packed*; *fans packed the stadium*; *packed with information*). **4** cover (a thing) with packaging. **5** be suitable for packing. **6** *colloq.* **a** carry (a gun etc.). **b** be capable of delivering (a forceful punch). **7** (of animals or Rugby forwards) form a pack. □ **pack in** *colloq.* stop, give up (*packed in his job*). **pack it in** (or **up**) *colloq.* end or stop it. **pack off** send (a person) away, esp. summarily. **pack them in** fill a theatre etc. with a capacity audience. **pack up** *colloq.* **1** stop functioning; break down. **2** retire from an activity, contest, etc. **send packing** *colloq.* dismiss summarily. [Low German or Dutch]

pack[2] *v.* select (a jury etc.) or fill (a meeting) so as to secure a decision in one's favour. [probably from PACT]

package *—n.* **1 a** bundle of things packed. **b** parcel, box, etc., in which things are packed. **2** (in full **package deal**) set of proposals or items offered or agreed to as a whole. **3** *Computing* piece of software suitable for a wide range of users. **4** *colloq.* = PACKAGE HOLIDAY. *—v.* **(-ging)** make up into or enclose in a package. □ **packager** *n.*

package holiday *n.* (also **package tour**) holiday (or tour) with travel, hotels, etc. at an inclusive price.

packaging *n.* **1** wrapping or container for goods. **2** process of packing goods.

packed lunch *n.* lunch of sandwiches etc. prepared and packed to be eaten away from home.

packed out *adj.* full, crowded.

packer *n.* person or thing that packs, esp. a dealer who prepares and packs food.

packet /ˈpækɪt/ *n.* **1** small package. **2** *colloq.* large sum of money won, lost, or spent. **3** (in full **packet-boat**) *hist.* mail-boat or passenger ship.

packhorse *n.* horse for carrying loads.

pack ice *n.* crowded floating ice in the sea.

packing *n.* material used to pack esp. fragile articles.

packthread *n.* stout thread for sewing or tying up packs.

pact *n.* agreement; treaty. [Latin *pactum*]

pad[1] *–n.* **1** thick piece of soft material used to protect, fill out hollows, hold or absorb liquid, etc. **2** sheets of blank paper fastened together at one edge, for writing or drawing on. **3** fleshy underpart of an animal's foot or of a human finger. **4** guard for the leg and ankle in sports. **5** flat surface for helicopter take-off or rocket-launching. **6** *slang* lodgings, flat, etc. **7** floating leaf of a water lily. *–v.* (**-dd-**) **1** provide with a pad or padding; stuff. **2** (foll. by *out*) lengthen or fill out (a book etc.) with unnecessary material. [probably Low German or Dutch]

pad[2] *–v.* (**-dd-**) **1** walk with a soft dull steady step. **2** travel, or tramp along (a road etc.), on foot. *–n.* sound of soft steady steps. [Low German *pad* PATH]

padded cell *n.* room with padded walls in a mental hospital.

padding *n.* soft material used to pad or stuff.

paddle[1] /ˈpæd(ə)l/ *–n.* **1** short broad-bladed oar used without a rowlock. **2** paddle-shaped instrument. **3** fin, flipper. **4** board on a paddle-wheel or mill-wheel. **5** action or spell of paddling. *–v.* (**-ling**) **1** move on water or propel a boat by paddles. **2** row gently. [origin unknown]

paddle[2] /ˈpæd(ə)l/ *–v.* (**-ling**) walk barefoot, or dabble the feet or hands, in shallow water. *–n.* act of paddling. [probably Low German or Dutch]

paddle-boat *n.* (also **paddle-steamer**) boat (or steamer) propelled by a paddle-wheel.

paddle-wheel *n.* wheel for propelling a ship, with boards round the circumference.

paddock /ˈpædək/ *n.* **1** small field, esp. for keeping horses in. **2** turf enclosure at a racecourse for horses or cars. [*parrock*, var. of PARK]

Paddy /ˈpædɪ/ *n.* (*pl.* **-ies**) *colloq.* often *offens.* Irishman. [Irish *Padraig* Patrick]

paddy[1] /ˈpædɪ/ *n.* (*pl.* **-ies**) **1** (in full **paddy-field**) field where rice is grown. **2** rice before threshing or in the husk. [Malay]

paddy[2] /ˈpædɪ/ *n.* (*pl.* **-ies**) *colloq.* rage; fit of temper. [from PADDY]

padlock /ˈpædlɒk/ *–n.* detachable lock hanging by a pivoted hook on the object fastened. *–v.* secure with a padlock. [origin unknown]

padre /ˈpɑːdrɪ/ *n.* chaplain in the army etc. [Italian, Spanish, and Portuguese, = father, priest]

paean /ˈpiːən/ *n.* (*US* **pean**) song of praise or triumph. [Latin from Greek]

paederast var. of PEDERAST.

paederasty var. of PEDERASTY.

paediatrics /ˌpiːdɪˈætrɪks/ *n.pl.* (treated as *sing.*) (*US* **pediatrics**) branch of medicine dealing with children and their diseases. □ **paediatric** *adj.* **paediatrician** /-əˈtrɪʃ(ə)n/ *n.* [from PAEDO-, Greek *iatros* physician]

paedo- *comb. form* (*US* **pedo-**) child. [Greek *pais paid-* child]

paedophile /ˈpiːdəˌfaɪl/ *n.* (*US* **pedophile**) person who displays paedophilia.

paedophilia /ˌpiːdəˈfɪlɪə/ *n.* (*US* **pedophilia**) sexual attraction felt towards children.

paella /paɪˈelə/ *n.* Spanish dish of rice, saffron, chicken, seafood, etc., cooked and served in a large shallow pan. [Latin PATELLA]

paeony var. of PEONY.

pagan /ˈpeɪgən/ *–n.* non-religious person, pantheist, or heathen, esp. in pre-Christian times. *–adj.* **1 a** of pagans. **b** irreligious. **2** pantheistic. □ **paganism** *n.* [Latin *paganus* from *pagus* country district]

page[1] *–n.* **1 a** leaf of a book, periodical, etc. **b** each side of this. **c** what is written or printed on this. **2** episode; memorable event. *–v.* (**-ging**) paginate. [Latin *pagina*]

page[2] *–n.* **1** liveried boy or man employed to run errands, attend to a door, etc. **2** boy as a personal attendant of a bride etc. *–v.* (**-ging**) **1** (in hotels, airports, etc.) summon, esp. by making an announcement. **2** summon by pager. [French]

pageant /ˈpædʒ(ə)nt/ *n.* **1 a** brilliant spectacle, esp. an elaborate parade. **b** spectacular procession or play illustrating historical events. **c** tableau etc. on a fixed stage or moving vehicle. **2** empty or specious show. [origin unknown]

pageantry *n.* (esp. on State occasions) spectacular show; pomp.

page-boy *n.* **1** = PAGE[2] *n.* 2. **2** woman's hairstyle with the hair bobbed and rolled under.

pager *n.* bleeping radio device, calling its wearer to the telephone etc.

paginate /'pædʒɪ,neɪt/ *v.* (**-ting**) assign numbers to the pages of (a book etc.). □ **pagination** /-'neɪʃ(ə)n/ *n.* [Latin: related to PAGE[1]]

pagoda /pə'gəʊdə/ *n.* **1** Hindu or Buddhist temple etc., esp. a many-tiered tower, in India and the Far East. **2** ornamental imitation of this. [Portuguese]

pah /pɑ:/ *int.* expressing disgust or contempt. [natural exclamation]

paid *past* and *past part.* of PAY[1].

paid-up *adj.* having paid one's subscription to a trade-union, club, etc., or having done what is required to be considered a full member of a particular group (*paid-up feminist*).

pail *n.* **1** bucket. **2** amount contained in this. □ **pailful** *n.* (*pl.* **-s**). [Old English]

pain *–n.* **1** any unpleasant bodily sensation produced by illness, accident, etc. **2** mental suffering. **3** (also **pain in the neck** or **arse**) *colloq.* troublesome person or thing; nuisance. *–v.* **1** cause pain to. **2** (as **pained** *adj.*) expressing pain (*pained expression*). □ **be at** (or **take**) **pains** take great care. **in pain** suffering pain. **on** (or **under**) **pain of** with (death etc.) as the penalty. [Latin *poena* penalty]

painful *adj.* **1** causing bodily or mental pain. **2** (esp. of part of the body) suffering pain. **3** causing trouble or difficulty; laborious (*painful climb*). □ **painfully** *adv.*

painkiller *n.* drug for alleviating pain. □ **painkilling** *adj.*

painless *adj.* not causing pain. □ **painlessly** *adv.*

painstaking /'peɪnz,teɪkɪŋ/ *adj.* careful, industrious, thorough. □ **painstakingly** *adv.*

paint *–n.* **1** pigment, esp. in liquid form, for colouring a surface. **2** this as a dried film or coating (*paint peeled off*). *–v.* **1 a** cover (a wall, object, etc.) with paint. **b** apply paint of a specified colour to (*paint the door green*). **2** depict (an object, scene, etc.) in paint; produce (a picture) thus. **3** describe vividly (*painted a gloomy picture*). **4** *joc.* or *archaic* **a** apply make-up to (the face, skin, etc.). **b** apply (a liquid to the skin etc.). □ **paint out** efface with paint. **paint the town red** *colloq.* enjoy oneself flamboyantly. [Latin *pingo pict-*]

paintbox *n.* box holding dry paints for painting pictures.

paintbrush *n.* brush for applying paint.

painted lady *n.* orange-red spotted butterfly.

painter[1] *n.* person who paints; artist or decorator.

painter[2] *n.* rope attached to the bow of a boat for tying it to a quay etc. [origin unknown]

painterly *adj.* **1** characteristic of a painter or paintings; artistic. **2** (of a painting) lacking clearly defined outlines.

painting *n.* **1** process or art of using paint. **2** painted picture.

paint shop *n.* part of a factory where cars etc. are sprayed or painted.

paintwork *n.* painted, esp. wooden, surface or area in a building etc.

painty *adj.* of or covered in paint (*painty smell*).

pair *–n.* **1** set of two people or things used together or regarded as a unit. **2** article (e.g. scissors, trousers, or pyjamas) consisting of two joined or corresponding parts. **3 a** engaged or married couple. **b** mated couple of animals. **4** two horses harnessed side by side (*coach and pair*). **5** member of a pair in relation to the other (*cannot find its pair*). **6** two playing-cards of the same denomination. **7** either or both of two MPs etc. on opposite sides agreeing not to vote on certain occasions. *–v.* **1** (often foll. by *off*) arrange or be arranged in couples. **2 a** join or be joined in marriage. **b** (of animals) mate. [Latin *paria*: related to PAR]

pair of scales *n.* simple balance.

Paisley /'peɪzlɪ/ *n.* (*pl.* **-s**) (often *attrib.*) **1** pattern of curved feather-shaped figures. **2** soft woollen shawl etc. having this pattern. [*Paisley* in Scotland]

pajamas *US* var. of PYJAMAS.

Paki /'pækɪ/ *n.* (*pl.* **-s**) *slang offens.* Pakistani. [abbreviation]

Pakistani /,pɑ:kɪ'stɑ:nɪ/ *–n.* (*pl.* **-s**) **1** native or national of Pakistan. **2** person of Pakistani descent. *–adj.* of Pakistan.

pal *–n. colloq.* friend, mate, comrade. *–v.* (**-ll-**) (usu. foll. by *up*) associate; form a friendship. [Romany]

palace /'pælɪs/ *n.* **1** official residence of a sovereign, president, archbishop, or bishop. **2** splendid or spacious building. [Latin *palatium*]

palace revolution *n.* (also **palace coup**) (usu. non-violent) overthrow of a sovereign, government, etc. by a bureaucracy.

palaeo- *comb. form* (*US* **paleo-**) ancient; prehistoric. [Greek *palaios*]

Palaeocene /'pælɪə,si:n/ (*US* **Paleocene**) *Geol.* *–adj.* of the earliest epoch of the Tertiary period. *–n.* this epoch or system. [from PALAEO-, Greek *kainos* new]

palaeography /ˌpælɪˈɒɡrəfɪ/ *n.* (*US* **paleography**) the study of ancient writing and documents. □ **palaeographer** *n.* [French: related to PALAEO-]

palaeolithic /ˌpælɪəʊˈlɪθɪk/ *adj.* (*US* **paleolithic**) of the early part of the Stone Age. [Greek *lithos* stone]

palaeontology /ˌpælɪɒnˈtɒlədʒɪ/ *n.* (*US* **paleontology**) the study of life in the geological past. □ **palaeontologist** *n.* [Greek *ōn ont-* being]

Palaeozoic /ˌpælɪəʊˈzəʊɪk/ (*US* **Paleozoic**) *–adj.* of an era of geological time marked by the appearance of plants and animals, esp. invertebrates. *–n.* this era. [Greek *zōion* animal]

palais /ˈpæleɪ/ *n. colloq.* public dance-hall. [French, = hall]

palanquin /ˌpælənˈkiːn/ *n.* (also **palankeen**) (in India and the East) covered litter for one. [Portuguese]

palatable /ˈpælətəb(ə)l/ *adj.* **1** pleasant to taste. **2** (of an idea etc.) acceptable, satisfactory.

palatal /ˈpælət(ə)l/ *–adj.* **1** of the palate. **2** (of a sound) made by placing the tongue against the hard palate (e.g. *y* in *yes*). *–n.* palatal sound.

palate /ˈpælət/ *n.* **1** structure closing the upper part of the mouth cavity in vertebrates. **2** sense of taste. **3** mental taste; liking. [Latin *palatum*]

palatial /pəˈleɪʃ(ə)l/ *adj.* (of a building) like a palace; spacious and splendid. □ **palatially** *adv.* [Latin: related to PALACE]

palatinate /pəˈlætɪˌneɪt/ *n.* territory under the jurisdiction of a Count Palatine.

palatine /ˈpæləˌtaɪn/ *adj.* (also **Palatine**) *hist.* **1** (of an official etc.) having local authority that elsewhere belongs only to a sovereign (*Count Palatine*). **2** (of a territory) subject to this authority. [Latin: related to PALACE]

palaver /pəˈlɑːvə(r)/ *n. colloq.* tedious fuss and bother. [Latin: related to PARABLE]

pale[1] *–adj.* **1** (of a person, colour, or complexion) light or faint; whitish, ashen. **2** of faint lustre; dim. *–v.* (**-ling**) **1** grow or make pale. **2** (often foll. by *before*, *beside*) seem feeble in comparison (with). □ **palely** *adv.* **paleness** *n.* **palish** *adj.* [Latin *pallidus*]

pale[2] *n.* **1** pointed piece of wood for fencing etc.; stake. **2** boundary. □ **beyond the pale** outside the bounds of acceptable behaviour. [Latin *palus*]

paleface *n.* name supposedly used by N. American Indians for the White man.

paleo- *comb. form US* var. of PALAEO-.

Paleocene *US* var. of PALAEOCENE.

paleography *US* var. of PALAEOGRAPHY.

paleolithic *US* var. of PALAEOLITHIC.

paleontology *US* var. of PALAEONTOLOGY.

Paleozoic *US* var. of PALAEOZOIC.

Palestinian /ˌpælɪˈstɪnɪən/ *–adj.* of Palestine. *–n.* **1** native of Palestine. **2** Arab, or a descendant of one, born or living in the area formerly called Palestine.

palette /ˈpælɪt/ *n.* **1** artist's thin board or slab for laying and mixing colours on. **2** range of colours used by an artist. [French from Latin *pala* spade]

palette-knife *n.* **1** thin flexible steel blade with a handle for mixing colours or applying or removing paint. **2** blunt round-ended flexible kitchen knife.

palimony /ˈpælɪmənɪ/ *n.* esp. *US colloq.* allowance paid by either partner of a separated unmarried couple to the other. [from PAL, ALIMONY]

palimpsest /ˈpælɪmpˌsest/ *n.* **1** writing-material or manuscript on which the original writing has been effaced for reuse. **2** monumental brass turned and re-engraved on the reverse side. [Greek *palin* again, *psēstos* rubbed]

palindrome /ˈpælɪnˌdrəʊm/ *n.* word or phrase reading the same backwards as forwards (e.g. *nurses run*). □ **palindromic** /-ˈdrɒmɪk/ *adj.* [Greek *palindromos* running back: related to PALIMPSEST, *drom-* run]

paling *n.* **1** fence of pales. **2** pale.

palisade /ˌpælɪˈseɪd/ *–n.* **1** fence of pales or iron railings. **2** strong pointed wooden stake. *–v.* (**-ding**) enclose or provide with a palisade. [French: related to PALE[2]]

pall[1] /pɔːl/ *n.* **1** cloth spread over a coffin etc. **2** shoulder-band with pendants, worn as an ecclesiastical vestment and sign of authority. **3** dark covering (*pall of darkness*). [Latin *pallium* cloak]

pall[2] /pɔːl/ *v.* (often foll. by *on*) become uninteresting (to). [from APPAL]

palladium /pəˈleɪdɪəm/ *n.* rare white metallic element used as a catalyst and in jewellery. [*Pallas*, name of an asteroid]

pallbearer *n.* person helping to carry or escort a coffin at a funeral.

pallet[1] /ˈpælɪt/ *n.* **1** straw mattress. **2** mean or makeshift bed. [Latin *palea* straw]

pallet[2] /ˈpælɪt/ *n.* portable platform for transporting and storing loads. [French: related to PALETTE]

palliasse /ˈpælɪˌæs/ *n.* straw mattress. [Latin: related to PALLET[1]]

palliate /ˈpælɪˌeɪt/ *v.* **(-ting) 1** alleviate (disease) without curing it. **2** excuse, extenuate. □ **palliative** /-ətɪv/ *n.* & *adj.* [Latin *pallio* cloak: related to PALL[1]]

pallid /ˈpælɪd/ *adj.* pale, esp. from illness. [Latin: related to PALE[1]]

pallor /ˈpælə(r)/ *n.* paleness. [Latin *palleo* be pale]

pally *adj.* **(-ier, -iest)** *colloq.* friendly.

palm[1] /pɑːm/ *n.* **1** (also **palm-tree**) (usu. tropical) tree-like plant with no branches and a mass of large leaves at the top. **2** leaf of this as a symbol of victory. [Latin *palma*]

palm[2] /pɑːm/ *–n.* **1** inner surface of the hand between the wrist and fingers. **2** part of a glove that covers this. *–v.* conceal in the hand. □ **palm off 1** (often foll. by *on*) impose fraudulently (on a person) (*palmed my old car off on him*). **2** (often foll. by *with*) cause (a person) to accept unwillingly or unknowingly (*palmed him off with my old car*). [Latin *palma*]

palmate /ˈpælmeɪt/ *adj.* **1** shaped like an open hand. **2** having lobes etc. like spread fingers. [Latin *palmatus*: related to PALM[2]]

palmetto /pælˈmetəʊ/ *n.* (*pl.* **-s**) small palm-tree. [Spanish *palmito* diminutive of *palma* PALM[1]]

palmistry /ˈpɑːmɪstrɪ/ *n.* fortune-telling from lines etc. on the palm of the hand. □ **palmist** *n.*

palm oil *n.* oil from various palms.

Palm Sunday *n.* Sunday before Easter, celebrating Christ's entry into Jerusalem.

palmy /ˈpɑːmɪ/ *adj.* **(-ier, -iest) 1** of, like, or abounding in palms. **2** triumphant, flourishing (*palmy days*).

palomino /ˌpæləˈmiːnəʊ/ *n.* (*pl.* **-s**) golden or cream-coloured horse with light-coloured mane and tail. [Latin *palumba* dove]

palpable /ˈpælpəb(ə)l/ *adj.* **1** able to be touched or felt. **2** readily perceived. □ **palpably** *adv.* [Latin *palpo* caress]

palpate /ˈpælpeɪt/ *v.* **(-ting)** examine (esp. medically) by touch. □ **palpation** /-ˈpeɪʃ(ə)n/ *n.*

palpitate /ˈpælpɪˌteɪt/ *v.* **(-ting)** pulsate, throb, tremble. [Latin *palpito* frequentative of *palpo* touch gently]

palpitation /ˌpælpɪˈteɪʃ(ə)n/ *n.* **1** throbbing, trembling. **2** (often in *pl.*) increased rate of heartbeat due to exertion, agitation, or disease.

palsy /ˈpɔːlzɪ/ *–n.* (*pl.* **-ies**) paralysis, esp. with involuntary tremors. *–v.* **(-ies, -ied)** affect with palsy. [French: related to PARALYSIS]

paltry /ˈpɔːltrɪ/ *adj.* **(-ier, -iest)** worthless, contemptible, trifling. □ **paltriness** *n.* [from *palt* rubbish]

pampas /ˈpæmpəs/ *n.pl.* large treeless plains in S. America. [Spanish from Quechua]

pampas-grass *n.* tall S. American ornamental grass.

pamper *v.* overindulge (a person, taste, etc.); spoil. [obsolete *pamp* cram]

pamphlet /ˈpæmflɪt/ *–n.* small usu. unbound booklet or leaflet. *–v.* **(-t-)** distribute pamphlets to. [*Pamphilus*, name of medieval poem]

pamphleteer /ˌpæmflɪˈtɪə(r)/ *n.* writer of (esp. political) pamphlets.

pan[1] *–n.* **1 a** broad usu. metal vessel used for cooking etc. **b** contents of this. **2** panlike vessel in which substances are heated etc. **3** similar shallow container, e.g. the bowl of a pair of scales. **4** lavatory bowl. **5** part of the lock in old guns. **6** hollow in the ground (*salt-pan*). *–v.* **(-nn-) 1** *colloq.* criticize severely. **2 a** (foll. by *off, out*) wash (gold-bearing gravel) in a pan. **b** search for gold thus. □ **pan out 1** (of an action etc.) turn out; work out well or in a specified way. **2** (of gravel) yield gold. □ **panful** *n.* (*pl.* **-s**). **panlike** *adj.* [Old English]

pan[2] *–v.* **(-nn-) 1** swing (a film camera) horizontally to give a panoramic effect or to follow a moving object. **2** (of a camera) be moved thus. *–n.* panning movement. [from PANORAMA]

pan- *comb. form* **1** all; the whole of. **2** relating to the whole of a continent, racial group, religion, etc. (*pan-American*). [Greek *pan*, neuter of *pas pantos* all]

panacea /ˌpænəˈsiːə/ *n.* universal remedy. [Greek: related to PAN-, *akos* remedy]

panache /pəˈnæʃ/ *n.* assertive flamboyance; confidence of style or manner. [French, = plume]

panama /ˈpænəˌmɑː/ *n.* straw hat with a brim and indented crown. [*Panama* in Central America]

panatella /ˌpænəˈtelə/ *n.* long thin cigar. [American Spanish, = long thin biscuit]

pancake *n.* **1** thin flat cake of fried batter usu. rolled up with a filling. **2** flat cake of make-up etc.

Pancake Day *n.* Shrove Tuesday (when pancakes are traditionally eaten).

pancake landing *n. colloq.* emergency aircraft landing with the undercarriage still retracted.

panchromatic /ˌpænkrəʊˈmætɪk/ *adj.* (of a film etc.) sensitive to all visible colours of the spectrum.

pancreas /'pæŋkrɪəs/ *n.* gland near the stomach supplying digestive fluid and secreting insulin. □ **pancreatic** /-'ætɪk/ *adj.* [Greek *kreas* flesh]

panda /'pændə/ *n.* **1** (also **giant panda**) large bearlike black and white mammal native to China and Tibet. **2** (also **red panda**) reddish-brown Himalayan racoon-like mammal. [Nepali]

panda car *n.* police patrol car.

pandemic /pæn'demɪk/ *adj.* (of a disease etc.) widespread; universal. [Greek *dēmos* people]

pandemonium /,pændɪ'məʊnɪəm/ *n.* **1** uproar; utter confusion. **2** scene of this. [place in hell in Milton's *Paradise Lost*: related to PAN-, DEMON]

pander /'pændə(r)/ *–v.* (foll. by *to*) gratify or indulge (a person or weakness etc.). *–n.* **1** procurer; pimp. **2** person who encourages coarse desires. [*Pandare*, name of a character in the story of Troilus and Cressida]

pandit var. of PUNDIT 1.

Pandora's box /pæn'dɔ:rəz/ *n.* process that once begun will generate many unmanageable problems. [a box in Greek mythology from which many ills were released on mankind]

p. & p. *abbr.* postage and packing.

pane *n.* single sheet of glass in a window or door. [Latin *pannus* a cloth]

panegyric /,pænɪ'dʒɪrɪk/ *n.* eulogy; speech or essay of praise. [Greek *agora* assembly]

panel /'pæn(ə)l/ *–n.* **1** distinct, usu. rectangular, section of a surface (e.g. of a wall, door, or vehicle). **2** strip of material in a garment. **3** team in a broadcast game, discussion, etc. **4 a** list of available jurors. **b** jury. *–v.* (**-ll-**; *US* **-l-**) fit, cover, or decorate with panels. [Latin diminutive of *pannus*: related to PANE]

panel-beater *n.* person who beats out the metal panels of vehicles.

panel game *n.* broadcast quiz etc. played by a panel.

panelling *n.* (*US* **paneling**) **1** panelled work. **2** wood for making panels.

panellist *n.* (*US* **panelist**) member of a panel.

pang *n.* (often in *pl.*) sudden sharp pain or painful emotion. [obsolete *pronge*]

pangolin /pæŋ'gəʊlɪn/ *n.* scaly Asian and African anteater. [Malay]

panic /'pænɪk/ *–n.* **1** sudden uncontrollable fear. **2** infectious fright, esp. in commercial dealings. *–v.* (**-ck-**) (often foll. by *into*) affect or be affected with panic (*was panicked into buying*). □ **panicky** *adj.* [Greek *Pan*, rural god]

panicle /'pænɪk(ə)l/ *n.* loose branching cluster of flowers, as in oats. [Latin *paniculum* diminutive of *panus* thread]

panic stations *n.pl. colloq.* state of emergency.

panic-stricken *adj.* (also **panic-struck**) affected with panic.

panjandrum /pæn'dʒændrəm/ *n.* **1** mock title for an important person. **2** pompous official etc. [invented word]

pannier /'pænɪə(r)/ *n.* basket, bag, or box, esp. one of a pair carried by a donkey etc., bicycle, or motor cycle. [Latin *panis* bread]

panoply /'pænəplɪ/ *n.* (*pl.* **-ies**) **1** complete or splendid array. **2** complete suit of armour. [Greek *hopla* arms]

panorama /,pænə'rɑ:mə/ *n.* **1** unbroken view of a surrounding region. **2** complete survey of a subject, series of events, etc. **3** picture or photograph containing a wide view. **4** continuous passing scene. □ **panoramic** /-'ræmɪk/ *adj.* [Greek *horama* view]

pan-pipes *n.pl.* musical instrument made of a series of short graduated pipes fixed together. [from *Pan*, Greek rural god]

pansy /'pænzɪ/ *n.* (*pl.* **-ies**) **1** cultivated plant with flowers of various rich colours. **2** *colloq. offens.* **a** effeminate man. **b** male homosexual. [French *pensée* thought, pansy]

pant *–v.* **1** breathe with short quick breaths. **2** (often foll. by *out*) utter breathlessly. **3** (usu. foll. by *for*) yearn, crave. **4** (of the heart etc.) throb violently. *–n.* **1** panting breath. **2** throb. [Greek: related to FANTASY]

pantaloons /,pæntə'lu:nz/ *n.pl.* (esp. women's) baggy trousers gathered at the ankles. [French from Italian]

pantechnicon /pæn'teknɪkən/ *n.* large furniture removal van. [from TECHNIC: originally as the name of a bazaar]

pantheism /'pænθɪ,ɪz(ə)m/ *n.* **1** belief that God is in all nature. **2** worship that admits or tolerates all gods. □ **pantheist** *n.* **pantheistic** /-'ɪstɪk/ *adj.* [Greek *theos* god]

pantheon /'pænθɪən/ *n.* **1** building in which illustrious dead are buried or have memorials. **2** the deities of a people collectively. **3** temple dedicated to all the gods. [Greek *theion* divine]

panther *n.* **1** leopard, esp. with black fur. **2** *US* puma. [Greek *panthēr*]

pantie-girdle *n.* woman's girdle with a crotch shaped like pants.

panties /'pæntɪz/ *n.pl. colloq.* short-legged or legless underpants worn by women and girls. [diminutive of PANTS]

pantihose /'pæntɪ,həʊz/ *n.* (usu. treated as *pl.*) *US* women's tights.

pantile /'pæntaɪl/ *n.* curved roof-tile. [from PAN¹]

panto /'pæntəʊ/ *n.* (*pl.* **-s**) *colloq.* = PANTOMIME[1]. [abbreviation]

pantograph /'pæntə,grɑ:f/ *n.* **1** instrument with jointed rods for copying a plan or drawing etc. on a different scale. **2** jointed framework conveying a current to an electric vehicle from overhead wires. [from PAN-, -GRAPH]

pantomime /'pæntə,maɪm/ *n.* **1** Christmas theatrical entertainment based on a fairy tale. **2** gestures and facial expression conveying meaning, esp. in drama and dance. **3** *colloq.* absurd or outrageous piece of behaviour. [Greek: related to PAN-, MIME]

pantry /'pæntrɪ/ *n.* (*pl.* **-ies**) **1** small room or cupboard in which crockery, cutlery, table linen, etc., are kept. **2** larder. [Latin *panis* bread]

pants *n.pl.* **1** underpants or knickers. **2** *US* trousers. □ **bore** (or **scare** etc.) **the pants off** *colloq.* bore, scare, etc., greatly. **with one's pants down** *colloq.* in an embarrassingly unprepared state. [abbreviation of PANTALOONS]

pap[1] *n.* **1** soft or semi-liquid food for infants or invalids. **2** light or trivial reading matter. [Low German or Dutch]

pap[2] *n. archaic* or *dial.* nipple. [Scandinavian]

papa /pə'pɑ:/ *n. archaic* father (esp. as a child's word). [Greek *papas*]

papacy /'peɪpəsɪ/ *n.* (*pl.* **-ies**) **1** pope's office or tenure. **2** papal system. [medieval Latin *papatia*: related to POPE]

papal /'peɪp(ə)l/ *adj.* of a pope or the papacy. [medieval Latin: related to POPE]

paparazzo /,pæpə'rætsəʊ/ *n.* (*pl.* **-zzi** /-tsɪ/) freelance photographer who pursues celebrities to photograph them. [Italian]

papaw var. of PAWPAW.

papaya var. of PAWPAW. [earlier form of PAWPAW]

paper –*n.* **1** material made in thin sheets from the pulp of wood etc., used for writing, drawing, or printing on, or as wrapping material etc. **2** (*attrib.*) **a** made of or using paper. **b** flimsy like paper. **3** = NEWSPAPER. **4 a** printed document. **b** (in *pl.*) identification etc. documents. **c** (in *pl.*) documents of a specified kind (*divorce papers*). **5** *Commerce* **a** negotiable documents, e.g. bills of exchange. **b** (*attrib.*) not actual; theoretical (*paper profits*). **6 a** set of printed questions in an examination. **b** written answers to these. **7** = WALLPAPER. **8** essay or dissertation. **9** piece of paper, esp. as a wrapper etc. –*v.* **1** decorate (a wall etc.) with wallpaper. **2** (foll. by *over*) **a** cover (a hole or blemish) with paper. **b** disguise or try to hide (a fault etc.). □ **on paper 1** in writing. **2** in theory; from written or printed evidence. [Latin PAPYRUS]

paperback *n.* (often *attrib.*) book bound in paper or card, not boards.

paper-boy *n.* (also **paper-girl**) boy or girl who delivers or sells newspapers.

paper-chase *n.* cross-country run following a trail of torn-up paper.

paper-clip *n.* clip of bent wire or plastic for fastening papers together.

paper-hanger *n.* person who hangs wallpaper, esp. for a living.

paper-knife *n.* blunt knife for opening letters etc.

paper-mill *n.* mill in which paper is made.

paper money *n.* banknotes.

paper round *n.* **1** job of regularly delivering newspapers. **2** route for this.

paper tiger *n.* apparently threatening, but ineffectual, person or thing.

paperweight *n.* small heavy object for keeping loose papers in place.

paperwork *n.* routine clerical or administrative work.

papery *adj.* like paper in thinness or texture.

papier mâché /,pæpɪeɪ 'mæʃeɪ/ *n.* paper pulp moulded into boxes, trays, etc. [French, = chewed paper]

papilla /pə'pɪlə/ *n.* (*pl.* **papillae** /-li:/) small nipple-like protuberance in or on the body, as that at the base of a hair, feather, etc. □ **papillary** *adj.* [Latin]

papist /'peɪpɪst/ *n.* often *derog.* **1** (often *attrib.*) Roman Catholic. **2** *hist.* advocate of papal supremacy. [related to POPE]

papoose /pə'pu:s/ *n.* N. American Indian young child. [Algonquian]

paprika /'pæprɪkə/ *n.* **1** red pepper. **2** condiment made from this. [Magyar]

Pap test *n.* cervical smear test. [*Papanicolaou*, name of a US scientist]

papyrus /pə'paɪərəs/ *n.* (*pl.* **papyri** /-raɪ/) **1** aquatic plant of N. Africa. **2 a** writing-material made in ancient Egypt from the pithy stem of this. **b** text written on this. [Latin from Greek]

par *n.* **1** average or normal amount, degree, condition, etc. (*feel below par*). **2** equality; equal status or footing (*on a par with*). **3** *Golf* number of strokes a first-class player should normally require for a hole or course. **4** face value of stocks and shares etc. (*at par*). **5** (in full **par of exchange**) recognized value of one country's currency in terms of another's. □ **par for the course** *colloq.* what is normal or to be expected. [Latin, = equal]

par- var. of PARA-[1] before a vowel or *h* (*parody*).

para /'pærə/ *n. colloq.* **1** paratrooper. **2** paragraph. [abbreviation]

para-[1] *prefix* (also **par-**) **1** beside (*paramilitary*). **2** beyond (*paranormal*). [Greek]

para-[2] *comb. form* protect, ward off (*parachute*; *parasol*). [Latin *paro* defend]

parable /'pærəb(ə)l/ *n.* **1** story used to illustrate a moral or spiritual lesson. **2** allegory. [Greek *parabolē* comparison]

parabola /pə'ræbələ/ *n.* open plane curve formed by the intersection of a cone with a plane parallel to its side. □ **parabolic** /,pærə'bɒlɪk/ *adj.* [Greek *parabolē* placing side by side: related to PARABLE]

paracetamol /,pærə'si:tə,mɒl/ *n.* **1** drug used to relieve pain and reduce fever. **2** tablet of this. [from *para-acetylamino*-phen*ol*]

parachute /'pærə,ʃu:t/ —*n.* rectangular or umbrella-shaped apparatus allowing a slow and safe descent esp. from an aircraft, or used to retard forward motion etc. (often *attrib.*: *parachute troops*). —*v.* (**-ting**) convey or descend by parachute. □ **parachutist** *n.* [French: related to PARA-[2], CHUTE[1]]

parade /pə'reɪd/ —*n.* **1** public procession. **2 a** ceremonial muster of troops for inspection. **b** = PARADE-GROUND. **3** ostentatious display (*made a parade of their wealth*). **4** public square, promenade, or row of shops. —*v.* (**-ding**) **1** march ceremonially. **2** assemble for parade. **3** display ostentatiously. **4** march through (streets etc.) in procession. □ **on parade 1** taking part in a parade. **2** on display. [Latin *paro* prepare]

parade-ground *n.* place for the muster and drilling of troops.

paradiddle /'pærə,dɪd(ə)l/ *n.* drum roll with alternate beating of sticks. [imitative]

paradigm /'pærə,daɪm/ *n.* example or pattern, esp. a set of noun or verb inflections. □ **paradigmatic** /-dɪg'mætɪk/ *adj.* [Latin from Greek]

paradise /'pærə,daɪs/ *n.* **1** (in some religions) heaven. **2** place or state of complete happiness. **3** (in full **earthly paradise**) abode of Adam and Eve; garden of Eden. □ **paradisaical** /-dɪ'seɪɪk(ə)l/ *adj.* **paradisal** /'pærə,daɪs(ə)l/ *adj.* **paradisiacal** /-dɪ'saɪək(ə)l/ *adj.* **paradisical** /-'dɪsɪk(ə)l/ *adj.* [Greek *paradeisos*]

paradox /'pærə,dɒks/ *n.* **1 a** seemingly absurd or contradictory though often true statement. **b** self-contradictory or absurd statement. **2** person or thing having contradictory qualities etc. **3** paradoxical quality. □ **paradoxical** /-'dɒksɪk(ə)l/ *adj.* **paradoxically** /-'dɒksɪkəlɪ/ *adv.* [Greek: related to PARA-[1], *doxa* opinion]

paraffin /'pærəfɪn/ *n.* **1** inflammable waxy or oily hydrocarbon distilled from petroleum or shale, used in liquid form (also **paraffin oil**) esp. as a fuel. **2** *Chem.* = ALKANE. [Latin, = having little affinity]

paraffin wax *n.* paraffin in its solid form.

paragon /'pærəgən/ *n.* (often foll. by *of*) model of excellence etc. [Greek *parakonē*]

paragraph /'pærə,grɑ:f/ —*n.* **1** distinct section of a piece of writing, beginning on a new often indented line. **2** symbol (usu. ¶) used to mark a new paragraph, or as a reference mark. **3** short item in a newspaper. —*v.* arrange (a piece of writing) in paragraphs. [Greek: related to PARA-[1], -GRAPH]

parakeet /'pærə,ki:t/ *n.* small usu. long-tailed parrot. [French: related to PARROT]

parallax /'pærə,læks/ *n.* **1** apparent difference in the position or direction of an object caused when the observer's position is changed. **2** angular amount of this. [Greek, = change]

parallel /'pærə,lel/ —*adj.* **1 a** (of lines or planes) continuously side by side and equidistant. **b** (foll. by *to, with*) (of a line or plane) having this relation (to or with another). **2** (of circumstances etc.) precisely similar, analogous, or corresponding. **3 a** (of processes etc.) occurring or performed simultaneously. **b** *Computing* involving the simultaneous performance of operations. —*n.* **1** person or thing precisely analogous to another. **2** comparison (*drew a parallel between them*). **3** (in full **parallel of latitude**) **a** each of the imaginary parallel circles of constant latitude on the earth's surface. **b** corresponding line on a map (*49th parallel*). **4** *Printing* two parallel lines (‖) as a reference mark. —*v.* (**-l-**) **1** be parallel, or correspond, to. **2** represent as similar; compare. **3** cite as a parallel instance. □ **in parallel** (of electric circuits) arranged so as to join at common points at each end. □ **parallelism** *n.* [Greek, = alongside one another]

parallel bars *n.pl.* pair of parallel rails on posts for gymnastics.

parallelepiped /,pærəle'lepɪ,ped, -lə'paɪpɪd/ *n.* solid body of which each face is a parallelogram. [Greek: related to PARALLEL, *epipedon* plane surface]

parallelogram /,pærə'lelə,græm/ *n.* four-sided plane rectilinear figure with opposite sides parallel.

paralyse /'pærə,laɪz/ *v.* (*US* **paralyze**) (**-sing** or **-zing**) **1** affect with paralysis. **2**

render powerless; cripple. [Greek: related to PARA-[1], *luō* loosen]

paralysis /pəˈrælɪsɪs/ *n.* **1** impairment or loss of esp. the motor function of the nerves, causing immobility. **2** powerlessness.

paralytic /ˌpærəˈlɪtɪk/ *–adj.* **1** affected by paralysis. **2** *slang* very drunk. *–n.* person affected by paralysis.

paramedic /ˌpærəˈmedɪk/ *n.* paramedical worker.

paramedical *adj.* (of services etc.) supplementing and assisting medical work.

parameter /pəˈræmɪtə(r)/ *n.* **1** *Math.* quantity constant in the case considered but varying in different cases. **2 a** (esp. measurable or quantifiable) characteristic or feature. **b** (loosely) limit or boundary, esp. of a subject for discussion. [Greek PARA-[1], -METER]

paramilitary /ˌpærəˈmɪlɪtərɪ/ *–adj.* (of forces) organized on military lines. *–n.* (*pl.* **-ies**) member of an unofficial paramilitary organization, esp. in N. Ireland.

paramount /ˈpærəˌmaʊnt/ *adj.* **1** supreme; most important. **2** in supreme authority. [Anglo-French *par* by, *amont* above: see AMOUNT]

paramour /ˈpærəˌmʊə(r)/ *n. archaic* or *derog.* illicit lover of a married person. [French *par amour* by love]

paranoia /ˌpærəˈnɔɪə/ *n.* **1** mental disorder with delusions of persecution and self-importance. **2** abnormal suspicion and mistrust. □ **paranoiac** *adj.* & *n.* **paranoiacally** *adv.* **paranoic** /-ˈnəʊɪk, -ˈnɔɪɪk/ *adj.* **paranoically** /-ˈnəʊɪkəlɪ, -ˈnɔɪɪkəlɪ/ *adv.* **paranoid** /ˈpærəˌnɔɪd/ *adj.* & *n.* [Greek: related to NOUS]

paranormal /ˌpærəˈnɔ:m(ə)l/ *adj.* beyond the scope of normal scientific investigation or explanation.

parapet /ˈpærəpɪt/ *n.* **1** low wall at the edge of a roof, balcony, bridge, etc. **2** defence of earth or stone. [French or Italian: related to PARA-[2], *petto* breast]

paraphernalia /ˌpærəfəˈneɪlɪə/ *n.pl.* (also treated as *sing.*) miscellaneous belongings, equipment, accessories, etc. [Greek: related to PARA-[1], *phernē* dower]

paraphrase /ˈpærəˌfreɪz/ *–n.* expression of a passage in other words. *–v.* (**-sing**) express the meaning of (a passage) thus. [Greek: related to PARA-[1]]

paraplegia /ˌpærəˈpli:dʒə/ *n.* paralysis below the waist. □ **paraplegic** *adj.* & *n.* [Greek: related to PARA-[1], *plēssō* strike]

parapsychology /ˌpærəsaɪˈkɒlədʒɪ/ *n.* the study of mental phenomena outside the sphere of ordinary psychology (hypnosis, telepathy, etc.).

paraquat /ˈpærəˌkwɒt/ *n.* a quick-acting highly toxic herbicide. [from PARA-[1], QUATERNARY]

parascending /ˈpærəˌsendɪŋ/ *n.* sport in which participants wearing open parachutes are towed behind a vehicle or motor boat to gain height before release for a conventional descent.

parasite /ˈpærəˌsaɪt/ *n.* **1** organism living in or on another and feeding on it. **2** person exploiting another or others. □ **parasitic** /-ˈsɪtɪk/ *adj.* **parasitically** /-ˈsɪtɪkəlɪ/ *adv.* **parasitism** *n.* [Greek: related to PARA-[1], *sitos* food]

parasol /ˈpærəˌsɒl/ *n.* light umbrella giving shade from the sun. [Italian: related to PARA-[2], *sole* sun]

paratrooper /ˈpærəˌtru:pə(r)/ *n.* member of a body of paratroops.

paratroops /ˈpærəˌtru:ps/ *n.pl.* parachute troops. [contraction]

paratyphoid /ˌpærəˈtaɪfɔɪd/ *n.* (often *attrib.*) fever resembling typhoid.

par avion /ˌpɑ:r æˈvjɔ̃/ *adv.* by airmail. [French, = by aeroplane]

parboil /ˈpɑ:bɔɪl/ *v.* boil until partly cooked. [Latin *par-* = PER-, confused with PART]

parcel /ˈpɑ:s(ə)l/ *–n.* **1** goods etc. wrapped up in a package for posting or carrying. **2** piece of land. **3** quantity dealt with in one commercial transaction. *–v.* (**-ll-**; *US* **-l-**) **1** (foll. by *up*) wrap as a parcel. **2** (foll. by *out*) divide into portions. [Latin: related to PARTICLE]

parch *v.* **1** make or become hot and dry. **2** roast (peas, corn, etc.) slightly. [origin unknown]

parchment /ˈpɑ:tʃmənt/ *n.* **1 a** skin, esp. of sheep or goat, prepared for writing or painting on. **b** manuscript written on this. **2** high-grade paper resembling parchment. [Latin *Pergamum*, now *Bergama* in Turkey]

pardon /ˈpɑ:d(ə)n/ *–n.* **1** forgiveness for an offence, error, etc. **2** (in full **free pardon**) remission of the legal consequences of a crime or conviction. *–v.* **1** forgive or excuse. **2** release from the legal consequences of an offence, error, etc. *–int.* (also **pardon me** or **I beg your pardon**) **1** formula of apology or disagreement. **2** request to repeat something said. □ **pardonable** *adj.* [Latin *perdono*: related to PER-, *dono* give]

pare /peə(r)/ *v.* (**-ring**) **1 a** trim or shave by cutting away the surface or edge. **b** (often foll. by *off, away*) cut off (the surface or edge). **2** (often foll. by *away, down*) diminish little by little. [Latin *paro* prepare]

parent /ˈpeərənt/ *–n.* **1** person who has or adopts a child; father or mother. **2** animal or plant from which others are derived. **3** (often *attrib.*) source, origin, etc. –*v.* (also *absol.*) be the parent of. □ **parental** /pə'rent(ə)l/ *adj.* **parenthood** *n.* [Latin *pario* bring forth]

parentage *n.* lineage; descent from or through parents.

parent company *n.* company of which others are subsidiaries.

parenthesis /pə'renθəsɪs/ *n.* (*pl.* **parentheses** /-ˌsiːz/) **1 a** explanatory or qualifying word, clause, or sentence inserted into a sentence etc., and usu. marked off by brackets, dashes, or commas. **b** (in *pl.*) round brackets ( ) used for this. **2** interlude or interval. □ **parenthetic** /ˌpærən'θetɪk/ *adj.* **parenthetically** /ˌpærən'θetɪkəlɪ/ *adv.* [Greek: related to PARA-[1], EN-, THESIS]

parenting *n.* (skill of) bringing up children.

parent-teacher association *n.* social and fund-raising organization of a school's parents and teachers.

par excellence /ˌpɑːr eksə'lɑ̃s/ *adv.* being the supreme example of its kind (*the short story par excellence*). [French]

parfait /'pɑːfeɪ/ *n.* **1** rich iced pudding of whipped cream, eggs, etc. **2** layers of ice-cream, meringue, etc., served in a tall glass. [French *parfait* PERFECT]

pariah /pə'raɪə/ *n.* **1** social outcast. **2** *hist.* member of a low caste or of no caste in S. India. [Tamil]

parietal /pə'raɪət(ə)l/ *adj.* of the wall of the body or any of its cavities. [Latin *paries* wall]

parietal bone *n.* either of a pair of bones in the skull.

paring *n.* strip or piece cut off.

parish /'pærɪʃ/ *n.* **1** area having its own church and clergyman. **2** (in full **civil parish**) local government district. **3** inhabitants of a parish. [Latin *parochia* from Greek *oikos* dwelling]

parish clerk *n.* official performing various duties for a church.

parish council *n.* administrative body in a civil parish.

parishioner /pə'rɪʃənə(r)/ *n.* inhabitant of a parish. [obsolete *parishen*: related to PARISH]

parish register *n.* book recording christenings, marriages, and burials, at a parish church.

parity /'pærɪtɪ/ *n.* **1** equality, equal status or pay. **2** parallelism or analogy (*parity of reasoning*). **3** equivalence of one currency with another; being at par. [Latin *paritas*: related to PAR]

park –*n.* **1** large public garden in a town, for recreation. **2** land attached to a country house etc. **3 a** large area of uncultivated land for public recreational use. **b** large enclosed area where wild animals are kept in captivity (*wildlife park*). **4** area for parking vehicles etc. (*car park*). **5** area for a specified purpose (*business park*). **6 a** *US* sports ground. **b** (usu. prec. by *the*) football pitch. –*v.* **1** (also *absol.*) leave (a vehicle) temporarily. **2** *colloq.* deposit and leave, usu. temporarily. □ **park oneself** *colloq.* sit down. [French from Germanic]

parka /'pɑːkə/ *n.* **1** long usu. green anorak with fur round the hood. **2** hooded skin jacket worn by Eskimos. [Aleutian]

parkin /'pɑːkɪn/ *n.* cake of ginger, oatmeal, treacle, etc. [origin uncertain]

parking-lot *n.* *US* outdoor car park.

parking-meter *n.* coin-operated meter allocating a length of time for which a vehicle may be parked in a street.

parking-ticket *n.* notice of a penalty imposed for parking illegally.

Parkinson's disease /'pɑːkɪns(ə)nz/ *n.* (also **Parkinsonism**) progressive disease of the nervous system with tremor, muscular rigidity, and emaciation. [*Parkinson*, name of a surgeon]

Parkinson's law /'pɑːkɪns(ə)nz/ *n.* notion that work expands to fill the time available for it. [*Parkinson*, name of a writer]

parkland *n.* open grassland with trees etc.

parky /'pɑːkɪ/ *adj.* (**-ier, -iest**) *colloq.* or *dial.* chilly. [origin unknown]

parlance /'pɑːləns/ *n.* vocabulary or idiom of a particular subject, group, etc. [French from *parler* speak]

parley /'pɑːlɪ/ –*n.* (*pl.* **-s**) conference of disputants, esp. to discuss peace terms etc. –*v.* (**-leys, -leyed**) (often foll. by *with*) hold a parley. [French *parler*: related to PARLANCE]

parliament /'pɑːləmənt/ *n.* **1** (**Parliament**) **a** (in the UK) highest legislature, consisting of the Sovereign, the House of Lords, and the House of Commons. **b** members of this for a particular period, esp. between elections. **2** similar legislature in other States. [French: related to PARLANCE]

parliamentarian /ˌpɑːləmen'teərɪən/ *n.* member of a parliament, esp. an expert in its procedures.

parliamentary /ˌpɑːlə'mentərɪ/ *adj.* **1** of a parliament. **2** enacted or established by a parliament. **3** (of language, behaviour, etc.) polite.

parlour /'pɑːlə(r)/ *n.* (*US* **parlor**) **1** *archaic* sitting-room in a private house. **2** esp. *US* shop providing specified goods or services (*beauty parlour*; *ice-cream* *parlour*). [Anglo-French: related to PARLEY]

parlour game *n.* indoor game, esp. a word-game.

parlous /'pɑ:ləs/ *adj. archaic* or *joc.* dangerous or difficult. [from PERILOUS]

Parmesan /ˌpɑ:mɪ'zæn/ *n.* hard dry cheese made orig. at Parma and usu. used grated. [Italian *parmegiano* of Parma]

parochial /pə'rəʊkɪəl/ *adj.* **1** of a parish. **2** (of affairs, views, etc.) merely local, narrow, or provincial. □ **parochialism** *n.* **parochially** *adv.* [Latin: related to PARISH]

parody /'pærədɪ/ *–n.* (*pl.* **-ies**) **1** humorous exaggerated imitation of an author, literary work, style, etc. **2** feeble imitation; travesty. *–v.* (**-ies, -ied**) **1** compose a parody of. **2** mimic humorously. □ **parodist** *n.* [Latin or Greek: related to PARA-[1], ODE]

parole /pə'rəʊl/ *–n.* **1** temporary or permanent release of a prisoner before the expiry of a sentence, on the promise of good behaviour. **2** such a promise. *–v.* (**-ling**) put (a prisoner) on parole. [French, = word: related to PARLANCE]

parotid /pə'rɒtɪd/ *–adj.* situated near the ear. *–n.* (in full **parotid gland**) salivary gland in front of the ear. [Greek: related to PARA-[1], *ous ōt-* ear]

paroxysm /'pærəkˌsɪz(ə)m/ *n.* **1** (often foll. by *of*) sudden attack or outburst (of rage, coughing, etc.). **2** fit of disease. □ **paroxysmal** /-'sɪzm(ə)l/ *adj.* [Greek *oxus* sharp]

parquet /'pɑ:keɪ/ *–n.* **1** flooring of wooden blocks arranged in a pattern. **2** *US* stalls of a theatre. *–v.* (**-eted** /-eɪd/; **-eting** /-eɪɪŋ/) floor (a room) thus. [French, diminutive of *parc* PARK]

parquetry /'pɑ:kɪtrɪ/ *n.* use of wooden blocks to make floors or inlay for furniture.

parr /pɑ:(r)/ *n.* young salmon. [origin unknown]

parricide /'pærɪˌsaɪd/ *n.* **1** murder of a near relative, esp. of a parent. **2** person who commits parricide. □ **parricidal** /-'saɪd(ə)l/ *adj.* [Latin: see PARENT, PATER, -CIDE]

parrot /'pærət/ *–n.* **1** mainly tropical bird with a short hooked bill, often vivid plumage, and the ability to mimic the human voice. **2** person who mechanically repeats another's words or actions. *–v.* (**-t-**) repeat mechanically. [French, diminutive of *Pierre* Peter]

parrot-fashion *adv.* (learning or repeating) mechanically, by rote.

parry /'pærɪ/ *–v.* (**-ies, -ied**) **1** avert or ward off (a weapon or attack), esp. with a countermove. **2** deal skilfully with (an awkward question etc.). *–n.* (*pl.* **-ies**) act of parrying. [Italian *parare* ward off]

parse /pɑ:z/ *v.* (**-sing**) **1** describe (a word in context) grammatically, stating its inflection, relation to the sentence, etc. **2** resolve (a sentence) into its component parts and describe them grammatically. [perhaps from French *pars* parts: related to PART]

parsec /'pɑ:sek/ *n.* unit of stellar distance, equal to about 3.25 light-years. [from PARALLAX, SECOND[2]]

parsimony /'pɑ:sɪmənɪ/ *n.* carefulness in the use of money etc.; stinginess. □ **parsimonious** /-'məʊnɪəs/ *adj.* [Latin *parco pars-* spare]

parsley /'pɑ:slɪ/ *n.* herb with crinkly aromatic leaves, used to season and garnish food. [Greek *petra* rock, *selinon* parsley]

parsnip /'pɑ:snɪp/ *n.* **1** plant with a pale-yellow tapering root. **2** this root eaten as a vegetable. [Latin *pastinaca*]

parson /'pɑ:s(ə)n/ *n.* **1** rector. **2** vicar; clergyman. [Latin: related to PERSON]

parsonage *n.* church house provided for a parson.

parson's nose *n.* fatty flesh at the rump of a cooked fowl.

part *–n.* **1** some but not all of a thing or group of things. **2** essential member, constituent, or component (*part of the family*; *spare parts*). **3** portion of a human or animal body. **4** division of a book, broadcast serial, etc., esp. issued or broadcast at one time. **5** each of several equal portions of a whole (*3 parts sugar to 2 parts flour*). **6 a** allotted share. **b** person's share in an action etc. (*had no part in it*). **c** duty (*not my part to interfere*). **7 a** character assigned to, or words spoken by, an actor on stage. **b** melody etc. assigned to a particular voice or instrument. **c** printed or written copy of an actor's or musician's part. **8** side in an agreement or dispute. **9** (in *pl.*) region or district (*am not from these parts*). **10** (in *pl.*) abilities (*man of many parts*). *–v.* **1** divide or separate into parts (*crowd parted*). **2 a** leave one another's company (*parted the best of friends*). **b** (foll. by *from*) say goodbye to. **3** (foll. by *with*) give up; hand over. **4** separate (hair of the head) to make a parting. *–adv.* in part; partly (*part iron and part wood*). □ **for the most part** see MOST. **for one's part** as far as one is concerned. **in part** (or **parts**) partly. **on the part of** made or done by (*no objection on my part*). **part and parcel** (usu. foll. by *of*) an essential part. **part company** see COMPANY. **play a part 1** be significant or contributory. **2** act deceitfully. **3** perform a theatrical role. **take**

in good part not be offended by. **take part** (often foll. by *in*) assist or have a share (in). **take the part of** support; side with. [Latin *pars part-*]

partake /pɑːˈteɪk/ *v.* (**-king**; *past* **partook**; *past part.* **partaken**) **1** (foll. by *of*, *in*) take a share or part. **2** (foll. by *of*) eat or drink some or *colloq.* all (of a thing). [back-formation from *partaker* = *part-taker*]

parterre /pɑːˈteə(r)/ *n.* **1** level space in a formal garden occupied by flower-beds. **2** *US* pit of a theatre. [French, = on the ground]

part-exchange *–n.* transaction in which goods are given as part of the payment. *–v.* give (goods) thus.

parthenogenesis /ˌpɑːθənəʊˈdʒenɪsɪs/ *n.* reproduction without fertilization, esp. in invertebrates and lower plants. [Greek *parthenos* virgin]

Parthian shot /ˈpɑːθɪən/ *n.* remark or glance etc. on leaving. [*Parthia*, ancient kingdom in W. Asia: from the custom of a retreating Parthian horseman firing a shot at the enemy]

partial /ˈpɑːʃ(ə)l/ *adj.* **1** not complete; forming only part. **2** biased. **3** (foll. by *to*) having a liking for. □ **partiality** /-ʃɪˈælɪtɪ/ *n.* **partially** *adv.* **partialness** *n.* [Latin: related to PART]

partial eclipse *n.* eclipse in which only part of the luminary is covered.

participant /pɑːˈtɪsɪpənt/ *n.* participator.

participate /pɑːˈtɪsɪˌpeɪt/ *v.* (**-ting**) (often foll. by *in*) take part or a share (in). □ **participation** /-ˈpeɪʃ(ə)n/ *n.* **participator** *n.* **participatory** *adj.* [Latin *particeps -cip-* taking PART]

participle /ˈpɑːtɪˌsɪp(ə)l/ *n.* word formed from a verb (e.g. *going*, *gone*, *being*, *been*) and used in compound verb-forms (e.g. *is going*, *has been*) or as an adjective (e.g. *working woman*, *burnt toast*). □ **participial** /-ˈsɪpɪəl/ *adj.* [Latin: related to PARTICIPATE]

particle /ˈpɑːtɪk(ə)l/ *n.* **1** minute portion of matter. **2** smallest possible amount (*particle of sense*). **3 a** minor part of speech, esp. a short undeclinable one. **b** common prefix or suffix such as *in-*, *-ness*. [Latin *particula* diminutive of *pars* PART]

particoloured /ˈpɑːtɪˌkʌləd/ *adj.* (*US* **-colored**) of more than one colour. [related to PART, COLOUR]

particular /pəˈtɪkjʊlə(r)/ *–adj.* **1** relating to or considered as one thing or person as distinct from others; individual (*in this particular case*). **2** more than is usual; special (*took particular care*). **3** scrupulously exact; fastidious. **4** detailed (*full and particular account*). *–n.* **1** detail; item. **2** (in *pl.*) information; detailed account. □ **in particular** especially, specifically. [Latin: related to PARTICLE]

particularity /pəˌtɪkjʊˈlærɪtɪ/ *n.* **1** quality of being individual or particular. **2** fullness or minuteness of detail.

particularize /pəˈtɪkjʊləˌraɪz/ *v.* (also **-ise**) (**-zing** or **-sing**) (also *absol.*) **1** name specially or one by one. **2** specify (items). □ **particularization** /-ˈzeɪʃ(ə)n/ *n.*

particularly /pəˈtɪkjʊləlɪ/ *adv.* **1** especially, very. **2** specifically (*particularly asked for you*). **3** in a particular or fastidious manner.

parting *n.* **1** leave-taking or departure (often *attrib.*: *parting words*). **2** dividing line of combed hair. **3** division; separating.

parting shot *n.* = PARTHIAN SHOT.

partisan /ˌpɑːtɪˈzæn/ (also **partizan**) *–n.* **1** strong, esp. unreasoning, supporter of a party, cause, etc. **2** guerrilla. *–adj.* **1** of partisans. **2** biased. □ **partisanship** *n.* [Italian: related to PART]

partition /pɑːˈtɪʃ(ə)n/ *–n.* **1** structure dividing a space, esp. a light interior wall. **2** division into parts, esp. *Polit.* of a country. *–v.* **1** divide into parts. **2** (foll. by *off*) separate (part of a room etc.) with a partition. [Latin *partior* divide]

partitive /ˈpɑːtɪtɪv/ *–adj.* (of a word, form, etc.) denoting part of a collective group or quantity. *–n.* partitive word (e.g. *some*, *any*) or form. [French or medieval Latin: related to PARTITION]

partizan var. of PARTISAN.

partly *adv.* **1** with respect to a part or parts. **2** to some extent.

partner /ˈpɑːtnə(r)/ *–n.* **1** person who shares or takes part with another or others, esp. in a business. **2** companion in dancing. **3** player (esp. one of two) on the same side in a game. **4** either member of a married or unmarried couple. *–v.* be the partner of. [alteration of *parcener* joint heir]

partnership *n.* **1** state of being a partner or partners. **2** joint business. **3** pair or group of partners.

part of speech *n.* grammatical class of words (in English noun, pronoun, adjective, adverb, verb, etc.).

partook *past* of PARTAKE.

partridge /ˈpɑːtrɪdʒ/ *n.* (*pl.* same or **-s**) game-bird, esp. European or Asian. [Greek *perdix*]

part-song *n.* song with three or more voice-parts, often unaccompanied.

part-time *–adj.* (esp. of a job) occupying less than the normal working week etc. *–adv.* (also **part time**) as a part-time activity (*works part time*).

part-timer *n.* person employed in part-time work.

parturient /pɑːˈtjʊərɪənt/ *adj. formal* about to give birth. [Latin *pario part-* bring forth]

parturition /ˌpɑːtjʊˈrɪʃ(ə)n/ *n. formal* giving birth.

party /ˈpɑːtɪ/ —*n.* (*pl.* **-ies**) **1** social gathering, usu. of invited guests. **2** people working or travelling together (*search party*). **3** political group putting forward candidates in elections and usu. organized on a national basis. **4** each side in an agreement or dispute. **5** (foll. by *to*) *Law* accessory (to an action). **6** *colloq.* person. —*v.* (**-ies, -ied**) attend a party; celebrate. [Romanic: related to PART]

party line *n.* **1** policy adopted by a political party etc. **2** shared telephone line.

party-wall *n.* wall common to adjoining buildings or rooms.

parvenu /ˈpɑːvəˌnjuː/ *n.* (*pl.* **-s**; *fem.* **parvenue**) (often *attrib.*) newly rich social climber; upstart. [Latin: related to PER-, *venio* come]

pas /pɑː/ *n.* (*pl.* same) step, esp. in ballet. [French, = step]

pascal *n.* **1** /ˈpæsk(ə)l/ SI unit of pressure. **2** (**Pascal** /pæsˈkɑːl/) *Computing* programming language used esp. in education. [*Pascal*, name of a scientist]

paschal /ˈpæsk(ə)l/ *adj.* **1** of the Jewish Passover. **2** of Easter. [Hebrew *pesaḥ*]

pas de deux /ˌpɑː də ˈdɜː/ *n.* dance for two. [French, = step for two]

pash *n. slang* brief infatuation. [abbreviation of PASSION]

pasha /ˈpɑːʃə/ *n. hist.* title (placed after the name) of a Turkish military commander, governor, etc. [Turkish]

Pashto /ˈpʌʃtəʊ/ —*n.* language of Afghanistan, parts of Pakistan, etc. —*adj.* of or in this language. [Pashto]

paso doble /ˌpæsəʊ ˈdəʊbleɪ/ *n.* Latin American ballroom dance. [Spanish, = double step]

pasque-flower /ˈpæsk/ *n.* a kind of anemone with bell-shaped purple flowers. [French *passe-fleur*]

pass[1] /pɑːs/ —*v.* **1** (often foll. by *along*, *by*, *down*, *on*, etc.) move onward, esp. past something. **2 a** go past; leave on one side or behind. **b** overtake, esp. in a vehicle. **3** (cause to) be transferred from one person or place to another (*title passes to his son*; *pass the butter*). **4** surpass; exceed (*passes all understanding*). **5** get through. **6 a** go unremarked or uncensured (*let the matter pass*). **b** (foll. by *as*, *for*) be accepted or known as. **7** move; cause to go (*passed her hand over her face*). **8 a** be successful or adequate, esp. in an examination. **b** be successful in (an examination). **c** (of an examiner) judge (a candidate) to be satisfactory. **9 a** (of a bill) be approved by (Parliament etc.). **b** cause or allow (a bill) to proceed. **c** (of a bill or proposal) be approved. **10** occur, elapse; happen (*time passes slowly*; *heard what passed*). **11** (cause to) circulate; be current. **12** spend (time or a period) (*passed the afternoon reading*). **13** (also *absol.*) (in field games) send (the ball) to a team-mate. **14 a** forgo one's turn or chance. **b** leave a quiz question etc. unanswered. **15** (foll. by *to*, *into*, *from*) change (from one form or state to another). **16** come to an end. **17** discharge (esp. faeces or urine) from the body. **18** (foll. by *on*, *upon*) utter (legal sentence, criticism) upon; adjudicate. —*n.* **1** act of passing. **2 a** success in an examination. **b** university degree without honours. **3 a** permit, esp. for admission, leave, etc. **b** ticket or permit giving free entry, access, travel, etc. **4** (in field games) transference of the ball to a team-mate. **5** desperate position (*come to a fine pass*). □ **in passing** in the course of conversation etc. **make a pass at** *colloq.* make sexual advances to. **pass away 1** *euphem.* die. **2** cease to exist. **pass by 1** go past. **2** disregard, omit. **pass muster** see MUSTER. **pass off 1** (of feelings etc.) disappear gradually. **2** (of proceedings) be carried through (in a specified way). **3** (foll. by *as*) misrepresent or disguise (a person or thing) as something else. **4** evade or lightly dismiss (an awkward remark etc.). **pass on 1** proceed. **2** *euphem.* die. **3** transmit to the next person in a series. **pass out 1** become unconscious. **2** complete military training. **pass over 1** omit, ignore, or disregard. **2** ignore the claims of (a person) to promotion etc. **3** *euphem.* die. **pass round 1** distribute. **2** give to one person after another. **pass the time of day** see TIME. **pass up** *colloq.* refuse or neglect (an opportunity etc.). **pass water** urinate. [Latin *passus* PACE[1]]

pass[2] /pɑːs/ *n.* narrow way through mountains. [var. of PACE[1]]

passable *adj.* **1** barely satisfactory; adequate. **2** (of a road, pass, etc.) that can be traversed. □ **passably** *adv.*

passage /ˈpæsɪdʒ/ *n.* **1** process or means of passing; transit. **2** = PASSAGEWAY. **3** liberty or right to pass through. **4** journey by sea or air. **5** transition from one state to another. **6** short extract from a book, piece of music, etc. **7** passing of a bill etc. into law. **8** duct etc. in the body. [French: related to PASS[1]]

passageway *n.* narrow path or way; corridor.

passbook *n.* book issued to an account-holder recording deposits and withdrawals.
passé /ˈpæseɪ/ *adj.* (*fem.* ***passée***) **1** old-fashioned. **2** past its prime. [French]
passenger /ˈpæsɪndʒə(r)/ *n.* **1** (often *attrib.*) traveller in or on a vehicle (other than the driver, pilot, crew, etc.) (*passenger seat*). **2** *colloq.* idle member of a team, crew, etc. [French *passager*: related to PASSAGE]
passer-by *n.* (*pl.* **passers-by**) person who goes past, esp. by chance.
passerine /ˈpæsə,ri:n/ *–n.* perching bird such as the sparrow and most land birds. *–adj.* of passerines. [Latin *passer* sparrow]
passim /ˈpæsɪm/ *adv.* throughout; at several points in a book, article, etc. [Latin]
passion /ˈpæʃ(ə)n/ *n.* **1** strong emotion. **2** outburst of anger (*flew into a passion*). **3** intense sexual love. **4 a** strong enthusiasm (*passion for football*). **b** object arousing this. **5** (**the Passion**) **a** suffering of Christ during his last days. **b** Gospel account of this. **c** musical setting of this. □ **passionless** *adj.* [Latin *patior pass-* suffer]
passionate /ˈpæʃənət/ *adj.* dominated, displaying, or caused by strong emotion. □ **passionately** *adv.*
passion-flower *n.* climbing plant with a flower supposedly suggestive of the instruments of the Crucifixion.
passion-fruit *n.* edible fruit of some species of passion-flower.
passion-play *n.* miracle play representing the Passion.
Passion Sunday *n.* fifth Sunday in Lent.
passive /ˈpæsɪv/ *adj.* **1** acted upon, not acting. **2** showing no interest or initiative; submissive. **3** *Chem.* not active; inert. **4** *Gram.* indicating that the subject undergoes the action of the verb (e.g. in *they were seen*). □ **passively** *adv.* **passivity** /-ˈsɪvɪtɪ/ *n.* [Latin: related to PASSION]
passive resistance *n.* non-violent refusal to cooperate.
passive smoking *n.* involuntary inhalation of others' cigarette smoke.
passkey *n.* **1** private key to a gate etc. **2** master-key.
passmark *n.* minimum mark needed to pass an examination.
Passover /ˈpɑ:s,əʊvə(r)/ *n.* Jewish spring festival commemorating the Exodus from Egypt. [from PASS[1], OVER]
passport *n.* **1** official document certifying the holder's identity and citizenship, and authorizing travel abroad. **2** (foll. by *to*) thing that ensures admission or attainment (*passport to success*). [French *passeport*: related to PASS[1], PORT[1]]
password *n.* prearranged selected word or phrase securing recognition, admission, etc.
past /pɑ:st/ *–adj.* **1** gone by in time (*in past years*; *the time is past*). **2** recently gone by (*the past month*). **3** of a former time (*past president*). **4** *Gram.* expressing a past action or state. *–n.* **1** (prec. by *the*) **a** past time. **b** past events (*cannot undo the past*). **2** person's past life, esp. if discreditable (*man with a past*). **3** past tense or form. *–prep.* **1** beyond in time or place (*is past two o'clock*; *lives just past the pub*). **2** beyond the range, duration, or compass of (*past endurance*). *–adv.* so as to pass by (*ran past*). □ **not put it past** believe it possible of (a person). **past it** *colloq.* old and useless. [from PASS[1]]
pasta /ˈpæstə/ *n.* dried flour paste in various shapes (e.g. lasagne or spaghetti). [Italian: related to PASTE]
paste /peɪst/ *–n.* **1** any moist fairly stiff mixture, esp. of powder and liquid. **2** dough of flour with fat, water, etc. **3** liquid adhesive used for sticking paper etc. **4** meat or fish spread (*anchovy paste*). **5** hard glasslike composition used for imitation gems. *–v.* (**-ting**) **1** fasten or coat with paste. **2** *slang* **a** beat or thrash. **b** bomb or bombard heavily. □ **pasting** *n.* (esp. in sense 2 of *v.*). [Latin *pasta* lozenge, from Greek]
pasteboard *n.* **1** stiff material made by pasting together sheets of paper. **2** (*attrib.*) flimsy, unsubstantial.
pastel /ˈpæst(ə)l/ *n.* **1** (often *attrib.*) light shade of a colour (*pastel blue*). **2** crayon of powdered pigments bound with a gum solution. **3** drawing in pastel. [French *pastel*, or Italian *pastello* diminutive of PASTA]
pastern /ˈpæst(ə)n/ *n.* part of a horse's foot between fetlock and hoof. [French from Latin]
paste-up *n.* document prepared for copying etc. by pasting sections on to a backing.
pasteurize /ˈpɑ:stʃə,raɪz/ *v.* (also **-ise**) (**-zing** or **-sing**) partially sterilize (milk etc.) by heating. □ **pasteurization** /-ˈzeɪʃ(ə)n/ *n.* [*Pasteur*, name of a chemist]
pastiche /pæˈsti:ʃ/ *n.* **1** picture or musical composition from or imitating various sources. **2** literary or other work composed in the style of a well-known author etc. [Latin *pasta* PASTE]
pastille /ˈpæstɪl/ *n.* small sweet or lozenge. [French from Latin]

pastime /'pɑːstaɪm/ *n.* recreation, hobby. [from PASS[1], TIME]

past master *n.* expert.

pastor /'pɑːstə(r)/ *n.* minister, esp. of a Nonconformist church. [Latin *pasco past-* feed]

pastoral /'pɑːstər(ə)l/ *–adj.* **1** of shepherds, flocks, or herds. **2** (of land) used for pasture. **3** (of a poem, picture, etc.) portraying (esp. romanticized) country life. **4** of a pastor. *–n.* **1** pastoral poem, play, picture, etc. **2** letter from a pastor (esp. a bishop) to the clergy or people. [Latin *pastoralis*: related to PASTOR]

pastorale /ˌpæstə'rɑːl/ *n.* (*pl.* **-s** or **-li** /-liː/) musical work with a rustic theme or atmosphere. [Italian: related to PASTORAL]

pastorate /'pɑːstərət/ *n.* **1** office or tenure of a pastor. **2** body of pastors.

pastrami /pæ'strɑːmɪ/ *n.* seasoned smoked beef. [Yiddish]

pastry /'peɪstrɪ/ *n.* (*pl.* **-ies**) **1** dough of flour, fat, and water used as a base and covering for pies etc. **2** cake etc. made wholly or partly of this. [from PASTE]

pastry-cook *n.* cook who specializes in pastry.

pasturage /'pɑːstʃərɪdʒ/ *n.* **1** land for pasture. **2** pasturing of cattle etc.

pasture /'pɑːstʃə(r)/ *–n.* **1** grassland suitable for grazing. **2** herbage for animals. *–v.* (**-ring**) **1** put (animals) to pasture. **2** (of animals) graze. [Latin: related to PASTOR]

pasty[1] /'pæstɪ/ *n.* (*pl.* **-ies**) pastry shaped around esp. a meat and vegetable filling. [Latin: related to PASTE]

pasty[2] /'peɪstɪ/ *adj.* (**-ier, -iest**) unhealthily pale (*pasty-faced*). □ **pastiness** *n.*

Pat. *abbr.* Patent.

pat[1] *–v.* (**-tt-**) **1** strike gently with a flat palm, esp. in affection, sympathy, etc. **2** flatten or mould by patting. *–n.* **1** light stroke or tap, esp. with the hand in affection etc. **2** sound made by this. **3** small mass (esp. of butter) formed by patting. □ **pat on the back** congratulatory gesture. [probably imitative]

pat[2] *–adj.* **1** prepared or known thoroughly. **2** apposite or opportune, esp. glibly so (*a pat answer*). *–adv.* **1** in a pat manner. **2** appositely. □ **have off pat** know or have memorized perfectly. [related to PAT[1]]

patch *–n.* **1** material used to mend a hole or as reinforcement. **2** shield protecting an injured eye. **3** large or irregular distinguishable area. **4** *colloq.* period of a specified, esp. unpleasant, kind (*went through a bad patch*). **5** piece of ground. **6** *colloq.* area assigned to, or patrolled by, esp. a police officer. **7** plants growing in one place (*cabbage patch*). **8** scrap, remnant. *–v.* **1** (often foll. by *up*) repair with a patch or patches. **2** (of material) serve as a patch to. **3** (often foll. by *up*) put together, esp. hastily. **4** (foll. by *up*) settle (a quarrel etc.), esp. hastily or temporarily. □ **not a patch on** *colloq.* greatly inferior to. [perhaps French, var. of PIECE]

patchboard *n.* board with electrical sockets linked by movable leads to enable changeable permutations of connection.

patchouli /pə'tʃuːlɪ/ *n.* **1** strongly scented E. Indian plant. **2** perfume from this. [native name in Madras]

patch pocket *n.* piece of cloth sewn on a garment as a pocket.

patch test *n.* test for allergy by applying patches of allergenic substances to the skin.

patchwork *n.* **1** (often. *attrib.*) stitching together of small pieces of variegated cloth to form a pattern (*patchwork quilt*). **2** thing composed of fragments etc.

patchy *adj.* (**-ier, -iest**) **1** uneven in quality. **2** having or existing in patches. □ **patchily** *adv.* **patchiness** *n.*

pate *n. archaic* or *colloq.* head. [origin unknown]

pâté /'pæteɪ/ *n.* paste of mashed and spiced meat or fish etc. [French, = PASTY[1]]

pâté de foie gras /ˌpæteɪ də fwɑː 'grɑː/ *n.* fatted goose liver pâté. [French]

patella /pə'telə/ *n.* (*pl.* **patellae** /-liː/) kneecap. □ **patellar** *adj.* [Latin, = pan, diminutive of *patina*: related to PATEN]

paten /'pæt(ə)n/ *n.* shallow dish for bread at the Eucharist. [Latin *patina*]

patent /'peɪt(ə)nt, 'pæt-/ *–n.* **1** official document conferring a right or title, esp. the sole right to make, use, or sell a specified invention. **2** invention or process so protected. *–adj.* **1** /'peɪt(ə)nt/ obvious, plain. **2** conferred or protected by patent. **3 a** proprietary. **b** to which one has a proprietary claim. *–v.* obtain a patent for (an invention). □ **patently** /'peɪtəntlɪ/ *adv.* (in sense 1 of *adj.*). [Latin *pateo* lie open]

patentee /ˌpeɪtən'tiː/ *n.* **1** person who takes out or holds a patent. **2** person entitled temporarily to the benefit of a patent.

patent leather *n.* glossy leather.

patent medicine *n.* proprietary medicine available without prescription.

patent office *n.* office issuing patents.

pater /'peɪtə(r)/ *n. colloq.* father. [Latin]

■ **Usage** *Pater* is now only found in jocular or affected use.

paterfamilias /ˌpeɪtəfəˈmɪlɪˌæs/ *n.* male head of a family or household. [Latin, = father of the family]
paternal /pəˈtɜ:n(ə)l/ *adj.* **1** of, like, or appropriate to a father; fatherly. **2** related through the father. **3** (of a government etc.) limiting freedom and responsibility by well-meant regulations. □ **paternally** *adv.* [Latin: related to PATER]
paternalism *n.* policy of governing or behaving in a paternal way. □ **paternalistic** /-ˈlɪstɪk/ *adj.*
paternity /pəˈtɜ:nɪtɪ/ *n.* **1** fatherhood. **2** one's paternal origin.
paternity suit *n.* lawsuit held to determine if a certain man is the father of a certain child.
paternoster /ˌpætəˈnɒstə(r)/ *n.* Lord's Prayer, esp. in Latin. [Latin *pater noster* our father]
path /pɑ:θ/ *n.* (*pl.* **paths** /pɑ:ðz/) **1** way or track made for or by walking. **2** line along which a person or thing moves (*flight path*). **3** course of action. [Old English]
pathetic /pəˈθetɪk/ *adj.* **1** arousing pity, sadness, or contempt. **2** *colloq.* miserably inadequate. □ **pathetically** *adv.* [Greek *pathos* from *paskhō* suffer]
pathetic fallacy *n.* attribution of human emotions to inanimate things, esp. in literature.
pathfinder *n.* explorer; pioneer.
pathogen /ˈpæθədʒ(ə)n/ *n.* agent causing disease. □ **pathogenic** /-ˈdʒenɪk/ *adj.* [Greek *pathos* suffering, -GEN]
pathological /ˌpæθəˈlɒdʒɪk(ə)l/ *adj.* **1** of pathology. **2** of or caused by physical or mental disorder (*pathological fear of spiders*). □ **pathologically** *adv.*
pathology /pəˈθɒlədʒɪ/ *n.* the study or symptoms of disease. □ **pathologist** *n.* [Greek *pathos*: related to PATHETIC]
pathos /ˈpeɪθɒs/ *n.* evocation of pity or sadness in speech, writing, etc. [Greek: related to PATHETIC]
pathway *n.* path or its course.
patience /ˈpeɪʃ(ə)ns/ *n.* **1** ability to endure delay, hardship, provocation, etc. **2** perseverance or forbearance. **3** solo card-game. [Latin: related to PASSION]
patient –*adj.* having or showing patience. –*n.* person receiving or registered to receive medical treatment. □ **patiently** *adv.*
patina /ˈpætɪnə/ *n.* (*pl.* **-s**) **1** film, usu. green, formed on old bronze. **2** similar film on other surfaces. **3** gloss produced by age on woodwork. [Latin: related to PATEN]
patio /ˈpætɪəʊ/ *n.* (*pl.* **-s**) **1** paved usu. roofless area adjoining a house. **2** inner roofless court in a Spanish or Spanish-American house. [Spanish]
patisserie /pəˈti:sərɪ/ *n.* **1** shop where pastries are made and sold. **2** pastries collectively. [Latin: related to PASTE]
Patna rice /ˈpætnə/ *n.* rice with long firm grains. [from *Patna* in India]
patois /ˈpætwɑ:/ *n.* (*pl.* same /-wɑ:z/) regional dialect, differing from the literary language. [French]
patriarch /ˈpeɪtrɪˌɑ:k/ *n.* **1** male head of a family or tribe. **2** (often in *pl.*) any of those regarded as fathers of the human race, esp. the sons of Jacob, or Abraham, Isaac, and Jacob, and their forefathers. **3** *Eccl.* **a** chief bishop in the Orthodox Church. **b** *RC Ch.* bishop ranking immediately below the pope. **4** venerable old man. □ **patriarchal** /-ˈɑ:k(ə)l/ *adj.* [Greek *patria* family, *arkhēs* ruler]
patriarchate /ˈpeɪtrɪˌɑ:kət/ *n.* **1** office, see, or residence of a Church patriarch. **2** rank of a tribal patriarch.
patriarchy /ˈpeɪtrɪˌɑ:kɪ/ *n.* (*pl.* **-ies**) male-dominated social system, with descent through the male line.
patrician /pəˈtrɪʃ(ə)n/ –*n. hist.* member of the nobility in ancient Rome. –*adj.* **1** aristocratic. **2** *hist.* of the ancient Roman nobility. [Latin *patricius*: related to PATER]
patricide /ˈpætrɪˌsaɪd/ *n.* = PARRICIDE (esp. with reference to the killing of one's father). □ **patricidal** /-ˈsaɪd(ə)l/ *adj.* [Latin, alteration of *parricida*]
patrimony /ˈpætrɪmənɪ/ *n.* (*pl.* **-ies**) **1** property inherited from one's father or ancestor. **2** heritage. □ **patrimonial** /-ˈməʊnɪəl/ *adj.* [Latin: related to PATER]
patriot /ˈpeɪtrɪət, ˈpæt-/ *n.* person devoted to and ready to defend his or her country. □ **patriotic** /-ˈɒtɪk/ *adj.* **patriotically** /-ˈɒtɪklɪ/ *adv.* **patriotism** *n.* [Greek *patris* fatherland]
patristic /pəˈtrɪstɪk/ *adj.* of the early Christian writers or their work. [Latin: related to PATER]
patrol /pəˈtrəʊl/ –*n.* **1** act of walking or travelling around an area, esp. regularly, for security or supervision. **2** guards, police, etc. sent out on patrol. **3** **a** troops sent out to reconnoitre. **b** such reconnaissance. **4** unit of six to eight Scouts or Guides. –*v.* (**-ll-**) **1** carry out a patrol of. **2** act as a patrol. [German *Patrolle* from French]
patrol car *n.* police car used for patrols.
patron /ˈpeɪtrən/ *n.* (*fem.* **patroness**) **1** person financially supporting a person, cause, etc. **2** customer of a shop etc. [Latin *patronus*: related to PATER]
patronage /ˈpætrənɪdʒ/ *n.* **1** patron's or customer's support. **2** right or control of

appointments to office, privileges, etc. **3** condescending manner.

patronize /ˈpætrəˌnaɪz/ *v.* (also **-ise**) (**-zing** or **-sing**) **1** treat condescendingly. **2** be a patron or customer of. □ **patronizing** *adj.* **patronizingly** *adv.*

patron saint *n.* saint regarded as protecting a person, place, activity, etc.

patronymic /ˌpætrəˈnɪmɪk/ *n.* name derived from the name of a father or ancestor (e.g. *Johnson, O'Brien, Ivanovich*). [Greek *patēr* father, *onoma* name]

patten /ˈpæt(ə)n/ *n. hist.* shoe or clog with a raised sole or set on an iron ring, for walking in mud etc. [French *patin*]

patter[1] *–n.* sound of quick light steps or taps. *–v.* make this sound (*rain pattering on the window-panes*). [from PAT[1]]

patter[2] *–n.* **1** rapid speech used by a comedian. **2** salesman's persuasive talk. *–v.* talk or say glibly or mechanically. [originally *pater*, = PATERNOSTER]

pattern /ˈpæt(ə)n/ *–n.* **1** repeated decorative design on wallpaper, cloth, etc. **2** regular or logical form, order, etc. (*behaviour pattern*). **3** model, design, or instructions for making something (*knitting pattern*). **4** excellent example, model (*pattern of elegance*). **5** wooden or metal shape from which a mould is made for a casting. **6** random combination of shapes or colours. *–v.* **1** (usu. foll. by *after*, *on*) model (a thing) on a design etc. **2** decorate with a pattern. [from PATRON]

patty /ˈpætɪ/ *n.* (*pl.* **-ies**) little pie or pasty. [French PÂTÉ, after PASTY[1]]

paucity /ˈpɔːsɪtɪ/ *n.* smallness of number or quantity. [Latin *paucus* few]

paunch /pɔːntʃ/ *n.* belly, stomach, esp. when protruding. □ **paunchy** *adj.* (**-ier**, **-iest**). [Anglo-French *pa(u)nche* from Latin *pantices* bowels]

pauper /ˈpɔːpə(r)/ *n.* poor person. □ **pauperism** *n.* [Latin, = poor]

pause /pɔːz/ *–n.* **1** temporary stop or silence. **2** *Mus.* mark (𝄐) over a note or rest that is to be lengthened. *–v.* (**-sing**) make a pause; wait. □ **give pause to** cause to hesitate. [Greek *pauō* stop]

pavane /pəˈvɑːn/ *n.* (also **pavan** /ˈpæv(ə)n/) *hist.* **1** a kind of stately dance. **2** music for this. [French from Spanish]

pave *v.* (**-ving**) cover (a street, floor, etc.) with a durable surface. □ **pave the way** (usu. foll. by *for*) make preparations. □ **paving** *n.* [Latin *pavio* ram (v.)]

pavement *n.* **1** paved path for pedestrians beside a road. **2** covering of a street, floor, etc., made of usu. rectangular stones. [Latin *pavimentum*: related to PAVE]

pavement artist *n.* artist who draws in chalk on paving-stones for tips.

pavilion /pəˈvɪljən/ *n.* **1** building at a sports ground for changing, refreshments, etc. **2** summerhouse or decorative shelter in a park. **3** large tent at a show, fair, etc. **4** building or stand for entertainments, at an exhibition, etc. [Latin *papilio* butterfly]

paving-stone *n.* large flat stone for paving.

pavlova /pævˈləʊvə/ *n.* meringue cake with cream and fruit. [*Pavlova*, name of a ballerina]

Pavlovian /pævˈləʊvɪən/ *adj.* **1** reacting predictably to a stimulus. **2** of such a stimulus or response. [*Pavlov*, name of a physiologist]

paw *–n.* **1** foot of an animal having claws or nails. **2** *colloq.* person's hand. *–v.* **1** strike or scrape with a paw or foot. **2** *colloq.* fondle awkwardly or indecently. [French *poue* from Germanic]

pawl *n.* **1** lever with a catch for the teeth of a wheel or bar. **2** *Naut.* short bar used to lock a capstan, windlass, etc. [Low German or Dutch]

pawn[1] *n.* **1** *Chess* piece of the smallest size and value. **2** person used by others for their own purposes. [French *poun* from Latin *pedo -onis* foot-soldier]

pawn[2] *–v.* **1** deposit (a thing) with a pawnbroker as security for money lent. **2** pledge or wager (one's life, honour, etc.). *–n.* object left in pawn. □ **in pawn** held as security. [French *pan* from Germanic]

pawnbroker *n.* person who lends money at interest on the security of personal property.

pawnshop *n.* pawnbroker's shop.

pawpaw /ˈpɔːpɔː/ *n.* (also **papaw** /pəˈpɔː/, **papaya** /pəˈpaɪə/) **1** elongated melon-shaped fruit with orange flesh. **2** tropical tree bearing this. [Spanish and Portuguese *papaya*]

pax *n.* **1** kiss of peace. **2** (as *int.*) *slang* call for a truce (used esp. by schoolchildren). [Latin, = peace]

pay *–v.* (*past* and *past part.* **paid**) **1** (also *absol.*) give (a person etc.) what is due for services done, goods received, debts incurred, etc. (*paid him in full*). **2 a** give (a usu. specified amount) for work done, a debt, etc. (*they pay £6 an hour*). **b** (foll. by *to*) hand over the amount of (a debt, wages, etc.) to (*paid the money to the assistant*). **3 a** give, bestow, or express (attention, a compliment, etc.) (*paid them no heed*). **b** make (a visit) (*paid a call on their uncle*). **4** (also *absol.*) (of a business, attitude, etc.) be profitable or advantageous to (a person etc.). **5** reward or punish (*shall pay you for that*).

6 (usu. as **paid** *adj.*) recompense (work, time, etc.) (*paid holiday*). **7** (usu. foll. by *out, away*) let out (a rope) by slackening it. *–n.* wages. □ **in the pay of** employed by. **pay back 1** repay. **2** punish or have revenge on. **pay for 1** hand over the money for. **2** bear the cost of. **3** suffer or be punished for (a fault etc.). **pay in** pay (money) into a bank etc. account. **pay its** (or **one's**) **way** cover costs. **pay one's last respects** attend a funeral to show respect. **pay off 1** dismiss (workers) with a final payment. **2** *colloq.* yield good results; succeed. **3** pay (a debt) in full. **pay one's respects** make a polite visit. **pay through the nose** *colloq.* pay much more than a fair price. **pay up** pay the full amount (of). **put paid to** *colloq.* **1** deal effectively with (a person). **2** terminate (hopes etc.). □ **payee** /peɪ'iː/ *n.* [Latin *paco* appease: related to PEACE]

payable *adj.* that must or may be paid; due (*payable in April*).

pay-as-you-earn *n.* deduction of income tax from wages at source.

pay-bed *n.* private hospital bed.

pay-claim *n.* (esp. a trade union's) demand for a pay increase.

pay-day *n.* day on which wages are paid.

PAYE *abbr.* pay-as-you-earn.

paying guest *n.* boarder.

payload *n.* **1** part of an aircraft's load yielding revenue. **2** explosive warhead carried by a rocket etc. **3** goods carried by a road vehicle.

paymaster *n.* **1** official who pays troops, workmen, etc. **2** usu. *derog.* person, organization, etc., to whom another owes loyalty because of payment given. **3** (in full **Paymaster General**) Treasury minister responsible for payments.

payment *n.* **1** paying. **2** amount paid. **3** reward, recompense.

pay-off *n. slang* **1** payment. **2** climax. **3** final reckoning.

payola /peɪ'əʊlə/ *n.* esp. *US slang* bribe offered for unofficial promotion of a product etc. in the media.

pay-packet *n.* envelope etc. containing an employee's wages.

pay phone *n.* coin-box telephone.

payroll *n.* list of employees receiving regular pay.

Pb *symb.* lead. [Latin *plumbum*]

PC *abbr.* **1** police constable. **2** Privy Councillor. **3** personal computer.

p.c. *abbr.* **1** per cent. **2** postcard.

PCB *abbr.* **1** polychlorinated biphenyl, any of several toxic aromatic compounds formed as waste in industrial processes. **2** *Computing* printed circuit board.

Pd *symb.* palladium.

pd. *abbr.* paid.

p.d.q. *abbr. colloq.* pretty damn quick.

PE *abbr.* physical education.

pea *n.* **1 a** hardy climbing plant with edible seeds growing in pods. **b** its seed. **2** similar plant (*sweet pea*; *chick-pea*). [from PEASE taken as a plural]

peace *n.* **1 a** quiet; tranquillity. **b** mental calm; serenity. **2 a** (often *attrib.*) freedom from or the cessation of war (*peace talks*). **b** (esp. **Peace**) treaty of peace between States etc. at war. **3** freedom from civil disorder. □ **at peace 1** in a state of friendliness. **2** serene. **3** *euphem.* dead. **hold one's peace** keep silent. **keep the peace** prevent, or refrain from, strife. **make one's peace** (often foll. by *with*) re-establish friendly relations. **make peace** agree to end a war or quarrel. [Latin *pax pac-*]

peaceable *adj.* **1** disposed to peace. **2** peaceful; tranquil. [Latin *placibilis* pleasing: related to PLEASE]

peace dividend *n.* public money which becomes available when defence spending is reduced.

peaceful *adj.* **1** characterized by peace; tranquil. **2** not infringing peace (*peaceful coexistence*). □ **peacefully** *adv.* **peacefulness** *n.*

peacemaker *n.* person who brings about peace. □ **peacemaking** *n.* & *adj.*

peace-offering *n.* propitiatory or conciliatory gift.

peace-pipe *n.* tobacco-pipe as a token of peace among N. American Indians.

peacetime *n.* period when a country is not at war.

peach[1] *n.* **1 a** round juicy fruit with downy yellow or pink skin. **b** tree bearing this. **2** yellowish-pink colour. **3** *colloq.* **a** person or thing of superlative quality. **b** attractive young woman. □ **peachy** *adj.* (**-ier**, **-iest**). [Latin *persica* Persian (apple)]

peach[2] *v.* (usu. foll. by *against, on*) *colloq.* turn informer; inform. [from obsolete *appeach*: related to IMPEACH]

peach Melba *n.* dish of peaches, ice-cream, and raspberry sauce.

peacock *n.* (*pl.* same or **-s**) male peafowl, with brilliant plumage and an erectile fanlike tail with eyelike markings. [from Latin *pavo* peacock, COCK[1]]

peacock blue *adj.* & *n.* (as adj. often hyphenated) lustrous greenish blue of a peacock's neck.

peacock butterfly *n.* butterfly with eyelike wing markings.

peafowl *n.* a kind of pheasant; peacock, peahen.

pea green *adj.* & *n.* (as adj. often hyphenated) bright green.

peahen *n.* female peafowl.

peak[1] –*n.* **1** projecting usu. pointed part, esp.: **a** the pointed top of a mountain. **b** a mountain with a peak. **c** a stiff brim at the front of a cap. **2 a** highest point of a curve, graph, etc. (*peak of the wave*). **b** time of greatest success, fitness, etc. **3** *attrib.* maximum, busiest (*peak viewing*; *peak hours*). –*v.* reach its highest value, quality, etc. (*output peaked*). □ **peaked** *adj.* [related to PICK[2]]

peak[2] *v.* **1** waste away. **2** (as **peaked** *adj.*) sharp-featured; pinched. [origin unknown]

peak-load *n.* maximum of electric power demand etc.

peaky *adj.* (**-ier, -iest**) **1** sickly; puny. **2** white-faced.

peal –*n.* **1 a** loud ringing of a bell or bells, esp. a series of changes. **b** set of bells. **2** loud repeated sound, esp. of thunder, laughter, etc. –*v.* **1** (cause to) sound in a peal. **2** utter sonorously. [from APPEAL]

pean *US* var. of PAEAN.

peanut *n.* **1** plant of the pea family bearing pods underground that contain seeds used for food and oil. **2** seed of this. **3** (in *pl.*) *colloq.* paltry thing or amount, esp. of money.

peanut butter *n.* paste of ground roasted peanuts.

pear /peə(r)/ *n.* **1** yellowish or greenish fleshy fruit, tapering towards the stalk. **2** tree bearing this. [Latin *pirum*]

pearl /pɜ:l/ –*n.* **1 a** (often *attrib.*) rounded usu. white or bluish-grey lustrous solid formed within the shell of certain oysters, highly prized as a gem. **b** imitation of this. **c** (in *pl.*) necklace of pearls. **2** precious thing; finest example. **3** thing like a pearl, e.g. a dewdrop or tear. –*v.* **1** *poet.* form or sprinkle with pearly drops. **2** reduce (barley etc.) to small rounded grains. **3** fish for pearls. [Italian *perla*, from Latin *perna* leg]

pearl barley *n.* barley ground to small rounded grains.

pearl bulb *n.* translucent electric light bulb.

pearl button *n.* mother-of-pearl button, or an imitation of it.

pearl-diver *n.* person who dives for pearl-oysters.

pearlite var. of PERLITE.

pearly –*adj.* (**-ier, -iest**) like, containing, or adorned with pearls; lustrous. –*n.* (*pl.* **-ies**) **1** pearly king or queen. **2** (in *pl.*) pearly king's or queen's clothes.

Pearly Gates *n.pl. colloq.* gates of Heaven.

pearly king *n.* (also **pearly queen**) London costermonger (or his wife) wearing clothes covered with pearl buttons.

pearly nautilus see NAUTILUS.

peasant /'pez(ə)nt/ *n.* **1** (in some rural agricultural countries) small farmer, agricultural worker. **2** *derog.* lout; boor. □ **peasantry** *n.* (*pl.* **-ies**). [Anglo-French *paisant* from *païs* country]

pease /pi:z/ *n.pl. archaic* peas. [Latin *pisa*]

pease-pudding *n.* boiled split peas (served esp. with boiled beef or ham).

peashooter *n.* small tube for blowing dried peas through as a toy.

pea-souper *n. colloq.* thick yellowish fog.

peat *n.* **1** partly carbonized vegetable matter used for fuel, in horticulture, etc. **2** cut piece of this. □ **peaty** *adj.* [perhaps Celtic: related to PIECE]

peatbog *n.* bog composed of peat.

pebble /'peb(ə)l/ *n.* small stone worn smooth esp. by the action of water. □ **pebbly** *adj.* [Old English]

pebble-dash *n.* mortar with stone chippings in it as a coating for external walls.

pecan /'pi:kən/ *n.* **1** pinkish-brown smooth nut with an edible kernel. **2** type of hickory producing this. [Algonquian]

peccadillo /,pekə'dɪləʊ/ *n.* (*pl.* **-es** or **-s**) trifling offence; venial sin. [Spanish *pecadillo*, from Latin *pecco* to sin (v.)]

peck[1] –*v.* **1** strike or bite with a beak. **2** kiss hastily or perfunctorily. **3 a** make (a hole) by pecking. **b** (foll. by *out, off*) remove or pluck out by pecking. **4** (also *absol.*) *colloq.* eat listlessly; nibble at. –*n.* **1** stroke, mark, or bite made by a beak. **2** hasty or perfunctory kiss. □ **peck at 1** eat (food) listlessly; nibble. **2** carp at; nag. **3** strike repeatedly with a beak. [probably Low German]

peck[2] *n.* measure of capacity for dry goods, equal to 2 gallons or 8 quarts. □ **a peck of** large number or amount of. [Anglo-French]

pecker *n. US coarse slang* penis. □ **keep your pecker up** *colloq.* remain cheerful.

pecking order *n.* social hierarchy, orig. as observed among hens.

peckish *adj. colloq.* hungry.

pectin /'pektɪn/ *n.* soluble gelatinous carbohydrate found in ripe fruits etc. and used as a setting agent in jams and jellies. □ **pectic** *adj.* [Greek *pēgnumi* make solid]

pectoral /'pektər(ə)l/ –*adj.* of or worn on the breast or chest (*pectoral fin*; *pectoral muscle*; *pectoral cross*). –*n.* pectoral muscle or fin. [Latin *pectus -tor-* chest]

peculate /ˈpekjʊˌleɪt/ *v.* (**-ting**) embezzle (money). □ **peculation** /-ˈleɪʃ(ə)n/ *n.* **peculator** *n.* [Latin: related to PECULIAR]

peculiar /pɪˈkjuːlɪə(r)/ *adj.* **1** strange; odd; unusual. **2 a** (usu. foll. by *to*) belonging exclusively (*peculiar to the time*). **b** belonging to the individual (*in their own peculiar way*). **3** particular; special (*point of peculiar interest*). [Latin *peculium* private property, from *pecu* cattle]

peculiarity /pɪˌkjuːlɪˈærɪtɪ/ *n.* (*pl.* **-ies**) **1** idiosyncrasy; oddity. **2** characteristic. **3** being peculiar.

peculiarly /pɪˈkjuːlɪəlɪ/ *adv.* **1** more than usually, especially (*peculiarly annoying*). **2** oddly.

pecuniary /pɪˈkjuːnɪərɪ/ *adj.* **1** of or concerning money. **2** (of an offence) entailing a money penalty. [Latin *pecunia* money, from *pecu* cattle]

pedagogue /ˈpedəˌgɒg/ *n. archaic* or *derog.* schoolmaster; teacher. □ **pedagogic** /-ˈgɒgɪk, -ˈgɒdʒɪk/ *adj.* **pedagogical** /-ˈgɒgɪk(ə)l, -ˈgɒdʒɪk(ə)l/ *adj.* [Greek *pais paid-* child, *agō* lead]

pedagogy /ˈpedəˌgɒdʒɪ, -ˌgɒgɪ/ *n.* science of teaching.

pedal *–n.* /ˈped(ə)l/ lever or key operated by foot, esp. in a vehicle, on a bicycle, or on some musical instruments (e.g. the organ). *–v.* /ˈped(ə)l/ (**-ll-**; *US* **-l-**) **1** operate the pedals of a bicycle, organ, etc. **2** propel (a bicycle etc.) with the pedals. *–adj.* /ˈpiːd(ə)l/ of the foot or feet. [Latin *pes ped-* foot]

pedalo /ˈpedəˌləʊ/ *n.* (*pl.* **-s**) pedal-operated pleasure-boat.

pedant /ˈped(ə)nt/ *n. derog.* person who insists on adherence to formal rules or literal meaning. □ **pedantic** /pɪˈdæntɪk/ *adj.* **pedantically** /pɪˈdæntɪkəlɪ/ *adv.* **pedantry** *n.* [French from Italian]

peddle /ˈped(ə)l/ *v.* (**-ling**) **1 a** sell (goods) as a pedlar. **b** advocate or promote. **2** sell (drugs) illegally. **3** engage in selling, esp. as a pedlar. [back-formation from PEDLAR]

peddler *n.* **1** person who sells drugs illegally. **2** *US* var. of PEDLAR.

pederast /ˈpedəˌræst/ *n.* (also **paederast**) man who engages in pederasty.

pederasty /ˈpedəˌræstɪ/ *n.* (also **paederasty**) anal intercourse between a man and a boy. [Greek *pais paid-* boy, *erastēs* lover]

pedestal /ˈpedɪst(ə)l/ *n.* **1** base supporting a column or pillar. **2** stone etc. base of a statue etc. □ **put on a pedestal** admire disproportionately, idolize. [Italian *piedestallo*, = foot of stall]

pedestrian /pɪˈdestrɪən/ *–n.* (often *attrib.*) person who is walking, esp. in a town. *–adj.* prosaic; dull; uninspired. □ **pedestrianize** *v.* (also **-ise**) (**-zing** or **-sing**). [Latin: related to PEDAL]

pedestrian crossing *n.* part of a road where crossing pedestrians have right of way.

pediatrics *US* var. of PAEDIATRICS.

pedicure /ˈpedɪˌkjʊə(r)/ *n.* **1** care or treatment of the feet, esp. the toenails. **2** person practising this for a living. [Latin *pes ped-* foot, *cura* care]

pedigree /ˈpedɪˌgriː/ *n.* **1** (often *attrib.*) recorded line of descent (esp. a distinguished one) of a person or pure-bred animal. **2** genealogical table. **3** *colloq.* 'life history' of a person, thing, idea, etc. □ **pedigreed** *adj.* [*pedegru* from French *pie de grue* (unrecorded) crane's foot, a mark denoting succession in pedigrees]

pediment /ˈpedɪmənt/ *n.* triangular part crowning the front of a building, esp. over a portico. [from *periment*, perhaps a corruption of PYRAMID]

pedlar /ˈpedlə(r)/ *n.* (*US* **peddler**) **1** travelling seller of small items. **2** (usu. foll. by *of*) retailer (of gossip etc.). [alteration of *pedder* from *ped* pannier]

pedo- *US* var. of PAEDO-.

pedometer /pɪˈdɒmɪtə(r)/ *n.* instrument for estimating distance walked by recording the number of steps taken. [Latin *pes ped-* foot: related to -METER]

pedophile *US* var. of PAEDOPHILE.

pedophilia *US* var. of PAEDOPHILIA.

peduncle /pɪˈdʌŋk(ə)l/ *n.* stalk of a flower, fruit, or cluster, esp. a main stalk bearing a solitary flower or subordinate stalks. □ **peduncular** /-kjʊlə(r)/ *adj.* [related to PEDOMETER, -UNCLE]

pee *colloq.* *–v.* (**pees, peed**) urinate. *–n.* **1** act of urinating. **2** urine. [from PISS]

peek *–v.* (usu. foll. by *in, out, at*) peep slyly, glance. *–n.* quick or sly look. [origin unknown]

peel *–v.* **1 a** strip the skin, rind, wrapping, etc. from. **b** (usu. foll. by *off*) strip (skin, peel, wrapping, etc.). **2 a** become bare of skin, paint, etc. **b** (often foll. by *off*) (of skin, paint, etc.) flake off. **3** (often foll. by *off*) *colloq.* (of a person) strip ready for exercise etc. *–n.* outer covering of a fruit, vegetable, etc.; rind. □ **peel off** veer away and detach oneself from a group etc. □ **peeler** *n.* [Old English from Latin *pilo* strip of hair]

peeling *n.* (usu. in *pl.*) stripped-off piece of peel.

peen *n.* wedge-shaped or thin or curved end of a hammer-head. [Latin *pinna* point]

peep[1] *–v.* **1** (usu. foll. by *at, in, out, into*) look through a narrow opening; look furtively. **2** (usu. foll. by *out*) come slowly into view; emerge. *–n.* **1** furtive or peering glance. **2** first appearance (*peep of day*). [origin unknown]

peep[2] –*v.* make a shrill feeble sound as of young birds, mice, etc. –*n.* **1** such a sound. **2** slight sound, utterance, or complaint (*not a peep out of them*). [imitative]

peep-hole *n.* small hole for peeping through.

peeping Tom *n.* furtive voyeur.

peep-show *n.* small exhibition of pictures etc. viewed through a lens or hole set into a box etc.

peer[1] *v.* (usu. foll. by *into*, *at*, etc.) look closely or with difficulty. [origin unknown]

peer[2] *n.* **1 a** (*fem.* **peeress**) member of one of the degrees of the nobility in Britain or Ireland, i.e. a duke, marquis, earl, viscount, or baron. **b** noble of any country. **2** person who is equal in ability, standing, rank, or value. [Latin *par* equal]

peerage *n.* **1** peers as a class; the nobility. **2** rank of peer or peeress.

peer group *n.* group of people of the same age, status, etc.

peerless *adj.* unequalled, superb.

peer of the realm *n.* peer entitled to sit in the House of Lords.

peeve *colloq.* –*v.* (**-ving**) (usu. as **peeved** *adj.*) irritate, annoy. –*n.* cause or state of irritation. [back-formation from PEEVISH]

peevish *adj.* irritable. □ **peevishly** *adv.* [origin unknown]

peewit /'pi:wɪt/ *n.* (also **pewit**) lapwing. [a sound imitative of its cry]

peg –*n.* **1** pin or bolt of wood, metal, etc., for holding things together, hanging garments on, holding up a tent, etc. **2** each of the pins used to tighten or loosen the strings of a violin etc. **3** pin for marking position, e.g. on a cribbage-board. **4** = CLOTHES-PEG. **5** occasion or pretext (*peg to hang an argument on*). **6** drink, esp. of spirits. –*v.* (**-gg-**) **1** (usu. foll. by *down*, *in*, *out*, etc.) fix (a thing) with a peg. **2** stabilize (prices, wages, etc.). **3** mark (the score) with pegs on a cribbage-board. □ **off the peg** (of clothes) ready-made. **peg away** (often foll. by *at*) work consistently. **peg out 1** *slang* die. **2** mark the boundaries of. **square peg in a round hole** misfit. **take a person down a peg or two** humble a person. [probably Low German or Dutch]

pegboard *n.* board with small holes for pegs, used for displays, games, etc.

peg-leg *n. colloq.* **1** artificial leg. **2** person with this.

pejorative /pɪ'dʒɒrətɪv/ –*adj.* derogatory. –*n.* derogatory word. [Latin *pejor* worse]

peke *n. colloq.* Pekingese. [abbreviation]

Pekingese /,pi:kɪ'ni:z/ *n.* (also **Pekinese**) (*pl.* same) lap-dog of a short-legged breed with long hair and a snub nose. [from *Peking* (Beijing) in China]

pelargonium /,pelə'gəʊnɪəm/ *n.* plant with red, pink, or white flowers and, often, fragrant leaves; geranium. [Greek *pelargos* stork]

pelf *n. derog.* or *joc.* money; wealth. [French: related to PILFER]

pelican /'pelɪkən/ *n.* large water-bird with a large bill and a pouch in its throat for storing fish. [Greek *pelekan*]

pelican crossing *n.* pedestrian crossing with traffic-lights operated by pedestrians.

pelisse /pɪ'li:s/ *n. hist.* **1** woman's long cloak with armholes or sleeves. **2** fur-lined cloak as part of a hussar's uniform. [Latin *pellicia* (garment) of fur, from *pellis* skin]

pellagra /pə'lægrə/ *n.* disease with cracking of the skin and often ending in insanity. [Italian *pelle* skin]

pellet /'pelɪt/ *n.* **1** small compressed ball of paper, bread, etc. **2** pill. **3** piece of small shot. [French *pelote* from Latin *pila* ball]

pellicle /'pelɪk(ə)l/ *n.* thin skin, membrane, or film. [Latin diminutive of *pellis* skin]

pell-mell /pel'mel/ *adv.* **1** headlong, recklessly. **2** in disorder or confusion. [French *pêle-mêle*]

pellucid /pɪ'lu:sɪd/ *adj.* **1** transparent. **2** (of style, speech, etc.) clear. [Latin: related to PER-]

pelmet /'pelmɪt/ *n.* narrow border of cloth, wood, etc. fitted esp. above a window to conceal the curtain rail. [probably French]

pelt[1] –*v.* **1** (usu. foll. by *with*) strike repeatedly with thrown objects. **2** (usu. foll. by *down*) (of rain etc.) fall quickly and torrentially. **3** run fast. –*n.* pelting. □ **at full pelt** as fast as possible. [origin unknown]

pelt[2] *n.* undressed skin, usu. of a fur-bearing mammal. [French, ultimately from Latin *pellis* skin]

pelvis /'pelvɪs/ *n.* basin-shaped cavity in most vertebrates, formed from the hip-bone with the sacrum and other vertebrae. □ **pelvic** *adj.* [Latin, = basin]

pen[1] –*n.* **1** instrument for writing etc. with ink. **2** (**the pen**) occupation of writing. –*v.* (**-nn-**) write. [Latin *penna* feather]

pen[2] *–n.* small enclosure for cows, sheep, poultry, etc. *–v.* (**-nn-**) (often foll. by *in, up*) enclose or shut up, esp. in a pen. [Old English]

pen[3] *n.* female swan. [origin unknown]

penal /ˈpiːn(ə)l/ *adj.* **1** of or concerning punishment or its infliction. **2** (of an offence) punishable, esp. by law. □ **penally** *adv.* [Latin *poena* PAIN]

penalize /ˈpiːnəˌlaɪz/ *v.* (also **-ise**) (**-zing** or **-sing**) **1** subject (a person) to a penalty or disadvantage. **2** make or declare (an action) penal.

penalty /ˈpenəltɪ/ *n.* (*pl.* **-ies**) **1** punishment for breaking a law, rule, or contract. **2** disadvantage, loss, etc., esp. as a result of one's own actions. **3** *Sport* disadvantage imposed for a breach of the rules etc. [medieval Latin: related to PENAL]

penalty area *n. Football* ground in front of the goal in which a foul by defenders involves the award of a penalty kick.

penalty kick *n. Football* free kick at the goal resulting from a foul in the penalty area.

penance /ˈpenəns/ *n.* **1** act of self-punishment as reparation for guilt. **2 a** (in the Roman Catholic and Orthodox Church) sacrament including confession of and absolution for sins. **b** penalty imposed, esp. by a priest, for a sin. □ **do penance** perform a penance. [related to PENITENT]

pence *pl.* of PENNY.

penchant /ˈpɑ̃ʃɑ̃/ *n.* (followed by *for*) inclination or liking. [French]

pencil /ˈpens(ə)l/ *–n.* **1** instrument for writing or drawing, usu. a thin rod of graphite etc. enclosed in a wooden cylinder or metal case. **2** (*attrib.*) resembling a pencil in shape (*pencil skirt*). *–v.* (**-ll-**; *US* **-l-**) **1** write, draw, or mark with a pencil. **2** (usu. foll. by *in*) write, note, or arrange provisionally. [Latin *penicillum* paintbrush]

pendant /ˈpend(ə)nt/ *n.* hanging jewel etc., esp. one attached to a necklace, bracelet, etc. [French *pendre* hang]

pendent /ˈpend(ə)nt/ *adj. formal* **1 a** hanging. **b** overhanging. **2** undecided, pending. □ **pendency** *n.*

pending *–predic. adj.* **1** awaiting decision or settlement, undecided. **2** about to come into existence (*patent pending*). *–prep.* **1** during (*pending further inquiries*). **2** until (*bailed pending trial*). [after French: see PENDANT]

pendulous /ˈpendjʊləs/ *adj.* hanging down; drooping and swinging. [Latin *pendulus* from *pendeo* hang]

pendulum /ˈpendjʊləm/ *n.* (*pl.* **-s**) weight suspended so as to swing freely, esp. a rod with a weighted end regulating a clock. [Latin neuter adjective: related to PENDULOUS]

penetrate /ˈpenɪˌtreɪt/ *v.* (**-ting**) **1 a** find access into or through. **b** (usu. foll. by *with*) imbue with; permeate. **2** see into, find out, or discern. **3** see through (darkness, fog, etc.). **4** be absorbed by the mind. **5** (as **penetrating** *adj.*) **a** having or suggesting sensitivity or insight. **b** (of a voice etc.) easily heard through or above other sounds; piercing. □ **penetrable** /-trəb(ə)l/ *adj.* **penetrability** /-trəˈbɪlɪtɪ/ *n.* **penetration** /-ˈtreɪʃ(ə)n/ *n.* **penetrative** /-trətɪv/ *adj.* [Latin]

pen-friend *n.* friend communicated with by letter only.

penguin /ˈpeŋgwɪn/ *n.* flightless black and white sea bird of the southern hemisphere, with wings developed into flippers for swimming underwater. [origin unknown]

penicillin /ˌpenɪˈsɪlɪn/ *n.* antibiotic, produced naturally by mould or synthetically. [Latin *penicillum*: related to PENCIL]

peninsula /pɪˈnɪnsjʊlə/ *n.* piece of land almost surrounded by water or projecting far into a sea etc. □ **peninsular** *adj.* [Latin *paene* almost, *insula* island]

penis /ˈpiːnɪs/ *n.* male organ of copulation and (in mammals) urination. [Latin]

penitent /ˈpenɪt(ə)nt/ *–adj.* repentant. *–n.* **1** repentant sinner. **2** person doing penance under the direction of a confessor. □ **penitence** *n.* **penitently** *adv.* [Latin *paeniteo* repent]

penitential /ˌpenɪˈtenʃ(ə)l/ *adj.* of penitence or penance.

penitentiary /ˌpenɪˈtenʃərɪ/ *–n.* (*pl.* **-ies**) *US* federal or State prison. *–adj.* **1** of penance. **2** of reformatory treatment. [Latin: related to PENITENT]

penknife *n.* small folding knife.

pen-name *n.* literary pseudonym.

pennant /ˈpenənt/ *n.* **1** tapering flag, esp. that flown at the masthead of a vessel in commission. **2** = PENNON. [blend of PENDANT and PENNON]

penniless /ˈpenɪlɪs/ *adj.* having no money; destitute.

pennon /ˈpenən/ *n.* **1** long narrow flag, triangular or swallow-tailed. **2** long pointed streamer on a ship. [Latin *penna* feather]

penny /ˈpenɪ/ *n.* (*pl.* for separate coins **-ies**, for a sum of money **pence** /pens/) **1** British coin and monetary unit equal to one-hundredth of a pound. **2** *hist.* British bronze coin and monetary unit equal to one-two-hundred-and-fortieth of a pound. □ **in for a penny, in for a pound** exhortation to total commitment to an undertaking. **pennies from heaven** unexpected benefits. **the penny drops** *colloq.* one understands at last. **penny wise and pound foolish** mean in small expenditures but wasteful of large amounts. **a pretty penny** a large sum of money. **two a penny** easily obtained and so almost worthless. [Old English]

penny black *n.* first adhesive postage stamp (1840, price one penny).

penny farthing *n.* early type of bicycle with a large front and small rear wheel.

penny-pinching –*n.* meanness. –*adj.* mean. □ **penny-pincher** *n.*

pennyroyal /ˌpenɪˈrɔɪəl/ *n.* creeping kind of mint. [Anglo-French *puliol real* royal thyme]

penny whistle *n.* tin pipe with six finger holes.

pennywort *n.* wild plant with rounded leaves, growing esp. in marshy places.

pennyworth *n.* as much as can be bought for a penny.

penology /piːˈnɒlədʒɪ/ *n.* the study of the punishment of crime and prison management. □ **penologist** *n.* [Latin *poena* penalty]

pen-pal *n. colloq.* = PEN-FRIEND.

pen-pushing *n. colloq. derog.* clerical work. □ **pen-pusher** *n.*

pension[1] /ˈpenʃ(ə)n/ –*n.* **1** regular payment made by a government to people above a specified age, to widows, or to the disabled. **2** similar payments made by an employer, private pension fund, etc. on the retirement of an employee. –*v.* grant a pension to. □ **pension off 1** dismiss with a pension. **2** cease to employ or use. [Latin *pendo pens-* pay]

pension[2] /pɑ̃ˈsjɔ̃/ *n.* European, esp. French, boarding-house. [French: related to PENSION[1]]

pensionable *adj.* **1** entitled to a pension. **2** (of a service, job, etc.) entitling an employee to a pension.

pensioner *n.* recipient of a pension, esp. the retirement pension. [French: related to PENSION[1]]

pensive /ˈpensɪv/ *adj.* deep in thought. □ **pensively** *adv.* [French *penser* think]

pent *adj.* (often foll. by *in*, *up*) closely confined; shut in (*pent-up feelings*). [from PEN[2]]

penta- *comb. form* five. [Greek *pente* five]

pentacle /ˈpentək(ə)l/ *n.* figure used as a symbol, esp. in magic, e.g. a pentagram. [medieval Latin *pentaculum*: related to PENTA-]

pentagon /ˈpentəgən/ *n.* **1** plane figure with five sides and angles. **2** (**the Pentagon**) **a** pentagonal Washington headquarters of the US forces. **b** leaders of the US forces. □ **pentagonal** /-ˈtægən(ə)l/ *adj.* [Greek *pentagōnon*: related to PENTA-]

pentagram /ˈpentəˌgræm/ *n.* five-pointed star. [Greek: see PENTA-, -GRAM]

pentameter /penˈtæmɪtə(r)/ *n.* line of verse with five metrical feet. [Greek: see PENTA-, -METER]

Pentateuch /ˈpentəˌtjuːk/ *n.* first five books of the Old Testament. [Greek *teukhos* book]

pentathlon /penˈtæθlən/ *n.* athletic event comprising five different events for each competitor. □ **pentathlete** /-ˈtæθliːt/ *n.* [Greek: see PENTA-, *athlon* contest]

pentatonic /ˌpentəˈtɒnɪk/ *adj.* consisting of five musical notes.

Pentecost /ˈpentɪˌkɒst/ *n.* **1** Whit Sunday. **2** Jewish harvest festival, on the fiftieth day after the second day of Passover. [Greek *pentēkostē* fiftieth (day)]

pentecostal /ˌpentɪˈkɒst(ə)l/ *adj.* (of a religious group) emphasizing the divine gifts, esp. the power to heal the sick, and often fundamentalist.

penthouse /ˈpenthaʊs/ *n.* (esp. luxurious) flat on the roof or top floor of a tall building. [Latin: related to APPEND]

penultimate /pɪˈnʌltɪmət/ *adj.* & *n.* last but one. [Latin *paenultimus* from *paene* almost, *ultimus* last]

penumbra /pɪˈnʌmbrə/ *n.* (*pl.* **-s** or **-brae** /-briː/) **1** partly shaded region around the shadow of an opaque body, esp. that around the shadow of the moon or earth in an eclipse. **2** partial shadow. □ **penumbral** *adj.* [Latin *paene* almost, UMBRA]

penurious /pɪˈnjʊərɪəs/ *adj.* **1** poor. **2** stingy; grudging. **3** scanty. [medieval Latin: related to PENURY]

penury /ˈpenjʊrɪ/ *n.* (*pl.* **-ies**) **1** destitution; poverty. **2** lack; scarcity. [Latin]

peon /ˈpiːən/ *n.* Spanish American day-labourer. [Portuguese and Spanish: related to PAWN[1]]

peony /ˈpiːənɪ/ *n.* (also **paeony**) (*pl.* **-ies**) plant with large globular red, pink, or white flowers. [Greek *paiōnia*]

people /ˈpiːp(ə)l/ –*n.pl.* except in sense 2. **1** persons in general or of a specified kind (*people don't like rudeness*; *famous people*). **2** persons composing a community, tribe, race, nation, etc. (*a warlike people*; *peoples of the Commonwealth*). **3** (**the people**) **a** the mass of people in a country etc. not having special rank or position. **b** these as an electorate. **4** parents or other relatives (*my people disapprove*). **5 a** subjects, armed followers, etc. **b** congregation of a parish priest etc. –*v.* (**-ling**) (usu. foll. by *with*) **1** fill with people, animals, etc.;

populate. **2** (esp. as **peopled** *adj.*) inhabit. [Latin *populus*]

PEP /pep/ *abbr.* Personal Equity Plan.

pep *colloq.* –*n.* vigour; spirit. –*v.* (**-pp-**) (usu. foll. by *up*) fill with vigour. [abbreviation of PEPPER]

pepper –*n.* **1** hot aromatic condiment from the dried berries of certain plants. **2** anything pungent. **3 a** capsicum plant, grown as a vegetable. **b** its fruit. –*v.* **1** sprinkle or treat with or as if with pepper. **2** pelt with missiles. [Sanskrit *pippalī*]

pepper-and-salt *adj.* with small patches of dark and light colour intermingled.

peppercorn *n.* **1** dried pepper berry. **2** (in full **peppercorn rent**) nominal rent.

pepper-mill *n.* device for grinding pepper by hand.

peppermint *n.* **1 a** mint plant grown for its strong-flavoured oil. **b** this oil. **2** sweet flavoured with peppermint.

pepperoni /ˌpepəˈrəʊnɪ/ *n.* beef and pork sausage seasoned with pepper. [Italian *peperone* chilli]

pepper-pot *n.* small container with a perforated lid for sprinkling pepper.

peppery *adj.* **1** of, like, or containing pepper. **2** hot-tempered. **3** pungent.

pep pill *n.* pill containing a stimulant drug.

peppy *adj.* (**-ier, -iest**) *colloq.* vigorous, energetic, bouncy.

pepsin /ˈpepsɪn/ *n.* enzyme contained in the gastric juice. [Greek *pepsis* digestion]

pep talk *n.* (usu. short) talk intended to enthuse, encourage, etc.

peptic /ˈpeptɪk/ *adj.* concerning or promoting digestion. [Greek *peptikos* able to digest]

peptic ulcer *n.* ulcer in the stomach or duodenum.

peptide /ˈpeptaɪd/ *n.* *Biochem.* compound consisting of two or more amino acids bonded in sequence. [Greek *peptos* cooked]

per *prep.* **1** for each (*two sweets per child; five miles per hour*). **2** by means of; by; through (*per post*). **3** (in full **as per**) in accordance with (*as per instructions*). □ **as per usual** *colloq.* as usual. [Latin]

per- *prefix* **1** through; all over (*pervade*). **2** completely; very (*perturb*). **3** to destruction; to the bad (*perdition; pervert*). [Latin *per-*: related to PER]

peradventure /pərədˈventʃə(r)/ *adv.* *archaic* or *joc.* perhaps. [French: related to PER, ADVENTURE]

perambulate /pəˈræmbjʊˌleɪt/ *v.* (**-ting**) **1** walk through, over, or about (streets, the country, etc.). **2** walk from place to place. □ **perambulation** /-ˈleɪʃ(ə)n/ *n.* [Latin *perambulo*: related to AMBLE]

perambulator *n. formal* = PRAM.

per annum /pər ˈænəm/ *adv.* for each year. [Latin]

percale /pəˈkeɪl/ *n.* closely woven cotton fabric. [French]

per capita /pə ˈkæpɪtə/ *adv.* & *adj.* (also **per caput** /ˈkæpʊt/) for each person. [Latin, = by heads]

perceive /pəˈsiːv/ *v.* (**-ving**) **1** apprehend, esp. through the sight; observe. **2** (usu. foll. by *that, how*, etc.) apprehend with the mind; understand; see or regard. □ **perceivable** *adj.* [Latin *percipio -cept-* seize, understand]

per cent /pə ˈsent/ (*US* **percent**) –*adv.* in every hundred. –*n.* **1** percentage. **2** one part in every hundred (*half a per cent*).

percentage *n.* **1** rate or proportion per cent. **2** proportion.

percentile /pəˈsentaɪl/ *n.* *Statistics* **1** each of 99 points at which a range of data is divided to make 100 groups of equal size. **2** each of these groups.

perceptible /pəˈseptɪb(ə)l/ *adj.* capable of being perceived by the senses or intellect. □ **perceptibility** /-ˈbɪlɪtɪ/ *n.* **perceptibly** *adv.* [Latin: related to PERCEIVE]

perception /pəˈsepʃ(ə)n/ *n.* **1** act or faculty of perceiving. **2** (often foll. by *of*) intuitive recognition of a truth, aesthetic quality, etc.; way of seeing, understanding. □ **perceptual** /pəˈseptʃʊəl/ *adj.*

perceptive /pəˈseptɪv/ *adj.* **1** sensitive; discerning. **2** capable of perceiving. □ **perceptively** *adv.* **perceptiveness** *n.* **perceptivity** /-ˈtɪvɪtɪ/ *n.*

perch[1] –*n.* **1** bar, branch, etc. used by a bird to rest on. **2** high place for a person or thing to rest on. **3** *hist.* measure of length, esp. for land, of $5^1/_2$ yards. –*v.* (usu. foll. by *on*) settle or rest on or as on a perch etc. [Latin *pertica* pole]

perch[2] *n.* (*pl.* same or **-es**) edible European spiny-finned freshwater fish. [Latin *perca* from Greek]

perchance /pəˈtʃɑːns/ *adv.* *archaic* or *poet.* **1** by chance. **2** maybe. [Anglo-French *par* by]

percipient /pəˈsɪpɪənt/ *adj.* able to perceive; conscious. □ **percipience** *n.* [Latin: related to PERCEIVE]

percolate /ˈpɜːkəˌleɪt/ *v.* (**-ting**) **1** (often foll. by *through*) **a** (of liquid etc.) filter or ooze gradually. **b** (of an idea etc.) permeate gradually. **2** prepare (coffee) in a percolator. **3** strain (a liquid, powder, etc.) through a fine mesh etc. □ **percolation** /-'leɪʃ(ə)n/ *n.* [Latin *colum* strainer]

percolator *n.* machine making coffee by circulating boiling water through ground beans.

percussion /pə'kʌʃ(ə)n/ *n.* **1 a** (often *attrib.*) playing of music by striking instruments with sticks etc. (*percussion instrument*). **b** such instruments collectively. **2** gentle tapping of the body in medical diagnosis. **3** forcible striking of one esp. solid body against another. □ **percussionist** *n.* **percussive** *adj.* [Latin *percutio -cuss-* strike]

percussion cap *n.* small amount of explosive powder contained in metal or paper and exploded by striking.

perdition /pə'dɪʃ(ə)n/ *n.* eternal death; damnation. [Latin *perdo -dit-* destroy]

peregrine /'perɪgrɪn/ *n.* (in full **peregrine falcon**) a kind of falcon much used for hawking. [Latin *peregrinus* foreign]

peremptory /pə'remptərɪ/ *adj.* **1** (of a statement or command) admitting no denial or refusal. **2** (of a person, manner, etc.) imperious; dictatorial. □ **peremptorily** *adv.* **peremptoriness** *n.* [Latin *peremptorius* deadly, decisive]

perennial /pə'renɪəl/ *–adj.* **1** lasting through a year or several years. **2** (of a plant) lasting several years. **3** lasting a long time or for ever. *–n.* perennial plant. □ **perennially** *adv.* [Latin *perennis* from *annus* year]

perestroika /,pere'strɔɪkə/ *n.* (in the former USSR) reform of the economic and political system. [Russian, = restructuring]

perfect /'pɜ:fɪkt/ *–adj.* **1** complete; not deficient. **2** faultless. **3** very enjoyable, excellent (*perfect evening*). **4** exact, precise (*perfect circle*). **5** entire, unqualified (*perfect stranger*). **6** *Gram.* (of a tense) denoting a completed action or event (e.g. *he has gone*). *–v.* /pə'fekt/ **1** make perfect. **2** complete. *–n. Gram.* the perfect tense. □ **perfectible** /pə'fektɪb(ə)l/ *adj.* **perfectibility** /pə,fektɪ'bɪlɪtɪ/ *n.* [Latin *perficere -fect-* complete (v.)]

perfection /pə'fekʃ(ə)n/ *n.* **1** making, becoming, or being perfect. **2** faultlessness. **3** perfect person, thing, or example. □ **to perfection** exactly; completely. [Latin: related to PERFECT]

perfectionism *n.* uncompromising pursuit of excellence. □ **perfectionist** *n.* & *adj.*

perfectly *adv.* **1** completely; quite. **2** in a perfect way.

perfect pitch *n.* = ABSOLUTE PITCH.

perfidy /'pɜ:fɪdɪ/ *n.* breach of faith; treachery. □ **perfidious** /-'fɪdɪəs/ *adj.* [Latin *perfidia* from *fides* faith]

perforate /'pɜ:fə,reɪt/ *v.* **(-ting) 1** make a hole or holes through; pierce. **2** make a row of small holes in (paper etc.) so that a part may be torn off easily. □ **perforation** /-'reɪʃ(ə)n/ *n.* [Latin *perforo* pierce through]

perforce /pə'fɔ:s/ *adv. archaic* unavoidably; necessarily. [French *par force* by FORCE[1]]

perform /pə'fɔ:m/ *v.* **1** (also *absol.*) carry into effect; do. **2** execute (a function, play, piece of music, etc.). **3** act in a play; play music, sing, etc.; execute tricks. **4** function. □ **performer** *n.* [Anglo-French: related to PER-, FURNISH]

performance /pə'fɔ:məns/ *n.* **1** (usu. foll. by *of*) **a** act, process, or manner of performing or functioning. **b** execution (of a duty etc.). **2** performing of a play, music, etc.; instance of this. **3** *colloq.* fuss; emotional scene.

performing arts *n.pl.* drama, music, dance, etc.

perfume /'pɜ:fju:m/ *–n.* **1** sweet smell. **2** fluid containing the essence of flowers etc.; scent. *–v.* /also pə'fju:m/ **(-ming)** impart a sweet scent to. [Italian *parfumare* smoke through]

perfumer /pə'fju:mə(r)/ *n.* maker or seller of perfumes. □ **perfumery** *n.* (*pl.* **-ies**).

perfunctory /pə'fʌŋktərɪ/ *adj.* done merely out of duty; superficial, careless. □ **perfunctorily** *adv.* **perfunctoriness** *n.* [Latin: related to FUNCTION]

pergola /'pɜ:gələ/ *n.* arbour or covered walk formed of growing plants trained over trellis-work. [Italian]

perhaps /pə'hæps/ *adv.* it may be; possibly.

peri- *prefix* round, about. [Greek]

perianth /'perɪ,ænθ/ *n.* outer part of a flower. [Greek *anthos* flower]

pericardium /,perɪ'kɑ:dɪəm/ *n.* (*pl.* **-dia**) membranous sac enclosing the heart. [Greek *kardia* heart]

perigee /'perɪ,dʒi:/ *n.* point of a planet's or comet's orbit where it is nearest the earth. [Greek *perigeion*]

perihelion /,perɪ'hi:lɪən/ *n.* (*pl.* **-lia**) point of a planet's or comet's orbit where it is nearest the sun's centre. [related to PERI-, Greek *hēlios* sun]

peril /'perɪl/ *n.* serious and immediate danger. □ **perilous** *adj.* **perilously** *adv.* [Latin *peric(u)lum*]

perimeter /pə'rɪmɪtə(r)/ *n.* **1 a** circumference or outline of a closed figure. **b** length of this. **2** outer boundary of an enclosed area. [Greek: related to -METER]

perineum /ˌperɪ'ni:əm/ *n.* (*pl.* **-nea**) region of the body between the anus and the scrotum or vulva. □ **perineal** *adj.* [Latin from Greek]

period /'pɪərɪəd/ *–n.* **1** length or portion of time. **2** distinct portion of history, a person's life, etc. **3** time forming part of a geological era. **4** interval between recurrences of an astronomical or other phenomenon. **5** time allowed for a lesson in school. **6** occurrence of menstruation (often *attrib.*: *period pains*). **7** complete sentence, esp. one consisting of several clauses. **8** esp. *US* **a** = FULL STOP 1. **b** *colloq.* used at the end of a statement to indicate finality (*I'm not going, period*). *–adj.* characteristic of some past period (*period furniture*). [Greek *hodos* way]

periodic /ˌpɪərɪ'ɒdɪk/ *adj.* appearing or occurring at intervals. □ **periodicity** /-rɪə'dɪsɪtɪ/ *n.*

periodical *–n.* newspaper, magazine, etc. issued at regular intervals. *–adj.* periodic. □ **periodically** *adv.*

periodic table *n.* arrangement of elements in order of increasing atomic number and in which elements of similar chemical properties appear at regular intervals.

periodontics /ˌperɪə'dɒntɪks/ *n.pl.* (treated as *sing.*) branch of dentistry concerned with the structures surrounding and supporting the teeth. [Greek *odous* tooth]

peripatetic /ˌperɪpə'tetɪk/ *–adj.* **1** (of a teacher) working in more than one school or college etc. **2** going from place to place; itinerant. *–n.* peripatetic person, esp. a teacher. [Greek *pateō* walk]

peripheral /pə'rɪfər(ə)l/ *–adj.* **1** of minor importance; marginal. **2** of the periphery. *–n.* any input, output, or storage device that can be controlled by a computer's central processing unit, e.g. a floppy disk or printer.

peripheral nervous system *n.* nervous system outside the brain and spinal cord.

periphery /pə'rɪfərɪ/ *n.* (*pl.* **-ies**) **1** boundary of an area or surface. **2** outer or surrounding region. [Greek *pherō* bear]

periphrasis /pə'rɪfrəsɪs/ *n.* (*pl.* **-phrases** /-ˌsi:z/) **1** roundabout way of speaking; circumlocution. **2** roundabout phrase. □ **periphrastic** /ˌperɪ'fræstɪk/ *adj.* [Greek: related to PHRASE]

periscope /'perɪˌskəʊp/ *n.* apparatus with a tube and mirrors or prisms, by which an observer in a trench, submerged submarine, or at the back of a crowd etc., can see things otherwise out of sight. □ **periscopic** /-'skɒpɪk/ *adj.*

perish /'perɪʃ/ *v.* **1** be destroyed; suffer death or ruin. **2 a** (esp. of rubber) lose its normal qualities; deteriorate, rot. **b** cause to rot or deteriorate. **3** (in *passive*) suffer from cold. [Latin *pereo*]

perishable *–adj.* liable to perish; subject to decay. *–n.* thing, esp. a foodstuff, subject to rapid decay.

perisher *n. slang* annoying person.

perishing *colloq.* *–adj.* **1** confounded. **2** freezing cold. *–adv.* confoundedly.

peristalsis /ˌperɪ'stælsɪs/ *n.* involuntary muscular wavelike movement by which the contents of the digestive tract are propelled along it. [Greek *peristellō* wrap around]

peritoneum /ˌperɪtə'ni:əm/ *n.* (*pl.* **-s** or **-nea**) membrane lining the cavity of the abdomen. □ **peritoneal** *adj.* [Greek *peritonos* stretched around]

peritonitis /ˌperɪtə'naɪtɪs/ *n.* inflammatory disease of the peritoneum.

periwig /'perɪwɪg/ *n.* esp. *hist.* wig. [alteration of PERUKE]

periwinkle[1] /'perɪˌwɪŋk(ə)l/ *n.* evergreen trailing plant with blue, purple, or white flowers. [Latin *pervinca*]

periwinkle[2] /'perɪˌwɪŋk(ə)l/ *n.* = WINKLE. [origin unknown]

perjure /'pɜ:dʒə(r)/ *v.refl.* (**-ring**) *Law* **1** wilfully tell a lie when on oath. **2** (as **perjured** *adj.*) guilty of or involving perjury. □ **perjurer** *n.* [French from Latin *juro* swear]

perjury *n.* (*pl.* **-ies**) *Law* act of wilfully telling a lie when on oath.

perk[1] *v.* □ **perk up 1** recover confidence, courage, life, or zest. **2** restore confidence, courage, or liveliness in. **3** smarten up. **4** raise (one's head etc.) briskly. [origin unknown]

perk[2] *n. colloq.* perquisite. [abbreviation]

perky *adj.* (**-ier, -iest**) lively; cheerful. □ **perkily** *adv.* **perkiness** *n.*

perlite /'pɜ:laɪt/ *n.* (also **pearlite**) glassy type of vermiculite used for insulation etc. [French *perle* pearl]

perm[1] *–n.* permanent wave. *–v.* give a permanent wave to. [abbreviation]

perm[2] *colloq.* *–n.* permutation. *–v.* make a permutation of. [abbreviation]

permafrost /'pɜ:məˌfrɒst/ *n.* subsoil which remains frozen all year, as in polar regions. [from PERMANENT, FROST]

permanent /'pɜ:mənənt/ *adj.* lasting, or intended to last or function, indefinitely. □ **permanence** *n.* **permanency** *n.* **permanently** *adv.* [Latin *permaneo* remain to the end]

permanent wave *n.* long-lasting artificial wave in the hair.

permeable /'pɜ:mɪəb(ə)l/ *adj.* capable of being permeated. □ **permeability** /-'bɪlɪtɪ/ *n.* [related to PERMEATE]

permeate /'pɜ:mɪ,eɪt/ *v.* (**-ting**) **1** penetrate throughout; pervade; saturate. **2** (usu. foll. by *through*, *among*, etc.) diffuse itself. □ **permeation** /-'eɪʃ(ə)n/ *n.* [Latin *permeo* pass through]

Permian /'pɜ:mɪən/ *–adj.* of the last period of the Palaeozoic era. *–n.* this period. [*Perm* in Russia]

permissible /pə'mɪsɪb(ə)l/ *adj.* allowable. □ **permissibility** /-'bɪlɪtɪ/ *n.* [French or medieval Latin: related to PERMIT]

permission /pə'mɪʃ(ə)n/ *n.* (often foll. by *to* + infin.) consent; authorization. [Latin *permissio*: related to PERMIT]

permissive /pə'mɪsɪv/ *adj.* **1** tolerant or liberal, esp. in sexual matters. **2** giving permission. □ **permissiveness** *n.* [French or medieval Latin: related to PERMIT]

permit *–v.* /pə'mɪt/ (**-tt-**) **1** give permission or consent to; authorize. **2 a** allow; give an opportunity to. **b** give an opportunity (*circumstances permitting*). **3** (foll. by *of*) admit. *–n.* /'pɜ:mɪt/ **1 a** document giving permission to act. **b** document etc. which allows entry. **2** *formal* permission. [Latin *permitto -miss-* allow]

permutation /,pɜ:mjʊ'teɪʃ(ə)n/ *n.* **1** one of the possible ordered arrangements or groupings of a set of things. **2** combination or selection of a specified number of things from a larger group, esp. matches in a football pool. [Latin *permuto* change thoroughly]

pernicious /pə'nɪʃəs/ *adj.* very harmful or destructive; deadly. [Latin *pernicies* ruin]

pernicious anaemia *n.* defective formation of red blood cells through lack of vitamin B.

pernickety /pə'nɪkɪtɪ/ *adj. colloq.* fastidious; over-precise. [origin unknown]

peroration /,perə'reɪʃ(ə)n/ *n.* concluding part of a speech. [Latin *oro* speak]

peroxide /pə'rɒksaɪd/ *–n.* **1 a** = HYDROGEN PEROXIDE. **b** (often *attrib.*) solution of hydrogen peroxide used esp. to bleach the hair. **2** compound of oxygen with another element containing the greatest possible proportion of oxygen. *–v.* (**-ding**) bleach (the hair) with peroxide. [from PER-, OXIDE]

perpendicular /,pɜ:pən'dɪkjʊlə(r)/ *–adj.* **1 a** (usu. foll. by *to*) at right angles (to a given line, plane, or surface). **b** at right angles to the plane of the horizon. **2** upright, vertical. **3** (of a slope etc.) very steep. **4** (**Perpendicular**) *Archit.* of the third stage of English Gothic (15th–16th c.) with vertical tracery in large windows. *–n.* **1** perpendicular line. **2** (prec. by *the*) perpendicular line or direction (*is out of the perpendicular*). □ **perpendicularity** /-'lærɪtɪ/ *n.* [Latin *perpendiculum* plumb-line]

perpetrate /'pɜ:pɪ,treɪt/ *v.* (**-ting**) commit (a crime, blunder, or anything outrageous). □ **perpetration** /-'treɪʃ(ə)n/ *n.* **perpetrator** *n.* [Latin *perpetro* perform]

perpetual /pə'petʃʊəl/ *adj.* **1** lasting for ever or indefinitely. **2** continuous, uninterrupted. **3** *colloq.* frequent (*perpetual interruptions*). □ **perpetually** *adv.* [Latin *perpetuus* continuous]

perpetual motion *n.* motion of a hypothetical machine which once set in motion would run for ever unless subject to an external force or to wear.

perpetuate /pə'petʃʊ,eɪt/ *v.* (**-ting**) **1** make perpetual. **2** preserve from oblivion. □ **perpetuation** /-'eɪʃ(ə)n/ *n.* **perpetuator** *n.* [Latin *perpetuo*]

perpetuity /,pɜ:pɪ'tju:ɪtɪ/ *n.* (*pl.* **-ies**) **1** state or quality of being perpetual. **2** perpetual annuity. **3** perpetual possession or position. □ **in perpetuity** for ever. [Latin: related to PERPETUAL]

perplex /pə'pleks/ *v.* **1** puzzle, bewilder, or disconcert. **2** complicate or confuse (a matter). □ **perplexedly** /-ɪdlɪ/ *adv.* **perplexing** *adj.* [Latin *perplexus* involved]

perplexity /pə'pleksɪtɪ/ *n.* (*pl.* **-ies**) **1** state of being perplexed. **2** thing that perplexes.

per pro. /pɜ: 'prəʊ/ *abbr.* through the agency of (used in signatures). [Latin *per procurationem*]

■ **Usage** The correct sequence is A *per pro.* B, where B is signing on behalf of A.

perquisite /'pɜ:kwɪzɪt/ *n.* **1** extra profit or allowance additional to a main income etc. **2** customary extra right or privilege. [Latin *perquiro -quisit-* search diligently for]

■ **Usage** *Perquisite* is sometimes confused with *prerequisite*, which means 'thing required as a precondition'.

perry /'perɪ/ *n.* (*pl.* **-ies**) drink made from fermented pear juice. [French *peré*: related to PEAR]

per se /pɜ: 'seɪ/ *adv.* by or in itself; intrinsically. [Latin]

persecute /'pɜ:sɪ,kju:t/ *v.* (**-ting**) **1** subject (a person etc.) to hostility or ill-treatment, esp. on grounds of political or religious belief. **2** harass, worry. □ **persecution** /-'kju:ʃ(ə)n/ *n.* **persecutor** *n.* [Latin *persequor -secut-* pursue]

persevere /ˌpɜːsɪˈvɪə(r)/ *v.* (**-ring**) (often foll. by *in, with*) continue steadfastly or determinedly; persist. □ **perseverance** *n.* [Latin: related to SEVERE]

Persian /ˈpɜːʃ(ə)n/ *–n.* **1** native or inhabitant of ancient or modern Persia (now Iran); person of Persian descent. **2** language of ancient Persia or modern Iran. **3** (in full **Persian cat**) cat of a breed with long silky hair. *–adj.* of or relating to Persia or its people or language.

■ **Usage** The preferred terms for the language (see sense 2 of the noun) are *Iranian* and *Farsi* respectively.

Persian lamb *n.* silky tightly curled fur of a young karakul, used in clothing.

persiflage /ˈpɜːsɪˌflɑːʒ/ *n.* light raillery, banter. [French]

persimmon /pɜːˈsɪmən/ *n.* **1** tropical evergreen tree. **2** its edible tomato-like fruit. [Algonquian]

persist /pəˈsɪst/ *v.* **1** (often foll. by *in*) continue firmly or obstinately (in an opinion or action) esp. despite obstacles, remonstrance, etc. **2** (of a phenomenon etc.) continue in existence; survive. □ **persistence** *n.* **persistent** *adj.* **persistently** *adv.* [Latin *sisto* stand]

person /ˈpɜːs(ə)n/ *n.* **1** individual human being. **2** living body of a human being (*found on my person*). **3** *Gram.* any of three classes of personal pronouns, verb-forms, etc.: the person speaking (**first person**); the person spoken to (**second person**); the person spoken of (**third person**). **4** (in *comb.*) used to replace *-man* in offices open to either sex (*salesperson*). **5** (in Christianity) God as Father, Son, or Holy Ghost. □ **in person** physically present. [Latin: related to PERSONA]

persona /pəˈsəʊnə/ *n.* (*pl.* **-nae** /-niː/) aspect of the personality as shown to or perceived by others. [Latin, = actor's mask]

personable *adj.* pleasing in appearance and behaviour.

personage *n.* person, esp. of rank or importance.

persona grata /pəˌsəʊnə ˈgrɑːtə/ *n.* (*pl.* ***personae gratae*** /-niː, -tiː/) person acceptable to certain others.

personal /ˈpɜːsən(ə)l/ *adj.* **1** one's own; individual; private. **2** done or made in person (*my personal attention*). **3** directed to or concerning an individual (*personal letter*). **4** referring (esp. in a hostile way) to an individual's private life or concerns (*personal remarks; no need to be personal*). **5** of the body and clothing (*personal hygiene*). **6** existing as a person (*a personal God*). **7** *Gram.* of or denoting one of the three persons (*personal pronoun*).

personal column *n.* part of a newspaper devoted to private advertisements and messages.

personal computer *n.* computer designed for use by a single individual.

personal equity plan *n.* scheme for tax-free personal investments through financial institutions.

personality /ˌpɜːsəˈnælɪtɪ/ *n.* (*pl.* **-ies**) **1** **a** person's distinctive character or qualities (*has a strong personality*). **b** socially attractive qualities (*was clever but had no personality*). **2** famous person (*TV personality*). **3** (in *pl.*) personal remarks.

personalize *v.* (also **-ise**) (**-zing** or **-sing**) **1** make personal, esp. by marking with one's name etc. **2** personify.

personally *adv.* **1** in person (*see to it personally*). **2** for one's own part (*speaking personally*). **3** in a personal manner (*took the criticism personally*).

personal organizer *n.* means of keeping track of personal affairs, esp. a loose-leaf notebook divided into sections.

personal pronoun *n.* pronoun replacing the subject, object, etc., of a clause etc., e.g. *I, we, you, them, us*.

personal property *n. Law* all one's property except land and those interests in land that pass to one's heirs.

personal stereo *n.* small portable cassette player, often with radio or CD player, used with lightweight headphones.

persona non grata /pəˌsəʊnə nɒn (or nəʊn) ˈgrɑːtə/ *n.* (*pl.* ***personae non gratae*** /-niː, -tiː/) unacceptable person.

personify /pəˈsɒnɪˌfaɪ/ *v.* (**-ies, -ied**) **1** represent (an abstraction or thing) as having human characteristics. **2** symbolize (a quality etc.) by a figure in human form. **3** (usu. as **personified** *adj.*) be a typical example of; embody (*she personifies youthful arrogance; he was niceness personified*). □ **personification** /-fɪˈkeɪʃ(ə)n/ *n.*

personnel /ˌpɜːsəˈnel/ *n.* staff of an organization, the armed forces, a public service, etc. [French, = personal]

personnel department *n.* part of an organization concerned with the appointment, training, and welfare of employees.

perspective /pəˈspektɪv/ *–n.* **1** **a** art of drawing solid objects on a two-dimensional surface so as to give the right impression of relative positions, size, etc. **b** picture so drawn. **2** apparent relation between visible objects as to position, distance, etc. **3** mental view of the relative importance of things. **4** view, esp. stretching into the distance. *–adj.* of or in perspective. □ **in** (or **out of**) **perspective 1** drawn or viewed according (or not according) to the rules of perspective. **2** correctly (or incorrectly) regarded in terms of relative importance. [Latin *perspicio -spect-* look at]

Perspex /'pɜ:speks/ *n. propr.* tough light transparent thermoplastic. [related to PERSPECTIVE]

perspicacious /,pɜ:spɪ'keɪʃəs/ *adj.* having mental penetration or discernment. □ **perspicacity** /-'kæsɪtɪ/ *n.* [Latin *perspicax*: related to PERSPECTIVE]

■ **Usage** *Perspicacious* is sometimes confused with *perspicuous*.

perspicuous /pə'spɪkjʊəs/ *adj.* **1** easily understood; clearly expressed. **2** expressing things clearly. □ **perspicuity** /-'kju:ɪtɪ/ *n.* [Latin: related to PERSPECTIVE]

■ **Usage** *Perspicuous* is sometimes confused with *perspicacious*.

perspiration /,pɜ:spɪ'reɪʃ(ə)n/ *n.* **1** sweat. **2** sweating. [French: related to PERSPIRE]

perspire /pə'spaɪə(r)/ *v.* (**-ring**) sweat. [Latin *spiro* breathe]

persuade /pə'sweɪd/ *v.* (**-ding**) **1** (often foll. by *of* or *that*) cause (another person or oneself) to believe; convince. **2** (often foll. by *to* + infin.) induce. □ **persuadable** *adj.* **persuasible** *adj.* [Latin *persuadeo -suas-* induce]

persuasion /pə'sweɪʒ(ə)n/ *n.* **1** persuading. **2** persuasiveness. **3** belief or conviction. **4** religious belief, or the group or sect holding it. [Latin: related to PERSUADE]

persuasive /pə'sweɪsɪv/ *adj.* good at persuading. □ **persuasively** *adv.* **persuasiveness** *n.* [French or medieval Latin: related to PERSUADE]

pert *adj.* **1** saucy, impudent. **2** jaunty. □ **pertly** *adv.* **pertness** *n.* [Latin *apertus* open]

pertain /pə'teɪn/ *v.* **1** (foll. by *to*) **a** relate or have reference to. **b** belong to as a part, appendage, or accessory. **2** (usu. foll. by *to*) be appropriate to. [Latin *pertineo* belong to]

pertinacious /,pɜ:tɪ'neɪʃəs/ *adj.* stubborn; persistent (in a course of action etc.). □ **pertinacity** /-'næsɪtɪ/ *n.* [Latin *pertinax*: related to PERTAIN]

pertinent /'pɜ:tɪnənt/ *adj.* (often foll. by *to*) relevant. □ **pertinence** *n.* **pertinency** *n.* [Latin: related to PERTAIN]

perturb /pə'tɜ:b/ *v.* **1** disturb mentally; agitate. **2** throw into confusion or disorder. □ **perturbation** /-'beɪʃ(ə)n/ *n.* [French from Latin]

peruke /pə'ru:k/ *n. hist.* wig. [French from Italian]

peruse /pə'ru:z/ *v.* (**-sing**) **1** read or study carefully. **2** *joc.* read or look at desultorily.□ **perusal** *n.* [originally = 'use up']

pervade /pə'veɪd/ *v.* (**-ding**) **1** spread throughout, permeate. **2** be rife among or through. □ **pervasion** *n.* **pervasive** *adj.* [Latin *pervado* penetrate]

perverse /pə'vɜ:s/ *adj.* **1** deliberately or stubbornly departing from what is reasonable or required. **2** intractable. □ **perversely** *adv.* **perversity** *n.* (*pl.* **-ies**). [Latin: related to PERVERT]

perversion /pə'vɜ:ʃ(ə)n/ *n.* **1** perverting or being perverted. **2** preference for an abnormal form of sexual activity. [Latin: related to PERVERT]

pervert *–v.* /pə'vɜ:t/ **1** turn (a person or thing) aside from its proper use or nature. **2** misapply (words etc.). **3** lead astray from right conduct or (esp. religious) beliefs; corrupt. **4** (as **perverted** *adj.*) showing perversion. *–n.* /'pɜ:vɜ:t/ perverted person, esp. sexually. [Latin *verto vers-* turn]

pervious /'pɜ:vɪəs/ *adj.* **1** permeable. **2** (usu. foll. by *to*) **a** affording passage. **b** accessible (to reason etc.). [Latin *via* road]

peseta /pə'seɪtə/ *n.* chief monetary unit of Spain. [Spanish]

pesky /'peskɪ/ *adj.* (**-ier, -iest**) esp. *US colloq.* troublesome; annoying. [origin unknown]

peso /'peɪsəʊ/ *n.* (*pl.* **-s**) chief monetary unit of several Latin American countries and of the Philippines. [Spanish]

pessary /'pesərɪ/ *n.* (*pl.* **-ies**) **1** device worn in the vagina to support the uterus or as a contraceptive. **2** vaginal suppository. [Latin from Greek]

pessimism /'pesɪ,mɪz(ə)m/ *n.* **1** tendency to be gloomy or expect the worst. **2** *Philos.* belief that this world is as bad as it could be or that all things tend to evil. □ **pessimist** *n.* **pessimistic** /-'mɪstɪk/ *adj.* **pessimistically** /-'mɪstɪkəlɪ/ *adv.* [Latin *pessimus* worst]

pest *n.* **1** troublesome or annoying person or thing. **2** destructive animal, esp. one which attacks food sources. [Latin *pestis* plague]

pester *v.* trouble or annoy, esp. with frequent or persistent requests. [probably French *empestrer* encumber: influenced by PEST]

pesticide /'pestɪ,saɪd/ *n.* substance for destroying pests, esp. insects.

pestilence /ˈpestɪləns/ *n.* fatal epidemic disease, esp. bubonic plague. [Latin *pestis* plague]

pestilent *adj.* **1** deadly. **2** harmful or morally destructive. **3** *colloq.* troublesome, annoying.

pestilential /ˌpestɪˈlenʃ(ə)l/ *adj.* **1** of or relating to pestilence. **2** pestilent.

pestle /ˈpes(ə)l/ *n.* club-shaped instrument for pounding substances in a mortar. [Latin *pistillum* from *pinso* pound]

pet[1] *–n.* **1** domestic or tamed animal kept for pleasure or companionship. **2** darling, favourite. *–attrib. adj.* **1** kept as a pet (*pet lamb*). **2** of or for pet animals (*pet food*). **3** often *joc.* favourite or particular (*pet hate*). **4** expressing fondness or familiarity (*pet name*). *–v.* **(-tt-) 1** fondle erotically. **2** treat as a pet; stroke, pat. [origin unknown]

pet[2] *n.* fit of ill-humour. [origin unknown]

petal /ˈpet(ə)l/ *n.* each of the parts of the corolla of a flower. □ **petalled** *adj.* [Greek *petalon* leaf]

petard /pɪˈtɑːd/ *n. hist.* small bomb used to blast down a door etc. [French]

peter /ˈpiːtə(r)/ *v.* □ **peter out** diminish, come to an end. [origin unknown]

Peter Pan *n.* person who remains youthful or is immature. [hero of J. M. Barrie's play (1904)]

petersham /ˈpiːtəʃəm/ *n.* thick corded silk ribbon. [Lord *Petersham*, name of an army officer]

pethidine /ˈpeθɪˌdiːn/ *n.* synthetic soluble analgesic used esp. in childbirth. [perhaps from the chemical *piperidine*]

petiole /ˈpetɪˌəʊl/ *n.* slender stalk joining a leaf to a stem. [French from Latin]

petit bourgeois /ˌpetɪ ˈbʊəʒwɑː/ *n.* (*pl.* **petits bourgeois** pronunc. same) member of the lower middle classes. [French]

petite /pəˈtiːt/ *adj.* (of a woman) of small and dainty build. [French, = little]

petit four /ˌpetɪ ˈfɔː(r)/ *n.* (*pl.* **petits fours** /ˈfɔːz/) very small fancy cake. [French, = small oven]

petition /pəˈtɪʃ(ə)n/ *–n.* **1** supplication, request. **2** formal written request, esp. one signed by many people, appealing to an authority. **3** *Law* application to a court for a writ etc. *–v.* **1** make or address a petition to. **2** (often foll. by *for*, *to*) appeal earnestly or humbly. [Latin *peto petit-* ask]

petit mal /ˌpetɪ ˈmæl/ *n.* mild form of epilepsy. [French, = little sickness]

petit point /ˌpetɪ ˈpwæ̃/ *n.* embroidery on canvas using small stitches. [French, = little point]

petrel /ˈpetr(ə)l/ *n.* sea bird, usu. flying far from land. [origin unknown]

Petri dish /ˈpiːtrɪ/ *n.* shallow covered dish used for the culture of bacteria etc. [*Petri*, name of a bacteriologist]

petrify /ˈpetrɪˌfaɪ/ *v.* **(-ies, -ied) 1** paralyse with fear, astonishment, etc. **2** change (organic matter) into a stony substance. **3** become like stone. □ **petrifaction** /-ˈfækʃ(ə)n/ *n.* [Latin *petra* rock, from Greek]

petrochemical /ˌpetrəʊˈkemɪk(ə)l/ *n.* substance industrially obtained from petroleum or natural gas.

petrodollar /ˈpetrəʊˌdɒlə(r)/ *n.* notional unit of currency earned by a petroleum-exporting country.

petrol /ˈpetr(ə)l/ *n.* **1** refined petroleum used as a fuel in motor vehicles, aircraft, etc. **2** (*attrib.*) concerned with the supply of petrol (*petrol pump*). [Latin: related to PETROLEUM]

petroleum /pɪˈtrəʊlɪəm/ *n.* hydrocarbon oil found in the upper strata of the earth, refined for use as fuel etc. [Latin *petra* rock, *oleum* oil]

petroleum jelly *n.* translucent solid mixture of hydrocarbons used as a lubricant, ointment, etc.

pet shop *n.* shop selling animals to be kept as pets.

petticoat /ˈpetɪˌkəʊt/ *n.* **1** woman's or girl's undergarment hanging from the waist or shoulders. **2** (*attrib.*) often *derog.* feminine. [*petty coat*]

pettifog /ˈpetɪˌfɒg/ *v.* **(-gg-) 1** practise legal trickery. **2** quibble or wrangle about trivial points. [origin unknown]

pettish *adj.* peevish, petulant; easily put out. [from PET[2]]

petty *adj.* **(-ier, -iest) 1** unimportant; trivial. **2** small-minded. **3** minor, inferior, on a small scale. **4** *Law* (of a crime) of lesser importance. □ **pettily** *adv.* **pettiness** *n.* [French *petit* small]

petty cash *n.* money from or for small items of receipt or expenditure.

petty officer *n.* naval NCO.

petulant /ˈpetjʊlənt/ *adj.* peevishly impatient or irritable. □ **petulance** *n.* **petulantly** *adv.* [Latin *peto* seek]

petunia /pɪˈtjuːnɪə/ *n.* cultivated plant with white, purple, red, etc., funnel-shaped flowers. [French *petun* tobacco]

pew *n.* **1** (in a church) long bench with a back; enclosed compartment. **2** *colloq.* seat (esp. *take a pew*). [Latin PODIUM]

pewit var. of PEEWIT.

pewter /ˈpjuːtə(r)/ *n.* **1** grey alloy of tin, antimony, and copper. **2** utensils made of this. [French *peutre*]

peyote /peɪˈəʊtɪ/ *n.* **1** Mexican cactus. **2** hallucinogenic drug prepared from this. [American Spanish from Nahuatl]

pfennig /ˈfenɪg/ *n.* one-hundredth of a Deutschmark. [German]

PG *abbr.* (of a film) classified as suitable for children subject to parental guidance.

pH /piː'eɪtʃ/ *n.* measure of the acidity or alkalinity of a solution. [German *Potenz* power, *H* (symbol for hydrogen)]

phagocyte /'fægə,saɪt/ *n.* leucocyte capable of engulfing and absorbing foreign matter. [Greek *phag-* eat, *kutos* cell]

phalanx /'fælæŋks/ *n.* (*pl.* **phalanxes** or **phalanges** /fə'lændʒiːz/) **1** *Gk Antiq.* line of battle, esp. a body of infantry drawn up in close order. **2** set of people etc. forming a compact mass, or banded for a common purpose. [Latin from Greek]

phallus /'fæləs/ *n.* (*pl.* **phalli** /-laɪ/ or **phalluses**) **1** (esp. erect) penis. **2** image of this as a symbol of natural generative power. □ **phallic** *adj.* [Latin from Greek]

phantasm /'fæn,tæz(ə)m/ *n.* illusion, phantom. □ **phantasmal** /-'tæzm(ə)l/ *adj.* [Latin: related to PHANTOM]

phantasmagoria /,fæntæzmə'gɔːrɪə/ *n.* shifting series of real or imaginary figures as seen in a dream. □ **phantasmagoric** /-'gɒrɪk/ *adj.* [probably from French *fantasmagorie*: related to PHANTASM]

phantom /'fæntəm/ —*n.* **1** ghost, apparition, spectre. **2** mental illusion. —*attrib. adj.* illusory. [Greek *phantasma*]

Pharaoh /'feərəʊ/ *n.* **1** ruler of ancient Egypt. **2** title of this ruler. [Old English from Church Latin *Pharao*, ultimately from Egyptian]

Pharisee /'færɪ,siː/ *n.* **1** member of an ancient Jewish sect, distinguished by strict observance of the traditional and written law. **2** self-righteous person; hypocrite. □ **Pharisaic** /-'seɪɪk/ *adj.* [Hebrew *pārûš*]

pharmaceutical /,fɑːmə'sjuːtɪk(ə)l/ *adj.* **1** of or engaged in pharmacy. **2** of the use or sale of medicinal drugs. [Latin from Greek *pharmakon* drug]

pharmaceutics *n.pl.* (usu. treated as *sing.*) = PHARMACY 1.

pharmacist /'fɑːməsɪst/ *n.* person qualified to prepare and dispense drugs.

pharmacology /,fɑːmə'kɒlədʒɪ/ *n.* the study of the action of drugs on the body. □ **pharmacological** /-kə'lɒdʒɪk(ə)l/ *adj.* **pharmacologist** *n.*

pharmacopoeia /,fɑːməkə'piːə/ *n.* **1** book, esp. one officially published, containing a list of drugs with directions for use. **2** stock of drugs. [Greek *pharmakopoios* drug-maker]

pharmacy /'fɑːməsɪ/ *n.* (*pl.* **-ies**) **1** preparation and (esp. medicinal) dispensing of drugs. **2** pharmacist's shop, dispensary.

pharynx /'færɪŋks/ *n.* (*pl.* **pharynges** /-rɪn,dʒiːz/ or **-xes**) cavity behind the nose and mouth. □ **pharyngeal** /-rɪn'dʒiːəl/ *adj.* **pharyngitis** /-rɪn'dʒaɪtɪs/ *n.* [Latin from Greek]

phase /feɪz/ —*n.* **1** stage in a process of change or development. **2** each of the aspects of the moon or a planet, according to the amount of its illumination. **3** *Physics* stage in a periodically recurring sequence, esp. the wave-form of alternating electric currents or light. —*v.* (**-sing**) carry out (a programme etc.) in phases or stages. □ **phase in** (or **out**) bring gradually into (or out of) use. [Greek *phasis* appearance]

Ph.D. *abbr.* Doctor of Philosophy. [Latin *philosophiae doctor*]

pheasant /'fez(ə)nt/ *n.* long-tailed gamebird. [Greek *Phasianos* of Phasis, name of a river associated with the bird]

phenobarbitone /,fiːnəʊ'bɑːbɪ,təʊn/ *n.* narcotic and sedative barbiturate drug used esp. to treat epilepsy. [from PHENOL, BARBITURATE]

phenol /'fiːnɒl/ *n.* **1** hydroxyl derivative of benzene. **2** any hydroxyl derivative of an aromatic hydrocarbon. [French]

phenomenal /fɪ'nɒmɪn(ə)l/ *adj.* **1** extraordinary, remarkable. **2** of the nature of a phenomenon. □ **phenomenally** *adv.*

phenomenon /fɪ'nɒmɪnən/ *n.* (*pl.* **-mena**) **1** fact or occurrence that appears or is perceived, esp. one of which the cause is in question. **2** remarkable person or thing. [Greek *phainō* show]

■ **Usage** The plural form of this word, *phenomena*, is often used mistakenly for the singular. This should be avoided.

pheromone /'ferə,məʊn/ *n.* substance secreted and released by an animal for detection and response by another usu. of the same species. [Greek *pherō* convey, HORMONE]

phew /fjuː/ *int.* expression of relief, astonishment, weariness, etc. [imitative]

phi /faɪ/ *n.* twenty-first letter of the Greek alphabet (Φ, φ). [Greek]

phial /'faɪəl/ *n.* small glass bottle, esp. for liquid medicine. [Greek *phiale* broad flat dish]

phil- var. of PHILO-.

-phil var. of -PHILE.

philadelphus /,fɪlə'delfəs/ *n.* flowering shrub, esp. the mock orange. [Latin from Greek]

philander /fɪ'lændə(r)/ *v.* flirt or have casual affairs with women. □ **philanderer** *n.* [Greek *anēr andr-* male person]

philanthropy /fɪ'lænθrəpɪ/ *n.* **1** love of mankind. **2** practical benevolence. □

philanthropic /-ˈθrɒpɪk/ *adj.* **philanthropist** *n.* [Greek *anthrōpos* human being]

philately /fɪˈlætəlɪ/ *n.* the study and collecting of postage stamps. □ **philatelist** *n.* [Greek *atelēs* tax-free]

-phile *comb. form* (also **-phil**) forming nouns and adjectives denoting fondness for what is specified (*bibliophile*). [Greek *philos* loving]

philharmonic /ˌfɪlhɑːˈmɒnɪk/ *adj.* fond of music (usu. in the names of orchestras etc.). [Italian: related to HARMONIC]

philippic /fɪˈlɪpɪk/ *n.* bitter verbal attack. [Greek from *Philip* II of Macedon]

Philistine /ˈfɪlɪˌstaɪn/ *–n.* **1** member of a people of ancient Palestine. **2** (usu. **philistine**) person who is hostile or indifferent to culture. *–adj.* (usu. **philistine**) hostile or indifferent to culture. □ **philistinism** /-stɪˌnɪz(ə)m/ *n.* [Hebrew *pᵉlištî*]

Phillips /ˈfɪlɪps/ *n.* (usu. *attrib.*) *propr.* denoting a screw with a cross-shaped slot, or a corresponding screwdriver. [name of the US manufacturer]

philo- *comb. form* (also **phil-** before a vowel or *h*) denoting a liking for what is specified. [Greek *philos* friend]

philodendron /ˌfɪləˈdendrən/ *n.* (*pl.* **-s** or **-dra**) tropical evergreen climber cultivated as a house-plant. [Greek *dendron* tree]

philology /fɪˈlɒlədʒɪ/ *n.* the study of language, esp. in its historical and comparative aspects. □ **philological** /-ləˈlɒdʒɪk(ə)l/ *adj.* **philologist** *n.* [French from Latin from Greek: related to PHILO-, -LOGY]

philosopher /fɪˈlɒsəfə(r)/ *n.* **1** expert in or student of philosophy. **2** person who lives by a philosophy or is wise.

philosophers' stone *n.* (also **philosopher's stone**) supreme object of alchemy, a substance supposed to change other metals into gold or silver.

philosophical /ˌfɪləˈsɒfɪk(ə)l/ *adj.* (also **philosophic**) **1** of or according to philosophy. **2** skilled in or devoted to philosophy. **3** calm in adversity. □ **philosophically** *adv.*

philosophize /fɪˈlɒsəˌfaɪz/ *v.* (also **-ise**) (**-zing** or **-sing**) **1** reason like a philosopher. **2** speculate; theorize. □ **philosophizer** *n.*

philosophy /fɪˈlɒsəfɪ/ *n.* (*pl.* **-ies**) **1** use of reason and argument in seeking truth and knowledge of reality, esp. knowledge of the causes and nature of things and of the principles governing existence. **2 a** particular system or set of beliefs reached by this. **b** personal rule of life. [Greek: related to PHILO-, *sophia* wisdom]

philtre /ˈfɪltə(r)/ *n.* (*US* **philter**) love-potion. [Greek *phileō* to love]

phlebitis /flɪˈbaɪtɪs/ *n.* inflammation of a vein. □ **phlebitic** /-ˈbɪtɪk/ *adj.* [Greek *phleps phleb-* vein]

phlegm /flem/ *n.* **1** thick viscous substance secreted by the mucous membranes of the respiratory passages, discharged by coughing. **2 a** calmness. **b** sluggishness. **3** *hist.* phlegm regarded as one of the four bodily humours. [Greek *phlegma*]

phlegmatic /flegˈmætɪk/ *adj.* calm, unexcitable. □ **phlegmatically** *adv.*

phloem /ˈfləʊem/ *n.* tissue conducting sap in plants. [Greek *phloos* bark]

phlox /flɒks/ *n.* (*pl.* same or **-es**) plant with scented clusters of esp. white, blue, or red flowers. [Greek *phlox*, name of a plant (literally 'flame')]

-phobe *comb. form* forming nouns denoting a person with a specified fear or aversion (*xenophobe*). [Greek *phobos* fear]

phobia /ˈfəʊbɪə/ *n.* abnormal or morbid fear or aversion. □ **phobic** *adj.* & *n.* [from -PHOBIA]

-phobia *comb. form* forming nouns denoting a specified fear or aversion (*agoraphobia*). □ **-phobic** *comb. form* forming adjectives.

phoenix /ˈfiːnɪks/ *n.* mythical bird, the only one of its kind, that burnt itself on a pyre and rose from the ashes to live again. [Greek *phoinix*]

phone *n.* & *v.* (**-ning**) *colloq.* = TELEPHONE. [abbreviation]

phone book *n.* = TELEPHONE DIRECTORY.

phonecard *n.* card containing prepaid units for use with a cardphone.

phone-in *n.* broadcast programme during which listeners or viewers telephone the studio and participate.

phoneme /ˈfəʊniːm/ *n.* unit of sound in a specified language that distinguishes one word from another (e.g. *p, b, d, t* as in pad, pat, bad, bat, in English). □ **phonemic** /-ˈniːmɪk/ *adj.* [Greek *phōneō* speak]

phonetic /fəˈnetɪk/ *adj.* **1** representing vocal sounds. **2** (of spelling etc.) corresponding to pronunciation. □ **phonetically** *adv.* [Greek: related to PHONEME]

phonetics *n.pl.* (usu. treated as *sing.*) **1** vocal sounds. **2** the study of these. □ **phonetician** /ˌfəʊnɪˈtɪʃ(ə)n/ *n.*

phoney /ˈfəʊnɪ/ (also **phony**) *colloq.* *–adj.* (**-ier, -iest**) **1** sham; counterfeit. **2** fictitious. *–n.* (*pl.* **-eys** or **-ies**) phoney person or thing. □ **phoniness** *n.* [origin unknown]

phonic /ˈfɒnɪk/ *adj.* of sound; of vocal sounds. [Greek *phōnē* voice]

phono- *comb. form* sound. [Greek *phōnē* voice, sound]

phonograph /'fəʊnə,grɑ:f/ *n.* **1** early form of gramophone. **2** *US* gramophone.

phonology /fə'nɒlədʒɪ/ *n.* the study of sounds in language or a particular language; a language's sound system. □ **phonological** /,fəʊnə'lɒdʒɪk(ə)l, ,fɒn-/ *adj.*

phony var. of PHONEY.

phosphate /'fɒsfeɪt/ *n.* salt or ester of phosphoric acid, esp. used as a fertilizer. [French: related to PHOSPHORUS]

phosphor /'fɒsfə(r)/ *n.* synthetic fluorescent or phosphorescent substance. [Latin PHOSPHORUS]

phosphorescence /,fɒsfə'res(ə)ns/ *n.* **1** radiation similar to fluorescence but detectable after excitation ceases. **2** emission of light without combustion or perceptible heat. □ **phosphoresce** *v.* (**-cing**). **phosphorescent** *adj.*

phosphorus /'fɒsfərəs/ *n. Chem.* nonmetallic element existing in allotropic forms, esp. as a whitish waxy substance burning slowly at ordinary temperatures and so luminous in the dark. □ **phosphoric** /-'fɒrɪk/ *adj.* **phosphorous** *adj.* [Greek *phōs* light, *-phoros* -bringing]

photo /'fəʊtəʊ/ *n.* (*pl.* **-s**) = PHOTOGRAPH *n.* [abbreviation]

photo- *comb. form* denoting: **1** light. **2** photography. [Greek *phōs phōt-* light]

photochemistry /,fəʊtəʊ'kemɪstrɪ/ *n.* the study of the chemical effects of light.

photocopier /'fəʊtəʊ,kɒpɪə(r)/ *n.* machine for producing photocopies.

photocopy *—n.* (*pl.* **-ies**) photographic copy of printed or written material. *—v.* (**-ies, -ied**) make a photocopy of.

photoelectric /,fəʊtəʊɪ'lektrɪk/ *adj.* marked by or using emissions of electrons from substances exposed to light. □ **photoelectricity** /-'trɪsɪtɪ/ *n.*

photoelectric cell *n.* device using the effect of light to generate current.

photo finish *n.* close finish of a race or contest, where the winner is distinguishable only on a photograph.

photofit *n.* reconstructed picture of a suspect made from composite photographs.

photogenic /,fəʊtəʊ'dʒenɪk/ *adj.* **1** looking attractive in photographs. **2** *Biol.* producing or emitting light.

photograph /'fəʊtə,grɑ:f/ *—n.* picture formed by means of the chemical action of light or other radiation on sensitive film. *—v.* (also *absol.*) take a photograph of (a person etc.). □ **photographer** /fə'tɒgrəfə(r)/ *n.* **photographic** /-'græfɪk/ *adj.* **photographically** /-'græfɪkəlɪ/ *adv.*

photography /fə'tɒgrəfɪ/ *n.* the taking and processing of photographs.

photogravure /,fəʊtəʊgrə'vjʊə(r)/ *n.* **1** image produced from a photographic negative transferred to a metal plate and etched in. **2** this process. [French *gravure* engraving]

photojournalism /,fəʊtəʊ'dʒɜ:nə,lɪz(ə)m/ *n.* the relating of news by photographs, esp. in magazines etc. □ **photojournalist** *n.*

photolithography /,fəʊtəʊlɪ'θɒgrəfɪ/ *n.* lithography using plates made photographically.

photometer /fəʊ'tɒmɪtə(r)/ *n.* instrument for measuring light. □ **photometric** /,fəʊtəʊ'metrɪk/ *adj.* **photometry** /-'tɒmɪtrɪ/ *n.*

photon /'fəʊtɒn/ *n.* quantum of electromagnetic radiation energy, proportional to the frequency of radiation. [after *electron*]

photo opportunity *n.* organized opportunity for the press etc. to photograph a celebrity.

photosensitive /,fəʊtəʊ'sensɪtɪv/ *adj.* reacting to light.

Photostat /'fəʊtəʊ,stæt/*—n. propr.* **1** type of photocopier. **2** copy made by it. *—v.* (**photostat**) (**-tt-**) make a Photostat of.

photosynthesis /,fəʊtəʊ'sɪnθɪsɪs/ *n.* process in which the energy of sunlight is used by organisms, esp. green plants, to synthesize carbohydrates from carbon dioxide and water. □ **photosynthesize** *v.* (also **-ise**) (**-zing** or **-sing**). **photosynthetic** /-'θetɪk/ *adj.*

phrase /freɪz/ *—n.* **1** group of words forming a conceptual unit, but not a sentence. **2** idiomatic or short pithy expression. **3** mode of expression. **4** *Mus.* group of notes forming a distinct unit within a melody. *—v.* (**-sing**) **1** express in words. **2** *Mus.* divide (music) into phrases, esp. in performance. □ **phrasal** *adj.* [Greek *phrasis* from *phrazō* tell]

phrase book *n.* book for travellers, listing useful expressions with their foreign equivalents.

phraseology /,freɪzɪ'ɒlədʒɪ/ *n.* (*pl.* **-ies**) **1** choice or arrangement of words. **2** mode of expression. □ **phraseological** /-zɪə'lɒdʒɪk(ə)l/ *adj.*

phrenetic var. of FRENETIC.

phrenology /frɪ'nɒlədʒɪ/ *n. hist.* the study of the shape and size of the cranium as a supposed indication of character and mental faculties. □ **phrenological** /-nə'lɒdʒɪk(ə)l/ *adj.* **phrenologist** *n.* [Greek *phrēn* mind]

phut /fʌt/ *n.* dull abrupt sound as of impact or an explosion. □ **go phut** *colloq.*

(esp. of a plan) collapse, break down. [perhaps from Hindi *phaṭnā* to burst]

phylactery /fɪ'læktərɪ/ *n.* (*pl.* **-ies**) small leather box containing Hebrew texts, worn by Jewish men at prayer. [Greek *phulassō* guard]

phyllo pastry var. of FILO PASTRY.

phylum /'faɪləm/ *n.* (*pl.* **phyla**) *Biol.* taxonomic rank below a kingdom, comprising a class or classes and subordinate taxa. [Greek *phulon* race]

physic /'fɪzɪk/ *n.* esp. *archaic.* **1** medicine. **2** art of healing. **3** medical profession. [Greek *phusikē* of nature]

physical /'fɪzɪk(ə)l/ *adj.* **1** of the body (*physical exercise*). **2** of matter; material. **3 a** of, or according to, the laws of nature. **b** of physics. **4** *US* medical examination. □ **physically** *adv.*

physical chemistry *n.* application of physics to the study of chemical behaviour.

physical geography *n.* branch of geography dealing with natural features.

physical jerks *n.pl. colloq.* physical exercises.

physical science *n.* science(s) used in the study of inanimate natural objects.

physician /fɪ'zɪʃ(ə)n/ *n.* doctor, esp. a specialist in medical diagnosis and treatment.

physicist /'fɪzɪsɪst/ *n.* person skilled in physics.

physics /'fɪzɪks/ *n.pl.* (treated as *sing.*) branch of science dealing with the properties and interactions of matter and energy. [Latin *physica* (pl.) from Greek: related to PHYSIC]

physio /'fɪzɪəʊ/ *n.* (*pl.* **-s**) *colloq.* **1** physiotherapy. **2** physiotherapist.

physiognomy /,fɪzɪ'ɒnəmɪ/ *n.* (*pl.* **-ies**) **1 a** cast or form of a person's features, expression, etc. **b** supposed art of judging character from facial characteristics etc. **2** external features of a landscape etc. [Greek: related to PHYSIC, GNOMON]

physiology /,fɪzɪ'ɒlədʒɪ/ *n.* **1** science of the functions of living organisms and their parts. **2** these functions. □ **physiological** /-zɪə'lɒdʒɪk(ə)l/ *adj.* **physiologist** *n.* [Latin: related to PHYSIC, -LOGY]

physiotherapy /,fɪzɪəʊ'θerəpɪ/ *n.* treatment of disease, injury, deformity, etc., by physical methods including massage, heat treatment, remedial exercise, etc. □ **physiotherapist** *n.* [related to PHYSIC, THERAPY]

physique /fɪ'zi:k/ *n.* bodily structure and development. [French: related to PHYSIC]

pi /paɪ/ *n.* **1** sixteenth letter of the Greek alphabet (Π, π). **2** (as π) the symbol of the ratio of the circumference of a circle to its diameter (approx. 3.14). [Greek]

pia mater /,paɪə 'meɪtə(r)/ *n.* delicate innermost membrane enveloping the brain and spinal cord. [Latin, = tender mother]

pianissimo /,pɪə'nɪsɪ,məʊ/ *Mus.* –*adj.* very soft. –*adv.* very softly. –*n.* (*pl.* **-s** or **-mi** /-mɪ/) very soft playing or passage. [Italian, superlative of PIANO²]

pianist /'pɪənɪst/ *n.* piano-player.

piano¹ /pɪ'ænəʊ/ *n.* (*pl.* **-s**) keyboard instrument with metal strings struck by hammers. [Italian, abbreviation of PIANOFORTE]

piano² /'pjɑ:nəʊ/ *Mus.* –*adj.* soft. –*adv.* softly. –*n.* (*pl.* **-s** or **-ni** /-nɪ/) soft playing or passage. [Latin *planus* flat, (of sound) soft]

piano-accordion *n.* accordion with a small keyboard like that of a piano.

pianoforte /,pɪænəʊ'fɔ:tɪ/ *n. formal* or *archaic* = PIANO¹. [Italian, earlier *piano e forte* soft and loud]

Pianola /pɪə'nəʊlə/ *n. propr.* a kind of automatic piano. [diminutive]

piazza /pɪ'ætsə/ *n.* public square or market-place. [Italian: related to PLACE]

pibroch /'pi:brɒk/ *n.* martial or funerary bagpipe music. [Gaelic]

pica /'paɪkə/ *n.* **1** unit of type-size (1/6 inch). **2** size of letters in typewriting (10 per inch). [Latin: related to PIE²]

picador /'pɪkə,dɔ:(r)/ *n.* mounted man with a lance in a bullfight. [Spanish]

picaresque /,pɪkə'resk/ *adj.* (of a style of fiction) dealing with the episodic adventures of rogues etc. [Spanish *pícaro* rogue]

■ **Usage** *Picaresque* is sometimes used to mean 'transitory' or 'roaming', but this is considered incorrect in standard English.

picayune /,pɪkə'ju:n/ *US colloq.* –*n.* **1** small coin. **2** insignificant person or thing. –*adj.* mean; contemptible; petty [French *picaillon*]

piccalilli /,pɪkə'lɪlɪ/ *n.* (*pl.* **-s**) pickle of chopped vegetables, mustard, and hot spices. [origin unknown]

piccaninny /,pɪkə'nɪnɪ/ *n.* (*US* **pickaninny**) (*pl.* **-ies**) often *offens.* small Black or Australian Aboriginal child. [West Indian Negro from Spanish *pequeño* or Portuguese *pequeno* little]

piccolo /'pɪkə,ləʊ/ *n.* (*pl.* **-s**) small flute sounding an octave higher than the ordinary one. [Italian, = small]

pick¹ –*v.* **1** (also *absol.*) choose carefully. **2** detach or pluck (a flower, fruit, etc.) from a stem, tree, etc. **3 a** probe with the finger, an instrument, etc. to remove unwanted matter. **b** clear (a bone, carcass, etc.) of scraps of meat etc. **4** (also *absol.*) (of a person) eat (food, a meal, etc.) in small bits. *–n.* **1** act of picking. **2 a** selection, choice. **b** right to select (*had first pick of the prizes*). **3** (usu. foll. by *of*) best (*the pick of the bunch*). □ **pick and choose** select fastidiously. **pick at 1** eat (food) without interest. **2** find fault with. **pick a person's brains** extract ideas, information, etc., from a person for one's own use. **pick holes in** find fault with (an idea etc.). **pick a lock** open a lock with an instrument other than the proper key, esp. with criminal intent. **pick off 1** pluck (leaves etc.) off. **2** shoot (people etc.) one by one without haste. **pick on 1** find fault with; nag at. **2** select. **pick out 1** take from a larger number. **2** distinguish from surrounding objects; identify. **3** play (a tune) by ear on the piano etc. **4** (often foll. by *in*, *with*) accentuate (decoration, a painting, etc.) with a contrasting colour. **pick over** select the best from. **pick a person's pockets** steal from a person's pockets. **pick a quarrel** start an argument deliberately. **pick to pieces** = *take to pieces* (see PIECE). **pick up 1** grasp and raise. **2 a** acquire by chance or without effort. **b** learn effortlessly. **3** stop for and take along with one. **4** become acquainted with (a person) casually, esp for sexual purposes. **5** (of one's health, the weather, share prices, etc.) recover, improve, etc. **6** (of an engine etc.) recover speed. **7** (of the police etc.) arrest. **8** detect by scrutiny or with a telescope, radio, etc. **9** accept the responsibility of paying (a bill etc.). **10** resume, take up anew (*picked up where we left off*). □ **picker** *n.* [from PIKE]

pick[2] *n.* **1** long-handled tool with a usu. curved iron bar pointed at one or both ends, used for breaking up hard ground etc. **2** *colloq.* plectrum. **3** any instrument for picking. [from PIKE]

pickaback var. of PIGGYBACK.

pickaninny *US* var. of PICCANINNY.

pickaxe /'pɪkæks/ *n.* (*US* **pickax**) = PICK[2] 1. [French: related to PIKE]

picket /'pɪkɪt/ *–n.* **1** one or more persons stationed outside a place of work to persuade others not to enter during a strike etc. **2** pointed stake driven into the ground. **3 a** small body of troops sent out to watch for the enemy. **b** group of sentries. *–v.* **(-t-) 1 a** station or act as a picket. **b** beset or guard with a picket or pickets. **2** secure (a place) with stakes. **3** tether (an animal). [French *piquer* prick]

picket line *n.* boundary established by workers on strike, esp. at the entrance to the place of work, which others are asked not to cross.

pickings *n.pl.* **1** profits or gains acquired easily or dishonestly. **2** left-overs.

pickle /'pɪk(ə)l/ *–n.* **1 a** (often in *pl.*) food, esp. vegetables, preserved in brine, vinegar, mustard, etc. **b** the liquid used for this. **2** *colloq.* plight (*in a pickle*). *–v.* **(-ling) 1** preserve in or treat with pickle. **2** (as **pickled** *adj.*) *slang* drunk. [Low German or Dutch *pekel*]

pick-me-up *n.* **1** tonic for the nerves etc. **2** a good experience that cheers.

pickpocket *n.* person who steals from people's pockets.

pick-up *n.* **1** *slang* person met casually, esp. for sexual purposes. **2** small open motor truck. **3** part of a record-player carrying the stylus. **4** device on an electric guitar etc. that converts string vibrations into electrical signals. **5** act of picking up.

picky *adj.* **(-ier, -iest)** *colloq.* excessively fastidious.

pick-your-own *adj.* (usu. *attrib.*) (of fruit and vegetables) dug or picked by the customer at the farm etc.

picnic /'pɪknɪk/ *–n.* **1** outing including an outdoor meal. **2** meal eaten out of doors. **3** (usu. with *neg.*) *colloq.* something agreeable or easily accomplished etc. *–v.* **(-ck-)** take part in a picnic. [French *pique-nique*]

pico- *comb. form* denoting a factor of 10^{-12} (*picometre*). [Spanish *pico* beak, peak, little bit]

Pict *n.* member of an ancient people of N. Britain. □ **Pictish** *adj.* [Latin]

pictograph /'pɪktə,grɑːf/ *n.* (also **pictogram** /-,græm/) **1** pictorial symbol for a word or phrase. **2** pictorial representation of statistics etc. □ **pictographic** /-'græfɪk/ *adj.* [Latin *pingo pict-* paint]

pictorial /pɪk'tɔːrɪəl/ *–adj.* **1** of or expressed in a picture or pictures. **2** illustrated. *–n.* periodical with pictures as the main feature. □ **pictorially** *adv.* [Latin *pictor* painter: related to PICTURE]

picture /'pɪktʃə(r)/ *–n.* **1 a** (often *attrib.*) painting, drawing, photograph, etc., esp. as a work of art. **b** portrait. **c** beautiful object. **2** total mental or visual impression produced; scene. **3 a** film. **b** (**the pictures**) cinema; cinema performance. *–v.* **(-ring) 1** (also *refl.*; often foll. by *to*) imagine (*pictured it to herself*). **2** represent in a picture. **3** describe graphically. □ **get the picture** *colloq.* grasp the drift of information etc. **in the picture** *colloq.* fully informed. [Latin *pingo pict-* paint]

picture postcard *n.* postcard with a picture on one side.

picturesque /ˌpɪktʃə'resk/ *adj.* **1** beautiful or striking to look at. **2** (of language etc.) strikingly graphic. [Italian *pittoresco*, assimilated to PICTURE]
picture window *n.* large window of one pane of glass.
piddle /'pɪd(ə)l/ *v.* (**-ling**) **1** *colloq.* urinate. **2** (as **piddling** *adj.*) *colloq.* trivial; trifling. **3** (foll. by *about, around*) work or act in a trifling way. [origin unknown]
pidgin /'pɪdʒɪn/ *n.* simplified language used between people not having a common language. [corruption of *business*]
pidgin English *n.* pidgin in which the chief language is English, used orig. between Chinese and Europeans.
pie[1] /paɪ/ *n.* **1** baked dish of meat, fish, fruit, etc., usu. with a top and base of pastry. **2** thing resembling a pie (*mud pie*). □ **easy as pie** very easy. [origin uncertain]
pie[2] /paɪ/ *n. archaic* magpie. [Latin *pica*]
piebald /'paɪbɔ:ld/ *–adj.* (esp. of a horse) having irregular patches of two colours, esp. black and white. *–n.* piebald animal. [from PIE[2], BALD]
piece /pi:s/ *–n.* **1 a** (often foll. by *of*) distinct portion forming part of or broken off from a larger object. **b** each of the parts of which a set or category is composed (*five-piece band*). **2** coin. **3** (usu. short) literary or musical composition; picture; play. **4** item or instance (*piece of news*; *piece of impudence*). **5 a** object used to make moves in a board-game. **b** chessman (strictly, other than a pawn). **6** definite quantity in which a thing is sold. **7** (often foll. by *of*) enclosed portion (of land etc.). **8** *slang derog.* woman. *–v.* (**-cing**) (usu. foll. by *together*) form into a whole; put together; join. □ **go to pieces** collapse emotionally. **in** (or **all in**) **one piece 1** unbroken. **2** unharmed. **of a piece** (often foll. by *with*) uniform, consistent. **a piece of cake** see CAKE. **a piece of one's mind** sharp rebuke or lecture. **say one's piece** give one's opinion or make a prepared statement. **take to pieces 1** break up or dismantle. **2** criticize harshly. [Anglo-French, probably from Celtic]
pièce de résistance /ˌpjes də reɪ'zi:stɑ̃s/ *n.* (*pl.* ***pièces de résistance*** pronunc. same) most important or remarkable item, esp. a dish at a meal. [French]
piecemeal *–adv.* piece by piece; gradually. *–adj.* gradual; unsystematic. [from PIECE, MEAL[1]]
piece-work *n.* work paid for according to the amount produced.
pie chart *n.* circle divided into sectors to represent relative quantities.
piecrust *n.* baked pastry crust of a pie.
pied /paɪd/ *adj.* particoloured. [from PIE[2]]
pied-à-terre /ˌpjeɪdɑ:'teə(r)/ *n.* (*pl.* ***pieds-à-terre*** pronunc. same) (usu. small) flat, house, etc. kept for occasional use. [French, literally 'foot to earth']
pie-eyed *adj. slang* drunk.
pie in the sky *n.* (used without an article) unrealistic prospect of future happiness.
pier *n.* **1 a** structure built out into the sea, a lake, etc., as a promenade and landing-stage. **b** breakwater. **2 a** support of an arch or of the span of a bridge; pillar. **b** solid masonry between windows etc. [Latin *pera*]
pierce *v.* (**-cing**) **1 a** (of a sharp instrument etc.) penetrate. **b** (often foll. by *with*) make a hole in or through with a sharp-pointed instrument. **c** make (a hole etc.). **2** (as **piercing** *adj.*) (of a glance, sound, light, pain, cold, etc.) keen, sharp, or unpleasantly penetrating. **3** (often foll. by *through, into*) force a way through or into, penetrate. [French *percer* from Latin *pertundo* bore through]
pier-glass *n.* large mirror, used orig. to fill wall-space between windows.
pierrot /'pɪərəʊ/ *n.* (*fem.* **pierrette** /pɪə'ret/) **1** white-faced entertainer in pier shows etc. with a loose white clown's costume. **2** French pantomime character so dressed. [French, diminutive of *Pierre* Peter]
pietà /ˌpɪe'tɑ:/ *n.* representation of the Virgin Mary holding the dead body of Christ on her lap. [Italian, = PIETY]
pietism /'paɪəˌtɪz(ə)m/ *n.* **1** pious sentiment. **2** exaggerated or affected piety. [German: related to PIETY]
piety /'paɪətɪ/ *n.* (*pl.* **-ies**) **1** quality of being pious. **2** pious act. [Latin: related to PIOUS]
piffle /'pɪf(ə)l/ *colloq.* *–n.* nonsense; empty speech. *–v.* (**-ling**) talk or act feebly; trifle. [imitative]
piffling *adj. colloq.* trivial; worthless.
pig *–n.* **1** omnivorous hoofed bristly broad-snouted mammal, esp. a domesticated kind. **2** its flesh as food. **3** *colloq.* greedy, dirty, or unpleasant person. **4** oblong mass of metal (esp. iron or lead) from a smelting-furnace. **5** *slang derog.* police officer. *–v.* (**-gg-**) *colloq.* eat (food) greedily. □ **buy a pig in a poke** acquire something without previous sight or knowledge of it. **pig it** *colloq.* live in a disorderly or filthy fashion. **pig out** (often foll. by *on*) esp. *US slang* eat gluttonously. [Old English]

pigeon /ˈpɪdʒ(ə)n/ *n.* bird of the dove family. [Latin *pipio -onis*]

pigeon-hole *–n.* each of a set of compartments on a wall etc. for papers, letters, etc. *–v.* **1** assign to a preconceived category. **2** deposit in a pigeonhole. **3** put aside for future consideration.

pigeon-toed *adj.* having the toes turned inwards.

piggery *n.* (*pl.* **-ies**) **1** pig farm. **2** = PIGSTY.

piggish *adj.* greedy; dirty; mean.

piggy *–n.* (*pl.* **-ies**) *colloq.* little pig. *–adj.* (**-ier, -iest**) **1** like a pig. **2** (of features etc.) like those of a pig.

piggyback (also **pickaback** /ˈpɪkə ˌbæk/) *–n.* ride on the back and shoulders of another person. *–adv.* on the back and shoulders of another person. [origin unknown]

piggy bank *n.* pig-shaped money box.

pigheaded *adj.* obstinate. □ **pigheadedness** *n.*

pig-iron *n.* crude iron from a smelting-furnace.

piglet /ˈpɪglɪt/ *n.* young pig.

pigment /ˈpɪgmənt/ *–n.* **1** colouring-matter used as paint or dye. **2** natural colouring-matter of animal or plant tissue. *–v.* colour with or as if with pigment. □ **pigmentary** *adj.* [Latin *pingo* paint]

pigmentation /ˌpɪgmənˈteɪʃ(ə)n/ *n.* **1** natural colouring of plants, animals, etc. **2** excessive colouring of tissue by the deposition of pigment.

pigmy var. of PYGMY.

pigskin *n.* **1** hide of a pig. **2** leather made from this.

pigsty *n.* (*pl.* **-ies**) **1** pen for pigs. **2** filthy house, room, etc.

pigswill /ˈpɪgswɪl/ *n.* kitchen refuse and scraps fed to pigs.

pigtail *n.* plait of hair hanging from the back of the head.

pike *n.* (*pl.* same or **-s**) **1** large voracious freshwater fish with a long narrow snout. **2** *hist.* weapon with a pointed metal head on a long wooden shaft. [Old English]

pikestaff *n.* wooden shaft of a pike. □ **plain as a pikestaff** quite plain or obvious.

pilaster /pɪˈlæstə(r)/ *n.* rectangular column projecting slightly from a wall. □ **pilastered** *adj.* [Latin *pila* pillar]

pilau /pɪˈlaʊ/ *n.* (also **pilaff, pilaf** /pɪˈlæf/) Middle Eastern or Indian dish of rice boiled with meat, vegetables, spices, etc. [Turkish]

pilchard /ˈpɪltʃəd/ *n.* small marine fish of the herring family. [origin unknown]

pile[1] *–n.* **1** heap of things laid upon one another. **2** large imposing building. **3** *colloq.* **a** large quantity. **b** large amount of money. **4 a** series of plates of dissimilar metals laid one on another alternately to produce an electric current. **b** = NUCLEAR REACTOR. **5** funeral pyre. *–v.* (**-ling**) **1 a** (often foll. by *up, on*) heap up. **b** (foll. by *with*) load. **2** (usu. foll. by *in, into, on, out of*, etc.) crowd hurriedly or tightly. □ **pile it on** *colloq.* exaggerate. **pile up 1** accumulate; heap up. **2** *colloq.* cause (a vehicle etc.) to crash. [Latin *pila*]

pile[2] *n.* **1** heavy beam driven vertically into the ground to support a bridge, the foundations of a house, etc. **2** pointed stake or post. [Latin *pilum* javelin]

pile[3] *n.* soft projecting surface on a carpet, velvet, etc. [Latin *pilus* hair]

pile-driver *n.* machine for driving piles into the ground.

piles *n.pl. colloq.* haemorrhoids. [Latin *pila* ball]

pile-up *n. colloq.* multiple crash of road vehicles.

pilfer /ˈpɪlfə(r)/ *v.* (also *absol.*) steal (objects), esp. in small quantities. [French *pelfre*]

pilgrim /ˈpɪlgrɪm/ *n.* **1** person who journeys to a sacred place for religious reasons. **2** traveller. [Latin: related to PEREGRINE]

pilgrimage *n.* **1** pilgrim's journey. **2** any journey taken for sentimental reasons.

Pilgrim Fathers *n.pl.* English Puritans who founded the colony of Plymouth, Massachusetts, in 1620.

pill *n.* **1 a** ball or flat disc of solid medicine for swallowing whole. **b** (usu. prec. by *the*) *colloq.* contraceptive pill. **2** unpleasant or painful necessity. [Latin *pila* ball]

pillage /ˈpɪlɪdʒ/ *–v.* (**-ging**) (also *absol.*) plunder, sack. *–n.* pillaging, esp. in war. [French *piller* plunder]

pillar /ˈpɪlə(r)/ *n.* **1** slender vertical structure of stone etc. used as a support or for ornament. **2** person regarded as a mainstay (*pillar of the faith*). **3** upright mass of air, water, rock, etc. □ **from pillar to post** (rushing etc.) from one place to another. [Latin *pila* pillar]

pillar-box *n.* public postbox shaped like a pillar.

pillar-box red *adj.* & *n.* bright red.

pillbox *n.* **1** shallow cylindrical box for holding pills. **2** hat of a similar shape. **3** *Mil.* small partly underground enclosed concrete fort.

pillion /ˈpɪljən/ *n.* seating for a passenger behind a motor cyclist. □ **ride**

pillion travel seated behind a motor cyclist. [Gaelic *pillean* small cushion]

pillory /'pɪlərɪ/ *–n.* (*pl.* **-ies**) *hist.* wooden framework with holes for the head and hands, holding a person and allowing him or her to be publicly ridiculed. *–v.* (**-ies, -ied**) **1** expose to ridicule. **2** *hist.* put in the pillory. [French]

pillow /'pɪləʊ/ *–n.* **1** soft support for the head, esp. in bed. **2** pillow-shaped block or support. *–v.* rest on or as if on a pillow. [Latin *pulvinus* cushion]

pillowcase *n.* (also **pillowslip**) washable cover for a pillow.

pilot /'paɪlət/ *–n.* **1** person who operates the controls of an aircraft. **2** person qualified to take charge of a ship entering or leaving harbour. **3** (usu. *attrib.*) experimental undertaking or test (*pilot scheme*). **4** guide. *–v.* (**-t-**) **1** act as a pilot of. **2** conduct or initiate as a pilot. [Greek *pēdon*]

pilot-light *n.* **1** small gas burner kept alight to light another. **2** electric indicator light or control light.

pilot officer *n.* lowest commissioned rank in the RAF.

pimento /pɪ'mentəʊ/ *n.* (*pl.* **-s**) **1** tree native to Jamaica. **2** berries of this, usu. crushed for culinary use; allspice. **3** = PIMIENTO. [Latin: related to PIGMENT]

pi meson var. of PION.

pimiento /ˌpɪmɪ'entəʊ/ *n.* (*pl.* **-s**) = SWEET PEPPER. [see PIMENTO]

pimp *–n.* man who lives off the earnings of a prostitute or a brothel. *–v.* act as a pimp. [origin unknown]

pimpernel /'pɪmpəˌnel/ *n.* = SCARLET PIMPERNEL. [Latin *piper* PEPPER]

pimple /'pɪmp(ə)l/ *n.* **1** small hard inflamed spot on the skin. **2** anything resembling a pimple. □ **pimply** *adj.* [Old English]

PIN /pɪn/ *abbr.* personal identification number (for use with a cashcard etc.).

pin *–n.* **1** small thin pointed piece of metal with a round or flattened head used (esp. in sewing) for holding things in place, attaching one thing to another, etc. **2** peg of wood or metal for various purposes. **3** (in idioms) something of small value (*not worth a pin*). **4** (in *pl.*) *colloq.* legs. *–v.* (**-nn-**) **1 a** (often foll. by *to, up, together*) fasten with a pin or pins. **b** transfix with a pin, lance, etc. **2** (usu. foll. by *on*) put (blame, responsibility, etc.) on (a person etc.). **3** (often foll. by *against, on,* etc.) seize and hold fast. □ **pin down 1** (often foll. by *to*) bind (a person etc.) to a promise, arrangement, etc. **2** force (a person) to declare his or her intentions. **3** restrict the actions of (an enemy etc.). **4** specify (a thing) precisely. **pin one's faith** (or **hopes** etc.) **on** rely implicitly on. [Latin *pinna* point etc.]

pina colada /ˌpi:nə kə'lɑ:də/ *n.* cocktail of pineapple juice, rum, and coconut. [Spanish]

pinafore /'pɪnəˌfɔ:(r)/ *n.* **1** apron, esp. with a bib. **2** (in full **pinafore dress**) collarless sleeveless dress worn over a blouse or jumper. [from PIN, AFORE]

pinball *n.* game in which small metal balls are shot across a board to strike pins.

pince-nez /pæns'neɪ/ *n.* (*pl.* same) pair of eyeglasses with a nose-clip. [French, = pinch-nose]

pincer movement *n.* movement by two wings of an army converging to surround an enemy.

pincers /'pɪnsəz/ *n.pl.* **1** (also **pair of pincers**) gripping-tool resembling scissors but with blunt jaws. **2** front claws of lobsters and some other crustaceans. [related to PINCH]

pinch *–v.* **1 a** squeeze tightly, esp. between finger and thumb. **b** (often *absol.*) (of a shoe etc.) constrict painfully. **2** (of cold, hunger, etc.) affect painfully. **3** *slang* **a** steal. **b** arrest. **4** (as **pinched** *adj.*) (of the features) drawn. **5 a** (usu. foll. by *in, of, for,* etc.) stint. **b** be niggardly. **6** (usu. foll. by *out, back, down*) remove (leaves, buds, etc.) to encourage bushy growth. *–n.* **1** act of pinching. **2** amount that can be taken up with fingers and thumb (*pinch of snuff*). **3** the stress caused by poverty etc. □ **at** (or **in**) **a pinch** in an emergency. [French *pincer*]

pinchbeck *–n.* goldlike alloy of copper and zinc used in cheap jewellery etc. *–adj.* counterfeit, sham. [*Pinchbeck*, name of a watchmaker]

pincushion *n.* small pad for holding pins.

pine[1] *n.* **1** evergreen coniferous tree with needle-shaped leaves growing in clusters. **2** its wood. **3** (*attrib.*) made of pine. [Latin *pinus*]

pine[2] *v.* (**-ning**) **1** (often foll. by *away*) decline or waste away from grief etc. **2** long eagerly. [Old English]

pineal /'pɪnɪəl/ *adj.* shaped like a pine cone. [Latin *pinea*: related to PINE[1]]

pineal body *n.* (also **pineal gland**) conical gland in the brain, secreting a hormone-like substance.

pineapple /'paɪnˌæp(ə)l/ *n.* **1** large juicy tropical fruit with yellow flesh and tough segmented skin. **2** plant bearing this. [from PINE[1], APPLE]

pine cone *n.* fruit of the pine.

pine nut *n.* edible seed of various pines.

ping –*n.* single short high ringing sound. –*v.* (cause to) make a ping. [imitative]

ping-pong *n. colloq.* = TABLE TENNIS. [imitative]

pinhead *n.* **1** head of a pin. **2** very small thing or spot. **3** *colloq.* stupid person.

pinhole *n.* **1** hole made by a pin. **2** hole into which a peg fits.

pinhole camera *n.* camera with a pinhole aperture and no lens.

pinion[1] /'pɪnjən/ –*n.* **1** outer part of a bird's wing. **2** *poet.* wing; flight-feather. –*v.* **1** cut off the pinion of (a wing or bird) to prevent flight. **2 a** bind the arms of (a person). **b** (often foll. by *to*) bind (the arms, a person, etc.) fast to a thing. [Latin *pinna*]

pinion[2] /'pɪnjən/ *n.* **1** small cog-wheel engaging with a larger one. **2** cogged spindle engaging with a wheel. [Latin *pinea* pine-cone: related to PINE[1]]

pink[1] –*n.* **1** pale red colour. **2** cultivated plant with fragrant flowers. **3** (prec. by *the*) the most perfect condition, the peak (*the pink of health*). **4** person with socialist tendencies. –*adj.* **1** of a pale red colour. **2** tending to socialism. □ **in the pink** *colloq.* in very good health. □ **pinkish** *adj.* **pinkness** *n.* **pinky** *adj.* [origin unknown]

pink[2] *v.* **1** pierce slightly. **2** cut a scalloped or zigzag edge on. [perhaps from Low German or Dutch]

pink[3] *v.* (of a vehicle engine) emit high-pitched explosive sounds caused by faulty combustion. [imitative]

pink gin *n.* gin flavoured with angostura bitters.

pinking shears *n.pl.* dressmaker's serrated shears for cutting a zigzag edge.

pinko /'pɪŋkəʊ/ *adj.* (*pl.* **-s**) esp. *US slang* socialist.

pin-money *n.* **1** *hist.* allowance to a woman from her husband. **2** very small sum of money.

pinnace /'pɪnɪs/ *n.* ship's small boat. [French]

pinnacle /'pɪnək(ə)l/ *n.* **1** culmination or climax. **2** natural peak. **3** small ornamental turret crowning a buttress, roof, etc. [Latin *pinna* PIN]

pinnate /'pɪneɪt/ *adj.* (of a compound leaf) having leaflets on either side of the leaf-stalk. [Latin *pinnatus* feathered: related to PINNACLE]

pinny /'pɪnɪ/ *n.* (*pl.* **-ies**) *colloq.* pinafore. [abbreviation]

pinpoint –*n.* **1** point of a pin. **2** something very small or sharp. **3** (*attrib.*) precise, accurate. –*v.* locate with precision.

pinprick *n.* trifling irritation.

pins and needles *n.pl.* tingling sensation in a limb recovering from numbness.

pinstripe *n.* **1** (often *attrib.*) narrow stripe in cloth (*pinstripe suit*). **2** (in *sing.* or *pl.*) pinstripe suit (*came wearing his pinstripes*). □ **pinstriped** *adj.*

pint /paɪnt/ *n.* **1** measure of capacity for liquids etc., $^1/_8$ gal. (0.57, *US* 0.47, litre). **2 a** *colloq.* pint of beer. **b** pint of a liquid, esp. milk. **3** measure of shellfish containable in a pint mug. [French]

pinta /'paɪntə/ *n. colloq.* pint of milk. [corruption of *pint of*]

pin-table *n.* table used in playing pinball.

pintail *n.* duck or grouse with a pointed tail.

pintle /'pɪnt(ə)l/ *n.* pin or bolt, esp. one on which some other part turns. [Old English]

pint-sized *adj. colloq.* very small.

pin-tuck *n.* very narrow ornamental tuck.

pin-up *n.* **1** photograph of a popular or sexually attractive person, hung on the wall. **2** person in such a photograph.

pin-wheel *n.* small Catherine wheel.

Pinyin /pɪn'jɪn/ *n.* system of romanized spelling for transliterating Chinese. [Chinese]

pion /'paɪɒn/ *n.* (also **pi meson**) subatomic particle having a mass many times greater than that of an electron. [from PI]

pioneer /,paɪə'nɪə(r)/ –*n.* **1** initiator of an enterprise; investigator of a subject etc. **2** explorer or settler; colonist. –*v.* **1** initiate (an enterprise etc.) for others to follow. **2** be a pioneer. [French *pionnier*: related to PAWN[1]]

pious /'paɪəs/ *adj.* **1** devout; religious. **2** sanctimonious. **3** dutiful. □ **piously** *adv.* **piousness** *n.* [Latin]

pip[1] –*n.* seed of an apple, pear, orange, grape, etc. –*v.* (**-pp-**) remove the pips from (fruit etc.). □ **pipless** *adj.* [abbreviation of PIPPIN]

pip[2] *n.* short high-pitched sound, usu. electronically produced, esp. as a time signal. [imitative]

pip[3] *v.* (**-pp-**) *colloq.* **1** hit with a shot. **2** (also **pip at the post**) defeat narrowly or at the last moment. [origin unknown]

pip[4] *n.* **1** any of the spots on a playing-card, dice, or domino. **2** star (1–3 according to rank) on the shoulder of an army officer's uniform. [origin unknown]

pip[5] *n.* **1** disease of poultry etc. **2** *colloq.* fit of disgust or bad temper (esp. *give one the pip*). [Low German or Dutch]

pipe –*n.* **1** tube of metal, plastic, etc., used to convey water, gas, etc. **2 a** narrow tube with a bowl at one end containing tobacco for smoking. **b** quantity of tobacco held by this. **3 a** wind instrument of a single tube. **b** any of the tubes by which sound is produced in an organ. **c** (in *pl.*) = BAGPIPES. **4** tubular organ, vessel, etc. in an animal's body. **5** high note or song, esp. of a bird. **6 a** boatswain's whistle. **b** sounding of this. **7** cask for wine, esp. as a measure, usu. = 105 gal. (about 477 litres). –*v.* (**-ping**) **1 a** convey (oil, water, gas, etc.) by pipes. **b** provide with pipes. **2** play (a tune etc.) on a pipe or pipes. **3** (esp. as **piped** *adj.*) transmit (recorded music etc.) by wire or cable. **4** (usu. foll. by *up*, *on*, *to*, etc.) *Naut.* **a** summon (a crew). **b** signal the arrival of (an officer etc.) on board. **5** utter in a shrill voice. **6** decorate or trim with piping. **7** lead or bring (a person etc.) by the sound of a pipe or pipes. □ **pipe down** *colloq.* be quiet or less insistent. **pipe up** begin to play, sing, speak, etc. □ **pipeful** *n.* (*pl.* **-s**). [Latin *pipo* chirp]

pipeclay *n.* fine white clay used for tobacco-pipes, whitening leather, etc.

pipe-cleaner *n.* piece of flexible tufted wire for cleaning a tobacco-pipe.

pipedream *n.* unattainable or fanciful hope or scheme. [originally as experienced when smoking an opium pipe]

pipeline *n.* **1** long, usu. underground, pipe for conveying esp. oil. **2** channel supplying goods, information, etc. □ **in the pipeline** being dealt with or prepared; under discussion, on the way.

piper *n.* person who plays a pipe, esp. the bagpipes.

pipette /pɪ'pet/ *n. Chem* slender tube for transferring or measuring small quantities of liquids. [French diminutive: related to PIPE]

piping *n.* **1** pipelike fold or cord for edging or decorating clothing, upholstery, etc. **2** ornamental lines of icing, cream, potato, etc. on a cake etc. **3** lengths of pipe, system of pipes. □ **piping hot** (of food, water, etc.) very hot.

pipit /'pɪpɪt/ *n.* small bird resembling a lark. [imitative]

pippin /'pɪpɪn/ *n.* **1** apple grown from seed. **2** red and yellow eating apple. [French]

pipsqueak *n. colloq.* insignificant or contemptible person or thing. [imitative]

piquant /'pi:kənt, -kɑ:nt/ *adj.* **1** agreeably pungent, sharp, or appetizing. **2** pleasantly stimulating to the mind. □ **piquancy** *n.* [French *piquer* prick]

pique /pi:k/ –*v.* (**piques, piqued, piquing**) **1** wound the pride of, irritate. **2** arouse (curiosity, interest, etc.). –*n.* resentment; hurt pride. [French: related to PIQUANT]

piquet /pɪ'ket/ *n.* card-game for two players with a pack of 32 cards. [French]

piracy /'paɪrəsɪ/ *n.* (*pl.* **-ies**) **1** robbery of ships at sea. **2** similar practice, esp. hijacking. **3** infringement of copyright etc. [related to PIRATE]

piranha /pɪ'rɑ:nə/ *n.* voracious S. American freshwater fish. [Portuguese]

pirate /'paɪərət/ –*n.* **1 a** seafaring robber attacking ships. **b** ship used by pirates. **2** (often *attrib.*) person who infringes another's copyright or business rights or who broadcasts without official authorization (*pirate radio station*). –*v.* (**-ting**) reproduce (a book etc.) or trade (goods) without permission. □ **piratical** /-'rætɪk(ə)l/ *adj.* [Latin *pirata* from Greek]

pirouette /,pɪrʊ'et/ –*n.* dancer's spin on one foot or the point of the toe. –*v.* (**-tting**) perform a pirouette. [French, = spinning-top]

piscatorial /,pɪskə'tɔ:rɪəl/ *adj.* of fishermen or fishing. □ **piscatorially** *adv.* [Latin *piscator* angler, from *piscis* fish]

Pisces /'paɪsi:z/ *n.* (*pl.* same) **1** constellation and twelfth sign of the zodiac (the Fish or Fishes). **2** person born when the sun is in this sign. [Latin, pl. of *piscis* fish]

piscina /pɪ'si:nə/ *n.* (*pl.* **-nae** /-ni:/ or **-s**) **1** stone basin near the altar in a church for draining water used in rinsing the chalice etc. **2** fish-pond. [Latin, from *piscis* fish]

piss *coarse slang* –*v.* **1** urinate. **2** discharge (blood etc.) with urine. **3** (as **pissed** *adj.*) drunk. –*n.* **1** urine. **2** act of urinating. □ **piss about** fool or mess about. **piss down** rain heavily. **piss off 1** go away. **2** (often as **pissed off** *adj.*) annoy; depress. **piss on** show utter contempt for (a person or thing). **take the piss** (often foll. by *out of*) mock; make fun of. [French, imitative]

piss artist *n.* **1** drunkard. **2** person who fools about.

piss-taking *n.* mockery. □ **piss-take** *n.* **piss-taker** *n.*

piss-up *n.* drinking spree.

pistachio /pɪ'stɑ:ʃɪəʊ/ *n.* (*pl.* **-s**) **1** edible pale-green nut. **2** tree yielding this. [Persian *pistah*]

piste /pi:st/ *n.* ski-run of compacted snow. [French, = racetrack]

pistil /'pɪstɪl/ *n.* female organs of a flower, comprising the stigma, style, and ovary. □ **pistillate** *adj.* [Latin: related to PESTLE]

pistol /'pɪst(ə)l/ *n.* small handgun. [Czech *pišt'al*]

piston /'pɪst(ə)n/ *n.* **1** sliding cylinder fitting closely in a tube in which it moves up and down, used in an internal-combustion engine to impart motion, or in a pump to receive motion. **2** sliding valve in a trumpet etc. [Italian: related to PESTLE]

piston-ring *n.* ring on a piston sealing the gap between piston and cylinder wall.

piston-rod *n.* rod or crankshaft by which a piston imparts motion.

pit[1] *–n.* **1 a** deep hole in the ground, usu. large. **b** coalmine. **c** covered hole as a trap for animals. **2** hollow on a surface, esp. an indentation of the skin. **3 a** = *orchestra pit* (see ORCHESTRA 2). **b** usu. *hist.* seating at the back of the stalls. **4** (**the pits**) *slang* worst imaginable place, situation, person, etc. **5 a** area at the side of a track where racing cars are serviced and refuelled. **b** sunken area in a workshop floor for access to a car's underside. *–v.* (**-tt-**) **1** (usu. foll. by *against*) set (one's wits, strength, etc.) in competition. **2** (usu. as **pitted** *adj.*) make pits, scars, craters, etc. in. **3** put into a pit. [Old English from Latin *puteus* well]

pit[2] *v.* (**-tt-**) (usu. as **pitted** *adj.*) remove stones from (fruit). [origin uncertain]

pita var. of PITTA.

pit-a-pat /'pɪtə,pæt/ (also **pitter-patter**) *–adv.* **1** with a sound like quick light steps. **2** falteringly (*heart went pit-a-pat*). *–n.* such a sound. [imitative]

pit bull terrier *n.* small American dog noted for ferocity.

pitch[1] *–v.* **1** erect and fix (a tent, camp, etc.). **2** throw. **3** fix in a definite position. **4** express in a particular style or at a particular level. **5** (often foll. by *against*, *into*, etc.) fall heavily, esp. headlong. **6** (of a ship etc.) plunge backwards and forwards in a lengthwise direction. **7** *Mus.* set at a particular pitch. **8** *Cricket* **a** cause (a bowled ball) to strike the ground at a specified point etc. **b** (of a ball) strike the ground thus. *–n.* **1** area of play in a field-game. **2** height, degree, intensity, etc. (*excitement had reached such a pitch*). **3** degree of slope, esp. of a roof. **4** *Mus.* quality of a sound governed by the rate of vibrations producing it; highness or lowness of a note. **5** act of throwing. **6** pitching motion of a ship etc. **7** *colloq.* salesman's persuasive talk. **8** place where a street vendor is stationed. **9** distance between successive points, lines, etc. (e.g. character spacing on a typewriter). □ **pitch in** *colloq.* set to work vigorously. **pitch into** *colloq.* **1** attack forcibly. **2** assail (food, work, etc.) vigorously. [origin uncertain]

pitch[2] *–n.* dark resinous substance from the distillation of tar or turpentine, used for making ships watertight etc. *–v.* coat with pitch. □ **pitchy** *adj.* (**-ier, -iest**). [Latin *pix pic-*]

pitch-black *adj.* (also **pitch-dark**) very or completely dark.

pitchblende /'pɪtʃblend/ *n.* uranium oxide occurring in pitchlike masses and yielding radium. [German: related to PITCH[2]]

pitched battle *n.* **1** vigorous argument etc. **2** planned battle between sides in prepared positions and on chosen ground.

pitched roof *n.* sloping roof.

pitcher[1] *n.* large jug with a lip and a handle. [related to BEAKER]

pitcher[2] *n.* player who delivers the ball in baseball.

pitchfork *–n.* long-handled two-pronged fork for pitching hay etc. *–v.* **1** throw with or as if with a pitchfork. **2** (usu. foll. by *into*) thrust (a person) forcibly into a position, office, etc.

pitch-pine *n.* pine-tree yielding much resin.

piteous /'pɪtɪəs/ *adj.* deserving or arousing pity; wretched. □ **piteously** *adv.* **piteousness** *n.* [Romanic: related to PITY]

pitfall *n.* **1** unsuspected danger or drawback. **2** covered pit for trapping animals.

pith *n.* **1** spongy white tissue lining the rind of an orange etc. **2** essential part. **3** spongy tissue in the stems and branches of plants. **4** strength; vigour; energy. [Old English]

pit-head *n.* **1** top of a mineshaft. **2** area surrounding this (also *attrib.*: *pit-head ballot*).

pith helmet *n.* protective sun-helmet made of dried pith from plants.

pithy *adj.* (**-ier, -iest**) **1** (of style, speech, etc.) terse and forcible. **2** of or like pith. □ **pithily** *adv.* **pithiness** *n.*

pitiable /'pɪtɪəb(ə)l/ *adj.* deserving or arousing pity or contempt. □ **pitiably** *adv.* [French: related to PITY]

pitiful /'pɪtɪ,fʊl/ *adj.* **1** causing pity. **2** contemptible. □ **pitifully** *adv.*

pitiless /'pɪtɪlɪs/ *adj.* showing no pity (*pitiless heat*). □ **pitilessly** *adv.*

pit of the stomach *n.* depression below the breastbone.

piton /'pi:tɒn/ *n.* peg driven into rock or a crack to support a climber or rope. [French]

pitta /'pɪtə/ *n.* (also **pita**) flat hollow unleavened bread which can be split and filled. [modern Greek, = a kind of cake]

pittance /'pɪt(ə)ns/ *n.* very small allowance or remuneration. [Romanic: related to PITY]

pitter-patter var. of PIT-A-PAT.

pituitary /pɪ'tju:ɪtərɪ/ *n.* (*pl.* **-ies**) (also **pituitary gland**) small ductless gland at the base of the brain. [Latin *pituita* phlegm]

pity /'pɪtɪ/ *–n.* **1** sorrow and compassion for another's suffering. **2** cause for regret (*what a pity!*). *–v.* (**-ies, -ied**) feel (often contemptuous) pity for. □ **take pity on** help out of pity for. □ **pitying** *adj.* **pityingly** *adv.* [Latin: related to PIETY]

pivot /'pɪvət/ *–n.* **1** shaft or pin on which something turns or oscillates. **2** crucial or essential person, point, etc. *–v.* (**-t-**) **1** turn on or as on a pivot. **2** provide with a pivot. □ **pivotal** *adj.* [French]

pixel /'pɪks(ə)l/ *n.* any of the minute areas of uniform illumination of which an image on a display screen is composed. [abbreviation of *picture element*]

pixie /'pɪksɪ/ *n.* (also **pixy**) (*pl.* **-ies**) fairy-like being. [origin unknown]

pizza /'pi:tsə/ *n.* Italian dish of a layer of dough baked with a topping of tomatoes, cheese, etc. [Italian, = pie]

pizzeria /,pi:tsə'ri:ə/ *n.* pizza restaurant.

pizzicato /,pɪtsɪ'kɑ:təʊ/ *Mus.* *–adv.* plucking. *–adj.* (of a note, passage, etc.) performed pizzicato. *–n.* (*pl.* **-s** or **-ti** /-tɪ/) note, passage, etc. played pizzicato. [Italian]

pl. *abbr.* **1** plural. **2** (usu. **Pl.**) place. **3** plate.

placable /'plækəb(ə)l/ *adj.* easily placated; mild; forgiving. □ **placability** /-'bɪlɪtɪ/ *n.* [Latin *placo* appease]

placard /'plækɑ:d/ *–n.* large notice for public display. *–v.* set up placards on (a wall etc.). [French from Dutch *placken* glue (v.)]

placate /plə'keɪt/ *v.* (**-ting**) pacify; conciliate. □ **placatory** /plə'keɪtərɪ/ *adj.* [Latin *placo* appease]

place *–n.* **1 a** particular portion of space. **b** portion of space occupied by a person or thing. **c** proper or natural position. **2** city, town, village, etc. **3** residence, home. **4** group of houses in a town etc., esp. a square. **5** (esp. large) country house. **6** rank or status. **7** space, esp. a seat, for a person. **8** building or area for a specific purpose (*place of work*). **9** point reached in a book etc. (*lost my place*). **10** particular spot on a surface, esp. of the skin (*sore place*). **11 a** employment or office. **b** duties or entitlements of office etc. (*not my place to criticize*). **12** position as a member of a team, student in a college, etc. **13** any of the first three (or four) positions in a race, esp. other than the winner. **14** position of a digit in a series indicated in decimal or similar notation. *–v.* (**-cing**) **1** put in a particular or proper place or state or order; arrange. **2** identify, classify, or remember correctly. **3** assign to a particular place, class, or rank; locate. **4** find employment or a living etc. for. **5** make or state (an order or bet etc.). **6** (often foll. by *in*, *on*, etc.) have (confidence etc.). **7** state the position of (any of the first three or four runners) in a race. **8** (as **placed** *adj.*) among the first three (or four) in a race. □ **give place to 1** make room for. **2** yield precedence to. **3** be succeeded by. **go places** *colloq.* be successful. **in place** in the right position; suitable. **in place of** in exchange for; instead of. **in places** at only some places or parts. **out of place 1** in the wrong position. **2** unsuitable. **put a person in his** (or **her**) **place** deflate a person. **take place** occur. **take the place of** be substituted for. □ **placement** *n.* [Latin *platea* broad way]

placebo /plə'si:bəʊ/ *n.* (*pl.* **-s**) **1** medicine with no physiological effect prescribed for psychological reasons. **2** dummy pill etc. used in a controlled trial. [Latin, = I shall be acceptable]

place-kick *n.* kick in football with the ball placed on the ground.

place-mat *n.* small table-mat for a person's plate.

place-name *n.* name of a town, village, etc.

placenta /plə'sentə/ *n.* (*pl.* **-tae** /-ti:/ or **-s**) organ in the uterus of pregnant mammals nourishing the foetus through the umbilical cord and expelled after birth. □ **placental** *adj.* [Greek, = flat cake]

placer /'pleɪsə(r)/ *n.* deposit of sand, gravel, etc. containing valuable minerals in particles. [American Spanish]

place-setting *n.* set of cutlery etc. for one person at a table.

placid /'plæsɪd/ *adj.* **1** calm; not easily excited or irritated. **2** tranquil, serene. □ **placidity** /plə'sɪdɪtɪ/ *n.* **placidly** *adv.* **placidness** *n.* [Latin *placeo* please]

placket /'plækɪt/ *n.* **1** opening or slit in a garment, for fastenings or access to a pocket. **2** flap of fabric under this. [var. of PLACARD]

plagiarize /'pleɪdʒə,raɪz/ *v.* (also **-ise**) (**-zing** or **-sing**) **1** (also *absol.*) take and pass off (another's thoughts, writings, etc.) as one's own. **2** pass off the thoughts etc. of (another person) as one's own. □ **plagiarism** *n.* **plagiarist** *n.* **plagiarizer** *n.* [Latin *plagiarius* kidnapper]

plague /pleɪg/ *–n.* **1** deadly contagious disease. **2** (foll. by *of*) *colloq.* infestation of a pest etc. **3** great trouble or affliction. **4** *colloq.* nuisance. *–v.* (**plagues, plagued, plaguing**) **1** *colloq.* pester, annoy. **2** affect as with plague; afflict, hinder (*plagued by back pain*). [Latin *plaga* stroke, infection]

plaice /pleɪs/ *n.* (*pl.* same) marine flatfish used as food. [Latin *platessa*]

plaid /plæd/ *n.* **1** (often *attrib.*) chequered or tartan, esp. woollen, twilled cloth (*plaid skirt*). **2** long piece of this worn over the shoulder in Highland Scottish costume. [Gaelic]

plain *–adj.* **1** clear, evident. **2** readily understood, simple. **3** (of food, decoration, etc.) simple. **4** not beautiful or distinguished-looking. **5** outspoken; straightforward. **6** unsophisticated; not luxurious (*a plain man; plain living*). *–adv.* **1** clearly. **2** simply. *–n.* **1** level tract of country. **2** basic knitting stitch. □ **plainly** *adv.* **plainness** *n.* [Latin *planus*]

plainchant *n.* = PLAINSONG.

plain chocolate *n.* dark chocolate without added milk.

plain clothes *n.pl.* ordinary clothes, not uniform (*plain-clothes police*).

plain dealing *n.* candour; straightforwardness.

plain flour *n.* flour containing no raising agent.

plain sailing *n.* uncomplicated situation or course of action.

plainsong *n.* unaccompanied church music sung in unison in medieval modes and in free rhythm corresponding to the accentuation of the words.

plain-spoken *adj.* frank.

plaint *n.* **1** *Law* accusation; charge. **2** *literary* complaint, lamentation. [French *plainte* from Latin *plango* lament]

plaintiff /'pleɪntɪf/ *n.* person who brings a case against another into court. [French *plaintif*: related to PLAINTIVE]

plaintive /'pleɪntɪv/ *adj.* expressing sorrow; mournful-sounding. □ **plaintively** *adv.* [French: related to PLAINT]

plait /plæt/ *–n.* length of hair, straw, etc., in three or more interlaced strands. *–v.* **1** weave (hair etc.) into a plait. **2** make by interlacing strands (*plaited belt*). [French *pleit* from Latin *plico* fold]

plan *–n.* **1** method or procedure for doing something; design, scheme, or intention. **2** drawing etc. of a building or structure, made by projection on to a horizontal plane. **3** map of a town or district. **4** scheme of an arrangement (*seating plan*). *–v.* (**-nn-**) **1** arrange (a procedure etc.) beforehand; form a plan; intend. **2** make a plan of or design for. **3** (as **planned** *adj.*) in accordance with a plan (*planned parenthood*). **4** make plans. □ **plan on** (often foll. by *pres. part.*) *colloq.* aim at; intend. □ **planning** *n.* [French]

planchette /plɑːn'ʃet/ *n.* small board on castors with a pencil, said to write spirit messages when a person's fingers rest lightly on it. [French diminutive: related to PLANK]

plane[1] *–n.* **1** flat surface such that a straight line joining any two points on it lies wholly in it. **2** level surface. **3** *colloq.* = AEROPLANE. **4** flat surface producing lift by the action of air or water over and under it (usu. in *comb.*: *hydroplane*). **5** (often foll. by *of*) level of attainment, knowledge, etc. *–adj.* **1** (of a surface etc.) perfectly level. **2** (of an angle, figure, etc.) lying in a plane. *–v.* (**-ning**) glide. [Latin *planus* PLAIN]

plane[2] *–n.* tool for smoothing a usu. wooden surface by paring shavings from it. *–v.* (**-ning**) **1** smooth with a plane. **2** (often foll. by *away, down*) pare with a plane. [Latin: related to PLANE[1]]

plane[3] *n.* tall tree with maple-like leaves and bark which peels in uneven patches. [Greek *platanos*]

planet /'plænɪt/ *n.* celestial body orbiting round a star. □ **planetary** *adj.* [Greek, = wanderer]

planetarium /,plænɪ'teərɪəm/ *n.* (*pl.* **-s** or **-ria**) **1** domed building in which images of stars, planets, constellations, etc. are projected. **2** device for such projection.

plangent /'plændʒ(ə)nt/ *adj. literary* **1** loud and reverberating. **2** plaintive. [Latin: related to PLAINT]

plank *–n.* **1** long flat piece of timber. **2** item in a political or other programme. *–v.* **1** provide or cover with planks. **2** (usu. foll. by *down*) *colloq.* **a** put down or deposit roughly or violently. **b** pay (money) on the spot. □ **walk the plank** *hist.* be made to walk blindfold along a plank over the side of a ship to one's death in the sea. [Latin *planca*]

planking *n.* planks as flooring etc.

plankton /'plæŋkt(ə)n/ *n.* chiefly microscopic organisms drifting in the sea or fresh water. [Greek, = wandering]

planner *n.* **1** person who plans new towns etc. **2** person who makes plans. **3** list, table, etc., with information helpful in planning.

planning permission *n.* formal permission for building etc., esp. from a local authority.

plant /plɑːnt/ *–n.* **1 a** organism usu. containing chlorophyll enabling it to live wholly on inorganic substances, and lacking the power of voluntary movement. **b** small organism of this kind, as distinguished from a shrub or tree. **2 a** machinery, fixtures, etc., used in industry. **b** factory. **3** *colloq.* something deliberately placed so as to incriminate another. *–v.* **1** place (seeds, plants, etc.) in soil for growing. **2** (often foll. by *in, on*, etc.) put or fix in position. **3** (often *refl.*) station (a person etc.), esp. as a spy. **4** cause (an idea etc.) to be established, esp. in another person's mind. **5** deliver (a blow, kiss, etc.) with a deliberate aim. **6** *colloq.* place (something incriminating) for later discovery. □ **plant out** transfer from a pot or frame to the open ground; set out (seedlings) at intervals. □ **plantlike** *adj.* [Latin *planta*]

plantain[1] /'plæntɪn/ *n.* plant with broad flat leaves spread close to the ground and seeds used as food for birds. [Latin *plantago*]

plantain[2] /'plæntɪn/ *n.* **1** a kind of banana plant, grown for its fruit. **2** banana-like fruit of this. [Spanish]

plantation /plɑ:n'teɪʃ(ə)n, plæn-/ *n.* **1** estate on which cotton, tobacco, etc. is cultivated. **2** area planted with trees etc. **3** *hist.* colony. [Latin: related to PLANT]

planter *n.* **1** manager or owner of a plantation. **2** container for houseplants.

plaque /plæk, plɑ:k/ *n.* **1** commemorative tablet, esp. fixed to a building. **2** deposit on teeth where bacteria proliferate. [Dutch *plak* tablet: related to PLACARD]

plasma /'plæzmə/ *n.* (also **plasm** /'plæz(ə)m/) **1 a** colourless fluid part of blood, lymph, or milk, in which corpuscles or fat-globules are suspended. **b** this taken from blood for transfusions. **2** = PROTOPLASM. **3** gas of positive ions and free electrons in about equal numbers. □ **plasmic** *adj.* [Greek *plassō* shape (v.)]

plaster /'plɑ:stə(r)/ *–n.* **1** soft mixture of lime, sand, and water etc. applied to walls, ceilings, etc., to dry into a smooth hard surface. **2** = STICKING-PLASTER. **3** = PLASTER OF PARIS. *–v.* **1** cover (a wall etc.) with plaster. **2** coat, daub, cover thickly. **3** stick or apply (a thing) thickly like plaster. **4** (often foll. by *down*) smooth (esp. hair) with water etc. **5** (as **plastered** *adj.*) *slang* drunk. □ **plasterer** *n.* [Greek *emplastron*]

plasterboard *n.* two boards with a filling of plaster for partitions, walls, etc.

plaster cast *n.* **1** bandage stiffened with plaster of Paris and applied to a broken limb etc. **2** statue or mould made of plaster.

plaster of Paris *n.* fine white gypsum plaster for plaster casts etc.

plastic /'plæstɪk/ *–n.* **1** synthetic resinous substance that can be given any shape. **2** (in full **plastic money**) *colloq.* credit card(s). *–adj.* **1** made of plastic. **2** capable of being moulded; pliant, supple. **3** giving form to clay, wax, etc. □ **plasticity** /-'tɪsɪtɪ/ *n.* **plasticize** /-,saɪz/ *v.* (also **-ise**) (**-zing** or **-sing**). **plasticizer** /-,saɪzə(r)/ *n.* (also **-iser**). **plasticky** *adj.* [Greek: related to PLASMA]

plastic arts *n.pl.* arts involving modelling or the representation of solid objects.

plastic bomb *n.* bomb containing plastic explosive.

plastic explosive *n.* putty-like explosive.

Plasticine /'plæstə,si:n/ *n. propr.* pliant material used for modelling.

plastic surgery *n.* reconstruction or repair of damaged or unsightly skin, muscle, etc., esp. by the transfer of tissue. □ **plastic surgeon** *n.*

plate *–n.* **1 a** shallow usu. circular vessel from which food is eaten or served. **b** contents of this. **2** similar vessel used for a collection in church etc. **3** (*collect.*) **a** utensils of silver, gold, or other metal. **b** objects of plated metal. **4** piece of metal with a name or inscription for affixing to a door etc. **5** illustration on special paper in a book. **6** thin sheet of metal, glass, etc., coated with a sensitive film for photography. **7** flat thin usu. rigid sheet of metal etc., often as part of a mechanism. **8 a** smooth piece of metal etc. for engraving. **b** impression from this. **9 a** silver or gold cup as a prize for a horse-race etc. **b** race with this as a prize. **10 a** thin piece of plastic material, moulded to the shape of the mouth, on which artificial teeth are mounted. **b** *colloq.* denture. **11** each of several rigid sheets of rock thought to form the earth's outer crust. **12** thin flat organic structure or formation. *–v.* (**-ting**) **1** apply a thin coat esp. of silver, gold, or tin to (another metal). **2** cover (esp. a ship) with plates of metal, for protection. □ **on a plate** *colloq.* available with little trouble to the recipient. **on one's plate** *colloq.* for one to deal with. □ **plateful** *n.* (*pl.* **-s**). [Latin *platta* from *plattus* flat]

plateau /'plætəʊ/ *–n.* (*pl.* **-x** or **-s** /-təʊz/) **1** area of fairly level high ground. **2** state of little variation after an increase. *–v.* (**plateauing, plateaus, plateaued**) (often foll. by *out*) reach a level or static state after an increase. [French: related to PLATE]

plate glass *n.* thick fine-quality glass for shop windows etc.

platelayer *n.* person employed in fixing and repairing railway rails.

platelet /'pleɪtlɪt/ *n.* small colourless disc of protoplasm found in blood and involved in clotting.

platen /'plæt(ə)n/ *n.* **1** plate in a printing-press which presses the paper against the type. **2** cylindrical roller in a typewriter etc. against which the paper is held. [French *platine*: related to PLATE]

plate-rack *n.* rack in which plates are placed to drain.

plate tectonics *n.pl.* (usu. treated as *sing.*) the study of the earth's surface based on the concept of moving 'plates' (see sense 11 of PLATE) forming its structure.

platform /'plætfɔ:m/ *n.* **1** raised level surface, esp. one from which a speaker addresses an audience or one alongside the line at a railway station. **2** floor area at the entrance to a bus etc. **3** thick sole of a shoe. **4** declared policy of a political party. [French: related to PLATE, FORM]

platinum /'plætɪnəm/ *n. Chem.* white heavy precious metallic element that does not tarnish. [earlier *platina* from Spanish, diminutive from *plata* silver]

platinum blonde (also **platinum blond**) *–adj.* silvery-blond. *–n.* person with such hair.

platitude /'plætɪ,tju:d/ *n.* commonplace remark, esp. one solemnly delivered. □ **platitudinous** /-'tju:dɪnəs/ *adj.* [French: related to PLATE]

Platonic /plə'tɒnɪk/ *adj.* **1** of Plato or his ideas. **2** (**platonic**) (of love or friendship) not sexual. [Greek *Platōn* (5th-4th c. BC), name of a Greek philosopher]

Platonism /'pleɪtə,nɪz(ə)m/ *n.* philosophy of Plato or his followers. □ **Platonist** *n.*

platoon /plə'tu:n/ *n.* **1** subdivision of a military company. **2** group of persons acting together. [French *peloton* diminutive of *pelote* PELLET]

platter *n.* large flat dish or plate. [Anglo-French *plater*: related to PLATE]

platypus /'plætɪpəs/ *n.* (*pl.* **-puses**) Australian aquatic egg-laying mammal, with a ducklike bill and flat tail. [Greek, = flat foot]

plaudit /'plɔ:dɪt/ *n.* (usu. in *pl.*) **1** round of applause. **2** expression of approval. [Latin *plaudite*, imperative of *plaudo plaus-* clap]

plausible /'plɔ:zɪb(ə)l/ *adj.* **1** (of a statement etc.) reasonable or probable. **2** (of a person) persuasive but deceptive. □ **plausibility** /-'bɪlɪtɪ/ *n.* **plausibly** *adv.* [Latin: related to PLAUDIT]

play *–v.* **1** (often foll. by *with*) occupy or amuse oneself pleasantly. **2** (foll. by *with*) act light-heartedly or flippantly with (a person's feelings etc.). **3 a** perform on or be able to perform on (a musical instrument). **b** perform (a piece of music etc.). **c** cause (a record, record-player, etc.) to produce sounds. **4 a** (foll. by *in*) perform a role in (a drama etc.). **b** perform (a drama or role) on stage etc. **c** give a dramatic performance at (a particular theatre or place). **5** act in real life the part of (*play truant*; *play the fool*). **6** (foll. by *on*) perform (a trick or joke etc.) on (a person). **7** *colloq.* cooperate; do what is wanted (*they won't play*). **8** gamble, gamble on. **9 a** take part in (a game or recreation). **b** compete with (another player or team) in a game. **c** occupy (a specified position) in a team for a game. **d** assign (a player) to a position. **10** move (a piece) or display (a playing-card) in one's turn in a game. **11** (also *absol.*) strike (a ball etc.) or execute (a stroke) in a game. **12** move about in a lively manner; flit, dart. **13** (often foll. by *on*) **a** touch gently. **b** emit light, water, etc. (*fountains gently playing*). **14** allow (a fish) to exhaust itself pulling against a line. **15** (often foll. by *at*) **a** engage half-heartedly (in an activity). **b** pretend to be. *–n.* **1** recreation, amusement, esp. as the spontaneous activity of children. **2 a** playing of a game. **b** action or manner of this. **3** dramatic piece for the stage etc. **4** activity or operation (*the play of fancy*). **5 a** freedom of movement. **b** space or scope for this. **6** brisk, light, or fitful movement. **7** gambling. □ **in** (or **out of**) **play** *Sport* (of the ball etc.) in (or not in) a position to be played according to the rules. **make a play for** *colloq.* make a conspicuous attempt to acquire. **make play with** use ostentatiously. **play about** (or **around**) behave irresponsibly. **play along** pretend to cooperate. **play back** play (sounds recently recorded). **play ball** *colloq.* cooperate. **play by ear 1** perform (music) without having seen it written down. **2** (also **play it by ear**) *colloq.* proceed step by step according to results. **play one's cards right** (or **well**) *colloq.* make good use of opportunities; act shrewdly. **play down** minimize the importance of. **played out** exhausted of energy or usefulness. **play fast and loose** act unreliably. **play the field** see FIELD. **play for time** seek to gain time by delaying. **play the game** observe the rules; behave honourably. **play havoc** (or **hell**) **with**

colloq. cause great confusion or difficulty to; disrupt. **play into a person's hands** act so as unwittingly to give a person an advantage. **play it cool** *colloq.* be relaxed or apparently indifferent. **play the market** speculate in stocks etc. **play off** (usu. foll. by *against*) **1** oppose (one person against another), esp. for one's own advantage. **2** play an extra match to decide a draw or tie. **play on 1** continue to play. **2** take advantage of (a person's feelings etc.). **play safe** (or **for safety**) avoid risks. **play up 1** behave mischievously. **2** annoy in this way. **3** cause trouble; be irritating. **play up to** flatter, esp. to win favour. **play with fire** take foolish risks. [Old English]

play-act *v.* **1** act in a play. **2** pretend; behave insincerely. □ **play-acting** *n.*

play-back *n.* playing back of a sound.

playbill *n.* poster advertising a play.

playboy *n.* wealthy pleasure-seeking man.

player *n.* **1** participant in a game. **2** person playing a musical instrument. **3** actor.

playfellow *n.* playmate.

playful *adj.* **1** fond of or inclined to play. **2** done in fun. □ **playfully** *adv.* **playfulness** *n.*

playgoer *n.* person who goes often to the theatre.

playground *n.* outdoor area for children to play in.

playgroup *n.* organized regular meeting of preschool children for supervised play.

playhouse *n.* theatre.

playing-card *n.* one of a set of usu. 52 oblong cards, divided into four suits and used in games.

playing-field *n.* field for outdoor games.

playlet *n.* short play.

playmate *n.* child's companion in play.

play-off *n.* match played to decide a draw or tie.

play on words *n.* pun.

play-pen *n.* portable enclosure for a young child to play in.

play school *n.* nursery school or kindergarten.

plaything *n.* **1** toy or other thing to play with. **2** person used merely as an object of amusement or pleasure.

playtime *n.* time for play or recreation.

playwright *n.* person who writes plays.

plc *abbr.* (also **PLC**) Public Limited Company.

plea *n.* **1** appeal, entreaty. **2** *Law* formal statement by or on behalf of a defendant. **3** excuse. [Latin *placitum* decree: related to PLEASE]

pleach *v.* entwine or interlace (esp. branches to form a hedge). [Latin: related to PLEXUS]

plead *v.* **1** (foll. by *with*) make an earnest appeal to. **2** (of an advocate) address a lawcourt. **3** maintain (a cause) in a lawcourt. **4** (foll. by *guilty* or *not guilty*) declare oneself to be guilty or not guilty of a charge. **5** allege as an excuse (*plead insanity*). **6** (often as **pleading** *adj.*) make an appeal or entreaty (*in a pleading voice*). [Anglo-French *pleder*: related to PLEA]

pleading *n.* (usu. in *pl.*) formal statement of the cause of an action or defence.

pleasant /'plez(ə)nt/ *adj.* (**-er, -est**) pleasing to the mind, feelings, or senses. □ **pleasantly** *adv.* [French: related to PLEASE]

pleasantry *n.* (*pl.* **-ies**) **1** amusing or polite remark. **2** humorous speech. **3** jocularity.

please /pli:z/ *v.* (**-sing**) **1** be agreeable to; make glad; give pleasure. **2** (in *passive*) **a** (foll. by *to* + infin.) be glad or willing to (*am pleased to help*). **b** (often foll. by *about, at, with*) derive pleasure or satisfaction (from). **3** (with *it* as subject) be the inclination or wish of (*it did not please him to attend*). **4** think fit (*take as many as you please*). **5** used in polite requests (*come in, please*). □ **if you please** if you are willing, esp. *iron.* to indicate unreasonableness (*then, if you please, we had to pay*). **please oneself** do as one likes. □ **pleased** *adj.* **pleasing** *adj.* [French *plaisir* from Latin *placeo*]

pleasurable /'pleʒərəb(ə)l/ *adj.* causing pleasure. □ **pleasurably** *adv.*

pleasure /'pleʒə(r)/ *n.* **1** feeling of satisfaction or joy. **2** enjoyment. **3** source of pleasure or gratification. **4** one's will or desire (*what is your pleasure?*). **5** sensual gratification. **6** (*attrib.*) done or used for pleasure. [French: related to PLEASE]

pleat –*n.* fold or crease, esp. a flattened fold in cloth doubled upon itself. –*v.* make a pleat or pleats in. [from PLAIT]

pleb *n. colloq.* usu. *derog.* = PLEBEIAN 2. □ **plebby** *adj.* [abbreviation of PLEBEIAN]

plebeian /plɪ'bi:ən/ –*n.* **1** commoner, esp. in ancient Rome. **2** working-class person, esp. an uncultured one. –*adj.* **1** of the common people. **2** uncultured, coarse. [Latin *plebs plebis* common people]

plebiscite /ˈplebɪ,saɪt/ *n.* referendum. [Latin *plebiscitum*: related to PLEBEIAN]
plectrum /ˈplektrəm/ *n.* (*pl.* **-s** or **-tra**) thin flat piece of plastic etc. for plucking the strings of a guitar etc. [Greek *plēssō* strike]
pledge *–n.* **1** solemn promise. **2** thing given as security against a debt etc. **3** thing put in pawn. **4** thing given as a token of favour etc., or of something to come. **5** drinking of a person's health, toast. **6** solemn promise to abstain from alcohol (*sign the pledge*). *–v.* (**-ging**) **1 a** deposit as security. **b** pawn. **2** promise solemnly by the pledge of (one's honour, word, etc.). **3** bind by a solemn promise. **4** drink to the health of. □ **pledge one's troth** see TROTH. [French *plege*]
Pleiades /ˈplaɪə,di:z/ *n.pl.* cluster of seven stars in the constellation Taurus. [Latin from Greek]
Pleistocene /ˈplaɪstə,si:n/ *Geol. –adj.* of the first epoch of the Quaternary period. *–n.* this epoch. [Greek *pleistos* most, *kainos* new]
plenary /ˈpli:nərɪ/ *adj.* **1** (of an assembly) to be attended by all members. **2** entire, unqualified (*plenary indulgence*). [Latin *plenus* full]
plenipotentiary /,plenɪpəˈtenʃərɪ/ *–n.* (*pl.* **-ies**) person (esp. a diplomat) invested with full authority to act. *–adj.* having this power. [Latin: related to PLENARY, POTENT]
plenitude /ˈplenɪ,tju:d/ *n. literary* **1** fullness, completeness. **2** abundance. [Latin: related to PLENARY]
plenteous /ˈplentɪəs/ *adj. literary* plentiful. [French *plentivous*: related to PLENTY]
plentiful /ˈplentɪ,fʊl/ *adj.* abundant, copious. □ **plentifully** *adv.*
plenty /ˈplentɪ/ *–n.* (often foll. by *of*) abundance, sufficient quantity or number (*we have plenty*; *plenty of time*; *a time of plenty*). *–adj. colloq.* plentiful. *–adv. colloq.* fully, quite. [Latin *plenitas*: related to PLENARY]
plenum /ˈpli:nəm/ *n.* full assembly of people or a committee etc. [Latin, neuter of *plenus* full]
pleonasm /ˈpli:ə,næz(ə)m/ *n.* use of more words than are needed (e.g. *see with one's eyes*). □ **pleonastic** /-ˈnæstɪk/ *adj.* [Greek *pleon* more]
plethora /ˈpleθərə/ *n.* over-abundance. [Greek, = fullness]
pleura /ˈplʊərə/ *n.* (*pl.* **-rae** /-ri:/) membrane enveloping the lungs. □ **pleural** *adj.* [Greek *pleura* rib]
pleurisy /ˈplʊərəsɪ/ *n.* inflammation of the pleura. □ **pleuritic** /-ˈrɪtɪk/ *adj.* [Greek: related to PLEURA]
plexus /ˈpleksəs/ *n.* (*pl.* same or **plexuses**) *Anat.* network of nerves or vessels (*solar plexus*). [Latin *plecto plex-* plait]
pliable /ˈplaɪəb(ə)l/ *adj.* **1** bending easily; supple. **2** yielding, compliant. □ **pliability** /-ˈbɪlɪtɪ/ *n.* [French: related to PLY[1]]
pliant /ˈplaɪənt/ *adj.* = PLIABLE 1. □ **pliancy** *n.*
pliers /ˈplaɪəz/ *n.pl.* pincers with parallel flat surfaces for holding small objects, bending wire, etc. [from dial. *ply* bend: related to PLIABLE]
plight[1] /plaɪt/ *n.* unfortunate condition or state. [Anglo-French *plit* PLAIT]
plight[2] /plaɪt/ *v. archaic* **1** pledge. **2** (foll. by *to*) engage (oneself) in marriage. □ **plight one's troth** see TROTH. [Old English]
plimsoll /ˈplɪms(ə)l/ *n.* (also **plimsole**) rubber-soled canvas sports shoe. [from PLIMSOLL LINE]
Plimsoll line /ˈplɪms(ə)l/ *n.* (also **Plimsoll mark**) marking on a ship's side showing the limit of legal submersion under various conditions. [*Plimsoll*, name of a politician]
plinth *n.* **1** lower square slab at the base of a column. **2** base supporting a vase or statue etc. [Greek, = tile]
Pliocene /ˈplaɪə,si:n/ *Geol. –adj.* of the last epoch of the Tertiary period. *–n.* this epoch. [Greek *pleiōn* more, *kainos* new]
PLO *abbr.* Palestine Liberation Organization.
plod *–v.* (**-dd-**) **1** walk doggedly or laboriously; trudge. **2** work slowly and steadily. *–n.* spell of plodding. □ **plodder** *n.* [probably imitative]
plonk[1] *–v.* **1** set down hurriedly or clumsily. **2** (usu. foll. by *down*) set down firmly. *–n.* heavy thud. [imitative]
plonk[2] *n. colloq.* cheap or inferior wine. [origin unknown]
plonker *n. coarse slang* **1** fool. **2** penis.
plop *–n.* sound as of a smooth object dropping into water without a splash. *–v.* (**-pp-**) fall or drop with a plop. *–adv.* with a plop. [imitative]
plosive /ˈpləʊsɪv/ *–adj.* pronounced with a sudden release of breath. *–n.* plosive sound. [from EXPLOSIVE]
plot *–n.* **1** defined and usu. small piece of land. **2** interrelationship of the main events in a play, novel, film, etc. **3** conspiracy or secret plan. *–v.* (**-tt-**) **1** make a plan or map of. **2** (also *absol.*) plan or contrive secretly (a crime etc.). **3** mark on a chart or diagram. **4** make (a curve etc.) by marking out a number of points. **5** provide (a play, novel, film,

etc.) with a plot. □ **plotter** *n.* [Old English and French *complot*]

plough /plaʊ/ (*US* **plow**) *–n.* **1** implement for cutting furrows in the soil and turning it up. **2** implement resembling this (*snowplough*). **3** (**the Plough**) the Great Bear (see BEAR²) or its seven bright stars. *–v.* **1** (also *absol.*) turn up (the earth) with a plough. **2** (foll. by *out, up*, etc.) turn or extract with a plough. **3** furrow or scratch (a surface) as with a plough. **4** produce (a furrow or line) thus. **5** (foll. by *through*) advance laboriously, esp. through work, a book, etc. **6** (foll. by *through, into*) move violently like a plough. **7** *colloq.* fail in an examination. □ **plough back 1** plough (grass etc.) into the soil to enrich it. **2** reinvest (profits) in the business producing them. [Old English]

ploughman *n.* (*US* **plowman**) person who uses a plough.

ploughman's lunch *n.* meal of bread and cheese with pickle and salad.

ploughshare *n.* (*US* **plowshare**) cutting blade of a plough.

plover /ˈplʌvə(r)/ *n.* plump-breasted wading bird, e.g. the lapwing. [Latin *pluvia* rain]

plow *US* var. of PLOUGH.

plowman *US* var. of PLOUGHMAN.

plowshare *US* var. of PLOUGHSHARE.

ploy *n.* cunning manoeuvre to gain advantage. [origin unknown]

PLR *abbr.* Public Lending Right.

pluck *–v.* **1** pick or pull out or away. **2** strip (a bird) of feathers. **3** pull at, twitch. **4** (foll. by *at*) tug or snatch at. **5** sound (the string of a musical instrument) with a finger or plectrum. **6** plunder. *–n.* **1** courage, spirit. **2** plucking; twitch. **3** animal's heart, liver, and lungs as food. □ **pluck up** summon up (one's courage etc.). [Old English]

plucky *adj.* (**-ier, -iest**) brave, spirited. □ **pluckily** *adv.* **pluckiness** *n.*

plug *–n.* **1** piece of solid material fitting tightly into a hole, used to fill a gap or cavity or act as a wedge or stopper. **2 a** device of metal pins in an insulated casing, fitting into holes in a socket for making an electrical connection. **b** *colloq.* electric socket. **3** = SPARK-PLUG. **4** *colloq.* piece of free publicity for an idea, product, etc. **5** cake or stick of tobacco; piece of this for chewing. *–v.* (**-gg-**) **1** (often foll. by *up*) stop (a hole etc.) with a plug. **2** *slang* shoot or hit (a person etc.). **3** *colloq.* seek to popularize (an idea, product, etc.) by constant recommendation. **4** *colloq.* (foll. by *away* (*at*)) work steadily (at). □ **plug in** connect electrically by inserting a plug into a socket. [Low German or Dutch]

plug-hole *n.* hole, esp. in a sink or bath, which can be closed by a plug.

plug-in *attrib. adj.* designed to be plugged into a socket.

plum *n.* **1 a** small sweet oval fleshy fruit with a flattish pointed stone. **b** tree bearing this. **2** reddish-purple colour. **3** raisin used in cooking. **4** *colloq.* something prized (often *attrib.*: *plum job*). □ **have a plum in one's mouth** have an affectedly rich voice. [Latin: related to PRUNE¹]

plumage /ˈpluːmɪdʒ/ *n.* bird's feathers. [French: related to PLUME]

plumb /plʌm/ *–n.* lead ball, esp. attached to the end of a line for finding the depth of water or testing whether a wall etc. is vertical. *–adv.* **1** exactly (*plumb in the centre*). **2** vertically. **3** *US slang* quite, utterly (*plumb crazy*). *–adj.* vertical. *–v.* **1 a** provide with plumbing. **b** (often foll. by *in*) fit as part of a plumbing system. **c** work as a plumber. **2** sound or test with a plumb. **3** reach or experience (an extreme feeling) (*plumb the depths of fear*). **4** learn in detail the facts about (a matter). □ **out of plumb** not vertical. [Latin *plumbum* lead]

plumber *n.* person who fits and repairs the apparatus of a water-supply, heating, etc.

plumbing *n.* **1** system or apparatus of water-supply etc. **2** work of a plumber. **3** *colloq.* lavatory installations.

plumb-line *n.* line with a plumb attached.

plume /pluːm/ *–n.* **1** feather, esp. a large one used for ornament. **2** ornament of feathers etc. worn on a helmet or hat or in the hair. **3** something resembling this (*plume of smoke*). *–v.* (**-ming**) **1** decorate or provide with a plume or plumes. **2** *refl.* (foll. by *on, upon*) pride (oneself on esp. something trivial). **3** (of a bird) preen (itself or its feathers). [Latin *pluma*]

plummet /ˈplʌmɪt/ *–n.* **1** plumb, plumb-line. **2** sounding-line. **3** weight attached to a fishing-line to keep the float upright. *–v.* (**-t-**) fall or plunge rapidly. [French: related to PLUMB]

plummy *adj.* (**-ier, -iest**) **1** abounding or rich in plums. **2** *colloq.* (of a voice) sounding affectedly rich in tone. **3** *colloq.* good, desirable.

plump¹ *–adj.* full or rounded in shape; fleshy. *–v.* (often foll. by *up, out*) make or become plump (*plumped up the cushion*). □ **plumpness** *n.* [Low German or Dutch *plomp* blunt]

plump² *–v.* **1** (foll. by *for*) decide on, choose. **2** (often foll. by *down*) drop or fall abruptly. *–n.* abrupt or heavy fall.

—*adv. colloq.* with a plump. [Low German or Dutch *plompen*, imitative]

plum pudding *n.* = CHRISTMAS PUDDING.

plumy /'plu:mɪ/ *adj.* (**-ier**, **-iest**) **1** plumelike, feathery. **2** adorned with plumes.

plunder /'plʌndə(r)/ —*v.* **1** rob or steal, esp. in wartime; loot. **2** exploit (another person's or common property) for one's own profit. —*n.* **1** activity of plundering. **2** property so acquired. [German *plündern*]

plunge —*v.* (**-ging**) **1** (usu. foll. by *in, into*) **a** thrust forcefully or abruptly. **b** dive. **c** (cause to) enter a condition or embark on a course impetuously (*they plunged into marriage*; *the room was plunged into darkness*). **2** immerse completely. **3 a** move suddenly and dramatically downward. **b** (foll. by *down, into*, etc.) move with a rush (*plunged down the stairs*). **4** *colloq.* run up gambling debts. —*n.* plunging action or movement; dive. □ **take the plunge** *colloq.* take a decisive step. [Romanic: related to PLUMB]

plunger *n.* **1** part of a mechanism that works with a plunging or thrusting movement. **2** rubber cup on a handle for clearing blocked pipes by a plunging and sucking action.

pluperfect /plu:'pɜ:fɪkt/ *Gram.* —*adj.* (of a tense) denoting an action completed prior to some past point of time (e.g. *he had gone by then*). —*n.* pluperfect tense. [Latin *plus quam perfectum* more than perfect]

plural /'plʊər(ə)l/ —*adj.* **1** more than one in number. **2** *Gram.* (of a word or form) denoting more than one. —*n. Gram.* **1** plural word or form. **2** plural number. [Latin: related to PLUS]

pluralism *n.* **1** form of society embracing many minority groups and cultural traditions. **2** the holding of more than one office at a time, esp. in the Church. □ **pluralist** *n.* **pluralistic** /-'lɪstɪk/ *adj.*

plurality /plʊə'rælɪtɪ/ *n.* (*pl.* **-ies**) **1** state of being plural. **2** = PLURALISM 2. **3** large number. **4** *US* majority that is not absolute.

pluralize /'plʊərə,laɪz/ *v.* (also **-ise**) (**-zing** or **-sing**) make plural, express in the plural.

plus —*prep.* **1** with the addition of (symbol +). **2** (of temperature) above zero (*plus 2°*). **3** *colloq.* with; having gained; newly possessing. —*adj.* **1** (after a number) at least (*fifteen plus*). **2** (after a grade etc.) rather better than (*beta plus*). **3** *Math.* positive. **4** having a positive electrical charge. **5** (*attrib.*) additional, extra. —*n.* **1** the symbol +. **2** additional or positive quantity. **3** advantage. —*conj. colloq.* also; and furthermore. [Latin, = more]

■ **Usage** The use of *plus* as a conjunction, as in *they arrived late, plus they wanted a meal*, is considered incorrect by some people.

plus-fours *n.pl.* men's long wide knickerbockers. [the length was increased by 4 inches to create an overhang]

plush —*n.* cloth of silk or cotton etc., with a long soft nap. —*adj.* **1** made of plush. **2** *colloq.* = PLUSHY. □ **plushly** *adv.* **plushness** *n.* [Latin: related to PILE[3]]

plushy *adj.* (**-ier**, **-iest**) *colloq.* stylish, luxurious. □ **plushiness** *n.*

Pluto /'plu:təʊ/ *n.* outermost known planet of the solar system. [Greek *Ploutōn*, god of the underworld]

plutocracy /plu:'tɒkrəsɪ/ *n.* (*pl.* **-ies**) **1 a** government by the wealthy. **b** State so governed. **2** wealthy élite. □ **plutocratic** /-tə'krætɪk/ *adj.* [Greek *ploutos* wealth]

plutocrat /'plu:tə,kræt/ *n.* **1** member of a plutocracy. **2** wealthy person.

plutonic /plu:'tɒnɪk/ *adj.* formed as igneous rock by solidification below the surface of the earth. [Latin *Pluto*, god of the underworld]

plutonium /plu:'təʊnɪəm/ *n.* radioactive metallic element. [*Pluto*, name of a planet]

pluvial /'plu:vɪəl/ *adj.* **1** of rain; rainy. **2** *Geol.* caused by rain. [Latin *pluvia* rain]

ply[1] /plaɪ/ *n.* (*pl.* **-ies**) **1** thickness or layer of cloth or wood etc. **2** strand of yarn or rope etc. [French *pli*: related to PLAIT]

ply[2] /plaɪ/ *v.* (**-ies**, **-ied**) **1** use or wield (a tool, weapon, etc.). **2** work steadily at (*ply one's trade*). **3** (foll. by *with*) **a** supply continuously (with food, drink, etc.). **b** approach repeatedly (with questions, etc.). **4 a** (often foll. by *between*) (of a vehicle etc.) travel regularly to and fro. **b** work (a route) thus. **5** (of a taxi-driver etc.) attend regularly for custom (*ply for hire*). [from APPLY]

Plymouth Brethren /'plɪməθ/ *n.pl.* Calvinistic religious body with no formal creed and no official order of ministers. [*Plymouth* in Devon]

plywood *n.* strong thin board made by gluing layers of wood with the direction of the grain alternating.

PM *abbr.* **1** prime minister. **2** post-mortem.

Pm *symb.* promethium.

p.m. *abbr.* after noon. [Latin *post meridiem*]

PMS *abbr.* premenstrual syndrome.

PMT *abbr.* premenstrual tension.

pneumatic /nju:'mætɪk/ *adj.* **1** filled with air or wind (*pneumatic tyre*). **2** operated by compressed air (*pneumatic drill*). [Greek *pneuma* wind]

pneumoconiosis /ˌnju:məʊˌkɒnɪ'əʊsɪs/ *n.* lung disease caused by the inhalation of dust or small particles. [Greek *pneumōn* lung, *konis* dust]

pneumonia /nju:'məʊnɪə/ *n.* inflammation of one or both lungs. [Greek *pneumōn* lung]

PO *abbr.* **1** Post Office. **2** postal order. **3** Petty Officer. **4** Pilot Officer.

Po *symb.* polonium.

po /pəʊ/ *n.* (*pl.* **-s**) *colloq.* chamber-pot. [from POT[1]]

poach[1] *v.* **1** cook (an egg) without its shell in or over boiling water. **2** cook (fish etc.) by simmering in a small amount of liquid. □ **poacher** *n.* [French *pochier*: related to POKE[2]]

poach[2] *v.* **1** (also *absol.*) catch (game or fish) illegally. **2** (often foll. by *on*) trespass or encroach on (another's property, territory, etc.). **3** appropriate (another's ideas, staff, etc.). □ **poacher** *n.* [earlier *poche*: related to POACH[1]]

pock *n.* (also **pock-mark**) small pus-filled spot on the skin, esp. caused by chickenpox or smallpox. □ **pock--marked** *adj.* [Old English]

pocket /'pɒkɪt/ *–n.* **1** small bag sewn into or on clothing, for carrying small articles. **2** pouchlike compartment in a suitcase, car door, etc. **3** one's financial resources (*beyond my pocket*). **4** isolated group or area (*pockets of resistance*). **5** cavity in the earth containing ore, esp. gold. **6** pouch at the corner or on the side of a billiard- or snooker-table into which balls are driven. **7** = AIR POCKET. **8** (*attrib.*) **a** small enough or intended for carrying in a pocket. **b** smaller than the usual size. *–v.* (**-t-**) **1** put into one's pocket. **2** appropriate, esp. dishonestly. **3** confine as in a pocket. **4** submit to (an injury or affront). **5** conceal or suppress (one's feelings). **6** *Billiards* etc. drive (a ball) into a pocket. □ **in pocket** having gained in a transaction. **in a person's pocket 1** under a person's control. **2** close to or intimate with a person. **out of pocket** having lost in a transaction. [Anglo-French diminutive: related to POKE[2]]

pocketbook *n.* **1** notebook. **2** folding case for papers or money carried in a pocket.

pocketful *n.* (*pl.* **-s**) as much as a pocket will hold.

pocket knife *n.* = PENKNIFE.

pocket money *n.* money for minor expenses, esp. given to children.

pod *–n.* long seed-vessel, esp. of a pea or bean. *–v.* (**-dd-**) **1** bear or form pods. **2** remove (peas etc.) from pods. [origin unknown]

podgy /'pɒdʒɪ/ *adj.* (**-ier, -iest**) **1** short and fat. **2** plump, fleshy. □ **podginess** *n.* [*podge* short fat person]

podium /'pəʊdɪəm/ *n.* (*pl.* **-s** or **podia**) rostrum. [Greek *podion* diminutive of *pous pod-* foot]

poem /'pəʊɪm/ *n.* **1** metrical composition, usu. concerned with feeling or imaginative description. **2** elevated composition in verse or prose. **3** something with poetic qualities (*a poem in stone*). [Greek *poieō* make]

poesy /'pəʊəzɪ/ *n.* *archaic* poetry. [French, ultimately as POEM]

poet /'pəʊɪt/ *n.* (*fem.* **poetess**) **1** writer of poems. **2** highly imaginative or expressive person. [Greek *poiētēs*: related to POEM]

poetaster /ˌpəʊɪ'tæstə(r)/ *n.* inferior poet. [from POET, Latin *-aster* derogatory suffix]

poetic /pəʊ'etɪk/ *adj.* (also **poetical**) of or like poetry or poets. □ **poetically** *adv.*

poetic justice *n.* very appropriate punishment or reward.

poetic licence *n.* writer's or artist's transgression of established rules for effect.

Poet Laureate *n.* (*pl.* **Poets Laureate**) poet appointed to write poems for State occasions.

poetry /'pəʊɪtrɪ/ *n.* **1** art or work of a poet. **2** poems collectively. **3** poetic or tenderly pleasing quality. [medieval Latin: related to POET]

po-faced *adj.* **1** solemn-faced, humourless. **2** smug. [perhaps from PO, influenced by *poker-faced*]

pogo /'pəʊgəʊ/ *n.* (*pl.* **-s**) (also **pogo stick**) stiltlike toy with a spring, used for jumping about on. [origin uncertain]

pogrom /'pɒgrəm/ *n.* organized massacre (orig. of Jews in Russia). [Russian]

poignant /'pɔɪnjənt/ *adj.* **1** painfully sharp to the emotions or senses; deeply moving. **2** arousing sympathy. **3** sharp or pungent in taste or smell. **4** pleasantly piquant. □ **poignance** *n.* **poignancy** *n.* **poignantly** *adv.* [Latin: related to POINT]

poinsettia /pɔɪn'setɪə/ *n.* plant with large scarlet bracts surrounding small yellow flowers. [*Poinsett*, name of a diplomat]

point *–n.* **1** sharp or tapered end of a tool, weapon, pencil, etc. **2** tip or extreme end. **3** that which in geometry has position but not magnitude. **4** particular place or position. **5** precise or critical moment (*when it came to the point, he refused*). **6** very small mark on a surface. **7** dot or other punctuation mark. **8** = DECIMAL POINT. **9** stage or degree in progress or increase (*abrupt to the point of rudeness*). **10** temperature at which a change of state occurs (*freezing point*). **11** single item or particular (*explained it point by point*). **12** unit of scoring in games or of measuring value etc. **13** significant or essential thing; what is intended or under discussion (*the point of my question; get to the point*). **14** sense, purpose; advantage, value (*saw no point in staying*). **15** characteristic (*tact is not his strong point*). **16 a** each of 32 directions marked at equal distances round a compass. **b** corresponding direction towards the horizon. **17** (usu. in *pl.*) pair of movable tapering rails that allow a train to pass from one line to another. **18** = POWER POINT. **19** (usu. in *pl.*) electrical contact in the distributor of a vehicle. **20** *Cricket* **a** fielder on the off side near the batsman. **b** this position. **21** tip of the toe in ballet. **22** promontory. **23** (usu. in *pl.*) extremities of a dog, horse, etc. –*v.* **1** (usu. foll. by *to, at*) **a** direct or aim (a finger, weapon, etc.). **b** direct attention. **2** (foll. by *at, towards*) aim or be directed to. **3** (foll. by *to*) indicate; be evidence of (*it all points to murder*). **4** give force to (words or actions). **5** fill the joints of (brickwork) with smoothed mortar or cement. **6** (also *absol.*) (of a dog) indicate the presence of (game) by acting as pointer. □ **at** (or **on**) **the point of** on the verge of. **beside the point** irrelevant. **in point** relevant (*the case in point*). **in point of fact** see FACT. **make a point of** insist on (doing etc.); treat or regard as essential; call particular attention to (an action). **point out** indicate; draw attention to. **point up** emphasize. **to the point** relevant; relevantly. **up to a point** to some extent but not completely. [Latin *pungo punct-* prick]

point-blank –*adj.* **1 a** (of a shot) aimed or fired at very close range. **b** (of a range) very close. **2** (of a remark etc.) blunt, direct. –*adv.* **1** at very close range. **2** directly, bluntly.

point-duty *n.* traffic control by a police officer, esp. at a road junction.

pointed *adj.* **1** sharpened or tapering to a point. **2** (of a remark etc.) having point; cutting. **3** emphasized. □ **pointedly** *adv.*

pointer *n.* **1** thing that points, e.g. the index hand of a gauge. **2** rod for pointing to features on a chart etc. **3** *colloq.* hint. **4** dog of a breed that on scenting game stands rigid looking towards it. **5** (in *pl.*) two stars in the Great Bear in line with the pole star.

pointillism /ˈpwæntɪˌlɪz(ə)m/ *n.* technique of impressionist painting using tiny dots of pure colour which become blended in the viewer's eye. □ **pointillist** *n.* & *adj.* [French *pointiller* mark with dots]

pointing *n.* **1** cement filling the joints of brickwork. **2** facing produced by this.

pointless *adj.* lacking purpose or meaning; ineffective, fruitless. □ **pointlessly** *adv.* **pointlessness** *n.*

point of honour *n.* thing of great importance to one's reputation or conscience.

point of no return *n.* point in a journey or enterprise at which it becomes essential or more practical to continue to the end.

point of order *n.* query in a debate etc. as to whether correct procedure is being followed.

point-of-sale *adj.* (usu. *attrib.*) of the place at which goods are retailed.

point of view *n.* **1** position from which a thing is viewed. **2** way of considering a matter.

point-to-point *n.* steeplechase for hunting horses.

poise /pɔɪz/ –*n.* **1** composure, self-possession. **2** equilibrium. **3** carriage (of the head etc.). –*v.* (**-sing**) **1** balance; hold suspended or supported. **2** be balanced or suspended. [Latin *pendo pens-* weigh]

poised *adj.* **1 a** composed, self-assured. **b** carrying oneself gracefully or with dignity. **2** (often foll. by *for*, or *to* + infin.) ready for action.

poison /ˈpɔɪz(ə)n/ –*n.* **1** substance that when introduced into or absorbed by a living organism causes death or injury, esp. one that kills by rapid action even in a small quantity. **2** *colloq.* harmful influence. –*v.* **1** administer poison to. **2** kill, injure, or infect with poison. **3** treat (a weapon) with poison. **4** corrupt or pervert (a person or mind). **5** spoil or destroy (a person's pleasure etc.). □ **poisoner** *n.* **poisonous** *adj.* [Latin: related to POTION]

poison ivy *n.* N. American climbing plant secreting an irritant oil from its leaves.

poison-pen letter *n.* malicious anonymous letter.

poke[1] –*v.* (**-king**) **1 a** thrust or push with the hand, a stick, etc. **b** (foll. by *out, up,* etc.) be thrust forward, protrude. **2** (foll. by *at* etc.) make thrusts. **3** thrust the end of a finger etc. against. **4** (foll. by *in*) produce (a hole etc. in a thing) by poking. **5** stir (a fire) with a poker. **6 a** (often foll. by *about, around*) potter. **b**

(foll. by *about*, *into*) pry; search. **7** *coarse slang* have sexual intercourse with. *–n.* **1** act of poking. **2** thrust, nudge. □ **poke fun at** ridicule. **poke one's nose into** *colloq.* pry or intrude into. [German or Dutch]

poke[2] *n. dial.* bag, sack. □ **buy a pig in a poke** see PIG. [French dial.]

poker[1] *n.* metal rod for stirring a fire.

poker[2] *n.* card-game in which bluff is used as players bet on the value of their hands. [origin unknown]

poker-face *n.* impassive countenance assumed by a poker-player. □ **poker-faced** *adj.*

poky /'pəʊkɪ/ *adj.* (**-ier, -iest**) (of a room etc.) small and cramped. □ **pokiness** *n.* [from POKE[1]]

polar /'pəʊlə(r)/ **1** *adj.* of or near a pole of the earth or of the celestial sphere. **2** having magnetic or electric polarity. **3** directly opposite in character. [Latin: related to POLE[2]]

polar bear *n.* large white bear living in the Arctic regions.

polar circle *n.* each of the circles parallel to the equator at 23° 27′ from either pole.

polarity /pə'lærɪtɪ/ *n.* (*pl.* **-ies**) **1** tendency of a magnet etc. to point with its extremities to the magnetic poles of the earth, or of a body to lie with its axis in a particular direction. **2** state of having two poles with contrary qualities. **3** state of having two opposite tendencies, opinions, etc. **4** electrical condition of a body (positive or negative). **5** attraction towards an object.

polarize /'pəʊlə,raɪz/ *v.* (also **-ise**) (**-zing** or **-sing**) **1** restrict the vibrations of (light-waves etc.) to one direction. **2** give magnetic or electric polarity to. **3** divide into two opposing groups. □ **polarization** /-'zeɪʃ(ə)n/ *n.*

Polaroid /'pəʊlə,rɔɪd/ *n. propr.* **1** material in thin sheets polarizing light passing through it. **2** camera with internal processing that produces a print rapidly after each exposure. **3** (in *pl.*) sunglasses with Polaroid lenses.

polder /'pəʊldə(r)/ *n.* piece of land reclaimed from the sea or a river, esp. in the Netherlands. [Dutch]

Pole *n.* **1** native or national of Poland. **2** person of Polish descent. [German from Polish]

pole[1] *n.* **1** long slender rounded piece of wood, metal, etc., esp. with the end placed in the ground as a support etc. **2** = PERCH[1] 3. □ **up the pole** *slang* **1** crazy. **2** in difficulty. [Latin *palus* stake]

pole[2] *n.* **1** (in full **north pole, south pole**) **a** each of the two points in the celestial sphere about which the stars appear to revolve. **b** each of the ends of the axis of rotation of the earth (*North Pole*; *South Pole*). **2** each of the two opposite points on the surface of a magnet at which magnetic forces are strongest. **3** each of two terminals (positive and negative) of an electric cell or battery etc. **4** each of two opposed principles. □ **be poles apart** differ greatly. [Greek, = axis]

■ **Usage** The spelling is *North Pole* and *South Pole* when used as geographical designations.

poleaxe /'pəʊlæks/ (*US* **-ax**) *–n.* **1** *hist.* = BATTLEAXE 1. **2** butcher's axe. *–v.* (**-xing**) **1** hit or kill with a poleaxe. **2** (esp. as **poleaxed** *adj.*) *colloq.* dumbfound, overwhelm. [Low German or Dutch: related to POLL, AXE]

polecat /'pəʊlkæt/ *n.* **1** small brownish-black mammal of the weasel family. **2** *US* skunk. [origin unknown]

pole-jump var. of POLE-VAULT.

polemic /pə'lemɪk/ *–n.* **1** forceful verbal or written controversy or argument. **2** (in *pl.*) art or practice of controversial discussion. *–adj.* (also **polemical**) involving dispute; controversial. □ **polemicist** /-sɪst/ *n.* [Greek *polemos* war]

pole star *n.* **1** star in the Little Bear, near the North Pole in the sky. **2** thing serving as a guide.

pole-vault (also **pole-jump**) *–n.* vault, or sport of vaulting, over a high bar with the aid of a pole held in the hands. *–v.* perform this. □ **pole-vaulter** *n.*

police /pə'li:s/ *–n.* (as *pl.*) **1** (usu. prec. by *the*) the civil force responsible for maintaining public order. **2** its members. **3** force with similar functions (*military police*). *–v.* (**-cing**) **1** keep (a place or people) in order by means of police or a similar body. **2** provide with police. **3** keep in order, administer, control (*problem of policing the new law*). [Latin: related to POLICY[1]]

police constable see CONSTABLE.

police dog *n.* dog, esp. an Alsatian, used in police work.

police force *n.* body of police of a country, district, or town.

policeman *n.* (*fem.* **policewoman**) member of a police force.

police officer *n.* member of a police force.

police State *n.* totalitarian State controlled by political police.

police station *n.* office of a local police force.

policy[1] /'pɒlɪsɪ/ *n.* (*pl.* **-ies**) **1** course of action adopted by a government, business, individual, etc. **2** prudent conduct; sagacity. [Latin *politia* POLITY]

■ **Usage** See note at *polity*.

policy[2] /ˈpɒlɪsɪ/ *n.* (*pl.* **-ies**) **1** contract of insurance. **2** document containing this. [French *police*, ultimately from Greek *apodeixis* proof]

policyholder *n.* person or body holding an insurance policy.

polio /ˈpəʊlɪəʊ/ *n.* = POLIOMYELITIS. [abbreviation]

poliomyelitis /ˌpəʊlɪəʊˌmaɪəˈlaɪtɪs/ *n.* infectious viral disease of the grey matter of the central nervous system with temporary or permanent paralysis. [Greek *polios* grey, *muelos* marrow]

Polish /ˈpəʊlɪʃ/ *–adj.* **1** of Poland. **2** of the Poles or their language. *–n.* language of Poland.

polish /ˈpɒlɪʃ/ *–v.* (often foll. by *up*) **1** make or become smooth or glossy by rubbing. **2** (esp. as **polished** *adj.*) refine or improve; add the finishing touches to. *–n.* **1** substance used for polishing. **2** smoothness or glossiness produced by friction. **3** refinement, elegance. □ **polish off** finish (esp. food) quickly. [Latin *polio*]

polite /pəˈlaɪt/ *adj.* (**politer, politest**) **1** having good manners; courteous. **2** cultivated, refined. □ **politely** *adv.* **politeness** *n.* [Latin *politus*: related to POLISH]

politic /ˈpɒlɪtɪk/ *–adj.* **1** (of an action) judicious, expedient. **2** (of a person) prudent, sagacious. **3** political (now only in *body politic*). *–v.* (**-ck-**) engage in politics. [Greek: related to POLITY]

political /pəˈlɪtɪk(ə)l/ *adj.* **1 a** of or concerning the State or its government, or public affairs generally. **b** of or engaged in politics. **2** taking or belonging to a side in politics. **3** concerned with seeking power, status, etc. (*political decision*). □ **politically** *adv.* [Latin: related to POLITIC]

political asylum *n.* State protection given to a political refugee from another country.

political economy *n.* the study of the economic aspects of government.

political geography *n.* geography dealing with boundaries and the possessions of States.

political prisoner *n.* person imprisoned for political reasons.

political science *n.* the study of political activity and systems of government.

politician /ˌpɒlɪˈtɪʃ(ə)n/ *n.* **1** person involved in politics, esp. professionally as an MP. **2** esp. *US derog.* person who manoeuvres; schemer, time-server.

politicize /pəˈlɪtɪˌsaɪz/ *v.* (also **-ise**) (**-zing** or **-sing**) **1 a** give a political character to. **b** make politically aware. **2** engage in or talk politics. □ **politicization** /-ˈzeɪʃ(ə)n/ *n.*

politico /pəˈlɪtɪˌkəʊ/ *n.* (*pl.* **-s**) *colloq.* politician or political enthusiast. [Spanish]

politics /ˈpɒlɪtɪks/ *n.pl.* **1** (treated as *sing.* or *pl.*) **a** art and science of government. **b** public life and affairs. **2** (usu. treated as *pl.*) political principles or practice (*what are his politics?*). **3** activities concerned with seeking power, status, etc.

polity /ˈpɒlɪtɪ/ *n.* (*pl.* **-ies**) **1** form or process of civil government. **2** organized society; State. [Greek *politēs* citizen, from *polis* city]

■ **Usage** This word is sometimes confused with *policy*.

polka /ˈpɒlkə/ *–n.* **1** lively dance of Bohemian origin. **2** music for this. *–v.* (**-kas, -kaed** /-kəd/ or **-ka'd, -kaing** /-kəɪŋ/) dance the polka. [Czech *půlka*]

polka dot *n.* round dot as one of many forming a regular pattern on a textile fabric etc.

poll /pəʊl/ *–n.* **1 a** (often in *pl.*) voting or the counting of votes at an election (*go to the polls*). **b** result of voting or number of votes recorded. **2** = OPINION POLL. **3** human head. *–v.* **1 a** take the vote or votes of. **b** receive (so many votes). **c** give (a vote). **2** record the opinion of (a person or group) in an opinion poll. **3** cut off the top of (a tree or plant), esp. make a pollard of. **4** (esp. as **polled** *adj.*) cut the horns off (cattle). [perhaps from Low German or Dutch]

pollack /ˈpɒlək/ *n.* (also **pollock**) (*pl.* same or **-s**) edible marine fish related to the cod. [origin unknown]

pollard /ˈpɒləd/ *–n.* **1** animal that has lost or cast its horns; ox, sheep, or goat of a hornless breed. **2** tree whose branches have been cut back to encourage the dense growth of young branches. *–v.* make (a tree) a pollard. [from POLL]

pollen /ˈpɒlən/ *n.* fine dustlike grains discharged from the male part of a flower, each containing the fertilizing element. [Latin]

pollen count *n.* index of the amount of pollen in the air, published as a warning to hay fever sufferers.

pollinate /ˈpɒlɪˌneɪt/ *v.* (**-ting**) (also *absol.*) convey pollen to or sprinkle (a stigma) with pollen. □ **pollination** /-ˈneɪʃ(ə)n/ *n.* **pollinator** *n.*

polling *n.* registering or casting of votes.

polling-booth *n.* compartment in which a voter stands to mark the ballot-paper.

polling-day *n.* election day.

polling-station *n.* building, often a school, used for voting at an election.

pollock var. of POLLACK.

pollster *n.* person who organizes an opinion poll.

poll tax *n.* **1** *informal* = COMMUNITY CHARGE. **2** *hist.* tax levied on every adult.

pollute /pə'lu:t/ *v.* **(-ting) 1** contaminate (the environment). **2** make foul or impure. □ **pollutant** *adj.* & *n.* **polluter** *n.* **pollution** *n.* [Latin *polluo -lut-*]

polo /'pəʊləʊ/ *n.* game like hockey played on horseback with a long-handled mallet. [Balti, = ball]

polonaise /,pɒlə'neɪz/ *n.* **1** slow dance of Polish origin. **2** music for this. [French: related to POLE]

polo-neck *n.* **1** high round turned-over collar. **2** sweater with this.

polonium /pə'ləʊnɪəm/ *n.* radioactive metallic element, occurring naturally in uranium ores. [medieval Latin *Polonia* Poland]

poltergeist /'pɒltə,gaɪst/ *n.* noisy mischievous ghost, esp. one causing physical damage. [German]

poltroon /pɒl'tru:n/ *n.* spiritless coward. □ **poltroonery** *n.* [Italian *poltro* sluggard]

poly /'pɒlɪ/ *n.* (*pl.* **-s**) *colloq.* polytechnic. [abbreviation]

poly- *comb. form* **1** many (*polygamy*). **2** polymerized (*polyunsaturated*; *polyester*). [Greek *polus* many]

polyandry /'pɒlɪ,ændrɪ/ *n.* polygamy in which a woman has more than one husband. □ **polyandrous** /-'ændrəs/ *adj.* [Greek *anēr andr-* male]

polyanthus /,pɒlɪ'ænθəs/ *n.* (*pl.* **-thuses**) flowering plant cultivated from hybridized primulas. [Greek *anthos* flower]

polychromatic /,pɒlɪkrəʊ'mætɪk/ *adj.* **1** many-coloured. **2** (of radiation) containing more than one wavelength. □ **polychromatism** /-'krəʊmə,tɪz(ə)m/ *n.*

polychrome /'pɒlɪ,krəʊm/ *—adj.* in many colours. *—n.* polychrome work of art. [Greek: related to POLY-, CHROME]

polyester /,pɒlɪ'estə(r)/ *n.* synthetic fibre or resin.

polyethene /'pɒlɪ,eθi:n/ *n.* = POLYTHENE.

polyethylene /,pɒlɪ'eθɪ,li:n/ *n.* = POLYTHENE.

polygamy /pə'lɪgəmɪ/ *n.* practice of having more than one wife or (less usu.) husband at once. □ **polygamist** *n.* **polygamous** *adj.* [Greek *gamos* marriage]

polyglot /'pɒlɪ,glɒt/ *—adj.* knowing, using, or written in several languages. *—n.* polyglot person. [Greek *glōtta* tongue]

polygon /'pɒlɪgən, -,gɒn/ *n.* figure with many (usu. five or more) sides and angles. □ **polygonal** /pə'lɪgən(ə)l/ *adj.* [Greek *-gōnos* angled]

polygraph /'pɒlɪ,grɑ:f/ *n.* machine for reading physiological characteristics (e.g. pulse-rate); lie-detector.

polygyny /pə'lɪdʒɪnɪ/ *n.* polygamy in which a man has more than one wife. □ **polygynous** /pə'lɪdʒɪnəs/ *adj.* [Greek *gunē* woman]

polyhedron /,pɒlɪ'hi:drən/ *n.* (*pl.* **-dra**) solid figure with many (usu. more than six) faces. □ **polyhedral** *adj.* [Greek *hedra* base]

polymath /'pɒlɪ,mæθ/ *n.* person of great or varied learning. [Greek *manthanō math-* learn]

polymer /'pɒlɪmə(r)/ *n.* compound of one or more large molecules formed from repeated units of smaller molecules. □ **polymeric** /-'merɪk/ *adj.* **polymerize** *v.* (also **-ise**) (**-zing** or **-sing**). **polymerization** /-'zeɪʃ(ə)n/ *n.* [Greek *polumeros* having many parts]

polymorphous /,pɒlɪ'mɔ:fəs/ *adj.* (also **polymorphic**) passing through various forms in successive stages of development.

polynomial /,pɒlɪ'nəʊmɪəl/ *—n.* expression of more than two algebraic terms. *—adj.* of or being a polynomial. [from POLY-, BINOMIAL]

polyp /'pɒlɪp/ *n.* **1** simple organism with a tube-shaped body. **2** small usu. benign growth on a mucous membrane. [Greek *pous* foot]

polyphony /pə'lɪfənɪ/ *n.* (*pl.* **-ies**) *Mus.* contrapuntal music. □ **polyphonic** /,pɒlɪ'fɒnɪk/ *adj.* [Greek *phōnē* sound]

polypropene /,pɒlɪ'prəʊpi:n/ *n.* = POLYPROPYLENE.

polypropylene /,pɒlɪ'prəʊpɪ,li:n/ *n.* any polymer of propylene, including thermoplastic materials used for films, fibres, or moulding materials.

polysaccharide /,pɒlɪ'sækə,raɪd/ *n.* any of a group of complex carbohydrates, e.g. starch. [see SACCHARIN]

polystyrene /,pɒlɪ'staɪə,ri:n/ *n.* a polymer of styrene, a kind of hard plastic, often foamed for packaging. [*styrene* from Greek *sturax* a resin]

polysyllabic /,pɒlɪsɪ'læbɪk/ *adj.* **1** having many syllables. **2** using words of many syllables. [medieval Latin from Greek]

polysyllable /'pɒlɪ,sɪləb(ə)l/ *n.* polysyllabic word.

polytechnic /,pɒlɪ'teknɪk/ *—n.* college offering courses in many (esp. vocational) subjects up to degree level. *—adj.*

giving instruction in various vocational or technical subjects. [Greek *tekhnē* art]

polytheism /ˈpɒlɪθiːˌɪz(ə)m/ *n.* belief in or worship of more than one god. □ **polytheist** *n.* **polytheistic** /-ˈɪstɪk/ *adj.* [Greek *theos* god]

polythene /ˈpɒlɪˌθiːn/ *n.* a tough light plastic. [from POLYETHYLENE]

polyunsaturated /ˌpɒlɪʌnˈsætʃəˌreɪtɪd/ *adj.* (of a fat or oil) having a chemical structure capable of further reaction and not contributing to the accumulation of cholesterol in the blood.

polyurethane /ˌpɒlɪˈjʊərəˌθeɪn/ *n.* synthetic resin or plastic used esp. in paints or foam. [related to UREA, ETHANE]

polyvinyl chloride /ˌpɒlɪˈvaɪnɪl/ *n.* a vinyl plastic used for electrical insulation or as a fabric etc.; PVC.

Pom *n. Austral. & NZ slang offens.* = POMMY. [abbreviation]

pomace /ˈpʌmɪs/ *n.* crushed apples in cider-making. [Latin *pomum* apple]

pomade /pəˈmɑːd/ *n.* scented ointment for the hair and head. [Italian: related to POMACE]

pomander /pəˈmændə(r)/ *n.* **1** ball of mixed aromatic substances. **2** container for this. [Anglo-French from medieval Latin]

pomegranate /ˈpɒmɪˌgrænɪt/ *n.* **1** tropical fruit with a tough rind, reddish pulp, and many seeds. **2** tree bearing this. [French *pome grenate* from Romanic, = many-seeded apple]

pomelo /ˈpʌməˌləʊ/ *n.* (*pl.* **-s**) **1** = SHADDOCK. **2** *US* = GRAPEFRUIT. [origin unknown]

pommel /ˈpʌm(ə)l/ —*n.* **1** knob, esp. at the end of a sword-hilt. **2** upward projecting front of a saddle. —*v.* (**-ll-**; *US* **-l-**) = PUMMEL. [Latin *pomum* apple]

Pommy /ˈpɒmɪ/ *n.* (also **pommie**) (*pl.* **-ies**) *Austral. & NZ slang offens.* British person, esp. a recent immigrant. [origin uncertain]

pomp *n.* **1** splendid display; splendour. **2** specious glory. [Latin from Greek *pompē*]

pom-pom /ˈpɒmpɒm/ *n.* automatic quick-firing gun. [imitative]

pompon /ˈpɒmpɒn/ *n.* (also **pompom**) **1** ornamental tuft or bobble on a hat, shoes, etc. **2** (often *attrib.*) dahlia etc. with small tightly-clustered petals. [French]

pompous /ˈpɒmpəs/ *n.* self-important, affectedly grand or solemn. □ **pomposity** /pɒmˈpɒsɪtɪ/ *n.* (*pl.* **-ies**). **pompously** *adv.* **pompousness** *n.* [Latin: related to POMP]

ponce *slang* —*n.* **1** man who lives off a prostitute's earnings; pimp. **2** *offens.* homosexual or effeminate man. —*v.* (**-cing**) act as a ponce. □ **ponce about** move about effeminately or ineffectually. [origin unknown]

poncho /ˈpɒntʃəʊ/ *n.* (*pl.* **-s**) cloak of a usu. blanket-like piece of cloth with a slit in the middle for the head. [South American Spanish]

pond *n.* small body of still water. [var. of POUND[3]]

ponder *v.* **1** think over; consider. **2** muse, be deep in thought. [Latin *pondero* weigh]

ponderable *adj. literary* having appreciable weight or significance. [Latin: related to PONDER]

ponderous /ˈpɒndərəs/ *adj.* **1** slow and awkward, esp. because of great weight. **2** (of style etc.) laborious; dull. □ **ponderously** *adv.* **ponderousness** *n.* [Latin *pondus -der-* weight]

pondweed *n.* aquatic plant growing in still water.

pong *v. colloq.* stink. □ **pongy** *adj.* (**-ier, -iest**). [origin unknown]

poniard /ˈpɒnjəd/ *n.* dagger. [French *poignard* from Latin *pugnus* fist]

pontiff /ˈpɒntɪf/ *n.* Pope. [Latin *pontifex -fic-* priest]

pontifical /pɒnˈtɪfɪk(ə)l/ *adj.* **1** papal. **2** pompously dogmatic. □ **pontifically** *adv.*

pontificate —*v.* /pɒnˈtɪfɪˌkeɪt/ (**-ting**) **1** be pompously dogmatic. **2** play the pontiff. —*n.* /pɒnˈtɪfɪkət/ **1** office of a bishop or pope. **2** period of this.

pontoon[1] /pɒnˈtuːn/ *n.* card-game in which players try to acquire cards with a face value totalling 21. [probably a corruption of VINGT-ET-UN]

pontoon[2] /pɒnˈtuːn/ *n.* **1** flat-bottomed boat. **2** each of several boats etc. used to support a temporary bridge. [Latin *ponto ponton-* punt]

pony /ˈpəʊnɪ/ *n.* (*pl.* **-ies**) horse of any small breed. [perhaps from French *poulenet* foal]

pony-tail *n.* hair drawn back, tied, and hanging down behind the head.

pony-trekking *n.* travelling across country on ponies for pleasure.

poodle /ˈpuːd(ə)l/ *n.* **1** dog of a breed with a curly coat that is usually clipped. **2** servile follower. [German *Pudel*]

poof /pʊf/ *n.* (also **poofter**) *slang offens.* effeminate or homosexual man. [origin unknown]

pooh /puː/ *int.* expressing impatience, contempt, or disgust at a bad smell. [imitative]

pooh-pooh /puːˈpuː/ *v.* express contempt for, ridicule. [reduplication of POOH]

pool[1] *n.* **1** small body of still water. **2** small shallow body of any liquid. **3** swimming-pool. **4** deep place in a river. [Old English]

pool[2] *–n.* **1 a** common supply of persons, vehicles, commodities, etc. for sharing by a group of people. **b** group of persons sharing duties etc. **2** common fund, e.g. of profits of separate firms or of players' stakes in gambling. **3** arrangement between competing parties to fix prices and share business. **4** *US* **a** game on a billiard-table with usu. 16 balls. **b** game on a billiard-table in which each player has a ball of a different colour with which he or she tries to pocket the others in fixed order, the winner taking all of the stakes. *–v.* **1** put into a common fund. **2** share in common. [French *poule*]

pools *n.pl.* (prec. by *the*) = FOOTBALL POOL.

poop *n.* stern of a ship; the deck which is furthest aft and highest. [Latin *puppis*]

poor *adj.* **1** without enough money to live comfortably. **2** (foll. by *in*) deficient in (a possession or quality). **3 a** scanty, inadequate. **b** less good than is usual or expected (*poor visibility*; *is a poor driver*). **c** paltry; inferior (*came a poor third*). **4** deserving pity or sympathy; unfortunate (*you poor thing*). **5** spiritless, despicable. □ **poor man's** inferior or cheaper substitute for. [Latin *pauper*]

poorhouse *n. hist.* = WORKHOUSE.

poor law *n. hist.* law concerning public support of the poor.

poorly *–adv.* in a poor manner, badly. *–predic. adj.* unwell.

poor relation *n.* inferior or subordinate member of a family etc.

pop[1] *–n.* **1** sudden sharp explosive sound as of a cork when drawn. **2** *colloq.* effervescent drink. *–v.* **(-pp-) 1** (cause to) make a pop. **2** (foll. by *in, out, up*, etc.) go, move, come, or put unexpectedly or abruptly (*pop out to the shop*). **3** *slang* pawn. *–adv.* with the sound of a pop (*go pop*). □ **pop off** *colloq.* die. **pop the question** *colloq.* propose marriage. [imitative]

pop[2] *n. colloq.* **1** (in full **pop music**) highly successful commercial music, esp. since the 1950s. **2** (*attrib.*) of or relating to pop music (*pop concert, group, song*). **3** pop record or song (*top of the pops*). [abbreviation]

pop[3] *n.* esp. *US colloq.* father. [from PAPA]

pop. *abbr.* population.

popadam var. of POPPADAM.

pop art *n.* art based on modern popular culture and the mass media.

popcorn *n.* maize which bursts open when heated.

pop culture *n.* commercial culture based on popular taste.

pope *n.* (also **Pope**) head of the Roman Catholic Church (*the Pope*; *we have a new pope*). [Greek *papas* patriarch]

popery /'pəʊpərɪ/ *n. derog.* papal system; Roman Catholicism.

pop-eyed *adj. colloq.* **1** having bulging eyes. **2** wide-eyed (with surprise etc.).

popgun *n.* child's toy gun shooting pellets etc. by the compression of air.

popinjay /'pɒpɪn,dʒeɪ/ *n.* fop, conceited person. [Arabic *babagha* parrot]

popish /'pəʊpɪʃ/ *adj. derog.* Roman Catholic.

poplar /'pɒplə(r)/ *n.* tall slender tree with a straight trunk and often tremulous leaves. [Latin *populus*]

poplin /'pɒplɪn/ *n.* plain-woven fabric usu. of cotton, with a corded surface. [French *papeline*]

poppadam /'pɒpədəm/ *n.* (also **poppadom, popadam**) *Ind.* thin, crisp, spiced bread eaten with curry etc. [Tamil]

popper *n.* **1** *colloq.* press-stud. **2** thing that pops (*party popper*).

poppet /'pɒpɪt/ *n. colloq.* (esp. as a term of endearment) small or dainty person. [Latin *pup(p)a* doll]

popping-crease *n. Cricket* line in front of and parallel to the wicket, within which the batsman stands. [from POP[1]]

poppy /'pɒpɪ/ *n.* (*pl.* **-ies**) **1** plant with showy esp. scarlet flowers and a milky sap. **2** artificial poppy worn on Remembrance Sunday. [Latin *papaver*]

poppycock *n. slang* nonsense. [Dutch *pappekak*]

Poppy Day *n.* = REMEMBRANCE SUNDAY.

populace /'pɒpjʊləs/ *n.* the common people. [Italian: related to POPULAR]

popular /'pɒpjʊlə(r)/ *adj.* **1** liked by many people. **2 a** of or for the general public. **b** prevalent among the general public (*popular fallacies*). **3** (sometimes *derog.*) adapted to the understanding, taste, or means of the people (*popular science*; *the popular press*). □ **popularity** /-'lærɪtɪ/ *n.* **popularly** *adv.* [Anglo-Latin *populus* PEOPLE]

popular front *n.* party or coalition combining left-wing groups.

popularize *v.* (also **-ise**) **(-zing** or **-sing) 1** make popular. **2** present (a difficult subject) in a readily understandable form. □ **popularization** /-'zeɪʃ(ə)n/ *n.*

popular music *n.* any music that appeals to a wide public.

populate /'pɒpjʊ,leɪt/ *v.* **(-ting) 1** inhabit, form the population of. **2** supply with inhabitants. [medieval Latin: related to PEOPLE]

population /ˌpɒpjʊˈleɪʃ(ə)n/ *n.* **1** inhabitants of a place, country, etc. **2** total number of these or any group of living things.

population explosion *n.* sudden large increase of population.

populist /ˈpɒpjʊlɪst/ *n.* politician claiming to represent the ordinary people. [Latin *populus* people]

populous /ˈpɒpjʊləs/ *adj.* thickly inhabited.

pop-up *adj.* involving parts that pop up automatically (*pop-up toaster*; *pop-up book*).

porcelain /ˈpɔːsəlɪn/ *n.* **1** hard fine translucent ceramic with a transparent glaze. **2** objects made of this. [Italian diminutive of *porca* sow]

porch *n.* covered entrance to a building. [Latin *porticus*]

porcine /ˈpɔːsaɪn/ *adj.* of or like pigs. [Latin: related to PORK]

porcupine /ˈpɔːkjʊˌpaɪn/ *n.* rodent with a body and tail covered with erectile spines. [Provençal: related to PORK, SPINE]

pore[1] *n.* esp. *Biol.* minute opening in a surface through which fluids etc. may pass. [Greek *poros*]

pore[2] *v.* (**-ring**) (foll. by *over*) **1** be absorbed in studying (a book etc.). **2** meditate on. [origin unknown]

pork *n.* flesh (esp. unsalted) of a pig, used as food. [Latin *porcus* pig]

porker *n.* pig raised for food.

pork pie *n.* pie of minced pork etc. eaten cold.

pork pie hat *n.* hat with a flat crown and a brim turned up all round.

porky *adj.* (**-ier, -iest**) **1** *colloq.* fat. **2** of or like pork.

porn (also **porno**) *—n. colloq.* pornography. *—attrib. adj.* pornographic. [abbreviation]

pornography /pɔːˈnɒɡrəfɪ/ *n.* **1** explicit representation of sexual activity in literature, films, etc., intended to stimulate erotic rather than aesthetic or emotional feelings. **2** literature etc. containing this. □ **pornographic** /-nəˈɡræfɪk/ *adj.* [Greek *pornē* prostitute]

porous /ˈpɔːrəs/ *adj.* **1** full of pores. **2** letting through air, water, etc. □ **porosity** /pɔːˈrɒsɪtɪ/ *n.* [Latin: related to PORE[1]]

porphyry /ˈpɔːrfɪrɪ/ *n.* (*pl.* **-ies**) hard rock composed of crystals of white or red feldspar in a red matrix. □ **porphyritic** /-ˈrɪtɪk/ *adj.* [Greek: related to PURPLE]

porpoise /ˈpɔːpəs/ *n.* sea mammal of the whale family, with a blunt rounded snout. [Latin *porcus* pig, *piscis* fish]

porridge /ˈpɒrɪdʒ/ *n.* **1** dish of oatmeal or cereal boiled in water or milk. **2** *slang* imprisonment. [alteration of POTTAGE]

porringer /ˈpɒrɪndʒə(r)/ *n.* small bowl, often with a handle, for soup etc. [French *potager*: related to POTTAGE]

port[1] *n.* **1** harbour. **2** town possessing a harbour. [Latin *portus*]

port[2] *n.* a kind of sweet fortified wine. [*Oporto* in Portugal]

port[3] *—n.* left-hand side of a ship or aircraft looking forward. *—v.* (also *absol.*) turn (the helm) to port. [probably originally the side turned to PORT[1]]

port[4] *n.* **1** opening in the side of a ship for entrance, loading, etc. **2** porthole. [Latin *porta* gate]

portable /ˈpɔːtəb(ə)l/ *—adj.* **1** easily movable, convenient for carrying. **2** (of a right, opinion, etc.) capable of being transferred or adapted in altered circumstances (*portable pension*). *—n.* portable version of an item, e.g. a television. □ **portability** /ˌpɔːtəˈbɪlɪtɪ/ *n.* [Latin *porto* carry]

portage /ˈpɔːtɪdʒ/ *—n.* **1** carrying of boats or goods overland between two navigable waters. **2** place where this is necessary. *—v.* (**-ging**) convey (a boat or goods) over a portage. [Latin *porto* carry]

Portakabin /ˈpɔːtəˌkæbɪn/ *n. propr.* prefabricated room or small building. [from PORTABLE, CABIN]

portal /ˈpɔːt(ə)l/ *n.* doorway or gate etc., esp. an elaborate one. [Latin: related to PORT[4]]

portcullis /pɔːtˈkʌlɪs/ *n.* strong heavy grating lowered to block a gateway in a fortress etc. [French, = sliding door]

portend /pɔːˈtend/ *v.* **1** foreshadow as an omen. **2** give warning of. [Latin *portendo*: related to PRO-[1], TEND[1]]

portent /ˈpɔːtent/ *n.* **1** omen, significant sign of something to come. **2** prodigy; marvellous thing. [Latin *portentum*: related to PORTEND]

portentous /pɔːˈtentəs/ *adj.* **1** like or being a portent. **2** pompously solemn.

porter[1] *n.* **1** person employed to carry luggage etc. **2** dark beer brewed from charred or browned malt. [Latin *porto* carry]

porter[2] *n.* gatekeeper or doorman, esp. of a large building. [Latin: related to PORT[4]]

porterage *n.* **1** hire of porters. **2** charge for this. [from PORTER[1]]

porterhouse steak *n.* choice cut of beef.

portfolio /pɔːtˈfəʊlɪəʊ/ *n.* (*pl.* **-s**) **1 a** folder for loose sheets of paper, drawings, etc. **b** samples of an artist's work. **2** range of investments held by a person, company, etc. **3** office of a minister of State (cf. MINISTER WITHOUT PORTFOLIO). [Italian *portafogli* sheet-carrier]

porthole *n.* aperture (esp. glazed) in a ship's side for letting in light.

portico /ˈpɔːtɪˌkəʊ/ *n.* (*pl.* **-es** or **-s**) colonnade; roof supported by columns at regular intervals, usu. attached as a porch to a building. [Latin *porticus* porch]

portion /ˈpɔːʃ(ə)n/ *–n.* **1** part or share. **2** amount of food allotted to one person. **3** one's destiny or lot. *–v.* **1** divide (a thing) into portions. **2** (foll. by *out*) distribute. [Latin *portio*]

Portland cement /ˈpɔːtlənd/ *n.* cement manufactured from chalk and clay. [Isle of *Portland* in Dorset]

Portland stone /ˈpɔːtlənd/ *n.* building limestone from the Isle of Portland.

portly /ˈpɔːtlɪ/ *adj.* (**-ier**, **-iest**) corpulent; stout. [Latin *porto* carry]

portmanteau /pɔːtˈmæntəʊ/ *n.* (*pl.* **-s** or **-x** /-təʊz/) trunk for clothes etc., opening into two equal parts. [Latin *porto* carry: related to MANTLE]

portmanteau word *n.* word combining the sounds and meanings of two others (e.g. *motel*, *Oxbridge*).

port of call *n.* place where a ship or a person stops on a journey.

portrait /ˈpɔːtrɪt/ *n.* **1** drawing, painting, photograph, etc. of a person or animal, esp. of the face. **2** description in words. □ **portraitist** *n.* [French: related to PORTRAY]

portraiture /ˈpɔːtrɪtʃə(r)/ *n.* **1** making portraits. **2** description in words. **3** portrait.

portray /pɔːˈtreɪ/ *v.* **1** make a likeness of. **2** describe in words. □ **portrayal** *n.* **portrayer** *n.* [French *portraire -trait* depict]

Portuguese /ˌpɔːtʃʊˈgiːz/ *–n.* (*pl.* same) **1 a** native or national of Portugal. **b** person of Portuguese descent. **2** language of Portugal. *–adj.* of Portugal, its people, or language. [medieval Latin]

Portuguese man-of-war *n.* (*pl.* **men-**) jellyfish with a large crest and poisonous sting.

pose /pəʊz/ *–v.* (**-sing**) **1** assume a certain attitude of the body, esp. when being photographed or painted. **2** (foll. by *as*) pretend to be (another person etc.) (*posing as a celebrity*). **3** behave affectedly to impress others. **4** put forward or present (a question etc.). **5** place (an artist's model etc.) in a certain attitude. *–n.* **1** attitude of body or mind. **2** affectation, pretence. [Latin *pauso* PAUSE, confused with Latin *pono* place]

poser *n.* **1** poseur. **2** *colloq.* puzzling question or problem.

poseur /pəʊˈzɜː(r)/ *n.* person who behaves affectedly. [French *poser* POSE]

posh *colloq. –adj.* smart; upper-class. *–adv.* in an upper-class way (*talk posh*). □ **posh up** smarten up. □ **poshly** *adv.* **poshness** *n.* [perhaps from slang *posh* a dandy, money]

posit /ˈpɒzɪt/ *v.* (**-t-**) assume as a fact, postulate. [Latin: related to POSITION]

position /pəˈzɪʃ(ə)n/ *–n.* **1** place occupied by a person or thing. **2** way in which a thing or its parts are placed or arranged. **3** proper place (*in position*). **4** advantage (*jockeying for position*). **5** attitude; view on a question. **6** situation in relation to others (*puts one in an awkward position*). **7** rank, status; social standing. **8** paid employment. **9** place where troops etc. are posted for strategical purposes. *–v.* place in position. □ **in a position to** able to. □ **positional** *adj.* [Latin *pono posit-* place]

positive /ˈpɒzɪtɪv/ *–adj.* **1** explicit; definite, unquestionable (*positive proof*). **2** (of a person) convinced, confident, or overconfident in an opinion. **3 a** absolute; not relative. **b** *Gram.* (of an adjective or adverb) expressing a simple quality without comparison. **4** *colloq.* downright (*it was a positive miracle*). **5** constructive (*positive thinking*). **6** marked by the presence and not absence of qualities (*positive reaction*). **7** esp. *Philos.* dealing only with matters of fact; practical. **8** tending in a direction naturally or arbitrarily taken as that of increase or progress. **9** greater than zero. **10** *Electr.* of, containing, or producing the kind of electrical charge produced by rubbing glass with silk; lacking electrons. **11** (of a photographic image) showing lights and shades or colours unreversed. *–n.* positive adjective, photograph, quantity, etc. □ **positively** *adv.* **positiveness** *n.* [Latin: related to POSITION]

positive discrimination *n.* practice of making distinctions in favour of groups considered to be underprivileged.

positive vetting *n.* inquiry into the background etc. of a candidate for a post involving national security.

positivism *n.* philosophical system recognizing only facts and observable phenomena. □ **positivist** *n.* & *adj.*

positron /ˈpɒzɪˌtrɒn/ *n.* *Physics* elementary particle with the same mass as but opposite (positive) charge to an electron. [*positive* elec*tron*]

posse /ˈpɒsɪ/ *n.* **1** strong force or company. **2** body of law-enforcers. [Latin, = be able]

possess /pə'zes/ *v.* **1** hold as property; own. **2** have (a faculty, quality, etc.). **3** occupy or dominate the mind of (*possessed by the devil*; *possessed by fear*). □ **be possessed of** own, have. **what possessed you?** an expression of incredulity. □ **possessor** *n.* [Latin *possideo possess-*]

possession /pə'zeʃ(ə)n/ *n.* **1** possessing or being possessed. **2** thing possessed. **3** holding or occupancy. **4** *Law* power or control similar to ownership but which may exist separately from it (*prosecuted for possession of drugs*). **5** (in *pl.*) property, wealth, subject territory, etc. **6** *Football* etc. control of the ball by a player.

possessive /pə'zesɪv/ *–adj.* **1** wanting to retain what one has, reluctant to share. **2** jealous and domineering. **3** *Gram.* indicating possession. *–n.* (in full **possessive case**) *Gram.* case of nouns and pronouns expressing possession. □ **possessiveness** *n.*

possibility /ˌpɒsɪ'bɪlɪtɪ/ *n.* (*pl.* **-ies**) **1** state or fact of being possible. **2** thing that may exist or happen. **3** (usu. in *pl.*) capability of being used; potential (*have possibilities*). [Latin *posse* be able]

possible /'pɒsɪb(ə)l/ *–adj.* **1** capable of existing, happening, being done, etc. **2** potential (*a possible way of doing it*). *–n.* **1** possible candidate, member of a team, etc. **2** highest possible score, esp. in shooting.

possibly *adv.* **1** perhaps. **2** in accordance with possibility (*cannot possibly go*).

possum /'pɒsəm/ *n.* **1** *colloq.* = OPOSSUM 1. **2** *Austral.* & *NZ colloq.* marsupial resembling an American opossum. □ **play possum** *colloq.* pretend to be unconscious; feign ignorance. [abbreviation]

post[1] /pəʊst/ *–n.* **1** long stout piece of timber or metal set upright in the ground etc. to support something, mark a position or boundary, etc. **2** pole etc. marking the start or finish of a race. *–v.* **1** (often foll. by *up*) attach (a notice etc.) in a prominent place. **2** announce or advertise by poster or list. [Latin *postis*]

post[2] /pəʊst/ *–n.* **1** official conveyance of parcels, letters, etc. (*send it by post*). **2** single collection or delivery of these; the letters etc. dispatched (*has the post arrived?*). **3** place where letters etc. are collected (*take it to the post*). *–v.* **1** put (a letter etc.) in the post. **2** (esp. as **posted** *adj.*) (often foll. by *up*) supply with information (*keep me posted*). **3 a** enter (an item) in a ledger. **b** (often foll. by *up*) complete (a ledger) in this way. [Latin: related to POSITION]

post[3] /pəʊst/ *–n.* **1** place where a soldier is stationed or which he or she patrols. **2** place of duty. **3 a** position taken up by a body of soldiers. **b** force occupying this. **c** fort. **4** job, paid employment. **5** = TRADING POST. *–v.* **1** place (soldiers, an employee, etc.). **2** appoint to a post or command. [French: related to POST[2]]

post- *prefix* after, behind. [Latin *post* (adv. and prep.)]

postage *n.* charge for sending a letter etc. by post.

postage stamp *n.* official stamp affixed to a letter etc., showing the amount of postage paid.

postal *adj.* of or by post. [French: related to POST[2]]

postal code *n.* = POSTCODE.

postal order *n.* money order issued by the Post Office.

postbag *n.* = MAILBAG.

postbox *n.* public box for posting mail.

postcard *n.* card for sending by post without an envelope.

postcode *n.* group of letters and figures in a postal address to assist sorting.

post-coital /pəʊst'kəʊɪt(ə)l/ *adj. formal* occurring after sexual intercourse.

postdate /pəʊst'deɪt/ *v.* (**-ting**) **1** give a date later than the actual one to (a document etc.). **2** follow in time.

poster *n.* **1** placard in a public place. **2** large printed picture.

poste restante /ˌpəʊst re'stɑ̃t/ *n.* department in a post office where letters are kept till called for. [French]

posterior /pɒ'stɪərɪə(r)/ *–adj.* **1** later; coming after. **2** at the back. *–n.* (in *sing.* or *pl.*) buttocks. [Latin, comparative of *posterus*: related to POST-]

posterity /pɒ'sterɪtɪ/ *n.* **1** succeeding generations. **2** person's descendants. [Latin: related to POSTERIOR]

postern /'pɒst(ə)n/ *n. archaic* back door; side way or entrance. [Latin: related to POSTERIOR]

poster paint *n.* gummy opaque paint.

post-free *adj.* & *adv.* carried by post free of charge, or with postage prepaid.

postgraduate /pəʊst'grædjʊət/ *–n.* person engaged in a course of study after taking a first degree. *–adj.* of or concerning postgraduates.

post-haste *adv.* with great speed.

posthumous /'pɒstjʊməs/ *adj.* **1** occurring after death. **2** (of a book etc.) published after the author's death. **3** (of a child) born after the death of its father. □ **posthumously** *adv.* [Latin *postumus* last]

postilion /pɒ'stɪljən/ *n.* (also **postillion**) person riding on the near horse of a team drawing a coach when there is no coachman. [Italian: related to POST[2]]

post-impressionism /ˌpəʊstɪm'preʃəˌnɪz(ə)m/ *n.* art intending to express the individual artist's conception of the objects represented. □ **post-impressionist** *n.* & *adj.*

post-industrial /ˌpəʊstɪn'dʌstrɪəl/ *adj.* of a society or economy which no longer relies on heavy industry.

postman *n.* (*fem.* **postwoman**) person employed to deliver and collect letters etc.

postmark –*n.* official mark on a letter, giving the place, date, etc., and cancelling the stamp. –*v.* mark (an envelope etc.) with this.

postmaster *n.* (*fem.* **postmistress**) official in charge of a post office.

post-modern /pəʊst'mɒd(ə)n/ *adj.* (in the arts etc.) of the movement reacting against modernism, esp. by drawing attention to former conventions. □ **post-modernism** *n.* **post-modernist** *n.* & *adj.*

post-mortem /pəʊst'mɔːtəm/ –*n.* **1** examination made after death, esp. to determine its cause. **2** *colloq.* discussion after a game, election, etc. –*adv.* & *adj.* after death. [Latin]

postnatal /pəʊst'neɪt(ə)l/ *adj.* of the period after childbirth.

Post Office *n.* **1** public department or corporation responsible for postal services. **2** (**post office**) room or building where postal business is carried on.

post-office box *n.* numbered place in a post office where letters are kept until called for.

post-paid *adj.* & *adv.* on which postage has been paid.

postpone /pəʊs'pəʊn/ *v.* (**-ning**) cause or arrange (an event etc.) to take place at a later time. □ **postponement** *n.* [Latin *pono* place]

postprandial /pəʊst'prændɪəl/ *adj. formal* or *joc.* after dinner or lunch. [Latin *prandium* a meal]

postscript /'pəʊstskrɪpt/ *n.* additional paragraph or remark, usu. at the end of a letter after the signature and introduced by 'PS'.

postulant /'pɒstjʊlənt/ *n.* candidate, esp. for admission to a religious order. [Latin: related to POSTULATE]

postulate –*v.* /'pɒstjʊˌleɪt/ (**-ting**) **1** (often foll. by *that*) assume as a necessary condition, esp. as a basis for reasoning; take for granted. **2** claim. –*n.* /'pɒstjʊlət/ **1** thing postulated. **2** prerequisite or condition. □ **postulation** /ˌpɒstjʊ'leɪʃ(ə)n/ *n.* [Latin *postulo*]

posture /'pɒstʃə(r)/ –*n.* **1** relative position of parts, esp. of the body; carriage, bearing. **2** mental attitude. **3** condition or state (of affairs etc.). –*v.* (**-ring**) **1** assume a mental or physical attitude, esp. for effect. **2** pose (a person). □ **postural** *adj.* [Latin: related to POSIT]

postwar /pəʊst'wɔː(r), 'pəʊst-/ *adj.* occurring or existing after a war.

posy /'pəʊzɪ/ *n.* (*pl.* **-ies**) small bunch of flowers. [alteration of POESY]

pot[1] –*n.* **1** rounded ceramic, metal, or glass vessel for holding liquids or solids or for cooking in. **2** flowerpot, teapot, etc. **3** contents of a pot. **4** chamber-pot; child's potty. **5** total amount bet in a game etc. **6** (usu. in *pl.*) *colloq.* large sum (*pots of money*). **7** *slang* silver cup etc. as a trophy. –*v.* (**-tt-**) **1** place in a pot. **2** (usu. as **potted** *adj.*) preserve in a sealed pot (*potted shrimps*). **3** pocket (a ball) in billiards etc. **4** abridge or epitomize. **5** shoot at, hit, or kill (an animal) with a pot-shot. **6** seize or secure. □ **go to pot** *colloq.* deteriorate; be ruined. □ **potful** *n.* (*pl.* **-s**). [Old English from Latin]

pot[2] *n. slang* marijuana. [Mexican Spanish *potiguaya*]

potable /'pəʊtəb(ə)l/ *adj.* drinkable. [Latin *poto* drink]

potage /pɒ'tɑːʒ/ *n.* thick soup. [French: related to POT[1]]

potash /'pɒtæʃ/ *n.* an alkaline potassium compound. [Dutch: related to POT[1], ASH[1]]

potassium /pə'tæsɪəm/ *n.* soft silver-white metallic element. [from POTASH]

potation /pə'teɪʃ(ə)n/ *n.* **1** a drink. **2** drinking. [Latin: related to POTION]

potato /pə'teɪtəʊ/ *n.* (*pl.* **-es**) **1** starchy plant tuber used for food. **2** plant bearing this. [Spanish *patata* from Taino *batata*]

potato crisp *n.* = CRISP.

pot-belly *n.* **1** protruding stomach. **2** person with this.

pot-boiler *n.* piece of art, writing, etc. done merely to earn money.

pot-bound *adj.* (of a plant) with roots filling the flowerpot, leaving no room to expand.

poteen /pɒ'tʃiːn/ *n. Ir.* illicit alcoholic spirit. [Irish *poitín* diminutive of *pota* POT[1]]

potent /'pəʊt(ə)nt/ *adj.* **1** powerful; strong. **2** (of a reason) cogent; forceful. **3** (of a male) capable of sexual erection or orgasm. □ **potency** *n.* [Latin *potens -ent-*: related to POSSE]

potentate /'pəʊtənˌteɪt/ *n.* monarch or ruler. [Latin: related to POTENT]

potential /pə'tenʃ(ə)l/ –*adj.* capable of coming into being or action; latent. –*n.* **1** capacity for use or development. **2** usable resources. **3** *Physics* quantity determining the energy of mass in a gravitational field or of charge in an electric field. □ **potentiality** /-ʃɪ'ælɪtɪ/

n. **potentially** *adv.* [Latin: related to POTENT]

potential difference *n.* difference of electric potential between two points.

pother /ˈpɒðə(r)/ *n. literary* noise, commotion, fuss. [origin unknown]

pot-herb *n.* herb grown in a kitchen garden.

pothole *n.* **1** deep hole or cave system in rock. **2** hole in a road surface. □ **potholer** *n.* **potholing** *n.*

pot-hook *n.* **1** hook over a hearth for hanging or lifting a pot. **2** curved stroke in handwriting.

pot-hunter *n.* **1** person who hunts for game at random. **2** person who competes merely for the prize.

potion /ˈpəʊʃ(ə)n/ *n.* dose of a liquid medicine, drug, poison, etc. [Latin *poto* drink]

pot luck *n.* whatever is available.

pot plant *n.* plant grown in a flowerpot.

pot-pourri /pəʊˈpʊərɪ/ *n.* (*pl.* **-s**) **1** scented mixture of dried petals and spices. **2** musical or literary medley. [French, = rotten pot]

pot roast *n.* piece of meat cooked slowly in a covered dish. □ **pot-roast** *v.*

potsherd /ˈpɒtʃɜːd/ *n.* esp. *Archaeol.* broken piece of ceramic material.

pot-shot *n.* **1** random shot. **2** casual attempt.

pottage /ˈpɒtɪdʒ/ *n. archaic* soup, stew. [French: related to POT[1]]

potter[1] *v.* (*US* **putter**) **1** (often foll. by *about, around*) work or occupy oneself in a desultory manner. **2** go slowly, dawdle, loiter (*pottered up to the pub*). [dial. *pote* push]

potter[2] *n.* maker of ceramic vessels. [Old English: related to POT[1]]

potter's wheel *n.* horizontal revolving disc to carry clay during moulding.

pottery *n.* (*pl.* **-ies**) **1** vessels etc. made of fired clay. **2** potter's work. **3** potter's workshop. [French: related to POTTER[2]]

potting shed *n.* shed in which plants are potted and tools etc. are stored.

potty[1] *adj.* (**-ier, -iest**) *slang* **1** foolish, crazy. **2** insignificant, trivial. □ **pottiness** *n.* [origin unknown]

potty[2] *n.* (*pl.* **-ies**) *colloq.* chamber-pot, esp. for a child.

pouch *–n.* **1** small bag or detachable outside pocket. **2** baggy area of skin under the eyes etc. **3 a** pocket-like receptacle of marsupials. **b** similar structure in various animals, e.g. in the cheeks of rodents. *–v.* **1** put or make into a pouch. **2** take possession of; pocket. [French: related to POKE[2]]

pouffe /puːf/ *n.* large firm cushion used as a low seat or footstool. [French]

poult /pəʊlt/ *n.* young domestic fowl, turkey, pheasant, etc. [contraction of PULLET]

poulterer /ˈpəʊltərə(r)/ *n.* dealer in poultry and usu. game. [*poulter*: related to POULT]

poultice /ˈpəʊltɪs/ *–n.* soft medicated usu. heated mass applied to the body and kept in place with muslin etc., to relieve soreness and inflammation. *–v.* (**-cing**) apply a poultice to. [Latin *puls* pottage]

poultry /ˈpəʊltrɪ/ *n.* domestic fowls (ducks, geese, turkeys, chickens, etc.), esp. as a source of food. [French: related to POULT]

pounce *–v.* (**-cing**) **1** spring or swoop, esp. as in capturing prey. **2** (often foll. by *on, upon*) **a** make a sudden attack. **b** seize eagerly upon a remark etc. *–n.* act of pouncing. [origin unknown]

pound[1] *n.* **1** unit of weight equal to 16 oz avoirdupois (0.4536 kg), 12 oz troy (0.3732 kg). **2** (in full **pound sterling**) (*pl.* same or **-s**) chief monetary unit of the UK etc. [Latin *pondo*]

pound[2] *v.* **1** crush or beat with repeated blows. **2** (foll. by *at, on*) deliver heavy blows or gunfire. **3** (foll. by *along* etc.) make one's way heavily or clumsily. **4** (of the heart) beat heavily. [Old English]

pound[3] *n.* enclosure where stray animals or officially removed vehicles are kept until claimed. [Old English]

poundage *n.* commission or fee of so much per pound sterling or weight.

pound coin *n.* (also **pound note**) coin or note worth one pound.

pounder *n.* (usu. in *comb.*) **1** thing or person weighing a specified number of pounds (*a five-pounder*). **2** gun firing a shell of a specified number of pounds.

pound of flesh *n.* any legal but morally offensive demand.

pour /pɔː(r)/ *v.* **1** (usu. foll. by *down, out, over*, etc.) flow or cause to flow esp. downwards in a stream or shower. **2** dispense (a drink) by pouring. **3** rain heavily. **4** (usu. foll. by *in, out*, etc.) come or go in profusion or rapid succession (*the crowd poured out*; *letters poured in*). **5** discharge or send freely. **6** (often foll. by *out*) utter at length or in a rush (*poured out their story*). [origin unknown]

pourboire /pʊəˈbwɑː(r)/ *n.* gratuity, tip. [French]

pout *–v.* **1** push the lips forward as a sign of displeasure or sulking. **2** (of the lips) be pushed forward. *–n.* this action. [origin unknown]

pouter *n.* a kind of pigeon that is able to inflate its crop.

poverty /'pɒvətɪ/ *n.* **1** being poor; want. **2** (often foll. by *of, in*) scarcity or lack. **3** inferiority, poorness. [Latin *pauper*]

poverty line *n.* minimum income needed for the necessities of life.

poverty-stricken *adj.* very poor.

poverty trap *n.* situation in which an increase of income incurs a loss of State benefits, making real improvement impossible.

POW *abbr.* prisoner of war.

pow *int.* expressing the sound of a blow or explosion. [imitative]

powder –*n.* **1** mass of fine dry particles. **2** medicine or cosmetic in this form. **3** = GUNPOWDER. –*v.* **1** apply powder to. **2** (esp. as **powdered** *adj.*) reduce to a fine powder (*powdered milk*). □ **powdery** *adj.* [Latin *pulvis -ver-* dust]

powder blue *adj.* & *n.* (as adj. often hyphenated) pale blue.

powder-puff *n.* soft pad for applying powder to the skin, esp. the face.

powder-room *n. euphem.* women's lavatory in a public building.

power –*n.* **1** ability to do or act. **2** particular faculty of body or mind. **3 a** influence, authority. **b** ascendancy, control (*the party in power*). **4** authorization; delegated authority. **5** influential person, body, or thing. **6** State having international influence. **7** vigour, energy. **8** active property or function (*heating power*). **9** *colloq.* large number or amount (*did me a power of good*). **10** capacity for exerting mechanical force or doing work (*horsepower*). **11** (often *attrib.*) mechanical or electrical energy as distinct from manual labour. **12 a** electricity supply. **b** particular source or form of energy (*hydroelectric power*). **13** *Physics* rate of energy output. **14** product obtained when a number is multiplied by itself a certain number of times (*2 to the power of 3 = 8*). **15** magnifying capacity of a lens. **16** deity. –*v.* **1** supply with mechanical or electrical energy. **2** (foll. by *up, down*) increase or decrease the power supplied to (a device); switch on or off. □ **the powers that be** those in authority. [Latin *posse* be able]

powerboat *n.* powerful motor boat.

power cut *n.* temporary withdrawal or failure of an electric power supply.

powerful *adj.* having much power or influence. □ **powerfully** *adv.* **powerfulness** *n.*

powerhouse *n.* **1** = POWER STATION. **2** person or thing of great energy.

powerless *adj.* **1** without power. **2** wholly unable. □ **powerlessness** *n.*

power line *n.* conductor supplying electrical power, esp. one supported by pylons or poles.

power of attorney *n.* authority to act for another person in legal and financial matters.

powerplant *n.* installation which provides power.

power point *n.* socket in a wall etc. for connecting an electrical device to the mains.

power-sharing *n.* coalition government, esp. as preferred on principle.

power station *n.* building where electrical power is generated for distribution.

powwow –*n.* meeting for discussion (orig. among N. American Indians). –*v.* hold a powwow. [Algonquian]

pox *n.* **1** virus disease leaving pockmarks. **2** *colloq.* = SYPHILIS. [alteration of *pocks* pl. of POCK]

poxy *adj.* (**-ier, -iest**) **1** infected by pox. **2** *slang* of poor quality; worthless.

pp *abbr.* pianissimo.

pp. *abbr.* pages.

p.p. *abbr.* (also **pp**) *per pro.*

ppm *abbr.* parts per million.

PPS *abbr.* **1** Parliamentary Private Secretary. **2** additional postscript. [sense 2 from *post-postscript*]

PR *abbr.* **1** public relations. **2** proportional representation.

Pr *symb.* praseodymium.

pr. *abbr.* pair.

practicable /'præktɪkəb(ə)l/ *adj.* **1** that can be done or used. **2** possible in practice. □ **practicability** /-'bɪlɪtɪ/ *n.* [French: related to PRACTICAL]

practical /'præktɪk(ə)l/ –*adj.* **1** of or concerned with practice rather than theory (*practical difficulties*). **2** suited to use; functional (*practical shoes*). **3** (of a person) good at making, organizing, or mending things. **4** sensible, realistic. **5** that is such in effect, virtual (*in practical control*). –*n.* practical examination or lesson. □ **practicality** /-'kælɪtɪ/ *n.* (*pl.* **-ies**). [Greek *praktikos* from *prassō* do]

practical joke *n.* humorous trick played on a person.

practically *adv.* **1** virtually, almost. **2** in a practical way.

practice /'præktɪs/ –*n.* **1** habitual action or performance. **2 a** repeated activity undertaken in order to improve a skill. **b** session of this. **3** action as opposed to theory. **4** the work, business, or place of business of a doctor, lawyer, etc. (*has a practice in town*). **5** procedure, esp. of a specified kind (*bad practice*). –*v.* *US* var. of PRACTISE. □ **in practice 1** when actually applied; in reality. **2** skilful from recent practice. **out of practice** lacking a former skill from lack of practice. [from PRACTISE]

practise /'præktɪs/ *v.* (*US* **practice**) (**-sing** or *US* **-cing**) **1** perform habitually; carry out in action. **2** do repeatedly as an exercise to improve a skill; exercise oneself in or on (an activity requiring skill). **3** (as **practised** *adj.*) experienced, expert. **4** (also *absol.*) be engaged in (a profession, religion, etc.). [Latin: related to PRACTICAL]

practitioner /præk'tɪʃənə(r)/ *n.* person practising a profession, esp. medicine.

praenomen /priː'nəʊmen/ *n.* ancient Roman's first or personal name (e.g. *Marcus* Tullius Cicero). [Latin: related to PRE-, NOMEN]

praesidium var. of PRESIDIUM.

praetor /'priːtə(r)/ *n.* ancient Roman magistrate below consul. [Latin]

praetorian guard /priː'tɔːrɪən/ *n.* bodyguard of the ancient Roman emperor.

pragmatic /præg'mætɪk/ *adj.* dealing with matters from a practical point of view. □ **pragmatically** *adv.* [Greek *pragma -mat-* deed]

pragmatism /'prægmə,tɪz(ə)m/ *n.* **1** pragmatic attitude or procedure. **2** philosophy that evaluates assertions solely by their practical consequences and bearing on human interests. □ **pragmatist** *n.* [Greek *pragma*: related to PRAGMATIC]

prairie /'preərɪ/ *n.* large area of treeless grassland, esp. in N. America. [Latin *pratum* meadow]

prairie dog *n.* N. American rodent making a barking sound.

prairie oyster *n.* seasoned raw egg, swallowed without breaking the yolk.

prairie wolf *n.* = COYOTE.

praise /preɪz/ *–v.* (**-sing**) **1** express warm approval or admiration of. **2** glorify (God) in words. *–n.* praising; commendation. [French *preisier* from Latin *pretium* price]

praiseworthy *adj.* worthy of praise.

praline /'prɑːliːn/ *n.* sweet made by browning nuts in boiling sugar. [French]

pram *n.* four-wheeled conveyance for a baby, pushed by a person on foot. [abbreviation of PERAMBULATOR]

prance /prɑːns/ *–v.* (**-cing**) **1** (of a horse) raise the forelegs and spring from the hind legs. **2** walk or behave in an elated or arrogant manner. *–n.* prancing, prancing movement. [origin unknown]

prang *slang –v.* **1** crash (an aircraft or vehicle). **2** damage by impact. **3** bomb (a target) successfully. *–n.* act of pranging. [imitative]

prank *n.* practical joke; piece of mischief. [origin unknown]

prankster *n.* practical joker.

praseodymium /,preɪzɪə'dɪmɪəm/ *n.* soft silvery metallic element of the lanthanide series. [Greek *prasios* green]

prat *n. slang* **1** fool. **2** buttocks. [origin unknown]

prate *–v.* (**-ting**) **1** chatter; talk too much. **2** talk foolishly or irrelevantly. *–n.* prating; idle talk. [Low German or Dutch]

prattle /'præt(ə)l/ *–v.* (**-ling**) chatter in a childish or inconsequential way. *–n.* childish or inconsequential chatter. [Low German *pratelen*: related to PRATE]

prawn *n.* edible shellfish like a large shrimp. [origin unknown]

pray *v.* (often foll. by *for* or *to* + infin. or *that* + clause) **1** say prayers; make devout supplication. **2 a** entreat. **b** ask earnestly (*prayed to be released*). **3** (as *imper.*) *archaic* please (*pray tell me*). [Latin *precor*]

prayer[1] /'preə(r)/ *n.* **1 a** request or thanksgiving to God or an object of worship. **b** formula used in praying (*the Lord's Prayer*). **c** act of praying. **d** religious service consisting largely of prayers (*morning prayer*). **2** entreaty to a person. [Latin: related to PRECARIOUS]

prayer[2] /'preɪə(r)/ *n.* person who prays.

prayer-book *n.* book of set prayers.

prayer-mat *n.* small carpet on which Muslims kneel to pray.

prayer-wheel *n.* revolving cylindrical box inscribed with or containing prayers, used esp. by Tibetan Buddhists.

praying mantis see MANTIS.

pre- *prefix* before (in time, place, order, degree, or importance). [Latin *prae* before]

preach *v.* **1** (also *absol.*) deliver (a sermon); proclaim or expound (the gospel etc.). **2** give moral advice in an obtrusive way. **3** advocate or inculcate (a quality or practice etc.). □ **preacher** *n.* [Latin *praedico* proclaim]

preamble /priː'æmb(ə)l/ *n.* **1** preliminary statement. **2** introductory part of a statute or deed etc. [Latin: related to AMBLE]

pre-amp /'priːæmp/ *n.* = PREAMPLIFIER. [abbreviation]

preamplifier /priː'æmplɪ,faɪə(r)/ *n.* electronic device that amplifies a weak signal (e.g. from a microphone or pickup) and transmits it to a main amplifier.

prearrange /,priːə'reɪndʒ/ *v.* (**-ging**) arrange beforehand. □ **prearrangement** *n.*

prebend /'prebənd/ *n.* **1** stipend of a canon or member of chapter. **2** portion of land or tithe from which this is drawn. □ **prebendal** /prɪ'bend(ə)l/ *adj.* [Latin *praebeo* grant]

prebendary /'prebəndərɪ/ *n.* (*pl.* **-ies**) holder of a prebend; honorary canon. [medieval Latin: related to PREBEND]

Precambrian /pri:'kæmbrɪən/ *Geol.* *–adj.* of the earliest geological era. *–n.* this era.

precarious /prɪ'keərɪəs/ *adj.* **1** uncertain; dependent on chance. **2** insecure, perilous. □ **precariously** *adv.* **precariousness** *n.* [Latin *precarius*: related to PRAY]

precast /pri:'kɑ:st/ *adj.* (of concrete) cast in its final shape before positioning.

precaution /prɪ'kɔ:ʃ(ə)n/ *n.* action taken beforehand to avoid risk or ensure a good result. □ **precautionary** *adj.* [Latin: related to CAUTION]

precede /prɪ'si:d/ *v.* (**-ding**) **1** come or go before in time, order, importance, etc. **2** (foll. by *by*) cause to be preceded. [Latin: related to CEDE]

precedence /'presɪd(ə)ns/ *n.* **1** priority in time, order, importance, etc. **2** right of preceding others. □ **take precedence** (often foll. by *over*, *of*) have priority (over).

precedent *–n.* /'presɪd(ə)nt/ previous case etc. taken as a guide for subsequent cases or as a justification. *–adj.* /prɪ'si:d(ə)nt, 'presɪ-/ preceding in time, order, importance, etc. [French: related to PRECEDE]

precentor /prɪ'sentə(r)/ *n.* person who leads the singing or (in a synagogue) the prayers of a congregation. [Latin *praecentor* from *cano* sing]

precept /'pri:sept/ *n.* **1** rule or guide, esp. for conduct. **2** lawful demand, esp. from one authority to another to levy rates. [Latin *praeceptum* maxim, order]

preceptor /prɪ'septə(r)/ *n.* teacher, instructor. □ **preceptorial** /ˌpri:sep'tɔ:rɪəl/ *adj.* [Latin: related to PRECEPT]

precession /prɪ'seʃ(ə)n/ *n.* slow movement of the axis of a spinning body around another axis. [Latin: related to PRECEDE]

precession of the equinoxes *n.* **1** slow retrograde motion of equinoctial points along the ecliptic. **2** resulting earlier occurrence of equinoxes in each successive sidereal year.

pre-Christian /pri:'krɪstʃ(ə)n/ *adj.* before Christianity.

precinct /'pri:sɪŋkt/ *n.* **1** enclosed area, e.g. around a cathedral, college, etc. **2** designated area in a town, esp. where traffic is excluded. **3** (in *pl.*) environs. [Latin *praecingo -cinct-* encircle]

preciosity /ˌpreʃɪ'ɒsɪtɪ/ *n.* affected refinement in art etc., esp. in the choice of words. [related to PRECIOUS]

precious /'preʃəs/ *–adj.* **1** of great value or worth. **2** beloved; much prized (*precious memories*). **3** affectedly refined. **4** *colloq.* often *iron.* **a** considerable (*a precious lot of good*). **b** expressing contempt or disdain (*keep your precious flowers!*). *–adv. colloq.* extremely, very (*had precious little left*). □ **preciousness** *n.* [Latin *pretium* price]

precious metals *n.pl.* gold, silver, and platinum.

precious stone *n.* piece of mineral of great value, esp. as used in jewellery.

precipice /'presɪpɪs/ *n.* **1** vertical or steep face of a rock, cliff, mountain, etc. **2** dangerous situation. [Latin *praeceps -cipit-* headlong]

precipitate *–v.* (**-ting**) /prɪ'sɪpɪˌteɪt/ **1** hasten the occurrence of; cause to occur prematurely. **2** (foll. by *into*) send rapidly into a certain state or condition (*was precipitated into war*). **3** throw down headlong. **4** *Chem.* cause (a substance) to be deposited in solid form from a solution. **5** *Physics* condense (vapour) into drops and so deposit it. *–adj.* /prɪ'sɪpɪtət/ **1** headlong; violently hurried (*precipitate departure*). **2** (of a person or act) hasty, rash. *–n.* /prɪ'sɪpɪtət/ **1** *Chem.* substance precipitated from a solution. **2** *Physics* moisture condensed from vapour, e.g. rain, dew.

precipitation /prɪˌsɪpɪ'teɪʃ(ə)n/ *n.* **1** precipitating or being precipitated. **2** rash haste. **3 a** rain or snow etc. falling to the ground. **b** quantity of this.

precipitous /prɪ'sɪpɪtəs/ *adj.* **1 a** of or like a precipice. **b** dangerously steep. **2** = PRECIPITATE *adj.*

précis /'preɪsi:/ *–n.* (*pl.* same /-si:z/) summary, abstract. *–v.* (**-cises** /-si:z/; **-cised** /-si:d/; **-cising** /-si:ɪŋ/) make a précis of. [French]

precise /prɪ'saɪs/ *adj.* **1 a** accurately expressed. **b** definite, exact. **2** punctilious; scrupulous in being exact. [Latin *praecido* cut short]

precisely *adv.* **1** in a precise manner; exactly. **2** (as a reply) quite so, as you say.

precision /prɪ'sɪʒ(ə)n/ *n.* **1** accuracy. **2** degree of refinement in measurement etc. **3** (*attrib.*) marked by or adapted for precision (*precision instruments*).

preclinical /pri:'klɪnɪk(ə)l/ *adj.* of the first, chiefly theoretical, stage of a medical education.

preclude /prɪ'klu:d/ *v.* **(-ding) 1** (foll. by *from*) prevent. **2** make impossible. [Latin *praecludo*: related to CLOSE[1]]

precocious /prɪ'kəʊʃəs/ *adj.* **1** often *derog.* (of esp. a child) prematurely developed in some respect. **2** (of an action etc.) indicating such development. □ **precociously** *adv.* **precociousness** *n.* **precocity** /-'kɒsɪtɪ/ *n.* [Latin *praecox -cocis* early ripe]

precognition /,pri:kɒg'nɪʃ(ə)n/ *n.* supposed foreknowledge, esp. of a supernatural kind.

preconceive /,pri:kən'si:v/ *v.* **(-ving)** form (an idea or opinion etc.) beforehand.

preconception /,pri:kən'sepʃ(ə)n/ *n.* preconceived idea, prejudice.

precondition /,pri:kən'dɪʃ(ə)n/ *n.* condition that must be fulfilled in advance.

precursor /pri:'kɜ:sə(r)/ *n.* **1 a** forerunner. **b** person who precedes in office etc. **2** harbinger. [Latin *praecurro -curs-* run before]

predate /pri:'deɪt/ *v.* **(-ting)** precede in time.

predator /'predətə(r)/ *n.* predatory animal. [Latin]

predatory *adj.* **1** (of an animal) preying naturally upon others. **2** plundering or exploiting others.

predecease /,pri:dɪ'si:s/ *v.* **(-sing)** die earlier than (another person).

predecessor /'pri:dɪ,sesə(r)/ *n.* **1** former holder of an office or position with respect to a later holder. **2** ancestor. **3** thing to which another has succeeded. [Latin *decessor*: related to DECEASE]

predestine /pri:'destɪn/ *v.* **(-ning) 1** determine beforehand. **2** ordain in advance by divine will or as if by fate. □ **predestination** /-'neɪʃ(ə)n/ *n.* [French or Church Latin: related to PRE-]

predetermine /,pri:dɪ'tɜ:mɪn/ *v.* **(-ning) 1** decree beforehand. **2** predestine.

predicament /prɪ'dɪkəmənt/ *n.* difficult or unpleasant situation. [Latin: related to PREDICATE]

predicant /'predɪkənt/ *hist.* —*adj.* (of a religious order) engaged in preaching. —*n.* predicant person, esp. a Dominican. [Latin: related to PREDICATE]

predicate —*v.* /'predɪ,keɪt/ **(-ting) 1** (also *absol.*) assert (something) about the subject of a proposition. **2** (foll. by *on*) found or base (a statement etc.) on. —*n.* /'predɪkət/ *Gram.* & *Logic* what is said about the subject of a sentence or proposition etc. (e.g. *went home* in *John went home*). □ **predicable** /'predɪkəb(ə)l/ *adj.* **predication** /-'keɪʃ(ə)n/ *n.* [Latin *praedico -dicat-* declare]

predicative /prɪ'dɪkətɪv/ *adj.* **1** *Gram.* (of an adjective or noun) forming or contained in the predicate, as *old* in *the dog is old*. **2** that predicates. [Latin: related to PREDICATE]

predict /prɪ'dɪkt/ *v.* (often foll. by *that*) foretell, prophesy. □ **predictor** *n.* [Latin *praedico -dict-* foretell]

predictable *adj.* that can be predicted or is to be expected. □ **predictability** /-'bɪlɪtɪ/ *n.* **predictably** *adv.*

prediction /prɪ'dɪkʃ(ə)n/ *n.* **1** predicting or being predicted. **2** thing predicted.

predilection /,pri:dɪ'lekʃ(ə)n/ *n.* (often foll. by *for*) preference or special liking. [Latin *praediligo* prefer]

predispose /,pri:dɪ'spəʊz/ *v.* **(-sing) 1** influence favourably in advance. **2** (foll. by *to*, or *to* + infin.) render liable or inclined beforehand. □ **predisposition** /-pə'zɪʃ(ə)n/ *n.*

predominant /prɪ'dɒmɪnənt/ *adj.* **1** predominating. **2** being the strongest or main element. □ **predominance** *n.* **predominantly** *adv.*

predominate /prɪ'dɒmɪ,neɪt/ *v.* **(-ting) 1** (foll. by *over*) have control. **2** be superior. **3** be the strongest or main element.

pre-echo /pri:'ekəʊ/ *n.* (*pl.* **-es**) **1** faint copy heard just before an actual sound in a recording, caused by the accidental transfer of signals. **2** foreshadowing.

pre-embryo /pri:'embrɪəʊ/ *n.* (*pl.* **-s**) potential human embryo in the first fourteen days after fertilization.

pre-eminent /pri:'emɪnənt/ *adj.* **1** excelling others. **2** outstanding. □ **pre-eminence** *n.* **pre-eminently** *adv.*

pre-empt /pri:'empt/ *v.* **1 a** forestall. **b** appropriate in advance. **2** obtain by pre-emption. [back-formation from PRE-EMPTION]

■ **Usage** *Pre-empt* is sometimes used to mean *prevent*, but this is considered incorrect in standard English.

pre-emption /pri:'empʃ(ə)n/ *n.* purchase or taking by one person or party before the opportunity is offered to others. [medieval Latin *emo empt-* buy]

pre-emptive /pri:'emptɪv/ *adj.* **1** pre-empting. **2** (of military action) intended to prevent attack by disabling the enemy.

preen *v.* **1** (of a bird) tidy (the feathers or itself) with its beak. **2** (of a person) smarten or admire (oneself, one's hair, clothes, etc.). **3** (often foll. by *on*) congratulate or pride (oneself). [origin unknown]

prefab /'pri:fæb/ *n. colloq.* prefabricated building. [abbreviation]

prefabricate /priːˈfæbrɪˌkeɪt/ *v.* (**-ting**) manufacture sections of (a building etc.) prior to their assembly on site.

preface /ˈprefəs/ —*n.* **1** introduction to a book stating its subject, scope, etc. **2** preliminary part of a speech. —*v.* (**-cing**) **1** (foll. by *with*) introduce or begin (a speech or event). **2** provide (a book etc.) with a preface. **3** (of an event etc.) lead up to (another). □ **prefatory** /-tərɪ/ *adj.* [Latin *praefatio*]

prefect /ˈpriːfekt/ *n.* **1** chief administrative officer of a district, esp. in France. **2** senior pupil in a school, helping to maintain discipline. [Latin *praeficio -fect-* set in authority over]

prefecture /ˈpriːfektʃə(r)/ *n.* **1** district under the government of a prefect. **2** prefect's office or tenure. [Latin: related to PREFECT]

prefer /prɪˈfɜː(r)/ *v.* (**-rr-**) **1** (often foll. by *to*, or *to* + infin.) like better (*prefers coffee to tea*). **2** submit (information, an accusation, etc.) for consideration. **3** promote or advance (a person). [Latin *praefero -lat-*]

preferable /ˈprefərəb(ə)l/ *adj.* to be preferred; more desirable. □ **preferably** *adv.*

preference /ˈprefərəns/ *n.* **1** preferring or being preferred. **2** thing preferred. **3** favouring of one person etc. before others. **4** prior right, esp. to the payment of debts. □ **in preference to** as a thing preferred over (another).

preference shares *n.pl.* (also **preference stock** *n.sing.*) shares or stock whose entitlement to dividend takes precedence over that of ordinary shares.

preferential /ˌprefəˈrenʃ(ə)l/ *adj.* **1** of or involving preference. **2** giving or receiving a favour. □ **preferentially** *adv.*

preferment /prɪˈfɜːmənt/ *n. formal* promotion to a higher office.

prefigure /priːˈfɪgə(r)/ *v. formal* (**-ring**) represent or imagine beforehand.

prefix /ˈpriːfɪks/ —*n.* **1** verbal element placed at the beginning of a word to qualify its meaning (e.g. *ex-*, *non-*). **2** title before a name (e.g. *Mr*). —*v.* (often foll. by *to*) **1** add as an introduction. **2** join (a word or element) as a prefix.

pregnant /ˈpregnənt/ *adj.* **1** having a child or young developing in the uterus. **2** full of meaning; significant; suggestive (*a pregnant pause*). □ **pregnancy** *n.* (*pl.* **-ies**). [Latin *praegnans*]

preheat /priːˈhiːt/ *v.* heat beforehand.

prehensile /priːˈhensaɪl/ *adj. Zool.* (of a tail or limb) capable of grasping. [Latin *prehendo -hens-* grasp]

prehistoric /ˌpriːhɪˈstɒrɪk/ *adj.* **1** of the period before written records. **2** *colloq.* utterly out of date. □ **prehistory** /-ˈhɪstərɪ/ *n.*

prejudge /priːˈdʒʌdʒ/ *v.* (**-ging**) form a premature judgement on (a person, issue, etc.).

prejudice /ˈpredʒʊdɪs/ —*n.* **1 a** preconceived opinion. **b** (foll. by *against*, *in favour of*) bias, partiality. **2** harm that results or may result from some action or judgement (*to the prejudice of*). —*v.* (**-cing**) **1** impair the validity or force of (a right, claim, statement, etc.). **2** (esp. as **prejudiced** *adj.*) cause (a person) to have a prejudice. □ **without prejudice** (often foll. by *to*) without detriment (to an existing right or claim). [Latin: related to JUDGE]

prejudicial /ˌpredʒʊˈdɪʃ(ə)l/ *adj.* (often foll. by *to*) causing prejudice; detrimental.

prelacy /ˈpreləsɪ/ *n.* (*pl.* **-ies**) **1** church government by prelates. **2** (prec. by *the*) prelates collectively. **3** office or rank of prelate. [Anglo-French from medieval Latin: related to PRELATE]

prelate /ˈprelət/ *n.* high ecclesiastical dignitary, e.g. a bishop. [Latin: related to PREFER]

prelim /ˈpriːlɪm/ *n. colloq.* **1** preliminary university examination. **2** (in *pl.*) pages preceding the main text of a book. [abbreviation]

preliminary /prɪˈlɪmɪnərɪ/ —*adj.* introductory, preparatory. —*n.* (*pl.* **-ies**) (usu. in *pl.*) **1** preliminary action or arrangement (*dispense with the preliminaries*). **2** preliminary trial or contest. [Latin *limen* threshold]

prelude /ˈpreljuːd/ —*n.* (often foll. by *to*) **1** action, event, or situation serving as an introduction. **2** introductory part of a poem etc. **3** *Mus.* **a** introductory piece to a fugue, suite, etc. **b** short piece of a similar type. —*v.* (**-ding**) **1** serve as a prelude to. **2** introduce with a prelude. [Latin *ludo lus-* play]

premarital /priːˈmærɪt(ə)l/ *adj.* existing or (esp. of sexual relations) occurring before marriage.

premature /ˈpremətʃə(r), -ˈtʃʊə(r)/ *adj.* **1 a** occurring or done before the usual or proper time (*a premature decision*). **b** too hasty. **2** (of a baby) born (esp. three or more weeks) before the end of gestation. □ **prematurely** *adv.* [Latin: related to PRE-, MATURE]

premed /priːˈmed/ *n. colloq.* = PREMEDICATION. [abbreviation]

premedication /ˌpriːmedɪˈkeɪʃ(ə)n/ *n.* medication to prepare for an operation etc.

premeditate /pri:'medɪ,teɪt/ *v.* (**-ting**) think out or plan beforehand (*premeditated murder*). □ **premeditation** /-'teɪʃ(ə)n/ *n.* [Latin: related to MEDITATE]

premenstrual /pri:'menstrʊəl/ *adj.* of the time shortly before each menstruation (*premenstrual tension*).

premier /'premɪə(r)/ *–n.* prime minister or other head of government. *–adj.* first in importance, order, or time. □ **premiership** *n.* [French, = first]

première /'premɪ,eə(r)/ *–n.* first performance or showing of a play or film. *–v.* (**-ring**) give a première of. [French feminine: related to PREMIER]

premise /'premɪs/ *n.* **1** *Logic* = PREMISS. **2** (in *pl.*) **a** house or other building with its grounds, outbuildings, etc. **b** *Law* houses, lands, or tenements previously specified in a document etc. □ **on the premises** in the building etc. concerned. [Latin *praemissa* set in front]

premiss /'premɪs/ *n. Logic* previous statement from which another is inferred. [var. of PREMISE]

premium /'pri:mɪəm/ *n.* **1** amount to be paid for a contract of insurance. **2** sum added to interest, wages, price, etc. **3** reward or prize. **4** (*attrib.*) (of a commodity) of the best quality and therefore more expensive. □ **at a premium 1** highly valued; above the usual or nominal price. **2** scarce and in demand. [Latin *praemium* reward]

Premium Bond *n.* (also **Premium Savings Bond**) government security without interest but with a draw for cash prizes.

premolar /pri:'məʊlə(r)/ *n.* (in full **premolar tooth**) tooth between the canines and molars.

premonition /,premə'nɪʃ(ə)n/ *n.* forewarning; presentiment. □ **premonitory** /prɪ'mɒnɪtərɪ/ *adj.* [Latin *moneo* warn]

prenatal /pri:'neɪt(ə)l/ *adj.* of the period before childbirth.

preoccupy /pri:'ɒkjʊ,paɪ/ *v.* (**-ies, -ied**) **1** (of a thought etc.) dominate the mind of (a person) to the exclusion of all else. **2** (as **preoccupied** *adj.*) otherwise engrossed; mentally distracted. □ **preoccupation** /-'peɪʃ(ə)n/ *n.* [Latin *praeoccupo* seize beforehand]

preordain /,pri:ɔ:'deɪn/ *v.* ordain or determine beforehand.

prep *n. colloq.* **1** homework, esp. in boarding-schools. **2** period when this is done. [abbreviation of PREPARATION]

prepack /pri:'pæk/ *v.* (also **pre-package** /-'pækɪdʒ/) pack (goods) on the site of production or before retail.

prepaid *past* and *past part.* of PREPAY.

preparation /,prepə'reɪʃ(ə)n/ *n.* **1** preparing or being prepared. **2** (often in *pl.*) something done to make ready. **3** specially prepared substance. **4** = PREP.

preparatory /prɪ'pærətərɪ/ *–adj.* (often foll. by *to*) serving to prepare; introductory. *–adv.* (often foll. by *to*) in a preparatory manner (*was packing preparatory to departure*).

preparatory school *n.* private primary school or *US* secondary school.

prepare /prɪ'peə(r)/ *v.* (**-ring**) **1** make or get ready for use, consideration, etc. **2** assemble (a meal etc.). **3 a** make (a person or oneself) ready or disposed in some way (*prepared them for a shock*). **b** get ready (*prepare to jump*). □ **be prepared** (often foll. by *for*, or *to* + infin.) be disposed or willing to. [Latin *paro* make ready]

preparedness /prɪ'peərɪdnɪs/ *n.* readiness, esp. for war.

prepay /pri:'peɪ/ *v.* (*past* and *past part.* **prepaid**) **1** pay (a charge) in advance. **2** pay postage on (a letter etc.) before posting. □ **prepayment** *n.*

preplan /pri:'plæn/ *v.* (**-nn-**) plan in advance.

preponderate /prɪ'pɒndə,reɪt/ *v.* (**-ting**) (often foll. by *over*) be greater in influence, quantity, or number; predominate. □ **preponderance** *n.* **preponderant** *adj.* [Latin *pondus -der-* weight]

preposition /,prepə'zɪʃ(ə)n/ *n. Gram.* word governing (and usu. preceding) a noun or pronoun and expressing a relation to another word, as in: 'the man *on* the platform', 'came *after* dinner', 'went *by* train'. □ **prepositional** *adj.* [Latin *praepono -posit-* place before]

prepossess /,pri:pə'zes/ *v.* **1** (usu. in *passive*) (of an idea, feeling, etc.) take possession of (a person). **2 a** prejudice (usu. favourably and spontaneously). **b** (as **prepossessing** *adj.*) attractive, appealing. □ **prepossession** /-'zeʃ(ə)n/ *n.*

preposterous /prɪ'pɒstərəs/ *adj.* **1** utterly absurd; outrageous. **2** contrary to nature, reason, or sense. □ **preposterously** *adv.* [Latin, = before behind]

preppy *n.* (*pl.* **-ies**) *US colloq.* student of an expensive private school or similar-looking person. [from PREP]

prepuce /'pri:pju:s/ *n.* **1** = FORESKIN. **2** fold of skin surrounding the clitoris. [Latin *praeputium*]

Pre-Raphaelite /pri:'ræfə,laɪt/ *–n.* member of a group of 19th-c. artists emulating Italian art before the time of Raphael. *–adj.* **1** of the Pre-Raphaelites. **2** (**pre-Raphaelite**) (esp. of a woman) like a type painted by the Pre-

Raphaelites (e.g. with long thick curly auburn hair).

pre-record /ˌpriːrɪ'kɔːd/ *v.* record (esp. material for broadcasting) in advance.

prerequisite /priː'rekwɪzɪt/ *–adj.* required as a precondition. *–n.* prerequisite thing.

■ **Usage** *Prerequisite* is sometimes confused with *perquisite* which means 'an extra profit, allowance, or right'.

---

prerogative /prɪ'rɒgətɪv/ *n.* right or privilege exclusive to an individual or class. [Latin *praerogo* ask first]

Pres. *abbr.* President.

presage /'presɪdʒ/ *–n.* **1** omen, portent. **2** presentiment, foreboding. *–v.* (**-ging**) **1** portend, foreshadow. **2** give warning of (an event etc.) by natural means. **3** (of a person) predict or have a presentiment of. [Latin *praesagium*]

presbyopia /ˌprezbɪ'əʊpɪə/ *n.* long-sightedness caused by loss of elasticity of the eye lens, occurring esp. in old age. □ **presbyopic** /-'ɒpɪk/*adj.* [Greek *presbus* old man, *ōps* eye]

presbyter /'prezbɪtə(r)/ *n.* **1** (in the Episcopal Church) minister of the second order; priest. **2** (in the Presbyterian Church) elder. [Church Latin from Greek, = elder]

Presbyterian /ˌprezbɪ'tɪərɪən/*–adj.* (of a church) governed by elders all of equal rank, esp. with ref. to the Church of Scotland. *–n.* member of a Presbyterian Church. □ **Presbyterianism** *n.*

presbytery /'prezbɪtərɪ/ *n.* (*pl.* **-ies**) **1** eastern part of a chancel. **2** body of presbyters, esp. a court next above a Kirk-session. **3** house of a Roman Catholic priest.

preschool /'priːskuːl/ *adj.* of the time before a child is old enough to go to school.

prescient /'presɪənt/ *adj.* having foreknowledge or foresight. □ **prescience** *n.* [Latin *praescio* know before]

prescribe /prɪ'skraɪb/ *v.* (**-bing**) **1 a** advise the use of (a medicine etc.). **b** recommend, esp. as a benefit. **2** lay down or impose authoritatively. [Latin *praescribo*]

■ **Usage** *Prescribe* is sometimes confused with *proscribe*.

---

prescript /'priːskrɪpt/ *n.* ordinance, law, command. [Latin: related to PRESCRIBE]

prescription /prɪ'skrɪpʃ(ə)n/ *n.* **1** act of prescribing. **2 a** doctor's (usu. written) instruction for the supply and use of a medicine. **b** medicine prescribed.

prescriptive /prɪ'skrɪptɪv/ *adj.* **1** prescribing, laying down rules. **2** arising from custom.

presence /'prez(ə)ns/ *n.* **1** being present. **2** place where a person is (*admitted to their presence*). **3** person's appearance or bearing, esp. when imposing. **4** person or spirit that is present (*the royal presence*; *aware of a presence in the room*). [Latin: related to PRESENT[1]]

presence of mind *n.* calmness and quick-wittedness in sudden difficulty etc.

present[1] /'prez(ə)nt/ *–adj.* **1** (usu. *predic.*) being in the place in question. **2 a** now existing, occurring, or being such. **b** now being considered etc. (*in the present case*). **3** *Gram.* expressing an action etc. now going on or habitually performed (*present participle*). *–n.* (prec. by *the*) **1** the time now passing (*no time like the present*). **2** *Gram.* present tense. □ **at present** now. **by these presents** *Law* by this document. **for the present** just now; for the time being. [Latin *praesens -ent-*]

present[2] /prɪ'zent/ *v.* **1** introduce, offer, or exhibit for attention or consideration. **2 a** (with a thing as object, foll. by *to*) offer or give as a gift (to a person). **b** (with a person as object, foll. by *with*) make available to; cause to have (*that presents us with a problem*). **3 a** (of a company, producer, etc.) put (a piece of entertainment) before the public. **b** (of a performer, compère, etc.) introduce. **4** introduce (a person) formally (*may I present my fiancé?*). **5 a** (of a circumstance) reveal (some quality etc.) (*this presents some difficulty*). **b** exhibit (an appearance etc.). **6** (of an idea etc.) offer or suggest itself. **7** deliver (a cheque, bill, etc.) for acceptance or payment. **8 a** (usu. foll. by *at*) aim (a weapon). **b** hold out (a weapon) in position for aiming. □ **present arms** hold a rifle etc. vertically in front of the body as a salute. □ **presenter** *n.* (in sense 3b). [Latin *praesento*: related to PRESENT[1]]

present[3] /'prez(ə)nt/ *n.* thing given, gift. [French: related to PRESENT[1]]

presentable /prɪ'zentəb(ə)l/ *adj.* of good appearance; fit to be presented. □ **presentability** /-'bɪlɪtɪ/ *n.* **presentably** *adv.*

presentation /ˌprezən'teɪʃ(ə)n/ *n.* **1 a** presenting or being presented. **b** thing presented. **2** manner or quality of presenting. **3** demonstration or display of materials, information, etc.; lecture.

present-day *attrib. adj.* of this time; modern.

presentiment /prɪ'zentɪmənt/ *n.* vague expectation; foreboding (esp. of misfortune).

presently *adv.* **1** soon; after a short time. **2** esp. *US* & *Scot.* at the present time; now.

preservative /prɪ'zɜ:vətɪv/ *–n.* substance for preserving perishable foodstuffs, wood, etc. *–adj.* tending to preserve.

preserve /prɪ'zɜ:v/ *–v.* **(-ving) 1** keep safe or free from decay etc. **2** maintain (a thing) in its existing state. **3** retain (a quality or condition). **4** treat (food) to prevent decomposition or fermentation. **5** keep (game etc.) undisturbed for private use. *–n.* (in *sing.* or *pl.*) **1** preserved fruit; jam. **2** place where game etc. is preserved. **3** sphere of activity regarded as a person's own. □ **preservation** /,prezə'veɪʃ(ə)n/ *n.* [Latin *servo* keep]

pre-set /pri:'set/ *v.* **(-tt-**; *past* and *past part.* **-set)** set or fix (a device) in advance of its operation.

preshrunk /pri:'ʃrʌŋk/ *adj.* (of fabric etc.) treated so that it shrinks during manufacture and not in use.

preside /prɪ'zaɪd/ *v.* **(-ding) 1** (often foll. by *at, over*) be chairperson or president of a meeting etc. **2** exercise control or authority. [Latin *sedeo* sit]

presidency /'prezɪdənsɪ/ *n.* (*pl.* **-ies**) **1** office of president. **2** period of this.

president /'prezɪd(ə)nt/ *n.* **1** head of a republican State. **2** head of a society or council etc. **3** head of certain colleges. **4** *US* head of a university, company, etc. **5** person in charge of a meeting. □ **presidential** /-'denʃ(ə)l/ *adj.*

presidium /prɪ'sɪdɪəm/ *n.* (also **praesidium**) standing committee in a Communist country. [Latin: related to PRESIDE]

press[1] *–v.* **1** apply steady force to (a thing in contact). **2 a** compress or squeeze a thing to flatten, shape, or smooth it. **b** squeeze (a fruit etc.) to extract its juice. **3** (foll. by *out of, from*, etc.) squeeze (juice etc.). **4** embrace or caress by squeezing (*pressed my hand*). **5** (foll. by *on, against*, etc.) exert pressure. **6** be urgent; demand immediate action. **7** (foll. by *for*) make an insistent demand. **8** (foll. by *up, round*, etc.) crowd. **9** (foll. by *on, forward*, etc.) hasten insistently. **10** (often in *passive*) (of an enemy etc.) bear heavily on. **11** (often foll. by *for*, or *to* + infin.) urge or entreat (*pressed me to stay*; *pressed me for an answer*). **12** (foll. by *on, upon*) **a** urge (an opinion, claim, or course of action). **b** force (an offer, a gift, etc.). **13** insist on (*did not press the point*). **14** manufacture (a gramophone record, car part, etc.) by using pressure to shape and extract from a sheet of material. *–n.* **1** act of pressing (*give it a press*). **2** device for compressing, flattening, shaping, extracting juice, etc. **3** = PRINTING-PRESS. **4** (prec. by *the*) **a** art or practice of printing. **b** newspapers etc. generally or collectively. **5** notice or publicity in newspapers etc. (*got a good press*). **6 (Press)** printing or publishing company. **7 a** crowding. **b** crowd (of people etc.). **8** the pressure of affairs. **9** esp. *Ir.* & *Scot.* large usu. shelved cupboard. □ **be pressed for** have barely enough (time etc.). **go** (or **send**) **to press** go or send to be printed. [Latin *premo press-*]

press[2] *v.* **1** *hist.* force to serve in the army or navy. **2** bring into use as a makeshift (*was pressed into service*). [obsolete *prest* from French, = loan]

press agent *n.* person employed to obtain advertising and press publicity.

press conference *n.* interview given to a number of journalists.

press gallery *n.* gallery for reporters, esp. in a legislative assembly.

press-gang *–n.* **1** *hist.* body of men employed to press men into army or navy service. **2** any group using coercive methods. *–v.* force into service.

pressie /'prezɪ/ *n.* (also **prezzie**) *colloq.* present, gift. [abbreviation]

pressing *–adj.* **1** urgent. **2** urging strongly (*pressing invitation*). *–n.* **1** thing made by pressing, e.g. a gramophone record. **2** series of these made at one time. **3** act of pressing (*all at one pressing*). □ **pressingly** *adv.*

press release *n.* statement issued to newspapers.

press-stud *n.* small fastening device engaged by pressing its two halves together.

press-up *n.* exercise in which the prone body is raised from the ground by placing the hands on the floor and straightening the arms.

pressure /'preʃə(r)/ *–n.* **1 a** exertion of continuous force on or against a body by another in contact with it. **b** force exerted. **c** amount of this (expressed by the force on a unit area) (*atmospheric pressure*). **2** urgency (*work under pressure*). **3** affliction or difficulty (*under financial pressure*). **4** constraining influence (*put pressure on us*). *–v.* **(-ring)** (often foll. by *into*) apply (esp. moral) pressure to; coerce; persuade. [Latin: related to PRESS[1]]

pressure-cooker *n.* airtight pan for cooking quickly under steam pressure. □ **pressure-cook** *v.*

pressure group *n.* group formed to influence public policy.

pressure point *n.* point where an artery can be pressed against a bone to inhibit bleeding.

pressurize *v.* (also **-ise**) (**-zing** or **-sing**) **1** (esp. as **pressurized** *adj.*) maintain normal atmospheric pressure in (an aircraft cabin etc.) at a high altitude. **2** raise to a high pressure. **3** pressure (a person). □ **pressurization** /-'zeɪʃ(ə)n/ *n.*

pressurized-water reactor *n.* nuclear reactor with water at high pressure as the coolant.

Prestel /'prestel/ *n. propr.* computerized visual information system operated by British Telecom. [from PRESS, TELECOMMUNICATION]

prestidigitator /ˌprestɪ'dɪdʒɪˌteɪtə(r)/ *n. formal* conjuror. □ **prestidigitation** /-'teɪʃ(ə)n/ *n.* [French: related to PRESTO, DIGIT]

prestige /pre'sti:ʒ/ *n.* **1** respect or reputation derived from achievements, power, associations, etc. **2** (*attrib.*) having or conferring prestige. □ **prestigious** /-'stɪdʒəs/ *adj.* [Latin *praestigiae* juggler's tricks]

presto /'prestəʊ/ *Mus.* *–adv.* & *adj.* in quick tempo. *–n.* (*pl.* **-s**) presto passage or movement. [Latin *praestus* quick]

prestressed /pri:'strest/ *adj.* (of concrete) strengthened by stretched wires within it.

presumably /prɪ'zju:məblɪ/ *adv.* as may reasonably be presumed.

presume /prɪ'zju:m/ *v.* (**-ming**) **1** (often foll. by *that*) suppose to be true; take for granted. **2** (often foll. by *to* + infin.) **a** take the liberty, be impudent enough (*presumed to question their authority*). **b** dare, venture (*may I presume to ask?*). **3** be presumptuous. **4** (foll. by *on*, *upon*) take advantage of or make unscrupulous use of (a person's good nature etc.). [Latin *praesumo*]

presumption /prɪ'zʌmpʃ(ə)n/ *n.* **1** arrogance, presumptuous behaviour. **2 a** presuming a thing to be true. **b** thing that is or may be presumed to be true. **3** ground for presuming. [Latin: related to PRESUME]

presumptive /prɪ'zʌmptɪv/ *adj.* giving grounds for presumption (*presumptive evidence*).

presumptuous /prɪ'zʌmptʃʊəs/ *adj.* unduly or overbearingly confident. □ **presumptuously** *adv.* **presumptuousness** *n.*

presuppose /ˌpri:sə'pəʊz/ *v.* (**-sing**) **1** assume beforehand. **2** imply. □ **presupposition** /-ˌsʌpə'zɪʃ(ə)n/ *n.*

pre-tax /pri:'tæks, 'pri:-/ *adj.* (of income etc.) before deduction of taxes.

pretence /prɪ'tens/ *n.* (*US* **pretense**) **1** pretending, make-believe. **2 a** pretext, excuse. **b** false show of intentions or motives. **3** (foll. by *to*) claim, esp. a false one (to merit etc.). **4** display; ostentation. [Anglo-Latin: related to PRETEND]

pretend /prɪ'tend/ *–v.* **1** claim or assert falsely so as to deceive (*pretend knowledge; pretended to be rich*). **2** imagine to oneself in play (*pretended it was night*). **3** (as **pretended** *adj.*) falsely claim to be such (*a pretended friend*). **4** (foll. by *to*) **a** lay claim to (a right or title etc.). **b** profess to have (a quality etc.). *–adj. colloq.* pretended; in pretence (*pretend money*). [Latin *praetendo*: related to TEND[1]]

pretender *n.* person who claims a throne, title, etc.

pretense *US* var. of PRETENCE.

pretension /prɪ'tenʃ(ə)n/ *n.* **1** (often foll. by *to*) **a** assertion of a claim. **b** justifiable claim. **2** pretentiousness. [medieval Latin: related to PRETEND]

pretentious /prɪ'tenʃəs/ *adj.* **1** making an excessive claim to merit or importance. **2** ostentatious. □ **pretentiously** *adv.* **pretentiousness** *n.*

preterite /'pretərɪt/ (*US* **preterit**) *Gram.* *–adj.* expressing a past action or state. *–n.* preterite tense or form. [Latin *praeteritum* past]

preternatural /ˌpri:tə'nætʃər(ə)l/ *adj.* extraordinary, exceptional; supernatural. [Latin *praeter* beyond]

pretext /'pri:tekst/ *n.* ostensible reason; excuse offered. [Latin *praetextus*: related to TEXT]

prettify /'prɪtɪˌfaɪ/ *v.* (**-ies, -ied**) make pretty, esp. in an affected way.

pretty /'prɪtɪ/ *–adj.* (**-ier, -iest**) **1** attractive in a delicate way (*pretty girl; pretty dress*). **2** fine or good of its kind. **3** *iron.* considerable, fine (*a pretty penny*). *–adv. colloq.* fairly, moderately. *–v.* (**-ies, -ied**) (often foll. by *up*) make pretty. □ **pretty much** (or **nearly** or **well**) *colloq.* almost; very nearly. □ **prettily** *adv.* **prettiness** *n.* [Old English]

pretty-pretty *adj. colloq.* too pretty.

pretzel /'prets(ə)l/ *n.* crisp knot-shaped salted biscuit. [German]

prevail /prɪ'veɪl/ *v.* **1** (often foll. by *against, over*) be victorious or gain mastery. **2** be the more usual or predominant. **3** exist or occur in general use or experience. **4** (foll. by *on, upon*) persuade. [Latin *praevaleo*: related to AVAIL]

prevalent /'prevələnt/ *adj.* **1** generally existing or occurring. **2** predominant. □ **prevalence** *n.* [related to PREVAIL]

prevaricate /prɪ'værɪ,keɪt/ *v.* (**-ting**) **1** speak or act evasively or misleadingly. **2** quibble, equivocate. □ **prevarication** /-'keɪʃ(ə)n/ *n.* **prevaricator** *n.* [Latin, = walk crookedly]

■ **Usage** *Prevaricate* is often confused with *procrastinate*, which means 'to defer or put off action'.

prevent /prɪ'vent/ *v.* (often foll. by *from* + verbal noun) stop from happening or doing something; hinder; make impossible (*the weather prevented me from going*). □ **preventable** *adj.* (also **preventible**). **prevention** *n.* [Latin *praevenio -vent-* hinder]

■ **Usage** The use of *prevent* without 'from' as in *prevented me going* is informal. An acceptable alternative is *prevented my going*.

preventative /prɪ'ventətɪv/ *adj.* & *n.* = PREVENTIVE.

preventive /prɪ'ventɪv/ *–adj.* serving to prevent, esp. disease. *–n.* preventive agent, measure, drug, etc.

preview /'pri:vju:/ *–n.* showing of a film, play, exhibition, etc., before it is seen by the general public. *–v.* see or show in advance.

previous /'pri:vɪəs/ *–adj.* **1** (often foll. by *to*) coming before in time or order. **2** *colloq.* hasty, premature. *–adv.* (foll. by *to*) before. □ **previously** *adv.* [Latin *praevius* from *via* way]

pre-war /pri:'wɔ:(r), 'pri:-/ *adj.* existing or occurring before a war.

prey /preɪ/ *–n.* **1** animal that is hunted or killed by another for food. **2** (often foll. by *to*) person or thing that is influenced by or vulnerable to (something undesirable) (*prey to morbid fears*). *–v.* (foll. by *on, upon*) **1** seek or take as prey. **2** (of a disease, emotion, etc.) exert a harmful influence (*it preyed on his mind*). [Latin *praeda*]

prezzie var. of PRESSIE.

price *–n.* **1** amount of money for which a thing is bought or sold. **2** what is or must be given, done, sacrificed, etc., to obtain or achieve something (*peace at any price*). **3** odds in betting. *–v.* (**-cing**) **1** fix or find the price of (a thing for sale). **2** estimate the value of. □ **at a price** at a high cost. **price on a person's head** reward for a person's capture or death. **what price ... ?** (often foll. by verbal noun) *colloq.* **1** what is the chance of ... ? (*what price your finishing the course?*). **2** *iron.* the much boasted ... proves disappointing (*what price your friendship now?*). [Latin *pretium*]

price-fixing *n.* maintaining of prices at a certain level by agreement between competing sellers.

priceless *adj.* **1** invaluable. **2** *colloq.* very amusing or absurd.

price tag *n.* **1** label on an item showing its price. **2** cost of an undertaking.

price war *n.* period of fierce competition among traders cutting prices.

pricey *adj.* (**-cier, -ciest**) *colloq.* expensive.

prick *–v.* **1** pierce slightly; make a small hole in. **2** (foll. by *off, out*) mark with small holes or dots. **3** trouble mentally (*my conscience pricked me*). **4** tingle. **5** (foll. by *out*) plant (seedlings etc.) in small holes pricked in the soil. *–n.* **1** act of pricking. **2** small hole or mark made by pricking. **3** pain caused as by pricking. **4** mental pain. **5** *coarse slang* **a** penis. **b** *derog.* contemptible man. □ **prick up one's ears 1** (of a dog etc.) make the ears erect when alert. **2** (of a person) become suddenly attentive. [Old English]

prickle /'prɪk(ə)l/ *–n.* **1** small thorn. **2** hard-pointed spine of a hedgehog etc. **3** prickling sensation. *–v.* (**-ling**) affect or be affected with a sensation of multiple pricking. [Old English]

prickly *adj.* (**-ier, -iest**) **1** having prickles. **2** (of a person) ready to take offence. **3** tingling. □ **prickliness** *n.*

prickly heat *n.* itchy inflammation of the skin, causing a tingling sensation and common in hot countries.

prickly pear *n.* **1** cactus with pear-shaped prickly fruit. **2** its fruit.

pride *–n.* **1 a** elation or satisfaction at one's achievements, qualities, possessions, etc. **b** object of this feeling; the flower or best. **2** high or overbearing opinion of one's worth or importance. **3** (in full **proper pride**) proper sense of what befits one's position; self-respect. **4** group (of certain animals, esp. lions). **5** best condition, prime. *–v.refl.* (**-ding**) (foll. by *on, upon*) be proud of. □ **take pride** (or **a pride**) **in 1** be proud of. **2** maintain in good condition or appearance. [Old English: related to PROUD]

pride of place *n.* most important or prominent position.

prie-dieu /pri:'djɜ:/ *n.* (*pl.* **prie-dieux** pronunc. same) kneeling-desk for prayer. [French, = pray God]

priest *n.* **1** ordained minister of the Roman Catholic or Orthodox Church, or of the Anglican Church (above a deacon and below a bishop). **2** (*fem.* **priestess**) official minister of a non-

Christian religion. □ **priesthood** *n.* **priestly** *adj.* [Latin PRESBYTER]

prig *n.* self-righteous or moralistic person. □ **priggish** *adj.* **priggishness** *n.* [origin unknown]

prim *adj.* (**primmer**, **primmest**) stiffly formal and precise; prudish. □ **primly** *adv.* **primness** *n.* [French: related to PRIME[1]]

prima ballerina /ˌpri:mə ˌbælə'ri:nə/ *n.* chief female dancer in a ballet. [Italian]

primacy /'praɪməsɪ/ *n.* (*pl.* **-ies**) **1** pre-eminence. **2** office of a primate. [Latin: related to PRIMATE]

prima donna /ˌpri:mə 'dɒnə/ *n.* (*pl.* **prima donnas**) **1** chief female singer in an opera. **2** temperamentally self-important person. □ **prima donna-ish** *adj.* [Italian]

prima facie /ˌpraɪmə 'feɪʃi:/ *–adv.* at first sight. *–adj.* (of evidence) based on the first impression. [Latin]

primal /'praɪm(ə)l/ *adj.* **1** primitive, primeval. **2** chief, fundamental. [Latin: related to PRIME[1]]

primary /'praɪmərɪ/ *–adj.* **1 a** of the first importance; chief. **b** fundamental, basic. **2** earliest, original; first in a series. **3** of the first rank in a series; not derived. **4** designating any of the colours red, green, and blue, or (for pigments) red, blue, and yellow, of which all other colours are mixtures. **5** (of education) for children below the age of 11. **6** (**Primary**) *Geol.* of the lowest series of strata. **7** *Biol.* of the first stage of development. *–n.* (*pl.* **-ies**) **1** thing that is primary. **2** (in full **primary election**) (in the US) preliminary election to appoint party conference delegates or to select candidates for a principal (esp. presidential) election. **3** = PRIMARY FEATHER. □ **primarily** /'praɪmərɪlɪ, -'meərɪlɪ/ *adv.* [Latin: related to PRIME[1]]

primary feather *n.* large flight-feather of a bird's wing.

primary school *n.* school for children below the age of 11.

primate /'praɪmeɪt/ *n.* **1** member of the highest order of mammals, including apes, monkeys, and man. **2** (also /-mət/) archbishop. [Latin *primas -at-* chief]

prime[1] *–adj.* **1** chief, most important. **2** first-rate, excellent. **3** primary, fundamental. **4** *Math.* **a** (of a number etc.) divisible only by itself and unity (e.g. 2, 3, 5, 7, 11). **b** (of numbers) having no common factor but unity. *–n.* **1** state of the highest perfection (*prime of life*). **2** (prec. by *the*; foll. by *of*) the best part. [Latin *primus* first]

prime[2] *v.* (**-ming**) **1** prepare (a thing) for use or action. **2** prepare (a gun) for firing or (an explosive) for detonation. **3** pour liquid into a (pump) to enable it to work. **4** prepare (wood etc.) for painting by applying a substance that prevents paint from being absorbed. **5** equip (a person) with information etc. **6** ply (a person) with food or drink in preparation for something. [origin unknown]

prime minister *n.* head of an elected government; principal minister.

primer[1] *n.* substance used to prime wood etc.

primer[2] *n.* **1** elementary textbook for teaching children to read. **2** introductory book. [Latin: related to PRIME[1]]

prime time *n.* (in broadcasting) time when audiences are largest.

primeval /praɪ'mi:v(ə)l/ *adj.* **1** of the first age of the world. **2** ancient, primitive. □ **primevally** *adv.* [Latin: related to PRIME[1], *aevum* age]

primitive /'prɪmɪtɪv/ *–adj.* **1** at an early stage of civilization (*primitive man*). **2** undeveloped, crude, simple (*primitive methods*). *–n.* **1** untutored painter with a direct naïve style. **2** picture by such a painter. □ **primitively** *adv.* **primitiveness** *n.* [Latin: related to PRIME[1]]

primogeniture /ˌpraɪməʊ'dʒenɪtʃə(r)/ *n.* **1** fact of being the first-born child. **2** (in full **right of primogeniture**) right of succession belonging to the first-born. [medieval Latin: related to PRIME[1], Latin *genitura* birth]

primordial /praɪ'mɔ:dɪəl/ *adj.* existing at or from the beginning, primeval. [Latin: related to PRIME[1], *ordior* begin]

primp *v.* **1** make (the hair, clothes, etc.) tidy. **2** *refl.* make (oneself) smart. [var. of PRIM]

primrose /'prɪmrəʊz/ *n.* **1 a** wild plant bearing pale yellow spring flowers. **b** its flower. **2** pale yellow colour. [French and medieval Latin, = first rose]

primrose path *n.* pursuit of pleasure.

primula /'prɪmjʊlə/ *n.* cultivated plant bearing primrose-like flowers in a wide variety of colours. [Latin diminutive: related to PRIME[1]]

Primus /'praɪməs/ *n. propr.* portable cooking stove burning vaporized oil. [Latin, = first]

prince *n.* (as a title usu. **Prince**) **1** male member of a royal family other than the reigning king. **2** ruler of a small State. **3** noble man in some countries. **4** (often foll. by *of*) chief or greatest (*the prince of novelists*). [Latin *princeps -cip-*]

Prince Consort *n.* (title conferred on) the husband of a reigning queen who is himself a prince.

princeling /'prɪnslɪŋ/ *n.* young or petty prince.

princely *adj.* (**-ier, iest**) **1** of or worthy of a prince. **2** sumptuous, generous, splendid.

Prince of Wales *n.* (title conferred on) the eldest son and heir apparent of the British monarch.

Prince Regent *n.* prince who acts as regent, esp. the future George IV.

princess /prɪn'ses/ *n.* (as a title usu. **Princess** /'prɪnses/) **1** wife of a prince. **2** female member of a royal family other than a queen. [French: related to PRINCE]

Princess Royal *n.* (title conferred on) the British monarch's eldest daughter.

principal /'prɪnsɪp(ə)l/ *–adj.* **1** (usu. *attrib.*) first in rank or importance; chief. **2** main, leading. *–n.* **1** chief person. **2** head of some schools, colleges, and universities. **3** leading performer in a concert, play, etc. **4** capital sum as distinct from interest or income. **5** person for whom another is agent etc. **6** civil servant of the grade below Secretary. **7** person directly responsible for a crime. □ **principally** *adv.* [Latin: related to PRINCE]

principal boy *n.* leading male role in a pantomime, usu. played by a woman.

principality /,prɪnsɪ'pælɪtɪ/ *n.* (*pl.* **-ies**) **1** State ruled by or government of a prince. **2** (**the Principality**) Wales.

principal parts *n.pl. Gram.* parts of a verb from which all other parts can be deduced.

principle /'prɪnsɪp(ə)l/ *n.* **1** fundamental truth or law as the basis of reasoning or action. **2 a** personal code of conduct (*person of high principle*). **b** (in *pl.*) personal rules of conduct (*has no principles*). **3** general law in physics etc. **4** law of nature forming the basis for the construction or working of a machine etc. **5** fundamental source; primary element. □ **in principle** in theory. **on principle** on the basis of a moral code. [Latin *principium* source]

principled *adj.* based on or having (esp. praiseworthy) principles of behaviour.

prink *v.* **1 a** (usu. *refl.*; often foll. by *up*) smarten (oneself) up. **b** dress oneself up. **2** (of a bird) preen. [origin unknown]

print *–v.* **1** produce or cause (a book, picture, etc.) to be produced by applying inked types, blocks, or plates, to paper, etc. **2** express or publish in print. **3 a** (often foll. by *on, in*) impress or stamp (a mark on a surface). **b** (often foll. by *with*) impress or stamp (a surface with a seal, die, etc.). **4** (often *absol.*) write (letters) without joining them up. **5** (often foll. by *off, out*) produce (a photograph) from a negative. **6** (usu. foll. by *out*) (of a computer etc.) produce output in printed form. **7** mark (a textile fabric) with a coloured design. **8** (foll. by *on*) impress (an idea, scene, etc. on the mind or memory). *–n.* **1** indentation or mark on a surface left by the pressure of a thing in contact with it. **2 a** printed lettering or writing. **b** words in printed form. **c** printed publication, esp. a newspaper. **3** picture or design printed from a block or plate. **4** photograph produced on paper from a negative. **5** printed cotton fabric. □ **in print 1** (of a book etc.) available from the publisher. **2** in printed form. **out of print** no longer available from the publisher. [Latin *premo*: related to PRESS[1]]

printed circuit *n.* electric circuit with thin strips of conductor printed on a flat insulating sheet.

printer *n.* **1** person who prints books etc. **2** owner of a printing business. **3** device that prints, esp. from a computer.

printing *n.* **1** production of printed books etc. **2** copies of a book printed at one time. **3** printed letters or writing imitating them.

printing-press *n.* machine for printing from types or plates etc.

printout *n.* computer output in printed form.

prior /'praɪə(r)/ *–adj.* **1** earlier. **2** (often foll. by *to*) coming before in time, order, or importance. *–adv.* (foll. by *to*) before (*left prior to his arrival*). *–n.* (*fem.* **prioress**) **1** superior of a religious house or order. **2** (in an abbey) deputy of an abbot. [Latin, = earlier]

priority /praɪ'ɒrɪtɪ/ *n.* (*pl.* **-ies**) **1** thing that is regarded as more important than others. **2** high(est) place among various things to be done (*gave priority to*). **3** right to do something before other people. **4** right to proceed ahead of other traffic. **5** (state of) being more important. □ **prioritize** *v.* (also **-ise**) (**-zing** or **-sing**). [medieval Latin: related to PRIOR]

priory /'praɪərɪ/ *n.* (*pl.* **-ies**) monastery governed by a prior or nunnery governed by a prioress. [Anglo-French and medieval Latin: related to PRIOR]

prise *v.* (also **prize**) (**-sing** or **-zing**) force open or out by leverage. [French: related to PRIZE[2]]

prism /'prɪz(ə)m/ *n.* **1** solid figure whose two ends are equal parallel rectilinear figures, and whose sides are parallelograms. **2** transparent body in this form, usu. triangular with refracting surfaces at an acute angle with each other, which separates white light into a spectrum of colours. [Greek *prisma -mat-* thing sawn]

prismatic /prɪz'mætɪk/ *adj.* **1** of, like, or using a prism. **2** (of colours) distributed (as if) by a transparent prism. [Greek: related to PRISM]

prison /'prɪz(ə)n/ *n.* **1** place of captivity, esp. a building to which persons are committed while awaiting trial or for punishment. **2** custody, confinement. [Latin *prehendo* seize]

prisoner /'prɪznə(r)/ *n.* **1** person kept in prison. **2** (in full **prisoner at the bar**) person in custody on a criminal charge and on trial. **3** person or thing confined by illness, another's grasp, etc. **4** (in full **prisoner of war**) person captured in war. □ **take prisoner** seize and hold as a prisoner. [Anglo-French: related to PRISON]

prisoner of conscience see CONSCIENCE.

prissy /'prɪsɪ/ *adj.* **(-ier, -iest)** prim, prudish. □ **prissily** *adv.* **prissiness** *n.* [perhaps from PRIM, SISSY]

pristine /'prɪsti:n/ *adj.* **1** in its original condition; unspoilt. **2** spotless; fresh as if new. **3** ancient, primitive. [Latin *pristinus* former]

■ **Usage** The use of *pristine* in sense 2 is considered incorrect by some people.

privacy /'prɪvəsɪ/ *n.* **1 a** being private and undisturbed. **b** right to this. **2** freedom from intrusion or public attention.

private /'praɪvət/ *–adj.* **1** belonging to an individual, one's own, personal (*private property*). **2** confidential, not to be disclosed to others (*private talks*). **3** kept or removed from public knowledge or observation. **4** not open to the public. **5** (of a place) secluded. **6** (of a person) not holding public office or an official position. **7** (of education or medical treatment) conducted outside the State system, at the individual's expense. *–n.* **1** private soldier. **2** (in *pl.*) *colloq.* genitals. □ **in private** privately. □ **privately** *adv.* [Latin *privo* deprive]

private bill *n.* parliamentary bill affecting an individual or corporation only.

private company *n.* company with restricted membership and no public share issue.

private detective *n.* detective engaged privately, outside an official police force.

private enterprise *n.* businesses not under State control.

privateer /,praɪvə'tɪə(r)/ *n.* **1** privately owned and officered warship holding a government commission. **2** its commander.

private eye *n. colloq.* private detective.

private hotel *n.* hotel not obliged to take all comers.

private means *n.pl.* income from investments etc., apart from earned income.

private member *n.* MP not holding government office.

private member's bill *n.* bill introduced by a private member, not part of government legislation.

private parts *n.pl. euphem.* genitals.

private sector *n.* the part of the economy free of direct State control.

private soldier *n.* ordinary soldier other than the officers.

private view *n.* viewing of an exhibition (esp. of paintings) before it opens to the public.

privation /praɪ'veɪʃ(ə)n/ *n.* lack of the comforts or necessities of life. [Latin: related to PRIVATE]

privative /'prɪvətɪv/ *adj.* **1** consisting in or marked by loss or absence. **2** *Gram.* expressing privation. [Latin: related to PRIVATE]

privatize /'praɪvə,taɪz/ *v.* (also **-ise**) **(-zing** or **-sing)** transfer (a business etc.) from State to private ownership. □ **privatization** /-'zeɪʃ(ə)n/ *n.*

privet /'prɪvɪt/ *n.* bushy evergreen shrub used for hedges. [origin unknown]

privilege /'prɪvɪlɪdʒ/ *–n.* **1** right, advantage, or immunity, belonging to a person, class, or office. **2** special benefit or honour (*a privilege to meet you*). *–v.* **(-ging)** invest with a privilege. □ **privileged** *adj.* [Latin: related to PRIVY, *lex leg-* law]

privy /'prɪvɪ/ *–adj.* **1** (foll. by *to*) sharing in the secret of (a person's plans etc.). **2** *archaic* hidden, secret. *–n.* (*pl.* **-ies**) lavatory, esp. an outside one. [French *privé* private place]

Privy Council *n.* body of advisers appointed by the sovereign (now chiefly honorary). □ **Privy Councillor** *n.* (also **Privy Counsellor**).

privy purse *n.* allowance from the public revenue for the monarch's private expenses.

privy seal *n.* seal formerly affixed to minor State documents.

prize[1] *–n.* **1** something that can be won in a competition, lottery, etc. **2** reward given as a symbol of victory or superiority. **3** something striven for or worth striving for. **4** (*attrib.*) **a** to which a prize is awarded (*prize poem*). **b** excellent of its kind. *–v.* **(-zing)** value highly (*a much prized possession*). [French: related to PRAISE]

prize[2] *n.* ship or property captured in naval warfare. [French *prise* from Latin *prehendo* seize]

prize[3] var. of PRISE.

prizefight *n.* boxing-match fought for a prize of money. □ **prizefighter** *n.*

prize-giving *n.* awarding of prizes, esp. formally at a school etc.

prizewinner *n.* winner of a prize. □ **prizewinning** *attrib. adj.*

PRO *abbr.* **1** Public Record Office. **2** public relations officer.

pro[1] /prəʊ/ *n.* (*pl.* **-s**) *colloq.* professional. [abbreviation]

pro[2] /prəʊ/ *–adj.* (of an argument or reason) for; in favour. *–n.* (*pl.* **-s**) reason in favour. *–prep.* in favour of. [Latin, = for, on behalf of]

pro-[1] *prefix* **1** favouring or supporting (*pro-government*). **2** acting as a substitute or deputy for (*proconsul*). **3** forwards (*produce*). **4** forwards and downwards (*prostrate*). **5** onwards (*progress*). **6** in front of (*protect*). [Latin *pro* in front (of)]

pro-[2] *prefix* before in time, place, order, etc. [Greek *pro* before]

proactive /prəʊ'æktɪv/ *adj.* (of a person, policy, etc.) taking the initiative. [from PRO-[2], after REACTIVE]

probability /,prɒbə'bɪlɪtɪ/ *n.* (*pl.* **-ies**) **1** being probable. **2** likelihood of something happening. **3** probable or most probable event. **4** *Math.* extent to which an event is likely to occur, measured by the ratio of the favourable cases to the total number of possible cases. □ **in all probability** most probably.

probable /'prɒbəb(ə)l/ *–adj.* (often foll. by *that*) that may be expected to happen or prove true; likely (*the probable explanation*; *it is probable that they forgot*). *–n.* probable candidate, member of a team, etc. □ **probably** *adv.* [Latin: related to PROVE]

probate /'prəʊbeɪt/ *n.* **1** official proving of a will. **2** verified copy of a will with a certificate as handed to executors. [Latin *probo* PROVE]

probation /prə'beɪʃ(ə)n/ *n.* **1** *Law* system of supervising and monitoring the behaviour of (esp. young) offenders, as an alternative to prison. **2** period of testing the character or abilities of esp. a new employee. □ **on probation** undergoing probation. □ **probationary** *adj.* [Latin: related to PROVE]

probationer *n.* person on probation.

probation officer *n.* official supervising offenders on probation.

probative /'prəʊbətɪv/ *adj. formal* affording proof. [Latin: related to PROVE]

probe *–n.* **1** penetrating investigation. **2** small device, esp. an electrode, for measuring, testing, etc. **3** blunt-ended surgical instrument for exploring a wound etc. **4** (in full **space probe**) unmanned exploratory spacecraft transmitting information about its environment. *–v.* (**-bing**) **1** examine or enquire into closely. **2** explore with a probe. [Latin *proba*: related to PROVE]

probity /'prəʊbɪtɪ/ *n.* uprightness, honesty. [Latin *probus* good]

problem /'prɒbləm/ *n.* **1** doubtful or difficult matter requiring a solution. **2** something hard to understand or accomplish. **3** (*attrib.*) causing problems (*problem child*). **4** puzzle or question for solution; exercise. [Greek *problēma -mat-*]

problematic /,prɒblə'mætɪk/ *adj.* (also **problematical**) attended by difficulty; doubtful or questionable. □ **problematically** *adv.* [Greek: related to PROBLEM]

proboscis /prəʊ'bɒsɪs/ *n.* (*pl.* **-sces**) **1** long flexible trunk or snout of some mammals, e.g. an elephant or tapir. **2** elongated mouth parts of some insects. [Greek *boskō* feed]

proboscis monkey *n.* monkey of Borneo, the male of which has a large pendulous nose.

procedure /prə'siːdʒə(r)/ *n.* **1** way of acting or advancing, esp. in business or legal action. **2** way of performing a task. **3** series of actions conducted in a certain order or manner. □ **procedural** *adj.* [French: related to PROCEED]

proceed /prə'siːd/ *v.* **1** (often foll. by *to*) go forward or on further; make one's way. **2** (often foll. by *with*, or *to* + infin.) continue with an activity; go on to do something (*proceeded with their work*; *proceeded to beg me*). **3** (of an action) be carried on or continued (*the case will now proceed*). **4** adopt a course of action (*how shall we proceed?*). **5** go on to say. **6** (foll. by *against*) start a lawsuit against (a person). **7** (often foll. by *from*) originate (*trouble proceeded from illness*). [Latin *cedo cess-* go]

proceeding *n.* **1** action or piece of conduct (*high-handed proceeding*). **2** (in *pl.*) (in full **legal proceedings**) lawsuit. **3** (in *pl.*) published report of discussions or a conference.

proceeds /'prəʊsiːdz/ *n.pl.* profits from sale etc. [pl. of obsolete *proceed* (n.) from PROCEED]

process[1] /'prəʊses/ *–n.* **1** course of action or proceeding, esp. a series of stages in manufacture etc. **2** progress or course (*in process of construction*). **3** natural or involuntary course or change (*process of growing old*). **4** action at law; summons or writ. **5** natural projection of a bone, stem, etc. **–***v.* **1** deal with by a particular process. **2** (as **processed** *adj.*) treat (food, esp. to prevent decay) (*processed cheese*). [Latin: related to PROCEED]

process[2] /prə'ses/ *v.* walk in procession. [back-formation from PROCESSION]

procession /prə'seʃ(ə)n/ *n.* **1** people or vehicles etc. advancing in orderly succession, esp. at a ceremony, demonstration, or festivity. **2** movement of such a group (*go in procession*). [Latin: related to PROCEED]

processional **–***adj.* **1** of processions. **2** used, carried, or sung in processions. **–***n. Eccl.* processional hymn or hymn book.

processor /'prəʊsesə(r)/ *n.* machine that processes things, esp.: **1** = CENTRAL PROCESSOR. **2** = FOOD PROCESSOR.

proclaim /prə'kleɪm/ *v.* **1** (often foll. by *that*) announce or declare publicly or officially. **2** declare to be (king, a traitor, etc.).□ **proclamation** /,prɒklə'meɪʃ(ə)n/ *n.* [Latin: related to CLAIM]

proclivity /prə'klɪvɪtɪ/ *n.* (*pl.* **-ies**) tendency, inclination. [Latin *clivus* slope]

procrastinate /prəʊ'kræstɪ,neɪt/ *v.* (**-ting**) defer action. □ **procrastination** /-'neɪʃ(ə)n/ *n.* **procrastinator** *n.* [Latin *cras* tomorrow]

■ **Usage** *Procrastinate* is often confused with *prevaricate*, which means 'to be evasive, quibble'.

procreate /'prəʊkrɪ,eɪt/ *v.* (**-ting**) (often *absol.*) produce (offspring) naturally. □ **procreation** /-'eɪʃ(ə)n/ *n.* **procreative** *adj.* [Latin: related to CREATE]

Procrustean /prəʊ'krʌstɪən/ *adj.* seeking to enforce uniformity ruthlessly or violently. [Greek *Prokroustēs*, name of a robber who fitted his victims to a bed by stretching them or cutting bits off them]

proctor /'prɒktə(r)/ *n.* disciplinary officer (usu. one of two) at certain universities. □ **proctorial** /-'tɔ:rɪəl/ *adj.* **proctorship** *n.* [from PROCURATOR]

procuration /,prɒkjʊ'reɪʃ(ə)n/ *n.* **1** *formal* act of procuring. **2** function or authorized action of an attorney. [Latin: related to PROCURE]

procurator /'prɒkjʊ,reɪtə(r)/ *n.* agent or proxy, esp. with power of attorney. [Latin *procurator* agent]

procurator fiscal *n.* (in Scotland) local coroner and public prosecutor.

procure /prə'kjʊə(r)/ *v.* (**-ring**) **1** obtain, esp. by care or effort; acquire (*managed to procure a copy*). **2** bring about (*procured their dismissal*). **3** (also *absol.*) obtain (women) for prostitution. □ **procurement** *n.* [Latin *curo* look after]

procurer *n.* (*fem.* **procuress**) person who obtains women for prostitution. [Latin PROCURATOR]

prod **–***v.* (**-dd-**) **1** poke with a finger, stick, etc. **2** stimulate to action. **3** (foll. by *at*) make a prodding motion. **–***n.* **1** poke, thrust. **2** stimulus to action. [origin unknown]

prodigal /'prɒdɪg(ə)l/ **–***adj.* **1** recklessly wasteful. **2** (foll. by *of*) lavish. **–***n.* **1** prodigal person. **2** (in full **prodigal son**) repentant wastrel, returned wanderer, etc. (Luke 15:11–32). □ **prodigality** /-'gælɪtɪ/ *n.* [Latin *prodigus* lavish]

prodigious /prə'dɪdʒəs/ *adj.* **1** marvellous or amazing. **2** enormous. **3** abnormal. [Latin: related to PRODIGY]

prodigy /'prɒdɪdʒɪ/ *n.* (*pl.* **-ies**) **1** exceptionally gifted or able person, esp. a precocious child. **2** marvellous, esp. extraordinary, thing. **3** (foll. by *of*) wonderful example (of a quality). [Latin *prodigium* portent]

produce **–***v.* /prə'dju:s/ (**-cing**) **1** manufacture or prepare (goods etc.). **2** bring forward for consideration, inspection, or use (*will produce evidence*). **3** bear, yield, or bring into existence (offspring, fruit, a harvest, etc.). **4** cause or bring about (a reaction, sensation, etc.). **5** *Geom.* extend or continue (a line). **6** supervise the production of (a play, film, broadcast, record, etc.). **–***n.* /'prɒdju:s/ **1 a** what is produced, esp. agricultural products collectively (*dairy produce*). **b** amount of this. **2** (often foll. by *of*) result (of labour, efforts, etc.). □ **producible** /prə'dju:sɪb(ə)l/ *adj.* [Latin *duco duct-* lead]

producer /prə'dju:sə(r)/ *n.* **1** person who produces goods etc. **2** person who supervises the production of a play, film, broadcast, etc.

product /'prɒdʌkt/ *n.* **1** thing or substance produced, esp. by manufacture. **2** result. **3** quantity obtained by multiplying. [Latin: related to PRODUCE]

production /prə'dʌkʃ(ə)n/ *n.* **1** producing or being produced, esp. in large quantities (*go into production*). **2** total yield. **3** thing produced, esp. a film, play, book, etc. [Latin: related to PRODUCE]

production line *n.* systematized sequence of operations involved in producing a commodity.

productive /prə'dʌktɪv/ *adj.* **1** of or engaged in the production of goods. **2** producing much (*productive writer*). **3** producing commodities of exchangeable value (*productive labour*). **4** (foll.

by *of*) producing or giving rise to (*productive of great annoyance*). □ **productively** *adv.* **productiveness** *n.* [Latin: related to PRODUCE]

productivity /ˌprɒdʌk'tɪvɪtɪ/ *n.* **1** being productive, capacity to produce. **2** amount produced by an industry, workforce, etc.

proem /'prəʊɪm/ *n.* preface etc. to a book or speech. [Latin from Greek]

Prof. *abbr.* Professor.

profane /prə'feɪn/ *–adj.* **1 a** irreverent, blasphemous. **b** (of language) obscene. **2** not sacred or biblical; secular. *–v.* (**-ning**) **1** treat (esp. a sacred thing) irreverently; disregard. **2** violate or pollute. □ **profanation** /ˌprɒfə'neɪʃ(ə)n/ *n.* [Latin *fanum* temple]

profanity /prə'fænɪtɪ/ *n.* (*pl.* **-ies**) **1** profane act or language; blasphemy. **2** swear-word.

profess /prə'fes/ *v.* **1** claim openly to have (a quality or feeling). **2** (often foll. by *to* + infin.) pretend, declare (*profess ignorance*). **3** affirm one's faith in or allegiance to. [Latin *profiteor -fess-* declare]

professed *adj.* **1** self-acknowledged (*professed Christian*). **2** alleged, ostensible. □ **professedly** /-sɪdlɪ/ *adv.*

profession /prə'feʃ(ə)n/ *n.* **1** vocation or calling, esp. learned or scientific (*medical profession*). **2** people in a profession. **3** declaration or avowal. □ **the oldest profession** *colloq.* prostitution.

professional *–adj.* **1** of, belonging to, or connected with a profession. **2 a** skilful, competent. **b** worthy of a professional (*professional conduct*). **3** engaged in a specified activity as one's main paid occupation (*professional boxer*). **4** *derog.* engaged in a specified activity, esp. fanatically (*professional agitator*). *–n.* professional person. □ **professionally** *adv.*

professionalism *n.* qualities associated with a profession, esp. competence, skill, etc.

professor /prə'fesə(r)/ *n.* **1 a** (often as a title) highest-ranking academic teaching in a university department; holder of a university chair. **b** *US* university teacher. **2** person who professes a religion etc. □ **professorial** /ˌprɒfɪ'sɔ:rɪəl/ *adj.* **professorship** *n.*

proffer *v.* offer. [French: related to PRO-[1], OFFER]

proficient /prə'fɪʃ(ə)nt/ *adj.* (often foll. by *in*, *at*) adept, expert. □ **proficiency** *n.* **proficiently** *adv.* [Latin *proficio -fect-* advance]

profile /'prəʊfaɪl/ *–n.* **1 a** outline, esp. of a human face, as seen from one side. **b** representation of this. **2** short biographical or character sketch. *–v.* (**-ling**) represent or describe by a profile. □ **keep a low profile** remain inconspicuous. [Italian *profilare* draw in outline]

profit /'prɒfɪt/ *–n.* **1** advantage or benefit. **2** financial gain; excess of returns over outlay. *–v.* (**-t-**) **1** (also *absol.*) be beneficial to. **2** obtain advantage or benefit (*profited by the experience*). □ **at a profit** with financial gain. [Latin *profectus*: related to PROFICIENT]

profitable *adj.* **1** yielding profit. **2** beneficial. □ **profitability** /-'bɪlɪtɪ/ *n.* **profitably** *adv.*

profit and loss account *n.* account showing net profit or loss at any time.

profiteer /ˌprɒfɪ'tɪə(r)/ *–v.* make or seek excessive profits, esp. illegally or on the black market. *–n.* person who profiteers.

profiterole /prə'fɪtəˌrəʊl/ *n.* small hollow choux bun, usu. filled with cream and covered with chocolate. [French diminutive: related to PROFIT]

profit margin *n.* profit after the deduction of costs.

profit-sharing *n.* sharing of profits, esp. between employer and employees.

profligate /'prɒflɪgət/ *–adj.* **1** recklessly extravagant. **2** licentious, dissolute. *–n.* profligate person. □ **profligacy** *n.* **profligately** *adv.* [Latin *profligo* ruin]

pro forma /prəʊ 'fɔ:mə/ *–adv. & adj.* as or being a matter of form. *–n.* (in full **pro-forma invoice**) invoice sent in advance of goods supplied. [Latin]

profound /prə'faʊnd/ *adj.* (**-er**, **-est**) **1** having or demanding great knowledge, study, or insight (*profound treatise*; *profound doctrines*). **2** intense, unqualified, thorough (*a profound sleep*; *profound indifference*). **3** deep (*profound crevasses*). □ **profoundly** *adv.* **profoundness** *n.* **profundity** /prə'fʌndɪtɪ/ *n.* (*pl.* **-ies**). [Latin *profundus*]

profuse /prə'fju:s/ *adj.* **1** (often foll. by *in*, *of*) lavish; extravagant. **2** exuberantly plentiful; copious (*profuse variety*). □ **profusely** *adv.* **profusion** /-'fju:ʒ(ə)n/ *n.* [Latin *fundo fus-* pour]

progenitor /prəʊ'dʒenɪtə(r)/ *n.* **1** ancestor. **2** predecessor. **3** original. [Latin *progigno* beget]

progeny /'prɒdʒɪnɪ/ *n.* **1** offspring; descendant(s). **2** outcome, issue. [Latin: related to PROGENITOR]

progesterone /prəʊ'dʒestəˌrəʊn/ *n.* a steroid hormone which stimulates the preparation of the uterus for pregnancy and maintains the uterus in the event of fertilization. [German: related to PRO-[1], GESTATION]

progestogen /prəʊ'dʒestədʒɪn/ *n.* **1** a steroid hormone (e.g. progesterone) maintaining pregnancy and preventing further ovulation. **2** similar synthetic hormone.

prognosis /prɒg'nəʊsɪs/ *n.* (*pl.* **-noses** /-si:z/) forecast, esp. of the course of a disease. [Greek *gignōskō* know]

prognostic /prɒg'nɒstɪk/ *–n.* **1** (often foll. by *of*) advance indication, esp. of the course of a disease. **2** prediction, forecast. *–adj.* (often foll. by *of*) foretelling, predictive. [Latin: related to PROGNOSIS]

prognosticate /prɒg'nɒstɪ,keɪt/ *v.* (**-ting**) **1** (often foll. by *that*) foretell, foresee, prophesy. **2** (of a thing) betoken, indicate. □ **prognostication** /-'keɪʃ(ə)n/ *n.* **prognosticator** *n.* [medieval Latin: related to PROGNOSTIC]

programme /'prəʊgræm/ (*US* **program**) *–n.* **1** list of events, performers, etc. at a public function etc. **2** radio or television broadcast. **3** plan of events (*programme is dinner and an early night*). **4** course or series of studies, lectures, etc. **5** (usu. **program**) series of coded instructions for a computer etc. *–v.* (**-mm-**; *US* **-m-**) **1** make a programme of. **2** (usu. **program**) express (a problem) or instruct (a computer) by means of a program. □ **programmable** *adj.* **programmatic** /-grə'mætɪk/ *adj.* **programmer** *n.* (in sense 5 of *n.*). [Greek *graphō* write]

progress *–n.* /'prəʊgres/ **1** forward or onward movement towards a destination. **2** advance or development; improvement (*made little progress*). **3** *hist.* State tour, esp. by royalty. *–v.* /prə'gres/ **1** move or be moved forward or onward; continue. **2** advance, develop, or improve (*science progresses*). □ **in progress** developing; going on. [Latin *progredior -gress-* go forward]

progression /prə'greʃ(ə)n/ *n.* **1** progressing. **2** succession; series. [Latin: related to PROGRESS]

progressive /prə'gresɪv/ *–adj.* **1** moving forward. **2** proceeding step by step; cumulative (*progressive drug use*). **3 a** favouring rapid political or social reform. **b** modern; efficient (*a progressive company*). **4** (of disease, violence, etc.) increasing in severity or extent. **5** (of taxation) increasing with the sum taxed. **6** (of a card-game, dance, etc.) with periodic changes of partners. **7** *Gram.* (of a tense) expressing action in progress, e.g. *am writing, was writing*. *–n.* (also **Progressive**) advocate of progressive political policies. □ **progressively** *adv.* [French or medieval Latin: related to PROGRESS]

prohibit /prə'hɪbɪt/ *v.* (**-t-**) (often foll. by *from* + verbal noun) **1** forbid. **2** prevent. □ **prohibitor** *n.* **prohibitory** *adj.* [Latin *prohibeo -hibit-*]

prohibited degrees var. of FORBIDDEN DEGREES.

prohibition /,prəʊhɪ'bɪʃ(ə)n, ,prəʊɪ'b-/ *n.* **1** forbidding or being forbidden. **2** edict or order that forbids. **3** (usu. **Prohibition**) legal ban on the manufacture and sale of alcohol, esp. in the US (1920–33). □ **prohibitionist** *n.* (in sense 3).

prohibitive /prə'hɪbɪtɪv/ *adj.* **1** prohibiting. **2** (of prices, taxes, etc.) extremely high (*prohibitive price*). □ **prohibitively** *adv.*

project *–n.* /'prɒdʒekt/ **1** plan; scheme. **2** extensive essay, piece of research, etc. by a student. *–v.* /prə'dʒekt/ **1** protrude; jut out. **2** throw; cast; impel. **3** extrapolate (results etc.) to a future time; forecast. **4** plan or contrive (a scheme etc.). **5** cause (light, shadow, images, etc.) to fall on a surface. **6** cause (a sound, esp. the voice) to be heard at a distance. **7** (often *refl.* or *absol.*) express or promote forcefully or effectively. **8** make a projection of (the earth, sky, etc.). **9 a** (also *absol.*) attribute (an emotion etc.) to an external object or person, esp. unconsciously. **b** (*refl.*) imagine (oneself) having another's feelings, being in the future, etc. [Latin *projicio -ject-* throw forth]

projectile /prə'dʒektaɪl/ *–n.* **1** missile, esp. fired by a rocket. **2** bullet, shell, etc. *–adj.* **1** capable of being projected by force, esp. from a gun. **2** projecting or impelling.

projection /prə'dʒekʃ(ə)n/ *n.* **1** projecting or being projected. **2** thing that projects or obtrudes. **3** presentation of an image etc. on a surface. **4** forecast or estimate (*projection of next year's profits*). **5 a** mental image viewed as an objective reality. **b** unconscious transfer of feelings etc. to external objects or persons. **6** representation on a plane surface of any part of the surface of the earth or a celestial sphere (*Mercator projection*). □ **projectionist** *n.* (in sense 3).

projector /prə'dʒektə(r)/ *n.* apparatus for projecting slides or film on to a screen.

prokaryote /prəʊ'kærɪət/ *n.* organism in which the chromosomes are not separated from the cytoplasm by a membrane; bacterium. [from PRO-[2], *karyo-* from Greek *karuon* kernel, *-ote* as in ZYGOTE]

prolactin /prəʊ'læktɪn/ *n.* hormone that stimulates milk production after childbirth. [from PRO-[1], LACTATION]

prolapse /ˈprəʊlæps/ *–n.* (also **prolapsus** /-ˈlæpsəs/) **1** forward or downward displacement of a part or organ. **2** prolapsed womb, rectum, etc. *–v.* (**-sing**) undergo prolapse. [Latin: related to LAPSE]

prolate /ˈprəʊleɪt/ *adj. Geom.* (of a spheroid) lengthened in the direction of a polar diameter. [Latin, = brought forward, prolonged]

prole *adj. & n. derog. colloq.* proletarian. [abbreviation]

prolegomenon /ˌprəʊlɪˈgɒmɪnən/ *n.* (*pl.* **-mena**) (usu. in *pl.*) preface to a book etc., esp. when critical or discursive. [Greek *legō* say]

proletarian /ˌprəʊlɪˈteərɪən/ *–adj.* of the proletariat. *–n.* member of the proletariat. [Latin *proles* offspring]

proletariat /ˌprəʊlɪˈteərɪət/ *n.* **1** wage-earners collectively. **2** esp. *derog.* lowest, esp. uneducated, class. [French: related to PROLETARIAN]

proliferate /prəˈlɪfəˌreɪt/ *v.* (**-ting**) **1** reproduce; produce (cells etc.) rapidly. **2** increase rapidly in numbers. □ **proliferation** /-ˈreɪʃ(ə)n/ *n.* [Latin *proles* offspring]

prolific /prəˈlɪfɪk/ *adj.* **1** producing many offspring or much output. **2** (often foll. by *of*) abundantly productive. **3** (often foll. by *in*) abounding, copious. □ **prolifically** *adv.* [medieval Latin: related to PROLIFERATE]

prolix /ˈprəʊlɪks/ *adj.* (of speech, writing, etc.) lengthy; tedious. □ **prolixity** /-ˈlɪksɪtɪ/ *n.* [Latin]

prologue /ˈprəʊlɒg/ *n.* **1** preliminary speech, poem, etc., esp. of a play. **2** (usu. foll. by *to*) introductory event. [Greek *logos* word]

prolong /prəˈlɒŋ/ *v.* **1** extend in time or space. **2** (as **prolonged** *adj.*) lengthy, esp. tediously so. □ **prolongation** /ˌprəʊlɒŋˈgeɪʃ(ə)n/ *n.* [Latin *longus* long]

prom *n. colloq.* **1** = PROMENADE *n.* 1. **2** = PROMENADE CONCERT. [abbreviation]

promenade /ˌprɒməˈnɑːd/ *–n.* **1** paved public walk, esp. along the sea front at a resort. **2** walk, ride, or drive, taken esp. for display or pleasure. *–v.* (**-ding**) **1** make a promenade (through). **2** lead (a person etc.) about, esp. for display. [French]

promenade concert *n.* concert with restricted seating and a large area for standing.

promenade deck *n.* upper deck on a passenger ship.

promenader *n.* **1** person who promenades. **2** regular attender at promenade concerts.

Promethean /prəˈmiːθɪən/ *adj.* daring or inventive. [*Prometheus*, a mortal punished by the Greek gods for stealing fire]

promethium /prəˈmiːθɪəm/ *n.* radioactive metallic element of the lanthanide series, found in nuclear waste. [*Prometheus*: see PROMETHEAN]

prominence /ˈprɒmɪnəns/ *n.* **1** being prominent. **2** jutting outcrop, mountain, etc. [Latin: related to PROMINENT]

prominent /ˈprɒmɪnənt/ *adj.* **1** jutting out, projecting. **2** conspicuous. **3** distinguished, important. [Latin *promineo* project]

promiscuous /prəˈmɪskjʊəs/ *adj.* **1** having frequent, esp. casual, sexual relationships. **2** mixed and indiscriminate. **3** *colloq.* carelessly irregular; casual. □ **promiscuity** /-ˈskjuːɪtɪ/ *n.* **promiscuously** *adv.* [Latin *misceo* mix]

promise /ˈprɒmɪs/ *–n.* **1** assurance that one will or will not undertake a certain action etc. (*promise of help*). **2** sign of future achievements, good results, etc. (*writer of great promise*). *–v.* (**-sing**) **1** (usu. foll. by *to* + infin., or *that* + clause; also *absol.*) make a promise (*promise not to be late*). **2** (often foll. by *to* + infin.) seem likely (to) (*promises to be a good book*). **3** *colloq.* assure (*I promise you, it will not be easy*). □ **promise well** (or **ill** etc.) hold out good (or bad etc.) prospects. [Latin *promissum* from *mitto miss-* send]

promised land *n.* (prec. by *the*) **1** *Bibl.* Canaan (Gen. 12:7 etc.). **2** any desired place, esp. heaven.

promising *adj.* likely to turn out well; hopeful, full of promise (*promising start*). □ **promisingly** *adv.*

promissory /ˈprɒmɪsərɪ/ *adj.* conveying or implying a promise. [medieval Latin: related to PROMISE]

promissory note *n.* signed document containing a written promise to pay a stated sum.

promo /ˈprəʊməʊ/ *n.* (*pl.* **-s**) *colloq.* **1** (often *attrib.*) promotion, advertising (*promo video*). **2** promotional video, trailer, etc. [abbreviation]

promontory /ˈprɒməntərɪ/ *n.* (*pl.* **-ies**) point of high land jutting out into the sea etc.; headland. [Latin]

promote /prəˈməʊt/ *v.* (**-ting**) **1** (often foll. by *to*) raise (a person) to a higher office, rank, etc. (*promoted to captain*). **2** help forward; encourage (a cause, process, etc.). **3** publicize and sell (a product). **4** *Chess* raise (a pawn) to the rank of queen etc. □ **promotion** /-ˈməʊʃ(ə)n/ *n.* **promotional**

/-'məʊʃən(ə)l/ *adj.* [Latin *promoveo -mot-*]

promoter *n.* **1** person who promotes, esp. a sporting event, theatrical production, etc. **2** (in full **company promoter**) person who promotes the formation of a joint-stock company. [medieval Latin: related to PROMOTE]

prompt *–adj.* acting, made, or done with alacrity; ready (*prompt reply*). *–adv.* punctually (*at six o'clock prompt*). *–v.* **1** (usu. foll. by *to*, or *to* + infin.) incite; urge (*prompted them to action*). **2 a** (also *absol.*) supply a forgotten word etc. to (an actor etc.). **b** assist (a hesitating speaker) with a suggestion. **3** give rise to; inspire (feeling, thought, action, etc.). *–n.* **1 a** act of prompting. **b** thing said to prompt an actor etc. **c** = PROMPTER. **2** *Computing* sign on a VDU screen to show that the system is waiting for input. □ **promptitude** *n.* **promptly** *adv.* **promptness** *n.* [Latin]

prompter *n.* person who prompts actors.

promulgate /'prɒməl,geɪt/ *v.* (**-ting**) **1** make known to the public; disseminate; promote. **2** proclaim (a decree, news, etc.). □ **promulgation** /-'geɪʃ(ə)n/ *n.* **promulgator** *n.* [Latin]

prone *adj.* **1 a** lying face downwards. **b** lying flat, prostrate. **c** having the front part downwards, esp. the palm. **2** (usu. foll. by *to*, or *to* + infin.) disposed or liable (*prone to bite his nails*). **3** (usu. in *comb.*) likely to suffer (*accident-prone*). □ **proneness** /'prəʊnnɪs/ *n.* [Latin]

prong *n.* each of two or more projecting pointed parts at the end of a fork etc. [origin unknown]

pronominal /prəʊ'nɒmɪn(ə)l/ *adj.* of, concerning, or being, a pronoun. [Latin: related to PRONOUN]

pronoun /'prəʊnaʊn/ *n.* word used instead of and to indicate a noun already mentioned or known, esp. to avoid repetition (e.g. *we, their, this, ourselves*). [from PRO-[1], NOUN]

pronounce /prə'naʊns/ *v.* (**-cing**) **1** (also *absol.*) utter or speak (words, sounds, etc.) in a certain, or esp. in the approved, way. **2** utter or proclaim (a judgement, sentence, etc.) officially, formally, or solemnly (*I pronounce you man and wife*). **3** state as one's opinion (*pronounced the beef excellent*). **4** (usu. foll. by *on, for, against, in favour of*) pass judgement (*pronounced for the defendant*). □ **pronounceable** *adj.* **pronouncement** *n.* [Latin *nuntio* announce]

pronounced *adj.* strongly marked; noticeable (*pronounced limp*).

pronto /'prɒntəʊ/ *adv. colloq.* promptly, quickly. [Latin: related to PROMPT]

pronunciation /prə,nʌnsɪ'eɪʃ(ə)n/ *n.* **1** pronouncing of a word, esp. with reference to a standard. **2** act of pronouncing. **3** way of pronouncing words etc. [Latin: related to PRONOUNCE]

proof *–n.* **1** facts, evidence, reasoning, etc. establishing or helping to establish a fact (*no proof that he was there*). **2** demonstration, proving (*not capable of proof*). **3** test, trial (*put them to the proof*). **4** standard of strength of distilled alcohol. **5** trial impression from type or film, for correcting before final printing. **6** step by step resolution of a mathematical or philosophical problem. **7** photographic print made for selection etc. *–adj.* **1** (often in *comb.*) impervious to penetration, ill effects, etc., esp. by a specified agent (*proof against corruption; childproof*). **2** being of proof alcoholic strength. *–v.* **1** make proof, esp. make (fabric) waterproof. **2** make a proof of (a printed work). [Latin *proba*: related to PROVE]

proofread *v.* (*past* and *past part.* **-read** /-red/) read and correct (printer's proofs). □ **proofreader** *n.*

prop[1] *–n.* **1** rigid, esp. separate, support. **2** person or thing that supports, comforts, etc. *–v.* (**-pp-**) (often foll. by *against, up*, etc.) support with or as if with a prop. [Low German or Dutch]

prop[2] *n. colloq.* = PROPERTY 3. [abbreviation]

prop[3] *n. colloq.* propeller. [abbreviation]

propaganda /,prɒpə'gændə/ *n.* **1** organized propagation of a doctrine by use of publicity, selected information, etc. **2** usu. *derog.* ideas etc. so propagated. □ **propagandist** *n.* & *adj.* **propagandize** *v.* (also **-ise**) (**-zing** or **-sing**). [Latin: related to PROPAGATE]

propagate /'prɒpə,geɪt/ *v.* (**-ting**) **1 a** breed (a plant, animal, etc.) from the parent stock. **b** (*refl.* or *absol.*) (of a plant, animal, etc.) reproduce itself. **2** disseminate (a belief, theory, etc.). **3** transmit (a vibration, earthquake, etc.). □ **propagation** /-'geɪʃ(ə)n/ *n.* [Latin *propago*]

propagator *n.* **1** person or thing that propagates. **2** small heated box for germinating seeds or raising seedlings.

propane /'prəʊpeɪn/ *n.* gaseous hydrocarbon used as bottled fuel. [*propionic acid*: related to PRO-[2], Greek *piōn* fat]

propanone /'prəʊpə,nəʊn/ *n. Chem.* = ACETONE. [from PROPANE]

propel /prə'pel/ *v.* (**-ll-**) drive or push forward; urge on. □ **propellant** *n.* & *adj.* [Latin *pello puls-* drive]

propeller *n.* revolving shaft with blades, esp. for propelling a ship or aircraft.

propene /'prəʊpi:n/ *n. Chem.* = PROPYLENE. [from PROPANE, ALKENE]

propensity /prə'pensɪtɪ/ *n.* (*pl.* **-ies**) inclination, tendency. [Latin *propensus* inclined]

proper /'prɒpə(r)/ *adj.* **1 a** accurate, correct (*gave him the proper amount*). **b** fit, suitable, right (*at the proper time*). **2** decent; respectable, esp. excessively so (*not quite proper*). **3** (usu. foll. by *to*) belonging or relating (*respect proper to them*). **4** (usu. placed after the noun) strictly so called; genuine (*this is the crypt, not the cathedral proper*). **5** *colloq.* thorough; complete (*a proper row*). [Latin *proprius* one's own]

proper fraction *n.* fraction less than unity, with the numerator less than the denominator.

properly *adv.* **1** fittingly, suitably (*do it properly*). **2** accurately, correctly (*properly speaking*). **3** rightly. **4** with decency; respectably (*behave properly*). **5** *colloq.* thoroughly (*properly puzzled*).

proper noun *n.* (also **proper name**) capitalized name for an individual person, place, animal, country, title, etc., e.g. 'Jane', 'Everest'.

propertied /'prɒpətɪd/ *adj.* having property, esp. land.

property /'prɒpətɪ/ *n.* (*pl.* **-ies**) **1** thing(s) owned; possession, esp. a house, land, etc. (*has money in property*). **2** attribute, quality, or characteristic (*property of dissolving grease*). **3** movable object used on a theatre stage or in a film. [Latin *proprietas*: related to PROPER]

prophecy /'prɒfɪsɪ/ *n.* (*pl.* **-ies**) **1 a** prophetic utterance, esp. biblical. **b** prediction of future events. **2** faculty, practice, etc. of prophesying (*gift of prophecy*). [Greek: related to PROPHET]

prophesy /'prɒfɪ,saɪ/ *v.* (**-ies, -ied**) **1** (usu. foll. by *that, who,* etc.) foretell (an event etc.). **2** speak as a prophet; foretell the future. [French *profecier*: related to PROPHECY]

prophet /'prɒfɪt/ *n.* (*fem.* **prophetess**) **1** teacher or interpreter of the supposed will of God. **2 a** person who foretells events. **b** spokesman; advocate (*prophet of the new order*). **3** (**the Prophet**) Muhammad. [Greek *prophētēs* spokesman]

prophetic /prə'fetɪk/ *adj.* **1** (often foll. by *of*) containing a prediction; predicting. **2** of a prophet. □ **prophetically** *adv.* [Latin: related to PROPHET]

prophylactic /,prɒfɪ'læktɪk/ *–adj.* tending to prevent disease etc. *–n.* **1** preventive medicine or action. **2** esp. *US* condom. [Greek, = keeping guard before]

prophylaxis /,prɒfɪ'læksɪs/ *n.* preventive treatment against disease. [from PRO-[2], Greek *phulaxis* guarding]

propinquity /prə'pɪŋkwɪtɪ/ *n.* **1** nearness in space; proximity. **2** close kinship. **3** similarity. [Latin *prope* near]

propitiate /prə'pɪʃɪ,eɪt/ *v.* (**-ting**) appease (an offended person etc.). □ **propitiable** *adj.* **propitiation** /-'eɪʃ(ə)n/ *n.* **propitiator** *n.* **propitiatory** /-ʃətərɪ/ *adj.* [Latin: related to PROPITIOUS]

propitious /prə'pɪʃəs/ *adj.* **1** (of an omen etc.) favourable, auspicious. **2** (often foll. by *for, to*) suitable, advantageous. [Latin *propitius*]

proponent /prə'pəʊnənt/ *n.* person advocating a motion, theory, or proposal. [Latin: related to PROPOSE]

proportion /prə'pɔ:ʃ(ə)n/ *–n.* **1 a** comparative part or share (*large proportion of the profits*). **b** comparative ratio (*proportion of births to deaths*). **2** correct or pleasing relation of things or parts of a thing (*has fine proportions; exaggerated out of all proportion*). **3** (in *pl.*) dimensions; size (*large proportions*). **4** *Math.* equality of ratios between two pairs of quantities, e.g. 3:5 and 9:15. *–v.* (usu. foll. by *to*) make proportionate (*proportion the punishment to the crime*). [Latin: related to PORTION]

proportional *adj.* in due proportion; comparable (*proportional increase in the expense*). □ **proportionally** *adv.*

proportional representation *n.* electoral system in which parties gain seats in proportion to the number of votes cast for them.

proportionate /prə'pɔ:ʃənət/ *adj.* = PROPORTIONAL. □ **proportionately** *adv.*

proposal /prə'pəʊz(ə)l/ *n.* **1 a** act of proposing something. **b** course of action etc. proposed. **2** offer of marriage.

propose /prə'pəʊz/ *v.* (**-sing**) **1** (also *absol.*) put forward for consideration or as a plan; suggest. **2** (usu. foll. by *to* + infin., or verbal noun) intend; purpose (*propose to open a café*). **3** (usu. foll. by *to*) offer oneself in marriage. **4** nominate (a person) as a member of a society, for an office, etc. □ **propose a toast** (or **somebody's health**) ask people to drink to someone's health. □ **proposer** *n.* [Latin *pono posit-* place]

proposition /,prɒpə'zɪʃ(ə)n/ *–n.* **1** statement, assertion. **2** scheme proposed, proposal. **3** *Logic* statement subject to proof or disproof. **4** *colloq.* problem, opponent, prospect, etc. for consideration (*difficult proposition*). **5** *Math.* formal statement of a theorem or problem, often including the demonstration. **6 a** likely commercial etc. enterprise etc. **b** person regarded similarly. **7** *colloq.* sexual proposal. *–v. colloq.* make a (esp. sexual) proposal to (*he propositioned her*). [Latin: related to PROPOSE]

propound /prə'paʊnd/ *v.* offer for consideration; propose. [*propo(u)ne* from Latin: related to PROPOSE]

proprietary /prə'praɪətərɪ/ *adj.* **1 a** of or holding property (*proprietary classes*). **b** of a proprietor (*proprietary rights*). **2** held in private ownership. [Latin *proprietarius*: related to PROPERTY]

proprietary medicine *n.* drug, medicine, etc. produced by a company, usu. under a patent.

proprietary name *n.* (also **proprietary term**) registered name of a product etc. as a trade mark.

proprietor /prə'praɪətə(r)/ *n.* (*fem.* **proprietress**) **1** holder of property. **2** owner of a business etc., esp. of a hotel. □ **proprietorial** /-'tɔ:rɪəl/ *adj.* [related to PROPRIETARY]

propriety /prə'praɪɪtɪ/ *n.* (*pl.* **-ies**) **1** fitness; rightness. **2** correctness of behaviour or morals. **3** (in *pl.*) details or rules of correct conduct. [French: related to PROPERTY]

propulsion /prə'pʌlʃ(ə)n/ *n.* **1** driving or pushing forward. **2** impelling influence. □ **propulsive** /-'pʌlsɪv/ *adj.* [related to PROPEL]

propylene /'prəʊpɪ,li:n/ *n.* gaseous hydrocarbon used in the manufacture of chemicals. [from *propyl*, a univalent radical of propane]

pro rata /prəʊ 'rɑ:tə/ *–adj.* proportional. *–adv.* proportionally. [Latin]

prorogue /prə'rəʊg/ *v.* (**-gues, -gued, -guing**) **1** discontinue the meetings of (a parliament etc.) without dissolving it. **2** (of a parliament etc.) be prorogued. □ **prorogation** /,prəʊrə'geɪʃ(ə)n/ *n.* [Latin *prorogo* extend]

prosaic /prə'zeɪɪk/ *adj.* **1** like prose, lacking poetic beauty. **2** unromantic; dull; commonplace. □ **prosaically** *adv.* [Latin: related to PROSE]

pros and cons *n.pl.* reasons or considerations for and against a proposition etc.

proscenium /prə'si:nɪəm/ *n.* (*pl.* **-s** or **-nia**) part of the stage in front of the curtain and the enclosing arch. [Greek: related to SCENE]

proscribe /prə'skraɪb/ *v.* (**-bing**) **1** forbid, esp. by law. **2** reject or denounce (a practice etc.). **3** outlaw (a person). □ **proscription** /-'skrɪpʃ(ə)n/ *n.* **proscriptive** /-'skrɪptɪv/ *adj.* [Latin, = publish in writing]

■ **Usage** *Proscribe* is sometimes confused with *prescribe*.

prose /prəʊz/ *–n.* **1** ordinary written or spoken language not in verse. **2** passage of prose, esp. for translation into a foreign language. **3** dull or matter-of-fact quality (*prose of existence*). *–v.* (**-sing**) talk tediously. [Latin *prosa* from *oratio* straightforward (discourse)]

prosecute /'prɒsɪ,kju:t/ *v.* (**-ting**) **1** (also *absol.*) institute legal proceedings against (a person), or with reference to (a claim, crime, etc.) (*decided not to prosecute*). **2** *formal* carry on (a trade, pursuit, etc.). □ **prosecutor** *n.* [Latin *prosequor -secut-* pursue]

prosecution /,prɒsɪ'kju:ʃ(ə)n/ *n.* **1 a** institution and continuation of (esp. criminal) legal proceedings. **b** prosecuting party in a court case. **2** prosecuting or being prosecuted (*in the prosecution of his hobby*).

proselyte /'prɒsə,laɪt/ *n.* **1** person converted, esp. recently, from one opinion, creed, party, etc., to another. **2** convert to Judaism. □ **proselytism** /-lə,tɪz(ə)m/ *n.* [Latin *proselytus* from Greek]

proselytize /'prɒsələ,taɪz/ *v.* (also **-ise**) (**-zing** or **-sing**) (also *absol.*) convert or seek to convert from one belief etc. to another.

prose poem *n.* piece of poetic writing in prose.

prosody /'prɒsədɪ/ *n.* **1** science of versification. **2** the study of speech-rhythms. □ **prosodic** /prə'sɒdɪk/ *adj.* **prosodist** *n.* [Greek *pros* to: related to ODE]

prospect *–n.* /'prɒspekt/ **1 a** (often in *pl.*) expectation, esp. of success in a career etc. (*job with no prospects*). **b** something one expects (*don't relish the prospect of meeting him*). **2** extensive view of landscape etc. (*striking prospect*). **3** mental picture. **4** possible or probable customer, subscriber, etc. *–v.* /prə'spekt/ (usu. foll. by *for*) explore, search (esp. a region) for gold etc. □ **prospector** /prə'spektə(r)/ *n.* [Latin: related to PROSPECTUS]

prospective /prə'spektɪv/ *adj.* some day to be; expected; future (*prospective bridegroom*). [Latin: related to PROSPECTUS]

prospectus /prə'spektəs/ *n.* (*pl.* **-tuses**) printed document advertising or describing a school, commercial enterprise, forthcoming book, etc. [Latin, = prospect, from *prospicio -spect-* look forward]

prosper *v.* be successful, thrive. [Latin *prospero*]
prosperity /prɒ'sperɪtɪ/ *n.* prosperous state; wealth; success.
prosperous /'prɒspərəs/ *adj.* **1** successful; rich; thriving. **2** auspicious (*prosperous wind*). □ **prosperously** *adv.* [French from Latin]
prostate /'prɒsteɪt/ *n.* (in full **prostate gland**) gland round the neck of the bladder in male mammals, releasing part of the semen. □ **prostatic** /-'stætɪk/ *adj.* [Greek *prostatēs* one who stands before]
prosthesis /prɒs'θi:sɪs/ *n.* (*pl.* **-theses** /-si:z/) **1** artificial leg etc; false tooth, breast, etc. **2** branch of surgery dealing with prostheses. □ **prosthetic** /-'θetɪk/ *adj.* [Greek, = placing in addition]
prostitute /'prɒstɪ,tju:t/ *–n.* **1** woman who engages in sexual activity for payment. **2** (usu. **male prostitute**) man or boy who engages in sexual activity, esp. with homosexual men, for payment. *–v.* (**-ting**) **1** (esp. *refl.*) make a prostitute of (esp. oneself). **2** misuse or offer (one's talents, skills, name, etc.) for money etc. □ **prostitution** /-'tju:ʃ(ə)n/ *n.* [Latin *prostituo -tut-* offer for sale]
prostrate *–adj.* /'prɒstreɪt/ **1 a** lying face downwards, esp. in submission. **b** lying horizontally. **2** overcome, esp. by grief, exhaustion, etc. **3** growing along the ground. *–v.* /prɒ'streɪt/ (**-ting**) **1** lay or throw (esp. a person) flat. **2** *refl.* throw (oneself) down in submission etc. **3** overcome; make weak. □ **prostration** /prɒ'streɪʃ(ə)n/ *n.* [Latin *prosterno -strat-* throw in front]
prosy /'prəʊzɪ/ *adj.* (**-ier, -iest**) tedious, commonplace, dull (*prosy talk*). □ **prosily** *adv.* **prosiness** *n.*
protactinium /,prəʊtæk'tɪnɪəm/ *n.* radioactive metallic element. [German: related to ACTINIUM]
protagonist /prə'tægənɪst/ *n.* **1** chief person in a drama, story, etc. **2** leading person in a contest etc.; principal performer. **3** (usu. foll. by *of, for*) advocate or champion of a cause etc. (*protagonist of women's rights*). [Greek: related to PROTO-, *agōnistēs* actor]

■ **Usage** The use of *protagonist* in sense 3 is considered incorrect by some people.

protean /prəʊ'ti:ən/ *adj.* variable, taking many forms; versatile. [*Proteus*, Greek sea-god who took various shapes]
protect /prə'tekt/ *v.* **1** (often foll. by *from, against*) keep (a person, thing, etc.) safe; defend, guard. **2** shield (home industry) from competition with import duties. [Latin *tego tect-* cover]
protection /prə'tekʃ(ə)n/ *n.* **1 a** protecting or being protected; defence. **b** thing, person, or animal that protects. **2** (also **protectionism**) theory or practice of protecting home industries. **3** *colloq.* **a** immunity from violence etc. obtained by payment to gangsters etc. **b** (in full **protection money**) money so paid. □ **protectionist** *n.* & *adj.*
protective /prə'tektɪv/ *adj.* protecting; intended or tending to protect. □ **protectively** *adv.* **protectiveness** *n.*
protective custody *n.* detention of a person for his or her own protection.
protector *n.* (*fem.* **protectress**) **1** person or thing that protects. **2** *hist.* regent ruling during the minority or absence of the sovereign. □ **protectorship** *n.*
protectorate /prə'tektərət/ *n.* **1 a** State that is controlled and protected by another. **b** this relation. **2** *hist.* **a** office of the protector of a kingdom or State. **b** period of this, esp. in England 1653–9.
protégé /'prɒtɪ,ʒeɪ/ *n.* (*fem.* **protégée** pronunc. same) person under the protection, patronage, tutelage, etc. of another. [French: related to PROTECT]
protein /'prəʊti:n/ *n.* any of a group of organic compounds composed of one or more chains of amino acids and forming an essential part of all living organisms. [Greek *prōtos* first]
pro tem /prəʊ 'tem/ *adj.* & *adv. colloq.* = PRO TEMPORE. [abbreviation]
pro tempore /prəʊ 'tempərɪ/ *adj.* & *adv.* for the time being. [Latin]
Proterozoic /,prəʊtərəʊ'zəʊɪk/ *Geol.* *–adj.* of the later part of the Precambrian era. *–n.* this time. [Greek *proteros* former, *zōē* life]
protest *–n.* /'prəʊtest/ **1** statement or act of dissent or disapproval. **2** *Law* written declaration that a bill has been presented and payment or acceptance refused. *–v.* /prə'test/ **1** (usu. foll. by *against, at, about*, etc.) make a protest. **2** affirm (one's innocence etc.) solemnly. **3** *Law* write or obtain a protest in regard to (a bill). **4** *US* object to (a decision etc.). □ **under protest** unwillingly. □ **protester** *n.* (also **protestor**). [Latin *protestor* declare formally]
Protestant /'prɒtɪst(ə)nt/ *–n.* member or follower of any of the Churches separating from the Roman Catholic Church after the Reformation. *–adj.* of the Protestant Churches or their members etc. □ **Protestantism** *n.* [related to PROTEST]
protestation /,prɒtɪ'steɪʃ(ə)n/ *n.* **1** strong affirmation. **2** protest. [Latin: related to PROTEST]
protium /'prəʊtɪəm/ *n.* ordinary isotope of hydrogen. [Latin: related to PROTO-]

proto- *comb. form* first. [Greek *prōtos*]

protocol /'prəʊtə,kɒl/ *–n.* **1** official formality and etiquette, esp. as observed on State occasions etc. **2** original draft of esp. the terms of a treaty. **3** formal statement of a transaction. *–v.* (**-ll-**) draw up or record in a protocol. [Greek *kolla* glue]

proton /'prəʊtɒn/ *n.* elementary particle with a positive electric charge equal to that of an electron, and occurring in all atomic nuclei. [Greek *prōtos* first]

protoplasm /'prəʊtə,plæz(ə)m/ *n.* material comprising the living part of a cell, consisting of a nucleus in membrane-enclosed cytoplasm. □ **protoplasmic** /-'plæzmɪk/ *adj.* [Greek: related to PROTO-, PLASMA]

prototype /'prəʊtə,taɪp/ *n.* **1** original as a pattern for imitations, improved forms, representations, etc. **2** trial model or preliminary version of a vehicle, machine, etc. □ **prototypic** /-'tɪpɪk/ *adj.* **prototypical** /-'tɪpɪk(ə)l/ *adj.* [Greek: related to PROTO-]

protozoan /,prəʊtə'zəʊən/ *–n.* (also **protozoon** /-'zəʊɒn/) (*pl.* **protozoa** /-'zəʊə/ or **-s**) unicellular microscopic organism, e.g. the amoeba. *–adj.* (also **protozoic** /-'zəʊɪk/) of this group. [from PROTO-, Greek *zōion* animal]

protract /prə'trækt/ *v.* (often as **protracted** *adj.*) prolong or lengthen. □ **protraction** *n.* [Latin *traho tract-* draw]

protractor *n.* instrument for measuring angles, usu. in the form of a graduated semicircle.

protrude /prə'tru:d/ *v.* (**-ding**) thrust forward; stick out; project. □ **protrusion** *n.* **protrusive** *adj.* [Latin *trudo trus-* thrust]

protuberant /prə'tju:bərənt/ *adj.* bulging out; prominent. □ **protuberance** *n.* [Latin: related to TUBER]

proud *adj.* **1** feeling greatly honoured or pleased (*proud to know him*). **2 a** (often foll. by *of*) haughty, arrogant (*too proud to speak to us*). **b** (often in *comb.*) having a proper pride; satisfied (*house-proud*; *proud of a job well done*). **3** (of an occasion, action, etc.) justly arousing or showing pride (*proud day*; *proud smile*). **4** imposing, splendid. **5** (often foll. by *of*) slightly projecting (*nail stood proud of the plank*). □ **do proud** *colloq.* treat with lavish generosity or honour (*did us proud*). □ **proudly** *adv.* [French *prud* valiant]

prove /pru:v/ *v.* (**-ving**; *past part.* **proved** or **proven** /'pru:v(ə)n, 'prəʊ-/) **1** (often foll. by *that*) demonstrate the truth of by evidence or argument. **2 a** (usu. foll. by *to* + infin.) be found (*it proved to be untrue*). **b** emerge as (*will prove the winner*). **3** test the accuracy of (a calculation). **4** establish the validity of (a will). **5** (of dough) rise in bread-making. □ **not proven** (in Scottish Law) verdict that there is insufficient evidence to establish guilt or innocence. **prove oneself** show one's abilities, courage, etc. □ **provable** *adj.* [Latin *probo* test, approve]

■ **Usage** The use of *proven* as the past participle is uncommon except in certain expressions, such as *of proven ability*. It is, however, standard in Scots and American English.

provenance /'prɒvɪnəns/ *n.* origin or place of origin; history. [French *provenir* from Latin]

Provençal /,prɒvɒn'sɑ:l/ *–adj.* of Provence. *–n.* native or language of Provence. [French: related to PROVINCE]

provender /'prɒvɪndə(r)/ *n.* **1** animal fodder. **2** *joc.* food. [Latin: related to PREBEND]

proverb /'prɒvɜ:b/ *n.* short pithy saying in general use, held to embody a general truth. [Latin *proverbium* from *verbum* word]

proverbial /prə'vɜ:bɪəl/ *adj.* **1** (esp. of a characteristic) well known; notorious (*his proverbial honesty*). **2** of or referred to in a proverb (*proverbial ill wind*). □ **proverbially** *adv.* [Latin: related to PROVERB]

provide /prə'vaɪd/ *v.* (**-ding**) **1** supply, furnish (*provided me with food*; *provided a chance*). **2 a** (usu. foll. by *for*, *against*) make due preparation. **b** (usu. foll. by *for*) take care of a person etc. with money, food, etc. (*provides for a large family*). **3** (usu. foll. by *that*) stipulate in a will, statute, etc. □ **provider** *n.* [Latin *provideo -vis-* foresee]

provided *conj.* (often foll. by *that*) on the condition or understanding that.

providence /'prɒvɪd(ə)ns/ *n.* **1** protective care of God or nature. **2** (**Providence**) God in this aspect. **3** foresight; thrift. [Latin: related to PROVIDE]

provident *adj.* having or showing foresight; thrifty. [Latin: related to PROVIDE]

providential /,prɒvɪ'denʃ(ə)l/ *adj.* **1** of or by divine foresight or interposition. **2** opportune, lucky. □ **providentially** *adv.*

Provident Society *n.* = FRIENDLY SOCIETY.

providing *conj.* = PROVIDED.

province /'prɒvɪns/ *n.* **1** principal administrative division of a country etc. **2** (**the provinces**) country outside a capital city, esp. regarded as uncultured or unsophisticated. **3** sphere of action;

business (*outside my province*). **4** branch of learning etc. (*in the province of aesthetics*). **5** district under an archbishop or metropolitan. **6** territory outside Italy under an ancient Roman governor. [Latin *provincia*]

provincial /prə'vɪnʃ(ə)l/ *–adj.* **1** of a province or provinces. **2** unsophisticated or uncultured. *–n.* **1** inhabitant of a province or the provinces. **2** unsophisticated or uncultured person. □ **provincialism** *n.*

provision /prə'vɪʒ(ə)n/ *–n.* **1 a** act of providing (*provision of nurseries*). **b** preparation, esp. for the future (*made provision for their old age*). **2** (in *pl.*) food, drink, etc., esp. for an expedition. **3** legal or formal stipulation or proviso. *–v.* supply with provisions. [Latin: related to PROVIDE]

provisional *–adj.* **1** providing for immediate needs only; temporary. **2** (**Provisional**) of the unofficial wing of the IRA, using terrorism. *–n.* (**Provisional**) member of the Provisional wing of the IRA. □ **provisionally** *adv.*

proviso /prə'vaɪzəʊ/ *n.* (*pl.* **-s**) **1** stipulation. **2** clause containing this. □ **provisory** *adj.* [Latin, = it being provided]

Provo /'prəʊvəʊ/ *n.* (*pl.* **-s**) *colloq.* Provisional. [abbreviation]

provocation /,prɒvə'keɪʃ(ə)n/ *n.* **1** provoking or being provoked (*did it under severe provocation*). **2** cause of annoyance.

provocative /prə'vɒkətɪv/ *adj.* **1** (usu. foll. by *of*) tending to provoke, esp. anger or sexual desire. **2** intentionally annoying or controversial. □ **provocatively** *adv.* **provocativeness** *n.*

provoke /prə'vəʊk/ *v.* (**-king**) **1** (often foll. by *to*, or *to* + infin.) rouse or incite (*provoked him to fury*). **2** call forth; instigate; cause (indignation, an inquiry, process, etc.). **3** (usu. foll. by *into* + verbal noun) irritate or stimulate (a person) (*provoked him into retaliating*). **4** tempt; allure. [Latin *provoco* call forth]

provost /'prɒvəst/ *n.* **1** head of some (esp. Oxbridge) colleges. **2** head of a cathedral chapter. **3** = PROVOST MARSHAL. [Latin *propositus* from *pono* place]

provost marshal /prə'vəʊ/ *n.* head of military police in camp or on active service.

prow *n.* **1** fore-part or bow of a ship. **2** pointed or projecting front part. [French *proue* from Greek *prōira*]

prowess /'praʊɪs/ *n.* **1** skill, expertise. **2** valour, gallantry. [French: related to PROUD]

prowl *–v.* (often foll. by *about*, *around*) roam (a place) esp. stealthily or restlessly or in search of prey, plunder, etc. *–n.* act of prowling. □ **on the prowl** prowling. □ **prowler** *n.* [origin unknown]

prox. *abbr.* proximo.

proximate /'prɒksɪmət/ *adj.* **1** nearest or next before or after (in place, order, time, causation, thought process, etc.). **2** approximate. [Latin *proximus* nearest]

proximity /prɒk'sɪmɪtɪ/ *n.* nearness in space, time, etc. (*in close proximity*). [Latin: related to PROXIMATE]

proximo /'prɒksɪ,məʊ/ *adj.* *Commerce* of next month (*the third proximo*). [Latin, = in the next (*mense* month)]

proxy /'prɒksɪ/ *n.* (*pl.* **-ies**) (also *attrib.*) **1** authorization given to a substitute or deputy (*proxy vote*; *married by proxy*). **2** person authorized to act thus. **3 a** written authorization for esp. proxy voting. **b** proxy vote. [obsolete *procuracy* procuration]

prude /pru:d/ *n.* excessively (often affectedly) squeamish or sexually modest person. □ **prudery** *n.* **prudish** *adj.* **prudishly** *adv.* **prudishness** *n.* [French: related to PROUD]

prudent /'pru:d(ə)nt/ *adj.* cautious; politic. □ **prudence** *n.* **prudently** *adv.* [Latin *prudens -ent-*: related to PROVIDENT]

prudential /pru:'denʃ(ə)l/ *adj.* of or showing prudence. □ **prudentially** *adv.*

prune[1] /pru:n/ *n.* dried plum. [Latin *prunum* from Greek]

prune[2] /pru:n/ *v.* (**-ning**) **1 a** (often foll. by *down*) trim (a bush etc.) by cutting away dead or overgrown branches etc. **b** (usu. foll. by *off*, *away*) lop (branches etc.) thus. **2** reduce (costs etc.) (*prune expenses*). **3 a** (often foll. by *of*) clear or remove superfluities from. **b** remove (superfluities). [French *prooignier* from Romanic: related to ROUND]

prurient /'prʊərɪənt/ *adj.* having or encouraging unhealthy sexual curiosity. □ **prurience** *n.* [Latin *prurio* itch]

Prussian /'prʌʃ(ə)n/ *–adj.* of Prussia, or esp. its rigidly militaristic tradition. *–n.* native of Prussia. [*Prussia*, former German state]

Prussian blue *n.* & *adj.* (as adj. often hyphenated) deep blue (pigment).

prussic acid /'prʌsɪk/ *n.* hydrocyanic acid. [French]

pry /praɪ/ *v.* (**pries, pried**) **1** (usu. foll. by *into*) inquire impertinently. **2** (usu. foll. by *into*, *about*, etc.) look or peer inquisitively. [origin unknown]

PS *abbr.* postscript.

psalm /sɑ:m/ *n.* **1** (also **Psalm**) sacred song, esp. from the Book of Psalms, esp. metrically chanted in a service. **2** (**the Psalms** or **the Book of Psalms**) Old Testament book containing the Psalms. [Latin *psalmus* from Greek]

psalmist *n.* composer of a psalm.

psalmody /'sɑ:mədɪ, 'sæl-/ *n.* practice or art of singing psalms, hymns, etc., esp. in public worship. [Greek: related to PSALM]

Psalter /'sɔ:ltə(r)/ *n.* **1** the Book of Psalms. **2** (**psalter**) version or copy of this. [Old English and French from Greek *psaltērion* stringed instrument]

psaltery /'sɔ:ltərɪ/ *n.* (*pl.* **-ies**) ancient and medieval instrument like a dulcimer but played by plucking the strings. [Latin: related to PSALTER]

psephology /sɪ'fɒlədʒɪ/ *n.* the statistical study of voting etc. □ **psephologist** *n.* [Greek *psēphos* pebble, vote]

pseud /sju:d/ *colloq.* –*adj.* (esp. intellectually) pretentious; not genuine. –*n.* such a person; poseur. [from PSEUDO-]

pseudo /'sju:dəʊ/ *adj.* & *n.* (*pl.* **-s**) = PSEUD.

pseudo- *comb. form* (also **pseud-** before a vowel) **1** false; not genuine (*pseudo-intellectual*). **2** resembling or imitating (*pseudo-acid*). [Greek *pseudēs* false]

pseudonym /'sju:dənɪm/ *n.* fictitious name, esp. of an author. [Greek: related to PSEUDO-, *onoma* name]

psi[1] /psaɪ, saɪ/ *n.* twenty-third letter of the Greek alphabet (Ψ, ψ). [Greek]

psi[2] *abbr.* pounds per square inch.

psittacosis /,sɪtə'kəʊsɪs/ *n.* contagious viral disease of esp. parrots, transmissible to human beings. [Greek *psittakos* parrot]

psoriasis /sə'raɪəsɪs/ *n.* skin disease marked by red scaly patches. [Greek *psōra* itch]

psst *int.* (also **pst**) whispered exclamation to attract a person's attention. [imitative]

PSV *abbr.* public service vehicle.

psych /saɪk/ *v. colloq.* **1** (usu. foll. by *up*; often *refl.*) prepare (oneself or another) mentally for an ordeal etc. **2** (often foll. by *out*) intimidate or frighten (a person), esp. for one's own advantage. **3** (usu. foll. by *out*) analyse (a person's motivation etc.) for one's own advantage (*can't psych him out*). [abbreviation]

psyche /'saɪkɪ/ *n.* the soul, spirit, or mind. [Latin from Greek]

psychedelia /,saɪkə'di:lɪə/ *n.pl.* **1** psychedelic phenomena. **2** subculture associated with these.

psychedelic /,saɪkə'delɪk/ *adj.* **1 a** expanding the mind's awareness etc., esp. with hallucinogenic drugs. **b** hallucinatory; bizarre. **c** (of a drug) producing hallucinations. **2** *colloq.* **a** producing a hallucinatory effect; vivid in colour or design etc. **b** (of colours, patterns, etc.) bright, bold, and often abstract. [Greek *psukhē* mind, *dēlos* clear]

psychiatry /saɪ'kaɪətrɪ/ *n.* the study and treatment of mental disease. □ **psychiatric** /-kɪ'ætrɪk/ *adj.* **psychiatrist** *n.* [from PSYCHO-, Greek *iatros* physician]

psychic /'saɪkɪk/ –*adj.* **1 a** (of a person) considered to have occult powers such as telepathy, clairvoyance, etc. **b** supernatural. **2** of the soul or mind. –*n.* person considered to have psychic powers; medium. [Greek *psukhē* soul, mind]

psychical *adj.* **1** concerning psychic phenomena or faculties (*psychical research*). **2** of the soul or mind. □ **psychically** *adv.*

psycho /'saɪkəʊ/ *colloq.* –*n.* (*pl.* **-s**) psychopath. –*adj.* psychopathic. [abbreviation]

psycho- *comb. form* of the mind or psychology. [Greek: related to PSYCHIC]

psychoanalysis /,saɪkəʊə'nælɪsɪs/ *n.* treatment of mental disorders by bringing repressed fears and conflicts into the conscious mind over a long course of interviews. □ **psychoanalyse** /-'ænə,laɪz/ *v.* (**-sing**). **psychoanalyst** /-'ænəlɪst/ *n.* **psychoanalytic** /-,ænə'lɪtɪk/ *adj.* **psychoanalytical** /-,ænə'lɪtɪk(ə)l/ *adj.*

psychokinesis /,saɪkəʊkɪ'ni:sɪs/ *n.* movement of objects supposedly by telepathy or mental effort.

psychological /,saɪkə'lɒdʒɪk(ə)l/ *adj.* **1** of or arising in the mind. **2** of psychology. **3** *colloq.* (of an ailment etc.) imaginary (*her cold is psychological*). □ **psychologically** *adv.*

psychological block *n.* mental inhibition caused by emotional factors.

psychological moment *n.* best time for achieving a particular effect or purpose.

psychological warfare *n.* campaign directed at reducing enemy morale.

psychology /saɪ'kɒlədʒɪ/ *n.* (*pl.* **-ies**) **1** the study of the human mind. **2** treatise on or theory of this. **3 a** mental characteristics etc. of a person or group. **b** mental aspects of an activity, situation, etc. (*psychology of crime*). □ **psychologist** *n.*

psychopath /'saɪkə,pæθ/ *n.* **1** mentally deranged person, esp. showing abnormal or violent social behaviour. **2** mentally or emotionally unstable person. □ **psychopathic** /-'pæθɪk/ *adj.*

psychopathology /ˌsaɪkəʊpəˈθɒlədʒɪ/ *n.* **1** the study of mental disorders. **2** mentally or behaviourally disordered state.

psychopathy /saɪˈkɒpəθɪ/ *n.* psychopathic or psychologically abnormal behaviour.

psychosis /saɪˈkəʊsɪs/ *n.* (*pl.* **-choses** /-siːz/) severe mental disorder with loss of contact with reality. [Greek: related to PSYCHE]

psychosomatic /ˌsaɪkəʊsəˈmætɪk/ *adj.* **1** (of a bodily disorder) mental, not physical, in origin. **2** of the mind and body together.

psychotherapy /ˌsaɪkəʊˈθerəpɪ/ *n.* treatment of mental disorder by psychological means. □ **psychotherapeutic** /-ˈpjuːtɪk/ *adj.* **psychotherapist** *n.*

psychotic /saɪˈkɒtɪk/ *–adj.* of or suffering from a psychosis. *–n.* psychotic person.

PT *abbr.* physical training.

Pt *symb.* platinum.

pt *abbr.* **1** part. **2** pint. **3** point. **4** *Naut.* port.

PTA *abbr.* parent-teacher association.

ptarmigan /ˈtɑːmɪgən/ *n.* game-bird with a grouselike appearance. [Gaelic]

Pte. *abbr.* Private (soldier).

pteridophyte /ˈterɪdəˌfaɪt/ *n.* flowerless plant, e.g. ferns, club-mosses, etc. [Greek *pteris* fern]

pterodactyl /ˌterəˈdæktɪl/ *n.* large extinct flying reptile. [Greek *pteron* wing, DACTYL]

pterosaur /ˈterəˌsɔː(r)/ *n.* flying reptile with large batlike wings. [Greek *pteron* wing, *saura* lizard]

PTO *abbr.* please turn over.

Ptolemaic /ˌtɒlɪˈmeɪɪk/ *adj. hist.* of Ptolemy or his theories. [Greek *Ptolemaios*, name of a 2nd-c. astronomer]

Ptolemaic system *n.* theory that the earth is the stationary centre of the universe.

ptomaine /ˈtəʊmeɪn/ *n.* any of various esp. toxic amine compounds in putrefying matter. [Greek *ptōma* corpse]

Pu *symb.* plutonium.

pub *n. colloq.* public house. [abbreviation]

pub-crawl *n. colloq.* drinking tour of several pubs.

puberty /ˈpjuːbətɪ/ *n.* period of sexual maturation. □ **pubertal** *adj.* [Latin *puber* adult]

pubes[1] /ˈpjuːbiːz/ *n.* (*pl.* same) **1** lower part of the abdomen at the front of the pelvis. **2** *colloq.* pubic hair. [Latin]

pubes[2] *pl.* of PUBIS.

pubescence /pjuːˈbes(ə)ns/ *n.* **1** beginning of puberty. **2** soft down on plants, or on animals, esp. insects. □ **pubescent** *adj.* [Latin: related to PUBES]

pubic /ˈpjuːbɪk/ *adj.* of the pubes or pubis.

pubis /ˈpjuːbɪs/ *n.* (*pl.* **pubes** /-biːz/) either of a pair of bones forming the two sides of the pelvis. [Latin *os pubis* bone of the PUBES]

public /ˈpʌblɪk/ *–adj.* **1** of the people as a whole (*public holiday*). **2** open to or shared by all (*public baths*). **3** done or existing openly (*public apology*). **4** (of a service, funds, etc.) provided by or concerning government (*public money*; *public records*). **5** of or involved in the affairs, esp. the government or entertainment, of the community (*distinguished public career*; *public figures*). *–n.* **1** (as *sing.* or *pl.*) community, or members of it, in general. **2** specified section of the community (*reading public*; *my public*). □ **go public 1** become a public company. **2** reveal one's plans etc. **in public** openly, publicly. □ **publicly** *adv.* [Latin]

public-address system *n.* set of loudspeakers, microphones, amplifiers, etc., used in addressing large audiences.

publican /ˈpʌblɪkən/ *n.* keeper of a public house. [Latin: related to PUBLIC]

publication /ˌpʌblɪˈkeɪʃ(ə)n/ *n.* **1 a** preparation and issuing of a book, newspaper, etc. to the public. **b** book etc. so issued. **2** making something publicly known. [Latin: related to PUBLIC]

public bar *n.* less or least expensive bar in a public house.

public company *n.* company that sells shares on the open market.

public convenience *n.* public lavatory.

public enemy *n.* notorious wanted criminal.

public figure *n.* famous person.

public health *n.* provision of adequate sanitation, drainage, etc. by government.

public house *n.* inn providing alcoholic drinks for consumption on the premises.

publicist /ˈpʌblɪsɪst/ *n.* publicity agent or public relations officer.

publicity /pʌbˈlɪsɪtɪ/ *n.* **1** public exposure. **2 a** advertising. **b** material used for this. [French: related to PUBLIC]

publicize /ˈpʌblɪˌsaɪz/ *v.* (also **-ise**) (**-zing** or **-sing**) advertise; make publicly known.

public lending right *n.* right of authors to payment when their books etc. are lent by public libraries.

public opinion *n.* views, esp. moral, that are generally prevalent.

public ownership *n.* State ownership of the means of production, distribution, or exchange.

public prosecutor *n.* law officer acting on behalf of the State or in the public interest.

public relations *n.pl.* (usu. treated as *sing.*) professional promotion of a favourable public image, esp. by a company, famous person, etc.

public relations officer *n.* person employed to promote public relations.

public school *n.* **1** private fee-paying secondary school, esp. for boarders. **2** *US, Austral.*, & *Scot.* non-fee-paying school.

public sector *n.* State-controlled part of an economy, industry, etc.

public servant *n.* State official.

public spirit *n.* willingness to engage in community action. □ **public-spirited** *adj.*

public transport *n.* buses, trains, etc., charging set fares and running on fixed routes.

public utility *n.* organization supplying water, gas, etc. to the community.

public works *n.pl.* building operations etc. done by or for the State.

publish /ˈpʌblɪʃ/ *v.* **1** (also *absol.*) prepare and issue (a book, newspaper, etc.) for public sale. **2** make generally known. **3** announce formally. [Latin: related to PUBLIC]

publisher *n.* person or (esp.) company that publishes books etc. for sale.

puce *adj.* & *n.* dark red or purple-brown. [Latin *pulex* flea]

puck[1] *n.* rubber disc used as a ball in ice hockey. [origin unknown]

puck[2] *n.* mischievous or evil sprite. □ **puckish** *adj.* **puckishly** *adv.* **puckishness** *n.* [Old English]

pucker –*v.* (often foll. by *up*) gather into wrinkles, folds, or bulges (*this seam is puckered up*). –*n.* such a wrinkle, bulge, fold, etc. [origin unknown]

pud /pʊd/ *n. colloq.* = PUDDING. [abbreviation]

pudding /ˈpʊdɪŋ/ *n.* **1 a** any of various sweet cooked dishes (*rice pudding*). **b** savoury dish containing flour, suet, etc. (*steak and kidney pudding*). **c** sweet course of a meal. **d** any of various sausages stuffed with oatmeal, spices, blood, etc. (*black pudding*). **2** *colloq.* plump, stupid, or lazy person. □ **puddingy** *adj.* [Latin *botellus* sausage]

puddle /ˈpʌd(ə)l/ –*n.* **1** small pool, esp. of rainwater. **2** clay and sand worked with water used as a watertight covering for embankments etc. –*v.* (**-ling**) **1** knead (clay and sand) into puddle. **2** stir (molten iron) to produce wrought iron by expelling carbon. □ **puddly** *adj.* [Old English]

pudendum /pju:ˈdendəm/ *n.* (*pl.* **pudenda**) (usu. in *pl.*) genitals, esp. of a woman. [Latin *pudeo* be ashamed]

pudgy /ˈpʌdʒɪ/ *adj.* (**-ier**, **-iest**) *colloq.* (esp. of a person) plump, podgy. □ **pudginess** *n.* [cf. PODGY]

puerile /ˈpjʊəraɪl/ *adj.* childish, immature. □ **puerility** /-ˈrɪlɪtɪ/ *n.* (*pl.* **-ies**). [Latin *puer* boy]

puerperal /pju:ˈɜ:pər(ə)l/ *adj.* of or caused by childbirth. [Latin *puer* boy, *pario* bear]

puerperal fever *n.* fever following childbirth and caused by uterine infection.

puff –*n.* **1 a** short quick blast of breath or wind. **b** sound of or like this. **c** small quantity of vapour, smoke, etc., emitted in one blast (*puff of smoke*). **2** light pastry cake containing jam, cream, etc. **3** gathered material in a dress etc. (*puff sleeve*). **4** extravagantly enthusiastic review, advertisement, etc., esp. in a newspaper. **5** = POWDER-PUFF. –*v.* **1** emit a puff of air or breath; blow with short blasts. **2** (usu. foll. by *away*, *out*, etc.) emit or move with puffs (*puffing away at his cigar*; *train puffed out*). **3** (usu. in *passive*; often foll. by *out*) *colloq.* put out of breath (*arrived puffed*). **4** breathe hard; pant. **5** (usu. foll. by *up*, *out*) inflate; swell (*his eye was puffed up*). **6** (usu. foll. by *out*, *up*, *away*) blow or emit (dust, smoke, etc.) with a puff. **7** smoke (a pipe etc.) in puffs. **8** (usu. as **puffed up** *adj.*) elate; make proud or boastful. **9** advertise or promote with exaggerated or false praise. □ **puff up** = sense 8 of *v.* [imitative]

puff-adder *n.* large venomous African viper which inflates the upper part of its body.

puffball *n.* ball-shaped fungus emitting clouds of spores.

puffin /ˈpʌfɪn/ *n.* N. Atlantic and N. Pacific sea bird with a large head and brightly coloured triangular bill. [origin unknown]

puff pastry *n.* leaved pastry made light and flaky by rolling and folding the dough many times.

puffy *adj.* (**-ier**, **-iest**) **1** swollen, puffed out. **2** *colloq.* short-winded. □ **puffily** *adv.* **puffiness** *n.*

pug *n.* (in full **pug-dog**) dog of a dwarf breed with a broad flat nose and wrinkled face. [origin unknown]

pugilist /ˈpju:dʒɪlɪst/ *n.* (esp. professional) boxer. □ **pugilism** *n.* **pugilistic** /-ˈlɪstɪk/ *adj.* [Latin *pugil* boxer]

pugnacious /pʌg'neɪʃəs/ *adj.* quarrelsome; disposed to fight. □ **pugnaciously** *adv.* **pugnacity** /-'næsɪtɪ/ *n.* [Latin *pugnax -acis* from *pugno* fight]

pug-nose *n.* short squat or snub nose. □ **pug-nosed** *adj.*

puissance /'pwi:sɔ̃s/ *n.* competitive jumping of large obstacles in show-jumping. [French: related to PUISSANT]

puissant /'pwi:sɒnt/ *adj. literary* or *archaic* powerful; mighty. [Romanic: related to POTENT]

puke *v.* & *n.* (**-king**) *slang* vomit. □ **pukey** *adj.* [imitative]

pukka /'pʌkə/ *adj. Anglo-Ind. colloq.* **1** genuine. **2** of good quality; reliable (*a pukka job*). [Hindi]

pulchritude /'pʌlkrɪ,tju:d/ *n. literary* beauty. □ **pulchritudinous** /-'tju:dɪnəs/ *adj.* [Latin *pulcher* beautiful]

pule *v.* (**-ling**) *literary* cry querulously or weakly; whimper. [imitative]

pull /pʊl/ —*v.* **1** exert force upon (a thing, person, etc.) to move it to oneself or the origin of the force (*pulled it nearer*). **2** exert a pulling force (*engine will not pull*). **3** extract (a cork or tooth) by pulling. **4** damage (a muscle etc.) by abnormal strain. **5 a** move (a boat) by pulling on the oars. **b** (of a boat etc.) be caused to move, esp. in a specified direction. **6** (often foll. by *up*) proceed with effort (up a hill etc.). **7** (foll. by *on*) bring out (a weapon) for use against (a person). **8** check the speed of (a horse), esp. to lose a race. **9** attract (custom or support). **10** draw (liquor) from a barrel etc. **11** (foll. by *at*) tear or pluck at. **12** (often foll. by *on*, *at*) inhale or drink deeply; draw or suck (on a pipe etc.). **13** (often foll. by *up*) remove (a plant) by the root. **14 a** *Cricket* strike (the ball) to the leg side. **b** *Golf* strike (the ball) widely to the left. **15** print (a proof etc.). **16** *slang* succeed in attracting sexually. —*n.* **1** act of pulling. **2** force exerted by this. **3** influence; advantage. **4** attraction or attention-getter. **5** deep draught of liquor. **6** prolonged effort, e.g. in going up a hill. **7** handle etc. for applying a pull. **8** printer's rough proof. **9** *Cricket* & *Golf* pulling stroke. **10** suck at a cigarette. □ **pull about 1** treat roughly. **2** pull from side to side. **pull apart** (or **to pieces**) = *take to pieces* (see PIECE). **pull back** (cause to) retreat. **pull down 1** demolish (esp. a building). **2** humiliate. **pull a face** distort the features, grimace. **pull a fast one** see FAST[1]. **pull in 1** (of a bus, train, etc.) arrive to take passengers. **2** (of a vehicle) move to the side of or off the road. **3** *colloq.* earn or acquire. **4** *colloq.* arrest. **pull a person's leg** deceive playfully. **pull off 1** remove by pulling. **2** succeed in achieving or winning. **pull oneself together** recover control of oneself. **pull the other one** *colloq.* expressing disbelief (with ref. to *pull a person's leg*). **pull out 1** take out by pulling. **2** depart. **3** withdraw from an undertaking. **4** (of a bus, train, etc.) leave a station, stop, etc. **5** (of a vehicle) move out from the side of the road, or to overtake. **pull over** (of a vehicle) pull in. **pull one's punches** avoid using one's full force. **pull the plug on** put an end to (by withdrawing resources etc.). **pull rank** take unfair advantage of one's seniority. **pull round** (or **through**) (cause to) recover from an illness. **pull strings** exert (esp. clandestine) influence. **pull together** work in harmony. **pull up 1** (cause to) stop moving. **2** pull out of the ground. **3** reprimand. **4** check oneself. **pull one's weight** (often *refl.*) do one's fair share of work. [Old English]

pullet /'pʊlɪt/ *n.* young hen, esp. one less than one year old. [Latin *pullus*]

pulley /'pʊlɪ/ *n.* (*pl.* **-s**) **1** grooved wheel or wheels for a cord etc. to pass over, set in a block and used for changing the direction of a force. **2** wheel or drum fixed on a shaft and turned by a belt, used esp. to increase speed or power. [French *polie*: related to POLE[2]]

pull-in *n.* roadside café or other stopping-place.

Pullman /'pʊlmən/ *n.* (*pl.* **-s**) **1** luxurious railway carriage or motor coach. **2** sleeping-car. [*Pullman*, name of the designer]

pull-out *n.* removable section of a magazine etc.

pullover *n.* knitted garment put on over the head and covering the top half of the body.

pullulate /'pʌljʊ,leɪt/ *v.* (**-ting**) **1** (of a seed, shoot, etc.) bud, sprout. **2** swarm, teem. **3** develop; spring up. **4** (foll. by *with*) abound. □ **pullulation** /-'leɪʃ(ə)n/ *n.* [Latin *pullulo* sprout]

pulmonary /'pʌlmənərɪ/ *adj.* **1** of the lungs. **2** having lungs or lunglike organs. **3** affected with or susceptible to lung disease. [Latin *pulmo -onis* lung]

pulp —*n.* **1** soft fleshy part of fruit etc. **2** soft thick wet mass, esp. from rags, wood, etc., used in paper-making. **3** (often *attrib.*) cheap fiction etc., orig. printed on rough paper. —*v.* reduce to or become pulp. □ **pulpy** *adj.* **pulpiness** *n.* [Latin]

pulpit /'pʊlpɪt/ *n.* **1** raised enclosed platform in a church etc. from which the preacher delivers a sermon. **2** (prec. by *the*) preachers collectively; preaching. [Latin *pulpitum* platform]

pulpwood *n.* timber suitable for making paper-pulp.

pulsar /'pʌlsɑ:(r)/ *n.* cosmic source of regular rapid pulses of radiation, e.g. a rotating neutron star. [from *puls*ating *star*, after *quasar*]

pulsate /pʌl'seɪt/ *v.* (**-ting**) **1** expand and contract rhythmically; throb. **2** vibrate, quiver, thrill. □ **pulsation** *n.* **pulsatory** *adj.* [Latin: related to PULSE[1]]

pulse[1] *–n.* **1 a** rhythmical throbbing of the arteries as blood is propelled through them, esp. in the wrists, temples, etc. **b** each beat of the arteries or heart. **2** throb or thrill of life or emotion. **3** general feeling or opinion. **4** single vibration of sound, electric current, light, etc., esp. as a signal. **5** rhythmical beat, esp. of music. *–v.* (**-sing**) pulsate. [Latin *pello puls-* drive, beat]

pulse[2] *n.* (as *sing.* or *pl.*) **1** edible seeds of various leguminous plants, e.g. chickpeas, lentils, beans, etc. **2** plant producing these. [Latin *puls*]

pulverize /'pʌlvə,raɪz/ *v.* (also **-ise**) (**-zing** or **-sing**) **1** reduce or crumble to fine particles or dust. **2** *colloq.* demolish, defeat utterly. □ **pulverization** /-'zeɪʃ(ə)n/ *n.* [Latin *pulvis -ver-* dust]

puma /'pju:mə/ *n.* wild American greyish-brown cat. [Spanish from Quechua]

pumice /'pʌmɪs/ *n.* (in full **pumicestone**) **1** light porous volcanic rock used in cleaning or polishing. **2** piece of this used for removing hard skin etc. [Latin *pumex pumic-*]

pummel /'pʌm(ə)l/ *v.* (**-ll-**; *US* **-l-**) strike repeatedly, esp. with the fists. [from POMMEL]

pump[1] *–n.* **1** machine or device for raising or moving liquids, compressing gases, inflating tyres, etc. **2** act of pumping; stroke of a pump. *–v.* **1** (often foll. by *in, out, into, up*, etc.) raise or remove (liquid, gas, etc.) with a pump. **2** (often foll. by *up*) fill (a tyre etc.) with air. **3** remove (water etc.) with a pump. **4** work a pump. **5** (often foll. by *out*) (cause to) move, pour forth, etc., as if by pumping. **6** persistently question (a person) to obtain information. **7 a** move vigorously up and down. **b** shake (a person's hand) effusively. □ **pump iron** *colloq.* exercise with weights. [origin uncertain]

pump[2] *n.* **1** plimsoll. **2** light dancing shoe. **3** *US* court shoe. [origin unknown]

pumpernickel /'pʌmpə,nɪk(ə)l/ *n.* German wholemeal rye bread. [German]

pumpkin /'pʌmpkɪn/ *n.* **1** large rounded yellow or orange fruit cooked as a vegetable. **2** large-leaved tendrilled plant bearing this. [Greek *pepōn* melon]

pun *–n.* humorous use of a word or words with two or more meanings; play on words. *–v.* (**-nn-**) (foll. by *on*; also *absol.*) make a pun or puns with (words). [origin unknown]

punch[1] *–v.* **1** strike, esp. with a closed fist. **2 a** pierce a hole in (metal, paper, etc.) as or with a punch. **b** pierce (a hole) thus. *–n.* **1** blow with a fist. **2** ability to deliver this. **3** *colloq.* vigour, momentum; effective force. **4** tool, machine, or device for punching holes or impressing a design in leather, metal, etc. □ **puncher** *n.* [var. of *pounce* emboss]

punch[2] *n.* drink of wine or spirits mixed with water, fruit juices, spices, etc., and usu. served hot. [origin unknown]

punch[3] *n.* (**Punch**) grotesque humpbacked puppet in *Punch and Judy* shows. □ **as pleased as Punch** extremely pleased. [abbreviation of *Punchinello*, name of the chief character in an Italian puppet-show]

punchball *n.* stuffed or inflated ball on a stand for punching as exercise or training.

punch-bowl *n.* **1** bowl for punch. **2** deep round hollow in a hill.

punch card *n.* (also **punched card** or **tape**) card etc. perforated according to a code, for conveying instructions or data to a data processor etc.

punch-drunk *adj.* stupefied from or as if from a series of heavy blows.

punch-line *n.* words giving the point of a joke or story.

punch-up *n. colloq.* fist-fight; brawl.

punchy *adj.* (**-ier, -iest**) vigorous; forceful.

punctilio /pʌŋk'tɪlɪəʊ/ *n.* (*pl.* **-s**) **1** delicate point of ceremony or honour. **2** etiquette of such points. **3** petty formality. [Italian and Spanish: related to POINT]

punctilious /pʌŋk'tɪlɪəs/ *adj.* **1** attentive to formality or etiquette. **2** precise in behaviour. □ **punctiliously** *adv.* **punctiliousness** *n.* [Italian: related to PUNCTILIO]

punctual /'pʌŋktʃʊəl/ *adj.* keeping to the appointed time; prompt. □ **punctuality** /-'ælɪtɪ/ *n.* **punctually** *adv.* [medieval Latin: related to POINT]

punctuate /'pʌŋktʃʊ,eɪt/ *v.* (**-ting**) **1** insert punctuation marks in. **2** interrupt at intervals (*punctuated his tale with heavy sighs*). [medieval Latin: related to PUNCTUAL]

punctuation /,pʌŋktʃʊ'eɪʃ(ə)n/ *n.* **1** system of marks used to punctuate a written passage. **2** use of, or skill in using, these.

punctuation mark *n.* any of the marks (e.g. full stop and comma) used in writing to separate sentences etc. and clarify meaning.

puncture /ˈpʌŋktʃə(r)/ *–n.* **1** prick or pricking, esp. the accidental piercing of a pneumatic tyre. **2** hole made in this way. *–v.* (**-ring**) **1** make or undergo a puncture (in). **2** prick, pierce, or deflate (pomposity etc.). [Latin *punctura*: related to POINT]

pundit /ˈpʌndɪt/ *n.* **1** (also **pandit**) learned Hindu. **2** often *iron.* expert. □ **punditry** *n.* [Hindustani from Sanskrit]

pungent /ˈpʌndʒ(ə)nt/ *adj.* **1** sharp or strong in taste or smell, esp. producing a smarting or pricking sensation. **2** (of remarks) penetrating, biting, caustic. **3** mentally stimulating. □ **pungency** *n.* [Latin: related to POINT]

punish /ˈpʌnɪʃ/ *v.* **1** inflict retribution on (an offender) or for (an offence). **2** *colloq.* inflict severe blows on (an opponent). **3** tax, abuse, or treat severely or improperly. □ **punishable** *adj.* **punishing** *adj.* [Latin *punio*]

punishment *n.* **1** punishing or being punished. **2** loss or suffering inflicted in this. **3** *colloq.* severe treatment or suffering.

punitive /ˈpju:nɪtɪv/ *adj.* **1** inflicting or intended to inflict punishment. **2** (of taxation etc.) extremely severe. [French or medieval Latin: related to PUNISH]

Punjabi /pʊnˈdʒɑ:bɪ/ *–n.* (*pl.* **-s**) **1** native of Punjab. **2** language of Punjab. *–adj.* of Punjab, its people, or language. [*Punjab*, State in India and province in Pakistan]

punk *n.* **1 a** (in full **punk rock**) anti-establishment and deliberately outrageous style of rock music. **b** (in full **punk rocker**) devotee of this. **2** esp. *US* young hooligan or petty criminal; lout. **3** soft crumbly fungus-infested wood used as tinder. [origin unknown]

punkah /ˈpʌŋkə/ *n.* large swinging cloth fan on a frame, worked by a cord or electrically. [Hindi]

punnet /ˈpʌnɪt/ *n.* small light basket or container for fruit or vegetables. [origin unknown]

punster *n.* person who makes puns, esp. habitually.

punt[1] *–n.* square-ended flat-bottomed pleasure boat propelled by a long pole. *–v.* **1** propel (a punt) with a pole. **2** travel or convey in a punt. □ **punter** *n.* [Low German or Dutch]

punt[2] *–v.* kick (a ball, esp. in Rugby) after it has dropped from the hands and before it reaches the ground. *–n.* such a kick. [origin unknown]

punt[3] *v.* **1** *colloq.* **a** bet on a horse etc. **b** speculate in shares etc. **2** (in some card-games) lay a stake against the bank. [French *ponter*]

punt[4] /pʊnt/ *n.* chief monetary unit of the Republic of Ireland. [Irish, = pound]

punter *n. colloq.* **1** person who gambles or lays a bet. **2 a** customer or client; member of an audience. **b** prostitute's client.

puny /ˈpju:nɪ/ *adj.* (**-ier, -iest**) **1** undersized. **2** weak, feeble. [French *puisné* born afterwards]

pup *–n.* young dog, wolf, rat, seal, etc. *–v.* (**-pp-**) (also *absol.*) (of a bitch etc.) bring forth (young). [from PUPPY]

pupa /ˈpju:pə/ *n.* (*pl.* **pupae** /-pi:/) insect in the stage between larva and imago. □ **pupal** *adj.* [Latin, = doll]

pupil[1] /ˈpju:pɪl/ *n.* person taught by another, esp. a schoolchild or student. [Latin *pupillus*, *-illa* diminutives of *pupus* boy, *pupa* girl]

pupil[2] *n.* dark circular opening in the centre of the iris of the eye. [related to PUPIL[1]]

puppet /ˈpʌpɪt/ *n.* **1** small figure moved esp. by strings as entertainment. **2** person controlled by another. □ **puppetry** *n.* [var. of POPPET]

puppet State *n.* country that is nominally independent but actually under the control of another power.

puppy /ˈpʌpɪ/ *n.* (*pl.* **-ies**) **1** young dog. **2** conceited or arrogant young man. [French: related to POPPET]

puppy-fat *n.* temporary fatness of a child or adolescent.

puppy love *n.* = CALF-LOVE.

purblind /ˈpɜ:blaɪnd/ *adj.* **1** partly blind; dim-sighted. **2** obtuse, dim-witted. □ **purblindness** *n.* [from *pur(e)* (= 'utterly') *blind*]

purchase /ˈpɜ:tʃəs/ *–v.* (**-sing**) **1** buy. **2** (often foll. by *with*) obtain or achieve at some cost. *–n.* **1** buying. **2** thing bought. **3 a** firm hold to prevent slipping; leverage. **b** device or tackle for moving heavy objects. **4** annual rent or return from land. □ **purchaser** *n.* [Anglo-French: related to PRO-[1], CHASE[1]]

purdah /ˈpɜ:də/ *n. Ind.* screening of women from strangers by a veil or curtain in some Muslim and Hindu societies. [Urdu]

pure *adj.* **1** unmixed, unadulterated (*pure white*; *pure malice*). **2** of unmixed origin or descent (*pure-blooded*). **3** chaste. **4** not morally corrupt. **5** guiltless. **6** sincere. **7** (of a sound) perfectly in tune. **8** (of a subject of study) abstract, not applied. □ **pureness** *n.* [Latin *purus*]

purée /ˈpjʊəreɪ/ *–n.* smooth pulp of vegetables or fruit etc. *–v.* **(-ées, -éed)** make a purée of. [French]

purely *adv.* **1** in a pure manner. **2** merely, solely, exclusively.

purgative /ˈpɜːgətɪv/ *–adj.* **1** serving to purify. **2** strongly laxative. *–n.* **1** purgative thing. **2** laxative. [Latin: related to PURGE]

purgatory /ˈpɜːgətərɪ/ *–n.* (*pl.* **-ies**) **1** *RC Ch.* supposed place or state of expiation of petty sins after death and before entering heaven. **2** place or state of temporary suffering or expiation. *–adj.* purifying. □ **purgatorial** /-ˈtɔːrɪəl/ *adj.* [medieval Latin: related to PURGE]

purge *–v.* **(-ging) 1** (often foll. by *of, from*) make physically or spiritually clean. **2** remove by cleansing. **3** rid (an organization, party, etc.) of unacceptable members. **4 a** empty (the bowels). **b** empty the bowels of (a person). **5** *Law* atone for (an offence, esp. contempt of court). *–n.* **1** act of purging. **2** purgative. [Latin *purgo* purify]

purify /ˈpjʊərɪ,faɪ/ *v.* **(-ies, -ied) 1** clear of extraneous elements; make pure. **2** (often foll. by *of, from*) make ceremonially pure or clean. □ **purification** /-fɪˈkeɪʃ(ə)n/ *n.* **purificatory** /-fɪ,keɪtərɪ/ *adj.* **purifier** *n.*

purist /ˈpjʊərɪst/ *n.* advocate of scrupulous purity, esp. in language or art. □ **purism** *n.* **puristic** /-ˈrɪstɪk/ *adj.*

puritan /ˈpjʊərɪt(ə)n/ *–n.* **1 (Puritan)** *hist.* member of a group of English Protestants who sought to simplify and regulate forms of worship after the Reformation. **2** purist member of any party. **3** strict observer of religion or morals. *–adj.* **1 (Puritan)** *hist.* of the Puritans. **2** scrupulous and austere in religion or morals. □ **puritanism** *n.* [Latin: related to PURE]

puritanical /,pjʊərɪˈtænɪk(ə)l/ *adj.* strictly religious or moral in behaviour. □ **puritanically** *adv.*

purity /ˈpjʊərɪtɪ/ *n.* pureness, cleanness.

purl[1] *–n.* **1** knitting stitch made by putting the needle through the front of the previous stitch and passing the yarn round the back of the needle. **2** chain of minute loops decorating the edges of lace etc. *–v.* (also *absol.*) knit with a purl stitch. [origin unknown]

purl[2] *v.* (of a brook etc.) flow with a babbling sound. [imitative]

purler *n. colloq.* headlong fall. [*purl* overturn]

purlieu /ˈpɜːljuː/ *n.* (*pl.* **-s**) **1** person's bounds, limits, or usual haunts. **2** *hist.* tract on the border of a forest. **3** (in *pl.*) outskirts, outlying region. [Anglo-French *puralé* from *aller* go]

purlin /ˈpɜːlɪn/ *n.* horizontal beam along the length of a roof. [Anglo-Latin *perlio*]

purloin /pəˈlɔɪn/ *v. formal* or *joc.* steal, pilfer. [Anglo-French *purloigner* from *loign* far]

purple /ˈpɜːp(ə)l/ *–n.* **1** colour between red and blue. **2** (in full **Tyrian purple**) crimson dye obtained from some molluscs. **3** purple robe, esp. of an emperor or senior magistrate. **4** scarlet official dress of a cardinal. **5** (prec. by *the*) position of rank, authority, or privilege. *–adj.* of a purple colour. *–v.* **(-ling)** make or become purple. □ **purplish** *adj.* [Greek *porphura*, a shellfish yielding dye]

purple heart *n. colloq.* heart-shaped stimulant tablet, esp. of amphetamine.

purple passage *n.* (also **purple patch**) ornate or elaborate literary passage.

purport *–v.* /pəˈpɔːt/ **1** profess; be intended to seem (*purports to be an officer*). **2** (often foll. by *that*) (of a document or speech) have as its meaning; state. *–n.* /ˈpɜːpɔːt/ **1** ostensible meaning. **2** sense or tenor (of a document or statement). □ **purportedly** /pəˈpɔːtɪdlɪ/ *adv.* [Latin: related to PRO-[1], *porto* carry]

purpose /ˈpɜːpəs/ *–n.* **1** object to be attained; thing intended. **2** intention to act. **3** resolution, determination. *–v.* **(-sing)** have as one's purpose; design, intend. □ **on purpose** intentionally. **to no purpose** with no result or effect. **to the purpose 1** relevant. **2** useful. [Latin *propono* PROPOSE]

purpose-built *adj.* (also **purpose-made**) built or made for a specific purpose.

purposeful *adj.* **1** having or indicating purpose. **2** intentional. **3** resolute. □ **purposefully** *adv.* **purposefulness** *n.*

purposeless *adj.* having no aim or plan.

purposely *adv.* on purpose.

purpose-made var. of PURPOSE-BUILT.

purposive /ˈpɜːpəsɪv/ *adj.* **1** having, serving, or done with a purpose. **2** purposeful; resolute.

purr /pɜː(r)/ *–v.* **1** (of a cat) make a low vibratory sound expressing contentment. **2** (of machinery etc.) run smoothly and quietly. **3** (of a person) express pleasure; utter purringly. *–n.* purring sound. [imitative]

purse *–n.* **1** small pouch for carrying money on the person. **2** *US* handbag. **3** money, funds. **4** sum as a present or prize in a contest. *–v.* **(-sing) 1** (often foll. by *up*) pucker or contract (the lips etc.). **2** become wrinkled. □ **hold the**

purse-strings have control of expenditure. [Greek, = leather bag]

purser *n.* officer on a ship who keeps the accounts, esp. the head steward in a passenger vessel.

pursuance /pə'sju:əns/ *n.* (foll. by *of*) carrying out or observance (of a plan, idea, etc.).

pursuant *adv.* (foll. by *to*) in accordance with. [French: related to PURSUE]

pursue /pə'sju:/ *v.* (**-sues, -sued, -suing**) **1** follow with intent to overtake, capture, or do harm to; go in pursuit. **2** continue or proceed along (a route or course of action). **3** follow or engage in (study or other activity). **4** proceed according to (a plan etc.). **5** seek after, aim at. **6** continue to investigate or discuss (a topic). **7** importune (a person) persistently. **8** (of misfortune etc.) persistently assail. □ **pursuer** *n.* [Latin *sequor* follow]

pursuit /pə'sju:t/ *n.* **1** act of pursuing. **2** occupation or activity pursued. □ **in pursuit of** pursuing. [French: related to SUIT]

pursuivant /'pɜ:sɪv(ə)nt/ *n.* officer of the College of Arms below a herald. [French: related to PURSUE]

purulent /'pjʊərʊlənt/ *adj.* of, containing, or discharging pus. □ **purulence** *n.* [Latin: related to PUS]

purvey /pə'veɪ/ *v.* provide or supply (food etc.) as one's business. □ **purveyor** *n.* [Latin: related to PROVIDE]

purview /'pɜ:vju:/ *n.* **1** scope or range of a document, scheme, etc. **2** range of physical or mental vision. [Anglo-French past part.: related to PURVEY]

pus *n.* thick yellowish or greenish liquid produced from infected tissue. [Latin *pus puris*]

push /pʊʃ/ **–***v.* **1** exert a force on (a thing) to move it or cause it to move away. **2** exert such a force (*do not push against the door*). **3 a** thrust forward or upward. **b** (cause to) project (*pushes out new roots*). **4** move forward or make (one's way) by force or persistence. **5** exert oneself, esp. to surpass others. **6** (often foll. by *to*, *into*, or *to* + infin.) urge, impel, or press (a person) hard; harass. **7** (often foll. by *for*) pursue or demand (a claim etc.) persistently. **8** promote, e.g. by advertising. **9** *colloq.* sell (a drug) illegally. **–***n.* **1** act of pushing; shove, thrust. **2** force exerted in this. **3** vigorous effort. **4** military attack in force. **5** enterprise, ambition. **6** use of influence to advance a person. □ **be pushed for** *colloq.* have very little of (esp. time). **give** (or **get**) **the push** *colloq.* dismiss or send (or be dismissed or sent) away. **push about** = *push around*. **push along** (often in *imper.*) *colloq.* depart, leave. **push around** *colloq.* bully. **push one's luck 1** take undue risks. **2** act presumptuously. **push off 1** push with an oar etc. to get a boat out into a river etc. **2** (often in *imper.*) *colloq.* go away. **push through** get (a scheme, proposal, etc.) completed or accepted quickly. [Latin: related to PULSATE]

push-bike *n. colloq.* bicycle.

push-button *n.* **1** button to be pushed, esp. to operate an electrical device. **2** (*attrib.*) operated thus.

pushchair *n.* folding chair on wheels, for pushing a young child along in.

pusher *n. colloq.* seller of illegal drugs.

pushful *adj.* pushy; arrogant. □ **pushfully** *adv.*

pushing *adj.* **1** pushy. **2** *predic colloq.* having nearly reached (a specified age).

pushover *n. colloq.* **1** something easily done. **2** person easily persuaded, defeated, etc.

push-start **–***n.* starting of a vehicle by pushing it to turn the engine. **–***v.* start (a vehicle) in this way.

Pushtu /'pʌʃtu:/ *n.* & *adj.* = PASHTO. [Persian]

push-up *n.* = PRESS-UP.

pushy *adj.* (**-ier, -iest**) *colloq.* excessively self-assertive. □ **pushily** *adv.* **pushiness** *n.*

pusillanimous /ˌpju:sɪ'lænɪməs/ *adj. formal* cowardly, timid. □ **pusillanimity** /-lə'nɪmɪtɪ/ *n.* [Church Latin *pusillanimis* from *pusillus* very small, *animus* mind]

puss /pʊs/ *n. colloq.* **1** cat (esp. as a form of address). **2** sly or coquettish girl. [Low German or Dutch]

pussy /'pʊsɪ/ *n.* (*pl.* **-ies**) **1** (also **pussy-cat**) *colloq.* cat. **2** *coarse slang* vulva.

pussyfoot *v. colloq.* **1** move stealthily. **2** equivocate; stall.

pussy willow *n.* willow with furry catkins.

pustulate /'pʌstjʊˌleɪt/ *v.* (**-ting**) form into pustules. [Latin: related to PUSTULE]

pustule /'pʌstju:l/ *n.* pimple containing pus. □ **pustular** *adj.* [Latin *pustula*]

put /pʊt/ **–***v.* (**-tt-**; *past* and *past part.* **put**) **1** move to or cause to be in a specified place or position (*put it in your pocket*; *put the children to bed*). **2** bring into a specified condition or state (*puts me in great difficulty*). **3** (often foll. by *on*, *to*) impose, enforce, assign, or apply (*put a tax on beer*; *where do you put the blame?*; *put a stop to it*; *put it to good use*). **4** place (a person) or (*refl.*) imagine (oneself) in a specified position (*put them at their ease*; *put yourself in my shoes*). **5** (foll. by *for*) substitute (one thing) for (another). **6** express in a specified way (*to put it mildly*). **7** (foll. by *at*) estimate (an amount etc.) at so much (*put the cost at £50*). **8** (foll. by *into*) express or translate in (words, or another language). **9** (foll. by *into*) invest (money in an asset, e.g. land). **10** (foll. by *on*) stake (money) on (a horse etc.). **11** (foll. by *to*) submit for attention (*put it to a vote*). **12** throw (esp. a shot or weight) as a sport. **13** (foll. by *back*, *off*, *out to sea*, etc.) (of a ship etc.) proceed in a specified direction. *–n.* throw of the shot etc. □ **put about 1** spread (information, a rumour, etc.). **2** *Naut.* turn round; put (a ship) on the opposite tack. **put across 1** communicate (an idea etc.) effectively. **2** (often in **put it** (or **one**) **across**) achieve by deceit. **put away 1** restore (a thing) to its usual or former place. **2** lay (money etc.) aside for future use. **3** imprison or commit to a home etc. **4** consume (food and drink), esp. in large quantities. **5** = *put down* 7. **put back 1** = *put away* 1. **2** change (a meeting etc.) to a later date or time. **3** move back the hands of (a clock or watch). **put a bold** etc. **face on it** see FACE. **put the boot in** see BOOT. **put by** = *put away* 2. **put down 1** suppress by force. **2** *colloq.* snub, humiliate. **3** record or enter in writing. **4** enter the name of (a person) on a list. **5** (foll. by *as*, *for*) account or reckon. **6** (foll. by *to*) attribute (*put it down to bad planning*). **7** put (an old or sick animal) to death. **8** pay as a deposit. **9** stop to let (passengers) get off. **put an end to** see END. **put one's foot down** see FOOT. **put one's foot in it** see FOOT. **put forth** (of a plant) send out (buds or leaves). **put forward 1** suggest or propose. **2** advance the time shown on (a clock or watch). **put in 1 a** enter or submit (a claim etc.). **b** (foll. by *for*) submit a claim for (a specified thing). **2** (foll. by *for*) be a candidate for (an appointment, election, etc.). **3** spend (time). **4** interpose (a remark, blow, etc.). **put it to a person** (often foll. by *that*) challenge a person to deny. **put off 1 a** postpone. **b** postpone an engagement with (a person). **2** (often foll. by *with*) evade (a person) with an excuse etc. **3** hinder, dissuade; offend, disconcert. **put on 1** clothe oneself with. **2** cause (a light etc.) to function. **3** cause (transport) to be available. **4** stage (a play, show, etc.). **5** advance the hands of (a clock or watch). **6 a** pretend to (an emotion). **b** assume, take on (a character or appearance). **c** (**put it on**) exaggerate one's feelings etc. **7** increase one's weight by (a specified amount). **8** (foll. by *to*) make aware of or put in touch with (*put us on to their new accountant*). **put on weight** increase one's weight. **put out 1 a** (often as **put out** *adj.*) disconcert or annoy. **b** (often *refl.*) inconvenience (*don't put yourself out*). **2** extinguish (a fire or light). **3** cause (a batsman or side) to be out. **4** allocate (work) to be done off the premises. **5** blind (a person's eyes). **put over** = *put across* 1. **put a sock in it** see SOCK[1]. **put through 1** carry out or complete. **2** (often foll. by *to*) connect by telephone. **put to flight** see FLIGHT[2]. **put together 1** assemble (a whole) from parts. **2** combine (parts) to form a whole. **put under** make unconscious. **put up 1** build, erect. **2** raise (a price etc.). **3** take or provide with accommodation (*put me up for the night*). **4** engage in (a defensive fight, struggle, etc.). **5** present (a proposal). **6** present oneself, or propose, for election. **7** provide (money) as a backer. **8** display (a notice). **9** offer for sale or competition. **put upon** (usu. in *passive*) *colloq.* take advantage of (a person) unfairly or excessively. **put a person up to** (usu. foll. by verbal noun) instigate a person to (*put them up to stealing*). **put up with** endure, tolerate. **put the wind up** see WIND[1]. **put a person wise** see WISE. **put words into a person's mouth** see MOUTH. [Old English]

putative /'pju:tətɪv/ *adj. formal* reputed, supposed (*his putative father*). [Latin *puto* think]

put-down *n. colloq.* snub.

put-on *n. colloq.* deception or hoax.

putrefy /'pju:trɪ,faɪ/ *v.* (**-ies, -ied**) **1** become or make putrid; go bad. **2** fester, suppurate. **3** become morally corrupt. □ **putrefaction** /-'fækʃ(ə)n/ *n.* **putrefactive** /-'fæktɪv/ *adj.* [Latin *puter putris* rotten]

putrescent /pju:'tres(ə)nt/ *adj.* rotting. □ **putrescence** *n.* [Latin: related to PUTRID]

putrid /'pju:trɪd/ *adj.* **1** decomposed, rotten. **2** foul, noxious. **3** corrupt. **4** *slang* of poor quality; contemptible; very unpleasant. □ **putridity** /-'trɪdɪtɪ/ *n.* [Latin *putreo* rot (v.)]

putsch /pʊtʃ/ *n.* attempt at political revolution; violent uprising. [Swiss German]

putt *–v.* (**-tt-**) strike (a golf ball) gently on a putting-green. *–n.* putting stroke. [from PUT]

puttee /'pʌtɪ/ *n. hist.* long strip of cloth wound round the leg from ankle to knee for protection and support, worn esp. by soldiers. [Hindi]

putter[1] *n.* golf club for putting.

putter[2] *US* var. of POTTER[1].

putting-green *n.* (in golf) smooth area of grass round a hole.

putty /'pʌtɪ/ *–n.* cement of whiting and linseed oil, used for fixing panes of glass, filling holes, etc. *–v.* (**-ies**, **-ied**) cover, fix, join, or fill with putty. [French *potée*: related to POT]

put-up job *n. colloq.* fraudulent scheme.

puzzle /'pʌz(ə)l/ *–n.* **1** difficult or confusing problem. **2** problem or toy designed to test knowledge or ingenuity. *–v.* (**-ling**) **1** confound or disconcert mentally. **2** (usu. foll. by *over* etc.) be perplexed (about). **3** (usu. as **puzzling** *adj.*) require much mental effort (*puzzling situation*). **4** (foll. by *out*) solve or understand by hard thought. □ **puzzlement** *n.* [origin unknown]

puzzler *n.* difficult question or problem.

PVC *abbr.* polyvinyl chloride.

PW *abbr.* policewoman.

PWR *abbr.* pressurized-water reactor.

pyaemia /paɪ'i:mɪə/ *n.* (*US* **pyemia**) blood-poisoning caused by pus-forming bacteria in the bloodstream. [Greek *puon* pus, *haima* blood]

pygmy /'pɪgmɪ/ *n.* (also **pigmy**) (*pl.* **-ies**) (often *attrib.*) **1** member of a dwarf people of esp. equatorial Africa. **2** very small person, animal, or thing. **3** insignificant person. [Latin from Greek]

pyjamas /pə'dʒɑ:məz/ *n.pl.* (*US* **pajamas**) **1** suit of loose trousers and jacket for sleeping in. **2** loose trousers worn by both sexes in some Asian countries. **3** (**pyjama**) (*attrib.*) of either part of a pair of pyjamas (*pyjama jacket*). [Urdu, = leg-clothing]

pylon /'paɪlən/ *n.* tall structure, esp. as a support for electric-power cables etc. [Greek *pulē* gate]

pyorrhoea /,paɪə'rɪə/ *n.* (*US* **pyorrhea**) **1** gum disease causing loosening of the teeth. **2** discharge of pus. [Greek *puon* pus, *rheō* flow]

pyracantha /,paɪərə'kænθə/ *n.* evergreen thorny shrub with white flowers and bright red or yellow berries. [Latin from Greek]

pyramid /'pɪrəmɪd/ *n.* **1** monumental, esp. stone, structure, with a square base and sloping triangular sides meeting at an apex, esp. an ancient Egyptian royal tomb. **2** solid of this shape with esp. a square or triangular base. **3** pyramid-shaped thing or pile of things. □ **pyramidal** /-'ræmɪd(ə)l/ *adj.* [Greek *puramis -mid-*]

pyramid selling *n.* system of selling goods in which agency rights are sold to an increasing number of distributors at successively lower levels.

pyre /'paɪə(r)/ *n.* heap of combustible material, esp. for burning a corpse. [Greek: related to PYRO-]

pyrethrum /paɪ'ri:θrəm/ *n.* **1** aromatic chrysanthemum. **2** insecticide from its dried flowers. [Latin from Greek]

pyretic /paɪ'retɪk/ *adj.* of, for, or producing fever. [Greek *puretos* fever]

Pyrex /'paɪəreks/ *n. propr.* hard heat-resistant glass, used esp. for ovenware. [invented word]

pyrexia /paɪ'reksɪə/ *n. Med.* = FEVER. [Greek *purexis*]

pyrites /paɪ'raɪti:z/ *n.* (in full **iron pyrites**) lustrous yellow mineral that is a sulphide of iron. [Greek: related to PYRE]

pyro- *comb. form* **1** denoting fire. **2** denoting a mineral etc. changed under the action of heat, or fiery in colour. [Greek *pur* fire]

pyromania /,paɪərəʊ'meɪnɪə/ *n.* obsessive desire to start fires. □ **pyromaniac** *n.* & *adj.*

pyrotechnics /,paɪərəʊ'teknɪks/ *n.pl.* **1** art of making fireworks. **2** display of fireworks. **3** any brilliant display. □ **pyrotechnic** *adj.*

pyrrhic /'pɪrɪk/ *adj.* (of a victory) won at too great a cost. [*Pyrrhus* of Epirus, who defeated the Romans in 279 BC, but suffered heavy losses]

Pythagoras' theorem /paɪ'θægərəs/ *n.* theorem that the square on the hypotenuse of a right-angled triangle is equal to the sum of the squares on the other two sides. [*Pythagoras* (6th c. BC), name of a Greek philosopher]

python /'paɪθ(ə)n/ *n.* large tropical constricting snake. [Greek *Puthōn*, name of a monster]

pyx /pɪks/ *n.* vessel for the consecrated bread of the Eucharist. [Greek *puxis* BOX[1]]

Q

Q[1] /kju:/ *n.* (also **q**) (*pl.* **Qs** or **Q's**) seventeenth letter of the alphabet.

Q[2] *abbr.* (also **Q.**) **1** Queen('s). **2** question.

QC *abbr.* Queen's Counsel.

QED *abbr.* which was to be proved. [Latin *quod erat demonstrandum*]

QM *abbr.* quartermaster.

qr. *abbr.* quarter(s).

qt *abbr.* quart(s).

qua /kwɑ:, kweɪ/ *conj.* in the capacity of. [Latin, = in the way in which]

quack[1] –*n.* harsh sound made by ducks. –*v.* utter this sound. [imitative]

quack[2] *n.* **1** unqualified practitioner, esp. of medicine; charlatan (often *attrib.*: *quack cure*). **2** *slang* any doctor. □ **quackery** *n.* [abbreviation of *quacksalver* from Dutch: probably related to QUACK[1], SALVE[1]]

quad[1] /kwɒd/ *n. colloq.* quadrangle. [abbreviation]

quad[2] /kwɒd/ *n. colloq.* quadruplet. [abbreviation]

quad[3] /kwɒd/ *colloq.* –*n.* quadraphonics. –*adj.* quadraphonic. [abbreviation]

Quadragesima /ˌkwɒdrəˈdʒesɪmə/ *n.* first Sunday in Lent. [Latin *quadragesimus* fortieth]

quadrangle /ˈkwɒdˌræŋg(ə)l/ *n.* **1** four-sided plane figure, esp. a square or rectangle. **2** four-sided court, esp. in colleges. □ **quadrangular** /-ˈræŋgjʊlə(r)/ *adj.* [Latin: related to QUADRI-, ANGLE[1]]

quadrant /ˈkwɒdrənt/ *n.* **1** quarter of a circle's circumference. **2** quarter of a circle enclosed by two radii at right angles. **3** quarter of a sphere etc. **4** any of four parts of a plane divided by two lines at right angles. **5 a** graduated quarter-circular strip of metal etc. **b** instrument graduated (esp. through an arc of 90°) for measuring angles. [Latin *quadrans -ant-*]

quadraphonic /ˌkwɒdrəˈfɒnɪk/ *adj.* (of sound reproduction) using four transmission channels. □ **quadraphonically** *adv.* **quadraphonics** *n.pl.* [from QUADRI-, STEREOPHONIC]

quadrate –*adj.* /ˈkwɒdrət/ esp. *Anat.* & *Zool.* square or rectangular. –*n.* /ˈkwɒdrət, -dreɪt/ rectangular object. –*v.* /kwɒˈdreɪt/ (**-ting**) make square. [Latin *quadro* make square]

quadratic /kwɒˈdrætɪk/ *Math.* –*adj.* involving the square (and no higher power) of an unknown quantity or variable (*quadratic equation*). –*n.* quadratic equation.

quadri- *comb. form* four. [Latin *quattuor* four]

quadriceps /ˈkwɒdrɪˌseps/ *n.* four-headed muscle at the front of the thigh. [from QUADRI-, BICEPS]

quadrilateral /ˌkwɒdrɪˈlætər(ə)l/ –*adj.* having four sides. –*n.* four-sided figure.

quadrille /kwɒˈdrɪl/ *n.* **1** a kind of square dance. **2** music for this. [French]

quadriplegia /ˌkwɒdrɪˈpli:dʒə/ *n.* paralysis of all four limbs. □ **quadriplegic** *adj.* & *n.* [from QUADRI-, Greek *plēgē* a blow]

quadruped /ˈkwɒdrʊˌped/ *n.* four-footed animal, esp. a mammal. [Latin: related to QUADRI-, *pes ped-* foot]

quadruple /ˈkwɒdrʊp(ə)l/ –*adj.* **1** fourfold; having four parts. **2** (of time in music) having four beats in a bar. –*n.* fourfold number or amount. –*v.* (**-ling**) multiply by four. [Latin: related to QUADRI-]

quadruplet /ˈkwɒdrʊplɪt/ *n.* each of four children born at one birth.

quadruplicate –*adj.* /kwɒˈdru:plɪkət/ **1** fourfold. **2** of which four copies are made. –*v.* /kwɒˈdru:plɪˌkeɪt/ (**-ting**) multiply by four.

quaff /kwɒf/ *v. literary* **1** drink deeply. **2** drain (a cup etc.) in long draughts. □ **quaffable** *adj.* [perhaps imitative]

quagmire /ˈkwɒgˌmaɪə(r), ˈkwæg-/ *n.* **1** muddy or boggy area. **2** hazardous situation. [from *quag* bog, MIRE]

quail[1] *n.* (*pl.* same or **-s**) small game-bird related to the partridge. [French *quaille*]

quail[2] *v.* flinch; show fear. [origin unknown]

quaint *adj.* attractively odd or old-fashioned. □ **quaintly** *adv.* **quaintness** *n.* [French *cointe* from Latin *cognosco* ascertain]

quake –*v.* (**-king**) shake, tremble. –*n. colloq.* earthquake. [Old English]

Quaker *n.* member of the Society of Friends. □ **Quakerism** *n.*

qualification /ˌkwɒlɪfɪˈkeɪʃ(ə)n/ *n.* **1** accomplishment fitting a person for a position or purpose. **2** thing that modifies or limits (*statement had many qualifications*). **3** qualifying or being qualified. □ **qualificatory** /ˈkwɒl-/

adj. [French or medieval Latin: related to QUALIFY]

qualify /ˈkwɒlɪˌfaɪ/ *v.* (**-ies, -ied**) **1** (often as **qualified** *adj.*) make competent or fit for a position or purpose. **2** make legally entitled. **3** (usu. foll. by *for*) (of a person) satisfy conditions or requirements. **4** modify or limit (a statement etc.) (*qualified approval*). **5** *Gram.* (of a word) attribute a quality to (esp. a noun). **6** moderate, mitigate; make less severe. **7** (foll. by *as*) be describable as, count as (*a grunt hardly qualifies as conversation*). □ **qualifier** *n.* [Latin *qualis* such as, of what kind]

qualitative /ˈkwɒlɪtətɪv/ *adj.* of quality as opposed to quantity. □ **qualitatively** *adv.* [Latin: related to QUALITY]

quality /ˈkwɒlɪtɪ/ *n.* (*pl.* **-ies**) **1** degree of excellence. **2 a** general excellence (*has quality*). **b** (*attrib.*) of high quality (*a quality product*). **3** attribute, faculty (*has many good qualities*). **4** relative nature or character. **5** timbre of a voice or sound. **6** *archaic* high social standing (*people of quality*). [Latin *qualis* such as, of what kind]

quality control *n.* maintaining of standards in products or services by inspection, testing samples, etc.

qualm /kwɑːm/ *n.* **1** misgiving; uneasy doubt. **2** scruple of conscience. **3** momentary faint or sick feeling. [origin uncertain]

quandary /ˈkwɒndərɪ/ *n.* (*pl.* **-ies**) **1** perplexed state. **2** practical dilemma. [origin uncertain]

quango /ˈkwæŋgəʊ/ *n.* (*pl.* **-s**) semi-public body with financial support from and senior appointments made by the government. [abbreviation of *qua*si (or *qua*si-*a*utonomous) *n*on-*g*overnmen-t(al) *o*rganization]

quanta *pl.* of QUANTUM.

quantify /ˈkwɒntɪˌfaɪ/ *v.* (**-ies, -ied**) **1** determine the quantity of. **2** express as a quantity. □ **quantifiable** *adj.* **quantification** /ˌkwɒntɪfɪˈkeɪʃ(ə)n/ *n.* [medieval Latin: related to QUANTITY]

quantitative /ˈkwɒntɪtətɪv/ *adj.* **1** of quantity as opposed to quality. **2** measured or measurable by quantity.

quantity /ˈkwɒntɪtɪ/ *n.* (*pl.* **-ies**) **1** property of things that is measurable. **2** size, extent, weight, amount, or number. **3** specified or considerable portion, number, or amount (*buys in quantity*; *small quantity of food*). **4** (in *pl.*) large amounts or numbers; an abundance. **5** length or shortness of a vowel sound or of a syllable. **6** *Math.* value, component, etc. that may be expressed in numbers. [Latin *quantus* how much]

quantity surveyor *n.* person who measures and prices building work.

quantum /ˈkwɒntəm/ *n.* (*pl.* **quanta**) **1** *Physics* discrete amount of energy proportional to the frequency of radiation it represents. **2** a required or allowed amount. [Latin *quantus* how much]

quantum jump *n.* (also **quantum leap**) **1** sudden large increase or advance. **2** *Physics* abrupt transition in an atom or molecule from one quantum state to another.

quantum mechanics *n.pl.* (usu. treated as *sing.*) (also **quantum theory**) *Physics* theory assuming that energy exists in discrete units.

quarantine /ˈkwɒrənˌtiːn/ —*n.* **1** isolation imposed on a person or animal to prevent infection or contagion. **2** period of this. —*v.* (**-ning**) put in quarantine. [Italian *quaranta* forty]

quark[1] /kwɑːk/ *n. Physics* component of elementary particles. [word used by Joyce in *Finnegans Wake* (1939)]

quark[2] /kwɑːk/ *n.* a kind of low-fat curd cheese. [German]

quarrel /ˈkwɒr(ə)l/ —*n.* **1** severe or angry dispute or contention. **2** break in friendly relations. **3** cause of complaint (*have no quarrel with him*). —*v.* (**-ll-**; *US* **-l-**) **1** (often foll. by *with*) find fault. **2** dispute; break off friendly relations. [Latin *querela* from *queror* complain]

quarrelsome *adj.* given to quarrelling.

quarry[1] /ˈkwɒrɪ/ —*n.* (*pl.* **-ies**) place from which stone etc. may be extracted. —*v.* (**-ies, -ied**) extract (stone) from a quarry. [Latin *quadrum* square]

quarry[2] /ˈkwɒrɪ/ *n.* (*pl.* **-ies**) **1** intended victim or prey. **2** object of pursuit. [Latin *cor* heart]

quarry tile *n.* unglazed floor-tile.

quart /kwɔːt/ *n.* liquid measure equal to a quarter of a gallon; two pints (0.946 litre). [Latin *quartus* fourth]

quarter /ˈkwɔːtə(r)/ —*n.* **1** each of four equal parts into which a thing is divided. **2** period of three months. **3** point of time 15 minutes before or after any hour. **4 a** 25 US or Canadian cents. **b** coin for this. **5** part of a town, esp. as occupied by a particular class (*residential quarter*). **6 a** point of the compass. **b** region at this. **7** direction, district, or source of supply (*help from any quarter*). **8** (in *pl.*) **a** lodgings. **b** accommodation of troops etc. **9 a** one fourth of a lunar month. **b** moon's position between the first and second (**first quarter**) or third and fourth (**last quarter**) of these. **10 a** each of the four parts into which a carcass is divided. **b** (in *pl.*) = HINDQUARTERS. **11** mercy towards an enemy etc. on condition of surrender. **12 a** grain measure equivalent to 8 bushels. **b** one-fourth of a hundredweight. **c** *colloq.* one-fourth of a pound weight. **13** each of four divisions on a shield. –*v.* **1** divide into quarters. **2** *hist.* divide (the body of an executed person) in this way. **3 a** put (troops etc.) into quarters. **b** provide with lodgings. **4** *Heraldry* place (coats of arms) on the four quarters of a shield. [Latin *quartarius*: related to QUART]

quarterback *n.* player in American football who directs attacking play.

quarter day *n.* one of four days on which quarterly payments are due, tenancies begin and end, etc.

quarterdeck *n.* part of a ship's upper deck near the stern, usu. reserved for officers.

quarter-final *n.* match or round preceding the semifinal.

quarter-hour *n.* **1** period of 15 minutes. **2** = QUARTER *n.* 3.

quarter-light *n.* small pivoted window in the side of a car, carriage, etc.

quarterly –*adj.* produced or occurring once every quarter of a year. –*adv.* once every quarter of a year. –*n.* (*pl.* **-ies**) quarterly journal.

quartermaster *n.* **1** regimental officer in charge of quartering, rations, etc. **2** naval petty officer in charge of steering, signals, etc.

quarter sessions *n.pl. hist.* court of limited criminal and civil jurisdiction, usu. held quarterly.

quarterstaff *n. hist.* stout pole 6–8 feet long, formerly used as a weapon.

quartet /kwɔː'tet/ *n.* **1** *Mus.* **a** composition for four performers. **b** the performers. **2** any group of four. [Latin *quartus*]

quarto /'kwɔːtəʊ/ *n.* (*pl.* **-s**) **1** size of a book or page given by folding a sheet of standard size twice to form four leaves. **2** book or sheet of this size. [Latin: related to QUART]

quartz /kwɔːts/ *n.* silica in various mineral forms. [German from Slavonic]

quartz clock *n.* (also **quartz watch**) clock or watch operated by vibrations of an electrically driven quartz crystal.

quasar /'kweɪzɑː(r)/ *n. Astron.* starlike object with a large redshift. [from *quas*i-stell*ar*]

quash /kwɒʃ/ *v.* **1** annul; reject as invalid, esp. by a legal procedure. **2** suppress, crush. [French *quasser* from Latin]

quasi- /'kweɪzaɪ/ *comb. form* **1** seemingly, not really. **2** almost. [Latin *quasi* as if]

quaternary /kwə'tɜːnərɪ/ –*adj.* **1** having four parts. **2** (**Quaternary**) *Geol.* of the most recent period in the Cenozoic era. –*n.* (**Quaternary**) *Geol.* this period. [Latin *quaterni* four each]

quatrain /'kwɒtreɪn/ *n.* four-line stanza. [French *quatre* four]

quatrefoil /'kætrəˌfɔɪl/ *n.* four-pointed or -leafed figure, esp. as an architectural ornament. [Anglo-French *quatre* four: related to FOIL[2]]

quattrocento /ˌkwætrəʊ'tʃentəʊ/ *n.* 15th-c. Italian art. [Italian, = 400, used for the years 1400–99]

quaver –*v.* **1** (esp. of a voice or sound) vibrate, shake, tremble. **2** sing or say with a quavering voice. –*n.* **1** *Mus.* note half as long as a crotchet. **2** trill in singing. **3** tremble in speech. □ **quavery** *adj.* [probably imitative]

quay /kiː/ *n.* artificial landing-place for loading and unloading ships. [French]

quayside *n.* land forming or near a quay.

queasy /'kwiːzɪ/ *adj.* (**-ier, -iest**) **1 a** (of a person) nauseous. **b** (of the stomach) easily upset, weak of digestion. **2** (of the conscience etc.) overscrupulous. □ **queasily** *adv.* **queasiness** *n.* [origin uncertain]

queen –*n.* **1** (as a title usu. **Queen**) female sovereign. **2** (in full **queen consort**) king's wife. **3** woman, country, or thing pre-eminent of its kind. **4** fertile female among ants, bees, etc. **5** most powerful piece in chess. **6** court-card depicting a queen. **7** (**the Queen**) national anthem when the sovereign is female. **8** *slang offens.* male homosexual. **9** belle or mock sovereign for some event (*Queen of the May*). –*v. Chess* convert (a pawn) into a queen when it reaches the opponent's side of the board. □ **queenly** *adj.* (**-ier, -iest**). **queenliness** *n.* [Old English]

Queen-Anne *n.* (often *attrib.*) style of English architecture, furniture, etc., in the early 18th c.

queen bee *n.* **1** fertile female bee. **2** woman who behaves as if she is the most important person in a group.

queen mother *n.* dowager who is mother of the sovereign.

Queen of the May *n.* = MAY QUEEN.

queen-post *n.* either of two upright timbers between the tie-beam and main rafters of a roof-truss.

Queensberry Rules /'kwiːnzbərɪ/ *n.pl.* standard rules, esp. of boxing. [from the name Marquis of *Queensberry*]

Queen's Counsel *n.* counsel to the Crown, taking precedence over other barristers.

Queen's English *n.* (prec. by *the*) English language correctly written or spoken.

Queen's evidence see EVIDENCE.

Queen's Guide *n.* Guide who has reached the highest rank of proficiency.

Queen's highway *n.* public road, regarded as being under the sovereign's protection.

Queen's Proctor *n.* official who has the right to intervene in probate, divorce, and nullity cases when collusion or the suppression of facts is alleged.

Queen's Scout *n.* Scout who has reached the highest standard of proficiency.

queer *–adj.* **1** strange, odd, eccentric. **2** shady, suspect, of questionable character. **3** slightly ill; faint. **4** *slang offens.* (esp. of a man) homosexual. *–n. slang offens.* homosexual. *–v. slang* spoil, put out of order. □ **in Queer Street** *slang* in difficulty, esp. in debt. **queer a person's pitch** *colloq.* spoil a person's chances. [origin uncertain]

quell *v.* **1** crush or put down (a rebellion etc.). **2** suppress (fear etc.). [Old English]

quench *v.* **1** satisfy (thirst) by drinking. **2** extinguish (a fire or light). **3** cool, esp. with water. **4** esp. *Metallurgy* cool (a hot substance) in cold water etc. **5** stifle or suppress (desire etc.). [Old English]

quern *n.* hand-mill for grinding corn. [Old English]

querulous /ˈkwerʊləs/ *adj.* complaining, peevish. □ **querulously** *adj.* [Latin *queror* complain]

query /ˈkwɪərɪ/ *–n.* (*pl.* **-ies**) **1** question. **2** question mark or the word *query* as a mark of interrogation. *–v.* (**-ies, -ied**) **1** ask or inquire. **2** call in question. **3** dispute the accuracy of. [Latin *quaere* imperative of *quaero* inquire]

quest *–n.* **1** search or seeking. **2** thing sought, esp. by a medieval knight. *–v.* (often foll. by *about*) go about in search of something (esp. of dogs seeking game). [Latin *quaero quaesit-* seek]

question /ˈkwestʃ(ə)n/ *–n.* **1** sentence worded or expressed so as to seek information or an answer. **2 a** doubt or dispute about a matter (*no question that he is dead*). **b** raising of such doubt etc. **3** matter to be discussed or decided. **4** problem requiring a solution. *–v.* **1** ask questions of; interrogate; subject (a person) to examination. **2** throw doubt upon; raise objections to. □ **be just a question of time** be certain to happen sooner or later. **be a question of** be at issue, be a problem (*it's a question of money*). **call in** (or **into**) **question** express doubts about. **in question** that is being discussed or referred to (*the person in question*). **out of the question** not worth discussing; impossible. □ **questioner** *n.* **questioning** *adj.* & *n.* **questioningly** *adv.* [Latin: related to QUEST]

questionable *adj.* doubtful as regards truth, quality, honesty, wisdom, etc.

question mark *n.* punctuation mark (?) indicating a question.

question-master *n.* person presiding over a quiz game etc.

questionnaire /ˌkwestʃəˈneə(r)/ *n.* formulated series of questions, esp. for statistical analysis. [French: related to QUESTION]

question time *n.* period in Parliament when MPs may question ministers.

queue /kjuː/ *–n.* line or sequence of persons, vehicles, etc. waiting their turn. *–v.* (**queues, queued, queuing** or **queueing**) (often foll. by *up*) form or join a queue. [Latin *cauda* tail]

queue-jump *v.* push forward out of turn in a queue.

quibble /ˈkwɪb(ə)l/ *–n.* **1** petty objection; trivial point of criticism. **2** evasion; argument relying on ambiguity. **3** *archaic* pun. *–v.* (**-ling**) use quibbles. □ **quibbling** *adj.* [origin uncertain]

quiche /kiːʃ/ *n.* savoury flan. [French]

quick *–adj.* **1** taking only a short time (*quick worker*). **2** arriving after a short time, prompt. **3** with only a short interval (*in quick succession*). **4** lively, intelligent, alert. **5** (of a temper) easily roused. **6** *archaic* alive (*the quick and the dead*). *–adv.* (also as *int.*) quickly. *–n.* **1** soft sensitive flesh, esp. below the nails. **2** seat of emotion (*cut to the quick*). □ **quickly** *adv.* [Old English]

quicken *v.* **1** make or become quicker; accelerate. **2** give life or vigour to; rouse. **3 a** (of a woman) reach a stage in pregnancy when movements of the foetus can be felt. **b** (of a foetus) begin to show signs of life.

quick-fire *attrib. adj.* rapid; in rapid succession.

quick-freeze *v.* freeze (food) rapidly so as to preserve its natural qualities.

quickie *n. colloq.* thing done or made quickly.

quicklime *n.* = LIME[1].

quick one *n. colloq.* drink (usu. alcoholic) taken quickly.

quicksand *n.* (often in *pl.*) **1** area of loose wet sand that sucks in anything placed on it. **2** treacherous situation etc.

quickset *–attrib. adj.* (of a hedge etc.) formed of cuttings, esp. hawthorn. *–n.* hedge formed in this way.

quicksilver *n.* mercury.

quickstep *n.* fast foxtrot.

quick-tempered *adj.* easily angered.

quick-witted *adj.* quick to grasp a situation, make repartee, etc. □ **quick-wittedness** *n.*

quid[1] *n.* (*pl.* same) *slang* one pound sterling. □ **quids in** *slang* in a position of profit. [probably from Latin *quid* what]

quid[2] *n.* lump of tobacco for chewing. [a dialect word, = CUD]

quiddity /'kwɪdɪtɪ/ *n.* (*pl.* **-ies**) **1** *Philos.* essence of a thing. **2** quibble; trivial objection. [Latin *quidditas* from *quid* what]

quid pro quo /ˌkwɪd prəʊ 'kwəʊ/ *n.* (*pl.* **quid pro quos**) return made (for a gift, favour, etc.). [Latin, = something for something]

quiescent /kwɪ'es(ə)nt/ *adj.* inert, dormant. □ **quiescence** *n.* [related to QUIET]

quiet /'kwaɪət/ **–***adj.* **1** with little or no sound or motion. **2** of gentle or peaceful disposition. **3** unobtrusive; not showy. **4** not overt; disguised. **5** undisturbed, uninterrupted; free or far from vigorous action. **6** informal (*quiet wedding*). **7** enjoyed in quiet (*quiet smoke*). **8** not anxious or remorseful. **9** not busy (*it is very quiet at work*). **10** peaceful (*all quiet on the frontier*). **–***n.* **1** silence; stillness. **2** undisturbed state; tranquillity. **–***v.* (often foll. by *down*) make or become quiet or calm. □ **be quiet** (esp. in *imper.*) cease talking etc. **keep quiet** (often foll. by *about*) say nothing. **on the quiet** secretly. □ **quietly** *adv.* **quietness** *n.* [Latin *quiesco* become calm]

quieten *v.* (often foll. by *down*) = QUIET *v.*

quietism *n.* passive contemplative attitude towards life, esp. as a form of mysticism. □ **quietist** *n.* & *adj.* [Italian: related to QUIET]

quietude /'kwaɪɪˌtju:d/ *n.* state of quiet.

quietus /kwaɪ'i:təs/ *n.* release from life; death, final riddance (*will get its quietus*). [medieval Latin: related to QUIET]

quiff *n.* **1** man's tuft of hair, brushed upward over the forehead. **2** curl plastered down on the forehead. [origin unknown]

quill *n.* **1** (in full **quill-feather**) large feather in a wing or tail. **2** hollow stem of this. **3** (in full **quill pen**) pen made of a quill. **4** (usu. in *pl.*) porcupine's spine. [probably Low German *quiele*]

quilt **–***n.* coverlet, esp. of quilted material. **–***v.* line a coverlet or garment with padding enclosed between layers of cloth by lines of stitching. □ **quilter** *n.* **quilting** *n.* [Latin *culcita* cushion]

quim *n. coarse slang* female genitals. [origin unknown]

quin *n. colloq.* quintuplet. [abbreviation]

quince *n.* **1** acid pear-shaped fruit used in jams etc. **2** tree bearing this. [originally a plural, from French *cooin*, from *Cydonia* in Crete]

quincentenary /ˌkwɪnsen'ti:nərɪ/ **–***n.* (*pl.* **-ies**) 500th anniversary; celebration of this. **–***adj.* of this anniversary. [Latin *quinque* five]

quincunx /'kwɪnkʌŋks/ *n.* five objects, esp. trees, at the corners and centre of a square or rectangle. [Latin, = five-twelfths]

quinine /'kwɪni:n/ *n.* bitter drug obtained from cinchona bark, used as a tonic and to reduce fever. [Spanish *quina* cinchona bark, from Quechua *kina* bark]

Quinquagesima /ˌkwɪŋkwə'dʒesɪmə/ *n.* Sunday before Lent. [Latin *quinquagesimus* fiftieth]

quinquennial /kwɪŋ'kwenɪəl/ *adj.* **1** lasting five years. **2** recurring every five years. □ **quinquennially** *adv.* [Latin *quinquennis* from *quinque* five, *annus* year]

quinquereme /'kwɪŋkwɪˌri:m/ *n.* ancient Roman galley with five files of oarsmen on each side. [Latin *quinque* five, *remus* oar]

quintessence /kwɪn'tes(ə)ns/ *n.* **1** (usu. foll. by *of*) purest and most perfect form, manifestation, or embodiment of a quality etc. **2** highly refined extract. □ **quintessential** /ˌkwɪntɪ'senʃ(ə)l/ *adj.* **quintessentially** /ˌkwɪntɪ'senʃəlɪ/ *adv.* [Latin *quinta essentia* fifth substance (underlying the four elements)]

quintet /kwɪn'tet/ *n.* **1** *Mus.* **a** composition for five performers. **b** the performers. **2** any group of five. [Latin *quintus*]

quintuple /'kwɪntjʊp(ə)l/ **–***adj.* fivefold; having five parts. **–***n.* fivefold number or amount. **–***v.* (**-ling**) multiply by five. [Latin *quintus* fifth]

quintuplet /'kwɪntjʊplɪt/ *n.* each of five children born at one birth.

quintuplicate **–***adj.* /kwɪn'tju:plɪkət/ **1** fivefold. **2** of which five copies are made. **–***v.* /kwɪn'tju:plɪˌkeɪt/ (**-ting**) multiply by five.

quip **–***n.* clever saying; epigram. **–***v.* (**-pp-**) make quips. [perhaps from Latin *quippe* forsooth]

quire *n.* 25 (formerly 24) sheets of paper. [Latin: related to QUATERNARY]

quirk *n.* **1** peculiar feature, peculiarity. **2** trick of fate. □ **quirky** *adj.* (**-ier, -iest**). [origin unknown]

quisling /'kwɪzlɪŋ/ *n.* traitor, collaborator. [*Quisling*, name of a Norwegian officer and collaborator with the Nazis]

quit –*v.* (**-tting**; *past* and *past part.* **quitted** or **quit**) **1** (also *absol.*) give up, let go, abandon (a task etc.). **2** *US* cease, stop (*quit grumbling*). **3** leave or depart from. –*predic. adj.* (foll. by *of*) rid of (*glad to be quit of the problem*). [Latin: related to QUIET]

quitch *n.* (in full **quitch-grass**) = COUCH². [Old English]

quite *adv.* **1** completely, entirely, wholly. **2** to some extent, rather. **3** (often foll. by *so*) said to indicate agreement. □ **quite a** (or **some**) a remarkable or outstanding (thing or person). **quite a few** *colloq.* a fairly large number (of). **quite something** *colloq.* remarkable thing or person. [var. of QUIT]

quits *predic. adj.* on even terms by retaliation or repayment. □ **call it quits** acknowledge that things are now even; agree to stop quarrelling. [probably related to QUIT]

quitter *n.* **1** person who gives up easily. **2** shirker.

quiver¹ /ˈkwɪvə(r)/ –*v.* tremble or vibrate with a slight rapid motion. –*n.* quivering motion or sound. [obsolete *quiver* nimble]

quiver² /ˈkwɪvə(r)/ *n.* case for arrows. [Anglo-French from Germanic]

quixotic /kwɪkˈsɒtɪk/ *adj.* extravagantly and romantically chivalrous. □ **quixotically** *adv.* [Don *Quixote*, in Cervantes' romance]

quiz –*n.* (*pl.* **quizzes**) **1** test of knowledge, esp. as entertainment. **2** interrogation, examination. –*v.* (**-zz-**) examine by questioning. [origin unknown]

quizzical /ˈkwɪzɪk(ə)l/ *adj.* expressing or done with mild or amused perplexity. □ **quizzically** *adv.*

quod *n. slang* prison. [origin unknown]

quoin /kɔɪn/ *n.* **1** external angle of a building. **2** cornerstone. **3** wedge used in printing and gunnery. [var. of COIN]

quoit /kɔɪt/ *n.* **1** ring thrown to encircle an iron peg. **2** (in *pl.*) game using these. [origin unknown]

quondam /ˈkwɒndæm/ *attrib. adj.* that once was, sometime, former. [Latin adv., = formerly]

quorate /ˈkwɔːreɪt/ *adj.* constituting or having a quorum. [from QUORUM]

quorum /ˈkwɔːrəm/ *n.* minimum number of members that must be present to constitute a valid meeting. [Latin, = of whom]

quota /ˈkwəʊtə/ *n.* **1** share to be contributed to, or received from, a total. **2** number of goods, people, etc., stipulated or permitted. [Latin *quotus* from *quot* how many]

quotable /ˈkwəʊtəb(ə)l/ *adj.* worth quoting.

quotation /kwəʊˈteɪʃ(ə)n/ *n.* **1** passage or remark quoted. **2** quoting or being quoted. **3** contractor's estimate. [medieval Latin: related to QUOTE]

quotation marks *n.pl.* inverted commas (' ' or " ") used at the beginning and end of a quotation etc.

quote –*v.* (**-ting**) **1** cite or appeal to (an author, book, etc.) in confirmation of some view. **2 a** repeat or copy out a passage from. **b** (foll. by *from*) cite (an author, book, etc.). **3** (foll. by *as*) cite (an author etc.) as proof, evidence, etc. **4 a** enclose (words) in quotation marks. **b** (as *int.*) verbal formula indicating opening quotation marks (*he said, quote, 'I shall stay'*). **5** (often foll. by *at*, also *absol.*) state the price of. –*n. colloq.* **1** passage quoted. **2** price quoted. **3** (usu. in *pl.*) quotation marks. [Latin *quoto* mark with numbers]

quoth /kwəʊθ/ *v.* (only in 1st and 3rd person) *archaic* said. [Old English]

quotidian /kwɒˈtɪdɪən/ *adj.* **1** occurring or recurring daily. **2** commonplace, trivial. [Latin *cotidie* daily]

quotient /ˈkwəʊʃ(ə)nt/ *n.* result of a division sum. [Latin *quotiens -ent-* how many times]

q.v. *abbr.* which see (in references). [Latin *quod vide*]

qwerty /ˈkwɜːtɪ/ *attrib. adj.* denoting the standard keyboard on English-language typewriters etc., with *q, w, e, r, t,* and *y* as the first keys on the top row of letters.

R

R[1] /ɑː(r)/ *n.* (also **r**) (*pl.* **Rs** or **R's**) eighteenth letter of the alphabet.

R[2] *abbr.* (also **R.**) **1** *Regina* (*Elizabeth R*). **2** *Rex.* **3** River. **4** (also ®) registered as a trade mark. **5** *Chess* rook.

r. *abbr.* (also **r**) **1** right. **2** radius.

RA *abbr.* **1 a** Royal Academy. **b** Royal Academician. **2** Royal Artillery.

Ra *symb.* radium.

rabbet /ˈræbɪt/ –*n.* step-shaped channel cut along the edge or face of a length of wood etc., usu. to receive the edge or tongue of another piece. –*v.* (**-t-**) **1** join or fix with a rabbet. **2** make a rabbet in. [French *rab(b)at*: related to REBATE[1]]

rabbi /ˈræbaɪ/ *n.* (*pl.* **-s**) **1** Jewish scholar or teacher, esp. of the law. **2** Jewish religious leader. □ **rabbinical** /rəˈbɪnɪk(ə)l/ *adj.* [Hebrew, = my master]

rabbit /ˈræbɪt/ –*n.* **1 a** burrowing plant-eating mammal of the hare family. **b** *US* hare. **2** its fur. –*v.* (**-t-**) **1** hunt rabbits. **2** (often foll. by *on*, *away*) *colloq.* talk pointlessly; chatter. [origin uncertain]

rabbit punch *n.* short chop with the edge of the hand to the nape of the neck.

rabble /ˈræb(ə)l/ *n.* **1** disorderly crowd, mob. **2** contemptible or inferior set of people. **3** (prec. by *the*) the lower or disorderly classes of the populace. [origin uncertain]

rabble-rouser *n.* person who stirs up the rabble or a crowd, esp. to agitate for social change.

Rabelaisian /ˌræbəˈleɪzɪən/ *adj.* **1** of or like the French satirist Rabelais or his writings. **2** marked by exuberant imagination and coarse humour.

rabid /ˈræbɪd, ˈreɪ-/ *adj.* **1** affected with rabies, mad. **2** violent, fanatical. □ **rabidity** /rəˈbɪdɪtɪ/ *n.* [Latin *rabio* rave]

rabies /ˈreɪbiːz/ *n.* contagious viral disease of esp. dogs, transmissible through saliva to humans etc. and causing madness; hydrophobia. [Latin: related to RABID]

RAC *abbr.* Royal Automobile Club.

raccoon var. of RACOON.

race[1] –*n.* **1** contest of speed between runners, horses, vehicles, ships, etc. **2** (in *pl.*) series of these for horses, dogs, etc., at a fixed time on a regular course. **3** contest between persons to be first to achieve something. **4 a** strong current in the sea or a river. **b** channel (*mill-race*). –*v.* (**-cing**) **1** take part in a race. **2** have a race with. **3** try to surpass in speed. **4** (foll. by *with*) compete in speed with. **5** cause to race. **6 a** go at full or excessive speed. **b** cause to do this. **7** (usu. as **racing** *adj.*) follow or take part in horse-racing (*a racing man*). [Old Norse]

race[2] *n.* **1** each of the major divisions of humankind, each having distinct physical characteristics. **2** fact or concept of division into races. **3** genus, species, breed, or variety of animals or plants. **4** group of persons, animals, or plants connected by common descent. **5** any great division of living creatures (*the human race*). [Italian *razza*]

racecourse *n.* ground for horse-racing.

racegoer *n.* person who frequents horse-races.

racehorse *n.* horse bred or kept for racing.

raceme /rəˈsiːm/ *n.* flower cluster with separate flowers attached by short stalks at equal distances along the stem. [Latin *racemus* grape-bunch]

race meeting *n.* sequence of horse-races at one place.

race relations *n.pl.* relations between members of different races in the same country.

race riot *n.* outbreak of violence due to racial antagonism.

racetrack *n.* **1** = RACECOURSE. **2** track for motor racing.

racial /ˈreɪʃ(ə)l/ *adj.* **1** of or concerning race. **2** on the grounds of or connected with difference in race. □ **racially** *adv.*

racialism /ˈreɪʃəˌlɪz(ə)m/ *n.* = RACISM. □ **racialist** *n.* & *adj.*

racing car *n.* motor car built for racing.

racing driver *n.* driver of a racing car.

racism *n.* **1** belief in the superiority of a particular race; prejudice based on this. **2** antagonism towards other races. □ **racist** *n.* & *adj.*

rack[1] –*n.* **1** framework, usu. with rails, bars, etc., for holding things. **2** cogged or toothed bar or rail engaging with a wheel or pinion etc. **3** *hist.* instrument of torture stretching the victim's joints. –*v.* **1** (of disease or pain) inflict suffering on. **2** *hist.* torture (a person) on the rack. **3** place in or on a rack. **4** shake violently. **5** injure by straining. □ **on the rack** suffering acute mental or physical pain. **rack one's brains** make a great mental effort. [Low German or Dutch]

rack[2] *n.* destruction (esp. *rack and ruin*). [from WRACK]

rack[3] *v.* (often foll. by *off*) draw off (wine, beer, etc.) from the lees. [Provençal *arracar* from *raca* stems and husks of grapes, dregs]

racket[1] /'rækɪt/ *n.* (also **racquet**) **1** bat with a round or oval frame strung with catgut, nylon, etc., used in tennis, squash, etc. **2** (in *pl.*) game like squash, played in a court of four plain walls. [French *raquette* from Arabic *rahat* palm of the hand]

racket[2] /'rækɪt/ *n.* **1** disturbance, uproar, din. **2** *slang* **a** scheme for obtaining money etc. by dishonest means. **b** dodge; sly game. **3** *colloq.* line of business. [perhaps imitative]

racketeer /ˌrækɪ'tɪə(r)/ *n.* person who operates a dishonest business. □ **racketeering** *n.*

rack-rent *n.* extortionate rent.

raconteur /ˌrækɒn'tɜ:(r)/ *n.* teller of anecdotes. [French: related to RECOUNT]

racoon /rə'ku:n/ *n.* (also **raccoon**) (*pl.* same or **-s**) **1** N. American mammal with a bushy tail and sharp snout. **2** its fur. [Algonquian]

racquet var. of RACKET[1].

racy *adj.* (**-ier, -iest**) **1** lively and vigorous in style. **2** risqué. **3** of distinctive quality (*a racy wine*). □ **raciness** *n.* [from RACE[2]]

rad *n.* unit of absorbed dose of ionizing radiation. [from *r*adiation *a*bsorbed *d*ose]

RADA /'rɑ:də/ *abbr.* Royal Academy of Dramatic Art.

radar /'reɪdɑ:(r)/ *n.* **1** system for detecting the direction, range, or presence of objects, by sending out pulses of high frequency electromagnetic waves which they reflect. **2** apparatus for this. [from *ra*dio *d*etection *a*nd *r*anging]

radar trap *n.* device using radar to detect speeding vehicles.

raddle /'ræd(ə)l/ *–n.* red ochre. *–v.* (**-ling**) **1** colour with raddle or too much rouge. **2** (as **raddled** *adj.*) worn out. [related to RUDDY]

radial /'reɪdɪəl/ *–adj.* **1** of or in rays. **2 a** arranged like rays or radii. **b** having spokes or radiating lines. **c** acting or moving along lines diverging from a centre. **3** (in full **radial-ply**) (of a tyre) having fabric layers arranged radially and the tread strengthened. *–n.* radial-ply tyre. □ **radially** *adv.* [medieval Latin: related to RADIUS]

radian /'reɪdɪən/ *n.* SI unit of angle, equal to an angle at the centre of a circle the arc of which is equal in length to the radius (1 radian is approx. 57°).

radiant /'reɪdɪənt/ *–adj.* **1** emitting rays of light. **2** (of eyes or looks) beaming with joy, hope, or love. **3** (of beauty) splendid or dazzling. **4** (of light) issuing in rays. *–n.* point or object from which light or heat radiates. □ **radiance** *n.* **radiantly** *adv.*

radiant heat *n.* heat transmitted by radiation.

radiate *–v.* /'reɪdɪˌeɪt/ (**-ting**) **1 a** emit rays of light, heat, etc. **b** (of light or heat) be emitted in rays. **2** emit (light, heat, etc.) from a centre. **3** transmit or demonstrate (joy etc.). **4** diverge or spread from a centre. *–adj.* /'reɪdɪət/ having divergent rays or parts radially arranged.

radiation /ˌreɪdɪ'eɪʃ(ə)n/ *n.* **1** radiating or being radiated. **2** *Physics* **a** emission of energy as electromagnetic waves or as moving particles. **b** energy transmitted in this way, esp. invisibly. **3** (in full **radiation therapy**) treatment of cancer etc. using radiation, e.g. X-rays or ultraviolet light.

radiation sickness *n.* sickness caused by exposure to radiation such as gamma rays.

radiator /'reɪdɪˌeɪtə(r)/ *n.* **1** device for heating a room etc., consisting of a metal case through which hot water or steam circulates. **2** engine-cooling device in a motor vehicle or aircraft.

radical /'rædɪk(ə)l/ *–adj.* **1** fundamental (*a radical error*). **2** far-reaching; thorough (*radical change*). **3** advocating thorough reform; holding extreme political views; revolutionary. **4** forming the basis; primary. **5** of the root of a number or quantity. **6** (of surgery etc.) seeking to ensure the removal of all diseased tissue. **7** of the roots of words. **8** *Bot.* of the root. *–n.* **1** person holding radical views or belonging to a radical party. **2** *Chem.* **a** = FREE RADICAL. **b** atom or a group of these normally forming part of a compound and remaining unaltered during the compound's ordinary chemical changes. **3** root of a word. **4** *Math.* quantity forming or expressed as the root of another. □ **radicalism** *n.* **radically** *adv.* [Latin: related to RADIX]

radicchio /rə'di:kɪəʊ/ *n.* (*pl.* **-s**) chicory with reddish-purple leaves. [Italian, = chicory]

radicle /'rædɪk(ə)l/ *n.* part of a plant embryo that develops into the primary root; rootlet. [Latin: related to RADIX]

radii *pl.* of RADIUS.

radio /'reɪdɪəʊ/ *–n.* (*pl.* **-s**) **1** (often *attrib.*) **a** transmission and reception of sound messages etc. by electromagnetic waves of radio frequency. **b** apparatus for receiving, broadcasting, or transmitting radio signals. **2 a** sound broadcasting (*prefers the radio*). **b** broadcasting station or channel (*Radio One*). —*v.* (**-es, -ed**) **1 a** send (a message) by radio. **b** send a message to (a person) by radio. **2** communicate or broadcast by radio. [short for *radio-telegraphy* etc.]

radio- *comb. form* **1** denoting radio or broadcasting. **2** connected with radioactivity. **3** connected with rays or radiation.

radioactive /ˌreɪdɪəʊˈæktɪv/ *adj.* of or exhibiting radioactivity.

radioactivity /ˌreɪdɪəʊækˈtɪvɪtɪ/ *n.* spontaneous disintegration of atomic nuclei, with the emission of usu. penetrating radiation or particles.

radiocarbon /ˌreɪdɪəʊˈkɑːbən/ *n.* radioactive isotope of carbon.

radio-controlled /ˌreɪdɪəʊkənˈtrəʊld/ *adj.* controlled from a distance by radio.

radio frequency *n.* (*pl.* **-ies**) frequency band of telecommunication, ranging from 10^4 to 10^{11} or 10^{12} Hz.

radiogram /ˈreɪdɪəʊˌgræm/ *n.* **1** combined radio and record-player. **2** picture obtained by X-rays etc. **3** telegram sent by radio.

radiograph /ˈreɪdɪəʊˌgrɑːf/ —*n.* **1** instrument recording the intensity of radiation. **2** = RADIOGRAM 2. —*v.* obtain a picture of by X-ray, gamma ray, etc. □ **radiographer** /-ˈɒgrəfə(r)/ *n.* **radiography** /-ˈɒgrəfɪ/ *n.*

radioisotope /ˌreɪdɪəʊˈaɪsəˌtəʊp/ *n.* radioactive isotope.

radiology /ˌreɪdɪˈɒlədʒɪ/ *n.* the study of X-rays and other high-energy radiation, esp. as used in medicine. □ **radiologist** *n.*

radiophonic /ˌreɪdɪəʊˈfɒnɪk/ *adj.* of or relating to electronically produced sound, esp. music.

radioscopy /ˌreɪdɪˈɒskəpɪ/ *n.* examination by X-rays etc. of objects opaque to light.

radio-telegraphy /ˌreɪdɪəʊtɪˈlegrəfɪ/ *n.* telegraphy using radio.

radio-telephony /ˌreɪdɪəʊtɪˈlefənɪ/ *n.* telephony using radio. □ **radio-telephone** /-ˈtelɪˌfəʊn/ *n.*

radio telescope *n.* directional aerial system for collecting and analysing radiation in the radio frequency range from stars etc.

radiotherapy /ˌreɪdɪəʊˈθerəpɪ/ *n.* treatment of disease by X-rays or other forms of radiation.

radish /ˈrædɪʃ/ *n.* **1** plant with a fleshy pungent root. **2** this root, eaten esp. raw. [Latin RADIX]

radium /ˈreɪdɪəm/ *n.* radioactive metallic element orig. obtained from pitchblende etc., used esp. in radiotherapy.

radius /ˈreɪdɪəs/ *n.* (*pl.* **radii** /-dɪˌaɪ/ or **radiuses**) **1 a** straight line from the centre to the circumference of a circle or sphere. **b** length of this. **2** distance from a centre (*within a radius of 20 miles*). **3 a** thicker and shorter of the two bones in the human forearm. **b** corresponding bone in a vertebrate's foreleg or a bird's wing. [Latin]

radix /ˈreɪdɪks/ *n.* (*pl.* **radices** /-dɪˌsiːz/) *Math.* number or symbol used as the basis of a numeration scale (e.g. ten in the decimal system). [Latin, = root]

radon /ˈreɪdɒn/ *n.* gaseous radioactive inert element arising from the disintegration of radium.

RAF *abbr.* /*colloq.* ræf/ Royal Air Force.

raffia /ˈræfɪə/ *n.* **1** palm-tree native to Madagascar. **2** fibre from its leaves, used for weaving and for tying plants etc. [Malagasy]

raffish /ˈræfɪʃ/ *adj.* **1** disreputable, rakish. **2** tawdry. [*raff* rubbish]

raffle /ˈræf(ə)l/ —*n.* fund-raising lottery with prizes. —*v.* (**-ling**) (often foll. by *off*) sell by means of a raffle. [French *raf(f)le*, a dice-game]

raft /rɑːft/ *n.* flat floating structure of timber or other materials for conveying persons or things. [Old Norse]

rafter /ˈrɑːftə(r)/ *n.* each of the sloping beams forming the framework of a roof. [Old English]

rag[1] *n.* **1** torn, frayed, or worn piece of woven material. **2** (in *pl.*) old or worn clothes. **3** (*collect.*) scraps of cloth used as material for paper, stuffing, etc. **4** *derog.* newspaper. □ **in rags** much torn. **rags to riches** poverty to affluence. [probably a back-formation from RAGGED]

rag[2] —*n.* **1** fund-raising programme of stunts, parades, and entertainment organized by students. **2** prank. **3 a** rowdy celebration. **b** noisy disorderly scene. —*v.* (**-gg-**) **1** tease; play rough jokes on. **2** engage in rough play; be noisy and riotous. [origin unknown]

rag[3] *n.* ragtime composition. [abbreviation]

ragamuffin /ˈrægəˌmʌfɪn/ *n.* child in ragged dirty clothes. [probably from RAG[1]]

rag-and-bone man *n.* itinerant dealer in old clothes, furniture, etc.

rag-bag *n.* **1** bag for scraps of fabric etc. **2** miscellaneous collection.

rag doll *n.* stuffed cloth doll.

rage —*n.* **1** fierce or violent anger. **2** fit of this. **3** violent action of a natural force. —*v.* (**-ging**) **1** be full of anger. **2** (often foll.

by *at, against*) speak furiously or madly. **3** (of wind, battle, etc.) be violent; be at its height. **4** (as **raging** *adj.*) extreme, very painful (*raging thirst*; *raging headache*). □ **all the rage** very popular, fashionable. [Latin RABIES]

ragged /'rægɪd/ *adj.* **1** torn; frayed. **2** in ragged clothes. **3** with a broken or jagged outline or surface. **4** faulty, imperfect; lacking finish, smoothness, or uniformity. [Old Norse]

ragged robin *n.* pink-flowered campion with tattered petals.

raglan /'ræglən/ *–adj.* (of a sleeve) running up to the neck of a garment. *–n.* (often *attrib.*) overcoat without shoulder seams, the sleeves running up to the neck. [Lord *Raglan*]

ragout /ræ'gu:/ *n.* meat stewed with vegetables and highly seasoned. [French]

ragtag *n.* (in full **ragtag and bobtail**) *derog.* rabble or common people. [from RAG[1]]

ragtime *n.* form of highly syncopated early jazz, esp. for the piano.

rag trade *n. colloq.* the clothing business.

ragwort /'rægwɜ:t/ *n.* yellow-flowered ragged-leaved plant.

raid *–n.* **1** rapid surprise attack, esp.: **a** in warfare. **b** in order to commit a crime, steal, or do harm. **2** surprise attack by police etc. to arrest suspected persons or seize illicit goods. *–v.* make a raid on. □ **raider** *n.* [Scots form of ROAD]

rail[1] *–n.* **1** level or sloping bar or series of bars: **a** used to hang things on. **b** as the top of banisters. **c** forming part of a fence or barrier as protection. **2** steel bar or continuous line of bars laid on the ground, usu. as a railway. **3** (often *attrib.*) railway. *–v.* **1** furnish with a rail or rails. **2** (usu. foll. by *in, off*) enclose with rails. □ **off the rails** disorganized; out of order; deranged. [French *reille* from Latin *regula* RULE]

rail[2] *v.* (often foll. by *at, against*) complain or protest strongly; rant. [French *railler*]

rail[3] *n.* wading bird often inhabiting marshes. [French]

railcar *n.* single powered railway coach.

railcard *n.* pass entitling the holder to reduced rail fares.

railing *n.* (usu. in *pl.*) fence or barrier made of rails.

raillery /'reɪlərɪ/ *n.* good-humoured ridicule. [French *raillerie*: related to RAIL[2]]

railman *n.* = RAILWAYMAN.

railroad *–n.* esp. *US* = RAILWAY. *–v.* (often foll. by *into, through*, etc.) coerce; rush (*railroaded into agreeing*; *railroaded through the Cabinet*).

railway *n.* **1** track or set of tracks of steel rails upon which trains run. **2** such a system worked by a single company. **3** organization and personnel required for its working.

railwayman *n.* railway employee.

raiment /'reɪmənt/ *n. archaic* clothing. [*arrayment*: related to ARRAY]

rain *–n.* **1 a** condensed atmospheric moisture falling in drops. **b** fall of such drops. **2** (in *pl.*) **a** (prec. by *the*) rainy season. **b** rainfalls. **3 a** falling liquid or solid particles or objects. **b** rainlike descent of these. *–v.* **1** (prec. by *it* as subject) rain falls. **2 a** fall like rain. **b** (prec. by *it* as subject) send in large quantities. **3** send down like rain; lavishly bestow (*rained blows upon him*). **4** (of the sky, clouds, etc.) send down rain. □ **rain off** (or *US* **out**) (esp. in *passive*) cause (an event etc.) to be cancelled because of rain. [Old English]

rainbow /'reɪmbəʊ/ *–n.* arch of colours formed in the sky by reflection, refraction, and dispersion of the sun's rays in falling rain or in spray or mist. *–adj.* many-coloured. [Old English: related to RAIN, BOW[1]]

rainbow trout *n.* large trout orig. of the Pacific coast of N. America.

rain check *n.* esp. *US* ticket given for later use when an outdoor event is interrupted or postponed by rain. □ **take a rain check on** reserve the right not to take up (an offer) until convenient.

raincoat *n.* waterproof or water-resistant coat.

raindrop *n.* single drop of rain.

rainfall *n.* **1** fall of rain. **2** quantity of rain falling within a given area in a given time.

rainforest *n.* luxuriant tropical forest with heavy rainfall.

rainproof *adj.* impervious to rain.

rainstorm *n.* storm with heavy rain.

rainwater *n.* water collected from fallen rain.

rainwear *n.* clothes for wearing in the rain.

rainy *adj.* (**-ier, -iest**) (of weather, a climate, day, etc.) in or on which rain is falling or much rain usually falls. [Old English: related to RAIN]

rainy day *n.* time of special need in the future.

raise /reɪz/ *–v.* (**-sing**) **1** put or take into a higher position. **2** (often foll. by *up*) cause to rise or stand up or be vertical. **3** increase the amount, value, or strength of. **4** (often foll. by *up*) construct or build up. **5** levy, collect, or bring together (*raise money*). **6** cause to be heard or considered (*raise an objection*). **7** set going or bring into being (*raise hopes*). **8** bring up, educate. **9** breed, grow. **10** promote to a higher rank. **11** (foll. by *to*) multiply a quantity to a power. **12** cause (bread) to rise. **13** *Cards* bet more than (another player). **14** end (a siege etc.). **15** remove (a barrier etc.). **16** cause (a ghost etc.) to appear. **17** *colloq.* get hold of, find. **18** rouse from sleep or death, or from a lair. –*n.* **1** *Cards* increase in a stake or bid. **2** esp. *US* increase in salary. □ **raise Cain** *colloq.* = *raise the roof.* **raise one's eyebrows** see EYEBROW. **raise from the dead** restore to life. **raise a laugh** cause others to laugh. **raise the roof** be very angry; cause an uproar. [Old Norse]

raisin /'reɪz(ə)n/ *n.* dried grape. [Latin: related to RACEME]

raison d'être /ˌreɪzɔ̃ 'detr/ *n.* (*pl.* ***raisons d'être*** pronunc. same) purpose or reason that accounts for, justifies, or originally caused a thing's existence. [French]

raj /rɑːdʒ/ *n.* (prec. by *the*) *hist.* British sovereignty in India. [Hindi]

raja /'rɑːdʒə/ *n.* (also **rajah**) *hist.* **1** Indian king or prince. **2** petty dignitary or noble in India. [Hindi from Sanskrit]

rake[1] –*n.* **1** implement consisting of a pole with a toothed crossbar at the end for drawing together hay etc. or smoothing loose soil or gravel. **2** similar implement used (e.g.) to draw in money at a gaming-table. –*v.* (**-king**) **1** collect or gather with or as with a rake. **2** make tidy or smooth with a rake. **3** use a rake. **4** search thoroughly, ransack. **5** direct gunfire along (a line) from end to end. **6** scratch or scrape. □ **rake in** *colloq.* amass (profits etc.). **rake up** revive the (unwelcome) memory of. [Old English]

rake[2] *n.* dissolute man of fashion. [*rakehell*: related to RAKE[1], HELL]

rake[3] –*v.* (**-king**) **1** set or be set at a sloping angle. **2** (of a mast or funnel) incline from the perpendicular towards the stern. –*n.* **1** raking position or build. **2** amount by which a thing rakes. [origin unknown]

rake-off *n. colloq.* commission or share.

rakish *adj.* **1** dashing; jaunty. **2** dissolute. □ **rakishly** *adv.* [from RAKE[2]]

rallentando /ˌrælən'tændəʊ/ *Mus.* –*adv.* & *adj.* with a gradual decrease of speed. –*n.* (*pl.* **-s** or **-di** /-dɪ/) passage to be performed in this way. [Italian]

rally[1] /'rælɪ/ –*v.* (**-ies, -ied**) **1** (often foll. by *round*) bring or come together as support or for action. **2** bring or come together again after a rout or dispersion. **3** recover after illness etc., revive. **4** revive (courage etc.). **5** (of share-prices etc.) increase after a fall. –*n.* (*pl.* **-ies**) **1** rallying or being rallied. **2** mass meeting of supporters or persons with a common interest. **3** competition for motor vehicles, mainly over public roads. **4** (in tennis etc.) extended exchange of strokes. [French *rallier*: related to RE-, ALLY]

rally[2] /'rælɪ/ *v.* (**-ies, -ied**) ridicule good-humouredly. [French *railler*: related to RAIL[2]]

rallycross *n.* motor racing over roads and cross-country.

RAM *abbr.* **1** Royal Academy of Music. **2** random-access memory.

ram –*n.* **1** uncastrated male sheep. **2** (**the Ram**) zodiacal sign or constellation Aries. **3** *hist.* = BATTERING-RAM. **4** falling weight of a pile-driving machine. **5** hydraulically operated water pump. –*v.* (**-mm-**) **1** force or squeeze into place by pressure. **2** (usu. foll. by *down, in,* etc.) beat down or drive in by heavy blows. **3** (of a ship, vehicle, etc.) strike violently, crash against. **4** (foll. by *against, into*) dash or violently impel. [Old English]

Ramadan /'ræməˌdæn/ *n.* ninth month of the Muslim year, with strict fasting from sunrise to sunset. [Arabic]

ramble /'ræmb(ə)l/ –*v.* (**-ling**) **1** walk for pleasure. **2** talk or write incoherently. –*n.* walk taken for pleasure. [Dutch *rammelen*]

rambler *n.* **1** person who rambles. **2** straggling or spreading rose.

rambling *adj.* **1** wandering. **2** disconnected, incoherent. **3** (of a house, street, etc.) irregularly arranged. **4** (of a plant) straggling, climbing.

rambutan /ræm'buːt(ə)n/ *n.* **1** red plum-sized prickly fruit. **2** E. Indian tree bearing this. [Malay]

RAMC *abbr.* Royal Army Medical Corps.

ramekin /'ræmɪkɪn/ *n.* **1** small dish for baking and serving an individual portion of food. **2** food served in this. [French *ramequin*]

ramification /ˌræmɪfɪ'keɪʃ(ə)n/ *n.* (usu. in *pl.*) **1** consequence. **2** subdivision of a complex structure or process. [French: related to RAMIFY]

ramify /'ræmɪˌfaɪ/ *v.* (**-ies, -ied**) (cause to) form branches, subdivisions, or offshoots; branch out. [Latin *ramus* branch]

ramp –*n.* **1** slope, esp. joining two levels of ground, floor, etc. **2** movable stairs for entering or leaving an aircraft. **3** transverse ridge in a road making vehicles slow down. –*v.* **1** furnish or build with a ramp. **2 a** assume a threatening posture.

b (often foll. by *about*) storm, rage. [French *ramper* crawl]

rampage –*v.* /ræm'peɪdʒ/ (**-ging**) **1** (often foll. by *about*) rush wildly or violently. **2** rage, storm. –*n.* /'ræmpeɪdʒ/ wild or violent behaviour. □ **on the rampage** rampaging. [perhaps from RAMP]

rampant /'ræmpənt/ *adj.* **1** unchecked, flourishing excessively. **2** rank, luxuriant. **3** (placed after the noun) *Heraldry* (of an animal) standing on its left hind foot with its forepaws in the air (*lion rampant*). **4** violent, fanatical. □ **rampancy** *n.* [French: related to RAMP]

rampart /'ræmpɑːt/ *n.* **1 a** defensive wall with a broad top and usu. a stone parapet. **b** walkway on top of this. **2** defence, protection. [French *remparer* fortify]

ramrod *n.* **1** rod for ramming down the charge of a muzzle-loading firearm. **2** thing that is very straight or rigid.

ramshackle /'ræm,ʃæk(ə)l/ *adj.* tumbledown, rickety. [related to RANSACK]

ran *past* of RUN.

ranch /rɑːntʃ/ –*n.* **1** cattle-breeding establishment, esp. in the US and Canada. **2** farm where other animals are bred (*mink ranch*). –*v.* farm on a ranch. □ **rancher** *n.* [Spanish *rancho* group of persons eating together]

rancid /'rænsɪd/ *adj.* smelling or tasting like rank stale fat. □ **rancidity** /-'sɪdɪtɪ/ *n.* [Latin *rancidus* stinking]

rancour /'ræŋkə(r)/ *n.* (*US* **rancor**) inveterate bitterness, malignant hate. □ **rancorous** *adj.* [Latin *rancor*: related to RANCID]

rand *n.* chief monetary unit of South Africa. [*the Rand*, gold-field district near Johannesburg]

R & B *abbr.* rhythm and blues.

R & D *abbr.* research and development.

random /'rændəm/ *adj.* made, done, etc., without method or conscious choice. □ **at random** without a particular aim. □ **randomize** *v.* (also **-ise**) (**-zing** or **-sing**). **randomization** /-'zeɪʃ(ə)n/ *n.* **randomly** *adv.* **randomness** *n.* [French *randon* from *randir* gallop]

random-access *adj. Computing* (of a memory or file) having all parts directly accessible, so that it need not read sequentially.

randy /'rændɪ/ *adj.* (**-ier, -iest**) eager for sexual gratification, lustful. □ **randily** *adv.* **randiness** *n.* [perhaps related to RANT]

ranee /'rɑːnɪ/ *n.* (also **rani**) (*pl.* **-s**) *hist.* raja's wife or widow. [Hindi]

rang *past* of RING[2].

range /reɪndʒ/ –*n.* **1 a** region between limits of variation, esp. scope of effective operation. **b** such limits. **2** area relevant to something. **3 a** distance attainable by a gun or projectile. **b** distance between a gun or projectile and its objective. **4** row, series, etc., esp. of mountains. **5** area with targets for shooting. **6** fireplace with ovens and hotplates for cooking. **7** area over which a thing is distributed. **8** distance that can be covered by a vehicle without refuelling. **9** distance between a camera and the subject to be photographed. **10** large area of open land for grazing or hunting. –*v.* (**-ging**) **1** reach; lie spread out; extend; be found over a specified district; vary between limits. **2** (usu. in *passive* or *refl.*) line up, arrange. **3** rove, wander. **4** traverse in all directions. [French: related to RANK[1]]

rangefinder *n.* instrument for estimating the distance of an object to be shot at or photographed.

ranger *n.* **1** keeper of a royal or national park, or of a forest. **2** member of a body of mounted soldiers. **3** (**Ranger**) senior Guide.

rangy /'reɪndʒɪ/ *adj.* (**-ier, -iest**) tall and slim.

rani var. of RANEE.

rank[1] –*n.* **1 a** position in a hierarchy, grade of advancement. **b** distinct social class; grade of dignity or achievement. **c** high social position. **d** place in a scale. **2** row or line. **3** single line of soldiers drawn up abreast. **4** place where taxis await customers. **5** order, array. –*v.* **1** have a rank or place. **2** classify, give a certain grade to. **3** arrange (esp. soldiers) in rank. □ **close ranks** maintain solidarity. **the ranks** common soldiers. [French *ranc*]

rank[2] *adj.* **1** luxuriant, coarse; choked with or apt to produce weeds or excessive foliage. **2 a** foul-smelling. **b** loathsome, corrupt. **3** flagrant, virulent, gross, complete (*rank outsider*). [Old English]

rank and file *n.* (usu. treated as *pl.*) ordinary members of an organization.

rankle /'ræŋk(ə)l/ *v.* (**-ling**) (of envy, disappointment, etc., or their cause) cause persistent annoyance or resentment. [French (*d*)*rancler* fester, from medieval Latin *dra*(*cu*)*nculus* little serpent]

ransack /'rænsæk/ *v.* **1** pillage or plunder (a house, country, etc.). **2** thoroughly search. [Old Norse *rannsaka* from *rann* house, *-saka* seek]

ransom /'rænsəm/ –*n.* **1** money demanded or paid for the release of a prisoner. **2** liberation of a prisoner in return for this. –*v.* **1** buy the freedom or restoration of; redeem. **2** = *hold to ransom* (see HOLD[1]). **3** release for a ransom. [Latin: related to REDEMPTION]

rant –*v.* speak loudly, bombastically, violently, or theatrically. –*n.* piece of ranting. □ **rant and rave** express anger noisily and forcefully. [Dutch]

ranunculus /rə'nʌŋkjʊləs/ *n.* (*pl.* **-luses** or **-li** /-,laɪ/) plant of the genus including buttercups. [Latin, diminutive of *rana* frog]

RAOC *abbr.* Royal Army Ordnance Corps.

rap[1] –*n.* **1** smart slight blow. **2** knock, sharp tapping sound. **3** *slang* blame, punishment. **4 a** rhythmic monologue recited to music. **b** (in full **rap music**) style of rock music with words recited. –*v.* (**-pp-**) **1** strike smartly. **2** knock; make a sharp tapping sound. **3** criticize adversely. **4** perform a rap. □ **take the rap** suffer the consequences. □ **rapper** *n.* [probably imitative]

rap[2] *n.* small amount, the least bit (*don't care a rap*). [Irish *ropaire* counterfeit coin]

rapacious /rə'peɪʃəs/ *adj.* grasping, extortionate, predatory. □ **rapacity** /rə'pæsɪtɪ/ *n.* [Latin *rapax*: related to RAPE[1]]

rape[1] –*n.* **1 a** act of forcing a woman or girl to have sexual intercourse against her will. **b** forcible sodomy. **2** (often foll. by *of*) violent assault or plunder, forcible interference. –*v.* (**-ping**) commit rape on. [Latin *rapio* seize]

rape[2] *n.* plant grown as fodder, and for its seed from which oil is extracted. [Latin *rapum*, *rapa* turnip]

rapid /'ræpɪd/ –*adj.* (**-er, -est**) **1** quick, swift. **2** acting or completed in a short time. **3** (of a slope) descending steeply. –*n.* (usu. in *pl.*) steep descent in a river-bed, with a swift current. □ **rapidity** /rə'pɪdɪtɪ/ *n.* **rapidly** *adv.* **rapidness** *n.* [Latin: related to RAPE[1]]

rapid eye movement *n.* type of jerky movement of the eyes during dreaming.

rapier /'reɪpɪə(r)/ *n.* **1** light slender sword for thrusting. **2** (*attrib.*) sharp (*rapier wit*). [French *rapière*]

rapine /'ræpaɪn/ *n. rhet.* plundering. [Latin: related to RAPE[1]]

rapist *n.* person who commits rape.

rapport /ræ'pɔ:(r)/ *n.* relationship or communication, esp. when useful and harmonious. [Latin *porto* carry]

rapprochement /ræ'prɒʃmɑ̃/ *n.* resumption of harmonious relations, esp. between States. [French: related to APPROACH]

rapscallion /ræp'skæljən/ *n. archaic* or *joc.* rascal. [perhaps from RASCAL]

rapt *adj.* **1** fully absorbed or intent, enraptured. **2** carried away with feeling or lofty thought. [Latin *raptus*: related to RAPE[1]]

rapture /'ræptʃə(r)/ *n.* **1** ecstatic delight. **2** (in *pl.*) great pleasure or enthusiasm or the expression of it. □ **rapturous** *adj.* [French or medieval Latin: related to RAPE[1]]

rare[1] *adj.* (**rarer, rarest**) **1** seldom done, found, or occurring; uncommon, unusual. **2** exceptionally good. **3** of less than the usual density. □ **rareness** *n.* [Latin *rarus*]

rare[2] *adj.* (**rarer, rarest**) (of meat) cooked so that the inside is still red and juicy; underdone. [Old English]

rarebit *n.* = WELSH RABBIT. [from RARE[1]]

rare earth *n.* lanthanide element.

rarefy /'reərɪ,faɪ/ *v.* (**-ies, -ied**) **1** make or become less dense or solid. **2** purify or refine (a person's nature etc.). **3** make (an idea etc.) subtle. □ **rarefaction** /-'fækʃ(ə)n/ *n.* [French or medieval Latin: related to RARE[1]]

rarely *adv.* **1** seldom, not often. **2** exceptionally.

raring /'reərɪŋ/ *adj. colloq.* enthusiastic, eager (*raring to go*). [participle of *rare*, dial. var. of ROAR or REAR[2]]

rarity /'reərətɪ/ *n.* (*pl.* **-ies**) **1** rareness. **2** uncommon thing. [Latin: related to RARE[1]]

rascal /'rɑ:sk(ə)l/ *n.* dishonest or mischievous person. □ **rascally** *adj.* [French *rascaille* rabble]

rase var. of RAZE.

rash[1] *adj.* reckless, impetuous, hasty. □ **rashly** *adv.* **rashness** *n.* [probably Old English]

rash[2] *n.* **1** eruption of the skin in spots or patches. **2** (usu. foll. by *of*) sudden widespread phenomenon (*rash of strikes*). [origin uncertain]

rasher *n.* thin slice of bacon or ham. [origin unknown]

rasp /rɑ:sp/ –*n.* **1** coarse kind of file having separate teeth. **2** grating noise or utterance. –*v.* **1 a** scrape with a rasp. **b** scrape roughly. **c** (foll. by *off*, *away*) remove by scraping. **2 a** make a grating sound. **b** say gratingly. **3** grate upon (a person or feelings). [French *raspe(r)*]

raspberry /'rɑ:zbərɪ/ *n.* (*pl.* **-ies**) **1 a** red blackberry-like fruit. **b** bramble bearing this. **2** *colloq.* sound made by blowing through the lips, expressing derision or disapproval. [origin unknown]

raspberry-cane *n.* raspberry plant.

Rastafarian /,ræstə'feərɪən/ (also **Rasta** /'ræstə/) –*n.* member of a Jamaican sect, often having dreadlocks and regarding Haile Selassie of Ethiopia as God.—*adj.* of this sect. [*Ras Tafari*, title of former Emperor Haile Selassie]

rat —*n.* **1 a** rodent like a large mouse. **b** similar rodent (*muskrat*; *water-rat*). **2** turncoat. **3** *colloq.* unpleasant or treacherous person. **4** (in *pl.*) *slang* exclamation of annoyance etc. —*v.* (**-tt-**) **1** hunt or kill rats. **2** (also foll. by *on*) inform (on); desert, betray. [Old English]

ratable var. of RATEABLE.

ratatat (also **rat-a-tat**) var. of RAT-TAT.

ratatouille /ˌrætəˈtuːɪ, -ˈtwiː/ *n.* dish of stewed onions, courgettes, tomatoes, aubergines, and peppers. [French dial.]

ratbag *n. slang* obnoxious person.

ratchet /ˈrætʃɪt/ *n.* **1** set of teeth on the edge of a bar or wheel with a catch ensuring motion in one direction only. **2** (in full **ratchet-wheel**) wheel with a rim so toothed. [French *rochet* lance-head]

rate[1] —*n.* **1** numerical proportion between two sets of things (*moving at a rate of 50 m.p.h.*) or as the basis of calculating an amount or value (*rate of interest*). **2** fixed or appropriate charge, cost, or value; measure of this (*postal rates*; *the rate for the job*). **3** pace of movement or change (*prices increasing at a great rate*). **4** (in *comb.*) class or rank (*first-rate*). **5** (in *pl.*) tax levied by local authorities on businesses (and formerly on private individuals) according to the value of buildings and land occupied. —*v.* (**-ting**) **1 a** estimate the worth or value of. **b** assign a value to. **2** consider, regard as. **3** (foll. by *as*) rank or be considered. **4 a** subject to the payment of a local rate. **b** value for the purpose of assessing rates. **5** be worthy of, deserve. □ **at any rate** in any case, whatever happens. **at this rate** if this example is typical. [Latin *rata*: related to RATIO]

■ **Usage** See note at *community charge*.

rate[2] *v.* (**-ting**) scold angrily. [origin unknown]

rateable *adj.* (also **ratable**) liable to rates.

rateable value *n.* value at which a business etc. is assessed for rates.

rate-capping *n. hist.* imposition of an upper limit on local authority rates. □ **rate-cap** *v.*

ratepayer *n.* person liable to pay rates.

rather /ˈrɑːðə(r)/ *adv.* **1** by preference (*would rather not go*). **2** (usu. foll. by *than*) more truly; as a more likely alternative (*is stupid rather than dishonest*). **3** more precisely (*a book, or rather, a pamphlet*). **4** slightly, to some extent (*became rather drunk*). **5** /rɑːˈðɜː(r)/ (as an emphatic response) assuredly (*Did you like it? – Rather!*). □ **had rather** would rather. [Old English comparative of *rathe* early]

ratify /ˈrætɪˌfaɪ/ *v.* (**-ies, -ied**) confirm or accept (an agreement made in one's name) by formal consent, signature, etc. □ **ratification** /-fɪˈkeɪʃ(ə)n/ *n.* [medieval Latin: related to RATE[1]]

rating *n.* **1** placing in a rank or class. **2** estimated standing of a person as regards credit etc. **3** non-commissioned sailor. **4** amount fixed as a local rate. **5** relative popularity of a broadcast programme as determined by the estimated size of the audience.

ratio /ˈreɪʃɪəʊ/ *n.* (*pl.* **-s**) quantitative relation between two similar magnitudes expressed as the number of times one contains the other (*in the ratio of three to two*). [Latin *reor rat-* reckon]

ratiocinate /ˌrætɪˈɒsɪˌneɪt/ *v.* (**-ting**) *literary* reason, esp. using syllogisms. □ **ratiocination** /-ˈneɪʃ(ə)n/ *n.* [Latin: related to RATIO]

ration /ˈræʃ(ə)n/ —*n.* **1** official allowance of food, clothing, etc., in a time of shortage. **2** (usu. in *pl.*) fixed daily allowance of food, esp. in the armed forces. —*v.* **1** limit (persons or provisions) to a fixed ration. **2** (usu. foll. by *out*) share out (food etc.) in fixed quantities. [Latin: related to RATIO]

rational /ˈræʃən(ə)l/ *adj.* **1** of or based on reason. **2** sensible. **3** endowed with reason. **4** rejecting what is unreasonable or cannot be tested by reason in religion or custom. **5** (of a quantity or ratio) expressible as a ratio of whole numbers. □ **rationality** /-ˈnælɪtɪ/ *n.* **rationally** *adv.* [Latin: related to RATION]

rationale /ˌræʃəˈnɑːl/ *n.* fundamental reason, logical basis. [neuter of Latin *rationalis*: related to RATIONAL]

rationalism /ˈræʃənəˌlɪz(ə)m/ *n.* practice of treating reason as the basis of belief and knowledge. □ **rationalist** *n.* & *adj.* **rationalistic** /-ˈlɪstɪk/ *adj.*

rationalize *v.* (also **-ise**) (**-zing** or **-sing**) **1** (often foll. by *away*) offer a rational but specious explanation of (one's behaviour or attitude). **2** make logical and consistent. **3** make (a business etc.) more efficient by reorganizing it to reduce or eliminate waste. □ **rationalization** /-ˈzeɪʃ(ə)n/ *n.*

ratline /ˈrætlɪn/ *n.* (also **ratlin**) (usu. in *pl.*) any of the small lines fastened across a sailing-ship's shrouds like ladder-rungs. [origin unknown]

rat race *n. colloq.* fiercely competitive struggle for position, power, etc.

ratsbane *n.* anything poisonous to rats, esp. a plant.

rattan /rə'tæn/ *n.* **1** climbing palm with long thin jointed pliable stems, used for furniture etc. **2** piece of rattan stem used as a walking-stick etc. [Malay]

rat-tat /ræt'tæt/ *n.* (also **rat-tat-tat** /ˌrættæt'tæt/, **ratatat**, **rat-a-tat** /ˌrætə'tæt/) rapping sound, esp. of a knocker. [imitative]

rattle /'ræt(ə)l/ *–v.* (**-ling**) **1 a** give out a rapid succession of short sharp hard sounds. **b** cause to do this. **c** cause such sounds by shaking something. **2** (often foll. by *along*) **a** move with a rattling noise. **b** move or travel briskly. **3 a** (usu. foll. by *off*) say or recite rapidly. **b** (usu. foll. by *on*) talk in a lively thoughtless way. **4** *colloq.* disconcert, alarm. *–n.* **1** rattling sound. **2** device or plaything made to rattle. □ **rattly** *adj.* [probably Low German or Dutch]

rattlesnake *n.* poisonous American snake with a rattling structure of horny rings on its tail.

rattling *–adj.* **1** that rattles. **2** brisk, vigorous (*rattling pace*). *–adv. colloq.* remarkably (*rattling good story*).

ratty *adj.* (**-ier**, **-iest**) **1** relating to or infested with rats. **2** *colloq.* irritable, bad-tempered. □ **rattily** *adv.* **rattiness** *n.*

raucous /'rɔ:kəs/ *adj.* harsh-sounding, loud and hoarse. □ **raucously** *adv.* **raucousness** *n.* [Latin]

raunchy /'rɔ:ntʃɪ/ *adj.* (**-ier**, **-iest**) *colloq.* coarse, earthy, sexually boisterous. □ **raunchily** *adv.* **raunchiness** *n.* [origin unknown]

ravage /'rævɪdʒ/ *–v.* (**-ging**) devastate, plunder. *–n.* **1** devastation. **2** (usu. in *pl.*; foll. by *of*) destructive effect. [French alteration from *ravine* rush of water]

rave *–v.* (**-ving**) **1** talk wildly or furiously in or as in delirium. **2** (usu. foll. by *about*, *over*) speak with rapturous admiration; go into raptures. **3** *colloq.* enjoy oneself freely (esp. *rave it up*). *–n.* **1** (usu. *attrib.*) *colloq.* highly enthusiastic review. **2** (also **rave-up**) *colloq.* lively party. **3** *slang* craze. [probably French dial. *raver*]

ravel /'ræv(ə)l/ *v.* (**-ll-**; *US* **-l-**) **1** entangle or become entangled. **2** fray out. **3** (often foll. by *out*) disentangle, unravel, separate into threads. [probably Dutch *ravelen*]

raven /'reɪv(ə)n/ *–n.* large glossy blue-black crow with a hoarse cry. *–adj.* glossy black. [Old English]

ravening /'rævənɪŋ/ *adj.* hungrily seeking prey; voracious. [French *raviner* from Latin: related to RAPINE]

ravenous /'rævənəs/ *adj.* **1** very hungry. **2** voracious. **3** rapacious. □ **ravenously** *adv.* [obsolete *raven* plunder, from French *raviner* ravage]

raver *n. colloq.* uninhibited pleasure-loving person.

ravine /rə'vi:n/ *n.* deep narrow gorge. [Latin: related to RAPINE]

raving *–n.* (usu. in *pl.*) wild or delirious talk. *–adj.* & *adv. colloq.* as an intensifier (*a raving beauty*; *raving mad*).

ravioli /ˌrævɪ'əʊlɪ/ *n.* small pasta envelopes containing minced meat etc. [Italian]

ravish /'rævɪʃ/ *v.* **1** *archaic* rape (a woman). **2** enrapture. □ **ravishment** *n.* [Latin: related to RAPE[1]]

ravishing *adj.* lovely, beautiful. □ **ravishingly** *adv.*

raw *adj.* **1** uncooked. **2** in the natural state; not processed or manufactured. **3** inexperienced, untrained. **4 a** stripped of skin; with the flesh exposed, unhealed. **b** sensitive to the touch from being so exposed. **5** (of the atmosphere, day, etc.) cold and damp. **6** crude in artistic quality; lacking finish. **7** (of the edge of cloth) without hem or selvage. □ **in the raw 1** in its natural state without mitigation (*life in the raw*). **2** naked. **touch on the raw** upset (a person) on a sensitive matter. [Old English]

raw-boned *adj.* gaunt.

raw deal *n.* harsh or unfair treatment.

rawhide *n.* **1** untanned hide. **2** rope or whip of this.

Rawlplug /'rɔ:lplʌg/ *n. propr.* cylindrical plug for holding a screw or nail in masonry. [*Raw*lings, name of the engineers who introduced it]

raw material *n.* material from which manufactured goods are made.

ray[1] *n.* **1** single line or narrow beam of light from a small or distant source. **2** straight line in which radiation travels to a given point. **3** (in *pl.*) radiation of a specified type (*X-rays*). **4** trace or beginning of an enlightening or cheering influence (*ray of hope*). **5** any of a set of radiating lines, parts, or things. **6** marginal floret of a composite flower, e.g. a daisy. [Latin RADIUS]

ray[2] *n.* large edible marine fish with a flat body and a long slender tail. [Latin *raia*]

ray[3] *n.* (also **re**) *Mus.* second note of a major scale. [Latin *resonare*, word arbitrarily taken]

rayon /'reɪɒn/ *n.* textile fibre or fabric made from cellulose. [from RAY[1]]

raze *v.* (also **rase**) (**-zing** or **-sing**) completely destroy; tear down (esp. *raze to the ground*). [Latin *rado ras-* scrape]

razor /'reɪzə(r)/ *n.* instrument with a sharp blade used in cutting hair, esp. shaving. [French *rasor*: related to RAZE]

razor-bill *n.* auk with a sharp-edged bill.

razor-blade *n.* flat piece of metal with a sharp edge, used in a safety razor.

razor-edge *n.* (also **razor's edge**) **1** keen edge. **2** sharp mountain-ridge. **3** critical situation. **4** sharp line of division.

razzle-dazzle /'ræzəl,dæz(ə)l/ *n.* (also **razzle**) *colloq.* **1 a** excitement; bustle. **b** spree (esp. *on the razzle*). **2** extravagant publicity. [reduplication of DAZZLE]

razzmatazz /,ræzmə'tæz/ *n. colloq.* **1** glamorous excitement, bustle. **2** spree. **3** insincere actions. [probably an alteration of RAZZLE-DAZZLE]

Rb *symb.* rubidium.

RC *abbr.* Roman Catholic.

Rd. *abbr.* Road.

RE *abbr.* **1** Religious Education. **2** Royal Engineers.

Re *symb.* rhenium.

re[1] /ri:/ *prep.* **1** in the matter of (as the first word in a heading). **2** *Commerce* about, concerning (in letters). [Latin, ablative of *res* thing]

re[2] var. of RAY[3].

re- *prefix* **1** attachable to almost any verb or its derivative, meaning: **a** once more; afresh, anew. **b** back; with return to a previous state. **2** (also **red-** before a vowel, as in *redolent*) in verbs and verbal derivatives denoting: **a** in return; mutually (*react*). **b** opposition (*resist*). **c** behind or after (*relic*). **d** retirement or secrecy (*recluse*). **e** off, away, down (*recede*; *relegate*; *repress*). **f** frequentative or intensive force (*redouble*; *resplendent*). **g** negative force (*recant*; *reveal*). [Latin]

■ **Usage** In sense 1, a hyphen is normally used when the word begins with *e* (*re-enact*), or to distinguish the compound from a more familiar one-word form (*re-cover* = cover again).

reach –*v.* **1** (often foll. by *out*) stretch out, extend. **2** (often foll. by *for*) stretch out the hand etc.; make a stretch or effort. **3** get as far as. **4** get to or attain. **5** make contact with the hand etc., or by telephone etc. (*could not be reached*). **6** hand, pass (*reach me that book*). **7** take with an outstretched hand. **8** *Naut.* sail with the wind abeam or abaft the beam. –*n.* **1** extent to which a hand etc. can be reached out, influence exerted, motion carried out, or mental powers used. **2** act of reaching out. **3** continuous extent, esp. of a river between two bends or of a canal between locks. **4** *Naut.* distance traversed in reaching. □ **reachable** *adj.* [Old English]

reach-me-down *n. colloq.* **1** ready-made garment. **2** = HAND-ME-DOWN.

reacquaint /,ri:ə'kweɪnt/ *v.* make acquainted again. □ **reacquaintance** *n.*

react /rɪ'ækt/ *v.* **1** (often foll. by *to*) respond to a stimulus; change or behave differently due to some influence (*reacted badly to the news*). **2** (often foll. by *against*) respond with repulsion to; tend in a reverse or contrary direction. **3** (foll. by *with*) (of a substance or particle) be the cause of chemical activity or interaction with another (*nitrous oxide reacts with the metal*). **4** (foll. by *with*) cause (a substance) to react with another.

reaction /rɪ'ækʃ(ə)n/ *n.* **1** reacting, response. **2** bad physical response to a drug etc. **3** occurrence of a condition after a period of its opposite. **4** tendency to oppose change or reform. **5** interaction of substances undergoing chemical change.

reactionary –*adj.* tending to oppose (esp. political) change or reform. –*n.* (*pl.* **-ies**) reactionary person.

reactivate /rɪ'æktɪ,veɪt/ *v.* (**-ting**) restore to a state of activity. □ **reactivation** /-'veɪʃ(ə)n/ *n.*

reactive /rɪ'æktɪv/ *adj.* **1** showing reaction. **2** reacting rather than taking the initiative. **3** susceptible to chemical reaction.

reactor *n.* **1** person or thing that reacts. **2** = NUCLEAR REACTOR.

read –*v.* (*past* and *past part.* **read** /red/) **1** (also *absol.*) reproduce mentally or (often foll. by *aloud*, *out*, *off*, etc.) vocally the written or printed words of (a book, author, etc.). **2** convert or be able to convert into the intended words or meaning (written or other symbols or the things expressed in this way) (*can't read music*). **3** understand by observing; interpret (*read me like a book*; *read his silence as consent*; *read my mind*; *reads tea-leaves*). **4** find (a thing) stated in print etc. (*read that you were leaving*). **5** (often foll. by *into*) assume as intended or deducible (*read too much into it*). **6** bring into a specified state by reading (*read myself to sleep*). **7 a** (of a recording instrument) show (a specified figure etc.). **b** interpret (a recording instrument) (*read the meter*). **8** convey meaning when read; have a certain wording (*it reads persuasively*; *reads from left to right*). **9** sound or affect a hearer or reader when read (*the book reads like a parody*). **10** study by reading (esp. a subject at university). **11** (as **read** /red/ *adj.*) versed in a subject (esp.

literature) by reading (*well-read person*). **12** (of a computer) copy or transfer (data). **13** hear and understand (over a radio) (*are you reading me?*). **14** replace a word etc. with (the correct one(s)) (*for 'this' read 'these'*). *–n.* **1** spell of reading. **2** *colloq.* book etc. as regards readability (*is a good read*). □ **read between the lines** look for or find hidden meaning. **read up** (often followed by *on*) make a special study of (a subject). **take as read** treat (a thing) as if it has been agreed. [Old English]

readable *adj.* **1** able to be read. **2** interesting to read. □ **readability** /-'bɪlɪtɪ/ *n.*

readdress /ˌri:ə'dres/ *v.* **1** change the address of (an item for posting). **2** address (a problem etc.) anew. **3** speak or write to anew.

reader *n.* **1** person who reads. **2** book intended to give reading practice, esp. in a foreign language. **3** device for producing an image that can be read from microfilm etc. **4** (also **Reader**) university lecturer of the highest grade below professor. **5** publisher's employee who reports on submitted manuscripts. **6** printer's proof-corrector. **7** person appointed to read aloud, esp. in church.

readership *n.* **1** readers of a newspaper etc. **2** (also **Readership**) position of Reader.

readily /'redɪlɪ/ *adv.* **1** without showing reluctance, willingly. **2** without difficulty.

readiness *n.* **1** ready or prepared state. **2** willingness. **3** facility; promptness in argument or action.

reading *n.* **1 a** act of reading (*reading of the will*). **b** matter to be read (*made exciting reading*). **2** (in *comb.*) used for reading (*reading-lamp*; *reading-room*). **3** literary knowledge. **4** entertainment at which a play, poems, etc., are read. **5** figure etc. shown by a recording instrument. **6** interpretation or view taken (*what is your reading of the facts?*). **7** interpretation made (of drama, music, etc.). **8** each of the successive occasions on which a bill must be presented to a legislature for acceptance. [Old English: related to READ]

readjust /ˌri:ə'dʒʌst/ *v.* adjust again or to a former state. □ **readjustment** *n.*

readmit /ˌri:əd'mɪt/ *v.* (**-tt-**) admit again. □ **readmission** *n.*

readopt /ˌri:ə'dɒpt/ *v.* adopt again. □ **readoption** *n.*

ready /'redɪ/ *–adj.* (**-ier**, **-iest**) (usu. *predic.*) **1** with preparations complete (*dinner is ready*). **2** in a fit state. **3** willing, inclined, or resolved (*he is always ready to complain*). **4** within reach; easily secured (*ready source of income*). **5** fit for immediate use. **6** immediate, unqualified (*found ready acceptance*). **7** prompt (*is always ready with excuses*). **8** (foll. by *to* + infin.) about to (*ready to burst*). **9** provided beforehand. *–adv.* (usu. in *comb.*) beforehand; so as not to require doing when the time comes for use etc. (*is ready packed*; *ready-mixed concrete*; *ready-made family*). *–n.* (*pl.* **-ies**) *slang* (prec. by *the*) = READY MONEY. *–v.* (**-ies**, **-ied**) make ready, prepare. □ **at the ready** ready for action. **make ready** prepare. [Old English]

ready-made *adj.* (also **ready-to-wear**) (esp. of clothes) made in a standard size, not to measure.

ready money *n.* **1** actual coin or notes. **2** payment on the spot.

ready reckoner *n.* book or table listing standard numerical calculations as used esp. in commerce.

reaffirm /ˌri:ə'fɜ:m/ *v.* affirm again. □ **reaffirmation** /-ˌæfə'meɪʃ(ə)n/ *n.*

reafforest /ˌri:ə'fɒrɪst/ *v.* replant (former forest land) with trees. □ **reafforestation** /-'steɪʃ(ə)n/ *n.*

reagent /ri:'eɪdʒ(ə)nt/ *n. Chem.* substance used to cause a reaction, esp. to detect another substance.

real[1] *–adj.* **1** actually existing or occurring. **2** genuine; rightly so called; not artificial. **3** *Law* consisting of immovable property such as land or houses (*real estate*). **4** appraised by purchasing power (*real value*). **5** *Math.* (of a quantity) having no imaginary part (see IMAGINARY 2). *–adv. Scot. & US colloq.* really, very. □ **for real** *colloq.* seriously, in earnest. **the real thing** (of an object or emotion) genuine, not inferior. [Anglo-French and Latin *realis* from *res* thing]

real[2] /reɪ'ɑ:l/ *n. hist.* coin and monetary unit in Spanish-speaking countries. [Spanish: related to ROYAL]

real ale *n.* beer regarded as brewed in a traditional way.

realign /ˌri:ə'laɪn/ *v.* **1** align again. **2** regroup in politics etc. □ **realignment** *n.*

realism *n.* **1** practice of regarding things in their true nature and dealing with them as they are. **2** fidelity to nature in representation; the showing of life etc. as it is. **3** *Philos.* doctrine that abstract concepts have an objective existence. □ **realist** *n.*

realistic /rɪə'lɪstɪk/ *adj.* **1** regarding things as they are; following a policy of realism. **2** based on facts rather than ideals. □ **realistically** *adv.*

reality /rɪ'ælɪtɪ/ *n.* (*pl.* **-ies**) **1** what is real or existent or underlies appearances. **2** (foll. by *of*) the real nature of. **3** real existence; state of being real. **4** resemblance to an original. □ **in reality** in fact. [medieval Latin or French: related to REAL[1]]

realize *v.* (also **-ise**) (**-zing** or **-sing**) **1** (often foll. by *that*) be fully aware of; conceive as real. **2** understand clearly. **3** present as real. **4** convert into actuality. **5 a** convert into money. **b** acquire (profit). **c** be sold for (a specified price). □ **realizable** *adj.* **realization** /-'zeɪʃ(ə)n/ *n.*

real life *n.* **1** life lived by actual people. **2** (*attrib.*) (**real-life**) actual, not fictional (*her real-life husband*).

reallocate /ri:'ælə,keɪt/ *v.* (**-ting**) allocate again or differently. □ **reallocation** /-'keɪʃ(ə)n/ *n.*

really /'rɪəlɪ/ *adv.* **1** in reality. **2** very (*really useful*). **3** indeed, I assure you. **4** expression of mild protest or surprise.

realm /relm/ *n.* **1** *formal* kingdom. **2** domain (*realm of myth*). [Latin REGIMEN]

real money *n.* current coin; cash.

real tennis *n.* original form of tennis played on an indoor court.

real time *n.* **1** actual time during which a process occurs. **2** (*attrib.*) (**real-time**) *Computing* (of a system) in which the response time is the actual time during which an event occurs.

realty /'ri:əltɪ/ *n.* real estate.

ream *n.* **1** twenty quires of paper. **2** (in *pl.*) large quantity of writing. [Arabic, = bundle]

reanimate /ri:'ænɪ,meɪt/ *v.* (**-ting**) **1** restore to life. **2** restore to activity or liveliness. □ **reanimation** /-'meɪʃ(ə)n/ *n.*

reap *v.* **1** cut or gather (esp. grain) as a harvest. **2** harvest the crop of (a field etc.). **3** receive as a result of one's own or others' actions. [Old English]

reaper *n.* **1** person who reaps. **2** reaping machine. **3** (**the Reaper** or **grim Reaper**) death personified.

reappear /,ri:ə'pɪə(r)/ *v.* appear again or as previously. □ **reappearance** *n.*

reapply /,ri:ə'plaɪ/ *v.* (**-ies, -ied**) apply again, esp. submit a further application (for a position etc.). □ **reapplication** /-æplɪ'keɪʃ(ə)n/ *n.*

reappoint /,ri:ə'pɔɪnt/ *v.* appoint to a position previously held. □ **reappointment** *n.*

reapportion /,ri:ə'pɔ:ʃ(ə)n/ *v.* apportion again or differently.

reappraise /,ri:ə'preɪz/ *v.* (**-sing**) appraise or assess again or differently. □ **reappraisal** *n.*

rear[1] **–***n.* **1** back part of anything. **2** space behind, or position at the back of, anything. **3** *colloq.* buttocks. **–***adj.* at the back. □ **bring up the rear** come last. [probably from REARWARD or REARGUARD]

rear[2] *v.* **1 a** bring up and educate (children). **b** breed and care for (animals). **c** cultivate (crops). **2** (of a horse etc.) raise itself on its hind legs. **3 a** set upright. **b** build. **c** hold upwards. **4** extend to a great height. [Old English]

rear admiral *n.* naval officer ranking below vice admiral.

rearguard *n.* body of troops detached to protect the rear, esp. in retreats. [French *rereguarde*]

rearguard action *n.* **1** engagement undertaken by a rearguard. **2** defensive stand or struggle, esp. when losing.

rear-lamp *n.* (also **rear-light**) usu. red light at the rear of a vehicle.

rearm /ri:'ɑ:m/ *v.* (also *absol.*) arm again, esp. with improved weapons. □ **rearmament** *n.*

rearmost *adj.* furthest back.

rearrange /,ri:ə'reɪndʒ/ *v.* (**-ging**) arrange again in a different way. □ **rearrangement** *n.*

rearrest /,ri:ə'rest/ **–***v.* arrest again. **–***n.* rearresting or being rearrested.

rearward /'rɪəwəd/ **–***n.* (esp. in prepositional phrases) rear (*to the rearward of*; *in the rearward*). **–***adj.* to the rear. **–***adv.* (also **rearwards**) towards the rear. [Anglo-French *rerewarde* = REARGUARD]

reason /'ri:z(ə)n/ **–***n.* **1** motive, cause, or justification. **2** fact adduced or serving as this. **3** intellectual faculty by which conclusions are drawn from premisses. **4** sanity (*lost his reason*). **5** sense; sensible conduct; what is right, practical, or practicable; moderation. **–***v.* **1** form or try to reach conclusions by connected thought. **2** (foll. by *with*) use argument with (a person) by way of persuasion. **3** (foll. by *that*) conclude or assert in argument. **4** (foll. by *into*, *out of*) persuade or move by argument. **5** (foll. by *out*) think out (consequences etc.). **6** (often as **reasoned** *adj.*) express in a logical way. **7** embody reason in (an amendment etc.). □ **by reason of** owing to. **in** (or **within**) **reason** within the bounds of moderation. **with reason** justifiably. [Latin *ratio*]

reasonable *adj.* **1** having sound judgement; moderate; ready to listen to reason. **2** not absurd. **3 a** not greatly less or more than might be expected. **b** inexpensive. **c** tolerable, fair. □ **reasonableness** *n.* **reasonably** *adv.*

reassemble /ˌriːəˈsemb(ə)l/ *v.* (**-ling**) assemble again or into a former state. □ **reassembly** *n.*
reassert /ˌriːəˈsɜːt/ *v.* assert again, esp. with renewed emphasis. □ **reassertion** *n.*
reassess /ˌriːəˈses/ *v.* assess again or differently. □ **reassessment** *n.*
reassign /ˌriːəˈsaɪn/ *v.* assign again or differently. □ **reassignment** *n.*
reassure /ˌriːəˈʃʊə(r)/ *v.* (**-ring**) **1** restore confidence to; dispel the apprehensions of. **2** confirm in an opinion or impression. □ **reassurance** *n.* **reassuring** *adj.*
reawaken /ˌriːəˈweɪkən/ *v.* awaken again.
rebate[1] /ˈriːbeɪt/ *n.* **1** partial refund. **2** deduction from a sum to be paid; discount. [French *rabattre*: related to RE-, ABATE]
rebate[2] /ˈriːbeɪt/ *n.* & *v.* (**-ting**) = RABBET.
rebel *–n.* /ˈreb(ə)l/ **1** person who fights against, resists, or refuses allegiance to, the established government. **2** person or thing that resists authority or control. *–attrib. adj.* /ˈreb(ə)l/ **1** rebellious. **2** of rebels. **3** in rebellion. *–v.* /rɪˈbel/ (**-ll-**; *US* **-l-**) (usu. foll. by *against*) **1** act as a rebel; revolt. **2** feel or display repugnance. [Latin: related to RE-, *bellum* war]
rebellion /rɪˈbeljən/ *n.* open resistance to authority, esp. organized armed resistance to an established government. [Latin: related to REBEL]
rebellious /rɪˈbeljəs/ *adj.* **1** tending to rebel. **2** in rebellion. **3** defying lawful authority. **4** (of a thing) unmanageable, refractory. □ **rebelliously** *adv.* **rebelliousness** *n.*
rebid /ˈriːbɪd/ *–v.* /also riːˈbɪd/ (**-dd-**; *past* and *past part.* **rebid**) bid again. *–n.* **1** act of rebidding. **2** bid so made.
rebind /riːˈbaɪnd/ *v.* (*past* and *past part.* **rebound**) bind (esp. a book) again or differently.
rebirth /riːˈbɜːθ/ *n.* **1** new incarnation. **2** spiritual enlightenment. **3** revival. □ **reborn** /riːˈbɔːn/ *adj.*
reboot /riːˈbuːt/ *v.* (often *absol.*) *Computing* boot up (a system) again.
rebound *–v.* /rɪˈbaʊnd/ **1** spring back after impact. **2** (foll. by *upon*) (of an action) have an adverse effect upon (the doer). *–n.* /ˈriːbaʊnd/ act of rebounding; recoil, reaction. □ **on the rebound** while still recovering from an emotional shock, esp. rejection by a lover. [French *rebonder*: related to BOUND[1]]
rebroadcast /riːˈbrɔːdkɑːst/ *–v.* (*past* **-cast** or **-casted**; *past part.* **-cast**) broadcast again. *–n.* repeat broadcast.
rebuff /rɪˈbʌf/ *–n.* **1** rejection of one who makes advances, proffers help, shows interest, makes a request, etc. **2** snub. *–v.* give a rebuff to. [French from Italian]
rebuild /riːˈbɪld/ *v.* (*past* and *past part.* **rebuilt**) build again or differently.
rebuke /rɪˈbjuːk/ *–v.* (**-king**) express sharp disapproval to (a person) for a fault; censure. *–n.* rebuking or being rebuked. [Anglo-French]
rebus /ˈriːbəs/ *n.* (*pl.* **rebuses**) representation of a word (esp. a name) by pictures etc. suggesting its parts. [Latin *rebus*, ablative pl. of *res* thing]
rebut /rɪˈbʌt/ *v.* (**-tt-**) **1** refute or disprove (evidence or a charge). **2** force or turn back; check. □ **rebuttal** *n.* [Anglo-French *rebuter*: related to BUTT[1]]
rec *n. colloq.* recreation ground. [abbreviation]
recalcitrant /rɪˈkælsɪtrənt/ *adj.* **1** obstinately disobedient. **2** objecting to restraint. □ **recalcitrance** *n.* [Latin *recalcitro* kick out, from *calx* heel]
recall /rɪˈkɔːl/ *–v.* **1** summon to return. **2** recollect, remember. **3** bring back to memory; serve as a reminder of. **4** revoke or annul (an action or decision). **5** revive, resuscitate. **6** take back (a gift). *–n.* /also ˈriːkɔl/ **1** summons to come back. **2** act of remembering. **3** ability to remember. **4** possibility of recalling, esp. in the sense of revoking (*beyond recall*).
recant /rɪˈkænt/ *v.* (also *absol.*) withdraw and renounce (a former belief or statement) as erroneous or heretical. □ **recantation** /ˌriːkænˈteɪʃ(ə)n/ *n.* [Latin: related to CHANT]
recap /ˈriːkæp/ *colloq.* *–v.* (**-pp-**) recapitulate. *–n.* recapitulation. [abbreviation]
recapitulate /ˌriːkəˈpɪtjʊˌleɪt/ *v.* (**-ting**) **1** go briefly through again; summarize. **2** go over the main points or headings of. [Latin: related to CAPITAL]
recapitulation /ˌriːkəˌpɪtjʊˈleɪʃ(ə)n/ *n.* **1** act of recapitulating. **2** *Mus.* part of a movement in which themes are restated. [Latin: related to RECAPITULATE]
recapture /riːˈkæptʃə(r)/ *–v.* (**-ring**) **1** capture again; recover by capture. **2** re-experience (a past emotion etc.). *–n.* act of recapturing.
recast /riːˈkɑːst/ *–v.* (*past* and *past part.* **recast**) **1** cast again (a play, net, votes, etc.). **2** put into a new form; improve the arrangement of. *–n.* **1** recasting. **2** recast form.
recce /ˈrekɪ/ *colloq.* *–n.* reconnaissance. *–v.* (**recced**, **recceing**) reconnoitre. [abbreviation]

recede /rɪ'siːd/ *v.* **(-ding) 1** go or shrink back or further off. **2** be left at an increasing distance by an observer's motion. **3** slope backwards (*a receding chin*). **4** decline in force or value. [Latin *recedere -cess-*: related to CEDE]

receipt /rɪ'siːt/ *–n.* **1** receiving or being received. **2** written acknowledgement of payment received. **3** (usu. in *pl.*) amount of money etc. received. **4** *archaic* recipe. *–v.* place a written or printed receipt on (a bill). □ **in receipt of** having received. [Anglo-French *receite*: related to RECEIVE]

receive /rɪ'siːv/ *v.* **(-ving) 1** take or accept (a thing offered, sent, or given). **2** acquire; be provided with. **3** have conferred or inflicted on one. **4** react to (news, a play, etc.) in a particular way. **5 a** stand the force or weight of. **b** bear up against; encounter with opposition. **6** consent to hear (a confession or oath) or consider (a petition). **7** (also *absol.*) accept (stolen goods knowingly). **8** admit; consent or prove able to hold; provide accommodation for. **9** (of a receptacle) be able to hold. **10** greet or welcome, esp. in a specified manner. **11** entertain as a guest etc. **12** admit to membership. **13** convert (broadcast signals) into sound or pictures. **14** (often as **received** *adj.*) give credit to; accept as authoritative or true. □ **be at** (or **on**) **the receiving end** *colloq.* bear the brunt of something unpleasant. [Latin *recipio -cept-* get back again]

received pronunciation *n.* the form of educated spoken English used in southern England.

receiver *n.* **1** person or thing that receives. **2** part of a machine or instrument that receives something (esp. the part of a telephone that contains the earpiece). **3** (in full **official receiver**) person appointed by a court to administer the property of a bankrupt or insane person, or property under litigation. **4** radio or television receiving apparatus. **5** person who receives stolen goods.

receivership *n.* **1** office of official receiver. **2** state of being dealt with by a receiver (esp. *in receivership*).

recent /'riːs(ə)nt/ *–adj.* **1** not long past; that happened, began to exist, or existed, lately. **2** not long established; lately begun; modern. **3** (**Recent**) *Geol.* of the most recent epoch of the Quaternary period. *–n.* (**Recent**) *Geol.* this epoch. □ **recently** *adv.* [Latin *recens -ent-*]

receptacle /rɪ'septək(ə)l/ *n.* **1** containing vessel, place, or space. **2** *Bot.* enlarged and modified area of the stem apex which bears the flower. [Latin: related to RECEIVE]

reception /rɪ'sepʃ(ə)n/ *n.* **1** receiving or being received. **2** way in which a person or thing is received (*cool reception*). **3** social occasion for receiving guests, esp. after a wedding. **4** place where guests or clients etc. report on arrival at a hotel, office, etc. **5 a** receiving of broadcast signals. **b** quality of this. [Latin: related to RECEIVE]

receptionist *n.* person employed to receive guests, clients, etc.

reception room *n.* room for receiving guests, clients, etc.

receptive /rɪ'septɪv/ *adj.* able or quick to receive impressions or ideas. □ **receptively** *adv.* **receptiveness** *n.* **receptivity** /ˌriːsep'tɪvɪtɪ/ *n.* [French or medieval Latin: related to RECEIVE]

recess /rɪ'ses, 'riːses/ *–n.* **1** space set back in a wall. **2** (often in *pl.*) remote or secret place. **3** temporary cessation from work, esp. of Parliament. *–v.* **1** make a recess in. **2** place in a recess. **3** *US* take a recess; adjourn. [Latin *recessus*: related to RECEDE]

recession /rɪ'seʃ(ə)n/ *n.* **1** temporary decline in economic activity or prosperity. **2** receding or withdrawal from a place or point. [Latin: related to RECESS]

recessional *–adj.* sung while the clergy and choir withdraw after a service. *–n.* recessional hymn.

recessive /rɪ'sesɪv/ *adj.* **1** tending to recede. **2** (of an inherited characteristic) appearing in offspring only when not masked by an inherited dominant characteristic.

recharge *–v.* /riː'tʃɑːdʒ/ **(-ging)** charge (a battery etc.) again or be recharged. *–n.* /'riːtʃɑːdʒ/ recharging or being recharged. □ **rechargeable** /riː'tʃɑːdʒəb(ə)l/ *adj.*

recheck *–v.* /riː'tʃek/ check again. *–n.* /'riːtʃek/ further check or inspection.

recherché /rə'ʃeəʃeɪ/ *adj.* **1** carefully sought out; rare or exotic. **2** far-fetched. [French]

rechristen /riː'krɪs(ə)n/ *v.* **1** christen again. **2** give a new name to.

recidivist /rɪ'sɪdɪvɪst/ *n.* person who relapses into crime. □ **recidivism** *n.* [Latin *recidivus* falling back: related to RECEDE]

recipe /'resɪpɪ/ *n.* **1** statement of the ingredients and procedure required for preparing a cooked dish. **2** (foll. by *for*) certain means to (an outcome) (*recipe for disaster*). [2nd sing. imperative of Latin *recipio* RECEIVE]

recipient /rɪ'sɪpɪənt/ *n.* person who receives something. [Italian or Latin: related to RECEIVE]

reciprocal /rɪ'sɪprək(ə)l/ *–adj.* **1** in return (*a reciprocal greeting*). **2** mutual. **3** *Gram.* (of a pronoun) expressing mutual relation (as in *each other*). *–n. Math.* expression or function so related to another that their product is unity ($^1/_2$ *is the reciprocal of 2*). □ **reciprocally** *adv.* [Latin *reciprocus* moving to and fro]

reciprocate /rɪ'sɪprə,keɪt/ *v.* **(-ting) 1** requite (affection etc.). **2** (foll. by *with*) give in return. **3** give and receive mutually; interchange. **4** (of a part of a machine) move backwards and forwards. □ **reciprocation** /-'keɪʃ(ə)n/ *n.*

reciprocity /,resɪ'prɒsɪtɪ/ *n.* **1** condition of being reciprocal. **2** mutual action. **3** give and take, esp. the interchange of privileges.

recital /rɪ'saɪt(ə)l/ *n.* **1** reciting or being recited. **2** concert of classical music given by a soloist or small group. **3** (foll. by *of*) detailed account of (connected things or facts); narrative.

recitation /,resɪ'teɪʃ(ə)n/ *n.* **1** reciting. **2** thing recited.

recitative /,resɪtə'ti:v/ *n.* musical declamation in the narrative and dialogue parts of opera and oratorio. [Italian *recitativo*: related to RECITE]

recite /rɪ'saɪt/ *v.* **(-ting) 1** repeat aloud or declaim (a poem or passage) from memory. **2** give a recitation. **3** enumerate. [Latin *recito* read out]

reckless /'reklɪs/ *adj.* disregarding the consequences or danger etc.; rash. □ **recklessly** *adv.* **recklessness** *n.* [Old English *reck* concern oneself]

reckon /'rekən/ *v.* **1** (often foll. by *that*) be of the considered opinion; think. **2** consider or regard (*reckoned to be the best*). **3** count or compute by calculation. **4** (foll. by *in*) count in or include in computation. **5** make calculations; add up an account or sum. **6** (foll. by *on*) rely on, count on, or base plans on. **7** (foll. by *with* or *without*) take (or fail to take) into account. [Old English]

reckoning *n.* **1** counting or calculating. **2** consideration or opinion. **3** settlement of an account.

reclaim /rɪ'kleɪm/ *v.* **1** seek the return of (one's property, rights, etc.). **2** bring (land) under cultivation, esp. from being under water. **3** win back or away from vice, error, or a waste condition. □ **reclaimable** *adj.* **reclamation** /,reklə'meɪʃ(ə)n/ *n.* [Latin *reclamare* cry out against]

reclassify /ri:'klæsɪ,faɪ/ *v.* **(-ies, -ied)** classify again or differently. □ **reclassification** /-fɪ'keɪʃ(ə)n/ *n.*

recline /rɪ'klaɪn/ *v.* **(-ning)** assume or be in a horizontal or relaxed leaning position. [Latin *reclino*]

reclothe /ri:'kləʊð/ *v.* **(-thing)** clothe again or differently.

recluse /rɪ'klu:s/ *n.* person given to or living in seclusion or isolation; hermit. □ **reclusive** *adj.* [Latin *recludo -clus-* shut away]

recognition /,rekəg'nɪʃ(ə)n/ *n.* recognizing or being recognized. [Latin: related to RECOGNIZE]

recognizance /rɪ'kɒgnɪz(ə)ns/ *n.* **1** bond by which a person undertakes before a court or magistrate to observe some condition, e.g. to appear when summoned. **2** sum pledged as surety for this. [French: related to RE-]

recognize /'rekəg,naɪz/ *v.* (also **-ise**) **(-zing** or **-sing) 1** identify as already known. **2** realize or discover the nature of. **3** (foll. by *that*) realize or admit. **4** acknowledge the existence, validity, character, or claims of. **5** show appreciation of; reward. **6** (foll. by *as, for*) treat. □ **recognizable** *adj.* [Latin *recognosco*]

recoil /rɪ'kɔɪl/ *–v.* **1** suddenly move or spring back in fear, horror, or disgust. **2** shrink mentally in this way. **3** rebound after an impact. **4** (foll. by *on, upon*) have an adverse reactive effect on (the originator). **5** (of a gun) be driven backwards by its discharge. *–n.* /also 'ri:kɔɪl/ act or sensation of recoiling. [French *reculer* from Latin *culus* buttocks]

recollect /,rekə'lekt/ *v.* **1** remember. **2** succeed in remembering; call to mind. [Latin *recolligo*: related to COLLECT[1]]

recollection /,rekə'lekʃ(ə)n/ *n.* **1** act or power of recollecting. **2** thing recollected. **3 a** person's memory. **b** time over which memory extends (*happened within my recollection*). [French or medieval Latin: related to RECOLLECT]

recolour /ri:'kʌlə(r)/ *v.* colour again or differently.

recombine /,ri:kəm'baɪn/ *v.* **(-ning)** combine again or differently.

recommence /,ri:kə'mens/ *v.* **(-cing)** begin again. □ **recommencement** *n.*

recommend /,rekə'mend/ *v.* **1** suggest as fit for some purpose or use. **2** advise as a course of action etc. **3** (of qualities, conduct, etc.) make acceptable or desirable. **4** (foll. by *to*) commend or entrust (to a person or a person's care). □ **recommendation** /-'deɪʃ(ə)n/ *n.* [medieval Latin: related to RE-]

recompense /'rekəm,pens/ *–v.* **(-sing) 1** make amends to (a person) or for (a loss etc.). **2** requite; reward or punish (a person or action). *–n.* **1** reward, requital. **2** retribution. [Latin: related to COMPENSATE]

reconcile /ˈrekən,saɪl/ *v.* (**-ling**) **1** make friendly again after an estrangement. **2** (usu. in *refl.* or *passive*; foll. by *to*) make acquiescent or contentedly submissive to (something disagreeable). **3** settle (a quarrel etc.). **4 a** harmonize, make compatible. **b** show the compatibility of by argument or in practice. □ **reconcilable** *adj.* **reconciliation** /-,sɪlɪˈeɪʃ(ə)n/ *n.* [Latin: related to CONCILIATE]

recondite /ˈrekən,daɪt/ *adj.* **1** (of a subject or knowledge) abstruse, out of the way, little known. **2** (of an author or style) dealing in abstruse knowledge or allusions, obscure. [Latin *recondo -dit-* put away]

recondition /,ri:kənˈdɪʃ(ə)n/ *v.* overhaul, renovate, make usable again.

reconnaissance /rɪˈkɒnɪs(ə)ns/ *n.* **1** survey of a region, esp. to locate an enemy or ascertain strategic features. **2** preliminary survey. [French: related to RECONNOITRE]

reconnect /,ri:kəˈnekt/ *v.* connect again. □ **reconnection** *n.*

reconnoitre /,rekəˈnɔɪtə(r)/ *v.* (*US* **reconnoiter**) (**-ring**) make a reconnaissance (of). [French: related to RECOGNIZE]

reconquer /ri:ˈkɒŋkə(r)/ *v.* conquer again. □ **reconquest** *n.*

reconsider /,ri:kənˈsɪdə(r)/ *v.* consider again, esp. for a possible change of decision. □ **reconsideration** /-ˈreɪʃ(ə)n/ *n.*

reconstitute /ri:ˈkɒnstɪ,tju:t/ *v.* (**-ting**) **1** reconstruct. **2** reorganize. **3** rehydrate (dried food etc.). □ **reconstitution** /-ˈtju:ʃ(ə)n/ *n.*

reconstruct /,ri:kənˈstrʌkt/ *v.* **1** build again. **2 a** form an impression of (past events) by assembling the evidence for them. **b** re-enact (a crime). **3** reorganize. □ **reconstruction** *n.*

reconvene /,ri:kənˈvi:n/ *v.* (**-ning**) convene again, esp. after a pause in proceedings.

reconvert /,ri:kənˈvɜ:t/ *v.* convert back to a former state. □ **reconversion** *n.*

recopy /ri:ˈkɒpɪ/ *v.* (**-ies, -ied**) copy again.

record –*n.* /ˈrekɔ:d/ **1 a** piece of evidence or information constituting an (esp. official) account of something that has occurred, been said, etc. **b** document etc. preserving this. **2** state of being set down or preserved in writing etc. **3** (in full **gramophone record**) disc carrying recorded sound in grooves on each surface, for reproduction by a record-player. **4** official report of the proceedings and judgement in a court of justice. **5 a** facts known about a person's past. **b** list of a person's previous criminal convictions. **6** (often *attrib.*) best performance (esp. in sport) or most remarkable event of its kind on record. **7** object serving as a memorial; portrait. –*v.* /rɪˈkɔ:d/ **1** set down in writing or some other permanent form for later reference. **2** convert (sound, a broadcast, etc.) into permanent form for later reproduction. □ **for the record** as an official statement etc. **go on record** state one's opinion openly, so that it is recorded. **have a record** have a recorded criminal conviction or convictions. **off the record** unofficially, confidentially. **on record** officially recorded; publicly known. [Latin *cor cordis* heart]

record-breaking *attrib. adj.* that breaks a record.

recorded delivery *n.* Post Office service in which the dispatch and receipt of an item are recorded.

recorder /rɪˈkɔ:də(r)/ *n.* **1** apparatus for recording, esp. a video or tape recorder. **2** (also **Recorder**) barrister or solicitor of at least ten years' standing, serving as a part-time judge. **3** wooden or plastic wind instrument with holes covered by the fingers. **4** keeper of records.

record-holder *n.* person who holds a record.

recording *n.* **1** process by which audio or video signals are recorded for later reproduction. **2** material or a programme recorded.

recordist *n.* person who records sound.

record-player *n.* apparatus for reproducing sound from gramophone records.

recount /rɪˈkaʊnt/ *v.* **1** narrate. **2** tell in detail. [Anglo-French *reconter*: related to RE-, COUNT[1]]

re-count –*v.* /ri:ˈkaʊnt/ count again. –*n.* /ˈri:kaʊnt/ re-counting, esp. of votes in an election.

recoup /rɪˈku:p/ *v.* **1** recover or regain (a loss). **2** compensate or reimburse for a loss. □ **recoupment** *n.* [French *recouper* cut back]

recourse /rɪˈkɔ:s, ˈri:-/ *n.* **1** resort to a possible source of help. **2** person or thing resorted to. □ **have recourse to** turn to (a person or thing) for help. [Latin: related to COURSE]

recover /rɪˈkʌvə(r)/ *v.* **1** regain possession, use, or control of. **2** return to health, consciousness, or to a normal state or position. **3** obtain or secure by legal process. **4** retrieve or make up for (a loss, setback, etc.). **5** *refl.* regain composure, consciousness, or control of one's limbs. **6** retrieve (reusable substances) from waste. □ **recoverable** *adj.* [Latin: related to RECUPERATE]

re-cover /ri:'kʌvə(r)/ *v.* **1** cover again. **2** provide (a chair etc.) with a new cover.

recovery *n.* (*pl.* **-ies**) recovering or being recovered. [Anglo-French *recoverie*: related to RECOVER]

recreant /'rekrɪənt/ *literary* –*adj.* craven, cowardly. –*n.* coward. [medieval Latin: related to CREED]

re-create /,ri:krɪ'eɪt/ *v.* (**-ting**) create over again, reproduce. □ **re-creation** *n.*

recreation /,rekrɪ'eɪʃ(ə)n/ *n.* **1** process or means of refreshing or entertaining oneself. **2** pleasurable activity. □ **recreational** *adj.* [Latin: related to CREATE]

recreation ground *n.* public land used for sports or games.

recriminate /rɪ'krɪmɪ,neɪt/ *v.* (**-ting**) make mutual or counter accusations. □ **recrimination** /-'neɪʃ(ə)n/ *n.* **recriminatory** /-nətərɪ/ *adj.* [medieval Latin: related to CRIME]

recross /ri:'krɒs/ *v.* cross again.

recrudesce /,ri:kru:'des/ *v.* (**-cing**) *formal* (of a disease, problem, etc.) break out again. □ **recrudescence** *n.* **recrudescent** *adj.* [Latin: related to CRUDE]

recruit /rɪ'kru:t/ –*n.* **1** newly enlisted serviceman or servicewoman. **2** new member of a society etc. **3** beginner. –*v.* **1** enlist (a person) as a recruit. **2** form (an army etc.) by enlisting recruits. **3** get or seek recruits. **4** replenish or reinvigorate (numbers, strength, etc.). □ **recruitment** *n.* [French dial. *recrute*: related to CREW[1]]

rectal /'rekt(ə)l/ *adj.* of or by means of the rectum.

rectangle /'rek,tæŋg(ə)l/ *n.* plane figure with four straight sides and four right angles, esp. other than a square. □ **rectangular** /-'tæŋgjʊlə(r)/ *adj.* [French or medieval Latin]

rectify /'rektɪ,faɪ/ *v.* (**-ies, -ied**) **1** adjust or make right. **2** purify or refine, esp. by repeated distillation. **3** convert (alternating current) to direct current. □ **rectifiable** *adj.* **rectification** /-fɪ'keɪʃ(ə)n/ *n.* **rectifier** *n.* [Latin *rectus* straight, right]

rectilinear /,rektɪ'lɪnɪə(r)/ *adj.* **1** bounded or characterized by straight lines. **2** in or forming a straight line. [Latin: related to RECTIFY]

rectitude /'rektɪ,tju:d/ *n.* **1** moral uprightness, righteousness. **2** correctness. [Latin *rectus* right]

recto /'rektəʊ/ *n.* (*pl.* **-s**) **1** right-hand page of an open book. **2** front of a printed leaf. [Latin, = on the right]

rector /'rektə(r)/ *n.* **1** (in the Church of England) incumbent of a parish where all tithes formerly passed to the incumbent (cf. VICAR). **2** *RC Ch.* priest in charge of a church or religious institution. **3** head of some universities and colleges. □ **rectorship** *n.* [Latin *rego rect-* rule]

rectory *n.* (*pl.* **-ies**) rector's house. [French or medieval Latin: related to RECTOR]

rectum /'rektəm/ *n.* (*pl.* **-s**) final section of the large intestine, terminating at the anus. [Latin, = straight]

recumbent /rɪ'kʌmbənt/ *adj.* lying down; reclining. [Latin *cumbo* lie]

recuperate /rɪ'ku:pə,reɪt/ *v.* (**-ting**) **1** recover from illness, exhaustion, loss, etc. **2** regain (health, a loss, etc.). □ **recuperation** /-'reɪʃ(ə)n/ *n.* **recuperative** /-rətɪv/ *adj.* [Latin *recupero*]

recur /rɪ'kɜ:(r)/ *v.* (**-rr-**) **1** occur again; be repeated. **2** (foll. by *to*) go back in thought or speech. **3** (as **recurring** *adj.*) (of a decimal fraction) with the same figure(s) repeated indefinitely (*1.6 recurring*). [Latin *curro* run]

recurrent /rɪ'kʌrənt/ *adj.* recurring; happening repeatedly. □ **recurrence** *n.*

recusant /'rekjʊz(ə)nt/ –*n.* person who refuses submission to an authority or compliance with a regulation, esp. *hist.* one who refused to attend services of the Church of England. –*adj.* of or being a recusant. □ **recusancy** *n.* [Latin *recuso* refuse]

recycle /ri:'saɪk(ə)l/ *v.* (**-ling**) convert (waste) to reusable material. □ **recyclable** *adj.*

red –*adj.* (**redder, reddest**) **1** of the colour ranging from that of blood to deep pink or orange. **2** flushed in the face with shame, anger, etc. **3** (of the eyes) bloodshot or red-rimmed. **4** (of the hair) reddish-brown, tawny. **5** having to do with bloodshed, burning, violence, or revolution. **6** *colloq.* Communist or socialist. **7** (**Red**) *hist.* Russian, Soviet. –*n.* **1** red colour or pigment. **2** red clothes or material. **3** *colloq.* Communist or socialist. □ **in the red** in debt or deficit. □ **reddish** *adj.* **redness** *n.* [Old English]

red admiral *n.* butterfly with red bands.

red-blooded *adj.* virile, vigorous.

redbreast *n. colloq.* robin.

redbrick *adj.* (of a university) founded in the 19th or early 20th c.

redcap *n.* member of the military police.

red card *n. Football* card shown by the referee to a player being sent off.

red carpet *n.* privileged treatment of an eminent visitor.

red cell *n.* (also **red corpuscle**) erythrocyte.

redcoat *n. hist.* British soldier.

Red Crescent *n.* equivalent of the Red Cross in Muslim countries.

Red Cross *n.* international organization bringing relief to victims of war or disaster.

redcurrant *n.* **1** small red edible berry. **2** shrub bearing this.

redden *v.* **1** make or become red. **2** blush.

redecorate /ri:'dekə,reɪt/ *v.* **(-ting)** decorate (a room etc.) again or differently. □ **redecoration** /-'reɪʃ(ə)n/ *n.*

redeem /rɪ'di:m/ *v.* **1** recover by expenditure of effort or by a stipulated payment. **2** make a single payment to cancel (a regular charge or obligation). **3** convert (tokens or bonds etc.) into goods or cash. **4** deliver from sin and damnation. **5** make up for; be a compensating factor in (*has one redeeming feature*). **6** (foll. by *from*) save from (a defect). **7** *refl.* save (oneself) from blame. **8** purchase the freedom of (a person). **9** save (a person's life) by ransom. **10** save or rescue or reclaim. **11** fulfil (a promise). □ **redeemable** *adj.* [Latin *emo* buy]

redeemer *n.* **1** person who redeems. **2** (**the Redeemer**) Christ.

redefine /,ri:dɪ'faɪn/ *v.* **(-ning)** define again or differently. □ **redefinition** /-defɪ'nɪʃ(ə)n/ *n.*

redemption /rɪ'dempʃ(ə)n/ *n.* **1** redeeming or being redeemed. **2** thing that redeems. [Latin: related to REDEEM]

redeploy /,ri:dɪ'plɔɪ/ *v.* send (troops, workers, etc.) to a new place or task. □ **redeployment** *n.*

redesign /,ri:dɪ'zaɪn/ *v.* design again or differently.

redevelop /,ri:dɪ'veləp/ *v.* replan or rebuild (esp. an urban area). □ **redevelopment** *n.*

red flag *n.* **1** symbol of socialist revolution. **2** warning of danger.

red-handed *adv.* in the act of committing a crime, doing wrong, etc.

red hat *n.* **1** cardinal's hat. **2** symbol of a cardinal's office.

redhead *n.* person with red hair.

red herring *n.* misleading clue; distraction.

red-hot *adj.* **1** heated until red. **2** *colloq.* highly exciting. **3** *colloq.* (of news) fresh; completely new. **4** intensely excited. **5** enraged.

red-hot poker *n.* cultivated plant with spikes of usu. red or yellow flowers.

redial /ri:'daɪəl/ *v.* **(-ll-;** *US* **-l-)** dial again.

rediffusion /,ri:dɪ'fju:ʒ(ə)n/ *n.* relaying of broadcast programmes, esp. by cable from a central receiver.

Red Indian *n. offens.* American Indian.

redirect /,ri:daɪ'rekt, -dɪ'rekt/ *v.* **1** direct again; send in a different direction. **2** readdress (a letter etc.).

rediscover /,ri:dɪ'skʌvə(r)/ *v.* discover again. □ **rediscovery** *n.* (*pl.* **-ies**).

redistribute /,ri:dɪ'strɪ,bju:t, ri:'dɪs-/ *v.* **(-ting)** distribute again or differently. □ **redistribution** /-'bju:ʃ(ə)n/ *n.*

■ **Usage** The second pronunciation given, with the stress on the second syllable, is considered incorrect by some people.

redivide /,ri:dɪ'vaɪd/ *v.* **(-ding)** divide again or differently.

red lead *n.* red form of lead oxide used as a pigment.

red-letter day *n.* day that is pleasantly noteworthy or memorable (orig. a festival marked in red on the calendar).

red light *n.* **1** signal to stop on a road, railway, etc. **2** warning.

red-light district *n.* district where many prostitutes work.

red meat *n.* meat that is red when raw (e.g. beef or lamb).

redneck *n. US* often *derog.* politically conservative working-class White in the southern US.

redo /ri:'du:/ *v.* (**redoing**; *3rd sing. present* **redoes**; *past* **redid**; *past part.* **redone**) **1** do again. **2** redecorate.

redolent /'redələnt/ *adj.* **1** (foll. by *of*, *with*) strongly reminiscent, suggestive, or smelling. **2** fragrant. □ **redolence** *n.* [Latin *oleo* smell]

redouble /ri:'dʌb(ə)l/ **–***v.* **(-ling) 1** make or grow greater or more intense or numerous. **2** *Bridge* double again a bid already doubled by an opponent. **–***n. Bridge* redoubling of a bid.

redoubt /rɪ'daʊt/ *n. Mil.* outwork or fieldwork without flanking defences. [French *redoute*: related to REDUCE]

redoubtable /rɪ'daʊtəb(ə)l/ *adj.* formidable.

redound /rɪ'daʊnd/ *v.* **1** (foll. by *to*) make a great contribution to (one's credit or advantage etc.). **2** (foll. by *upon*, *on*) come back or recoil upon. [Latin *unda* wave]

red pepper *n.* **1** cayenne pepper. **2** ripe red fruit of the capsicum plant.

redpoll *n.* finch with a red forehead, similar to a linnet.

redraft /ri:'drɑ:ft/ *v.* draft (a text) again, usu. differently.

red rag *n.* something that excites a person's rage.

redraw /riːˈdrɔː/ *v.* (*past* **redrew**; *past part.* **redrawn**) draw again or differently.

redress /rɪˈdres/ –*v.* **1** remedy or rectify (a wrong or grievance etc.). **2** readjust, set straight again. –*n.* **1** reparation for a wrong. **2** (foll. by *of*) redressing (a grievance etc.). □ **redress the balance** restore equality. [French: related to DRESS]

red rose *n.* emblem of Lancashire or the Lancastrians.

redshank *n.* sandpiper with bright-red legs.

redshift *n.* displacement of the spectrum to longer wavelengths in the light coming from receding galaxies etc.

redskin *n. colloq. offens.* American Indian.

red squirrel *n.* native British squirrel with reddish fur.

redstart *n.* red-tailed songbird. [from RED, obsolete *steort* tail]

red tape *n.* excessive bureaucracy or formality, esp. in public business.

reduce /rɪˈdjuːs/ *v.* (**-cing**) **1** make or become smaller or less. **2** (foll. by *to*) bring by force or necessity (to some undesirable state or action) (*reduced them to tears*; *reduced to begging*). **3** convert to another (esp. simpler) form (*reduced it to a powder*). **4** convert (a fraction) to the form with the lowest terms. **5** (foll. by *to*) bring, simplify, or adapt by classification or analysis (*the dispute may be reduced to three issues*). **6** make lower in status or rank. **7** lower the price of. **8** lessen one's weight or size. **9** weaken (*is in a very reduced state*). **10** impoverish. **11** subdue, bring back to obedience. **12** *Chem.* **a** (cause to) combine with hydrogen. **b** (cause to) undergo addition of electrons. **13 a** (in surgery) restore (a dislocated etc. part) to its proper position. **b** remedy (a dislocation etc.) in this way. □ **reducible** *adj.* [Latin *duco* bring]

reduced circumstances *n.pl.* poverty after relative prosperity.

reductio ad absurdum /rɪˌdʌktɪəʊ æd æbˈzɜːdəm/ *n.* proof of the falsity of a premiss by showing that its logical consequence is absurd. [Latin, = reduction to the absurd]

reduction /rɪˈdʌkʃ(ə)n/ *n.* **1** reducing or being reduced. **2** amount by which prices etc. are reduced. **3** smaller copy of a picture etc. □ **reductive** *adj.*

redundant /rɪˈdʌnd(ə)nt/ *adj.* **1** superfluous. **2** that can be omitted without any loss of significance. **3** (of a person) no longer needed at work and therefore unemployed. □ **redundancy** *n.* (*pl.* **-ies**). [Latin: related to REDOUND]

reduplicate /rɪˈdjuːplɪˌkeɪt/ *v.* (**-ting**) **1** make double. **2** repeat. **3** repeat (a letter or syllable or word) exactly or with a slight change (e.g. *hurly-burly*, *see-saw*). □ **reduplication** /-ˈkeɪʃ(ə)n/ *n.*

redwing *n.* thrush with red underwings.

redwood *n.* very large Californian conifer yielding red wood.

re-echo /riːˈekəʊ/ *v.* (**-es**, **-ed**) echo repeatedly; resound.

reed *n.* **1 a** water or marsh plant with a firm stem. **b** tall straight stalk of this. **2 a** strip of cane etc. vibrating to produce the sound in some wind instruments. **b** (esp. in *pl.*) such an instrument. □ **reeded** *adj.* [Old English]

reed-bed *n.* bed or growth of reeds.

re-educate /riːˈedjʊˌkeɪt/ *v.* (**-ting**) educate again, esp. to change a person's views. □ **re-education** /-ˈkeɪʃ(ə)n/ *n.*

reedy *adj.* (**-ier**, **-iest**) **1** full of reeds. **2** like a reed. **3** (of a voice) like a reed instrument in tone. □ **reediness** *n.*

reef[1] *n.* **1** ridge of rock or coral etc. at or near the surface of the sea. **2 a** lode of ore. **b** bedrock surrounding this. [Old Norse *rif*]

reef[2] –*n.* each of several strips across a sail, for taking it in or rolling it up to reduce its surface area in a high wind. –*v.* take in a reef or reefs of (a sail). [Dutch from Old Norse]

reefer *n.* **1** *slang* marijuana cigarette. **2** thick double-breasted jacket. [from REEF[2]]

reef-knot *n.* symmetrical double knot.

reek –*v.* (often foll. by *of*) **1** smell strongly and unpleasantly. **2** have unpleasant or suspicious associations (*reeks of corruption*). –*n.* **1** foul or stale smell. **2** esp. *Scot.* smoke. **3** vapour, visible exhalation. [Old English]

reel –*n.* **1** cylindrical device on which thread, silk, yarn, paper, film, wire, etc., are wound. **2** quantity of thread etc. wound on a reel. **3** device for winding and unwinding a line as required, esp. in fishing. **4** revolving part in various machines. **5 a** lively folk or Scottish dance. **b** music for this. –*v.* **1** wind (thread, fishing-line, etc.) on a reel. **2** (foll. by *in*, *up*) draw (fish etc.) in or up with a reel. **3** stand, walk, or run unsteadily. **4** be shaken mentally or physically. **5** rock from side to side, or swing violently. **6** dance a reel. □ **reel off** say or recite very rapidly and without apparent effort. [Old English]

re-elect /ˌriːɪˈlekt/ *v.* elect again, esp. to a further term of office. □ **re-election** /-ɪˈlekʃ(ə)n/ *n.*

re-embark /ˌriːɪmˈbɑːk/ *v.* go or put on board ship again.

re-emerge /ˌriːɪˈmɜːdʒ/ *v.* (**-ging**) emerge again; come back out. □ **re-emergence** *n.*
re-emphasize /riːˈemfəˌsaɪz/ *v.* (also **-ise**) (**-zing** or **-sing**) place renewed emphasis on.
re-employ /ˌriːɪmˈplɔɪ/ *v.* employ again. □ **re-employment** *n.*
re-enact /ˌriːɪˈnækt/ *v.* act out (a past event). □ **re-enactment** *n.*
re-engage /ˌriːɪnˈɡeɪdʒ/ *v.* (**-ging**) engage again.
re-enlist /ˌriːɪnˈlɪst/ *v.* enlist again, esp. in the armed services.
re-enter /riːˈentə(r)/ *v.* enter again; go back in.
re-entrant /riːˈentrənt/ *adj.* (of an angle) pointing inwards, reflex.
re-entry /riːˈentrɪ/ *n.* (*pl.* **-ies**) act of entering again, esp. (of a spacecraft, missile, etc.) re-entering the earth's atmosphere.
re-equip /ˌriːɪˈkwɪp/ *v.* (**-pp-**) provide or be provided with new equipment.
re-establish /ˌriːɪˈstæblɪʃ/ *v.* establish again or anew. □ **re-establishment** *n.*
reeve[1] *n. hist.* **1** chief magistrate of a town or district. **2** official supervising a landowner's estate. [Old English]
reeve[2] *v.* (*past* **rove** or **reeved**) *Naut.* **1** (usu. foll. by *through*) thread (a rope or rod etc.) through a ring or other aperture. **2** fasten (a rope or block) in this way. [probably Dutch *reven*]
reeve[3] *n.* female ruff. [origin unknown]
re-examine /ˌriːɪɡˈzæmɪn/ *v.* (**-ning**) examine again or further. □ **re-examination** /-ˈneɪʃ(ə)n/ *n.*
ref[1] *n. colloq.* referee in sports. [abbreviation]
ref[2] *n. Commerce* reference. [abbreviation]
reface /riːˈfeɪs/ *v.* (**-cing**) put a new facing on (a building).
refashion /riːˈfæʃ(ə)n/ *v.* fashion again or differently.
refectory /rɪˈfektərɪ/ *n.* (*pl.* **-ies**) dining-room, esp. in a monastery or college. [Latin *reficio* renew]
refectory table *n.* long narrow table.
refer /rɪˈfɜː(r)/ *v.* (**-rr-**) (usu. foll. by *to*) **1** make an appeal or have recourse to (some authority or source of information) (*referred to his notes*). **2** send on or direct (a person, or a question for decision). **3** (of a person speaking) make an allusion or direct the hearer's or reader's attention (*did not refer to our problems*). **4** (of a statement etc.) be relevant; relate (*these figures refer to last year*). **5** send (a person) to a medical specialist etc. **6** (foll. by *back to*) **a** return (a document etc.) to its sender for clarification. **b** send (a proposal etc.) back to (a lower body, court, etc.). **7** fail (a candidate in an examination). □ **referable** /rɪˈfɜːrəb(ə)l/ *adj.* [Latin *refero relat-* carry back]
referee /ˌrefəˈriː/ —*n.* **1** umpire, esp. in football or boxing. **2** person referred to for a decision in a dispute etc. **3** person willing to testify to the character of an applicant for employment etc. —*v.* (**-rees, -reed**) act as referee (for).
reference /ˈrefərəns/ *n.* **1** referring of a matter for decision, settlement, or consideration to some authority. **2** scope given to this authority. **3** (foll. by *to*) **a** relation, respect, or correspondence. **b** allusion. **c** direction to a book etc. (or a passage in it) where information may be found. **d** book or passage so cited. **4** act of looking up a passage etc., or referring to a book or person for information. **5 a** written testimonial supporting an applicant for employment etc. **b** person giving this. □ **with** (or **in**) **reference to** regarding; as regards; about. □ **referential** /-ˈrenʃ(ə)l/ *adj.*
reference book *n.* book intended to be consulted for occasional information rather than to be read continuously.
referendum /ˌrefəˈrendəm/ *n.* (*pl.* **-s** or **-da**) vote on an important political question open to all the electors of a State. [Latin: related to REFER]
referral /rɪˈfɜːr(ə)l/ *n.* referring of a person to a medical specialist etc.
referred pain *n.* pain felt in a part of the body other than its actual source.
refill —*v.* /riːˈfɪl/ fill again. —*n.* /ˈriːfɪl/ **1** thing that refills, esp. another drink. **2** act of refilling. □ **refillable** /-ˈfɪləb(ə)l/ *adj.*
refine /rɪˈfaɪn/ *v.* (**-ning**) **1** free from impurities or defects. **2** make or become more polished, elegant, or cultured.
refined *adj.* polished, elegant, cultured.
refinement *n.* **1** refining or being refined. **2** fineness of feeling or taste. **3** polish or elegance in behaviour or manner. **4** added development or improvement (*car with several refinements*). **5** subtle reasoning; fine distinction.
refiner *n.* person or firm whose business is to refine crude oil, metal, sugar, etc.
refinery *n.* (*pl.* **-ies**) place where oil, sugar, etc. is refined.
refit —*v.* /riːˈfɪt/ (**-tt-**) esp. *Naut.* make or become serviceable again by repairs, renewals, etc. —*n.* /ˈriːfɪt/ refitting.
reflate /riːˈfleɪt/ *v.* (**-ting**) cause reflation of (a currency or economy etc.). [from RE-, after *inflate, deflate*]
reflation /riːˈfleɪʃ(ə)n/ *n.* inflation of a financial system to restore its previous

condition after deflation. □ **reflationary** *adj.* [from RE-, after *inflation, deflation*]

reflect /rɪ'flekt/ *v.* **1** (of a surface or body) throw back (heat, light, sound, etc.). **2** (of a mirror) show an image of; reproduce to the eye or mind. **3** correspond in appearance or effect to (*their behaviour reflects their upbringing*). **4 a** (of an action, result, etc.) show or bring (credit, discredit, etc.). **b** (*absol.*; usu. foll. by *on, upon*) bring discredit on. **5 a** (often foll. by *on, upon*) meditate on; think about. **b** (foll. by *that, how*, etc.) consider; remind oneself. [Latin *flecto flex-* bend]

reflection /rɪ'flekʃ(ə)n/ *n.* (also **reflexion**) **1** reflecting or being reflected. **2 a** reflected light, heat, or colour. **b** reflected image. **3** reconsideration (*on reflection*). **4** (often foll. by *on*) discredit or thing bringing discredit. **5** (often foll. by *on, upon*) idea arising in the mind; comment.

reflective *adj.* **1** (of a surface etc.) reflecting. **2** (of mental faculties) concerned in reflection or thought. **3** (of a person or mood etc.) thoughtful; given to meditation. □ **reflectively** *adv.* **reflectiveness** *n.*

reflector *n.* **1** piece of glass or metal etc. for reflecting light in a required direction, e.g. a red one on the back of a motor vehicle or bicycle. **2 a** telescope etc. using a mirror to produce images. **b** the mirror itself.

reflex /'ri:fleks/ *–adj.* **1** (of an action) independent of the will, as an automatic response to the stimulation of a nerve. **2** (of an angle) exceeding 180°. *–n.* **1** reflex action. **2** sign or secondary manifestation (*law is a reflex of public opinion*). **3** reflected light or image. [Latin: related to REFLECT]

reflex camera *n.* camera in which the viewed image is formed by a mirror, enabling the scene to be correctly composed and focused.

reflexion var. of REFLECTION.

reflexive /rɪ'fleksɪv/ *Gram. –adj.* **1** (of a word or form, esp. of a pronoun) referring back to the subject of a sentence (e.g. *myself*). **2** (of a verb) having a reflexive pronoun as its object (as in *to wash oneself*). *–n.* reflexive word or form, esp. a pronoun (e.g. *myself*).

reflexology /,ri:flek'sɒlədʒɪ/ *n.* massage through points on the feet, hands, and head, to relieve tension and treat illness. □ **reflexologist** *n.*

refloat /ri:'fləʊt/ *v.* set (a stranded ship) afloat again.

refocus /ri:'fəʊkəs/ *v.* (**-s-** or **-ss-**) focus again or anew.

reforest /ri:'fɒrɪst/ *v.* = REAFFOREST. □ **reforestation** /-'steɪʃ(ə)n/ *n.*

reforge /ri:'fɔ:dʒ/ *v.* (**-ging**) forge again or differently.

reform /rɪ'fɔ:m/ *–v.* **1** make or become better by the removal of faults and errors. **2** abolish or cure (an abuse or malpractice). *–n.* **1** removal of faults or abuses, esp. moral, political, or social. **2** improvement made or suggested. □ **reformative** *adj.*

re-form /ri:'fɔ:m/ *v.* form again. □ **re-formation** /-'meɪʃ(ə)n/ *n.*

reformat /ri:'fɔ:mæt/ *v.* (**-tt-**) format anew.

reformation /,refə'meɪʃ(ə)n/ *n.* **1** reforming or being reformed, esp. a radical change for the better in political, religious, or social affairs. **2** (**the Reformation**) *hist.* 16th-c. movement for the reform of abuses in the Roman Church ending in the establishment of the Reformed or Protestant Churches.

reformatory /rɪ'fɔ:mətərɪ/ *–n.* (*pl.* **-ies**) *US* & *hist.* institution for the reform of young offenders. *–adj.* producing reform.

Reformed Church *n.* a Protestant (esp. Calvinist) Church.

reformer *n.* person who advocates or brings about (esp. political or social) reform.

reformism /rɪ'fɔ:mɪz(ə)m/ *n.* policy of reform rather than abolition or revolution. □ **reformist** *n.* & *adj.*

reformulate /ri:'fɔ:mjʊ,leɪt/ *v.* (**-ting**) formulate again or differently. □ **reformulation** /-'leɪʃ(ə)n/ *n.*

refract /rɪ'frækt/ *v.* (of water, air, glass, etc.) deflect (a ray of light etc.) at a certain angle when it enters obliquely from another medium. □ **refraction** *n.* **refractive** *adj.* [Latin *refringo -fract-* break open]

refractor *n.* **1** refracting medium or lens. **2** telescope using a lens to produce an image.

refractory /rɪ'fræktərɪ/ *adj.* **1** stubborn, unmanageable, rebellious. **2** (of a wound, disease, etc.) not yielding to treatment. **3** (of a substance) hard to fuse or work. [Latin: related to REFRACT]

refrain[1] /rɪ'freɪn/ *v.* (foll. by *from*) avoid doing (an action) (*refrain from smoking*). [Latin *frenum* bridle]

refrain[2] /rɪ'freɪn/ *n.* **1** recurring phrase or lines, esp. at the ends of stanzas. **2** music accompanying this. [Latin: related to REFRACT]

refrangible /rɪ'frændʒɪb(ə)l/ *adj.* that can be refracted. [Latin: related to REFRACT]

refreeze /ri:'fri:z/ *v.* (**-zing**; *past* **refroze**; *past part.* **refrozen**) freeze again.

refresh /rɪ'freʃ/ *v.* **1** give new spirit or vigour to. **2** revive (the memory), esp. by consulting the source of one's information. □ **refreshing** *adj.* **refreshingly** *adv.* [French: related to FRESH]

refresher *n.* **1** something that refreshes, esp. a drink. **2** *Law* extra fee payable to counsel in a prolonged case.

refresher course *n.* course reviewing or updating previous studies.

refreshment *n.* **1** refreshing or being refreshed. **2** (usu. in *pl.*) food or drink.

refrigerant /rɪ'frɪdʒərənt/ *–n.* substance used for refrigeration. *–adj.* cooling. [Latin: related to REFRIGERATE]

refrigerate /rɪ'frɪdʒə,reɪt/ *v.* (**-ting**) **1** make or become cool or cold. **2** subject (food etc.) to cold in order to freeze or preserve it. □ **refrigeration** /-'reɪʃ(ə)n/ *n.* [Latin *refrigero* from *frigus* cold]

refrigerator *n.* cabinet or room in which food etc. is kept cold.

refroze *past* of REFREEZE.

refrozen *past part.* of REFREEZE.

refuel /ri:'fju:əl/ *v.* (**-ll-**; *US* **-l-**) replenish a fuel supply; supply with more fuel.

refuge /'refju:dʒ/ *n.* **1** shelter from pursuit, danger, or trouble. **2** person or place etc. offering this. [Latin *refugium* from *fugio* flee]

refugee /,refjʊ'dʒi:/ *n.* person taking refuge, esp. in a foreign country, from war, persecution, or natural disaster. [French *réfugié*: related to REFUGE]

refulgent /rɪ'fʌldʒ(ə)nt/ *adj. literary* shining, gloriously bright. □ **refulgence** *n.* [Latin *refulgeo* shine brightly]

refund *–v.* /rɪ'fʌnd/ (also *absol.*) **1** pay back (money or expenses). **2** reimburse (a person). *–n.* /'ri:fʌnd/ **1** act of refunding. **2** sum refunded. □ **refundable** /rɪ'fʌndəb(ə)l/ *adj.* [Latin *fundo* pour]

refurbish /ri:'fɜ:bɪʃ/ *v.* **1** brighten up. **2** restore and redecorate. □ **refurbishment** *n.*

refurnish /ri:'fɜ:nɪʃ/ *v.* furnish again or differently.

refusal /rɪ'fju:z(ə)l/ *n.* **1** refusing or being refused. **2** (in full **first refusal**) right or privilege of deciding to take or leave a thing before it is offered to others.

refuse[1] /rɪ'fju:z/ *v.* (**-sing**) **1** withhold acceptance of or consent to (*refuse an offer, orders*). **2** (often foll. by *to* + infin.) indicate unwillingness or inability (*I refuse to go*; *car refuses to start*; *I refuse!*). **3** (often with double object) not grant (a request) made by (a person). **4** (also *absol.*) (of a horse) be unwilling to jump (a fence etc.). [French *refuser*]

refuse[2] /'refju:s/ *n.* items rejected as worthless; waste. [French: related to REFUSE[1]]

refusenik /rɪ'fju:znɪk/ *n. hist.* Soviet Jew who has been refused permission to emigrate to Israel.

refute /rɪ'fju:t/ *v.* (**-ting**) **1** prove the falsity or error of (a statement etc. or the person advancing it). **2** rebut by argument. **3** deny or contradict (without argument). □ **refutation** /,refjʊ'teɪʃ(ə)n/ *n.* [Latin *refuto*]

■ **Usage** The use of *refute* in sense 3 is considered incorrect by some people. It is often confused in this sense with *repudiate*.

reg /redʒ/ *n. colloq.* = REGISTRATION MARK. [abbreviation]

regain /rɪ'geɪn/ *v.* obtain possession or use of after loss (*regain consciousness*).

regal /'ri:g(ə)l/ *adj.* **1** of or by a monarch or monarchs. **2** fit for a monarch; magnificent. □ **regality** /rɪ'gælɪtɪ/ *n.* **regally** *adv.* [Latin *rex reg-* king]

regale /rɪ'geɪl/ *v.* (**-ling**) **1** entertain lavishly with feasting. **2** (foll. by *with*) entertain with (talk etc.). [French *régaler*: related to GALLANT]

regalia /rɪ'geɪlɪə/ *n.pl.* **1** insignia of royalty used at coronations. **2** insignia of an order or of civic dignity. [medieval Latin: related to REGAL]

regard /rɪ'gɑ:d/ *–v.* **1** gaze on steadily (usu. in a specified way) (*regarded them suspiciously*). **2** heed; take into account. **3** look upon or think of in a specified way (*regard it as an insult*). *–n.* **1** gaze; steady or significant look. **2** (foll. by *to, for*) attention or care. **3** (foll. by *for*) esteem; kindly feeling; respectful opinion. **4** respect; point attended to (*in this regard*). **5** (in *pl.*) expression of friendliness in a letter etc.; compliments. □ **as regards** about, concerning; in respect of. **in** (or **with**) **regard to** as concerns; in respect of. [French *regard(er)*: related to GUARD]

regardful *adj.* (foll. by *of*) mindful of.

regarding *prep.* about, concerning; in respect of.

regardless *–adj.* (foll. by *of*) without regard or consideration for. *–adv.* without paying attention.

regatta /rɪ'gætə/ *n.* event consisting of rowing or yacht races. [Italian]

regency /'ri:dʒənsɪ/ *n.* (*pl.* **-ies**) **1** office of regent. **2** commission acting as regent. **3 a** period of office of a regent or regency commission. **b** (**Regency**) (in the UK) 1811 to 1820. [medieval Latin *regentia*: related to REGENT]

regenerate *–v.* /rɪ'dʒenə,reɪt/ (**-ting**) **1** bring or come into renewed existence;

generate again. **2** improve the moral condition of. **3** impart new, more vigorous, or spiritually higher life or nature to. **4** *Biol.* regrow or cause (new tissue) to regrow. *–adj.* /rɪ'dʒenərət/ spiritually born again, reformed. □ **regeneration** /-'reɪʃ(ə)n/ *n.* **regenerative** /-rətɪv/ *adj.*

regent /'ri:dʒ(ə)nt/ *–n.* person appointed to administer a State because the monarch is a minor or is absent or incapacitated. *–adj.* (after the noun) acting as regent (*Prince Regent*). [Latin *rego* rule]

reggae /'regeɪ/ *n.* W. Indian style of music with a strongly accented subsidiary beat. [origin unknown]

regicide /'redʒɪ,saɪd/ *n.* **1** person who kills or helps to kill a king. **2** killing of a king. [Latin *rex reg-* king, -CIDE]

regime /reɪ'ʒi:m/ *n.* (also **régime**) **1** method or system of government. **2** prevailing order or system of things. **3** regimen. [French: related to REGIMEN]

regimen /'redʒɪmən/ *n.* prescribed course of exercise, way of life, or diet. [Latin *rego* rule]

regiment *–n.* /'redʒɪmənt/ **1 a** permanent unit of an army, usu. commanded by a colonel and divided into several companies, troops, or batteries. **b** operational unit of artillery etc. **2** (usu. foll. by *of*) large or formidable array or number. *–v.* /'redʒɪ,ment/ **1** organize (esp. oppressively) in groups or according to a system. **2** form into a regiment or regiments. □ **regimentation** /-'teɪʃ(ə)n/ *n.* [Latin: related to REGIMEN]

regimental /,redʒɪ'ment(ə)l/ *–adj.* of a regiment. *–n.* (in *pl.*) military uniform, esp. of a particular regiment. □ **regimentally** *adv.*

Regina /rɪ'dʒaɪnə/ *n.* **1** (after the name) reigning queen (*Elizabeth Regina*). **2** *Law* the Crown (*Regina v. Jones*). [Latin, = queen: related to REX]

region /'ri:dʒ(ə)n/ *n.* **1** geographical area or division, having definable boundaries or characteristics (*fertile region*). **2** administrative area, esp. in Scotland. **3** part of the body (*lumbar region*). **4** sphere or realm (*region of metaphysics*). □ **in the region of** approximately. □ **regional** *adj.* **regionally** *adv.* [Latin *rego* rule]

register /'redʒɪstə(r)/ *–n.* **1** official list, e.g. of births, marriages, and deaths, of children in a class, of shipping, of professionally qualified persons, or of qualified voters in a constituency. **2** book in which items are recorded for reference. **3** device recording speed, force, etc. **4 a** compass of a voice or instrument. **b** part of this compass (*lower register*). **5** adjustable plate for widening or narrowing an opening and regulating a draught, esp. in a fire-grate. **6 a** set of organ pipes. **b** sliding device controlling this. **7** = CASH REGISTER. **8** form of a language (colloquial, literary, etc.) used in particular circumstances. **9** *Computing* a memory location having specific properties and quick access time. *–v.* **1** set down (a name, fact, complaint, etc.) formally; record in writing. **2** enter or cause to be entered in a particular register. **3** commit (a letter etc.) to registered post. **4** (of an instrument) record automatically; indicate. **5 a** express (an emotion) facially or by gesture (*registered surprise*). **b** (of an emotion) show in a person's face or gestures. **6** make an impression on a person's mind. [Latin *regero -gest-* transcribe, record]

registered nurse *n.* nurse with a State certificate of competence.

registered post *n.* postal procedure with special precautions for safety and for compensation in case of loss.

register office *n.* State office where civil marriages are conducted.

■ **Usage** *Register office* is the official name, although *registry office* is often heard in colloquial usage.

registrar /,redʒɪ'strɑ:(r)/ *n.* **1** official responsible for keeping a register. **2** chief administrator in a university, college, etc. **3** hospital doctor training as a specialist. [medieval Latin: related to REGISTER]

registration /,redʒɪ'streɪʃ(ə)n/ *n.* registering or being registered. [French or medieval Latin: related to REGISTER]

registration mark *n.* (also **registration number**) combination of letters and numbers identifying a vehicle etc.

registry /'redʒɪstrɪ/ *n.* (*pl.* **-ies**) place where registers or records are kept. [medieval Latin: related to REGISTER]

registry office *n.* = REGISTER OFFICE.

Regius professor /'ri:dʒɪəs/ *n.* holder of a chair founded by a sovereign (esp. one at Oxford or Cambridge instituted by Henry VIII) or filled by Crown appointment. [Latin *regius* royal]

regrade /ri:'greɪd/ *v.* (**-ding**) grade again or differently.

regress *–v.* /rɪ'gres/ **1** move backwards; return to a former, esp. worse, state. **2** *Psychol.* (cause to) return mentally to a former stage of life. *–n.* /'ri:gres/ act of regressing. □ **regression** /rɪ'greʃ(ə)n/ *n.* **regressive** /rɪ'gresɪv/ *adj.* [Latin *regredior -gress-* go back]

regret /rɪ'gret/ *–v.* (**-tt-**) **1** feel or express sorrow, repentance, or distress over (an action or loss etc.). **2** acknowledge with sorrow or remorse (*regret to say*). **–***n.* feeling of sorrow, repentance, etc., over an action or loss etc. □ **give** (or **send**) **one's regrets** formally decline an invitation. [French *regretter*]

regretful *adj.* feeling or showing regret. □ **regretfully** *adv.*

regrettable *adj.* (of events or conduct) undesirable, unwelcome; deserving censure. □ **regrettably** *adv.*

regroup /ri:'gru:p/ *v.* **1** group or arrange again or differently. **2** *Mil.* prepare for a fresh attack.

regrow /ri:'grəʊ/ *v.* grow again, esp. after an interval. □ **regrowth** *n.*

regular /'regjʊlə(r)/ **–***adj.* **1** acting, done, or recurring uniformly or calculably in time or manner; habitual, constant, orderly. **2** conforming to a rule or principle; systematic. **3** harmonious, symmetrical. **4** conforming to a standard of etiquette or procedure. **5** properly constituted or qualified; pursuing an occupation as one's main pursuit (*regular soldier*). **6** *Gram.* (of a noun, verb, etc.) following the normal type of inflection. **7** *colloq.* thorough, absolute (*a regular hero*). **8** (before or after the noun) bound by religious rule; belonging to a religious or monastic order (*canon regular*). **9** (of a person) defecating or menstruating at predictable times. **–***n.* **1** regular soldier. **2** *colloq.* regular customer, visitor, etc. **3** one of the regular clergy. □ **regularity** /-'lærɪtɪ/ *n.* **regularize** *v.* (also **-ise**) (**-zing** or **-sing**). **regularly** *adv.* [Latin *regula* rule]

regulate /'regjʊ,leɪt/ *v.* (**-ting**) **1** control by rule. **2** subject to restrictions. **3** adapt to requirements. **4** alter the speed of (a machine or clock) so that it works accurately. □ **regulator** *n.* **regulatory** /-lətərɪ/ *adj.* [Latin: related to REGULAR]

regulation /,regjʊ'leɪʃ(ə)n/ *n.* **1** regulating or being regulated. **2** prescribed rule. **3** (*attrib.*) **a** in accordance with regulations; of the correct type etc. **b** *colloq.* usual.

regulo /'regjʊ,ləʊ/ *n.* (usu. foll. by a numeral) each of the numbers of a scale denoting temperature in a gas oven (*cook at regulo 6*). [*Regulo*, propr. term for a thermostatic gas oven control]

regurgitate /rɪ'gɜ:dʒɪ,teɪt/ *v.* (**-ting**) **1** bring (swallowed food) up again to the mouth. **2** reproduce, rehash (information etc.). □ **regurgitation** /-'teɪʃ(ə)n/ *n.* [Latin *gurges -git-* whirlpool]

rehabilitate /,ri:hə'bɪlɪ,teɪt/ *v.* (**-ting**) **1** restore to effectiveness or normal life by training etc., esp. after imprisonment or illness. **2** restore to former privileges or reputation or a proper condition. □ **rehabilitation** /-'teɪʃ(ə)n/ *n.* [medieval Latin: related to RE-, ABILITY]

rehang /ri:'hæŋ/ *v.* (*past* and *past part.* **rehung**) hang again or differently.

rehash **–***v.* /ri:'hæʃ/ put (old material) into a new form without significant change or improvement. **–***n.* /'ri:hæʃ/ **1** material rehashed. **2** rehashing.

rehear /ri:'hɪə(r)/ *v.* (*past* and *past part.* **reheard** /-'hɜ:d/) hear (esp. a judicial case) again.

rehearsal /rɪ'hɜ:s(ə)l/ *n.* **1** trial performance or practice of a play, music, etc. **2** process of rehearsing.

rehearse /rɪ'hɜ:s/ *v.* (**-sing**) **1** practise (a play, music, etc.) for later public performance. **2** hold a rehearsal. **3** train (a person) by rehearsal. **4** recite or say over. **5** give a list of, enumerate. [Anglo-French: related to HEARSE]

reheat /ri:'hi:t/ *v.* heat again.

rehouse /ri:'haʊz/ *v.* (**-sing**) house elsewhere.

rehung *past* and *past part.* of REHANG.

Reich /raɪx/ *n.* the former German State, esp. the Third Reich. [German, = empire]

reign /reɪn/ **–***v.* **1** be king or queen. **2** prevail (*confusion reigns*). **3** (as **reigning** *attrib. adj.*) (of a winner, champion, etc.) currently holding the title etc. **–***n.* **1** sovereignty, rule. **2** period during which a sovereign rules. [Latin *regnum*: related to REX]

reimburse /,ri:ɪm'bɜ:s/ *v.* (**-sing**) **1** repay (a person who has expended money). **2** repay (a person's expenses). □ **reimbursement** *n.*

reimpose /,ri:ɪm'pəʊz/ *v.* (**-sing**) impose again, esp. after a lapse.

rein /reɪn/ **–***n.* (in *sing.* or *pl.*) **1** long narrow strap with each end attached to the bit, used to guide or check a horse etc. **2** similar device used to restrain a child. **3** means of control. **–***v.* **1** check or manage with reins. **2** (foll. by *up*, *back*) pull up or back with reins. **3** (foll. by *in*) hold in as with reins. **4** govern, restrain, control. □ **give free rein to** allow freedom of action or expression. **keep a tight rein on** allow little freedom to. [French *rene* from Latin *retinēre* RETAIN]

reincarnation /,ri:ɪnkɑ:'neɪʃ(ə)n/ *n.* rebirth of a soul in a new body. □ **reincarnate** /-'kɑ:neɪt/ *v.* (**-ting**). **reincarnate** /-'kɑ:nət/ *adj.*

reindeer /'reɪndɪə(r)/ *n.* (*pl.* same or **-s**) subarctic deer with large antlers. [Old Norse]

reinforce /,ri:ɪn'fɔ:s/ *v.* (**-cing**) strengthen or support, esp. with additional personnel or material, or by an increase of numbers, quantity, or size etc. [French *renforcer*]

reinforced concrete *n.* concrete with metal bars or wire etc. embedded to increase its strength.

reinforcement *n.* **1** reinforcing or being reinforced. **2** thing that reinforces. **3** (in *pl.*) reinforcing personnel or equipment etc.

reinsert /ˌri:ɪn'sɜ:t/ *v.* insert again.

reinstate /ˌri:ɪn'steɪt/ *v.* (**-ting**) **1** replace in a former position. **2** restore (a person etc.) to former privileges. □ **reinstatement** *n.*

reinsure /ˌri:ɪn'ʃʊə(r)/ *v.* (**-ring**) insure again (esp. of an insurer transferring risk to another insurer). □ **reinsurance** *n.*

reinterpret /ˌri:ɪn'tɜ:prɪt/ *v.* (**-t-**) interpret again or differently. □ **reinterpretation** /-'teɪʃ(ə)n/ *n.*

reintroduce /ˌri:ɪntrə'dju:s/ *v.* (**-cing**) introduce again. □ **reintroduction** /-'dʌkʃ(ə)n/ *n.*

reinvest /ˌri:ɪn'vest/ *v.* invest again (esp. proceeds or interest). □ **reinvestment** *n.*

reissue /ri:'ɪʃu:/ *–v.* (**-ues**, **-ued**, **-uing**) issue again or in a different form. *–n.* new issue, esp. of a previously published book.

reiterate /ri:'ɪtəˌreɪt/ *v.* (**-ting**) say or do again or repeatedly. □ **reiteration** /-'reɪʃ(ə)n/ *n.*

reject *–v.* /rɪ'dʒekt/ **1** put aside or send back as not to be used, done, or complied with etc. **2** refuse to accept or believe in. **3** rebuff or withhold affection from (a person). **4** show an immune response to (a transplant) so that it fails. *–n.* /'ri:dʒekt/ thing or person rejected as unfit or below standard. □ **rejection** /rɪ'dʒekʃ(ə)n/ *n.* [Latin *rejicio -ject-* throw back]

rejig /ri:'dʒɪg/ *v.* (**-gg-**) **1** re-equip (a factory etc.) for a new kind of work. **2** rearrange.

rejoice /rɪ'dʒɔɪs/ *v.* (**-cing**) **1** feel great joy. **2** be glad. **3** (foll. by *in*, *at*) take delight. [French *rejoir*: related to JOY]

rejoin[1] /ri:'dʒɔɪn/ *v.* **1** join together again; reunite. **2** join (a companion etc.) again.

rejoin[2] /rɪ'dʒɔɪn/ *v.* **1** say in answer, retort. **2** reply to a charge or pleading in a lawsuit. [French *rejoindre*: related to JOIN]

rejoinder /rɪ'dʒɔɪndə(r)/ *n.* what is said in reply; retort. [Anglo-French: related to REJOIN[2]]

rejuvenate /rɪ'dʒu:vəˌneɪt/ *v.* (**-ting**) make (as if) young again. □ **rejuvenation** /-'neɪʃ(ə)n/ *n.* [Latin *juvenis* young]

rekindle /ri:'kɪnd(ə)l/ *v.* (**-ling**) kindle again.

relabel /ri:'leɪb(ə)l/ *v.* (**-ll-**; *US* **-l-**) label (esp. a commodity) again or differently.

relapse /rɪ'læps/ *–v.* (**-sing**) (usu. foll. by *into*) fall back or sink again (into a worse state after improvement). *–n.* /also 'ri:-/ relapsing, esp. a deterioration in a patient's condition after partial recovery. [Latin *labor laps-* slip]

relate /rɪ'leɪt/ *v.* (**-ting**) **1** narrate or recount. **2** (usu. foll. by *to*, *with*) connect (two things) in thought or meaning; associate. **3** (foll. by *to*) have reference to. **4** (foll. by *to*) feel connected or sympathetic to. [Latin: related to REFER]

related *adj.* connected, esp. by blood or marriage.

relation /rɪ'leɪʃ(ə)n/ *n.* **1 a** the way in which one person or thing is related or connected to another. **b** connection, correspondence, contrast, or feeling prevailing between persons or things (*bears no relation to the facts*; *enjoyed good relations for many years*). **2** relative. **3** (in *pl.*) **a** (foll. by *with*) dealings (with others). **b** sexual intercourse. **4** = RELATIONSHIP. **5 a** narration (*his relation of the events*). **b** narrative. □ **in relation to** as regards. [Latin: related to REFER]

relationship *n.* **1** state or instance of being related. **2 a** connection or association (*good working relationship*). **b** *colloq.* emotional (esp. sexual) association between two people.

relative /'relətɪv/ *–adj.* **1** considered in relation to something else (*relative velocity*). **2** (foll. by *to*) proportioned to (something else) (*growth is relative to input*). **3** implying comparison or contextual relation (*'heat' is a relative word*). **4** comparative (*their relative merits*). **5** having mutual relations; corresponding in some way; related to each other. **6** (foll. by *to*) having reference or relating to (*the facts relative to the issue*). **7** *Gram.* **a** (of a word, esp. a pronoun) referring to an expressed or implied antecedent and attaching a subordinate clause to it, e.g. *which*, *who*. **b** (of a clause) attached to an antecedent by a relative word. *–n.* **1** person connected by blood or marriage. **2** species related to another by common origin. **3** *Gram.* relative word, esp. a pronoun. □ **relatively** *adv.* [Latin: related to REFER]

relative atomic mass *n.* the ratio of the average mass of one atom of an element to one-twelfth of the mass of an atom of carbon-12.

relative density *n.* the ratio between the mass of a substance and that of the same volume of a substance used as a standard (usu. water or air).

relative molecular mass *n.* the ratio of the average mass of one molecule of an element or compound to one-twelfth of the mass of an atom of carbon-12.

relativity /ˌreləˈtɪvɪtɪ/ *n.* **1** being relative. **2** *Physics* **a** (**special theory of relativity**) theory based on the principle that all motion is relative and that light has a constant velocity. **b** (**general theory of relativity**) theory extending this to gravitation and accelerated motion.

relax /rɪˈlæks/ *v.* **1** make or become less stiff, rigid, or tense. **2** make or become less formal or strict (*rules were relaxed*). **3** reduce or abate (one's attention, efforts, etc.). **4** cease work or effort. **5** (as **relaxed** *adj.*) at ease; unperturbed. [Latin *relaxo*: related to LAX]

relaxation /ˌri:lækˈseɪʃ(ə)n/ *n.* **1** relaxing or being relaxed. **2** recreation.

relay /ˈri:leɪ/ –*n.* **1** fresh set of people etc. substituted for tired ones. **2** supply of material similarly used. **3** = RELAY RACE. **4** device activating an electric circuit etc. in response to changes affecting itself. **5 a** device to receive, reinforce, and transmit a message, broadcast, etc. **b** relayed message or transmission. –*v.* /also rɪˈleɪ/ receive (a message, broadcast, etc.) and transmit it to others. [French *relai* from Latin *laxo*: see LAX]

re-lay /ri:ˈleɪ/ *v.* (*past* and *past part.* **re-laid**) lay again or differently.

relay race *n.* race between teams of which each member in turn covers part of the distance.

relearn /ri:ˈlɜ:n/ *v.* learn again.

release /rɪˈli:s/ –*v.* (**-sing**) **1** (often foll. by *from*) set free; liberate, unfasten. **2** allow to move from a fixed position. **3 a** make (information, a recording, etc.) publicly available. **b** issue (a film etc.) for general exhibition. –*n.* **1** liberation from a restriction, duty, or difficulty. **2** handle or catch that releases part of a mechanism. **3** news item etc. made available for publication (*press release*). **4 a** film or record etc. that is released. **b** releasing or being released in this way. [French *relesser* from Latin *relaxo* RELAX]

relegate /ˈrelɪˌgeɪt/ *v.* (**-ting**) **1** consign or dismiss to an inferior position. **2** transfer (a sports team) to a lower division of a league etc. **3** banish. □ **relegation** /-ˈgeɪʃ(ə)n/ *n.* [Latin *relego* send away]

relent /rɪˈlent/ *v.* relax severity, abandon a harsh intention, yield to compassion. [medieval Latin *lentus* flexible]

relentless *adj.* unrelenting, oppressively constant. □ **relentlessly** *adv.*

re-let /ri:ˈlet/ –*v.* (**-tt-**; *past* and *past part.* **-let**) let (a property) for a further period or to a new tenant. –*n.* re-let property.

relevant /ˈrelɪv(ə)nt/ *adj.* (often foll. by *to*) bearing on or having reference to the matter in hand. □ **relevance** *n.* [Latin *relevo*: related to RELIEVE]

reliable /rɪˈlaɪəb(ə)l/ *adj.* of consistently good character or quality; dependable. □ **reliability** /-ˈbɪlɪtɪ/ *n.* **reliably** *adv.*

reliance /rɪˈlaɪəns/ *n.* (foll. by *in*, *on*) trust, confidence. □ **reliant** *adj.*

relic /ˈrelɪk/ *n.* **1** object that is interesting because of its age or association. **2** part of a dead holy person's body or belongings kept as an object of reverence. **3** surviving custom or belief etc. from a past age. **4** memento or souvenir. **5** (in *pl.*) what has survived. **6** (in *pl.*) dead body or remains of a person. [Latin *reliquiae* remains: related to RELINQUISH]

relict /ˈrelɪkt/ *n.* object surviving in its primitive form. [French *relicte*: related to RELIC]

relief /rɪˈli:f/ *n.* **1 a** alleviation of or deliverance from pain, distress, anxiety, etc. **b** feeling accompanying such deliverance. **2** feature etc. that diversifies monotony or relaxes tension. **3** assistance (esp. financial) given to those in special need or difficulty. **4 a** replacing of a person or persons on duty by another or others. **b** person or persons replacing others in this way. **5** (usu. *attrib.*) thing supplementing another in some service (*relief bus*). **6 a** method of moulding, carving, or stamping in which the design stands out from the surface. **b** piece of sculpture etc. in relief. **c** representation of relief given by an arrangement of line, colour, or shading. **7** vividness, distinctness (*brings the facts out in sharp relief*). **8** (foll. by *of*) reinforcement (esp. the raising of a siege) of a place. **9** esp. *Law* redress of a hardship or grievance. [French and Italian: related to RELIEVE]

relief map *n.* map indicating hills and valleys by shading etc. rather than by contour lines alone.

relief road *n.* road taking traffic around a congested area.

relieve /rɪˈli:v/ *v.* (**-ving**) **1** bring or give relief to. **2** mitigate the tedium or monotony of. **3** release (a person) from a duty by acting as or providing a substitute. **4** (foll. by *of*) take (esp. a burden or duty) away from (a person). □ **relieve one's**

feelings use strong language or vigorous behaviour when annoyed. **relieve oneself** urinate or defecate. □ **relieved** *adj.* [Latin *relevo* raise again, alleviate]

relievo /rɪˈliːvəʊ/ *n.* (*pl.* **-s**) = RELIEF 6. [Italian *rilievo*: related to RELIEF]

relight /riːˈlaɪt/ *v.* (*past* and *past part.* **-lit**) light (a fire etc.) again.

religion /rɪˈlɪdʒ(ə)n/ *n.* **1** belief in a superhuman controlling power, esp. in a personal God or gods entitled to obedience and worship. **2** expression of this in worship. **3** particular system of faith and worship. **4** life under monastic vows. **5** thing that one is devoted to. [Latin *religio* bond]

religiosity /rɪˌlɪdʒɪˈɒsɪtɪ/ *n.* state of being religious or too religious. [Latin: related to RELIGIOUS]

religious /rɪˈlɪdʒəs/ *–adj.* **1** devoted to religion; pious, devout. **2** of or concerned with religion. **3** of or belonging to a monastic order. **4** scrupulous, conscientious. *–n.* (*pl.* same) person bound by monastic vows. □ **religiously** *adv.* [Latin *religiosus*: related to RELIGION]

reline /riːˈlaɪn/ *v.* (**-ning**) put a new lining in (a garment etc.).

relinquish /rɪˈlɪŋkwɪʃ/ *v.* **1** surrender or resign (a right or possession). **2** give up or cease from (a habit, plan, belief, etc.). **3** relax hold of. □ **relinquishment** *n.* [Latin *relinquo -lict-* leave behind]

reliquary /ˈrelɪkwərɪ/ *n.* (*pl.* **-ies**) esp. *Relig.* receptacle for a relic or relics. [French *reliquaire*: related to RELIC]

relish /ˈrelɪʃ/ *–n.* **1** (often foll. by *for*) great liking or enjoyment. **2 a** appetizing flavour. **b** attractive quality. **3** condiment eaten with plainer food to add flavour. **4** (foll. by *of*) distinctive taste or tinge. *–v.* **1** get pleasure out of; enjoy greatly. **2** anticipate with pleasure. [French *reles* remainder: related to RELEASE]

relive /riːˈlɪv/ *v.* (**-ving**) live (an experience etc.) over again, esp. in the imagination.

reload /riːˈləʊd/ *v.* (also *absol.*) load (esp. a gun) again.

relocate /ˌriːləʊˈkeɪt/ *v.* (**-ting**) **1** locate in a new place. **2** move to a new place (esp. to live or work). □ **relocation** /-ˈkeɪʃ(ə)n/ *n.*

reluctant /rɪˈlʌkt(ə)nt/ *adj.* (often foll. by *to* + infin.) unwilling or disinclined. □ **reluctance** *n.* **reluctantly** *adv.* [Latin *luctor* struggle]

rely /rɪˈlaɪ/ *v.* (**-ies, -ied**) (foll. by *on, upon*) **1** depend on with confidence or assurance. **2** be dependent on. [Latin *religo* bind closely]

REM *abbr.* rapid eye movement.

remade *past* and *past part.* of REMAKE.

remain /rɪˈmeɪn/ *v.* **1** be left over after others or other parts have been removed, used, or dealt with. **2** be in the same place or condition during further time; stay (*remained at home*). **3** (foll. by compl.) continue to be (*remained calm; remains President*). [Latin *remaneo*]

remainder *–n.* **1** residue. **2** remaining persons or things. **3** number left after division or subtraction. **4** copy or copies of a book left unsold when demand has almost ceased. *–v.* dispose of a remainder of (books) at a reduced price. [Anglo-French: related to REMAIN]

remains *n.pl.* **1** what remains after other parts have been removed or used etc. **2** relics of antiquity, esp. of buildings. **3** dead body.

remake *–v.* /riːˈmeɪk/ (**-king**; *past* and *past part.* **remade**) make again or differently. *–n.* /ˈriːmeɪk/ thing that has been remade, esp. a cinema film.

remand /rɪˈmɑːnd/ *–v.* return (a prisoner) to custody, esp. to allow further inquiry. *–n.* recommittal to custody. □ **on remand** in custody pending trial. [Latin *remando*]

remand centre *n.* institution to which accused persons are remanded.

remark /rɪˈmɑːk/ *–v.* **1** (often foll. by *that*) **a** say by way of comment. **b** *archaic* take notice of; regard with attention. **2** (usu. foll. by *on, upon*) make a comment. *–n.* **1** written or spoken comment; anything said. **2 a** noticing (*worthy of remark*). **b** commenting (*let it pass without remark*). [French *remarquer*: related to MARK[1]]

remarkable *adj.* worth notice; exceptional; striking. □ **remarkably** *adv.* [French *remarquable*: related to REMARK]

remarry /riːˈmærɪ/ *v.* (**-ies, -ied**) marry again. □ **remarriage** *n.*

REME /ˈriːmiː/ *abbr.* Royal Electrical and Mechanical Engineers.

remeasure /riːˈmeʒə(r)/ *v.* (**-ring**) measure again.

remedial /rɪˈmiːdɪəl/ *adj.* **1** affording or intended as a remedy. **2** (of teaching etc.) for slow or disadvantaged pupils. [Latin: related to REMEDY]

remedy /ˈremɪdɪ/ *–n.* (*pl.* **-ies**) (often foll. by *for, against*) **1** medicine or treatment. **2** means of counteracting or removing anything undesirable. **3** redress; legal or other reparation. *–v.* (**-ies, -ied**) rectify; make good. □ **remediable** /rɪˈmiːdɪəb(ə)l/ *adj.* [Latin *remedium* from *medeor* heal]

remember /rɪˈmembə(r)/ *v.* **1** (often foll. by *to* + infin. or *that* + clause) keep in the memory; not forget. **2** (also *absol.*)

bring back into one's thoughts. **3** think of or acknowledge (a person), esp. in making a gift etc. **4** (foll. by *to*) convey greetings from (one person) to (another) (*remember me to John*). [Latin: related to MEMORY]

remembrance /rɪ'membrəns/ *n.* **1** remembering or being remembered. **2** a memory or recollection. **3** keepsake, souvenir. **4** (in *pl.*) greetings conveyed through a third person. [French: related to REMEMBER]

Remembrance Day *n.* **1** = REMEMBRANCE SUNDAY. **2** *hist.* Armistice Day.

Remembrance Sunday *n.* Sunday nearest 11 Nov., when those killed in the wars of 1914–18 and 1939–45 and later conflicts are commemorated.

remind /rɪ'maɪnd/ *v.* (usu. foll. by *of* or *to* + infin. or *that* + clause) cause (a person) to remember or think of (*reminds me of her father*; *reminded them of the time*).

reminder *n.* **1** thing that reminds, esp. a repeat letter or bill. **2** (often foll. by *of*) memento.

reminisce /ˌremɪ'nɪs/ *v.* (**-cing**) indulge in reminiscence.

reminiscence /ˌremɪ'nɪs(ə)ns/ *n.* **1** remembering things past. **2** (in *pl.*) collection in literary form of incidents and experiences remembered. [Latin *reminiscor* remember]

reminiscent *adj.* **1** (foll. by *of*) reminding or suggestive of. **2** concerned with reminiscence.

remiss /rɪ'mɪs/ *adj.* careless of duty; lax, negligent. [Latin: related to REMIT]

remission /rɪ'mɪʃ(ə)n/ *n.* **1** reduction of a prison sentence on account of good behaviour. **2** remitting of a debt or penalty etc. **3** diminution of force, effect, or degree (esp. of disease or pain). **4** (often foll. by *of*) forgiveness (of sins etc.). [Latin: related to REMIT]

remit –*v.* /rɪ'mɪt/ (**-tt-**) **1** cancel or refrain from exacting or inflicting (a debt, punishment, etc.). **2** abate or slacken; cease partly or entirely. **3** send (money etc.) in payment. **4 a** (foll. by *to*) refer (a matter for decision etc.) to some authority. **b** send back (a case) to a lower court. **5** postpone or defer. **6** pardon (sins etc.). –*n.* /'ri:mɪt/ **1** terms of reference of a committee etc. **2** item remitted for consideration. [Latin *remitto -miss-*]

remittance *n.* **1** money sent, esp. by post. **2** sending of money.

remittent *adj.* (of a fever or disease) abating at intervals.

remix –*v.* /ri:'mɪks/ mix again. –*n.* /'ri:mɪks/ remixed recording.

remnant /'remnənt/ *n.* **1** small remaining quantity. **2** piece of cloth etc. left when the greater part has been used or sold. [French: related to REMAIN]

remodel /ri:'mɒd(ə)l/ *v.* (**-ll-**; *US* **-l-**) **1** model again or differently. **2** reconstruct.

remold *US* var. of REMOULD.

remonstrate /'remən,streɪt/ *v.* (**-ting**) (foll. by *with*) make a protest; argue forcibly. □ **remonstrance** /rɪ'mɒnstrəns/ *n.* **remonstration** /-'streɪʃ(ə)n/ *n.* [medieval Latin *monstro* show]

remorse /rɪ'mɔ:s/ *n.* **1** deep regret for a wrong committed. **2** compunction; compassion, mercy (*without remorse*). [medieval Latin *mordeo mors-* bite]

remorseful *adj.* filled with repentance. □ **remorsefully** *adv.*

remorseless *adj.* without compassion. □ **remorselessly** *adv.*

remortgage /ri:'mɔ:gɪdʒ/ –*v.* (**-ging**) (also *absol.*) mortgage again; revise the terms of an existing mortgage on (a property). –*n.* different or altered mortgage.

remote /rɪ'məʊt/ *adj.* (**remoter, remotest**) **1** far away, far apart, distant. **2** isolated; secluded. **3** distantly related (*remote ancestor*). **4** slight, faint (*a remote hope*; *not the remotest chance*). **5** aloof; not friendly. □ **remotely** *adv.* **remoteness** *n.* [Latin *remotus*: related to REMOVE]

remote control *n.* **1** control of an apparatus from a distance by means of signals transmitted from a radio or electronic device. **2** such a device.

remould (*US* **remold**) –*v.* /ri:'məʊld/ **1** mould again; refashion. **2** re-form the tread of (a tyre). –*n.* /'ri:məʊld/ remoulded tyre.

removal /rɪ'mu:v(ə)l/ *n.* **1** removing or being removed. **2** transfer of furniture etc. on moving house.

remove /rɪ'mu:v/ –*v.* (**-ving**) **1** take off or away from the place occupied. **2 a** convey to another place; change the situation of. **b** get rid of; dismiss. **3** cause to be no longer present or available; take away (*privileges were removed*). **4** (in *passive*; foll. by *from*) distant or remote in condition (*country is not far removed from anarchy*). **5** (as **removed** *adj.*) (esp. of cousins) separated by a specified number of steps of descent (*a first cousin twice removed* = a grandchild of a first cousin). –*n.* **1** degree of remoteness; distance. **2** stage in a gradation; degree (*several removes from what I expected*). **3** form or division in some schools. □ **removable** *adj.* [Latin *removeo -mot-*]

remunerate /rɪ'mju:nə,reɪt/ *v.* (**-ting**) **1** reward; pay for services rendered. **2** serve as or provide recompense for (work etc.) or to (a person). □ **remuneration** /-'reɪʃ(ə)n/ *n.* **remunerative** /-rətɪv/ *adj.* [Latin *munus -ner-* gift]

Renaissance /rɪ'neɪs(ə)ns/ *n.* **1** revival of art and literature in the 14th–16th c. **2** period of this. **3** (often *attrib.*) style of art, architecture, etc. developed during this era. **4** (**renaissance**) any similar revival. [French *naissance* birth]

Renaissance man *n.* person with many talents or pursuits, esp. in the humanities.

renal /'ri:n(ə)l/ *adj.* of the kidneys. [Latin *renes* kidneys]

rename /ri:'neɪm/ *v.* (**-ming**) name again; give a new name to.

renascent /rɪ'næs(ə)nt/ *adj.* springing up anew; being reborn. □ **renascence** *n.*

renationalize /ri:'næʃənə,laɪz/ *v.* (also **-ise**) (**-zing** or **-sing**) nationalize again (an originally nationalized and more recently privatized industry etc.). □ **renationalization** /-'zeɪʃ(ə)n/ *n.*

rend *v.* (*past* and *past part.* **rent**) *archaic* tear or wrench forcibly. [Old English]

render *v.* **1** cause to be or become (*rendered us helpless*). **2** give or pay (money, service, etc.), esp. in return or as a thing due. **3** (often foll. by *to*) **a** give (assistance). **b** show (obedience etc.). **c** do (a service etc.). **4** submit; send in; present (an account, reason, etc.). **5 a** represent or portray. **b** act (a role). **c** *Mus.* perform; execute. **6** translate. **7** (often foll. by *down*) melt down (fat etc.). **8** cover (stone or brick) with a coat of plaster. □ **rendering** *n.* (esp. in senses 5, 6, and 8). [Latin *reddo* give back]

rendezvous /'rɒndɪ,vu:/ —*n.* (*pl.* same /-,vu:z/) **1** agreed or regular meeting-place. **2** meeting by arrangement. —*v.* (**rendezvouses** /-,vu:z/; **rendezvoused** /-,vu:d/; **rendezvousing** /-,vu:ɪŋ/) meet at a rendezvous. [French, = present yourselves]

rendition /ren'dɪʃ(ə)n/ *n.* interpretation or rendering of a dramatic role, piece of music, etc. [French: related to RENDER]

renegade /'renɪ,geɪd/ *n.* person who deserts a party or principles. [medieval Latin: related to RENEGE]

renege /rɪ'ni:g, -'neɪg/ *v.* (**-ging**) (often foll. by *on*) go back on (one's word etc.). [Latin *nego* deny]

renegotiate /,ri:nɪ'gəʊʃɪ,eɪt/ *v.* (**-ting**) (also *absol.*) negotiate again or on different terms. □ **renegotiation** /-'eɪʃ(ə)n/ *n.*

renew /rɪ'nju:/ *v.* **1** revive; make new again; restore to the original state. **2** reinforce; resupply; replace. **3** repeat or re-establish, resume after an interruption (*renewed our acquaintance*). **4** (also *absol.*) grant or be granted continuation of (a licence, subscription, lease, etc.). **5** recover (strength etc.). □ **renewable** *adj.* **renewal** *n.*

rennet /'renɪt/ *n.* **1** curdled milk found in the stomach of an unweaned calf. **2** preparation made from the stomach-membrane of a calf or from certain fungi, used in making cheese etc. [probably Old English: related to RUN]

renounce /rɪ'naʊns/ *v.* (**-cing**) **1** consent formally to abandon (a claim, right, etc.). **2** repudiate; refuse to recognize any longer. **3** decline further association or disclaim relationship with. [Latin *nuntio* announce]

renovate /'renə,veɪt/ *v.* (**-ting**) restore to good condition; repair. □ **renovation** /-'veɪʃ(ə)n/ *n.* **renovator** *n.* [Latin *novus* new]

renown /rɪ'naʊn/ *n.* fame, high distinction. [French *renomer* make famous]

renowned *adj.* famous, celebrated.

rent[1] —*n.* **1** tenant's periodical payment to an owner for the use of land or premises. **2** payment for the use of equipment etc. —*v.* **1** (often foll. by *from*) take, occupy, or use at a rent. **2** (often foll. by *out*) let or hire (a thing) for rent. **3** (foll. by *at*) be let at a specified rate. [French *rente*: related to RENDER]

rent[2] *n.* **1** large tear in a garment etc. **2** opening in clouds etc. [from REND]

rent[3] *past* and *past part.* of REND.

rental /'rent(ə)l/ *n.* **1** amount paid or received as rent. **2** act of renting. [Anglo-French or Anglo-Latin: related to RENT[1]]

rent-boy *n.* young male prostitute.

rentier /'rɑ̃tɪ,eɪ/ *n.* person living on income from property, investments, etc. [French]

renumber /ri:'nʌmbə(r)/ *v.* change the number or numbers given or allocated to.

renunciation /rɪ,nʌnsɪ'eɪʃ(ə)n/ *n.* **1** renouncing or giving up. **2** self-denial.

reoccupy /ri:'ɒkjʊ,paɪ/ *v.* (**-ies, -ied**) occupy again. □ **reoccupation** /-'peɪʃ(ə)n/ *n.*

reoccur /,ri:ə'kɜ:(r)/ *v.* (**-rr-**) occur again or habitually. □ **reoccurrence** /-'kʌrəns/ *n.*

reopen /ri:'əʊpən/ *v.* open again.

reorder /ri:'ɔ:də(r)/ —*v.* **1** order again. **2** put into a new order. —*n.* renewed or repeated order for goods.

reorganize /riː'ɔːgə,naɪz/ *v.* (also **-ise**) (**-zing** or **-sing**) organize differently. □ **reorganization** /-'zeɪʃ(ə)n/ *n.*

reorient /riː'ɔːrɪ,ent/ *v.* **1** give a new direction or outlook to (ideas, a person, etc.). **2** help (a person) find his or her bearings again. **3** (refl., often foll. by *to*) adjust oneself to or come to terms with something.

reorientate /riː'ɔːrɪən,teɪt/ *v.* (**-ting**) = REORIENT. □ **reorientation** /-'teɪʃ(ə)n/ *n.*

rep[1] *n. colloq.* representative, esp. a commercial traveller. [abbreviation]

rep[2] *n. colloq.* **1** repertory. **2** repertory theatre or company. [abbreviation]

repack /riː'pæk/ *v.* pack again.

repackage /riː'pækɪdʒ/ *v.* (**-ging**) **1** package again or differently. **2** present in a new form.

repaid *past* and *past part.* of REPAY.

repaint *–v.* /riː'peɪnt/ **1** paint again or differently. **2** restore the paint or colouring of. *–n.* /'riːpeɪnt/ act of repainting.

repair[1] /rɪ'peə(r)/ *–v.* **1** restore to good condition after damage or wear. **2** set right or make amends for (a loss, wrong, error, etc.). *–n.* **1** restoring to sound condition (*in need of repair*). **2** result of this (*the repair hardly shows*). **3** good or relative condition for working or using (*in bad repair*). □ **repairable** *adj.* **repairer** *n.* [Latin *paro* make ready]

repair[2] /rɪ'peə(r)/ *v.* (foll. by *to*) resort; have recourse; go. [Latin: related to REPATRIATE]

repaper /riː'peɪpə(r)/ *v.* paper (a wall etc.) again.

reparable /'repərəb(ə)l/ *adj.* (of a loss etc.) that can be made good. [Latin: related to REPAIR[1]]

reparation /,repə'reɪʃ(ə)n/ *n.* **1** making amends. **2** (esp. in *pl.*) compensation for war damages.

repartee /,repɑː'tiː/ *n.* **1** practice or skill of making witty retorts. **2** conversation characterized by such retorts. [French *repartie* from *repartir* reply promptly: related to PART]

repast /rɪ'pɑːst/ *n. formal* **1** meal. **2** food and drink for this. [Latin *repasco -past-* feed]

repatriate *–v.* /riː'pætrɪ,eɪt/ (**-ting**) return (a person) to his or her native land. *–n.* /riː'pætrɪət/ repatriated person. □ **repatriation** /-'eɪʃ(ə)n/ *n.* [Latin *repatrio* go back home, from *patria* native land]

repay /riː'peɪ/ *v.* (*past* and *past part.* **repaid**) **1** pay back (money). **2** make repayment to (a person). **3** requite, reward (a service, action, etc.) (*repaid their kindness*; *book repays study*). □ **repayable** *adj.* **repayment** *n.*

repeal /rɪ'piːl/ *–v.* revoke or annul (a law etc.). *–n.* repealing. [French: related to APPEAL]

repeat /rɪ'piːt/ *–v.* **1** say or do over again. **2** recite, rehearse, or report (something learnt or heard). **3** recur; appear again. **4** (of food) be tasted after being swallowed due to belching. *–n.* **1** **a** repeating. **b** thing repeated (often *attrib.*: *repeat prescription*). **2** repeated broadcast. **3** *Mus.* **a** passage intended to be repeated. **b** mark indicating this. **4** pattern repeated in wallpaper etc. □ **repeat itself** recur in the same form. **repeat oneself** say or do the same thing over again. □ **repeatable** *adj.* **repeatedly** *adv.* [Latin *peto* seek]

repeater *n.* **1** person or thing that repeats. **2** firearm which fires several shots without reloading. **3** watch or clock which repeats its last strike when required. **4** device for the retransmission of an electrical message.

repel /rɪ'pel/ *v.* (**-ll-**) **1** drive back; ward off (*repel an attacker*). **2** refuse to accept (*repelled offers of help*). **3** be repulsive or distasteful to. **4** resist mixing with or admitting (*oil and water repel each other*; *surface repels moisture*). **5** (of a magnetic pole) push away from itself (*like poles repel*). □ **repellent** *adj.* & *n.* [Latin *repello -puls-*]

repent /rɪ'pent/ *v.* **1** (often foll. by *of*) feel deep sorrow about one's actions etc. **2** (also *absol.*) wish one had not done; resolve not to continue (a wrongdoing etc.). □ **repentance** *n.* **repentant** *adj.* [Latin *paeniteo*]

repercussion /,riːpə'kʌʃ(ə)n/ *n.* **1** indirect effect or reaction following an event or act. **2** recoil after impact. **3** echo. [Latin: related to RE-]

repertoire /'repə,twɑː(r)/ *n.* **1** stock of works that a performer etc. knows or is prepared to perform. **2** stock of techniques etc. (*repertoire of excuses*). [Latin: related to REPERTORY]

repertory /'repətərɪ/ *n.* (*pl.* **-ies**) **1** performance of various plays for short periods by one company. **2** repertory theatres collectively. **3** store or collection, esp. of information, instances, etc. **4** = REPERTOIRE. [Latin *reperio* find]

repertory company *n.* theatrical company that performs plays from a repertoire.

repetition /,repə'tɪʃ(ə)n/ *n.* **1** **a** repeating or being repeated. **b** thing repeated. **2** copy. □ **repetitious** *adj.* **repetitive** /rɪ'petətɪv/ *adj.*

repetitive strain injury *n.* painful esp. hand or arm condition resulting from prolonged repetitive movements.
rephrase /ri:'freɪz/ *v.* (**-sing**) express differently.
repine /rɪ'paɪn/ *v.* (**-ning**) (often foll. by *at, against*) fret; be discontented. [from PINE[2], after *repent*]
replace /rɪ'pleɪs/ *v.* (**-cing**) **1** put back in place. **2** take the place of; succeed; be substituted for. **3** find or provide a substitute for. **4** (often foll. by *with, by*) fill up the place of.
replacement *n.* **1** replacing or being replaced. **2** person or thing that replaces another.
replant /ri:'plɑ:nt/ *v.* **1** transfer (a plant etc.). **2** plant (ground) again.
replay *–v.* /ri:'pleɪ/ play (a match, recording, etc.) again. *–n.* /'ri:pleɪ/ replaying of a match, recorded incident in a game, etc.
replenish /rɪ'plenɪʃ/ *v.* **1** (often foll. by *with*) fill up again. **2** renew (a supply etc.). □ **replenishment** *n.* [French *plenir* from *plein* full]
replete /rɪ'pli:t/ *adj.* (often foll. by *with*) **1** well-fed, gorged. **2** filled or well-supplied. □ **repletion** *n.* [Latin *pleo* fill]
replica /'replɪkə/ *n.* **1** exact copy, esp. a duplicate of a work, made by the original artist. **2** copy or model, esp. on a smaller scale. [Italian *replicare* REPLY]
replicate /'replɪkeɪt/ *v.* repeat (an experiment etc.); make a replica of. □ **replication** /-'keɪʃ(ə)n/ *n.* [Latin *replico* fold back]
reply /rɪ'plaɪ/ *–v.* (**-ies, -ied**) **1** (often foll. by *to*) make an answer, respond in word or action. **2** say in answer. *–n.* (*pl.* **-ies**) **1** replying (*what did they say in reply?*). **2** what is replied; response. [Latin *replico* fold back]
repoint /ri:'pɔɪnt/ *v.* point (esp. brickwork) again.
repopulate /ri:'pɒpjʊ,leɪt/ *v.* (**-ting**) populate again or increase the population of.
report /rɪ'pɔ:t/ *–v.* **1 a** bring back or give an account of. **b** state as fact or news; narrate, describe, or repeat, esp. as an eyewitness or hearer etc. **c** relate as spoken by another. **2** make an official or formal statement about. **3** (often foll. by *to*) bring (an offender or offence) to the attention of the authorities. **4** (often foll. by *to*) present oneself to a person as having returned or arrived. **5** (also *absol.*) take down word for word, summarize, or write a description of for publication. **6** make or send in a report. **7** (foll. by *to*) be responsible to (a superior etc.). *–n.* **1** account given or opinion formally expressed after investigation or consideration. **2** description, summary, or reproduction of a scene, speech, law case, etc., esp. for newspaper publication or broadcast. **3** common talk; rumour. **4** way a person or thing is spoken of (*hear a good report of you*). **5** periodical statement on (esp. a school pupil's) work, conduct, etc. **6** sound of a gunshot etc. □ **reportedly** *adv.* [Latin *porto* bring]
reportage /,repɔ:'tɑ:ʒ/ *n.* **1** reporting of news for the media. **2** typical style of this. **3** factual journalistic material in a book etc. [from REPORT, after French]
reported speech *n.* speaker's words with the person, tense, etc. adapted, e.g. *he said that he would go*.
reporter *n.* person employed to report news etc. for the media.
repose[1] /rɪ'pəʊz/ *–n.* **1** cessation of activity, excitement, or toil. **2** sleep. **3** peaceful or quiescent state; tranquillity. *–v.* (**-sing**) **1** (also *refl.*) lie down in rest. **2** (often foll. by *in, on*) lie, be lying or laid, esp. in sleep or death. [Latin: related to PAUSE]
repose[2] /rɪ'pəʊz/ *v.* (**-sing**) (foll. by *in*) place (trust etc.) in. [from RE-, POSE]
reposeful *adj.* showing or inducing repose. □ **reposefully** *adv.* [from REPOSE[1]]
reposition /,ri:pə'zɪʃ(ə)n/ *v.* **1** move or place in a different position. **2** alter one's position.
repository /rɪ'pɒzɪtərɪ/ *n.* (*pl.* **-ies**) **1** place where things are stored or may be found, esp. a warehouse or museum. **2** receptacle. **3** (often foll. by *of*) **a** book, person, etc. regarded as a store of information etc. **b** recipient of secrets etc. [Latin: related to REPOSE[2]]
repossess /,ri:pə'zes/ *v.* regain possession of (esp. property on which payment is in arrears). □ **repossession** *n.*
repot /ri:'pɒt/ *v.* (**-tt-**) move (a plant) to another, esp. larger, pot.
reprehensible /,reprɪ'hensɪb(ə)l/ *adj.* blameworthy. [Latin *prehendo* seize]
represent[1] /,reprɪ'zent/ *v.* **1** stand for or correspond to. **2** (often in *passive*) be a specimen of. **3** embody; symbolize. **4** place a likeness of before the mind or senses. **5** (often foll. by *as, to be*) describe or depict as; declare. **6** (foll. by *that*) allege. **7** show, or play the part of, on stage. **8** be a substitute or deputy for; be entitled to act or speak for. **9** be elected as a member of a legislature etc. by. [Latin: related to PRESENT[2]]
represent[2] /,ri:prɪ'zent/ *v.* submit (a cheque etc.) again for payment.
representation /,reprɪzen'teɪʃ(ə)n/ *n.* **1** representing or being represented. **2**

thing that represents another. **3** (esp. in *pl.*) statement made of allegations or opinions.

representational *adj. Art* depicting a subject as it appears to the eye.

representative /ˌreprɪ'zentətɪv/ *–adj.* **1** typical of a class. **2** containing typical specimens of all or many classes (*representative sample*). **3 a** consisting of elected deputies etc. **b** based on representation by these (*representative government*). **4** (foll. by *of*) serving as a portrayal or symbol of. *–n.* **1** (foll. by *of*) sample, specimen, or typical embodiment of. **2 a** agent of a person or society. **b** commercial traveller. **3** delegate; substitute. **4** deputy etc. in a representative assembly. [French or medieval Latin: related to REPRESENT[1]]

repress /rɪ'pres/ *v.* **1 a** keep under; quell. **b** suppress; prevent from sounding, rioting, or bursting out. **2** *Psychol.* actively exclude (an unwelcome thought) from conscious awareness. **3** (usu. as **repressed** *adj.*) subject (a person) to the suppression of his or her thoughts or impulses. □ **repression** *n.* **repressive** *adj.* [Latin: related to PRESS[1]]

reprice /riː'praɪs/ *v.* (**-cing**) price again or differently.

reprieve /rɪ'priːv/ *–v.* (**-ving**) **1** remit or postpone the execution of (a condemned person). **2** give respite to. *–n.* **1 a** reprieving or being reprieved. **b** warrant for this. **2** respite. [*repry* from French *reprendre -pris* take back]

reprimand /'reprɪˌmɑːnd/ *–n.* (esp. official) rebuke. *–v.* administer this to. [Latin: related to REPRESS]

reprint *–v.* /riː'prɪnt/ print again. *–n.* /'riːprɪnt/ **1** reprinting of a book etc. **2** book etc. reprinted. **3** quantity reprinted.

reprisal /rɪ'praɪz(ə)l/ *n.* act of retaliation. [medieval Latin: related to REPREHEND]

reprise /rɪ'priːz/ *n.* **1** repeated passage in music. **2** repeated item in a musical programme. [French: related to REPRIEVE]

repro /'riːprəʊ/ *n.* (*pl.* **-s**) (often *attrib.*) *colloq.* reproduction or copy. [abbreviation]

reproach /rɪ'prəʊtʃ/ *–v.* express disapproval to (a person or oneself) for a fault. *–n.* **1** rebuke or censure. **2** (often foll. by *to*) thing that brings disgrace or discredit. **3** state of disgrace or discredit. □ **above** (or **beyond**) **reproach** perfect, blameless. [French *reprochier*]

reproachful *adj.* full of or expressing reproach. □ **reproachfully** *adv.*

reprobate /'reprəˌbeɪt/ *n.* unprincipled or immoral person. [Latin: related to PROVE]

reprocess /riː'prəʊses/ *v.* process again or differently.

reproduce /ˌriːprə'djuːs/ *v.* (**-cing**) **1** produce a copy or representation of. **2** cause to be seen or heard etc. again (*tried to reproduce the sound exactly*). **3** produce further members of the same species by natural means. **4** *refl.* produce offspring. □ **reproducible** *adj.*

reproduction /ˌriːprə'dʌkʃ(ə)n/ *n.* **1** reproducing or being reproduced, esp. the production of further members of the same species. **2** copy of a work of art. **3** (*attrib.*) (of furniture etc.) imitating an earlier style. **4** quality of reproduced sound. □ **reproductive** *adj.*

reprogram /riː'prəʊgræm/ *v.* (also **reprogramme**) (**-mm-**; *US* **-m-**) program (esp. a computer) again or differently. □ **reprogramable** *adj.* (also **reprogrammable**).

reproof /rɪ'pruːf/ *n. formal* **1** blame (*glance of reproof*). **2** rebuke. [French *reprove*: related to REPROVE]

reprove /rɪ'pruːv/ *v.* (**-ving**) *formal* rebuke (a person, conduct, etc.). [Latin: related to REPROBATE]

reptile /'reptaɪl/ *n.* **1** cold-blooded scaly animal of a class including snakes, lizards, crocodiles, turtles, tortoises, etc. **2** mean, grovelling, or repulsive person. □ **reptilian** /-'tɪlɪən/ *adj.* & *n.* [Latin *repo rept-* creep]

republic /rɪ'pʌblɪk/ *n.* State in which supreme power is held by the people or their elected representatives or by an elected or nominated president, not by a monarch etc. [Latin *res* concern: related to PUBLIC]

republican *–adj.* **1** of or constituted as a republic. **2** characteristic of a republic. **3** advocating or supporting republican government. *–n.* **1** person advocating or supporting republican government. **2** (**Republican**) **a** *US* supporter of the Republican Party. **b** *Ir.* supporter of the IRA or Sinn Féin. □ **republicanism** *n.*

republish /riː'pʌblɪʃ/ *v.* publish again or in a new edition etc. □ **republication** /-'keɪʃ(ə)n/ *n.*

repudiate /rɪ'pjuːdɪˌeɪt/ *v.* (**-ting**) **1 a** disown, disavow, reject. **b** refuse dealings with. **c** deny. **2** refuse to recognize or obey (authority or a treaty). **3** refuse to discharge (an obligation or debt). □ **repudiation** /-'eɪʃ(ə)n/ *n.* [Latin *repudium* divorce]

■ **Usage** See note at *refute*.

repugnance /rɪ'pʌgnəns/ *n.* **1** antipathy; aversion. **2** inconsistency or incompatibility of ideas etc. [Latin *pugno* fight]
repugnant *adj.* **1** extremely distasteful. **2** contradictory.
repulse /rɪ'pʌls/ *–v.* (**-sing**) **1** drive back by force of arms. **2 a** rebuff. **b** refuse. *–n.* **1** repulsing or being repulsed. **2** rebuff. [Latin: related to REPEL]
repulsion /rɪ'pʌlʃ(ə)n/ *n.* **1** aversion, disgust. **2** *Physics* tendency of bodies to repel each other.
repulsive /rɪ'pʌlsɪv/ *adj.* causing aversion or loathing; disgusting. □ **repulsively** *adv.* [French *répulsif* or REPULSE]
repurchase /ri:'pɜ:tʃɪs/ *–v.* (**-sing**) purchase again. *–n.* act of purchasing again.
reputable /'repjʊtəb(ə)l/ *adj.* of good repute; respectable. [French or medieval Latin: related to REPUTE]
reputation /,repjʊ'teɪʃ(ə)n/ *n.* **1** what is generally said or believed about a person's or thing's character (*reputation for honesty*; *reputation of being a crook*). **2** state of being well thought of; respectability (*lost its reputation*). [Latin: related to REPUTE]
repute /rɪ'pju:t/ *–n.* reputation. *–v.* (as **reputed** *adj.*) **1** be generally considered (*is reputed to be the best*). **2** passing as, but probably not (*his reputed father*). □ **reputedly** *adv.* [Latin *puto* think]
request /rɪ'kwest/ *–n.* **1** act of asking for something (*came at his request*). **2** thing asked for. *–v.* **1** ask to be given, allowed, or favoured with. **2** (foll. by *to* + infin.) ask (a person) to do something. **3** (foll. by *that*) ask that. □ **by** (or **on**) **request** in response to an expressed wish. [Latin: related to REQUIRE]
request stop *n.* bus-stop at which a bus stops only if requested.
requiem /'rekwɪəm/ *n.* **1** (**Requiem**) (also *attrib.*) chiefly *RC Ch.* mass for the repose of the souls of the dead. **2** music for this. [Latin, = rest]
require /rɪ'kwaɪə(r)/ *v.* (**-ring**) **1** need; depend on for success or fulfilment (*the work requires patience*). **2** lay down as an imperative (*required by law*). **3** command; instruct (a person etc.). **4** order; insist on (an action or measure). □ **requirement** *n.* [Latin *requiro -quisit-* seek]
requisite /'rekwɪzɪt/ *–adj.* required by circumstances; necessary to success etc. *–n.* (often foll. by *for*) thing needed (for some purpose). [Latin: related to REQUIRE]
requisition /,rekwɪ'zɪʃ(ə)n/ *–n.* **1** official order laying claim to the use of property or materials. **2** formal written demand that some duty should be performed. **3** being called or put into service. *–v.* demand the use or supply of, esp. by requisition order. [Latin: related to REQUIRE]
requite /rɪ'kwaɪt/ *v.* (**-ting**) **1** make return for (a service). **2** reward or avenge (a favour or injury). **3** (often foll. by *for*) make return to (a person). **4** reciprocate (love etc.). □ **requital** *n.* [from RE-, *quite* = QUIT]
reran *past* of RERUN.
reread /ri:'ri:d/ *v.* (*past* and *past part.* **reread** /-'red/) read again.
rerecord /,ri:rɪ'kɔ:d/ *v.* record again.
reredos /'rɪədɒs/ *n.* ornamental screen covering the wall at the back of an altar. [Anglo-French: related to ARREARS, *dos* back]
re-release /,ri:rɪ'li:s/ *–v.* (**-sing**) release (a record, film, etc.) again. *–n.* re-released record, film, etc.
re-route /ri:'ru:t/ *v.* (**-teing**) send or carry by a different route.
rerun *–v.* /ri:'rʌn/ (**-nn-**; *past* **reran**; *past part.* **rerun**) **1** run (a race, film, etc.) again. **2** repeat (a course of action). *–n.* /'ri:rʌn/ **1** act of rerunning. **2** film etc. shown again. **3** repetition (of events).
resale /ri:'seɪl/ *n.* sale of a thing previously bought.
resat *past* and *past part.* of RESIT.
reschedule /ri:'ʃedju:l, -'skedʒʊəl/ *v.* (**-ling**) alter the schedule of; replan.
rescind /rɪ'sɪnd/ *v.* abrogate, revoke, cancel. □ **rescission** /-'sɪʒ(ə)n/ *n.* [Latin *rescindo -sciss-* cut off]
rescript /ri:'skrɪpt/ *n.* **1** Roman emperor's or Pope's written reply to an appeal for a decision. **2** official edict or announcement. [Latin *rescribo -script-* reply in writing]
rescue /'reskju:/ *–v.* (**-ues, -ued, -uing**) (often foll. by *from*) save or set free from danger or harm. *–n.* rescuing or being rescued. □ **rescuer** *n.* [Romanic: related to RE-, EX-[1], QUASH]
reseal /ri:'si:l/ *v.* seal again. □ **resealable** *adj.*
research /rɪ'sɜ:tʃ, 'ri:sɜ:tʃ/ *–n.* (often *attrib.*) systematic investigation and study of materials, sources, etc., in order to establish facts and reach conclusions. *–v.* do research into or for. □ **researcher** *n.* [French: related to SEARCH]

■ **Usage** The second pronunciation, with the stress on the first syllable, is considered incorrect by some people.

research and development *n.* work directed towards the innovation, introduction, and improvement of products and processes.

resell /riː'sel/ *v.* (*past* and *past part.* **resold**) sell (an object etc.) after buying it.

resemblance /rɪ'zembləns/ *n.* likeness or similarity. [Anglo-French: related to RESEMBLE]

resemble /rɪ'zemb(ə)l/ *v.* (**-ling**) be like; have a similarity to, or the same appearance as. [French *sembler* seem]

resent /rɪ'zent/ *v.* feel indignation at; be aggrieved by (a circumstance, action, or person). [Latin *sentio* feel]

resentful *adj.* feeling resentment. □ **resentfully** *adv.*

resentment *n.* indignant or bitter feelings. [Italian or French: related to RESENT]

reservation /ˌrezə'veɪʃ(ə)n/ *n.* **1** reserving or being reserved. **2** thing booked, e.g. a room in a hotel. **3** spoken or unspoken limitation or exception to an agreement etc. **4** (in full **central reservation**) strip of land between the carriageways of a road. **5** area of land reserved for occupation by American Indians etc. [Latin: related to RESERVE]

reserve /rɪ'zɜːv/ *—v.* (**-ving**) **1** put aside, keep back for a later occasion or special use. **2** order to be specially retained or allocated for a particular person or at a particular time. **3** retain or secure (*reserve the right to*). *—n.* **1** thing reserved for future use; extra amount. **2** limitation or exception attached to something. **3** self-restraint; reticence; lack of cordiality. **4** company's profit added to capital. **5** (in *sing.* or *pl.*) assets kept readily available. **6** (in *sing.* or *pl.*) **a** troops withheld from action to reinforce or protect others. **b** forces in addition to the regular army etc., but available in an emergency. **7** member of the military reserve. **8** extra player chosen as a possible substitute in a team. **9** land reserved for special use, esp. as a habitat (*nature reserve*). □ **in reserve** unused and available if required. **reserve judgement** postpone giving one's opinion. [Latin *servo* keep]

reserved *adj.* **1** reticent; slow to reveal emotion or opinions; uncommunicative. **2** set apart, destined for a particular use.

reserve price *n.* lowest acceptable price stipulated for an item sold at auction.

reservist *n.* member of the military reserve.

reservoir /'rezəˌvwɑː(r)/ *n.* **1** large natural or artificial lake as a source of water supply. **2** receptacle for fluid. **3** supply of information etc. [French: related to RESERVE]

reset /riː'set/ *v.* (**-tt-**; *past* and *past part.* **reset**) set (a bone, gems, a clock etc.) again or differently.

resettle /riː'set(ə)l/ *v.* (**-ling**) settle again or elsewhere. □ **resettlement** *n.*

reshape /riː'ʃeɪp/ *v.* (**-ping**) shape or form again or differently.

reshuffle /riː'ʃʌf(ə)l/ *—v.* (**-ling**) **1** shuffle (cards) again. **2** change the posts of (government ministers etc.). *—n.* act of reshuffling.

reside /rɪ'zaɪd/ *v.* (**-ding**) **1** have one's home, dwell permanently. **2** (foll. by *in*) (of power, a right, etc.) be vested in. **3** (foll. by *in*) (of a quality) be present or inherent in. [Latin *sedeo* sit]

residence /'rezɪd(ə)ns/ *n.* **1** process of residing or being resident. **2 a** place where a person resides. **b** house, esp. one of pretension. □ **in residence** living or working at a specified place, esp. for the performance of duties (*artist in residence*).

residency /'rezɪdənsɪ/ *n.* (*pl.* **-ies**) **1** = RESIDENCE 1, 2a. **2** permanent or regular engagement of a musician, artist, etc., in one place.

resident *—n.* **1** (often foll. by *of*) **a** permanent inhabitant. **b** non-migratory species of bird. **2** guest in a hotel etc. staying overnight. *—adj.* **1** residing; in residence. **2** having quarters at one's workplace etc. (*resident housekeeper*). **3** located in. **4** (of birds etc.) non-migratory.

residential /ˌrezɪ'denʃ(ə)l/ *adj.* **1** suitable for or occupied by dwellings (*residential area*). **2** used as a residence (*residential hotel*). **3** based on or connected with residence (*residential course*).

residual /rɪ'zɪdjʊəl/ *—adj.* left as a residue or residuum. *—n.* residual quantity.

residuary /rɪ'zɪdjʊərɪ/ *adj.* **1** of the residue of an estate (*residuary bequest*). **2** residual.

residue /'rezɪˌdjuː/ *n.* **1** what is left over or remains; remainder. **2** what remains of an estate after the payment of charges, debts, and bequests. [Latin *residuum*: related to RESIDUUM]

residuum /rɪ'zɪdjʊəm/ *n.* (*pl.* **-dua**) **1** substance left after combustion or evaporation. **2** residue. [Latin: related to RESIDE]

resign /rɪ'zaɪn/ *v.* **1** (often foll. by *from*) give up office, one's employment, etc. **2** relinquish, surrender (a right, task, etc.). **3** *refl.* (usu. foll. by *to*) reconcile (oneself etc.) to the inevitable. [Latin *signo* sign]

re-sign /riːˈsaɪn/ *v.* sign again.

resignation /ˌrezɪɡˈneɪʃ(ə)n/ *n.* **1** resigning, esp. from one's job or office. **2** letter etc. conveying this. **3** reluctant acceptance of the inevitable. [medieval Latin: related to RESIGN]

resigned *adj.* **1** (often foll. by *to*) having resigned oneself; resolved to endure. **2** indicative of this □ **resignedly** /-nɪdlɪ/ *adv.*

resilient /rɪˈzɪlɪənt/ *adj.* **1** resuming its original shape after compression etc. **2** readily recovering from a setback. □ **resilience** *n.* [Latin: related to SALIENT]

resin /ˈrezɪn/ *–n.* **1** adhesive substance secreted by some plants and trees. **2** (in full **synthetic resin**) organic compound made by polymerization etc. and used in plastics. *–v.* (**-n-**) rub or treat with resin. □ **resinous** *adj.* [Latin]

resist /rɪˈzɪst/ *–v.* **1** withstand the action or effect of. **2** stop the course or progress of. **3** abstain from (pleasure, temptation, etc.). **4** strive against; try to impede; refuse to comply with (*resist arrest*). **5** offer opposition; refuse to comply. *–n.* protective coating of a resistant substance. □ **resistible** *adj.* [Latin *sisto* stop]

resistance *n.* **1** resisting; refusal to comply. **2** power of resisting. **3** ability to withstand disease. **4** impeding or stopping effect exerted by one thing on another. **5** *Physics* property of hindering the conduction of electricity, heat, etc. **6** resistor. **7** secret organization resisting a régime, esp. in an occupied country. □ **resistant** *adj.* [Latin: related to RESIST]

resistor *n.* device having resistance to the passage of an electric current.

resit *–v.* /riːˈsɪt/ (**-tt-**; *past* and *past part.* **resat**) sit (an examination) again after failing. *–n.* /ˈriːsɪt/ **1** resitting of an examination. **2** examination specifically for this.

resold *past* and *past part.* of RESELL.

resoluble /rɪˈzɒljʊb(ə)l/ *adj.* **1** that can be resolved. **2** (foll. by *into*) analysable into. [Latin: related to RESOLVE]

resolute /ˈrezəˌluːt/ *adj.* determined, decided, firm of purpose. □ **resolutely** *adv.* [Latin: related to RESOLVE]

resolution /ˌrezəˈluːʃ(ə)n/ *n.* **1** resolute temper or character. **2** thing resolved on; intention. **3** formal expression of opinion or intention by a legislative body or public meeting. **4** (usu. foll. by *of*) solving of a doubt, problem, or question. **5** separation into components. **6** (foll. by *into*) conversion into another form. **7** *Mus.* causing discord to pass into concord. **8 a** smallest interval measurable by a scientific instrument. **b** resolving power.

resolve /rɪˈzɒlv/ *–v.* (**-ving**) **1** make up one's mind; decide firmly (*resolved to leave, on leaving*). **2** cause (a person) to do this (*events resolved him to leave*). **3** solve, explain, or settle (a doubt, argument, etc.). **4** (foll. by *that*) (of an assembly or meeting) pass a resolution by vote. **5** (often foll. by *into*) (cause to) separate into constituent parts; analyse. **6** (foll. by *into*) reduce by mental analysis into. **7** *Mus.* convert or be converted into concord. *–n.* firm mental decision or intention; determination. [Latin: related to SOLVE]

resolved *adj.* resolute, determined.

resonant /ˈrezənənt/ *adj.* **1** (of sound) echoing, resounding; continuing to sound; reinforced or prolonged by reflection or vibration. **2** (of a body, room, etc.) tending to reinforce or prolong sounds, esp. by vibration. **3** (often foll. by *with*) (of a place) resounding. □ **resonance** *n.* [Latin: related to RESOUND]

resonate /ˈrezəˌneɪt/ *v.* (**-ting**) produce or show resonance; resound. □ **resonator** *n.* [Latin: related to RESONANT]

resort /rɪˈzɔːt/ *–n.* **1** place frequented esp. for holidays or for a specified purpose or quality (*seaside resort*; *health resort*). **2 a** thing to which one has recourse; expedient, measure. **b** (foll. by *to*) recourse to; use of (*without resort to violence*). *–v.* **1** (foll. by *to*) turn to as an expedient (*resorted to force*). **2** (foll. by *to*) go often or in large numbers to. □ **in the** (or **as a**) **last resort** when all else has failed. [French *sortir* go out]

re-sort /riːˈsɔːt/ *v.* sort again or differently.

resound /rɪˈzaʊnd/ *v.* **1** (often foll. by *with*) (of a place) ring or echo. **2** (of a voice, instrument, sound, etc.) produce echoes; go on sounding; fill a place with sound. **3 a** (of a reputation etc.) be much talked of. **b** (foll. by *through*) produce a sensation. **4** (of a place) re-echo (a sound). [Latin: related to SOUND[1]]

resounding *adj.* **1** ringing, echoing. **2** notable, emphatic (*a resounding success*).

resource /rɪˈzɔːs/ *–n.* **1** expedient or device. **2** (often in *pl.*) means available; stock or supply that can be drawn on; asset. **3** (in *pl.*) country's collective wealth. **4** skill in devising expedients (*person of great resource*). **5** (in *pl.*) one's inner strength, ingenuity, etc. *–v.* (**-cing**) provide with resources. □ **resourceful** *adj.* (in sense 4). **resourcefully** *adv.* **resourcefulness** *n.* [French: related to SOURCE]

respect /rɪ'spekt/ *–n.* **1** deferential esteem felt or shown towards a person or quality. **2** (foll. by *of, for*) heed or regard. **3** aspect, detail, etc. (*correct in all respects*). **4** reference, relation (*with respect to*). **5** (in *pl.*) polite messages or attentions (*give her my respects*). *–v.* **1** regard with deference or esteem. **2 a** avoid interfering with or harming. **b** treat with consideration. **c** refrain from offending (a person, feelings, etc.). □ **in respect of** (or **with respect to**) as concerns. □ **respecter** *n.* [Latin *respicio -spect-* look back at]

respectable *adj.* **1** of acceptable social standing; decent and proper in appearance or behaviour. **2** fairly competent (*a respectable try*). **3** reasonably good in condition, appearance, number, size, etc. □ **respectability** /-'bɪlɪtɪ/ *n.* **respectably** *adv.*

respectful *adj.* showing deference. □ **respectfully** *adv.*

respecting *prep.* with regard to; concerning.

respective *adj.* of or relating to each of several individually (*go to your respective seats*). [French or medieval Latin: related to RESPECT]

respectively *adv.* for each separately or in turn, and in the order mentioned (*she and I gave £10 and £1 respectively*).

respell /riː'spel/ *v.* (*past* and *past part.* **respelt** or **respelled**) spell again or differently, esp. phonetically.

respiration /ˌrespə'reɪʃ(ə)n/ *n.* **1 a** breathing. **b** single breath in or out. **2** *Biol.* (in living organisms) the absorption of oxygen and the release of energy and carbon dioxide. [Latin *spiro* breathe]

respirator /'respəˌreɪtə(r)/ *n.* **1** apparatus worn over the face to warm, filter, or purify inhaled air. **2** apparatus for maintaining artificial respiration.

respire /rɪ'spaɪə(r)/ *v.* (**-ring**) **1** (also *absol.*) breathe (air etc.); inhale and exhale. **2** (of a plant) carry out respiration. □ **respiratory** /rɪ'spɪrətərɪ/ *adj.*

respite /'respaɪt/ *n.* **1** interval of rest or relief. **2** delay permitted before the discharge of an obligation or the suffering of a penalty. [Latin: related to RESPECT]

resplendent /rɪ'splend(ə)nt/ *adj.* brilliant, dazzlingly or gloriously bright. □ **resplendence** *n.* [Latin *resplendeo* shine]

respond /rɪ'spɒnd/ *v.* **1** answer, reply. **2** act or behave in a corresponding manner. **3** (usu. foll. by *to*) show sensitiveness to by behaviour or change (*does not respond to kindness*). **4** (of a congregation) make set answers to a priest etc. [Latin *respondeo -spons-*]

respondent *–n.* defendant, esp. in an appeal or divorce case. *–adj.* in the position of defendant.

response /rɪ'spɒns/ *n.* **1** answer given in a word or act; reply. **2** feeling, movement, or change caused by a stimulus or influence. **3** (often in *pl.*) any part of the liturgy said or sung in answer to the priest. [Latin: related to RESPOND]

responsibility /rɪˌspɒnsə'bɪlɪtɪ/ *n.* (*pl.* **-ies**) **1 a** (often foll. by *for, of*) being responsible. **b** authority; managerial freedom (*job with more responsibility*). **2** person or thing for which one is responsible; duty, commitment. **3** capacity for rational conduct (*diminished responsibility*).

responsible /rɪ'spɒnsəb(ə)l/ *adj.* **1** (often foll. by *to, for*) liable to be called to account (to a person or for a thing). **2** morally accountable for one's actions; capable of rational conduct. **3** of good credit, position, or repute; respectable; evidently trustworthy. **4** (often foll. by *for*) being the primary cause. **5** involving responsibility. □ **responsibly** *adv.*

responsive /rɪ'spɒnsɪv/ *adj.* **1** (often foll. by *to*) responding readily (to some influence). **2** sympathetic. **3 a** answering. **b** by way of answer. □ **responsiveness** *n.*

respray *–v.* /riː'spreɪ/ spray again (esp. a vehicle with paint). *–n.* /'riːspreɪ/ act of respraying.

rest[1] *–v.* **1** cease from exertion, action, etc. **2** be still or asleep, esp. to refresh oneself or recover strength. **3** give relief or repose to; allow to rest. **4** (foll. by *on, upon, against*) lie on; be supported by. **5** (foll. by *on, upon*) depend or be based on. **6** (foll. by *on, upon*) (of a look) alight or be steadily directed on. **7** (foll. by *on, upon*) place for support or foundation on. **8** (of a problem or subject) be left without further investigation or discussion (*let the matter rest*). **9 a** lie in death. **b** (foll. by *in*) lie buried in (a churchyard etc.). **10** (as **rested** *adj.*) refreshed by resting. *–n.* **1** repose or sleep. **2** cessation of exertion, activity, etc. **3** period of resting. **4** support for holding or steadying something. **5** *Mus.* **a** interval of silence. **b** sign denoting this. □ **at rest** not moving; not agitated or troubled; dead. **be resting** *euphem.* (of an actor) be out of work. **rest one's case** conclude one's argument etc. **rest on one's laurels** not seek further success. **rest on one's oars** relax one's efforts. **set at rest** settle or relieve (a question, a person's mind, etc.). [Old English]

rest[2] *–n.* (prec. by *the*) the remaining part or parts; the others; the remainder of some quantity or number. *–v.* **1** remain in a specified state (*rest assured*). **2** (foll. by *with*) be left in the hands or charge of (*the final arrangements rest with you*). □ **for the rest** as regards anything else. [French *rester* remain]

restart *–v.* /riːˈstɑːt/ start again. *–n.* /ˈriːstɑːt/ act of restarting.

restate /riːˈsteɪt/ *v.* (**-ting**) express again or differently, esp. for emphasis. □ **restatement** *n.*

restaurant /ˈrestəˌrɒnt/ *n.* public premises where meals may be bought and eaten. [French from *restaurer* RESTORE]

restaurant car *n.* dining-car.

restaurateur /ˌrestərəˈtɜː(r)/ *n.* restaurant-keeper.

rest-cure *n.* rest usu. of some weeks as a medical treatment.

restful *adj.* giving rest or a feeling of rest; quiet, undisturbed. □ **restfully** *adv.* **restfulness** *n.*

rest home *n.* place where old or convalescent people are cared for.

restitution /ˌrestɪˈtjuːʃ(ə)n/ *n.* **1** restoring of a thing to its proper owner. **2** reparation for an injury (esp. *make restitution*). [Latin]

restive /ˈrestɪv/ *adj.* **1** fidgety; restless. **2** (of a horse) jibbing; refractory. **3** (of a person) resisting control. □ **restively** *adv.* **restiveness** *n.* [French: related to REST[2]]

restless *adj.* **1** without rest or sleep. **2** uneasy; agitated. **3** constantly in motion, fidgeting, etc. □ **restlessly** *adv.* **restlessness** *n.* [Old English: related to REST[1]]

restock /riːˈstɒk/ *v.* (also *absol.*) stock again or differently.

restoration /ˌrestəˈreɪʃ(ə)n/ *n.* **1** restoring or being restored. **2** model or representation of the supposed original form of a thing. **3** (**Restoration**) *hist.* **a** (prec. by *the*) re-establishment of the British monarchy in 1660. **b** (often *attrib.*) literary period following this (*Restoration comedy*).

restorative /rɪˈstɒrətɪv/ *–adj.* tending to restore health or strength. *–n.* restorative medicine, food, etc.

restore /rɪˈstɔː(r)/ *v.* (**-ring**) **1** bring back to the original state by rebuilding, repairing, etc. **2** bring back to health etc. **3** give back to the original owner etc. **4** reinstate. **5** replace; put back; bring back to a former condition. **6** make a representation of the supposed original state of (a ruin, extinct animal, etc.). □ **restorer** *n.* [Latin *restauro*]

restrain /rɪˈstreɪn/ *v.* **1** (often *refl.*, usu. foll. by *from*) check or hold in; keep in check, under control, or within bounds. **2** repress, keep down. **3** confine, imprison. [Latin *restringo -strict-*]

restraint *n.* **1** restraining or being restrained. **2** restraining agency or influence. **3** moderation; self-control. **4** reserve of manner. **5** confinement, esp. because of insanity.

restrict /rɪˈstrɪkt/ *v.* **1** confine, limit. **2** withhold from general circulation or disclosure. □ **restriction** *n.* [Latin: related to RESTRAIN]

restrictive /rɪˈstrɪktɪv/ *adj.* restricting. [French or medieval Latin: related to RESTRICT]

restrictive practice *n.* agreement that limits competition or output in industry.

rest room *n.* esp. *US* public lavatory.

restructure /riːˈstrʌktʃə(r)/ *v.* (**-ring**) give a new structure to; rebuild; rearrange.

restyle /riːˈstaɪl/ *v.* (**-ling**) reshape; remake in a new style.

result /rɪˈzʌlt/ *–n.* **1** consequence, issue, or outcome of something. **2** satisfactory outcome (*gets results*). **3** end product of calculation. **4** (in *pl.*) list of scores or winners etc. in examinations or sporting events. *–v.* **1** (often foll. by *from*) arise as the actual, or follow as a logical, consequence. **2** (often foll. by *in*) have a specified end or outcome (*resulted in a large profit*). [Latin *resulto* spring back]

resultant *–adj.* resulting, esp. as the total outcome of more or less opposed forces. *–n.* force etc. equivalent to two or more acting in different directions at the same point.

resume /rɪˈzjuːm/ *v.* (**-ming**) **1** begin again or continue after an interruption. **2** begin to speak, work, or use again; recommence. **3** get back; take back (*resume one's seat*). [Latin *sumo sumpt-* take]

résumé /ˈrezjʊˌmeɪ/ *n.* summary. [French: related to RESUME]

resumption /rɪˈzʌmpʃ(ə)n/ *n.* resuming. □ **resumptive** *adj.* [Latin: related to RESUME]

resurface /riːˈsɜːfɪs/ *v.* (**-cing**) **1** lay a new surface on (a road etc.). **2** return to the surface. **3** turn up again.

resurgent /rɪˈsɜːdʒ(ə)nt/ *adj.* rising or arising again. □ **resurgence** *n.* [Latin *resurgo -surrect-* rise again]

resurrect /ˌrezəˈrekt/ *v.* **1** *colloq.* revive the practice, use, or memory of. **2** raise or rise from the dead. [back-formation from RESURRECTION]

resurrection /ˌrezəˈrekʃ(ə)n/ *n.* **1** rising from the dead. **2** (**Resurrection**) Christ's rising from the dead. **3** revival after disuse, inactivity, or decay. [Latin: related to RESURGENT]

resuscitate /rɪ'sʌsɪ,teɪt/ *v.* (**-ting**) **1** revive from unconsciousness or apparent death. **2** revive, restore. □ **resuscitation** /-'teɪʃ(ə)n/ *n.* [Latin *suscito* raise]

retail /'ri:teɪl/ *–n.* sale of goods in small quantities to the public, and usu. not for resale. *–adj.* & *adv.* by retail; at a retail price. *–v.* **1** sell (goods) by retail. **2** (often foll. by *at*, *of*) (of goods) be sold in this way (esp. for a specified price). **3** /also rɪ'teɪl/ recount; relate details of. □ **retailer** *n.* [French *taillier* cut: related to TALLY]

retain /rɪ'teɪn/ *v.* **1 a** keep possession of; not lose; continue to have. **b** not abolish, discard, or alter. **2** keep in one's memory. **3** keep in place; hold fixed. **4** secure the services of (a person, esp. a barrister) with a preliminary payment. [Latin *retineo -tent-*]

retainer *n.* **1** fee for securing a person's services. **2** faithful servant (esp. *old retainer*). **3** reduced rent paid to retain unoccupied accommodation. **4** person or thing that retains.

retake *–v.* /ri:'teɪk/ (**-king**; *past* **retook**; *past part.* **retaken**) **1** take (a photograph, exam, etc.) again. **2** recapture. *–n.* /'ri:teɪk/ **1** act of filming a scene or recording music etc. again. **2** film or recording obtained in this way. **3** act of taking an exam etc. again.

retaliate /rɪ'tælɪ,eɪt/ *v.* (**-ting**) repay an injury, insult, etc. in kind; attack in return. □ **retaliation** /-'eɪʃ(ə)n/ *n.* **retaliatory** /-'tæljətərɪ/ *adj.* [Latin *talis* such]

retard /rɪ'tɑ:d/ *v.* **1** make slow or late. **2** delay the progress or accomplishment of. □ **retardant** *adj.* & *n.* **retardation** /,ri:tɑ:'deɪʃ(ə)n/ *n.* [Latin *tardus* slow]

retarded *adj.* backward in mental or physical development.

retch *v.* make a motion of vomiting, esp. involuntarily and without effect. [Old English]

retell /ri:'tel/ *v.* (*past* and *past part.* **retold**) tell again or differently.

retention /rɪ'tenʃ(ə)n/ *n.* **1** retaining or being retained. **2** condition of retaining bodily fluid (esp. urine) normally evacuated. [Latin: related to RETAIN]

retentive /rɪ'tentɪv/ *adj.* **1** tending to retain. **2** (of memory etc.) not forgetful. [French or medieval Latin: related to RETAIN]

retexture /ri:'tekstʃə(r)/ *v.* (**-ring**) treat (material, a garment, etc.) so as to restore its original texture.

rethink *–v.* /ri:'θɪŋk/ (*past* and *past part.* **rethought**) consider again, esp. with a view to making changes. *–n.* /'ri:θɪŋk/ reassessment; rethinking.

reticence /'retɪs(ə)ns/ *n.* **1** avoidance of saying all one knows or feels, or more than is necessary. **2** disposition to silence; taciturnity. □ **reticent** *adj.* [Latin *reticeo* keep silent]

reticulate *–v.* /rɪ'tɪkjʊ,leɪt/ (**-ting**) divide or be divided in fact or appearance into a network. *–adj.* /rɪ'tɪkjʊlət/ reticulated. □ **reticulation** /-'leɪʃ(ə)n/ *n.* [Latin *reticulum* diminutive of *rete* net]

retie /ri:'taɪ/ *v.* (**retying**) tie again.

retina /'retɪnə/ *n.* (*pl.* **-s** or **-nae** /-,ni:/) layer at the back of the eyeball sensitive to light. □ **retinal** *adj.* [Latin *rete* net]

retinue /'retɪ,nju:/ *n.* body of attendants accompanying an important person. [French: related to RETAIN]

retire /rɪ'taɪə(r)/ *v.* (**-ring**) **1 a** leave office or employment, esp. because of age. **b** cause (a person) to retire from work. **2** withdraw, go away, retreat. **3** seek seclusion or shelter. **4** go to bed. **5** withdraw (troops). **6** *Cricket* (of a batsman) voluntarily end or be compelled to suspend one's innings. □ **retire into oneself** become uncommunicative or unsociable. [French *tirer* draw]

retired *adj.* **1** having retired from employment. **2** withdrawn from society or observation; secluded.

retirement *n.* **1 a** retiring. **b** period of one's life as a retired person. **2** seclusion.

retirement pension *n.* pension paid by the State to retired people above a certain age.

retiring *adj.* shy; fond of seclusion.

retold *past* and *past part.* of RETELL.

retook *past* of RETAKE.

retort[1] /rɪ'tɔ:t/ *–n.* incisive, witty, or angry reply. *–v.* **1 a** say by way of a retort. **b** make a retort. **2** repay (an insult or attack) in kind. [Latin *retorqueo -tort-* twist]

retort[2] /rɪ'tɔ:t/ *–n.* **1** vessel with a long neck turned downwards, used in distilling liquids. **2** vessel for heating coal to generate gas. *–v.* purify (mercury) by heating in a retort. [medieval Latin: related to RETORT[1]]

retouch /ri:'tʌtʃ/ *v.* improve (a picture, photograph, etc.) by minor alterations.

retrace /rɪ'treɪs/ *v.* (**-cing**) **1** go back over (one's steps etc.). **2** trace back to a source or beginning. **3** recall the course of (a thing) in one's memory.

retract /rɪ'trækt/ *v.* **1** withdraw (a statement or undertaking). **2** draw or be drawn back or in. □ **retractable** *adj.* **retraction** *n.* [Latin *retraho -tract-* draw back]

retractile /rɪ'træktaɪl/ *adj.* capable of being retracted.

retrain /riː'treɪn/ *v.* train again or further, esp. for new work.

retread –*v.* /riː'tred/ **1** (*past* **retrod**; *past part.* **retrodden**) tread (a path etc.) again. **2** (*past, past part.* **retreaded**) put a fresh tread on (a tyre). –*n.* /'riːtred/ retreaded tyre.

retreat /rɪ'triːt/ –*v.* **1** (esp. of military forces) go back, retire; relinquish a position. **2** recede. –*n.* **1 a** act of retreating. **b** *Mil.* signal for this. **2** withdrawal into privacy or security. **3** place of shelter or seclusion. **4** period of seclusion for prayer and meditation. **5** *Mil.* bugle-call at sunset. [Latin: related to RETRACT]

retrench /rɪ'trentʃ/ *v.* **1** cut down expenses; introduce economies. **2** reduce the amount of (costs). □ **retrenchment** *n.* [French: related to TRENCH]

retrial /riː'traɪəl/ *n.* second or further (judicial) trial.

retribution /ˌretrɪ'bjuːʃ(ə)n/ *n.* requital, usu. for evil done; vengeance. □ **retributive** /rɪ'trɪbjʊtɪv/ *adj.* [Latin: related to TRIBUTE]

retrieve /rɪ'triːv/ –*v.* (**-ving**) **1 a** regain possession of. **b** recover by investigation or effort of memory. **2** obtain (information stored in a computer etc.). **3** (of a dog) find and bring in (killed or wounded game etc.). **4** (foll. by *from*) rescue (esp. from a bad state). **5** restore to a flourishing state; revive. **6** repair or set right (a loss or error etc.) (*managed to retrieve the situation*). –*n.* possibility of recovery (*beyond retrieve*). □ **retrievable** *adj.* **retrieval** *n.* [French *trouver* find]

retriever *n.* dog of a breed used for retrieving game.

retro /'retrəʊ/ *slang* –*adj.* reviving or harking back to the past. –*n.* retro fashion or style.

retro- *comb. form* **1** denoting action back or in return. **2** *Anat. & Med.* denoting location behind. [Latin]

retroactive /ˌretrəʊ'æktɪv/ *adj.* (esp. of legislation) effective from a past date.

retrod *past* of RETREAD.

retrodden *past part.* of RETREAD.

retrograde /'retrəˌgreɪd/ –*adj.* **1** directed backwards. **2** reverting, esp. to an inferior state; declining. **3** reversed (*retrograde order*). –*v.* **1** move backwards; recede. **2** decline, revert. [Latin *retrogradior -gress-* move backwards]

retrogress /ˌretrə'gres/ *v.* **1** move backwards. **2** deteriorate. □ **retrogression** /-'greʃ(ə)n/ *n.* **retrogressive** *adj.*

retrorocket /'retrəʊˌrɒkɪt/ *n.* auxiliary rocket for slowing down a spacecraft etc.

retrospect /'retrəˌspekt/ *n.* □ **in retrospect** when looking back. [from RETRO-, PROSPECT]

retrospection /ˌretrə'spekʃ(ə)n/ *n.* looking back into the past.

retrospective /ˌretrə'spektɪv/ –*adj.* **1** looking back on or dealing with the past. **2** (of a statute etc.) applying to the past as well as the future. –*n.* exhibition, recital, etc. showing an artist's development over his or her lifetime. □ **retrospectively** *adv.*

retroussé /rə'truːseɪ/ *adj.* (of the nose) turned up at the tip. [French]

retroverted /'retrəʊˌvɜːtɪd/ *adj.* (of the womb) inclined backwards. [Latin: related to RETRO-, *verto* turn]

retrovirus /'retrəʊˌvaɪərəs/ *n.* any of a group of RNA viruses which form DNA during the replication of their RNA, and so transfer genetic material into the DNA of host cells. [from the initial letters of *reverse transcriptase* + VIRUS]

retry /riː'traɪ/ *v.* (**-ies, -ied**) try (a defendant or lawsuit) a second or further time.

retsina /ret'siːnə/ *n.* Greek white wine flavoured with resin. [modern Greek]

retune /riː'tjuːn/ *v.* (**-ning**) **1** tune (a musical instrument) again or differently. **2** tune (a radio etc.) to a different frequency.

return /rɪ'tɜːn/ –*v.* **1** come or go back. **2** bring, put, or send back. **3** pay back or reciprocate; give in response. **4** yield (a profit). **5** say in reply; retort. **6** (in cricket or tennis etc.) hit or send (the ball) back. **7** state, mention, or describe officially, esp. in answer to a writ or formal demand. **8** (of an electorate) elect as an MP, government, etc. –*n.* **1** coming or going back. **2 a** giving, sending, putting, or paying back. **b** thing given or sent back. **3** (in full **return ticket**) ticket for a journey to a place and back to the starting-point. **4** (in *sing.* or *pl.*) **a** proceeds or profit of an undertaking. **b** acquisition of these. **5** formal statement compiled or submitted by order (*income-tax return*). **6** (in full **return match** or **game**) second match etc. between the same opponents. **7 a** person's election as an MP etc. **b** returning officer's announcement of this. □ **by return (of post)** by the next available post in the return direction. **in return** as an exchange or reciprocal action. **many happy returns (of the day)** greeting on a birthday. □ **returnable** *adj.* [Romanic: related to TURN]

returnee /ˌrɪtɜː'niː/ *n.* person who returns home from abroad, esp. after war service.

returning officer *n.* official conducting an election in a constituency and announcing the results.

retying *pres. part.* of RETIE.

retype /riːˈtaɪp/ *v.* (**-ping**) type again, esp. to correct errors.

reunify /riːˈjuːnɪˌfaɪ/ *v.* (**-ies, -ied**) restore (esp. separated territories) to a political unity. □ **reunification** /-fɪˈkeɪʃ(ə)n/ *n.*

reunion /riːˈjuːnjən/ *n.* **1** reuniting or being reunited. **2** social gathering, esp. of people formerly associated.

reunite /ˌriːjuːˈnaɪt/ *v.* (**-ting**) (cause to) come together again.

reupholster /ˌriːʌpˈhəʊlstə(r)/ *v.* upholster anew.

reuse *–v.* /riːˈjuːz/ (**-sing**) use again. *–n.* /riːˈjuːs/ second or further use. □ **reusable** /-ˈjuːzəb(ə)l/ *adj.*

Rev. *abbr.* Reverend.

rev *colloq. –n.* (in *pl.*) number of revolutions of an engine per minute. *–v.* (**-vv-**) **1** (of an engine) revolve; turn over. **2** (also *absol.*; often foll. by *up*) cause (an engine) to run quickly. [abbreviation]

revalue /riːˈvæljuː/ *v.* (**-ues, -ued, -uing**) give a different, esp. higher, value to (a currency etc.). □ **revaluation** /-ˈeɪʃ(ə)n/ *n.*

revamp /riːˈvæmp/ *v.* **1** renovate, revise, improve. **2** patch up.

Revd *abbr.* Reverend.

reveal /rɪˈviːl/ *v.* **1** display or show; allow to appear. **2** (often as **revealing** *adj.*) disclose, divulge, betray (*revealing remark*). **3** (in *refl.* or *passive*) come to sight or knowledge. [Latin *velum* veil]

reveille /rɪˈvælɪ/ *n.* military waking-signal. [French *réveillez* wake up]

revel /ˈrev(ə)l/ *–v.* (**-ll-**; *US* **-l-**) **1** have a good time; be extravagantly festive. **2** (foll. by *in*) take keen delight in. *–n.* (in *sing.* or *pl.*) revelling. □ **reveller** *n.* **revelry** *n.* (*pl.* **-ies**). [Latin: related to REBEL]

revelation /ˌrevəˈleɪʃ(ə)n/ *n.* **1 a** revealing, esp. the supposed disclosure of knowledge to man by a divine or supernatural agency. **b** knowledge disclosed in this way. **2** striking disclosure. **3** (**Revelation** or *colloq.* **Revelations**) (in full **the Revelation of St John the Divine**) last book of the New Testament.

revenge /rɪˈvendʒ/ *–n.* **1** retaliation for an offence or injury. **2** act of retaliation. **3** desire for this; vindictive feeling. **4** (in games) win after an earlier defeat. *–v.* (**-ging**) **1** (in *refl.* or *passive*; often foll. by *on, upon*) inflict retaliation for (an offence). **2** avenge (a person). [Latin: related to VINDICATE]

revengeful *adj.* eager for revenge. □ **revengefully** *adv.*

revenue /ˈrevəˌnjuː/ *n.* **1 a** income, esp. a substantial one. **b** (in *pl.*) items constituting this. **2** State's annual income from which public expenses are met. **3** department of the civil service collecting this. [French *revenu* from Latin *revenio* return]

reverberate /rɪˈvɜːbəˌreɪt/ *v.* (**-ting**) **1** (of sound, light, or heat) be returned, echoed, or reflected repeatedly. **2** return (a sound etc.) in this way. **3** (of an event etc.) produce a continuing effect, shock, etc. □ **reverberant** *adj.* **reverberation** /-ˈreɪʃ(ə)n/ *n.* **reverberative** /-rətɪv/ *adj.* [Latin *verbero* beat]

revere /rɪˈvɪə(r)/ *v.* (**-ring**) hold in deep and usu. affectionate or religious respect. [Latin *vereor* fear]

reverence /ˈrevərəns/ *–n.* **1** revering or being revered. **2** capacity for revering. *–v.* (**-cing**) regard or treat with reverence. [Latin: related to REVERE]

reverend /ˈrevərənd/ *adj.* (esp. as the title of a clergyman) deserving reverence. [Latin *reverendus*: related to REVERE]

Reverend Mother *n.* Mother Superior of a convent.

reverent /ˈrevərənt/ *adj.* feeling or showing reverence. □ **reverently** *adv.* [Latin: related to REVERE]

reverential /ˌrevəˈrenʃ(ə)l/ *adj.* of the nature of, due to, or characterized by reverence. □ **reverentially** *adv.* [medieval Latin: related to REVERENCE]

reverie /ˈrevərɪ/ *n.* fit of abstracted musing, day-dream. [French]

revers /rɪˈvɪə(r)/ *n.* (*pl.* same /-ˈvɪəz/) **1** turned-back edge of a garment revealing the undersurface. **2** material on this surface. [French: related to REVERSE]

reverse /rɪˈvɜːs/ *–v.* (**-sing**) **1** turn the other way round or up or inside out. **2** change to the opposite character or effect. **3** (cause to) travel backwards. **4** make (an engine etc.) work in a contrary direction. **5** revoke or annul (a decree, act, etc.). *–adj.* **1** backwards or upside down. **2** opposite or contrary in character or order; inverted. *–n.* **1** opposite or contrary (*the reverse is the case*). **2** contrary of the usual manner (*printed in reverse*). **3** piece of misfortune; disaster; defeat. **4** reverse gear or motion. **5** reverse side. **6** side of a coin etc. bearing the secondary design. **7** verso of a printed leaf. □ **reverse arms** hold a rifle with the butt upwards. **reverse the charges** have the recipient of a telephone call pay for it. □ **reversal** *n.* **reversible** *adj.* [Latin *verto vers-* turn]

reverse gear *n.* gear used to make a vehicle etc. go backwards.

reversing light *n.* white light at the rear of a vehicle showing that it is in reverse gear.

reversion /rɪ'vɜːʃ(ə)n/ *n.* **1** return to a previous state, habit, etc. **2** *Biol.* return to ancestral type. **3** legal right (esp. of the original owner, or his or her heirs) to possess or succeed to property on the death of the present possessor. [Latin: related to REVERSE]

revert /rɪ'vɜːt/ *v.* **1** (foll. by *to*) return to a former state, practice, opinion, etc. **2** (of property, an office, etc.) return by reversion. □ **revertible** *adj.* (in sense 2).

review /rɪ'vjuː/ —*n.* **1** general survey or assessment of a subject or thing. **2** survey of the past. **3** revision or reconsideration (*is under review*). **4** display and formal inspection of troops etc. **5** published criticism of a book, play, etc. **6** periodical with critical articles on current events, the arts, etc. —*v.* **1** survey or look back on. **2** reconsider or revise. **3** hold a review of (troops etc.). **4** write a review of (a book, play, etc.). □ **reviewer** *n.* [French *revoir*: related to VIEW]

revile /rɪ'vaɪl/ *v.* (**-ling**) abuse verbally. [French: related to VILE]

revise /rɪ'vaɪz/ *v.* (**-sing**) **1** examine or re-examine and improve or amend (esp. written or printed matter). **2** consider and alter (an opinion etc.). **3** (also *absol.*) go over (work learnt or done) again, esp. for an examination. □ **revisory** *adj.* [Latin *reviso* from *video vis-* see]

Revised Standard Version *n.* revision published in 1946–57 of the American Standard Version of the Bible (itself based on the English RV).

Revised Version *n.* revision published in 1881–95 of the Authorized Version of the Bible.

revision /rɪ'vɪʒ(ə)n/ *n.* **1** revising or being revised. **2** revised edition or form. [Latin: related to REVISE]

revisionism *n.* often *derog.* revision or modification of an orthodoxy, esp. of Marxism. □ **revisionist** *n.* & *adj.*

revisit /riː'vɪzɪt/ *v.* (**-t-**) visit again.

revitalize /riː'vaɪtəˌlaɪz/ *v.* (also **-ise**) (**-zing** or **-sing**) imbue with new life and vitality.

revival /rɪ'vaɪv(ə)l/ *n.* **1** reviving or being revived. **2** new production of an old play etc. **3** revived use of an old practice, style, etc. **4 a** reawakening of religious fervour. **b** campaign to promote this.

revivalism *n.* promotion of a revival, esp. of religious fervour. □ **revivalist** *n.* & *adj.*

revive /rɪ'vaɪv/ *v.* (**-ving**) **1** come or bring back to consciousness, life, or strength. **2** come or bring back to existence, or to use or notice etc. [Latin *vivo* live]

revivify /rɪ'vɪvɪˌfaɪ/ *v.* (**-ies, -ied**) restore to animation, vigour, or life. □ **revivification** /-fɪ'keɪʃ(ə)n/ *n.* [Latin: related to VIVIFY]

revoke /rɪ'vəʊk/ —*v.* (**-king**) **1** rescind, withdraw, or cancel. **2** *Cards* fail to follow suit when able to do so. —*n. Cards* revoking. □ **revocable** /'revəkəb(ə)l/ *adj.* **revocation** /ˌrevə'keɪʃ(ə)n/ *n.* [Latin *voco* call]

revolt /rɪ'vəʊlt/ —*v.* **1** rise in rebellion. **2 a** affect with strong disgust. **b** (often foll. by *at*, *against*) feel strong disgust. —*n.* **1** act of rebelling. **2** state of insurrection. **3** sense of disgust. **4** mood of protest or defiance. [Italian: related to REVOLVE]

revolting *adj.* disgusting, horrible. □ **revoltingly** *adv.*

revolution /ˌrevə'luːʃ(ə)n/ *n.* **1** forcible overthrow of a government or social order. **2** any fundamental change or reversal of conditions. **3** revolving. **4 a** single completion of an orbit or rotation. **b** time taken for this. **5** cyclic recurrence. [Latin: related to REVOLVE]

revolutionary —*adj.* **1** involving great and often violent change. **2** of or causing political revolution. —*n.* (*pl.* **-ies**) instigator or supporter of political revolution.

revolutionize *v.* (also **-ise**) (**-zing** or **-sing**) change fundamentally.

revolve /rɪ'vɒlv/ *v.* (**-ving**) **1** (cause to) turn round, esp. on an axis; rotate. **2** move in a circular orbit. **3** ponder (a problem etc.) in the mind. **4** (foll. by *around*) have as its chief concern; be centred upon (*his life revolves around his job*). [Latin *revolvo -volut-*]

revolver *n.* pistol with revolving chambers enabling several shots to be fired without reloading.

revolving door *n.* door with usu. four partitions turning round a central axis.

revue /rɪ'vjuː/ *n.* entertainment of short usu. satirical sketches and songs. [French: related to REVIEW]

revulsion /rɪ'vʌlʃ(ə)n/ *n.* **1** abhorrence. **2** sudden violent change of feeling. [Latin *vello vuls-* pull]

reward /rɪ'wɔːd/ —*n.* **1 a** return or recompense for service or merit. **b** requital for good or evil. **2** sum offered for the detection of a criminal, restoration of lost property, etc. —*v.* give a reward to (a person) or for (a service etc.). [Anglo-French *reward(er)* REGARD]

rewarding *adj.* (of an activity etc.) worthwhile; satisfying.

rewind /riː'waɪnd/ *v.* (*past* and *past part.* **rewound**) wind (a film or tape etc.) back.

rewire /riː'waɪə(r)/ *v.* (**-ring**) provide with new electrical wiring.

reword /riː'wɜːd/ *v.* express in different words.

rework /riː'wɜːk/ *v.* revise; refashion; remake. □ **reworking** *n.*

rewrite *–v.* /riː'raɪt/ (**-ting**; *past* **rewrote**; *past part.* **rewritten**) write again or differently. *–n.* /'riːraɪt/ **1** rewriting. **2** thing rewritten.

Rex *n.* **1** (after the name) reigning king (*George Rex*). **2** *Law* the Crown (*Rex v. Jones*). [Latin]

Rf *symb.* rutherfordium.

RFC *abbr.* Rugby Football Club.

Rh *symb.* rhodium.

r.h. *abbr.* right hand.

rhapsodize /'ræpsə,daɪz/ *v.* (also **-ise**) (**-zing** or **-sing**) talk or write rhapsodies.

rhapsody /'ræpsədɪ/ *n.* (*pl.* **-ies**) **1** enthusiastic or extravagant speech or composition. **2** piece of music in one movement, often based on national, folk, or popular melodies. □ **rhapsodic** /ræp'sɒdɪk/ *adj.* [Greek *rhaptō* stitch: related to ODE]

rhea /'riːə/ *n.* S. American flightless ostrich-like bird. [Greek *Rhea* mother name of Zeus]

rhenium /'riːnɪəm/ *n.* rare metallic element occurring naturally in molybdenum ores. [Latin *Rhenus* Rhine]

rheostat /'riːə,stæt/ *n.* instrument used to control an electric current by varying the resistance. [Greek *rheos* stream]

rhesus /'riːsəs/ *n.* (in full **rhesus monkey**) small N. Indian monkey. [*Rhesus*, mythical king of Thrace]

rhesus factor *n.* antigen occurring on the red blood cells of most humans and some other primates.

rhesus negative *adj.* lacking the rhesus factor.

rhesus positive *adj.* having the rhesus factor.

rhetoric /'retərɪk/ *n.* **1** art of effective or persuasive speaking or writing. **2** language designed to persuade or impress (esp. seen as overblown and meaningless). [Greek *rhētōr* orator]

rhetorical /rɪ'tɒrɪk(ə)l/ *adj.* **1** expressed artificially or extravagantly. **2** of the nature or art of rhetoric. □ **rhetorically** *adv.* [Greek: related to RHETORIC]

rhetorical question *n.* question used for effect but not seeking an answer (e.g. *who cares?* for *nobody cares*).

rheumatic /ruː'mætɪk/ *–adj.* of, suffering from, producing, or produced by rheumatism. *–n.* person suffering from rheumatism. □ **rheumatically** *adv.* **rheumaticky** *adj. colloq.* [Greek *rheuma* stream]

rheumatic fever *n.* fever with inflammation and pain in the joints.

rheumatics *n.pl.* (treated as *sing.*; often prec. by *the*) *colloq.* rheumatism.

rheumatism /'ruːmə,tɪz(ə)m/ *n.* disease marked by inflammation and pain in the joints, muscles, or fibrous tissue, esp. rheumatoid arthritis.

rheumatoid /'ruːmə,tɔɪd/ *adj.* having the character of rheumatism.

rheumatoid arthritis *n.* chronic progressive disease causing inflammation and stiffening of the joints.

rhinestone *n.* imitation diamond. [river *Rhine* in Germany]

rhino /'raɪnəʊ/ *n.* (*pl.* same or **-s**) *colloq.* rhinoceros. [abbreviation]

rhinoceros /raɪ'nɒsərəs/ *n.* (*pl.* same or **-roses**) large thick-skinned mammal with usu. one horn on its nose. [Greek *rhis rhin-* nose, *keras* horn]

rhizome /'raɪzəʊm/ *n.* underground rootlike stem bearing both roots and shoots. [Greek *rhizoma*]

rho /rəʊ/ *n.* seventeenth letter of the Greek alphabet (Ρ, ρ). [Greek]

rhodium /'rəʊdɪəm/ *n.* hard white metallic element used in making alloys and plating jewellery. [Greek *rhodon* rose]

rhododendron /,rəʊdə'dendrən/ *n.* (*pl.* **-s** or **-dra**) evergreen shrub with large clusters of bell-shaped flowers. [Greek *rhodon* rose, *dendron* tree]

rhomboid /'rɒmbɔɪd/ *–adj.* (also **rhomboidal** /-'bɔɪd(ə)l/) like a rhombus. *–n.* quadrilateral of which only the opposite sides and angles are equal. [Greek: related to RHOMBUS]

rhombus /'rɒmbəs/ *n.* (*pl.* **-buses** or **-bi** /-baɪ/) *Geom.* parallelogram with oblique angles and equal sides. [Greek *rhombos*]

RHS *abbr.* Royal Horticultural Society.

rhubarb /'ruːbɑːb/ *n.* **1 a** plant with long fleshy dark-red leaf-stalks cooked as a dessert. **b** these stalks. **2 a** *colloq.* indistinct conversation or noise, from the repeated use of the word 'rhubarb' by a crowd. **b** *slang* nonsense. [Greek *rha* rhubarb, *barbaros* foreign]

rhyme /raɪm/ *–n.* **1** identity of sound between words or their endings, esp. in verse. **2** (in *sing.* or *pl.*) verse or a poem having rhymes. **3** use of rhyme. **4** word providing a rhyme. *–v.* (**-ming**) **1 a** (of words or lines) produce a rhyme. **b** (foll. by *with*) act as or treat (a word) as a rhyme (with another). **2** make or write rhymes. **3** put or make (a story etc.) into rhyme. □ **rhyme or reason** sense, logic. [Latin: related to RHYTHM]

rhymester *n.* writer of (esp. simple) rhymes.

rhyming slang *n.* slang that replaces words by rhyming words or phrases, e.g. *suit* by *whistle and flute*.

rhythm /'rɪð(ə)m/ *n.* **1 a** periodical accent and the duration of notes in music, esp. as beats in a bar. **b** type of structure formed by this (*samba rhythm*). **2** measured regular flow of verse or prose determined by the length of and stress on syllables. **3** *Physiol.* pattern of successive strong and weak movements. **4** regularly recurring sequence of events. □ **rhythmic** *adj.* **rhythmical** *adj.* **rhythmically** *adv.* [Greek *rhuthmos*]

rhythm and blues *n.* popular music with blues themes and a strong rhythm.

rhythm method *n.* abstention from sexual intercourse near the time of ovulation, as a method of birth control.

rhythm section *n.* piano (or guitar etc.), bass, and drums in a dance or jazz band.

rib –*n.* **1** each of the curved bones joined to the spine in pairs and protecting the chest. **2** joint of meat from this part of an animal. **3** supporting ridge, timber, rod, etc. across a surface or through a structure. **4** *Knitting* combination of plain and purl stitches producing a ribbed design. –*v.* (**-bb-**) **1** provide with ribs; act as the ribs of. **2** *colloq.* make fun of; tease. **3** mark with ridges. [Old English]

ribald /'rɪb(ə)ld/ *adj.* coarsely or disrespectfully humorous; obscene. [French *riber* be licentious]

ribaldry *n.* ribald talk or behaviour.

riband /'rɪbənd/ *n.* ribbon. [French *riban*]

ribbed *adj.* having ribs or riblike markings.

ribbing *n.* **1** ribs or a riblike structure. **2** *colloq.* teasing.

ribbon /'rɪbən/ *n.* **1 a** narrow strip or band of fabric, used esp. for trimming or decoration. **b** material in this form. **2** ribbon worn to indicate some honour or membership of a sports team etc. **3** long narrow strip of anything (*typewriter ribbon*). **4** (in *pl.*) ragged strips (*torn to ribbons*). [var. of RIBAND]

ribbon development *n.* building of houses one house deep along a road leading out of a town or village.

ribcage *n.* wall of bones formed by the ribs round the chest.

riboflavin /,raɪbəʊ'fleɪvɪn/ *n.* (also **riboflavine** /-vi:n/) vitamin of the B complex, found in liver, milk, and eggs. [*ribose* sugar, Latin *flavus* yellow]

ribonucleic acid /,raɪbənju:'kli:ɪk/ *n.* nucleic acid in living cells, involved in protein synthesis. [*ribose* sugar]

rib-tickler *n.* something amusing, joke.

rice *n.* **1** swamp grass cultivated in esp. Asian marshes. **2** grains of this, used as food. [French *ris* ultimately from Greek *oruza*]

rice-paper *n.* edible paper made from the bark of an oriental tree and used for painting and in cookery.

rich *adj.* **1** having much wealth. **2** splendid, costly, elaborate. **3** valuable (*rich offerings*). **4** copious, abundant, ample (*rich supply of ideas*). **5** (often foll. by *in*, *with*) (of soil or a region etc.) fertile; abundant in resources etc. (*rich in nutrients*). **6** (of food or diet) containing much fat or spice etc. **7** (of the mixture in an internal-combustion engine) containing a high proportion of fuel. **8** (of colour, sound, or smell) mellow and deep, strong and full. **9** highly amusing or ludicrous; outrageous. □ **richness** *n.* [Old English and French]

riches *n.pl.* abundant means; valuable possessions. [French *richeise*: related to RICH]

richly *adv.* **1** in a rich way. **2** fully, thoroughly (*richly deserves success*).

Richter scale /'rɪktə/ *n.* scale of 0–10 for representing the strength of an earthquake. [*Richter*, name of a seismologist]

rick[1] *n.* stack of hay etc. [Old English]

rick[2] (also **wrick**) –*n.* slight sprain or strain. –*v.* sprain or strain slightly. [Low German *wricken*]

rickets /'rɪkɪts/ *n.* (treated as *sing.* or *pl.*) deficiency disease of children with softening of the bones. [origin uncertain]

rickety /'rɪkɪtɪ/ *adj.* **1** insecure, shaky. **2** suffering from rickets. □ **ricketiness** *n.*

rickrack var. of RICRAC.

rickshaw /'rɪkʃɔ:/ *n.* (also **ricksha** /-ʃə/) light two-wheeled hooded vehicle drawn by one or more persons. [abbreviation of *jinrickshaw* from Japanese]

ricochet /'rɪkə,ʃeɪ/ –*n.* **1** rebounding of esp. a shell or bullet off a surface. **2** hit made after this. –*v.* (**-cheted** /-,ʃeɪd/; **-cheting** /-,ʃeɪɪŋ/ or **-chetted** /-,ʃetɪd/; **-chetting** /-,ʃetɪŋ/) (of a projectile) make a ricochet. [French]

ricotta /rɪ'kɒtə/ *n.* soft Italian cheese. [Latin: related to RE-, *coquo* cook]

ricrac /'rɪkræk/ *n.* (also **rickrack**) zigzag braided trimming for garments. [from RACK[1]]

rid *v.* (**-dd-**; *past* and *past part.* **rid**) (foll. by *of*) free (a person or place) of something unwanted. □ **be** (or **get**) **rid of** be freed or relieved of; dispose of. [Old Norse]

riddance /ˈrɪd(ə)ns/ *n.* getting rid of something. □ **good riddance** expression of relief at getting rid of something.

ridden *past part.* of RIDE.

riddle[1] /ˈrɪd(ə)l/ *–n.* **1** verbal puzzle or test, often with a trick answer. **2** puzzling fact, thing, or person. *–v.* **(-ling)** speak in riddles. [Old English: related to READ]

riddle[2] /ˈrɪd(ə)l/ *–v.* **(-ling)** (usu. foll. by *with*) **1** make many holes in, esp. with gunshot. **2** (in *passive*) fill; permeate (*riddled with errors*). **3** pass through a riddle. *–n.* coarse sieve. [Old English]

ride *–v.* **(-ding**; *past* **rode**; *past part.* **ridden** /ˈrɪd(ə)n/) **1** (often foll. by *on, in*) travel or be carried on (a bicycle etc.) or esp. *US* in (a vehicle); be conveyed (*rode her bike*; *rode on her bike*; *rode the tram*). **2** (often foll. by *on*; also *absol.*) be carried by (a horse etc.). **3** be carried or supported by (*ship rides the waves*). **4** traverse or take part in on horseback etc. (*ride 50 miles*; *rode the prairie*). **5 a** lie at anchor; float buoyantly. **b** (of the moon) seem to float. **6** yield to (a blow) so as to reduce its impact. **7** give a ride to; cause to ride (*rode me home*). **8** (of a rider) cause (a horse etc.) to move forward (*rode their horses at the fence*). **9** (as **ridden** *adj.*) (foll. by *by*, *with*, or in *comb.*) be dominated by; be infested with (*ridden with guilt*; *rat-ridden cellar*). *–n.* **1** journey or spell of riding in a vehicle, or on a horse, bicycle, person's back, etc. **2** path (esp. through woods) for riding on. **3** specified kind of ride (*bumpy ride*). **4** amusement for riding on at a fairground etc. □ **let a thing ride** leave it undisturbed. **ride again** reappear as strong etc. as ever. **ride high** be elated or successful. **ride out** come safely through (a storm, danger, etc.). **ride roughshod over** see ROUGHSHOD. **ride up** (of a garment) work upwards out of place. **take for a ride** *colloq.* hoax or deceive. [Old English]

rider *n.* **1** person who rides (esp. a horse). **2** additional remark following a statement, verdict, etc. □ **riderless** *adj.*

ridge *–n.* **1** line of the junction of two surfaces sloping upwards towards each other (*ridge of a roof*). **2** long narrow hilltop, mountain range, or watershed. **3** any narrow elevation across a surface. **4** elongated region of high barometric pressure. **5** raised strip of esp. ploughed land. *–v.* **(-ging)** mark with ridges. □ **ridgy** *adj.* [Old English]

ridge-pole *n.* horizontal roof pole of a long tent.

ridgeway *n.* road or track along a ridge.

ridicule /ˈrɪdɪˌkjuːl/ *–n.* derision, mockery. *–v.* **(-ling)** make fun of; mock; laugh at. [Latin *rideo* laugh]

ridiculous /rɪˈdɪkjʊləs/ *adj.* **1** deserving or inviting ridicule. **2** unreasonable. □ **ridiculously** *adv.* **ridiculousness** *n.*

riding[1] /ˈraɪdɪŋ/ *n.* sport or pastime of travelling on horseback.

riding[2] /ˈraɪdɪŋ/ *n. hist.* former administrative division **(East, North, West Riding)** of Yorkshire. [Old English from Old Norse, = third part]

riding-light *n.* light shown by a ship at anchor.

riding-school *n.* establishment teaching horsemanship.

Riesling /ˈriːzlɪŋ/ *n.* **1** a kind of grape. **2** white wine made from this. [German]

rife *predic. adj.* **1** of common occurrence; widespread. **2** (foll. by *with*) abounding in. [Old English, probably from Old Norse]

riff *n.* short repeated phrase in jazz etc. [abbreviation of RIFFLE]

riffle /ˈrɪf(ə)l/ *–v.* **(-ling) 1** (often foll. by *through*) leaf quickly through (pages). **2 a** turn (pages) in quick succession. **b** shuffle (playing-cards), esp. by flexing and combining the two halves of a pack. *–n.* **1** act of riffling. **2** *US* **a** shallow disturbed part of a stream. **b** patch of waves or ripples. [perhaps var. of RUFFLE]

riff-raff /ˈrɪfræf/ *n.* (often prec. by *the*) rabble; disreputable people. [French *rif et raf*]

rifle[1] /ˈraɪf(ə)l/ *–n.* **1** gun with a long rifled barrel, esp. one fired from the shoulder. **2** (in *pl.*) riflemen. *–v.* **(-ling)** make spiral grooves in (a gun, its barrel, or its bore) to make a projectile spin. [French]

rifle[2] /ˈraɪf(ə)l/ *v.* **(-ling)** (often foll. by *through*) **1** search and rob. **2** carry off as booty. [French]

rifleman *n.* soldier armed with a rifle.

rifle-range *n.* place for rifle-practice.

rifle-shot *n.* **1** shot fired with a rifle. **2** distance coverable by this.

rifling *n.* arrangement of grooves on the inside of a gun's barrel.

rift *–n.* **1** crack, split; break (in cloud etc.). **2** disagreement; breach. **3** cleft in earth or rock. *–v.* tear or burst apart. [Scandinavian: related to RIVEN]

rift-valley *n.* steep-sided valley formed by subsidence between nearly parallel faults.

rig[1] *–v.* **(-gg-) 1** provide (a ship) with sails, rigging, etc. **2** (often foll. by *out, up*) fit with clothes or other equipment. **3** (foll. by *up*) set up hastily or as a makeshift. **4** assemble and adjust the parts of (an aircraft). *–n.* **1** arrangement of a ship's masts, sails, etc. **2** equipment for a special purpose, e.g. a radio transmitter. **3** = OIL RIG. **4** *colloq.* style of dress; uniform (*in full rig*). □ **rigged** *adj.* (also in *comb.*). [perhaps from Scandinavian]

rig[2] *–v.* (**-gg-**) manage or fix (a result etc.) fraudulently (*rigged the election*). *–n.* trick, dodge, or way of swindling. □ **rig the market** cause an artificial rise or fall in prices. □ **rigger** *n.* [origin unknown]

rigger *n.* **1** worker on an oil rig. **2** person who rigs or who arranges rigging.

rigging *n.* ship's spars, ropes, etc.

right /raɪt/ *–adj.* **1** (of conduct etc.) just, morally or socially correct (*do the right thing*). **2** true, correct (*which is the right way?*). **3** suitable or preferable (*right person for the job*). **4** sound or normal; healthy; satisfactory (*engine doesn't sound right*). **5** on or towards the east side of the human body, or of any object etc., when facing north. **6** (of a side of fabric etc.) meant for display or use. **7** *colloq.* real; complete (*made a right mess of it*). **8** (also **Right**) *Polit.* of the Right. *–n.* **1** that which is correct or just; fair treatment (often in *pl.*: *rights and wrongs of the case*). **2** justification or fair claim (*has no right to speak*). **3** legal or moral entitlement; authority to act (*human rights*; *right of reply*). **4** right-hand part, region, or direction. **5** *Boxing* **a** right hand. **b** blow with this. **6** (often **Right**) **a** conservative political group or section. **b** conservatives collectively. **7** side of a stage to the right of a person facing the audience. *–v.* **1** (often *refl.*) restore to a proper, straight, or vertical position. **2** correct or avenge (mistakes, wrongs, etc.); set in order; make reparation. *–adv.* **1** straight (*go right on*). **2** *colloq.* immediately (*do it right now*). **3 a** (foll. by *to*, *round*, *through*, etc.) all the way (*sank right to the bottom*). **b** (foll. by *off*, *out*, etc.) completely (*came right off its hinges*). **4** exactly, quite (*right in the middle*). **5** justly, properly, correctly, truly, satisfactorily (*not holding it right*; *if I remember right*). **6** on or to the right side. *–int. colloq.* expressing agreement or assent. □ **by right** (or **rights**) if right were done. **do right by** act dutifully towards (a person). **in one's own right** through one's own position or effort etc. **in the right** having justice or truth on one's side. **in one's right mind** sane. **of** (or **as of**) **right** having legal or moral etc. entitlement. **on the right side of** *colloq.* **1** in the favour of (a person etc.). **2** somewhat less than (a specified age). **put** (or **set**) **right 1** restore to order, health, etc. **2** correct the mistaken impression etc. of (a person). **put** (or **set**) **to rights** make correct or well ordered. **right away** (or **off**) immediately. **right oh!** (or **ho!**) = RIGHTO. **right on!** *slang* expression of strong approval or encouragement. **a right one** *colloq.* foolish or funny person. **right you are!** *colloq.* exclamation of assent. **too right** *slang* expression of agreement. □ **rightness** *n.* [Old English]

right angle *n.* angle of 90°.

right arm *n.* one's most reliable helper.

right bank *n.* bank of a river on the right facing downstream.

righten *v.* make right or correct.

righteous /'raɪtʃəs/ *adj.* (of a person or conduct) morally right; virtuous, law-abiding. □ **righteously** *adv.* **righteousness** *n.* [Old English]

rightful *adj.* **1 a** (of a person) legitimately entitled to (a position etc.) (*rightful heir*). **b** (of status or property etc.) that one is entitled to. **2** (of an action etc.) equitable, fair. □ **rightfully** *adv.* [Old English]

right hand *n.* = RIGHT-HAND MAN.

right-hand *attrib. adj.* **1** on or towards the right side of a person or thing. **2** done with the right hand. **3** (of a screw) = RIGHT-HANDED 4b.

right-handed *adj.* **1** naturally using the right hand for writing etc. **2** (of a tool etc.) for use by the right hand. **3** (of a blow) struck with the right hand. **4 a** turning to the right. **b** (of a screw) turned clockwise to tighten. □ **right-handedly** *adv.* **right-handedness** *n.*

right-hander *n.* **1** right-handed person. **2** right-handed blow.

right-hand man *n.* indispensable or chief assistant.

Right Honourable *n.* title given to certain high officials, e.g. Privy Counsellors.

rightism *n.* political conservatism. □ **rightist** *n.* & *adj.*

rightly *adv.* justly, properly, correctly, justifiably.

right-minded *adj.* (also **right-thinking**) having sound views and principles.

rightmost *adj.* furthest to the right.

righto /'raɪtəʊ/ *int. colloq.* expressing agreement or assent.

right of way *n.* **1** right established by usage to pass over another's ground. **2** path subject to such a right. **3** right of a vehicle to precedence.

Right Reverend *n.* bishop's title.

right turn *n.* turn of 90 degrees to the right.

rightward /ˈraɪtwəd/ *–adv.* (also **rightwards**) towards the right. *–adj.* going towards or facing the right.

right wing *–n.* **1** more conservative section of a political party or system. **2** right side of a football etc. team on the field. *–adj.* (**right-wing**) conservative or reactionary. □ **right-winger** *n.*

rigid /ˈrɪdʒɪd/ *adj.* **1** not flexible; unbendable. **2** (of a person, conduct, etc.) inflexible, unbending, harsh. □ **rigidity** /-ˈdʒɪdɪtɪ/ *n.* **rigidly** *adv.* **rigidness** *n.* [Latin *rigidus* from *rigeo* be stiff]

rigmarole /ˈrɪgmə,rəʊl/ *n.* **1** lengthy and complicated procedure. **2** rambling or meaningless talk or tale. [originally *ragman roll* catalogue]

rigor[1] /ˈrɪgə(r), ˈraɪgɔ:(r)/ *n.* feeling of cold with shivering and a rise in temperature, preceding a fever etc. [Latin *rigeo* be stiff]

rigor[2] *US* var. of RIGOUR.

rigor mortis /,rɪgə ˈmɔ:tɪs/ *n.* stiffening of the body after death.

rigorous /ˈrɪgərəs/ *adj.* **1** firm; strict, severe. **2** strictly exact or accurate. □ **rigorously** *adv.* **rigorousness** *n.* [related to RIGOUR]

rigour /ˈrɪgə(r)/ *n.* (*US* **rigor**) **1 a** severity, strictness, harshness. **b** (in *pl.*) harsh measures or conditions. **2** logical exactitude. **3** strict enforcement of rules etc. (*utmost rigour of the law*). **4** austerity of life. [Latin: related to RIGOR[1]]

rig-out *n. colloq.* outfit of clothes.

rile *v.* (**-ling**) *colloq.* anger, irritate. [French from Latin]

rill *n.* small stream. [probably Low German or Dutch]

rim *n.* **1** edge or border, esp. of something circular. **2** outer edge of a wheel, holding the tyre. **3** part of spectacle frames around the lens. □ **rimless** *adj.* **rimmed** *adj.* (also in *comb.*). [Old English]

rime[1] *–n.* **1** frost. **2** hoar-frost. *–v.* (**-ming**) cover with rime. [Old English]

rime[2] *archaic* var. of RHYME.

rind /raɪnd/ *n.* tough outer layer or covering of fruit and vegetables, cheese, bacon, etc. [Old English]

ring[1] *–n.* **1** circular band, usu. of metal, worn on a finger. **2** circular band of any material. **3** rim of a cylindrical or circular object, or a line or band round it. **4** mark etc. resembling a ring (*rings round his eyes*; *smoke rings*). **5** ring in the cross-section of a tree, produced by one year's growth. **6 a** enclosure for a circus performance, boxing, betting at races, showing of cattle, etc. **b** (prec. by *the*) bookmakers collectively. **7 a** people or things in a circle. **b** such an arrangement. **8** traders, spies, politicians, etc., combined illicitly for profit etc. **9** circular or spiral course. **10** = GAS RING. **11 a** thin disc of particles etc. round a planet. **b** halo round the moon. *–v.* **1** (often foll. by *round*, *about*, *in*) make or draw a circle round; encircle. **2** put a ring on (a bird etc.) or through the nose of (a pig, bull, etc.). □ **run** (or **make**) **rings round** *colloq.* outclass or outwit (another person). [Old English]

ring[2] *–v.* (*past* **rang**; *past part.* **rung**) **1** (often foll. by *out* etc.) give a clear resonant or vibrating sound of or as of a bell. **2 a** make (esp. a bell) ring. **b** (*absol.*) call by ringing a bell (*you rang, sir?*). **3** (also *absol.*; often foll. by *up*) call by telephone (*will ring you*). **4** (usu. foll. by *with*, *to*) (of a place) resound with a sound, fame, etc. (*theatre rang with applause*). **5** (of the ears) be filled with a sensation of ringing. **6 a** sound (a peal etc.) on bells. **b** (of a bell) sound (the hour etc.). **7** (foll. by *in*, *out*) usher in or out with bell-ringing (*rang out the Old Year*). **8** convey a specified impression (*words rang true*). *–n.* **1** ringing sound or tone. **2** act or sound of ringing a bell. **3** *colloq.* telephone call (*give me a ring*). **4** specified feeling conveyed by words etc. (*had a melancholy ring*). **5** set of esp. church bells. □ **ring back** make a return telephone call to. **ring a bell** *colloq.* begin to revive a memory. **ring down** (or **up**) **the curtain 1** cause the curtain to be lowered or raised. **2** (foll. by *on*) mark the end or the beginning of (an enterprise etc.). **ring in** report or make contact by telephone. **ring off** end a telephone call. **ring round** telephone several people. **ring up 1** call by telephone. **2** record (an amount etc.) on a cash register. [Old English]

ring-binder *n.* loose-leaf binder with ring-shaped clasps.

ring-dove *n.* woodpigeon.

ringer *n.* bell-ringer. □ **be a ringer** (or **dead ringer**) **for** *slang* resemble (a person) exactly.

ring-fence *v.* (**-cing**) protect or guarantee (funds).

ring finger *n.* third finger, esp. of the left hand, on which a wedding ring is usu. worn.

ringing tone *n.* sound heard after dialling an unengaged number.

ringleader *n.* leading instigator of a crime, mischief, etc.

ringlet /ˈrɪŋlɪt/ *n.* curly lock of esp. long hair. □ **ringleted** *adj.*

ringmaster *n.* person directing a circus performance.

ring-pull *attrib. adj.* (of a tin) having a ring for pulling to break its seal.

ring road *n.* bypass encircling a town.

ringside *n.* area immediately beside a boxing or circus ring etc. (often *attrib.*: *ringside view*).

ringworm *n.* fungal skin infection causing circular inflamed patches, esp. on the scalp.

rink *n.* **1** area of ice for skating or curling etc. **2** enclosed area for roller-skating. **3** building containing either of these. **4** strip of bowling-green. **5** team in bowls or curling. [apparently from French *renc* RANK[1]]

rinse –*v.* (**-sing**) (often foll. by *through*, *out*) **1** wash or treat with clean water etc. **2** wash lightly. **3** put (clothes etc.) through clean water after washing. **4** (foll. by *out*, *away*) clear (impurities) by rinsing. –*n.* **1** rinsing (*give it a rinse*). **2** temporary hair tint (*blue rinse*). [French *rincer*]

riot /'raɪət/ –*n.* **1 a** violent disturbance by a crowd of people. **b** (*attrib.*) involved in suppressing riots (*riot police*). **2** loud uncontrolled revelry. **3** (foll. by *of*) lavish display or sensation (*riot of colour and sound*). **4** *colloq.* very amusing thing or person. –*v.* make or engage in a riot. □ **read the Riot Act** act firmly to suppress insubordination; give warning. **run riot 1** throw off all restraint. **2** (of plants) grow or spread uncontrolled. □ **rioter** *n.* **riotous** *adj.* [French]

RIP *abbr.* may he, she, or they rest in peace. [Latin *requiesca(n)t in pace*]

rip[1] –*v.* (**-pp-**) **1** tear or cut (a thing) quickly or forcibly away or apart (*ripped out the lining*). **2 a** make (a hole etc.) by ripping. **b** make a long tear or cut in. **3** come violently apart; split. **4** rush along. –*n.* **1** long tear or cut. **2** act of ripping. □ **let rip** *colloq.* **1** (allow to) proceed or act without restraint or interference. **2** speak violently. **rip into** *colloq.* attack (a person) verbally. **rip off** *colloq.* **1** swindle. **2** steal. [origin unknown]

rip[2] *n.* stretch of rough water caused by meeting currents. [origin uncertain]

rip[3] *n.* **1** dissolute person; rascal. **2** worthless horse. [origin uncertain]

riparian /raɪ'peərɪən/ *adj.* of or on a river-bank (*riparian rights*). [Latin *ripa* bank]

rip-cord *n.* cord for releasing a parachute from its pack.

ripe *adj.* **1** (of grain, fruit, cheese, etc.) ready to be reaped, picked, or eaten. **2** mature, fully developed (*ripe in judgement*). **3** (of a person's age) advanced. **4** (often foll. by *for*) fit or ready (*ripe for development*). □ **ripeness** *n.* [Old English]

ripen *v.* make or become ripe.

rip-off *n. colloq.* swindle, financial exploitation.

riposte /rɪ'pɒst/ –*n.* **1** quick retort. **2** quick return thrust in fencing. –*v.* (**-ting**) deliver a riposte. [Italian: related to RESPOND]

ripper *n.* **1** person or thing that rips. **2** murderer who mutilates the victims' bodies.

ripple /'rɪp(ə)l/ –*n.* **1** ruffling of the water's surface, small wave or waves. **2** gentle lively sound, e.g. of laughter or applause. **3** wavy appearance in hair, material, etc. **4** slight variation in the strength of a current etc. **5** ice-cream with veins of syrup (*raspberry ripple*). –*v.* (**-ling**) **1** (cause to) form or flow in ripples. **2** show or sound like ripples. □ **ripply** *adj.* [origin unknown]

rip-roaring *adj.* **1** wildly noisy or boisterous. **2** excellent, first-rate.

ripsaw *n.* coarse saw for sawing wood along the grain.

rise /raɪz/ –*v.* (**-sing**; *past* **rose** /rəʊz/; *past part.* **risen** /'rɪz(ə)n/) **1** come or go up. **2** grow, project, expand, or incline upwards; become higher. **3** appear or be visible above the horizon. **4** get up from lying, sitting, kneeling, or from bed; become erect. **5** (of a meeting etc.) adjourn. **6** reach a higher position, level, amount, intensity, etc. **7** make progress socially etc. (*rose from the ranks*). **8 a** come to the surface of liquid. **b** (of a person) react to provocation (*rise to the bait*). **9** come to life again. **10** (of dough) swell by the action of yeast etc. **11** (often foll. by *up*) rebel (*rise up against them*). **12** originate (*river rises in the mountains*). **13** (of wind) start to blow. **14** (of a person's spirits) become cheerful. –*n.* **1** rising. **2** upward slope, hill, or movement (*house stood on a rise*). **3 a** increase in amount, extent, sound, pitch, etc. (*rise in unemployment*). **b** increase in salary. **4** increase in status or power; upward progress. **5** movement of fish to the surface. **6** origin. **7 a** vertical height of a step, arch, incline, etc. **b** = RISER 2. □ **get** (or **take**) **a rise out of** *colloq.* provoke a reaction from (a person), esp. by teasing. **on the rise** on the increase. **rise above** be superior to (petty feelings, difficulties, etc.). **rise to** develop powers equal to (an occasion). [Old English]

riser *n.* **1** person who rises from bed (*early riser*). **2** vertical section between the treads of a staircase.

risible /'rɪzɪb(ə)l/ *adj.* laughable, ludicrous. [Latin *rideo ris-* laugh]

rising –*adj.* **1** advancing to maturity or high standing (*rising young lawyer*). **2** approaching a specified age (*rising* *five*). **3** (of ground) sloping upwards. *–n.* revolt or insurrection.

rising damp *n.* moisture absorbed from the ground into a wall.

risk *–n.* **1** chance or possibility of danger, loss, injury, etc. (*health risk*; *risk of fire*). **2** person or thing causing a risk or regarded in relation to risk (*is a poor risk*). *–v.* **1** expose to risk. **2** accept the chance of (*risk getting wet*). **3** venture on. □ **at risk** exposed to danger. **at one's (own) risk** accepting responsibility, agreeing to make no claims. **at the risk of** with the possibility of (an adverse consequence). **put at risk** expose to danger. **run a** (or **the**) **risk** (often foll. by *of*) expose oneself to danger or loss etc. **take a risk** (or **risks**) chance the possibility of danger etc. [French *risque(r)* from Italian]

risky *adj.* (**-ier**, **-iest**) **1** involving risk. **2** = RISQUÉ. □ **riskily** *adv.* **riskiness** *n.*

risotto /rɪ'zɒtəʊ/ *n.* (*pl.* **-s**) Italian savoury rice dish cooked in stock. [Italian]

risqué /'rɪskeɪ, -'keɪ/ *adj.* (of a story etc.) slightly indecent. [French: related to RISK]

rissole /'rɪsəʊl/ *n.* cake of spiced minced meat, coated in breadcrumbs and fried. [French]

rit. *abbr. Mus.* ritardando.

ritardando /ˌrɪtɑː'dændəʊ/ *adv.* & *n.* (*pl.* **-s** or **-di** /-dɪ/) *Mus.* = RALLENTANDO. [Italian]

rite *n.* **1** religious or solemn observance, act, or procedure (*burial rites*). **2** body of customary observances characteristic of a Church etc. (*Latin rite*). [Latin *ritus*]

rite of passage *n.* (often in *pl.*) event marking a change or stage in life, e.g. marriage.

ritual /'rɪtʃʊəl/ *–n.* **1 a** prescribed order of a ceremony etc. **b** solemn or colourful pageantry etc. **2** procedure regularly followed. *–adj.* of or done as a ritual or rite (*ritual murder*). □ **ritually** *adv.* [Latin: related to RITE]

ritualism *n.* regular or excessive practice of ritual. □ **ritualist** *n.* **ritualistic** /-'lɪstɪk/ *adj.* **ritualistically** /-'lɪstɪkəlɪ/ *adv.*

ritzy /'rɪtzɪ/ *adj.* (**-ier**, **-iest**) *colloq.* high-class, luxurious, showily smart. [from *Ritz*, name of luxury hotels]

rival /'raɪv(ə)l/ *–n.* (often *attrib.*) **1** person competing with another. **2** person or thing that equals another in quality. *–v.* (**-ll-**; *US* **-l-**) be, seem, or claim to be the rival of or comparable to. [Latin *rivus* stream]

rivalry *n.* (*pl.* **-ies**) being rivals; competition.

riven /'rɪv(ə)n/ *adj. literary* split, torn. [past part. of *rive* from Old Norse]

river /'rɪvə(r)/ *n.* **1** copious natural stream of water flowing to the sea or a lake etc. **2** copious flow (*rivers of blood*). □ **sell down the river** *colloq.* betray or let down. [Latin *ripa* bank]

riverside *n.* (often *attrib.*) ground along a river-bank.

rivet /'rɪvɪt/ *–n.* nail or bolt for joining metal plates etc., with the headless end beaten out when in place. *–v.* (**-t-**) **1 a** join or fasten with rivets. **b** beat out or press down the end of (a nail or bolt). **c** fix, make immovable. **2 a** (foll. by *on*, *upon*) direct intently (one's eyes or attention etc.). **b** (esp. as **riveting** *adj.*) engross (a person or the attention). [French *river* fasten]

riviera /ˌrɪvɪ'eərə/ *n.* coastal subtropical region, esp. that of SE France and NW Italy. [Italian, = sea-shore]

rivulet /'rɪvjʊlɪt/ *n.* small stream. [Latin *rivus* stream]

RM *abbr.* Royal Marines.

rm. *abbr.* room.

RMA *abbr.* Royal Military Academy.

RN *abbr.* Royal Navy.

Rn *symb.* radon.

RNA *abbr.* ribonucleic acid.

RNLI *abbr.* Royal National Lifeboat Institution.

roach *n.* (*pl.* same or **-es**) small freshwater fish of the carp family. [French]

road *n.* **1 a** way with a prepared surface, for vehicles, pedestrians, etc. **b** part of this for vehicles only (*step out into the road*). **2** one's way or route. **3** (usu. in *pl.*) piece of water near the shore in which ships can ride at anchor. □ **any road** *dial.* = ANYWAY 2, 3. **get out of the** (or **my** etc.) **road** *dial.* stop obstructing a person. **in the** (or **one's**) **road** *dial.* forming an obstruction. **one for the road** *colloq.* final (esp. alcoholic) drink before departure. **on the road** travelling, esp. as a firm's representative, itinerant performer, or vagrant. **the road to** way of getting to or achieving (*road to London*; *road to ruin*). [Old English: related to RIDE]

roadbed *n.* **1** foundation structure of a railway. **2** foundation material for a road. **3** *US* part of a road on which vehicles travel.

roadblock *n.* barrier set up on a road in order to stop and examine traffic.

road fund licence *n.* disc displayed on a vehicle certifying payment of road tax.

road-hog *n. colloq.* reckless or inconsiderate road-user.

road-holding *n.* stability of a moving vehicle.

road-house *n.* inn or club on a major road.

roadie *n. colloq.* assistant of a touring band etc., erecting and maintaining equipment.

road-metal *n.* broken stone used in road-making etc.

road sense *n.* capacity for safe behaviour in traffic etc.

roadshow *n.* **1** television or radio series broadcasting each programme from a different venue. **2** any touring political or advertising campaign or touring entertainment.

roadside *n.* (often *attrib.*) strip of land beside a road.

road sign *n.* sign giving information or instructions to road users.

roadstead *n.* = ROAD 3.

roadster *n.* open car without rear seats.

road tax *n.* periodic tax payable on road vehicles.

road test –*n.* test of a vehicle's roadworthiness. –*v.* (**road-test**) test (a vehicle) on the road.

roadway *n.* **1** road. **2** part of a road intended for vehicles.

roadworks *n.pl.* construction, repair, etc. of roads.

roadworthy *adj.* fit to be used on the road. □ **roadworthiness** *n.*

roam –*v.* **1** ramble, wander. **2** travel unsystematically over, through, or about. –*n.* act of roaming; ramble. □ **roamer** *n.* [origin unknown]

roan –*adj.* (of esp. a horse) having a coat thickly interspersed with hairs of another colour. –*n.* roan animal. [French]

roar –*n.* **1 a** loud deep hoarse sound, as made by a lion. **b** similar sound. **2** loud laugh. –*v.* **1** (often foll. by *out*) utter loudly or make a roar, roaring laugh, etc. **2** travel in a vehicle at high speed, esp. with the engine roaring. [Old English]

roaring drunk *predic. adj.* very drunk and noisy.

roaring forties *n.pl.* stormy ocean tracts between lat. 40° and 50° S.

roaring success *n.* great success.

roaring trade *n.* (also **roaring business**) very brisk trade or business.

roaring twenties *n.pl.* decade of the 1920s.

roast –*v.* **1 a** cook (food, esp. meat) or (of food) be cooked in an oven or by open heat (*roast chestnuts*). **b** heat (coffee beans) before grinding. **2** *refl.* expose (oneself etc.) to fire or heat. **3** criticize severely, denounce. –*attrib. adj.* roasted (*roast beef*). –*n.* **1 a** roast meat. **b** dish of this. **c** piece of meat for roasting. **2** process of roasting. [French *rost(ir)* from Germanic]

roaster *n.* **1** oven, dish, apparatus, etc. for roasting. **2** fowl, potato, etc. for roasting.

roasting –*adj.* very hot. –*n.* severe criticism or denunciation.

rob *v.* (**-bb-**) (often foll. by *of*) **1** (also *absol.*) take unlawfully from, esp. by force or threat (*robbed the safe*; *robbed her of her jewels*). **2** deprive of what is due or normal (*robbed of sleep*). □ **robber** *n.* [French *rob(b)er* from Germanic]

robbery *n.* (*pl.* **-ies**) **1** act of robbing. **2** *colloq.* excessive charge or cost.

robe –*n.* **1 a** long loose outer garment. **b** (often in *pl.*) this worn as an indication of rank, office, profession, etc. **2** esp. *US* dressing-gown. –*v.* (**-bing**) clothe in a robe; dress. [French]

robin /'rɒbɪn/ *n.* **1** (also **robin redbreast**) small brown red-breasted bird. **2** *US* red-breasted thrush. [pet form of *Robert*]

Robin Hood *n.* person who steals from the rich to give to the poor.

robinia /rə'bɪnɪə/ *n.* any of various N. American trees or shrubs, e.g. a locust tree or false acacia. [*Robin*, name of a French gardener]

robot /'rəʊbɒt/ *n.* **1** machine resembling or functioning like a human. **2** machine automatically completing a mechanical process. **3** person who acts mechanically. □ **robotic** /-'bɒtɪk/ *adj.* **robotize** *v.* (also **-ise**) (**-zing** or **-sing**). [Czech]

robotics /rəʊ'bɒtɪks/ *n.pl.* (usu. treated as *sing.*) art, science, or study of robot design and operation.

robust /rəʊ'bʌst/ *adj.* (**-er**, **-est**) **1** strong and sturdy, esp. in physique or construction. **2** (of exercise, discipline, etc.) vigorous, requiring strength. **3** (of mental attitude, argument, etc.) straightforward, vigorous. **4** (of a statement, reply, etc.) bold, firm, unyielding. □ **robustly** *adv.* **robustness** *n.* [Latin *robur* strength]

roc *n.* gigantic bird of Eastern legend. [Spanish from Arabic]

rochet /'rɒtʃɪt/ *n.* surplice-like vestment of a bishop or abbot. [French from Germanic]

rock[1] *n.* **1 a** hard material of the earth's crust, often exposed on the surface. **b** similar material on other planets. **2** *Geol.* any natural material, hard or soft (e.g. clay), consisting of one or more minerals. **3 a** projecting rock forming a hill, cliff, reef, etc. **b** (**the Rock**) Gibraltar. **4** large detached stone. **5** *US* stone of any size. **6** firm and dependable support or protection. **7** hard sweet usu. in the form of a peppermint-flavoured stick. **8** *slang* precious stone, esp. a diamond. □ **get one's rocks off** *coarse slang* achieve (esp. sexual) satisfaction. **on the rocks** *colloq.* **1** short of money. **2** (of a marriage etc.) broken down. **3** (of a drink) served neat with ice-cubes. [French *roque*, *roche*]

rock[2] *–v.* **1** move gently to and fro; set, maintain, or be in, such motion. **2** (cause to) sway; shake, oscillate, reel. **3** distress, perturb (*rocked by the news*). *–n.* **1** rocking movement. **2** spell of this. **3 a** = ROCK AND ROLL. **b** rock and roll-influenced popular music. □ **rock the boat** *colloq.* disturb a stable situation. [Old English]

rockabilly /ˈrɒkəˌbɪlɪ/ *n.* rock and roll combined with hill-billy music.

rock and roll *n.* (also **rock 'n' roll**) popular dance-music originating in the 1950s with a heavy beat and often a blues element.

rock-bottom *–adj.* (of prices etc.) the very lowest. *–n.* very lowest level.

rock-cake *n.* small rough-surfaced spicy currant bun.

rock-crystal *n.* transparent colourless quartz, usu. in hexagonal prisms.

rocker *n.* **1** curved bar etc. on which something can rock. **2** rocking-chair. **3** devotee of rock music, esp. a leather-clad motor cyclist. **4 a** device for rocking. **b** pivoted switch operating between 'on' and 'off' positions. □ **off one's rocker** *slang* crazy.

rockery *n.* (*pl.* **-ies**) construction of stones with soil between them for growing rock-plants on.

rocket /ˈrɒkɪt/ *–n.* **1** cylindrical firework or signal etc. propelled to a great height after ignition. **2** engine operating on the same principle, providing thrust but not dependent on air intake. **3** rocket-propelled missile, spacecraft, etc. **4** *slang* severe reprimand. *–v.* (**-t-**) **1 a** move rapidly upwards or away. **b** increase rapidly (*prices rocketed*). **2** bombard with rockets. [French *roquette* from Italian]

rocketry *n.* science or practice of rocket propulsion.

rock-face *n.* vertical surface of natural rock.

rockfall *n.* descent or mass of loose fallen rocks.

rock-garden *n.* = ROCKERY.

rocking-chair *n.* chair mounted on rockers or springs for gently rocking in.

rocking-horse *n.* toy horse on rockers or springs.

rock-plant *n.* plant growing on or among rocks.

rock-salmon *n.* any of several fishes, esp. the catfish and dogfish.

rock-salt *n.* common salt as a solid mineral.

rocky[1] *adj.* (**-ier, -iest**) of, like, or full of rock or rocks. □ **rockiness** *n.*

rocky[2] *adj.* (**-ier, -iest**) *colloq.* unsteady, tottering, unstable. □ **rockiness** *n.*

rococo /rəˈkəʊkəʊ/ *–adj.* **1** of a late baroque style of 18th-c. decoration. **2** (of literature, music, architecture, etc.) highly ornate. *–n.* this style. [French]

rod *n.* **1** slender straight cylindrical bar or stick. **2 a** cane for flogging. **b** (prec. by *the*) use of this. **3** = FISHING-ROD. **4** *hist.* (as a measure) perch or square perch (see PERCH[1]). □ **make a rod for one's own back** make trouble for oneself. [Old English]

rode *past* of RIDE.

rodent /ˈrəʊd(ə)nt/ *n.* mammal with strong incisors and no canine teeth, e.g. the rat, mouse, squirrel, beaver, and porcupine. [Latin *rodo* gnaw]

rodeo /ˈrəʊdɪəʊ, rəˈdeɪəʊ/ *n.* (*pl.* **-s**) **1** exhibition of cowboys' skills in handling animals. **2** round-up of cattle on a ranch for branding etc. [Spanish]

rodomontade /ˌrɒdəmɒnˈteɪd/ *n.* boastful talk or behaviour. [French from Italian]

roe[1] /rəʊ/ *n.* **1** (also **hard roe**) mass of eggs in a female fish's ovary. **2** (also **soft roe**) milt of a male fish. [Low German or Dutch]

roe[2] /rəʊ/ *n.* (*pl.* same or **-s**) (also **roe-deer**) small kind of deer. [Old English]

roebuck *n.* male roe-deer.

roentgen /ˈrʌntjən/ *n.* (also **röntgen**) unit of ionizing radiation. [*Röntgen*, name of a physicist]

rogation /rəʊˈgeɪʃ(ə)n/ *n.* (usu. in *pl.*) litany of the saints chanted on the three days before Ascension day. [Latin *rogo* ask]

Rogation Days *n.pl.* the three days before Ascension Day.

roger /ˈrɒdʒə(r)/ *int.* **1** your message has been received and understood (used in radio communication etc.). **2** *slang* I agree. [from the name, code for *R*]

rogue /rəʊg/ *n.* **1** dishonest or unprincipled person. **2** *joc.* mischievous person, esp. a child. **3** (usu. *attrib.*) wild fierce animal driven away or living apart from others (*rogue elephant*). **4** (often *attrib.*) inexplicably aberrant result or phenomenon; inferior or defective specimen. [origin unknown]

roguery /ˈrəʊgərɪ/ *n.* (*pl.* **-ies**) conduct or action characteristic of rogues.

rogues' gallery *n. colloq.* collection of photographs of known criminals etc., used for identification.

roguish *adj.* **1** playfully mischievous. **2** characteristic of rogues. □ **roguishly** *adv.* **roguishness** *n.*

roister *v.* (esp. as **roistering** *adj.*) revel noisily; be uproarious. □ **roisterer** *n.* [Latin: related to RUSTIC]

role *n.* (also **rôle**) **1** actor's part in a play, film, etc. **2** person's or thing's function. [French: related to ROLL]

role model *n.* person on whom others model themselves.

role-playing *n.* (also **role-play**) acting of characters or situations as an aid in psychotherapy, language-teaching, etc. □ **role-play** *v.*

roll /rəʊl/ –*v.* **1** (cause to) move or go in some direction by turning on an axis (*ball rolled under the table*; *rolled the barrel into the cellar*). **2 a** make cylindrical or spherical by revolving between two surfaces or over on itself (*rolled a newspaper*). **b** make thus (*rolled a cigarette*). **c** gather into a mass or shape (*rolled the dough into a ball*; *rolled himself into a ball*). **3** (often foll. by *along*, *by*, etc.) (cause to) move, advance, or be conveyed on or (of time etc.) as if on wheels etc. (*bus rolled past*; *rolled the tea trolley*; *years rolled by*; *rolled by in his car*). **4** flatten or form by passing a roller etc. over or by passing between rollers (*roll the lawn*; *roll pastry*). **5** rotate (*his eyes rolled*; *he rolled his eyes*). **6 a** wallow (*dog rolled in the dust*). **b** (of a horse etc.) lie on its back and kick about. **7** (of a moving ship, aircraft, vehicle, or person) sway to and fro sideways or walk unsteadily (*rolled out of the pub*). **8 a** undulate (*rolling hills*; *rolling mist*). **b** carry or propel with undulations (*river rolls its waters to the sea*). **9** (cause to) start functioning or moving (*cameras rolled*). **10** sound or utter with vibrations or a trill (*thunder rolled; rolls his* rs). –*n.* **1** rolling motion or gait; undulation (*roll of the hills*). **2 a** spell of rolling (*roll in the mud*). **b** gymnastic exercise in which the body rolls in a forward or backward circle. **c** (esp. **a roll in the hay**) *colloq.* sexual intercourse etc. **3** rhythmic rumbling sound of thunder etc. **4** complete revolution of an aircraft about its longitudinal axis. **5** anything forming a cylinder by being turned over on itself without folding (*roll of carpet*; *sausage roll*). **6 a** small portion of bread individually baked. **b** this with a specified filling (*ham roll*). **7** thing cylindrical in shape (*rolls of fat*; *roll of hair*). **8 a** official list or register (*electoral roll*). **b** total numbers on this. □ **be rolling in** *colloq.* have plenty of (esp. money). **rolled into one** combined in one person or thing. **roll in** arrive in great numbers or quantity. **roll on** *v.* **1** put on or apply by rolling. **2** (in *imper.*) *colloq.* come quickly (*roll on Friday!*). **roll up 1** *colloq.* arrive in a vehicle; appear on the scene. **2** make into or form a roll. **strike off the rolls** debar (esp. a solicitor) from practising. [Latin *rotulus* diminutive: related to ROTA]

roll-call *n.* calling out a list of names to establish who is present.

rolled gold *n.* thin coating of gold applied to a base metal by rolling.

rolled oats *n.pl.* husked and crushed oats.

roller *n.* **1 a** revolving cylinder for smoothing, spreading, crushing, stamping, hanging a towel on, etc., used alone or in a machine. **b** cylinder for diminishing friction when moving a heavy object. **2** small cylinder on which hair is rolled for setting. **3** long swelling wave.

roller bearing *n.* bearing like a ball-bearing but with small cylinders instead of balls.

roller blind *n.* blind on a roller.

roller-coaster *n.* **1** switchback at a fair etc. **2** (*attrib.*) (of emotions etc.) uncontrollable, unstable.

roller-skate –*n.* metal frame with small wheels, fitted to shoes for riding on a hard surface. –*v.* (**-ting**) move on roller-skates. □ **roller-skater** *n.*

roller towel *n.* towel with the ends joined, hung on a roller.

rollicking /'rɒlɪkɪŋ/ *adj.* jovial, exuberant. [origin unknown]

rolling drunk *predic. adj.* swaying or staggering from drunkenness.

rolling-mill *n.* machine or factory for rolling metal into shape.

rolling-pin *n.* cylinder for rolling out pastry, dough, etc.

rolling-stock *n.* **1** locomotives, carriages, etc. used on a railway. **2** *US* road vehicles of a company.

rolling stone *n.* unsettled rootless person.

rollmop *n.* rolled uncooked pickled herring fillet. [German *Rollmops*]

roll-neck *adj.* (of a garment) having a high loosely turned-over neck.

roll of honour *n.* list of those honoured, esp. the dead in war.

roll-on –*attrib. adj.* (of deodorant etc.) applied by means of a rotating ball in the neck of the container. –*n.* light elastic corset.

roll-on roll-off *adj.* (of a ship, etc.) in which vehicles are driven directly on and off.

roll-top desk *n.* desk with a flexible cover sliding in curved grooves.

roll-up *n.* (also **roll-your-own**) hand-rolled cigarette.

roly-poly /ˌrəʊlɪ'pəʊlɪ/ *–n.* (*pl.* **-ies**) (also **roly-poly pudding**) pudding made of a rolled strip of suet pastry covered with jam etc. and boiled or baked. *–adj.* podgy, plump. [probably ROLL]

ROM *n. Computing* read-only memory. [abbreviation]

rom. *abbr.* roman (type).

Roman /'rəʊmən/ *–adj.* **1** of ancient Rome, its territory, people, etc. **2** of medieval or modern Rome. **3** = ROMAN CATHOLIC. **4** (**roman**) (of type) plain and upright, used in ordinary print. **5** (of the alphabet etc.) based on the ancient Roman system with letters A–Z. *–n.* **1** citizen or soldier of the ancient Roman Republic or Empire. **2** citizen of modern Rome. **3** = ROMAN CATHOLIC. **4** (**roman**) roman type. [Latin]

Roman candle *n.* firework discharging flaming coloured balls.

Roman Catholic *–adj.* of the part of the Christian Church acknowledging the Pope as its head. *–n.* member of this Church. □ **Roman Catholicism** *n.*

romance /rəʊ'mæns/ *–n.* /also 'rəʊ-/ **1** idealized, poetic, or unworldly atmosphere or tendency. **2 a** love affair. **b** mutual attraction in this. **c** sentimental or idealized love. **3 a** literary genre concerning romantic love, stirring action, etc. **b** work of this genre. **4** medieval, esp. verse, tale of chivalry, common in the Romance languages. **5 a** exaggeration, lies. **b** instance of this. **6** (**Romance**) (often *attrib.*) languages descended from Latin. **7** *Mus.* short informal piece. *–v.* (**-cing**) **1** exaggerate, distort the truth, fantasize. **2** court, woo. [Romanic: related to ROMANIC]

■ **Usage** The alternative pronunciation given for the noun, with the stress on the first syllable, is considered incorrect by some people.

Roman Empire *n. hist.* that established by Augustus in 27 BC and divided by Theodosius in AD 395.

Romanesque /ˌrəʊmə'nesk/ *–n.* style of European architecture *c.* 900–1200, with massive vaulting and round arches. *–adj.* of this style.

Romanian /rəʊ'meɪnɪən/ (also **Rumanian** /ru:-/) *–n.* **1 a** native or national of Romania. **b** person of Romanian descent. **2** language of Romania. *–adj.* of Romania, its people, or language.

Romanic /rəʊ'mænɪk/ *–n.* = ROMANCE *n.* 6. *–adj.* **1 a** of Romance. **b** Romance-speaking. **2** descended from, or inheriting the civilization etc. of, the ancient Romans. [Latin *Romanicus*: related to ROMAN]

romanize /'rəʊməˌnaɪz/ *v.* (also **-ise**) (**-zing** or **-sing**) **1** make Roman or Roman Catholic in character. **2** put into the Roman alphabet or roman type. □ **romanization** /-'zeɪʃ(ə)n/ *n.*

Roman law *n.* law-code of ancient Rome, forming the basis of many modern codes.

Roman nose *n.* aquiline high-bridged nose.

roman numeral *n.* any of the Roman letters representing numbers: I = 1, V = 5, X = 10, L = 50, C = 100, D = 500, M = 1000.

Romano- *comb. form* Roman; Roman and (*Romano-British*).

romantic /rəʊ'mæntɪk/ *–adj.* **1** of, characterized by, or suggestive of romance (*romantic picture*). **2** inclined towards or suggestive of romance in love (*romantic evening*; *romantic words*). **3** (of a person) imaginative, visionary, idealistic. **4 a** (of style in art, music, etc.) concerned more with feeling and emotion than with form and aesthetic qualities. **b** (also **Romantic**) of the 18th–19th-c. romantic movement or style in the European arts. **5** (of a project etc.) unpractical, fantastic. *–n.* **1** romantic person. **2** romanticist. □ **romantically** *adv.* [French: related to ROMANCE]

romanticism *n.* (also **Romanticism**) adherence to a romantic style in art, music, etc.

romanticist *n.* (also **Romanticist**) writer or artist of the romantic school.

romanticize *v.* (also **-ise**) (**-zing** or **-sing**) **1** make romantic; exaggerate (*romanticized account*). **2** indulge in romantic thoughts or actions.

Romany /'rɒmənɪ/ *–n.* (*pl.* **-ies**) **1** Gypsy. **2** language of the Gypsies. *–adj.* of Gypsies or the Romany language. [Romany *Rom* Gypsy]

Romeo /'rəʊmɪəʊ/ *n.* (*pl.* **-s**) passionate male lover or seducer. [name of a character in Shakespeare]

romp *–v.* **1** play roughly and energetically. **2** (foll. by *along*, *past*, etc.) *colloq.* proceed without effort. *–n.* spell of romping. □ **romp in** (or **home**) *colloq.* win easily. [perhaps from RAMP]

rompers *n. pl.* (also **romper suit**) young child's one-piece garment covering the trunk and usu. the legs.

rondeau /'rɒndəʊ/ *n.* (*pl.* **rondeaux** pronunc. same or /-əʊz/) poem of ten or thirteen lines with only two rhymes throughout and with the opening words used twice as a refrain. [French: related to RONDEL]

rondel /'rɒnd(ə)l/ *n.* rondeau, esp. one of special form. [French: related to ROUND: cf. ROUNDEL]

rondo /'rɒndəʊ/ *n.* (*pl.* **-s**) musical form with a recurring leading theme. [French RONDEAU]

röntgen var. of ROENTGEN.

rood *n.* **1** crucifix, esp. one raised on a rood-screen. **2** quarter of an acre. [Old English]

rood-screen *n.* carved screen separating nave and chancel.

roof *–n.* (*pl.* **-s**) **1 a** upper covering of a building. **b** top of a covered vehicle. **c** top inner surface of an oven, refrigerator, etc. **2** overhead rock in a cave or mine etc. *–v.* **1** (often foll. by *in*, *over*) cover with or as with a roof. **2** be the roof of. □ **go through the roof** *colloq.* (of prices etc.) rise dramatically. **hit** (or **go through**) **the roof** *colloq.* become very angry. [Old English]

roof-garden *n.* garden on the flat roof of a building.

roofing *n.* material for a roof.

roof of the mouth *n.* palate.

roof-rack *n.* framework for luggage on top of a vehicle.

rooftop *n.* **1** outer surface of a roof. **2** (in *pl.*) tops of houses etc. □ **shout it from the rooftops** make a thing embarrassingly public.

roof-tree *n.* ridge-piece of a roof.

rook[1] /rʊk/ *–n.* black bird of the crow family nesting in colonies. *–v.* **1** *colloq.* charge (a customer) extortionately. **2** win money at cards etc., esp. by swindling. [Old English]

rook[2] /rʊk/ *n.* chess piece with a battlement-shaped top. [French from Arabic]

rookery *n.* (*pl.* **-ies**) colony of rooks, penguins, or seals.

rookie /'rʊkɪ/ *n. slang* new recruit. [corruption of *recruit*]

room /ru:m/ *–n.* **1** space for, or occupied by, something; capacity (*takes up too much room*; *room for improvement*). **2 a** part of a building enclosed by walls, floor, and ceiling. **b** (in *pl.*) apartments or lodgings. **c** people in a room (*room fell silent*). *–v. US* have room(s); lodge, board. [Old English]

rooming-house *n.* lodging house.

room-mate *n.* person sharing a room.

room service *n.* provision of food etc. in a hotel bedroom.

roomy *adj.* (**-ier**, **-iest**) having much room, spacious. □ **roominess** *n.*

roost *–n.* branch or perch for a bird, esp. to sleep. *–v.* settle for rest or sleep. □ **come home to roost** (of a scheme etc.) recoil unfavourably. [Old English *hrōst*]

rooster *n.* domestic cock.

root[1] *–n.* **1 a** part of a plant normally below the ground, conveying nourishment from the soil. **b** (in *pl.*) branches or fibres of this. **c** small plant with a root for transplanting. **2 a** plant with an edible root. **b** such a root. **3** (in *pl.*) emotional attachment or family ties to a place or community. **4 a** embedded part of a hair, tooth, nail, etc. **b** part of a thing attaching it to a greater whole. **5** (often *attrib.*) basic cause, source, nature, or origin (*root of all evil*; *roots in the distant past*; *root cause*; *the root of things*). **6 a** number that when multiplied by itself a usu. specified number of times gives a specified number or quantity (*cube root of eight is two*). **b** square root. **c** value of an unknown quantity satisfying a given equation. **7** core of a word, without prefixes, suffixes, etc. *–v.* **1** (cause to) take root; grow roots (*root them firmly*). **2** (esp. as **rooted** *adj.*) fix firmly; establish (*rooted objection to*; *reaction rooted in fear*). **3** (usu. foll. by *out*, *up*) drag or dig up by the roots. □ **root and branch** thorough(ly), radical(ly). **root out** find and get rid of. **strike** (or **take**) **root 1** begin to grow and draw nourishment from the soil. **2** become established. □ **rootless** *adj.* [Old English]

root[2] *v.* **1** (also *absol.*) (often foll. by *up*) turn up (the ground) with the snout, beak, etc., in search of food. **2 a** (foll. by *around*, *in*, etc.) rummage. **b** (foll. by *out* or *up*) find or extract by rummaging. **3** (foll. by *for*) *US slang* encourage by applause or support. [Old English and Old Norse]

rootstock *n.* **1** rhizome. **2** plant into which a graft is inserted. **3** primary form from which offshoots have arisen.

rope *–n.* **1 a** stout cord made by twisting together strands of hemp, wire, etc. **b** piece of this. **2** (foll. by *of*) quantity of onions, pearls, etc. strung together. **3** (prec. by *the*) **a** halter for hanging a person. **b** execution by hanging. *–v.* (**-ping**) **1** fasten, secure, or catch with rope. **2** (usu. foll. by *off*, *in*) enclose with rope. **3** *Mountaineering* connect with or attach to a rope. □ **know** (or **learn** or **show**) **the ropes** know (or learn or show) how to do a thing properly. **rope in** persuade to take part. **rope into** persuade to take part in (*roped into washing up*). [Old English]

rope-ladder *n.* two ropes with cross-pieces, used as a ladder.

ropy *adj.* (also **ropey**) (**-ier**, **-iest**) *colloq.* poor in quality. □ **ropiness** *n.*

Roquefort /'rɒkfɔ:(r)/ *n. propr.* soft blue cheese made from ewes' milk. [*Roquefort* in France]

ro-ro /'rəʊrəʊ/ *attrib. adj.* roll-on roll-off. [abbreviation]

rorqual /'rɔ:kw(ə)l/ *n.* whale with a dorsal fin. [French from Norwegian]

Rorschach test /'rɔ:ʃɑ:k/ *n.* personality test based on the subject's interpretation of a standard set of ink-blots. [*Rorschach*, name of a psychiatrist]

rosaceous /rəʊ'zeɪʃəs/ *adj.* of a large plant family including the rose. [Latin: related to ROSE[1]]

rosary /'rəʊzərɪ/ *n.* (*pl.* **-ies**) **1** *RC Ch.* repeated sequence of prayers. **2** string of beads for keeping count in this. [Latin *rosarium* rose-garden]

rose[1] /rəʊz/ *–n.* **1** prickly bush or shrub bearing usu. fragrant red, pink, yellow, or white flowers. **2** this flower. **3** flowering plant resembling this (*Christmas rose*). **4 a** pinkish-red colour. **b** (usu. in *pl.*) rosy complexion (*roses in her cheeks*). **5** sprinkling-nozzle of a watering-can etc. **6** circular electric light mounting on a ceiling. **7 a** representation of a rose in heraldry etc. **b** rose-shaped design. **8** (in *pl.*) used to express luck, ease, success, etc. (*roses all the way*; *everything's roses*). *–adj.* = ROSE-COLOURED 1. [Latin *rosa*]

rose[2] *past* of RISE.

rosé /'rəʊzeɪ/ *n.* light pink wine. [French]

rosebowl *n.* bowl for cut roses, esp. as a prize in a competition.

rosebud *n.* **1** bud of a rose. **2** pretty young woman.

rose-bush *n.* rose plant.

rose-coloured *adj.* **1** pinkish-red. **2** optimistic, cheerful (*wears rose-coloured glasses*).

rose-hip *n.* = HIP[2].

rosemary /'rəʊzmərɪ/ *n.* evergreen fragrant shrub used as a herb. [*rosmarine* from Latin *ros* dew: related to MARINE]

rosette /rəʊ'zet/ *n.* **1** rose-shaped ornament of ribbon etc., esp. as a supporter's badge or as a prize in a competition. **2** rose-shaped carving. [French diminutive: related to ROSE[1]]

rose-water *n.* perfume made from roses.

rose-window *n.* circular window with roselike tracery.

rosewood *n.* any of several fragrant close-grained woods used in making furniture.

rosin /'rɒzɪn/ *–n.* resin, esp. in solid form. *–v.* (**-n-**) rub (esp. a violin bow etc.) with rosin. [alteration of RESIN]

RoSPA /'rɒspə/ *abbr.* Royal Society for the Prevention of Accidents.

roster /'rɒstə(r)/ *–n.* list or plan of turns of duty etc. *–v.* place on a roster. [Dutch *rooster*, literally 'gridiron']

rostrum /'rɒstrəm/ *n.* (*pl.* **rostra** or **-s**) platform for public speaking, an orchestral conductor, etc. [Latin]

rosy /'rəʊzɪ/ *adj.* (**-ier**, **-iest**) **1** pink or red. **2** optimistic, hopeful (*rosy future*). □ **rosily** *adv.* **rosiness** *n.*

rot *–v.* (**-tt-**) **1** (of animal or vegetable matter) lose its original form by the chemical action of bacteria, fungi, etc.; decay. **2** gradually perish or waste away (*left to rot in prison*). **3** cause to rot, make rotten. *–n.* **1** rotting; decay. **2** *slang* nonsense (*talks rot*). **3** decline in standards etc. (*rot set in*). *–int.* expressing incredulity or ridicule. [Old English]

rota /'rəʊtə/ *n.* list of duties to be done or names of people to do them in turn; roster. [Latin, = wheel]

Rotarian /rəʊ'teərɪən/ *–n.* member of Rotary. *–adj.* of Rotary.

rotary /'rəʊtərɪ/ *–adj.* acting by rotation (*rotary drill*). *–n.* (*pl.* **-ies**) **1** rotary machine. **2** (**Rotary**) (in full **Rotary International**) worldwide charitable society of businessmen, orig. entertaining in rotation. [medieval Latin: related to ROTA]

Rotary Club *n.* local branch of Rotary.

rotate /rəʊ'teɪt/ *v.* (**-ting**) **1** move round an axis or centre, revolve. **2** take or arrange (esp. crops) in rotation. **3** act or take place in rotation (*chairmanship will rotate*). □ **rotatable** *adj.* **rotatory** /'rəʊtətərɪ, -'teɪtərɪ/ *adj.* [Latin: related to ROTA]

rotation /rəʊ'teɪʃ(ə)n/ *n.* **1** rotating or being rotated. **2** recurrence; recurrent series or period; regular succession. **3** the growing of different crops in regular order to avoid exhausting the soil. □ **rotational** *adj.*

Rotavator /'rəʊtə,veɪtə(r)/ *n.* (also **Rotovator**) *propr.* machine with a rotating blade for breaking up or tilling the soil. [from ROTARY, CULTIVATOR]

rote *n.* (usu. prec. by *by*; also *attrib.*) mechanical or habitual repetition (in order to memorize) (*rote learning*). [origin unknown]

rot-gut *n. slang* cheap harmful alcohol.

rotisserie /rəʊ'tɪsərɪ/ *n.* **1** restaurant etc. where meat is roasted or barbecued. **2** rotating spit for roasting or barbecuing meat. [French: related to ROAST]

rotor /'rəʊtə(r)/ *n.* **1** rotary part of a machine. **2** rotary aerofoil on a helicopter, providing lift. [related to ROTATE]

Rotovator var. of ROTAVATOR.

rotten /'rɒt(ə)n/ *adj.* (**-er**, **-est**) **1** rotting or rotted; fragile from age or use. **2** morally or politically corrupt. **3** *slang* **a** disagreeable, unpleasant, bad (*had a rotten time*). **b** worthless (*rotten idea*). **c**

ill (*feel rotten*). □ **rottenly** *adv.* **rottenness** *n.* [Old Norse: related to ROT]

rotten borough *n. hist.* (before 1832) English borough electing an MP though having very few voters.

rotter *n. slang* nasty or contemptible person. [from ROT]

Rottweiler /ˈrɒtˌvaɪlə(r), -ˌwaɪlə(r)/ *n.* black-and-tan dog noted for ferocity. [*Rottweil* in Germany]

rotund /rəʊˈtʌnd/ *adj.* **1** plump, podgy. **2** (of speech etc.) sonorous, grandiloquent. □ **rotundity** *n.* [Latin *rotundus*: related to ROTA]

rotunda /rəʊˈtʌndə/ *n.* circular building, hall, or room, esp. domed. [Italian *rotonda*: related to ROTUND]

rouble /ˈruːb(ə)l/ *n.* (also **ruble**) chief monetary unit of Russia etc. [French from Russian]

roué /ˈruːeɪ/ *n.* (esp. elderly) debauchee. [French]

rouge /ruːʒ/ –*n.* red cosmetic for colouring the cheeks. –*v.* (**-ging**) **1** colour with or apply rouge. **2** become red, blush. [Latin *rubeus* red]

rough /rʌf/ –*adj.* **1** uneven or bumpy, not smooth, level, or polished. **2** shaggy or coarse-haired. **3** boisterous, coarse; violent, not mild, quiet, or gentle (*rough fellow*; *rough play*; *rough sea*). **4** (of wine etc.) sharp or harsh in taste. **5** harsh, insensitive (*rough words*; *rough treatment*). **6 a** unpleasant, severe, demanding (*had a rough time*). **b** unfortunate; undeserved (*had rough luck*). **c** (often foll. by *on*) hard or unfair (towards). **7** lacking finish etc. **8** incomplete, rudimentary, approximate (*rough attempt*; *rough sketch*; *rough estimate*). **9** (of stationery etc.) used for rough notes etc. **10** *colloq.* unwell; depressed (*feeling rough*). –*adv.* in a rough manner (*play rough*). –*n.* **1** (usu. prec. by *the*) hardship (*take the rough with the smooth*). **2** rough ground, esp. on a golf-course (*ball went into the rough*). **3** violent person (*bunch of roughs*). **4** unfinished or natural state (*written it in rough*). –*v.* **1** (foll. by *up*) ruffle (feathers, hair, etc.), esp. by rubbing. **2** (foll. by *out, in*) shape, plan, or sketch roughly. □ **rough it** *colloq.* do without basic comforts. **rough up** *slang* attack violently. □ **roughish** *adj.* **roughness** *n.* [Old English]

roughage *n.* coarse fibrous material in food, stimulating intestinal action.

rough-and-ready *adj.* crude but effective; not over-particular.

rough-and-tumble –*adj.* irregular, scrambling, disorderly. –*n.* disorderly fight; scuffle.

roughcast –*n.* (often *attrib.*) plaster of lime and gravel, used on outside walls. –*adj.* (of a plan etc.) roughly formed, preliminary. –*v.* (*past* and *past part.* **-cast**) **1** coat with roughcast. **2** prepare in outline.

rough diamond *n.* **1** uncut diamond. **2** rough-mannered but honest person.

rough-dry *v.* dry (clothes) without ironing.

roughen *v.* make or become rough.

rough-hewn *adj.* uncouth, unrefined.

rough house *n. slang* disturbance or row; boisterous play.

rough justice *n.* **1** treatment that is approximately fair. **2** unjust treatment.

roughly *adv.* **1** in a rough manner. **2** approximately (*roughly 20 people*). □ **roughly speaking** approximately.

roughneck *n. colloq.* **1** worker on an oil rig. **2** rough or rowdy person.

rough-rider *n.* person who breaks in or rides unbroken horses.

roughshod *adj.* (of a horse) having shoes with nail-heads projecting to prevent slipping. □ **ride roughshod over** treat inconsiderately or arrogantly.

roulade /ruːˈlɑːd/ *n.* **1** rolled piece of meat, sponge, etc. with a filling. **2** quick succession of notes, usu. sung to one syllable. [French *rouler* roll]

roulette /ruːˈlet/ *n.* gambling game in which a ball is dropped on to a revolving numbered wheel. [French, = little wheel]

round –*adj.* **1** shaped like a circle, sphere, or cylinder; convex; circular, curved, not angular. **2** done with or involving circular motion. **3** entire, continuous, complete (*round dozen*). **4** candid, outspoken. **5** (usu. *attrib.*) (of a number) expressed for brevity as a complete number (*£297.32, or in round figures £300*). **6** (of a voice, style, etc.) flowing, sonorous. –*n.* **1** round object or form. **2 a** revolving motion or course (*yearly round*). **b** recurring series of activities, meetings, etc. (*continuous round of pleasure*; *round of talks*). **3 a** fixed route for deliveries (*milk round*). **b** route etc. for supervision or inspection (*watchman's round*; *doctor's rounds*). **4** drinks etc. for all members of a group. **5 a** one bullet, shell, etc. **b** act of firing this. **6 a** slice from a loaf of bread. **b** sandwich made from two slices. **c** joint of beef from the haunch. **7** set, series, or sequence of actions in turn, esp.: **a** one spell of play in a game etc. **b** one stage in a competition. **8** *Golf* playing of all the holes in a course once. **9** song for unaccompanied voices overlapping at intervals. **10** rung of a ladder. **11** (foll. by *of*) circumference or extent of (*in all the* *round of Nature*). –*adv.* **1** with circular motion (*wheels go round*). **2** with return to the starting-point or an earlier state (*summer soon comes round*). **3** with change to an opposite position, opinion, etc. (*turned round to look*; *soon won them round*). **4** to, at, or affecting a circumference, area, group, etc. (*tea was handed round*; *may I look round?*). **5** in every direction within a radius (*spread destruction round*). **6** circuitously (*go the long way round*). **7** to a person's house, a convenient place, etc. (*ask him round*; *will be round soon*; *brought the car round*). **8** measuring (a specified distance) in girth. –*prep.* **1** so as to encircle or enclose (*a blanket round him*). **2** at or to points on the circumference of (*sat round the table*). **3** with successive visits to (*hawks them round the cafés*). **4** within a radius of (*towns round Birmingham*). **5** having as an axis or central point (*planned a book round the War*). **6 a** so as to pass in a curved course (*go round the corner*). **b** having so passed (*be round the corner*). **c** in the resulting position (*find them round the corner*). –*v.* **1** give or take a round shape. **2** pass round (a corner, cape, etc.). **3** (usu. foll. by *up, down*) express (a number) approximately, for brevity. □ **go the round** (or **rounds**) (of news etc.) be passed on. **in the round 1** with all features shown; all things considered. **2** with the audience on at least three sides of the stage. **3** (of sculpture) with all sides shown. **round about 1** all round; on all sides (of). **2** approximately (*round about £50*). **round and round** several times round. **round the bend** see BEND[1]. **round off** make complete or less angular. **round on** attack unexpectedly, esp. verbally. **round out 1** provide with more details. **2** complete, finish. **round the twist** see TWIST. **round up 1** collect or bring together. **2** = sense 3 of *v.* □ **roundish** *adj.* **roundness** *n.* [Latin: related to ROTUND]

roundabout –*n.* **1** road junction at which traffic circulates in one direction round a central island. **2 a** large revolving device for children to ride on in a playground. **b** = MERRY-GO-ROUND 1. –*adj.* circuitous.

round brackets *n.pl.* brackets of the form ( ).

round dance *n.* dance in which couples move in circles or dancers form one large circle.

roundel /'raʊnd(ə)l/ *n.* **1** circular mark, esp. identifying military aircraft. **2** small disc, esp. a medallion. [French *rondel(le)*: related to ROUND]

roundelay /'raʊndɪ,leɪ/ *n.* short simple song with a refrain. [alteration of French *rondelet* diminutive: related to ROUNDEL]

rounder *n.* **1** (in *pl.*; treated as *sing.*) ball game in which players hit the ball and run through a round of bases. **2** complete run as a unit of scoring in rounders.

Roundhead *n. hist.* member of the Parliamentary party in the English Civil War.

roundly *adv.* bluntly, severely (*told them roundly*).

round robin *n.* **1** petition, esp. with signatures in a circle to conceal the order of writing. **2** *US* tournament in which each competitor plays every other.

round-shouldered *adj.* having shoulders bent forward and a rounded back.

roundsman *n.* tradesman's employee delivering goods.

Round Table *n.* **1** international charitable association. **2** (**round table**) assembly for discussion, esp. at a conference (often *attrib.*: *round-table talks*).

round trip *n.* trip to one or more places and back again.

round-up *n.* **1** systematic rounding up. **2** summary or résumé.

roundworm *n.* worm with a rounded body.

rouse /raʊz/ *v.* (**-sing**) **1** (cause to) wake. **2** (often foll. by *up*, often *refl.*) stir up, make or become active or excited (*was roused to protest*). **3** anger (*terrible when roused*). **4** evoke (feelings). [origin unknown]

rousing *adj.* exciting, stirring (*rousing song*).

roustabout /'raʊstə,baʊt/ *n.* **1** labourer on an oil rig. **2** unskilled or casual labourer. [*roust* rout out, rouse]

rout[1] –*n.* **1** disorderly retreat of defeated troops (*put them to rout*). **2** overthrow, defeat. –*v.* put to flight, defeat. [French: related to ROUTE]

rout[2] *v.* = ROOT[2]. [var. of ROOT[2]]

route /ru:t/ –*n.* way or course taken (esp. regularly) from one place to another. –*v.* (**-teing**) send, forward, or direct by a particular route. [French *route* road, from Latin *rupta* (*via*)]

route march *n.* training-march for troops.

routine /ru:'ti:n/ –*n.* **1** regular course or procedure, unvarying performance of certain acts. **2** set sequence in a dance, comedy act, etc. **3** *Computing* sequence of instructions for a particular task. –*adj.* **1** performed as part of a routine (*routine duties*). **2** of a customary or standard kind. □ **routinely** *adv.* [French: related to ROUTE]

roux /ru:/ *n.* (*pl.* same) mixture of fat and flour used in sauces etc. [French]

rove[1] *v.* (**-ving**) **1** wander without settling; roam, ramble. **2** (of eyes) look about. [probably Scandinavian]

rove[2] *past* of REEVE[2].

rover[1] *n.* wanderer.

rover[2] *n.* pirate. [Low German or Dutch]

roving eye *n.* tendency to infidelity.

row[1] /rəʊ/ *n.* **1** line of persons or things. **2** line of seats across a theatre etc. **3** street with houses along one or each side. □ **in a row 1** forming a row. **2** *colloq.* in succession (*two days in a row*). [Old English]

row[2] /rəʊ/ *–v.* **1** (often *absol.*) propel (a boat) with oars. **2** convey (a passenger) thus. *–n.* **1** spell of rowing. **2** trip in a rowing-boat. □ **rower** *n.* [Old English]

row[3] /raʊ/ *colloq.* *–n.* **1** loud noise or commotion. **2** fierce quarrel or dispute. **3** severe reprimand. *–v.* **1** make or engage in a row. **2** reprimand. [origin unknown]

rowan /ˈrəʊən/ *n.* (in full **rowan-tree**) **1** *Scot.* & *N.Engl.* mountain ash. **2** (in full **rowan-berry**) its scarlet berry. [Scandinavian]

row-boat *n.* *US* = ROWING-BOAT.

rowdy /ˈraʊdɪ/ *–adj.* (**-ier**, **-iest**) noisy and disorderly. *–n.* (*pl.* **-ies**) rowdy person. □ **rowdily** *adv.* **rowdiness** *n.* **rowdyism** *n.* [origin unknown]

rowel /ˈraʊəl/ *n.* spiked revolving disc at the end of a spur. [Latin *rotella* diminutive: related to ROTA]

rowing-boat *n.* small boat propelled by oars.

rowlock /ˈrɒlək/ *n.* device on a boat's side for holding an oar in place. [*oarlock* from Old English: related to OAR, LOCK[1]]

royal /ˈrɔɪəl/ *–adj.* **1** of, suited to, or worthy of a king or queen. **2** in the service or under the patronage of a king or queen. **3** of the family of a king or queen. **4** majestic, splendid. **5** exceptional, first-rate (*had a royal time*). *–n.* *colloq.* member of the royal family. □ **royally** *adv.* [Latin: related to REGAL]

royal blue *adj.* & *n.* (as adj. often hyphenated) deep vivid blue.

Royal British Legion *n.* national association of ex-members of the armed forces, founded in 1921.

Royal Commission *n.* commission of inquiry appointed by the Crown at the request of Government.

royal family *n.* family of a sovereign.

royal flush *n.* straight poker flush headed by an ace.

royal icing *n.* hard white icing for cakes.

royalist *n.* supporter of monarchy, or *hist.* of the royal side in the English Civil War. □ **royalism** *n.*

royal jelly *n.* substance secreted by worker bees and fed by them to future queen bees.

Royal Marine *n.* British marine (see MARINE *n.* 1).

Royal Navy *n.* British navy.

royalty *n.* (*pl.* **-ies**) **1** royal office, dignity, or power; being royal. **2 a** royal persons. **b** member of a royal family. **3** percentage of profit from a book, public performance, patent, etc. paid to the author etc. **4 a** royal right (now esp. over minerals) granted by the sovereign. **b** payment made by a producer of minerals etc. to the owner of the site etc. [French: related to ROYAL]

royal warrant *n.* warrant authorizing a tradesperson to supply goods to a specified royal person.

royal 'we' *n.* use of 'we' instead of 'I' by a single person.

RP *abbr.* received pronunciation.

RPI *abbr.* retail price index.

rpm *abbr.* revolutions per minute.

RPO *abbr.* Royal Philharmonic Orchestra.

RSA *abbr.* **1** Royal Society of Arts. **2** Royal Scottish Academy; Royal Scottish Academician.

RSC *abbr.* Royal Shakespeare Company.

RSI *abbr.* repetitive strain injury.

RSJ *abbr.* rolled steel joist.

RSM *abbr.* Regimental Sergeant-Major.

RSPB *abbr.* Royal Society for the Protection of Birds.

RSPCA *abbr.* Royal Society for the Prevention of Cruelty to Animals.

RSV *abbr.* Revised Standard Version (of the Bible).

RSVP *abbr.* (in an invitation etc.) please answer. [French *répondez s'il vous plaît*]

rt. *abbr.* right.

Rt. Hon. *abbr.* Right Honourable.

Rt. Revd. *abbr.* (also **Rt. Rev.**) Right Reverend.

RU *abbr.* Rugby Union.

Ru *symb.* ruthenium.

rub *–v.* (**-bb-**) **1** move something, esp. one's hand, with firm pressure over the surface of. **2** (usu. foll. by *against*, *in*, *on*, *over*) apply (one's hand etc.) in this way. **3** clean, polish, chafe, or make dry, sore, or bare by rubbing. **4** (foll. by *in*, *into*, *through*, *over*) apply (polish etc.) by rubbing. **5** (often foll. by *together*, *against*, *on*) move with contact or friction or slide (objects) against each other. **6** (of cloth, skin, etc.) become frayed, worn, sore, or bare with friction.

–n. **1** act or spell of rubbing (*give it a rub*). **2** impediment or difficulty (*there's the rub*). □ **rub along** *colloq.* cope or manage routinely. **rub down** dry, smooth, or clean by rubbing. **rub it in** (or **rub a person's nose in it**) emphasize or repeat an embarrassing fact etc. **rub off 1** (usu. foll. by *on*) be transferred by contact, be transmitted (*his attitudes have rubbed off on me*). **2** remove by rubbing. **rub out** erase with a rubber. **rub shoulders with** associate with. **rub up 1** polish. **2** brush up (a subject or one's memory). **rub up the wrong way** irritate. [Low German]

rubato /ru:'bɑ:təʊ/ *n. Mus.* (*pl.* **-s** or **-ti** /-tɪ/) temporary disregarding of strict tempo. [Italian, = robbed]

rubber[1] *n.* **1** tough elastic substance made from the latex of plants or synthetically. **2** piece of this or a similar substance for erasing esp. pencil marks. **3** *colloq.* condom. **4** (in *pl.*) *US* galoshes. □ **rubbery** *adj.* **rubberiness** *n.* [from RUB]

rubber[2] *n.* match of esp. three successive games between the same sides or persons at whist, bridge, cricket, etc. [origin unknown]

rubber band *n.* loop of rubber for holding papers etc. together.

rubberize *v.* (also **-ise**) (**-zing** or **-sing**) treat or coat with rubber.

rubberneck *colloq.* *–n.* inquisitive person, esp. a tourist or sightseer. *–v.* behave like a rubberneck.

rubber plant *n.* **1** evergreen tropical plant often cultivated as a house-plant. **2** (also **rubber tree**) tropical tree yielding latex.

rubber stamp *–n.* **1** device for inking and imprinting on a surface. **2 a** person who mechanically copies or endorses others' actions. **b** indication of such endorsement. *–v.* (**rubber-stamp**) approve automatically.

rubbing *n.* impression or copy made by rubbing.

rubbish /'rʌbɪʃ/ *–n.* **1** waste material; refuse, litter. **2** worthless material; trash. **3** (often as *int.*) nonsense. *–v.* *colloq.* criticize contemptuously. □ **rubbishy** *adj.* [Anglo-French *rubbous*]

rubble /'rʌb(ə)l/ *n.* rough fragments of stone, brick, etc., esp. from a demolished building. [French *robe* spoils]

rub-down *n.* rubbing down.

rubella /ru:'belə/ *n. formal* German measles. [Latin *rubellus* reddish]

Rubicon /'ru:bɪ,kɒn/ *n.* boundary; point from which there is no going back. [*Rubicon*, river on an ancient frontier of Italy]

rubicund /'ru:bɪ,kʌnd/ *adj.* (of a face, complexion, etc.) ruddy, high-coloured. [Latin *rubeo* be red]

rubidium /ru:'bɪdɪəm/ *n.* soft silvery metallic element. [Latin *rubidus* red]

Rubik's cube /'ru:bɪks/ *n.* cube-shaped puzzle in which composite faces must be restored to single colours by rotation. [*Rubik*, name of its inventor]

ruble var. of ROUBLE.

rubric /'ru:brɪk/ *n.* **1** heading or passage in red or special lettering. **2** explanatory words. **3** established custom or rule. **4** direction for the conduct of divine service in a liturgical book. [Latin *ruber* red]

ruby /'ru:bɪ/ *–n.* (*pl.* **-ies**) **1** rare precious stone varying in colour from deep crimson to pale rose. **2** deep red colour. *–adj.* of this colour. [Latin *rubeus* red]

ruby wedding *n.* fortieth wedding anniversary.

RUC *abbr.* Royal Ulster Constabulary.

ruche /ru:ʃ/ *n.* frill or gathering of lace etc. □ **ruched** *adj.* [French, = beehive]

ruck[1] *n.* **1** (prec. by *the*) main body of competitors not likely to overtake the leaders. **2** undistinguished crowd or group. **3** *Rugby* loose scrum. [apparently Scandinavian]

ruck[2] *–v.* (often foll. by *up*) make or become creased or wrinkled. *–n.* crease or wrinkle. [Old Norse]

rucksack /'rʌksæk/ *n.* bag carried on the back, esp. by hikers. [German]

ruckus /'rʌkəs/ *n.* esp. *US informal* row, commotion. [perhaps from RUCTION or RUMPUS]

ruction /'rʌkʃ(ə)n/ *n. colloq.* **1** disturbance or tumult. **2** (in *pl.*) row, heated arguments. [origin unknown]

rudder *n.* flat piece hinged vertically to the stern of a ship or on the tailplane of an aircraft etc., for steering. □ **rudderless** *adj.* [Old English]

ruddy /'rʌdɪ/ *adj.* (**-ier, -iest**) **1** (of a person, complexion, etc.) freshly or healthily red. **2** reddish. **3** *colloq.* bloody, damnable. □ **ruddily** *adv.* **ruddiness** *n.* [Old English]

rude *adj.* **1** impolite or offensive. **2** roughly made or done; crude (*rude plough*). **3** primitive or uneducated (*rude simplicity*). **4** abrupt, sudden, startling (*rude awakening*). **5** *colloq.* indecent, lewd (*rude joke*). **6** vigorous or hearty (*rude health*). □ **rudely** *adv.* **rudeness** *n.* [Latin *rudis*]

rudiment /'ru:dɪmənt/ *n.* **1** (in *pl.*) elements or first principles of a subject. **2** (in *pl.*) imperfect beginning of something undeveloped or yet to develop. **3** vestigial or undeveloped part or organ.

□ **rudimentary** /-ˈmentərɪ/ *adj.* [Latin: related to RUDE]

rue[1] *v.* (**rues, rued, rueing** or **ruing**) repent of; wish to be undone or non-existent (esp. *rue the day*). [Old English]

rue[2] *n.* evergreen shrub with bitter strong-scented leaves. [Greek *rhutē*]

rueful *adj.* genuinely or humorously sorrowful. □ **ruefully** *adv.* **ruefulness** *n.* [from RUE[1]]

ruff[1] *n.* **1** projecting starched frill worn round the neck, esp. in the 16th c. **2** projecting or coloured ring of feathers or hair round a bird's or animal's neck. **3** domestic pigeon. **4** (*fem.* **reeve** /riːv/) wading bird with a ruff. [perhaps = ROUGH]

ruff[2] *–v.* trump at cards. *–n.* trumping. [French *ro(u)ffle*]

ruffian /ˈrʌfɪən/ *n.* violent lawless person. [Italian *ruffiano*]

ruffle /ˈrʌf(ə)l/ *–v.* (**-ling**) **1 a** disturb the smoothness or tranquillity of. **b** undergo this. **2** gather (lace etc.) into a ruffle. **3** (often foll. by *up*) (of a bird) erect (its feathers) in anger, display, etc. *–n.* frill of lace etc., esp. round the wrist or neck. [origin unknown]

rufous /ˈruːfəs/ *adj.* (esp. of animals) reddish-brown. [Latin *rufus*]

rug *n.* **1** thick floor covering, usu. smaller than a carpet. **2** thick woollen coverlet or wrap. □ **pull the rug from under** deprive of support; weaken, unsettle. [probably Scandinavian]

Rugby /ˈrʌgbɪ/ *n.* (in full **Rugby football**) team game played with an oval ball that may be kicked or carried. [*Rugby school*, where it was first played]

Rugby League *n.* partly professional Rugby with teams of 13.

Rugby Union *n.* amateur Rugby with teams of 15.

rugged /ˈrʌgɪd/ *adj.* **1** (esp. of ground) rough, uneven. **2** (of features) wrinkled, furrowed, irregular. **3 a** unpolished; lacking refinement (*rugged grandeur*). **b** harsh in sound. **4** robust, hardy. □ **ruggedly** *adv.* **ruggedness** *n.* [probably Scandinavian]

rugger /ˈrʌgə(r)/ *n. colloq.* Rugby.

ruin /ˈruːɪn/ *–n.* **1** destroyed, wrecked, or spoiled state. **2** downfall or elimination (*ruin of my hopes*). **3** complete loss of one's property or position (*bring to ruin*). **4** (in *sing.* or *pl.*) remains of a building etc. that has suffered ruin. **5** cause of ruin (*the ruin of us*). *–v.* **1 a** bring to ruin (*extravagance has ruined me*). **b** spoil, damage. **2** (esp. as **ruined** *adj.*) reduce to ruins. □ **in ruins** completely wrecked (*hopes were in ruins*). [Latin *ruo* fall]

ruination /ˌruːɪˈneɪʃ(ə)n/ *n.* **1** bringing to ruin. **2** ruining or being ruined.

ruinous *adj.* **1** bringing ruin, disastrous (*ruinous expense*). **2** dilapidated. □ **ruinously** *adv.*

rule *–n.* **1** compulsory principle governing action. **2** prevailing custom or standard; normal state of things. **3** government or dominion (*under British rule*). **4** graduated straight measure; ruler. **5** code of discipline of a religious order. **6** order made by a judge or court with reference to a particular case only. **7** *Printing* thin line or dash. *–v.* (**-ling**) **1** dominate; keep under control. **2** (often foll. by *over*) have sovereign control of (*rules over a vast kingdom*). **3** (often foll. by *that*) pronounce authoritatively. **4 a** make parallel lines across (paper). **b** make (a straight line) with a ruler etc. □ **as a rule** usually. **rule out** exclude; pronounce irrelevant or ineligible. **rule the roost** be in control. [Latin *regula*]

rule of thumb *n.* rule based on experience or practice rather than theory.

ruler *n.* **1** person exercising government or dominion. **2** straight usu. graduated strip of wood, metal, or plastic used to draw or measure.

ruling *n.* authoritative pronouncement.

rum[1] *n.* spirit distilled from sugar-cane or molasses. [origin unknown]

rum[2] *adj.* (**rummer, rummest**) *colloq.* odd, strange, queer. [origin unknown]

Rumanian var. of ROMANIAN.

rumba /ˈrʌmbə/ *n.* **1** Latin American ballroom dance orig. from Cuba. **2** music for this. [American Spanish]

rum baba *n.* sponge cake soaked in rum syrup.

rumble /ˈrʌmb(ə)l/ *–v.* (**-ling**) **1** make a continuous deep resonant sound as of distant thunder. **2** (foll. by *along, by, past*, etc.) (esp. of a vehicle) move with a rumbling noise. **3** (often *absol.*) *slang* find out the esp. discreditable truth about. *–n.* rumbling sound. [probably Dutch *rommelen*]

rumbustious /rʌmˈbʌstʃəs/ *adj. colloq.* boisterous, noisy, uproarious. [probably var. of *robustious* from ROBUST]

ruminant /ˈruːmɪnənt/ *–n.* animal that chews the cud. *–adj.* **1** of ruminants. **2** meditative. [related to RUMINATE]

ruminate /ˈruːmɪˌneɪt/ *v.* (**-ting**) **1** meditate, ponder. **2** chew the cud. □ **rumination** /-ˈneɪʃ(ə)n/ *n.* **ruminative** /-nətɪv/ *adj.* [Latin *rumen* throat]

rummage /ˈrʌmɪdʒ/ *–v.* (**-ging**) **1** search, esp. unsystematically. **2** (foll. by *out, up*) find among other things. *–n.* rummaging. [French *arrumage* from *arrumer* stow cargo]

rummage sale *n.* esp. *US* jumble sale.
rummy /'rʌmɪ/ *n.* card-game played usu. with two packs. [origin unknown]
rumour /'ru:mə(r)/ (*US* **rumor**) –*n.* (often foll. by *of* or *that*) general talk, assertion, or hearsay of doubtful accuracy (*heard a rumour that you are leaving*). –*v.* (usu. in *passive*) report by way of rumour (*it is rumoured that you are leaving*). [Latin *rumor* noise]
rump *n.* **1** hind part of a mammal or bird, esp. the buttocks. **2** remnant of a parliament etc. [probably Scandinavian]
rumple /'rʌmp(ə)l/ *v.* (**-ling**) crease, ruffle. [Dutch *rompelen*]
rump steak *n.* cut of beef from the rump.
rumpus /'rʌmpəs/ *n. colloq.* disturbance, brawl, row, or uproar. [origin unknown]
run –*v.* (**-nn-**; *past* **ran**; *past part.* **run**) **1** go with quick steps, never having both or all feet on the ground at once. **2** flee, abscond. **3** go or travel hurriedly or briefly (*I'll just run down to the shops*). **4 a** advance by or as by rolling or on wheels, or smoothly or easily. **b** (cause to) be in action or operation or go in a specified way (*left the engine running*; *ran the car into a tree*). **5** be current or operative (*lease runs for 99 years*). **6** travel on its route (*train is running late*). **7** (of a play etc.) be staged or presented (*now running at the Apollo*). **8** extend; have a course, order, or tendency (*road runs by the coast*; *prices are running high*). **9 a** (often *absol.*) compete in (a race). **b** finish a race in a specified position. **10** (often foll. by *for*) seek election (*ran for president*). **11** flow (with) or be wet; drip (with) (*walls running with condensation*). **12 a** cause (water etc.) to flow. **b** fill (a bath) thus. **13** spread rapidly (*ink ran over the table*). **14** traverse (a course, race, or distance). **15** perform (an errand). **16** publish (an article etc.) in a newspaper etc. **17** direct or manage (a business etc.). **18** own and use (a vehicle) regularly. **19** transport in a private vehicle (*ran me to the station*). **20** enter (a horse etc.) for a race. **21** smuggle (guns etc.). **22** chase or hunt. **23** allow (an account) to accumulate before paying. **24** (of a dyed colour) spread from the dyed parts. **25 a** (of a thought, the eye, the memory, etc.) pass quickly (*ideas ran through my mind*). **b** pass (one's eye) quickly (*ran my eye down the page*). **26** (of tights etc.) ladder. **27** (of esp. the eyes or nose) exude liquid. –*n.* **1** running. **2** short excursion. **3** distance travelled. **4** general tendency. **5** regular route. **6** continuous stretch, spell, or course (*run of bad luck*). **7** (often foll. by *on*) high general demand (*run on the dollar*). **8** quantity produced at one time (*print run*). **9** average type or class (*general run of customers*). **10** point scored in cricket or baseball. **11** (foll. by *of*) free use of or access to (*run of the house*). **12 a** animal's regular track. **b** enclosure for fowls etc. **c** range of pasture. **13** ladder in tights etc. **14** *Mus.* rapid scale passage. **15** (in full **the runs**) *colloq.* diarrhoea. □ **on the run** fleeing. **run about 1** bustle, hurry. **2** (esp. of children) play freely. **run across** happen to meet or find. **run after 1** pursue at a run. **2** pursue, esp. sexually. **run along** *colloq.* depart. **run around 1** take from place to place by car etc. **2** (often foll. by *with*) *slang* engage in esp. promiscuous sexual relations. **run away 1** (often foll. by *from*) flee, abscond. **2** mentally evade (a problem etc.). **run away with 1** carry off. **2** win easily. **3** deprive of self-control, carry away. **4** consume (money etc.). **5** (of a horse) bolt with (a rider etc.). **6** leave home to have a relationship with (esp. another person's husband or wife). **run down 1** knock down. **2** reduce the numbers etc. of. **3** (of an unwound clock etc.) stop. **4** discover after a search. **5** *colloq.* disparage. **run dry 1** cease to flow. **2** = *run out* 1. **run for it** seek safety by fleeing. **run** (or **good run**) **for one's money 1** vigorous or close competition. **2** some return for outlay or effort. **run the gauntlet** see GAUNTLET[2]. **run high** (of feelings) be strong. **run in 1** run (an engine or vehicle) carefully when new. **2** *colloq.* arrest. **run in the family** (of a trait) be common in a family. **run into 1** collide with. **2** encounter. **3** reach as many as (a usu. high figure). **run into the ground** *colloq.* bring (a person) to exhaustion etc. **run low** (or **short**) become depleted, have too little. **run off 1** flee. **2** produce (copies etc.) on a machine. **3** decide (a race etc.) after heats or a tie. **4** (cause to) flow away. **5** write or recite fluently. **run off with 1** steal. **2** = *run away with* 6. **run on 1** continue in operation. **2** speak volubly or incessantly. **3** continue on the same line as the preceding matter. **run out 1** come to an end. **2** (foll. by *of*) exhaust one's stock of. **3** put down the wicket of (a running batsman). **run out on** *colloq.* desert (a person). **run over 1** (of a vehicle etc.) knock down or crush. **2** overflow. **3** study or repeat quickly. **run ragged** exhaust (a person). **run rings round** see RING[1]. **run riot** see RIOT. **run a** (or **the**) **risk** see RISK. **run through 1** examine or rehearse briefly. **2** peruse. **3**

deal successively with. **4** spend money rapidly or recklessly. **5** pervade. **6** pierce with a sword etc. **run to 1** have the money, resources, or ability for. **2** reach (an amount or number). **3** (of a person) show a tendency to (*runs to fat*). **run to earth** see EARTH. **run to seed** see SEED. **run up 1** accumulate (a debt etc.). **2** build or make hurriedly. **3** raise (a flag). **run up against** meet with (a difficulty etc.). [Old English]

runabout *n.* light car or aircraft.

run-around *n.* (esp. in phr. **give a person the run-around**) *colloq.* deceit or evasion.

runaway *n.* **1** fugitive. **2** bolting animal, vehicle out of control. **3** (*attrib.*) that is running away or out of control (*runaway slave*; *runaway inflation*).

run-down –*n.* **1** reduction in numbers. **2** detailed analysis. –*adj.* **1** decayed, dilapidated. **2** exhausted (from overwork, illness, etc.).

rune *n.* **1** letter of the earliest Germanic alphabet. **2** similar mark of mysterious or magic significance. □ **runic** *adj.* [Old Norse]

rung[1] *n.* **1** step of a ladder. **2** strengthening crosspiece in a chair etc. [Old English]

rung[2] *past part.* of RING[2].

run-in *n.* **1** approach to an action or event. **2** *colloq.* quarrel.

runnel /ˈrʌn(ə)l/ *n.* **1** brook. **2** gutter. [Old English]

runner *n.* **1** person, horse, etc. that runs, esp. in a race. **2** creeping rooting plant-stem. **3** rod, groove, roller, or blade on which a thing, e.g. a sledge, slides. **4** sliding ring on a rod etc. **5** messenger. **6** (in full **runner bean**) twining bean plant with long flat green edible seed pods. **7** long narrow ornamental cloth or rug. □ **do a runner** *slang* abscond, leave hastily; flee.

runner-up *n.* (*pl.* **runners-up** or **runner-ups**) competitor or team taking second place.

running –*n.* **1** action of runners in a race etc. **2** way a race etc. proceeds. –*adj.* **1** continuous (*running battle*). **2** consecutive (*three days running*). **3** done with a run (*running jump*). □ **in** (or **out of**) **the running** (of a competitor) with a good (or poor) chance of success. **make** (or **take up**) **the running** take the lead; set the pace. **take a running jump** (esp. as *int.*) *slang* go away.

running-board *n.* footboard on either side of a vehicle.

running commentary *n.* verbal description of an esp. sporting event.

running knot *n.* knot that slips along a rope etc. to allow tightening etc.

running mate *n. US* **1** candidate for vice-president etc. **2** horse intended to set the pace for another horse in a race.

running repairs *n.pl.* minor or temporary repairs etc.

running sore *n.* suppurating sore; festering situation etc.

running water *n.* flowing water, esp. on tap.

runny *adj.* (**-ier, -iest**) **1** tending to run or flow. **2** excessively fluid.

run-off *n.* additional election, race, etc., after a tie.

run-of-the-mill *adj.* ordinary, undistinguished.

run-out *n.* dismissal of a batsman by being run out.

runt *n.* **1** smallest pig etc. in a litter. **2** weakling; undersized person. [origin unknown]

run-through *n.* **1** rehearsal. **2** brief survey.

run-up *n.* (often foll. by *to*) preparatory period.

runway *n.* specially prepared surface for aircraft taking off and landing.

rupee /ruːˈpiː/ *n.* chief monetary unit of India, Pakistan, etc. [Hindustani]

rupiah /ruːˈpiːə/ *n.* chief monetary unit of Indonesia. [related to RUPEE]

rupture /ˈrʌptʃə(r)/ –*n.* **1** breaking; breach. **2** breach in a relationship; disagreement and parting. **3** abdominal hernia. –*v.* (**-ring**) **1** burst (a cell or membrane etc.). **2** sever (a connection). **3** affect with or suffer a hernia. [Latin *rumpo rupt-* break]

rural /ˈrʊər(ə)l/ *adj.* in, of, or suggesting the country (*rural seclusion*). [Latin *rus rur-* the country]

rural dean see DEAN[1].

rural district *n. hist.* group of country parishes with an elected council.

ruse /ruːz/ *n.* stratagem, trick. [French]

rush[1] –*v.* **1** go, move, flow, or act precipitately or with great speed. **2** move or transport with great haste (*was rushed to hospital*). **3** (foll. by *at*) **a** move suddenly towards. **b** begin or attack impetuously. **4** perform or deal with hurriedly (*don't rush your dinner*). **5** force or induce (a person) to act hastily. **6** attack or capture by sudden assault. **7** *slang* overcharge (a customer). –*n.* **1 a** rushing; violent or speedy advance or attack. **b** sudden flow, flood. **2** period of great activity. **3** (*attrib.*) done with great haste or speed (*a rush job*). **4** sudden migration of large numbers. **5** (foll. by *on, for*) sudden strong demand for a commodity. **6** (in *pl.*) *colloq.* first uncut prints of a film. [French *ruser*: related to RUSE]

rush[2] *n.* **1** marsh plant with slender tapering pith-filled stems, used for making chair-bottoms, baskets, etc. **2** stem of this. □ **rushy** *adj.* [Old English]

rush candle *n.* candle made of rush pith dipped in tallow.

rush hour *n.* (often hyphenated when *attrib.*) time(s) each day when traffic is heaviest.

rushlight *n.* rush candle.

rusk *n.* slice of bread rebaked as a light biscuit, esp. as baby food. [Spanish or Portuguese *rosca* twist]

russet /'rʌsɪt/ *–adj.* reddish-brown. *–n.* **1** russet colour. **2** rough-skinned russet-coloured apple. [Latin *russus*]

Russian /'rʌʃ(ə)n/ *–n.* **1 a** native or national of Russia or (loosely) the former Soviet Union. **b** person of Russian descent. **2** language of Russia. *–adj.* **1** of Russia or (loosely) the former Soviet Union or its people. **2** of or in Russian.

Russian roulette *n.* firing of a revolver, with one chamber loaded, at one's head, after spinning the chamber.

Russian salad *n.* salad of mixed diced vegetables with mayonnaise.

Russo- *comb. form* Russian; Russian and.

rust *–n.* **1** reddish corrosive coating formed on iron, steel, etc. by oxidation, esp. when wet. **2** fungal plant-disease with rust-coloured spots. **3** impaired state due to disuse or inactivity. **4** reddish-brown. *–v.* **1** affect or be affected with rust. **2** become impaired through disuse. [Old English]

rustic /'rʌstɪk/ *–adj.* **1** of or like country people or country life. **2** unsophisticated. **3** of rude or rough workmanship. **4** made of untrimmed branches or rough timber (*rustic bench*). **5** *Archit.* with a roughened or rough-hewn surface. *–n.* country person, peasant. □ **rusticity** /-'tɪsɪtɪ/ *n.* [Latin *rus* the country]

rusticate /'rʌstɪ,keɪt/ *v.* **(-ting) 1** send down (a student) temporarily from university. **2** retire to or live in the country. **3** make rustic. □ **rustication** /-'keɪʃ(ə)n/ *n.*

rustle /'rʌs(ə)l/ *–v.* **(-ling) 1** (cause to) make a gentle sound as of dry blown leaves. **2** (also *absol.*) steal (cattle or horses). *–n.* rustling sound. □ **rustle up** *colloq.* produce at short notice. □ **rustler** *n.* (esp. in sense 2 of *v.*). [imitative]

rustproof *–adj.* not susceptible to corrosion by rust. *–v.* make rustproof.

rusty *adj.* **(-ier, -iest) 1** rusted or affected by rust. **2** stiff with age or disuse. **3** (of knowledge etc.) impaired, esp. by neglect (*my French is rusty*). **4** rust-coloured. **5** (of black clothes) discoloured by age. □ **rustiness** *n.*

rut[1] *–n.* **1** deep track made by the passage of wheels. **2** established (esp. tedious) practice or routine (*in a rut*). *–v.* **(-tt-)** mark with ruts. [probably French: related to ROUTE]

rut[2] *–n.* periodic sexual excitement of a male deer etc. *–v.* **(-tt-)** be affected with rut. [Latin *rugio* roar]

ruthenium /ru:'θi:nɪəm/ *n.* rare hard white metallic element from platinum ores. [medieval Latin *Ruthenia* Russia]

rutherfordium /,rʌðə'fɔ:dɪəm/ *n.* artificial metallic element. [*Rutherford*, name of a physicist]

ruthless /'ru:θlɪs/ *adj.* having no pity or compassion. □ **ruthlessly** *adv.* **ruthlessness** *n.* [*ruth* pity, from RUE[1]]

RV *abbr.* Revised Version (of the Bible).

-ry *suffix* = -ERY (*infantry*; *rivalry*).

rye /raɪ/ *n.* **1 a** cereal plant. **b** grain of this used for bread and fodder. **2** (in full **rye whisky**) whisky distilled from fermented rye. [Old English]

ryegrass *n.* forage or coarse lawn grass. [alteration of *ray-grass*]

S

S[1] /es/ *n.* (also **s**) (*pl.* **Ss** or **S's**) **1** nineteenth letter of the alphabet. **2** S-shaped thing.
S[2] *abbr.* (also **S.**) **1** Saint. **2** South, Southern.
S[3] *symb.* sulphur.
s. *abbr.* **1** second(s). **2** *hist.* shilling(s). **3** son. [sense 2 originally from Latin *solidus*]
-s' *suffix* denoting the possessive case of plural nouns and sometimes of singular nouns ending in *s* (*the boys' shoes*; *Charles' book*). [Old English inflection]
's *abbr.* **1** is; has (*he's*; *she's got it*; *John's*; *Charles's*). **2** us (*let's*).
-'s *suffix* denoting the possessive case of singular nouns and of plural nouns not ending in *-s* (*John's book*; *book's cover*; *children's shoes*).
SA *abbr.* **1** Salvation Army. **2** South Africa. **3** South Australia.
sabbath /'sæbəθ/ *n.* religious day of rest kept by Christians on Sunday and Jews on Saturday. [Hebrew, = rest]
sabbatical /sə'bætɪk(ə)l/ *—adj.* (of leave) granted at intervals to a university teacher for study or travel. *—n.* period of sabbatical leave. [Greek: related to SABBATH]
saber *US* var. of SABRE.
sable /'seɪb(ə)l/ *—n.* (*pl.* same or -s) **1** small brown-furred mammal of N. Europe and N. Asia. **2** its skin or fur. *—adj.* **1** (usu. placed after noun) *Heraldry* black. **2** esp. *poet.* gloomy. [Slavonic]
sabot /'sæbəʊ/ *n.* **1** shoe carved from wood. **2** wooden-soled shoe. [French]
sabotage /'sæbə,tɑ:ʒ/ *—n.* deliberate damage to productive capacity, esp. as a political act. *—v.* (**-ging**) **1** commit sabotage on. **2** destroy, spoil. [French: related to SABOT]
saboteur /,sæbə'tɜ:(r)/ *n.* person who commits sabotage. [French]
sabre /'seɪbə(r)/ *n.* (*US* **saber**) **1** curved cavalry sword. **2** light tapering fencing-sword. [French from German *Sabel*]
sabre-rattling *n.* display or threat of military force.
sac *n.* membranous bag in an animal or plant. [Latin: related to SACK[1]]
saccharin /'sækərɪn/ *n.* a sugar substitute. [medieval Latin *saccharum* sugar]
saccharine /'sækə,ri:n/ *adj.* excessively sentimental or sweet.
sacerdotal /,sækə'dəʊt(ə)l/ *adj.* of priests or priestly office. [Latin *sacerdos -dot-* priest]
sachet /'sæʃeɪ/ *n.* **1** small bag or packet containing shampoo etc. **2** small scented bag for perfuming drawers etc. [French diminutive: related to SAC]
sack[1] *—n.* **1 a** large strong bag for storage or conveyance. **b** quantity contained in a sack. **2** (prec. by *the*) *colloq.* dismissal from employment. **3** (prec. by *the*) *US slang* bed. *—v.* **1** put into a sack or sacks. **2** *colloq.* dismiss from employment. [Latin *saccus*]
sack[2] *—v.* plunder and destroy (a captured town etc.). *—n.* such sacking. [French *mettre à sac* put in a sack]
sack[3] *n. hist.* white wine from Spain and the Canaries. [French *vin sec* dry wine]
sackbut /'sækbʌt/ *n.* early form of trombone. [French]
sackcloth *n.* **1** coarse fabric of flax or hemp used for sacks. **2** clothing for penance or mourning (esp. *sackcloth and ashes*).
sacking *n.* material for making sacks; sackcloth.
sacral /'seɪkr(ə)l/ *adj.* **1** *Anat.* of the sacrum. **2** of or for sacred rites. [Latin *sacrum* sacred]
sacrament /'sækrəmənt/ *n.* **1** symbolic Christian ceremony, e.g. baptism and Eucharist. **2** (also **Blessed** or **Holy Sacrament**) (prec. by *the*) Eucharist. **3** sacred thing. □ **sacramental** /-'ment(ə)l/ *adj.* [Latin: related to SACRED]
sacred /'seɪkrɪd/ *adj.* **1 a** (often foll. by *to*) dedicated to a god. **b** connected with religion (*sacred music*). **2** safeguarded or required esp. by tradition; inviolable. [Latin *sacer* holy]
sacred cow *n. colloq.* traditionally hallowed idea or institution.
sacrifice /'sækrɪ,faɪs/ *—n.* **1 a** voluntary relinquishing of something valued. **b** thing so relinquished. **c** the loss entailed. **2 a** slaughter of an animal or person or surrender of a possession, as an offering to a deity. **b** animal, person, or thing so offered. *—v.* (**-cing**) **1** give up (a thing) as a sacrifice. **2** (foll. by *to*) devote or give over to. **3** (also *absol.*) offer or kill as a sacrifice. □ **sacrificial** /-'fɪʃ(ə)l/ *adj.* [Latin: related to SACRED]

sacrilege /ˈsækrɪlɪdʒ/ *n.* violation of what is regarded as sacred. □ **sacrilegious** /-ˈlɪdʒəs/ *adj.* [Latin: related to SACRED, *lego* take]

sacristan /ˈsækrɪst(ə)n/ *n.* person in charge of a sacristy and church contents. [medieval Latin: related to SACRED]

sacristy /ˈsækrɪstɪ/ *n.* (*pl.* **-ies**) room in a church where vestments, sacred vessels, etc., are kept. [medieval Latin: related to SACRED]

sacrosanct /ˈsækrəʊ,sæŋkt/ *adj.* most sacred; inviolable. □ **sacrosanctity** /-ˈsæŋkt-/ *n.* [Latin: related to SACRED, SAINT]

sacrum /ˈseɪkrəm/ *n.* (*pl.* **sacra** or **-s**) triangular bone between the two hipbones. [Latin *os sacrum* sacred bone]

sad *adj.* (**sadder, saddest**) **1** unhappy. **2** causing sorrow. **3** regrettable. **4** shameful, deplorable. □ **sadden** *v.* **sadly** *adv.* **sadness** *n.* [Old English]

saddle /ˈsæd(ə)l/ **–***n.* **1** seat of leather etc. strapped on a horse etc. for riding. **2** bicycle etc. seat. **3** joint of meat consisting of the two loins. **4** ridge rising to a summit at each end. **–***v.* (**-ling**) **1** put a saddle on (a horse etc.). **2** (foll. by *with*) burden (a person) with a task etc. □ **in the saddle 1** mounted. **2** in office or control. [Old English]

saddleback *n.* **1** roof of a tower with two opposite gables. **2** hill with a concave upper outline. **3** black pig with a white stripe across the back. □ **saddlebacked** *adj.*

saddle-bag *n.* **1** each of a pair of bags laid across the back of a horse etc. **2** bag attached to a bicycle saddle etc.

saddler *n.* maker of or dealer in saddles etc.

saddlery /ˈsædlərɪ/ *n.* (*pl.* **-ies**) saddler's goods, trade, or premises.

Sadducee /ˈsædjʊ,siː/ *n.* member of a Jewish sect of the time of Christ that denied the resurrection of the dead. [Hebrew]

sadhu /ˈsɑːduː/ *n.* (in India) holy man, sage, or ascetic. [Sanskrit]

sadism /ˈseɪdɪz(ə)m/ *n.* **1** *colloq.* enjoyment of cruelty to others. **2** sexual perversion characterized by this. □ **sadist** *n.* **sadistic** /səˈdɪstɪk/ *adj.* **sadistically** /səˈdɪstɪkəlɪ/ *adv.* [de *Sade*, name of an author]

sado-masochism /,seɪdəʊˈmæsə,kɪz(ə)m/ *n.* sadism and masochism in one person. □ **sado-masochist** *n.* **sado-masochistic** /-ˈkɪstɪk/ *adj.*

s.a.e. *abbr.* stamped addressed envelope.

safari /səˈfɑːrɪ/ *n.* (*pl.* **-s**) expedition, esp. in Africa, to observe or hunt animals (*go on safari*). [Swahili from Arabic *safara* to travel]

safari park *n.* park where wild animals are kept in the open for viewing from vehicles.

safe –*adj.* **1** free of danger or injury. **2** secure, not risky (*in a safe place*). **3** reliable, certain. **4** prevented from escaping or doing harm (*have got him safe*). **5** (also **safe and sound**) uninjured; with no harm done. **6** cautious, unenterprising. **–***n.* **1** strong lockable cabinet etc. for valuables. **2** = MEAT SAFE. □ **on the safe side** with a margin for error. □ **safely** *adv.* [French *sauf* from Latin *salvus*]

safe conduct *n.* **1** immunity given from arrest or harm. **2** document securing this.

safe deposit *n.* building containing strongrooms and safes for hire.

safeguard –*n.* protecting proviso, circumstance, etc. **–***v.* guard or protect (rights etc.).

safe house *n.* place of refuge etc. for spies, terrorists, etc.

safe keeping *n.* preservation in a safe place.

safe period *n.* time during the month when conception is least likely.

safe sex *n.* sexual activity in which precautions are taken against sexually transmitted diseases, esp. Aids.

safety *n.* being safe; freedom from danger or risk.

safety-belt *n.* **1** = SEAT-BELT. **2** belt or strap worn to prevent injury.

safety-catch *n.* device preventing a gun-trigger or machinery from being operated accidentally.

safety curtain *n.* fireproof curtain between a stage and auditorium.

safety lamp *n.* miner's lamp so protected as not to ignite firedamp.

safety match *n.* match igniting only on a specially prepared surface.

safety net *n.* net placed to catch an acrobat etc. in case of a fall.

safety pin *n.* pin with a guarded point.

safety razor *n.* razor with a guard to prevent cutting the skin.

safety-valve *n.* **1** (in a steam boiler) automatic valve relieving excess pressure. **2** means of venting excitement etc. harmlessly.

saffron /ˈsæfrən/ **–***n.* **1** deep yellow food colouring and flavouring made from dried crocus stigmas. **2** colour of this. **–***adj.* deep yellow. [French from Arabic]

sag –*v.* (**-gg-**) **1** sink or subside, esp. unevenly. **2** have a downward bulge or curve in the middle. **3** fall in price. **–***n.* state or extent of sagging. □ **saggy** *adj.* [Low German or Dutch]

saga /ˈsɑːgə/ *n.* **1** long heroic story, esp. medieval Icelandic or Norwegian. **2** series of connected novels concerning a family's history etc. **3** long involved story. [Old Norse: related to SAW[3]]

sagacious /səˈgeɪʃ(ə)s/ *adj.* showing insight or good judgement. □ **sagacity** /səˈgæsɪtɪ/ *n.* [Latin *sagax -acis*]

sage[1] *n.* culinary herb with dull greyish-green leaves. [French from Latin SALVIA]

sage[2] *–n.* often *iron.* wise man. *–adj.* wise, judicious, experienced. □ **sagely** *adv.* [French from Latin *sapio* be wise]

sagebrush *n.* growth of shrubby aromatic plants in some semi-arid regions of western N. America.

Sagittarius /ˌsædʒɪˈteərɪəs/ *n.* (*pl.* **-es**) **1** constellation and ninth sign of the zodiac (the Archer). **2** person born when the sun is in this sign. □ **Sagittarian** *adj.* & *n.* [Latin, = archer]

sago /ˈseɪgəʊ/ *n.* (*pl.* **-s**) **1** a starch used in puddings etc. **2** (in full **sago palm**) any of several tropical palms and cycads yielding this. [Malay]

sahib /sɑːb/ *n. hist.* (in India) form of address to European man. [Arabic, = lord]

said *past* and *past part.* of SAY.

sail *–n.* **1** piece of material extended on rigging to catch the wind and propel a boat or ship. **2** ship's sails collectively. **3** voyage or excursion in a sailing-boat. **4** ship, esp. as discerned from its sails. **5** wind-catching apparatus of a windmill. *–v.* **1** travel on water by the use of sails or engine-power. **2** begin a voyage (*sails at nine*). **3 a** navigate (a ship etc.). **b** travel on (a sea). **4** set (a toy boat) afloat. **5** glide or move smoothly or in a stately manner. **6** (often foll. by *through*) *colloq.* succeed easily (*sailed through the exams*). □ **sail close to the wind 1** sail as nearly against the wind as possible. **2** come close to indecency or dishonesty. **sail into** *colloq.* attack physically or verbally. **under sail** with sails set. [Old English]

sailboard *n.* board with a mast and sail, used in windsurfing. □ **sailboarder** *n.* **sailboarding** *n.*

sailcloth *n.* **1** material used for sails. **2** canvas-like dress material.

sailing-boat *n.* (also **sailing-ship**) vessel driven by sails.

sailor *n.* **1** member of a ship's crew, esp. one below the rank of officer. **2** person considered with regard to seasickness (*a good sailor*). [originally *sailer*: see -ER[1]]

sailplane *n.* glider designed for sustained flight.

sainfoin /ˈsænfɔɪn/ *n.* pink-flowered fodder-plant. [Latin *sanctus* holy, *foenum* hay]

saint /seɪnt, before a name usu. sənt/ *–n.* (*abbr.* **St** or **S**; *pl.* **Sts** or **SS**) **1** holy or (in some Churches) formally canonized person regarded as worthy of special veneration. **2** very virtuous person. *–v.* (as **sainted** *adj.*) saintly. □ **sainthood** *n.* **saintlike** *adj.* [Latin *sanctus* holy]

St Bernard /ˈbɜːnəd/ *n.* (in full **St Bernard dog**) very large dog of a breed orig. kept in the Alps to rescue travellers.

St John's wort /dʒɒnz/ *n.* yellow-flowered plant.

St Leger /ˈledʒə(r)/ *n.* horse-race at Doncaster for three-year-olds. [from the name of the founder]

saintly *adj.* (**-ier, -iest**) very holy or virtuous. □ **saintliness** *n.*

St Vitus's dance /ˈvaɪtəsɪz/ *n.* disease producing involuntary convulsive movements of the body.

sake[1] *n.* □ **for Christ's** (or **God's** or **goodness'** or **Heaven's** or **Pete's** etc.) **sake** expression of impatience, supplication, anger, etc. **for the sake of** (or **for one's sake**) out of consideration for; in the interest of; because of; in order to please, honour, get, or keep. [Old English]

sake[2] /ˈsɑːkɪ/ *n.* Japanese rice wine. [Japanese]

salaam /səˈlɑːm/ *–n.* **1** (chiefly as a Muslim greeting) Peace! **2** Muslim low bow with the right palm on the forehead. **3** (in *pl.*) respectful compliments. *–v.* make a salaam (to). [Arabic]

salacious /səˈleɪʃəs/ *adj.* **1** indecently erotic. **2** lecherous. □ **salaciousness** *n.* **salacity** /səˈlæsɪtɪ/ *n.* [Latin *salax -acis*: related to SALIENT]

salad /ˈsæləd/ *n.* cold mixture of usu. raw vegetables, often with a dressing. [French *salade* from Latin *sal* salt]

salad cream *n.* creamy salad-dressing.

salad days *n.* period of youthful inexperience.

salad-dressing *n.* = DRESSING 2a.

salamander /ˈsæləˌmændə(r)/ *n.* **1** tailed newtlike amphibian once thought able to endure fire. **2** similar mythical creature. [Greek *salamandra*]

salami /səˈlɑːmɪ/ *n.* (*pl.* **-s**) highly-seasoned orig. Italian sausage. [Italian]

sal ammoniac /ˌsæl əˈməʊnɪˌæk/ *n.* ammonium chloride, a white crystalline salt. [Latin *sal* salt, *ammoniacus* of Jupiter Ammon]

salary /ˈsælərɪ/ *–n.* (*pl.* **-ies**) fixed regular wages, usu. monthly or quarterly, esp. for white-collar work. *–v.* (**-ies, -ied**) (usu. as **salaried** *adj.*) pay a salary to. [Latin *salarium* money for buying salt]

sale *n.* **1** exchange of a commodity for money etc.; act or instance of selling. **2** amount sold (*sales were enormous*). **3** temporary offering of goods at reduced prices. **4 a** event at which goods are sold. **b** public auction. □ **on** (or **for**) **sale** offered for purchase. [Old English]

saleable *adj.* fit or likely to be sold. □ **saleability** /-'bɪlɪtɪ/ *n.*

sale of work *n.* sale of home-made goods etc. for charity.

sale or return *n.* arrangement by which a purchaser may return surplus goods to the supplier without payment.

saleroom *n.* room where auctions are held.

salesman *n.* **1** man employed to sell goods. **2** *US* commercial traveller.

salesmanship *n.* skill in selling.

salesperson *n.* salesman or saleswoman.

sales talk *n.* persuasive talk promoting goods or an idea etc.

saleswoman *n.* woman employed to sell goods.

salicylic acid /ˌsælɪ'sɪlɪk/ *n.* chemical used as a fungicide and in aspirin and dyes. □ **salicylate** /sə'lɪsɪˌleɪt/ *n.* [Latin *salix* willow]

salient /'seɪlɪənt/ *–adj.* **1** prominent, conspicuous. **2** (of an angle, esp. in fortification) pointing outwards. *–n.* salient angle or part of a fortification; outward bulge in a military line. [Latin *salio* leap]

saline /'seɪlaɪn/ *–adj.* **1** containing salt or salts. **2** tasting of salt. **3** of chemical salts. **4** of the nature of a salt. *–n.* **1** salt lake, spring, etc. **2** saline solution. □ **salinity** /sə'lɪnɪtɪ/ *n.* **salinization** /ˌsælɪnaɪ'zeɪʃ(ə)n/ *n.* [Latin *sal* salt]

saliva /sə'laɪvə/ *n.* colourless liquid secreted into the mouth by glands. □ **salivary** /sə'laɪ-, 'sælɪ-/ *adj.* [Latin]

salivate /'sælɪˌveɪt/ *v.* (**-ting**) secrete saliva, esp. in excess. □ **salivation** /-'veɪʃ(ə)n/ *n.* [Latin *salivare*: related to SALIVA]

sallow[1] /'sæləʊ/ *adj.* (**-er, -est**) (esp. of the skin) yellowish. [Old English]

sallow[2] /'sæləʊ/ *n.* **1** low-growing willow. **2** a shoot or the wood of this. [Old English]

sally /'sælɪ/ (*pl.* **-ies**) *–n.* **1** sudden military charge; sortie. **2** excursion. **3** witticism. *–v.* (**-ies, -ied**) **1** (usu. foll. by *out, forth*) set out on a walk, journey, etc. **2** (usu. foll. by *out*) make a military sally. [French *saillie* from Latin *salio* leap]

salmon /'sæmən/ *–n.* (*pl.* usu. same or **-s**) large expensive edible fish with orange-pink flesh. *–adj.* salmon-pink. [Latin *salmo*]

salmonella /ˌsælmə'nelə/ *n.* (*pl.* **-llae** /-liː/) **1** bacterium causing food poisoning. **2** such food poisoning. [*Salmon*, name of a veterinary surgeon]

salmon pink *adj.* & *n.* (as adj. often hyphenated) orange-pink colour of salmon flesh.

salmon trout *n.* large silver-coloured trout.

salon /'sælɒn/ *n.* **1** room or establishment of a hairdresser, beautician, etc. **2** *hist.* meeting of eminent people in the home of a lady of fashion. **3** reception room, esp. of a continental house. [French: related to SALOON]

saloon /sə'luːn/ *n.* **1 a** large room or hall on a ship, in a hotel, etc. **b** public room for a specified purpose (*billiard-saloon*). **2** (in full **saloon car**) (usu. four-seater) car with the body closed off from the luggage area. **3** *US* drinking-bar. **4** (in full **saloon bar**) more comfortable bar in a public house. [French *salon*]

salsa /'sælsə/ *n.* a kind of dance music of Cuban origin, with jazz and rock elements. [Spanish: related to SAUCE]

salsify /'sælsɪfɪ/ *n.* (*pl.* **-ies**) plant with long fleshy edible roots. [French from Italian]

SALT /sɔːlt, sɒlt/ *abbr.* Strategic Arms Limitation Talks (or Treaty).

salt /sɔːlt, sɒlt/ *–n.* **1** (also **common salt**) sodium chloride, esp. mined or evaporated from sea water, and used for seasoning or preserving food. **2** chemical compound formed from the reaction of an acid with a base. **3** piquancy; wit. **4** (in *sing.* or *pl.*) **a** substance resembling salt in taste, form, etc. (*bath salts*). **b** (esp. in *pl.*) substance used as a laxative. **5** (also **old salt**) experienced sailor. **6** = SALT-CELLAR. *–adj.* containing, tasting of, or preserved with salt. *–v.* **1** cure, preserve, or season with salt or brine. **2** sprinkle (a road etc.) with salt. □ **salt away** (or **down**) *slang* put (money etc.) by. **the salt of the earth** most admirable or honest person or people (Matt. 5:13). **take with a pinch** (or **grain**) **of salt** regard sceptically. **worth one's salt** efficient, capable. [Old English]

salt-cellar *n.* container for salt at table. [earlier *salt saler* from French *salier* salt-box]

salting *n.* (esp. in *pl.*) marsh overflowed by the sea.

saltire /'sɔːlˌtaɪə(r)/ *n.* X-shaped cross dividing a shield in four. [French *sautoir* stile]

salt-lick *n.* place where animals lick salt from the ground.

salt-mine *n.* mine yielding rock-salt.

salt-pan *n.* vessel, or depression near the sea, used for getting salt by evaporation.

saltpetre /ˌsɒlt'piːtə(r), ˌsɔːlt-/ *n.* (*US* **saltpeter**) white crystalline salty substance used in preserving meat and in gunpowder. [Latin *sal petrae*, = salt of rock]

salt-water *adj.* of or living in the sea.

salty *adj.* (-**ier**, -**iest**) **1** tasting of or containing salt. **2** (of wit etc.) piquant. □ **saltiness** *n.*

salubrious /sə'luːbrɪəs/ *adj.* health-giving; healthy. □ **salubrity** *n.* [Latin *salus* health]

saluki /sə'luːkɪ/ *n.* (*pl.* **-s**) dog of a tall slender silky-coated breed. [Arabic]

salutary /'sæljʊtərɪ/ *adj.* having a good effect. [Latin: related to SALUTE]

salutation /ˌsæljuː'teɪʃ(ə)n/ *n. formal* sign or expression of greeting.

salute /sə'luːt/ –*n.* **1** gesture of respect, homage, greeting etc. **2** *Mil. & Naut.* prescribed gesture or use of weapons or flags as a sign of respect etc. **3** ceremonial discharge of a gun or guns. –*v.* (**-ting**) **1 a** make a salute to. **b** (often foll. by *to*) perform a salute. **2** greet. **3** commend. [Latin *salus -ut-* health]

salvage /'sælvɪdʒ/ –*n.* **1** rescue of property from the sea, a fire, etc. **2** property etc. so saved. **3 a** saving and use of waste materials. **b** materials salvaged. –*v.* (**-ging**) **1** save from a wreck etc. **2** retrieve from a disaster etc. (*salvaged her pride*). □ **salvageable** *adj.* [Latin: related to SAVE[1]]

salvation /sæl'veɪʃ(ə)n/ *n.* **1** saving or being saved. **2** deliverance from sin and damnation. **3** religious conversion. **4** person or thing that saves. □ **salvationist** *n.* (esp. with ref. to the Salvation Army). [Latin: related to SAVE[1]]

Salvation Army *n.* worldwide evangelical Christian quasi-military organization helping the poor.

salve[1] –*n.* **1** healing ointment. **2** (often foll. by *for*) thing that soothes or consoles. –*v.* (**-ving**) soothe. [Old English]

salve[2] *v.* (**-ving**) save from wreck or fire etc. □ **salvable** *adj.* [back-formation from SALVAGE]

salver *n.* tray, esp. silver, for drinks, letters, etc. [Spanish *salva* assaying of food]

salvia /'sælvɪə/ *n.* garden plant of the sage family with red or blue flowers. [Latin, = SAGE[1]]

salvo /'sælvəʊ/ *n.* (*pl.* **-es** or **-s**) **1** simultaneous discharge of guns etc. **2** round of applause. [Italian *salva*]

sal volatile /ˌsæl və'lætɪlɪ/ *n.* solution of ammonium carbonate used as smelling-salts. [Latin, = volatile salt]

SAM *abbr.* surface-to-air missile.

Samaritan /sə'mærɪt(ə)n/ *n.* **1** (in full **good Samaritan**) charitable or helpful person (Luke 10:33 etc.). **2** member of a counselling organization. [originally = inhabitant of ancient Samaria]

samarium /sə'meərɪəm/ *n.* metallic element of the lanthanide series. [ultimately from *Samarski*, name of an official]

samba /'sæmbə/ –*n.* **1** ballroom dance of Brazilian origin. **2** music for this. –*v.* (**-bas**, **-baed** or **-ba'd** /-bəd/, **-baing** /-bəɪŋ/) dance the samba. [Portuguese]

same –*adj.* **1** (often prec. by *the*) identical; not different (*on the same bus*). **2** unvarying (*same old story*). **3** (usu. prec. by *this, these, that, those*) just mentioned (*this same man later died*). –*pron.* (prec. by *the*) **1** the same person or thing. **2** *Law* or *archaic* the person or thing just mentioned. –*adv.* (usu. prec. by *the*) similarly; in the same way (*feel the same*). □ **all** (or **just**) **the same 1** nevertheless. **2** emphatically the same. **at the same time 1** simultaneously. **2** notwithstanding. **be all** (or **just**) **the same to** make no difference to. **same here** *colloq.* the same applies to me. □ **sameness** *n.* [Old Norse]

samizdat /ˌsæmɪz'dæt/ *n.* clandestine publication of banned literature. [Russian]

samosa /sə'məʊsə/ *n.* fried triangular pastry containing spiced vegetables or meat. [Hindustani]

samovar /'sæməˌvɑː(r)/ *n.* Russian tea-urn. [Russian]

Samoyed /'sæməˌjed/ *n.* **1** member of a people of northern Siberia. **2** (also **samoyed**) dog of a white Arctic breed. [Russian]

sampan /'sæmpæn/ *n.* small boat used in the Far East. [Chinese]

samphire /'sæmˌfaɪə(r)/ *n.* edible maritime rock-plant. [French, = St Peter('s herb)]

sample /'sɑːmp(ə)l/ –*n.* **1** small representative part or quantity. **2** specimen. **3** illustrative or typical example. –*v.* (**-ling**) **1** take or give samples of. **2** try the qualities of. **3** experience briefly. [Anglo-French: related to EXAMPLE]

sampler[1] *n.* piece of embroidery using various stitches as a specimen of proficiency. [French: related to EXEMPLAR]

sampler[2] *n.* **1** person or thing that samples. **2** *US* collection of representative items etc.

sampling *n.* technique of digitally encoding a piece of sound and re-using it as part of a composition or recording.

Samson /'sæms(ə)n/ *n.* person of great strength. [*Samson* in the Old Testament]

samurai /'sæmʊ,raɪ/ *n.* (*pl.* same) **1** Japanese army officer. **2** *hist.* member of a Japanese military caste. [Japanese]

sanatorium /,sænə'tɔ:rɪəm/ *n.* (*pl.* **-s** or **-ria**) **1** residential clinic, esp. for convalescents and the chronically sick. **2** room etc. for sick people in a school etc. [Latin *sano* heal]

sanctify /'sæŋktɪ,faɪ/ *v.* (**-ies, -ied**) **1** consecrate; treat as holy. **2** free from sin. **3** justify; sanction. □ **sanctification** /-fɪ'keɪʃ(ə)n/ *n.* [Latin *sanctus* holy]

sanctimonious /,sæŋktɪ'məʊnɪəs/ *adj.* ostentatiously pious. □ **sanctimoniously** *adv.* **sanctimoniousness** *n.* **sanctimony** /'sæŋktɪmənɪ/ *n.* [Latin *sanctimonia* sanctity]

sanction /'sæŋkʃ(ə)n/ —*n.* **1** approval by custom or tradition; express permission. **2** confirmation of a law etc. **3** penalty for disobeying a law or rule, or a reward for obeying it. **4** *Ethics* moral force encouraging obedience to any rule of conduct. **5** (esp. in *pl.*) (esp. economic) action by a State against another to abide by an international agreement etc. —*v.* **1** authorize or agree to (an action etc.). **2** ratify; make (a law etc.) binding. [Latin *sancio sanct-* make sacred]

sanctity /'sæŋktɪtɪ/ *n.* holiness, sacredness; inviolability. [Latin *sanctus* holy]

sanctuary /'sæŋktʃʊərɪ/ *n.* (*pl.* **-ies**) **1** holy place. **2 a** holiest part of a temple etc. **b** chancel. **3** place where birds, wild animals, etc., are bred and protected. **4** place of refuge.

sanctum /'sæŋktəm/ *n.* (*pl.* **-s**) **1** holy place. **2** *colloq.* study, den.

sand —*n.* **1** fine loose grains resulting from the erosion of esp. siliceous rocks and forming the seashore, deserts, etc. **2** (in *pl.*) **a** grains of sand. **b** expanse of sand. **c** sandbank. —*v.* smooth with sandpaper or sand. [Old English]

sandal[1] /'sænd(ə)l/ *n.* shoe with an openwork upper or no upper, usu. fastened by straps. [Latin from Greek]

sandal[2] /'sænd(ə)l/ *n.* = SANDALWOOD. [Sanskrit *candana*]

sandal-tree *n.* tree yielding sandalwood.

sandalwood *n.* **1** scented wood of a sandal-tree. **2** perfume from this.

sandbag —*n.* bag filled with sand, used for temporary defences etc. —*v.* (**-gg-**) defend or hit with sandbag(s).

sandbank *n.* sand forming a shallow place in the sea or a river.

sandblast —*v.* roughen, treat, or clean with a jet of sand driven by compressed air or steam. —*n.* this jet. □ **sandblaster** *n.*

sandboy *n.* □ **happy as a sandboy** extremely happy or carefree. [probably = a boy hawking sand for sale]

sandcastle *n.* model castle made of sand at the seashore.

sand-dune *n.* (also **sand-hill**) = DUNE.

sander *n.* power tool for sanding.

sandman *n.* imaginary person causing tiredness in children.

sand-martin *n.* bird nesting in sandy banks.

sandpaper —*n.* paper with an abrasive coating for smoothing or polishing. —*v.* rub with this.

sandpiper *n.* wading bird frequenting wet sandy areas.

sandpit *n.* pit containing sand, for children to play in.

sandstone *n.* sedimentary rock of compressed sand.

sandstorm *n.* storm with clouds of sand raised by the wind.

sandwich /'sænwɪdʒ/ —*n.* **1** two or more slices of bread with a filling. **2** layered cake with jam or cream. —*v.* **1** put (a thing, statement, etc.) between two of another character. **2** squeeze in between others (*sat sandwiched in the middle*). [from the Earl of *Sandwich*]

sandwich-board *n.* each of two boards worn front and back to carry advertisements.

sandwich course *n.* course with alternate periods of study and work experience.

sandy *adj.* (**-ier, -iest**) **1** having much sand. **2 a** (of hair) reddish. **b** sand-coloured. □ **sandiness** *n.*

sane *adj.* **1** of sound mind; not mad. **2** (of views etc.) moderate, sensible. [Latin *sanus* healthy]

sang *past* of SING.

sang-froid /sɑ̃'frwɑ:/ *n.* calmness in danger or difficulty. [French, = cold blood]

sangria /sæŋ'gri:ə/ *n.* Spanish drink of red wine with lemonade, fruit, etc. [Spanish, = bleeding]

sanguinary /'sæŋgwɪnərɪ/ *adj.* **1** bloody. **2** bloodthirsty. [Latin *sanguis -guin-* blood]

sanguine /'sæŋgwɪn/ *adj.* **1** optimistic, confident. **2** (of the complexion) florid, ruddy.

Sanhedrin /'sænɪdrɪn/ *n.* highest court of justice and the supreme council in ancient Jerusalem. [Greek *sunedrion* council]

sanitarium /,sænɪ'teərɪəm/ *n.* (*pl.* **-s** or **-ria**) *US* = SANATORIUM. [related to SANITARY]

sanitary /ˈsænɪtərɪ/ *adj.* **1** (of conditions etc.) affecting health. **2** hygienic. □ **sanitariness** *n.* [Latin *sanitas*: related to SANE]

sanitary towel *n.* (*US* **sanitary napkin**) absorbent pad used during menstruation.

sanitation /ˌsænɪˈteɪʃ(ə)n/ *n.* **1** sanitary conditions. **2** maintenance etc. of these. **3** disposal of sewage and refuse etc.

sanitize /ˈsænɪˌtaɪz/ *v.* (also **-ise**) (**-zing** or **-sing**) **1** make sanitary; disinfect. **2** *colloq.* censor (information etc.) to make it more acceptable.

sanity /ˈsænɪtɪ/ *n.* **1** being sane. **2** moderation. [Latin *sanitas*: related to SANE]

sank *past* of SINK.

sansculotte /ˌsænzkjʊˈlɒt/ *n.* (esp. in the French Revolution) extreme republican. [French, literally = 'without knee-breeches']

sanserif /sænˈserɪf/ *n.* (also **sans-serif**) form of type without serifs. [apparently from *sans* without, SERIF]

Sanskrit /ˈsænskrɪt/ *–n.* ancient and sacred language of the Hindus in India. *–adj.* of or in this language. [Sanskrit, = composed]

Santa Claus /ˈsæntə ˌklɔːz/ *n.* person said to bring children presents on Christmas Eve. [Dutch, = St Nicholas]

sap[1] *–n.* **1** vital juice circulating in plants. **2** vigour, vitality. **3** *slang* foolish person. *–v.* (**-pp-**) **1** drain or dry (wood) of sap. **2** weaken. [Old English]

sap[2] *–n.* tunnel or trench dug to get nearer to the enemy. *–v.* (**-pp-**) **1** dig saps. **2** undermine. [French *sappe* or Italian *zappa* spade]

sapient /ˈseɪpɪənt/ *adj. literary* **1** wise. **2** aping wisdom. □ **sapience** *n.* [Latin *sapio* be wise]

sapling *n.* young tree. [from SAP[1]]

sapper *n.* **1** person who digs saps. **2** soldier of the Royal Engineers (esp. as the official term for a private).

Sapphic /ˈsæfɪk/ *adj.* **1** of Sappho or her poetry. **2** lesbian. [Greek *Sappho*, poetess of Lesbos]

sapphire /ˈsæfaɪə(r)/ *–n.* **1** transparent blue precious stone. **2** its bright blue colour. *–adj.* (also **sapphire blue**) bright blue. [Greek *sappheiros* lapis lazuli]

sappy *adj.* (**-ier, -iest**) **1** full of sap. **2** young and vigorous.

saprophyte /ˈsæprəˌfaɪt/ *n.* plant or micro-organism living on dead or decayed organic matter. [Greek *sapros* rotten, *phuō* grow]

saraband /ˈsærəˌbænd/ *n.* **1** slow stately Spanish dance. **2** music for this. [Spanish *zarabanda*]

Saracen /ˈsærəs(ə)n/ *n. hist.* Arab or Muslim at the time of the Crusades. [Greek *sarakēnos*]

sarcasm /ˈsɑːˌkæz(ə)m/ *n.* ironically scornful language. □ **sarcastic** /sɑːˈkæstɪk/ *adj.* **sarcastically** /sɑːˈkæstɪkəlɪ/ *adv.* [Greek *sarkazō* speak bitterly]

sarcoma /sɑːˈkəʊmə/ *n.* (*pl.* **-s** or **-mata**) malignant tumour of connective tissue. [Greek *sarx sark-* flesh]

sarcophagus /sɑːˈkɒfəgəs/ *n.* (*pl.* **-phagi** /-ˌgaɪ/) stone coffin. [Greek, = flesh-consumer]

sardine /sɑːˈdiːn/ *n.* (*pl.* same or **-s**) young pilchard etc. sold in closely packed tins. □ **like sardines** crowded close together. [French from Latin]

sardonic /sɑːˈdɒnɪk/ *adj.* bitterly mocking or cynical. □ **sardonically** *adv.* [Greek *sardonios* Sardinian]

sardonyx /ˈsɑːdənɪks/ *n.* onyx in which white layers alternate with yellow or orange ones. [Greek *sardonux*]

sargasso /sɑːˈgæsəʊ/ *n.* (*pl.* **-s** or **-es**) (also **sargassum**) (*pl.* **-gassa**) seaweed with berry-like air-vessels. [Portuguese]

sarge *n. slang* sergeant. [abbreviation]

sari /ˈsɑriː/ *n.* (*pl.* **-s**) length of cloth draped round the body, traditionally worn by women of the Indian subcontinent. [Hindi]

sarky /ˈsɑːkɪ/ *adj.* (**-ier, -iest**) *slang* sarcastic. [abbreviation]

sarnie /ˈsɑːnɪ/ *n. colloq.* sandwich. [abbreviation]

sarong /səˈrɒŋ/ *n.* Malay and Javanese garment of a long strip of cloth tucked round the waist or under the armpits. [Malay]

sarsaparilla /ˌsɑːsəpəˈrɪlə/ *n.* **1** preparation of the dried roots of various plants, esp. smilax, used to flavour some drinks and medicines and formerly as a tonic. **2** plant yielding this. [Spanish]

sarsen /ˈsɑːs(ə)n/ *n.* sandstone boulder carried by ice during a glacial period. [from SARACEN]

sarsenet /ˈsɑːsnɪt/ *n.* soft silk material used esp. for linings. [Anglo-French from *sarzin* SARACEN]

sartorial /sɑːˈtɔːrɪəl/ *adj.* of men's clothes or tailoring. □ **sartorially** *adv.* [Latin *sartor* tailor]

SAS *abbr.* Special Air Service.

sash[1] *n.* strip or loop of cloth etc. worn over one shoulder or round the waist. [Arabic, = muslin]

sash[2] *n.* frame holding the glass in a sash-window. [from CHASSIS]

sashay /ˈsæʃeɪ/ *v.* esp. *US colloq.* walk or move ostentatiously, casually, or diagonally. [French *chassé*]

sash-cord *n.* strong cord attaching the sash-weights to a window sash.

sash-weight *n.* weight attached to each end of a window sash.

sash-window *n.* window sliding up and down in grooves.

sass *US colloq.* —*n.* impudence, cheek. —*v.* be impudent to. [var. of SAUCE]

sassafras /'sæsə,fræs/ *n.* **1** small N. American tree. **2** medicinal preparation from its leaves or bark. [Spanish or Portuguese]

Sassenach /'sæsə,næk/ *n. Scot.* & *Ir.* usu. *derog.* English person. [Gaelic *Sasunnoch*]

sassy *adj.* (**-ier, -iest**) esp. *US colloq.* impudent, cheeky. [var. of SAUCY]

SAT /sæt/ *abbr.* standard assessment task.

Sat. *abbr.* Saturday.

sat *past* and *past part.* of SIT.

Satan /'seɪt(ə)n/ *n.* the Devil; Lucifer. [Hebrew, = enemy]

satanic /sə'tænɪk/ *adj.* of or like Satan; hellish; evil. □ **satanically** *adv.*

Satanism /'seɪtə,nɪz(ə)m/ *n.* **1** worship of Satan. **2** pursuit of evil. □ **Satanist** *n.* & *adj.*

satchel /'sætʃ(ə)l/ *n.* small shoulder-bag for carrying school-books etc. [Latin: related to SACK[1]]

sate *v.* (**-ting**) *formal* gratify fully; surfeit. [probably dial. *sade* satisfy]

sateen /sæ'ti:n/ *n.* glossy cotton fabric like satin. [*satin* after *velveteen*]

satellite /'sætə,laɪt/ —*n.* **1** celestial or artificial body orbiting the earth or another planet. **2** (in full **satellite State**) small country controlled by another. —*attrib. adj.* transmitted by satellite (*satellite television*). [Latin *satelles -lit-* attendant]

satellite dish *n.* dish-shaped aerial for receiving satellite television.

satiate /'seɪʃɪ,eɪt/ *v.* (**-ting**) = SATE. □ **satiable** /-ʃəb(ə)l/ *adj.* **satiation** /-'eɪʃ(ə)n/ *n.* [Latin *satis* enough]

satiety /sə'taɪɪtɪ/ *n. formal* being sated. [Latin: related to SATIATE]

satin /'sætɪn/ —*n.* silk etc. fabric glossy on one side. —*adj.* smooth as satin. □ **satiny** *adj.* [Arabic *zaitūnī*]

satinwood *n.* a kind of yellow glossy timber.

satire /'sætaɪə(r)/ *n.* **1** ridicule, irony, etc., used to expose folly or vice etc. **2** work using this. □ **satirical** /sə'tɪrɪk(ə)l/ *adj.* **satirically** /sə'tɪrɪkəlɪ/ *adv.* [Latin *satira* medley]

satirist /'sætərɪst/ *n.* **1** writer of satires. **2** satirical person.

satirize /'sætə,raɪz/ *v.* (also **-ise**) (**-zing** or **-sing**) attack or describe with satire.

satisfaction /,sætɪs'fækʃ(ə)n/ *n.* **1** satisfying or being satisfied (*derived great satisfaction*). **2** thing that satisfies (*is a great satisfaction to me*). **3** (foll. by *for*) atonement; compensation (*demanded satisfaction*).

satisfactory /,sætɪs'fæktərɪ/ *adj.* adequate; giving satisfaction. □ **satisfactorily** *adv.*

satisfy /'sætɪs,faɪ/ *v.* (**-ies, -ied**) **1 a** meet the expectations or desires of. **b** be adequate. **2** meet (an appetite or want). **3** rid (a person) of such an appetite or want. **4** pay (a debt or creditor). **5** adequately fulfil or comply with (conditions etc.). **6** (often foll. by *of*, *that*) convince, esp. with proof etc. □ **satisfy oneself** (often foll. by *that*) become certain. [Latin *satisfacio*]

satrap /'sætræp/ *n.* **1** provincial governor in the ancient Persian empire. **2** subordinate ruler. [Persian, = protector of the land]

satsuma /sæt'su:mə/ *n.* variety of tangerine. [*Satsuma*, province in Japan]

saturate /'sætʃə,reɪt/ *v.* (**-ting**) **1** fill with moisture. **2** (often foll. by *with*) fill to capacity. **3** cause (a substance etc.) to absorb, hold, etc. as much as possible of another substance etc. **4** supply (a market) beyond demand. **5** (as **saturated** *adj.*) (of fat molecules) containing the greatest number of hydrogen atoms. [Latin *satur* full]

saturation /,sætʃə'reɪʃ(ə)n/ *n.* saturating or being saturated.

saturation point *n.* stage beyond which no more can be absorbed or accepted.

Saturday /'sætə,deɪ/ —*n.* day of the week following Friday. —*adv. colloq.* **1** on Saturday. **2** (**Saturdays**) on Saturdays; each Saturday. [Latin: related to SATURNALIA]

saturnalia /,sætə'neɪlɪə/ *n.* (*pl.* same or **-s**) **1** (usu. **Saturnalia**) *Rom. Hist.* festival of Saturn in December, the predecessor of Christmas. **2** (as *sing.* or *pl.*) scene of wild revelry. [Latin, pl. from *Saturnus* Roman god]

saturnine /'sætə,naɪn/ *adj.* of gloomy temperament or appearance.

satyr /'sætə(r)/ *n.* **1** (in Greek and Roman mythology) woodland god with some horselike or goatlike features. **2** lecherous man. [Greek *saturos*]

sauce /sɔ:s/ —*n.* **1** liquid or viscous accompaniment to a dish. **2** something adding piquancy or excitement. **3** *colloq.* impudence, impertinence, cheek. —*v.* (**-cing**) *colloq.* be impudent to; cheek. [Latin *salsus* salted]

sauce-boat *n.* jug or dish for serving sauces etc.

saucepan *n.* cooking pan, usu. round with a lid and a projecting handle, used on a hob.

saucer *n.* **1** shallow circular dish for standing a cup on. **2** thing of this shape. □ **saucerful** *n.* (*pl.* **-s**). [French *saussier*]

saucy *adj.* (**-ier**, **-iest**) impudent, cheeky. □ **saucily** *adv.* **sauciness** *n.*

sauerkraut /'saʊə,kraʊt/ *n.* German dish of pickled cabbage. [German]

sauna /'sɔ:nə/ *n.* **1** period spent in a special room heated very hot, to clean the body. **2** such a room. [Finnish]

saunter /'sɔ:ntə(r)/ *–v.* walk slowly; stroll. *–n.* leisurely walk. [origin unknown]

saurian /'sɔ:rɪən/ *adj.* of or like a lizard. [Greek *saura* lizard]

sausage /'sɒsɪdʒ/ *n.* **1 a** seasoned minced meat etc. in a cylindrical edible skin. **b** piece of this. **2** sausage-shaped object. □ **not a sausage** *colloq.* nothing at all. [French *saussiche*]

sausage meat *n.* minced meat used in sausages etc.

sausage roll *n.* sausage meat in a pastry roll.

sauté /'səʊteɪ/ *–attrib. adj.* (esp. of potatoes) fried quickly in a little fat. *–n.* food so cooked. *–v.* (**sautéd** or **sautéed**) cook in this way. [French *sauter* jump]

Sauternes /səʊ'tɜ:n/ *n.* sweet white wine from Sauternes in the Bordeaux region of France. [*Sauternes* in France]

savage /'sævɪdʒ/ *–adj.* **1** fierce; cruel. **2** wild; primitive. *–n.* **1** *derog.* member of a primitive tribe. **2** cruel or barbarous person. *–v.* (**-ging**) **1** attack and maul. **2** attack verbally. □ **savagely** *adv.* **savagery** *n.* (*pl.* **-ies**). [French from Latin *silva* a wood]

savannah /sə'vænə/ *n.* (also **savanna**) grassy plain in tropical and subtropical regions. [Spanish]

savant /'sæv(ə)nt, sæ'vɑ̃/ *n.* (*fem.* **savante** /'sæv(ə)nt/, /sæ'vɑ̃t/) learned person. [French]

save[1] *–v.* (**-ving**) **1** (often foll. by *from*) rescue or keep from danger, harm, etc. **2** (often foll. by *up*) keep (esp. money) for future use. **3 a** (often *refl.*) relieve (another or oneself) from spending (money, time, trouble, etc.); prevent exposure to (annoyance etc.). **b** obviate the need for. **4** preserve from damnation; convert. **5 a** avoid losing (a game, match, etc.). **b** prevent (a goal etc.) from being scored. *–n. Football* etc. prevention of a goal etc. □ **savable** *adj.* (also **saveable**). [Latin *salvo* from *salvus* safe]

save[2] *archaic* or *poet.* *–prep.* except; but. *–conj.* (often foll. by *for*) except; but. [Latin *salvo*, *salva*, ablative sing. of *salvus* safe]

save-as-you-earn *n.* saving by regular deduction from earnings at source.

saveloy /'sævə,lɔɪ/ *n.* seasoned dried smoked sausage. [Italian *cervellata*]

saver *n.* **1** person who saves esp. money. **2** (often in *comb.*) thing that saves (time etc.). **3** cheap (esp. off-peak) fare.

saving *–adj.* (often in *comb.*) making economical use of (*labour-saving*). *–n.* **1** anything that is saved. **2** an economy (*a saving in expenses*). **3** (usu. in *pl.*) money saved. **4** act of preserving or rescuing. *–prep.* **1** except. **2** without offence to (*saving your presence*).

saving grace *n.* redeeming quality.

savings bank *n.* bank paying interest on small deposits.

savings certificate *n.* interest-bearing Government certificate issued to savers.

saviour /'seɪvjə(r)/ *n.* (*US* **savior**) **1** person who saves from danger etc. **2** (**Saviour**) (prec. by *the*, *our*) Christ. [Latin: related to SAVE[1]]

savoir faire /,sævwɑ: 'feə(r)/ *n.* ability to behave appropriately; tact. [French]

savor *US* var. of SAVOUR.

savory[1] /'seɪvərɪ/ *n.* (*pl.* **-ies**) aromatic herb used esp. in cookery. [Latin *satureia*]

savory[2] *US* var. of SAVOURY.

savour /'seɪvə(r)/ (*US* **savor**) *–n.* **1** characteristic taste, flavour, etc. **2** hint of a different quality etc. in something. *–v.* **1** appreciate and enjoy (food, an experience, etc.). **2** (foll. by *of*) imply or suggest (a specified quality). [Latin *sapor*]

savoury /'seɪvərɪ/ (*US* **savory**) *–adj.* **1** having an appetizing taste or smell. **2** (of food) salty or piquant, not sweet. **3** pleasant; acceptable. *–n.* (*pl.* **-ies**) savoury dish. □ **savouriness** *n.*

savoy /sə'vɔɪ/ *n.* cabbage with wrinkled leaves. [*Savoy* in SE France]

savvy /'sævɪ/ *slang* *–v.* (**-ies**, **-ied**) know. *–n.* knowingness; understanding. *–adj.* (**-ier**, **-iest**) *US* knowing; wise. [Pidgin alteration of Spanish *sabe usted* you know]

saw[1] *–n.* **1** hand tool with a toothed blade used to cut esp. wood with a to-and-fro movement. **2** power tool with a toothed rotating disk or moving band, for cutting. *–v.* (*past part.* **sawn** or **sawed**) **1** cut (wood etc.) or make (boards etc.) with a saw. **2** use a saw. **3 a** move with a sawing motion (*sawing away on his violin*). **b** divide (the air etc.) with gesticulations. [Old English]

saw[2] *past* of SEE[1].

saw[3] *n.* proverb; maxim. [Old English: related to SAY]
sawdust *n.* powdery wood particles produced in sawing.
sawfish *n.* (*pl.* same or **-es**) large marine fish with a toothed flat snout.
sawmill *n.* factory for sawing planks.
sawn *past part.* of SAW[1].
sawn-off *adj.* (*US* **sawed-off**) (of a shotgun) with part of the barrel sawn off.
sawtooth *adj.* (also **sawtoothed**) serrated.
sawyer *n.* person who saws timber, esp. for a living.
sax *n. colloq.* saxophone. [abbreviation]
saxe /sæks/ *n.* & *adj.* (in full **saxe blue**; as adj. often hyphenated) light greyish-blue colour. [French, = Saxony]
saxifrage /'sæksɪ,freɪdʒ/ *n.* rock-plant with small white, yellow, or red flowers. [Latin *saxum* rock, *frango* break]
Saxon /'sæks(ə)n/ *–n.* **1** *hist.* **a** member of the Germanic people that conquered parts of England in 5th–6th c. **b** (usu. **Old Saxon**) language of the Saxons. **2** = ANGLO-SAXON. *–adj.* **1** *hist.* of the Saxons. **2** = ANGLO-SAXON. [Latin *Saxo -onis*]
saxophone /'sæksə,fəʊn/ *n.* metal woodwind reed instrument used esp. in jazz. □ **saxophonist** /-'sɒfənɪst/ *n.* [*Sax*, name of the maker]
say *–v.* (*3rd sing. present* **says** /sez/; *past* and *past part.* **said** /sed/) **1** (often foll. by *that*) **a** utter (specified words); remark. **b** express (*say what you feel*). **2** (often foll. by *that*) **a** state; promise or prophesy. **b** have specified wording; indicate (*clock says ten to six*). **3** (in *passive*; usu. foll. by *to* + infin.) be asserted (*is said to be old*). **4** (foll. by *to* + infin.) *colloq.* tell to do (*he said to hurry*). **5** convey (information) (*spoke, but said little*). **6** offer as an argument or excuse (*much to be said in favour of it*). **7** (often *absol.*) give an opinion or decision as to (*hard to say*). **8** take as an example or as near enough (*paid, say, £20*). **9** recite or repeat (prayers, Mass, tables, a lesson, etc.). **10** convey (inner meaning etc.) (*what is the poem saying?*). **11** (**the said**) *Law* or *joc.* the previously mentioned. *–n.* **1** opportunity to express a view (*let him have his say*). **2** share in a decision (*had no say in it*). □ **I'll say** *colloq.* yes indeed. **I say!** exclamation of surprise etc. or drawing attention. **that is to say** in other words, more explicitly. [Old English]
SAYE *abbr.* save-as-you-earn.
saying *n.* maxim, proverb, etc. □ **go without saying** be too obvious to need mention.
say-so *n. colloq.* **1** power of decision. **2** mere assertion (*his say-so is not enough*).
Sb *symb.* antimony. [Latin *stibium*]
S-bend *n.* S-shaped bend in a road or pipe.
Sc *symb.* scandium.
sc. *abbr.* scilicet.
s.c. *abbr.* small capitals.
scab *–n.* **1** crust over a healing cut, sore, etc. **2** (often *attrib.*) *colloq. derog.* blackleg. **3** skin disease, esp. in animals. **4** fungous plant disease. *–v.* (**-bb-**) **1** *colloq. derog.* act as a blackleg. **2** form a scab, heal over. □ **scabby** *adj.* (**-ier, -iest**). [Old Norse: cf. SHABBY]
scabbard /'skæbəd/ *n. hist.* sheath of a sword etc. [Anglo-French]
scabies /'skeɪbi:z/ *n.* contagious skin disease causing itching. [Latin]
scabious /'skeɪbɪəs/ *n.* plant with esp. blue pincushion-shaped flowers. [medieval Latin *scabiosa* (*herba*) named as curing scabies]
scabrous /'skeɪbrəs/ *adj.* **1** rough, scaly. **2** indecent, salacious. [Latin]
scaffold /'skæfəʊld/ *n.* **1** *hist.* platform for the execution of criminals. **2** = SCAFFOLDING. [Romanic: related to EX-[1], CATAFALQUE]
scaffolding *n.* **1 a** temporary structure of poles, planks, etc., for building work. **b** materials for this. **2** any temporary framework.
scalar /'skeɪlə(r)/ *Math.* & *Physics* *–adj.* (of a quantity) having only magnitude, not direction. *–n.* scalar quantity. [Latin: related to SCALE[3]]
scalawag var. of SCALLYWAG.
scald /skɔ:ld, skɒld/ *–v.* **1** burn (the skin etc.) with hot liquid or steam. **2** heat (esp. milk) to near boiling point. **3** (usu. foll. by *out*) clean with boiling water. *–n.* burn etc. caused by scalding. [Latin *excaldo* from *calidus* hot]
scale[1] *–n.* **1** each of the thin horny plates protecting the skin of fish and reptiles. **2** something resembling this. **3** white deposit formed in a kettle etc. by hard water. **4** tartar formed on teeth. *–v.* (**-ling**) **1** remove scale(s) from. **2** form or come off in scales. □ **scaly** *adj.* (**-ier, -iest**). [French *escale*]
scale[2] *n.* **1 a** (often in *pl.*) weighing machine. **b** (also **scale-pan**) each of the dishes on a simple balance. **2** (**the Scales**) zodiacal sign or constellation Libra. □ **tip** (or **turn**) **the scales 1** be the decisive factor. **2** (usu. foll. by *at*) weigh. [Old Norse *skál* bowl]
scale[3] *–n.* **1** graded classification system (*high on the social scale*). **2 a** (often *attrib.*) ratio of reduction or enlargement in a map, model, picture, etc. (*on a*

scale of one inch to the mile; *a scale model*). **b** relative dimensions. **3** *Mus.* set of notes at fixed intervals, arranged in order of pitch. **4 a** set of marks on a line used in measuring etc. **b** rule determining the distances between these. **c** rod etc. on which these are marked. —*v.* (**-ling**) **1 a** climb (a wall, height, etc.). **b** climb (the social scale, heights of ambition, etc.). **2** represent proportionally; reduce to a common scale. □ **in scale** in proportion. **scale down** (or **up**) make or become smaller (or larger) in proportion. **to scale** uniformly in proportion. [Latin *scala* ladder]

scalene /'skeɪli:n/ *adj.* (esp. of a triangle) having unequal sides. [Greek *skalēnos* unequal]

scallion /'skæljən/ *n.* esp. *US* shallot; spring onion etc. [Latin from *Ascalon* in ancient Palestine]

scallop /'skæləp, 'skɒl-/ (also **scollop** /'skɒl-/) —*n.* **1** edible mollusc with two fan-shaped ridged shells. **2** (in full **scallop shell**) single shell of a scallop, often used for cooking or serving food in. **3** (in *pl.*) ornamental edging of semicircular curves. —*v.* (**-p-**) ornament with scallops. □ **scalloping** *n.* (in sense 3 of *n.*). [French ESCALOPE]

scallywag /'skælɪ,wæg/ *n.* (also **scalawag** /'skælə-/) scamp, rascal. [origin unknown]

scalp —*n.* **1** skin on the head, with the hair etc. attached. **2** *hist.* this cut off as a trophy by an American Indian. —*v.* **1** *hist.* take the scalp of (an enemy). **2** *US colloq.* resell (shares etc.) at a high or quick profit. [probably Scandinavian]

scalpel /'skælp(ə)l/ *n.* surgeon's small sharp knife. [Latin *scalpo* scratch]

scam *n. US slang* trick, fraud. [origin unknown]

scamp *n. colloq.* rascal; rogue. [probably Dutch]

scamper —*v.* run and skip. —*n.* act of scampering. [perhaps from SCAMP]

scampi /'skæmpɪ/ *n.pl.* large prawns. [Italian]

scan —*v.* (**-nn-**) **1** look at intently or quickly. **2** (of a verse etc.) be metrically correct. **3 a** examine (a surface etc.) to detect radioactivity etc. **b** traverse (a particular region) with a radar etc. beam. **4** resolve (a picture) into its elements of light and shade for esp. television transmission. **5** analyse the metrical structure of (verse). **6** obtain an image of (part of the body) using a scanner. —*n.* **1** scanning. **2** image obtained by scanning. [Latin *scando* climb, scan]

scandal /'skænd(ə)l/ *n.* **1** cause of public outrage. **2** outrage etc. so caused. **3** malicious gossip. □ **scandalous** *adj.* **scandalously** *adv.* [Greek *skandalon*, = snare]

scandalize *v.* (also **-ise**) (**-zing** or **-sing**) offend morally; shock.

scandalmonger *n.* person who habitually spreads scandal.

Scandinavian /,skændɪ'neɪvɪən/ —*n.* **1 a** native or inhabitant of Scandinavia (Denmark, Norway, Sweden, and Iceland). **b** person of Scandinavian descent. **2** family of languages of Scandinavia. —*adj.* of Scandinavia. [Latin]

scandium /'skændɪəm/ *n.* metallic element occurring naturally in lanthanide ores. [Latin *Scandia* Scandinavia]

scanner *n.* **1** device for scanning or systematically examining all the parts of something. **2** machine for measuring radiation, ultrasound reflections, etc., from the body as a diagnostic aid.

scansion /'skænʃ(ə)n/ *n.* metrical scanning of verse. [Latin: related to SCAN]

scant *adj.* barely sufficient; deficient. [Old Norse]

scanty *adj.* (**-ier, -iest**) **1** of small extent or amount. **2** barely sufficient. □ **scantily** *adv.* **scantiness** *n.*

scapegoat /'skeɪpgəʊt/ *n.* person blamed for others' shortcomings (with ref. to Lev. 16). [obsolete *scape* escape]

scapula /'skæpjʊlə/ *n.* (*pl.* **-lae** /-,li:/ or **-s**) shoulder-blade. [Latin]

scapular /'skæpjʊlə(r)/ —*adj.* of the shoulder or shoulder-blade. —*n.* short monastic cloak.

scar[1] —*n.* **1** usu. permanent mark on the skin from a wound etc. **2** emotional damage from grief etc. **3** sign of damage. **4** mark left on a plant by the loss of a leaf etc. —*v.* (**-rr-**) **1** (esp. as **scarred** *adj.*) mark with a scar or scars (*scarred for life*). **2** form a scar. [French *eschar(r)e*]

scar[2] *n.* (also **scaur**) steep craggy outcrop of a mountain or cliff. [Old Norse, = reef]

scarab /'skærəb/ *n.* **1 a** sacred dung-beetle of ancient Egypt. **b** a kind of beetle. **2** ancient Egyptian gem cut in the form of a beetle. [Latin *scarabaeus* from Greek]

scarce /skeəs/ —*adj.* **1** (usu. *predic.*) (esp. of food, money, etc.) in short supply. **2** rare. —*adv. archaic* or *literary* scarcely. □ **make oneself scarce** *colloq.* keep out of the way; surreptitiously disappear. [French *scars* Latin *excerpto* EXCERPT]

scarcely *adv.* **1** hardly, only just (*had scarcely arrived*). **2** surely not (*can scarcely have said so*). **3** esp. *iron.* not (*scarcely expected to be insulted*).

scarcity *n.* (*pl.* **-ies**) (often foll. by *of*) lack or shortage, esp. of food.

scare /skeə(r)/ –*v.* (**-ring**) **1** frighten, esp. suddenly. **2** (as **scared** *adj.*) (usu. foll. by *of*, or *to* + infin.) frightened; terrified. **3** (usu. foll. by *away*, *off*, *up*, etc.) drive away by frightening. **4** become scared (*they don't scare easily*). –*n.* **1** sudden attack of fright. **2** alarm caused by rumour etc. (*a measles scare*). [Old Norse]

scarecrow *n.* **1** human figure dressed in old clothes and set up in a field to scare birds away. **2** *colloq.* badly-dressed, grotesque-looking, or very thin person.

scaremonger *n.* person who spreads alarming rumours. □ **scaremongering** *n.*

scarf[1] *n.* (*pl.* **scarves** /skɑ:vz/ or **-s**) piece of material worn esp. round the neck or over the head, for warmth or ornament. [French *escarpe*]

scarf[2] –*v.* join the ends of (timber etc.) by bevelling or notching them to fit and then bolting them etc. –*n.* (*pl.* **-s**) joint made by scarfing. [probably French *escarf*]

scarify[1] /ˈskærɪˌfaɪ/ *v.* (**-ies, -ied**) **1 a** make slight incisions in. **b** cut off skin from. **2** hurt by severe criticism etc. **3** loosen (soil). □ **scarification** /-fɪˈkeɪʃ(ə)n/ *n.* [Greek *skariphos* stylus]

scarify[2] /ˈskeərɪˌfaɪ/ *v.* (**-ies, -ied**) *colloq.* scare.

scarlatina /ˌskɑ:ləˈti:nə/ *n.* = SCARLET FEVER. [Italian: related to SCARLET]

scarlet /ˈskɑ:lət/ –*adj.* of brilliant red tinged with orange. –*n.* **1** scarlet colour or pigment. **2** scarlet clothes or material (*dressed in scarlet*). [French *escarlate*]

scarlet fever *n.* infectious bacterial fever with a scarlet rash.

scarlet pimpernel *n.* wild plant with small esp. scarlet flowers.

scarlet woman *n. derog.* promiscuous woman, prostitute.

scarp –*n.* steep slope, esp. the inner side of a ditch in a fortification. –*v.* make perpendicular or steep. [Italian *scarpa*]

scarper *v. slang* run away, escape. [probably Italian *scappare* escape]

scarves *pl.* of SCARF[1].

scary /ˈskeərɪ/ *adj.* (**-ier, -iest**) *colloq.* frightening.

scat[1] *v.* (**-tt-**) (usu. in *imper.*) *colloq.* depart quickly. [perhaps an abbreviation of SCATTER]

scat[2] –*n.* wordless jazz singing. –*v.* (**-tt-**) sing scat. [probably imitative]

scathing /ˈskeɪðɪŋ/ *adj.* witheringly scornful. □ **scathingly** *adv.* [Old Norse]

scatology /skæˈtɒlədʒɪ/ *n.* excessive interest in excrement or obscenity. □ **scatological** /-təˈlɒdʒɪk(ə)l/ *adj.* [Greek *skōr skat-* dung]

scatter –*v.* **1 a** throw about; strew. **b** cover by scattering. **2 a** (cause to) move in flight etc.; disperse. **b** disperse or cause (hopes, clouds, etc.) to disperse. **3** (as **scattered** *adj.*) wide apart or sporadic (*scattered villages*). **4** *Physics* deflect or diffuse (light, particles, etc.). –*n.* **1** act of scattering. **2** small amount scattered. **3** extent of distribution. [probably var. of SHATTER]

scatterbrain *n.* person lacking concentration. □ **scatterbrained** *adj.*

scatty /ˈskætɪ/ *adj.* (**-ier, -iest**) *colloq.* scatterbrained. □ **scattily** *adv.* **scattiness** *n.*

scaur var. of SCAR[2].

scavenge /ˈskævɪndʒ/ *v.* (**-ging**) (usu. foll. by *for*; also *absol.*) search for and collect (discarded items). [back-formation from SCAVENGER]

scavenger *n.* **1** person who scavenges. **2** animal feeding on carrion, refuse, etc. [Anglo-French *scawager*: related to SHOW]

Sc.D. *abbr.* Doctor of Science. [Latin *scientiae doctor*]

SCE *abbr.* Scottish Certificate of Education.

scenario /sɪˈnɑ:rɪəʊ/ *n.* (*pl.* **-s**) **1** outline of the plot of a play, film, etc. **2** postulated sequence of future events. [Italian]

■ **Usage** *Scenario* should not be used in standard English to mean 'situation', as in *it was an unpleasant scenario*.

scene /si:n/ *n.* **1** place in which events, real or fictional, occur. **2 a** incident, real or fictional. **b** description of this. **3** public display of emotion, temper, etc. (*made a scene in the restaurant*). **4 a** continuous portion of a play in a fixed setting; subdivision of an act. **b** similar section of a film, book, etc. **5 a** piece of scenery used in a play. **b** these collectively. **6** landscape or view. **7** *colloq.* **a** area of interest (*not my scene*). **b** milieu (*well-known on the jazz scene*). □ **behind the scenes 1** offstage. **2** secret; secretly. **set the scene** describe the location of events. [Greek *skēnē* tent, stage]

scenery *n.* **1** natural features of a landscape, esp. when picturesque. **2** painted backcloths, props, etc., used as the background in a play etc. [Italian: related to SCENARIO]

scene-shifter *n.* person who moves scenery in a theatre.

scenic *adj.* **1 a** picturesque. **b** of natural scenery. **2** of or on the stage. □ **scenically** *adv.*

scent /sent/ *–n.* **1** distinctive, esp. pleasant, smell. **2** = PERFUME 2. **3 a** perceptible smell left by an animal. **b** clues etc. leading to a discovery. **c** power of detecting esp. smells. *–v.* **1 a** discern by scent. **b** sense (*scented danger*). **2** (esp. as **scented** *adj.*) make fragrant (*scented soap*). □ **put** (or **throw**) **off the scent** deceive by false clues etc. **scent out** discover by smelling or searching. [French *sentir* perceive]

scepter *US* var. of SCEPTRE.

sceptic /'skeptɪk/ *n.* (*US* **skeptic**) **1** person inclined to doubt accepted opinions. **2** person who doubts the truth of religions. **3** philosopher who questions the possibility of knowledge. □ **scepticism** /-ˌsɪz(ə)m/ *n.* [Greek *skeptomai* observe]

sceptical *adj.* (*US* **skeptical**) inclined to doubt accepted opinions; critical; incredulous. □ **sceptically** *adv.*

sceptre /'septə(r)/ *n.* (*US* **scepter**) staff as a symbol of sovereignty. [Greek *skēptō* lean on]

schadenfreude /'ʃɑ:dənˌfrɔɪdə/ *n.* malicious enjoyment of another's misfortunes. [German *Schaden* harm, *Freude* joy]

schedule /'ʃedju:l/ *–n.* **1 a** list of intended events, times, etc. **b** plan of work. **2** list of rates or prices. **3** *US* timetable. **4** tabulated list. *–v.* (**-ling**) **1** include in a schedule. **2** make a schedule of. **3** list (a building) for preservation. □ **according to** (or **on**) **schedule** as planned; on time. [Latin *schedula* slip of paper]

scheduled flight *n.* (also **scheduled service** etc.) regular public flight, service, etc.

schema /'ski:mə/ *n.* (*pl.* **schemata** or **-s**) synopsis, outline, or diagram. [Greek *skhēma -at-* form, figure]

schematic /skɪ'mætɪk/ *–adj.* of or as a scheme or schema; diagrammatic. *–n.* diagram, esp. of an electronic circuit. □ **schematically** *adv.*

schematize /'ski:məˌtaɪz/ *v.* (also **-ise**) (**-zing** or **-sing**) put in schematic form.

scheme /ski:m/ *–n.* **1** systematic plan or arrangement (*colour scheme*). **2** artful plot. **3** timetable, outline, syllabus, etc. *–v.* (**-ming**) plan, esp. secretly or deceitfully. □ **scheming** *adj.* [Greek: related to SCHEMA]

scherzo /'skeəˌtsəʊ/ *n.* (*pl.* **-s**) *Mus.* vigorous, often playful, piece, esp. as part of a larger work. [Italian, = jest]

schism /'skɪz(ə)m/ *n.* division of a group (esp. religious) into sects etc., usu. over doctrine. □ **schismatic** /-'mætɪk/ *adj.* & *n.* [Greek *skhizō* to split]

schist /ʃɪst/ *n.* layered crystalline rock. [Greek *skhizō* to split]

schizo /'skɪtsəʊ/ *colloq. –adj.* schizophrenic. *–n.* (*pl.* **-s**) schizophrenic person. [abbreviation]

schizoid /'skɪtsɔɪd/ *–adj.* tending to schizophrenia but usu. without delusions. *–n.* schizoid person.

schizophrenia /ˌskɪtsə'fri:nɪə/ *n.* mental disease marked by a breakdown in the relation between thoughts, feelings, and actions, and often with delusions and retreat from social life. □ **schizophrenic** /-'frenɪk/ *adj.* & *n.* [Greek *skhizō* to split, *phrēn* mind]

schlock /ʃlɒk/ *n. US colloq.* trash. [Yiddish *shlak* a blow]

schmaltz /ʃmɔ:lts/ *n. colloq.* esp. *US* sentimentality, esp. in music, drama, etc. □ **schmaltzy** *adj.* [Yiddish]

schmuck /ʃmʌk/ *n. slang* esp. *US* foolish or contemptible person. [Yiddish]

schnapps /ʃnæps/ *n.* any of various spirits drunk in N. Europe. [German]

schnitzel /'ʃnɪtz(ə)l/ *n.* escalope of veal. [German]

scholar /'skɒlə(r)/ *n.* **1** learned person, academic. **2** holder of a scholarship. **3** person of specified academic ability (*poor scholar*). □ **scholarly** *adj.* [Latin: related to SCHOOL[1]]

scholarship *n.* **1 a** academic achievement, esp. of a high level. **b** standards of a good scholar (*shows great scholarship*). **2** financial award for a student etc., given for scholarly achievement.

scholastic /skə'læstɪk/ *adj.* **1** of schools, education, etc.; academic. **2** *hist.* of scholasticism. [Greek: related to SCHOOL[1]]

scholasticism /ˌskə'læstɪˌsɪz(ə)m/ *n. hist.* medieval western Church philosophy.

school[1] /sku:l/ *–n.* **1 a** educational institution for pupils up to 19 years of age, or (*US*) including college or university level. **b** (*attrib.*) of or for use in school (*school dinners*). **2 a** school buildings, pupils, staff, etc. **b** time of teaching; the teaching itself (*no school today*). **3** university department or faculty. **4 a** group of similar artists etc., esp. followers of an artist etc. **b** group of like-minded people (*belongs to the old school*). **5** group of card-players etc. **6** *colloq.* instructive circumstances etc. (*school of adversity*). *–v.* **1** send to school; educate. **2** (often foll. by *to*) discipline, train, control. **3** (as **schooled** *adj.*) (foll. by *in*) educated or trained (*schooled in humility*). □ **at** (*US* **in**) **school** attending lessons etc. **go to school** attend lessons. [Greek *skholē*]

school[2] /sku:l/ *n.* (often foll. by *of*) shoal of fish, whales, etc. [Low German or Dutch]

school age *n.* age-range of school attendance.
schoolboy *n.* boy attending school.
schoolchild *n.* child attending school.
schoolgirl *n.* girl attending school.
schoolhouse *n.* school building, esp. in a village.
schooling *n.* education, esp. at school.
school-leaver *n.* person finishing secondary school (esp. considered as joining the job market).
schoolmaster *n.* head or assistant male teacher.
schoolmistress *n.* head or assistant female teacher.
schoolroom *n.* room used for lessons.
schoolteacher *n.* teacher in a school.
school year *n.* period from September to July.
schooner /'sku:nə(r)/ *n.* **1** fore-and-aft rigged ship with two or more masts. **2 a** measure or glass for esp. sherry. **b** *US* & *Austral.* tall beer-glass. [origin uncertain]
schottische /ʃɒ'ti:ʃ/ *n.* **1** a kind of slow polka. **2** music for this. [German, = Scottish]
sciatic /saɪ'ætɪk/ *adj.* **1** of the hip. **2** of the sciatic nerve. **3** suffering from or liable to sciatica. [Greek *iskhion* hip]
sciatica /saɪ'ætɪkə/ *n.* neuralgia of the hip and leg. [Latin: related to SCIATIC]
sciatic nerve *n.* largest nerve, running from pelvis to thigh.
science /'saɪəns/ *n.* **1** branch of knowledge involving systematized observation and experiment. **2 a** knowledge so gained, or on a specific subject. **b** pursuit or principles of this. **3** skilful technique. [Latin *scio* know]
science fiction *n.* fiction with a scientific theme, esp. concerned with the future, space, other worlds, etc.
science park *n.* area containing science-based industries.
scientific /,saɪən'tɪfɪk/ *adj.* **1 a** following the systematic methods of science. **b** systematic, accurate. **2** of, used in, or engaged in science. □ **scientifically** *adv.*
scientist /'saɪəntɪst/ *n.* student or expert in science.
Scientology /,saɪən'tɒlədʒɪ/ *n.* system of religious philosophy based on self-improvement and graded courses of study and training. □ **Scientologist** *n.* & *adj.* [Latin *scientia* knowledge]
sci-fi /'saɪfaɪ/ *n.* (often *attrib.*) *colloq.* science fiction. [abbreviation]
scilicet /'saɪlɪ,set/ *adv.* that is to say (used esp. in explanation of an ambiguity). [Latin]
scimitar /'sɪmɪtə(r)/ *n.* curved oriental sword. [French and Italian]
scintilla /sɪn'tɪlə/ *n.* trace. [Latin, = spark]
scintillate /'sɪntɪ,leɪt/ *v.* **(-ting) 1** (esp. as **scintillating** *adj.*) talk cleverly; be brilliant. **2** sparkle; twinkle. □ **scintillation** /-'leɪʃ(ə)n/ *n.* [Latin: related to SCINTILLA]
scion /'saɪən/ *n.* **1** shoot of a plant etc., esp. one cut for grafting or planting. **2** descendant; younger member of (esp. a noble) family. [French]
scirocco var. of SIROCCO.
scissors /'sɪzəz/ *n.pl.* (also **pair of scissors** *sing.*) hand-held cutting instrument with two pivoted blades opening and closing. [Latin *caedo* cut: related to CHISEL]
sclerosis /sklə'rəʊsɪs/ *n.* **1** abnormal hardening of body tissue. **2** (in full **multiple** or **disseminated sclerosis**) serious progressive disease of the nervous system. □ **sclerotic** /-'rɒtɪk/ *adj.* [Greek *sklēros* hard]
scoff[1] *–v.* (usu. foll. by *at*) speak scornfully; mock. *–n.* mocking words; taunt. [perhaps from Scandinavian]
scoff[2] *colloq. –v.* eat greedily. *–n.* food; a meal. [Afrikaans *schoff* from Dutch]
scold /skəʊld/ *–v.* **1** rebuke (esp. a child). **2** find fault noisily. *–n. archaic* nagging woman. □ **scolding** *n.* [probably Old Norse]
scollop var. of SCALLOP.
sconce *n.* wall-bracket for a candlestick or light-fitting. [Latin *(ab)sconsa* covered (light)]
scone /skɒn, skəʊn/ *n.* small cake of flour, fat, and milk, baked quickly. [origin uncertain]
scoop *–n.* **1** spoon-shaped object, esp.: **a** a short-handled deep shovel for loose materials. **b** a large long-handled ladle for liquids. **c** the excavating part of a digging-machine etc. **d** an instrument for serving ice-cream etc. **2** quantity taken up by a scoop. **3** scooping movement. **4** exclusive news item. **5** large profit made quickly. *–v.* **1** (usu. foll. by *out*) hollow out (as if) with a scoop. **2** (usu. foll. by *up*) lift (as if) with a scoop. **3** forestall (a rival newspaper etc.) with a scoop. **4** secure (a large profit etc.), esp. suddenly. [Low German or Dutch]
scoot *v.* (esp. in *imper.*) *colloq.* depart quickly, flee. [origin unknown]
scooter *n.* **1** child's toy with a footboard on two wheels and a long steering-handle. **2** (in full **motor scooter**) low-powered motor cycle with a shieldlike protective front.
scope *n.* **1** range or opportunity (*beyond the scope of our research*). **2** extent of mental ability, outlook, etc. (*intellect limited in its scope*). [Greek, = target]

-scope *comb. form* forming nouns denoting: **1** device looked at or through (*telescope*). **2** instrument for observing or showing (*oscilloscope*). □ **-scopic** /'skɒpɪk/ *comb. form* forming adjectives. [Greek *skopeō* look at]

-scopy *comb. form* indicating viewing or observation, usu. with an instrument ending in *-scope* (*microscopy*).

scorbutic /skɔː'bjuːtɪk/ *adj.* of, like, or affected with scurvy. [Latin *scorbutus* scurvy]

scorch *–v.* **1** burn or discolour the surface of with dry heat. **2** become so discoloured etc. **3** (as **scorching** *adj.*) *colloq.* **a** (of the weather) very hot. **b** (of criticism etc.) stringent; harsh. *–n.* mark made by scorching. [origin unknown]

scorched earth policy *n.* policy of destroying anything that might be of use to an invading enemy.

scorcher *n. colloq.* very hot day.

score *–n.* **1 a** number of points, goals, runs, etc., made by a player or side in some games. **b** respective numbers of points etc. at the end of a game (*score was five–nil*). **c** act of gaining esp. a goal. **2** (*pl.* same or **-s**) twenty or a set of twenty. **3** (in *pl.*) a great many (*scores of people*). **4** reason or motive (*rejected on that score*). **5** *Mus.* **a** copy of a composition showing all the vocal and instrumental parts arranged one below the other. **b** music for a film or play, esp. for a musical. **6** notch, line, etc. cut or scratched into a surface. **7** record of money owing. *–v.* **(-ring) 1 a** win or gain (a goal, points, success, etc.). **b** count for (points in a game etc.) (*a boundary scores six*). **2 a** make a score in a game (*failed to score*). **b** keep score in a game. **3** mark with notches etc. **4** have an advantage (*that is where he scores*). **5** *Mus.* (often foll. by *for*) orchestrate or arrange (a piece of music). **6** *slang* **a** obtain drugs illegally. **b** make a sexual conquest. □ **keep score** (or **the score**) register scores as they are made. **know the score** *colloq.* be aware of the essential facts. **on that score** so far as that is concerned. **score off** (or **score points off**) *colloq.* humiliate, esp. verbally. **score out** delete. □ **scorer** *n.* [Old Norse: related to SHEAR]

scoreboard *n.* large board for displaying the score in a game or match.

score-book *n.* (also **score-card** or **-sheet**) printed book etc. for entering esp. cricket scores in.

scoria /'skɔːrɪə/ *n.* (*pl.* **scoriae** /-rɪ,iː/) **1** cellular lava, or fragments of it. **2** slag or dross of metals. □ **scoriaceous** /-'eɪʃəs/ *adj.* [Greek *skōria* refuse]

scorn *–n.* disdain, contempt, derision. *–v.* **1** hold in contempt. **2** reject or refuse to do as unworthy. [French *escarnir*]

scornful *adj.* (often foll. by *of*) contemptuous. □ **scornfully** *adv.*

Scorpio /'skɔːpɪəʊ/ *n.* (*pl.* **-s**) **1** constellation and eighth sign of the zodiac (the Scorpion). **2** person born when the sun is in this sign. [Greek *skorpios* scorpion]

scorpion /'skɔːpɪən/ *n.* **1** arachnid with pincers and a jointed stinging tail. **2 (the Scorpion)** zodiacal sign or constellation Scorpio.

Scot *n.* **1** native of Scotland. **2** person of Scottish descent. [Latin *Scottus*]

Scotch *–adj.* var. of SCOTTISH or SCOTS. *–n.* **1** var. of SCOTTISH or SCOTS. **2** Scotch whisky. [from SCOTTISH]

■ **Usage** *Scots* or *Scottish* is preferred to *Scotch* in Scotland, except in the compound nouns *Scotch broth*, *egg*, *fir*, *mist*, *terrier*, and *whisky*.

scotch *v.* **1** put an end to; frustrate. **2** *archaic* wound without killing. [origin unknown]

Scotch broth *n.* meat soup with pearl barley etc.

Scotch egg *n.* hard-boiled egg in sausage meat.

Scotch fir *n.* (also **Scots fir**) = SCOTS PINE.

Scotch mist *n.* thick drizzly mist.

Scotch terrier *n.* (also **Scottish terrier**) small rough-haired terrier.

Scotch whisky *n.* whisky distilled in Scotland.

scot-free *adv.* unharmed, unpunished. [obsolete *scot* tax]

Scots (also **Scotch**) esp. *Scot. –adj.* **1** = SCOTTISH *adj.* **2** in the dialect, accent, etc., of (esp. Lowlands) Scotland. *–n.* **1** = SCOTTISH *n.* **2** form of English spoken in (esp. Lowlands) Scotland. [var. of SCOTTISH]

Scots fir var. of SCOTCH FIR.

Scotsman *n.* (*fem.* **Scotswoman**) = SCOT.

Scots pine *n.* (also **Scottish pine**) a kind of pine tree.

Scottie *n.* (also **Scottie dog**) *colloq.* Scotch terrier.

Scottish (also **Scotch**) *–adj.* of Scotland or its inhabitants. *–n.* (prec. by *the*; treated as *pl.*) people of Scotland.

Scottish pine var. of SCOTS PINE.

Scottish terrier var. of SCOTCH TERRIER.

scoundrel /'skaʊndr(ə)l/ *n.* unscrupulous villain; rogue. [origin unknown]

scour[1] *–v.* **1 a** cleanse by rubbing. **b** (usu. foll. by *away*, *off*, etc.) clear (rust, stains, etc.) by rubbing etc. **2** clear out (a pipe, channel, etc.) by flushing through.

–*n.* scouring or being scoured. □ **scourer** *n.* [French *escurer*]

scour[2] *v.* search thoroughly, esp. by scanning (*scoured the streets for him; scoured the newspaper*). [origin unknown]

scourge /skɜːdʒ/ –*n.* **1** person or thing seen as causing suffering. **2** whip. –*v.* (**-ging**) **1** whip. **2** punish, oppress. [Latin *corrigia* whip]

Scouse *colloq.* –*n.* **1** Liverpool dialect. **2** (also **Scouser**) native of Liverpool. –*adj.* of Liverpool. [from LOBSCOUSE]

scout –*n.* **1** soldier etc. sent ahead to get esp. military intelligence. **2** search for this. **3** = TALENT-SCOUT. **4** (also **Scout**) member of the Scout Association, an (orig. boys') association intended to develop character. **5** domestic worker at an Oxford college. –*v.* **1** (often foll. by *for*) go about searching for information etc. **2** (foll. by *about*, *around*) make a search. **3** (often foll. by *out*) *colloq.* explore to get information about (territory etc.). □ **scouting** *n.* [French *escoute(r)* from Latin *ausculto* listen]

Scouter *n.* adult leader of Scouts.

Scoutmaster *n.* person in charge of a group of Scouts.

scow *n.* esp. *US* flat-bottomed boat. [Dutch]

scowl –*n.* severe frowning or sullen expression. –*v.* make a scowl. [Scandinavian]

scrabble /ˈskræb(ə)l/ –*v.* (**-ling**) scratch or grope about, esp. in search of something. –*n.* **1** act of scrabbling. **2** (**Scrabble**) *propr.* game in which players build up words from letter-blocks on a board. [Dutch]

scrag –*n.* **1** (also **scrag-end**) inferior end of a neck of mutton. **2** skinny person or animal. –*v.* (**-gg-**) *slang* **1** strangle, hang. **2** handle roughly, beat up. [origin uncertain]

scraggy *adj.* (**-ier**, **-iest**) thin and bony. □ **scragginess** *n.*

scram *v.* (**-mm-**) (esp. in *imper.*) *colloq.* go away. [perhaps from SCRAMBLE]

scramble /ˈskræmb(ə)l/ –*v.* (**-ling**) **1** clamber, crawl, climb, etc., esp. hurriedly or anxiously. **2** (foll. by *for*, *at*) struggle with competitors (for a thing or share). **3** mix together indiscriminately. **4** cook (eggs) by stirring them in a pan over heat. **5** change the speech frequency of (a broadcast transmission or telephone conversation) so as to make it unintelligible without a decoding device. **6** (of fighter aircraft or pilots) take off quickly in an emergency or for action. –*n.* **1** act of scrambling. **2** difficult climb or walk. **3** (foll. by *for*) eager struggle or competition. **4** motor-cycle race over rough ground. **5** emergency take-off by fighter aircraft. [imitative]

scrambler *n.* device for scrambling telephone conversations.

scrap[1] –*n.* **1** small detached piece; fragment. **2** rubbish or waste material. **3** discarded metal for reprocessing (often *attrib.*: *scrap metal*). **4** (with *neg.*) smallest piece or amount. **5** (in *pl.*) **a** odds and ends. **b** uneaten food. –*v.* (**-pp-**) discard as useless. [Old Norse: related to SCRAPE]

scrap[2] *colloq.* –*n.* fight or rough quarrel. –*v.* (**-pp-**) have a scrap. [perhaps from SCRAPE]

scrapbook *n.* blank book for sticking cuttings, drawings, etc., in.

scrape –*v.* (**-ping**) **1 a** move a hard or sharp edge across (a surface), esp. to make smooth. **b** apply (a hard or sharp edge) in this way. **2** (foll. by *away*, *off*, etc.) remove by scraping. **3 a** rub (a surface) harshly against another. **b** scratch or damage by scraping. **4** make (a hollow) by scraping. **5 a** draw or move with a scraping sound. **b** make such a sound. **c** produce such a sound from. **6** (often foll. by *along*, *by*, *through*, etc.) move almost touching surrounding obstacles etc. (*scraped through the gap*). **7** narrowly achieve (a living, an examination pass, etc.). **8** (often foll. by *by*, *through*) **a** barely manage. **b** pass an examination etc. with difficulty. **9** (foll. by *together*, *up*) bring, provide, or amass with difficulty. **10** be economical. **11** draw back a foot in making a clumsy bow. **12** (foll. by *back*) draw (the hair) tightly back. –*n.* **1** act or sound of scraping. **2** scraped place; graze. **3** *colloq.* predicament caused by rashness etc. □ **scrape the barrel** *colloq.* be reduced to a limited choice etc. [Old Norse]

scraper *n.* device for scraping, esp. paint etc. from a surface.

scrap heap *n.* **1** pile of scrap. **2** state of being discarded as useless.

scrapie /ˈskreɪpɪ/ *n.* viral disease of sheep, characterized by lack of coordination.

scraping *n.* (esp. in *pl.*) fragment produced by scraping.

scrap merchant *n.* dealer in scrap.

scrappy *adj.* (**-ier**, **-iest**) **1** consisting of scraps. **2** incomplete; carelessly arranged or put together.

scrapyard *n.* place where (esp. metal) scrap is collected for reuse.

scratch –*v.* **1** score, mark, or wound superficially, esp. with a sharp object. **2** (also *absol.*) scrape, esp. with the nails to relieve itching. **3** make or form by scratching. **4** (foll. by *together*, *up*, etc.)

= SCRAPE 9. **5** (foll. by *out*, *off*, *through*) strike (out) (writing etc.). **6** (also *absol.*) withdraw (a competitor, oneself, etc.) from a race or competition. **7** (often foll. by *about*, *around*, etc.) **a** scratch the ground etc. in search. **b** search haphazardly (*scratching about for evidence*). –*n.* **1** mark or wound made by scratching. **2** sound of scratching. **3** spell of scratching oneself. **4** *colloq.* superficial wound. **5** line from which competitors in a race (esp. those not receiving a handicap) start. –*attrib.adj.* **1** collected by chance. **2** collected or made from whatever is available; heterogeneous. **3** with no handicap given (*scratch race*). □ **from scratch 1** from the beginning. **2** without help. **scratch one's head** be perplexed. **scratch the surface** deal with a matter only superficially. **up to scratch** up to the required standard. [origin uncertain]

scratchy *adj.* (**-ier, -iest**) **1** tending to make scratches or a scratching noise. **2** causing itching. **3** (of a drawing etc.) untidy, careless. □ **scratchily** *adv.* **scratchiness** *n.*

scrawl –*v.* **1** write or make (marks) in a hurried untidy way. **2** (foll. by *out*) cross out by scrawling over. –*n.* **1** hurried untidy manner of writing. **2** example of this. □ **scrawly** *adj.* [origin uncertain]

scrawny *adj.* (**-ier, -iest**) lean, scraggy. [dial.]

scream –*n.* **1** loud high-pitched cry of fear, pain, etc. **2** similar sound or cry. **3** *colloq.* hilarious occurrence or person. –*v.* **1** emit a scream. **2** speak or sing (words etc.) in a screaming tone. **3** make or move with a screaming sound. **4** laugh uncontrollably. **5** be blatantly obvious. [Old English]

scree *n.* (in *sing.* or *pl.*) **1** small loose stones. **2** mountain slope covered with these. [Old Norse, = landslip]

screech –*n.* harsh piercing scream. –*v.* utter with or make a screech. □ **screechy** *adj.* (**-ier, -iest**). [Old English (imitative)]

screech-owl *n.* owl that screeches, esp. a barn-owl.

screed *n.* **1** long usu. tiresome piece of writing or speech. **2** layer of cement etc. applied to level a surface. [probably from SHRED]

screen –*n.* **1** fixed or movable upright partition for separating, concealing, or protecting from heat etc. **2** thing used to conceal or shelter. **3 a** concealing stratagem. **b** protection thus given. **4 a** blank surface on which a photographic image is projected. **b** (prec. by *the*) the cinema industry; films collectively. **5** surface of a cathode-ray tube etc., esp. of a television, on which images appear. **6** = SIGHT-SCREEN. **7** = WINDSCREEN. **8** frame with fine netting to keep out insects etc. **9** large sieve or riddle. **10** system of checking for disease, an ability, attribute, etc. –*v.* **1** (often foll. by *from*) **a** shelter; hide. **b** protect from detection, censure, etc. **2** (foll. by *off*) conceal behind a screen. **3** show (a film, television programme, etc.). **4** prevent from causing, or protect from, electrical interference. **5** test or check (a person or group) for a disease, reliability, loyalty, etc. **6** sieve. [French]

screenplay *n.* film script.

screen printing *n.* printing process with ink forced through a prepared sheet of fine material.

screen test *n.* audition for a part in a film.

screenwriter *n.* person who writes for the cinema.

screw /skru:/ –*n.* **1** thin cylinder or cone with a spiral ridge or thread running round the outside (**male screw**) or the inside (**female screw**). **2** (in full **wood-screw**) metal male screw with a slotted head and a sharp point. **3** (in full **screw-bolt**) blunt metal male screw on which a nut is threaded to bolt things together. **4** straight screw used to exert pressure. **5** (in *sing.* or *pl.*) instrument of torture acting in this way. **6** (in full **screw-propeller**) propeller with twisted blades acting like a screw on the water or air. **7** one turn of a screw. **8** (foll. by *of*) small twisted-up paper (of tobacco etc.). **9** (in billiards etc.) an oblique curling motion of the ball. **10** *slang* prison warder. **11** *coarse slang* **a** act of sexual intercourse. **b** partner in this. –*v.* **1** fasten or tighten with a screw or screws. **2** turn (a screw). **3** twist or turn round like a screw. **4** (of a ball etc.) swerve. **5** (foll. by *out of*) extort (consent, money, etc.) from. **6** (also *absol.*) *coarse slang* have sexual intercourse with. **7** swindle. □ **have a screw loose** *colloq.* be slightly crazy. **put the screws on** *colloq.* pressurize, intimidate. **screw up 1** contract or contort (one's face etc.). **2** contract and crush (a piece of paper etc.) into a tight mass. **3** summon up (one's courage etc.). **4** *slang* **a** bungle. **b** spoil (an event, opportunity, etc.). **c** upset, disturb mentally. [French *escroue*]

screwball *n. US slang* crazy or eccentric person.

screwdriver *n.* tool with a tip that fits into the head of a screw to turn it.

screw top *n.* (also (with hyphen) *attrib.*) screwed-on cap or lid.

screw-up *n. slang* bungle, mess.

screwy *adj.* (**-ier, -iest**) *slang* **1** crazy or eccentric. **2** absurd. □ **screwiness** *n.*

scribble /'skrɪb(ə)l/ *–v.* (**-ling**) **1** write or draw carelessly or hurriedly. **2** *joc.* be an author or writer. *–n.* **1** scrawl. **2** hasty note etc. [Latin *scribillo* diminutive: related to SCRIBE]

scribe *–n.* **1** ancient or medieval copyist of manuscripts. **2** ancient Jewish record-keeper or professional theologian and jurist. **3** pointed instrument for making marks on wood etc. **4** *colloq.* writer, esp. a journalist. *–v.* (**-bing**) mark with a scribe. □ **scribal** *adj.* [Latin *scriba* from *scribo* write]

scrim *n.* open-weave fabric for lining or upholstery etc. [origin unknown]

scrimmage /'skrɪmɪdʒ/ *–n.* tussle; brawl. *–v.* (**-ging**) engage in this. [from SKIRMISH]

scrimp *v.* skimp. [origin unknown]

scrip *n.* **1** provisional certificate of money subscribed, entitling the holder to dividends. **2** (*collect.*) such certificates. **3** extra share or shares instead of a dividend. [abbreviation of *subscription receipt*]

script *–n.* **1** text of a play, film, or broadcast. **2** handwriting; written characters. **3** type imitating handwriting. **4** alphabet or system of writing. **5** examinee's written answers. *–v.* write a script for (a film etc.). [Latin *scriptum* from *scribo* write]

scripture /'skrɪptʃə(r)/ *n.* **1** sacred writings. **2** (**Scripture** or **the Scriptures**) the Bible. □ **scriptural** *adj.* [Latin: related to SCRIPT]

scriptwriter *n.* person who writes scripts for films, TV, etc. □ **scriptwriting** *n.*

scrivener /'skrɪvənə(r)/ *n. hist.* **1** copyist or drafter of documents. **2** notary. [French *escrivein*]

scrofula /'skrɒfjʊlə/ *n.* disease with glandular swellings, probably a form of tuberculosis. □ **scrofulous** *adj.* [Latin *scrofa* a sow]

scroll /skrəʊl/ *–n.* **1** roll of parchment or paper, esp. written on. **2** book in the ancient roll form. **3** ornamental design imitating a roll of parchment. *–v.* (often foll. by *down*, *up*) move (a display on a VDU screen) to view earlier or later material. [originally (*sc*)*rowle* ROLL]

scrolled *adj.* having a scroll ornament.

Scrooge /skru:dʒ/ *n.* miser. [name of a character in Dickens]

scrotum /'skrəʊtəm/ *n.* (*pl.* **scrota** or **-s**) pouch of skin containing the testicles. □ **scrotal** *adj.* [Latin]

scrounge *v.* (**-ging**) (also *absol.*) obtain by cadging. □ **on the scrounge** scrounging. □ **scrounger** *n.* [dial. *scrunge* steal]

scrub[1] *–v.* (**-bb-**) **1** clean by rubbing, esp. with a hard brush and water. **2** (often foll. by *up*) (of a surgeon etc.) clean and disinfect the hands and arms before operating. **3** *colloq.* scrap or cancel. **4** use water to remove impurities from (gases etc.). *–n.* scrubbing or being scrubbed. [Low German or Dutch]

scrub[2] *n.* **1 a** brushwood or stunted forest growth. **b** land covered with this. **2** (*attrib.*) small or dwarf variety (*scrub pine*). □ **scrubby** *adj.* [from SHRUB]

scrubber *n.* **1** *slang* promiscuous woman. **2** apparatus for purifying gases etc.

scruff[1] *n.* back of the neck (esp. *scruff of the neck*). [perhaps from Old Norse *skoft* hair]

scruff[2] *n. colloq.* scruffy person. [origin uncertain]

scruffy *adj.* (**-ier, -iest**) *colloq.* shabby, slovenly, untidy. □ **scruffily** *adv.* **scruffiness** *n.* [*scruff* = SCURF]

scrum *n.* **1** scrummage. **2** *colloq.* = SCRIMMAGE. [abbreviation]

scrum-half *n.* half-back who puts the ball into the scrum.

scrummage *n. Rugby* massed forwards on each side pushing to gain possession of the ball thrown on the ground between them. [related to SCRIMMAGE]

scrump *v. colloq.* steal from an orchard or garden. [related to SCRUMPY]

scrumptious /'skrʌmpʃəs/ *adj. colloq.* **1** delicious. **2** delightful. [origin unknown]

scrumpy /'skrʌmpɪ/ *n. colloq.* rough cider. [dial. *scrump* small apple]

scrunch *–v.* **1** (usu. foll. by *up*) crumple. **2** crunch. *–n.* crunching sound. [var. of CRUNCH]

scruple /'skru:p(ə)l/ *–n.* **1** (often in *pl.*) moral concern. **2** doubt caused by this. *–v.* (**-ling**) (foll. by *to* + infin.; usu. with *neg.*) hesitate because of scruples. [Latin]

scrupulous /'skru:pjʊləs/ *adj.* **1** conscientious, thorough. **2** careful to avoid doing wrong. **3** punctilious; over-attentive to details. □ **scrupulously** *adv.* [Latin: related to SCRUPLE]

scrutineer /,skru:tɪ'nɪə(r)/ *n.* person who scrutinizes ballot-papers.

scrutinize /'skru:tɪ,naɪz/ *v.* (also **-ise**) (**-zing** or **-sing**) subject to scrutiny.

scrutiny /'skru:tɪnɪ/ *n.* (*pl.* **-ies**) **1** critical gaze. **2** close investigation. **3** official examination of ballot-papers. [Latin *scrutinium* from *scrutor* examine]

scuba /'sku:bə/ *n.* (*pl.* **-s**) aqualung. [acronym of *s*elf-*c*ontained *u*nderwater *b*reathing *a*pparatus]

scuba-diving *n.* swimming underwater using a scuba. □ **scuba-dive** *v.* **scuba-diver** *n.*

scud –*v.* (**-dd-**) **1** move straight and fast; skim along (*scudding clouds*). **2** *Naut.* run before the wind. –*n.* **1** spell of scudding. **2** scudding motion. **3** vapoury driving clouds or shower. [perhaps an alteration of SCUT]

scuff –*v.* **1** graze or brush against. **2** mark or wear out (shoes) in this way. **3** shuffle or drag the feet. –*n.* mark of scuffing. [imitative]

scuffle /'skʌf(ə)l/ –*n.* confused struggle or fight at close quarters. –*v.* (**-ling**) engage in a scuffle. [probably Scandinavian: related to SHOVE]

scull –*n.* **1** either of a pair of small oars. **2** oar over the stern of a boat to propel it, usu. by a twisting motion. **3** (in *pl.*) sculling race. –*v.* (often *absol.*) propel (a boat) with sculls. [origin unknown]

sculler *n.* **1** user of sculls. **2** boat for sculling.

scullery /'skʌlərɪ/ *n.* (*pl.* **-ies**) back kitchen; room for washing dishes etc. [Anglo-French *squillerie*]

scullion /'skʌljən/ *n. archaic* **1** cook's boy. **2** person who washes dishes etc. [origin unknown]

sculpt *v.* sculpture. [shortening of SCULPTOR]

sculptor *n.* (*fem.* **sculptress**) artist who sculptures. [Latin: related to SCULPTURE]

sculpture /'skʌlptʃə(r)/ –*n.* **1** art of making three-dimensional or relief forms, by chiselling, carving, modelling, casting, etc. **2** work of sculpture. –*v.* (**-ring**) **1** represent in or adorn with sculpture. **2** practise sculpture. □ **sculptural** *adj.* [Latin *sculpo sculpt-* carve]

scum –*n.* **1** layer of dirt, froth, etc. at the top of liquid. **2** *derog.* worst part, person, or group (*scum of the earth*). –*v.* (**-mm-**) **1** remove scum from. **2** form a scum (on). □ **scummy** *adj.* (**-ier, -iest**). [Low German or Dutch]

scumbag *n. slang* contemptible person.

scupper[1] *n.* hole in a ship's side to drain water from the deck. [French *escopir* to spit]

scupper[2] *v. slang* **1** sink (a ship or its crew). **2** defeat or ruin (a plan etc.). **3** kill. [origin unknown]

scurf *n.* dandruff. □ **scurfy** *adj.* [Old English]

scurrilous /'skʌrɪləs/ *adj.* grossly or indecently abusive. □ **scurrility** /skə'rɪlɪtɪ/ *n.* (*pl.* **-ies**). **scurrilously** *adv.* **scurrilousness** *n.* [Latin *scurra* buffoon]

scurry /'skʌrɪ/ –*v.* (**-ies, -ied**) run or move hurriedly, esp. with short quick steps; scamper. –*n.* (*pl.* **-ies**) **1** act or sound of scurrying. **2** flurry of rain or snow. [abbreviation of *hurry-scurry* reduplication of HURRY]

scurvy /'skɜ:vɪ/ –*n.* disease caused by a deficiency of vitamin C. –*adj.* (**-ier, -iest**) paltry, contemptible. □ **scurvily** *adv.* [from SCURF]

scut *n.* short tail, esp. of a hare, rabbit, or deer. [origin unknown]

scutter *v.* & *n. colloq.* scurry. [perhaps an alteration of SCUTTLE[2]]

scuttle[1] /'skʌt(ə)l/ *n.* **1** = COAL-SCUTTLE. **2** part of a car body between the windscreen and the bonnet. [Old Norse from Latin *scutella* dish]

scuttle[2] /'skʌt(ə)l/ –*v.* (**-ling**) scurry; flee from danger etc. –*n.* hurried gait; precipitate flight. [perhaps related to dial. *scuddle* frequentative of SCUD]

scuttle[3] /'skʌt(ə)l/ –*n.* hole with a lid in a ship's deck or side. –*v.* let water into (a ship) to sink it. [Spanish *escotilla* hatchway]

Scylla and Charybdis /ˌsɪlə, kə'rɪbdɪs/ *n.pl.* two dangers or extremes such that one can be avoided only by approaching the other. [names of a monster and a whirlpool in Greek mythology]

scythe /saɪð/ –*n.* mowing and reaping implement with a long handle and curved blade swung over the ground. –*v.* (**-thing**) cut with a scythe. [Old English]

SDI *abbr.* strategic defence initiative.

SDLP *abbr.* (in N. Ireland) Social Democratic and Labour Party.

SDP *abbr. hist.* Social Democratic Party.

SE *abbr.* **1** south-east. **2** south-eastern.

Se *symb.* selenium.

sea *n.* **1** expanse of salt water that covers most of the earth's surface. **2** any part of this. **3** named tract of this partly or wholly enclosed by land (*North Sea*). **4** large inland lake (*Sea of Galilee*). **5** waves of the sea; their motion or state (*choppy sea*). **6** (foll. by *of*) vast quantity or expanse. **7** (*attrib.*) living or used in, on, or near the sea (often prefixed to the name of a marine animal, plant, etc., having a superficial resemblance to what it is named after) (*sea lettuce*). □ **at sea 1** in a ship on the sea. **2** perplexed, confused. **by sea** in a ship or ships. **go to sea** become a sailor. **on the sea 1** = *at sea* 1. **2** on the coast. [Old English]

sea anchor *n.* bag to retard the drifting of a ship.

sea anemone *n.* marine animal with tube-shaped body and petal-like tentacles.

seabed *n.* ocean floor.

sea bird *n.* bird living near the sea.

seaboard *n.* **1** seashore or coastline. **2** coastal region.

seaborne *adj.* transported by sea.

sea change *n.* notable or unexpected transformation.

sea cow *n.* **1** sirenian. **2** walrus.

sea dog *n.* old sailor.

seafarer *n.* **1** sailor. **2** traveller by sea. □ **seafaring** *adj.* & *n.*

seafood *n.* (often *attrib.*) edible sea fish or shellfish (*seafood restaurant*).

sea front *n.* part of a town directly facing the sea.

seagoing *adj.* (of ships) fit for crossing the sea.

sea green *adj.* & *n.* (as adj. often hyphenated) bluish-green.

seagull *n.* = GULL[1].

sea horse *n.* **1** small upright fish with a head like a horse's. **2** mythical creature with a horse's head and fish's tail.

seakale *n.* plant with young shoots used as a vegetable.

seal[1] *–n.* **1** piece of stamped wax, lead, paper, etc., attached to a document or to a receptacle, envelope, etc., to guarantee authenticity or security. **2** engraved piece of metal etc. for stamping a design on a seal. **3** substance or device used to close a gap etc. **4** anything regarded as a confirmation or guarantee (*seal of approval*). **5** decorative adhesive stamp. *–v.* **1** close securely or hermetically. **2** stamp, fasten, or fix with a seal. **3** certify as correct with a seal or stamp. **4** (often foll. by *up*) confine securely. **5** settle or decide (*their fate is sealed*). **6** (foll. by *off*) prevent entry to or exit from (an area). □ **set one's seal to** (or **on**) authorize or confirm. [Latin *sigillum*]

seal[2] *–n.* fish-eating amphibious marine mammal with flippers. *–v.* hunt for seals. [Old English]

sealant *n.* material for sealing, esp. to make airtight or watertight.

sea legs *n.pl.* ability to keep one's balance and avoid seasickness at sea.

sea level *n.* mean level of the sea's surface, used in reckoning the height of hills etc. and as a barometric standard.

sealing-wax *n.* mixture softened by heating and used to make seals.

sea lion *n.* large, eared seal.

Sea Lord *n.* naval member of the Admiralty Board.

sealskin *n.* **1** skin or prepared fur of a seal. **2** (often *attrib.*) garment made from this.

seals of office *n.pl.* seals held, esp. by the Lord Chancellor or a Secretary of State.

seam *–n.* **1** line where two edges join, esp. of cloth or boards. **2** fissure between parallel edges. **3** wrinkle. **4** stratum of coal etc. *–v.* **1** join with a seam. **2** (esp. as **seamed** *adj.*) mark or score with a seam. □ **seamless** *adj.* [Old English]

seaman *n.* **1** person whose work is at sea. **2** sailor, esp. one below the rank of officer.

seamanship *n.* skill in managing a ship or boat.

seam bowler *n. Cricket* bowler who makes the ball deviate by bouncing it off its seam.

sea mile *n.* = unit varying between approx. 2,014 yards (1,842 metres) and 2,035 yards (1,861 metres).

seamstress /'semstrɪs/ *n.* (also **sempstress**) woman who sews, esp. for a living. [Old English: related to SEAM]

seamy *adj.* (**-ier, -iest**) **1** disreputable or sordid (esp. *the seamy side*). **2** marked with or showing seams. □ **seaminess** *n.*

seance /'seɪɑ̃s/ *n.* meeting at which a spiritualist attempts to make contact with the dead. [French]

sea pink *n.* maritime plant with bright pink flowers.

seaplane *n.* aircraft designed to take off from and land on water.

seaport *n.* town with a harbour.

sear *v.* **1** scorch, cauterize. **2** cause anguish to. **3** brown (meat) quickly at a high temperature to retain its juices in cooking. [Old English]

search /sɜːtʃ/ *–v.* **1** (also *absol.*) look through or go over thoroughly to find something. **2** examine or feel over (a person) to find anything concealed. **3** probe (*search one's conscience*). **4** (foll. by *for*) look thoroughly in order to find. **5** (as **searching** *adj.*) (of an examination) thorough; keenly questioning (*searching gaze*). **6** (foll. by *out*) look for; seek out. *–n.* **1** act of searching. **2** investigation. □ **in search of** trying to find. **search me!** *colloq.* I do not know. □ **searcher** *n.* **searchingly** *adv.* [Anglo-French *cerchier*]

searchlight *n.* **1** powerful outdoor electric light with a concentrated beam that can be turned in any direction. **2** light or beam from this.

search-party *n.* group of people conducting an organized search.

search warrant *n.* official authorization to enter and search a building.

sea room *n.* space at sea for a ship to turn etc.

sea salt *n.* salt produced by evaporating sea water.

seascape *n.* picture or view of the sea.

Sea Scout *n.* member of the maritime branch of the Scout Association.

seashell *n.* shell of a salt-water mollusc.

seashore *n.* land next to the sea.

seasick *adj.* nauseous from the motion of a ship at sea. □ **seasickness** *n.*

seaside *n.* sea-coast, esp. as a holiday resort.

season /'si:z(ə)n/ –*n.* **1** each of the climatic divisions of the year (spring, summer, autumn, winter). **2** proper or suitable time. **3** time when something is plentiful, active, etc. **4** (usu. prec. by *the*) = HIGH SEASON. **5** time of year for an activity or for social life generally (*football season*; *London in the season*). **6** indefinite period. **7** *colloq.* = SEASON TICKET. –*v.* **1** flavour (food) with salt, herbs, etc. **2** enhance with wit etc. **3** moderate. **4** (esp. as **seasoned** *adj.*) make or become suitable by exposure to the weather or experience (*seasoned wood*; *seasoned campaigner*). □ **in season 1** (of food) plentiful and good. **2** (of an animal) on heat. [Latin *satio* sowing]

seasonable *adj.* **1** suitable or usual to the season. **2** opportune. **3** apt.

■ **Usage** *Seasonable* is sometimes confused with *seasonal*.

seasonal *adj.* of, depending on, or varying with the season. □ **seasonally** *adv.*

seasoning *n.* salt, herbs, etc. added to food to enhance its flavour.

season ticket *n.* ticket entitling the holder to unlimited travel, access, etc., in a given period.

seat –*n.* **1** thing made or used for sitting on. **2 a** buttocks. **b** part of a garment covering them. **3** part of a chair etc. on which the buttocks rest. **4** place for one person in a theatre etc. **5** position as an MP, committee member, etc., or the right to occupy it. **6** supporting or guiding part of a machine. **7** location (*seat of learning*). **8** country mansion. **9** manner of sitting on a horse etc. –*v.* **1** cause to sit. **2** provide sitting accommodation for (*bus seats 50*). **3** (as **seated** *adj.*) sitting. **4** put or fit in position. □ **be seated** sit down. **by the seat of one's pants** *colloq.* by instinct rather than knowledge. **take a seat** sit down. [Old Norse: related to SIT]

seat-belt *n.* belt securing motor vehicle or aircraft passengers.

-seater *comb. form* having a specified number of seats.

seating *n.* **1** seats collectively. **2** sitting accommodation.

sea urchin *n.* small marine animal with a spiny shell.

sea wall *n.* wall built to stop flooding or erosion by the sea.

seaward /'si:wəd/ –*adv.* (also **seawards**) towards the sea. –*adj.* going or facing towards the sea.

seaway *n.* **1** inland waterway open to seagoing ships. **2** ship's progress. **3** ship's path across the sea.

seaweed *n.* plant growing in the sea or on rocks on a shore.

seaworthy *adj.* fit to put to sea. □ **seaworthiness** *n.*

sebaceous /sɪ'beɪʃəs/ *adj.* fatty; secreting oily matter. [Latin *sebum* tallow]

Sec. *abbr.* (also **sec.**) secretary.

sec[1] *abbr.* secant.

sec[2] *n. colloq.* (in phrases) second, moment (*wait a sec*). [abbreviation]

sec. *abbr.* second(s).

sec *adj.* (of wine) dry. [French]

secant /'si:kənt/ *n. Math.* **1** ratio of the hypotenuse to the shorter side adjacent to an acute angle (in a right-angled triangle). **2** line cutting a curve at one or more points. [French]

secateurs /ˌsekə'tɜ:z/ *n.pl.* pruning clippers used with one hand. [French]

secede /sɪ'si:d/ *v.* (**-ding**) withdraw formally from a political federation or religious body. [Latin *secedo -cess-*]

secession /sɪ'seʃ(ə)n/ *n.* act of seceding. □ **secessionist** *n.* & *adj.* [Latin: related to SECEDE]

seclude /sɪ'klu:d/ *v.* (**-ding**) (also *refl.*) **1** keep (a person or place) apart from others. **2** (esp. as **secluded** *adj.*) screen from view. [Latin *secludo -clus-*]

seclusion /sɪ'klu:ʒ(ə)n/ *n.* secluded state or place.

second[1] /'sekənd/ –*adj.* **1** next after first. **2** additional (*ate a second cake*). **3** subordinate; inferior. **4** *Mus.* performing a lower or subordinate part (*second violins*). **5** such as to be comparable to (*a second Callas*). –*n.* **1** runner-up. **2** person or thing besides the first or previously mentioned one. **3** second gear. **4** (in *pl.*) inferior goods. **5** (in *pl.*) *colloq.* second helping or course. **6** assistant to a duellist, boxer, etc. –*v.* **1** support; back up. **2** formally support (a nomination, resolution, or its proposer). □ **at second hand** indirectly. □ **seconder** *n.* (esp. in sense 2 of *v.*). [Latin *secundus* from *sequor* follow]

second[2] /'sekənd/ *n.* **1** sixtieth of a minute of time or of an angle. **2** *colloq.* very short time (*wait a second*). [medieval Latin *secunda* (*minuta*) secondary (minute)]

second[3] /sɪ'kɒnd/ *v.* transfer (a person) temporarily to another department etc. □ **secondment** *n.* [French *en second* in the second rank]

secondary /'sekəndərɪ/ *–adj.* **1** coming after or next below what is primary. **2** derived from or supplementing what is primary. **3** (of education, a school, etc.) following primary, esp. from the age of 11. *–n.* (*pl.* **-ies**) secondary thing. □ **secondarily** *adv.* [Latin: related to SECOND[1]]

secondary colour *n.* result of mixing two primary colours.

secondary picketing *n.* picketing of premises of a firm not directly involved in an industrial dispute.

second-best *adj.* & *n.* next after best.

second chamber *n.* upper house of a parliament.

second class *–n.* second-best group, category, postal service, or accommodation. *–adj.* & *adv.* (**second-class**) of or by the second class (*second-class citizens*; *travelled second-class*).

second cousin *n.* son or daughter of one's parent's cousin.

second-degree *adj.* denoting burns that cause blistering but not permanent scars.

second fiddle see FIDDLE.

second-guess *v. colloq.* **1** anticipate by guesswork. **2** criticize with hindsight.

second-hand *–adj.* **1 a** having had a previous owner; not new. **b** (*attrib.*) (of a shop etc.) where such goods can be bought. **2** (of information etc.) indirect, not from one's own observation etc. *–adv.* **1** on a second-hand basis. **2** indirectly.

second lieutenant *n.* army officer next below lieutenant.

secondly *adv.* **1** furthermore. **2** as a second item.

second nature *n.* acquired tendency that has become instinctive.

second officer *n.* assistant mate on a merchant ship.

second person see PERSON.

second-rate *adj.* mediocre; inferior.

second sight *n.* clairvoyance.

second string *n.* alternative course of action etc.

second thoughts *n.pl.* revised opinion or resolution.

second wind *n.* **1** recovery of normal breathing during exercise after initial breathlessness. **2** renewed energy to continue.

secrecy /'si:krəsɪ/ *n.* state of being secret; habit or faculty of keeping secrets (*done in secrecy*).

secret /'si:krɪt/ *–adj.* **1** kept or meant to be kept private, unknown, or hidden. **2** acting or operating secretly. **3** fond of secrecy. *–n.* **1** thing kept or meant to be kept secret. **2** mystery. **3** effective but not generally known method (*what's their secret?*; *the secret of success*). □ **in secret** secretly. □ **secretly** *adv.* [Latin *secerno secret-* separate]

secret agent *n.* spy.

secretaire /ˌsekrɪ'teə(r)/ *n.* escritoire. [French: related to SECRETARY]

secretariat /ˌsekrɪ'teərɪət/ *n.* **1** administrative office or department. **2** its members or premises. [medieval Latin: related to SECRETARY]

secretary /'sekrɪtərɪ/ *n.* (*pl.* **-ies**) **1** employee who assists with correspondence, records, making appointments, etc. **2** official of a society or company who writes letters, organizes business, etc. **3** principal assistant of a government minister, ambassador, etc. □ **secretarial** /-'teərɪəl/ *adj.* **secretaryship** *n.* [Latin *secretarius*: related to SECRET]

secretary bird *n.* long-legged crested African bird.

Secretary-General *n.* principal administrator of an organization.

Secretary of State *n.* **1** head of a major government department. **2** *US* = FOREIGN MINISTER.

secret ballot *n.* ballot in which votes are cast in secret.

secrete /sɪ'kri:t/ *v.* (**-ting**) **1** (of a cell, organ, etc.) produce and discharge (a substance). **2** conceal. □ **secretory** *adj.* [from SECRET]

secretion /sɪ'kri:ʃ(ə)n/ *n.* **1 a** process of secreting. **b** secreted substance. **2** act of concealing. [Latin: related to SECRET]

secretive /'si:krətɪv/ *adj.* inclined to make or keep secrets; uncommunicative. □ **secretively** *adv.* **secretiveness** *n.*

secret police *n.* police force operating secretly for political ends.

secret service *n.* government department concerned with espionage.

secret society *n.* society whose members are sworn to secrecy about it.

sect *n.* **1** group sharing (usu. unorthodox) religious, political, or philosophical doctrines. **2** (esp. exclusive) religious denomination. [Latin *sequor* follow]

sectarian /sek'teərɪən/ *–adj.* **1** of a sect. **2** devoted, esp. narrow-mindedly, to one's sect. *–n.* member of a sect. □ **sectarianism** *n.* [medieval Latin *sectarius* adherent]

section /'sekʃ(ə)n/ *–n.* **1** each of the parts of a thing or out of which a thing can be fitted together. **2** part cut off. **3** subdivision. **4** *US* **a** area of land. **b** district of a town. **5** act of cutting or separating surgically. **6 a** cutting of a solid by a plane. **b** resulting figure or area. *–v.* **1** arrange in or divide into sections. **2** compulsorily commit to a psychiatric hospital. [Latin *seco sect-* cut]

sectional /'sekʃən(ə)l/ *adj.* **1 a** of a social group (*sectional interests*). **b** partisan. **2** made in sections. **3** local rather than general. □ **sectionally** *adv.*

sector /'sektə(r)/ *n.* **1** distinct part of an enterprise, society, the economy, etc. **2** military subdivision of an area. **3** plane figure enclosed by two radii of a circle, ellipse, etc., and the arc between them. [Latin: related to SECTION]

secular /'sekjʊlə(r)/ *adj.* **1** not concerned with religion; not sacred; worldly (*secular education*; *secular music*). **2** (of clerics) not monastic. □ **secularism** *n.* **secularize** *v.* (also **-ise**) (**-zing** or **-sing**). **secularization** /-'zeɪʃ(ə)n/ *n.* [Latin *saeculum* an age]

secure /sɪ'kjʊə(r)/ *—adj.* **1** untroubled by danger or fear. **2** safe. **3** reliable; stable; fixed. *—v.* (**-ring**) **1** make secure or safe. **2** fasten or close securely. **3** succeed in obtaining. □ **securely** *adv.* [Latin *se* without, *cura* care]

security *n.* (*pl.* **-ies**) **1** secure condition or feeling. **2** thing that guards or guarantees. **3 a** safety against espionage, theft, etc. **b** organization for ensuring this. **4** thing deposited as a guarantee of an undertaking or loan, to be forfeited in case of default. **5** (often in *pl.*) document as evidence of a loan, certificate of stock, bonds, etc.

security risk *n.* person or thing that threatens security.

sedan /sɪ'dæn/ *n.* **1** (in full **sedan chair**) *hist.* enclosed chair for one, carried on poles by two men. **2** *US* enclosed car with four or more seats. [origin uncertain]

sedate /sɪ'deɪt/ *—adj.* tranquil and dignified; serious. *—v.* (**-ting**) put under sedation. □ **sedately** *adv.* **sedateness** *n.* [Latin *sedo* settle, calm]

sedation /sɪ'deɪʃ(ə)n/ *n.* act of calming, esp. by sedatives. [Latin: related to SEDATE]

sedative /'sedətɪv/ *—n.* calming drug or influence. *—adj.* calming, soothing. [medieval Latin: related to SEDATE]

sedentary /'sedəntərɪ/ *adj.* **1** sitting. **2** (of work etc.) done while sitting. **3** (of a person) disinclined to exercise. [Latin *sedeo* sit]

sedge *n.* waterside or marsh plant resembling coarse grass. □ **sedgy** *adj.* [Old English]

sediment /'sedɪmənt/ *n.* **1** grounds; dregs. **2** matter deposited on the land by water or wind. □ **sedimentary** /-'mentərɪ/ *adj.* **sedimentation** /-'teɪʃ(ə)n/ *n.* [Latin *sedeo* sit]

sedition /sɪ'dɪʃ(ə)n/ *n.* conduct or speech inciting to rebellion. □ **seditious** *adj.* [Latin *seditio*]

seduce /sɪ'dju:s/ *v.* (**-cing**) **1** entice into sexual activity or wrongdoing. **2** coax or lead astray. □ **seducer** *n.* [Latin *se-* away, *duco duct-* lead]

seduction /sɪ'dʌkʃ(ə)n/ *n.* **1** seducing or being seduced. **2** thing that tempts or attracts.

seductive /sɪ'dʌktɪv/ *adj.* alluring, enticing. □ **seductively** *adv.* **seductiveness** *n.*

seductress /sɪ'dʌktrɪs/ *n.* female seducer. [obsolete *seductor* male seducer: related to SEDUCE]

sedulous /'sedjʊləs/ *adj.* persevering, diligent, painstaking. □ **sedulity** /sɪ'dju:lɪtɪ/ *n.* **sedulously** *adv.* [Latin *sedulus* zealous]

sedum /'si:dəm/ *n.* fleshy-leaved plant with yellow, pink, or white flowers, e.g. the stonecrop. [Latin, = houseleek]

see[1] *v.* (*past* **saw**; *past part.* **seen**) **1** perceive with the eyes. **2** have or use this power. **3** discern mentally; understand. **4** watch (a film, game, etc.). **5** ascertain, learn (*will see if he's here*). **6** imagine, foresee (*see trouble ahead*). **7** look at for information (*see page 15*). **8** meet and recognize (*I saw your mother in town*). **9 a** meet socially or on business; visit or be visited by (*is too ill to see anyone*; *must see a doctor*). **b** meet regularly as a boyfriend or girlfriend. **10** reflect, wait for clarification (*we shall have to see*). **11** experience (*I never thought to see it*). **12** find attractive (*can't think what she sees in him*). **13** escort, conduct (*saw them home*). **14** witness (an event etc.) (*see the New Year in*). **15** ensure (*see that it is done*). **16 a** (in poker etc.) equal (a bet). **b** equal the bet of (a player). □ **see about 1** attend to. **2** consider. **see the back of** *colloq.* be rid of. **see fit** see FIT[1]. **see the light 1** realize one's mistakes etc. **2** undergo religious conversion. **see off 1** be present at the departure of (a person). **2** *colloq.* ward off, get the better of. **see out 1** accompany out of a building etc. **2** finish (a project etc.) completely. **3** survive (a period etc.). **see over** inspect; tour. **see red** *colloq.* become enraged. **see stars** *colloq.* see lights as a result of a blow on the head. **see things** *colloq.* have hallucinations. **see through** detect the truth or true nature of. **see a person through** support a person during a difficult time. **see a thing through** finish it completely. **see to it** (foll. by *that*) ensure. [Old English]

see[2] *n.* **1** area under the authority of a bishop or archbishop. **2** his office or jurisdiction. [Latin *sedes* seat]

seed –*n.* **1 a** part of a plant capable of developing into another such plant. **b** seeds collectively, esp. for sowing. **2** semen. **3** prime cause, beginning. **4** offspring, descendants. **5** (in tennis etc.) seeded player. –*v.* **1 a** place seeds in. **b** sprinkle (as) with seed. **2** sow seeds. **3** produce or drop seed. **4** remove seeds from (fruit etc.). **5** place a crystal etc. in (a cloud) to produce rain. **6** *Sport* **a** so position (a strong competitor in a knockout competition) that he or she will not meet other strong competitors in early rounds. **b** arrange (the order of play) in this way. □ **go** (or **run**) **to seed 1** cease flowering as seed develops. **2** become degenerate, unkempt, etc. □ **seedless** *adj.* [Old English]

seed-bed *n.* **1** bed prepared for sowing. **2** place of development.

seedling *n.* young plant raised from seed rather than from a cutting etc.

seed-pearl *n.* very small pearl.

seed-potato *n.* potato kept for seed.

seedsman *n.* dealer in seeds.

seedy *adj.* (**-ier**, **-iest**) **1** shabby, unkempt. **2** *colloq.* unwell. **3** full of or going to seed. □ **seediness** *n.*

seeing *conj.* (usu. foll. by *that*) considering that, inasmuch as, because.

seek *v.* (*past* and *past part.* **sought** /sɔːt/) **1** (often foll. by *for*, *after*) search or inquire. **2 a** try or want to find or get or reach (*sought my hand*). **b** request (*sought help*). **3** endeavour (*seek to please*). □ **seek out 1** search for and find. **2** single out as a friend etc. □ **seeker** *n.* [Old English]

seem *v.* (often foll. by *to* + infin.) appear or feel (*seems ridiculous*). □ **I** etc. **can't seem to** I etc. appear unable to (*can't seem to manage it*). **it seems** (or **would seem**) (often foll. by *that*) it appears to be the case. [Old Norse]

seeming *adj.* apparent but perhaps doubtful (*his seeming interest*). □ **seemingly** *adv.*

seemly *adj.* (**-ier**, **-iest**) in good taste; decorous. □ **seemliness** *n.* [Old Norse: related to SEEM]

seen *past part.* of SEE[1].

See of Rome *n.* the papacy.

seep *v.* ooze out; percolate. [Old English]

seepage *n.* **1** act of seeping. **2** quantity that seeps out.

seer *n.* **1** person who sees. **2** prophet; visionary.

seersucker /ˈsɪəˌsʌkə(r)/ *n.* linen, cotton, etc. fabric with a puckered surface. [Persian]

see-saw /ˈsiːsɔː/ –*n.* **1 a** long plank balanced on a central support, for children to sit on at each end and move up and down alternately. **b** this game. **2** up-and-down or to-and-fro motion. **3** close contest with alternating advantage. –*v.* **1** play on a see-saw. **2** move up and down. **3** vacillate in policy, emotion, etc. –*adj.* & *adv.* with up-and-down or backward-and-forward motion. [reduplication of SAW[1]]

seethe /siːð/ *v.* (**-thing**) **1** boil, bubble over. **2** be very angry, resentful, etc. [Old English]

see-through *adj.* (esp. of clothing) translucent.

segment /ˈsegmənt/ –*n.* **1** each part into which a thing is or can be divided. **2** part of a circle or sphere etc. cut off by an intersecting line or plane. –*v.* usu. /-ˈment/ divide into segments. □ **segmental** /-ˈment(ə)l/ *adj.* **segmentation** /-ˈteɪʃ(ə)n/ *n.* [Latin *seco* cut]

segregate /ˈsegrɪˌgeɪt/ *v.* (**-ting**) **1** put apart; isolate. **2** separate (esp. an ethnic group) from the rest of the community. [Latin *grex greg-* flock]

segregation /ˌsegrɪˈgeɪʃ(ə)n/ *n.* **1** enforced separation of ethnic groups in a community etc. **2** segregating or being segregated. □ **segregationist** *n.* & *adj.*

seigneur /seɪˈnjɜː(r)/ *n.* feudal lord. □ **seigneurial** *adj.* [French from Latin *senior* SENIOR]

seine /seɪn/ –*n.* fishing-net with floats at the top and weights at the bottom edge. –*v.* (**-ning**) fish or catch with a seine. [Old English *segne*]

seise var. of SEIZE 6.

seismic /ˈsaɪzmɪk/ *adj.* of earthquakes. [Greek *seismos* earthquake]

seismogram /ˈsaɪzməˌgræm/ *n.* record given by a seismograph.

seismograph /ˈsaɪzməˌgrɑːf/ *n.* instrument that records the force, direction, etc., of earthquakes. □ **seismographic** /-ˈgræfɪk/ *adj.*

seismology /saɪzˈmɒlədʒɪ/ *n.* the study of earthquakes. □ **seismological** /-məˈlɒdʒɪk(ə)l/ *adj.* **seismologist** *n.*

seize /siːz/ *v.* (**-zing**) **1** (often foll. by *on*, *upon*) take hold of forcibly or suddenly. **2** take possession of forcibly or by legal power. **3** affect suddenly (*panic seized us*). **4** (often foll. by *on*, *upon*) take advantage of (an opportunity etc.). **5** (often foll. by *on*, *upon*) comprehend quickly or clearly. **6** (also **seise**) (usu. foll. by *of*) *Law* put in possession of. □ **seized** (or **seised**) **of 1** possessing legally. **2** aware or informed of. **seize up 1** (of a mechanism) become jammed. **2** (of part of the body etc.) become stiff. [French *saisir*]

seizure /'si:ʒə(r)/ *n.* **1** seizing or being seized. **2** sudden attack, esp. of epilepsy or apoplexy.

seldom /'seldəm/ *adv.* rarely, not often. [Old English]

select /sɪ'lekt/ *–v.* choose, esp. with care. *–adj.* **1** chosen for excellence or suitability. **2** (of a society etc.) exclusive. [Latin *seligo -lect-*]

select committee *n.* small parliamentary committee conducting a special inquiry.

selection /sɪ'lekʃ(ə)n/ *n.* **1** selecting or being selected. **2** selected person or thing. **3** things from which a choice may be made. **4** evolutionary process by which some species thrive better than others.

selective *adj.* **1** of or using selection (*selective schools*). **2** able to select. **3** (of memory etc.) selecting what is convenient. □ **selectively** *adv.* **selectivity** /-'tɪvɪtɪ/ *n.*

selector *n.* **1** person who selects, esp. a team. **2** device in a vehicle, machinery, etc. that selects the required gear etc.

selenium /sɪ'li:nɪəm/ *n.* non-metallic element occurring naturally in various metallic sulphide ores. [Greek *selēnē* moon]

self *n.* (*pl.* **selves**) **1** individuality, personality, or essence (*showed his true self*; *is her old self again*). **2** object of introspection or reflexive action. **3 a** one's own interests or pleasure. **b** concentration on these. **4** *Commerce* or *colloq.* myself, yourself, etc. (*cheque drawn to self*). [Old English]

self- *comb. form* expressing reflexive action: **1** of or by oneself or itself (*self-locking*). **2** on, in, for, or of oneself or itself (*self-absorbed*).

self-abasement /,selfə'beɪsmənt/ *n.* self-humiliation; cringing.

self-absorption /,selfəb'zɔ:pʃ(ə)n/ *n.* absorption in oneself. □ **self-absorbed** *adj.*

self-abuse /,selfə'bju:s/ *n. archaic* masturbation.

self-addressed /,selfə'drest/ *adj.* (of an envelope) bearing one's own address for a reply.

self-adhesive /,selfəd'hi:sɪv/ *adj.* (of an envelope, label, etc.) adhesive, esp. without wetting.

self-advancement /,selfəd'vɑ:nsmənt/ *n.* advancement of oneself.

self-aggrandizement /,selfə'grændɪzmənt/ *n.* process of enriching oneself or making oneself powerful. □ **self-aggrandizing** /-'grændaɪzɪŋ/ *adj.*

self-analysis /,selfə'næləsɪs/ *n.* analysis of oneself, one's motives, character, etc.

self-appointed /,selfə'pɔɪntɪd/ *adj.* designated so by oneself, not by others (*self-appointed critic*).

self-assembly /,selfə'semblɪ/ *adj.* assembled by the buyer from a kit.

self-assertive /,selfə'sɜ:tɪv/ *adj.* confident or aggressive in promoting oneself, one's rights, etc. □ **self-assertion** *n.*

self-assured /,selfə'ʃʊəd/ *n.* self-confident. □ **self-assurance** *n.*

self-aware /,selfə'weə(r)/ *adj.* conscious of one's character, feelings, motives, etc. □ **self-awareness** *n.*

self-catering /self'keɪtərɪŋ/ *adj.* (of a holiday, accommodation etc.) with cooking facilities provided, but no food.

self-censorship /self'sensəʃɪp/ *n.* censoring of oneself.

self-centred /self'sentəd/ *adj.* (*US* **-centered**) preoccupied with oneself; selfish. □ **self-centredly** *adv.* **self-centredness** *n.*

self-cleaning /self'kli:nɪŋ/ *adj.* (esp. of an oven) cleaning itself when heated.

self-conceit /,selfkən'si:t/ *n.* high or exaggerated opinion of oneself.

self-confessed /,selfkən'fest/ *adj.* openly admitting oneself to be.

self-confident /self'kɒnfɪd(ə)nt/ *adj.* having confidence in oneself. □ **self-confidence** *n.* **self-confidently** *adv.*

self-congratulatory /,selfkəŋ'grætjʊ,lətərɪ/ *adj.* = SELF-SATISFIED. □ **self-congratulation** /-'leɪʃ(ə)n/ *n.*

self-conscious /self'kɒnʃəs/ *adj.* nervous, shy, or embarrassed. □ **self-consciously** *adv.* **self-consciousness** *n.*

self-consistent /,selfkən'sɪst(ə)nt/ *adj.* (of parts of the same whole etc.) consistent; not conflicting. □ **self-consistency** *n.*

self-contained /,selfkən'teɪnd/ *adj.* **1** (of a person) uncommunicative; independent. **2** (of accommodation) complete in itself, having no shared entrance or facilities.

self-control /,selfkən'trəʊl/ *n.* power of controlling one's behaviour, emotions, etc. □ **self-controlled** *adj.*

self-critical /self'krɪtɪk(ə)l/ *adj.* critical of oneself, one's abilities, etc. □ **self-criticism** /-,sɪz(ə)m/ *n.*

self-deception /,selfdɪ'sepʃ(ə)n/ *n.* deceiving of oneself, esp. about one's motives or feelings. □ **self-deceit** /-dɪ'si:t/ *n.*

self-defeating /,selfdɪ'fi:tɪŋ/ *adj.* (of an action etc.) doomed to failure because of internal inconsistencies; achieving the opposite of what is intended.

self-defence /,selfdɪ'fens/ *n.* (*US* **-defense**) physical or verbal defence of one's body, property, rights, reputation, etc.

self-delusion /ˌselfdɪ'lu:ʒ(ə)n/ *n.* act of deluding oneself.

self-denial /ˌselfdɪ'naɪəl/ *n.* asceticism, esp. to discipline oneself. □ **self-denying** *adj.*

self-deprecation /ˌselfdeprɪ'keɪʃ(ə)n/ *n.* belittling of oneself. □ **self-deprecating** /-'deprɪˌkeɪtɪŋ/ *adj.*

self-destruct /ˌselfdɪ'strʌkt/ *—v.* (of a spacecraft, bomb, etc.) explode or disintegrate automatically, esp. when pre-set to do so. *—attrib. adj.* enabling a thing to self-destruct (*self-destruct device*).

self-destruction /ˌselfdɪ'strʌkʃ(ə)n/ *n.* **1** destroying of itself or oneself or one's chances, happiness, etc. **2** act of self-destructing. □ **self-destructive** *adj.*

self-determination /ˌselfdɪˌtɜ:mɪ'neɪʃ(ə)n/ *n.* **1** nation's right to determine its own government etc. **2** ability to act with free will.

self-discipline /self'dɪsɪplɪn/ *n.* **1** ability to apply oneself. **2** self-control. □ **self-disciplined** *adj.*

self-discovery /ˌselfdɪ'skʌvərɪ/ *n.* process of acquiring insight into one's character, desires, etc.

self-doubt /self'daʊt/ *n.* lack of confidence in oneself.

self-drive /self'draɪv/ *adj.* (of a hired vehicle) driven by the hirer.

self-educated /self'edjʊˌkeɪtɪd/ *adj.* educated by one's own reading etc., without formal instruction.

self-effacing /ˌselfɪ'feɪsɪŋ/ *adj.* retiring; modest. □ **self-effacement** *n.*

self-employed /ˌselfɪm'plɔɪd/ *adj.* working as a freelance or for one's own business etc. □ **self-employment** *n.*

self-esteem /ˌselfɪ'sti:m/ *n.* good opinion of oneself.

self-evident /self'evɪd(ə)nt/ *adj.* obvious; without the need of proof or further explanation. □ **self-evidence** *n.* **self-evidently** *adv.*

self-examination /ˌselfɪgˌzæmɪ'neɪʃ(ə)n/ *n.* **1** the study of one's own conduct etc. **2** examining of one's own body for signs of illness.

self-explanatory /ˌselfɪk'splænətərɪ/ *adj.* not needing explanation.

self-expression /ˌselfɪk'spreʃ(ə)n/ *n.* artistic or free expression.

self-financing /self'faɪnænsɪŋ, -'nænsɪŋ/ *adj.* (of an institution or undertaking) that pays for itself without subsidy.

self-fulfilling /ˌselffʊl'fɪlɪŋ/ *adj.* (of a prophecy etc.) bound to come true as a result of its being made.

self-fulfilment /ˌselffʊl'fɪlmənt/ *n.* fulfilment of one's ambitions etc.

self-governing /self'gʌvənɪŋ/ *adj.* governing itself or oneself. □ **self-government** *n.*

self-help /self'help/ *n.* (often *attrib.*) use of one's own abilities, resources, etc. to solve one's problems etc. (*formed a self-help group*).

self-image /self'ɪmɪdʒ/ *n.* one's conception of oneself.

self-important /ˌselfɪm'pɔ:t(ə)nt/ *adj.* conceited; pompous. □ **self-importance** *n.*

self-imposed /ˌselfɪm'pəʊzd/ *adj.* (of a task etc.) imposed on and by oneself.

self-improvement /ˌselfɪm'pru:vmənt/ *n.* improvement of oneself or one's life etc. by one's own efforts.

self-induced /ˌselfɪn'dju:st/ *adj.* induced by oneself or itself.

self-indulgent /ˌselfɪn'dʌldʒ(ə)nt/ *adj.* **1** indulging in one's own pleasure, feelings, etc. **2** (of a work of art etc.) lacking economy and control. □ **self-indulgence** *n.*

self-inflicted /ˌselfɪn'flɪktɪd/ *adj.* inflicted by and on oneself.

self-interest /self'ɪntrəst/ *n.* one's personal interest or advantage. □ **self-interested** *adj.*

selfish *adj.* concerned chiefly with one's own interests or pleasure; actuated by or appealing to self-interest. □ **selfishly** *adv.* **selfishness** *n.*

self-justification /ˌselfdʒʌstɪfɪ'keɪʃ(ə)n/ *n.* justification or excusing of oneself.

self-knowledge /self'nɒlɪdʒ/ *n.* understanding of oneself.

selfless *adj.* unselfish. □ **selflessly** *adv.* **selflessness** *n.*

self-made *adj.* successful or rich by one's own effort.

self-opinionated /ˌselfə'pɪnjəˌneɪtɪd/ *adj.* stubbornly adhering to one's opinions.

self-perpetuating /ˌselfpə'petju:ˌeɪtɪŋ/ *adj.* perpetuating itself or oneself without external agency.

self-pity /self'pɪtɪ/ *n.* feeling sorry for oneself. □ **self-pitying** *adj.*

self-pollination /ˌselfpɒlɪ'neɪʃ(ə)n/ *n.* pollination of a flower by pollen from the same plant. □ **self-pollinating** *adj.*

self-portrait /self'pɔ:trɪt/ *n.* portrait or description of oneself by oneself.

self-possessed /ˌselfpə'zest/ *adj.* calm and composed. □ **self-possession** /-'zeʃ(ə)n/ *n.*

self-preservation /ˌselfprezə'veɪʃ(ə)n/ *n.* **1** keeping oneself safe. **2** instinct for this.

self-proclaimed /ˌselfprəˈkleɪmd/ *adj.* proclaimed by oneself or itself to be such.

self-propelled /ˌselfprəˈpeld/ *adj.* (of a vehicle etc.) propelled by its own power. □ **self-propelling** *adj.*

self-raising /selfˈreɪzɪŋ/ *adj.* (of flour) containing a raising agent.

self-realization /ˌselfrɪəlaɪˈzeɪʃ(ə)n/ *n.* development of one's abilities etc.

self-regard /ˌselfrɪˈgɑːd/ *n.* proper regard for oneself.

self-regulating /selfˈregjʊˌleɪtɪŋ/ *adj.* regulating oneself or itself without intervention. □ **self-regulation** /-ˈleɪʃ(ə)n/ *n.* **self-regulatory** /-lətərɪ/ *adj.*

self-reliance /ˌselfrɪˈlaɪəns/ *n.* reliance on one's own resources etc.; independence. □ **self-reliant** *adj.*

self-reproach /ˌselfrɪˈprəʊtʃ/ *n.* reproach directed at oneself.

self-respect /ˌselfrɪˈspekt/ *n.* respect for oneself. □ **self-respecting** *adj.*

self-restraint /ˌselfrɪˈstreɪnt/ *n.* self-control.

self-righteous /selfˈraɪtʃəs/ *adj.* smugly sure of one's rightness. □ **self-righteously** *adv.* **self-righteousness** *n.*

self-rule /selfˈruːl/ *n.* self-government.

self-sacrifice /selfˈsækrɪˌfaɪs/ *n.* selflessness; self-denial. □ **self-sacrificing** *adj.*

selfsame *adj.* (prec. by *the*) very same, identical.

self-satisfied /selfˈsætɪsˌfaɪd/ *adj.* complacent; self-righteous. □ **self-satisfaction** /-ˈfækʃ(ə)n/ *n.*

self-sealing /selfˈsiːlɪŋ/ *adj.* **1** (of a tyre etc.) automatically able to seal small punctures. **2** (of an envelope) self-adhesive.

self-seed /selfˈsiːd/ *v.* (of a plant) propagate itself by seed. □ **self-seeder** *n.*

self-seeking /selfˈsiːkɪŋ/ *adj.* & *n.* selfish.

self-service /selfˈsɜːvɪs/ *–adj.* (often *attrib.*) (of a shop, restaurant, etc.) with customers serving themselves and paying at a checkout etc. *–n. colloq.* self-service restaurant etc.

self-starter /selfˈstɑːtə(r)/ *n.* **1** electrical appliance for starting an engine. **2** ambitious person with initiative.

self-styled *adj.* called so by oneself.

self-sufficient /ˌselfsəˈfɪʃ(ə)nt/ *adj.* able to supply one's own needs; independent. □ **self-sufficiency** *n.*

self-supporting /ˌselfsəˈpɔːtɪŋ/ *adj.* financially self-sufficient.

self-taught /selfˈtɔːt/ *adj.* self-educated.

self-willed /selfˈwɪld/ *adj.* obstinately pursuing one's own wishes.

self-worth /selfˈwɜːθ/ *n.* = SELF-ESTEEM.

sell *–v.* (*past* and *past part.* **sold** /səʊld/) **1** exchange or be exchanged for money (*these sell well*). **2** stock for sale (*do you sell eggs?*). **3** (foll. by *at*, *for*) have a specified price (*sells at £5*). **4** (also *refl.*) betray or prostitute for money etc. **5** (also *refl.*) advertise or publicize (a product, oneself, etc.). **6** cause to be sold (*name alone will sell it*). **7** *colloq.* make (a person) enthusiastic about (an idea etc.). *–n. colloq.* **1** manner of selling (*soft sell*). **2** deception; disappointment. □ **sell down the river** see RIVER. **sell off** sell at reduced prices. **sell out 1** (also *absol.*) sell (all one's stock, shares, etc.). **2** betray; be treacherous or disloyal. **sell short** disparage, underestimate. **sell up** sell one's business, house, etc. [Old English]

sell-by date *n.* latest recommended date of sale.

seller *n.* **1** person who sells. **2** thing that sells well or badly.

seller's market *n.* (also **sellers' market**) trading conditions favourable to the seller.

selling-point *n.* advantageous feature.

Sellotape /ˈseləˌteɪp/ *–n. propr.* adhesive usu. transparent tape. *–v.* (**sellotape**) (**-ping**) fix with Sellotape. [from CELLULOSE]

sell-out *n.* **1** commercial success, esp. the selling of all tickets for a show. **2** betrayal.

selvage /ˈselvɪdʒ/ *n.* (also **selvedge**) fabric edging woven to prevent cloth from fraying. [from SELF, EDGE]

selves *pl.* of SELF.

semantic /sɪˈmæntɪk/ *adj.* of meaning in language. □ **semantically** *adv.* [Greek *sēmainō* to mean]

semantics *n.pl.* (usu. treated as *sing.*) branch of linguistics concerned with meaning.

semaphore /ˈseməˌfɔː(r)/ *–n.* **1** system of signalling with the arms or two flags. **2** railway signalling apparatus consisting of a post with a movable arm or arms etc. *–v.* (**-ring**) signal or send by semaphore. [Greek *sēma* sign, *pherō* bear]

semblance /ˈsembləns/ *n.* (foll. by *of*) appearance; show (*a semblance of anger*). [French *sembler* resemble]

semen /ˈsiːmən/ *n.* reproductive fluid of males. [Latin *semen semin-* seed]

semester /sɪˈmestə(r)/ *n.* half-year course or term in (esp. US) universities. [Latin *semestris* from *sex* six, *mensis* month]

semi /ˈsemɪ/ *n.* (*pl.* **-s**) *colloq.* semi-detached house. [abbreviation]

semi- *prefix* **1** half. **2** partly. [Latin]

semibreve /'semɪ,bri:v/ *n. Mus.* note equal to four crochets.

semicircle /'semɪ,sɜ:k(ə)l/ *n.* half of a circle or of its circumference. □ **semicircular** /-'sɜ:kjʊlə(r)/ *adj.*

semicolon /,semɪ'kəʊlən/ *n.* punctuation mark (;) of intermediate value between a comma and full stop.

semiconductor /,semɪkən'dʌktə(r)/ *n.* substance that in certain conditions has electrical conductivity intermediate between insulators and metals.

semi-conscious /,semɪ'kɒnʃəs/ *adj.* partly or imperfectly conscious.

semi-detached /,semɪdɪ'tætʃt/ *–adj.* (of a house) joined to one other on one side only. *–n.* such a house.

semifinal /,semɪ'faɪn(ə)l/ *n.* match or round preceding the final. □ **semifinalist** *n.*

seminal /'semɪn(ə)l/ *adj.* **1** of seed, semen, or reproduction; germinal. **2** (of ideas etc.) forming a basis for future development. [Latin: related to SEMEN]

seminar /'semɪ,nɑ:(r)/ *n.* **1** small discussion class at a university etc. **2** short intensive course of study. **3** conference of specialists. [German: related to SEMINARY]

seminary /'semɪnərɪ/ *n.* (*pl.* **-ies**) training-college for priests or rabbis etc. □ **seminarist** *n.* [Latin: related to SEMEN]

semiotics /,semɪ'ɒtɪks/ *n.* the study of signs and symbols and their use, esp. in language. □ **semiotic** *adj.* [Greek *sēmeiōtikos* of signs]

semi-permeable /,semɪ'pɜ:mɪəb(ə)l/ *adj.* (of a membrane etc.) allowing small molecules to pass through.

semiprecious /,semɪ'preʃəs/ *adj.* (of a gem) less valuable than a precious stone.

semi-professional /,semɪprə'feʃən(ə)l/ *–adj.* **1** (of a footballer, musician, etc.) paid for an activity but not relying on it for a living. **2** of semi-professionals. *–n.* semi-professional person.

semiquaver /'semɪ,kweɪvə(r)/ *n. Mus.* note equal to half a quaver.

semi-skilled /,semɪ'skɪld/ *adj.* (of work or a worker) needing or having some training.

semi-skimmed /,semɪ'skɪmd/ *adj.* (of milk) from which some of the cream has been skimmed.

Semite /'si:maɪt/ *n.* member of the peoples said to be descended from Shem (Gen. 10), including esp. the Jews and Arabs. [Greek *Sēm* Shem]

Semitic /sɪ'mɪtɪk/ *adj.* **1** of the Semites, esp. the Jews. **2** of languages of the family including Hebrew and Arabic.

semitone /'semɪ,təʊn/ *n.* half a tone in the musical scale.

semitropical /,semɪ'trɒpɪk(ə)l/ *adj.* = SUBTROPICAL.

semivowel /'semɪ,vaʊəl/ *n.* **1** sound intermediate between a vowel and a consonant. **2** letter representing this. (e.g. *w*, *y*)

semolina /,semə'li:nə/ *n.* **1** hard grains left after the milling of flour, used in milk puddings etc. **2** pudding of this. [Italian *semolino*]

sempstress var. of SEAMSTRESS.

Semtex *n. propr.* malleable odourless plastic explosive. [from *Semtín* in Czechoslovakia, where it was originally made]

SEN *abbr.* State Enrolled Nurse.

Sen. *abbr.* **1** Senior. **2** Senator.

senate /'senɪt/ *n.* **1** legislative body, esp. the upper and smaller assembly in the US, France, etc. **2** governing body of a university or (*US*) a college. **3** ancient Roman State council. [Latin *senatus* from *senex* old man]

senator /'senətə(r)/ *n.* member of a senate. □ **senatorial** /-'tɔ:rɪəl/ *adj.* [Latin: related to SENATE]

send *v.* (*past* and *past part.* **sent**) **1 a** order or cause to go or be conveyed. **b** propel (*sent him flying*). **c** cause to become (*sent me mad*). **2** send a message etc. (*he sent to warn me*). **3** (of God, etc.) grant, bestow, or inflict; bring about; cause to be. **4** *slang* put into ecstasy. □ **send away for** order (goods) by post. **send down 1** rusticate or expel from a university. **2** send to prison. **send for 1** summon. **2** order by post. **send in 1** cause to go in. **2** submit (an entry etc.) for a competition etc. **send off 1** dispatch (a letter, parcel, etc.). **2** attend the departure of (a person) as a sign of respect etc. **3** *Sport* (of a referee) order (a player) to leave the field. **send off for** = *send away for*. **send on** transmit further or in advance of oneself. **send up 1** cause to go up. **2** transmit to a higher authority. **3** *colloq.* ridicule by mimicking. **send word** send information. □ **sender** *n.* [Old English]

send-off *n.* party etc. at the departure of a person, start of a project, etc.

send-up *n. colloq.* satire, parody.

senescent /sɪ'nes(ə)nt/ *adj.* growing old. □ **senescence** *n.* [Latin *senex* old]

seneschal /'senɪʃ(ə)l/ *n.* steward of a medieval great house. [French, = old servant]

senile /'si:naɪl/ *adj.* **1** of old age. **2** mentally or physically infirm because of old age. □ **senility** /sɪ'nɪlɪtɪ/ *n.* [Latin: related to SENESCENT]

senile dementia *n.* illness of old people with loss of memory and control of bodily functions etc.
senior /'si:nɪə(r)/ *–adj.* **1** more or most advanced in age, standing, or position. **2** (placed after a person's name) senior to a relative of the same name. *–n.* **1** senior person. **2** one's elder or superior. □ **seniority** /-'ɒrɪtɪ/ *n.* [Latin comparative of *senex* old]
senior citizen *n.* old-age pensioner.
senior nursing officer *n.* person in charge of nursing services in a hospital.
senior school *n.* school for children esp. over the age of 11.
senior service *n.* Royal Navy.
senna /'senə/ *n.* **1** cassia. **2** laxative from the dried pod of this. [Arabic]
señor /sen'jɔ:(r)/ *n.* (*pl.* **señores** /-rez/) title used of or to a Spanish-speaking man. [Spanish from Latin *senior* SENIOR]
señora /sen'jɔ:rə/ *n.* title used of or to a Spanish-speaking esp. married woman.
señorita /,senjə'ri:tə/ *n.* title used of or to a young esp. unmarried Spanish-speaking woman.
sensation /sen'seɪʃ(ə)n/ *n.* **1** feeling in one's body (*sensation of warmth*). **2** awareness, impression (*sensation of being watched*). **3 a** intense interest, shock, etc. felt among a large group. **b** person, event, etc., causing this. **4** sense of touch. [medieval Latin: related to SENSE]
sensational *adj.* **1** causing or intended to cause great public excitement etc. **2** dazzling; wonderful (*you look sensational*). □ **sensationalize** *v.* (also **-ise**) (**-zing** or **-sing**). **sensationally** *adv.*
sensationalism *n.* use of or interest in the sensational. □ **sensationalist** *n.* & *adj.*
sense *–n.* **1 a** any of the five bodily faculties transmitting sensation. **b** sensitiveness of all or any of these (*good sense of smell*). **2** ability to perceive or feel. **3** (foll. by *of*) consciousness; awareness (*sense of guilt*). **4** quick or accurate appreciation, understanding, or instinct (*sense of humour*). **5** practical wisdom, common sense. **6 a** meaning of a word etc. **b** intelligibility or coherence. **7** prevailing opinion (*sense of the meeting*). **8** (in *pl.*) sanity, ability to think. *–v.* (**-sing**) **1** perceive by a sense or senses. **2** be vaguely aware of. **3** realize. **4** (of a machine etc.) detect. □ **come to one's senses 1** regain consciousness. **2** regain common sense. **in a** (or **one**) **sense** if the statement etc. is understood in a particular way. **make sense** be intelligible or practicable. **make sense of** show or find the meaning of. **take leave of one's senses** go mad. [Latin *sensus* from *sentio sens-* feel]
senseless *adj.* **1** pointless; foolish. **2** unconscious. □ **senselessly** *adv.* **senselessness** *n.*
sense-organ *n.* bodily organ conveying external stimuli to the sensory system.
sensibility /,sensɪ'bɪlɪtɪ/ *n.* (*pl.* **-ies**) **1** capacity to feel. **2 a** sensitiveness. **b** exceptional degree of this. **3** (in *pl.*) tendency to feel offended etc.

■ **Usage** *Sensibility* should not be used in standard English to mean 'possession of good sense'.

sensible /'sensɪb(ə)l/ *adj.* **1** having or showing wisdom or common sense. **2 a** perceptible by the senses. **b** great enough to be perceived. **3** (of clothing etc.) practical. **4** (foll. by *of*) aware. □ **sensibly** *adv.*
sensitive /'sensɪtɪv/ *adj.* **1** (often foll. by *to*) acutely susceptible to external stimuli or impressions; having sensibility. **2** easily offended or hurt. **3** (often foll. by *to*) (of an instrument etc.) responsive to or recording slight changes. **4** (of photographic materials) responding (esp. rapidly) to light. **5** (of a topic etc.) requiring tactful treatment or secrecy. □ **sensitively** *adv.* **sensitiveness** *n.* **sensitivity** /-'tɪvɪtɪ/ *n.*
sensitize /'sensɪ,taɪz/ *v.* (also **-ise**) (**-zing** or **-sing**) make sensitive. □ **sensitization** /-'zeɪʃ(ə)n/ *n.*
sensor *n.* device for detecting or measuring a physical property. [from SENSORY]
sensory *adj.* of sensation or the senses. [Latin *sentio sens-* feel]
sensual /'sensjʊəl/ *adj.* **1 a** of physical, esp. sexual, pleasure. **b** enjoying or giving this, voluptuous. **2** showing sensuality (*sensual lips*). □ **sensualism** *n.* **sensually** *adv.* [Latin: related to SENSE]

■ **Usage** *Sensual* is sometimes confused with *sensuous*, which dces not have the sexual overtones of *sensual*.

sensuality /,sensjʊ'ælɪtɪ/ *n.* (esp. sexual) gratification of the senses.
sensuous /'sensjʊəs/ *adj.* of or affecting the senses, esp. aesthetically. □ **sensuously** *adv.* **sensuousness** *n.* [Latin: related to SENSE]

■ **Usage** See note at *sensual*.

sent *past* and *past part.* of SEND.
sentence /'sent(ə)ns/ *–n.* **1** statement, question, exclamation, or command containing or implying a subject and predicate (e.g. *I went*; *come here!*). **2 a**

decision of a lawcourt, esp. the punishment allotted to a convicted criminal. **b** declaration of this. –*v.* (**-cing**) **1** declare the sentence of (a convicted criminal etc.). **2** (foll. by *to*) declare (such a person) to be condemned to (a punishment). [Latin *sententia* from *sentio* consider]

sententious /sen'tenʃəs/ *adj.* **1** pompously moralizing. **2** affectedly formal in style. **3** aphoristic; using maxims. □ **sententiousness** *n.* [Latin: related to SENTENCE]

sentient /'senʃ(ə)nt/ *adj.* capable of perception and feeling. □ **sentience** *n.* **sentiency** *n.* **sentiently** *adv.* [Latin *sentio* feel]

sentiment /'sentɪmənt/ *n.* **1** mental feeling. **2** (often in *pl.*) what one feels, opinion. **3** opinion or feeling, as distinct from its expression (*the sentiment is good*). **4** emotional or irrational view. **5** such views collectively, esp. as an influence. **6** tendency to be swayed by feeling. **7 a** mawkish or exaggerated emotion. **b** display of this.

sentimental /ˌsentɪ'ment(ə)l/ *adj.* **1** of or showing sentiment. **2** showing or affected by emotion rather than reason. □ **sentimentalism** *n.* **sentimentalist** *n.* **sentimentality** /-'tælɪtɪ/ *n.* **sentimentalize** *v.* (also **-ise**) (**-zing** or **-sing**). **sentimentally** *adv.*

sentimental value *n.* value given to a thing because of its associations.

sentinel /'sentɪn(ə)l/ *n.* sentry or lookout. [French from Italian]

sentry /'sentrɪ/ *n.* (*pl.* **-ies**) soldier etc. stationed to keep guard. [perhaps from obsolete *centrinel*, var. of SENTINEL]

sentry-box *n.* cabin for sheltering a standing sentry.

sepal /'sep(ə)l/ *n.* division or leaf of a calyx. [perhaps from SEPARATE, PETAL]

separable /'sepərəb(ə)l/ *adj.* able to be separated. □ **separability** /-'bɪlɪtɪ/ *n.* [Latin: related to SEPARATE]

separate –*adj.* /'sepərət/ forming a unit by itself, existing apart; disconnected, distinct, or individual. –*n.* /'sepərət/ (in *pl.*) trousers, skirts, etc. that are not parts of suits. –*v.* /'sepəˌreɪt/ (**-ting**) **1** make separate, sever. **2** prevent union or contact of. **3** go different ways. **4** (esp. as **separated** *adj.*) cease to live with one's spouse. **5** (foll. by *from*) secede. **6 a** divide or sort into parts or sizes. **b** (often foll. by *out*) extract or remove (an ingredient etc.). □ **separately** *adv.* **separateness** *n.* [Latin *separo* (v.)]

separation /ˌsepə'reɪʃ(ə)n/ *n.* **1** separating or being separated. **2** (in full **judicial** or **legal separation**) legal arrangement by which a couple remain married but live apart. [Latin: related to SEPARATE]

separatist /'sepərətɪst/ *n.* person who favours separation, esp. political independence. □ **separatism** *n.*

separator /'sepəˌreɪtə(r)/ *n.* machine for separating, e.g. cream from milk.

Sephardi /sɪ'fɑːdɪ/ *n.* (*pl.* **Sephardim**) Jew of Spanish or Portuguese descent. □ **Sephardic** *adj.* [Hebrew, = Spaniard]

sepia /'siːpɪə/ *n.* **1** dark reddish-brown colour or paint. **2** brown tint used in photography. [Greek, = cuttlefish]

sepoy /'siːpɔɪ/ *n. hist.* native Indian soldier under European, esp. British, discipline. [Persian *sipāhī* soldier]

sepsis /'sepsɪs/ *n.* septic condition. [Greek: related to SEPTIC]

Sept. *abbr.* September.

sept *n.* clan, esp. in Ireland. [alteration of SECT]

September /sep'tembə(r)/ *n.* ninth month of the year. [Latin *septem* seven, originally the 7th month of the Roman year]

septennial /sep'tenɪəl/ *adj.* **1** lasting for seven years. **2** recurring every seven years.

septet /sep'tet/ *n.* **1** *Mus.* **a** composition for seven performers. **b** the performers. **2** any group of seven. [Latin *septem* seven]

septic /'septɪk/ *adj.* contaminated with bacteria, putrefying. [Greek *sēpō* rot]

septicaemia /ˌseptɪ'siːmɪə/ *n.* (*US* **septicemia**) blood-poisoning. □ **septicaemic** *adj.* [from SEPTIC, Greek *haima* blood]

septic tank *n.* tank in which sewage is disintegrated through bacterial activity.

septuagenarian /ˌseptjʊədʒɪ'neərɪən/ *n.* person from 70 to 79 years old. [Latin *septuageni* 70 each]

Septuagesima /ˌseptjʊə'dʒesɪmə/ *n.* Sunday before Sexagesima. [Latin, = seventieth]

Septuagint /'septjʊəˌdʒɪnt/ *n.* Greek version of the Old Testament including the Apocrypha. [Latin *septuaginta* seventy]

septum /'septəm/ *n.* (*pl.* **septa**) partition such as that between the nostrils or the chambers of a poppy-fruit or of a shell. [Latin *s(a)eptum* from *saepio* enclose]

septuple /'septjʊp(ə)l/ –*adj.* **1** sevenfold, having seven parts. **2** being seven times as many or as much. –*n.* sevenfold number or amount. [Latin *septem* seven]

sepulchral /sɪ'pʌlkr(ə)l/ *adj.* **1** of a tomb or interment. **2** funereal, gloomy. [Latin: related to SEPULCHRE]

sepulchre /'sepəlkə(r)/ (*US* **sepulcher**) *–n.* tomb, esp. cut in rock or built of stone or brick. *–v.* (**-ring**) **1** place in a sepulchre. **2** serve as a sepulchre for. [Latin *sepelio* bury]

sepulture /'sepəltʃə(r)/ *n.* burying, interment. [Latin: related to SEPULCHRE]

sequel /'si:kw(ə)l/ *n.* **1** what follows (esp. as a result). **2** novel, film, etc., that continues the story of an earlier one. [Latin *sequor* follow]

sequence /'si:kwəns/ *n.* **1** succession. **2** order of succession. **3** set of things belonging next to one another; unbroken series. **4** part of a film dealing with one scene or topic. [Latin: related to SEQUEL]

sequencer *n.* programmable electronic device for storing sequences of musical notes, chords, etc., and transmitting them when required to an electronic musical instrument. □ **sequencing** *n.*

sequential /sɪ'kwenʃ(ə)l/ *adj.* forming a sequence or consequence. □ **sequentially** *adv.* [from SEQUENCE]

sequester /sɪ'kwestə(r)/ *v.* **1** (esp. as **sequestered** *adj.*) seclude, isolate. **2** = SEQUESTRATE. [Latin *sequester* trustee]

sequestrate /'si:kwɪ,streɪt, sɪ'kwe-/ *v.* (**-ting**) **1** confiscate. **2** take temporary possession of (a debtor's estate etc.). □ **sequestration** /,si:kwɪ'streɪʃ(ə)n/ *n.* **sequestrator** /'si:kwɪ,streɪtə(r)/ *n.* [Latin: related to SEQUESTER]

sequin /'si:kwɪn/ *n.* circular spangle, esp. sewn on to clothing. □ **sequinned** *adj.* (also **sequined**). [Italian *zecchino* a gold coin]

sequoia /sɪ'kwɔɪə/ *n.* extremely tall Californian evergreen conifer. [*Sequoiah*, name of a Cherokee]

seraglio /sə'rɑ:lɪəʊ/ *n.* (*pl.* **-s**) **1** harem. **2** *hist.* Turkish palace. [Italian *serraglio* from Turkish]

seraph /'serəf/ *n.* (*pl.* **-im** or **-s**) angelic being of the highest order of the celestial hierarchy. □ **seraphic** /sə'ræfɪk/ *adj.* [Hebrew]

Serb *–n.* **1** native of Serbia in SE Europe. **2** person of Serbian descent. *–adj.* = SERBIAN. [Serbian *Srb*]

Serbian /'sɜ:bɪən/ *–n.* **1** dialect of the Serbs. **2** = SERB. *–adj.* of Serbia.

Serbo-Croat /,sɜ:bəʊ'krəʊæt/ (also **Serbo-Croatian** /-krəʊ'eɪʃ(ə)n/) *–n.* main official language of Yugoslavia, combining Serbian and Croatian. *–adj.* of this language.

serenade /,serə'neɪd/ *–n.* **1** piece of music performed at night, esp. beneath a lover's window. **2** orchestral suite for a small ensemble. *–v.* (**-ding**) perform a serenade to. □ **serenader** *n.* [Italian: related to SERENE]

serendipity /,serən'dɪpɪtɪ/ *n.* faculty of making happy discoveries by accident. □ **serendipitous** *adj.* [coined by Horace Walpole]

serene /sɪ'ri:n/ *adj.* (**-ner, -nest**) **1** clear and calm. **2** tranquil, unperturbed. □ **serenely** *adv.* **sereneness** *n.* **serenity** /sɪ'renɪtɪ/ *n.* [Latin]

serf *n.* **1** *hist.* labourer who was not allowed to leave the land on which he worked. **2** oppressed person, drudge. □ **serfdom** *n.* [Latin *servus* slave]

serge *n.* durable twilled worsted etc. fabric. [French *sarge*, *serge*]

sergeant /'sɑ:dʒ(ə)nt/ *n.* **1** non-commissioned Army or RAF officer next below warrant-officer. **2** police officer below inspector. [French *sergent* from Latin *serviens -ent-* servant]

sergeant-major *n.* (in full **regimental sergeant-major**) warrant-officer assisting the adjutant of a regiment or battalion.

serial /'sɪərɪəl/ *–n.* (also *attrib.*) story etc. published, broadcast, or shown in instalments. *–adj.* **1** of, in, or forming a series. **2** *Mus.* using transformations of a fixed series of notes (see SERIES 4). □ **serially** *adv.* [from SERIES]

serialize *v.* (also **-ise**) (**-zing** or **-sing**) publish or produce in instalments. □ **serialization** /-'zeɪʃ(ə)n/ *n.*

serial killer *n.* person who murders continually with no apparent motive.

serial number *n.* number identifying an item in a series.

series /'sɪəri:z, -rɪz/ *n.* (*pl.* same) **1** number of similar or related things, events, etc.; succession, row, or set. **2** set of related but individual programmes. **3** set of related geological strata. **4** arrangement of the twelve notes of the chromatic scale as a basis for serial music. **5** set of electrical circuits or components arranged so that the same current passes through each successively. □ **in series** in ordered succession. [Latin *sero* join]

serif /'serɪf/ *n.* slight projection at the extremities of a printed letter (as in T contrasted with T) (cf. SANSERIF). [origin uncertain]

serio-comic /,sɪərɪəʊ'kɒmɪk/ *adj.* combining the serious and the comic.

serious /'sɪərɪəs/ *adj.* **1** thoughtful, earnest. **2** important, demanding consideration. **3** not negligible; dangerous, frightening (*serious injury*). **4** sincere, in earnest, not frivolous. **5** (of music, literature, etc.) intellectual in content or appeal; not popular. □ **seriously** *adv.* **seriousness** *n.* [Latin *seriosus*]

serjeant /ˈsɑːdʒ(ə)nt/ *n.* (in full **serjeant-at-law**, *pl.* **serjeants-at-law**) *hist.* barrister of the highest rank. [var. of SERGEANT]

serjeant-at-arms *n.* (*pl.* **serjeants-at--arms**) official of a court, city, or parliament, with ceremonial duties.

sermon /ˈsɜːmən/ *n.* **1** spoken or written discourse on religion or morals etc., esp. delivered in church. **2** admonition, reproof. [Latin *sermo -onis* speech]

sermonize *v.* (also **-ise**) (**-zing** or **-sing**) moralize (to).

serous /ˈsɪərəs/ *adj.* **1** of or like serum; watery. **2** (of a gland or membrane) having a serous secretion. □ **serosity** /-ˈrɒsɪtɪ/ *n.* [related to SERUM]

serpent /ˈsɜːpənt/ *n.* **1** snake, esp. large. **2** sly or treacherous person. [Latin *serpo* creep]

serpentine /ˈsɜːpənˌtaɪn/ *—adj.* **1** of or like a serpent. **2** coiling, meandering. **3** cunning, treacherous. *—n.* soft usu. dark-green rock, sometimes mottled.

SERPS *abbr.* State earnings-related pension scheme.

serrated /səˈreɪtɪd/ *adj.* with a sawlike edge. □ **serration** *n.* [Latin *serra* saw]

serried /ˈserɪd/ *adj.* (of ranks of soldiers etc.) close together. [French *serrer* to close]

serum /ˈsɪərəm/ *n.* (*pl.* **sera** or **-s**) **1** liquid that separates from a clot when blood coagulates, esp. used for inoculation. **2** watery fluid in animal bodies. [Latin, = whey]

servant /ˈsɜːv(ə)nt/ *n.* **1** person employed to do domestic duties, esp. in a wealthy household. **2** devoted follower or helper. [French: related to SERVE]

serve *—v.* (**-ving**) **1** do a service for (a person, community, etc.). **2** be a servant to. **3** carry out duties (*served on six committees*). **4** (foll. by *in*) be employed in (esp. the armed forces) (*served in the navy*). **5 a** be useful to or serviceable for. **b** meet requirements; perform a function. **6 a** go through a due period of (apprenticeship, a prison sentence, etc.). **b** go through (a due period) of imprisonment etc. **7** present (food) to eat. **8** (in full **serve at table**) act as a waiter. **9 a** attend to (a customer etc.). **b** (foll. by *with*) supply with (goods). **10** treat (a person) in a specified way. **11 a** (often foll. by *on*) deliver (a writ etc.). **b** (foll. by *with*) deliver a writ etc. to. **12** (also *absol.*) (in tennis etc.) deliver (a ball etc.) to begin or resume play. **13** (of an animal) copulate with (a female). *—n.* = SERVICE *n.* 16a, b. □ **serve a person right** be a person's deserved punishment etc. **serve up** *derog.* offer (*served up the same old excuses*). [Latin *servio*]

server *n.* **1 a** person who serves. **b** utensil for serving food. **2** celebrant's assistant at a mass etc.

servery *n.* (*pl.* **-ies**) room or counter from which meals etc. are served.

service /ˈsɜːvɪs/ *—n.* **1** work, or the doing of work, for another or for a community etc. (often in *pl.*: *the services of a lawyer*). **2** work done by a machine etc. (*has given good service*). **3** assistance or benefit given. **4** provision or supplying of a public need, e.g. transport, or (often in *pl.*) of water, gas, electricity, etc. **5** employment as a servant. **6** state or period of employment (*resigned after 15 years' service*). **7** public or Crown department or organization (*civil service*). **8** (in *pl.*) the armed forces. **9** (*attrib.*) of the kind issued to the armed forces (*service revolver*). **10 a** ceremony of worship. **b** form of liturgy for this. **11 a** provision for the maintenance of a machine etc. **b** routine maintenance of a vehicle etc. **12** assistance given to customers. **13 a** serving of food, drinks, etc.; quality of this. **b** extra charge nominally made for this. **14** (in *pl.*) = SERVICE AREA (*motorway services*). **15** set of dishes, plates, etc., for serving meals (*dinner service*). **16 a** act of serving in tennis etc. **b** person's turn to serve. **c** (in full **service game**) game in which a specified player serves. *—v.* (**-cing**) **1** maintain or repair (a car, machine, etc.). **2** supply with a service. □ **at a person's service** ready to serve a person. **be of service** be helpful or useful. **in service 1** employed as a servant. **2** in use. **out of service** not available for use, not working. [Latin *servitium* from *servus* slave]

serviceable *adj.* **1** useful or usable; able to render service. **2** durable but plain. □ **serviceability** /-ˈbɪlɪtɪ/ *n.*

service area *n.* area beside a major road providing petrol, refreshments, toilet facilities, etc.

service charge *n.* additional charge for service in a restaurant etc.

service flat *n.* flat in which domestic service and sometimes meals are provided for an extra fee.

service industry *n.* industry providing services, not goods.

serviceman *n.* **1** man in the armed forces. **2** man providing service or maintenance.

service road *n.* road serving houses, shops, etc., lying back from the main road.

service station *n.* = GARAGE *n.* 2.

servicewoman *n.* woman in the armed forces.

serviette /ˌsɜːvɪ'et/ *n.* table-napkin. [French: related to SERVE]

servile /'sɜːvaɪl/ *adj.* **1** of or like a slave. **2** fawning; subservient. □ **servility** /-'vɪlɪtɪ/ *n.* [Latin *servus* slave]

serving *n.* quantity of food for one person.

servitor /'sɜːvɪt(ə)r/ *n. archaic* servant, attendant. [Latin: related to SERVE]

servitude /'sɜːvɪˌtjuːd/ *n.* slavery, subjection. [Latin *servus* slave]

servo /'sɜːvəʊ/ *n.* (*pl.* **-s**) **1** powered mechanism producing motion at a higher level of energy than the input level. **2** (in *comb.*) involving this. [Latin *servus* slave]

sesame /'sesəmɪ/ *n.* **1** E. Indian plant with oil-yielding seeds. **2** its seeds. □ **open sesame** magic phrase for opening a locked door or gaining access. [Greek]

sesqui- /'seskwɪ/ *comb. form* denoting $1^1/_2$ (*sesquicentennial*). [Latin]

sessile /'sesaɪl/ *adj.* **1** (of a flower, leaf, eye, etc.) attached directly by its base without a stalk or peduncle. **2** fixed in one position; immobile. [Latin: related to SESSION]

session /'seʃ(ə)n/ *n.* **1** period devoted to an activity (*recording session*). **2** assembly of a parliament, court, etc. **3** single meeting for this. **4** period during which these are regularly held. **5** academic year. □ **in session** assembled for business; not on vacation. □ **sessional** *adj.* [Latin *sedeo sess-* sit]

sestet /ses'tet/ *n.* **1** last six lines of a sonnet. **2** sextet. [Italian *sesto* sixth]

set[1] *v.* (**-tt-**; *past* and *past part.* **set**) **1** put, lay, or stand in a certain position etc. **2** apply (one thing) to (another) (*set pen to paper*). **3 a** fix ready or in position. **b** dispose suitably for use, action, or display. **4 a** adjust (a clock or watch) to show the right time. **b** adjust (an alarm clock) to sound at the required time. **5 a** fix, arrange, or mount. **b** insert (a jewel) in a ring etc. **6** make (a device) ready to operate. **7** lay (a table) for a meal. **8** style (the hair) while damp. **9** (foll. by *with*) ornament or provide (a surface). **10** make or bring into a specified state; cause to be (*set things in motion*; *set it on fire*). **11** (of jelly, cement, etc.) harden or solidify. **12** (of the sun, moon, etc.) move towards or below the earth's horizon. **13** show (a story etc.) as happening in a certain time or place. **14 a** (foll. by *to* + infin.) cause (a person or oneself) to do a specified thing. **b** (foll. by pres. part.) start (a person or thing) doing something. **15** give as work to be done or a matter to be dealt with. **16** exhibit as a model etc. (*set an example*). **17** initiate; lead (*set the fashion*; *set the pace*). **18** establish (a record etc.). **19** determine or decide. **20** appoint or establish (*set them in authority*). **21 a** put parts of (a broken or dislocated bone, limb, etc.) together for healing. **b** deal with (a fracture etc.) in this way. **22** (in full **set to music**) provide (words etc.) with music for singing. **23 a** compose (type etc.). **b** compose the type etc. for (a book etc.). **24** (of a tide, current, etc.) have a certain motion or direction. **25** (of a face) assume a hard expression. **26 a** cause (a hen) to sit on eggs. **b** place (eggs) for a hen to sit on. **27** (of eyes etc.) become motionless. **28** feel or show a certain tendency (*opinion is setting against it*). **29 a** (of blossom) form into fruit. **b** (of fruit) develop from blossom. **c** (of a tree) develop fruit. **30** (of a dancer) take a position facing one's partner. **31** (of a hunting dog) take a rigid attitude indicating the presence of game. **32** *dial.* or *slang* sit. □ **set about 1** begin or take steps towards. **2** *colloq.* attack. **set against 1** consider or reckon as a balance or compensation for. **2** cause to oppose. **set apart** separate, reserve, differentiate. **set aside 1** put to one side. **2** keep for future use. **3** disregard or reject. **set back 1** place further back in place or time. **2** impede or reverse the progress of. **3** *colloq.* cost (a person) a specified amount. **set down 1** record in writing. **2** allow to alight. **3** (foll. by *to*) attribute to. **4** (foll. by *as*) explain as. **set eyes on** see EYE. **set foot on** (or **in**) enter or go to (a place etc.). **set forth 1** begin a journey. **2** expound. **set in 1** (of weather etc.) begin, become established. **2** insert. **set off 1** begin a journey. **2** detonate (a bomb etc.). **3** initiate, stimulate. **4** cause (a person) to start laughing, talking, etc. **5** adorn; enhance. **6** (foll. by *against*) use as a compensating item. **set on** (or **upon**) **1** attack violently. **2** cause or urge to attack. **set oneself up as** pretend or claim to be. **set out 1** begin a journey. **2** (foll. by *to* + infin.) intend. **3** demonstrate, arrange, or exhibit. **4** mark out. **5** declare. **set sail** hoist the sails, begin a voyage. **set to** begin vigorously, esp. fighting, arguing, or eating. **set up 1** place in position or view. **2** start (a business etc.). **3** establish in some capacity. **4** supply the needs of. **5** begin making (a loud sound). **6** cause (a condition or situation). **7** prepare (a task etc. for another). **8** restore the health of (a person). **9** establish (a record). **10** *colloq.* frame or cause (a person) to look foolish; cheat. [Old English]

set[2] *n.* **1** group of linked or similar things or persons. **2** section of society. **3** collection of objects for a specified purpose (*cricket set*; *teaset*). **4** radio or television receiver. **5** (in tennis etc.) group of games counting as a unit towards winning a match. **6** *Math.* collection of things sharing a property. **7** direction or position in which something sets or is set. **8** slip, shoot, bulb, etc., for planting. **9** setting, stage furniture, etc., for a play or film etc. **10** styling of the hair while damp. **11** (also **sett**) badger's burrow. **12** (also **sett**) granite paving-block. [senses 1–6 from French *sette*; senses 7–12 from SET[1]]

set[3] *adj.* **1** prescribed or determined in advance; fixed (*a set meal*). **2** (of a phrase or speech etc.) having invariable or predetermined wording; not extempore. **3** prepared for action. **4** (foll. by *on*, *upon*) determined to get or achieve etc. [past part. of SET[1]]

set-back *n.* reversal or arrest of progress; relapse.

set piece *n.* **1** formal or elaborate arrangement, esp. in art or literature. **2** fireworks arranged on scaffolding etc.

set square *n.* right-angled triangular plate for drawing lines, esp. at 90°, 45°, 60°, or 30°.

sett var. of SET[2] 11, 12.

settee /se'ti:/ *n.* = SOFA. [origin uncertain]

setter *n.* dog of a long-haired breed trained to stand rigid when scenting game.

set theory *n.* the study or use of sets in mathematics.

setting *n.* **1** position or manner in which a thing is set. **2** immediate surroundings of a house etc. **3** period, place, etc., of a story, drama, etc. **4** frame etc. for a jewel. **5** music to which words are set. **6** cutlery etc. for one person at a table. **7** level at which a machine is set to operate (*on a high setting*).

settle[1] /'set(ə)l/ *v.* (**-ling**) **1** (often foll. by *down*, *in*) establish or become established in an abode or lifestyle. **2** (often foll. by *down*) **a** regain calm after disturbance; come to rest. **b** adopt a regular or secure style of life. **c** (foll. by *to*) apply oneself (*settled down to work*). **3** (cause to) sit, alight, or come down to stay for some time. **4** make or become composed, certain, quiet, or fixed. **5** determine, decide, or agree upon. **6** resolve (a dispute, matter, etc.). **7** agree to terminate (a lawsuit). **8** (foll. by *for*) accept or agree to (esp. a less desirable alternative). **9** (also *absol.*) pay (a debt, account, etc.). **10** (as **settled** *adj.*) established (*settled weather*). **11** calm (nerves, the stomach, etc.). **12 a** colonize. **b** establish colonists in. **13** subside; fall to the bottom or on to a surface. □ **settle up** (also *absol.*) pay (an account, debt, etc.). **settle with 1** pay (a creditor). **2** get revenge on. [Old English: related to SIT]

settle[2] /'set(ə)l/ *n.* high-backed wooden bench, often with a box below the seat. [Old English]

settlement *n.* **1** settling or being settled. **2 a** place occupied by settlers. **b** small village. **3 a** political or financial etc. agreement. **b** arrangement ending a dispute. **4 a** terms on which property is given to a person. **b** deed stating these. **c** amount or property given.

settler *n.* person who settles abroad.

set-to *n.* (*pl.* **-tos**) *colloq.* fight, argument.

set-up *n.* **1** arrangement or organization. **2** manner, structure, or position of this. **3** instance of setting a person up (see *set up* 10).

seven /'sev(ə)n/ *adj.* & *n.* **1** one more than six. **2** symbol for this (7, vii, VII). **3** size etc. denoted by seven. **4** seven o'clock. [Old English]

sevenfold *adj.* & *adv.* **1** seven times as much or as many. **2** consisting of seven parts.

seven seas *n.* (prec. by *the*) the oceans of the world.

seventeen /,sevən'ti:n/ *adj.* & *n.* **1** one more than sixteen. **2** symbol for this (17, xvii, XVII). **3** size etc. denoted by seventeen. □ **seventeenth** *adj.* & *n.* [Old English]

seventh *adj.* & *n.* **1** next after sixth. **2** one of seven equal parts of a thing. □ **seventhly** *adv.*

Seventh-Day Adventists *n.pl.* sect of Adventists observing the sabbath on Saturday.

seventh heaven *n.* state of intense joy.

seventy /'sevəntɪ/ *adj.* & *n.* (*pl.* **-ies**) **1** seven times ten. **2** symbol for this (70, lxx, LXX). **3** (in *pl.*) numbers from 70 to 79, esp. the years of a century or of a person's life. □ **seventieth** *adj.* & *n.* [Old English]

sever /'sevə(r)/ *v.* divide, break, or make separate, esp. by cutting (*severed artery*). [Anglo-French *severer* from Latin *separo*]

several /'sevr(ə)l/ *–adj.* & *pron.* more than two but not many; a few. *–adj. formal* separate or respective (*went their several ways*). □ **severally** *adv.* [Latin *separ* distinct]

severance *n.* **1** act of severing. **2** severed state.

severance pay *n.* payment made to an employee on termination of a contract.

severe /sɪ'vɪə(r)/ *adj.* **1** rigorous and harsh (*severe critic*). **2** serious (*severe shortage*). **3** forceful (*severe storm*). **4** extreme (*severe winter*). **5** exacting (*severe competition*). **6** plain in style. □ **severely** *adv.* **severity** /-'verɪtɪ/ *n.* [Latin *severus*]

Seville orange /'sevɪl, sə'vɪl/ *n.* bitter orange used for marmalade. [*Seville* in Spain]

sew /səʊ/ *v.* (*past part.* **sewn** or **sewed**) fasten, join, etc., with a needle and thread or a sewing-machine. □ **sew up 1** join or enclose by sewing. **2** (esp. in *passive*) *colloq.* satisfactorily arrange or finish; gain control of. [Old English]

sewage /'su:ɪdʒ/ *n.* waste matter conveyed in sewers. [from SEWER]

sewage farm *n.* (also **sewage works**) place where sewage is treated.

sewer /'su:ə(r)/ *n.* conduit, usu. underground, for carrying off drainage water and sewage. [Anglo-French *sever(e)*: related to EX-[1], *aqua* water]

sewerage /'su:ərɪdʒ/ *n.* system of, or drainage by, sewers.

sewing /'səʊɪŋ/ *n.* material or work to be sewn.

sewing-machine *n.* machine for sewing or stitching.

sewn *past part.* of SEW.

sex –*n.* **1** each of the main groups (male and female) into which living things are categorized on the basis of their reproductive functions (*what sex is your dog?*). **2** sexual instincts, desires, etc., or their manifestation. **3** *colloq.* sexual intercourse. **4** (*attrib.*) of or relating to sex or sexual differences. –*v.* **1** determine the sex of. **2** (as **sexed** *adj.*) having a specified sexual appetite (*highly sexed*). [Latin *sexus*]

sexagenarian /,seksədʒɪ'neərɪən/ *n.* person from 60 to 69 years old. [Latin *-arius* from *sexaginta* sixty]

Sexagesima /,seksə'dʒesɪmə/ *n.* Sunday before Quinquagesima. [Latin, = sixtieth]

sex appeal *n.* sexual attractiveness.

sex change *n.* apparent change of sex by hormone treatment and surgery.

sex chromosome *n.* chromosome determining the sex of an organism.

sexism *n.* prejudice or discrimination, esp. against women, on the grounds of sex. □ **sexist** *adj.* & *n.*

sexless *adj.* **1** neither male nor female. **2** lacking sexual desire or attractiveness.

sex life *n.* person's sexual activity.

sex maniac *n. colloq.* person obsessed with sex.

sex object *n.* person regarded as an object of sexual gratification.

sex offender *n.* person who commits a sexual crime.

sexology /sek'sɒlədʒɪ/ *n.* the study of sexual relationships or practices. □ **sexologist** *n.*

sex symbol *n.* person widely noted for sex appeal.

sextant /'sekst(ə)nt/ *n.* instrument with a graduated arc of 60°, used in navigation and surveying for measuring the angular distance of objects by means of mirrors. [Latin *sextans -ntis* sixth part]

sextet /sek'stet/ *n.* **1** *Mus.* **a** composition for six performers. **b** the performers. **2** any group of six. [alteration of SESTET after Latin *sex* six]

sexton /'sekst(ə)n/ *n.* person who looks after a church and churchyard, often acting as bell-ringer and gravedigger. [French *segerstein* from Latin *sacristanus*]

sextuple /'seks,tju:p(ə)l/ –*adj.* **1** sixfold. **2** having six parts. **3** being six times as many or as much. –*n.* sixfold number or amount. [medieval Latin from Latin *sex* six]

sextuplet /'seks,tjʊplɪt, -'tju:plɪt/ *n.* each of six children born at one birth.

sexual /'sekʃʊəl/ *adj.* of sex, the sexes, or relations between them. □ **sexuality** /-'ælɪtɪ/ *n.* **sexually** *adv.*

sexual intercourse *n.* method of reproduction involving insertion of the penis into the vagina, usu. followed by ejaculation.

sexy *adj.* (**-ier**, **-iest**) **1** sexually attractive, stimulating, or aroused. **2** *colloq.* (of a project etc.) exciting, trendy. □ **sexily** *adv.* **sexiness** *n.*

SF *abbr.* science fiction.

sf *abbr.* sforzando.

sforzando /sfɔ:'tsændəʊ/ *Mus.* –*adj.* & *adv.* with sudden emphasis. –*n.* (*pl.* **-s** or **-di** /-dɪ/) **1** suddenly emphasized note or group of notes. **2** increase in emphasis and loudness. [Italian]

Sgt. *abbr.* Sergeant.

sh *int.* = HUSH.

shabby /'ʃæbɪ/ *adj.* (**-ier**, **-iest**) **1** faded and worn, dingy, dilapidated. **2** contemptible (*a shabby trick*). □ **shabbily** *adv.* **shabbiness** *n.* [related to SCAB]

shack –*n.* roughly built hut or cabin. –*v.* (foll. by *up*) *slang* cohabit, esp. as lovers. [perhaps from Mexican *jacal* wooden hut]

shackle /'ʃæk(ə)l/ –*n.* **1** metal loop or link, closed by a bolt, used to connect chains etc. **2** fetter for the ankle or wrist. **3** (usu. in *pl.*) restraint, impediment. –*v.* (**-ling**) fetter, impede, restrain. [Old English]

shad *n.* (*pl.* same or **-s**) large edible marine fish. [Old English]

shaddock /'ʃædək/ *n.* **1** largest citrus fruit, with a thick yellow skin and bitter pulp. **2** tree bearing these. [Capt. *Shaddock*, who introduced it to the W. Indies in the 17th c.]

shade –*n.* **1** comparative darkness (and usu. coolness) given by shelter from direct light and heat. **2** area so sheltered. **3** darker part of a picture etc. **4** colour, esp. as darker or lighter than one similar. **5** comparative obscurity. **6** slight amount (*a shade better*). **7** lampshade. **8** screen against the light. **9** (in *pl.*) esp. *US colloq.* sunglasses. **10** *literary* ghost. **11** (in *pl.*; foll. by *of*) reminder of, suggesting (esp. something undesirable) (*shades of Hitler!*). –*v.* (**-ding**) **1** screen from light. **2** cover, moderate, or exclude the light of. **3** darken, esp. with parallel lines to show shadow etc. **4** (often foll. by *away*, *off*, *into*) pass or change gradually. [Old English]

shading *n.* light and shade shown on a map or drawing by parallel lines etc.

shadow /'ʃædəʊ/ –*n.* **1** shade; patch of shade. **2** dark shape projected by a body intercepting rays of light. **3** inseparable attendant or companion. **4** person secretly following another. **5** slightest trace (*not a shadow of doubt*). **6** weak or insubstantial remnant (*a shadow of his former self*). **7** (*attrib.*) denoting members of an Opposition party holding posts parallel to those of the government (*shadow Cabinet*). **8** shaded part of a picture. **9** gloom or sadness. –*v.* **1** cast a shadow over. **2** secretly follow and watch. [Old English: related to SHADE]

shadow-boxing *n.* boxing with an imaginary opponent as training.

shadowy *adj.* **1** like or having a shadow. **2** vague, indistinct.

shady /'ʃeɪdɪ/ *adj.* (**-ier, -iest**) **1** giving shade. **2** situated in shade. **3** disreputable; of doubtful honesty. □ **shadily** *adv.* **shadiness** *n.*

shaft /ʃɑ:ft/ *n.* **1** narrow usu. vertical space, for access to a mine, or (in a building) for a lift, ventilation, etc. **2** (foll. by *of*) **a** ray (of light). **b** bolt (of lightning). **3** stem or handle of a tool etc. **4** long narrow part supporting, connecting, or driving thicker part(s) etc. **5 a** *archaic* arrow, spear. **b** its long slender stem. **6** hurtful or provocative remark (*shafts of wit*). **7** each of the pair of poles between which a horse is harnessed to a vehicle. **8** central stem of a feather. **9** column, esp. between the base and capital. [Old English]

shag[1] *n.* **1** coarse kind of cut tobacco. **2 a** rough mass of hair etc. **b** (*attrib.*) (of a carpet) with a long rough pile **3** cormorant, esp. the crested cormorant. [Old English]

shag[2] *v.* (**-gg**) *coarse slang* **1** have sexual intercourse with. **2** (usu. in *passive*; often foll. by *out*) exhaust, tire out. [origin unknown]

shaggy *adj.* (**-ier, -iest**) **1** hairy, rough-haired. **2** unkempt. □ **shagginess** *n.*

shaggy-dog story *n.* long rambling joke, amusing only by its pointlessness.

shagreen /ʃæ'gri:n/ *n.* **1** a kind of untanned granulated leather. **2** sharkskin. [var. of CHAGRIN]

shah *n. hist.* former monarch of Iran. [Persian]

shake –*v.* (**-king**; *past* **shook** /ʃʊk/; *past part.* **shaken**) **1** move forcefully or quickly up and down or to and fro. **2** (cause to) tremble or vibrate. **3** agitate, shock, or upset the composure of. **4** weaken or impair in courage, effectiveness, etc. **5** (of a voice, note, etc.) tremble; trill. **6** gesture with (one's fist, a stick, etc.). **7** *colloq.* shake hands (*they shook on the deal*). –*n.* **1** shaking or being shaken. **2** jerk or shock. **3** (in *pl.*; prec. by *the*) *colloq.* fit of trembling. **4** *Mus.* trill. **5** = MILK SHAKE. □ **no great shakes** *colloq.* mediocre, poor. **shake down 1** settle or cause to fall by shaking. **2** settle down; become established. **shake hands** (often foll. by *with*) clasp hands as a greeting, farewell, in congratulation, as confirmation of a deal, etc. **shake one's head** turn one's head from side to side in refusal, denial, disapproval, or concern. **shake off** get rid of or evade (a person or thing). **shake out 1** empty by shaking. **2** open (a sail, flag, etc.) by shaking. **shake up 1** mix by shaking. **2** restore to shape by shaking. **3** disturb or make uncomfortable; rouse from apathy, conventionality, etc. [Old English]

shaker *n.* **1** person or thing that shakes. **2** container for shaking together the ingredients of cocktails etc.

Shakespearian /ʃeɪk'spɪərɪən/ *adj.* (also **Shakespearean**) of Shakespeare.

shake-up *n.* upheaval or drastic reorganization.

shako /'ʃækəʊ/ *n.* (*pl.* **-s**) cylindrical plumed peaked military hat. [Hungarian *csákó*]

shaky *adj.* (**-ier, -iest**) **1** unsteady; trembling. **2** unsound, infirm. **3** unreliable. □ **shakily** *adv.* **shakiness** *n.*

shale *n.* soft rock of consolidated mud or clay that splits easily. □ **shaly** *adj.* [German: related to SCALE[2]]

shall /ʃæl, ʃ(ə)l/ *v.aux.* (*3rd sing. present* **shall**; *archaic 2nd sing. present* **shalt**; *past* **should** /ʃʊd, ʃəd/) (foll. by infin. without *to*, or *absol.*; present and past only in use) **1** (in the 1st person) expressing the future tense or (with *shall* stressed) emphatic intention (*I shall return soon*). **2** (in the 2nd and 3rd persons) expressing a strong assertion, command, or duty (*they shall go to the party*; *thou shalt not steal*; *they shall obey*). **3** (in 2nd-person questions) expressing an enquiry, esp. to avoid the form of a request (*shall you go to France?*). □ **shall I?** (or **we**) do you want me (or us) to? [Old English]

shallot /ʃə'lɒt/ *n.* onion-like plant with a cluster of small bulbs. [French: related to SCALLION]

shallow /'ʃæləʊ/ *–adj.* **1** of little depth. **2** superficial, trivial. *–n.* (often in *pl.*) shallow place. □ **shallowness** *n.* [Old English]

shalom /ʃə'lɒm/ *n.* & *int.* Jewish salutation at meeting or parting. [Hebrew]

shalt *archaic 2nd person sing.* of SHALL.

sham *–v.* (**-mm-**) **1** feign, pretend. **2** pretend to be. *–n.* **1** imposture, pretence. **2** bogus or false person or thing. *–adj.* pretended, counterfeit. [origin unknown]

shaman /'ʃæmən/ *n.* witch-doctor or priest claiming to communicate with gods etc. □ **shamanism** *n.* [Russian]

shamble /'ʃæmb(ə)l/ *–v.* (**-ling**) walk or run awkwardly, dragging the feet. *–n.* shambling gait. [perhaps related to SHAMBLES]

shambles *n.pl.* (usu. treated as *sing.*) **1** *colloq.* mess, muddle. **2** butcher's slaughterhouse. **3** scene of carnage. [pl. of *shamble* table for selling meat]

shambolic /ʃæm'bɒlɪk/ *adj. colloq.* chaotic, unorganized. [from SHAMBLES after *symbolic*]

shame *–n.* **1** distress or humiliation caused by consciousness of one's guilt, dishonour, or folly. **2** capacity for feeling this. **3** state of disgrace or discredit. **4 a** person or thing that brings disgrace etc. **b** thing that is wrong or regrettable. *–v.* (**-ming**) **1** bring shame on; make ashamed; put to shame. **2** (foll. by *into*, *out of*) force by shame (*shamed into confessing*). □ **for shame!** reproof to a shameless person. **put to shame** humiliate by being greatly superior. [Old English]

shamefaced *adj.* **1** showing shame. **2** bashful, shy. □ **shamefacedly** /also -sɪdlɪ/ *adv.*

shameful *adj.* disgraceful, scandalous. □ **shamefully** *adv.* **shamefulness** *n.*

shameless *adj.* **1** having or showing no shame. **2** impudent. □ **shamelessly** *adv.*

shammy /'ʃæmɪ/ *n.* (*pl.* **-ies**) (in full **shammy leather**) *colloq.* = CHAMOIS 2. [representing corrupted pronunciation]

shampoo /ʃæm'pu:/ *–n.* **1** liquid for washing the hair. **2** similar substance for washing cars, carpets, etc. *–v.* (**-poos**, **-pooed**) wash with shampoo. [Hindustani]

shamrock /'ʃæmrɒk/ *n.* trefoil, used as an emblem of Ireland. [Irish]

shandy /'ʃændɪ/ *n.* (*pl.* **-ies**) beer with lemonade or ginger beer. [origin unknown]

shanghai /ʃæŋ'haɪ/ *v.* (**-hais**, **-haied**, **-haiing**) **1** *colloq.* trick or force someone into doing something. **2** trick or force (a person) into serving as a sailor. [*Shanghai* in China]

shank *n.* **1 a** leg. **b** lower part of the leg. **c** shin-bone. **2** shaft or stem, esp. the part of a tool etc. joining the handle to the working end. [Old English]

shanks's mare *n.* (also **shanks's pony**) one's own legs as transport.

shan't /ʃɑ:nt/ *contr.* shall not.

shantung /ʃæn'tʌŋ/ *n.* soft undressed Chinese silk. [*Shantung*, Chinese province]

shanty[1] /'ʃæntɪ/ *n.* (*pl.* **-ies**) **1** hut or cabin. **2** shack. [origin unknown]

shanty[2] /'ʃæntɪ/ *n.* (*pl.* **-ies**) (in full **sea shanty**) sailors' work song. [probably French *chanter*: related to CHANT]

shanty town *n.* area with makeshift housing.

shape *–n.* **1** effect produced by a thing's outline. **2** external form or appearance. **3** specific form or guise (*in the shape of an excuse*). **4** good or specified condition (*back in shape*; *in poor shape*). **5** person or thing seen in outline or indistinctly. **6** mould or pattern. **7** moulded jelly etc. **8** piece of material, paper, etc., made or cut in a particular form. *–v.* (**-ping**) **1** give a certain shape or form to; fashion, create. **2** influence (one's life, course, etc.). **3** (usu. foll. by *up*) show signs of developing; show promise. **4** (foll. by *to*) adapt or make conform. □ **in any shape or form** in any form at all (*don't like jazz in any shape or form*). **take shape** take on a definite form. [Old English]

shapeless *adj.* lacking definite or attractive shape. □ **shapelessness** *n.*

shapely *adj.* (**-ier**, **-iest**) pleasing in appearance, elegant, well-proportioned. □ **shapeliness** *n.*

shard *n.* broken piece of pottery or glass etc. [Old English]

share[1] *–n.* **1** portion of a whole allotted to or taken from a person. **2** each of the equal parts into which a company's capital is divided, entitling its owner to a proportion of the profits. *–v.* (**-ring**) **1** (also *absol.*) have or use with another or others; get, have, or give a share of (*we shared a room*; *refused to share*; *shared his food*). **2** (foll. by *in*) participate. **3** (often foll. by *out*) divide and distribute (*let's share the last cake*). **4**

have in common (*shared the same beliefs*). [Old English: related to SHEAR]

share[2] *n.* = PLOUGHSHARE. [Old English: related to SHARE[1]]

shareholder *n.* owner of shares in a company.

share-out *n.* act of sharing out, distribution.

shark[1] *n.* large voracious marine fish. [origin unknown]

shark[2] *n. colloq.* swindler, profiteer. [origin unknown]

sharkskin *n.* **1** skin of a shark. **2** smooth slightly shiny fabric.

sharp *–adj.* **1** having an edge or point able to cut or pierce. **2** tapering to a point or edge. **3** abrupt, steep, angular. **4** well-defined, clean-cut. **5 a** severe or intense. **b** (of food etc.) pungent, acid. **6** (of a voice etc.) shrill and piercing. **7** (of words or temper etc.) harsh. **8** acute; quick to understand. **9** artful, unscrupulous. **10** vigorous or brisk. **11** *Mus.* above the normal pitch; a semitone higher than a specified pitch (*C sharp*). *–n.* **1** *Mus.* **a** note a semitone above natural pitch. **b** sign (♯) indicating this. **2** *colloq.* swindler, cheat. *–adv.* **1** punctually (*at nine o'clock sharp*). **2** suddenly (*pulled up sharp*). **3** at a sharp angle. **4** *Mus.* above true pitch (*sings sharp*). □ **sharply** *adv.* **sharpness** *n.* [Old English]

sharpen *v.* make or become sharp. □ **sharpener** *n.*

sharper *n.* swindler, esp. at cards.

sharpish *colloq. –adj.* fairly sharp. *–adv.* **1** fairly sharply. **2** quite quickly.

sharp practice *n.* dishonest or dubious dealings.

sharpshooter *n.* skilled marksman.

sharp-witted *adj.* keenly perceptive or intelligent.

shat *past* and *past. part.* of SHIT.

shatter *v.* **1** break suddenly in pieces. **2** severely damage or destroy. **3** (esp. in *passive*) greatly upset or discompose. **4** (usu. as **shattered** *adj.*) *colloq.* exhaust. [origin unknown]

shave *–v.* (**-ving**; *past part.* **shaved** or (as *adj.*) **shaven**) **1** remove (bristles or hair) with a razor. **2** (also *absol.*) remove bristles or hair with a razor from (a person, face, leg, etc.). **3** reduce by a small amount. **4** pare (wood etc.) to shape it. **5** miss or pass narrowly. *–n.* **1** shaving or being shaved. **2** narrow miss or escape. **3** tool for shaving wood etc. [Old English]

shaver *n.* **1** thing that shaves. **2** electric razor. **3** *colloq.* young lad.

Shavian /ˈʃeɪvɪən/ *–adj.* of or like the writings of G. B. Shaw. *–n.* admirer of Shaw. [*Shavius*, Latinized form of *Shaw*]

shaving *n.* thin strip cut off wood etc.

shawl *n.* large usu. rectangular piece of fabric worn over the shoulders or head, or wrapped round a baby. [Urdu from Persian *shāl*]

she /ʃiː/ *–pron.* (*obj.* **her**; *poss.* **her**; *pl.* **they**) the woman, girl, female animal, ship, or country, etc. previously named or in question. *–n.* **1** female; woman. **2** (in *comb.*) female (*she-goat*). [Old English]

s/he *pron.* written representation of 'he or she' used to indicate either sex.

sheaf *–n.* (*pl.* **sheaves**) bundle of things laid lengthways together and usu. tied, esp. reaped corn or a collection of papers. *–v.* make into sheaves. [Old English]

shear *–v.* (*past* **sheared**; *past part.* **shorn** or **sheared**) **1** (also *absol.*) clip the wool off (a sheep etc.). **2** remove or take off by cutting. **3** cut with scissors or shears etc. **4** (foll. by *of*) **a** strip bare. **b** deprive. **5** (often foll. by *off*) distort, be distorted, or break, from structural strain. *–n.* **1** strain produced by pressure in the structure of a substance. **2** (in *pl.*) (also **pair of shears** *sing.*) large scissor-shaped clipping or cutting instrument. □ **shearer** *n.* [Old English]

sheath *n.* (*pl.* **-s** /ʃiːðz, ʃiːθs/) **1** close-fitting cover, esp. for the blade of a knife or sword. **2** condom. **3** enclosing case, covering, or tissue. **4** woman's close-fitting dress. [Old English]

sheathe /ʃiːð/ *v.* (**-thing**) **1** put into a sheath. **2** encase; protect with a sheath.

sheath knife *n.* dagger-like knife carried in a sheath.

sheave *v.* (**-ving**) make into sheaves.

sheaves *pl.* of SHEAF.

shebeen /ʃɪˈbiːn/ *n.* esp. *Ir.* unlicensed drinking place. [Irish]

shed[1] *n.* one-storeyed usu. wooden structure for storage or shelter, or as a workshop. [from SHADE]

shed[2] *v.* (**-dd-**; *past* and *past part.* **shed**) **1** let, or cause to, fall off (*trees shed their leaves*). **2** take off (clothes). **3** reduce (an electrical power load) by disconnection etc. **4** cause to fall or flow (*shed blood*; *shed tears*). **5** disperse, diffuse, radiate (*shed light*). **6** get rid of (*IBM are shedding 200 jobs*; *shed your inhibitions*). □ **shed light on** help to explain. [Old English]

she'd /ʃiːd/ *contr.* **1** she had. **2** she would.

sheen *n.* **1** gloss or lustre. **2** brightness. □ **sheeny** *adj.* [Old English, = beautiful]

sheep *n.* (*pl.* same) **1** mammal with a thick woolly coat, esp. kept for its wool or meat. **2** timid, silly, or easily-led person. **3** (usu. in *pl.*) member of a minister's congregation. [Old English]

sheep-dip *n.* preparation or place for cleansing sheep of vermin by dipping.

sheepdog *n.* **1** dog trained to guard and herd sheep. **2** dog of a breed suitable for this.

sheepfold *n.* pen for sheep.

sheepish *adj.* embarrassed or shy; ashamed. □ **sheepishly** *adv.*

sheepshank *n.* knot for shortening a rope temporarily.

sheepskin *n.* **1** (often *attrib.*) sheep's skin with the wool on. **2** leather from sheep's skin.

sheer[1] *–adj.* **1** mere, complete (*sheer luck*). **2** (of a cliff etc.) perpendicular. **3** (of a textile) diaphanous. *–adv.* directly, perpendicularly. [Old English]

sheer[2] *v.* **1** esp. Naut. swerve or change course. **2** (foll. by *away*, *off*) turn away, esp. from a person or topic one dislikes or fears. [origin unknown]

sheet[1] *–n.* **1** large rectangle of cotton etc. used esp. in pairs as inner bedclothes. **2** broad usu. thin flat piece of paper, metal, etc. **3** wide expanse of water, ice, flame, falling rain, etc. **4** page of unseparated postage stamps. **5** *derog.* newspaper. *–v.* **1** provide or cover with sheets. **2** form into sheets. **3** (of rain etc.) fall in sheets. [Old English]

sheet[2] *n.* rope or chain attached to the lower corner of a sail to hold or control it. [Old English: related to SHEET[1]]

sheet anchor *n.* **1** emergency reserve anchor. **2** person or thing depended on in the last resort.

sheeting *n.* material for making bed linen.

sheet metal *n.* metal rolled or hammered etc. into thin sheets.

sheet music *n.* music published in sheets, not bound.

sheikh /ʃeɪk/ *n.* **1** chief or head of an Arab tribe, family, or village. **2** Muslim leader. □ **sheikhdom** *n.* [Arabic]

sheila /'ʃi:lə/ *n. Austral.* & *NZ slang* girl, young woman. [origin uncertain]

shekel /'ʃek(ə)l/ *n.* **1** chief monetary unit of modern Israel. **2** *hist.* silver coin and unit of weight in ancient Israel etc. **3** (in *pl.*) *colloq.* money; riches. [Hebrew]

shelduck /'ʃeldʌk/ *n.* (*pl.* same or **-s**; *masc.* **sheldrake**, *pl.* same or **-s**) bright-plumaged wild duck. [probably from dial. *sheld* pied, DUCK[1]]

shelf *n.* (*pl.* **shelves**) **1** wooden etc. board projecting from a wall, or as part of a unit, used to store things. **2 a** projecting horizontal ledge in a cliff face etc. **b** reef or sandbank. □ **on the shelf 1** (of a woman) regarded as too old to hope for marriage. **2** (esp. of a retired person) put aside as if no longer useful. [Low German]

shelf-life *n.* time for which a stored item remains usable.

shelf-mark *n.* code on a library book showing where it is kept.

shell *–n.* **1 a** hard outer case of many molluscs, the tortoise, etc. **b** hard but fragile case of an egg. **c** hard outer case of a nut-kernel, seed, etc. **2 a** explosive projectile for use in a big gun etc. **b** hollow container for fireworks, cartridges, etc. **3** shell-like thing, esp.: **a** a light racing-boat. **b** the metal framework of a vehicle etc. **c** the walls of an unfinished or gutted building, ship, etc. *–v.* **1** remove the shell or pod from. **2** bombard with shells. □ **come out of one's shell** become less shy. **shell out** (also *absol.*) *colloq.* pay (money). □ **shell-less** *adj.* **shell-like** *adj.* [Old English]

she'll /ʃi:l/ *contr.* she will; she shall.

shellac /ʃə'læk/ *–n.* resin used for making varnish. *–v.* (**-ck-**) varnish with shellac. [from SHELL, LAC]

shelled *adj.* **1** having a shell. **2** with its shell removed.

shellfish *n.* (*pl.* same) **1** aquatic mollusc with a shell. **2** crustacean.

shell-shock *n.* nervous breakdown caused by warfare. □ **shell-shocked** *adj.*

Shelta /'ʃeltə/ *n.* ancient hybrid secret language used by Irish tinkers, Gypsies, etc.

shelter *–n.* **1** protection from danger, bad weather, etc. **2** place giving shelter or refuge. *–v.* **1** act or serve as a shelter to; protect; conceal; defend. **2** find refuge; take cover. [origin unknown]

shelve *v.* (**-ving**) **1** put aside, esp. temporarily. **2** put (books etc.) on a shelf. **3** fit with shelves. **4** (of ground etc.) slope. □ **shelving** *n.*

shelves *pl.* of SHELF.

shemozzle /ʃɪ'mɒz(ə)l/ *n. slang* **1** brawl or commotion. **2** muddle. [Yiddish]

shenanigan /ʃɪ'nænɪgən/ *n.* (esp. in *pl.*) *colloq.* mischievous or dubious behaviour, carryings-on. [origin unknown]

shepherd /'ʃepəd/ *–n.* **1** (*fem.* **shepherdess**) person employed to tend sheep. **2** member of the clergy in charge of a congregation. *–v.* **1 a** tend (sheep etc.). **b** guide (followers etc.). **2** marshal or drive (a crowd etc.) like sheep. [Old English: related to SHEEP, HERD]

shepherd's pie *n.* = COTTAGE PIE.

Sheraton /'ʃerət(ə)n/ *n.* (often *attrib.*) style of English furniture *c.*1790. [name of a furniture-maker]

sherbet /'ʃɜ:bət/ *n.* **1** flavoured sweet effervescent powder or drink. **2** drink of sweet diluted fruit juices. [Turkish and Persian from Arabic]

sherd *n.* = POTSHERD. [Old English]

sheriff /'ʃerɪf/ *n.* **1 a** (also **High Sheriff**) chief executive officer of the Crown in a county, administering justice etc. **b** honorary officer elected annually in some towns. **2** *US* elected chief law-enforcing officer in a county. **3** (also **sheriff-depute**) *Scot.* chief judge of a county or district. [Old English: related to SHIRE, REEVE[1]]

Sherpa *n.* (*pl.* same or **-s**) member of a Himalayan people living on the borders of Nepal and Tibet. [native name]

sherry /'ʃerɪ/ *n.* (*pl.* **-ies**) **1** fortified wine orig. from S. Spain. **2** glass of this. [*Xeres* in Andalusia]

she's /ʃi:z, ʃɪz/ *contr.* **1** she is. **2** she has.

Shetland pony /'ʃetlənd/ *n.* pony of a small hardy rough-coated breed. [*Shetland Islands*, NNE of Scotland]

shew *archaic* var. of SHOW.

shiatsu /ʃɪ'ætsu:/ *n.* Japanese therapy in which pressure is applied, chiefly with fingers and hands, to specific points on the body. [Japanese, = finger pressure]

shibboleth /'ʃɪbə,leθ/ *n.* long-standing formula, doctrine, or phrase, etc., held to be true by a party or sect. [Hebrew (Judg. 12:6)]

shied *past* & *past part.* of SHY[2].

shield *–n.* **1 a** piece of armour held in front of the body for protection when fighting. **b** person or thing giving protection. **2** shield-shaped trophy. **3** protective plate or screen in machinery etc. **4** *Heraldry* stylized representation of a shield for displaying a coat of arms etc. *–v.* protect or screen. [Old English]

shier *compar.* of SHY[1].

shiest *superl.* of SHY[1].

shift *–v.* **1** (cause to) change or move from one position to another. **2** remove, esp. with effort. **3** *slang* **a** hurry. **b** consume (food or drink). **4** *US* change (gear) in a vehicle. *–n.* **1** act of shifting. **2 a** relay of workers. **b** time for which they work. **3 a** device, stratagem, or expedient. **b** trick or evasion. **4** woman's straight unwaisted dress or petticoat. **5** *Physics* displacement of a spectral line. **6** key on a keyboard used to switch between lower and upper case etc. **7** *US* **a** gear lever in a vehicle. **b** mechanism for this. □ **make shift** manage; get along somehow. **shift for oneself** rely on one's own efforts. **shift one's ground** take up a new position in an argument etc. [Old English]

shiftless *adj.* lacking resourcefulness; lazy.

shifty *adj. colloq.* (**-ier**, **-iest**) evasive; deceitful. □ **shiftily** *adv.* **shiftiness** *n.*

Shiite /'ʃi:aɪt/ *–n.* adherent of the branch of Islam rejecting the first three Sunni caliphs. *–adj.* of this branch. [Arabic *Shiah*, = party]

shillelagh /ʃɪ'leɪlə, -lɪ/ *n.* Irish cudgel. [*Shillelagh* in Ireland]

shilling /'ʃɪlɪŋ/ *n.* **1** *hist.* former British coin and monetary unit worth one-twentieth of a pound. **2** monetary unit in Kenya, Tanzania, and Uganda. [Old English]

shilly-shally /'ʃɪlɪ,ʃælɪ/ *v.* (**-ies**, **-ied**) be undecided; vacillate. [from *shall I?*]

shim *–n.* thin wedge in machinery etc. to make parts fit. *–v.* (**-mm-**) fit or fill up with a shim. [origin unknown]

shimmer *–v.* shine tremulously or faintly. *–n.* tremulous or faint light. [Old English]

shin *–n.* **1** front of the leg below the knee. **2** cut of beef from this part. *–v.* (**-nn-**) (usu. foll. by *up*, *down*) climb quickly by clinging with the arms and legs. [Old English]

shin-bone *n.* = TIBIA.

shindig /'ʃɪndɪg/ *n. colloq.* **1** lively noisy party. **2** = SHINDY 1. [probably from SHINDY]

shindy /'ʃɪndɪ/ *n.* (*pl.* **-ies**) *colloq.* **1** brawl, disturbance, or noise. **2** = SHINDIG 1. [perhaps an alteration of SHINTY]

shine *–v.* (**-ning**; *past* and *past part.* **shone** /ʃɒn/ or **shined**) **1** emit or reflect light; be bright; glow. **2** (of the sun, a star, etc.) be visible. **3** cause (a lamp etc.) to shine. **4** (*past* and *past part.* **shined**) polish. **5** be brilliant; excel. *–n.* **1** light; brightness. **2** high polish; lustre. □ **take a shine to** *colloq.* take a fancy to. [Old English]

shiner *n. colloq.* black eye.

shingle[1] /'ʃɪŋg(ə)l/ *n.* small smooth pebbles, esp. on the sea-shore. □ **shingly** *adj.* [origin uncertain]

shingle[2] /'ʃɪŋg(ə)l/ *–n.* **1** rectangular wooden tile used on roofs etc. **2** *archaic* **a** shingled hair. **b** shingling of hair. *–v.* (**-ling**) **1** roof with shingles. **2** *archaic* **a** cut (a woman's hair) short. **b** cut the hair of (a person or head) in this way. [Latin *scindula*]

shingles /'ʃɪŋg(ə)lz/ *n.pl.* (usu. treated as *sing.*) acute painful viral inflammation of the nerve ganglia, with a rash often encircling the body. [Latin *cingulum* girdle]

Shinto /'ʃɪntəʊ/ *n.* Japanese religion with the worship of ancestors and nature-spirits. □ **Shintoism** *n.* **Shintoist** *n.* [Chinese, = way of the gods]

shinty /ˈʃɪntɪ/ *n.* (*pl.* **-ies**) **1** game like hockey, but with taller goalposts. **2** stick or ball used in this. [origin uncertain]

shiny *adj.* (**-ier, -iest**) **1** having a shine. **2** (of clothing) with the nap worn off. □ **shininess** *n.*

ship –*n.* **1** large seagoing vessel. **2** *US* aircraft. **3** spaceship. –*v.* (**-pp-**) **1** put, take, or send away in a ship. **2 a** take in (water) over a ship's side etc. **b** lay (oars) at the bottom of a boat. **c** fix (a rudder etc.) in place. **3 a** embark. **b** (of a sailor) take service on a ship. **4** deliver (goods) to an agent for forwarding. □ **ship off** send away. **when a person's ship comes home** (or **in**) when a person's fortune is made. [Old English]

-ship *suffix* forming nouns denoting: **1** quality or condition (*friendship*; *hardship*). **2** status, office, etc. (*authorship*; *lordship*). **3** tenure of office (*chairmanship*). **4** specific skill (*workmanship*). **5** members of a group (*readership*). [Old English]

shipboard *attrib. adj.* used or occurring on board a ship.

shipbuilder *n.* person, company, etc., that constructs ships. □ **shipbuilding** *n.*

ship-canal *n.* canal large enough for ships.

shipload *n.* as many goods or passengers as a ship can hold.

shipmate *n.* fellow member of a ship's crew.

shipment *n.* **1** amount of goods shipped. **2** act of shipping goods etc.

shipowner *n.* owner of a ship, ships, or shares in ships.

shipper *n.* person or company that ships goods. [Old English]

shipping *n.* **1** transport of goods etc. **2** ships, esp. a navy.

ship's boat *n.* small boat carried on board a ship.

shipshape *adv.* & *predic.adj.* trim, neat, tidy.

shipwreck –*n.* **1 a** destruction of a ship by a storm, foundering, etc. **b** ship so destroyed. **2** (often foll. by *of*) ruin of hopes, dreams, etc. –*v.* **1** inflict shipwreck on. **2** suffer shipwreck.

shipwright *n.* **1** shipbuilder. **2** ship's carpenter.

shipyard *n.* place where ships are built etc.

shire *n.* county. [Old English]

shire-horse *n.* heavy powerful draught-horse.

shirk *v.* (also *absol.*) avoid (duty, work, etc.). □ **shirker** *n.* [German *Schurke* scoundrel]

shirr –*n.* elasticated gathered threads in a garment etc. forming smocking. –*v.* gather (material) with parallel threads. □ **shirring** *n.* [origin unknown]

shirt *n.* upper-body garment of cotton etc., usu. front-opening. □ **keep one's shirt on** *colloq.* keep one's temper. **put one's shirt on** *colloq.* bet all one has on. □ **shirting** *n.* **shirtless** *adj.* [Old English]

shirtsleeve *n.* (usu. in *pl.*) sleeve of a shirt. □ **in shirtsleeves** without one's jacket on.

shirt-tail *n.* curved part of a shirt below the waist.

shirtwaister *n.* woman's dress with a bodice like a shirt.

shirty *adj.* (**-ier, -iest**) *colloq.* angry; annoyed. □ **shirtily** *adv.* **shirtiness** *n.*

shish kebab /ˌʃɪʃ kɪˈbæb/ *n.* pieces of meat and vegetables grilled on skewers. [Turkish: related to KEBAB]

shit *coarse slang* –*n.* **1** faeces. **2** act of defecating. **3** contemptible person. **4** nonsense. –*int.* exclamation of anger etc. –*v.* (**-tt-**; *past* and *past part.* **shitted, shat** or **shit**) defecate or cause the defecation of (faeces etc.). [Old English]

shitty *adj.* (**-ier, -iest**) *coarse slang* **1** disgusting, contemptible. **2** covered with excrement.

shiver[1] /ˈʃɪvə(r)/ –*v.* tremble with cold, fear, etc. –*n.* **1** momentary shivering movement. **2** (in *pl.*, prec. by *the*) attack of shivering. □ **shivery** *adj.* [origin uncertain]

shiver[2] /ˈʃɪvə(r)/ –*n.* (esp. in *pl.*) small fragment or splinter. –*v.* break into shivers. [related to dial. *shive* slice]

shoal[1] –*n.* multitude, esp of fish swimming together. –*v.* (of fish) form shoals. [Dutch: cf. SCHOOL[2]]

shoal[2] –*n.* **1 a** area of shallow water. **b** submerged sandbank visible at low water. **2** (esp. in *pl.*) hidden danger. –*v.* (of water) get shallower. [Old English]

shock[1] –*n.* **1** violent collision, impact, tremor, etc. **2** sudden and disturbing effect on the emotions etc. **3** acute prostration following a wound, pain, etc. **4** = ELECTRIC SHOCK. **5** disturbance in the stability of an organization etc. –*v.* **1 a** horrify; outrage. **b** (*absol.*) cause shock. **2** affect with an electric or pathological shock. [French *choc*, *choquer*]

shock[2] –*n.* group of corn-sheaves in a field. –*v.* arrange (corn) in shocks. [origin uncertain]

shock[3] *n.* unkempt or shaggy mass of hair. [origin unknown]

shock absorber *n.* device on a vehicle etc. for absorbing shocks, vibrations, etc.

shocker *n. colloq.* **1** shocking person or thing. **2** sensational novel etc.

shocking *adj.* **1** causing shock; scandalous. **2** *colloq.* very bad. □ **shockingly** *adv.*

shocking pink *adj.* & *n.* (as adj. often hyphenated) vibrant shade of pink.

shockproof *adj.* resistant to the effects of (esp. physical) shock.

shock therapy *n.* (also **shock treatment**) treatment of depressive patients by electric shock etc.

shock troops *n.pl.* troops specially trained for assault.

shock wave *n.* **1** moving region of high air pressure caused by an explosion or by a supersonic body. **2** wave of emotional shock (*the news sent shock waves throughout the region*).

shod *past* and *past part.* of SHOE.

shoddy /ˈʃɒdɪ/ *adj.* (**-ier, -iest**) **1** poorly made. **2** counterfeit. □ **shoddily** *adv.* **shoddiness** *n.* [origin unknown]

shoe /ʃuː/ *–n.* **1** protective foot-covering of leather etc., esp. one not reaching above the ankle. **2** protective metal rim for a horse's hoof. **3** thing like a shoe in shape or use. **4** = BRAKE SHOE. *–v.* (**shoes, shoeing**; *past* and *past part.* **shod**) **1** fit (esp. a horse etc.) with a shoe or shoes. **2** (as **shod** *adj.*) (in *comb.*) having shoes etc. of a specified kind (*roughshod*). □ **be in a person's shoes** be in his or her situation, difficulty, etc. [Old English]

shoehorn *n.* curved implement for easing the heel into a shoe.

shoelace *n.* cord for lacing up shoes.

shoemaker *n.* maker of boots and shoes. □ **shoemaking** *n.*

shoestring *n.* **1** shoelace. **2** *colloq.* small esp. inadequate amount of money.

shoe-tree *n.* shaped block for keeping a shoe in shape.

shone *past* and *past part.* of SHINE.

shoo *–int.* exclamation used to frighten away animals etc. *–v.* (**shoos, shooed**) **1** utter the word 'shoo!'. **2** (usu. foll. by *away*) drive away by shooing. [imitative]

shook *past* of SHAKE.

shoot *–v.* (*past* and *past part.* **shot**) **1 a** (also *absol.*) cause (a weapon) to fire. **b** kill or wound with a bullet, arrow, etc. **2** send out, discharge, etc., esp. swiftly. **3** (often foll. by *out, along, forth*, etc.) come or go swiftly or vigorously. **4 a** (of a plant etc.) put forth buds etc. **b** (of a bud etc.) appear. **5** hunt game etc. with a gun. **6** film or photograph. **7** (also *absol.*) esp. Football **a** score (a goal). **b** take a shot at (the goal). **8** (of a boat) sweep swiftly down or under (a bridge, rapids, etc.). **9** (usu. foll. by *through, up*, etc.) (of a pain) seem to stab. **10** (often foll. by *up*; also *absol.*) *slang* inject (a drug). *–n.* **1 a** young branch or sucker. **b** new growth of a plant. **2 a** hunting party, expedition, etc. **b** land shot over for game. **3** = CHUTE[1]. *–int. colloq.* invitation to ask questions etc. □ **shoot down 1** kill by shooting. **2** cause (an aircraft etc.) to crash by shooting. **3** argue effectively against. **shoot one's bolt** *colloq.* do all that is in one's power. **shoot one's mouth off** *slang* talk too much or indiscreetly. **shoot up 1** grow rapidly. **2** rise suddenly. **3** terrorize by indiscriminate shooting. **the whole shoot** (or **the whole shooting match**) *colloq.* everything. [Old English]

shooting-brake *n. archaic* estate car.

shooting star *n.* small rapidly moving meteor.

shooting-stick *n.* walking-stick with a foldable seat.

shop *–n.* **1** place for the retail sale of goods or services. **2** act of going shopping (*did a big shop*). **3** place for manufacture or repair (*engineering-shop*). **4** one's profession etc. as a subject of conversation (*talk shop*). **5** *colloq.* institution, place of business, etc. *–v.* (**-pp-**) **1** go to a shop or shops to buy goods. **2** *slang* inform against (a criminal etc.). □ **all over the shop** *colloq.* **1** in disorder. **2** everywhere. **shop around** look for the best bargain. □ **shopper** *n.* [French *eschoppe*]

shop assistant *n.* person serving in a shop.

shop-floor *n.* **1** production area in a factory etc. **2** workers as distinct from management.

shopkeeper *n.* owner or manager of a shop.

shoplift *v.* steal goods while appearing to shop. □ **shoplifter** *n.*

shopping *n.* **1** (often *attrib.*) purchase of goods etc. **2** goods purchased.

shopping centre *n.* area or complex of shops.

shop-soiled *adj.* soiled or faded by display in a shop.

shop steward *n.* elected representative of workers in a factory etc.

shopwalker *n.* supervisor in a large shop.

shore[1] *n.* **1** land adjoining the sea, a lake etc. **2** (usu. in *pl.*) country (*foreign shores*). □ **on shore** ashore. [Low German or Dutch]

shore[2] *–n.* prop or beam set against a ship, wall, etc., as a support. *–v.* (**-ring**) (often foll. by *up*) support (as if) with a shore or shores; hold up. [Low German or Dutch]

shoreline *n.* line where shore and water meet.

shorn *past part.* of SHEAR.

short –*adj.* **1 a** measuring little from head to foot, top to bottom, or end to end; not long. **b** not long in duration. **c** seeming short (*a few short years of happiness*). **2 a** (usu. foll. by *of, on*) deficient; scanty (*short of spoons*). **b** not far-reaching; acting or being near at hand (*short range*). **3 a** concise; brief. **b** curt; uncivil. **4** (of the memory) unable to remember distant events. **5** (of a vowel or syllable) having the lesser of the two recognized durations. **6** (of pastry) easily crumbled. **7** (of stocks etc.) sold or selling when the amount is not in hand, with reliance on getting the deficit at a lower price in time for delivery. **8** (of a drink of spirits) undiluted. **9** (of odds or a chance) nearly even. –*adv.* **1** before the natural or expected time or place; abruptly. **2** rudely. –*n.* **1** short circuit. **2** *colloq.* short drink. **3** short film. –*v.* short-circuit. □ **be caught** (or **taken**) **short 1** be put at a disadvantage. **2** *colloq.* urgently need to use the lavatory. **be short for** be an abbreviation for. **come short of** = *fall short of.* **for short** as a short name (*Tom for short*). **in short** briefly. **short of 1** see sense 2a of *adj.* **2** less than (*nothing short of a miracle*). **3** distant from (*two miles short of home*). **4** without going so far as (*did everything short of resigning*). **short on** *colloq.* see sense 2a of *adj.* □ **shortish** *adj.* **shortness** *n.* [Old English]

shortage *n.* (often foll. by *of*) deficiency; lack.

short back and sides *n.* short simple haircut.

shortbread *n.* rich biscuit of butter, flour, and sugar.

shortcake *n.* **1** = SHORTBREAD. **2** cake of short pastry filled with fruit and cream.

short-change *v.* cheat, esp. by giving insufficient change.

short circuit –*n.* electric circuit through small resistance, esp. instead of the resistance of a normal circuit. –*v.* (**short-circuit**) **1** cause a short circuit (in). **2** shorten or avoid by taking a more direct route etc.

shortcoming *n.* deficiency; defect.

shortcrust *n.* (in full **shortcrust pastry**) a type of crumbly pastry.

short cut *n.* **1** route shorter than the usual one. **2** quick method.

shorten *v.* become or make shorter or short.

shortening *n.* fat for pastry.

shortfall *n.* deficit.

shorthand *n.* **1** (often *attrib.*) system of rapid writing using special symbols. **2** abbreviated or symbolic mode of expression.

short-handed *adj.* understaffed.

shorthand typist *n.* typist qualified in shorthand.

shorthorn *n.* animal of a breed of cattle with short horns.

shortie var. of SHORTY.

short list –*n.* list of selected candidates from which a final choice is made. –*v.* (**short-list**) put on a short list.

short-lived *adj.* ephemeral.

shortly *adv.* **1** (often foll. by *before, after*) soon. **2** in a few words; curtly. [Old English]

short-range *adj.* **1** having a short range. **2** relating to the immediate future.

shorts *n.pl.* **1** trousers reaching to the knees or higher. **2** *US* underpants.

short shrift *n.* curt or dismissive treatment. [Old English *shrift* confession: related to SHRIVE]

short sight *n.* inability to focus on distant objects.

short-sighted *adj.* **1** having short sight. **2** lacking imagination or foresight. □ **short-sightedly** *adv.* **short-sightedness** *n.*

short-staffed *adj.* understaffed.

short temper *n.* temper easily lost. □ **short-tempered** *adj.*

short-term *adj.* of or for a short period of time.

short wave *n.* radio wave of frequency greater than 3 MHz.

short weight *n.* weight less than it is alleged to be.

short-winded *adj.* easily becoming breathless.

shorty *n.* (also **shortie**) (*pl.* **-ies**) *colloq.* person or garment shorter than average.

shot[1] *n.* **1** firing of a gun, cannon, etc. (*heard a shot*). **2** attempt to hit by shooting or throwing etc. **3 a** single non-explosive missile for a gun etc. **b** (*pl.* same or **-s**) small lead pellet used in quantity in a single charge. **c** (as *pl.*) these collectively. **4 a** photograph. **b** continuous film sequence. **5 a** stroke or a kick in a ball game. **b** *colloq.* attempt, guess (*had a shot at it*). **6** *colloq.* person of specified shooting skill (*a good shot*). **7** ball thrown by a shot-putter. **8** launch of a space rocket. **9** range etc. to or at which a thing will carry or act. **10** *colloq.* **a** drink of esp. spirits. **b** injection of a drug etc. □ **like a shot** *colloq.* without hesitation; willingly. [Old English]

shot[2] *past* and *past part.* of SHOOT. –*adj.* (of coloured material) woven so as to show different colours at different angles. □ **shot through** (usu. foll. by *with*) permeated or suffused.

shotgun *n.* gun for firing small shot at short range.

shotgun wedding *n. colloq.* wedding enforced because of the bride's pregnancy.

shot in the arm *n. colloq.* stimulus or encouragement.

shot in the dark *n.* mere guess.

shot-put *n.* athletic contest in which a shot is thrown. □ **shot-putter** *n.*

should /ʃʊd, ʃəd/ *v.aux.* (*3rd sing.* **should**) *past* of SHALL, used esp.: **1** in reported speech (*I said I should be home soon*). **2 a** to express obligation or likelihood (*I should tell you; you should have read it; they should have arrived by now*). **b** to express a tentative suggestion (*I should like to add*). **3 a** expressing the conditional mood in the 1st person (*I should have been killed if I had gone*). **b** forming a conditional clause (*if you should see him*).

shoulder /'ʃəʊldə(r)/ –*n.* **1** part of the body at which the arm, foreleg, or wing is attached. **2** either of the two projections below the neck. **3** upper foreleg of an animal as meat. **4** (often in *pl.*) shoulder regarded as supportive, comforting, etc. (*a shoulder to cry on; has broad shoulders*). **5** strip of land next to a road. **6** the part of a garment covering the shoulder. –*v.* **1 a** push with the shoulder. **b** make one's way thus. **2** take on (a burden etc.). □ **put one's shoulder to the wheel** make a great effort. **shoulder arms** hold a rifle with the barrel against the shoulder and the butt in the hand. **shoulder to shoulder 1** side by side. **2** with united effort. [Old English]

shoulder bag *n.* bag hung from the shoulder by a strap.

shoulder-blade *n.* either of the large flat bones of the upper back.

shoulder-length *adj.* (of hair etc.) reaching to the shoulders.

shoulder-pad *n.* pad in a garment to bulk out the shoulder.

shoulder-strap *n.* **1** strip of cloth going over the shoulder from front to back of a garment. **2** strap suspending a bag etc. from the shoulder. **3** strip of cloth from shoulder to collar, esp. on a military uniform.

shouldn't /'ʃʊd(ə)nt/ *contr.* should not.

shout –*v.* **1** speak or cry loudly. **2** say or express loudly. –*n.* **1** loud cry of joy etc., or calling attention. **2** *colloq.* one's turn to buy a round of drinks etc. □ **shout down** reduce to silence by shouting. [perhaps related to SHOOT]

shove /ʃʌv/ –*v.* (**-ving**) **1** (also *absol.*) push vigorously. **2** *colloq.* put casually (*shoved it in a drawer*). –*n.* act of shoving. □ **shove off 1** start from the shore in a boat. **2** *slang* depart. [Old English]

shove-halfpenny *n.* form of shovelboard played with coins etc. on a table.

shovel /'ʃʌv(ə)l/ –*n.* **1** spadelike tool with raised sides, for shifting coal etc. **2** (part of) a machine with a similar form or function. –*v.* (**-ll-**; *US* **-l-**) **1** move (as if) with a shovel. **2** *colloq.* move in large quantities or roughly (*shovelled peas into his mouth*). □ **shovelful** *n.* (*pl.* **-s**). [Old English]

shovelboard *n.* game played esp. on a ship's deck by pushing discs over a marked surface.

shoveller /'ʃʌvələ(r)/ *n.* (also **shoveler**) duck with a shovel-like beak.

show /ʃəʊ/ –*v.* (*past part.* **shown** or **showed**) **1** be, allow, or cause to be, visible; manifest (*buds are beginning to show; white shows the dirt*). **2** (often foll. by *to*) offer for scrutiny etc. (*show your tickets please*). **3 a** indicate (one's feelings) (*showed his anger*). **b** accord, grant (favour, mercy, etc.). **4** (of feelings etc.) be manifest (*his dislike shows*). **5 a** demonstrate; point out; prove (*showed it to be false; showed his competence*). **b** (usu. foll. by *how to* + infin.) instruct by example (*showed them how to knit*). **6** (*refl.*) exhibit oneself (as being) (*showed herself to be fair*). **7** exhibit in a show. **8** (often foll. by *in, out, up, round*, etc.) conduct or lead (*showed them to their rooms*). **9** *colloq.* = *show up* 3 (*he didn't show*). –*n.* **1** showing. **2** spectacle, display, exhibition, etc. **3** public entertainment or performance. **4 a** outward appearance or display. **b** empty appearance; mere display. **5** *colloq.* undertaking, business, etc. **6** *Med.* discharge of blood etc. at the onset of childbirth. □ **good** (or **bad** or **poor**) **show!** *colloq.* that was well (or badly) done. **on show** being exhibited. **show one's hand** disclose one's plans. **show off 1** display to advantage. **2** *colloq.* act pretentiously. **show up 1** make or be conspicuous or clearly visible. **2** expose or humiliate. **3** *colloq.* appear; arrive. **show willing** show a willingness to help etc. [Old English]

showbiz *n. colloq.* = SHOW BUSINESS.

show business *n. colloq.* theatrical profession.

showcase –*n.* **1** glass case for exhibiting goods etc. **2** event etc. designed to exhibit someone or something to advantage. –*v.* (**-sing**) display in or as if in a showcase.

showdown *n.* final test or confrontation.

shower –*n.* **1** brief fall of rain, snow, etc. **2 a** brisk flurry of bullets, dust, etc. **b** sudden copious arrival of gifts, honours, etc. **3** (in full **shower-bath**) **a** cubicle, bath, etc. in which one stands under a spray of water. **b** apparatus etc. used for this. **c** act of bathing in a shower. **4** *US* party for giving presents to a prospective bride etc. **5** *slang* contemptible person or group. –*v.* **1** discharge (water, missiles, etc.) in a shower. **2** take a shower. **3** (usu. foll. by *on, upon*) lavishly bestow (gifts etc.). **4** descend in a shower. □ **showery** *adj.* [Old English]

showerproof *adj.* resistant to light rain.

showgirl *n.* female singer and dancer in musicals, variety shows, etc.

show house *n.* (also **show flat**) furnished and decorated new house etc., on show to prospective buyers.

showing *n.* **1** display, performance. **2** quality of performance. **3** presentation of a case; evidence.

showjumping *n.* sport of riding horses competitively over a course of fences etc. □ **showjumper** *n.*

showman *n.* **1** proprietor or manager of a circus etc. **2** person skilled in publicity, esp. self-advertisement. □ **showmanship** *n.*

shown *past part.* of SHOW.

show-off *n. colloq.* person who shows off.

show of hands *n.* raised hands indicating a vote for or against.

show-piece *n.* **1** item presented for display. **2** outstanding specimen.

show-place *n.* tourist attraction.

showroom *n.* room used to display goods for sale.

show-stopper *n. colloq.* act in a show receiving prolonged applause.

show trial *n.* judicial trial designed to frighten or impress the public.

showy *adj.* (**-ier, -iest**) **1** brilliant; gaudy. **2** striking. □ **showily** *adv.* **showiness** *n.*

shrank *past* of SHRINK.

shrapnel /ˈʃræpn(ə)l/ *n.* **1** fragments of an exploded bomb etc. **2** shell containing pieces of metal etc., timed to burst short of impact. [*Shrapnel*, name of the inventor of the shell]

shred –*n.* **1** scrap or fragment. **2** least amount (*not a shred of evidence*). –*v.* (**-dd-**) tear or cut into shreds. □ **shredder** *n.* [Old English]

shrew /ʃru:/ *n.* **1** small mouselike long-nosed mammal. **2** bad-tempered or scolding woman. □ **shrewish** *adj.* (in sense 2). [Old English]

shrewd /ʃru:d/ *adj.* astute; clever and judicious. □ **shrewdly** *adv.* **shrewdness** *n.* [perhaps from obsolete *shrew* to curse, from SHREW]

shriek –*n.* shrill scream or sound. –*v.* make or utter in a shriek. [Old Norse]

shrike *n.* bird with a strong hooked and toothed bill. [Old English]

shrill –*adj.* **1** piercing and high-pitched in sound. **2** *derog.* sharp, unrestrained. –*v.* utter with or make a shrill sound. □ **shrillness** *n.* **shrilly** *adv.* [origin uncertain]

shrimp –*n.* **1** (*pl.* same or **-s**) small edible crustacean, turning pink when boiled. **2** *colloq.* very small person. –*v.* try to catch shrimps. [origin uncertain]

shrine *n.* **1** esp. *RC Ch.* **a** place for special worship or devotion. **b** tomb or reliquary. **2** place hallowed by some memory or association. [Latin *scrinium* bookcase]

shrink –*v.* (*past* **shrank**; *past part.* **shrunk** or (esp. as *adj.*) **shrunken**) **1** make or become smaller, esp. from moisture, heat, or cold. **2** (usu. foll. by *from*) recoil; flinch. –*n.* **1** act of shrinking. **2** *slang* psychiatrist. [Old English]

shrinkage *n.* **1** process or degree of shrinking. **2** allowance made by a shop etc. for loss by wastage, theft, etc.

shrink-wrap *v.* enclose (an article) in film that shrinks tightly on to it.

shrive *v.* (**-ving**; *past* **shrove**; *past part.* **shriven**) *RC Ch. archaic* **1** (of a priest) hear and absolve (a penitent). **2** (*refl.*) submit oneself to a priest for confession etc. [Old English *scrīfan* impose as penance]

shrivel /ˈʃrɪv(ə)l/ *v.* (**-ll-**; *US* **-l-**) contract into a wrinkled or dried-up state. [perhaps from Old Norse]

shroud –*n.* **1** wrapping for a corpse. **2** thing that conceals. **3** (in *pl.*) ropes supporting a mast. –*v.* **1** clothe (a body) for burial. **2** cover or conceal. [Old English, = garment]

shrove *past* of SHRIVE.

Shrovetide *n.* Shrove Tuesday and the two days preceding it.

Shrove Tuesday *n.* day before Ash Wednesday.

shrub *n.* any woody plant smaller than a tree and with branches near the ground. □ **shrubby** *adj.* [Old English]

shrubbery *n.* (*pl.* **-ies**) area planted with shrubs.

shrug –*v.* (**-gg-**) (often *absol.*) slightly and momentarily raise (the shoulders) to express indifference, doubt, etc. –*n.* act of shrugging. □ **shrug off** dismiss as unimportant. [origin unknown]

shrunk (also **shrunken**) *past part.* of SHRINK.

shudder –*v.* **1** shiver, esp. convulsively, from fear, cold, etc. **2** feel strong repugnance, fear, etc. (*shudder at the thought*). **3** vibrate. –*n.* **1** act of shuddering. **2** (in *pl.*; prec. by *the*) *colloq.* state of shuddering. [Low German or Dutch]

shuffle /'ʃʌf(ə)l/ –*v.* (**-ling**) **1** (also *absol.*) drag (the feet) in walking etc. **2** (also *absol.*) rearrange or intermingle (esp. cards or papers). **3 a** prevaricate, be evasive. **b** keep shifting one's position. –*n.* **1** act of shuffling; shuffling walk or movement. **2** change of relative positions. **3** shuffling dance. □ **shuffle off** remove, get rid of. [Low German]

shufti /'ʃʊftɪ/ *n.* (*pl.* **-s**) *colloq.* look, glimpse. [Arabic *šaffa* try to see]

shun *v.* (**-nn-**) avoid; keep clear of. [Old English]

shunt –*v.* **1** move (a train) between sidings etc.; (of a train) be shunted. **2** move or put aside; redirect. –*n.* **1** shunting or being shunted. **2** *Electr.* conductor joining two points of a circuit, through which current may be diverted. **3** *Surgery* alternative path for the circulation of the blood. **4** *slang* collision of vehicles, esp. one behind another. [perhaps from SHUN]

shush /ʃʊʃ, ʃʌʃ/ –*int.* hush! –*v.* **1** quieten (a person or people) by saying "shush". **2** fall silent. [imitative]

shut *v.* (**-tt-**; *past* and *past part.* **shut**) **1 a** move (a door, window, lid, etc.) into position to block an opening. **b** close or seal (a room, box, eye, etc.) by moving a door etc. **2** become or be capable of being closed or sealed. **3** become or make closed for trade. **4** fold or contract (a book, telescope, etc.). **5** (usu. foll. by *in*, *out*) keep in or out of a room etc. **6** (usu. foll. by *in*) catch (a finger, dress, etc.) by shutting something on it. **7** bar access to. □ **be** (or **get**) **shut of** *slang* be (or get) rid of. **shut down 1** stop (a factory etc.) from operating. **2** (of a factory etc.) stop operating. **shut off 1** stop the flow of (water, gas, etc.). **2** separate from society etc. **shut out 1** exclude. **2** screen from view. **3** prevent. **4** block from the mind. **shut up 1** close all doors and windows of. **2** imprison. **3** put (a thing) away in a box etc. **4** (esp. in *imper.*) *colloq.* stop talking. **shut up shop** close a business, shop, etc., temporarily or permanently. [Old English]

shut-down *n.* closure of a factory etc.

shut-eye *n. colloq.* sleep.

shutter –*n.* **1** movable hinged cover for a window. **2** device that exposes the film in a camera. –*v.* provide with shutters.

shuttle /'ʃʌt(ə)l/ –*n.* **1 a** (in a loom) instrument pulling the weft-thread between the warp-threads. **b** (in a sewing-machine) bobbin carrying the lower thread. **2** train, bus, etc. used in a shuttle service. **3** = SPACE SHUTTLE. –*v.* (**-ling**) (cause to) move to and fro like a shuttle. [Old English: related to SHOOT]

shuttlecock *n.* cork with a ring of feathers, or a similar plastic device, struck to and fro in badminton.

shuttle diplomacy *n.* negotiations conducted by a mediator travelling between disputing parties.

shuttle service *n.* transport service operating to and fro over a short route.

shy[1] /ʃaɪ/ –*adj.* (**shyer**, **shyest** or **shier**, **shiest**) **1 a** timid and nervous in company; self-conscious. **b** (of animals etc.) easily startled. **2** (in *comb.*) disliking or fearing (*work-shy*). –*v.* (**shies**, **shied**) **1** (usu. foll. by *at*) (esp. of a horse) turn suddenly aside in fright. **2** (usu. foll. by *away from*, *at*) avoid involvement in. –*n.* sudden startled movement. □ **shyly** *adv.* (also **shily**). **shyness** *n.* [Old English]

shy[2] –*v.* (**shies**, **shied**) (also *absol.*) fling, throw. –*n.* (*pl.* **shies**) fling, throw. [origin unknown]

Shylock /'ʃaɪlɒk/ *n.* hard-hearted money-lender. [name of a character in a play by Shakespeare]

shyster /'ʃaɪstə(r)/ *n.* esp. *US colloq.* unscrupulous or unprofessional person. [origin uncertain]

SI *abbr.* the international system of units of measurement. [French *Système International*]

Si *symb.* silicon.

si /si:/ *n.* = TE. [French from Italian]

Siamese /,saɪə'mi:z/ –*n.* (*pl.* same) **1** native or language of Siam (now Thailand) in Asia. **2** (in full **Siamese cat**) cat of a cream-coloured short-haired breed with dark markings and blue eyes. –*adj.* of Siam, its people, or language.

Siamese twins *n.pl.* **1** twins joined at some part of the body. **2** any closely associated pair.

sibilant /'sɪbɪlənt/ –*adj.* **1** sounded with a hiss. **2** hissing. –*n.* sibilant letter or sound. □ **sibilance** *n.* **sibilancy** *n.* [Latin]

sibling *n.* each of two or more children having one or both parents in common. [Old English, = akin]

sibyl /'sɪbɪl/ *n.* pagan prophetess. [Greek *sibulla*]

sibylline /'sɪbɪ,laɪn/ *adj.* **1** of or from a sibyl. **2** oracular; prophetic. [Latin: related to SIBYL]

sic /sɪk/ *adv.* (usu. in brackets) used, spelt, etc., as written (confirming, or emphasizing, the quoted or copied words). [Latin, = so]

sick –*adj.* **1** esp. *US* unwell, ill. **2** vomiting or likely to vomit. **3** (often foll. by *of*) *colloq.* **a** disgusted; surfeited. **b** angry, esp. because of surfeit. **4** *colloq.* (of a joke etc.) cruel, morbid, perverted, offensive. **5 a** mentally disordered. **b** (esp. in *comb.*) pining (*lovesick*). –*n. colloq.* vomit. –*v.* (usu. foll. by *up*) *colloq.* vomit. □ **take** (or **fall**) **sick** *colloq.* be taken ill. [Old English]

sickbay *n.* room, cabin, etc. for those who are sick.

sickbed *n.* invalid's bed.

sicken *v.* **1** affect with disgust etc. **2 a** (often foll. by *for*) show symptoms of illness. **b** (often foll. by *at*, or *to* + infin.) feel nausea or disgust. **3** (as **sickening** *adj.*) **a** disgusting. **b** *colloq.* very annoying. □ **sickeningly** *adv.*

sickle /ˈsɪk(ə)l/ *n.* short-handled tool with a semicircular blade, used for reaping etc. [Old English]

sick-leave *n.* leave granted because of illness.

sickle-cell *n.* sickle-shaped blood cell, esp. as found in a type of severe hereditary anaemia.

sickly *adj.* (**-ier**, **-iest**) **1 a** weak; apt to be ill. **b** languid, faint, or pale. **2** causing ill health. **3** sentimental or mawkish. **4** of or inducing nausea. [related to SICK]

sickness *n.* **1** being ill; disease. **2** vomiting or a tendency to vomit.

sick-pay *n.* pay given during sick-leave.

side –*n.* **1 a** each of the surfaces bounding an object. **b** vertical inner or outer surface. **c** such a surface as distinct from the top or bottom, front or back. **2 a** right or left part of a person or animal, esp. of the torso. **b** left or right half or a specified part of a thing. **c** (often in *comb.*) adjoining position (*seaside*; *stood at my side*). **d** direction (*from all sides*). **3 a** either surface of a thing regarded as having two surfaces. **b** writing filling one side of a sheet of paper. **4** aspect of a question, character, etc. (*look on the bright side*). **5 a** each of two competing groups in war, politics, games, etc. **b** cause etc. regarded as being in conflict with another. **6 a** part or region near the edge. **b** (*attrib.*) subordinate, peripheral, or detached part (*side-road*; *side-table*). **7** *colloq.* television channel. **8** each of the bounding lines of a plane rectilinear figure. **9** position nearer or farther than, or right or left of, a given dividing line. **10** line of descent through one parent. **11** (in full **side spin**) spin given to a billiard-ball etc. by hitting it on one side. **12** *slang* cheek; pretensions (*has no side about him*). –*v.* (**-ding**) (usu. foll. by *with*) take a side in a conflict etc. □ **by the side of 1** close to. **2** compared with. **let the side down** embarrass or fail one's colleagues. **on one side 1** not in the main or central position. **2** aside. **on the . . . side** somewhat (*on the high side*). **on the side 1** as a sideline. **2** illicitly. **3** *US* as a side dish. **side by side** standing close together, esp. for mutual support. **take sides** support one or other cause etc. [Old English]

sideboard *n.* table or esp. a flat-topped cupboard for dishes, table linen, etc.

sideboards *n.pl. colloq.* hair grown by a man down the sides of his face.

sideburns *n.pl.* = SIDEBOARDS. [earlier *burnsides*, after General *Burnside* (d. 1881)]

side-car *n.* passenger compartment attached to the side of a motor cycle.

sided *adj.* **1** having sides. **2** (in *comb.*) having a specified number or type of sides.

side-door *n.* **1** door at the side of a building. **2** indirect means of access.

side-drum *n.* small double-headed drum.

side-effect *n.* secondary (usu. undesirable) effect.

sidekick *n. colloq.* friend, associate; henchman.

sidelight *n.* **1** light from the side. **2** small light at the side of the front of a vehicle. **3** *Naut.* light on the side of a moving ship.

sideline *n.* **1** work etc. done in addition to one's main activity. **2** (usu. in *pl.*) **a** line bounding the side of a hockey-pitch etc. **b** space next to these where spectators etc. sit. □ **on the sidelines** not directly concerned.

sidelong –*adj.* (esp. of a glance) oblique. –*adv.* obliquely.

sidereal /saɪˈdɪərɪəl/ *adj.* of the constellations or fixed stars. [Latin *sidus sider-* star]

sidereal day *n.* time between successive meridional transits of a star etc.

side-road *n.* minor road, esp. branching from a main road.

side-saddle –*n.* saddle for a woman riding with both legs on the same side of the horse. –*adv.* riding in this position.

sideshow *n.* **1** small show or stall in an exhibition, fair, etc. **2** minor incident or issue.

sidesman *n.* assistant churchwarden who takes the collection etc.

side-splitting *adj.* causing violent laughter.

sidestep –*n.* step to the side. –*v.* (**-pp-**) **1** avoid by stepping sideways. **2** evade.

side-swipe –*n.* **1** glancing blow on or from the side. **2** incidental criticism etc. –*v.* hit (as if) with a side-swipe.
sidetrack *v.* divert or diverge from the main course or issue.
sidewalk *n. US* pavement.
sideways –*adv.* **1** to or from a side. **2** with one side facing forward. –*adj.* to or from a side.
side-whiskers *n.pl.* whiskers on the cheeks.
side wind *n.* wind from the side.
siding *n.* short track at the side of a railway line, used for shunting.
sidle /'saɪd(ə)l/ *v.* (**-ling**) (usu. foll. by *along*, *up*) walk timidly or furtively. [shortening of SIDELONG]
SIDS *abbr.* sudden infant death syndrome; cot-death.
siege *n.* **1** surrounding and blockading of a town, castle, etc. **2** similar operation by police etc. to force an armed person out of a building. □ **lay siege to** conduct the siege of. **raise the siege of** abandon, or cause the abandonment of, an attempted siege of. [French *sege* seat]
siemens /'si:mənz/ *n.* (*pl.* same) SI unit of conductance, equal to one reciprocal ohm. [von *Siemens*, name of an engineer]
sienna /sɪ'enə/ *n.* **1** a kind of earth used as a pigment. **2** its colour of yellowish-brown (**raw sienna**) or reddish-brown (**burnt sienna**). [*Siena* in Tuscany]
sierra /sɪ'erə/ *n.* long jagged mountain chain, esp. in Spain or Spanish America. [Spanish from Latin *serra* saw]
siesta /sɪ'estə/ *n.* afternoon sleep or rest, esp. in hot countries. [Spanish from Latin *sexta* (*hora*) sixth hour]
sieve /sɪv/ –*n.* perforated or meshed utensil for separating solids or coarse material from liquids or fine particles, or for pulping. –*v.* (**-ving**) sift. [Old English]
sift *v.* **1** put through a sieve. **2** (usu. foll. by *from*, *out*) separate (finer or coarser parts) from material. **3** sprinkle (esp. sugar) from a perforated container. **4** examine (evidence, facts, etc.). **5** (of snow, light, etc.) fall as if from a sieve. [Old English]
sigh /saɪ/ –*v.* **1** emit an audible breath in sadness, weariness, relief, etc. **2** (foll. by *for*) yearn for. **3** express with sighs. **4** make a sighing sound. –*n.* **1** act of sighing. **2** sound made in sighing. [Old English]
sight /saɪt/ –*n.* **1 a** faculty of seeing. **b** act of seeing or being seen. **2** thing seen. **3** opinion (*in my sight*). **4** range of vision (*out of sight*). **5** (usu. in *pl.*) noteworthy features of a town etc. **6 a** device on a gun, telescope, etc., for assisting aim or observation. **b** aim or observation so gained. **7** *colloq.* unsightly person or thing (*looked a sight*). **8** *colloq.* great deal (*a sight too clever*). –*v.* **1** get sight of, observe the presence of (*they sighted land*). **2** aim (a gun etc.) with a sight. □ **at first sight** on first glimpse or impression. **at** (or **on**) **sight** as soon as a person or a thing has been seen. **catch** (or **lose**) **sight of** begin (or cease) to see or be aware of. **in sight 1** visible. **2** near at hand. **set one's sights on** aim at. [Old English: related to SEE[1]]
sighted *adj.* **1** not blind. **2** (in *comb.*) having specified vision (*long-sighted*).
sight for sore eyes *n. colloq.* welcome person or thing.
sightless *adj.* blind.
sightly *adj.* attractive to look at.
sight-read *v.* read (music) at sight.
sight-screen *n. Cricket* large white screen placed near the boundary in line with the wicket to help the batsman see the ball.
sightseer *n.* person visiting the sights of a place. □ **sightseeing** *n.*
sight unseen *adv.* without previous inspection.
sigma /'sɪgmə/ *n.* eighteenth letter of the Greek alphabet (Σ, σ, or, when final, ς). [Latin from Greek]
sign /saɪn/ –*n.* **1** thing indicating a quality, state, future event, etc. (*sign of weakness*). **2** mark, symbol, etc. **3** gesture or action conveying an order etc. **4** signboard; signpost. **5** each of the twelve divisions of the zodiac. –*v.* **1 a** (also *absol.*) write (one's name) on a document etc. as authorization. **b** sign (a document) as authorization. **2** communicate by gesture (*signed to me to come*). **3** engage or be engaged by signing a contract etc. (see also *sign on*, *sign up*). □ **sign away** relinquish (property etc.) by signing. **sign in 1** sign a register on arrival. **2** get (a person) admitted by signing a register. **sign off 1** end work, broadcasting, etc. **2** withdraw one's claim to unemployment benefit after finding work. **sign on 1** agree to a contract etc. **2** employ (a person). **3** register as unemployed. **sign out** sign a register on departing. **sign up 1** engage (a person). **2** enlist in the armed forces. **3** enrol. [Latin *signum*]
signal[1] /'sɪgn(ə)l/ –*n.* **1 a** sign (usu. prearranged) conveying information etc. **b** message of such signs. **2** immediate cause of action etc. (*her death was a signal for hope*). **3 a** electrical impulse or impulses or radio waves transmitted as a signal. **b** sequence of these. **4** device on a railway giving instructions or warnings to train-drivers etc. –*v.* (**-ll-**; *US* **-l-**) **1** make signals. **2 a** (often foll. by

to + infin.) make signals to; direct. **b** transmit or express by signal; announce. □ **signaller** *n.* [Latin: *signum* sign]

signal[2] /ˈsɪgn(ə)l/ *attrib. adj.* remarkable, noteworthy. □ **signally** *adv.* [French *signalé*: related to SIGNAL[1]]

signal-box *n.* building beside a railway track from which signals are controlled.

signalize *v.* (also **-ise**) (**-zing** or **-sing**) **1** make noteworthy or remarkable. **2** indicate.

signalman *n.* railway signal operator.

signatory /ˈsɪgnətərɪ/ *–n.* (*pl.* **-ies**) party that has signed an agreement, esp. a treaty. *–adj.* having signed such an agreement etc. [Latin: related to SIGN]

signature /ˈsɪgnətʃə(r)/ *n.* **1 a** person's name, initials, etc. used in signing. **b** act of signing. **2** *Mus.* **a** = KEY SIGNATURE. **b** = TIME SIGNATURE. **3** *Printing* section of a book made from one sheet folded and cut. [medieval Latin: related to SIGNATORY]

signature tune *n.* tune used regularly to introduce a particular broadcast or performer.

signboard *n.* board displaying a name or symbol etc. outside a shop or hotel etc.

signet /ˈsɪgnɪt/ *n.* small seal. [French or medieval Latin: related to SIGN]

signet-ring *n.* ring with a seal set in it.

significance /sɪgˈnɪfɪkəns/ *n.* **1** importance. **2** meaning. **3** being significant. **4** extent to which a result deviates from a hypothesis such that the difference is due to more than errors in sampling. [Latin: related to SIGNIFY]

significant *adj.* **1** having a meaning; indicative. **2** noteworthy; important. □ **significantly** *adv.* [Latin: related to SIGNIFY]

significant figure *n.* digit conveying information about a number containing it.

signify /ˈsɪgnɪˌfaɪ/ *v.* (**-ies, -ied**) **1** be a sign or indication of. **2** mean; symbolize. **3** make known. **4** be of importance; matter. □ **signification** /-fɪˈkeɪʃ(ə)n/ *n.* [Latin: related to SIGN]

sign language *n.* system of communication by gestures, used esp. by the deaf.

sign of the cross *n.* Christian sign made by tracing a cross with the hand.

signor /ˈsiːnjɔː(r)/ *n.* (*pl.* **-nori** /-ˈnjɔːriː/) title used of or to an Italian-speaking man. [Latin *senior* SENIOR]

signora /siːˈnjɔːrə/ *n.* title used of or to an Italian-speaking esp. married woman.

signorina /ˌsiːnjəˈriːnə/ *n.* title used of or to an Italian-speaking esp. unmarried woman.

signpost *–n.* **1** post on a road etc. indicating direction etc. **2** indication, guide. *–v.* provide with a signpost or signposts.

signwriter *n.* person who paints signboards etc.

Sikh /siːk, sɪk/ *n.* member of an Indian monotheistic sect. [Hindi, = disciple]

silage /ˈsaɪlɪdʒ/ *n.* **1** green fodder stored in a silo. **2** storage in a silo. [alteration of ENSILAGE after *silo*]

silence /ˈsaɪləns/ *–n.* **1** absence of sound. **2** abstinence from speech or noise. **3** avoidance of mentioning a thing, betraying a secret, etc. *–v.* (**-cing**) make silent, esp. by force or superior argument. □ **in silence** without speech or other sound. [Latin: related to SILENT]

silencer *n.* device for reducing the noise of a vehicle's exhaust, a gun, etc.

silent *adj.* not speaking; not making or accompanied by any sound. □ **silently** *adv.* [Latin *sileo* be silent]

silent majority *n.* the mass of allegedly moderate people who rarely express an opinion.

silhouette /ˌsɪluːˈet/ *–n.* **1** picture showing the outline only, usu. in black on white or cut from paper. **2** dark shadow or outline against a lighter background. *–v.* (**-ting**) represent or (usu. in *passive*) show in silhouette. [*Silhouette*, name of a politician]

silica /ˈsɪlɪkə/ *n.* silicon dioxide, occurring as quartz etc. and as a main constituent of sandstone and other rocks. □ **siliceous** /-ˈlɪʃəs/ *adj.* [Latin *silex -lic-* flint]

silica gel *n.* hydrated silica in a hard granular form used as a drying agent.

silicate /ˈsɪlɪˌkeɪt/ *n.* compound of a metal with silicon and oxygen.

silicon /ˈsɪlɪkən/ *n. Chem.* non-metallic element occurring widely in silica and silicates.

silicon chip *n.* silicon microchip.

silicone /ˈsɪlɪˌkəʊn/ *n.* any organic compound of silicon, with high resistance to cold, heat, water, etc.

silicosis /ˌsɪlɪˈkəʊsɪs/ *n.* lung fibrosis caused by inhaling dust containing silica.

silk *n.* **1** fine soft lustrous fibre produced by silkworms. **2** (often *attrib.*) thread or cloth from this. **3** (in *pl.*) cloth or garments of silk, esp. as worn by a jockey. **4** *colloq.* Queen's (or King's) Counsel, as having the right to wear a silk gown. **5** fine soft thread (*embroidery silk*). □ **take silk** become a Queen's (or King's) Counsel. [Old English *sioloc*]

silken *adj.* **1** made of silk. **2** soft or lustrous.

silk-screen printing *n.* = SCREEN PRINTING.

silkworm *n.* caterpillar that spins a cocoon of silk.

silky *adj.* (**-ier, -iest**) **1** soft and smooth like silk. **2** suave. □ **silkily** *adv.* **silkiness** *n.*

sill *n.* slab of stone, wood, or metal at the foot of a window or doorway. [Old English]

sillabub var. of SYLLABUB.

silly /'sɪlɪ/ *–adj.* (**-ier, -iest**) **1** foolish, imprudent. **2** weak-minded. **3** *Cricket* (of a fielder or position) very close to the batsman. *–n.* (*pl.* **-ies**) *colloq.* foolish person. □ **sillily** *adv.* **silliness** *n.* [Old English, = happy]

silo /'saɪləʊ/ *n.* (*pl.* **-s**) **1** pit or airtight barn etc. in which green crops are kept for fodder. **2** pit or tower for storing grain, cement, etc. **3** underground storage chamber for a guided missile. [Spanish from Latin]

silt *–n.* sediment in a channel, harbour, etc. *–v.* (often foll. by *up*) choke or be choked with silt. [perhaps Scandinavian]

Silurian /saɪ'ljʊərɪən/ *Geol. –adj.* of the third period of the Palaeozoic era. *–n.* this period. [*Silures*, people of ancient Wales]

silvan var. of SYLVAN.

silver *–n.* **1** greyish-white lustrous precious metallic element. **2** colour of this. **3** silver or cupro-nickel coins. **4** household cutlery. **5** = SILVER MEDAL. *–adj.* of or coloured like silver. *–v.* **1** coat or plate with silver. **2** provide (a mirror-glass) with a backing of tin amalgam etc. **3** make silvery. **4** turn grey or white. [Old English]

silver band *n.* band playing silver-plated instruments.

silver birch *n.* common birch with silver-coloured bark.

silverfish *n.* (*pl.* same or **-es**) **1** small silvery wingless insect. **2** silver-coloured fish.

silver jubilee *n.* 25th anniversary.

silver lining *n.* consolation or hope in misfortune.

silver medal *n.* medal of silver, usu. awarded as second prize.

silver paper *n.* aluminium foil.

silver plate *n.* vessels, cutlery, etc., plated with silver. □ **silver-plated** *adj.*

silver sand *n.* fine pure sand used in gardening.

silver screen *n.* (usu. prec. by *the*) cinema films collectively.

silverside *n.* upper side of a round of beef.

silversmith *n.* worker in silver.

silver tongue *n.* eloquence.

silverware *n.* articles of or plated with silver.

silver wedding *n.* 25th anniversary of a wedding.

silvery *adj.* **1** like silver in colour or appearance. **2** having a clear gentle ringing sound.

silviculture /'sɪlvɪ,kʌltʃə(r)/ *n.* (also **sylviculture**) cultivation of forest trees. [Latin *silva* a wood: related to CULTURE]

simian /'sɪmɪən/ *–adj.* **1** of the anthropoid apes. **2** like an ape or monkey. *–n.* ape or monkey. [Latin *simia* ape]

similar /'sɪmɪlə(r)/ *adj.* **1** like, alike. **2** (often foll. by *to*) having a resemblance. **3** *Geom.* shaped alike. □ **similarity** /-'lærɪtɪ/ *n.* (*pl.* **-ies**). **similarly** *adv.* [Latin *similis* like]

simile /'sɪmɪlɪ/ *n.* **1** esp. poetical comparison of one thing with another using the words 'like' or 'as' (e.g. *as brave as a lion*). **2** use of this. [Latin, neuter of *similis* like]

similitude /sɪ'mɪlɪ,tju:d/ *n.* **1** guise, appearance. **2** comparison; expression of a comparison. [Latin: related to SIMILE]

simmer *–v.* **1** bubble or boil gently. **2** be in a state of suppressed anger or excitement. *–n.* simmering condition. □ **simmer down** become less agitated. [perhaps imitative]

simnel cake /'sɪmn(ə)l/ *n.* rich fruit cake, usu. with a marzipan layer and decoration, eaten esp. at Easter. [Latin *simila* fine flour]

simony /'saɪmənɪ/ *n.* buying or selling of ecclesiastical privileges. [from *Simon* Magus (Acts 8:18)]

simoom /sɪ'mu:m/ *n.* hot dry dust-laden desert wind. [Arabic]

simper *–v.* **1** smile in a silly or affected way. **2** express by or with simpering. *–n.* such a smile. [origin unknown]

simple /'sɪmp(ə)l/ *adj.* (**simpler, simplest**) **1** understood or done easily and without difficulty. **2** not complicated or elaborate; plain. **3** not compound or complex. **4** absolute, unqualified, straightforward (*the simple truth*). **5** foolish; gullible, feeble-minded. □ **simpleness** *n.* [Latin *simplus*]

simple fracture *n.* fracture of the bone only without a wound.

simple interest *n.* interest payable on a capital sum only.

simple-minded *adj.* foolish; feeble-minded. □ **simple-mindedness** *n.*

simpleton *n.* gullible or halfwitted person.

simplicity /sɪm'plɪsɪtɪ/ *n.* fact or condition of being simple.

simplify /'sɪmplɪ,faɪ/ *v.* **(-ies, -ied)** make simple or simpler. □ **simplification** /-fɪ'keɪʃ(ə)n/ *n.*

simplistic /sɪm'plɪstɪk/ *adj.* excessively or affectedly simple. □ **simplistically** *adv.*

simply *adv.* **1** in a simple manner. **2** absolutely (*simply astonishing*). **3** merely (*was simply trying to please*).

simulate /'sɪmjʊ,leɪt/ *v.* **(-ting) 1** pretend to be, have, or feel. **2** imitate or counterfeit. **3** reproduce the conditions of (a situation etc.), e.g. for training. **4** produce a computer model of (a process). □ **simulation** /-'leɪʃ(ə)n/ *n.* **simulator** *n.* [Latin: related to SIMILAR]

simultaneous /,sɪməl'teɪnɪəs/ *adj.* (often foll. by *with*) occurring or operating at the same time. □ **simultaneity** /-tə'neɪɪtɪ/ *n.* **simultaneously** *adv.* [Latin *simul* at the same time]

sin[1] *–n.* **1 a** breaking of divine or moral law, esp. deliberately. **b** such an act. **2** offence against good taste or propriety etc. *–v.* **(-nn-) 1** commit a sin. **2** (foll. by *against*) offend. [Old English]

sin[2] /saɪn/ *abbr.* sine.

sin bin *n. colloq. Ice Hockey* penalty box.

since *–prep.* throughout or during the period after (*has been here since June*; *happened since yesterday*). *–conj.* **1** during or in the time after (*what have you done since we met?*). **2** because. *–adv.* **1** from that time or event until now (*has not seen him since*). **2** ago (*many years since*). [Old English, = after that]

sincere /sɪn'sɪə(r)/ *adj.* **(sincerer, sincerest) 1** free from pretence. **2** genuine, honest, frank. □ **sincerity** /-'serɪtɪ/ *n.* [Latin]

sincerely *adv.* in a sincere manner. □ **yours sincerely** formula for ending an informal letter.

sine *n.* ratio of the side opposite a given angle (in a right-angled triangle) to the hypotenuse. [Latin SINUS]

sinecure /'saɪnɪ,kjʊə(r), 'sɪn-/ *n.* profitable or prestigious position requiring little or no work. [Latin *sine cura* without care]

sine die /,saɪnɪ 'daɪɪ, ,sɪneɪ 'di:eɪ/ *adv. formal* indefinitely (*postponed sine die*). [Latin]

sine qua non /,sɪneɪ kwɑ: 'nəʊn/ *n.* indispensable condition or qualification. [Latin, = without which not]

sinew /'sɪnju:/ *n.* **1** tough fibrous tissue uniting muscle to bone; a tendon. **2** (in *pl.*) muscles; bodily strength. **3** (in *pl.*) strength or framework of a thing. □ **sinewy** *adj.* [Old English]

sinful *adj.* committing or involving sin. □ **sinfully** *adv.* **sinfulness** *n.*

sing *–v.* (*past* **sang**; *past part.* **sung**) **1** utter musical sounds, esp. words with a set tune. **2** utter or produce by singing. **3** (of the wind, a kettle, etc.) hum, buzz, or whistle. **4** (of the ears) hear a humming sound. **5** *slang* turn informer. **6** (foll. by *of*) *literary* celebrate in verse. *–n.* act or spell of singing. □ **sing out** shout. **sing the praises of** praise enthusiastically. □ **singer** *n.* [Old English]

singe /sɪndʒ/ *–v.* **(-geing) 1** burn superficially; scorch. **2** burn off the tips of (hair). *–n.* superficial burn. [Old English]

singer-songwriter *n.* person who sings and writes songs.

Singhalese var. of SINHALESE.

single /'sɪŋg(ə)l/ *–adj.* **1** one only, not double or multiple. **2** united or undivided. **3** for or done by one person etc. **4** one by itself (*a single tree*). **5** regarded separately (*every single thing*). **6** not married. **7** (with *neg.* or *interrog.*) even one (*not a single car*). **8** (of a flower) having only one circle of petals. *–n.* **1** single thing, esp. a single room in a hotel. **2** (in full **single ticket**) ticket valid for an outward journey only. **3** pop record with one item on each side. **4** *Cricket* hit for one run. **5** (usu. in *pl.*) game with one player on each side. **6** (in *pl.*) unmarried people. *–v.* (foll. by *out*) choose for special attention etc. □ **singly** *adv.* [Latin *singulus*]

single-breasted *adj.* (of a coat etc.) having only one vertical row of buttons and overlapping little down the front.

single combat *n.* duel.

single cream *n.* thin cream with a relatively low fat content.

single-decker *n.* bus with only one deck.

single file *–n.* line of people one behind another. *–adv.* one behind the other.

single-handed *adv.* without help. □ **single-handedly** *adv.*

single-minded *adj.* having or intent on only one aim. □ **single-mindedly** *adv.* **single-mindedness** *n.*

single parent *n.* person bringing up a child or children alone.

singlet /'sɪŋglɪt/ *n.* sleeveless vest. [after *doublet*]

singleton /'sɪŋg(ə)lt(ə)n/ *n.* **1** one card only of a suit in a player's hand. **2** single person or thing. [after *simpleton*]

singsong *–n.* informal singing party. *–adj.* monotonously rising and falling. [from SING, SONG]

singular /'sɪŋgjʊlə(r)/ *–adj.* **1** unique; outstanding; extraordinary, strange. **2** *Gram.* (of a word or form) denoting a single person or thing. *–n. Gram.* **1** singular word or form. **2** the singular number. □ **singularity** /-'lærɪtɪ/ *n.* **singularly** *adv.* [Latin: related to SINGLE]

sinh /ʃaɪn, saɪ'neɪtʃ/ *abbr. Math.* hyperbolic sine. [*sine, hyperbolic*]

Sinhalese /ˌsɪnhə'li:z, ˌsɪnə'li:z/ (also **Singhalese** /ˌsɪŋg-/) *–n.* (*pl.* same) **1** member of a N. Indian people now forming the majority of the population of Sri Lanka. **2** their language. *–adj.* of this people or language. [Sanskrit]

sinister /'sɪnɪstə(r)/ *adj.* **1** evil or villainous in appearance or manner. **2** wicked, criminal. **3** ominous. **4** *Heraldry* of or on the left-hand side of a shield etc. (i.e. to the observer's right). [Latin, = left]

sink *–v.* (*past* **sank** or **sunk**; *past part.* **sunk** or as *adj.* **sunken**) **1** fall or come slowly downwards. **2** disappear below the horizon. **3 a** go or penetrate below the surface esp. of a liquid. **b** (of a ship) go to the bottom of the sea etc. **4** settle comfortably. **5 a** decline in strength etc. **b** (of the voice) descend in pitch or volume. **6** cause or allow to sink or penetrate. **7** cause (a plan, person, etc.) to fail. **8** dig (a well) or bore (a shaft). **9** engrave (a die). **10** invest (money). **11 a** knock (a ball) into a pocket or hole in billiards, golf, etc. **b** achieve this by (a stroke). **12** overlook or forget (*sank their differences*). *–n.* **1** plumbed-in basin, esp. in a kitchen. **2** place where foul liquid collects. **3** place of vice. □ **sink in 1** penetrate or permeate. **2** become understood. [Old English]

sinker *n.* weight used to sink a fishing-line or sounding-line.

sinking fund *n.* money set aside gradually for the eventual repayment of a debt.

sinner *n.* person who sins, esp. habitually.

Sinn Fein /ʃɪn 'feɪn/ *n.* political wing of the IRA. [Irish, = we ourselves]

Sino- *comb. form* Chinese; Chinese and (*Sino-American*). [Greek *Sinai* the Chinese]

sinology /saɪ'nɒlədʒɪ/ *n.* the study of the Chinese language, Chinese history, etc. □ **sinologist** *n.*

sinuous /'sɪnjʊəs/ *adj.* having many curves; undulating. □ **sinuosity** /-'ɒsɪtɪ/ *n.* [Latin: related to SINUS]

sinus /'saɪnəs/ *n.* cavity of bone or tissue, esp. in the skull connecting with the nostrils. [Latin, = bosom, recess]

sinusitis /ˌsaɪnə'saɪtɪs/ *n.* inflammation of a sinus.

-sion see -ION.

sip *–v.* (**-pp-**) drink in small mouthfuls. *–n.* **1** small mouthful of liquid. **2** act of taking this. [perhaps var. of SUP[1]]

siphon /'saɪf(ə)n/ *–n.* **1** tube shaped like an inverted V or U with unequal legs, used to convey liquid from a container to a lower level by atmospheric pressure. **2** bottle from which aerated water is forced by the pressure of gas. *–v.* (often foll. by *off*) **1** (cause to) flow through a siphon. **2** divert or set aside (funds etc.). [Greek, = pipe]

sir *n.* **1** polite form of address or reference to a man. **2** (**Sir**) title prefixed to the forename of a knight or baronet. [from SIRE]

sire *–n.* **1** male parent of an animal, esp. a stallion. **2** *archaic* form of address to a king. **3** *archaic* father or male ancestor. *–v.* (**-ring**) (esp. of an animal) beget. [French from Latin *senior* SENIOR]

siren /'saɪərən/ *n.* **1 a** device for making a loud wailing or warning sound. **b** this sound. **2** (in Greek mythology) woman or winged creature whose singing lured unwary sailors on to rocks. **3** (often *attrib.*) temptress; seductress. [Greek *seirēn*]

sirenian /sɑɪ'ri:nɪən/ *n.* any one of an order of large aquatic plant-eating mammals.

sirloin /'sɜ:lɔɪn/ *n.* upper and choicer part of a loin of beef. [French: related to SUR-[1], LOIN]

sirocco /sɪ'rɒkəʊ/ *n.* (also **scirocco**) (*pl.* **-s**) **1** Saharan simoom. **2** warm sultry wind in S. Europe. [Arabic *sharūk*]

sirup *US* var. of SYRUP.

sis *n. colloq.* sister. [abbreviation]

sisal /'saɪs(ə)l/ *n.* **1** fibre made from a Mexican agave. **2** this plant. [*Sisal*, the port of Yucatan]

siskin /'sɪskɪn/ *n.* yellowish-green songbird. [Dutch]

sissy /'sɪsɪ/ (also **cissy**) *colloq. –n.* (*pl.* **-ies**) effeminate or cowardly person. *–adj.* (**-ier, -iest**) effeminate; cowardly. [from SIS]

sister *n.* **1** woman or girl in relation to her siblings. **2** female fellow member of a trade union, feminist group, etc. **3** senior female nurse. **4** member of a female religious order. **5** (often *attrib.*) of the same type, design, or origin etc. (*sister ship*; *prose, the younger sister of verse*). □ **sisterly** *adj.* [Old English]

sisterhood *n.* **1** relationship between or as between sisters. **2** society of esp. religious or charitable women. **3** community of feeling between women.

sister-in-law *n.* (*pl.* **sisters-in-law**) **1** sister of one's wife or husband. **2** wife of one's brother.

Sisyphean /ˌsɪsɪ'fi:ən/ *adj.* (of toil) endless and fruitless like that of Sisyphus (who endlessly pushed a stone uphill in Hades). [Latin from Greek]

sit *v.* (**-tt-**; *past* and *past part.* **sat**) **1** support the body by resting the buttocks on the ground or a seat etc. **2** cause to sit; place in a sitting position. **3 a** (of a bird) perch or warm the eggs in its nest. **b** (of an animal) rest with the hind legs bent and the buttocks on the ground. **4** (of a committee etc.) be in session. **5** (usu. foll. by *for*) pose (for a portrait). **6** (foll. by *for*) be a Member of Parliament for (a constituency). **7** (often foll. by *for*) take (an examination). **8** be in a more or less permanent position or condition (*left sitting in Rome*; *parcel sitting on the doorstep*). **9** (of clothes etc.) fit or hang in a certain way. **10** babysit. □ **be sitting pretty** be comfortably placed. **sit at a person's feet** be a person's pupil. **sit back** relax one's efforts. **sit down 1** sit after standing. **2** cause to sit. **3** (foll. by *under*) submit tamely to (an insult etc.). **sit in 1** occupy a place as a protest. **2** (foll. by *for*) take the place of. **3** (foll. by *on*) be present as a guest or observer at (a meeting etc.). **sit in judgement** be censorious or self-righteous. **sit on 1** be a member of (a committee etc.). **2** hold a session or inquiry concerning. **3** *colloq.* delay action about. **4** *colloq.* repress, rebuke, or snub. **sit on the fence** remain neutral or undecided. **sit out 1** take no part in (a dance etc.). **2** stay till the end of (esp. an ordeal). **3** sit outdoors. **sit tight** *colloq.* **1** remain firmly in one's place. **2** not yield. **sit up 1** rise from lying to sitting. **2** sit firmly upright. **3** go to bed late. **4** *colloq.* become interested or aroused etc. **sit well on** suit or fit. [Old English]

sitar /'sɪtɑ:(r), sɪ'tɑ:(r)/ *n.* long-necked Indian lute. [Hindi]

sitcom *n. colloq.* situation comedy. [abbreviation]

sit-down –*attrib. adj.* **1** (of a meal) eaten sitting at a table. **2** (of a protest etc.) with demonstrators occupying their workplace or sitting down on the ground in a public place. –*n.* **1** spell of sitting. **2** sit-down protest etc.

site –*n.* **1** ground chosen or used for a town or building. **2** place of or for some activity (*camping site*). –*v.* (**-ting**) locate, place. [Latin *situs*]

sit-in *n.* protest involving sitting in.

Sitka /'sɪtkə/ *n.* (in full **Sitka spruce**) fast-growing spruce yielding timber. [*Sitka* in Alaska]

sits vac /sɪts 'væk/ *abbr.* situations vacant.

sitter *n.* **1** person who sits, esp. for a portrait. **2** = BABYSITTER (see BABYSIT). **3** *colloq.* easy catch or shot.

sitting –*n.* **1** continuous period spent engaged in an activity (*finished the book in one sitting*). **2** time during which an assembly is engaged in business. **3** session in which a meal is served. –*adj.* **1** having sat down. **2** (of an animal or bird) still. **3** (of an MP etc.) current.

sitting duck *n.* (also **sitting target**) *colloq.* easy target.

sitting-room *n.* room for relaxed sitting in.

sitting tenant *n.* tenant occupying premises.

situate /'sɪtjʊˌeɪt/ *v.* (**-ting**) (usu. in *passive*) **1** put in a certain position or circumstances. **2** establish or indicate the place of; put in a context. [Latin *situo*: related to SITE]

situation /ˌsɪtjʊ'eɪʃ(ə)n/ *n.* **1** place and its surroundings. **2** circumstances; position; state of affairs. **3** *formal* paid job. □ **situational** *adj.*

situation comedy *n.* broadcast comedy based on characters dealing with awkward domestic situations.

sit-up *n.* physical exercise of sitting up from a supine position without using the arms or hands.

sit-upon *n. colloq.* buttocks.

six *adj.* & *n.* **1** one more than five. **2** symbol for this (6, vi, VI). **3** size etc. denoted by six. **4** *Cricket* hit scoring six runs. **5** six o'clock. □ **at sixes and sevens** in confusion or disagreement. **knock** (or **hit**) **for six** *colloq.* utterly surprise or overcome. [Old English]

sixer *n.* **1** *Cricket* hit for six runs. **2** Brownie or Cub in charge of a group of six.

sixfold *adj.* & *adv.* **1** six times as much or as many. **2** consisting of six parts.

sixpence /'sɪkspəns/ *n.* **1** sum of six esp. old pence. **2** *hist.* coin worth this.

sixpenny *adj.* costing or worth sixpence, esp. before decimalization.

six-shooter *n.* (also **six-gun**) revolver with six chambers.

sixteen /ˌsɪks'ti:n, 'sɪks-/ *adj.* & *n.* **1** one more than fifteen. **2** symbol for this (16, xvi, XVI). **3** size etc. denoted by sixteen. □ **sixteenth** *adj.* & *n.* [Old English]

sixth *adj.* & *n.* **1** next after fifth. **2** any of six equal parts of a thing. □ **sixthly** *adv.*

sixth form *n.* form in a secondary school for pupils over 16.

sixth-form college *n.* separate college for pupils over 16.

sixth-former *n.* sixth-form pupil.

sixth sense *n.* supposed intuitive or extrasensory faculty.

sixty /'sɪkstɪ/ *adj.* & *n.* (*pl.* **-ies**) **1** six times ten. **2** symbol for this (60, lx, LX). **3** (in *pl.*) numbers from 60 to 69, esp. the years of a century or of a person's life. □ **sixtieth** *adj.* & *n.* [Old English]

sizable var. of SIZEABLE.

size[1] *–n.* **1** relative dimensions, magnitude. **2** each of the classes into which similar things are divided according to size. *–v.* (**-zing**) sort in sizes or according to size. □ **the size of it** *colloq.* the truth of the matter. **size up** *colloq.* form a judgement of. □ **sized** *adj.* (also in *comb.*). [French *sise*]

size[2] *–n.* sticky solution used in glazing paper, stiffening textiles, etc. *–v.* (**-zing**) treat with size. [perhaps = SIZE[1]]

sizeable *adj.* (also **sizable**) large or fairly large.

sizzle /'sɪz(ə)l/ *–v.* (**-ling**) **1** sputter or hiss, esp. in frying. **2** *colloq.* be very hot or excited etc. *–n.* sizzling sound. □ **sizzling** *adj.* & *adv.* [imitative]

SJ *abbr.* Society of Jesus.

ska /skɑ:/ *n.* a kind of fast orig. Jamaican pop music. [origin unknown]

skate[1] *–n.* **1** boot with a blade attached for gliding on ice; this blade. **2** = ROLLER-SKATE. *–v.* (**-ting**) **1 a** move on skates. **b** perform (a specified figure) on skates. **2** (foll. by *over*) refer fleetingly to, disregard. □ **get one's skates on** *slang* make haste. **skate on thin ice** *colloq.* behave rashly, risk danger. □ **skater** *n.* [Dutch *schaats* from French]

skate[2] *n.* (*pl.* same or **-s**) large flat marine fish used as food. [Old Norse]

skateboard *–n.* short narrow board on two wheeled trucks, for riding on while standing. *–v.* ride on a skateboard. □ **skateboarder** *n.*

skedaddle /skɪ'dæd(ə)l/ *v.* (**-ling**) *colloq.* depart quickly, flee. [origin unknown]

skein /skeɪn/ *n.* **1** loosely-coiled bundle of yarn or thread. **2** flock of wild geese etc. in flight. [French *escaigne*]

skeleton /'skelɪt(ə)n/ *n.* **1** hard framework of bones etc. of an animal. **2** supporting framework or structure of a thing. **3** very thin person or animal. **4** useless or dead remnant. **5** outline sketch, epitome. **6** (*attrib.*) having only the essential or minimum number of persons, parts, etc. (*skeleton staff*). □ **skeletal** *adj.* [Greek *skellō* dry up]

skeleton in the cupboard *n.* discreditable or embarrassing secret.

skeleton key *n.* key designed to fit many locks.

skeptic *US* var. of SCEPTIC.

skeptical *US* var. of SCEPTICAL.

skerry /'skerɪ/ *n.* (*pl.* **-ies**) *Scot.* reef, rocky island. [Old Norse]

sketch *–n.* **1** rough or unfinished drawing or painting. **2** rough draft or general outline. **3** short usu. humorous play. **4** short descriptive essay etc. *–v.* **1** make or give a sketch of. **2** draw sketches. **3** (often foll. by *in*, *out*) outline briefly. [Greek *skhēdios* extempore]

sketch-book *n.* (also **sketch-block**) pad of drawing-paper for sketching.

sketch-map *n.* roughly-drawn map with few details.

sketchy *adj.* (**-ier**, **-iest**) **1** giving only a rough outline, like a sketch. **2** *colloq.* unsubstantial or imperfect, esp. through haste. □ **sketchily** *adv.* **sketchiness** *n.*

skew *–adj.* oblique, slanting, set askew. *–n.* slant. *–v.* **1** make skew. **2** distort. **3** move obliquely. □ **on the skew** askew. [French: related to ESCHEW]

skewbald /'skju:bɔ:ld/ *–adj.* (esp. of a horse) with irregular patches of white and another colour. *–n.* skewbald animal. [origin uncertain]

skewer /'skju:ə(r)/ *–n.* long pin designed for holding meat together while cooking. *–v.* fasten together or pierce (as) with a skewer. [origin uncertain]

skew-whiff *adj.* & *adv. colloq.* askew.

ski /ski:/ *–n.* (*pl.* **-s**) **1** each of a pair of long narrow pieces of wood etc., fastened under the feet for travelling over snow. **2** similar device under a vehicle or aircraft. *–v.* (**skis**, **ski'd** or **skied** /ski:d/, **skiing**) travel on skis. □ **skier** *n.* [Norwegian from Old Norse]

skid *–v.* (**-dd-**) **1** (of a vehicle etc.) slide on slippery ground, esp. sideways or obliquely. **2** cause (a vehicle) to skid. *–n.* **1** act of skidding. **2** runner beneath an aircraft for use when landing. □ **on the skids** *colloq.* about to be discarded or defeated. **put the skids under** *colloq.* hasten the downfall or failure of. [origin unknown]

skid-pan *n.* slippery surface for drivers to practise control of skidding.

skid row *n. US slang* part of a town frequented by vagrants etc.

skiff *n.* light rowing- or sculling-boat. [French *esquif*: related to SHIP]

ski-jump *n.* steep slope levelling off before a sharp drop to allow a skier to leap through the air. □ **ski-jumping** *n.*

skilful *adj.* (*US* **skillful**) (often foll. by *at*, *in*) having or showing skill. □ **skilfully** *adv.*

ski-lift *n.* device for carrying skiers up a slope, usu. a cable with hanging seats.

skill *n.* (often foll. by *in*) ability to do something well; technique, expertise. [Old Norse, = difference]

skilled *adj.* **1** (often foll. by *in*) skilful. **2** (of work or a worker) requiring or having skill or special training.

skillet /'skɪlɪt/ *n.* **1** small long-handled metal cooking-pot. **2** *US* frying-pan. [French]

skillful *US* var. of SKILFUL.

skim –*v.* (**-mm-**) **1 a** take a floating layer from the surface of (a liquid). **b** take (cream etc.) from the surface of a liquid. **2 a** barely touch (a surface) in passing over. **b** (often followed by *over*) deal with or treat (a matter) superficially. **3** (often foll. by *over, along*) go or glide lightly. **4** (often followed by *through*) read or look over cursorily. –*n.* skimming. [French: related to SCUM]

skimmia /'skɪmɪə/ *n.* evergreen shrub with red berries. [Japanese]

skim milk *n.* (also **skimmed milk**) milk from which the cream has been removed.

skimp *v.* **1** (often followed by *on*) economize; use a meagre or insufficient amount of, stint. **2** (often foll. by *in*) supply (a person etc.) meagrely with food etc. **3** do hastily or carelessly. [cf. SCRIMP]

skimpy *adj.* (**-ier, -iest**) meagre; insufficient. □ **skimpiness** *n.*

skin –*n.* **1** flexible covering of a body. **2 a** skin of a flayed animal with or without the hair etc. **b** material prepared from skins. **3** complexion of the skin. **4** outer layer or covering, esp. of a fruit, sausage, etc. **5** film like skin on a liquid etc. **6** container for liquid, made of an animal's skin. **7** *slang* skinhead. –*v.* (**-nn-**) **1** remove the skin from. **2** graze (part of the body). **3** *slang* swindle. □ **be skin and bone** be very thin. **by** (or **with**) **the skin of one's teeth** by a very narrow margin. **get under a person's skin** *colloq.* interest or annoy a person intensely. **have a thick** (or **thin**) **skin** be insensitive (or sensitive). **no skin off one's nose** *colloq.* of no consequence to one. □ **skinless** *adj.* [Old Norse]

skin-deep *adj.* superficial.

skin-diver *n.* underwater swimmer without a diving-suit, usu. with aqualung and flippers. □ **skin-diving** *n.*

skinflint *n.* miser.

skinful *n. colloq.* enough alcohol to make one drunk.

skin-graft *n.* **1** surgical transplanting of skin. **2** skin transferred in this way.

skinhead *n.* youth with a shaven head, esp. one of an aggressive gang.

skinny *adj.* (**-ier, -iest**) thin or emaciated. □ **skinniness** *n.*

skint *adj. slang* having no money left. [= *skinned*]

skin-tight *adj.* (of a garment) very close-fitting.

skip[1] –*v.* (**-pp-**) **1 a** move along lightly, esp. with alternate hops. **b** jump lightly, esp. over a skipping-rope. **c** gambol, caper, frisk. **2** (often foll. by *from, off, to*) move quickly from one point, subject, etc. to another. **3** (also *absol.*) omit parts of (a text, subject, etc.). **4** *colloq.* miss intentionally, not attend. **5** *colloq.* leave hurriedly. –*n.* skipping movement or action. □ **skip it** *colloq.* abandon a topic etc. [probably Scandinavian]

skip[2] *n.* **1** large container for building refuse etc. **2** container for transporting or raising materials in mining etc. [Old Norse]

skipjack *n.* (in full **skipjack tuna**) (*pl.* same or **-s**) small striped Pacific tuna used as food. [from SKIP[1], JACK]

skipper –*n.* **1** captain of a ship or aircraft. **2** captain of a sporting team. –*v.* be captain of. [Low German or Dutch *schipper*]

skipping-rope *n.* length of rope turned over the head and under the feet while jumping it as a game or exercise.

skirl –*n.* shrill sound, esp. of bagpipes. –*v.* make a skirl. [probably Scandinavian]

skirmish /'skɜːmɪʃ/ –*n.* **1** minor battle. **2** short argument or contest of wit etc. –*v.* engage in a skirmish. [French from Germanic]

skirt –*n.* **1** woman's garment hanging from the waist. **2** the part of a coat etc. hanging below the waist. **3** hanging part at the base of a hovercraft. **4** (in *sing.* or *pl.*) edge, border, extreme part. **5** (also **bit of skirt**) *slang offens.* woman. **6** (in full **skirt of beef** etc.) cut of meat from the flank or diaphragm. –*v.* (often foll. by *around*) **1** go or lie along or round the edge of. **2** avoid dealing with (an issue etc.). [Old Norse: related to SHIRT]

skirting-board *n.* narrow board etc. along the bottom of a room-wall.

ski-run *n.* slope prepared for skiing.

skit *n.* light, usu. short, piece of satire or burlesque. [perhaps from Old Norse: related to SHOOT]

skittish *adj.* **1** lively, playful. **2** (of a horse etc.) nervous, inclined to shy. [perhaps related to SKIT]

skittle /'skɪt(ə)l/ *n.* **1** pin used in skittles. **2** (in *pl.*; usu. treated as *sing.*) game of trying to bowl down usu. nine wooden pins. [origin unknown]

skive *v.* (**-ving**) (often followed by *off*) *slang* evade work; play truant. □ **skiver** *n.* [Old Norse]

skivvy /'skɪvɪ/ –*n.* (*pl.* **-ies**) *colloq. derog.* female domestic servant. –*v.*

(-ies, -ied) work as a skivvy. [origin unknown]
skua /'skju:ə/ *n.* large predatory sea bird. [Old Norse]
skulduggery /skʌl'dʌgərɪ/ *n.* trickery; unscrupulous behaviour. [origin unknown]
skulk *v.* move stealthily; lurk, hide. [Scandinavian]
skull *n.* **1** bony case of the brain of a vertebrate. **2** bony skeleton of the head. **3** head as the seat of intelligence. [origin unknown]
skull and crossbones *n.pl.* representation of a skull with two crossed thigh-bones as an emblem of piracy or death.
skullcap *n.* peakless cap covering the crown only.
skunk *n.* (*pl.* same or **-s**) **1** black and white striped mammal emitting a powerful stench when attacked. **2** *colloq.* contemptible person. [American Indian]
sky /skaɪ/ *–n.* (*pl.* **skies**) (in *sing.* or *pl.*) atmosphere and outer space as seen from the earth. *–v.* **(skies, skied)** *Cricket* etc. hit (a ball) high. □ **to the skies** without reserve (*praise to the skies*). [Old Norse, = cloud]
sky blue *adj.* & *n.* (as adj. often hyphenated) bright clear blue.
skydiving *n.* sport of performing acrobatic manoeuvres under free fall before opening a parachute. □ **skydiver** *n.*
sky-high *adv.* & *adj.* very high.
skyjack *v. slang* hijack (an aircraft).
skylark *–n.* lark that sings while soaring. *–v.* play tricks, frolic.
skylight *n.* window in a roof.
skyline *n.* outline of hills, buildings, etc. against the sky.
sky-rocket *–n.* = ROCKET 1. *–v.* (esp. of prices) rise very rapidly.
skyscraper *n.* very tall building.
skyward /'skaɪwəd/ *–adv.* (also **skywards**) towards the sky. *–adj.* moving skyward.
sky-writing *n.* writing in aeroplane smoke-trails.
slab *n.* **1** flat thick esp. rectangular piece of solid material, esp. stone. **2** mortuary table. [origin unknown]
slack[1] *–adj.* **1** (of rope etc.) not taut. **2** inactive or sluggish. **3** negligent, remiss. **4** (of tide etc.) neither ebbing nor flowing. *–n.* **1** slack part of a rope (*haul in the slack*). **2** slack period. **3** (in *pl.*) informal trousers. *–v.* **1** loosen (rope etc.). **2** *colloq.* take a rest, be lazy. □ **slack off 1** loosen. **2** (also **slack up**) reduce one's level of activity; reduce speed. □ **slackness** *n.* [Old English]
slack[2] *n.* coal-dust or fragments of coal. [probably Low German or Dutch]
slacken *v.* make or become slack. □ **slacken off** = *slack off* (see SLACK[1]).
slacker *n.* shirker.
slag *–n.* **1** refuse left after smelting etc. **2** *slang derog.* prostitute; promiscuous woman. *–v.* **(-gg-) 1** form slag. **2** (often foll. by *off*) *slang* insult, slander. □ **slaggy** *adj.* [Low German]
slag-heap *n.* hill of refuse from a coal-mine, steelworks, etc.
slain *past part.* of SLAY[1].
slake *v.* **(-king) 1** assuage or satisfy (thirst, a desire, etc.). **2** temper (quick-lime) by combination with water. [Old English: related to SLACK[1]]
slalom /'slɑ:ləm/ *n.* **1** ski-race down a zigzag obstacle course. **2** obstacle race in canoes etc. [Norwegian]
slam[1] *–v.* **(-mm-) 1** shut forcefully and loudly. **2** put down loudly. **3** put or do suddenly (*slam the brakes on*; *car slammed to a halt*). **4** *slang* criticize severely. **5** *slang* hit. **6** *slang* conquer easily. *–n.* sound or action of slamming. [probably Scandinavian]
slam[2] *n. Cards* winning of every trick in a game. [origin uncertain]
slander *–n.* **1** false and damaging utterance about a person. **2** uttering of this. *–v.* utter slander about. □ **slanderous** *adj.* [French *esclandre*: related to SCANDAL]
slang *–n.* very informal words, phrases, or meanings, not regarded as standard and often used by a specific profession, class, etc. *–v.* use abusive language (to). □ **slangy** *adj.* [origin unknown]
slanging-match *n.* prolonged exchange of insults.
slant /slɑ:nt/ *–v.* **1** slope; lie or (cause to) go obliquely. **2** (often as **slanted** *adj.*) present (information) in a biased or particular way. *–n.* **1** slope; oblique position. **2** point of view, esp. a biased one. *–adj.* sloping, oblique. □ **on a** (or **the**) **slant** aslant. [Scandinavian]
slantwise *adv.* aslant.
slap *–v.* **(-pp-) 1** strike with the palm or a flat object, or so as to make a similar noise. **2** lay forcefully (*slapped it down*). **3** put hastily or carelessly (*slap paint on*). **4** (often foll. by *down*) *colloq.* reprimand or snub. *–n.* **1** blow with the palm or a flat object. **2** slapping sound. *–adv.* suddenly, fully, directly (*ran slap into him*). [Low German, imitative]
slap and tickle *n. colloq.* sexual horseplay.
slap-bang *adv. colloq.* violently, headlong.
slapdash *–adj.* hasty and careless. *–adv.* in this manner.

slap-happy *adj. colloq.* cheerfully casual or flippant.
slap in the face *n.* rebuff or affront.
slap on the back *n.* congratulations.
slapstick *n.* boisterous comedy.
slap-up *attrib. adj. colloq.* excellent, lavish.
slash –*v.* **1** cut or gash with a knife etc. **2** (often foll. by *at*) deliver or aim cutting blows. **3** reduce (prices etc.) drastically. **4** censure vigorously. –*n.* **1** slashing cut or stroke. **2** *Printing* oblique stroke; solidus. **3** *slang* act of urinating. [origin unknown]
slat *n.* thin narrow piece of wood, plastic, or metal, esp. as in a fence or Venetian blind. [French *esclat* splinter]
slate –*n.* **1** (esp. bluish-grey) metamorphic rock easily split into flat smooth plates. **2** piece of this as a tile or *hist.* for writing on. **3** bluish-grey colour of slate. **4** list of nominees for office etc. –*v.* (**-ting**) **1** roof with slates. **2** *colloq.* criticize severely. **3** *US* make arrangements for (an event etc.). **4** *US* nominate for office etc. –*adj.* of slate or the colour of slate. □ **on the slate** on (usu. informal) credit. □ **slating** *n.* **slaty** *adj.* [French *esclate*, feminine of *esclat*: related to SLAT]
slattern /'slæt(ə)n/ *n.* slovenly woman. □ **slatternly** *adj.* [origin uncertain]
slaughter /'slɔ:tə(r)/ –*v.* **1** kill (animals) for food or skins or because of disease. **2** kill (people) ruthlessly or on a great scale. **3** *colloq.* defeat utterly. –*n.* act of slaughtering. □ **slaughterer** *n.* [Old Norse: related to SLAY]
slaughterhouse *n.* place for the slaughter of animals as food.
Slav /slɑ:v/ –*n.* member of a group of peoples in central and eastern Europe speaking Slavonic languages. –*adj.* of the Slavs. [Latin *Sclavus*, ethnic name]
slave –*n.* **1** person who is owned by and has to serve another. **2** drudge, hard worker. **3** (foll. by *of*, *to*) obsessive devotee (*slave of fashion*). **4** machine, or part of one, directly controlled by another. –*v.* (**-ving**) (often foll. by *at*, *over*) work very hard. [French *esclave* from Latin *Sclavus* SLAV (captive)]
slave-driver *n.* **1** overseer of slaves. **2** demanding boss.
slave labour *n.* forced labour.
slaver[1] *n. hist.* ship or person engaged in the slave-trade.
slaver[2] /'slævə(r)/ –*v.* **1** dribble. **2** (foll. by *over*) drool over. –*n.* **1** dribbling saliva. **2 a** fulsome flattery. **b** drivel, nonsense. [Low German or Dutch]
slavery /'sleɪvərɪ/ *n.* **1** condition of a slave. **2** drudgery. **3** practice of having slaves.
slave-trade *n. hist.* dealing in slaves, esp. African Blacks.
Slavic /'slɑ:vɪk/ *adj.* & *n.* = SLAVONIC.
slavish *adj.* **1** like slaves. **2** without originality. □ **slavishly** *adv.*
Slavonic /slə'vɒnɪk/ –*adj.* **1** of the group of languages including Russian, Polish, and Czech. **2** of the Slavs. –*n.* Slavonic language-group. [related to SLAV]
slay *v.* (*past* **slew** /slu:/; *past part.* **slain**) **1** *literary* = KILL 1. **2** = KILL 4. □ **slayer** *n.* [Old English]
sleaze *n. colloq.* sleaziness. [back-formation from SLEAZY]
sleazy *adj.* (**-ier**, **-iest**) squalid, tawdry. □ **sleazily** *adv.* **sleaziness** *n.* [origin unknown]
sled *US* –*n.* sledge. –*v.* (**-dd-**) ride on a sledge. [Low German]
sledge –*n.* vehicle on runners for use on snow. –*v.* (**-ging**) travel or convey by sledge. [Dutch *sleedse*]
sledgehammer /'sledʒ,hæmə(r)/ *n.* **1** large heavy long-handled hammer used to break stone etc. **2** (*attrib.*) heavy or powerful (*sledgehammer blow*). [Old English *slecg*: related to SLAY]
sleek –*adj.* **1** (of hair, skin, etc.) smooth and glossy. **2** looking well-fed and comfortable. –*v.* make sleek. □ **sleekly** *adv.* **sleekness** *n.* [var. of SLICK]
sleep –*n.* **1** natural recurring condition of suspended consciousness, with the eyes closed and the muscles relaxed. **2** period of sleep (*had a sleep*). **3** state like sleep; rest, quiet, death. –*v.* (*past* and *past part.* **slept**) **1 a** be in a state of sleep. **b** fall asleep. **2** (foll. by *at*, *in*, etc.) spend the night. **3** provide beds etc. for (*house sleeps six*). **4** (foll. by *with*, *together*) have sexual intercourse, esp. in bed. **5** (foll. by *on*) put off (a decision) until the next day. **6** (foll. by *through*) fail to be woken by. **7** be inactive or dead. **8** (foll. by *off*) remedy by sleeping. □ **get to sleep** manage to fall asleep. **go to sleep** **1** begin to sleep. **2** (of a limb) become numb. **put to sleep 1** anaesthetize. **2** put down (an animal). **sleep around** *colloq.* be sexually promiscuous. **sleep in** sleep later than usual in the morning. [Old English]
sleeper *n.* **1** person or animal that sleeps. **2** horizontal beam supporting a railway track. **3 a** sleeping-car. **b** berth in this. **4** ring or stud worn in a pierced ear to keep the hole open.
sleeping-bag *n.* padded bag to sleep in when camping etc.
sleeping-car *n.* (also **sleeping-carriage**) railway coach with berths.
sleeping partner *n.* partner not sharing in the actual work of a firm.

sleeping-pill *n.* pill to induce sleep.

sleeping policeman *n.* ramp etc. in the road to make traffic slow down.

sleeping sickness *n.* tropical disease causing extreme lethargy.

sleepless *adj.* **1** lacking sleep (*sleepless night*). **2** unable to sleep. **3** continually active. □ **sleeplessness** *n.*

sleepwalk *v.* walk about while asleep. □ **sleepwalker** *n.*

sleepy *adj.* (**-ier, -iest**) **1** drowsy. **2** quiet, inactive (*sleepy town*). □ **sleepily** *adv.* **sleepiness** *n.*

sleet *–n.* **1** snow and rain falling together. **2** hail or snow melting as it falls. *–v.* (prec. by *it* as subject) sleet falls (*it is sleeting*). □ **sleety** *adj.* [Old English]

sleeve *n.* **1** part of a garment that encloses an arm. **2** cover of a gramophone record. **3** tube enclosing a rod etc. □ **up one's sleeve** in reserve. □ **sleeved** *adj.* (also in *comb.*). **sleeveless** *adj.* [Old English]

sleigh /sleɪ/ *–n.* sledge, esp. for riding on. *–v.* travel on a sleigh. [Dutch *slee*: related to SLEDGE]

sleight of hand /slaɪt/ *n.* dexterity, esp. in conjuring. [Old Norse: related to SLY]

slender *adj.* (**-er, -est**) **1 a** of small girth or breadth. **b** gracefully thin. **2** relatively small, scanty, inadequate. [origin unknown]

slept *past* and *past part.* of SLEEP.

sleuth /slu:θ/ *colloq. –n.* detective. *–v.* investigate crime etc. [Old Norse]

slew[1] /slu:/ (also **slue**) *–v.* (often foll. by *round*) turn or swing forcibly to a new position. *–n.* such a turn. [origin unknown]

slew[2] *past* of SLAY.

slice *–n.* **1** thin flat piece or wedge of esp. food cut off or out. **2** share; part. **3** long-handled kitchen utensil with a broad flat perforated blade. **4** *Sport* stroke that sends the ball obliquely. *–v.* (**-cing**) **1** (often foll. by *up*) cut into slices. **2** (foll. by *off*) cut off. **3** (foll. by *into, through*) cut (as) with a knife. **4** strike (a ball) with a slice. [French *esclice* from Germanic]

slick *–adj. colloq.* **1 a** skilful or efficient. **b** superficially or pretentiously smooth and dexterous; glib. **2** sleek, smooth. *–n.* large patch of oil etc., esp. on the sea. *–v. colloq.* **1** (usu. foll. by *back, down*) flatten (one's hair etc.). **2** (usu. foll. by *up*) make sleek or smart. □ **slickly** *adv.* **slickness** *n.* [Old English]

slide *–v.* (*past* and *past part.* **slid**) **1** move along a smooth surface with continuous contact on the same part of the thing moving. **2** move quietly or smoothly; glide. **3** glide over ice without skates. **4** (foll. by *over*) barely touch upon (a delicate subject etc.). **5** (often foll. by *into*) move quietly or unobtrusively. *–n.* **1** act of sliding. **2** rapid decline. **3** inclined plane down which children, goods, etc., slide. **4** track made by or for sliding, esp. on ice. **5** part of a machine or instrument that slides. **6 a** mounted transparency viewed with a projector. **b** piece of glass holding an object for a microscope. **7** = HAIR-SLIDE. □ **let things slide** be negligent; allow deterioration. [Old English]

slide-rule *n.* ruler with a sliding central strip, graduated logarithmically for making rapid calculations.

sliding scale *n.* scale of fees, taxes, wages, etc., that varies according to some other factor.

slight /slaɪt/ *–adj.* **1 a** small; insignificant. **b** inadequate. **2** slender, frail-looking. **3** (in *superl.*) any whatever (*if there were the slightest chance*). *–v.* treat disrespectfully; ignore. *–n.* act of slighting. □ **slightly** *adv.* **slightness** *n.* [Old Norse]

slim *–adj.* (**slimmer, slimmest**) **1** not fat, slender. **2** small, insufficient (*slim chance*). *–v.* (**-mm-**) (often foll. by *down*) **1** become slimmer by dieting, exercise, etc. **2** make smaller (*slimmed it down to 40 pages*). □ **slimmer** *n.* **slimming** *n.* & *adj.* **slimmish** *adj.* [Low German or Dutch]

slime *n.* thick slippery mud or sticky substance produced by an animal or plant. [Old English]

slimline *adj.* **1** of slender design. **2** (of a drink) not fattening.

slimy *adj.* (**-ier, -iest**) **1** like, covered with, or full of slime. **2** *colloq.* disgustingly obsequious. □ **sliminess** *n.*

sling[1] *–n.* **1** strap etc. used to support or raise a thing. **2** bandage supporting an injured arm from the neck. **3** strap etc. for firing a stone etc. by hand. *–v.* (*past* and *past part.* **slung**) **1** *colloq.* throw. **2** suspend with a sling. □ **sling one's hook** *slang* go away. [Old Norse or Low German or Dutch]

sling[2] *n.* sweetened drink of spirits (esp. gin) and water. [origin unknown]

sling-back *n.* shoe held in place by a strap above the heel.

slink *v.* (*past* and *past part.* **slunk**) (often foll. by *off, away, by*) move in a stealthy or guilty manner. [Old English]

slinky *adj.* (**-ier, -iest**) (of a garment) close-fitting and sinuous.

slip[1] *–v.* (**-pp-**) **1** slide unintentionally or momentarily; lose one's footing or balance. **2** go or move with a sliding motion. **3** escape or fall from being slippery or not being held properly. **4**

(often foll. by *in, out, away*) go unobserved or quietly. **5 a** make a careless or slight error. **b** fall below standard. **6** place or slide stealthily or casually (*slipped a coin to him*). **7** release from restraint or connection. **8** move (a stitch) to the other needle without knitting it. **9** (foll. by *on, off*) pull (a garment) easily or hastily on or off. **10** escape from; evade (*dog slipped its collar; slipped my mind*). –*n.* **1** act of slipping. **2** careless or slight error. **3 a** pillowcase. **b** petticoat. **4** (in *sing.* or *pl.*) = SLIPWAY. **5** *Cricket* **a** fielder stationed for balls glancing off the bat to the off side. **b** (in *sing.* or *pl.*) this position. □ **give a person the slip** escape from; evade. **let slip 1** utter inadvertently. **2** miss (an opportunity). **3** release, esp. from a leash. **slip up** *colloq.* make a mistake. [probably from Low German *slippen*]

slip[2] *n.* **1** small piece of paper, esp. for writing on. **2** piece cut from a plant for grafting or planting. □ **slip of a** small and slim (*slip of a girl*). [Low German or Dutch]

slip[3] *n.* clay and water mixture for decorating earthenware. [Old English, = slime]

slip-knot *n.* **1** knot that can be undone by a pull. **2** running knot.

slip of the pen *n.* (also **slip of the tongue**) small written (or spoken) mistake.

slip-on –*attrib. adj.* easily slipped on and off. –*n.* slip-on shoe or garment.

slippage *n.* act or an instance of slipping.

slipped disc *n.* displaced disc between vertebrae causing lumbar pain.

slipper *n.* light loose soft indoor shoe.

slippery *adj.* **1** difficult to grasp, stand on, etc. because smooth or wet. **2** unreliable, unscrupulous. □ **slipperiness** *n.* [Old English]

slippery slope *n.* course leading eventually to disaster.

slippy *adj.* (**-ier, -iest**) *colloq.* slippery. □ **look** (or **be**) **slippy** make haste.

slip-road *n.* road for entering or leaving a motorway etc.

slipshod *adj.* careless, slovenly.

slipstream *n.* current of air or water driven back by a revolving propeller or a moving vehicle.

slip-up *n. colloq.* mistake.

slipway *n.* ramp for building ships or landing boats.

slit –*n.* straight narrow incision or opening. –*v.* (**-tt-**; *past* and *past part.* **slit**) **1** make a slit in. **2** cut into strips. [Old English]

slither /'slɪðə(r)/ –*v.* slide unsteadily. –*n.* act of slithering. □ **slithery** *adj.* [var. of *slidder*: related to SLIDE]

sliver /'slɪvə(r)/ –*n.* long thin piece cut or split off. –*v.* **1** break off as a sliver. **2** break or form into slivers. [Old English]

Sloane /sləʊn/ *n.* (in full **Sloane Ranger**) *slang* fashionable and conventional upper-class young person. □ **Sloaney** *adj.* [*Sloane* Square in London, and Lone *Ranger*, cowboy hero]

slob *n. colloq. derog.* lazy, untidy, or fat person. [Irish *slab* mud]

slobber –*v.* **1** dribble. **2** (foll. by *over*) drool over. –*n.* dribbling saliva. □ **slobbery** *adj.* [Dutch]

sloe /sləʊ/ *n.* **1** = BLACKTHORN. **2** its small sour bluish-black fruit. [Old English]

slog –*v.* (**-gg-**) **1** hit hard and usu. wildly. **2** work or walk doggedly. –*n.* **1** hard random hit. **2 a** hard steady work or walk. **b** spell of this. [origin unknown]

slogan /'sləʊgən/ *n.* **1** catchy phrase used in advertising etc. **2** party cry; watchword. [Gaelic, = war cry]

sloop *n.* small one-masted fore-and-aft rigged vessel. [Dutch *sloep*]

slop –*v.* (**-pp-**) **1** (often foll. by *over*) spill over the edge of a vessel. **2** wet (the floor etc.) by slopping. –*n.* **1** liquid spilled or splashed. **2** sloppy language. **3** (in *pl.*) dirty waste water or wine etc. from a kitchen, bedroom, or prison vessels. **4** (in *sing.* or *pl.*) unappetizing weak liquid food. □ **slop about** move about in a slovenly manner. **slop out** carry slops out (in prison etc.). [Old English]

slope –*n.* **1** inclined position, direction, or state. **2** piece of rising or falling ground. **3** difference in level between the two ends or sides of a thing. **4** place for skiing on a mountain etc. –*v.* (**-ping**) **1** have or take a slope, slant. **2** cause to slope. □ **slope arms** place one's rifle in a sloping position against one's shoulder. **slope off** *slang* go away, esp. to evade work etc. [*aslope* crosswise]

sloppy *adj.* (**-ier, -iest**) **1** wet, watery, too liquid. **2** careless, untidy. **3** foolishly sentimental. □ **sloppily** *adv.* **sloppiness** *n.*

slosh –*v.* **1** (often foll. by *about*) splash or flounder. **2** *slang* hit, esp. heavily. **3** *colloq.* **a** pour (liquid) clumsily. **b** pour liquid on. –*n.* **1** slush. **2** act or sound of splashing. **3** *slang* heavy blow. [var. of SLUSH]

sloshed *predic. adj. slang* drunk.

slot –*n.* **1** slit in a machine etc. for a thing, esp. a coin, to be inserted. **2** slit, groove, etc. for a thing. **3** allotted place in a schedule, esp. in broadcasting. –*v.* (**-tt-**) **1** (often foll. by *in, into*) place or be placed (as if) into a slot. **2** provide with slots. [French *esclot* hollow of breast]

sloth /sləʊθ/ *n.* **1** laziness, indolence. **2** slow-moving S. American mammal that hangs upside down in trees. [from SLOW]

slothful *adj.* lazy. □ **slothfully** *adv.*

slot-machine *n.* machine worked by the insertion of a coin, esp. selling small items or providing amusement.

slouch *–v.* stand, move, or sit in a drooping fashion. *–n.* **1** slouching posture or movement. **2** *slang* incompetent or slovenly worker etc. [origin unknown]

slouch hat *n.* hat with a wide flexible brim.

slough[1] /slaʊ/ *n.* swamp, miry place. [Old English]

slough[2] /slʌf/ *–n.* part that an animal casts or moults, esp. a snake's cast skin. *–v.* (often foll. by *off*) cast or drop off as a slough. [origin unknown]

Slough of Despond *n.* state of hopeless depression.

Slovak /ˈsləʊvæk/ *–n.* **1** native of Slovakia in Czechoslovakia. **2** language of Slovakia, one of the two official languages of Czechoslovakia. *–adj.* of the Slovaks or their language. [native name]

sloven /ˈslʌv(ə)n/ *n.* untidy or careless person. [origin uncertain]

slovenly *–adj.* careless and untidy; unmethodical. *–adv.* in a slovenly manner. □ **slovenliness** *n.*

slow /sləʊ/ *–adj.* **1 a** taking a relatively long time to do a thing (also foll. by *of*: *slow of speech*). **b** acting, moving, or done without speed, not quick. **2** not conducive to speed (*slow route*). **3** (of a clock etc.) showing a time earlier than is correct. **4** (of a person) not understanding or learning readily. **5** dull, tedious. **6** slack, sluggish (*business is slow*). **7** (of a fire or oven) giving little heat. **8** *Photog.* (of a film) needing long exposure. **9** reluctant; not hasty (*slow to anger*). *–adv.* slowly (also in *comb.*: *slow-moving traffic*). *–v.* (usu. foll. by *down*, *up*) **1** reduce one's speed or the speed of (a vehicle etc.). **2** reduce one's pace of life. □ **slowish** *adj.* **slowly** *adv.* **slowness** *n.* [Old English]

slowcoach *n. colloq.* slow person.

slow-down *n.* action of slowing down.

slow motion *n.* **1** speed of a film at which actions etc. appear much slower than usual. **2** simulation of this in real action.

slow-worm *n.* small European legless lizard. [Old English *slow* of uncertain origin]

sludge *n.* **1** thick greasy mud or sediment. **2** sewage. □ **sludgy** *adj.* [cf. SLUSH]

slue var. of SLEW[1].

slug[1] *n.* **1** small shell-less mollusc often destroying plants. **2 a** bullet, esp. of irregular shape. **b** missile for an airgun. **3** *Printing* metal bar used in spacing. **4** mouthful of drink (esp. spirits). [Scandinavian]

slug[2] *US –v.* (**-gg-**) hit hard. *–n.* hard blow. □ **slug it out** fight it out. [origin unknown]

sluggard /ˈslʌɡəd/ *n.* lazy person. [related to SLUG[1]]

sluggish *adj.* inert; slow-moving. □ **sluggishly** *adv.* **sluggishness** *n.*

sluice /slu:s/ *–n.* **1** (also **sluice-gate**, **sluice-valve**) sliding gate or other contrivance for regulating the volume or flow of water. **2** water so regulated. **3** (**sluice-way**) artificial water-channel, esp. for washing ore. **4** place for rinsing. **5** act of rinsing. *–v.* (**-cing**) **1** provide or wash with a sluice or sluices. **2** rinse, esp. with running water. **3** (foll. by *out*, *away*) wash out or away with a flow of water. **4** (of water) rush out (as if) from a sluice. [French *escluse*]

slum *–n.* **1** house unfit for human habitation. **2** (often in *pl.*) overcrowded and squalid district in a city. *–v.* (**-mm-**) visit slums, esp. out of curiosity. □ **slum it** *colloq.* put up with conditions less comfortable than usual. □ **slummy** *adj.* [originally cant]

slumber *v. & n. poet.* or *joc.* sleep. [Old English]

slump *–n.* sudden severe or prolonged fall in prices and trade, usu. bringing widespread unemployment. *–v.* **1** undergo a slump. **2** sit or fall heavily or limply. [imitative]

slung *past* and *past part.* of SLING[1].

slunk *past* and *past part.* of SLINK.

slur *–v.* (**-rr-**) **1** pronounce indistinctly with sounds running into one another. **2** *Mus.* perform (notes) legato. **3** *archaic* or *US* put a slur on (a person or a person's character). **4** (usu. foll. by *over*) pass over (a fact, fault, etc.) lightly. *–n.* **1** imputation of wrongdoing. **2** act of slurring. **3** *Mus.* curved line joining notes to be slurred. [origin unknown]

slurp *colloq. –v.* eat or esp. drink noisily. *–n.* sound of this. [Dutch]

slurry /ˈslʌrɪ/ *n.* thin semi-liquid cement, mud, manure, etc. [related to dial. *slur* thin mud]

slush *n.* **1** thawing muddy snow. **2** silly sentimentality. □ **slushy** *adj.* (**-ier**, **-iest**). [origin unknown]

slush fund *n.* reserve fund, esp. for political bribery.

slut *n. derog.* slovenly or promiscuous woman. □ **sluttish** *adj.* [origin unknown]
sly /slaɪ/ *adj.* **(slyer, slyest) 1** cunning, crafty, wily. **2** secretive. **3** knowing; insinuating. □ **on the sly** secretly. □ **slyly** *adv.* **slyness** *n.* [Old Norse: related to SLAY]
Sm *symb.* samarium.
smack[1] *–n.* **1** sharp slap or blow. **2** hard hit at cricket etc. **3** loud kiss. **4** loud sharp sound. *–v.* **1** slap. **2** part (one's lips) noisily in anticipation of food. **3** move, hit, etc., with a smack. *–adv. colloq.* **1** with a smack. **2** suddenly; directly; violently. **3** exactly (*smack in the centre*). □ **a smack in the eye** (or **face**) *colloq.* rebuff; setback. [imitative]
smack[2] (foll. by *of*) *–v.* **1** have a flavour of; taste of. **2** suggest (*smacks of nepotism*). *–n.* **1** flavour. **2** barely discernible quality. [Old English]
smack[3] *n.* single-masted sailing-boat. [Low German or Dutch]
smack[4] *n. slang* heroin or other hard drug. [probably alteration of Yiddish *schmeck* sniff]
smacker *n. slang* **1** loud kiss. **2 a** £1. **b** *US* $1.
small /smɔːl/ *–adj.* **1** not large or big. **2** not great in importance, amount, number, power, etc. **3** not much; little (*paid small attention*). **4** insignificant (*from small beginnings*). **5** of small particles (*small shot*). **6** on a small scale (*small farmer*). **7** poor or humble. **8** mean; ungenerous. **9** young (*small child*). *–n.* **1** slenderest part of a thing, esp. of the back. **2** (in *pl.*) *colloq.* underwear, esp. as laundry. *–adv.* into small pieces (*chop it small*). □ **feel** (or **look**) **small** be humiliated or ashamed. □ **smallish** *adj.* **smallness** *n.* [Old English]
small arms *n.pl.* portable firearms.
small beer *n.* trifling thing.
small change *n.* coins, not notes.
small fry *n.* unimportant people; children.
smallholder *n.* farmer of a smallholding.
smallholding *n.* agricultural holding smaller than a farm.
small hours *n.pl.* period soon after midnight.
small-minded *adj.* petty; narrow in outlook.
smallpox *n. hist.* acute contagious disease with fever and pustules, usu. leaving scars.
small print *n.* unfavourable clauses etc. in a contract, usu. printed small.
small-scale *adj.* made or occurring on a small scale.
small talk *n.* light social conversation.
small-time *adj. colloq.* unimportant, petty.
smarm *–v. colloq.* **1** (often foll. by *down*) smooth, plaster flat (hair etc.). **2** be ingratiating. *–n. colloq.* obsequiousness. [dial.]
smarmy *adj.* **(-ier, -iest)** *colloq.* ingratiating. □ **smarmily** *adv.* **smarminess** *n.*
smart *–adj.* **1** well-groomed, neat. **2** brightly coloured, newly painted, etc. **3** stylish, fashionable. **4** (esp. *US*) clever, ingenious, quickwitted. **5** quick, brisk. **6** painfully severe; sharp, vigorous. *–v.* **1** feel or give pain. **2** rankle. **3** (foll. by *for*) suffer the consequences of. *–n.* sharp pain; stinging sensation. *–adv.* smartly. □ **smartish** *adj.* & *adv.* **smartly** *adv.* **smartness** *n.* [Old English]
smart alec *n.* (also **smart aleck**) *colloq.* conceited know-all.
smarten *v.* (usu. foll. by *up*) make or become smart.
smart money *n.* money invested by people with expert knowledge.
smash *–v.* **1** (often foll. by *up*) **a** break into pieces; shatter. **b** bring or come to sudden destruction, defeat, or disaster. **2** (foll. by *into, through*) move with great force. **3** (foll. by *in*) break with a crushing blow. **4** hit (a ball etc.) with great force, esp. downwards. *–n.* **1** act of smashing, collision. **2** sound of this. **3** (in full **smash hit**) very successful play, song, performer, etc. *–adv.* with a smash. [imitative]
smash-and-grab *n.* robbery in which a shop-window is smashed and goods seized.
smasher *n. colloq.* beautiful or pleasing person or thing.
smashing *adj. colloq.* excellent, wonderful.
smash-up *n.* violent collision.
smattering *n.* slight superficial knowledge of a language etc. [origin unknown]
smear *–v.* **1** daub or mark with grease etc. **2** smudge. **3** defame. *–n.* **1** act of smearing. **2** *Med.* **a** material smeared on a microscopic slide etc. for examination. **b** specimen of this. □ **smeary** *adj.* [Old English]
smear test *n.* = CERVICAL SMEAR.
smell *–n.* **1** faculty of perceiving odours. **2** quality in substances that is perceived by this. **3** unpleasant odour. **4** act of inhaling to ascertain smell. *–v.* (*past* and *past part.* **smelt** or **smelled**) **1** perceive or examine by smell. **2** emit an odour; stink. **3** seem by smell to be (*smells sour*). **4** (foll. by *of*) **a** emit the odour of (*smells of fish*). **b** be suggestive

of (*smells of dishonesty*). **5** perceive; detect (*smell a bargain*). **6** have or use a sense of smell. □ **smell a rat** suspect trickery etc. **smell out** detect by smell or investigation. [Old English]

smelling-salts *n.pl.* sharp-smelling substances sniffed to relieve faintness etc.

smelly *adj.* (**-ier, -iest**) having a strong or unpleasant smell. □ **smelliness** *n.*

smelt[1] *v.* **1** extract metal from (ore) by melting. **2** extract (metal) in this way. □ **smelter** *n.* [Low German or Dutch *smelten*]

smelt[2] *past* and *past part.* of SMELL.

smelt[3] *n.* (*pl.* same or **-s**) small edible green and silver fish. [Old English]

smidgen /'smɪdʒ(ə)n/ *n.* (also **smidgin**) *colloq.* small bit or amount. [perhaps from *smitch* in the same sense]

smilax /'smaɪlæks/ *n.* any of several climbing shrubs. [Greek, = bindweed]

smile *–v.* (**-ling**) **1** have or assume a happy, kind, or amused expression, with the corners of the mouth turned up. **2** express by smiling (*smiled a welcome*). **3** give (a smile) of a specified kind (*smiled a sardonic smile*). **4** (foll. by *on*, *upon*) favour (*fortune smiled on me*). *–n.* **1** act of smiling. **2** smiling expression or aspect. [perhaps from Scandinavian]

smirch *–v.* soil; discredit. *–n.* spot, stain. [origin unknown]

smirk *–n.* conceited or silly smile. *–v.* give a smirk. [Old English]

smite *v.* (**-ting**; *past* **smote**; *past part.* **smitten** /'smɪt(ə)n/) **1** *archaic* or *literary* **a** hit. **b** chastise; defeat. **2** (in *passive*) affect strongly; seize (*smitten with regret*; *smitten by her beauty*). [Old English]

smith *n.* **1** blacksmith. **2** (esp. in *comb.*) worker in metal (*goldsmith*). **3** (esp. in *comb.*) craftsman (*wordsmith*). [Old English]

smithereens /,smɪðə'ri:nz/ *n.pl.* small fragments. [dial. *smithers*]

smithy /'smɪðɪ/ *n.* (*pl.* **-ies**) blacksmith's workshop, forge. [related to SMITH]

smitten *past part.* of SMITE.

smock *–n.* **1** loose shirtlike garment often ornamented with smocking. **2** loose overall. *–v.* adorn with smocking. [Old English]

smocking *n.* ornamental effect on cloth made by gathering it tightly with stitches.

smog *n.* smoke-laden fog. □ **smoggy** *adj.* (**-ier, -iest**). [portmanteau word]

smoke *–n.* **1** visible vapour from a burning substance. **2** act of smoking tobacco. **3** *colloq.* cigarette or cigar. *–v.* (**-king**) **1 a** inhale and exhale the smoke of (a cigarette etc.). **b** do this habitually. **2** emit smoke or visible vapour. **3** darken or preserve with smoke (*smoked salmon*). □ **go up in smoke** *colloq.* come to nothing. **smoke out 1** drive out by means of smoke. **2** drive out of hiding etc. [Old English]

smoke bomb *n.* bomb that emits dense smoke on exploding.

smoke-free *adj.* **1** free from smoke. **2** where smoking is not permitted.

smokeless *adj.* producing little or no smoke; free from smoke.

smokeless zone *n.* district where only smokeless fuel may be used.

smoker *n.* **1** person who habitually smokes. **2** compartment on a train where smoking is allowed.

smokescreen *n.* **1** cloud of smoke concealing (esp. military) operations. **2** ruse for disguising one's activities.

smokestack *n.* **1** chimney or funnel of a locomotive or steamer. **2** tall chimney.

smoky *adj.* (**-ier, -iest**) **1** emitting, filled with, or obscured by, smoke. **2** stained with or coloured like smoke. **3** having the flavour of smoked food. □ **smokiness** *n.*

smolder *US* var. of SMOULDER.

smooch *colloq.* *–n.* **1** period of slow close dancing. **2** period of kissing and caressing. *–v.* engage in a smooch. □ **smoochy** *adj.* [imitative]

smooth /smu:ð/ *–adj.* **1** having an even surface; free from projections, dents, and roughness. **2** that can be traversed without check. **3** (of the sea etc.) calm, flat. **4** (of a journey etc.) easy. **5** not harsh in sound or taste. **6** suave, conciliatory; slick. **7** not jerky. *–v.* **1** (often foll. by *out*, *down*) make or become smooth. **2** (often foll. by *out*, *down*, *over*, *away*) reduce or get rid of (differences, faults, difficulties, etc.) in fact or appearance. *–n.* smoothing touch or stroke. *–adv.* smoothly. □ **smoothly** *adv.* **smoothness** *n.* [Old English]

smoothie /'smu:ðɪ/ *n. colloq.*, often *derog.* smooth person.

smooth-tongued *adj.* insincerely flattering.

smorgasbord /'smɔ:gəs,bɔ:d/ *n.* various esp. savoury dishes as hors d'œuvres or a buffet meal. [Swedish]

smote *past* of SMITE.

smother /'smʌðə(r)/ *v.* **1** suffocate, stifle. **2** (foll. by *in, with*) overwhelm or cover with (kisses, gifts, kindness, etc.). **3** extinguish (a fire) by covering it. **4 a** die of suffocation. **b** have difficulty breathing. **5** (often foll. by *up*) suppress or conceal. [Old English]

smoulder /'sməʊldə(r)/ (*US* **smolder**) *–v.* **1** burn slowly without flame or internally. **2** (of emotions) be fierce but suppressed. **3** (of a person) show silent emotion. –*n.* smouldering. [origin unknown]

smudge –*n.* blurred or smeared line, mark, blot, etc. –*v.* (**-ging**) **1** make a smudge on or of. **2** become smeared or blurred. □ **smudgy** *adj.* [origin unknown]

smug *adj.* (**smugger**, **smuggest**) self-satisfied. □ **smugly** *adv.* **smugness** *n.* [Low German *smuk* pretty]

smuggle /'smʌg(ə)l/ *v.* (**-ling**) **1** (also *absol.*) import or export illegally, esp. without paying duties. **2** (foll. by *in, out*) convey secretly. □ **smuggler** *n.* **smuggling** *n.* [Low German]

smut –*n.* **1** small flake of soot etc. **2** spot or smudge made by this. **3** obscene talk, pictures, or stories. **4** fungous disease of cereals. –*v.* (**-tt-**) mark with smuts. □ **smutty** *adj.* (**-ier**, **-iest**). [origin unknown]

Sn *symb.* tin. [Latin *stannum*]

snack *n.* **1** light, casual, or hurried meal. **2** small amount of food eaten between meals. [Dutch]

snack bar *n.* place where snacks are sold.

snaffle /'snæf(ə)l/ –*n.* (in full **snaffle-bit**) simple bridle-bit without a curb. –*v.* (**-ling**) *colloq.* steal; seize. [Low German or Dutch perhaps from *snavel* beak]

snafu /snæ'fu:/ *slang* –*adj.* in utter confusion. –*n.* this state. [acronym of '*s*ituation *n*ormal: *a*ll *f*ouled (or *f*ucked) *u*p']

snag –*n.* **1** unexpected obstacle or drawback. **2** jagged projection. **3** tear in material etc. –*v.* (**-gg-**) catch or tear on a snag. [probably Scandinavian]

snail *n.* slow-moving gastropod mollusc with a spiral shell. [Old English]

snail's pace *n.* very slow movement.

snake –*n.* **1** long limbless reptile. **2** (also **snake in the grass**) traitor; secret enemy. –*v.* (**-king**) move or twist like a snake. [Old English]

snake-charmer *n.* person appearing to make snakes move by music etc.

snakes and ladders *n.pl.* (usu. treated as *sing.*) board-game with counters moved up 'ladders' and down 'snakes'.

snakeskin –*n.* skin of a snake. –*adj.* made of snakeskin.

snaky *adj.* **1** of or like a snake. **2** winding, sinuous. **3** cunning, treacherous.

snap –*v.* (**-pp-**) **1** break suddenly or with a cracking sound. **2** (cause to) emit a sudden sharp crack. **3** open or close with a snapping sound. **4** speak or say irritably. **5** (often foll. by *at*) make a sudden audible bite. **6** move quickly (*snap into action*). **7** photograph. –*n.* **1** act or sound of snapping. **2** crisp biscuit (*brandy snap*). **3** snapshot. **4** (in full **cold snap**) sudden brief spell of cold weather. **5 a** card-game in which players call 'snap' when two similar cards are exposed. **b** (as *int.*) on noticing an (often unexpected) similarity. **6** vigour, liveliness. –*adv.* with a snap (*heard it go snap*). –*adj.* done without forethought (*snap decision*). □ **snap out of** *slang* get rid of (a mood, etc.) by a sudden effort. **snap up** accept (an offer etc.) quickly or eagerly. [Low German or Dutch *snappen* seize]

snapdragon *n.* plant with a two-lipped flower.

snap-fastener *n.* = PRESS-STUD.

snapper *n.* any of several edible marine fish.

snappish *adj.* **1** curt; ill-tempered; sharp. **2** inclined to snap.

snappy *adj.* (**-ier**, **-iest**) *colloq.* **1** brisk, lively. **2** neat and elegant (*snappy dresser*). **3** snappish. □ **make it snappy** be quick. □ **snappily** *adv.*

snapshot *n.* casual or informal photograph.

snare /sneə(r)/ –*n.* **1** trap, esp. with a noose, for birds or animals. **2** trap, trick, or temptation. **3** (in *sing.* or *pl.*) twisted strings of gut, hide, or wire stretched across the lower head of a side-drum to produce a rattle. **4** (in full **snare drum**) drum fitted with snares. –*v.* (**-ring**) catch in a snare; trap. [Old Norse]

snarl[1] –*v.* **1** growl with bared teeth. **2** speak, say, or express angrily. –*n.* act or sound of snarling. [*snar* from Low German]

snarl[2] –*v.* (often foll. by *up*) twist; entangle; hamper the movement of (traffic etc.); become entangled or congested. –*n.* knot, tangle. [from SNARE]

snarl-up *n. colloq.* traffic jam; muddle.

snatch –*v.* **1** (often foll. by *away, from*) seize or remove quickly, eagerly, or unexpectedly. **2 a** steal (a handbag etc.) by grabbing. **b** *slang* kidnap. **3** secure with difficulty. **4** (foll. by *at*) **a** try to seize. **b** take (an offer etc.) eagerly. –*n.* **1** act of snatching. **2** fragment of a song or talk etc. **3** *US slang* kidnapping. **4** short spell of activity etc. [related to SNACK]

snazzy *adj.* (**-ier**, **-iest**) *slang* smart, stylish, showy. □ **snazzily** *adv.* **snazziness** *n.* [origin unknown]

sneak –*v.* **1** (foll. by *in, out, past, away,* etc.) go or convey furtively. **2** *slang* steal unobserved. **3** *slang* tell tales; turn informer. **4** (as **sneaking** *adj.*) **a** furtive (*sneaking affection*). **b** persistent and puzzling (*sneaking feeling*). –*n.* **1** mean-spirited underhand person. **2** *slang* tell-tale; informer. *—adj.* acting or done without warning; secret. □ **sneaky** *adj.* (**-ier**, **-iest**). [origin uncertain]

sneaker *n. slang* soft-soled canvas shoe.

sneak-thief *n.* thief who steals without breaking in.

sneer *—n.* contemptuous smile or remark. *—v.* **1** (often foll. by *at*) smile or speak derisively. **2** say with a sneer. □ **sneering** *adj.* **sneeringly** *adv.* [origin unknown]

sneeze *—n.* sudden loud involuntary expulsion of air from the nose and mouth caused by irritation of the nostrils. *—v.* (**-zing**) make a sneeze. □ **not to be sneezed at** *colloq.* worth having or considering. [Old English]

snick *—v.* **1** make a small notch or incision in. **2** *Cricket* deflect (the ball) slightly with the bat. *—n.* **1** small notch or cut. **2** *Cricket* slight deflection of the ball. [*snickersnee* long knife, ultimately from Dutch]

snicker *n.* & *v.* = SNIGGER. [imitative]

snide *adj.* sneering; slyly derogatory. [origin unknown]

sniff *—v.* **1** inhale air audibly through the nose. **2** (often foll. by *up*) draw in through the nose. **3** smell the scent of by sniffing. *—n.* **1** act or sound of sniffing. **2** amount of air etc. sniffed up. □ **sniff at** show contempt for. **sniff out** = *smell out*. [imitative]

sniffer *n.* person who sniffs, esp. a drug etc. (often in *comb.*: *glue-sniffer*).

sniffer-dog *n. colloq.* dog trained to sniff out drugs or explosives.

sniffle /ˈsnɪf(ə)l/ *—v.* (**-ling**) sniff slightly or repeatedly. *—n.* **1** act of sniffling. **2** (in *sing.* or *pl.*) cold in the head causing sniffling. [imitative: cf. SNIVEL]

sniffy *adj. colloq.* (**-ier**, **-iest**) disdainful. □ **sniffily** *adv.* **sniffiness** *n.*

snifter *n. slang* small alcoholic drink. [dial. *snift* sniff]

snigger *—n.* half-suppressed laugh. *—v.* utter this. [var. of SNICKER]

snip *—v.* (**-pp-**) (also *absol.*) cut with scissors etc., esp. in small quick strokes. *—n.* **1** act of snipping. **2** piece snipped off. **3** *slang* **a** something easily done. **b** bargain. [Low German or Dutch *snippen*]

snipe *—n.* (*pl.* same or **-s**) wading bird with a long straight bill. *—v.* (**-ping**) **1** fire shots from hiding, usu. at long range. **2** (often foll. by *at*) make a sly critical attack. □ **sniper** *n.* (in sense 1 of *v.*). [probably Scandinavian]

snippet /ˈsnɪpɪt/ *n.* **1** small piece cut off. **2** (usu. in *pl.*) **a** scrap of information etc. **b** short extract from a book etc.

snitch *slang —v.* **1** steal. **2** (often foll. by *on*) inform on a person. *—n.* informer. [origin unknown]

snivel /ˈsnɪv(ə)l/ *—v.* (**-ll-**; *US* **-l-**) **1** weep with sniffling. **2** run at the nose; sniffle. **3** show weak or tearful sentiment. *—n.* act of snivelling. [Old English]

snob *n.* person who despises those inferior in social position, wealth, intellect, taste, etc. (*intellectual snob*). □ **snobbery** *n.* **snobbish** *adj.* **snobby** *adj.* (**-ier**, **-iest**). [origin unknown]

snog *slang —v.* (**-gg-**) engage in kissing and caressing. *—n.* period of this. [origin unknown]

snood *n.* ornamental hairnet, worn usu. at the back of the head. [Old English]

snook *n. slang* contemptuous gesture with the thumb to the nose and the fingers spread. □ **cock a snook** (often foll. by *at*) **1** make this gesture. **2** register one's contempt. [origin unknown]

snooker *—n.* **1** game played on an oblong cloth-covered table with a cue-ball, 15 red, and 6 coloured balls. **2** position in this game in which a direct shot would lose points. *—v.* **1** (also *refl.*) subject (oneself or an opponent) to a snooker. **2** (esp. as **snookered** *adj.*) *slang* thwart, defeat. [origin unknown]

snoop *colloq. —v.* **1** pry into another's affairs. **2** (often foll. by *about*, *around*) investigate transgressions of rules, the law, etc. *—n.* act of snooping. □ **snooper** *n.* **snoopy** *adj.* [Dutch]

snooty *adj.* (**-ier**, **-iest**) *colloq.* supercilious; conceited; snobbish. □ **snootily** *adv.* [origin unknown]

snooze *colloq. —n.* short sleep, nap. *—v.* (**-zing**) take a snooze. [origin unknown]

snore *—n.* snorting or grunting sound of breathing during sleep. *—v.* (**-ring**) make this sound. [imitative]

snorkel /ˈsnɔːk(ə)l/ *—n.* **1** breathing-tube for an underwater swimmer. **2** device for supplying air to a submerged submarine. *—v.* (**-ll-**; *US* **-l-**) use a snorkel. [German *Schnorchel*]

snort *—n.* **1** explosive sound made esp. by horses by the sudden forcing of breath through the nose. **2** similar human sound showing contempt, incredulity, etc. **3** *colloq.* small drink of liquor. **4** *slang* inhaled dose of powdered cocaine etc. *—v.* **1** make a snort. **2** (also *absol.*) *slang* inhale (esp. cocaine). **3** express or utter with a snort. [imitative]

snot *n. slang* nasal mucus. [probably Low German or Dutch: related to SNOUT]

snotty *adj.* (**-ier**, **-iest**) *slang* **1** running or covered with nasal mucus. **2** snooty. **3** mean, contemptible. □ **snottily** *adv.* **snottiness** *n.*

snout *n.* **1** projecting nose and mouth of an animal. **2** *derog.* person's nose. **3** pointed front of a thing. [Low German or Dutch]

snow /snəʊ/ *–n.* **1** frozen atmospheric vapour falling to earth in light white flakes. **2** fall or layer of this. **3** thing resembling snow in whiteness or texture etc. **4** *slang* cocaine. *–v.* **1** (prec. by *it* as subject) snow falls (*it is snowing*; *if it snows*). **2** (foll. by *in*, *over*, *up*, etc.) confine or block with snow. □ **be snowed under** be overwhelmed, esp. with work. [Old English]

snowball *–n.* ball of compressed snow for throwing in play. *–v.* **1** throw or pelt with snowballs. **2** increase rapidly.

snowball-tree *n.* guelder rose.

snowberry *n.* (*pl.* **-ies**) shrub with white berries.

snow-blind *adj.* temporarily blinded by the glare from snow.

snowblower *n.* machine that clears snow by blowing.

snowbound *adj.* prevented by snow from going out or travelling.

snowcap *n.* snow-covered mountain peak. □ **snowcapped** *adj.*

snowdrift *n.* bank of snow heaped up by the wind.

snowdrop *n.* early spring plant with white drooping flowers.

snowfall *n.* **1** fall of snow. **2** amount of this.

snowflake *n.* each of the flakes in which snow falls.

snow goose *n.* white Arctic goose.

snowline *n.* level above which snow never melts entirely.

snowman *n.* figure resembling a human, made of compressed snow.

snowmobile /ˈsnəʊmə,bi:l/ *n.* motor vehicle, esp. with runners or Caterpillar tracks, for travel over snow.

snowplough *n.* (*US* **snowplow**) device or vehicle for clearing roads of thick snow.

snowshoe *n.* racket-shaped attachment to a boot for walking on snow without sinking in.

snowstorm *n.* heavy fall of snow, esp. with a high wind.

snow white *adj.* & *n.* (as adj. often hyphenated) pure white.

snowy *adj.* (**-ier, -iest**) **1** of or like snow. **2** (of the weather etc.) with much snow.

snowy owl *n.* large white Arctic owl.

SNP *abbr.* Scottish National Party.

Snr. *abbr.* Senior.

snub *–v.* (**-bb-**) rebuff or humiliate with sharp words or coldness. *–n.* act of snubbing. *–adj.* short and blunt in shape. [Old Norse, = chide]

snub nose *n.* short turned-up nose. □ **snub-nosed** *adj.*

snuff[1] *–n.* charred part of a candle-wick. *–v.* trim the snuff from (a candle). □ **snuff it** *slang* die. **snuff out 1** extinguish (a candle flame). **2** put an end to (hopes etc.). [origin unknown]

snuff[2] *–n.* powdered tobacco or medicine taken by sniffing. *–v.* take snuff. [Dutch]

snuffbox *n.* small box for holding snuff.

snuffer *n.* device for snuffing or extinguishing a candle.

snuffle /ˈsnʌf(ə)l/ *–v.* (**-ling**) **1** make sniffing sounds. **2** speak or say nasally or whiningly. **3** breathe noisily, esp. with a blocked nose. *–n.* snuffling sound or tone. □ **snuffly** *adj.* [Low German or Dutch *snuffelen*]

snug *–adj.* (**snugger, snuggest**) **1** cosy, comfortable, sheltered. **2** close-fitting. *–n.* small room in a pub. □ **snugly** *adv.* [probably Low German or Dutch]

snuggery *n.* (*pl.* **-ies**) snug place, den.

snuggle /ˈsnʌg(ə)l/ *v.* (**-ling**) settle or draw into a warm comfortable position.

so[1] /səʊ/ *–adv.* **1** to such an extent (*stop complaining so*; *so small as to be invisible*; *not so late as I expected*). **2** in this or that way; in the manner, position, or state described or implied (*place your feet so*; *am not cold but may become so*). **3** also (*he went and so did I*). **4** indeed, actually (*you said it was good, and so it is*). **5** very (*I am so glad*). **6** (with verbs of saying or thinking etc.) thus, this, that (*I think so*; *so he said*). *–conj.* (often foll. by *that*) **1** consequently (*was ill, so couldn't come*). **2** in order that (*came early so that I could see you*). **3** and then; as the next step (*so then I gave up*; *and so to bed*). **4** (introducing a question) then; after that (*so what did you do?*). □ **and so on** (or **forth**) **1** and others of the same kind. **2** and in other similar ways. **or so** approximately (*50 or so*). **so as to** in order to. **so be it** expression of acceptance or resignation. **so long!** *colloq.* goodbye. **so much 1** a certain amount (of). **2** nothing but (*so much nonsense*). **so much for** that is all that need be done or said about (a thing). **so so** *adj.* & *adv. colloq.* only moderately good or well. **so what?** *colloq.* that is not significant. [Old English]

so[2] var. of SOH.

-so *comb. form* = -SOEVER.

soak *–v.* **1** make or become thoroughly wet through saturation. **2** (of rain etc.) drench. **3** (foll. by *in*, *up*) absorb (liquid, knowledge, etc.). **4** *refl.* (often foll. by *in*) steep (oneself) in a subject etc. **5** (foll. by *in, into, through*) (of liquid) go or penetrate by saturation. **6** *colloq.* extort money from. **7** *colloq.* drink heavily. **—***n.* **1** act of soaking; prolonged spell in a bath. **2** *colloq.* hard drinker. [Old English]

soakaway *n.* pit into which liquids may flow and then percolate slowly into the subsoil.

soaking *adj.* (in full **soaking wet**) wet through.

so-and-so /'səʊən,səʊ/ *n.* (*pl.* **-so's**) **1** particular but unspecified person or thing. **2** *colloq.* objectionable person.

soap **—***n.* **1** cleansing agent yielding lather when rubbed in water. **2** *colloq.* = SOAP OPERA. **—***v.* apply soap to. [Old English]

soapbox *n.* makeshift stand for a speaker in the street etc.

soap flakes *n.pl.* thin flakes of soap for washing clothes etc.

soap opera *n.* broadcast drama serial with domestic themes (orig. sponsored in the US by soap manufacturers).

soap powder *n.* powdered soap, esp. with additives, for washing clothes etc.

soapstone *n.* steatite.

soapsuds *n.pl.* = SUDS.

soapwort *n.* plant with pink or white flowers, and leaves yielding a soapy substance.

soapy *adj.* (**-ier, -iest**) **1** of or like soap. **2** containing or smeared with soap. **3** unctuous, flattering. □ **soapily** *adv.* **soapiness** *n.*

soar *v.* **1** fly or rise high. **2** reach a high level or standard. **3** fly without flapping the wings or using power. [French *essorer*]

sob **—***v.* (**-bb-**) **1** inhale convulsively, usu. with weeping. **2** utter with sobs. **—***n.* act or sound of sobbing. [imitative]

sober **—***adj.* (**soberer, soberest**) **1** not drunk. **2** not given to drink. **3** moderate, tranquil, sedate, serious. **4** not exaggerated. **5** (of a colour etc.) quiet; dull. **—***v.* (often foll. by *down, up*) make or become sober. □ **soberly** *adv.* [French from Latin]

sobriety /sə'braɪɪtɪ/ *n.* being sober. [Latin: related to SOBER]

sobriquet /'səʊbrɪ,keɪ/ *n.* (also **soubriquet** /'su:-/) nickname. [French]

sob story *n. colloq.* story or explanation appealing for sympathy.

Soc. *abbr.* **1** Socialist. **2** Society.

so-called *adj.* commonly called, often incorrectly.

soccer /'sɒkə(r)/ *n.* Association football. [from Assoc.]

sociable /'səʊʃəb(ə)l/ *adj.* liking company, gregarious; friendly. □ **sociability** /-'bɪlɪtɪ/ *n.* **sociably** *adv.* [Latin *socius* companion]

social /'səʊʃ(ə)l/ **—***adj.* **1** of society or its organization, esp. of the relations of people or classes of people. **2** living in organized communities. **3** needing companionship; gregarious. **—***n.* social gathering, esp. of a club. □ **socially** *adv.* [Latin: related to SOCIABLE]

social climber *n.* person anxious to gain a higher social status.

social contract *n.* agreement between the State and population for mutual advantage.

social democracy *n.* political system favouring a mixed economy and democratic social change. □ **social democrat** *n.*

socialism *n.* **1** political and economic theory advocating State ownership and control of the means of production, distribution, and exchange. **2** social system based on this. □ **socialist** *n.* & *adj.* **socialistic** /-'lɪstɪk/ *adj.* [French: related to SOCIAL]

socialite /'səʊʃə,laɪt/ *n.* person moving in fashionable society.

socialize /'səʊʃə,laɪz/ *v.* (also **-ise**) (**-zing** or **-sing**) **1** mix socially. **2** make social. **3** organize on socialistic principles. □ **socialization** /-'zeɪʃ(ə)n/ *n.*

social science *n.* the study of society and social relationships. □ **social scientist** *n.*

social security *n.* State assistance to the poor and unemployed etc.

social services *n.pl.* welfare services provided by the State, esp. education, health, and housing.

social work *n.* professional or voluntary work with disadvantaged groups. □ **social worker** *n.*

society /sə'saɪətɪ/ *n.* (*pl.* **-ies**) **1** organized and interdependent community. **2** system and organization of this. **3** aristocratic part of this; its members (*polite society; society would not approve*). **4** mixing with others; companionship, company. **5** club, association (*music society; building society*). □ **societal** *adj.* [Latin *societas*]

Society of Friends *n.* pacifist Christian sect with no written creed or ordained ministers; Quakers.

Society of Jesus see JESUIT.

socio- *comb. form* of society or sociology (and) (*socio-economic*). [Latin: related to SOCIAL]

sociology /,səʊsɪ'ɒlədʒɪ/ *n.* the study of society and social problems. □ **sociological** /-ə'lɒdʒɪk(ə)l/ *adj.* **sociologist** *n.* [French: related to SOCIAL]

sock[1] *n.* **1** knitted covering for the foot and lower leg. **2** insole. □ **pull one's socks up** *colloq.* make an effort to improve. **put a sock in it** *slang* be quiet. [Old English *socc* from Greek *sukkhos* slipper]

sock[2] *colloq.* –*v.* hit hard. –*n.* hard blow. □ **sock it to** attack or address (a person or people) vigorously. [origin unknown]

socket /'sɒkɪt/ *n.* hollow for something to fit into etc., esp. a device receiving an electric plug, light-bulb, etc. [Anglo-French]

Socratic /sə'krætɪk/ *adj.* of Socrates or his philosophy.

Socratic irony *n.* pose of ignorance to entice others into refutable statements.

Socratic method *n.* dialectic, procedure by question and answer.

sod[1] *n.* **1** turf, piece of turf. **2** surface of the ground. [Low German or Dutch]

sod[2] *coarse slang* –*n.* **1** unpleasant or awkward person or thing. **2** fellow (*lucky sod*). –*v.* **(-dd-) 1** damn (*sod them!*). **2** (as **sodding** *adj.*) damned. □ **sod off** go away. [abbreviation of SODOMITE]

soda /'səʊdə/ *n.* **1** compound of sodium in common use. **2** (in full **soda water**) effervescent water used esp. with spirits etc. as a drink. [perhaps from Latin *sodanum* from Arabic]

soda bread *n.* bread leavened with baking-soda.

soda fountain *n.* **1** device supplying soda water. **2** shop or counter with this.

sodden /'sɒd(ə)n/ *adj.* **1** saturated; soaked through. **2** stupid or dull etc. with drunkenness. [archaic past part. of SEETHE]

sodium /'səʊdɪəm/ *n.* soft silver-white metallic element. [from SODA]

sodium bicarbonate *n.* white crystalline compound used in baking-powder.

sodium chloride *n.* common salt.

sodium hydroxide *n.* strongly alkaline compound used in soap etc.; caustic soda.

sodium lamp *n.* lamp using sodium vapour and giving a yellow light.

sodium nitrate *n.* white powdery compound used in fertilizers etc.

sodomite /'sɒdə,maɪt/ *n.* person who practises sodomy. [Greek: related to SODOMY]

sodomy /'sɒdəmɪ/ *n.* = BUGGERY. □ **sodomize** *v.* (also **-ise**) **(-zing** or **-sing)**. [Latin from *Sodom*: Gen. 18,19]

Sod's Law *n.* = MURPHY'S LAW.

soever /səʊ'evə(r)/ *adv. literary* of any kind; to any extent (*how great soever it may be*).

-soever *comb. form* of any kind; to any extent (*whatsoever*; *howsoever*).

sofa /'səʊfə/ *n.* long upholstered seat with a back and arms. [Arabic *shuffa*]

sofa bed *n.* sofa that can be converted into a bed.

soffit /'sɒfɪt/ *n.* undersurface of an arch, lintel, etc. [French *soffite*, Italian *soffitta*]

soft –*adj.* **1** not hard; easily cut or dented; malleable. **2** (of cloth etc.) smooth; fine; not rough. **3** (of wind etc.) mild, gentle. **4** (of water) low in mineral salts and lathering easily. **5** (of light or colour etc.) not brilliant or glaring. **6** (of sound) gentle, not loud. **7** (of a consonant) sibilant (as *c* in *ice*, *s* in *pleasure*). **8** (of an outline etc.) vague, blurred. **9** gentle, conciliatory. **10** compassionate, sympathetic. **11** feeble, half-witted, silly, sentimental. **12** *colloq.* (of a job etc.) easy. **13** (of drugs) not highly addictive. **14** (also **soft-core**) (of pornography) not highly obscene. **15** (of currency) likely to fall in value; not readily exchangeable into other currencies. –*adv.* softly. □ **be soft on** *colloq.* **1** be lenient towards. **2** be infatuated with. **have a soft spot for** be fond of. □ **softish** *adj.* **softly** *adv.* **softness** *n.* [Old English]

softball *n.* form of baseball using a softer and larger ball.

soft-boiled *adj.* (of an egg) boiled leaving the yolk soft.

soft-centred *adj.* **1** (of a sweet) having a soft centre. **2** soft-hearted; sentimental.

soft drink *n.* non-alcoholic drink.

soften /'sɒf(ə)n/ *v.* **1** make or become soft or softer. **2** (often foll. by *up*) **a** make weaker by preliminary attack. **b** make (a person) more receptive to persuasion. □ **softener** *n.*

soft fruit *n.* small stoneless fruit (a strawberry or currant).

soft furnishings *n.pl.* curtains, rugs, etc.

soft-hearted *adj.* tender, compassionate. □ **soft-heartedness** *n.*

softie *n.* (also **softy**) (*pl.* **-ies**) *colloq.* weak, silly, or soft-hearted person.

softly-softly *adj.* (also **softly, softly**) (of strategy) cautious and cunning.

soft option *n.* easier alternative.

soft palate *n.* rear part of the palate.

soft pedal –*n.* piano pedal that softens the tone. –*v.* **(soft-pedal) (-ll-;** *US* **-l-)** refrain from emphasizing; be restrained.

soft roe see ROE[1].

soft sell *n.* restrained salesmanship.

soft soap –*n. colloq.* persuasive flattery. –*v.* **(soft-soap)** *colloq.* persuade with flattery.

soft-spoken *adj.* having a gentle voice.

soft target *n.* vulnerable person or thing.

soft touch *n. colloq.* gullible person, esp. over money.
software *n.* programs for a computer.
softwood *n.* easily sawn wood of coniferous trees.
softy var. of SOFTIE.
soggy /'sɒɡɪ/ *adj.* (**-ier**, **-iest**) sodden, saturated; too moist (*soggy bread*). □ **sogginess** *n.* [dial. *sog* marsh]
soh /səʊ/ *n.* (also **so**) *Mus.* fifth note of a major scale. [Latin *solve*, word arbitrarily taken]
soigné /'swɑːnjeɪ/ *adj.* (*fem.* ***soignée*** pronunc. same) well-groomed. [French]
soil[1] *n.* **1** upper layer of earth in which plants grow. **2** ground belonging to a nation; territory (*on French soil*). [Latin *solium* seat, *solum* ground]
soil[2] *–v.* **1** make dirty; smear or stain. **2** defile; discredit. *–n.* **1** dirty mark. **2** filth; refuse. [French *soill(i)er*]
soil pipe *n.* discharge-pipe of a lavatory.
soirée /'swɑːreɪ/ *n.* evening party, usu. for conversation or music. [French]
soixante-neuf /ˌswɑːsɑ̃t'nɜːf/ *n. slang* mutual oral stimulation of the genitals. [French, = sixty-nine]
sojourn /'sɒdʒ(ə)n/ *–n.* temporary stay. *–v.* stay temporarily. [French *sojorner*]
sola /'səʊlə/ *n.* pithy-stemmed E. Indian swamp plant. [Urdu]
solace /'sɒləs/ *–n.* comfort in sadness, disappointment, or tedium. *–v.* (**-cing**) give solace to. [Latin *solatium*]
solan /'səʊlən/ *n.* (in full **solan goose**) large gooselike gannet. [Old Norse]
solar /'səʊlə(r)/ *adj.* of or reckoned by the sun. [Latin *sol* sun]
solar battery *n.* (also **solar cell**) device converting solar radiation into electricity.
solar day *n.* interval between meridian transits of the sun.
solarium /sə'leərɪəm/ *n.* (*pl.* **-ria**) room with sun-lamps or a glass roof etc. [Latin: related to SOLAR]
solar panel *n.* panel that absorbs the sun's rays as an energy source.
solar plexus *n.* complex of nerves at the pit of the stomach.
solar system *n.* the sun and the celestial bodies whose motion it governs.
solar year *n.* time taken for the earth to travel once round the sun.
sola topi *n.* sun-helmet made from the pith of the sola plant.
sold *past* and *past part.* of SELL. *–adj.* (foll. by *on*) *colloq.* enthusiastic about.
solder /'sɒldə(r), 'səʊ-/ *–n.* fusible alloy used to join metals or wires etc. *–v.* join with solder. [Latin: related to SOLID]
soldering iron *n.* heated tool for melting and applying solder.
soldier /'səʊldʒə(r)/ *–n.* **1** member of an army. **2** (in full **common soldier**) private or NCO in an army. **3** *colloq.* finger of bread for dipping into egg. *–v.* serve as a soldier. □ **soldier on** *colloq.* persevere doggedly. □ **soldierly** *adj.* [French *soulde*, originally = soldier's pay]
soldier of fortune *n.* mercenary.
soldiery *n.* soldiers, esp. of a specified character.
sole[1] *–n.* **1** undersurface of the foot. **2** part of a shoe, sock, etc., under the foot, esp. other than the heel. **3** lower surface or base of a plough, golf-club head, etc. *–v.* (**-ling**) provide (a shoe etc.) with a sole. □ **-soled** *adj.* (in *comb.*). [Latin *solea* sandal]
sole[2] *n.* (*pl.* same or **-s**) flat-fish used as food. [Latin *solea* sandal, which the shape of fish resembles]
sole[3] *adj.* one and only; single, exclusive. [French from Latin *solus*]
solecism /'sɒlɪˌsɪz(ə)m/ *n.* **1** mistake of grammar or idiom. **2** offence against etiquette. □ **solecistic** /-'sɪstɪk/ *adj.* [Greek *soloikos* speaking incorrectly]
solely /'səʊllɪ/ *adv.* **1** alone (*solely responsible*). **2** only (*did it solely out of duty*).
solemn /'sɒləm/ *adj.* **1** serious and dignified. **2** formal. **3** awe-inspiring. **4** (of a person) serious or cheerless in manner. **5** grave, sober (*solemn promise*). □ **solemnly** *adv.* **solemness** *n.* [Latin *solemnis*]
solemnity /sə'lemnɪtɪ/ *n.* (*pl.* **-ies**) **1** being solemn. **2** rite, ceremony.
solemnize /'sɒləmˌnaɪz/ *v.* (also **-ise**) (**-zing** or **-sing**) **1** duly perform (esp. a marriage ceremony). **2** make solemn. □ **solemnization** /-'zeɪʃ(ə)n/ *n.*
solenoid /'səʊləˌnɔɪd/ *n.* cylindrical coil of wire acting as a magnet when carrying electric current. [French from Greek *sōlēn* tube]
sol-fa /'sɒlfɑː/ *n.* system of syllables representing musical notes. [*sol* var. of SOH, FA]
soli *pl.* of SOLO.
solicit /sə'lɪsɪt/ *v.* (**-t-**) **1** seek (esp. business) repeatedly or earnestly. **2** (also *absol.*) accost as a prostitute. □ **solicitation** /-'teɪʃ(ə)n/ *n.* [Latin *sollicitus* anxious]
solicitor *n.* lawyer qualified to advise clients and instruct barristers. [French: related to SOLICIT]
Solicitor-General *n.* (*pl.* **Solicitors-General**) law officer below the Attorney-General or the Lord Advocate.
solicitous *adj.* **1** showing interest or concern. **2** (foll. by *to* + infin.) eager,

anxious. □ **solicitously** *adv.* [Latin: related to SOLICIT]

solicitude *n.* being solicitous. [Latin: related to SOLICITOUS]

solid /ˈsɒlɪd/ *–adj.* (**-er, -est**) **1** firm and stable in shape; not liquid or fluid. **2** of such material throughout, not hollow. **3** of the same substance throughout (*solid silver*). **4** sturdily built; not flimsy or slender. **5 a** three-dimensional. **b** of solids (*solid geometry*). **6 a** sound, reliable (*solid arguments*). **b** dependable (*solid friend*). **7** sound but unexciting (*solid piece of work*). **8** financially sound. **9** uninterrupted (*four solid hours*). **10** unanimous, undivided. **11** (of printing) without spaces. *–n.* **1** solid substance or body. **2** (in *pl.*) solid food. **3** *Geom.* three-dimensional body or magnitude. *–adv.* solidly (*jammed solid*). □ **solidly** *adv.* **solidness** *n.* [Latin *solidus*]

solidarity /ˌsɒlɪˈdærɪtɪ/ *n.* **1** unity, esp. political or in an industrial dispute. **2** mutual dependence. [French: related to SOLID]

solidify /səˈlɪdɪˌfaɪ/ *v.* (**-ies, -ied**) make or become solid. □ **solidification** /-fɪˈkeɪʃ(ə)n/ *n.*

solidity /səˈlɪdɪtɪ/ *n.* being solid; firmness.

solid-state *adj.* using the electronic properties of solids (e.g. a semiconductor) to replace those of valves.

solidus /ˈsɒlɪdəs/ *n.* (*pl.* **solidi** /-ˌdaɪ/) oblique stroke (/). [Latin: related to SOLID]

soliloquy /səˈlɪləkwɪ/ *n.* (*pl.* **-quies**) **1** talking without or regardless of hearers, esp. in a play. **2** this part of a play. □ **soliloquist** *n.* **soliloquize** *v.* (also **-ise**) (**-zing** or **-sing**). [Latin *solus* alone, *loquor* speak]

solipsism /ˈsɒlɪpˌsɪz(ə)m/ *n.* philosophical theory that the self is all that exists or can be known. □ **solipsist** *n.* [Latin *solus* alone, *ipse* self]

solitaire /ˌsɒlɪˈteə(r)/ *n.* **1** jewel set by itself. **2** ring etc. with this. **3** game for one player in which pegs etc. are removed from a board by jumping others over them. **4** *US* = PATIENCE 3. [French: see SOLITARY]

solitary /ˈsɒlɪtərɪ/ *–adj.* **1** living or being alone; not gregarious; lonely. **2** secluded. **3** single, sole. *–n.* (*pl.* **-ies**) **1** recluse. **2** *colloq.* = SOLITARY CONFINEMENT. □ **solitariness** *n.* [Latin *solitarius* from *solus* alone]

solitary confinement *n.* isolation in a separate prison cell.

solitude /ˈsɒlɪˌtju:d/ *n.* **1** being solitary. **2** lonely place. [Latin *solitudo*: related to SOLITARY]

solo /ˈsəʊləʊ/ *–n.* (*pl.* **-s**) **1** (*pl.* **-s** or **soli** /-lɪ/) musical piece or passage, or a dance, performed by one person. **2** thing done by one person, esp. an unaccompanied flight. **3** (in full **solo whist**) type of whist in which one player may oppose the others. *–v.* (**-es, -ed**) perform a solo. *–adv.* unaccompanied, alone. [Italian from Latin: related to SOLE[3]]

soloist /ˈsəʊləʊɪst/ *n.* performer of a solo, esp. in music.

Solomon's seal *n.* flowering plant with drooping green and white flowers. [*Solomon*, king of Israel]

solstice /ˈsɒlstɪs/ *n.* either of the times when the sun is furthest from the equator. [Latin *solstitium* 'the sun standing still']

soluble /ˈsɒljʊb(ə)l/ *adj.* **1** that can be dissolved, esp. in water. **2** solvable. □ **solubility** /-ˈbɪlɪtɪ/ *n.* [Latin *solvo solut-* release]

solute /ˈsɒlju:t/ *n.* dissolved substance.

solution /səˈlu:ʃ(ə)n/ *n.* **1** solving or means of solving a problem. **2 a** conversion of a solid or gas into a liquid by mixture with a liquid. **b** state resulting from this. **3** dissolving or being dissolved.

solve *v.* (**-ving**) answer, remove, or effectively deal with (a problem). □ **solvable** *adj.*

solvent *–adj.* **1** able to pay one's debts; not in debt. **2** able to dissolve or form a solution with something. *–n.* solvent liquid etc. □ **solvency** *n.* (in sense 2 of *adj.*).

somatic /səˈmætɪk/ *adj.* of the body, not of the mind. □ **somatically** *adv.* [Greek *sōma -mat-* body]

sombre /ˈsɒmbə(r)/ *adj.* (also *US* **somber**) dark, gloomy, dismal. □ **sombrely** *adv.* **sombreness** *n.* [Latin *sub ombra* under shade]

sombrero /sɒmˈbreərəʊ/ *n.* (*pl.* **-s**) broad-brimmed hat worn esp. in Latin America. [Spanish: related to SOMBRE]

some /sʌm/ *–adj.* **1** unspecified amount or number of (*some water*; *some apples*; *some of them*). **2** unknown or unspecified (*some day*; *some fool broke it*). **3** approximately (*some ten days*). **4** considerable (*went to some trouble*; *at some cost*). **5** (usu. stressed) **a** at least a modicum of (*have some consideration*). **b** such up to a point (*that is some help*). **c** *colloq.* remarkable (*I call that some story*). *–pron.* some people or things, some number or amount (*I have some already*). *–adv. colloq.* to some extent (*do it some more*). [Old English]

-some[1] *suffix* forming adjectives meaning: **1** producing (*fearsome*). **2** characterized by being (*gladsome*). **3** apt to (*tiresome*; *meddlesome*). **4** suitable for (*cuddlesome*). [Old English]

-some[2] *suffix* forming nouns from numerals, meaning 'a group of' (*foursome*). [Old English]

somebody –*pron.* some person. –*n.* (*pl.* **-ies**) important person.

some day *adv.* at some time in the future.

somehow *adv.* **1** for some reason or other (*somehow I don't trust him*). **2** in some way; by some means.

someone *n.* & *pron.* = SOMEBODY.

someplace *adv. US colloq.* = SOMEWHERE.

somersault /'sʌmə,sɒlt/ –*n.* leap or roll with the body turning through a circle. –*v.* perform this. [French *sobre* above, *saut* jump]

something *n.* & *pron.* **1** unspecified or unknown thing (*something has happened*). **2** unexpressed or intangible quantity, quality, or extent (*something strange about it*). **3** *colloq.* notable person or thing. □ **something else** *colloq.* something exceptional. **something like** approximately. **something of** to some extent (*something of an expert*). [Old English: related to SOME, THING]

sometime –*adv.* **1** at some time. **2** formerly. –*attrib. adj.* former.

sometimes *adv.* occasionally.

somewhat *adv.* to some extent.

somewhen *adv. colloq.* at some time.

somewhere –*adv.* in or to some place. –*pron.* some unspecified place.

somnambulism /sɒm'næmbjʊ,lɪz(ə)m/ *n.* sleepwalking. □ **somnambulant** *adj.* **somnambulist** *n.* [Latin *somnus* sleep, *ambulo* walk]

somnolent /'sɒmnələnt/ *adj.* **1** sleepy, drowsy. **2** inducing drowsiness. □ **somnolence** *n.* [Latin: related to SOMNAMBULISM]

son /sʌn/ *n.* **1** boy or man in relation to his parent(s). **2** male descendant. **3** (foll. by *of*) male member of a family, etc. **4** male descendent or inheritor of a quality etc. (*sons of freedom*). **5** form of address, esp. to a boy. [Old English]

sonar /'səʊnɑ:(r)/ *n.* **1** system for the underwater detection of objects by reflected sound. **2** apparatus for this. [*so*und *na*vigation and *r*anging]

sonata /sə'nɑ:tə/ *n.* composition for one or two instruments, usu. in three or four movements. [Italian, = sounded]

sonatina /,sɒnə'ti:nə/ *n.* simple or short sonata. [Italian, diminutive of SONATA]

son et lumière /,sɒneɪ'lu:mjeə(r)/ *n.* entertainment by night at a historic building etc., using lighting effects and recorded sound to give a dramatic narrative of its history. [French, = sound and light]

song *n.* **1** words set to music or meant to be sung. **2** vocal music. **3** musical composition suggestive of a song. **4** cry of some birds. □ **for a song** *colloq.* very cheaply. [Old English: related to SING]

song and dance *n. colloq.* fuss, commotion.

songbird *n.* bird with a musical call.

songbook *n.* book of song lyrics and music.

song cycle *n.* set of linked songs.

songster *n.* (*fem.* **songstress**) **1** singer. **2** songbird.

song thrush *n.* common thrush, noted for singing.

songwriter *n.* writer of songs or the music for them.

sonic /'sɒnɪk/ *adj.* of or using sound or sound waves. [Latin *sonus* sound]

sonic bang *n.* (also **sonic boom**) noise made when an aircraft passes the speed of sound.

sonic barrier *n.* = SOUND BARRIER.

son-in-law *n.* (*pl.* **sons-in-law**) daughter's husband.

sonnet /'sɒnɪt/ *n.* poem of 14 lines with a fixed rhyme-scheme and, in English, usu. ten syllables per line. [French *sonnet* or Italian *sonetto*]

sonny /'sʌnɪ/ *n. colloq.* familiar form of address to a young boy.

sonorous /'sɒnərəs/ *adj.* **1** having a loud, full, or deep sound; resonant. **2** (of language, style, etc.) imposing. □ **sonority** /sə'nɒrɪtɪ/ *n.* [Latin]

soon *adv.* **1** in a short time (*shall soon know*). **2** relatively early (*must you go so soon?*). **3** readily or willingly (*would sooner go*; *would as soon stay*). □ **as** (or **so**) **soon as** at the moment that; not later than; as early as (*came as soon as I could*). **sooner or later** at some future time; eventually. □ **soonish** *adv.* [Old English]

soot /sʊt/ *n.* black powdery deposit from smoke. [Old English]

sooth *n. archaic* truth. [Old English]

soothe /su:ð/ *v.* (**-thing**) **1** calm (a person, feelings, etc). **2** soften or mitigate (pain etc.). [Old English]

soothsayer /'su:θ,seɪə(r)/ *n.* seer, prophet.

sooty *adj.* (**-ier**, **-iest**) **1** covered with soot. **2** black or brownish-black.

sop –*n.* **1** thing given or done to pacify or bribe. **2** piece of bread etc. dipped in gravy etc. –*v.* (**-pp-**) **1** (as **sopping** *adj.*) drenched (*came home sopping*; *sopping wet clothes*). **2** (foll. by *up*) soak or mop up. [Old English]

sophism /'sɒfɪz(ə)m/ *n.* false argument, esp. one intended to deceive. [Greek *sophos* wise]

sophist /'sɒfɪst/ *n.* captious or clever but fallacious reasoner. □ **sophistic** /sə'fɪstɪk/ *adj.* [Greek: related to SOPHISM]

sophisticate /sə'fɪstɪkət/ *n.* sophisticated person. [medieval Latin: related to SOPHISM]

sophisticated /sə'fɪstɪ,keɪtɪd/ *adj.* **1** (of a person) worldly-wise; cultured; elegant. **2** (of a thing, idea, etc.) highly developed and complex. □ **sophistication** /-'keɪʃ(ə)n/ *n.*

sophistry /'sɒfɪstrɪ/ *n.* (*pl.* **-ies**) **1** use of sophisms. **2** a sophism.

sophomore /'sɒfə,mɔ:(r)/ *n. US* second-year university or high-school student. [*sophum*, obsolete var. of SOPHISM]

soporific /,sɒpə'rɪfɪk/ *–adj.* inducing sleep. *–n.* soporific drug or influence. □ **soporifically** *adv.* [Latin *sopor* sleep]

sopping see SOP.

soppy *adj.* (**-ier**, **-iest**) *colloq.* mawkishly sentimental; silly; infatuated. □ **soppily** *adv.* **soppiness** *n.* [from SOP]

soprano /sə'prɑ:nəʊ/ *n.* (*pl.* **-s**) **1 a** highest singing-voice. **b** female or boy singer with this voice. **2** instrument of a high or the highest pitch in its family. [Italian *sopra* above]

sorbet /'sɔ:beɪ/ *n.* **1** water-ice. **2** sherbet. [Arabic *sharba* to drink]

sorcerer /'sɔ:sərə(r)/ *n.* (*fem.* **sorceress**) magician, wizard. □ **sorcery** *n.* (*pl.* **-ies**). [French *sourcier*: related to SORT]

sordid /'sɔ:dɪd/ *adj.* **1** dirty, squalid. **2** ignoble, mercenary. □ **sordidly** *adv.* **sordidness** *n.* [Latin *sordidus*]

sore *–adj.* **1** (of a part of the body) painful. **2** suffering pain. **3** aggrieved, vexed. **4** *archaic* grievous or severe (*in sore need*). *–n.* **1** inflamed place on the skin or flesh. **2** source of distress or annoyance. *–adv. archaic* grievously, severely. □ **soreness** *n.* [Old English]

sorely *adv.* extremely (*sorely tempted*; *sorely vexed*).

sore point *n.* subject causing distress or annoyance.

sorghum /'sɔ:gəm/ *n.* tropical cereal grass. [Italian *sorgo*]

sorority /sə'rɒrɪtɪ/ *n.* (*pl.* **-ies**) *US* female students' society in a university or college. [Latin *soror* sister]

sorrel[1] /'sɒr(ə)l/ *n.* sour-leaved herb. [Germanic: related to SOUR]

sorrel[2] /'sɒr(ə)l/ *–adj.* of a light reddish-brown colour. *–n.* **1** this colour. **2** sorrel animal, esp. a horse. [French]

sorrow /'sɒrəʊ/ *–n.* **1** mental distress caused by loss or disappointment etc. **2** cause of sorrow. *–v.* feel sorrow, mourn. [Old English]

sorrowful *adj.* feeling, causing, or showing sorrow. □ **sorrowfully** *adv.*

sorry /'sɒrɪ/ *–adj.* (**-ier**, **-iest**) **1** pained, regretful, penitent (*sorry about the mess*). **2** (foll. by *for*) feeling pity or sympathy for. **3** (*attrib.*) wretched (*a sorry sight*). *–int.* expression of apology. □ **sorry for oneself** dejected. [Old English: related to SORE]

sort *–n.* **1** group of similar things etc.; class or kind. **2** *colloq.* person of a specified kind (*a good sort*). *–v.* (often foll. by *out*, *over*) arrange systematically; put in order. □ **of a sort** (or **of sorts**) *colloq.* barely deserving the name (*a holiday of sorts*). **out of sorts** slightly unwell; in low spirits. **sort of** *colloq.* as it were; to some extent. **sort out 1** separate into sorts. **2** select from a varied group. **3** disentangle or put into order. **4** solve. **5** *colloq.* deal with or reprimand. [Latin *sors sort-* lot]

sortie /'sɔ:tɪ/ *–n.* **1** sally, esp. from a besieged garrison. **2** operational military flight. *–v.* (**-ties**, **-tied**, **-tieing**) make a sortie. [French]

SOS /,esəʊ'es/ *n.* (*pl.* **SOSs**) **1** international code-signal of extreme distress. **2** urgent appeal for help. [letters easily recognized in Morse]

sostenuto /,sɒstə'nu:təʊ/ *Mus. –adv.* & *adj.* in a sustained or prolonged manner. *–n.* (*pl.* **-s**) passage to be played in this way. [Italian]

sot *n.* habitual drunkard. □ **sottish** *adj.* [Old English]

sotto voce /,sɒtəʊ 'vəʊtʃɪ/ *adv.* in an undertone. [Italian]

sou /su:/ *n.* **1** *colloq.* very small sum of money. **2** *hist.* former French coin of low value. [French from Latin: related to SOLID]

soubrette /su:'bret/ *n.* **1** pert maidservant etc. in a comedy. **2** actress taking this part. [French]

soubriquet var. of SOBRIQUET.

soufflé /'su:fleɪ/ *n.* light spongy sweet or savoury dish usu. made with stiffly beaten egg-whites and gelatine. [French, = blown]

sough /saʊ, sʌf/ *–v.* moan or whisper like the wind in trees etc. *–n.* this sound. [Old English]

sought *past* and *past part.* of SEEK.

sought-after *adj.* generally desired.

souk /su:k/ *n.* market-place in Muslim countries. [Arabic]

soul /səʊl/ *n.* **1** spiritual or immaterial part of a person, often regarded as immortal. **2** moral, emotional, or intellectual nature of a person. **3** personification or pattern (*the very soul of discretion*). **4** an individual (*not a soul in sight*). **5** person regarded with familiarity or pity etc. (*the poor soul; a good soul*). **6** person regarded as an animating or essential part (*life and soul*). **7** energy or intensity, esp. in a work of art. **8** = SOUL MUSIC. □ **upon my soul** exclamation of surprise. [Old English]

soul-destroying *adj.* (of an activity etc.) tedious, monotonous.

soul food *n.* traditional food of American Blacks.

soulful *adj.* having, expressing, or evoking deep feeling. □ **soulfully** *adv.*

soulless *adj.* **1** lacking sensitivity or noble qualities. **2** undistinguished or uninteresting.

soul mate *n.* person ideally suited to another.

soul music *n.* Black American music with rhythm and blues, gospel, and rock elements.

soul-searching *n.* introspection.

sound[1] *–n.* **1** sensation caused in the ear by the vibration of the surrounding air or other medium. **2** vibrations causing this sensation. **3** what is or may be heard. **4** idea or impression conveyed by words (*don't like the sound of that*). **5** mere words. *–v.* **1** (cause to) emit sound. **2** utter, pronounce (*sound a warning note*). **3** convey an impression when heard (*sounds worried*). **4** give an audible signal for (an alarm etc.). **5** test (the lungs etc.) by the sound produced. □ **sound off** talk loudly or express one's opinions forcefully. □ **soundless** *adj.* [Latin *sonus*]

sound[2] *–adj.* **1** healthy; not diseased, injured, or rotten. **2** (of an opinion, policy, etc.) correct, well-founded. **3** financially secure. **4** undisturbed (*sound sleeper*). **5** thorough (*sound thrashing*). *–adv.* soundly (*sound asleep*). □ **soundly** *adv.* **soundness** *n.* [Old English]

sound[3] *v.* **1** test the depth or quality of the bottom of (the sea or a river etc.). **2** (often foll. by *out*) inquire (esp. discreetly) into the opinions or feelings of (a person). [French *sonder* from Latin *sub unda* under the wave]

sound[4] *n.* strait (of water). [Old English, = swimming]

sound barrier *n.* high resistance of air to objects moving at speeds near that of sound.

sound bite *n.* short pithy extract from an interview, speech, etc., as part of a news etc. broadcast.

soundbox *n.* the hollow body of a stringed musical instrument, providing resonance.

sound effect *n.* sound other than speech or music made artificially for a film, broadcast, etc.

sounding *n.* **1** measurement of the depth of water. **2** (in *pl.*) region close enough to the shore for sounding. **3** (in *pl.*) cautious investigation.

sounding-balloon *n.* balloon used to obtain information about the upper atmosphere.

sounding-board *n.* **1 a** person etc. used to test opinion. **b** means of disseminating opinions etc. **2** canopy directing sound towards an audience.

sounding-line *n.* line used in sounding.

sounding-rod *n.* rod used in sounding water in a ship's hold.

soundproof *–adj.* impervious to sound. *–v.* make soundproof.

sound system *n.* equipment for sound reproduction.

soundtrack *n.* **1** the sound element of a film or videotape. **2** recording of this made available separately. **3** any single track in a multi-track recording.

sound wave *n.* wave of compression and rarefaction, by which sound is transmitted in the air etc.

soup /su:p/ *–n.* liquid food made by boiling meat, fish, or vegetables. *–v.* (usu. foll. by *up*) *colloq.* **1** increase the power of (an engine). **2** enliven (*a souped-up version of the original*). □ **in the soup** *colloq.* in difficulties. [French]

soupçon /'su:psɔ̃/ *n.* small quantity; trace. [French: related to SUSPICION]

soup-kitchen *n.* place dispensing soup etc. to the poor.

soup-plate *n.* deep wide-rimmed plate.

soup-spoon *n.* large round-bowled spoon.

soupy *adj.* (**-ier, -iest**) **1** like soup. **2** sentimental.

sour *–adj.* **1** acid in taste or smell, esp. because unripe or fermented. **2** morose; bitter. **3** (of a thing) unpleasant; distasteful. **4** (of the soil) dank. *–v.* make or become sour. □ **go** (or **turn**) **sour 1** turn out badly. **2** lose one's keenness. □ **sourly** *adv.* **sourness** *n.* [Old English]

source /sɔ:s/ *n.* **1** place from which a river or stream issues. **2** place of origination. **3** person or document etc. providing information. □ **at source** at the point of origin or issue. [French: related to SURGE]

sour grapes *n.pl.* resentful disparagement of something one covets.

sourpuss *n. colloq.* sour-tempered person.

souse /saʊs/ *–v.* (**-sing**) **1** immerse in pickle or other liquid. **2** (as **soused** *adj.*) *colloq.* drunk. **3** (usu. foll. by *in*) soak (a thing) in liquid. *–n.* **1 a** pickle made with salt. **b** *US* food in pickle. **2** plunge or drench in water. [French *sous*]

soutane /suːˈtɑːn/ *n.* cassock worn by a Roman Catholic priest. [French from Italian *sotto* under]

south *–n.* **1** point of the horizon 90° clockwise from east. **2** compass point corresponding to this. **3** direction in which this lies. **4** (usu. **the South**) part of the world, a country, or a town to the south. *–adj.* **1** towards, at, near, or facing the south. **2** from the south (*south wind*). *–adv.* **1** towards, at, or near the south. **2** (foll. by *of*) further south than. □ **to the south** (often foll. by *of*) in a southerly direction. [Old English]

South African *–adj.* of the republic of South Africa. *–n.* **1** native or national of South Africa. **2** person of South African descent.

South American *–adj.* of South America. *–n.* native or national of a South American country.

southbound *adj.* travelling or leading southwards.

South-East *n.* part of a country or town to the south-east.

south-east *–n.* **1** point of the horizon midway between south and east. **2** direction in which this lies. *–adj.* of, towards, or coming from the south-east. *–adv.* towards, at, or near the south-east.

southeaster /saʊθˈiːstə(r)/ *n.* south-east wind.

south-easterly *adj.* & *adv.* = SOUTH-EAST.

south-eastern *adj.* on the south-east side.

southerly /ˈsʌðəlɪ/ *–adj.* & *adv.* **1** in a southern position or direction. **2** (of a wind) from the south. *–n.* (*pl.* **-ies**) such a wind.

southern /ˈsʌð(ə)n/ *adj.* of or in the south. □ **southernmost** *adj.*

Southern Cross *n.* southern constellation in the shape of a cross.

southerner *n.* native or inhabitant of the south.

Southern hemisphere *n.* the half of the earth south of the equator.

southern lights *n.pl.* aurora australis.

southpaw *colloq.* *–n.* left-handed person, esp. in boxing. *–adj.* left-handed.

south pole see POLE[2].

South Sea *n.* (also **South Seas**) southern Pacific Ocean.

south-south-east *n.* point or direction midway between south and south-east.

south-south-west *n.* point or direction midway between south and south-west.

southward /ˈsaʊθwəd/ *–adj.* & *adv.* (also **southwards**) towards the south. *–n.* southward direction or region.

South-West *n.* part of a country or town to the south-west.

south-west *–n.* **1** point of the horizon midway between south and west. **2** direction in which this lies. *–adj.* of, towards, or coming from the south-west. *–adv.* towards, at, or near the south-west.

southwester /saʊθˈwestə(r)/ *n.* south-west wind.

south-westerly *adj.* & *adv.* = SOUTH-WEST.

south-western *adj.* on the south-west side.

souvenir /ˌsuːvəˈnɪə(r)/ *n.* memento of an occasion, place, etc. [French]

sou'wester /saʊˈwestə(r)/ *n.* **1** waterproof hat with a broad flap covering the neck. **2** south-west wind. [from SOUTH-WESTER]

sovereign /ˈsɒvrɪn/ *–n.* **1** supreme ruler, esp. a monarch. **2** *hist.* British gold coin nominally worth £1. *–adj.* **1** supreme (*sovereign power*). **2** self-governing (*sovereign State*). **3** royal (*our sovereign lord*). **4** excellent; effective (*sovereign remedy*). **5** unmitigated (*sovereign contempt*). [French *so(u)verain*: *-g-* by association with *reign*]

sovereignty *n.* (*pl.* **-ies**) **1** supremacy. **2 a** self-government. **b** self-governing State.

Soviet /ˈsəʊvɪət, ˈsɒ-/ *hist.* *–adj.* of the USSR or its people. *–n.* **1** citizen of the USSR. **2** (**soviet**) elected council in the USSR. **3** revolutionary council of workers, peasants, etc. [Russian]

sow[1] /səʊ/ *v.* (*past* **sowed**; *past part.* **sown** or **sowed**) **1** (also *absol.*) **a** scatter (seed) on or in the earth. **b** (often foll. by *with*) plant with seed. **2** initiate (*sow hatred*). □ **sow one's wild oats** indulge in youthful excess or promiscuity. [Old English]

sow[2] /saʊ/ *n.* adult female pig. [Old English]

soy *n.* **1** (in full **soy sauce**) sauce from pickled soya beans. **2** (in full **soy bean**) = SOYA 1. [Japanese]

soya /ˈsɔɪə/ *n.* **1** (in full **soya bean**) **a** leguminous plant yielding edible oil and flour and used to replace animal protein. **b** seed of this. **2** (in full **soya sauce**) = SOY 1. [Malay: related to SOY]

sozzled /ˈsɒz(ə)ld/ *adj.* *colloq.* very drunk. [dial. *sozzle* mix sloppily, imitative]

spa /spɑː/ *n.* **1** curative mineral spring. **2** resort with this. [*Spa* in Belgium]

space *–n.* **1 a** continuous expanse in which things exist and move. **b** amount of this taken by a thing or available. **2** interval between points or objects. **3** empty area (*make a space*). **4 a** outdoor urban recreation area (*open space*). **b** large unoccupied region (*wide open spaces*). **5** = OUTER SPACE. **6** interval of time (*in the space of an hour*). **7** amount of paper used in writing, available for advertising, etc. **8 a** blank between printed, typed, or written words, etc. **b** piece of metal providing this. **9** freedom to think, be oneself, etc. (*need my own space*). *–v.* (**-cing**) **1** set or arrange at intervals. **2** put spaces between. **3** (as **spaced** *adj.*) (often foll. by *out*) *slang* euphoric, esp. from taking drugs. □ **space out** spread out (more) widely. □ **spacer** *n.* [Latin *spatium*]

space age *–n.* era of space travel. *–attrib. adj.* (**space-age**) very modern.

spacecraft *n.* vehicle for travelling in outer space.

Space Invaders *n.pl.* computer game in which players combat aliens.

spaceman *n.* (*fem.* **spacewoman**) astronaut.

space-saving *adj.* occupying little space or helping to save space.

spaceship *n.* spacecraft.

space shuttle *n.* spacecraft for repeated use, esp. between the earth and a space station.

space station *n.* artificial satellite as a base for operations in outer space.

spacesuit *n.* sealed pressurized suit for an astronaut in outer space.

space–time *n.* fusion of the concepts of space and time as a four-dimensional continuum.

spacious /ˈspeɪʃəs/ *adj.* having ample space; roomy. □ **spaciously** *adv.* **spaciousness** *n.* [Latin: related to SPACE]

spade[1] *n.* long-handled digging tool with a broad sharp-edged metal blade. □ **call a spade a spade** speak bluntly. □ **spadeful** *n.* (*pl.* **-s**). [Old English]

spade[2] *n.* **1 a** playing-card of a suit denoted by black inverted heart-shaped figures with short stalks. **b** (in *pl.*) this suit. **2** *slang offens.* Black person. [Italian *spada* sword: related to SPADE[1]]

spadework *n.* hard preparatory work.

spaghetti /spəˈgetɪ/ *n.* pasta in long thin strands. [Italian]

spaghetti Bolognese /ˌbɒləˈneɪz/ *n.* spaghetti with a meat and tomato sauce.

spaghetti junction *n.* multi-level road junction, esp. on a motorway.

spaghetti western *n.* cowboy film made by Italians, esp. cheaply.

Spam *n. propr.* tinned meat made from ham. [*sp*iced h*am*]

span[1] *–n.* **1** full extent from end to end. **2** each part of a bridge between supports. **3** maximum lateral extent of an aeroplane or its wing, or a bird's wing, etc. **4 a** maximum distance between the tips of the thumb and little finger. **b** this as a measure of 9 in. *–v.* (**-nn-**) **1** stretch from side to side of; extend across. **2** bridge (a river etc.). [Old English]

span[2] see SPICK AND SPAN.

spandrel /ˈspændrɪl/ *n.* space between the curve of an arch and the surrounding rectangular moulding, or between the curves of adjoining arches and the moulding above. [origin uncertain]

spangle /ˈspæŋg(ə)l/ *–n.* small piece of glittering material, esp. one of many used to ornament a dress etc.; sequin. *–v.* (**-ling**) (esp. as **spangled** *adj.*) cover with or as with spangles (*star-spangled*). [obsolete *spang* from Dutch]

Spaniard /ˈspænjəd/ *n.* **1** native or national of Spain. **2** person of Spanish descent. [French *Espaigne* Spain]

spaniel /ˈspænj(ə)l/ *n.* dog of a breed with a long silky coat and drooping ears. [French *espaigneul* Spanish (dog)]

Spanish /ˈspænɪʃ/ *–adj.* of Spain, its people, or language. *–n.* **1** the language of Spain and Spanish America. **2** (**the Spanish**) (*pl.*) the people of Spain. [*Spain* in Europe]

Spanish Main *n. hist.* NE coast of S. America and adjoining parts of the Caribbean Sea.

Spanish omelette *n.* omelette with chopped vegetables in the mix.

Spanish onion *n.* large mild-flavoured onion.

spank *–v.* **1** slap, esp. on the buttocks as punishment. **2** (of a horse etc.) move briskly. *–n.* slap, esp. on the buttocks. [imitative]

spanker *n. Naut.* fore-and-aft sail set on the after side of the mizen-mast.

spanking *–adj.* **1** brisk. **2** *colloq.* striking; excellent. *–adv. colloq.* very (*spanking new*). *–n.* slapping on the buttocks.

spanner *n.* tool for turning a nut on a bolt etc. [German]

spanner in the works *n. colloq.* impediment.

spar[1] *n.* **1** stout pole, esp. as a ship's mast etc. **2** main longitudinal beam of an aeroplane wing. [Old Norse *sperra* or French *esparre*]

spar[2] *–v.* (**-rr-**) **1** make the motions of boxing without heavy blows. **2** argue. *–n.* **1** sparring motion. **2** boxing-match. [Old English]

spar[3] *n.* easily split crystalline mineral. [Low German]

spare –*adj.* **1 a** not required for normal or immediate use; extra. **b** for emergency or occasional use. **2** lean; thin. **3** frugal. –*n.* spare part. –*v.* (**-ring**) **1** afford to give, do without; dispense with (*spared me ten minutes*). **2 a** refrain from killing, hurting, etc. **b** abstain from inflicting (*spare me this task*). **3** be frugal or grudging of (*no expense spared*). □ **go spare** *colloq.* become very angry or distraught. **not spare oneself** exert one's utmost efforts. **spare a person's life** not kill him or her. **to spare** left over; additional (*an hour to spare*). □ **sparely** *adv.* **spareness** *n.* [Old English]

spare part *n.* duplicate, esp. as a replacement.

spare-rib *n.* closely-trimmed ribs of esp. pork. [Low German *ribbesper*, associated with SPARE]

spare time *n.* leisure.

spare tyre *n. colloq.* roll of fat round the waist.

sparing *adj.* **1** frugal; economical. **2** restrained. □ **sparingly** *adv.*

spark –*n.* **1** fiery particle thrown from a fire, alight in ashes, or produced by a flint, match, etc. **2** (often foll. by *of*) small amount (*spark of interest*). **3 a** flash of light between electric conductors etc. **b** this serving to ignite the explosive mixture in an internal-combustion engine. **4 a** flash of wit etc. **b** (also **bright spark**) witty or lively person. –*v.* **1** emit a spark or sparks. **2** (often foll. by *off*) stir into activity; initiate. □ **sparky** *adj.* [Old English]

sparkle /'spɑ:k(ə)l/ –*v.* (**-ling**) **1 a** emit or seem to emit sparks; glitter, glisten. **b** be witty; scintillate. **2** (of wine etc.) effervesce. –*n.* **1** glitter. **2** lively quality (*the song lacks sparkle*). □ **sparkly** *adj.*

sparkler *n.* **1** hand-held sparkling firework. **2** *colloq.* diamond.

spark-plug *n.* (also **sparking-plug**) device for making a spark in an internal-combustion engine.

sparring partner *n.* **1** boxer employed to spar with another as training. **2** person with whom one enjoys arguing.

sparrow /'spærəʊ/ *n.* small brownish-grey bird. [Old English]

sparrowhawk *n.* small hawk.

sparse *adj.* thinly dispersed or scattered. □ **sparsely** *adv.* **sparseness** *n.* **sparsity** *n.* [Latin *spargo spars-* scatter]

Spartan /'spɑ:t(ə)n/ –*adj.* **1** of Sparta in ancient Greece. **2** austere, rigorous, frugal. –*n.* citizen of Sparta. [Latin]

spasm /'spæz(ə)m/ *n.* **1** sudden involuntary muscular contraction. **2** convulsive movement or emotion etc. **3** (usu. foll. by *of*) *colloq.* brief spell. [Greek *spasma* from *spaō* pull]

spasmodic /spæz'mɒdɪk/ *adj.* of or in spasms, intermittent. □ **spasmodically** *adv.* [Greek: related to SPASM]

spastic /'spæstɪk/ –*adj.* of or having cerebral palsy. –*n.* **1** spastic person. **2** *slang offens.* stupid or incompetent person. [Greek: related to SPASM]

spat[1] *past* and *past part.* of SPIT[1].

spat[2] *n.* (usu. in *pl.*) *hist.* short gaiter covering a shoe. [abbreviation of *spatterdash*: related to SPATTER]

spat[3] *n. colloq.* petty or brief quarrel. [probably imitative]

spat[4] *n.* spawn of shellfish, esp. the oyster. [Anglo-French, of unknown origin]

spate *n.* **1** river-flood (*river in spate*). **2** large or excessive number of similar events (*spate of car thefts*). [origin unknown]

spathe /speɪð/ *n.* large bract(s) enveloping a flower-cluster. [Greek *spathē* broad blade]

spatial /'speɪʃ(ə)l/ *adj.* of space. □ **spatially** *adv.* [Latin: related to SPACE]

spatter –*v.* splash or scatter in drips. –*n.* **1** splash. **2** pattering. [imitative]

spatula /'spætjʊlə/ *n.* broad-bladed flexible implement used for spreading, stirring, mixing paints, etc. [Latin diminutive: related to SPATHE]

spawn –*v.* **1 a** (of a fish, frog, etc.) produce (eggs). **b** be produced as eggs or young. **2** produce or generate in large numbers. –*n.* **1** eggs of fish, frogs, etc. **2** mycelium of mushrooms or other fungi. [Anglo-French *espaundre*: related to EXPAND]

spay *v.* sterilize (a female animal) by removing the ovaries. [Anglo-French: related to ÉPÉE]

speak *v.* (*past* **spoke**; *past part.* **spoken**) **1** utter words in an ordinary voice. **2** utter (words, the truth, etc.). **3 a** converse; talk (*spoke to her earlier*; *had to speak to the children about rudeness*). **b** (foll. by *of*, *about*) mention in writing etc. **c** (foll. by *for*) act as spokesman for. **4** (foll. by *to*) speak with reference to; support in words (*spoke to the resolution*). **5** make a speech. **6** use or be able to use (a specified language). **7 a** convey an idea (*actions speak louder than words*). **b** (usu. foll. by *to*) communicate feeling etc.; affect, touch (*the sunset spoke to her*). □ **generally** (or **strictly** etc.) **speaking** in the general (or strict etc.) sense. **not** (or **nothing**) **to**

speak of not (or nothing) worth mentioning. **on speaking terms** friendly enough to converse. **speak for itself** be sufficient evidence. **speak out** (often followed by *against*) give one's opinion courageously. **speak up 1** speak loudly or freely; speak louder. **2** (followed by *for*) defend. **speak volumes** be very significant. [Old English]

-speak *comb. form* jargon (*Newspeak*; *computer speak*).

speakeasy *n.* (*pl.* **-ies**) *US hist. slang* place where alcoholic liquor was sold illicitly.

speaker *n.* **1** person who speaks, esp. in public. **2** person who speaks a specified language (esp. in *comb.*: *a French-speaker*). **3** (**Speaker**) presiding officer in a legislative assembly, esp. the House of Commons. **4** = LOUDSPEAKER.

speaking clock *n.* telephone service announcing the correct time.

spear —*n.* **1** thrusting or throwing weapon with a long shaft and a pointed usu. steel tip. **2 a** tip and stem of asparagus, broccoli, etc. **b** blade of grass etc. —*v.* pierce or strike (as) with a spear. [Old English]

spearhead —*n.* **1** point of a spear. **2** person or group leading an attack etc. —*v.* act as the spearhead of (an attack etc.).

spearmint *n.* common garden mint, used in cookery and to flavour chewing-gum etc.

spearwort *n.* aquatic plant with narrow spear-shaped leaves and yellow flowers.

spec[1] *n. colloq.* speculation. □ **on spec** as a gamble. [abbreviation]

spec[2] *n. colloq.* detailed working description; specification. [abbreviation of SPECIFICATION]

special /ˈspeʃ(ə)l/ —*adj.* **1 a** exceptional. **b** peculiar; specific. **2** for a particular purpose. **3** for children with special needs (*special school*). —*n.* special constable, train, edition of a newspaper, dish on a menu, etc. □ **specially** *adv.* **specialness** *n.* [Latin: related to SPECIES]

Special Branch *n.* police department dealing with political security.

special constable *n.* person trained to assist the police in routine duties or in an emergency.

special correspondent *n.* journalist writing on special events or a special subject.

special delivery *n.* delivery of mail outside the normal delivery schedule.

special edition *n.* extra late edition of a newspaper.

special effects *n.pl.* illusions created by props, camera-work, etc.

specialist *n.* **1** person trained in a particular branch of a profession, esp. medicine. **2** person who specially studies a subject or area.

speciality /ˌspeʃɪˈælɪtɪ/ *n.* (*pl.* **-ies**) **1** special subject, product, activity, etc. **2** special feature or skill.

specialize /ˈspeʃəˌlaɪz/ *v.* (also **-ise**) (**-zing** or **-sing**) **1** (often foll. by *in*) **a** be or become a specialist. **b** devote oneself to an interest, skill, etc. (*specializes in insulting people*). **2** (esp. in *passive*) adapt for a particular purpose (*specialized organs*). **3** (as **specialized** *adj.*) of a specialist (*specialized work*). □ **specialization** /-ˈzeɪʃ(ə)n/ *n.* [French: related to SPECIAL]

special licence *n.* licence allowing immediate marriage without banns.

special pleading *n.* biased reasoning.

specialty /ˈspeʃəltɪ/ *n.* (*pl.* **-ies**) esp. *US* = SPECIALITY.

specie /ˈspiːʃiː, -ʃɪ/ *n.* coin as opposed to paper money. [related to SPECIES]

species /ˈspiːʃɪz, -ʃiːz, -siːz/ *n.* (*pl.* same) **1** class of things having some common characteristics. **2** group of animals or plants within a genus, differing only slightly from others and capable of interbreeding. **3** kind, sort. [Latin *specio* look]

specific /spəˈsɪfɪk/ —*adj.* **1** clearly defined (*a specific purpose*). **2** relating to a particular subject; peculiar. **3** exact, giving full details (*was specific about his wishes*). **4** *archaic* (of medicine etc.) having a distinct effect in curing a certain disease. —*n.* **1** *archaic* specific medicine or remedy. **2** specific aspect or factor (*discussed specifics*; *from the general to the specific*). □ **specifically** *adv.* **specificity** /-ˈfɪsɪtɪ/ *n.* [Latin: related to SPECIES]

specification /ˌspesɪfɪˈkeɪʃ(ə)n/ *n.* **1** act of specifying. **2** (esp. in *pl.*) detail of the design and materials etc. of work done or to be done. [medieval Latin: related to SPECIFY]

specific gravity *n.* = RELATIVE DENSITY.

specific heat capacity *n.* heat required to raise the temperature of the unit mass of a given substance by a given amount (usu. one degree).

specify /ˈspesɪˌfaɪ/ *v.* (**-ies, -ied**) **1** (also *absol.*) name or mention expressly or as a condition (*specified a two-hour limit*). **2** include in specifications. [Latin: related to SPECIFIC]

specimen /ˈspesɪmɪn/ *n.* **1** individual or sample taken as an example of a class or whole, esp. in experiments etc. **2** sample of urine for testing. **3** *colloq.* usu. *derog.* person of a specified sort. [Latin *specio* look]

specious /'spi:ʃəs/ *adj.* plausible but wrong (*specious argument*). [Latin: related to SPECIES]

speck –*n.* **1** small spot or stain. **2** particle. –*v.* (esp. as **specked** *adj.*) marked with specks. [Old English]

speckle /'spek(ə)l/ –*n.* speck, esp. one of many markings. –*v.* (**-ling**) (esp. as **speckled** *adj.*) mark with speckles. [Dutch *spekkel*]

specs *n.pl. colloq.* spectacles. [abbreviation]

spectacle /'spektək(ə)l/ *n.* **1** striking, impressive, or ridiculous sight. **2** public show. **3** object of public attention. [Latin *specio spect-* look]

spectacled *adj.* wearing spectacles.

spectacles *n.pl.* pair of lenses in a frame resting on the nose and ears, used to correct defective eyesight.

spectacular /spek'tækjʊlə(r)/ –*adj.* striking, impressive, lavish. –*n.* spectacular show. □ **spectacularly** *adv.*

spectator *n.* person who watches a show, game, incident, etc. □ **spectate** *v.* (**-ting**) *informal.* [Latin: related to SPECTACLE]

spectator sport *n.* sport attracting many spectators.

specter *US* var. of SPECTRE.

spectra *pl.* of SPECTRUM.

spectral /'spektr(ə)l/ *adj.* **1** of or like a spectre; ghostly. **2** of the spectrum or spectra. □ **spectrally** *adv.*

spectre /'spektə(r)/ *n.* (*US* **specter**) **1** ghost. **2** haunting presentiment (*spectre of war*). [Latin *spectrum* from *specio* look]

spectrometer /spek'trɒmɪtə(r)/ *n.* instrument for measuring observed spectra.

spectroscope /'spektrə,skəʊp/ *n.* instrument for recording spectra for examination. □ **spectroscopic** /-'skɒpɪk/ *adj.* **spectroscopy** /-'trɒ skəpɪ/ *n.*

spectrum /'spektrəm/ *n.* (*pl.* **-tra**) **1** band of colours as seen in a rainbow etc. **2** entire or wide range of a subject, emotion, etc. **3** distribution of electromagnetic radiation in which the parts are arranged according to wavelength. [Latin *specio* look]

specula *pl.* of SPECULUM.

speculate /'spekjʊ,leɪt/ *v.* (**-ting**) **1** (usu. foll. by *on, upon, about*) theorize, conjecture. **2** deal in a commodity or asset in the hope of profiting from fluctuating prices. □ **speculation** /-'leɪʃ(ə)n/ *n.* **speculative** /-lətɪv/ *adj.* **speculator** *n.* [Latin *specula* watch-tower, from *specio* look]

speculum /'spekjʊləm/ *n.* (*pl.* **-la**) **1** instrument for dilating orifices of the body. **2** mirror of polished metal in a telescope. [Latin, = mirror]

sped *past* and *past part.* of SPEED.

speech *n.* **1** faculty, act, or manner of speaking. **2** formal public address. **3** language of a nation, group, etc. [Old English: related to SPEAK]

speech day *n.* school celebration with speeches, prize-giving, etc.

speechify /'spi:tʃɪ,faɪ/ *v.* (**-ies, -ied**) *joc.* make esp. boring or long speeches.

speechless *adj.* temporarily silent because of emotion etc.

speech therapy *n.* treatment for defective speech.

speed –*n.* **1** rapidity of movement. **2** rate of progress or motion. **3** gear appropriate to a range of speeds of a bicycle. **4** *Photog.* **a** sensitivity of film to light. **b** light-gathering power of a lens. **c** duration of an exposure. **5** *slang* amphetamine drug. **6** *archaic* success, prosperity. –*v.* (*past* and *past part.* **sped**) **1** go or send quickly. **2** (*past* and *past part.* **speeded**) travel at an illegal or dangerous speed. **3** *archaic* be or make prosperous or successful. □ **at speed** moving quickly. **speed up** move or work faster. □ **speeder** *n.* [Old English]

speedboat *n.* high-speed motor boat.

speed limit *n.* maximum permitted speed on a road etc.

speedo /'spi:dəʊ/ *n.* (*pl.* **-s**) *colloq.* = SPEEDOMETER. [abbreviation]

speedometer /spi:'dɒmɪtə(r)/ *n.* instrument on a vehicle indicating its speed.

speedway *n.* **1 a** motor-cycle racing. **b** arena for this. **2** *US* road or track for fast traffic.

speedwell *n.* small plant with bright blue flowers. [from SPEED, WELL[1]]

speedy *adj.* (**-ier, -iest**) **1** rapid. **2** done without delay; prompt. □ **speedily** *adv.* **speediness** *n.*

speleology /,spi:lɪ'ɒlədʒɪ/ *n.* the study of caves. [Greek *spēlaion* cave]

spell[1] *v.* (*past* and *past part.* **spelt** or **spelled**) **1** (also *absol.*) write or name correctly the letters of (a word etc.). **2 a** (of letters) form (a word etc.). **b** result in (*spell ruin*). □ **spell out 1** make out (words etc.) letter by letter. **2** explain in detail. □ **speller** *n.* [French *espeller* related to SPELL[2]]

spell[2] *n.* **1** words used as a charm or incantation etc. **2** effect of these. **3** fascination exercised by a person, activity, etc. [Old English]

spell[3] *n.* **1** short or fairly short period (*a cold spell*). **2** period of some activity or work. [Old English, = substitute]
spellbind /'spelbaɪnd/ *v.* (*past* and *past part.* **spellbound**) **1** (esp. as **spellbinding** *adj.*) hold the attention as if with a spell; entrance. **2** (as **spellbound** *adj.*) entranced, fascinated.
spelling *n.* **1** way a word is spelt. **2** ability to spell.
spelt[1] *past* and *past part.* of SPELL[1].
spelt[2] *n.* a kind of wheat giving very fine flour. [Old English]
spend *v.* (*past* and *past part.* **spent**) **1** pay out (money). **2 a** use or consume (time or energy). **b** use up (material etc.). **3** (as **spent** *adj.*) having lost its original force or strength; exhausted. □ **spend a penny** *colloq.* go to the lavatory. □ **spender** *n.* [Latin: related to EXPEND]
spendthrift –*n.* extravagant person. –*adj.* extravagant.
sperm *n.* (*pl.* same or **-s**) **1** = SPERMATOZOON. **2** semen. [Greek *sperma -mat-*]
spermaceti /ˌspɜːmə'setɪ/ *n.* white waxy substance from the sperm whale, used for ointments etc. [medieval Latin, = whale sperm]
spermatozoon /ˌspɜːmətəʊ'zəʊən/ *n.* (*pl.* **-zoa**) mature sex cell in semen. [from SPERM, Greek *zōion* animal]
sperm bank *n.* store of semen for artificial insemination.
sperm count *n.* number of spermatozoa in one ejaculation or a measured amount of semen.
spermicide /'spɜːmɪˌsaɪd/ *n.* substance able to kill spermatozoa. □ **spermicidal** /-'saɪd(ə)l/ *adj.*
sperm whale *n.* large whale yielding spermaceti.
spew *v.* (also **spue**) **1** (often foll. by *up*) vomit. **2** (often foll. by *out*) (cause to) gush. [Old English]
sphagnum /'sfægnəm/ *n.* (*pl.* **-na**) (in full **sphagnum moss**) moss growing in bogs, used as packing etc. [Greek *sphagnos*]
sphere *n.* **1** solid figure with every point on its surface equidistant from its centre; its surface. **2** ball, globe. **3 a** field of action, influence, etc. **b** social class. **4** *hist.* each of the revolving shells in which celestial bodies were thought to be set. [Greek *sphaira* ball]
spherical /'sferɪk(ə)l/ *adj.* **1** shaped like a sphere. **2** of spheres. □ **spherically** *adv.*
spheroid /'sfɪərɔɪd/ *n.* spherelike but not perfectly spherical body. □ **spheroidal** /-'rɔɪd(ə)l/ *adj.*
sphincter /'sfɪŋktə(r)/ *n.* ring of muscle surrounding and closing an opening in the body. [Greek *sphiggō* bind tight]
sphinx /sfɪŋks/ *n.* **1** (**Sphinx**) (in Greek mythology) winged monster with a woman's head and a lion's body, whose riddle Oedipus guessed. **2** *Antiq.* **a** ancient Egyptian stone figure with a lion's body and a human or animal head. **b** (**the Sphinx**) huge sphinx near the Pyramids at Giza. **3** inscrutable person. [Greek]
spice –*n.* **1** aromatic or pungent vegetable substance used to flavour food. **2** spices collectively. **3 a** piquant quality. **b** (foll. by *of*) slight flavour or suggestion. –*v.* (**-cing**) **1** flavour with spice. **2** (foll. by *with*) enhance (*spiced with wit*). [French *espice*]
spick and span *adj.* **1** neat and clean. **2** smart and new. [earlier *span and span new*, fresh and new like a shaved chip]
spicy *adj.* (**-ier, -iest**) **1** of or flavoured with spice. **2** piquant; sensational, improper. □ **spiciness** *n.*
spider /'spaɪdə(r)/ *n.* eight-legged arthropod of which many species spin webs esp. to capture insects as food. [Old English: related to SPIN]
spider crab *n.* crab with long thin legs.
spider monkey *n.* monkey with long limbs and a prehensile tail.
spider plant *n.* house plant with long narrow striped leaves.
spidery *adj.* elongated and thin (*spidery handwriting*).
spiel /ʃpiːl/ *n. slang* glib speech or story; sales pitch. [German, = game]
spigot /'spɪgət/ *n.* **1** small peg or plug, esp. in a cask. **2** device for controlling the flow of liquid in a tap. [related to SPIKE[2]]
spike[1] –*n.* **1 a** sharp point. **b** pointed piece of metal, esp. the top of an iron railing. **2 a** metal point in the sole of a running-shoe to prevent slipping. **b** (in *pl.*) spiked running-shoes. **3** pointed metal rod used for filing rejected news items. **4** large nail. –*v.* (**-king**) **1** put spikes on or into. **2** fix on a spike. **3** *colloq.* **a** lace (a drink) with alcohol etc. **b** contaminate with something added. **4** *colloq.* reject (a newspaper story). □ **spike a person's guns** spoil his or her plans. [Low German or Dutch: related to SPOKE[1]]
spike[2] *n.* cluster of flower-heads on a long stem. [Latin *spica*]
spikenard /'spaɪknɑːd/ *n.* **1** tall sweet-smelling Indian plant. **2** *hist.* perfumed ointment formerly made from this. [medieval Latin *spica nardi*]
spiky *adj.* (**-ier, -iest**) **1** like a spike; having or sticking up in spikes. **2** *colloq.*

touchy, irritable. □ **spikily** *adv.* **spikiness** *n.*

spill[1] –*v.* (*past* and *past part.* **spilt** or **spilled**) **1** fall or run or cause (liquid, powder, etc.) to fall or run out of a container, esp. accidentally. **2 a** throw from a vehicle, saddle, etc. **b** (foll. by *into, out* etc.) (esp. of a crowd) leave a place quickly. **3** *slang* disclose (information etc.). **4** shed (blood). –*n.* **1** spilling or being spilt. **2** tumble, esp. from a horse, bicycle, etc. □ **spill the beans** *colloq.* divulge information etc. **spill over** overflow. □ **spillage** *n.* [Old English]

spill[2] *n.* thin strip of wood or paper etc. for lighting a fire, pipe, etc. [Low German or Dutch]

spillikin /ˈspɪlɪkɪn/ *n.* **1** splinter of wood etc. **2** (in *pl.*) game in which thin rods are removed one at a time from a heap without moving the others. [from SPILL[2]]

spillway *n.* passage for surplus water from a dam.

spin –*v.* (**-nn-**; *past* and *past part.* **spun**) **1** (cause to) turn or whirl round quickly. **2** (also *absol.*) **a** draw out and twist (wool, cotton, etc.) into threads. **b** make (yarn) in this way. **3** (of a spider, silkworm, etc.) make (a web, cocoon, etc.) by extruding a fine viscous thread. **4** (esp. of the head) be dizzy through excitement etc. **5** tell or write (a story etc.). **6** (as **spun** *adj.*) made into threads (*spun glass*; *spun gold*). **7** toss (a coin). **8** = SPIN-DRY. –*n.* **1** spinning motion; whirl. **2** rotating dive of an aircraft. **3** secondary twisting motion, e.g. of a ball in flight. **4** *colloq.* brief drive, esp. in a car. □ **spin out** prolong. **spin a yarn** tell a story. [Old English]

spina bifida /ˌspaɪnə ˈbɪfɪdə/ *n.* congenital spinal defect in which part of the spinal cord protrudes. [Latin, = cleft spine]

spinach /ˈspɪnɪdʒ/ *n.* green vegetable with edible leaves. [French *espinache*]

spinal /ˈspaɪn(ə)l/ *adj.* of the spine. [Latin: related to SPINE]

spinal column *n.* spine.

spinal cord *n.* cylindrical nervous structure within the spine.

spin bowler *n. Cricket* bowler who imparts spin to a ball.

spindle /ˈspɪnd(ə)l/ *n.* **1** slender rod or bar, often tapered, for twisting and winding thread. **2** pin or axis that revolves or on which something revolves. **3** turned piece of wood used as a banister, chair leg, etc. [Old English: related to SPIN]

spindle tree *n.* tree with hard wood used for spindles.

spindly *adj.* (**-ier, -iest**) long or tall and thin; thin and weak.

spin-drier *n.* (also **spin-dryer**) machine for drying clothes by spinning them in a rapidly revolving drum. □ **spin-dry** *v.*

spindrift /ˈspɪndrɪft/ *n.* spray on the surface of the sea. [Scots var. of *spoondrift* from obsolete *spoon* scud]

spine *n.* **1** vertebrae extending from the skull to the coccyx; backbone. **2** needle-like outgrowth of an animal or plant. **3** part of a book enclosing the page-fastening. **4** sharp ridge or projection. [Latin *spina*]

spine-chiller *n.* frightening and usu. exciting story, film, etc. □ **spine-chilling** *adj.*

spineless *adj.* **1** having no spine; invertebrate. **2** lacking resolve, feeble.

spinet /spɪˈnet/ *n. hist.* small harpsichord with oblique strings. [Italian *spinetta*]

spinnaker /ˈspɪnəkə(r)/ *n.* large triangular sail opposite the mainsail of a racing-yacht. [*Sphinx*, name of the yacht first using it]

spinner *n.* **1** spin bowler. **2** person or thing that spins, esp. a manufacturer engaged in cotton-spinning. **3** revolving bait.

spinneret /ˈspɪnəˌret/ *n.* **1** spinning-organ in a spider etc. **2** device for forming synthetic fibre.

spinney /ˈspɪnɪ/ *n.* (*pl.* **-s**) small wood; thicket. [Latin *spinetum* from *spina* thorn]

spinning-jenny *n. hist.* machine for spinning fibres with more than one spindle at a time.

spinning wheel *n.* household device for spinning yarn or thread, with a spindle driven by a wheel with a crank or treadle.

spin-off *n.* incidental result or benefit, esp. from technology.

spinster *n.* **1** *formal* unmarried woman. **2** woman, esp. elderly, thought unlikely to marry. □ **spinsterish** *adj.* [originally = woman who spins]

spiny *adj.* (**-ier, -iest**) having many spines.

spiny anteater *n.* = ECHIDNA.

spiraea /ˌspaɪˈrɪə/ *n.* (*US* **spirea**) shrub with clusters of small white or pink flowers. [Greek: related to SPIRAL]

spiral /ˈspaɪər(ə)l/ –*adj.* **1** coiled in a plane or as round a cylinder or cone. **2** having this shape. –*n.* **1** spiral curve or thing (*spiral of smoke*). **2** progressive rise or fall of two or more quantities alternately because each depends on the other(s). –*v.* (**-ll-**; *US* **-l-**) **1** move in a spiral course. **2** (of prices, wages, etc.) rise or fall continuously. □ **spirally** *adv.* [Greek *speira* coil]

spiral staircase *n.* circular staircase round a central axis.

spirant /'spaɪərənt/ *–adj.* uttered with a continuous expulsion of breath. *–n.* such a consonant. [Latin *spiro* breathe]

spire *n.* **1** tapering structure, esp. on a church tower. **2** any tapering thing. [Old English]

spirea *US* var. of SPIRAEA.

spirit /'spɪrɪt/ *–n.* **1** person's essence or intelligence; soul. **2 a** rational or intelligent being without a material body. **b** ghost. **3 a** person's character (*an unbending spirit*). **b** attitude (*took it in the wrong spirit*). **c** type of person (*is a free spirit; a kindred spirit*). **d** prevailing tendency (*spirit of the age*). **4 a** (usu. in *pl.*) strong distilled liquor, e.g. whisky or gin. **b** distilled volatile liquid (*wood spirit*). **c** purified alcohol (*methylated spirit*). **5 a** courage, vivacity. **b** (in *pl.*) state of mind, mood (*in high spirits; his spirits were dashed*). **6** essential as opposed to formal meaning (*the spirit of the law*). *–v.* (**-t-**) (usu. foll. by *away, off*, etc.) convey rapidly or mysteriously. □ **in spirit** inwardly. [Latin *spiritus*: related to SPIRANT]

spirited *adj.* **1** lively, courageous. **2** (in *comb.*) in a specified mood (*high-spirited*). □ **spiritedly** *adv.*

spirit gum *n.* quick-drying gum for attaching false hair.

spirit-lamp *n.* lamp burning methylated spirit etc. instead of oil.

spiritless *adj.* lacking vigour.

spirit-level *n.* device with a glass tube nearly filled with alcohol, used to test horizontality.

spiritual /'spɪrɪtʃʊəl/ *–adj.* **1** of the spirit or soul (*spiritual relationship; spiritual home*). **2** religious, divine, inspired. **3** refined, sensitive. *–n.* (also **Negro spiritual**) religious song orig. of American Blacks. □ **spirituality** /-'ælɪtɪ/ *n.* **spiritually** *adv.*

spiritualism *n.* belief in, and supposed practice of, communication with the dead, esp. through mediums. □ **spiritualist** *n.* **spiritualistic** /-'lɪstɪk/ *adj.*

spirituous /'spɪrɪtʃʊəs/ *adj.* **1** very alcoholic. **2** distilled as well as fermented.

spirochaete /'spaɪərəʊˌki:t/ *n.* any of various flexible spiral-shaped bacteria. [Latin from Greek *speira* coil, *khaitē* long hair]

spirogyra /ˌspaɪərəʊ'dʒaɪərə/ *n.* freshwater alga containing spiral bands of chlorophyll. [Greek *speira* coil, *guros* round]

spit[1] *–v.* (**-tt-**; *past* and *past part.* **spat** or **spit**) **1 a** (also *absol.*) eject (esp. saliva) from the mouth. **b** do this in contempt or anger. **2** utter vehemently. **3** (of a fire, gun, etc.) throw out with an explosion. **4** (of rain) fall lightly. **5** make a spitting noise. *–n.* **1** spittle. **2** act of spitting. □ **spit it out** *colloq.* say it quickly and concisely. [Old English]

spit[2] *–n.* **1** rod for skewering meat for roasting on a fire etc. **2** point of land projecting into the sea. *–v.* (**-tt-**) pierce (as) with a spit. [Old English]

spit and polish *n. colloq.* esp. military cleaning and polishing.

spite *–n.* ill will, malice. *–v.* (**-ting**) hurt, harm, or frustrate (a person) through spite. □ **in spite of** notwithstanding. [French: related to DESPITE]

spiteful *adj.* malicious. □ **spitefully** *adv.*

spitfire *n.* person of fiery temper.

spit-roast *v.* roast on a spit.

spitting distance *n. colloq.* very short distance.

spitting image *n.* (foll. by *of*) *colloq.* double of (a person).

spittle /'spɪt(ə)l/ *n.* saliva. [related to SPIT[1]]

spittoon /spɪ'tu:n/ *n.* vessel to spit into.

spiv *n. colloq.* man, esp. a flashily-dressed one, living from shady dealings. □ **spivvish** *adj.* **spivvy** *adj.* [origin unknown]

splash *–v.* **1** scatter or cause (liquid) to scatter in drops. **2** wet with spattered liquid etc. **3 a** (usu. foll. by *across, along, about*, etc.) move while spattering liquid etc. **b** jump or fall into water etc. with a splash. **4** display (news) prominently. **5** decorate with scattered colour. **6** spend (money) ostentatiously. *–n.* **1** act or noise of splashing. **2** quantity of liquid splashed. **3** mark etc. made by splashing. **4** prominent news feature, display, etc. **5** patch of colour. **6** *colloq.* small quantity of soda water etc. (in drink). □ **make a splash** attract attention. **splash out** *colloq.* spend money freely. □ **splashy** *adj.* (**-ier, -iest**). [imitative]

splashback *n.* panel behind a sink etc. to protect the wall from splashes.

splashdown *n.* landing of a spacecraft on the sea. □ **splash down** *v.*

splat *colloq.* *–n.* sharp splattering sound. *–adv.* with a splat. *–v.* (**-tt-**) fall or hit with a splat. [abbreviation of SPLATTER]

splatter *–v.* splash esp. with a continuous noisy action; spatter. *–n.* noisy splashing sound. [imitative]

splay *–v.* **1** spread apart. **2** (of an opening) have its sides diverging. **3** construct (an opening) with divergent sides. *–n.* surface at an oblique angle to another. *–adj.* splayed. [from DISPLAY]

spleen *n.* **1** abdominal organ regulating the quality of the blood. **2** moroseness, irritability (from the earlier belief that the spleen was the seat of such feelings). [Greek *splēn*]

spleenwort *n.* evergreen fern formerly used as a remedy.

splendid /'splendɪd/ *adj.* **1** magnificent, sumptuous. **2** impressive, glorious, dignified (*splendid isolation*). **3** excellent; fine. □ **splendidly** *adv.* [Latin: related to SPLENDOUR]

splendiferous /splen'dɪfərəs/ *adj. colloq.* splendid. [from SPLENDOUR]

splendour /'splendə(r)/ *n.* (*US* **splendor**) dazzling brightness; magnificence. [Latin *splendeo* shine]

splenetic /splɪ'netɪk/ *adj.* bad-tempered; peevish. □ **splenetically** *adv.* [Latin: related to SPLEEN]

splenic /'splenɪk, 'spli:-/ *adj.* of or in the spleen. [Latin from Greek: related to SPLEEN]

splice –*v.* (**-cing**) **1** join (ropes) by interweaving strands. **2** join (pieces of wood or tape etc.) by overlapping. **3** (esp. as **spliced** *adj.*) *colloq.* join in marriage. –*n.* join made by splicing. □ **splice the main brace** *Naut. hist. slang* issue an extra tot of rum. [probably Dutch *splissen*]

splint –*n.* strip of wood etc. bound to a broken limb while it sets. –*v.* secure with a splint. [Low German or Dutch]

splinter –*n.* small sharp fragment of wood, stone, glass, etc. –*v.* break into splinters; shatter. □ **splintery** *adj.* [Dutch: related to SPLINT]

splinter group *n.* breakaway political group.

split –*v.* (**-tt-**; *past* and *past part.* **split**) **1 a** break, esp. with the grain or into halves; break forcibly. **b** (often foll. by *up*) divide into parts, esp. equal shares (*they split the money*). **2** (often foll. by *off, away*) remove or be removed by breaking or dividing. **3 a** (usu. foll. by *on, over*, etc.) divide into disagreeing or hostile parties (*split on the question of picketing*). **b** (foll. by *with*) quarrel or cease association with. **4** cause the fission of (an atom). **5** *slang* leave, esp. suddenly. **6** (usu. foll. by *on*) *colloq.* inform. **7 a** (as **splitting** *adj.*) (of a headache) severe. **b** (of the head) suffer from a severe headache, noise, etc. –*n.* **1** act or result of splitting. **2** disagreement; schism. **3** (in *pl.*) feat of leaping in the air or sitting down with the legs at right angles to the body in front and behind or on either side. **4** dish of split bananas etc. with ice-cream. □ **split the difference** take the average of two proposed amounts. **split hairs** make insignificant distinctions. **split one's sides** laugh uncontrollably. **split up** separate, end a relationship. [Dutch]

split infinitive *n.* infinitive with an adverb etc. inserted between *to* and the verb.

split-level *adj.* (of a room etc.) with more than one level.

split pea *n.* pea dried and split in half for cooking.

split personality *n.* condition in which a person seems to have two alternating personalities.

split pin *n.* metal cotter passed through a hole and held by the pressing back of the two ends.

split-screen *n.* screen on which two or more separate images are displayed.

split second –*n.* **1** very brief moment. **2** (of timing) very accurate. –*attrib. adj.* (**split-second**) **1** very rapid. **2** (of timing) very accurate.

splodge *colloq.* –*n.* daub, blot, or smear. –*v.* (**-ging**) make a splodge on. □ **splodgy** *adj.* [alteration of SPLOTCH]

splosh *colloq.* –*v.* move with a splashing sound. –*n.* **1** splashing sound. **2** splash of water etc. [imitative]

splotch *n.* & *v.* = SPLODGE. □ **splotchy** *adj.* [origin uncertain]

splurge *colloq.* –*n.* **1** sudden extravagance. **2** ostentatious display or effort. –*v.* (**-ging**) (usu. foll. by *on*) spend large sums of money or make a great effort. [probably imitative]

splutter –*v.* **1 a** speak, say, or express in a choking manner. **b** emit spitting sounds. **2** speak rapidly or incoherently. –*n.* spluttering speech or sound. [from SPUTTER]

spoil –*v.* (*past* and *past part.* **spoilt** or **spoiled**) **1 a** make or become useless or unsatisfactory. **b** reduce the enjoyment etc. of (*the news spoiled his dinner*). **2** make (esp. a child) unpleasant by over-indulgence. **3** (of food) go bad. **4** render (a ballot-paper) invalid by improper marking. –*n.* (usu. in *pl.*) **1** plunder, stolen goods. **2** profit or advantage from success or position. □ **be spoiling for** aggressively seek (a fight etc.). **spoilt for choice** having so many choices that it is difficult to choose. [Latin *spolio*]

spoilage *n.* **1** paper spoilt in printing. **2** spoiling of food etc. by decay.

spoiler *n.* **1** retarding device on an aircraft, interrupting the air flow. **2** similar device on a vehicle to increase contact with the ground at speed.

spoilsport *n.* person who spoils others' enjoyment.

spoilt *past* and *past part.* of SPOIL.

spoke[1] *n.* each of the rods running from the hub to the rim of a wheel. □ **put a spoke in a person's wheel** thwart or hinder a person. □ **spoked** *adj.* [Old English]

spoke[2] *past* of SPEAK.

spoken *past part.* of SPEAK. *adj.* (in *comb.*) speaking in a specified way (*well-spoken*). □ **spoken for** claimed (*this seat is spoken for*).

spokeshave *n.* tool for planing curved surfaces. [from SPOKE[1]]

spokesman *n.* (*fem.* **spokeswoman**) person speaking for a group etc. [from SPOKE[2]]

spokesperson *n.* (*pl.* **-s** or **-people**) spokesman or spokeswoman.

spoliation /ˌspəʊlɪˈeɪʃ(ə)n/ *n.* plundering, pillage. [Latin: related to SPOIL]

spondee /ˈspɒndiː/ *n.* metrical foot consisting of two long syllables (––). □ **spondaic** /-ˈdeɪɪk/ *adj.* [Greek *spondē* libation, with which songs in this metre were associated]

sponge /spʌndʒ/ –*n.* **1** sea animal with a porous body wall and a rigid internal skeleton. **2** this skeleton or a piece of porous rubber etc. used in bathing, cleaning, etc. **3** thing like a sponge in consistency etc., esp. a sponge cake. **4** act of sponging. –*v.* (**-ging**) **1** wipe or cleanse with a sponge. **2** (often foll. by *out*, *away*, etc.) wipe off or efface (as) with a sponge. **3** (often foll. by *up*) absorb (as) with a sponge. **4** (often foll. by *on*, *off*) live as a parasite. □ **spongiform** *adj.* (esp. in senses 1, 2 of the *n.*). [Latin *spongia*]

sponge bag *n.* waterproof bag for toilet articles.

sponge cake *n.* light spongy cake.

sponge pudding *n.* light spongy pudding.

sponger *n.* parasitic person.

sponge rubber *n.* porous rubber.

spongy *adj.* (**-ier, -iest**) like a sponge, porous, elastic, absorbent. □ **sponginess** *n.*

sponsor /ˈspɒnsə(r)/ –*n.* **1** person who pledges money to a charity etc. in return for another person fulfilling a sporting etc. challenge. **2 a** patron of an artistic or sporting activity etc. **b** company etc. supporting a broadcast in return for advertising time. **3** person who introduces legislation. **4** godparent at a baptism or (esp. *RC Ch.*) person who presents a candidate for baptism. –*v.* be a sponsor for. □ **sponsorial** /-ˈsɔːrɪəl/ *adj.* **sponsorship** *n.* [Latin *spondeo spons-* pledge]

spontaneous /spɒnˈteɪnɪəs/ *adj.* **1** acting, done, or occurring without external cause. **2** instinctive, automatic, natural. **3** (of style or manner) gracefully natural. □ **spontaneity** /ˌspɒntəˈneɪətɪ, -ˈniːɪtɪ/ *n.* **spontaneously** *adv.* [Latin *sponte* of one's own accord]

spontaneous combustion *n.* ignition of a substance from internal heat.

spoof *n.* & *v. colloq.* **1** parody. **2** hoax, swindle. [invented word]

spook –*n. colloq.* ghost. –*v.* esp. *US* frighten, unnerve. [Low German or Dutch]

spooky *adj.* (**-ier, -iest**) *colloq.* ghostly, eerie. □ **spookily** *adv.* **spookiness** *n.*

spool –*n.* **1** reel for winding magnetic tape, yarn, etc., on. **2** revolving cylinder of an angler's reel. –*v.* wind on a spool. [French *espole* or Germanic *spole*]

spoon –*n.* **1 a** utensil with a bowl and a handle for lifting food to the mouth, stirring, etc. **b** spoonful, esp. of sugar. **2** spoon-shaped thing, esp. (in full **spoon-bait**) a revolving metal fish-lure. –*v.* **1** (often foll. by *up*, *out*) take (liquid etc.) with a spoon. **2** hit (a ball) feebly upwards. **3** *colloq.* kiss and cuddle. □ **spoonful** *n.* (*pl.* **-s**). [Old English]

spoonbill *n.* wading bird with a broad flat-tipped bill.

spoonerism /ˈspuːnəˌrɪz(ə)m/ *n.* (usu. accidental) transposition of the initial letters etc. of two or more words. [*Spooner*, name of a scholar]

spoonfeed *v.* (*past* and *past part.* **-fed**) **1** feed with a spoon. **2** give such extensive help etc. to (a person) that he or she need make no effort.

spoor *n.* animal's track or scent. [Dutch]

sporadic /spəˈrædɪk/ *adj.* occurring only sparsely or occasionally. □ **sporadically** *adv.* [Greek *sporas -ad-* scattered]

spore *n.* reproductive cell of many plants and micro-organisms. [Greek *spora* seed]

sporran /ˈspɒrən/ *n.* pouch worn in front of the kilt. [Gaelic *sporan*]

sport –*n.* **1 a** game or competitive activity, usu. played outdoors and involving physical exertion, e.g. cricket, football, racing. **b** these collectively. **2** (in *pl.*) meeting for competing in sports, esp. athletics. **3** amusement, fun. **4** *colloq.* **a** fair, generous, or sporting person. **b** person with a specified attitude to games, rules, etc. **5** animal or plant deviating from the normal type. –*v.* **1** amuse oneself, play about. **2** wear or exhibit, esp. ostentatiously. □ **in sport** jestingly. **make sport of** ridicule. [from DISPORT]

sporting *adj.* **1** interested or concerned in sport. **2** generous, fair. □ **a sporting**

chance some possibility of success. □ **sportingly** *adv.*

sportive *adj.* playful.

sports car *n.* low-built fast car.

sports coat *n.* (also **sports jacket**) man's informal jacket.

sports ground *n.* piece of land used for sports.

sportsman *n.* (*fem.* **sportswoman**) **1** person who takes part in sport, esp. professionally. **2** fair and generous person. □ **sportsmanlike** *adj.* **sportsmanship** *n.*

sportswear *n.* clothes for sports or informal wear.

sporty *adj.* (**-ier**, **-iest**) *colloq.* **1** fond of sport. **2** rakish, showy. □ **sportily** *adv.* **sportiness** *n.*

spot –*n.* **1** small roundish area or mark differing in colour, texture, etc., from the surface it is on. **2** pimple or blemish. **3** moral blemish or stain. **4** particular place, locality. **5** particular part of one's body or aspect of one's character. **6** *colloq.* one's esp. regular position in an organization, programme, etc. **7 a** *colloq.* small quantity (*spot of trouble*). **b** drop (*spot of rain*). **8** = SPOTLIGHT. **9** (usu. *attrib.*) money paid or goods delivered immediately after a sale (*spot cash*). –*v.* (**-tt-**) **1** *colloq.* pick out, recognize, catch sight of. **2** watch for and take note of (trains, talent, etc.). **3** (as **spotted** *adj.*) marked or decorated with spots. **4** make spots, rain slightly. □ **in a spot** (or **in a tight** etc. **spot**) *colloq.* in difficulty. **on the spot 1** at the scene of an event. **2** *colloq.* in a position demanding response or action. **3** without delay. **4** without moving forwards or backwards (*running on the spot*). [perhaps from Low German or Dutch]

spot check *n.* sudden or random check.

spotless *adj.* absolutely clean or pure. □ **spotlessly** *adv.*

spotlight –*n.* **1** beam of light directed on a small area. **2** lamp projecting this. **3** full publicity. –*v.* (*past* and *past part.* **-lighted** or **-lit**) **1** direct a spotlight on. **2** draw attention to.

spot on *adj. colloq.* precise; on target.

spotted dick *n.* suet pudding containing currants.

spotter *n.* **1** (often in *comb.*) person who spots people or things (*train-spotter*). **2** (in full **spotter plane**) aircraft used to locate enemy positions etc.

spotty *adj.* (**-ier**, **-iest**) **1** marked with spots. **2** patchy, irregular. □ **spottiness** *n.*

spot-weld *v.* join (two metal surfaces) by welding at discrete points. □ **spot weld** *n.* **spot welder** *n.* **spot welding** *n.*

spouse /spaʊz/ *n.* husband or wife. [Latin *sponsus sponsa* betrothed]

spout –*n.* **1** projecting tube or lip used for pouring from a teapot, kettle, jug, etc., or on a fountain, roof-gutter, etc. **2** jet or column of liquid etc. –*v.* **1** discharge or issue forcibly in a jet. **2** utter or speak at length or pompously. □ **up the spout** *slang* **1** useless, ruined, broken down. **2** pregnant. [Dutch]

sprain –*v.* wrench (an ankle, wrist, etc.), causing pain or swelling. –*n.* such a wrench. [origin unknown]

sprang *past* of SPRING.

sprat *n.* small edible marine fish. [Old English]

sprawl –*v.* **1 a** sit, lie, or fall with limbs flung out untidily. **b** spread (one's limbs) thus. **2** (of writing, a plant, a town, etc.) be irregular or straggling. –*n.* **1** sprawling movement, position, or mass. **2** straggling urban expansion. [Old English]

spray[1] –*n.* **1** water etc. flying in small drops. **2** liquid sprayed with an aerosol etc. **3** device for this. –*v.* **1** (also *absol.*) throw (liquid) as spray. **2** (also *absol.*) sprinkle (an object) thus, esp. with insecticide. **3** (of a tom-cat) mark its environment with urine, to attract females. □ **sprayer** *n.* [origin uncertain]

spray[2] *n.* **1** sprig of flowers or leaves, or a small branch; decoratively arranged bunch of flowers. **2** ornament in a similar form. [Old English]

spray-gun *n.* device for spraying paint etc.

spread /spred/ –*v.* (*past* and *past part.* **spread**) **1** (often foll. by *out*) **a** open, extend, or unfold. **b** cause to cover a surface or larger area. **c** display thus. **2** (often foll. by *out*) have a wide, specified, or increasing extent. **3** become or make widely known, felt, etc. (*rumours are spreading*). **4 a** cover (*spread the wall with paint*). **b** lay (a table). –*n.* **1** act of spreading. **2** capability or extent of spreading (*has a large spread*). **3** diffusion (*spread of learning*). **4** breadth. **5** increased girth (*middle-aged spread*). **6** difference between two rates, prices, etc. **7** *colloq.* elaborate meal. **8** paste for spreading on bread etc. **9** bedspread. **10** printed matter spread across more than one column. □ **spread oneself** be lavish or discursive. **spread one's wings** develop one's powers fully. [Old English]

spread eagle –*n.* figure of an eagle with legs and wings extended as an emblem. –*v.* (**spread-eagle**) **1** (usu. as **spread-eagled** *adj.*) place (a person) with arms and legs spread out. **2** defeat utterly.

spreadsheet *n.* computer program for the manipulation and retrieval of esp. tabulated figures, esp. for accounting.

spree *n. colloq.* **1** extravagant outing (*shopping spree*). **2** bout of fun or drinking etc. [origin unknown]

sprig –*n.* **1** small branch or shoot. **2** ornament resembling this, esp. on fabric. –*v.* (**-gg-**) ornament with sprigs (*sprigged muslin*). [Low German *sprick*]

sprightly /'spraɪtlɪ/ *adj.* (**-ier**, **-iest**) vivacious, lively, brisk. □ **sprightliness** *n.* [from *spright*, var. of SPRITE]

spring –*v.* (*past* **sprang**; *past part.* **sprung**) **1** rise rapidly or suddenly, leap, jump. **2** move rapidly by or as by the action of a spring. **3** (usu. foll. by *from*) originate (from ancestors, a source, etc.). **4** act or appear suddenly or unexpectedly (*a breeze sprang up*; *spring to mind*; *spring to life*). **5** (often foll. by *on*) present (a thing or circumstance etc.) suddenly or unexpectedly (*sprang it on me*). **6** *slang* contrive the escape of (a person from prison etc.). **7** rouse (game) from a covert etc. **8** (usu. as **sprung** *adj.*) provide (a mattress etc.) with springs. –*n.* **1** jump, leap. **2** recoil. **3** elasticity. **4** elastic device, usu. of coiled metal, used esp. to drive clockwork or for cushioning in furniture or vehicles. **5 a** (often *attrib.*) the first season of the year, in which new vegetation begins to appear. **b** (often foll. by *of*) early stage of life etc. **6** place where water, oil, etc., wells up from the earth; basin or flow so formed. **7** motive for or origin of an action, custom, etc. □ **spring a leak** develop a leak. **spring up** come into being, appear. □ **springlike** *adj.* [Old English]

spring balance *n.* balance that measures weight by the tension of a spring.

springboard *n.* **1** flexible board for leaping or diving from. **2** source of impetus.

springbok /'sprɪŋbɒk/ *n.* (*pl.* same or **-s**) S. African gazelle. [Afrikaans]

spring chicken *n.* **1** young fowl for eating. **2** youthful person.

spring-clean –*n.* (also **spring-cleaning**) thorough cleaning of a house, esp. in spring. –*v.* clean (a house) thus.

spring equinox *n.* (also **vernal equinox**) equinox about 20 March.

springer *n.* small spaniel of a breed used to spring game.

spring fever *n.* restlessness or lethargy associated with spring.

spring greens *n.pl.* young cabbage leaves.

spring onion *n.* young onion eaten raw.

spring roll *n.* Chinese fried pancake filled with vegetables.

spring tide *n.* tide just after the new and the full moon when there is the greatest difference between high and low water.

springtime *n.* season of spring.

springy *adj.* (**-ier**, **-iest**) springing back quickly when squeezed, bent, or stretched; elastic. □ **springiness** *n.*

sprinkle /'sprɪŋk(ə)l/ –*v.* (**-ling**) **1** scatter in small drops or particles. **2** (often foll. by *with*) subject to sprinkling with liquid etc. **3** (of liquid etc.) fall on in this way. **4** distribute in small amounts. –*n.* (usu. foll. by *of*) **1** light shower. **2** = SPRINKLING. [origin uncertain]

sprinkler *n.* device for sprinkling a lawn or extinguishing fires.

sprinkling *n.* small sparse number or amount.

sprint –*v.* **1** run a short distance at full speed. **2** run (a specified distance) thus. –*n.* **1** such a run. **2** short burst in cycling, swimming, etc. □ **sprinter** *n.* [Old Norse]

sprit *n.* small diagonal spar from the mast to the upper outer corner of a sail. [Old English]

sprite *n.* elf, fairy. [*sprit*, contraction of SPIRIT]

spritsail /'sprɪts(ə)l/ *n.* sail extended by a sprit.

spritzer /'sprɪtsə(r)/ *n.* drink of wine with soda water. [German, = a splash]

sprocket /'sprɒkɪt/ *n.* each of several teeth on a wheel engaging with links of a chain. [origin unknown]

sprout –*v.* **1** put forth (shoots, hair, etc.). **2** begin to grow. –*n.* **1** shoot of a plant. **2** = BRUSSELS SPROUT. [Old English]

spruce[1] /spru:s/ –*adj.* neatly dressed etc.; smart. –*v.* (**-cing**) (usu. foll. by *up*) make or become smart. □ **sprucely** *adv.* **spruceness** *n.* [perhaps from SPRUCE[2]]

spruce[2] /spru:s/ *n.* **1** conifer with dense conical foliage. **2** its wood. [obsolete *Pruce* Prussia]

sprung see SPRING.

spry /spraɪ/ *adj.* (**spryer**, **spryest**) lively, nimble. □ **spryly** *adv.* [origin unknown]

spud –*n.* **1** *colloq.* potato. **2** small narrow spade for weeding. –*v.* (**-dd-**) (foll. by *up*, *out*) remove with a spud. [origin unknown]

spue var. of SPEW.

spumante /spu:'mæntɪ/ *n.* Italian sparkling white wine. [Italian, = sparkling]

spume *n.* & *v.* (**-ming**) froth, foam. □ **spumy** *adj.* (**-ier**, **-iest**). [Latin *spuma*]
spun *past* and *past part.* of SPIN.
spunk *n.* **1** *colloq.* courage, mettle, spirit. **2** *coarse slang* semen. **3** touchwood. [origin unknown]
spunky *adj.* (**-ier**, **-iest**) *colloq.* brave, spirited.
spun silk *n.* cheap material containing waste silk.
spur —*n.* **1** small spike or spiked wheel worn on a rider's heel for urging on a horse. **2** stimulus, incentive. **3** spur-shaped thing, esp.: **a** a projection from a mountain or mountain range. **b** a branch road or railway. **c** a hard projection on a cock's leg. —*v.* (**-rr-**) **1** prick (a horse) with spurs. **2** incite or stimulate. □ **on the spur of the moment** on impulse. [Old English]
spurge *n.* plant with an acrid milky juice. [Latin *expurgare* to clean out]
spurious /ˈspjʊərɪəs/ *adj.* not genuine, fake. [Latin]
spurn *v.* reject with disdain or contempt. [Old English]
spurt —*v.* **1** (cause to) gush out in a jet or stream. **2** make a sudden effort. —*n.* **1** sudden gushing out, jet. **2** short burst of speed, growth, etc. [origin unknown]
sputnik /ˈspʊtnɪk, ˈspʌt-/ *n.* Russian artificial satellite orbiting the earth. [Russian]
sputter —*v.* make a series of quick explosive sounds, splutter. —*n.* this sound. [Dutch (imitative)]
sputum /ˈspju:təm/ *n.* (*pl.* **sputa**) **1** saliva. **2** expectorated matter, used esp. in diagnosis. [Latin]
spy /spaɪ/ —*n.* (*pl.* **spies**) **1** person who secretly collects and reports information for a government, company, etc. **2** person watching others secretly. —*v.* (**spies, spied**) **1** discern, see. **2** (often foll. by *on*) act as a spy. **3** (often foll. by *into*) pry. □ **spy out** explore or discover, esp. secretly. [French *espie*, *espier*]
spyglass *n.* small telescope.
spyhole *n.* peep-hole.
sq. *abbr.* square.
Sqn. Ldr. *abbr.* squadron leader.
squab /skwɒb/ —*n.* **1** young (esp. unfledged) pigeon or other bird. **2** short fat person. **3** stuffed cushion, esp. as part of a car-seat. **4** sofa, ottoman. —*adj.* short and fat, squat. [perhaps from Scandinavian]
squabble /ˈskwɒb(ə)l/ —*n.* petty or noisy quarrel. —*v.* (**-ling**) engage in this. [probably imitative]
squad /skwɒd/ *n.* **1** small group sharing a task etc., esp. of soldiers or police officers (*drug squad*). **2** *Sport* team. [French *escouade*]
squad car *n.* police car.
squaddie *n.* (also **squaddy**) (*pl.* **-ies**) *slang* recruit; private.
squadron /ˈskwɒdrən/ *n.* **1** unit of the RAF with 10–18 aircraft. **2** detachment of warships employed on a particular duty. **3** organized group etc., esp. a cavalry division of two troops. [Italian *squadrone*: related to SQUAD]
squadron leader *n.* commander of an RAF squadron, next below Wing Commander.
squalid /ˈskwɒlɪd/ *adj.* **1** filthy, dirty. **2** mean or poor in appearance. [Latin]
squall /skwɔ:l/ —*n.* **1** sudden or violent wind, esp. with rain, snow, or sleet. **2** discordant cry; scream (esp. of a baby). —*v.* **1** utter a squall; scream. **2** utter with a squall. □ **squally** *adj.* [probably alteration of SQUEAL after BAWL]
squalor /ˈskwɒlə(r)/ *n.* filthy or squalid state. [Latin]
squander /ˈskwɒndə(r)/ *v.* spend wastefully. [origin unknown]
square /skweə(r)/ —*n.* **1** rectangle with four equal sides. **2** object of (approximately) this shape. **3** open (usu. four-sided) area surrounded by buildings. **4** product of a number multiplied by itself (*16 is the square of 4*). **5** L- or T-shaped instrument for obtaining or testing right angles. **6** *slang* conventional or old-fashioned person. —*adj.* **1** square-shaped. **2** having or in the form of a right angle (*square corner*). **3** angular, not round. **4** designating a unit of measure equal to the area of a square whose side is one of the unit specified (*square metre*). **5** (often foll. by *with*) level, parallel. **6** (usu. foll. by *to*) at right angles. **7** sturdy, squat (*a man of square frame*). **8** arranged; settled (*get things square*). **9** (also **all square**) **a** with no money owed. **b** (of scores) equal. **10** fair and honest. **11** direct (*met with a square refusal*). **12** *slang* conventional or old-fashioned. —*adv.* **1** squarely (*hit me square on the jaw*). **2** fairly, honestly. —*v.* (**-ring**) **1** make square. **2** multiply (a number) by itself. **3** (usu. foll. by *to*, *with*) adjust; make or be suitable or consistent; reconcile. **4** mark out in squares. **5** settle or pay (a bill etc.). **6** place (one's shoulders etc.) squarely facing forwards. **7** *colloq.* pay or bribe (a person). **8** (also *absol.*) make the scores of (a match etc.) equal. □ **back to square one** *colloq.* back to the starting-point with no progress made. **out of square** not at right angles. **square the circle 1** construct a square equal in area to a given circle. **2** do what is impossible. **square peg in a round hole** see PEG. **square up** settle an account etc. **square**

up to 1 move threateningly towards (a person). **2** face and tackle (a difficulty etc.) resolutely. □ **squarely** *adv.* **squareness** *n.* **squarish** *adj.* [French *esquare*, Latin *quadra*]

square-bashing *n. slang* military drill on a barrack-square.

square brackets *n.pl.* brackets of the form [ ].

square dance *n.* dance with usu. four couples facing inwards from four sides.

square deal *n.* fair bargain or treatment.

square leg *n.* fielding position in cricket at some distance on the batsman's leg side and nearly opposite the stumps.

square meal *n.* substantial meal.

square measure *n.* measure expressed in square units.

square-rigged *adj.* with the principal sails at right angles to the length of the ship.

square root *n.* number that multiplied by itself gives a specified number.

squash[1] /skwɒʃ/ –*v.* **1** crush or squeeze, esp. flat or into pulp. **2** (often foll. by *into*) *colloq.* put or make one's way by squeezing. **3** belittle, bully (a person). **4** suppress (a proposal, allegation, etc.). –*n.* **1** crowd; crowded state. **2** drink made of crushed fruit. **3** (in full **squash rackets**) game played with rackets and a small ball in a closed court. □ **squashy** *adj.* (**-ier, -iest**). [French *esquasser*: related to EX-[1], QUASH]

squash[2] /skwɒʃ/ *n.* (*pl.* same or **-es**) **1** trailing annual plant. **2** edible gourd of this. [Narragansett]

squat /skwɒt/ –*v.* (**-tt-**) **1** sit on one's heels or on the ground with the knees drawn up. **2** *colloq.* sit down. **3** occupy a building as a squatter. –*adj.* (**squatter, squattest**) short and thick, dumpy. –*n.* **1** squatting posture. **2** place occupied by squatters. [French *esquatir* flatten]

squatter *n.* person who inhabits unoccupied premises without permission.

squaw *n.* N. American Indian woman or wife. [Narragansett]

squawk –*n.* **1** loud harsh cry, esp. of a bird. **2** complaint. –*v.* utter a squawk. [imitative]

squeak –*n.* **1** short high-pitched cry or sound. **2** (also **narrow squeak**) narrow escape. –*v.* **1** make a squeak. **2** utter (words) shrilly. **3** (foll. by *by, through*) *colloq.* pass narrowly. **4** *slang* turn informer. [imitative: related to SQUEAL, SHRIEK]

squeaky *adj.* (**-ier, -iest**) making a squeaking sound. □ **squeakily** *adv.* **squeakiness** *n.*

squeaky clean *adj.* (usu. hyphenated when *attrib.*) *colloq.* **1** completely clean. **2** above criticism.

squeal –*n.* prolonged shrill sound or cry. –*v.* **1** make, or utter with, a squeal. **2** *slang* turn informer. **3** *colloq.* protest vociferously. [imitative]

squeamish /'skwi:mɪʃ/ *adj.* **1** easily nauseated or disgusted. **2** fastidious. □ **squeamishly** *adv.* **squeamishness** *n.* [Anglo-French *escoymos*]

squeegee /'skwi:dʒi:/ *n.* rubber-edged implement on a handle, for cleaning windows etc. [*squeege*, alteration of SQUEEZE]

squeeze –*v.* (**-zing**) **1** (often foll. by *out*) **a** exert pressure on, esp. to extract moisture etc. **b** extract (moisture) by squeezing. **2** reduce in size or alter in shape by squeezing. **3** force or push into or through a small or narrow space. **4 a** harass or pressure (a person). **b** (usu. foll. by *out of*) obtain by extortion, entreaty, etc. **5** press (a person's hand) in sympathy etc. –*n.* **1** squeezing or being squeezed. **2** close embrace. **3** crowd, crowded state. **4** small quantity produced by squeezing (*squeeze of lemon*). **5** restriction on borrowing, investment, etc., in a financial crisis. □ **put the squeeze on** *colloq.* coerce or pressure. [origin unknown]

squeeze-box *n. colloq.* accordion or concertina.

squelch –*v.* **1 a** make a sucking sound as of treading in thick mud. **b** move with a squelching sound. **2** disconcert, silence. –*n.* act or sound of squelching. □ **squelchy** *adj.* [imitative]

squib *n.* **1** small hissing firework that finally explodes. **2** satirical essay. [perhaps imitative]

squid *n.* (*pl.* same or **-s**) ten-armed marine cephalopod used as food. [origin unknown]

squidgy /'skwɪdʒɪ/ *adj.* (**-ier, -iest**) *colloq.* squashy, soggy. [imitative]

squiffy /'skwɪfɪ/ *adj.* (**-ier, -iest**) *slang* slightly drunk. [origin unknown]

squiggle /'skwɪg(ə)l/ *n.* short curly line, esp. in handwriting. □ **squiggly** *adj.* [imitative]

squill *n.* bulbous plant resembling a bluebell. [Latin *squilla*]

squint –*v.* **1** have eyes that do not move together but look in different directions. **2** (often foll. by *at*) look obliquely or with half-closed eyes. –*n.* **1** condition causing squinting. **2** stealthy or sidelong glance. **3** *colloq.* glance, look. **4** oblique opening in a church wall affording a view of the altar. [obsolete *asquint*, perhaps from Dutch *schuinte* slant]

squire –*n.* **1** country gentleman, esp. the chief landowner of a district. **2** *hist.* knight's attendant. –*v.* (**-ring**) (of a man) attend or escort (a woman). [related to ESQUIRE]

squirearchy /'skwaɪə,rɑ:kɪ/ *n.* (*pl.* **-ies**) landowners collectively.

squirm –*v.* **1** wriggle, writhe. **2** show or feel embarrassment. –*n.* squirming movement. [imitative]

squirrel /'skwɪr(ə)l/ –*n.* **1** bushy-tailed usu. tree-living rodent. **2** its fur. **3** hoarder. –*v.* (**-ll-**; *US* **-l-**) **1** (often foll. by *away*) hoard. **2** (often foll. by *around*) bustle about. [Greek *skiouros*, from *skia* shade, *oura* tail]

squirt –*v.* **1** eject (liquid etc.) in a jet. **2** be ejected in this way. **3** splash with a squirted substance. –*n.* **1 a** jet of water etc. **b** small quantity squirted. **2** syringe. **3** *colloq.* insignificant but self-assertive person. [imitative]

squish *colloq.* –*n.* slight squelching sound. –*v.* move with a squish. □ **squishy** *adj.* (**-ier**, **-iest**). [imitative]

Sr *symb.* strontium.

Sr. *abbr.* **1** Senior. **2** Señor. **3** Signor.

SRN *abbr.* State Registered Nurse.

SS *abbr.* **1** steamship. **2** *hist.* Nazi special police force. **3** Saints. [sense 2 from German *Schutz-Staffel*]

SSE *abbr.* south-south-east.

SSW *abbr.* south-south-west.

St *abbr.* Saint.

St. *abbr.* Street.

st. *abbr.* stone (in weight).

stab –*v.* (**-bb-**) **1** pierce or wound with a knife etc. **2** (often foll. by *at*) aim a blow with such a weapon. **3** cause a sensation like being stabbed (*stabbing pain*). **4** hurt or distress (a person, feelings, etc.). –*n.* **1** act of stabbing. **2** wound from this. **3** *colloq.* attempt. [origin unknown]

stability /stə'bɪlɪtɪ/ *n.* being stable. [Latin: related to STABLE]

stabilize /'steɪbɪ,laɪz/ *v.* (also **-ise**) (**-zing** or **-sing**) make or become stable. □ **stabilization** /-'zeɪʃ(ə)n/ *n.*

stabilizer *n.* (also **-iser**) **1** device used to keep esp. a ship, aircraft, or (in *pl.*) child's bicycle stable. **2** food additive for preserving texture.

stab in the back –*n.* treacherous attack. –*v.* betray.

stable /'steɪb(ə)l/ –*adj.* (**-bler**, **-blest**) **1** firmly fixed or established; not likely to move or change. **2** (of a person) not easily upset or disturbed. –*n.* **1** building for keeping horses. **2** establishment for training racehorses. **3** racehorses from one stable. **4** persons, products, etc., having a common origin or affiliation. **5** such an origin or affiliation. –*v.* (**-ling**) put or keep in a stable. □ **stably** *adv.* [Latin *stabilis* from *sto* to stand]

stable-companion *n.* (also **stable-mate**) **1** horse of the same stable. **2** member of the same organization.

stabling *n.* accommodation for horses.

staccato /stə'kɑ:təʊ/ esp. Mus. –*adv.* & *adj.* with each sound or note sharply distinct. –*n.* (*pl.* **-s**) staccato passage or delivery. [Italian]

stack –*n.* **1** (esp. orderly) pile or heap. **2** = HAYSTACK. **3** *colloq.* large quantity (*a stack of work*; *stacks of money*). **4 a** = CHIMNEY-STACK. **b** = SMOKESTACK. **c** tall factory chimney. **5** stacked group of aircraft. **6** part of a library where books are compactly stored. **7** high detached rock, esp. off the coast of Scotland. –*v.* **1** pile in a stack or stacks. **2 a** arrange (cards) secretly for cheating. **b** manipulate (circumstances etc.) to suit one. **3** cause (aircraft) to fly in circles while waiting to land. [Old Norse]

stadium /'steɪdɪəm/ *n.* (*pl.* **-s**) athletic or sports ground with tiered seats for spectators. [Greek *stadion*]

staff /stɑ:f/ –*n.* **1 a** stick or pole for use in walking or as a weapon. **b** stick or rod as a sign of office etc. **c** person or thing that supports. **2 a** people employed in a business etc. **b** those in authority in a school etc. **c** body of officers assisting an officer in high command (*general staff*). **3** (*pl.* **-s** or **staves**) *Mus.* set of usu. five parallel lines on or between which notes are placed to indicate their pitch. –*v.* provide (an institution etc.) with staff. [Old English]

staff college *n.* college where officers are trained for staff duties.

staff nurse *n.* nurse ranking just below a sister.

staff sergeant *n.* senior sergeant of a non-infantry company.

stag *n.* **1** adult male deer. **2** *Stock Exch. slang* person who applies for new shares, intending to sell at once for a profit. [Old English]

stag beetle *n.* beetle with branched mandibles like antlers.

stage –*n.* **1** point or period in a process or development. **2 a** raised platform, esp. for performing plays etc. on. **b** (prec. by *the*) theatrical profession, drama. **c** scene of action. **3 a** regular stopping-place on a route. **b** distance between two of these. **4** *Astronaut.* section of a rocket with a separate engine. –*v.* (**-ging**) **1** present (a play etc.) on stage. **2** arrange, organize (*staged a demonstration*). [French *estage*, ultimately from Latin *sto* stand]

stagecoach *n. hist.* large closed horse-drawn coach running on a regular route by stages.
stagecraft *n.* theatrical skill or experience.
stage direction *n.* instruction in a play as to actors' movements, sound effects, etc.
stage fright *n.* performer's fear of an audience.
stage-hand *n.* person moving stage scenery etc.
stage-manage *v.* **1** be the stage-manager of. **2** arrange and control for effect.
stage-manager *n.* person responsible for lighting and mechanical arrangements etc. on stage.
stage-struck *adj.* obsessed with becoming an actor.
stage whisper *n.* **1** an aside. **2** loud whisper meant to be overheard.
stagey var. of STAGY.
stagflation /stæg'fleɪʃ(ə)n/ *n. Econ.* state of inflation without a corresponding increase of demand and employment. [blend of *stagnation*, *inflation*]
stagger –*v.* **1** (cause to) walk unsteadily. **2** shock, confuse. **3** arrange (events etc.) so that they do not coincide. **4** arrange (objects) so that they are not in line. –*n.* **1** tottering movement. **2** (in *pl.*) disease, esp. of horses and cattle, causing staggering. [Old Norse]
staggering *adj.* astonishing, bewildering. □ **staggeringly** *adv.*
staghound *n.* large dog used for hunting deer.
staging /'steɪdʒɪŋ/ *n.* **1** presentation of a play etc. **2 a** platform or support, esp. temporary. **b** shelves for plants in a greenhouse.
staging post *n.* regular stopping-place, esp. on an air route.
stagnant /'stægnənt/ *adj.* **1** (of liquid) motionless, having no current. **2** dull, sluggish. □ **stagnancy** *n.* [Latin *stagnum* pool]
stagnate /stæg'neɪt/ *v.* **(-ting)** be or become stagnant. □ **stagnation** *n.*
stag-party *n. colloq.* all-male celebration held esp. for a man about to marry.
stagy /'steɪdʒɪ/ *adj.* (also **stagey**) **(-ier, -iest)** theatrical, artificial, exaggerated.
staid *adj.* of quiet and steady character; sedate. [= stayed, past part. of STAY[1]]
stain –*v.* **1** discolour or be discoloured by the action of liquid sinking in. **2** spoil, damage (a reputation, character, etc.). **3** colour (wood, glass, etc.) with a penetrating substance. **4** impregnate (a specimen) with a colouring agent for microscopic examination. –*n.* **1** discoloration; spot, mark. **2** blot, blemish; damage to a reputation etc. **3** substance used in staining. [earlier *distain* from French *desteindre*]
stained glass *n.* coloured glass in a leaded window etc.
stainless *adj.* **1** without stains. **2** not liable to stain.
stainless steel *n.* chrome steel resisting rust and tarnish.
stair *n.* **1** each of a set of fixed indoor steps. **2** (usu. in *pl.*) set of such steps. [Old English]
staircase *n.* flight of stairs and the supporting structure.
stair-rod *n.* rod securing a carpet between two steps.
stairway *n.* = STAIRCASE.
stairwell *n.* shaft for a staircase.
stake[1] –*n.* **1** stout sharpened stick driven into the ground as a support, boundary mark, etc. **2** *hist.* **a** post to which a condemned person was tied to be burnt alive. **b** (prec. by *the*) such death as a punishment. –*v.* **(-king) 1** secure or support with a stake or stakes. **2** (foll. by *off*, *out*) mark off (an area) with stakes. **3** establish (a claim). □ **stake out** *colloq.* place under surveillance. [Old English]
stake[2] –*n.* **1** sum of money etc. wagered on an event. **2** (often foll. by *in*) interest or concern, esp. financial. **3** (in *pl.*) **a** prize-money, esp. in a horse-race. **b** such a race. –*v.* **1** wager. **2** *US colloq.* support, esp. financially. □ **at stake** risked, to be won or lost. [Old English]
stakeholder *n.* independent party with whom money etc. wagered is deposited.
stake-out *n.* esp. *US colloq.* period of surveillance.
Stakhanovite /stə'kɑːnə,vaɪt/ *n.* (often *attrib.*) exceptionally productive worker. [*Stakhanov*, name of a Russian coalminer]
stalactite /'stælək,taɪt/ *n.* icicle-like deposit of calcium carbonate hanging from the roof of a cave etc. [Greek *stalaktos* dripping]
stalagmite /'stæləg,maɪt/ *n.* icicle-like deposit of calcium carbonate rising from the floor of a cave etc. [Greek *stalagma* a drop]
stale –*adj.* **1 a** not fresh. **b** musty, insipid, or otherwise the worse for age or use. **2** trite, unoriginal (*stale joke*). **3** (of an athlete or performer) impaired by excessive training. –*v.* **(-ling)** make or become stale. □ **staleness** *n.* [Anglo-French *estaler* halt]
stalemate –*n.* **1** *Chess* position counting as a draw, in which a player cannot move except into check. **2** deadlock. –*v.* **(-ting) 1** *Chess* bring (a player) to a

stalemate. **2** bring to a deadlock. [obsolete *stale*: related to STALE, MATE[2]]

Stalinism /ˈstɑːlɪˌnɪz(ə)m/ *n.* centralized authoritarian form of socialism associated with Stalin. □ **Stalinist** *n.* & *adj.* [*Stalin*, a Soviet statesman]

stalk[1] /stɔːk/ *n.* **1** main stem of a herbaceous plant. **2** slender attachment or support of a leaf, flower, fruit, etc. **3** similar support for an organ etc. in an animal. [diminutive of (now dial.) *stale* rung]

stalk[2] /stɔːk/ *–v.* **1** pursue (game or an enemy) stealthily. **2** stride, walk in a haughty manner. **3** *formal* or *rhet.* move silently or threateningly through (a place) (*fear stalked the land*). *–n.* **1** stalking of game. **2** haughty gait. [Old English: related to STEAL]

stalking-horse *n.* **1** horse concealing a hunter. **2** pretext concealing one's real intentions or actions. **3** weak political candidate forcing an election in the hope of a more serious contender coming forward.

stall[1] /stɔːl/ *–n.* **1** trader's booth or table in a market etc. **2** compartment for one animal in a stable or cowhouse. **3** fixed, usu. partly enclosed, seat in the choir or chancel of a church. **4** (usu. in *pl.*) each of the seats on the ground floor of a theatre. **5 a** compartment for one person in a shower-bath etc. **b** compartment for one horse at the start of a race. **6 a** stalling of an engine or aircraft. **b** condition resulting from this. *–v.* **1** (of a vehicle or its engine) stop because of an overload on the engine or an inadequate supply of fuel to it. **2** (of an aircraft or its pilot) lose control because the speed is too low. **3** cause to stall. [Old English]

stall[2] /stɔːl/ *v.* **1** play for time when being questioned etc. **2** delay, obstruct. [*stall* 'decoy': probably related to STALL[1]]

stallholder *n.* person in charge of a stall at a market etc.

stallion /ˈstæljən/ *n.* uncastrated adult male horse. [French *estalon*]

stalwart /ˈstɔːlwət/ *–adj.* **1** strong, sturdy. **2** courageous, resolute, reliable. *–n.* stalwart person, esp. a loyal comrade. [Old English, = place, WORTH]

stamen /ˈsteɪmən/ *n.* organ producing pollen in a flower. [Latin, = warp, thread]

stamina /ˈstæmɪnə/ *n.* physical or mental endurance. [Latin, pl. of STAMEN]

stammer *–v.* **1** speak haltingly, esp. with pauses or rapid repetitions of the same syllable. **2** (often foll. by *out*) utter (words) in this way. *–n.* **1** tendency to stammer. **2** instance of stammering. [Old English]

stamp *–v.* **1 a** bring down (one's foot) heavily, esp. on the ground. **b** (often foll. by *on*) crush or flatten in this way. **c** walk heavily. **2 a** impress (a design, mark, etc.) on a surface. **b** impress (a surface) with a pattern etc. **3** affix a postage or other stamp to. **4** assign a specific character to; mark out. *–n.* **1** instrument for stamping. **2 a** mark or design made by this. **b** impression of an official mark required to be made on deeds, bills of exchange, etc., as evidence of payment of tax. **3** small adhesive piece of paper indicating that payment has been made, esp. a postage stamp. **4** mark or label etc. on a commodity as evidence of quality etc. **5** act or sound of stamping the foot. **6** characteristic mark or quality. □ **stamp on 1** impress (an idea etc.) on (the memory etc.). **2** suppress. **stamp out 1** produce by cutting out with a die etc. **2** put an end to, destroy. [Old English]

stamp-collector *n.* philatelist.

stamp-duty *n.* duty imposed on certain legal documents.

stampede /stæmˈpiːd/ *–n.* **1** sudden flight or hurried movement of animals or people. **2** response of many persons at once to a common impulse. *–v.* (**-ding**) (cause to) take part in a stampede. [Spanish *estampida* crash, uproar]

stamping-ground *n. colloq.* favourite haunt.

stance /stɑːns, stæns/ *n.* **1** standpoint; attitude. **2** attitude or position of the body, esp. when hitting a ball etc. [Italian *stanza* standing]

stanch /stɑːntʃ, stɔːntʃ/ *v.* (also **staunch**) **1** restrain the flow of (esp. blood). **2** restrain the flow from (esp. a wound). [French *estanchier*]

stanchion /ˈstɑːnʃ(ə)n/ *n.* **1** upright post or support. **2** upright bar or frame for confining cattle in a stall. [Anglo-French]

stand *–v.* (*past* and *past part.* **stood** /stʊd/) **1** have, take, or maintain an upright position, esp. on the feet or a base. **2** be situated (*here once stood a village*). **3** be of a specified height. **4** be in a specified state (*stands accused; it stands as follows*). **5** set in an upright or specified position (*stood it against the wall*). **6 a** move to and remain in a specified position (*stand aside*). **b** take a specified attitude (*stand aloof*). **7** maintain a position; avoid falling, moving, or being moved. **8** assume a stationary position; cease to move. **9** remain valid or unaltered. **10** *Naut.* hold a specified course. **11** endure, tolerate. **12** provide at one's own expense (*stood him a* *drink*). **13** (often foll. by *for*) be a candidate (for office etc.) (*stood for Parliament*). **14** act in a specified capacity (*stood proxy*). **15** undergo (trial). *–n.* **1** cessation from progress, stoppage. **2 a** *Mil.* halt made to repel an attack. **b** resistance to attack or compulsion (esp. *make a stand*). **c** *Cricket* prolonged period at the wicket by two batsmen. **3** position taken up; attitude adopted. **4** rack, set of shelves, etc. for storage. **5** open-fronted stall or structure for a trader, exhibitor, etc. **6** standing-place for vehicles. **7 a** raised structure to sit or stand on. **b** *US* witness-box. **8** each halt made for a performance on a tour. **9** group of growing plants (*stand of trees*). □ **as it stands 1** in its present condition. **2** in the present circumstances. **stand by 1** stand nearby; look on without interfering. **2** uphold, support (a person). **3** adhere to (a promise etc.). **4** be ready for action. **stand a chance** see CHANCE. **stand corrected** accept correction. **stand down** withdraw from a position or candidacy. **stand for 1** represent, signify, imply. **2** *colloq.* endure, tolerate. **stand one's ground** not yield. **stand in** (usu. foll. by *for*) deputize. **stand off 1** move or keep away. **2** temporarily dismiss (an employee). **stand on** insist on, observe scrupulously. **stand on one's own feet** (or **own two feet**) be self-reliant or independent. **stand out 1** be prominent or outstanding. **2** (usu. foll. by *against, for*) persist in opposition or support. **stand to 1** *Mil.* stand ready for an attack. **2** abide by. **3** be likely or certain to. **stand to reason** be obvious. **stand up 1 a** rise to one's feet. **b** come to, remain in, or place in a standing position. **2** (of an argument etc.) be valid. **3** *colloq.* fail to keep an appointment with. **stand up for** support, side with. **stand up to 1** face (an opponent) courageously. **2** be resistant to (wear, use, etc.). **take one's stand on** base one's argument etc. on, rely on. [Old English]

standard /'stændəd/ *–n.* **1** object, quality, or measure serving as a basis, example, or principle to which others conform or should conform or by which others are judged. **2 a** level of excellence etc. required or specified (*not up to standard*). **b** average quality (*of a low standard*). **3** ordinary procedure etc. **4** distinctive flag. **5 a** upright support. **b** upright pipe. **6 a** tree or shrub that stands without support. **b** shrub grafted on an upright stem and trained in tree form. **7** tune or song of established popularity. *–adj.* **1** serving or used as a standard. **2** of a normal or prescribed quality, type, or size. **3** of recognized and permanent value; authoritative (*standard book on jazz*). **4** (of language) conforming to established educated usage. [Anglo-French: related to EXTEND, and in senses 5 and 6 of *n.* influenced by STAND]

standard assessment task *n.* standard test given to schoolchildren.

standard-bearer *n.* **1** soldier who carries a standard. **2** prominent leader in a cause.

standardize *v.* (also **-ise**) (**-zing** or **-sing**) cause to conform to a standard. □ **standardization** /-'zeɪʃ(ə)n/ *n.*

standard lamp *n.* lamp on a tall upright with a base.

standard of living *n.* degree of material comfort of a person or group.

standard time *n.* uniform time for places in approximately the same longitude, established in a country or region by law or custom.

stand-by *n.* (*pl.* **-bys**) **1** (often *attrib.*) person or thing ready if needed in an emergency etc. **2** readiness for duty (*on stand-by*).

stand-in *n.* deputy or substitute.

standing *–n.* **1** esteem or repute, esp. high; status. **2** duration (*of long standing*). *–adj.* **1** that stands, upright. **2** established, permanent (*a standing rule; a standing army*). **3** (of a jump, start, etc.) performed with no run-up. **4** (of water) stagnant.

standing committee *n.* committee that is permanent during the existence of the appointing body.

standing joke *n.* object of permanent ridicule.

standing order *n.* instruction to a banker to make regular payments, or to a retailer for a regular supply of goods.

standing orders *n.pl.* rules governing procedure in a parliament, council, etc.

standing ovation *n.* prolonged applause from an audience which has risen to its feet.

standing-room *n.* space to stand in.

stand-off half *n. Rugby* half-back forming a link between the scrum-half and the three-quarters.

standoffish /stænd'ɒfɪʃ/ *adj.* cold or distant in manner.

standpipe *n.* vertical pipe extending from a water supply, esp. one connecting a temporary tap to the mains.

standpoint *n.* point of view.

standstill *n.* stoppage; inability to proceed.

stand-up *attrib. adj.* **1** (of a meal) eaten standing. **2** (of a fight) violent and thorough. **3** (of a collar) not turned down. **4**

(of a comedian) telling jokes to an audience.

stank *past* of STINK.

stanza /'stænzə/ *n.* basic metrical unit of a poem etc., typically of four to twelve rhymed lines. [Italian]

staphylococcus /,stæfɪlə'kɒkəs/ *n.* (*pl.* **-cocci** /-kaɪ/) bacterium sometimes forming pus. □ **staphylococcal** *adj.* [Greek *staphulē* bunch of grapes, *kokkos* berry]

staple[1] /'steɪp(ə)l/ *–n.* U-shaped metal bar or piece of wire with pointed ends for driving into and holding papers together, or holding an electrical wire in place, etc. *–v.* (**-ling**) fasten or provide with a staple. □ **stapler** *n.* [Old English]

staple[2] /'steɪp(ə)l/ *–n.* **1** principal or important article of commerce (*staples of British industry*). **2** chief element or main component. **3** fibre of cotton or wool etc. with regard to its quality (*cotton of fine staple*). *–attrib. adj.* **1** main or principal (*staple diet*). **2** important as a product or export. [French *estaple* market]

star /stɑ:(r)/ *–n.* **1** celestial body appearing as a luminous point in the night sky. **2** large naturally luminous gaseous body such as the sun. **3** celestial body regarded as influencing fortunes etc. **4** thing like a star in shape or appearance. **5** decoration or mark of rank or excellence etc., usu. with radiating points. **6 a** famous or brilliant person; principal performer (*star of the show*). **b** (*attrib.*) outstanding (*star pupil*). *–v.* (**-rr-**) **1** appear or present as principal performer(s). **2** (esp. as **starred** *adj.*) mark, set, or adorn with a star or stars. □ **stardom** *n.* [Old English]

starboard /'stɑ:bəd, -bɔ:d/ *–n.* right-hand side of a ship or aircraft looking forward. *–v.* (also *absol.*) turn (the helm) to starboard. [Old English, = *steer board*]

starch *–n.* **1** polysaccharide obtained chiefly from cereals and potatoes. **2** preparation of this for stiffening fabric. **3** stiffness of manner; formality. *–v.* stiffen (clothing) with starch. [Old English: related to STARK]

starchy *adj.* (**-ier, -iest**) **1** of, like, or containing starch. **2** prim, formal. □ **starchily** *adv.* **starchiness** *n.*

stardust *n.* **1** multitude of stars looking like dust. **2** romance, magic feeling.

stare /steə(r)/ *–v.* (**-ring**) **1** (usu. foll. by *at*) look fixedly, esp. in curiosity, surprise, horror, etc. **2** reduce (a person) to a specified condition by staring (*stared me into silence*). *–n.* staring gaze. □ **stare a person in the face** be evident or imminent. **stare a person out** stare at a person until he or she looks away. [Old English]

starfish *n.* (*pl.* same or **-es**) echinoderm with five or more radiating arms.

star-gazer *n. colloq.* usu. *derog.* or *joc.* astronomer or astrologer.

stark *–adj.* **1** sharply evident (*in stark contrast*). **2** desolate, bare. **3** absolute (*stark madness*). *–adv.* completely, wholly (*stark naked*). □ **starkly** *adv.* **starkness** *n.* [Old English]

starkers /'stɑ:kəz/ *predic. adj. slang* stark naked.

starlet /'stɑ:lɪt/ *n.* promising young performer, esp. a film actress.

starlight *n.* light of the stars.

starling /'stɑ:lɪŋ/ *n.* gregarious bird with blackish speckled lustrous plumage. [Old English]

starlit *adj.* **1** lit by stars. **2** with stars visible.

Star of David *n.* two interlaced equilateral triangles used as a Jewish and Israeli symbol.

starry *adj.* (**-ier, -iest**) **1** full of stars. **2** like a star.

starry-eyed *adj. colloq.* **1** enthusiastic but impractical. **2** euphoric.

Stars and Stripes *n.pl.* national flag of the US.

star-studded *adj.* covered with stars; featuring many famous performers.

START /stɑ:t/ *abbr.* Strategic Arms Reduction Treaty (or Talks).

start *–v.* **1** begin. **2** set in motion or action (*started a fire*). **3** set oneself in motion or action. **4** begin a journey etc. **5** (often foll. by *up*) (cause to) begin operating. **6 a** cause or enable (a person) to make a beginning (*started me in business*). **b** (foll. by pres. part.) cause (a person) to begin (*started me coughing*). **7** (often foll. by *up*) establish. **8** give a signal to (competitors) to start in a race. **9** (often foll. by *up, from,* etc.) jump in surprise, pain, etc. **10** spring out, up, etc. **11** conceive (a baby). **12** rouse (game etc.). **13 a** (of timbers etc.) spring out; give way. **b** cause (timbers etc.) to do this. *–n.* **1** beginning. **2** place from which a race etc. begins. **3** advantage given at the beginning of a race etc. **4** advantageous initial position in life, business, etc. **5** sudden movement of surprise, pain, etc. □ **for a start** *colloq.* as a beginning. **start off** begin; begin to move. **start out** begin a journey. **start up** arise; occur. [Old English]

starter *n.* **1** device for starting a vehicle engine etc. **2** first course of a meal. **3** person giving the signal for the start of a race. **4** horse or competitor starting in a race. □ **for starters** *colloq.* to start with.

starting-block *n.* shaped block for a runner's feet at the start of a race.

starting price *n.* odds ruling at the start of a horse-race.

startle /'stɑ:t(ə)l/ *v.* (**-ling**) shock or surprise. [Old English]

star turn *n.* main item in an entertainment etc.

starve *v.* (**-ving**) **1** (cause to) die of hunger or suffer from malnourishment. **2** *colloq.* feel very hungry (*I'm starving*). **3 a** suffer from mental or spiritual want. **b** (foll. by *for*) feel a strong craving for. **4** (foll. by *of*) deprive of. **5** compel by starving (*starved into surrender*). □ **starvation** /-'veɪʃ(ə)n/ *n.* [Old English, = die]

starveling /'stɑ:vlɪŋ/ *n. archaic* starving person or animal.

Star Wars *n.pl. colloq.* strategic defence initiative.

stash *colloq.* –*v.* (often foll. by *away*) **1** conceal; put in a safe place. **2** hoard. –*n.* **1** hiding-place. **2** thing hidden. [origin unknown]

stasis /'steɪsɪs/ *n.* (*pl.* **stases** /-si:z/) **1** inactivity; stagnation. **2** stoppage of circulation. [Greek]

state –*n.* **1** existing condition or position of a person or thing. **2** *colloq.* **a** excited or agitated mental condition (esp. *in a state*). **b** untidy condition. **3** (usu. **State**) **a** political community under one government. **b** this as part of a federal republic. **4** (usu. **State**) (*attrib.*) **a** of, for, or concerned with the State. **b** reserved for or done on occasions of ceremony. **5** (usu. **State**) civil government. **6** pomp. **7** (**the States**) USA. –*v.* (**-ting**) **1** express in speech or writing. **2** fix, specify. **3** *Mus.* play (a theme etc.), esp. for the first time. □ **in state** with all due ceremony. **lie in state** be laid in a public place of honour before burial. [partly from ESTATE, partly from Latin STATUS]

stateless *adj.* having no nationality or citizenship.

stately *adj.* (**-ier, -iest**) dignified; imposing. □ **stateliness** *n.*

stately home *n.* large historic house, esp. one open to the public.

statement *n.* **1** stating or being stated; expression in words. **2** thing stated. **3** formal account of facts. **4** record of transactions in a bank account etc. **5** notification of the amount due to a tradesman etc.

state of emergency *n.* condition of danger or disaster in a country, with normal constitutional procedures suspended.

state of the art –*n.* current stage of esp. technological development. –*attrib. adj.* (usu. **state-of-the-art**) absolutely up-to-date (*state-of-the-art weaponry*).

stateroom *n.* **1** state apartment. **2** large private cabin in a passenger ship.

State school *n.* school largely managed and funded by the public authorities.

statesman *n.* (*fem.* **stateswoman**) distinguished and capable politician or diplomat. □ **statesmanlike** *adj.* **statesmanship** *n.*

static /'stætɪk/ –*adj.* **1** stationary; not acting or changing. **2** *Physics* concerned with bodies at rest or forces in equilibrium. –*n.* **1** static electricity. **2** atmospherics. □ **statically** *adv.* [Greek *statikos* from *sta-* stand]

static electricity *n.* electricity not flowing as a current.

statics *n.pl.* (usu. treated as *sing.*) **1** science of bodies at rest or of forces in equilibrium. **2** = STATIC.

station /'steɪʃ(ə)n/ –*n.* **1 a** regular stopping-place on a railway line. **b** buildings of this. **c** (in *comb.*) centre where vehicles of a specified type depart and arrive (*coach station*). **2** person or thing's allotted place or building etc. **3** centre for a particular service or activity. **4** establishment involved in broadcasting. **5 a** military or naval base. **b** inhabitants of this. **6** position in life; rank, status. **7** *Austral.* & *NZ* large sheep or cattle farm. –*v.* **1** assign a station to. **2** put in position. [Latin *statio* from *sto stat-* stand]

stationary *adj.* **1** not moving. **2** not meant to be moved. **3** unchanging. [Latin: related to STATION]

stationer *n.* dealer in stationery.

stationery *n.* writing-materials, office supplies, etc.

Stationery Office *n.* the Government's publishing house.

stationmaster *n.* official in charge of a railway station.

station of the cross *n. RC Ch.* each of a series of images representing the events in Christ's Passion before which prayers are said.

station-wagon *n.* esp. *US* estate car.

statistic /stə'tɪstɪk/ *n.* statistical fact or item. [German: related to STATE]

statistical *adj.* of statistics. □ **statistically** *adv.*

statistics *n.pl.* **1** (usu. treated as *sing.*) science of collecting and analysing significant numerical data. **2** such analysed data. □ **statistician** /,stætɪ'stɪʃ(ə)n/ *n.*

statuary /'stætʃʊərɪ/ –*adj.* of or for statues (*statuary art*). –*n.* (*pl.* **-ies**) **1** statues collectively. **2** making statues. **3** sculptor. [Latin: related to STATUE]

statue /'stætʃu:/ *n.* sculptured figure of a person or animal, esp. life-size or larger. [Latin *statua*]
statuesque /,stætʃʊ'esk/ *adj.* like, or having the dignity or beauty of, a statue.
statuette /,stætʃʊ'et/ *n.* small statue.
stature /'stætʃə(r)/ *n.* **1** height of a (esp. human) body. **2** calibre, esp. moral; eminence. [Latin *statura*]
status /'steɪtəs/ *n.* **1** rank, social position, relative importance. **2** superior social etc. position. [Latin: related to STATURE]
status quo /,steɪtəs 'kwəʊ/ *n.* existing state of affairs. [Latin]
status symbol *n.* a possession etc. intended to indicate the owner's superiority.
statute /'stætʃu:t/ *n.* **1** written law passed by a legislative body. **2** rule of a corporation, founder, etc., intended to be permanent. [Latin *statutum* from *statuo* set up]
statute-book *n.* **1** book(s) containing the statute law. **2** body of a country's statutes.
statute law *n.* **1** (*collect.*) body of principles and rules of law laid down in statutes. **2** a statute.
statute mile see MILE 1.
statutory /'statʃʊtərɪ/ *adj.* required or enacted by statute. □ **statutorily** *adv.*
staunch[1] /stɔ:ntʃ/ *adj.* **1** loyal. **2** (of a ship, joint, etc.) strong, watertight, airtight, etc. □ **staunchly** *adv.* [French *estanche*]
staunch[2] var. of STANCH.
stave *–n.* **1** each of the curved slats forming the sides of a cask, pail, etc. **2** = STAFF *n.* 3. **3** stanza or verse. *–v.* (**-ving**; *past* and *past part.* **stove** or **staved**) (usu. foll. by *in*) break a hole in, damage, crush by forcing inwards. □ **stave off** avert or defer (danger etc.). [from STAFF]
stay[1] *–v.* **1** continue in the same place or condition; not depart or change. **2** (often foll. by *at, in, with*) reside temporarily. **3** *archaic* or *literary* **a** stop or check. **b** (esp. in *imper.*) pause. **4** postpone (judgement etc.). **5** assuage (hunger etc.), esp. temporarily. *–n.* **1** act or period of staying. **2** suspension or postponement of a sentence, judgement, etc. **3** prop, support. **4** (in *pl.*) *hist.* (esp. boned) corset. □ **stay the course** endure to the end. **stay in** remain indoors. **stay the night** remain overnight. **stay put** *colloq.* remain where it is placed or where one is. **stay up** not go to bed (until late). [Anglo-French from Latin *sto* stand: sense 3 of *n.* from French, formed as STAY[2]]
stay[2] *n.* **1** *Naut.* rope or guy supporting a mast, flagstaff, etc. **2** supporting cable on an aircraft etc. [Old English from Germanic]
stay-at-home *–attrib. adj.* rarely going out. *–n.* such a person.
stayer *n.* person or animal with great endurance.
staying power *n.* endurance.
staysail /'steɪseɪl, -s(ə)l/ *n.* sail extended on a stay.
STD *abbr.* subscriber trunk dialling.
stead /sted/ *n.* □ **in a person's** (or **thing's**) **stead** as a substitute; in a person's (or thing's) place. **stand a person in good stead** be advantageous or useful to him or her. [Old English, = place]
steadfast /'stedfɑ:st/ *adj.* constant, firm, unwavering. □ **steadfastly** *adv.* **steadfastness** *n.* [Old English: related to STEAD]
steady /'stedɪ/ *–adj.* (**-ier, -iest**) **1** firmly fixed or supported; unwavering. **2** uniform and regular (*steady pace*; *steady increase*). **3 a** constant. **b** persistent. **4** (of a person) serious and dependable. **5** regular, established (*steady girlfriend*). *–v.* (**-ies, -ied**) make or become steady. *–adv.* steadily. *–n.* (*pl.* **-ies**) *colloq.* regular boyfriend or girlfriend. □ **go steady** (often foll. by *with*) *colloq.* have as a regular boyfriend or girlfriend. **steady on!** be careful! □ **steadily** *adv.* **steadiness** *n.* [from STEAD]
steady state *n.* unvarying condition, esp. in a physical process.
steak /steɪk/ *n.* **1** thick slice of meat (esp. beef) or fish, usu. grilled or fried. **2** beef cut for stewing or braising. [Old Norse]
steak-house *n.* restaurant specializing in beefsteaks.
steal *–v.* (*past* **stole**; *past part.* **stolen**) **1** (also *absol.*) take (another's property) illegally or without right or permission, esp. in secret. **2** obtain surreptitiously, insidiously, or artfully (*stole a kiss*). **3** (foll. by *in, out, away, up*, etc.) move, esp. silently or stealthily. *–n.* **1** *US colloq.* act of stealing or theft. **2** *colloq.* easy task or good bargain. □ **steal a march on** get an advantage over by surreptitious means. **steal the show** outshine other performers, esp. unexpectedly. **steal a person's thunder** take away the attention due to someone else by using his or her words, ideas, etc. [Old English]
stealth /stelθ/ *n.* secrecy, secret behaviour. [Old English: related to STEAL]
stealthy *adj.* (**-ier, -iest**) done or moving with stealth; furtive. □ **stealthily** *adv.* **stealthiness** *n.*
steam *–n.* **1 a** gas into which water is changed by boiling. **b** condensed vapour formed from this. **2 a** power obtained

from steam. **b** *colloq.* power or energy. —*v.* **1 a** cook (food) in steam. **b** treat with steam. **2** give off steam. **3 a** move under steam power. **b** (foll. by *ahead*, *away*, etc.) *colloq.* proceed or travel fast or with vigour. **4** (usu. foll. by *up*) **a** cover or become covered with condensed steam. **b** (as **steamed up** *adj.*) *colloq.* angry or excited. [Old English]

steamboat *n.* steam-driven boat.

steam engine *n.* **1** engine which uses steam to generate power. **2** locomotive powered by this.

steamer *n.* **1** steamship. **2** vessel for steaming food in.

steam hammer *n.* forging-hammer powered by steam.

steam iron *n.* electric iron that emits steam.

steamroller —*n.* **1** heavy slow-moving vehicle with a roller, used to flatten new-made roads. **2** a crushing power or force. —*v.* crush or move forcibly or indiscriminately; force.

steamship *n.* steam-driven ship.

steam train *n.* train pulled by a steam engine.

steamy *adj.* (**-ier**, **-iest**) **1** like or full of steam. **2** *colloq.* erotic. □ **steamily** *adv.* **steaminess** *n.*

steatite /'stɪətaɪt/ *n.* impure form of talc, esp. soapstone. [Greek *stear steat-* tallow]

steed *n. archaic* or *poet.* horse. [Old English]

steel —*n.* **1** strong malleable alloy of iron and carbon, used esp. for making tools, weapons, etc. **2** strength, firmness (*nerves of steel*). **3** steel rod for sharpening knives. —*adj.* of or like steel. —*v.* (also *refl.*) harden or make resolute. [Old English]

steel band *n.* band playing chiefly calypso-style music on percussion instruments made from oil drums.

steel wool *n.* abrasive substance consisting of a mass of fine steel shavings.

steelworks *n.pl.* (usu. treated as *sing.*) factory producing steel. □ **steelworker** *n.*

steely *adj.* (**-ier**, **-iest**) **1** of or like steel. **2** severe; resolute. □ **steeliness** *n.*

steelyard *n.* balance with a graduated arm along which a weight is moved.

steep[1] —*adj.* **1** sloping sharply. **2** (of a rise or fall) rapid. **3** (*predic.*) *colloq.* **a** exorbitant; unreasonable. **b** exaggerated; incredible. —*n.* steep slope; precipice. □ **steepen** *v.* **steepish** *adj.* **steeply** *adv.* **steepness** *n.* [Old English]

steep[2] —*v.* soak or bathe in liquid. —*n.* **1** act of steeping. **2** liquid for steeping. □ **steep in 1** pervade or imbue with. **2** make deeply acquainted with (a subject etc.). [Old English]

steeple /'sti:p(ə)l/ *n.* tall tower, esp. with a spire, above the roof of a church. [Old English: related to STEEP[1]]

steeplechase *n.* **1** horse-race with ditches, hedges, etc., to jump. **2** cross-country foot-race. □ **steeplechasing** *n.*

steeplejack *n.* repairer of tall chimneys, steeples, etc.

steer[1] *v.* **1** (also *absol.*) guide (a vehicle, ship, etc.) with a wheel or rudder etc. **2** direct or guide (one's course, other people, a conversation, etc.) in a specified direction. □ **steer clear of** avoid. □ **steering** *n.* [Old English]

steer[2] *n.* = BULLOCK. [Old English]

steerage *n.* **1** act of steering. **2** *archaic* cheapest part of a ship's accommodation.

steering-column *n.* column on which a steering-wheel is mounted.

steering committee *n.* committee deciding the order of business, the course of operations, etc.

steering-wheel *n.* wheel by which a vehicle etc. is steered.

steersman *n.* person who steers a ship.

stegosaurus /,stegə'sɔ:rəs/ *n.* (*pl.* **-ruses**) plant-eating dinosaur with a double row of bony plates along the spine. [Greek *stegē* covering, *sauros* lizard]

stela /'sti:lə/ *n.* (*pl.* **stelae** /-li:/) (also **stele** /sti:l, 'sti:lɪ/) *Archaeol.* upright slab or pillar usu. inscribed and sculpted, esp. as a gravestone. [Latin and Greek]

stellar /'stelə(r)/ *adj.* of a star or stars. [Latin *stella* star]

stem[1] —*n.* **1** main body or stalk of a plant. **2** stalk of a fruit, flower, or leaf. **3** stem-shaped part, as: **a** the slender part of a wineglass. **b** the tube of a tobacco-pipe. **c** a vertical stroke in a letter or musical note. **4** *Gram.* root or main part of a noun, verb, etc., to which inflections are added. **5** main upright timber at the bow of a ship (*from stem to stern*). —*v.* (**-mm-**) (foll. by *from*) spring or originate from. [Old English]

stem[2] *v.* (**-mm-**) check or stop. [Old Norse]

stench *n.* foul smell. [Old English: related to STINK]

stencil /'stensɪl/ —*n.* **1** (in full **stencil-plate**) thin sheet in which a pattern is cut, placed on a surface and printed or inked over etc. to reproduce the pattern. **2** pattern so produced. **3** waxed sheet etc. from which a stencil is made by a typewriter. —*v.* (**-ll-**; *US* **-l-**) **1** (often foll. by *on*) produce (a pattern) with a stencil. **2** mark (a surface) in this way. [French *estanceler* sparkle, from Latin *scintilla* spark]

Sten gun *n.* lightweight sub-machine-gun. [*S* and *T* (initials of its inventors' surnames) + *-en* after BREN]

stenographer /ste'nɒgrəfə(r)/ *n.* esp. *US* shorthand typist. [Greek *stenos* narrow]

stentorian /sten'tɔ:rɪən/ *adj.* loud and powerful. [*Stentor*, name of a herald in Homer's *Iliad*]

step *–n.* **1 a** complete movement of one leg in walking or running. **b** distance so covered. **2** unit of movement in dancing. **3** measure taken, esp. one of several in a course of action. **4** surface of a stair, stepladder, etc.; tread. **5** short distance. **6** sound or mark made by a foot in walking etc. **7** manner of walking etc. **8** degree in the scale of promotion or precedence etc. **9 a** stepping in unison or to music (esp. *in* or *out of step*). **b** state of conforming (*refuses to keep step with the team*). **10** (in *pl.*) (also **pair of steps**) = STEPLADDER. *–v.* (**-pp-**) **1** lift and set down one's foot or alternate feet in walking. **2** come or go in a specified direction by stepping. **3** make progress in a specified way (*stepped into a new job*). **4** (foll. by *off*, *out*) measure (distance) by stepping. **5** perform (a dance). □ **mind** (or **watch**) **one's step** be careful. **step by step** gradually; cautiously. **step down** resign. **step in 1** enter. **2** intervene. **step on it** *colloq.* accelerate; hurry up. **step out 1** be active socially. **2** take large steps. **step out of line** behave inappropriately or disobediently. **step up** increase, intensify. [Old English]

step- *comb. form* denoting a relationship resulting from a parent's later marriage. [Old English, = orphaned]

stepbrother *n.* son of one's step-parent by a previous partner.

stepchild *n.* one's husband's or wife's child by a previous partner.

stepdaughter *n.* female stepchild.

stepfather *n.* male step-parent.

stephanotis /ˌstefə'nəʊtɪs/ *n.* fragrant tropical climbing plant. [Greek]

stepladder *n.* short folding ladder with flat steps.

stepmother *n.* female step-parent.

step-parent *n.* mother's or father's spouse who is not one's own parent.

steppe *n.* level grassy unforested plain. [Russian]

stepping-stone *n.* **1** large stone in a stream etc. helping one to cross. **2** means of progress.

stepsister *n.* daughter of one's step-parent by a previous partner.

stepson *n.* male stepchild.

-ster *suffix* denoting a person engaged in or associated with a particular activity or quality (*brewster*; *gangster*; *youngster*). [Old English]

stereo /'sterɪəʊ, 'stɪə-/ *–n.* (*pl.* **-s**) **1 a** stereophonic record-player etc. **b** stereophonic sound reproduction (see STEREOPHONIC). **2** = STEREOSCOPE. *–adj.* **1** = STEREOPHONIC. **2** = STEREOSCOPIC (see STEREOSCOPE). [abbreviation]

stereo- *comb. form* solid; having three dimensions. [Greek *stereos* solid]

stereophonic /ˌsterɪəʊ'fɒnɪk, ˌstɪə-/ *adj.* using two or more channels, giving the effect of naturally distributed sound.

stereoscope /'sterɪəˌskəʊp, 'stɪə-/ *n.* device for producing a three-dimensional effect by viewing two slightly different photographs together. □ **stereoscopic** /-'skɒpɪk/ *adj.*

stereotype /'sterɪəʊˌtaɪp, 'stɪə-/ *–n.* **1 a** person or thing seeming to conform to a widely accepted type. **b** such a type, idea, or attitude. **2** printing-plate cast from a mould of composed type. *–v.* (**-ping**) **1** (esp. as **stereotyped** *adj.*) cause to conform to a type; standardize. **2 a** print from a stereotype. **b** make a stereotype of. [French: related to STEREO-]

sterile /'steraɪl/ *adj.* **1** unable to produce a crop, fruit, or young; barren. **2** unproductive (*sterile discussion*). **3** free from living micro-organisms etc. □ **sterility** /stə'rɪlɪtɪ/ *n.* [Latin]

sterilize /'sterɪˌlaɪz/ *v.* (also **-ise**) (**-zing** or **-sing**) **1** make sterile. **2** deprive of reproductive powers. □ **sterilization** /-'zeɪʃ(ə)n/ *n.*

sterling /'stɜ:lɪŋ/ *–adj.* **1** of or in British money (*pound sterling*). **2** (of a coin or precious metal) genuine; of standard value or purity. **3** (of a person etc.) genuine, reliable. *–n.* British money. [Old English, = penny]

sterling silver *n.* silver of $92^1/_2$% purity.

stern[1] *adj.* severe, grim; authoritarian. □ **sternly** *adv.* **sternness** *n.* [Old English]

stern[2] *n.* rear part, esp. of a ship or boat. [Old Norse: related to STEER[1]]

sternum /'stɜ:nəm/ *n.* (*pl.* **-na** or **-nums**) breastbone. [Greek *sternon* chest]

steroid /'stɪərɔɪd, 'ste-/ *n.* any of a group of organic compounds including many hormones, alkaloids, and vitamins. [from STEROL]

sterol /'sterɒl/ *n.* naturally occurring steroid alcohol. [from CHOLESTEROL, etc.]

stertorous /'stɜ:tərəs/ *adj.* (of breathing etc.) laboured and noisy. [Latin *sterto* snore]

stet *v.* (**-tt-**) (usu. written on a proof-sheet etc.) ignore or cancel (the alteration); let the original stand. [Latin, = let it stand]

stethoscope /'steθə,skəʊp/ *n.* instrument used in listening to the heart, lungs, etc. [Greek *stēthos* breast]

stetson /'stets(ə)n/ *n.* slouch hat with a very wide brim and high crown. [*Stetson*, name of a hat-maker]

stevedore /'sti:və,dɔ:(r)/ *n.* person employed in loading and unloading ships. [Spanish *estivador*]

stew *–v.* **1** cook by long simmering in a closed vessel. **2** fret, be anxious. **3** *colloq.* swelter. **4** (of tea etc.) become bitter or strong from infusing too long. **5** (as **stewed** *adj.*) *colloq.* drunk. *–n.* **1** dish of stewed meat etc. **2** *colloq.* agitated or angry state. □ **stew in one's own juice** suffer the consequences of one's actions. [French *estuver*]

steward /'stju:əd/ *–n.* **1** passengers' attendant on a ship, aircraft, or train. **2** official supervising a meeting, show, etc. **3** person responsible for supplies of food etc. for a college or club etc. **4** property manager. *–v.* act as a steward (of). □ **stewardship** *n.* [Old English, = house-warden]

stewardess /,stju:ə'des, 'stju:ədɪs/ *n.* female steward, esp. on a ship or aircraft.

stick[1] *n.* **1 a** short slender length of wood. **b** this as a support or weapon. **2** thin rod of wood etc. for a particular purpose (*cocktail stick*). **3** implement used to propel the ball in hockey or polo etc. **4** gear lever. **5** conductor's baton. **6** sticklike piece of celery, dynamite, etc. **7** (often prec. by *the*) punishment, esp. by beating. **8** *colloq.* adverse criticism. **9** *colloq.* piece of wood as part of a house or furniture. **10** *colloq.* person, esp. when dull or unsociable. [Old English]

stick[2] *v.* (*past* and *past part.* **stuck**) **1** (foll. by *in, into, through*) insert or thrust (a thing or its point). **2** stab. **3** (foll. by *in, into, on,* etc.) **a** fix or be fixed on a pointed thing. **b** fix or be fixed (as) by a pointed end. **4** fix or be fixed (as) by adhesive etc. **5** remain (in the mind). **6** lose or be deprived of movement or action through adhesion, jamming, etc. **7** *colloq.* **a** put in a specified position or place. **b** remain in a place. **8** *colloq.* (of an accusation etc.) be convincing or regarded as valid. **9** *colloq.* endure, tolerate. **10** (foll. by *at*) *colloq.* persevere with. □ **be stuck for** be at a loss for or in need of. **be stuck on** *colloq.* be infatuated with. **be stuck with** *colloq.* be unable to get rid of. **get stuck in** (or **into** a thing) *slang* begin in earnest. **stick around** *colloq.* linger; remain. **stick at nothing** be absolutely ruthless. **stick by** (or **with**) stay loyal or close to. **stick in one's throat** be against one's principles. **stick it out** *colloq.* endure a burden etc. to the end. **stick one's neck out** be rashly bold. **stick out** (cause to) protrude. **stick out for** persist in demanding. **stick to 1** remain fixed on or to. **2** remain loyal to. **3** keep to (a subject etc.). **stick together** *colloq.* remain united or mutually loyal. **stick up 1** be or make erect or protruding upwards. **2** fasten to an upright surface. **3** *colloq.* rob or threaten with a gun. **stick up for** support or defend. [Old English]

sticker *n.* **1** adhesive label. **2** persistent person.

sticking-plaster *n.* adhesive plaster for wounds etc.

stick insect *n.* insect with a twiglike body.

stick-in-the-mud *n. colloq.* unprogressive or old-fashioned person.

stickleback /'stɪk(ə)l,bæk/ *n.* small spiny-backed fish. [Old English, = thorn-back]

stickler *n.* (foll. by *for*) person who insists on something (*stickler for accuracy*). [obsolete *stickle* be umpire]

stick-up *n. colloq.* robbery using a gun.

sticky *adj.* (**-ier, -iest**) **1** tending or intended to stick or adhere. **2** glutinous, viscous. **3** humid. **4** *colloq.* difficult, awkward; unpleasant, painful (*sticky problem*). □ **stickily** *adv.* **stickiness** *n.*

sticky wicket *n. colloq.* difficult circumstances.

stiff *–adj.* **1** rigid; inflexible. **2** hard to bend, move, or turn etc. **3** hard to cope with; needing strength or effort (*stiff climb*). **4** severe or strong (*stiff penalty*). **5** formal, constrained. **6** (of a muscle, person, etc.) aching owing to exertion, injury, etc. **7** (of esp. an alcoholic drink) strong. **8** (foll. by *with*) *colloq.* abounding in. *–adv. colloq.* utterly, extremely (*bored stiff; worried stiff*). *–n. slang* **1** corpse. **2** foolish or useless person. □ **stiffish** *adj.* **stiffly** *adv.* **stiffness** *n.* [Old English]

stiffen *v.* make or become stiff. □ **stiffening** *n.*

stiff-necked *adj.* obstinate; haughty.

stiff upper lip *n.* appearance of being calm in adversity.

stifle /'staɪf(ə)l/ *v.* (**-ling**) **1** suppress. **2** feel or make unable to breathe easily; suffocate. **3** kill by suffocating. □ **stifling** *adj.* & *adv.* [origin uncertain]

stigma /'stɪgmə/ *n.* (*pl.* **-s** or, esp. in sense 3, **-mata** /-mətə, -'mɑ:tə/) **1** shame, disgrace. **2** part of the pistil that receives the pollen in pollination. **3** (in *pl.*) (in Christian belief) marks like those on Christ's body after the Crucifixion, appearing on the bodies of certain saints etc. [Greek *stigma -mat-* brand, dot]

stigmatize /'stɪgmə,taɪz/ *v.* (also **-ise**) (**-zing** or **-sing**) (often foll. by *as*) brand as unworthy or disgraceful. [Greek *stigmatizō*: related to STIGMA]

stile *n.* steps allowing people but not animals to climb over a fence or wall. [Old English]

stiletto /stɪ'letəʊ/ *n.* (*pl.* **-s**) **1** short dagger. **2** (in full **stiletto heel**) **a** long tapering heel of a shoe. **b** shoe with such a heel. **3** pointed instrument for making eyelets etc. [Italian diminutive: related to STYLE]

still[1] *–adj.* **1** not or hardly moving. **2** with little or no sound; calm and tranquil. **3** (of a drink) not effervescing. *–n.* **1** deep silence (*still of the night*). **2** static photograph (as opposed to a motion picture), esp. a single shot from a cinema film. *–adv.* **1** without moving (*sit still*). **2** even now or at a particular time (*is he still here?*). **3** nevertheless. **4** (with *compar.*) even, yet, increasingly (*still greater efforts*). *–v.* make or become still; quieten. □ **stillness** *n.* [Old English]

still[2] *n.* apparatus for distilling spirits etc. [obsolete *still* (v.) = DISTIL]

stillbirth *n.* birth of a dead child.

stillborn *adj.* **1** born dead. **2** abortive.

still life *n.* (*pl.* **lifes**) painting or drawing of inanimate objects, e.g. fruit or flowers.

still-room *n.* **1** room for distilling. **2** housekeeper's storeroom or pantry.

stilt *n.* **1** either of a pair of poles with foot supports for walking at a distance above the ground. **2** each of a set of piles or posts supporting a building etc. [Low German or Dutch]

stilted *adj.* **1** (of literary style etc.) stiff and unnatural; bombastic. **2** standing on stilts.

Stilton /'stɪlt(ə)n/ *n. propr.* strong rich esp. blue-veined cheese. [*Stilton* in England]

stimulant /'stɪmjʊlənt/ *–adj.* stimulating, esp. bodily or mental activity. *–n.* stimulant substance or influence. [Latin: related to STIMULATE]

stimulate /'stɪmjʊ,leɪt/ *v.* (**-ting**) **1** act as a stimulus to. **2** animate, excite, arouse. □ **stimulation** /-'leɪʃ(ə)n/ *n.* **stimulative** /-lətɪv/ *adj.* **stimulator** *n.* [Latin: related to STIMULUS]

stimulus /'stɪmjʊləs/ *n.* (*pl.* **-li** /-,laɪ/) thing that rouses to activity. [Latin, = goad]

sting *–n.* **1** sharp wounding organ of an insect, snake, nettle, etc. **2 a** act of inflicting a wound with this. **b** the wound itself or the pain caused by it. **3** painful quality or effect. **4** pungency, vigour. **5** *slang* swindle. *–v.* (*past* and *past part.* **stung**) **1 a** wound or pierce with a sting. **b** be able to sting. **2** feel or give a tingling physical or sharp mental pain. **3** (foll. by *into*) incite, esp. painfully (*stung into replying*). **4** *slang* swindle, charge exorbitantly. □ **sting in the tail** unexpected final pain or difficulty. [Old English]

stinger *n.* stinging animal or thing, esp. a sharp blow.

stinging-nettle *n.* nettle with stinging hairs.

stingray *n.* broad flat-fish with a poisonous spine at the base of its tail.

stingy /'stɪndʒɪ/ *adj.* (**-ier, -iest**) *colloq.* niggardly, mean. □ **stingily** *adv.* **stinginess** *n.* [perhaps from STING]

stink *–v.* (*past* **stank** or **stunk**; *past part.* **stunk**) **1** emit a strong offensive smell. **2** (often foll. by *out*) fill (a place) with a stink. **3** (foll. by *out* etc.) drive (a person) out etc. by a stink. **4** *colloq.* be or seem very unpleasant. *–n.* **1** strong or offensive smell. **2** *colloq.* row or fuss. [Old English]

stink bomb *n.* device emitting a stink when opened.

stinker *n. slang* objectionable or difficult person or thing.

stinking *–adj.* **1** that stinks. **2** *slang* very objectionable. *–adv. slang* extremely and usu. objectionably (*stinking rich*).

stint *–v.* **1** supply (food or aid etc.) meanly or grudgingly. **2** (often *refl.*) supply (a person etc.) in this way. *–n.* **1** limitation of supply or effort (*without stint*). **2** allotted amount of work (*do one's stint*). **3** small sandpiper. [Old English]

stipend /'staɪpend/ *n.* salary, esp. of a clergyman. [Latin *stipendium*]

stipendiary /staɪ'pendjərɪ, stɪ-/ *–adj.* receiving a stipend. *–n.* (*pl.* **-ies**) person receiving a stipend. [Latin: related to STIPEND]

stipendiary magistrate *n.* paid professional magistrate.

stipple /'stɪp(ə)l/ *–v.* (**-ling**) **1** draw or paint or engrave etc. with dots instead of lines. **2** roughen the surface of (paint, cement, etc.). *–n.* **1** stippling. **2** effect of stippling. [Dutch]

stipulate /'stɪpjʊ,leɪt/ *v.* (**-ting**) demand or specify as part of a bargain etc. □

stipulation /-ˈleɪʃ(ə)n/ *n.* [Latin *stipulari*]

stir[1] –*v.* (**-rr-**) **1** move a spoon etc. round and round in (a liquid etc.), esp. to mix ingredients. **2 a** cause to move, esp. slightly. **b** be or begin to be in motion. **3** rise from sleep. **4** arouse, inspire, or excite (the emotions, a person, etc.). **5** *colloq.* cause trouble between people by gossiping etc. –*n.* **1** act of stirring. **2** commotion, excitement. □ **stir in** add (an ingredient) by stirring. **stir up 1** mix thoroughly by stirring. **2** stimulate, excite. □ **stirrer** *n.* [Old English]

stir[2] *n. slang* prison. [origin unknown]

stir-fry –*v.* fry rapidly while stirring. –*n.* stir-fried dish.

stirrup /ˈstɪrəp/ *n.* metal loop supporting a horse-rider's foot. [Old English, = climbing-rope]

stirrup-cup *n.* cup of wine etc. offered to a departing traveller, orig. a rider.

stirrup-leather *n.* (also **stirrup-strap**) strap attaching a stirrup to a saddle.

stirrup-pump *n.* hand-operated water-pump with a foot-rest, used to extinguish small fires.

stitch –*n.* **1 a** (in sewing, knitting, or crocheting) single pass of a needle, or the resulting thread or loop etc. **b** particular method of sewing etc. **2** least bit of clothing (*hadn't a stitch on*). **3** sharp pain in the side induced by running etc. –*v.* sew; make stitches (in). □ **in stitches** *colloq.* laughing uncontrollably. **stitch up 1** join or mend by sewing. **2** *slang* trick, cheat, betray. [Old English: related to STICK[2]]

stitch in time *n.* timely remedy.

stoat *n.* mammal of the weasel family with brown fur turning mainly white in the winter. [origin unknown]

stock –*n.* **1** store of goods etc. ready for sale or distribution etc. **2** supply or quantity of anything for use. **3** equipment or raw material for manufacture or trade etc. (*rolling-stock*). **4** farm animals or equipment. **5 a** capital of a business. **b** shares in this. **6** reputation or popularity (*his stock is rising*). **7 a** money lent to a government at fixed interest. **b** right to receive such interest. **8** line of ancestry (*comes of Cornish stock*). **9** liquid basis for soup etc. made by stewing bones, vegetables, etc. **10** fragrant-flowered cruciferous cultivated plant. **11** plant into which a graft is inserted. **12** main trunk of a tree etc. **13** (in *pl.*) *hist.* timber frame with holes for the feet in which offenders were locked as a public punishment. **14** base, support, or handle for an implement or machine. **15** butt of a rifle etc. **16** (in *pl.*) supports for a ship during building or repair. **17** band of cloth worn round the neck. –*attrib. adj.* **1** kept in stock and so regularly available. **2** hackneyed, conventional. –*v.* **1** have (goods) in stock. **2** provide (a shop or a farm etc.) with goods, livestock, etc. **3** fit (a gun etc.) with a stock. □ **in** (or **out of**) **stock** available (or not available) immediately for sale etc. **stock up** (often foll. by *with*) provide with or get stocks or supplies (of). **take stock 1** make an inventory of one's stock. **2** (often foll. by *of*) review (a situation etc.). [Old English]

stockade /stɒˈkeɪd/ –*n.* line or enclosure of upright stakes. –*v.* (**-ding**) fortify with this. [Spanish *estacada*]

stockbreeder *n.* livestock farmer.

stockbroker *n.* = BROKER 2. □ **stockbroking** *n.*

stock-car *n.* specially strengthened car for use in racing with deliberate bumping.

Stock Exchange *n.* **1** place for dealing in stocks and shares. **2** dealers working there.

stockholder *n.* owner of stocks or shares.

stockinet /ˌstɒkɪˈnet/ *n.* (also **stockinette**) elastic knitted fabric. [probably from *stocking-net*]

stocking *n.* **1** long knitted covering for the leg and foot, of nylon, wool, silk, etc. **2** differently-coloured lower leg of a horse etc. □ **in one's stocking** (or **stockinged**) **feet** without shoes. [from STOCK]

stocking-stitch *n.* alternate rows of plain and purl.

stock-in-trade *n.* **1** requisite(s) of a trade or profession. **2** characteristic or essential product; characteristic behaviour, actions, etc.

stockist *n.* dealer in specified types of goods.

stockjobber *n.* = JOBBER 2.

stock market *n.* **1** = STOCK EXCHANGE. **2** transactions on this.

stockpile –*n.* accumulated stock of goods etc. held in reserve. –*v.* (**-ling**) accumulate a stockpile of.

stockpot *n.* pot for making soup stock.

stockroom *n.* room for storing goods.

stock-still *adj.* motionless.

stocktaking *n.* **1** making an inventory of stock. **2** review of one's position etc.

stocky *adj.* (**-ier**, **-iest**) short and sturdy. □ **stockily** *adv.* **stockiness** *n.*

stockyard *n.* enclosure for the sorting or temporary keeping of cattle.

stodge /stɒdʒ/ *n. colloq.* heavy fattening food. [imitative, after *stuff* and *podge*]

stodgy *adj.* (**-ier, -iest**) **1** (of food) heavy and glutinous. **2** dull and uninteresting. □ **stodgily** *adv.* **stodginess** *n.*

Stoic /'stəʊɪk/ *—n.* **1** member of the ancient Greek school of philosophy which sought virtue as the greatest good and taught control of one's feelings and passions. **2** (**stoic**) stoical person. *—adj.* **1** of or like the Stoics. **2** (**stoic**) = STOICAL. [Greek *stoa* portico]

stoical *adj.* having or showing great self-control in adversity. □ **stoically** *adv.*

Stoicism /'stəʊɪˌsɪz(ə)m/ *n.* **1** philosophy of the Stoics. **2** (**stoicism**) stoical attitude.

stoke *v.* (**-king**) (often foll. by *up*) **1** feed and tend (a fire or furnace etc.). **2** *colloq.* fill oneself with food. [back-formation from STOKER]

stokehold *n.* compartment in a steamship containing its boilers and furnace.

stokehole *n.* space for stokers in front of a furnace.

stoker *n.* person who tends a furnace, esp. on a steamship. [Dutch]

STOL *abbr.* short take-off and landing.

stole[1] *n.* **1** woman's garment like a long wide scarf, worn over the shoulders. **2** strip of silk etc. worn similarly by a priest. [Greek *stolē* equipment, clothing]

stole[2] *past* of STEAL.

stolen *past part.* of STEAL.

stolid /'stɒlɪd/ *adj.* not easily excited or moved; impassive, unemotional. □ **stolidity** /-'lɪdɪtɪ/ *n.* **stolidly** *adv.* [Latin]

stoma /'stəʊmə/ *n.* (*pl.* **-s** or **stomata**) **1** minute pore in the epidermis of a leaf. **2** small mouthlike artificial orifice made in the stomach. [Greek *stoma* mouth]

stomach /'stʌmək/ *—n.* **1 a** internal organ in which digestion occurs. **b** any of several such organs in animals. **2** lower front of the body. **3** (usu. foll. by *for*) **a** appetite. **b** inclination. *—v.* **1** find palatable. **2** endure (usu. with *neg.*: *cannot stomach it*). [Greek *stoma* mouth]

stomach-ache *n.* pain in the belly or bowels.

stomacher *n. hist.* pointed bodice of a dress, often jewelled or embroidered. [probably French: related to STOMACH]

stomach-pump *n.* syringe for forcing liquid etc. into or out of the stomach.

stomach upset *n.* temporary digestive disorder.

stomp *—v.* tread or stamp heavily. *—n.* lively jazz dance with heavy stamping. [var. of STAMP]

stone *—n.* **1 a** solid non-metallic mineral matter; rock. **b** small piece of this. **2** (often in *comb.*) piece of stone of a definite shape or for a particular purpose. **3 a** thing resembling stone, e.g. the hard case of the kernel in some fruits. **b** (often in *pl.*) hard morbid concretion in the body. **4** (*pl.* same) unit of weight equal to 14 lb. **5** = PRECIOUS STONE. **6** (*attrib.*) made of stone. *—v.* (**-ning**) **1** pelt with stones. **2** remove the stones from (fruit). □ **cast** (or **throw**) **stones** speak ill of a person. **leave no stone unturned** try all possible means. **a stone's throw** a short distance. [Old English]

Stone Age *n.* prehistoric period when weapons and tools were made of stone.

stonechat *n.* small brown bird with black and white markings.

stone-cold *adj.* completely cold.

stone-cold sober *predic. adj.* completely sober.

stonecrop *n.* succulent rock-plant.

stoned *adj. slang* drunk or drugged.

stone-dead *adj.* completely dead.

stone-deaf *adj.* completely deaf.

stone-fruit *n.* fruit with flesh enclosing a stone.

stoneground *adj.* (of flour) ground with millstones.

stonemason *n.* person who cuts, prepares, and builds with stone.

stonewall *v.* **1** obstruct (a discussion or investigation) with evasive answers etc. **2** *Cricket* bat with excessive caution.

stoneware *n.* ceramic ware which is impermeable and partly vitrified but opaque.

stonewashed *adj.* (esp. of denim) washed with abrasives to give a worn or faded look.

stonework *n.* masonry.

stonker /'stɒŋkə(r)/ *n. slang* excellent person or thing. □ **stonking** *adj.* [20th c.: origin unknown]

stony *adj.* (**-ier, -iest**) **1** full of stones. **2 a** hard, rigid. **b** unfeeling, uncompromising. □ **stonily** *adv.* **stoniness** *n.*

stony-broke *adj. slang* entirely without money.

stood *past* and *past part.* of STAND.

stooge *colloq.* *—n.* **1** butt or foil, esp. for a comedian. **2** assistant or subordinate, esp. for routine or unpleasant work. *—v.* (**-ging**) **1** (foll. by *for*) act as a stooge for. **2** (foll. by *about*, *around*, etc.) move about aimlessly. [origin unknown]

stook /stu:k, stʊk/ *—n.* group of sheaves of grain stood on end in a field. *—v.* arrange in stooks. [related to Low German *stūke*]

stool *n.* **1** single seat without a back or arms. **2** = FOOTSTOOL. **3** (usu. in *pl.*) = FAECES. [Old English]

stoolball *n.* team game with pairs of batters scoring runs between two bases.

stool-pigeon *n.* **1** person acting as a decoy. **2** police informer.

stoop[1] *–v.* **1** lower the body, sometimes bending the knee; bend down. **2** stand or walk with the shoulders habitually bent forward. **3** (foll. by *to* + infin.) condescend. **4** (foll. by *to*) descend to (some conduct). *–n.* stooping posture. [Old English]

stoop[2] *n. US* porch, small veranda, or steps in front of a house. [Dutch *stoep*]

stop *–v.* (**-pp-**) **1 a** put an end to the progress, motion, or operation of. **b** effectively hinder or prevent. **c** discontinue (*stop playing*). **2** come to an end (*supplies suddenly stopped*). **3** cease from motion, speaking, or action. **4** defeat. **5** *slang* receive (a blow etc.). **6** remain; stay for a short time. **7** (often foll. by *up*) block or close up (a hole, leak, etc.). **8** not permit or supply as usual (*stop their wages*). **9** (in full **stop payment of** or **on**) instruct a bank to withhold payment on (a cheque). **10** fill (a tooth). **11** press (a violin etc. string) to obtain the required pitch. *–n.* **1** stopping or being stopped. **2** designated stopping-place for a bus or train etc. **3** = FULL STOP. **4** device for stopping motion at a particular point. **5** change of pitch effected by stopping a string. **6 a** (in an organ) row of pipes of one character. **b** knob etc. operating these. **7** *Optics & Photog.* = DIAPHRAGM 3a. **8 a** effective diameter of a lens. **b** device for reducing this. **9** (of sound) = PLOSIVE. □ **pull out all the stops** make extreme effort. **put a stop to** cause to end. **stop at nothing** be ruthless. **stop off** (or **over**) break one's journey. [Old English]

stopcock *n.* externally operated valve regulating the flow through a pipe etc.

stopgap *n.* temporary substitute.

stop-go *n.* alternate stopping and restarting, esp. of the economy.

stopoff *n.* break in a journey.

stopover *n.* break in a journey, esp. overnight.

stoppage *n.* **1** interruption of work owing to a strike etc. **2** (in *pl.*) sum deducted from pay, for tax, national insurance, etc. **3** condition of being blocked or stopped.

stopper *–n.* plug for closing a bottle etc. *–v.* close with this.

stop press *n.* (often *attrib.*) late news inserted in a newspaper after printing has begun.

stopwatch *n.* watch that can be stopped and started, used to time races etc.

storage /ˈstɔːrɪdʒ/ *n.* **1 a** storing of goods etc. **b** method of or space for storing. **2** cost of storing. **3** storing of data in a computer etc.

storage battery *n.* (also **storage cell**) battery (or cell) for storing electricity.

storage heater *n.* electric heater releasing heat stored outside peak hours.

store *–n.* **1** quantity of something kept available for use. **2** (in *pl.*) **a** articles gathered for a particular purpose. **b** supply of, or place for keeping, these. **3 a** = DEPARTMENT STORE. **b** esp. *US* shop. **c** (often in *pl.*) shop selling basic necessities. **4** warehouse for keeping furniture etc. temporarily. **5** device in a computer for keeping retrievable data. *–v.* (**-ring**) **1** (often foll. by *up, away*) accumulate for future use. **2** put (furniture etc.) in a store. **3** stock or provide with something useful. **4** keep (data) for retrieval. □ **in store 1** kept in readiness. **2** coming in the future. **3** (foll. by *for*) awaiting. **set store by** consider important. [French *estore(r)* from Latin *instauro* renew]

storehouse *n.* storage place.

storekeeper *n.* **1** storeman. **2** *US* shopkeeper.

storeman *n.* person in charge of a store of goods.

storeroom *n.* storage room.

storey /ˈstɔːrɪ/ *n.* (*pl.* **-s**) **1** = FLOOR *n.* 3. **2** thing forming a horizontal division. □ **-storeyed** *adj.* (in *comb.*). [Anglo-Latin: related to HISTORY, perhaps originally meaning a tier of painted windows]

storied /ˈstɔːrɪd/ *adj. literary* celebrated in or associated with stories or legends.

stork *n.* long-legged usu. white wading bird. [Old English]

storm *–n.* **1** violent atmospheric disturbance with strong winds and usu. thunder, rain, or snow. **2** violent political etc. disturbance. **3** (foll. by *of*) **a** violent shower of missiles or blows. **b** outbreak of applause, hisses, etc. **4 a** direct assault by troops on a fortified place. **b** capture by such an assault. *–v.* **1** attack or capture by storm. **2** (usu. foll. by *in, out of,* etc.) move violently or angrily (*stormed out*). **3** (often foll. by *at, away*) talk violently, rage, bluster. □ **take by storm 1** capture by direct assault. **2** rapidly captivate. [Old English]

storm centre *n.* **1** point to which the wind spirals inward in a cyclonic storm. **2** centre of controversy etc.

storm cloud *n.* **1** heavy rain-cloud. **2** threatening situation.

storm-door *n.* additional outer door.

storm in a teacup *n.* great excitement over a trivial matter.

storm petrel *n.* (also **stormy petrel**) **1** small black and white N. Atlantic petrel. **2** person causing unrest.

storm trooper *n.* member of the storm troops.

storm troops *n.pl.* **1** = SHOCK TROOPS. **2** *hist.* Nazi political militia.

stormy *adj.* (**-ier, -iest**) **1** of or affected by storms. **2** (of a wind etc.) violent. **3** full of angry feeling or outbursts (*stormy meeting*). □ **stormily** *adv.* **storminess** *n.*

stormy petrel var. of STORM PETREL.

story /'stɔ:rɪ/ *n.* (*pl.* **-ies**) **1** account of imaginary or past events; tale, anecdote. **2** history of a person or institution etc. **3** (in full **story-line**) narrative or plot of a novel, play, etc. **4** facts or experiences worthy of narration. **5** *colloq.* fib. [Anglo-French *estorie* from Latin: related to HISTORY]

storyteller *n.* **1** person who tells stories. **2** *colloq.* liar. □ **storytelling** *n.* & *adj.*

stoup /stu:p/ *n.* **1** basin for holy-water. **2** *archaic* flagon, beaker. [Old Norse]

stout *–adj.* **1** rather fat, corpulent, bulky. **2** thick or strong. **3** brave, resolute. *–n.* strong dark beer. □ **stoutly** *adv.* **stoutness** *n.* [Anglo-French from Germanic]

stout-hearted *adj.* courageous.

stove[1] *n.* closed apparatus burning fuel or using electricity for heating or cooking. [Low German or Dutch]

stove[2] *past* and *past part.* of STAVE *v.*

stove-pipe *n.* pipe carrying smoke and gases from a stove to a chimney.

stow /stəʊ/ *v.* pack (goods, cargo, etc.) tidily and compactly. □ **stow away 1** place (a thing) out of the way. **2** be a stowaway on a ship etc. [from BESTOW]

stowage *n.* **1** stowing. **2** place for this.

stowaway *n.* person who hides on a ship or aircraft etc. to travel free.

strabismus /strə'bɪzməs/ *n. Med.* squinting, squint. [Greek *strabos* squinting]

straddle /'stræd(ə)l/ *v.* (**-ling**) **1 a** sit or stand across (a thing) with the legs spread. **b** be situated on both sides of. **2** part (one's legs) widely. [from STRIDE]

strafe /strɑ:f, streɪf/ *v.* (**-fing**) bombard; attack with gunfire. [German, = punish]

straggle /'stræg(ə)l/ *–v.* (**-ling**) **1** lack compactness or tidiness. **2** be dispersed or sporadic. **3** trail behind in a race etc. *–n.* straggling or scattered group. □ **straggler** *n.* **straggly** *adj.* (**-ier, -iest**). [origin uncertain]

straight /streɪt/ *–adj.* **1** extending uniformly in the same direction; not bent or curved. **2** successive, uninterrupted (*three straight wins*). **3** ordered; level; tidy (*put things straight*). **4** honest, candid. **5** (of thinking etc.) logical. **6** (of theatre, music, etc.) serious, classical, not popular or comic. **7 a** unmodified. **b** (of a drink) undiluted. **8** *colloq.* **a** (of a person etc.) conventional, respectable. **b** heterosexual. **9** direct, undeviating. *–n.* **1** straight part, esp. the concluding stretch of a racetrack. **2** straight condition. **3** sequence of five cards in poker. **4** *colloq.* conventional person; heterosexual. *–adv.* **1** in a straight line; direct. **2** in the right direction. **3** correctly. □ **go straight** (of a criminal) become honest. **straight away** immediately. **straight off** *colloq.* without hesitation. □ **straightish** *adj.* **straightness** *n.* [originally a past part. of STRETCH]

straightaway /,streɪtə'weɪ/ *adv.* = *straight away*.

straighten /'streɪt(ə)n/ *v.* **1** (often foll. by *out*) make or become straight. **2** (foll. by *up*) stand erect after bending.

straight eye *n.* ability to detect deviation from the straight.

straight face *n.* intentionally expressionless face. □ **straight-faced** *adj.*

straight fight *n. Polit.* contest between two candidates only.

straight flush *n.* flush in numerical sequence.

straightforward /streɪt'fɔ:wəd/ *adj.* **1** honest or frank. **2** (of a task etc.) simple.

straight man *n.* comedian's stooge.

strain[1] *–v.* **1** stretch tightly; make or become taut or tense. **2** injure by overuse or excessive demands. **3** exercise (oneself, one's senses, a thing, etc.) intensely; press to extremes. **4** strive intensively. **5** (foll. by *at*) tug, pull. **6** distort from the true intention or meaning. **7 a** clear (a liquid) of solid matter by passing it through a sieve etc. **b** (foll. by *out*) filter (solids) out from a liquid. *–n.* **1 a** act of straining. **b** force exerted in this. **2** injury caused by straining a muscle etc. **3** severe mental or physical demand or exertion (*suffering from strain*). **4** snatch of music or poetry. **5** tone or tendency in speech or writing (*more in the same strain*). [French *estrei(g)n-* from Latin *stringo*]

strain[2] *n.* **1** breed or stock of animals, plants, etc. **2** tendency; characteristic. [Old English, = begetting]

strained *adj.* **1** constrained, artificial. **2** (of a relationship) mutually distrustful or tense.

strainer *n.* device for straining liquids etc.

strait *n.* **1** (in *sing.* or *pl.*) narrow channel connecting two large bodies of

water. **2** (usu. in *pl.*) difficulty or distress. [French *estreit* from Latin *strictus* narrow]
straitened /'streɪt(ə)nd/ *adj.* of or marked by poverty.
strait-jacket *–n.* **1** strong garment with long sleeves for confining a violent prisoner etc. **2** restrictive measures. *–v.* (**-t-**) **1** restrain with a strait-jacket. **2** severely restrict.
strait-laced *adj.* puritanical.
strand[1] *–v.* **1** run aground. **2** (as **stranded** *adj.*) in difficulties, esp. without money or transport. *–n.* foreshore; beach. [Old English]
strand[2] *n.* **1** each of the twisted threads or wires making a rope or cable etc. **2** single thread or strip of fibre. **3** lock of hair. **4** element; component. [origin unknown]
strange /streɪndʒ/ *adj.* **1** unusual, peculiar, surprising, eccentric. **2** (often foll. by *to*) unfamiliar, foreign. **3** (foll. by *to*) unaccustomed. **4** not at ease. □ **strangely** *adv.* **strangeness** *n.* [French *estrange* from Latin *extraneus*]
stranger *n.* **1** person new to a particular place or company. **2** (often foll. by *to*) person one does not know. **3** (foll. by *to*) person unaccustomed to (*no stranger to controversy*).
strangle /'stræŋg(ə)l/ *v.* (**-ling**) **1** squeeze the windpipe or neck of, esp. so as to kill. **2** hamper, suppress. □ **strangler** *n.* [Latin *strangulo*]
stranglehold *n.* **1** throttling hold in wrestling. **2** deadly grip. **3** complete control.
strangulate /'stræŋgjʊ,leɪt/ *v.* (**-ting**) compress (a vein, intestine, etc.), preventing circulation. [Latin: related to STRANGLE]
strangulation /,stræŋgjʊ'leɪʃ(ə)n/ *n.* **1** strangling or being strangled. **2** strangulating.
strap *–n.* **1** strip of leather etc., often with a buckle, for holding things together etc. **2** narrow strip of fabric worn over the shoulders as part of a garment. **3** loop for grasping to steady oneself in a moving vehicle. **4** (**the strap**) punishment by beating with a leather strap. *–v.* (**-pp-**) **1** (often foll. by *down*, *up*, etc.) secure or bind with a strap. **2** beat with a strap. □ **strapless** *adj.* [dial., = STROP]
straphanger *n. slang* standing passenger in a bus, train, etc. □ **straphang** *v.*
strapping *adj.* large and sturdy.
strata *pl.* of STRATUM.

■ **Usage** It is incorrect to use *strata* as the singular noun instead of *stratum*.

stratagem /'strætədʒəm/ *n.* **1** cunning plan or scheme. **2** trickery. [Greek *stratēgos* a general]
strategic /strə'ti:dʒɪk/ *adj.* **1** of or promoting strategy. **2** (of materials) essential in war. **3** (of bombing or weapons) done or for use as a longer-term military objective. □ **strategically** *adv.*
strategy /'strætɪdʒɪ/ *n.* (*pl.* **-ies**) **1** long-term plan or policy (*economic strategy*). **2** art of war. **3** art of moving troops, ships, aircraft, etc. into favourable positions. □ **strategist** *n.*
strathspey /stræθ'speɪ/ *n.* **1** slow Scottish dance. **2** music for this. [*Strathspey*, valley of the river Spey]
stratify /'strætɪ,faɪ/ *v.* (**-ies**, **-ied**) (esp. as **stratified** *adj.*) arrange in strata or grades etc. □ **stratification** /-fɪ'keɪʃ(ə)n/ *n.* [French: related to STRATUM]
stratigraphy /strə'tɪgrəfɪ/ *n. Geol.* & *Archaeol.* **1** relative position of strata. **2** the study of this. □ **stratigraphic** /,strætɪ'græfɪk/ *adj.* [from STRATUM]
stratosphere /'strætə,sfɪə(r)/ *n.* layer of atmosphere above the troposphere, extending to about 50 km from the earth's surface. □ **stratospheric** /-'sferɪk/ *adj.* [from STRATUM]
stratum /'strɑ:təm/ *n.* (*pl.* **strata**) **1** layer or set of layers of any deposited substance, esp. of rock. **2** atmospheric layer. **3** social class. [Latin *sterno* strew]
straw *n.* **1** dry cut stalks of grain as fodder or material for bedding, packing, etc. **2** single stalk of straw. **3** thin tube for sucking drink through. **4** insignificant thing. **5** pale yellow colour. □ **clutch at straws** try any remedy in desperation. **straw in the wind** indication of future developments. [Old English]
strawberry /'strɔ:bərɪ/ *n.* (*pl.* **-ies**) **1** pulpy red fruit with a seed-studded surface. **2** plant with runners and white flowers bearing this. [Old English: related to STRAW, for unknown reason]
strawberry mark *n.* reddish birthmark.
straw vote *n.* (also **straw poll**) unofficial ballot as a test of opinion.
stray *–v.* **1** wander from the right place or from one's companions; go astray. **2** deviate morally or mentally. *–n.* strayed person, animal, or thing. *–adj.* **1** strayed, lost. **2** isolated, occasional. **3** *Physics* wasted or unwanted. [Anglo-French *strey*: related to ASTRAY]
streak *–n.* **1** long thin usu. irregular line or band, esp. of colour. **2** strain in a person's character. **3** spell or series (*winning streak*). *–v.* **1** mark with

streaks. **2** move very rapidly. **3** *colloq.* run naked in public. □ **streaker** *n.* [Old English, = pen-stroke]

streaky *adj.* (**-ier**, **-iest**) **1** full of streaks. **2** (of bacon) with streaks of fat.

stream –*n.* **1** flowing body of water, esp. a small river. **2** flow of a fluid or of a mass of people. **3** current or direction in which things are moving or tending (*against the stream*). **4** group of schoolchildren of similar ability taught together. –*v.* **1** move as a stream. **2** run with liquid. **3** be blown in the wind. **4** emit a stream of (blood etc.). **5** arrange (schoolchildren) in streams. □ **on stream** in operation or production. [Old English]

streamer *n.* **1** long narrow strip of ribbon or paper. **2** long narrow flag. **3** banner headline.

streamline *v.* (**-ning**) **1** give (a vehicle etc.) the form which presents the least resistance to motion. **2** make simple or more efficient.

street *n.* **1 a** public road in a city, town, or village. **b** this with the houses etc. on each side. **2** people who live or work in a particular street. □ **on the streets** living by prostitution. **streets ahead** (often foll. by *of*) *colloq.* much superior (to). **up** (or **right up**) **one's street** *colloq.* what one likes, knows about, etc. [Old English]

streetcar *n. US* tram.

street credibility *n.* (also **street cred**) *slang* familiarity with a fashionable urban subculture.

streetwalker *n.* prostitute seeking customers in the street.

streetwise *adj.* knowing how to survive modern urban life.

strength *n.* **1** being strong; degree or manner of this. **2 a** person or thing giving strength. **b** positive attribute. **3** number of people present or available; full number. □ **from strength to strength** with ever-increasing success. **in strength** in large numbers. **on the strength of** on the basis of. [Old English: related to STRONG]

strengthen /'streŋθ(ə)n/ *v.* make or become stronger.

strenuous /'strenjʊəs/ *adj.* **1** requiring or using great effort. **2** energetic. □ **strenuously** *adv.* [Latin]

streptococcus /ˌstreptə'kɒkəs/ *n.* (*pl.* **-cocci** /-kaɪ/) bacterium of a type often causing infectious diseases. □ **streptococcal** *adj.* [Greek *streptos* twisted, *kokkos* berry]

streptomycin /ˌstreptəʊ'maɪsɪn/ *n.* antibiotic effective against many disease-producing bacteria. [Greek *streptos* twisted, *mukēs* fungus]

stress –*n.* **1 a** pressure or tension. **b** quantity measuring this. **2 a** physical or mental strain. **b** distress caused by this. **3 a** emphasis. **b** emphasis on a syllable or word. –*v.* **1** emphasize. **2** subject to stress. □ **lay stress on** emphasize. [shortening of DISTRESS]

stressful *adj.* causing stress.

stretch –*v.* **1** draw, be drawn, or be able to be drawn out in length or size. **2** make or become taut. **3** place or lie at full length or spread out. **4** (also *absol.*) **a** extend (a limb etc.). **b** thrust out one's limbs and tighten one's muscles after being relaxed. **5** have a specified length or extension; extend. **6** strain or exert extremely; exaggerate (*stretch the truth*). –*n.* **1** continuous extent, expanse, or period. **2** stretching or being stretched. **3** (*attrib.*) elastic (*stretch fabric*). **4** *colloq.* period of imprisonment etc. **5** *US* straight side of a racetrack. □ **at a stretch** in one period. **stretch one's legs** exercise oneself by walking. **stretch out 1** extend (a limb etc.). **2** last; prolong. **stretch a point** agree to something not normally allowed. □ **stretchy** *adj.* (**-ier**, **-iest**). [Old English]

stretcher *n.* **1** two poles with canvas etc. between, for carrying a person in a lying position. **2** brick etc. laid along the face of a wall.

strew /stru:/ *v.* (*past part.* **strewn** or **strewed**) **1** scatter or spread about over a surface. **2** (usu. foll. by *with*) spread (a surface) with scattered things. [Old English: related to STRAW]

'strewth var. of 'STRUTH.

stria /'straɪə/ *n.* (*pl.* **striae** /-i:/) slight ridge or furrow. [Latin]

striate –*adj.* /'straɪɪt/ (also **striated** /-eɪtɪd/) marked with striae. –*v.* /'straɪeɪt/ (**-ting**) mark with slight ridges. □ **striation** /straɪ'eɪʃ(ə)n/ *n.*

stricken /'strɪkən/ *adj.* overcome with illness or misfortune etc. [archaic past part. of STRIKE]

strict *adj.* **1** precisely limited or defined; undeviating (*strict diet*). **2** requiring complete obedience or exact performance. □ **strictly speaking** applying words or rules in their strict sense. □ **strictly** *adv.* **strictness** *n.* [Latin *stringo strict-* draw tight]

stricture /'strɪktʃə(r)/ *n.* (usu. in *pl.*; often foll. by *on*, *upon*) critical or censorious remark. [Latin: related to STRICT]

stride –*v.* (**-ding**; *past* **strode**; *past part.* **stridden** /'strɪd(ə)n/) **1** walk with long firm steps. **2** cross with one step. **3** bestride. –*n.* **1 a** single long step. **b** length of this. **2** gait as determined by the length of stride. **3** (usu. in *pl.*) progress (*great strides*). **4** steady progress (*get into one's stride*). □ **take in one's stride** manage easily. [Old English]

strident /'straɪd(ə)nt/ *adj.* loud and harsh. □ **stridency** *n.* **stridently** *adv.* [Latin *strido* creak]

strife *n.* conflict; struggle. [French *estrif*: related to STRIVE]

strike –*v.* (**-king**; *past* **struck**; *past part.* **struck** or *archaic* **stricken**) **1** deliver (a blow) or inflict a blow on; hit. **2** come or bring sharply into contact with (*ship struck a rock*). **3** propel or divert with a blow. **4** (cause to) penetrate (*struck terror into him*). **5** ignite (a match) or produce (sparks etc.) by friction. **6** make (a coin) by stamping. **7** produce (a musical note) by striking. **8 a** (also *absol.*) (of a clock) indicate (the time) with a chime etc. **b** (of time) be so indicated. **9 a** attack suddenly. **b** (of a disease) afflict. **10** cause to become suddenly (*struck dumb*). **11** reach or achieve (*strike a balance*). **12** agree on (a bargain). **13** assume (an attitude) suddenly and dramatically. **14** discover or find (oil etc.) by drilling etc. **15** occur to or appear to (*strikes me as silly*). **16** (of employees) engage in a strike. **17** lower or take down (a flag or tent etc.). **18** take a specified direction. **19** (also *absol.*) secure a hook in the mouth of (a fish) by jerking the tackle. –*n.* **1** act of striking. **2 a** organized refusal to work until a grievance is remedied. **b** similar refusal to participate. **3** sudden find or success. **4** attack, esp. from the air. □ **on strike** taking part in an industrial etc. strike. **strike home 1** deal an effective blow. **2** have the intended effect. **strike off 1** remove with a stroke. **2** delete (a name etc.) from a list, esp. a professional register. **strike out 1** hit out. **2** act vigorously. **3** delete (an item or name etc.). **4** set off (*struck out eastwards*). **strike up 1** start (an acquaintance, conversation, etc.), esp. casually. **2** (also *absol.*) begin playing (a tune etc.). **struck on** *colloq.* infatuated with. [Old English, = go, stroke]

strikebreaker *n.* person working or employed in place of strikers.

strike pay *n.* allowance paid to strikers by their union.

striker *n.* **1** employee on strike. **2** *Football* attacking player positioned forward.

striking *adj.* impressive; attracting attention. □ **strikingly** *adv.*

Strine *n.* **1** comic transliteration of Australian pronunciation. **2** (esp. uneducated) Australian English. [= *Australian* in Strine]

string –*n.* **1** twine or narrow cord. **2** piece of this or of similar material used for tying or holding together, pulling, forming the head of a racket, etc. **3** length of catgut or wire etc. on a musical instrument, producing a note by vibration. **4 a** (in *pl.*) stringed instruments in an orchestra etc. **b** (*attrib.*) of stringed instruments (*string quartet*). **5** (in *pl.*) condition or complication (*no strings attached*). **6** set of things strung together; series or line. **7** tough side of a bean-pod etc. –*v.* (*past* and *past part.* **strung**) **1** fit (a racket, violin, archer's bow, etc.) with a string or strings, or (a violin etc. bow) with horsehairs etc. **2** tie with string. **3** thread on a string. **4** arrange in or as a string. **5** remove the strings from (a bean). □ **on a string** under one's control. **string along** *colloq.* **1** deceive. **2** (often foll. by *with*) keep company (with). **string out** extend; prolong. **string up 1** hang up on strings etc. **2** kill by hanging. **3** (usu. as **strung up** *adj.*) make tense. [Old English]

string-course *n.* raised horizontal band of bricks etc. on a building.

stringed *adj.* (of musical instruments) having strings.

stringent /'strɪndʒ(ə)nt/ *adj.* (of rules etc.) strict, precise; leaving no loophole for discretion. □ **stringency** *n.* **stringently** *adv.* [Latin: related to STRICT]

stringer *n.* **1** longitudinal structural member in a framework, esp. of a ship or aircraft. **2** *colloq.* freelance newspaper correspondent.

string vest *n.* vest with large meshes.

stringy *adj.* (**-ier**, **-iest**) like string, fibrous. □ **stringiness** *n.*

strip[1] –*v.* (**-pp-**) **1** (often foll. by *of*) remove the clothes or covering from. **2** (often foll. by *off*) undress oneself. **3** (often foll. by *of*) deprive (a person) of property or titles. **4** leave bare. **5** (often foll. by *down*) remove the accessory fittings of or take apart (a machine etc.). **6** damage the thread of (a screw) or the teeth of (a gearwheel). **7** remove (paint) or remove paint from (a surface) with solvent. **8** (often foll. by *from*) pull (a covering etc.) off (*stripped the masks from their faces*). –*n.* **1** act of stripping, esp. in striptease. **2** *colloq.* distinctive outfit worn by a sports team. [Old English]

strip[2] *n.* long narrow piece. □ **tear a person off a strip** *colloq.* rebuke a person. [Low German *strippe* strap]

strip cartoon *n.* = COMIC STRIP.

strip club *n.* club at which striptease is performed.

stripe *n.* **1** long narrow band or strip differing in colour or texture from the surface on either side of it. **2** *Mil.* chevron etc. denoting military rank. [perhaps from Low German or Dutch]

striped *adj.* marked with stripes.

strip light *n.* tubular fluorescent lamp.

stripling *n.* youth not yet fully grown. [from STRIP[2]]

stripper *n.* **1** person or thing that strips something. **2** device or solvent for removing paint etc. **3** striptease performer.

strip-search *–n.* search involving the removal of all a person's clothes. *–v.* search in this way.

striptease /'strɪpti:z/ *n.* entertainment in which the performer slowly and erotically undresses.

stripy *adj.* (**-ier, -iest**) striped.

strive *v.* (**-ving**; *past* **strove**; *past part.* **striven** /'strɪv(ə)n/) **1** try hard (*strive to succeed*). **2** (often foll. by *with, against*) struggle. [French *estriver*]

strobe *n. colloq.* stroboscope. [abbreviation]

stroboscope /'strəʊbəˌskəʊp/ *n.* **1** *Physics* instrument for determining speeds of rotation etc. by shining a bright light at intervals so that a rotating object appears stationary. **2** lamp made to flash intermittently, esp. for this purpose. □ **stroboscopic** /-'skɒpɪk/ *adj.* [Greek *strobos* whirling]

strode *past* of STRIDE.

stroke *–n.* **1** act of striking; blow, hit. **2** sudden disabling attack caused esp. by thrombosis; apoplexy. **3 a** action or movement, esp. as one of a series. **b** slightest action (*stroke of work*). **4** single complete motion of a wing, oar, etc. **5** (in rowing) the mode or action of moving the oar (*row a fast stroke*). **6** whole motion of a piston in either direction. **7** specified mode of swimming. **8** specially successful or skilful effort (*a stroke of diplomacy*). **9** mark made by a single movement of a pen, paintbrush, etc. **10** detail contributing to the general effect. **11** sound of a striking clock. **12** (in full **stroke oar**) oar or oarsman nearest the stern, setting the time of the stroke. **13** act or spell of stroking. *–v.* (**-king**) **1** pass one's hand gently along the surface of (hair or fur etc.). **2** act as the stroke of (a boat or crew). □ **at a stroke** by a single action. **on the stroke (of)** punctually (at). [Old English: related to STRIKE]

stroll /strəʊl/ *–v.* walk in a leisurely way. *–n.* short leisurely walk. [probably from German *Strolch* vagabond]

strolling players *n.pl. hist.* travelling actors etc.

strong *–adj.* (**stronger** /'strɒŋgə(r)/; **strongest** /'strɒŋgɪst/) **1** able to resist; not easily damaged, overcome, or disturbed. **2** healthy. **3** capable of exerting great force or of doing much; muscular, powerful. **4** forceful in effect (*strong wind*). **5** firmly held (*strong suspicion*). **6** (of an argument etc.) convincing. **7** intense (*strong light*). **8** formidable (*strong candidate*). **9** (of a solution or drink etc.) not very diluted. **10** of a specified number (*200 strong*). **11** *Gram.* (of a verb) forming inflections by a change of vowel within the stem (e.g. *swim, swam*). *–adv.* strongly. □ **come on strong** behave aggressively. **going strong** *colloq.* continuing vigorously; in good health etc. □ **strongish** *adj.* **strongly** *adv.* [Old English]

strong-arm *attrib. adj.* using force (*strong-arm tactics*).

strongbox *n.* small strongly made chest for valuables.

stronghold *n.* **1** fortified place. **2** secure refuge. **3** centre of support for a cause etc.

strong language *n.* swearing.

strong-minded *adj.* determined.

strong point *n.* (also **strong suit**) thing at which one excels.

strongroom *n.* room, esp. in a bank, for keeping valuables safe from fire and theft.

strontium /'strɒntɪəm/ *n.* soft silver-white metallic element. [*Strontian* in Scotland]

strontium-90 *n.* radioactive isotope of strontium found in nuclear fallout and concentrated in bones and teeth when ingested.

strop *–n.* device, esp. a strip of leather, for sharpening razors. *–v.* (**-pp-**) sharpen on a strop. [Low German or Dutch]

stroppy /'strɒpɪ/ *adj.* (**-ier, -iest**) *colloq.* bad-tempered; awkward to deal with. [origin uncertain]

strove *past* of STRIVE.

struck *past* and *past part.* of STRIKE.

structural /'strʌktʃər(ə)l/ *adj.* of a structure. □ **structurally** *adv.*

structuralism *n.* doctrine that structure rather than function is important. □ **structuralist** *n.* & *adj.*

structure /'strʌktʃə(r)/ *–n.* **1 a** constructed unit, esp. a building. **b** way in which a building etc. is constructed. **2** framework (*new wages structure*). *–v.* (**-ring**) give structure to; organize. [Latin *struo struct-* build]

strudel /'stru:d(ə)l/ *n.* thin leaved pastry rolled round a filling and baked. [German]

struggle /'strʌg(ə)l/ *–v.* (**-ling**) **1** violently try to get free of restraint. **2** (often foll. by *for*, or *to* + infin.) try hard under difficulties (*struggled for power*; *struggled to win*). **3** (foll. by *with*, *against*) contend; fight. **4** (foll. by *along*, *up*, etc.) progress with difficulty. **5** (esp. as **struggling** *adj.*) have difficulty in gaining recognition or a living (*struggling artist*). –*n.* **1** act or spell of struggling. **2** hard or confused contest. [origin uncertain]

strum –*v.* (**-mm-**) **1** (often foll. by *on*; also *absol.*) play on (a guitar, piano, etc.), esp. carelessly or unskilfully. **2** play (a tune etc.) in this way. –*n.* sound or spell of strumming. [imitative: cf. THRUM[1]]

strumpet /'strʌmpɪt/ *n. archaic* or *rhet.* prostitute. [origin unknown]

strung *past* and *past part.* of STRING.

strut –*n.* **1** bar in a framework, designed to resist compression. **2** strutting gait. –*v.* (**-tt-**) **1** walk stiffly and pompously. **2** brace with struts. [Old English]

'struth /stru:θ/ *int.* (also **'strewth**) *colloq.* exclamation of surprise. [*God's truth*]

strychnine /'strɪkni:n/ *n.* highly poisonous alkaloid used in small doses as a stimulant. [Greek *strukhnos* nightshade]

Sts *abbr.* Saints.

stub –*n.* **1** remnant of a pencil or cigarette etc. **2** counterfoil of a cheque or receipt etc. **3** stump. –*v.* (**-bb-**) **1** strike (one's toe) against something. **2** (usu. foll. by *out*) extinguish (a cigarette) by pressure. [Old English]

stubble /'stʌb(ə)l/ *n.* **1** stalks of corn etc. left in the ground after the harvest. **2** short stiff hair or bristles. □ **stubbly** *adj.* [Latin *stupula*]

stubborn /'stʌbən/ *adj.* obstinate, inflexible. □ **stubbornly** *adj.* **stubbornness** *n.* [origin unknown]

stubby *adj.* (**-ier**, **-est**) short and thick.

stucco /'stʌkəʊ/ –*n.* (*pl.* **-es**) plaster or cement for coating walls or moulding into decorations. –*v.* (**-es**, **-ed**) coat with stucco. [Italian]

stuck *past* and *past part.* of STICK[2].

stuck-up *adj.* conceited, snobbish. [STICK[2]]

stud[1] –*n.* **1** large-headed projecting nail, boss, or knob, esp. for ornament. **2** double button, esp. for use with two buttonholes in a shirt-front. –*v.* (**-dd-**) **1** set with or as with studs. **2** (as **studded** *adj.*) (foll. by *with*) thickly set or strewn with. [Old English]

stud[2] *n.* **1 a** number of horses kept for breeding etc. **b** place where these are kept. **2** stallion. **3** *colloq.* young man, esp. one noted for sexual prowess. **4** (in full **stud poker**) form of poker with betting after the dealing of cards face up. □ **at stud** (of a stallion) hired out for breeding. [Old English]

stud-book *n.* book containing the pedigrees of horses.

studding-sail /'stʌns(ə)l/ *n.* extra sail set in light winds. [Low German or Dutch]

student /'stju:d(ə)nt/ *n.* **1** person who is studying, esp. at a place of higher or further education. **2** (*attrib.*) studying in order to become (*student nurse*). □ **studentship** *n.* [Latin: related to STUDY]

stud-farm *n.* place where horses are bred.

studio /'stju:dɪəʊ/ *n.* (*pl.* **-s**) **1** workroom of a painter, photographer, etc. **2** place for making films, recordings, or broadcast programmes. [Italian]

studio couch *n.* couch convertible into a bed.

studio flat *n.* one-roomed flat.

studious /'stju:dɪəs/ *adj.* **1** assiduous in study. **2** painstaking. □ **studiously** *adv.* [Latin: related to STUDY]

study /'stʌdɪ/ –*n.* (*pl.* **-ies**) **1** acquisition of knowledge, esp. from books. **2** (in *pl.*) pursuit of academic knowledge. **3** private room used for reading, writing, etc. **4** piece of work, esp. a drawing, done for practice or as an experiment. **5** portrayal in literature etc. of behaviour or character etc. **6** musical composition designed to develop a player's skill. **7** thing worth observing (*his face was a study*). **8** thing that is or deserves to be investigated. –*v.* (**-ies**, **-ied**) **1** make a study of; investigate (a subject) (*study law*). **2** (often foll. by *for*) apply oneself to study. **3** scrutinize closely (a visible object). **4** learn (one's role etc.). **5** take pains to achieve (a result) or pay regard to (a subject or principle etc.). **6** (as **studied** *adj.*) deliberate, affected (*studied politeness*). [Latin *studium*]

stuff –*n.* **1** material; fabric. **2** substance or things not needing to be specified (*lot of stuff on the news*). **3** particular knowledge or activity (*know one's stuff*). **4** woollen fabric. **5** trash, nonsense. **6** (prec. by *the*) **a** *colloq.* supply, esp. of drink or drugs. **b** *slang* money. –*v.* **1** pack (a receptacle) tightly (*stuff a cushion with feathers*). **2** (foll. by *in*, *into*) force or cram (a thing). **3** fill out the skin of (an animal etc.) with material to restore the original shape. **4** fill (food, esp. poultry) with a mixture, esp. before cooking. **5** (also *refl.*) fill with food; eat greedily. **6** push, esp. hastily or clumsily. **7** (usu. in *passive*; foll. by *up*) block up (the nose etc.). **8** *slang* (expressing contempt) dispose of (*you can stuff the job*). **9** *coarse slang offens.* have sexual intercourse with (a woman). □ **get**

stuffed *slang* exclamation of dismissal, contempt, etc. **stuff and nonsense** exclamation of incredulity or ridicule. [French *estoffe*]

stuffed shirt *n. colloq.* pompous person.

stuffing *n.* **1** padding for cushions etc. **2** mixture used to stuff food, esp. before cooking.

stuffy *adj.* **(-ier, -iest) 1** (of a room etc.) lacking fresh air. **2** dull or uninteresting. **3** (of the nose etc.) stuffed up. **4** dull and conventional. □ **stuffily** *adv.* **stuffiness** *n.*

stultify /'stʌltɪ,faɪ/ *v.* **(-ies, -ied)** make ineffective or useless, esp. by routine. □ **stultification** /-fɪ'keɪʃ(ə)n/ *n.* [Latin *stultus* foolish]

stumble /'stʌmb(ə)l/ *–v.* **(-ling) 1** involuntarily lurch forward or almost fall. **2** (often foll. by *along*) walk with repeated stumbles. **3** speak haltingly. **4** (foll. by *on, upon, across*) find by chance. *–n.* act of stumbling. [related to STAMMER]

stumbling-block *n.* obstacle.

stump *–n.* **1** part of a cut or fallen tree still in the ground. **2** similar part (e.g. of a branch or limb) cut off or worn down. **3** *Cricket* each of the three uprights of a wicket. **4** (in *pl.*) *joc.* legs. *–v.* **1** (of a question etc.) be too hard for; baffle. **2** (as **stumped** *adj.*) at a loss, baffled. **3** *Cricket* put (a batsman) out by touching the stumps with the ball while he is out of the crease. **4** walk stiffly or noisily. **5** (also *absol.*) *US* traverse (a district) making political speeches. □ **stump up** *colloq.* pay or produce (the money required). [Low German or Dutch]

stumpy *adj.* **(-ier, -iest)** short and thick. □ **stumpiness** *n.*

stun *v.* **(-nn-) 1** knock senseless; stupefy. **2** bewilder, shock. [French: related to ASTONISH]

stung *past* and *past part.* of STING.

stunk *past* and *past part.* of STINK.

stunner *n. colloq.* stunning person or thing.

stunning *adj. colloq.* extremely attractive or impressive. □ **stunningly** *adv.*

stunt[1] *v.* retard the growth or development of. [obsolete *stunt* foolish, short]

stunt[2] *n.* **1** something unusual done for publicity. **2** trick or daring feat. [origin unknown]

stunt man *n.* man employed to perform dangerous stunts in place of an actor.

stupefy /'stju:pɪ,faɪ/ *v.* **(-ies, -ied) 1** make stupid or insensible. **2** astonish, amaze. □ **stupefaction** /-'fækʃ(ə)n/ *n.* [French from Latin *stupeo* be amazed]

stupendous /stju:'pendəs/ *adj.* amazing or prodigious, esp. in size. □ **stupendously** *adv.* [Latin: related to STUPEFY]

stupid /'stju:pɪd/ *adj.* **(stupider, stupidest) 1** unintelligent, foolish (*a stupid fellow*). **2** typical of stupid persons (*stupid mistake*). **3** uninteresting, boring. **4** in a stupor. □ **stupidity** /-'pɪdɪtɪ/ *n.* (*pl.* **-ies**). **stupidly** *adv.* [Latin: related to STUPENDOUS]

stupor /'stju:pə(r)/ *n.* dazed, torpid, or helplessly amazed state. [Latin: related to STUPEFY]

sturdy /'stɜ:dɪ/ *adj.* **(-ier, -iest) 1** robust; strongly built. **2** vigorous (*sturdy resistance*). □ **sturdily** *adv.* **sturdiness** *n.* [French *esturdi*]

sturgeon /'stɜ:dʒ(ə)n/ *n.* (*pl.* same or **-s**) large sharklike fish yielding caviare. [Anglo-French from Germanic]

stutter *–v.* **1** stammer, esp. by involuntary repetition of the initial consonants of words. **2** (often foll. by *out*) utter (words) in this way. *–n.* act or habit of stuttering. [dial. *stut*]

sty[1] /staɪ/ *n.* (*pl.* **sties**) = PIGSTY. [Old English]

sty[2] /staɪ/ *n.* (also **stye**) (*pl.* **sties** or **styes**) inflamed swelling on the edge of an eyelid. [Old English]

Stygian /'stɪdʒɪən/ *adj. literary* dark, gloomy. [literally = of the *Styx*, a river round Hades in Greek mythology]

style /staɪl/ *–n.* **1** kind or sort, esp. in regard to appearance and form (*elegant style of house*). **2** manner of writing, speaking, or performing. **3** distinctive manner of a person, artistic school, or period. **4** correct way of designating a person or thing. **5** superior quality or manner (*do it in style*). **6** fashion in dress etc. **7** pointed tool for scratching or engraving. **8** *Bot.* narrow extension of the ovary supporting the stigma. *–v.* **(-ling) 1** design or make etc. in a particular (esp. fashionable) style. **2** designate in a specified way. [Latin *stilus*]

stylish *adj.* **1** fashionable; elegant. **2** superior. □ **stylishly** *adv.* **stylishness** *n.*

stylist /'staɪlɪst/ *n.* **1 a** designer of fashionable styles etc. **b** hairdresser. **2** stylish writer or performer.

stylistic /staɪ'lɪstɪk/ *adj.* of esp. literary style. □ **stylistically** *adv.*

stylized /'staɪlaɪzd/ *adj.* (also **-ised**) painted, drawn, etc. in a conventional non-realistic style.

stylus /'staɪləs/ *n.* (*pl.* **-luses**) **1** sharp needle following a groove in a gramophone record and transmitting the recorded sound for reproduction. **2** pointed writing tool. [Latin: related to STYLE]

stymie /'staɪmɪ/ (also **stimy**) *–n.* (*pl.* **-ies**) **1** *Golf* situation where an opponent's ball lies between one's ball and the hole. **2** difficult situation. *–v.* (**-mies, -mied, -mying** or **-mieing**) **1** obstruct; thwart. **2** *Golf* block with a stymie. [origin unknown]

styptic /'stɪptɪk/ *–adj.* checking bleeding. *–n.* styptic substance. [Greek *stuphō* contract]

styrene /'staɪəri:n/ *n.* liquid hydrocarbon easily polymerized and used in making plastics etc. [Greek *sturax* a resin]

suasion /'sweɪʒ(ə)n/ *n. formal* persuasion (*moral suasion*). [Latin *suadeo suas-* urge]

suave /swɑ:v/ *adj.* smooth; polite; sophisticated. □ **suavely** *adv.* **suavity** /-vɪtɪ/ *n.* [Latin *suavis*]

sub *colloq. –n.* **1** submarine. **2** subscription. **3** substitute. **4** sub-editor. *–v.* (**-bb-**) **1** (usu. foll. by *for*) act as a substitute. **2** sub-edit. [abbreviation]

sub- *prefix* **1** at, to, or from a lower position (*subordinate*; *submerge*; *subtract*). **2** secondary or inferior position (*subclass*; *subtotal*). **3** nearly; more or less (*subarctic*). [Latin]

subaltern /'sʌbəlt(ə)n/ *n.* officer below the rank of captain, esp. a second lieutenant. [Latin: related to ALTERNATE]

sub-aqua /sʌb'ækwə/ *adj.* of underwater swimming or diving.

subaquatic /,sʌbə'kwætɪk/ *adj.* underwater.

subatomic /,sʌbə'tɒmɪk/ *adj.* occurring in, or smaller than, an atom.

subcommittee *n.* committee formed from a main committee for a special purpose.

subconscious /sʌb'kɒnʃəs/ *–adj.* of the part of the mind which is not fully conscious but influences actions etc. *–n.* this part of the mind. □ **subconsciously** *adv.*

subcontinent /'sʌb,kɒntɪnənt/ *n.* large land mass, smaller than a continent.

subcontract *–v.* /,sʌbkən'trækt/ **1** employ another contractor to do (work) as part of a larger project. **2** make or carry out a subcontract. *–n.* /sʌb'kɒntrækt/ secondary contract. □ **subcontractor** /-'træktə(r)/ *n.*

subculture /'sʌb,kʌltʃə(r)/ *n.* distinct cultural group within a larger culture.

subcutaneous /,sʌbkju:'teɪnɪəs/ *adj.* under the skin.

subdivide /,sʌbdɪ'vaɪd/ *v.* (**-ding**) divide again after a first division. □ **subdivision** /-,vɪʒ(ə)n/ *n.*

subdue /səb'dju:/ *v.* (**-dues, -dued, -duing**) **1** conquer, subjugate, or tame. **2** (as **subdued** *adj.*) softened; lacking in intensity; toned down. [Latin *subduco*]

sub-editor /sʌb'edɪtə(r)/ *n.* **1** assistant editor. **2** person who edits material for printing. □ **sub-edit** *v.* (**-t-**).

subfusc /'sʌbfʌsk/ *–adj. formal* dull; dusky. *–n.* formal clothing at some universities. [Latin *fuscus* dark brown]

subgroup /'sʌbgru:p/ *n.* subset of a group.

subheading /'sʌb,hedɪŋ/ *n.* subordinate heading or title.

subhuman /sʌb'hju:mən/ *adj.* (of behaviour, intelligence, etc.) less than human.

subject *–n.* /'sʌbdʒɪkt/ **1 a** matter, theme, etc. to be discussed, described, represented, etc. **b** (foll. by *for*) person, circumstance, etc., giving rise to a specified feeling, action, etc. (*subject for congratulation*). **2** field of study. **3** *Logic & Gram.* noun or its equivalent about which a sentence is predicated and with which the verb agrees. **4** any person, except a monarch, living under a government. **5** *Philos.* **a** thinking or feeling entity; the conscious mind esp. as opposed to anything external to it. **b** central substance of a thing as opposed to its attributes. **6** *Mus.* theme; leading phrase or motif. **7** person of specified tendencies (*a hysterical subject*). *–adj.* /'sʌbdʒɪkt/ **1** (foll. by *to*) conditional upon (*subject to your approval*). **2** (foll. by *to*) liable or exposed to (*subject to infection*). **3** (often foll. by *to*) owing obedience to a government etc.; in subjection. *–adv.* /'sʌbdʒɪkt/ (foll. by *to*) conditionally upon (*subject to your consent, I shall go*). *–v.* /səb'dʒekt/ **1** (foll. by *to*) make liable; expose (*subjected us to hours of waiting*). **2** (usu. foll. by *to*) subdue to one's sway etc. □ **subjection** /səb'dʒekʃ(ə)n/ *n.* [Latin *subjectus* placed under]

subjective /səb'dʒektɪv/ *adj.* **1** (of art, written history, an opinion, etc.) not impartial or literal; personal. **2** esp. *Philos.* of the individual consciousness or perception; imaginary, partial, or distorted. **3** *Gram.* of the subject. □ **subjectively** *adv.* **subjectivity** /,sʌbdʒek'tɪvɪtɪ/ *n.* [Latin: related to SUBJECT]

subjoin /sʌb'dʒɔɪn/ *v.* add (an illustration, anecdote, etc.) at the end. [Latin *subjungo -junct-*]

sub judice /sʌb 'dʒu:dɪsɪ/ *adj. Law* under judicial consideration and therefore prohibited from public discussion elsewhere. [Latin]

subjugate /'sʌbdʒʊ,geɪt/ *v.* (**-ting**) bring into subjection; vanquish. □ **subjugation** /-'geɪʃ(ə)n/ *n.* **subjugator** *n.* [Latin *jugum* yoke]

subjunctive /səb'dʒʌŋktɪv/ *Gram.* *–adj.* (of a mood) expressing what is imagined, wished, or possible (e.g. *if I were you*; *be that as it may*). *–n.* this mood or form. [Latin: related to SUBJOIN]

sublease *–n.* /'sʌbli:s/ lease granted by a tenant to a subtenant. *–v.* /sʌb'li:s/ (**-sing**) lease to a subtenant.

sublet *–n.* /'sʌblet/ = SUBLEASE *n.* *–v.* /sʌb'let/ (**-tt-**; *past* and *past part.* **-let**) = SUBLEASE *v.*

sub-lieutenant /ˌsʌblef'tenənt/ *n.* officer ranking next below lieutenant.

sublimate *–v.* /'sʌblɪˌmeɪt/ (**-ting**) **1** divert (esp. sexual energy) into socially more acceptable activity. **2** convert (a substance) from the solid state directly to vapour by heat, and usu. allow it to solidify again. **3** refine; purify; idealize. *–n.* /'sʌblɪmət/ sublimated substance. □ **sublimation** /-'meɪʃ(ə)n/ *n.* [Latin: related to SUBLIME]

sublime /sə'blaɪm/ *–adj.* (**sublimer**, **sublimest**) **1** of the most exalted or noble kind; awe-inspiring. **2** arrogantly unruffled (*sublime indifference*). *–v.* (**-ming**) **1** = SUBLIMATE *v.* 2. **2** purify or elevate by or as if by sublimation; make sublime. **3** become pure (as if) by sublimation. □ **sublimely** *adv.* **sublimity** /-'lɪmɪtɪ/ *n.* [Latin *sublimis*]

subliminal /səb'lɪmɪn(ə)l/ *adj. Psychol.* (of a stimulus etc.) below the threshold of sensation or consciousness. □ **subliminally** *adv.* [Latin *limen -min-* threshold]

sub-machine-gun /ˌsʌbmə'ʃi:ngʌn/ *n.* hand-held lightweight machine-gun.

submarine /ˌsʌbmə'ri:n, 'sʌb-/ *–n.* vessel, esp. an armed warship, capable of operating under water. *–attrib. adj.* existing, occurring, done, or used under the sea. □ **submariner** /-'mærɪnə(r)/ *n.*

submerge /səb'mɜ:dʒ/ *v.* (**-ging**) **1** place, go, or dive under water. **2** inundate with work, problems, etc. □ **submergence** *n.* **submersion** /-'mɜ:ʃ(ə)n/ *n.* [Latin *mergo mers-* dip]

submersible /səb'mɜ:sɪb(ə)l/ *–n.* submarine operating under water for short periods. *–adj.* capable of submerging.

submicroscopic /sʌbˌmaɪkrə'skɒpɪk/ *adj.* too small to be seen by an ordinary microscope.

submission /səb'mɪʃ(ə)n/ *n.* **1 a** submitting or being submitted. **b** thing submitted. **2** submissiveness. [Latin *submissio*: related to SUBMIT]

submissive /səb'mɪsɪv/ *adj.* humble, obedient. □ **submissively** *adv.* **submissiveness** *n.*

submit /səb'mɪt/ *v.* (**-tt-**) **1** (usu. foll. by *to*) **a** cease resistance; yield. **b** *refl.* surrender (oneself) to the control of another etc. **2** present for consideration. **3** (usu. foll. by *to*) subject (a person or thing) to a process, treatment, etc. [Latin *mitto miss-* send]

subnormal /sʌb'nɔ:m(ə)l/ *adj.* below or less than normal, esp. in intelligence.

suborder *n.* taxonomic category between an order and a family.

subordinate *–adj.* /sə'bɔ:dɪnət/ (usu. foll. by *to*) of inferior importance or rank; secondary, subservient. *–n.* /sə'bɔ:dɪnət/ person working under another. *–v.* /sə'bɔ:dɪˌneɪt/ (**-ting**) (usu. foll. by *to*) make or treat as subordinate. □ **subordination** /-'neɪʃ(ə)n/ *n.* [Latin: related to ORDAIN]

subordinate clause *n.* clause serving as an adjective, adverb, or noun in a main sentence.

suborn /sə'bɔ:n/ *v.* induce by bribery etc. to commit perjury etc. [Latin *orno* equip]

sub-plot *n.* secondary plot in a play etc.

subpoena /sə'pi:nə/ *–n.* writ ordering a person to attend a lawcourt. *–v.* (*past* and *past part.* **-naed** or **-na'd**) serve a subpoena on. [Latin, = under penalty]

sub rosa /sʌb 'rəʊzə/ *adj.* & *adv.* in secrecy or confidence. [Latin, = under the rose]

subroutine *n. Computing* routine designed to perform a frequently used operation within a program.

subscribe /səb'skraɪb/ *v.* (**-bing**) **1** (usu. foll. by *to, for*) **a** pay (a specified sum), esp. regularly, for membership of an organization, receipt of a publication, etc. **b** contribute money to a fund, for a cause, etc. **2** (usu. foll. by *to*) agree with an opinion etc. (*I subscribe to that*). □ **subscribe to** arrange to receive (a periodical etc.) regularly. [Latin *scribo script-* write]

subscriber *n.* **1** person who subscribes. **2** person hiring a telephone line.

subscriber trunk dialling *n.* automatic connection of trunk calls by dialling.

subscript /'sʌbskrɪpt/ *–adj.* written or printed below the line. *–n.* subscript number etc.

subscription /səb'skrɪpʃ(ə)n/ *–n.* **1 a** act of subscribing. **b** money subscribed. **2** membership fee, esp. paid regularly. *–attrib. adj.* paid for mainly by advance sales of tickets (*subscription concert*).

subsection *n.* division of a section.

subsequent /'sʌbsɪkwənt/ *adj.* (usu. foll. by *to*) following, esp. as a consequence. □ **subsequently** *adv.* [Latin *sequor* follow]

subservient /səb'sɜ:vɪənt/ *adj.* **1** servile. **2** (usu. foll. by *to*) instrumental. **3**

(usu. foll. by *to*) subordinate. □ **subservience** *n.* [Latin *subservio*]

subset *n.* set of which all the elements are contained in another set.

subside /səb'saɪd/ *v.* (**-ding**) **1** become tranquil; abate (*excitement subsided*). **2** (of water etc.) sink. **3** (of the ground) cave in; sink. □ **subsidence** /-'saɪd(ə)ns, 'sʌbsɪd(ə)ns/ *n.* [Latin *subsido*]

subsidiary /səb'sɪdɪərɪ/ *–adj.* **1** supplementary; auxiliary. **2** (of a company) controlled by another. *–n.* (*pl.* **-ies**) subsidiary thing, person, or company. [Latin: related to SUBSIDY]

subsidize /'sʌbsɪ,daɪz/ *v.* (also **-ise**) (**-zing** or **-sing**) **1** pay a subsidy to. **2** partially pay for by subsidy.

subsidy /'sʌbsɪdɪ/ *n.* (*pl.* **-ies**) **1** money granted esp. by the State to keep down the price of commodities etc. **2** any monetary grant. [Latin *subsidium* help]

subsist /səb'sɪst/ *v.* **1** (often foll. by *on*) keep oneself alive; be kept alive. **2** remain in being; exist. [Latin *subsisto*]

subsistence /səb'sɪst(ə)ns/ *n.* **1** state or instance of subsisting. **2 a** means of support; livelihood. **b** (often *attrib.*) minimal level of existence or income.

subsistence farming *n.* farming which supports the farmer's household but produces no surplus.

subsoil /'sʌbsɔɪl/ *n.* soil immediately under the surface soil.

subsonic /sʌb'sɒnɪk/ *adj.* of speeds less than that of sound.

substance /'sʌbst(ə)ns/ *n.* **1** particular kind of material having uniform properties. **2** reality; solidity. **3** content or essence as opposed to form etc. (*substance of his remarks*). **4** wealth and possessions (*woman of substance*). □ **in substance** generally; essentially. [Latin *substantia*]

substandard /sʌb'stændəd/ *adj.* of less than the required or normal quality or size.

substantial /səb'stænʃ(ə)l/ *adj.* **1 a** of real importance or value. **b** large in size or amount. **2** solid; sturdy. **3** commercially successful; wealthy. **4** essential; largely true. **5** real; existing. □ **substantially** *adv.* [Latin: related to SUBSTANCE]

substantiate /səb'stænʃɪ,eɪt/ *v.* (**-ting**) prove the truth of (a charge, claim, etc.). □ **substantiation** /-'eɪʃ(ə)n/ *n.*

substantive /'sʌbstəntɪv/ *–adj.* /also səb'stæntɪv/ **1** genuine, actual, real. **2** not slight; substantial. *–n. Gram.* = NOUN. □ **substantively** *adv.*

substitute /'sʌbstɪ,tju:t/ *–n.* **1** (also *attrib.*) person or thing acting or used in place of another. **2** artificial alternative to a food etc. *–v.* (**-ting**) (often foll. by *for*) (cause to) act as a substitute. □ **substitution** /-'tju:ʃ(ə)n/ *n.* [Latin *substituo -tut-*]

substratum /'sʌb,strɑ:təm/ *n.* (*pl.* **-ta**) underlying layer or substance.

substructure *n.* underlying or supporting structure.

subsume /səb'sju:m/ *v.* (**-ming**) (usu. foll. by *under*) include (an instance, idea, category, etc.) in a rule, class, etc. [Latin *sumo* take]

subtenant /'sʌb,tenənt/ *n.* person renting a room etc. from its tenant. □ **subtenancy** *n.* (*pl.* **-ies**).

subtend /sʌb'tend/ *v.* (of a line) be opposite (an angle or arc). [Latin: related to TEND[1]]

subterfuge /'sʌbtə,fju:dʒ/ *n.* **1** attempt to avoid blame or defeat esp. by lying or deceit. **2** statement etc. used for such a purpose. [Latin]

subterranean /,sʌbtə'reɪnɪən/ *adj.* underground. [Latin *terra* land]

subtext *n.* underlying theme.

subtitle /'sʌb,taɪt(ə)l/ *–n.* **1** secondary or additional title of a book etc. **2** caption on a film etc., esp. translating dialogue. *–v.* (**-ling**) provide with a subtitle or subtitles.

subtle /'sʌt(ə)l/ *adj.* (**subtler, subtlest**) **1** elusive, mysterious; hard to grasp. **2** (of scent, colour, etc.) faint, delicate. **3 a** perceptive (*subtle intellect*). **b** ingenious (*subtle device*). □ **subtlety** *n.* (*pl.* **-ies**). **subtly** *adv.* [Latin *subtilis*]

subtotal *n.* total of one part of a group of figures to be added.

subtract /səb'trækt/ *v.* (often foll. by *from*) deduct (a number etc.) from another. □ **subtraction** /-'trækʃ(ə)n/ *n.* [Latin *subtraho* draw away]

subtropics /sʌb'trɒpɪks/ *n.pl.* regions adjacent to the tropics. □ **subtropical** *adj.*

suburb /'sʌbɜ:b/ *n.* outlying district of a city. [Latin *urbs* city]

suburban /sə'bɜ:bən/ *adj.* **1** of or characteristic of suburbs. **2** *derog.* provincial in outlook. □ **suburbanite** *n.*

suburbia /sə'bɜ:bɪə/ *n.* often *derog.* suburbs, their inhabitants, and their way of life.

subvention /səb'venʃ(ə)n/ *n.* subsidy. [Latin *subvenio* assist]

subversive /səb'vɜ:sɪv/ *–adj.* seeking to subvert (esp. a government). *–n.* subversive person. □ **subversion** *n.* **subversively** *adv.* **subversiveness** *n.* [medieval Latin *subversivus*: related to SUBVERT]

subvert /səb'vɜ:t/ *v.* overthrow or weaken (a government etc.). [Latin *verto vers-* turn]

subway /'sʌbweɪ/ *n.* **1** pedestrian tunnel beneath a road etc. **2** esp. *US* underground railway.

subzero /sʌb'zɪərəʊ/ *adj.* (esp. of temperature) lower than zero.

suc- *prefix* assim. form of SUB- before *c*.

succeed /sək'si:d/ *v.* **1 a** (often foll. by *in*) have success. **b** be successful. **2** follow; come next after. **3** (often foll. by *to*) come into an inheritance, office, title, or property (*succeeded to the throne*). [Latin *succedo -cess-* come after]

success /sək'ses/ *n.* **1** accomplishment of an aim; favourable outcome. **2** attainment of wealth, fame, or position. **3** successful thing or person. [Latin: related to SUCCEED]

successful *adj.* having success; prosperous. □ **successfully** *adv.*

succession /sək'seʃ(ə)n/ *n.* **1 a** process of following in order; succeeding. **b** series of things or people one after another. **2 a** right of succeeding to the throne, an office, inheritance, etc. **b** act or process of so succeeding. **c** those having such a right. □ **in succession** one after another. **in succession to** as the successor of.

successive /sək'sesɪv/ *adj.* following one after another; consecutive. □ **successively** *adv.*

successor /sək'sesə(r)/ *n.* (often foll. by *to*) person or thing that succeeds another.

succinct /sək'sɪŋkt/ *adj.* brief; concise. □ **succinctly** *adv.* **succinctness** *n.* [Latin *cingo cinct-* gird]

succour /'sʌkə(r)/ (*US* **succor**) *–n.* aid, esp. in time of need. *–v.* give succour to. [Latin *succurro* run to help]

succubus /'sʌkjʊbəs/ *n.* (*pl.* **-buses** or **-bi** /-ˌbaɪ/) female demon formerly believed to have sexual intercourse with sleeping men. [Latin, = prostitute]

succulent /'sʌkjʊlənt/ *–adj.* **1** juicy; palatable. **2** *Bot.* (of a plant, its leaves, or stems) thick and fleshy. *–n. Bot.* succulent plant. □ **succulence** *n.* [Latin *succus* juice]

succumb /sə'kʌm/ *v.* (usu. foll. by *to*) **1** surrender (*succumbed to temptation*). **2** die (from) (*succumbed to his injuries*). [Latin *cumbo* lie]

such *–adj.* **1** (often foll. by *as*) of the kind or degree indicated (*such people; people such as these*). **2** so great or extreme (*not such a fool as that*). **3** of a more than normal kind or degree (*such awful food*). *–pron.* such a person or persons; such a thing or things. □ **as such** as being what has been indicated or named; in itself (*there is no theatre as such*). **such as** for example. [Old English, = so like]

such-and-such *–attrib. adj.* of a particular kind but not needing to be specified. *–n.* such a person or thing.

suchlike *colloq. –attrib. adj.* of such a kind. *–n.* things, people, etc. of such a kind.

suck *–v.* **1** draw (a fluid) into the mouth by suction. **2** (also *absol.*) draw fluid from (a thing) in this way. **3** roll the tongue round (a sweet etc.). **4** make a sucking action or sound. **5** (usu. foll. by *down, in*) engulf or drown in a sucking movement. *–n.* act or period of sucking. □ **suck dry** exhaust the contents of by sucking. **suck in 1** absorb. **2** involve (a person) esp. against his or her will. **suck up 1** (often foll. by *to*) *colloq.* behave obsequiously. **2** absorb. [Old English]

sucker /'sʌkə(r)/ *n.* **1 a** gullible person. **b** (foll. by *for*) person susceptible to. **2 a** rubber cup etc. adhering by suction. **b** similar organ of an organism. **3** shoot springing from a root or stem below ground.

suckle /'sʌk(ə)l/ *v.* (**-ling**) **1** feed (young) from the breast or udder. **2** feed by sucking the breast etc.

suckling *n.* unweaned child or animal.

sucrose /'su:krəʊz/ *n.* sugar from sugar cane, sugar beet, etc. [French *sucre* SUGAR]

suction /'sʌkʃ(ə)n/ *n.* **1** act of sucking. **2 a** production of a partial vacuum by the removal of air etc. so that liquid etc. is forced in or adhesion is procured. **b** force so produced. [Latin *sugo suct-* suck]

Sudanese /ˌsu:də'ni:z/ *–adj.* of Sudan. *–n.* (*pl.* same) **1** native, national, or inhabitant of Sudan. **2** person of Sudanese descent. [*Sudan* in NE Africa]

sudden /'sʌd(ə)n/ *adj.* done or occurring unexpectedly or abruptly. □ **all of a sudden** suddenly. □ **suddenly** *adv.* **suddenness** *n.* [Latin *subitaneus*]

sudden death *n. colloq.* decision in a tied game etc. dependent on one move, card, etc.

sudden infant death syndrome *n.* = COT-DEATH.

sudorific /ˌsu:də'rɪfɪk/ *–adj.* causing sweating. *–n.* sudorific drug. [Latin *sudor* sweat]

suds *n.pl.* froth of soap and water. □ **sudsy** *adj.* [Low German *sudde* or Dutch *sudse* marsh, bog]

sue /su:, sju:/ *v.* (**sues, sued, suing**) **1** (also *absol.*) begin a law suit against. **2** (often foll. by *to, for*) make application to a lawcourt for redress. **3** (often foll. by *to, for*) make entreaty to a person for a favour. [Anglo-French *suer* from Latin *sequor* follow]

suede /sweɪd/ *n.* (often *attrib.*) **1** leather with the flesh side rubbed to a nap. **2** cloth imitating it. [French, = Sweden]

suet /'su:ɪt/ *n.* hard white fat on the kidneys or loins of oxen, sheep, etc. □ **suety** *adj.* [Anglo-French *seu*, from Latin *sebum*]

suf- *prefix* assim. form of SUB- before *f*.

suffer /'sʌfə(r)/ *v.* **1** undergo pain, grief, damage, etc. **2** undergo, experience, or be subjected to (pain, loss, grief, defeat, change, etc.). **3** tolerate (*does not suffer fools gladly*). **4** (usu. foll. by *to* + infin.) *archaic* allow. □ **sufferer** *n.* [Latin *suffero*]

sufferance *n.* tacit consent. □ **on sufferance** tolerated but not encouraged. [Latin: related to SUFFER]

suffice /sə'faɪs/ *v.* (**-cing**) **1** (often foll. by *for*, or *to* + infin.) be adequate. **2** satisfy. □ **suffice it to say** I shall say only this. [Latin *sufficio*]

sufficiency /sə'fɪʃənsɪ/ *n.* (*pl.* **-ies**) (often foll. by *of*) adequate amount.

sufficient *adj.* sufficing, adequate. □ **sufficiently** *adv.*

suffix /'sʌfɪks/ —*n.* letter(s) added at the end of a word to form a derivative. —*v.* append, esp. as a suffix. [Latin *figo fix-* fasten]

suffocate /'sʌfə,keɪt/ *v.* (**-ting**) **1** choke or kill by stopping breathing, esp. by pressure, fumes, etc. **2** (often foll. by *by*, *with*) produce a choking or breathlessness in. **3** be or feel suffocated. □ **suffocating** *adj.* **suffocation** /-'keɪʃ(ə)n/ *n.* [Latin *suffoco* from *fauces* throat]

suffragan /'sʌfrəgən/ *n.* **1** bishop assisting a diocesan bishop. **2** bishop in relation to his archbishop or metropolitan. [medieval Latin *suffraganeus*]

suffrage /'sʌfrɪdʒ/ *n.* right of voting in political elections. [Latin *suffragium*]

suffragette /,sʌfrə'dʒet/ *n. hist.* woman seeking suffrage by organized protest.

suffuse /sə'fju:z/ *v.* (**-sing**) (of colour, moisture, etc.) spread throughout from within. □ **suffusion** /-'fju:ʒ(ə)n/ *n.* [Latin *suffundo* pour over]

Sufi /'su:fɪ/ *n.* (*pl.* **-s**) Muslim mystic. □ **Sufic** *adj.* **Sufism** *n.* [Arabic]

sug- *prefix* assim. form of SUB- before *g*.

sugar /'ʃʊgə(r)/ —*n.* **1** sweet crystalline substance esp. from sugar cane and sugar beet, used in cookery etc.; sucrose. **2** *Chem.* soluble usu. sweet crystalline carbohydrate, e.g. glucose. **3** esp. *US colloq.* darling (as a term of address). —*v.* sweeten or coat with sugar. [French *sukere*, from Arabic *sukkar*]

sugar beet *n.* beet yielding sugar.

sugar cane *n.* tropical grass yielding sugar.

sugar-daddy *n. slang* elderly man who lavishes gifts on a young woman.

sugar loaf *n.* conical moulded mass of sugar.

sugar soap *n.* alkaline compound for cleaning or removing paint.

sugary *adj.* **1** containing or like sugar. **2** excessively sweet or esp. sentimental. □ **sugariness** *n.*

suggest /sə'dʒest/ *v.* **1** (often foll. by *that*) propose (a theory, plan, etc.). **2 a** evoke (an idea etc.). **b** hint at. □ **suggest itself** (of an idea etc.) come into the mind. [Latin *suggero -gest-*]

suggestible *adj.* **1** easily influenced. **2** capable of being suggested. □ **suggestibility** /-'bɪlɪtɪ/ *n.*

suggestion /sə'dʒestʃ(ə)n/ *n.* **1** suggesting or being suggested. **2** theory, plan, etc., suggested. **3** slight trace, hint. **4** *Psychol.* insinuation of a belief etc. into the mind. [Latin: related to SUGGEST]

suggestive /sə'dʒestɪv/ *adj.* **1** (usu. foll. by *of*) hinting (at). **2** (of a remark, joke, etc.) indecent. □ **suggestively** *adv.*

suicidal /,su:ɪ'saɪd(ə)l/ *adj.* **1** inclined to commit suicide. **2** of suicide. **3** self-destructive; rash. □ **suicidally** *adv.*

suicide /'su:ɪ,saɪd/ *n.* **1 a** intentional killing of oneself. **b** person who commits suicide. **2** self-destructive action or course (*political suicide*). [Latin *sui* of oneself, -CIDE]

sui generis /,sju:aɪ 'dʒenərɪs, ,su:ɪ 'gen-/ *adj.* of its own kind; unique. [Latin]

suit /su:t, sju:t/ —*n.* **1** set of matching clothes, usu. a jacket and trousers or skirt. **2** (esp. in *comb.*) clothes for a special purpose (*swimsuit*). **3** any of the four sets (spades, hearts, diamonds, clubs) making up a pack of cards. **4** lawsuit. **5 a** petition, esp. to a person in authority. **b** *archaic* courting a woman (*paid suit to her*). —*v.* **1** go well with (a person's appearance etc.). **2** (also *absol.*) meet the demands or requirements of; satisfy; agree with. **3** make fitting; accommodate; adapt. **4** (as **suited** *adj.*) appropriate; well-fitted (*not suited to be a nurse*). □ **suit oneself** do as one chooses. [Anglo-French *siute*]

suitable *adj.* (usu. foll. by *to*, *for*) well-fitted; appropriate. □ **suitability** /-'bɪlɪtɪ/ *n.* **suitably** *adv.*

suitcase *n.* case for carrying clothes etc., with a handle and a flat hinged lid.

suite /swi:t/ *n.* **1** set, esp. of rooms in a hotel etc. or a sofa and armchairs. **2** *Mus.* set of instrumental pieces performed as a unit. [French: related to SUIT]

suitor /'su:tə(r), 'sju:-/ *n.* **1** man wooing a woman. **2** plaintiff or petitioner in a lawsuit. [Anglo-French from Latin]

sulfa *US* var. of SULPHA.
sulfate *US* var. of SULPHATE.
sulfide *US* var. of SULPHIDE.
sulfite *US* var. of SULPHITE.
sulfonamide *US* var. of SULPHONAMIDE.
sulfur *US* var. of SULPHUR
sulfuric *US* var. of SULPHURIC.
sulfurous *US* var. of SULPHUROUS.
sulk –*v*. be sulky. –*n*. (also in *pl*., prec. by *the*) period of sullen silence. [perhaps a back-formation from SULKY]
sulky *adj*. (**-ier, -iest**) sullen or silent, esp. from resentment or bad temper. □ **sulkily** *adv*. **sulkiness** *n*. [perhaps from obsolete *sulke* hard to dispose of]
sullen /'sʌlən/ *adj*. passively resentful, sulky, morose. □ **sullenly** *adv*. **sullenness** *n*. [Anglo-French *sol* SOLE[3]]
sully /'sʌlɪ/ *v*. (**-ies, -ied**) disgrace or tarnish (a reputation etc.). [French *souiller*: related to SOIL[2]]
sulpha /'sʌlfə/ *n*. (*US* **sulfa**) any of various sulphonamides (often *attrib*.: *sulpha drug*). [abbreviation]
sulphate /'sʌlfeɪt/ *n*. (*US* **sulfate**) salt or ester of sulphuric acid. [Latin SULPHUR]
sulphide /'sʌlfaɪd/ *n*. (*US* **sulfide**) binary compound of sulphur.
sulphite /'sʌlfaɪt/ *n*. (*US* **sulfite**) salt or ester of sulphurous acid. [French: related to SULPHATE]
sulphonamide /sʌl'fɒnə,maɪd/ *n*. (*US* **sulfonamide**) any of a class of antibiotic drugs containing sulphur. [German *Sulfon* (related to SULPHUR), *amide* a derivative of AMMONIA]
sulphur /'sʌlfə(r)/ *n*. (*US* **sulfur**) **1** pale-yellow non-metallic element burning with a blue flame and a suffocating smell. **2** pale greenish-yellow colour. [Anglo-French from Latin]
sulphur dioxide *n*. colourless pungent gas formed by burning sulphur in air and dissolving it in water.
sulphureous /sʌl'fjʊərɪəs/ *adj*. (*US* **sulfureous**) of or like sulphur.
sulphuric /sʌl'fjʊərɪk/ *adj*. (*US* **sulfuric**) *Chem*. containing sulphur with a valency of six.
sulphuric acid *n*. dense oily highly corrosive acid.
sulphurous /'sʌlfərəs/ *adj*. (*US* **sulfurous**) **1** of or like sulphur. **2** *Chem*. containing sulphur with a valency of four.
sulphurous acid *n*. weak acid used as a reducing and bleaching acid.
sultan /'sʌlt(ə)n/ *n*. Muslim sovereign. □ **sultanate** *n*. [Arabic]
sultana /sʌl'tɑ:nə/ *n*. **1** seedless raisin. **2** sultan's mother, wife, concubine, or daughter. [Italian]
sultry /'sʌltrɪ/ *adj*. (**-ier, -iest**) **1** (of weather etc.) hot and close. **2** (of a person etc.) passionate, sensual. □ **sultrily** *adv*. **sultriness** *n*. [obsolete *sulter* (v.): related to SWELTER]
sum –*n*. **1** total resulting from addition. **2** amount of money (*a large sum*). **3 a** arithmetical problem. **b** (esp. *pl*.) *colloq*. arithmetic work, esp. elementary. –*v*. (**-mm-**) find the sum of. □ **in sum** in brief. **sum up 1** (esp. of a judge) give a summing-up. **2** form or express an opinion of (a person, situation, etc.). **3** summarize. [Latin *summa*]
sumac /'su:mæk, 'ʃu:-/ *n*. (also **sumach**) **1** shrub with reddish conical fruits used as a spice. **2** dried and ground leaves of this used in tanning and dyeing. [French from Arabic]
summarize /'sʌmə,raɪz/ *v*. (also **-ise**) (**-zing** or **-sing**) make or be a summary of.
summary /'sʌmərɪ/ –*n*. (*pl*. **-ies**) brief account. –*adj*. without details or formalities; brief. □ **summarily** *adv*. [Latin: related to SUM]
summation /sə'meɪʃ(ə)n/ *n*. **1** finding of a total. **2** a summing-up.
summer /'sʌmə(r)/ *n*. **1** (often *attrib*.) warmest season of the year. **2** (often foll. by *of*) mature stage of life etc. □ **summery** *adj*. [Old English]
summer-house *n*. light building in a garden etc. for sitting in in fine weather.
summer pudding *n*. pudding of soft fruit pressed in a bread case.
summer school *n*. course of summer lectures etc. held esp. at a university.
summer solstice *n*. solstice about 21 June.
summertime *n*. season or period of summer.
summer time *n*. period from March to October when clocks are advanced an hour.
summing-up *n*. **1** judge's review of evidence given to a jury. **2** recapitulation of the main points of an argument etc.
summit /'sʌmɪt/ *n*. **1** highest point, top. **2** highest degree of power, ambition, etc. **3** (in full **summit meeting, talks**, etc.) conference of heads of government. [Latin *summus* highest]
summon /'sʌmən/ *v*. **1** order to come or appear, esp. in a lawcourt. **2** (usu. foll. by *to* + infin.) call upon (*summoned her to assist*). **3** call together. **4** (often foll. by *up*) gather (courage, spirits, resources, etc.). [Latin *summoneo*]
summons –*n*. (*pl*. **summonses**) authoritative call to attend or do something, esp. to appear in court. –*v*. esp. *Law* serve with a summons.

sumo /'su:məʊ/ *n.* Japanese wrestling in which a wrestler is defeated by touching the ground with any part of the body except the soles of the feet or by moving outside the ring. [Japanese]

sump *n.* **1** casing holding the oil in an internal-combustion engine. **2** pit, well, hole, etc. in which superfluous liquid collects. [Low German or Dutch]

sumptuary /'sʌmptʃʊərɪ/ *adj. Law* regulating (esp. private) expenditure. [Latin *sumptus* cost]

sumptuous /'sʌmptʃʊəs/ *adj.* rich, lavish, costly. □ **sumptuously** *adv.* **sumptuousness** *n.* [Latin: related to SUMPTUARY]

Sun. *abbr.* Sunday.

sun –*n.* **1 a** the star round which the earth orbits and from which it receives light and warmth. **b** this light or warmth. **2** any star. –*v.* (**-nn-**) *refl.* bask in the sun. □ **under the sun** anywhere in the world. □ **sunless** *adj.* [Old English]

sunbathe *v.* (**-thing**) bask in the sun, esp. to tan the body. □ **sunbather** *n.*

sunbeam *n.* ray of sunlight.

sunbed *n.* **1** long lightweight, usu. folding, chair for sunbathing. **2** bed for lying on under a sun-lamp.

sunblock *n.* lotion protecting the skin from the sun.

sunburn *n.* inflammation and tanning of the skin from exposure to the sun. □ **sunburnt** *adj.* (also **sunburned**).

sundae /'sʌndeɪ/ *n.* ice-cream with fruit, nuts, syrup, etc. [perhaps from SUNDAY]

Sunday /'sʌndeɪ/ –*n.* **1** first day of the week, a Christian holiday and day of worship. **2** *colloq.* newspaper published on Sundays. –*adv. colloq.* **1** on Sunday. **2** (**Sundays**) on Sundays; each Sunday. [Old English]

Sunday best *n. joc.* person's best clothes, esp. for Sunday use.

Sunday school *n.* religious class on Sundays for children.

sunder *v. archaic* or *literary* separate. [Old English: cf. ASUNDER]

sundew *n.* small insect-consuming bog-plant.

sundial *n.* instrument showing the time by the shadow of a pointer in sunlight.

sundown *n.* sunset.

sundry /'sʌndrɪ/ –*adj.* various; several. –*n.* (*pl.* **-ies**) (in *pl.*) items or oddments not mentioned individually. [Old English: related to SUNDER]

sunfish *n.* (*pl.* same or **-es**) any of various almost spherical fish.

sunflower *n.* tall plant with large golden-rayed flowers.

sung *past part.* of SING.

sun-glasses *n.pl.* glasses tinted to protect the eyes from sunlight or glare.

sunk *past* and *past part.* of SINK.

sunken *adj.* **1** at a lower level; submerged. **2** (of the cheeks etc.) hollow, depressed. [past part. of SINK]

sun-lamp *n.* lamp giving ultraviolet rays for therapy, to tan, etc.

sunlight *n.* light from the sun.

sunlit *adj.* illuminated by sunlight.

sun lounge *n.* room with large windows to receive sunlight.

Sunni /'sʌnɪ/ –*n.* (*pl.* same or **-s**) **1** one of the two main branches of Islam, accepting law based not only on the Koran, but on Muhammad's words and acts. **2** adherent of this branch. –*adj.* (also **Sunnite**) of or relating to Sunni. [Arabic *Sunna* = way, rule]

sunny *adj.* (**-ier**, **-iest**) **1** bright with or warmed by sunlight. **2** cheery, bright. □ **sunnily** *adv.* **sunniness** *n.*

sunrise *n.* **1** sun's rising. **2** time of this.

sun-roof *n.* panel in a car's roof that can be opened.

sunset *n.* **1** sun's setting. **2** time of this.

sunshade *n.* parasol; awning.

sunshine *n.* **1 a** light of the sun. **b** area lit by the sun. **2** fine weather. **3** cheerfulness. **4** *colloq.* form of address.

sunspot *n.* dark patch on the sun's surface.

sunstroke *n.* acute prostration from excessive exposure to the sun.

suntan *n.* brownish skin colour caused by exposure to the sun. □ **suntanned** *adj.*

suntrap *n.* sunny, esp. sheltered, place.

sun-up *n.* esp. *US* sunrise.

sup¹ –*v.* (**-pp-**) **1** take by sips or spoonfuls. **2** esp. *N.Engl. colloq.* drink (alcohol). –*n.* sip of liquid. [Old English]

sup² *v.* (**-pp-**) *archaic* take supper. [French]

sup- *prefix* assim. form of SUB- before *p*.

super /'su:pə(r)/ –*adj.* (also as *int.*) *colloq.* excellent; splendid. –*n. colloq.* **1** superintendent. **2** supernumerary. [shortening of words beginning *super-*]

super- *comb. form* forming nouns, adjectives, and verbs, meaning: **1** above, beyond, or over (*superstructure*; *supernormal*). **2** to an extreme degree (*superabundant*). **3** extra good or large of its kind (*supertanker*). **4** of a higher kind (*superintendent*). [Latin]

superabundant /,su:pərə'bʌnd(ə)nt/ *adj.* abounding beyond what is normal or right. □ **superabundance** *n.* [Latin: related to SUPER-, ABOUND]

superannuate /,su:pər'ænjʊ,eɪt/ *v.* (**-ting**) **1** pension (a person) off. **2** dismiss or discard as too old. **3** (as **super-**

annuated *adj.*) too old for work or use. [Latin *annus* year]

superannuation /ˌsu:pərˌænjʊ'eɪʃ(ə)n/ *n.* **1** pension. **2** payment towards this.

superb /su:'pɜ:b/ *adj.* **1** *colloq.* excellent. **2** magnificent. □ **superbly** *adv.* [Latin, = proud]

supercargo /'su:pəˌkɑ:gəʊ/ *n.* (*pl.* **-es**) officer in a merchant ship managing sales etc. of cargo. [Spanish *sobrecargo*]

supercharge /'su:pəˌtʃɑ:dʒ/ *v.* (**-ging**) **1** (usu. foll. by *with*) charge (the atmosphere etc.) with energy, emotion, etc. **2** use a supercharger on.

supercharger *n.* device supplying air or fuel to an internal-combustion engine at above atmospheric pressure to increase efficiency.

supercilious /ˌsu:pə'sɪlɪəs/ *adj.* contemptuous; haughty. □ **superciliously** *adv.* **superciliousness** *n.* [Latin *supercilium* eyebrow]

supercomputer /'su:pəkəmˌpju:tə(r)/ *n.* powerful computer capable of dealing with complex mathematical problems.

superconductivity /ˌsu:pəˌkɒndʌk'tɪvɪtɪ/ *n. Physics* property of zero electrical resistance in some substances at very low absolute temperatures. □ **superconducting** /-kən'dʌktɪŋ/ *adj.*

superconductor /ˌsu:pəkən'dʌktə(r)/ *n. Physics* substance having superconductivity.

superego /ˌsu:pər'i:gəʊ/ *n.* (*pl.* **-s**) *Psychol.* part of the mind that acts as a conscience and responds to social rules.

supererogation /ˌsu:pərˌerə'geɪʃ(ə)n/ *n.* doing more than duty requires. [Latin *supererogo* pay in addition]

superficial /ˌsu:pə'fɪʃ(ə)l/ *adj.* **1** of or on the surface; lacking depth. **2** swift or cursory (*superficial examination*). **3** apparent but not real (*superficial resemblance*). **4** (esp. of a person) shallow. □ **superficiality** /-ʃɪ'ælɪtɪ/ *n.* **superficially** *adv.* [Latin: related to FACE]

superfine /'su:pəˌfaɪn/ *adj. Commerce* of extra quality. [Latin: related to FINE[1]]

superfluity /ˌsu:pə'flu:ɪtɪ/ *n.* (*pl.* **-ies**) **1** state of being superfluous. **2** superfluous amount or thing. [Latin *fluo* to flow]

superfluous /su:'pɜ:fluəs/ *adj.* more than is needed or wanted; useless. [Latin *fluo* to flow]

superglue /'su:pəˌglu:/ *n.* exceptionally strong glue.

supergrass /'su:pəˌgrɑ:s/ *n. colloq.* police informer implicating many people.

superhuman /ˌsu:pə'hju:mən/ *adj.* exceeding normal human capability.

superimpose /ˌsu:pərɪm'pəʊz/ *v.* (**-sing**) (usu. foll. by *on*) lay (a thing) on something else. □ **superimposition** /-pə'zɪʃ(ə)n/ *n.*

superintend /ˌsu:pərɪn'tend/ *v.* supervise, direct. □ **superintendence** *n.*

superintendent /ˌsu:pərɪn'tend(ə)nt/ *n.* **1** police officer above the rank of chief inspector. **2 a** person who superintends. **b** director of an institution etc.

superior /su:'pɪərɪə(r)/ —*adj.* **1** in a higher position; of higher rank. **2 a** high-quality (*superior leather*). **b** supercilious (*had a superior air*). **3** (often foll. by *to*) better or greater in some respect. **4** written or printed above the line. —*n.* **1** person superior to another esp. in rank. **2** head of a monastery etc. (*Mother Superior*). □ **superiority** /-'ɒrɪtɪ/ *n.* [Latin comparative of *superus* above]

superlative /su:'pɜ:lətɪv/ —*adj.* **1** of the highest quality or degree; excellent. **2** *Gram.* (of an adjective or adverb) expressing the highest degree of a quality (e.g. *bravest, most fiercely*). —*n.* **1** *Gram.* superlative form of an adjective or adverb. **2** (in *pl.*) high praise; exaggerated language. [French from Latin]

superman /'su:pəˌmæn/ *n.* **1** *colloq.* man of exceptional strength or ability. **2** *Philos.* ideal person not subject to conventional morality etc.

supermarket /'su:pəˌmɑ:kɪt/ *n.* large self-service store selling food, household goods, etc.

supernatural /ˌsu:pə'nætʃər(ə)l/ —*adj.* not attributable to, or explicable by, the laws of nature; magical; mystical. —*n.* (prec. by *the*) supernatural forces, effects, etc. □ **supernaturally** *adv.*

supernova /ˌsu:pə'nəʊvə/ *n.* (*pl.* **-vae** /-vi:/ or **-s**) star increasing suddenly in brightness.

supernumerary /ˌsu:pə'nju:mərərɪ/ —*adj.* **1** in excess of the normal number; extra. **2** engaged for extra work. **3** (of an actor) appearing on stage but not speaking. —*n.* (*pl.* **-ies**) supernumerary person or thing. [Latin: related to NUMBER]

superphosphate /ˌsu:pə'fɒsfeɪt/ *n.* fertilizer made from phosphate rock.

superpower /'su:pəˌpaʊə(r)/ *n.* extremely powerful nation.

superscript /'su:pəskrɪpt/ —*adj.* written or printed above. —*n.* superscript number or symbol. [Latin *scribo* write]

supersede /ˌsu:pə'si:d/ *v.* (**-ding**) **1** take the place of. **2** replace with another person or thing. □ **supersession** /-'seʃ(ə)n/ *n.* [Latin *supersedeo*]

supersonic /ˌsu:pə'sɒnɪk/ *adj.* of or having a speed greater than that of sound. □ **supersonically** *adv.*

superstar /ˈsuːpəˌstɑː(r)/ *n.* extremely famous or renowned actor, musician, etc.

superstition /ˌsuːpəˈstɪʃ(ə)n/ *n.* **1** belief in the supernatural; irrational fear of the unknown. **2** practice, belief, or religion based on this. □ **superstitious** *adj.* **superstitiously** *adv.* [Latin]

superstore /ˈsuːpəˌstɔː(r)/ *n.* large supermarket.

superstructure /ˈsuːpəˌstrʌktʃə(r)/ *n.* structure built on top of another.

supertanker /ˈsuːpəˌtæŋkə(r)/ *n.* very large tanker ship.

supertax /ˈsuːpəˌtæks/ *n.* additional tax on incomes above a certain level.

supervene /ˌsuːpəˈviːn/ *v.* **(-ning)** *formal* occur as an interruption or change. □ **supervention** /-ˈvenʃ(ə)n/ *n.* [Latin *supervenio*]

supervise /ˈsuːpəˌvaɪz/ *v.* **(-sing)** superintend, oversee. □ **supervision** /-ˈvɪʒ(ə)n/ *n.* **supervisor** *n.* **supervisory** *adj.* [Latin *supervideo -vis-*]

superwoman /ˈsuːpəˌwʊmən/ *n. colloq.* woman of exceptional strength or ability.

supine /ˈsuːpaɪn/ —*adj.* **1** lying face upwards. **2** inert, indolent. —*n.* Latin verbal noun used only in the accusative and ablative. [Latin]

supper *n.* **1** late evening snack. **2** evening meal, esp. light. [French *souper*]

supplant /səˈplɑːnt/ *v.* take the place of, esp. by underhand means. [Latin *supplanto* trip up]

supple /ˈsʌp(ə)l/ *adj.* **(suppler, supplest)** flexible, pliant. □ **suppleness** *n.* [Latin *supplex*]

supplement —*n.* /ˈsʌplɪmənt/ **1** thing or part added to improve or provide further information. **2** separate section, esp. a colour magazine, of a newspaper etc. —*v.* /ˈsʌplɪmənt, -ˌment/ provide a supplement for. □ **supplemental** /-ˈment(ə)l/ *adj.* **supplementary** /-ˈmentərɪ/ *adj.* **supplementation** /-ˈteɪʃ(ə)n/ *n.* [Latin *suppleo* supply]

suppliant /ˈsʌplɪənt/ —*adj.* supplicating. —*n.* supplicating person. [Latin: related to SUPPLICATE]

supplicate /ˈsʌplɪˌkeɪt/ *v.* **(-ting)** *literary* **1** petition humbly to (a person) or for (a thing). **2** (foll. by *to, for*) make a petition. □ **supplicant** *adj.* & *n.* **supplication** /-ˈkeɪʃ(ə)n/ *n.* **supplicatory** /-kətərɪ/ *adj.* [Latin *supplico*]

supply /səˈplaɪ/ —*v.* **(-ies, -ied) 1** provide (a thing needed). **2** (often foll. by *with*) provide (a person etc. with a thing). **3** meet or make up for (a deficiency or need etc.). —*n.* (*pl.* **-ies**) **1** providing of what is needed. **2** stock, store, amount, etc., of something provided or obtainable. **3** (in *pl.*) provisions and equipment for an army, expedition, etc. **4** (often *attrib.*) schoolteacher etc. acting as a temporary substitute for another. □ **in short supply** scarce. **supply and demand** *Econ.* quantities available and required, as factors regulating price. □ **supplier** *n.* [Latin *suppleo* fill up]

supply-side *attrib. adj. Econ.* denoting a policy of low taxation etc. to encourage production and investment.

support /səˈpɔːt/ —*v.* **1** carry all or part of the weight of; keep from falling, sinking, or failing. **2** provide for (a family etc.). **3** strengthen, encourage. **4** bear out; tend to substantiate. **5** give help or approval to (a person, team, sport, etc.); further (a cause etc.). **6** speak in favour of (a resolution etc.). **7** (also *absol.*) take a secondary part to (a principal actor etc.); perform a secondary act to (the main act) at a pop concert etc. —*n.* **1** supporting or being supported. **2** person or thing that supports. **3** secondary act at a pop concert etc. □ **in support of** so as to support. [Latin *porto* carry]

supporter *n.* person or thing that supports a cause, team, etc.

supporting film *n.* (also **supporting picture** etc.) less important film in a cinema programme.

supportive *adj.* providing (esp. emotional) support or encouragement. □ **supportively** *adv.* **supportiveness** *n.*

suppose /səˈpəʊz/ *v.* **(-sing)** (often foll. by *that*) **1** assume; be inclined to think. **2** take as a possibility or hypothesis (*suppose you are right; supposing you are right*). **3** (in *imper.*) as a formula of proposal (*suppose we try again*). **4** (of a theory or result etc.) require as a condition (*that supposes we're on time*). **5** (in *imper.* or *pres. part.* forming a question) in the circumstances that; if (*suppose he won't let you?*). **6** (as **supposed** *adj.*) presumed (*his supposed brother*). **7** (in *passive*; foll. by *to* + infin.) **a** be expected or required (*was supposed to write to you*). **b** (with *neg.*) ought not; not be allowed to (*you are not supposed to go in there*). □ **I suppose so** expression of hesitant agreement. [French: related to POSE]

supposedly /səˈpəʊzɪdlɪ/ *adv.* allegedly; as is generally believed.

supposition /ˌsʌpəˈzɪʃ(ə)n/ *n.* **1** thing supposed. **2** act of supposing.

suppositious /ˌsʌpəˈzɪʃəs/ *adj.* hypothetical.

suppository /sə'pɒzɪtərɪ/ *n.* (*pl.* **-ies**) medical preparation melting in the rectum or vagina. [Latin *suppositorius* placed underneath]

suppress /sə'pres/ *v.* **1** put an end to, esp. forcibly. **2** prevent (information, feelings, a reaction, etc.) from being seen, heard, or known. **3 a** partly or wholly eliminate (electrical interference etc.). **b** equip (a device) to reduce the interference caused by it. □ **suppressible** *adj.* **suppression** *n.* **suppressor** *n.* [Latin: related to PRESS[1]]

suppurate /'sʌpjʊ,reɪt/ *v.* (**-ting**) **1** form pus. **2** fester. □ **suppuration** /-'reɪʃ(ə)n/ *n.* [Latin: related to PUS]

supra /'su:prə/ *adv.* above or earlier (in a book etc.). [Latin]

supra- *prefix* above.

supranational /,su:prə'næʃən(ə)l/ *adj.* transcending national limits.

supremacy /su:'preməsɪ/ *n.* (*pl.* **-ies**) **1** being supreme. **2** highest authority.

supreme /su:'pri:m/ *adj.* **1** highest in authority or rank. **2** greatest; most important. **3** (of a penalty or sacrifice etc.) involving death. □ **supremely** *adv.* [Latin]

Supreme Court *n.* highest judicial court in a State etc.

supremo /su:'pri:məʊ/ *n.* (*pl.* **-s**) person in overall charge. [Spanish, = SUPREME]

sur-[1] *prefix* = SUPER- (*surcharge*; *surrealism*). [French]

sur-[2] *prefix* assim. form of SUB- before *r*.

surcease /sɜ:'si:s/ *literary* –*n.* cessation. –*v.* (**-sing**) cease. [French *sursis* delayed, omitted]

surcharge –*n.* /'sɜ:tʃɑ:dʒ/ additional charge or payment. –*v.* /'sɜ:tʃɑ:dʒ, -'tʃɑ:dʒ/ (**-ging**) exact a surcharge from. [French: related to SUR-[1]]

surd *Math* –*adj.* (of a number) irrational. –*n.* surd number, esp. the root of an integer. [Latin, = deaf]

sure /ʃʊə(r), ʃɔ:(r)/ –*adj.* **1** (often foll. by *of* or *that*) convinced. **2** having adequate reason for a belief or assertion. **3** (foll. by *of*) confident in anticipation or knowledge of. **4** reliable or unfailing. **5** (foll. by *to* + infin.) certain. **6** undoubtedly true or truthful. –*adv. colloq.* certainly. □ **be sure** (in *imper.* or *infin.*; foll. by *that* + clause or *to* + infin.) take care to; not fail to. **for sure** *colloq.* certainly. **make sure** make or become certain; ensure. **sure enough** *colloq.* in fact; certainly. **to be sure** admittedly; indeed, certainly. □ **sureness** *n.* [French from Latin *securus*]

sure-fire *attrib. adj. colloq.* certain to succeed.

sure-footed *adj.* never stumbling or making a mistake.

surely *adv.* **1** with certainty or safety (*slowly but surely*). **2** as an appeal to likelihood or reason (*surely that can't be right*).

surety /'ʃʊərətɪ, 'ʃɔ:-/ *n.* (*pl.* **-ies**) **1** money given as a guarantee of performance etc. **2** (esp. in phr. **stand surety for**) person who takes responsibility for another's debt, obligation, etc. [French from Latin]

surf –*n.* foam of the sea breaking on the shore or reefs. –*v.* practise surfing. □ **surfer** *n.* [origin unknown]

surface /'sɜ:fɪs/ –*n.* **1 a** the outside of a thing. **b** area of this. **2** any of the limits of a solid. **3** top of a liquid or of the ground etc. **4** outward or superficial aspect. **5** *Geom.* set of points with length and breadth but no thickness. **6** (*attrib.*) **a** of or on the surface. **b** superficial. –*v.* (**-cing**) **1** give the required surface to (a road, paper, etc.). **2** rise or bring to the surface. **3** become visible or known. **4** *colloq.* wake up; get up. □ **come to the surface** become perceptible. [French: related to SUR-[1]]

surface mail *n.* mail carried by land or sea.

surface tension *n.* tension of the surface-film of a liquid, tending to minimize its surface area.

surfboard *n.* long narrow board used in surfing.

surfeit /'sɜ:fɪt/ –*n.* **1** an excess, esp. in eating or drinking. **2** resulting fullness. –*v.* (**-t-**) **1** overfeed. **2** (foll. by *with*) (cause to) be wearied through excess. [French: related to SUR-[1], FEAT]

surfing *n.* sport of riding the surf on a board.

surge –*n.* **1** sudden rush. **2** heavy forward or upward motion. **3** sudden increase in price, activity, etc. **4** sudden increase in voltage of an electric current. **5** swell of the sea. –*v.* (**-ging**) **1** move suddenly and powerfully forwards. **2** (of an electric current etc.) increase suddenly. **3** (of the sea etc.) swell. [Latin *surgo* rise]

surgeon /'sɜ:dʒ(ə)n/ *n.* **1** medical practitioner qualified in surgery. **2** naval or military medical officer.

surgery /'sɜ:dʒərɪ/ *n.* (*pl.* **-ies**) **1** treatment of bodily injuries or disorders by incision or manipulation etc. as opposed to drugs. **2** place where or time when a doctor, dentist, etc., treats patients, or an MP, lawyer, etc., holds consultations. [Latin *chirurgia*, from Greek *kheir* hand, *ergō* work]

surgical /'sɜ:dʒɪk(ə)l/ *adj.* **1** of or by surgeons or surgery. **2 a** used in surgery. **b** worn to correct a deformity etc. **3**

(esp. of military action) swift and precise. □ **surgically** *adv.*

surgical spirit *n.* methylated spirit used for cleansing etc.

surly /'sɜ:lɪ/ *adj.* (**-ier**, **-iest**) bad-tempered; unfriendly. □ **surliness** *n.* [obsolete *sirly* haughty: related to SIR]

surmise /sə'maɪz/ *—n.* conjecture. *—v.* (**-sing**) (often foll. by *that*) infer doubtfully; guess; suppose. [Latin *supermitto -miss-* accuse]

surmount /sə'maʊnt/ *v.* **1** overcome (a difficulty or obstacle). **2** (usu. in *passive*) cap or crown. □ **surmountable** *adj.* [French: related to SUR-[1]]

surname /'sɜ:neɪm/ *n.* family name, usu. inherited or acquired by marriage. [obsolete *surnoun* from Anglo-French: related to SUR-[1]]

surpass /sə'pɑ:s/ *v.* **1** be greater or better than, outdo. **2** (as **surpassing** *adj.*) pre-eminent. [French: related to SUR-[1]]

surplice /'sɜ:plɪs/ *n.* loose white vestment worn by clergy and choristers. [Anglo-French *surplis*]

surplus /'sɜ:pləs/ *—n.* **1** amount left over. **2** excess of revenue over expenditure. *—adj.* exceeding what is needed or used. [Anglo-French]

surprise /sə'praɪz/ *—n.* **1** unexpected or astonishing thing. **2** emotion caused by this. **3** catching or being caught unawares. **4** (*attrib.*) unexpected; made or done etc. without warning. *—v.* (**-sing**) **1** affect with surprise; turn out contrary to the expectations of. **2** (usu. in *passive*; foll. by *at*) shock, scandalize. **3** capture or attack by surprise. **4** come upon (a person) unawares. **5** (foll. by *into*) startle (a person) into an action etc. □ **take by surprise** affect with surprise, esp. by an unexpected encounter or statement. □ **surprising** *adj.* **surprisingly** *adv.* [French]

surreal /sə'rɪəl/ *adj.* unreal; dreamlike; bizarre. [back-formation from SURREALISM]

surrealism *n.* 20th-c. movement in art and literature, attempting to express the subconscious mind by dream imagery, bizarre juxtapositions, etc. □ **surrealist** *n.* & *adj.* **surrealistic** /-'lɪstɪk/ *adj.* **surrealistically** /-'lɪstɪkəlɪ/ *adv.* [French: related to SUR-[1], REAL[1]]

surrender /sə'rendə(r)/ *—v.* **1** hand over; relinquish. **2** submit, esp. to an enemy. **3** *refl.* (foll. by *to*) yield to a habit, emotion, influence, etc. **4** give up rights under (a life-insurance policy) in return for a smaller sum received immediately. **5** abandon (hope etc.). *—n.* act of surrendering. □ **surrender to bail** duly appear in court after release on bail. [Anglo-French: related to SUR-[1]]

surreptitious /,sʌrəp'tɪʃəs/ *adj.* done by stealth; clandestine. □ **surreptitiously** *adv.* [Latin *surripio* seize secretly]

surrogate /'sʌrəgət/ *n.* **1** substitute. **2** deputy, esp. of a bishop in granting marriage licences. □ **surrogacy** *n.* [Latin *rogo* ask]

surrogate mother *n.* woman who bears a child on behalf of another woman, usu. by artificial insemination of her own egg by the other woman's partner.

surround /sə'raʊnd/ *—v.* come or be all round; encircle, enclose. *—n.* **1** border or edging, esp. an area of floor between the walls and carpet of a room. **2** surrounding area or substance. [Latin: related to SUR-[1], *unda* wave]

surroundings *n.pl.* objects or conditions around or affecting a person or thing; environment.

surtax /'sɜ:tæks/ *n.* additional tax, esp. on high incomes. [French: related to SUR-[1]]

surtitle /'sɜ:,taɪt(ə)l/ *n.* explanatory caption projected on to a screen above the stage during an opera.

surveillance /sɜ:'veɪləns/ *n.* close observation undertaken by the police etc. [French: related to SUR-[1], *veiller* watch]

survey *—v.* /sə'veɪ/ **1** view or consider as a whole. **2** examine the condition of (a building etc.). **3** determine the boundaries, extent, ownership, etc. of (a district etc.). *—n.* /'sɜ:veɪ/ **1** general view or consideration. **2 a** act of surveying property. **b** statement etc. resulting from this. **3** investigation of public opinion etc. **4** map or plan made by surveying. [Latin: related to SUPER-, *video* see]

surveyor /sə'veɪə(r)/ *n.* person who surveys land and buildings, esp. for a living.

survival /sə'vaɪv(ə)l/ *n.* **1** surviving. **2** relic.

survive /sə'vaɪv/ *v.* (**-ving**) **1** continue to live or exist. **2** live or exist longer than. **3** remain alive after or continue to exist in spite of (a danger, accident, etc.). □ **survivor** *n.* [Anglo-French *survivre* from Latin *supervivo*]

sus var. of SUSS.

sus- *prefix* assim. form of SUB- before *c*, *p*, *t*.

susceptibility /sə,septə'bɪlɪtɪ/ *n.* (*pl.* **-ies**) **1** being susceptible. **2** (in *pl.*) person's feelings.

susceptible /sə'septəb(ə)l/ *adj.* **1** impressionable, sensitive, emotional. **2** (*predic.*) **a** (foll. by *to*) liable or vulnerable to. **b** (foll. by *of*) allowing; admitting of (proof etc.). □ **susceptibly** *adv.* [Latin *suscipio -cept-* take up]

sushi /'su:ʃɪ/ *n.* Japanese dish of balls of cold rice topped with raw fish etc. [Japanese]

suspect –*v.* /sə'spekt/ **1** be inclined to think. **2** have an impression of the existence or presence of. **3** (often foll. by *of*) mentally accuse. **4** doubt the genuineness or truth of. –*n.* /'sʌspekt/ suspected person. –*adj.* /'sʌspekt/ subject to or deserving suspicion. [Latin *suspicio -spect-*]

suspend /sə'spend/ *v.* **1** hang up. **2** keep inoperative or undecided for a time. **3** debar temporarily from a function, office, etc. **4** (as **suspended** *adj.*) (of particles or a body in a fluid) floating between the top and bottom. [Latin *suspendo -pens-*]

suspended animation *n.* temporary deathlike condition.

suspended sentence *n.* judicial sentence left unenforced subject to good behaviour during a specified period.

suspender *n.* **1** attachment to hold up a stocking or sock by its top. **2** (in *pl.*) *US* braces.

suspender belt *n.* woman's undergarment with suspenders.

suspense /sə'spens/ *n.* state of anxious uncertainty or expectation. □ **suspenseful** *adj.* [French, = delay]

suspension /sə'spenʃ(ə)n/ *n.* **1** suspending or being suspended. **2** springs etc. supporting a vehicle on its axles. **3** substance consisting of particles suspended in a medium.

suspension bridge *n.* bridge with a roadway suspended from cables supported by towers.

suspicion /sə'spɪʃ(ə)n/ *n.* **1** unconfirmed belief; distrust. **2** suspecting or being suspected. **3** (foll. by *of*) slight trace of. □ **above suspicion** too obviously good etc. to be suspected. **under suspicion** suspected. [Latin: related to SUSPECT]

suspicious /sə'spɪʃəs/ *adj.* **1** prone to or feeling suspicion. **2** causing suspicion. □ **suspiciously** *adv.*

suss *v. slang* (also **sus**) (**-ss-**) (usu. foll. by *out*) **1** investigate, inspect. **2** work out; realize. □ **on suss** on suspicion (of having committed a crime). [abbreviation]

sustain /sə'steɪn/ *v.* **1** support, bear the weight of, esp. for a long period. **2** encourage, support. **3** (of food) nourish. **4** endure, stand. **5** suffer (defeat or injury etc.). **6** (of a court etc.) uphold or decide in favour of (an objection etc.). **7** corroborate (a statement or charge). **8** maintain (effort etc.). □ **sustainable** *adj.* [Latin *sustineo* keep up]

sustenance /'sʌstɪnəns/ *n.* **1** nourishment, food. **2** means of support. [Anglo-French: related to SUSTAIN]

suttee /sʌ'ti:, 'sʌtɪ/ *n.* esp. *hist.* **1** Hindu custom of a widow's suicide on her husband's funeral pyre. **2** widow undergoing this. [Sanskrit *satī* faithful wife]

suture /'su:tʃə(r)/ –*n.* **1** stitching of the edges of a wound or incision. **2** thread or wire used for this. –*v.* (**-ring**) stitch (a wound or incision). [Latin *suo sut-* sew]

suzerain /'su:zərən/ *n.* **1** *hist.* feudal overlord. **2** *archaic* sovereign or State partially controlling another State that is internally autonomous. □ **suzerainty** *n.* [French]

svelte /svelt/ *adj.* slender, lissom, graceful. [French from Italian]

SW *abbr.* **1** south-west. **2** south-western.

swab /swɒb/ –*n.* **1 a** absorbent pad used in surgery. **b** specimen of a secretion taken for examination. **2** mop etc. for cleaning or mopping up. –*v.* (**-bb-**) **1** clean with a swab. **2** (foll. by *up*) absorb (moisture) with a swab. **3** mop clean (a ship's deck) [Dutch]

swaddle /'swɒd(ə)l/ *v.* (**-ling**) wrap (esp. a baby) tightly. [from SWATHE]

swaddling-clothes *n.pl.* narrow bandages formerly used to wrap and restrain a baby.

swag *n.* **1** *slang* booty of burglars etc. **2** *Austral.* & *NZ* traveller's bundle. **3** festoon of flowers, foliage, drapery, etc. □ **swagged** *adj.* [probably Scandinavian]

swagger –*v.* walk or behave arrogantly. –*n.* swaggering gait or manner. [from SWAG]

swagger stick *n.* short cane carried by a military officer.

Swahili /swə'hi:lɪ/ *n.* (*pl.* same) **1** member of a Bantu people of Zanzibar and adjacent coasts. **2** their language. [Arabic]

swain *n.* **1** *archaic* country youth. **2** *poet.* young lover or suitor. [Old Norse, = lad]

swallow[1] /'swɒləʊ/ –*v.* **1** cause or allow (food etc.) to pass down the throat. **2** perform the muscular movement required to do this. **3** accept meekly or credulously. **4** repress (a feeling etc.) (*swallow one's pride*). **5** articulate (words etc.) indistinctly. **6** (often foll. by *up*) engulf or absorb; exhaust. –*n.* **1** act of swallowing. **2** amount swallowed. [Old English]

swallow[2] /'swɒləʊ/ *n.* migratory swift-flying bird with a forked tail. [Old English]

swallow-dive *n.* & *v.* dive with the arms outspread until close to the water.

swallow-tail *n.* **1** deeply forked tail. **2** butterfly etc. with this.

swam *past* of SWIM.

swami /'swɑːmɪ/ *n.* (*pl.* **-s**) Hindu male religious teacher. [Hindi *svami*]

swamp /swɒmp/ *–n.* (area of) waterlogged ground. *–v.* **1** overwhelm, flood, or soak with water. **2** overwhelm or make invisible etc. with an excess or large amount of something. □ **swampy** *adj.* (**-ier**, **-iest**). [origin uncertain]

swan /swɒn/ *–n.* large usu. white waterbird with a long flexible neck. *–v.* (**-nn-**) (usu. foll. by *about*, *off*, etc.) *colloq.* move or go aimlessly, casually, or with a superior air. [Old English]

swank *colloq.* *–n.* ostentation, swagger. *–v.* show off. □ **swanky** *adj* (**-ier**, **-iest**). [origin uncertain]

swansong *n.* person's last work or act before death or retirement etc.

swap /swɒp/ (also **swop**) *–v.* (**-pp-**) exchange or barter. *–n.* **1** act of swapping. **2** thing for swapping or swapped. [originally = 'hit', imitative]

SWAPO /'swɑːpəʊ/ *abbr.* (also **Swapo**) South West Africa People's Organization.

sward /swɔːd/ *n. literary* expanse of turf. [Old English, = skin]

swarf /swɔːf/ *n.* fine chips or filings of stone, metal, etc. [Old Norse]

swarm[1] /swɔːm/ *–n.* **1** cluster of bees leaving the hive with the queen to establish a new colony. **2** large cluster of insects, birds, or people. **3** (in *pl.*; foll. by *of*) great numbers. *–v.* **1** move in or form a swarm. **2** (foll. by *with*) (of a place) be overrun, crowded, or infested with. [Old English]

swarm[2] /swɔːm/ *v.* (foll. by *up*) climb (a rope or tree etc.) by clinging with the hands and knees etc. [origin unknown]

swarthy /'swɔːðɪ/ *adj.* (**-ier**, **-iest**) dark, dark-complexioned. [obsolete *swarty* from *swart* black, from Old English]

swashbuckler /'swɒʃ,bʌklə(r)/ *n.* swaggering adventurer. □ **swashbuckling** *adj.* & *n.* [*swash* strike noisily, BUCKLER]

swastika /'swɒstɪkə/ *n.* **1** ancient symbol formed by an equal-armed cross with each arm continued at a right angle. **2** this with clockwise continuations as the symbol of Nazi Germany. [Sanskrit]

swat /swɒt/ *–v.* (**-tt-**) **1** crush (a fly etc.) with a sharp blow. **2** hit hard and abruptly. *–n.* swatting blow. [dial. var. of SQUAT]

swatch /swɒtʃ/ *n.* **1** sample, esp. of cloth. **2** collection of samples. [origin unknown]

swath /swɔːθ/ *n.* (also **swathe** /sweɪð/) (*pl.* **-s** /swɔːθs, swɔːðz, sweɪðz/) **1** ridge of cut grass or corn etc. **2** space left clear by a mower etc. **3** broad strip. [Old English]

swathe /sweɪð/ *–v.* (**-thing**) bind or wrap in bandages or garments etc. *–n.* bandage or wrapping. [Old English]

sway *–v.* **1** (cause to) lean or move unsteadily from side to side. **2** oscillate; waver. **3 a** control the motion or direction of. **b** influence; rule over. *–n.* **1** rule, influence, or government (*hold sway*). **2** swaying motion. [origin uncertain]

swear /sweə(r)/ *–v.* (*past* **swore**; *past part.* **sworn**) **1 a** (often foll. by *to* + infin. or *that* + clause) state or promise solemnly or on oath. **b** (cause to) take (an oath) (*swore them to secrecy*). **2** *colloq.* insist (*swore he was fit*). **3** (often foll. by *at*) use profane or obscene language. **4** (foll. by *by*) **a** appeal to as a witness in taking an oath (*swear by Almighty God*). **b** *colloq.* have great confidence in (*swears by yoga*). **5** (foll. by *to*; usu. in *neg.*) say certainly (*could not swear to it*). *–n.* spell of swearing. □ **swear blind** *colloq.* affirm emphatically. **swear in** induct into office etc. with an oath. **swear off** *colloq.* promise to abstain from (drink etc.). [Old English]

swear-word *n.* profane or indecent word.

sweat /swet/ *–n.* **1** moisture exuded through the pores, esp. from heat or nervousness. **2** state or period of sweating. **3** *colloq.* state of anxiety (*in a sweat*). **4** *colloq.* **a** drudgery, effort. **b** laborious task. **5** condensed moisture on a surface. *–v.* (*past* and *past part.* **sweated** or *US* **sweat**) **1** exude sweat. **2** be terrified or suffer. **3** (of a wall etc.) exhibit surface moisture. **4** (cause to) drudge or toil. **5** emit like sweat. **6** make (a horse, athlete, etc.) sweat by exercise. **7** (as **sweated** *adj.*) (of goods, labour, etc.) produced by or subjected to exploitation. □ **no sweat** *colloq.* no bother, no trouble. **sweat blood** *colloq.* **1** work strenuously. **2** be very anxious. **sweat it out** *colloq.* endure a difficult experience to the end. □ **sweaty** *adj.* (**-ier**, **-iest**). [Old English]

sweat-band *n.* band fitted inside a hat or worn round a wrist etc. to absorb sweat.

sweater *n.* jersey or pullover.

sweatshirt *n.* sleeved cotton sweater.

sweatshop *n.* factory where sweated labour is used.

Swede *n.* **1 a** native or national of Sweden. **b** person of Swedish descent. **2** (**swede**) large yellow-fleshed turnip orig. from Sweden. [Low German or Dutch]

Swedish –*adj.* of Sweden, its people, or language. –*n.* language of Sweden.

sweep –*v.* (*past* and *past part.* **swept**) **1** clean or clear (a room or area etc.) (as) with a broom. **2** (often foll. by *up*) clean a room etc. in this way. **3** (often foll. by *up*) collect or remove (dirt etc.) by sweeping. **4** (foll. by *aside, away*, etc.) **a** push (as) with a broom. **b** dismiss abruptly. **5** (foll. by *along, down*, etc.) carry or drive along with force. **6** (foll. by *off, away*, etc.) remove or clear forcefully. **7** traverse swiftly or lightly. **8** impart a sweeping motion to. **9** swiftly cover or affect. **10 a** glide swiftly; speed along. **b** go majestically. **11** (of landscape etc.) be rolling or spacious. –*n.* **1** act or motion of sweeping. **2** curve in the road, sweeping line of a hill, etc. **3** range or scope. **4** = CHIMNEY-SWEEP. **5** sortie by aircraft. **6** *colloq.* = SWEEPSTAKE. □ **make a clean sweep of 1** completely abolish or expel. **2** win all the prizes etc. in (a competition etc.). **sweep away** abolish swiftly. **sweep the board 1** win all the money at stake. **2** win all possible prizes etc. **sweep under the carpet** see CARPET. [Old English]

sweeper *n.* **1** person who cleans by sweeping. **2** manual device for sweeping carpets etc. **3** *Football* defensive player positioned close to the goalkeeper.

sweeping –*adj.* **1** wide in range or effect (*sweeping changes*). **2** generalized, arbitrary (*sweeping statement*). –*n.* (in *pl.*) dirt etc. collected by sweeping.

sweepstake *n.* **1** form of gambling in which all stakes are pooled and paid to the winners. **2** race with betting of this kind. **3** prize(s) won in a sweepstake.

sweet –*adj.* **1** tasting of sugar. **2** smelling pleasant like roses or perfume etc.; fragrant. **3** (of sound etc.) melodious or harmonious. **4** fresh; not salt, sour, or bitter. **5** gratifying or attractive. **6** amiable, pleasant. **7** *colloq.* pretty, charming. **8** (foll. by *on*) *colloq.* fond of; in love with. –*n.* **1** small shaped piece of sweet substance, usu. made with sugar or chocolate. **2** sweet dish or course of a meal. □ **sweetish** *adj.* **sweetly** *adv.* [Old English]

sweet-and-sour *attrib. adj.* cooked in a sauce containing sugar and vinegar or lemon etc.

sweetbread *n.* pancreas or thymus of an animal, esp. as food.

sweet-brier *n.* wild rose with small fragrant leaves.

sweetcorn *n.* sweet-flavoured maize kernels.

sweeten *v.* **1** make or become sweet or sweeter. **2** make agreeable or less painful. □ **sweetening** *n.*

sweetener *n.* **1** substance used to sweeten food or drink. **2** *colloq.* bribe or inducement.

sweetheart *n.* **1** lover or darling. **2** term of endearment.

sweetie *n. colloq.* **1** = SWEET 1. **2** sweetheart.

sweetmeal *n.* sweetened wholemeal.

sweetmeat *n.* **1** = SWEET 1. **2** small fancy cake.

sweetness *n.* being sweet; fragrance. □ **sweetness and light** (esp. uncharacteristic) mildness and reason.

sweet pea *n.* climbing plant with fragrant flowers.

sweet pepper *n.* mild pepper.

sweet potato *n.* **1** tropical climbing plant with sweet tuberous roots used for food. **2** root of this.

sweetshop *n.* confectioner's shop.

sweet talk *colloq.* –*n.* flattery, blandishment. –*v.* (**sweet-talk**) flatter in order to persuade.

sweet tooth *n.* liking for sweet-tasting things.

sweet william *n.* cultivated plant with clusters of vivid fragrant flowers.

swell –*v.* (*past part.* **swollen** /'swəulən/ or **swelled**) **1** (cause to) grow bigger, louder, or more intense. **2** (often foll. by *up*) rise or raise up from the surrounding surface. **3** (foll. by *out*) bulge. **4** (of the heart etc.) feel full of joy, pride, relief, etc. **5** (foll. by *with*) be hardly able to restrain (pride etc.). –*n.* **1** act or state of swelling. **2** heaving of the sea with unbreaking waves. **3 a** crescendo. **b** mechanism in an organ etc. for producing a crescendo or diminuendo. **4** *colloq.* dandy. **5** protuberance. –*adj. colloq.* **1** esp. *US* fine, excellent. **2** smart, fashionable. □ **have** (or **get**) **a swelled** (or **swollen**) **head** be (or become) conceited. [Old English]

swelling *n.* abnormal bodily protuberance.

swelter –*v.* be uncomfortably hot. –*n.* sweltering condition. [Old English]

swept *past* and *past part.* of SWEEP.

swerve –*v.* (**-ving**) (cause to) change direction, esp. abruptly. –*n.* swerving movement. [Old English, = scour]

swift –*adj.* **1** quick, rapid. **2** prompt. –*n.* swift-flying migratory bird with long wings. □ **swiftly** *adv.* **swiftness** *n.* [Old English]

swig –*v.* (**-gg-**) *colloq.* drink in large draughts. –*n.* swallow of drink, esp. large. [origin unknown]

swill –*v.* **1** (often foll. by *out*) rinse or flush. **2** drink greedily. –*n.* **1** act of rinsing. **2** mainly liquid refuse as pig-food. [Old English]

swim –*v.* (**-mm-**; *past* **swam**; *past part.* **swum**) **1** propel the body through water with limbs, fins, or tail. **2** traverse (a stretch of water or distance) by swimming. **3** perform (a stroke) by swimming. **4** float on a liquid. **5** appear to undulate, reel, or whirl. **6** feel dizzy (*my head swam*). **7** (foll. by *in*, *with*) be flooded. –*n.* period or act of swimming. □ **in the swim** *colloq.* involved in or aware of what is going on. □ **swimmer** *n.* [Old English]

swimming-bath *n.* (also **swimming-pool**) artificial pool for swimming.

swimming-costume *n.* = BATHING-COSTUME.

swimmingly *adv. colloq.* smoothly, without impediment.

swimsuit *n.* swimming-costume, esp. one-piece for women and girls.

swimwear *n.* clothing for swimming in.

swindle /'swɪnd(ə)l/ –*v.* (**-ling**) (often foll. by *out of*) **1** cheat of money etc. **2** cheat a person of (money etc.) (*swindled £200 out of him*). –*n.* **1** act of swindling. **2** fraudulent person or thing. □ **swindler** *n.* [back-formation from *swindler* from German]

swine *n.* (*pl.* same) **1** *formal* or *US* pig. **2** *colloq.* (*pl.* same or **-s**) **a** contemptible person. **b** unpleasant or difficult thing. □ **swinish** *adj.* [Old English]

swing –*v.* (*past* and *past part.* **swung**) **1 a** (cause to) move with a to-and-fro or curving motion, as of an object attached at one end and hanging free at the other; sway. **b** hang so as to be free to swing. **2** oscillate or revolve. **3** move by gripping something and leaping etc. (*swung from tree to tree*). **4** walk with a swing. **5** (foll. by *round*) move to face the opposite direction. **6** change one's opinion or mood. **7** (foll. by *at*) attempt to hit. **8** (also **swing it**) play (music) with a swing rhythm. **9** *colloq.* (of a party etc.) be lively etc. **10** have a decisive influence on (voting etc.). **11** *colloq.* achieve, manage. **12** *colloq.* be executed by hanging. –*n.* **1** act, motion, or extent of swinging. **2** swinging or smooth gait, rhythm, or action. **3 a** seat slung by ropes etc. for swinging on or in. **b** period of swinging on this. **4 a** jazz or dance music with an easy flowing rhythm. **b** rhythmic feeling or drive of this. **5** discernible change, esp. in votes or points scored etc. □ **swings and roundabouts** situation affording equal gain and loss. □ **swinger** *n.* [Old English]

swing-boat *n.* boat-shaped swing at fairs.

swing-bridge *n.* bridge that can be swung aside to let ships pass.

swing-door *n.* self-closing door opening both ways.

swingeing /'swɪndʒɪŋ/ *adj.* **1** (of a blow) forcible. **2** huge or far-reaching (*swingeing economies*). [archaic *swinge* strike hard, from Old English]

swing-wing *n.* aircraft wing that can move from a right-angled to a swept-back position.

swipe *colloq.* –*v.* (**-ping**) **1** (often foll. by *at*) hit hard and recklessly. **2** steal. –*n.* reckless hard hit or attempted hit. [perhaps var. of SWEEP]

swirl –*v.* move, flow, or carry along with a whirling motion. –*n.* **1** swirling motion. **2** twist or curl. □ **swirly** *adj.* [perhaps from Low German or Dutch]

swish –*v.* **1** swing (a thing) audibly through the air, grass, etc. **2** move with or make a swishing sound. –*n.* swishing action or sound. –*adj. colloq.* smart, fashionable. [imitative]

Swiss –*adj.* of Switzerland or its people. –*n.* (*pl.* same) **1** native or national of Switzerland. **2** person of Swiss descent. [French *Suisse*]

Swiss roll *n.* cylindrical sponge cake with a jam etc. filling.

switch –*n.* **1** device for completing and breaking an electric circuit. **2 a** transfer, change-over, or deviation. **b** exchange. **3** flexible shoot cut from a tree. **4** light tapering rod. **5** *US* railway points. –*v.* **1** (foll. by *on*, *off*) turn (an electrical device) on or off. **2** change or transfer. **3** exchange. **4** whip or flick with a switch. □ **switch off** *colloq.* cease to pay attention. [Low German]

switchback *n.* **1** ride at a fair etc., with extremely steep ascents and descents. **2** (often *attrib.*) such a railway or road.

switchboard *n.* apparatus for making connections between electric circuits, esp. in telephony.

switched-on *adj. colloq.* **1** up to date; aware of what is going on. **2** excited; under the influence of drugs.

swivel /'swɪv(ə)l/ –*n.* coupling between two parts enabling one to revolve without turning the other. –*v.* (**-ll-**; *US* **-l-**) turn on or as on a swivel. [Old English]

swivel chair *n.* chair with a revolving seat.

swizz *n.* (also **swiz**) *colloq.* **1** something unfair or disappointing. **2** swindle. [origin unknown]

swizzle /'swɪz(ə)l/ *n.* **1** *colloq.* frothy mixed alcoholic drink esp. of rum or gin and bitters. **2** *slang* = SWIZZ. [origin unknown]

swizzle-stick *n.* stick used for frothing or flattening drinks.

swollen *past part.* of SWELL.

swoon *v.* & *n. literary* faint. [Old English]

swoop –*v.* **1** (often foll. by *down*) descend rapidly like a bird of prey. **2** (often foll. by *on*) make a sudden attack. –*n.* swooping movement or action. [Old English]

swop var. of SWAP.

sword /sɔːd/ *n.* **1** weapon with a long blade and hilt with a handguard. **2** (prec. by *the*) **a** war. **b** military power. □ **put to the sword** kill. [Old English]

sword dance *n.* dance with the brandishing of swords or with swords laid on the ground.

swordfish *n.* (*pl.* same or **-es**) large marine fish with swordlike upper jaw.

sword of Damocles /'dæmə,kliːz/ *n.* an immediate danger. [from *Damokles*, who had a sword hung by a hair over him]

swordplay *n.* **1** fencing. **2** repartee; lively argument.

swordsman *n.* person of (usu. specified) skill with a sword. □ **swordsmanship** *n.*

swordstick *n.* hollow walking-stick containing a blade that can be used as a sword.

swore *past* of SWEAR.

sworn *past part.* of SWEAR. –*attrib. adj.* bound (as) by an oath (*sworn enemies*).

swot *colloq.* –*v.* (**-tt-**) **1** study hard. **2** (usu. foll. by *up, up on*) study (a subject) hard or hurriedly. –*n.* usu. *derog.* person who swots. [dial. var. of SWEAT]

swum *past part.* of SWIM.

swung *past* and *past part.* of SWING.

sybarite /'sɪbə,raɪt/ *n.* self-indulgent or voluptuous person. □ **sybaritic** /-'rɪtɪk/ *adj.* [*Sybaris*, ancient city in S. Italy]

sycamore /'sɪkə,mɔː(r)/ *n.* **1** large maple or its wood. **2** *US* plane-tree or its wood. [Greek *sukomoros*]

sycophant /'sɪkə,fænt/ *n.* flatterer; toady. □ **sycophancy** *n.* **sycophantic** /-'fæntɪk/ *adj.* [Greek *sukophantēs*]

syl- *prefix* assim. form of SYN- before *l*.

syllabary /'sɪləbərɪ/ *n.* (*pl.* **-ies**) list of characters representing syllables. [related to SYLLABLE]

syllabic /sɪ'læbɪk/ *adj.* of or in syllables. □ **syllabically** *adv.*

syllable /'sɪləb(ə)l/ *n.* **1** unit of pronunciation forming the whole or part of a word and usu. having one vowel sound often with consonant(s) before or after (e.g. *water* has two, *inferno* three). **2** character(s) representing a syllable. **3** the least amount of speech or writing. □ **in words of one syllable** plainly, bluntly. [Greek *sullabē*]

syllabub /'sɪlə,bʌb/ *n.* (also **sillabub**) dessert of flavoured, sweetened, and whipped cream or milk. [origin unknown]

syllabus /'sɪləbəs/ *n.* (*pl.* **-buses** or **-bi** /-,baɪ/) programme or outline of a course of study, teaching, etc. [misreading of Greek *sittuba* label]

syllepsis /sɪ'lepsɪs/ *n.* (*pl.* **syllepses** /-siːz/) figure of speech in which a word is applied to two others in different senses (e.g. *caught the train and a cold*) or to two others of which it grammatically suits one only (e.g. *neither you nor he knows*) (cf. ZEUGMA). [Greek: related to SYLLABLE]

syllogism /'sɪlə,dʒɪz(ə)m/ *n.* reasoning in which a conclusion is drawn from two given or assumed propositions. □ **syllogistic** /-'dʒɪstɪk/ *adj.* [Greek *logos* reason]

sylph /sɪlf/ *n.* **1** elemental spirit of the air. **2** slender graceful woman or girl. □ **sylphlike** *adj.* [Latin]

sylvan /'sɪlv(ə)n/ *adj.* (also **silvan**) **1 a** of the woods. **b** having woods. **2** rural. [Latin *silva* a wood]

sylviculture var. of SILVICULTURE.

sym- *prefix* assim. form of SYN- before *b, m, p*.

symbiosis /,sɪmbaɪ'əʊsɪs, ,sɪmbɪ-/ *n.* (*pl.* **-bioses** /-siːz/) **1** interaction between two different organisms living in close physical association, usu. to the advantage of both. **2** mutually advantageous association between persons. □ **symbiotic** /-'ɒtɪk/ *adj.* [Greek, = living together]

symbol /'sɪmb(ə)l/ *n.* **1** thing regarded as typifying or representing something (*white is a symbol of purity*). **2** mark, sign, etc. representing an object, idea, function, or process; logo. □ **symbolic** /-'bɒlɪk/ *adj.* **symbolically** /-'bɒlɪkəlɪ/ *adv.* [Greek *sumbolon*]

symbolism *n.* **1 a** use of symbols. **b** symbols collectively. **2** artistic and poetic movement or style using symbols to express ideas, emotions, etc. □ **symbolist** *n.*

symbolize *v.* (also **-ise**) (**-zing** or **-sing**) **1** be a symbol of. **2** represent by symbols. [French: related to SYMBOL]

symmetry /'sɪmɪtrɪ/ *n.* (*pl.* **-ies**) **1 a** correct proportion of parts. **b** beauty resulting from this. **2 a** structure allowing an object to be divided into parts of an equal shape and size. **b** possession of such a structure. **3** repetition of exactly similar parts facing each other or a centre. □ **symmetrical** /-'metrɪk(ə)l/ *adj.* **symmetrically** /-'metrɪkəlɪ/ *adv.* [Greek *summetria*]

sympathetic /ˌsɪmpəˈθetɪk/ *adj.* **1** of or expressing sympathy. **2** pleasant, likeable. **3** (foll. by *to*) favouring (a proposal etc.). □ **sympathetically** *adv.*

sympathize /ˈsɪmpəˌθaɪz/ *v.* (also **-ise**) (**-zing** or **-sing**) (often foll. by *with*) **1** feel or express sympathy. **2** agree. □ **sympathizer** *n.*

sympathy /ˈsɪmpəθɪ/ *n.* (*pl.* **-ies**) **1 a** sharing of another's feelings. **b** capacity for this. **2 a** (often foll. by *with*) sharing or tendency to share (with a person etc.) in an emotion, sensation, or condition. **b** (in *sing.* or *pl.*) compassion or commiseration; condolences. **3** (often foll. by *for*) approval. **4** (in *sing.* or *pl.*; often foll. by *with*) agreement (with a person etc.) in opinion or desire. □ **in sympathy** (often foll. by *with*) having, showing, or resulting from sympathy. [Greek, = fellow-feeling]

symphony /ˈsɪmfənɪ/ *n.* (*pl.* **-ies**) **1** large-scale composition for full orchestra in several movements. **2** instrumental interlude in a large-scale vocal work. **3** = SYMPHONY ORCHESTRA. □ **symphonic** /-ˈfɒnɪk/ *adj.* [from SYN-, Greek *phōnē* sound]

symphony orchestra *n.* large orchestra suitable for playing symphonies etc.

symposium /sɪmˈpəʊzɪəm/ *n.* (*pl.* **-sia**) **1** conference, or collection of essays, on a particular subject. **2** philosophical or other friendly discussion. [Greek *sumpotēs* fellow-drinker]

symptom /ˈsɪmptəm/ *n.* **1** physical or mental sign of disease. **2** sign of the existence of something. □ **symptomatic** /-ˈmætɪk/ *adj.* [Greek *piptō* fall]

syn- *prefix* with, together, alike. [Greek *sun* with]

synagogue /ˈsɪnəˌgɒg/ *n.* **1** building for Jewish religious observance and instruction. **2** Jewish congregation. [Greek, = assembly]

synapse /ˈsaɪnæps, ˈsɪn-/ *n. Anat.* junction of two nerve-cells. [Greek *haptō* join]

sync /sɪŋk/ (also **synch**) *colloq.* —*n.* synchronization. —*v.* synchronize. □ **in** (or **out of**) **sync** (often foll. by *with*) according or agreeing well (or badly). [abbreviation]

synchromesh /ˈsɪŋkrəʊˌmeʃ/ *n.* (often *attrib.*) system of gear-changing, esp. in vehicles, in which the gearwheels revolve at the same speed during engagement. [abbreviation of *synchronized mesh*]

synchronic /sɪŋˈkrɒnɪk/ *adj.* concerned with a subject as it exists at one point in time. □ **synchronically** *adv.* [from SYN-, Greek *khronos* time]

synchronism /ˈsɪŋkrəˌnɪz(ə)m/ *n.* **1** being or treating as synchronic or synchronous. **2** process of synchronizing sound and picture.

synchronize /ˈsɪŋkrəˌnaɪz/ *v.* (also **-ise**) (**-zing** or **-sing**) **1** (often foll. by *with*) make or be synchronous (with). **2** make the sound and picture of (a film etc.) coincide. **3** cause (clocks etc.) to show the same time. □ **synchronization** /-ˈzeɪʃ(ə)n/ *n.*

■ **Usage** *Synchronize* should not be used in standard English to mean 'coordinate' or 'combine'.

synchronous /ˈsɪŋkrənəs/ *adj.* (often foll. by *with*) existing or occurring at the same time.

syncopate /ˈsɪŋkəˌpeɪt/ *v.* (**-ting**) **1** displace the beats or accents in (music). **2** shorten (a word) by dropping interior letters. □ **syncopation** /-ˈpeɪʃ(ə)n/ *n.* [Latin: related to SYNCOPE]

syncope /ˈsɪŋkəpɪ/ *n.* **1** *Gram.* syncopation. **2** fainting through a fall in blood pressure. [Greek *sunkopē* cutting off]

syncretize /ˈsɪŋkrəˌtaɪz/ *v.* (also **-ise**) (**-zing** or **-sing**) attempt, esp. inconsistently, to unify or reconcile differing schools of thought. □ **syncretic** /-ˈkretɪk/ *adj.* **syncretism** *n.* [Greek]

syndic /ˈsɪndɪk/ *n.* any of various university or government officials. [Greek *sundikos*, = advocate]

syndicalism /ˈsɪndɪkəˌlɪz(ə)m/ *n. hist.* movement for transferring industrial ownership and control to workers' unions. □ **syndicalist** *n.* [French: related to SYNDIC]

syndicate —*n.* /ˈsɪndɪkət/ **1** combination of individuals or businesses to promote a common interest. **2** agency supplying material simultaneously to a number of newspapers etc. **3** group of people who gamble, organize crime, etc. **4** committee of syndics. —*v.* /ˈsɪndɪˌkeɪt/ (**-ting**) **1** form into a syndicate. **2** publish (material) through a syndicate. □ **syndication** /-ˈkeɪʃ(ə)n/ *n.* [Latin: related to SYNDIC]

syndrome /ˈsɪndrəʊm/ *n.* **1** group of concurrent symptoms of a disease. **2** characteristic combination of opinions, emotions, behaviour, etc. [Greek *sundromē* running together]

synecdoche /sɪˈnekdəkɪ/ *n.* figure of speech in which a part is made to represent the whole or vice versa (e.g. *new faces at the club*; *England lost to India*). [Greek, = taking together]

synod /ˈsɪnəd/ *n.* Church council of delegated clergy and sometimes laity. [Greek, = meeting]

synonym /ˈsɪnənɪm/ *n.* word or phrase that means the same as another (e.g. *shut* and *close*). [Greek *onoma* name]

synonymous /sɪˈnɒnɪməs/ *adj.* (often foll. by *with*) **1** having the same meaning. **2** suggestive of; associated with (*his name is synonymous with terror*).

synopsis /sɪˈnɒpsɪs/ *n.* (*pl.* **synopses** /-siːz/) summary or outline. [Greek *opsis* view]

synoptic /sɪˈnɒptɪk/ *adj.* of or giving a synopsis. [Greek: related to SYNOPSIS]

Synoptic Gospels *n.pl.* Gospels of Matthew, Mark, and Luke.

synovia /saɪˈnəʊvɪə, sɪ-/ *n. Physiol.* viscous fluid lubricating joints etc. □ **synovial** *adj.* [medieval Latin]

syntax /ˈsɪntæks/ *n.* **1** grammatical arrangement of words. **2** rules or analysis of this. □ **syntactic** /-ˈtæktɪk/ *adj.* **syntactically** /-ˈtæktɪkəlɪ/ *adv.* [Greek, = arrangement]

synth /sɪnθ/ *n. colloq.* = SYNTHESIZER.

synthesis /ˈsɪnθəsɪs/ *n.* (*pl.* **-theses** /-ˌsiːz/) **1 a** combining of elements into a whole. **b** result of this. **2** *Chem.* artificial production of compounds from their constituents as distinct from extraction from plants etc. [Greek, = placing together]

synthesize /ˈsɪnθəˌsaɪz/ *v.* (also **-ise**) (**-zing** or **-sing**) make a synthesis of.

synthesizer *n.* electronic, usu. keyboard, instrument producing a wide variety of sounds.

synthetic /sɪnˈθetɪk/ *–adj.* **1** made by chemical synthesis, esp. to imitate a natural product. **2** affected, insincere. *–n.* synthetic substance. □ **synthetically** *adv.*

syphilis /ˈsɪfəlɪs/ *n.* contagious venereal disease. □ **syphilitic** /-ˈlɪtɪk/ *adj.* [*Syphilus*, name of a character in a poem of 1530]

Syriac /ˈsɪrɪˌæk/ *–n.* language of ancient Syria, western Aramaic. *–adj.* of or in Syriac.

Syrian /ˈsɪrɪən/ *–n.* **1** native or national of Syria. **2** person of Syrian descent. *–adj.* of Syria.

syringa /sɪˈrɪŋgə/ *n.* **1** = MOCK ORANGE. **2** lilac or similar related plant. [related to SYRINGE]

syringe /sɪˈrɪndʒ, ˈsɪ-/ *–n.* device for sucking in and ejecting liquid in a fine stream. *–v.* (**-ging**) sluice or spray with a syringe. [Greek *surigx* pipe]

syrup /ˈsɪrəp/ *n.* (*US* **sirup**) **1 a** sweet sauce of sugar dissolved in boiling water. **b** similar fluid as a drink, medicine, etc. **2** condensed sugar-cane juice; molasses, treacle. **3** excessive sweetness of manner or style. □ **syrupy** *adj.* [Arabic *sharab*]

system /ˈsɪstəm/ *n.* **1** complex whole; set of connected things or parts; organized body of things. **2 a** set of organs in the body with a common structure or function. **b** human or animal body as a whole. **3** method; scheme of action, procedure, or classification. **4** orderliness. **5** (prec. by *the*) prevailing political or social order, esp. regarded as oppressive. □ **get a thing out of one's system** *colloq.* get rid of a preoccupation or anxiety. [Greek *sustēma -mat-*]

systematic /ˌsɪstəˈmætɪk/ *adj.* **1** methodical; according to a system. **2** regular, deliberate. □ **systematically** *adv.*

systematize /ˈsɪstəməˌtaɪz/ *v.* (also **-ise**) (**-zing** or **-sing**) make systematic. □ **systematization** /-ˈzeɪʃ(ə)n/ *n.*

systemic /sɪˈstemɪk/ *adj.* **1** *Physiol.* of the whole body. **2** (of an insecticide etc.) entering the plant via the roots or shoots and freely transported within its tissues. □ **systemically** *adv.*

systems analysis *n.* analysis of a complex process etc. in order to improve its efficiency esp. by using a computer. □ **systems analyst** *n.*

T

T[1] /ti:/ *n.* (also **t**) (*pl.* **Ts** or **T's**) **1** twentieth letter of the alphabet. **2** T-shaped thing (esp. *attrib.*: *T-joint*). □ **to a T** exactly; to a nicety.

T[2] *symb.* tritium.

t. (also **t**) *abbr.* **1** ton(s). **2** tonne(s).

TA *abbr.* Territorial Army.

Ta *symb.* tantalum.

ta /tɑ:/ *int. colloq.* thank you. [infantile form]

tab[1] —*n.* **1** small flap or strip of material attached for grasping, fastening, or hanging up, or for identification. **2** *US colloq.* bill (*picked up the tab*). **3** distinguishing mark on a staff officer's collar. —*v.* (**-bb-**) provide with a tab or tabs. □ **keep tabs** (or **a tab**) **on** *colloq.* **1** keep account of. **2** have under observation or in check. [probably dial.]

tab[2] *n.* = TABULATOR 2. [abbreviation]

tabard /'tæbəd/ *n.* **1** herald's official coat emblazoned with royal arms. **2** woman's or girl's sleeveless jerkin. **3** *hist.* knight's short emblazoned garment worn over armour. [French]

tabasco /tə'bæskəʊ/ *n.* **1** pungent pepper. **2** (**Tabasco**) *propr.* sauce made from this. [*Tabasco* in Mexico]

tabby /'tæbɪ/ *n.* (*pl.* **-ies**) **1** grey or brownish cat with dark stripes. **2** a kind of watered silk. [French from Arabic]

tabernacle /'tæbə,næk(ə)l/ *n.* **1** *hist.* tent used as a sanctuary by the Israelites during the Exodus. **2** niche or receptacle, esp. for the Eucharistic elements. **3** Nonconformist meeting-house. [Latin: related to TAVERN]

tabla /'tæblə/ *n.* pair of small drums played with the hands, esp. in Indian music. [Arabic, = drum]

table /'teɪb(ə)l/ —*n.* **1** flat surface on a leg or legs, used for eating, working at, etc. **2 a** food provided in a household (*keeps a good table*). **b** group seated for dinner etc. **3 a** set of facts or figures in columns etc. (*table of contents*). **b** matter contained in this. **c** = MULTIPLICATION TABLE. —*v.* (**-ling**) **1** bring forward for discussion etc. at a meeting. **2** esp. *US* postpone consideration of (a matter). □ **at table** taking a meal at a table. **on the table** offered for discussion. **turn the tables** (often foll. by *on*) reverse circumstances to one's advantage (against). **under the table** *colloq.* **1** very drunk. **2** = *under the counter* (see COUNTER[1]). [Latin *tabula* board]

tableau /'tæbləʊ/ *n.* (*pl.* **-x** /-əʊz/) **1** picturesque presentation. **2** group of silent motionless people representing a scene on stage. [French, = picture, diminutive of TABLE]

tablecloth *n.* cloth spread over a table, esp. for meals.

table d'hôte /,tɑ:b(ə)l 'dəʊt/ *n.* (often *attrib.*) meal from a set menu at a fixed price. [French, = host's table]

tableland *n.* elevated plateau.

table licence *n.* licence to serve alcoholic drinks with meals only.

table linen *n.* tablecloths, napkins, etc.

tablespoon *n.* **1** large spoon for serving food. **2** amount held by this. □ **tablespoonful** *n.* (*pl.* **-s**).

tablet /'tæblɪt/ *n.* **1** small solid dose of a medicine etc. **2** bar of soap etc. **3** flat slab of esp. stone, usu. inscribed. **4** *US* writing-pad. [Latin diminutive: related to TABLE]

table talk *n.* informal talk at table.

table tennis *n.* indoor ball game played with small bats on a table divided by a net.

tabletop *n.* surface of a table.

tableware *n.* dishes, plates, etc., for meals.

table wine *n.* wine of ordinary quality.

tabloid /'tæblɔɪd/ *n.* small-sized, often popular or sensational, newspaper. [from TABLET]

taboo /tə'bu:/ (also **tabu**) —*n.* (*pl.* **-s**) **1** ritual isolation of a person or thing as sacred or accursed. **2** prohibition imposed by social custom. —*adj.* avoided or prohibited, esp. by social custom (*taboo words*). —*v.* (**-oos**, **-ooed** or **-us**, **-ued**) **1** put under taboo. **2** exclude or prohibit, esp. socially. [Tongan]

tabor /'teɪbə(r)/ *n. hist.* small drum, esp. used to accompany a pipe. [French]

tabu var. of TABOO.

tabular /'tæbjʊlə(r)/ *adj.* of or arranged in tables or lists. [Latin: related to TABLE]

tabulate /'tæbjʊ,leɪt/ *v.* (**-ting**) arrange (figures or facts) in tabular form. □ **tabulation** /-'leɪʃ(ə)n/ *n.*

tabulator *n.* **1** person or thing that tabulates. **2** device on a typewriter etc. for advancing to a sequence of set positions in tabular work.

tacho /'tækəʊ/ *n.* (*pl.* **-s**) *colloq.* = TACHOMETER. [abbreviation]

tachograph /ˈtækəˌgrɑːf/ *n.* device in a vehicle recording speed and travel time. [Greek *takhos* speed]

tachometer /təˈkɒmɪtə(r)/ *n.* instrument measuring velocity or rate of rotation of a shaft (esp. in a vehicle).

tacit /ˈtæsɪt/ *adj.* understood or implied without being stated (*tacit consent*). □ **tacitly** *adv.* [Latin *taceo* be silent]

taciturn /ˈtæsɪˌtɜːn/ *adj.* saying little; uncommunicative. □ **taciturnity** /-ˈtɜːnɪtɪ/ *n.* [Latin: related to TACIT]

tack[1] *–n.* **1** small sharp broad-headed nail. **2** *US* drawing-pin. **3** long stitch for joining fabrics etc. lightly or temporarily together. **4** (in sailing) direction, or temporary change of direction, esp. taking advantage of a side wind (*starboard tack*). **5** course of action or policy (*change tack*). **6** sticky condition of varnish etc. *–v.* **1** (often foll. by *down* etc.) fasten with tacks. **2** stitch lightly together. **3** (foll. by *to*, *on*, *on to*) add or append. **4 a** change a ship's course by turning its head to the wind. **b** make a series of such tacks. [probably related to French *tache* clasp, nail]

tack[2] *n.* saddle, bridle, etc., of a horse. [from TACKLE]

tack[3] *n. colloq.* cheap or shoddy material; tat, kitsch. [back formation from TACKY[2]]

tackle /ˈtæk(ə)l/ *–n.* **1** equipment for a task or sport. **2** mechanism, esp. of ropes, pulley-blocks, hooks, etc., for lifting weights, managing sails, etc. **3** windlass with its ropes and hooks. **4** act of tackling in football etc. *–v.* **(-ling) 1** try to deal with (a problem or difficulty). **2** grapple with (an opponent). **3** confront (a person) in discussion or argument. **4** intercept or stop (a player running with the ball). □ **tackler** *n.* [Low German]

tackle-block *n.* pulley over which a rope runs.

tacky[1] *adj.* **(-ier, -iest)** slightly sticky. □ **tackiness** *n.* [from TACK[1]]

tacky[2] *adj.* **(-ier, -iest)** *colloq.* **1** in poor taste, cheap. **2** tatty, shabby. □ **tackiness** *n.* [origin unknown]

taco /ˈtækəʊ/ *n.* (*pl.* **-s**) Mexican dish of meat etc. in a folded tortilla. [Mexican Spanish]

tact *n.* **1** skill in dealing with others, esp. in delicate situations. **2** intuitive perception of the right thing to do or say. [Latin *tango tact-* touch]

tactful *adj.* having or showing tact. □ **tactfully** *adv.*

tactic /ˈtæktɪk/ *n.* **1** tactical manoeuvre. **2** = TACTICS. [Greek from *tasso* arrange]

tactical *adj.* **1** of tactics (*tactical retreat*). **2** (of bombing etc.) done in direct support of military or naval operations. **3** adroitly planning or adroitly planned. □ **tactically** *adv.*

tactics *n.pl.* **1** (also treated as *sing.*) disposition of armed forces, esp. in warfare. **2** short-term procedure adopted in carrying out a scheme or achieving an end. □ **tactician** /tækˈtɪʃ(ə)n/ *n.*

tactile /ˈtæktaɪl/ *adj.* **1** of the sense of touch. **2** perceived by touch; tangible. □ **tactility** /-ˈtɪlɪtɪ/ *n.* [Latin: related to TACT]

tactless *adj.* having or showing no tact. □ **tactlessly** *adv.*

tadpole *n.* larva, esp. of a frog, toad, or newt. [related to TOAD, POLL]

taffeta /ˈtæfɪtə/ *n.* fine lustrous silk or silklike fabric. [French or medieval Latin from Persian]

taffrail /ˈtæfreɪl/ *n.* rail round a ship's stern. [Dutch *taffereel* panel]

Taffy /ˈtæfɪ/ *n.* (*pl.* **-ies**) *colloq.* often *offens.* Welshman. [a supposed pronunciation of *Davy* = *David*]

tag[1] *–n.* **1** label, esp. on an object to show its address, price, etc. **2** metal etc. point on a shoelace etc. **3** loop or flap for handling or hanging a thing. **4** loose or ragged end. **5** trite quotation or stock phrase. *–v.* **(-gg-) 1** provide with a tag or tags. **2** (often foll. by *on*, *on to*) join or attach. □ **tag along** (often foll. by *with*) go along, accompany passively. [origin unknown]

tag[2] *–n.* children's chasing game. *–v.* **(-gg-)** touch in a game of tag. [origin unknown]

tag end *n.* esp. *US* last remnant.

tagliatelle /ˌtæljəˈtelɪ/ *n.* narrow ribbon-shaped pasta. [Italian]

t'ai chi /taɪ ˈtʃiː/ *n.* (in full **t'ai chi ch'uan** /ˈtʃwɑːn/) Chinese martial art and system of callisthenics with slow controlled movements. [Chinese, = great ultimate boxing]

tail[1] *–n.* **1** hindmost part of an animal, esp. extending beyond the body. **2 a** thing like a tail, esp. an extension at the rear. **b** rear of a procession etc. **3** rear part of an aeroplane, vehicle, or rocket. **4** luminous trail following a comet. **5** inferior, weaker, or last part of anything. **6** part of a shirt or coat below the waist at the back. **7** (in *pl.*) *colloq.* **a** tailcoat. **b** evening dress including this. **8** (in *pl.*) reverse of a coin as a choice when tossing. **9** *colloq.* person following another. *–v.* **1** remove the stalks of (fruit). **2** (often foll. by *after*) *colloq.* follow closely. □ **on a person's tail** closely following a person. **tail off** (or **away**) gradually decrease or diminish; end inconclusively. **with one's tail between one's legs** dejected, humiliated. □ **tailless** *adj.* [Old English]

tail[2] *Law* –*n.* limitation of ownership, esp. of an estate limited to a person and that person's heirs. –*adj.* so limited (*estate tail*). □ **in tail** under such a limitation. [French *taillier* cut: related to TALLY]

tailback *n.* long line of traffic caused by an obstruction.

tailboard *n.* hinged or removable flap at the rear of a lorry etc.

tailcoat *n.* man's coat with a long divided flap at the back, worn as part of formal dress.

tail-end *n.* hindmost, lowest, or last part.

tailgate *n.* **1** esp. *US* = TAILBOARD. **2** rear door of an estate car or hatchback.

tail-light *n.* (also **tail-lamp**) *US* rear light on a vehicle etc.

tailor /ˈteɪlə(r)/ –*n.* maker of clothes, esp. men's outer garments to measure. –*v.* **1** make (clothes) as a tailor. **2** make or adapt for a special purpose. **3** work as or be a tailor. [Anglo-French *taillour*: related to TAIL[2]]

tailored *adj.* **1** (of clothing) well or closely fitted. **2** = TAILOR-MADE.

tailor-made *adj.* **1** made to order by a tailor. **2** made or suited for a particular purpose.

tailpiece *n.* **1** rear appendage. **2** final part of a thing. **3** decoration in a blank space at the end of a chapter etc.

tailpipe *n.* rear section of an exhaust-pipe.

tailplane *n.* horizontal aerofoil at the tail of an aircraft.

tailspin *n.* **1** spin by an aircraft with the tail spiralling. **2** state of chaos or panic.

tail wind *n.* wind blowing in the direction of travel.

taint –*n.* **1** spot or trace of decay, infection, corruption, etc. **2** corrupt condition or infection. –*v.* **1** affect with a taint; become tainted. **2** (foll. by *with*) affect slightly. [Latin: related to TINGE]

take –*v.* (**-king**; *past* **took** /tʊk/; *past part.* **taken**) **1** lay hold of; get into one's hands. **2** acquire, capture, earn, or win. **3** get by purchase, hire, or formal agreement (*take lodgings*; *took a taxi*). **4** (in a recipe) use. **5** regularly buy (a newspaper etc.). **6** obtain after qualifying (*take a degree*). **7** occupy (*take a chair*). **8** make use of (*take the next turning on the left*; *take the bus*). **9** consume (food or medicine). **10 a** be effective (*inoculation did not take*). **b** (of a plant, seed, etc.) begin to grow. **11** require or use up (*will only take a minute*). **12** carry or accompany (*take the book home*; *bus will take you*). **13** remove; steal (*someone has taken my pen*). **14** catch or be infected with (fire or fever etc.). **15 a** experience, seek, or be affected by (*take fright*; *take pleasure*). **b** exert (*take no notice*). **16** find out and note (*took his address*; *took her temperature*). **17** understand; assume (*I took you to mean yes*). **18** treat, deal with, or regard in a specified way (*took it badly*; *took the corner too fast*). **19** (foll. by *for*) regard as being (*do you take me for an idiot?*). **20 a** accept, receive (*take the offer*; *take a call*; *takes boarders*). **b** hold (*takes 3 pints*). **c** submit to; tolerate (*take a joke*). **21** wear (*takes size 10*). **22** choose or assume (*took a job*; *took the initiative*). **23** derive (*takes its name from the inventor*). **24** (foll. by *from*) subtract (*take 3 from 9*). **25** perform or effect (*take notes*; *take an oath*; *take a look*). **26** occupy or engage oneself in (*take a rest*). **27** conduct (*took prayers*). **28** teach, be taught, or be examined in (a subject). **29 a** make (a photograph). **b** photograph (a person etc.). **30** (in *imper.*) use as an example (*take Napoleon*). **31** *Gram.* have or require as part of a construction (*this verb takes an object*). **32** have sexual intercourse with (a woman). **33** (in *passive*; foll. by *by*, *with*) be attracted or charmed by. –*n.* **1** amount taken or caught at a time etc. **2** scene or film sequence photographed continuously at one time. □ **be taken ill** become ill, esp. suddenly. **have what it takes** *colloq.* have the necessary qualities etc. for success. **take account of** see ACCOUNT. **take advantage of** see ADVANTAGE. **take after** resemble (a parent etc.). **take against** begin to dislike. **take aim** see AIM. **take apart 1** dismantle. **2** *colloq.* beat or defeat. **3** *colloq.* criticize severely. **take away 1** remove or carry elsewhere. **2** subtract. **3** buy (hot food etc.) for eating elsewhere. **take back 1** retract (a statement). **2** convey to an original position. **3** carry in thought to a past time. **4 a** return (goods) to a shop. **b** (of a shop) accept such goods. **5** accept (a person) back into one's affections, into employment, etc. **take the biscuit** (or **bun** or **cake**) *colloq.* be the most remarkable. **take down 1** write down (spoken words). **2** remove or dismantle. **3** lower (a garment worn below the waist). **take effect** see EFFECT. **take for granted** see GRANT. **take fright** see FRIGHT. **take heart** be encouraged. **take in 1** receive as a lodger etc. **2** undertake (work) at home. **3** make (a garment etc.) smaller. **4** understand; observe (*did you take that in?*). **5** cheat. **6** include. **7** *colloq.* visit (a place) on the way to another (*took in Bath*). **8** absorb into the body. **take in hand 1** undertake; start doing or dealing with. **2** undertake to control or reform (a person). **take into account** see ACCOUNT. **take it 1** (often foll. by *that*) assume. **2** *colloq.* endure in a specified way (*took it badly*). **take it easy** see EASY. **take it into one's head** see HEAD. **take it on one** (or **oneself**) (foll. by *to* + infin.) venture or presume. **take it or leave it** (esp. in *imper.*) accept it or not. **take it out of 1** exhaust the strength of. **2** have revenge on. **take it out on** relieve one's frustration by treating aggressively. **take off 1 a** remove (clothing) from the body. **b** remove or lead away. **c** withdraw (transport, a show, etc.). **2** deduct. **3** depart, esp. hastily. **4** *colloq.* mimic humorously. **5** begin a jump. **6** become airborne. **7** (of a scheme, enterprise, etc.) become successful. **8** have (a period) away from work. **take oneself off** go away. **take on 1** undertake (work etc.). **2** engage (an employee). **3** be willing or ready to meet (an opponent etc.). **4** acquire (new meaning etc.). **5** *colloq.* show strong emotion. **take out 1** remove; extract. **2** escort on an outing or as a sexual partner. **3** get (a licence, summons, etc.) issued. **4** *US* = *take away* 3. **5** *slang* murder or destroy. **take a person out of himself** or **herself** make a person forget his or her worries. **take over 1** succeed to the management or ownership of. **2** take control. **take part** see PART. **take place** see PLACE. **take shape** assume a distinct form; develop. **take one's time** not hurry. **take to 1** begin or fall into the habit of (*took to smoking*). **2** have recourse to. **3** adapt oneself to. **4** form a liking for. **take up 1** become interested or engaged in (a pursuit). **2** adopt as a protégé. **3** occupy (time or space). **4** begin (residence etc.). **5** resume after an interruption. **6** (often foll. by *on*) interrupt or question (a speaker) (on a point). **7** accept (an offer etc.). **8** shorten (a garment). **9** lift up. **10** absorb. **11** pursue (a matter etc.) further, esp. with those in authority. **take a person up on** accept (a person's offer etc.). **take up with** begin to associate with. [Old English from Old Norse]

take-away –*attrib. adj.* (of food) bought cooked for eating elsewhere. –*n.* **1** this food. **2** establishment selling this.

take-home pay *n.* employee's pay after the deduction of tax etc.

take-off *n.* **1** act of becoming airborne. **2** act of mimicking.

take-over *n.* assumption of control (esp. of a business); buying-out.

taker *n.* person who takes a bet, accepts an offer, etc.

take-up *n.* acceptance of a thing offered.

taking –*adj.* attractive, captivating. –*n.* (in *pl.*) amount of money taken at a show, in a shop, etc.

talc *n.* **1** talcum powder. **2** magnesium silicate formed as soft flat plates, used as a lubricator etc. [Arabic from Persian *talk*]

talcum /'tælkəm/ *n.* **1** = TALC 2. **2** (in full **talcum powder**) powdered talc applied to the skin, usu. perfumed. [medieval Latin: see TALC]

tale *n.* **1** (usu. fictitious) narrative or story. **2** allegation, often malicious or in breach of confidence. [Old English]

talent /'tælənt/ *n.* **1** special aptitude or faculty (*talent for music*). **2** high mental ability. **3 a** person or persons of talent. **b** *colloq.* attractive members of the opposite sex (*plenty of local talent*). **4** ancient esp. Greek weight and unit of currency. □ **talented** *adj.* [Greek *talanton*]

talent-scout *n.* (also **talent-spotter**) person seeking new talent, esp. in sport or entertainment.

talisman /'tælɪzmən/ *n.* (*pl.* **-s**) ring, stone, etc. thought to have magic powers, esp. to bring good luck. □ **talismanic** /-'mænɪk/ *adj.* [French and Spanish from Greek]

talk /tɔːk/ –*v.* **1** (often foll. by *to, with*) converse or communicate verbally. **2** have the power of speech. **3** (often foll. by *about*) **a** discuss; express; utter (*talked cricket*; *talking nonsense*). **b** (in *imper.*) *colloq.* as an emphatic statement (*talk about expense!*). **4** use (a language) in speech (*talking Spanish*). **5** (foll. by *at*) address pompously. **6** (usu. foll. by *into, out of*) bring into a specified condition etc. by talking (*talked himself hoarse*; *did you talk them into it?*). **7** betray secrets. **8** gossip (*people will talk*). **9** have influence (*money talks*). –*n.* **1** conversation, talking. **2** particular mode of speech (*baby-talk*). **3** informal address or lecture. **4 a** rumour or gossip (*talk of a merger*). **b** its theme (*the talk was all babies*). **5** empty promises; boasting. **6** (often in *pl.*) discussions or negotiations. □ **now you're talking** *colloq.* I like what you say, suggest, etc. **talk back** reply defiantly. **talk down to** speak condescendingly to. **talk a person** etc. **down 1** silence by loudness or persistence. **2** bring (a pilot or aircraft) to landing by radio. **talk of 1** discuss or mention. **2** (often foll. by verbal noun) express some intention of (*talked of moving to London*). **talk out** block (a bill in Parliament) by prolonging discussion to the time of adjournment. **talk over** discuss at length. **talk a person over** (or **round**) gain agreement by talking. **talk shop** talk about one's occupation etc. **talk to** rebuke, scold. □ **talker** *n.* [from TALE or TELL]

talkative /'tɔ:kətɪv/ *adj.* fond of or given to talking.

talkback *n.* (often *attrib.*) system of two-way communication by loudspeaker.

talkie *n. colloq.* (esp. early) film with a soundtrack.

talking *–adj.* **1** that talks, or is able to talk (*talking parrot*). **2** expressive (*talking eyes*). *–n.* in senses of TALK *v.* □ **talking of** as we are discussing.

talking book *n.* recorded reading of a book, esp. for the blind.

talking-point *n.* topic for discussion.

talking-shop *n. derog.* arena or opportunity for empty talk.

talking-to *n. colloq.* reproof, reprimand.

tall /tɔ:l/ *–adj.* **1** of more than average height. **2** of a specified height (*about six feet tall*). **3** higher than the surrounding objects (*tall building*). *–adv.* as if tall; proudly (*sit tall*). □ **tallish** *adj.* **tallness** *n.* [Old English, = swift]

tallboy *n.* tall chest of drawers.

tall order *n.* unreasonable demand.

tallow /'tæləʊ/ *n.* hard (esp. animal) fat melted down to make candles, soap, etc. □ **tallowy** *adj.* [Low German]

tall ship *n.* sailing-ship with a high mast.

tall story *n. colloq.* extravagant story that is difficult to believe.

tally /'tælɪ/ *–n.* (*pl.* **-ies**) **1** reckoning of a debt or score. **2** total score or amount. **3** mark registering the number of objects delivered or received. **4** *hist.* **a** piece of notched wood for keeping account. **b** account kept thus. **5** identification ticket or label. **6** corresponding thing, counterpart, or duplicate. *–v.* (**-ies, -ied**) (often foll. by *with*) agree or correspond. [Latin *talea* rod]

tally-ho /,tælɪ'həʊ/ *–int.* huntsman's cry on sighting a fox. *–n.* (*pl.* **-s**) cry of this. *–v.* (**-hoes, -hoed**) **1** utter a cry of 'tally-ho'. **2** indicate (a fox) or urge (hounds) with this cry. [cf. French *taïaut*]

Talmud /'tælmʊd, -məd/ *n.* body of Jewish civil and ceremonial law and legend. □ **Talmudic** /-'mʊdɪk/ *adj.* **Talmudist** *n.* [Hebrew, = instruction]

talon /'tælən/ *n.* claw, esp. of a bird of prey. [Latin *talus* ankle]

talus /'teɪləs/ *n.* (*pl.* **tali** /-laɪ/) ankle-bone supporting the tibia. [Latin, = ankle]

tamarind /'tæmərɪnd/ *n.* **1** tropical evergreen tree. **2** fruit pulp from this used as food and in drinks. [Arabic, = Indian date]

tamarisk /'tæmərɪsk/ *n.* seashore shrub usu. with small pink or white flowers. [Latin]

tambour /'tæmbʊə(r)/ *n.* **1** drum. **2** circular frame holding fabric taut for embroidering. [French: related to TABOR]

tambourine /,tæmbə'ri:n/ *n.* small shallow drum with jingling discs in its rim, shaken or banged as an accompaniment. [French, diminutive of TAMBOUR]

tame *–adj.* **1** (of an animal) domesticated; not wild or shy. **2** insipid; dull (*tame entertainment*). **3** (of a person) amenable. *–v.* (**-ming**) **1** make tame; domesticate. **2** subdue, curb. □ **tameable** *adj.* **tamely** *adv.* **tameness** *n.* **tamer** *n.* (also in *comb.*). [Old English]

Tamil /'tæmɪl/ *–n.* **1** member of a people of South India and Sri Lanka. **2** language of this people. *–adj.* of this people or language. [native name]

tam-o'-shanter /,tæmə'ʃæntə(r)/ *n.* floppy round esp. woollen beret, of Scottish origin. [hero of a poem by Burns]

tamp *v.* ram down hard or tightly. [*tampion* stopper for gun-muzzle, from French *tampon*]

tamper *v.* (foll. by *with*) **1** meddle with or change illicitly. **2** exert a secret or corrupt influence upon; bribe. [var. of TEMPER]

tampon /'tæmpɒn/ *n.* plug of soft material used esp. to absorb menstrual blood. [French: related to TAMP]

tam-tam /'tæmtæm/ *n.* large metal gong. [Hindi]

tan[1] *–n.* **1** = SUNTAN. **2** yellowish-brown colour. **3** bark, esp. of oak, used to tan hides. *–adj.* yellowish-brown. *–v.* (**-nn-**) **1** make or become brown by exposure to sunlight. **2** convert (raw hide) into leather. **3** *slang* beat, thrash. [medieval Latin *tanno*, perhaps from Celtic]

tan[2] *abbr.* tangent.

tandem /'tændəm/ *–n.* **1** bicycle with two or more seats one behind another. **2** group of two people etc. with one behind or following the other. **3** carriage driven tandem. *–adv.* with two or more horses harnessed one behind another (*drive tandem*). □ **in tandem 1** one behind another. **2** alongside each other; together. [Latin, = at length]

tandoor /'tændʊə(r)/ *n.* clay oven. [Hindustani]

tandoori /tæn'dʊərɪ/ *n.* food spiced and cooked over charcoal in a tandoor (often *attrib.*: *tandoori chicken*). [Hindustani]

tang *n.* **1** strong taste or smell. **2** characteristic quality. **3** projection on the blade of esp. a knife, by which it is held firm in the handle. [Old Norse *tange* point]

tangent /'tændʒ(ə)nt/ *n.* **1** (often *attrib.*) straight line, curve, or surface that meets a curve at a point, but does not intersect it. **2** ratio of two sides (other than the hypotenuse) opposite and adjacent to an acute angle in a right-angled triangle. □ **at a tangent** diverging from a previous course or from what is relevant or central (*go off at a tangent*). [Latin *tango tact-* touch]

tangential /tæn'dʒenʃ(ə)l/ *adj.* **1** of or along a tangent. **2** divergent. **3** peripheral. □ **tangentially** *adv.*

tangerine /ˌtændʒə'ri:n/ *n.* **1** small sweet thin-skinned citrus fruit like an orange; mandarin. **2** deep orange-yellow colour. [*Tangier* in Morocco]

tangible /'tændʒɪb(ə)l/ *adj.* **1** perceptible by touch. **2** definite; clearly intelligible; not elusive (*tangible proof*). □ **tangibility** /-'bɪlɪtɪ/ *n.* **tangibleness** *n.* **tangibly** *adv.* [Latin: related to TANGENT]

tangle /'tæŋg(ə)l/ –*v.* (**-ling**) **1** intertwine (threads or hairs etc.) or become entwined in a confused mass; entangle. **2** (foll. by *with*) *colloq.* become involved (esp. in conflict) with (*don't tangle with me*). **3** complicate (*tangled affair*). –*n.* **1** confused mass of intertwined threads etc. **2** confused state. [origin uncertain]

tangly *adj.* (**-ier**, **-iest**) tangled.

tango /'tæŋgəʊ/ –*n.* (*pl.* **-s**) **1** slow S. American ballroom dance. **2** music for this. –*v.* (**-goes**, **-goed**) dance the tango. [American Spanish]

tangy *adj.* (**-ier**, **-iest**) having a strong usu. acid tang.

tanh /θæn, tænʃ, tæn'eɪtʃ/ *abbr.* hyperbolic tangent.

tank –*n.* **1** large container, usu. for liquid or gas. **2** heavy armoured fighting vehicle moving on continuous tracks. –*v.* (usu. foll. by *up*) fill the tank of (a vehicle etc.) with fuel. □ **tankful** *n.* (*pl.* **-s**). [originally Indian, = pond, from Gujarati]

tankard /'tæŋkəd/ *n.* **1** tall beer mug with a handle. **2** contents of or amount held by this (*drank a tankard of ale*). [probably Dutch *tankaert*]

tanked up *predic. adj. colloq.* drunk.

tank engine *n.* steam engine with integral fuel and water containers.

tanker *n.* ship, aircraft, or road vehicle for carrying liquids, esp. oil, in bulk.

tanner *n.* person who tans hides.

tannery *n.* (*pl.* **-ies**) place where hides are tanned.

tannic /'tænɪk/ *adj.* of tan (sense 3). [French *tannique*: related to TANNIN]

tannic acid *n.* natural yellowish organic compound used as a mordant and astringent.

tannin /'tænɪn/ *n.* any of various organic compounds found in tree-barks and oak-galls, used in leather production. [French *tanin*: related to TAN[1]]

Tannoy /'tænɔɪ/ *n. propr.* type of public-address system. [origin uncertain]

tansy /'tænzɪ/ *n.* (*pl.* **-ies**) plant with yellow flowers and aromatic leaves. [Greek *athanasia* immortality]

tantalize /'tæntəˌlaɪz/ *v.* (also **-ise**) (**-zing** or **-sing**) **1** torment or tease by the sight or promise of the unobtainable. **2** raise and then dash the hopes of. □ **tantalization** /-'zeɪʃ(ə)n/ *n.* [*Tantalus*, mythical king punished in Hades with sight of water and fruit which drew back whenever he tried to reach them]

tantalum /'tæntələm/ *n.* rare hard white metallic element. □ **tantalic** *adj.* [related to TANTALIZE]

tantalus /'tæntələs/ *n.* stand in which spirit-decanters may be locked up but visible. [see TANTALIZE]

tantamount /'tæntəˌmaʊnt/ *predic. adj.* (foll. by *to*) equivalent to. [Italian *tanto montare* amount to so much]

tantra /'tæntrə/ *n.* any of a class of Hindu or Buddhist mystical or magical writings. [Sanskrit, = doctrine]

tantrum /'tæntrəm/ *n.* (esp. child's) outburst of bad temper or petulance. [origin unknown]

Taoiseach /'ti:ʃəx/ *n.* prime minister of the Irish Republic. [Irish, = chief, leader]

Taoism /'taʊɪz(ə)m, 'tɑ:əʊ-/ *n.* Chinese philosophy advocating humility and religious piety. □ **Taoist** *n.* [Chinese *dao* right way]

tap[1] –*n.* **1** device by which a flow of liquid or gas from a pipe or vessel can be controlled. **2** tapping of a telephone etc. **3** taproom. –*v.* (**-pp-**) **1** provide (a cask) or let out (liquid) with a tap. **2** draw sap from (a tree) by cutting into it. **3** obtain information or supplies from. **4** extract or obtain; discover and exploit (*mineral wealth waiting to be tapped*; *tap skills of young people*). **5** connect a listening device to (a telephone etc.). □ **on tap 1** ready to be drawn off by tap. **2** *colloq.* freely available. [Old English]

tap[2] –*v.* (**-pp-**) **1** (foll. by *at*, *on*) strike a gentle but audible blow. **2** (often foll. by *against*, *on*, etc.) strike or cause (a thing) to strike lightly (*tapped me on the* *shoulder*). **3** (often foll. by *out*) make by a tap or taps (*tapped out the rhythm*). **4** tap-dance. –*n.* **1 a** light blow; rap. **b** sound of this. **2 a** tap-dancing. **b** metal attachment on a tap-dancer's shoe. [imitative]

tapas /'tæpæs/ *n.pl.* (often *attrib.*) small savoury esp. Spanish dishes. [Spanish]

tap-dance –*n.* rhythmic dance performed with shoes with metal taps. –*v.* perform a tap-dance. □ **tap-dancer** *n.* **tap-dancing** *n.*

tape –*n.* **1** narrow strip of woven material for tying up, fastening, etc. **2** this across the finishing line of a race. **3** (in full **adhesive tape**) strip of adhesive plastic etc. for fastening, masking, insulating, etc. **4 a** = MAGNETIC TAPE. **b** reel or cassette containing this. **c** tape recording. **5** = TAPE-MEASURE. –*v.* (**-ping**) **1 a** fasten or join etc. with tape. **b** apply tape to. **2** (foll. by *off*) seal or mark off with tape. **3** record on magnetic tape. **4** measure with tape. □ **have** (or **get**) **a person** or **thing taped** *colloq.* understand (him, it, etc.) fully. [Old English]

tape deck *n.* machine for using audiotape (separate from the amplifier, speakers, etc.).

tape machine *n.* **1** machine for recording telegraph messages. **2** = TAPE RECORDER.

tape-measure *n.* strip of marked tape or flexible metal for measuring.

taper –*n.* **1** wick coated with wax etc. for conveying a flame. **2** slender candle. –*v.* (often foll. by *off*) **1** diminish or reduce in thickness towards one end. **2** make or become gradually less. [Old English]

tape recorder *n.* apparatus for recording and replaying sounds on magnetic tape. □ **tape-record** *v.* **tape recording** *n.*

tapestry /'tæpɪstrɪ/ *n.* (*pl.* **-ies**) **1 a** thick fabric in which coloured weft threads are woven to form pictures or designs. **b** (usu. wool) embroidery imitating this. **c** piece of this. **2** events or circumstances etc. seen as interwoven etc. (*life's rich tapestry*). □ **tapestried** *adj.* [*tapissery* from French *tapis* carpet]

tapeworm *n.* parasitic intestinal flatworm with a segmented body.

tapioca /,tæpɪ'əʊkə/ *n.* starchy substance in hard white grains, obtained from cassava and used for puddings etc. [Tupi-Guarani]

tapir /'teɪpə(r), -pɪə(r)/ *n.* nocturnal Central and S. American or Malaysian hoofed mammal with a short flexible snout. [Tupi]

tappet /'tæpɪt/ *n.* lever or projecting part in machinery giving intermittent motion. [from TAP[2]]

taproom *n.* room in a pub serving drinks on tap.

tap root *n.* tapering root growing vertically downwards.

tar[1] –*n.* **1** dark thick inflammable liquid distilled from wood or coal etc., used as a preservative of wood and iron, in making roads, as an antiseptic, etc. **2** similar substance formed in the combustion of tobacco etc. –*v.* (**-rr-**) cover with tar. □ **tar and feather** smear with tar and then cover with feathers as a punishment. **tarred with the same brush** having the same faults. [Old English]

tar[2] *n. colloq.* sailor. [from TARPAULIN]

taramasalata /,tærəməsə'lɑ:tə/ *n.* (also **taramosalata**) pâté made from roe with olive oil, seasoning, etc. [Greek *taramas* roe, *salata* SALAD]

tarantella /,tærən'telə/ *n.* **1** whirling S. Italian dance. **2** music for this. [Italian from *Taranto* in Italy]

tarantula /tə'ræntjʊlə/ *n.* **1** large hairy tropical spider. **2** large black S. European spider. [medieval Latin: related to TARANTELLA]

tarboosh /tɑ:'bu:ʃ/ *n.* cap like a fez. [Arabic from Persian]

tardy /'tɑ:dɪ/ *adj.* (**-ier, -iest**) **1** slow to act, come, or happen. **2** delaying or delayed. □ **tardily** *adv.* **tardiness** *n.* [Latin *tardus* slow]

tare[1] *n.* **1** vetch, esp. as a cornfield weed or fodder. **2** (in *pl.*) *Bibl.* an injurious cornfield weed (Matt. 13:24-30). [origin unknown]

tare[2] *n.* **1** allowance made for the weight of packing or wrapping around goods. **2** weight of a vehicle without fuel or load. [Arabic *tarha*]

target /'tɑ:gɪt/ –*n.* **1** mark fired or aimed at, esp. a round object marked with concentric circles. **2** person or thing aimed or fired at etc. (*an easy target*). **3** objective or result aimed at. **4** butt for criticism, abuse, etc. –*v.* (**-t-**) **1** identify or single out as a target. **2** aim or direct (*missiles targeted on major cities*). [French *targe* shield]

tariff /'tærɪf/ *n.* **1** table of fixed charges (*hotel tariff*). **2 a** duty on a particular class of goods. **b** list of duties or customs due. [Arabic, = notification]

tarlatan /'tɑ:lət(ə)n/ *n.* thin stiff open-weave muslin. [French; probably originally Indian]

Tarmac /'tɑ:mæk/ –*n. propr.* **1** = TARMACADAM. **2** runway etc. made of this. –*v.* (**tarmac**) (**-ck-**) apply tarmacadam to. [abbreviation]

tarmacadam /ˌtɑːməˈkædəm/ *n.* stone or slag bound with bitumen, used in paving roads etc. [from TAR[1], MACADAM]

tarn *n.* small mountain lake. [Old Norse]

tarnish /ˈtɑːnɪʃ/ *–v.* **1** (cause to) lose lustre. **2** impair (one's reputation etc.). *–n.* **1** loss of lustre, esp. as a film on a metal's surface. **2** blemish, stain. [French *ternir* from *terne* dark]

taro /ˈtɑːrəʊ/ *n.* (*pl.* **-s**) tropical plant with tuberous roots used as food. [Polynesian]

tarot /ˈtærəʊ/ *n.* (often *attrib.*) **1** (in *sing.* or *pl.*) **a** pack of mainly picture cards used in fortune-telling. **b** any game played with a similar pack of 78 cards. **2** any card from a tarot pack. [French]

tarpaulin /tɑːˈpɔːlɪn/ *n.* **1** heavy-duty cloth waterproofed esp. with tar. **2** sheet or covering of this. [from TAR[1], PALL[1]]

tarragon /ˈtærəgən/ *n.* bushy herb used in salads, stuffings, vinegar, etc. [medieval Latin from Greek]

tarry[1] /ˈtɑːrɪ/ *adj.* (**-ier, -iest**) of, like, or smeared with tar.

tarry[2] /ˈtærɪ/ *v.* (**-ies, -ied**) *archaic* linger, stay, wait. [origin unknown]

tarsal /ˈtɑːs(ə)l/ *–adj.* of the ankle-bones. *–n.* tarsal bone. [from TARSUS]

tarsus /ˈtɑːsəs/ *n.* (*pl.* **tarsi** /-saɪ/) **1** bones of the ankle and upper foot. **2** shank of a bird's leg. [Greek]

tart[1] *n.* **1** open pastry case containing jam etc. **2** pie with a fruit or other sweet filling. □ **tartlet** *n.* [French *tarte*]

tart[2] *–n. slang* **1** prostitute; promiscuous woman. **2** *slang offens.* girl or woman. *–v.* (foll. by *up*) *colloq.* (usu. *refl.*) smarten or dress up, esp. gaudily. [probably abbreviation of SWEETHEART]

tart[3] *adj.* **1** sharp or acid in taste. **2** (of a remark etc.) cutting, bitter. □ **tartly** *adv.* **tartness** *n.* [Old English]

tartan /ˈtɑːt(ə)n/ *n.* **1** pattern of coloured stripes crossing at right angles, esp. denoting a Scottish Highland clan. **2** woollen cloth woven in this pattern (often *attrib.*: *tartan scarf*). [origin uncertain]

Tartar /ˈtɑːtə(r)/ *–n.* **1 a** member of a group of Central Asian peoples including Mongols and Turks. **b** Turkic language of these peoples. **2** (**tartar**) harsh or formidable person. *–adj.* **1** of Tartars. **2** of Central Asia east of the Caspian Sea. [French or medieval Latin]

tartar /ˈtɑːtə(r)/ *n.* **1** hard deposit that forms on the teeth. **2** deposit that forms a hard crust in wine. [medieval Latin from Greek]

tartare /tɑːˈtɑː(r)/ *adj.* (in phr. **sauce tartare**) = TARTAR SAUCE. [French]

tartaric /tɑːˈtærɪk/ *adj.* of or from tartar.

tartaric acid *n.* natural acid found esp. in unripe grapes, used in baking powders etc.

tartar sauce *n.* sauce of mayonnaise and chopped gherkins, capers, etc. [from TARTAR]

tartrazine /ˈtɑːtrəˌziːn/ *n.* brilliant yellow dye from tartaric acid, used to colour food etc.

tarty *adj. colloq.* (**-ier, -iest**) (esp. of a woman) vulgar, gaudy; promiscuous. [from TART[2]]

Tarzan /ˈtɑːz(ə)n/ *n. colloq.* agile muscular man. [name of a character in stories by E. R. Burroughs]

task /tɑːsk/ *–n.* piece of work to be done. *–v.* make great demands on (a person's powers etc.). □ **take to task** rebuke, scold. [medieval Latin *tasca*, probably = *taxa* TAX]

task force *n.* (also **task group**) armed force or other group organized for a specific operation or task.

taskmaster *n.* (*fem.* **taskmistress**) person who makes others work hard.

Tass *n.* official Russian news agency. [Russian]

tassel /ˈtæs(ə)l/ *n.* **1** tuft of loosely hanging threads or cords etc. as decoration. **2** tassel-like flower-head of some plants, esp. maize. □ **tasselled** *adj.* (*US* **taseled**). [French *tas(s)el* clasp]

taste /teɪst/ *–n.* **1 a** sensation caused in the mouth by contact with a soluble substance. **b** faculty of perceiving this (*bitter to the taste*). **2** small sample of food or drink. **3** slight experience (*taste of success*). **4** (often foll. by *for*) liking or predilection (*expensive tastes*). **5** aesthetic discernment in art, clothes, conduct, etc. (*in poor taste*). *–v.* (**-ting**) **1** sample the flavour of (food etc.) by taking it into the mouth. **2** (also *absol.*) perceive the flavour of (*cannot taste with a cold*). **3** (esp. with *neg.*) eat or drink a small portion of (*had not tasted food for days*). **4** experience (*never tasted failure*). **5** (often foll. by *of*) have a specified flavour (*tastes of onions*). □ **to one's taste** pleasing, suitable. [French from Romanic]

taste bud *n.* cell or nerve-ending on the surface of the tongue by which things are tasted.

tasteful *adj.* having, or done in, good taste. □ **tastefully** *adv.* **tastefulness** *n.*

tasteless *adj.* **1** lacking flavour. **2** having, or done in, bad taste. □ **tastelessly** *adv.* **tastelessness** *n.*

taster *n.* **1** person employed to test food or drink by tasting. **2** small sample.

tasting *n.* gathering at which food or drink is tasted and evaluated.
tasty *adj.* (**-ier, -iest**) **1** pleasing in flavour; appetizing. **2** *colloq.* attractive. □ **tastily** *adv.* **tastiness** *n.*
tat[1] *n. colloq.* tatty things; rubbish, junk. [back-formation from TATTY]
tat[2] *v.* (**-tt-**) do, or make by, tatting. [origin unknown]
tat[3] see TIT[2].
ta-ta /tæ'tɑː/ *int. colloq.* goodbye. [origin unknown]
tatter *n.* (usu. in *pl.*) rag; irregularly torn cloth or paper etc. □ **in tatters** *colloq.* **1** torn in many places. **2** destroyed, ruined. [Old Norse]
tattered *adj.* in tatters.
tatting /'tætɪŋ/ *n.* **1** a kind of handmade knotted lace used for trimming etc. **2** process of making this. [origin unknown]
tattle /'tæt(ə)l/ *–v.* (**-ling**) prattle, chatter, gossip. *–n.* gossip; idle talk. [Flemish *tatelen*, imitative]
tattoo[1] /tə'tuː, tæ-/ *n.* **1** evening drum or bugle signal recalling soldiers to quarters. **2** elaboration of this with music and marching as an entertainment. **3** rhythmic tapping or drumming. [earlier *tap-too* from Dutch *taptoe*, literally 'close the tap' (of the cask)]
tattoo[2] /tə'tuː, tæ-/ *–v.* (**-oos, -ooed**) **1** mark (skin) indelibly by puncturing it and inserting pigment. **2** make (a design) in this way. *–n.* such a design. □ **tattooer** *n.* **tattooist** *n.* [Polynesian]
tatty /'tætɪ/ *adj.* (**-ier, -iest**) *colloq.* **1** tattered; shabby. **2** inferior. **3** tawdry. □ **tattily** *adv.* **tattiness** *n.* [originally Scots, = shaggy, apparently related to TATTER]
tau /taʊ, tɔː/ *n.* nineteenth letter of the Greek alphabet (Τ, τ). [Greek]
taught *past* and *past part.* of TEACH.
taunt /tɔːnt/ *–n.* insult; provocation. *–v.* insult; provoke contemptuously. [French *tant pour tant* tit for tat, smart rejoinder]
taupe /təʊp/ *adj.* & *n.* grey tinged with esp. brown. [French, = MOLE[1]]
Taurus /'tɔːrəs/ *n.* (*pl.* **-es**) **1** constellation and second sign of the zodiac (the Bull). **2** person born when the sun is in this sign. □ **Taurean** *adj.* & *n.* [Latin, = bull]
taut /tɔːt/ *adj.* **1** (of a rope etc.) tight; not slack. **2** (of nerves etc.) tense. **3** (of a ship etc.) in good condition. □ **tauten** *v.* **tautly** *adv.* **tautness** *n.* [perhaps = TOUGH]
tautology /tɔː'tɒlədʒɪ/ *n.* (*pl.* **-ies**) repetition using different words, esp. as a fault of style (e.g. *arrived one after the other in succession*). □ **tautological** /-tə'lɒdʒɪk(ə)l/ *adj.* **tautologous** /-ləgəs/ *adj.* [Greek *tauto* the same]
tavern /'tæv(ə)n/ *n. archaic* or *literary* inn, pub. [Latin *taberna*]
taverna /tə'vɜːnə/ *n.* Greek restaurant. [modern Greek: related to TAVERN]
tawdry /'tɔːdrɪ/ *adj.* (**-ier, -iest**) showy but worthless; gaudy. □ **tawdrily** *adv.* **tawdriness** *n.* [*tawdry lace* from *St Audrey's lace*]
tawny /'tɔːnɪ/ *adj.* (**-ier, -iest**) orange-brown or yellow-brown. [Anglo-French *tauné*: related to TAN[1]]
tawny owl *n.* reddish-brown European owl.
tax *–n.* **1** money compulsorily levied by the State or local authorities on individuals, property, or businesses. **2** (usu. foll. by *on, upon*) strain, heavy demand, or burdensome obligation. *–v.* **1** impose a tax on. **2** deduct tax from (income etc.). **3** make heavy demands on (*taxes my patience*). **4** (often foll. by *with*) confront (a person) with a fault etc; call to account. □ **taxable** *adj.* [Latin *taxo* censure, compute]
taxa *pl.* of TAXON.
taxation /tæk'seɪʃ(ə)n/ *n.* imposition or payment of tax. [Latin: related to TAX]
tax avoidance *n.* minimizing payment of tax by financial manoeuvring.
tax-deductible *adj.* (of expenditure) legally deductible from income before tax assessment.
tax disc *n.* road tax receipt displayed on the windscreen of a vehicle.
tax evasion *n.* illegal non-payment or underpayment of tax.
tax-free *adj.* exempt from tax.
tax haven *n.* country etc. where taxes are low.
taxi /'tæksɪ/ *–n.* (*pl.* **-s**) (in full **taxi-cab**) car licensed to ply for hire and usu. fitted with a taximeter. *–v.* (**-xis, -xied, -xiing** or **-xying**) **1** (of an aircraft or pilot) drive on the ground before take-off or after landing. **2** go or convey in a taxi. [abbreviation of *taximeter cab*]
taxidermy /'tæksɪˌdɜːmɪ/ *n.* art of preparing, stuffing, and mounting the skins of animals. □ **taxidermist** *n.* [Greek *taxis* arrangement, *derma* skin]
taximeter /'tæksɪˌmiːtə(r)/ *n.* automatic fare-indicator fitted to a taxi. [French: related to TAX]
taxi rank *n.* (*US* **taxi stand**) place where taxis wait to be hired.
taxman *n. colloq.* inspector or collector of taxes.
taxon /'tæks(ə)n/ *n.* (*pl.* **taxa**) any taxonomic group. [back-formation from TAXONOMY]
taxonomy /tæk'sɒnəmɪ/ *n.* classification of living and extinct organisms. □

taxonomic /-sə'nɒmɪk/ *adj.* **taxonomical** /-sə'nɒmɪk(ə)l/ *adj.* **taxonomically** /-sə'nɒmɪkəlɪ/ *adv.* **taxonomist** *n.* [Greek *taxis* arrangement, *-nomia* distribution]

taxpayer *n.* person who pays taxes.

tax return *n.* declaration of income for taxation purposes.

tayberry /'teɪbərɪ/ *n.* (*pl.* **-ies**) hybrid fruit between the blackberry and raspberry. [River *Tay* in Scotland]

TB *abbr.* **1** tubercle bacillus. **2** tuberculosis.

Tb *symb.* terbium.

t.b.a. *abbr.* to be announced.

T-bone /'ti:bəʊn/ *n.* T-shaped bone, esp. in steak from the thin end of a loin.

tbsp. *abbr.* tablespoonful.

Tc *symb.* technetium.

TCP *abbr. propr.* a disinfectant and germicide. [*t*richloro*p*henylmethyliodasalicyl]

Te *symb.* tellurium.

te /ti:/ *n.* (also **ti**) seventh note of a major scale. [earlier *si*: French from Italian]

tea *n.* **1 a** (in full **tea plant**) Asian evergreen shrub or small tree. **b** its dried leaves. **2** drink made by infusing tea-leaves in boiling water. **3** infusion of other leaves etc. (*camomile tea*; *beef tea*). **4 a** light afternoon meal of tea, bread, cakes, etc. **b** = HIGH TEA. [probably Dutch *tee* from Chinese]

tea bag *n.* small perforated bag of tea for infusion.

tea break *n.* pause in work etc. to drink tea.

tea caddy *n.* container for tea-leaves.

teacake *n.* light usu. toasted sweet bun eaten at tea.

teach *v.* (*past* and *past part.* **taught** /tɔ:t/) **1 a** give systematic information, instruction, or training to (a person) or about (a subject or skill) (*taught me to swim*). **b** (*absol.*) practise this professionally. **c** communicate, instruct in (*suffering taught me patience*). **2** advocate as a moral etc. principle (*taught forgiveness*). **3** (foll. by *to* + infin.) **a** instruct (a person) by example or punishment (*that will teach you not to disobey*). **b** *colloq.* discourage (a person) from (*that will teach you to laugh*). □ **teachable** *adj.* [Old English]

teacher *n.* person who teaches, esp. in a school.

tea chest *n.* light metal-lined plywood box for transporting tea-leaves.

teaching *n.* **1** profession of a teacher. **2** (often in *pl.*) what is taught; doctrine.

tea cloth *n.* = TEA TOWEL.

tea cosy *n.* cover to keep a teapot warm.

teacup *n.* **1** cup from which tea is drunk. **2** amount held by this. □ **teacupful** *n.* (*pl.* **-s**).

tea dance *n.* afternoon tea with dancing.

teak *n.* **1** a hard durable timber. **2** large Indian or SE Asian deciduous tree yielding this. [Portuguese from Malayalam]

teal *n.* (*pl.* same) small freshwater duck. [origin unknown]

tea lady *n.* woman employed to make tea in offices etc.

tea-leaf *n.* **1** dried leaf of tea. **2** (esp. in *pl.*) these as dregs. **3** *rhyming slang* thief.

team –*n.* **1** set of players forming one side in a game. **2** two or more people working together. **3** set of draught animals. –*v.* **1** (usu. foll. by *up*) join in a team or in common action (*teamed up with them*). **2** (foll. by *with*) match or coordinate (clothes). [Old English]

team-mate *n.* fellow-member of a team.

team spirit *n.* willingness to act for the communal good.

teamster *n.* **1** *US* lorry-driver. **2** driver of a team of animals.

teamwork *n.* combined action; co-operation.

tea-planter *n.* proprietor or cultivator of a tea plantation.

teapot *n.* pot with a handle, spout, and lid, for brewing and then pouring tea.

tear[1] /teə(r)/ –*v.* (*past* **tore**; *past part.* **torn**) **1** (often foll. by *up*) pull apart or to pieces with some force (*tore up the letter*). **2 a** make a hole or rent in this way; undergo this (*have torn my coat*; *curtain tore*). **b** make (a hole or rent). **3** (foll. by *away*, *off*, *at*, etc.) pull violently (*tore off the cover*; *tore down the notice*). **4** violently disrupt or divide (*torn by guilt*). **5** *colloq.* go hurriedly (*tore across the road*). –*n.* **1** hole etc. caused by tearing. **2** torn part of cloth etc. □ **be torn between** have difficulty in choosing between. **tear apart 1** search (a place) exhaustively. **2** criticize forcefully. **3** destroy; divide utterly; distress greatly. **tear one's hair out** *colloq.* behave with extreme desperation. **tear into** *colloq.* **1** severely reprimand. **2** start (an activity) vigorously. **tear oneself away** leave reluctantly. **tear to shreds** *colloq.* refute or criticize thoroughly. **that's torn it** *colloq.* that has spoiled things etc. [Old English]

tear[2] /tɪə(r)/ *n.* **1** drop of clear salty liquid secreted by glands from the eye, and shed esp. in grief. **2** tearlike thing; drop. □ **in tears** crying. [Old English]

tearaway *n. colloq.* unruly young person.

tear-drop *n.* single tear.

tear-duct *n.* drain for carrying tears to or from the eye.
tearful *adj.* **1** crying or inclined to cry. **2** sad (*tearful event*). □ **tearfully** *adv.*
tear-gas *n.* gas causing severe irritation to the eyes.
tearing hurry *n. colloq.* great hurry.
tear-jerker *n. colloq.* sentimental story, film, etc.
tearoom *n.* small unlicensed café serving tea etc.
tea rose *n.* hybrid shrub with a tealike scent.
tease /ti:z/ –*v.* (**-sing**) (also *absol.*) **1 a** make fun of playfully, unkindly, or annoyingly; irritate. **b** allure, esp. sexually, while withholding satisfaction. **2** pick (wool etc.) into separate fibres. **3** dress (cloth) esp. with teasels. –*n.* **1** *colloq.* person fond of teasing. **2** act of teasing (*only a tease*). □ **tease out** separate by disentangling. [Old English]
teasel /'ti:z(ə)l/ *n.* (also **teazel, teazle**) **1** plant with large prickly heads that are dried and used to raise the nap on woven cloth. **2** other device used for this purpose. □ **teaseler** *n.* [Old English: related to TEASE]
teaser *n.* **1** person who teases. **2** *colloq.* hard question or task.
teaset *n.* set of crockery for serving tea.
teashop *n.* = TEAROOM.
teaspoon *n.* **1** small spoon for stirring tea. **2** amount held by this. □ **teaspoonful** *n.* (*pl.* **-s**).
teat *n.* **1** mammary nipple, esp. of an animal. **2** rubber nipple for sucking from a bottle. [French from Germanic]
teatime *n.* time in the afternoon when tea is served.
tea towel *n.* towel for drying washed crockery etc.
tea trolley *n.* (*US* **tea wagon**) small trolley from which tea is served.
teazel (also **teazle**) var. of TEASEL.
TEC /tek/ *abbr.* Training and Enterprise Council.
tec *n. colloq.* detective. [abbreviation]
tech *n.* (also **tec**) *colloq.* technical college. [abbreviation]
technetium /tek'ni:ʃ(ə)m/ *n.* artificially produced radioactive metallic element. [Greek *tekhnētos* artificial]
technic /'teknɪk/ *n.* **1** (usu. in *pl.*) **a** technology. **b** technical terms, details, methods, etc. **2** technique. [Greek *tekhnē* art]
technical *adj.* **1** of the mechanical arts and applied sciences (*technical college*). **2** of a particular subject or craft etc. or its techniques (*technical terms*). **3** (of a book or discourse etc.) using technical language; specialized. **4** due to mechanical failure (*technical hitch*). **5** strictly or legally interpreted (*lost on a technical point*). □ **technically** *adv.*
technicality /,teknɪ'kælɪtɪ/ *n.* (*pl.* **-ies**) **1** being technical. **2** technical expression. **3** technical point or detail (*acquitted on a technicality*).
technical knockout *n.* ruling by the referee that a boxer has lost because he is not fit to continue.
technician /tek'nɪʃ(ə)n/ *n.* **1** person doing practical or maintenance work in a laboratory etc. **2** person skilled in artistic etc. technique. **3** expert in practical science.
Technicolor /'teknɪ,kʌlə(r)/ *n.* (often *attrib.*) **1** *propr.* process of colour cinematography. **2** (usu. **technicolor**) *colloq.* **a** vivid colour. **b** artificial brilliance.
technique /tek'ni:k/ *n.* **1** mechanical skill in art. **2** skilful manipulation of a situation, people, etc. **3** manner of artistic execution in music, painting, etc. [French: related to TECHNIC]
technocracy /tek'nɒkrəsɪ/ *n.* (*pl.* **-ies**) **1** rule or control by technical experts. **2** instance or application of this. [Greek *tekhnē* art]
technocrat /'teknə,kræt/ *n.* exponent or advocate of technocracy. □ **technocratic** /-'krætɪk/ *adj.*
technology /tek'nɒlədʒɪ/ *n.* (*pl.* **-ies**) **1** knowledge or use of the mechanical arts and applied sciences (*lacked the technology*). **2** these subjects collectively. □ **technological** /-nə'lɒdʒɪk(ə)l/ *adj.* **technologically** /-nə'lɒdʒɪkəlɪ/ *adv.* **technologist** *n.* [Greek *tekhnologia* systematic treatment, from *tekhnē* art]
tectonic /tek'tɒnɪk/ *adj.* **1** of building or construction. **2** of the deformation and subsequent structural changes of the earth's crust (see PLATE TECTONICS). [Greek *tektōn* craftsman]
tectonics *n.pl.* (usu. treated as *sing.*) study of the earth's large-scale structural features (see PLATE TECTONICS).
Ted *n.* (also **ted**) *colloq.* Teddy boy. [abbreviation]
teddy /'tedɪ/ *n.* (also **Teddy**) (*pl.* **-ies**) (in full **teddy bear**) soft toy bear. [*Teddy*, pet form of *Theodore* Roosevelt]
Teddy boy /'tedɪ/ *n. colloq.* youth, esp. of the 1950s, wearing Edwardian-style clothes, hairstyle, etc. [*Teddy*, pet form of *Edward*]
tedious /'ti:dɪəs/ *adj.* tiresomely long; wearisome. □ **tediously** *adv.* **tediousness** *n.* [Latin: related to TEDIUM]
tedium /'ti:dɪəm/ *n.* tediousness. [Latin *taedium* from *taedet* it bores]
tee[1] *n.* = T[1]. [phonetic spelling]

tee[2] /ti:/ *–n.* **1 a** cleared space from which the golf ball is struck at the start of play for each hole. **b** small wooden or plastic support for a golf ball used then. **2** mark aimed at in bowls, quoits, curling, etc. *–v.* **(tees, teed)** (often foll. by *up*) place (a ball) on a golf tee. □ **tee off 1** play a ball from a tee. **2** *colloq.* start, begin. [origin unknown]

tee-hee /ti:'hi:/ (also **te-hee**) *–int.* expressing esp. derisive amusement. *–n.* titter, giggle. *–v.* **(-hees, -heed)** titter, giggle. [imitative]

teem[1] *v.* **1** be abundant. **2** (foll. by *with*) be full of or swarming with (*teeming with ideas*). [Old English, = give birth to]

teem[2] *v.* (often foll. by *down*) (of water etc.) flow copiously; pour (*teeming with rain*). [Old Norse]

teen *attrib. adj.* = TEENAGE. [abbreviation]

-teen *suffix* forming numerals from 13 to 19. [Old English]

teenage *attrib. adj.* of or characteristic of teenagers. □ **teenaged** *adj.*

teenager *n.* person from 13 to 19 years of age.

teens /ti:nz/ *n.pl.* years of one's age from 13 to 19 (*in his teens*).

teensy /'ti:nzɪ/ *adj.* **(-ier, -iest)** *colloq.* = TEENY.

teeny /'ti:nɪ/ *adj.* **(-ier, -iest)** *colloq.* tiny. [var. of TINY]

teeny-bopper *n. colloq.* young teenager, usu. a girl, who follows the latest fashions.

teeny-weeny *adj.* (also **teensy-weensy**) very tiny.

teepee var. of TEPEE.

teeter *v.* totter; move unsteadily. [dial. *titter*]

teeth *pl.* of TOOTH.

teethe /ti:ð/ *v.* **(-thing)** grow or cut teeth, esp. milk teeth.

teething-ring *n.* ring for an infant to bite on while teething.

teething troubles *n.pl.* initial difficulties in an enterprise etc.

teetotal /ti:'təʊt(ə)l/ *adj.* of or advocating total abstinence from alcohol. □ **teetotalism** *n.* **teetotaller** *n.* [reduplication of TOTAL]

teff *n.* an African cereal. [Amharic]

TEFL /'tef(ə)l/ *abbr.* teaching of English as a foreign language.

Teflon /'teflɒn/ *n. propr.* non-stick coating for kitchen utensils. [from *te*tra-, *fl*uor-, *-on*]

te-hee var. of TEE-HEE.

Tel. *abbr.* (also **tel.**) telephone.

tele- *comb. form* **1** at or to a distance (*telekinesis, telescope*). **2** television (*telecast*). **3** by telephone (*telesales*). [Greek *tēle* far off]

tele-ad /'telɪ,æd/ *n.* advertisement telephoned to a newspaper etc.

telecast *–n.* television broadcast. *–v.* transmit by television. □ **telecaster** *n.*

telecommunication /,telɪkə,mju:nɪ'keɪʃ(ə)n/ *n.* **1** communication over a distance by circuits using cable, fibre optics, satellites, radio etc. **2** (usu. in *pl.*) technology of this.

teleconference /'telɪ,kɒnfərəns/ *n.* conference with participants linked by telephone etc. □ **teleconferencing** *n.*

telefax /'telɪ,fæks/ *n.* = FAX. [abbreviation of *telefacsimile*]

telegram /'telɪ,græm/ *n.* message sent by telegraph and delivered in printed form.

■ **Usage** Since 1981 *telegram* has not been in UK official use, except for international messages. See also *telemessage.*

---

telegraph /'telɪ,grɑ:f/ *–n.* (often *attrib.*) device or system for transmitting messages or signals to a distance, esp. by making and breaking an electrical connection (*telegraph wire*). *–v.* **1** (often followed by *to*) send a message by telegraph to. **2** send or communicate by telegraph (*telegraphed my concern*). **3** give advance indication of (*telegraphed his punch*). □ **telegraphist** /tɪ'legrəfɪst/ *n.*

telegraphic /,telɪ'græfɪk/ *adj.* **1** of or by telegraphs or telegrams. **2** economically worded. □ **telegraphically** *adv.*

telegraphy /tɪ'legrəfɪ/ *n.* communication by telegraph.

telekinesis /,telɪkaɪ'ni:sɪs, -kɪ'ni:sɪs/ *n.* supposed paranormal force moving objects at a distance. □ **telekinetic** /-'netɪk/ *adj.* [Greek *kineō* move]

telemarketing /'telɪ,mɑ:kɪtɪŋ/ *n.* marketing of goods etc. by unsolicited telephone calls.

telemessage /'telɪ,mesɪdʒ/ *n.* message sent by telephone or telex and delivered in printed form.

■ **Usage** *Telemessage* has been in UK official use since 1981 for inland messages, replacing *telegram.*

---

telemetry /tɪ'lemətrɪ/ *n.* process of recording the readings of an instrument and transmitting them by radio. □ **telemeter** /tɪ'lemɪtə(r)/ *n.*

teleology /,ti:lɪ'ɒlədʒɪ, ,te-/ *n.* (*pl.* **-ies**) *Philos.* **1** explanation of phenomena by the purpose they serve. **2** *Theol.* doctrine of design and purpose in the material world. □ **teleological** /-ə'lɒdʒɪk(ə)l/ *adj.* [Greek *telos* end]

telepathy /tɪ'lepəθɪ/ *n.* supposed paranormal communication of thoughts. □ **telepathic** /ˌtelɪ'pæθɪk/ *adj.* **telepathically** /ˌtelɪ'pæθɪkəlɪ/ *adv.*

telephone /'telɪˌfəʊn/ —*n.* **1** apparatus for transmitting sound (esp. speech) to a distance, esp. by using optical or electrical signals. **2** handset etc. used in this. **3** system of communication using a network of telephones. —*v.* **(-ning) 1** speak to or send (a message) by telephone. **2** make a telephone call. □ **on the telephone** having or using a telephone. **over the telephone** using the telephone. □ **telephonic** /-'fɒnɪk/ *adj.* **telephonically** /-'fɒnɪkəlɪ/ *adv.*

telephone book *n.* = TELEPHONE DIRECTORY.

telephone booth *n.* (also **telephone kiosk, telephone box**) booth etc. with a telephone for public use.

telephone directory *n.* book listing telephone subscribers and numbers.

telephone number *n.* number used to call a particular telephone.

telephonist /tɪ'lefənɪst/ *n.* operator in a telephone exchange or at a switchboard.

telephony /tɪ'lefənɪ/ *n.* transmission of sound by telephone.

telephoto /ˌtelɪ'fəʊtəʊ/ *n.* (*pl.* **-s**) (in full **telephoto lens**) lens used in telephotography.

telephotography /ˌtelɪfə'tɒgrəfɪ/ *n.* photographing of distant objects with a system of lenses giving a large image. □ **telephotographic** /-ˌfəʊtə'græfɪk/ *adj.*

teleprinter /'telɪˌprɪntə(r)/ *n.* device for transmitting, receiving, and printing telegraph messages.

teleprompter /'telɪˌprɒmptə(r)/ *n.* device beside a television or cinema camera that slowly unrolls a speaker's script out of sight of the audience.

telesales /'telɪˌseɪlz/ *n.pl.* selling by telephone.

telescope /'telɪˌskəʊp/ —*n.* **1** optical instrument using lenses or mirrors to magnify distant objects. **2** = RADIO TELESCOPE. —*v.* **(-ping) 1** press or drive (sections of a tube, colliding vehicles, etc.) together so that one slides into another. **2** close or be capable of closing in this way. **3** compress so as to occupy less space or time.

telescopic /ˌtelɪ'skɒpɪk/ *adj.* **1** of or made with a telescope (*telescopic observations*). **2** (esp. of a lens) able to focus on and magnify distant objects. **3** consisting of sections that telescope. □ **telescopically** *adv.*

telescopic sight *n.* telescope on a rifle etc. used for sighting.

teletext /'telɪˌtekst/ *n.* computerized news and information service transmitted to the televisions of subscribers.

telethon /'telɪˌθɒn/ *n.* exceptionally long television programme, esp. to raise money for charity. [from TELE-, MARATHON]

Teletype /'telɪˌtaɪp/ *n. propr.* a kind of teleprinter.

televise /'telɪˌvaɪz/ *v.* **(-sing)** broadcast on television.

television /'telɪˌvɪʒ(ə)n, -'vɪʒ(ə)n/ *n.* **1** system for reproducing on a screen visual images transmitted (usu. with sound) by radio signals or cable. **2** (in full **television set**) device with a screen for receiving these signals. **3** television broadcasting. □ **televisual** /-'vɪʒʊəl/ *adj.*

telex /'teleks/ (also **Telex**) —*n.* international system of telegraphy by teleprinters using the public telecommunications network. —*v.* send, or communicate with, by telex. [from TELEPRINTER, EXCHANGE]

tell *v.* (*past* and *past part.* **told** /təʊld/) **1** relate in speech or writing (*tell me a story*). **2** make known; express in words (*tell me your name*). **3** reveal or signify to (a person) (*your face tells me everything*). **4** utter (*tell lies*). **5 a** (often foll. by *of, about*) divulge information etc.; reveal a secret, the truth etc. (*told her about Venice*; *book tells you how to cook*; *promise you won't tell*; *time will tell*). **b** (foll. by *on*) *colloq.* inform against. **6** (foll. by *to* + infin.) direct; order (*tell them to wait*). **7** assure (*it's true, I tell you*). **8** decide, determine, distinguish (*tell one from the other*). **9** (often foll. by *on*) produce a noticeable effect or influence (*strain told on me*; *evidence tells against you*). **10** (often *absol.*) count (votes) at a meeting, election, etc. □ **tell apart** distinguish between (*could not tell them apart*). **tell off** *colloq.* scold. **tell tales** make known another person's faults etc. **tell the time** read the time from a clock or watch. **you're telling me** *colloq.* I agree wholeheartedly. [Old English: related to TALE]

teller *n.* **1** person working at the counter of a bank etc. **2** person who counts votes. **3** person who tells esp. stories (*teller of tales*).

telling *adj.* having a marked effect; striking; impressive. □ **tellingly** *adv.*

telling-off *n.* (*pl.* **tellings-off**) *colloq.* scolding.

tell-tale *n.* **1** person who reveals secrets about another. **2** (*attrib.*) that reveals or betrays (*tell-tale smile*). **3** automatic monitoring or registering device.

tellurium /te'ljʊərɪəm/ *n.* rare lustrous silver-white element used in semiconductors. □ **telluric** *adj.* [Latin *tellus -ur-* earth]

telly /'telɪ/ *n.* (*pl.* **-ies**) *colloq.* **1** television. **2** television set. [abbreviation]

temerity /tɪ'merɪtɪ/ *n.* rashness; audacity. [Latin *temere* rashly]

temp *colloq.* —*n.* temporary employee, esp. a secretary. —*v.* work as a temp. [abbreviation]

temper —*n.* **1** mental disposition, mood (*placid temper*). **2** irritation or anger (*fit of temper*). **3** tendency to lose one's temper (*have a temper*). **4** composure, calmness (*lose one's temper*). **5** hardness or elasticity of metal. —*v.* **1** bring (metal or clay) to a proper hardness or consistency. **2** (foll. by *with*) moderate, mitigate (*temper justice with mercy*). □ **in a bad** (or **out of**) **temper** irritable, angry. **in a good temper** amicable, happy. [Latin *tempero* mingle]

tempera /'tempərə/ *n.* **1** method of painting using an emulsion, e.g. of pigment with egg-yolk and water, esp. on canvas. **2** this emulsion. [Italian]

temperament /'temprəmənt/ *n.* person's or animal's nature and character (*nervous temperament*). [Latin: related to TEMPER]

temperamental /ˌtemprə'ment(ə)l/ *adj.* **1** of temperament. **2 a** (of a person) unreliable; moody. **b** *colloq.* (of esp. a machine) unreliable, unpredictable. □ **temperamentally** *adv.*

temperance /'tempərəns/ *n.* **1** moderation, esp. in eating and drinking. **2** (often *attrib.*) abstinence, esp. total, from alcohol (*temperance hotel*). [Latin: related to TEMPER]

temperate /'tempərət/ *adj.* **1** avoiding excess. **2** moderate. **3** (of a region or climate) mild. [Latin: related to TEMPER]

temperature /'temprɪtʃə(r)/ *n.* **1** measured or perceived degree of heat or cold of a thing, region, etc. **2** *colloq.* body temperature above the normal (*have a temperature*). **3** degree of excitement in a discussion etc. [Latin: related to TEMPER]

tempest /'tempɪst/ *n.* violent storm. [Latin *tempus* time]

tempestuous /tem'pestʃʊəs/ *adj.* stormy; turbulent. □ **tempestuously** *adv.*

tempi *pl.* of TEMPO.

template /'templɪt, -pleɪt/ *n.* piece of thin board or metal plate etc., used as a pattern in cutting or drilling etc. [originally *templet,* diminutive of *temple,* device in a loom to keep the cloth stretched]

temple[1] /'temp(ə)l/ *n.* building for the worship, or seen as the dwelling-place, of a god or gods etc. [Latin *templum*]

temple[2] /'temp(ə)l/ *n.* flat part of either side of the head between the forehead and the ear. [French from Latin]

tempo /'tempəʊ/ *n.* (*pl.* **-s** or **-pi** /-piː/) **1** speed at which music is or should be played. **2** speed or pace. [Latin *tempus -por-* time]

temporal /'tempər(ə)l/ *adj.* **1** worldly as opposed to spiritual; secular. **2** of time. **3** *Gram.* denoting time or tense (*temporal conjunction*). **4** of the temples of the head (*temporal artery*). [Latin *tempus -por-* time]

temporary /'tempərərɪ/ —*adj.* lasting or meant to last only for a limited time. —*n.* (*pl.* **-ies**) person employed temporarily. □ **temporarily** *adv.* **temporariness** *n.*

temporize /'tempəˌraɪz/ *v.* (also **-ise**) (**-zing** or **-sing**) **1** avoid committing oneself so as to gain time; procrastinate. **2** comply temporarily; adopt a time-serving policy.

tempt *v.* **1** entice or incite (a person) to do what is wrong or forbidden (*tempted him to steal it*). **2** allure, attract. **3** risk provoking (fate etc.). □ **be tempted to** be strongly disposed to. □ **tempter** *n.* **temptress** *n.* [Latin *tempto, tento* try, test]

temptation /temp'teɪʃ(ə)n/ *n.* **1** tempting or being tempted; incitement, esp. to wrongdoing. **2** attractive thing or course of action. **3** *archaic* putting to the test.

tempting *adj.* attractive, inviting. □ **temptingly** *adv.*

tempura /tem'pʊərə/ *n.* Japanese dish of fish, shellfish, etc., fried in batter. [Japanese]

ten *adj.* & *n.* **1** one more than nine. **2** symbol for this (10, x, X). **3** size etc. denoted by ten. **4** ten o'clock. □ **ten to one** very probably. [Old English]

tenable /'tenəb(ə)l/ *adj.* **1** maintainable or defensible against attack or objection (*tenable position*). **2** (foll. by *for, by*) (of an office etc.) that can be held for (a specified period) or by (a specified class of person). □ **tenability** /-'bɪlɪtɪ/ *n.* [French *tenir* hold]

tenacious /tɪ'neɪʃəs/ *adj.* **1** (often foll. by *of*) keeping a firm hold. **2** persistent, resolute. **3** (of memory) retentive. □ **tenaciously** *adv.* **tenacity** /tɪ'næsɪtɪ/ *n.* [Latin *tenax -acis* from *teneo* hold]

tenancy /'tenənsɪ/ *n.* (*pl.* **-ies**) **1** status of or possession as a tenant. **2** duration of this.

tenant /'tenənt/ *n.* **1** person who rents land or property from a landlord. **2**

(often foll. by *of*) occupant of a place. [French: related to TENABLE]

tenant farmer *n.* person who farms rented land.

tenantry *n.* tenants of an estate etc.

tench *n.* (*pl.* same) European freshwater fish of the carp family. [Latin *tinca*]

Ten Commandments *n.pl.* (prec. by *the*) rules of conduct given by God to Moses (Exod. 20:1–17).

tend[1] *v.* **1** (often foll. by *to*) be apt or inclined (*tends to lose his temper*; *tends to fat*). **2** be moving; hold a course (*tends in our direction*). [Latin *tendo tens-* or *tent-* stretch]

tend[2] *v.* take care of, look after (an invalid, sheep, a machine etc.). [from ATTEND]

tendency /'tendənsɪ/ *n.* (*pl.* **-ies**) (often foll. by *to, towards*) leaning or inclination. [medieval Latin: related to TEND[1]]

tendentious /ten'denʃəs/ *adj. derog.* calculated to promote a particular cause or viewpoint; biased; controversial. □ **tendentiously** *adv.* **tendentiousness** *n.*

tender[1] *adj.* (**tenderer, tenderest**) **1** easily cut or chewed, not tough (*tender steak*). **2** susceptible to pain or grief; vulnerable; compassionate (*tender heart*). **3** sensitive; fragile; delicate (*tender skin*; *tender reputation*). **4** loving, affectionate. **5** requiring tact (*tender subject*). **6** (of age) early, immature (*of tender years*). □ **tenderly** *adv.* **tenderness** *n.* [Latin *tener*]

tender[2] –*v.* **1** offer, present (one's services, resignation, money as payment, etc.). **2** (often foll. by *for*) offer a tender. –*n.* offer, esp. in writing, to execute work or supply goods at a stated price. □ **put out to tender** seek competitive tenders for (work etc.). □ **tenderer** *n.* [French: related to TEND[1]]

tender[3] *n.* **1** person who looks after people or things. **2** supply ship attending a larger one etc. **3** truck coupled to a steam locomotive to carry fuel and water. [from TEND[2]]

tenderfoot *n.* (*pl* **-s** or **-feet**) newcomer, novice.

tender-hearted /,tendə'hɑ:tɪd/ *adj.* easily moved; compassionate. □ **tender-heartedness** *n.*

tenderize /'tendə,raɪz/ *v.* (also **-ise**) (**-zing** or **-sing**) make (esp. meat) tender by beating, hanging, marinading, etc. □ **tenderizer** *n.*

tenderloin *n.* **1** middle part of pork loin. **2** *US* undercut of sirloin.

tender mercies *n.pl. iron.* harsh treatment.

tender spot *n.* subject on which a person is touchy.

tendon /'tend(ə)n/ *n.* cord of strong connective tissue attaching a muscle to a bone etc. □ **tendinitis** /-'naɪtɪs/ *n.* [Latin *tendo* stretch]

tendril /'tendrɪl/ *n.* slender leafless shoot by which some climbing plants cling. [probably from French *tendrillon*]

tenebrous /'tenɪbrəs/ *adj. literary* dark, gloomy. [Latin *tenebrosus*]

tenement /'tenɪmənt/ *n.* **1** room or flat within a house or block of flats. **2** (also **tenement-house** or **-block**) house or block so divided. [Latin *teneo* hold]

tenet /'tenɪt/ *n.* doctrine, principle. [Latin, = he holds]

tenfold *adj.* & *adv.* **1** ten times as much or as many. **2** consisting of ten parts.

ten-gallon hat *n.* cowboy's large broad-brimmed hat.

tenner *n. colloq.* ten-pound or ten-dollar note.

tennis /'tenɪs/ *n.* game in which two or four players strike a ball with rackets over a net stretched across a court. [probably French *tenez* take! (as a server's call)]

tennis elbow *n.* sprain caused by overuse of forearm muscles.

tenon /'tenən/ *n.* wooden projection made for insertion into a cavity, esp. a mortise, in another piece. [Latin: related to TENOR]

tenor /'tenə(r)/ *n.* **1 a** male singing-voice between baritone and alto or countertenor. **b** singer with this voice. **2** (often *attrib.*) instrument with a similar range. **3** (usu. foll. by *of*) general meaning. **4** (usu. foll. by *of*) prevailing course, esp. of a person's life or habits. [Latin *teneo* hold]

tenosynovitis /,tenəʊ,saɪnəʊ'vaɪtɪs/ *n.* injury of esp. a wrist tendon resulting from repetitive strain. [Greek *tenōn* tendon, SYNOVIA]

tenpin *n.* pin used in tenpin bowling.

tenpin bowling *n.* game in which ten pins or skittles are bowled at in an alley.

tense[1] –*adj.* **1** stretched tight, strained. **2** causing tenseness (*tense moment*). –*v.* (**-sing**) make or become tense. □ **tense up** become tense. □ **tensely** *adv.* **tenseness** *n.* [Latin *tensus*: related to TEND[1]]

tense[2] *n.* **1** form of a verb indicating the time (also the continuance or completeness) of the action etc. **2** set of such forms as a paradigm. [Latin *tempus* time]

tensile /'tensaɪl/ *adj.* **1** of tension. **2** capable of being stretched. □ **tensility** /-'sɪlɪtɪ/ *n.* [medieval Latin: related to TENSE[1]]

tensile strength *n.* resistance to breaking under tension.

tension /'tenʃ(ə)n/ *–n.* **1** stretching or being stretched; tenseness. **2** mental strain or excitement. **3** strained (political, social, etc.) state or relationship. **4** stress produced by forces pulling apart. **5** degree of tightness of stitches in knitting and machine sewing. **6** voltage (*high tension*; *low tension*). *–v.* subject to tension. □ **tensional** *adj.* [Latin: related to TEND[1]]

tent *n.* **1** portable canvas etc. shelter or dwelling supported by poles and cords attached to pegs driven into the ground. **2** tentlike enclosure, e.g. supplying oxygen to a patient. [Latin: related to TEND[1]]

tentacle /'tentək(ə)l/ *n.* **1** long slender flexible appendage of an (esp. invertebrate) animal, used for feeling, grasping, or moving. **2** channel for gathering information, exercising influence, etc. □ **tentacled** *adj.* [Latin: related to TEMPT]

tentative /'tentətɪv/ *adj.* **1** experimental. **2** hesitant, not definite (*tentative suggestion*). □ **tentatively** *adv.* **tentativeness** *n.* [medieval Latin: related to TEMPT]

tenter *n.* machine for stretching cloth to dry in shape. [medieval Latin *tentorium*: related to TEND[1]]

tenterhook *n.* hook to which cloth is fastened on a tenter. □ **on tenterhooks** in a state of suspense or agitation due to uncertainty.

tenth *adj.* & *n.* **1** next after ninth. **2** any of ten equal parts of a thing. □ **tenthly** *adv.*

tent-stitch *n.* **1** series of parallel diagonal stitches. **2** such a stitch.

tenuous /'tenjʊəs/ *adj.* **1** slight, insubstantial (*tenuous connection*). **2** (of a distinction etc.) oversubtle. **3** thin, slender, small. **4** rarefied. □ **tenuity** /-'ju:ɪtɪ/ *n.* **tenuously** *adv.* [Latin *tenuis*]

tenure /'tenjə(r)/ *n.* **1** condition, or form of right or title, under which (esp. real) property is held. **2** (often foll. by *of*) **a** holding or possession of an office or property. **b** period of this. **3** guaranteed permanent employment, esp. as a teacher or lecturer. □ **tenured** *adj.* [Latin *teneo*]

tepee /'ti:pi:/ *n.* (also **teepee**) N. American Indian's conical tent. [Dakota]

tepid /'tepɪd/ *adj.* **1** lukewarm. **2** unenthusiastic. □ **tepidity** /tɪ'pɪdɪtɪ/ *n.* **tepidly** *adv.* [Latin]

tequila /tɪ'ki:lə/ *n.* Mexican liquor made from an agave. [*Tequila* in Mexico]

tera- *comb. form* denoting a factor of 10^{12}. [Greek *teras* monster]

terbium /'tɜ:bɪəm/ *n.* silvery metallic element of the lanthanide series. [*Ytterby* in Sweden]

tercel /'tɜ:s(ə)l/ *n.* (also **tiercel** /'tɪəs(ə)l/) male hawk, esp. a peregrine or goshawk. [Latin *tertius* third]

tercentenary /,tɜ:sen'ti:nərɪ/ *n.* (*pl.* **-ies**) **1** three-hundredth anniversary. **2** celebration of this. [Latin *ter*, = three times]

teredo /tə'ri:dəʊ/ *n.* (*pl.* **-s**) bivalve mollusc that bores into submerged timbers of ships etc. [Latin from Greek]

tergiversate /'tɜ:dʒɪvə,seɪt/ *v.* (**-ting**) **1** change one's party or principles; apostatize. **2** make conflicting or evasive statements. □ **tergiversation** /-'seɪʃ(ə)n/ *n.* **tergiversator** *n.* [Latin *tergum* back, *verto* turn]

term *–n.* **1** word for a definite concept, esp. specialized (*technical term*). **2** (in *pl.*) language used; mode of expression (*in no uncertain terms*). **3** (in *pl.*) relation, footing (*on good terms*). **4** (in *pl.*) **a** stipulations (*accepts your terms*). **b** charge or price (*reasonable terms*). **5 a** limited, usu. specified, period (*term of five years*; *in the short term*). **b** period of weeks during which instruction is given or during which a lawcourt holds sessions. **6** *Logic* word or words that may be the subject or predicate of a proposition. **7** *Math.* **a** each of the quantities in a ratio or series. **b** part of an algebraic expression. **8** completion of a normal length of pregnancy. *–v.* call, name (*was termed a bigot*). □ **bring to terms** cause to accept conditions. **come to terms** yield, give way. **come to terms with** reconcile oneself to (a difficulty etc.). **in terms of** in the language peculiar to; referring to. □ **termly** *adj.* & *adv.* [Latin TERMINUS]

termagant /'tɜ:məgənt/ *n.* overbearing woman; virago. [French *Tervagan* from Italian]

terminable /'tɜ:mɪnəb(ə)l/ *adj.* able to be terminated.

terminal /'tɜ:mɪn(ə)l/ *–adj.* **1 a** (of a condition or disease) fatal. **b** (of a patient) dying. **2** of or forming a limit or terminus (*terminal station*). *–n.* **1** terminating thing; extremity. **2** terminus for trains or long-distance buses. **3** = AIR TERMINAL. **4** point of connection for closing an electric circuit. **5** apparatus for the transmission of messages to and from a computer, communications system, etc. □ **terminally** *adv.* [Latin: related to TERMINUS]

terminate /'tɜ:mɪ,neɪt/ *v.* (**-ting**) **1** bring or come to an end. **2** (foll. by *in*) (of a word) end in (a specified letter etc.).

termination /ˌtɜːmɪˈneɪʃ(ə)n/ *n.* **1** terminating or being terminated. **2** induced abortion. **3** ending or result. **4** word's final syllable or letter.

termini *pl.* of TERMINUS.

terminology /ˌtɜːmɪˈnɒlədʒɪ/ *n.* (*pl.* **-ies**) **1** system of specialized terms. **2** science of the use of terms. □ **terminological** /-nəˈlɒdʒɪk(ə)l/ *adj.* [German: related to TERMINUS]

terminus /ˈtɜːmɪnəs/ *n.* (*pl.* **-ni** /-ˌnaɪ/ or **-nuses**) **1** station at the end of a railway or bus route. **2** point at the end of a pipeline etc. [Latin, = end, limit, boundary]

termite /ˈtɜːmaɪt/ *n.* small tropical ant-like social insect destructive to timber. [Latin *termes -mitis*]

terms of reference *n.pl.* scope of an inquiry etc.; definition of this.

tern *n.* marine gull-like bird with a long forked tail. [Scandinavian]

ternary /ˈtɜːnərɪ/ *adj.* composed of three parts. [Latin *terni*, = three each]

terrace /ˈterəs/ *–n.* **1** flat area made on a slope for cultivation. **2** level paved area next to a house. **3** row of houses built in one block of uniform style. **4** tiered standing accommodation for spectators at a sports ground. *–v.* (**-cing**) form into or provide with a terrace or terraces. [Latin *terra* earth]

terrace house *n.* (also **terraced house**) house in a terrace.

terracotta /ˌterəˈkɒtə/ *n.* **1 a** unglazed usu. brownish-red earthenware. **b** statuette of this. **2** its colour. [Italian, = baked earth]

terra firma /ˌterə ˈfɜːmə/ *n.* dry land, firm ground. [Latin]

terrain /təˈreɪn/ *n.* tract of land, esp. in geographical or military contexts. [Latin: related to TERRENE]

terra incognita /ˌterə ɪŋˈkɒgnɪtə, ˌɪnkɒgˈniːtə/ *n.* unexplored region. [Latin, = unknown land]

terrapin /ˈterəpɪn/ *n.* **1** N. American edible freshwater turtle. **2** (**Terrapin**) *propr.* type of prefabricated one-storey building. [Algonquian]

terrarium /təˈreərɪəm/ *n.* (*pl.* **-s** or **-ria**) **1** place for keeping small land animals. **2** sealed transparent globe etc. containing growing plants. [Latin *terra* earth, after *aquarium*]

terrazzo /teˈrætsəʊ/ *n.* (*pl.* **-s**) smooth flooring-material of stone chips set in concrete. [Italian, = terrace]

terrene /teˈriːn/ *adj.* **1** of the earth; worldly. **2** of earth, earthy. **3** terrestrial. [Latin *terrenus* from *terra* earth]

terrestrial /təˈrestrɪəl/ *adj.* **1** of or on the earth; earthly. **2** of or on dry land. [Latin *terrestris*: related to TERRENE]

terrible /ˈterɪb(ə)l/ *adj.* **1** *colloq.* very great or bad (*terrible bore*). **2** *colloq.* very incompetent (*terrible at maths*). **3** causing or likely to cause terror; dreadful, formidable. [Latin *terreo* frighten]

terribly *adv.* **1** *colloq.* very, extremely (*terribly nice*). **2** in a terrible manner.

terrier /ˈterɪə(r)/ *n.* small dog of various breeds originally used for digging out foxes etc. [French *chien terrier* dog that chases to earth]

terrific /təˈrɪfɪk/ *adj.* **1** *colloq.* **a** huge; intense (*terrific noise*). **b** excellent (*did a terrific job*). **2** causing terror. □ **terrifically** *adv.* [Latin: related to TERRIBLE]

terrify /ˈterɪˌfaɪ/ *v.* (**-ies**, **-ied**) fill with terror (*terrified of dogs*). □ **terrifying** *adj.* **terrifyingly** *adv.*

terrine /təˈriːn/ *n.* **1** pâté or similar food. **2** earthenware vessel, esp. for pâté. [Latin *terra* earth]

territorial /ˌterɪˈtɔːrɪəl/ *–adj.* **1** of territory or a district (*territorial possessions*; *territorial right*). **2** tending to defend one's territory. *–n.* (**Territorial**) member of the Territorial Army. □ **territorially** *adv.* [Latin: related to TERRITORY]

Territorial Army *n.* local volunteer reserve force.

territorial waters *n.pl.* area of sea under the jurisdiction of a State, esp. within a stated distance of the shore.

territory /ˈterɪtərɪ/ *n.* (*pl.* **-ies**) **1** extent of the land under the jurisdiction of a ruler, State, etc. **2** (**Territory**) organized division of a country, esp. one not yet admitted to the full rights of a State. **3** sphere of action etc.; province. **4** commercial traveller's sales area. **5** animal's or human's defended space or area. **6** area defended by a team or player in a game. [Latin *terra* land]

terror /ˈterə(r)/ *n.* **1** extreme fear. **2 a** terrifying person or thing. **b** *colloq.* formidable or troublesome person or thing, esp. a child. **3** organized intimidation; terrorism. [Latin *terreo* frighten]

terrorist *n.* (often *attrib.*) person using esp. organized violence against a government etc. □ **terrorism** *n.* [French: related to TERROR]

terrorize /ˈterəˌraɪz/ *v.* (also **-ise**) (**-zing** or **-sing**) **1** fill with terror. **2** use terrorism against. □ **terrorization** /-ˈzeɪʃ(ə)n/ *n.*

terror-stricken *adj.* (also **terror-struck**) affected with terror.

terry /ˈterɪ/ *n.* (often *attrib.*) looped pile fabric used esp. for towels and nappies. [origin unknown]

terse *adj.* (**terser**, **tersest**) **1** brief, concise. **2** curt, abrupt. □ **tersely** *adv.* **terseness** *n.* [Latin *tergo ters-* wipe]

tertiary /'tɜːʃərɪ/ –*adj.* **1** third in order or rank etc. **2** (**Tertiary**) of the first period in the Cenozoic era. –*n.* (**Tertiary**) Tertiary period. [Latin *tertius* third]

tervalent /'tɜːvələnt, -'veɪlənt/ *adj.* having a valency of three. [from TERCENTENARY, VALENCE[1]]

Terylene /'terɪˌliːn/ *n. propr.* synthetic textile fibre of polyester. [from *terephthalic* acid, ETHYLENE]

TESL /'tes(ə)l/ *abbr.* teaching of English as a second language.

tesla /'teslə/ *n.* SI unit of magnetic induction. [*Tesla*, name of a scientist]

TESSA /'tesə/ *abbr.* tax exempt special savings account.

tessellated /'tesəˌleɪtɪd/ *adj.* **1** of or resembling a mosaic. **2** regularly chequered. [Latin *tessella* diminutive of TESSERA]

tessellation /ˌtesə'leɪʃ(ə)n/ *n.* close arrangement of polygons, esp. in a repeated pattern.

tessera /'tesərə/ *n.* (*pl.* **tesserae** /-ˌriː/) small square block used in mosaic. [Latin from Greek]

tessitura /ˌtesɪ'tʊərə/ *n.* range of a singing voice or vocal part. [Italian, = TEXTURE]

test[1] –*n.* **1** critical examination or trial of a person's or thing's qualities. **2** means, procedure, or standard for so doing. **3** minor examination, esp. in school (*spelling test*). **4** *colloq.* test match. –*v.* **1** put to the test. **2** try or tax severely. **3** examine by means of a reagent. □ **put to the test** cause to undergo a test. **test out** put to a practical test. □ **testable** *adj.* [Latin *testu(m)* earthen pot: related to TEST[2]]

test[2] *n.* shell of some invertebrates. [Latin *testa* pot, tile, shell]

testa /'testə/ *n.* (*pl.* **testae** /-tiː/) seed's protective outer covering. [Latin: related to TEST[2]]

testaceous /te'steɪʃəs/ *adj.* having a hard continuous shell.

testament /'testəmənt/ *n.* **1** will (esp. *last will and testament*). **2** (usu. foll. by *to*) evidence, proof (*is testament to his loyalty*). **3** *Bibl.* **a** covenant, dispensation. **b** (**Testament**) division of the Bible (see OLD TESTAMENT, NEW TESTAMENT). [Latin *testamentum* will: related to TESTATE]

testamentary /ˌtestə'mentərɪ/ *adj.* of, by, or in a will.

testate /'testeɪt/ –*adj.* having left a valid will at death. –*n.* testate person. □ **testacy** *n.* (*pl.* **-ies**). [Latin *testor* testify, from *testis* witness]

testator /te'steɪtə(r)/ *n.* (*fem.* **testatrix** /te'steɪtrɪks/) (esp. deceased) person who has made a will. [Latin: related to TESTATE]

test card *n.* still television picture outside normal programme hours used for adjusting brightness, definition, etc.

test case *n. Law* case setting a precedent for other similar cases.

test drive *n.* drive taken to judge the performance of a vehicle. □ **test-drive** *v.*

tester *n.* **1** person or thing that tests. **2** bottle etc. containing a cosmetic for trial in a shop.

testes *pl.* of TESTIS.

test flight *n.* aircraft flight for evaluation purposes. □ **test-fly** *v.*

testicle /'testɪk(ə)l/ *n.* male organ that produces spermatozoa etc., esp. one of a pair in the scrotum in man and most mammals. [Latin, diminutive of *testis* witness]

testify /'testɪˌfaɪ/ *v.* (**-ies, -ied**) **1** (often foll. by *to*) (of a person or thing) bear witness; be evidence of (*testified to the facts*). **2** give evidence. **3** affirm or declare. [Latin *testificor* from *testis* witness]

testimonial /ˌtestɪ'məʊnɪəl/ *n.* **1** certificate of character, conduct, or qualifications. **2** gift presented to a person (esp. in public) as a mark of esteem etc. [French: related to TESTIMONY]

testimony /'testɪmənɪ/ *n.* (*pl.* **-ies**) **1** witness's statement under oath etc. **2** declaration or statement of fact. **3** evidence, demonstration (*produce testimony*). [Latin *testimonium* from *testis* witness]

testis /'testɪs/ *n.* (*pl.* **testes** /-tiːz/) *Anat. & Zool.* testicle. [Latin, = witness (cf. TESTICLE)]

test match *n.* international cricket or Rugby match, usu. in a series.

testosterone /te'stɒstəˌrəʊn/ *n.* male sex hormone formed in the testicles. [from TESTIS, STEROL]

test paper *n.* **1** minor examination paper. **2** paper impregnated with a substance changing colour under known conditions.

test pilot *n.* pilot who test-flies aircraft.

test-tube *n.* thin glass tube closed at one end, used for chemical tests etc.

test-tube baby *n. colloq.* baby conceived by *in vitro* fertilization.

testy *adj.* (**-ier, -iest**) irritable, touchy. □ **testily** *adv.* **testiness** *n.* [French *teste* head: related to TEST[2]]

tetanus /'tetənəs/ *n.* bacterial disease causing painful spasm of the voluntary muscles. [Greek *teinō* stretch]

tetchy /'tetʃɪ/ *adj.* (**-ier, -iest**) peevish, irritable. □ **tetchily** *adv.* **tetchiness** *n.* [*teche* blemish, fault]

tête-à-tête /ˌteɪtɑːˈteɪt/ *–n.* (often *attrib.*) private conversation between two persons. *–adv.* privately without a third person (*dined tête-à-tête*). [French, literally 'head-to-head']

tether /ˈteðə(r)/ *–n.* rope etc. confining a grazing animal. *–v.* tie with a tether. □ **at the end of one's tether** at the limit of one's patience, resources, etc. [Old Norse]

tetra- *comb. form* four. [Greek *tettares* four]

tetrad /ˈtetræd/ *n.* group of four. [Greek: related to TETRA-]

tetragon /ˈtetrəˌgɒn/ *n.* plane figure with four angles and sides. □ **tetragonal** /tɪˈtrægən(ə)l/ *adj.* [Greek *-gōnos* -angled]

tetrahedron /ˌtetrəˈhiːdrən/ *n.* (*pl.* **-dra** or **-s**) four-sided solid; triangular pyramid. □ **tetrahedral** *adj.* [Greek *hedra* base]

tetralogy /teˈtrælədʒɪ/ *n.* (*pl.* **-ies**) group of four related novels, plays, operas, etc.

tetrameter /teˈtræmɪtə(r)/ *n. Prosody* verse of four measures.

Teuton /ˈtjuːt(ə)n/ *n.* member of a Teutonic nation, esp. a German. [Latin *Teutones*, ancient tribe of N. Europe]

Teutonic /tjuːˈtɒnɪk/ *adj.* **1** of the Germanic peoples or languages. **2** German. [Latin: related to TEUTON]

text *n.* **1** main body of a book as distinct from notes etc. **2** original book or document, esp. as distinct from a paraphrase etc. **3** passage from Scripture, esp. as the subject of a sermon. **4** subject, theme. **5** (in *pl.*) books prescribed for study. **6** data in textual form, esp. as stored, processed, or displayed in a word processor etc. [Latin *texo text-* weave]

textbook *–n.* book for use in studying, esp. a standard account of a subject. *–attrib. adj.* **1** exemplary, accurate. **2** instructively typical.

text editor *n. Computing* system or program allowing the user to enter and edit text.

textile /ˈtekstaɪl/ *–n.* **1** (often in *pl.*) fabric, cloth, or fibrous material, esp. woven. **2** fibre, yarn. *–adj.* **1** of weaving or cloth (*textile industry*). **2** woven (*textile fabrics*). [Latin: related to TEXT]

text processing *n. Computing* manipulation of text, esp. transforming it from one format to another.

textual /ˈtekstʃʊəl/ *adj.* of, in, or concerning a text. □ **textually** *adv.*

texture /ˈtekstʃə(r)/ *–n.* **1** feel or appearance of a surface or substance. **2** arrangement of threads etc. in textile fabric. *–v.* (**-ring**) (usu. as **textured** *adj.*) **1** provide with a texture. **2** (of vegetable protein) provide with a texture resembling meat. □ **textural** *adj.* [Latin: related to TEXT]

Th *symb.* thorium.

-th *suffix* (also **-eth**) forming ordinal and fractional numbers from *four* onwards. [Old English]

Thai /taɪ/ *–n.* (*pl.* same or **-s**) **1 a** native or national of Thailand. **b** person of Thai descent. **2** language of Thailand. *–adj.* of Thailand. [Thai, = free]

thalidomide /θəˈlɪdəˌmaɪd/ *n.* sedative drug found in 1961 to cause foetal malformation when taken early in pregnancy. [from ph*thali*mido*glutari*mide]

thallium /ˈθælɪəm/ *n.* rare soft white metallic element. [Greek *thallos* green shoot]

than /ðən, ðæn/ *conj.* introducing a comparison (*plays better than he did before*; *more bread than meat in these sausages*; *cost more than £100*; *you are older than he*). [Old English, originally = THEN]

■ **Usage** With reference to the last example, it is also legitimate to say *you are older than him*, with *than* treated as a preposition, esp. in less formal contexts.

---

thane *n. hist.* **1** man who held land from an English king or other superior by military service. **2** man who held land from a Scottish king and ranked with an earl's son; chief of a clan. [Old English]

thank *–v.* **1** express gratitude to (*thanked him for the present*). **2** hold responsible (*you can thank yourself for that*). *–n.* (in *pl.*) **1** gratitude. **2** expression of gratitude. **3** (as a formula) thank you (*thanks for your help*). □ **thank goodness** (or **God** or **heavens** etc.) *colloq.* expression of relief etc. **thanks to** as the result of (*thanks to my foresight*; *thanks to your obstinacy*). **thank you** polite formula expressing gratitude. [Old English]

thankful *adj.* **1** grateful, pleased. **2** expressive of thanks.

thankfully *adv.* **1** in a thankful manner. **2** let us be thankful (that) (*thankfully, it didn't rain*).

■ **Usage** The use of *thankfully* in sense 2 is common, but is considered incorrect by some people.

---

thankless *adj.* **1** not expressing or feeling gratitude. **2** (of a task etc.) giving no pleasure or profit; unappreciated.

thanksgiving *n.* **1** expression of gratitude, esp. to God. **2** (**Thanksgiving** or **Thanksgiving Day**) fourth Thursday in November (a national holiday in the US).

that /ðæt/ *–demons. pron.* (*pl.* **those** /ðəuz/) **1** person or thing indicated, named, or understood (*I heard that; who is that in the garden?*). **2** contrasted with *this* (*this is much better than that*). **3** (esp. in relative constructions) the one, the person, etc. (*a table like that described above*). **4** /ðət/ (*pl.* **that**) used instead of *which* or *whom* to introduce a defining clause (*the book that you sent me; there is nothing here that matters*). *–demons. adj.* (*pl.* **those** /ðəuz/) designating the person or thing indicated, named, understood, etc. (cf. sense 1 of *pron.*). *–adv.* **1** to such a degree; so (*have done that much*). **2** *colloq.* very (*not that good*). *–conj.* /ðət/ introducing a subordinate clause indicating: **1** statement or hypothesis (*they say that he is better*). **2** purpose (*we eat that we may live*). **3** result (*am so sleepy that I cannot work*). □ **all that** very (*not all that good*). **that is** (or **that is to say**) formula introducing or following an explanation of a preceding word or words. **that's that** formula indicating conclusion or completion. [Old English]

■ **Usage** In sense 4 of the pronoun, *that* usually specifies or identifies something referred to, whereas *who* or *which* need not: compare *the book that you sent me is lost* with *the book, which I gave you, is lost*. *That* is often omitted in senses 1 and 3 of the conjunction: *they say he is ill*.

thatch *–n.* **1** roof-covering of straw, reeds, etc. **2** *colloq.* hair of the head. *–v.* (also *absol.*) cover with thatch. □ **thatcher** *n.* [Old English]

thaw *–v.* **1** (often foll. by *out*) pass from a frozen into a liquid or unfrozen state. **2** (usu. prec. by *it* as subject) (of the weather) become warm enough to melt ice etc. **3** become warm enough to lose numbness etc. **4** become or make genial. **5** (often foll. by *out*) cause to thaw. *–n.* **1** thawing. **2** warmth of weather that thaws. [Old English]

the /before a vowel ðɪ, before a consonant ðə, when stressed ðiː/ *–adj.* (called the definite article) **1** denoting person(s) or thing(s) already mentioned, under discussion, implied, or familiar (*gave the man a wave*). **2** describing as unique (*the Thames*). **3 a** (foll. by defining adj.) which is, who are, etc. (*Edward the Seventh*). **b** (foll. by adj. used *absol.*) denoting a class described (*from the sublime to the ridiculous*). **4** best known or best entitled to the name (with *the* stressed: *do you mean* the *Kipling?*). **5** indicating a following defining clause or phrase (*the book that you borrowed*). **6 a** indicating that a singular noun represents a species etc. (*the cat is a mammal*). **b** used with a noun which figuratively represents an occupation etc. (*went on the stage*). **c** (foll. by the name of a unit) a, per (*5p in the pound*). *–adv.* (preceding comparatives in expressions of proportional variation) in or by that (or such a) degree; on that account (*the more the merrier; the more he has the more he wants*). [Old English]

theatre /ˈθɪətə(r)/ *n.* (*US* **theater**) **1** building or outdoor area for dramatic performances. **2** writing and production of plays. **3** room or hall for lectures etc. with seats in tiers. **4** operating theatre. **5 a** scene or field of action (*the theatre of war*). **b** (*attrib.*) designating weapons intermediate between tactical and strategic. [Greek *theatron*]

theatrical /θɪˈætrɪk(ə)l/ *–adj.* **1** of or for the theatre or acting. **2** (of a manner or person etc.) calculated for effect; showy. *–n.* (in *pl.*) dramatic performances (*amateur theatricals*). □ **theatricality** /-ˈkælɪtɪ/ *n.* **theatrically** *adv.*

thee *objective case* of THOU[1].

theft *n.* act of stealing. [Old English: related to THIEF]

their /ðeə(r)/ *poss. pron.* (*attrib.*) of or belonging to them. [Old Norse]

theirs /ðeəz/ *poss. pron.* the one or ones of or belonging to them (*it is theirs; theirs are over here*). □ **of theirs** of or belonging to them (*a friend of theirs*).

theism /ˈθiːɪz(ə)m/ *n.* belief in gods or a god, esp. a god supernaturally revealed to man. □ **theist** *n.* **theistic** /-ˈɪstɪk/ *adj.* [Greek *theos* god]

them /ð(ə)m, or, when stressed, ðem/ *–pron.* **1** *objective case* of THEY. **2** *colloq.* they (*it's them again*). *–demons. adj. slang* or *dial.* those. [Old Norse]

theme *n.* **1** subject or topic of a talk, book, etc. **2** *Mus.* prominent melody in a composition. **3** *US* school exercise on a given subject. □ **thematic** /θɪˈmætɪk/ *adj.* **thematically** /θɪˈmætɪkəlɪ/ *adv.* [Greek *thema -mat-*]

theme park *n.* amusement park organized round a unifying idea.

theme song *n.* (also **theme tune**) **1** recurrent melody in a musical play or film. **2** signature tune.

themselves /ðəmˈselvz/ *pron.* **1** *emphat. form* of THEY or THEM. **2** *refl. form* of THEM. □ **be themselves** act in their normal, unconstrained manner. **by themselves** see *by oneself*.

then /ðen/ *–adv.* **1** at that time. **2 a** next; after that. **b** and also. **3 a** in that case (*then you should have said so*). **b** implying grudging or impatient concession (*all right then, if you must*). **c** used parenthetically to resume a narrative etc. (*the policeman, then, knocked on the door*). —*attrib. adj.* such at the time in question (*the then King*). —*n.* that time (*until then*). □ **then and there** immediately and on the spot. [Old English]

thence /ðens/ *adv.* (also **from thence**) *archaic* or *literary* **1** from that place. **2** for that reason. [Old English]

thenceforth /ðens'fɔ:θ/ *adv.* (also **thenceforward** /-'fɔ:wəd/) *archaic* or *literary* from that time onward.

theo- *comb. form* God or god(s). [Greek *theos* god]

theocracy /θɪ'ɒkrəsɪ/ *n.* (*pl.* **-ies**) form of government by God or a god directly, or through a priestly order etc. □ **theocratic** /θɪə'krætɪk/ *adj.*

theodolite /θɪ'ɒdə,laɪt/ *n.* surveying-instrument for measuring horizontal and vertical angles with a rotating telescope. [origin unknown]

theologian /θɪə'ləʊdʒɪən, -dʒ(ə)n/ *n.* expert in theology. [French: related to THEOLOGY]

theology /θɪ'ɒlədʒɪ/ *n.* (*pl.* **-ies**) the study or a system of theistic (esp. Christian) religion. □ **theological** /θɪə'lɒdʒɪk(ə)l/ *adj.* **theologically** /θɪə'lɒdʒɪkəlɪ/ *adv.* [Greek: related to THEO-]

theorem /'θɪərəm/ *n.* esp. *Math.* **1** general proposition that is not self-evident but is proved by reasoning. **2** rule in algebra etc., esp. one expressed by symbols or formulae. [Greek *theōreō* look at]

theoretical /θɪə'retɪk(ə)l/ *adj.* **1** concerned with knowledge but not with its practical application. **2** based on theory rather than experience. □ **theoretically** *adv.*

theoretician /,θɪərə'tɪʃ(ə)n/ *n.* person concerned with the theoretical aspects of a subject.

theorist /'θɪərɪst/ *n.* holder or inventor of a theory.

theorize /'θɪəraɪz/ *v.* (also **-ise**) (**-zing** or **-sing**) evolve or indulge in theories.

theory /'θɪərɪ/ *n.* (*pl.* **-ies**) **1** supposition or system of ideas explaining something, esp. one based on general principles independent of the particular things to be explained (*atomic theory; theory of evolution*). **2** speculative (esp. fanciful) view (*one of my pet theories*). **3** abstract knowledge or speculative thought (*all very well in theory*). **4** exposition of the principles of a science etc. (*the theory of music*). **5** collection of propositions to illustrate the principles of a mathematical subject (*probability theory*). [Greek: related to THEOREM]

theosophy /θɪ'ɒsəfɪ/ *n.* (*pl.* **-ies**) any of various philosophies professing to achieve knowledge of God by spiritual ecstasy, direct intuition, or special individual relations, esp. a modern movement following Hindu and Buddhist teachings and seeking universal brotherhood. □ **theosophical** /θɪə'sɒfɪk(ə)l/ *adj.* **theosophist** *n.* [Greek *theosophos* wise concerning God]

therapeutic /,θerə'pju:tɪk/ *adj.* **1** of, for, or contributing to, the cure of disease. **2** soothing, conducive to well-being. □ **therapeutically** *adv.* [Greek *therapeuō* wait on, cure]

therapeutics *n.pl.* (usu. treated as *sing.*) branch of medicine concerned with cures and remedies.

therapy /'θerəpɪ/ *n.* (*pl.* **-ies**) non-surgical treatment of disease or disability. □ **therapist** *n.* [Greek *therapeia* healing]

there /ðeə(r)/ —*adv.* **1** in, at, or to that place or position (*lived there for a year; goes there daily*). **2** at that point (in speech, performance, writing, etc.). **3** in that respect (*I agree with you there*). **4** used for emphasis in calling attention (*you there!*). **5** used to indicate the fact or existence of something (*there is a house on the corner*). —*n.* that place (*lives near there*). —*int.* **1** expressing confirmation, triumph, etc. (*there! what did I tell you?*). **2** used to soothe a child etc. (*there, there, never mind*). □ **there and then** = *then and there*. [Old English]

thereabouts *adv.* (also **thereabout**) **1** near that place. **2** near that number, quantity, etc.

thereafter *adv. formal* after that.

thereby *adv.* by that means, as a result of that. □ **thereby hangs a tale** much could be said about that.

therefore /'ðeəfɔ:(r)/ *adv.* for that reason; accordingly, consequently.

therein *adv. formal* **1** in that place etc. **2** in that respect.

thereof *adv. formal* of that or it.

thereto *adv. formal* **1** to that or it. **2** in addition.

thereupon *adv.* **1** in consequence of that. **2** immediately after that.

therm *n.* unit of heat, esp. as the statutory unit of gas supplied, equivalent to 100,000 British thermal units (1.055 x 10^8 joules). [Greek *thermē* heat]

thermal /'θɜ:m(ə)l/ —*adj.* **1** of, for, or producing heat. **2** promoting the retention of heat (*thermal underwear*). —*n.* rising current of warm air (used by gliders etc. to gain height). □ **thermally** *adv.* [French: related to THERM]

thermal unit *n.* unit for measuring heat.

thermionic /ˌθɜ:mɪ'ɒnɪk/ *adj.* of electrons emitted from a very hot substance. [from THERMO-, ION]

thermionic valve *n.* device giving a flow of thermionic electrons in one direction, used esp. in the rectification of a current and in radio reception.

thermo- *comb. form* heat. [Greek]

thermocouple /'θɜ:məʊˌkʌp(ə)l/ *n.* device for measuring temperatures by means of a pair of different metals in contact at a point and generating a thermoelectric voltage.

thermodynamics /ˌθɜ:məʊdaɪ'næmɪks/ *n.pl.* (usu. treated as *sing.*) science of the relations between heat and other forms of energy. □ **thermodynamic** *adj.*

thermoelectric /ˌθɜ:məʊɪ'lektrɪk/ *adj.* producing electricity by a difference of temperatures.

thermometer /θə'mɒmɪtə(r)/ *n.* instrument for measuring temperature, esp. a graduated glass tube containing mercury or alcohol. [French: related to THERMO-, -METER]

thermonuclear /ˌθɜ:məʊ'nju:klɪə(r)/ *adj.* **1** relating to nuclear reactions that occur only at very high temperatures. **2** (of weapons) using thermonuclear reactions.

thermoplastic /ˌθɜ:məʊ'plæstɪk/ *–adj.* that becomes plastic on heating and hardens on cooling. *–n.* thermoplastic substance.

Thermos /'θɜ:məs/ *n.* (in full **Thermos flask**) *propr.* vacuum flask. [Greek: related to THERMO-]

thermosetting /'θɜ:məʊˌsetɪŋ/ *adj.* (of plastics) setting permanently when heated.

thermosphere /'θɜ:məˌsfɪə(r)/ *n.* region of the atmosphere beyond the mesosphere.

thermostat /'θɜ:məˌstæt/ *n.* device that automatically regulates or responds to temperature. □ **thermostatic** /-'stætɪk/ *adj.* **thermostatically** /-'stætɪkəlɪ/ *adv.* [from THERMO-, Greek *statos* standing]

thesaurus /θɪ'sɔ:rəs/ *n.* (*pl.* **-ri** /-raɪ/ or **-ruses**) book that lists words in groups of synonyms and related concepts. [Greek: related to TREASURE]

these *pl.* of THIS.

thesis /'θi:sɪs/ *n.* (*pl.* **theses** /-si:z/) **1** proposition to be maintained or proved. **2** dissertation, esp. by a candidate for a higher degree. [Greek, = putting]

Thespian /'θespɪən/ *–adj.* of drama. *–n.* actor or actress. [Greek *Thespis*, name of a Greek tragedian]

theta /'θi:tə/ *n.* eighth letter of the Greek alphabet (Θ, θ). [Greek]

they /ðeɪ/ *pron.* (*obj.* **them**; *poss.* **their, theirs**) **1** *pl.* of HE, SHE, IT. **2** people in general (*so they say*). **3** those in authority (*they have raised taxes*). [Old Norse]

they'd /ðeɪd/ *contr.* **1** they had. **2** they would.

they'll /'ðeɪəl, ðeəl/ *contr.* **1** they will. **2** they shall.

they're /ðeə(r), ðe(r),/ *contr.* they are.

they've /ðeɪv/ *contr.* they have.

thiamine /'θaɪəmɪn, -ˌmi:n/ *n.* (also **thiamin**) B vitamin found in unrefined cereals, beans, and liver, a deficiency of which causes beriberi. [Greek *theion* sulphur, *amin* from VITAMIN]

thick *–adj.* **1** of great or specified extent between opposite surfaces. **2** (of a line etc.) broad; not fine. **3** arranged closely; crowded together; dense. **4** (usu. foll. by *with*) densely covered or filled (*air thick with smoke*). **5 a** firm in consistency; containing much solid matter. **b** made of thick material (*a thick coat*). **6 a** muddy, cloudy; impenetrable by sight. **b** (of one's head) suffering from a hangover, headache, etc. **7** *colloq.* stupid. **8 a** (of a voice) indistinct. **b** (of an accent) very marked. **9** *colloq.* intimate, very friendly. *–n.* thick part of anything. *–adv.* thickly (*snow was falling thick*). □ **a bit thick** *colloq.* unreasonable or intolerable. **in the thick of** at the busiest part of. **through thick and thin** under all conditions; in spite of all difficulties. □ **thickish** *adj.* **thickly** *adv.* [Old English]

thicken *v.* **1** make or become thick or thicker. **2** become more complicated (*plot thickens*). □ **thickener** *n.*

thickening *n.* **1** becoming thick or thicker. **2** substance used to thicken liquid. **3** thickened part.

thicket /'θɪkɪt/ *n.* tangle of shrubs or trees. [Old English: related to THICK]

thickhead *n. colloq.* stupid person. □ **thickheaded** *adj.*

thickness *n.* **1** being thick. **2** extent of this. **3** layer of material (*use three thicknesses*).

thickset *adj.* **1** heavily or solidly built. **2** set or growing close together.

thick-skinned *adj.* not sensitive to criticism.

thief *n.* (*pl.* **thieves** /θi:vz/) person who steals, esp. secretly. [Old English]

thieve *v.* (**-ving**) **1** be a thief. **2** steal (a thing). [Old English: related to THIEF]

thievery *n.* stealing.

thievish *adj.* given to stealing.

thigh /θaɪ/ *n.* part of the leg between the hip and the knee. [Old English]

thigh-bone *n.* = FEMUR.

thimble *n.* metal or plastic cap worn to protect the finger and push the needle in sewing. [Old English: related to THUMB]

thimbleful *n.* (*pl.* **-s**) small quantity, esp. of drink.

thin *–adj.* (**thinner**, **thinnest**) **1** having opposite surfaces close together; of small thickness or diameter. **2** (of a line) narrow or fine. **3** made of thin material (*thin dress*). **4** lean; not plump. **5** not dense or copious (*thin hair*). **6** of slight consistency. **7** weak; lacking an important ingredient (*thin blood; a thin voice*). **8** (of an excuse etc.) flimsy or transparent. *–adv.* thinly (*cut the bread very thin*). *–v.* (**-nn-**) **1** (often foll. by *down*) make or become thin or thinner. **2** (often foll. by *out*) make or become less dense or crowded or numerous. □ **have a thin time** *colloq.* have a wretched or uncomfortable time. **thin on the ground** few in number. **thin on top** balding. □ **thinly** *adv.* **thinness** *n.* **thinnish** *adj.* [Old English]

thine /ðaɪn/ *poss. pron. archaic* **1** (*predic.* or *absol.*) of or belonging to thee. **2** (*attrib.* before a vowel) = THY. [Old English]

thin end of the wedge see WEDGE.

thing *n.* **1** entity, idea, action, etc., that exists or may be thought about or perceived. **2** inanimate material object (*take that thing away*). **3** unspecified item (*a few things to buy*). **4** act, idea, or utterance (*silly thing to do*). **5** event (*unfortunate thing to happen*). **6** quality (*patience is a useful thing*). **7** person regarded with pity, contempt, or affection (*poor thing!*). **8** specimen or type (*latest thing in hats*). **9** *colloq.* one's special interest (*not my thing*). **10** *colloq.* something remarkable (*there's a thing!*). **11** (prec. by *the*) *colloq.* **a** what is proper or fashionable. **b** what is needed (*just the thing*). **c** what is to be considered (*the thing is, shall we go or not?*). **d** what is important. **12** (in *pl.*) personal belongings or clothing (*where are my things?*). **13** (in *pl.*) equipment (*painting things*). **14** (in *pl.*) affairs in general (*not in the nature of things*). **15** (in *pl.*) circumstances, conditions (*things look good*). **16** (in *pl.* with a following adjective) all that is so describable (*things Greek*). □ **do one's own thing** *colloq.* pursue one's own interests or inclinations. **have a thing about** *colloq.* be obsessed or prejudiced about. **make a thing of** *colloq.* **1** regard as essential. **2** cause a fuss about. [Old English]

thingummy /ˈθɪŋəmɪ/ *n.* (*pl.* **-ies**) (also **thingumabob** /-məˌbɒb/, **thingumajig** /-məˌdʒɪg/) *colloq.* person or thing whose name one has forgotten or does not know.

think *–v.* (*past* and *past part.* **thought** /θɔːt/) **1** be of the opinion (*think that they will come*). **2** judge or consider (*is thought to be a fraud*). **3** exercise the mind (*let me think for a moment*). **4** (foll. by *of* or *about*) **a** consider; be or become aware of. **b** form or entertain the idea of; imagine. **5** have a half-formed intention (*I think I'll stay*). **6** form a conception of. **7** recognize the presence or existence of (*thought no harm in it*). *–n. colloq.* act of thinking (*have a think*). □ **think again** revise one's plans or opinions. **think aloud** utter one's thoughts as soon as they occur. **think better of** change one's mind about (an intention) after reconsideration. **think fit** see FIT[1]. **think little** (or **nothing**) **of** consider to be insignificant. **think much** (or **a lot** or **highly**) **of** have a high opinion of. **think out 1** consider carefully. **2** produce (an idea etc.) by thinking. **think over** reflect upon in order to reach a decision. **think through** reflect fully upon (a problem etc.). **think twice** use careful consideration, avoid hasty action, etc. **think up** *colloq.* devise. [Old English]

thinker *n.* **1** person who thinks, esp. in a specified way (*an original thinker*). **2** person with a skilled or powerful mind.

thinking *–attrib. adj.* intelligent, rational. *–n.* opinion, judgement.

think-tank *n. colloq.* body of experts providing advice and ideas on national or commercial problems.

thinner *n.* solvent for diluting paint etc.

thin-skinned *adj.* sensitive to criticism.

thiosulphate /ˌθaɪəʊˈsʌlfeɪt/ *n.* sulphate in which one oxygen atom is replaced by sulphur. [Greek *theion* sulphur]

third *adj.* & *n.* **1** next after second. **2** each of three equal parts of a thing. □ **thirdly** *adv.* [Old English: related to THREE]

third degree *–n.* long and severe questioning, esp. by police to obtain information or a confession. *–adj.* (**third-degree**) denoting burns of the most severe kind, affecting lower layers of tissue.

third man *n.* fielder positioned near the boundary behind the slips.

third party *–n.* **1** another party besides the two principals. **2** bystander etc. *–adj.* (**third-party**) (of insurance) covering damage or injury suffered by a person other than the insured.

third person *n.* **1** = THIRD PARTY. **2** *Gram.* see PERSON.

third-rate *adj.* inferior; very poor.

third reading *n.* third presentation of a bill to a legislative assembly.
Third Reich *n.* Nazi regime, 1933–45.
Third World *n.* (usu. prec. by *the*) developing countries of Asia, Africa, and Latin America.
thirst –*n.* **1** need to drink; discomfort caused by this. **2** desire, craving. –*v.* (often foll. by *for* or *after*) **1** feel thirst. **2** have a strong desire. [Old English]
thirsty *adj.* (**-ier, -iest**) **1** feeling thirst. **2** (of land, a season, etc.) dry or parched. **3** (often foll. by *for* or *after*) eager. **4** *colloq.* causing thirst (*thirsty work*). □ **thirstily** *adv.* **thirstiness** *n.* [Old English: related to THIRST]
thirteen /θɜːˈtiːn/ *adj.* & *n.* **1** one more than twelve. **2** symbol for this (13, xiii, XIII). **3** size etc. denoted by thirteen. □ **thirteenth** *adj.* & *n.* [Old English: related to THREE]
thirty /ˈθɜːtɪ/ *adj.* & *n.* (*pl.* **-ies**) **1** three times ten. **2** symbol for this (30, xxx, XXX). **3** (in *pl.*) numbers from 30 to 39, esp. the years of a century or of a person's life. □ **thirtieth** *adj.* & *n.* [Old English: related to THREE]
Thirty-nine Articles *n.pl.* points of doctrine assented to by those taking orders in the Church of England.
this /ðɪs/ –*demons. pron.* (*pl.* **these** /ðiːz/) **1** person or thing close at hand or indicated or already named or understood (*can you see this?*; *this is my cousin*). **2** (contrasted with *that*) the person or thing nearer to hand or more immediately in mind. –*demons. adj.* (*pl.* **these** /ðiːz/) **1** designating the person or thing close at hand etc. (cf. senses 1, 2 of *pron.*). **2** (of time) the present or current (*am busy all this week*). **3** *colloq.* (in narrative) designating a person or thing previously unmentioned (*then up came this policeman*). –*adv.* to the degree or extent indicated (*knew him when he was this high*). □ **this and that** *colloq.* various unspecified things. [Old English]
thistle /ˈθɪs(ə)l/ *n.* **1** prickly plant, usu. with globular heads of purple flowers. **2** this as the Scottish national emblem. [Old English]
thistledown *n.* light down containing thistle-seeds and blown about in the wind.
thistly *adj.* overgrown with thistles.
thither /ˈðɪðə(r)/ *adv. archaic* or *formal* to or towards that place. [Old English]
tho' (also **tho**) var. of THOUGH.
thole *n.* (in full **thole-pin**) **1** pin in the gunwale of a boat as the fulcrum for an oar. **2** each of two such pins forming a rowlock. [Old English]
thong *n.* narrow strip of hide or leather. [Old English]
thorax /ˈθɔːræks/ *n.* (*pl.* **-races** /-rəˌsiːz/ or **-raxes**) *Anat.* & *Zool.* part of the trunk between the neck and the abdomen. □ **thoracic** /θɔːˈræsɪk/ *adj.* [Latin from Greek]
thorium /ˈθɔːrɪəm/ *n. Chem.* radioactive metallic element. [*Thor*, name of Scandinavian god of thunder]
thorn *n.* **1** sharp-pointed projection on a plant. **2** thorn-bearing shrub or tree. □ **thorn in one's flesh** (or **side**) constant nuisance. □ **thornless** *adj.* [Old English]
thorny *adj.* (**-ier, -iest**) **1** having many thorns. **2** problematic, causing disagreement. □ **thornily** *adv.* **thorniness** *n.* [Old English: related to THORN]
thorough /ˈθʌrə/ *adj.* **1** complete and unqualified; not superficial. **2** acting or done with great care and completeness. **3** absolute (*thorough nuisance*). □ **thoroughly** *adv.* **thoroughness** *n.* [related to THROUGH]
thoroughbred –*adj.* **1** of pure breed. **2** high-spirited. –*n.* thoroughbred animal, esp. a horse.
thoroughfare *n.* road or path open at both ends, esp. for traffic.
thoroughgoing *attrib. adj.* thorough; complete.
those *pl.* of THAT.
thou[1] /ðaʊ/ *pron.* (*obj.* **thee** /ðiː/; *poss.* **thy** or **thine**; *pl.* **ye** or **you**) *archaic* second person singular pronoun. [Old English]

■ **Usage** *Thou* has now been replaced by *you* except in some formal, liturgical, dialect, and poetic uses.

thou[2] /θaʊ/ *n.* (*pl.* same or **-s**) *colloq.* **1** thousand. **2** one thousandth. [abbreviation]
though /ðəʊ/ (also **tho'**) –*conj.* **1** despite the fact that; in spite of being (*though it was early we left*; *though annoyed, I agreed*). **2** (introducing a possibility) even if (*ask him though he may refuse*). **3** and yet; nevertheless. –*adv. colloq.* however; all the same. [Old Norse]
thought[1] /θɔːt/ *n.* **1** process or power of thinking; faculty of reason. **2** way of thinking associated with a particular time, group, etc. **3** sober reflection or consideration. **4** idea or piece of reasoning produced by thinking. **5** (foll. by *of* + verbal noun or *to* + infin.) partly formed intention (*had no thought to go*). **6** (usu. in *pl.*) what one is thinking; one's opinion. **7** (prec. by *a*) somewhat (*a thought arrogant*). □ **in thought** meditating. [Old English: related to THINK]

thought[2] *past* and *past part.* of THINK.

thoughtful *adj.* **1** engaged in or given to meditation. **2** (of a book, writer, etc.) giving signs of serious thought. **3** (often foll. by *of*) (of a person or conduct) considerate. □ **thoughtfully** *adv.* **thoughtfulness** *n.*

thoughtless *adj.* **1** careless of consequences or of others' feelings. **2** due to lack of thought. □ **thoughtlessly** *adv.* **thoughtlessness** *n.*

thought-reader *n.* person supposedly able to perceive another's thoughts.

thousand /'θaʊz(ə)nd/ *adj.* & *n.* (*pl.* **thousands** or (in sense 1) **thousand**) (in *sing.* prec. by *a* or *one*) **1** ten hundred. **2** symbol for this (1,000, m, M). **3** (in *sing.* or *pl.*) *colloq.* large number. □ **thousandfold** *adj.* & *adv.* **thousandth** *adj.* & *n.* [Old English]

thrall /θrɔːl/ *n. literary* **1** (often foll. by *of*, *to*) slave (of a person, or of a power or influence). **2** slavery (*in thrall*). □ **thraldom** *n.* [Old English from Old Norse]

thrash –*v.* **1** beat or whip severely. **2** defeat thoroughly. **3** deliver repeated blows. **4** (foll. by *about*, *around*) move or fling (esp. the limbs) about violently. **5** = THRESH 1. –*n* **1** act of thrashing. **2** *slang* (esp. lavish) party. □ **thrash out** discuss to a conclusion. [Old English]

thread /θred/ –*n.* **1 a** spun-out cotton, silk, or glass etc.; yarn. **b** length of this. **2** thin cord of twisted yarns used esp. in sewing and weaving. **3** continuous aspect of a thing (*the thread of life*; *thread of his argument*). **4** spiral ridge of a screw. –*v.* **1** pass a thread through (a needle). **2** put (beads) on a thread. **3** insert (a strip of material, e.g. film or magnetic tape) into equipment. **4** make (one's way) carefully through a crowded place, over a difficult route, etc. [Old English: related to THROW]

threadbare *adj.* **1** (of cloth) with the nap worn away and the thread visible. **2** (of a person) wearing such clothes. **3** hackneyed.

threadworm *n.* parasitic threadlike worm.

threat /θret/ *n.* **1** declaration of an intention to punish or hurt if an order etc. is not obeyed. **2** indication of something undesirable coming (*threat of war*). **3** person or thing as a likely cause of harm etc. [Old English]

threaten *v.* **1** make a threat or threats against. **2** be a sign of (something undesirable). **3** (foll. by *to* + infin.) announce one's intention to do an undesirable thing. **4** (also *absol.*) warn of the infliction of (harm etc.). **5** (as **threatened** *adj.*) (of a species etc.) likely to become extinct. [Old English]

three *adj.* & *n.* **1 a** one more than two. **b** symbol for this (3, iii, III). **2** size etc. denoted by three. [Old English]

three-cornered *adj.* **1** triangular. **2** (of a contest etc.) between three parties.

three-decker *n.* **1** warship with three gun-decks. **2** thing with three levels or divisions.

three-dimensional *adj.* having or appearing to have length, breadth, and depth.

threefold *adj.* & *adv.* **1** three times as much or as many. **2** consisting of three parts.

three-legged race *n.* running-race between pairs, one member of each pair having the left leg tied to the right leg of the other.

three-line whip *n.* written notice to MPs from their leader insisting on attendance at a debate and voting a certain way.

threepence /'θrepəns, 'θrʊp-/ *n.* sum of three pence.

threepenny /'θrepənɪ, 'θrʊp-/ *attrib. adj.* costing three pence.

three-piece –*n.* three-piece suit or suite. –*attrib. adj.* (esp. of a suit or suite) consisting of three items.

three-ply –*adj.* of three strands or layers etc. –*n.* **1** three-ply wool. **2** three-ply wood.

three-point turn *n.* method of turning a vehicle round in a narrow space by moving forwards, backwards, and forwards again.

three-quarter *n.* (also **three-quarter back**) *Rugby* any of three or four players just behind the half-backs.

three-quarters *n.pl.* three parts out of four.

three Rs *n.pl.* (prec. by *the*) reading, writing, and arithmetic.

threescore *n.* & *adj. archaic* sixty.

threesome *n.* group of three persons.

three-way *adj.* involving three directions or participants.

threnody /'θrenədɪ/ *n.* (*pl.* **-ies**) song of lamentation or mourning. [Greek]

thresh *v.* **1** beat out or separate grain from (corn etc.). **2** = THRASH *v.* 4. □ **thresher** *n.* [Old English]

threshing-floor *n.* hard level floor for threshing esp. with flails.

threshold /'θreʃəʊld/ *n.* **1** strip of wood or stone forming the bottom of a doorway and crossed in entering a house etc. **2** point of entry or beginning. **3** limit below which a stimulus causes no reaction. [Old English: related to THRASH in the sense 'tread']

threw *past* of THROW.

thrice *adv. archaic* or *literary* **1** three times. **2** (esp. in *comb.*) highly (*thrice-blessed*). [related to THREE]

thrift *n.* **1** frugality; careful use of money etc. **2** the sea pink. [Old Norse: related to THRIVE]

thriftless *adj.* wasteful.

thrifty *adj.* (**-ier, -iest**) economical. □ **thriftily** *adv.* **thriftiness** *n.*

thrill *–n.* **1** wave or nervous tremor of emotion or sensation (*a thrill of joy*). **2** throb, pulsation. *–v.* **1** (cause to) feel a thrill. **2** quiver or throb with or as with emotion. [Old English, = pierce: related to THROUGH]

thriller *n.* exciting or sensational story or play etc., esp. about crime or espionage.

thrips *n.* (*pl.* same) an insect harmful to plants. [Greek, = woodworm]

thrive *v.* (**-ving**; *past* **throve** or **thrived**; *past part.* **thriven** /'θrɪv(ə)n/ or **thrived**) **1** prosper, flourish. **2** grow rich. **3** (of a child, animal, or plant) grow vigorously. [Old Norse]

thro' var. of THROUGH.

throat *n.* **1 a** windpipe or gullet. **b** front part of the neck containing this. **2** *literary* narrow passage, entrance, or exit. □ **cut one's own throat** harm oneself or one's interests. **ram** (or **thrust**) **down a person's throat** force on a person's attention. [Old English]

throaty *adj.* (**-ier, -iest**) (of a voice) hoarsely resonant. □ **throatily** *adv.* **throatiness** *n.*

throb *–v.* (**-bb-**) **1** pulsate, esp. with more than the usual force or rapidity. **2** vibrate with a persistent rhythm or with emotion. *–n.* **1** throbbing. **2** (esp. violent) pulsation. [imitative]

throe *n.* (usu. in *pl.*) violent pang, esp. of childbirth or death. □ **in the throes of** struggling with the task of. [Old English, alteration of original *throwe*, perhaps by association with *woe*]

thrombosis /θrɒm'bəʊsɪs/ *n.* (*pl.* **-boses** /-si:z/) coagulation of the blood in a blood-vessel or organ. [Greek, = curdling]

throne *–n.* **1** chair of State for a sovereign or bishop etc. **2** sovereign power (*came to the throne*). *–v.* (**-ning**) enthrone. [Greek *thronos*]

throng *–n.* (often foll. by *of*) crowd, esp. of people. *–v.* **1** come in great numbers (*crowds thronged to the stadium*). **2** flock into or crowd round; fill with or as with a crowd. [Old English]

throstle /'θrɒs(ə)l/ *n.* song thrush. [Old English]

throttle /'θrɒt(ə)l/ *–n.* **1 a** valve controlling the flow of fuel or steam etc. in an engine. **b** (in full **throttle-lever**) lever or pedal operating this valve. **2** throat, gullet, or windpipe. *–v.* (**-ling**) **1** choke or strangle. **2** prevent the utterance etc. of. **3** control (an engine or steam etc.) with a throttle. □ **throttle back** (or **down**) reduce the speed of (an engine or vehicle) by throttling. [perhaps from THROAT]

through /θru:/ (also **thro'**, *US* **thru**) *–prep.* **1 a** from end to end or side to side of. **b** going in one side or end and out the other of. **2** between or among (*swam through the waves*). **3** from beginning to end of (*read through the letter; went through many difficulties*). **4** because of; by the agency, means, or fault of (*lost it through carelessness*). **5** *US* up to and including (*Monday through Friday*). *–adv.* **1** through a thing; from side to side, end to end, or beginning to end. **2** so as to be connected by telephone (*will put you through*). *–attrib. adj.* **1** (of a journey, route, etc.) done without a change of line or vehicle etc. or with one ticket. **2** (of traffic) going through a place to its destination. **3** (of a road) open at both ends. □ **be through** *colloq.* **1** (often foll. by *with*) have finished. **2** (often foll. by *with*) cease to have dealings. **3** have no further prospects. **through and through** thoroughly, completely. [Old English]

throughout /θru:'aʊt/ *–prep.* right through; from end to end of. *–adv.* in every part or respect.

throughput *n.* amount of material put through a process, esp. in manufacturing or computing.

throve *past* of THRIVE.

throw /θrəʊ/ *–v.* (*past* **threw** /θru:/; *past part.* **thrown**) **1** propel with force through the air. **2** force violently into, or compel to be in, a specified position or state (*thrown on the rocks; threw themselves down; thrown out of work*). **3** turn or move (part of the body) quickly or suddenly (*threw an arm out*). **4** project or cast (light, a shadow, etc.). **5 a** bring to the ground in wrestling. **b** (of a horse) unseat (its rider). **6** *colloq.* disconcert (*the question threw me*). **7** (foll. by *on, off*, etc.) put (clothes etc.) hastily on or off etc. **8 a** cause (dice) to fall on a table etc. **b** obtain (a specified number) by throwing dice. **9** cause to pass or extend suddenly to another state or position (*threw a bridge across the river*). **10** operate (a switch or lever). **11** form on a potter's wheel. **12** have (a fit or tantrum etc.). **13** give (a party). *–n.* **1** act of throwing or being thrown. **2** distance a thing is or may be thrown. **3** (prec. by *a*) *slang* each; per item (*sold at £10 a throw*). □ **throw away 1** discard as useless or unwanted. **2** waste or fail to make use of (an opportunity etc.). **throw back 1** revert to ancestral character. **2** (usu. in *passive*; foll. by *on*) compel to rely on. **throw in 1** interpose (a word or remark). **2** include at no extra cost. **3** throw (a football) from the edge of the pitch where it has gone out of play. **throw in the towel** (or **sponge**) admit defeat. **throw off 1** discard; contrive to get rid of. **2** write or utter in an offhand manner. **throw oneself at** seek blatantly as a sexual partner. **throw oneself into** engage vigorously in. **throw oneself on** (or **upon**) rely completely on. **throw open** (often foll. by *to*) **1** cause to be suddenly or widely open. **2** make accessible. **throw out 1** put out forcibly or suddenly. **2** discard as unwanted. **3** reject (a proposal). **throw over** desert, abandon. **throw together 1** assemble hastily. **2** bring into casual contact. **throw up 1** abandon. **2** resign from. **3** *colloq.* vomit. **4** erect hastily. **5** bring to notice. [Old English, = twist]

throw-away *attrib. adj.* **1** meant to be thrown away after (one) use. **2** spoken in a deliberately casual way. **3** disposed to throwing things away (*throw-away society*).

throwback *n.* **1** reversion to ancestral character. **2** instance of this.

throw-in *n.* throwing in of a football during play.

thrown *past part.* of THROW.

thru *US* var. of THROUGH.

thrum[1] *–v.* (**-mm-**) **1** play (a stringed instrument) monotonously or unskilfully. **2** (often foll. by *on*) drum idly. *–n.* **1** such playing. **2** resulting sound. [imitative]

thrum[2] *n.* **1** unwoven end of a warp-thread, or the whole of such ends, left when the finished web is cut away. **2** any short loose thread. [Old English]

thrush[1] *n.* any of various songbirds, esp. the song thrush and mistle thrush. [Old English]

thrush[2] *n.* **1** fungous disease, esp. of children, affecting the mouth and throat. **2** similar disease of the vagina. [origin unknown]

thrust *–v.* (*past* and *past part.* **thrust**) **1** push with a sudden impulse or with force. **2** (foll. by *on*) impose (a thing) forcibly; enforce acceptance of (a thing). **3** (foll. by *at, through*) pierce, stab; lunge suddenly. **4** make (one's way) forcibly. **5** (as **thrusting** *adj.*) aggressive, ambitious. *–n.* **1** sudden or forcible push or lunge. **2** propulsive force produced by a jet or rocket engine. **3** strong attempt to penetrate an enemy's line or territory. **4** remark aimed at a person. **5** stress between the parts of an arch etc. **6** (often foll. by *of*) chief theme or gist of remarks etc. [Old Norse]

thud *–n.* low dull sound as of a blow on a non-resonant surface. *–v.* (**-dd-**) make or fall with a thud. [probably Old English]

thug *n.* **1** violent ruffian. **2** (**Thug**) *hist.* member of a religious organization of robbers and assassins in India. □ **thuggery** *n.* **thuggish** *adj.* [Hindi]

thulium /ˈθjuːlɪəm/ *n.* metallic element of the lanthanide series. [Latin *Thule*, name of a region in the remote north]

thumb /θʌm/ *–n.* **1** short thicker finger on the human hand, set apart from the other four. **2** part of a glove etc. for a thumb. *–v.* **1** wear or soil (pages etc.) with a thumb. **2** turn over pages with or as with a thumb (*thumbed through the directory*). **3** request or get (a lift) by signalling with a raised thumb. **4** use the thumb in a gesture. □ **thumb one's nose** = *cock a snook* (see SNOOK). **thumbs down** indication of rejection. **thumbs up** indication of satisfaction or approval. **under a person's thumb** completely dominated by a person. [Old English]

thumb index *n.* set of lettered grooves cut down the side of a book for easy reference.

thumbnail *n.* **1** nail of a thumb. **2** (*attrib.*) concise (*thumbnail sketch*).

thumbprint *n.* impression of a thumb esp. for identification.

thumbscrew *n.* instrument of torture for crushing the thumbs.

thump *–v.* **1** beat or strike heavily, esp. with the fist. **2** throb strongly. **3** (foll. by *at, on*, etc.) knock loudly. *–n.* **1** heavy blow. **2** dull sound of this. [imitative]

thumping *adj. colloq.* (esp. as an intensifier) huge (*a thumping lie; a thumping great house*).

thunder /ˈθʌndə(r)/ *–n.* **1** loud noise caused by lightning and due to the expansion of rapidly heated air. **2** resounding loud deep noise (*thunders of applause*). **3** strong censure or denunciation. *–v.* **1** (prec. by *it* as subject) thunder sounds (*it is thundering; if it thunders*). **2** make or proceed with a noise like thunder. **3** utter (threats, compliments, etc.) loudly. **4** (foll. by *against* etc.) make violent threats etc. against. □ **steal a person's thunder** see STEAL. □ **thundery** *adj.* [Old English]

thunderbolt *n.* **1** flash of lightning with a simultaneous crash of thunder. **2** unexpected occurrence or announcement. **3** supposed bolt or shaft as a

destructive agent, esp. as an attribute of a god.
thunderclap *n.* **1** crash of thunder. **2** something startling or unexpected.
thundercloud *n.* cumulus cloud charged with electricity and producing thunder and lightning.
thunder-fly *n.* = THRIPS.
thundering *adj. colloq.* (esp. as an intensifier) huge (*a thundering nuisance; a thundering great bruise*).
thunderous *adj.* **1** like thunder. **2** very loud.
thunderstorm *n.* storm with thunder and lightning and usu. heavy rain or hail.
thunderstruck *predic. adj.* amazed.
Thur. *abbr.* (also **Thurs.**) Thursday.
thurible /'θjʊərɪb(ə)l/ *n.* censer. [Latin *thus thur-* incense]
Thursday /'θɜːzdeɪ/ *–n.* day of the week following Wednesday. *–adv. colloq.* **1** on Thursday. **2** (**Thursdays**) on Thursdays; each Thursday. [Old English]
thus /ðʌs/ *adv. formal* **1 a** in this way. **b** as indicated. **2 a** accordingly. **b** as a result or inference. **3** to this extent; so (*thus far*; *thus much*). [Old English]
thwack *–v.* hit with a heavy blow. *–n.* heavy blow. [imitative]
thwart /θwɔːt/ *–v.* frustrate or foil (a person, plan, etc.). *–n.* rower's seat. [Old Norse, = across]
thy /ðaɪ/ *poss. pron.* (*attrib.*) (also **thine** *predic.* or before a vowel) *archaic* of or belonging to thee. [from THINE]

■ **Usage** *Thy* has now been replaced by *your* except in some formal, liturgical, dialect, and poetic uses.

thyme /taɪm/ *n.* any of several herbs with aromatic leaves. [Greek *thumon*]
thymol /'θaɪmɒl/ *n.* antiseptic obtained from oil of thyme.
thymus /'θaɪməs/ *n.* (*pl.* **thymi** /-maɪ/) lymphoid organ situated in the neck of vertebrates. [Greek]
thyroid /'θaɪrɔɪd/ *n.* (in full **thyroid gland**) **1** large ductless gland in the neck of vertebrates, secreting a hormone which regulates growth and development. **2** extract prepared from the thyroid gland of animals and used in treating goitre etc. [Greek *thureos* oblong shield]
thyroid cartilage *n.* large cartilage of the larynx, forming the Adam's apple.
thyself *pron. archaic* **1** emphat. form of THOU[1], THEE. **2** refl. form of THEE.
Ti *symb.* titanium.
ti var. of TE.
tiara /tɪ'ɑːrə/ *n.* **1** jewelled ornamental band worn on the front of a woman's hair. **2** three-crowned diadem worn by a pope. □ **tiaraed** *adj.* [Latin from Greek]
tibia /'tɪbɪə/ *n.* (*pl.* **tibiae** /-bɪ,iː/) *Anat.* inner of two bones extending from the knee to the ankle. □ **tibial** *adj.* [Latin]
tic *n.* (in full **nervous tic**) occasional involuntary contraction of the muscles, esp. of the face. [French from Italian]
tick[1] *–n.* **1** slight recurring click, esp. that of a watch or clock. **2** *colloq.* moment. **3** mark (✓) to denote correctness, check items in a list, etc. *–v.* **1** (of a clock etc.) make ticks. **2 a** mark with a tick. **b** (often foll. by *off*) mark (an item) with a tick in checking. □ **tick off** *colloq.* reprimand. **tick over 1** (of an engine etc.) idle. **2** (of a person, project, etc.) be functioning at a basic level. **what makes a person tick** *colloq.* person's motivation. [probably imitative]
tick[2] *n.* **1** parasitic arachnid on the skin of dogs, cattle, etc. **2** parasitic insect on sheep and birds etc. [Old English]
tick[3] *n. colloq.* credit (*buy goods on tick*). [apparently an abbreviation of TICKET in *on the ticket*]
tick[4] *n.* **1** cover of a mattress or pillow. **2** = TICKING. [Greek *thēkē* case]
ticker *n. colloq.* **1** heart. **2** watch. **3** *US* = TAPE MACHINE 1.
ticker-tape *n.* **1** paper strip from a tape machine. **2** this or similar material thrown from windows etc. to greet a celebrity.
ticket /'tɪkɪt/ *–n.* **1** written or printed piece of paper or card entitling the holder to enter a place, participate in an event, travel by public transport, etc. **2** notification of a traffic offence etc. (*parking ticket*). **3** certificate of discharge from the army. **4** certificate of qualification as a ship's master, pilot, etc. **5** price etc. label. **6** esp. *US* **a** list of candidates put forward by one group, esp. a political party. **b** principles of a party. **7** (prec. by *the*) *colloq.* what is correct or needed. *–v.* (**-t-**) attach a ticket to. [obsolete French *étiquet*]
ticking *n.* stout usu. striped material used to cover mattresses etc. [from TICK[4]]
tickle /'tɪk(ə)l/ *–v.* (**-ling**) **1 a** touch or stroke (a person etc.) playfully or lightly so as to produce laughter and spasmodic movement. **b** produce this sensation. **2** excite agreeably; amuse. **3** catch (a trout etc.) by rubbing it so that it moves backwards into the hand. *–n.* **1** act of tickling. **2** tickling sensation. □ **tickled pink** (or **to death**) *colloq.* extremely amused or pleased. □ **tickly** *adj.* [probably frequentative of TICK[1]]
ticklish *adj.* **1** sensitive to tickling. **2** (of a matter or person) difficult to handle.

tick-tack *n.* a kind of manual semaphore used by racecourse bookmakers.

tick-tock *n.* ticking of a large clock etc.

tidal /ˈtaɪd(ə)l/ *adj.* relating to, like, or affected by tides. □ **tidally** *adv.*

tidal wave *n.* **1** exceptionally large ocean wave, esp. one caused by an underwater earthquake. **2** widespread manifestation of feeling etc.

tidbit *US* var. of TITBIT.

tiddler *n. colloq.* **1** small fish, esp. a stickleback or minnow. **2** unusually small thing. [perhaps related to TIDDLY[2] and *tittlebat*, a childish form of *stickleback*]

tiddly[1] *adj.* (**-ier**, **-iest**) *colloq.* slightly drunk. [origin unknown]

tiddly[2] *adj.* (**-ier**, **-iest**) *colloq.* little. [origin unknown]

tiddly-wink /ˈtɪdlɪwɪŋk/ *n.* **1** counter flicked with another into a cup etc. **2** (in *pl.*) this game. [perhaps related to TIDDLY[1]]

tide *n.* **1 a** periodic rise and fall of the sea due to the attraction of the moon and sun. **b** water as affected by this. **2** time or season (usu. in *comb.*: *Whitsuntide*). **3** marked trend of opinion, fortune, or events. □ **tide (-ding) over** provide (a person) with what is needed during a difficult period. [Old English, = TIME]

tidemark *n.* **1** mark made by the tide at high water. **2 a** line left round a bath by the dirty water. **b** *colloq.* line between washed and unwashed parts of a person's body.

tidetable *n.* table indicating the times of high and low tides.

tideway *n.* tidal part of a river.

tidings /ˈtaɪdɪŋz/ *n.* (as *sing.* or *pl.*) *archaic* or *joc.* news. [Old English, probably from Old Norse]

tidy *–adj.* (**-ier**, **-iest**) **1** neat, orderly. **2** (of a person) methodical. **3** *colloq.* considerable (*a tidy sum*). *–n.* (*pl.* **-ies**) **1** receptacle for holding small objects etc. **2** esp. *US* cover for a chair-back etc. *–v.* (**-ies**, **-ied**) (also *absol.*; often foll. by *up*) put in good order; make (oneself, a room, etc.) tidy. □ **tidily** *adv.* **tidiness** *n.* [originally = timely etc., from TIDE]

tie /taɪ/ *–v.* (**tying**) **1** attach or fasten with string or cord etc. **2 a** form (a string, ribbon, shoelace, necktie, etc.) into a knot or bow. **b** form (a knot or bow) in this way. **3** (often foll. by *down*) restrict (a person) in some way (*is tied to his job*). **4** (often foll. by *with*) achieve the same score or place as another competitor (*tied with her for first place*). **5** hold (rafters etc.) together by a crosspiece etc. **6** *Mus.* unite (written notes) by a tie. *–n.* **1** cord or wire etc. used for fastening. **2** strip of material worn round the collar and tied in a knot at the front. **3** thing that unites or restricts persons (*family ties*). **4** draw, dead heat, or equality of score among competitors. **5** match between any pair from a group of competing players or teams. **6** (also **tie-beam** etc.) rod or beam holding parts of a structure together. **7** *Mus.* curved line above or below two notes of the same pitch indicating that they are to be played without a break between them. □ **tie in** (foll. by *with*) bring into or have a close association or agreement. **tie up 1** bind securely with cord etc. **2** invest or reserve (capital etc.) so that it is not immediately available for use. **3** (often foll. by *with*) = *tie in*. **4** (usu. in *passive*) fully occupy (a person). **5** bring to a satisfactory conclusion. [Old English]

tie-break *n.* (also **tie-breaker**) means of deciding a winner from competitors who have tied.

tied *attrib. adj.* **1** (of a house) occupied subject to the tenant's working for its owner. **2** (of a public house etc.) bound to supply the products of a particular brewery only.

tie-dye *n.* (also **tie and dye**) method of producing dyed patterns by tying string etc. to keep the dye away from parts of the fabric.

tie-in *n.* **1** connection or association. **2** joint promotion of related commodities etc. (e.g. a book and a film).

tie-pin *n.* ornamental pin for holding a tie in place.

tier *n.* row, rank, or unit of a structure, as one of several placed one above another (*tiers of seats*). □ **tiered** *adj.* [French *tire* from *tirer* draw, elongate]

tiercel var. of TERCEL.

tie-up *n.* connection, association.

tiff *n.* slight or petty quarrel. [origin unknown]

tiffin /ˈtɪfɪn/ *n. Ind.* light meal, esp. lunch. [apparently from *tiffing* sipping]

tiger /ˈtaɪgə(r)/ *n.* **1** large Asian animal of the cat family, with a yellow-brown coat with black stripes. **2** fierce, energetic, or formidable person. [Greek *tigris*]

tiger-cat *n.* any moderate-sized feline resembling the tiger, e.g. the ocelot.

tiger lily *n.* tall garden lily with dark-spotted orange flowers.

tiger moth *n.* moth with richly spotted and streaked wings.

tight /taɪt/ *–adj.* **1** closely held, drawn, fastened, fitting, etc. (*tight hold*; *tight skirt*). **2** too closely fitting. **3** impermeable, impervious, esp. (in *comb.*) to a specified thing (*watertight*). **4** tense; stretched. **5** *colloq.* drunk. **6** *colloq.*

stingy. **7** (of money or materials) not easily obtainable. **8 a** (of precautions, a programme, etc.) stringent, demanding. **b** presenting difficulties (*tight situation*). **9** produced by or requiring great exertion or pressure (*tight squeeze*). *–adv.* tightly (*hold tight!*). □ **tightly** *adv.* **tightness** *n.* [Old Norse]

tight corner *n.* (also **tight place** or **spot**) difficult situation.

tighten *v.* make or become tighter.

tight-fisted *adj.* stingy.

tight-lipped *adj.* with or as with the lips compressed to restrain emotion or speech; determinedly reticent.

tightrope *n.* rope stretched tightly high above the ground, on which acrobats perform.

tights *n.pl.* **1** thin close-fitting wool or nylon etc. garment covering the legs, feet, and the lower part of the torso, worn by women and girls. **2** similar garment worn by a dancer, acrobat, etc.

tigress /'taɪgrɪs/ *n.* female tiger.

tike var. of TYKE.

tilde /'tɪldə/ *n.* mark (~) put over a letter, e.g. over a Spanish *n* when pronounced *ny* (as in *señor*). [Latin: related to TITLE]

tile *–n.* **1** thin slab of concrete or baked clay etc. used for roofing or paving etc. **2** similar slab of glazed pottery, cork, linoleum, etc., for covering a wall, floor, etc. **3** thin flat piece used in a game (esp. in mah-jong). *–v.* (**-ling**) cover with tiles. □ **on the tiles** *colloq.* having a spree. □ **tiler** *n.* [Latin *tegula*]

tiling *n.* **1** process of fixing tiles. **2** area of tiles.

till[1] *–prep.* **1** up to or as late as (*wait till six o'clock*). **2** up to the time of (*faithful till death*). *–conj.* **1** up to the time when (*wait till I return*). **2** so long that (*laughed till I cried*). [Old Norse: related to TILL[3]]

■ **Usage** In all senses, *till* can be replaced by *until* which is more formal in style.

---

till[2] *n.* drawer for money in a shop or bank etc., esp. with a device recording the amount of each purchase. [origin unknown]

till[3] *v.* cultivate (land). □ **tiller** *n.* [Old English, = strive for]

tillage *n.* **1** preparation of land for growing crops. **2** tilled land.

tiller *n.* bar fitted to a boat's rudder to turn it in steering. [Anglo-French *telier* weaver's beam]

tilt *–v.* **1** (cause to) assume a sloping position; heel over. **2** (foll. by *at*) strike, thrust, or run at, with a weapon. **3** (foll. by *with*) engage in a contest. *–n.* **1** tilting. **2** sloping position. **3** (of medieval knights etc.) charging with a lance against an opponent or at a mark. **4** attack, esp. with argument or satire (*have a tilt at*). □ **full** (or **at full**) **tilt 1** at full speed. **2** with full force. **tilt at windmills** see WINDMILL. [Old English, = unsteady]

tilth *n.* **1** tillage, cultivation. **2** tilled soil. [Old English: related to TILL[3]]

timber *n.* **1** wood prepared for building, carpentry, etc. **2** piece of wood or beam, esp. as the rib of a vessel. **3** large standing trees. **4** (esp. as *int.*) warning cry that a tree is about to fall. [Old English, = building]

timbered *adj.* **1** made wholly or partly of timber. **2** (of country) wooded.

timberline *n.* line or level above which no trees grow.

timbre /'tæmbə(r), 'tæ̃mbrə/ *n.* distinctive character of a musical sound or voice apart from its pitch and volume. [Greek: related to TYMPANUM]

timbrel /'tɪmbr(ə)l/ *n. archaic* tambourine. [French: related to TIMBRE]

time *–n.* **1** indefinite continued progress of existence, events, etc., in the past, present, and future, regarded as a whole. **2** progress of this as affecting persons or things. **3** portion of time belonging to particular events or circumstances (*the time of the Plague*; *prehistoric times*). **4** allotted or available portion of time (*had no time to eat*). **5** point of time, esp. in hours and minutes (*the time is 7.30*). **6** (prec. by *a*) indefinite period. **7** time or an amount of time as reckoned by a conventional standard (*eight o'clock New York time*; *the time allowed is one hour*). **8** occasion (*last time*). **9** moment etc. suitable for a purpose etc. (*the time to act*). **10** (in *pl.*) expressing multiplication (*five times six is thirty*). **11** lifetime (*will last my time*). **12** (in *sing.* or *pl.*) conditions of life or of a period (*hard times*). **13** *slang* prison sentence (*is doing time*). **14** apprenticeship (*served his time*). **15** period of gestation. **16** date or expected date of childbirth or death. **17** measured time spent in work. **18 a** any of several rhythmic patterns of music. **b** duration of a note. *–v.* (**-ming**) **1** choose the time for. **2** do at a chosen or correct time. **3** arrange the time of arrival of. **4** ascertain the time taken by. □ **against time** with utmost speed, so as to finish by a specified time. **ahead of time** earlier than expected. **all the time 1** during the whole of the time referred to (often despite some contrary expectation etc.). **2** constantly. **at one time 1** in a known but unspecified past period. **2** simultaneously. **at the same time 1** simultaneously. **2** nevertheless. **at times** intermittently. **for the time being** until some other arrangement is made. **half the time** *colloq.* as often as not. **have no time for 1** be unable or unwilling to spend time on. **2** dislike. **have a time of it** undergo trouble or difficulty. **in no time 1** very soon. **2** very quickly. **in time 1** not late, punctual. **2** eventually. **3** in accordance with a given rhythm. **keep time** move or sing etc. in time. **pass the time of day** *colloq.* exchange a greeting or casual remarks. **time after time 1** on many occasions. **2** in many instances. **time and** (or **time and time**) **again** on many occasions. **the time of one's life** period of exceptional enjoyment. **time out of mind** a longer time than anyone can remember. **time was** there was a time. [Old English]

time and a half *n.* one and a half times the normal rate of payment.

time-and-motion *adj.* (usu. *attrib.*) measuring the efficiency of industrial and other operations.

time bomb *n.* bomb designed to explode at a pre-set time.

time capsule *n.* box etc. containing objects typical of the present time, buried for future discovery.

time clock *n.* clock with a device for recording workers' hours of work.

time exposure *n.* exposure of photographic film for longer than the slowest normal shutter setting.

time-honoured *adj.* esteemed by tradition or through custom.

timekeeper *n.* **1** person who records time, esp. of workers or in a game. **2 a** watch or clock as regards accuracy (*a good timekeeper*). **b** person as regards punctuality. □ **timekeeping** *n.*

time-lag *n.* interval of time between a cause and effect.

timeless *adj.* not affected by the passage of time. □ **timelessly** *adv.* **timelessness** *n.*

time-limit *n.* limit of time within which a task must be done.

timely *adj.* (**-ier, -iest**) opportune; coming at the right time. □ **timeliness** *n.*

timepiece *n.* clock or watch.

timer *n.* person or device that measures or records time taken.

time-served *adj.* having completed a period of apprenticeship or training.

time-server *n. derog.* person who changes his or her view to suit the prevailing circumstances, fashion, etc. □ **time-serving** *adj.*

time-share *n.* share in a property under a time-sharing scheme.

time-sharing *n.* **1** use of a holiday home at contractually agreed different times by several joint owners. **2** operation of a computer system by several users for different operations at the same time.

time sheet *n.* sheet of paper for recording hours of work etc.

time-shift –*v.* move from one time to another, esp. record (a television programme) for later viewing. –*n.* movement from one time to another (*the continual time-shifts make the plot difficult to follow*).

time signal *n.* audible signal of the exact time of day.

time signature *n. Mus.* indication of tempo following a clef.

time switch *n.* switch acting automatically at a pre-set time.

timetable –*n.* list of times at which events are scheduled to take place, esp. the arrival and departure of transport or a sequence of lessons. –*v.* (**-ling**) include in or arrange to a timetable; schedule.

time zone *n.* range of longitudes where a common standard time is used.

timid *adj.* (**timider, timidest**) easily frightened; apprehensive. □ **timidity** /-ˈmɪdɪtɪ/ *n.* **timidly** *adv.* [Latin *timeo* fear]

timing *n.* **1** way an action or process is timed. **2** regulation of the opening and closing of valves in an internal-combustion engine.

timorous /ˈtɪmərəs/ *adj.* **1** timid. **2** frightened. □ **timorously** *adv.* [medieval Latin: related to TIMID]

timpani /ˈtɪmpənɪ/ *n.pl.* (also **tympani**) kettledrums. □ **timpanist** *n.* [Italian, pl. of *timpano* = TYMPANUM]

tin –*n.* **1** silvery-white metallic element, used esp. in alloys and in making tin plate. **2** container made of tin or tinned iron, esp. airtight for preserving food. **3** = TIN PLATE. –*v.* (**-nn-**) **1** seal (food) in a tin for preservation. **2** cover or coat with tin. [Old English]

tin can *n.* tin container, esp. an empty one.

tincture /ˈtɪŋktʃə(r)/ –*n.* (often foll. by *of*) **1** slight flavour or trace. **2** tinge (of a colour). **3** medicinal solution (of a drug) in alcohol (*tincture of quinine*). –*v.* (**-ring**) **1** colour slightly; tinge, flavour. **2** (often foll. by *with*) affect slightly (with a quality). [Latin: related to TINGE]

tinder *n.* dry substance that readily catches fire from a spark. □ **tindery** *adj.* [Old English]

tinder-box *n. hist.* box containing tinder, flint, and steel, formerly used for kindling fires.

tine *n.* prong, tooth, or point of a fork, comb, antler, etc. [Old English]
tin foil *n.* foil made of tin, aluminium, or tin alloy, used for wrapping food.
ting –*n.* tinkling sound as of a bell. –*v.* (cause to) emit this sound. [imitative]
tinge –*v.* (**-ging**) (often foll. by *with*; often in *passive*) **1** colour slightly. **2** affect slightly. –*n.* **1** tendency towards or trace of some colour. **2** slight admixture of a feeling or quality. [Latin *tingo tinct-* dye]
tingle /ˈtɪŋg(ə)l/ –*v.* (**-ling**) **1** feel a slight prickling, stinging, or throbbing sensation. **2** cause this (*the reply tingled in my ears*). –*n.* tingling sensation. □ **tingly** *adj.* [probably from TINKLE]
tin hat *n. colloq.* military steel helmet.
tinker –*n.* **1** itinerant mender of kettles and pans etc. **2** *Scot.* & *Ir.* Gypsy. **3** *colloq.* mischievous person or animal. **4** spell of tinkering. –*v.* **1** (foll. by *at*, *with*) work in an amateurish or desultory way. **2** work as a tinker. [origin unknown]
tinkle /ˈtɪŋk(ə)l/ –*v.* (**-ling**) (cause to) make a succession of short light ringing sounds. –*n.* **1** tinkling sound. **2** *colloq.* telephone call. □ **tinkly** *adj.* [imitative]
tinnitus /tɪˈnaɪtəs/ *n. Med.* condition with ringing in the ears. [Latin *tinnio tinnit-* ring, tinkle]
tinny *adj.* (**-ier**, **-iest**) **1** of or like tin. **2** flimsy, insubstantial. **3** (of sound) thin and metallic.
tin-opener *n.* tool for opening tins.
tin-pan alley *n.* world of composers and publishers of popular music.
tin plate *n.* sheet iron or sheet steel coated with tin.
tinpot *attrib. adj.* cheap, inferior.
tinsel /ˈtɪns(ə)l/ *n.* **1** glittering metallic strips, threads, etc., used as decoration. **2** superficial brilliance or splendour. **3** (*attrib.*) gaudy, flashy. □ **tinselled** *adj.* **tinselly** *adj.* [Latin *scintilla* spark]
tinsmith *n.* worker in tin and tin plate.
tinsnips *n.* clippers for cutting sheet metal.
tint –*n.* **1** variety of a colour, esp. made by adding white. **2** tendency towards or admixture of a different colour (*red with a blue tint*). **3** faint colour spread over a surface. –*v.* apply a tint to; colour. [*tinct*: related to TINGE]
tin-tack *n.* iron tack.
tintinnabulation /ˌtɪntɪˌnæbjʊˈleɪʃ(ə)n/ *n.* ringing or tinkling of bells. [Latin *tintinnabulum* bell]
tin whistle *n.* = PENNY WHISTLE.
tiny /ˈtaɪnɪ/ *adj.* (**-ier**, **-iest**) very small or slight. □ **tinily** *adv.* **tininess** *n.* [origin unknown]
-tion see -ION.
tip[1] –*n.* **1** extremity or end, esp. of a small or tapering thing. **2** small piece or part attached to the end of a thing. **3** leaf-bud of tea. –*v.* (**-pp-**) provide with a tip. □ **on the tip of one's tongue** about to be said or remembered. **tip of the iceberg** small evident part of something much larger. [Old Norse]
tip[2] –*v.* (**-pp-**) **1** (often foll. by *over*, *up*) **a** lean or slant. **b** cause to do this. **2** (foll. by *into* etc.) **a** overturn or cause to overbalance. **b** discharge the contents of (a container etc.) in this way. –*n.* **1 a** slight push or tilt. **b** light stroke. **2** place where material (esp. refuse) is tipped. □ **tip the scales** see SCALE[2]. [origin uncertain]
tip[3] –*v.* (**-pp-**) **1** make a small present of money to, esp. for a service given. **2** name as the likely winner of a race or contest etc. **3** strike or touch lightly. –*n.* **1** small money present, esp. for a service given. **2** piece of private or special information, esp. regarding betting or investment. **3** small or casual piece of advice. □ **tip off** give (a person) a hint or piece of special information or warning. **tip a person the wink** give a person private information. [origin uncertain]
tip-off *n.* hint or warning etc.
tipper *n.* (often *attrib.*) road haulage vehicle that tips at the back to discharge its load.
tippet /ˈtɪpɪt/ *n.* **1** long piece of fur etc. worn by a woman round the shoulders. **2** similar garment worn by judges, clergy, etc. [probably from TIP[1]]
tipple /ˈtɪp(ə)l/ –*v.* (**-ling**) **1** drink intoxicating liquor habitually. **2** drink (liquor) repeatedly in small amounts. –*n. colloq.* alcoholic drink. □ **tippler** *n.* [origin unknown]
tipstaff *n.* **1** sheriff's officer. **2** metal-tipped staff carried as a symbol of office. [from TIP[1]]
tipster *n.* person who gives tips, esp. about betting at horse-races.
tipsy /ˈtɪpsɪ/ *adj.* (**-ier**, **-iest**) **1** slightly drunk. **2** caused by or showing intoxication (*a tipsy lurch*). □ **tipsily** *adv.* **tipsiness** *n.* [from TIP[2]]
tiptoe –*n.* the tips of the toes. –*v.* (**-toes**, **-toed**, **-toeing**) walk on tiptoe, or very stealthily. –*adv.* (also **on tiptoe**) with the heels off the ground.
tiptop *colloq.* –*adj.* highest in excellence. –*n.* highest point of excellence. –*adv.* most excellently.
tip-up *attrib. adj.* able to be tipped, e.g. of a theatre seat.
TIR *abbr.* international road transport. [French *transport international routier*]

tirade /taɪ'reɪd/ *n.* long vehement denunciation or declamation. [French from Italian]

tire[1] *v.* (**-ring**) **1** make or grow weary. **2** exhaust the patience or interest of; bore. **3** (in *passive*; foll. by *of*) have had enough of; be fed up with. [Old English]

tire[2] *n.* **1** band of metal placed round the rim of a wheel to strengthen it. **2** *US* var. of TYRE. [perhaps = archaic *tire* 'head-dress']

tired *adj.* **1** weary; ready for sleep. **2** (of an idea etc.) hackneyed. □ **tiredly** *adv.* **tiredness** *n.*

tireless *adj.* not tiring easily, energetic. □ **tirelessly** *adv.* **tirelessness** *n.*

tiresome *adj.* **1** wearisome, tedious. **2** *colloq.* annoying. □ **tiresomely** *adv.* **tiresomeness** *n.*

tiro /'taɪə,rəʊ/ *n.* (also **tyro**) (*pl.* **-s**) beginner, novice. [Latin, = recruit]

'tis /tɪz/ *archaic* it is. [contraction]

tissue /'tɪʃuː, 'tɪsjuː/ *n.* **1** any of the coherent collections of specialized cells of which animals or plants are made (*muscular tissue*). **2** = TISSUE-PAPER. **3** disposable piece of thin soft absorbent paper for wiping, drying, etc. **4** fine woven esp. gauzy fabric. **5** (foll. by *of*) connected series (*tissue of lies*). [French *tissu* woven cloth]

tissue-paper *n.* thin soft paper for wrapping etc.

tit[1] *n.* any of various small birds. [probably from Scandinavian]

tit[2] *n.* □ **tit for tat** blow for blow; retaliation. [= earlier *tip* in *tip for tap*: see TIP[2]]

tit[3] *n.* **1** *coarse slang* woman's breast. **2** *colloq.* nipple. [Old English]

Titan /'taɪt(ə)n/ *n.* (often **titan**) person of very great strength, intellect, or importance. [Greek, = member of a race of giants]

titanic /taɪ'tænɪk/ *adj.* gigantic, colossal. □ **titanically** *adv.* [Greek: related to TITAN]

titanium /taɪ'teɪnɪəm, tɪ-/ *n.* grey metallic element. [Greek: related to TITAN]

titbit *n.* (*US* **tidbit**) **1** dainty morsel. **2** piquant item of news etc. [perhaps from dial. *tid* tender]

titchy *adj.* (**-ier, -iest**) *colloq.* very small. [*titch* small person, from *Tich*, name of a comedian]

titfer *n. slang* hat. [abbreviation of *tit for tat*, rhyming slang]

tithe /taɪð/ –*n.* **1** one-tenth of the annual produce of land or labour, formerly taken as a tax for the Church. **2** tenth part. –*v.* (**-thing**) **1** subject to tithes. **2** pay tithes. [Old English, = tenth]

tithe barn *n.* barn built to hold tithes paid in kind.

Titian /'tɪʃ(ə)n/ *adj.* (of hair) bright auburn. [*Titian*, name of a painter]

titillate /'tɪtɪ,leɪt/ *v.* (**-ting**) **1** excite, esp. sexually. **2** tickle. □ **titillation** /-'leɪʃ(ə)n/ *n.* [Latin]

titivate /'tɪtɪ,veɪt/ *v.* (**-ting**) (often *refl.*) *colloq.* smarten up; put the finishing touches to. □ **titivation** /-'veɪʃ(ə)n/ *n.* [earlier *tidivate*, perhaps from TIDY after *cultivate*]

title /'taɪt(ə)l/ *n.* **1** name of a book, work of art, etc. **2** heading of a chapter, document, etc. **3 a** = TITLE-PAGE. **b** book, magazine, etc., in terms of its title (*brought out two new titles*). **4** (usu. in *pl.*) caption or credit in a film etc. **5** name indicating a person's status (e.g. *queen*, *professor*) or used as a form of address or reference (e.g. *Lord*, *Mr*, *Your Grace*). **6** championship in sport. **7** *Law* **a** right to ownership of property with or without possession. **b** facts constituting this. **c** (foll. by *to*) just or recognized claim. [Latin *titulus*]

titled *adj.* having a title of nobility or rank.

title-deed *n.* legal instrument as evidence of a right.

title-holder *n.* person who holds a title, esp. a sporting champion.

title-page *n.* page at the beginning of a book giving the title, author, etc.

title role *n.* part in a play etc. that gives it its name (e.g. *Othello*).

titmouse *n.* (*pl.* **titmice**) small active tit. [Old English *tit* little, *māse* titmouse, assimilated to MOUSE]

titrate /taɪ'treɪt/ *v.* (**-ting**) *Chem.* ascertain the amount of a constituent in (a solution) by reaction with a known concentration of reagent. □ **titration** /-'treɪʃ(ə)n/ *n.* [French *titre* title]

titter –*v.* laugh covertly; giggle. –*n.* covert laugh. [imitative]

tittle /'tɪt(ə)l/ *n.* **1** small written or printed stroke or dot. **2** particle; whit (*not one jot or tittle*). [Latin: related to TITLE]

tittle-tattle /'tɪt(ə)l,tæt(ə)l/ –*n.* petty gossip. –*v.* (**-ling**) gossip, chatter. [reduplication of TATTLE]

tittup /'tɪtəp/ –*v.* (**-p-** or **-pp-**) go about friskily or jerkily; bob up and down; canter. –*n.* such a gait or movement. [perhaps imitative]

titular /'tɪtjʊlə(r)/ *adj.* **1** of or relating to a title. **2** existing, or being, in name or title only (*titular ruler*). [French: related to TITLE]

tizzy /'tɪzɪ/ *n.* (*pl.* **-ies**) *colloq.* state of agitation (*in a tizzy*). [origin unknown]

T-junction *n.* road junction at which one road joins another at right angles without crossing it.

Tl *symb.* thallium.
Tm *symb.* thulium.
TNT *abbr.* trinitrotoluene, a high explosive formed from toluene.
to /tə/, *before a vowel* /tʊ/, when stressed /tu:/ *–prep.* **1** introducing a noun expressing: **a** what is reached, approached, or touched (*fell to the ground*; *went to Paris*; *five minutes to six*). **b** what is aimed at (*throw it to me*). **c** as far as (*went on to the end*). **d** what is followed (*made to order*). **e** what is considered or affected (*am used to that*; *that is nothing to me*). **f** what is caused or produced (*turn to stone*). **g** what is compared (*nothing to what it once was*; *equal to the occasion*). **h** what is increased (*add it to mine*). **i** what is involved or composed as specified (*there is nothing to it*). **2** introducing the infinitive: **a** as a verbal noun (*to get there is the priority*). **b** expressing purpose, consequence, or cause (*we eat to live*; *left him to starve*; *I'm sorry to hear that*). **c** as a substitute for *to* + infinitive (*wanted to come but was unable to*). *–adv.* **1** in the normal or required position or condition (*come to*; *heave to*). **2** (of a door) in a nearly closed position. □ **to and fro 1** backwards and forwards. **2** repeatedly between the same points. [Old English]
toad *n.* **1** froglike amphibian breeding in water but living chiefly on land. **2** repulsive person. [Old English]
toadflax *n.* plant with yellow or purple flowers.
toad-in-the-hole *n.* sausages baked in batter.
toadstool *n.* fungus, usu. poisonous, with a round top and slender stalk.
toady *–n.* (*pl.* **-ies**) sycophant. *–v.* (foll. by *to*) (**-ies**, **-ied**) behave servilely to; fawn upon. □ **toadyism** *n.* [contraction of *toad-eater*]
toast *–n.* **1** sliced bread browned on both sides by radiant heat. **2 a** person or thing in whose honour a company is requested to drink. **b** call to drink or an instance of drinking in this way. *–v.* **1** brown by radiant heat. **2** warm (one's feet, oneself, etc.) at a fire etc. **3** drink to the health or in honour of (a person or thing). [French *toster* roast]
toaster *n.* electrical device for making toast.
toasting-fork *n.* long-handled fork for making toast.
toastmaster *n.* (*fem.* **toastmistress**) person responsible for announcing toasts at a public occasion.
toast rack *n.* rack for holding slices of toast at table.
tobacco /tə'bækəʊ/ *n.* (*pl.* **-s**) **1** plant of American origin with narcotic leaves used for smoking, chewing, or snuff. **2** its leaves, esp. as prepared for smoking. [Spanish *tabaco*, of American Indian origin]
tobacconist /tə'bækənɪst/ *n.* dealer in tobacco, cigarettes, etc.
toboggan /tə'bɒgən/ *–n.* long light narrow sledge for sliding downhill over snow or ice. *–v.* ride on a toboggan. [Canadian French from Algonquian]
toby jug /'təʊbɪ/ *n.* jug or mug in the form of a stout man wearing a three-cornered hat. [familiar form of the name *Tobias*]
toccata /tə'kɑ:tə/ *n.* musical composition for a keyboard instrument, designed to exhibit the performer's touch and technique. [Italian, = touched]
tocsin /'tɒksɪn/ *n.* alarm bell or signal. [Provençal *tocasenh*]
tod *n. slang* □ **on one's tod** alone; on one's own. [rhyming slang *on one's Tod Sloan*]
today /tə'deɪ/ *–adv.* **1** on this present day. **2** nowadays. *–n.* **1** this present day. **2** modern times. [Old English]
toddle /'tɒd(ə)l/ *–v.* (**-ling**) **1** walk with short unsteady steps like a small child. **2** *colloq.* **a** walk, stroll. **b** (usu. foll. by *off* or *along*) depart. *–n.* act of toddling. [origin unknown]
toddler *n.* child who is just learning to walk.
toddy /'tɒdɪ/ *n.* (*pl.* **-ies**) drink of spirits with hot water and sugar etc. [Hindustani *tār* palm]
to-do /tə'du:/ *n.* (*pl.* **-s**) commotion or fuss.
toe *–n.* **1** any of the five terminal projections of the foot. **2** corresponding part of an animal. **3** part of a shoe etc. that covers the toes. **4** lower end or tip of an implement etc. *–v.* (**toes, toed, toeing**) touch (a starting-line etc.) with the toes. □ **on one's toes** alert. **toe the line** conform, esp. under pressure. [Old English]
toecap *n.* (usu. strengthened) outer covering of the toe of a boot or shoe.
toe-hold *n.* **1** small foothold. **2** small beginning or advantage.
toenail *n.* nail of each toe.
toff *n. slang* upper-class person. [perhaps from *tuft*, = titled undergraduate]
toffee /'tɒfɪ/ *n.* **1** firm or hard sweet made by boiling sugar, butter, etc. together. **2** this substance. □ **for toffee** *slang* (prec. by *can't* etc.) (denoting incompetence) at all (*they couldn't sing for toffee*). [origin unknown]

toffee-apple *n.* apple with a coating of toffee.

toffee-nosed *adj. slang* snobbish, superior.

tofu /'təʊfu:, 'tɒfu:/ *n.* curd of mashed soya beans. [Japanese]

tog[1] *colloq.* –*n.* (usu. in *pl.*) item of clothing. –*v.* (**-gg-**) (foll. by *out, up*) dress. [apparently originally cant: ultimately related to Latin TOGA]

tog[2] *n.* unit of thermal resistance used to express the insulating properties of clothes and quilts. [arbitrary, probably from TOG[1]]

toga /'təʊgə/ *n. hist.* ancient Roman citizen's loose flowing outer garment. □ **togaed** *adj.* (also **toga'd**). [Latin]

together /tə'geðə(r)/ –*adv.* **1** in company or conjunction (*walking together; were at school together*). **2** simultaneously (*both shouted together*). **3** one with another (*talking together*). **4** into conjunction; so as to unite (*tied them together; put two and two together*). **5** into company or companionship. **6** uninterruptedly (*he could talk for three hours together*). –*adj. colloq.* well-organized; self-assured; emotionally stable. □ **together with** as well as. [Old English: related to TO, GATHER]

togetherness *n.* **1** being together. **2** feeling of comfort from this.

toggle /'tɒg(ə)l/ *n.* **1** fastener for a garment consisting of a crosspiece which passes through a hole or loop. **2** *Computing* switch action that is operated the same way but with opposite effect on successive occasions. [origin unknown]

toggle switch *n.* electric switch with a lever to be moved usu. up and down.

toil –*v.* **1** work laboriously or incessantly. **2** make slow painful progress. –*n.* intensive labour; drudgery. [Anglo-French *toil(er)* dispute]

toilet /'tɔɪlɪt/ *n.* **1** = LAVATORY. **2** process of washing oneself, dressing, etc. (*at one's toilet*). [French *toilette* diminutive of *toile* cloth]

toilet paper *n.* paper for cleaning oneself after excreting.

toilet roll *n.* roll of toilet paper.

toiletry /'tɔɪlɪtrɪ/ *n.* (*pl.* **-ies**) (usu. in *pl.*) article or cosmetic used in washing, dressing, etc.

toilet soap *n.* soap for washing oneself.

toilette /twɑ:'let/ *n.* = TOILET 2. [French]

toilet-training *n.* training of a young child to use the lavatory. □ **toilet-train** *v.*

toilet water *n.* dilute perfume used after washing.

toils /tɔɪlz/ *n.pl.* net, snare. [*toil* from French: related to TOILET]

toilsome /'tɔɪlsəm/ *adj.* involving toil.

toing and froing /,tu:ɪŋ ənd 'frəʊɪŋ/ *n.* constant movement to and fro; bustle; dispersed activity. [from TO, FRO]

Tokay /tə'keɪ/ *n.* a sweet Hungarian wine. [*Tokaj* in Hungary]

token /'təʊkən/ *n.* **1** thing serving as a symbol, reminder, or mark (*as a token of affection; in token of my esteem*). **2** voucher. **3** thing equivalent to something else, esp. money. **4** (*attrib.*) **a** perfunctory (*token effort*). **b** conducted briefly to demonstrate strength of feeling (*token strike*). **c** chosen by tokenism to represent a group (*token woman*). □ **by this** (or **the same**) **token 1** similarly. **2** moreover. [Old English]

tokenism *n.* **1** granting of minimum concessions, esp. to minority groups. **2** making of only a token effort.

told *past* and *past part.* of TELL[1].

tolerable /'tɒlərəb(ə)l/ *adj.* **1** endurable. **2** fairly good. □ **tolerably** *adv.* [Latin: related to TOLERATE]

tolerance /'tɒlərəns/ *n.* **1** willingness or ability to tolerate; forbearance. **2** allowable variation in any measurable property.

tolerant *adj.* **1** disposed to tolerate others or their acts or opinions. **2** (foll. by *of*) enduring or patient.

tolerate /'tɒlə,reɪt/ *v.* (**-ting**) **1** allow the existence or occurrence of without authoritative interference. **2** endure (suffering etc.). **3** find or treat as endurable. **4** be able to take or undergo (drugs, treatment, etc.) without adverse effects. [Latin *tolero*]

toleration /,tɒlə'reɪʃ(ə)n/ *n.* tolerating or being tolerated, esp. the allowing of religious differences without discrimination. [Latin: related to TOLERATE]

toll[1] /təʊl, tɒl/ *n.* **1** charge to use a bridge, road, etc. **2** cost or damage caused by a disaster etc. □ **take its toll** be accompanied by loss, injury, etc.

toll[2] /təʊl/ –*v.* **1 a** (of a bell) sound with slow uniform strokes. **b** ring (a bell) in this way. **c** (of a bell) announce or mark (a death etc.) in this way. **2** strike (the hour). –*n.* **1** tolling. **2** stroke of a bell. [(now dial.) *toll* entice, pull, from an Old English root]

toll-bridge *n.* bridge at which a toll is charged.

toll-gate *n.* gate preventing passage until a toll is paid.

toll-road *n.* road maintained by the tolls collected on it.

toluene /'tɒljʊ,i:n/ *n.* colourless aromatic liquid hydrocarbon derivative of benzene, used in the manufacture of explosives etc. [*Tolu* in Colombia]

tom *n.* (in full **tom-cat**) male cat. [abbreviation of the name *Thomas*]

tomahawk /'tɒmə,hɔ:k/ *n.* N. American Indian war-axe. [Renape]

tomato /tə'mɑ:təʊ/ *n.* (*pl.* **-es**) **1** glossy red or yellow pulpy edible fruit. **2** plant bearing this. [ultimately from Mexican *tomatl*]

tomb /tu:m/ *n.* **1** burial-vault. **2** grave. **3** sepulchral monument. [Greek *tumbos*]

tombola /tɒm'bəʊlə/ *n.* lottery with tickets drawn from a drum for immediate prizes. [French or Italian]

tomboy *n.* boisterous girl who enjoys activities traditionally associated with boys. □ **tomboyish** *adj.* [from TOM]

tombstone *n.* memorial stone over a grave, usu. with an epitaph.

Tom, Dick, and Harry *n.* (also **Tom, Dick, or Harry**) (usu. prec. by *any* or *every*) person taken at random (*any Tom, Dick, or Harry can walk in*).

tome *n.* heavy book or volume. [Greek *temnō* cut]

tomfool *–n.* foolish person. *–attrib. adj.* silly, foolish.

tomfoolery /tɒm'fu:lərɪ/ *n.* foolish behaviour.

Tommy /'tɒmɪ/ *n.* (*pl.* **-ies**) *colloq.* British private soldier. [*Tommy Atkins*, name used in specimens of completed official forms]

tommy-gun *n.* sub-machine-gun. [*Thompson*, name of its co-inventor]

tommy-rot *n. slang* nonsense. [from TOM]

tomography /tə'mɒgrəfɪ/ *n.* method of radiography displaying details in a selected plane within the body. [Greek *tomē* a cutting]

tomorrow /tə'mɒrəʊ/ *–adv.* **1** on the day after today. **2** at some future time. *–n.* **1** the day after today. **2** the near future. [from TO, MORROW]

tomtit *n.* tit, esp. a blue tit.

tom-tom /'tɒmtɒm/ *n.* **1** primitive drum beaten with the hands. **2** tall drum used in jazz bands etc. [Hindi *tamtam*, imitative]

ton /tʌn/ *n.* **1** (in full **long ton**) unit of weight equal to 2,240 lb (1016.05 kg). **2** (in full **short ton**) unit of weight equal to 2,000 lb (907.19 kg). **3** = METRIC TON. **4** (in full **displacement ton**) unit of measurement of a ship's weight or volume. **5** (usu. in *pl.*) *colloq.* large number or amount (*tons of people*). **6** *slang* **a** speed of 100 m.p.h. **b** £100. □ **weigh a ton** *colloq.* be very heavy. [originally the same word as TUN]

tonal /'təʊn(ə)l/ *adj.* of or relating to tone or tonality. □ **tonally** *adv.* [medieval Latin: related to TONE]

tonality /tə'nælɪtɪ/ *n.* (*pl.* **-ies**) **1** *Mus.* **a** relationship between the tones of a musical scale. **b** observance of a single tonic key as the basis of a composition. **2** colour scheme of a picture.

tone *–n.* **1** musical or vocal sound, esp. with reference to its pitch, quality, and strength. **2** (often in *pl.*) modulation of the voice expressing a particular feeling or mood (*a cheerful tone*). **3** manner of expression in writing or speaking. **4** *Mus.* **a** musical sound, esp. of a definite pitch and character. **b** interval of a major second, e.g. C–D. **5 a** general effect of colour or of light and shade in a picture. **b** tint or shade of a colour. **6** prevailing character of the morals and sentiments etc. in a group. **7** proper firmness of the body. **8** state of good or specified health. *–v.* (**-ning**) **1** give the desired tone to. **2** modify the tone of. **3** (often foll. by *to*) attune. **4** (foll. by *with*) (esp. of colour) be in harmony with. □ **tone down** make or become softer in tone. **tone up** make or become stronger in tone. □ **toneless** *adj.* **tonelessly** *adv.* **toner** *n.* [Greek *tonos* from *teinō* stretch]

tone-deaf *adj.* unable to perceive differences of musical pitch accurately.

tone poem *n.* orchestral composition with a descriptive or rhapsodic theme.

tongs *n.pl.* implement with two arms for grasping coal, sugar, etc. [Old English]

tongue /tʌŋ/ *–n.* **1** fleshy muscular organ in the mouth used in tasting, licking, and swallowing, and (in man) for speech. **2** tongue of an ox etc. as food. **3** faculty of or tendency in speech (*a sharp tongue*). **4** particular language (*the German tongue*). **5** thing like a tongue in shape or position, esp.: **a** a long low promontory. **b** a strip of leather etc. under the laces in a shoe. **c** the clapper of a bell. **d** the pin of a buckle. **e** a projecting strip on a board etc. fitting into the groove of another. *–v.* (**-guing**) use the tongue to articulate (notes) in playing a wind instrument. □ **find** (or **lose**) **one's tongue** be able (or unable) to express oneself after a shock etc. **hold one's tongue** see HOLD[1]. **with one's tongue in one's cheek** insincerely or ironically. [Old English]

tongue-and-groove *–n.* (often *attrib.*) planking etc. with a projecting strip down one side and a groove down the other. *–v.* **1** panel with tongue-and-groove. **2** (as **tongued and grooved** *adj.*) having a tongue-and-groove joint.

tongue-in-cheek *–adj.* ironic. *–adv.* insincerely or ironically.

tongue-tie *n.* speech impediment due to a malformation of the tongue.

tongue-tied *adj.* **1** too shy or embarrassed to speak. **2** having a tongue-tie.

tongue-twister *n.* sequence of words difficult to pronounce quickly and correctly.

tonic /'tɒnɪk/ *–n.* **1** invigorating medicine. **2** anything serving to invigorate. **3** = TONIC WATER. **4** *Mus.* keynote. *–adj.* invigorating. [Greek: related to TONE]

tonic sol-fa *n. Mus.* system of notation used esp. in teaching singing.

tonic water *n.* carbonated water flavoured with quinine.

tonight /tə'naɪt/ *–adv.* on the present or approaching evening or night. *–n.* the evening or night of the present day. [Old English]

tonnage /'tʌnɪdʒ/ *n.* **1** ship's internal cubic capacity or freight-carrying capacity. **2** charge per ton on freight or cargo. [related to TON]

tonne /tʌn/ *n.* = METRIC TON. [French: related to TON]

tonsil /'tɒns(ə)l/ *n.* either of two small organs, one on each side of the root of the tongue. [Latin]

tonsillectomy /ˌtɒnsə'lektəmɪ/ *n.* (*pl.* **-ies**) surgical removal of the tonsils.

tonsillitis /ˌtɒnsə'laɪtɪs/ *n.* inflammation of the tonsils.

tonsorial /tɒn'sɔːrɪəl/ *adj.* usu. *joc.* of a hairdresser or hairdressing. [Latin *tondeo tons-* shave]

tonsure /'tɒnʃə(r)/ *–n.* **1** shaving of the crown of the head or the entire head, esp. of a person entering the priesthood or a monastic order. **2** bare patch made in this way. *–v.* (**-ring**) give a tonsure to. [Latin: related to TONSORIAL]

ton-up *attrib. adj. slang* (of a motor cyclist) achieving a speed of 100 m.p.h., esp. habitually.

too *adv.* **1** to a greater extent than is desirable or permissible (*too large*). **2** *colloq.* very (*not too sure*). **3** in addition (*I'm coming too*). **4** moreover (*food was bad, and expensive too*). □ **none too** rather less than (*feeling none too good*). **too bad** see BAD. **too much** intolerable. **too much for 1** more than a match for. **2** beyond what is endurable by. **too right** see RIGHT. [stressed form of TO]

took *past* of TAKE.

tool *–n.* **1** implement used to carry out mechanical functions by hand or by machine. **2** thing used in an occupation or pursuit (*tools of one's trade*). **3** person merely used by another. **4** *coarse slang* penis. *–v.* **1** dress (stone) with a chisel. **2** impress a design on (leather). **3** (foll. by *along*, *around*, etc.) *slang* drive or ride, esp. in a casual or leisurely manner. [Old English]

toolmaker *n.* person who makes precision tools. □ **toolmaking** *n.*

tool-pusher *n.* worker directing the drilling on an oil rig.

toot *–n.* short sharp sound as made by a trumpet. *–v.* **1** sound (a trumpet etc.) with a short sharp sound. **2** give out such a sound. [probably imitative]

tooth *n.* (*pl.* **teeth**) **1** each of a set of hard bony enamel-coated structures in the jaws of most vertebrates, used for biting and chewing. **2** toothlike part or projection, e.g. the cog of a gearwheel, the point of a saw or comb, etc. **3** (often foll. by *for*) taste; appetite. **4** (in *pl.*) force, effectiveness. □ **armed to the teeth** completely and elaborately armed. **fight tooth and nail** fight very fiercely. **get one's teeth into** devote oneself seriously to. **in the teeth of 1** in spite of (opposition or difficulty etc.). **2** contrary to (instructions etc.). **3** directly against (the wind etc.). □ **toothed** *adj.* (also in *comb.*). **toothless** *adj.* [Old English]

toothache *n.* pain in a tooth or teeth.

toothbrush *n.* brush for cleaning the teeth.

tooth-comb *n.* = FINE-TOOTH COMB.

toothpaste *n.* paste for cleaning the teeth.

toothpick *n.* small sharp stick for removing food lodged between the teeth.

tooth powder *n.* powder for cleaning the teeth.

toothsome *adj.* (of food) delicious.

toothy *adj.* (**-ier**, **-iest**) having large, numerous, or prominent teeth.

tootle /'tuːt(ə)l/ *v.* (**-ling**) **1** toot gently or repeatedly. **2** (usu. foll. by *along*, *around*, etc.) *colloq.* move casually.

tootsy /'tʊtsɪ/ *n.* (*pl.* **-ies**) *slang* usu. *joc.* foot. [origin uncertain]

top[1] *–n.* **1** highest point or part. **2 a** highest rank or place. **b** person occupying this. **c** upper end or head (*top of the table*). **3** upper surface or part of a thing. **4** stopper of a bottle, lid of a jar, etc. **5** garment for the upper part of the body. **6** utmost degree; height (*at the top of his voice*). **7** (in *pl.*) *colloq.* person or thing of the best quality. **8** (esp. in *pl.*) leaves etc. of a plant grown esp. for its root (*turnip-tops*). **9** *Naut.* platform round the head of the lower mast. **10** = TOP GEAR (*climbed the hill in top*). *–attrib. adj.* **1** highest in position. **2** highest in degree or importance. *–v.* (**-pp-**) **1** provide with a top, cap, etc. **2** be higher or better than; surpass; be at the top of (*topped the list*). **3** reach the top of (a hill etc.). **4** *slang* kill. **5** *Golf* hit (a ball) above the centre. □ **off the top of one's head** see HEAD. **on top** in a superior position; above. **on top of 1** fully in command of. **2** in close proximity to. **3** in addition to. **on top of**

the world *colloq.* exuberant. **over the top 1** over the parapet of a trench (and into battle). **2** beyond what is normally acceptable (*that joke was over the top*). **top off** (or **up**) put an end or the finishing touch to. **top up 1** complete (an amount or number). **2** fill up (a partly full container). **3** top up something for (a person) (*may I top you up with sherry?*). □ **topmost** *adj.* [Old English]

top[2] *n.* toy spinning on a point when set in motion. [Old English]

topaz /'təʊpæz/ *n.* transparent mineral, usu. yellow, used as a gem. [Greek *topazos*]

top brass *n. colloq.* highest-ranking officers.

topcoat *n.* **1** overcoat. **2** outer coat of paint etc.

top dog *n. colloq.* victor; master.

top drawer *n. colloq.* high social position or origin.

top-dress *v.* apply fertilizer on the top of (earth) instead of ploughing it in. □ **top-dressing** *n.*

tope *v.* (**-ping**) *archaic* or *literary* drink alcohol to excess, esp. habitually. □ **toper** *n.* [origin uncertain]

topee var. of TOPI.

top-flight *adj.* of the highest rank of achievement.

topgallant /tɒp'gælənt/ *n.* mast, sail, yard, or rigging immediately above the topmast and topsail.

top gear *n.* highest gear.

top hat *n.* tall silk hat.

top-heavy *adj.* disproportionately heavy at the top.

topi /'təʊpɪ/ *n.* (also **topee**) (*pl.* **-s**) hat, esp. a sola topi. [Hindi, = hat]

topiary /'təʊpɪərɪ/ *–adj.* concerned with or formed by clipping shrubs, trees, etc. into ornamental shapes. *–n.* topiary art. [Greek *topos* place]

topic /'tɒpɪk/ *n.* subject of a discourse, conversation, or argument. [Greek *topos* place, commonplace]

topical *adj.* dealing with the news, current affairs, etc. □ **topicality** /-'kælɪtɪ/ *n.* **topically** *adv.*

topknot *n.* knot, tuft, crest, or bow of ribbon, worn or growing on the head.

topless *adj.* **1** without a top. **2 a** (of clothes) having no upper part. **b** (of esp. a woman) bare-breasted. **c** (of a place) where women go topless; employing bare-breasted women.

top-level *adj.* of the highest level of importance, prestige, etc.

topmast *n.* mast next above the lower mast.

top-notch *adj. colloq.* first-rate.

topography /tə'pɒgrəfɪ/ *n.* **1** detailed description, representation on a map, etc., of the features of a town, district, etc. **2** such features. □ **topographer** *n.* **topographical** /,tɒpə'græfɪk(ə)l/ *adj.* [Greek *topos* place]

topology /tə'pɒlədʒɪ/ *n.* the study of geometrical properties unaffected by changes of shape or size. □ **topological** /,tɒpə'lɒdʒɪk(ə)l/ *adj.* [Greek *topos* place]

topper *n. colloq.* = TOP HAT.

topping *–adj. archaic slang* excellent. *–n.* thing that tops another thing, esp. sauce on a dessert etc.

topple /'tɒp(ə)l/ *v.* (**-ling**) **1** (often foll. by *over*, *down*) (cause to) fall as if top-heavy. **2** overthrow. [from TOP[1]]

topsail /'tɒpseɪl, -s(ə)l/ *n.* square sail next above the lowest; fore-and-aft sail on a gaff.

top secret *adj.* of the highest secrecy.

topside *n.* **1** outer side of a round of beef. **2** side of a ship above the water-line.

topsoil *n.* top layer of soil.

topspin *n.* spinning motion imparted to a ball in tennis etc. by hitting it forward and upward.

topsy-turvy /,tɒpsɪ'tɜːvɪ/ *adv.* & *adj.* **1** upside down. **2** in utter confusion. [from TOP[1], obsolete *terve* overturn]

top-up *n.* addition; something that serves to top up.

toque /təʊk/ *n.* woman's small brimless hat. [French]

tor *n.* hill or rocky peak. [Old English]

torch *n.* **1** portable battery-powered electric lamp. **2** thing lit for illumination. **3** source of heat, illumination, or enlightenment. □ **carry a torch for** suffer from unrequited love for. [Latin: related to TORT]

torchlight *n.* light of a torch or torches. □ **torchlit** *adj.*

torch song *n.* popular song of unrequited love.

tore *past* of TEAR[1].

toreador /'tɒrɪə,dɔː(r)/ *n.* bullfighter, esp. on horseback. [Latin *taurus*]

torment *–n.* /'tɔːment/ **1** severe physical or mental suffering. **2** cause of this. *–v.* /tɔː'ment/ **1** subject to torment. **2** tease or worry excessively. □ **tormentor** /-'mentə(r)/ *n.* [Latin *tormentum*: related to TORT]

tormentil /'tɔːməntɪl/ *n.* low-growing plant with bright yellow flowers. [French from medieval Latin]

torn *past part.* of TEAR[1].

tornado /tɔː'neɪdəʊ/ *n.* (*pl.* **-es**) violent storm of small extent with whirling winds. [Spanish *tronada* thunderstorm]

torpedo /tɔː'piːdəʊ/ *–n.* (*pl.* **-es**) **1** cigar-shaped self-propelled underwater missile that explodes on impact with a ship.

2 similar device dropped from an aircraft. —*v.* (**-es, -ed**) **1** destroy or attack with a torpedo. **2** destroy or damage (a policy, institution, plan, etc.). [Latin, = electric ray: related to TORPOR]

torpedo-boat *n.* small fast warship armed with torpedoes.

torpid /'tɔ:pɪd/ *adj.* **1** sluggish, inactive, apathetic. **2** numb. **3** (of a hibernating animal) dormant. □ **torpidity** /-'pɪdɪtɪ/ *n.* [Latin: related to TORPOR]

torpor /'tɔ:pə(r)/ *n.* torpid condition. [Latin *torpeo* be sluggish]

torque /tɔ:k/ *n.* **1** *Mech.* twisting or rotating force, esp. in a machine. **2** *hist.* necklace of twisted metal, esp. of the ancient Gauls and Britons. [Latin: related to TORT]

torr /tɔ:(r)/ *n.* (*pl.* same) unit of pressure equal to 133.32 pascals ($^1/_{760}$ of one atmosphere). [*Torricelli*, name of a physicist]

torrent /'tɒrənt/ *n.* **1** rushing stream of liquid. **2** (in *pl.*) great downpour of rain. **3** (usu. foll. by *of*) violent or copious flow (*torrent of abuse*). □ **torrential** /tə'renʃ(ə)l/ *adj.* [French from Italian]

torrid /'tɒrɪd/ *adj.* **1 a** (of the weather) very hot and dry. **b** (of land etc.) parched by such weather. **2** passionate, intense. [Latin *torreo tost-* parch]

torrid zone *n.* the part of the earth between the Tropics of Cancer and Capricorn.

torsion /'tɔ:ʃ(ə)n/ *n.* twisting, esp. of one end of a body while the other is held fixed. □ **torsional** *adj.* [Latin: related to TORT]

torso /'tɔ:səʊ/ *n.* (*pl.* **-s**) **1** trunk of the human body. **2** statue of this. [Latin *thyrsus* rod]

tort *n. Law* breach of duty (other than under contract) leading to liability for damages. □ **tortious** /'tɔ:ʃəs/ *adj.* [Latin *torqueo tort-* twist]

tortilla /tɔ:'ti:jə/ *n.* thin flat orig. Mexican maize cake eaten hot. [Spanish diminutive of *torta* cake]

tortoise /'tɔ:təs/ *n.* slow-moving reptile with a horny domed shell. [medieval Latin *tortuca*]

tortoiseshell /'tɔ:tə,ʃel/ —*n.* **1** yellowish-brown mottled or clouded outer shell of some turtles. **2 a** = TORTOISESHELL CAT. **b** = TORTOISESHELL BUTTERFLY. —*adj.* having the colouring or appearance of tortoiseshell.

tortoiseshell butterfly *n.* butterfly with wings mottled like tortoiseshell.

tortoiseshell cat *n.* domestic cat with markings resembling tortoiseshell.

tortuous /'tɔ:tʃʊəs/ *adj.* **1** full of twists and turns. **2** devious, circuitous. □ **tortuously** *adv.* [Latin: related to TORT]

■ **Usage** *Tortuous* should not be confused with *torturous* which means 'involving torture, excruciating'.

---

torture /'tɔ:tʃə(r)/ —*n.* **1** infliction of severe bodily pain, esp. as a punishment or means of persuasion. **2** severe physical or mental suffering. —*v.* (**-ring**) subject to torture. □ **torturer** *n.* **torturous** *adj.* [Latin *tortura* twisting: related to TORT]

Tory /'tɔ:rɪ/ —*n.* (*pl.* **-ies**) **1** *colloq.* = CONSERVATIVE *n.* 2. **2** *hist.* member of the party that gave rise to the Conservative party (opp. WHIG). —*adj. colloq.* = CONSERVATIVE *adj.* 3. □ **Toryism** *n.* [originally = Irish outlaw]

tosa /'təʊsə/ *n.* dog of a breed of mastiff, orig. kept for dog-fighting. [Japanese]

tosh *n. colloq.* rubbish, nonsense. [origin unknown]

toss —*v.* **1** throw up (a ball etc.), esp. with the hand. **2** roll about, throw, or be thrown, restlessly or from side to side. **3** (usu. foll. by *to, away, aside, out,* etc.) throw (a thing) lightly or carelessly. **4 a** throw (a coin) into the air to decide a choice etc. by the side on which it lands. **b** (also *absol.*; often foll. by *for*) settle a question or dispute with (a person) in this way. **5** (of a bull etc.) throw (a person etc.) up with the horns. **6** coat (food) with dressing etc. by shaking it. —*n.* **1** act of tossing (a coin, the head, etc.). **2** fall, esp. from a horse. □ **toss one's head** throw it back esp. in anger, impatience, etc. **toss off 1** drink off at a draught. **2** dispatch (work) rapidly or without effort. **3** *coarse slang* masturbate. **toss up** toss a coin. [origin unknown]

toss-up *n.* **1** doubtful matter. **2** tossing of a coin.

tot[1] *n.* **1** small child. **2** dram of liquor. [originally dial.]

tot[2] *v.* (**-tt-**) **1** (usu. foll. by *up*) add (figures etc.). **2** (foll. by *up*) (of items) mount up. □ **tot up to** amount to. [abbreviation of TOTAL or of Latin *totum* the whole]

total /'təʊt(ə)l/ —*adj.* **1** complete, comprising the whole (*total number of votes*). **2** absolute, unqualified (*in total ignorance*). —*n.* total number or amount. —*v.* (**-ll-**; *US* **-l-**) **1 a** amount in number to. **b** find the total of. **2** (foll. by *to, up to*) amount to. [medieval Latin *totus* entire]

totalitarian /təʊ,tælɪ'teərɪən/ *adj.* of a one-party form of government requiring complete subservience to the State. □ **totalitarianism** *n.*

totality /təʊ'tælɪtɪ/ *n.* **1** complete amount. **2** time during which an eclipse is total.

totalizator /'təʊtəlaɪ,zeɪtə(r)/ *n.* (also **totalisator**) **1** device showing the number and amount of bets staked on a race, to facilitate the division of the total among those backing the winner. **2** system of betting based on this.

totalize /'təʊtə,laɪz/ *v.* (also **-ise**) (**-zing** or **-sing**) collect into a total; find the total of.

totally *adv.* completely.

tote[1] *n. slang* totalizator. [abbreviation]

tote[2] *v.* (**-ting**) esp. *US colloq.* carry, convey (*toting a gun*). [originally US, probably of dial. origin]

tote bag *n.* woman's large bag for shopping etc.

totem /'təʊtəm/ *n.* **1** natural object, esp. an animal, adopted esp. by N. American Indians as an emblem of a clan or individual. **2** image of this. □ **totemic** /-'temɪk/ *adj.* [Algonquian]

totem-pole *n.* pole on which totems are carved or hung.

t'other /'tʌðə(r)/ *adj.* & *pron. dial.* or *joc.* the other. [*thet other* 'that other']

totter –*v.* **1** stand or walk unsteadily or feebly. **2 a** (of a building etc.) shake as if about to collapse. **b** (of a system of government etc.) be about to fall. –*n.* unsteady or shaky movement or gait. □ **tottery** *adj.* [Dutch]

totting-up *n.* **1** adding of separate items. **2** adding up of convictions for driving offences, possibly resulting in disqualification.

toucan /'tu:kən/ *n.* tropical American fruit-eating bird with an immense beak. [Tupi]

touch /tʌtʃ/ –*v.* **1** come into or be in physical contact with (a thing, each other, etc.). **2** (often foll. by *with*) bring the hand etc. into contact with. **3** bring (two things) into mutual contact. **4** rouse tender or painful feelings in. **5** strike lightly. **6** (usu. with *neg.*) **a** disturb, harm, or affect. **b** have any dealings with. **c** consume, use (*I don't touch alcohol*). **7** concern. **8 a** reach as far as, esp. momentarily. **b** (usu. with *neg.*) approach in excellence etc. (*can't touch him for style*). **9** modify (*pity touched with fear*). **10** (as **touched** *adj.*) *colloq.* slightly mad. **11** (usu. foll. by *for*) *slang* ask for and get money etc. from (a person) (*touched him for £5*). –*n.* **1** act of touching (*felt a touch on my arm*). **2** faculty of perception through physical contact, esp. with the fingers. **3 a** small amount; slight trace. **b** (prec. by *a*) slightly (*a touch too salty*). **4 a** manner of playing keys or strings. **b** response of the keys or strings. **c** style of workmanship, writing, etc. **5** distinguishing manner or detail (*a professional touch*). **6** special skill (*have lost my touch*). **7** (esp. in *pl.*) light stroke with a pencil etc. **8** *slang* **a** act of getting money etc. from a person by asking. **b** = SOFT TOUCH. **9** *Football* part of the field outside the side limits. □ **in touch** (often foll. by *with*) **1** in communication. **2** aware. **lose touch** (often foll. by *with*) **1** cease to be informed. **2** cease to be in contact. **out of touch** (often foll. by *with*) **1** not in correspondence. **2** not up to date. **3** lacking in awareness. **touch at** (of a ship) call at (a port etc.). **touch bottom** **1** reach the bottom of water with one's feet. **2** be at the lowest or worst point. **touch down** (of an aircraft) make contact with the ground in landing. **touch off** **1** explode by touching with a match etc. **2** initiate (a process) suddenly. **touch on** (or **upon**) **1** refer to or mention briefly or casually. **2** verge on. **touch up** **1** give finishing touches to or retouch. **2** *slang* sexually molest. **touch wood** touch something wooden to avert ill luck. [French *tochier*]

touch-and-go *adj.* critical, risky.

touchdown *n.* act of touching down by an aircraft.

touché /tu:'ʃeɪ/ *int.* **1** acknowledgement of a justified accusation or retort. **2** acknowledgement of a hit by a fencing-opponent. [French, = *touched*]

touching –*adj.* moving; pathetic. –*prep. literary* concerning. □ **touchingly** *adv.*

touch-line *n.* (in various sports) either of the lines marking the side boundaries of the pitch.

touch-paper *n.* paper impregnated with nitre, for igniting fireworks etc.

touchstone *n.* **1** dark schist or jasper used for testing alloys by marking it with them. **2** criterion.

touch-type *v.* type without looking at the keys. □ **touch-typist** *n.*

touchwood *n.* readily inflammable wood etc., esp. when made soft by fungi.

touchy *adj.* (**-ier**, **-iest**) apt to take offence; over-sensitive. □ **touchily** *adv.* **touchiness** *n.*

tough /tʌf/ –*adj.* **1** hard to break, cut, tear, or chew. **2** able to endure hardship; hardy. **3** unyielding, stubborn, difficult (*it was a tough job*). **4** *colloq.* **a** acting sternly; hard (*get tough with*). **b** (of circumstances, luck, etc.) severe, hard. **5** *colloq.* criminal or violent. –*n.* tough person, esp. a ruffian. □ **toughen** *v.* **toughness** *n.* [Old English]

toupee /'tu:peɪ/ *n.* hairpiece to cover a bald spot. [French]

tour /tʊə(r)/ *–n.* **1 a** journey from place to place as a holiday; sightseeing excursion. **b** a walk round; inspection (*made a tour of the garden*). **2** spell of duty on military or diplomatic service. **3** series of performances, matches, etc., at different places. *–v.* **1** (usu. foll. by *through*) make a tour. **2** make a tour of (a country etc.). □ **on tour** (esp. of a team, theatre company, etc.) touring. [Latin: related to TURN]

tour de force /ˌtʊə də 'fɔ:s/ *n.* (*pl.* ***tours de force***) outstanding feat or performance. [French]

tourer *n.* car or caravan for touring in.

tourism *n.* commercial organization and operation of holidays.

tourist *n.* **1** holiday-maker, esp. abroad (often *attrib.*: *tourist season*). **2** member of a touring sports team.

tourist class *n.* lowest class of passenger accommodation in a ship, aircraft, etc.

touristy *adj.* usu. *derog.* appealing to or visited by many tourists.

tourmaline /'tʊərməˌli:n/ *n.* mineral of various colours used as a gemstone. [French from Sinhalese]

tournament /'tʊənəmənt/ *n.* **1** large contest of many rounds (*chess tournament*). **2** display of military exercises etc. (*Royal Tournament*). **3** *hist.* pageant with jousting. [French: related to TOURNEY]

tournedos /'tʊənəˌdəʊ/ *n.* (*pl.* same /-ˌdəʊz/) small round thick cut from a fillet of beef. [French]

tourney /'tʊənɪ/ *–n.* (*pl.* **-s**) tournament. *–v.* (**-eys, -eyed**) take part in a tournament. [French: related to TURN]

tourniquet /'tʊənɪˌkeɪ/ *n.* device for stopping the flow of blood through an artery by constriction. [French]

tour operator *n.* travel agent specializing in package holidays.

tousle /'taʊz(ə)l/ *v.* (**-ling**) **1** make (esp. the hair) untidy. **2** handle roughly. [dial. *touse*]

tout *–v.* **1** (usu. foll. by *for*) solicit custom persistently; pester customers. **2** solicit the custom of (a person) or for (a thing), esp. sell (tickets) at a price higher than the official one. **3** spy out the movements and condition of racehorses in training. *–n.* person who touts, esp. tickets. [Old English, = peep]

tow¹ /təʊ/ *–v.* pull (a boat, vehicle, etc.) along by a rope etc. *–n.* towing or being towed. □ **have in** (or **on**) **tow 1** be towing. **2** be accompanied by and often in charge of (a person). **on tow** being towed. [Old English]

tow² *n.* coarse part of flax or hemp prepared for spinning. [Low German *touw*]

toward /tə'wɔ:d/ *prep.* = TOWARDS.

towards /tə'wɔ:dz/ *prep.* **1** in the direction of (*set out towards town*). **2** as regards; in relation to (*attitude towards death*). **3** as a contribution to; for (*put it towards her holiday*). **4** near (*towards the end of our journey*). [Old English, = future: related to TO, -WARD]

tow-bar *n.* bar for towing esp. a caravan.

towel /'taʊəl/ *–n.* absorbent cloth or paper etc. used for drying after washing. *–v.* (**-ll-**; *US* **-l-**) (often *refl.*) wipe or dry with a towel. [French *toail(l)e* from Germanic]

towelling *n.* thick soft absorbent cloth, used esp. for towels.

tower *–n.* **1** tall structure, often part of a church, castle, etc. **2** fortress etc. with a tower. **3** tall structure housing machinery etc. (*cooling tower*; *control tower*). *–v.* **1** (usu. foll. by *above*, *up*) reach or be high or above; be superior. **2** (as **towering** *adj.*) **a** high, lofty (*towering intellect*). **b** violent (*towering rage*). [Greek *turris*]

tower block *n.* tall building containing offices or flats.

tower of strength *n.* person who gives strong emotional support.

tow-headed *adj.* having very light or unkempt hair.

town *n.* **1 a** densely populated built-up defined area, between a city and a village in size. **b** densely populated area, esp. as opposed to the country. **2 a** London or the chief city or town in an area (*went up to town*). **b** central business area in a neighbourhood. □ **go to town** *colloq.* do something with great energy or enthusiasm. **on the town** *colloq.* enjoying night-life in a town. [Old English]

town clerk *n.* *US* & *hist.* official in charge of the records etc. of a town.

town crier *n.* = CRIER 2.

townee var. of TOWNIE.

town gas *n.* manufactured gas for domestic and commercial use.

town hall *n.* headquarters of local government, with public meeting rooms etc.

town house *n.* **1** town residence, esp. of a person with a house in the country. **2** terrace house. **3** house in a planned group in a town.

townie /'taʊnɪ/ *n.* (also **townee** /-'ni:/) *derog.* town inhabitant ignorant of country life.

town planning *n.* planning of the construction and growth of towns. □ **town planner** *n.*
townscape *n.* **1** visual appearance of a town or towns. **2** picture of a town.
townsfolk *n.* inhabitants of a town or towns.
township *n.* **1** *S.Afr.* urban area set aside for Black occupation. **2** *US & Can.* **a** division of a county. **b** district six miles square. **3** *hist.* small town or village forming part of a large parish. **4** *Austral. & NZ* small town.
townsman *n.* (*fem.* **townswoman**) inhabitant of a town.
townspeople *n.pl.* people of a town.
tow-path *n.* path by a river or canal, orig. used for towing a boat by horse.
toxaemia /tɒk'siːmɪə/ *n.* (*US* **toxemia**) **1** blood-poisoning. **2** increased blood pressure in pregnancy. [related to TOXIC, Greek *haima* blood]
toxic /'tɒksɪk/ *adj.* **1** poisonous. **2** of poison. □ **toxicity** /-'sɪsɪtɪ/ *n.* [Greek *toxikon* poison for arrows]
toxicology /ˌtɒksɪ'kɒlədʒɪ/ *n.* the study of poisons. □ **toxicological** /-kə'lɒdʒɪk(ə)l/ *adj.* **toxicologist** *n.*
toxin /'tɒksɪn/ *n.* poison produced by a living organism.
toxocara /ˌtɒksə'kɑːrə/ *n.* parasitic worm in dogs and cats.
toxocariasis /ˌtɒksəkə'raɪəsɪs/ *n.* disease resulting from infection by the toxocara.
toy –*n.* **1** plaything. **2** thing regarded as providing amusement. **3** (usu. *attrib.*) diminutive breed of dog etc. –*v.* (usu. foll. by *with*) **1** trifle, amuse oneself, flirt. **2** move a thing idly. [origin unknown]
toy boy *n. colloq.* woman's much younger boyfriend.
toyshop *n.* shop selling toys.
trace[1] –*v.* (**-cing**) **1 a** observe or find vestiges or signs of by investigation. **b** (often foll. by *along, through, to,* etc.) follow or mark the track or position of. **c** (often foll. by *back*) follow to its origins. **2** copy (a drawing etc.) by drawing over its lines on superimposed translucent paper. **3** (often foll. by *out*) mark out, delineate, sketch, or write, esp. laboriously. **4** make one's way along (a path etc.). –*n.* **1 a** indication of something having existed; vestige. **b** very small quantity. **2** track or footprint. **3** track left by the moving pen of an instrument etc. □ **traceable** *adj.* [Latin *traho* draw]
trace[2] *n.* each of the two side-straps, chains, or ropes by which a horse draws a vehicle. □ **kick over the traces** become insubordinate or reckless. [French *trais*, pl. of TRAIT]
trace element *n.* chemical element required only in minute amounts by living organisms for normal growth.
tracer *n.* **1** bullet etc. that is visible in flight because of flames etc. emitted. **2** artificial radioactive isotope which can be followed through the body by the radiation it produces.
tracery *n.* (*pl.* **-ies**) **1** ornamental stone openwork, esp. in the upper part of a Gothic window. **2** fine decorative pattern.
trachea /trə'kiːə/ *n.* (*pl.* **-cheae** /-'kiːiː/) windpipe. [Latin from Greek]
tracheotomy /ˌtrækɪ'ɒtəmɪ/ *n.* (*pl.* **-ies**) incision of the trachea to relieve an obstruction.
tracing *n.* **1** traced copy of a drawing etc. **2** act of tracing.
tracing-paper *n.* translucent paper for making tracings.
track[1] –*n.* **1 a** mark(s) left by a person, animal, vehicle, etc. **b** (in *pl.*) such marks, esp. footprints. **2** rough path, esp. one beaten by use. **3** continuous railway line. **4 a** racecourse; circuit. **b** prepared course for runners etc. **5 a** groove on a gramophone record. **b** section of a record, CD, or magnetic tape containing one song etc. **c** lengthwise strip of magnetic tape containing a single sequence of signals. **6** line of travel (*track of the comet*). **7** band round the wheels of a tank etc. **8** line of thought or action. –*v.* **1** follow the track of. **2** trace (a course, development, etc.) by vestiges. **3** (often foll. by *back, in,* etc.) (of a film or television camera) move in relation to the subject being filmed. □ **in one's tracks** *colloq.* where one stands, instantly (*stopped him in his tracks*). **keep** (or **lose**) **track of** follow (or fail to follow) the course of. **make tracks** *colloq.* depart. **make tracks for** *colloq.* go in pursuit of or towards. **off the track** away from the subject. **track down** reach or capture by tracking. □ **tracker** *n.* [French *trac*]
tracker dog *n.* police dog tracking by scent.
track events *n.pl.* running-races as opposed to jumping etc.
track record *n.* person's past performance.
track shoe *n.* runner's spiked shoe.
track suit *n.* loose warm suit worn for exercising etc.
tract[1] *n.* **1** stretch or extent of territory, esp. large. **2** bodily organ or system (*digestive tract*). [Latin *traho tract-* pull]

tract[2] *n.* pamphlet, esp. propagandist. [apparently Latin *tractatus* from *tracto* handle]

tractable *adj.* (of a person or material) easily handled; manageable. □ **tractability** /-'bılıtı/ *n.* [Latin *tracto* handle]

traction /'trækʃ(ə)n/ *n.* **1** act of hauling or pulling a thing over a surface. **2** sustained therapeutic pulling on a limb etc. with pulleys, weights, etc. [French or medieval Latin: related to TRACT[1]]

traction-engine *n.* steam or diesel engine for drawing heavy loads on roads, fields, etc.

tractor /'træktə(r)/ *n.* **1** vehicle used for pulling farm machinery etc. **2** traction-engine. [related to TRACTION]

trad *colloq.* –*n.* traditional jazz. –*adj.* traditional. [abbreviation]

trade –*n.* **1 a** buying and selling. **b** this between nations etc. **c** business conducted for profit (esp. as distinct from a profession). **d** business of a specified nature or time (*Christmas trade*; *tourist trade*). **2** skilled craft practised professionally. **3** (usu. prec. by *the*) people engaged in a specific trade (*the trade will never agree*). **4** *US* transaction, esp. a swap. **5** (usu. in *pl.*) trade wind. –*v.* (**-ding**) **1** (often foll. by *in*, *with*) engage in trade; buy and sell. **2 a** exchange in commerce. **b** exchange (insults, blows, etc.). **c** *US* swap. **3** (usu. foll. by *with*, *for*) have a transaction with a person for a thing. □ **trade in** (often foll. by *for*) exchange (esp. a used car) in part payment for another. **trade off** exchange, esp. as a compromise. **trade on** take advantage of. □ **tradable** *adj.* **tradeable** *adj.* [Low German, = track: related to TREAD]

trade-in *n.* thing given in part exchange for another.

trade mark *n.* **1** device or name secured by law or custom as representing a company, product, etc. **2** distinctive characteristic etc.

trade name *n.* **1** name by which a thing is called in a trade. **2** name given to a product. **3** name under which a business trades.

trade-off *n.* balance, compromise.

trade price *n.* price charged to the retailer.

trader *n.* **1** person engaged in trade. **2** merchant ship.

tradescantia /,trædı'skæntıə/ *n.* (usu. trailing) plant with large blue, white, or pink flowers. [*Tradescant*, name of a naturalist]

trade secret *n.* **1** secret device or technique used esp. in a trade. **2** *joc.* any secret.

tradesman *n.* (*fem.* **tradeswoman**) person engaged in trade, esp. a shopkeeper.

tradespeople *n.pl.* people engaged in trade.

Trades Union Congress *n.* official representative body of British trade unions.

trade union *n.* (also **trades union**) organized association of workers in a trade, profession, etc., formed to protect and further their rights and interests. □ **trade-unionism** *n.* **trade-unionist** *n.*

trade wind *n.* wind blowing continually towards the equator and deflected westward.

trading *n.* act of engaging in trade.

trading estate *n.* specially-designed industrial and commercial area.

trading post *n.* store etc. in a remote or unsettled region.

trading-stamp *n.* stamp given to customers by some shops and exchangeable in large numbers for goods or cash.

tradition /trə'dıʃ(ə)n/ *n.* **1 a** custom, opinion, or belief handed down to posterity. **b** this process of handing down. **2** artistic, literary, etc. principles based on experience and practice; any one of these. [Latin *trado -dit-* hand on, betray]

traditional *adj.* **1** of, based on, or obtained by tradition. **2** (of jazz) in the style of the early 20th c. □ **traditionally** *adv.*

traditionalism *n.* respect or support for tradition. □ **traditionalist** *n.* & *adj.*

traduce /trə'dju:s/ *v.* (**-cing**) speak ill of; misrepresent. □ **traducement** *n.* **traducer** *n.* [Latin *traduco*, = disgrace]

traffic /'træfık/ –*n.* **1** vehicles moving on a public highway or in the air or at sea. **2** (usu. foll. by *in*) trade, esp. illegal (*drugs traffic*). **3** coming and going of people or goods by road, rail, air, sea, etc. **4** dealings between people etc. (*had no traffic with them*). **5** messages etc. transmitted through a communications system; volume of this. –*v.* (**-ck-**) **1** (usu. foll. by *in*) deal in something, esp. illegally. **2** deal in; barter. □ **trafficker** *n.* [French from Italian]

traffic island *n.* raised area in a road to divide traffic streams and for pedestrians to use in crossing.

traffic jam *n.* traffic at a standstill because of roadworks, an accident, etc.

traffic-light *n.* (also **traffic-lights** *n.pl.*) signal controlling road traffic by coloured lights.

traffic warden *n.* official employed to help control road traffic and esp. parking.

tragedian /trə'dʒi:dıən/ *n.* **1** writer of tragedies. **2** (*fem.* **tragedienne** /-dı'en/)

actor in tragedy. [French: related to TRAGEDY]

tragedy /'trædʒɪdɪ/ *n.* (*pl.* **-ies**) **1** serious accident, disaster, etc.; sad event. **2 a** play dealing with tragic events and ending unhappily, esp. with the downfall of the protagonist. **b** such plays as a genre. [Greek *tragōidia*]

tragic /'trædʒɪk/ *adj.* **1** disastrous; greatly distressing; very sad. **2** of tragedy. □ **tragically** *adv.*

tragicomedy /,trædʒɪ'kɒmədɪ/ *n.* (*pl.* **-ies**) play or situation with a mixture of comedy and tragedy. □ **tragicomic** *adj.*

trail –*n.* **1** track or scent left by a moving thing, person, etc. **2** beaten path, esp. through a wild region. **3** long line of people or things following behind something. **4** part dragging behind a thing or person. –*v.* **1** draw or be drawn along behind, esp. on the ground. **2** (often foll. by *behind*) walk wearily. **3** follow the trail of; pursue. **4** be losing in a contest (*trailing by three points*). **5** (usu. foll. by *away*, *off*) peter out; tail off. **6 a** (of a plant etc.) grow or hang over a wall, along the ground, etc. **b** hang loosely. **7** (often *refl.*) drag (oneself, one's limbs, etc.) along wearily etc. [French or Low German]

trail-blazer *n.* **1** person who marks a new track through wild country. **2** pioneer. □ **trail-blazing** *n.*

trailer *n.* **1** set of brief extracts from a film etc., used to advertise it in advance. **2** vehicle towed by another, esp.: **a** the rear section of an articulated lorry. **b** an open cart. **c** a platform for transporting a boat etc. **d** *US* a caravan.

trailing edge *n.* rear edge of an aircraft's wing etc.

train –*v.* **1 a** (often foll. by *to* + infin.) teach (a person, animal, oneself, etc.) a specified skill, esp. by practice. **b** undergo this process (*trained as a teacher*). **2** bring or come to physical efficiency by exercise, diet, etc. **3** (often foll. by *along, up*) guide the growth of (a plant). **4** (usu. as **trained** *adj.*) make (the mind, eye, etc.) discerning through practice etc. **5** (often foll. by *on*) point or aim (a gun, camera, etc.) at an object etc. –*n.* **1** series of railway carriages or trucks drawn by an engine. **2** thing dragged along behind or forming the back part of a dress, robe, etc. **3** succession or series of people, things, events, etc. (*train of thought*). **4** body of followers; retinue. □ **in train** properly arranged or directed. □ **trainee** /-'ni:/ *n.* [Latin *traho* draw]

train-bearer *n.* person holding up the train of a robe etc.

trainer *n.* **1** person who trains horses, athletes, footballers, etc. **2** aircraft or simulator used to train pilots. **3** soft running shoe.

training *n.* process of teaching or learning a skill etc.

train-spotter *n.* person who collects locomotive numbers as a hobby. □ **train-spotting** *n.*

traipse *colloq.* –*v.* (**-sing**) tramp or trudge wearily. –*n.* tedious journey on foot. [origin unknown]

trait /treɪ, treɪt/ *n.* characteristic. [Latin *tractus*: related to TRACT[1]]

traitor /'treɪtə(r)/ *n.* (*fem.* **traitress**) (often foll. by *to*) person who is treacherous or disloyal, esp. to his or her country. □ **traitorous** *adj.* [Latin *traditor*: related to TRADITION]

trajectory /trə'dʒektərɪ/ *n.* (*pl.* **-ies**) path of an object moving under given forces. [Latin *traicio -ject-* throw across]

tram *n.* **1** (also **tramcar**) electrically-powered passenger road vehicle running on rails. **2** four-wheeled vehicle used in coalmines. [Low German and Dutch *trame* beam]

tramlines *n.pl.* **1** rails for a tramcar. **2** *colloq.* pair of long parallel lines at the sides of a tennis or badminton court.

trammel /'træm(ə)l/ –*n.* **1** (usu. in *pl.*) impediment; hindrance (*trammels of domesticity*). **2** triple drag-net for fishing. –*v.* (**-ll-**; *US* **-l-**) hamper. [medieval Latin *tremaculum*]

tramp –*v.* **1 a** walk heavily and firmly. **b** go on foot, esp. a distance. **2 a** cross on foot, esp. wearily or reluctantly. **b** cover (a distance) in this way. **3** (often foll. by *down*) tread on; trample; stamp on. **4** live as a tramp. –*n.* **1** itinerant vagrant or beggar. **2** sound of a person, or esp. people, walking, marching, etc. **3** long walk. **4** *slang derog.* promiscuous woman. [Germanic]

trample /'træmp(ə)l/ *v.* (**-ling**) **1** tread under foot. **2** press down or crush in this way. □ **trample on 1** tread heavily on. **2** treat roughly or with contempt. [from TRAMP]

trampoline /'træmpə,li:n/ –*n.* strong fabric sheet connected by springs to a horizontal frame, used for gymnastic jumping. –*v.* (**-ning**) use a trampoline. [Italian *trampolino*]

tramway *n.* rails for a tram.

trance /trɑ:ns/ *n.* **1 a** sleeplike state without response to stimuli. **b** hypnotic or cataleptic state. **2** such a state as entered into by a medium. **3** rapture, ecstasy. [Latin *transeo* pass over]

tranny /'trænɪ/ *n.* (*pl.* **-ies**) *colloq.* transistor radio. [abbreviation]

tranquil /'træŋkwɪl/ *adj.* calm, serene, undisturbed. □ **tranquillity** /-'kwɪlɪtɪ/ *n.* **tranquilly** *adv.* [Latin]

tranquillize *v.* (*US* **tranquilize**, **-ise**) (**-zing** or **-sing**) make tranquil, esp. by a drug etc.

tranquillizer *n.* (*US* **tranquilizer**, **-iser**) drug used to diminish anxiety.

trans- *prefix* **1** across, beyond. **2** on or to the other side of. **3** through. [Latin]

transact /træn'zækt/ *v.* perform or carry through (business). [Latin: related to ACT]

transaction /træn'zækʃ(ə)n/ *n.* **1 a** piece of esp. commercial business done. **b** transacting of business etc. **2** (in *pl.*) published reports of discussions, papers read, etc., at the meetings of a learned society.

transalpine /trænz'ælpaɪn/ *adj.* on the north side of the Alps. [Latin]

transatlantic /ˌtrænzət'læntɪk/ *adj.* **1** beyond the Atlantic, esp.: **a** American. **b** *US* European. **2** crossing the Atlantic.

transceiver /træn'si:və(r)/ *n.* combined radio transmitter and receiver.

transcend /træn'send/ *v.* **1** be beyond the range or grasp of (human experience, reason, belief, etc.). **2** excel; surpass. [Latin *scando* climb]

transcendent *adj.* **1** excelling, surpassing. **2** transcending human experience. **3** (esp. of God) existing apart from, not subject to the limitations of, the material universe. □ **transcendence** *n.* **transcendency** *n.*

transcendental /ˌtrænsen'dent(ə)l/ *adj.* **1** *Philos.* a priori, not based on experience; intuitively accepted; innate in the mind. **2 a** visionary, abstract. **b** vague, obscure. □ **transcendentally** *adv.*

transcendentalism *n.* transcendental philosophy. □ **transcendentalist** *n.*

Transcendental Meditation *n.* method of detaching oneself from problems, anxiety, etc., by silent meditation and repetition of a mantra.

transcontinental /ˌtrænzkɒntɪ'nent(ə)l/ *adj.* extending across a continent.

transcribe /træn'skraɪb/ *v.* (**-bing**) **1** copy out. **2** write out (shorthand, notes, etc.) in full. **3** record for subsequent reproduction. **4** arrange (music) for a different instrument etc. □ **transcriber** *n.* **transcription** /-'skrɪpʃ(ə)n/ *n.* [Latin *transcribo -script-*]

transcript /'trænskrɪpt/ *n.* written copy.

transducer /trænz'dju:sə(r)/ *n.* any device for converting a non-electrical signal into an electrical one, e.g. pressure into voltage. [Latin: related to DUCT]

transept /'trænsept/ *n.* **1** part of a cross-shaped church at right angles to the nave. **2** either arm of this. [Latin: related to SEPTUM]

transexual var. of TRANSSEXUAL.

transfer –*v.* /træns'fɜ:(r)/ (**-rr-**) **1** (often foll. by *to*) **a** convey, remove, or hand over (a thing etc.). **b** make over the possession of (property, a ticket, rights, etc.) to a person. **2** change or move to another group, club, department, etc. **3** change from one station, route, etc., to another on a journey. **4** convey (a design) from one surface to another. **5** change (meaning) by extension or metaphor. –*n.* /'trænsfɜ:(r)/ **1** transferring or being transferred. **2** design etc. conveyed or to be conveyed from one surface to another. **3** football player etc. who is transferred. **4** document effecting conveyance of property, a right, etc. □ **transferable** /-'fɜ:rəb(ə)l/ *adj.* [Latin *fero lat-* bear]

transference /'trænsfərəns/ *n.* **1** transferring or being transferred. **2** *Psychol.* redirection of childhood emotions to a new object, esp. to a psychoanalyst.

transfiguration /ˌtrænsfɪgə'reɪʃ(ə)n/ *n.* **1** change of form or appearance. **2 a** Christ's appearance in radiant glory to three of his disciples (Matt. 17:2, Mark 9:2–3). **b** (**Transfiguration**) festival of Christ's transfiguration, 6 August. [Latin: related to TRANSFIGURE]

transfigure /træns'fɪgə(r)/ *v.* (**-ring**) change in form or appearance, esp. so as to elevate or idealize. [Latin]

transfix /træns'fɪks/ *v.* **1** paralyse with horror or astonishment. **2** pierce with a sharp implement or weapon. [Latin: related to FIX]

transform /træns'fɔ:m/ *v.* **1** make a thorough or dramatic change in the form, appearance, character, etc., of. **2** change the voltage etc. of (an alternating current). □ **transformation** /-fə'meɪʃ(ə)n/ *n.* [Latin]

transformer *n.* apparatus for reducing or increasing the voltage of an alternating current.

transfuse /træns'fju:z/ *v.* (**-sing**) **1 a** transfer (blood) from one person or animal to another. **b** inject (liquid) into a blood-vessel to replace lost fluid. **2** permeate. □ **transfusion** *n.* [Latin: related to FOUND[3]]

transgress /trænz'gres/ *v.* (also *absol.*) go beyond the bounds or limits set by (a commandment, law, etc.); sin. □ **transgression** *n.* **transgressor** *n.* [Latin *transgredior -gress-*]

transient /'trænzɪənt/ *adj.* of short duration; passing. □ **transience** *n.* [Latin: related to TRANCE]

transistor /træn'zɪstə(r)/ *n.* **1** semiconductor device with three connections, capable of amplification in addition to rectification. **2** (in full **transistor radio**) portable radio with transistors. [from TRANSFER, RESISTOR]

transistorize *v.* (also **-ise**) (**-zing** or **-sing**) equip with transistors (rather than valves).

transit /'trænzɪt/ *n.* **1** going, conveying, or being conveyed, esp. over a distance. **2** passage or route. **3** apparent passage of a celestial body across the meridian of a place, or across the sun or a planet. □ **in transit** while going or being conveyed. [Latin: related to TRANCE]

transit camp *n.* camp for the temporary accommodation of soldiers, refugees, etc.

transition /træn'zɪʃ(ə)n/ *n.* **1** passing or change from one place, state, condition, etc., to another. **2** *Art* change from one style to another, esp. *Archit.* from Norman to Early English. □ **transitional** *adj.* **transitionally** *adv.* [Latin: related to TRANSIT]

transitive /'trænsɪtɪv/ *adj.* (of a verb) taking a direct object (whether expressed or implied), e.g. *saw* in *saw the donkey, saw that she was ill.* [Latin: related to TRANSIT]

transitory /'trænsɪtərɪ/ *adj.* not permanent; brief, transient. □ **transitorily** *adv.* **transitoriness** *n.* [Latin: related to TRANSIT]

translate /træn'sleɪt/ *v.* (**-ting**) **1** (also *absol.*) (often foll. by *into*) express the sense of (a word, text, etc.) in another language or in another, esp. simpler, form. **2** be translatable, bear translation (*does not translate well*). **3** interpret (*translated his silence as dissent*). **4** move or change, esp. from one person, place, or condition, to another. □ **translatable** *adj.* **translation** *n.* **translator** *n.* [Latin: related to TRANSFER]

transliterate /trænz'lɪtə,reɪt/ *v.* (**-ting**) represent (a word etc.) in the closest corresponding letters of a different script. □ **transliteration** /-'reɪʃ(ə)n/ *n.* [Latin *littera* letter]

translucent /træns'lu:s(ə)nt/ *adj.* allowing light to pass through; semi-transparent. □ **translucence** *n.* **translucency** *n.* [Latin *luceo* shine]

transmigrate /,trænzmaɪ'greɪt/ *v.* (**-ting**) **1** (of the soul) pass into a different body. **2** migrate. □ **transmigration** /-'greɪʃ(ə)n/ *n.* [Latin]

transmission /trænz'mɪʃ(ə)n/ *n.* **1** transmitting or being transmitted. **2** broadcast programme. **3** mechanism transmitting power from the engine to the axle in a vehicle.

transmit /trænz'mɪt/ *v.* (**-tt-**) **1 a** pass or hand on; transfer (*transmitted the message; how diseases are transmitted*). **b** communicate (ideas, emotions, etc.). **2 a** allow (heat, light, sound, electricity, etc.) to pass through. **b** be a medium for (ideas, emotions, etc.) (*his message transmits hope*). **3** broadcast (a radio or television programme). □ **transmissible** *adj.* **transmittable** *adj.* [Latin *mitto miss-* send]

transmitter *n.* **1** person or thing that transmits. **2** equipment used to transmit radio or other electronic signals.

transmogrify /trænz'mɒgrɪ,faɪ/ *v.* (**-ies, -ied**) *joc.* transform, esp. in a magical or surprising manner. □ **transmogrification** /-fɪ'keɪʃ(ə)n/ *n.* [origin unknown]

transmute /trænz'mju:t/ *v.* (**-ting**) **1** change the form, nature, or substance of. **2** *hist.* change (base metals) into gold. □ **transmutation** /-'teɪʃ(ə)n/ *n.* [Latin *muto* change]

transoceanic /,trænzəʊʃɪ'ænɪk/ *adj.* **1** beyond the ocean. **2** crossing the ocean.

transom /'trænsəm/ *n.* **1** horizontal bar of wood or stone across a window or the top of a door. **2** *US* = TRANSOM WINDOW. [French *traversin*: related to TRAVERSE]

transom window *n.* window above a transom.

transparency /træns'pærənsɪ/ *n.* (*pl.* **-ies**) **1** being transparent. **2** picture, esp. a photograph, to be viewed by light passing through it. [medieval Latin: related to TRANSPARENT]

transparent /træns'pærənt/ *adj.* **1** allowing light to pass through so that bodies can be distinctly seen. **2 a** (of a disguise, pretext, etc.) easily seen through. **b** (of a quality etc.) evident; obvious. **3** easily understood; frank. □ **transparently** *adv.* [Latin *pareo* appear]

transpire /træn'spaɪə(r)/ *v.* (**-ring**) **1** (usu. prec. by *it* as subject) (of a secret or fact) come to be known; turn out; prove to be the case (*it transpired he knew nothing about it*). **2** occur; happen. **3** emit (vapour or moisture), or be emitted, through the skin, lungs, or leaves; perspire. □ **transpiration** /-spɪ'reɪʃ(ə)n/ *n.* (in sense 3). [Latin *spiro* breathe]

■ **Usage** The use of *transpire* in sense 2 is considered incorrect by some people.

transplant –*v.* /træns'plɑ:nt/ **1** plant in another place (*transplanted the daffodils*). **2** transfer (living tissue or an organ) to another part of the body or to another body. –*n.* /'trænsplɑ:nt/ **1 a** transplanting of an organ or tissue. **b** such an organ etc. **2** thing, esp. a plant, transplanted. □ **transplantation** /-'teɪʃ(ə)n/ *n.* [Latin]

transponder /træn'spɒndə(r)/ *n.* device for receiving a radio signal and automatically transmitting a different signal. [from TRANSMIT, RESPOND]

transport –*v.* /træns'pɔ:t/ **1** take or carry (a person, goods, etc.) to another place. **2** *hist.* deport (a criminal) to a penal colony. **3** (as **transported** *adj.*) (usu. foll. by *with*) affected with strong emotion. –*n.* /'trænspɔ:t/ **1 a** system of conveying people, goods, etc., from place to place. **b** means of this (*our transport has arrived*). **2** ship, aircraft, etc. used to carry soldiers, stores, etc. **3** (esp. in *pl.*) vehement emotion (*transports of joy*). □ **transportable** /-'pɔ:təb(ə)l/ *adj.* [Latin *porto* carry]

transportation /,trænspɔ:'teɪʃ(ə)n/ *n.* **1** conveying or being conveyed. **2 a** system of conveying. **b** esp. *US* means of this. **3** *hist.* deportation of convicts.

transport café *n.* roadside café for (esp. commercial) drivers.

transporter *n.* vehicle used to transport other vehicles or heavy machinery etc. by road.

transporter bridge *n.* bridge carrying vehicles etc. across water on a suspended moving platform.

transpose /træns'pəʊz/ *v.* (**-sing**) **1 a** cause (two or more things) to change places. **b** change the position of (a thing) in a series. **2** change the order or position of (words or a word) in a sentence. **3** put (music) into a different key. □ **transposition** /-pə'zɪʃ(ə)n/ *n.* [French: related to POSE]

transputer /træns'pju:tə(r)/ *n.* microprocessor with integral memory designed for parallel processing. [from TRANSISTOR, COMPUTER]

transsexual /træn'sekʃʊəl/ (also **transexual**) –*adj.* having the physical characteristics of one sex and an overwhelming psychological identification with the other. –*n.* **1** transsexual person. **2** person who has had a sex change.

transship /træns'ʃɪp/ *v.* (**-pp-**) transfer from one ship or form of transport to another. □ **transshipment** *n.*

transubstantiation /,trænsəb,stænʃɪ'eɪʃ(ə)n/ *n. RC Ch.* conversion of the Eucharistic elements wholly into the body and blood of Christ. [medieval Latin: related to TRANS-, SUBSTANCE]

transuranic /,trænsjʊə'rænɪk/ *adj.* (of a chemical element) having a higher atomic number than uranium.

transverse /'trænzvɜ:s/ *adj.* situated, arranged, or acting in a crosswise direction. □ **transversely** *adv.* [Latin *transverto -vers-* turn across]

transvestite /trænz'vestaɪt/ *n.* man deriving esp. sexual pleasure from dressing in women's clothes. □ **transvestism** *n.* [Latin *vestio* clothe]

trap[1] –*n.* **1** device, often baited, for catching animals. **2** trick betraying a person into speech or an act. **3** arrangement to catch an unsuspecting person. **4** device for hurling an object, e.g. a clay pigeon, into the air to be shot at. **5** compartment from which a greyhound is released at the start of a race. **6** device that sends a ball into the air. **7** curve in a downpipe etc. that fills with liquid and forms a seal against the return of gases. **8** two-wheeled carriage (*pony and trap*). **9** = TRAPDOOR. **10** *slang* mouth (esp. *shut one's trap*). –*v.* (**-pp-**) **1** catch (an animal) in a trap. **2** catch or catch out (a person) by means of a trick etc. **3** stop and retain in or as in a trap. **4** provide (a place) with traps. [Old English]

trap[2] *n.* (in full **trap-rock**) dark-coloured igneous rock. [Swedish]

trapdoor *n.* door in a floor, ceiling, or roof.

trapeze /trə'pi:z/ *n.* crossbar suspended by ropes as a swing for acrobatics etc. [Latin: related to TRAPEZIUM]

trapezium /trə'pi:zɪəm/ *n.* (*pl.* **-s** or **-zia**) **1** quadrilateral with only one pair of sides parallel. **2** *US* = TRAPEZOID 1. [Greek *trapezion*]

trapezoid /'træpɪ,zɔɪd/ *n.* **1** quadrilateral with no two sides parallel. **2** *US* = TRAPEZIUM 1. [Greek: related to TRAPEZIUM]

trapper *n.* person who traps wild animals, esp. for their fur.

trappings *n.pl.* **1** ornamental accessories. **2** harness of a horse, esp. when ornamental. [*trap* from French *drap* cloth]

Trappist –*n.* monk of an order vowed to silence. –*adj.* of this order. [*La Trappe* in Normandy]

trash –*n.* **1** esp. *US* worthless or waste stuff; rubbish. **2** worthless person or persons. –*v. slang* wreck, vandalize. □ **trashy** *adj.* (**-ier**, **-iest**). [origin unknown]

trash can *n. US* dustbin.

trattoria /ˌtrætəˈriːə/ *n.* Italian restaurant. [Italian]

trauma /ˈtrɔːmə/ *n.* (*pl.* **traumata** /-mətə/ or **-s**) **1** profound emotional shock. **2** physical injury. **3** physical shock syndrome following this. □ **traumatize** *v.* (also **-ise**) (**-zing** or **-sing**). [Greek, = wound]

traumatic /trɔːˈmætɪk/ *adj.* **1** of or causing trauma. **2** *colloq.* distressing (*traumatic experience*). □ **traumatically** *adv.* [Greek: related to TRAUMA]

travail /ˈtræveɪl/ *literary* —*n.* **1** painful effort. **2** pangs of childbirth. —*v.* make a painful effort, esp. in childbirth. [French *travaillier*]

travel /ˈtræv(ə)l/ —*v.* (**-ll-**; *US* **-l-**) **1** go from one place to another; make a journey, esp. a long one or abroad. **2 a** journey along or through (a country). **b** cover (a distance) in travelling. **3** *colloq.* withstand a long journey (*wines that do not travel*). **4** go from place to place as a salesman. **5** move or proceed as specified (*light travels faster than sound*). **6** *colloq.* move quickly. **7** pass, esp. in a deliberate manner, from point to point (*her eye travelled over the scene*). **8** (of a machine or part) move or operate in a specified way. —*n.* **1 a** travelling, esp. in foreign countries. **b** (often in *pl.*) spell of this. **2** range, rate, or mode of motion of a part in machinery. [originally = TRAVAIL]

travel agency *n.* agency that makes the necessary arrangements for travellers. □ **travel agent** *n.*

travelled *adj.* (*US* **traveled**) experienced in travelling (also in *comb.*: *much-travelled*).

traveller *n.* (*US* **traveler**) **1** person who travels or is travelling. **2** travelling salesman. **3** Gypsy.

traveller's cheque *n.* cheque for a fixed amount that may be cashed on signature abroad.

traveller's joy *n.* wild clematis.

traveller's tale *n.* incredible and probably untrue story.

travelling salesman *n.* = COMMERCIAL TRAVELLER.

travelogue /ˈtrævəˌlɒg/ *n.* film or illustrated lecture about travel. [from TRAVEL, after *monologue*]

travel-sick *adj.* suffering from nausea caused by motion in travelling.

traverse —*v.* /trəˈvɜːs/ (**-sing**) **1** travel or lie across (*traversed the country*; *pit traversed by a beam*). **2** consider or discuss the whole extent of (a subject). —*n.* /ˈtrævəs/ **1** sideways movement. **2** traversing. **3** thing that crosses another. □ **traversal** *n.* [French: related to TRANSVERSE]

travesty /ˈtrævɪstɪ/ —*n.* (*pl.* **-ies**) grotesque misrepresentation or imitation (*travesty of justice*). —*v.* (**-ies, -ied**) make or be a travesty of. [French *travestir* disguise, from Italian]

trawl —*v.* **1** fish with a trawl or seine. **2 a** catch by trawling. **b** (often foll. by *through*) search thoroughly (*trawled her memory for their names*). —*n.* **1** act of trawling. **2** (in full **trawl-net**) large wide-mouthed fishing-net dragged by a boat along the sea bottom. [probably Dutch *traghel* drag-net]

trawler *n.* boat used for trawling.

tray *n.* **1** flat board, usu. with a raised rim, for carrying dishes. **2** shallow lidless box for papers or small articles, sometimes forming a drawer in a cabinet etc. [Old English]

treacherous /ˈtretʃərəs/ *adj.* **1** guilty of or involving treachery. **2** (of the weather, ice, the memory, etc.) likely to fail or give way. □ **treacherously** *adv.* [French from *trichier* cheat: related to TRICK]

treachery /ˈtretʃərɪ/ *n.* (*pl.* **-ies**) violation of faith or trust; betrayal.

treacle /ˈtriːk(ə)l/ *n.* **1** syrup produced in refining sugar. **2** molasses. □ **treacly** *adj.* [French from Latin *theriaca* antidote against a snake-bite, from *thērion* wild animal]

tread /tred/ —*v.* (*past* **trod**; *past part.* **trodden** or **trod**) **1** (often foll. by *on*) set down one's foot; walk, step. **2 a** walk on. **b** (often foll. by *down*) press or crush with the feet. **3** perform (steps etc.) by walking. **4** (often foll. by *in, into*) press down into the ground with the feet (*trod dirt into the carpet*). —*n.* **1** manner or sound of walking. **2** top surface of a step or stair. **3** thick moulded part of a vehicle tyre for gripping the road. **4 a** part of a wheel that touches the ground or rail. **b** part of a rail that the wheels touch. **5** part of the sole of a shoe that rests on the ground. □ **tread the boards** be an actor. **tread on air** feel elated. **tread on a person's toes** offend a person; encroach on a person's privileges etc. **tread water** maintain an upright position in water by moving the feet and hands. [Old English]

treadle /ˈtred(ə)l/ *n.* lever worked by the foot and imparting motion to a machine. [Old English: related to TREAD]

treadmill *n.* **1** device for producing motion by the weight of persons or animals stepping on steps attached to a revolving upright wheel. **2** similar device used for exercise. **3** monotonous routine work.

treadwheel *n.* = TREADMILL 1, 2.

treason /ˈtri:z(ə)n/ *n.* violation by a subject of allegiance to the sovereign or State. [Latin: related to TRADITION]

■ **Usage** The crime of *petty treason* was abolished in 1828. This is why *high treason*, originally distinguished from *petty treason*, now means the same as *treason*.

treasonable *adj.* involving or guilty of treason.

treasure /ˈtreʒə(r)/ *–n.* **1 a** precious metals or gems. **b** hoard of these. **c** accumulated wealth. **2** thing valued for its rarity, workmanship, associations, etc. (*art treasures*). **3** *colloq.* much loved or highly valued person. *–v.* (**-ring**) **1** value highly. **2** (often foll. by *up*) store up as valuable. [Greek *thēsauros*]

treasure hunt *n.* **1** search for treasure. **2** game in which players seek a hidden object from a series of clues.

treasurer *n.* person in charge of the funds of a society etc.

treasure trove *n.* treasure of unknown ownership found hidden.

treasury *n.* (*pl.* **-ies**) **1** place or building where treasure is stored. **2** funds or revenue of a State, institution, or society. **3** (**Treasury**) **a** department managing the public revenue of a country. **b** offices and officers of this.

Treasury bench *n.* front bench in the House of Commons occupied by Cabinet ministers etc.

treasury bill *n.* bill of exchange issued by the government to raise money for temporary needs.

treat *–v.* **1** act or behave towards or deal with (a person or thing) in a certain way (*treated me kindly*; *treat it as a joke*). **2** apply a process to (*treat it with acid*). **3** apply medical care or attention to. **4** present or deal with (a subject) in literature or art. **5** (often foll. by *to*) provide with food, drink, or entertainment at one's own expense (*treated us to dinner*). **6** (often foll. by *with*) negotiate terms (with a person). **7** (often foll. by *of*) give a spoken or written exposition. *–n.* **1** event or circumstance (esp. when unexpected or unusual) that gives great pleasure. **2** meal, entertainment, etc., designed to do this. **3** (prec. by *a*) extremely good or well (*they looked a treat*; *has come on a treat*). □ **treatable** *adj.* [Latin *tracto* handle]

treatise /ˈtri:tɪz/ *n.* a written work dealing formally and systematically with a subject. [Anglo-French: related to TREAT]

treatment *n.* **1** process or manner of behaving towards or dealing with a person or thing. **2** medical care or attention. **3** manner of treating a subject in literature or art. **4** (prec. by *the*) *colloq.* the customary way of dealing with a person, situation, etc. (*got the full treatment*).

treaty /ˈtri:tɪ/ *n.* (*pl.* **-ies**) **1** formal agreement between States. **2** agreement between parties, esp. for the purchase of property. [Latin: related to TREAT]

treble /ˈtreb(ə)l/ *–adj.* **1 a** threefold. **b** triple. **c** three times as much or many (*treble the amount*). **2** high-pitched. *–n.* **1** treble quantity or thing. **2** hit on the narrow band between the two middle circles of a dartboard, scoring treble. **3 a** *Mus.* = SOPRANO (esp. a boy's voice or part, or an instrument). **b** high-pitched voice. **4** high-frequency output of a radio, record-player, etc. *–v.* (**-ling**) make or become three times as much or many; increase threefold; multiply by three. □ **trebly** *adv.* [Latin: related to TRIPLE]

treble chance *n.* method of competing in a football pool in which the chances of winning depend on the number of draws and home and away wins predicted by the competitors.

treble clef *n.* clef placing the G above middle C on the second lowest line of the staff.

tree *–n.* **1** perennial plant with a woody self-supporting main stem or trunk and usu. unbranched for some distance above the ground. **2** piece or frame of wood etc. for various purposes (*shoetree*). **3** = FAMILY TREE. *–v.* (**trees**; **treed**) force to take refuge in a tree. □ **grow on trees** (usu. with *neg.*) be plentiful. □ **treeless** *adj.* [Old English]

treecreeper *n.* small creeping bird feeding on insects in tree-bark.

tree-fern *n.* large fern with an upright trunklike stem.

tree line *n.* = TIMBERLINE.

tree ring *n.* ring in a cross-section of a tree, from one year's growth.

tree surgeon *n.* person who treats decayed trees in order to preserve them.

treetop *n.* topmost part of a tree.

trefoil /ˈtrefɔɪl/ *n.* **1** leguminous plant with leaves of three leaflets, esp. clover. **2** three-lobed ornamentation, esp. in tracery windows. [Anglo-French: related to TRI-, FOIL[2]]

trek orig. *S.Afr.* *–v.* (**-kk-**) **1** travel or make one's way arduously. **2** esp. *hist.* migrate or journey with one's belongings by ox-wagon. *–n.* **1 a** long or arduous journey or walk (*quite a trek to the launderette*). **b** each stage of this. **2** organized migration of a body of people. □ **trekker** *n.* [Dutch, = draw]

trellis /'trelɪs/ *n.* (in full **trellis-work**) lattice of light wooden or metal bars, esp. as a support for climbing plants. [French *trelis*]

trematode /'tremə,təʊd/ *n.* a kind of parasitic flatworm. [Greek *trēma* hole]

tremble /'tremb(ə)l/ *–v.* (**-ling**) **1** shake involuntarily from emotion, weakness, etc. **2** be in a state of extreme apprehension. **3** quiver (*leaves trembled in the breeze*). *–n.* trembling; quiver (*tremble in his voice*). [medieval Latin: related to TREMULOUS]

trembler *n.* automatic vibrator for making and breaking an electrical circuit.

trembly *adj.* (**-ier**, **-iest**) *colloq.* trembling.

tremendous /trɪ'mendəs/ *adj.* **1** *colloq.* remarkable, considerable, excellent. **2** awe-inspiring, overpowering. □ **tremendously** *adv.* [Latin *tremendus* to be trembled at: related to TREMOR]

tremolo /'tremə,ləʊ/ *n.* (*pl.* **-s**) tremulous effect in music. [Italian: related to TREMULOUS]

tremor /'tremə(r)/ *n.* **1** shaking, quivering. **2** thrill (of fear, exultation, etc.). **3** (in full **earth tremor**) slight earthquake. [Latin *tremo* tremble]

tremulous /'tremjʊləs/ *adj.* trembling. □ **tremulously** *adv.* [Latin *tremulus*: related to TREMOR]

trench *–n.* **1** long narrow usu. deep ditch. **2** *Mil.* this dug by troops as a shelter from enemy fire. *–v.* **1** dig a trench or trenches in (the ground). **2** turn over the earth of (a field, garden, etc.) by digging a succession of ditches. [French *trenche*, *-ier* cut]

trenchant /'trentʃ(ə)nt/ *adj.* (of style or language etc.) incisive, terse, vigorous. □ **trenchancy** *n.* **trenchantly** *adv.* [French: related to TRENCH]

trench coat *n.* **1** soldier's lined or padded waterproof coat. **2** loose belted raincoat.

trencher *n. hist.* wooden or earthenware platter for serving food. [Anglo-French: related to TRENCH]

trencherman *n.* person who eats well, or in a specified manner.

trench warfare *n.* war carried on from trenches.

trend *–n.* general direction and tendency (esp. of events, fashion, or opinion). *–v.* **1** bend or turn away in a specified direction. **2** have a general tendency. [Old English]

trend-setter *n.* person who leads the way in fashion etc.

trendy *colloq.*; often *derog.* *–adj.* (**-ier**, **-iest**) fashionable. *–n.* (*pl.* **-ies**) fashionable person. □ **trendily** *adv.* **trendiness** *n.*

trepan /trɪ'pæn/ *–n.* cylindrical saw formerly used by surgeons for removing part of the skull. *–v.* (**-nn-**) perforate (the skull) with a trepan. [Greek *trupanon* auger]

trepidation /,trepɪ'deɪʃ(ə)n/ *n.* fear, anxiety. [Latin *trepidus* flurried]

trespass /'trespəs/ *–v.* **1** (usu. foll. by *on*, *upon*) make an unlawful or unauthorized intrusion (esp. on land or property). **2** (foll. by *on*) make unjustifiable claims on; encroach on (*trespass on your hospitality*). *–n.* **1** *Law* act of trespassing. **2** *archaic* sin, offence. □ **trespasser** *n.* [medieval Latin: related to TRANS-, PASS[1]]

tress *n.* **1** long lock of human (esp. female) hair. **2** (in *pl.*) woman's or girl's head of hair. [French]

trestle /'tres(ə)l/ *n.* **1** supporting structure for a table etc., consisting of two frames fixed at an angle or hinged, or of a bar with two divergent pairs of legs. **2** (in full **trestle-table**) table of a board or boards on trestles etc. **3** (in full **trestle-work**) open braced framework to support a bridge etc. [Latin *transtrum* cross-beam]

trews /tru:z/ *n.pl.* close-fitting usu. tartan trousers. [Irish and Gaelic: related to TROUSERS]

tri- *comb. form* three or three times. [Latin and Greek]

triad /'traɪæd/ *n.* **1** group of three (esp. notes in a chord). **2** the number three. **3** (also **Triad**) Chinese secret society, usu. criminal. □ **triadic** /-'ædɪk/ *adj.* [Latin from Greek]

trial /'traɪəl/ *n.* **1** judicial examination and determination of issues between parties by a judge with or without a jury. **2** test (*will give you a trial*). **3** trying thing or person (*trials of old age*). **4** match held to select players for a team. **5** (often in *pl.*) contest involving performance by horses, dogs, motor cycles, etc. □ **on trial 1** being tried in a court of law. **2** being tested; to be chosen or retained only if suitable. [Anglo-French: related to TRY]

trial and error *n.* repeated (usu. unsystematic) attempts continued until successful.

trial run *n.* preliminary operational test.

triangle /'traɪ,æŋg(ə)l/ *n.* **1** plane figure with three sides and angles. **2** any three things not in a straight line, with imaginary lines joining them. **3** implement of this shape. **4** musical instrument consisting of a steel rod bent into a triangle, struck with a small steel rod. **5** situation, esp. an emotional relationship, involving three people. □ **triangular** /-'æŋgjʊlə(r)/ *adj.* [Latin: related to TRI-]

triangulate /traɪ'æŋgjʊ,leɪt/ *v.* (**-ting**) measure and map out (an area) by dividing it into triangles. □ **triangulation** /-'leɪʃ(ə)n/ *n.*

Triassic /traɪ'æsɪk/ *Geol.* –*adj.* of the earliest period of the Mesozoic era. –*n.* this period. [related to TRIAD]

triathlon /traɪ'æθlən/ *n.* athletic contest of three events for all competitors. [from TRI- after DECATHLON]

tribe *n.* **1** group of (esp. primitive) families or communities, linked by social, religious, or blood ties, and usu. having a common culture and dialect and a recognized leader. **2** any similar natural or political division. **3** usu. *derog.* set or number of persons, esp. of one profession etc. or family. □ **tribal** *adj.* **tribalism** *n.* [Latin *tribus*]

tribesman *n.* (*fem.* **-woman**) member of a tribe.

tribology /traɪ'bɒlədʒɪ/ *n.* the study of friction, wear, lubrication, and the design of bearings. [Greek *tribō* rub]

tribulation /,trɪbjʊ'leɪʃ(ə)n/ *n.* great affliction. [Latin *tribulum* threshing-sledge]

tribunal /traɪ'bju:n(ə)l/ *n.* **1** board appointed to adjudicate in some matter. **2** court of justice. **3** seat or bench for a judge or judges. [Latin: related to TRIBUNE]

tribune /'trɪbju:n/ *n.* **1** popular leader or demagogue. **2** (in full **tribune of the people**) official in ancient Rome chosen by the people to protect their interests. [Latin *tribunus*: related to TRIBE]

tributary /'trɪbjʊtərɪ/ –*n.* (*pl.* **-ies**) **1** river or stream flowing into a larger river or lake. **2** *hist.* person or State paying or subject to tribute. –*adj.* **1** (of a river etc.) that is a tributary. **2** *hist.* **a** paying tribute. **b** serving as tribute. [Latin: related to TRIBUTE]

tribute /'trɪbju:t/ *n.* **1** thing said or done or given as a mark of respect or affection etc. **2** (foll. by *to*) indication of (some praiseworthy quality) (*their success is a tribute to their perseverance*). **3** *hist.* **a** periodic payment by one State or ruler to another, esp. as a sign of dependence. **b** obligation to pay this. [Latin *tributum* neuter past part. of *tribuo -ut-* assign, originally divide between TRIBES]

trice *n.* □ **in a trice** in an instant. [*trice* haul up, from Low German and Dutch]

triceps /'traɪseps/ *n.* muscle (esp. in the upper arm) with three points of attachment at one end. [Latin *caput* head]

triceratops /traɪ'serə,tɒps/ *n.* dinosaur with three sharp horns on the forehead and a wavy-edged collar round the neck. [Greek, = three-horned face]

trichinosis /,trɪkɪ'nəʊsɪs/ *n.* disease caused by hairlike worms usu. ingested in meat. [Greek *thrix trikh-* hair]

trichology /trɪ'kɒlədʒɪ/ *n.* the study of hair. □ **trichologist** *n.*

trichromatic /,traɪkrə'mætɪk/ *adj.* **1** having or using three colours. **2** (of vision) having the normal three colour-sensations, i.e. red, green, and purple.

trick –*n.* **1** action or scheme undertaken to deceive or outwit. **2** illusion (*trick of the light*). **3** special technique; knack. **4** **a** feat of skill or dexterity. **b** unusual action (e.g. begging) learned by an animal. **5** foolish or discreditable act; practical joke (*a mean trick to play*). **6** idiosyncrasy (*has a trick of repeating himself*). **7** **a** cards played in one round of a card-game. **b** point gained in this. **8** (*attrib.*) done to deceive or mystify (*trick photography*; *trick question*). –*v.* **1** deceive by a trick; outwit. **2** (often foll. by *out of*) swindle (*tricked out of his savings*). **3** (foll. by *into*) cause to do something by trickery (*tricked into marriage*; *tricked me into agreeing*). **4** foil, baffle; take by surprise. □ **do the trick** *colloq.* achieve the required result. **how's tricks?** *colloq.* how are you? **trick or treat** esp. *US* children's custom of calling at houses at Hallowe'en with the threat of pranks if they are not given a small gift. **trick out** (or **up**) dress or deck out. [French]

trickery *n.* deception, use of tricks.

trickle /'trɪk(ə)l/ –*v.* (**-ling**) **1** (cause to) flow in drops or a small stream. **2** come or go slowly or gradually (*information trickles out*). –*n.* trickling flow. [probably imitative]

trickle charger *n.* electrical charger for batteries that works at a steady slow rate.

trickster *n.* deceiver, rogue.

tricksy *adj.* (**-ier**, **-iest**) full of tricks; playful.

tricky *adj.* (**-ier**, **-iest**) **1** requiring care and adroitness (*tricky job*). **2** crafty, deceitful. □ **trickily** *adv.* **trickiness** *n.*

tricolour /'trɪkələ(r)/ *n.* (*US* **tricolor**) flag of three bands of different colours, esp. the French or Irish national flags. [French: related to TRI-]

tricot /'trɪkəʊ/ *n.* knitted fabric. [French]

tricycle /'traɪsɪk(ə)l/ *n.* three-wheeled pedal-driven vehicle similar to a bicycle.

trident /'traɪd(ə)nt/ *n.* three-pronged spear. [Latin *dens dent-* tooth]

Tridentine /traɪ'dentaɪn/ *adj.* of the Council of Trent, held at Trento in Italy 1545–63, esp. as the basis of Roman Catholic orthodoxy. [medieval Latin *Tridentum* Trento]

tried *past* and *past part.* of TRY.

triennial /traɪ'enɪəl/ *adj.* lasting, or recurring every, three years. [Latin *annus* year]

trier /'traɪə(r)/ *n.* **1** person who perseveres. **2** tester, esp. of foodstuffs.

trifle /'traɪf(ə)l/ –*n.* **1** thing of slight value or importance. **2 a** small amount, esp. of money. **b** (prec. by *a*) somewhat (*a trifle annoyed*). **3** dessert of sponge cake with custard, jelly, fruit, cream, etc. –*v.* (**-ling**) **1** talk or act frivolously. **2** (foll. by *with*) treat or deal with frivolously; flirt heartlessly with. [originally *trufle* from French = *truf(f)e* deceit]

trifling *adj.* **1** unimportant, petty. **2** frivolous.

triforium /traɪ'fɔ:rɪəm/ *n.* (*pl.* **-ria**) gallery or arcade above the arches of the nave, choir, and transepts of a church. [Anglo-Latin]

trig *n. colloq.* trigonometry. [abbreviation]

trigger –*n.* **1** movable device for releasing a spring or catch and so setting off a mechanism (esp. that of a gun). **2** event, occurrence, etc., that sets off a chain reaction. –*v.* (often foll. by *off*) set (an action or process) in motion; precipitate. □ **quick on the trigger** quick to respond. [*tricker* from Dutch *trekker* from *trekken* pull]

trigger-happy *adj.* apt to shoot on the slightest provocation.

trigonometry /,trɪgə'nɒmətrɪ/ *n.* branch of mathematics dealing with the relations of the sides and angles of triangles and with the relevant functions of any angles. □ **trigonometric** /-nə'metrɪk/ *adj.* **trigonometrical** /-nə'metrɪk(ə)l/ *adj.* [Greek *trigōnon* triangle]

trig point *n.* reference point on high ground, used in triangulation.

trike *n. colloq.* tricycle. [abbreviation]

trilateral /traɪ'lætər(ə)l/ *adj.* **1** of, on, or with three sides. **2** involving three parties. [Latin: related to TRI-]

trilby /'trɪlbɪ/ *n.* (*pl.* **-ies**) soft felt hat with a narrow brim and indented crown. [*Trilby*, name of a character in a novel by G. du Maurier]

trilingual /traɪ'lɪŋgw(ə)l/ *adj.* **1** able to speak three languages. **2** spoken or written in three languages.

trill –*n.* **1** quavering sound, esp. a rapid alternation of sung or played notes. **2** bird's warbling. **3** pronunciation of *r* with vibration of the tongue. –*v.* **1** produce a trill. **2** warble (a song) or pronounce (*r* etc.) with a trill. [Italian]

trillion /'trɪljən/ *n.* (*pl.* same) **1** a million million (10^{12}). **2** (now less often) a million million million (10^{18}). □ **trillionth** *adj.* & *n.* [French or Italian: related to TRI-, MILLION, after *billion*]

trilobite /'traɪlə,baɪt/ *n.* a kind of fossil marine arthropod. [from TRI-, Greek *lobos* lobe]

trilogy /'trɪlədʒɪ/ *n.* (*pl.* **-ies**) group of three related novels, plays, operas, etc.

trim –*v.* (**-mm-**) **1 a** make neat or of the required size or form, esp. by cutting away irregular or unwanted parts. **b** set in good order. **2** (foll. by *off, away*) cut off (unwanted parts). **3** ornament, decorate. **4** adjust the balance of (a ship or aircraft) by arranging its cargo etc. **5** arrange (sails) to suit the wind. **6 a** associate oneself with currently prevailing views, esp. to advance oneself. **b** hold a middle course in politics or opinion. **7** *colloq.* **a** rebuke sharply. **b** thrash. **c** get the better of in a bargain etc. –*n.* **1** state of readiness or fitness (*in perfect trim*). **2** ornament or decorative material. **3** trimming of a person's hair. –*adj.* (**trimmer**, **trimmest**) **1** neat or spruce. **2** in good order; well arranged or equipped. [Old English, = make firm]

trimaran /'traɪmə,ræn/ *n.* vessel like a catamaran, with three hulls side by side. [from CATAMARAN]

trimeter /'trɪmɪtə(r)/ *n. Prosody* line of verse of three measures. [Greek: see TRI-, -METER]

trimming *n.* **1** ornamentation or decoration, esp. for clothing. **2** (in *pl.*) *colloq.* usual accompaniments, esp. of the main course of a meal.

Trinitarian /,trɪnɪ'teərɪən/ –*n.* believer in the Trinity. –*adj.* of this belief. □ **Trinitarianism** *n.*

trinitrotoluene /traɪ,naɪtrə'tɒljʊ,i:n/ *n.* (also **trinitrotoluol** /-'tɒljʊ,ɒl/) = TNT.

trinity /'trɪnɪtɪ/ *n.* (*pl.* **-ies**) **1** state of being three. **2** group of three. **3** (**the Trinity** or **Holy Trinity**) *Theol.* the three persons of the Christian Godhead (Father, Son, and Holy Spirit). [Latin *trinitas* from *trinus* threefold]

Trinity Sunday *n.* Sunday next after Whit Sunday.

Trinity term *n.* university and law term beginning after Easter.

trinket /'trɪŋkɪt/ *n.* trifling ornament, esp. a piece of jewellery. □ **trinketry** *n.* [origin unknown]

trio /'tri:əʊ/ *n.* (*pl.* **-s**) **1** group of three. **2** *Mus.* **a** composition for three performers. **b** the performers. [French and Italian from Latin]

trip –*v.* (**-pp-**) **1 a** (often foll. by *up*) (cause to) stumble, esp. by catching the feet. **b** (foll. by *up*) (cause to) make a slip or blunder. **2 a** move with quick light steps. **b** (of a rhythm etc.) run lightly. **3** make an excursion to a place. **4 a** operate (a mechanism) suddenly by knocking aside a catch etc. **b** automatically cut out. **5** *slang* have a hallucinatory experience caused by a drug. –*n.* **1** journey or excursion, esp. for pleasure. **2 a** stumble or blunder. **b** tripping or being tripped up. **3** nimble step. **4** *slang* drug-induced hallucinatory experience. **5** device for tripping a mechanism etc. [Dutch *trippen* skip, hop]

tripartite /traɪ'pɑ:taɪt/ *adj.* **1** consisting of three parts. **2** shared by or involving three parties. [Latin *partior* divide]

tripe *n.* **1** first or second stomach of a ruminant, esp. an ox, as food. **2** *colloq.* nonsense, rubbish. [French]

triple /'trɪp(ə)l/ –*adj.* **1** consisting of three usu. equal parts or things; threefold. **2** involving three parties. **3** three times as much or many. –*n.* **1** threefold number or amount. **2** set of three. –*v.* (**-ling**) multiply by three. □ **triply** *adv.* [Latin *triplus* from Greek]

triple crown *n.* winning of all three of a group of sporting events, esp. in Rugby.

triple jump *n.* athletic contest comprising a hop, step, and jump.

triplet /'trɪplɪt/ *n.* **1** each of three children or animals born at one birth. **2** set of three things, esp. three equal notes played in the time of two of the same value.

triplex /'trɪpleks/ *adj.* triple, threefold. [Latin]

triplicate –*adj.* /'trɪplɪkət/ **1** existing in three examples or copies. **2** having three corresponding parts. **3** tripled. –*n.* /'trɪplɪkət/ each of a set of three copies or corresponding parts. –*v.* /'trɪplɪˌkeɪt/ (**-ting**) **1** make in three copies. **2** multiply by three. □ **in triplicate** in three copies. □ **triplication** /-'keɪʃ(ə)n/ *n.* [Latin: related to TRIPLEX]

tripod /'traɪpɒd/ *n.* **1** three-legged stand for a camera etc. **2** stool, table, or utensil resting on three feet or legs. [Greek, = three-footed]

tripos /'traɪpɒs/ *n.* (at Cambridge University) honours examinations for primary degrees. [related to TRIPOD]

tripper *n.* person who goes on a pleasure trip.

triptych /'trɪptɪk/ *n.* picture or relief carving on three panels, usu. hinged together at the sides. [after DIPTYCH]

trip-wire *n.* wire stretched close to the ground to trip up an intruder or to operate an alarm or other device when disturbed.

trireme /'traɪri:m/ *n.* ancient Greek warship, with three files of oarsmen on each side. [Latin *remus* oar]

trisect /traɪ'sekt/ *v.* divide into three (usu. equal) parts. □ **trisection** *n.* [Latin *seco sect-* cut]

trite *adj.* (of a phrase, observation, etc.) hackneyed. □ **tritely** *adv.* **triteness** *n.* [Latin *tero trit-* rub]

tritium /'trɪtɪəm/ *n.* radioactive isotope of hydrogen with a mass about three times that of ordinary hydrogen. [Greek *tritos* third]

triumph /'traɪʌmf/ –*n.* **1 a** state of victory or success (*returned in triumph*). **b** a great success or achievement. **2** supreme example (*a triumph of engineering*). **3** joy at success; exultation (*triumph in her face*). **4** processional entry of a victorious general into ancient Rome. –*v.* **1** (often foll. by *over*) gain a victory; be successful. **2** (of an ancient Roman general) ride in triumph. **3** (often foll. by *over*) exult. □ **triumphal** /-'ʌmf(ə)l/ *adj.* [French from Latin]

■ **Usage** *Triumphal*, meaning 'of or used in celebrating a triumph' as in *triumphal arch* should not be confused with *triumphant* meaning 'victorious' or 'exultant'.

triumphalism /traɪ'ʌmfəˌlɪz(ə)m/ *n.* excessive exultation over the victories of one's own party etc. □ **triumphalist** *adj.* & *n.*

triumphant /traɪ'ʌmf(ə)nt/ *adj.* **1** victorious, successful. **2** exultant. □ **triumphantly** *adv.*

■ **Usage** See note at *triumph*.

triumvirate /traɪ'ʌmvərət/ *n.* ruling group of three men, esp. in ancient Rome. [Latin *tres* three, *vir* man]

trivalent /traɪ'veɪlənt/ *adj.* *Chem.* having a valency of three. □ **trivalency** *n.*

trivet /'trɪvɪt/ *n.* iron tripod or bracket for a pot or kettle to stand on. [apparently from Latin *tripes* three-footed]

trivia /'trɪvɪə/ *n.pl.* trifles or trivialities.

trivial /'trɪvɪəl/ *adj.* **1** of small value or importance; trifling. **2** (of a person etc.) concerned only with trivial things. □ **triviality** /-'ælɪtɪ/ *n.* (*pl.* **-ies**). **trivially** *adv.* [Latin *trivialis* commonplace, from *trivium* three-way street corner]

trivialize *v.* (also **-ise**) (**-zing** or **-sing**) make or treat as trivial; minimize. □ **trivialization** /-'zeɪʃ(ə)n/ *n.*

trochee /'trəʊki:/ *n. Prosody* metrical foot consisting of one long followed by one short syllable (–∪). □ **trochaic** /trə'keɪɪk/ *adj.* [Greek, = running]

trod *past* and *past part.* of TREAD.

trodden *past part.* of TREAD.

troglodyte /'trɒglə,daɪt/ *n.* cave-dweller. [Greek *trōglē* hole]

troika /'trɔɪkə/ *n.* **1 a** Russian vehicle with a team of three horses abreast. **b** this team. **2** group of three people, esp. as an administrative council. [Russian]

Trojan /'trəʊdʒ(ə)n/ *–adj.* of ancient Troy in Asia Minor. *–n.* **1** native or inhabitant of Troy. **2** person who works, fights, etc. courageously. [Latin *Troia* Troy]

Trojan Horse *n.* **1** hollow wooden horse used by the Greeks to enter Troy. **2** person or device planted to bring about an enemy's downfall.

troll[1] *n.* (in Scandinavian folklore) fabulous being, esp. a giant or dwarf dwelling in a cave. [Old Norse]

troll[2] *v.* fish by drawing bait along in the water. [perhaps related to French *troller* to quest]

trolley /'trɒlɪ/ *n.* (*pl.* **-s**) **1** table, stand, or basket on wheels or castors for serving food, transporting luggage etc., gathering purchases in a supermarket, etc. **2** low truck running on rails. **3** (in full **trolley-wheel**) wheel attached to a pole etc. used for collecting current from an overhead electric wire to drive a vehicle. [dial., perhaps from TROLL[2]]

trolley bus *n.* electric bus using a trolley-wheel.

trollop /'trɒləp/ *n.* disreputable girl or woman. [perhaps related to archaic *trull* prostitute]

trombone /trɒm'bəʊn/ *n.* brass wind instrument with a sliding tube. □ **trombonist** *n.* [French or Italian *tromba* TRUMPET]

trompe-l'œil /trɔ̃p'lɜ:ɪ/ *n.* (often *attrib.*) painting etc. designed to give an illusion of reality. [French, literally 'deceives the eye']

-tron *suffix Physics* forming nouns denoting: **1** elementary particle (*positron*). **2** particle accelerator. [from ELECTRON]

troop *–n.* **1** assembled company; assemblage of people or animals. **2** (in *pl.*) soldiers, armed forces. **3** cavalry unit under a captain. **4** unit of artillery or armoured vehicles. **5** grouping of three or more Scout patrols. *–v.* (foll. by *in, out, off*, etc.) come together or move in large numbers. □ **troop the colour** transfer a flag ceremonially at a public mounting of garrison guards. [French *troupe*]

trooper *n.* **1** private soldier in a cavalry or armoured unit. **2** *Austral.* & *US* mounted or State police officer. **3** cavalry horse. **4** troop-ship.

troop-ship *n.* ship used for transporting troops.

trope *n.* figurative use of a word. [Greek *tropos* from *trepō* turn]

trophy /'trəʊfɪ/ *n.* (*pl.* **-ies**) **1** cup etc. as a prize in a contest. **2** memento or souvenir of success in hunting, war, etc. [Greek *tropaion*]

tropic /'trɒpɪk/ *–n.* **1** parallel of latitude 23°27´ north (**tropic of Cancer**) or south (**tropic of Capricorn**) of the Equator. **2** each of two corresponding circles on the celestial sphere where the sun appears to turn when at its greatest declination. **3** (**the Tropics**) region between the tropics of Cancer and Capricorn. *–adj.* = TROPICAL. [Greek *tropē* turn]

tropical *adj.* of or typical of the Tropics.

troposphere /'trɒpə,sfɪə(r)/ *n.* lowest layer of atmosphere extending about 6–10 km upwards from the earth's surface. [Greek *tropos* turn]

Trot *n. colloq.* usu. *derog.* Trotskyite. [abbreviation]

trot *–v.* (**-tt-**) **1** (of a person) run at a moderate pace. **2** (of a horse) proceed at a steady pace faster than a walk, lifting each diagonal pair of legs alternately. **3** *colloq.* walk, go. **4** cause (a horse or person) to trot. **5** traverse (a distance) at a trot. *–n.* **1** action or exercise of trotting (*proceed at a trot; went for a trot*). **2** (**the trots**) *slang* diarrhoea. □ **on the trot** *colloq.* **1** in succession (*six days on the trot*). **2** continually busy (*kept me on the trot*). **trot out 1** *colloq.* introduce (an opinion etc.) tediously or repeatedly. **2** cause (a horse) to trot to show his paces. [French]

troth /trəʊθ/ *n. archaic* **1** faith, loyalty. **2** truth. □ **pledge** (or **plight**) **one's troth** pledge one's word, esp. in marriage or betrothal. [Old English: related to TRUTH]

Trotskyism /'trɒtskɪ,ɪz(ə)m/ *n.* political principles of L. Trotsky, esp. as urging worldwide socialist revolution. □ **Trotskyist** *n.* **Trotskyite** *n. derog.*

trotter *n.* **1** (usu. in *pl.*) animal's foot as food. **2** horse bred or trained for trotting.

troubadour /'tru:bə,dʊə(r)/ *n.* **1** singer or poet. **2** French medieval lyric poet singing of courtly love. [Provençal *trobar* find, compose]

trouble /'trʌb(ə)l/ *–n.* **1** difficulty or distress; vexation, affliction (*had trouble with my car*). **2 a** inconvenience; unpleasant exertion; bother. **b** cause of this (*she was no trouble*). **3** perceived failing (*the trouble with me is that I can't say no*). **4** dysfunction (*kidney trouble*; *engine trouble*). **5 a** disturbance (*crowd trouble*; *don't want any trouble*). **b** (in *pl.*) political or social unrest, public disturbances, esp. (**the Troubles**) in N. Ireland. *–v.* (**-ling**) **1** cause distress or anxiety to; disturb. **2** be disturbed or worried (*don't trouble about it*). **3** afflict; cause pain etc. to. **4** (often *refl.*) subject or be subjected to inconvenience or unpleasant exertion (*sorry to trouble you*; *don't trouble yourself*). □ **ask** (or **look**) **for trouble** *colloq.* invite trouble by one's actions, behaviour, etc.; be rash or indiscreet. **in trouble 1** involved in a matter likely to bring censure or punishment. **2** *colloq.* pregnant while unmarried. [Latin: related to TURBID]

troublemaker *n.* person habitually causing trouble. □ **troublemaking** *n.*

troubleshooter *n.* **1** mediator in a dispute. **2** person who traces and corrects faults in machinery or in an organization etc. □ **troubleshooting** *n.*

troublesome *adj.* causing trouble, annoying.

trough /trɒf/ *n.* **1** long narrow open receptacle for water, animal feed, etc. **2** channel or hollow like this. **3** elongated region of low barometric pressure. [Old English]

trounce *v.* (**-cing**) **1** defeat heavily. **2** beat, thrash. **3** punish severely. [origin unknown]

troupe /tru:p/ *n.* company or band, esp. of artistes. [French, = TROOP]

trouper *n.* **1** member of a theatrical troupe. **2** staunch colleague.

trousers /'traʊzəz/ *n.pl.* **1** two-legged outer garment reaching from the waist usu. to the ankles. **2** (**trouser**) (*attrib.*) designating part of this (*trouser leg*). □ **wear the trousers** (esp. of a wife) dominate in a marriage. □ **trousered** *adj.* [in pl. after *drawers*: Irish and Gaelic *triubhas* trews]

trouser suit *n.* woman's suit of trousers and jacket.

trousseau /'tru:səʊ/ *n.* (*pl.* **-s** or **-x** /-səʊz/) bride's collection of clothes etc. [French: related to TRUSS]

trout *n.* (*pl.* same or **-s**) fish related to the salmon, valued as food. [Latin *tructa*]

trove *n.* = TREASURE TROVE. [Anglo-French *trové* from *trover* find]

trowel /'traʊəl/ *n.* **1** small flat-bladed tool for spreading mortar etc. **2** scoop for lifting small plants or earth. [Latin *trulla*]

troy *n.* (in full **troy weight**) system of weights used for precious metals and gems, with a pound of 12 ounces or 5,760 grains; 1 oz troy = 31.1035 g. [probably *Troyes* in France]

truant /'tru:ənt/ *–n.* **1** child who stays away from school. **2** person who avoids work etc. *–adj.* shirking, idle, wandering. *–v.* (also **play truant**) be a truant. □ **truancy** *n.* (*pl.* **-ies**). [French, probably from Celtic]

truce /tru:s/ *n.* temporary agreement to cease hostilities. [originally *trewes* pl.: Old English, = covenant: related to TRUE]

truck[1] *n.* **1** lorry. **2** open railway wagon for freight. **3** axle unit on a skateboard. [perhaps from TRUCKLE]

truck[2] *n.* □ **have no truck with** avoid dealing with. [French *troquer*]

trucker *n.* esp. *US* long-distance lorry-driver.

truckle /'trʌk(ə)l/ *–n.* (in full **truckle-bed**) low bed on wheels, stored under a larger bed. *–v.* (**-ling**) (foll. by *to*) submit obsequiously. [Latin *trochlea* pulley]

truculent /'trʌkjʊlənt/ *adj.* aggressively defiant. □ **truculence** *n.* **truculently** *adv.* [Latin *trux truc-* fierce]

trudge *–v.* (**-ging**) **1** go on foot, esp. laboriously. **2** traverse (a distance) in this way. *–n.* trudging walk. [origin unknown]

true /tru:/ *–adj.* (**truer**, **truest**) **1** in accordance with fact or reality (*a true story*). **2** genuine; rightly or strictly so called. **3** (often foll. by *to*) loyal, faithful. **4** (foll. by *to*) accurately conforming to (a type or standard) (*true to form*). **5** correctly positioned or balanced; upright, level. **6** exact, accurate (*a true copy*). *–adv.* **1** *archaic* truly (*tell me true*). **2** accurately (*aim true*). **3** without variation (*breed true*). □ **come true** actually happen. **out of true** out of alignment. **true to life** accurately representing reality. [Old English]

true-blue *–adj.* extremely loyal or orthodox. *–n.* such a person, esp. a Conservative.

true-love *n.* sweetheart.

true north *n.* north according to the earth's axis, not magnetic north.

truffle /'trʌf(ə)l/ *n.* **1** edible rich-flavoured underground fungus. **2** sweet made of a chocolate mixture covered with cocoa etc. [probably Dutch from French]

trug *n.* shallow oblong garden-basket usu. of wood strips. [perhaps a dial. var. of TROUGH]

truism /'tru:ɪz(ə)m/ *n.* statement too hackneyed to be worth making, e.g. 'Nothing lasts for ever'.

truly /'tru:lɪ/ *adv.* **1** sincerely (*am truly grateful*). **2** really, indeed (*truly, I do not know*). **3** loyally (*served them truly*). **4**

accurately (*is not truly depicted*). **5** properly (*well and truly*). [Old English: related to TRUE]

trump[1] –*n.* **1 a** playing-card of a suit temporarily ranking above the others. **b** (in *pl.*) this suit (*hearts are trumps*). **2** *colloq.* generous or loyal person. –*v.* **1** defeat (a card or its player) with a trump. **2** *colloq.* outdo. □ **come** (or **turn**) **up trumps** *colloq.* **1** turn out better than expected. **2** be greatly successful or helpful. **trump up** fabricate or invent (an accusation etc.) (*trumped-up charge*). [corruption of TRIUMPH in the same (now obsolete) sense]

trump[2] *n. archaic* trumpet-blast. [French *trompe*]

trump card *n.* **1** card belonging to, or turned up to determine, a trump suit. **2** *colloq.* valuable resource, esp. kept in reserve.

trumpery /'trʌmpərɪ/ –*n.* (*pl.* **-ies**) **1** worthless finery. **2** worthless thing; rubbish. –*adj.* showy but worthless; trashy; shallow. [French *tromperie* deceit]

trumpet /'trʌmpɪt/ –*n.* **1** brass instrument with a flared bell and bright penetrating tone. **2** trumpet-shaped thing (*ear-trumpet*). **3** sound of or like a trumpet. –*v.* (**-t-**) **1 a** blow a trumpet. **b** (of an enraged elephant etc.) make a trumpet-like cry. **2** proclaim loudly. □ **trumpeter** *n.* [French diminutive: related to TRUMP[2]]

trumpet-call *n.* urgent summons to action.

truncate /trʌŋ'keɪt/ *v.* (**-ting**) cut the top or the end from; shorten. □ **truncation** /-'keɪʃ(ə)n/ *n.* [Latin: related to TRUNK]

truncheon /'trʌntʃ(ə)n/ *n.* short club carried by a police officer. [French *tronchon* stump: related to TRUNK]

trundle /'trʌnd(ə)l/ *v.* (**-ling**) roll or move, esp. heavily or noisily. [var. of obsolete or dial. *trendle*: related to TREND]

trunk *n.* **1** main stem of a tree. **2** body without the limbs and head. **3** large box with a hinged lid for luggage, storage, etc. **4** *US* boot of a car. **5** elephant's elongated prehensile nose. **6** (in *pl.*) men's close-fitting shorts worn for swimming etc. [Latin *truncus* cut short]

trunk call *n.* long-distance telephone call.

trunk line *n.* main line of a railway, telephone system, etc.

trunk road *n.* important main road.

truss –*n.* **1** framework supporting a roof, bridge, etc. **2** surgical appliance worn to support a hernia. **3** bundle of hay or straw. **4** compact terminal cluster of flowers or fruit. –*v.* **1** tie up (a fowl) for cooking. **2** (often foll. by *up*) tie (a person) up with the arms to the sides. **3** support (a roof or bridge etc.) with a truss or trusses. [French]

trust –*n.* **1** firm belief in the reliability, truth, or strength etc. of a person or thing. **2** confident expectation. **3** responsibility (*position of great trust*). **4** commercial credit (*obtained goods on trust*). **5** *Law* **a** arrangement whereby a person or group manages property on another's behalf. **b** property so held. **c** body of trustees. **6** association of companies for reducing competition etc. –*v.* **1** place trust in; believe in; rely on the character or behaviour of. **2** (foll. by *with*) allow (a person) to have or use (a thing) from confidence in its careful use (*trusted her with my car*). **3** (often foll. by *that*) have faith, confidence, or hope that a thing will take place (*I trust you will come*). **4** (foll. by *to*) consign (a thing) to (a person) with trust. **5** (foll. by *in*) place reliance in (*we trust in you*). **6** (foll. by *to*) place (esp. undue) reliance on (*trust to luck*). □ **in trust** (of property) managed by one or more persons on behalf of another. **take on trust** accept (an assertion etc.) without evidence or investigation. [Old Norse]

trustee /trʌs'tiː/ *n.* person or member of a board managing property in trust with a legal obligation to administer it solely for the purposes specified. □ **trusteeship** *n.*

trustful *adj.* full of trust or confidence. □ **trustfully** *adv.*

trusting *adj.* having trust; trustful. □ **trustingly** *adv.*

trustworthy *adj.* deserving of trust; reliable. □ **trustworthiness** *n.*

trusty –*adj.* (**-ier**, **-iest**) *archaic* or *joc.* trustworthy (*a trusty steed*). –*n.* (*pl.* **-ies**) prisoner given special privileges for good behaviour.

truth /truːθ/ *n.* (*pl.* **truths** /truːðz, truːθs/) **1** quality or state of being true. **2** what is true. □ **in truth** *literary* truly, really. [Old English: related to TRUE]

truthful *adj.* **1** habitually speaking the truth. **2** (of a story etc.) true. □ **truthfully** *adv.* **truthfulness** *n.*

try /traɪ/ –*v.* (**-ies**, **-ied**) **1** make an effort with a view to success (often foll. by *to* + infin.; *colloq.* foll. by *and* + infin.: *tried to be on time*; *try and be early*). **2** make an effort to achieve (*tried my best*). **3 a** test by use or experiment. **b** test the qualities of. **4** make severe demands on (*tries my patience*). **5** examine the effectiveness of for a purpose (*try cold water*; *have you tried kicking it?*). **6** ascertain the state of fastening of (a door, window, etc.). **7 a** investigate and decide (a case or issue) judicially. **b** (often foll. by *for*) subject (a person) to trial (*tried for murder*). **8** (foll. by *for*) apply or compete for; seek to reach or attain (*try for a gold medal*). –*n.* (*pl.* **-ies**) **1** effort to accomplish something. **2** *Rugby* touching-down of the ball behind the opposing goal-line, scoring points and entitling the scoring side to a kick at the goal. □ **try one's hand** test how skilful one is, esp. at the first attempt. **try it on** *colloq.* try to get away with an unreasonable request etc. **try on** put on (clothes etc.) to see if they fit etc. **try out** put to the test, test thoroughly. [originally = separate, distinguish, from French *trier* sift]

■ **Usage** Use of the verb *try* with *and* (see sense 1) is uncommon in negative contexts (except in the imperative, e.g. *don't try and get the better of me*) and in the past tense.

trying *adj.* annoying, vexatious; hard to endure.

try-on *n. colloq.* **1** act of trying it on or trying on (clothes etc.). **2** attempt to deceive.

try-out *n.* experimental test.

tryst /trɪst/ *n. archaic* meeting, esp. of lovers. [French]

tsar /zɑː(r)/ *n.* (also **czar**) (*fem.* **tsarina** /-ˈriːnə/) *hist.* title of the former emperors of Russia. □ **tsarist** *n.* (usu. *attrib.*). [Latin *Caesar*]

tsetse /ˈtsetsɪ, ˈtetsɪ/ *n.* African fly feeding on blood and transmitting esp. sleeping-sickness. [Tswana]

T-shirt /ˈtiːʃɜːt/ *n.* short-sleeved casual top having the form of a T when spread out.

tsp. *abbr.* (*pl.* **tsps.**) teaspoonful.

T-square /ˈtiːskweə(r)/ *n.* T-shaped instrument for drawing right angles.

tsunami /tsuːˈnɑːmɪ/ *n.* (*pl.* **-s**) long high sea wave caused by underwater earthquakes etc. [Japanese]

TT *abbr.* **1** Tourist Trophy. **2** tuberculin-tested. **3 a** teetotal. **b** teetotaller.

tub –*n.* **1** open flat-bottomed usu. round vessel. **2** tub-shaped (usu. plastic) carton. **3** *colloq.* bath. **4** *colloq.* clumsy slow boat. –*v.* (**-bb-**) plant, bathe, or wash in a tub. [probably Low German or Dutch]

tuba /ˈtjuːbə/ *n.* (*pl.* **-s**) low-pitched brass wind instrument. [Latin, = trumpet]

tubby /ˈtʌbɪ/ *adj.* (**-ier, -iest**) short and fat. □ **tubbiness** *n.*

tube –*n.* **1** long hollow cylinder. **2** soft metal or plastic cylinder sealed at one end and holding a semi-liquid substance (*tube of toothpaste*). **3** hollow cylindrical organ in the body. **4** (often prec. by *the*) *colloq.* London underground (*went by tube*). **5 a** cathode-ray tube, esp. in a television set. **b** (prec. by *the*) esp. *US colloq.* television. **6** *US* thermionic valve. **7** = INNER TUBE. **8** *Austral. slang* can of beer. –*v.* (**-bing**) **1** equip with tubes. **2** enclose in a tube. [Latin]

tuber *n.* **1** thick rounded part of a stem or rhizome, usu. found underground and covered with modified buds, e.g. in a potato. **2** similar root of a dahlia etc. [Latin, = hump, swelling]

tubercle /ˈtjuːbək(ə)l/ *n.* small rounded swelling on the body or in an organ, esp. as characteristic of tuberculosis. □ **tuberculous** /-ˈbɜːkjʊləs/ *adj.* [Latin *tuberculum*, diminutive of TUBER]

tubercle bacillus *n.* bacterium causing tuberculosis.

tubercular /tjʊˈbɜːkjʊlə(r)/ *adj.* of or having tubercles or tuberculosis.

tuberculin /tjʊˈbɜːkjʊlɪn/ *n.* sterile liquid from cultures of tubercle bacillus, used in the diagnosis and treatment of tuberculosis.

tuberculin-tested *adj.* (of milk) from cows shown to be free of tuberculosis.

tuberculosis /tjʊˌbɜːkjʊˈləʊsɪs/ *n.* infectious bacterial disease marked by tubercles, esp. in the lungs.

tuberose /ˈtjuːbəˌrəʊz/ *n.* plant with scented white funnel-like flowers.

tuberous /ˈtjuːbərəs/ *adj.* having tubers; of or like a tuber.

tubing *n.* length of tube or quantity of tubes.

tub-thumper *n. colloq.* ranting preacher or orator.

tubular /ˈtjuːbjʊlə(r)/ *adj.* **1** tube-shaped. **2** having or consisting of tubes. **3** (of furniture etc.) having a tubular framework.

tubular bells *n.pl.* orchestral instrument of vertically suspended brass tubes struck with a hammer.

tubule /ˈtjuːbjuːl/ *n.* small tube in a plant or animal body. [Latin *tubulus*, diminutive: related to TUBE]

TUC *abbr.* Trades Union Congress.

tuck –*v.* **1** (often foll. by *in*, *up*) **a** draw, fold, or turn the outer or end parts of (cloth or clothes etc.) close together so as to be held; push in the edge of (a thing) so as to confine it (*tucked his shirt into his trousers*). **b** push in the edges of bedclothes around (a person) (*came to tuck me in*). **2** draw together into a small space (*tucked its head under its wing*). **3** stow (a thing) away in a specified place or way (*tucked it in a corner*; *tucked it out of sight*). **4** make a stitched fold in (cloth etc.). –*n.* **1** flattened usu. stitched fold in cloth etc. **2** *colloq.* food, esp. cakes and sweets (also *attrib.*: *tuck box*). □ **tuck in** *colloq.* eat heartily. **tuck into** (or **away**) *colloq.* eat (food) heartily (*tucked into* *their dinner; could really tuck it away*). [Low German or Dutch]

tucker –*n.* **1** *hist.* piece of lace or linen etc. in or on a woman's bodice. **2** *Austral. & NZ slang* food. –*v.* (esp. in *passive*; often foll. by *out*) *US & Austral. colloq.* tire.

tuck-in *n. colloq.* large meal.

tuck shop *n.* small shop selling sweets etc. to schoolchildren.

-tude *suffix* forming abstract nouns (*altitude*; *solitude*). [Latin *-tudo*]

Tudor /'tju:də(r)/ *adj.* **1** of the royal family of England 1485–1603 or this period. **2** of the architectural style of this period, esp. with half-timbering. [Owen *Tudor*, name of the grandfather of Henry VII]

Tues. *abbr.* (also **Tue.**) Tuesday.

Tuesday /'tju:zdeɪ/ –*n.* day of the week following Monday. –*adv.* **1** *colloq.* on Tuesday. **2** (**Tuesdays**) on Tuesdays; each Tuesday. [Old English]

tufa /'tju:fə/ *n.* **1** porous limestone rock formed round mineral springs. **2** = TUFF. [Italian: related to TUFF]

tuff *n.* rock formed from volcanic ash. [Latin *tofus*]

tuffet /'tʌfɪt/ *n.* clump of grass; small mound. [var. of TUFT]

tuft *n.* bunch or collection of threads, grass, feathers, hair, etc., held or growing together at the base. □ **tufted** *adj.* **tufty** *adj.* [probably French *tofe*]

tug –*v.* (**-gg-**) **1** (often foll. by *at*) pull hard or violently; jerk. **2** tow (a ship etc.) by a tugboat. –*n.* **1** hard, violent, or jerky pull. **2** sudden strong emotion. **3** small powerful boat for towing ships. [related to TOW[1]]

tugboat *n.* = TUG *n.* 3.

tug of love *n. colloq.* dispute over the custody of a child.

tug of war *n.* **1** trial of strength between two sides pulling opposite ways on a rope. **2** decisive or severe contest.

tuition /tju:'ɪʃ(ə)n/ *n.* **1** teaching, esp. if paid for. **2** fee for this. [Latin *tueor tuit-* look after]

tulip /'tju:lɪp/ *n.* **1** bulbous spring-flowering plant with showy cup-shaped flowers. **2** its flower. [Turkish *tul(i)band* TURBAN (from its shape), from Persian]

tulip-tree *n.* tree producing tulip-like flowers.

tulle /tju:l/ *n.* soft fine silk etc. net for veils and dresses. [*Tulle* in France]

tum *n. colloq.* stomach. [abbreviation of TUMMY]

tumble /'tʌmb(ə)l/ –*v.* (**-ling**) **1** (cause to) fall suddenly, clumsily, or headlong. **2** fall rapidly in amount etc. (*prices tumbled*). **3** (often foll. by *about, around*) roll or toss to and fro. **4** move or rush in a headlong or blundering manner. **5** (often foll. by *to*) *colloq.* grasp the meaning behind an idea, circumstance, etc. (*he quickly tumbled to our plan*). **6** overturn; fling or push roughly or carelessly. **7** perform acrobatic feats, esp. somersaults. **8** rumple or disarrange. –*n.* **1** sudden or headlong fall. **2** somersault or other acrobatic feat. **3** untidy or confused state. [Low German *tummeln*]

tumbledown *adj.* falling or fallen into ruin; dilapidated.

tumble-drier *n.* (also **tumble-dryer**) machine for drying washing in a heated rotating drum. □ **tumble-dry** *v.*

tumbler *n.* **1** drinking-glass with no handle or foot. **2** acrobat. **3** part of a lock that holds the bolt until lifted by a key. **4** a kind of pigeon that turns over backwards in flight.

tumbrel /'tʌmbr(ə)l/ *n.* (also **tumbril**) *hist.* open cart in which condemned persons were taken to the guillotine in the French Revolution. [French *tomber* fall]

tumescent /tjʊ'mes(ə)nt/ *adj.* swelling. □ **tumescence** *n.* [Latin: related to TUMOUR]

tumid /'tju:mɪd/ *adj.* **1** swollen, inflated. **2** (of style etc.) inflated, bombastic. □ **tumidity** /-'mɪdɪtɪ/ *n.*

tummy /'tʌmɪ/ *n.* (*pl.* **-ies**) *colloq.* stomach. [a childish pronunciation]

tummy-button *n.* navel.

tumour /'tju:mə(r)/ *n.* (*US* **tumor**) a swelling, esp. from an abnormal growth of tissue. □ **tumorous** *adj.* [Latin *tumeo* swell]

tumult /'tju:mʌlt/ *n.* **1** uproar or din, esp. of a disorderly crowd. **2** angry demonstration by a mob; riot. **3** conflict of emotions in the mind. [Latin: related to TUMOUR]

tumultuous /tjʊ'mʌltʃʊəs/ *adj.* noisy; turbulent; violent.

tumulus /'tju:mjʊləs/ *n.* (*pl.* **-li** /-ˌlaɪ/) ancient burial mound. [Latin: related to TUMOUR]

tun *n.* **1** large beer or wine cask. **2** brewer's fermenting-vat. [Old English]

tuna /'tju:nə/ *n.* (*pl.* same or **-s**) **1** large edible marine fish. **2** (in full **tuna-fish**) its flesh as food. [American Spanish]

tundra /'tʌndrə/ *n.* vast level treeless Arctic region with underlying permafrost. [Lappish]

tune –*n.* melody. –*v.* (**-ning**) **1** put (a musical instrument) in tune. **2 a** adjust (a radio etc.) to the frequency of a signal. **b** (foll. by *in*) adjust a radio receiver to the required signal. **3** adjust (an engine etc.) to run efficiently. □ **in** (or **out of**) **tune 1** having (or not having) the correct pitch or intonation (*sings in tune*). **2**

(usu. foll. by *with*) harmonizing (or clashing) with one's company, surroundings, etc. **to the tune of** *colloq.* to the considerable sum of. **tuned in** (often foll. by *to*) *colloq.* acquainted; in rapport; up to date. **tune up 1** bring one's instrument to the proper pitch. **2** bring to the most efficient condition. [var. of TONE]

tuneful *adj.* melodious, musical. □ **tunefully** *adv.*

tuneless *adj.* unmelodious, unmusical. □ **tunelessly** *adv.*

tuner *n.* **1** person who tunes musical instruments, esp. pianos. **2 a** part of a radio or television receiver for tuning. **b** radio receiver as a separate unit in a high-fi system. **3** electronic device for tuning a guitar etc.

tungsten /'tʌŋst(ə)n/ *n.* dense metallic element with a very high melting point. [Swedish, = heavy stone]

tunic /'tju:nɪk/ *n.* **1** close-fitting short coat of police or military etc. uniform. **2** loose often sleeveless garment reaching to the knees. [Latin]

tuning-fork *n.* two-pronged steel fork giving a particular note when struck.

tunnel /'tʌn(ə)l/ *—n.* **1** underground passage dug through a hill or under a road, river, etc., esp. for a railway or road. **2** underground passage dug by an animal. *—v.* (**-ll-**; *US* **-l-**) **1** (foll. by *through*, *into*, etc.) make a tunnel through. **2** make (one's way) by tunnelling. [French diminutive of *tonne* TUN]

tunnel vision *n.* **1** vision which is poor or lost outside the centre of the normal field of vision. **2** *colloq.* inability to grasp a situation's wider implications.

tunny /'tʌnɪ/ *n.* (*pl.* same or **-ies**) = TUNA. [Greek *thunnos*]

tup *—n.* ram. *—v.* (**-pp-**) (of a ram) copulate with (a ewe). [origin unknown]

Tupi /'tu:pɪ/ *—n.* (*pl.* same or **-s**) **1** member of an American Indian people of the Amazon valley. **2** their language. *—adj.* of this people or language. [Tupi]

tuppence /'tʌpəns/ *n.* = TWOPENCE. [phonetic spelling]

tuppenny /'tʌpənɪ/ *adj.* = TWOPENNY. [phonetic spelling]

Tupperware /'tʌpə,weə(r)/ *n. propr.* range of plastic containers for storing food. [*Tupper*, name of the manufacturer]

turban /'tɜ:bən/ *n.* **1** man's headdress of fabric wound round a cap or the head, worn esp. by Muslims and Sikhs. **2** woman's hat resembling this. □ **turbaned** *adj.* [Persian: cf. TULIP]

turbid /'tɜ:bɪd/ *adj.* **1** (of a liquid or colour) muddy, thick; not clear. **2** (of style etc.) confused, disordered. □ **turbidity** /-'bɪdɪtɪ/ *n.* [Latin *turba* crowd]

■ **Usage** *Turbid* is sometimes confused with *turgid* which means 'swollen, inflated; pompous'.

turbine /'tɜ:baɪn/ *n.* rotary motor driven by a flow of water, steam, gas, wind, etc. [Latin *turbo -in-* spinning-top, whirlwind]

turbo /'tɜ:bəʊ/ *n.* (*pl.* **-s**) = TURBOCHARGER.

turbo- *comb. form* turbine.

turbocharger *n.* supercharger driven by a turbine powered by the engine's exhaust gases.

turbofan *n.* jet engine in which a turbine-driven fan provides additional thrust.

turbojet *n.* **1** jet engine in which the jet also operates a turbine-driven air-compressor. **2** aircraft powered by this.

turboprop *n.* **1** jet engine in which a turbine is used as in a turbojet and also to drive a propeller. **2** aircraft powered by this. [from PROP[3]]

turbot /'tɜ:bət/ *n.* (*pl.* same or **-s**) large European flat-fish prized as food. [French from Swedish]

turbulent /'tɜ:bjʊlənt/ *adj.* **1** disturbed; in commotion. **2** (of a flow of air etc.) varying irregularly. **3** restless; riotous. □ **turbulence** *n.* **turbulently** *adv.* [Latin *turba* crowd]

Turco- *comb. form* (also **Turko-**) Turkish; Turkish and. [medieval Latin: related to TURK]

turd *n. coarse slang* **1** lump of excrement. **2** contemptible person. [Old English]

tureen /tjʊə'ri:n/ *n.* deep covered dish for soup. [from TERRINE]

turf *—n.* (*pl.* **-s** or **turves**) **1 a** layer of grass etc. with earth and matted roots as the surface of grassland. **b** piece of this cut from the ground. **2** slab of peat for fuel. **3** (prec. by *the*) **a** horse-racing generally. **b** general term for race-courses. *—v.* **1** cover (ground) with turf. **2** (foll. by *out*) *colloq.* expel or eject (a person or thing). □ **turfy** *adj.* [Old English]

turf accountant *n.* bookmaker.

turgescent /tɜ:'dʒes(ə)nt/ *adj.* becoming turgid. □ **turgescence** *n.* [Latin: related to TURGID]

turgid /'tɜ:dʒɪd/ *adj.* **1** swollen, inflated. **2** (of language) pompous, bombastic. □ **turgidity** /-'dʒɪdɪtɪ/ *n.* [Latin *turgeo* swell]

■ **Usage** *Turgid* is sometimes confused with *turbid* which means 'muddy, not clear; confused'.

Turk *n.* **1 a** native or national of Turkey. **b** person of Turkish descent. **2** member of a Central Asian people from whom the Ottomans derived, speaking a Turkic language. **3** *offens.* ferocious or wild person. [origin unknown]

turkey /'tɜ:kɪ/ *n.* (*pl.* **-s**) **1** large orig. American bird bred for food. **2** its flesh as food. **3** *US slang* theatrical failure; flop. □ **talk turkey** *US colloq.* talk frankly; get down to business. [originally of the guinea-fowl, imported from *Turkey*]

turkeycock *n.* male turkey.

Turki /'tɜ:kɪ/ **–***adj.* of a group of languages and peoples including Turkish. **–***n.* this group. □ **Turkic** *adj.* [Persian: related to TURK]

Turkish **–***adj.* of Turkey, the Turks, or their language. **–***n.* this language.

Turkish bath *n.* **1** hot-air or steam bath followed by washing, massage, etc. **2** (in *sing.* or *pl.*) building for this.

Turkish carpet *n.* wool carpet with a thick pile and traditional bold design.

Turkish coffee *n.* strong black coffee.

Turkish delight *n.* sweet of lumps of flavoured gelatine coated in powdered sugar.

Turkish towel *n.* towel made of cotton terry.

Turko- var. of TURCO-.

Turk's head *n.* turban-like ornamental knot.

turmeric /'tɜ:mərɪk/ *n.* **1** E. Indian plant of the ginger family. **2** its powdered rhizome used as a spice in curry etc. or for yellow dye. [perhaps from French *terre mérite*]

turmoil /'tɜ:mɔɪl/ *n.* **1** violent confusion; agitation. **2** din and bustle. [origin unknown]

turn /tɜ:n/ **–***v.* **1** move around a point or axis; give or receive a rotary motion (*turned the wheel*; *the wheel turns*). **2** change in position so that a different side, end, or part becomes outermost or uppermost etc.; invert or reverse (*it turned inside out*; *turned it upside down*). **3 a** give a new direction to (*turn your face this way*). **b** take a new direction (*turn left here*). **4** aim in a certain way (*turned the hose on them*). **5** (foll. by *into*) change in nature, form, or condition to (*turned into a frog*; *turned the book into a play*). **6** (foll. by *to*) **a** set about (*turned to doing the ironing*). **b** have recourse to (*turned to drink*; *turned to me for help*). **c** go on to consider next (*let us now turn to your report*). **7** become (*turned nasty*). **8 a** (foll. by *against*) make or become hostile to (*has turned her against us*). **b** (foll. by *on*, *upon*) become hostile to; attack (*suddenly turned on them*). **9** (of hair or leaves) change colour. **10** (of milk) become sour. **11** (of the stomach) be nauseated. **12** cause (milk) to become sour or (the stomach) to be nauseated. **13** (of the head) become giddy. **14** translate (*turn it into French*). **15** move to the other side of; go round (*turned the corner*). **16** pass the age or time of (*he has turned 40*; *it has turned 4 o'clock*). **17** (foll. by *on*) depend on; be determined by. **18** send or put; cause to go (*was turned loose*; *turned the water out into a basin*). **19** perform (a somersault etc.). **20** remake (esp. a sheet) putting the less worn outer side on the inside. **21** make (a profit). **22** divert (a bullet). **23** blunt (a knife etc.). **24** shape (an object) on a lathe. **25** give an (esp. elegant) form to (*turn a compliment*). **26** (of the tide) change direction. **–***n.* **1** turning; rotary motion. **2** changed or a change of direction or tendency (*took a sudden turn to the left*). **3** point at which a turning or change occurs. **4** turning of a road. **5** change of direction of the tide. **6** change in the course of events (*a turn for the worse*). **7** tendency or disposition; facility of forming (*is of a mechanical turn of mind*; *has a neat turn of phrase*). **8** opportunity or obligation etc. that comes successively to each of several persons etc. (*my turn to pay*). **9** short walk or ride (*took a turn in the park*). **10** short performance, variety act. **11** service of a specified kind (*did me a good turn*). **12** purpose (*served my turn*). **13** *colloq.* momentary nervous shock (*gave me a turn*). **14** *Mus.* ornament consisting of the principal note with those above and below it. □ **at every turn** continually. **by turns** in rotation; alternately. **in turn** in succession. **in one's turn** when one's turn comes. **not know which way** (or **where**) **to turn** be at a loss, unsure how to act, etc. **out of turn 1** when it is not one's turn. **2** inappropriately (*did I speak out of turn?*). **take turns** (or **take it in turns**) act alternately. **to a turn** (esp. cooked) perfectly. **turn about** move so as to face in a new direction. **turn and turn about** alternately. **turn around** esp. *US* = *turn round*. **turn away 1** turn to face in another direction. **2** reject. **3** send away. **turn back 1** begin or cause to retrace one's steps. **2** fold back. **turn down 1** reject (a proposal etc.). **2** reduce the volume or strength of (sound, heat, etc.) by turning a knob etc. **3** fold down. **turn one's hand to** see HAND. **turn a person's head** see HEAD. **turn in 1** hand in or return. **2** achieve or register (a performance, score, etc.). **3** *colloq.* go to bed for the night. **4** incline inwards. **5**

hand over (a suspect etc.) to the authorities. **6** *colloq.* abandon (a plan etc.). **turn off 1 a** stop the flow or operation of (water, electricity, etc.) by a tap, switch, etc. **b** operate (a tap, switch, etc.) to achieve this. **2** enter a side-road. **3** *colloq.* cause to lose interest. **turn on 1 a** start the flow or operation of (water, electricity, etc.) by means of a tap, switch, etc. **b** operate (a tap, switch, etc.) to achieve this. **2** *colloq.* excite; stimulate, esp. sexually. **turn on one's heel** see HEEL[1]. **turn out 1** expel. **2** extinguish (an electric light etc.). **3** dress or equip (*well turned out*). **4** produce (goods etc.). **5** empty or clean out (a room etc.). **6** empty (a pocket). **7** *colloq.* assemble; attend a meeting etc. **8** (often foll. by *to* + infin. or *that* + clause) prove to be the case; result (*turned out to be true*; *see how things turn out*). **turn over 1** reverse the position of (*turn over the page*). **2 a** cause (an engine) to run. **b** (of an engine) start running. **3** consider; ponder. **4** (foll. by *to*) **a** transfer the care or conduct of (a person or thing) to (a person) (*shall turn it all over to my deputy*). **b** = *turn in* 5. **5** do business to the gross value of (*turns over £5000 a week*). **turn over a new leaf** reform one's conduct. **turn round 1** turn so as to face in a new direction. **2 a** unload and reload (a ship etc.). **b** receive, process, and send out again; cause to progress through a system. **3** adopt new opinions or policy. **turn the tables** see TABLE. **turn tail** turn one's back; run away. **turn to** set about one's work. **turn turtle** see TURTLE. **turn up 1** increase the volume or strength of by turning a knob etc. **2** discover or reveal. **3** be found, esp. by chance. **4** happen or present itself; (of a person) arrive (*people turned up late*). **5** shorten (a garment) by raising its hem. **6** fold over or upwards. [Old English *tyruan*, from Greek *tornos* lathe]

turn-about *n.* **1** turning about. **2** abrupt change of policy etc.

turn-buckle *n.* threaded device for tightly connecting parts of a metal rod or wire.

turncoat *n.* person who changes sides.

turner *n.* person who works with a lathe.

turnery *n.* **1** objects made on a lathe. **2** work with a lathe.

turning *n.* **1 a** road that branches off another. **b** place where this occurs. **2 a** use of a lathe. **b** (in *pl.*) chips or shavings from a lathe.

turning-circle *n.* smallest circle in which a vehicle can turn without reversing.

turning-point *n.* point at which a decisive change occurs.

turnip /'tɜ:nɪp/ *n.* **1** plant with a globular root. **2** its root as a vegetable. □ **turnipy** *adj.* [dial. *neep* (Old English from Latin *napu*)]

turnip-top *n.* turnip leaves as a vegetable.

turnkey *n.* (*pl.* **-s**) *archaic* jailer.

turn-off *n.* **1** turning off a main road. **2** *colloq.* something that repels or causes a loss of interest.

turn-on *n.* *colloq.* person or thing that causes (esp. sexual) excitement.

turnout *n.* **1** number of people attending a meeting, voting at an election, etc. **2** set or display of equipment, clothes, etc.

turnover *n.* **1** act of turning over. **2** gross amount of money taken in a business. **3** rate at which goods are sold and replaced in a shop. **4** rate at which people enter and leave employment etc. **5** small pie made by folding pastry over a filling.

turnpike *n.* **1** *hist.* **a** toll-gate. **b** road on which a toll was charged. **2** *US* motorway on which a toll is charged.

turn-round *n.* **1 a** unloading and reloading between trips. **b** receiving, processing, and sending out again; progress through a system. **2** reversal of an opinion or tendency.

turnstile *n.* gate with revolving arms allowing people through singly.

turntable *n.* **1** circular revolving plate on which records are played. **2** circular revolving platform for turning a railway locomotive.

turn-up *n.* **1** turned up end of a trouser leg. **2** *colloq.* unexpected happening.

turpentine /'tɜ:pən,taɪn/ *n.* resin from any of various trees. [Latin *terebinthina*]

turpentine substitute *n.* = WHITE SPIRIT.

turpitude /'tɜ:pɪ,tju:d/ *n.* *formal* depravity, wickedness. [Latin *turpis* disgraceful]

turps *n.* *colloq.* oil of turpentine. [abbreviation]

turquoise /'tɜ:kwɔɪz/ —*n.* **1** semiprecious stone, usu. opaque and greenish- or sky-blue. **2** greenish-blue colour. —*adj.* of this colour. [French, = Turkish]

turret /'tʌrɪt/ *n.* **1** small tower, esp. decorating a building. **2** low flat usu. revolving armoured tower for a gun and gunners in a ship, aircraft, fort, or tank. **3** rotating holder for tools in a lathe etc. □ **turreted** *adj.* [French diminutive: related to TOWER]

turtle /'tɜ:t(ə)l/ *n.* **1** aquatic reptile with flippers and a horny shell. **2** its flesh, used for soup. □ **turn turtle** capsize. [alteration of earlier *tortue*: related to TORTOISE]
turtle-dove /'tɜt(ə)l,dʌv/ *n.* wild dove noted for its soft cooing and affection for its mate. [Latin *turtur*]
turtle-neck *n.* high close-fitting neck on a knitted garment.
Tuscan /'tʌskən/ *–n.* **1** inhabitant of Tuscany. **2** form of Italian spoken in Tuscany; standard Italian. *–adj.* **1** of Tuscany or the Tuscans. **2** *Archit.* of the plainest of the classical orders. [Latin]
tusk *n.* long pointed tooth, esp. protruding from a closed mouth, as in the elephant, walrus, etc. □ **tusked** *adj.* [Old English]
tussle /'tʌs(ə)l/ *–n.* struggle, scuffle. *–v.* (**-ling**) engage in a tussle. [originally Scots and Northern English, perhaps diminutive of *touse*: related to TOUSLE]
tussock /'tʌsək/ *n.* clump of grass etc. □ **tussocky** *adj.* [perhaps from dial. *tusk* tuft]
tut var. of TUT-TUT.
tutelage /'tju:tɪlɪdʒ/ *n.* **1** guardianship. **2** being under this. **3** tuition. [Latin *tutela*: related to TUTOR]
tutelary /'tju:tɪlərɪ/ *adj.* **1 a** serving as guardian. **b** of a guardian. **2** giving protection. [Latin: related to TUTELAGE]
tutor /'tju:tə(r)/ *–n.* **1** private teacher. **2** university teacher supervising the studies or welfare of assigned undergraduates. *–v.* **1** act as tutor to. **2** work as a tutor. □ **tutorship** *n.* [Latin *tueor tut-* watch]
tutorial /tju:'tɔ:rɪəl/ *–adj.* of a tutor or tuition. *–n.* period of undergraduate tuition individually or in a small group. [Latin *tutorius*: related to TUTOR]
tutti /'tʊtɪ/ *Mus. –adj.* & *adv.* with all voices or instruments together. *–n.* (*pl.* **-s**) such a passage. [Italian, pl. of *tutto* all]
tutti-frutti /,tu:tɪ'fru:tɪ/ *n.* (*pl.* **-s**) ice-cream containing small pieces of mixed glacé fruit. [Italian, = all fruits]
tut-tut /tʌt'tʌt/ (also **tut**) *–int.* expressing disapproval or impatience. *–n.* such an exclamation. *–v.* (**-tt-**) exclaim this. [imitative of a click of the tongue]
tutu /'tu:tu:/ *n.* ballet dancer's short skirt of stiffened frills. [French]
tu-whit, tu-whoo /tʊ,wɪt tʊ'wu:/ *n.* representation of the cry of an owl. [imitative]
tux *n. US colloq.* = TUXEDO. [abbreviation]
tuxedo /tʌk'si:dəʊ/ *n.* (*pl.* **-s** or **-es**) *US* **1** dinner-jacket. **2** suit of clothes including this. [*Tuxedo* Park in US]
TV *abbr.* television.
TVEI *abbr.* Technical and Vocational Educational Initiative.
twaddle /'twɒd(ə)l/ *n.* silly writing or talk; nonsense. [earlier *twattle*, alteration of TATTLE]
twain *adj.* & *n. archaic* two. [Old English, masculine form of TWO]
twang *–n.* **1** sound made by a plucked string or released bowstring. **2** nasal quality of a voice. *–v.* (cause to) emit this sound. □ **twangy** *adj.* [imitative]
'twas /twɒz/ *archaic* it was. [contraction]
twat /twɒt/ *n. coarse slang* **1** female genitals. **2** contemptible person. [origin unknown]
tweak *–v.* **1** pinch and twist sharply; jerk. **2** make fine adjustments to (a mechanism). *–n.* act of tweaking. [probably dial. *twick*, TWITCH]
twee *adj.* (**tweer** /'twi:ə(r)/; **tweest** /'twi:ɪst/) *derog.* affectedly dainty or quaint. [a childish pronunciation of SWEET]
tweed *n.* **1** rough-surfaced woollen cloth, usu. of mixed flecked colours. **2** (in *pl.*) clothes made of tweed. [alteration of *tweel* (Scots var. of TWILL)]
tweedy *adj.* (**-ier, -iest**) **1** of or dressed in tweed. **2** characteristic of country gentry; heartily informal.
'tween *prep. archaic* = BETWEEN. [abbreviation]
tweet *–n.* chirp of a small bird. *–v.* make this noise. [imitative]
tweeter *n.* loudspeaker for high frequencies.
tweezers /'twi:zəz/ *n.pl.* small pair of pincers for taking up small objects, plucking out hairs, etc. [originally *tweezes* pl. of obsolete *tweeze*, a case for small instruments]
twelfth *adj.* & *n.* **1** next after eleventh. **2** each of twelve equal parts of a thing. [Old English: related to TWELVE]
Twelfth Night *n.* 5 Jan., eve of Epiphany.
twelve *adj.* & *n.* **1** one more than eleven. **2** symbol for this (12, xii, XII). **3** size etc. denoted by twelve. **4** twelve o'clock. **5** (**the Twelve**) the apostles. [Old English]
twelvefold *adj.* & *adv.* **1** twelve times as much or as many. **2** consisting of twelve parts.
twelvemonth *n. archaic* year.
twenty /'twentɪ/ *adj.* & *n.* (*pl.* **-ies**) **1** product of two and ten. **2** symbol for this (20, xx, XX). **3** (in *pl.*) numbers from 20 to 29, esp. the years of a century or of a person's life. □ **twentieth** *adj.* & *n.* [Old English]

twenty-twenty vision *n.* (also **20/20 vision**) **1** vision of normal acuity. **2** *colloq.* good eyesight.

'twere /twɜː(r)/ *archaic* it were. [contraction]

twerp *n.* (also **twirp**) *slang* stupid or objectionable person. [origin unknown]

twice *adv.* **1** two times; on two occasions. **2** in double degree or quantity (*twice as good*). [Old English: related to TWO]

twiddle /ˈtwɪd(ə)l/ *–v.* (**-ling**) twirl, adjust, or play randomly or idly. *–n.* act of twiddling. □ **twiddle one's thumbs 1** make them rotate round each other. **2** have nothing to do. □ **twiddly** *adj.* [probably imitative]

twig[1] *n.* very small thin branch of a tree or shrub. □ **twiggy** *adj.* [Old English]

twig[2] *v.* (**-gg-**) *colloq.* understand; realize. [origin unknown]

twilight /ˈtwaɪlaɪt/ *n.* **1** light from the sky when the sun is below the horizon, esp. in the evening. **2** period of this. **3** faint light. **4** period of decline or destruction. [from TWO, LIGHT[1]]

twilight zone *n.* **1** decrepit urban area. **2** undefined or intermediate zone or area.

twilit /ˈtwaɪlɪt/ *adj.* dimly illuminated by twilight.

twill *n.* fabric so woven as to have a surface of diagonal parallel ridges. □ **twilled** *adj.* [Old English, = two-thread]

'twill *archaic* it will. [contraction]

twin *–n.* **1** each of a closely related or associated pair, esp. of children or animals born at a birth. **2** exact counterpart of a person or thing. **3 (the Twins)** zodiacal sign or constellation Gemini. *–adj.* forming, or being one of, such a pair (*twin brothers*). *–v.* (**-nn-**) **1 a** join intimately together. **b** (foll. by *with*) pair. **2** bear twins. **3** link (a town) with one in a different country, for friendship and cultural exchange. □ **twinning** *n.* [Old English: related to TWO]

twin bed *n.* each of a pair of single beds. □ **twin-bedded** *adj.*

twine *–n.* **1** strong coarse string of twisted strands of fibre. **2** coil, twist. *–v.* (**-ning**) **1** form (a string etc.) by twisting strands. **2** weave (a garland etc.). **3** (often foll. by *with*) garland (a brow etc.). **4** (often foll. by *round*, *about*) coil or wind. **5** *refl.* (of a plant) grow in this way. [Old English]

twin-engined *adj.* having two engines.

twinge /twɪndʒ/ *n.* sharp momentary local pain or pang. [Old English]

twinkle /ˈtwɪŋk(ə)l/ *–v.* (**-ling**) **1** (of a star or light etc.) shine with rapidly intermittent gleams. **2** (of the eyes) sparkle. **3** (of the feet) move lightly and rapidly. *–n.* **1** sparkle or gleam of the eyes. **2** twinkling light. **3** light rapid movement. □ **in a twinkle** (or **a twinkling** or **the twinkling of an eye**) in an instant. □ **twinkly** *adj.* [Old English]

twin set *n.* woman's matching cardigan and jumper.

twin town *n.* town twinned with another.

twirl *–v.* spin, swing, or twist quickly and lightly round. *–n.* **1** twirling motion. **2** flourish made with a pen. [origin uncertain]

twirp var. of TWERP.

twist *–v.* **1 a** change the form of by rotating one end and not the other or the two ends in opposite directions. **b** undergo such a change. **c** wrench or pull out of shape with a twisting action (*twisted my ankle*). **2 a** wind (strands etc.) about each other. **b** form (a rope etc.) in this way. **3 a** give a spiral form to. **b** take a spiral form. **4** (foll. by *off*) break off by twisting. **5** misrepresent the meaning of (words). **6 a** take a winding course. **b** make (one's way) in a winding manner. **7** *colloq.* cheat. **8** (as **twisted** *adj.*) *derog.* (of a person or mind) neurotic; perverted. **9** dance the twist. *–n.* **1** act of twisting. **2** twisted state. **3** thing formed by twisting. **4** point at which a thing twists or bends. **5** usu. *derog.* peculiar tendency of mind or character etc. **6** unexpected development of events, esp. in a story etc. **7** (prec. by *the*) popular 1960s dance with a twisting movement of the hips. □ **round the twist** *slang* crazy. **twist a person's arm** *colloq.* coerce, esp. using moral pressure. **twist round one's finger** easily persuade or dominate (a person). □ **twisty** *adj.* (**-ier**, **-iest**). [related to TWIN, TWINE]

twister *n. colloq.* swindler.

twit[1] *n. slang* foolish person. [originally dial., perhaps from TWIT[2]]

twit[2] *v.* (**-tt-**) reproach or taunt, usu. good-humouredly. [Old English]

twitch *–v.* **1** (of features, muscles, etc.) move or contract spasmodically. **2** pull sharply at. *–n.* **1** sudden involuntary contraction or movement. **2** sudden pull or jerk. **3** *colloq.* state of nervousness. □ **twitchy** *adj.* (**-ier**, **-iest**) (in sense 3 of *n.*). [probably Old English]

twitcher *n. colloq.* bird-watcher seeking sightings of rare birds.

twitter *–v.* **1** (esp. of a bird) emit a succession of light tremulous sounds. **2** utter or express in this way. *–n.* **1** act of twittering. **2** *colloq.* tremulously excited state. □ **twittery** *adj.* [imitative]

'twixt *prep. archaic* = BETWIXT. [contraction]

two /tu:/ *adj.* & *n.* **1** one more than one. **2** symbol for this (2, ii, II). **3** size etc. denoted by two. **4** two o'clock. □ **in two** in or into two pieces. **put two and two together** infer from known facts. [Old English]

two-bit *attrib. adj. US colloq.* cheap, petty.

two-dimensional *adj.* **1** having or appearing to have length and breadth but no depth. **2** lacking substance; superficial.

two-edged *adj.* double-edged.

two-faced *adj.* insincere; deceitful.

twofold *adj.* & *adv.* **1** twice as much or as many. **2** consisting of two parts.

two-handed *adj.* **1** having, using, or requiring the use of two hands. **2** (of a card-game) for two players.

twopence /'tʌpəns/ *n.* **1** sum of two pence. **2** (esp. with *neg.*) *colloq.* thing of little value (*don't care twopence*).

twopenny /'tʌpənɪ/ *attrib. adj.* **1** costing two pence. **2** *colloq.* cheap, worthless.

twopenny-halfpenny /ˌtʌpnɪ'heɪpnɪ/ *attrib. adj.* cheap, insignificant.

two-piece –*adj.* (of a suit etc.) consisting of two matching items. –*n.* two-piece suit etc.

two-ply –*adj.* of two strands or layers etc. –*n.* **1** two-ply wool. **2** two-ply wood.

twosome *n.* two persons together.

two-step *n.* dance in march or polka time.

two-stroke –*attrib. adj.* (of an internal-combustion engine) having its power cycle completed in one up-and-down movement of the piston. –*n.* two-stroke engine.

two-time *v. colloq.* **1** be unfaithful to (a lover). **2** swindle. □ **two-timer** *n.*

two-tone *adj.* having two colours or sounds.

'twould /twʊd/ *archaic* it would. [contraction]

two-way *adj.* **1** involving two directions or participants. **2** (of a radio) capable of transmitting and receiving signals.

two-way mirror *n.* panel of glass that can be seen through from one side and is a mirror on the other.

-ty[1] *suffix* forming nouns denoting quality or condition (*cruelty*; *plenty*). [French from Latin *-tas -tatis*]

-ty[2] *suffix* denoting tens (*ninety*). [Old English *-tig*]

tycoon /taɪ'ku:n/ *n.* business magnate. [Japanese, = great lord]

tying *pres. part.* of TIE.

tyke /taɪk/ *n.* (also **tike**) **1** unpleasant or coarse man. **2** small child. [Old Norse]

tympani var. of TIMPANI.

tympanum /'tɪmpənəm/ *n.* (*pl.* **-s** or **-na**) **1** middle ear. **2** eardrum. **3** *Archit.* **a** vertical triangular space forming the centre of a pediment. **b** similar space over a door between the lintel and the arch. [Greek *tumpanon* drum]

Tynwald /'tɪnwɒld/ *n.* parliament of the Isle of Man. [Old Norse, = assembly-field]

type /taɪp/ –*n.* **1** sort, class, or kind. **2** person, thing, or event exemplifying a class or group. **3** (in *comb.*) made of, resembling, or functioning as (*ceramic-type material*; *Cheddar-type cheese*). **4** *colloq.* person, esp. of a specified character (*a quiet type*; *not my type*). **5** object, conception, or work of art, serving as a model for subsequent artists. **6** *Printing* **a** piece of metal etc. with a raised letter or character on its upper surface for printing. **b** kind or size of such pieces (*printed in large type*). **c** set or supply of these (*ran short of type*). –*v.* (**-ping**) **1** write with a typewriter. **2** typecast. **3** esp. *Biol.* & *Med.* assign to a type; classify. [Greek *tupos* impression]

typecast *v.* (*past* and *past part.* **-cast**) assign (an actor or actress) repeatedly to the same type of role.

typeface *n. Printing* **1** inked surface of type. **2** set of characters in one design.

typescript *n.* typewritten document.

typesetter *n. Printing* **1** person who composes type. **2** composing-machine. □ **typesetting** *n.*

typewriter *n.* machine with keys for producing printlike characters one at a time on paper inserted round a roller.

typewritten *adj.* produced on a typewriter.

typhoid /'taɪfɔɪd/ *n.* (in full **typhoid fever**) infectious bacterial fever attacking the intestines.

typhoon /taɪ'fu:n/ *n.* violent hurricane in E. Asian seas. [Chinese, = great wind, and Arabic]

typhus /'taɪfəs/ *n.* infectious fever with a purple rash, headaches, and usu. delirium. [Greek, = stupor]

typical /'tɪpɪk(ə)l/ *adj.* **1** serving as a characteristic example; representative (*a typical English pub*). **2** (often foll. by *of*) characteristic of a particular person, thing, or type (*typical of him to refuse*). □ **typicality** /-'kælɪtɪ/ *n.* **typically** *adv.* [medieval Latin: related to TYPE]

typify /'tɪpɪˌfaɪ/ *v.* (**-ies, -ied**) **1** be typical of. **2** represent by or as a type or symbol. □ **typification** /-fɪ'keɪʃ(ə)n/ *n.* [Latin: related to TYPE]

typist /'taɪpɪst/ *n.* person who types, esp. for a living.

typo /ˈtaɪpəʊ/ *n.* (*pl.* **-s**) *colloq.* typographical error. [abbreviation]

typography /taɪˈpɒgrəfɪ/ *n.* **1** printing as an art. **2** style and appearance of printed matter. □ **typographer** *n.* **typographical** /-pəˈgræfɪk(ə)l/ *adj.* **typographically** /-pəˈgræfɪkəlɪ/ *adv.* [French: related to TYPE]

tyrannical /tɪˈrænɪk(ə)l/ *adj.* despotic; unjustly severe. □ **tyrannically** *adv.* [Greek: related to TYRANT]

tyrannize /ˈtɪrəˌnaɪz/ *v.* (also **-ise**) (**-zing** or **-sing**) (often foll. by *over*) treat despotically or cruelly. [French: related to TYRANT]

tyrannosaurus /tɪˌrænəˈsɔːrəs/ *n.* (*pl.* **-ruses**) (also **tyrannosaur**) dinosaur with very short front legs and a long well-developed tail. [from TYRANT, after *dinosaur*]

tyranny /ˈtɪrənɪ/ *n.* (*pl.* **-ies**) **1** cruel and arbitrary use of authority. **2 a** rule by a tyrant. **b** period of this. **c** State ruled by a tyrant. □ **tyrannous** *adj.* [Greek: related to TYRANT]

tyrant /ˈtaɪərənt/ *n.* **1** oppressive or cruel ruler. **2** person exercising power arbitrarily or cruelly. [Greek *turannos*]

tyre /ˈtaɪə(r)/ *n.* (*US* **tire**) rubber covering, usu. inflated, placed round a wheel to form a soft contact with the road. [var. of TIRE[2]]

Tyrian /ˈtɪrɪən/ *–adj.* of ancient Tyre in Phoenicia. *–n.* native or citizen of Tyre. [Latin *Tyrus* Tyre]

Tyrian purple see PURPLE *n.* 2.

tyro var. of TIRO.

tzatziki /tsætˈsiːkɪ/ *n.* Greek side dish of yoghurt with cucumber. [modern Greek]

U

U[1] /ju:/ *n.* (also **u**) (*pl.* **Us** or **U's**) **1** twenty-first letter of the alphabet. **2** U-shaped object or curve.

U[2] /ju:/ *adj. colloq.* upper class or supposedly upper class. [abbreviation]

U[3] *abbr.* (also **U.**) universal (of films classified as suitable for all).

U[4] *symb.* uranium.

UB40 *abbr.* **1** card issued to people claiming unemployment benefit. **2** *colloq.* unemployed person. [*u*nemployment *b*enefit]

ubiquitous /ju:'bɪkwɪtəs/ *adj.* **1** (seemingly) present everywhere simultaneously. **2** often encountered. □ **ubiquity** *n.* [Latin *ubique* everywhere]

U-boat /'ju:bəʊt/ *n. hist.* German submarine. [German *Untersee* undersea]

u.c. *abbr.* upper case.

UCCA /'ʌkə/ *abbr.* Universities Central Council on Admissions.

UDA *abbr.* Ulster Defence Association (a loyalist paramilitary organization).

udder *n.* baglike mammary organ of cattle etc., with several teats. [Old English]

UDI *abbr.* unilateral declaration of independence.

UDR *abbr.* Ulster Defence Regiment.

UEFA /ju:'eɪfə/ *abbr.* Union of European Football Associations.

UFO /'ju:fəʊ/ *n.* (also **ufo**) (*pl.* **-s**) unidentified flying object.

ugh /əx, ʌg/ *int.* **1** expressing disgust etc. **2** sound of a cough or grunt. [imitative]

Ugli /'ʌglɪ/ *n.* (*pl.* **-lis** or **-lies**) *propr.* mottled green and yellow hybrid of a grapefruit and a tangerine. [from UGLY]

uglify /'ʌglɪ,faɪ/ *v.* (**-ies, -ied**) make ugly.

ugly /'ʌglɪ/ *adj.* (**-lier, -liest**) **1** unpleasant to the eye, ear, or mind etc. (*ugly scar*; *ugly snarl*). **2** unpleasantly suggestive; discreditable (*ugly rumours*). **3** threatening, dangerous (*an ugly look*). **4** morally repulsive (*ugly vices*). □ **ugliness** *n.* [Old Norse]

ugly customer *n.* threatening or violent person.

ugly duckling *n.* person lacking early promise but blossoming later.

UHF *abbr.* ultrahigh frequency.

uh-huh /'ʌhʌ/ *int. colloq.* yes; indeed. [imitative]

UHT *abbr.* ultra heat treated (esp. of milk, for long keeping).

UK *abbr.* United Kingdom.

Ukrainian /ju:'kreɪnɪən/ **–***n.* **1** native or language of Ukraine. **2** person of Ukrainian descent. **–***adj.* of Ukraine, its people, or language. [*Ukraine* in eastern Europe]

ukulele /ju:kə'leɪlɪ/ *n.* small four-stringed Hawaiian guitar. [Hawaiian]

ulcer *n.* **1** open sore on or in the body, often forming pus. **2** corrupting influence etc. □ **ulcerous** *adj.* [Latin *ulcus -cer-*]

ulcerate /'ʌlsə,reɪt/ *v.* (**-ting**) form into or affect with an ulcer. □ **ulceration** /-'reɪʃ(ə)n/ *n.*

-ule *suffix* forming diminutive nouns (*globule*). [Latin *-ulus*]

ullage /'ʌlɪdʒ/ *n.* **1** amount by which a cask etc. falls short of being full. **2** loss by evaporation or leakage. [French from Latin]

ulna /'ʌlnə/ *n.* (*pl.* **ulnae** /-ni:/) **1** thinner and longer bone in the forearm, opposite to the thumb. **2** corresponding bone in an animal's foreleg or a bird's wing. □ **ulnar** *adj.* [Latin]

ulster *n.* long loose overcoat of rough cloth. [*Ulster* in Ireland]

Ulsterman *n.* (*fem.* **Ulsterwoman**) native of Ulster.

ult. *abbr.* ultimo.

ulterior /ʌl'tɪərɪə(r)/ *adj.* not evident or admitted; hidden, secret (esp. *ulterior motive*). [Latin, = further]

ultimate /'ʌltɪmət/ **–***adj.* **1** last or last possible, final. **2** fundamental, primary, basic (*ultimate truths*). **–***n.* **1** (prec. by *the*) best achievable or imaginable. **2** final or fundamental fact or principle. □ **ultimately** *adv.* [Latin *ultimus* last]

ultimatum /,ʌltɪ'meɪtəm/ *n.* (*pl.* **-s**) final statement of terms, the rejection of which could cause hostility etc. [Latin: related to ULTIMATE]

ultimo /'ʌltɪ,məʊ/ *adj. Commerce* of last month (*the 28th ultimo*). [Latin, = in the last (*mense* month)]

ultra /'ʌltrə/ **–***adj.* extreme, esp. in religion or politics. **–***n.* extremist. [see ULTRA-]

ultra- *comb. form* **1** extreme(ly), excessive(ly) (*ultra-modern*). **2** beyond. [Latin *ultra* beyond]

ultrahigh /,ʌltrə'haɪ/ *adj.* (of a frequency) in the range 300 to 3000 megahertz.

ultramarine /,ʌltrəmə'ri:n/ **–***n.* **1** brilliant blue pigment orig. from lapis lazuli. **2** colour of this. *–adj.* of this colour. [Italian and medieval Latin, = beyond the sea, from where lapis lazuli was brought]

ultramicroscopic /ˌʌltrəˌmaɪkrəˈskɒpɪk/ *adj.* too small to be seen by an ordinary optical microscope.

ultramontane /ˌʌltrəˈmɒnteɪn/ *–adj.* **1** situated beyond the Alps. **2** advocating supreme papal authority. *–n.* **1** person living beyond the Alps. **2** advocate of supreme papal authority. [medieval Latin: related to MOUNTAIN]

ultrasonic /ˌʌltrəˈsɒnɪk/ *adj.* of or using sound waves pitched above the range of human hearing. □ **ultrasonically** *adv.*

ultrasonics *n.pl.* (usu. treated as *sing.*) science of ultrasonic waves.

ultrasound /ˈʌltrəˌsaʊnd/ *n.* ultrasonic waves.

ultraviolet /ˌʌltrəˈvaɪələt/ *adj.* of or using radiation with a frequency just beyond that of the violet end of the visible spectrum.

ultra vires /ˌʌltrə ˈvaɪəˌriːz/ *adv.* & *predic.adj.* beyond one's legal power or authority. [Latin]

ululate /ˈjuːlʊˌleɪt/ *v.* (**-ting**) howl, wail. □ **ululation** /-ˈleɪʃ(ə)n/ *n.* [Latin]

um *int.* expressing hesitation or a pause in speech. [imitative]

umbel /ˈʌmb(ə)l/ *n.* flower-cluster with stalks springing from a common centre and forming a flat or curved surface. □ **umbellate** *adj.* [Latin *umbella* sunshade]

umbelliferous /ˌʌmbəˈlɪfərəs/ *adj.* (of a plant) bearing umbels, such as parsley and carrot.

umber /ˈʌmbə(r)/ *–n.* **1** natural pigment like ochre but darker and browner. **2** colour of this. *–adj.* of this colour. [Latin *umbra* shadow]

umbilical /ʌmˈbɪlɪk(ə)l/ *adj.* of the navel. [from UMBILICUS]

umbilical cord *n.* cordlike structure attaching a foetus to the placenta.

umbilicus /ʌmˈbɪlɪkəs/ *n.* (*pl.* **-ci** /-ˌsaɪ/ or **-cuses**) navel. [Latin]

umbra /ˈʌmbrə/ *n.* (*pl.* **-s** or **-brae** /-briː/) total shadow, esp. that cast on the earth by the moon during a solar eclipse. [Latin, = shadow]

umbrage /ˈʌmbrɪdʒ/ *n.* offence taken (esp. *take umbrage at*). [Latin: related to UMBRA]

umbrella /ʌmˈbrelə/ *n.* **1** collapsible cloth canopy on a central stick, used against rain, strong sun, etc. **2** protection, patronage. **3** (often *attrib.*) coordinating agency (*umbrella organization*). [Italian diminutive: related to UMBRA]

umlaut /ˈʊmlaʊt/ *n.* **1** mark (¨) used over a vowel, esp. in Germanic languages, to indicate a vowel change. **2** such a vowel change, e.g. German *Mann, Männer* /ˈmenə(r)/, English *man, men*. [German]

umpire /ˈʌmpaɪə(r)/ *–n.* person enforcing rules and settling disputes in esp. cricket or between disputants. *–v.* (**-ring**) (often foll. by *for, in*, etc.) act as umpire (in). [French *nonper* not equal: related to PEER[2]]

umpteen /ˈʌmptiːn, -ˈtiːn/ *colloq. –adj.* indefinitely many; a lot of. *–pron.* indefinitely many. □ **umpteenth** *adj.* [jocular formation on -TEEN]

UN *abbr.* United Nations.

un-[1] *prefix* **1** added to adjectives and participles and their derivative nouns and adverbs, meaning: **a** not (*unusable*). **b** reverse of (esp. with implied approval etc.) (*unselfish*; *unsociable*). **2** (less often) added to nouns, meaning 'a lack of', 'the reverse of' (*unrest*; *untruth*). [Old English]

■ **Usage** The number of words that can be formed with this prefix (and with *un-*[2]) is virtually unlimited; consequently only a selection can be given here.

un-[2] *prefix* added to verbs and (less often) nouns, forming verbs denoting: **1** reversal (*undress*; *unsettle*). **2** deprivation (*unmask*). **3** release from (*unburden*; *uncage*). **4** causing to be no longer (*unman*). [Old English]

■ **Usage** See note at *un-*[1].

un-[3] *prefix Chem.* denoting 'one', combined with other numerical roots *nil* (= 0), *un* (= 1), *bi* (= 2), etc., to form the names of elements based on the atomic number, and terminated with *-ium*, e.g. *unnilquadium* = 104, *ununbium* = 112. [Latin *unus* one]

unabashed /ˌʌnəˈbæʃt/ *adj.* not abashed.

unabated /ˌʌnəˈbeɪtɪd/ *adj.* not abated; undiminished.

unable /ʌnˈeɪb(ə)l/ *predic. adj.* (usu. foll. by *to* + infin.) not able.

unabridged /ˌʌnəˈbrɪdʒd/ *adj.* complete; not abridged.

unacademic /ˌʌnækəˈdemɪk/ *adj.* (of a person, book, etc.) not academic.

unacceptable /ˌʌnəkˈseptəb(ə)l/ *adj.* not acceptable. □ **unacceptably** *adv.*

unaccompanied /ˌʌnəˈkʌmpənɪd/ *adj.* **1** not accompanied. **2** *Mus.* without accompaniment.

unaccomplished /ˌʌnəˈkʌmplɪʃt/ *adj.* **1** uncompleted. **2** lacking accomplishments.

unaccountable /ˌʌnəˈkaʊntəb(ə)l/ *adj.* **1** without explanation; strange. **2** not answerable for one's actions. □ **unaccountably** *adv.*
unaccounted /ˌʌnəˈkaʊntɪd/ *adj.* (often foll. by *for*) unexplained; excluded.
unaccustomed /ˌʌnəˈkʌstəmd/ *adj.* **1** (usu. foll. by *to*) not accustomed. **2** unusual (*unaccustomed silence*).
unacknowledged /ˌʌnəkˈnɒlɪdʒd/ *adj.* not acknowledged.
unacquainted /ˌʌnəˈkweɪntɪd/ *adj.* (usu. foll. by *with*) not acquainted.
unadopted /ˌʌnəˈdɒptɪd/ *adj.* (of a road) not maintained by a local authority.
unadorned /ˌʌnəˈdɔːnd/ *adj.* plain.
unadulterated /ˌʌnəˈdʌltəˌreɪtɪd/ *adj.* **1** pure. **2** complete, utter.
unadventurous /ˌʌnədˈventʃərəs/ *adj.* not adventurous.
unadvised /ˌʌnədˈvaɪzd/ *adj.* **1** indiscreet; rash. **2** without advice. □ **unadvisedly** /-zɪdlɪ/ *adv.*
unaffected /ˌʌnəˈfektɪd/ *adj.* **1** (usu. foll. by *by*) not affected. **2** free from affectation. □ **unaffectedly** *adv.*
unaffiliated /ˌʌnəˈfɪlɪˌeɪtɪd/ *adj.* not affiliated.
unafraid /ˌʌnəˈfreɪd/ *adj.* not afraid.
unaided /ʌnˈeɪdɪd/ *adj.* without help.
unalike /ˌʌnəˈlaɪk/ *adj.* not alike; different.
unalloyed /ˌʌnəˈlɔɪd/ *adj.* **1** complete; utter (*unalloyed joy*). **2** pure.
unalterable /ʌnˈɔːltərəb(ə)l/ *adj.* not alterable.
unaltered /ʌnˈɔːltəd/ *adj.* not altered; remaining the same.
unambiguous /ˌʌnæmˈbɪgjʊəs/ *adj.* not ambiguous; clear or definite in meaning. □ **unambiguously** *adv.*
unambitious /ˌʌnæmˈbɪʃəs/ *adj.* not ambitious.
un-American /ˌʌnəˈmerɪkən/ *adj.* **1** uncharacteristic of Americans. **2** contrary to US interests, treasonable.
unamused /ˌʌnəˈmjuːzd/ *adj.* not amused.
unanimous /juːˈnænɪməs/ *adj.* **1** all in agreement (*committee was unanimous*). **2** (of an opinion, vote, etc.) by all without exception (*unanimous choice*). □ **unanimity** /-nəˈnɪmɪtɪ/ *n.* **unanimously** *adv.* [Latin *unus* one, *animus* mind]
unannounced /ˌʌnəˈnaʊnst/ *adj.* not announced; without warning (of arrival etc.).
unanswerable /ʌnˈɑːnsərəb(ə)l/ *adj.* **1** irrefutable (*unanswerable case*). **2** unable to be answered (*unanswerable question*).
unanswered /ʌnˈɑːnsəd/ *adj.* not answered.
unanticipated /ˌʌnænˈtɪsɪˌpeɪtɪd/ *adj.* not anticipated.
unappealing /ˌʌnəˈpiːlɪŋ/ *adj.* unattractive.
unappetizing /ʌnˈæpɪˌtaɪzɪŋ/ *adj.* not appetizing.
unappreciated /ˌʌnəˈpriːʃɪˌeɪtɪd/ *adj.* not appreciated.
unappreciative /ˌʌnəˈpriːʃətɪv/ *adj.* not appreciative.
unapproachable /ˌʌnəˈprəʊtʃəb(ə)l/ *adj.* **1** inaccessible. **2** (of a person) unfriendly.
unarmed /ʌnˈɑːmd/ *adj.* not armed; without weapons.
unashamed /ˌʌnəˈʃeɪmd/ *adj.* **1** feeling no guilt. **2** blatant; bold. □ **unashamedly** /-mɪdlɪ/ *adv.*
unassailable /ˌʌnəˈseɪləb(ə)l/ *adj.* unable to be attacked; impregnable.
unassuming /ˌʌnəˈsjuːmɪŋ/ *adj.* not pretentious; modest.
unattached /ˌʌnəˈtætʃt/ *adj.* **1** not engaged, married, etc. **2** (often foll. by *to*) not attached, esp. to a particular organization etc.
unattainable /ˌʌnəˈteɪnəb(ə)l/ *adj.* not attainable.
unattended /ˌʌnəˈtendɪd/ *adj.* **1** (usu. foll. by *to*) not attended. **2** (of a person, vehicle, etc.) alone.
unattractive /ˌʌnəˈtræktɪv/ *adj.* not attractive. □ **unattractively** *adv.*
unattributable /ˌʌnəˈtrɪbjʊtəb(ə)l/ *adj.* (esp. of published information) not attributed to a source etc.
unauthorized /ʌnˈɔːθəˌraɪzd/ *adj.* (also **-ised**) not authorized.
unavailable /ˌʌnəˈveɪləb(ə)l/ *adj.* not available. □ **unavailability** /-ˈbɪlɪtɪ/ *n.*
unavailing /ˌʌnəˈveɪlɪŋ/ *adj.* achieving nothing. □ **unavailingly** *adv.*
unavoidable /ˌʌnəˈvɔɪdəb(ə)l/ *adj.* inevitable. □ **unavoidably** *adv.*
unaware /ˌʌnəˈweə(r)/ —*adj.* **1** (usu. foll. by *of* or *that*) not aware. **2** unperceptive. —*adv.* = UNAWARES. □ **unawareness** *n.*
unawares *adv.* **1** unexpectedly. **2** inadvertently.
unbalanced /ʌnˈbælənst/ *adj.* **1** emotionally unstable. **2** biased (*unbalanced report*).
unban /ʌnˈbæn/ *v.* (**-nn-**) remove prohibited status from; allow.
unbar /ʌnˈbɑː(r)/ *v.* (**-rr-**) **1** unlock, open. **2** remove a bar from (a gate etc.).
unbearable /ʌnˈbeərəb(ə)l/ *adj.* unendurable. □ **unbearably** *adv.*
unbeatable /ʌnˈbiːtəb(ə)l/ *adj.* not beatable; excelling.
unbeaten /ʌnˈbiːt(ə)n/ *adj.* **1** not beaten. **2** (of a record etc.) not surpassed.
unbecoming /ˌʌnbɪˈkʌmɪŋ/ *adj.* **1** unflattering (*unbecoming hat*). **2** (usu. foll.

by *to*, *for*) not fitting; indecorous. □ **unbecomingly** *adv.*

unbeknown /ˌʌnbɪˈnəʊn/ *adj.* (also **unbeknownst** /-ˈnəʊnst/) (foll. by *to*) without the knowledge of (*unbeknown to us*).

unbelief /ˌʌnbɪˈliːf/ *n.* lack of esp. religious belief. □ **unbeliever** *n.* **unbelieving** *adj.*

unbelievable /ˌʌnbɪˈliːvəb(ə)l/ *adj.* not believable; incredible. □ **unbelievably** *adv.*

unbend /ʌnˈbend/ *v.* (*past* and *past part.* **unbent**) **1** straighten. **2** relax; become affable.

unbending *adj.* **1** inflexible. **2** firm; austere.

unbiased /ʌnˈbaɪəst/ *adj.* (also **unbiassed**) impartial.

unbidden /ʌnˈbɪd(ə)n/ *adj.* not commanded or invited (*arrived unbidden*).

unbind /ʌnˈbaɪnd/ *v.* (*past* and *past part.* **unbound**) release; unfasten, untie.

unbleached /ʌnˈbliːtʃt/ *adj.* not bleached.

unblemished /ʌnˈblemɪʃt/ *adj.* not blemished.

unblinking /ʌnˈblɪŋkɪŋ/ *adj.* **1** not blinking. **2** steadfast; stolid.

unblock /ʌnˈblɒk/ *v.* remove an obstruction from.

unblushing /ʌnˈblʌʃɪŋ/ *adj.* **1** shameless. **2** frank.

unbolt /ʌnˈbəʊlt/ *v.* release the bolt of (a door etc.).

unborn /ʌnˈbɔːn/ *adj.* not yet, or never to be, born (*unborn child*; *unborn hopes*).

unbosom /ʌnˈbʊz(ə)m/ *v.* **1** disclose (thoughts etc.). **2** (*refl.*; often foll. by *of*) unburden oneself (of thoughts etc.).

unbothered /ʌnˈbɒðəd/ *predic. adj.* not bothered; unconcerned.

unbound[1] /ʌnˈbaʊnd/ *adj.* **1** not bound. **2** unconstrained. **3 a** (of a book) without a binding. **b** having paper covers.

unbound[2] *past* and *past part.* of UNBIND.

unbounded /ʌnˈbaʊndɪd/ *adj.* infinite (*unbounded optimism*).

unbreakable /ʌnˈbreɪkəb(ə)l/ *adj.* not breakable.

unbridgeable /ʌnˈbrɪdʒəb(ə)l/ *adj.* unable to be bridged.

unbridle /ʌnˈbraɪd(ə)l/ *v.* (**-ling**) remove a bridle, constraints, etc., from (a horse, one's tongue, etc.) (*unbridled insolence*).

unbroken /ʌnˈbrəʊkən/ *adj.* **1** not broken. **2** untamed (*unbroken horse*). **3** uninterrupted (*unbroken sleep*). **4** unsurpassed (*unbroken record*).

unbuckle /ʌnˈbʌk(ə)l/ *v.* (**-ling**) release the buckle of (a strap, shoe, etc.).

unburden /ʌnˈbɜːd(ə)n/ *v.* (often *refl.*; often followed by *to*) relieve (oneself, one's conscience, etc.) by confession etc.

unbusinesslike /ʌnˈbɪznɪsˌlaɪk/ *adj.* not businesslike.

unbutton /ʌnˈbʌt(ə)n/ *v.* **1** unfasten the buttons of (a garment, person, etc.). **2** (*absol.*) *colloq.* relax.

uncalled-for /ʌnˈkɔːldfɔː(r)/ *adj.* (of a remark, action, etc.) rude, unnecessary.

uncanny /ʌnˈkænɪ/ *adj.* (**-ier**, **-iest**) seemingly supernatural; mysterious. □ **uncannily** *adv.* **uncanniness** *n.*

uncapped /ʌnˈkæpt/ *adj. Sport* (of a player) not yet awarded his cap or never having been selected to represent his country.

uncared-for /ʌnˈkeədfɔː(r)/ *adj.* disregarded; neglected.

uncaring /ʌnˈkeərɪŋ/ *adj.* neglectful, lacking compassion.

unceasing /ʌnˈsiːsɪŋ/ *adj.* not ceasing; continuous (*unceasing effort*).

uncensored /ʌnˈsensəd/ *adj.* not censored.

unceremonious /ˌʌnserɪˈməʊnɪəs/ *adj.* **1** abrupt; discourteous. **2** informal. □ **unceremoniously** *adv.*

uncertain /ʌnˈsɜːt(ə)n/ *adj.* **1** not certainly knowing or known (*result is uncertain*). **2** unreliable. **3** changeable, erratic (*uncertain weather*). □ **in no uncertain terms** clearly and forcefully. □ **uncertainly** *adv.* **uncertainty** *n.* (*pl.* **-ies**).

unchain /ʌnˈtʃeɪn/ *v.* remove the chain(s) from; release.

unchallengeable /ʌnˈtʃælɪndʒəb(ə)l/ *adj.* not challengeable; unassailable.

unchallenged /ʌnˈtʃælɪndʒd/ *adj.* not challenged.

unchangeable /ʌnˈtʃeɪndʒəb(ə)l/ *adj.* unable to be changed.

unchanged /ʌnˈtʃeɪndʒd/ *adj.* not changed; unaltered.

unchanging /ʌnˈtʃeɪndʒɪŋ/ *adj.* not changing; remaining the same.

unchaperoned /ʌnˈʃæpəˌrəʊnd/ *adj.* without a chaperone.

uncharacteristic /ˌʌnkærɪktəˈrɪstɪk/ *adj.* not characteristic. □ **uncharacteristically** *adv.*

uncharitable /ʌnˈtʃærɪtəb(ə)l/ *adj.* censorious, severe in judgement. □ **uncharitably** *adv.*

uncharted /ʌnˈtʃɑːtɪd/ *adj.* not mapped or surveyed.

unchecked /ʌnˈtʃekt/ *adj.* **1** not checked. **2** unrestrained (*unchecked violence*).

unchivalrous /ʌnˈʃɪvəlrəs/ *adj.* not chivalrous. □ **unchivalrously** *adv.*

unchristian /ʌn'krɪstʃ(ə)n/ *adj.* contrary to Christian principles, esp. uncaring or selfish.

uncial /'ʌnsɪəl/ *–adj.* of or written in rounded unjoined letters similar to capitals, found in manuscripts of the 4th–8th c. *–n.* uncial letter, style, or MS. [Latin *uncia* inch]

uncircumcised /ʌn'sɜ:kəm,saɪzd/ *adj.* not circumcised.

uncivil /ʌn'sɪvɪl/ *adj.* ill-mannered; impolite. □ **uncivilly** *adv.*

uncivilized /ʌn'sɪvɪ,laɪzd/ *adj.* (also **-ised**) **1** not civilized. **2** rough; uncultured.

unclaimed /ʌn'kleɪmd/ *adj.* not claimed.

unclasp /ʌn'klɑ:sp/ *v.* **1** loosen the clasp(s) of. **2** release the grip of (a hand etc.).

unclassified /ʌn'klæsɪ,faɪd/ *adj.* **1** not classified. **2** (of State information) not secret.

uncle /'ʌŋk(ə)l/ *n.* **1 a** brother of one's father or mother. **b** aunt's husband. **2** *colloq.* (form of address by a child to) parent's male friend. **3** *slang* esp. *hist.* pawnbroker. [Latin *avunculus*]

-uncle *suffix* forming nouns, usu. diminutives (*carbuncle*). [Latin *-unculus*]

unclean /ʌn'kli:n/ *adj.* **1** not clean. **2** unchaste. **3** religiously impure; forbidden.

unclear /ʌn'klɪə(r)/ *adj.* **1** not clear or easy to understand. **2** (of a person) uncertain (*I'm unclear as to what you mean*).

unclench /ʌn'klentʃ/ *v.* **1** release (clenched hands etc.). **2** (of hands etc.) become relaxed or open.

Uncle Sam *n. colloq.* US government.

unclothe /ʌn'kləʊð/ *v.* (**-thing**) **1** remove clothes, leaves, etc. from. **2** expose, reveal.

unclouded /ʌn'klaʊdɪd/ *adj.* **1** clear; bright. **2** untroubled (*unclouded serenity*).

uncluttered /ʌn'klʌtəd/ *adj.* not cluttered; austere, simple.

uncoil /ʌn'kɔɪl/ *v.* unwind.

uncoloured /ʌn'kʌləd/ *adj.* **1** having no colour. **2** not influenced; impartial.

uncombed /ʌn'kəʊmd/ *adj.* (of hair or a person) not combed.

uncomfortable /ʌn'kʌmftəb(ə)l/ *adj.* **1** not comfortable. **2** uneasy; disquieting (*uncomfortable silence*). □ **uncomfortably** *adv.*

uncommercial /,ʌnkə'mɜ:ʃ(ə)l/ *adj.* not commercial.

uncommitted /,ʌnkə'mɪtɪd/ *adj.* **1** not committed. **2** not politically attached.

uncommon /ʌn'kɒmən/ *adj.* **1** unusual. **2** remarkably great etc. (*uncommon appetite*). □ **uncommonly** *adv.* **uncommonness** *n.*

uncommunicative /,ʌnkə'mju:nɪkətɪv/ *adj.* taciturn.

uncompetitive /,ʌnkəm'petɪtɪv/ *adj.* not competitive.

uncomplaining /,ʌnkəm'pleɪnɪŋ/ *adj.* not complaining; resigned. □ **uncomplainingly** *adv.*

uncompleted /,ʌnkəm'pli:tɪd/ *adj.* not completed; incomplete.

uncomplicated /ʌn'kɒmplɪ,keɪtɪd/ *adj.* simple; straightforward.

uncomplimentary /,ʌnkɒmplɪ'mentərɪ/ *adj.* insulting.

uncomprehending /,ʌnkɒmprɪ'hendɪŋ/ *adj.* not comprehending.

uncompromising /ʌn'kɒmprə,maɪzɪŋ/ *adj.* stubborn; unyielding. □ **uncompromisingly** *adv.*

unconcealed /,ʌnkən'si:ld/ *adj.* not concealed; obvious.

unconcern /,ʌnkən'sɜ:n/ *n.* calmness; indifference; apathy. □ **unconcerned** *adj.* **unconcernedly** /-nɪdlɪ/ *adv.*

unconditional /,ʌnkən'dɪʃən(ə)l/ *adj.* not subject to conditions; complete (*unconditional surrender*). □ **unconditionally** *adv.*

unconditioned reflex *n.* instinctive response to a stimulus.

unconfined /,ʌnkən'faɪnd/ *adj.* not confined; boundless.

unconfirmed /,ʌnkən'fɜ:md/ *adj.* not confirmed.

uncongenial /,ʌnkən'dʒi:nɪəl/ *adj.* not congenial.

unconnected /,ʌnkə'nektɪd/ *adj.* **1** not physically joined. **2** not connected or associated. **3** disconnected (*unconnected ideas*).

unconquerable /ʌn'kɒŋkərəb(ə)l/ *adj.* not conquerable.

unconscionable /ʌn'kɒnʃənəb(ə)l/ *adj.* **1** without or contrary to conscience. **2** excessive (*unconscionable waste*). □ **unconscionably** *adv.* [from UN-[1], CONSCIENCE]

unconscious /ʌn'kɒnʃəs/ *–adj.* not conscious (*fell unconscious*; *unconscious prejudice*). *–n.* normally inaccessible part of the mind affecting the emotions etc. □ **unconsciously** *adv.* **unconsciousness** *n.*

unconsidered /,ʌnkən'sɪdəd/ *adj.* **1** not considered; disregarded. **2** not premeditated.

unconstitutional /,ʌnkɒnstɪ'tju:ʃən(ə)l/ *adj.* in breach of a political constitution or procedural rules. □ **unconstitutionally** *adv.*

unconstrained /,ʌnkən'streɪnd/ *adj.* not constrained or compelled.

uncontaminated /ˌʌnkənˈtæmɪˌneɪtɪd/ *adj.* not contaminated.
uncontested /ˌʌnkənˈtestɪd/ *adj.* not contested.
uncontrollable /ˌʌnkənˈtrəʊləb(ə)l/ *adj.* not controllable. □ **uncontrollably** *adv.*
uncontrolled /ˌʌnkənˈtrəʊld/ *adj.* not controlled; unrestrained.
uncontroversial /ˌʌnkɒntrəˈvɜːʃ(ə)l/ *adj.* not controversial.
unconventional /ˌʌnkənˈvenʃən(ə)l/ *adj.* unusual; unorthodox. □ **unconventionality** /-ˈnælɪtɪ/ *n.* **unconventionally** *adv.*
unconvinced /ˌʌnkənˈvɪnst/ *adj.* not convinced.
unconvincing /ˌʌnkənˈvɪnsɪŋ/ *adj.* not convincing. □ **unconvincingly** *adv.*
uncooked /ʌnˈkʊkt/ *adj.* not cooked; raw.
uncooperative /ˌʌnkəʊˈɒpərətɪv/ *adj.* not cooperative.
uncoordinated /ˌʌnkəʊˈɔːdɪˌneɪtɪd/ *adj.* **1** not coordinated. **2** clumsy.
uncork /ʌnˈkɔːk/ *v.* **1** draw the cork from (a bottle). **2** vent (feelings etc.).
uncorroborated /ˌʌnkəˈrɒbəˌreɪtɪd/ *adj.* (esp. of evidence etc.) not corroborated.
uncountable /ʌnˈkaʊntəb(ə)l/ *adj.* **1** inestimable, immense (*uncountable wealth*). **2** (of a noun) not used in the plural or with the indefinite article (e.g. *happiness, milk*).
uncouple /ʌnˈkʌp(ə)l/ *v.* **(-ling)** release from couplings or couples.
uncouth /ʌnˈkuːθ/ *adj.* uncultured, rough. [Old English, = unknown]
uncover /ʌnˈkʌvə(r)/ *v.* **1** remove a cover or covering from. **2** disclose (*uncovered the truth*).
uncritical /ʌnˈkrɪtɪk(ə)l/ *adj.* **1** not critical; complacently accepting. **2** not in accordance with the principles of criticism. □ **uncritically** *adv.*
uncross /ʌnˈkrɒs/ *v.* **1** remove from a crossed position. **2** (as **uncrossed** *adj.*) (of a cheque) not crossed.
uncrown /ʌnˈkraʊn/ *v.* **1** deprive of a crown, a position, etc. **2** (as **uncrowned** *adj.*) **a** not crowned. **b** having the status but not the name of (*uncrowned king of boxing*).
unction /ˈʌŋkʃ(ə)n/ *n.* **1 a** anointing with oil etc. as a religious rite or medical treatment. **b** oil, ointment, etc. so used. **2 a** soothing words or thought. **b** excessive or insincere flattery. **3 a** emotional fervency. **b** pretence of this. [Latin *ungo unct-* anoint]
unctuous /ˈʌŋktʃʊəs/ *adj.* **1** unpleasantly flattering; oily. **2** greasy or soapy. □ **unctuously** *adv.* [medieval Latin: related to UNCTION]
uncultivated /ʌnˈkʌltɪˌveɪtɪd/ *adj.* not cultivated.
uncured /ʌnˈkjʊəd/ *adj.* **1** not cured. **2** (of pork etc.) not salted or smoked.
uncurl /ʌnˈkɜːl/ *v.* straighten out, untwist.
uncut /ʌnˈkʌt/ *adj.* **1** not cut. **2** (of a book) with the pages sealed or untrimmed. **3** (of a book, film, etc.) complete; uncensored. **4** (of esp. a diamond) not shaped. **5** (of fabric) with a looped pile.
undamaged /ʌnˈdæmɪdʒd/ *adj.* intact.
undated /ʌnˈdeɪtɪd/ *adj.* without a date.
undaunted /ʌnˈdɔːntɪd/ *adj.* not daunted.
undeceive /ˌʌndɪˈsiːv/ *v.* **(-ving)** (often foll. by *of*) free (a person) from a misconception, deception, or error.
undecided /ˌʌndɪˈsaɪdɪd/ *adj.* **1** not settled. **2** irresolute.
undeclared /ˌʌndɪˈkleəd/ *adj.* not declared.
undefeated /ˌʌndɪˈfiːtɪd/ *adj.* not defeated.
undefended /ˌʌndɪˈfendɪd/ *adj.* not defended.
undefined /ˌʌndɪˈfaɪnd/ *adj.* not defined; vague, indefinite.
undemanding /ˌʌndɪˈmɑːndɪŋ/ *adj.* not demanding; easily done or satisfied (*undemanding reading*).
undemocratic /ˌʌndeməˈkrætɪk/ *adj.* not democratic.
undemonstrative /ˌʌndɪˈmɒnstrətɪv/ *adj.* not emotionally expressive; reserved.
undeniable /ˌʌndɪˈnaɪəb(ə)l/ *adj.* indisputable; certain. □ **undeniably** *adv.*
under *–prep.* **1 a** in or to a position lower than; below; beneath (*under the table*). **b** on the inside of (*vest under his shirt*). **2** inferior to; less than (*no-one under a major*; *is under 18*; *was under £20*). **3 a** subject to; controlled by (*under constraint*; *born under Saturn*; *prospered under him*). **b** undergoing (*is under repair*). **c** classified or subsumed in (*under two headings*). **4** at the foot of or sheltered by (*under the cliff*). **5** planted with (a crop). **6** powered by (sail, steam, etc.). *–adv.* **1** in or to a lower position or condition (*kept him under*). **2** *colloq.* in or into unconsciousness (*put him under*). *–adj.* lower (*under jaw*). □ **under arms** see ARM[2]. **under one's belt** see BELT. **under one's breath** see BREATH. **under a cloud** see CLOUD. **under control** see CONTROL. **under the counter** see COUNTER[1]. **under fire** see FIRE. **under a person's nose** see NOSE. **under the sun** anywhere in the world.

under way in motion; in progress. **under the weather** see WEATHER. □ **undermost** *adj.* [Old English]

under- *prefix* in senses of UNDER: **1** below, beneath (*underground*). **2** lower; subordinate (*under-secretary*). **3** insufficiently, incompletely (*undercook*; *underdeveloped*).

underachieve /ˌʌndərəˈtʃiːv/ *v.* (**-ving**) do less well than might be expected (esp. academically). □ **underachiever** *n.*

underact /ˌʌndərˈækt/ *v. Theatr.* act with insufficient force.

under-age *adj.* (also **under age**) not old enough.

underarm –*adj.* & *adv. Sport*, esp. *Cricket* with the arm below shoulder-level. –*attrib. adj.* **1** under the arm (*underarm seam*). **2** in the armpit.

underbelly *n.* (*pl.* **-ies**) undersurface of an animal, vehicle, etc., esp. as vulnerable to attack.

underbid –*v.* /ˌʌndəˈbɪd/ (**-dd-**; *past* and *past part.* **-bid**) **1** make a lower bid than. **2** (also *absol.*) *Bridge* etc. bid less on (one's hand) than warranted. –*n.* /ˈʌndəˌbɪd/ such a bid.

undercarriage *n.* **1** wheeled retractable structure beneath an aircraft, used for landing etc. **2** supporting frame of a vehicle.

undercharge /ˌʌndəˈtʃɑːdʒ/ *v.* (**-ging**) **1** charge too little to (a person). **2** give too little charge to (a gun, electric battery, etc.).

underclothes *n.pl.* clothes worn under others, esp. next to the skin.

underclothing *n.* underclothes collectively.

undercoat *n.* **1 a** layer of paint under a topcoat. **b** paint for this. **2** animal's under layer of hair etc.

undercook /ˌʌndəˈkʊk/ *v.* cook insufficiently.

undercover /ˌʌndəˈkʌvə(r)/ *adj.* (usu. *attrib.*) **1** surreptitious. **2** spying incognito, esp. by infiltration (*undercover agent*).

undercroft *n.* crypt. [obsolete *croft* from Latin]

undercurrent *n.* **1** current below the surface. **2** underlying often contrary feeling, influence, etc. (*undercurrent of protest*).

undercut –*v.* /ˌʌndəˈkʌt/ (**-tt-**; *past* and *past part.* **-cut**) **1** sell or work at a lower price than. **2** strike (a ball) to make it rise high. **3** cut away the part below. **4** undermine. –*n.* /ˈʌndəˌkʌt/ underside of sirloin.

underdeveloped /ˌʌndədɪˈveləpt/ *adj.* **1** not fully developed; immature. **2** (of a country etc.) with unexploited potential. □ **underdevelopment** *n.*

underdog *n.* **1** oppressed person. **2** loser in a fight.

underdone /ˌʌndəˈdʌn/ *adj.* undercooked.

underemployed /ˌʌndərɪmˈplɔɪd/ *adj.* not fully occupied. □ **underemployment** *n.*

underestimate –*v.* /ˌʌndərˈestɪˌmeɪt/ (**-ting**) form too low an estimate of. –*n.* /ˌʌndərˈestɪmət/ estimate that is too low. □ **underestimation** /-ˈmeɪʃ(ə)n/ *n.*

underexpose /ˌʌndərɪkˈspəʊz/ *v.* (**-sing**) expose (film) for too short a time etc. □ **underexposure** *n.*

underfed /ˌʌndəˈfed/ *adj.* malnourished.

underfelt *n.* felt laid under a carpet.

underfloor *attrib. adj.* beneath the floor (*underfloor heating*).

underfoot /ˌʌndəˈfʊt/ *adv.* (also **under foot**) **1** under one's feet. **2** on the ground.

underfunded *adj.* provided with insufficient money.

undergarment *n.* piece of underclothing.

undergo /ˌʌndəˈgəʊ/ *v.* (*3rd sing. present* **-goes**; *past* **-went**; *past part.* **-gone**) be subjected to; suffer; endure.

undergraduate /ˌʌndəˈgrædjʊət/ *n.* person studying for a first degree.

underground –*adv.* /ˌʌndəˈgraʊnd/ **1** beneath the ground. **2** in or into secrecy or hiding. –*adj.* /ˈʌndəˌgraʊnd/ **1** situated underground. **2** secret, subversive. **3** unconventional (*underground literature*). –*n.* /ˈʌndəˌgraʊnd/ **1** underground railway. **2** secret subversive group or activity.

undergrowth *n.* dense shrubs etc., esp. in a wood.

underhand *adj.* **1** deceitful; crafty; secret. **2** *Sport*, esp. *Cricket* underarm.

underlay[1] –*v.* /ˌʌndəˈleɪ/ (*past* and *past part.* **-laid**) lay something under (a thing) to support or raise it. –*n.* /ˈʌndəˌlaɪ/ thing so laid (esp. under a carpet).

underlay[2] *past* of UNDERLIE.

underlie /ˌʌndəˈlaɪ/ *v.* (**-lying**; *past* **-lay**; *past part.* **-lain**) **1** (also *absol.*) lie under (a stratum etc.). **2** (also *absol.*) (esp. as **underlying** *adj.*) be the basis of (a doctrine, conduct, etc.). **3** exist beneath the superficial aspect of.

underline /ˌʌndəˈlaɪn/ *v.* (**-ning**) **1** draw a line under (a word etc.) to give emphasis, indicate italic type, etc. **2** emphasize, stress.

underling /ˈʌndəlɪŋ/ *n.* usu. *derog.* subordinate.

underlying *pres. part.* of UNDERLIE.

undermanned /ˌʌndəˈmænd/ *adj.* having an insufficient crew or staff.

undermentioned /ˌʌndəˈmenʃ(ə)nd/ *adj.* mentioned later in a book etc.

undermine /ˌʌndəˈmaɪn/ *v.* (**-ning**) **1** injure (a person, reputation, health, etc.) secretly or insidiously. **2** wear away the base of (*banks were undermined*). **3** make an excavation under.

underneath /ˌʌndəˈni:θ/ *–prep.* **1** at or to a lower place than, below. **2** on the inside of. *–adv.* **1** at or to a lower place. **2** inside. *–n.* lower surface or part. *–adj.* lower. [Old English: related to NETHER]

undernourished /ˌʌndəˈnʌrɪʃt/ *adj.* insufficiently nourished. □ **undernourishment** *n.*

underpaid *past* and *past part.* of UNDERPAY.

underpants *n.pl.* undergarment, esp. men's, covering the genitals and buttocks.

underpart *n.* lower or subordinate part.

underpass *n.* **1** road etc. passing under another. **2** subway.

underpay /ˌʌndəˈpeɪ/ *v.* (*past* and *past part.* **-paid**) pay too little to (a person) or for (a thing). □ **underpayment** *n.*

underpin /ˌʌndəˈpɪn/ *v.* (**-nn-**) **1** support from below with masonry etc. **2** support, strengthen.

underplay /ˌʌndəˈpleɪ/ *v.* **1** make little of. **2** *Theatr.* underact.

underpopulated /ˌʌndəˈpɒpjʊˌleɪtɪd/ *adj.* having an insufficient or very small population.

underprice /ˌʌndəˈpraɪs/ *v.* (**-cing**) price lower than what is usual or appropriate.

underprivileged /ˌʌndəˈprɪvɪlɪdʒd/ *adj.* less privileged than others; having below average income, rights, etc.

underrate /ˌʌndəˈreɪt/ *v.* (**-ting**) have too low an opinion of.

underscore /ˌʌndəˈskɔ:(r)/ *v.* (**-ring**) = UNDERLINE.

undersea *adj.* below the sea or its surface.

underseal *–v.* seal the underpart of (esp. a vehicle against rust etc.). *–n.* protective coating for this.

under-secretary /ˌʌndəˈsekrətərɪ/ *n.* (*pl.* **-ies**) subordinate official, esp. a junior minister or senior civil servant.

undersell /ˌʌndəˈsel/ *v.* (*past* and *past part.* **-sold**) sell at a lower price than (another seller).

undersexed /ˌʌndəˈsekst/ *adj.* having unusually weak sexual desires.

undershirt *n.* esp. *US* man's or boy's vest.

undershoot /ˌʌndəˈʃu:t/ *v.* (*past* and *past part.* **-shot**) land short of (a runway etc.).

undershot *adj.* **1** (of a water-wheel) turned by water flowing under it. **2** (of a lower jaw) projecting beyond the upper jaw.

underside *n.* lower or under side or surface.

undersigned /ˌʌndəˈsaɪnd/ *adj.* (usu. *absol.*) whose signature is appended (*we, the undersigned*).

undersized /ˌʌndəˈsaɪzd/ *adj.* smaller than average.

underskirt *n.* petticoat.

underslung /ˌʌndəˈslʌŋ/ *adj.* supported from above.

undersold *past* and *past part.* of UNDERSELL.

underspend /ˌʌndəˈspend/ *v.* (*past* and *past part.* **-spent**) (usu. *absol.*) spend less than (the expected amount), or too little.

understaffed /ˌʌndəˈstɑ:ft/ *adj.* having too few staff.

understand /ˌʌndəˈstænd/ *v.* (*past* and *past part.* **-stood**) **1** perceive the meaning of (words, a person, a language, a subject, etc.) (*understood you perfectly*; *cannot understand algebra*). **2** perceive the significance or cause of (*do not understand why he came*). **3** (often *absol.*) sympathize with, know how to deal with (*quite understand your difficulty*; *ask her, she understands*). **4** (often foll. by *that* or *absol.*) infer, take as implied (*am I to understand that you refuse?*; *he is old, I understand*). **5** supply (an implied missing word) mentally. □ **understand each other 1** know each other's views. **2** agree or collude. □ **understandable** *adj.* **understandably** *adv.* [Old English: related to STAND]

understanding *–n.* **1** ability to understand or think; intelligence. **2** individual's perception of a situation etc. **3** agreement, esp. informal (*had an understanding*). **4** sympathy; tolerance. *–adj.* **1** having understanding or insight. **2** sympathetic. □ **understandingly** *adv.*

understate /ˌʌndəˈsteɪt/ *v.* (**-ting**) **1** express mildly or in a restrained way. **2** represent as less than it actually is. □ **understatement** *n.*

understeer *n.* tendency of a vehicle not to turn sharply enough.

understood *past* and *past part.* of UNDERSTAND.

understudy esp. *Theatr.* *–n.* (*pl.* **-ies**) person ready to take on another's role etc. when required. *–v.* (**-ies, -ied**) **1** study (a role etc.) thus. **2** act as an understudy to.

undersubscribed /ˌʌndəsəb'skraɪbd/ *adj.* without sufficient subscribers, participants, etc.

undersurface *n.* lower or under surface.

undertake /ˌʌndə'teɪk/ *v.* (**-king**; *past* **-took**; *past part.* **-taken**) **1** agree to perform or be responsible for; engage in, enter upon (work, a responsibility, etc.). **2** (usu. foll. by *to* + infin.) promise. **3** guarantee (*undertake that he is innocent*).

undertaker /'ʌndəˌteɪkə(r)/ *n.* professional funeral organizer.

undertaking /ˌʌndə'teɪkɪŋ/ *n.* **1** work etc. undertaken, enterprise (*serious undertaking*). **2** promise. **3** /'ʌn-/ professional funeral management.

underthings *n.pl.* underclothes.

undertone *n.* **1** subdued tone or colour. **2** underlying quality or feeling.

undertook *past* of UNDERTAKE.

undertow *n.* current below the surface of the sea contrary to the surface current.

underused /ˌʌndə'ju:zd/ *adj.* not used to capacity.

undervalue /ˌʌndə'vælju:/ *v.* (**-ues**, **-ued**, **-uing**) **1** value insufficiently. **2** underestimate.

undervest *n.* vest.

underwater /ˌʌndə'wɔ:tə(r)/ *–adj.* situated or done under water. *–adv.* under water.

underwear *n.* underclothes.

underweight *–adj.* /ˌʌndə'weɪt/ below normal weight. *–n.* /'ʌndəˌweɪt/ insufficient weight.

underwent *past* of UNDERGO.

underwhelm /ˌʌndə'welm/ *v.joc.* fail to impress. [alteration of OVERWHELM]

underworld *n.* **1** those who live by organized crime and vice. **2** mythical abode of the dead under the earth.

underwrite /ˌʌndə'raɪt/ *v.* (**-ting**; *past* **-wrote**; *past part.* **-written**) **1 a** sign and accept liability under (an insurance policy, esp. on shipping etc.). **b** accept (liability) in this way. **2** undertake to finance or support. **3** engage to buy all the unsold stock in (a company etc.). □ **underwriter** /'ʌn-/ *n.*

undescended /ˌʌndɪ'sendɪd/ *adj.* (of a testicle) not descending normally into the scrotum.

undeserved /ˌʌndɪ'zɜ:vd/ *adj.* not deserved. □ **undeservedly** /-vɪdlɪ/ *adv.*

undeserving /ˌʌndɪ'zɜ:vɪŋ/ *adj.* not deserving.

undesigned /ˌʌndɪ'zaɪnd/ *adj.* unintentional.

undesirable /ˌʌndɪ'zaɪərəb(ə)l/ *–adj.* objectionable, unpleasant. *–n.* undesirable person. □ **undesirability** /-'bɪlɪtɪ/ *n.*

undetectable /ˌʌndɪ'tektəb(ə)l/ *adj.* not detectable.

undetected /ˌʌndɪ'tektɪd/ *adj.* not detected.

undetermined /ˌʌndɪ'tɜ:mɪnd/ *adj.* = UNDECIDED.

undeterred /ˌʌndɪ'tɜ:d/ *adj.* not deterred.

undeveloped /ˌʌndɪ'veləpt/ *adj.* not developed.

undid *past* of UNDO.

undies /'ʌndɪz/ *n.pl. colloq.* (esp. women's) underclothes. [abbreviation]

undifferentiated /ˌʌndɪfə'renʃɪˌeɪtɪd/ *adj.* not differentiated; amorphous.

undigested /ˌʌndaɪ'dʒestɪd/ *adj.* **1** not digested. **2** (of facts etc.) not properly arranged or considered.

undignified /ʌn'dɪgnɪˌfaɪd/ *adj.* lacking dignity.

undiluted /ˌʌndaɪ'lju:tɪd/ *adj.* **1** not diluted. **2** complete, utter.

undiminished /ˌʌndɪ'mɪnɪʃt/ *adj.* not diminished or lessened.

undine /'ʌndi:n/ *n.* female water-spirit. [Latin *unda* wave]

undiplomatic /ˌʌndɪplə'mætɪk/ *adj.* tactless.

undisciplined /ʌn'dɪsɪplɪnd/ *adj.* lacking discipline; not disciplined.

undisclosed /ˌʌndɪs'kləʊzd/ *adj.* not revealed or made known.

undiscovered /ˌʌndɪ'skʌvəd/ *adj.* not discovered.

undiscriminating /ˌʌndɪ'skrɪmɪˌneɪtɪŋ/ *adj.* lacking good judgement.

undisguised /ˌʌndɪs'gaɪzd/ *adj.* not disguised; open.

undismayed /ˌʌndɪs'meɪd/ *adj.* not dismayed.

undisputed /ˌʌndɪ'spju:tɪd/ *adj.* not disputed or called in question.

undistinguished /ˌʌndɪ'stɪŋgwɪʃt/ *adj.* not distinguished; mediocre.

undisturbed /ˌʌndɪ'stɜ:bd/ *adj.* not disturbed or interfered with.

undivided /ˌʌndɪ'vaɪdɪd/ *adj.* not divided or shared; whole, entire (*undivided attention*).

undo /ʌn'du:/ *v.* (*3rd sing. present* **-does**; *past* **-did**; *past part.* **-done**; *pres. part.* **-doing**) **1** unfasten (a coat, button, parcel, etc.), or the clothing of (a person). **2** annul, cancel (*cannot undo the past*). **3** ruin the prospects, reputation, or morals of.

undoing *n.* **1** ruin or cause of ruin. **2** reversing of an action etc. **3** opening or unfastening.

undone /ʌn'dʌn/ *adj.* **1** not done. **2** not fastened. **3** *archaic* ruined.

undoubted /ʌn'daʊtɪd/ *adj.* certain, not questioned. □ **undoubtedly** *adv.*

undreamed /ʌn'dri:md, ʌn'dremt/ *adj.* (also **undreamt** /ʌn'dremt/) (often foll. by *of*) not dreamed, thought, or imagined.

undress /ʌn'dres/ —*v.* **1** take off one's clothes. **2** take the clothes off (a person). —*n.* **1** ordinary or casual dress, esp. as opposed to full dress or uniform. **2** naked or partially clothed state.

undressed /ʌn'drest/ *adj.* **1** not, or no longer, dressed. **2** (of food) without a dressing. **3** (of leather etc.) not treated.

undrinkable /ʌn'drɪŋkəb(ə)l/ *adj.* unfit for drinking.

undue /ʌn'dju:/ *adj.* excessive, disproportionate. □ **unduly** *adv.*

undulate /'ʌndjʊ,leɪt/ *v.* (**-ting**) (cause to) have a wavy motion or look. □ **undulation** /-'leɪʃ(ə)n/ *n.* [Latin *unda* wave]

undying /ʌn'daɪɪŋ/ *adj.* immortal; never-ending (*undying love*).

unearned /ʌn'ɜ:nd/ *adj.* not earned.

unearned income *n.* income from investments etc. rather than from working.

unearth /ʌn'ɜ:θ/ *v.* discover by searching, digging, or rummaging.

unearthly /ʌn'ɜ:θlɪ/ *adj.* **1** supernatural, mysterious. **2** *colloq.* absurdly early or inconvenient (*unearthly hour*). □ **unearthliness** *n.*

unease /ʌn'i:z/ *n.* nervousness, anxiety.

uneasy /ʌn'i:zɪ/ *adj.* (**-ier, -iest**) **1** nervous, anxious. **2** disturbing (*uneasy suspicion*). □ **uneasily** *adv.* **uneasiness** *n.*

uneatable /ʌn'i:təb(ə)l/ *adj.* not able to be eaten (cf. INEDIBLE).

uneaten /ʌn'i:t(ə)n/ *adj.* left not eaten.

uneconomic /,ʌni:kə'nɒmɪk, ,ʌnek-/ *adj.* not economic; unprofitable.

uneconomical *adj.* not economical; wasteful.

unedifying /ʌn'edɪ,faɪɪŋ/ *adj.* distasteful, degrading.

unedited /ʌn'edɪtɪd/ *adj.* not edited.

uneducated /ʌn'edjʊ,keɪtɪd/ *adj.* not educated.

unembarrassed /,ʌnɪm'bærəst/ *adj.* not embarrassed.

unemotional /,ʌnɪ'məʊʃən(ə)l/ *adj.* not emotional; lacking emotion.

unemphatic /,ʌnɪm'fætɪk/ *adj.* not emphatic.

unemployable /,ʌnɪm'plɔɪəb(ə)l/ *adj.* unfit for paid employment. □ **unemployability** /-'bɪlɪtɪ/ *n.*

unemployed /,ʌnɪm'plɔɪd/ *adj.* **1** out of work. **2** not in use.

unemployment /,ʌnɪm'plɔɪmənt/ *n.* **1** being unemployed. **2** lack of employment in a country etc.

unemployment benefit *n.* State payment made to an unemployed person.

unencumbered /,ʌnɪn'kʌmbəd/ *adj.* **1** (of an estate) not having liabilities (e.g. a mortgage). **2** free; not burdened.

unending /ʌn'endɪŋ/ *adj.* endless or seemingly endless.

unendurable /,ʌnɪn'djʊərəb(ə)l/ *adj.* too bad to be borne.

unenlightened /,ʌnɪn'laɪt(ə)nd/ *adj.* not enlightened.

unenterprising /ʌn'entə,praɪzɪŋ/ *adj.* not enterprising.

unenthusiastic /,ʌnɪn,θju:zɪ'æstɪk/ *adj.* not enthusiastic. □ **unenthusiastically** *adv.*

unenviable /ʌn'envɪəb(ə)l/ *adj.* not enviable.

unequal /ʌn'i:kw(ə)l/ *adj.* **1** (often foll. by *to*) not equal. **2** of varying quality. **3** unfair (*unequal contest*). □ **unequally** *adv.*

unequalled *adj.* (*US* **-aled**) superior to all others.

unequivocal /,ʌnɪ'kwɪvək(ə)l/ *adj.* not ambiguous, plain, unmistakable. □ **unequivocally** *adv.*

unerring /ʌn'ɜ:rɪŋ/ *adj.* not erring; true, certain. □ **unerringly** *adv.*

UNESCO /ju:'neskəʊ/ *abbr.* (also **Unesco**) United Nations Educational, Scientific, and Cultural Organization.

unethical /ʌn'eθɪk(ə)l/ *adj.* not ethical, esp. unscrupulous or unprofessional. □ **unethically** *adv.*

uneven /ʌn'i:v(ə)n/ *adj.* **1** not level or smooth. **2** of variable quality etc. **3** (of a contest) unequal. □ **unevenly** *adv.* **unevenness** *n.*

uneventful /,ʌnɪ'ventfʊl/ *adj.* not eventful. □ **uneventfully** *adv.*

unexampled /,ʌnɪg'zɑ:mp(ə)ld/ *adj.* without precedent.

unexceptionable /,ʌnɪk'sepʃənəb(ə)l/ *adj.* entirely satisfactory.

■ **Usage** See note at *exceptionable.*

unexceptional /,ʌnɪk'sepʃən(ə)l/ *adj.* usual, normal, ordinary.

unexciting /,ʌnɪk'saɪtɪŋ/ *adj.* not exciting; dull.

unexpected /,ʌnɪk'spektɪd/ *adj.* not expected; surprising. □ **unexpectedly** *adv.* **unexpectedness** *n.*

unexplained /,ʌnɪk'spleɪnd/ *adj.* not explained.

unexplored /,ʌnɪk'splɔ:d/ *adj.* not explored.

unexposed /,ʌnɪk'spəʊzd/ *adj.* not exposed.

unexpressed /ˌʌnɪk'sprest/ *adj.* not expressed or made known (*unexpressed fears*).

unexpurgated /ʌn'ekspəˌgeɪtɪd/ *adj.* (esp. of a text etc.) complete.

unfading /ʌn'feɪdɪŋ/ *adj.* never fading.

unfailing /ʌn'feɪlɪŋ/ *adj.* not failing or dwindling; constant; reliable. □ **unfailingly** *adv.*

unfair /ʌn'feə(r)/ *adj.* not fair, just, or impartial. □ **unfairly** *adv.* **unfairness** *n.*

unfaithful /ʌn'feɪθfʊl/ *adj.* **1** not faithful, esp. adulterous. **2** treacherous; disloyal. □ **unfaithfully** *adv.* **unfaithfulness** *n.*

unfamiliar /ˌʌnfə'mɪljə(r)/ *adj.* not familiar. □ **unfamiliarity** /-lɪ'ærɪtɪ/ *n.*

unfashionable /ʌn'fæʃənəb(ə)l/ *adj.* not fashionable. □ **unfashionably** *adv.*

unfasten /ʌn'fɑːs(ə)n/ *v.* **1** make or become loose. **2** open the fastening(s) of. **3** detach.

unfathomable /ʌn'fæðəməb(ə)l/ *adj.* incapable of being fathomed.

unfavourable /ʌn'feɪvərəb(ə)l/ *adj.* (*US* **unfavorable**) not favourable; adverse, hostile. □ **unfavourably** *adv.*

unfeasible /ʌn'fiːzɪb(ə)l/ *adj.* not feasible; impractical.

unfeeling /ʌn'fiːlɪŋ/ *adj.* unsympathetic, harsh.

unfeigned /ʌn'feɪnd/ *adj.* genuine, sincere.

unfertilized /ʌn'fɜːtɪˌlaɪzd/ *adj.* (also **-ised**) not fertilized.

unfetter /ʌn'fetə(r)/ *v.* release from fetters.

unfilled /ʌn'fɪld/ *adj.* not filled.

unfinished /ʌn'fɪnɪʃt/ *adj.* not finished; incomplete.

unfit /ʌn'fɪt/ *adj.* (often foll. by *for*, or *to* + infin.) not fit.

unfitted /ʌn'fɪtɪd/ *adj.* **1** not fit. **2** not fitted or suited. **3** having no fittings.

unfitting /ʌn'fɪtɪŋ/ *adj.* not suitable, unbecoming.

unfix /ʌn'fɪks/ *v.* release, loosen, or detach.

unflagging /ʌn'flægɪŋ/ *adj.* tireless, persistent.

unflappable /ʌn'flæpəb(ə)l/ *adj. colloq.* imperturbable; calm. □ **unflappability** /-'bɪlɪtɪ/ *n.*

unflattering /ʌn'flætərɪŋ/ *adj.* not flattering. □ **unflatteringly** *adv.*

unfledged /ʌn'fledʒd/ *adj.* **1** (of a person) inexperienced. **2** (of a bird) not yet fledged.

unflinching /ʌn'flɪntʃɪŋ/ *adj.* not flinching. □ **unflinchingly** *adv.*

unfold /ʌn'fəʊld/ *v.* **1** open the fold or folds of, spread out. **2** reveal (thoughts etc.). **3** become opened out. **4** develop.

unforced /ʌn'fɔːst/ *adj.* **1** easy, natural. **2** not compelled or constrained.

unforeseeable /ˌʌnfɔː'siːəb(ə)l/ *adj.* not foreseeable.

unforeseen /ˌʌnfɔː'siːn/ *adj.* not foreseen.

unforgettable /ˌʌnfə'getəb(ə)l/ *adj.* that cannot be forgotten; memorable, wonderful.

unforgivable /ˌʌnfə'gɪvəb(ə)l/ *adj.* that cannot be forgiven.

unforgiving /ˌʌnfə'gɪvɪŋ/ *adj.* not forgiving.

unformed /ʌn'fɔːmd/ *adj.* **1** not formed; undeveloped. **2** shapeless.

unforthcoming /ˌʌnfɔːθ'kʌmɪŋ/ *adj.* not forthcoming.

unfortunate /ʌn'fɔːtʃənət/ *–adj.* **1** unlucky. **2** unhappy. **3** regrettable. *–n.* unfortunate person.

unfortunately *adv.* **1** (qualifying a sentence) it is unfortunate that. **2** in an unfortunate manner.

unfounded /ʌn'faʊndɪd/ *adj.* without foundation (*unfounded rumour*).

unfreeze /ʌn'friːz/ *v.* (**-zing**; *past* **unfroze**; *past part.* **unfrozen**) **1** (cause to) thaw. **2** derestrict (assets, credits, etc.).

unfrequented /ˌʌnfrɪ'kwentɪd/ *adj.* not frequented.

unfriendly /ʌn'frendlɪ/ *adj.* (**-ier, -iest**) not friendly; hostile.

unfrock /ʌn'frɒk/ *v.* = DEFROCK.

unfroze *past* of UNFREEZE.

unfrozen *past part.* of UNFREEZE.

unfulfilled /ˌʌnfʊl'fɪld/ *adj.* not fulfilled.

unfunny /ʌn'fʌnɪ/ *adj.* (**-ier, -iest**) failing to amuse.

unfurl /ʌn'fɜːl/ *v.* **1** unroll, spread out (a sail, umbrella, etc.). **2** become unrolled.

unfurnished /ʌn'fɜːnɪʃt/ *adj.* **1** (usu. foll. by *with*) not supplied. **2** without furniture.

ungainly /ʌn'geɪnlɪ/ *adj.* awkward, clumsy. □ **ungainliness** *n.* [obsolete *gain* straight, from Old Norse]

ungenerous /ʌn'dʒenərəs/ *adj.* mean. □ **ungenerously** *adv.*

ungentlemanly /ʌn'dʒentəlmənlɪ/ *adj.* not gentlemanly.

unget-at-able /ˌʌnget'ætəb(ə)l/ *adj. colloq.* inaccessible.

ungird /ʌn'gɜːd/ *v.* release the girdle, belt, etc. of.

ungodly /ʌn'gɒdlɪ/ *adj.* **1** impious, wicked. **2** *colloq.* outrageous (*ungodly hour*).

ungovernable /ʌn'gʌvənəb(ə)l/ *adj.* uncontrollable, violent.

ungraceful /ʌn'greɪsfʊl/ *adj.* lacking grace or elegance. □ **ungracefully** *adv.*

ungracious /ʌn'greɪʃəs/ *adj.* discourteous; grudging. □ **ungraciously** *adv.*

ungrammatical /ˌʌngrə'mætɪk(ə)l/ *adj.* contrary to the rules of grammar. □ **ungrammatically** *adv.*

ungrateful /ʌn'greɪtfʊl/ *adj.* not feeling or showing gratitude. □ **ungratefully** *adv.*

ungreen /ʌn'gri:n/ *adj.* not concerned with the protection of the environment; harmful to the environment.

ungrudging /ʌn'grʌdʒɪŋ/ *adj.* not grudging.

unguarded /ʌn'gɑ:dɪd/ *adj.* **1** incautious, thoughtless (*unguarded remark*). **2** not guarded.

unguent /'ʌŋgwənt/ *n.* soft ointment or lubricant. [Latin *unguo* anoint]

ungulate /'ʌŋgjʊlət/ *–adj.* hoofed. *–n.* hoofed mammal. [Latin *ungula* hoof, claw]

unhallowed /ʌn'hæləʊd/ *adj.* **1** not consecrated. **2** not sacred, wicked.

unhampered /ʌn'hæmpəd/ *adj.* not hampered.

unhand /ʌn'hænd/ *v. rhet.* or *joc.* take one's hands off (a person); release.

unhappy /ʌn'hæpɪ/ *adj.* (**-ier, -iest**) **1** miserable. **2** unfortunate. **3** disastrous. □ **unhappily** *adv.* **unhappiness** *n.*

unharmed /ʌn'hɑ:md/ *adj.* not harmed.

unharness /ʌn'hɑ:nɪs/ *v.* remove a harness from.

unhealthy /ʌn'helθɪ/ *adj.* (**-ier, -iest**) **1** in poor health. **2 a** harmful to health. **b** unwholesome. **c** *slang* dangerous. □ **unhealthily** *adv.* **unhealthiness** *n.*

unheard /ʌn'hɜ:d/ *adj.* **1** not heard. **2** (usu. **unheard-of**) unprecedented.

unheeded /ʌn'hi:dɪd/ *adj.* disregarded.

unhelpful /ʌn'helpfʊl/ *adj.* not helpful. □ **unhelpfully** *adv.*

unhesitating /ʌn'hezɪˌteɪtɪŋ/ *adj.* without hesitation. □ **unhesitatingly** *adv.*

unhindered /ʌn'hɪndəd/ *adj.* not hindered.

unhinge /ʌn'hɪndʒ/ *v.* (**-ging**) **1** take (a door etc.) off its hinges. **2** (esp. as **unhinged** *adj.*) make mad or crazy.

unhistorical /ˌʌnhɪ'stɒrɪk(ə)l/ *adj.* not historical.

unhitch /ʌn'hɪtʃ/ *v.* **1** release from a hitched state. **2** unhook, unfasten.

unholy /ʌn'həʊlɪ/ *adj.* (**-ier, -iest**) **1** impious, wicked. **2** *colloq.* dreadful; outrageous (*unholy row*).

unhook /ʌn'hʊk/ *v.* **1** remove from a hook or hooks. **2** unfasten the hook(s) of.

unhoped-for /ʌn'həʊptfɔ:(r)/ *adj.* not hoped for or expected.

unhorse /ʌn'hɔ:s/ *v.* (**-sing**) throw (a rider) from a horse.

unhurried /ʌn'hʌrɪd/ *adj.* not hurried.

unhurt /ʌn'hɜ:t/ *adj.* not hurt.

unhygienic /ˌʌnhaɪ'dʒi:nɪk/ *adj.* not hygienic.

uni /'ju:nɪ/ *n.* (*pl.* **-s**) esp. *Austral.* & *NZ colloq.* university. [abbreviation]

uni- *comb. form* one; having or consisting of one. [Latin *unus* one]

Uniat /'ju:nɪˌæt/ (also **Uniate** /-ət/) *–adj.* of the Church in E. Europe or the Near East, acknowledging papal supremacy but retaining its own liturgy etc. *–n.* member of such a Church. [Latin *unio* UNION]

unicameral /ju:nɪ'kæmər(ə)l/ *adj.* having a single legislative chamber. [related to CHAMBER]

UNICEF /'ju:nɪˌsef/ *abbr.* United Nations Children's (orig. International Children's Emergency) Fund.

unicellular /ju:nɪ'seljʊlə(r)/ *adj.* (of an organism etc.) consisting of a single cell.

unicorn /'ju:nɪˌkɔ:n/ *n.* mythical horse with a single straight horn. [Latin *cornu* horn]

unicycle /'ju:nɪˌsaɪk(ə)l/ *n.* single-wheeled cycle, esp. as used by acrobats. □ **unicyclist** *n.*

unidentified /ˌʌnaɪ'dentɪˌfaɪd/ *adj.* not identified.

unification /ju:nɪfɪ'keɪʃ(ə)n/ *n.* unifying or being unified. □ **unificatory** *adj.*

Unification Church *n.* religious organization founded by Sun Myung Moon.

uniform /'ju:nɪˌfɔ:m/ *–adj.* **1** unvarying (*uniform appearance*). **2** conforming to the same standard, rules, etc. **3** constant over a period (*uniform acceleration*). *–n.* distinctive clothing worn by soldiers, police, schoolchildren, etc. □ **uniformed** *adj.* **uniformity** /-'fɔ:mɪtɪ/ *n.* **uniformly** *adv.* [Latin: related to FORM]

unify /'ju:nɪˌfaɪ/ *v.* (**-ies, -ied**) make or become united or uniform. [Latin: related to UNI-]

unilateral /ju:nɪ'lætər(ə)l/ *adj.* done by or affecting only one person or party (*unilateral disarmament*). □ **unilaterally** *adv.*

unilateralism *n.* unilateral disarmament. □ **unilateralist** *n.* & *adj.*

unimaginable /ˌʌnɪ'mædʒɪnəb(ə)l/ *adj.* impossible to imagine.

unimaginative /ˌʌnɪ'mædʒɪnətɪv/ *adj.* lacking imagination; stolid, dull. □ **unimaginatively** *adv.*

unimpaired /ˌʌnɪm'peəd/ *adj.* not impaired.

unimpeachable /ˌʌnɪm'pi:tʃəb(ə)l/ *adj.* beyond reproach or question.

unimpeded /ˌʌnɪm'pi:dɪd/ *adj.* not impeded.

unimportant /ˌʌnɪmˈpɔːt(ə)nt/ *adj.* not important.
unimpressed /ˌʌnɪmˈprest/ *adj.* not impressed.
unimpressive /ˌʌnɪmˈpresɪv/ *adj.* not impressive.
uninformed /ˌʌnɪnˈfɔːmd/ *adj.* not informed; ignorant.
uninhabitable /ˌʌnɪnˈhæbɪtəb(ə)l/ *adj.* unfit for habitation.
uninhabited /ˌʌnɪnˈhæbɪtɪd/ *adj.* not inhabited.
uninhibited /ˌʌnɪnˈhɪbɪtɪd/ *adj.* not inhibited.
uninitiated /ˌʌnɪˈnɪʃɪˌeɪtɪd/ *adj.* not initiated, admitted, or instructed.
uninjured /ʌnˈɪndʒəd/ *adj.* not injured.
uninspired /ˌʌnɪnˈspaɪəd/ *adj.* not inspired; commonplace, pedestrian.
uninspiring /ˌʌnɪnˈspaɪərɪŋ/ *adj.* not inspiring.
unintelligent /ˌʌnɪnˈtelɪdʒ(ə)nt/ *adj.* not intelligent.
unintelligible /ˌʌnɪnˈtelɪdʒɪb(ə)l/ *adj.* not intelligible.
unintended /ˌʌnɪnˈtendɪd/ *adj.* not intended.
unintentional /ˌʌnɪnˈtenʃən(ə)l/ *adj.* not intentional. □ **unintentionally** *adv.*
uninterested /ʌnˈɪntrəstɪd/ *adj.* not interested; indifferent.
uninteresting /ʌnˈɪntrəstɪŋ/ *adj.* not interesting.
uninterrupted /ˌʌnɪntəˈrʌptɪd/ *adj.* not interrupted.
uninvited /ˌʌnɪnˈvaɪtɪd/ *adj.* not invited.
uninviting /ˌʌnɪnˈvaɪtɪŋ/ *adj.* unattractive, repellent.
union /ˈjuːnjən/ *n.* **1** uniting or being united. **2 a** whole formed from parts or members. **b** political unit so formed. **3** = TRADE UNION. **4** marriage. **5** concord (*perfect union*). **6 (Union) a** university social club and (at Oxbridge) debating society. **b** buildings of this. **7** *Math.* totality of the members of two or more sets. **8** mixed fabric, e.g. cotton with linen or silk. [Latin *unus* one]
union-bashing *n. colloq.* media or government campaign against trade unions.
unionist *n.* **1 a** member of a trade union. **b** advocate of trade unions. **2** (usu. **Unionist**) member of a party advocating continued union between Great Britain and Northern Ireland. □ **unionism** *n.*
unionize *v.* (also **-ise**) (**-zing** or **-sing**) organize in or into a trade union. □ **unionization** /-ˈzeɪʃ(ə)n/ *n.*
Union Jack *n.* (also **Union flag**) national ensign of the United Kingdom.
unique /juːˈniːk/ *adj.* **1** being the only one of its kind; having no like, equal, or parallel. **2** remarkable (*unique opportunity*). □ **uniquely** *adv.* [Latin *unicus* from *unus* one]

■ **Usage** In sense 1, *unique* cannot be qualified by adverbs such as *absolutely*, *most*, and *quite*. The use of *unique* in sense 2 is regarded as incorrect by some people.

unisex /ˈjuːnɪˌseks/ *adj.* (of clothing, hairstyles, etc.) designed for both sexes.
unison /ˈjuːnɪs(ə)n/ *n.* **1** concord (*acted in perfect unison*). **2** coincidence in pitch of sounds or notes (*sung in unison*). [Latin *sonus* SOUND[1]]
unit /ˈjuːnɪt/ *n.* **1 a** individual thing, person, or group, esp. for calculation. **b** smallest component of a complex whole. **2** quantity as a standard of measurement (*unit of heat*; *SI unit*). **3** smallest share in a unit trust. **4** part of a mechanism with a specified function. **5** fitted item of furniture, esp. as part of a set. **6** subgroup with a special function. **7** group of buildings, wards, etc., in a hospital. **8 a** single-digit number. **b** the number 'one'. [Latin *unus* one]
Unitarian /juːnɪˈteərɪən/ —*n.* **1** person who believes that God is one, not a Trinity. **2** member of a religious body so believing. —*adj.* of Unitarians. □ **Unitarianism** *n.* [Latin *unitas* UNITY]
unitary /ˈjuːnɪtərɪ/ *adj.* **1** of a unit or units. **2** marked by unity or uniformity. [from UNIT OR UNITY]
unit cost *n.* cost of producing one item.
unite /jʊˈnaɪt/ *v.* (**-ting**) **1** join together; combine, esp. for a common purpose or action (*united in their struggle*). **2** join in marriage. **3** (cause to) form a physical or chemical whole (*oil will not unite with water*). [Latin *unio -it-* from *unus* one]
United Kingdom *n.* Great Britain and Northern Ireland.
United Nations *n.pl.* (as *sing.* or *pl.*) supranational peace-seeking organization.
United Reformed Church *n.* Church formed in 1972 from the English Presbyterian and Congregational Churches.
United States *n.* (in full **United States of America**) federal republic of 50 States, mostly in N. America and including Alaska and Hawaii.
unit price *n.* price charged for each unit of goods supplied.
unit trust *n.* company investing contributions from many persons in various securities and paying proportional dividends.

unity /'ju:nɪtɪ/ *n.* (*pl.* **-ies**) **1 a** oneness; being one; interconnected parts constituting a whole (*national unity*). **b** such a complex whole (*person regarded as a unity*). **2 a** being united; solidarity. **b** harmony (*lived together in unity*). **3** the number 'one'. [Latin *unus* one]

univalent /ˌju:nɪ'veɪlənt/ *adj.* having a valency of one. [from UNI-, VALENCE[1]]

univalve /'ju:nɪˌvælv/ *Zool.* –*adj.* having one valve. –*n.* univalve mollusc.

universal /ˌju:nɪ'vɜ:s(ə)l/ –*adj.* of, belonging to, or done etc. by all; applicable to all cases. –*n.* term, characteristic, or concept of general application. □ **universality** /-'sælɪtɪ/ *n.* **universally** *adv.* [Latin: related to UNIVERSE]

universal coupling *n.* (also **universal joint**) coupling or joint which can transmit rotary power by a shaft at any angle.

universal time *n.* = GREENWICH MEAN TIME.

universe /'ju:nɪˌvɜ:s/ *n.* **1** all existing things; Creation. **2** all mankind. **3** *Statistics* & *Logic* all the objects under consideration. [Latin *universus* combined into one]

university /ˌju:nɪ'vɜ:sɪtɪ/ *n.* (*pl.* **-ies**) **1** educational institution of advanced learning and research conferring degrees. **2** members of this. [Latin: related to UNIVERSE]

unjust /ʌn'dʒʌst/ *adj.* not just, not fair. □ **unjustly** *adv.* **unjustness** *n.*

unjustifiable /ʌn'dʒʌstɪˌfaɪəb(ə)l/ *adj.* not justifiable. □ **unjustifiably** *adv.*

unjustified /ʌn'dʒʌstɪˌfaɪd/ *adj.* not justified.

unkempt /ʌn'kempt/ *adj.* untidy, dishevelled. [= *uncombed*]

unkind /ʌn'kaɪnd/ *adj.* not kind; harsh, cruel. □ **unkindly** *adv.* **unkindness** *n.*

unknot /ʌn'nɒt/ *v.* (**-tt-**) release the knot(s) of, untie.

unknowable /ʌn'nəʊəb(ə)l/ –*adj.* that cannot be known. –*n.* **1** unknowable thing. **2** (**the Unknowable**) the postulated absolute or ultimate reality.

unknowing /ʌn'nəʊɪŋ/ *adj.* (often foll. by *of*) not knowing; ignorant, unconscious. □ **unknowingly** *adv.*

unknown /ʌn'nəʊn/ –*adj.* (often foll. by *to*) not known, unfamiliar. –*n.* unknown thing, person, or quantity. □ **unknown to** without the knowledge of (*did it unknown to me*).

unknown quantity *n.* mysterious or obscure person or thing.

Unknown Soldier *n.* unidentified soldier etc. symbolizing a nation's dead in war.

Unknown Warrior *n.* = UNKNOWN SOLDIER.

unlabelled /ʌn'leɪb(ə)ld/ *adj.* (*US* **unlabeled**) not labelled; without a label.

unlace /ʌn'leɪs/ *v.* (**-cing**) **1** undo the lace(s) of. **2** unfasten or loosen in this way.

unladen /ʌn'leɪd(ə)n/ *adj.* not laden.

unladen weight *n.* weight of a vehicle etc. when not loaded.

unladylike /ʌn'leɪdɪˌlaɪk/ *adj.* not ladylike.

unlatch /ʌn'lætʃ/ *v.* **1** release the latch of. **2** open in this way.

unlawful /ʌn'lɔ:fʊl/ *adj.* illegal, not permissible. □ **unlawfully** *adv.*

unleaded /ʌn'ledɪd/ *adj.* (of petrol etc.) without added lead.

unlearn /ʌn'lɜ:n/ *v.* (*past* and *past part.* **unlearned** or **unlearnt**) **1** forget deliberately. **2** rid oneself of (a habit, false information, etc.).

unlearned[1] /ʌn'lɜ:nɪd/ *adj.* not well educated; ignorant.

unlearned[2] /ʌn'lɜ:nd/ *adj.* (also **unlearnt** /-'lɜ:nt/) not learnt.

unleash /ʌn'li:ʃ/ *v.* **1** release from a leash or restraint. **2** set free to engage in pursuit or attack.

unleavened /ʌn'lev(ə)nd/ *adj.* not leavened; made without yeast etc.

unless /ʌn'les/ *conj.* if not; except when (*shall go unless I hear from you*). [= *on less*]

unlettered /ʌn'letəd/ *adj.* illiterate; not well educated.

unlicensed /ʌn'laɪs(ə)nst/ *adj.* not licensed, esp. to sell alcohol.

unlike /ʌn'laɪk/ –*adj.* **1** not like; different from. **2** uncharacteristic of (*greed is unlike her*). **3** dissimilar, different. –*prep.* differently from (*acts quite unlike anyone else*).

unlikely /ʌn'laɪklɪ/ *adj.* (**-ier, -iest**) **1** improbable (*unlikely tale*). **2** (foll. by *to* + infin.) not expected (*unlikely to die*). **3** unpromising (*unlikely candidate*). □ **unlikeliness** *n.*

unlike signs *n.pl. Math.* plus and minus.

unlimited /ʌn'lɪmɪtɪd/ *adj.* unrestricted; enormous (*unlimited expanse*).

unlined[1] /ʌn'laɪnd/ *adj.* without lines or wrinkles.

unlined[2] /ʌn'laɪnd/ *adj.* without a lining.

unlisted /ʌn'lɪstɪd/ *adj.* not in a published list, esp. of Stock Exchange prices or telephone numbers.

unlit /ʌn'lɪt/ *adj.* not lit.

unload /ʌn'ləʊd/ *v.* **1** (also *absol.*) remove a load from (a vehicle etc.). **2** remove (a load) from a vehicle etc. **3** remove the ammunition from (a gun etc.). **4** *colloq.* get rid of.

unlock /ʌn'lɒk/ *v.* **1 a** release the lock of (a door, box, etc.). **b** release or disclose by unlocking. **2** release thoughts, feelings, etc. from (one's mind etc.).

unlooked-for /ʌn'lʊktfɔ:(r)/ *adj.* unexpected.

unloose /ʌn'lu:s/ *v.* (**-sing**) (also **unloosen**) unfasten, loose; set free.

unlovable /ʌn'lʌvəb(ə)l/ *adj.* not lovable.

unloved /ʌn'lʌvd/ *adj.* not loved.

unlovely /ʌn'lʌvlɪ/ *adj.* not attractive; unpleasant.

unloving /ʌn'lʌvɪŋ/ *adj.* not loving.

unlucky /ʌn'lʌkɪ/ *adj.* (**-ier, -iest**) **1** not fortunate or successful. **2** wretched. **3** bringing bad luck. **4** ill-judged. □ **unluckily** *adv.*

unmade /ʌn'meɪd/ *adj.* (esp. of a bed) not made.

unmake /ʌn'meɪk/ *v.* (**-king**; *past* and *past part.* **unmade**) undo; destroy, depose, annul.

unman /ʌn'mæn/ *v.* (**-nn-**) make weak, cowardly, etc.; cause to weep etc.

unmanageable /ʌn'mænɪdʒəb(ə)l/ *adj.* not easily managed or controlled.

unmanly /ʌn'mænlɪ/ *adj.* not manly.

unmanned /ʌn'mænd/ *adj.* **1** not manned. **2** overcome by emotion etc.

unmannerly /ʌn'mænəlɪ/ *adj.* ill-mannered. □ **unmannerliness** *n.*

unmarked /ʌn'mɑ:kt/ *adj.* **1** not marked. **2** not noticed.

unmarried /ʌn'mærɪd/ *adj.* not married, single.

unmask /ʌn'mɑ:sk/ *v.* **1 a** remove the mask from. **b** expose the true character of. **2** remove one's mask.

unmatched /ʌn'mætʃt/ *adj.* not matched or equalled.

unmentionable /ʌn'menʃənəb(ə)l/ *–adj.* unsuitable for polite conversation. *–n.* (in *pl.*) *joc.* undergarments.

unmerciful /ʌn'mɜ:sɪ,fʊl/ *adj.* merciless. □ **unmercifully** *adv.*

unmerited /ʌn'merɪtɪd/ *adj.* not merited.

unmet /ʌn'met/ *adj.* (of a demand, goal, etc.) not achieved or fulfilled.

unmethodical /,ʌnmɪ'θɒdɪk(ə)l/ *adj.* not methodical.

unmindful /ʌn'maɪndfʊl/ *adj.* (often foll. by *of*) not mindful.

unmissable /ʌn'mɪsəb(ə)l/ *adj.* that cannot or should not be missed.

unmistakable /,ʌnmɪ'steɪkəb(ə)l/ *adj.* clear, obvious, plain. □ **unmistakably** *adv.*

unmitigated /ʌn'mɪtɪ,geɪtɪd/ *adj.* not mitigated; absolute (*unmitigated disaster*).

unmixed /ʌn'mɪkst/ *adj.* not mixed.

unmodified /ʌn'mɒdɪ,faɪd/ *adj.* not modified.

unmoral /ʌn'mɒr(ə)l/ *adj.* not concerned with morality (cf. IMMORAL). □ **unmorality** /,ʌnmə'rælɪtɪ/ *n.*

unmoved /ʌn'mu:vd/ *adj.* **1** not moved. **2** constant in purpose. **3** unemotional.

unmusical /ʌn'mju:zɪk(ə)l/ *adj.* **1** discordant. **2** unskilled in or indifferent to music.

unnameable /ʌn'neɪməb(ə)l/ *adj.* too bad to be named or mentioned.

unnamed /ʌn'neɪmd/ *adj.* not named.

unnatural /ʌn'nætʃər(ə)l/ *adj.* **1** contrary to nature; not normal. **2** lacking natural feelings, esp. cruel or wicked. **3** artificial. **4** affected. □ **unnaturally** *adv.*

unnecessary /ʌn'nesəsərɪ/ *adj.* **1** not necessary. **2** superfluous. □ **unnecessarily** *adv.*

unneeded /ʌn'ni:dɪd/ *adj.* not needed.

unnerve /ʌn'nɜ:v/ *v.* (**-ving**) deprive of confidence etc.

unnoticeable /ʌn'nəʊtɪsəb(ə)l/ *adj.* not easily seen or noticed.

unnoticed /ʌn'nəʊtɪst/ *adj.* not noticed.

unnumbered /ʌn'nʌmbəd/ *adj.* **1** without a number. **2** not counted. **3** countless.

unobjectionable /,ʌnəb'dʒekʃənəb(ə)l/ *adj.* not objectionable; acceptable.

unobservant /,ʌnəb'zɜ:v(ə)nt/ *adj.* not observant.

unobserved /,ʌnəb'zɜ:vd/ *adj.* not observed.

unobtainable /,ʌnəb'teɪnəb(ə)l/ *adj.* that cannot be obtained.

unobtrusive /,ʌnəb'tru:sɪv/ *adj.* not making oneself or itself noticed. □ **unobtrusively** *adv.*

unoccupied /ʌn'ɒkjʊ,paɪd/ *adj.* not occupied.

unofficial /,ʌnə'fɪʃ(ə)l/ *adj.* not officially authorized or confirmed. □ **unofficially** *adv.*

unofficial strike *n.* strike not ratified by the strikers' trade union.

unopened /ʌn'əʊpənd/ *adj.* not opened.

unopposed /,ʌnə'pəʊzd/ *adj.* not opposed.

unorganized /ʌn'ɔ:gə,naɪzd/ *adj.* (also **-ised**) not organized.

unoriginal /,ʌnə'rɪdʒɪn(ə)l/ *adj.* lacking originality; derivative.

unorthodox /ʌn'ɔ:θə,dɒks/ *adj.* not orthodox.

unpack /ʌn'pæk/ *v.* **1** (also *absol.*) open and empty (a package, luggage, etc.). **2** take (a thing) from a package etc.

unpaid /ʌn'peɪd/ *adj.* (of a debt or a person) not paid.

unpainted /ʌn'peɪntɪd/ *adj.* not painted.

unpaired /ʌn'peəd/ *adj.* **1** not being one of a pair. **2** not united or arranged in pairs.

unpalatable /ʌn'pælətəb(ə)l/ *adj.* (of food, an idea, suggestion, etc.) disagreeable, distasteful.

unparalleled /ʌn'pærə,leld/ *adj.* unequalled.

unpardonable /ʌn'pɑ:dənəb(ə)l/ *adj.* that cannot be pardoned. □ **unpardonably** *adv.*

unparliamentary /,ʌnpɑ:lə'mentərɪ/ *adj.* contrary to proper parliamentary usage.

unparliamentary language *n.* oaths or abuse.

unpasteurized /ʌn'pɑ:stʃə,raɪzd/ *adj.* (also **-ised**) not pasteurized.

unpatriotic /,ʌnpætrɪ'ɒtɪk/ *adj.* not patriotic.

unperson /'ʌn,pɜ:s(ə)n/ *n.* person said not to exist, esp. by the State.

unperturbed /,ʌnpə'tɜ:bd/ *adj.* not perturbed.

unpick /ʌn'pɪk/ *v.* undo the sewing of (stitches, a garment, etc.).

unpin /ʌn'pɪn/ *v.* (**-nn-**) unfasten or detach by removing or opening a pin or pins.

unplaced /ʌn'pleɪst/ *adj.* not placed, esp. not one of the first three in a race etc.

unplanned /ʌn'plænd/ *adj.* not planned.

unplayable /ʌn'pleɪəb(ə)l/ *adj.* **1** *Sport* (of a ball) too fast etc. to be returned. **2** that cannot be played.

unpleasant /ʌn'plez(ə)nt/ *adj.* not pleasant, disagreeable. □ **unpleasantly** *adv.* **unpleasantness** *n.*

unpleasing /ʌn'pli:zɪŋ/ *adj.* not pleasing.

unplug /ʌn'plʌg/ *v.* (**-gg-**) **1** disconnect (an electrical device) by removing its plug from the socket. **2** unstop.

unplumbed /ʌn'plʌmd/ *adj.* **1** not plumbed. **2** not fully explored or understood.

unpointed /ʌn'pɔɪntɪd/ *adj.* **1** having no point or points. **2** not punctuated. **3** (of brickwork etc.) not pointed.

unpolished /ʌn'pɒlɪʃt/ *adj.* not polished or refined; rough.

unpolitical /,ʌnpə'lɪtɪk(ə)l/ *adj.* not concerned with politics.

unpopular /ʌn'pɒpjʊlə(r)/ *adj.* not popular; disliked. □ **unpopularity** /-'lærɪtɪ/ *n.*

unpopulated /ʌn'pɒpjʊ,leɪtɪd/ *adj.* not populated.

unpractical /ʌn'præktɪk(ə)l/ *adj.* **1** not practical. **2** (of a person) without practical skill.

unpractised /ʌn'præktɪst/ *adj.* (*US* **unpracticed**) **1** not experienced or skilled. **2** not put into practice.

unprecedented /ʌn'presɪ,dentɪd/ *adj.* having no precedent; unparalleled. □ **unprecedentedly** *adv.*

unpredictable /,ʌnprɪ'dɪktəb(ə)l/ *adj.* that cannot be predicted. □ **unpredictability** /-'bɪlɪtɪ/ *n.* **unpredictably** *adv.*

unprejudiced /ʌn'predʒʊdɪst/ *adj.* not prejudiced.

unpremeditated /,ʌnpri:'medɪ,teɪtɪd/ *adj.* not deliberately planned, unintentional.

unprepared /,ʌnprɪ'peəd/ *adj.* not prepared; not ready.

unprepossessing /,ʌnpri:pə'zesɪŋ/ *adj.* unattractive.

unpretentious /,ʌnprɪ'tenʃəs/ *adj.* simple, modest, unassuming.

unpriced /ʌn'praɪst/ *adj.* not having a price fixed, marked, or stated.

unprincipled /ʌn'prɪnsɪp(ə)ld/ *adj.* lacking or not based on moral principles.

unprintable /ʌn'prɪntəb(ə)l/ *adj.* too offensive or indecent to be printed.

unproductive /,ʌnprə'dʌktɪv/ *adj.* not productive.

unprofessional /,ʌnprə'feʃən(ə)l/ *adj.* **1** contrary to professional standards. **2** unskilled, amateurish. □ **unprofessionally** *adv.*

unprofitable /ʌn'prɒfɪtəb(ə)l/ *adj.* not profitable.

unprogressive /,ʌnprə'gresɪv/ *adj.* not progressive, old-fashioned.

unpromising /ʌn'prɒmɪsɪŋ/ *adj.* not likely to turn out well.

unprompted /ʌn'prɒmptɪd/ *adj.* spontaneous.

unpronounceable /,ʌnprə'naʊnsəb(ə)l/ *adj.* that cannot be pronounced.

unpropitious /,ʌnprə'pɪʃəs/ *adj.* not propitious.

unprotected /,ʌnprə'tektɪd/ *adj.* not protected.

unprovable /ʌn'pru:vəb(ə)l/ *adj.* that cannot be proved.

unproved /ʌn'pru:vd/ *adj.* (also **unproven** /-v(ə)n/) not proved.

unprovoked /,ʌnprə'vəʊkt/ *adj.* without provocation.

unpublished /ʌn'pʌblɪʃt/ *adj.* not published.

unpunctual /ʌn'pʌŋktʃʊəl/ *adj.* not punctual.

unpunished /ʌn'pʌnɪʃt/ *adj.* not punished.

unputdownable /ˌʌnpʊtˈdaʊnəb(ə)l/ *adj. colloq.* (of a book) compulsively readable.

unqualified /ʌnˈkwɒlɪˌfaɪd/ *adj.* **1** not legally or officially qualified. **2** complete (*unqualified success*). **3** not competent (*unqualified to say*).

unquenchable /ʌnˈkwentʃəb(ə)l/ *adj.* that cannot be quenched.

unquestionable /ʌnˈkwestʃənəb(ə)l/ *adj.* that cannot be disputed or doubted. □ **unquestionably** *adv.*

unquestioned /ʌnˈkwestʃ(ə)nd/ *adj.* not disputed or doubted; definite, certain.

unquestioning /ʌnˈkwestʃənɪŋ/ *adj.* **1** asking no questions. **2** (of obedience etc.) absolute. □ **unquestioningly** *adv.*

unquiet /ʌnˈkwaɪət/ *adj.* **1** restless, agitated. **2** anxious.

unquote /ʌnˈkwəʊt/ *v.* (as *int.*) verbal formula indicating closing quotation marks.

unravel /ʌnˈræv(ə)l/ *v.* (**-ll-**; *US* **-l-**) **1** make or become disentangled, unknitted, unknotted, etc. **2** probe and solve (a mystery etc.). **3** undo (esp. knitted fabric).

unread /ʌnˈred/ *adj.* **1** (of a book etc.) not read. **2** (of a person) not well-read.

unreadable /ʌnˈriːdəb(ə)l/ *adj.* too dull, bad, or difficult to read.

unready /ʌnˈredɪ/ *adj.* **1** not ready. **2** hesitant.

unreal /ʌnˈrɪəl/ *adj.* **1** not real. **2** imaginary. **3** *slang* incredible. □ **unreality** /-ɪˈælɪtɪ/ *n.*

unrealistic /ˌʌnrɪəˈlɪstɪk/ *adj.* not realistic. □ **unrealistically** *adv.*

unrealizable /ʌnˈrɪəlaɪzəb(ə)l/ *adj.* (also **-isable**) that cannot be realized.

unrealized /ʌnˈrɪəlaɪzd/ *adj.* (also **-ised**) not realized.

unreason /ʌnˈriːz(ə)n/ *n.* madness; chaos; disorder.

unreasonable /ʌnˈriːzənəb(ə)l/ *adj.* **1** excessive (*unreasonable demands*). **2** not heeding reason. □ **unreasonably** *adv.*

unreasoning /ʌnˈriːzənɪŋ/ *adj.* not reasoning.

unrecognizable /ʌnˈrekəgˌnaɪzəb(ə)l/ *adj.* (also **-isable**) that cannot be recognized.

unrecognized /ʌnˈrekəgˌnaɪzd/ *adj.* (also **-ised**) not acknowledged.

unrecorded /ˌʌnrɪˈkɔːdɪd/ *adj.* not recorded.

unredeemed /ˌʌnrɪˈdiːmd/ *adj.* not redeemed.

unreel /ʌnˈriːl/ *v.* unwind from a reel.

unrefined /ˌʌnrɪˈfaɪnd/ *adj.* not refined.

unreflecting /ˌʌnrɪˈflektɪŋ/ *adj.* not thoughtful.

unreformed /ˌʌnrɪˈfɔːmd/ *adj.* not reformed.

unregenerate /ˌʌnrɪˈdʒenərət/ *adj.* obstinately wrong or bad.

unregistered /ʌnˈredʒɪstəd/ *adj.* not registered.

unregulated /ʌnˈregjʊˌleɪtɪd/ *adj.* not regulated.

unrehearsed /ˌʌnrɪˈhɜːst/ *adj.* not rehearsed.

unrelated /ˌʌnrɪˈleɪtɪd/ *adj.* not related.

unrelenting /ˌʌnrɪˈlentɪŋ/ *adj.* not abating, yielding, or relaxing; unmerciful. □ **unrelentingly** *adv.*

unreliable /ˌʌnrɪˈlaɪəb(ə)l/ *adj.* not reliable; erratic. □ **unreliability** /-ˈbɪlɪtɪ/ *n.*

unrelieved /ˌʌnrɪˈliːvd/ *adj.* not relieved; monotonously uniform.

unremarkable /ˌʌnrɪˈmɑːkəb(ə)l/ *adj.* not remarkable; uninteresting, ordinary.

unremarked /ˌʌnrɪˈmɑːkt/ *adj.* not mentioned or remarked upon.

unremitting /ˌʌnrɪˈmɪtɪŋ/ *adj.* incessant. □ **unremittingly** *adv.*

unremunerative /ˌʌnrɪˈmjuːnərətɪv/ *adj.* not, or not very, profitable.

unrepeatable /ˌʌnrɪˈpiːtəb(ə)l/ *adj.* **1** that cannot be done, made, or said again. **2** too indecent to repeat.

unrepentant /ˌʌnrɪˈpent(ə)nt/ *adj.* not repentant, impenitent. □ **unrepentantly** *adv.*

unrepresentative /ˌʌnreprɪˈzentətɪv/ *adj.* not representative.

unrepresented /ˌʌnreprɪˈzentɪd/ *adj.* not represented.

unrequited /ˌʌnrɪˈkwaɪtɪd/ *adj.* (of love etc.) not returned.

unreserved /ˌʌnrɪˈzɜːvd/ *adj.* **1** not reserved. **2** total; without reservation. □ **unreservedly** /-vɪdlɪ/ *adv.*

unresisting /ˌʌnrɪˈzɪstɪŋ/ *adj.* not resisting.

unresolved /ˌʌnrɪˈzɒlvd/ *adj.* **1** irresolute, undecided. **2** (of questions etc.) undetermined.

unresponsive /ˌʌnrɪˈspɒnsɪv/ *adj.* not responsive.

unrest /ʌnˈrest/ *n.* disturbed or dissatisfied state (*industrial unrest*).

unrestrained /ˌʌnrɪˈstreɪnd/ *adj.* not restrained.

unrestricted /ˌʌnrɪˈstrɪktɪd/ *adj.* not restricted.

unrewarded /ˌʌnrɪˈwɔːdɪd/ *adj.* not rewarded.

unrewarding /ˌʌnrɪˈwɔːdɪŋ/ *adj.* not rewarding or satisfying.

unrighteous /ʌnˈraɪtʃəs/ *adj.* wicked.

unripe /ʌnˈraɪp/ *adj.* not ripe.

unrivalled /ʌnˈraɪv(ə)ld/ *adj.* (*US* **unrivaled**) having no equal.

unroll /ʌn'rəʊl/ *v.* **1** open out from a rolled-up state. **2** display or be displayed like this.

unromantic /,ʌnrə'mæntɪk/ *adj.* not romantic.

unruffled /ʌn'rʌf(ə)ld/ *adj.* calm.

unruly /ʌn'ru:lɪ/ *adj.* (**-ier, -iest**) undisciplined, disorderly. □ **unruliness** *n.* [related to RULE]

unsaddle /ʌn'sæd(ə)l/ *v.* (**-ling**) **1** remove the saddle from. **2** unhorse.

unsafe /ʌn'seɪf/ *adj.* not safe.

unsaid /ʌn'sed/ *adj.* not uttered or expressed (*left it unsaid*).

unsaleable /ʌn'seɪləb(ə)l/ *adj.* not saleable.

unsalted /ʌn'sɔ:ltɪd/ *adj.* not salted.

unsatisfactory /,ʌnsætɪs'fæktərɪ/ *adj.* poor, unacceptable.

unsatisfied /ʌn'sætɪs,faɪd/ *adj.* not satisfied.

unsatisfying /ʌn'sætɪs,faɪɪŋ/ *adj.* not satisfying.

unsaturated /ʌn'sætʃə,reɪtɪd/ *adj. Chem.* (of esp. a fat or oil) having double or triple bonds in its molecule and therefore capable of further reaction.

unsavoury /ʌn'seɪvərɪ/ *adj.* (*US* **unsavory**) **1** disgusting, unpleasant. **2** morally offensive.

unsay /ʌn'seɪ/ *v.* (*past* and *past part.* **unsaid**) retract (a statement).

unscalable /ʌn'skeɪləb(ə)l/ *adj.* that cannot be scaled.

unscarred /ʌn'skɑ:d/ *adj.* not scarred or damaged.

unscathed /ʌn'skeɪðd/ *adj.* without injury.

unscheduled /ʌn'ʃedju:ld/ *adj.* not scheduled.

unschooled /ʌn'sku:ld/ *adj.* uneducated, untrained.

unscientific /,ʌnsaɪən'tɪfɪk/ *adj.* not scientific in method etc. □ **unscientifically** *adv.*

unscramble /ʌn'skræmb(ə)l/ *v.* (**-ling**) make plain, decode, interpret (a scrambled transmission etc.).

unscreened /ʌn'skri:nd/ *adj.* **1 a** (esp. of coal) not passed through a screen or sieve. **b** not checked, esp. for security or medical problems. **2** not having a screen. **3** not shown on a screen.

unscrew /ʌn'skru:/ *v.* **1** unfasten by removing a screw or screws. **2** loosen (a screw or screw-top).

unscripted /ʌn'skrɪptɪd/ *adj.* (of a speech etc.) delivered impromptu.

unscrupulous /ʌn'skru:pjʊləs/ *adj.* having no scruples, unprincipled. □ **unscrupulously** *adv.* **unscrupulousness** *n.*

unseal /ʌn'si:l/ *v.* break the seal of; open (a letter, receptacle, etc.).

unseasonable /ʌn'si:zənəb(ə)l/ *adj.* **1** not seasonable. **2** untimely, inopportune. □ **unseasonably** *adv.*

unseasonal /ʌn'si:zən(ə)l/ *adj.* not typical of, or appropriate to, the time or season. □ **unseasonally** *adv.*

unseat /ʌn'si:t/ *v.* **1** remove from (esp. a parliamentary) seat. **2** dislodge from a seat, esp. on horseback.

unseeded /ʌn'si:dɪd/ *adj. Sport* (of a player) not seeded.

unseeing /ʌn'si:ɪŋ/ *adj.* **1** unobservant. **2** blind. □ **unseeingly** *adv.*

unseemly /ʌn'si:mlɪ/ *adj.* (**-ier, -iest**) **1** indecent. **2** unbecoming. □ **unseemliness** *n.*

unseen /ʌn'si:n/ *—adj.* **1** not seen. **2** invisible. **3** (of a translation) to be done without preparation. *—n.* unseen translation.

unselfconscious /,ʌnself'kɒnʃəs/ *adj.* not self-conscious. □ **unselfconsciously** *adv.* **unselfconsciousness** *n.*

unselfish /ʌn'selfɪʃ/ *adj.* concerned about others; sharing. □ **unselfishly** *adv.* **unselfishness** *n.*

unsentimental /,ʌnsentɪ'ment(ə)l/ *adj.* not sentimental.

unsettle /ʌn'set(ə)l/ *v.* (**-ling**) **1** disturb; discompose. **2** derange.

unsettled /ʌn'set(ə)ld/ *adj.* **1** restless, disturbed; unpredictable, changeable. **2** open to change or further discussion. **3** (of a bill etc.) unpaid.

unsex /ʌn'seks/ *v.* deprive (a person, esp. a woman) of the qualities of her or his sex.

unshackle /ʌn'ʃæk(ə)l/ *v.* (**-ling**) **1** release from shackles. **2** set free.

unshakeable /ʌn'ʃeɪkəb(ə)l/ *adj.* firm; obstinate. □ **unshakeably** *adv.*

unshaken /ʌn'ʃeɪkən/ *adj.* not shaken.

unshaven /ʌn'ʃeɪv(ə)n/ *adj.* not shaved.

unsheathe /ʌn'ʃi:ð/ *v.* (**-thing**) remove (a knife etc.) from a sheath.

unshockable /ʌn'ʃɒkəb(ə)l/ *adj.* unable to be shocked.

unshrinking /ʌn'ʃrɪŋkɪŋ/ *adj.* unhesitating, fearless.

unsighted /ʌn'saɪtɪd/ *adj.* **1** not sighted or seen. **2** prevented from seeing.

unsightly /ʌn'saɪtlɪ/ *adj.* ugly. □ **unsightliness** *n.*

unsigned /ʌn'saɪnd/ *adj.* not signed.

unsinkable /ʌn'sɪŋkəb(ə)l/ *adj.* unable to be sunk.

unskilful /ʌn'skɪlfʊl/ *adj.* (*US* **unskillful**) not skilful.

unskilled /ʌn'skɪld/ *adj.* lacking, or (of work) not needing, special skill or training.

unsliced /ʌn'slaɪst/ *adj.* (esp. of a loaf of bread) not sliced.

unsmiling /ʌn'smaɪlɪŋ/ *adj.* not smiling.

unsmoked /ʌn'sməʊkt/ *adj.* not cured by smoking (*unsmoked bacon*).
unsociable /ʌn'səʊʃəb(ə)l/ *adj.* not sociable, disliking company.

■ **Usage** See note at *unsocial.*

unsocial /ʌn'səʊʃ(ə)l/ *adj.* **1** not social; not suitable for or seeking society. **2** outside the normal working day (*unsocial hours*). **3** antisocial.

■ **Usage** *Unsocial* is sometimes confused with *unsociable.*

unsoiled /ʌn'sɔɪld/ *adj.* not soiled or dirtied.
unsold /ʌn'səʊld/ *adj.* not sold.
unsolicited /ˌʌnsə'lɪsɪtɪd/ *adj.* not asked for; voluntary.
unsolved /ʌn'sɒlvd/ *adj.* not solved.
unsophisticated /ˌʌnsə'fɪstɪˌkeɪtɪd/ *adj.* artless, simple, natural.
unsorted /ʌn'sɔ:tɪd/ *adj.* not sorted.
unsought /ʌn'sɔ:t/ *adj.* **1** not sought for. **2** without being requested.
unsound /ʌn'saʊnd/ *adj.* **1** unhealthy, not sound. **2** rotten, weak; unreliable. **3** ill-founded. □ **of unsound mind** insane. □ **unsoundness** *n.*
unsparing /ʌn'speərɪŋ/ *adj.* **1** lavish. **2** merciless.
unspeakable /ʌn'spi:kəb(ə)l/ *adj.* **1** that cannot be expressed in words. **2** indescribably bad. □ **unspeakably** *adv.*
unspecific /ˌʌnspə'sɪfɪk/ *adj.* not specific; general, inexact.
unspecified /ʌn'spesɪˌfaɪd/ *adj.* not specified.
unspectacular /ˌʌnspek'tækjʊlə(r)/ *adj.* not spectacular; dull.
unspoiled /ʌn'spɔɪld/ *adj.* (also **unspoilt**) not spoilt.
unspoken /ʌn'spəʊkən/ *adj.* **1** not expressed in speech. **2** not uttered as speech.
unsporting /ʌn'spɔ:tɪŋ/ *adj.* not fair or generous.
unsportsmanlike /ʌn'spɔ:tsmənˌlaɪk/ *adj.* unsporting.
unstable /ʌn'steɪb(ə)l/ *adj.* (**unstabler, unstablest**) **1** not stable; likely to fall. **2** not stable emotionally. **3** changeable. □ **unstably** *adv.*
unstained /ʌn'steɪnd/ *adj.* not stained.
unstated /ʌn'steɪtɪd/ *adj.* not stated or declared.
unsteady /ʌn'stedɪ/ *adj.* (**-ier, -iest**) **1** not steady or firm. **2** changeable. **3** not uniform or regular. □ **unsteadily** *adv.* **unsteadiness** *n.*
unstick /ʌn'stɪk/ *v.* (*past* and *past part.* **unstuck**) separate (a thing stuck to another). □ **come unstuck** *colloq.* come to grief, fail.
unstinted /ʌn'stɪntɪd/ *adj.* not stinted.
unstinting /ʌn'stɪntɪŋ/ *adj.* lavish; limitless. □ **unstintingly** *adv.*
unstitch /ʌn'stɪtʃ/ *v.* undo the stitches of.
unstop /ʌn'stɒp/ *v.* (**-pp-**) **1** unblock. **2** remove the stopper from.
unstoppable /ʌn'stɒpəb(ə)l/ *adj.* that cannot be stopped or prevented.
unstrap /ʌn'stræp/ *v.* (**-pp-**) undo the strap(s) of.
unstressed /ʌn'strest/ *adj.* not pronounced with stress.
unstring /ʌn'strɪŋ/ *v.* (*past* and *past part.* **unstrung**) **1** remove or relax the string(s) of (a bow, harp, etc.). **2** remove (beads etc.) from a string. **3** (esp. as **unstrung** *adj.*) unnerve.
unstructured /ʌn'strʌktʃəd/ *adj.* **1** not structured. **2** informal.
unstuck *past* and *past part.* of UNSTICK.
unstudied /ʌn'stʌdɪd/ *adj.* easy, natural, spontaneous.
unsubstantial /ˌʌnsəb'stænʃ(ə)l/ *adj.* = INSUBSTANTIAL.
unsubstantiated /ˌʌnsəb'stænʃɪˌeɪtɪd/ *adj.* not substantiated.
unsubtle /ʌn'sʌt(ə)l/ *adj.* not subtle; obvious; clumsy.
unsuccessful /ˌʌnsək'sesfʊl/ *adj.* not successful. □ **unsuccessfully** *adv.*
unsuitable /ʌn'su:təb(ə)l, ʌn'sju:t-/ *adj.* not suitable. □ **unsuitability** /-'bɪlɪtɪ/ *n.* **unsuitably** *adv.*
unsuited /ʌn'su:tɪd, ʌn'sju:t-/ *adj.* **1** (usu. foll. by *for*) not fit. **2** (usu. foll. by *to*) not adapted.
unsullied /ʌn'sʌlɪd/ *adj.* not sullied.
unsung /ʌn'sʌŋ/ *adj.* not celebrated, unrecognized (*unsung heroes*).
unsupervised /ʌn'su:pəˌvaɪzd/ *adj.* not supervised.
unsupported /ˌʌnsə'pɔ:tɪd/ *adj.* not supported.
unsure /ʌn'ʃʊə(r)/ *adj.* not sure.
unsurpassed /ˌʌnsə'pɑ:st/ *adj.* not surpassed.
unsurprised /'ʌnsə'praɪzd/ *adj.* not surprised.
unsurprising /ˌʌnsə'praɪzɪŋ/ *adj.* not surprising. □ **unsurprisingly** *adv.*
unsuspecting /ˌʌnsə'spektɪŋ/ *adj.* not suspecting. □ **unsuspected** *adj.*
unsustainable /ˌʌnsə'steɪnəb(ə)l/ *adj.* that cannot be sustained.
unsweetened /ʌn'swi:t(ə)nd/ *adj.* not sweetened.
unswept /ʌn'swept/ *adj.* not swept.
unswerving /ʌn'swɜ:vɪŋ/ *adj.* steady, constant. □ **unswervingly** *adv.*
unsymmetrical /ˌʌnsɪ'metrɪk(ə)l/ *adj.* not symmetrical.
unsympathetic /ˌʌnsɪmpə'θetɪk/ *adj.* not sympathetic. □ **unsympathetically** *adv.*

unsystematic /ˌʌnsɪstəˈmætɪk/ *adj.* not systematic. □ **unsystematically** *adv.*
untainted /ʌnˈteɪntɪd/ *adj.* not tainted.
untalented /ʌnˈtæləntɪd/ *adj.* not talented.
untameable /ʌnˈteɪməb(ə)l/ *adj.* that cannot be tamed.
untamed /ʌnˈteɪmd/ *adj.* not tamed, wild.
untangle /ʌnˈtæŋg(ə)l/ *v.* (**-ling**) disentangle.
untapped /ʌnˈtæpt/ *adj.* not (yet) tapped or used (*untapped resources*).
untarnished /ʌnˈtɑːnɪʃt/ *adj.* not tarnished.
untaught /ʌnˈtɔːt/ *adj.* (of a person, knowledge, etc.) not taught.
untaxed /ʌnˈtækst/ *adj.* (of a person, commodity, etc.) not taxed.
unteachable /ʌnˈtiːtʃəb(ə)l/ *adj.* (of a person, subject, etc.) incapable of being taught.
untenable /ʌnˈtenəb(ə)l/ *adj.* (of a theory etc.) not tenable.
untested /ʌnˈtestɪd/ *adj.* not tested or proved.
untether /ʌnˈteðə(r)/ *v.* release (an animal) from a tether.
unthinkable /ʌnˈθɪŋkəb(ə)l/ *adj.* **1** unimaginable, inconceivable. **2** *colloq.* highly unlikely or undesirable. □ **unthinkably** *adv.*
unthinking /ʌnˈθɪŋkɪŋ/ *adj.* **1** thoughtless. **2** unintentional, inadvertent. □ **unthinkingly** *adv.*
unthread /ʌnˈθred/ *v.* take the thread out of (a needle etc.).
unthrone /ʌnˈθrəʊn/ *v.* (**-ning**) dethrone.
untidy /ʌnˈtaɪdɪ/ *adj.* (**-ier, -iest**) not neat or orderly. □ **untidily** *adv.* **untidiness** *n.*
untie /ʌnˈtaɪ/ *v.* (**untying**) **1** undo (a knot, package, etc.). **2** release from bonds or attachment.
until /ənˈtɪl/ *prep.* & *conj.* = TILL[1]. [earlier *untill*: *un* from Old Norse *und* as far as]

■ **Usage** *Until*, as opposed to *till*, is used esp. at the beginning of a sentence and in formal style, e.g. *until you told me, I had no idea*; *he resided there until his decease.*

untimely /ʌnˈtaɪmlɪ/ *adj.* **1** inopportune. **2** (of death) premature. □ **untimeliness** *n.*
untiring /ʌnˈtaɪərɪŋ/ *adj.* tireless. □ **untiringly** *adv.*
untitled /ʌnˈtaɪt(ə)ld/ *adj.* having no title.
unto /ˈʌntʊ/ *prep. archaic* = TO (in all uses except signalling the infinitive). [from UNTIL, with *to* replacing *til*]
untold /ʌnˈtəʊld/ *adj.* **1** not told. **2** immeasurable (*untold misery*).
untouchable /ʌnˈtʌtʃəb(ə)l/ —*adj.* that may not be touched. —*n.* member of a hereditary Hindu group held to defile members of higher castes on contact. □ **untouchability** /-ˈbɪlɪtɪ/ *n.*

■ **Usage** The use of this term, and social restrictions accompanying it, were declared illegal under the Indian constitution in 1949.

untouched /ʌnˈtʌtʃt/ *adj.* **1** not touched. **2** not affected physically, emotionally, etc. **3** not discussed.
untoward /ˌʌntəˈwɔːd/ *adj.* **1** inconvenient, unlucky. **2** awkward. **3** perverse, refractory. **4** unseemly.
untraceable /ʌnˈtreɪsəb(ə)l/ *adj.* that cannot be traced.
untrained /ʌnˈtreɪnd/ *adj.* not trained.
untrammelled /ʌnˈtræm(ə)ld/ *adj.* not trammelled, unhampered.
untranslatable /ˌʌntrænsˈleɪtəb(ə)l/ *adj.* that cannot be translated satisfactorily.
untreated /ʌnˈtriːtɪd/ *adj.* not treated.
untried /ʌnˈtraɪd/ *adj.* **1** not tried or tested. **2** inexperienced.
untroubled /ʌnˈtrʌb(ə)ld/ *adj.* calm, tranquil.
untrue /ʌnˈtruː/ *adj.* **1** not true. **2** (often foll. by *to*) not faithful or loyal. **3** deviating from an accepted standard.
untrustworthy /ʌnˈtrʌstˌwɜːðɪ/ *adj.* not trustworthy. □ **untrustworthiness** *n.*
untruth /ʌnˈtruːθ/ *n.* **1** being untrue. **2** lie.
untruthful /ʌnˈtruːθfʊl/ *adj.* not truthful. □ **untruthfully** *adv.*
untuck /ʌnˈtʌk/ *v.* free (bedclothes etc.) from being tucked in or up.
unturned /ʌnˈtɜːnd/ *adj.* **1** not turned over, round, away, etc. **2** not shaped by turning.
untutored /ʌnˈtjuːtəd/ *adj.* uneducated, untaught.
untwine /ʌnˈtwaɪn/ *v.* (**-ning**) untwist, unwind.
untwist /ʌnˈtwɪst/ *v.* open from a twisted or spiralled state.
unusable /ʌnˈjuːzəb(ə)l/ *adj.* not usable.
unused *adj.* **1** /ʌnˈjuːzd/ **a** not in use. **b** never having been used. **2** /ʌnˈjuːst/ (foll. by *to*) not accustomed.
unusual /ʌnˈjuːʒʊəl/ *adj.* **1** not usual. **2** remarkable. □ **unusually** *adv.*
unutterable /ʌnˈʌtərəb(ə)l/ *adj.* inexpressible; beyond description. □ **unutterably** *adv.*
unvarnished /ʌnˈvɑːnɪʃt/ *adj.* **1** not varnished. **2** plain and straightforward (*the unvarnished truth*).
unvarying /ʌnˈveərɪɪŋ/ *adj.* not varying.
unveil /ʌnˈveɪl/ *v.* **1** uncover (a statue etc.) ceremonially. **2** reveal. **3** remove a veil from; remove one's veil.

unverified /ʌn'verɪ,faɪd/ *adj.* not verified.

unversed /ʌn'vɜ:st/ *adj.* (usu. foll. by *in*) not experienced or skilled.

unviable /ʌn'vaɪəb(ə)l/ *adj.* not viable.

unvoiced /ʌn'vɔɪst/ *adj.* **1** not spoken. **2** (of a consonant etc.) not voiced.

unwaged /ʌn'weɪdʒd/ *adj.* not receiving a wage; unemployed.

unwanted /ʌn'wɒntɪd/ *adj.* not wanted.

unwarrantable /ʌn'wɒrəntəb(ə)l/ *adj.* unjustifiable. □ **unwarrantably** *adv.*

unwarranted /ʌn'wɒrəntɪd/ *adj.* **1** unauthorized. **2** unjustified.

unwary /ʌn'weərɪ/ *adj.* (often foll. by *of*) not cautious. □ **unwarily** *adv.* **unwariness** *n.*

unwashed /ʌn'wɒʃt/ *adj.* not washed or clean. □ **the great unwashed** *colloq.* the rabble.

unwavering /ʌn'weɪvərɪŋ/ *adj.* not wavering. □ **unwaveringly** *adv.*

unweaned /ʌn'wi:nd/ *adj.* not yet weaned.

unwearying /ʌn'wɪərɪɪŋ/ *adj.* persistent.

unwelcome /ʌn'welkəm/ *adj.* not welcome or acceptable.

unwell /ʌn'wel/ *adj.* ill.

unwholesome /ʌn'həʊlsəm/ *adj.* **1** detrimental to physical or moral health. **2** unhealthy-looking.

unwieldy /ʌn'wi:ldɪ/ *adj.* (**-ier, -iest**) cumbersome or hard to manage, owing to size, shape, etc. □ **unwieldily** *adv.* **unwieldiness** *n.* [*wieldy* active, from WIELD]

unwilling /ʌn'wɪlɪŋ/ *adj.* not willing or inclined; reluctant. □ **unwillingly** *adv.* **unwillingness** *n.*

unwind /ʌn'waɪnd/ *v.* (*past* and *past part.* **unwound**) **1** draw out or become drawn out after having been wound. **2** *colloq.* relax.

unwinking /ʌn'wɪŋkɪŋ/ *adj.* **1** not winking. **2** vigilant.

unwise /ʌn'waɪz/ *adj.* foolish, imprudent. □ **unwisely** *adv.*

unwished /ʌn'wɪʃt/ *adj.* (usu. foll. by *for*) not wished for.

unwitting /ʌn'wɪtɪŋ/ *adj.* **1** not knowing or aware (*an unwitting offender*). **2** unintentional. □ **unwittingly** *adv.* [Old English: related to WIT]

unwonted /ʌn'wəʊntɪd/ *adj.* not customary or usual.

unworkable /ʌn'wɜ:kəb(ə)l/ *adj.* not workable; impracticable.

unworkmanlike /ʌn'wɜ:kmən,laɪk/ *adj.* badly done or made.

unworldly /ʌn'wɜ:ldlɪ/ *adj.* spiritual; naïve. □ **unworldliness** *n.*

unworn /ʌn'wɔ:n/ *adj.* not worn or impaired by wear.

unworried /ʌn'wʌrɪd/ *adj.* not worried; calm.

unworthy /ʌn'wɜ:ðɪ/ *adj.* (**-ier, -iest**) **1** (often foll. by *of*) not worthy of or befitting a person etc. **2** discreditable, unseemly. □ **unworthily** *adv.* **unworthiness** *n.*

unwound *past* and *past part.* of UNWIND.

unwrap /ʌn'ræp/ *v.* (**-pp-**) **1** remove the wrapping from. **2** open, unfold. **3** become unwrapped.

unwritten /ʌn'rɪt(ə)n/ *adj.* **1** not written. **2** (of a law etc.) based on custom or judicial decision, not on statute.

unyielding /ʌn'ji:ldɪŋ/ *adj.* **1** not yielding. **2** firm, obstinate.

unzip /ʌn'zɪp/ *v.* (**-pp-**) unfasten the zip of.

up –*adv.* **1** at, in, or towards a higher place or a place regarded as higher, e.g. the north, a capital or a university (*up in the air*; *up in Scotland*; *went up to London*; *came up in 1989*). **2 a** to or in an erect or required position or condition (*stood it up*; *wound up the watch*). **b** in or into an active condition (*stirred up trouble*; *the hunt is up*). **3** in a stronger or leading position (*three goals up*; *am £10 up*; *is well up in class*). **4** to a specified place, person, or time (*a child came up to me*; *fine up till now*). **5** higher in price or value (*our costs are up*; *shares are up*). **6 a** completely (*burn up*; *eat up*). **b** more loudly or clearly (*speak up*). **7** completed (*time is up*). **8** into a compact, accumulated, or secure state (*pack up*; *save up*; *tie up*). **9** out of bed, having risen (*are you up yet?*; *sun is up*). **10** happening, esp. unusually (*something is up*). **11** (usu. foll. by *before*) appearing for trial etc. (*up before the magistrate*). **12** (of a road etc.) being repaired. **13** (of a jockey) in the saddle. –*prep.* **1** upwards and along, through, or into (*climbed up the ladder*; *went up the road*). **2** from the bottom to the top of. **3 a** at or in a higher part of (*is up the street*). **b** towards the source of (a river). –*adj.* **1** directed upwards (*up stroke*). **2** of travel towards a capital or centre (*the up train*). –*n.* spell of good fortune. –*v.* (**-pp-**) **1** *colloq.* start, esp. abruptly, to speak or act (*upped and hit him*). **2** raise (*upped their prices*). □ **be all up with** be hopeless for (a person). **on the up** (or **up and up**) *colloq.* steadily improving. **up against 1** close to. **2** in or into contact with. **3** *colloq.* confronted with (a problem etc.). **up and about** (or **doing**) having risen from bed; active. **up and down 1** to and fro (along). **2** *colloq.* in varying health or spirits. **up for** available for or standing for (office etc.) (*up* *for sale*). **up to 1** until. **2** below or equal to. **3** incumbent on (*it is up to you to say*). **4** capable of. **5** occupied or busy with. **up to date** see DATE[1]. [Old English]

up- *prefix* in senses of UP, added: **1** as an adverb to verbs and verbal derivations, = 'upwards' (*upcurved*; *update*). **2** as a preposition to nouns forming adverbs and adjectives (*up-country*; *uphill*). **3** as an adjective to nouns (*upland*; *upstroke*).

up-and-coming *adj. colloq.* (of a person) promising; progressing.

up-and-over *adj.* (of a door) opening by being raised and pushed back into a horizontal position.

upbeat *–n.* unaccented beat in music. *–adj. colloq.* optimistic, cheerful.

upbraid /ʌp'breɪd/ *v.* (often foll. by *with, for*) chide, reproach. [Old English: related to BRAID = brandish]

upbringing *n.* rearing of a child. [obsolete *upbring* to rear]

up-country *adv.* & *adj.* inland.

upcurved /'ʌpkɜ:vd/ *adj.* curved upwards.

update *–v.* /ʌp'deɪt/ (**-ting**) bring up to date. *–n.* /'ʌpdeɪt/ **1** updating. **2** updated information etc.

up-end /ʌp'end/ *v.* set or rise up on end.

upfield *adv.* in or to a position nearer to the opponents' end of a field.

upfront /ʌp'frʌnt/ *colloq. –adv.* (usu. **up front**) **1** at the front; in front. **2** (of payments) in advance. *–adj.* **1** honest, frank, direct. **2** (of payments) made in advance.

upgrade *v.* /ʌp'greɪd/ (**-ding**) **1** raise in rank etc. **2** improve (equipment etc.).

upheaval /ʌp'hi:v(ə)l/ *n.* violent or sudden change or disruption. [from *upheave*, = heave or lift up]

uphill *–adv.* /ʌp'hɪl/ up a slope. *–adj.* /'ʌphɪl/ **1** sloping up; ascending. **2** arduous.

uphold /ʌp'həʊld/ *v.* (*past* and *past part.* **upheld**) **1** confirm (a decision etc.). **2** support, maintain (a custom etc.). □ **upholder** *n.*

upholster /ʌp'həʊlstə(r)/ *v.* provide (furniture) with upholstery. [back formation from UPHOLSTERER]

upholsterer *n.* person who upholsters, esp. for a living. [obsolete *upholster* from UPHOLD in sense 'keep in repair']

upholstery *n.* **1** covering, padding, springs, etc. for furniture. **2** upholsterer's work.

upkeep *n.* **1** maintenance in good condition. **2** cost or means of this.

upland /'ʌplənd/ *–n.* (usu. in *pl.*) higher or inland parts of a country. *–adj.* of these parts.

uplift *–v.* /ʌp'lɪft/ **1** raise. **2** (esp. as **uplifting** *adj.*) elevate morally or emotionally. *–n.* /'ʌplɪft/ **1** *colloq.* elevating influence. **2** support for the bust etc.

up-market *adj.* & *adv.* of or directed at the upper end of the market; classy.

upmost var. of UPPERMOST.

upon /ə'pɒn/ *prep.* = ON. [from *up on*]

■ **Usage** *Upon* is sometimes more formal than *on*, but is standard in *once upon a time* and *upon my word*.

upper[1] *–attrib. adj.* **1** higher in place; situated above another part. **2** higher in rank etc. (*upper class*). *–n.* part of a boot or shoe above the sole. □ **on one's uppers** *colloq.* very short of money.

upper[2] *n. slang* amphetamine or other stimulant.

upper case *n.* capital letters.

upper crust *n. colloq.* (prec. by *the*) the aristocracy.

upper-cut *–n.* upwards blow delivered with the arm bent. *–v.* hit upwards with the arm bent.

upper hand *n.* (prec. by *the*) dominance, control.

Upper House *n.* higher house in a legislature, esp. the House of Lords.

uppermost *–adj.* (also **upmost**) **1** highest. **2** predominant. *–adv.* at or to the uppermost position.

uppish *adj. colloq.* uppity.

uppity /'ʌpɪtɪ/ *adj. colloq.* self-assertive, arrogant.

upright *–adj.* **1** erect, vertical. **2** (of a piano) with vertical strings. **3** honourable or honest. *–n.* **1** upright post or rod, esp. as a structural support. **2** upright piano. [Old English]

uprising *n.* insurrection.

uproar *n.* tumult; violent disturbance. [Dutch, = commotion]

uproarious /ʌp'rɔ:rɪəs/ *adj.* **1** very noisy. **2** provoking loud laughter; very funny. □ **uproariously** *adv.*

uproot /ʌp'ru:t/ *v.* **1** pull (a plant etc.) up from the ground. **2** displace (a person). **3** eradicate.

uprush *n.* upward rush.

ups-a-daisy var. of UPSY-DAISY.

ups and downs *n.pl.* **1** rises and falls. **2** mixed fortune.

upset *–v.* /ʌp'set/ (**-tt-**; *past* and *past part.* **upset**) **1** overturn. **2** disturb the composure or digestion of. **3** disrupt. *–n.* /'ʌpset/ **1** emotional or physical disturbance. **2** surprising result. *–adj.* /ʌp'set, 'ʌp-/ disturbed (*upset stomach*).

upshot *n.* outcome, conclusion.

upside down /ˌʌpsaɪd 'daʊn/ *adv.* & *adj.* **1** with the upper and lower parts reversed; inverted. **2** in or into total disorder. [from *up so down*, perhaps = 'up as if down']

upsilon /'ju:psɪ,lɒn, ʌp'saɪlən/ *n.* twentieth letter of the Greek alphabet (Υ, υ). [Greek, = slender U, from *psilos* slender, with ref. to its later coincidence in sound with Greek *oi*]

upstage /ʌp'steɪdʒ/ *–adj.* & *adv.* nearer the back of a theatre stage. *–v.* (**-ging**) **1** move upstage to make (another actor) face away from the audience. **2** divert attention from (a person) to oneself.

upstairs *–adv.* /ʌp'steəz/ to or on an upper floor. *–attrib. adj.* /'ʌpsteəz/ situated upstairs. *–n.* /ʌp'steəz/ upper floor.

upstanding /ʌp'stændɪŋ/ *adj.* **1** standing up. **2** strong and healthy. **3** honest.

upstart *–n.* newly successful, esp. arrogant, person. *–adj.* **1** that is an upstart. **2** of upstarts.

upstate *US –n.* provincial, esp. northern, part of a State. *–attrib. adj.* of this part. *–adv.* in or to this part.

upstream *adv.* & *adj.* in the direction contrary to the flow of a stream etc.

upstroke *n.* upwards stroke.

upsurge *n.* upward surge.

upswept *adj.* (of hair) combed to the top of the head.

upswing *n.* upward movement or trend.

upsy-daisy /'ʌpsɪ,deɪzɪ/ *int.* (also **ups-a--daisy**) expressing encouragement to a child who is being lifted or has fallen. [earlier *up-a-daisy*]

uptake *n.* **1** *colloq.* understanding (esp. *quick* or *slow on the uptake*). **2** taking up (of an offer etc.).

upthrust *n.* **1** upward thrust. **2** upward displacement of part of the earth's crust.

uptight /ʌp'taɪt/ *adj. colloq.* **1** nervously tense or angry. **2** rigidly conventional.

uptown *US –attrib. adj.* of the residential part of a town or city. *–adv.* in or into this part. *–n.* this part.

upturn *–n.* /'ʌptɜ:n/ upward trend; improvement. *–v.* /ʌp'tɜ:n/ turn up or upside down.

UPVC *abbr.* unplasticized polyvinyl chloride.

upward /'ʌpwəd/ *–adv.* (also **upwards**) towards what is higher, more important, etc. *–adj.* moving or extending upwards. □ **upwards of** more than (*upwards of forty*).

upwardly *adv.* in an upward direction.

upwardly mobile *adj.* aspiring to advance socially or professionally.

upwind /ʌp'wɪnd/ *adj.* & *adv.* in the direction from which the wind is blowing.

uranium /jʊ'reɪnɪəm/ *n.* radioactive grey dense metallic element, capable of nuclear fission and used as a source of nuclear energy. [*Uranus*, name of a planet]

urban /'ɜ:bən/ *adj.* of, living in, or situated in a town or city. [Latin *urbs* city]

urbane /ɜ:'beɪn/ *adj.* suave; elegant. □ **urbanity** /-'bænɪtɪ/ *n.* [Latin: related to URBAN]

urban guerrilla *n.* terrorist operating in an urban area.

urbanize /'ɜ:bə,naɪz/ *v.* (also **-ise**) (**-zing** or **-sing**) make urban, esp. by destroying the rural quality of (a district). □ **urbanization** /-'zeɪʃ(ə)n/ *n.*

urchin /'ɜ:tʃɪn/ *n.* **1** mischievous, esp. ragged, child. **2** = SEA URCHIN. [Latin *ericius* hedgehog]

Urdu /'ʊədu:, 'ɜ:-/ *n.* language related to Hindi but with many Persian words, used esp. in Pakistan. [Hindustani]

-ure *suffix* forming: **1** nouns of action (*seizure*). **2** nouns of result (*creature*). **3** collective nouns (*nature*). [Latin *-ura*]

urea /jʊə'ri:ə/ *n.* soluble nitrogenous compound contained esp. in urine. [French *urée* from Greek *ouron* urine]

ureter /jʊə'ri:tə(r)/ *n.* duct conveying urine from the kidney to the bladder. [Greek *oureō* urinate]

urethra /jʊə'ri:θrə/ *n.* (*pl.* **-s**) duct conveying urine from the bladder. [Greek: related to URETER]

urge *–v.* (**-ging**) **1** (often foll. by *on*) drive forcibly; hasten. **2** encourage or entreat earnestly or persistently. **3** (often foll. by *on*, *upon*) advocate (an action or argument etc.) emphatically (to a person). *–n.* **1** urging impulse or tendency. **2** strong desire. [Latin *urgeo*]

urgent *adj.* **1** requiring immediate action or attention. **2** importunate. □ **urgency** *n.* **urgently** *adv.* [French: related to URGE]

uric /'jʊərɪk/ *adj.* of urine. [French *urique*: related to URINE]

uric acid *n.* constituent of urine.

urinal /jʊə'raɪn(ə)l/ *n.* place or receptacle for urination by men. [Latin: related to URINE]

urinary /'jʊərɪnərɪ/ *adj.* of or relating to urine.

urinate /'jʊərɪ,neɪt/ *v.* (**-ting**) discharge urine. □ **urination** /-'neɪʃ(ə)n/ *n.*

urine /'jʊərɪn/ *n.* waste fluid secreted by the kidneys and discharged from the bladder. [Latin *urina*]

urn *n.* **1** vase with a foot and usu. a rounded body, used esp. for the ashes of the dead. **2** large vessel with a tap, in which tea or coffee etc. is made or kept hot. [Latin *urna*]

urogenital /jʊərə'dʒenɪt(ə)l/ *adj.* of the urinary and reproductive systems. [Greek *ouron* urine]

urology /jʊə'rɒlədʒɪ/ *n.* the study of the urinary system. □ **urological** /-rə'lɒdʒɪk(ə)l/ *adj.*

Ursa Major /,ɜ:sə/ *n.* = *Great Bear* (see BEAR²). [Latin]

Ursa Minor *n.* = *Little Bear* (see BEAR²). [Latin]

ursine /'ɜ:saɪn/ *adj.* of or like a bear. [Latin *ursus* bear]

US *abbr.* United States.

us /əs, ʌs/ *pron.* **1** *objective case* of WE (*they saw us*). **2** *colloq.* = WE (*it's us again*). **3** *colloq.* = ME¹ (*give us a kiss*). [Old English]

USA *abbr.* United States of America.

usable /'ju:zəb(ə)l/ *adj.* that can be used.

USAF *abbr.* United States Air Force.

usage /'ju:sɪdʒ/ *n.* **1** use, treatment (*damaged by rough usage*). **2** customary practice, esp. in the use of a language or as creating a precedent in law.

use –*v.* /ju:z/ (**using**) **1** cause to act or serve for a purpose; bring into service. **2** treat in a specified manner (*used him shamefully*). **3** exploit for one's own ends. **4** /ju:s/ did or had habitually (*I used to drink*; *it used not* (or *did not use*) *to rain so often*). **5** (as **used** *adj.*) second-hand. **6** (as **used** /ju:st/ *predic. adj.*) (foll. by *to*) familiar by habit; accustomed (*used to hard work*). –*n.* /ju:s/ **1** using or being used. **2** right or power of using (*lost the use of his legs*). **3** benefit, advantage (*a torch would be of use*; *it's no use talking*). **4** custom or usage (*established by long use*). □ **have no use for 1** not need. **2** dislike, be contemptuous of. **in use** being used. **make use of 1** use. **2** benefit from. **out of use** not being used. **use up 1** consume completely. **2** find a use for (leftovers etc.). [French *us*, *user*, ultimately from Latin *utor us-*]

useful *adj.* **1** that can be used to advantage; helpful; beneficial. **2** *colloq.* creditable, efficient (*useful footballer*). □ **make oneself useful** help. □ **usefully** *adv.* **usefulness** *n.*

useless *adj.* **1** serving no purpose; unavailing. **2** *colloq.* feeble or ineffectual (*useless at swimming*). □ **uselessly** *adv.* **uselessness** *n.*

user *n.* person who uses a thing.

user-friendly *adj.* (of a computer etc.) easy to use.

usher –*n.* **1** person who shows people to their seats in a cinema, church, etc. **2** doorkeeper at a court etc. –*v.* **1** act as usher to. **2** (usu. foll. by *in*) announce, herald, or show in. [Latin *ostium* door]

usherette /,ʌʃə'ret/ *n.* female usher, esp. in a cinema.

USSR *abbr. hist.* Union of Soviet Socialist Republics.

usual /'ju:ʒʊəl/ *adj.* **1** customary, habitual (*the usual time*). **2** (*absol.*, prec. by *the*, *my*, etc.) *colloq.* person's usual drink etc. □ **as usual** as is (or was) usual. □ **usually** *adv.* [Latin: related to USE]

usurer /'ju:ʒərə(r)/ *n.* person who practises usury.

usurp /jʊ'zɜ:p/ *v.* seize (a throne or power etc.) wrongfully. □ **usurpation** /ju:zə'peɪʃ(ə)n/ *n.* **usurper** *n.* [French from Latin]

usury /'ju:ʒərɪ/ *n.* **1** lending of money at interest, esp. at an exorbitant or illegal rate. **2** interest at this rate. □ **usurious** /-'ʒʊərɪəs/ *adj.* [Anglo-French or medieval Latin: related to USE]

utensil /ju:'tens(ə)l/ *n.* implement or vessel, esp. for kitchen use. [medieval Latin: related to USE]

uterine /'ju:tə,raɪn/ *adj.* of the uterus.

uterus /'ju:tərəs/ *n.* (*pl.* **uteri** /-,raɪ/) womb. [Latin]

utilitarian /,ju:tɪlɪ'teərɪən/ –*adj.* **1** designed to be useful rather than attractive; severely practical. **2** of utilitarianism. –*n.* adherent of utilitarianism.

utilitarianism *n.* doctrine that actions are right if they are useful or benefit a majority.

utility /ju:'tɪlɪtɪ/ *n.* (*pl.* **-ies**) **1** usefulness. **2** useful thing. **3** = PUBLIC UTILITY. **4** (*attrib.*) basic and standardized (*utility furniture*). [Latin *utilis* useful: related to USE]

utility room *n.* room for domestic appliances, e.g. a washing-machine, boiler, etc.

utility vehicle *n.* vehicle serving various functions.

utilize /'ju:tɪ,laɪz/ *v.* (also **-ise**) (**-zing** or **-sing**) use; turn to account. □ **utilization** /-'zeɪʃ(ə)n/ *n.* [Italian: related to UTILITY]

utmost /'ʌtməʊst/ –*attrib. adj.* furthest, extreme, greatest. –*n.* utmost point or degree etc. □ **do one's utmost** do all that one can. [Old English, = *outmost*]

Utopia /ju:'təʊpɪə/ *n.* imagined perfect place or state of things. □ **Utopian** *adj.* (also **utopian**). [title of a book by Thomas More, from Greek *ou* not, *topos* place]

utter¹ *attrib. adj.* complete, absolute. □ **utterly** *adv.* [Old English, comparative of OUT]

utter[2] *v.* **1** emit audibly. **2** express in words. **3** *Law* put (esp. forged money) into circulation. [Dutch]
utterance *n.* **1** act of uttering. **2** thing spoken. **3** power or manner of speaking.
uttermost *attrib. adj.* utmost.
U-turn /'ju:tɜ:n/ *n.* **1** U-shaped turn of a vehicle so as to face in the opposite direction. **2** abrupt reversal of policy.
UV *abbr.* ultraviolet.
uvula /'ju:vjʊlə/ *n.* (*pl.* **uvulae** /-,li:/) fleshy part of the soft palate hanging above the throat. □ **uvular** *adj.* [Latin diminutive of *uva* grape]
uxorial /ʌk'sɔ:rɪəl/ *adj.* of a wife. [Latin *uxor* wife]
uxorious /ʌk'sɔ:rɪəs/ *adj.* greatly or excessively fond of one's wife.

V

V[1] /viː/ *n.* (also **v**) (*pl.* **Vs** or **V's**) **1** twenty-second letter of the alphabet. **2** V-shaped thing. **3** (as a roman numeral) 5.
V[2] *abbr.* volt(s).
V[3] *symb.* vanadium.
v. *abbr.* **1** verse. **2** versus. **3** very. **4** *vide.*
vac *n. colloq.* **1** vacation. **2** vacuum cleaner. [abbreviation]
vacancy /ˈveɪkənsɪ/ *n.* (*pl.* **-ies**) **1** being vacant. **2** unoccupied job. **3** available room in a hotel etc.
vacant *adj.* **1** not filled or occupied. **2** not mentally active; showing no interest. □ **vacantly** *adv.* [Latin: related to VACATE]
vacant possession *n.* ownership of an unoccupied house etc.
vacate /vəˈkeɪt/ *v.* (**-ting**) leave vacant, cease to occupy (a house, post, etc.). [Latin *vaco* be empty]
vacation /vəˈkeɪʃ(ə)n/ **—***n.* **1** fixed holiday period, esp. in universities and law-courts. **2** *US* holiday. **3** vacating or being vacated. **—***v. US* take a holiday. [Latin: related to VACATE]
vaccinate /ˈvæksɪˌneɪt/ *v.* (**-ting**) inoculate with a vaccine to immunize against a disease. □ **vaccination** /-ˈneɪʃ(ə)n/ *n.* **vaccinator** *n.*
vaccine /ˈvæksiːn/ *n.* preparation, orig. cowpox virus, used in vaccination. [Latin *vacca* cow]
vacillate /ˈvæsɪˌleɪt/ *v.* (**-ting**) be irresolute; fluctuate. □ **vacillation** /-ˈleɪʃ(ə)n/ *n.* **vacillator** *n.* [Latin]
vacuole /ˈvækjʊˌəʊl/ *n.* tiny space in an organ or cell, containing air, fluid, etc. [Latin *vacuus* empty]
vacuous /ˈvækjʊəs/ *adj.* **1** expressionless. **2** showing absence of thought or intelligence, inane. □ **vacuity** /vəˈkjuːɪtɪ/ *n.* **vacuously** *adv.* [Latin *vacuus* empty]
vacuum /ˈvækjʊəm/ **—***n.* (*pl.* **-s** or **-cua**) **1** space entirely devoid of matter. **2** space or vessel from which all or some of the air has been pumped out. **3** absence of the normal or previous content, activities, etc. **4** (*pl.* **-s**) *colloq.* vacuum cleaner. **—***v. colloq.* clean with a vacuum cleaner. [Latin *vacuus* empty]
vacuum brake *n.* brake worked by the exhaustion of air.
vacuum cleaner *n.* machine for removing dust etc. by suction. □ **vacuum-clean** *v.*
vacuum flask *n.* vessel with a double wall enclosing a vacuum, ensuring that the contents remain hot or cold.
vacuum-packed *adj.* sealed after the partial removal of air.
vacuum tube *n.* tube with a near-vacuum for the free passage of electric current.
vade-mecum /ˌvɑːdɪˈmeɪkəm/ *n.* handbook etc. used constantly. [Latin, = go with me]
vagabond /ˈvægəˌbɒnd/ **—***n.* wanderer, esp. an idle one. **—***attrib. adj.* wandering, roving. □ **vagabondage** *n.* [Latin *vagor* wander]
vagary /ˈveɪgərɪ/ *n.* (*pl.* **-ies**) caprice, whim. [Latin *vagor* wander]
vagina /vəˈdʒaɪnə/ *n.* (*pl.* **-s** or **-nae** /-niː/) canal from the uterus to the vulva in female mammals. □ **vaginal** *adj.* [Latin, = sheath]
vagrant /ˈveɪgrənt/ **—***n.* unemployed itinerant. **—***adj.* wandering, roving. □ **vagrancy** *n.* [Anglo-French]
vague /veɪg/ *adj.* **1** uncertain or ill-defined. **2** (of a person or mind) imprecise; inexact in thought, expression, or understanding. □ **vaguely** *adv.* **vagueness** *n.* [Latin *vagus* wandering]
vain *adj.* **1** having too high an opinion of one's looks, abilities, etc. **2** empty, trivial (*vain triumphs*). **3** useless; futile (*in the vain hope of finding it*). □ **in vain 1** without success. **2** lightly or profanely (*take his name in vain*). □ **vainly** *adv.* [Latin *vanus*]
vainglory /veɪnˈglɔːrɪ/ *n.* boastfulness; extreme vanity. □ **vainglorious** *adj.* [French *vaine gloire*]
valance /ˈvæləns/ *n.* (also **valence**) short curtain round the frame or canopy of a bedstead, above a window, etc. [Anglo-French *valer* descend]
vale *n.* (*archaic* except in place-names) valley. [Latin *vallis*]
valediction /ˌvælɪˈdɪkʃ(ə)n/ *n. formal* **1** bidding farewell. **2** words used in this. □ **valedictory** *adj.* & *n.* (*pl.* **-ies**). [Latin *vale* farewell]
valence[1] /ˈveɪləns/ *n.* = VALENCY.
valence[2] var. of VALANCE.
valency /ˈveɪlənsɪ/ *n.* (*pl.* **-ies**) combining power of an atom measured by the number of hydrogen atoms it can displace or combine with. [Latin *valentia* power]

valentine /'vælən,taɪn/ *n.* **1** card sent, often anonymously, as a mark of love on St Valentine's Day (14 Feb.). **2** sweetheart chosen on this day. [*Valentine*, name of two saints]

valerian /və'lɪərɪən/ *n.* any of various flowering herbs, esp. used as a sedative. [French from medieval Latin]

valet /'vælɪt, -leɪ/ *–n.* gentleman's personal servant. *–v.* (**-t-**) **1** work as a valet (for). **2** clean or clean out (a car). [French *va(s)let*, related to VARLET, VASSAL]

valetudinarian /,vælɪ,tju:dɪ'neərɪən/ *–n.* person of poor health or who is unduly anxious about health. *–adj.* of a valetudinarian. □ **valetudinarianism** *n.* [Latin *valetudo* health]

valiant /'væljənt/ *adj.* brave. □ **valiantly** *adv.* [Latin *valeo* be strong]

valid /'vælɪd/ *adj.* **1** (of a reason, objection, etc.) sound, defensible. **2 a** executed with the proper formalities, legally acceptable (*valid contract*; *valid passport*). **b** not yet expired. □ **validity** /və'lɪdɪtɪ/ *n.* [Latin *validus* strong: related to VALIANT]

validate /'vælɪ,deɪt/ *v.* (**-ting**) make valid; ratify. □ **validation** /-'deɪʃ(ə)n/ *n.*

valise /və'li:z/ *n. US* small portmanteau. [French from Italian]

Valium /'vælɪəm/ *n. propr.* drug diazepam used as a tranquillizer. [origin uncertain]

valley /'vælɪ/ *n.* (*pl.* **-s**) low area between hills, usu. with a stream or river flowing through it. [French: related to VALE]

valour /'vælə(r)/ *n.* (*US* **valor**) courage, esp. in battle. □ **valorous** *adj.* [Latin *valeo* be strong]

valuable /'væljʊəb(ə)l/ *–adj.* of great value, price, or worth. *–n.* (usu. in *pl.*) valuable thing. □ **valuably** *adv.*

valuation /,væljʊ'eɪʃ(ə)n/ *n.* **1** estimation (esp. professional) of a thing's worth. **2** worth so estimated.

value /'vælju:/ *–n.* **1** worth, desirability, or utility, or the qualities on which these depend. **2** worth as estimated (*set a high value on my time*). **3** amount for which a thing can be exchanged in the open market. **4** equivalent of a thing. **5** (in full **value for money**) something well worth the money spent. **6** effectiveness (*news value*). **7** (in *pl.*) one's principles, priorities, or standards. **8** *Mus.* duration of a note. **9** *Math.* amount denoted by an algebraic term. *–v.* (**-ues, -ued, -uing**) **1** estimate the value of, esp. professionally. **2** have a high or specified opinion of. □ **valueless** *adj.* **valuer** *n.* [French past part. of *valoir* be worth, from Latin *valeo*]

value added tax *n.* tax levied on the rise in value of services and goods at each stage of production.

value judgement *n.* subjective estimate of worth etc.

valve *n.* **1** device controlling flow through a pipe etc., esp. allowing movement in one direction only. **2** structure in an organ etc. allowing a flow of blood etc. in one direction only. **3** = THERMIONIC VALVE. **4** device to vary the effective length of the tube in a trumpet etc. **5** half-shell of an oyster, mussel, etc. □ **valvular** /'vælvjʊlə(r)/ *adj.* [Latin *valva* leaf of a folding door]

vamoose /və'mu:s/ *v. US slang* depart hurriedly. [Spanish *vamos* let us go]

vamp[1] *–n.* upper front part of a boot or shoe. *–v.* **1** (often foll. by *up*) repair or furbish. **2** (foll. by *up*) make by patching or from odds and ends. **3** improvise a musical accompaniment. [French *avantpié* front of the foot]

vamp[2] *colloq. –n.* woman who uses sexual attraction to exploit men. *–v.* allure and exploit (a man). [abbreviation of VAMPIRE]

vampire /'væmpaɪə(r)/ *n.* **1** supposed ghost or reanimated corpse sucking the blood of sleeping persons. **2** person who preys ruthlessly on others. **3** (in full **vampire bat**) tropical (esp. South American) bloodsucking bat. [French or German from Magyar]

van[1] *n.* **1** small covered goods vehicle. **2** railway carriage for luggage and for the guard. [abbreviation of CARAVAN]

van[2] *n.* vanguard, forefront. [abbreviation]

vanadium /və'neɪdɪəm/ *n.* hard grey metallic element used to strengthen steel. [Old Norse *Vanadís* name of the Scandinavian goddess Freyja]

vandal /'vænd(ə)l/ *n.* person who wilfully or maliciously damages property. □ **vandalism** *n.* [*Vandals*, name of a Germanic people that sacked Rome and destroyed works of art in the 5th c.: Latin from Germanic]

vandalize /'vændə,laɪz/ *v.* (also **-ise**) (**-zing** or **-sing**) wilfully or maliciously destroy or damage (esp. public property).

vane *n.* **1** weather-vane. **2** blade of a screw propeller or windmill etc. [dial. var. of obsolete *fane* banner]

vanguard /'vængɑ:d/ *n.* **1** foremost part of an advancing army etc. **2** leaders of a movement etc. [French *avan(t)garde* from *avant* before: related to GUARD]

vanilla /və'nɪlə/ *n.* **1 a** tropical fragrant climbing orchid. **b** (in full **vanilla-pod**) fruit of this. **2** extract from the vanilla-pod, or a synthetic substance, used as flavouring. [Spanish diminutive of *vaina* pod]

vanish /'vænɪʃ/ *v.* **1** disappear. **2** cease to exist. [Latin: related to VAIN]

vanishing cream *n.* skin ointment that leaves no visible trace.

vanishing-point *n.* **1** point at which receding parallel lines appear to meet. **2** stage of complete disappearance.

vanity /'vænɪtɪ/ *n.* (*pl.* **-ies**) **1** conceit about one's appearance or attainments. **2** futility, unsubstantiality, unreal thing (*the vanity of human achievement*). **3** ostentatious display. [Latin: related to VAIN]

vanity bag *n.* (also **vanity case**) woman's make-up bag or case.

vanity unit *n.* wash-basin set into a unit with cupboards beneath.

vanquish /'væŋkwɪʃ/ *v. literary* conquer, overcome. [Latin *vinco*]

vantage /'vɑ:ntɪdʒ/ *n.* **1** (also **vantage point**) place giving a good view. **2** *Tennis* = ADVANTAGE. [French: related to ADVANTAGE]

vapid /'væpɪd/ *adj.* insipid; dull; flat. □ **vapidity** /və'pɪdɪtɪ/ *n.* [Latin *vapidus*]

vapor *US* var. of VAPOUR.

vaporize /'veɪpə,raɪz/ *v.* (also **-ise**) (**-zing** or **-sing**) change into vapour. □ **vaporization** /-'zeɪʃ(ə)n/ *n.*

vapour /'veɪpə(r)/ *n.* (*US* **vapor**) **1** moisture or other substance diffused or suspended in air, e.g. mist, smoke. **2** gaseous form of a substance. **3** medicinal inhalant. □ **vaporous** *adj.* **vapoury** *adj.* [Latin *vapor* steam]

vapour trail *n.* trail of condensed water from an aircraft etc.

variable /'veərɪəb(ə)l/ *–adj.* **1** changeable, adaptable. **2** apt to vary; not constant. **3** *Math.* (of a quantity) indeterminate; able to assume different numerical values. *–n.* variable thing or quantity. □ **variability** /-'bɪlɪtɪ/ *n.* **variably** *adv.*

variance /'veərɪəns/ *n.* **1** (usu. prec. by *at*) difference of opinion; dispute (*we were at variance*). **2** discrepancy.

variant *–adj.* **1** differing in form or details from a standard (*variant spelling*). **2** having different forms (*forty variant types*). *–n.* variant form, spelling, type, etc.

variation /,veərɪ'eɪʃ(ə)n/ *n.* **1** varying. **2** departure from the normal kind, amount, a standard, etc. (*prices are subject to variation*). **3** extent of this. **4** variant thing. **5** *Mus.* theme in a changed or elaborated form.

varicoloured /'veərɪ,kʌləd/ *adj.* (*US* **varicolored**) **1** variegated in colour. **2** of various colours. [Latin *varius* VARIOUS]

varicose /'værɪ,kəʊs/ *adj.* (esp. of a vein etc.) permanently and abnormally dilated. [Latin *varix* varicose vein]

varied /'veərɪd/ *adj.* showing variety.

variegated /'veərɪ,geɪtɪd/ *adj.* **1** with irregular patches of different colours. **2** having leaves of two or more colours. □ **variegation** /-'geɪʃ(ə)n/ *n.* [Latin: related to VARIOUS]

variety /və'raɪətɪ/ *n.* (*pl.* **-ies**) **1** diversity; absence of uniformity; many-sidedness. **2** quantity or collection of different things (*for a variety of reasons*). **3 a** class of things that differ from the rest in the same general class. **b** member of such a class. **4** (foll. by *of*) different form of a thing, quality, etc. **5** *Biol.* subdivision of a species. **6** series of dances, songs, comedy acts, etc. (*variety show*). [Latin: related to VARIOUS]

various /'veərɪəs/ *adj.* **1** different, diverse (*from various backgrounds*). **2** several (*for various reasons*). □ **variously** *adv.* [Latin *varius*]

■ **Usage** *Various* (unlike *several*) cannot be used with *of*, as (wrongly) in *various of the guests arrived late.*

varlet /'vɑ:lɪt/ *n. archaic* menial; rascal. [French var. of *vaslet* VALET]

varnish /'vɑ:nɪʃ/ *–n.* **1** resinous solution used to give a hard shiny transparent coating. **2** similar preparation (*nail varnish*). **3** deceptive outward appearance or show. *–v.* **1** apply varnish to. **2** give a deceptively attractive appearance to. [French *vernis*, probably ultimately from *Berenice* in Cyrenaica]

varsity /'vɑ:sɪtɪ/ *n.* (*pl.* **-ies**) *colloq.* (esp. with ref. to sports) university. [abbreviation]

vary /'veərɪ/ *v.* (**-ies, -ied**) **1** be or become different; be of different kinds; change. **2** make different; modify. [Latin *vario*: related to VARIOUS]

vas *n.* (*pl.* **vasa** /'veɪsə/) vessel or duct. [Latin, = vessel]

vascular /'væskjʊlə(r)/ *adj.* of or containing vessels for conveying blood, sap, etc. [Latin *vasculum* diminutive of VAS]

vas deferens /væs 'defə,renz/ *n.* (*pl.* **vasa deferentia** /,veɪsə ,defə'renʃɪə/) sperm duct of the testicle.

vase /vɑ:z/ *n.* vessel used as an ornament or container for flowers. [Latin: related to VAS]

vasectomy /və'sektəmɪ/ *n.* (*pl.* **-ies**) removal of part of each vas deferens, esp. for sterilization.

Vaseline /'væsɪ,li:n/ *n. propr.* type of petroleum jelly used as an ointment etc. [German *Wasser* water, Greek *elaion* oil]

vassal /'væs(ə)l/ *n.* **1** *hist.* feudal tenant of land. **2** humble dependant. □ **vassalage** *n.* [medieval Latin *vassallus* retainer]

vast /vɑ:st/ *adj.* immense, huge. □ **vastly** *adv.* **vastness** *n.* [Latin]

VAT /,vi:eɪ'ti:, væt/ *abbr.* value added tax.

vat *n.* tank, esp. for holding liquids in brewing, distilling, food manufacture, dyeing, and tanning. [dial. var. of *fat*, from Old English]

Vatican /'vætɪkən/ *n.* palace or government of the Pope in Rome. [name of a hill in Rome]

vaudeville /'vɔ:dəvɪl/ *n.* esp. *US* **1** variety entertainment. **2** light stage play with interspersed songs. □ **vaudevillian** /-'vɪlɪən/ *adj.* & *n.* [French]

vault /vɔ:lt/ –*n.* **1** arched roof. **2** vault-like covering (*vault of heaven*). **3** underground storage chamber or place of interment beneath a church or in a cemetery etc. **4** act of vaulting. –*v.* **1** leap, esp. using the hands or a pole. **2** spring over in this way. **3** (esp. as **vaulted**) **a** make in the form of a vault. **b** provide with a vault or vaults. [Latin *volvo* roll]

vaulting *n.* arched work in a vaulted roof or ceiling.

vaulting-horse *n.* wooden box for vaulting over.

vaunt /vɔ:nt/ *v.* & *n.* *literary* boast. [Latin: related to VAIN]

VC *abbr.* Victoria Cross.

VCR *abbr.* video cassette recorder.

VD *abbr.* venereal disease.

VDU *abbr.* visual display unit.

VE *abbr.* Victory in Europe (in 1945).

've *abbr.* (usu. after pronouns) have (*I've*; *they've*).

veal *n.* calf's flesh as food. [French from Latin *vitulus* calf]

vector /'vektə(r)/ *n.* **1** *Math.* & *Physics* quantity having direction as well as magnitude. **2** carrier of disease. [Latin *veho vect-* convey]

Veda /'veɪdə/ *n.* (in *sing.* or *pl.*) oldest Hindu scriptures. □ **Vedic** *adj.* [Sanskrit, = knowledge]

VE day *n.* 8 May, the day marking Victory in Europe (in 1945).

veer /vɪə(r)/ –*v.* **1** change direction, esp. (of the wind) clockwise. **2** change in course or opinion etc. –*n.* change of direction. [French *virer*]

veg /vedʒ/ *n. colloq.* vegetable(s). [abbreviation]

vegan /'vi:gən/ –*n.* person who does not eat animals or animal products. –*adj.* using or containing no animal products. [shortening of VEGETARIAN]

vegeburger var. of VEGGIE BURGER.

vegetable /'vedʒtəb(ə)l/ –*n.* **1** plant, esp. a herbaceous plant used for food, e.g. a cabbage, potato, or bean. **2** *colloq. derog.* **a** *offens.* person who is severely mentally incapacitated, esp. through brain injury etc. **b** dull or inactive person. –*adj.* of, derived from, or relating to plant life or vegetables as food. [Latin: related to VEGETATE]

vegetable marrow see MARROW 1.

vegetal /'vedʒɪt(ə)l/ *adj.* of or like plants. [medieval Latin: related to VEGETATE]

vegetarian /,vedʒɪ'teərɪən/ –*n.* person who does not eat meat or fish. –*adj.* excluding animal food, esp. meat (*vegetarian diet*). □ **vegetarianism** *n.* [from VEGETABLE]

vegetate /'vedʒɪ,teɪt/ *v.* (**-ting**) **1** live an uneventful or monotonous life. **2** grow as plants do. [Latin *vegeto* animate]

vegetation /,vedʒɪ'teɪʃ(ə)n/ *n.* plants collectively; plant life. [medieval Latin: related to VEGETATE]

vegetative /'vedʒɪtətɪv/ *adj.* **1** concerned with growth and development as distinct from sexual reproduction. **2** of vegetation. [French or medieval Latin: related to VEGETATE]

veggie /'vedʒɪ/ *n.* (also **vegie**) *colloq.* vegetarian. [abbreviation]

veggie burger *n.* (also **vegeburger**) flat cake like a hamburger but containing vegetables or soya protein instead of meat.

vehement /'vi:əmənt/ *adj.* showing or caused by strong feeling; ardent (*vehement protest*). □ **vehemence** *n.* **vehemently** *adv.* [Latin]

vehicle /'vi:ɪk(ə)l/ *n.* **1** conveyance used on land or in space. **2** thing or person as a medium for expression or action. **3** liquid etc. as a medium for suspending pigments, drugs, etc. □ **vehicular** /vɪ'hɪkjʊlə(r)/ *adj.* [Latin *veho* carry]

veil /veɪl/ –*n.* **1** piece of usu. transparent fabric attached to a woman's hat etc., esp. to conceal or protect the face. **2** piece of linen etc. as part of a nun's headdress. **3** thing that hides or disguises (*a veil of silence*). –*v.* **1** cover with a veil. **2** (esp. as **veiled** *adj.*) partly conceal (*veiled threats*). □ **beyond the veil** in the unknown state of life after death. **draw a veil over** avoid discussing; hush up. **take the veil** become a nun. [Latin *velum*]

vein /veɪn/ *n.* **1 a** any of the tubes conveying blood to the heart. **b** (in general use) any blood-vessel. **2** rib of an insect's wing or leaf. **3** streak of a different colour in wood, marble, cheese, etc. **4** fissure in rock filled with ore. **5** specified character or tendency; mood

(*spoke in a sarcastic vein*). □ **veined** *adj.* **veiny** *adj.* (**-ier, -iest**). [Latin *vena*]
Velcro /'velkrəʊ/ *n. propr.* fastener consisting of two strips of fabric which cling when pressed together. [French *velours croché* hooked velvet]
veld /velt/ *n.* (also **veldt**) *S.Afr.* open country. [Afrikaans: related to FIELD]
veleta /və'li:tə/ *n.* ballroom dance in triple time. [Spanish, = weather-vane]
vellum /'veləm/ *n.* **1 a** fine parchment, orig. calfskin. **b** manuscript on this. **2** smooth writing-paper imitating vellum. [French *velin*: related to VEAL]
velocity /vɪ'lɒsɪtɪ/ *n.* (*pl.* **-ies**) speed, esp. of inanimate things (*wind velocity*; *velocity of light*). [Latin *velox* swift]
velodrome /'velə,drəʊm/ *n.* place or building with a track for cycle-racing. [French *vélo* bicycle]
velour /və'lʊə(r)/ *n.* (also **velours** pronunc. same) plushlike fabric. [French]
velvet /'velvɪt/ *–n.* **1** soft fabric with a thick short pile on one side. **2** furry skin on a growing antler. *–adj.* of, like, or soft as velvet. □ **on velvet** in an advantageous or prosperous position. □ **velvety** *adj.* [Latin *villus* tuft, down]
velveteen /,velvɪ'ti:n/ *n.* cotton fabric with a pile like velvet.
velvet glove *n.* outward gentleness, esp. cloaking firmness.
Ven. *abbr.* Venerable (as the title of an archdeacon).
venal /'vi:n(ə)l/ *adj.* corrupt; able to be bribed; involving bribery. □ **venality** /-'nælɪtɪ/ *n.* **venally** *adv.* [Latin *venum* thing for sale]

■ **Usage** *Venal* is sometimes confused with *venial*, which means 'pardonable'.

---

vend *v.* offer (small wares) for sale. □ **vendible** *adj.* [Latin *vendo* sell]
vendetta /ven'detə/ *n.* **1** blood feud. **2** prolonged bitter quarrel. [Latin: related to VINDICTIVE]
vending-machine *n.* slot-machine selling small items.
vendor *n. Law* seller, esp. of property. [Anglo-French: related to VEND]
veneer /vɪ'nɪə(r)/ *–n.* **1** thin covering of fine wood etc. **2** (often foll. by *of*) deceptively pleasing appearance. *–v.* **1** apply a veneer to (wood etc.). **2** disguise. [German *furnieren* to furnish]
venerable /'venərəb(ə)l/ *adj.* **1** entitled to deep respect on account of character, age, associations, etc. (*venerable priest*; *venerable relics*). **2** title of an archdeacon in the Church of England. [Latin: related to VENERATE]
venerate /'venə,reɪt/ *v.* (**-ting**) respect deeply. □ **veneration** /-'reɪʃ(ə)n/ *n.* **venerator** *n.* [Latin *veneror* revere]
venereal /vɪ'nɪərɪəl/ *adj.* **1** of sexual desire or intercourse. **2** of venereal disease. [Latin *venus veneris* sexual love]
venereal disease *n.* disease contracted by sexual intercourse with an infected person or congenitally.
Venetian /vɪ'ni:ʃ(ə)n/ *–n.* native, citizen, or dialect of Venice. *–adj.* of Venice. [from French or medieval Latin *Venetia* Venice]
venetian blind *n.* window-blind of adjustable horizontal slats.
vengeance /'vendʒ(ə)ns/ *n.* punishment inflicted for wrong to oneself or one's cause. □ **with a vengeance** to a high or excessive degree (*punctuality with a vengeance*). [French *venger* from Latin *vindico* avenge]
vengeful *adj.* vindictive; seeking vengeance. □ **vengefully** *adv.* [obsolete *venge* avenge: related to VENGEANCE]
venial /'vi:nɪəl/ *adj.* (of a sin or fault) pardonable; not mortal. □ **veniality** /-'ælɪtɪ/ *n.* **venially** *adv.* [Latin *venia* forgiveness]

■ **Usage** *Venial* is sometimes confused with *venal*, which means 'corrupt'.

---

venison /'venɪs(ə)n/ *n.* deer's flesh as food. [Latin *venatio* hunting]
Venn diagram *n.* diagram using overlapping and intersecting circles etc. to show the relationships between mathematical sets. [*Venn*, name of a logician]
venom /'venəm/ *n.* **1** poisonous fluid of esp. snakes. **2** malignity; virulence. □ **venomous** *adj.* **venomously** *adv.* [Latin *venenum*]
venous /'vi:nəs/ *adj.* of, full of, or contained in, veins. [Latin: related to VEIN]
vent[1] *–n.* **1** opening allowing the passage of air etc. **2** outlet; free expression (*gave vent to my anger*). **3** anus, esp. of a lower animal. *–v.* **1** make a vent in (a cask etc.). **2** give free expression to. □ **vent one's spleen on** scold or ill-treat without cause. [Latin *ventus* wind]
vent[2] *n.* slit in a garment, esp. in the lower edge of the back of a jacket. [French *fente* from Latin *findo* cleave]
ventilate /'ventɪ,leɪt/ *v.* (**-ting**) **1** cause air to circulate freely in (a room etc.). **2** air (a question, grievance, etc.). **3** *Med.* **a** oxygenate (the blood). **b** admit or force air into (the lungs). □ **ventilation** /-'leɪʃ(ə)n/ *n.* [Latin *ventilo* blow, winnow: related to VENT[1]]
ventilator *n.* **1** appliance or aperture for ventilating a room etc. **2** *Med.* = RESPIRATOR 2.
ventral /'ventr(ə)l/ *adj.* of or on the abdomen. [*venter* abdomen, from Latin]
ventricle /'ventrɪk(ə)l/ *n.* **1** cavity in the body. **2** hollow part of an organ, esp. the

brain or heart. □ **ventricular** /-'trɪkjʊlə(r)/ *adj.* [Latin *ventriculus* diminutive of *venter* belly]

ventriloquism /ven'trɪlə,kwɪz(ə)m/ *n.* (also **ventriloquy**) skill of speaking without moving the lips, esp. as entertainment with a dummy. □ **ventriloquist** *n.* [Latin *venter* belly, *loquor* speak]

venture /'ventʃə(r)/ –*n.* **1** risky undertaking. **2** commercial speculation. –*v.* (**-ring**) **1** dare; not be afraid. **2** dare to go, make, or put forward (*venture out*; *venture an opinion*). **3 a** expose to risk; stake. **b** take risks. [from ADVENTURE]

Venture Scout *n.* senior Scout.

venturesome *adj.* **1** disposed to take risks. **2** risky.

venue /'venju:/ *n.* place for a match, meeting, concert, etc. [French, from *venir* come]

Venus fly-trap /'vi:nəs/ *n.* insectivorous plant. [Latin *Venus* goddess of love]

veracious /və'reɪʃəs/ *adj. formal* **1** truthful by nature. **2** (of a statement etc.) true. □ **veracity** /və'ræsɪtɪ/ *n.* [Latin *verax* from *verus* true]

veranda /və'rændə/ *n.* (usu. covered) platform along the side of a house. [Hindi from Portuguese *varanda*]

verb *n.* word used to indicate action, a state, or an occurrence (e.g. *hear*, *be*, *happen*). [Latin *verbum* word]

verbal –*adj.* **1** of words. **2** oral, not written. **3** of a verb. **4** (of a translation) literal. **5** talkative. –*n.* **1** *slang* verbal statement to the police. **2** *slang* stream of abuse. □ **verbally** *adv.* [Latin: related to VERB]

■ **Usage** Some people reject sense 2 of *verbal* as illogical, and prefer *oral*. However, *verbal* is the usual term in expressions such as *verbal communication*, *verbal contract*, and *verbal evidence*.

verbalize *v.* (also **-ise**) (**-zing** or **-sing**) put into words.

verbal noun *n.* noun derived from a verb (e.g. *smoking* in *smoking is forbidden*): see -ING[1].

verbatim /vɜ:'beɪtɪm/ *adv.* & *adj.* in exactly the same words. [medieval Latin: related to VERB]

verbena /vɜ:'bi:nə/ *n.* (*pl.* same) plant of a genus of usu. annual or biennial plants with clusters of fragrant flowers. [Latin]

verbiage /'vɜ:bɪɪdʒ/ *n. derog.* too many words or unnecessarily difficult words. [French: related to VERB]

verbose /vɜ:'bəʊs/ *adj.* using more words than are needed. □ **verbosity** /-'bɒsɪtɪ/ *n.* [Latin *verbosus* from *verbum* word]

verdant /'vɜ:d(ə)nt/ *adj.* **1** (of grass, a field, etc.) green, lush. **2** (of a person) unsophisticated, green. □ **verdancy** *n.* [perhaps from French *verdeant* from *viridis* green]

verdict /'vɜ:dɪkt/ *n.* **1** decision of a jury in a civil or criminal case. **2** decision; judgement. [Anglo-French *verdit* from *ver* true, *dit* saying]

verdigris /'vɜ:dɪ,gri:/ *n.* greenish-blue substance that forms on copper or brass. [French, = green of Greece]

verdure /'vɜ:djə(r)/ *n. literary* green vegetation or its colour. [French *verd* green]

verge[1] *n.* **1** edge or border. **2** brink (*on the verge of tears*). **3** grass edging of a road etc. [Latin *virga* rod]

verge[2] *v.* (**-ging**) **1** (foll. by *on*) border on. **2** incline downwards or in a specified direction. [Latin *vergo* bend]

verger *n.* **1** church caretaker and attendant. **2** officer preceding a bishop etc. with a staff. [Anglo-French: related to VERGE[1]]

verify /'verɪ,faɪ/ *v.* (**-ies, -ied**) **1** establish the truth, correctness, or validity of by examination etc. (*verified my figures*). **2** (of an event etc.) bear out (a prediction or promise). □ **verifiable** *adj.* **verification** /-fɪ'keɪʃ(ə)n/ *n.* [medieval Latin: related to VERY]

verily /'verɪlɪ/ *adv. archaic* really, truly. [from VERY]

verisimilitude /,verɪsɪ'mɪlɪ,tju:d/ *n.* appearance of being true or real. [Latin *verus* true, *similis* like]

veritable /'verɪtəb(ə)l/ *adj.* real; rightly so called (*a veritable feast*). □ **veritably** *adv.* [French: related to VERITY]

verity /'verɪtɪ/ *n.* (*pl.* **-ies**) **1** a fundamental truth. **2** *archaic* truth. [Latin *veritas* truth]

vermicelli /,vɜ:mɪ'tʃelɪ/ *n.* **1** pasta in long slender threads. **2** shreds of chocolate as cake decoration etc. [Latin *vermis* worm]

vermicide /'vɜ:mɪ,saɪd/ *n.* drug that kills intestinal worms. [Latin *vermis* worm]

vermiculite /və'mɪkjʊ,laɪt/ *n.* a hydrous silicate mineral used esp. as a moisture-holding medium for plant growth. [Latin *vermiculatus* worm-eaten, from *vermis* worm]

vermiform /'vɜ:mɪ,fɔ:m/ *adj.* worm-shaped. [medieval Latin: related to VERMICIDE]

vermiform appendix *n.* small blind tube extending from the caecum in man and some other mammals.

vermilion /və'mɪljən/ –*n.* **1** cinnabar. **2 a** brilliant red pigment made esp. from this. **b** colour of this. –*adj.* of this colour.

[Latin *vermiculus* diminutive of *vermis* worm]

vermin /ˈvɜːmɪn/ *n.* (usu. treated as *pl.*) **1** mammals and birds harmful to game, crops, etc., e.g. foxes and rats. **2** parasitic worms or insects. **3** vile people. □ **verminous** *adj.* [Latin *vermis* worm]

vermouth /ˈvɜːməθ/ *n.* wine flavoured with aromatic herbs. [German: related to WORMWOOD]

vernacular /vəˈnækjʊlə(r)/ *–n.* **1** language or dialect of a particular country. **2** language of a particular class or group. **3** homely speech. *–adj.* (of language) native; not foreign or formal. [Latin *vernaculus* native]

vernal /ˈvɜːn(ə)l/ *adj.* of or in spring. [Latin *ver* spring]

vernal equinox var. of SPRING EQUINOX.

vernier /ˈvɜːnɪə(r)/ *n.* small movable graduated scale for obtaining fractional parts of subdivisions on a fixed scale. [*Vernier*, name of a mathematician]

veronal /ˈverən(ə)l/ *n.* sedative drug. [German from *Verona* in Italy]

veronica /vəˈrɒnɪkə/ *n.* speedwell. [medieval Latin, probably from St *Veronica*]

verruca /vəˈruːkə/ *n.* (*pl.* **verrucae** /-siː/ or **-s**) wart or similar growth, esp. on the foot. [Latin]

versatile /ˈvɜːsəˌtaɪl/ *adj.* **1** adapting easily to different subjects or occupations; skilled in many subjects or occupations. **2** having many uses. □ **versatility** /-ˈtɪlɪtɪ/ *n.* [Latin *verto vers-* turn]

verse *n.* **1** poetry. **2** stanza of a poem or song. **3** each of the short numbered divisions of the Bible. **4** poem. [Latin *versus*: related to VERSATILE]

versed /vɜːst/ *adj.* (foll. by *in*) experienced or skilled in. [Latin *versor* be engaged in]

versicle /ˈvɜːsɪk(ə)l/ *n.* each of a priest's short sentences in a liturgy, answered by the congregation. [Latin diminutive: related to VERSE]

versify /ˈvɜːsɪˌfaɪ/ *v.* (**-ies**, **-ied**) **1** turn into or express in verse. **2** compose verses. □ **versification** /-fɪˈkeɪʃ(ə)n/ *n.* **versifier** *n.*

version /ˈvɜːʃ(ə)n/ *n.* **1** account of a matter from a particular point of view. **2** book etc. in a particular edition or translation (*Authorized Version*). **3** form or variant. [Latin *verto vers-* turn]

verso /ˈvɜːsəʊ/ *n.* (*pl.* **-s**) **1** left-hand page of an open book. **2** back of a printed leaf. [Latin *verso* (*folio*) on the turned (leaf)]

versus /ˈvɜːsəs/ *prep.* against (esp. in law and sport). [Latin: related to VERSE]

vertebra /ˈvɜːtɪbrə/ *n.* (*pl.* **-brae** /-ˌbriː/) each segment of a backbone. □ **vertebral** *adj.* [Latin *verto* turn]

vertebrate /ˈvɜːtɪbrət/ *–adj.* (of an animal) having a backbone. *–n.* vertebrate animal. [Latin *vertebratus* jointed: related to VERTEBRA]

vertex /ˈvɜːteks/ *n.* (*pl.* **-tices** /-tɪˌsiːz/ or **-texes**) **1** highest point; top, apex. **2 a** each angular point of a triangle, polygon, etc. **b** meeting-point of lines that form an angle. [Latin, = whirlpool, crown of a head, from *verto* turn]

vertical /ˈvɜːtɪk(ə)l/ *–adj.* **1** at right angles to a horizontal plane. **2** in a direction from top to bottom of a picture etc. **3** of or at the vertex. *–n.* vertical line or plane. □ **vertically** *adv.* [Latin: related to VERTEX]

vertical take-off *n.* take-off of an aircraft directly upwards.

vertiginous /vɜːˈtɪdʒɪnəs/ *adj.* of or causing vertigo. [Latin: related to VERTIGO]

vertigo /ˈvɜːtɪˌgəʊ/ *n.* dizziness caused esp. by heights. [Latin, = whirling, from *verto* turn]

vervain /ˈvɜːveɪn/ *n.* any of several verbenas, esp. one with small blue, white, or purple flowers. [Latin: related to VERBENA]

verve *n.* enthusiasm, vigour, spirit. [French]

very /ˈverɪ/ *–adv.* **1** in a high degree (*did it very easily*). **2** in the fullest sense (foll. by *own* or superl. adj.: *do your very best*; *my very own room*). *–adj.* actual; truly such (*the very thing we need*; *his very words*; *the very same*). □ **not very** in a low degree, far from being. **very good** (or **well**) formula of consent or approval. **very high frequency** (in radio) 30–300 megahertz. **Very Reverend** title of a dean. [Latin *verus* true]

Very light /ˈvɪərɪ/ *n.* flare projected from a pistol for signalling or illuminating part of a battlefield etc. [*Very*, name of its inventor]

vesicle /ˈvesɪk(ə)l/ *n.* small bladder, bubble, or blister. [Latin]

vespers *n.pl.* evensong. [Latin *vesper* evening]

vessel /ˈves(ə)l/ *n.* **1** hollow receptacle, esp. for liquid. **2** ship or boat, esp. a large one. **3** duct or canal etc. holding or conveying blood or sap, etc., esp. = BLOOD-VESSEL. [Latin diminutive: related to VAS]

vest *–n.* **1** undergarment worn on the trunk. **2** *US & Austral.* waistcoat. *–v.* **1** (foll. by *with*) bestow (powers, authority, etc.) on. **2** (foll. by *in*) confer (property or power) on (a person) with an immediate fixed right of future possession. **3** clothe (oneself), esp. in vestments. [Latin *vestis* garment]

vestal virgin *n. Rom. Antiq.* virgin consecrated to Vesta and vowed to chastity. [*Vesta*, Roman goddess of the hearth and home]

vested interest *n.* **1** personal interest in a state of affairs, usu. with an expectation of gain. **2** *Law* interest (usu. in land or money held in trust) recognized as belonging to a person.

vestibule /'vestɪ,bju:l/ *n.* **1** hall or lobby of a building. **2** *US* enclosed space between railway-carriages. [Latin]

vestige /'vestɪdʒ/ *n.* **1** trace; sign. **2** slight amount; particle. **3** atrophied part or organ of an animal or plant that was well developed in ancestors. □ **vestigial** /-'tɪdʒɪəl/ *adj.* [Latin *vestigium* footprint]

vestment /'vestmənt/ *n.* ceremonial garment, esp. a chasuble. [Latin: related to VEST]

vestry /'vestrɪ/ *n.* (*pl.* **-ies**) church room or building for keeping vestments etc. in.

vet[1] –*n. colloq.* veterinary surgeon. –*v.* (**-tt-**) make a careful and critical examination of (a scheme, work, candidate, etc.). [abbreviation]

vet[2] *n. US* veteran. [abbreviation]

vetch *n.* plant of the pea family used largely for fodder. [Latin *vicia*]

veteran /'vetərən/ *n.* **1** (often *attrib.*) old soldier or long-serving member of any group (*war veteran*; *veteran actor*). **2** *US* ex-serviceman or servicewoman. [Latin *vetus -er-* old]

veteran car *n.* car made before 1916, or (strictly) before 1905.

veterinarian /,vetərɪ'neərɪən/ *n. formal* veterinary surgeon.

veterinary /'vetə,rɪnərɪ/ –*adj.* of or for the diseases and injuries of animals. –*n.* (*pl.* **-ies**) veterinary surgeon. [Latin *veterinae* cattle]

veterinary surgeon *n.* person qualified to treat animals.

veto /'vi:təʊ/ –*n.* (*pl.* **-es**) **1** right to reject a measure, resolution, etc. unilaterally. **2** rejection, prohibition. –*v.* (**-oes, -oed**) **1** reject (a measure etc.). **2** forbid, prohibit. [Latin, = I forbid]

vex *v.* **1** anger, irritate. **2** *archaic* grieve, afflict. [Latin *vexo* afflict]

vexation /vek'seɪʃ(ə)n/ *n.* **1** vexing or being vexed. **2** annoying or distressing thing.

vexatious /vek'seɪʃ(ə)s/ *adj.* **1** causing vexation. **2** *Law* (of litigation) lacking sufficient grounds and seeking only to annoy the defendant.

vexed *adj.* (of a question) much discussed; problematic.

v.g.c. *abbr.* very good condition.

VHF *abbr.* very high frequency.

via /'vaɪə/ *prep.* through (*London to Rome via Paris*; *send it via your son*). [Latin, ablative of *via* way]

viable /'vaɪəb(ə)l/ *adj.* **1** (of a plan etc.) feasible, esp. economically. **2** (esp. of a foetus) capable of developing and surviving independently. □ **viability** /-'bɪlɪtɪ/ *n.* [French *vie* life]

viaduct /'vaɪə,dʌkt/ *n.* long bridge, esp. a series of arches, carrying a road or railway across a valley or hollow. [Latin *via* way, after AQUEDUCT]

vial /'vaɪəl/ *n.* small (usu. cylindrical glass) vessel, esp. for holding medicines. [related to PHIAL]

viand /'vaɪənd/ *n. formal* (usu. in *pl.*) article of food. [Latin *vivo* live]

viaticum /vaɪ'ætɪkəm/ *n.* (*pl.* **-ca**) Eucharist given to a dying person. [Latin *via* road]

vibes *n.pl. colloq.* **1** vibrations, esp. feelings communicated. **2** = VIBRAPHONE. [abbreviation]

vibrant /'vaɪbrənt/ *adj.* **1** vibrating. **2** (often foll. by *with*) thrilling, lively. **3** (of sound) resonant. **4** (of colours) bright and striking. □ **vibrancy** *n.* **vibrantly** *adv.* [Latin: related to VIBRATE]

vibraphone /'vaɪbrə,fəʊn/ *n.* instrument like a xylophone but with motor-driven resonators under the metal bars giving a vibrato effect. [from VIBRATO]

vibrate /vaɪ'breɪt/ *v.* (**-ting**) **1** move rapidly to and fro. **2** (of a sound) throb; resonate. **3** (foll. by *with*) quiver, thrill. **4** swing to and fro, oscillate. [Latin *vibro* shake]

vibration /vaɪ'breɪʃ(ə)n/ *n.* **1** vibrating. **2** (in *pl.*) **a** mental, esp. occult, influence. **b** atmosphere or feeling communicated.

vibrato /vɪ'brɑ:təʊ/ *n.* rapid slight variation in musical pitch producing a tremulous effect. [Italian: related to VIBRATE]

vibrator /vaɪ'breɪtə(r)/ *n.* device that vibrates, esp. an instrument for massage or sexual stimulation. □ **vibratory** *adj.*

viburnum /vaɪ'bɜ:nəm/ *n.* a shrub, usu. with white flowers. [Latin, = wayfaring-tree]

vicar /'vɪkə(r)/ *n.* clergyman of a Church of England parish where he formerly received a stipend rather than tithes: cf. RECTOR 1. [Latin *vicarius* substitute: related to VICE[3]]

vicarage *n.* vicar's house.

vicarious /vɪ'keərɪəs/ *adj.* **1** experienced indirectly or second-hand. **2** acting or done for another. **3** deputed, delegated. □ **vicariously** *adv.* [Latin: related to VICAR]

vice[1] *n.* **1** immoral conduct. **2** form of this (*the vice of gluttony*). **3** weakness; indulgence (*brandy is my one vice*). [Latin *vitium*]

vice[2] *n.* (*US* **vise**) clamp with two jaws holding an object so as to leave the hands free to work on it. [*vis* screw, from Latin *vitis* vine]

vice[3] /'vaɪsɪ/ *prep.* in the place of; succeeding. [Latin, ablative of (*vix*) *vicis* change]

vice- *comb. form* forming nouns meaning: **1** substitute, deputy (*vice-president*). **2** next in rank to (*vice admiral*). [related to VICE[3]]

vice-chancellor /vaɪs'tʃɑ:nsələ(r)/ *n.* deputy chancellor (esp. administrator of a university).

vice-president /vaɪs'prezɪd(ə)nt/ *n.* official ranking below and deputizing for a president. □ **vice-presidency** *n.* (*pl.* **-ies**). **vice-presidential** /-'denʃ(ə)l/ *adj.*

viceregal /vaɪs'ri:g(ə)l/ *adj.* of a viceroy.

vicereine /'vaɪsreɪn/ *n.* **1** viceroy's wife. **2** woman viceroy. [French: related to VICE-, *reine* queen]

vice ring *n.* group of criminals organizing prostitution.

viceroy /'vaɪsrɔɪ/ *n.* sovereign's deputy ruler in a colony, province, etc. [French: related to VICE-, *roy* king]

vice squad *n.* police department concerned with prostitution etc.

vice versa /ˌvaɪsɪ 'vɜ:sə, vaɪs 'vɜ:sə/ *adj.* with the order of the terms changed; the other way round. [Latin, = the position being reversed]

vichyssoise /ˌvi:ʃi:'swɑ:z/ *n.* (usu. chilled) creamy soup of leeks and potatoes. [French, = of Vichy]

Vichy water /'vi:ʃi:/ *n.* effervescent mineral water from Vichy in France.

vicinity /vɪ'sɪnɪtɪ/ *n.* (*pl.* **-ies**) **1** surrounding district. **2** (foll. by *to*) nearness. □ **in the vicinity** (often foll. by *of*) near (to). [Latin *vicinus* neighbour]

vicious /'vɪʃəs/ *adj.* **1** bad-tempered, spiteful (*vicious dog, remark*). **2** violent (*vicious attack*). **3** corrupt, depraved. **4** (of reasoning etc.) faulty, unsound. □ **viciously** *adv.* **viciousness** *n.* [Latin: related to VICE[1]]

vicious circle *n.* self-perpetuating, harmful sequence of cause and effect.

vicious spiral *n.* vicious circle, esp. as causing inflation.

vicissitude /vɪ'sɪsɪˌtju:d/ *n. literary* change, esp. of fortune. [Latin: related to VICE[3]]

victim /'vɪktɪm/ *n.* **1** person or thing injured or destroyed (*road victim*; *victim of greed*). **2** prey; dupe (*fell victim to his charm*). **3** creature sacrificed to a deity or in a religious rite. [Latin]

victimize *v.* (also **-ise**) (**-zing** or **-sing**) **1** single out for punishment or discrimination. **2** make (a person etc.) a victim. □ **victimization** /-'zeɪʃ(ə)n/ *n.*

victor /'vɪktə(r)/ *n.* winner in a battle or contest. [Latin *vinco vict-* conquer]

Victoria Cross /vɪk'tɔ:rɪə/ *n.* highest decoration for conspicuous bravery in the armed services. [Queen *Victoria*]

Victorian *–adj.* **1** of the time of Queen Victoria. **2** prudish; strict. *–n.* person of this time.

Victoriana /vɪkˌtɔ:rɪ'ɑ:nə/ *n.pl.* articles, esp. collectors' items, of the Victorian period.

Victoria sponge /vɪk'tɔ:rɪə/ *n.* sandwich sponge cake with a jam filling.

victorious /vɪk'tɔ:rɪəs/ *adj.* **1** conquering, triumphant. **2** marked by victory. □ **victoriously** *adv.* [Latin: related to VICTOR]

victory /'vɪktərɪ/ *n.* (*pl.* **-ies**) defeat of an enemy or opponent.

victual /'vɪt(ə)l/ *–n.* (usu. in *pl.*) food, provisions. *–v.* (**-ll-**; *US* **-l-**) **1** supply with victuals. **2** obtain stores. **3** eat victuals. [Latin *victus* food]

victualler /'vɪtlə(r)/ *n.* (*US* **victualer**) **1** person etc. who supplies victuals. **2** (in full **licensed victualler**) publican etc. licensed to sell alcohol.

vicuña /vɪ'kju:nə/ *n.* **1** S. American mammal like a llama, with fine silky wool. **2 a** cloth from its wool. **b** imitation of this. [Spanish from Quechua]

vide /'vi:deɪ/ *v.* (in *imper.*) see, consult (a passage in a book etc.). [Latin *video* see]

videlicet /vɪ'delɪˌset/ *adv.* = VIZ. [Latin *video* see, *licet* allowed]

video /'vɪdɪəʊ/ *–adj.* **1** of the recording (or reproduction) of moving pictures on magnetic tape. **2** of the broadcasting of television pictures. *–n.* (*pl.* **-s**) **1** such recording or broadcasting. **2** *colloq.* = VIDEO RECORDER. **3** *colloq.* a film on videotape. *–v.* (**-oes**, **-oed**) record on videotape. [Latin, = I see]

video cassette *n.* cassette of videotape.

videodisc *n.* disc for recording moving pictures and sound.

video game *n.* computer game played on a television screen.

video nasty *n. colloq.* horrific or pornographic video film.

video recorder *n.* (also **video cassette recorder**) apparatus for recording and playing videotapes.

video shop *n.* shop hiring out or selling video films etc.

videotape *–n.* magnetic tape for recording moving pictures and sound. *–v.* (**-ping**) record on this.
videotape recorder *n.* = VIDEO RECORDER.
videotex /ˈvɪdɪəʊˌteks/ *n.* (also **videotext**) any electronic information system, esp. teletext or viewdata.
vie /vaɪ/ *v.* (**vies; vied; vying**) (often foll. by *with*) compete; strive for superiority. [probably French: related to ENVY]
Vietnamese /ˌvɪetnəˈmiːz/ *–adj.* of Vietnam. *–n.* (*pl.* same) native or language of Vietnam.
view /vjuː/ *–n.* **1** range of vision (*came into view*). **2 a** what is seen; prospect, scene, etc. **b** picture etc. of this. **3 a** opinion. **b** manner of considering a thing (*took a long-term view*). **4** inspection by the eye or mind (*private view*). *–v.* **1** look at; inspect with the idea of purchasing; survey visually or mentally. **2** form a mental impression or opinion of; consider. **3** watch television. □ **have in view 1** have as one's object. **2** bear (a circumstance) in mind. **in view of** considering. **on view** being shown or exhibited. **with a view to** with the hope or intention of. [Latin *video* see]
viewdata *n.* news and information service from a computer source, connected to a television screen by a telephone link.
viewer *n.* **1** person who views, esp. television. **2** device for looking at film transparencies etc.
viewfinder *n.* device on a camera showing the borders of the proposed photograph.
viewpoint *n.* point of view.
vigil /ˈvɪdʒɪl/ *n.* **1** keeping awake during the night etc., esp. to keep watch or pray. **2** eve of a festival or holy day. [Latin *vigilia*]
vigilance *n.* watchfulness, caution. □ **vigilant** *adj.* [Latin: related to VIGIL]
vigilante /ˌvɪdʒɪˈlæntɪ/ *n.* member of a self-appointed group maintaining order etc. [Spanish, = vigilant]
vignette /viːˈnjet/ *n.* **1** short description, character sketch. **2** book illustration not in a definite border. **3** photograph etc. with the background shaded off. [French, diminutive: related to VINE]
vigour /ˈvɪgə(r)/ *n.* (*US* **vigor**) **1** physical or mental strength or energy. **2** healthy growth. **3** forcefulness; trenchancy, animation. □ **vigorous** *adj.* **vigorously** *adv.* [French from Latin *vigeo* be lively]
Viking /ˈvaɪkɪŋ/ *n.* Scandinavian pirate and raider of the 8th–11th c. [Old Norse]
vile *adj.* **1** disgusting. **2** depraved. **3** *colloq.* abominable (*vile weather*). □ **vilely** *adv.* **vileness** *n.* [Latin *vilis* cheap, base]
vilify /ˈvɪlɪˌfaɪ/ *v.* (**-ies, -ied**) defame; malign. □ **vilification** /-fɪˈkeɪʃ(ə)n/ *n.* [Latin: related to VILE]
villa /ˈvɪlə/ *n.* **1** country house; mansion. **2** rented holiday home, esp. abroad. **3** (usu. as part of an address) detached or semi-detached house in a residential district. [Italian and Latin]
village /ˈvɪlɪdʒ/ *n.* **1** country settlement, larger than a hamlet and smaller than a town. **2** self-contained village-like community within a city etc. (*Greenwich village*; *Olympic village*). □ **villager** *n.* [Latin: related to VILLA]
villain /ˈvɪlən/ *n.* **1** wicked person. **2** chief evil character in a play, story, etc. **3** *colloq.* professional criminal. **4** *colloq.* rascal. [Latin: related to VILLA]
villainous *adj.* wicked.
villainy *n.* (*pl.* **-ies**) wicked behaviour or act. [French: related to VILLAIN]
villein /ˈvɪlɪn/ *n. hist.* feudal tenant entirely subject to a lord or attached to a manor. □ **villeinage** *n.* [var. of VILLAIN]
vim *n. colloq.* vigour. [perhaps from Latin, accusative of *vis* energy]
vinaigrette /ˌvɪnɪˈgret/ *n.* **1** salad dressing of oil, wine vinegar, and seasoning. **2** small bottle for smelling-salts. [French, diminutive: related to VINEGAR]
vindicate /ˈvɪndɪˌkeɪt/ *v.* (**-ting**) **1** clear of blame or suspicion. **2** establish the existence, merits, or justice of (something disputed etc.). **3** justify by evidence or argument. □ **vindication** /-ˈkeɪʃ(ə)n/ *n.* **vindicator** *n.* **vindicatory** *adj.* [Latin *vindico* claim]
vindictive /vɪnˈdɪktɪv/ *adj.* vengeful. □ **vindictively** *adv.* **vindictiveness** *n.* [Latin *vindicta* vengeance: related to VINDICATE]
vine *n.* **1** climbing or trailing plant with a woody stem, esp. bearing grapes. **2** stem of this. [Latin *vinea* vineyard]
vinegar /ˈvɪnɪgə(r)/ *n.* sour liquid got from malt, wine, cider, etc., by fermentation and used as a condiment or for pickling. □ **vinegary** *adj.* [French, = sour wine: related to EAGER]
vineyard /ˈvɪnjəd/ *n.* plantation of grapevines, esp. for wine-making.
vingt-et-un /ˌvæteɪˈɜː/ *n.* = PONTOON[1]. [French, = twenty-one]
vino /ˈviːnəʊ/ *n. slang* wine, esp. of an inferior kind. [Italian, = wine]
vinous /ˈvaɪnəs/ *adj.* **1** of, like, or due to wine. **2** addicted to wine. [Latin *vinum* wine]
vintage /ˈvɪntɪdʒ/ *–n.* **1 a** season's produce of grapes. **b** wine from this. **2 a** gathering of grapes for wine-making. **b** season of this. **3** wine of high quality

from a particular year and district. **4 a** year etc. when a thing was made etc. **b** thing made etc. in a particular year etc. —*adj.* **1** of high or peak quality. **2** of a past season. [Latin *vinum* wine]

vintage car *n.* car made 1917–1930.

vintner *n.* wine-merchant. [Anglo-Latin from French, ultimately from Latin *vinetum* vineyard, from *vinum* wine]

vinyl /'vaɪnɪl/ *n.* plastic made by polymerization, esp. polyvinyl chloride. [Latin *vinum* wine]

viol /'vaɪəl/ *n.* medieval stringed instrument of various sizes, like a violin but held vertically. [French from Provençal]

viola[1] /vɪ'əʊlə/ *n.* instrument larger than the violin and of lower pitch. [Italian and Spanish: related to VIOL]

viola[2] /'vaɪələ/ *n.* any plant of the genus including the pansy and violet, esp. a cultivated hybrid. [Latin, = violet]

viola da gamba /vɪˌəʊlə də 'gæmbə/ *n.* viol held between the player's legs.

violate /'vaɪəˌleɪt/ *v.* **(-ting) 1** disregard; break (an oath, treaty, law, etc.). **2** treat (a sanctuary etc.) profanely; disrespect. **3** disturb (a person's privacy etc.). **4** rape. □ **violable** *adj.* **violation** /-'leɪʃ(ə)n/ *n.* **violator** *n.* [Latin *violo*]

violence /'vaɪələns/ *n.* **1** being violent. **2** violent conduct or treatment. **3** unlawful use of force. □ **do violence to** act contrary to; outrage. [Latin: related to VIOLENT]

violent /'vaɪələnt/ *adj.* **1** involving or using great physical force (*violent person*; *violent storm*). **2 a** intense, vehement (*violent pain*; *violent dislike*). **b** lurid (*violent colours*). **3** (of death) resulting from violence or poison. □ **violently** *adv.* [French from Latin]

violet /'vaɪələt/ —*n.* **1** sweet-scented plant with usu. purple, blue, or white flowers. **2** bluish-purple colour at the end of the spectrum furthest from red. **3** pigment or clothes or material of this colour. —*adj.* of this colour. [French diminutive of *viole* VIOLA[2]]

violin /ˌvaɪə'lɪn/ *n.* high-pitched stringed instrument played with a bow. □ **violinist** *n.* [Italian diminutive of VIOLA[1]]

violist /'vaɪəlɪst/ *n.* viol- or viola-player.

violoncello /ˌvaɪələn'tʃeləʊ/ *n.* (*pl.* **-s**) *formal* = CELLO. [Italian, diminutive of *violone* bass viol]

VIP *abbr.* very important person.

viper /'vaɪpə(r)/ *n.* **1** small venomous snake. **2** malignant or treacherous person. [Latin]

virago /vɪ'rɑːgəʊ/ *n.* (*pl.* **-s**) fierce or abusive woman. [Latin, = female warrior]

viral /'vaɪər(ə)l/ *adj.* of or caused by a virus.

virgin /'vɜːdʒɪn/ —*n.* **1** person who has never had sexual intercourse. **2 (the Virgin)** Christ's mother Mary. **3 (the Virgin)** sign or constellation Virgo. —*adj.* **1** not yet used etc. **2** virginal. [Latin *virgo -gin-*]

virginal —*adj.* of or befitting a virgin. —*n.* (usu. in *pl.*) *Mus.* legless spinet in a box. [Latin: related to VIRGIN]

virgin birth *n.* **1** (usu. preceded by *the*) doctrine of Christ's birth from a virgin mother. **2** parthenogenesis.

Virginia creeper /və'dʒɪnɪə/ *n.* ornamental vine. [*Virginia* in US]

virginity /və'dʒɪnɪtɪ/ *n.* state of being a virgin.

Virgo /'vɜːgəʊ/ *n.* (*pl.* **-s**) **1** constellation and sixth sign of the zodiac (the Virgin). **2** person born when the sun is in this sign. [Latin: related to VIRGIN]

virile /'vɪraɪl/ *adj.* **1** (of a man) vigorous or strong. **2** sexually potent. **3** of a man as distinct from a woman or child. □ **virility** /vɪ'rɪlɪtɪ/ *n.* [Latin *vir* man]

virology /vaɪ'rɒlədʒɪ/ *n.* the study of viruses. □ **virologist** *n.*

virtual /'vɜːtʃʊəl/ *adj.* being so in practice though not strictly or in name (*the virtual manager*; *a virtual promise*). [medieval Latin: related to VIRTUE]

virtually *adv.* in effect, nearly, almost.

virtual reality *n.* simulation of the real world by a computer.

virtue /'vɜːtʃuː/ *n.* **1** moral excellence; goodness. **2** particular form of this. **3** (esp. female) chastity. **4** good quality (*has the virtue of speed*). **5** efficacy (*no virtue in such drugs*). □ **by** (or **in**) **virtue of** on account of, because of. [Latin: related to VIRILE]

virtuoso /ˌvɜːtʃʊ'əʊsəʊ/ *n.* (*pl.* **-si** /-siː/ or **-s**) (often *attrib.*) highly skilled artist, esp. a musician (*virtuoso performance*). □ **virtuosic** /-'ɒsɪk/ *adj.* **virtuosity** /-'ɒsɪtɪ/ *n.* [Italian: related to VIRTUOUS]

virtuous /'vɜːtʃʊəs/ *adj.* **1** morally good. **2** *archaic* chaste. □ **virtuously** *adv.* [Latin: related to VIRTUE]

virulent /'vɪrʊlənt/ *adj.* **1** strongly poisonous. **2** (of a disease) violent. **3** bitterly hostile. □ **virulence** *n.* **virulently** *adv.* [Latin: related to VIRUS]

virus /'vaɪərəs/ *n.* **1** microscopic organism often causing diseases. **2** = COMPUTER VIRUS. [Latin, = poison]

visa /'viːzə/ *n.* endorsement on a passport etc., esp. allowing entrance to or exit from a country. [Latin, = seen]

visage /ˈvɪzɪdʒ/ *n. literary* face. [Latin *visus* sight]

vis-à-vis /ˌviːzəˈviː/ *–prep.* **1** in relation to. **2** in comparison with. *–adv.* opposite. [French, = face to face: related to VISAGE]

viscera /ˈvɪsərə/ *n.pl.* internal organs of the body. [Latin]

visceral *adj.* **1** of the viscera. **2** of feelings rather than reason.

viscid /ˈvɪsɪd/ *adj.* glutinous, sticky. [Latin: related to VISCOUS]

viscose /ˈvɪskəʊz/ *n.* **1** cellulose in a highly viscous state, used for making rayon etc. **2** fabric made from this. [Latin: related to VISCOUS]

viscount /ˈvaɪkaʊnt/ *n.* British nobleman ranking between an earl and a baron. □ **viscountcy** *n.* (*pl.* **-ies**). [Anglo-French: related to VICE-, COUNT²]

viscountess *n.* **1** viscount's wife or widow. **2** woman holding the rank of viscount.

viscous /ˈvɪskəs/ *adj.* **1** glutinous, sticky. **2** semifluid. **3** not flowing freely. □ **viscosity** /-ˈkɒsɪtɪ/ *n.* (*pl.* **-ies**). [Latin *viscum* birdlime]

vise *US* var. of VICE².

visibility /ˌvɪzɪˈbɪlɪtɪ/ *n.* **1** being visible. **2** range or possibility of vision as determined by the light and weather.

visible /ˈvɪzɪb(ə)l/ *adj.* **1** able to be seen, perceived, or ascertained. **2** (of exports etc.) consisting of actual goods. □ **visibly** *adv.* [Latin: related to VISION]

vision /ˈvɪʒ(ə)n/ *n.* **1** act or faculty of seeing, sight. **2** thing or person seen in a dream or trance. **3** mental picture (*visions of hot toast*). **4** imaginative insight. **5** statesmanlike foresight. **6** beautiful person etc. **7** television or cinema picture, esp. of specified quality (*poor vision*). [Latin *video vis-* see]

visionary *–adj.* **1** given to seeing visions or to fanciful theories. **2** having vision or foresight. **3** not real, imaginary. **4** not practicable. *–n.* (*pl.* **-ies**) visionary person.

visit /ˈvɪzɪt/ *–v.* (**-t-**) **1** (also *absol.*) go or come to see or inspect (a person, place, etc.). **2** stay temporarily with (a person) or at (a place). **3** (of a disease, calamity, etc.) attack. **4 a** (foll. by *with*) punish (a person). **b** (often foll. by *upon*) inflict punishment for (a sin). *–n.* **1 a** act of visiting. **b** temporary stay, esp. as a guest. **2** (foll. by *to*) occasion of going to a doctor etc. **3** formal or official call. [Latin: related to VISION]

visitant /ˈvɪzɪt(ə)nt/ *n.* **1** visitor, esp. a ghost etc. **2** migratory bird resting temporarily in an area.

visitation /ˌvɪzɪˈteɪʃ(ə)n/ *n.* **1** official visit of inspection. **2** trouble etc. seen as divine punishment. **3** (**Visitation**) **a** visit of the Virgin Mary to Elizabeth. **b** festival of this.

visitor *n.* **1** person who visits. **2** migrant bird staying for part of the year.

visitors' book *n.* book for visitors to a hotel, church, etc., to sign, make remarks in, etc.

visor /ˈvaɪzə(r)/ *n.* (also **vizor**) **1** movable part of a helmet covering the face. **2** shield for the eyes, esp. one at the top of a vehicle windscreen. [Anglo-French *viser*: related to VISAGE]

vista /ˈvɪstə/ *n.* **1** long narrow view as between rows of trees. **2** mental view of a long series of events. [Italian]

visual /ˈvɪʒʊəl/ *adj.* of or used in seeing. □ **visually** *adv.* [Latin *visus* sight]

visual aid *n.* film etc. as a teaching aid.

visual display unit *n. Computing* device displaying data on a screen.

visualize *v.* (also **-ise**) (**-zing** or **-sing**) imagine visually. □ **visualization** /-ˈzeɪʃ(ə)n/ *n.*

vital /ˈvaɪt(ə)l/ *–adj.* **1** of or essential to organic life (*vital functions*). **2** essential, indispensable (*of vital importance*). **3** full of life or activity. **4** fatal (*vital error*). *–n.* (in *pl.*) the body's vital organs, e.g. the heart and brain. □ **vitally** *adv.* [Latin *vita* life]

vitality /vaɪˈtælɪtɪ/ *n.* **1** liveliness, animation. **2** ability to survive or endure. [Latin: related to VITAL]

vitalize /ˈvaɪtəˌlaɪz/ *v.* (also **-ise**) (**-zing** or **-sing**) **1** endow with life. **2** make lively or vigorous. □ **vitalization** /-ˈzeɪʃ(ə)n/ *n.*

vital statistics *n.pl.* **1** *joc.* measurements of a woman's bust, waist, and hips. **2** the number of births, marriages, deaths, etc.

vitamin /ˈvɪtəmɪn/ *n.* any of various substances present in many foods and essential to health and growth (*vitamin A, B, C,* etc.). [Latin *vita* life, AMINE]

vitamin B complex *n.* any of a group of vitamins often found together in foods.

vitaminize /ˈvɪtəmɪˌnaɪz/ *v.* (also **-ise**) (**-zing** or **-sing**) add vitamins to.

vitiate /ˈvɪʃɪˌeɪt/ *v.* (**-ting**) **1** impair, debase. **2** make invalid or ineffectual. □ **vitiation** /-ˈeɪʃ(ə)n/ *n.* [Latin: related to VICE¹]

viticulture /ˈvɪtɪˌkʌltʃə(r)/ *n.* cultivation of grapes. [Latin *vitis* vine]

vitreous /ˈvɪtrɪəs/ *adj.* of or like glass. [Latin *vitrum* glass]

vitreous humour *n.* clear fluid in the eye between the lens and the retina.

vitrify /ˈvɪtrɪˌfaɪ/ *v.* (**-ies, -ied**) change into glass or a glasslike substance, esp. by heat. □ **vitrifaction** /-ˈfækʃ(ə)n/ *n.*

vitrification /-frˈkeɪʃ(ə)n/ *n.* [French or medieval Latin: related to VITREOUS]
vitriol /ˈvɪtrɪəl/ *n.* **1** sulphuric acid or a sulphate. **2** caustic or hostile speech or criticism. [Latin *vitrum*]
vitriolic /ˌvɪtrɪˈɒlɪk/ *adj.* caustic, hostile.
vituperate /vaɪˈtjuːpəˌreɪt/ *v.* (**-ting**) criticize abusively. □ **vituperation** /-ˈreɪʃ(ə)n/ *n.* **vituperative** /-rətɪv/ *adj.* [Latin]
viva[1] /ˈvaɪvə/ *colloq.* —*n.* (*pl.* **-s**) = VIVA VOCE. —*v.* (**vivas, vivaed, vivaing**) = VIVA-VOCE. [abbreviation]
viva[2] /ˈviːvə/ —*int.* long live. —*n.* cry of this as a salute etc. [Italian, = let live]
vivace /vɪˈvɑːtʃɪ/ *adv. Mus.* in a lively manner. [Latin: related to VIVACIOUS]
vivacious /vɪˈveɪʃəs/ *adj.* lively, animated. □ **vivacity** /vɪˈvæsɪtɪ/ *n.* [Latin *vivax* from *vivo* live]
vivarium /vaɪˈveərɪəm/ *n.* (*pl.* **-ria** or **-s**) **1** glass bowl etc. for keeping animals for scientific study. **2** enclosure for keeping animals in (nearly) their natural state. [Latin]
viva voce /ˌvaɪvə ˈvəʊtʃɪ/ —*adj.* oral. —*adv.* orally. —*n.* oral examination. —*v.* (**viva-voce**) (**-vocees, -voceed, -voceing**) examine orally. [medieval Latin, = with the living voice]
vivid /ˈvɪvɪd/ *adj.* **1** (of light or colour) strong, intense. **2** (of a memory, description, the imagination, etc.) clear, lively, graphic. □ **vividly** *adv.* **vividness** *n.* [Latin]
vivify /ˈvɪvɪˌfaɪ/ *v.* (**-ies, -ied**) enliven, animate, give life to. [French from Latin]
viviparous /vɪˈvɪpərəs/ *adj. Zool.* bringing forth young alive. [Latin *vivus* alive, *pario* produce]
vivisect /ˈvɪvɪˌsekt/ *v.* perform vivisection on.
vivisection /ˌvɪvɪˈsekʃ(ə)n/ *n.* surgical experimentation on living animals for scientific research. □ **vivisectional** *adj.* **vivisectionist** *n.* & *adj.* **vivisector** /ˈvɪvɪˌsektə(r)/ *n.* [Latin *vivus* living, DISSECTION]
vixen /ˈvɪks(ə)n/ *n.* **1** female fox. **2** spiteful woman. [Old English: related to FOX]
viz. /vɪz, or by substitution ˈneɪmlɪ/ *adv.* namely; that is to say; in other words. [abbreviation of VIDELICET, *z* = medieval Latin symbol for abbreviation of *-et*]
vizier /vɪˈzɪə(r)/ *n. hist.* high official in some Muslim countries. [ultimately from Arabic]
vizor var. of VISOR.
V-neck *n.* (often *attrib.*) V-shaped neckline on a pullover etc.
vocable /ˈvəʊkəb(ə)l/ *n.* word, esp. with reference to form not meaning. [Latin *voco* call]
vocabulary /vəˈkæbjʊlərɪ/ *n.* (*pl.* **-ies**) **1** words used by a particular language, book, branch of science, or author. **2** list of these, in alphabetical order with definitions or translations. **3** individual's stock of words (*limited vocabulary*). **4** set of artistic or stylistic forms or techniques. [medieval Latin: related to VOCABLE]
vocal /ˈvəʊk(ə)l/ —*adj.* **1** of or uttered by the voice. **2** outspoken (*very vocal about his rights*). —*n.* (in *sing.* or *pl.*) sung part or piece of music. □ **vocally** *adv.* [Latin: related to VOICE]
vocal cords *n.* voice-producing part of the larynx.
vocalist *n.* singer.
vocalize *v.* (also **-ise**) (**-zing** or **-sing**) **1** form (a sound) or utter (a word) with the voice. **2** articulate, express. □ **vocalization** /-ˈzeɪʃ(ə)n/ *n.*
vocation /vəʊˈkeɪʃ(ə)n/ *n.* **1 a** strong feeling of suitability for a particular career. **b** this regarded as a divine call to a career in the Church. **2** employment, trade, profession. □ **vocational** *adj.* [Latin *voco* call]
vocative /ˈvɒkətɪv/ *Gram.* —*n.* case of a noun used in addressing a person or thing. —*adj.* of or in this case.
vociferate /vəˈsɪfəˌreɪt/ *v.* (**-ting**) **1** utter noisily. **2** shout, bawl. □ **vociferation** /-ˈreɪʃ(ə)n/ *n.* **vociferator** *n.* [Latin: related to VOICE, *fero* bear]
vociferous /vəˈsɪfərəs/ *adj.* **1** noisy, clamorous. **2** insistently and forcibly outspoken. □ **vociferously** *adv.*
vodka /ˈvɒdkə/ *n.* alcoholic spirit distilled esp. in Russia from rye etc. [Russian]
vogue /vəʊg/ *n.* **1** (prec. by *the*) prevailing fashion. **2** (often *attrib.*) popular use (*had a great vogue*). □ **in vogue** in fashion. □ **voguish** *adj.* [French from Italian]
voice —*n.* **1 a** sound formed in the larynx and uttered by the mouth, esp. by a person speaking, singing, etc. **b** power of this (*lost her voice*). **2 a** use of the voice; spoken or written expression (esp. *give voice*). **b** opinion so expressed. **c** right to express an opinion. **d** medium for expression. **3** *Gram.* set of verbal forms showing whether a verb is active or passive. —*v.* (**-cing**) **1** express. **2** (esp. as **voiced** *adj.*) utter with vibration of the vocal cords (e.g. *b*, *d*). □ **in good voice** singing or speaking well or easily. **with one voice** unanimously. [Latin *vox voc-*]
voice-box *n.* larynx.

voice in the wilderness *n.* unheeded advocate of reform.
voiceless *adj.* **1** dumb, speechless. **2** uttered without vibration of the vocal cords (e.g. *f, p*).
voice-over *n.* commentary in a film etc. by an unseen narrator.
void *–adj.* **1** empty, vacant. **2** (of a contract etc.) invalid, not legally binding (*null and void*). *–n.* empty space, vacuum. *–v.* **1** render void. **2** excrete; empty (the bowels etc.). □ **void of** lacking, free from. [French]
voile /vɔɪl, vwɑ:l/ *n.* fine semi-transparent fabric. [French, = VEIL]
vol. *abbr.* volume.
volatile /'vɒlə,taɪl/ *adj.* **1** changeable in mood; fickle. **2** (of trading conditions etc.) unstable. **3** (of a political situation etc.) likely to erupt in violence. **4** *Chem.* evaporating rapidly. □ **volatility** /-'tɪlɪtɪ/ *n.* [Latin *volo* fly]
volatilize /və'lætɪ,laɪz/ *v.* (also **-ise**) (**-zing** or **-sing**) turn into vapour. □ **volatilization** /-'zeɪʃ(ə)n/ *n.*
vol-au-vent /'vɒləʊ,vɑ̃/ *n.* small puff pastry case with a savoury filling. [French, literally 'flight in the wind']
volcanic /vɒl'kænɪk/ *adj.* of, like, or from a volcano. □ **volcanically** *adv.*
volcano /vɒl'keɪnəʊ/ *n.* (*pl.* **-es**) **1** mountain or hill from which lava, steam, etc. escape through openings in the earth's crust. **2** volatile situation. [Latin *Volcanus* Vulcan, Roman god of fire]
vole *n.* small plant-eating rodent. [originally *vole-mouse* from Norwegian *voll* field]
volition /və'lɪʃ(ə)n/ *n.* act or power of willing. □ **of one's own volition** voluntarily. □ **volitional** *adj.* [Latin *volo* wish]
volley /'vɒlɪ/ *–n.* (*pl.* **-s**) **1 a** simultaneous firing of a number of weapons. **b** bullets etc. so fired. **2** (usu. foll. by *of*) torrent (of abuse etc.). **3** playing of a ball in tennis, football, etc., before it touches the ground. *–v.* (**-eys**, **-eyed**) return or send by or in a volley. [French *volée* from Latin *volo* fly]
volleyball *n.* game for two teams of six hitting a large ball by hand over a high net.
volt /vəʊlt/ *n.* SI unit of electromotive force, the difference of potential that would carry one ampere of current against one ohm resistance. [*Volta*, name of a physicist]
voltage *n.* electromotive force expressed in volts.
volte-face /vɒlt'fɑ:s/ *n.* sudden reversal of one's attitude or opinion. [French from Italian]
voltmeter *n.* instrument measuring electric potential in volts.
voluble /'vɒljʊb(ə)l/ *adj.* speaking or spoken fluently or at length. □ **volubility** /-'bɪlɪtɪ/ *n.* **volubly** *adv.* [Latin *volvo* roll]
volume /'vɒlju:m/ *n.* **1** single book forming part or all of a work. **2 a** solid content, bulk. **b** space occupied by a gas or liquid. **c** (foll. by *of*) amount or quantity. **3** strength of sound, loudness. **4** (foll. by *of*) **a** moving mass of water etc. **b** (usu. in *pl.*) mass of smoke etc. [Latin *volumen*: related to VOLUBLE, ancient books being in roll form]
volumetric /,vɒljʊ'metrɪk/ *adj.* of measurement by volume. □ **volumetrically** *adv.* [from VOLUME, METRIC]
voluminous /və'lu:mɪnəs/ *adj.* **1** (of drapery etc.) loose and ample. **2** written or writing at great length. [Latin: related to VOLUME]
voluntary /'vɒləntrɪ/ *–adj.* **1** acting, done, or given willingly; not compulsory; intentional. **2** unpaid (*voluntary work*). **3** (of an institution) supported by charity. **4** (of a school) built by a charity but maintained by a local education authority. **5** brought about by voluntary action. **6** (of a movement, muscle, or limb) controlled by the will. *–n.* (*pl.* **-ies**) organ solo played before or after a church service. □ **voluntarily** *adv.* **voluntariness** *n.* [Latin *voluntas* will]
volunteer /,vɒlən'tɪə(r)/ *–n.* person who voluntarily undertakes a task or enters military etc. service. *–v.* **1** (often foll. by *to* + infin.) undertake or offer (one's services, a remark, etc.) voluntarily. **2** (often foll. by *for*) be a volunteer. [French: related to VOLUNTARY]
voluptuary /və'lʌptʃʊərɪ/ *n.* (*pl.* **-ies**) person who seeks luxury and sensual pleasure. [Latin: related to VOLUPTUOUS]
voluptuous /və'lʌptʃʊəs/ *adj.* **1** of, tending to, occupied with, or derived from, sensuous or sensual pleasure. **2** (of a woman) curvaceous and sexually desirable. □ **voluptuously** *adv.* [Latin *voluptas* pleasure]
volute /və'lu:t/ *n.* spiral stonework scroll as an ornament of esp. Ionic capitals. [Latin *volvo -ut-* roll]
vomit /'vɒmɪt/ *–v.* (**-t-**) **1** eject (contents of the stomach) through the mouth; be sick. **2** (of a volcano, chimney, etc.) eject violently, belch forth. *–n.* matter vomited from the stomach. [Latin]
voodoo /'vu:du:/ *–n.* religious witchcraft as practised esp. in the W. Indies. *–v.* (**-doos**, **-dooed**) affect by voodoo; bewitch. [Dahomey]
voracious /və'reɪʃəs/ *adj.* **1** gluttonous, ravenous. **2** very eager (*voracious*

reader). □ **voraciously** *adv.* **voracity** /və'ræsɪtɪ/ *n.* [Latin *vorax* from *voro* devour]

vortex /'vɔːteks/ *n.* (*pl.* **-texes** or **-tices** /-tɪ,siːz/) **1** whirlpool, whirlwind. **2** whirling motion or mass. **3** thing viewed as destructive or devouring (*the vortex of society*). □ **vortical** *adj.* [Latin: related to VERTEX]

vorticist /'vɔːtɪsɪst/ *n.* futuristic painter, writer, etc., of a school based on the so-called 'vortices' of modern civilization. □ **vorticism** *n.*

votary /'vəʊtərɪ/ *n.* (*pl.* **-ies**; *fem.* **votaress**) (usu. foll. by *of*) **1** person dedicated to the service of a god or cult. **2** devotee of a person, occupation, etc. [Latin: related to VOTE]

vote –*n.* **1** formal expression of choice or opinion by a ballot, show of hands, etc., in an election etc. **2** (usu. prec. by *the*) right to vote, esp. in a State election. **3** opinion expressed by a vote (*vote of no confidence*). **4** votes given by or for a particular group (*the Welsh vote*; *the Labour vote*). –*v.* (**-ting**) **1** (often foll. by *for*, *against*) give a vote. **2 a** enact or resolve by a majority of votes. **b** grant (a sum of money) by vote. **3** *colloq.* pronounce by general consent. **4** (often foll. by *that*) suggest, urge. □ **vote down** defeat (a proposal etc.) in a vote. **vote in** elect by voting. **vote off** dismiss from (a committee etc.) by voting. **vote out** dismiss from office etc. by voting. **vote with one's feet** *colloq.* indicate an opinion by one's presence or absence. [Latin *votum* from *voveo vot-* vow]

voter *n.* person voting or with the right to vote at an election.

votive /'vəʊtɪv/ *adj.* offered or consecrated in fulfilment of a vow (*votive offering*). [Latin: related to VOTE]

vouch *v.* (foll. by *for*) answer for, be surety for (*will vouch for the truth of this*; *can vouch for him*). [French *vo(u)cher* summon, invoke]

voucher *n.* **1** document exchangeable for goods or services. **2** receipt. [from Anglo-French, or from VOUCH]

vouchsafe /vaʊtʃ'seɪf/ *v.* (**-fing**) *formal* **1** condescend to grant. **2** (foll. by *to* + infin.) condescend.

vow –*n.* solemn, esp. religious, promise (*monastic vows*; *marriage vows*). –*v.* **1** promise solemnly. **2** *archaic* declare solemnly. [French *vou(er)*: related to VOTE]

vowel /'vaʊəl/ *n.* **1** speech-sound made with vibration of the vocal cords but without audible friction. **2** letter(s) representing this, as *a*, *e*, *i*, *o*, *u*, *aw*, *ah*. [Latin: related to VOCAL]

vox pop *n.* (often *attrib.*) *colloq.* popular opinion as represented by informal comments from the public. [abbreviation of VOX POPULI]

vox populi /,vɒks 'pɒpjʊ,laɪ/ *n.* public opinion, popular belief. [Latin, = the people's voice]

voyage /'vɔɪɪdʒ/ –*n.* journey, esp. a long one by sea or in space. –*v.* (**-ging**) make a voyage. □ **voyager** *n.* [Latin VIATICUM]

voyeur /vwɑː'jɜː(r)/ *n.* **1** person who derives sexual pleasure from secretly observing others' sexual activity or organs. **2** (esp. covert) spectator. □ **voyeurism** *n.* **voyeuristic** /-'rɪstɪk/ *adj.* [French *voir* see]

vs. *abbr.* versus.

V-sign /'viːsaɪn/ *n.* **1** sign of the letter V made with the first two fingers pointing up and the back of the hand facing outwards, as a gesture of abuse etc. **2** similar sign made with the palm of the hand facing outwards, as a symbol of victory.

VSO *abbr.* Voluntary Service Overseas.

VSOP *abbr.* Very Special Old Pale (brandy).

VTO *abbr.* vertical take-off.

VTOL /'viːtɒl/ *abbr.* vertical take-off and landing.

VTR *abbr.* videotape recorder.

vulcanite /'vʌlkə,naɪt/ *n.* hard black vulcanized rubber. [related to VULCANIZE]

vulcanize /'vʌlkə,naɪz/ *v.* (also **-ise**) (**-zing** or **-sing**) treat (rubber etc.) with sulphur at a high temperature to strengthen it. □ **vulcanization** /-'zeɪʃ(ə)n/ *n.* [*Vulcan*: related to VOLCANO]

vulcanology /,vʌlkə'nɒlədʒɪ/ *n.* the study of volcanoes. □ **vulcanological** /-nə'lɒdʒɪk(ə)l/ *adj.* **vulcanologist** *n.*

vulgar /'vʌlgə(r)/ *adj.* **1 a** coarse; indecent; tasteless. **b** of or characteristic of the common people. **2** common; prevalent (*vulgar errors*). □ **vulgarly** *adv.* [Latin *vulgus* common people]

vulgar fraction *n.* fraction expressed by numerator and denominator, not decimally.

vulgarian /vʌl'geərɪən/ *n.* vulgar (esp. rich) person.

vulgarism *n.* vulgar word, expression, action, or habit.

vulgarity /vʌl'gærɪtɪ/ *n.* (*pl.* **-ies**) vulgar act, expression, or state.

vulgarize /'vʌlgə,raɪz/ *v.* (also **-ise**) (**-zing** or **-sing**) **1** make vulgar. **2** spoil by popularizing. □ **vulgarization** /-'zeɪʃ(ə)n/ *n.*

vulgar tongue *n.* (prec. by *the*) national or vernacular language.

Vulgate /ˈvʌlgeɪt/ *n.* 4th-c. Latin version of the Bible. [Latin: related to VULGAR]
vulnerable /ˈvʌlnərəb(ə)l/ *adj.* **1** easily wounded or harmed. **2** (foll. by *to*) exposed to damage, temptation, etc. □ **vulnerability** /-ˈbɪlɪtɪ/ *n.* **vulnerably** *adv.* [Latin *vulnus -er-* wound]
vulpine /ˈvʌlpaɪn/ *adj.* **1** of or like a fox. **2** crafty, cunning. [Latin *vulpes* fox]
vulture /ˈvʌltʃə(r)/ *n.* **1** large carrion-eating bird of prey, reputed to gather with others in anticipation of a death. **2** rapacious person. [Anglo-French from Latin]
vulva /ˈvʌlvə/ *n.* (*pl.* **-s**) external female genitals. [Latin]
vv. *abbr.* **1** verses. **2** volumes.
vying *pres. part.* of VIE.

W

W[1] /ˈdʌb(ə)ljuː/ *n.* (also **w**) (*pl.* **Ws** or **W's**) twenty-third letter of the alphabet.

W[2] *abbr.* (also **W.**) **1** watt(s). **2** West; Western.

W[3] *symb.* tungsten. [*wolframium*, Latinized name]

w. *abbr.* **1** wicket(s). **2** wide(s). **3** with.

WA *abbr.* Western Australia.

wacky *adj.* (**-ier**, **-iest**) *slang* crazy. [originally dial., = left-handed]

wad /wɒd/ *–n.* **1** lump of soft material used esp. to keep things apart or in place or to block a hole. **2** bundle of banknotes or documents. *–v.* (**-dd-**) **1** stop up or keep in place with a wad. **2** line, stuff, or protect with wadding. [origin uncertain]

wadding *n.* soft fibrous material used in quilt-making etc., or to pack fragile articles.

waddle /ˈwɒd(ə)l/ *–v.* (**-ling**) walk with short steps and a swaying motion. *–n.* waddling gait. [from WADE]

wade *–v.* (**-ding**) **1** walk through water, mud, etc., esp. with difficulty. **2** (foll. by *through*) go through (a tedious task, book, etc.). **3** (foll. by *into*) *colloq.* attack (a person or task) vigorously. *–n.* spell of wading. □ **wade in** *colloq.* make a vigorous attack or intervention. [Old English]

wader *n.* **1** long-legged water-bird that wades. **2** (in *pl.*) high waterproof boots.

wadi /ˈwɒdɪ/ *n.* (*pl.* **-s**) rocky watercourse in N. Africa etc., dry except in the rainy season. [Arabic]

wafer *n.* **1** very thin light crisp sweet biscuit. **2** disc of unleavened bread used in the Eucharist. **3** disc of red paper stuck on a legal document instead of a seal. [Anglo-French *wafre* from Germanic]

wafer-thin *adj.* very thin.

waffle[1] /ˈwɒf(ə)l/ *colloq.* *–n.* verbose but aimless or ignorant talk or writing. *–v.* (**-ling**) indulge in waffle. [dial., = yelp]

waffle[2] /ˈwɒf(ə)l/ *n.* small crisp batter cake. [Dutch]

waffle-iron *n.* utensil, usu. of two shallow metal pans hinged together, for baking waffles.

waft /wɒft/ *–v.* convey or travel easily and smoothly as through air or over water. *–n.* (usu. foll. by *of*) whiff or scent. [originally 'convoy (ship etc.)' from Dutch or Low German *wachter* from *wachten* to guard]

wag[1] *–v.* (**-gg-**) shake or wave to and fro. *–n.* single wagging motion (*with a wag of his tail*). □ **tongues wag** there is talk. [Old English]

wag[2] *n.* facetious person. [Old English]

wage *–n.* (in *sing.* or *pl.*) fixed regular payment to an employee, esp. a manual worker. *–v.* (**-ging**) carry on (a war etc.). [Anglo-French from Germanic]

waged *adj.* in regular paid employment.

wage-earner *n.* person who works for wages.

wager *n.* & *v.* = BET. [Anglo-French: related to WAGE]

waggish *adj.* playful, facetious. □ **waggishly** *adv.* **waggishness** *n.*

waggle /ˈwæg(ə)l/ *v.* (**-ling**) *colloq.* wag.

waggly *adj.* unsteady; waggling.

wagon /ˈwægən/ *n.* (also **waggon**) **1** four-wheeled vehicle for heavy loads. **2** railway vehicle, esp. an open truck. **3** tea trolley. □ **on the wagon** (or **water-wagon**) *slang* teetotal. [Dutch: related to WAIN]

wagoner *n.* (also **waggoner**) driver of a wagon.

wagon-load *n.* as much as a wagon can carry.

wagtail *n.* small bird with a long tail in frequent motion.

waif *n.* **1** homeless and helpless person, esp. an abandoned child. **2** ownerless object or animal. [Anglo-French, probably from Scandinavian]

waifs and strays *n.pl.* **1** homeless or neglected children. **2** odds and ends.

wail *–n.* **1** prolonged plaintive high-pitched cry of pain, grief, etc. **2** sound like this. *–v.* **1** utter a wail. **2** lament or complain persistently or bitterly. [Old Norse]

wain *n.* *archaic* wagon. [Old English]

wainscot /ˈweɪnskət/ *n.* boarding or wooden panelling on the lower part of a room-wall. [Low German *wagenschot* from *wagen* WAGON]

wainscoting *n.* **1** wainscot. **2** material for this.

waist *n.* **1 a** part of the human body below the ribs and above the hips; narrower middle part of the normal human figure. **b** circumference of this. **2** narrow middle of a violin, wasp, etc. **3 a** part of a garment encircling the waist. **b** *US* blouse, bodice. **4** part of a ship between the forecastle and the quarterdeck. □

waisted *adj.* (also in *comb.*). [Old English: related to WAX[2]]
waistband *n.* strip of cloth forming the waist of a garment.
waistcoat *n.* close-fitting waist-length garment without sleeves or collar, worn usu. over a shirt and under a jacket.
waist-deep *adj.* & *adv.* (also **waist-high**) up to the waist.
waistline *n.* outline or size of a person's body at the waist.
wait –*v.* **1 a** defer action or departure for a specified time or until some event occurs (*wait a minute*; *wait till I come*; *wait for a fine day*). **b** be expectant. **2** await (an opportunity, one's turn, etc.). **3** defer (a meal etc.) until a person's arrival. **4** (usu. as **waiting** *n.*) park a vehicle for a short time. **5** act as a waiter or attendant. **6** (foll. by *on, upon*) **a** await the convenience of. **b** serve as an attendant to. **c** pay a respectful visit to. –*n.* **1** period of waiting. **2** (usu. foll. by *for*) watching for an enemy (*lie in wait*). **3** (in *pl.*) *archaic* street singers of Christmas carols. □ **wait and see** await the progress of events. **wait up** (often foll. by *for*) not go to bed until a person arrives or an event happens. **you wait!** used to imply a threat, warning, etc. [Germanic: related to WAKE[1]]
waiter *n.* man who serves at table in a hotel or restaurant etc.
waiting game *n.* the delaying of action in order to have a greater effect later.
waiting-list *n.* list of people waiting for a thing not immediately available.
waiting-room *n.* room for people to wait in, esp. to see a doctor etc. or at a station.
waitress *n.* woman who serves at table in a hotel or restaurant etc.
waive *v.* (**-ving**) refrain from insisting on or using (a right, claim, opportunity, etc.). [Anglo-French *weyver*: related to WAIF]
waiver *n. Law* **1** waiving of a legal right etc. **2** document recording this.
wake[1] –*v.* (**-king**; *past* **woke** or **waked**; *past part.* **woken** or **waked**) **1** (often foll. by *up*) (cause to) cease to sleep. **2** (often foll. by *up*) (cause to) become alert or attentive. **3** *archaic* (except as **waking** *adj.* & *n.*) be awake (*waking hours*). **4** disturb with noise. **5** evoke (an echo). –*n.* **1** watch beside a corpse before burial; attendant lamentation and (less often) merrymaking. **2** (usu. in *pl.*) annual holiday in (industrial) northern England. [Old English]
wake[2] *n.* **1** track left on the water's surface by a moving ship. **2** turbulent air left behind a moving aircraft etc. □ **in the wake of** following, as a result of. [Low German from Old Norse]
wakeful *adj.* **1** unable to sleep. **2** (of a night etc.) sleepless. **3** vigilant. □ **wakefully** *adv.* **wakefulness** *n.*
waken *v.* make or become awake. [Old Norse]
wale *n.* **1** = WEAL[1]. **2** ridge on corduroy etc. **3** *Naut.* a broad thick timber along a ship's side. [Old English]
walk /wɔ:k/ –*v.* **1 a** progress by lifting and setting down each foot in turn, never having both feet off the ground at once. **b** (of a quadruped) go with the slowest gait. **2 a** travel or go on foot. **b** take exercise in this way. **3** traverse on foot at walking speed, tread the floor or surface of. **4** cause to walk with one (*walk the dog*). –*n.* **1 a** act of walking, the ordinary human gait. **b** slowest gait of an animal. **c** person's manner of walking. **2 a** distance which can be walked in a (usu. specified) time (*ten minutes' walk from here*). **b** excursion on foot. **3** place or track intended or suitable for walking. □ **walk all over** *colloq.* **1** defeat easily. **2** take advantage of. **walk away from 1** easily outdistance. **2** refuse to become involved with. **walk away with** *colloq.* = *walk off with*. **walk into** *colloq.* **1** encounter through unwariness. **2** get (a job) easily. **walk off with** *colloq.* **1** steal. **2** win easily. **walk on air** feel elated. **walk out 1** depart suddenly or angrily. **2** stop work in protest. **walk out on** desert, abandon. **walk the streets** be a prostitute. □ **walkable** *adj.* [Old English]
walkabout *n.* **1** informal stroll among a crowd by a visiting dignitary. **2** period of wandering in the bush by an Australian Aboriginal.
walker *n.* **1** person or animal that walks. **2 a** framework in which a baby can walk unaided. **b** = WALKING FRAME.
walkie-talkie /ˌwɔ:kɪˈtɔ:kɪ/ *n.* two-way radio carried on the person.
walk-in *attrib. adj.* (of a storage area) large enough to walk into.
walking *n.* & *adj.* in senses of WALK *n.*
walking frame *n.* tubular metal frame for disabled or old people to help them walk.
walking-stick *n.* stick carried for support when walking.
Walkman *n.* (*pl.* **-s**) *propr.* type of personal stereo.
walk of life *n.* occupation, profession.
walk-on *n.* **1** (in full **walk-on part**) non-speaking dramatic role. **2** player of this.
walk-out *n.* sudden angry departure, esp. as a protest or strike.
walk-over *n.* easy victory.

walkway *n.* passage or path (esp. raised) for walking along.

wall /wɔːl/ *–n.* **1** continuous vertical narrow structure of usu. brick or stone, esp. enclosing or dividing a space or supporting a roof. **2** thing like a wall, esp.: **a** a steep side of a mountain. **b** *Anat.* the outermost layer or enclosing membrane etc. of an organ etc. *–v.* **1** (esp. as **walled** *adj.*) surround with a wall. **2 a** (usu. foll. by *up*, *off*) block (a space etc.) with a wall. **b** (foll. by *up*) enclose within a sealed space. □ **go to the wall** be defeated or pushed aside. **up the wall** *colloq.* crazy or furious. **walls have ears** beware of eavesdroppers. □ **wall-less** *adj.* [Latin *vallum* rampart]

wallaby /'wɒləbɪ/ *n.* (*pl.* **-ies**) marsupial similar to but smaller than a kangaroo. [Aboriginal]

wallah /'wɒlə/ *n. slang* person concerned with or in charge of a usu. specified thing, business, etc. [Hindi]

wall bar *n.* one of a set of parallel bars attached to the wall of a gymnasium, on which exercises are performed.

wallet /'wɒlɪt/ *n.* small flat esp. leather case for holding banknotes etc. [Anglo-French]

wall-eye /'wɔːlaɪ/ *n.* **1** eye with a streaked or opaque white iris. **2** eye squinting outwards. □ **wall-eyed** *adj.* [Old Norse]

wallflower *n.* **1** fragrant spring garden plant. **2** *colloq.* woman sitting out a dance for lack of partners.

wall game *n.* form of football played at Eton.

Walloon /wɒ'luːn/ *–n.* **1** member of a people inhabiting S. and E. Belgium and neighbouring France. **2** French dialect spoken by this people. *–adj.* of or concerning the Walloons or their language. [medieval Latin *Wallo -onis*]

wallop /'wɒləp/ *colloq. –v.* (**-p-**) **1** thrash; beat. **2** (as **walloping** *adj.*) huge. *–n.* **1** heavy blow. **2** beer. [earlier senses 'gallop', 'boil', from French *waloper* from Germanic: cf. GALLOP]

wallow /'wɒləʊ/ *–v.* **1** (esp. of an animal) roll about in mud etc. **2** (usu. foll. by *in*) indulge in unrestrained pleasure, misery, etc. *–n.* **1** act of wallowing. **2** place used by buffalo etc. for wallowing. [Old English]

wallpaper *–n.* **1** paper for pasting on to interior walls as decoration. **2** usu. *derog.* trivial background noise, music, etc. *–v.* decorate with wallpaper.

wall-to-wall *adj.* **1** (of a carpet) fitted to cover a whole room etc. **2** *colloq.* ubiquitous (*wall-to-wall pop music*).

wally /'wɒlɪ/ *n.* (*pl.* **-ies**) *slang* foolish or inept person. [origin uncertain]

walnut /'wɔːlnʌt/ *n.* **1** tree with aromatic leaves and drooping catkins. **2** nut of this tree. **3** its timber. [Old English, = foreign nut]

walrus /'wɔːlrəs/ *n.* (*pl.* same or **-es**) large amphibious long-tusked arctic mammal. [Dutch]

walrus moustache *n.* long thick drooping moustache.

waltz /wɔːls/ *–n.* **1** ballroom dance in triple time performed by couples revolving with sliding steps. **2** music for this. *–v.* **1** dance a waltz. **2** (often foll. by *in*, *out*, *round*, etc.) *colloq.* move easily, lightly, casually, etc. □ **waltz off with** *colloq.* **1** steal. **2** win easily. [German *Walzer* from *walzen* revolve]

wampum /'wɒmpəm/ *n.* beads made from shells and strung together for use as money, decoration, etc. by N. American Indians. [Algonquian]

wan /wɒn/ *adj.* (**wanner**, **wannest**) pale; exhausted-looking. □ **wanly** *adv.* **wanness** *n.* [Old English, = dark]

wand /wɒnd/ *n.* **1** supposedly magic stick used by a fairy, magician, etc. **2** staff as a symbol of office. **3** *colloq.* conductor's baton. [Old Norse]

wander /'wɒndə(r)/ *v.* **1** (often foll. by *in*, *off*, etc.) go about from place to place aimlessly. **2 a** wind about; meander. **b** stray from a path etc. **3** talk or think incoherently; be inattentive or delirious. □ **wanderer** *n.* [Old English: related to WEND]

wandering Jew *n.* person who never settles down.

wanderlust *n.* eagerness for travelling or wandering; restlessness. [German]

wane *–v.* (**-ning**) **1** (of the moon) decrease in apparent size. **2** decrease in power, vigour, importance, size, etc. *–n.* process of waning. □ **on the wane** waning; declining. [Old English]

wangle /'wæŋg(ə)l/ *colloq. –v.* (**-ling**) (often *refl.*) contrive to obtain (a favour etc.). *–n.* act of wangling. [origin unknown]

wank *coarse slang –v.* masturbate. *–n.* act of masturbating. [origin unknown]

Wankel engine /'wæŋk(ə)l/ *n.* internal-combustion engine with a continuously rotated and eccentrically pivoted shaft. [*Wankel*, name of an engineer]

wanker *n. coarse slang* contemptible or ineffectual person.

wannabe /'wɒnəbɪ/ *n. slang* **1** avid fan who tries to emulate the person he or she admires. **2** anybody who would like to be someone else. [corruption of *want to be*]

want /wɒnt/ *–v.* **1 a** (often foll. by *to* + infin.) desire; wish for possession of; need (*wants a drink*; *wants it done immediately*). **b** require to be attended to; need (*garden wants weeding*). **c** (foll. by *to* + infin.) *colloq.* ought; should (*you want to be careful*). **2** (usu. foll. by *for*) lack; be deficient. **3** be without or fall short by. **4** (as **wanted** *adj.*) (of a suspected criminal etc.) sought by the police. *–n.* **1** (often foll. by *of*) lack, absence, or deficiency (*could not go for want of time*). **2** poverty; need. [Old Norse]

wanting *adj.* **1** lacking (in quality or quantity); not equal to requirements. **2** absent, not supplied.

wanton /'wɒnt(ə)n/ *–adj.* **1** licentious; sexually promiscuous. **2** capricious; arbitrary; motiveless (*wanton wind*; *wanton destruction*). **3** luxuriant; unrestrained (*wanton profusion*). *–n. literary* licentious person. □ **wantonly** *adv.* [from obsolete *wantowen*, = undisciplined]

wapiti /'wɒpɪtɪ/ *n.* (*pl.* **-s**) N. American deer. [a Cree word]

war /wɔ:(r)/ *–n.* **1 a** armed hostilities between esp. nations; conflict. **b** specific instance or period of this. **c** suspension of international law etc. during this. **2** hostility or contention between people, groups, etc. **3** (often foll. by *on*) sustained campaign against crime, poverty, etc. *–v.* (**-rr-**) **1** (as **warring** *adj.*) rival; fighting. **2** make war. □ **at war** (often foll. by *with*) engaged in a war. **go to war** declare or begin a war. **have been in the wars** *colloq.* appear injured etc. [Anglo-French from Germanic]

warble /'wɔ:b(ə)l/ *–v.* (**-ling**) **1** sing in a gentle trilling manner. **2** speak in a warbling manner. *–n.* warbled song or utterance. [French *werble(r)*]

warbler *n.* bird that warbles.

war crime *n.* crime violating the international laws of war. □ **war criminal** *n.*

war cry *n.* **1** phrase or name shouted to rally one's troops. **2** party slogan etc.

ward /wɔ:d/ *n.* **1** separate part of a hospital or room for a particular group of patients. **2** administrative division of a constituency. **3 a** minor under the care of a guardian or court. **b** (in full **ward of court**) minor or mentally deficient person placed under the protection of a court. **4** (in *pl.*) the corresponding notches and projections in a key and a lock. **5** *archaic* guardian's control. □ **ward off 1** parry (a blow). **2** avert (danger etc.). [Old English]

-ward *suffix* (also **-wards**) added to nouns of place or destination and to adverbs of direction and forming: **1** adverbs (usu. **-wards**) meaning 'towards' (*backwards*; *homewards*). **2** adjectives (usu. **-ward**) meaning 'turned or tending towards' (*downward*; *onward*). **3** (less commonly) nouns meaning 'the region towards or about' (*look to the eastward*). [Old English]

war dance *n.* dance performed by primitive peoples etc. before a battle or to celebrate victory.

warden /'wɔ:d(ə)n/ *n.* **1** (often in *comb.*) supervising official (*traffic warden*). **2** president or governor of a college, hospital, etc. [Anglo-French and French: related to GUARDIAN]

warder /'wɔ:də(r)/ *n.* (*fem.* **wardress**) prison officer. [French: related to GUARD]

wardrobe /'wɔ:drəʊb/ *n.* **1** large cupboard for storing clothes. **2** person's stock of clothes. **3** costume department of a theatre etc. [French]

wardrobe mistress *n.* (*masc.* **wardrobe master**) person in charge of a theatrical wardrobe.

wardroom *n.* mess in a warship for commissioned officers.

-wards var. of -WARD.

wardship *n.* tutelage.

ware *n.* **1** (esp. in *comb.*) things of a specified kind made usu. for sale (*chinaware*; *hardware*). **2** (usu. in *pl.*) articles for sale. **3** ceramics etc. of a specified kind (*Delft ware*). [Old English]

warehouse *–n.* **1** building in which goods are stored. **2** wholesale or large retail store. *–v.* /also -haʊz/ (**-sing**) store temporarily in a repository.

warfare *n.* waging war, campaigning.

war-game *n.* **1** military training exercise. **2** battle etc. conducted with toy soldiers.

warhead *n.* explosive head of a missile.

warhorse *n.* **1** *hist.* trooper's powerful horse. **2** *colloq.* veteran soldier, politician, etc.

warlike *adj.* **1** hostile. **2** soldierly. **3** military.

warlock /'wɔ:lɒk/ *n. archaic* sorcerer. [Old English, = traitor]

warlord *n.* military commander or commander-in-chief.

warm /wɔ:m/ *–adj.* **1** of or at a fairly high temperature. **2** (of clothes etc.) affording warmth. **3 a** sympathetic, friendly, loving. **b** hearty, enthusiastic. **4** *colloq. iron.* dangerous, difficult, hostile. **5** *colloq.* **a** (in a game) close to the object etc. sought. **b** near to guessing. **6** (of a colour etc.) reddish or yellowish;

suggestive of warmth. **7** *Hunting* (of a scent) fresh and strong. *—v.* **1** make warm. **2 a** (often foll. by *up*) warm oneself. **b** (often foll. by *to*) become animated or sympathetic. *—n.* **1** act of warming. **2** warmth of the atmosphere etc. □ **warm up 1** make or become warm. **2** prepare for a performance etc. by practising. **3** reach a temperature for efficient working. **4** reheat (food). □ **warmly** *adv.* **warmth** *n.* [Old English]

warm-blooded *adj.* **1** having blood temperature well above that of the environment. **2** ardent.

war memorial *n.* monument to those killed in a war.

warm-hearted *adj.* kind, friendly. □ **warm-heartedness** *n.*

warming-pan *n. hist.* container for live coals with a flat body and a long handle, used for warming a bed.

warmonger *n.* person who promotes war. □ **warmongering** *n.* & *adj.*

warm-up *n.* period of preparatory exercise.

warm work *n.* **1** work etc. that makes one warm through exertion. **2** dangerous conflict etc.

warn /wɔːn/ *v.* **1** (also *absol.*) **a** (often foll. by *of* or *that*) inform of danger, unknown circumstances, etc. **b** (foll. by *to* + infin.) advise (a person) to take certain action. **c** (often foll. by *against*) inform (a person etc.) about a specific danger. **2** (usu. with *neg.*) admonish. □ **warn off** tell (a person) to keep away (from). [Old English]

warning *n.* **1** in senses of WARN. **2** thing that warns. [Old English]

war of nerves *n.* attempt to wear down an opponent psychologically.

warp /wɔːp/ *—v.* **1 a** make or become distorted, esp. through heat, damp, etc. **b** make or become perverted or strange (*warped sense of humour*). **2** haul (a ship) by a rope attached to a fixed point. *—n.* **1 a** warped state, esp. of timber. **b** perversion of the mind. **2** lengthwise threads in a loom. **3** rope used in warping a ship. [Old English]

warpaint *n.* **1** paint used to adorn the body before battle, esp. by N. American Indians. **2** *colloq.* make-up.

warpath *n.* □ **on the warpath 1** (of N. American Indians) going to war. **2** *colloq.* seeking a confrontation.

warrant /ˈwɒrənt/ *—n.* **1** thing that authorizes an action. **2 a** written authorization, money voucher, etc. **b** written authorization allowing police to search premises, arrest a suspect, etc. **3** certificate of service rank held by a warrant-officer. *—v.* **1** serve as a warrant for; justify. **2** guarantee or attest to esp. the genuineness of. □ **I** (or **I'll**) **warrant** I am certain; no doubt. [French *warant*, from Germanic]

warrant-officer *n.* officer ranking between commissioned officers and NCOs.

warranty *n.* (*pl.* **-ies**) **1** undertaking as to the ownership or quality of a thing sold etc., often accepting responsibility for defects or repairs over a specified period. **2** (usu. foll. by *for* + verbal noun) authority or justification. [Anglo-French *warantie*: related to WARRANT]

warren /ˈwɒrən/ *n.* **1** network of rabbit burrows. **2** densely populated or labyrinthine building or district. [Anglo-French *warenne* from Germanic]

warring see WAR *v.*

warrior /ˈwɒrɪə(r)/ *n.* **1** person experienced or distinguished in fighting. **2** fighting man, esp. of primitive peoples. **3** (*attrib.*) martial (*warrior nation*). [French *werreior*: related to WAR]

warship *n.* ship used in war.

wart /wɔːt/ *n.* **1** small hard round growth on the skin. **2** protuberance on the skin of an animal, surface of a plant, etc. □ **warts and all** *colloq.* with no attempt to conceal blemishes. □ **warty** *adj.* [Old English]

wart-hog *n.* African wild pig.

wartime *n.* period during which a war is being waged.

wary /ˈweərɪ/ *adj.* (**-ier**, **-iest**) **1** on one's guard; circumspect. **2** (foll. by *of*) cautious. **3** showing caution. □ **warily** *adv.* **wariness** *n.* [*ware* look out for, avoid]

was *1st* & *3rd sing. past* of BE.

wash /wɒʃ/ *—v.* **1** cleanse with liquid, esp. water. **2** (foll. by *out*, *off*, *away*, etc.) **a** remove (a stain) in this way. **b** (of a stain etc.) be removed by washing. **3** wash oneself or one's hands and face. **4** wash clothes, dishes, etc. **5** (of fabric or dye) bear washing without damage. **6** (of an argument etc.) stand scrutiny; be believed or acceptable. **7** (of a river, waters, etc.) touch. **8** (of liquid) carry along in a specified direction (*was washed overboard*; *washed up on the shore*). **9** (foll. by *over*, *along*, etc.) sweep, move, or splash. **10** (foll. by *over*) occur all around without greatly affecting (a person). **11** sift (ore) by the action of water. **12** brush watery paint or ink over. **13** *poet.* moisten, water. *—n.* **1 a** washing or being washed. **b** (prec. by *the*) a laundry etc. (*sent them to the wash*). **2** clothes etc. for washing or just washed. **3** motion of agitated water or air, esp. from the passage of a ship etc. or aircraft. **4** kitchen slops given to pigs. **5 a** thin, weak, or inferior liquid food. **b** liquid food for animals. **6** liquid to spread over a surface to cleanse, heal, or colour. **7** thin coating of water-colour. □ **come out in the wash** *colloq.* be resolved in the course of time. **wash one's dirty linen in public** let private quarrels or difficulties become generally known. **wash down 1** wash completely. **2** (usu. foll. by *with*) accompany or follow (food) with a drink. **wash one's hands of** renounce responsibility for. **wash out 1** clean the inside of by washing. **2** clean (a garment etc.) by brief washing. **3** *colloq.* rain off. **4** = sense 2 of *v.* **wash up 1** (also *absol.*) wash (dishes etc.) after use. **2** *US* wash one's face and hands. **3** carry on to a shore. □ **washable** *adj.* [Old English]

wash-basin *n.* plumbed-in basin for washing one's hands etc.

washboard *n.* **1** ribbed board on which clothes are scrubbed. **2** this as a percussion instrument.

washed out *adj.* (also **washed-out**) **1** faded; pale. **2** *colloq.* pale, exhausted.

washed up *adj.* (also **washed-up**) esp. *US slang* defeated, having failed.

washer *n.* **1** person or machine that washes. **2** flat ring inserted at a joint to tighten it and prevent leakage or under the head of a screw etc., or under a nut, to disperse its pressure.

washer-up *n.* (*pl.* **washers-up**) (also **washer-upper**) person who washes up dishes etc.

washerwoman *n.* laundress.

washing *n.* clothes etc. for washing or just washed.

washing-machine *n.* machine for washing clothes.

washing-powder *n.* soap powder or detergent for washing clothes.

washing-soda *n.* sodium carbonate, used dissolved in water for washing and cleaning.

washing-up *n.* **1** process of washing dishes etc. **2** used dishes etc. for washing.

wash-out *n.* *colloq.* complete failure, non-event.

washroom *n.* esp. *US* public toilet.

washstand *n.* piece of furniture to hold a basin, jug, soap, etc.

washy *adj.* (**-ier, -iest**) **1** too watery or weak. **2** lacking vigour or intensity. □ **washily** *adv.* **washiness** *n.*

wasn't /'wɒz(ə)nt/ *contr.* was not.

Wasp /wɒsp/ *n.* (also **WASP**) *US* usu. *derog.* middle-class American White Protestant. [*W*hite *A*nglo-*S*axon *P*rotestant]

wasp /wɒsp/ *n.* stinging insect with black and yellow stripes. [Old English]

waspish *adj.* irritable, snappish.

wasp-waist *n.* very slender waist.

wassail /'wɒseɪl, -s(ə)l/ *archaic* —*n.* festive occasion; drinking-bout. —*v.* make merry. [Old Norse *ves heill* be in health: related to WHOLE]

wastage /'weɪstɪdʒ/ *n.* **1** amount wasted. **2** loss by use, wear, or leakage. **3** (also **natural wastage**) loss of employees other than by redundancy.

waste —*v.* (**-ting**) **1** use to no purpose or with inadequate result or extravagantly. **2** fail to use (esp. an opportunity). **3** (often foll. by *on*) **a** give (advice etc.) without effect. **b** (often in *passive*) fail to be appreciated or used properly (*she was wasted on him*; *feel wasted in this job*). **4** wear gradually away; make or become weak. **5** devastate. —*adj.* **1** superfluous; no longer needed. **2** not inhabited or cultivated. —*n.* **1** act of wasting. **2** waste material. **3** waste region. **4** being used up; diminution by wear. **5** = WASTE PIPE. □ **go** (or **run**) **to waste** be wasted. [Latin: related to VAST]

waste disposal unit *n.* device fitted to a sink etc. for disposing of household waste.

wasteful *adj.* **1** extravagant. **2** causing or showing waste. □ **wastefully** *adv.*

wasteland *n.* **1** unproductive or useless area of land. **2** place or time considered spiritually or intellectually barren.

waste paper *n.* used or valueless paper.

waste-paper basket *n.* receptacle for waste paper.

waste pipe *n.* pipe to carry off waste material.

waste product *n.* useless by-product of manufacture or of an organism.

waster *n.* **1** wasteful person. **2** *colloq.* wastrel.

wastrel /'weɪstr(ə)l/ *n.* good-for-nothing person.

watch /wɒtʃ/ —*v.* **1** keep the eyes fixed on. **2** keep under observation; follow observantly. **3** (often foll. by *for*) be in an alert state; be vigilant. **4** (foll. by *over*) look after; take care of. —*n.* **1** small portable timepiece for carrying on the wrist or in a pocket. **2** state of alert or constant observation or attention. **3** *Naut.* **a** usu. four-hour spell of duty. **b** (in full **starboard** or **port watch**) each of the halves into which a ship's crew is divided to take alternate watches. **4** *hist.* watchman or watchmen. □ **on the watch for** waiting for (an anticipated occurrence). **watch it** (or **oneself**) *colloq.* be careful. **watch out** (often foll. by *for*) be on one's guard. □ **watcher** *n.* (also in *comb.*). [Old English: related to WAKE[1]]

watchdog *n.* **1** dog guarding property etc. **2** person or body monitoring others' rights etc.
watchful *adj.* **1** accustomed to watching, alert. **2** on the watch. □ **watchfully** *adv.* **watchfulness** *n.*
watching brief *n.* brief of a barrister who follows a case for a client not directly concerned.
watchmaker *n.* person who makes and repairs watches and clocks.
watchman *n.* man employed to look after an empty building etc. at night.
watch-night service *n.* religious service held on the last night of the year.
watch-tower *n.* tower for keeping watch from.
watchword *n.* phrase summarizing a guiding principle.
water /'wɔ:tə(r)/ —*n.* **1** colourless transparent liquid compound of oxygen and hydrogen. **2** liquid consisting chiefly of this and found in seas and rivers, in rain, and in secretions of organisms. **3** expanse of water; a sea, lake, river, etc. **4** (in *pl.*) part of a sea or river. **5** (often as **the waters**) mineral water at a spa etc. **6** state of a tide. **7** solution of a specified substance in water (*lavender-water*). **8** transparency and brilliance of a gem. **9** (*attrib.*) **a** found in or near water. **b** of, for, or worked by water. **c** involving, using, or yielding water. **10** (usu. in *pl.*) amniotic fluid, released during labour. —*v.* **1** sprinkle or soak with water. **2** supply (a plant) with water. **3** give water to (an animal). **4** secrete water. **5** (as **watered** *adj.*) (of silk etc.) having irregular wavy glossy markings. **6** take in a supply of water. □ **by water** using a ship etc. for transport. **like water** in great quantity, profusely. **make one's mouth water** cause one's saliva to flow, stimulate one's appetite or anticipation. **of the first water** of the finest quality or extreme degree. **water down 1** dilute. **2** make less forceful or horrifying. **water under the bridge** past events accepted as irrevocable. [Old English]
Water-bearer var. of WATER-CARRIER.
water-bed *n.* mattress filled with water.
water-biscuit *n.* thin crisp unsweetened biscuit.
water-buffalo *n.* common domestic Indian buffalo.
water bus *n.* boat carrying passengers on a regular run on a river, lake, etc.
water-cannon *n.* device using a jet of water to disperse a crowd etc.
Water-carrier *n.* (also **Water-bearer**) (prec. by *the*) zodiacal sign or constellation Aquarius.
water chestnut *n.* corm from a sedge, used in Chinese cookery.
water-clock *n.* clock measuring time by the flow of water.
water-closet *n.* lavatory that can be flushed.
water-colour *n.* (*US* **water-color**) **1** artists' paint made of pigment to be diluted with water and not oil. **2** picture painted with this. **3** art of painting with water-colours. □ **water-colourist** *n.*
water-cooled *adj.* cooled by the circulation of water.
watercourse *n.* **1** brook or stream. **2** bed of this.
watercress *n.* pungent cress growing in running water and used in salad.
water-diviner *n.* dowser.
waterfall *n.* stream flowing over a precipice or down a steep hillside.
waterfowl *n.* (usu. collect. as *pl.*) birds frequenting water.
waterfront *n.* part of a town adjoining a river etc.
water-glass *n.* solution of sodium or potassium silicate used esp. for preserving eggs.
water-hammer *n.* knocking noise in a water-pipe when a tap is suddenly turned off.
water-hole *n.* shallow depression in which water collects.
water-ice *n.* flavoured and frozen water and sugar etc.
watering-can *n.* portable container with a long spout, for watering plants.
watering-hole *n.* **1** = WATERING-PLACE 1. **2** *slang* bar.
watering-place *n.* **1** pool from which animals regularly drink. **2** spa or seaside resort.
water jump *n.* jump over water in a steeplechase etc.
water-level *n.* **1 a** surface of the water in a reservoir etc. **b** height of this. **2** level below which the ground is saturated with water. **3** level using water to determine the horizontal.
water lily *n.* aquatic plant with floating leaves and flowers.
water-line *n.* line along which the surface of water touches a ship's side.
waterlogged *adj.* saturated or filled with water.
Waterloo /,wɔ:tə'lu:/ *n.* decisive defeat or contest. [*Waterloo* in Belgium, where Napoleon was defeated]
water main *n.* main pipe in a water-supply system.
waterman *n.* **1** boatman plying for hire. **2** oarsman as regards skill in keeping the boat balanced.

watermark –*n.* faint design in some paper identifying the maker etc. –*v.* mark with this.

water-meadow *n.* meadow periodically flooded by a stream.

water melon *n.* large dark-green melon with red pulp and watery juice.

water-mill *n.* mill worked by a water-wheel.

water-pistol *n.* toy pistol shooting a jet of water.

water polo *n.* game played by swimmers, with a ball like a football.

water-power *n.* mechanical force derived from the weight or motion of water.

waterproof –*adj.* impervious to water. –*n.* waterproof garment or material. –*v.* make waterproof.

water-rat *n.* = WATER-VOLE.

water-rate *n.* charge made for the use of the public water-supply.

watershed *n.* **1** line of separation between waters flowing to different rivers, basins, etc. **2** turning-point in affairs. [from *shed* ridge]

waterside *n.* edge of a sea, lake, or river.

water-ski –*n.* each of a pair of skis for skimming the surface of the water when towed by a motor boat. –*v.* travel on water-skis. □ **water-skier** *n.*

water-softener *n.* apparatus for softening hard water.

waterspout *n.* gyrating column of water and spray between sea and cloud.

water table *n.* = WATER-LEVEL 2.

watertight *adj.* **1** closely fastened or fitted so as to prevent the passage of water. **2** (of an argument etc.) unassailable.

water-tower *n.* tower with an elevated tank to give pressure for distributing water.

water-vole *n.* aquatic vole.

waterway *n.* navigable channel.

water-wheel *n.* wheel driven by water to work machinery, or to raise water.

water-wings *n.pl.* inflated floats fixed on the arms of a person learning to swim.

waterworks *n.* **1** establishment for managing a water-supply. **2** *colloq.* shedding of tears. **3** *colloq.* urinary system.

watery *adj.* **1** containing too much water. **2** too thin in consistency. **3** of or consisting of water. **4** vapid, uninteresting. **5** (of colour) pale. **6** (of the sun, moon, or sky) rainy-looking. **7** (of eyes) moist; tearful. □ **wateriness** *n.*

watt /wɒt/ *n.* SI unit of power, equivalent to one joule per second, corresponding to the rate of energy in an electric circuit where the potential difference is one volt and the current one ampere. [*Watt*, name of an engineer]

wattage *n.* amount of electrical power expressed in watts.

watt-hour *n.* energy used when one watt is applied for one hour.

wattle[1] /ˈwɒt(ə)l/ *n.* **1** structure of interlaced rods and sticks used for fences etc. **2** Australian acacia with pliant branches and golden flowers used as the national emblem. [Old English]

wattle[2] /ˈwɒt(ə)l/ *n.* fleshy appendage on the head or throat of a turkey or other birds. [origin unknown]

wattle and daub *n.* network of rods and twigs plastered with clay or mud as a building material.

wave –*v.* (**-ving**) **1 a** (often foll. by *to*) move a hand etc. to and fro in greeting or as a signal. **b** move (a hand etc.) in this way. **2 a** show a sinuous or sweeping motion as of a flag, tree, corn, etc. in the wind. **b** impart a waving motion to. **3** direct (a person) by waving (*waved them away*; *waved them to follow*). **4** express (a greeting etc.) by waving. **5** give an undulating form to (hair etc.). **6** (of hair etc.) have such a form. –*n.* **1** ridge of water between two depressions. **2** long body of water curling into an arch and breaking on the shore. **3** thing compared to this, e.g. a body of persons in one of successive advancing groups. **4** gesture of waving. **5 a** process of waving the hair. **b** undulating form produced by this. **6** temporary occurrence or increase of a condition or influence (*wave of enthusiasm*; *heat wave*). **7** *Physics* **a** disturbance of the particles of esp. a fluid medium for the propagation or direction of motion, heat, light, sound, etc. **b** single curve in this motion. **8** undulating line or outline. □ **make waves** *colloq.* cause trouble. **wave aside** dismiss as intrusive or irrelevant. **wave down** wave to (a vehicle or driver) to stop. [Old English]

waveband *n.* range of radio wavelengths between certain limits.

wave-form *n.* *Physics* curve showing the shape of a wave at a given time.

wavelength *n.* **1** distance between successive crests of a wave. **2** this as a distinctive feature of radio waves from a transmitter. **3** *colloq.* particular mode or range of thought.

wavelet *n.* small wave.

wave machine *n.* device at a swimming-pool producing waves.

waver *v.* **1** be or become unsteady; begin to give way. **2** be irresolute. **3** (of a light) flicker. [Old Norse: related to WAVE]

wavy *adj.* (**-ier, -iest**) having waves or alternate contrary curves. □ **waviness** *n.*

wax[1] –*n.* **1** sticky plastic yellowish substance secreted by bees as the material of honeycomb. **2** this bleached and purified, used for candles, modelling, etc. **3** any similar substance, e.g. the yellow substance secreted by the ear. –*v.* **1** cover or treat with wax. **2** remove unwanted hair from (legs etc.) using wax. □ **waxy** *adj.* (**-ier, -iest**). [Old English]

wax[2] *v.* **1** (of the moon) increase in apparent size. **2** become larger or stronger. **3** pass into a specified state or mood (*wax lyrical*). □ **wax and wane** undergo alternate increases and decreases. [Old English]

waxen *adj.* **1** smooth or pale like wax. **2** *archaic* made of wax.

waxwing *n.* any of various birds with tips like red sealing-wax to some wing-feathers.

waxwork *n.* **1** object, esp. a lifelike dummy, modelled in wax. **2** (in *pl.*) exhibition of wax dummies.

way –*n.* **1** road, track, path, etc., for passing along. **2** course or route for reaching a place (*asked the way to London*; *the way out*). **3** method or plan for attaining an object. **4** style, manner (*I like the way you dress*). **5** person's chosen or habitual course of action. **6** normal course of events (*that is always the way*). **7** travelling distance; length traversed or to be traversed. **8** unimpeded opportunity or space to advance (*make way*). **9** advance in some direction; impetus, progress (*under way*). **10** being engaged in movement from place to place; time spent in this (*on the way home*). **11** specified direction (*step this way*). **12** *colloq.* scope or range. **13** line of occupation or business. **14** specified condition or state (*things are in a bad way*). **15** respect (*is useful in some ways*). **16** (in *pl.*) part into which a thing is divided (*split it three ways*). **17** (in *pl.*) structure of timber etc. down which a new ship is launched. –*adv. colloq.* far (*way off*). □ **by the way** incidentally. **by way of 1** by means of. **2** as a form of. **3** passing through. **come one's way** become available to one. **get out of the** (or **my** etc.) **way** stop obstructing a person. **go out of one's way** make a special effort. **in a way** to some extent. **in the** (or **one's**) **way** forming an obstruction. **lead the way** act as guide or leader. **look the other way** ignore what one should notice. **on the** (or **one's**) **way 1** in the course of a journey etc. **2** having progressed. **3** *colloq.* (of a child) conceived but not yet born. **on the way out** *colloq.* going out of fashion or favour. **out of the way 1** no longer an obstacle. **2** disposed of. **3** unusual. **4** (of a place) remote. [Old English]

way back *adv. colloq.* long ago.

waybill *n.* list of passengers or parcels on a vehicle.

wayfarer *n.* traveller, esp. on foot. □ **wayfaring** *n.* & *adj.*

waylay *v.* (*past* and *past part.* **waylaid**) **1** lie in wait for. **2** stop to talk to or rob.

way of life *n.* principles or habits governing all one's actions etc.

way-out *adj. colloq.* unusual; eccentric.

-ways *suffix* forming adjectives and adverbs of direction or manner (*sideways*).

ways and means *n.pl.* **1** methods of achieving something. **2** methods of raising government revenue.

wayside *n.* **1** side of a road. **2** land at the side of a road.

wayward *adj.* childishly self-willed; capricious. □ **waywardness** *n.* [from AWAY, -WARD]

Wb *abbr.* weber(s).

WC *abbr.* **1** water-closet. **2** West Central.

W/Cdr. *abbr.* wing commander.

we /wi:, wɪ/ *pron.* (*obj.* **us**; *poss.* **our, ours**) **1** *pl.* of I[2]. **2** used for or by a royal person in a proclamation etc. or by an editor etc. in a formal context. [Old English]

WEA *abbr.* Workers' Educational Association.

weak *adj.* **1** deficient in strength, power, vigour, resolution, or number. **2** unconvincing. **3** *Gram.* (of a verb) forming inflections by the addition of a suffix to the stem. □ **weakish** *adj.* [Old Norse]

weaken *v.* make or become weak or weaker.

weak-kneed *adj. colloq.* lacking resolution.

weakling *n.* feeble person or animal.

weakly –*adv.* in a weak manner. –*adj.* (**-ier, -iest**) sickly, not robust.

weak-minded *adj.* **1** mentally deficient. **2** lacking in resolution.

weak moment *n.* time when one is unusually compliant or susceptible.

weakness *n.* **1** being weak. **2** weak point. **3** (foll. by *for*) self-indulgent liking (*weakness for chocolate*).

weak point *n.* (also **weak spot**) **1** place where defences are assailable. **2** flaw in an argument or character or in resistance to temptation.

weal[1] –*n.* ridge raised on the flesh by a stroke of a rod or whip. –*v.* mark with a weal. [var. of WALE]

weal[2] *n. literary* welfare. [Old English]

wealth /welθ/ *n.* **1** riches. **2** being rich. **3** (foll. by *of*) abundance. [Old English]

wealthy *adj.* (**-ier, -iest**) having an abundance, esp. of money.

wean *v.* **1** accustom (an infant or other young mammal) to food other than (esp. its mother's) milk. **2** (often foll. by *from, away from*) disengage (from a habit etc.) by enforced discontinuance. [Old English, = accustom]

weapon /'wepən/ *n.* **1** thing designed, used, or usable for inflicting bodily harm. **2** means for gaining the advantage in a conflict. [Old English]

weaponry *n.* weapons collectively.

wear /weə(r)/ *–v.* (*past* **wore**; *past part.* **worn**) **1** have on one's person as clothing or an ornament etc. **2** exhibit or present (a facial expression etc.) (*wore a frown*). **3** *colloq.* (usu. with *neg.*) tolerate. **4** (often foll. by *away, down*) **a** injure the surface of, or partly obliterate or alter, by rubbing, stress, or use. **b** undergo such injury or change. **5** (foll. by *off, away*) rub or be rubbed off. **6** make (a hole etc.) by constant rubbing or dripping etc. **7** (often foll. by *out*) exhaust. **8** (foll. by *down*) overcome by persistence. **9** (foll. by *well* etc.) endure continued use or life. **10** (of time) pass, esp. tediously. **11** (of a ship) fly (a flag). *–n.* **1** wearing or being worn. **2** things worn; fashionable or suitable clothing (*sportswear; footwear*). **3** (in full **wear and tear**) damage from continuous use. □ **wear one's heart on one's sleeve** show one's feelings openly. **wear off** lose effectiveness or intensity. **wear out** **1** use or be used until useless. **2** tire or be tired out. **wear thin** (of patience, excuses, etc.) begin to fail. **wear the trousers** see TROUSERS. □ **wearer** *n.* [Old English]

wearisome /'wɪərɪsəm/ *adj.* tedious; tiring by monotony or length.

weary /'wɪərɪ/ *–adj.* (**-ier, -iest**) **1** very tired after exertion or endurance. **2** (foll. by *of*) no longer interested in, tired of. **3** tiring, tedious. *–v.* (**-ies, -ied**) make or grow weary. □ **wearily** *adv.* **weariness** *n.* [Old English]

weasel /'wi:z(ə)l/ *n.* small flesh-eating mammal related to the stoat and ferret. [Old English]

weasel word *n.* (usu. in *pl.*) word that is intentionally ambiguous or misleading.

weather /'weðə(r)/ *–n.* **1** state of the atmosphere at a place and time as regards heat, cloudiness, dryness, sunshine, wind, and rain etc. **2** (*attrib.*) *Naut.* windward. *–v.* **1** expose to or affect by atmospheric changes; season (wood). **2** be discoloured or worn in this way. **3 a** come safely through (a storm). **b** survive (a difficult period etc.). **4** get to the windward of (a cape etc.). □ **keep a weather eye open** be watchful. **make heavy weather of** *colloq.* exaggerate the difficulty presented by. **under the weather** *colloq.* indisposed. [Old English]

weather-beaten *adj.* affected by exposure to the weather.

weatherboard *n.* **1** sloping board attached to the bottom of an outside door to keep out the rain etc. **2** each of a series of overlapping horizontal boards on a wall. □ **weatherboarding** *n.* (in sense 2 of *n.*).

weathercock *n.* **1** weather-vane in the form of a cock. **2** inconstant person.

weather forecast *n.* assessment of likely weather.

weatherman *n.* meteorologist, esp. one who broadcasts a weather forecast.

weatherproof *adj.* resistant to the effects of bad weather, esp. rain.

weather-vane *n.* **1** revolving pointer on a church spire etc. to show the direction of the wind. **2** inconstant person.

weave[1] *–v.* (**-ving**; *past* **wove**; *past part.* **woven** or **wove**) **1 a** form (fabric) by interlacing long threads in two directions. **b** form (thread) into fabric in this way. **2** make fabric in this way. **3 a** (foll. by *into*) make (facts etc.) into a story or connected whole. **b** make (a story) in this way. *–n.* style of weaving. [Old English]

weave[2] *v.* (**-ving**) move repeatedly from side to side; take an intricate course to avoid obstructions. □ **get weaving** *slang* begin action; hurry. [Old Norse: related to WAVE]

weaver *n.* **1** person who weaves fabric. **2** (in full **weaver-bird**) tropical bird building elaborately woven nests.

web *n.* **1 a** woven fabric. **b** amount woven in one piece. **2** complex series (*web of lies*). **3** cobweb, gossamer, or a similar product of a spinning creature. **4** membrane between the toes of a swimming animal or bird. **5** large roll of paper used in printing. **6** thin flat connecting part in machinery etc. □ **webbed** *adj.* [Old English]

webbing *n.* strong narrow closely-woven fabric used for belts etc.

weber /'veɪbə(r)/ *n.* the SI unit of magnetic flux. [*Weber*, name of a physicist]

web-footed *adj.* having the toes connected by webs.

Wed. *abbr.* (also **Weds.**) Wednesday.

wed *v.* (**-dd-**; *past* and *past part.* **wedded** or **wed**) **1** usu. *formal* or *literary* marry. **2** unite. **3** (as **wedded** *adj.*) of or in marriage (*wedded bliss*). **4** (as **wedded** *adj.*) (foll. by *to*) obstinately attached or devoted to (a pursuit etc.). [Old English, = pledge]

we'd /wi:d, wɪd/ *contr.* **1** we had. **2** we should; we would.

wedding /'wedɪŋ/ *n.* marriage ceremony. [Old English: related to WED]

wedding breakfast *n.* meal etc. between a wedding and departure for the honeymoon.

wedding cake *n.* rich iced cake served at a wedding reception.

wedding ring *n.* ring worn by a married person.

wedge –*n.* **1** piece of tapering wood or metal etc. driven between two objects or parts to secure or separate them. **2** anything resembling a wedge. **3** golf club with a wedge-shaped head. –*v.* (**-ging**) **1** secure or fasten with a wedge. **2** force open or apart with a wedge. **3** (foll. by *in, into*) pack or thrust (a thing or oneself) tightly in or into. □ **thin end of the wedge** *colloq.* thing of little importance in itself, but likely to lead to more serious developments. [Old English]

Wedgwood /'wedʒwʊd/ *n. propr.* **1** a kind of fine stoneware usu. with a white cameo design. **2** its characteristic blue colour. [*Wedgwood*, name of a potter]

wedlock /'wedlɒk/ *n.* the married state. □ **born in** (or **out of**) **wedlock** born of married (or unmarried) parents. [Old English, = marriage vow]

Wednesday /'wenzdeɪ/ –*n.* day of the week following Tuesday. –*adv. colloq.* **1** on Wednesday. **2** (**Wednesdays**) on Wednesdays; each Wednesday. [Old English]

Weds. *abbr.* var. of WED.

wee[1] *adj.* (**weer** /'wi:ə(r)/; **weest** /'wi:ɪst/) **1** esp. *Scot.* little. **2** *colloq.* tiny. [Old English]

wee[2] *n. colloq.* = WEE-WEE.

weed –*n.* **1** wild plant growing where it is not wanted. **2** thin weak-looking person or horse. **3** (prec. by *the*) *slang* **a** marijuana. **b** tobacco. –*v.* **1 a** clear (an area) of weeds. **b** remove unwanted parts from. **2** (foll. by *out*) **a** sort out and remove (inferior or unwanted parts etc.). **b** rid of inferior parts, unwanted members, etc. **3** cut off or uproot weeds. [Old English]

weed-killer *n.* chemical used to destroy weeds.

weeds /wi:dz/ *n.pl.* (in full **widow's weeds**) *archaic* deep mourning worn by a widow. [Old English, = garment]

weedy *adj.* (**-ier, -iest**) **1** weak, feeble. **2** having many weeds.

week *n.* **1** period of seven days reckoned usu. from midnight on Saturday. **2** any period of seven days. **3** the six days between Sundays. **4 a** the five days Monday to Friday. **b** time spent working in this period (*35-hour week*; *three-day week*). [Old English]

weekday *n.* day other than Sunday or Saturday and Sunday.

weekend *n.* **1** Sunday and Saturday or part of Saturday. **2** this period extended slightly esp. for a holiday or visit etc.

weekender *n.* person who spends the weekend away from home; weekend visitor.

weekly –*adj.* done, produced, or occurring once a week. –*adv.* once a week. –*n.* (*pl.* **-ies**) weekly newspaper or periodical.

weeny /'wi:nɪ/ *adj.* (**-ier, -iest**) *colloq.* tiny. [from WEE[1]]

weep –*v.* (*past* and *past part.* **wept**) **1** shed tears. **2** (often foll. by *for*) bewail, lament over. **3 a** be covered with or send forth drops. **b** come or send forth in drops; exude liquid. **4** (as **weeping** *adj.*) (of a tree) having drooping branches. –*n.* spell of weeping. [Old English]

weepie *n. colloq.* sentimental or emotional film, play, etc.

weepy *adj.* (**-ier, -iest**) *colloq.* inclined to weep; tearful.

weevil /'wi:vɪl/ *n.* destructive beetle feeding esp. on grain. [Low German]

wee-wee /'wi:wi:/ *colloq.* –*n.* **1** act of urinating. **2** urine. –*v.* (**-wees, -weed**) urinate. [origin unknown]

weft *n.* **1** threads woven across a warp to make fabric. **2** yarn for these. **3** thing woven. [Old English: related to WEAVE[1]]

weigh /weɪ/ *v.* **1** find the weight of. **2** balance in the hands to guess or as if to guess the weight of. **3** (often foll. by *out*) take a definite weight of (a substance); measure out (a specified weight) (*weigh out the flour*; *weigh out 6 oz*). **4 a** estimate the relative value, importance, or desirability of. **b** (foll. by *with, against*) compare. **5** be equal to (a specified weight). **6** have (esp. a specified) importance; exert an influence. **7** (often foll. by *on*) be heavy or burdensome (to); be depressing (to). □ **weigh down 1** bring down by exerting weight. **2** be oppressive to. **weigh in** (of a boxer before a contest, or a jockey after a race) be weighed. **weigh in with** *colloq.* advance (an argument etc.) boldly. **weigh out** (of a jockey) be weighed before a race. **weigh up** *colloq.* form an estimate of. **weigh one's words** carefully choose the way one expresses something. [Old English, = carry]

weighbridge *n.* weighing-machine for vehicles.

weigh-in *n.* weighing of a boxer before a fight.

weight /weɪt/ *–n.* **1** force experienced by a body as a result of the earth's gravitation. **2** heaviness of a body regarded as a property of it. **3 a** quantitative expression of a body's weight. **b** scale of such weights (*troy weight*). **4** body of a known weight for use in weighing or weight training. **5** heavy body, esp. as used in a mechanism etc. **6** load or burden. **7** influence, importance. **8** *Athletics* = SHOT[1] 7. *–v.* **1 a** attach a weight to. **b** hold down with a weight. **2** (foll. by *with*) impede or burden. □ **throw one's weight about** (or **around**) *colloq.* be unpleasantly self-assertive. **worth one's weight in gold** very useful or helpful. [Old English]

weighting *n.* extra allowance paid in special cases.

weightless *adj.* (of a body, esp. in an orbiting spacecraft etc.) not apparently acted on by gravity. □ **weightlessness** *n.*

weightlifting *n.* sport of lifting heavy weights. □ **weightlifter** *n.*

weight training *n.* physical training using weights.

weighty *adj.* (**-ier, -iest**) **1** heavy. **2** momentous. **3** (of utterances etc.) deserving consideration. **4** influential, authoritative. □ **weightily** *adv.* **weightiness** *n.*

weir /wɪə(r)/ *n.* dam across a river to raise the level of water upstream or regulate its flow. [Old English]

weird *adj.* **1** uncanny, supernatural. **2** *colloq.* queer, incomprehensible. □ **weirdly** *adv.* **weirdness** *n.* [Old English *wyrd* destiny]

weirdo /'wɪədəʊ/ *n.* (*pl.* **-s**) *colloq.* odd or eccentric person.

welch var. of WELSH.

welcome /'welkəm/ *–n.* act of greeting or receiving gladly; kind or glad reception. *–int.* expressing such a greeting. *–v.* (**-ming**) receive with a welcome. *–adj.* **1** that one receives with pleasure (*welcome guest; welcome news*). **2** (foll. by *to*, or *to* + infin.) cordially allowed or invited (*you are welcome to use my car*). □ **make welcome** receive hospitably. **outstay one's welcome** stay too long as a visitor etc. **you are welcome** there is no need for thanks. [Old English]

weld *–v.* **1 a** hammer or press (pieces of iron or other metal usu. heated but not melted) into one piece. **b** join by fusion with an electric arc etc. **c** form by welding into some article. **2** fashion into an effectual or homogeneous whole. *–n.* welded joint. □ **welder** *n.* [alteration of WELL[2], probably influenced by the form *welled*]

welfare /'welfeə(r)/ *n.* **1** well-being, happiness; health and prosperity (of a person or community etc.). **2** (**Welfare**) **a** welfare centre or office. **b** financial support given by the State. [from WELL[1], FARE]

welfare state *n.* **1** system whereby the State undertakes to protect the health and well-being of its citizens by means of grants, pensions, etc. **2** country practising this system.

welfare work *n.* organized effort for the welfare of the poor, disabled, etc.

welkin *n. poet.* sky. [Old English, = cloud]

well[1] *–adv.* (**better, best**) **1** in a satisfactory way (*works well*). **2** with some distinction (*plays the piano well*). **3** in a kind way (*treated me well*). **4** thoroughly, carefully (*polish it well*). **5** with heartiness or approval (*speak well of*). **6** probably, reasonably (*you may well be right*). **7** to a considerable extent (*is well over forty*). *–adj.* (**better, best**) **1** (usu. *predic.*) in good health. **2** (*predic.*) **a** in a satisfactory state or position. **b** advisable (*it would be well to enquire*). *–int.* expressing surprise, resignation, etc., or used to introduce speech. □ **leave well alone** avoid needless change or disturbance. **well and truly** decisively, completely. **well away 1** having made considerable progress. **2** *colloq.* fast asleep or drunk. **well done!** expressing praise for something done. **well worth** certainly worth. [Old English]

well[2] *–n.* **1** shaft sunk into the ground to obtain water, oil, etc. **2** enclosed space like a well-shaft, e.g. in the middle of a building for stairs or a lift, or for light or ventilation. **3** (foll. by *of*) source. **4** (in *pl.*) spa. **5** = INK-WELL. **6** *archaic* water-spring. **7** railed space in a lawcourt. *–v.* (foll. by *out, up*) spring as from a fountain. [Old English]

we'll /wi:l, wɪl/ *contr.* we shall; we will.

well-adjusted *adj.* **1** mentally and emotionally stable. **2** in a good state of adjustment.

well-advised *adj.* (usu. foll. by *to* + infin.) prudent.

well-appointed *adj.* having all the necessary equipment.

well-attended *adj.* attended by a large number of people.

well-balanced *adj.* sane, sensible.

well-behaved *adj.* habitually behaving well.

well-being *n.* state of being contented, healthy, etc.

well-born *adj.* of noble family.

well-bred *adj.* having or showing good breeding or manners.
well-built *adj.* big, strong, and well-proportioned.
well-connected *adj.* associated, esp. by birth, with persons of good social position.
well-disposed *adj.* (often foll. by *towards*) friendly or sympathetic.
well-dressed *adj.* fashionably smart.
well-earned *adj.* fully deserved.
well-founded *adj.* (of suspicions etc.) based on good evidence.
well-groomed *adj.* with carefully tended hair, clothes, etc.
well-head *n.* source.
well-heeled *adj. colloq.* wealthy.
wellies /'welɪz/ *n.pl. colloq.* wellingtons. [abbreviation]
well-informed *adj.* having much knowledge or information about a subject.
wellington /'welɪŋt(ə)n/ *n.* (in full **wellington boot**) waterproof boot usu. reaching the knee. [Duke of *Wellington*]
well-intentioned *adj.* having or showing good intentions.
well-judged *adj.* opportunely, skilfully, or discreetly done.
well-kept *adj.* kept in good order or condition.
well-known *adj.* known to many.
well-made *adj.* **1** strongly manufactured. **2** having a good build.
well-mannered *adj.* having good manners.
well-meaning *adj.* (also **well-meant**) well-intentioned (but ineffective).
wellnigh *adv.* almost (*wellnigh impossible*).
well off *adj.* (also **well-off**) **1** having plenty of money. **2** in a fortunate situation.
well-oiled *adj. colloq.* very drunk.
well-paid *adj.* **1** (of a job) that pays well. **2** (of a person) amply rewarded for a job.
well-preserved *adj.* **1** in good condition. **2** (of an old person) showing little sign of age.
well-read *adj.* knowledgeable through much reading.
well-received *adj.* welcomed; favourably received.
well-rounded *adj.* complete and symmetrical.
well-spoken *adj.* articulate or refined in speech.
well-spring *n.* = WELL-HEAD.
well-to-do *adj.* prosperous.
well-tried *adj.* often tested with good results.
well-trodden *adj.* much frequented.
well-wisher *n.* person who wishes one well.
well-worn *adj.* **1** much worn by use. **2** (of a phrase etc.) trite.
Welsh –*adj.* of or relating to Wales or its people or language. –*n.* **1** the Celtic language of Wales. **2** (prec. by *the*; treated as *pl.*) the people of Wales. [Old English, ultimately from Latin *Volcae*, name of a Celtic people]
welsh *v.* (also **welch**) **1** (of a loser of a bet, esp. a bookmaker) decamp without paying. **2** evade an obligation. **3** (foll. by *on*) **a** fail to carry out a promise to (a person). **b** fail to honour (an obligation). [origin unknown]
Welshman *n.* man who is Welsh by birth or descent.
Welsh rabbit *n.* (also, by folk etymology, **Welsh rarebit**) dish of melted cheese etc. on toast.
Welshwoman *n.* woman who is Welsh by birth or descent.
welt –*n.* **1** leather rim sewn round the edge of a shoe-upper for the sole to be attached to. **2** = WEAL[1]. **3** ribbed or reinforced border of a garment. **4** heavy blow. –*v.* **1** provide with a welt. **2** raise weals on; thrash. [origin unknown]
welter[1] –*v.* **1** roll, wallow. **2** (foll. by *in*) lie prostrate or be soaked in. –*n.* **1** general confusion. **2** (foll. by *of*) disorderly mixture or contrast. [Low German or Dutch]
welter[2] *n.* heavy rider or boxer. [origin unknown]
welterweight *n.* **1** weight in certain sports intermediate between lightweight and middleweight, in the amateur boxing scale 63.5–67 kg. **2** sportsman of this weight.
wen *n.* benign tumour on the skin, esp. on the scalp. [Old English]
wench *n. joc.* girl or young woman. [abbreviation of *wenchel*, from Old English, = child]
wend *v.* □ **wend one's way** make one's way. [Old English, = turn]
Wendy house /'wendɪ/ *n.* children's small houselike tent or structure for playing in. [*Wendy*, name of a character in Barrie's *Peter Pan*]
went *past* of GO[1].
wept *past* of WEEP.
were *2nd sing. past, pl. past*, and *past subjunctive* of BE.
we're /wɪə(r)/ *contr.* we are.
weren't /wɜːnt/ *contr.* were not.
werewolf /'weəwʊlf/ *n.* (*pl.* **-wolves**) mythical being who at times changes from a person to a wolf. [Old English]
Wesleyan /'wezlɪən/ –*adj.* of or relating to a Protestant denomination founded by John Wesley. –*n.* member of this denomination.

west *–n.* **1 a** point of the horizon where the sun sets at the equinoxes. **b** compass point corresponding to this. **c** direction in which this lies. **2** (usu. **the West**) **a** European civilization. **b** States of western Europe and N. America. **c** western part of a country, town, etc. *–adj.* **1** towards, at, near, or facing the west. **2** from the west (*west wind*). *–adv.* **1** towards, at, or near the west. **2** (foll. by *of*) further west than. □ **go west** *slang* be killed or destroyed etc. **to the west** (often followed by *of*) in a westerly direction. [Old English]

westbound *adj.* travelling or leading westwards.

West Country *n.* south-western England.

West End *n.* main entertainment and shopping area of London.

westering *adj.* (of the sun) nearing the west.

westerly *–adj.* & *adv.* **1** in a western position or direction. **2** (of a wind) from the west. *–n.* (*pl.* **-ies**) such a wind.

western *–adj.* of or in the west. *–n.* film or novel about cowboys in western North America. □ **westernmost** *adj.*

westerner *n.* native or inhabitant of the west.

westernize *v.* (also **-ise**) (**-zing** or **-sing**) influence with, or convert to, the ideas and customs etc. of the West.

West Indian *n.* **1** native or national of the West Indies. **2** person of West Indian descent.

west-north-west *n.* point or direction midway between west and north-west.

West Side *n. US* western part of Manhattan.

west-south-west *n.* point or direction midway between west and south-west.

westward /ˈwestwəd/ *–adj.* & *adv.* (also **westwards**) towards the west. *–n.* westward direction or region.

wet *–adj.* (**wetter, wettest**) **1** soaked or covered with water or other liquid. **2** (of the weather etc.) rainy. **3** (of paint etc.) not yet dried. **4** used with water (*wet shampoo*). **5** *colloq.* feeble, inept. *–v.* (**-tt-**; *past* and *past part.* **wet** or **wetted**) **1** make wet. **2 a** urinate in or on (*wet the bed*). **b** *refl.* urinate involuntarily. *–n.* **1** liquid that wets something. **2** rainy weather. **3** *colloq.* feeble or inept person. **4** *colloq.* Conservative with liberal tendencies. **5** *colloq.* drink. □ **wet behind the ears** *colloq.* immature, inexperienced. **wet through** (or **to the skin**) with one's clothes soaked. □ **wetly** *adv.* **wetness** *n.* [Old English]

wet blanket *n. colloq.* gloomy person hindering others' enjoyment.

wet dream *n.* erotic dream with the involuntary ejaculation of semen.

wether /ˈweðə(r)/ *n.* castrated ram. [Old English]

wetland *n.* (often in *pl.*) a swamp or other unusually damp area of land.

wet-nurse *–n.* woman employed to suckle another's child. *–v.* **1** act as a wet-nurse to. **2** *colloq.* treat as if helpless.

wet suit *n.* rubber garment worn by skin-divers etc. to keep warm.

we've /wiːv/ *contr.* we have.

Wg. Cdr. *abbr.* wing commander.

whack *colloq.* *–v.* **1** strike or beat forcefully. **2** (as **whacked** *adj.*) tired out. *–n.* **1** sharp or resounding blow. **2** *slang* share. □ **have a whack at** *slang* attempt. [imitative]

whacking *colloq.* *–adj.* very large. *–adv.* very.

whale *–n.* (*pl.* same or **-s**) very large marine mammal with a streamlined body and horizontal tail. *–v.* (**-ling**) hunt whales. □ **a whale of a** *colloq.* an exceedingly good or fine etc. [Old English]

whalebone *n.* elastic horny substance in the upper jaw of some whales.

whale-oil *n.* oil from the blubber of whales.

whaler *n.* whaling ship or seaman.

wham *int. colloq.* expressing forcible impact. [imitative]

wharf /wɔːf/ *–n.* (*pl.* **wharves** /wɔːvz/ or **-s**) quayside area to which a ship may be moored to load and unload. *–v.* **1** moor (a ship) at a wharf. **2** store (goods) on a wharf. [Old English]

wharfage *n.* **1** accommodation at a wharf. **2** fee for this.

what /wɒt/ *–interrog. adj.* **1** asking for a choice from an indefinite number or for a statement of amount, number, or kind (*what books have you read?*). **2** *colloq.* = WHICH *interrog.adj.* (*what book have you chosen?*). *–adj.* (usu. in an exclamation) how great or remarkable (*what luck!*). *–rel. adj.* the or any ... that (*will give you what help I can*). *–pron.* (corresponding to the functions of the *adj.*) **1** what thing or things? (*what is your name?*; *I don't know what you mean*). **2** (asking for a remark to be repeated) = what did you say? **3** how much (*what you must have suffered!*). **4** (as *rel. pron.*) that or those which; a or the or any thing which (*what followed was worse*; *tell me what you think*). *–adv.* to what extent (*what does it matter?*). □ **what about** what is the news or your opinion of. **what-d'you-call-it** *colloq.* substitute for a name not recalled. **what ever** what at all or in any way (*what *ever do you mean?*) (see also WHATEVER). **what for** *colloq.* **1** for what reason? **2** severe reprimand (esp. *give a person what for*). **what have you** (prec. by *or* or *and*) *colloq.* anything else similar. **what not** (prec. by *and*) other similar things. **what's-his** (or **-her** or **-its**) **-name** *colloq.* substitute for a name not recalled. **what's what** *colloq.* what is useful or important etc. **what with** *colloq.* because of (usu. several things). [Old English]

whatever /wɒt'evə(r)/ *adj.* & *pron.* **1** = WHAT (in relative uses) with the emphasis on indefiniteness (*lend me whatever you can*; *whatever money you have*). **2** though anything (*we are safe whatever happens*). **3** (with *neg.* or *interrog.*) at all; of any kind (*there is no doubt whatever*).

whatnot *n. colloq.* indefinite or trivial thing.

whatsoever /ˌwɒtsəʊ'evə(r)/ *adj.* & *pron.* = WHATEVER.

wheat *n.* **1** cereal plant bearing dense four-sided seed-spikes. **2** its grain, used in making flour etc. [Old English]

wheatear *n.* small migratory bird. [related to WHITE, ARSE]

wheaten *adj.* made of wheat.

wheat germ *n.* embryo of the wheat grain, extracted as a source of vitamins.

wheatmeal *n.* flour made from wheat with some of the bran and germ removed.

wheedle /'wi:d(ə)l/ *v.* **(-ling) 1** coax by flattery or endearments. **2** (foll. by *out*) get (a thing) out of a person or cheat (a person) out of a thing by wheedling. [origin uncertain]

wheel *–n.* **1** circular frame or disc which revolves on an axle and is used for vehicular or other mechanical motion. **2** wheel-like thing. **3** motion as of a wheel, esp. the movement of a line of soldiers with one end as a pivot. **4** (in *pl.*) *slang* car. **5** = STEERING-WHEEL. *–v.* **1 a** turn on an axis or pivot. **b** swing round in line with one end as a pivot. **2 a** (often foll. by *about*, *round*) change direction or face another way. **b** cause to do this. **3** push or pull (a wheeled thing, or its load or occupant). **4** go in circles or curves. □ **at the wheel 1** driving a vehicle. **2** directing a ship. **3** in control. **on wheels** (or **oiled wheels**) smoothly. **wheel and deal** engage in political or commercial scheming. **wheels within wheels 1** intricate machinery. **2** *colloq.* indirect or secret agencies. □ **wheeled** *adj.* (also in *comb.*). [Old English]

wheelbarrow *n.* small handcart with one wheel and two shafts.

wheelbase *n.* distance between the axles of a vehicle.

wheelchair *n.* chair on wheels for an invalid or disabled person.

wheel-clamp *n.* = CLAMP[1] *n.* 2.

-wheeler *comb. form* vehicle with a specified number of wheels (*three-wheeler*).

wheeler-dealer *n.* person who wheels and deals.

wheel-house *n.* steersman's shelter.

wheelie *n. slang* stunt of riding a bicycle or motor cycle with the front wheel off the ground.

wheel-spin *n.* rotation of a vehicle's wheels without traction.

wheelwright *n.* person who makes or repairs wheels.

wheeze *–v.* **(-zing) 1** breathe with an audible whistling sound. **2** utter with this sound. *–n.* **1** sound of wheezing. **2** *colloq.* clever scheme. □ **wheezy** *adj.* **(-ier, -iest)**. **wheezily** *adv.* **wheeziness** *n.* [probably from Old Norse, = hiss]

whelk *n.* marine mollusc with a spiral shell. [Old English]

whelm *v. poet.* **1** engulf. **2** crush with weight. [Old English]

whelp *–n.* **1** young dog; puppy. **2** *archaic* cub. **3** ill-mannered child or youth. *–v.* (also *absol.*) give birth to (a whelp or whelps or (*derog.*) a child). [Old English]

when *–interrog. adv.* **1** at what time? **2** on what occasion? **3** how soon? *–rel. adv.* (prec. by *time* etc.) at or on which (*there are times when I could cry*). *–conj.* **1** at the or any time that; as soon as (*come when you like*; *come when ready*). **2** although (*why stand when you could sit?*). **3** after which; and then; but just then (*was nearly asleep when the bell rang*). *–pron.* what time?; which time (*till when can you stay?*; *since when it has improved*). *–n.* time, occasion (*fixed the where and when*). [Old English]

whence *formal –interrog. adv.* from what place? *–conj.* **1** to the place from which (*return whence you came*). **2** (often prec. by *place* etc.) from which. **3** and thence (*whence it follows that*). [Old English: related to WHEN]

■ **Usage** The use of *from whence* rather than simply *whence* (as in *the place from whence they came*), though common, is generally considered incorrect.

whenever /wen'evə(r)/ *conj.* & *adv.* **1** at whatever time; on whatever occasion. **2** every time that.

whensoever /ˌwensəʊ'evə(r)/ *conj.* & *adv. formal* = WHENEVER.

where /weə(r)/ *–interrog. adv.* **1** in or to what place or position? **2** in what respect? (*where does it concern us?*). *–rel. adv.* (prec. by *place* etc.) in or to which (*places where they meet*). *–conj.* **1** in or to the or any place, direction, or respect in which (*go where you like; tick where applicable*). **2** and there (*reached Crewe, where the car broke down*). *–pron.* what place? (*where do you come from?*). *–n.* place; scene of something (see WHEN *n.*). [Old English]

whereabouts *–interrog. adv.* /ˌweərə'baʊts/ approximately where? *–n.* /'weərəˌbaʊts/ (as *sing.* or *pl.*) person's or thing's location.

whereas /weər'æz/ *conj.* **1** in contrast or comparison with the fact that. **2** (esp. in legal preambles) taking into consideration the fact that.

whereby /weə'baɪ/ *conj.* by what or which means.

wherefore *–adv. archaic* **1** for what reason? **2** for which reason. *–n.* see WHY.

wherein /weər'ɪn/ *conj. formal* in what or which place or respect.

whereof /weər'ɒv/ *conj. formal* of what or which.

whereupon /ˌweərə'pɒn/ *conj.* immediately after which.

wherever /weər'evə(r)/ *–adv.* in or to whatever place. *–conj.* in every place that.

wherewithal /'weəwɪˌðɔːl/ *n. colloq.* money etc. needed for a purpose.

wherry /'werɪ/ *n.* (*pl.* **-ies**) **1** light rowing-boat usu. for carrying passengers. **2** large light barge. [origin unknown]

whet *v.* (**-tt-**) **1** sharpen (a tool). **2** stimulate (the appetite or a desire etc.). [Old English]

whether /'weðə(r)/ *conj.* introducing the first or both of alternative possibilities (*I doubt whether it matters; I do not know whether they have arrived or not*). □ **whether or no** whether it is so or not. [Old English]

whetstone *n.* tapered stone used with water to sharpen tools.

whew /hwjuː/ *int.* expressing surprise, consternation, or relief. [imitative]

whey /weɪ/ *n.* watery liquid left when milk forms curds. [Old English]

which *–interrog. adj.* asking for choice from a definite set of alternatives (*which John do you mean?; say which book you prefer*). *–rel. adj.* being the one just referred to; and this or these (*ten years, during which time they admitted nothing*). *–interrog. pron.* **1** which person or persons? (*which of you is responsible?*). **2** which thing or things? (*say which you prefer*). *–rel. pron.* (*poss.* **of which, whose** /huːz/) **1** which thing or things, usu. introducing a clause not essential for identification (*the house, which is empty, has been damaged*). **2** used in place of *that* after *in* or *that* (*there is the house in which I was born; that which you have just seen*). [Old English]

whichever /wɪtʃ'evə(r)/ *adj. & pron.* any which (*take whichever you like*).

whiff *n.* **1** puff or breath of air, smoke, etc. **2** smell. **3** (foll. by *of*) trace of scandal etc. **4** small cigar. [imitative]

Whig *n. hist.* member of the British reforming and constitutional party succeeded in the 19th c. by the Liberal Party. □ **Whiggery** *n.* **Whiggish** *adj.* **Whiggism** *n.* [*whiggamer, -more*, nickname of 17th-c. Scots rebels]

while *–n.* period of time (*a long while ago; waited a while; all this while*). *–conj.* **1** during the time that; for as long as; at the same time as (*while I was away, the house was burgled; fell asleep while reading*). **2** in spite of the fact that; whereas (*while I want to believe it, I cannot*). *–v.* (**-ling**) (foll. by *away*) pass (time etc.) in a leisurely or interesting way. *–rel. adv.* (prec. by *time* etc.) during which (*the summer while I was abroad*). □ **between whiles** in the intervals. **for a while** for some time. **in a while** soon. **the while** in the meantime. **worth while** (or **worth one's while**) worth the time or effort spent. [Old English]

■ **Usage** *Worth while* (two words) is used only predicatively, as in *thought it worth while to ring the police*, whereas *worthwhile* is used both predicatively and attributively.

whilst /waɪlst/ *adv. & conj.* while. [from WHILE]

whim *n.* **1** sudden fancy; caprice. **2** capriciousness. [origin unknown]

whimper *–v.* make feeble, querulous, or frightened sounds. *–n.* such a sound. [imitative]

whimsical /'wɪmzɪk(ə)l/ *adj.* capricious, fantastic. □ **whimsicality** /-'kælɪtɪ/ *n.* **whimsically** *adv.*

whimsy /'wɪmzɪ/ *n.* (*pl.* **-ies**) = WHIM. [origin uncertain]

whin *n.* (in *sing.* or *pl.*) gorse. [Scandinavian]

whinchat *n.* small songbird.

whine *–n.* **1** complaining long-drawn wail as of a dog. **2** similar shrill prolonged sound. **3** querulous tone or complaint. *–v.* (**-ning**) emit or utter a whine; complain. [Old English]

whinge /wɪndʒ/ *v.* (**-geing** or **-ging**) *colloq.* whine; grumble peevishly. [Old English]

whinny /ˈwɪnɪ/ *–n.* (*pl.* **-ies**) gentle or joyful neigh. *–v.* (**-ies**, **-ied**) give a whinny. [imitative]

whip *–n.* **1** lash attached to a stick for urging on animals or punishing etc. **2 a** member of a political party in Parliament appointed to control its discipline and tactics. **b** whips' written notice requesting or requiring attendance for voting at a division etc., variously underlined according to the degree of urgency (*three-line whip*). **c** (prec. by *the*) party discipline and instructions (*asked for the Labour whip*). **3** dessert made with whipped cream etc. **4** = WHIPPER-IN. *–v.* (**-pp-**) **1** beat or urge on with a whip. **2** beat (cream or eggs etc.) into a froth. **3** take or move suddenly, unexpectedly, or rapidly (*whipped out a knife*; *whipped behind the door*). **4** *slang* steal. **5** *slang* **a** excel. **b** defeat. **6** bind with spirally wound twine. **7** sew with overcast stitches. □ **whip in** bring (hounds) together. **whip on** urge into action. **whip up** excite or stir up. [Low German or Dutch]

whipcord *n.* tightly twisted cord.

whip hand *n.* **1** hand that holds the whip (in riding etc.). **2** (usu. prec. by *the*) advantage or control in a situation.

whiplash *n.* flexible end of a whip.

whiplash injury *n.* injury to the neck caused by a jerk of the head, esp. as in a motor accident.

whipper-in /ˌwɪpəˈrɪn/ *n.* (*pl.* **whippers-in**) huntsman's assistant who manages the hounds.

whippersnapper /ˈwɪpəˌsnæpə(r)/ *n.* **1** small child. **2** insignificant but presumptuous person.

whippet /ˈwɪpɪt/ *n.* crossbred dog of the greyhound type used for racing. [probably from obsolete *whippet* move briskly, from *whip it*]

whipping boy *n.* scapegoat.

whipping-top *n.* top kept spinning by blows of a lash.

whippoorwill /ˈwɪpʊəwɪl/ *n.* American nightjar. [imitative]

whip-round *n. colloq.* informal collection of money among a group of people.

whipstock *n.* handle of a whip.

whirl *–v.* **1** swing round and round; revolve rapidly. **2** (foll. by *away*) convey or go rapidly in a vehicle etc. **3** send or travel swiftly in an orbit or a curve. **4** (of the brain etc.) seem to spin round. *–n.* **1** whirling movement. **2** state of intense activity (*the social whirl*). **3** state of confusion (*in a whirl*). □ **give it a whirl** *colloq.* attempt it. [Old Norse, and Low German or Dutch]

whirligig /ˈwɜːlɪgɪg/ *n.* **1** spinning or whirling toy. **2** merry-go-round. **3** revolving motion.

whirlpool *n.* powerful circular eddy of water.

whirlwind *n.* **1** rapidly whirling mass or column of air. **2** (*attrib.*) very rapid.

whirr *–n.* continuous rapid buzz or soft clicking sound. *–v.* (**-rr-**) make this sound. [Scandinavian]

whisk *–v.* **1** (foll. by *away*, *off*) **a** brush with a sweeping movement. **b** take suddenly. **2** whip (cream, eggs, etc.). **3** convey or go (esp. out of sight) lightly or quickly. **4** wave or lightly brandish. *–n.* **1** whisking action or motion. **2** utensil for whisking eggs or cream etc. **3** bunch of grass, twigs, bristles, etc., for removing dust or flies. [Scandinavian]

whisker *n.* **1** (usu. in *pl.*) hair growing on a man's face, esp. on the cheek. **2** each of the bristles on the face of a cat etc. **3** *colloq.* small distance (*within a whisker of*). □ **whiskered** *adj.* **whiskery** *adj.* [from WHISK]

whisky /ˈwɪskɪ/ *n.* (*Ir.* & *US* **whiskey**) (*pl.* **-ies** or **-eys**) spirit distilled esp. from malted grain, esp. barley or rye. [abbreviation of *usquebaugh* from Gaelic, = water of life]

whisper *–v.* **1 a** speak very softly without vibration of the vocal cords. **b** talk or say in a barely audible tone or in a secret or confidential way. **2** rustle or murmur. *–n.* **1** whispering speech or sound. **2** thing whispered. □ **it is whispered** there is a rumour. [Old English]

whist *n.* card-game usu. for two pairs of players. [earlier *whisk*, perhaps from WHISK (with ref. to whisking away the tricks): perhaps associated with *whist*! (= silence)]

whist drive *n.* social occasion with the playing of progressive whist.

whistle /ˈwɪs(ə)l/ *–n.* **1** clear shrill sound made by forcing breath through a small hole between nearly closed lips. **2** similar sound made by a bird, the wind, a missile, etc. **3** instrument used to produce such a sound. *–v.* (**-ling**) **1** emit a whistle. **2 a** give a signal or express surprise or derision by whistling. **b** (often foll. by *up*) summon or give a signal to (a dog etc.) by whistling. **3** (also *absol.*) produce (a tune) by whistling. **4** (foll. by *for*) vainly seek or desire. [Old English]

whistle-stop *n.* **1** *US* small unimportant town on a railway. **2** politician's brief pause for an electioneering speech on tour.

Whit –*n.* = WHITSUNTIDE. –*attrib. adj.* of Whitsuntide or Whit Sunday. [Old English, = white]

whit *n.* particle; least possible amount (*not a whit better*). [apparently = WIGHT]

white –*adj.* **1** resembling a surface reflecting sunlight without absorbing any of the visible rays; of the colour of milk or snow. **2** nearly this colour; pale, esp. in the face. **3 (White) a** of the human group having light-coloured skin. **b** of or relating to White people. **4** albino (*white mouse*). **5** (of hair) having lost its colour, esp. in old age. **6** (of coffee) with milk or cream. –*n.* **1** white colour or pigment. **2 a** white clothes or material. **b** (in *pl.*) white garments as worn in cricket, tennis, etc. **3 a** (in a game or sport) white piece, ball, etc. **b** player using these. **4** = EGG-WHITE. **5** whitish part of the eyeball round the iris. **6 (White)** member of a light-skinned race. □ **bleed white** drain of wealth etc. □ **whiteness** *n.* **whitish** *adj.* [Old English]

white ant *n.* termite.

whitebait *n.* (*pl.* same) (usu. in *pl.*) small silvery-white young of herrings and sprats, esp. as food.

white cell *n.* leucocyte.

white-collar *attrib. adj.* (of a worker or work) non-manual; clerical, professional.

white corpuscle *n.* = WHITE CELL.

white elephant *n.* useless possession.

white feather *n.* symbol of cowardice.

white flag *n.* symbol of surrender.

White Friar *n.* Carmelite.

white goods *n.pl.* large domestic electrical equipment.

whitehead *n. colloq.* white or white-topped skin-pustule.

white heat *n.* **1** temperature at which metal emits white light. **2** state of intense passion or activity. □ **white-hot** *adj.*

white hope *n.* person expected to achieve much.

white horses *n.pl.* white-crested waves at sea.

white lead *n.* mixture of lead carbonate and hydrated lead oxide used as pigment.

white lie *n.* harmless or trivial untruth.

white light *n.* apparently colourless light, e.g. ordinary daylight.

white magic *n.* magic used for beneficent purposes.

white meat *n.* poultry, veal, rabbit, and pork.

whiten *v.* make or become white. □ **whitener** *n.*

white noise *n.* noise containing many frequencies with equal intensities.

white-out *n.* dense blizzard, esp. in polar regions.

White Paper *n.* Government report giving information.

white pepper *n.* pepper made by grinding a ripe or husked berry.

White Russian *n.* = BYELORUSSIAN.

white sauce *n.* sauce of flour, melted butter, and milk or cream.

White slave *n.* woman tricked or forced into prostitution.

white spirit *n.* light petroleum as a solvent.

white sugar *n.* purified sugar.

white tie *n.* man's white bow-tie as part of full evening dress.

whitewash –*n.* **1** solution of quicklime or whiting for whitening walls etc. **2** means employed to conceal mistakes or faults. –*v.* **1** cover with whitewash. **2** attempt to clear the reputation of by concealing facts.

white wedding *n.* wedding at which the bride wears a formal white wedding dress.

white whale *n.* northern cetacean, white when adult.

whitewood *n.* pale wood, esp. prepared for staining etc.

whither /'wɪðə(r)/ *archaic* –*adv.* **1** to what place or state? **2** (prec. by *place* etc.) to which. –*conj.* **1** to the or any place to which (*go whither you will*). **2** and thither. [Old English]

whiting[1] *n.* (*pl.* same) small white-fleshed fish used as food. [Dutch: related to WHITE]

whiting[2] *n.* ground chalk used in whitewashing etc.

whitlow /'wɪtləʊ/ *n.* inflammation near a fingernail or toenail. [originally *white* FLAW[1]]

Whitsun /'wɪts(ə)n/ –*n.* = WHITSUNTIDE. –*adj.* = WHIT. [*Whitsun Day* = Whit Sunday]

Whit Sunday *n.* seventh Sunday after Easter, commemorating Pentecost.

Whitsuntide *n.* weekend or week including Whit Sunday.

whittle /'wɪt(ə)l/ *v.* **(-ling) 1** (often foll. by *at*) pare (wood etc.) with repeated slicing with a knife. **2** (often foll. by *away*, *down*) reduce by repeated subtractions. [dial. *thwittle*]

whiz (also **whizz**) –*n.* sound made by a body moving through the air at great speed. –*v.* **(-zz-)** move with or make a whiz. [imitative]

whiz-kid *n. colloq.* brilliant or highly successful young person.

WHO *abbr.* World Health Organization.

who /hu:/ *pron.* (*obj.* **whom** /hu:m/ or *colloq.* **who**; *poss.* **whose** /hu:z/) **1 a** what or which person or persons? (*who* *called?*; *you know who it was*). **b** what sort of person or persons? (*who am I to object?*). **2** (a person) that (*anyone who wishes can come*; *the woman whom you met*; *the man who you saw*). **3** and or but he, they, etc. (*gave it to Tom, who sold it to Jim*). [Old English]

■ **Usage** In the last two examples of sense 2 *whom* is correct, but *who* is common in less formal contexts.

whoa /wəʊ/ *int.* used to stop or slow a horse etc. [var. of HO]

who'd /hu:d/ *contr.* **1** who had. **2** who would.

whodunit /hu:'dʌnɪt/ *n.* (also **whodunnit**) *colloq.* detective story, play, or film. [= *who done* (illiterate for *did*) *it?*]

whoever /hu:'evə(r)/ *pron.* (*obj.* **whoever** or *formal* **whomever** /hu:m-/; *poss.* **whosever** /hu:z-/) **1** the or any person or persons who (*whoever comes is welcome*). **2** though anyone (*whoever else objects, I do not*).

whole /həʊl/ —*adj.* **1** uninjured, unbroken, intact, or undiminished. **2** not less than; all there is of. **3** (of blood or milk etc.) with no part removed. —*n.* **1** thing complete in itself. **2** all there is of a thing. **3** (foll. by *of*) all members etc. of (*the whole of London knows it*). □ **as a whole** as a unity; not as separate parts. **on the whole** taking everything relevant into account. **whole lot** see LOT. □ **wholeness** *n.* [Old English]

wholefood *n.* food which has not been unnecessarily processed or refined.

wholegrain *attrib. adj.* made with or containing whole grains (*wholegrain rice*).

wholehearted *adj.* **1** completely devoted. **2** done with all possible effort or sincerity. □ **wholeheartedly** *adv.*

wholemeal *n.* (usu. *attrib.*) meal or flour with none of the bran or germ removed.

whole number *n.* number without fractions; integer.

wholesale —*n.* selling of goods in large quantities to be retailed by others. —*adj.* & *adv.* **1** by wholesale. **2** on a large scale. —*v.* (**-ling**) sell wholesale. □ **wholesaler** *n.* [*by whole sale*]

wholesome *adj.* **1** promoting physical, mental, or moral health. **2** prudent (*wholesome respect*). [Old English: related to WHOLE]

wholewheat *n.* (usu. *attrib.*) wheat with none of the bran or germ removed.

wholism var. of HOLISM.

wholly /'həʊllɪ/ *adv.* **1** entirely; without limitation. **2** purely.

whom *objective case* of WHO.

whomever *objective case* of WHOEVER.

whomsoever *objective case* of WHOSOEVER.

whoop /hu:p, wu:p/ —*n.* **1** loud cry of or as of excitement etc. **2** long rasping indrawn breath in whooping cough. —*v.* utter a whoop. □ **whoop it up** *colloq.* **1** engage in revelry. **2** *US* make a stir. [imitative]

whoopee /wʊ'pi:/ *int.* expressing exuberant joy. □ **make whoopee** /'wʊpɪ/ *colloq.* **1** have fun, make merry. **2** make love. [imitative]

whooping cough /'hu:pɪŋ/ *n.* infectious bacterial disease, esp. of children, with a series of short violent coughs followed by a whoop.

whoops /wʊps/ *int. colloq.* expressing surprise or apology, esp. on losing balance or making an obvious mistake. [var. of OOPS]

whop *v.* (**-pp-**) *slang* **1** thrash. **2** defeat. [origin unknown]

whopper *n. slang* **1** something big of its kind. **2** great lie.

whopping *adj. colloq.* (esp. as an intensifier) huge (*a whopping success*; *a whopping great lie*).

whore /hɔ:(r)/ *n.* **1** prostitute. **2** *derog.* promiscuous woman. [Old English]

whore-house *n.* brothel.

whorl /wɜ:l/ *n.* **1** ring of leaves etc. round a stem. **2** one turn of a spiral. [apparently var. of WHIRL]

whortleberry /'wɜ:t(ə)l,berɪ/ *n.* (*pl.* **-ies**) bilberry. [origin unknown]

whose /hu:z/ —*interrog. pron.* of or belonging to which person (*whose is this book?*). —*interrog. adj.* of whom or which (*whose book is this?*). —*rel. pron.* of whom; of which (*the man, whose name was Tim*; *the house whose roof was damaged*).

whosoever /,hu:səʊ'evə(r)/ *pron.* (*obj.* **whomsoever** /,hu:m-/; *poss.* **whosesoever** /,hu:z-/) *archaic* = WHOEVER.

who's who *n.* **1** who or what each person is (*know who's who*). **2** list with facts about notable persons.

why /waɪ/ —*adv.* **1** for what reason or purpose (*why did you do it?*; *I do not know why you came*). **2** (prec. by *reason* etc.) for which (*the reasons why I did it*). —*int.* expressing: **1** surprised discovery or recognition (*why, it's you!*). **2** impatience (*why, of course I do!*). **3** reflection (*why, yes, I think so*). **4** objection (*why, what is wrong with it?*). □ **whys and wherefores** reasons; explanation. [Old English: related to WHAT]

WI *abbr.* **1** West Indies. **2** Women's Institute.

wick *n.* strip or thread feeding a flame with fuel. □ **get on a person's wick** *colloq.* annoy a person. [Old English]

wicked /'wɪkɪd/ *adj.* (**-er**, **-est**) **1** sinful, iniquitous, immoral. **2** spiteful. **3** playfully malicious. **4** *colloq.* very bad. **5** *slang* excellent. □ **wickedly** *adv.* **wickedness** *n.* [origin uncertain]

wicker *n.* plaited osiers etc. as material for baskets etc. [Scandinavian]

wickerwork *n.* **1** wicker. **2** things made of wicker.

wicket /'wɪkɪt/ *n.* **1** *Cricket* **a** three stumps with the bails in position defended by a batsman. **b** ground between two wickets. **c** state of this. **d** instance of a batsman being got out (*bowler has taken four wickets*). **2** (in full **wicket-door** or **-gate**) small door or gate, esp. beside or in a larger one or closing the lower part only of a doorway. [Anglo-French *wiket* = French *guichet*]

wicket-keeper *n.* fieldsman stationed close behind a batsman's wicket.

widdershins /'wɪdəʃɪnz/ *adv.* (also **withershins** /'wɪð-/) esp. *Scot.* **1** in a direction contrary to the sun's course (considered unlucky). **2** anticlockwise. [German, = contrary]

wide *–adj.* **1** having sides far apart, broad, not narrow (*wide river*; *wide sleeve*; *wide angle*). **2** (following a measurement) in width (*a metre wide*). **3 a** extending far (*wide range*; *wide experience*). **b** considerable (*wide margin*). **4** not restricted (*a wide public*). **5 a** liberal; unprejudiced (*takes wide views*). **b** not specialized; general. **6** open to the full extent (*wide eyes*). **7** (foll. by *of*) not within a reasonable distance of, far from (*wide shot*; *wide of the target*). **8** (in *comb.*) extending over the whole of (*nationwide*). *–adv.* **1** widely. **2** to the full extent. **3** far from the target etc. (*shooting wide*). *–n.* = WIDE BALL. □ **give a wide berth to** see BERTH. **wide of the mark** see MARK[1]. **wide open** (often foll. by *to*) exposed (to attack etc.). **the wide world** all the world, great as it is. [Old English]

wide awake *adj.* **1** fully awake. **2** *colloq.* wary, knowing.

wide ball *n. Cricket* ball judged to be beyond the batsman's reach, so scoring a run.

wide-eyed *adj.* surprised; naïve.

widely *adv.* **1** to a wide extent; far apart. **2** extensively. **3** by many people (*it is widely thought that*). **4** considerably; to a large degree (*holds a widely different view*).

widen *v.* make or become wider.

widespread *adj.* widely distributed.

widgeon /'wɪdʒ(ə)n/ *n.* (also **wigeon**) a kind of wild duck. [origin uncertain]

widow /'wɪdəʊ/ *–n.* **1** woman who has lost her husband by death and not married again. **2** woman whose husband is often away on a specified activity (*golf widow*). *–v.* **1** make into a widow or widower. **2** (as **widowed** *adj.*) bereft by the death of a spouse. □ **widowhood** *n.* [Old English]

widower *n.* man who has lost his wife by death and not married again.

widow's peak *n.* V-shaped growth of hair towards the centre of the forehead.

width *n.* **1** measurement from side to side. **2** large extent. **3** liberality of views etc. **4** strip of material of full width. □ **widthways** *adv.* [from WIDE]

wield *v.* hold and use; command, exert (a weapon, tool, power, etc.). [Old English]

Wiener schnitzel /'vi:nə ˌʃnɪts(ə)l/ *n.* veal cutlet breaded, fried, and garnished. [German]

wife *n.* (*pl.* **wives**) **1** married woman, esp. in relation to her husband. **2** *archaic* woman. □ **wifely** *adj.* [Old English, = woman]

wig *n.* artificial head of hair. [abbreviation of PERIWIG]

wigeon var. of WIDGEON.

wigging *n. colloq.* reprimand. [origin uncertain]

wiggle /'wɪg(ə)l/ *colloq. –v.* (**-ling**) move from side to side etc. *–n.* act of wiggling; kink in a line etc. □ **wiggly** *adj.* (**-ier**, **-iest**). [Low German or Dutch *wiggelen*]

wight /waɪt/ *n. archaic* person. [Old English, = thing, creature]

wigwam /'wɪgwæm/ *n.* N. American Indian's hut or tent. [Ojibwa]

wilco /'wɪlkəʊ/ *int. colloq.* expressing compliance or agreement. [abbreviation of *will comply*]

wild /waɪld/ *–adj.* **1** in its original natural state; not domesticated, cultivated, or civilized (*wild cat*; *wild strawberry*). **2** unrestrained, disorderly, uncontrolled (*wild youth*; *wild hair*). **3** tempestuous (*wild night*). **4** intensely eager, frantic (*wild excitement*; *wild delight*). **5** (foll. by *about*) *colloq.* enthusiastically devoted to. **6** *colloq.* infuriated. **7** haphazard, ill-aimed, rash (*wild guess*; *wild venture*). **8** *colloq.* exciting, delightful. *–adv.* in a wild manner. *–n.* **1** wild tract of land. **2** desert. □ **in the wild** in an uncultivated etc. state. **in the wilds** *colloq.* far from towns etc. **run wild** grow or stray unchecked or undisciplined. □ **wildly** *adv.* **wildness** *n.* [Old English]

wild card *n.* **1** card having any rank chosen by the player holding it. **2** *Computing* character that will match any character or combination of characters. **3** person or thing that can be used in several different ways.

wildcat *–n.* **1** hot-tempered or violent person. **2** exploratory oil well. *–adj.* (*attrib.*) **1** (of a strike) sudden and unofficial. **2** reckless; financially unsound.

wildebeest /'wɪldəˌbiːst/ *n.* (*pl.* same or **-s**) = GNU. [Afrikaans: related to WILD, BEAST]

wilderness /'wɪldənɪs/ *n.* **1** desert; uncultivated region or garden area. **2** (foll. by *of*) confused assemblage. [Old English: related to WILD, DEER]

wildfire *n. hist.* combustible liquid used in war. □ **spread like wildfire** spread with great speed.

wildfowl *n.* (*pl.* same) game-bird.

wild-goose chase *n.* foolish or hopeless quest.

wild hyacinth *n.* = BLUEBELL.

wildlife *n.* wild animals collectively.

Wild West *n.* western US before the establishment of law and order.

wile *–n.* (usu. in *pl.*) stratagem, trick. *–v.* (**-ling**) (foll. by *away*, *into*, etc.) lure. [perhaps from Scandinavian]

wilful /'wɪlfʊl/ *adj.* (*US* **willful**) **1** intentional, deliberate (*wilful murder*; *wilful neglect*). **2** obstinate. □ **wilfully** *adv.* [from WILL[2]]

will[1] *v.aux.* (*3rd sing. present* **will**; *past* **would** /wʊd/) **1** (strictly only in the 2nd and 3rd persons: see SHALL) expressing a future statement, command, etc. (*you will regret this*; *they will leave at once*). **2** expressing the speaker's intention (*I will return soon*). **3** wish or desire (*will you have a drink?*; *come when you will*). **4** expressing a request as a question (*will you please open the window?*). **5** be able to (*the jar will hold a kilo*). **6** have a habit or tendency to (*accidents will happen*; *will sit there for hours*). **7** expressing probability or expectation (*that will be my wife*). [Old English]

will[2] *–n.* **1** faculty by which a person decides what to do. **2** strong desire or intention (*will to live*). **3** determination, will-power (*has a strong will*). **4** legal written directions for the disposal of one's property after death. **5** disposition towards others (*good will*). **6** *archaic* what one desires or ordains. *–v.* **1** try to cause by will-power (*willed her to win*). **2** intend; desire. **3** bequeath by a will. □ **at will** whenever one wishes. **with a will** energetically or resolutely. [Old English]

willful *US* var. of WILFUL.

willie var. of WILLY.

willies /'wɪlɪz/ *n.pl. colloq.* nervous discomfort (*gives me the willies*). [origin unknown]

willing *adj.* **1** ready to consent or undertake. **2** given or done etc. by a willing person. □ **willingly** *adv.* **willingness** *n.*

will-o'-the-wisp /ˌwɪləðə'wɪsp/ *n.* **1** phosphorescent light seen on marshy ground. **2** elusive person. [= *William of the torch*]

willow /'wɪləʊ/ *n.* tree with pliant branches yielding osiers and timber for cricket-bats etc., usu. growing near water. [Old English]

willow-herb *n.* plant with leaves like a willow.

willow-pattern *n.* conventional Chinese design of blue on white porcelain etc.

willow-warbler *n.* small woodland bird with a tuneful song.

willowy *adj.* **1** lithe and slender. **2** having willows.

will-power *n.* control by deliberate purpose over impulse.

willy /'wɪlɪ/ *n.* (also **willie**) (*pl.* **-ies**) *colloq.* penis. [diminutive of *William*]

willy-nilly /ˌwɪlɪ'nɪlɪ/ *adv.* whether one likes it or not. [later spelling of *will I, nill I* I am willing, I am unwilling]

wilt *–v.* **1** wither, droop. **2** lose energy, flag. *–n.* plant-disease causing wilting. [originally dial.]

wily /'waɪlɪ/ *adj.* (**-ier, -iest**) crafty, cunning. □ **wiliness** *n.*

wimp *n. colloq.* feeble or ineffectual person. □ **wimpish** *adj.* [origin uncertain]

wimple /'wɪmp(ə)l/ *n.* headdress also covering the neck and the sides of the face, worn by some nuns. [Old English]

win *–v.* (**-nn-**; *past* and *past part.* **won** /wʌn/) **1** secure as a result of a fight, contest, bet, effort, etc. **2** be the victor; be victorious in. *–n.* victory in a game etc. □ **win the day** be victorious in battle, argument, etc. **win over** persuade, gain the support of. **win one's spurs** *colloq.* gain distinction or fame. **win through** (or **out**) overcome obstacles. **you can't win** *colloq.* there is no way to succeed or to please. □ **winnable** *adj.* [Old English, = toil]

wince *–n.* start or involuntary shrinking movement of the face, showing pain or distress. *–v.* (**-cing**) give a wince. [Germanic: related to WINK]

wincey /'wɪnsɪ/ *n.* (*pl.* **-s**) lightweight fabric of wool and cotton or linen. [apparently an alteration of *woolsey* in LINSEY-WOOLSEY]

winceyette /ˌwɪnsɪ'et/ *n.* lightweight flannelette.

winch –*n.* **1** crank of a wheel or axle. **2** windlass. –*v.* lift with a winch. [Old English]

wind[1] –*n.* **1** air in natural motion, esp. a current of this. **2 a** breath, esp. as needed in exercise or playing a wind instrument. **b** power of breathing easily. **3** empty talk. **4** gas generated in the bowels etc. **5** wind instruments of an orchestra etc. **6** scent carried by the wind. –*v.* **1** cause to be out of breath by exertion or a blow. **2** make (a baby) bring up wind after feeding. **3** detect the presence of by a scent. □ **get wind of** begin to suspect the existence of. **get** (or **have**) **the wind up** *colloq.* be alarmed or frightened. **in the wind** about to happen. **like the wind** swiftly. **put the wind up** *colloq.* alarm, frighten. **take the wind out of a person's sails** frustrate a person by anticipating an action or remark etc. □ **windless** *adj.* [Old English]

wind[2] /waɪnd/ –*v.* (*past* and *past part.* **wound**) **1** (often as **winding** *adj.*) go in a spiral, curved, or crooked course. **2** make (one's way) thus. **3** wrap closely; coil. **4 a** provide with a coiled thread etc. **b** surround with or as with a coil. **5** wind up (a clock etc.). –*n.* **1** bend or turn in a course. **2** single turn when winding. □ **wind down 1** lower by winding. **2** unwind. **3** draw gradually to a close. **wind off** unwind. **wind up 1** coil the whole of. **2** tighten the coiling or coiled spring of (esp. a clock). **3** *colloq.* **a** increase the intensity of (feelings etc.), excite. **b** provoke (a person) to anger etc. **4** bring to a conclusion; end. **5 a** arrange the affairs of and dissolve (a company). **b** cease business and go into liquidation. **6** *colloq.* arrive finally. [Old English]

windbag *n. colloq.* person who talks a lot but says little of any value.

wind-break *n.* thing serving to break the force of the wind.

windburn *n.* inflammation of the skin caused by exposure to the wind.

windcheater *n.* wind-resistant jacket.

wind-cone *n.* = WIND-SOCK.

wind-down *n. colloq.* gradual lessening of excitement or activity.

winder *n.* winding mechanism, esp. of a clock or watch.

windfall *n.* **1** fruit, esp. an apple, blown to the ground by the wind. **2** unexpected good fortune, esp. a legacy.

winding-sheet *n.* sheet in which a corpse is wrapped for burial.

wind instrument *n.* musical instrument sounded by an air-current, esp. the breath.

wind-jammer *n.* merchant sailing-ship.

windlass /'wɪndləs/ *n.* machine with a horizontal axle for hauling or hoisting. [Old Norse, = winding-pole]

windmill *n.* **1** mill worked by the wind acting on its sails. **2** toy consisting of a stick with curved vanes that revolve in a wind. □ **tilt at windmills** attack an imaginary enemy.

window /'wɪndəʊ/ *n.* **1 a** opening in a wall etc., usu. with glass to admit light etc. **b** the glass itself. **2** space for display behind the window of a shop. **3** window-like opening. **4** opportunity to learn from observation. **5** transparent part in an envelope showing an address. **6** VDU display showing a particular part of the data. □ **windowless** *adj.* [Old Norse, = wind-eye]

window-box *n.* box placed outside a window for growing flowers.

window-dressing *n.* **1** art of arranging a display in a shop-window etc. **2** adroit presentation of facts etc. to give a deceptively favourable impression.

window-pane *n.* pane of glass in a window.

window-seat *n.* **1** seat below a window, esp. in an alcove. **2** seat next to a window in an aircraft, train, etc.

window-shop *v.* look at goods displayed in shop-windows, without buying anything.

window-sill *n.* sill below a window.

windpipe *n.* air-passage from the throat to the lungs.

windscreen *n.* screen of glass at the front of a motor vehicle.

windscreen wiper *n.* blade moving in an arc to keep a windscreen clear of rain etc.

windshield *n. US* = WINDSCREEN.

wind-sock *n.* canvas cylinder or cone on a mast to show the direction of the wind at an airfield etc.

windsurfing *n.* sport of riding on water on a sailboard. □ **windsurf** *v.* **windsurfer** *n.*

windswept *adj.* exposed to or swept back by the wind.

wind-tunnel *n.* tunnel-like device producing an air-stream past models of aircraft etc. for the study of aerodynamics.

wind-up /'waɪndʌp/ –*n.* **1** conclusion; finish. **2** *colloq.* attempt to provoke. –*attrib. adj.* (of a mechanism) operating by being wound up.

windward /'wɪndwəd/ –*adj.* & *adv.* on the side from which the wind is blowing. –*n.* windward direction.

windy *adj.* (**-ier, -iest**) **1** stormy with or exposed to wind. **2** generating or characterized by flatulence. **3** *colloq.* wordy. **4**

colloq. nervous, frightened. □ **windiness** *n.* [Old English: related to WIND[1]]

wine *–n.* **1** fermented grape juice as an alcoholic drink. **2** fermented drink resembling this made from other fruits etc. **3** dark-red colour of red wine. *–v.* **(-ning)** (esp. in phr. **wine and dine**) **1** drink wine. **2** entertain with wine. [Old English]

wine bar *n.* bar or small restaurant where wine is the main drink available.

winebibber *n.* tippler.

wine cellar *n.* **1** cellar for storing wine. **2** its contents.

wineglass *n.* glass for wine, usu. with a stem and foot.

wine list *n.* list of wines available in a restaurant etc.

winepress *n.* press in which grapes are squeezed in making wine.

wine vinegar *n.* vinegar made from wine as distinct from malt etc.

wine waiter *n.* waiter responsible for serving wine.

wing *–n.* **1** each of the limbs or organs by which a bird etc. is able to fly. **2** winglike structure supporting an aircraft. **3** part of a building etc. extended in a certain direction. **4 a** forward player at either end of a line in football, hockey, etc. **b** side part of a playing-area. **5** (in *pl.*) sides of a theatre stage. **6** polarized section of a political party in terms of its views. **7** flank of a battle array. **8** the part of a vehicle over a wheel. **9** air-force unit of several squadrons or groups. *–v.* **1** travel or traverse on wings. **2** wound in a wing or an arm. **3** equip with wings. **4** enable to fly; send in flight. □ **on the wing** flying, in flight. **take under one's wing** treat as a protégé. **take wing** fly away. □ **winged** *adj.* **winglike** *adj.* [Old Norse]

wing-case *n.* horny cover of an insect's wing.

wing-chair *n.* chair with side-pieces at the top of a high back.

wing-collar *n.* man's high stiff collar with turned-down corners.

wing commander *n.* RAF officer next below group captain.

winger *n.* **1** (in football etc.) wing player. **2** (in *comb.*) member of a specified political wing.

wing-nut *n.* nut with projections for the fingers to turn it.

wing-span *n.* (also **wing-spread**) measurement right across the wings.

wink *–v.* **1** (often foll. by *at*) close and open one eye quickly, esp. as a signal. **2** close and open (one or both eyes) quickly. **3** (of a light etc.) twinkle; (of an indicator) flash on and off. *–n.* **1** act of winking. **2** *colloq.* short sleep. □ **in a wink** very quickly. **wink at** purposely avoid seeing; pretend not to notice. □ **winker** *n.* (in sense 3 of *v.*). [Old English]

winkle /ˈwɪŋk(ə)l/ *–n.* small edible sea snail. *–v.* **(-ling)** (foll. by *out*) extract with difficulty. [abbreviation of PERIWINKLE[2]]

winkle-picker *n. slang* long pointed shoe.

winner *n.* **1** person etc. that wins. **2** *colloq.* successful or highly promising idea etc.

winning *–adj.* **1** having or bringing victory. **2** attractive (*winning smile*). *–n.* (in *pl.*) money won. □ **winningly** *adv.*

winning-post *n.* post marking the end of a race.

winnow /ˈwɪnəʊ/ *v.* **1** blow (grain) free of chaff etc. by an air-current. **2** (foll. by *out*, *away*, *from*, etc.) get rid of (chaff etc.) from grain. **3** sift, examine (evidence etc.). [Old English: related to WIND[1]]

wino /ˈwaɪnəʊ/ *n.* (*pl.* **-s**) *slang* alcoholic.

winsome *adj.* attractive, engaging. □ **winsomely** *adv.* **winsomeness** *n.* [Old English, = joyous]

winter *–n.* **1** coldest and last season of the year. **2** (*attrib.*) characteristic of or fit for winter. *–v.* (usu. foll. by *at*, *in*) pass the winter. [Old English]

winter garden *n.* garden or conservatory of plants flourishing in winter.

wintergreen *n.* a kind of plant remaining green all winter.

winter jasmine *n.* jasmine with yellow flowers in winter.

winter solstice *n.* about 22 Dec.

winter sports *n.pl.* sports performed on snow or ice.

wintertime *n.* season or period of winter.

wintry *adj.* **(-ier, -iest) 1** characteristic of winter. **2** lacking warmth; unfriendly. □ **wintriness** *n.*

winy *adj.* **(-ier, -iest)** wine-flavoured.

wipe *–v.* **(-ping) 1** clean or dry the surface of by rubbing. **2** rub (a cloth) over a surface. **3** spread (a liquid etc.) over a surface by rubbing. **4** (often foll. by *away*, *off*, etc.) **a** clear or remove by wiping. **b** erase or eliminate completely. *–n.* **1** act of wiping. **2** piece of specially treated material for wiping (*antiseptic wipes*). □ **wipe down** clean (a wall etc.) by wiping. **wipe the floor with** *colloq.* inflict a humiliating defeat on. **wipe off** annul (a debt etc.). **wipe out 1** destroy, annihilate, obliterate. **2** clean the inside of. **wipe up 1** dry (dishes etc.). **2** take up (a liquid etc.) by wiping. [Old English]

wiper *n.* = WINDSCREEN WIPER.

wire –*n.* **1 a** metal drawn out into a thread or thin flexible rod. **b** piece of this. **c** (*attrib.*) made of wire. **2** length of this for fencing or to carry an electric current etc. **3** *colloq.* telegram. –*v.* (**-ring**) **1** provide, fasten, strengthen, etc., with wire. **2** (often foll. by *up*) install electrical circuits in (a building, equipment, etc.). **3** *colloq.* telegraph. □ **get one's wires crossed** become confused and misunderstood. [Old English]

wire-haired *adj.* (esp. of a dog) with stiff or wiry hair.

wireless *n.* radio; radio receiving set.

wire netting *n.* netting of meshed wire.

wire-tapping *n.* tapping of telephone lines to eavesdrop.

wire wool *n.* mass of fine wire for scouring or rubbing down.

wireworm *n.* destructive larva of a kind of beetle.

wiring *n.* system or installation of wires providing electrical circuits.

wiry *adj.* (**-ier, -iest**) **1** sinewy, untiring. **2** like wire; tough, coarse. □ **wiriness** *n.*

wisdom /'wɪzdəm/ *n.* **1** experience and knowledge together with the power of applying them. **2** prudence; common sense. **3** wise sayings. [Old English: related to WISE[1]]

wisdom tooth *n.* hindmost molar usu. cut at about 20 years of age.

wise[1] /waɪz/ *adj.* **1** having, showing, or dictated by wisdom. **2** prudent, sensible. **3** having knowledge (often in *comb.*: *streetwise*; *worldly-wise*). **4** suggestive of wisdom. **5** *US colloq.* alert, crafty. □ **be** (or **get**) **wise to** *colloq.* be (or become) aware of. **none the wiser** knowing no more than before. **put wise** (often foll. by *to*) *colloq.* inform (of). **wise up** esp. *US colloq.* put or get wise. □ **wisely** *adv.* [Old English]

wise[2] /waɪz/ *n. archaic* way, manner, or degree. □ **in no wise** not at all. [Old English]

-wise *suffix* forming adjectives and adverbs of manner (*clockwise*; *lengthwise*) or respect (*moneywise*).

■ **Usage** More fanciful phrase-based combinations, such as *employment-wise* (= as regards employment) are restricted to informal contexts.

wiseacre /'waɪz,eɪkə(r)/ *n.* person who affects a wise manner. [Dutch *wijsseggher* soothsayer]

wisecrack *colloq.* –*n.* smart pithy remark. –*v.* make a wisecrack.

wise guy *n. colloq.* know-all.

wise man *n.* wizard, esp. one of the Magi.

wisent /'wi:z(ə)nt/ *n.* European bison. [German: cf. BISON]

wish –*v.* **1** (often foll. by *for*) have or express a desire or aspiration for (*wish for happiness*). **2** have as a desire or aspiration (*I wish I could sing*). **3** want or demand (*I wish to go*; *I wish you to do it*). **4** express one's hopes for (*wish you success*). **5** (foll. by *on, upon*) *colloq.* foist on. –*n.* **1 a** desire, request. **b** expression of this. **2** thing desired. □ **best** (or **good**) **wishes** hopes felt or expressed for another's happiness etc. [Old English]

wishbone *n.* forked bone between the neck and breast of a fowl often broken between two people, the longer portion entitling the holder to make a wish.

wishful *adj.* (often foll. by *to* + infin.) desiring. □ **wishfully** *adv.*

wish-fulfilment *n.* tendency for subconscious desire to be satisfied in fantasy.

wishful thinking *n.* belief founded on wishes rather than facts.

wishing-well *n.* well into which coins are dropped and a wish is made.

wishy-washy /'wɪʃɪ,wɒʃɪ/ *adj. colloq.* **1** feeble in quality or character. **2** weak, watery. [from WASH]

wisp *n.* **1** small bundle or twist of straw etc. **2** small separate quantity of smoke, hair, etc. **3** small thin person etc. □ **wispy** *adj.* (**-ier, -iest**). [origin uncertain]

wisteria /wɪ'stɪərɪə/ *n.* (also **wistaria**) climbing plant with blue, purple, or white hanging flowers. [*Wistar*, name of an anatomist]

wistful *adj.* yearning, mournfully expectant or wishful. □ **wistfully** *adv.* **wistfulness** *n.* [apparently an assimilation of obsolete *wistly* 'intently' to *wishful*]

wit *n.* **1** (in *sing.* or *pl.*) intelligence; quick understanding. **2 a** unexpected combining or contrasting of ideas or expressions. **b** power of giving pleasure by this. **3** person possessing such power. □ **at one's wit's** (or **wits'**) **end** utterly at a loss or in despair. **have** (or **keep**) **one's wits about one** be alert. **live by one's wits** live by ingenious or crafty expedients, without a settled occupation. **out of one's wits** mad. **to wit** that is to say, namely. [Old English]

witch *n.* **1** sorceress, woman supposed to have dealings with the Devil or evil spirits. **2** old hag. **3** fascinating girl or woman. [Old English]

witchcraft *n.* **1** use of magic. **2** bewitching charm.

witch-doctor *n.* tribal magician of primitive people.

witchery *n.* = WITCHCRAFT.

witches' sabbath *n.* supposed midnight orgy of the Devil and witches.

witch-hazel *n.* (also **wych-hazel**) **1** American shrub with bark yielding an astringent lotion. **2** this lotion.

witch-hunt *n.* campaign against persons suspected of unpopular or unorthodox views, esp. Communists.

with /wɪð/ *prep.* expressing: **1** instrument or means used (*cut with a knife*). **2** **a** association or company (*lives with his mother*; *works with Shell*). **b** parting of company (*dispense with*). **3** cause (*shiver with fear*). **4** possession (*man with dark hair*; *filled with water*). **5** circumstances (*sleep with the window open*). **6** manner (*handle with care*). **7** agreement (*sympathize with*). **8** disagreement, antagonism (*incompatible with*; *quarrel with*). **9** understanding (*are you with me?*). **10** reference or regard (*be patient with them*; *how are things with you?*). □ **away** (or **in** or **out** etc.) **with** (as *int.*) take, send, or put (a person or thing) away (or in or out etc.). **with it** *colloq.* **1** up to date. **2** alert and comprehending. **with that** thereupon. [Old English]

withdraw /wɪð'drɔː/ *v.* (*past* **withdrew**; *past part.* **withdrawn**) **1** pull or take aside or back. **2** discontinue, cancel, retract. **3** remove; take away. **4** take (money) out of an account. **5** retire or move apart. **6** (as **withdrawn** *adj.*) abnormally shy and unsociable; mentally detached. [from WITH = away]

withdrawal *n.* **1** withdrawing or being withdrawn. **2** process of ceasing to take an addictive drug etc., often with an unpleasant reaction (*withdrawal symptoms*). **3** = COITUS INTERRUPTUS.

withe /wɪθ, wɪð, waɪð/ *n.* (also **withy** /'wɪðɪ/) (*pl.* **withes** or **withies**) tough flexible shoot, esp. of willow, used for binding, basketwork, etc. [Old English]

wither /'wɪðə(r)/ *v.* **1** (often foll. by *up*) make or become dry and shrivelled. **2** (often foll. by *away*) deprive of or lose vigour or freshness. **3** (esp. as **withering** *adj.*) blight with scorn etc. □ **witheringly** *adv.* [apparently var. of WEATHER]

withers /'wɪðəz/ *n.pl.* ridge between a horse's shoulder-blades. [obsolete *wither* against (the collar)]

withershins var. of WIDDERSHINS.

withhold /wɪð'həʊld/ *v.* (*past* and *past part.* **-held**) **1** hold back; restrain. **2** refuse to give, grant, or allow. [from WITH = away]

within /wɪ'ðɪn/ *–adv.* **1** inside. **2** indoors. **3** in spirit (*pure within*). *–prep.* **1** inside. **2** **a** not beyond or out of. **b** not transgressing or exceeding. **3** not further off than (*within three miles*; *within ten days*). □ **within one's grasp** close enough to be obtained. **within reach** (or **sight**) **of** near enough to be reached or seen. [Old English: related to WITH, IN]

without /wɪ'ðaʊt/ *–prep.* **1** not having or feeling or showing. **2** with freedom from. **3** in the absence of. **4** with neglect or avoidance of. **5** *archaic* outside. *–adv. archaic* or *literary* **1** outside. **2** out of doors. [Old English: related to WITH, OUT]

withstand /wɪð'stænd/ *v.* (*past* and *past part.* **-stood**) oppose, hold out against. [Old English: related to WITH, STAND]

withy var. of WITHE.

witless *adj.* foolish, crazy. [Old English: related to WIT]

witness /'wɪtnɪs/ *–n.* **1** = EYEWITNESS. **2** **a** person giving sworn testimony. **b** person attesting another's signature to a document. **3** (foll. by *to*, *of*) person or thing whose existence etc. attests or proves something. **4** testimony, evidence, confirmation. *–v.* **1** be an eyewitness of. **2** be witness to the authenticity of (a signature etc.). **3** serve as evidence or an indication of. **4** (foll. by *against*, *for*, *to*) give or serve as evidence. □ **bear witness to** (or **of**) **1** attest the truth of. **2** state one's belief in. **call to witness** appeal to for confirmation etc. [Old English: related to WIT]

witness-box *n.* (*US* **witness-stand**) enclosure in a lawcourt from which witnesses give evidence.

witter /'wɪtə(r)/ *v.* (often foll. by *on*) *colloq.* chatter annoyingly or on trivial matters. [origin unknown]

witticism /'wɪtɪˌsɪz(ə)m/ *n.* witty remark. [from WITTY]

wittingly /'wɪtɪŋlɪ/ *adv.* aware of what one is doing; intentionally. [from WIT]

witty *adj.* (**-ier, -iest**) showing esp. verbal wit. □ **wittily** *adv.* **wittiness** *n.* [Old English: related to WIT]

wives *pl.* of WIFE.

wizard /'wɪzəd/ *–n.* **1** sorcerer; magician. **2** person of remarkable powers, genius. *–adj. slang* wonderful. □ **wizardry** *n.* [from WISE[1]]

wizened /'wɪz(ə)nd/ *adj.* shrivelled-looking. [Old English]

WNW *abbr.* west-north-west.

WO *abbr.* Warrant-Officer.

woad *n.* **1** plant yielding a blue dye. **2** dye from this. [Old English]

wobble /'wɒb(ə)l/ *–v.* (**-ling**) **1** sway from side to side. **2** stand or go unsteadily; stagger. **3** waver, vacillate. *–n.* state or instance of wobbling. [cf. Low German *wabbeln*]

wobbly *adj.* (**-ier, -iest**) **1** tending to wobble. **2** wavy (*wobbly line*). **3** weak after illness. **4** wavering, insecure (*the economy was wobbly*). □ **throw a wobbly** *slang* have a tantrum or fit of nerves.

wodge *n. colloq.* chunk, lump. [alteration of WEDGE]

woe /wəʊ/ *n.* **1** affliction; bitter grief. **2** (in *pl.*) calamities. □ **woe betide** see BETIDE. **woe is me** alas. [Old English]

woebegone /ˈwəʊbɪˌgɒn/ *adj.* dismal-looking. [from WOE, *begone* = surrounded]

woeful *adj.* **1** sorrowful. **2** causing or feeling affliction. **3** very bad. □ **woefully** *adv.*

wog *n. slang offens.* foreigner, esp. a non-White one. [origin unknown]

woggle /ˈwɒg(ə)l/ *n.* leather etc. ring through which the ends of a Scout's neckerchief are passed at the neck. [origin unknown]

wok *n.* bowl-shaped frying-pan used in esp. Chinese cookery. [Chinese]

woke *past* of WAKE[1].

woken *past part.* of WAKE[1].

wold /wəʊld/ *n.* high open uncultivated land or moor. [Old English]

wolf /wʊlf/ *–n.* (*pl.* **wolves** /wʊlvz/) **1** wild animal related to the dog, usu. hunting in packs. **2** *slang* man who seduces women. *–v.* (often foll. by *down*) devour greedily. □ **cry wolf** raise false alarms. **keep the wolf from the door** avert starvation. □ **wolfish** *adj.* [Old English]

wolfhound *n.* dog of a kind used orig. to hunt wolves.

wolf in sheep's clothing *n.* hostile person who pretends friendship.

wolfram /ˈwʊlfrəm/ *n.* **1** tungsten. **2** tungsten ore. [German]

wolfsbane *n.* aconite.

wolf-whistle *n.* whistle made by a man to a sexually attractive woman.

wolverine /ˈwʊlvəˌriːn/ *n.* N. American animal of the weasel family. [related to WOLF]

wolves *pl.* of WOLF.

woman /ˈwʊmən/ *n.* (*pl.* **women** /ˈwɪmɪn/) **1** adult human female. **2** the female sex. **3** *colloq.* wife or girlfriend. **4** (prec. by *the*) feminine characteristics (*brought out the woman in him*). **5** (*attrib.*) female (*woman doctor*). **6** (in *comb.*) woman of a specified nationality, skill, etc. (*Englishwoman*; *horsewoman*). **7** *colloq.* charwoman. [Old English]

womanhood *n.* **1** female maturity. **2** womanly instinct. **3** womankind.

womanish *adj. derog.* effeminate, unmanly.

womanize *v.* (also **-ise**) (**-zing** or **-sing**) chase after women; philander. □ **womanizer** *n.*

womankind *n.* (also **womenkind**) women in general.

womanly *adj.* having or showing qualities associated with women. □ **womanliness** *n.*

womb /wuːm/ *n.* organ of conception and gestation in a woman and other female mammals. [Old English]

wombat /ˈwɒmbæt/ *n.* burrowing plant-eating Australian marsupial. [Aboriginal]

women *pl.* of WOMAN.

womenfolk *n.* **1** women in general. **2** the women in a family.

womenkind var. of WOMANKIND.

women's libber *n. colloq.* supporter of women's liberation.

women's liberation *n.* (also **women's lib**) *colloq.* movement urging the liberation of women from domestic duties and subservient status.

women's rights *n.pl.* position of legal and social equality with men.

won *past* and *past part.* of WIN.

wonder /ˈwʌndə(r)/ *–n.* **1** emotion, esp. admiration, excited by what is unexpected, unfamiliar, or inexplicable. **2** strange or remarkable thing, specimen, event, etc. **3** (*attrib.*) having marvellous or amazing properties etc. (*wonder drug*; *wonder woman*). *–v.* **1** be filled with wonder or great surprise. **2** (foll. by *that*) be surprised to find. **3** desire or be curious to know (*I wonder what the time is*). □ **I shouldn't wonder** *colloq.* I think it likely. **no** (or **small**) **wonder** one cannot be surprised. **work** (or **do**) **wonders 1** do miracles. **2** be remarkably effective. [Old English]

wonderful *adj.* very remarkable or admirable. □ **wonderfully** *adv.* [Old English]

wonderland *n.* **1** fairyland. **2** land of surprises or marvels.

wonderment *n.* surprise, awe.

wondrous /ˈwʌndrəs/ *poet.* *–adj.* wonderful. *–adv.* wonderfully (*wondrous kind*).

wonky /ˈwɒŋkɪ/ *adj.* (**-ier, -iest**) *slang* **1** crooked, askew. **2** loose, unsteady. **3** unreliable. [fanciful]

wont /wəʊnt/ *–predic. adj. archaic* or *literary* (foll. by *to* + infin.) accustomed. *–n. formal* or *joc.* what is customary, one's habit. [Old English]

won't /wəʊnt/ *contr.* will not.

wonted /ˈwəʊntɪd/ *attrib. adj.* habitual, usual.

woo *v.* (**woos, wooed**) **1** court; seek the hand or love of. **2** try to win (fame, fortune, etc.). **3** seek the favour or support of. **4** coax or importune. □ **wooer** *n.* [Old English]

wood /wʊd/ *n.* **1 a** hard fibrous substance of the trunk or branches of a tree or shrub. **b** this for timber or fuel. **2** (in *sing.* or *pl.*) growing trees densely occupying a tract of land. **3** wooden cask for wine etc. **4** wooden-headed golf club. **5** = BOWL[2] *n.* 1. □ **not see the wood for the trees** fail to grasp the main issue from over-attention to details. **out of the wood** (or **woods**) out of danger or difficulty. [Old English]

wood anemone *n.* a wild spring-flowering anemone.

woodbine *n.* honeysuckle.

woodchuck *n.* N. American marmot. [American Indian name]

woodcock *n.* game-bird related to the snipe.

woodcraft *n.* **1** knowledge of woodland, esp. in camping etc. **2** skill in woodwork.

woodcut *n.* **1** relief cut on wood. **2** print made from this.

woodcutter *n.* person who cuts timber.

wooded *adj.* having woods or many trees.

wooden /'wʊd(ə)n/ *adj.* **1** made of wood. **2** like wood. **3 a** stiff, clumsy. **b** expressionless. □ **woodenly** *adv.* **woodenness** *n.*

woodland *n.* (often *attrib.*) wooded country, woods.

woodlouse *n.* (*pl.* **-lice**) small land crustacean with many legs.

woodman *n.* forester.

woodpecker *n.* bird that taps tree-trunks in search of insects.

woodpigeon *n.* dove with white patches like a ring round its neck.

woodpile *n.* pile of wood, esp. for fuel.

wood pulp *n.* wood-fibre prepared for paper-making.

woodruff *n.* white-flowered plant with fragrant leaves.

woodshed *n.* shed where wood for fuel is stored. □ **something nasty in the woodshed** *colloq.* shocking thing kept secret.

woodwind *n.* **1** wind instruments that were (mostly) orig. made of wood, e.g. the flute, clarinet, oboe, and saxophone. **2** one such instrument.

woodwork *n.* **1** making of things in wood. **2** things made of wood. □ **crawl out of the woodwork** *colloq.* (of something distasteful) appear.

woodworm *n.* **1** wood-boring larva of a kind of beetle. **2** condition of wood affected by this.

woody *adj.* (**-ier, -iest**) **1** wooded. **2** like or of wood. □ **woodiness** *n.*

woodyard *n.* yard where wood is used or stored.

woody nightshade *n.* a kind of nightshade with poisonous red berries.

woof[1] /wʊf/ *–n.* gruff bark of a dog. *–v.* give a woof. [imitative]

woof[2] /wu:f/ *n.* = WEFT 1. [Old English: related to WEB]

woofer /'wu:fə(r)/ *n.* loudspeaker for low frequencies. [from WOOF[1]]

wool /wʊl/ *n.* **1** fine soft wavy hair from the fleece of sheep etc. **2** woollen yarn or cloth or clothing. **3** wool-like substance (*steel wool*). □ **pull the wool over a person's eyes** deceive a person. [Old English]

wool-gathering *n.* absent-mindedness.

woollen (*US* **woolen**) *–adj.* made wholly or partly of wool. *–n.* **1** woollen fabric. **2** (in *pl.*) woollen garments. [Old English]

woolly *–adj.* (**-ier, -iest**) **1** bearing wool. **2** like wool. **3** woollen (*a woolly cardigan*). **4** (of a sound) indistinct. **5** (of thought) vague or confused. *–n.* (*pl.* **-ies**) *colloq.* woollen garment, esp. a pullover. □ **woolliness** *n.*

Woolsack *n.* **1** Lord Chancellor's wool-stuffed seat in the House of Lords. **2** his position.

woozy *adj.* (**-ier, -iest**) *colloq.* **1** dizzy or unsteady. **2** slightly drunk. □ **woozily** *adv.* **wooziness** *n.* [origin unknown]

wop *n. slang offens.* Italian or other S. European. [origin uncertain]

Worcester sauce /'wʊstə/ *n.* a pungent sauce. [*Worcester* in England]

word /wɜ:d/ *–n.* **1** meaningful element of speech, usu. shown with a space on either side of it when written or printed. **2** speech, esp. as distinct from action. **3** one's promise or assurance. **4** (in *sing.* or *pl.*) thing said, remark, conversation. **5** (in *pl.*) text of a song or an actor's part. **6** (in *pl.*) angry talk (*have words*). **7** news, message (*send word*). **8** command (*gave the word to begin*). *–v.* put into words; select words to express. □ **in other words** expressing the same thing differently. **in so many words** in those very words; explicitly. **in a** (or **one**) **word** briefly. **my** (or **upon my**) **word** exclamation of surprise etc. **take a person at his** or **her word** interpret a person's words literally. **take a person's word for it** believe a person's statement without investigation etc. **the Word** (or **Word of God**) the Bible. **word for word** in exactly the same or (of translation) corresponding words. □ **wordless** *adj.* [Old English]

word-blindness *n.* = DYSLEXIA.

word-game *n.* game involving the making or selection etc. of words.

wording *n.* form of words used.

word of mouth *n.* speech (only).

word-perfect *adj.* knowing one's part etc. by heart.

wordplay *n.* witty use of words, esp. punning.

word processor *n.* computer program, or device incorporating a computer, used for storing text entered from a keyboard, making corrections, and providing a printout. □ **word-process** *v.* **word processing** *n.*

wordy *adj.* (-**ier**, -**iest**) using or expressed in too many words. □ **wordily** *adv.* **wordiness** *n.*

wore *past* of WEAR.

work /wɜːk/ –*n.* **1** application of mental or physical effort to a purpose; use of energy. **2** task to be undertaken. **3** thing done or made by work; result of an action. **4** employment or occupation etc., esp. as a means of earning income. **5** literary or musical composition. **6** actions or experiences of a specified kind (*nice work!*). **7** (in *comb.*) things made of a specified material or with specified tools etc. (*ironwork*; *needlework*). **8** (in *pl.*) operative part of a clock or machine. **9** *Physics* the exertion of force overcoming resistance or producing molecular change. **10** (in full **the works**) *colloq.* **a** all that is available or needed. **b** full, esp. harsh, treatment. **11** (in *pl.*) operations of building or repair (*road works*). **12** (in *pl.*; often treated as *sing.*) factory. **13** (usu. in *pl.*) *Theol.* meritorious act. **14** (usu. in *pl.* or in *comb.*) defensive structure (*earthworks*). –*v.* **1** do work; be engaged in bodily or mental activity. **2** be employed in certain work (*works in industry*). **3** make efforts (*works for peace*). **4** (foll. by *in*) be a craftsman in (a material). **5** operate or function, esp. effectively (*how does this machine work?*; *your idea will not work*). **6** operate, manage, control (*cannot work the machine*). **7 a** put or keep in operation or at work; cause to toil (*works the staff hard*). **b** cultivate (land). **8 a** bring about; produce as a result (*worked miracles*). **b** *colloq.* arrange (matters) (*worked it so that we could go*; *can you work things for us?*). **9** knead, hammer; bring to a desired shape or consistency. **10** do, or make by, needlework etc. **11** (cause to) progress or penetrate, or make (one's way), gradually or with difficulty in a specified way (*worked the peg into the hole*; *worked our way through the crowd*). **12** (foll. by *loose* etc.) gradually become (loose etc.) by constant movement. **13** artificially excite (*worked themselves into a rage*). **14 a** purchase with one's labour instead of money (*work one's passage*). **b** obtain by labour the money for (*worked my way through university*). **15** (foll. by *on*, *upon*) have influence. **16** be in motion or agitated; ferment. □ **at work** in action, engaged in work. **get worked up** become angry, excited, or tense. **have one's work cut out** be faced with a hard task. **work in** find a place for. **work off** get rid of by work or activity. **work out 1 a** solve (a sum) or find (an amount) by calculation. **b** solve, understand (a problem, person, etc.). **2** (foll. by *at*) be calculated. **3** give a definite result (*this sum will not work out*). **4** have a result (*the plan worked out well*). **5** provide for the details of (*has worked out a scheme*). **6** engage in physical exercise or training. **work over 1** examine thoroughly. **2** *colloq.* treat with violence. **work through** arrive at an understanding of (a problem) etc. **work to rule** (as a protest) follow official working rules exactly in order to reduce output. **work up 1** bring gradually to an efficient or (of a painting etc.) advanced state. **2** (foll. by *to*) advance gradually to a climax etc. **3** elaborate or excite by degrees. **4** mingle (ingredients). **5** learn (a subject) by study. **work wonders** see WONDER. [Old English]

workable *adj.* that can be worked, will work, or is worth working. □ **workability** /-'bɪlɪtɪ/ *n.*

workaday *adj.* ordinary, everyday, practical.

workaholic /ˌwɜːkə'hɒlɪk/ *n. colloq.* person addicted to working.

work-basket *n.* basket for sewing materials.

workbench *n.* bench for manual work, esp. carpentry.

workbook *n.* student's book with exercises.

workbox *n.* box for tools, needlework, etc.

work camp *n.* camp at which community work is done, esp. by young volunteers.

workday *n.* day on which work is usually done.

worker *n.* **1** person who works, esp. for an employer. **2** neuter bee or ant. **3** person who works hard.

work experience *n.* scheme intended to give young people temporary experience of employment.

workforce *n.* **1** workers engaged or available. **2** number of these.

workhouse *n. hist.* public institution for the poor of a parish.

working –*attrib. adj.* **1 a** engaged in work (*working mother*; *working man*).

b while so engaged (*all his working life*; *in working hours*). **2** functioning or able to function (*working model*). —*n.* **1** activity of work. **2** functioning. **3** mine or quarry. **4** (usu. in *pl.*) machinery, mechanism.

working capital *n.* capital actually used in a business.

working class *n.* social class employed, esp. in manual or industrial work, for wages. □ **working-class** *adj.*

working day *n.* **1** workday. **2** part of the day devoted to work.

working hypothesis *n.* hypothesis as a basis for action.

working knowledge *n.* knowledge adequate to work with.

working lunch *n.* lunch at which business is conducted.

working order *n.* condition in which a machine works.

working party *n.* group of people appointed to study and advise on a particular problem.

workload *n.* amount of work to be done.

workman *n.* **1** man employed to do manual labour. **2** person with regard to skill in a job (*a good workman*).

workmanlike *adj.* competent, showing practised skill.

workmanship *n.* degree of skill in doing a task or of finish in the product made.

workmate *n.* person working alongside another.

work of art *n.* fine picture, poem, building, etc.

workout *n.* session of physical exercise or training.

workpiece *n.* thing worked on with a tool or machine.

workplace *n.* place at which a person works.

workroom *n.* room for working in.

worksheet *n.* **1** paper for recording work done or in progress. **2** paper listing questions or activities for students etc. to work through.

workshop *n.* **1** room or building in which goods are manufactured. **2** place or meeting for concerted discussion or activity (*dance workshop*).

work-shy *adj.* disinclined to work.

workstation *n.* **1** location of a stage in a manufacturing process. **2** computer terminal or the desk etc. where this is located.

work study *n.* assessment of methods of working so as to achieve maximum productivity.

work table *n.* table for working at.

worktop *n.* flat surface for working on, esp. in a kitchen.

work-to-rule *n.* working to rule.

world /wɜːld/ *n.* **1 a** the earth, or a planetary body like it. **b** its countries and people. **2** the universe, all that exists. **3 a** the time, state, or scene of human existence. **b** (prec. by *the*, *this*) mortal life. **4** secular interests and affairs. **5** human affairs; active life. **6** average, respectable, or fashionable people or their customs or opinions. **7** all that concerns or all who belong to a specified class or sphere of activity (*the world of sport*). **8** (foll. by *of*) vast amount. **9** (*attrib.*) affecting many nations, of all nations (*world politics*; *world champion*). □ **bring** (or **come**) **into the world** give birth (or be born). **for all the world** (foll. by *like*, *as if*) precisely. **in the world** of all; at all (*what in the world is it?*). **man** (or **woman**) **of the world** person experienced and practical in human affairs. **out of this world** *colloq.* extremely good etc. **think the world of** have a very high regard for. [Old English]

world-beater *n.* person or thing surpassing all others.

world-class *adj.* of a quality or standard regarded as high throughout the world.

World Cup *n.* competition between sporting teams from various countries.

world-famous *adj.* known throughout the world.

worldly *adj.* (**-ier**, **-iest**) **1** of the affairs of the world, temporal, earthly (*worldly goods*). **2** experienced in life, sophisticated, practical. □ **worldliness** *n.*

worldly-wise *adj.* prudent or shrewd in one's dealings with the world.

world music *n.* pop music that incorporates local or ethnic elements (esp. from the developing world).

world war *n.* war involving many major nations.

world-weary *adj.* bored with human affairs. □ **world-weariness** *n.*

worldwide —*adj.* occurring in or known in all parts of the world. —*adv.* throughout the world.

worm /wɜːm/ —*n.* **1** any of various types of creeping invertebrate animals with long slender bodies and no limbs. **2** larva of an insect, esp. in fruit or wood. **3** (in *pl.*) intestinal parasites. **4** insignificant or contemptible person. **5** spiral part of a screw. —*v.* **1** (often *refl.*) move with a crawling motion. **2** *refl.* (foll. by *into*) insinuate oneself into favour etc. **3** (foll. by *out*) obtain (a secret etc.) by cunning persistence. **4** rid (a dog etc.) of worms. [Old English]

worm-cast *n.* convoluted mass of earth left on the surface by a burrowing earthworm.

wormeaten *adj.* **1** eaten into by worms; decayed. **2** old and dilapidated.

worm-hole *n.* hole left by the passage of a worm.

worm's-eye view *n.* view from below or from a humble position.

wormwood /'wɜ:mwʊd/ *n.* **1** plant with a bitter aromatic taste. **2** bitter mortification; source of this. [Old English: cf. VERMOUTH]

wormy *adj.* (**-ier, -iest**) **1** full of worms. **2** wormeaten. □ **worminess** *n.*

worn /wɔ:n/ *past part.* of WEAR[1]. *–adj.* **1** damaged by use or wear. **2** looking tired and exhausted.

worrisome /'wʌrɪsəm/ *adj.* causing worry.

worry /'wʌrɪ/ *–v.* (**-ies, -ied**) **1** give way to anxiety. **2** harass, importune; be a trouble or anxiety to. **3** (of a dog etc.) shake or pull about with the teeth. **4** (as **worried** *adj.*) uneasy. *–n.* (*pl.* **-ies**) **1** thing that causes anxiety or disturbs tranquility. **2** disturbed state of mind; anxiety. □ **worrier** *n.* [Old English, = strangle]

worry beads *n.pl.* string of beads manipulated with the fingers to occupy or calm oneself.

worse /wɜ:s/ *–adj.* **1** more bad. **2** (*predic.*) in or into worse health or a worse condition (*is getting worse*). *–adv.* more badly; more ill. *–n.* **1** worse thing or things (*you might do worse than accept*). **2** (prec. by *the*) worse condition (*a change for the worse*). □ **none the worse** (often foll. by *for*) not adversely affected (by). **the worse for wear 1** damaged by use. **2** injured. **worse luck** unfortunately. **worse off** in a worse (esp. financial) position. [Old English]

worsen *v.* make or become worse.

worship /'wɜ:ʃɪp/ *–n.* **1 a** homage or service to a deity. **b** acts, rites, or ceremonies of this. **2** adoration, devotion. **3** (**Worship**) (prec. by *His, Her, Your*) forms of description or address for a mayor, certain magistrates, etc. *–v.* (**-pp-**; *US* **-p-**) **1** adore as divine; honour with religious rites. **2** idolize or regard with adoration. **3** attend public worship. **4** be full of adoration. □ **worshipper** *n.* [Old English: related to WORTH, -SHIP]

worshipful *adj.* (also **Worshipful**) *archaic* (esp. in old titles of companies or officers) honourable, distinguished.

worst /wɜ:st/ *–adj.* most bad. *–adv.* most badly. *–n.* worst part or possibility (*prepare for the worst*). *–v.* get the better of; defeat. □ **at its** etc. **worst** in the worst state. **at worst** (or **the worst**) in the worst possible case. **do your worst** expression of defiance. **get the worst of it** be defeated. **if the worst comes to the worst** if the worst happens. [Old English: related to WORSE]

worsted /'wʊstɪd/ *n.* **1** fine woollen yarn. **2** fabric made from this. [*Worste(a)d* in Norfolk]

wort /wɜ:t/ *n.* **1** *archaic* (except in names) plant (*liverwort*). **2** infusion of malt before it is fermented into beer. [Old English]

worth /wɜ:θ/ *–predic. adj.* (*used like a preposition*) **1** of a value equivalent to (*is worth £50*; *is worth very little*). **2** such as to justify or repay (*worth doing*; *not worth the trouble*). **3** possessing or having property amounting to (*is worth a million pounds*). *–n.* **1** what a person or thing is worth; the (usu. high) merit of (*of great worth*). **2** equivalent of money in a commodity (*ten pounds' worth of petrol*). □ **for all one is worth** *colloq.* with one's utmost efforts. **for what it is worth** without a guarantee of its truth or value. **worth it** *colloq.* worth while. **worth one's salt** see SALT. **worth one's weight in gold** see WEIGHT. **worth while** (or **one's while**) see WHILE. [Old English]

worthless *adj.* without value or merit. □ **worthlessness** *n.*

worthwhile *adj.* that is worth the time, effort, or money spent.

■ **Usage** See note at *while.*

---

worthy /'wɜ:ðɪ/ *–adj.* (**-ier, -iest**) **1** deserving respect, estimable (*lived a worthy life*). **2** entitled to (esp. condescending) recognition (*a worthy old couple*). **3 a** (foll. by *of* or *to* + infin.) deserving (*worthy of a mention*). **b** (foll. by *of*) adequate or suitable to the dignity etc. of (*words worthy of the occasion*). *–n.* (*pl.* **-ies**) **1** worthy person. **2** person of some distinction. □ **worthily** *adv.* **worthiness** *n.*

-worthy *comb. form* forming adjectives meaning: **1** deserving of (*noteworthy*). **2** suitable for (*roadworthy*).

would /wʊd, wəd/ *v.aux.* (*3rd sing.* **would**) *past* of WILL[1], used esp.: **1** in reported speech (*he said he would be home by evening*). **2** to express a condition (*they would have been killed if they had gone*). **3** to express habitual action (*would wait every evening*). **4** to express a question or polite request (*would they like it?*; *would you come in, please?*). **5** to express probability (*she would be over fifty by now*). **6** to express consent (*they would not help*).

would-be *attrib. adj.* desiring or aspiring to be.

wouldn't /'wʊd(ə)nt/ *contr.* would not.

wound[1] /wu:nd/ *–n.* **1** injury done to living tissue by a deep cut or heavy blow etc. **2** pain inflicted on one's feelings; injury to one's reputation. *–v.* inflict a wound on. [Old English]

wound[2] *past* and *past part.* of WIND[2].

wound up *adj.* excited; tense; angry.

wove *past* of WEAVE[1].

woven *past part.* of WEAVE[1].

wow[1] /waʊ/ *–int.* expressing astonishment or admiration. *–n. slang* sensational success. *–v. slang* impress greatly. [imitative]

wow[2] /waʊ/ *n.* slow pitch-fluctuation in sound-reproduction, perceptible in long notes. [imitative]

WP *abbr.* word processor.

WPC *abbr.* woman police constable.

w.p.m. *abbr.* words per minute.

WRAC *abbr.* Women's Royal Army Corps.

wrack *n.* **1** seaweed cast up or growing on the shore. **2** destruction. [Low German or Dutch *wrak*: cf. WRECK]

WRAF *abbr.* Women's Royal Air Force.

wraith *n.* **1** ghost. **2** spectral appearance of a living person supposed to portend that person's death. [origin unknown]

wrangle /'ræŋg(ə)l/ *–n.* noisy argument or dispute. *–v.* (**-ling**) engage in a wrangle. [Low German or Dutch]

wrap *–v.* (**-pp-**) **1** (often foll. by *up*) envelop in folded or soft encircling material. **2** (foll. by *round, about*) arrange or draw (a pliant covering) round (a person). **3** (foll. by *round*) *slang* crash (a vehicle) into (a stationary object). *–n.* **1** shawl, scarf, etc. **2** esp. *US* wrapping material. □ **take the wraps off** disclose. **under wraps** in secrecy. **wrapped up in** engrossed or absorbed in. **wrap up 1** *colloq.* finish off (a matter). **2** put on warm clothes (*wrap up well*). **3** (in *imper.*) *slang* be quiet. [origin unknown]

wraparound *adj.* (also **wrapround**) **1** (esp. of clothing) designed to wrap round. **2** curving or extending round at the edges.

wrap-over *–attrib. adj.* (of a garment) overlapping when worn. *–n.* such a garment.

wrapper *n.* **1** cover for a sweet, book, posted newspaper, etc. **2** loose enveloping robe or gown.

wrapping *n.* (esp. in *pl.*) material used to wrap; wraps, wrappers.

wrapping paper *n.* strong or decorative paper for wrapping parcels.

wrapround var. of WRAPAROUND.

wrasse /ræs/ *n.* bright-coloured marine fish. [Cornish *wrach*]

wrath /rɒθ/ *n. literary* extreme anger. [Old English: related to WROTH]

wrathful *adj. literary* extremely angry. □ **wrathfully** *adv.*

wreak *v.* **1** (usu. foll. by *upon*) give play to (vengeance or one's anger etc.). **2** cause (damage etc.) (*wreak havoc*). [Old English, = avenge]

wreath /ri:θ/ *n.* (*pl.* **-s** /ri:ðz/) **1** flowers or leaves fastened in a ring, esp. as an ornament for the head or for laying on a grave etc. **2** curl or ring of smoke, cloud, or soft fabric. [Old English: related to WRITHE]

wreathe /ri:ð/ *v.* (**-thing**) **1** encircle or cover as, with, or like a wreath. **2** (foll. by *round*) wind (one's arms etc.) round (a person etc.). **3** (of smoke etc.) move in wreaths.

wreck *–n.* **1** the sinking or running aground of a ship. **2** ship that has suffered a wreck. **3** greatly damaged building, thing, or person. **4** (foll. by *of*) wretched remnant. *–v.* **1 a** seriously damage (a vehicle etc.). **b** ruin (hopes, a life, etc.). **2** cause the wreck of (a ship). [Anglo-French *wrec* from Germanic]

wreckage *n.* **1** wrecked material. **2** remnants of a wreck. **3** act of wrecking.

wrecker *n.* **1** person or thing that wrecks or destroys. **2** esp. *US* person employed in demolition or breaking up damaged vehicles. **3** esp. *hist.* person on the shore who tries to bring about a shipwreck for plunder or profit.

Wren *n.* member of the Women's Royal Naval Service. [from the abbreviation WRNS]

wren *n.* small usu. brown short-winged songbird with an erect tail. [Old English]

wrench *–n.* **1** violent twist or oblique pull or tearing off. **2** tool like a spanner for gripping and turning nuts etc. **3** painful uprooting or parting. *–v.* **1** twist or pull violently round or sideways. **2** (often foll. by *off, away*, etc.) pull off with a wrench. [Old English]

wrest *v.* **1** wrench away from a person's grasp. **2** (foll. by *from*) obtain by effort or with difficulty. [Old English]

wrestle /'res(ə)l/ *–n.* **1** contest in which two opponents grapple and try to throw each other to the ground, esp. as an athletic sport. **2** hard struggle. *–v.* (**-ling**) **1** (often foll. by *with*) take part or fight in a wrestle. **2 a** (foll. by *with, against*) struggle. **b** (foll. by *with*) do one's utmost to deal with (a task, difficulty, etc.). □ **wrestler** *n.* **wrestling** *n.* [Old English]

wretch *n.* **1** unfortunate or pitiable person. **2** (often as a playful term of depreciation) reprehensible person. [Old English, = outcast]

wretched /ˈretʃɪd/ *adj.* (**wretcheder, wretchedest**) **1** unhappy, miserable; unwell. **2** of bad quality; contemptible. **3** displeasing, hateful. □ **wretchedly** *adv.* **wretchedness** *n.*

wrick var. of RICK[2].

wriggle /ˈrɪg(ə)l/ –*v.* (**-ling**) **1** (of a worm etc.) twist or turn its body with short writhing movements. **2** make wriggling motions. **3** (foll. by *along*, *through*, etc.) go thus (*wriggled through the gap*). **4** be evasive. –*n.* act of wriggling. □ **wriggle out of** *colloq.* avoid on a pretext. □ **wriggly** *adj.* [Low German *wriggelen*]

wright /raɪt/ *n.* maker or builder (usu. in *comb.*: *playwright*; *shipwright*). [Old English: related to WORK]

wring –*v.* (*past* and *past part.* **wrung**) **1 a** squeeze tightly. **b** (often foll. by *out*) squeeze and twist, esp. to remove liquid. **2** break by twisting. **3** distress, torture. **4** extract by squeezing. **5** (foll. by *out*, *from*) obtain by pressure or importunity; extort. –*n.* act of wringing. □ **wring one's hands** clasp them as a gesture of distress. **wring the neck of** kill (a chicken etc.) by twisting its neck. [Old English]

wringer *n.* device for wringing water from washed clothes etc.

wringing *adj.* (in full **wringing wet**) so wet that water can be wrung out.

wrinkle /ˈrɪŋk(ə)l/ –*n.* **1** crease in the skin, esp. caused by age. **2** similar mark in another flexible surface. **3** *colloq.* useful tip or clever expedient. –*v.* (**-ling**) **1** make wrinkles in. **2** form wrinkles. [probably related to Old English *gewrinclod* sinuous]

wrinkly –*adj.* (-**ier**, -**iest**) having wrinkles. –*n. slang offens.* old or middle-aged person.

wrist *n.* **1** joint connecting the hand with the arm. **2** part of a garment covering this. [Old English]

wristlet *n.* band or ring to strengthen, guard, or adorn the wrist.

wrist-watch *n.* small watch worn on a strap etc. round the wrist.

writ[1] *n.* form of written command to act or not act in some way. [Old English: related to WRITE]

writ[2] *archaic past part.* of WRITE. □ **writ large** in magnified or emphasized form.

write *v.* (**-ting**; *past* **wrote**; *past part.* **written**) **1** mark paper or some other surface with symbols, letters, or words. **2** form or mark (such symbols etc.). **3** form or mark the symbols of (a word or sentence, or document etc.). **4** fill or complete (a sheet, cheque, etc.) with writing. **5** transfer (data) into a computer store. **6** (esp. in *passive*) indicate (a quality or condition) by one's or its appearance (*guilt was written on his face*). **7** compose for written or printed reproduction or publication. **8** (usu. foll. by *to*) write and send a letter (to a person). **9** *US* write and send a letter to (a person). **10** convey (news etc.) by letter. **11** state in a book etc. **12** (foll. by *into*, *out of*) include or exclude (a character or episode) in, or from, a story by changing the text. □ **write down** record in writing. **write in** send a suggestion, query, etc., in writing to esp. a broadcasting station. **write off 1** (foll. by *for*) = *send away for* (see SEND). **2** cancel the record of (a bad debt etc.); acknowledge the loss of (an asset). **3** completely destroy (a vehicle etc.). **4** dismiss as insignificant. **write out** write in full or in finished form. **write up 1** write a full account of; bring (a diary etc.) up to date. **2** praise in writing. [Old English]

write-off *n.* thing written off, esp. a vehicle too badly damaged to be repaired.

writer *n.* **1** person who writes or has written something. **2** person who writes books, author.

writer's cramp *n.* muscular spasm due to excessive writing.

write-up *n.* written or published account, review.

writhe /raɪð/ *v.* (**-thing**) **1** twist or roll oneself about in or as in acute pain. **2** suffer mental torture or embarrassment (*writhed with shame*). [Old English]

writing *n.* **1** written words etc. **2** handwriting. **3** (usu. in *pl.*) author's works. □ **in writing** in written form. **the writing on the wall** ominously significant event etc.

writing-desk *n.* desk for writing at, esp. with compartments for papers etc.

writing-paper *n.* paper for writing (esp. letters) on.

written *past part.* of WRITE.

WRNS *abbr.* Women's Royal Naval Service.

wrong –*adj.* **1** mistaken; not true; in error. **2** unsuitable; less or least desirable (*the wrong road*; *a wrong decision*). **3** contrary to law or morality (*it is wrong to steal*). **4** amiss; out of order, in a bad or abnormal condition (*something wrong with my heart*; *has gone wrong*). –*adv.* (usually placed last) in a wrong manner or direction; with an incorrect result (*guessed wrong*). –*n.* **1** what is morally wrong. **2** unjust action (*suffer a wrong*). –*v.* **1** treat unjustly. **2** mistakenly attribute bad motives to. □

do wrong sin. **do wrong to** malign or mistreat (a person). **get wrong 1** misunderstand (a person etc.). **2** obtain an incorrect answer to. **get** (or **get hold of**) **the wrong end of the stick** misunderstand completely. **go wrong 1** take the wrong path. **2** stop functioning properly. **3** depart from virtuous behaviour. **in the wrong** responsible for a quarrel, mistake, or offence. **on the wrong side of 1** out of favour with (a person). **2** somewhat more than (a stated age). **wrong side out** inside out. **wrong way round** in the opposite or reverse of the normal or desirable orientation or sequence etc. □ **wrongly** *adv.* **wrongness** *n.* [Old English]

wrongdoer *n.* person who behaves immorally or illegally. □ **wrongdoing** *n.*

wrong-foot *v. colloq.* **1** (in tennis, football, etc.) catch (an opponent) off balance. **2** disconcert; catch unprepared.

wrongful *adj.* unwarranted, unjustified (*wrongful arrest*). □ **wrongfully** *adv.*

wrong-headed *adj.* perverse and obstinate.

wrong side *n.* worse or undesired or unusable side of esp. fabric.

wrote *past* of WRITE.

wroth /rəʊθ/ *predic. adj. archaic* angry. [Old English]

wrought /rɔːt/ *archaic past* and *past part.* of WORK. –*adj.* (of metals) beaten out or shaped by hammering.

wrought iron *n.* tough malleable form of iron suitable for forging or rolling, not cast.

wrung *past* and *past part.* of WRING.

WRVS *abbr.* Women's Royal Voluntary Service.

wry /raɪ/ *adj.* (**wryer, wryest** or **wrier, wriest**) **1** distorted or turned to one side. **2** (of a face, smile, etc.) contorted in disgust, disappointment, or mockery. **3** (of humour) dry and mocking. □ **wryly** *adv.* **wryness** *n.* [Old English]

wryneck *n.* small woodpecker able to turn its head over its shoulder.

WSW *abbr.* west-south-west.

wt *abbr.* weight.

wych- *comb. form* in names of trees with pliant branches (*wych-alder*; *wych-elm*). [Old English, = bending]

wych-hazel var. of WITCH-HAZEL.

Wykehamist /ˈwɪkəmɪst/ –*adj.* of Winchester College. –*n.* past or present member of Winchester College. [William of *Wykeham*, name of the founder]

WYSIWYG /ˈwɪzɪwɪg/ *adj.* (also **wysiwyg**) *Computing* denoting a form of text onscreen exactly corresponding to its printout. [acronym of *w*hat *y*ou *s*ee *i*s *w*hat *y*ou *g*et]

X

X[1] /eks/ *n.* (also **x**) (*pl.* **Xs** or **X's**) **1** twenty-fourth letter of the alphabet. **2** (as a roman numeral) ten. **3** (usu. **x**) *Algebra* first unknown quantity. **4** unknown or unspecified number or person etc. **5** cross-shaped symbol used esp. to indicate position (*X marks the spot*) or incorrectness, or to symbolize a kiss or a vote, or as the signature of a person who cannot write.

X[2] *symb.* (of films) classified as suitable for adults only.

■ **Usage** This symbol was superseded in the UK in 1983 by *18*, but it is still used in the US.

X chromosome /'eks,krəʊmə,səʊm/ *n.* (in humans and some other mammals) sex chromosome of which the number in female cells is twice that in male cells. [*X* as an arbitrary label]

Xe *symb.* xenon.

xenon /'zenɒn, 'zi:-/ *n.* heavy inert gaseous element. [Greek, neuter of *xenos* strange]

xenophobia /,zenə'fəʊbɪə/ *n.* hatred or fear of foreigners. □ **xenophobic** *adj.* [Greek *xenos* strange, stranger]

xerography /zɪə'rɒgrəfɪ/ *n.* dry copying process in which powder adheres to areas remaining electrically charged after exposure of the surface to light from an image of the document to be copied. [Greek *xēros* dry]

Xerox /'zɪərɒks/ *–n. propr.* **1** machine for copying by xerography. **2** copy thus made. *–v.* (**xerox**) reproduce by this process.

xi /saɪ, ksaɪ/ *n.* fourteenth letter of the Greek alphabet (Ξ, ξ). [Greek]

-xion see -ION.

Xmas /'krɪsməs, 'eksməs/ *n. colloq.* = CHRISTMAS. [abbreviation, with X for the initial chi of Greek *Khristos* Christ]

X-ray /'eksreɪ/ *–n.* **1** (in *pl.*) electromagnetic radiation of short wavelength, able to pass through opaque bodies. **2** photograph made by X-rays, esp. showing the position of bones etc. by their greater absorption of the rays. *–v.* photograph, examine, or treat with X-rays. [X, originally with ref. to the unknown nature of the rays]

xylem /'zaɪləm/ *n. Bot.* woody tissue. [Greek]

xylophone /'zaɪlə,fəʊn/ *n.* musical instrument of graduated wooden or metal bars struck with small wooden hammers. □ **xylophonist** *n.* [Greek *xulon* wood]

Y

Y[1] /waɪ/ *n.* (also **y**) (*pl.* **Ys** or **Y's**) **1** twenty-fifth letter of the alphabet. **2** (usu. **y**) *Algebra* second unknown quantity. **3** Y-shaped thing.

Y[2] *symb.* yttrium.

-y[1] *suffix* forming adjectives: **1** from nouns and adjectives, meaning: **a** full of; having the quality of (*messy*). **b** addicted to (*boozy*). **2** from verbs, meaning 'inclined to', 'apt to' (*sticky*). [Old English]

-y[2] *suffix* (also **-ey**, **-ie**) forming diminutive nouns, pet names, etc. (*granny*; *Sally*; *nightie*). [originally *Scottish*]

-y[3] *suffix* forming nouns denoting state, condition, or quality (*orthodoxy*). [Latin *-ia*, Greek *-eia*]

yacht /jɒt/ *–n.* **1** light sailing-vessel. **2** larger usu. power-driven vessel for cruising. *–v.* race or cruise in a yacht. [Dutch *jaghtschip*, literally 'pursuit-ship']

yachtsman *n.* (*fem.* **yachtswoman**) person who sails yachts.

yack *slang –n.* trivial or unduly persistent conversation. *–v.* engage in this. [imitative]

yah /jɑː/ *int.* (also **yah boo**) expressing derision or defiance. [imitative]

yahoo /jɑːˈhuː/ *n.* bestial person. [name of a race of brutes in *Gulliver's Travels*]

Yahweh /ˈjɑːweɪ/ *n.* (also **Yahveh** /-veɪ/) = JEHOVAH.

yak *n.* long-haired Tibetan ox. [Tibetan]

Yale lock *n. propr.* type of lock with a revolving barrel, used for doors etc. [*Yale*, name of its inventor]

yam *n.* **1 a** tropical or subtropical climbing plant. **b** edible starchy tuber of this. **2** *US* sweet potato. [Portuguese or Spanish]

yammer /ˈjæmə(r)/ *colloq.* or *dial. –n.* **1** lament, wail, grumble. **2** voluble talk. *–v.* utter a yammer. [Old English]

yang *n.* (in Chinese philosophy) the active male principle of the universe (cf. YIN).

Yank *n. colloq.* often *derog.* American. [abbreviation of YANKEE]

yank *v.* & *n. colloq.* pull with a jerk. [origin unknown]

Yankee /ˈjæŋkɪ/ *n. colloq.* **1** often *derog.* = YANK. **2** *US* inhabitant of New England or of the northern States. [origin uncertain: perhaps from Dutch *Janke*, diminutive of *Jan* John, as a nickname]

yap *–v.* (**-pp-**) **1** bark shrilly or fussily. **2** *colloq.* talk noisily, foolishly, or complainingly. *–n.* sound of yapping. □ **yappy** *adj.* (**-ier**, **-iest**) in sense 1 of *v.* [imitative]

yarborough /ˈjɑːbərə/ *n.* whist or bridge hand with no card above a 9. [Earl of *Yarborough*, said to have betted against it]

yard[1] *n.* **1** unit of linear measure (3 ft, 0.9144 metre). **2** this length of material. **3** square or cubic yard. **4** spar slung across a mast for a sail to hang from. **5** (in *pl.*; foll. by *of*) *colloq.* a great length. [Old English, = stick]

yard[2] *n.* **1** piece of enclosed ground, esp. attached to a building or used for a particular purpose. **2** *US* & *Austral.* garden of a house. [Old English, = enclosure]

yardage *n.* number of yards of material etc.

yard-arm *n.* either end of a ship's yard.

Yardie /ˈjɑːdɪ/ *n. slang* member of a Jamaican or W. Indian gang engaging in organized crime, esp. drug-trafficking. [Jamaican English, = house, home]

yardstick *n.* **1** standard of comparison. **2** measuring rod a yard long, usu. divided into inches etc.

yarmulke /ˈjɑːməlkə/ *n.* (also **yarmulka**) skullcap worn by Jewish men. [Yiddish]

yarn *–n.* **1** spun thread, esp. for knitting, weaving, etc. **2** *colloq.* story, traveller's tale, anecdote. *–v. colloq.* tell yarns. [Old English]

yarrow /ˈjærəʊ/ *n.* perennial plant, esp. milfoil. [Old English]

yashmak /ˈjæʃmæk/ *n.* veil concealing the face except the eyes, worn by some Muslim women. [Arabic]

yaw *–v.* (of a ship or aircraft etc.) fail to hold a straight course; go unsteadily. *–n.* yawing of a ship etc. from its course. [origin unknown]

yawl *n.* a kind of ship's boat or sailing- or fishing-boat. [Low German *jolle* or Dutch *jol*]

yawn *–v.* **1** open the mouth wide and inhale, esp. when sleepy or bored. **2** gape, be wide open. *–n.* **1** act of yawning. **2** *colloq.* boring idea, activity, etc. [Old English]

yaws /jɔːz/ *n.pl.* (usu. treated as *sing.*) contagious tropical skin-disease with large red swellings. [origin unknown]

Yb *symb.* ytterbium.

Y chromosome *n.* /ˈwaɪ,krəʊmə,səʊm/ (in humans and some other mammals) sex chromosome occurring only in male cells. [*Y* as an arbitrary label]

yd *abbr.* (*pl.* **yds**) yard (measure).

ye[1] /jiː/ *pron. archaic pl.* of THOU[1] (*ye gods!*).

ye[2] /jiː/ *adj. pseudo-archaic* = THE (*ye olde tea-shoppe*). [from the obsolete *y*-shaped letter for *th*]

yea /jeɪ/ *archaic* –*adv.* **1** yes. **2** indeed (*ready, yea eager*). –*n.* utterance of 'yea'; 'yes' vote. [Old English]

yeah /jeə/ *adv. colloq.* yes. [a casual pronunciation of YES]

year /jɪə(r)/ *n.* **1** time occupied by the earth in one revolution round the sun, approx. 365¼ days. **2** = CALENDAR YEAR. **3** period of twelve months, starting at any point (*four years ago*; *tax year*). **4** (in *pl.*) age, time of life (*young for his years*). **5** (usu. in *pl.*) *colloq.* very long time. **6** group of students entering college etc. in the same academic year. [Old English]

yearbook *n.* annual publication dealing with events or aspects of the (usu. preceding) year.

yearling *n.* animal between one and two years old.

yearly –*adj.* **1** done, produced, or occurring once a year. **2** of or lasting a year. –*adv.* once a year.

yearn /jɜːn/ *v.* be filled with longing, compassion, or tenderness. □ **yearning** *n.* & *adj.* [Old English]

yeast *n.* greyish-yellow fungus obtained esp. from fermenting malt liquors and used as a fermenting agent, to raise bread, etc. [Old English]

yeasty *adj.* (**-ier, -iest**) **1** of, like, or tasting of yeast; frothy. **2** in a ferment. **3** working like yeast. **4** (of talk etc.) light and superficial.

yell –*n.* loud sharp cry; shout. –*v.* cry, shout. [Old English]

yellow /ˈjeləʊ/ –*adj.* **1** of the colour of buttercups, lemons, egg-yolks, etc. **2** having a yellow skin or complexion. **3** *colloq.* cowardly. –*n.* **1** yellow colour or pigment. **2** yellow clothes or material. –*v.* turn yellow. □ **yellowish** *adj.* **yellowness** *n.* **yellowy** *adj.* [Old English: related to GOLD]

yellow-belly *n. colloq.* coward.

yellow card *n.* card shown by the referee to a football-player being cautioned.

yellow fever *n.* tropical virus disease with fever and jaundice.

yellow flag *n.* flag displayed by a ship in quarantine.

yellowhammer *n.* bunting of which the male has a yellow head, neck, and breast.

Yellow Pages *n.pl. propr.* telephone directory on yellow paper, listing and classifying business subscribers.

yellow pepper *n.* ripe yellow fruit of the capsicum plant.

yellow spot *n.* point of acutest vision in the retina.

yellow streak *n. colloq.* trait of cowardice.

yelp –*n.* sharp shrill cry as of a dog in pain or excitement. –*v.* utter a yelp. [Old English]

yen[1] *n.* (*pl.* same) chief monetary unit of Japan. [Japanese from Chinese]

yen[2] *colloq.* –*n.* longing or yearning. –*v.* (**-nn-**) feel a longing. [Chinese]

yeoman /ˈjəʊmən/ *n.* **1** esp. *hist.* man holding and cultivating a small landed estate. **2** member of the yeomanry force. □ **yeomanly** *adj.* [from earlier *yoman*, *yeman*, etc., probably = young man]

Yeoman of the Guard *n.* member of the sovereign's bodyguard.

yeomanry *n.* (*pl.* **-ies**) **1** body or class of yeomen. **2** *hist.* volunteer cavalry force raised from the yeoman class.

Yeoman Warder *n.* (correct term for) a "beefeater" at the Tower of London.

yep *adv.* & *n.* (also **yup**) *US colloq.* = YES.

yes –*adv.* **1** indicating that the answer to the question is affirmative, the statement etc. made is correct, the request or command will be complied with, or the person summoned or addressed is present. **2** (**yes?**) **a** indeed? is that so? **b** what do you want? –*n.* **1** utterance of the word *yes*. **2** affirmation or assent. **3** 'yes' vote. □ **say yes** grant a request, confirm a statement. [Old English, = *yea let it be*]

yes-man *n. colloq.* weakly acquiescent person.

yesterday /ˈjestə,deɪ/ –*adv.* **1** on the day before today. **2** in the recent past. –*n.* **1** the day before today. **2** the recent past. [Old English]

yesteryear /ˈjestə,jɪə(r)/ *n. archaic* or *rhet.* **1** last year. **2** the recent past. [Old English *yester*- that is last past, YEAR]

yet –*adv.* **1** as late as, or until, now or then (*there is yet time*; *your best work yet*). **2** (with *neg.* or *interrog.*) so soon as, or by, now or then (*it is not time yet*; *have you finished yet?*). **3** again; in addition (*more and yet more*). **4** in the remaining time available (*I will do it yet*). **5** (foll. by *compar.*) even (*a yet more difficult task*). **6** nevertheless; and or but in spite of that. –*conj.* but at the same time; but nevertheless. [Old English]

yeti /ˈjetɪ/ *n.* = ABOMINABLE SNOWMAN. [Tibetan]

yew *n.* **1** dark-leaved evergreen tree bearing berry-like cones. **2** its wood. [Old English]

Y-fronts /ˈwaɪfrʌnts/ *n. propr.* men's or boys' briefs with a Y-shaped seam at the front.

YHA *abbr.* Youth Hostels Association.

Yid *n. slang offens.* Jew. [back-formation from YIDDISH]

Yiddish /ˈjɪdɪʃ/ *–n.* language used by Jews in or from Europe, orig. a German dialect with words from Hebrew etc. *–adj.* of this language. [German *jüdisch* Jewish]

yield *–v.* **1** produce or return as a fruit, profit, or result. **2** give up; surrender, concede. **3 a** (often foll. by *to*) surrender; submit; defer to. **b** (as **yielding** *adj.*) compliant; submissive; soft and pliable. **4** (foll. by *to*) give right of way to (other traffic). **5** (foll. by *to*) be inferior or confess inferiority to (*I yield to none in this matter*). *–n.* amount yielded or produced. [Old English, = pay]

yin *n.* (in Chinese philosophy) the passive female principle of the universe (cf. YANG).

yippee /jɪˈpiː/ *int.* expressing delight or excitement. [natural exclamation]

YMCA *abbr.* Young Men's Christian Association.

yob /jɒb/ *n. slang* lout, hooligan. □ **yobbish** *adj.* [back slang for BOY]

yobbo /ˈjɒbəʊ/ *n.* (*pl.* **-s**) *slang* = YOB.

yodel /ˈjəʊd(ə)l/ *–v.* (**-ll-**; *US* **-l-**) sing with melodious inarticulate sounds and frequent changes between falsetto and normal voice in the manner of Swiss mountain-dwellers. *–n.* yodelling cry. □ **yodeller** *n.* [German]

yoga /ˈjəʊgə/ *n.* **1** Hindu system of meditation and asceticism designed to effect reunion with the universal spirit. **2** system of physical exercises and breathing control used in yoga. [Sanskrit, = union]

yoghurt /ˈjɒgət/ *n.* (also **yogurt**) semi-solid sourish food made from milk fermented by added bacteria. [Turkish]

yogi /ˈjəʊgɪ/ *n.* (*pl.* **-s**) devotee of yoga. [Hindustani: related to YOGA]

yoicks *int.* cry used by fox-hunters to urge on the hounds. [origin unknown]

yoke *–n.* **1** wooden crosspiece fastened over the necks of two oxen etc. and attached to the plough or wagon to be pulled. **2** (*pl.* same or **-s**) pair (of oxen etc.). **3** object like a yoke in form or function, e.g. a wooden shoulder-piece for carrying a pair of pails, the top section of a garment from which the rest hangs. **4** sway, dominion, or servitude. **5** bond of union, esp. of marriage. *–v.* (**-king**) **1** put a yoke on. **2** couple or unite (a pair). **3** (foll. by *to*) link (one thing) to (another). **4** match or work together. [Old English]

yokel /ˈjəʊk(ə)l/ *n.* rustic; country bumpkin. [perhaps dial.]

yolk /jəʊk/ *n.* yellow inner part of an egg. [Old English: related to YELLOW]

Yom Kippur /jɒm ˈkɪpə(r)/ *n.* most solemn religious fast day of the Jewish year, Day of Atonement. [Hebrew]

yon *adj.* & *adv. literary* & *dial.* yonder. [Old English]

yonder *–adv.* over there; at some distance in that direction; in the place indicated. *–adj.* situated yonder.

yonks *n.pl. slang* a long time (*yonks ago*). [origin unknown]

yoo-hoo /ˈjuːhuː/ *int.* used to attract a person's attention. [natural exclamation]

yore *n.* □ **of yore** a long time ago. [Old English, = long ago]

york *v. Cricket* bowl out with a yorker. [back-formation from YORKER]

yorker *n. Cricket* ball that pitches immediately under the bat. [probably with ref. to the practice of Yorkshire cricketers]

Yorkist /ˈjɔːkɪst/ *–n. hist.* follower of the House of York, esp. in the Wars of the Roses. *–adj.* of the House of York.

Yorkshire pudding /ˈjɔːkʃə/ *n.* baked batter eaten with roast beef.

Yorkshire terrier /ˈjɔːkʃə/ *n.* small long-haired blue and tan kind of terrier.

you /juː/ *pron.* (*obj.* **you**; *poss.* **your**, **yours**) **1** the person or persons addressed. **2** (as *int.* with a noun) in an exclamatory statement (*you fools!*). **3** (in general statements) one, a person, people (*you get used to it*). □ **you and yours** you and your family, property, etc. [Old English, originally objective case of YE[1]]

you'd /juːd/ *contr.* **1** you had. **2** you would.

you'll /juːl/ *contr.* you will; you shall.

young /jʌŋ/ *–adj.* (**younger** /ˈjʌŋgə(r)/; **youngest** /ˈjʌŋgɪst/) **1** not far advanced in life, development, or existence; not yet old. **2 a** immature, inexperienced. **b** youthful. **3** of or characteristic of youth (*young love*). **4** representing young people (*Young Farmers*). **5 a** distinguishing a son from his father (*young George*). **b** as a familiar or condescending form of address (*listen, young lady*). **6** (**younger**) distinguishing one person from another of the same name (*the younger Pitt*). *–n.* (*collect.*) offspring, esp. of animals. □ **youngish** *adj.* [Old English]

young person *n. Law* person aged between 14 and 17 years.

youngster *n.* child, young person.

your /jɔ:(r)/ *poss. pron.* **1** of or belonging to you. **2** *colloq.* often *derog.* much talked of; well known (*your typical professor*). [Old English]

you're /jɔ:(r)/ *contr.* you are.

yours /jɔ:z/ *poss. pron.* **1** the one or ones belonging to you (*it is yours; yours are over there*). **2** your letter (*yours of the 10th*). **3** introducing a formula ending a letter (*yours ever; yours truly*). □ **of yours** of or belonging to you (*friend of yours*).

yourself /jɔ:'self/ *pron.* (*pl.* **yourselves**) **1 a** *emphat. form* of YOU. **b** *refl. form* of YOU. **2** in your normal state of body or mind (*are quite yourself again*). □ **be yourself** see ONESELF.

youth /ju:θ/ *n.* (*pl.* **-s** /ju:ðz/) **1** being young; period between childhood and adult age. **2** vigour, enthusiasm, inexperience, or other characteristic of this period. **3** young man. **4** (as *pl.*) young people collectively (*the youth of the country*). [Old English: related to YOUNG]

youth club *n.* place for young people's leisure activities.

youthful *adj.* young or still having the characteristics of youth. □ **youthfully** *adv.* **youthfulness** *n.*

youth hostel *n.* any of a chain of cheap lodgings for holiday-makers, esp. walkers and cyclists.

you've /ju:v/ *contr.* you have.

yowl –*n.* loud wailing cry of or as of a cat or dog in distress. –*v.* utter a yowl. [imitative]

yo-yo /'jəʊjəʊ/ *n.* (*pl.* **yo-yos**) *propr.* **1** toy consisting of a pair of discs with a deep groove between them in which string is attached and wound, and which can be made to fall and rise. **2** thing that repeatedly falls and rises. [origin unknown]

yr. *abbr.* **1** year(s). **2** younger. **3** your.

yrs. *abbr.* **1** years. **2** yours.

YTS *abbr.* Youth Training Scheme.

ytterbium /ɪ'tɜ:bɪəm/ *n.* metallic element of the lanthanide series. [*Ytterby* in Sweden]

yttrium /'ɪtrɪəm/ *n.* metallic element resembling the lanthanides. [related to YTTERBIUM]

yuan /ju:'ɑ:n/ *n.* (*pl.* same) chief monetary unit of China. [Chinese]

yucca /'jʌkə/ *n.* subtropical white-flowered plant with swordlike leaves, often grown as a house-plant. [Carib]

yuck /jʌk/ *int.* (also **yuk**) *slang* expression of strong distaste. [imitative]

yucky *adj.* (also **yukky**) (**-ier, -iest**) *slang* **1** messy, repellent. **2** sickly, sentimental.

Yugoslav /'ju:gə,slɑ:v/ (also **Jugoslav**) –*n.* **1** native or national of Yugoslavia. **2** person of Yugoslav descent. –*adj.* of Yugoslavia. □ **Yugoslavian** /-'slɑ:vɪən/ *adj.* & *n.* [Serbo-Croat *jug* south: related to SLAV]

yuk var. of YUCK.

yukky var. of YUCKY.

yule *n.* (in full **yule-tide**) *archaic* the Christmas festival. [Old English]

yule-log *n.* **1** large log traditionally burnt on Christmas Eve. **2** log-shaped chocolate cake eaten at Christmas.

yummy /'jʌmɪ/ *adj.* (**-ier, -iest**) *colloq.* tasty, delicious. [from YUM-YUM]

yum-yum *int.* expressing pleasure from eating or the prospect of eating. [natural exclamation]

yup var. of YEP.

yuppie /'jʌpɪ/ *n.* (also **yuppy**) (*pl.* **-ies**) (often *attrib.*) *colloq.*, usu. *derog.* young ambitious professional person working in a city. [from *y*oung *u*rban *p*rofessional]

YWCA *abbr.* Young Women's Christian Association.

Z

Z /zed/ *n.* (also **z**) (*pl.* **Zs** or **Z's**) **1** twenty-sixth letter of the alphabet. **2** (usu. **z**) *Algebra* third unknown quantity.

zabaglione /ˌzæbəˈljəʊnɪ/ *n.* Italian dessert of whipped and heated egg-yolks, sugar, and wine. [Italian]

zany /ˈzeɪnɪ/ *adj.* (**-ier, -iest**) comically idiotic; crazily ridiculous. [French or Italian]

zap *slang* —*v.* (**-pp-**) **1 a** kill or destroy; attack. **b** hit hard (*zapped the ball over the net*). **2** move quickly. **3** overwhelm emotionally. —*int.* expressing the sound or impact of a bullet, ray gun, etc., or any sudden event. [imitative]

zappy *adj.* (**-ier, -iest**) *colloq.* lively, energetic.

Zarathustrian var. of ZOROASTRIAN.

zeal *n.* earnestness or fervour; hearty persistent endeavour. [Greek *zēlos*]

zealot /ˈzelət/ *n.* extreme partisan; fanatic. □ **zealotry** *n.*

zealous /ˈzeləs/ *adj.* full of zeal; enthusiastic. □ **zealously** *adv.*

zebra /ˈzebrə, ˈzi:-/ *n.* (*pl.* same or **-s**) black-and-white striped African animal of the family including the ass and horse. [Italian or Portuguese from Congolese]

zebra crossing *n.* striped street-crossing where pedestrians have precedence.

zebu /ˈzi:bu:/ *n.* (*pl.* same or **-s**) humped ox of Asia and Africa. [French]

zed *n.* letter Z. [Greek ZETA]

zee /zi:/ *n.* *US* letter Z. [var. of ZED]

Zeitgeist /ˈtsaɪtgaɪst/ *n.* the spirit of the times. [German]

Zen *n.* form of Buddhism emphasizing meditation and intuition. [Japanese, = meditation]

Zend *n.* an interpretation of the Avesta. [Persian]

Zend-Avesta *n.* Zoroastrian sacred writings of the Avesta (or text) and Zend (or commentary).

zenith /ˈzenɪθ/ *n.* **1** point of the heavens directly above an observer. **2** highest point (of power or prosperity etc.). [Latin from Arabic]

zephyr /ˈzefə(r)/ *n.* *literary* mild gentle breeze. [Greek, = west wind]

Zeppelin /ˈzepəlɪn/ *n.* large dirigible German airship of the early 20th c. [Count F. von *Zeppelin*, name of an airman]

zero /ˈzɪərəʊ/ *n.* (*pl.* **-s**) **1** figure 0; nought; nil. **2** point on the scale of a thermometer etc. from which a positive or negative quantity is reckoned. **3** (*attrib.*) no, not any (*zero growth*). **4** (in full **zero-hour**) **a** hour at which a planned, esp. military, operation is timed to begin. **b** crucial moment. **5** lowest or earliest point (*down to zero*; *the year zero*). □ **zero in on** (**-oes, -oed**) **1** take aim at. **2** focus one's attention on. [Arabic: related to CIPHER]

zero option *n.* disarmament proposal for the total removal of certain types of weapons on both sides.

zero-rated *adj.* on which no VAT is charged.

zest *n.* **1** piquancy; stimulating flavour or quality. **2 a** keen enjoyment or interest. **b** (often foll. by *for*) relish. **c** gusto. **3** scraping of orange or lemon peel as flavouring. □ **zestful** *adj.* **zestfully** *adv.* [French]

zeta /ˈzi:tə/ *n.* sixth letter of the Greek alphabet (Ζ, ζ). [Greek]

zeugma /ˈzju:gmə/ *n.* figure of speech using a verb or adjective with two nouns, to one of which it is strictly applicable while the word appropriate to the other is not used (e.g. *with weeping eyes and* [sc. *grieving*] *hearts*) (cf. SYLLEPSIS). [Greek, = a yoking, from *zugon* yoke]

ziggurat /ˈzɪgəˌræt/ *n.* rectangular stepped tower in ancient Mesopotamia, surmounted by a temple. [Assyrian]

zigzag /ˈzɪgzæg/ —*adj.* with abrupt alternate right and left turns (*zigzag line*). —*n.* zigzag line; thing having the form of a zigzag or having sharp turns. —*adv.* with a zigzag course. —*v.* (**-gg-**) move in a zigzag course. [French from German]

zilch *n.* esp. *US slang* nothing. [origin uncertain]

zillion /ˈzɪljən/ *n.* *colloq.* indefinite large number. [probably after *million*]

Zimmer frame *n.* *propr.* a kind of walking-frame. [*Zimmer*, name of the maker]

zinc *n.* greyish-white metallic element used as a component of brass and in galvanizing sheet iron. [German *Zink*]

zing *colloq.* —*n.* vigour, energy. —*v.* move swiftly, esp. with a high-pitched ringing sound. [imitative]

zinnia /'zɪnɪə/ *n.* garden plant with showy flowers. [*Zinn*, name of a physician and botanist]

Zion /'zaɪən/ *n.* **1** ancient Jerusalem; its holy hill. **2 a** the Jewish people or religion. **b** the Christian Church. **3** the Kingdom of Heaven. [Hebrew *ṣîyôn*]

Zionism *n.* movement for the re-establishment and development of a Jewish nation in what is now Israel. □ **Zionist** *n.* & *adj.*

zip –*n.* **1** light fast sound. **2** energy, vigour. **3 a** (in full **zip-fastener**) fastening device of two flexible strips with interlocking projections, closed or opened by sliding a clip along them. **b** (*attrib.*) having a zip-fastener (*zip bag*). –*v.* (**-pp-**) **1** (often foll. by *up*) fasten with a zip-fastener. **2** move with zip or at high speed. [imitative]

Zip code *n. US* postcode. [zone *i*mprovement *p*lan]

zipper *n.* esp. *US* = ZIP 3a.

zippy *adj.* (**-ier**, **-iest**) *colloq.* lively, speedy.

zircon /'zɜːkən/ *n.* zirconium silicate of which some translucent varieties are cut into gems. [German *Zirkon*]

zirconium /zə'kəʊnɪəm/ *n.* grey metallic element.

zit *n.* esp. *US slang* pimple. [origin unknown]

zither /'zɪðə(r)/ *n.* stringed instrument with a flat soundbox, placed horizontally and played with the fingers and a plectrum. [Latin: related to GUITAR]

zloty /'zlɒtɪ/ *n.* (*pl.* same or **-s**) chief monetary unit of Poland. [Polish]

Zn *symb.* zinc.

zodiac /'zəʊdɪ,æk/ *n.* **1** belt of the heavens including all apparent positions of the sun, moon, and planets as known to ancient astronomers, and divided into twelve equal parts (**signs of the zodiac**). **2** diagram of these signs. □ **zodiacal** /zə'daɪək(ə)l/ *adj.* [Greek *zōion* animal]

zombie /'zɒmbɪ/ *n.* **1** *colloq.* person who acts mechanically or lifelessly. **2** corpse said to have been revived by witchcraft. [West African]

zone –*n.* **1** area having particular features, properties, purpose, or use (*danger zone*; *smokeless zone*). **2** well-defined region of more or less beltlike form. **3** area between two concentric circles. **4** encircling band of colour etc. **5** *archaic* belt, girdle. –*v.* (**-ning**) **1** encircle as or with a zone. **2** arrange or distribute by zones. **3** assign as or to a particular area. □ **zonal** *adj.* [Greek *zōnē* girdle]

zonked /zɒŋkt/ *adj. slang* (often foll. by *out*) exhausted; intoxicated. [*zonk* hit]

zoo *n.* zoological garden. [abbreviation]

zoological /,zəʊə'lɒdʒɪk(ə)l, ,zuːə-/ *adj.* of zoology.

■ **Usage** See note at *zoology*.

zoological garden *n.* (also **zoological gardens** *n.pl.*) public garden or park with a collection of animals for exhibition and study.

zoology /zəʊ'ɒlədʒɪ, ,zuː'ɒl-/ *n.* the study of animals. □ **zoologist** *n.* [Greek *zōion* animal]

■ **Usage** The second pronunciation given for *zoology*, *zoological*, and *zoologist*, with the first syllable pronounced as in *zoo*, although extremely common, is considered incorrect by some people.

zoom –*v.* **1** move quickly, esp. with a buzzing sound. **2** cause an aeroplane to mount at high speed and a steep angle. **3** (often foll. by *in* or *in on*) (of a camera) change rapidly from a long shot to a close-up (of). **4** (of prices etc.) rise sharply. –*n.* **1** aeroplane's steep climb. **2** zooming camera shot. [imitative]

zoom lens *n.* lens allowing a camera to zoom by varying the focal length.

zoophyte /'zəʊə,faɪt/ *n.* plantlike animal, esp. a coral, sea anemone, or sponge. [Greek *zōion* animal, *phuton* plant]

Zoroastrian /,zɒrəʊ'æstrɪən/ (also **Zarathustrian** /,zærə'θʊstrɪən/) –*adj.* of Zoroaster (or Zarathustra) or the dualistic religious system taught by him. –*n.* follower of Zoroaster. □ **Zoroastrianism** *n.* [*Zoroaster*, Persian founder of the religion]

Zr *symb.* zirconium.

zucchini /zuː'kiːnɪ/ *n.* (*pl.* same or **-s**) esp. *US & Austral.* courgette. [Italian, pl. of *zucchino*, diminutive of *zucca* gourd]

Zulu /'zuːluː/ –*n.* (*pl.* **-s**) **1** member of a S. African Bantu people. **2** their language. –*adj.* of this people or language. [native name]

zygote /'zaɪgəʊt/ *n. Biol.* cell formed by the union of two gametes. [Greek *zugōtos* yoked: related to ZEUGMA]

APPENDIX

Better English Guide

The aim of this guide is to give a short description of accepted conventions in written standard English. It does not cover other forms and varieties of English.

A. PUNCTUATION MARKS

Punctuation is a complicated subject, and only the main principles can be discussed here. The explanations are based on practice in British English; usage in American English differs in some instances. The main headings are as follows:

1. General remarks
2. Capital letter
3. Full stop
4. Comma
5. Semicolon
6. Colon
7. Question mark
8. Exclamation mark
9. Apostrophe
10. Quotation marks
11. Brackets (Parentheses)
12. Dash
13. Hyphen

1. General remarks

The purpose of punctuation is to mark out strings of words into manageable groups and help clarify their meaning (or in some cases to prevent a wrong meaning being deduced). The marks most commonly used to divide a piece of prose or other writing are the full stop, the semicolon, and the comma, with the strength of the dividing or separating role diminishing from the full stop to the comma. The full stop therefore marks the main division into sentences; the semicolon joins sentences (as in this sentence); and the comma (which is the most flexible in use and causes most problems) separates smaller elements with the least loss of continuity. Brackets and dashes also serve as separators—often more strikingly than commas, as in this sentence.

2. Capital letter

2.1.1 This is used for the first letter of the word beginning a sentence in most cases:

He decided not to come. Later he changed his mind.

2.1.2 A sentence or clause playing a subordinate or parenthetic role within a larger one does not normally begin with a capital letter:

I have written several letters (there are many to be written) and hope to finish them tomorrow.

2.1.3 In the following, however, the sentence is a separate one and therefore does begin with a capital letter:

There is more than one possibility. (You have said this often before.) So we should think carefully before acting.

2.1.4 A capital letter also begins sentences that form quoted speech:

The assistant turned and replied, 'There are no more left.'

2.2 The use of capital letters for proper names, titles, etc. is discussed in section C of this appendix (p. 1087).

3. Full stop ■

3.1 This is used to mark the end of a sentence when it is a statement (that is, not a question or exclamation). In prose, sentences marked by full stops normally represent a discrete or distinct statement; more closely connected or complementary statements are joined by a semicolon (as here).

3.2.1 Full stops are used to mark abbreviations (*Weds., Gen., p.m.*). They are often omitted in abbreviations that are familiar or very common (*Dr, Mr, Mrs*, etc.), in abbreviations that consist entirely of capital letters (*BBC, GMT*, etc.), and in acronyms that are pronounced as a word rather than a sequence of letters (*Intelsat*, *Ernie*, etc.).

3.2.2 If the full stop following an abbreviation comes right at the end of a sentence, another full stop is not added:

They have a collection of many animals, including dogs, cats, tortoises, snakes, etc.

but note this example with two full stops:

They have a collection of many animals (dogs, cats, tortoises, snakes, etc.).

3.3 A sequence of three full stops is used to mark an ellipsis or omission in a sequence of words, especially when forming an incomplete quotation.

He left the room, banged the door, . . . and went out.

When the omission occurs at the end of a sentence, a fourth point is added as the full stop of the whole sentence:

The report said: 'There are many issues to be considered, of which the chief are money, time, and personnel. . . . Let us consider personnel first.'

3.4 A full stop is used as a decimal point (*10.5%*; *£1.65*), and to divide hours and minutes in giving time (*6.15 p.m.*), although a colon is usual in American use (*6:15 p.m.*).

4. Comma ,

4.1 Use of the comma is more difficult to describe than the use of other punctuation marks, and there is much variation in practice. Essentially, its role is to give detail to the structure of sentences, especially longer ones, and make their meaning clear. Too many commas can be distracting; too few can make a piece of writing difficult to read or, worse, difficult to understand.

4.2.1 The comma is widely used to separate the main clauses of a compound sentence when they are not sufficiently close in meaning or content to form a continuous unpunctuated sentence, and are not distinct enough to warrant a semicolon. A conjunction such as *and*, *but*, *yet*, etc., is normally used:

The road runs through a beautiful wooded valley, and the railway line follows it closely.

The bus was late, so we had to wait.

4.2.2 It is considered incorrect to join the clauses of a compound sentence without a conjunction. In the following sentence, the comma should either be replaced by a semicolon, or be retained and followed by *and*:

incorrect *I like swimming very much, I go to the pool every week.*

4.2.3 It is also considered incorrect to separate a subject from its verb with a comma:

incorrect *Those with the smallest incomes and no other means, should get most support.*

4.3.1 Commas are usually inserted between adjectives coming before a noun:

An enterprising, ambitious person.

A cold, damp, badly heated room.

4.3.2 But the comma is omitted when the last adjective has a closer relation to the noun than the others:

A distinguished foreign politician.

A little old lady.

4.4 An important role of the comma is to prevent ambiguity or (momentary) misunderstanding, especially after a verb used intransitively (without a noun as direct object) where it might otherwise be taken to be transitive (with a direct object):

With the police pursuing, the people shouted loudly.

Other examples follow:

He did not want to leave, from a feeling of loyalty.

In the valley below, the houses appeared very small.

However, much as I should like to I cannot agree.
(compare *However much I should like to I cannot agree.*)

4.5.1 Commas are used in pairs to separate elements in a sentence that are not part of the main statement:

I should like you all, ladies and gentlemen, to raise your glasses.

There is no sense, as far as I can see, in this suggestion.

It appears, however, that we were wrong.

4.5.2 The comma is also used to separate a relative clause from its antecedent when the clause is not serving an identifying function:

The book, which was on the table, was a present.

In the above sentence, the information in the *which* clause is incidental to the main statement; without the comma, it would form an essential part of it in identifying which book is being referred to (and could be replaced by *that*):

The book which/that was on the table was a present.

4.6.1 Commas are used to separate items in a list or sequence. Usage varies as to the inclusion of a comma before *and* and the last item; the practice in this dictionary is to include it:

The following will report at 9.30 sharp: Jones, Smith, Thompson, and Williams.

4.6.2 A final comma before *and*, when used regularly and consistently, has the advantage of clarifying the grouping at a composite name occurring at the end of a list:

We shall go to Smiths, Boots, Woolworths, and Marks and Spencer.

4.7 A comma is used in numbers of four or more figures, to separate each group of three consecutive figures starting from the right (e.g. *10,135,793*).

5. Semicolon ;

5.1.1 The main role of the semicolon is to unite sentences that are closely associated or that complement or parallel each other in some way, as in the following:

In the north of the city there is a large industrial area with little private housing; further east is the university.

To err is human; to forgive, divine.

5.1.2 It is often used as a stronger division in a sentence that is already divided by commas:

He came out of the house, which lay back from the road, and saw her at the end of the path; but instead of continuing towards her, he hid until she had gone.

5.2 It is used in a similar way in lists of names or other items, to indicate a stronger division:

I should like to thank the managing director, Stephen Jones; my secretary, Mary Cartwright; and my assistant, Kenneth Sloane.

6. Colon

6.1 The main role of the colon is to separate main clauses when there is a step forward from the first to the second, especially from introduction to main point, from general statement to example, from cause to effect, and from premiss to conclusion:

There is something I want to say: I should like you all to know how grateful I am to you.

It was not easy: to begin with I had to find the right house.

The weather was bad: so we decided to stay at home.
(In this example, a comma could be used, but the emphasis on cause and effect would be much reduced.)

6.2 It also introduces a list of items. In this use a dash should not be added:

The following will be needed: a pen, pencil, rubber, piece of paper, and ruler.

6.3 It is used to introduce, more formally and emphatically than a comma would, speech or quoted material:

I told them last week: 'Do not in any circumstances open this door.'

7. Question mark ?

7.1.1 This is used in place of the full stop to show that the preceding sentence is a question:

Do you want another piece of cake?

He really is her husband?

7.1.2 It is not used when the question is implied by indirect speech:

I asked you whether you wanted another piece of cake.

7.2 It is used (often in brackets) to express doubt or uncertainty about a word or phrase immediately following or preceding it:

Julius Caesar, born (?) 100 BC.

They were then seen boarding a bus (to London?).

8. Exclamation mark !

This is used after an exclamatory word, phrase, or sentence expressing any of the following:

8.1 Absurdity:

What an idea!

8.2 Command or warning:

Go to your room!

Be careful!

8.3 Contempt or disgust:

They are revolting!

8.4 Emotion or pain:

I hate you!

That really hurts!

Ouch!

8.5 Enthusiasm:

I'd love to come!

8.6 Wish or regret:

Let me come!

If only I could swim!

8.7 Wonder, admiration, or surprise:

What a good idea!

Aren't they beautiful!

9. Apostrophe ’

9.1.1 The main use is to indicate the possessive case, as in *John's book*, *the girls' mother*, etc. It comes before the *s* in singular and plural nouns not ending in *s*, as in *the boy's games* and *the women's games*. It comes after the *s* in plural nouns ending in *s*, as in *the boys' games*.

9.1.2 In singular nouns ending in *s* practice differs between (for example) *Charles'* and *Charles's*; in some cases the shorter form is preferable for reasons of sound, as in *Xerxes' fleet*.

9.1.3 It is also used to indicate a place or business, e.g. *the butcher's*. In this use it is often omitted in some names, e.g. *Smiths*, *Lloyds Bank*.

9.2 It is used to indicate a contraction, e.g. *he's*, *wouldn't*, *o'clock*.

9.3 It is sometimes used to form a plural of individual letters or numbers, although this use is diminishing. It is helpful in *cross your t's* but unnecessary in *MPs* and *1940s*.

9.4 For its use as a quotation mark, see section 10.

10. Quotation marks ‘ ’

10.1 The main use is to indicate direct speech and quotations. A single turned comma (‘) is normally used at the beginning, and a single apostrophe (’) at the end of the quoted matter:

She said, ‘I have something to ask you.’

10.2 The closing quotation mark should come after any punctuation mark which is part of the quoted matter, but before any mark which is not:

They were described as ‘an unruly bunch’.

Did I hear you say ‘go away!’?

10.3 Punctuation dividing a sentence of quoted speech is put inside the quotation marks:

‘Go away,’ he said, ‘and don't ever come back.’

10.4 Quotation marks are also used for cited words and phrases:

What does ‘integrated circuit’ mean?

10.5 A quotation within a quotation is put in double quotation marks:

'Have you any idea,' he said, 'what "integrated circuit" means?'

Many publications use single within double quotations:

"Have you any idea," he said, "what 'integrated circuit' means?"

11. Brackets

11.1 The types of brackets used in normal punctuation are round brackets ( ) and square brackets [ ].

The name 'parentheses' is also used, especially for round brackets.

11.2 The main use of round brackets is to enclose explanations and extra information or comment:

He is (as he always was) a rebel.

Zimbabwe (formerly Rhodesia).

They talked about Machtpolitik *(power politics).*

11.3 They are used to give references and citations:

Thomas Carlyle (1795–1881).

A discussion of integrated circuits (see p. 38).

11.4 They are used to enclose reference letters or figures, e.g. *(1), (a).*

11.5 They are used to enclose optional words:

There are many (apparent) difficulties.
(In this example, the difficulties may or may not be only apparent.)

11.6.1 Square brackets are used less often. The main use is to enclose extra information attributable to someone (normally an editor) other than the writer of the surrounding text:

The man walked in, and his sister [Sarah] greeted him.

11.6.2 They are used in some contexts to convey special kinds of information, especially when round brackets are used also for other purposes: for example, in this dictionary they are used to give the etymologies at the end of entries.

12. Dash —

12.1 A single dash is used to indicate a pause, whether a hesitation in speech or to introduce an explanation or expansion of what comes before it:

'I think you should have—told me,' he replied.

We then saw the reptiles—snakes, crocodiles, that sort of thing.

12.2 A pair of dashes is used to indicate asides and parentheses, in the same way as commas, explained at 4.5.1 above, but forming a more distinct break:

People in the north are more friendly—and helpful—than those in the south.

There is nothing to be gained—unless you want a more active social life—in moving to the city.

12.3 A single dash is sometimes used to indicate an omitted word, for example a coarse word in reported speech:

'— you all,' he said.

13. Hyphen -

13.1 The hyphen has two main functions: to link words or elements of words into longer words and compounds, and to mark the division of a word at the end of a line in print or writing.

13.2.1 The use of the hyphen to connect words to form compound words is diminishing in English, especially when the elements are of one syllable as in *birdsong*, *eardrum*, and *playgroup*, and also in some longer formations such as *figurehead* and *nationwide*. The hyphen is used more often in routine and occasional couplings, especially when reference to the sense of the separate elements is considered important or unavoidable, as in *ankle-bone*. It is often retained to avoid awkward collisions of letters, as in *fast-talk*.

13.2.2 The hyphen serves to connect words that have a syntactic link, as in *hard-covered books* and *French-speaking people*, where the reference is to books with hard covers and people who speak French, rather than hard books with covers and French people who can speak (which would be the sense conveyed if the hyphens were omitted). It is also used to avoid more extreme kinds of ambiguity, as in *twenty-odd people*.

13.2.3 A particularly important use of the hyphen is to link compounds and phrases used attributively, as in a *well-known man* (but *the man is well known*), and *Christmas-tree lights* (but *the lights on the Christmas tree*).

13.2.4 It is also used to connect elements to form words in cases such as *re-enact* (where the collision of two *e*s would be awkward), *re-form* (= to form again, to distinguish it from *reform*), and some other prefixed words such as those in *anti-*, *non-*, *over-*, and *post-*. Usage varies in this regard, and much depends on how well established and clearly recognizable the resulting formation is. When the second element is a name, a hyphen is usual (as in *anti-Darwinian*).

13.2.5 It is used to indicate a common second element in all but the last of a list, e.g. *two-, three-, or fourfold.*

13.3 The hyphen used to divide a word at the end of a line is a different matter, because it is not a permanent feature of the spelling. It is more common in print, where the text has to be accurately spaced and the margin justified; in handwritten and typed or word-processed material it can be avoided altogether. In print, words need to be divided carefully and consistently, taking account of the appearance and structure of the word. Detailed guidance on word division may be found in the *Oxford Spelling Dictionary* (2nd edition, 1995).

B. SPELLING RULES

1. General remarks

British spelling was largely standardized by the middle of the 18th century, and American variants established by the early 19th, but many spelling conventions were fixed by printers as early as 1500, and since various changes in pronunciation have occurred in the ensuing centuries, present-day pronunciation and spelling are often at variance. Also, the 'neutral' vowel sound of unstressed syllables gives no guidance as to spelling, which is usually determined by the origin of the word, and care must be taken with words containing unstressed syllables such as *de-*, *di-*, *en-*, *in-*, *-par-*, *-per-*. These notes cover a few of the more common difficulties: for other individual points of uncertainty, the main part of the dictionary should be consulted, e.g. for pairs of words distinguished by meaning, such as *affect/effect*, *amend/emend*, *complement/compliment*, *enquire/inquire*, *its/it's*, *loath/loathe*, *stationary/stationery*. The following words may be difficult to find in a dictionary if the spelling is not known: *diphtheria*, *dissect*, *eczema*, *fuchsia*, *guerrilla*, *minuscule*, *necessary*,*ophthalmic*, *pejorative*, *semantics*. Note that silent letters occur especially in the combinations *gn-*, *kn-*, *mn-*, *pn-*, *ps-*, *pt-*, *rh-*, and that words ending in vowels other than *e* often have irregular inflections. (For a discussion of hyphenation see the section.)

2. *i* before *e*

2.1 For words pronounced with an 'ee' (/iː/) sound, the traditional rule '*i* before *e* except after *c*' is fairly reliable. The exceptions are (*a*) *seize* (and *seise*), (*b*) *either* and *neither* (if you pronounce them that way; also *heinous, inveigle*), (*c*) Latin words such as *prima facie* and *species*, and (*d*) words in which a stem ending in *-e-* is followed by a suffix beginning with *-i-*, e.g. *caffeine, casein, codeine, plebeian, protein*. Note that the syllable *-feit* is so spelt, e.g. in *counterfeit, forfeit, surfeit*, and that *mischief* is spelt like *chief*.

2.2 Words pronounced with an 'ay' (/eɪ/) or long 'i' (/aɪ/) sound generally have *-ei-*: e.g. *beige, heinous, reign, veil, eiderdown, height, kaleidoscope.* Words with other sounds follow no rules and must simply become familiar to the eye, e.g. *foreign* (related to *reign*), *friend, heifer, leisure, Madeira, sieve, sovereign* (like *foreign*), *their, view, weir, weird.*

3. Doubling consonants

3.1 When a suffix beginning with a vowel (such as *-able, -ed, -er, -ing,* or *-ish*) is added to a word ending in a consonant, the consonant is usually doubled if it is a single consonant preceded by a single vowel, and comes at the end of a stressed syllable. So *controllable, dropped, permitted, bigger, abetter, trekking, beginning, transferring, reddish, forgotten*, but *sweated, sweeter, appealing, greenish* (more than one vowel), *planting* (more than one consonant), *balloted, happened, preferable, profiting, rocketing* (not ending a stressed syllable). A secondary stress (not generally marked in this dictionary) is often sufficient to elicit a doubled consonant, e.g. *caravanned, confabbed, diagrammed, formatted, humbugged, programmed, zigzagged*, and (in British use) *kidnapped, worshipped,* though note *invalided* and (in British use) *benefited.* Other variable or exceptional verbs include *canvas, coif, curvet, ricochet, target, tittup,* and *wainscot.* Verbs ending in a vowel followed by *-c* generally form inflections in *-cked, -cking*, e.g. *bivouac, mimic, picnic.*

3.2 Derivative verbs formed by the addition of prefixes follow the pattern of the root verb, as in *inputting, leapfrogging, outcropped, outfitting*; note that *benefit* is not derived from *fit*, and the forms *benefitted* and *benefitting* are standard only in American English.

3.3 In British English, the letter *l* is doubled if it follows a single vowel, regardless of stress, e.g *labelled, travelling, jeweller*, but *heeled, airmailed, coolish* (more than one vowel). In American English the double *l* occurs only if ending a stressed syllable, e.g. *labeled*, *traveling*, *jeweler* in American use, but *dispelled*, *gelled* in both British and American use (the double *l* may be retained in the present tense in American use, e.g. *appall*, *enthrall*). Exceptions retaining single *l*: *paralleled*, *devilish*; exceptions having double *l* (in British use): *woollen*, *woolly*; note variability of *cruel(l)er*, *cruel(l)est*.

3.4 The letter *s* is not usually doubled before the suffix *-es*, either in plural nouns, e.g. *focuses*, *gases*, *pluses*, *yeses*, or in the present tense of verbs, e.g. *focuses*, *gases*. However, verbal forms in *-s(s)ed*, *-s(s)ing* are variable, and doubling only after stressed syllables is often preferable, e.g. *gassing*, *nonplussed*, but *biased*, *focused*, *focusing*, Variants are common: e.g. see BUS. See also 'Forming plurals' below.

3.5 The consonants *h*, *w*, *x*, and *y* are never doubled: *hurrahed*, *guffawed*, *mower*, *boxing*, *stayed*. Silent consonants are also never doubled: *crocheting*, *précising*.

4. Dropping silent e

A final silent *e* is usually dropped when adding a suffix beginning with a vowel, e.g. *bluish*, *bravest*, *continuous*, *queued*, *refusal*, *writing*. Exceptions are noted below:

4.1 **before *-ing*** The *e* is retained in *dyeing*, *singeing*, *swingeing*, and (usually) *routeing*, to distinguish them from *dying*, *singing*, *swinging*, and *routing*. It is commonly retained in *ageing*, *bingeing*, *blueing*, *clueing*, *cueing*, *twingeing*, *whingeing*, and sometimes in *glu(e)ing*, *hing(e)ing*, *ru(e)ing*, *spong(e)ing*, *ting(e)ing*. It is also retained for words ending in *-ee*, *-oe*, *-ye*, e.g. *canoeing*, *eyeing*, *fleeing*, *hoeing*, *shoeing*, *tiptoeing*. Otherwise it is dropped: *charging*, *icing*, *lunging*, *staging*, etc.

4.2 **words ending in *-ce* or *-ge*** The *e* is retained to preserve the sound of the consonant, e.g. *advantageous*, *courageous*, *knowledgeable*, *noticeable*, *manageable*, *peaceable*.

4.3 **before *-able*** The dropping of *e* before *-able* is very unpredictable, and the first (or only) spelling given in the main part of the dictionary should be preferred. The endings *-ceable* and *-geable* are usual, and no letter is dropped in *agreeable*, *foreseeable*. The *e* is retained in *probeable* to distinguish it from *probable*. The *e* is more often dropped in American English.

4.4 **before *-age*** The *e* is usually dropped: *cleavage*, *dosage*, *wastage*, Exceptions: *acreage*, *metreage* (always), *mil(e)age* (optional). Note also that *linage* and *lineage* are different words.

4.5 **before -*y*** The *e* is usually dropped: *bony*, *icy*, *grimy*. Exceptions: (*a*) after *u* (*gluey*); (*b*) after *g* (*cottagey*, *villagey*, but optional in *cag(e)y*, *stag(e)y*); (*c*) after *c* (usual in *dicey*, optional in *pric(e)y* and *spac(e)y*, occasionally seen in *pacy* and *spicy*, but otherwise dropped, e.g. *bouncy*, *chancy*, *fleecy*, *lacy*, etc.). The *e* is retained in *holey* to distinguish it from *holy*, and an extra *e* is added to separate two *y*s, e.g. *clayey*. It may be retained or added for clarity in more unusual words, e.g. *chocolatey*, *echoey*.

A silent *e* is not usually dropped when adding a suffix beginning with a consonant, e.g. *useful*, *homeless*, *safely*, *movement*, *whiteness*, *lifelike*, *awesome*. Exceptions: *argument*, *awful*, *duly*, *ninth*, *truly*, *wholly*. When such a suffix is added to words ending in *-dge*, American English tends to drop the *e*, e.g. *acknowledgment*, *fledgling*, and this practice is sometimes seen in British English (notably in *judgment*, which is usual in legal contexts).

5. Forming plurals

5.1 **Simple nouns** Regular plurals are formed by adding *s*, or after *s*, *sh*, *ss*, *z*, *x*, *ch* (unless pronounced 'hard') by adding *es*: *books*, *boxes*, *pizzas*, *queues*, *arches*, *stomachs*. An apostrophe should not be used. Nouns ending in *-y* preceded by a consonant (or *-quy*) form plurals ending in *-ies*, e.g. *rubies*, *soliloquies*, but *boys*, *monkeys*. Exceptions: *laybys*, *stand-bys*, most names (e.g. *the Kennedys*). Nouns ending in *-f* or *-fe* (not *-ff*, *-ffe*) may form plurals in *-ves*, either always (e.g. *halves*, *leaves*) or optionally (e.g. *hooves*, *scarves*), or may always have regular plurals (e.g. *beliefs*, *chiefs*); these should be checked in the main part of the dictionary. Nouns ending in *-o* or *-i* are variable and should be checked in the main part of the dictionary; a number of long-established English words have only plurals in *-oes* (e.g. *heroes*, *potatoes*, *tomatoes*) but plurals in *-os* are common, and are usual among words which are less naturalized (e.g. *arpeggios*), or are formed by abbreviation (e.g. *kilos*), or have a vowel preceding the *-o* (e.g. *radios*). Nouns ending in *-ful* form regular plurals in *-fuls* (see Usage Note at CUPFUL). Only the letter *z* is regularly doubled in forming plurals: *fezzes*, *quizzes*, but *gases*, *yeses* (see 'Doubling consonants' above). Nouns ending in *-man* form plurals in *-men*, e.g. *chairmen*, *postmen*, *spokeswomen*, etc., but note *caymans*, *dragomans*, *talismans*, *Turcomans*. Other irregular plurals are noted in the main text of the dictionary.

5.2 **Compound nouns** Most compound nouns pluralize the last element: *break-ins*, *forget-me-nots*, *major generals*, *man-hours*, *ne'er-do-wells*, *round-ups*, *sergeant majors*, *vice-chancellors*. Exceptions include: (i) nouns followed by prepositional phrases, e.g. *Chancellors of the Exchequer*, *commanders-in-chief*, *daughters-in-law*, *ladies-in-*

waiting, *men-of-war*, *rights of way*; (ii) nouns denoting persons, followed by adverbs, e.g. *hangers-on*, *passers-by*, *runners-up*; (iii) nouns followed by adjectives, e.g. *battles royal*, *cousins german*, *heirs presumptive*, *notaries public*, *Governors-General* (though terms in common use, especially if hyphenated, may not follow this rule, e.g. *Secretary-Generals*); (iv) nouns denoting persons and containing *man* or *woman*, which pluralize both elements, e.g. *women doctors*, *menservants*, *gentlemen farmers*.

5.3 **Foreign and classical plurals** Words adopted into English generally form regular English plurals, but words not fully naturalized may form the plural as in the language of origin, e.g. *bureaux*, *cherubim*, *lire*, *virtuosi*. Many words of Greek and Latin origin retain classical plurals, though they may be used only in technical contexts, e.g. *formulae*, *indices*, *stadia*. In general, in forming classical plurals, *-us* becomes *-i* (occasionally *-era* or *-ora*); *-a* becomes *-ae*; *-um* and *-on* become *-a*; *-ex* and *-ix* become *-ices*; *-nx* becomes *-nges*; *-is* becomes *-es* or *-ides*; and *-os* becomes *-oi*. Note that many nouns regularly form only English plurals, e.g. *agendas*, *censuses*, *irises*, *octopuses*, *omnibuses*, *phoenixes*, *thermoses*. Care should be taken with words ending in *-a*, e.g. *addenda*, *bacteria*, *criteria*, *phenomena*, and *strata* are plural, but *nebula* and *vertebra* are singular.

6. Common suffixes

Several common suffixes occur in different forms, which may cause spelling difficulties: users of the dictionary should be careful to check if unsure of accepted usage. The most frequent sources of uncertainty are as follows:

6.1 **-able/-ible** The suffix *-ible* is found only in a number of long-established words taken directly from Latin or modelled on these. Modern formations on English roots use *-able* (see also 'Dropping silent *e*' above).

6.2 **-ance/-ence (and -ant/-ent)** These endings are largely dependent on the source of the word in Latin, and must be checked in the main part of the dictionary. Note especially *currant/current* and *dependant/dependent*.

6.3 **-cede/-ceed** The suffix *-ceed* occurs only in *exceed*, *proceed*, *succeed*; otherwise *concede*, *intercede*, *precede*, *recede*, etc. (note also *supersede*).

6.4 **-ction/-xion** The *-x-* is recommended in *complexion* and *crucifixion*, but not in other words in this dictionary.

6.5 **-er/-or** The ending *-or* is found mainly in words of classical or French origin, especially in the combinations *-ator*, *-ctor*, *-essor.* It is also retained in legal use where *-er* is more usual, e.g. *divisor.* See also Usage Note at ADVISER.

6.6 **-er/-re** American spelling often uses *-er* for *-re* in words such as *centre, fibre, theatre,* etc., but not in *acre, cadre, lucre, massacre, mediocre, ogre,* and *wiseacre.* Note also the usage indicated in the entries for *meter* and *metre* in the main part of the dictionary.

6.7 **-ice/-ise** In standard British use, *licence* and *practice* are nouns, *license* and *practise* are verbs; in American use, the *-ise* form is used for both noun and verb. Note also the distinction between *prophesy* and *prophecy.*

6.8 **-ise/-ize/-yse** The verbal ending *-ize* has been in general use since the 16th century; it is favoured in American English and in much British writing, and remains the current preferred style of Oxford University Press in academic and general books published in Britain. However, the alternative spelling *-ise* is now widespread (partly under the influence of French), especially in Britain, and may be adopted provided that its use is consistent. A number of verbs always end in *-ise* in British use, notably *advertise, chastise, despise, disguise, franchise, merchandise, surmise,* and all verbs ending in *-cise, -prise, -vise* (including *comprise, excise, prise* (= open), *supervise, surprise, televise,* etc.), but *-ize* is always used in *prize* (= value), *capsize, size.* Spellings with *-yze* (*analyze, paralyze*) are acceptable only in American use.

6.9 **-our/-or** Most words ending in *-our* in British use are spelt with *-or* in American use. However, British spelling often uses *-or* (e.g. *error, stupor, tremor*), and the *u* is dropped before some suffixes (e.g. *coloration, honorary, vaporize,* but note *colourist, honourable, savoury*). It is advisable to check such spellings in the dictionary.

7. ae and oe

The use of the printed ligatures *æ* and *œ* is becoming rare, and there is a trend in favour of replacing *ae* and *oe* with simple *e*, especially in American and in scientific use. The main part of the dictionary should be checked for individual words.

C. CAPITALIZATION

The use of capital letters in punctuating sentences has been discussed above; their use to distinguish proper nouns or 'names' from ordinary words is subject to wide variation in practice. The standard

OUP style is outlined below, but the most important criterion is consistency within a single piece of writing.

1. Capital letters are used for the names of people and places (*John Smith*, *Paris*, *Oxford Street*, *New South Wales*, *the Black Sea*, *the Iron Duke*); the names of peoples and languages and derived words directly relating to them (*Englishman*, *Austrian*, *French*, *Swahili*, *Americanize*); the names of institutions and institutional groups (*the Crown*, *the Government*, *the British Museum*, *the House of Representatives*, *the Department of Trade*); the names of religious institutions and denominations and their adherents (*Judaism*, *Nonconformism*, *Methodist*, *Protestants*) and of societies and organizations (*the Royal Society*); the names of months and days (*Tuesday*, *March*, *Easter Day*); abstract qualities personified (*the face of Nature*, *O Death!*) or used as sobriquets (*a Blue* in university sport, *a Red* = communist); and names of other non-personal things (*the Flying Scotsman*).

Note that *the Baptist Church* is an institution, but *the Baptist church* is a building; a *Democrat* belongs to a political party, but a *democrat* simply supports democracy; *Northern Ireland* is a name with recognized status, but *northern England* is not.

2. A capital letter is used for words derived from a proper name, if the connection with the name is direct, or felt to be continuing (*Christian*, *Homeric*, *Marxism*), but not if it is more remote or conventional (*chauvinistic*, *quixotic*, *guillotine*).

3. A capital letter is used by convention in many names that are trade marks (*Elastoplast*, *Filofax*, *Hoover*, *Xerox*) or are otherwise associated with a particular manufacturer etc. (*Jaguar*, *Spitfire*). Some proprietary terms are now conventionally spelt with a lower case initial (*baby buggy*, *biro*, *cellophane*, *jeep*), and this is generally true of established verbs derived from proprietary terms (*to hoover*, *to xerox*).

4. Capital letters are used in titles of courtesy or rank, including compound titles (*His Royal Highness the Prince of Wales*, *President Carter*, *Sir John Smith*, *Lord Chief Justice*, *Lieutenant-Colonel*, *Vice-President*, *Your Grace*, *His Excellency*).

5. A capital letter is used for the personal pronoun *I* and for the interjection *O*.

6. A capital letter is used for the deity (*God*, *Father*, *Allah*, *Almighty*). However, the use of capitals in possessive determiners and possessive pronouns (*in His name*) is now generally considered old-fashioned.

7. Capital letters are used for the first and other important words in titles of books, newspapers, plays, films, television programmes, etc., and in headings and captions (*The Merchant of Venice*, *Pride and Prejudice*, *Book of Common Prayer*, *New Testament*, *Talmud*, *Guide to the Use of the Dictionary*).

8. Capital letters are used for historical events and periods (*the Dark Ages*, *Early Minoan*, *Perpendicular*, *the Renaissance*, *the First World War*); also for geological time divisions, but not for certain archaeological periods (*Devonian*, *Palaeozoic*, but *neolithic*).

9. Capital letters are frequently used in abbreviations, with or without full stops (*BBC*, *DoE*, *M.Litt.*).

10. A capital letter is used for a compass direction when abbreviated (*N*, *NNE*, *NE*) or when denoting a region (*unemployment in the North*).

11. A capital letter is frequently used to begin a line of English verse.

12. The use of a capital letter elsewhere than at the beginning of a word is seen in certain names (*MacDonald*, *O'Reilly*) and in some trade marks, and is conventional in some foreign languages.

D. ITALICIZATION

Italic type makes a word or phrase stand out from its context. It is used especially in the following ways:

1. For titles of books, plays, major musical works, works of art, long poems, periodical publications, and individual ships, trains, aircraft, etc.: *Jane Eyre*, *Henry V*, *The Magic Flute*, Michelangelo's *David*, *Paradise Lost*, the *Daily Telegraph*, the *Marie Celeste*, HMS *Dreadnought*.

The words *The* or *A* may or may not be part of a full title: the *Oxford Times*, *The Times*, *The Economist*, the *Messiah* by Handel, *A London Symphony*). Unless the exact title is to be cited, the article may be omitted if the work is well known or has already been cited: Darwin's *Origin of Species*.

Strictly, inflections are printed in roman: the *Marie Celeste*'s crew, a pile of *New Yorker*s.

2. For foreign words and phrases, when still perceived as foreign. When a foreign word becomes sufficiently naturalized, it is printed in roman. Headwords in this dictionary use italic or roman based on the current frequency of use in English. Words which would nor-

mally be printed in roman are sometimes italicized for consistency when other related words are being used, or when an English word exists with the same spelling, as with *pension* for a Continental boarding house.

3. For the Latin names of genera and species, e.g. *Malus*, *Homo sapiens*.

4. For distinguishing a word or phrase from the surrounding text, especially to emphasize it, or when mentioning a technical word for the first time. Italics may be used, for example, to distinguish stage directions in plays; in dictionaries, they typically distinguish markers for parts of speech, labels concerning register or restriction of use, and example sentences.

5. Italic type is *not* used for the following:

5.1 Titles of chapters in books, articles in periodicals, shorter poems, television and radio programmes; these may be referred to in quotation marks: an article on 'Oral Tradition' in the *Journal of Theology*, an episode of 'Neighbours', 'Sonnet VI' in *Selected Poems*.

5.2 The names of sacred texts or their subdivisions; quotation marks are not used: the Koran, Genesis, Epistle to the Romans.

5.3 Musical works identified by a description (Beethoven's Fifth Symphony).

5.4 Names of buildings or of types of vehicle (the Red Lion, the Colosseum, a Ford Cortina).

5.5 Most short abbreviations, including units of measurement (cf., e.g., ibid., i.e., km, op., pro tem, q.v.).

6. If a piece of text is already printed in italics, then the function of italicization is taken by roman type: She was reading *On the Use of* Verfremdungseffekt *in Brecht's Plays*.

E. REFERENCES TO PEOPLE

Names should normally be printed in the form by which the bearer is most commonly known, or is known to prefer: e.g. Arthur C. Clarke (not 'A. C. Clarke'), T. S. Eliot (not 'Thomas S. Eliot'), R. Vaughan Williams (not 'R. V. Williams'). Forenames should not be abbreviated (George not Geo., William not Wm.) unless reproducing a signature or manuscript or a commercial style.

Titles should have capital letters; they are frequently abbreviated for convenience. 'Mr' is applicable to any male; the use of 'Master' for boys is old-fashioned. 'Mrs' usually designates a married woman who has adopted her husband's surname. Both 'Miss' and 'Ms' are used by unmarried women, and by married women (retaining their maiden name) who object to having attention drawn to their marital status. 'Dr' is appropriate for those holding university doctorates, and is also given as a courtesy title to medical doctors (but not surgeons); it should not be combined with any of the above titles, nor used together with the letters indicating the doctorate (e.g. D.Phil.). Married professional women often practise under their maiden names. 'Esq.' after a surname is used by professionals as an alternative to 'Mr' (no longer restricted to professionals and holders of a bachelor's degree). 'Reverend' is used for ministers of religion; it should not be used with a surname alone: e.g. the Rev. J. Brown (or the Rev. Mr Brown if the name or initial is unknown), but not 'the Reverend Brown'. The abbreviation 'Revd' is sometimes preferred.

Traditionally, a woman may adopt her husband's name on marriage, and the couple are then jointly called 'Mr and Mrs John Smith'. Strictly, the wife is correctly referred to as 'Mrs John Smith', and to give her own name ('Mrs Mary Smith') would at one time have indicated a divorcee. However, this distinction is now made only in the most formal circumstances. Divorced women may retain either their maiden or married names, with Mrs, Miss, or Ms according to preference, and 'Mrs Mary Smith' is generally acceptable as a form for married women. The maiden or unmarried name of a man or woman may be indicated by 'né(e)': e.g. Mary Seymour-Smith (née Seymour), John Seymour-Smith (né Smith).

A title should not be used in a signature, though it may be placed after it in brackets for information: e.g. Robin Smith (Miss), M.T. Brown (Mrs).

A person's former or alternative name or title is indicated only if necessary or to avoid confusion: Michael (now Sir Michael) Tippett, Laurence (later Lord) Olivier, Lord Home of the Hirsel (then Sir Alec Douglas-Home). (Further discussion of titles of rank and nobility is beyond the scope of this book.) The owner of an **adopted name** uses it for all purposes, and may have adopted it legally. Though a former name may be indicated, it should not be referred to as the person's 'real name': e.g. Woody Allen (born Allen Stuart Konigsberg), Mohammed Ali (formerly Cassius Clay).

A **pseudonym** is only used for a specific purpose, e.g. as a pen-name or stage name, though a person may be best known by it: e.g. George Eliot (pseudonym of Mary Ann, later Marian, Evans). It may appear in quotation marks: e.g. 'Lewis Carroll' (Charles L. Dodgson). A **nickname** supplements the owner's name and is often placed in quotation marks (Charlie 'Bird' Parker), though it may replace the original name altogether (Fats Waller). An **alias** is generally a false name assumed with intent to deceive.

F. OFFENSIVE LANGUAGE AND SEXISM

In general, terms should be avoided which convey an impression of over-generalization, describing people as though they were merely instances of a particular feature, or especially imposing on them a depreciatory stereotype. One should certainly avoid abbreviated colloquial forms referring to race (e.g. *Paki*, *Jap*); the term *race* is itself best avoided, except in strictly anthropological contexts, in favour of *nation*, *people*, *ethnic group*, *community*. Words referring to racial type or to physical or mental handicap which have been used as terms of abuse, or have been associated with discrimination, are frequently therefore avoided even in their original neutral senses. The use of adjectives rather than nouns in describing groups is usually preferable (e.g. *the Hispanic community* not *the Hispanics*; *disabled people* not *the disabled*).

See also Usage Notes at ESKIMO, MUHAMMADAN, NEGRESS, NEGRO, SCOTCH, UNTOUCHABLE, and at MONGOLISM and GAY.

The replacement of offensive or potentially offensive vocabulary with 'politically correct' euphemistic phrases, while often well-intentioned, can create confusion unless the replacements are familiar to the intended audience (e.g. 'learning difficulties': educational problems or mental handicap?), and frequently offers a target for ridicule.

There is a widespread tendency to replace terms for occupations or titles which are unnecessarily marked for gender (e.g. *flight attendant* for *stewardess*), and to substitute *-person* for *-man* in words such as *chairman*, *salesman*, and *spokesman*. Opinions vary very widely concerning the desirability of such substitutions. A balance needs to be struck between the desire to avoid sexist language and the common sense of one's audience; sensitivity to context is needed to determine the borderline between sensible accommodation and absurdity. The extending of this tendency to cover words with only tenuous etymological links with sex (e.g. *masterpiece*, *manhandle*, *manhole*) is not generally accepted, and extreme forms such as *herstory* for *history* have little place outside specifically feminist writing.

The English language lacks a third person singular pronoun or possessive adjective applying neutrally to both sexes. The older convention was to use *he*, *him*, *his* for both sexes (e.g. *Each member must pay his subscription*), but this is now often felt to exclude women and girls. Acceptable alternatives include (i) rephrasing in the plural (e.g. *All members must pay their subscriptions*); and (ii) using both pronouns or possessives (e.g. *Each member must pay his or her subscription*), though this is often cumbersome; *his/her* and *he/she* (or even *s/he*) are awkward to read aloud. The use of *they* and *their* in the singular is common in informal speech (e.g. *Each member must pay their subscription*), but is still considered ungrammatical and should be avoided in formal speech and writing.